AF762067

World Checklist of Cyperaceae

Sedges

Dedicated to the memory of our co-author Tatyana Egorova, 1930–2007

World Checklist of

Cyperaceae

Sedges

Rafaël Govaerts and David A. Simpson

with

Jeremy Bruhl, Tatyana Egorova, Paul Goetghebeur & Karen Wilson

Kew Publishing

Royal Botanic Gardens, Kew

Kew

PLANTS PEOPLE
POSSIBILITIES

First published in 2007 by
Royal Botanic Gardens, Kew
Richmond, Surrey, TW9 3AB, UK
www.kew.org

ISBN 978-1-84246-199-0

British Library Cataloguing in Publication Data
A catalogue record for this book is available from the British Library

Typesetting and page layout: Christine Beard
Design by Media Resources,
Royal Botanic Gardens, Kew

Printed in the United Kingdom by Lightning Source

Contents

Introduction

Cyperaceae (sedges) are the third largest family in the Monocotyledons and seventh largest family in the angiosperms. They are more-or-less cosmopolitan in distribution with hotspots of diversity in north-eastern South America, eastern and southern Africa, South and Southeast Asia and Australasia. Sedges have economic importance going back into antiquity, highlighted by the role of papyrus (*Cyperus papyrus*) in early Egyptian society. Today their importance is wide-ranging, from the problems they present as weeds (*Cyperus rotundus* is known as the world's worst weed) to their considerable number of uses – providing materials for basketry, matting, construction (particularly thatch), perfumery, medicine and fuel, as well as food and animal fodder. The numbers of species of horticultural interest has increased, coinciding with a general interest in 'ornamental grasses'. Their importance has often been under-estimated, but there is no doubt that they play a significant, sometimes crucial role in many local economies, particularly in the tropics. Sedges are also important indicators of environmental damage, especially to aquatic habitats. They are particularly sensitive to the lowering of water tables and/or water pollution and there are observable declines in many species. Some genera are threatened by forest destruction and the prevalence of narrow endemic taxa within these genera may mean that some taxa are now extinct.

Cyperaceae have long been recognised as a taxonomically difficult group due to their reduced, but nevertheless complex, floral and vegetative structure. As a result there has been little consensus on tribal and generic boundaries; moreover there have been comparatively few specialists undertaking systematic studies on the family over the years. Recent DNA studies are beginning to shed light on relationships within the family but there is much work still to be done.

There is a real need for a nomenclator that provides a reliable guide to accepted names. This is particularly important as a basic framework for exchanging information on species-rich groups such as the genus *Carex* and in providing a framework for addressing conservation status and needs.

Systematic research will always give rise to name changes, new species will inevitably be found and overlooked names will continue to be discovered. We do not pretend that this list is complete or entirely correct, so we solicit input and constructive criticism. Please send comments by email to **r.govaerts@kew.org** and **d.simpson@kew.org** or by regular post to the authors at The Herbarium, Royal Botanic Gardens, Kew, Richmond, Surrey TW9 3AB, UK. The Checklist is also available online as a searchable database (**www.kew.org/wcsp/**). The database is live and changes based on any comments made on this or the online version can be incorporated. Of course, as soon as this checklist is printed it will become out of date, but the database which underpins it will constantly be updated. Any changes will appear on the online version at the URL above, new names will be added after a minimum of one year after publication to allow for an editing phase.

Currently the checklist includes 5387 accepted species in Cyperaceae in 106 genera and 26,150 names in total.

How to use the Checklist

Structure

The checklist is derived from a database encompassing 24 fields and complying with the data standards proposed by the Taxonomic Databases Working Group (TDWG). Its compilation was effected using Microsoft Access 2002 for Windows, within which editing was also carried out. The arrangement of genera being alphabetical with accepted and synonymous genera intercalated.

Names

Names of accepted genera and their species and infraspecific taxa are listed alphabetically. Synonymised genera (and species) are intercalated. For each accepted taxon, associated synonyms are listed chronologically if heterotypic, with any homotypic synonyms following in a given lead; in addition, all synonyms in an accepted genus are listed alphabetically at the end of that genus. Place and date of publication of all names are given. Citation of authors follows *Authors of Plant Names* (Brummitt & Powell 1992; updated and available electronically at http://www.ipni.org/ipni/query_author.html); for book abbreviations, the standard is *Taxonomic Literature*, 2nd edn. (Stafleu & Cowan 1976-88; supplements, 1992-2000) also available electronically (http://tl2.idcpublishers.info/); and periodicals are abbreviated according to *Botanico-Periodicum-Huntianum/ 2* (Bridson 2004). A question mark (?) following a name and author indicates that a place of publication has yet to be established. Names of hybrids are preceded by a multiplication sign (?), with the place of publication being followed by the names of the parents if known. Basionyms or replaced synonyms of accepted names are designated by an asterisk (*). For genera, the number of accepted species and the geographical distribution are furnished together with general comments and the suprageneric taxa as used at Kew.

Acceptance of taxa

Acceptance of species and infraspecific taxa is based not only on assessments of literature but also, where possible, by reference to specialist advice and (where necessary) to the herbarium or living collections. Generic limits follow Goetghebeur (1998).

New names and combinations

A few necessary new names and combinations have been made in the text to bring known taxa into line with current practice

Geographical distribution

Distributions of **species** and taxa of lower rank are furnished in two ways: firstly by a generalised statement in narrative form, and secondly as TDWG geographical codes (Brummitt 2001) also available electronically (http://www.nhm.ac.uk/hosted_sites/tdwg/geogrphy.html) expressed to that system's third level. Examples of the former include:

E. & C. U.S.A.
Texas to C. America
Mexico (Veracruz)
Europe to Iran
E. Himalaya, Tibet, China (W. Yunnan)
Philippines (Luzon)
S. Trop. America

When the presence of a taxon in a given region or location is not certainly known, a question mark is used, e.g. New Ireland ?; when an exact location within a country is not known, a question mark within brackets is used, e.g. Mexico (?). Distributions of **genera** are furnished in a relatively simplified form, any special features being given within brackets.

With respect to the TDWG codes, the **region** is indicated by the two-digit number (representative of the first two levels), the first digit also indicating the continent. The letter codes following the digits, when given, represent the third-level **unit** (a country, state or other comparable area). They usually are the first three letters of a given unit's name, but sometimes are contractions. If the country code is not known, '+' is used. For taxa that are known or appear to be extinct in a given region, '†' is used after the country code. Naturalisation is expressed by putting the third-level codes in lower case and, if in a second-level region all occurrences are the result of naturalisation, the code number for the region is placed in brackets. The application of question marks is as indicated above for geographical regions. Examples include:

12 SPA	[SW. Europe: Spain]
32 +	[C. Asia (more exact distribution not known)]
36 CHN? 38 JAP KOR	[Doubtful in China and Eastern Asia; China, Japan, Korea]
51 NZN NZS	[New Zealand: North and South Islands]
76 ARI 77 NWM TEX	[SW. & SC. U.S.A., Mexico & C. America:
79 ALL 80 GUA HON	Arizona, New Mexico, Texas, Mexico, Guatemala and Honduras]
77 TEX†	[SC. U.S.A.: Texas, where extinct]

Life-forms

The terminology for *life-forms*, definitions of which follow, is based on the system of Raunkiær (1934, especially chapters 1 and 2) with modifications derived from *Flora van België, het Groothertogdom Luxemburg, Noord-Frankrijk en de aangrenzende gebieden* (De Langhe *et al.* 1983: pp. xvii-xviii, 869 (fig. 16)).

Main Categories:

nanophanerophyte (*nanophan.*)
stems: woody and persisting for several years
buds: above soil level but normally below 3 m
e.g.: shrubs

chamerophyte (*cham.*)
stems: herbaceous and/or woody and persisting for several years
buds: on or just above soil level, never above 50 cm
e.g.: *Alyssum, Acaena, Acantholimon, Saxifraga*

hemicryptophyte (*hemicr.*)
stems: herbaceous, often dying back after the growing season, with shoots at soil level surviving
buds: just on or below soil level
e.g.: *Isolepis prolifera*

geophyte
hemicryptophytes that survive unfavourable seasons in the form of a rhizome, bulb, tuber or rootbud.
e.g.: *Carex arenaria* is a rhizome geophyte.

helophyte (*hel.*)
hemicryptophytes that grow in soil saturated with water or in water. leaf and flower bearing shoots rise above water
e.g.: *Cyperus papyrus*

hydrophyte
stems: vegetative shoots sunk in water
buds: permanently or temporarily on the bottom of the water

therophyte (*ther.*)
plants that survive unfavourable seasons in the form of seeds and complete their life-history during the favourable season.
e.g. annuals

Additional information:
climbing (*cl.*)
succulent (*succ.*)
parasitic (*par.*)

Abbreviations

A .... alpine/arctic
Agg. .... aggregate
al. .... alii: others
Arch. .... archipelago
app. .... approaching, close to
auct. .... of author
C. .... Central
cham. .... chamaephyte
cit. .... citatus: cited
cl. .... climbing
Co. .... county
comb. .... combinatio: combination
cons. .... conservandus: to be conserved
cppo...................... centre page pull-out
cult. .... cultus: cultivated
cv. .... cultivarietas: cultivar
descr. .... description
Distr. .... district
DT .... dry tropical (desert/steppe)
E. .... East(ern)
etc. .... et cetera: and the rest
e.g. .... exampli gratia: for example
G .... temperate
hel. .... helophyte
hort. .... hortorum: of gardens
I./Is .... island(s)
ICBN .... International Code of Botanical Nomenclature
i.e. .... id est: that is
ign. .... ignotus: unknown
in litt. .... in litteris: in correspondence
ined. .... ineditus: unpublished, provisional name
inq. .... inquilinus: naturalised
i.q. .... idem quod: the same as
Kep. .... kepulauan (islands)
Medit. .... mediterranean
MT .... monsoon tropical (savanna)
Mt./Mts .... mountain(s)
N. .... North(ern)
nanophan. .... nanophanerophyte
No. .... numero: number
noh .... new orchid hybrids
nom. cons. .... nomen conservandum: name conserved in ICBN

nom. illeg. .... nomen illegitimum: illegitimate name
nom. inval. .... invalid name
nom. nud. .... nomen nudum: name without a description
nom. rejic. .... nomen rejiciendum: name rejected in ICBN
nom. superfl. .... nomen superfluum: name superfluous when published
nov. .... novus: new
orth. var. .... orthographic variant
par. .... parasitic
Pen. .... peninsula(r)
phan. .... phanerophyte
p.p. .... pro parte: partly
Prov. .... province
q.e. .... quod est: which is
q.v. .... quod vide: which see
Reg. .... region
Rep. .... republic
S .... Subtropical
S. .... South(ern)
seq. .... sequens: following
s.l. .... sensu lato: in the broad sense
sp. .... species
s.p. .... without page number
sphalm. .... sphalmate: by mistake
s.s. .... sensu stricto: in the narrow sense
st. .... status
subtrop. .... subtropical
syn. .... synonymon: synonym
T .... tropical
temp. .... temperate
ther. .... therophyte
trop. .... tropical
vol. .... volume: volume
viz. .... videlicet: namely
W. .... West(ern)
WT .... wet tropical
? .... not known, doubtful
† .... extinct

Acknowledgements

We would particularly like to thank Nick Black for writing and maintaining the computer programs and reports used to create the checklist. Thanks are also due to Simon Owens, Keeper of the Herbarium at Kew for providing the facilities to allow the work to come to fruition, the staff at the Kew library and Phillip Cribb for initiating and susbequently managing the World Checklist of Monocotyledons project of which this work forms a part. We are also grateful to correspondents who have commented on entries in the online version of the checklist.

Co-authors

Jeremy Bruhl is Associate Professor of Botany and Director of the N.C.W. Beadle Herbarium at the University of New England, New South Wales, Australia.

Tatyana Egorova is a botanist at the V. L. Komarov Botanical Institute, St Petersburg, Russia.

Paul Goetghebeur is Director of the Botanical Garden at Gent University in Belgium.

Karen Wilson is a Special Botanist and Manager, Plant Diversity at the Royal Botanic Gardens, Sydney, Australia.

References

Bridson, G., comp. & ed. (2004). Botanico-Periodicum Huntianum/Second Edition. Pittsburgh: Hunt Institute for Botanical Documentation

Brummitt, R. K. (2001). World Geographical Scheme for Recording Plant Distributions, ed 2. xvi, 138 pp. Hunt Institute for Botanical Documentation, Carnegie-Mellon University, Pittsburgh, Penna. (for the International Working Group on Taxonomic Databases for Plant Sciences). (Plant Taxonomic Database Standards, 2: version 1.0.) [http://www.tdwg.org/geo2.htm]

Brummitt, R. K. & Powell, C. E., (1992). Authors of Plant Names. 732 pp. Kew: Royal Botanic Gardens. [http://www.ipni.org/ipni/query_author.html]

De Langhe, J. E. *et al.* (1983). Flora van België, het Groothertogdom Luxemburg, Noord-Frankrijk en de aangrenzende gebieden. civ, 970 pp., illus., map. Patrimonium, Nationale Plantentuin van België, Meise.

Goetghebeur, P. (1998). Cyperaceae. In: K. Kubitzki (ed.), the families and genera of vascular plants, vol. 4. Flowering plants, monocotyledons: Alismatanae and Commelinanae (except Gramineae): pp. 141–190. Springer-Verlag, Berlin, Germany.

Greuter, W. *et al.* (1993). Names in Current Use for Extant Plant Genera (Names in current use, 3). xxvii, 1464 pp. Koeltz, Koenigstein. (Regnum Vegetabile.)

Raunkiær, C. (1934). The Life Forms of Plants and Statistical Plant Geography. xvi, 632 pp., illus. Oxford University Press, London.

Stafleu, F. & Cowan, R. S. (1976–88). Taxonomic Literature: A Selective Guide to Botanical Publications and Collections with Dates, Commentaries and Types. 2nd edn. 7 vols. Utrecht: Bohn, Scheltema & Holkema. (Regnum Vegetabilie 94, 98, 105, 110, 112, 115, 116.) Continued as Stafleu, F. *et al.* (1992–2000). Taxonomic Literature, Supplement. Vols. 1-6. Koenigstein, Germany: Koeltz. (Regnum Vegetabile 125, passim. As of 2000 six volumes published.) [http://tl2.idcpublishers.info/]

Cyperaceae

Acorus [*Acoraceae*]

Acorus brasiliensius Schott ex Nees = ***Scleria ciliata*** Michx. var. ***ciliata***

Abildgaardia

Abildgaardia Vahl = ***Fimbristylis*** Vahl
Abildgaardia abbreviata Lye = [26 ZAM]
Abildgaardia abortiva (Steud.) Lye = ***Bulbostylis abortiva*** (Steud.) C.B.Clarke
Abildgaardia acutispicata Lye = [26 ZAM]
Abildgaardia afro-orientalis Lye = ***Bulbostylis afro-orientalis*** (Lye) R.W.Haines
Abildgaardia angustespicata Lye = [25 TAN]
Abildgaardia aphylla Kunth = ***Fimbristylis bahiensis*** Steud.
Abildgaardia argentobrunnea (C.B.Clarke) Lye = ***Bulbostylis argentobrunnea*** C.B.Clarke
Abildgaardia baeothryon A.St.-Hil. = ***Fimbristylis bahiensis*** Steud.
Abildgaardia barbata (Rottb.) P.Beauv. = ***Bulbostylis barbata*** (Rottb.) C.B.Clarke
Abildgaardia boeckeleriana (Schweinf.) Lye = ***Bulbostylis boeckeleriana*** (Schweinf.) Beetle
Abildgaardia boeckeleriana var. *transiens* (K.Schum.) Lye = ***Bulbostylis boeckeleriana*** var. ***transiens*** (K.Schum.) R.W.Haines
Abildgaardia brevifolia Steud. = ***Fimbristylis acicularis*** R.Br.
Abildgaardia buchananii (C.B.Clarke) Lye = ***Bulbostylis buchananii*** C.B.Clarke
Abildgaardia burchellii (Ficalho & Hiern) Lye = ***Bulbostylis burchellii*** (Ficalho & Hiern) C.B.Clarke
Abildgaardia capillaris (L.) Lye = ***Bulbostylis capillaris*** (L.) Kunth ex C.B.Clarke
Abildgaardia capitata Lye = [26 ZAM]
Abildgaardia cardiocarpoides (Cherm.) Lye = ***Bulbostylis cardiocarpoides*** Cherm.
Abildgaardia chinensis Gand. = ?
Abildgaardia cinnamometorum (Vahl) Thwaites = ***Fimbristylis cinnamometorum*** (Vahl) Kunth
Abildgaardia clarkeana (Hutch. ex Bodard) Lye = ***Bulbostylis clarkeana*** Hutch. ex Bodard
Abildgaardia coleotricha (Hochst. ex A.Rich.) Lye = ***Bulbostylis coleotricha*** (Hochst. ex A.Rich.) C.B.Clarke
Abildgaardia coleotricha var. *miegei* (Bodard) Lye = ***Bulbostylis coleotricha*** var. ***miegei*** (Bodard) R.W.Haines
Abildgaardia collina (Ridl.) Lye = ***Bulbostylis scabricaulis*** Cherm.
Abildgaardia compressa J.Presl & C.Presl = ***Fimbristylis ovata*** (Burm.f.) J.Kern
Abildgaardia congolensis (De Wild.) Lye = ***Bulbostylis congolensis*** De Wild.
Abildgaardia contexta (Nees) Lye = ***Bulbostylis contexta*** (Nees) Bodard
Abildgaardia cruciformis Lye = ***Bulbostylis cruciformis*** (Lye) R.W.Haines
Abildgaardia cyperoides (J.Presl & C.Presl) Nees & Meyen = ***Fimbristylis fusca*** (Nees) Benth. ex C.B.Clarke
Abildgaardia densa (Wall.) Lye = ***Bulbostylis densa*** (Wall.) Hand.-Mazz.
Abildgaardia densa subsp. *afromontana* Lye = ***Bulbostylis densa*** subsp. ***afromontana*** (Lye) R.W.Haines
Abildgaardia densicaespitosa Lye = ***Bulbostylis densicaespitosa*** (Lye) R.W.Haines
Abildgaardia densiflora Lye = [26 ZAM]
Abildgaardia disticha Lye = ***Bulbostylis distichoides*** Lye
Abildgaardia elegantissima Lye = ***Bulbostylis elegantissima*** (Lye) R.W.Haines
Abildgaardia eragrostis Nees = ***Fimbristylis eragrostis*** (Nees) Hance
Abildgaardia erratica (Hook.f.) Lye = ***Bulbostylis erratica*** (Hook.f.) C.B.Clarke
Abildgaardia erratica subsp. *schoenoides* (Kunth) Lye = ***Bulbostylis schoenoides*** (Kunth) C.B.Clarke
Abildgaardia festucoides (Poir.) Lye = ***Bulbostylis festucoides*** (Poir.) C.B.Clarke
Abildgaardia filamentosa (Vahl) Lye = ***Bulbostylis filamentosa*** (Vahl) C.B.Clarke
Abildgaardia filamentosa var. *holubii* (C.B.Clarke) Lye = ***Bulbostylis scabricaulis*** Cherm.
Abildgaardia filamentosa var. *metralis* (Cherm.) Lye = ***Bulbostylis filamentosa*** (Vahl) C.B.Clarke
Abildgaardia fimbristyloides F.Muell. = ***Fimbristylis fimbristyloides*** (F.Muell.) Druce
Abildgaardia fulvescens Thwaites = ***Fimbristylis fulvescens*** (Thwaites) Thwaites
Abildgaardia fusca Nees = ***Fimbristylis fusca*** (Nees) Benth. ex C.B.Clarke
Abildgaardia fusca var. *longifolia* Boeck. = ***Fimbristylis fulvescens*** (Thwaites) Thwaites
Abildgaardia glaberrima (Kük.) Lye = ***Bulbostylis glaberrima*** Kük.
Abildgaardia hensii (C.B.Clarke) Lye = ***Bulbostylis hensii*** (C.B.Clarke) R.W.Haines
Abildgaardia heterostachya (Cherm.) Lye = ***Bulbostylis heterostachya*** Cherm.
Abildgaardia hispidula (Vahl) Lye = ***Bulbostylis hispidula*** (Vahl) R.W.Haines
Abildgaardia hispidula subsp. *brachyphylla* (Cherm.) Lye = ***Bulbostylis hensii*** (C.B.Clarke) R.W.Haines
Abildgaardia hispidula var. *filiformis* (C.B.Clarke) Lye = ***Bulbostylis hispidula*** subsp. ***filiformis*** (C.B.Clarke) R.W.Haines
Abildgaardia hispidula subsp. *filiformis* (C.B.Clarke) Lye = ***Bulbostylis hispidula*** subsp. ***filiformis*** (C.B.Clarke) R.W.Haines
Abildgaardia hispidula subsp. *halophila* Lye = ***Bulbostylis hispidula*** subsp. ***halophila*** (Lye) R.W.Haines
Abildgaardia hispidula subsp. *intermedia* Lye = ***Bulbostylis hispidula*** subsp. ***intermedia*** (Lye) R.W.Haines
Abildgaardia hispidula subsp. *longispicata* (Lye) J.-P.Lebrun & Stork = ***Bulbostylis hispidula*** subsp. ***longispicata*** Lye

Abildgaardia hispidula subsp. *macroglumis* (Lye) J.-P.Lebrun & Stork = ***Bulbostylis hispidula*** subsp. ***macroglumis*** Lye
Abildgaardia hispidula var. *oligostachys* (Hochst. ex A.Rich.) Lye = ***Bulbostylis oligostachys*** (Hochst. ex A.Rich.) C.B.Clarke
Abildgaardia hispidula var. *pyriformis* Lye = ***Bulbostylis hispidula*** subsp. ***pyriformis*** (Lye) R.W.Haines
Abildgaardia hispidula subsp. *pyriformis* (Lye) Lye = ***Bulbostylis hispidula*** subsp. ***pyriformis*** (Lye) R.W.Haines
Abildgaardia hispidula subsp. *senegalensis* (Cherm.) J.-P.Lebrun & Stork = ***Bulbostylis hispidula*** subsp. ***senegalensis*** (Cherm.) Van den Berghen
Abildgaardia humilis (Kunth) Lye = ***Bulbostylis humilis*** (Kunth) C.B.Clarke
Abildgaardia hygrophila (Gordon-Gray) Lye = ***Fimbristylis hygrophila*** Gordon-Gray
Abildgaardia igneotonsa (Raymond) Kornas = ***Bulbostylis igneotonsa*** Raymond
Abildgaardia indica (Rich. ex Pers.) Nees = ***Fimbristylis ovata*** (Burm.f.) J.Kern
Abildgaardia javana Nees = ***Fimbristylis polytrichoides*** (Retz.) R.Br.
Abildgaardia javanensis Gand. = ***Fimbristylis ovata*** (Burm.f.) J.Kern
Abildgaardia javanica Steud. = ***Fimbristylis polytrichoides*** (Retz.) R.Br.
Abildgaardia johnstonii (C.B.Clarke) Lye = ***Bulbostylis johnstonii*** C.B.Clarke
Abildgaardia lacunosa Lye = [26 ZAM]
Abildgaardia lanceolata Schumach. = ***Fimbristylis tristachya*** R.Br. var. ***tristachya***
Abildgaardia lanifera (Boeck.) Lye = ***Bulbostylis lanifera*** (Boeck.) Kük.
Abildgaardia laxispicata Lye = [26 ZAM]
Abildgaardia leiolepis (Kük.) Lye = ***Bulbostylis leiolepis*** (Kük.) R.W.Haines
Abildgaardia longispicata Lye = [24 ETH]
Abildgaardia macra (Ridl.) Lye = ***Bulbostylis macra*** (Ridl.) C.B.Clarke
Abildgaardia macroanthela Lye = [26 ZAM]
Abildgaardia macrostachya Lye = ***Bulbostylis macrostachya*** (Lye) R.W.Haines
Abildgaardia malawiensis Lye = [26 MLW]
Abildgaardia megastachys (Ridl.) Lye = ***Bulbostylis megastachys*** (Ridl.) C.B.Clarke
Abildgaardia metalliphila Lye = [23 ZAI]
Abildgaardia mexicana (Palla) Kral = ***Fimbristylis mexicana*** Palla
Abildgaardia microcarpa Lye = ***Bulbostylis microcarpa*** (Lye) R.W.Haines
Abildgaardia microcephala Lye = [26 ZAM]
Abildgaardia microelegans Lye = ***Bulbostylis microelegans*** (Lye) R.W.Haines
Abildgaardia microrotundata Lye = [25 TAN]
Abildgaardia miegei (Bodard) Lye = ***Bulbostylis coleotricha*** var. ***miegei*** (Bodard) R.W.Haines
Abildgaardia monostachya (L.) Vahl = ***Fimbristylis ovata*** (Burm.f.) J.Kern
Abildgaardia natalensis Gand. = [27 NAT]
Abildgaardia nervosa J.Presl & C.Presl = ***Fimbristylis schoenoides*** (Retz.) Vahl
Abildgaardia nudiuscula Lye = [26 ZAM]
Abildgaardia oligostachys (Hochst. ex A.Rich.) Lye = ***Bulbostylis oligostachys*** (Hochst. ex A.Rich.) C.B.Clarke
Abildgaardia oritrephes (Ridl.) Lye = ***Bulbostylis oritrephes*** (Ridl.) C.B.Clarke
Abildgaardia ovata (Burm.f.) Kral = ***Fimbristylis ovata*** (Burm.f.) J.Kern
Abildgaardia pallescens Lye = ***Bulbostylis pallescens*** (Lye) R.W.Haines
Abildgaardia papillosa Kral & M.T.Strong = [84 BZE]
Abildgaardia paradoxa (Spreng.) Lye = ***Bulbostylis paradoxa*** (Spreng.) Lindm.
Abildgaardia parva (Ridl.) Lye = ***Bulbostylis pusilla*** (Hochst. ex A.Rich.) C.B.Clarke
Abildgaardia parvinux (C.B.Clarke) Lye = ***Bulbostylis parvinux*** C.B.Clarke
Abildgaardia pauciflora (Brongn.) Kunth = ***Fimbristylis fusca*** (Nees) Benth. ex C.B.Clarke
Abildgaardia pilosa (Willd.) Nees = ***Bulbostylis pilosa*** (Willd.) Cherm.
Abildgaardia pluricephala Lye = [26 ZAM]
Abildgaardia polycephala Brongn. = ***Androtrichum trigynum*** (Spreng.) H.Pfeiff.
Abildgaardia polygama Spreng. = [84]
Abildgaardia polystachya Spreng. = (Poaceae)
Abildgaardia pubescens C.Presl = ***Bulbostylis hispidula*** (Vahl) R.W.Haines subsp. ***hispidula***
Abildgaardia pusilla (Hochst. ex A.Rich.) Lye = ***Bulbostylis pusilla*** (Hochst. ex A.Rich.) C.B.Clarke
Abildgaardia pusilla subsp. *congolensis* (De Wild.) Lye = ***Bulbostylis congolensis*** De Wild.
Abildgaardia pusilla subsp. *yalingensis* (Cherm.) Lye = ***Bulbostylis pusilla*** (Hochst. ex A.Rich.) C.B.Clarke
Abildgaardia renschii (Boeck.) Lye = ***Bulbostylis renschii*** (Boeck.) C.B.Clarke
Abildgaardia rhizomatosa Lye = ***Bulbostylis rhizomatosa*** (Lye) R.W.Haines
Abildgaardia rottboelliana Nees ex Hook. & Arn. = ***Fimbristylis ovata*** (Burm.f.) J.Kern
Abildgaardia rotundata (Kük.) Lye = ***Bulbostylis rotundata*** (Kük.) R.W.Haines
Abildgaardia schimperana (Hochst. ex A.Rich.) Lye = ***Bulbostylis schimperiana*** (Hochst. ex A.Rich.) C.B.Clarke
Abildgaardia schlechteri (C.B.Clarke) Lye = ***Bulbostylis schlechteri*** C.B.Clarke
Abildgaardia schoenoides R.Br. = ***Fimbristylis squarrulosa*** F.Muell.
Abildgaardia scirpoides Nees = ***Fimbristylis bahiensis*** Steud.
Abildgaardia scleropus (C.B.Clarke) Lye = ***Bulbostylis scleropus*** C.B.Clarke
Abildgaardia scrobiculata Lye = [26 ZAM]
Abildgaardia setacea Griseb. = ***Bulbostylis setacea*** (Griseb.) Svenson
Abildgaardia setifolia (A.Rich.) Lye = ***Bulbostylis atrosanguinea*** (Boeck.) C.B.Clarke
Abildgaardia sphaerocarpa (Boeck.) Lye = ***Bulbostylis sphaerocarpa*** (Boeck.) C.B.Clarke
Abildgaardia squarrosa Lye = [25 KEN]
Abildgaardia striatella (C.B.Clarke) Lye = ***Bulbostylis humilis*** (Kunth) C.B.Clarke
Abildgaardia subumbellata Lye = ***Bulbostylis subumbellata*** (Lye) R.W.Haines
Abildgaardia tanzaniae Lye = ***Bulbostylis tanzaniae*** (Lye) R.W.Haines
Abildgaardia taylorii (K.Schum.) Lye = ***Bulbostylis taylorii*** (K.Schum.) C.B.Clarke
Abildgaardia trabeculata (C.B.Clarke) Lye = ***Bulbostylis trabeculata*** C.B.Clarke
Abildgaardia trabeculata var. *microglumis* Lye = ***Bulbostylis trabeculata*** var. ***microglumis*** (Lye) R.W.Haines

Abildgaardia trichobasis (Baker) Lye = ***Bulbostylis trichobasis*** (Baker) C.B.Clarke
Abildgaardia triflora (L.) Lye = ***Fimbristylis triflora*** (L.) K.Schum. ex Engl.
Abildgaardia tristachya Vahl = ***Fimbristylis triflora*** (L.) K.Schum. ex Engl.
Abildgaardia ugandensis Lye = ***Bulbostylis ugandensis*** (Lye) R.W.Haines
Abildgaardia vaginata R.Br. = ***Fimbristylis vaginata*** (R.Br.) Domin
Abildgaardia vanderystii (Cherm.) Lye = ***Bulbostylis vanderystii*** Cherm.
Abildgaardia variegata (Gordon-Gray) Lye = ***Fimbristylis variegata*** Gordon-Gray
Abildgaardia wallichiana (Schult.) Lye = ***Bulbostylis barbata*** (Rottb.) C.B.Clarke subsp. ***barbata***
Abildgaardia willdenowii Lye = ***Bulbostylis barbata*** (Rottb.) C.B.Clarke subsp. ***barbata***
Abildgaardia wombaliensis (De Wild.) Lye = ***Bulbostylis wombaliensis*** (De Wild.) R.W.Haines
Abildgaardia yalingensis (Cherm.) Lye = ***Bulbostylis pusilla*** (Hochst. ex A.Rich.) C.B.Clarke

Unplaced Names:
Abildgaardia abbreviata Lye, Lidia 1: 35 (1985). = [26 ZAM]
Abildgaardia acutispicata Lye, Lidia 1: 33 (1985). = [26 ZAM]
Abildgaardia angustespicata Lye, Lidia 1: 33 (1985). = [25 TAN]
Abildgaardia capitata Lye, Lidia 1: 37 (1985). = [26 ZAM]
Abildgaardia chinensis Gand., Bull. Soc. Bot. France 66: 296 (1919 publ. 1920). = ?
Abildgaardia densiflora Lye, Lidia 1: 31 (1985). = [26 ZAM]
Abildgaardia lacunosa Lye, Lidia 1: 33 (1985). = [26 ZAM]
Abildgaardia laxispicata Lye, Lidia 1: 37 (1985). = [26 ZAM]
Abildgaardia longispicata Lye, Lidia 1: 35 (1985). = [24 ETH]
Abildgaardia macroanthela Lye, Lidia 1: 35 (1985). = [26 ZAM]
Abildgaardia malawiensis Lye, Lidia 1: 35 (1985). = [26 MLW]
Abildgaardia metalliphila Lye, Lidia 1: 37 (1985). = [23 ZAI]
Abildgaardia microcephala Lye, Lidia 1: 33 (1985). = [26 ZAM]
Abildgaardia microrotundata Lye, Lidia 1: 33 (1985). = [25 TAN]
Abildgaardia natalensis Gand., Bull. Soc. Bot. France 66: 296 (1919 publ. 1920). = [27 NAT]
Abildgaardia nudiuscula Lye, Lidia 1: 37 (1985). = [26 ZAM]
Abildgaardia papillosa Kral & M.T.Strong, Sida 18: 842 (1999). = [84 BZE]
Abildgaardia pluricephala Lye, Lidia 1: 35 (1985). = [26 ZAM]
Abildgaardia polygama Spreng., Novi Provent.: 45 (1818). = [84]
Abildgaardia scrobiculata Lye, Lidia 1: 37 (1985). = [26 ZAM]
Abildgaardia squarrosa Lye, Lidia 1: 31 (1985). = [25 KEN]

Abildgardia

Abildgardia Rchb. = ***Fimbristylis*** Vahl

Acorellus

Acorellus Palla ex Kneuck. = ***Cyperus*** L.
Acorellus distachyus (All.) Palla = ***Cyperus laevigatus*** subsp. ***distachyos*** (All.) Maire & Weiller
Acorellus laevigatus (L.) Palla = ***Cyperus laevigatus*** L.
Acorellus laevigatus subsp. *distachyos* (All.) Holub = ***Cyperus laevigatus*** subsp. ***distachyos*** (All.) Maire & Weiller
Acorellus pallae Kneuck. = ***Cyperus laevigatus*** L. subsp. ***laevigatus***
Acorellus pannonicus (Jacq.) Palla = ***Cyperus pannonicus*** Jacq.

Acriulus

Acriulus Ridl. = ***Scleria*** P.J.Bergius
Acriulus griegiifolius Ridl. = ***Scleria griegiifolia*** (Ridl.) C.B.Clarke
Acriulus madagascariensis Ridl. = ***Scleria griegiifolia*** (Ridl.) C.B.Clarke
Acriulus titan C.B.Clarke = [2]

Unplaced Names:
Acriulus titan C.B.Clarke, Kew Bull., Addit. Ser. 8: 62 (1908). = [2]

Acrocarpus

Acrocarpus Nees = ***Lagenocarpus*** Nees
Acrocarpus campestris Nees = ***Lagenocarpus minarum*** (Nees) Kuntze
Acrocarpus densifolius Nees = ***Lagenocarpus minarum*** (Nees) Kuntze
Acrocarpus humilis Nees = ***Lagenocarpus humilis*** (Nees) Kuntze
Acrocarpus junciformis (Kunth) Nees = ***Lagenocarpus junciformis*** (Kunth) Kuntze
Acrocarpus minarum Nees = ***Lagenocarpus minarum*** (Nees) Kuntze
Acrocarpus polyphyllus Nees = ***Lagenocarpus polyphyllus*** (Nees) Kuntze
Acrocarpus stellatus Nees ex Boeck. = ***Lagenocarpus stellatus*** (Boeck.) Kuntze

Acrolepis

Acrolepis Schrad. = ***Ficinia*** Schrad.
Acrolepis ferruginea Nees ex Boeck. = ***Ficinia trichodes*** (Schrad.) Benth. & Hook.f.
Acrolepis ramosissima (Kunth) Boeck. = ***Ficinia ramosissima*** Kunth
Acrolepis trichodes Schrad. = ***Ficinia trichodes*** (Schrad.) Benth. & Hook.f.

Actaeogeton

Actaeogeton Steud. = ***Scirpus*** Tourn. ex L.

Actinoschoenus

Actinoschoenus Benth., Hooker's Icon. Pl. 14: 33 (1881).
Gabon, Zambia, W. Indian Ocean, Sri Lanka to S. China and N. Australia. 23 EQG GAB 26 ZAM 29 MAU MDG 36 CHH CHS 40 IND SRL 41 CBD THA VIE 42 BOR MLY PHI SUM 50 NTA.
2 Species

Actinoschoenus repens J.Raynal, Adansonia, n.s., 7: 92 (1967).
Zambia. 26 ZAM. Hemicr.

Actinoschoenus thouarsii (Kunth) Benth., Hooker's Icon. Pl. 14: 33 (1881).
WC. Trop. Africa, W. Indian Ocean, Sri Lanka, S. China, Indo-China to W. & C. Malesia, N. Australia. 23 EQG GAB 29 MAU MDG 36 CHH CHS 40 IND SRL 41 CBD THA VIE 42 BOR MLY PHI SUM 50 NTA. Hemicr.
Schoenus aphyllus Vahl, Enum. Pl. 2: 209 (1805). *Arthrostylis aphylla* (Vahl) Boeck., Linnaea 37: 524 (1872), nom. illeg.
**Arthrostylis thouarsii* Kunth, Enum. Pl. 2: 284 (1837). *Fimbristylis thouarsii* (Kunth) Merr., Enum. Philipp. Fl. Pl. 1: 127 (1923).
Rhynchospora echinata Willd. ex Kunth, Enum. Pl. 2: 284 (1837).
Rhynchospora pauciflora Willd. ex Kunth, Enum. Pl. 2: 284 (1837).
Arthrostylis chinensis Benth., Fl. Hongk.: 397 (1861). *Actinoschoenus chinensis* (Benth.) Benth., Hooker's Icon. Pl. 14: 33 (1881). *Fimbristylis chinensis* (Benth.) Tang & F.T.Wang, in Fl. Reipubl. Popul. Sin. 11: 106 (1961).
Arthrostylis filiformis Thwaites, Enum. Pl. Zeyl.: 352 (1864). *Actinoschoenus filiformis* (Thwaites) Benth., Hooker's Icon. Pl. 14: 33 (1881). *Fimbristylis actinoschoenus* C.B.Clarke in J.D.Hooker, Fl. Brit. India 6: 650 (1893). *Fimbristylis filiformis* (Thwaites) Druce, Bot. Soc. Exch. Club Brit. Isles 1916: 623 (1917), nom. illeg.
Fimbristylis laxiflora C.B.Clarke in T.A.Durand & H.Schinz, Consp. Fl. Afric. 5: 607 (1894), nom. inval.
Actinoschoenus humbertii Cherm., Mém. Acad. Malagache 10: 39 (1931). *Arthrostylis humbertii* (Cherm.) Kük., Feddes Repert. Spec. Nov. Regni Veg. 53: 200 (1944).

Synonyms:
Actinoschoenus chinensis (Benth.) Benth. = ***Actinoschoenus thouarsii*** (Kunth) Benth.
Actinoschoenus erinaceus (Ridl.) Raymond = ***Sphaerocyperus erinaceus*** (Ridl.) Lye
Actinoschoenus filiformis (Thwaites) Benth. = ***Actinoschoenus thouarsii*** (Kunth) Benth.
Actinoschoenus humbertii Cherm. = ***Actinoschoenus thouarsii*** (Kunth) Benth.

Actinoscirpus

Actinoscirpus (Ohwi) R.W.Haines & Lye, Bot. Not. 124: 481 (1971).
Trop. & Subtrop. Asia to N. Australia. 36 CHS 38 OGA TAI 40 ASS EHM IND NEP PAK SRL WHM 41 THA VIE 42 BOR JAW MLY PHI SUL SUM 43 BIS NWG SOL 50 NTA QLD WAU.
1 Species
Hymenochaeta P.Beauv. ex Lestib., Essai Cypér.: 43 (1819), nom. rejic.

Actinoscirpus grossus (L.f.) Goetgh. & D.A.Simpson, Kew Bull. 46: 171 (1991).
Trop. & Subtrop. Asia to N. Australia. 36 CHS 38 OGA TAI 40 ASS EHM IND NEP PAK SRL WHM 41 THA VIE 42 BOR JAW MLY PHI SUL SUM 43 BIS NWG SOL 50 NTA QLD WAU. Hemicr. or tuber geophyte.
**Scirpus grossus* L.f., Suppl. Pl.: 104 (1782). *Hymenochaeta grossa* (L.f.) Nees in R.Wight, Contr. Bot. India: 119 (1834). *Schoenoplectus grossus* (L.f.) Palla, Allg. Bot. Z. Syst. 17(Beih.): 3 (1911).

var. ***grossus***
Trop. & Subtrop. Asia to N. Australia. 36 CHS 38 OGA TAI 40 ASS EHM IND NEP PAK SRL WHM 41 THA VIE 42 BOR JAW MLY PHI SUL SUM 43 BIS NWG SOL 50 NTA QLD WAU. Hemicr. or tuber geophyte.
Scirpus maximus Roxb., Fl. Ind. 1: 231 (1820). *Isolepis maxima* A.Dietr., Sp. Pl. 2: 130 (1832). *Hymenochaeta maxima* (Roxb.) Nees, Linnaea 9: 293 (1834).
Hymenochaeta haematodes Nees & Meyen, Linnaea 9: 293 (1834), nom. inval.
Scirpus aemulans Steud., Syn. Pl. Glumac. 2: 87 (1855).
Scirpus canaliculatotriqueter Steud., Syn. Pl. Glumac. 2: 318 (1855).
Scirpus griffithii Boeck., Flora 41: 595 (1858).
Scirpus scaberrimus Boeck., Flora 41: 595 (1858).

var. ***kysoor*** (Roxb.) Noltie, Edinburgh J. Bot. 51: 173 (1994).
Indian subcontinent. 40 IND NEP. Hemicr. or rhizome geophyte.
**Scirpus kysoor* Roxb., Fl. Ind. 1: 230 (1820). *Rhynchospora kysoor* (Roxb.) A.Dietr., Sp. Pl. 2: 74 (1832). *Scirpus grossus* f. *kysoor* (Roxb.) Beetle, Amer. J. Bot. 33: 661 (1946).

Adamantogeton

Adamantogeton Schrad. ex Nees = ***Lagenocarpus*** Nees

Adupla

Adupla Bosc = ***Cyperus*** L.

Afrotrilepis

Afrotrilepis (Gilg) J.Raynal, Adansonia 3: 258 (1963).
W. & WC. Trop. Africa. 22 BEN BKN GHA GUI IVO LBR MLI NGA SEN SIE TOG 23 CMN EQG GAB.
2 Species
Catagyna Hutch. & Dalziel, Fl. W. Trop. Afr. 2: 490 (1936), nom. illeg.

Afrotrilepis jaegeri J.Raynal, Adansonia, n.s., 3: 259 (1963).
Sierra Leone. 22 SIE. Hemicr. or cham.

Afrotrilepis pilosa (Boeck.) J.Raynal, Adansonia, n.s., 3: 258 (1963).
W. & WC. Trop. Africa. 22 BEN BKN GHA GUI IVO LBR MLI NGA SEN SIE TOG 23 CMN EQG GAB. Hemicr. or cham.
**Trilepis pilosa* Boeck., Linnaea 29: 10 (1875). *Eriospora pilosa* (Boeck.) Benth., Hooker's Icon. Pl. 14: 30 (1880). *Catagyna pilosa* (Boeck.) Hutch. in J.Hutchinson & J.M.Dalziel, Fl. W. Trop. Afr. 2: 490 (1936).

Agistron

Agistron Raf. = ***Uncinia*** Pers.
Agistron erinacea (Cav.) Raf. = ***Uncinia erinacea*** (Cav.) Pers.
Agistron phleoides (Cav.) Raf. = ***Uncinia phleoides*** (Cav.) Pers.

Agylla

Agylla Phil. = ***Machaerina*** Vahl
Agylla ensifolia Phil. = ***Machaerina scirpoidea*** (Steud.) T.Koyama

Albikia

Albikia J.Presl & C.Presl = ***Hypolytrum*** Rich. ex Pers.
Albikia schoenoides C.Presl = ***Hypolytrum nemorum*** (Vahl) Spreng. subsp. ***nemorum***
Albikia scirpoides C.Presl = ***Hypolytrum nemorum*** (Vahl) Spreng. subsp. ***nemorum***

Aliniella

Aliniella J.Raynal = ***Alinula*** J.Raynal
Aliniella lipocarphioides (Kük.) J.Raynal = ***Alinula lipocarphioides*** (Kük.) J.Raynal

Alinula

Alinula J.Raynal, Adansonia, n.s., 17: 43 (1977).
Ethiopia to S. Africa, Madagascar. 23 ZAI 24 ETH 25 KEN TAN UGA 26 MLW MOZ ZAM 27 NAM NAT TVL 29 MDG.
4 Species
Aliniella J.Raynal, Adansonia, n.s., 13: 157 (1973), nom. illeg. *Raynalia* Soják, Cas. Nár. Mus., Odd. Prír. 148: 193 (1979 publ. 1980).
Marisculus Goetgh., Bull. Jard. Bot. Natl. Belg. 47: 444 (1977).

Alinula lipocarphioides (Kük.) J.Raynal, Adansonia, n.s., 17: 43 (1977).
Ethiopia to Zambia. 23 ZAI 24 ETH 25 KEN TAN UGA 26 ZAM. Ther.
**Ficinia lipocarphioides* Kük., Repert. Spec. Nov. Regni Veg. Beih. 40(1): 125 (1936). *Aliniella lipocarphioides* (Kük.) J.Raynal, Adansonia, n.s., 13: 157 (1973). *Raynalia lipocarphioides* (Kük.) Soják, Cas. Nár. Mus., Odd. Prír. 148: 193 (1979 publ. 1980). *Cyperus lipocarphioides* (Kük.) Lye, Nordic J. Bot. 3: 230 (1983).

Alinula malawica (J.Raynal) Goetgh. & Vorster, Bull. Jard. Bot. Natl. Belg. 58: 462 (1988).
Zambia to Malawi. 26 MLW ZAM. Ther.
**Mariscus malawicus* J.Raynal, Adansonia, n.s., 13: 159 (1973). *Cyperus malawicus* (J.Raynal) Lye, Nordic J. Bot. 3: 231 (1983).

Alinula paradoxa (Cherm.) Goetgh. & Vorster, Bull. Jard. Bot. Natl. Belg. 58: 461 (1988).
Kenya to S. Africa, W. Madagascar. 25 KEN TAN 26 MOZ 27 NAM NAT TVL 29 MDG. Ther.
**Lipocarpha paradoxa* Cherm., Bull. Soc. Bot. France 68: 425 (1921 publ. 1922). *Mariscus paradoxus* (Cherm.) Cherm., Bull. Soc. Bot. France 72: 169 (1925). *Cyperus subparadoxus* Kük. in H.G.A.Engler (ed.), Pflanzenr., IV, 20(101): 525 (1936). *Cyperus fimbristyloides* T.Koyama, Bot. Mag. (Tokyo) 73: 438 (1960).

Alinula peteri (Kük.) Goetgh. & Vorster, Bull. Jard. Bot. Natl. Belg. 58: 464 (1988).
Ethiopia to Zambia. 24 ETH 25 KEN TAN 26 MLW ZAM. Ther.
**Ascolepis peteri* Kük., Repert. Spec. Nov. Regni Veg. Beih. 40(1): 384 Anh.: 124 (1936). *Marisculus peteri* (Kük.) Goetgh., Bull. Jard. Bot. Natl. Belg. 47: 444 (1977). *Cyperus microaureus* Lye, Lidia 3: 132 (1994).

Ammorrhiza

Ammorrhiza Ehrh. = ***Carex*** L.

Amphiscirpus

Amphiscirpus Oteng-Yeb., Notes Roy. Bot. Gard. Edinburgh 33: 308 (1974).
W. Canada to W. & NC. U.S.A., Chile to W. & S. Argentina. 71 ABT BRC SAS 73 COL MNT ORE WAS WYO 74 NDA NEB 76 CAL NEV UTA 79 MXC 85 AGS AGW CLS.
1 Species

Amphiscirpus nevadensis (S.Watson) Oteng-Yeb., Notes Roy. Bot. Gard. Edinburgh 33: 308 (1974).
W. Canada to W. & NC. U.S.A., C. Mexico, Chile to W. & S. Argentina. 71 ABT BRC SAS 73 COL MNT ORE WAS WYO 74 NDA NEB 76 CAL NEV UTA 79 MXC 85 AGS AGW CLS. Hemicr.
**Scirpus nevadensis* S.Watson, Botany [Fortieth Parallel]: 306 (1871). *Schoenoplectus nevadensis* (S.Watson) Soják, Cas. Nár. Mus., Odd. Prír. 140: 127 (1972).
Scirpus remireoides Griseb., Abh. Königl. Ges. Wiss. Göttingen 19: 266 (1874). *Scirpus nevadensis* var. *remireoides* (Griseb.) Beetle, Amer. J. Bot. 33: 663 (1946).
Isolepis oreophila Phil., Anales Mus. Nac. Santiago de Chile 1: 79 (1891).
Scirpus chubutensis C.B.Clarke, Kew Bull., Addit. Ser. 9: 89 (1908).

Androcoma

Androcoma Nees = ***Androtrichum*** (Brongn.) Brongn.
Androcoma speciosa Nees = ***Androtrichum giganteum*** (Kunth) H.Pfeiff.

Androtrichum

Androtrichum (Brongn.) Brongn. in L.I.Duperrey, Voy. Monde, Phan.: 177 (1834).
Brazil to NE. Argentina. 84 BZS 85 AGE PAR URU.
2 Species
Comostemum Nees, Linnaea 9: 283 (1834).
Androcoma Nees, J. Bot. (Hooker) 2: 396 (1840).
Megarrhena Schrad. ex Nees in C.F.P.von Martius & auct. sucÏ. (eds.), Fl. Bras. 2(1): 5 (1842), pro syn.

Androtrichum giganteum (Kunth) H.Pfeiff., Revista Sudamer. Bot. 6: 185 (1940).
Brazil to NE. Argentina. 84 BZS 85 AGE PAR URU. Hemicr. or rhizome geophyte.
**Scirpus giganteus* Kunth, Enum. Pl. 2: 172 (1837).
Androcoma speciosa Nees, J. Bot. (Hooker) 2: 397 (1840). *Androtrichum speciosum* (Nees) Herter, Revista Sudamer. Bot. 9: 163 (1953).

Androtrichum trigynum (Spreng.) H.Pfeiff., Repert. Spec. Nov. Regni Veg. 42: 10 (1937).
Brazil to NE. Argentina. 84 BZS 85 AGE URU. Hemicr. or rhizome geophyte.
**Cyperus trigynus* Spreng., Syst. Veg. 1: 216 (1824).
Eriophorum montevidense Link, Hort. Berol. 1: 331 (1827). *Trichophorum montevidense* (Link) G.Don in J.C.Loudon, Hort. Brit.: 468 (1830). *Comostemum montevidense* (Link) Nees, Linnaea 9: 283 (1834). *Androtrichum montevidense* (Link) Schrad., Index Seminum (GOET): 1 (1835). *Scirpus montevidensis* (Link) C.B.Clarke, Kew Bull., Addit. Ser. 8: 89 (1908), nom. illeg.
Cyperus globuliferus J.Presl & C.Presl, Reliq. Haenk. 1: 165 (1828). *Comostemum globuliferum* (J.Presl & C.Presl) Nees in C.F.P.von Martius & auct. suc. (eds.), Fl. Bras. 2(1): 5 (1842).

Cyperus preslii A.Dietr., Sp. Pl. 2: 211 (1832).
Abildgaardia polycephala Brongn. in L.I.Duperrey, Voy. Monde, Phan.: 176 (1833). *Androtrichum polycephalum* (Brongn.) Kunth, Enum. Pl. 2: 250 (1837).

***Synonyms*:**
Androtrichum montevidense (Link) Schrad. = ***Androtrichum trigynum*** (Spreng.) H.Pfeiff.
Androtrichum polycephalum (Brongn.) Kunth = ***Androtrichum trigynum*** (Spreng.) H.Pfeiff.
Androtrichum speciosum (Nees) Herter = ***Androtrichum giganteum*** (Kunth) H.Pfeiff.

Anerma

Anerma Schrad. ex Nees = ***Scleria*** P.J.Bergius
Anerma hispidula (Hochst. ex A.Rich.) Schrad. ex Nees = ***Scleria hispidula*** Hochst. ex A.Rich.

Anithista

Anithista Raf. = ***Carex*** L.
Anithista clavata (Thunb.) Raf. = ***Carex clavata*** Thunb.
Anithista flava (L.) Raf. = ***Carex flava*** L.
Anithista lacustris (Willd.) Raf. = ***Carex lacustris*** Willd.
Anithista latifolia (Moench) Raf. = ***Carex plantaginea*** Lam.
Anithista littorea (Labill.) Raf. = ***Carex pumila*** Thunb.
Anithista patula Raf. = ***Carex pauciflora*** Lightf.
Anithista riparia Raf. = ***Carex lacustris*** Willd.
Anithista triticea (Wahlenb. ex Steud.) Raf. = ***Carex clavata*** Thunb.

Anogyna

Anogyna Nees = ***Lagenocarpus*** Nees
Anogyna tremula Nees = ***Lagenocarpus rigidus*** (Kunth) Nees subsp. ***rigidus***

Anosporum

Anosporum Nees = ***Cyperus*** L.
Anosporum ablepharum (Griseb.) Maury ex Micheli = ***Oxycaryum cubense*** (Poepp. & Kunth) Palla
Anosporum cephalotes (Vahl) Kurz = ***Cyperus cephalotes*** Vahl
Anosporum colymbetes (Kotschy & Peyr.) Boeck. = ***Cyperus colymbetes*** Kotschy & Peyr.
Anosporum cubense (Poepp. & Kunth) Boeck. = ***Oxycaryum cubense*** (Poepp. & Kunth) Palla
Anosporum cubense var. *gracile* Boeck. = ***Oxycaryum cubense*** (Poepp. & Kunth) Palla
Anosporum monocephalum (Roxb.) Nees = ***Cyperus cephalotes*** Vahl
Anosporum nudicaule (Poir.) Boeck. = ***Cyperus pectinatus*** Vahl
Anosporum pallidum Boeck. = ***Cyperus platystylis*** R.Br.
Anosporum paraguayense Maury = ***Oxycaryum cubense*** (Poepp. & Kunth) Palla
Anosporum pectinatum (Vahl) Lye = ***Cyperus pectinatus*** Vahl
Anosporum piliferum Maury = ***Oxycaryum cubense*** (Poepp. & Kunth) Palla
Anosporum schinzii Boeck. = ***Oxycaryum cubense*** (Poepp. & Kunth) Palla

Anthophyllum

Anthophyllum Steud. = ***Desmoschoenus*** Hook.f.
Anthophyllum urvillei Steud. = ***Desmoschoenus spiralis*** (A.Rich.) Hook.f.

Antrolepis

Antrolepis Welw. = ***Ascolepis*** Nees ex Steud.

Apartea

Apartea Pellegr. = ***Mapania*** Aubl.
Apartea le-testui Pellegr. = ***Mapania amplivaginata*** K.Schum.

Aplostemon

Aplostemon Raf. = ***Fimbristylis*** Vahl
Aplostemon atropurpureum (Retz.) Raf. = ***Eleocharis atropurpurea*** (Retz.) Kunth
Aplostemon bracteatum (Bigelow) Raf. = ***Trichophorum cespitosum*** (L.) Hartm. subsp. ***cespitosum***
Aplostemon compressum (L.) Raf. = ***Blysmus compressus*** (L.) Panz. ex Link
Aplostemon polytrichoides (Retz.) Raf. = ***Fimbristylis polytrichoides*** (Retz.) R.Br.
Aplostemon triquetrum Raf. = ***Fimbristylis argentea*** (Rottb.) Vahl

Archaeocarex

Archaeocarex Börner = ***Schoenoxiphium*** Nees
Archaeocarex dregeana (Kunth) Pissjauk. = ***Schoenoxiphium dregeanum*** Kunth
Archaeocarex ecklonii (Nees) Pissjauk. = ***Schoenoxiphium ecklonii*** Nees
Archaeocarex kunthiana (Kük.) Pissjauk. = ***Schoenoxiphium sparteum*** (Wahlenb.) C.B.Clarke
Archaeocarex rufus (Nees) Fedde & J.Schust. = ***Schoenoxiphium rufum*** Nees
Archaeocarex spartea (Wahlenb.) Pissjauk. = ***Schoenoxiphium sparteum*** (Wahlenb.) C.B.Clarke
Archaeocarex thunbergii (Nees) Pissjauk. = ***Schoenoxiphium thunbergii*** Nees

Arthrostylis

Arthrostylis R.Br., Prodr.: 229 (1810).
N. Australia. 50 NSW QLD WAU.
1 Species

Arthrostylis aphylla R.Br., Prodr.: 229 (1810).
Fimbristylis aphylla (R.Br.) F.Muell., Fragm. 9: 9 (1875).
N. Australia. 50 NSW QLD WAU. Hemicr. or rhizome geophyte.

***Synonyms*:**
Arthrostylis aphylla (Vahl) Boeck. = ***Actinoschoenus thouarsii*** (Kunth) Benth.
Arthrostylis chinensis Benth. = ***Actinoschoenus thouarsii*** (Kunth) Benth.
Arthrostylis filiformis Thwaites = ***Actinoschoenus thouarsii*** (Kunth) Benth.
Arthrostylis humbertii (Cherm.) Kük. = ***Actinoschoenus thouarsii*** (Kunth) Benth.
Arthrostylis kennyi F.M.Bailey = ***Schoenus kennyi*** (F.M.Bailey) S.T.Blake
Arthrostylis thouarsii Kunth = ***Actinoschoenus thouarsii*** (Kunth) Benth.

Ascolepis

Ascolepis Nees ex Steud., Syn. Pl. Glumac. 2: 105 (1855).
Trop. & Subtrop. 22 BEN BKN GHA GUI IVO MLI NGA SEN SIE TOG 23 BUR CAF CMN CON GAB ZAI 24 CHA ETH SUD 25 KEN TAN UGA

26 ANG MLW ZAM ZIM 27 CPP NAM NAT OFS SWZ TVL 29 MDG 41 LAO THA VIE 80 PAN 82 GUY SUR VEN 83 BOL 84 BZC BZN 85 AGE PAR URU.

22 Species

Platylepis Kunth, Enum. Pl. 2: 269 (1837), nom. illeg.

Pterachne Schrad. ex Nees in C.F.P.von Martius & auct. suc. (eds.), Fl. Bras. 2(1): 62 (1842), nom. inval.

Pterogyne Schrad. ex Nees in C.F.P.von Martius & auct. suc. (eds.), Fl. Bras. 2(1): nota (1842), nom. inval.

Antrolepis Welw., Apont.: 578 (1859), nom. inval.

Ascolepis ampullacea J.Raynal, Adansonia, n.s., 13: 159 (1973).
Zambia. 26 ZAM. Ther.

Ascolepis brasiliensis (Kunth) Benth. ex C.B.Clarke in T.A.Durand & H.Schinz, Consp. Fl. Afric. 5: 651 (1894).
Trop. Africa & America. 22 GUI IVO NGA SEN 23 CMN 29 MDG 80 PAN 82 GUY SUR VEN 83 BOL 84 BZC BZN 85 AGE PAR URU. Hemicr.

**Platylepis brasiliensis* Kunth, Enum. Pl. 2: 269 (1837).

Platylepis guianensis Nees in C.F.P.von Martius & auct. suc. (eds.), Fl. Bras. 2(1): 63 (1842).

Platylepis leucocephala Nees in C.F.P.von Martius & auct. suc. (eds.), Fl. Bras. 2(1): 63 (1842). *Ascolepis leucocephala* (Nees) L.T.Eiten, Simpos. Cerrado Univ. S. Paulo: 221 (1963).

Platylepis xanthocephala Nees in C.F.P.von Martius & auct. suc. (eds.), Fl. Bras. 2(1): 62 (1842).

Kyllinga decora Steud., Syn. Pl. Glumac. 2: 317 (1855).

Ascolepis capensis (Kunth) Ridl., Trans. Linn. Soc. London, Bot. 2: 164 (1884).
Trop. & S. Africa. 22 BEN BKN GHA GUI IVO MLI NGA SIE TOG 23 CMN GAB 24 ETH 25 KEN TAN UGA 26 ZAM 27 CPP NAT OFS SWZ TVL. Hemicr. or rhizome geophyte.

**Platylepis capensis* Kunth, Enum. Pl. 2: 269 (1837).

Platylepis dioica Steud., Syn. Pl. Glumac. 2: 131 (1855).

Ascolepis capensis var. *lacera* C.B.Clarke in T.A.Durand & H.Schinz, Consp. Fl. Afric. 5: 651 (1894), nom. inval.

Ascolepis densa Goetgh., Adansonia, n.s., 19: 285 (1980).
S. Tanzania to S. Trop. Africa. 23 ZAI 25 TAN 26 ANG ZAM ZIM. Hemicr. or tuber geophyte.

Ascolepis dipsacoides (Schumach.) J.Raynal, Adansonia, n.s., 8: 99 (1968).
W. & WC. Trop. Africa, Indo-China. 22 BEN GHA IVO NGA SEN TOG 23 CMN 41 LAO THA. Ther.

**Kyllinga dipsacoides* Schumach., Beskr. Guin. Pl.: 41 (1827).

subsp. ***dipsacoides***
W. & WC. Trop. Africa. 22 BEN GHA IVO NGA SEN TOG 23 CMN. Ther.

Ascolepis setigera Hutch. in J.Hutchinson & J.M.Dalziel, Fl. W. Trop. Afr. 2: 472 (1936), no latin descr.

subsp. ***siamensis*** (C.B.Clarke) J.Raynal, Adansonia, n.s., 8: 99 (1968).
Indo-China. 41 LAO THA. Ther.

**Scirpus squarrosus* var. *siamensis* C.B.Clarke, Beih. Bot. Centralbl. 27(2): 460 (1910). *Scirpus siamensis* (C.B.Clarke) J.Kern, Blumea 9: 219 (1958).

Ascolepis gracilis Turrill, Hooker's Icon. Pl. 31: t. 3020 (1915).

Ascolepis elata Welw., Trans. Linn. Soc. London 27: 79 (1869).
Nigeria to Sudan and Angola. 22 NGA 23 BUR 24 CHA SUD 26 ANG ZIM. Hemicr.

Ascolepis eriocauloides (Steud.) Nees ex Steud., Syn. Pl. Glumac. 2: 105 (1855).
Ethiopia. 24 ETH. Hemicr. or tuber geophyte.

**Kyllinga eriocauloides* Steud., Flora 25: 597 (1842).

Isolepis ascolepis A.Rich., Tent. Fl. Abyss. 2: 501 (1850).

Ascolepis erythrocephala S.S.Hooper, Kew Bull. 37: 605 (1983).
Tanzania to Zimbabwe. 25 TAN 26 ZIM. Ther.

Ascolepis fibrillosa Goetgh., Bull. Jard. Bot. Natl. Belg. 47: 439 (1977).
S. Zaïre to Angola. 23 ZAI 26 ANG ZAM. Hemicr. or rhizome geophyte.

Ascolepis hemisphaerica Peter ex Goethg., Adansonia, n.s., 19: 283 (1980).
Burundi, Tanzania. 23 BUR 25 TAN. Hemicr.

Ascolepis lineariglumis Lye, Nordic J. Bot. 2: 561 (1983).
Sudan to Zimbabwe. 24 SUD 25 KEN TAN UGA 26 ZAM ZIM. Ther. or hemicr.

Ascolepis protea var. *splendida* K.Schum. in O.Warburg (ed.), Kunene-Sambesi Exped.: 177 (1903).

Ascolepis lineariglumis var. *pulcherrima* Lye, Nordic J. Bot. 2: 562 (1983).

Ascolepis majestuosus P.A.Duvign. & G.Léonard, Bull. Soc. Roy. Bot. Belgique 90: 188 (1958).
Zaïre. 23 ZAI. Hemicr. or rhizome geophyte.

Ascolepis menonguensis Meneses, Garcia de Orta 4: 259 (1956).
Angola. 26 ANG. Hemicr. or rhizome geophyte.

Ascolepis metallorum P.A.Duvign. & G.Léonard, Bull. Soc. Roy. Bot. Belgique 90: 268 (1958).
S. Zaïre. 23 ZAI. Hemicr.

Ascolepis neglecta Goetgh., Bull. Jard. Bot. Natl. Belg. 47: 441 (1977).
Zaïre. 23 ZAI. Hemicr. or rhizome geophyte.

Ascolepis pinguis C.B.Clarke, Ann. Mus. Congo Belge, Bot., II, 1(2): 69 (1900).
Burundi to Zambia. 23 BUR CON ZAI 25 TAN 26 ZAM. Hemicr.

Ascolepis protea Welw., Trans. Linn. Soc. London 27: 75 (1868).
Trop. Africa. 22 BEN BKN GHA GUI IVO MLI NGA SEN SIE TOG 23 CAF CMN ZAI 24 CHA ETH 25 TAN 26 ANG MLW ZAM ZIM. Hemicr.

subsp. ***anthemiflora*** (Welw.) Lye, Nordic J. Bot. 2: 566 (1983).
Tanzania to S. Trop. Africa. 23 ZAI 25 TAN 26 ANG MLW ZAM. Hemicr.

**Ascolepis anthemiflora* Welw., Trans. Linn. Soc. London 27: 77 (1869).

subsp. ***atropurpurea*** Lye, Nordic J. Bot. 2: 563 (1983).
Zambia. 26 ZAM. Hemicr. or bulb geophyte.

subsp. ***bellidiflora*** (Welw.) Lye, Nordic J. Bot. 2: 566 (1983).
Trop. Africa. 22 NGA 24 ETH 25 TAN 26 ANG ZIM. Hemicr.

Ascolepis protea var. *bellidiflora* Welw., Trans. Linn. Soc. London 27: 76 (1869). *Ascolepis bellidiflora* (Welw.) Cherm., Arch. Bot. Mém. 4(7): 29 (1931).
Ascolepis protea var. *transiens* Kük., Repert. Spec. Nov. Regni Veg. Beih. 40(1): 123 (1936).
Ascolepis protea var. *tuberosa* Kük., Repert. Spec. Nov. Regni Veg. Beih. 40(1): 124 (1936).

subsp. ***chrysocephala*** Lye, Nordic J. Bot. 2: 564 (1983).
S. Tanzania. 25 TAN. Hemicr. or bulb geophyte.

subsp. ***protea***
Trop. Africa. 22 BKN GHA GUI IVO MLI NGA SEN SIE TOG 23 CAF CMN ZAI 24 CHA 25 TAN 26 ANG. Hemicr.
Ascolepis speciosa var. *ochracea* Meneses, Garcia de Orta 4(2): 260 (1957). *Ascolepis protea* var. *ochracea* (Meneses) Goetgh., Adansonia, n.s., 19: 277 (1980).
Ascolepis protea var. *floribunda* Goetgh., Adansonia, n.s., 19: 277 (1980).
Ascolepis protea var. *stellata* Goetgh., Adansonia, n.s., 19: 279 (1980).

subsp. ***rhizomatosa*** Lye, Nordic J. Bot. 2: 562 (1983).
S. Tanzania to Zambia. 25 TAN 26 ZAM. Hemicr. or rhizome geophyte.

var. ***santolinoides*** Welw., Trans. Linn. Soc. London 27: 76 (1869).
Tanzania to Angola. 25 TAN 26 ANG ZAM. Hemicr. or bulb geophyte.

Ascolepis pseudopeteri Goetgh., Adansonia, n.s., 19: 286 (1980).
Zambia. 26 ZAM. Ther.

Ascolepis pusilla Ridl., Trans. Linn. Soc. London, Bot. 2: 164 (1884).
Trop. & S. Africa, Madagascar, Indo-China. 22 GHA GUI MLI NGA SEN TOG 23 CAF CMN 24 CHA 25 TAN UGA 26 ZAM ZIM 27 NAM 29 MDG 41 THA VIE. Ther.

var. ***cylindrica*** S.S.Hooper, Kew Bull. 37: 608 (1983).
Tanzania (Iringa). 25 TAN. Ther.

var. ***echinata*** S.S.Hooper, Kew Bull. 37: 607 (1983).
Tanzania to Zambia. 25 TAN 26 ZAM. Ther.

var. ***miscocuspis*** Lye, Nordic J. Bot. 2: 564 (1983).
S. Tanzania to Zimbabwe. 25 TAN 26 ZAM ZIM. Ther.

var. ***pusilla***
Trop. & S. Africa, Madagascar, Indo-China. 22 GHA GUI MLI NGA SEN TOG 23 CAF CMN 24 CHA 25 TAN UGA 26 ZIM 27 NAM 29 MDG 41 THA VIE. Ther.
Lipocarpha multibracteata C.B.Clarke in D.Oliver & auct. suc. (eds.), Fl. Trop. Afr. 8: 472 (1902).

Ascolepis speciosa Welw., Trans. Linn. Soc. London 27: 78 (1869).
Angola. 26 ANG. Hemicr. or rhizome geophyte.
Ascolepis oliveri Vatke & Höpfner ex C.B.Clarke in T.A.Durand & H.Schinz, Consp. Fl. Afric. 5: 652 (1894).
Ascolepis vatkeana Boeck. ex C.B.Clarke in T.A.Durand & H.Schinz, Consp. Fl. Afric. 5: 652 (1894).

Ascolepis spinulosa Goetgh., Bull. Jard. Bot. Natl. Belg. 47: 438 (1977).
Zaïre. 23 ZAI. Hemicr. or rhizome geophyte.

Ascolepis trigona Goetgh., Adansonia, n.s., 19: 286 (1980).
S. Zaïre to Zambia. 23 ZAI. Hemicr. or rhizome geophyte.

Synonyms:
Ascolepis anthemiflora Welw. = ***Ascolepis protea*** subsp. ***anthemiflora*** (Welw.) Lye
Ascolepis bellidiflora (Welw.) Cherm. = ***Ascolepis protea*** subsp. ***bellidiflora*** (Welw.) Lye
Ascolepis capensis var. *lacera* C.B.Clarke = ***Ascolepis capensis*** (Kunth) Ridl.
Ascolepis gracilis Turrill = ***Ascolepis dipsacoides*** subsp. ***siamensis*** (C.B.Clarke) J.Raynal
Ascolepis kyllingioides Steud. = ***Lipocarpha microcephala*** (R.Br.) Kunth
Ascolepis leucocephala (Nees) L.T.Eiten = ***Ascolepis brasiliensis*** (Kunth) Benth. ex C.B.Clarke
Ascolepis peariglumis var. *pulcherrima* Lye = ***Ascolepis lineariglumis*** Lye
Ascolepis oliveri Vatke & Höpfner ex C.B.Clarke = ***Ascolepis speciosa*** Welw.
Ascolepis peteri Kük. = ***Alinula peteri*** (Kük.) Goetgh. & Vorster
Ascolepis protea var. *bellidiflora* Welw. = ***Ascolepis protea*** subsp. ***bellidiflora*** (Welw.) Lye
Ascolepis protea var. *floribunda* Goetgh. = ***Ascolepis protea*** Welw. subsp. ***protea***
Ascolepis protea var. *ochracea* (Meneses) Goetgh. = ***Ascolepis protea*** Welw. subsp. ***protea***
Ascolepis protea var. *splendida* K.Schum. = ***Ascolepis lineariglumis*** Lye
Ascolepis protea var. *stellata* Goetgh. = ***Ascolepis protea*** Welw. subsp. ***protea***
Ascolepis protea var. *transiens* Kük. = ***Ascolepis protea*** subsp. ***bellidiflora*** (Welw.) Lye
Ascolepis protea var. *tuberosa* Kük. = ***Ascolepis protea*** subsp. ***bellidiflora*** (Welw.) Lye
Ascolepis setigera Hutch. = ***Ascolepis dipsacoides*** (Schumach.) J.Raynal subsp. ***dipsacoides***
Ascolepis speciosa var. *ochracea* Meneses = ***Ascolepis protea*** Welw. subsp. ***protea***
Ascolepis tenuior Steud. = ***Lipocarpha squarrosa*** (L.) Goetgh.
Ascolepis vatkeana Boeck. ex C.B.Clarke = ***Ascolepis speciosa*** Welw.
Ascolepis venezuelensis Schnee = ***Lipocarpha maculata*** (Michx.) Torr.

Ascopholis

Ascopholis C.E.C.Fisch. = ***Cyperus*** L.
Ascopholis gamblei C.E.C.Fisch. = ***Cyperus mollipes*** (C.B.Clarke) K.Schum.

Asterochaete

Asterochaete Nees = ***Carpha*** Banks & Sol. ex R.Br.
Asterochaete angustifolia Nees = ***Carpha glomerata*** Nees
Asterochaete arundinacea (Sol. ex Vahl) Kunth = ***Costularia arundinacea*** (Sol. ex Vahl) Kük.
Asterochaete capitellata Nees = ***Carpha capitellata*** (Nees) Boeck.
Asterochaete elongata Kunth = ***Costularia elongata*** (Kunth) Kük.
Asterochaete glomerata (Nees) Nees = ***Carpha glomerata*** Nees
Asterochaete ludwigii Hochst. = ***Carpha capitellata*** (Nees) Boeck.
Asterochaete nitens Kunth = ***Carpha nitens*** (Kunth) Kük.
Asterochaete tenuis Kunth = ***Carpha glomerata*** Nees

Asteroschoenus

Asteroschoenus Nees = ***Rhynchospora*** Vahl
Asteroschoenus consanguineus (Kunth) Nees = ***Rhynchospora consanguinea*** (Kunth) Boeck.
Asteroschoenus consanguineus var. *angustifolius* Nees = ***Rhynchospora consanguinea*** (Kunth) Boeck.
Asteroschoenus rigidus (Kunth) Nees = ***Rhynchospora consanguinea*** (Kunth) Boeck.
Asteroschoenus speciosus (Kunth) Nees = ***Rhynchospora consanguinea*** (Kunth) Boeck.

Astroschoenus

Astroschoenus Lindl. = ***Rhynchospora*** Vahl

Atomostylis

Atomostylis Steud. = ***Cyperus*** L.
Atomostylis cyperiformis Steud. = ***Cyperus pectinatus*** Vahl
Atomostylis flavescens Steud. = ***Cyperus pectinatus*** Vahl

Aulacorhynchus

Aulacorhynchus Nees = ***Tetraria*** P.Beauv.
Aulacorhynchus crinifolius Nees = ***Tetraria crinifolia*** (Nees) C.B.Clarke

Baeochortus

Baeochortus Ehrh. = ***Carex*** L.

Baeothryon

Baeothryon Ehrh. ex A.Dietr. = ***Eleocharis*** R.Br.
Baeothryon alpinum (L.) T.V.Egorova = ***Trichophorum alpinum*** (L.) Pers.
Baeothryon atacamense (Phil.) Soják = ***Zameioscirpus atacamensis*** (Phil.) Dhooge & Goetgh.
Baeothryon cespitosum (L.) A.Dietr. = ***Trichophorum cespitosum*** (L.) Hartm.
Baeothryon cespitosum subsp. *delicatulum* (Fernald) Á.Löve & D.Löve = ***Trichophorum cespitosum*** (L.) Hartm. subsp. ***cespitosum***
Baeothryon clementis (M.E.Jones) Á.Löve & D.Löve = ***Oreobolopsis clementis*** (M.E.Jones) Dhooge & Goetgh.
Baeothryon clintonii (A.Gray) Á.Löve & D.Löve = ***Trichophorum clintonii*** (A.Gray) S.G.Sm.
Baeothryon confervoides (Poir.) A.Dietr. = ***Websteria confervoides*** (Poir.) S.S.Hooper
Baeothryon crinitum (Poir.) A.Dietr. = ***Ficinia crinita*** (Poir.) B.L Burtt
Baeothryon distigmaticum (Kük.) T.V.Egorova = ***Trichophorum pumilum*** (Vahl) Schinz & Thell. subsp. ***pumilum***
Baeothryon dolichocarpum (Zakirov) T.V.Egorova = ***Trichophorum dolichocarpum*** Zakirov
Baeothryon festucoides (Poir.) A.Dietr. = ***Bulbostylis festucoides*** (Poir.) C.B.Clarke
Baeothryon fistulosum (Link) A.Dietr. = ***Eleocharis acutangula*** (Roxb.) Schult.
Baeothryon flavescens (Poir.) A.Dietr. = ***Eleocharis flavescens*** (Poir.) Urb.
Baeothryon germanicum (Palla) Holub = ***Trichophorum cespitosum*** subsp. ***germanicum*** (Palla) Hegi
Baeothryon halleri (Vill.) T.Nees = ***Eleocharis quinqueflora*** (Hartmann) O.Schwarz
Baeothryon hudsonianum (Michx.) Soják = ***Trichophorum alpinum*** (L.) Pers.
Baeothryon mattfeldianum (Kük.) Á.Löve & D.Löve = ***Scirpus mattfeldianus*** Kük.
Baeothryon nanum (Spreng.) A.Dietr. = ***Eleocharis parvula*** (Roem. & Schult.) Palla
Baeothryon pauciflorum (Lightf.) A.Dietr. = ***Eleocharis quinqueflora*** (Hartmann) O.Schwarz
Baeothryon planifolium (Spreng.) Soják = ***Trichophorum planifolium*** (Spreng.) Palla
Baeothryon pumilum (Vahl) Á.Löve & D.Löve = ***Trichophorum pumilum*** (Vahl) Schinz & Thell.
Baeothryon pusillum (Vahl) A.Dietr. = ***Eleocharis nana*** Kunth
Baeothryon pygmaeum (Lam.) A.Dietr. = ***Eleocharis retroflexa*** subsp. ***chaetaria*** (Roem. & Schult.) T.Koyama
Baeothryon retroflexum (Poir.) A.Dietr. = ***Eleocharis retroflexa*** (Poir.) Urb.
Baeothryon rigidum (Steud.) Soják = ***Trichophorum rigidum*** (Steud.) Goetgh., Muasya & D.A.Simpson
Baeothryon subcapitatum (Thwaites & Hook.) Á.Löve & D.Löve = ***Trichophorum subcapitatum*** (Thwaites & Hook.) D.A.Simpson
Baeothryon subcapitatum subsp. *celebicum* (J.Kern) Soják = ***Trichophorum subcapitatum*** (Thwaites & Hook.) D.A.Simpson
Baeothryon subcapitatum subsp. *pulogense* (Merr.) Soják = ***Trichophorum subcapitatum*** (Thwaites & Hook.) D.A.Simpson
Baeothryon tenuiculum (Schrad. ex Schult.) A.Dietr. = ***Eleocharis filiculmis*** Kunth
Baeothryon trichoides (Kunth) A.Dietr. = ***Eleocharis acicularis*** (L.) Roem. & Schult. subsp. ***acicularis***
Baeothryon uniflorum (Trautv.) T.V.Egorova = ***Trichophorum uniflorum*** (Trautv.) Malyschev & Lukitsch.
Baeothryon variegatum (Poir.) A.Dietr. = ***Eleocharis variegata*** (Poir.) C.Presl
Baeothryon verecundum (Fernald) Á.Löve & D.Löve = ***Trichophorum planifolium*** (Spreng.) Palla

Baumea

Baumea Gaudich. = ***Machaerina*** Vahl
Baumea acuminata Regel = ?
Baumea acuta (Labill.) Palla = ***Machaerina acuta*** (Labill.) J.Kern
Baumea arthrophylla (Nees) Boeck. = ***Machaerina arthrophylla*** (Nees) T.Koyama
Baumea articulata (R.Br.) S.T.Blake = ***Machaerina articulata*** (R.Br.) T.Koyama
Baumea aspericaulis (Kük.) S.T.Blake = ***Machaerina aspericaulis*** (Kük.) T.Koyama
Baumea australis (Nees) S.T.Blake = ***Gahnia australis*** (Nees) K.L.Wilson
Baumea brownii Boeck. = ***Machaerina rubiginosa*** (Spreng.) T.Koyama
Baumea complanata (Berggr.) S.T.Blake = ***Machaerina complanata*** (Berggr.) T.Koyama
Baumea crassa Thwaites = ***Machaerina rubiginosa*** (Spreng.) T.Koyama
Baumea deplanchei Boeck. = ***Machaerina deplanchei*** (Boeck.) T.Koyama
Baumea disticha (C.B.Clarke) S.T.Blake = ***Machaerina disticha*** (C.B.Clarke) T.Koyama
Baumea ensifolia (Boeck.) Palla = ***Machaerina iridifolia*** (Bory) T.Koyama
Baumea ensigera (Hance) S.T.Blake = ***Machaerina ensigera*** (Hance) T.Koyama
Baumea falcata Nees = ***Machaerina falcata*** (Nees) T.Koyama
Baumea flexuosa Boeck. = ***Machaerina flexuosa*** (Boeck.) J.Kern

Baumea glomerata Gaudich. = ***Machaerina glomerata*** (Gaudich.) T.Koyama
Baumea gunnii (Hook.f.) S.T.Blake = ***Machaerina gunnii*** (Hook.f.) J.Kern
Baumea huttonii (Kirk) S.T.Blake = ***Machaerina huttonii*** (Kirk) T.Koyama
Baumea iridifolia (Nees) Boeck. = ***Machaerina flexuosa*** (Boeck.) J.Kern subsp. ***flexuosa***
Baumea iridifolia subsp. *laevinux* J.Raynal = ***Machaerina flexuosa*** subsp. ***laevinux*** (J.Raynal) Lye
Baumea johnsonii K.L.Wilson = [50 NSW]
Baumea juncea (R.Br.) Palla = ***Machaerina juncea*** (R.Br.) T.Koyama
Baumea laxa (Nees) Boeck. = ***Machaerina laxa*** (Nees) T.Koyama
Baumea loculata Boeck. = ***Machaerina articulata*** (R.Br.) T.Koyama
Baumea longifolia Boeck. = ***Gahnia filum*** (Labill.) F.Muell.
Baumea mariscoides Gaudich. = ***Machaerina mariscoides*** (Gaudich.) J.Kern
Baumea meyenii Kunth = ***Machaerina mariscoides*** (Gaudich.) J.Kern
Baumea milnei (C.B.Clarke) S.T.Blake = ***Machaerina milnei*** (C.B.Clarke) T.Koyama
Baumea montana J.Raynal = ***Machaerina montana*** (J.Raynal) Lye
Baumea muelleri (C.B.Clarke) S.T.Blake = ***Machaerina muelleri*** (C.B.Clarke) T.Koyama
Baumea nuda (Steud.) S.T.Blake = ***Machaerina nuda*** (Steud.) J.Kern
Baumea planifolia (Benth.) K.L.Wilson =
Baumea preissii Nees = ***Machaerina preissii*** (Nees) L.A.S.Johnson & Koyama
Baumea riparia (Nees) Boeck. = ***Machaerina riparia*** (Nees) T.Koyama
Baumea rubiginosa (Spreng.) Boeck. = ***Machaerina rubiginosa*** (Spreng.) T.Koyama
Baumea schoenoides (R.Br.) Boeck. = ***Machaerina acuta*** (Labill.) J.Kern
Baumea tenax (Hook.f.) S.T.Blake = ***Machaerina tenax*** (Hook.f.) T.Koyama
Baumea teretifolia (R.Br.) Palla = ***Machaerina teretifolia*** (R.Br.) T.Koyama
Baumea tetragona (Labill.) S.T.Blake = ***Machaerina tetragona*** (Labill.) T.Koyama
Baumea vaginalis (Benth.) S.T.Blake = ***Machaerina vaginalis*** (Benth.) T.Koyama
Baumea veillonis J.Raynal = [60 NWC]

Unplaced Names:
Baumea acuminata Regel, Cat. Pl. Hort. Aksakov.: 15 (1860). = ?
Baumea johnsonii K.L.Wilson, Telopea 1: 457 (1980). = [50 NSW]
Baumea veillonis J.Raynal, Adansonia, n.s., 13: 469 (1973 publ. 1974). = [60 NWC]

Becquerelia

Becquerelia Brongn. in L.I.Duperrey, Voy. Monde, Phan.: 161 (1833).
Trop. America. 80 BLZ COS NIC PAN 81 TRT 82 FRG GUY SUR VEN 83 CLM ECU PER 84 BZC BZE BZL BZN BZS.
8 Species

Becquerelia bullata C.B.Clarke & H.Pfeiff., Repert. Spec. Nov. Regni Veg. 18: 382 (1922).
N. Brazil. 84 BZN. Hemicr.

Becquerelia clarkei T.Koyama, Mem. New York Bot. Gard. 16: 49 (1967).
Brazil (Minas Gerais, Bahia). 84 BZE BZL. Hemicr. or rhizome geophyte.
**Hoppia bicolor* C.B.Clarke, Kew Bull., Addit. Ser. 8: 62 (1908). *Bisboeckelera bicolor* (C.B.Clarke) Standl., Contr. U. S. Natl. Herb. 17: 429 (1914).
Bisboeckelera irrigua var. *major* H.Pfeiff., Repert. Spec. Nov. Regni Veg. 42: 255 (1937).

Becquerelia cymosa Brongn. in L.I.Duperrey, Voy. Monde, Phan.: 161 (1833). *Scleria cymosa* (Brongn.) Boeck., Vidensk. Meddel. Naturhist. Foren. Kjøbenhavn 1869: 155 (1869).
Trinidad-Tobago, C. & S. Trop. America. 80 COS NIC PAN 81 TRT 82 FRG GUY SUR VEN 83 CLM ECU PER 84 BZC BZE BZL BZS. Hemicr.
Becquerelia verruculosa Nees in C.F.P.von Martius & auct. suc. (eds.), Fl. Bras. 2(1): 191 (1842).
Becquerelia verruculosa var. *major* Nees in C.F.P.von Martius & auct. suc. (eds.), Fl. Bras. 2(1): 191 (1842).
Becquerelia verruculosa var. *minor* Nees in C.F.P.von Martius & auct. suc. (eds.), Fl. Bras. 2(1): 191 (1842).

Becquerelia discolor Kunth, Enum. Pl. 2: 362 (1837). *Scleria discolor* (Kunth) Boeck., Linnaea 38: 541 (1874).
NE. Brazil. 84 BZE. Hemicr. or rhizome geophyte.

Becquerelia divaricata (Nees) Palla, Akad. Wiss. Mainz, Abh. Math.-Naturwiss. Kl. 79: 198 (1908).
Costa Rica to S. Trop. America. 80 COS 82 FRG GUY SUR VEN 83 CLM PER 84 BZE. Hemicr.
**Scleria divaricata* Nees, Linnaea 9: 303 (1834), nom. inval.
Becquerelia cymosa Kunth, Enum. Pl. 2: 361 (1837), nom. illeg.
Becquerelia merkeliana Nees in C.F.P.von Martius & auct. suc. (eds.), Fl. Bras. 2(1): 191 (1842). *Scleria merkeliana* (Nees) C.B.Clarke, Ill. Cyper.: t. 133 (1909). *Becquerelia cymosa* subsp. *merkeliana* (Nees) T.Koyama, Mem. New York Bot. Gard. 17(1): 29 (1967).

Becquerelia martii (Nees) H.Pfeiff., Repert. Spec. Nov. Regni Veg. 18: 381 (1922).
N. Brazil. 84 BZN. Hemicr.
**Calyptrocarya martii* Nees in C.F.P.von Martius & auct. suc. (eds.), Fl. Bras. 2(1): 195 (1842).
Becquerelia martii f. *typica* H.Pfeiff., Repert. Spec. Nov. Regni Veg. 18: 381 (1922), nom. inval.
Becquerelia martii f. *laxa* H.Pfeiff., Repert. Spec. Nov. Regni Veg. 18: 381 (1922).

Becquerelia muricata (Boeck.) Nees in C.F.P.von Martius & auct. suc. (eds.), Fl. Bras. 2(1): 192 (1842).
SE. & S. Brazil. 84 BZL BZS. Hemicr.
**Scleria muricata* Boeck., Linnaea 38: 540 (1874).

Becquerelia tuberculata (Boeck.) H.Pfeiff., Repert. Spec. Nov. Regni Veg. 18: 379 (1922).
Trinidad, C. America to Guyana. 80 BLZ 81 TRT 82 GUY SUR VEN 83 CLM 84 BZN. Hemicr.
**Scleria tuberculata* Boeck., Linnaea 38: 540 (1874).

Synonyms:
Becquerelia bicolor H.Pfeiff. = ***Calyptrocarya bicolor*** (H.Pfeiff.) T.Koyama
Becquerelia bicolor f. *humilis* H.Pfeiff. = ***Calyptrocarya bicolor*** (H.Pfeiff.) T.Koyama

Becquerelia bicolor f. *ramosissima* H.Pfeiff. = ***Calyptrocarya bicolor*** (H.Pfeiff.) T.Koyama
Becquerelia bicolor f. *verticillata* H.Pfeiff. = ***Calyptrocarya bicolor*** (H.Pfeiff.) T.Koyama
Becquerelia cymosa Kunth = ***Becquerelia divaricata*** (Nees) Palla
Becquerelia cymosa subsp. *merkeliana* (Nees) T.Koyama = ***Becquerelia divaricata*** (Nees) Palla
Becquerelia gigantea Hochst. ex Steud. = ?
Becquerelia glomerata Kunth = ***Calyptrocarya glomerulata*** (Brongn.) Urb.
Becquerelia glomerulata Brongn. = ***Calyptrocarya glomerulata*** (Brongn.) Urb.
Becquerelia glomerulata var. *angustifolia* (Lindl. & Nees ex Kunth) H.Pfeiff. = ***Calyptrocarya glomerulata*** (Brongn.) Urb.
Becquerelia glomerulata var. *euglomerulata* H.Pfeiff. = ***Calyptrocarya glomerulata*** (Brongn.) Urb.
Becquerelia glomerulata f. *minor* H.Pfeiff. = ***Calyptrocarya glomerulata*** (Brongn.) Urb.
Becquerelia glomerulata f. *minuta* H.Pfeiff. = ***Calyptrocarya glomerulata*** (Brongn.) Urb.
Becquerelia glomerulata f. *rubra* H.Pfeiff. = ***Calyptrocarya glomerulata*** (Brongn.) Urb.
Becquerelia glomerulata f. *scabrinux* H.Pfeiff. = ***Calyptrocarya glomerulata*** (Brongn.) Urb.
Becquerelia glomerulata f. *typica* H.Pfeiff. = ***Calyptrocarya glomerulata*** (Brongn.) Urb.
Becquerelia martii f. *laxa* H.Pfeiff. = ***Becquerelia martii*** (Nees) H.Pfeiff.
Becquerelia martii f. *typica* H.Pfeiff. = ***Becquerelia martii*** (Nees) H.Pfeiff.
Becquerelia merkeliana Nees = ***Becquerelia divaricata*** (Nees) Palla
Becquerelia verruculosa Nees = ***Becquerelia cymosa*** Brongn.
Becquerelia verruculosa var. *major* Nees = ***Becquerelia cymosa*** Brongn.
Becquerelia verruculosa var. *minor* Nees = ***Becquerelia cymosa*** Brongn.

***Unplaced Names*:**
Becquerelia gigantea Hochst. ex Steud., Syn. Pl. Glumac. 2: 180 (1855). = ?

Beera

Beera P.Beauv. = ***Hypolytrum*** Rich. ex Pers.
Beera caespitosa Schrad. ex Kunth = ***Hypolytrum strictum*** Poepp. ex Kunth
Beera laxa Schrad. ex Kunth = ***Hypolytrum laxum*** Kunth
Beera silvatica Schrad. ex Nees = ***Hypolytrum longifolium*** (Rich.) Nees subsp. ***longifolium***

Bisboeckelera

Bisboeckelera Kuntze, Revis. Gen. Pl. 2: 747 (1891).
S. Trop. America. 82 FRG GUY SUR VEN 83 CLM PER 84 BZN.
4 Species
Hoppia Nees in C.F.P.von Martius & auct. suc. (eds.), Fl. Bras. 2(1): 199 (1842).

Bisboeckelera irrigua (Nees) Kuntze, Revis. Gen. Pl. 2: 747 (1891).
SE. Colombia to SW. Guyana and N. Brazil. 82 GUY VEN 83 CLM PER 84 BZN. Hemicr.
**Hoppia irrigua* Nees in C.F.P.von Martius & auct. suc. (eds.), Fl. Bras. 2(1): 199 (1842).

Bisboeckelera longifolia (Rudge) Kuntze, Revis. Gen. Pl. 2: 747 (1891).
N. South America to N. Brazil. 82 FRG GUY SUR 84 BZN. Hemicr.
**Schoenus longifolius* Rudge, Pl. Guian.: 14 (1805). *Dichromena longifolia* (Rudge) Spreng., Syst. Veg. 1: 203 (1824). *Calyptrocarya longifolia* (Rudge) Kunth, Enum. Pl. 2: 365 (1837).
Hoppia angustifolia Boeck., Flora 54: 37 (1871). *Bisboeckelera angustifolia* (Boeck.) Kuntze, Revis. Gen. Pl. 2: 747 (1891). *Bisboeckelera longifolia* var. *angustifolia* (Boeck.) H.Pfeiff., Repert. Spec. Nov. Regni Veg. 42: 256 (1937).

Bisboeckelera microcephala (Boeck.) T.Koyama, Mem. New York Bot. Gard. 16: 50 (1967).
Guianas. 82 FRG GUY SUR. Hemicr.
**Hoppia microcephala* Boeck., Flora 54: 37 (1871).

Bisboeckelera vinacea Standl., Contr. U. S. Natl. Herb. 17: 429 (1914).
Colombia. 83 CLM. Hemicr.
Bisboeckelera irrigua var. *laevinux* H.Pfeiff., Repert. Spec. Nov. Regni Veg. 42: 255 (1937).

***Synonyms*:**
Bisboeckelera angustifolia (Boeck.) Kuntze = ***Bisboeckelera longifolia*** (Rudge) Kuntze
Bisboeckelera berroi (C.B.Clarke) Standl. = ***Cyperus berroi*** (C.B.Clarke) Barros
Bisboeckelera bicolor (C.B.Clarke) Standl. = ***Becquerelia clarkei*** T.Koyama
Bisboeckelera irrigua var. *laevinux* H.Pfeiff. = ***Bisboeckelera vinacea*** Standl.
Bisboeckelera irrigua var. *major* H.Pfeiff. = ***Becquerelia clarkei*** T.Koyama
Bisboeckelera longifolia var. *angustifolia* (Boeck.) H.Pfeiff. = ***Bisboeckelera longifolia*** (Rudge) Kuntze
Bisboeckelera paporiensis Suess. = ***Diplacrum capitatum*** (Willd.) Boeck.

Bitteria

Bitteria Börner = ***Carex*** L.
Bitteria digitata (L.) Fedde & J.Schust. = ***Carex digitata*** L.
Bitteria pediformis (C.A.Mey.) Fedde & J.Schust. = ***Carex pediformis*** C.A.Mey.
Bitteria pilulifera (L.) Fedde & J.Schust. = ***Carex pilulifera*** L.

Blepharolepis

Blepharolepis Nees = ***Scirpus*** Tourn. ex L.
Blepharolepis leucoblepharis Nees = [84]

***Unplaced Names*:**
Blepharolepis leucoblepharis Nees in C.F.P.von Martius & auct. suc. (eds.), Fl. Bras. 2(1): 91 (1842). = [84]

Blysmocarex

Blysmocarex N.A.Ivanova = ***Kobresia*** Willd.
Blysmocarex macrantha (Boeck.) K.V.Ivanova = ***Kobresia macrantha*** Boeck.
Blysmocarex macrantha subsp. *nudicarpa* (Y.C.Yang) D.S.Deng = ***Kobresia macrantha*** Boeck.
Blysmocarex macrantha subsp. *stolonifera* (Y.C.Tang ex P.C.Li) D.S.Deng = ***Kobresia stolonifera*** Y.C.Tang ex P.C.Li
Blysmocarex nudicarpa Y.C.Yang = ***Kobresia macrantha*** Boeck.

Blysmopsis

Blysmopsis Oteng-Yeb. = ***Blysmus*** Panz. ex Schult.
Blysmopsis rufa (Huds.) Oteng-Yeb. = ***Blysmus rufus*** (Huds.) Link

Blysmus

Blysmus Panz. ex Schult. in J.J.Roemer & J.A.Schultes, Mant. 2: 41 (1824).
Temp. Northern Hemisphere. 10 DEN FIN GRB IRE NOR SWE 11 AUT BGM CZE GER HUN NET POL SAR SWI 12 FRA 13 ALB BUL GRC ITA ROM YUG 14 BLR BLT KRY RUC RUE RUN RUS RUW UKR 20 MOR 30 ALT WSB YAK 32 KAZ KGZ TKM TZK UZB 33 NCS TCS 34 AFG IRN IRQ TUR 36 CHN CHT CHX 37 MON 40 EHM IND NEP PAK WHM 70 ASK NUN NWT YUK 71 MAN SAS 72 NBR NFL NSC ONT PEI QUE.
3 Species
Nomochloa P.Beauv. in T.G.Lestiboudois, Essai Cypér.: 37 (1819).
Leptolepis Boeck., Beitr. Cyper. 1: 31 (1888).
Blysmopsis Oteng-Yeb., Notes Roy. Bot. Gard. Edinburgh 33: 309 (1974).

Blysmus compressus (L.) Panz. ex Link, Hort. Berol. 1: 278 (1827).
Europe to Himalaya. 10 DEN FIN GRB NOR SWE 11 AUT BGM CZE GER HUN NET POL SAR SWI 12 FRA 13 ALB BUL GRC ITA ROM YUG 14 BLR BLT KRY RUC RUE RUN RUS RUW UKR 20 MOR 30 32 KAZ KGZ TKM TZK UZB 33 NCS TCS 34 AFG IRN IRQ TUR 36 CHN CHT CHX 37 MON 40 EHM IND NEP PAK WHM. Hemicr.
**Schoenus compressus* L., Sp. Pl.: 43 (1753). *Scirpus compressus* (L.) Pers., Syn. Pl. 1: 66 (1805), nom. illeg. *Aplostemon compressum* (L.) Raf., J. Phys. Chim. Hist. Nat. Arts 89: 105 (1819). *Chaetospora compressa* (L.) Gray, Nat. Arr. Brit. Pl. 2: 71 (1821). *Heleogiton compressa* (L.) Peterm., Fl. Bienitz: 22 (1841). *Nomochloa compressa* (L.) Beetle, Amer. J. Bot. 31: 262 (1944).

var. ***brevifolius*** (Decne.) Karth. in S.Karthikeyan & al., Fl. Ind. Enumerat. – Monocot.: 32 (1989).
S. Siberia to Himalaya. 30 32 KGZ TZK 34 AFG 36 CHN CHT CHX 37 MON 40 EHM IND PAK WHM. Hemicr.
Blysmus brevifolius Decne. in V.Jacquemont, Voy. Inde 4: 173 (1844).
**Scirpus brevifolius* Decne. in V.Jacquemont, Voy. Inde 4: 173 (1844). *Scirpus caricis* var. *brevifolius* (Decne.) C.B.Clarke in J.D.Hooker, Fl. Brit. India 6: 660 (1893). *Blysmus compressus* subsp. *brevifolius* (Decne.) Kukkonen, Ann. Bot. Fenn. 32: 154 (1995).
Leptolepis tibetica Boeck., Beitr. Cyper. 1: 31 (1888).
Scirpus caricis var. *dissitus* C.B.Clarke in J.D.Hooker, Fl. Brit. India 6: 661 (1893).
Scirpus caricis var. *sikkimensis* C.B.Clarke in J.D.Hooker, Fl. Brit. India 6: 661 (1893). *Blysmus compressus* var. *sikkimensis* (C.B.Clarke) Karth. in S.Karthikeyan & al., Fl. Ind. Enumerat. – Monocot.: 32 (1989).
Blysmus sinocompressus Tang & F.T.Wang, in Fl. Reipubl. Popul. Sin. 11: 41, 224 (1961).

subsp. ***compressus***
Europe to Himalaya. 10 DEN FIN GRB NOR SWE 11 AUT BGM CZE GER HUN NET POL SAR SWI 12 FRA 13 ALB BUL GRC ITA ROM YUG 14 BLR BLT KRY RUC RUE RUN RUS RUW UKR 20 MOR 32 KAZ TKM TZK UZB 33 NCS TCS 34 AFG IRN IRQ TUR 36 CHX 40 EHM NEP WHM. Hemicr.
Carex uliginosa L., Sp. Pl.: 973 (1753).
Scirpus planifolius Grimm, Nova Acta Phys.-Med. Acad. Caes. Leop.-Carol. Nat. Cur. 3(App.): 259 (1767).
Scirpus caricis Retz., Fl. Scand. Prodr.: 11 (1779), nom. illeg.
Cyperus horizontalis Salisb., Prodr. Stirp. Chap. Allerton: 30 (1796).
Scirpus cariciformis Vest, Man. Bot.: 287 (1805).
Scirpus caricinus Schrad., Fl. Germ. 1: 132 (1806), nom. illeg.
Scirpus distichus Peterm., Flora 27: 343 (1844). *Blysmus distichus* (Peterm.) Montandon, Bull. Murith. Soc. Valais. Sci. Nat.: 346 (1868).
Scirpus compressus var. *gracilis* Zapal., Consp. Fl. Gallic. Crit. 1: 124 (1906). *Blysmus compressus* f. *gracilis* (Zapal.) Soó, Acta Bot. Acad. Sci. Hung. 16: 368 (1970 publ. 1971).
Blysmus tenuis Gilli, Repert. Spec. Nov. Regni Veg. 46: 48 (1939). *Scirpus tenuis* (Gilli) Parsa, Fl. Iran 5: 435 (1951).

var. ***dissitus*** (Duthie) Karth. in S.Karthikeyan & al., Fl. Ind. Enumerat. – Monocot.: 32 (1989).
W. Himalaya. 40 WHM. Hemicr.
**Scirpus dissitus* Duthie, Saharunpore Report: s.p. (1885).

subsp. ***subulifolius*** A.P.Khokhr., Byull. Moskovsk. Obshch. Isp. Prir., Otd. Biol. 102: 46 (1997).
Turkey. 34 TUR. Hemicr.

Blysmus mongolicola Kitag., J. Jap. Bot. 44: 274 (1969).
Mongolia. 37 MON. Hemicr.

Blysmus rufus (Huds.) Link, Hort. Berol. 1: 278 (1827).
N., C. & E. Europe to Mongolia, Subarctic America to Canada. 10 DEN FIN GRB IRE NOR SWE 11 CZE? GER NET POL 14 RUN RUW 30 ALT WSB YAK 32 KAZ KGZ TZK UZB 36 CHX 37 MON 40 WHM 70 ASK NUN NWT YUK 71 MAN SAS 72 NBR NFL NSC ONT PEI QUE. Hemicr.
Scirpus rufus subsp. *exilis* Printz in ?. *Blysmus exilis* (Printz) N.A.Ivanova, Bot. Zhurn. S.S.S.R. 24: 502 (1939).
Schoenus ferrugineus Huds., Fl. Angl.: 14 (1762), nom. illeg.
**Schoenus rufus* Huds., Fl. Angl., ed. 2: 15 (1778). *Scirpus rufus* (Huds.) Schrad., Fl. Germ. 1: 133 (1806). *Chaetospora rufa* (Huds.) Gray, Nat. Arr. Brit. Pl. 2: 71 (1821). *Cyperus rufus* (Huds.) Missbach & E.H.L.Krause in J.Sturm, Deutschl. Fl., ed. 2, 2: 37 (1900), nom. illeg. *Nomochloa rufa* (Huds.) Beetle, Amer. J. Bot. 31: 262 (1944). *Blysmopsis rufa* (Huds.) Oteng-Yeb., Notes Roy. Bot. Gard. Edinburgh 33: 310 (1974).
Schoenus brunneus Sol. ex Roem. & Schult., Syst. Veg. 2: 65 (1817).
Scirpus bifolius Wallr., Sched. Crit.: 27 (1822). *Blysmus rufus* var. *bifolius* (Wallr.) Nyman, Consp. Fl. Eur., Suppl. 2(1): 318 (1889).
Scirpus wallrothii Bernh., Flora 8: 607 (1825).

Scirpus littoralis Flüggé ex Rchb., Fl. Germ. Excurs.: 74 (1830).
Scirpus rufus var. *neogaeus* Fernald, Rhodora 45: 287 (1943). *Nomochloa rufa* var. *neogaea* (Fernald) Beetle, Amer. J. Bot. 31: 262 (1944).

***Synonyms*:**

Blysmus bonannii (C.Presl) Steud. = ***Schoenus nigricans*** L.
Blysmus brevifolius Decne. = ***Blysmus compressus*** var. ***brevifolius*** (Decne.) Karth.
Blysmus compressus subsp. *brevifolius* (Decne.) Kukkonen = ***Blysmus compressus*** var. ***brevifolius*** (Decne.) Karth.
Blysmus compressus f. *gracilis* (Zapal.) Soó = ***Blysmus compressus*** (L.) Panz. ex Link subsp. ***compressus***
Blysmus compressus var. *sikkimensis* (C.B.Clarke) Karth. = ***Blysmus compressus*** var. ***brevifolius*** (Decne.) Karth.
Blysmus distichus (Peterm.) Montandon = ***Blysmus compressus*** (L.) Panz. ex Link subsp. ***compressus***
Blysmus exilis (Printz) N.A.Ivanova = ***Blysmus rufus*** (Huds.) Link
Blysmus rufus var. *bifolius* (Wallr.) Nyman = ***Blysmus rufus*** (Huds.) Link
Blysmus sinocompressus Tang & F.T.Wang = ***Blysmus compressus*** var. ***brevifolius*** (Decne.) Karth.
Blysmus tenuis Gilli = ***Blysmus compressus*** (L.) Panz. ex Link subsp. ***compressus***

Boeckeleria

Boeckeleria T.Durand = ***Tetraria*** P.Beauv.
Boeckeleria nigrovaginata (Nees) H.Pfiriff. = ***Tetraria nigrovaginata*** (Nees) C.B.Clarke

Bolboschoenus

Bolboschoenus (Asch.) Palla in W.D.J.Koch, Syn. Deut. Schweiz. Fl., ed. 3: 2531 (1904).
Cosmopolitan. 10 DEN FIN GRB IRE NOR SWE 11 AUT BGM CZE GER HUN NET POL SWI 12 BAL COR FRA POR SAR SPA 13 ALB BLT BUL GRC ITA KRI ROM RUC RUE SIC TUE UKR YUG 14 BLR BLT KRY RUC RUE RUN RUS RUW UKR 20 ALG EGY LBY MOR TUN 21 AZO? CNY CVI MDR 22 MLI MTN SEN 24 DJI SOM 25 KEN TAN 26 ANG 27 CPP NAM 29 MAU MDG 30 ALT BRY CTA IRK KRA TVA WSB YAK 31 AMU KAM KHA PRM SAK 32 KAZ KGZ TKM TZK UZB 33 NCS TCS 34 AFG CYP IRN IRQ LBS PAL SIN TUR 35 GST SAU 36 CHM CHN CHQ CHS CHX 37 MON 38 JAP KOR NNS TAI 40 ASS IND PAK 41 THA VIE 42 PHI 43 NWG 50 NFK NSW NTA QLD SOA TAS VIC WAU 51 NZN NZS 60 NWC 63 HAW 70 ASK NWT 71 ABT BRC MAN SAS 72 NBR NSC ONT PEI QUE 73 COL IDA MNT ORE WAS WYO 74 ILL IOW KAN MIN MSO NDA NEB OKL SDA WIS 75 CNT INI MAI MAS MIC NWH NWJ NWY OHI PEN RHO VER 76 ARI CAL NEV UTA 77 NWM TEX 78 ALA ARK DEL FLA GEO LOU MRY MSI NCA SCA TEN VRG 79 MXC MXE MXN MXS MXT 81 BER CUB 82 FRG? GUY SUR VEN 83 PER 84 85 AGE AGS AGW PAR URU.
12 Species
Scirpocyperus Ség., Pl. Veron. 3: 71 (1754).
Reigera Opiz, Seznam: 83 (1852).

Bolboschoenus caldwellii (V.J.Cook) Soják, Cas. Nár. Mus., Odd. Prír. 141: 62 (1972).
Australia, New Zealand. 50 NSW NTA QLD SOA TAS VIC WAU 51 NZN NZS. Hemicr.
**Scirpus caldwellii* V.J.Cook, Trans. & Proc. Roy. Soc. New Zealand 76: 568 (1947).

Bolboschoenus capensis (Burm.f.) Holub, Folia Geobot. Phytotax. 18: 204 (1983).
Cape Prov. 27 CPP. Hemicr.
**Scirpus capensis* Burm.f., Fl. Indica, Prodr. Fl. Cap.: 3 (1768).

Bolboschoenus fluviatilis (Torr.) Soják, Cas. Nár. Mus., Odd. Prír. 141: 62 (1972).
Russian Far East to Indo-China, N. America to NE. Mexico. 31+ 36 CHM CHN 38 JAP KOR 41 VIE 50 NFK NSW QLD 51 NZN 60 NWC 71 ABT BRC MAN SAS 72 NBR ONT QUE 73 COL IDA MNT ORE WAS 74 ILL IOW KAN MIN MSO NDA NEB OKL SDA WIS 75 CNT INI MAI MAS MIC NWH NWJ NWY OHI PEN VER 76 ARI CAL UTA 77 NWM 78 ALA DEL MRY TEN VRG 79 MXE. Hemicr.
**Scirpus maritimus* var. *fluviatilis* Torr., Ann. Lyceum Nat. Hist. New York 3: 325 (1836). *Scirpus fluviatilis* (Torr.) A.Gray, Manual, ed. 2: 500 (1856). *Bolboschoenus maritimus* subsp. *fluviatilis* (Torr.) Á.Löve & D.Löve, Taxon 30: 845 (1981). *Schoenoplectus fluviatilis* (Torr.) M.T.Strong, Novon 3: 203 (1993).
Scirpus yagara Ohwi, Mem. Coll. Sci. Kyoto Imp. Univ., Ser. B, Biol. 18(1): 110 (1944). *Bolboschoenus fluviatilis* subsp. *yagara* (Ohwi) T.Koyama, Acta Phytotax. Geobot. 31: 140 (1980).
Scirpus perviridis V.J.Cook, Trans. & Proc. Roy. Soc. New Zealand 76: 570 (1947).

Bolboschoenus glaucus (Lam.) S.G.Sm., Novon 5: 101 (1995).
Senegal, N. Africa, E. Europe to S. Pakistan, introduced in America. 14 KRY RUS UKR 20 EGY 22 SEN 32 KAZ 33 NCS TCS 34 IRN TUR 40 PAK (73) ida ore (75) nwj nwy (76) cal. Hemicr.
**Scirpus glaucus* Lam., Tabl. Encycl. 1: 142 (1791).
Scirpus tuberosus Desf., Fl. Atlant. 1: 50 (1798). *Scirpus maritimus* var. *tuberosus* (Desf.) Roem. & Schult., Syst. Veg. 2: 139 (1817). *Bolboschoenus maritimus* subsp. *tuberosus* (Desf.) T.Koyama, Acta Phytotax. Geobot. 31: 148 (1980).
Scirpus macrostachys Willd., Enum. Pl.: 78 (1809). *Scirpus maritimus* var. *macrostachys* (Willd.) Dumort., Fl. Belg.: 144 (1827). *Bolboschoenus macrostachys* (Willd.) Grossh., Fl. Kavkaza 1: 145 (1928).

Bolboschoenus laticarpus Marhold, Hroudová, Duchá?ek & Zákr., Phyton (Horn) 44: 7 (2004).
Europe. 10 DEN 11 AUT CZE GER HUN NET POL 12 FRA 13 BLT BUL ROM RUC RUE UKR. Hemicr.

Bolboschoenus maritimus (L.) Palla in W.D.J.Koch, Syn. Deut. Schweiz. Fl., ed. 3: 2532 (1904).
Temp. & Subtrop. 10 DEN FIN GRB IRE NOR SWE 11 AUT BGM CZE GER HUN NET POL SWI 12 BAL COR FRA POR SAR SPA 13 ALB BUL GRC ITA KRI ROM SIC TUE YUG 14 BLR BLT KRY RUC RUE RUN RUS RUW UKR 20 ALG EGY LBY MOR TUN 21 AZO? CNY CVI MDR 22 MLI MTN SEN 24 DJI SOM 25 KEN TAN 29 MAU MDG 30 ALT 31 32 KAZ KGZ TKM UZB 33 NCS TCS 34 AFG CYP IRN IRQ LBS PAL SIN TUR 35 GST SAU 36 CHQ CHX 37 MON 40 ASS

IND PAK 41 THA VIE 63 HAW 70 ASK NWT 71 ABT BRC MAN SAS 72 NBR NSC ONT PEI QUE 73 COL IDA MNT ORE WAS WYO 74 ILL IOW KAN MIN MSO NDA NEB OKL SDA 75 CNT MAI MAS MIC NWH NWJ NWY RHO 76 ARI CAL NEV UTA 77 NWM TEX 79 MXC MXN MXS 81 CUB 82 SUR 83 PER 84 85 AGE AGS AGW. Hemicr. or hel.

**Scirpus maritimus* L., Sp. Pl.: 51 (1753). *Reigera maritima* (L.) Opiz, Seznam: 83 (1852). *Schoenoplectus maritimus* (L.) Lye, Blyttia 29: 145 (1971).

Scirpus maritimus var. *agonus* Beetle, Ann. Lyceum Nat. Hist. New York 3: 325 (1836). *Scirpus maritimus* f. *agonus* (Beetle) Fernald, Rhodora 45: 288 (1943).

Scirpus maritimus var. *ovatus* Nees in C.F.P.von Martius & auct. suc. (eds.), Fl. Bras. 2(1): 106 (1842).

subsp. ***affinis*** (Roth) T.Koyama, Brittonia 31: 284 (1979).

SE. European Russia to Indo-China. 14 RUS 30 ALT 32 KAZ UZB 33 NCS TCS 34 AFG IRN 36 CHQ 37 MON 40 ASS IND PAK 41 THA VIE. Hemicr.

Scirpus strobilinus Roxb., Fl. Ind. 1: 219 (1820). *Bolboschoenus strobilinus* (Roxb.) V.I.Krecz., in Fl. Tadzhikskoi SSR 2: 47 (1963).

**Scirpus affinis* Roth, Nov. Pl. Sp.: 30 (1821). *Bolboschoenus affinis* (Roth) Drobow, Trudy Bot. Muz. Imp. Akad. Nauk 16: 139 (1916). *Scirpus maritimus* subsp. *affinis* (Roth) Norl., Bot. Not. 125: 404 (1972). *Bolboschoenus maritimus* var. *affinis* (Roth) Parmar in B.V.Shetty & V.Singh, Fl. Rajasthan 3: 888 (1993).

Scirpus vulpinicolor T.Koyama, Acta Phytotax. Geobot. 17: 50 (1957).

Bolboschoenus popovii E.M.Egorova, Trudy Bot. Inst. Akad. Nauk S.S.S.R., Ser. 1, Fl. Sist. Vyssh. Rast. 3: 21 (1967).

var. ***cymosus*** (Rchb.) Kit Tan & Oteng-Yeb. in P.H. Davis (ed.), Fl. Turkey 9: 64 (1985).

Europe to Medit. 11 GER 13 ROM 14 KRY RUC RUS UKR 20 ALG 33 NCS TCS 34 TUR. Hemicr.

**Scirpus maritimus* var. *cymosus* Rchb., Fl. Germ. Excurs.: 79 (1830). *Bolboschoenus maritimus* subsp. *cymosus* (Rchb.) Soják, Cas. Nár. Mus., Odd. Prír. 141: 62 (1972).

subsp. ***maritimus***

Temp. & Subtrop. 10 DEN FIN GRB IRE NOR SWE 11 AUT BGM CZE GER HUN NET POL SWI 12 BAL COR FRA POR SAR SPA 13 ALB BUL GRC ITA KRI ROM SIC TUE YUG 14 BLR BLT KRY RUC RUE RUN RUS RUW UKR 20 ALG EGY LBY MOR TUN 21 AZO? CNY CVI MDR 22 MLI MTN SEN 24 DJI SOM 25 KEN TAN 29 MAU MDG 30 31 32 KAZ KGZ TKM UZB 33 NCS TCS 34 AFG CYP IRN IRQ LBS PAL SIN TUR 35 GST SAU 36 CHX 40 IND PAK. Hemicr. or hel.

Bolboschoenus maritimus var. *desoulavii* Drobow in ?. *Bolboschoenus desoulavii* (Drobow) A.E.Kozhevn., Sosud. Rast. Sovet. Dal'nego Vostoka 3: 188 (1988).

Scirpus corymbosus Forssk., Fl. Aegypt.-Arab.: 14 (1775).

Scirpus mucronatus Pollich, Hist. Pl. Palat. 1: 44 (1776), nom. illeg.

Scirpus cyperoides Lam., Fl. Franç. 3: 553 (1778), nom. illeg.

Scirpus compactus Hoffm., Deutschl. Fl., ed. 2, 1: t. 25 (1791). *Bolboschoenus compactus* (Hoffm.) Drobow, Trudy Bot. Muz. Imp. Akad. Nauk 11: 92 (1913). *Bolboschoenus maritimus* var. *compactus* (Hoffm.) T.V.Egorova, in Fl. Severo-Vostoka Evropeiskoi Chasti SSSR 2: 18 (1976).

Scirpus macrostachyos Lam., Tabl. Encycl. 1: 142 (1791).

Scirpus aegyptiacus Poir. in J.B.A.M.de Lamarck, Encycl. 6: 770 (1805).

Scirpus riparius Pers., Syn. Pl. 1: 68 (1805).

Scirpus tridentatus Roxb., Fl. Ind. 1: 225 (1820).

Scirpus cephalotes B.Heyne ex Roth, Nov. Pl. Sp.: 30 (1821), nom. inval.

Scirpus maritimus var. *monostachyus* Lej. & Courtois, Comp. Fl. Belg. 1: 44 (1828).

Scirpus vulgaris Mazziari, Ionios Anthol. 2: 460 (1834).

Scirpus decumanus Willd. ex Kunth, Enum. Pl. 2: 168 (1837).

Scirpus lucidus Less. ex Kunth, Enum. Pl. 2: 168 (1837).

Scirpus swampianus Bosc ex Kunth, Enum. Pl. 2: 168 (1837).

Scirpus salinus Schmidt ex Steud., Nomencl. Bot., ed. 2, 2: 541 (1841).

Scirpus maritimus var. *digynus* Godr., Fl. Lorraine 3: 91 (1844).

Schoenus macrostachyus Noë ex Rchb., Icon. Fl. Germ. Helv. 8: 34 (1846).

Scirpus maritimus f. *longifolius* Peterm., Anal. Pfl.-Schlüss.: 481 (1846), nom. illeg. *Bolboschoenus maritimus* f. *longifolius* (Peterm.) Soó, Acta Bot. Acad. Sci. Hung. 16: 368 (1970 publ. 1971).

Isolepis grandispica Steud., Syn. Pl. Glumac. 2: 318 (1855). *Scirpus grandispicus* (Steud.) Berhaut, Bull. Soc. Bot. France 100: 176 (1953). *Bolboschoenus grandispicus* (Steud.) Lewej. & Lobin, Courier Forschungsinst. Senckenberg 52: 267 (1982).

Scirpus megastachyus Steud., Syn. Pl. Glumac. 2: 87 (1855). *Scirpus maritimus* var. *megastachyus* (Steud.) Nyman, Consp. Fl. Eur.: 763 (1882).

Scirpus squarrosulus Steud., Syn. Pl. Glumac. 2: 88 (1855).

Scirpus strobiliferus Steud., Syn. Pl. Glumac. 2: 317 (1855).

Scirpus maritimus f. *subumbellatus* Schur, Enum. Pl. Transsilv.: 693 (1866), nom. illeg. *Bolboschoenus maritimus* f. *subumbellatus* (Schur) Soó, Acta Bot. Acad. Sci. Hung. 16: 368 (1970 publ. 1971).

Scirpocyperus septentrionalis Montandon in F.Friche-Joset, Syn. Fl. Jura, ed. 2: 346 (1868).

Scirpus auronitens Nees & Ehrenb. ex Boeck., Linnaea 36: 723 (1870).

Scirpus laciniatus Nees & Ehrenb. ex Boeck., Linnaea 36: 723 (1870).

Scirpus hyalinolepis Steud. ex Jard., Bull. Soc. Linn. Normandie, II, 9: 280 (1875).

Scirpus fernaldii E.P.Bicknell, Torreya 1: 96 (1901). *Bolboschoenus fernaldii* (E.P.Bicknell) Soó ex Govaerts, World Checklist Seed Pl. 2(1): 14 (1996).

Scirpus polyopsis Bubani, Fl. Pyren. 4: 197 (1902).

Scirpus maritimus f. *digynus* Zapal., Consp. Fl. Gallic. Crit. 1: 123 (1906), nom. illeg. *Bolboschoenus maritimus* f. *zapalowiczii* Soó, Acta Bot. Acad. Sci. Hung. 16: 368 (1970 publ. 1971), nom. illeg.

Scirpus maritimus f. *sarmaticus* Zapal., Consp. Fl. Gallic. Crit. 1: 124 (1906), nom. illeg. *Bolboschoenus maritimus* f. *sarmaticus* (Zapal.) Soó, Acta Bot. Acad. Sci. Hung. 16: 368 (1970 publ. 1971), without basionym date.

Scirpus boraeanus Tourlet, Cat. Pl. Vasc. Indre-et-Loire: 524 (1908), nom. inval.

subsp. ***paludosus*** (A.Nelson) T.Koyama, Acta Phytotax. Geobot. 31: 148 (1980).

Hawaiian Is., N. America to N. Argentina. 63 HAW 70 ASK NWT 71 ABT BRC MAN SAS 72 NBR NSC ONT PEI QUE 73 COL IDA MNT ORE WAS WYO 74 ILL IOW KAN MIN MSO NDA NEB OKL SDA 75 CNT MAI MAS MIC NWH NWJ NWY RHO 76 ARI CAL NEV UTA 77 NWM TEX 79 MXC MXN MXS 81 CUB 82 SUR 83 PER 84 85 AGE AGS AGW. Hemicr.

Scirpus campestris Britton in N.L.Britton & A.Brown, Ill. Fl. N. U.S. 1: 267 (1896), nom. illeg. *Scirpus robustus* var. *campestris* Fernald, Rhodora 2: 241 (1900). *Scirpus interior* Britton, Man. Fl. N. States, ed. 2: 178 (1905).

**Scirpus paludosus* A.Nelson, Bull. Torrey Bot. Club 26: 5 (1899). *Scirpus robustus* var. *paludosus* (A.Nelson) Fernald, Rhodora 2: 241 (1900). *Scirpus campestris* var. *paludosus* (A.Nelson) Fernald, Rhodora 8: 162 (1906). *Scirpus maritimus* var. *paludosus* (A.Nelson) Kük., Repert. Spec. Nov. Regni Veg. 23: 200 (1926). *Bolboschoenus paludosus* (A.Nelson) Soó, Acta Bot. Acad. Sci. Hung. 16: 368 (1970 publ. 1971).

Scirpus pacificus Britton ex Parish, Bull. S. Calif. Acad. Sci. 4: 132 (1904).

Scirpus brittonianus Piper, Contr. U. S. Natl. Herb. 11: 157 (1906).

Scirpus paludosus var. *atlanticus* Fernald, Rhodora 45: 291 (1943).

Bolboschoenus medianus (V.J.Cook) Soják, Cas. Nár. Mus., Odd. Prír. 141: 63 (1972).

SE. Australia, New Zealand. 50 NSW SOA TAS VIC 51 NZN NZS. Hemicr.

**Scirpus medianus* V.J.Cook, Trans. & Proc. Roy. Soc. New Zealand 76: 569 (1947).

Bolboschoenus nobilis (Ridl.) Goetgh. & D.A.Simpson, Kew Bull. 46: 173 (1991).

S. Angola to Namibia. 26 ANG 27 NAM. Hemicr.

**Scirpus nobilis* Ridl., Trans. Linn. Soc. London, Bot. 2: 159 (1884). *Schoenoplectus nobilis* (Ridl.) Lye, Lidia 3: 144 (1995).

Bolboschoenus planiculmis (F.Schmidt) E.M.Egorova, Trudy Bot. Inst. Akad. Nauk S.S.S.R., Ser. 1, Fl. Sist. Vyssh. Rast. 3: 20 (1967).

C. Europe to Asia. 11 CZE 14 KRY RUS UKR 30 ALT BRY CTA IRK KRA TVA WSB YAK 31 KAM KHA PRM SAK 32 KAZ KGZ TZK 33 NCS TCS 36 CHS 37 MON 38 JAP KOR NNS TAI 40 IND 42 PHI 43 NWG. Hemicr.

**Scirpus planiculmis* F.Schmidt, Reis. Amur-Land., Bot.: 190 (1868).

Scirpus koshewnikowii Litv., Wild Pl. Tombov Gov.: 142 (1888). *Bolboschoenus koshewnikowii* (Litv.) A.E.Kozhevn., Sosud. Rast. Sovet. Dal'nego Vostoka 3: 189 (1988).

Scirpus compactus var. *orientalis* Litv., Sched. Herb. Fl. Ross. 8: 2792 (1922).

Scirpus biconcavus Ohwi, Mem. Coll. Sci. Kyoto Imp. Univ., Ser. B, Biol. 18(1): 109 (1944).

Bolboschoenus robustus (Pursh) Soják, Cas. Nár. Mus., Odd. Prír. 141: 63 (1972).

S. & E. U.S.A. to Mexico, Caribbean to S. Trop. America. 75 CNT MAI MAS NWH NWJ NWY PEN RHO 76 CAL 77 TEX 78 ALA ARK DEL FLA GEO LOU MRY MSI NCA SCA VRG 79 MXN MXT 81 BER 82 FRG? GUY SUR VEN 85 AGE PAR URU. Hemicr.

Scirpus maritimus var. *macrostachyus* Michx., Fl. Bor.-Amer. 1: 32 (1803). **Scirpus robustus* Pursh, Fl. Amer. Sept. 1: 56 (1813). *Scirpus maritimus* var. *robustus* (Pursh) Kük., Repert. Spec. Nov. Regni Veg. 23: 200 (1926). *Schoenoplectus robustus* (Pursh) M.T.Strong, Novon 3: 203 (1993). *Bolboschoenus maritimus* subsp. *macrostachyus* (Michx.) Soják, Cas. Nár. Mus., Odd. Prír. 152: 19 (1983).

Scirpus robustus var. *compactus* Burtt Davy, Fl. W. Calif.: 88 (1901).

Scirpus robustus f. *protrusus* Fernald, Rhodora 45: 392 (1943).

Bolboschoenus stagnicola (Raymond) Soják, Cas. Nár. Mus., Odd. Prír. 148: 193 (1979 publ. 1980).

Pakistan. 40 PAK. Hemicr.

**Scirpus stagnicola* Raymond, Naturaliste Canad. 99: 237 (1972).

Bolboschoenus yagara (Ohwi) Y.C.Yang & M.Zhan, Acta Biol. Plateau Sin. 7: 14 (1987 publ. 1988).

Temp. Eurasia. 11 CZE 31 AMU 38 JAP KOR TAI. Hemicr.

Synonyms:

Bolboschoenus affinis (Roth) Drobow = ***Bolboschoenus maritimus*** subsp. ***affinis*** (Roth) T.Koyama

Bolboschoenus compactus (Hoffm.) Drobow = ***Bolboschoenus maritimus*** (L.) Palla subsp. ***maritimus***

Bolboschoenus desoulavii (Drobow) A.E.Kozhevn. = ***Bolboschoenus maritimus*** (L.) Palla subsp. ***maritimus***

Bolboschoenus fernaldii (E.P.Bicknell) Soó ex Govaerts = ***Bolboschoenus maritimus*** (L.) Palla subsp. ***maritimus***

Bolboschoenus fluviatilis subsp. *yagara* (Ohwi) T.Koyama = ***Bolboschoenus fluviatilis*** (Torr.) Soják

Bolboschoenus grandispicus (Steud.) Lewej. & Lobin = ***Bolboschoenus maritimus*** (L.) Palla subsp. ***maritimus***

Bolboschoenus koshewnikowii (Litv.) A.E.Kozhevn. = ***Bolboschoenus planiculmis*** (F.Schmidt) E.M.Egorova

Bolboschoenus macrostachys (Willd.) Grossh. = ***Bolboschoenus glaucus*** S.G.Sm.

Bolboschoenus maritimus var. *affinis* (Roth) Parmar = ***Bolboschoenus maritimus*** subsp. ***affinis*** (Roth) T.Koyama

Bolboschoenus maritimus subsp. *compactus* (Hoffm.) Hejn? = ***Bolboschoenus maritimus*** (L.) Palla subsp. ***maritimus***

Bolboschoenus maritimus var. *compactus* (Hoffm.) T.V.Egorova = ***Bolboschoenus maritimus*** (L.) Palla subsp. ***maritimus***

Bolboschoenus maritimus subsp. *cymosus* (Rchb.) Soják = ***Bolboschoenus maritimus*** var. ***cymosus*** (Rchb.) Kit Tan & Oteng-Yeb.
Bolboschoenus maritimus var. *desoulavii* Drobow = ***Bolboschoenus maritimus*** (L.) Palla subsp. ***maritimus***
Bolboschoenus maritimus subsp. *fluviatilis* (Torr.) Á.Löve & D.Löve = ***Bolboschoenus fluviatilis*** (Torr.) Soják
Bolboschoenus maritimus f. *longifolius* (Peterm.) Soó = ***Bolboschoenus maritimus*** (L.) Palla subsp. ***maritimus***
Bolboschoenus maritimus subsp. *macrostachyus* (Michx.) Soják = ***Bolboschoenus robustus*** (Pursh) Soják
Bolboschoenus maritimus f. *sarmaticus* (Zapal.) Soó = ***Bolboschoenus maritimus*** (L.) Palla subsp. ***maritimus***
Bolboschoenus maritimus f. *subumbellatus* (Schur) Soó = ***Bolboschoenus maritimus*** (L.) Palla subsp. ***maritimus***
Bolboschoenus maritimus subsp. *tuberosus* (Desf.) T.Koyama = ***Bolboschoenus glaucus*** S.G.Sm.
Bolboschoenus maritimus f. *zapalowiczii* Soó = ***Bolboschoenus maritimus*** (L.) Palla subsp. ***maritimus***
Bolboschoenus novae-angliae (Britton) S.G.Sm. = ***Schoenoplectus novae-angliae*** (Britton) M.T.Strong
Bolboschoenus paludosus (A.Nelson) Soó = ***Bolboschoenus maritimus*** subsp. ***paludosus*** (A.Nelson) T.Koyama
Bolboschoenus popovii E.M.Egorova = ***Bolboschoenus maritimus*** subsp. ***affinis*** (Roth) T.Koyama
Bolboschoenus schmidii (Raymond) Holub = ***Bolboschoenus maritimus*** subsp. ***affinis*** (Roth) T.Koyama
Bolboschoenus strobilinus (Roxb.) V.I.Krecz. = ***Bolboschoenus maritimus*** subsp. ***affinis*** (Roth) T.Koyama

Borabora

Borabora Steud. = ***Cyperus*** L.
Borabora cyperoidea Steud. = ***Cyperus macrophyllus*** (Brongn.) Boeck.

Buekia

Buekia Nees = ***Neesenbeckia*** Levyns
Buekia punctoria (Vahl) Nees = ***Neesenbeckia punctoria*** (Vahl) Levyns

Bulbostylis

Bulbostylis Kunth, Enum. Pl. 2: 205 (1837), nom. cons.
Trop. & Subtrop. to C. Asia. (12) fra (13) ita† 20 MOR 21 CVI 22 BEN BKN GAM GHA GNB GUI IVO LBR MLI MTN NGA NGR SEN SIE TOG 23 BUR CAF CMN CON EQG GAB GGI RWA ZAI 24 CHA ETH SOC SOM SUD 25 KEN TAN UGA 26+ ANG MLW MOZ ZAM ZIM 27 BOT CPP LES NAM NAT OFS SWZ TVL 28 STH 29 ALD COM MAU MDG REU SEY 31 PRM 33 TCS 34 CHA TUR 35 YEM 36 CHS 38 JAP KZN NNS TAI 40 ASS BAN EHM IND NEP PAK SRL WHM 41 THA VIE 42 BOR JAW LSI MLY MOL PHI SUL SUM 43 NWG 50 NFK NSW NTA QLD SOA WAU 60 NWC (63) haw 72 NBR NSC ONT QUE 74 ILL IOW KAN MIN MSO NEB OKL WIS 75 CNT INI MAI MAS MIC NWH NWJ NWY OHI PEN RHO VER WVA 76 ARI CAL 77 NWM TEX 78 ALA ARK DEL FLA GEO KTY LOU MRY MSI NCA SCA TEN VRG WDC 79+ MXC MXE MXG MXI MXN MXS MXT 80 BLZ COS ELS GUA HON NIC PAN 81 ARU BAH CUB DOM HAI JAM LEE NLA PUE TRT WIN 82 FRG GUY SUR VEN 83 BOL CLM ECU PER 84 BZC BZE BZL BZN BZS 85 AGE AGW PAR PER URU.
203 Species
Stenophyllus Raf., Neogenyton: 4 (1825).
Oncostylis Nees in C.F.P.von Martius & auct. suc. (eds.), Fl. Bras. 2(1): 80 (1842).

Bulbostylis abortiva (Steud.) C.B.Clarke in J.Buchan, Nyassaland: 21 (1889).
Trop. Africa, Comores, Madagascar. 22 BEN BKN GHA IVO MLI NGA NGR SEN SIE TOG 23 BUR EQG GAB ZAI 24 CHA 25 KEN TAN UGA 26 ANG MLW MOZ 29 COM MDG. Ther. or hemicr.
**Fimbristylis abortiva* Steud., Syn. Pl. Glumac. 2: 111 (1855). *Abildgaardia abortiva* (Steud.) Lye, Bot. Not. 127: 495 (1974).
Scirpus schweinfurthianus Boeck., Linnaea 36: 757 (1870).
Bulbostylis camporum K.Schum. ex C.B.Clarke in D.Oliver & auct. suc. (eds.), Fl. Trop. Afr. 8: 441 (1902), pro syn.
Bulbostylis classensii De Wild., Pl. Bequaert. 4: 192 (1927).

Bulbostylis afro-orientalis (Lye) R.W.Haines in R.W.Haines & K.A.Lye, Sedges & Rushes E. Afr., App. 3: 1 (1983).
E. Kenya. 25 KEN. Ther.
**Abildgaardia afro-orientalis* Lye, Nordic J. Bot. 3: 233 (1983).

Bulbostylis alpestris Urb., Symb. Antill. 7: 168 (1912). *Stenophyllus alpestris* (Urb.) Britton, Bull. Torrey Bot. Club 43: 446 (1916). *Fimbristylis alpestris* (Urb.) Alain, Bull. Torrey Bot. Club 92: 290 (1965).
Dominican Rep. 81 DOM. Hemicr. or rhizome geophyte.

Bulbostylis amambayensis Barros, Darwiniana 6: 122 (1942).
Paraguay. 85 PAR. Hemicr. or rhizome geophyte.

Bulbostylis andongensis (Ridl.) C.B.Clarke in T.A.Durand & H.Schinz, Consp. Fl. Afric. 5: 611 (1894).
WC. Trop. Africa to Angola. 23 CON GAB ZAI 26 ANG. Ther.
**Fimbristylis andongensis* Ridl., Trans. Linn. Soc. London, Bot. 2: 153 (1884).

Bulbostylis andringitrensis Cherm., Bull. Soc. Bot. France 69: 720 (1922 publ. 1923).
C. Madagascar. 29 MDG. Ther.

Bulbostylis aphylla Barros, Darwiniana 5: 180 (1941).
Paraguay to Argentina (Misiones). 85 AGE PAR URU. Hemicr. or rhizome geophyte.

Bulbostylis arcuata Kral, Sida 4: 91 (1971).
Mexico (S. Baja California to Jalisco). 79 MXE MXN MXS. Ther.

Bulbostylis argentobrunnea C.B.Clarke in D.Oliver & auct. suc. (eds.), Fl. Trop. Afr. 8: 440 (1902). *Abildgaardia argentobrunnea* (C.B.Clarke) Lye, Bot. Not. 127: 495 (1974).
E. Kenya. 25 KEN. Hemicr.

Bulbostylis atrosanguinea (Boeck.) C.B.Clarke in T.A.Durand & H.Schinz, Consp. Fl. Afric. 5: 611 (1894).
Ethiopia to Angola, Yemen. 23 BUR RWA ZAI 24 ETH 25 KEN TAN UGA 26 ANG MLW 35 YEM. Hemicr.

Fimbristylis setifolia A.Rich., Tent. Fl. Abyss. 2: 507 (1850). *Bulbostylis setifolia* (A.Rich.) Bodard, Ann. Fac. Sci. Univ. Dakar 9: 66 (1963), nom. illeg. *Abildgaardia setifolia* (A.Rich.) Lye, Bot. Not. 127: 497 (1974).
**Scirpus atrosanguineus* Boeck., Bot. Jahrb. Syst. 7: 276 (1886). *Fimbristylis atrosanguinea* (Boeck.) K.Schum. in H.G.A.Engler, Pflanzenw. Ost-Afrikas, C: 125 (1895).

Bulbostylis aturensis (Maury) C.B.Clarke, Bull. Misc. Inform. Kew, Addit. Ser. 8: 26 (1908).
SW. Venezuela. 82 VEN. Ther.
**Scirpus aturensis* Maury, J. Bot. (Morot) 3: 199 (1889). *Bulbostylis leucostachya* var. *aturensis* (Maury) H.Pfeiff., Repert. Spec. Nov. Regni Veg. 31: 323 (1933).

Bulbostylis barbata (Rottb.) C.B.Clarke in J.D.Hooker, Fl. Brit. India 6: 651 (1893).
Cape Verde to Australia. (12) fra 21 CVI 22 BEN BKN GAM GHA GNB IVO LBR MLI NGA NGR SEN TOG 23 ZAI 24 CHA ETH SOM 29 MAU MDG REU SEY 36 CHS 38 JAP KZN NNS TAI 40 ASS EHM IND NEP PAK SRL WHM 41 THA VIE 42 BOR JAW LSI MLY MOL PHI SUL SUM 43 NWG 50 NSW NTA QLD SOA WAU (77) tex (78) ala fla geo lou msi nca sca ten (79) mxn (80) cos gua (81) win (82) ven. Ther.
**Scirpus barbatus* Rottb., Descr. Icon. Rar. Pl.: 52 (1773). *Isolepis barbata* (Rottb.) R.Br., Prodr.: 222 (1810). *Abildgaardia barbata* (Rottb.) P.Beauv., Fl. Oware 2: 47 (1816). *Cyperus barbata* (Rottb.) Poir. in J.B.A.M.de Lamarck, Encycl., Suppl. 5: 186 (1817). *Fimbristylis barbata* (Rottb.) Benth., Fl. Austral. 7: 321 (1878). *Iria barbata* (Rottb.) Kuntze, Revis. Gen. Pl. 2: 753 (1891). *Stenophyllus barbatus* (Rottb.) T.Cooke, Fl. Bombay 2(2): 887 (1908). *Stenophyllus barbatus* (Rottb.) Fyson, Bot. Bull. Pres. Coll. Madras: t. 64 (1919), nom. illeg.
Scirpus barbatus var. *americanus* Boeck., Linnaea 36: 752 (1870).
Stenophyllus barbatus f. *oligostachyus* Domin, Biblioth. Bot. 85: 465 (1915).

subsp. ***barbata***
Cape Verde to Australia. (12) fra 21 CVI 22 BEN BKN GAM GHA GNB IVO LBR MLI NGA NGR SEN TOG 23 ZAI 24 CHA ETH SOM 29 MAU MDG REU SEY 36 CHS 38 JAP KZN NNS TAI 40 ASS EHM IND NEP PAK SRL WHM 41 THA VIE 42 BOR JAW LSI MLY MOL PHI SUL SUM 43 NWG 50 NSW NTA QLD SOA WAU (77) tex (78) ala fla geo lou msi nca sca ten (79) mxn (80) gua (81) win (82) ven. Ther.
Scirpus antarcticus Vahl, Enum. Pl. 2: 261 (1805), nom. illeg.
Scirpus minimus Vahl, Enum. Pl. 2: 253 (1805).
Scirpus monander Roxb., Fl. Ind. 1: 222 (1820), nom. illeg.
Fimbristylis monandra Schult. in J.J.Roemer & J.A.Schultes, Mant. 2: 59 (1824).
Isolepis wallichiana Schult. in J.J.Roemer & J.A.Schultes, Mant. 2: 533 (1824). *Scirpus wallichianus* (Schult.) Spreng., Syst. Veg. 4(2): 27 (1827). *Bulbostylis wallichiana* (Schult.) Beetle, Leafl. W. Bot. 4: 45 (1944). *Abildgaardia wallichiana* (Schult.) Lye, Nordic J. Bot. 3: 239 (1983).
Scirpus muehlenbergii Spreng., Syst. Veg. 1: 207 (1824).
Isolepis fascicularis Sieber ex Kunth, Enum. Pl. 2: 209 (1837), nom. inval.
Isolepis willdenowii Kunth, Enum. Pl. 2: 210 (1837), nom. illeg. *Abildgaardia willdenowii* Lye, Nordic J. Bot. 1: 758 (1981 publ. 1982).
Scirpus brachyphyllus Willd. ex Kunth, Enum. Pl. 2: 212 (1837).
Scirpus brevifolius Bernh. ex Kunth, Enum. Pl. 2: 212 (1837).
Scirpus tristachyos Bojer, Hortus Maurit.: 383 (1837), nom. inval.
Fimbristylis tenuissima Steud., Syn. Pl. Glumac. 2: 109 (1855).
Isolepis cumingii Steud., Syn. Pl. Glumac. 2: 101 (1855). *Fimbristylis cumingii* (Steud.) Náves in F.M.Blanco, Fl. Filip., ed. 3, 4(13A): 308 (1880). *Scirpus mindorensis* Beetle, Leafl. W. Bot. 4: 46 (1944).
Isolepis involucellata Steud., Syn. Pl. Glumac. 2: 101 (1855).
Isolepis armerioides Miq., Fl. Ned. Ind. 3: 310 (1856). *Fimbristylis armerioides* (Miq.) Beetle, Leafl. W. Bot. 4: 45 (1944).
Isolepis foenicularis Schrad. ex Boeck., Linnaea 35: 752 (1870).
Isolepis subtristachya Hochst. ex Boeck., Linnaea 36: 752 (1870).
Scirpus dussii Boeck., Beitr. Cyper. 2: 38 (1890).
Isolepis subtristachya Hochst. ex C.B.Clarke in T.A.Durand & H.Schinz, Consp. Fl. Afric. 5: 611 (1894), nom. inval.
Stenophyllus floridanus Nash, Bull. Torrey Bot. Club 22: 161 (1895). *Bulbostylis floridana* (Nash) Fernald, Rhodora 40: 392 (1938).
Scirpus dussianus Boeck. ex Duss, Ann. Inst. Bot.-Géol. Colon. Marseille 3: 548 (1896 publ. 1897).
Bulbostylis willdenowii C.B.Clarke in D.Oliver & auct. suc. (eds.), Fl. Trop. Afr. 8: 431 (1902), pro syn.
Scirpus sterilis Maiden & Betche, Proc. Linn. Soc. New South Wales 33: 315 (1908).

subsp. ***pulchella*** (Thwaites) T.Koyama, Bot. Mag. (Tokyo) 93(1032): 341 (1980).
S. India, Sri Lanka, Indo-China. 40 IND SRL 41 VIE. Ther.
**Isolepis pulchella* Thwaites, Enum. Pl. Zeyl.: 350 (1864). *Fimbristylis pulchella* (Thwaites) Trimen, Syst. Cat. Fl. Pl. Ceylon: 102 (1885). *Bulbostylis barbata* var. *pulchella* (Thwaites) C.B.Clarke in J.D.Hooker, Fl. Brit. India 6: 652 (1893).

Bulbostylis basalis Fosberg, Kew Bull. 31: 829 (1977).
Aldabra. 29 ALD. Ther.

Bulbostylis bathiei Cherm., Bull. Soc. Bot. France 72: 617 (1925).
E. Madagascar. 29 MDG. Hemicr.

Bulbostylis bodardii S.S.Hooper, Kew Bull. 26: 582 (1972).
W. Trop. Africa. 22 GUI SEN. Hemicr. or rhizome geophyte.
**Bulbostylis zambesica* var. *occidentalis* Bodard, Bull. Soc. Bot. France 108: 309 (1961).

Bulbostylis boeckeleriana (Schweinf.) Beetle, Amer. Midl. Naturalist 41: 458 (1949).
Ethiopia to S. Africa. 23 RWA 24 ETH 25 KEN TAN 26 ZIM 27 CPP NAT TVL. Hemicr.

Scirpus boeckelerianus Schweinf., Bull. Herb. Boissier, App. 2: 50 (1894). *Abildgaardia boeckeleriana* (Schweinf.) Lye, Bot. Not. 126: 327 (1973).

var. ***boeckeleriana***
Ethiopia to S. Africa. 23 RWA 24 ETH 25 KEN TAN 26 ZIM 27 CPP NAT TVL. Hemicr.

var. ***transiens*** (K.Schum.) R.W.Haines in R.W.Haines & K.A.Lye, Sedges & Rushes E. Afr., App. 3: 1 (1983).
E. Trop. Africa. 25 KEN TAN. Hemicr.
**Fimbristylis transiens* K.Schum. in H.G.A.Engler, Pflanzenw. Ost-Afrikas, C: 124 (1895). *Bulbostylis transiens* (K.Schum.) C.B.Clarke in D.Oliver & auct. suc. (eds.), Fl. Trop. Afr. 8: 444 (1902). *Abildgaardia boeckeleriana* var. *transiens* (K.Schum.) Lye, Nordic J. Bot. 3: 239 (1983).

Bulbostylis boivinii C.B.Clarke ex Cherm., Notul. Syst. (Paris) 5: 275 (1936).
Comoros. 29 COM. Hemicr. or rhizome geophyte.

Bulbostylis boliviana Palla, Oesterr. Bot. Z. 59: 191 (1909).
Bolivia. 83 BOL. Hemicr. or rhizome geophyte.

Bulbostylis bozumensis Cherm., Arch. Bot. Mém. 3: 11 (1936).
Central African Rep., N. Zaïre. 23 CAF ZAI. Hemicr. or rhizome geophyte.

Bulbostylis brasiliensis Palla, Denkschr. Kaiserl. Akad. Wiss., Wien. Math.-Naturwiss. Kl. 79: 178 (1908).
Brazil. 84+. Hemicr. or rhizome geophyte.

Bulbostylis brevifolia Palla, Denkschr. Kaiserl. Akad. Wiss., Wien. Math.-Naturwiss. Kl. 79: 178 (1908).
Brazil. 84+. Hemicr. or rhizome geophyte.

Bulbostylis buchananii C.B.Clarke in D.Oliver & auct. suc. (eds.), Fl. Trop. Afr. 8: 437 (1902). *Abildgaardia buchananii* (C.B.Clarke) Lye, Bot. Not. 127: 495 (1974).
Kenya to Zimbabwe. 23 BUR ZAI 25 KEN TAN 26 MLW ZAM ZIM. Ther. or hemicr.

Bulbostylis burbidgeae K.L.Wilson, Telopea 1: 458 (1980).
Western Australia (Fortescue). 50 WAU. Hemicr. or rhizome geophyte.

Bulbostylis burchellii (Ficalho & Hiern) C.B.Clarke in T.A.Durand & H.Schinz, Consp. Fl. Afric. 5: 612 (1894).
Tanzania to S. Africa. 25 TAN 27 BOT CPP NAM NAT OFS SWZ TVL. Hemicr.
**Fimbristylis burchellii* Ficalho & Hiern, Trans. Linn. Soc. London, Bot. 2: 28 (1881). *Abildgaardia burchellii* (Ficalho & Hiern) Lye, Bot. Not. 127: 495 (1974).

Bulbostylis capillaris (L.) Kunth ex C.B.Clarke in J.D.Hooker, Fl. Brit. India 6: 652 (1885).
SE. Canada to C. America, Caribbean. (63) haw 72 NBR NSC ONT QUE 74 ILL IOW KAN MIN MSO NEB OKL WIS 75 CNT INI MAI MAS MIC NWH NWJ NWY OHI PEN RHO VER WVA 76 ARI CAL 77 NWM TEX 78 ALA ARK DEL FLA GEO KTY LOU MRY MSI NCA SCA TEN VRG WDC 79 MXC MXE MXG MXN MXS MXT 80 BLZ GUA HON NIC 81 CUB DOM PUE TRT WIN 82 FRG GUY SUR VEN 83 BOL ECU PER 84 BZC 85 AGE AGW PAR URU. Ther. or hemicr.
**Scirpus capillaris* L., Sp. Pl.: 49 (1753). *Isolepis capillaris* (L.) Roem. & Schult., Syst. Veg. 2: 118 (1817). *Trichelostylis capillaris* (L.) Alph.Wood, Class-book Bot.: 742 (1861). *Fimbristylis capillaris* (L.) A.Gray, Manual, ed. 5: 567 (1867). *Iria capillaris* (L.) Kuntze, Revis. Gen. Pl. 2: 753 (1891). *Stenophyllus capillaris* (L.) Britton, Bull. Torrey Bot. Club 21: 30 (1894). *Bulbostylis capillaris* var. *genuina* Osten, Anales Mus. Hist. Nat. Montevideo, II, 3: 192 (1931), nom. inval. *Abildgaardia capillaris* (L.) Lye, Bot. Not. 127: 495 (1974).

subsp. ***antillana*** (Britton) T.Koyama, J. Arnold Arbor. 60: 322 (1979).
Dominican Rep. to Lesser Antilles. 81 DOM PUE WIN. Ther. or hemicr.
**Stenophyllus antillanus* Britton, Bull. Torrey Bot. Club 43: 447 (1916). *Bulbostylis antillana* (Britton) Fernald, Rhodora 40: 392 (1938).

subsp. ***capillaris***
SE. Canada to C. America, Caribbean. (63) haw 72 NBR NSC ONT QUE 74 ILL IOW KAN MIN MSO NEB OKL WIS 75 CNT INI MAI MAS MIC NWH NWJ NWY OHI PEN RHO VER WVA 76 ARI CAL 77 NWM TEX 78 ALA ARK DEL FLA GEO KTY LOU MRY MSI NCA SCA TEN VRG WDC 79 MXC MXE MXG MXN MXS MXT 80 BLZ GUA HON NIC 81 CUB DOM PUE TRT WIN 82 FRG GUY SUR VEN 83 BOL ECU PER 84 BZC 85 AGE AGW PAR URU. Ther.
Cyperus minimus L., Sp. Pl.: 44 (1753).
Scirpus castaneus Muhl., Descr. Gram.: 38 (1817), nom. illeg.
Isolepis fusca Link, Jahrb. Gewächsk. 1(3): 79 (1820).
Isolepis muehlenbergii A.Dietr., Sp. Pl. 2: 106 (1832).
Fimbristylis coarctata Schwein. in L.C.Beck, Bot. North. Middle States: 423 (1833).
Isolepis coarctata Torr., Ann. Lyceum Nat. Hist. New York 3: 102 (1836).
Isolepis capillaris f. *borealiamericanae* Kunth, Enum. Pl. 2: 212 (1837).
Isolepis capillaris f. *europaea* Kunth, Enum. Pl. 2: 212 (1837).
Isolepis capillaris f. *indicae* Kunth, Enum. Pl. 2: 212 (1837).
Isolepis capillaris f. *meridionaliamericanae* Kunth, Enum. Pl. 2: 212 (1837).
Oncostylis ciliata Nees in C.F.P.von Martius & auct. suc. (eds.), Fl. Bras. 2(1): 84 (1842).
Oncostylis dubia Nees in C.F.P.von Martius & auct. suc. (eds.), Fl. Bras. 2(1): 84 (1842).
Cyperus pubescens Steud., Syn. Pl. Glumac. 2: 50 (1854), nom. illeg.
Isolepis radiciflora Steud., Syn. Pl. Glumac. 2: 318 (1855).
Scirpus trifidus Hance, J. Bot. 16: 112 (1878).
Scirpus capillaris var. *elatior* Griseb., Abh. Königl. Ges. Wiss. Göttingen 34: 317 (1879). *Bulbostylis capillaris* var. *elatior* (Griseb.) Osten, Anales Mus. Nac. Montevideo, II, 3: 193 (1931).
Fimbristylis capillaris f. *major* Maury, J. Bot. (Morot) 3: 197 (1889). *Bulbostylis capillaris* var. *major* (Maury) R.Knuth, Repert. Spec. Nov. Regni Veg. Beih. 43: 144 (1926).
Fimbristylis capillaris f. *minima* Maury, J. Bot. (Morot) 3: 197 (1889). *Bulbostylis capillaris*

f. *minima* (Maury) R.Knuth, Repert. Spec. Nov. Regni Veg. Beih. 43: 144 (1926).
Scirpus microstachys Boeck., Beitr. Cyper. 2: 14 (1890). *Bulbostylis capillaris* var. *microstachya* (Boeck.) Barros, Anales Mus. Argent. Ci. Nat. "Bernardino Rivadavia" 41: 358 (1945).
Isolepis gracillima Hochst. ex C.B.Clarke in T.A.Durand & H.Schinz, Consp. Fl. Afric. 5: 612 (1894), nom. inval.
Scirpus capillaris var. *sphaerolepis* Boeck., Allg. Bot. Z. Syst. 2: 55 (1896).
Bulbostylis capillaris var. *pyriformis* C.B.Clarke ex Lindm., Bih. Kongl. Svenska Vetensk.-Akad. Handl. 26(3; 9): 20 (1900).
Stenophyllus capillaris var. *trifidus* (Kunth) Domin, Biblioth. Bot. 85: 464 (1915).
Stenophyllus capillaris var. *cryptostachys* Fernald, Rhodora 19: 154 (1917).
Bulbostylis capillaris f. *contracta* Kük. ex Osten, Anales Mus. Nac. Montevideo, II, 3: 192 (1931).
Bulbostylis capillaris f. *fasciculata* Osten, Anales Mus. Hist. Nat. Montevideo, II, 3: 192 (1931).
Bulbostylis capillaris var. *major* Uittien, Recueil Trav. Bot. Néerl. 30: 188 (1933), nom. illeg.
Bulbostylis capillaris var. *crebra* Fernald, Rhodora 40: 395 (1938).
Bulbostylis capillaris var. *isopoda* Fernald, Rhodora 40: 395 (1938).
Bulbostylis capillaris f. *stenantha* Kük. ex Barros, Darwiniana 5: 181 (1941).

subsp. **insulana** M.T.Strong, Contr. U. S. Natl. Herb. 52: 243 (2005).
Cuba to Puerto Rico. 81 CUB DOM PUE. Hemicr. or rhizome geophyte

Bulbostylis carajana Kral & M.T.Strong, Sida 18: 844 (1999).
Brazil (Pará). 84 BZN. Ther.

Bulbostylis cardiocarpoides Cherm., Rev. Zool. Bot. Africaines 24: 298 (1934). *Abildgaardia cardiocarpoides* (Cherm.) Lye, Bot. Not. 127: 495 (1974).
WC. Trop. Africa to Uganda. 23 BUR GAB ZAI 25 UGA. Hemicr.

Bulbostylis ciliata (C.Presl) C.Presl ex Nees in C.F.P.von Martius & auct. suc. (eds.), Fl. Bras. 2(1): 84 (1842).
Mexico to Brazil. 79+ 80 PAN 83 PER 84 BZE BZN. Ther.
**Isolepis ciliata* C.Presl, Reliq. Haenk. 1: 188 (1828). *Stenophyllus capillaris* var. *ciliatus* (C.Presl) J.F.Macbr., Publ. Field Mus. Nat. Hist., Bot. Ser. 4: 167 (1929).
Isolepis langsdorffiana Kunth, Enum. Pl. 2: 214 (1837). *Bulbostylis langsdorffiana* (Kunth) C.B.Clarke in I.Urban, Symb. Antill. 2: 89 (1900).
Cyperus pubifolius Steud., Syn. Pl. Glumac. 2: 50 (1854).

Bulbostylis ciliatifolia (Elliott) Fernald, Rhodora 40: 391 (1938).
SE. U.S.A. to Texas, Cuba. 77 TEX 78 ALA FLA GEO LOU MSI NCA SCA TEN VRG 81 CUB. Ther.
**Scirpus ciliatifolius* Elliott, Sketch Bot. S. Carolina 1: 82 (1816). *Isolepis ciliatifolia* (Elliott) Torr., Ann. Lyceum Nat. Hist. New York 3: 350 (1836). *Trichelostylis ciliatifolia* (Elliott) Alph.Wood, Amer. Bot. Fl.: 264 (1870). *Stenophyllus ciliatifolius* (Elliott) C.Mohr, Bull. Torrey Bot. Club 24: 22 (1897).

var. ***ciliatifolia***.
SE. U.S.A. 78 ALA FLA GEO MSI NCA SCA VRG. Ther.

var. ***coarctata*** (Elliott) Kral, Sida 4: 96 (1971).
SE. U.S.A. to Texas, Cuba. 77 TEX 78 ALA FLA GEO LOU MSI NCA SCA TEN VRG 81 CUB. Hemicr.
**Scirpus coarctatus* Elliott, Sketch Bot. S. Carolina 1: 83 (1816). *Trichelostylis coarctata* (Elliott) Alph.Wood, Class-book Bot.: 742 (1861). *Fimbristylis capillaris* var. *coarctata* (Elliott) Britton, Bull. Torrey Bot. Club 15: 102 (1888). *Stenophyllus coarctatus* (Elliott) Britton in J.K.Small, Fl. S.E. U.S.: 189 (1903). *Bulbostylis coarctata* (Elliott) Fernald, Rhodora 40: 392 (1938).
Stenophyllus carteri Britton ex Small, Bull. New York Bot. Gard. 3: 420 (1903). *Fimbristylis carteri* (Britton ex Small) Alain, Bull. Torrey Bot. Club 92: 290 (1965).

Bulbostylis cinnamomea (Boeck.) C.B.Clarke in J.Buchan, Nyassaland: 21 (1889).
Zaïre to S. Africa, Mauritius, Madagascar. 23 ZAI 26 MLW 27 TVL 29 MAU MDG. Hemicr.
**Scirpus cinnamomeus* Boeck., Bot. Jahrb. Syst. 5: 505 (1884). *Fimbristylis cinnamomea* (Boeck.) K.Schum. in H.G.A.Engler, Pflanzenw. Ost-Afrikas, C: 125 (1895).
Scirpus pretorianus Boeck. ex C.B.Clarke in T.A. Durand & H.Schinz, Consp. Fl. Afric. 5: 613 (1894), nom. inval.
Scirpus transvaalensis Boeck. ex C.B.Clarke. in T.A. Durand & H.Schinz, Consp. Fl. Afric. 5: 163 (1894), nom. inval.

Bulbostylis cioniana (Savi) Lye, Mitt. Bot. Staatssamml. München 10: 547 (1971).
Sahara to NC. Africa. (13) ita† 20 MOR 22 BEN BKN GHA GUI MLI NGA NGR SEN SIE 23 GAB ZAI 24 CHA. Ther.
**Fimbristylis cioniana* Savi, Mem. Vald. 3: 98 (1842).
Fimbristylis vermoesenii De Wild., Pl. Bequaert. 4: 205 (1927).

Bulbostylis circinata (Kunth) C.B.Clarke, Bull. Misc. Inform. Kew, Addit. Ser. 8: 28 (1908).
N. South America to Brazil. 82 FRG GUY SUR 84 BZS. Hemicr. or rhizome geophyte.
**Isolepis circinata* Kunth, Enum. Pl. 2: 214 (1837). *Oncostylis circinata* (Kunth) Nees in C.F.P.von Martius & auct. suc. (eds.), Fl. Bras. 2(1): 82 (1842). *Scirpus circinatus* (Kunth) Boeck., Linnaea 36: 762 (1870). *Stenophyllus circinatus* (Kunth) H.Pfeiff., Bot. Arch. 6: 190 (1924).
Isolepis crinita Schrad. ex Nees in C.F.P.von Martius & auct. suc. (eds.), Fl. Bras. 2(1): 82 (1842).

Bulbostylis clarkeana Hutch. ex Bodard, Bull. Soc. Bot. France 108: 308 (1962). *Abildgaardia clarkeana* (Hutch. ex Bodard) Lye, Bot. Not. 127: 495 (1974).
W. Trop. Africa to Kenya. 22+ 23 BUR 25 KEN TAN UGA. Hemicr.

Bulbostylis closii Barros, Darwiniana 5: 181 (1941).
Brazil to NE. Argentina. 84 BZS 85 AGE PER. Hemicr. or rhizome geophyte.

Bulbostylis coleotricha (Hochst. ex A.Rich.) C.B.Clarke in T.A.Durand & H.Schinz, Consp. Fl. Afric. 5: 613 (1894).

Trop. Africa. 22 BEN BKN GHA GUI IVO MLI NGA NGR SEN SIE TOG 23 BUR CMN EQG GAB ZAI 24 CHA ETH 25 KEN UGA. Ther.

**Fimbristylis coleotricha* Hochst. ex A.Rich., Tent. Fl. Abyss. 2: 503 (1850). *Scirpus coleotrichus* (Hochst. ex A.Rich.) Boeck., Linnaea 36: 763 (1870). *Abildgaardia coleotricha* (Hochst. ex A.Rich.) Lye, Bot. Not. 127: 495 (1974).

var. ***coleotricha***

Trop. Africa. 22 BEN BKN GHA GUI IVO MLI NGA NGR SEN SIE TOG 23 BUR CMN EQG GAB ZAI 24 CHA ETH 25 KEN UGA. Ther.

Bulbostylis seretii De Wild., Pl. Bequaert. 4: 197 (1927).

var. ***miegei*** (Bodard) R.W.Haines in R.W.Haines & K.A.Lye, Sedges & Rushes E. Afr., App. 3: 1 (1983).

Uganda, Mali. 22 MLI 25 UGA. Ther.

**Bulbostylis miegei* Bodard, Bull. Soc. Bot. France 108: 307 (1962). *Abildgaardia miegei* (Bodard) Lye, Bot. Not. 127: 497 (1974). *Abildgaardia coleotricha* var. *miegei* (Bodard) Lye, Nordic J. Bot. 3: 239 (1983).

Bulbostylis comorensis C.B.Clarke, Bull. Misc. Inform. Kew, Addit. Ser. 8: 26 (1908).

Comoros. 29 COM. Hemicr. or rhizome geophyte.

Bulbostylis congolensis De Wild., Pl. Bequaert. 4: 194 (1927). *Abildgaardia congolensis* (De Wild.) Lye, Bot. Not. 127: 495 (1974). *Abildgaardia pusilla* subsp. *congolensis* (De Wild.) Lye, Nordic J. Bot. 3: 239 (1983). *Bulbostylis pusilla* subsp. *congolensis* (De Wild.) R.W.Haines in R.W.Haines & K.A.Lye, Sedges & Rushes E. Afr., App. 3: 1 (1983).

W. Trop. Africa to Kenya. 22 BEN BKN GHA GUI IVO LBR MLI NGA SIE TOG 23 BUR GAB ZAI 25 KEN TAN UGA. Ther.

Bulbostylis polytricha Cherm., Arch. Bot. Mém. 7: 39 (1931).

Bulbostylis holotricha Peter, Repert. Spec. Nov. Regni Veg. Beih. 40(1): 127 (1936).

Bulbostylis conifera (Kunth) Beetle, Leafl. W. Bot. 4: 45 (1944).

S. Trop. America. 82 FRG GUY SUR VEN 83 BOL 84 BZC BZE BZL BZN BZS 85 URU. Hemicr.

Fimbristylis conifera Rchb. ex Kunth, Enum. Pl. 2: 206 (1837).

**Isolepis conifera* Kunth, Enum. Pl. 2: 206 (1837). *Oncostylis conifera* (Kunth) Nees in C.F.P.von Martius & auct. suc. (eds.), Fl. Bras. 2(1): 81 (1842). *Stenophyllus coniferus* (Kunth) Britton, Bull. Torrey Bot. Club 43: 445 (1916).

Isolepis ciliifolia Steud., Syn. Pl. Glumac. 2: 99 (1855).

Scirpus reichenbachii Boeck., Linnaea 36: 737 (1870).

Bulbostylis conostachya (Boeck.) Beetle, Leafl. W. Bot. 4: 45 (1944).

Brazil. 84+. Hemicr. or rhizome geophyte.

**Isolepis conostachya* Boeck., Flora 41: 597 (1858). *Scirpus conostachys* (Boeck.) Boeck., Linnaea 36: 756 (1870). *Bulbostylis junciformis* var. *conostachya* (Boeck.) Lindm., Bih. Kongl. Svenska Vetensk.-Akad. Handl. 26(3; 9): 19 (1900).

Bulbostylis consanguinea (Kunth) C.B.Clarke, Bull. Misc. Inform. Kew 1908: 110 (1908).

Brazil to NE. Argentina. 83 BOL 84 85 AGE PAR URU. Hemicr. or rhizome geophyte.

**Isolepis consanguinea* Kunth, Enum. Pl. 2: 211 (1837). *Scirpus consanguineus* (Kunth) Boeck., Vidensk. Meddel. Naturhist. Foren. Kjøbenhavn 1869: 138 (1860).

Oncostylis kunthiana Nees in C.F.P.von Martius & auct. suc. (eds.), Fl. Bras. 2(1): 87 (1842).

Oncostylis kunthiana var. *laxa* Nees in C.F.P.von Martius & auct. suc. (eds.), Fl. Bras. 2(1): 87 (1842).

Bulbostylis conspicua (Boeck.) H.Pfeiff., Repert. Spec. Nov. Regni Veg. 27: 90 (1929).

Brazil (Minas Gerais). 84 BZL. Hemicr. or rhizome geophyte.

**Fimbristylis conspicua* Boeck., Allg. Bot. Z. Syst. 2: 56 (1896). *Stenophyllus conspicuus* (Boeck.) H.Pfeiff., Repert. Spec. Nov. Regni Veg. 27: 90 (1929).

Bulbostylis contexta (Nees) Bodard, Ann. Fac. Sci. Univ. Dakar 9: 77 (1963).

Ethiopia to S. Africa, Madagascar. 23 BUR 24 ETH 25 TAN 26+ 27 BOT CPP LES NAM NAT SWZ TVL. Hemicr.

**Trichelostylis contexta* Nees, Linnaea 10: 146 (1836). *Fimbristylis contexta* (Nees) Kunth, Enum. Pl. 2: 245 (1837). *Abildgaardia contexta* (Nees) Lye, Bot. Not. 127: 495 (1974).

Isolepis collina Kunth, Enum. Pl. 2: 208 (1837). *Bulbostylis collina* (Kunth) C.B.Clarke in T.A. Durand & H.Schinz, Consp. Fl. Afric. 5: 613 (1894). *Stenophyllus collinus* (Kunth) Chiov., Miss. Biol. Borana, Racc. Bot., Angiosp.-Gymnosp.: 299 (1939).

Scirpus collinus Boeck., Linnaea 26: 746 (1870).

Scirpus zeyheri Boeck., Linnaea 36: 752 (1870). *Bulbostylis zeyheri* (Boeck.) C.B.Clarke in T.A. Durand & H.Schinz, Consp. Fl. Afric. 5: 616 (1894).

Scirpus trispiculatus Boeck. ex C.B.Clarke in T.A.Durand & H.Schinz, Consp. Fl. Afric. 5: 613 (1894), nom. inval.

Bulbostylis burkei C.B.Clarke in W.H.Harvey & auct. suc. (eds.), Fl. Cap. 7: 208 (1898). *Scirpus burkei* (C.B.Clarke) K.Schum., Just's Bot. Jahresber. 26(1): 329 (1900).

Bulbostylis craspedota Chiov., Fl. Somala 2: 437 (1932). *Stenophyllus craspedotus* (Chiov.) Chiov., Fl. Somala 2: 437 (1932), nom. inval.

S. Somalia to N. Kenya. 24 SOM 25 KEN. Hemicr. or rhizome geophyte.

Bulbostylis cruciformis (Lye) R.W.Haines in R.W. Haines & K.A.Lye, Sedges & Rushes E. Afr., App. 3: 1 (1983).

Kenya. 25 KEN. Ther.

**Abildgaardia cruciformis* Lye, Nordic J. Bot. 1: 752 (1981 publ. 1982).

Bulbostylis cupricola Goetgh., Bull. Jard. Bot. Natl. Belg. 54: 92 (1984).

Zaïre. 23 ZAI. Ther.

Bulbostylis curassavica (Britton) Kük. & Ekman, Ark. Bot. 22A(17): 6 (1929).

SW. Mexico, Caribbean. 79 MXS 81 ARU HAI NLA PUE. Ther.

Bulbostylis floccosa var. *pumilio* C.B.Clarke in I.Urban, Symb. Antill. 5: 290 (1907). **Stenophyllus curassavicus* Britton, Bull. Torrey Bot. Club 43: 445 (1916). *Fimbristylis curassavica* (Britton) Alain, Bull. Torrey Bot. Club 92: 290 (1965).

Bulbostylis curassavica var. *pallescens* Kük. & Ekman, Ark. Bot. 22A(16): 8 (1929).

Bulbostylis cylindrica C.B.Clarke in D.Oliver & auct. suc. (eds.), Fl. Trop. Afr. 8: 446 (1902).
Angola. 26 ANG. Hemicr. or rhizome geophyte.

Bulbostylis decaryi Cherm., Bull. Soc. Bot. France 83: 72 (1936).
Madagascar. 29 MDG. Ther.

Bulbostylis densa (Wall.) Hand.-Mazz., Vegetationsbilder 20(7): 16 (1930).
Trop. & Subtrop. Old World. 22 GUI MLI NGR SEN SIE TOG 23 BUR CMN EQG GAB GGI RWA ZAI 24 CHA ETH SUD 25 KEN TAN UGA 26 ANG MLW 31 PRM 36 CHS 38 JAP KZN NNS TAI 40 ASS BAN EHM IND NEP SRL WHM 41 THA VIE 42 JAW LSI MLY PHI SUL SUM 43 NWG 50 NFK NSW QLD. Ther.
**Scirpus densus* Wall. in W.Roxburgh, Fl. Ind. 1: 231 (1820). *Isolepis densa* (Wall.) Schult. in J.J.Roemer & J.A.Schultes, Mant. 2: 70 (1824). *Fimbristylis densa* (Wall.) T.Koyama & T.I.Chuang, Quart. J. Taiwan Mus. 13: 229 (1960), nom. illeg. *Abildgaardia densa* (Wall.) Lye, Bot. Not. 127: 496 (1974). *Isolepis tenuissima* D.Don, Prodr. Fl. Nepal.: 40 (1825), nom. illeg. *Bulbostylis tenuissima* Nakai, Bull. Natl. Sci. Mus. Tokyo 31: 129 (1952), nom. illeg.

subsp. ***afromontana*** (Lye) R.W.Haines in R.W.Haines & K.A.Lye, Sedges & Rushes E. Afr., App. 3: 1 (1983).
Trop. Africa. 22 GUI MLI NGR SEN SIE TOG 23 BUR CMN GAB GGI RWA ZAI 24 CHA ETH SUD 25 KEN TAN UGA 26 ANG MLW. Ther.
Isolepis trifida Nees in R.Wight, Contr. Bot. India: 108 (1834). *Fimbristylis trifida* (Nees) Trin., Syst. Cat. Fl. Pl. Ceylon: 102 (1885). *Bulbostylis capillaris* var. *trifida* (Nees) C.B.Clarke in D.Oliver & auct. suc. (eds.), Fl. Trop. Afr. 8: 441 (1902). *Bulbostylis trifida* (Nees) Nelmes, Kew Bull. 5: 209 (1950).
Bulbostylis trifida var. *biegensis* Cherm., Bull. Soc. Bot. France 82: 341 (1935).
Fimbristylis minutissima Maire, Bull. Soc. Hist. Nat. Afrique N. 34: 139 (1943).
Fimbristylis marrana Miré & Quézel, J. Agric. Trop. Bot. Appl. 8: 112 (1961), nom. inval.
**Abildgaardia densa* subsp. *afromontana* Lye, Nordic J. Bot. 3: 237 (1983).

var. ***cameroonensis*** S.S.Hooper, Kew Bull. 26: 583 (1972).
Cameroon (E. Mt. Cameroon). 23 CMN. Ther.

var. ***capitata*** (Miq.) Ohwi, Mem. Coll. Sci. Kyoto Imp. Univ., Ser. B, Biol. 18: 40 (1944).
China to Temp. E. Asia. 36 CHS 38 JAP KZN NNS TAI. Ther.
**Isolepis capillaris* var. *capitata* Miq., Prolus. Fl. Jap.: 75 (1866).
Bulbostylis japonica C.B.Clarke, Bull. Misc. Inform. Kew, Addit. Ser. 8: 27 (1908).

subsp. ***densa***
Trop. Asia, E. Australia. 31 PRM 40 ASS BAN EHM IND NEP SRL WHM 41 THA VIE 42 JAW LSI MLY PHI SUL SUM 43 NWG 50 NFK NSW QLD. Ther.
Fimbristylis capillacea Hochst. ex Steud., Syn. Pl. Glumac. 2: 111 (1855).
Isolepis trichokolea Steud., Syn. Pl. Glumac. 2: 96 (1855). *Bulbostylis trichokolea* (Steud.) Beetle, Leafl. W. Bot. 4: 45 (1944).

Bulbostylis densicaespitosa (Lye) R.W.Haines in R.W.Haines & K.A.Lye, Sedges & Rushes E. Afr., App. 3: 1 (1983).
Kenya (Tane River). 25 KEN. Hemicr.
**Abildgaardia densicaespitosa* Lye, Nordic J. Bot. 1: 754 (1981 publ. 1982).

Bulbostylis disticha Ohwi & T.Koyama, Bull. Natl. Sci. Mus. Tokyo, n.s., 3(38): 18 (1956).
China (Zhejiang). 36 CHS. Hemicr. or rhizome geophyte.

Bulbostylis distichoides Lye, Lidia 5: 156 (2001).
Brazil (Bahia). 84 BZE. Ther.
**Abildgaardia disticha* Lye, Nordic J. Bot. 6: 765 (1986).

Bulbostylis edwalliana (Boeck.) Prata & M.G.López, Kew Bull. 57: 234 (2002).
WC. & SE. Brazil. 84 BZC BZL. Hemicr. or rhizome geophyte.
**Scirpus edwallianus* Boeck., Vidensk. Meddel. Naturhist. Foren. Kjøbenhavn 1894: 238 (1894 publ. 1895).
Bulbostylis francaensis Palla, Denkschr. Kaiserl. Akad. Wiss., Wien. Math.-Naturwiss. Kl. 79: 177 (1908). *Stenophyllus francaensis* (Palla) H.Pfeiff., Bot. Arch. 6: 191 (1924).

Bulbostylis elegantissima (Lye) R.W.Haines in R.W.Haines & K.A.Lye, Sedges & Rushes E. Afr., App. 3: 1 (1983).
Tanzania. 25 TAN. Ther.
**Abildgaardia elegantissima* Lye, Nordic J. Bot. 1: 755 (1981 publ. 1982).

Bulbostylis eleocharoides Kral & M.T.Strong, Sida 18: 846 (1999).
Bolivia to Paraguay. 83 BOL 85 PAR. Hemicr. or rhizome geophyte.

Bulbostylis emmerichiae T.Koyama, Bradea 1: 429 (1974).
Brazil. 84 BZC. Hemicr. or rhizome geophyte.

Bulbostylis eriophora (Steud.) H.Pfeiff., Revista Sudamer. Bot. 7: 129 (1943).
SE. China. 36 CHS. Hemicr. or rhizome geophyte.
**Cyperus eriophorus* Steud., Syn. Pl. Glumac. 2: 50 (1854).

Bulbostylis erratica (Hook.f.) C.B.Clarke in D.Oliver & auct. suc. (eds.), Fl. Trop. Afr. 8: 434 (1902).
W. & WC. Trop. Africa. 22 NGA SIE 23 CMN GGI. Hemicr. or rhizome geophyte.
**Schoenus erraticus* Hook.f., J. Proc. Linn. Soc., Bot. 6: 22 (1862). *Abildgaardia erratica* (Hook.f.) Lye, Nordic J. Bot. 3: 239 (1983). *Bulbostylis schoenoides* subsp. *erratica* (Hook.f.) Lye in R.W.Haines & K.A.Lye, Sedges & Rushes E. Afr.: 101 (1983).
Scirpus kamerunensis K.Schum. ex C.B.Clarke in D.Oliver & auct. suc. (eds.), Fl. Trop. Afr. 8: 435 (1902).

Bulbostylis eustachii J.M.Black ex Eardley in J.M.Black, Fl. S. Austral., ed. 2: 945 (1957).
South Australia. 50 SOA. Hemicr. or rhizome geophyte.

Bulbostylis fasciculata Uittien, Recueil Trav. Bot. Néerl. 22: 338 (1925).
N. South America to Brazil. 82 FRG GUY SUR 84 BZC. Hemicr. or rhizome geophyte.

Bulbostylis festucoides (Poir.) C.B.Clarke in T.A.Durand & H.Schinz, Consp. Fl. Afric. 5: 613 (1894).
Mascarenes. 22 GUI? 29 MAU REU. Hemicr. or rhizome geophyte.
**Scirpus festucoides* Poir. in J.B.A.M.de Lamarck, Encycl. 6: 752 (1805). *Baeothryon festucoides* (Poir.) A.Dietr., Sp. Pl. 2: 94 (1832). *Isolepis festucoides* (Poir.) Kunth, Enum. Pl. 2: 207 (1837). *Schoenus festucoides* (Poir.) Willd. ex Kunth, Enum. Pl. 2: 20 (1837). *Abildgaardia festucoides* (Poir.) Lye, Nordic J. Bot. 1: 757 (1981 publ. 1982).

Bulbostylis filamentosa (Vahl) C.B.Clarke in T.A. Durand & H.Schinz, Consp. Fl. Afric. 5: 613 (1894).
Trop. & S. Africa. 22 BEN BKN GHA GUI IVO MLI NGA NGR SIE TOG 23 BUR CAF CON GAB RWA ZAI 24 CHA 25 KEN TAN UGA 26 ANG MLW 27 BOT CPP NAM NAT SWZ TVL. Hemicr.
**Scirpus filamentosus* Vahl, Enum. Pl. 2: 262 (1805). *Isolepis filamentosa* (Vahl) Roem. & Schult., Syst. Veg. 2: 113 (1817). *Fimbristylis filamentosa* (Vahl) K.Schum. in H.G.A.Engler, Pflanzenw. Ost-Afrikas, C: 125 (1895). *Bulbostylis junciformis* var. *filamentosa* (Vahl) H.Pfeiff., Revista Sudamer. Bot. 7: 129 (1943). *Abildgaardia filamentosa* (Vahl) Lye, Bot. Not. 127: 496 (1974).
Fimbristylis cardiocarpa Ridl., Trans. Linn. Soc. London, Bot. 2: 154 (1884), nom. illeg. *Bulbostylis cardiocarpa* C.B.Clarke in T.A.Durand & H.Schinz, Consp. Fl. Afric. 5: 612 (1894).
Bulbostylis rehmannii C.B.Clarke in T.A.Durand & H.Schinz, Consp. Fl. Afric. 5: 615 (1894), nom. inval.
Bulbostylis metralis Cherm., Arch. Bot. Mém. 7: 34 (1931). *Abildgaardia filamentosa* var. *metralis* (Cherm.) Lye, Bot. Not. 127: 496 (1974). *Bulbostylis filamentosa* var. *metralis* (Cherm.) R.W.Haines in R.W.Haines & K.A.Lye, Sedges & Rushes E. Afr., App. 3: 1 (1983).

Bulbostylis fimbriata (Nees) C.B.Clarke in I.Urban, Symb. Antill. 2: 86 (1900).
Trop. America. 81 WIN 83 PER 84 BZL BZN. Ther.
**Oncostylis fimbriata* Nees in C.F.P.von Martius & auct. suc. (eds.), Fl. Bras. 2(1): 88 (1842). *Scirpus fimbriatus* (Nees) Boeck., Vidensk. Meddel. Naturhist. Foren. Kjøbenhavn 1869: 138 (1869), nom. illeg. *Isolepis fimbriata* (Nees) Schrad. ex B.D.Jacks., Index Kew. 1: 1237 (1893). *Stenophyllus fimbriatus* (Nees) Britton, Bull. Torrey Bot. Club 43: 446 (1916).

Bulbostylis fimbristyloides C.B.Clarke, Bull. Soc. Bot. France 54(8): 28 (1907).
W. Trop. Africa to Zambia. 22 BKN GUI MLI 23 CAF ZAI 24 CHA 26 ZAM. Hemicr. or rhizome geophyte.
Scirpus fimbristyloides K.Schum. ex C.B.Clarke, Bull. Soc. Bot. France 54(8): 28 (1907).
Bulbostylis cyrtathera Cherm., Arch. Bot. Mém. 7: 35 (1931).

Bulbostylis firingalavensis Cherm., Bull. Soc. Bot. France 68: 419 (1921 publ. 1922).
W. Madagascar. 29 MDG. Ther.

Bulbostylis flexuosa (Ridl.) Goetgh., Bull. Jard. Bot. Natl. Belg. 55: 241 (1985).
S. Zaïre to Angola. 23 ZAI 26 ANG ZAM. Hemicr. or rhizome geophyte.
**Fimbristylis flexuosa* Ridl., Trans. Linn. Soc. London, Bot. 2: 155 (1884).

Bulbostylis floccosa (Griseb.) C.B.Clarke in I.Urban, Symb. Antill. 2: 86 (1900).
Cuba. 81 CUB. Hemicr. or rhizome geophyte.
**Scirpus floccosus* Griseb., Cat. Pl. Cub.: 241 (1866). *Fimbristylis floccosa* (Griseb.) Alain, Bull. Torrey Bot. Club 92: 290 (1965).

Bulbostylis fluviatilis Kral & Davidse, Ann. Missouri Bot. Gard. 74: 437 (1987).
S. Trop. America. 82 VEN 83 BOL CLM 84 BZC BZN. Ther.

Bulbostylis funckii (Steud.) C.B.Clarke in I.Urban, Symb. Antill. 5: 290 (1907).
Trop. & Subtrop. America. 76 ARI 77 NWM 79 MXC MXE MXG MXS 80 COS ELS GUA HON 81 PUE 82 VEN 83 BOL 85 AGE. Ther.
**Isolepis funckii* Steud., Syn. Pl. Glumac. 2: 91 (1855). *Stenophyllus funckii* (Steud.) Britton, Bull. Torrey Bot. Club 21: 30 (1894).
Scirpus tenuispicatus Boeck., Linnaea 36: 740 (1870). *Bulbostylis tenuispicata* (Boeck.) Barros, Anales Mus. Argent. Ci. Nat. "Bernardino Rivadavia" 41: 338 (1945).
Scirpus heterocarpus S.Watson, Proc. Amer. Acad. Arts 18: 171 (1883).

Bulbostylis fusiformis Goetgh., Bull. Jard. Bot. Natl. Belg. 55: 238 (1985).
Zaïre. 23 ZAI. Ther.

Bulbostylis gardneriana (Steud.) Beetle, Leafl. W. Bot. 4: 45 (1944).
Brazil. 84+. Hemicr. or rhizome geophyte.
**Isolepis gardneriana* Steud., Syn. Pl. Glumac. 2: 99 (1855).

Bulbostylis glaberrima Kük., Notizbl. Bot. Gart. Berlin-Dahlem 9: 308 (1925). *Abildgaardia glaberrima* (Kük.) Lye, Bot. Not. 127: 496 (1974).
E. Trop. Africa. 25 KEN UGA. Hemicr. or rhizome geophyte.

Bulbostylis graminifolia C.B.Clarke, Bull. Misc. Inform. Kew, Addit. Ser. 8: 27 (1908).
Brazil. 84+. Hemicr. or rhizome geophyte.

Bulbostylis guineensis Cherm. ex Bodard, Bull. Soc. Bot. France 108: 308 (1962).
Guinea. 22 GUI. Hemicr. or rhizome geophyte.

Bulbostylis haitiensis Kük., Repert. Spec. Nov. Regni Veg. 23: 199 (1926).
Haiti. 81 HAI. Hemicr. or rhizome geophyte.

Bulbostylis hensii (C.B.Clarke) R.W.Haines in R.W.Haines & K.A.Lye, Sedges & Rushes E. Afr., App. 3: 1 (1983).
Trop. Africa. 22 BEN GHA GUI IVO LBR NGA SEN SIE TOG 23 BUR CAF CMN EQG GAB GGI RWA ZAI 25 KEN TAN UGA. Ther. or hemicr.
**Fimbristylis hensii* C.B.Clarke in D.Oliver & auct. suc. (eds.), Fl. Trop. Afr. 8: 419 (1902). *Abildgaardia hensii* (C.B.Clarke) Lye in R.W.Haines & K.A.Lye, Sedges & Rushes E. Afr., Suppl. 3: 2 (1983).
Fimbristylis exilis var. *brachyphylla* Cherm., Arch. Bot. Bull. Mens. 4(7): 32 (1931). *Fimbristylis hispidula* subsp. *brachyphylla* (Cherm.) Napper, Kew Bull. 25: 440 (1971). *Abildgaardia hispidula* subsp. *brachyphylla* (Cherm.) Lye, Nordic J. Bot. 3: 239 (1983). *Bulbostylis hispidula* subsp. *brachyphylla* (Cherm.) R.W.Haines in R.W.Haines & K.A.Lye, Sedges & Rushes E. Afr., App. 3: 1 (1983).

Bulbostylis heterostachya Cherm., Bull. Soc. Bot. France 72: 20 (1925). *Abildgaardia heterostachya* (Cherm.) Lye, Nordic J. Bot. 3: 239 (1983).
C. Madagascar. 29 MDG. Ther.

Bulbostylis hispaniolica Kük. & Ekman, Ark. Bot. 22A(17): 6 (1929). *Fimbristylis hispaniolica* (Kük. & Ekman) Alain, Bull. Torrey Bot. Club 92: 290 (1965).
Haiti. 81 HAI. Hemicr. or rhizome geophyte.

Bulbostylis hispida (Steud.) Beetle, Amer. Midl. Naturalist 34: 728 (1945).
S. America. 8+. Hemicr. or rhizome geophyte.
**Isolepis hispida* Steud., Syn. Pl. Glumac. 2: 104 (1855).

Bulbostylis hispidula (Vahl) R.W.Haines in R.W.Haines & K.A.Lye, Sedges & Rushes E. Afr., App. 3: 1 (1983).
Trop. & Subtrop. to Caucasus. 22 BEN BKN GAM GHA GNB GUI LBR MLI MTN NGA NGR SEN TOG 23 BUR CMN GAB ZAI 24 CHA SOC SOM 25 KEN TAN UGA 27 BOT CPP NAM NAT SWZ TVL 29 MDG SEY 33 TCS 34 TUR 42 JAW LSI PHI (78) mry 79 MXC MXE MXG MXN MXS MXT 80 BLZ COS ELS GUA HON NIC PAN 81 CUB PUE 82 GUY VEN 83 BOL CLM 84 BZC BZE BZN. Ther. or hemicr.
**Scirpus hispidulus* Vahl, Enum. Pl. 2: 276 (1805). *Fimbristylis hispidula* (Vahl) Kunth, Enum. Pl. 2: 277 (1837). *Abildgaardia hispidula* (Vahl) Lye, Bot. Not. 127: 496 (1974).

subsp. ***filiformis*** (C.B.Clarke) R.W.Haines in R.W.Haines & K.A.Lye, Sedges & Rushes E. Afr., App. 3: 1 (1983).
E. Trop. Africa. 25 KEN UGA. Ther.
**Bulbostylis filiformis* C.B.Clarke in D.Oliver & auct. suc. (eds.), Fl. Trop. Afr. 8: 441 (1902). *Stenophyllus filiformis* (C.B.Clarke) H.Pfeiff., Bot. Arch. 6: 191 (1924). *Abildgaardia hispidula* var. *filiformis* (C.B.Clarke) Lye, Bot. Not. 127: 496 (1974). *Abildgaardia hispidula* subsp. *filiformis* (C.B.Clarke) Lye, Nordic J. Bot. 3: 239 (1983).

subsp. ***halophila*** (Lye) R.W.Haines in R.W.Haines & K.A.Lye, Sedges & Rushes E. Afr., App. 3: 1 (1983).
E. Tanzania. 25 TAN. Ther.
**Abildgaardia hispidula* subsp. *halophila* Lye, Nordic J. Bot. 3: 235 (1983).

subsp. ***hispidula***
Trop. & Subtrop. to Caucasus. 22 BEN BKN GAM GHA GNB GUI LBR MLI NGA SEN TOG 23 BUR CMN GAB ZAI 24 CHA SOC SOM 25 KEN TAN UGA 27 BOT CPP NAM NAT SWZ TVL 29 MDG SEY 33 TCS 34 TUR 42 JAW LSI PHI (78) mry 79 MXE MXG MXN MXS MXT 80 BLZ COS ELS GUA HON NIC PAN 81 CUB PUE 82 GUY VEN 83 BOL CLM 84 BZC BZE BZN. Ther. or hemicr.
Isolepis exilis Kunth in F.W.H.von Humboldt, A.J.A.Bonpland & C.S.Kunth, Nov. Gen. Sp. 1: 224 (1816). *Fimbristylis exilis* (Kunth) Roem. & Schult. in J.J.Roemer & J.A.Schultes, Syst. Veg. 2: 98 (1817). *Scirpus exilis* (Kunth) Poir. in J.B.A.M.de Lamarck, Encycl., Suppl. 5: 105 (1817). *Bulbostylis exilis* (Kunth) Lye, Mitt. Bot. Staatssamml. München 10: 547 (1971).
Abildgaardia pubescens C.Presl, Reliq. Haenk. 1: 180 (1828). *Fimbristylis preslii* Kunth, Enum. Pl. 2: 228 (1837). *Bulbostylis pubescens* (C.Presl) Svenson in N.L.Britton & al. (eds.), N. Amer. Fl. 18: 542 (1957).
Chaetospora distachya Nees, Linnaea 10: 192 (1836).
Isolepis pubigera Rchb. ex Kunth, Enum. Pl. 2: 227 (1837).
Fimbristylis vestita Steud., Syn. Pl. Glumac. 2: 112 (1855).
Isolepis consocialis Steud., Syn. Pl. Glumac. 2: 318 (1855).
Isolepis perrottetii Steud., Syn. Pl. Glumac. 2: 318 (1855).
Isolepis pubiculmis Boeck., Linnaea 37: 28 (1871).
Scirpus chirigota C.Wright, Anales Acad. Ci. Méd. Habana 8: 81 (1871). *Fimbristylis chirigota* (C.Wright) Alain, Bull. Torrey Bot. Club 92: 290 (1965).
Trichelostylis pilosula Nees ex Boeck., Linnaea 37: 28 (1871).
Isolepis schweinfurthiana Oliv., Trans. Linn. Soc. London 29: 167 (1875).
Scirpus hildebrandtii Boeck., Flora 58: 263 (1875). *Fimbristylis hildebrandtii* (Boeck.) Ridl., Trans. Linn. Soc. London, Bot. 2: 155 (1884).
Scirpus granulatohirtellus Boeck. ex Schinz, Verh. Bot. Vereins Prov. Brandenburg 31: 179 (1890).
Fimbristylis trichophora Steud. ex C.B.Clarke in T.A.Durand & H.Schinz, Consp. Fl. Afric. 5: 605 (1894), nom. inval.
Isolepis abyssinica Hochst. ex C.B.Clarke in T.A.Durand & H.Schinz, Consp. Fl. Afric. 5: 605 (1894), nom. inval.
Fimbristylis clavinux C.B.Clarke, Ill. Cyper.: t. 41 (1909).
Bulbostylis woronowii Palla, Vestn. Tiflissk. Bot. Sada 21: 20 (1912).
Fimbristylis exilis var. *hirsutissima* Cherm., Mém. Acad. Malagache 10: 34 (1931).

subsp. ***intermedia*** (Lye) R.W.Haines in R.W.Haines & K.A.Lye, Sedges & Rushes E. Afr., App. 3: 1 (1983).
E. Kenya. 25 KEN. Ther.
**Abildgaardia hispidula* subsp. *intermedia* Lye, Nordic J. Bot. 3: 236 (1983).

subsp. ***longispicata*** Lye, Lidia 3: 207 (1996). *Abildgaardia hispidula* subsp. *longispicata* (Lye) J.-P.Lebrun & Stork, Énum. Pl. Fl. Afr. Trop. 4: 597 (1997).
Ethiopia. 24 ETH. Ther.

subsp. ***macroglumis*** Lye, Willdenowia 25: 596 (1996). *Abildgaardia hispidula* subsp. *macroglumis* (Lye) J.-P.Lebrun & Stork, Énum. Pl. Fl. Afr. Trop. 4: 597 (1997).
S. Somalia. 24 SOM. Ther.

subsp. ***pyriformis*** (Lye) R.W.Haines in R.W.Haines & K.A.Lye, Sedges & Rushes E. Afr., App. 3: 1 (1983).
Somalia to Uganda. 24 SOM 25 KEN UGA. Ther.
**Abildgaardia hispidula* var. *pyriformis* Lye, Bot. Not. 127: 494 (1974). *Abildgaardia hispidula* subsp. *pyriformis* (Lye) Lye, Nordic J. Bot. 3: 239 (1983).

subsp. ***senegalensis*** (Cherm.) Van den Berghen, in Fl. Illustr. Senegal 9: 159 (1988).
W. Trop. Africa. 22 GHA GUI MLI MTN NGA NGR SEN. Ther.

**Fimbristylis exilis* var. *senegalensis* Cherm., Arch. Bot. (Forlì) 4(7): 23 (1936). *Fimbristylis hispidula* subsp. *senegalensis* (Cherm.) Napper, Kew Bull. 25: 440 (1971). *Abildgaardia hispidula* subsp. *senegalensis* (Cherm.) J.-P.Lebrun & Stork, Énum. Pl. Fl. Afr. Trop. 3: 166 (1995).

Bulbostylis humilis (Kunth) C.B.Clarke in T.A.Durand & H.Schinz, Consp. Fl. Afric. 5: 614 (1894).
Yemen, Mozambique to S. Africa. 26 MOZ 27 TVL 35 YEM. Ther.
Fimbristylis arenaria Nees, Linnaea 10: 146 (1836). *Scirpus arenarius* (Nees) Boeck., Linnaea 36: 741 (1870).
Isolepis breviculmis Kunth, Enum. Pl. 2: 207 (1837). *Scirpus breviculmis* (Kunth) Boeck., Linnaea 36: 742 (1870). *Bulbostylis breviculmis* (Kunth) C.B.Clarke in T.A.Durand & H.Schinz, Consp. Fl. Afric. 5: 612 (1894).
**Isolepis humilis* Kunth, Enum. Pl. 2: 207 (1837). *Abildgaardia humilis* (Kunth) Lye, Bot. Not. 127: 496 (1974).
Isolepis humillima Hochst. ex C.B.Clarke in T.A.Durand & H.Schinz, Consp. Fl. Afric. 5: 616 (1894), nom. inval.
Bulbostylis striatella C.B.Clarke in W.H.Harvey & auct. suc. (eds.), Fl. Cap. 7: 305 (1898). *Abildgaardia striatella* (C.B.Clarke) Lye, Nordic J. Bot. 3: 239 (1983).

Bulbostylis humpatensis Meneses, Garcia de Orta 4: 253 (1956).
Angola. 26 ANG. Hemicr. or rhizome geophyte.

Bulbostylis igneotonsa Raymond, Naturaliste Canad. 99: 29 (1972). *Abildgaardia igneotonsa* (Raymond) Kornas, Flora 176: 64 (1985).
Zambia. 26 ZAM. Hemicr. or rhizome geophyte.

Bulbostylis jacobinae (Steud.) Lindm., Bih. Kongl. Svenska Vetensk.-Akad. Handl. 26(9): 18 (1901).
Brazil. 84 BZC BZE. Hemicr. or rhizome geophyte.
**Fimbristylis jacobinae* Steud., Syn. Pl. Glumac. 2: 111 (1855).

Bulbostylis johnstonii C.B.Clarke in D.Oliver & auct. suc. (eds.), Fl. Trop. Afr. 8: 442 (1902). *Abildgaardia johnstonii* (C.B.Clarke) Lye, Bot. Not. 127: 496 (1974).
Tanzania (Moshi). 25 TAN. Ther. or hemicr.

Bulbostylis junciformis (Kunth) C.B.Clarke, Trans. Linn. Soc. London, Bot. 4: 512 (1895).
Mexico to Trop. America. 79 MXC MXE MXG MXS MXT 80 BLZ COS ELS GUA HON NIC PAN 81 CUB HAI JAM TRT 82 FRG GUY SUR VEN 83 BOL ECU PER 84 BZC BZE BZL BZN BZS 85 AGE AGW PAR URU. Hemicr.
**Isolepis junciformis* Kunth in F.W.H.von Humboldt, A.J.A.Bonpland & C.S.Kunth, Nov. Gen. Sp. 1: 222 (1816). *Scirpus junciformis* (Kunth) Poir. in J.B.A.M.de Lamarck, Encycl., Suppl. 5: 105 (1817), nom. illeg. *Oncostylis junciformis* (Kunth) Nees in C.F.P.von Martius & auct. suc. (eds.), Fl. Bras. 2(1): 85 (1842). *Fimbristylis junciformis* (Kunth) Maury, J. Bot. (Morot) 3: 197 (1889), nom. illeg. *Stenophyllus junciformis* (Kunth) Britton, Bull. Torrey Bot. Club 43: 442 (1916).
Isolepis congesta Schrad. ex Schult. in J.J.Roemer & J.A.Schultes, Mant. 2: 70 (1824). *Oncostylis junciformis* var. *congesta* (Schrad. ex Schult.) Nees in C.F.P.von Martius & auct. suc. (eds.), Fl. Bras. 2(1): 86 (1842).
Scirpus humboldtii Spreng., Syst. Veg. 1: 213 (1824).
Scirpus congestus Rchb. ex Kunth, Enum. Pl. 2: 211 (1837).
Scirpus juncoides Willd. ex Kunth, Enum. Pl. 2: 211 (1837), nom. inval.
Scirpus lusulinus Rchb. ex Kunth, Enum. Pl. 2: 211 (1837).
Oncostylis junciformis var. *ambigua* Nees in C.F.P.von Martius & auct. suc. (eds.), Fl. Bras. 2(1): 86 (1842).
Oncostylis junciformis var. *humboldtiana* Nees in C.F.P.von Martius & auct. suc. (eds.), Fl. Bras. 2(1): 86 (1842).
Oncostylis junciformis f. *laxa* Nees in C.F.P.von Martius & auct. suc. (eds.), Fl. Bras. 2(1): 86 (1842).
Oncostylis junciformis var. *speciosior* Nees in C.F.P.von Martius & auct. suc. (eds.), Fl. Bras. 2(1): 86 (1842).
Oncostylis junciformis var. *stricta* Nees in C.F.P.von Martius & auct. suc. (eds.), Fl. Bras. 2(1): 86 (1842).
Isolepis berlandieri Steud., Syn. Pl. Glumac. 2: 103 (1855).
Scirpus berlandieri Steud., Syn. Pl. Glumac. 2: 103 (1855).
Scirpus junciformis f. *robustus* Kuntze, Revis. Gen. Pl. 3(2): 337 (1898).
Bulbostylis junciformis var. *laxiuscula* Kük., Repert. Spec. Nov. Regni Veg. 23: 198 (1926).
Bulbostylis junciformis var. *fulvicoma* Kük., Ostenia: 176 (1933).
Bulbostylis junciformis var. *ambigua* Steyerm., Fieldiana, Bot. 28(1): 37 (1951).

Bulbostylis juncoides (Vahl) Kük. ex Herter, Estud. Bot. Reg. Uruguay 4: 40 (1930 publ. 1931).
Arizona to Texas and Trop. America. 76 ARI 77 NWM TEX 79 MXC MXE MXG MXN MXS MXT 80 BLZ COS ELS GUA HON NIC PAN 81 JAM PUE 82 FRG GUY VEN 83 BOL ECU PER 84 BZC BZE BZL 85 AGE AGW PAR URU. Hemicr. or bulb geophyte.
**Schoenus juncoides* Vahl, Enum. Pl. 2: 211 (1805). *Bulbostylis juncoides* var. *genuina* Osten, Anales Mus. Hist. Nat. Montevideo, II, 3: 187 (1931), nom. inval. *Fimbristylis juncoides* (Vahl) Alain, Bull. Torrey Bot. Club 92: 290 (1965).
Isolepis hirtella Schrad. ex Schult. in J.J.Roemer & J.A.Schultes, Mant. 2: 70 (1824). *Bulbostylis hirtella* (Schrad. ex Schult.) Nees in C.F.P.von Martius & auct. suc. (eds.), Fl. Bras. 2(1): 85 (1842). *Oncostylis hirtella* (Schrad. ex Schult.) Nees in C.F.P.von Martius & auct. suc. (eds.), Fl. Bras. 2(1): 85 (1842). *Stenophyllus hirtellus* (Schrad. ex Schult.) H.Pfeiff., Bot. Arch. 6: 190 (1924).
Isolepis arenaria Nees, Linnaea 9: 291 (1834), nom. inval. *Stenophyllus arenarius* (Nees) J.F.Macbr., Publ. Field Columbian Mus., Bot. Ser. 4: 167 (1929).
Oncostylis arenaria Nees in C.F.P.von Martius & auct. suc. (eds.), Fl. Bras. 2(1): 87 (1842). *Bulbostylis arenaria* (Nees) Lindm., Bih. Kongl. Svenska Vetensk.-Akad. Handl. 26(9): 19 (1901). *Fimbristylis savannarum* Alain, Bull. Torrey Bot. Club 92: 289 (1965).
Oncostylis hirtella var. *hispidula* Nees in C.F.P.von Martius & auct. suc. (eds.), Fl. Bras. 2(1): 85 (1842).
Scirpus glaziovii Boeck., Vidensk. Meddel. Naturhist. Foren. Kjøbenhavn 1871: 152 (1871). *Bulbostylis glaziovii* (Boeck.) C.B.Clarke, Bull. Misc. Inform. Kew, Addit. Ser. 8: 26 (1908).

Scirpus juncoides var. *nanus* Griseb., Abh. Königl. Ges. Wiss. Göttingen 19: 267 (1874). *Bulbostylis juncoides* var. *nana* (Griseb.) Barros, Anales Mus. Argent. Ci. Nat. "Bernardino Rivadavia" 41: 342 (1945).
Scirpus lorentzii Boeck., Linnaea 38: 378 (1874). *Bulbostylis juncoides* var. *lorentzii* (Boeck.) Kük. ex Osten, Anales Mus. Nac. Montevideo, II, 3: 187 (1931).
Fimbristylis capillaris var. *pilosa* Britton, Bull. Torrey Bot. Club 15: 102 (1888).
Isolepis beyrichii Schrad. ex B.D.Jacks., Index Kew. 1: 1236 (1893).
Bulbostylis argentina Palla, Oesterr. Bot. Z. 57: 258 (1907). *Fimbristylis argentina* (Palla) Molfino, Physis (Buenos Aires) 8: 109 (1925). *Stenophyllus argentinus* (Palla) Killip, J. Wash. Acad. Sci. 21: 347 (1931).
Bulbostylis fendleri C.B.Clarke, Bull. Misc. Inform. Kew, Addit. Ser. 8: 27 (1908).
Stenophyllus hirtellus var. *elatior* H.Pfeiff., Bot. Arch. 6: 191 (1924).
Stenophyllus hirtellus var. *humilis* H.Pfeiff., Bot. Arch. 6: 191 (1924).
Bulbostylis juncoides var. *ampliceps* Kük. ex Osten, Anales Mus. Nac. Montevideo, II, 3: 188 (1931).
Bulbostylis juncoides f. *hirtella* Osten, Anales Mus. Nac. Montevideo, II, 3: 188 (1931).
Bulbostylis juncoides f. *subfimbriata* Kük. ex Osten, Anales Mus. Nac. Montevideo, II, 3: 188 (1931).
Bulbostylis juncoides f. *leucocarpa* Barros, Anales Mus. Argent. Ci. Nat. "Bernardino Rivadavia" 41: 345 (1945).

Bulbostylis kanongaensis Goetgh., Bull. Jard. Bot. Natl. Belg. 54: 94 (1984).
Zaïre. 23 ZAI. Ther.

Bulbostylis kirkii C.B.Clarke in W.H.Harvey & auct. suc. (eds.), Fl. Cap. 7: 209 (1898). *Scirpus kirkii* (C.B.Clarke) K.Schum., Just's Bot. Jahresber. 26(1): 329 (1900).
KwaZulu-Natal. 27 NAT. Hemicr. or rhizome geophyte.
Scirpus macrolepis Boeck. ex C.B.Clarke in T.A.Durand & H.Schinz, Consp. Fl. Afric. 5: 614 (1894), nom. inval.

Bulbostylis labillardierei (Steud.) Beetle, Leafl. W. Bot. 4: 45 (1944).
New Caledonia. 60 NWC. Hemicr. or rhizome geophyte.
**Isolepis labillardierei* Steud., Syn. Pl. Glumac. 2: 96 (1855).

Bulbostylis laeta C.B.Clarke, Bull. Misc. Inform. Kew, Addit. Ser. 8: 28 (1908).
Brazil (Minas Gerais). 84 BZL. Hemicr. or rhizome geophyte.

Bulbostylis lagoensis (Boeck.) Prata & M.G.López, Kew Bull. 56: 1008 (2001).
WC. & SE. Brazil. 84 BZC BZL. Hemicr. or rhizome geophyte.
**Scirpus lagoensis* Boeck., Beitr. Cyper. 2: 15 (1890).

Bulbostylis lanata (Kunth) Lindm., Bih. Kongl. Svenska Vetensk.-Akad. Handl. 26(9): 18 (1900), nom. cons.
S. Trop. America. 82 FRG GUY SUR VEN 83 CLM 84 BZE BZN. Hemicr. or cham.
**Isolepis lanata* Kunth in F.W.H.von Humboldt, A.J.A.Bonpland & C.S.Kunth, Nov. Gen. Sp. 1: 220 (1816). *Scirpus lanatus* (Kunth) Poir. in J.B.A.M.de Lamarck, Encycl., Suppl. 5: 104 (1817). *Schoenus lanatus* (Kunth) Willd. ex Kunth, Enum. Pl. 2: 206 (1837), nom. illeg. *Stenophyllus lanatus* (Kunth) Britton, Bull. Torrey Bot. Club 58: 318 (1931).
Bulbostylis lanata var. *xyrioides* Lindm., Bih. Kongl. Svenska Vetensk.-Akad. Handl. 26(3; 9): 18 (1900).

Bulbostylis laniceps C.B.Clarke in D.Oliver & auct. suc. (eds.), Fl. Trop. Afr. 8: 433 (1902). *Fimbristylis laniceps* (C.B.Clarke) K.Schum. in H.G.A.Engler, Pflanzenw. Ost-Afrikas, C: 125 (1895).
Trop. Africa. 22 GHA GUI IVO LBR NGA SIE TOG 23 CAF CMN EQG GAB ZAI 24 CHA 26 ANG ZAM. Hemicr.
Bulbostylis wittei Cherm., Rev. Zool. Bot. Africaines 22: 69 (1932).

Bulbostylis lanifera (Boeck.) Kük., Repert. Spec. Nov. Regni Veg. Beih. 40(1): 412 (1937).
Senegal to Chad. 22 BKN GHA IVO NGA SEN TOG 23 ZAI 34 CHA. Hemicr.
**Scirpus laniferus* Boeck., Linnaea 36: 768 (1870). *Fimbristylis lanifera* (Boeck.) K.Schum. in H.G.A.Engler, Pflanzenw. Ost-Afrikas, C: 125 (1895). *Abildgaardia lanifera* (Boeck.) Lye, Nordic J. Bot. 1: 757 (1981 publ. 1982).
Bulbostylis togoensis Cherm., Bull. Soc. Bot. France 81: 267 (1934).

Bulbostylis lateritica Cherm., Bull. Soc. Bot. France 73: 553 (1926 publ. 1927).
SW. Madagascar. 29 MDG. Ther.

Bulbostylis latifolia Kral & M.T.Strong, Sida 18: 848 (1999).
Brazil (Goiás). 84 BZC. Hemicr.

Bulbostylis leiolepis (Kük.) R.W.Haines in R.W.Haines & K.A.Lye, Sedges & Rushes E. Afr., App. 3: 1 (1983).
Tanzania (Uha). 25 TAN. Ther.
**Bulbostylis schimperiana* var. *leiolepis* Kük., Repert. Spec. Nov. Regni Veg. Beih. 40(1): 126 (1939). *Abildgaardia leiolepis* (Kük.) Lye, Nordic J. Bot. 1: 757 (1981 publ. 1982).

Bulbostylis leucostachya (Kunth) Kunth, Enum. Pl. 2: 205 (1837).
SE. Colombia to N. Brazil. 82 VEN 83 CLM 84 BZN. Hemicr. or cham.
**Isolepis leucostachya* Kunth in F.W.H.von Humboldt, A.J.A.Bonpland & C.S.Kunth, Nov. Gen. Sp. 1: 220 (1816). *Scirpus leucostachyus* (Kunth) Poir. in J.B.A.M.de Lamarck, Encycl., Suppl. 5: 104 (1817). *Stenophyllus leucostachyus* (Kunth) Britton, Bull. Torrey Bot. Club 43: 445 (1916).

Bulbostylis lichtensteiniana (Kunth) C.B.Clarke in T.A.Durand & H.Schinz, Consp. Fl. Afric. 5: 614 (1894).
St. Helena. 28 STH. Hemicr.
**Isolepis lichtensteiniana* Kunth, Enum. Pl. 2: 207 (1837). *Scirpus lichtensteinianus* (Kunth) Boeck., Linnaea 36: 745 (1870). *Fimbristylis lichtensteiniana* (Kunth) Hemsl., Rep. Challenger, Bot. 1(2): 86 (1884).

Bulbostylis lineolata Goetgh., Bull. Jard. Bot. Natl. Belg. 55: 248 (1985).
Tanzania to Angola. 23 BUR ZAI 25 TAN 26 ANG ZIM. Ther.

Bulbostylis loefgrenii (Boeck.) Prata & M.G.López, Kew Bull. 56: 1008 (2001).
Brazil (São Paulo). 84 BZL. Hemicr. or rhizome geophyte.
**Scirpus loefgrenii* Boeck., Beitr. Cyper. 2: 16 (1890).

Bulbostylis lombardii Kral & M.T.Strong, Sida 18: 850 (1999).
Brazil (Minas Gerais). 84 BZL. Hemicr. or rhizome geophyte.

Bulbostylis longiradiata Goetgh., Bull. Jard. Bot. Natl. Belg. 55: 219 (1985).
Zaïre, to Burundi. 23 BUR ZAI. Ther.

Bulbostylis macra (Ridl.) C.B.Clarke in T.A.Durand & H.Schinz, Consp. Fl. Afric. 5: 614 (1894).
S. Tanzania to Angola. 23 BUR ZAI 25 TAN 26 ANG MLW. Hemicr.
**Fimbristylis macra* Ridl., Trans. Linn. Soc. London, Bot. 2: 150 (1884). *Abildgaardia macra* (Ridl.) Lye, Bot. Not. 127: 497 (1974).
Fimbristylis zambesica K.Schum. in H.G.A.Engler, Pflanzenw. Ost-Afrikas, C: 125 (1895). *Bulbostylis zambesica* (K.Schum.) C.B.Clarke in D.Oliver & auct. suc. (eds.), Fl. Trop. Afr. 8: 430 (1902).

Bulbostylis macrostachya (Lye) R.W.Haines in R.W.Haines & K.A.Lye, Sedges & Rushes E. Afr., App. 3: 1 (1983).
Tanzania (Mbeya Reg.). 25 TAN. Hemicr.
**Abildgaardia macrostachya* Lye, Nordic J. Bot. 1: 749 (1981 publ. 1982).

Bulbostylis mahafalensis Cherm., Bull. Soc. Bot. France 68: 421 (1921 publ. 1922).
SW. Madagascar. 29 MDG. Ther.

Bulbostylis major Palla, Denkschr. Kaiserl. Akad. Wiss., Wien. Math.-Naturwiss. Kl. 79: 177 (1908).
Brazil. 84+. Hemicr. or rhizome geophyte.

Bulbostylis megastachys (Ridl.) C.B.Clarke in T.A. Durand & H.Schinz, Consp. Fl. Afric. 5: 614 (1894).
Zaïre, Angola, Zimbabwe. 23 ZAI 26 ANG ZIM. Hemicr. or rhizome geophyte.
**Fimbristylis megastachys* Ridl., Trans. Linn. Soc. London, Bot. 2: 156 (1884). *Abildgaardia megastachys* (Ridl.) Lye, Nordic J. Bot. 1: 757 (1981 publ. 1982).
Bulbostylis stricta Turrill, Bull. Misc. Inform. Kew 1925: 70 (1925).

Bulbostylis melanocephala (Ridl.) C.B.Clarke in T.A. Durand & H.Schinz, Consp. Fl. Afric. 5: 615 (1894).
Angola, Northern Prov. 26 ANG 27 TVL. Hemicr. or rhizome geophyte.
**Fimbristylis melanocephala* Ridl., Trans. Linn. Soc. London, Bot. 2: 151 (1884).

Bulbostylis micans Lindm., Bih. Kongl. Svenska Vetensk.-Akad. Handl. 26(9): 17 (1901).
Brazil. 84+. Hemicr. or rhizome geophyte.

Bulbostylis micranthera Cherm., Bull. Soc. Bot. France 68: 421 (1921 publ. 1922).
Madagascar. 29 MDG. Ther.

Bulbostylis microcarpa (Lye) R.W.Haines in R.W.Haines & K.A.Lye, Sedges & Rushes E. Afr., App. 3: 1 (1983).
Zaïre to Tanzania. 23 ZAI 25 TAN. Ther.
**Abildgaardia microcarpa* Lye, Nordic J. Bot. 3: 234 (1983).

Bulbostylis microelegans (Lye) R.W.Haines in R.W.Haines & K.A.Lye, Sedges & Rushes E. Afr., App. 3: 1 (1983).
WC. & E. Trop. Africa. 23 BUR RWA 25 KEN TAN UGA. Ther.
**Abildgaardia microelegans* Lye, Nordic J. Bot. 1: 756 (1981 publ. 1982).

Bulbostylis micromucronata Goetgh., Bull. Jard. Bot. Natl. Belg. 54: 94 (1984).
Zaïre. 23 ZAI. Ther.

Bulbostylis moggii Schönland & Turrill, Bull. Misc. Inform. Kew 1925: 70 (1925).
Botswana. 27 BOT. Hemicr. or rhizome geophyte.

Bulbostylis mozambica Raymond, Naturaliste Canad. 99: 30 (1972).
Mozambique. 26 MOZ. Hemicr. or rhizome geophyte.

Bulbostylis mucronata C.B.Clarke, Bot. Jahrb. Syst. 38: 135 (1906).
Namibia. 27 NAM. Hemicr. or rhizome geophyte.

Bulbostylis multispiculata Cherm., Arch. Bot. Mém. 7: 38 (1931).
Central African Rep. 23 CAF. Hemicr. or rhizome geophyte.

Bulbostylis neglecta (Hemsl.) C.B.Clarke in T.A.Durand & H.Schinz, Consp. Fl. Afric. 5: 615 (1894).
St. Helena. 28 STH†. Ther. Last coll. in 1885.
**Fimbristylis neglecta* Hemsl., Rep. Challenger, Bot. 1(2): 86 (1884).

Bulbostylis nemoides Goetgh., Bull. Jard. Bot. Natl. Belg. 54: 98 (1984).
Burundi. 23 BUR. Ther.

Bulbostylis nesiotica (I.M.Johnst.) Fernald, Rhodora 40: 392 (1938).
Revellagigedo Is. (Benedicto I.). 79 MXI. Hemicr. or rhizome geophyte.
**Stenophyllus nesioticus* I.M.Johnst., Univ. Calif. Publ. Bot. 7: 438 (1922).

Bulbostylis nesiotis (Hemsl.) C.B.Clarke, Bull. Misc. Inform. Kew, Addit. Ser. 8: 27 (1908).
Brazil (Trinidade). 84 BZL. Hemicr. or rhizome geophyte.
**Fimbristylis nesiotis* Hemsl., Rep. Challenger, Bot. 1(2): 130 (1884).

Bulbostylis oligostachys (Hochst. ex A.Rich.) C.B.Clarke in T.A.Durand & H.Schinz, Consp. Fl. Afric. 5: 615 (1894).
Ethiopia to Angola. 23 ZAI 24 ETH 25 KEN TAN UGA 26 ANG. Ther.
**Fimbristylis oligostachys* Hochst. ex A.Rich., Tent. Fl. Abyss. 2: 593 (1850). *Scirpus oligostachys* (Hochst. ex A.Rich.) Boeck., Linnaea 36: 743 (1870). *Fimbristylis oligostachya* (Hochst. ex A.Rich.) K.Schum. in H.G.A.Engler, Pflanzenw. Ost-Afrikas, C: 125 (1895), nom. illeg. *Abildgaardia hispidula* var. *oligostachys* (Hochst. ex A.Rich.) Lye, Bot. Not. 127: 496 (1974). *Abildgaardia oligostachys* (Hochst. ex A.Rich.) Lye, Nordic J. Bot. 3: 239 (1983).
Fimbristylis minima Hochst. ex Boeck., Linnaea 36: 744 (1870), nom. inval.
Scirpus purpureoater Boeck., Linnaea 38: 279 (1874). *Fimbristylis purpureoatra* (Boeck.) C.B.Clarke ex Engl., Abh. Königl. Akad. Wiss. Berlin 2: 149 (1891 publ. 1892).

Bulbostylis oritrephes (Ridl.) C.B.Clarke, Trans. Linn. Soc. London, Bot. 4: 54 (1894).
Trop. Africa & S. Africa. 22 GUI NGA SIE 23 BUR CMN GAB RWA ZAI 25 KEN TAN UGA 26 ANG MLW 27 CPP LES NAT OFS SWZ TVL. Hemicr.
**Fimbristylis oritrephes* Ridl., Trans. Linn. Soc. London, Bot. 2: 155 (1884). *Abildgaardia oritrephes* (Ridl.) Lye, Bot. Not. 127: 497 (1974).

subsp. ***australis*** B.L.Burtt, Notes Roy. Bot. Gard. Edinburgh 43: 353 (1986).
S. Africa. 27 CPP LES NAT OFS SWZ TVL. Hemicr. or rhizome geophyte.

subsp. ***oritrephes***
Trop. Africa. 22 GUI NGA SIE 23 BUR CMN GAB RWA ZAI 25 KEN TAN UGA 26 ANG MLW. Hemicr.
Bulbostylis trichobasis var. *leptocaulis* C.B.Clarke, Bull. Herb. Boissier, II, 1: 58 (1900).
Bulbostylis trichobasis var. *uniseriata* C.B.Clarke, Bull. Soc. Roy. Bot. Belgique 39: 37 (1900).
Bulbostylis caespitosa Peter, Repert. Spec. Nov. Regni Veg. Beih. 40(1): 127 (1936).

Bulbostylis ottonis (Boeck.) Beetle, Amer. Midl. Naturalist 34: 730 (1945).
Brazil. 84+. Hemicr. or rhizome geophyte.
**Isolepis ottonis* Boeck., Flora 41: 419 (1858).

Bulbostylis pachypoda Kral & M.T.Strong, Sida 18: 852 (1999).
Brazil (Minas Gerais to Mato Grosso). 84 BZC BZL. Hemicr.

Bulbostylis pallescens (Lye) R.W.Haines in R.W.Haines & K.A.Lye, Sedges & Rushes E. Afr., App. 3: 1 (1983).
S. Somalia to Kenya (Garissa). 24 SOM 25 KEN. Ther.
**Abildgaardia pallescens* Lye, Nordic J. Bot. 1: 751 (1981 publ. 1982).

Bulbostylis paradoxa (Spreng.) Lindm., Bih. Kongl. Svenska Vetensk.-Akad. Handl. 26(3; 9): 17 (1900).
Trop. America. 79 MXG MXS MXT 80 BLZ COS HON NIC PAN 81 CUB 82 FRG GUY SUR VEN 83 BOL CLM 84 BZC BZE BZL BZN BZS. Hemicr.
Schoenus spadiceus Kunth in F.W.H.von Humboldt, A.J.A.Bonpland & C.S.Kunth, Nov. Gen. Sp. 1: 227 (1816), nom. illeg. *Bulbostylis spadicea* Kük., Repert. Spec. Nov. Regni Veg. 23: 197 (1926).
**Schoenus paradoxus* Spreng., Syst. Veg. 1: 190 (1824). *Isolepis paradoxa* (Spreng.) Kunth, Enum. Pl. 2: 206 (1837), nom. illeg. *Oncostylis paradoxa* (Spreng.) Nees in C.F.P.von Martius & auct. suc. (eds.), Fl. Bras. 2(1): 81 (1842). *Scirpus paradoxus* (Spreng.) Boeck., Vidensk. Meddel. Naturhist. Foren. Kjøbenhavn 1869: 136 (1869). *Stenophyllus paradoxus* (Spreng.) Standl., Contr. U. S. Natl. Herb. 18: 88 (1916). *Fimbristylis paradoxa* (Spreng.) Alain, Bull. Torrey Bot. Club 92: 290 (1965). *Abildgaardia paradoxa* (Spreng.) Lye, Nordic J. Bot. 1: 758 (1981 publ. 1982).
Eriocaulon sphacelatum Willd. ex Kunth, Enum. Pl. 2: 206 (1837).
Rhynchospora perrigida Boeck., Allg. Bot. Z. Syst. 2: 93 (1896).

Bulbostylis paraensis C.B.Clarke, Bull. Misc. Inform. Kew, Addit. Ser. 8: 28 (1908). *Bulbostylis capillaris* var. *paraensis* (C.B.Clarke) Kük., Bot. Jahrb. Syst. 56(125): 15 (1921).
Brazil. 84 BZL BZN. Ther.
Isolepis paraensis Schrad. ex Nees in C.F.P.von Martius & auct. suc. (eds.), Fl. Bras. 2(1): 86 (1842), nom. inval.

Bulbostylis parvinux C.B.Clarke in W.H.Harvey & auct. suc. (eds.), Fl. Cap. 7: 207 (1898). *Scirpus parvinux* (C.B.Clarke) K.Schum., Just's Bot. Jahresber. 26(1): 329 (1900). *Abildgaardia parvinux* (C.B.Clarke) Lye, Mitt. Bot. Staatssamml. München 10: 545 (1971).
KwaZulu-Natal. 27 NAT. Hemicr. or rhizome geophyte.

Bulbostylis pauciflora (Liebm.) C.B.Clarke, Bull. Misc. Inform. Kew, Addit. Ser. 8: 26 (1908), nom. cons.
Caribbean. 81 ARU CUB DOM HAI LEE NLA PUE. Hemicr.
**Oncostylis pauciflora* Liebm., Mexic. Neldeagt. Pl., V, 2: 241 (1851). *Isolepis pauciflora* (Liebm.) Steud., Syn. Pl. Glumac. 2: 104 (1855). *Stenophyllus pauciflorus* (Liebm.) Britton, Sci. Surv. Porto Rico & Virgin Islands 6: 526 (1930).
Isolepis disticha Boeck., Flora 41: 415 (1858). *Scirpus subdistichus* Boeck., Linnaea 36: 490 (1870). *Bulbostylis subdisticha* (Boeck.) Beetle, Leafl. W. Bot. 4: 45 (1944).
Stenophyllus portoricensis Britton, Torreya 13: 216 (1913). *Bulbostylis portoricensis* (Britton) Fernald, Rhodora 40: 392 (1938). *Fimbristylis portoricensis* (Britton) Alain, Bull. Torrey Bot. Club 92: 290 (1965).
Bulbostylis ekmanii Kük., Repert. Spec. Nov. Regni Veg. 23: 197 (1926).

Bulbostylis perrieri Cherm., Bull. Soc. Bot. France 68: 420 (1921 publ. 1922).
W. Madagascar. 29 MDG. Ther.

Bulbostylis pilosa (Willd.) Cherm., Bull. Soc. Bot. France 81: 266 (1934).
Trop. Africa. 22 BEN GHA IVO MLI NGA SEN TOG 23 BUR CMN GAB ZAI 25 KEN TAN UGA 26 ANG ZIM. Hemicr.
**Schoenus pilosus* Willd., Phytographia: 3 (1794). *Abildgaardia pilosa* (Willd.) Nees, Linnaea 9: 289 (1834). *Isolepis pilosa* (Willd.) Steud., Syn. Pl. Glumac. 2: 97 (1855). *Fimbristylis africana* C.B.Clarke in T.A.Durand & H.Schinz, Consp. Fl. Afric. 5: 691 (1894). *Fimbristylis pilosa* (Willd.) K.Schum. in H.G.A.Engler, Pflanzenw. Ost-Afrikas, C: 124 (1895), nom. illeg.
Fimbristylis aphyllanthoides Welw. ex Ridl., Trans. Linn. Soc. London, Bot. 2: 151 (1884). *Bulbostylis aphyllanthoides* (Welw. ex Ridl.) C.B.Clarke in T.A.Durand & H.Schinz, Consp. Fl. Afric. 5: 611 (1894).

Bulbostylis pohliana (Boeck.) Beetle, Amer. Midl. Naturalist 34: 731 (1945).
SE. Brazil. 84 BZL. Hemicr. or rhizome geophyte.
**Isolepis pohliana* Boeck., Flora 41: 596 (1858).

Bulbostylis psammophila Cherm., Bull. Soc. Bot. France 68: 422 (1921 publ. 1922).
W. Madagascar. 29 MDG. Hemicr. or rhizome geophyte.

Bulbostylis pseudocollina Cherm., Bull. Soc. Bot. France 72: 174 (1925).
W. Madagascar. 29 MDG. Ther.

Bulbostylis pseudojunciformis (Boeck.) Beetle, Amer. Midl. Naturalist 34: 731 (1945).
Mexico. 79+. Hemicr. or rhizome geophyte.
**Isolepis pseudojunciformis* Boeck., Flora 40: 35 (1857).

Bulbostylis pseudoperennis Goetgh., Bull. Jard. Bot. Natl. Belg. 55: 235 (1985).
Zaïre. 23 ZAI. Ther. or hemicr.

Bulbostylis puberula C.B.Clarke in J.D.Hooker, Fl. Brit. India 6: 652 (1893).
C. Africa, Mascarenes, S. India to W. Malesia. 23 CMN ZAI 26 MLW 29 MDG REU 40 IND SRL 41 THA VIE 42 BOR JAW MLY SUM. Ther.
**Scirpus puberulus* Poir. in J.B.A.M.de Lamarck, Encycl. 6: 767 (1805), nom. illeg. *Scirpus thouarsii* Roem. & Schult., Syst. Veg. 2: 134 (1817). *Isolepis thouarsii* (Roem. & Schult.) Nees, Linnaea 9: 291 (1834), nom. illeg. *Isolepis puberula* (Poir.) Kunth, Enum. Pl. 2: 213 (1837). *Stenophyllus puberulus* (Poir.) Killip, J. Siam Soc., Nat. Hist. Suppl. 7: 57 (1927).
Isolepis sieberi Schrad., Anal. Fl. Cap. 1: 23 (1832).
Isolepis gracilis Nees in R.Wight, Contr. Bot. India: 109 (1834), nom. illeg.
Bulbostylis puberula var. *gracilis* C.E.C.Fisch. in J.S.Gamble, Fl. Madras: 1662 (1931).

Bulbostylis pusilla (Hochst. ex A.Rich.) C.B.Clarke in T.A.Durand & H.Schinz, Consp. Fl. Afric. 5: 615 (1894).
Trop. Africa. 22 BEN BKN MLI NGA NGR SIE 23 CAF CMN EQG GAB ZAI 24 CHA ETH 25 KEN TAN UGA 26 ANG MLW ZIM. Ther.
**Fimbristylis pusilla* Hochst. ex A.Rich., Tent. Fl. Abyss. 2: 506 (1850). *Abildgaardia pusilla* (Hochst. ex A.Rich.) Lye, Bot. Not. 127: 497 (1974).
Fimbristylis parva Ridl., Trans. Linn. Soc. London, Bot. 2: 153 (1884). *Bulbostylis parva* (Ridl.) C.B.Clarke in T.A.Durand & H.Schinz, Consp. Fl. Afric. 5: 615 (1894). *Abildgaardia parva* (Ridl.) Lye, Nordic J. Bot. 1: 758 (1981 publ. 1982).
Bulbostylis elegans Cherm., Arch. Bot. Mém. 7: 37 (1931), nom. illeg.
Bulbostylis yalingensis Cherm., Arch. Bot. Mém. 7: 40 (1931). *Abildgaardia yalingensis* (Cherm.) Lye, Nordic J. Bot. 1: 758 (1981 publ. 1982). *Abildgaardia pusilla* subsp. *yalingensis* (Cherm.) Lye, Nordic J. Bot. 3: 239 (1983). *Bulbostylis pusilla* subsp. *yalingensis* (Cherm.) R.W.Haines in R.W.Haines & K.A.Lye, Sedges & Rushes E. Afr., App. 3: 1 (1983).

Bulbostylis pyriformis S.T.Blake, Proc. Roy. Soc. Queensland 52: 55 (1941).
N. & E. Australia. 50 NSW NTA QLD. Hemicr. or rhizome geophyte.

Bulbostylis quaternella (Ridl.) Goetgh., Bull. Jard. Bot. Natl. Belg. 55: 223 (1985).
Zaïre to Angola. 23 ZAI 26 ANG. Hemicr.
**Fimbristylis quaternella* Ridl., Trans. Linn. Soc. London, Bot. 2: 152 (1884).

Bulbostylis radiciflora (Maury) Beetle, Amer. Midl. Naturalist 41: 484 (1949).
S. America. 8+. Hemicr. or rhizome geophyte.
**Scirpus radiciflorus* Maury, J. Bot. (Morot) 3: 199 (1889).

Bulbostylis rarissima (Steud.) C.B.Clarke in T.A.Durand & H.Schinz, Consp. Fl. Afric. 5: 651 (1894).
Ethiopia. 24 ETH. Hemicr. or rhizome geophyte.
**Cyperus rarissimus* Steud., Syn. Pl. Glumac. 2: 16 (1854).

Bulbostylis renschii (Boeck.) C.B.Clarke in T.A.Durand & H.Schinz, Consp. Fl. Afric. 5: 615 (1894).
Madagascar. 29 MDG. Hemicr. or rhizome geophyte.
**Scirpus renschii* Boeck., Bot. Jahrb. Syst. 5: 504 (1884). *Abildgaardia renschii* (Boeck.) Lye, Mitt. Bot. Staatssamml. München 10: 547 (1971).
Bulbostylis renschii var. *flaccidula* Cherm., Bull. Soc. Bot. France 68: 419 (1921).

Bulbostylis rhizomatosa (Lye) R.W.Haines in R.W.Haines & K.A.Lye, Sedges & Rushes E. Afr., App. 3: 1 (1983).
S. Zaïre to Zambia. 23 ZAI 26 ZAM. Hemicr.
**Abildgaardia rhizomatosa* Lye, Nordic J. Bot. 1: 749 (1981 publ. 1982).

Bulbostylis rotundata (Kük.) R.W.Haines in R.W.Haines & K.A.Lye, Sedges & Rushes E. Afr., App. 3: 1 (1983).
Tanzania. 25 TAN. Ther.
**Fimbristylis rotundata* Kük., Repert. Spec. Nov. Regni Veg. Beih. 40(1): 126 (1936). *Abildgaardia rotundata* (Kük.) Lye, Nordic J. Bot. 1: 758 (1981 publ. 1982).

Bulbostylis rufescens (Boeck.) Beetle, Amer. Midl. Naturalist 41: 485 (1949).
Brazil. 84+. Hemicr.
**Scirpus rufescens* Boeck., Vidensk. Meddel. Naturhist. Foren. Kjøbenhavn 1869: 138 (1869).

Bulbostylis rugosa M.G.López, Bonplandia (Corrientes) 9: 29 (1996).
Paraguay to Argentina (Corrientes). 85 AGE PAR. Hemicr. or rhizome geophyte.

Bulbostylis rumokensis Cherm. ex Goetgh., Bull. Jard. Bot. Natl. Belg. 54: 100 (1984).
Zaïre. 23 ZAI. Ther.

Bulbostylis scabra (J.Presl & C.Presl) C.B.Clarke, Bull. Herb. Boissier 6(App. 1): 21 (1898).
S. Trop. America. 83 PER 84 BZC BZE BZL BZN BZS 85 AGE PAR URU. Hemicr.
**Isolepis scabra* J.Presl & C.Presl, Reliq. Haenk. 1: 188 (1828). *Scirpus scaber* (J.Presl & C.Presl) Kuntze, Revis. Gen. Pl. 3(2): 337 (1898). *Fimbristylis scabra* (J.Presl & C.Presl) Hauman & Vanderveken, Anales Mus. Nac. Buenos Aires 29: 207 (1917), nom. illeg. *Stenophyllus scaber* (J.Presl & C.Presl) H.Pfeiff., Bot. Arch. 6: 190 (1924).
Scirpus humilis Willd. ex Kunth, Enum. Pl. 2: 212 (1837), nom. inval.
Fimbristylis urvilleana Steud., Syn. Pl. Glumac. 2: 112 (1855).
Scirpus sphaerolepis Boeck., Vidensk. Meddel. Naturhist. Foren. Kjøbenhavn 1869: 104 (1869). *Stenophyllus sphaerolepis* (Boeck.) Rusby, Mem. Torrey Bot. Club 4: 270 (1895). *Bulbostylis sphaerolepis* (Boeck.) Beetle, Amer. Midl. Naturalist 41: 487 (1949).
Scirpus niederleinianus Boeck., Beitr. Cyper. 2: 16 (1890).
Bulbostylis scabra f. *evolutior* Lindm., Bih. Kongl. Svenska Vetensk.-Akad. Handl. 26(3; 9): 18 (1901).

Bulbostylis scabricaulis Cherm., Bull. Soc. Bot. France 68: 419 (1921 publ. 1922).
Trop. & S. Africa, Madagascar. 22 BEN BKN GHA GUI IVO LBR MLI NGA SEN SIE TOG 23 BUR GAB RWA ZAI 25 KEN TAN UGA 26 ANG 27 TVL 29 MDG. Hemicr.
Fimbristylis collina Ridl., Trans. Linn. Soc. London, Bot. 2: 154 (1884). *Abildgaardia collina* (Ridl.) Lye, Nordic J. Bot. 1: 757 (1981 publ. 1982).

Bulbostylis cardiocarpa var. *holubii* C.B.Clarke in D.Oliver & auct. suc. (eds.), Fl. Trop. Afr. 8: 434 (1902). *Abildgaardia filamentosa* var. *holubii* (C.B.Clarke) Lye, Bot. Not. 127: 496 (1974).

Bulbostylis schaffneri (Boeck.) C.B.Clarke, Bull. Misc. Inform. Kew, Addit. Ser. 8: 26 (1908).
SC. U.S.A. to NE. Mexico. 77 NWM 79 MXE. Hemicr.
**Scirpus schaffneri* Boeck., Bot. Jahrb. Syst. 7: 275 (1886).
Scirpus pringlei Britton, Bull. Torrey Bot. Club 15: 103 (1888). *Bulbostylis pringlei* (Britton) Beetle, Leafl. W. Bot. 4: 45 (1944).

Bulbostylis schimperiana (Hochst. ex A.Rich.) C.B.Clarke in T.A.Durand & H.Schinz, Consp. Fl. Afric. 5: 616 (1894).
Ethiopia to Tanzania. 24 ETH 25 KEN TAN UGA. Ther.
**Isolepis schimperiana* Hochst. ex A.Rich., Tent. Fl. Abyss. 2: 501 (1851). *Scirpus schimperianus* (Hochst. ex A.Rich.) Boeck., Linnaea 36: 750 (1870). *Trichelostylis schimperiana* (Hochst. ex A.Rich.) Hochst. ex Boeck., Linnaea 37: 28 (1871). *Fimbristylis schimperiana* (Hochst. ex A.Rich.) K.Schum in H.G.A.Engler, Pflanzenw. Ost-Afrikas, C: 125 (1895). *Abildgaardia schimperana* (Hochst. ex A.Rich.) Lye, Bot. Not. 127: 497 (1974).
Fimbristylis subumbellata K.Schum. in H.G.A.Engler, Pflanzenw. Ost-Afrikas, C: 125 (1895).
Scirpus fimbristyloides K.Schum. ex C.B.Clarke in D.Oliver & auct. suc. (eds.), Fl. Trop. Afr. 8: 436 (1902).
Fimbristylis humilis Peter, Repert. Spec. Nov. Regni Veg. Beih. 40(1): 126 (1936).

Bulbostylis schlechteri C.B.Clarke, Bull. Herb. Boissier, II, 4: 995 (1904). *Abildgaardia schlechteri* (C.B.Clarke) Lye, Nordic J. Bot. 1: 758 (1981 publ. 1982).
Northern Prov. 27 TVL. Hemicr. or rhizome geophyte.

Bulbostylis schoenoides (Kunth) C.B.Clarke, Trans. Linn. Soc. London, Bot. 4: 54 (1894).
Uganda to S. Africa. 25 KEN TAN UGA 26 MLW 27 CPP LES NAT SWZ TVL. Hemicr.
**Isolepis schoenoides* Kunth, Enum. Pl. 2: 208 (1837). *Fimbristylis schoenoides* (Kunth) K.Schum. in H.G.A.Engler, Pflanzenw. Ost-Afrikas, C: 125 (1895), nom. illeg. *Abildgaardia erratica* subsp. *schoenoides* (Kunth) Lye, Nordic J. Bot. 3: 239 (1983).

Bulbostylis schomburgkiana (Steud.) M.T.Strong, Brittonia 45: 162 (1993).
N. South America to N. Brazil. 82 FRG VEN 84 BZN. Hemicr.
**Fimbristylis schomburgkiana* Steud., Syn. Pl. Glumac. 2: 111 (1855).
Fimbristylis strictior Steud., Syn. Pl. Glumac. 2: 119 (1855).

Bulbostylis schraderiana (Steud.) Beetle, Leafl. W. Bot. 4: 45 (1944).
Brazil. 84+. Hemicr. or rhizome geophyte.
**Isolepis schraderiana* Steud., Syn. Pl. Glumac. 2: 103 (1855).

Bulbostylis scirpoides Kral & M.T.Strong, Sida 18: 855 (1999).
Brazil (Brasília D.F., Goiás). 84 BZC. Hemicr. or tuber geophyte.

Bulbostylis scleropus C.B.Clarke in W.H.Harvey & auct. suc. (eds.), Fl. Cap. 7: 207 (1898). *Scirpus scleropus* (C.B.Clarke) K.Schum., Just's Bot. Jahresber. 26(1): 329 (1900). *Abildgaardia scleropus* (C.B.Clarke) Lye, Bot. Not. 127: 497 (1974).
Burundi to S. Africa. 23 BUR ZAI 26 MLW ZAM 27 CPP LES NAT OFS TVL. Hemicr.
Bulbostylis grandibulbosa Kük. in R.E.Fries, Wiss. Erg. Schwed. Rhod.-Kongo Exped. 1: 6 (1921).

Bulbostylis sellowiana (Kunth) Palla, Denkschr. Kaiserl. Akad. Wiss., Wien. Math.-Naturwiss. Kl. 79: 179 (1908).
S. Brazil. 84 BZS. Hemicr. or rhizome geophyte.
**Isolepis sellowiana* Kunth, Enum. Pl. 2: 208 (1837). *Scirpus sellowianus* (Kunth) Boeck., Vidensk. Meddel. Naturhist. Foren. Kjøbenhavn 1869: 138 (1869). *Fimbristylis sellowiana* (Kunth) Hauman & Vanderveken, Anales Mus. Nac. Buenos Aires 29: 207 (1917). *Stenophyllus sellowianus* (Kunth) I.M.Johnst., Univ. Calif. Publ. Bot. 7: 438 (1922).

Bulbostylis sepiacea Kral, Sida 4: 86 (1971).
Revillagigedo Is. (Socorro I.). 79 MXI. Hemicr. or rhizome geophyte.

Bulbostylis setacea (Griseb.) Svenson, Contr. Ocas. Mus. Hist. Nat. Colegio "De La Salle" 4: 10 (1946).
Caribbean. 81 CUB NLA. Ther.
**Abildgaardia setacea* Griseb., Cat. Pl. Cub.: 238 (1866). *Fimbristylis grisebachii* Greenm. ex Combs, Trans. Acad. Sci. St. Louis 7: 474 (1897). *Bulbostylis grisebachii* (Greenm. ex Combs) Kük., Repert. Spec. Nov. Regni Veg. 23: 197 (1926).
Fimbristylis cubensis C.B.Clarke, Ill. Cyper.: t. 43 (1909).

Bulbostylis smithii Barros, Darwiniana 11: 761 (1959).
Brazil (Minas Gerais). 84 BZL. Hemicr.

Bulbostylis somaliensis Lye, Edinburgh J. Bot. 53: 223 (1996).
Somalia. 24 SOM. Ther.

subsp. ***confusa*** Lye, Edinburgh J. Bot. 53: 225 (1996).
S. Somalia. 24 SOM. Ther.

subsp. ***microcarpa*** (Chiov.) Lye in M.Thulin (ed.), Fl. Somalia 4: 111 (1995).
S. Somalia (Bay Reg.). 24 SOM. Ther.
**Fimbristylis cioniana* var. *microcarpa* Chiov., Result. Sci. Miss. Stefan.-Paoli Somal. Ital.: 180 (1916).

subsp. ***somaliensis***
C. Somalia. 24 SOM. Ther.

Bulbostylis spectabilis Kral & M.T.Strong, Sida 18: 857 (1999).
Brazil (Paraná). 84 BZS. Hemicr. or rhizome geophyte.

Bulbostylis sphaerocarpa (Boeck.) C.B.Clarke in T.A. Durand & H.Schinz, Consp. Fl. Afric. 5: 616 (1894).
S. Tanzania. 25 TAN. Ther.
**Scirpus sphaerocarpus* Boeck., Linnaea 36: 741 (1870). *Fimbristylis sphaerocarpa* (Boeck.) K.Schum. in H.G.A.Engler, Pflanzenw. Ost-Afrikas, C: 125 (1895). *Abildgaardia sphaerocarpa* (Boeck.) Lye, Bot. Not. 127: 497 (1974).

Bulbostylis sphaerocephala (Boeck.) Lindm., Bih. Kongl. Svenska Vetensk.-Akad. Handl. 26(9): 18 (1901).
Brazil to Argentina (Misiones). 83 BOL 84 BZC 85 PAR. Hemicr. or rhizome geophyte.

**Scirpus sphaerocephalus* Boeck., Vidensk. Meddel. Naturhist. Foren. Kjøbenhavn 1869: 137 (1869). *Fimbristylis sphaerocephala* (Boeck.) Hauman & Vanderveken, Anales Mus. Nac. Buenos Aires 29: 207 (1917), nom. illeg.
Scirpus brunneovaginatus Boeck., Vidensk. Meddel. Naturhist. Foren. Kjøbenhavn 41–42: 25 (1879). *Bulbostylis sphaerocephala* var. *brunneovaginata* (Boeck.) C.B.Clarke, Bull. Herb. Boissier, II, 3: 1018 (1903). *Fimbristylis brunneovaginata* (Boeck.) H.Pfeiff., Revista Sudamer. Bot. 7: 128 (1943).
Scirpus setifolius Boeck., Flora 62: 160 (1879). *Bulbostylis setifolia* (Boeck.) Beetle, Amer. Midl. Naturalist 41: 486 (1949).
Scirpus sphaerocephalus var. *macrocephalus* Maury, Mém. Soc. Phys. Genève 31(1): 141 (1889). *Bulbostylis sphaerocephala* var. *macrocephala* (Maury) Barros, Anales Mus. Nac. Montevideo, II, 3: 186 (1931).
Scirpus schwackeanus Boeck., Beitr. Cyper. 2: 15 (1890). *Bulbostylis sphaerocephala* f. *schwackeana* (Boeck.) Lindm., Bih. Kongl. Svenska Vetensk.-Akad. Handl. 26(3; 9): 18 (1900).

Bulbostylis splendens M.T.Strong, Brittonia 48: 96 (1996).
Brazil (Rondônia). 84 BZN. Hemicr. or rhizome geophyte.

Bulbostylis stenocarpa Kük., Bot. Jahrb. Syst. 56(125): 15 (1921).
Trop. America. 79 MXT 81 CUB PUE 82 FRG GUY SUR VEN 83 BOL PER 84 BZC BZN. Hemicr. or rhizome geophyte.
Bulbostylis papillosa Kük., Repert. Spec. Nov. Regni Veg. 23: 198 (1926).

Bulbostylis stenophylla (Elliott) C.B.Clarke, Bull. Misc. Inform. Kew, Addit. Ser. 8: 26 (1908).
SE. U.S.A., Cuba. 78 FLA GEO NCA SCA 81 CUB. Ther.
**Scirpus stenophyllus* Elliott, Sketch Bot. S. Carolina 1: 83 (1816). *Isolepis stenophylla* (Elliott) Torr., Ann. Lyceum Nat. Hist. New York 3: 853 (1836). *Trichelostylis stenophylla* (Elliott) Alph.Wood, Class-book Bot.: 742 (1861). *Stenophyllus stenophyllus* (Elliott) Britton, Bull. Torrey Bot. Club 21: 30 (1890). *Fimbristylis stenophylla* (Elliott) Alain, Bull. Torrey Bot. Club 92: 290 (1965).
Dichromena caespitosum Muhl., Descr. Gram.: 14 (1817). *Dichromena caespitosa* (Muhl.) Spreng., Syst. Veg. 1: 202 (1824).
Stenophyllus caespitosus Raf., Neogenyton: 4 (1825).

Bulbostylis striata (Ruiz & Pav.) H.Pfeiff., Revista Sudamer. Bot. 7: 129 (1943).
Peru. 83 PER. Hemicr. or rhizome geophyte.
**Cyperus striatus* Ruiz & Pav., Fl. Peruv. 1: 47 (1798).
Cyperus atropurpureus Pers., Syn. Pl. 1: 60 (1805).
Cyperus collinus Vahl, Enum. Pl. 2: 301 (1805).

Bulbostylis subaphylla C.B.Clarke in I.Urban, Symb. Antill. 2: 86 (1900). *Stenophyllus subaphyllus* (C.B.Clarke) Britton, Bull. Dept. Agric. Jamaica 5(Suppl. 1): 12 (1907).
Greater Antilles. 81 CUB HAI JAM PUE. Hemicr. or rhizome geophyte.
Stenophyllus harrisii Britton, Torreya 20: 83 (1920). *Fimbristylis harrisii* (Britton) C.D.Adams, Phytologia 21: 66 (1971).
Bulbostylis subaphylla var. *longiglumis* Kük., Repert. Spec. Nov. Regni Veg. 23: 198 (1926).
Bulbostylis subaphylla var. *rigida* Kük., Repert. Spec. Nov. Regni Veg. 23: 199 (1926).
Bulbostylis subefimbriata Kük., Ark. Bot. 22A(17): 7 (1929).

Bulbostylis subsphaerocephala E.G.Camus, Notul. Syst. (Paris) 1: 292 (1910).
Vietnam. 41 VIE. Hemicr. or rhizome geophyte.

Bulbostylis subspinescens C.B.Clarke in J.D.Hooker, Fl. Brit. India 6: 652 (1893).
India (Orissa). 40 IND. Hemicr. or rhizome geophyte.

Bulbostylis subumbellata (Lye) R.W.Haines in R.W.Haines & K.A.Lye, Sedges & Rushes E. Afr., App. 3: 1 (1983).
Uganda. 25 UGA. Hemicr. or rhizome geophyte.
**Abildgaardia subumbellata* Lye, Nordic J. Bot. 1: 756 (1981 publ. 1982).

Bulbostylis surinamensis Uittien, Recueil Trav. Bot. Néerl. 22: 338 (1925).
Guyana to Suriname. 82 FRG? GUY SUR. Hemicr. or rhizome geophyte.

Bulbostylis svensoniana Steyerm., Fieldiana, Bot. 27: 37 (1951).
Venezuela, Paraguay. 82 VEN 85 PAR. Hemicr.

Bulbostylis swamyi Govind., Proc. Indian Acad. Sci., Pl. Sci. 94: 11 (1985).
S. India. 40 IND. Hemicr. or rhizome geophyte.

Bulbostylis tanzaniae (Lye) R.W.Haines in R.W.Haines & K.A.Lye, Sedges & Rushes E. Afr., App. 3: 1 (1983).
Zaïre, Burundi, S. Tanzania, Zambia. 23 BUR ZAI 25 TAN 26 ZAM. Ther.
**Abildgaardia tanzaniae* Lye, Nordic J. Bot. 1: 753 (1981 publ. 1082).

Bulbostylis taylorii (K.Schum.) C.B.Clarke in T.A.Durand & H.Schinz, Consp. Fl. Afric. 5: 616 (1894).
Tanzania (Zanzibar). 25 TAN. Ther.
**Fimbristylis taylorii* K.Schum. in H.G.A.Engler, Pflanzenw. Ost-Afrikas, C: 125 (1895). *Abildgaardia taylorii* (K.Schum.) Lye, Bot. Not. 127: 497 (1974).

Bulbostylis tenella (Link) C.B.Clarke, Bull. Misc. Inform. Kew, Addit. Ser. 8: 28 (1908).
N. Brazil. 84 BZN. Hemicr. or rhizome geophyte.
**Scirpus tenellus* Link, Jahrb. Gewächsk. 1(3): 80 (1820). *Fimbristylis tenella* (Link) Schult. in J.J.Roemer & J.A.Schultes, Mant. 2: 59 (1824). *Isolepis tenella* (Link) Kunth, Enum. Pl. 2: 213 (1837). *Oncostylis tenella* (Link) Nees in C.F.P.von Martius & auct. suc. (eds.), Fl. Bras. 2(1): 82 (1842). *Stenophyllus tenellus* (Link) H.Pfeiff., Bot. Arch. 6: 191 (1924).

Bulbostylis tenerrima (Fisch. & C.A.Mey. ex Ledeb.) Palla, Trudy Tiflissk. Bot. Sada 12(App. 1): 8 (1912).
NE. Turkey to W. Transcaucasus. 33 TCS 34 TUR. Ther.
**Isolepis tenerrima* Fisch. & C.A.Mey. ex Ledeb., Fl. Ross. 4: 258 (1852).

Bulbostylis tenuifolia (Rudge) J.F.Macbr., Publ. Field Mus. Nat. Hist., Bot. Ser. 11: 5 (1931).
Mexico to Brazil. 79 MXC MXE MXG MXS MXT 80 BLZ COS ELS GUA HON NIC PAN 81 CUB JAM TRT 82 FRG GUY SUR VEN 83 BOL CLM PER 84 BZL BZS. Ther.
**Scirpus tenuifolius* Rudge, Pl. Guian.: 18 (1805). *Isolepis tenuifolia* (Rudge) A.Dietr., Sp. Pl. 2: 131

(1832). *Oncostylis tenuifolia* (Rudge) Nees in C.F.P.von Martius & auct. suc. (eds.), Fl. Bras. 2(1): 83 (1842). *Bulbostylis capillaris* var. *tenuifolia* (Rudge) C.B.Clarke in I.Urban, Symb. Antill. 2: 89 (1900). *Scirpus capillaris* var. *tenuifolius* (Rudge) Pilg., Bot. Jahrb. Syst. 30: 142 (1901). *Stenophyllus tenuifolius* (Rudge) Britton, Bull. Torrey Bot. Club 43: 448 (1916).
Isolepis asperula Kunth in F.W.H.von Humboldt, A.J.A.Bonpland & C.S.Kunth, Nov. Gen. Sp. 1: 221 (1816). *Scirpus asperulus* (Kunth) Spreng., Syst. Veg. 1: 213 (1824). *Bulbostylis asperula* (Kunth) C.B.Clarke, Bull. Misc. Inform. Kew, Addit. Ser. 8: 29 (1908). *Stenophyllus tenuifolius* var. *asperula* (Kunth) H.Pfeiff., Bot. Arch. 6: 189 (1924).
Isolepis bufonia Kunth in F.W.H.von Humboldt, A.J.A.Bonpland & C.S.Kunth, Nov. Gen. Sp. 1: 222 (1816). *Scirpus bufonius* (Kunth) Spreng., Syst. Veg. 1: 218 (1824). *Fimbristylis bufonia* (Kunth) Alain, Bull. Torrey Bot. Club 92: 291 (1965).
Scirpus humilis Link, Jahrb. Gewächsk. 1: 78 (1820), nom. illeg.
Isolepis asperiuscula Kunth, Enum. Pl. 2: 212 (1837).
Scirpus fuscus Willd. ex Kunth, Enum. Pl. 2: 212 (1837).
Stenophyllus tenuifolius var. *latesquamata* H.Pfeiff., Bot. Arch. 6: 189 (1924).
Stenophyllus tenuifolius var. *longinux* H.Pfeiff., Bot. Arch. 6: 189 (1924).

Bulbostylis trabeculata C.B.Clarke in T.A.Durand & H.Schinz, Consp. Fl. Afric. 5: 616 (1894). *Abildgaardia trabeculata* (C.B.Clarke) Lye, Nordic J. Bot. 1: 758 (1981 publ. 1982).
E. Trop. Africa to Namibia. 25 KEN UGA 26 ANG 27 BOT NAM. Ther.

var. ***microglumis*** (Lye) R.W.Haines in R.W.Haines & K.A.Lye, Sedges & Rushes E. Afr., App. 3: 1 (1983).
E. Trop. Africa. 25 KEN UGA. Ther.
**Abildgaardia trabeculata* var. *microglumis* Lye, Nordic J. Bot. 1: 755 (1981 publ. 1982).

var. ***trabeculata***
Angola, Namibia, Botswana. 26 ANG 27 BOT NAM. Ther.

Bulbostylis trichobasis (Baker) C.B.Clarke in T.A.Durand & H.Schinz, Consp. Fl. Afric. 5: 616 (1894).
Madagascar. 29 MDG. Hemicr. or rhizome geophyte.
Fimbristylis cinerea Ridl., J. Linn. Soc., Bot. 20: 335 (1883).
**Scirpus trichobasis* Baker, J. Linn. Soc., Bot. 20: 298 (1893). *Abildgaardia trichobasis* (Baker) Lye, Nordic J. Bot. 1: 758 (1981 publ. 1982).
Bulbostylis trichobasis var. *elatior* Cherm., Bull. Soc. Bot. France 68: 421 (1921).

Bulbostylis trichoides (Schrad. ex Schult.) Beetle, Leafl. W. Bot. 4: 45 (1944).
Brazil. 84+. Hemicr. or rhizome geophyte.
**Isolepis trichoides* Schrad. ex Schult. in J.J.Roemer & J.A.Schultes, Mant. 2: 64 (1824).

Bulbostylis trilobata Kral, Sida 4: 90 (1971).
S. Mexico to C. America. 79 MXG MXS MXT 80 GUA HON. Ther.

Bulbostylis trullata Goetgh., Bull. Jard. Bot. Natl. Belg. 55: 247 (1985).
Zaïre. 23 ZAI. Ther.

Bulbostylis truncata (Nees) M.T.Strong, Brittonia 45: 165 (1993).
S. Trop. America. 82 GUY SUR VEN 83 BOL CLM 84 BZC BZE BZL BZN. Hemicr.
**Oncostylis truncata* Nees in C.F.P.von Martius & auct. suc. (eds.), Fl. Bras. 2(1): 83 (1842).
Bulbostylis capillaris var. *glabra* Uittien, Recueil Trav. Bot. Néerl. 30: 188 (1933).

Bulbostylis tuerckheimii Urb., Symb. Antill. 7: 169 (1912). *Stenophyllus tuerckheimii* (Urb.) Britton, Bull. Torrey Bot. Club 43: 446 (1916).
Dominican Rep. 81 DOM. Hemicr. or rhizome geophyte.

Bulbostylis turbinata S.T.Blake, Proc. Roy. Soc. Queensland 52: 56 (1941).
Australia. 50 NTA QLD SOA? WAU. Hemicr. or rhizome geophyte.

Bulbostylis ugandensis (Lye) R.W.Haines in R.W.Haines & K.A.Lye, Sedges & Rushes E. Afr., App. 3: 1 (1983).
Uganda, Rwanda, Burundi, Tanzania. 23 BUR RWA 25 TAN UGA. Ther. or hemicr.
**Abildgaardia ugandensis* Lye, Bot. Not. 127: 493 (1974).

Bulbostylis vaginosa Kük., Notizbl. Bot. Gart. Berlin-Dahlem 9: 308 (1925).
Kenya. 25 KEN. Hemicr. or rhizome geophyte.

Bulbostylis vanderystii Cherm., Rev. Zool. Bot. Africaines 24: 299 (1934). *Abildgaardia vanderystii* (Cherm.) Lye, Nordic J. Bot. 1: 758 (1981 publ. 1982).
Zaïre. 23 ZAI. Hemicr.

Bulbostylis vestita (Kunth) C.B.Clarke in I.Urban, Symb. Antill. 2: 87 (1900).
Mexico to Trop. America. 79 MXC MXG MXI MXS MXT 80 BLZ COS GUA HON NIC 81 CUB PUE TRT 82 FRG SUR VEN 83 BOL PER 84 BZE BZN 85 PAR. Ther. or hemicr.
**Isolepis vestita* Kunth, Enum. Pl. 2: 210 (1837). *Oncostylis vestita* (Kunth) Nees in C.F.P.von Martius & auct. suc. (eds.), Fl. Bras. 2(1): 88 (1842). *Scirpus vestitus* (Kunth) Rchb. ex Boeck., Linnaea 36: 753 (1870). *Fimbristylis vestita* (Kunth) Hemsl., Biol. Cent.-Amer., Bot. 3: 460 (1885). *Stenophyllus vestitus* (Kunth) Britton, Bull. Torrey Bot. Club 43: 446 (1916).
Scirpus barbatus Vahl ex Kunth, Enum. Pl. 2: 211 (1837), nom. inval.
Scirpus hirsutus Willd. ex Kunth, Enum. Pl. 2: 210 (1837).
Oncostylis hispida Liebm., Mexic. Neldeagt. Pl., V, 2: 240 (1851).
Scirpus hirtus Griseb., Cat. Pl. Cub.: 241 (1866), nom. illeg.

Bulbostylis viguieri Cherm., Bull. Soc. Bot. France 68: 419 (1921 publ. 1922).
C. Madagascar. 29 MDG. Ther.

Bulbostylis viridicarinata (De Wild.) Goetgh., Bull. Jard. Bot. Natl. Belg. 54: 104 (1984).
Trop. Africa. 22 GHA GUI IVO NGA TOG 23 BUR CAF ZAI 26 MOZ ZAM. Ther.
**Fimbristylis viridicarinata* De Wild., Pl. Bequaert. 4: 207] (1927).
Fimbristylis tisserantii Cherm., Arch. Bot. Bull. Mens. 4(7): 32 (1931). *Bulbostylis tisserantii* (Cherm.) Lye, Mitt. Bot. Staatssamml. München 10: 547 (1971), nom. illeg.

Bulbostylis warei (Torr.) C.B.Clarke, Bull. Misc. Inform. Kew, Addit. Ser. 8: 26 (1908).
SE. U.S.A. 78 ALA FLA GEO NCA SCA. Hemicr.
**Isolepis warei* Torr., Ann. Lyceum Nat. Hist. New York 3: 354 (1836). *Trichelostylis warei* (Torr.) Alph.Wood, Class-book Bot.: 742 (1861). *Stenophyllus warei* (Torr.) Britton, Bull. Torrey Bot. Club 21: 30 (1894).

Bulbostylis wilsonii (Britton) Urb., Ark. Bot. 22A(17): 7 (1929).
Bahamas. 81 BAH. Hemicr. or rhizome geophyte.
**Stenophyllus wilsonii* Britton, Torreya 13: 215 (1913).

Bulbostylis wombaliensis (De Wild.) R.W.Haines in R.W.Haines & K.A.Lye, Sedges & Rushes E. Afr., App. 3: 1 (1983).
Uganda, Zaïre, Zambia. 23 ZAI 25 UGA 26 ZAM. Ther.
**Fimbristylis wombaliensis* De Wild., Pl. Bequaert. 4: 208 (1927). *Abildgaardia wombaliensis* (De Wild.) Lye, Bot. Not. 127: 497 (1974).

Bulbostylis xerophila Cherm., Bull. Soc. Bot. France 68: 420 (1921 publ. 1922).
Madagascar. 29 MDG. Ther.

***Synonyms*:**

Bulbostylis Steven = ***Eleocharis*** R.Br.
Bulbostylis albescens (Miq.) Beetle = ***Rhynchospora albescens*** (Miq.) Kük.
Bulbostylis antillana (Britton) Fernald = ***Bulbostylis capillaris*** subsp. ***antillana*** (Britton) T.Koyama
Bulbostylis aphyllanthoides (Welw. ex Ridl.) C.B.Clarke = ***Bulbostylis pilosa*** (Willd.) Cherm.
Bulbostylis arenaria (Nees) Lindm. = ***Bulbostylis juncoides*** (Vahl) Kük. ex Herter
Bulbostylis argentina Palla = ***Bulbostylis juncoides*** (Vahl) Kük. ex Herter
Bulbostylis asperula (Kunth) C.B.Clarke = ***Bulbostylis tenuifolia*** (Rudge) J.F.Macbr.
Bulbostylis barbata var. *pulchella* (Thwaites) C.B.Clarke = ***Bulbostylis barbata*** subsp. ***pulchella*** (Thwaites) T.Koyama
Bulbostylis breviculmis (Kunth) C.B.Clarke = ***Bulbostylis humilis*** (Kunth) C.B.Clarke
Bulbostylis burkei C.B.Clarke = ***Bulbostylis contexta*** (Nees) Bodard
Bulbostylis caespitosa Peter = ***Bulbostylis oritrephes*** (Ridl.) C.B.Clarke subsp. ***oritrephes***
Bulbostylis camporum K.Schum. ex C.B.Clarke = ***Bulbostylis abortiva*** (Steud.) C.B.Clarke
Bulbostylis capillaris subsp. *capitata* (Miq.) T.Koyama = ***Bulbostylis densa*** var. ***capitata*** (Miq.) Ohwi
Bulbostylis capillaris var. *coarctata* (Elliott) J.F.Macbr. = ***Bulbostylis capillaris*** (L.) Kunth ex C.B.Clarke subsp. ***capillaris***
Bulbostylis capillaris f. *contracta* Kük. ex Osten = ***Bulbostylis capillaris*** (L.) Kunth ex C.B.Clarke subsp. ***capillaris***
Bulbostylis capillaris var. *crebra* Fernald = ***Bulbostylis capillaris*** (L.) Kunth ex C.B.Clarke subsp. ***capillaris***
Bulbostylis capillaris var. *elatior* (Griseb.) Osten = ***Bulbostylis capillaris*** (L.) Kunth ex C.B.Clarke subsp. ***capillaris***
Bulbostylis capillaris f. *fasciculata* Osten = ***Bulbostylis capillaris*** (L.) Kunth ex C.B.Clarke subsp. ***capillaris***
Bulbostylis capillaris var. *genuina* Osten = ***Bulbostylis capillaris*** (L.) Kunth ex C.B.Clarke
Bulbostylis capillaris var. *glabra* Uittien = ***Bulbostylis truncata*** (Nees) M.T.Strong
Bulbostylis capillaris var. *isopoda* Fernald = ***Bulbostylis capillaris*** (L.) Kunth ex C.B.Clarke subsp. ***capillaris***
Bulbostylis capillaris var. *major* (Maury) R.Knuth = ***Bulbostylis capillaris*** (L.) Kunth ex C.B.Clarke subsp. ***capillaris***
Bulbostylis capillaris var. *major* Uittien = ***Bulbostylis capillaris*** (L.) Kunth ex C.B.Clarke subsp. ***capillaris***
Bulbostylis capillaris var. *microstachya* (Boeck.) Barros = ***Bulbostylis capillaris*** (L.) Kunth ex C.B.Clarke subsp. ***capillaris***
Bulbostylis capillaris f. *minima* (Maury) R.Knuth = ***Bulbostylis capillaris*** (L.) Kunth ex C.B.Clarke subsp. ***capillaris***
Bulbostylis capillaris var. *paraensis* (C.B.Clarke) Kük. = ***Bulbostylis paraensis*** C.B.Clarke
Bulbostylis capillaris var. *pyriformis* C.B.Clarke ex Lindm. = ***Bulbostylis capillaris*** (L.) Kunth ex C.B.Clarke subsp. ***capillaris***
Bulbostylis capillaris f. *stenantha* Kük. ex Barros = ***Bulbostylis capillaris*** (L.) Kunth ex C.B.Clarke subsp. ***capillaris***
Bulbostylis capillaris var. *tenuifolia* (Rudge) C.B.Clarke = ***Bulbostylis tenuifolia*** (Rudge) J.F.Macbr.
Bulbostylis capillaris var. *trifida* (Nees) C.B.Clarke = ***Bulbostylis densa*** subsp. ***afromontana*** (Lye) R.W.Haines
Bulbostylis capitata (L.) Steven = ***Eleocharis geniculata*** (L.) Roem. & Schult.
Bulbostylis cardiocarpa C.B.Clarke = ***Bulbostylis filamentosa*** (Vahl) C.B.Clarke
Bulbostylis cardiocarpa var. *holubii* C.B.Clarke = ***Bulbostylis scabricaulis*** Cherm.
Bulbostylis classensii De Wild. = ***Bulbostylis abortiva*** (Steud.) C.B.Clarke
Bulbostylis coarctata (Elliott) Fernald = ***Bulbostylis ciliatifolia*** var. ***coarctata*** (Elliott) Kral
Bulbostylis collina (Kunth) C.B.Clarke = ***Bulbostylis contexta*** (Nees) Bodard
Bulbostylis curassavica var. *pallescens* Kük. & Ekman = ***Bulbostylis curassavica*** (Britton) Kük. & Ekman
Bulbostylis cyrtathera Cherm. = ***Bulbostylis fimbristyloides*** C.B.Clarke
Bulbostylis ekmanii Kük. = ***Bulbostylis pauciflora*** (Liebm.) C.B.Clarke
Bulbostylis elegans Cherm. = ***Bulbostylis pusilla*** (Hochst. ex A.Rich.) C.B.Clarke
Bulbostylis equitans (Kük.) Raymond = ***Nemum equitans*** (Kük.) J.Raynal
Bulbostylis exilis (Kunth) Lye = ***Bulbostylis hispidula*** (Vahl) R.W.Haines subsp. ***hispidula***
Bulbostylis fasciculata Cherm. =
Bulbostylis fendleri C.B.Clarke = ***Bulbostylis juncoides*** (Vahl) Kük. ex Herter
Bulbostylis filamentosa var. *metralis* (Cherm.) R.W. Haines = ***Bulbostylis filamentosa*** (Vahl) C.B.Clarke
Bulbostylis filiformis C.B.Clarke = ***Bulbostylis hispidula*** subsp. ***filiformis*** (C.B.Clarke) R.W.Haines
Bulbostylis floccosa var. *pumilio* C.B.Clarke = ***Bulbostylis curassavica*** (Britton) Kük. & Ekman
Bulbostylis floridana (Nash) Fernald = ***Bulbostylis barbata*** (Rottb.) C.B.Clarke subsp. ***barbata***
Bulbostylis francaensis Palla = ***Bulbostylis edwalliana*** (Boeck.) Prata & M.G.López
Bulbostylis geniculatus (L.) Steven = ***Eleocharis geniculata*** (L.) Roem. & Schult.
Bulbostylis glaziovii (Boeck.) C.B.Clarke = ***Bulbostylis juncoides*** (Vahl) Kük. ex Herter
Bulbostylis grandibulbosa Kük. = ***Bulbostylis scleropus*** C.B.Clarke

Bulbostylis grisebachii (Greenm. ex Combs) Kük. = ***Bulbostylis setacea*** (Griseb.) Svenson
Bulbostylis hirta (Thunb.) Svenson = ***Fimbristylis squarrosa*** Vahl var. ***squarrosa***
Bulbostylis hirtella (Schrad. ex Schult.) Nees = ***Bulbostylis juncoides*** (Vahl) Kük. ex Herter
Bulbostylis hispidula subsp. *brachyphylla* (Cherm.) R.W. Haines = ***Bulbostylis hensii*** (C.B.Clarke) R.W.Haines
Bulbostylis holotricha Peter = ***Bulbostylis congolensis*** De Wild.
Bulbostylis japonica C.B.Clarke = ***Bulbostylis densa*** var. ***capitata*** (Miq.) Ohwi
Bulbostylis junciformis var. *ambigua* Steyerm. = ***Bulbostylis junciformis*** (Kunth) C.B.Clarke
Bulbostylis junciformis var. *congesta* (Schrad. ex Schult.) Steyerm. = ***Bulbostylis consanguinea*** (Kunth) C.B.Clarke
Bulbostylis junciformis var. *conostachya* (Boeck.) Lindm. = ***Bulbostylis conostachya*** (Boeck.) Beetle
Bulbostylis junciformis var. *filamentosa* (Vahl) H.Pfeiff. = ***Bulbostylis filamentosa*** (Vahl) C.B.Clarke
Bulbostylis junciformis var. *fulvicoma* Kük. = ***Bulbostylis junciformis*** (Kunth) C.B.Clarke
Bulbostylis junciformis var. *laxiuscula* Kük. = ***Bulbostylis junciformis*** (Kunth) C.B.Clarke
Bulbostylis juncoides var. *ampliceps* Kük. ex Osten = ***Bulbostylis juncoides*** (Vahl) Kük. ex Herter
Bulbostylis juncoides var. *genuina* Osten = ***Bulbostylis juncoides*** (Vahl) Kük. ex Herter
Bulbostylis juncoides f. *hirtella* Osten = ***Bulbostylis juncoides*** (Vahl) Kük. ex Herter
Bulbostylis juncoides f. *leucocarpa* Barros = ***Bulbostylis juncoides*** (Vahl) Kük. ex Herter
Bulbostylis juncoides var. *lorentzii* (Boeck.) Kük. ex Osten = ***Bulbostylis juncoides*** (Vahl) Kük. ex Herter
Bulbostylis juncoides var. *nana* (Griseb.) Barros = ***Bulbostylis juncoides*** (Vahl) Kük. ex Herter
Bulbostylis juncoides f. *subfimbriata* Kük. ex Osten = ***Bulbostylis juncoides*** (Vahl) Kük. ex Herter
Bulbostylis lanata var. *xyrioides* Lindm. = ***Bulbostylis lanata*** (Kunth) Lindm.
Bulbostylis langsdorffiana (Kunth) C.B.Clarke = ***Bulbostylis ciliata*** (C.Presl) C.Presl ex Nees
Bulbostylis leucostachya var. *aturensis* (Maury) H.Pfeiff. = ***Bulbostylis aturensis*** (Maury) C.B.Clarke
Bulbostylis maculosus Steven = ***Eleocharis maculosa*** (Vahl) Roem. & Schult.
Bulbostylis metralis Cherm. = ***Bulbostylis filamentosa*** (Vahl) C.B.Clarke
Bulbostylis miegei Bodard = ***Bulbostylis coleotricha*** var. ***miegei*** (Bodard) R.W.Haines
Bulbostylis ovatus (Roth) Steven = ***Eleocharis ovata*** (Roth) Roem. & Schult.
Bulbostylis palustris (L.) Steven = ***Eleocharis palustris*** (L.) Roem. & Schult.
Bulbostylis papillosa Kük. = ***Bulbostylis stenocarpa*** Kük.
Bulbostylis parva (Ridl.) C.B.Clarke = ***Bulbostylis pusilla*** (Hochst. ex A.Rich.) C.B.Clarke
Bulbostylis polytricha Cherm. = ***Bulbostylis congolensis*** De Wild.
Bulbostylis portoricensis (Britton) Fernald = ***Bulbostylis pauciflora*** (Liebm.) C.B.Clarke
Bulbostylis pringlei (Britton) Beetle = ***Bulbostylis schaffneri*** (Boeck.) C.B.Clarke
Bulbostylis puberula var. *gracilis* C.E.C.Fisch. = ***Bulbostylis puberula*** C.B.Clarke
Bulbostylis pubescens (C.Presl) Svenson = ***Bulbostylis hispidula*** (Vahl) R.W.Haines subsp. ***hispidula***
Bulbostylis pusilla subsp. *congolensis* (De Wild.) R.W.Haines = ***Bulbostylis congolensis*** De Wild.
Bulbostylis pusilla subsp. *yalingensis* (Cherm.) R.W.Haines = ***Bulbostylis pusilla*** (Hochst. ex A.Rich.) C.B.Clarke
Bulbostylis rehmannii C.B.Clarke = ***Bulbostylis filamentosa*** (Vahl) C.B.Clarke
Bulbostylis renschii var. *flaccidula* Cherm. = ***Bulbostylis renschii*** (Boeck.) C.B.Clarke
Bulbostylis scabra f. *evolutior* Lindm. = ***Bulbostylis scabra*** (J.Presl & C.Presl) C.B.Clarke
Bulbostylis schimperiana var. *leiolepis* Kük. = ***Bulbostylis leiolepis*** (Kük.) R.W.Haines
Bulbostylis schoenoides subsp. *erratica* (Hook.f.) Lye = ***Bulbostylis erratica*** (Hook.f.) C.B.Clarke
Bulbostylis seretii De Wild. = ***Bulbostylis coleotricha*** (Hochst. ex A.Rich.) C.B.Clarke var. ***coleotricha***
Bulbostylis setifolia (A.Rich.) Bodard = ***Bulbostylis atrosanguinea*** (Boeck.) C.B.Clarke
Bulbostylis setifolia (Boeck.) Beetle = ***Bulbostylis sphaerocephala*** (Boeck.) Lindm.
Bulbostylis spadicea Kük. = ***Bulbostylis paradoxa*** (Spreng.) Lindm.
Bulbostylis sphaerocephala var. *brunneovaginata* (Boeck.) C.B.Clarke = ***Bulbostylis sphaerocephala*** (Boeck.) Lindm.
Bulbostylis sphaerocephala var. *macrocephala* (Maury) Barros = ***Bulbostylis sphaerocephala*** (Boeck.) Lindm.
Bulbostylis sphaerocephala f. *schwackeana* (Boeck.) Lindm. = ***Bulbostylis sphaerocephala*** (Boeck.) Lindm.
Bulbostylis sphaerolepis (Boeck.) Beetle = ***Bulbostylis scabra*** (J.Presl & C.Presl) C.B.Clarke
Bulbostylis striatella C.B.Clarke = ***Bulbostylis humilis*** (Kunth) C.B.Clarke
Bulbostylis stricta Turrill = ***Bulbostylis megastachys*** (Ridl.) C.B.Clarke
Bulbostylis subaphylla var. *longiglumis* Kük. = ***Bulbostylis subaphylla*** C.B.Clarke
Bulbostylis subaphylla var. *rigida* Kük. = ***Bulbostylis subaphylla*** C.B.Clarke
Bulbostylis subdisticha (Boeck.) Beetle = ***Bulbostylis pauciflora*** (Liebm.) C.B.Clarke
Bulbostylis subefimbriata Kük. = ***Bulbostylis subaphylla*** C.B.Clarke
Bulbostylis tenuispicata (Boeck.) Barros = ***Bulbostylis funckii*** (Steud.) C.B.Clarke
Bulbostylis tenuissima Nakai = ***Bulbostylis densa*** (Wall.) Hand.-Mazz. subsp. ***densa***
Bulbostylis tisserantii (Cherm.) Lye = ***Bulbostylis viridicarinata*** (De Wild.) Goetgh.
Bulbostylis tisserantii Cherm. = [23 CAF]
Bulbostylis togoensis Cherm. = ***Bulbostylis lanifera*** (Boeck.) Kük.
Bulbostylis transiens (K.Schum.) C.B.Clarke = ***Bulbostylis boeckeleriana*** var. ***transiens*** (K.Schum.) R.W.Haines
Bulbostylis trichobasis var. *elatior* Cherm. = ***Bulbostylis trichobasis*** (Baker) C.B.Clarke
Bulbostylis trichobasis var. *leptocaulis* C.B.Clarke = ***Bulbostylis oritrephes*** (Ridl.) C.B.Clarke subsp. ***oritrephes***
Bulbostylis trichobasis var. *uniseriata* C.B.Clarke = ***Bulbostylis oritrephes*** (Ridl.) C.B.Clarke subsp. ***oritrephes***
Bulbostylis trichokolea (Steud.) Beetle = ***Bulbostylis densa*** (Wall.) Hand.-Mazz. subsp. ***densa***
Bulbostylis trifida (Nees) Nelmes = ***Bulbostylis densa*** subsp. ***afromontana*** (Lye) R.W.Haines

Bulbostylis trifida var. *biegensis* Cherm. = ***Bulbostylis densa*** subsp. ***afromontana*** (Lye) R.W.Haines
Bulbostylis tuberculosa (Michx.) Steven = ***Eleocharis tuberculosa*** (Michx.) Roem. & Schult.
Bulbostylis wallichiana (Schult.) Beetle = ***Bulbostylis barbata*** (Rottb.) C.B.Clarke subsp. ***barbata***
Bulbostylis willdenowii C.B.Clarke = ***Bulbostylis barbata*** (Rottb.) C.B.Clarke subsp. ***barbata***
Bulbostylis wittei Cherm. = ***Bulbostylis laniceps*** C.B.Clarke
Bulbostylis woronowii Palla = ***Bulbostylis hispidula*** (Vahl) R.W.Haines subsp. ***hispidula***
Bulbostylis yalingensis Cherm. = ***Bulbostylis pusilla*** (Hochst. ex A.Rich.) C.B.Clarke
Bulbostylis zambesica (K.Schum.) C.B.Clarke = ***Bulbostylis macra*** (Ridl.) C.B.Clarke
Bulbostylis zambesica var. *occidentalis* Bodard = ***Bulbostylis bodardii*** S.S.Hooper
Bulbostylis zeyheri (Boeck.) C.B.Clarke = ***Bulbostylis contexta*** (Nees) Bodard

***Unplaced Names*:**
Bulbostylis tisserantii Cherm., Arch. Bot. Mém. 3: 11 (1936). = [23 CAF]

Calisto

Calisto Néraud ex Gaudich. = [29 MAU]

***Unplaced Names*:**
Calisto Néraud ex Gaudich., Voy. Uranie: 26 (1826). = [29 MAU]

Callistachys

Callistachys Heuff. = ***Carex*** L.
Callistachys pyrenaica (Wahlenb.) Heuff. = ***Carex pyrenaica*** Wahlenb.

Calyptrocarya

Calyptrocarya Nees, Linnaea 9: 304 (1834).
Mexico to Trop. America. 79 MXT 80 BLZ COS GUA HON NIC PAN 81 TRT 82 FRG GUY SUR VEN 83 BOL CLM ECU GAL PER 84 BZC BZL BZN 85 PAR.
8 Species

Calyptrocarya bicolor (H.Pfeiff.) T.Koyama, Mem. New York Bot. Gard. 17(1): 43 (1967).
S. Trop. America. 82 FRG GUY SUR VEN 83 CLM ECU PER 84 BZC BZN. Hemicr.
**Becquerelia bicolor* H.Pfeiff., Repert. Spec. Nov. Regni Veg. 18: 381 (1922).
Becquerelia bicolor f. *humilis* H.Pfeiff., Repert. Spec. Nov. Regni Veg. 18: 382 (1922).
Becquerelia bicolor f. *ramosissima* H.Pfeiff., Repert. Spec. Nov. Regni Veg. 18: 382 (1922).
Becquerelia bicolor f. *verticillata* H.Pfeiff., Repert. Spec. Nov. Regni Veg. 18: 382 (1922).
Calyptrocarya poeppigiana f. *grandifolia* R.Gross ex J.F.Macbr., Publ. Field Mus. Nat. Hist., Bot. Ser. 11: 41 (1931).

Calyptrocarya delascioi Davidse & Kral, Ann. Missouri Bot. Gard. 75: 855 (1988).
Venezuela. 82 VEN. Hemicr.

Calyptrocarya glomerulata (Brongn.) Urb., Symb. Antill. 2: 169 (1900).
SE. Mexico to Trop. America. 79 MXT 80 BLZ COS GUA HON NIC PAN 81 TRT 82 FRG GUY SUR VEN 83 BOL CLM ECU GAL PER 84 BZC BZL BZN 85 PAR. Hemicr.
Dichromena fragifera A.Dietr., Sp. Pl. 2: 169 (1832). *Calyptrocarya fragifera* (A.Dietr.) Nees ex Kunth, Enum. Pl. 2: 364 (1837).
**Becquerelia glomerulata* Brongn. in L.I.Duperrey, Voy. Monde, Phan.: 163 (1833). *Becquerelia glomerulata* f. *typica* H.Pfeiff., Repert. Spec. Nov. Regni Veg. 18: 381 (1922), nom. inval.
Scleria palmetto Rchb. ex Core, Brittonia 2: 104 (1836), nom. inval.
Becquerelia glomerata Kunth, Enum. Pl. 2: 364 (1837), nom. illeg.
Calyptrocarya angustifolia Lindl. & Nees ex Kunth, Enum. Pl. 2: 364 (1837). *Becquerelia glomerulata* var. *angustifolia* (Lindl. & Nees ex Kunth) H.Pfeiff., Repert. Spec. Nov. Regni Veg. 18: 381 (1922). *Calyptrocarya fragifera* var. *angustifolia* (Lindl. & Nees ex Kunth) A.Gross & Brade, Arq. Mus. Nac. Rio de Janeiro 34: 53 (1932).
Calyptrocarya brevicaulis Nees ex Kunth, Enum. Pl. 2: 365 (1837). *Calyptrocarya angustifolia* var. *brevicaulis* (Nees ex Kunth) Kuntze, Revis. Gen. Pl. 2: 747 (1891).
Calyptrocarya palmetto Nees in C.F.P.von Martius & auct. suc. (eds.), Fl. Bras. 2(1): 195 (1842), nom. inval. *Becquerelia glomerulata* f. *rubra* H.Pfeiff., Repert. Spec. Nov. Regni Veg. 18: 381 (1922).
Hypolytrum multinerve Hochst. ex Steud., Syn. Pl. Glumac. 2: 133 (1855).
Calyptrocarya longifolia Boeck., Linnaea 38: 431 (1874), nom. illeg.
Calyptrocarya intermedia C.B.Clarke, Kew Bull., Addit. Ser. 8: 66 (1908).
Becquerelia glomerulata var. *euglomerulata* H.Pfeiff., Repert. Spec. Nov. Regni Veg. 18: 380 (1922).
Becquerelia glomerulata f. *minor* H.Pfeiff., Repert. Spec. Nov. Regni Veg. 18: 381 (1922).
Becquerelia glomerulata f. *minuta* H.Pfeiff., Repert. Spec. Nov. Regni Veg. 18: 380 (1922).
Becquerelia glomerulata f. *scabrinux* H.Pfeiff., Repert. Spec. Nov. Regni Veg. 18: 381 (1922).
Scleria clandestina Rchb. ex Core, Brittonia 2: 103 (1936), pro syn.

Calyptrocarya irwiniana T.Koyama, Jap. J. Bot. 20: 123 (1969).
WC. Brazil. 84 BZC. Hemicr.

Calyptrocarya luzuliformis T.Koyama, Mem. New York Bot. Gard. 17(1): 44 (1967).
Costa Rica to S. Trop. America. 80 COS 82 GUY SUR VEN 83 BOL CLM ECU PER 84. Hemicr.

Calyptrocarya monocephala Hochst. ex Steud., Syn. Pl. Glumac. 2: 181 (1855). *Diplacrum monocephalum* (Hochst. ex Steud.) H.Pfeiff., Repert. Spec. Nov. Regni Veg. 18: 383 (1922).
N. South America. 82 FRG GUY SUR VEN. Hemicr.
Calyptrocarya monocephala var. *triceps* Kük., Bot. Jahrb. Syst. 56(125): 24 (1921).

Calyptrocarya montesii Davidse & Kral, Ann. Missouri Bot. Gard. 75: 853 (1988).
Venezuela. 82 VEN. Ther.

Calyptrocarya poeppigiana Kunth, Enum. Pl. 2: 364 (1837).
Costa Rica to S. Trop. America. 80 COS PAN 82 FRG GUY SUR VEN 83 BOL CLM ECU PER 84 BZN. Hemicr.

Calyptrocarya schottmuelleriana Boeck., Bot. Jahrb. Syst. 5: 509 (1884).

***Synonyms*:**
Calyptrocarya angustifolia Lindl. & Nees ex Kunth = ***Calyptrocarya glomerulata*** (Brongn.) Urb.
Calyptrocarya angustifolia var. *brevicaulis* (Nees ex Kunth) Kuntze = ***Calyptrocarya glomerulata*** (Brongn.) Urb.
Calyptrocarya brevicaulis Nees ex Kunth = ***Calyptrocarya glomerulata*** (Brongn.) Urb.
Calyptrocarya bullata C.B.Clarke ex H.Pfeiff. = ***Hypolytrum bullatum*** C.B.Clarke
Calyptrocarya fragifera (A.Dietr.) Nees ex Kunth = ***Calyptrocarya glomerulata*** (Brongn.) Urb.
Calyptrocarya fragifera var. *angustifolia* (Lindl. & Nees ex Kunth) A.Gross & Brade = ***Calyptrocarya glomerulata*** (Brongn.) Urb.
Calyptrocarya intermedia C.B.Clarke = ***Calyptrocarya glomerulata*** (Brongn.) Urb.
Calyptrocarya longifolia Boeck. = ***Calyptrocarya glomerulata*** (Brongn.) Urb.
Calyptrocarya longifolia (Rudge) Kunth = ***Bisboeckelera longifolia*** (Rudge) Kuntze
Calyptrocarya martii Nees = ***Becquerelia martii*** (Nees) H.Pfeiff.
Calyptrocarya monocephala var. *triceps* Kük. = ***Calyptrocarya monocephala*** Hochst. ex Steud.
Calyptrocarya palmetto Nees = ***Calyptrocarya glomerulata*** (Brongn.) Urb.
Calyptrocarya poeppigiana f. *grandifolia* R.Gross ex J.F. Macbr. = ***Calyptrocarya bicolor*** (H.Pfeiff.) T.Koyama
Calyptrocarya schottmuelleriana Boeck. = ***Calyptrocarya poeppigiana*** Kunth

Calyptrolepis

Calyptrolepis Steud. = ***Rhynchospora*** Vahl
Calyptrolepis junciformis (Steud.) Steud. = ***Rhynchospora oligantha*** A.Gray var. ***oligantha***

Calyptrostylis

Calyptrostylis Nees = ***Rhynchospora*** Vahl
Calyptrostylis articulata (Roxb.) Nees = ***Rhynchospora corymbosa*** (L.) Britton
Calyptrostylis asperula Nees = ***Rhynchospora asperula*** (Nees) Steud.
Calyptrostylis asperula var. *minor* Nees = ***Rhynchospora corymbosa*** (L.) Britton
Calyptrostylis capitata Nees = ***Rhynchospora terminalis*** (Nees) Steud.
Calyptrostylis ceylanica (Nees) Nees = ***Rhynchospora triflora*** Vahl
Calyptrostylis divergens Nees = ***Rhynchospora corymbosa*** (L.) Britton
Calyptrostylis fascicularis Nees = ***Rhynchospora corymbosa*** (L.) Britton
Calyptrostylis florida (Rudge) Nees = ***Rhynchospora corymbosa*** (L.) Britton
Calyptrostylis gaudichaudii Nees = ***Rhynchospora corymbosa*** (L.) Britton
Calyptrostylis gigantea (Link) Nees = ***Rhynchospora gigantea*** Link
Calyptrostylis longirostris Nees = ***Rhynchospora amazonica*** subsp. ***guianensis*** (Kük.) T.Koyama
Calyptrostylis paniculata Liebm. = ***Rhynchospora schiedeana*** (Schltdl.) Kunth
Calyptrostylis rudgei Nees = ***Rhynchospora corymbosa*** (L.) Britton
Calyptrostylis schiedeana (Kunth) Liebm. = ***Rhynchospora schiedeana*** (Schltdl.) Kunth
Calyptrostylis terminalis Nees = ***Rhynchospora terminalis*** (Nees) Steud.
Calyptrostylis terminalis var. *pallida* Nees = ***Rhynchospora terminalis*** (Nees) Steud.

Campylostachys

Campylostachys E.Mey. = ***Fimbristylis*** Vahl
Campylostachys abbreviata E.Mey. = ***Fimbristylis ferruginea*** (L.) Vahl var. ***ferruginea***

Capeobolus

Capeobolus Browning, S. African J. Bot. 65: 218 (1999).
Cape Prov. 27 CPP.
1 Species

Capeobolus brevicaulis (C.B.Clarke) Browning, S. African J. Bot. 65: 218 (1999).
Cape Prov. 27 CPP. Hemicr.
Tetraria brevicaulis C.B.Clarke in T.A.Durand & H.Schinz, Consp. Fl. Afric. 5: 659 (1894), nom. nud. **Costularia brevicaulis* C.B.Clarke in W.H.Harvey & auct. suc. (eds.), Fl. Cap. 7: 274 (1898).

Capitularia

Capitularia J.V.Suringar = ***Capitularina*** J.Kern
Capitularia foliata Uittien = ***Capitularina involucrata*** (J.V.Suringar) J.Kern
Capitularia foliata var. *archboldii* Uittien = ***Capitularina involucrata*** (J.V.Suringar) J.Kern
Capitularia involucrata J.V.Suringar = ***Capitularina involucrata*** (J.V.Suringar) J.Kern

Capitularina

Capitularina J.Kern, in Fl. Males. 7: 458 (1974).
Papuasia. 43 NWG SOL.
1 Species
**Capitularia* J.V.Suringar, Nova Guinea 8: 711 (1912).

Capitularina involucrata (J.V.Suringar) J.Kern, in Fl. Males. 7: 460 (1974).
New Guinea to Solomon Is. 43 NWG SOL. Rhizome geophyte.
**Capitularia involucrata* J.V.Suringar, Nova Guinea 8: 711 (1912). *Chorizandra involucrata* (J.V.Suringar) Ridl., Trans. Linn. Soc. London, Bot. 9: 244 (1916).
Capitularia foliata Uittien, Recueil Trav. Bot. Néerl. 33: 289 (1936).
Capitularia foliata var. *archboldii* Uittien, J. Arnold Arbor. 20: 214 (1939).

Carex

Carex L., Sp. Pl.: 972 (1753).
Cosmopolitan. 10 AUT BOR DEN FIN FOR GRB ICE IRE NOR SVA SWE 11 AUT BGM CZE GER HUN NET NOR POL SVA SWI 12 BAL COR FRA ITA POR SAR SOC SPA SWI 13 ALB ALG BLR BLT BUK BUL GRC ITA KRI LBY MOR ROM RUC RUE RUN RUS RUW SIC TUE TUN UKR YUG 14 BLR BLT KRY RUC RUE RUN RUS RUW UKR 20 ALG EGY LBY MOR TUN 21 AZO CNY CVI MDR 22 GHA GUI MLI NGA SIE 23 BUR CAF CMN GGI RWA ZAI 24 ETH SOM SUD 25 KEN TAN UGA 26 ANG MLW MOZ ZAM ZIM 27 BOT CPP LES NAT

OFS SWZ TVL 28 STH 29 COM MAU MDG REU ROD 30 ALT BRY CTA IRK KRA TVA WSB WSC YAK 31 AMU KAM KHA KOR KUR MAG PRM SAK 32 KAZ KGZ TKM TZK UZB 33+ NCS TCS 34 AFG CYP EAI IRN IRQ LBS PAL SIN TUR 35 SAU YEM 36 CHC CHH CHI CHM CHN CHQ CHS CHT CHX 37 MON 38 JAP KOR KZN NNS OGA TAI 40 ASS BAN EHM IND NEP PAK SRL WHM 41 AMU CBD LAO MYA NCB SCS THA VIE 42 BOR CKI JAW LSI MLY MOL PHI SUL SUM XMS 43 BIS NWG SOL 50 NFK NSW NTA QLD SOA TAS VIC WAU 51 ATP CTM KER NZN NZS 60 FIJ NWC SAM VAN 61 COO MRQ SCI TUB 62 CRL MRN 63 HAW 70 ALU ASK GNL LAB NFL NUN NWT QUE YUK 71 ABR ABT BRC MAN SAS 72 BRC IDA LAB MNT NBR NFL NSC nsw ONT ORE PEI QUE WAS 73 COL IDA IOW MNT ORE VOL WAS WYO 74 ILL IOW KAN MIN MSO NDA NEB OKL SDA WIC WIS 75 CNT ill INI MAI MAS MIC MIN MWH NWH NWJ NWY OHI PEN RHO VER VRG WVA 76 ARI CAL NEV UTA 77 NWM TEX 78 ALA ARK ARL DEL FLA GEO KTY LOU MRY MSI NCA SCA TEN VRG WDC 79 MXC MXE MXG MXN MXS MXT 80 BLZ COS ELS GUA HON NIC PAN 81 BER CUB DOM HAI JAM LEE PUE WIN 82 GUY VEN 83 BOL CLM ECU PER 84 BZC BZL BZN BZS 85 AGE AGS AGW CHC CHS CLC CLN CLS JNF PAR URU 90 FAL MAQ SGE TDC.

1781 Species

Cyperoides Ség., Pl. Veron. 3: 73 (1754).
Mondo Adans., Fam. Pl. 2: 496 (1763).
Ulva Adans., Fam. Pl. 2: 496 (1763).
Ammorrhiza Ehrh., Beitr. Naturk. 4: 146 (1789).
Baeochortus Ehrh., Beitr. Naturk. 4: 149 (1789).
Caricella Ehrh., Beitr. Naturk. 4: 146 (1789).
Chordorrhiza Ehrh., Beitr. Naturk. 4: 148 (1789).
Drymeia Ehrh., Beitr. Naturk. 4: 148 (1789).
Heleonastes Ehrh., Beitr. Naturk. 4: 147 (1789).
Leptostachys Ehrh., Beitr. Naturk. 4: 147 (1789).
Limonaetes Ehrh., Beitr. Naturk. 4: 148 (1789).
Phaeolorum Ehrh., Beitr. Naturk. 4: 149 (1789).
Polyglochin Ehrh., Beitr. Naturk. 4: 146 (1789).
Psyllophora Ehrh., Beitr. Naturk. 4: 146 (1789).
Ptacoseia Ehrh., Beitr. Naturk. 4: 147 (1789).
Physiglochis Neck., Elem. Bot. 3: 245 (1790).
Schelhammeria Moench, Suppl. Meth.: 119 (1802).
Scuria Raf., J. Phys. Chim. Hist. Nat. Arts 89: 106 (1819).
Triodus Raf., J. Phys. Chim. Hist. Nat. Arts 89: 106 (1819).
Triplima Raf., Amer. Monthly Mag. & Crit. Rev. 4: 195 (1819).
Vignea P.Beauv. in T.G.Lestiboudois, Essai Cypér.: 22 (1819).
Trasus Gray, Nat. Arr. Brit. Pl. 2: 53 (1821).
Phyllostachys Torr., Ann. Lyceum Nat. Hist. New York 3: 404 (1836).
Anithista Raf., Good Book: 26 (1840).
Deweya Raf., Good Book: 26 (1840).
Diemisa Raf., Good Book: 27 (1840).
Edritria Raf., Good Book: 26 (1840).
Facolos Raf., Good Book: 26 (1840).
Forexeta Raf., Good Book: 28 (1840).
Itheta Raf., Good Book: 28 (1840).
Kolerma Raf., Good Book: 27 (1840).
Loncoperis Raf., Good Book: 27 (1840).
Loxanisa Raf., Good Book: 25 (1840).
Loxotrema Raf., Good Book: 25 (1840).
Meltrema Raf., Good Book: 26 (1840).
Neskiza Raf., Good Book: 27 (1840).
Olamblis Raf., Good Book: 26 (1840).
Olotrema Raf., Good Book: 25 (1840).
Onkerma Raf., Good Book: 27 (1840).
Osculisa Raf., Good Book: 27 (1840).
Temnemis Raf., Good Book: 27 (1840).
Callistachys Heuff., Flora 27: 528 (1844).
Cryptoglochin Heuff., Flora 27: 527 (1844).
Genersichia Heuff., Flora 27(2): 528 (1844).
Leucoglochin Heuffl., Flora 27: 528 (1844).
Maukschia Heuff., Flora 27: 527 (1844).
Heuffelia Opiz, Ost & West, Beibl. 98: 392 (1845).
Pseudocarex Miq., Ann. Mus. Bot. Lugduno-Batavi 2: 146 (1865).
Dornera Heuff. ex Schur, Enum. Pl. Transsilv.: 697 (1866).
Vignantha Schur, Enum. Pl. Transsilv.: 705 (1866).
Coleachyron J.Gay ex Boiss., Fl. Orient. 5: 408 (1882).
Neilreichia Kotula, Spraw. Komis. Fizjogr. 17: 136 (1883).
Homalostachys Boeck., Beitr. Cyper. 1: 37 (1888).
Caricina St.-Lag. in A.Cariot, Étude Fl., ed. 8, 2: 880 (1889).
Caricinella St.-Lag. in A.Cariot, Étude Fl., ed. 8, 2: 880 (1889).
Bitteria Börner, Abh. Naturwiss. Vereine Bremen 21: 269 (1913).
Chionanthula Börner, Abh. Naturwiss. Vereine Bremen 21: 275 (1913).
Dapedostachys Börner, Abh. Naturwiss. Vereine Bremen 21: 265 (1913).
Desmiograstis Börner, Abh. Naturwiss. Vereine Bremen 21: 274 (1913).
Echinochlaenia Börner, Abh. Naturwiss. Vereine Bremen 21: 264 (1913).
Kuekenthalia Börner, Abh. Naturwiss. Vereine Bremen 21: 266 (1913).
Lamprochlaenia Börner, Abh. Naturwiss. Vereine Bremen 21: 272 (1913).
Leptovignea Börner, Abh. Naturwiss. Vereine Bremen 21: 273 (1913).
Limivasculum Börner, Abh. Naturwiss. Vereine Bremen 21: 268 (1913).
Manochlaenia Börner, Abh. Naturwiss. Vereine Bremen 21: 271 (1913).
Proteocarpus Börner, Abh. Naturwiss. Vereine Bremen 21: 266 (1913).
Rhaptocalymma Borrer, Abh. Naturwiss. Vereine Bremen 21: 272 (1913).
Rhynchopera Börner, Abh. Naturwiss. Vereine Bremen 21: 272 (1913).
Thysanocarex Börner, Abh. Naturwiss. Vereine Bremen 21: 274 (1913).
Vignidula Börner, Abh. Naturwiss. Vereine Bremen 21: 275 (1913).
Diplocarex Hayata, Icon. Pl. Formosan. 10: 70 (1921).
Vesicarex Steyerm., Fieldiana, Bot. 28(1): 63 (1951).

Carex* ×*abitibiana Lepage, Naturaliste Canad. 86: 71 (1959). *C. aquatilis* × *C. stricta*.
Québec. 72 QUE. Hemicr. or rhizome geophyte.

Carex aboriginum M.E.Jones, Bull. Montana State Univ., Biol. Ser. 15: 69 (1910).
Montana. 73 MNT. Hemicr. or rhizome geophyte.

Carex* ×*abortiva E.Holmb., Bot. Not. 1929: 15 (1929). *C. brunnescens* × *C. cinerea*.

N. European Russia to Siberia. 14 RUN 30 WSB. Hemicr. or rhizome geophyte.

Carex abrupta Mack., Bull. Torrey Bot. Club 43: 618 (1916 publ. 1917).
Idaho to California. 73 IDA ORE 76 CAL NEV. Hemicr. or rhizome geophyte.

Carex abscondita Mack., Bull. Torrey Bot. Club 37: 244 (1910).
C. & E. U.S.A. 74 MSO OKL 75 CNT MAS NWJ NWY PEN RHO 77 TEX 78 ALA ARK DEL FLA GEO KTY LOU MRY MSI NCA SCA TEN VRG WDC. Hemicr. or rhizome geophyte.
**Carex ptychocarpa* Steud., Syn. Pl. Glumac. 2: 234 (1855), nom. illeg.
Carex digitalis var. *glauca* Chapm., Fl. South. U.S.: 541 (1860). *Carex abscondita* var. *glauca* (Chapm.) Fernald, Rhodora 37: 406 (1935).
Carex ptychocarpa var. *macrophylla* L.H.Bailey, Mem. Torrey Bot. Club 1: 48 (1889).
Carex magnifolia Mack. in J.K.Small, Fl. S.E. U.S., ed. 2: 1325 (1913).
Carex abscondita var. *rostellata* Fernald, Rhodora 44: 386 (1942).

Carex acaulis d'Urv., Mém. Soc. Linn. Paris 4: 599 (1826).
S. Argentina, Falkland Is. 85 AGS 90 FAL. Hemicr. or rhizome geophyte.

Carex accrescens Ohwi, Mem. Coll. Sci. Kyoto Imp. Univ., Ser. B, Biol. 6: 255 (1931).
Siberia to Japan. 30 YAK 31 KAM KUR MAG PRM SAK 36 CHI CHM 37 MON 38 JAP KOR. Hemicr. or rhizome geophyte.
Carex pallida C.A.Mey., Mém. Sav. Étr. Acad. St. Pétersbourg 1: 215 (1831), nom. illeg.
Carex pallida var. *papillosa* Kük. in H.G.A.Engler (ed.), Pflanzenr., IV, 20(38): 134 (1909).
Carex pallida var. *angustifolia* Y.L.Chang, in Fl. Plant. Herb. Chinae Bor.-Or. 11: 207 (1976). *Carex pallida* f. *angustifolia* (Y.L.Chang) Y.L.Chou, Bull. Bot. Lab. N. E. Forest. Inst., Harbin 4: 40 (1979).
Carex pallida var. *lithophiloides* A.E.Kozhevn., Sosud. Rast. Sovet. Dal'nego Vostoka 3: 381 (1988).

Carex acicularis Boott in J.D.Hooker, Fl. Nov.-Zel. 1: 280 (1853).
New Zealand. 51 NZN NZS. Hemicr. or rhizome geophyte.
Carex pyrenaica F.Muell., Fragm. 8: 251 (1874), nom. illeg.
Carex inconspicua Colenso, Trans. & Proc. New Zealand Inst. 28: 612 (1895).

Carex acidicola Naczi, Novon 12: 510 (2002).
SE. U.S.A. 78 ALA GEO. Hemicr. or rhizome geophyte.

Carex acocksii C.Archer, S. African J. Bot. 63: 342 (1997 publ. 1998).
Cape Prov. (Calvinia Div.). 27 CPP. Hemicr. or rhizome geophyte.

Carex acuta L., Sp. Pl.: 978 (1753). *Vignea acuta* (L.) Rchb. in J.C.Mössler, Handb. Gewächsk. ed. 2, 3: 1626 (1830). *Osculisa acuta* (L.) Raf., Good Book: 27 (1840). *Vignantha acuta* (L.) Schur, Enum. Pl. Transsilv.: 706 (1866).
NW. Africa, Europe to C. Asia. 10 DEN FIN GRB IRE NOR SWE 11 AUT BGM CZE GER HUN NET POL SWI 12 BAL FRA POR SAR SPA 13 ALB BUL GRC ITA ROM SIC YUG 14 BLR BLT RUC RUE RUN RUS RUW UKR 20 ALG MOR TUN 21 AZO 30 ALT WSB YAK 32 KAZ KGZ 33 NCS TCS 34 LBS TUR 37 MON. Hemicr. or rhizome geophyte.
Carex gracilis Curtis, Fl. Londin. 4: 282 (1782). *Carex acuta* f. *gracilis* (Curtis) Sanio, Bot. Centralbl. 6: 441 (1881). *Carex gracilis* subsp. *eugracilis* Kük., Allg. Bot. Z. Syst. 3: 171 (1897), nom. inval. *Carex cespitosa* var. *gracilis* (Curtis) Fiori, Nuov. Fl. Italia 1: 188 (1923).
Carex virens Thuill., Fl. Env. Paris: 489 (1790).
Carex ambigua Moench, Methodus: 324 (1794).
Carex pratensis Hosé, Ann. Bot. (Usteri) 21: 35 (1797).
Carex virginiana Woods in A.Rees, Cycl. 6: 100 (1806).
Carex moenchiana Wender., Flora 6: 718 (1823). *Vignea moenchiana* (Wender.) Rchb. in J.C.Mössler, Handb. Gewächsk. ed. 2, 3: 1626 (1830). *Vignantha moenchiana* (Wender.) Schur, Enum. Pl. Transsilv.: 706 (1866). *Carex acuta* var. *moenchiana* (Wender.) Nyman, Consp. Fl. Eur.: 777 (1882).
Carex acuta var. *sparsiflora* Dewey, Amer. J. Sci. Arts 10: 265 (1826).
Carex recurva Willd. ex Kunth, Enum. Pl. 2: 413 (1837), nom. illeg.
Carex riparia Moench ex Steud., Nomencl. Bot., ed. 2, 1: 295 (1840), nom. inval.
Carex corynophora Peterm., Flora 27: 335 (1844). *Carex gracilis* f. *corynophora* (Peterm.) Kük. in H.G.A.Engler (ed.), Pflanzenr., IV, 20(38): 321 (1909).
Carex prolixa Fr., Novit. Fl. Suec. Mant. 3: 228 (1845). *Vignantha prolixa* (Fr.) Fourr., Ann. Soc. Linn. Lyon, n.s., 17: 176 (1869). *Carex acuta* subsp. *prolixa* (Fr.) Nyman, Consp. Fl. Eur.: 777 (1878). *Carex acuta* var. *prolixa* (Fr.) Hartm., Handb. Skand. Fl., ed. 11: 469 (1879).
Carex tricostata Fr., Novit. Fl. Suec. Mant. 3: 152 (1845). *Carex gracilis* var. *tricostata* (Fr.) Asch., Fl. Brandenburg 1: 776 (1864). *Carex acuta* var. *tricostata* (Fr.) Crép., Man. Fl. Belgique, ed. 2: 322 (1866).
Carex acuta amblylepis Peterm., Anal. Pfl.-Schlüss.: 501 (1846). *Carex acuta* var. *amblylepis* (Peterm.) Nyman, Consp. Fl. Eur.: 777 (1882). *Carex gracilis* f. *amblylepis* (Peterm.) Asch. & Graebn., Syn. Mitteleur. Fl. 2(2): 94 (1902).
Carex acuta dioica Peterm., Anal. Pfl.-Schlüss.: 501 (1846). *Carex gracilis* lus. *dioica* (Peterm.) Soó, Acta Bot. Acad. Sci. Hung. 16: 370 (1970 publ. 1971).
Carex acuta f. *pedicellata* Peterm., Anal. Pfl.-Schlüss.: 501 (1846). *Carex gracilis* f. *pedicellata* (Peterm.) Soó, Acta Bot. Acad. Sci. Hung. 16: 370 (1970 publ. 1971).
Carex acuta f. *vaginata* Peterm., Anal. Pfl.-Schlüss.: 501 (1846). *Carex gracilis* f. *vaginata* (Peterm.) Soó, Acta Bot. Acad. Sci. Hung. 16: 370 (1970 publ. 1971).
Carex touranginiana Boreau, Fl. Centre France, ed. 2: 532 (1849). *Vignantha touranginiana* (Boreau) Fourr., Ann. Soc. Linn. Lyon, n.s., 17: 176 (1869). *Carex acuta* f. *tourangiana* (Boreau) Sanio, Bot. Cantralbl. 6: 442 (1881).
Carex rudis Wimm., Flora 33: 619 (1850). *Carex gracilis* f. *rudis* (Wimm.) Kük., Allg. Bot. Z. Syst. 3: 171 (1897). *Carex gracilis* var. *rudis* (Wimm.) Kük. in H.G.A.Engler (ed.), Pflanzenr., IV, 20(38): 322 (1909).

Carex mauritanica Boiss. & Reut., Pugill. Pl. Afr. Bor. Hispan.: 116 (1852). *Carex elata* subsp. *mauritanica* (Boiss. & Reut.) A.Galán, Lagascalia 18: 57 (1995).
Vignea strictifolia Opiz, Seznam: 103 (1852).
Carex virginica Steud., Syn. Pl. Glumac. 2: 217 (1855).
Carex wimmeri Steud., Syn. Pl. Glumac. 2: 215 (1855).
Carex schummelii Siegert, Verh. Bot. Vereins Prov. Brandenburg 2: 112 (1860). *Carex tricostata* var. *schummelii* (Siegert) Nyman, Consp. Fl. Eur., Suppl. 2: 323 (1890).
Carex pulchella Nyl. ex Trevir, Bull. Soc. Imp. Naturalistes Moscou 36(1): 537 (1863).
Carex acuta var. *intermedia* Èelak., Prodr. Fl. Böhmen: 63 (1867). *Carex gracilis* subsp. *intermedia* (Èelak.) Soó, Acta Bot. Acad. Sci. Hung. 16: 370 (1970 publ. 1971).
Carex cunninghamii Boott, Ill. Carex 4: 171 (1867).
Carex strictifolia Opiz ex Nyman, Consp. Fl. Eur.: 777 (1882).
Carex rufa var. *seminuda* Beck, Fl. Nieder-Österreich: 136 (1890). *Carex gracilis* f. *seminuda* (Beck) Kük. in H.G.A.Engler (ed.), Pflanzenr., IV, 20(38): 322 (1909).
Carex acuta f. *graciliflora* Legrand, Fl. Anal. Berry, ed. 2: 339 (1894). *Carex gracilis* f. *graciliflora* (Legrand) Soó, Acta Bot. Acad. Sci. Hung. 16: 370 (1970 publ. 1971).
Carex gracilis var. *angustifolia* Kük., Allg. Bot. Z. Syst. 3: 171 (1897).
Carex gracilis f. *brachystachya* Kük., Allg. Bot. Z. Syst. 3: 172 (1897).
Carex gracilis subsp. *erecta* Kük., Allg. Bot. Z. Syst. 3: 171 (1897).
Carex gracilis var. *humilis* Kük., Allg. Bot. Z. Syst. 3: 171 (1897).
Carex gracilis subsp. *corynophora* Asch. & Graebn., Syn. Mitteleur. Fl. 2(2): 90 (1902).
Carex fuscovaginata Kük., Bull. Herb. Boissier, II, 4: 56 (1904).
Carex gracilis var. *approximata* Kük. ex Junge, Verh. Naturwiss. Vereins Hamburg 3(12): 10 (1904).
Carex gracilis var. *libanotica* Kük. in H.G.A.Engler (ed.), Pflanzenr., IV, 20(38): 323 (1909).
Carex lilibaea Lojac., Fl. Sicul. 3: 223 (1909).
Carex graciliformis V.I.Krecz. in P.F.Majevski, Fl. Sredn. Rossii, ed. 6: 199 (1933).
Carex podperae Otruba, Sched. Fl. Exsicc. Reip. Bohem. Slov. Cent. 8: 22 (1933), pro syn.
Carex sareptana V.I.Krecz. in P.F.Majevski, Fl. Centr. Russ., ed. 6: 190 (1933).
Carex dichroandra V.I.Krecz. in V.L.Komarov (ed.), Fl. URSS 3: 596 (1935).
Carex gracilis f. *podperae* Soó, Acta Bot. Acad. Sci. Hung. 16: 370 (1970 publ. 1971).

Carex acutata Boott, Proc. Linn. Soc. London 1: 287 (1846).
NW. Venezuela to Chile. 82 VEN 83 CLM ECU PER 85 AGW CLC CLS. Hemicr. or rhizome geophyte.
Carex multispicata Kunze in C.Schkuhr, Beschr. Riedgräs., Suppl.: 42 (1840). *Carex acutata* var. *multispicata* (Kunze) Kük. in H.G.A.Engler (ed.), Pflanzenr., IV, 20(38): 702 (1909).
Carex physocarpa Nees ex Boott, Proc. Linn. Soc. London 1: 287 (1846), nom. illeg.
Carex foeminea Steud., Syn. Pl. Glumac. 2: 203 (1855).
Carex feminea Steud. ex Boott, Ill. Carex 4: 138 (1867).
Carex purpurea Boott, Ill. Carex 4: 157 (1867).
Carex anwandteri Phil., Anales Univ. Chile 93: 498 (1896).
Carex acutata var. *ciliata* Kük., Bot. Jahrb. Syst. 27: 549 (1899).

Carex acutiformis Ehrh., Beitr. Naturk. 4: 43 (1789).
Carex riparia var. *acutiformis* (Ehrh.) Fiori, Nuov. Fl. Italia 1: 201 (1923).
Europe, Medit. to Kashmir. 10 DEN FIN GRB IRE NOR SWE 11 AUT BGM CZE GER HUN NET POL SWI 12 COR FRA SAR SPA 13 ALB BUL GRC? ITA ROM SIC YUG 14 BLR BLT KRY RUC RUE RUN RUS RUW UKR 20 ALG MOR 25 TAN UGA 27 CPP LES NAT OFS TVL 30 ALT 32 KAZ KGZ TKM TZK UZB 33 TCS 34 AFG IRN IRQ PAL TUR 40 PAK WHM (72) ont (75) cnt ini mas mic nwy (78) mry. Hemicr.
Carex rufa Schrank, Baier. Fl. 1: 305 (1789), nom. illeg.
Carex rigens Thuill., Fl. Env. Paris: 488 (1790).
Carex scheuchzeri J.F.Gmel., Syst. Nat. 2: 144 (1791).
Carex spadicea Roth, Tent. Fl. Germ. 2(2): 461 (1793), nom. illeg.
Carex paludosa Gooden., Trans. Linn. Soc. London 2: 202 (1794). *Trasus paludosus* (Gooden.) Gray, Nat. Arr. Brit. Pl. 2: 64 (1821). *Edritria paludosa* (Gooden.) Raf., Good Book: 26 (1840).
Carex dubia Hosé, Ann. Bot. (Usteri) 21: 34 (1797).
Carex riparia Geners., Elench.: 874 (1798), sensu auct.
Carex intermedia Suter, Helvet. Fl. 2: 261 (1802), nom. illeg.
Carex palustris Suter, Helvet. Fl. 2: 261 (1802).
Carex kochiana DC., Cat. Pl. Horti Monsp.: 8 (1813). *Carex spadicea* var. *kochiana* (DC.) Crép., Man. Fl. Belgique, ed. 2: 325 (1866).
Carex acutiformis f. *minor* Peterm., Anal. Pfl.-Schlüss.: 516 (1846).
Carex paludosa f. *depauperata* Lange, Bot. Not. 1849: 165 (1849). *Carex acutiformis* f. *depauperata* (Lange) Kük. in H.G.A.Engler (ed.), Pflanzenr., IV, 20(38): 734 (1909).
Carex spadicea f. *maxima* Urb., Verh. Bot. Vereins Prov. Brandenburg 22: 54 (1881). *Carex acutiformis* f. *maxima* (Urb.) Kük. in H.G.A.Engler (ed.), Pflanzenr., IV, 20(38): 734 (1909).
Carex olgae Regel in A.P.Fedchenko, Putesh. Turkestan 3: 83 (1882).
Carex paludosa var. *spadicea* Nyman, Consp. Fl. Eur.: 768 (1882). *Carex acutiformis* f. *spadicea* (Nyman) Asch. & Graebn., Syn. Mitteleur. Fl. 2(2): 215 (1903).
Carex paludosa var. *longiglumis* St.-Lag. in A.Cariot, Étude Fl., ed. 8, 2: 865 (1889).
Carex acutiformis var. *abbreviata* Beck, Fl. Nieder-Österreich: 143 (1890).
Carex acutiformis f. *gracilior* Kük., Allg. Bot. Z. Syst. 5(Beih. 1): 64 (1899).
Carex acutatiformis H.E.Hess, Ber. Schweiz. Bot. Ges. 63: 355 (1953).

Carex adelostoma V.I.Krecz. in V.L.Komarov (ed.), Fl. URSS 3: 603 (1935).
Subarctic. 10 FIN ICE 14 RUN 30 WSB YAK 70 ASK NUN NWT 71 MAN 72 LAB NFL. Hemicr. or rhizome geophyte.
Carex buxbaumii var. *alpicola* Hartm., Handb. Skand. Fl.: 41 (1820).
Carex buxbaumii var. *mutica* Hartm., Handb. Skand. Fl.: 40 (1820). *Carex buxbaumii* subsp. *mutica* (Hartm.) Isov., Ann. Bot. Fenn. 14: 208 (1977).
Carex morrisseyi A.E.Porsild, Sargentia 4: 21 (1943).

Carex adrienii E.G.Camus in H.Lecomte, Fl. Indo-Chine 7: 186 (1912).
S. China to Vietnam. 36 CHC CHS 41 VIE. Hemicr. or rhizome geophyte.
Carex scaposa var. *baviensis* Franch., Nouv. Arch. Mus. Hist. Nat., III, 8: 255 (1896).
Carex chienii F.T.Wang & Tang, Contr. Inst. Bot. Natl. Acad. Peiping 6: 60 (1949).
Carex kwangtungensis F.T.Wang & Tang, Contr. Inst. Bot. Natl. Acad. Peiping 6: 59 (1949).

Carex adusta Boott in W.J.Hooker, Fl. Bor.-Amer. 2: 215 (1839). *Carex mirabilis* var. *adusta* (Boott) Prov., Fl. Canad. 2: 648 (1863).
Canada to N. U.S.A. 71 ABT BRC MAN SAS 72 NFL NSC ONT QUE 74 MIN WIS 75 MAI MIC NWY. Hemicr. or rhizome geophyte.
Carex adusta var. *glomerata* L.H.Bailey, Carex Cat.: 2 (1884).
Carex albolutescens var. *glomerata* Olney ex L.H. Bailey, Proc. Amer. Acad. Arts 22: 149 (1886).
Carex pinguis L.H.Bailey, Bull. Geol. Nat. Hist. Surv. 3: 22 (1887).

Carex aematorrhyncha Desv. in C.Gay, Fl. Chil. 6: 224 (1854). *Carex filiformis* var. *aematorrhyncha* (Desv.) W.Boott in S.Watson & al., Bot. California 2: 250 (1880). *Carex filiformis* subsp. *aematorrhyncha* (Desv.) Kük., Bot. Jahrb. Syst. 27: 556 (1899).
Ecuador to Falkland Is. 83 ECU 85 AGS CLC CLS 90 FAL. Hemicr. or rhizome geophyte.

var. ***aematorrhyncha***
Ecuador to Falkland Is. 83 ECU 85 AGS CLS 90 FAL. Hemicr. or rhizome geophyte.
Carex campyloxys Steud., Syn. Pl. Glumac. 2: 213 (1855).
Carex potamostauros Steud. in W.Lechler, Berberid. Amer. Austral.: 52 (1857).
Carex filiformis var. *gracilis* Kurtz ex Kük., Bot. Jahrb. Syst. 27: 556 (1899).

var. ***corralensis*** (Phil.) Kük. in H.G.A.Engler (ed.), Pflanzenr., IV, 20(38): 748 (1909).
Chile to W. Argentina. 85 AGS CLC. Hemicr. or rhizome geophyte.
**Carex corralensis* Phil., Linnaea 33: 273 (1865).

Carex aequialta Kük. in H.G.A.Engler (ed.), Pflanzenr., IV, 20(38): 354 (1909).
E. China, Japan. 36 CHS 38 JAP. Hemicr. or rhizome geophyte.

Carex aequivoca V.I.Krecz. in V.L.Komarov (ed.), Fl. URSS 3: 600 (1935).
E. Turkey to NW. Iran. 33 TCS 34 IRN IRQ TUR. Hemicr. or rhizome geophyte.

Carex ×aestivaliformis Mack., Bull. Torrey Bot. Club 37: 238 (1910). *C. aestivalis* × *C. gracillima*.
E. U.S.A. 75 MAS NWH NWJ NWY PEN 78 DEL GEO KTY VRG. Hemicr. or rhizome geophyte.

Carex aestivalis M.A.Curtis ex A.Gray, Amer. J. Sci. Arts 42: 28 (1842).
E. U.S.A. 75 NWY PEN VER WVA 78 ALA GEO KTY MRY NCA SCA TEN VRG. Hemicr. or rhizome geophyte.
Carex darlingtonii Schwein. ex M.A.Curtis, Amer. J. Sci. Arts, II, 7: 410 (1849).
Carex tabularia M.A.Curtis ex Boott, Ill. Carex 1: 54 (1858).

Carex aethiopica Schkuhr, Beschr. Riedgräs. 1: 107 (1801).
Cape Prov. 27 CPP. Hemicr. or rhizome geophyte.
Carex laevigata Wahlenb., Kongl. Vetensk. Acad. Nya Handl. 1803: 153 (1803), nom. illeg.
Carex iridifolia Kunth, Enum. Pl. 2: 492 (1837). *Carex aethiopica* var. *iridifolia* (Kunth) C.B.Clarke in T.A.Durand & H.Schinz, Consp. Fl. Afric. 5: 679 (1894).
Carex aethiopica var. *latispica* C.B.Clarke in W.H. Harvey & auct. suc. (eds.), Fl. Cap. 7: 308 (1898).

Carex agglomerata C.B.Clarke, J. Linn. Soc., Bot. 36: 269 (1903).
Qinghai to C. China. 36 CHC CHN CHQ. Hemicr. or rhizome geophyte.

var. ***agglomerata***
Qinghai to C. China. 36 CHC CHN CHQ. Hemicr. or rhizome geophyte.

var. ***rhizomata*** Y.C.Yang, Acta Biol. Plateau Sin. 3: 89 (1984).
China (Gansu). 36 CHN. Hemicr. or rhizome geophyte.

Carex aggregata Mack., Bull. Torrey Bot. Club 37: 246 (1910). *Carex sparganioides* var. *aggregata* (Mack.) Gleason, Phytologia 4: 22 (1952).
SE. Canada to NC. & E. U.S.A. 72 ONT 74 ILL IOW MIN MSO NEB OKL SDA WIS 75 INI NWJ NWY OHI PEN WVA 78 ALA DEL KTY MRY NCA SCA TEN VRG WDC. Hemicr. or rhizome geophyte.
**Carex agglomerata* Mack., Bull. Torrey Bot. Club 33: 442 (1906), nom. illeg.

Carex ×akitaensis Fujiw., J. Phytogeogr. Taxon. 35: 81 (1987). *C. middendorffii* × *C. sodoensis*.
Japan. 38 JAP. Hemicr. or rhizome geophyte.

Carex ×akiyamana Ohwi, Mem. Coll. Sci. Kyoto Imp. Univ., Ser. B, Biol. 11(5): 511 (1936). *C. flabellata* × *C. shimidzensis*.
Japan. 38 JAP. Hemicr. or rhizome geophyte.

Carex aladagensis Kük. ex J.Schultze-Motel, Willdenowia 3: 92 (1962).
Turkey. 34 TUR. Hemicr. or rhizome geophyte.

Carex alajica Litv., Trudy Bot. Muz. Imp. Akad. Nauk 7: 99 (1910).
C. Asia to NW. China. 32 KGZ TKM 36 CHX. Hemicr. or rhizome geophyte.

Carex alata Torr., Ann. Lyceum Nat. Hist. New York 3: 396 (1836). *Kolerma alata* (Torr.) Raf., Good Book: 27 (1840). *Carex straminea* var. *alata* (Torr.) Boeck., Linnaea 39: 117 (1875). *Carex albolutescens* var. *alata* (Torr.) Kük. in H.G.A.Engler (ed.), Pflanzenr., IV, 20(38): 209 (1909).
SE. Canada to C. & E. U.S.A. 72 ONT 74 ILL MSO OKL 75 CNT INI MAS MIC NWH NWJ NWY OHI PEN RHO 77 TEX 78 ALA ARK DEL FLA GEO KTY LOU MRY NCA SCA TEN VRG WDC. Hemicr. or rhizome geophyte.

Carex alba Scop., Fl. Carniol., ed. 2, 2: 216 (1772).
E. Pyrenees to Russian Far East. 11 AUT CZE GER HUN POL SWI 12 FRA SPA 13 ITA ROM YUG 14 RUE RUN RUS UKR 30 WSB 31+ 32 KAZ 33 NCS 36 CHX 37 MON. Hemicr. or rhizome geophyte.
Carex nemorosa Schrank, Baier. Fl. 1: 291 (1789).

Carex argentea J.F.Gmel., Syst. Nat. 2: 143 (1791).
Carex inclusa Turcz. ex Boott, Ill. Carex 4: 177 (1867).
Carex alba subsp. *ussuriensis* Kom., Trudy Imp. S.-Peterburgsk. Bot. Sada 18: 443 (1901).
Carex alba f. *aprica* Kük. in H.G.A.Engler (ed.), Pflanzenr., IV, 20(38): 500 (1909).
Carex ajanensis Vorosch., Izv. Bot. Sada Akad. Nauk S.S.S.R. 60: 35 (1965).

Carex albata Boott ex Franch., Nouv. Arch. Mus. Hist. Nat., III, 8: 216 (1896). *Carex nubigena* var. *albata* (Boott ex Franch.) Kük. in H.G.A.Engler (ed.), Pflanzenr., IV, 20(38): 146 (1909). *Carex nubigena* subsp. *albata* (Boott ex Franch.) T.Koyama, Bot. Mag. (Tokyo) 72: 306 (1959).
Russian Far East, Japan (SW. Hokkaido, N. & C. Honshu). 31+ 36 CHC 38 JAP. Hemicr. or rhizome geophyte.

Carex albicans Willd. ex Spreng., Syst. Veg. 3: 818 (1826).
SE. Canada to Mexico. 72 NBR NSC ONT QUE 74 ILL IOW KAN MIN MSO NEB OKL 75 CNT INI MAI MAS MIC NWH NWJ NWY OHI PEN RHO VER WVA 77 TEX 78 ALA ARK DEL FLA GEO KTY LOU MRY MSI NCA SCA TEN VRG 79 MXC MXE MXG. Hemicr. or rhizome geophyte.

var. ***albicans***
SE. Canada to NC. & E. U.S.A. 72 NSC ONT QUE 74 ILL IOW KAN MIN MSO NEB OKL 75 CNT INI MAI MAS MIC NWH NWJ NWY OHI PEN RHO VER WVA 77 TEX 78 ALA ARK DEL GEO KTY LOU MRY MSI NCA SCA TEN VRG. Hemicr. or rhizome geophyte.
Carex varia Muhl. ex Wahlenb., Kongl. Vetensk. Acad. Nya Handl. 24: 159 (1803), nom. illeg.
Carex alpestris Dewey, Amer. J. Sci. Arts 7: 268 (1824), nom. illeg. *Loxotrema alpestris* Raf., Good Book: 25 (1840).
Carex collecta Dewey, Amer. J. Sci. Arts 11: 314 (1826). *Carex novae-angliae* var. *collecta* (Dewey) Dewey, Class-book Bot.: 421 (1845).
Carex pensylvanica Torr., Ann. Lyceum Nat. Hist. New York 3: 410 (1836), nom. illeg.
Carex pensylvanica var. *muehlenbergii* A.Gray ex Torr., Ann. Lyceum Nat. Hist. New York 3: 410 (1836). *Carex nigromarginata* var. *muehlenbergii* (A.Gray ex Torr.) Gleason, Phytologia 4: 22 (1952). *Carex emmonsii* var. *muehlenbergii* (A.Gray ex Torr.) Rettig, Sida 13: 451 (1989).
Carex alpestris var. *tripla* Dewey, Amer. J. Sci. Arts, II, 35: 59 (1863).
Carex pilulifera f. *americana* Boeck., Linnaea 41: 216 (1877).
Carex varia var. *colorata* L.H.Bailey, Mem. Torrey Bot. Club 1: 40 (1889). *Carex artitecta* f. *colorata* (L.H.Bailey) D.S.Carp., Fl. Vermont, ed. 3: 57 (1937).
Carex varia f. *colorata* (L.H.Bailey) Kük. in H.G.A.Engler (ed.), Pflanzenr. 1: 41 (1889).
Carex davisii f. *glabrescens* Kük. in H.G.A.Engler (ed.), Pflanzenr., IV, 20(38): 588 (1909).
Carex varia f. *distincta* Howe, Rep. (Annual) Regents Univ. State New York New York State Mus., IV, 20(38): 449 (1909).
Carex artitecta Mack. in N.L.Britton & al. (eds.), N. Amer. Fl. 18: 189 (1935).
Carex artitecta var. *subtilirostris* F.J.Herm., Rhodora 40: 79 (1938).

var. ***australis*** (L.H.Bailey) Rettig, Sida 14: 133 (1990).
SE. U.S.A. to Mexico. 74 KAN 77 TEX 78 ALA ARK FLA GEO LOU MSI NCA SCA VRG 79 MXC MXE MXG. Hemicr. or rhizome geophyte.
Carex rhynchophysa Liebm., Mexic. Halvgr.: 264 (1850), nom. illeg. *Carex physorhyncha* Liebm. ex Steud., Syn. Pl. Glumac. 2: 219 (1855).
Carex nigromarginata f. *minor* Boeck., Bot. Jahrb. Syst. 1: 364 (1881).
**Carex varia* var. *australis* L.H.Bailey, Bot. Gaz. 17: 153 (1892). *Carex varia* f. *australis* (L.H.Bailey) Kük. in H.G.A.Engler (ed.), Pflanzenr., IV, 20(38): 449 (1909). *Carex emmonsii* var. *australis* (L.H.Bailey) Rettig, Sida 13: 452 (1989).

var. ***emmonsii*** (Dewey ex Torr.) Rettig, Sida 14: 133 (1990).
E. Canada to NC. & E. U.S.A. 72 NBR NSC ONT QUE 74 ILL MSO WIS 75 CNT INI MAI MAS MIC NWH NWJ NWY OHI PEN RHO VER WVA 78 ARK DEL FLA GEO KTY MRY NCA SCA TEN VRG. Hemicr. or rhizome geophyte.
Carex davisii Dewey, Amer. J. Sci. Arts 10: 279 (1826), nom. illeg.
**Carex emmonsii* Dewey ex Torr., Ann. Lyceum Nat. Hist. New York 3: 411 (1836). *Carex novae-angliae* var. *emmonsii* (Dewey ex Torr.) Carey in A.Gray, Manual: 556 (1848). *Carex lucorum* var. *emmonsii* (Dewey ex Torr.) Chapm., Fl. South. U.S.: 539 (1860).
Carex varia var. *minor* Boott in W.J.Hooker, Fl. Bor.-Amer. 2: 223 (1839). *Carex nigromarginata* var. *minor* (Boott) Gleason, Phytologia 4: 22 (1952).
Carex emmonsii var. *colorata* Howe, Rep. (Annual) Regents Univ. State New York State Mus. 48: 176 (1896).
Carex emmonsii var. *distincta* Howe, Rep. (Annual) Regents Univ. State New York State Mus. 48: 176 (1896).

Carex albida L.H.Bailey, Mem. Torrey Bot. Club 1: 9 (1889). *Carex luzulifolia* f. *albida* (L.H.Bailey) Kük. in H.G.A.Engler (ed.), Pflanzenr., IV, 20(38): 558 (1909).
California (Sonoma Co.). 76 CAL. Hemicr. or rhizome geophyte.
Carex sonomensis Stacey, Leafl. W. Bot. 2: 63 (1937).

Carex albidibasis T.Koyama, J. Jap. Bot. 30: 136 (1955).
Japan. 38 JAP. Hemicr. or rhizome geophyte.

Carex albolutescens Schwein., Ann. Lyceum Nat. Hist. New York 1: 66 (1828).
C. & E. U.S.A., Mexico (Chiapas). 74 ILL MSO OKL 75 CNT INI MAS MIC NWJ NWY OHI PEN RHO WVA 77 TEX 78 ALA ARK DEL GEO KTY LOU MRY MSI NCA SCA TEN VRG WDC 79 MXT. Hemicr. or rhizome geophyte.
Carex foenea Muhl. ex Torr., Ann. Lyceum Nat. Hist. New York 3: 315 (1836), nom. illeg. *Carex straminea* var. *foenea* Torr., Ann. Lyceum Nat. Hist. New York 3: 395 (1836).
Carex straminea var. *intermedia* Gay, Ann. Sci. Nat., Bot., II, 10: 364 (1838).
Carex foenea var. *ferruginea* A.Gray, Manual, ed. 5: 580 (1867). *Carex straminea* var. *ferruginea* (A.Gray) L.H.Bailey, Bull. Torrey Bot. Club 20: 421 (1893). *Carex alata* var. *ferruginea* (A.Gray)

Fernald, Proc. Amer. Acad. Arts 37: 477 (1902). *Carex albolutescens* var. *ferruginea* (A.Gray) Kük. in H.G.A.Engler (ed.), Pflanzenr., IV, 20(38): 209 (1909). *Carex festucacea* var. *ferruginea* (A.Gray) Farw., Pap. Michigan Acad. Sci. 2: 17 (1923).
Carex straminea var. *chlorostachys* Boeck., Linnaea 39: 118 (1875).

Carex albonigra Mack. in P.A.Rydberg, Fl. Rocky Mts.: 137 (1917).
Subarctic America to W. U.S.A. 70 ASK NWT YUK 71 ABT BRC 73 COL IDA MNT WAS WYO 76 ARI CAL UTA 77 NWM. Hemicr.

Carex albowii Leskov, Zhurn. Russk. Bot. Obshch. 16: 72 (1931).
Caucasus. 33 TCS. Hemicr. or rhizome geophyte.

Carex albula Allan, Trans. & Proc. Roy. Soc. New Zealand 76: 589 (1947).
New Zealand South I. 51 NZS. Hemicr. or rhizome geophyte.

Carex albursina E.Sheld., Bull. Torrey Bot. Club 20: 284 (1893).
E. Canada to NC. & E. U.S.A. 72 ONT QUE 74 ILL IOW MIN MSO OKL WIS 75 CNT INI MAS MIC NWH NWJ NWY OHI PEN VER WVA 78 ALA ARK DEL GEO KTY MRY NCA SCA TEN VRG WDC. Hemicr. or rhizome geophyte.
**Carex laxiflora* var. *latifolia* Boott, Ill. Carex 1: 38 (1858). *Carex lasiocarpa* var. *latifolia* (Boott) Gilly, Iowa State Coll. J. Sci 21: 125 (1946).

Carex allanii Hamlin, Trans. Roy. Soc. New Zealand, Bot. 1: 272 (1962).
New Zealand South I. 51 NZS. Hemicr. or rhizome geophyte.

Carex alligata Boott, Ill. Carex 4: 129 (1867).
Hawaiian Is. 63 HAW. Hemicr. or rhizome geophyte.
Carex alligata var. *degeneri* R.W.Krauss, Pacific Sci. 4: 287 (1950).
Carex pluvia R.W.Krauss, Pacific Sci. 4: 271 (1950).
Carex pluvia var. *koolauensis* R.W.Krauss, Pacific Sci. 4: 271 (1950).

Carex alliiformis C.B.Clarke, J. Linn. Soc., Bot. 36: 270 (1903).
Japan (S. Kyushu) to Taiwan, SE. China, Vietnam. 36 CHC CHH CHS 38 JAP NNS TAI 41 VIE. Hemicr. or rhizome geophyte.
Carex purpurascens Kük., J. Coll. Sci. Imp. Univ. Tokyo 22: 496 (1906).

Carex allivescens V.I.Krecz., Bot. Mater. Gerb. Bot. Inst. Komarova Akad. Nauk S.S.S.R. 9: 190 (1946).
China (Gansu). 36 CHN. Hemicr. or rhizome geophyte.

Carex alma L.H.Bailey, Mem. Torrey Bot. Club 1: 50 (1889).
S. U.S.A. to C. Mexico. 76 ARI CAL NEV 77 NWM TEX 79 MXC MXE MXN. Hemicr. or rhizome geophyte.
Carex vitrea T.Holm, Amer. J. Sci. 166: 454 (1903).
Carex agrostoides Mack., Bull. Torrey Bot. Club 34: 607 (1907 publ. 1908).
Carex arizonensis C.B.Clarke, Kew Bull., Addit. Ser. 8: 68 (1908), nom. illeg.

Carex ×almii E.Holmb., Bot. Not. 1929: 28 (1929). *C. dioica* × *C. tenuiflora.*
N. & NE. Europe. 10 SWE 14 RUN. Hemicr. or rhizome geophyte.

Carex alopecoidea Tuck., Enum. Meth. Caric.: 18 (1843).
S. Canada to N. & EC. U.S.A. 71 MAN SAS 72 NSC ONT QUE 74 ILL IOW MIN NDA WIS 75 CNT INI MAI MAS MIC NWJ NWY OHI PEN VER 78 DEL KTY MRY WDC. Hemicr. or rhizome geophyte.
Carex alopecoidea f. *sparsispicata* Dewey, Amer. J. Sci. Arts, II, 8: 350 (1849).

Carex alopecuroides D.Don ex Tilloch & Taylor, Philos. Mag. J. 62: 455 (Dec. 1823). *Carex japonica* var. *alopecuroides* (D.Don ex Tilloch & Taylor) C.B.Clarke in J.D.Hooker, Fl. Brit. India 6: 737 (1894).
Trop. & Subtrop. Asia. 36 CHC CHS 38 JAP NNS TAI 40 ASS EHM NEP WHM 41 VIE 42 JAW PHI SUL SUM 43 NWG. Hemicr. or rhizome geophyte.
Carex chlorostachys D.Don ex Tilloch & Taylor, Philos. Mag. J. 62: 455 (1823), nom. illeg. *Carex doniana* Spreng., Syst. Veg. 3: 825 (1826). *Carex alopecuroides* var. *chlorostachys* C.B.Clarke, J. Linn. Soc., Bot. 36: 271 (1903). *Carex japonica* var. *chlorostachys* (C.B.Clarke) Kük. in H.G.A.Engler (ed.), Pflanzenr., IV, 20(38): 620 (1909). *Carex japonica* subsp. *chlorostachys* (C.B.Clarke) T.Koyama in K.Hara, Fl. E. Himal.: 382 (1966).
Carex emodorum Spreng., Syst. Veg. 3: 818 (1826).
Carex consocialis Steud., Syn. Pl. Glumac. 2: 222 (1855).
Carex zollingeri Kunze ex Steud., Syn. Pl. Glumac. 2: 221 (1855).
Carex phaenocarpa Franch., Bull. Soc. Philom. Paris, VIII, 7: 90 (1895). *Carex japonica* var. *phaenocarpa* (Franch.) Kük. in H.G.A.Engler (ed.), Pflanzenr., IV, 20(38): 621 (1909).
Carex subtransversa C.B.Clarke, Philipp. J. Sci., C 2: 108 (1907). *Carex alopecuroides* subsp. *subtransversa* (C.B.Clarke) T.Koyama, in Fl. Taiwan 5: 354 (1978).
Carex kawakamii Hayata, J. Coll. Sci. Imp. Univ. Tokyo 30(1): 385 (1911).
Carex pseudojaponica Hayata, J. Coll. Sci. Imp. Univ. Tokyo 30(1): 392 (1911), nom. illeg.
Carex sasakii Hayata, J. Coll. Sci. Imp. Univ. Tokyo 30(1): 395 (1911).
Carex hayatana Honda, Bot. Mag. (Tokyo) 43: 191 (1929).

Carex ×alsatica Zahn, Oesterr. Bot. Z. 40: 363 (1890). *C. flava* × *C. viridula* subsp. *brachyrrhyncha.*
Europe, E. Canada to NE. U.S.A. 10 GRB IRE 14+ 72 NFL ONT QUE 75 MIC. Hemicr. or rhizome geophyte.

nothosubsp. ***alsatica***. *C. flava* × *C. viridula* subsp. *brachyrrhyncha.*
Europe. 10 GRB IRE 14+. Hemicr. or rhizome geophyte.
Carex ×mixta Corb., Nouv. Fl. Normandie: 607 (1893), nom. illeg.

nothosubsp. ***subviridula*** (Fernald) Govaerts in R.H.A.Govaerts & D.A.Simpson, World Checklist of Cyperaceae: 41 (2007). *C. flava* × *C. viridula.*
E. Canada to NE. U.S.A. 72 NFL ONT QUE 75 MIC. Hemicr. or rhizome geophyte.
**Carex ×subviridula* Fernald, Rhodora 35: 231 (1933).

Carex alsophila F.Muell., Fragm. 8: 257 (1874).
Victoria. 50 VIC. Hemicr. or rhizome geophyte.

Carex amgunensis F.Schmidt, Reis. Amur-Land., Bot.: 69 (1868).
N. & E. European Russia to N. China. 14 RUE RUN 30 WSB 31+ 36 CHC CHM 37 MON. Hemicr. or rhizome geophyte.

Carex amicta Boott, Ill. Carex 4: 131 (1867).
Colombia to Ecuador. 83 CLM ECU. Hemicr. or rhizome geophyte.

Carex amphibola Steud., Syn. Pl. Glumac. 2: 234 (1855). *Carex grisea* var. *amphibola* (Steud.) Kük. in H.G.A.Engler (ed.), Pflanzenr., IV, 20(38): 520 (1909).
SE. Canada to C. & E. U.S.A. 72 ONT 74 ILL MSO OKL 75 CNT INI MAS MIC NWJ NWY OHI PEN WVA 77 TEX 78 ALA ARK DEL GEO KTY LOU MRY MSI NCA SCA TEN VRG WDC. Hemicr. or rhizome geophyte.
Carex laxiflora Elliott, Sketch Bot. S. Carolina 2: 549 (1824), nom. illeg.
Carex grisea var. *minor* Olney in E.Hall, Pl. Tex.: 25 (1873).

Carex amplectens Mack., Bull. Torrey Bot. Club 43: 611 (1916 publ. 1917).
California. 76 CAL. Hemicr. or rhizome geophyte.

Carex amplifolia Boott in W.J.Hooker, Fl. Bor.-Amer. 2: 228 (1839).
SW. Canada to W. U.S.A., New Mexico. 71 BRC 73 IDA MNT ORE WAS 76 CAL 77 NWM. Hemicr. or rhizome geophyte.

Carex andersonii Boott in J.D.Hooker, Fl. Antarct.: 364 (1846).
S. Chile to S. Argentina. 85 AGS CLS. Hemicr. or rhizome geophyte.
Carex eleopsammodes Steud., Syn. Pl. Glumac. 2: 213 (1855).

Carex andicola G.A.Wheeler, Hickenia 2: 196 (1996).
W. Argentina to C. Chile. 85 AGS AGW CLC. Hemicr. or rhizome geophyte.

Carex andina Phil., Anales Univ. Chile 93: 485 (1896).
C. Chile to SW. Argentina. 85 AGS CLC. Hemicr. or rhizome geophyte.
Carex andina var. *subabscondita* Kük., Repert. Spec. Nov. Regni Veg. 8: 7 (1910).

Carex andringitrensis Cherm., Bull. Soc. Bot. France 70: 410 (1923).
C. Madagascar. 29 MDG. Hemicr. or rhizome geophyte.

Carex angolensis Nelmes, Bull. Misc. Inform. Kew 1940: 162 (1940).
Angola. 26 ANG. Hemicr. or rhizome geophyte.

Carex angustata Boott in W.J.Hooker, Fl. Bor.-Amer. 2: 218 (1839). *Carex stricta* var. *angustata* (Boott) L.H.Bailey in A.Gray, Manual, ed. 6: 600 (1890). *Carex stricta* f. *angustata* (Boott) Kük. in H.G.A.Engler (ed.), Pflanzenr., IV, 20(38): 330 (1909). *Carex strictior* f. *angustata* (Boott) House, Bull. New York State Mus. Nat. Hist. 243–244: 60 (1923).
NW. U.S.A. to Mexico (Baja California). 73 IDA ORE WAS 76 CAL 79 MXN. Hemicr. or rhizome geophyte.
Carex acuta var. *erecta* Dewey, Amer. J. Sci. Arts 10: 265 (1825). *Carex stricta* var. *erecta* (Dewey) C.F.Reed, Phytologia 67: 453 (1989).
Carex angustata var. *strictior* Dudley, Cornell Univ. Sci. Bull. 2: 113 (1886).
Carex auriculata L.H.Bailey, Mem. Torrey Bot. Club 1: 19 (1889), nom. illeg.
Carex eurycarpa T.Holm, Amer. J. Sci. 170: 303 (1905).
Carex oxycarpa T.Holm, Amer. J. Sci. 170: 303 (1905). *Carex eurycarpa* var. *oxycarpa* (T.Holm) Kük. in H.G.A.Engler (ed.), Pflanzenr., IV, 20(38): 339 (1909).
Carex egregia Mack., Bull. Torrey Bot. Club 42: 414 (1915).
Carex eurycarpa var. *attenuata* Kük., Repert. Spec. Nov. Regni Veg. 26: 254 (1929).

Carex angustifructus (Kük.) Nelmes, Bull. Misc. Inform. Kew 1939: 271 (1939).
Himalaya to Tibet. 36 CHT 40 EHM WHM. Hemicr. or rhizome geophyte.
**Carex atrofusca* var. *angustifructus* Kük. in H.G.A.Engler (ed.), Pflanzenr., IV, 20(38): 554 (1909).

Carex angustispica Reznicek & S.González, Contr. Univ. Michigan Herb. 23: 346 (2001).
Mexico (Oaxaca). 79 MXS. Hemicr.

Carex angustisquama Franch., Bull. Soc. Philom. Paris, VIII, 7: 39 (1895).
Japan (N. Honshu). 38 JAP. Hemicr. or rhizome geophyte.

Carex angustiutricula F.T.Wang & Tang ex L.K.Dai, Acta Phytotax. Sin. 37: 182 (1999).
China (Sichuan). 36 CHC. Hemicr. or rhizome geophyte.

Carex ×aniaiensis Fujiw. & Y.Matsuda, J. Phytogeogr. Taxon. 40: 77 (1992). *C. podogyna* × *C. sadoensis*.
Japan. 38 JAP. Hemicr. or rhizome geophyte.

Carex anisoneura V.I.Krecz. in V.L.Komarov (ed.), Fl. URSS 3: 410 (1935).
C. Asia. 32 KGZ UZB. Hemicr. or rhizome geophyte.

Carex anisostachys Liebm., Mexic. Halvgr.: 266 (1850).
Mexico to Venezuela. 79 MXC MXE MXG MXS MXT 80 GUA 82 VEN. Hemicr. or rhizome geophyte.
Carex chlorocarpos Liebm., Mexic. Halvgr.: 265 (1850), nom. illeg. *Carex liebmannii* Walp., Ann. Bot. Syst. 3: 705 (1852). *Carex anisostachys* var. *chlorocarpos* (Liebm.) L.H.Bailey, Mem. Torrey Bot. Club 1: 66 (1889).
Carex olivacea Liebm., Mexic. Halvgr.: 267 (1850), nom. illeg. *Carex monticola* Boeck., Bot. Jahrb. Syst. 1: 364 (1881), nom. illeg. *Carex anisostachys* var. *monticola* Kük. in H.G.A.Engler (ed.), Pflanzenr., IV, 20(38): 580 (1909). *Carex androgyna* L.H.Bailey, Proc. Amer. Acad. Arts 22: 101 (1887).
Carex oaxacana L.H.Bailey, Bot. Gaz. 25: 271 (1898).
Carex felipensis C.B.Clarke, Kew Bull., Addit. Ser. 8: 84 (1908).

Carex annectens (E.P.Bicknell) E.P.Bicknell, Bull. Torrey Bot. Club 35: 492 (1908).
SE. Canada to C. & E. U.S.A. 72 ONT QUE 74 ILL IOW KAN MIN MSO NEB OKL WIS 75 CNT INI MAI MAS MIC NWH NWJ NWY OHI PEN RHO VER WVA 78 ARK DEL KTY LOU MRY MSI NCA SCA TEN VRG WDC. Hemicr. or rhizome geophyte.
Carex vulpinoidea var. *ambigua* Barratt, N. Amer. Caric.: 62 (1840). *Carex setacea* var. *ambigua* (Barratt) Fernald, Rhodora 8: 167 (1906). *Carex*

annectens var. *ambigua* (Barratt) Gleason, Phytologia 4: 22 (1952).
Carex biformis var. *sterilis* F.W.Schultz, Flora 24: 55 (1841).
Carex xanthocarpa E.P.Bicknell, Bull. Torrey Bot. Club 23: 21 (1896), nom. illeg. *Carex vulpinoidea* var. *xanthocarpa* Kük. in H.G.A.Engler (ed.), Pflanzenr., IV, 20(38): 148 (1909). *Carex annectens* var. *xanthocarpa* (Kük.) Wiegand, Rhodora 24: 74 (1922). *Carex bicknellii* E.G.Camus, Notul. Syst. (Paris) 1: 239 (1910), nom. illeg.
**Carex xanthocarpa* var. *annectens* E.P.Bicknell, Bull. Torrey Bot. Club 23: 22 (1896). *Carex vulpinoidea* f. *annectens* (E.P.Bicknell) Kük. in H.G.A.Engler (ed.), Pflanzenr., IV, 20(38): 148 (1909). *Carex vulpinoidea* var. *annectens* (E.P.Bicknell) Farw., Pap. Michigan Acad. Sci. 1: 91 (1923).
Carex brachyglossa Mack., Bull. Torrey Bot. Club 50: 355 (1923).

Carex anningensis F.T.Wang & Tang ex P.C.Li, Acta Phytotax. Sin. 37: 171 (1999).
China (Yunnan). 36 CHC. Hemicr. or rhizome geophyte.

Carex anomoea Hand.-Mazz., Symb. Sin. 7: 1267 (1936).
Nepal to Sikkim. 40 EHM NEP. Hemicr. or rhizome geophyte.
**Carex inaequalis* Boott ex C.B.Clarke in J.D.Hooker, Fl. Brit. India 6: 726 (1894), nom. illeg. *Carex inaequabilis* V.I.Krecz., Bot. Mater. Gerb. Bot. Inst. Komarova Akad. Nauk S.S.S.R. 9: 192 (1946), nom. illeg.

Carex anthoxanthea C.Presl, Reliq. Haenk. 1: 203 (1828).
Russian Far East to NW. U.S.A. 31 MAG 70 ALU ASK YUK 71 BRC 73 WAS. Hemicr. or rhizome geophyte.
Carex leiocarpa C.A.Mey., Mém. Sav. Étr. Acad. St. Pétersbourg 1: 208 (1831). *Carex anthoxanthea* var. *leiocarpa* (C.A.Mey.) Kük. in H.G.A.Engler (ed.), Pflanzenr., IV, 20(38): 97 (1909).

Carex* ×*anticostensis (Fernald) Lepage, Naturaliste Canad. 83: 108 (1956). *C. rostrata* × *C. saxatilis* var. *militaris*.
E. Canada. 72 NFL QUE. Hemicr. or rhizome geophyte.
Carex inflata var. *anticostensis* Fernald, Rhodora 44: 329 (1942).
Carex rostrata var. *anticostensis* (Fernald) Fernald, Rhodora 48: 146 (1946).

nothovar. ***anticostensis***. *C. rostrata* × *C. saxatilis* var. *militaris*.
E. Canada. 72 NFL QUE. Hemicr. or rhizome geophyte.
Carex ×*anticostensis* nothovar. *minor* Lepage, Naturaliste Canad. 83: 111 (1956).

nothovar. ***inflatior*** Lepage, Naturaliste Canad. 83: 110 (1956). *C. rostrata* × *C. saxatilis* var. *rhomalea*.
E. Canada. 72 QUE. Hemicr. or rhizome geophyte.
Carex ×*anticostensis* nothovar. *longidens* Lepage, Naturaliste Canad. 83: 110 (1956).

Carex antoniensis A.Chev., Rev. Bot. Appl. Agric. Trop. 15: 1032 (1935).
Cape Verde (Santo Antão). 21 CVI. Hemicr. or rhizome geophyte.

Carex aperta Boott in W.J.Hooker, Fl. Bor.-Amer. 2: 218 (1839).
W. Canada to NW. U.S.A. 36 CHM? 71 ABT BRC 73 IDA MNT ORE WAS. Hemicr. or rhizome geophyte.
Carex stylosa var. *virens* L.H.Bailey, Proc. Amer. Acad. Arts 22: 79 (1886).
Carex acutina var. *tenuior* L.H.Bailey, Mem. Torrey Bot. Club 1: 53 (1889).
Carex turgidula L.H.Bailey, Bot. Gaz. 25: 271 (1898).
Carex accedens T.Holm, Amer. J. Sci. 166: 457 (1903).
Carex acutina var. *petrophila* T.Holm, Amer. J. Sci., IV, 16: 32 (1903).
Carex bovina Howell, Fl. N. W. Amer. 1: 702 (1903).
Carex aperta f. *concinnula* T.Holm, Amer. J. Sci., V, 2: 290 (1921).
Carex aperta f. *hydroessa* T.Holm, Amer. J. Sci., V, 2: 290 (1921).
Carex aperta f. *mimetica* T.Holm, Amer. J. Sci., V, 2: 290 (1921).
Carex aperta var. *umbrosa* Kük., Repert. Spec. Nov. Regni Veg. 26: 254 (1929).
Carex aperta var. *viridans* Kük., Repert. Spec. Nov. Regni Veg. 26: 254 (1929).

Carex aphanolepis Franch. & Sav., Enum. Pl. Jap. 2: 580 (1878). *Carex japonica* var. *aphanolepis* (Franch. & Sav.) Kük. in H.G.A.Engler (ed.), Pflanzenr., IV, 20(38): 620 (1909).
Russian Far East to E. China, Japan, Vietnam. 31 PRM 36 CHS 38 JAP KOR 41 VIE. Hemicr. or rhizome geophyte.
Carex japonica var. *humilis* Franch., Nouv. Arch. Mus. Hist. Nat., III, 10: 77 (1898).
Carex vernicosa Franch. ex C.B.Clarke, Kew Bull., Addit. Ser. 8: 85 (1908).

Carex aphylla Kunth, Enum. Pl. 2: 421 (1837).
Chile, S. Argentina. 85 AGS CLC. Hemicr. or rhizome geophyte.
Schoenus marginatus Kunze ex Kunth, Enum. Pl. 2: 421 (1837).
Carex berteroana Desv. in C.Gay, Fl. Chil. 6: 198 (1854).
Carex ebracteata Phil., Linnaea 33: 271 (1865).

Carex aphyllopus Kük., Bull. Herb. Boissier, II, 4: 53 (1904).
C. Japan. 38 JAP. Hemicr. or rhizome geophyte.
Carex paludicola (Ohwi) Ohwi, Mem. Coll. Sci. Kyoto Imp. Univ., Ser. B, Biol. 11(5): 512 (1936).

Carex apiahyensis Palla, Denkschr. Kaiserl. Akad. Wiss., Wien. Math.-Naturwiss. Kl. 79: 199 (1908).
Brazil. 84+. Hemicr. or rhizome geophyte.

Carex apoiensis Akiyama, Caric. Far East. Reg. As.: 117 (1955).
Japan. 38 JAP. Hemicr. or rhizome geophyte.

Carex appalachica J.M.Webber & P.W.Ball, Taxon 28: 614 (1979).
SE. Canada to NE. & EC. U.S.A. 72 ONT QUE 75 CNT MAI MAS NWH NWJ NWY OHI PEN VER WVA 78 GEO KTY MRY NCA SCA TEN VRG WDC. Hemicr. or rhizome geophyte.

Carex appressa R.Br., Prodr.: 242 (1810). *Carex paniculata* var. *appressa* (R.Br.) Cheeseman, Trans. & Proc. New Zealand Inst. 16: 427 (1884). *Carex appressa* var. *typica* Domin, Biblioth. Bot. 89: 493 (1915), nom. inval.

New Guinea, Australia, New Zealand, New Caledonia. 43 NWG 50 NSW NTA QLD SOA TAS VIC WAU 51 CTM NZN NZS 60 NWC. Hemicr. or rhizome geophyte.

var. ***appressa***

New Guinea, Australia, New Zealand North I., New Caledonia. 43 NWG 50 NSW NTA QLD SOA TAS VIC WAU 51 NZN 60 NWC. Hemicr. or rhizome geophyte.

Carex collata Boott, London J. Bot. 3: 417 (1844).

Carex diaphana Boott, Ill. Carex 4: 156 (1867). *Carex appressa* var. *diaphana* (Boott) Kük. in H.G.A.Engler (ed.), Pflanzenr., IV, 20(38): 179 (1909).

Carex virgata var. *abbreviata* Boeck., Linnaea 39: 99 (1875).

Carex chlorantha var. *composita* F.Muell. in G.Bentham, Fl. Austral. 7: 440 (1878).

Carex paniculata var. *subdiaphana* F.Muell. in G.Bentham, Fl. Austral. 7: 440 (1878).

Carex appressa f. *minor* Kük. in H.G.A.Engler (ed.), Pflanzenr., IV, 20(38): 179 (1909).

var. ***virgata*** (Sol. ex Boott) Kük. in H.G.A.Engler (ed.), Pflanzenr., IV, 20(38): 179 (1909).

Tasmania, New Zealand, Chatham Is. 50 TAS 51 CTM NZN NZS. Hemicr. or rhizome geophyte.

**Carex virgata* Sol. ex Boott in J.D.Hooker, Fl. Nov.-Zel. 1: 282 (1853). *Carex paniculata* var. *virgata* (Sol. ex Boott) Cheeseman, Trans. & Proc. New Zealand Inst. 16: 427 (1884).

Carex discolor Reinw. ex de Vriese, Pl. Ind. Bat. Orient.: 143 (1857).

Carex appropinquata Schumach., Enum. Pl. 1: 266 (1801). *Vignea appropinquata* (Schumach.) Soják, Cas. Nár. Mus., Odd. Prír. 148: 194 (1979 publ. 1980).

Europe to Caucasus. 10 DEN FIN GRB IRE NOR SWE 11 AUT BGM CZE GER HUN NET POL SWI 12 FRA 13 GRC ITA ROM YUG 14 BLR BLT RUC RUE RUN RUS RUW UKR 30 KRA WSB 32 KAZ 33 NCS TCS 34 TUR. Hemicr. or rhizome geophyte.

Carex paradoxa Willd., Schriften Ges. Naturf. Freunde Berlin 1794: 30 (1794), nom. illeg. *Vignea paradoxa* Rchb. in J.C.Mössler, Handb. Gewächsk. ed. 2, 3: 1621 (1830). *Caricina paradoxa* (Rchb.) St.-Lag. in A.Cariot, Étude Fl., ed. 8, 2: 876 (1889). *Carex paniculata* var. *paradoxa* (Rchb.) Fiori, Nuov. Fl. Italia 1: 185 (1923).

Carex paradoxa f. *sparsiflora* Lange, Haandb. Danske Fl., ed. 4: 122 (1886).

Carex paradoxa f. *brachystachya* Schatz, Verh. Gesch. der Baar 8: 14 (1893).

Carex paradoxa var. *gracilis* Asch. & Graebn., Syn. Mitteleur. Fl. 2(2): 45 (1902).

Carex aquatilis Wahlenb., Kongl. Vetensk. Acad. Nya Handl. 1803: 165 (1803). *Vignea aquatilis* (Wahlenb.) Rchb. in J.C.Mössler, Handb. Gewächsk. ed. 2, 3: 1624 (1830). *Neskiza aquatilis* (Wahlenb.) Raf., Good Book: 27 (1840). *Carex aquatilis* var. *genuina* Syme in J.E.Smith, Engl. Bot., ed. 3: 113 (1870), nom. inval.

N., C. & E. Europe to Siberia, E. Canada to U.S.A. 10 FIN GRB IRE NOR SWE 11 GER NET 14 BLT RUC RUE RUN RUS RUW 30 KRA WSB YAK 31 KAM KHA MAG 70 ASK GNL NUN NWT YUK 71 ABT BRC MAN SAS 72 LAB NBR NFL NSC ONT PEI QUE 73 COL IDA MNT ORE WAS WYO 74 ILL IOW KAN MIN MSO NDA NEB SDA WIS 75 CNT INI MAI MAS MIC NWJ NWY OHI PEN VER 76 ARI CAL NEV UTA 77 NWM 78 VRG (90) sge. Hemicr. or rhizome geophyte.

var. ***aquatilis***

N., C. & E. Europe to Siberia, E. Canada to U.S.A. 10 FIN GRB IRE NOR SWE 11 GER NET 14 BLT RUC RUE RUN RUS RUW 30 KRA WSB YAK 70 ASK NUN NWT YUK 71 ABT BRC MAN SAS 72 LAB NBR NFL NSC ONT PEI QUE 73 COL IDA MNT ORE WAS WYO 74 ILL IOW KAN MIN MSO NDA NEB SDA WIS 75 CNT INI MAI MAS MIC NWJ NWY OHI (90) sge. Hemicr. or rhizome geophyte.

Carex nardifolia Wahlenb., Kongl. Vetensk. Acad. Nya Handl. 1803: 165 (1803).

Carex aquatilis var. *elatior* Bab., Man. Brit. Bot.: 341 (1843).

Carex borealis Láng, Flora 26: 142 (1843).

Carex aquatilis var. *planifolia* F.Nyl., Spic. Fl. Fenn. 2: 23 (1844).

Carex aquatilis var. *sphagnophila* Fr., Summa Veg. Scand.: 72 (1845).

Carex aquatilis f. *cuspidata* Laest. ex Andersson, Cyper. Scand.: 46 (1849).

Carex aquatilis f. *virescens* Andersson, Cyper. Scand.: 46 (1849).

Carex arctophila Nyl. ex Fr., Summa Veg. Scand.: 561 (1849).

Carex aquatilis var. *watsonii* Syme in J.E.Smith, Engl. Bot., ed. 3: 113 (1870).

Carex acutinella var. *congesta* Boott in S.Watson & al., Bot. California 2: 238 (1889).

Carex variabilis L.H.Bailey, Mem. Torrey Bot. Club 1: 18 (1889).

Carex aquatilis f. *epigaea* Kurtz, Bot. Jahrb. Syst. 19: 478 (1894).

Carex chionophila T.Holm, Amer. J. Sci. 166: 457 (1903).

Carex limnocharis T.Holm, Amer. J. Sci. 166: 457 (1903).

Carex rhomboidea T.Holm, Amer. J. Sci. 166: 35 (1903). *Carex aquatilis* var. *rhomboidea* (T.Holm) Kelso, Biol. Leafl. 66: 34 (1953).

Carex sphacelata T.Holm, Amer. J. Sci. 166: 457 (1903).

Carex variabilis var. *sciaphila* T.Holm, Amer. J. Sci., IV, 16: 22 (1903). *Carex aquatilis* f. *sciaphila* (T.Holm) Kük. in H.G.A.Engler (ed.), Pflanzenr., IV, 20(38): 311 (1909). *Carex aquatilis* var. *sciaphila* (T.Holm) Kelso, Biol. Leafl. 66: 34 (1953).

Carex aquatilis f. *laxa* Kük. in H.G.A.Engler (ed.), Pflanzenr., IV, 20(38): 311 (1909).

Carex suksdorfii Kük., Repert. Spec. Nov. Regni Veg. 16: 434 (1920).

Carex suksdorfii var. *ovalis* Kük., Repert. Spec. Nov. Regni Veg. 16: 435 (1920).

Carex acutinella Mack. in N.L.Britton & al. (eds.), N. Amer. Fl. 18: 407 (1935).

Carex interimus Maguire, Brittonia 5: 200 (1944).

Carex aquatilis var. *plumbeophila* Kelso, Biol. Leafl. 51: 2 (1950).

var. ***dives*** (T.Holm) Kük. in H.G.A.Engler (ed.), Pflanzenr., IV, 20(38): 311 (1909).

S. Alaska to W. U.S.A. 70 ASK 71 BRC 73 MNT ORE WAS 76 CAL. Hemicr. or rhizome geophyte.

Carex sitchensis Prescott ex Bong., Veg. Sitcha: 51 (1833).
Carex wilkesii Olney, U.S. Expl. Exped., Phan. 2: 477 (1854).
Carex howellii L.H.Bailey, Mem. Torrey Bot. Club 1: 45 (1889).
**Carex dives* T.Holm, Amer. J. Sci. 167: 312 (1904).
Carex pachystoma T.Holm, Amer. J. Sci. 170: 302 (1905). *Carex aquatilis* var. *pachystoma* (T.Holm) Kük. in H.G.A.Engler (ed.), Pflanzenr., IV, 20(38): 311 (1909).
Carex panda C.B.Clarke, Kew Bull., Addit. Ser. 8: 71 (1908).
Carex aquatilis f. *angustata* Kük. in H.G.A.Engler (ed.), Pflanzenr., IV, 20(38): 309 (1909).
Carex aquatilis var. *sitchensis* (J.D.Prescott ex Bong.) Kelso, Biol. Leafl. 66: 34 (1953).

var. ***minor*** Boott, Ill. Carex 4: 163 (1867).
Subarctic. 10 FIN 14 RUE RUN 30 KRA WSB YAK 31 KAM KHA MAG 70 ASK GNL NUN NWT YUK 72 QUE. Hemicr. or rhizome geophyte.
Carex concolor R.Br., Chlor. Melvill.: 283 (1823). *Diemisa concolor* (R.Br.) Raf., Good Book: 27 (1840). *Carex rigida* var. *concolor* (R.Br.) Kük. in H.G.A.Engler (ed.), Pflanzenr., IV, 20(38): 301 (1909).
Carex washingtonia Eaton, Man. Bot.: 156 (1829).
Carex stans Drejer, Rev. Car. Bor.: 40 (1841). *Carex aquatilis* var. *stans* (Drejer) Boott, Ill. Carex 4: 163 (1867). *Carex aquatilis* subsp. *stans* (Drejer) Hultén, Circumpol. Pl. 1: 74 (1962).
Carex uzoni Kom., Repert. Spec. Nov. Regni Veg. 13: 165 (1914).
Carex concolor var. *minuscula* Kuvaev in V.B.Kuvaev & al., Rast. Pokrov Ostrova Sibiryakova: 52 (1994). *Carex minuscula* (Kuvaev) Rebrist., Bot. Zhurn. (Moscow & Leningrad) 82(7): 33 (1997).

var. ***substricta*** Kük. in H.G.A.Engler (ed.), Pflanzenr., IV, 20(38): 309 (1909). *Carex substricta* (Kük.) Mack. in P.A.Rydberg, Fl. Rocky Mts.: 139 (1917).
S. Canada to NC. & NE. U.S.A. 71 MAN 72 NBR NSC ONT QUE 74 ILL IOW KAN MIN MSO NDA NEB SDA WIS 75 CNT INI MAI MAS MIC NWJ NWY OHI PEN VER 78 VRG. Hemicr. or rhizome geophyte.

Carex aquilonalis Akiyama, J. Jap. Bot. 13: 651 (1937).
Japan. 38 JAP. Hemicr. or rhizome geophyte.

Carex ×arakanei T.Koyama, J. Jap. Bot. 33: 255 (1958).
Japan. 38 JAP. Hemicr. or rhizome geophyte.

Carex arapahoensis Clokey, Rhodora 21: 83 (1919).
WC. U.S.A. 73 COL WYO 76 UTA. Hemicr. or rhizome geophyte.

Carex archeri Boott in J.D.Hooker, Fl. Tasman. 2: 98 (1858).
Victoria, Tasmania. 50 TAS VIC. Hemicr. or rhizome geophyte.

Carex arcta Boott, Ill. Carex 4: 155 (1867).
Yukon to N. & W. U.S.A. 70 YUK 71 ABT BRC MAN 72 NBR ONT QUE 73 IDA MNT ORE WAS 74 MIN WIS 75 MAI MAS MIC NWH NWY VER 76 CAL. Hemicr. or rhizome geophyte.
Carex canescens var. *polystachya* Boott in J.Richardson, Arct. Search. Exped. 2: 344 (1851).
Carex kunzei Olney ex A.Gray, Proc. Amer. Acad. Arts 8: 406 (1872).
Carex canescens var. *oregana* L.H.Bailey, Mem. Torrey Bot. Club 1: 75 (1889). *Carex arcta* var. *oregana* (L.H.Bailey) B.Boivin, Phytologia 43: 49 (1979).
Carex heleonastes var. *scabriuscula* Kük. in H.G.A. Engler (ed.), Pflanzenr., IV, 20(38): 215 (1909).

Carex arctata Boott in W.J.Hooker, Fl. Bor.-Amer. 2: 227 (1839).
C. & E. Canada to NC. & NE. U.S.A. 71 MAN 72 LAB NBR NFL NSC ONT PEI QUE 74 MIN WIS 75 CNT INI MAI MAS MIC NWH NWY OHI PEN RHO VER. Hemicr. or rhizome geophyte.
Carex sylvatica Dewey, Amer. J. Sci. Arts 10: 40 (1826), nom. illeg.
Carex arctata var. *faxonii* L.H.Bailey, Bot. Gaz. 13: 87 (1888).

Carex arctiformis Mack. in N.L.Britton & al. (eds.), N. Amer. Fl. 18: 97 (1931). *Carex canescens* subsp. *arctiformis* (Mack.) Calder & R.L.Taylor, Canad. J. Bot. 43: 1389 (1963).
SE. Alaska to British Columbia. 70 ASK 71 BRC. Hemicr. or rhizome geophyte.
Carex heleonastes var. *robustior* Kük. in H.G.A.Engler (ed.), Pflanzenr., IV, 20(38): 215 (1909). *Carex cinerea* var. *robustior* (Kük.) B.Boivin, Provancheria 25: 98 (1992).

Carex arenaria L., Sp. Pl.: 973 (1753). *Vignea arenaria* (L.) Rchb. in J.C.Mössler, Handb. Gewächsk. ed. 2, 3: 1618 (1830).
SW. Europe to C. Russia. 10 DEN FIN GRB IRE NOR SWE 11 BGM GER NET POL 12 FRA POR SPA 14 BLT RUW (73) ore (78) del mry nca vrg. Hemicr. or rhizome geophyte.
Carex spadicea Gilib., Excerc. Phyt. 2: 546 (1792), nom. illeg.
Carex witheringii Gray, Nat. Arr. Brit. Pl. 2: 48 (1821).
Carex intermedia d'Urv., Mém. Soc. Linn. Paris 1: 378 (1822), nom. illeg.
Carex schreberi Desv., Fl. Anjou: 73 (1827), nom. illeg.
Carex ohmuelleriana O.Lang, Flora 26: 240 (1843). *Vignea ohmuelleriana* (O.Lang) Pritz., Icon. Bot. Index: t. 1161 (1855). *Carex brizoides* subsp. *ohmuelleriana* (O.Lang) Nyman, Consp. Fl. Eur.: 779 (1882). *Carex schreberi* subsp. *ohmuelleriana* (O. Lang) Nyman, Consp. Fl. Eur., Suppl. 2: 324 (1890).
Carex sabuletorum M.Bieb. ex Boott, Ill. Carex 4: 178 (1867).
Carex arenaria var. *pseudoarenaria* T.Marsson, Fl. Neu-Vorpommern: 523 (1869).
Carex arenaria f. *remota* T.Marsson, Fl. Neu-Vorpommern: 523 (1869).
Carex arenaria f. *pallida* Lackow., Fl. Berl., ed. 12: 21 (1901).
Carex arenaria f. *pumila* Lackow., Allg. Bot. Z. Syst. 7: 171 (1901).
Carex arenaria var. *spiralis* Asch. & Graebn., Syn. Mitteleur. Fl. 2(2): 30 (1902).
Carex arenaria var. *tenuis* Asch. & Graebn., Syn. Mitteleur. Fl. 2(2): 30 (1902).
Carex arenaria var. *prostrata* De Langhe, Bull. Soc. Roy. Bot. Belgique 104: 15 (1971).

Carex argentina Barros, Darwiniana 8: 409 (1948).
W. Argentina. 85 AGS AGW. Hemicr. or rhizome geophyte.

Carex argunensis Turcz. ex Ledeb., Fl. Ross. 4: 267 (1852). *Carex rupestris* subsp. *argunensis* (Turcz. ex Ledeb.) Vorosch. in A.K.Skvortsov (ed.), Florist. Issl. Razn. Raionakh SSSR: 156 (1985).
Siberia to NE. China. 30 TVA 31+ 36 CHM 37 MON. Hemicr. or rhizome geophyte.

subsp. ***alticola*** (Popl. ex Sukaczev) Malyschev, in Fl. Sibir., Cyperac. 3: 64 (1990).
Siberia, Russian Far East. 30+ 31+. Hemicr. or rhizome geophyte.
**Carex alticola* Popl. ex Sukaczev, Arbeiten Amur-Exped., Bot. 1: 277 (1912). *Carex rupestris* subsp. *alticola* (Popl. ex Sukaczev) Vorosch. in A.K.Skvortsov (ed.), Florist. Issl. Razn. Raionakh SSSR: 156 (1985).

subsp. ***argunensis***
Siberia to NE. China. 30 TVA 31+ 36 CHM 37 MON. Hemicr. or rhizome geophyte.
Carex rupestris Turcz. ex Ledeb., Fl. Ross. 4: 267 (1852), nom. illeg.

Carex argyi H.Lév. & Vaniot, Bull. Soc. Agric. Sarthe 60: 78 (1905).
China (Hubei to Jiangsu). 36 CHC CHS. Hemicr. or rhizome geophyte.

Carex argyrantha Tuck. ex Boott, Ill. Carex 3: 119 (1862).
E. Canada to NE. & EC. U.S.A. 72 NBR NSC ONT QUE 75 CNT MAI MAS MIC NWH NWJ NWY OHI PEN RHO VER WVA 78 MRY NCA TEN VRG. Hemicr. or rhizome geophyte.
Carex foenea var. *perplexa* L.H.Bailey, Mem. Torrey Bot. Club 1: 27 (1889).

Carex aridula V.I.Krecz., Bot. Mater. Gerb. Bot. Inst. Komarova Akad. Nauk S.S.S.R. 9: 191 (1946).
Tibet to C. China. 36 CHC CHI CHN CHQ CHT. Hemicr. or rhizome geophyte.

Carex arisanensis Hayata, J. Coll. Sci. Imp. Univ. Tokyo 30(1): 378 (1911).
S. Nansei-shoto, Taiwan, China (Guangxi). 36 CHS 38 NNS TAI. Hemicr. or rhizome geophyte.

Carex aristatisquamata Tang & F.T.Wang ex L.K.Dai, Acta Phytotax. Sin. 32: 179 (1994).
China (Sichuan). 36 CHC. Hemicr. or rhizome geophyte.

Carex aristulifera P.C.Li, Acta Bot. Yunnan. 12: 141 (1990).
SC. China. 36 CHC. Hemicr. or rhizome geophyte.

Carex arkansana (L.H.Bailey) L.H.Bailey, Bot. Gaz. 21: 6 (1896).
C. & EC. U.S.A. 74 ILL KAN MSO OKL 77 TEX 78 ARK. Hemicr. or rhizome geophyte.
**Carex rosea* var. *arkansana* L.H.Bailey, Bot. Gaz. 13: 87 (1888).

Carex arnellii H.Christ in N.J.W.Scheutz, Pl. Vasc. Jenis.: 177 (1888).
European Russia to Russian Far East. 14 RUC RUE RUS 30 WSB 31+ 32 KAZ 36 CHI CHM CHN 37 MON. Hemicr. or rhizome geophyte.
Carex maximowiczii Boeck., Linnaea 41: 237 (1877), nom. illeg.
Carex turczaninowiana Meinsh. ex Korsh., Trudy Imp. S.-Peterburgsk. Bot. Sada 13: 411 (1892).
Carex subconcolor Kitag., Bot. Mag. (Tokyo) 48: 20 (1934).

Carex arnottiana Nees ex Drejer, Symb. Caric.: 16 (1844).
Sri Lanka. 40 SRL. Hemicr. or rhizome geophyte.
Carex socia Boott, Proc. Linn. Soc. London 1: 254 (1845).

Carex arsenei Kük., Repert. Spec. Nov. Regni Veg. 8: 326 (1910).
C. & SW. Mexico. 79 MXC MXS. Hemicr. or rhizome geophyte.

Carex* ×*arthuriana Beckm. & Figert, Ber. Deutsch. Bot. Ges. 7: 30 (1889). *C. canescens* × *C. remota*. *Vignea* × *arthuriana* (Beckm. & Figert) Dostál, Folia Mus. Rerum Nat. Bohemiae Occid., Bot. 21: 16 (1984).
C. & E. Europe. 11 GER 14+. Hemicr. or rhizome geophyte.

Carex asperifructus Kük., Acta Horti Gothob. 5: 111 (1930).
Qinghai, China (Shanxi). 36 CHN CHQ. Hemicr. or rhizome geophyte.

Carex asraoi D.M.Verma, J. Econ. Taxon. Bot. 7: 606 (1985 publ. 1986).
Assam. 40 ASS. Hemicr. or rhizome geophyte.

Carex assiniboinensis Boott, Bot. Gaz. 9: 91 (1884).
C. & E. Canada, NC. U.S.A. to Michigan. 71 MAN SAS 72 ONT 74 IOW MIN NDA SDA WIS 75 MIC. Hemicr. or rhizome geophyte.
Carex assiniboinensis f. *ambulans* J.-P.Bernard, Naturaliste Canad. 86: 16 (1959).

Carex asynchrona Naczi, Sida 13: 487 (1989).
Mexico (Tamaulipas). 79 MXE. Hemicr. or rhizome geophyte.

Carex atherodes Spreng., Syst. Veg. 3: 828 (1826).
Subarctic and Temp. Northern Hemisphere. 10 FIN 11 GER POL 14 BLT RUC RUE RUN 30 WSB 31+ 32 KAZ 33 TCS 34 TUR 36 CHI CHM CHN CHX 37 MON 70 ASK NWT YUK 71 ABT BRC MAN SAS 72 ONT QUE 73 COL IDA MNT ORE WAS WYO 74 ILL IOW MIN MSO NDA NEB SDA WIS 75 INI MAI MAI MIC NWH NWY OHI PEN VER VER WVA 76 ARI CAL NEV UTA 77 NWM 78 VRG. Hemicr. or rhizome geophyte.
Carex aristata var. *cujavica* Asch. & Sprib. in ?. *Carex atherodes* var. *cujavica* (Asch. & Sprib.) Rauschert, Feddes Repert. 93: 16 (1982).
Carex aristata var. *lanceisquama* Hand.-Mazz. in ?. *Carex lanceisquama* (Hand.-Mazz.) V.I.Krecz., Bot. Mater. Gerb. Bot. Inst. Komarova Akad. Nauk S.S.S.R. 9: 187 (1946).
Carex aristata var. *vix-vaginans* Kük. in ?. *Carex atherodes* var. *vix-vaginans* (Kük.) A.E.Kozhevn., Sosud. Rast. Sovet. Dal'nego Vostoka 3: 342 (1988).
Carex aristata var. *kirschsteinii* Asch. & Graebn., Vossische Z.: 485 (15 Oct. 1899).
Carex aristata R.Br. in J.Richardson, Bot. App.: 751 (1823), nom. illeg. *Carex trichocarpa* var. *aristata* L.H.Bailey, Bot. Gaz. 10: 294 (1885).
Carex aristata subsp. *orostachys* C.A.Mey. in C.F.von Ledebour, Fl. Altaic. 4: 231 (1833). *Carex trichocarpa* var. *orostachys* (C.A.Mey.) Kük., Bot. Centralbl. 77: 92 (1899). *Carex aristata* var. *orostachys* (C.A.Mey.) C.B.Clarke, J. Linn. Soc., Bot. 36: 272 (1903).
Carex orthostachys C.A.Mey. in C.F.von Ledebour, Fl. Altaic. 4: 231 (1833). *Carex atherodes* var. *orthostachys* (C.A.Mey.) A.E.Kozhevn., Sosud. Rast. Sovet. Dal'nego Vostoka 3: 342 (1988).

Carex frankii Steud., Nomencl. Bot., ed. 2, 1: 290 (1840), nom. illeg.
Carex aristata var. *longilanceolata* Dewey, Amer. J. Sci. Arts 18: 102 (1854). *Carex atherodes* var. *longilanceolata* (Dewey) Gilly, Iowa State Coll. J. Sci. 21: 128 (1946).
Carex mirata Dewey, Amer. J. Sci. Arts, II, 39: 71 (1865).
Carex mirata var. *minor* Dewey, Amer. J. Sci. Arts, II, 39: 73 (1865).
Carex aristata var. *glabra* Uechtr., Verh. Bot. Vereins Prov. Brandenburg 8: 103 (1866). *Carex atherodes* f. *glabra* (Uechtr.) Lepage, Naturaliste Canad. 96: 769 (1969).
Carex aristata var. *imberbis* A.Gray, Manual, ed. 5: 597 (1867). *Carex atherodes* f. *imberbis* (A.Gray) B.Boivin, Phytologia 43: 99 (1979).
Carex siegertiana Uechtr. ex Garcke, Fl. N. Mitt.-Deutschland, ed. 8: 435 (1867). *Carex hirta* subsp. *siegertiana* (Uechtr. ex Garcke) Nyman, Consp. Fl. Eur.: 769 (1882). *Carex atherodes* var. *siegertiana* (Uechtr. ex Garcke) Rauschert, Feddes Repert. 93: 16 (1982).
Carex siegertiana var. *glabra* Uechtr. ex Garcke, Fl. N. Mitt.-Deutschland, ed. 8: 435 (1867). *Carex atherodes* var. *glabra* (Uechtr. ex Garcke) Rauschert, Feddes Repert. 93: 16 (1982).
Carex trichocarpa var. *imberbis* A.Gray, Manual, ed. 5: 597 (1867).
Carex trichocarpa var. *deweyi* L.H.Bailey, Bot. Gaz. 10: 293 (1885).
Carex trichocarpa var. *laeviconica* Hitchc., Trans. Acad. Sci. St. Louis 5: 524 (1891).
Carex fuscifructus C.B.Clarke in J.D.Hooker, Fl. Brit. India 6: 742 (1894).
Carex amurensis var. *mandschurica* Kük., Bot. Centralbl. 77: 95 (1899).
Carex aristata var. *kirschsteiniana* Asch., Graebn. & Kük., Allg. Bot. Z. Syst. 1899(11): 183 (1899).
Carex trichocarpa var. *maxima* Kük., Bot. Centralbl. 77: 93 (1899). *Carex aristata* var. *maxima* (Kük.) Kük. in H.G.A.Engler (ed.), Pflanzenr., IV, 20(38): 754 (1909). *Carex atherodes* var. *maxima* (Kük.) A.E.Kozhevn., Sosud. Rast. Sovet. Dal'nego Vostoka 3: 342 (1988).
Carex trichocarpa var. *turbinata* Dewey, Amer. J. Sci., IV, 11: 159 (1899).
Carex glaberrima Meinsh., Trudy Imp. S.-Peterburgsk. Bot. Sada 18: 369 (1900).
Carex aristata var. *hirtiformis* H.Lév. & Vaniot, Bull. Acad. Int. Géogr. Bot. 11: 68 (1902).
Carex aristata var. *subaristata* Kük. in H.G.A.Engler (ed.), Pflanzenr., IV, 20(38): 755 (1909).
Carex pergrandis V.I.Krecz. & Luchnik, Trudy Dal'nevost. Fil. Akad. Nauk S.S.S.R., Ser. Bot. 2: 894 (1937).
Carex orthostachys var. *spuria* Y.L.Chang & Y.L.Yang, in Fl. Plant. Herb. Chinae Bor.-Or. 11: 204 (1976).
Carex atherodes var. *kirschsteinii* (Asch., Graebn. & Kük.) Rauschert, Feddes Repert. 93: 16 (1982).

Carex athrostachya Olney, Proc. Amer. Acad. Arts 8: 393 (1872).
Alaska to W. & WC. U.S.A., Mexico (N. Baja California). 70 ASK YUK 71 ABT BRC MAN SAS 73 COL IDA MNT ORE WAS WYO 74 NDA SDA 76 ARI CAL NEV UTA 77 TEX 79 MXN. Hemicr. or rhizome geophyte.

Carex atlantica L.H.Bailey, Bull. Torrey Bot. Club 20: 425 (1893).
E. Canada to C. & E. U.S.A. 72 NSC ONT QUE 74 ILL MSO 75 CNT INI MAI MAS MIC NWH NWJ NWY OHI PEN RHO VER WVA 77 TEX 78 ALA ARK DEL FLA GEO KTY LOU MRY MSI NCA SCA TEN VRG WDC. Hemicr. or rhizome geophyte.

subsp. ***atlantica***
E. Canada to C. & E. U.S.A. 72 NSC 74 ILL MSO 75 CNT INI MAI MAS MIC NWH NWJ NWY OHI PEN RHO WVA 77 TEX 78 ALA ARK DEL FLA GEO KTY LOU MRY MSI NCA SCA TEN VRG WDC. Hemicr. or rhizome geophyte.
Carex stellulata var. *conferta* Chapm., Fl. South. U.S.: 534 (1860). *Carex echinata* var. *conferta* (Chapm.) L.H.Bailey, Carex Cat.: 2 (1884).
Carex incomperta E.P.Bicknell, Bull. Torrey Bot. Club 35: 494 (1908).
Carex atlantica var. *incomperta* (E.P.Bicknell) F.J.Herm., Rhodora 67: 362 (1965).

subsp. ***capillacea*** (L.H.Bailey) Reznicek, Contr. Univ. Michigan Herb. 14: 191 (1980).
E. Canada to E. U.S.A. 72 NSC ONT QUE 74 ILL 75 CNT INI MAI MAS MIC NWH NWJ NWY OHI PEN RHO VER 77 TEX 78 ALA ARK DEL FLA GEO KTY LOU MRY MSI NCA SCA TEN VRG WDC. Hemicr. or rhizome geophyte.
Carex howei var. *capillacea* (L.H.Bailey) Fernald in ?
**Carex interior* var. *capillacea* L.H.Bailey, Bull. Torrey Bot. Club 20: 426 (1893). *Carex scirpoides* var. *capillacea* (L.H.Bailey) Fernald, Rhodora 10: 47 (1908). *Carex stellulata* f. *capillacea* (L.H.Bailey) Kük. in H.G.A.Engler (ed.), Pflanzenr., IV, 20(38): 231 (1909). *Carex atlantica* var. *capillacea* (L.H.Bailey) Cronquist in H.A.Gleason & A.Cronquist, Man. Vasc. Pl. Northeast. U.S. & Adjacent Canada, ed. 2: 863 (1991).
Carex howei Mack., Bull. Torrey Bot. Club 37: 245 (1910).
Carex mohriana Mack. in N.L.Britton & al. (eds.), N. Amer. Fl. 18: 106 (1931).
Carex delicatula E.P.Bicknell, Bull. Torrey Bot. Club 35: 495 (Nov. 1908), nom. illeg.

Carex atlasica (H.Lindb.) Tattou, Bocconea 8: 198 (1998).
Morocco. 20 MOR. Hemicr. or rhizome geophyte.
**Carex leporina* subsp. *atlasica* H.Lindb., Acta Soc. Sci. Fenn., Ser. B, Opera Biol. 1(2): 29 (1932).

Carex atractodes F.J.Herm., J. Wash. Acad. Sci. 40: 283 (1950).
C. & S. Mexico. 79 MXC MXT. Hemicr. or rhizome geophyte.

Carex atrata L., Sp. Pl.: 976 (1753). *Trasus atratus* (L.) Gray, Nat. Arr. Brit. Pl. 2: 56 (1821). *Loxanisa atrata* (L.) Raf., Good Book: 26 (1840). *Carex atrata* var. *typica* Beck, Fl. Nieder-Österreich 1: 137 (1890), nom. inval.
Temp. Eurasia, Greenland. 10 FIN FOR GRB ICE NOR SWE 11 AUT CZE GER POL SWI 12 FRA SPA 13 BUL GRC ITA ROM YUG 14 BLR? RUC? RUE RUN UKR 30 ALT WSB 31 KHA SAK 32 KGZ TZK 33 NCS TCS 34 AFG IRN TUR 36 CHC CHM CHN CHT CHX 37 MON 38 JAP KOR TAI 40 EHM NEP 70 GNL. Hemicr. or rhizome geophyte.

subsp. ***apodostachya*** (Ohwi) T.Koyama, J. Fac. Sci. Univ. Tokyo, Sect. 3, Bot. 8: 149 (1962).
Taiwan. 38 TAI. Hemicr. or rhizome geophyte.
**Carex apodostachya* Ohwi, Jap. J. Bot. 7: 188 (1934).

subsp. ***aterrima*** (Hoppe) Èelak., Prodr. Fl. Böhmen: 67 (1867).
Mts. of Europe and Asia. 11 AUT CZE GER POL SWI 12 FRA SPA 13 BUL GRC ITA ROM YUG 14 RUE UKR 30 ALT WSB 31 KHA SAK 32 KAZ KGZ 33 NCS TCS 34 TUR 36 CHX 37 MON. Hemicr. or rhizome geophyte.
Carex nigra Schkuhr, Beschr. Riedgräs. 2: 38 (1806), nom. illeg.
**Carex aterrima* Hoppe, Denkschr. Königl.-Baier. Bot. Ges. Regensburg 1: 3 (1815). *Carex atrata* var. *aterrima* (Hoppe) Hartm., Sv. Norsk Exc.-Fl.: 131 (1846).
Carex perfusca V.I.Krecz. in V.L.Komarov (ed.), Fl. URSS 3: 600 (1935).

subsp. ***atrata***
Europe to Caucasus, Greenland, NE. China to C. Japan. 10 FIN FOR GRB ICE NOR SWE 11 AUT CZE GER POL SWI 12 FRA SPA 13 GRC ITA ROM YUG 14 BLR? RUC? RUN UKR 30 WSB 33 NCS TCS 34 TUR 36 CHM 38 JAP KOR 70 GNL. Hemicr. or rhizome geophyte.
Carex atrata var. *dubia* Gaudin, Etrennes Fl.: 178 (1804).
Carex atrata var. *varia* Gaudin, Etrennes Fl.: 178 (1804).
Carex dubia Gaudin, Etrennes Fl.: 178 (1804), nom. illeg.
Carex distachya Willd., Sp. Pl. 4: 220 (1805), nom. illeg. *Carex nigra* var. *distachya* Nyman, Consp. Fl. Eur.: 776 (1882).
Carex aethostachya Schkuhr, Beschr. Riedgräs. 2: 53 (1806).
Carex atrofusca Steven, Mém. Soc. Imp. Naturalistes Moscou 4: 67 (1813), nom. illeg.
Carex atrata var. *rectiuscula* Hartm., Handb. Skand. Fl.: 41 (1820).
Carex sudetica Opiz, Naturalientausch 11: 413 (1826). *Osculisa sudetica* (Opiz) Raf., Good Book: 27 (1840).
Carex aterrima W.D.J.Koch, Syn. Fl. Germ. Helv.: 874 (1837).
Carex cylindrica Miel. ex Kunth, Enum. Pl. 2: 484 (1837).
Carex atrata var. *brunnescens* Andersson, Cyper. Scand.: 37 (1849).
Carex castanea Miel., Flora 32: 665 (1849), nom. illeg. *Carex atrata* var. *castanea* Nyman, Consp. Fl. Eur.: 776 (1882). *Carex atrata* f. *castanea* (Nyman) K.Richt., Pl. Eur. 1: 157 (1890). *Carex capillaris* var. *castanea* Nyman, Consp. Fl. Eur.: 769 (1882).
Carex atrata var. *spadicea* Beurl., Bot. Not. 1853: 36 (1853).
Carex atrata var. *laxa* Neilr., Fl. Nied.-Oesterr.: 107 (1857).
Carex atrata var. *ovata* Boott, Ill. Carex 3: 114 (1862).
Carex atropurpurea Fisch. ex Trevir., Bull. Soc. Imp. Naturalistes Moscou 36(1): 538 (1863).
Carex atrata f. *gelida* Schur, Enum. Pl. Transsilv.: 708 (1866).
Carex atrata var. *bicolor* Èelak., Prodr. Fl. Böhmen: 67 (1867).
Carex tauricola Boiss., Fl. Orient. 5: 415 (1882).
Carex atrata var. *heterostachya* Almq., Öfvers. Kongl. Vetensk.-Akad. Förh. 41(7): 70 (1884).
Carex aterrima var. *wolfii* Kneuck., Bull. Murith. Soc. Valais. Sci. Nat. 19–20: 80 (1890). *Carex atrata* f. *wolfii* (Kneuck.) Kük. in H.G.A.Engler (ed.), Pflanzenr., IV, 20(38): 399 (1909).
Carex atrata var. *spiculosior* Norman, Forh. Vidensk.-Selsk. Kristiania 1893: 50 (1893).
Carex pseudoatrata Meinsh., Trudy Imp. S.-Peterburgsk. Bot. Sada 18: 346 (1900).
Carex pseudocilicica Hausskn., Mitth. Thüring. Bot. Vereins, n.s., 15: 6 (1900), nom. inval. *Carex atrata* var. *pseudocilicica* Kük. in H.G.A.Engler (ed.), Pflanzenr., IV, 20(38): 400 (1909).
Carex atrata f. *decolorans* Neuman, Sver. Fl.: 698 (1901).
Carex atrata var. *longistolonifera* Kük. in H.G.A.Engler (ed.), Pflanzenr., IV, 20(38): 400 (1909). *Carex atrata* subsp. *longistolonifera* (Kük.) S.Yun Liang, in Fl. Reipubl. Popul. Sin. 12: 114 (2000).
Carex chalciolepis var. *larimerana* Kelso, Biol. Leafl. 30: 2 (1945). *Carex atrata* var. *larimerana* (Kelso) Kelso, Biol. Leafl. 66: 32 (1953).
Carex atrata var. *japonoalpina* T.Koyama, J. Jap. Bot. 30: 313 (1955). *Carex japonoalpina* (T.Koyama) T.Koyama, Acta Phytotax. Geobot. 16: 154 (1956). *Carex perfusca* var. *japonoalpina* (T.Koyama) Kitag., Neo-Lineam. Fl. Manshur.: 134 (1979).

subsp. ***caucasica*** (Steven) Kük. in H.G.A.Engler (ed.), Pflanzenr., IV, 20(38): 400 (1909).
Russia (Ural'skiy Chrebet), Caucasus, W. & C. Asia, Taiwan. 14 RUE RUN 30 WSB 32 KAZ KGZ TZK 33 NCS TCS 34 AFG IRN TUR. Hemicr. or rhizome geophyte.
**Carex caucasica* Steven, Mém. Soc. Imp. Naturalistes Moscou 4: 68 (1813). *Carex atrata* var. *caucasica* (Steven) Boott, Ill. Carex 3: 114 (1862).
Carex urbis-malorum Popov, Bull. Soc. Imp. Naturalistes Moscou, n.s., 44: 131 (1935).
Carex karacolica Polozhij, Sist. Zametki Mater. Gerb. Krylova Tomsk. Gosud. Univ. Kuybysheva 79–80: 11 (1956).

subsp. ***pullata*** (Boott) Kük. in H.G.A.Engler (ed.), Pflanzenr., IV, 20(38): 400 (1909).
C. & E. Himalaya to C. China. 36 CHC CHN CHT 38 JAP? 40 EHM NEP. Hemicr. or rhizome geophyte.
Carex atrata var. *glacialis* Boott, Ill. Carex 3: 114 (1862).
**Carex atrata* var. *pullata* Boott, Ill. Carex 3: 114 (1862).
Carex duthiei C.B.Clarke in J.D.Hooker, Fl. Brit. India 6: 831 (1894).
Carex atrata var. *sinensis* Kük. in H.G.A.Engler (ed.), Pflanzenr., IV, 20(38): 400 (1909).

Carex atratiformis Britton, Bull. Torrey Bot. Club 22: 222 (1895). *Carex atrata* subsp. *atratiformis* (Britton) Kük. in H.G.A.Engler (ed.), Pflanzenr., IV, 20(38): 399 (1909).
Subarctic America to NE. U.S.A. 70 ASK NUN NWT YUK 71 ABT BRC MAN SAS 72 LAB NBR NFL

NSC ONT QUE 75 MAI MIC NWH NWY VER. Hemicr. or rhizome geophyte.
Carex ovata Rudge, Trans. Linn. Soc. London 7: 96 (1804), nom. illeg.
Carex raymondii Calder, Rhodora 54: 246 (1952). *Carex atratiformis* subsp. *raymondii* (Calder) Scoggan, Fl. Canada 1: 51 (1978).

Carex atrivaginata Nelmes, Mém. Mus. Natl. Hist. Nat., Sér. B, Bot. 4: 106 (1955).
Vietnam. 41 VIE. Hemicr. or rhizome geophyte.

Carex atrofusca Schkuhr, Beschr. Riedgräs. 1: 106 (1801). *Proteocarpus atrofuscus* (Schkuhr) Fedde & J.Schust., Just's Bot. Jahresber. 41(2): 11 (1913 publ. 1918).
Subarctic to Temp. Northern Hemisphere. 10 FIN GRB NOR SWE 11 AUT CZE SWI 12 FRA 13 BLR? ITA ROM RUN YUG 30 ALT KRA WSB YAK 31 KHA MAG 32 KAZ KGZ TZK UZB 34 AFG 36 CHC CHI CHM CHN CHQ CHT CHX 37 MON 40 EHM NEP WHM 70 ASK GNL NUN NWT YUK 71 MAN 72 LAB NFL ONT QUE. Hemicr. or rhizome geophyte.

subsp. ***atrofusca***
Subarctic to Temp. Eurasia. 10 FIN GRB NOR SWE 11 AUT CZE SWI 12 FRA 13 BLR? ITA ROM RUN YUG 30 ALT KRA WSB YAK 31 KHA MAG 32 KGZ TZK UZB 34 AFG 36 CHC CHT 37 MON 40 WHM 70 ASK GNL NUN NWT YUK 71 MAN 72 LAB NFL ONT QUE. Hemicr. or rhizome geophyte.
Carex ustulata Wahlenb., Kongl. Vetensk. Acad. Nya Handl. 1803: 156 (1803). *Trasus ustulatus* (Wahlenb.) Gray, Nat. Arr. Brit. Pl. 2: 66 (1821).
Carex maculata Turcz. ex Ledeb., Fl. Ross. 4: 295 (1852), nom. illeg.
Carex ustulata var. *pallisa* Meinsh., Trudy Imp. S.-Peterburgsk. Bot. Sada 18: 357 (1900).
Carex atrofusca f. *flavescens* Kük. in H.G.A.Engler (ed.), Pflanzenr., IV, 20(38): 554 (1909).
Carex stilbophaea V.I.Krecz. in V.L.Komarov (ed.), Fl. URSS 3: 605 (1935).
Carex atrofusca var. *decolorata* A.E.Porsild, Sargentia 4: 20 (1943). *Carex atrofusca* f. *decolorata* (A.E.Porsild) B.Boivin, Naturaliste Canad. 75: 208 (1948).
Carex atrofusca var. *nortoniana* B.Boivin, Naturaliste Canad. 75: 208 (1948).
Carex sisukensis Akiyama, Caric. Far East. Reg. As.: 125 (1955).
Carex taskanensis A.P.Khokhr., Byull. Moskovsk. Obshch. Isp. Prir., Otd. Biol. 88(5): 95 (1983).

var. ***major*** (Boeck.) Raymond, Canad. Field-Naturalist 66: 98 (1952).
Siberia to W. North America. 30 WSB 31+ 32+ 36 CHI CHM CHN CHQ 37 MON 70 ASK GNL NUN NWT YUK 71 MAN. Hemicr. or rhizome geophyte.
Carex coriophora Fisch. & C.A.Mey. ex Kunth, Enum. Pl. 2: 463 (1837). *Carex atrofusca* var. *coriophora* (Fisch. & C.A.Mey. ex Kunth) Kük. in H.G.A.Engler (ed.), Pflanzenr., IV, 20(38): 554 (1909).
Carex maculata Turcz. ex Steud., Nomencl. Bot., ed. 2, 1: 292 (1840), nom. inval.
**Carex ustulata* var. *major* Boeck., Linnaea 41: 260 (1877).

subsp. ***minor*** (Boott) T.Koyama, in Fl. E. Himal., 3rd. Rep.: 122 (1975).
C. Asia to Himalaya. 32 KAZ KGZ TZK UZB 36 CHC CHN CHQ CHT CHX 40 EHM NEP. Hemicr. or rhizome geophyte.
**Carex ustulata* var. *minor* Boott, Ill. Carex 1: 71 (1858). *Carex atrofusca* var. *minor* (Boott) Kük. in H.G.A.Engler (ed.), Pflanzenr., IV, 20(38): 554 (1909).
Carex oxyleuca V.I.Krecz. in V.L.Komarov (ed.), Fl. URSS 3: 605 (1935).

Carex atrofuscoides K.T.Fu, in Fl. Tsinlingensis 1(1): 446 (1976).
Tibet to C. China. 36 CHC CHN CHQ CHT. Hemicr. or rhizome geophyte.

Carex atropicta Steud., Syn. Pl. Glumac. 2: 204 (1855). *Limivasculum atropictum* (Steud.) Fedde & J.Schust., Just's Bot. Jahresber. 41(2): 11 (1913 publ. 1918).
C. & S. Chile, Argentina, Falkland Is. 85 AGE AGS CLC CLS 90 FAL. Hemicr. or rhizome geophyte.
Carex lechleri Phil., Linnaea 29: 83 (1857), nom. illeg.
Carex fuegina Phil., Anales Univ. Chile 93: 491 (1896).
Carex peraltae Phil., Anales Univ. Chile 93: 490 (1896).
Carex atropicta var. *platycarpa* Phil. ex Kük., Bot. Jahrb. Syst. 27: 520 (1899).

Carex atrosquama Mack., Proc. Biol. Soc. Wash. 25: 51 (1912). *Carex atrata* subsp. *atrosquama* (Mack.) Hultén, Acta Univ. Lund., n.s., 38: 354 (1942). *Carex atrata* var. *atrosquama* (Mack.) Cronquist, Vasc. Pl. Pacific Northw. 1: 243 (1969).
Alaska to WC. U.S.A. 70 ASK NWT YUK 71 ABT BRC 73 COL IDA MNT ORE WAS 76 UTA. Hemicr. or rhizome geophyte.
Carex apoda Clokey, Amer. J. Sci., V, 3: 88 (1922).
Carex viridior Mack. in L.Abrams, Ill. Fl. Pacific States 1: 331 (1923).

Carex atroviridis Ohwi, Mem. Coll. Sci. Kyoto Imp. Univ., Ser. B, Biol. 6: 241 (1931).
Japan (Yakushima). 38 JAP. Hemicr. or rhizome geophyte.
Carex yakusimensis Masam., J. Soc. Trop. Agric. 4: 77 (1932).

Carex aueri Kalela, Ann. Acad. Sci. Fenn., Ser. A 54(5): 60 (1940).
Chile. 85+. Hemicr. or rhizome geophyte.

Carex augustini Tuyama, Bot. Mag. (Tokyo) 52: 463 (1938).
Kazan-retto (Minami-iwo-jima). 38 KZN. Hemicr. or rhizome geophyte.

Carex augustinowiczii Meinsh., Trudy Imp. S.-Peterburgsk. Bot. Sada 12: 411 (1892).
NE. China, Korea, Sakhalin, N. & C. Japan. 31 SAK 36 CHM CHN 38 JAP KOR. Hemicr. or rhizome geophyte.

var. ***augustinowiczii***
NE. China, Korea, Sakhalin, N. & C. Japan. 31 SAK 36 CHM CHN 38 JAP KOR. Hemicr. or rhizome geophyte.
Carex eleusinoides var. *flaccidior* F.Schmidt, Reis. Amur-Land., Bot.: 196 (1868). *Carex flaccidior* (F.Schmidt) Miyabe & Kudô, Trans. Sapporo Nat. Hist. Soc. 7: 28 (1918).
Carex bidentula Franch., Bull. Soc. Philom. Paris, VIII, 7: 41 (1895).

Carex soyaeensis Kük., Bull. Herb. Boissier, II, 4: 53 (1904). *Carex augustinowiczii* subsp. *soyaeensis* (Kük.) T.V.Egorova, Bot. Zhurn. (Moscow & Leningrad) 75: 863 (1990).
Carex curtiglumis C.B.Clarke, Kew Bull., Addit. Ser. 8: 77 (1908).
Carex infirma C.B.Clarke, Kew Bull., Addit. Ser. 8: 78 (1908).

var. ***sharensis*** (Franch.) Ohwi, Mem. Coll. Sci. Kyoto Imp. Univ., Ser. B, Biol. 11: 313 (1936).
Japan (Hokkaido). 38 JAP. Hemicr. or rhizome geophyte.
**Carex sharensis* Franch., Bull. Soc. Philom. Paris, VIII, 7: 90 (1895).

Carex aurea Nutt., Gen. N. Amer. Pl. 2: 205 (1818). *Neskiza aurea* (Nutt.) Raf., Good Book: 27 (1840).
Subarctic and W. North America to N. & C. U.S.A. 70 ASK NUN NWT YUK 71 ABT BRC MAN SAS 72 NBR NFL NSC ONT PEI QUE 73 COL IDA MNT ORE WAS WYO 74 MIN NDA NEB SDA WIS 75 CNT INI MAI MAS MIC NWH NWY OHI RHO VER 76 CAL NEV UTA 77 NWM TEX. Hemicr. or rhizome geophyte.
Carex mutica R.Br. in J.Richardson, Bot. App.: 751 (1823).
Carex pyriformis F.W.Schultz, Pollichia 15: 122 (1855).
Carex aurea f. *colorata* Farw., Rep. (Annual) Michigan Acad. Sci. 21: 362 (1920).

Carex aureolensis Steud., Syn. Pl. Glumac. 2: 223 (1855).
C. & SE. U.S.A. to NE. Mexico, Brazil to NE. Argentina. 74 ILL KAN MSO NEB OKL 77 NWM TEX 78 ALA ARK FLA GEO KTY LOU MSI NCA SCA TEN VRG 79 MXE 84 BZS 85 AGE PAR URU. Hemicr. or rhizome geophyte.
Carex involucrata Boeck., Flora 38: 594 (1855).
Carex atrovirens Boeck., Beitr. Cyper. 1: 52 (1888).

Carex austrina Mack., Bull. Torrey Bot. Club 34: 151 (1907). *Carex muehlenbergii* var. *australis* Olney in E.Hall, Pl. Tex.: 25 (1873). *Carex muehlenbergii* var. *austrina* (Mack.) Small, Fl. S.E. U.S.: 218 (1903), nom. illeg.
C. & SE. U.S.A. 74 IOW KAN MSO NEB OKL 77 TEX 78 ALA ARK KTY LOU MSI NCA TEN. Hemicr. or rhizome geophyte.
Carex varia var. *pedicellata* Dewey, Amer. J. Sci. Arts 11: 163 (1826). *Carex pedicellata* (Dewey) Britton, Mem. Torrey Bot. Club 5: 87 (1894).
Carex communis var. *wheeleri* L.H.Bailey, Mem. Torrey Bot. Club 1: 41 (1889). *Carex pedicellata* var. *wheeleri* (L.H.Bailey) Britton, Mem. Torrey Bot. Club 5: 88 (1894).

Carex austro-occidentalis F.T.Wang & Tang ex Y.C.Tang, in Fl. Reipubl. Popul. Sin. 12: 527 (2000).
China (Sichuan). 36 CHC. Hemicr.

Carex austroafricana (Kük.) Raymond, Naturaliste Canad. 91: 126 (1964).
S. Africa. 27 CPP LES NAT OFS SWZ TVL. Hemicr. or rhizome geophyte.
**Carex cernua* var. *austroafricana* Kük. in H.G.A.Engler (ed.), Pflanzenr., IV, 20(38): 354 (1909).

Carex austroamericana G.A.Wheeler, Brittonia 38: 319 (1986).
S. Chile to S. Argentina. 85 AGS CLS. Hemicr. or rhizome geophyte.

Carex austrocaroliniana L.H.Bailey, Bull. Torrey Bot. Club 20: 428 (1893).
SE. U.S.A. 78 ALA GEO KTY NCA SCA TEN. Hemicr. or rhizome geophyte.
**Carex caroliniana* Buckley, Amer. J. Sci. Arts 45: 178 (1843), nom. illeg.

Carex austrokoreensis Ohwi, Acta Phytotax. Geobot. 5: 54 (1936).
S. Korea. 38 KOR. Hemicr. or rhizome geophyte.

Carex austromexicana Reznicek, Syst. Bot. 7: 340 (1982).
Mexico (Oaxaca). 79 MXS. Hemicr. or rhizome geophyte.

Carex austrosinensis Tang & F.T.Wang ex S.Y.Liang, Acta Phytotax. Sin. 33: 488 (1995).
China (Guangdong). 36 CHS. Hemicr. or rhizome geophyte.

Carex austroussuriensis A.E.Kozhevn., Byull. Moskovsk. Obshch. Isp. Prir., Otd. Biol. 92: 135 (1987).
S. Russian Far East. 31 PRM. Hemicr. or rhizome geophyte.

Carex austrozhejiangensis C.Z.Zheng & X.F.Jin, Acta Phytotax. Sin. 42: 546 (2004).
China (Zhejiang). 36 CHS. Hemicr. or rhizome geophyte.

Carex autumnalis Ohwi, Mem. Coll. Sci. Kyoto Imp. Univ., Ser. B, Biol. 5(3): 251 (1930).
E. China, Japan (Honshu, Shikoku), Nansei-shoto. 36 CHS 38 JAP NNS. Hemicr. or rhizome geophyte.

Carex aztecica Mack. in N.L.Britton & al. (eds.), N. Amer. Fl. 18: 229 (1935).
C. Mexico to Guatemala. 79 MXC MXS MXT 80 GUA. Hemicr. or rhizome geophyte.
**Carex fuscotincta* Mack., Bull. Torrey Bot. Club 36: 478 (1909), nom. illeg.

Carex azuayae Steyerm., Phytologia 9: 337 (1963).
Ecuador (Azuay). 83 ECU. Hemicr. or rhizome geophyte.

Carex baccans Nees in R.Wight, Contr. Bot. India: 122 (1834). *Carex baccans* f. *normalis* Kuntze, Revis. Gen. Pl. 2: 747 (1891), nom. inval.
S. China, Taiwan, Indo-China, Malesia. 36 CHC CHH CHS 38 TAI 40 ASS EHM IND NEP SRL 41 THA VIE 42 JAW MLY PHI SUL SUM 43 NWG. Hemicr. or rhizome geophyte.
Carex curvirostris Kunze in C.Schkuhr, Beschr. Riedgräs., Suppl.: 79 (1842).
Carex recurvirostra Kunze in C.Schkuhr, Beschr. Riedgräs., Suppl.: 79 (1842).
Carex javanica Boeck., Beitr. Cyper. 1: 43 (1888).
Carex baccans f. *aurantiaca* Kuntze, Revis. Gen. Pl. 2: 747 (1891).
Carex baccans f. *nigra* Kuntze, Revis. Gen. Pl. 2: 747 (1891).
Carex baccans f. *recurvirostris* Kuntze, Revis. Gen. Pl. 2: 747 (1891).
Carex baccans var. *siccifructus* C.B.Clarke in J.D.Hooker, Fl. Brit. India 6: 723 (1894).

Carex backii Boott in W.J.Hooker, Fl. Bor.-Amer. 2: 210 (1839).
Canada to N. U.S.A. 71 ABT BRC MAN SAS 72 NBR ONT QUE 73 COL MNT WYO 74 MIN NDA SDA WIS 75 MAI MIC NWH NWJ NWY

PEN VER. Hemicr. or rhizome geophyte.
Carex durifolia L.H.Bailey, Bull. Torrey Bot. Club 20: 428 (1893).

Carex baileyi Britton, Bull. Torrey Bot. Club 22: 220 (1895).
SE. Canada to E. U.S.A. 72 QUE 75 MAI MAS NWH NWY PEN VER WVA 78 KTY MRY NCA TEN VRG. Hemicr. or rhizome geophyte.

Carex baimaensis S.W.Su, Bull. Bot. Res., Harbin 10(4): 6 (1990).
China (Anhui). 36 CHS. Hemicr. or rhizome geophyte.

Carex baiposhanensis P.C.Li, Acta Bot. Yunnan. 12: 145 (1990).
China (Sichuan). 36 CHC. Hemicr. or rhizome geophyte.

Carex ×bakkerana D.T.E.Ploeg & Rudolphy, Gorteria 10: 178 (1981). *C. acutiformis* × *C. rostrata*.
WC. Europe. 11 GER NET. Hemicr. or rhizome geophyte.

Carex balansae Franch., Nouv. Arch. Mus. Hist. Nat., III, 8: 258 (1896).
Vietnam. 41 VIE. Hemicr. or rhizome geophyte.

Carex baldensis L., Cent. Pl. II: 32 (1756). *Chionanthula baldensis* (L.) Fedde & J.Schust., Just's Bot. Jahresber. 41(2): 9 (1913 publ. 1918).
Alps. 11 AUT GER SWI 12 FRA 13 ITA. Hemicr. or rhizome geophyte.

Carex balfourii Kük. in H.G.A.Engler (ed.), Pflanzenr., IV, 20(38): 129 (1909).
Réunion. 29 REU. Hemicr. or rhizome geophyte.

Carex ballsii Nelmes, Bull. Misc. Inform. Kew 1940: 134 (1940).
Mexico (Veracruz). 79 MXG. Hemicr. or rhizome geophyte.

Carex baltzellii Chapm., Amer. J. Sci. Arts, II, 3(9): 355 (1847).
SE. U.S.A. 78 ALA FLA GEO MSI. Hemicr. or rhizome geophyte.
Carex chapmanii Torr. ex Boott, Ill. Carex 1: 16 (1858), nom. illeg.

Carex bambusetorum Merr., Philipp. J. Sci., C 13: 132 (1918).
China (Guangdong). 36 CHS. Hemicr. or rhizome geophyte.

Carex banksii Boott, Trans. Linn. Soc. London 20: 119 (1846).
Juan Fernández Is., Chile, Argentina, Falkland Is. 85 AGS AGW CLC CLS JNF 90 FAL. Hemicr. or rhizome geophyte.

var. ***abbreviata*** Kük., Bot. Jahrb. Syst. 27: 522 (1899).
Argentina (Neuquén). 85 AGS. Hemicr. or rhizome geophyte.

var. ***banksii***
Juan Fernández Is., Chile, Argentina, Falkland Is. 85 AGS AGW CLC CLS JNF 90 FAL. Hemicr. or rhizome geophyte.
Carex atrata Banks ex Boott, Proc. Linn. Soc. London 1: 257 (1845), nom. illeg.
Carex magellanica Banks ex Boott, Proc. Linn. Soc. London 1: 257 (1845), nom. illeg.
Carex germana Boott in J.D.Hooker, Fl. Antarct.: 466 (1847).
Carex pterygocarpa Steud., Syn. Pl. Glumac. 2: 222 (1855).
Carex leucocarpa Phil., Linnaea 29: 82 (1857).
Carex gilliesii Phil., Anales Univ. Chile 1873: 556 (1873). *Carex banksii* var. *cilliesii* (Phil.) Kük. in H.G.A.Engler (ed.), Pflanzenr., IV, 20(38): 555 (1909).
Carex dessaueri Phil., Anales Univ. Chile 93: 496 (1896).
Carex ochrostachya Phil., Anales Univ. Chile 93: 500 (1896). *Carex banksii* f. *ochrostachys* (Phil.) Kük. in H.G.A.Engler (ed.), Pflanzenr., IV, 20(38): 555 (1909).
Carex promaucana Phil., Anales Univ. Chile 93: 496 (1896).
Carex trifida var. *franchetii* Kurtz, Revista Fac. Agron. Univ. Nac. La Plata 7: 192 (1896).
Carex banksii f. *brevifolia* Kük., Bot. Jahrb. Syst. 27: 522 (1899).
Carex banksii var. *distans* Kük., Bot. Jahrb. Syst. 27: 521 (1899).
Carex banksii f. *pallescens* Kük., Bot. Jahrb. Syst. 27: 522 (1899).

var. ***fonkii*** (Phil.) Kük., Bot. Jahrb. Syst. 27: 522 (1899).
S. Chile to S. Argentina. 85 AGS CLS. Hemicr. or rhizome geophyte.
**Carex fonkii* Phil., Linnaea 29: 83 (1857).

var. ***odontolepis*** (Phil.) Kük. in H.G.A.Engler (ed.), Pflanzenr., IV, 20(38): 554 (1909).
S. Chile to S. Argentina. 85 AGS CLS. Hemicr. or rhizome geophyte.
**Carex odontolepis* Phil., Linnaea 29: 82 (1857).

Carex baohuashanica Tang & F.T.Wang ex L.K.Dai, Acta Phytotax. Sin. 32: 182 (1994).
SE. China. 36 CHS. Hemicr. or rhizome geophyte.

Carex barbarae Dewey in W.H.Emory, Rep. U.S. Mex. Bound. 2(1): 231 (1858).
W. U.S.A. 73 ORE 76 CAL. Hemicr. or rhizome geophyte.
Carex laciniata Boott, Ill. Carex 4: 175 (1867).
Carex lacunarum T.Holm, Amer. J. Sci. 166: 458 (1903). *Carex magnifica* var. *lacunarum* (T.Holm) Kük. in H.G.A.Engler (ed.), Pflanzenr., IV, 20(38): 366 (1909).

Carex barbata Boott, Ill. Carex 1: 68 (1858). *Carex gunniana* var. *barbata* (Boott) Kük. in H.G.A.Engler (ed.), Pflanzenr., IV, 20(38): 663 (1909).
Tasmania. 50 TAS. Hemicr. or rhizome geophyte.

Carex baronii Baker, J. Linn. Soc., Bot. 21: 451 (1885).
Madagascar. 29 MDG. Hemicr. or rhizome geophyte.

Carex barrattii Torr. ex Schwein., Ann. Lyceum Nat. Hist. New York 1: 361 (1824). *Olamblis barrattii* (Torr. ex Schwein.) Raf., Good Book: 26 (1840) (1840).
E. U.S.A. 75 CNT NWJ NWY PEN 78 ALA DEL GEO MRY NCA SCA TEN VRG. Hemicr. or rhizome geophyte.
Carex littoralis Schwein., Ann. Lyceum Nat. Hist. New York 1: 70 (1824).
Carex variegata Scheele, Linnaea 23: 565 (1850), nom. illeg.
Carex jacca Carey ex Boott, Ill. Carex 1: 69 (1858).

Carex barrosii Nelmes, Kew Bull. 10: 87 (1955).
Chile, C. & S. Argentina, Falkland Is. 85 AGE AGS CLS 90 FAL. Hemicr. or rhizome geophyte.

Carex basiantha Steud., Syn. Pl. Glumac. 2: 232 (1855).
C. & SE. U.S.A. 74 OKL 77 TEX 78 ALA ARK FLA GEO LOU MSI NCA SCA TEN. Hemicr. or rhizome geophyte.
Carex willdenowii var. *pauciflora* Olney ex L.H.Bailey, Contr. U. S. Natl. Herb. 2: 482 (1894).

Carex bathiei H.Lév., Bull. Acad. Int. Géogr. Bot. 27: 5 (1917).
C. Madagascar. 29 MDG. Hemicr. or rhizome geophyte.

Carex bavicola Raymond, Mém. Jard. Bot. Montréal 53: 28 (1959).
Vietnam. 41 VIE. Hemicr. or rhizome geophyte.

Carex bebbii Olney ex Britton, Mem. Torrey Bot. Club 1: 55 (1889). *Carex tribuloides* var. *bebbii* (Olney ex Britton) L.H.Bailey, Mem. Torrey Bot. Club 1: 55 (1889).
Subarctic America to N. U.S.A. 70 ASK NWT 71 ABT BRC MAN SAS 72 LAB NBR NFL NSC ONT PEI QUE 73 COL IDA MNT ORE WAS WYO 74 ILL IOW MIN NDA NEB SDA WIS 75 CNT INI MAI MAS MIC NWH NWJ NWY OHI PEN VER 76 UTA. Hemicr. or rhizome geophyte.

Carex beckii G.A.Wheeler, Darwiniana 40: 201 (2002).
Bolivia. 83 BOL. Hemicr. or rhizome geophyte.

Carex* ×*beckmanniana Figert, Deutsche Bot. Monatsschr. 7: 185 (1889). *C. riparia* × *C. rostrata*.
Europe. 11 GER. Hemicr. or rhizome geophyte.

Carex bella L.H.Bailey, Bot. Gaz. 17: 152 (1892).
WC. U.S.A. to Mexico (Nuevo León). 73 COL WYO 74 SDA 76 ARI CAL UTA 77 NWM 79 MXE. Hemicr. or rhizome geophyte.
**Carex atrata* var. *discolor* L.H.Bailey, J. Bot. 26: 321 (1888).
Carex uncompahgre Kelso, Biol. Leafl. 38: 1 (1947).

Carex bequaertii De Wild., Pl. Bequaert. 4: 246 (1927).
NE. Zaïre to E. Trop. Africa. 23 RWA ZAI 24 ETH 25 KEN TAN UGA. Hemicr.

var. ***bequaertii***
NE. Zaïre to E. Trop. Africa. 23 RWA ZAI 24 ETH 25 KEN TAN UGA. Hemicr.

var. ***maxima*** Lye, Nordic J. Bot. 3: 244 (1983).
Uganda (Mt. Morungole). 25 UGA. Hemicr.

Carex berggrenii Petrie, Trans. & Proc. New Zealand Inst. 18: 297 (1885 publ. 1886).
New Zealand. 51 NZN NZS. Hemicr. or rhizome geophyte.

Carex bermudiana Hemsl., J. Bot. 21: 260 (1883).
Bermuda. 81 BER. Hemicr. or rhizome geophyte.

Carex berteroniana Steud., Flora 25: 604 (1842).
Juan Fernández Is. (Alejandro Selkirk I.). 85 JNF. Hemicr. or rhizome geophyte.
Carex paleata Boott, Trans. Linn. Soc. London 20: 127 (1846).

Carex bichenoviana Boott in J.D.Hooker, Fl. Tasman. 2: 101 (1858). *Carex pumila* var. *bichenoviana* (Boott) Kük. in H.G.A.Engler (ed.), Pflanzenr., IV, 20(38): 740 (1909).
SE. Australia. 50 NSW SOA TAS VIC (51) nzn. Hemicr. or rhizome geophyte.

Carex bicknellii Britton & A.Br., Ill. Fl. N. U.S. 1: 360 (1896).
S. Canada, NC. & NE. U.S.A. to South Carolina. 71 MAN 72 ONT 74 ILL IOW KAN MIN MSO NDA NEB SDA WIS 75 CNT INI MAI MAS MIC NWJ NWY OHI PEN RHO 78 SCA. Hemicr. or rhizome geophyte.
Carex straminea var. *crawei* Boott, Ill. Carex 3: 21 (1862). *Carex brevior* var. *crawei* (Boott) B.Boivin, Provancheria 25: 127 (1992).

Carex bicolor Bellardi ex All., Fl. Pedem. 2: 267 (1785). *Vignea bicolor* (Bellardi ex All.) Rchb. in J.C.Mössler, Handb. Gewächsk. ed. 2, 3: 1624 (1830).
Europe, N. Asia, Subarctic America to Canada. 10 ICE NOR SWE 11 AUT SWI 12 FRA 13 ITA ROM YUG 14 RUE RUN UKR 30 IRK KRA WSB YAK 31 KAM 70 ASK GNL NUN NWT YUK 71 BRC MAN SAS 72 NFL ONT QUE. Hemicr. or rhizome geophyte.
Carex androgyna Balb., Elenco: 97 (1801).
Carex cenisia Balb., Arch. Bot. (Leipzig) 3: 187 (1803).
Carex pusilla Pers., Syn. Pl. 2: 536 (1807).
Carex minima Boullu, Ann. Soc. Bot. Lyon 5: 88 (1878). *Carex bicolor* var. *minima* (Boullu) Nyman, Consp. Fl. Eur., Suppl. 2: 324 (1890).

Carex biegensis Cherm., Bull. Soc. Bot. France 82: 343 (1935).
Zaïre. 23 ZAI. Hemicr. or rhizome geophyte.

Carex bigelowii Torr. ex Schwein., Ann. Lyceum Nat. Hist. New York 1: 67 (1824). *Carex saxatilis* var. *bigelowii* (Torr. ex Schwein.) Torr., Ann. Lyceum Nat. Hist. New York 2: 297 (1836). *Carex rigida* var. *bigelowii* (Torr. ex Schwein.) Tuck., Enum. Meth. Caric.: 19 (1843).
Subarctic to N. U.S.A., Himalaya. 10 FIN FOR GRB ICE IRE NOR SVA SWE 11 AUT CZE GER POL 14 RUE RUN 30 ALT BRY IRK KRA TVA WSB YAK 31 KAM MAG 36 CHN CHX 37 MON 38 JAP KOR 40 NEP 70 ASK GNL NUN NWT YUK 71 BRC MAN SAS 72 LAB NBR NFL ONT QUE 75 MAI NWH NWY VER. Hemicr. or rhizome geophyte.

subsp. ***arctisibirica*** (Jurtzev) Á.Löve & D.Löve, Univ. Colorado Stud. Ser. Biol. 24: 22 (1966).
Subarctic Eurasia. 11 NOR SVA 14 RUN 30 KRA WSB YAK. Hemicr. or rhizome geophyte.
**Carex ensifolia* subsp. *arctisibirica* Jurtzev, Novosti Sist. Vyssh. Rast. 2: 308 (1965). *Carex arctisibirica* (Jurtzev) Czern., Sosud. Rast. SSSR: 194 (1981). *Carex bigelowii* var. *arctisibirica* (Jurtzev) Malyschev, in Fl. Sibir., Cyperac. 3: 163 (1990).

subsp. ***bigelowii***
Subarctic to N. U.S.A., Himalaya. 10 FIN FOR GRB ICE IRE NOR SWE 11 AUT CZE GER POL 14 RUN 31 KAM 38 JAP KOR 40 NEP 70 ASK GNL NUN NWT 71 MAN SAS 72 LAB NBR NFL ONT QUE 75 MAI NWH NWY VER. Hemicr. or rhizome geophyte.
Carex rigida Gooden., Trans. Linn. Soc. London 2: 193 (1794), nom. illeg. *Onkerma rigida* Raf., Good Book: 27 (1840). *Vignantha rigida* (Raf.) Schur, Enum. Pl. Transsilv.: 705 (1866). *Carex cespitosa* var. *rigida* (Raf.) Benth. & Hook.f., Handb. Brit. Fl., ed. 6: 496 (1896). *Carex bigelowii* subsp. *rigida* (Raf.) W.Schultze-Motel, Willdenowia 4: 326 (1968).
Carex washingtoniana Dewey, Amer. J. Sci. Arts 10: 272 (1826).

Neskiza saxatilis Raf., Good Book: 27 (1840).
Carex hyperborea Drejer, Rev. Car. Bor.: 43 (1841).
Carex salina var. *hyperborea* (Drejer) Nyman, Consp. Fl. Eur.: 776 (1882).
Carex saxatilis f. *infuscata* Drejer, Rev. Car. Bor.: 43 (1841). *Carex rigida* f. *infuscata* (Drejer) Kük. in H.G.A.Engler (ed.), Pflanzenr., IV, 20(38): 301 (1909).
Carex saxatilis f. *luteosa* Drejer, Rev. Car. Bor.: 43 (1841). *Carex rigida* f. *luteosa* (Drejer) Kük. in H.G.A.Engler (ed.), Pflanzenr., IV, 20(38): 301 (1909).
Carex saxatilis f. *pudica* Drejer, Rev. Car. Bor.: 43 (1841). *Carex rigida* f. *pudica* (Drejer) Kük. in H.G.A.Engler (ed.), Pflanzenr., IV, 20(38): 301 (1909).
Carex rigida f. *glacialis* Fr., Novit. Fl. Suec. Mant. 3: 148 (1845).
Carex rigida var. *saxatilis* Fr., Novit. Fl. Suec. Mant. 3: 148 (1845).
Carex hyperborea var. *latifolia* Andersson, Cyper. Scand.: 52 (1849). *Carex rigida* f. *latifolia* (Andersson) Kük. in H.G.A.Engler (ed.), Pflanzenr., IV, 20(38): 302 (1909).
Carex hyperborea var. *paradoxa* Drejer in G.C.Oeder & al. (eds.), Fl. Dan.: t. 2483 (1849). *Carex rigida* f. *paradoxa* (Drejer) Kük. in H.G.A.Engler (ed.), Pflanzenr., IV, 20(38): 302 (1909).
Carex saxatilis var. *pallidior* A.Blytt, Norges Fl. 1: 210 (1861).
Carex dubitata Dewey ex Boott, Ill. Carex 4: 167 (1867).
Carex vulgaris var. *alpina* Boott, Ill. Carex 4: 167 (1867).
Carex vulgaris var. *hyperborea* Boott, Ill. Carex 4: 167 (1867).
Carex anguillata Drejer in G.C.Oeder & al. (eds.), Fl. Dan.: t. 2846 (1871). *Carex aquatilis* var. *anguillata* (Drejer) Nyman, Consp. Fl. Eur.: 777 (1882). *Carex rigida* f. *anguillata* (Drejer) Kük. in H.G.A.Engler (ed.), Pflanzenr., IV, 20(38): 302 (1909). *Carex bigelowii* f. *anguillata* (Drejer) Fernald, Rhodora 44: 298 (1942).
Carex drejeriana Lange in G.C.Oeder & al. (eds.), Fl. Dan.: t. 2975 (1880). *Carex rigida* f. *drejeriana* (Lange) Kük. in H.G.A.Engler (ed.), Pflanzenr., IV, 20(38): 302 (1909).
Carex spiralis Ewing, Trans. Nat. Hist. Soc. Glasgow 2: 110 (1886–1887 publ. 1888).
Carex fyllae T.Holm, Bot. Jahrb. Syst. 8: 294 (1887).
Carex hyperborea f. *pumila* Lange, Meddel. Grønland 3: 290 (1887).
Carex warmingii Holm ex Lange, Consp. Fl. Groenland. 2: 290 (1887).
Carex drejeriana f. *cuspidata* Rosenv., Meddel. Grønland 3: 722 (1892).
Carex hyperborea f. *cuspidata* Rosenv., Till. Gronl. Fanerog. 2: 723 (1892). *Carex rigida* f. *cuspidata* (Rosenv.) Kük. in H.G.A.Engler (ed.), Pflanzenr., IV, 20(38): 302 (1909).
Carex rigida f. *aggregata* Almq. ex Neuman, Sver. Fl.: 707 (1901).
Carex rigida f. *juncelliformis* Almq. ex Neuman, Sver. Fl.: 707 (1901).
Carex rigida var. *hesperia* Piper, Contr. U. S. Natl. Herb. 11: 173 (1906).
Carex hartzii Gand., Bull. Soc. Bot. France 66: 296 (1919 publ. 1920).
Carex rigida var. *inferalpina* Laest., Nova Acta Regiae Soc. Sci. Upsal. 11: 287 (1939).
Carex bigelowii f. *glacialis* (Fr.) Raymond, Naturaliste Canad. 77: 57 (1950).

subsp. ***ensifolia*** (Turcz. ex Gorodkov) Holub., Folia Geobot. Phytotax. 3: 190 (1968).
European Russia (S. Ural'skiy Chrebet) to Mongolia. 14 RUE 30 ALT BRY IRK TVA WSB 36 CHN CHX 37 MON. Hemicr. or rhizome geophyte.
**Carex rigida* var. *ensifolia* Turcz. ex Gorodkov, Zhurn. Russk. Bot. Obshch. 15: 182 (1930). *Carex ensifolia* (Turcz. ex Gorodkov) V.I.Krecz. in V.L.Komarov (ed.), Fl. URSS 3: 227 (1935).

subsp. ***lugens*** (T.Holm) T.V.Egorova, Novosti Sist. Vyssh. Rast. 10: 104 (1973).
Siberia to Subarctic America. 30 YAK 31 KAM MAG 70 ASK NUN NWT YUK 71 BRC. Hemicr. or rhizome geophyte.
**Carex lugens* T.Holm, Amer. J. Sci., IV, 10: 269 (1900). *Carex ensifolia* subsp. *lugens* (T.Holm) Vorosch. in A.K.Skvortsov (ed.), Florist. Issl. Razn. Raionakh SSSR: 154 (1985).
Carex yukonensis Britton, Bull. New York Bot. Gard. 2: 159 (1901).
Carex consimilis T.Holm, Amer. J. Sci. 166: 457 (1903).
Carex cyclocarpa T.Holm, Amer. J. Sci. 166: 457 (1903).

Carex* ×*biharica Simonk., Enum. Fl. Transsilv.: 548 (1887). *C. canescens* × *C. echinata*. *Vignea* × *behurica* (Simonk.) Soják, Cas. Nár. Mus., Odd. Prír. 148: 194 (1979 publ. 1980).
Romania to Siberia. 13 ROM 14+ 30+. Hemicr. or rhizome geophyte.
Carex ×*tetrastachya* Traunst., Flora 33: 366 (1850), nom. illeg.

Carex bilateralis Hayata, J. Coll. Sci. Imp. Univ. Tokyo 30(1): 380 (1911).
Taiwan, New Guinea. 38 TAI 43 NWG. Hemicr. or rhizome geophyte.
Carex kotoensis Hayata ex Ohwi, Jap. J. Bot. 7: 189 (1934), pro syn.
Carex subteinogyna Ohwi, Mem. Coll. Sci. Kyoto Imp. Univ., Ser. B, Biol. 11(5): 469 (1936).
Carex spathaceobracteata Kük., Bot. Jahrb. Syst. 70: 466 (1940).
Carex acrophila S.T.Blake, J. Arnold Arbor. 28: 114 (1947).
Carex asperinervis T.Koyama, Acta Phytotax. Geobot. 16: 6 (1955).

Carex biltmoreana Mack., Bull. Torrey Bot. Club 37: 234 (1910).
EC. U.S.A. (S. Appalachian Mts.). 78 GEO NCA SCA. Hemicr. or rhizome geophyte.

Carex binervis Sm., Trans. Linn. Soc. London 5: 268 (1800). *Trasus binervis* (Sm.) Gray, Nat. Arr. Brit. Pl. 2: 63 (1821).
W. Europe, NW. Morocco. 10 FIN GRB IRE NOR 11 BGM GER 12 FRA POR SPA 20 MOR. Hemicr. or rhizome geophyte.
Carex distans Lightf., Fl. Scot. 2: 561 (1777), nom. illeg.
Carex fulva Hornem. in G.C.Oeder & al. (eds.), Fl. Dan.: t. 1768 (1823), nom. illeg.
Carex versicolor Dumort., Fl. Belg.: 148 (1827).

Carex multinervis Krock. ex Steud., Nomencl. Bot., ed. 2, 1: 293 (1840).
Carex binervis var. *alpina* Drejer, Rev. Car. Bor.: 56 (1841).
Carex binervis f. *nigrescens* Druce, J. Bot. 28: 44 (1890).
Carex sadleri E.F.Linton, J. Bot. 36: 43 (1898).

Carex bitchuensis T.Hoshino & H.Ikeda, J. Jap. Bot. 78: 24 (2003).
Japan (Honshu). 38 JAP. Hemicr. or rhizome geophyte.

Carex blakei Nelmes, Bull. Misc. Inform. Kew 1939: 311 (1939).
SE. Australia. 50 NSW VIC. Hemicr. or rhizome geophyte.

Carex blanda Dewey, Amer. J. Sci. Arts 10: 45 (1826). *Carex anceps* var. *blanda* (Dewey) Boott in W.J.Hooker, Fl. Bor.-Amer. 2: 226 (1839). *Deweya blanda* (Dewey) Raf., Good Book: 26 (1840). *Carex laxiflora* var. *blanda* (Dewey) Boott, Ill. Carex 1: 37 (1858).
SE. Canada to C. & E. U.S.A. 72 ONT QUE 73 COL WYO 74 ILL IOW KAN MSO NDA NEB OKL SDA WIS 75 CNT INI MAI MAS MIC NWH NWJ NWY OHI PEN RHO VER WVA 77 NWM TEX 78 ALA ARK DEL FLA GEO KTY LOU MRY MSI NCA SCA TEN VRG WDC. Hemicr. or rhizome geophyte.

Carex blepharicarpa Franch., Bull. Soc. Philom. Paris, VIII, 7: 46 (1895).
Russian Far East, Korea, Japan. 31 KUR SAK 38 JAP KOR. Hemicr. or rhizome geophyte.

var. ***blepharicarpa***
Japan. 38 JAP. Hemicr. or rhizome geophyte.
Carex hayatae H.Lév., Repert. Spec. Nov. Regni Veg. 5: 240 (1908).
Carex sadae H.Lév. & Vaniot, Bull. Acad. Int. Géogr. Bot. 19: 33 (1909).
Carex yezo-montana Akiyama, J. Fac. Sci. Hokkaido Univ., Ser. 5, Bot. 1: 62 (1931).

var. ***duneensis*** (Meinsh.) Akiyama in ?
Russian Far East, Korea, Japan. 31 KUR SAK 38 JAP KOR. Hemicr. or rhizome geophyte.
Carex dueensis Meinsh., Trudy Imp. S.-Peterburgsk. Bot. Sada 18: 387 (1900).
**Carex duneensis* Meinsh., Trudy Imp. S.-Peterburgsk. Bot. Sada 18: 387 (1900).
Carex duneensis var. *rostrata* Meinsh., Trudy Imp. S.-Peterburgsk. Bot. Sada 18: 387 (1900).

var. ***hirtifructus*** (Kük.) Ohwi, Mem. Coll. Sci. Kyoto Imp. Univ., Ser. B, Biol. 6: 247 (1931).
Japan. 38 JAP. Hemicr. or rhizome geophyte.
**Carex hirtifructus* Kük., Bull. Herb. Boissier, II, 5: 1163 (1905).
Carex blepharicarpa var. *stenocarpa* Ohwi, Mem. Coll. Sci. Kyoto Imp. Univ., Ser. B, Biol. 5: 269 (1930).

Carex blinii H.Lév. & Vaniot, Bull. Soc. Bot. France 53: 316 (1906).
China (Guizhou) to Indo-China. 36 CHC 41 THA VIE. Hemicr. or rhizome geophyte.

Carex bodinieri Franch., Bull. Soc. Philom. Paris, VIII, 7: 85 (1895).
SE. China. 36 CHS. Hemicr. or rhizome geophyte.

Carex boecheriana Á.Löve, D.Löve & Raymond, Canad. J. Bot. 35: 748 (1957).
Greenland. 70 GNL. Hemicr. or rhizome geophyte.
**Carex capillaris* subsp. *robustior* Drejer ex Böcher, Meddel. Grønland: 147 (1952).

Carex boelckeiana Barros, Colecç. Ci. Inst. Nac. Tecnol. Agropecu 8: 70 (1969).
WC. Argentina. 85 AGS. Hemicr. or rhizome geophyte.

Carex ×boenninghausiana Weihe, Flora 9: 759 (1826). *C. paniculata* × *C. remota*. *Vignea* ×*boenninghauseniana* (Weihe) Rchb. in J.C.Mössler, Handb. Gewächsk. ed. 2, 3: 1616 (1830). *Carex axillaris* subsp. *boenninghausiana* (Weihe) Nyman, Consp. Fl. Eur.: 780 (1882).
Europe. 10 NOR 11 GER 13 ITA SIC. Hemicr. or rhizome geophyte.
Carex ×*boenninghauseniana* Kunth, Enum. Pl. 2: 404 (1837).

Carex ×bogstadensis Kük. in ?. *C. rhynchophysa* × C. *vesicaria*.
European Russia to Siberia. 14+ 30+. Hemicr. or rhizome geophyte.

Carex bohemica Schreb., Beschr. Gräs. 2: 52 (1772). *Vignea bohemica* (Schreb.) Soják, Cas. Nár. Mus., Odd. Prír. 148: 194 (1979 publ. 1980).
Azores, Europe, Temp. Asia. 10 DEN FIN 11 AUT BGM CZE GER HUN POL SWI 12 FRA POR 13 ITA ROM YUG 14 BLR RUC RUE RUN RUS RUW 21 AZO 30 WSB 31+ 32 KAZ 33 NCS 36 CHM 38 JAP. Hemicr. or rhizome geophyte.
Carex cyperoides L., Syst. Veg. ed. 13: 703 (1774). *Schelhammeria cyperoides* (L.) Dumort., Fl. Belg.: 145 (1827). *Vignea cyperoides* (L.) Peterm., Fl. Lips. Excurs.: 49 (1838). *Caricina cyperoides* (L.) St.-Lag. in A.Cariot, Étude Fl., ed. 8, 2: 879 (1889). *Thysanocarex cyperoides* (L.) Fedde & J.Schust., Just's Bot. Jahresber. 41(2): 12 (1918).
Carex cyperoidea Houtt., Nat. Hist., II, 13: 517 (1782).
Schelhammeria capitata Moench, Suppl. Meth.: 119 (1802).
Carex cyperoides f. *aggregata* Domin, Sitzungsber. Königl. Böhm. Ges. Wiss. Prag, Math.-Naturwiss. Cl. 18: 77 (1904). *Carex bohemica* f. *aggregata* (Domin) Soó, Acta Bot. Acad. Sci. Hung. 16: 369 (1970 publ. 1971).

Carex bolanderi Olney, Proc. Amer. Acad. Arts 7: 393 (1868). *Carex deweyana* var. *bolanderi* (Olney) Boott in S.Watson & al., Bot. California 2: 236 (1880).
W. Canada to W. U.S.A., Mexico (Chihuahua, Durango). 71 BRC 73 IDA MNT ORE WAS 76 ARI CAL NEV UTA 77 NWM 79 MXE. Hemicr. or rhizome geophyte.
Carex bolanderi var. *artemisiana* Kelso, Biol. Leafl. 66: 37 (1953).

Carex boliviensis Van Heurck & Müll.Arg., Observ. Bot. Descript. Pl. Nov. Herb. Van Heurckiani 1: 32 (1870).
C. & S. Mexico, Peru to N. Argentina. 79 MXC MXE MXS 83 BOL PER 85 AGW. Hemicr. or rhizome geophyte.

subsp. ***boliviensis***
C. & S. Mexico, Peru to N. Argentina. 79 MXC MXE MXS 83 BOL PER 85 AGW. Hemicr. or rhizome geophyte.
Carex lorentziana Griseb., Abh. Königl. Ges. Wiss. Göttingen 19: 271 (1874).
Carex galbana L.H.Bailey, Bot. Gaz. 25: 271 (1898).

subsp. ***occidentalis*** Reznicek & S.González, Contr. Univ. Michigan Herb. 23: 343 (2001).
Mexico. 79 MXE MXS. Hemicr.

Carex bonanzensis Britton, Bull. New York Bot. Gard. 2: 160 (1901). *Vignea bonanzensis* (Britton) Soják, Cas. Nár. Mus., Odd. Prír. 148: 194 (1979 publ. 1980).
Siberia to Subarctic America. 30 WSB 31 MAG 70 ASK NWT YUK. Hemicr. or rhizome geophyte.
Carex cajanderi Kük., Oefvers. Förh. Finska Vetensk.-Soc. 45(8): 3 (1903).

Carex bonariensis Desf. ex Poir. in J.B.A.M.de Lamarck, Encycl., Suppl. 3: 250 (1813).
S. Ecuador to Argentina. 83 ECU PER 84 BZL BZS 85 AGE AGS AGW CLC CLS PAR URU. Hemicr. or rhizome geophyte.
Carex bonariensis var. *tolimensis* Maury, J. Bot. (Morot) 2: 423 (1888).
Carex bonariensis f. *pumila* Osten, Anales Mus. Hist. Nat. Montevideo, II, 3: 237 (1931).

var. ***bonariensis***
S. Ecuador to Argentina. 83 ECU PER 84 BZL BZS 85 AGE AGS AGW CLC CLS PAR URU. Hemicr. or rhizome geophyte.
Carex compressa Vahl ex Kunth, Enum. Pl. 2: 378 (1837).
Carex papillosa Nees, J. Bot. (Hooker) 2: 398 (1840).
Carex tuberculata Nees ex Boott, Ill. Carex 2: 76 (1860), nom. illeg.

var. ***glabrescens*** Kurtz ex Kük., Bot. Jahrb. Syst. 27: 512 (1899).
Argentina (Córdoba, Tucumán). 85 AGE AGW. Hemicr. or rhizome geophyte.

Carex bonplandii Kunth, Enum. Pl. 2: 380 (1837).
Arizona to S. Trop. America. 76 ARI 79 MXG MXS 80 COS GUA PAN 82 VEN 83 BOL CLM ECU PER 84 BZN. Hemicr. or rhizome geophyte.
Carex bonplandii var. *angustifolia* Boott, Ill. Carex 3: 115 (1862).
Carex heptastachya Boeck., Linnaea 39: 114 (1875).
Carex durandii Boeck., Allg. Bot. Z. Syst. 2: 189 (1896).
Carex bonplandii var. *humilior* Kük. in H.G.A.Engler (ed.), Pflanzenr., IV, 20(38): 232 (1909).
Carex bonplandii var. *robustior* Kük. in H.G.A.Engler (ed.), Pflanzenr., IV, 20(38): 232 (1909).

Carex boottiana Hook. & Arn., Bot. Beechey Voy.: 273 (1838). *Carex oahuensis* var. *boottiana* (Hook. & Arn.) Kük. in H.G.A.Engler (ed.), Pflanzenr., IV, 20(38): 632 (1909).
SE. China to Temp. E. Asia. 36 CHS 38 JAP KOR KZN NNS OGA TAI. Hemicr. or rhizome geophyte.
Carex bongardii Boott, Proc. Linn. Soc. London 1: 260 (1845).
Carex bongardii var. *robusta* Franch. & Sav., Enum. Pl. Jap. 2: 561 (1878). *Carex wahuensis* subsp. *robusta* (Franch. & Sav.) T.Koyama, in Fl. Taiwan 5: 326 (1978).
Carex stupenda H.Lév. & Vaniot, Bull. Acad. Int. Géogr. Bot. 10: 195 (1901).
Carex wilfordii C.B.Clarke, J. Linn. Soc., Bot. 36: 318 (1904).
Carex reflexistyla Hayata, J. Coll. Sci. Imp. Univ. Tokyo 30(1): 393 (1911).
Carex boninensis Koidz., Bot. Mag. (Tokyo) 32: 55 (1918).

Carex borbonica Lam., Encycl. 3: 387 (1792).
Mascarenes. 29 MAU REU. Hemicr. or rhizome geophyte.
Carex typhoides Bory, Voy. îles Afrique 3: 24 (1804).
Carex giraudiana Steud., Syn. Pl. Glumac. 2: 209 (1855).
Carex boryana var. *rigidifolia* Boeck., Linnaea 41: 285 (1877).
Carex borbonica var. *pallidior* Kük. in H.G.A.Engler (ed.), Pflanzenr., IV, 20(38): 655 (1909).

Carex borealihinganica Y.L.Chang & Y.L.Yang, in Fl. Plant. Herb. Chinae Bor.-Or. 11: 206 (1976).
NE. China. 36 CHM. Hemicr. or rhizome geophyte.

Carex borii Nelmes, Kew Bull. 4: 38 (1949).
Pakistan. 40 PAK. Hemicr. or rhizome geophyte.
Carex borii f. *lutea* R.R.Stewart, Annot. Cat. Vasc. Pl. W. Pakistan & Kashmir: 74 (1972).

Carex boryana Schkuhr, Beschr. Riedgräs. 2: 43 (1806).
Gulf of Guinea Is., Uganda, W. Indian Ocean. 23 GGI 25 UGA 29 MAU MDG REU. Hemicr. or rhizome geophyte.
Carex subulata Wahlenb., Kongl. Vetensk. Acad. Nya Handl. 24: 152 (1803), nom. illeg.
Carex boryana var. *minor* Boott, Ill. Carex 3: 111 (1862).

Carex bostrychostigma Maxim., Bull. Acad. Imp. Sci. Saint-Pétersbourg 31: 118 (1887).
Primorye, NE. China, Korea, C. & S. Japan. 31 PRM 36 CHM CHN CHS 38 JAP KOR. Hemicr. or rhizome geophyte.
Carex explens Kük., Bull. Herb. Boissier, II, 2: 1017 (1902).

Carex brachyanthera Ohwi, Jap. J. Bot. 7: 190 (1934).
Taiwan, Nansei-shoto, New Guinea. 38 NNS TAI 43 NWG. Hemicr. or rhizome geophyte.
Carex tricuspidata Kük., Bot. Jahrb. Syst. 70: 466 (1940).

Carex brachycalama Griseb., Abh. Königl. Ges. Wiss. Göttingen 24: 315 (1879).
C. Mexico, Ecuador to Argentina (Córdoba). 79 MXC 83 BOL ECU PER 85 AGE. Hemicr. or rhizome geophyte.

Carex brachystachys Schrank, Baier. Fl. 1: 294 (1789).
Mts. of C. & S. Europe. 11 AUT CZE GER POL SWI 12 FRA SPA 13 ITA ROM YUG. Hemicr. or rhizome geophyte.
Carex strigosa All., Fl. Pedem. 2: 270 (1785), nom. illeg.
Carex valesiaca Suter, Helvet. Fl. 1: 259 (1802).
Carex compressa Kit. ex Willd., Sp. Pl. 4: 276 (1805), nom. illeg.
Carex tenuis Host, Icon. Descr. Gram. Austriac. 4: 51 (1809), nom. illeg.
Carex mielichhoferi Willd. ex Kunth, Enum. Pl. 2: 465 (1837), nom. inval.
Carex linearis Clairv. ex Boott, Ill. Carex 4: 204 (1867), nom. illeg.
Carex scheuchzeri Gaudin ex Boott, Ill. Carex 4: 204 (1867).

Carex bracteosa Kunze in C.Schkuhr, Beschr. Riedgräs., Suppl.: 12 (1840). *Vignea bracteosa* (Kunze) Rchb. in J.C.Mössler, Handb. Gewächsk. ed. 2, 3: 1619 (1830).
C. & S. Chile. 85 CLC CLS. Hemicr. or rhizome geophyte.
Carex luteola Nees ex Boott, Ill. Carex 4: 181 (1867).

Carex bradei R.Gross, Repert. Spec. Nov. Regni Veg. 42: 176 (1937).
Brazil (Rio de Janeiro). 84 BZL. Hemicr. or rhizome geophyte.

Carex brainerdii Mack., Bull. Torrey Bot. Club 40: 534 (1913).
Oregon to California. 73 ORE 76 CAL. Hemicr. or rhizome geophyte.

Carex brasiliensis A.St.-Hil., Voy. Distr. Diam. 1: 369 (1833). *Carex procera* var. *brasiliensis* (A.St.-Hil.) Boott, Ill. Carex 4: 138 (1867).
Brazil to NE. Argentina. 84 BZS 85 AGE PAR. Hemicr. or rhizome geophyte.
Carex procera Kunth, Enum. Pl. 2: 491 (1837).
Carex tetrasticha Boeck., Beitr. Cyper. 1: 35 (1888).
Carex brasiliensis f. *robusta* C.B.Clarke ex Kük., Verh. Bot. Vereins Prov. Brandenburg 47: 210 (1905 publ. 1906).

Carex brassii Nelmes, Mem. New York Bot. Gard. 9: 100 (1954).
Malawi. 26 MLW. Hemicr. or rhizome geophyte.

Carex breviaristata K.T.Fu, in Fl. Tsinlingensis 1(1): 444 (1976).
China. 36 CHN CHS. Hemicr. or rhizome geophyte.

Carex brevicollis DC. in J.B.A.M.de Lamarck & A.P.de Candolle, Fl. Franç., ed. 3, 6: 295 (1815).
S. & EC. Europe to Caucasus. 11 CZE HUN 12 FRA SPA 13 BUL ROM YUG 14 UKR 33 NCS TCS 34 TUR. Hemicr. or rhizome geophyte.
Carex rhynchocarpa Heuff., Flora 16: 365 (1833).

Carex breviculmis R.Br., Prodr.: 242 (1810). *Carex breviculmis* var. *typica* Domin, Biblioth. Bot. 89: 496 (1915), nom. inval.
Trop. & Subtrop. Asia, Australia, New Zealand. 36 CHC CHM CHN CHS 38 JAP KOR NNS TAI 40 EHM IND NEP 41 MYA VIE 42 BOR JAW PHI SUL 43 NWG 50 NFK NSW QLD SOA TAS VIC 51 NZN. Hemicr. or rhizome geophyte.

var. ***breviculmis***
Trop. & Subtrop. Asia, Australia, New Zealand. 36 CHC CHM CHN CHS 38 JAP KOR TAI 40 EHM IND NEP 41 MYA VIE 42 BOR JAW PHI SUL 43 NWG 50 NFK NSW QLD SOA TAS VIC 51 NZN. Hemicr. or rhizome geophyte.
Carex breviculmis subsp. *royleana* Nees in R.Wight, Contr. Bot. India: 127 (1834).
Carex leucochlora Bunge, Mém. Sav. Étr. Acad. St. Pétersbourg 2: 142 (1835).
Carex langsdorffii Boott, Proc. Linn. Soc. London 1: 259 (1845).
Carex eggytera Steud., Syn. Pl. Glumac. 2: 220 (1855).
Carex filiculmis Franch. & Sav., Enum. Pl. Jap. 2: 563 (1878). *Carex breviculmis* f. *filiculmis* (Franch. & Sav.) Kük. in H.G.A.Engler (ed.), Pflanzenr., IV, 20(38): 470 (1909).
Carex boniana Boeck., Allg. Bot. Z. Syst. 2: 190 (1896).
Carex candolleana H.Lév. & Vaniot, Bull. Acad. Int. Géogr. Bot. 10: 273 (1901).
Carex wardiana H.Lév. & Vaniot, Bull. Acad. Int. Géogr. Bot. 10: 276 (1901). *Carex breviculmis* f. *wardiana* (H.Lév. & Vaniot) Kük. in H.G.A. Engler (ed.), Pflanzenr., IV, 20(38): 470 (1909).
Carex hypochlora Freyn, Oesterr. Bot. Z. 53: 26 (1903). *Carex umbrosa* subsp. *hypochlora* (Freyn) T.V.Egorova, Novosti Sist. Vyssh. Rast. 10: 103 (1973).
Carex breviculmis f. *longiaristata* Kük. in H.G.A. Engler (ed.), Pflanzenr., IV, 20(38): 470 (1909).
Carex breviculmis var. *stipitata* Kük. in H.G.A. Engler (ed.), Pflanzenr., IV, 20(38): 469 (1909).
Carex morrisonicola Hayata, J. Coll. Sci. Imp. Univ. Tokyo 30(1): 387 (1911). *Carex leucochlora* var. *morrisonicola* (Hayata) Katsuy., Bull. Kanagawa Pref. Mus., Nat. Sci. 22: 58 (1993).
Carex lonchophora Ohwi, Mem. Coll. Sci. Kyoto Imp. Univ., Ser. B, Biol. 3: 274 (1930).
Carex leucochlora var. *petrogena* Kitag., Bot. Mag. (Tokyo) 48: 11 (1934).
Carex conorrhyncha Nelmes, Kew Bull. 11: 182 (1956).
Carex horikawae K.Okamoto, Bull. Okayama Coll. Sci. 1: 56 (1965). *Carex leucochlora* var. *horikawae* (K.Okamoto) Katsuy., Bull. Kanagawa Pref. Mus., Nat. Sci. 22: 59 (1993).

var. ***cupulifera*** (Hayata) Y.C.Tang & S.Yun Liang, in Fl. Reipubl. Popul. Sin. 12: 163 (2000).
Taiwan. 38 TAI. Hemicr.
**Carex orthostemon* var. *cupulifera* Hayata, J. Coll. Sci. Imp. Univ. Tokyo 30(1): 389 (1911).

var. ***discoidea*** (Boott) Boott, Ill. Carex 4: 181 (1867).
Japan (Honshu, Kyushu), Nansei-shoto. 38 JAP NNS. Hemicr. or rhizome geophyte.
**Carex discoidea* Boott, Mem. Amer. Acad. Arts, n.s., 6: 419 (1859). *Carex breviculmis* f. *discoidea* (Boott) Kük. in H.G.A.Engler (ed.), Pflanzenr., IV, 20(38): 470 (1909).
Carex aphanandra Franch. & Sav., Enum. Pl. Jap. 2: 564 (1878). *Carex breviculmis* f. *aphanandra* (Franch. & Sav.) Kük. in H.G.A.Engler (ed.), Pflanzenr., IV, 20(38): 470 (1909).
Carex perangusta Ohwi, Mem. Coll. Sci. Kyoto Imp. Univ., Ser. B, Biol. 5(3): 273 (1930).

var. ***montivaga*** (S.T.Blake) Noot., in Fl. Males. 9(1): 151 (1979).
New Guinea. 43 NWG. Hemicr. or rhizome geophyte.
Carex bulbostylis Kük., Bot. Jahrb. Syst. 69: 264 (1938), nom. illeg.
**Carex montivaga* S.T.Blake, J. Arnold Arbor. 28: 109 (1947).

var. ***perciliata*** Kük. in H.G.A.Engler (ed.), Pflanzenr., IV, 20(38): 469 (1909). *Carex perciliata* (Kük.) Nelmes, Kew Bull. 1: 26 (1946).
Malesia to New Guinea. 42 BOR SUL 43 NWG. Hemicr. or rhizome geophyte.
Carex brevis S.T.Blake, J. Arnold Arbor. 28: 111 (1947).
Carex tricholoma S.T.Blake, J. Arnold Arbor. 28: 110 (1947).

var. ***pluricostata*** Kük. in H.G.A.Engler (ed.), Pflanzenr., IV, 20(38): 470 (1909).
China, S. Korea, C. Japan to Taiwan. 36 CHN CHS 38 JAP KOR NNS TAI. Hemicr. or rhizome geophyte.
Carex fibrillosa Franch. & Sav., Enum. Pl. Jap. 2: 564 (1878). *Carex breviculmis* f. *fibrillosa* (Franch. & Sav.) Kük. in H.G.A.Engler (ed.), Pflanzenr., IV, 20(38): 470 (1909). *Carex leucochlora* f. *fibrillosa* (Franch. & Sav.) K.T.Fu,

in Fl. Tsinlingensis 1(1): 249 (1976). *Carex breviculmis* subsp. *fibrillosa* (Franch. & Sav.) T.Koyama, in Fl. Taiwan 5: 336 (1978).

Carex tosaensis Akiyama, J. Fac. Sci. Hokkaido Univ., Ser. 5, Bot. 3: 267 (1935).

Carex pallens C.P.Wang, J. Nanjing Univ. (Biology) 1962(2): 49 (1962).

var. ***puberula*** (Boott) T.Koyama in ?

N. & C. Japan. 38 JAP. Hemicr. or rhizome geophyte.

**Carex puberula* Boott in M.C.Perry, Jap. Exped. 2: 324 (1856).

Carex brevicuspis C.B.Clarke, J. Linn. Soc., Bot. 36: 277 (1903).

China, Taiwan. 36 CHC CHN CHS 38 TAI. Hemicr. or rhizome geophyte.

var. ***basiflora*** (C.B.Clarke) Kük. in H.G.A.Engler (ed.), Pflanzenr., IV, 20(38): 630 (1909).

China (Shaanxi, Hubei). 36 CHC CHN. Hemicr. or rhizome geophyte.

**Carex basiflora* C.B.Clarke, J. Linn. Soc., Bot. 36: 274 (1903).

var. ***brevicuspis***

SE. China, Taiwan. 36 CHS 38 TAI. Hemicr. or rhizome geophyte.

Carex brevior (Dewey) Mack. ex Lunell, Amer. Midl. Naturalist 4: 235 (1915).

Canada to U.S.A., Mexico (Tamaulipas). 71 ABT BRC MAN SAS 72 ONT QUE 73 COL IDA MNT WAS WYO 74 ILL IOW KAN MIN MSO NDA NEB OKL SDA WIS 75 CNT INI MAI MAS MIC NWH NWJ NWY OHI PEN RHO VER WVA 76 ARI 77 NWM TEX 78 ARK DEL KTY MRY MSI TEN VRG WDC 79 MXE. Hemicr. or rhizome geophyte.

**Carex straminea* var. *brevior* Dewey, Amer. J. Sci. Arts 11: 158 (1826). *Carex festucacea* var. *brevior* (Dewey) Fernald, Proc. Amer. Acad. Arts 37: 477 (1902).

Carex breviscapa C.B.Clarke in J.D.Hooker, Fl. Brit. India 6: 736 (1894).

Sri Lanka to Queensland. 36 CHH CHS 38 NNS TAI 40 SRL 41 THA VIE 42 BOR JAW MLY PHI SUM 43 NWG 50 QLD. Hemicr. or rhizome geophyte.

Carex jackiana var. *breviculmis* Thwaites & Hook.f., Enum. Pl. Zeyl.: 356 (1864).

Carex curtisii Ridl., Mat. Fl. Malay. Penins. 3: 117 (1907).

Carex lutchuensis Ohwi, Mem. Coll. Sci. Kyoto Imp. Univ., Ser. B, Biol. 5(3): 270 (1930).

Carex laxisquamata Tang & F.T.Wang, in Fl. Hainan. 4: 535 (1977).

Carex breweri Boott, Ill. Carex 4: 142 (1867).

W. U.S.A. 73 ORE WAS 76 CAL NEV. Hemicr. or rhizome geophyte.

Carex brizoides L., Cent. Pl. I: 31 (1755). *Vignea brizoides* (L.) Rchb. in J.C.Mössler, Handb. Gewächsk. ed. 2, 3: 1611 (1830). *Carex brizoides* var. *genuina* Èelak., Prodr. Fl. Böhmen: 62 (1867), nom. inval. *Caricina brizoidea* (L.) St.-Lag. in A.Cariot, Étude Fl., ed. 8, 2: 877 (1889).

C. & S. Europe to C. Ukraine, NW. Turkey. 11 AUT BGM CZE GER HUN NET POL SWI 12 FRA SWI 13 ALB ITA ROM TUE YUG 14 BLR BLT UKR 34 TUR. Hemicr. or rhizome geophyte.

Carex praecox J.Jundz., Opisan. Rósl.: 889 (1830), nom. illeg.

Carex brizoides var. *nemoralis* Wimm., Fl. Schles., ed. 2, 1: 401 (1844).

Carex brizoides f. *remota* Peterm., Flora 27: 330 (1844).

Carex psammophila Schur, Enum. Pl. Transsilv.: 700 (1866).

Vignea psammophila Schur, Enum. Pl. Transsilv.: 700 (1866).

Vignea pseudoschreberi Schur, Enum. Pl. Transsilv.: 700 (1866).

Carex enervis Dulac, Fl. Hautes -Pyrénées: 53 (1867), nom. illeg.

Carex brizoides f. *brunnescens* Kük., Mitth. Thüring. Bot. Vereins, n.s., 2: 45 (1892).

Carex brizoides f. *subheterostachya* Asch. & Graebn., Syn. Mitteleur. Fl. 2(2): 34 (1902).

Carex bromoides Willd., Sp. Pl. 4: 258 (1805). *Loncoperis bromoides* (Willd.) Raf., Good Book: 27 (1840).

E. Canada to NC. & E. U.S.A., Mexico (Hidalgo). 72 NBR NSC ONT QUE 74 ILL MIN MSO WIS 75 CNT INI MAI MAS MIC NWH NWJ NWY OHI PEN RHO VER WVA 77 TEX 78 ALA ARK DEL FLA GEO KTY LOU MRY MSI NCA SCA TEN VRG WDC 79 MXE. Hemicr. or rhizome geophyte.

subsp. ***bromoides***

E. Canada to NC. & E. U.S.A., Mexico (Hidalgo). 72 NBR NSC ONT QUE 74 ILL MIN MSO WIS 75 CNT INI MAI MAS MIC NWH NWJ NWY OHI PEN RHO VER WVA 77 TEX 78 ALA ARK DEL FLA GEO KTY LOU MRY MSI NCA SCA TEN VRG WDC 79 MXE. Hemicr. or rhizome geophyte.

subsp. ***montana*** Naczi, Contr. Univ. Michigan Herb. 17: 220 (1990).

Virginia to South Carolina. 78 NCA SCA VRG. Hemicr. or rhizome geophyte.

Carex brongniartii Kunth, Enum. Pl. 2: 380 (1837).

C. Chile, N. Argentina, Uruguay. 85 AGE AGW CLC URU. Hemicr. or rhizome geophyte.

Carex muricata Schltdl., Linnaea 6: 29 (1831), nom. illeg.

**Carex muehlenbergii* Brongn. in L.I.Duperrey, Voy. Monde, Phan.: 151 (1833), nom. illeg.

Carex hypoxanthos Steud., Syn. Pl. Glumac. 2: 193 (1855).

Carex brownii Tuck., Enum. Meth. Caric.: 21 (1843).

China to SE. Australia. 36 CHC CHN CHS 38 JAP KOR NNS TAI 43 NWG 50 NSW VIC (51) nzn. Hemicr. or rhizome geophyte.

subsp. ***brownii***

C. China to SE. Australia. 36 CHC CHN CHS 38 JAP KOR TAI 43 NWG 50 NSW VIC (51) nzn. Hemicr. or rhizome geophyte.

Carex striata R.Br., Prodr.: 243 (1810), nom. illeg. *Vignea striata* Rchb. in J.C.Mössler, Handb. Gewächsk. ed. 2, 3: 1625 (1830).

Carex rigens Boott, Mem. Amer. Acad. Arts, n.s., 6: 419 (1859), nom. illeg.

Carex transversa Miq., Ann. Mus. Bot. Lugduno-Batavi 2: 148 (1866), nom. illeg.

Carex nipposinica Ohwi, Acta Phytotax. Geobot. 11: 255 (1942).

subsp. ***transversa*** (Boott) J.Kern, in Fl. Males. 9(1): 163 (1979).
SE. China, Temp. E. Asia, New Guinea. 36 CHS 38 JAP KOR NNS 43 NWG. Hemicr. or rhizome geophyte.
**Carex transversa* Boott in W.J.Hooker, Fl. Bor.-Amer. 2: 215 (1839). *Carex brownii* var. *transversa* (Boott) Kük. in H.G.A.Engler (ed.), Pflanzenr., IV, 20(38): 614 (1909).

Carex brunnea Thunb. in J.A.Murray, Syst. Veg. ed. 14: 844 (1784). *Olotrema brunnea* (Thunb.) Raf., Good Book: 25 (1840).
NE. Trop. Africa, W. Indian Ocean, Subtrop. Asia, E. Australia to Lord Howe I. 24 ETH SOM 29 MAU MDG REU ROD 35 YEM 36 CHC CHN CHS 38 JAP KOR NNS OGA TAI 41 VIE 50 NFK NSW QLD (75) mas. Hemicr. or rhizome geophyte.

var. ***abscondita*** T.Koyama in ?
Japan (Honshu: Mikawa Prov.). 38 JAP. Hemicr. or rhizome geophyte.

var. ***arabica*** S.S.Hooper, Kew Bull. 39: 748 (1984).
Yemen. 35 YEM. Hemicr. or rhizome geophyte.

subsp. ***brunnea***
W. Indian Ocean, Subtrop. Asia, E. Australia to Lord Howe I. 29 MAU MDG REU ROD 36 CHC CHN CHS 38 JAP KOR NNS OGA TAI 41 VIE 50 NFK NSW QLD (75) mas. Hemicr. or rhizome geophyte.
Carex acuta var. *rufa* L., Sp. Pl.: 978 (1753).
Carex acuta var. *personata* Fr., Novit. Fl. Suec. Alt.: 281 (1828). *Carex gracilis* var. *personata* (Fr.) Kük. in H.G.A.Engler (ed.), Pflanzenr., IV, 20(38): 322 (1909).
Carex commersoniana Sieber ex Kunth, Enum. Pl. 2: 391 (1837).
Carex conjugata Willd. ex Kunth, Enum. Pl. 2: 391 (1837), nom. inval.
Carex acuta var. *fluviatilis* Hartm., Handb. Skand. Fl., ed. 3: 219 (1838). *Carex gracilis* var. *fluviatilis* (Hartm.) Kük. in H.G.A.Engler (ed.), Pflanzenr., IV, 20(38): 322 (1909).
Carex aemulans Liebm. & Drejer, Flora 22: 142 (1839). *Carex paludosa* var. *aemulans* (Liebm. & Drejer) Nyman, Consp. Fl. Eur., Suppl. 2: 320 (1890).
Carex acuta var. *colorata* F.Nyl., Spic. Fl. Fenn. 2: 20 (1844).
Carex albescens F.Nyl., Spic. Fl. Fenn. 2: 90 (1844).
Carex prolixa var. *protensa* Fr., Novit. Fl. Suec. Mant. 3: 150 (1845). *Carex gracilis* f. *protensa* (Fr.) Kük. in H.G.A.Engler (ed.), Pflanzenr., IV, 20(38): 322 (1909).
Carex acuta f. *zygostachya* Rchb., Icon. Fl. Germ. Helv. 8: t. 232 (1846). *Carex gracilis* f. *zygostachya* (Rchb.) Kük. in H.G.A.Engler (ed.), Pflanzenr., IV, 20(38): 321 (1909).
Carex compacta Wimm., Flora 33: 618 (1850), nom. illeg. *Caricina compacta* St.-Lag. in A.Cariot, Étude Fl., ed. 8, 2: 874 (1889). *Carex gracilis* var. *compacta* (St.-Lag.) Kük., Allg. Bot. Z. Syst. 3: 171 (1897).
Carex acuta var. *minor* Trevir. in C.F.von Ledebour, Fl. Ross. 4: 313 (1852), nom. illeg.
Carex asperifolia Steud., Syn. Pl. Glumac. 2: 195 (1855).
Carex assimilis Steud., Syn. Pl. Glumac. 2: 195 (1855).
Carex boissieri Steud., Syn. Pl. Glumac. 2: 215 (1855).
Carex ischnantha Steud., Syn. Pl. Glumac. 2: 204 (1855).
Carex mauritanica Steud., Syn. Pl. Glumac. 2: 204 (1855).
Carex tenuispica Steud., Syn. Pl. Glumac. 2: 319 (1855).
Carex gracilis var. *minor* Boott, Ill. Carex 1: 59 (1858). *Carex brunnea* var. *minor* (Boott) Kük. in H.G.A.Engler (ed.), Pflanzenr., IV, 20(38): 601 (1909).
Carex acuta var. *nutans* Èelak., Prodr. Fl. Böhmen: 63 (1867).
Carex acuta var. *staminifera* Boott, Ill. Carex 4: 166 (1867).
Carex vulgaris var. *tricostata* Boott, Ill. Carex 4: 168 (1867).
Carex fabri Hance, J. Linn. Soc., Bot. 13: 90 (1873).
Carex brunnea var. *scabrifolia* Boeck., Linnaea 39: 146 (1875).
Carex acuta var. *chlorotica* Èelak., Prodr. Fl. Böhmen 4: 734 (1881). *Carex gracilis* var. *chlorotica* (Èelak.) Kük. in H.G.A.Engler (ed.), Pflanzenr., IV, 20(38): 323 (1909).
Carex acuta f. *rigida* Sanio, Bot. Centralbl. 6: 440 (1881).
Carex acuta var. *sphaerocarpa* Uechtr. in E.Fiek, Fl. Schlesien: 483 (1881). *Carex gracilis* var. *sphaerocarpa* (Uechtr.) Kük. in H.G.A.Engler (ed.), Pflanzenr., IV, 20(38): 323 (1909).
Carex acuta var. *longiglumis* Cariot, Étude Fl., ed. 8, 2: 856 (1889).
Carex acuta var. *pseudaquatilis* Appel, Deutsche Bot. Monatsschr. 10: 191 (1892).
Carex mosoynensis Franch., Bull. Soc. Philom. Paris, VIII, 7: 31 (1895).
Carex brunnea f. *simplex* Kük., Bot. Jahrb. Syst. 36(82): 8 (1905).
Carex gentilis var. *oshimensis* Kük. in H.G.A. Engler (ed.), Pflanzenr., IV, 20(38): 603 (1909).
Carex nakaharae Hayata, J. Coll. Sci. Imp. Univ. Tokyo 30(1): 387 (1911). *Carex gentilis* subsp. *nakaharae* (Hayata) T.Koyama, in Fl. Taiwan 5: 322 (1978).
Carex masoalensis Cherm., Bull. Soc. Bot. France 70: 414 (1923).
Carex amami-oshimensis Akiyama, J. Fac. Sci. Hokkaido Univ., Ser. 5, Bot. 2: 38, 186 (1932).
Carex hattoriana Nakai ex Tuyama, Bot. Mag. (Tokyo) 49: 508 (1935).
Carex nakiri Ohwi, Mem. Coll. Sci. Kyoto Imp. Univ., Ser. B, Biol. 11(5): 467 (1936), pro syn.
Carex spadiceovaginata Ohwi, Bot. Mag. (Tokyo) 56: 215 (1942).
Carex buruensis Nelmes, Kew Bull. 5: 201 (1950).
Carex megacarpa T.Koyama, Bot. Mag. (Tokyo) 69: 210 (1956).

subsp. ***occidentalis*** Lye, Bull. Mus. Natl. Hist. Nat., B, Adansonia 18: 235 (1996).
Ethiopia to N. Somalia. 24 ETH SOM. Hemicr.

var. ***stipitinux*** (C.B.Clarke ex Franch.) Kük. in H.G.A. Engler (ed.), Pflanzenr., IV, 20(38): 602 (1909).
C. & S. China. 36 CHC CHN CHS. Hemicr. or rhizome geophyte.
**Carex stipitinux* C.B.Clarke ex Franch., Bull. Soc. Philom. Paris, VIII, 7: 31 (1894).

Carex brunnescens (Pers.) Poir. in J.B.A.M.de Lamarck, Encycl., Suppl. 3: 286 (1813).
Temp. Northern Hemisphere. 10 FIN ICE NOR SWE 11 AUT CZE GER POL SWI 12 FRA 13 ITA ROM YUG 14 BLR BLT RUC RUE RUN RUS RUW 30 ALT KRA WSB 31 KAM KHA KUR MAG SAK 32 KAZ 37 MON 38 JAP KOR 70 ASK GNL NUN NWT YUK 71 ABT BRC MAN SAS 72 LAB NBR NFL NSC ONT PEI QUE 73 COL IDA MNT ORE WAS WYO 74 MIN WIS 75 CNT INI MAI MAS MIC NWH NWJ NWY OHI PEN RHO VER WVA 76 CAL UTA 78 DEL GEO MRY NCA SCA TEN VRG. Hemicr. or rhizome geophyte.
**Carex curta* var. *brunnescens* Pers., Syn. Pl. 2: 539 (1807). *Carex canescens* var. *brunnescens* (Pers.) W.D.J.Koch, Syn. Fl. Germ. Helv.: 754 (1837). *Facolos brunnescens* (Pers.) Raf., Good Book: 26 (1840). *Carex canescens* subsp. *brunnescens* (Pers.) Asch. & Graebn., Syn. Mitteleur. Fl. 2(2): 61 (1902). *Vignea brunnescens* (Pers.) Soják, Cas. Nár. Mus., Odd. Prír. 148: 194 (1979 publ. 1980).

subsp. ***brunnescens***
Temp. Northern Hemisphere. 10 FIN ICE NOR SWE 11 AUT CZE GER POL SWI 12 FRA 13 ITA ROM YUG 14 BLR BLT RUC RUE RUN RUS RUW 30 ALT KRA WSB 31 KAM KHA KUR MAG SAK 32+ 37 MON 38 JAP KOR 70 ASK GNL NUN NWT YUK 71 ABT BRC MAN SAS 72 LAB NBR NFL NSC ONT PEI QUE 73 COL IDA MNT ORE WAS WYO 74 MIN WIS 75 CNT INI MAI MAS MIC NWH NWJ NWY OHI PEN RHO VER WVA 76 CAL. Hemicr. or rhizome geophyte.
Carex canescens var. *alpicola* Wahlenb., Kongl. Vetensk. Acad. Nya Handl. 24: 147 (1803).
Carex gebhardii Hoppe, Caricolog. Germ.: t. 30 (1826), nom. illeg.
Carex persoonii Sieber ex W.D.J.Koch, Syn. Fl. Germ. Helv., ed. 2: 871 (1844), nom. illeg. *Carex canescens* var. *persoonii* H.Christ, Bull. Soc. Roy. Bot. Belgique 24(2): 18 (1885), nom. illeg.
Carex vitilis var. *sylvatica* Meinsh., Fl. Ingr.: 402 (1878).
Carex brunnescens subsp. *alaskana* Kalela, Ann. Bot. Fenn. 2: 207 (1965). *Vignea brunnescens* subsp. *alaskana* (Kalela) Soják, Cas. Nár. Mus., Odd. Prír. 148: 194 (1979 publ. 1980).
Carex brunnescens subsp. *pacifica* Kalela, Ann. Bot. Fenn. 2: 211 (1965). *Vignea brunnescens* subsp. *pacifica* (Kalela) Soják, Cas. Nár. Mus., Odd. Prír. 148: 194 (1979 publ. 1980).

subsp. ***sphaerostachya*** (Tuck.) Kalela, Ann. Bot. Fenn. 2: 203 (1965).
Subarctic to E. U.S.A. 10 FIN 14 BLT RUC 30 WSB 31+ 32+ 38 JAP 71 MAN 72 LAB NBR NFL NSC ONT PEI QUE 74 MIN WIS 75 CNT INI MAI MAS MIC NWH NWJ NWY OHI PEN RHO VER WVA 78 DEL GEO MRY NCA SCA TEN VRG. Hemicr. or rhizome geophyte.
**Carex canescens* var. *sphaerostachya* Tuck., Enum. Meth. Caric.: 10 (1843). *Carex sphaerostachya* (Tuck.) Dewey, Amer. J. Sci. Arts 49: 44 (1845). *Carex brunnescens* var. *sphaerostachya* (Tuck.) Kük. in H.G.A.Engler (ed.), Pflanzenr., IV, 20(38): 220 (1909).
Carex buckleyi Dewey, Amer. J. Sci. Arts 48: 143 (1845).
Carex vitilis Fr., Novit. Fl. Suec. Mant. 3: 137 (1845). *Carex canescens* var. *vitilis* (Fr.) J.Carey in A.Gray, Manual, ed. 2: 514 (1856). *Carex persoonii* subsp. *vitilis* (Fr.) Nyman, Consp. Fl. Eur.: 780 (1882). *Vignea brunnescens* subsp. *vitilis* (Fr.) Soják, Cas. Nár. Mus., Odd. Prír. 148: 194 (1979 publ. 1980).
Carex brunnescens var. *gracilior* Britton in N.L. Britton & A.Brown, Ill. Fl. N. U.S. 1: 351 (1896).

Carex brunnipes Reznicek, Syst. Bot. 11: 82 (1986).
C. Mexico to Guatemala. 79 MXC MXE MXT 80 GUA. Hemicr. or rhizome geophyte.

Carex brysonii Naczi, Contr. Univ. Michigan Herb. 19: 195 (1993).
Alabama (Black Warrior River). 78 ALA. Hemicr. or rhizome geophyte.

Carex buchananii Berggr., J. Bot. 18: 104 (1880). *Carex lucida* var. *buchananii* (Berggr.) Kük. in H.G.A. Engler (ed.), Pflanzenr., IV, 20(38): 686 (1909).
New Zealand North I. 51 NZN. Hemicr.

Carex bucharica Kük., Repert. Spec. Nov. Regni Veg. 16: 433 (1920). *Kobresia bucharica* (Kük.) K.V. Ivanova, Bot. Zhurn. S.S.S.R. 24: 491 (1939).
C. Asia to NE. Afghanistan. 32 TZK UZB 34 AFG. Hemicr. or rhizome geophyte.

Carex buekii Wimm., Fl. Schles., ed. 3: 81 (1857). *Carex cespitosa* var. *buekii* (Wimm.) Fiori, Nuov. Fl. Italia 1: 188 (1923).
C. Europe to Caucasus. 11 AUT CZE GER HUN POL 13 BUL ITA? ROM YUG 14 RUC RUS UKR 32 KAZ 33 NCS. Hemicr. or rhizome geophyte.
Carex banatica Heuff., Verh. K. K. Zool.-Bot. Ges. Wien 8: 222 (1858).

Carex bulbostylis Mack., Bull. Torrey Bot. Club 42: 617 (1915).
Tennessee to Texas. 74 OKL 77 TEX 78 ARK LOU MSI TEN. Hemicr. or rhizome geophyte.
Carex grisea var. *globosa* L.H.Bailey in A.Gray, Manual: 605 (1848). *Carex amphibola* var. *globosa* (L.H.Bailey) L.H.Bailey, Mem. Torrey Bot. Club 5: 72 (1894).

Carex bullata Willd., Sp. Pl. 4: 309 (1805).
W. Nova Scotia to E. U.S.A. 72 NSC 75 CNT MAI MAS NWJ NWY RHO 78 ARK DEL GEO MRY NCA TEN VRG. Hemicr. or rhizome geophyte.
Carex cylindrica Schwein., Ann. Lyceum Nat. Hist. New York 1: 71 (1824).
Carex greenii Boeck., Flora 41: 649 (1858). *Carex bullata* var. *greenii* (Boeck.) Fernald, Rhodora 8: 202 (1906).
Carex physema Dewey, Amer. J. Sci. Arts, II, 29: 347 (1860).
Carex bullata var. *olneyi* Fernald, Rhodora 3: 52 (1901).

Carex burchelliana Boeck., Linnaea 41: 234 (1877).
NE. Cape Prov., Northern Prov. 27 CPP TVL. Hemicr. or rhizome geophyte.

Carex burttii Noltie, Edinburgh J. Bot. 50: 186 (1993).
E. Himalaya to SE. Tibet. 36 CHT 40 EHM. Hemicr. or rhizome geophyte.

Carex bushii Mack., Bull. Torrey Bot. Club 37: 241 (1910).
C. & E. U.S.A. 74 ILL KAN MSO NEB OKL 75 CNT

INI MAS MIC NWJ NWY OHI PEN RHO WVA 77 TEX 78 ARK DEL GEO MRY MSI NCA TEN VRG WDC. Hemicr. or rhizome geophyte.

Carex hirsuta var. *cuspidata* Dewey in A.Wood, Class-book Bot.: 758 (1860). *Carex caroliniana* var. *cuspidata* (Dewey) Shinners, Field & Lab. 22: 30 (1954).

Carex buxbaumii Wahlenb., Kongl. Vetensk. Acad. Nya Handl. 1803: 163 (1803). *Physiglochis buxbaumii* (Wahlenb.) Raf., Good Book: 27 (1840).

Temp. Northern Hemisphere, SE. Australia. 10 DEN FIN GRB ICE IRE† NOR SWE 11 AUT CZE GER NET POL SWI 12 FRA 13 BUL ITA ROM YUG 14 BLR BLT RUC RUE RUN RUS RUW UKR 30 WSB 31 KUR 32 KAZ 33 NCS 38 JAP KOR 50 VIC 70 ASK GNL NWT YUK 71 ABT BRC MAN SAS 72 LAB NBR NFL NSC ONT QUE 73 COL IDA MNT ORE WAS WYO 74 ILL IOW KAN MIN MSO NDA NEB SDA WIS 75 CNT INI MAI MAS MIC NWH NWY OHI PEN RHO VER WVA 76 UTA 78 ARK DEL MRY NCA SCA VRG. Hemicr. or rhizome geophyte.

Carex polygama Schkuhr, Beschr. Riedgräs. 1: 84 (1801), nom. illeg.

Carex subulata Schumach., Enum. Pl. 1: 270 (1801), nom. illeg.

Carex buxbaumii var. *australis* Andersson, Cyper. Scand.: 39 (1849).

Carex buxbaumii f. *heterostachya* Andersson, Cyper. Scand.: 39 (1849). *Carex polygama* var. *heterostachya* (Andersson) Farw., Rep. (Annual) Michigan Acad. Sci. 19: 254 (1917).

Carex oligandra F.Muell. ex Boott, Ill. Carex 4: 136 (1867).

Carex tubulata K.Schum. ex Boott, Ill. Carex 4: 136 (1867).

Carex pseudobuxbaumii M.Winkl., Oesterr. Bot. Z. 18: 72 (1868).

Carex buxbaumii f. *oenipontana* Grembl. ex Appel, Mitth. Thüring. Bot. Vereins, n.s., 8: 43 (1890).

Carex buxbaumii f. *mitis* Norman ex Christian, Forh. Vidensk.-Selsk. Kristiania 1893: 50 (1893).

Carex buxbaumii f. *virescens* Norman ex Christian, Forh. Vidensk.-Selsk. Kristiania 1893: 50 (1893).

Carex picea Franch., Bull. Soc. Philom. Paris, VIII, 7: 39 (1895).

Carex buxbaumii f. *dilutior* Kük. in H.G.A.Engler (ed.), Pflanzenr., IV, 20(38): 394 (1909).

Carex holmiana Mack., Bull. Torrey Bot. Club 36: 481 (1909).

Carex buxbaumii var. *anticostensis* Raymond, Naturaliste Canad. 77: 58 (1950).

Carex buxbaumii f. *pedunculata* Raymond, Naturaliste Canad. 77: 58 (1950).

Carex caduca Boott, Ill. Carex 4: 157 (1867).

SC. & S. Chile, S. Argentina, Falkland Is. 85 AGS CLC CLS 90 FAL. Hemicr. or rhizome geophyte.

Carex ortegae Phil., Anales Univ. Chile 93: 488 (1896). *Carex caduca* var. *ortegae* (Phil.) Kük., Bot. Jahrb. Syst. 27: 497 (1899).

Carex caeligena Reznicek, Contr. Univ. Michigan Herb. 17: 280 (1990).

Mexico (Tamaulipas). 79 MXE. Hemicr. or rhizome geophyte.

Carex ×caesariensis Mack. in N.L.Britton & al. (eds.), N. Amer. Fl. 18: 440 (1935). *C. pellita* × *C. trichocarpa*.

NE. U.S.A. 75 NWJ PEN. Hemicr. or rhizome geophyte.

Carex caespititia Nees in R.Wight, Contr. Bot. India: 127 (1834).

Assam to SC. China. 36 CHC CHT 40 ASS. Hemicr. or rhizome geophyte.

Carex calcicola Tang & F.T.Wang, Acta Phytotax. Sin. 24: 244 (1986).

China (Guizhou). 36 CHC. Hemicr. or rhizome geophyte.

Carex calcifugens Naczi, Novon 12: 512 (2002).

SE. U.S.A. 78 FLA GEO NCA SCA. Hemicr. or rhizome geophyte.

Carex californica L.H.Bailey, Mem. Torrey Bot. Club 1: 9 (1889). *Carex polymorpha* var. *californica* (L.H.Bailey) Kük. in H.G.A.Engler (ed.), Pflanzenr., IV, 20(38): 515 (1909).

W. U.S.A. 73 IDA ORE WAS 76 CAL. Hemicr. or rhizome geophyte.

Carex callista Nelmes, Mém. Mus. Natl. Hist. Nat., Sér. B, Bot. 4: 119 (1955).

Myanmar. 41 MYA. Hemicr. or rhizome geophyte.

Carex callitrichos V.I.Krecz. in V.L.Komarov (ed.), Fl. URSS 3: 613 (1935). *Carex nanella* var. *callitrichos* (V.I.Krecz.) Vorosch. in A.K.Skvortsov (ed.), Florist. Issl. Razn. Raionakh SSSR: 155 (1985).

Siberia to Japan (Kyushu). 30+ 31 KUR PRM 36 CHM 38 JAP KOR. Hemicr. or rhizome geophyte.

Carex cambodiensis Nelmes, Kew Bull. 1: 11 (1946).

Cambodia. 41 CBD. Hemicr. or rhizome geophyte.

Carex camposii Boiss. & Reut., Pugill. Pl. Afr. Bor. Hispan.: 117 (1852).

SE. Portugal, S. Spain. 12 POR SPA. Hemicr. or rhizome geophyte.

Carex laevigata Boiss., Voy. Bot. Espagne: 631 (1842), nom. illeg.

Carex campylorhina V.I.Krecz. in V.L.Komarov (ed.), Fl. URSS 3: 612 (1935).

Russian Far East to NE. China. 31+ 36 CHM. Hemicr. or rhizome geophyte.

**Carex pilosa* f. *gracilior* Meinsh., Trudy Imp. S.-Peterburgsk. Bot. Sada 18(3): 385 (1901).

Carex canaliculata P.C.Li, Acta Bot. Yunnan. 12: 148 (1990).

SC. China. 36 CHC. Hemicr. or rhizome geophyte.

Carex canariensis Kük., Allg. Bot. Z. Syst. 6: 235 (1900). *Vignea canariensis* (Kük.) Soják, Cas. Nár. Mus., Odd. Prír. 148: 195 (1979 publ. 1980).

Canary Is. 21 CNY. Hemicr. or rhizome geophyte.

Carex ×candriana Kneuck., Allg. Bot. Z. Syst. 5: 195 (1899). *C. atrata* × *C. norvegica*.

N. Europe. 10 FIN. Hemicr. or rhizome geophyte.

Carex canescens L., Sp. Pl.: 974 (1753). *Vignea canescens* (L.) Rchb. in J.C.Mössler, Handb. Gewächsk. ed. 2, 3: 1612 (1830). *Caricina canescens* (L.) St.-Lag. in A.Cariot, Étude Fl., ed. 8, 2: 877 (1889).

Temp. 10 DEN FIN GRB ICE IRE NOR SWE 11 AUT BGM CZE GER HUN NET POL SWI 12 SPA 13 BUL ITA ROM YUG 14 BLR BLT RUC RUE RUN RUS RUW UKR 30 ALT KRA WSB YAK 31 KAM KHA MAG PRM SAK 32 KAZ KGZ UZB 33 TCS 34 TUR 36 CHI CHM CHX 37

MON 38 JAP KOR 40 WHM 43 NWG 50 NSW TAS VIC 70 ASK GNL NUN NWT YUK 71 ABT BRC MAN SAS 72 LAB NBR NFL NSC ONT PEI QUE 73 COL IDA MNT ORE WAS WYO 74 ILL MIN SDA WIS 75 CNT INI MAI MAS MIC NWH NWJ NWY OHI PEN RHO VER 76 ARI CAL NEV UTA 77 NWM 78 DEL MRY NCA SCA VRG WDC 85 AGE AGS CLC CLS 90 FAL. Hemicr. or rhizome geophyte.

subsp. ***canescens***

Temp. Northern Hemisphere, SE. Australia. 10 DEN FIN GRB ICE IRE NOR SWE 11 AUT BGM CZE GER HUN NET POL SWI 12 SPA 13 BUL ITA ROM YUG 14 BLR BLT RUC RUE RUN RUS RUW UKR 30 ALT KRA WSB YAK 31 KAM KHA MAG PRM SAK 32 KGZ UZB 33 TCS 34 TUR 36 CHI CHM CHX 37 MON 38 JAP KOR 40 WHM 43 NWG 50 NSW TAS VIC 70 ASK GNL NUN NWT YUK 71 ABT BRC MAN SAS 72 LAB NBR NFL NSC ONT PEI QUE 73 COL IDA MNT ORE WAS WYO 74 ILL MIN SDA WIS 75 CNT INI MAI MAS MIC NWH NWJ NWY OHI PEN RHO VER 76 ARI CAL NEV UTA 77 NWM 78 DEL MRY NCA SCA VRG WDC. Hemicr. or rhizome geophyte.

Carex brizoides Huds., Fl. Angl.: 349 (1762), nom. illeg.

Carex cinerea Pollich, Hist. Pl. Palat. 2: 571 (1777). *Vignea cinerea* (Pollich) Dostál, Folia Mus. Rerum Nat. Bohemiae Occid., Bot. 21: 16 (1984).

Carex richardii Thuill., Fl. Env. Paris: 482 (1790).

Carex tenella Ehrh., Beitr. Naturk. 6: 83 (1791), nom. illeg.

Carex curta Gooden., Trans. Linn. Soc. London 2: 145 (1794). *Facolos curta* (Gooden.) Raf., Good Book: 26 (1840). *Vignea curta* (Gooden.) Opiz, Seznam: 103 (1852).

Carex compressa Hosé, Ann. Bot. (Usteri) 21: 33 (1797). *Edritria compressa* (Hosé) Raf., Good Book: 26 (1840).

Carex badia Pers., Syn. Pl. 2: 541 (1807). *Onkerma badia* (Pers.) Raf., Good Book: 27 (1840).

Carex canescens f. *bracteata* Klett & Richt., Fl. Leipzig: 752 (1830).

Carex persoonii Láng, Flora 25: 748 (1842). *Vignea persoonii* (Láng) Schur, Verh. Mitth. Siebenbürg. Vereins Naturwiss. Hermannstadt 4: 78 (1853).

Carex canescens var. *tenuis* Láng, Linnaea 24: 538 (1851).

Carex salina Bong. ex Ledeb., Fl. Ross. 4: 281 (1852), nom. illeg.

Carex canescens var. *robusta* A.Blytt, Norges Fl. 1: 201 (1861).

Carex canescens var. *laetevirens* Asch., Fl. Brandenburg 1: 770 (1864).

Carex canescens var. *gracillima* Schur, Enum. Pl. Transsilv.: 702 (1866).

Carex breutelii Schnizl. ex Lange, Consp. Fl. Groenland. 1: 136 (1881).

Carex canescens var. *longibracteata* Porcius, Magyar Növényt. Lapok 9: 131 (1885).

Carex kanitzii Porcius, Magyar Növényt. Lapok 9: 131 (1885).

Carex canescens var. *brunnea* Macoun, Cat. Canad. Pl. 2: 377 (1890).

Carex canescens var. *robustina* Macoun, Cat. Canad. Pl. 2: 376 (1890).

Carex subsabulosa Norman, Forh. Vidensk.-Selsk. Kristiania 1893: 48 (1893). *Carex canescens* f. *subsabulosa* (Norman) Toivonen, Acta Bot. Fenn. 116: 17 (1981).

Carex canescens var. *subvitilis* Neuman, Sver. Fl.: 710 (1901).

Carex canescens var. *congesta* Domin, Sitzungsber. Königl. Böhm. Ges. Wiss. Prag, Math.-Naturwiss. Cl. 58: 47 (1903).

Carex disjuncta E.P.Bicknell, Bull. Torrey Bot. Club 35: 495 (1908).

Carex canescens var. *subtenella* Kük. in H.G.A. Engler (ed.), Pflanzenr., IV, 20(38): 218 (1909).

Carex canescens var. *heterostachya* Farw., Rep. (Annual) Michigan Acad. Sci. 19: 247 (1917).

Carex hylaea V.I.Krecz. in V.L.Komarov (ed.), Fl. URSS 3: 594 (1935).

subsp. ***disjuncta*** (Fernald) Toivonen, Ann. Bot. Fenn. 18: 94 (1981).

E. Canada to NC. & E. U.S.A. 72 LAB NBR NFL NSC ONT PEI QUE 74 ILL MIN WIS 75 CNT INI MAI MAS MIC NWH NWJ NWY OHI PEN RHO VER 78 DEL MRY NCA SCA VRG WDC. Hemicr. or rhizome geophyte.

**Carex canescens* var. *disjuncta* Fernald, Proc. Amer. Acad. Arts 37: 488 (1902).

var. ***robustior*** Blytt ex Andersson, Cyper. Scand.: 57 (1849). *Carex curta* var. *robustior* (Blytt ex Andersson) D.M.Moore & Chater, Bot. Not. 124: 326 (1971).

S. South America, Falkland Is. 85 AGE AGS CLC CLS 90 FAL. Hemicr. or rhizome geophyte.

Carex similis d'Urv., Mém. Soc. Linn. Paris 4: 599 (1826).

Carex canescens var. *alpicola* Kurtz, Revista Fac. Agron. Univ. Nac. La Plata 7: 353 (1896).

Carex canescens var. *dubia* Kük., Bot. Jahrb. Syst. 27: 509 (1899), nom. illeg.

Carex canescens var. *fallax* Kurtz ex Kük. in P.F.A. Ascherson & K.O.R.Graebner, Syn. Mitteleur. Fl. 2(2): 61 (1902).

Carex skottsbergii Gand., Bull. Soc. Bot. France 60: 27 (1913), nom. illeg.

Carex canina Dunn, J. Bot. 48: 325 (1910).

SE. China. 36 CHS. Hemicr. or rhizome geophyte.

Carex capillacea Boott, Ill. Carex 1: 44 (1858). *Carex rara* subsp. *capillacea* (Boott) Kük. in H.G.A.Engler (ed.), Pflanzenr., IV, 20(38): 102 (1909).

Assam to Russian Far East and New Guinea, SE. Australia, New Zealand. 31 KUR PRM SAK 36 CHC CHS CHT 38 JAP KOR TAI 40 ASS 41 VIE 42 BOR JAW PHI SUL SUM 43 NWG 50 NSW TAS VIC 51 NZS. Hemicr. or rhizome geophyte.

var. ***capillacea***

Assam to Japan and New Guinea, SE. Australia, New Zealand. 36 CHC CHS CHT 38 JAP KOR TAI 40 ASS 41 VIE 42 BOR JAW PHI SUL SUM 43 NWG 50 NSW TAS VIC 51 NZS. Hemicr. or rhizome geophyte.

Carex ontakensis Franch. & Sav., Enum. Pl. Jap. 1: 550 (1875).

Carex simplicissima F.Muell., Fragm. 9: 191 (1875).

Carex capillacea var. *yunnanensis* Franch., Nouv. Arch. Mus. Hist. Nat., III, 8: 198 (1896).

Carex wallii Petrie, Trans. & Proc. New Zealand Inst. 53: 371 (1921).

Carex capillacea var. *linzensis* Y.C.Yang, in Fl. Xizangica 5: 442 (1987).

var. ***sachalinensis*** (F.Schmidt) Ohwi, Mem. Coll. Sci. Kyoto Imp. Univ., Ser. B, Biol. 11: 442 (1936).
Russian Far East to New Guinea. 31 KUR PRM SAK 38 JAP KOR 42 BOR PHI 43 NWG. Hemicr. or rhizome geophyte.
Carex nana Boott, Mem. Amer. Acad. Arts, n.s., 6: 418 (1859), nom. illeg. *Carex rara* var. *nana* Kük. in H.G.A.Engler (ed.), Pflanzenr., IV, 20(38): 103 (1909).
**Carex uda* var. *sachalinensis* F.Schmidt, Reis. Amur-Land., Bot. 1: 353 (1868).
Carex aomorensis Franch., Nouv. Arch. Mus. Hist. Nat., III, 8: 198 (1896). *Carex capillacea* subsp. *aomorensis* (Franch.) T.V.Egorova, Novosti Sist. Vyssh. Rast. 22: 58 (1985).

Carex capillaris L., Sp. Pl.: 977 (1753). *Trasus capillaris* (L.) Gray, Nat. Arr. Brit. Pl. 2: 66 (1821). *Loxotrema capillaris* (L.) Raf., Good Book: 25 (1840). *Carex capillaris* f. *typica* B.Boivin, Naturaliste Canad. 75: 203 (1948), nom. inval.
Subarctic & Temp. Northern Hemisphere. 10 FIN FOR GRB ICE NOR SVA SWE 11 AUT CZE GER POL SWI 12 FRA SPA 13 ITA ROM YUG 14 BLR BLT RUC RUE RUN RUS RUW UKR 20 MOR 30 KRA WSB YAK 31 KUR MAG SAK 32 KAZ KGZ UZB 33 NCS 34 TUR 36 CHI CHN CHQ CHT CHX 38 JAP KOR 40 PAK 70 ASK GNL NUN NWT YUK 71 ABT BRC MAN SAS 72 LAB NBR NFL NSC ONT QUE 73 COL IDA MNT ORE WAS WYO 74 MIN NDA SDA WIS 75 MAI MIC NWH NWY 76 CAL NEV UTA 77 NWM. Hemicr. or rhizome geophyte.

subsp. ***capillaris***
Temp. Northern Hemisphere. 10 FIN FOR GRB ICE NOR SVA SWE 11 AUT CZE GER POL SWI 12 FRA SPA 13 ITA ROM YUG 14 BLR BLT RUC RUE RUN RUS RUW UKR 20 MOR 30 KRA WSB YAK 31 KUR SAK 32 KAZ KGZ UZB 33 NCS 34 TUR 36 CHI CHN CHQ CHT CHX 38 JAP KOR 40 PAK 70 ASK GNL NUN NWT YUK 71 ABT BRC MAN SAS 72 LAB NBR NFL NSC ONT QUE 73 COL IDA MNT ORE WAS WYO 74 MIN NDA SDA WIS 75 MAI MIC NWH NWY 76 CAL NEV UTA 77 NWM. Hemicr. or rhizome geophyte.
Carex pendula Geners., Elench.: 871 (1798), nom. illeg.
Carex plena Clairv., Man. Herbor. Suisse: 292 (1811).
Carex chlorostachys Steven, Mém. Soc. Imp. Naturalistes Moscou 4: 68 (1813). *Carex capillaris* subsp. *chlorostachys* (Steven) Á.Löve, D.Löve & Raymond, Canad. J. Bot. 35: 749 (1957).
Carex capillaris f. *major* Drejer, Rev. Car. Bor.: 54 (1841).
Carex capillaris var. *tenuior* Drejer, Rev. Car. Bor.: 53 (1841).
Carex nana Cham. ex Steud., Syn. Pl. Glumac. 2: 228 (1855). *Carex capillaris* var. *nana* (Cham. ex Steud.) Kük. in H.G.A.Engler (ed.), Pflanzenr., IV, 20(38): 591 (1909).
Carex capillaris var. *paludosa* F.Schmidt, Reis. Amur-Land., Bot.: 69 (1868).
Carex saskatschewana Boeck., Linnaea 41: 159 (1877). *Carex capillaris* f. *saskatschewana* (Boeck.) L.H.Bailey, Mem. Torrey Bot. Club 1: 70 (1889).
Carex capillaris f. *minima* Beck, Fl. Nieder-Österreich: 144 (1890).
Carex capillaris var. *elongata* Olney ex Fernald, Proc. Amer. Acad. Arts 37: 509 (1902).
Carex tiogana D.W.Taylor & J.D.Mastrog., Novon 9: 120 (1999).
Carex chlorostachys var. *conferta* Tang & F.T.Wang ex L.K.Dai, in Fl. Reipubl. Popul. Sin. 12: 526 (2000).

subsp. ***fuscidula*** (V.I.Krecz. ex Egorova) Á.Löve & D.Löve, Bot. Not. 128: 504 (1976).
Subarctic. 14 RUE RUN 30 KRA WSB YAK 31 MAG 70 ASK. Hemicr. or rhizome geophyte.
**Carex fuscidula* V.I.Krecz. ex Egorova, Novosti Sist. Vyssh. Rast. 1: 36 (1964).

Carex capilliculmis S.R.Zhang, Novon 15: 498 (2005).
Qinghai to C. China. 36 CHC CHN CHQ. Hemicr. or rhizome geophyte.
**Carex filamentosa* K.T.Fu, in Fl. Tsinlingensis 1(1): 445 (1976), nom. illeg.

Carex capilliformis Franch., Bull. Soc. Philom. Paris, VIII, 7: 89 (1895).
C. China. 36 CHC CHN. Hemicr. or rhizome geophyte.
Carex capilliformis var. *major* Kük., Bot. Jahrb. Syst. 36(82): 9 (1905).

Carex capitata L., Syst. Nat. ed. 10, 2: 1261 (1759). *Diemisa capitata* (L.) Raf., Good Book: 27 (1840). *Psyllophora capitata* (L.) Schur, Enum. Pl. Transsilv.: 697 (1866). *Vignea capitata* (L.) Soják, Cas. Nár. Mus., Odd. Přír. 148: 195 (1979 publ. 1980).
Alps, N. & E. Europe to Russian Far East, Subarctic America to W. U.S.A., S. South America. 10 FIN ICE NOR SWE 11 AUT 12 FRA† 13 ITA 14 RUE RUN 30 ALT BRY CTA IRK KRA WSB YAK 31 KHA MAG 33 NCS TCS 37 MON 70 ASK GNL NUN NWT YUK 71 ABT BRC MAN SAS 72 LAB NFL ONT QUE 73 COL MNT WYO 75 NWH 76 CAL NEV 79 MXE MXN 85 AGS AGW CLC. Hemicr. or rhizome geophyte.
Carex capitata f. *alpicola* Andersson, Nya Bot. Not. Stockh.: 2 (1849).
Carex arctogena Harry Sm., Acta Phytogeogr. Suec. 13: 193 (1940). *Carex capitata* f. *arctogena* (Harry Sm.) Raymond, Contr. Inst. Bot. Univ. Montréal 64: 38 (1949). *Carex capitata* subsp. *arctogena* (Harry Sm.) Böcher ex Á.Löve, D.Löve & B.M.Kapoor, Arctic Alpine Res. 3: 144 (1971).
Carex antarctogena Roiv., Ann. Bot. Soc. Zool.-Bot. Fenn. "Vanamo" 28(2): 197 (1954).

Carex capitellata Boiss. & Balansa in P.E.Boissier, Fl. Orient. 5: 399 (1882).
N. Turkey to Transcaucasus. 33 TCS 34 TUR. Hemicr. or rhizome geophyte.

Carex capricornis Meinsh. ex Maxim., Bull. Acad. Imp. Sci. Saint-Pétersbourg 31: 119 (1887).
Siberia to N. & C. Japan. 30+ 31 PRM 36 CHM CHS 38 JAP KOR. Hemicr. or rhizome geophyte.
Carex pseudocyperus var. *brachystachya* Regel & Maack, Tent. Fl. Ussur.: 165 (1861). *Carex brachystachya* (Regel & Maack) Akiyama, J. Fac. Sci. Hokkaido Univ., Ser. 5, Bot. 2: 219 (1932), nom. illeg.

Carex cardiolepis Nees in R.Wight, Contr. Bot. India: 127 (1834).
Afghanistan to SC. China. 34 AFG 36 CHC CHQ CHT 40 NEP PAK WHM. Hemicr. or rhizome geophyte.

Carex careyana Torr. ex Dewey, Amer. J. Sci. Arts 30: 60 (1836).
SE. Canada to NC. & E. U.S.A. 72 ONT 74 ILL IOW MIN MSO 75 INI MIC NWY OHI PEN WVA 78 ALA ARK GEO KTY TEN VRG WDC. Hemicr. or rhizome geophyte.
Carex heterostachya Torr., Amer. J. Sci. Arts, II, 2: 248 (1846), nom. illeg.

Carex ×cariei Aubin, Bull. Mens. Soc. Linn. Lyon 56: 18 (1987). *C. pallescens* × *C. panicea*.
France. 12 FRA. Hemicr. or rhizome geophyte.

Carex caroliniana Schwein., Ann. Lyceum Nat. Hist. New York 1: 67 (1824).
C. & E. U.S.A. 74 KAN MSO OKL 75 INI NWJ OHI PEN WVA 77 TEX 78 ARK DEL GEO MRY MSI NCA TEN VRG WDC. Hemicr. or rhizome geophyte.

Carex carsei Petrie, Trans. & Proc. New Zealand Inst. 54: 570 (1923).
New Zealand North I. 51 NZN. Hemicr. or rhizome geophyte.

Carex caryophyllea Latourr., Chlor. Lugd.: 27 (1785).
Temp. Old World. 10 DEN FIN GRB IRE NOR SWE 11 AUT BGM CZE GER HUN NET POL SWI 12 COR FRA POR SAR SPA 13 ALB BUL GRC ITA ROM SIC TUE YUG 14 BLR BLT KRY RUC RUE RUS RUW UKR 30 WSB 31 KAM KUR SAK 32 KAZ TKM 33 NCS TCS 34 TUR 37 MON 38 JAP KOR (75) mai mas (78) wdc. Hemicr. or rhizome geophyte.

var. **caryophyllea**
Temp. Old World. 10 DEN FIN GRB IRE NOR SWE 11 AUT BGM CZE GER HUN NET POL SWI 12 COR FRA POR SAR SPA 13 ALB BUL GRC ITA ROM SIC TUE YUG 14 BLR BLT KRY RUC RUE RUS RUW UKR 30 WSB 32 KAZ TKM 33 NCS TCS 34 TUR 37 MON 38 JAP (75) mai mas (78) wdc. Hemicr. or rhizome geophyte.
Carex praecox Jacq., Fl. Austriac. 5: 23 (1778), nom. illeg.
Carex verna Chaix & Vill., Hist. Pl. Dauphiné 1: 312 (1786), nom. illeg. *Olamblis verna* (Chaix & Vill.) Raf., Good Book: 26 (1840).
Carex refracta Roth, Tent. Fl. Germ. 2(2): 451 (1793). *Carex praecox* var. *reflexa* (Roth) Rchb., Icon. Fl. Germ. Helv. 12: t. 635 (1850). *Carex caryophyllea* f. *refracta* (Roth) Kük. in H.G.A.Engler (ed.), Pflanzenr., IV, 20(38): 465 (1909).
Carex mollis Host, Icon. Descr. Gram. Austriac. 4: 46 (1809), nom. illeg. *Carex caryophyllea* var. *mollis* Asch. & Graebn., Syn. Mitteleur. Fl. 1: 126 (1896).
Carex praecox var. *caespitosa* J.Fleisch., Riedgr. Wurtemb.: 18 (1832). *Carex caryophyllea* var. *caespitosa* (J.Fleisch.) Kük. in H.G.A.Engler (ed.), Pflanzenr., IV, 20(38): 465 (1909).
Carex praecox f. *pygmaea* J.Fleisch., Riedgr. Wurtemb.: 18 (1832). *Carex caryophyllea* f. *pygmaea* (J.Fleisch.) Kük. in H.G.A.Engler (ed.), Pflanzenr., IV, 20(38): 464 (1909).
Carex praecox f. *umbrosiformis* J.Fleisch., Riedgr. Wurtemb.: 18 (1832). *Carex caryophyllea* f. *umbrosiformis* (J.Fleisch.) Kük. in H.G.A.Engler (ed.), Pflanzenr., IV, 20(38): 464 (1909).
Carex praecox var. *elatior* Bogenh., Taschenb. Fl. Jena: 378 (1850).
Carex trachyantha Dorner ex Heuff., Verh. K. K. Zool.-Bot. Ges. Wien 8: 219 (1858). *Carex praecox* var. *trachyantha* (Dorner ex Heuff.) Nyman, Consp. Fl. Eur.: 773 (1882).
Carex anomala Janka, Linnaea 30: 609 (1860), nom. illeg.
Carex praecox var. *dumetorum* A.Blytt, Norges Fl. 1: 225 (1861).
Carex praecox var. *magna* Schur, Enum. Pl. Transsilv.: 716 (1866).
Carex pediformis var. *rostrata* F.Schmidt, Reis. Amur-Land., Bot.: 68 (1868), nom. illeg.
Carex setosa Franch. & Sav., Enum. Pl. Jap. 1: 142 (1873), nom. illeg.
Carex caryophyllea subsp. *nervata* Franch. & Sav., Enum. Pl. Jap. 2: 141 (1877).
Carex praecox var. *vidalii* Franch. & Sav., Enum. Pl. Jap. 2: 565 (1878). *Carex caryophyllea* f. *vidalii* (Franch. & Sav.) Kük. in H.G.A.Engler (ed.), Pflanzenr., IV, 20(38): 466 (1909).
Carex praecox var. *insularis* H.Christ in W.Barbey, Fl. Sard. Comp.: 64 (1885). *Carex insulana* (H.Christ) Rouy in G.Rouy & J.Foucaud, Fl. France 13: 443 (1912). *Carex caryophyllea* subsp. *insularis* (H.Christ) Arrigoni, Boll. Soc. Sarda Sci. Nat. 23: 220 (1984).
Carex praecox var. *diastachya* Callmé, Deutsche Bot. Monatsschr. 6: 2 (1888).
Carex verna f. *longibracteata* Beck, Fl. Nieder-Österreich: 238 (1890). *Carex caryophyllea* f. *longibracteata* (Beck) Kük. in H.G.A.Engler (ed.), Pflanzenr., IV, 20(38): 465 (1909).
Carex verna var. *minor* Beck, Fl. Nieder-Österreich: 139 (1890).
Carex verna f. *pedunculata* Beck, Fl. Nieder-Österreich: 138 (1890). *Carex caryophyllea* f. *pedunculata* (Beck) Kük. in H.G.A.Engler (ed.), Pflanzenr., IV, 20(38): 465 (1909).
Carex verna var. *caespitiformis* Waisb., Oesterr. Bot. Z. 43: 281 (1893).
Carex caryophyllea var. *angustifolia* Asch. & Graebn., Syn. Mitteleur. Fl. 1: 126 (1896).
Carex caryophyllea f. *pallescens* Kneuck., Allg. Bot. Z. Syst. 2: 104 (1896).
Carex praecox f. *fissa* Kük., Mitth. Thüring. Bot. Vereins, n.s., 10: 37 (1897).
Carex praecox var. *longifrons* Waisb. & Kük., Mitth. Thüring. Bot. Vereins, n.s., 10: 37 (1897).
Carex praecox f. *longivaginata* Kük., Mitth. Thüring. Bot. Vereins, n.s., 10: 37 (1897). *Carex caryophyllea* f. *longivaginata* (Kük.) Kük. in H.G.A.Engler (ed.), Pflanzenr., IV, 20(38): 465 (1909).
Carex praecox f. *luxurians* Kük., Mitth. Thüring. Bot. Vereins, n.s., 10: 37 (1897). *Carex caryophyllea* f. *luxurians* (Kük.) Kük. in H.G.A.Engler (ed.), Pflanzenr., IV, 20(38): 464 (1909).
Carex praecox f. *subpolyrrhiza* Kük., Mitth. Thüring. Bot. Vereins, n.s., 10: 37 (1897). *Carex caryophyllea* f. *subpolyrrhiza* (Kük.) Kük. in H.G.A.Engler (ed.), Pflanzenr., IV, 20(38): 465 (1909).
Carex verna f. *argyrata* Waisb., Oesterr. Bot. Z. 47: 430 (1897). *Carex caryophyllea* f. *argyrata* (Waisb.) Kük. in H.G.A.Engler (ed.), Pflanzenr., IV, 20(38): 465 (1909).

Carex verna f. *chlorantha* Waisb., Oesterr. Bot. Z. 47: 430 (1897). *Carex caryophyllea* f. *chlorantha* (Waisb.) Kük. in H.G.A.Engler (ed.), Pflanzenr., IV, 20(38): 465 (1909).
Carex verna f. *globuliflora* Waisb., Oesterr. Bot. Z. 47: 430 (1897). *Carex caryophyllea* f. *globuliflora* (Waisb.) Kük. in H.G.A.Engler (ed.), Pflanzenr., IV, 20(38): 465 (1909).
Carex verna f. *glomerata* Waisb., Oesterr. Bot. Z. 47: 4 (1897). *Carex caryophyllea* f. *glomerata* (Waisb.) Kük. in H.G.A.Engler (ed.), Pflanzenr., IV, 20(38): 465 (1909).
Carex verna f. *laxiflora* Waisb., Oesterr. Bot. Z. 47: 430 (1897). *Carex caryophyllea* f. *laxiflora* (Waisb.) Kük. in H.G.A.Engler (ed.), Pflanzenr., IV, 20(38): 465 (1909).
Carex verna var. *oxycarpa* Waisb., Oesterr. Bot. Z. 47: 430 (1897). *Carex caryophyllea* var. *oxycarpa* (Waisb.) Kük. in H.G.A.Engler (ed.), Pflanzenr., IV, 20(38): 465 (1909).
Carex praecox var. *umbrosa* De Wild. & T.Durand, Prodr. Fl. Belg. 3: 64 (1899).
Carex heribaudiana H.Lév. & Vaniot, Bull. Acad. Int. Géogr. Bot. 10: 275 (1901).
Carex multipes H.Lév. & Vaniot, Bull. Acad. Int. Géogr. Bot. 10: 271 (1901).
Carex caryophyllea f. *dissita* Franch. ex Kük. in H.G.A.Engler (ed.), Pflanzenr., IV, 20(38): 466 (1909).
Carex caryophyllea f. *leiocarpa* Franch. ex Kük. in H.G.A.Engler (ed.), Pflanzenr., IV, 20(38): 466 (1909).
Carex fuscotincta Merino, Fl. Galicia 3: 179 (1909).
Carex ruthenica V.I.Krecz. in V.L.Komarov (ed.), Fl. URSS 3: 610 (1935).
Carex scabricuspis V.I.Krecz. in V.L.Komarov (ed.), Fl. URSS 3: 609 (1935).

var. ***microtricha*** (Franch.) Kük. in H.G.A.Engler (ed.), Pflanzenr., IV, 20(38): 466 (1909).
Russian Far East, Korea, N. & C. Japan. 31 KAM KUR SAK 38 JAP KOR. Hemicr. or rhizome geophyte.
**Carex microtricha* Franch., Nouv. Arch. Mus. Hist. Nat., III, 9: 189 (1897).
Carex squamoidea Akiyama, J. Fac. Sci. Hokkaido Univ., Ser. 5, Bot. 1: 60 (1931).

Carex castanea Wahlenb., Kongl. Vetensk. Acad. Nya Handl. 24: 155 (1803). *Loxotrema castanea* (Wahlenb.) Raf., Good Book: 25 (1840).
C. & E. Canada to NC. & NE. U.S.A. 71 MAN 72 LAB NBR NFL NSC ONT QUE 74 MIN WIS 75 CNT MAI MIC MWH NWY VER 78 TEN. Hemicr. or rhizome geophyte.
Carex flexilis Rudge, Trans. Linn. Soc. London 7: 98 (1804).
Carex blepharophora A.Gray, Ann. Lyceum Nat. Hist. New York 3: 235 (1835).

Carex castanostachya K.Schum. ex Kük. in H.G.A.Engler (ed.), Pflanzenr., IV, 20(38): 276 (1909).
Kenya (Taita Hills), Tanzania. 25 KEN TAN. Hemicr.

Carex catamarcensis C.B.Clarke ex Kük., Bot. Jahrb. Syst. 27: 518 (1899).
Argentina (Catamarca, Tucumán), N. Chile. 85 AGW CLN. Hemicr. or rhizome geophyte.
Carex latibracteolata Kük., Bot. Jahrb. Syst. 27: 518 (1899).

Carex cataphyllodes Nelmes, Kew Bull. 1: 13 (1946).
SW. Thailand. 41 THA. Hemicr. or rhizome geophyte.

Carex cataractae R.Br., Prodr.: 242 (1810). *Carex oederi* var. *cataractae* (R.Br.) Kük. in H.G.A.Engler (ed.), Pflanzenr., IV, 20(38): 675 (1909).
SE. Australia, New Zealand, S. South America. 50 TAS 51 NZS 85 AGS CLC. Hemicr. or rhizome geophyte.

Carex caudispicata F.T.Wang & Tang ex P.C.Li, Acta Phytotax. Sin. 37: 170 (1999).
China (Yunnan). 36 CHC. Hemicr. or rhizome geophyte.

Carex cavaleriensis H.Lév. & Vaniot, Repert. Spec. Nov. Regni Veg. 3: 349 (1907).
China. 36+. Hemicr. or rhizome geophyte.

Carex caxinensis F.J.Herm., Brittonia 23: 146 (1971).
Guatemala. 80 GUA. Hemicr. or rhizome geophyte.

Carex celebica Kük., Bot. Jahrb. Syst. 70: 465 (1940).
Malesia to New Guinea. 42 SUL SUM 43 NWG. Hemicr. or rhizome geophyte.
Carex constricta S.T.Blake, J. Arnold Arbor. 28: 112 (1947).

Carex cenantha A.E.Kozhevn., Byull. Moskovsk. Obshch. Isp. Prir., Otd. Biol. 92(6): 129 (1987).
Russian Far East. 31+. Hemicr. or rhizome geophyte.

Carex cephaloidea (Dewey) Dewey ex Boott, Ill. Carex 3: 123 (1862).
E. Canada to NC. & NE. U.S.A. 72 NBR ONT QUE 74 ILL IOW MIN WIS 75 CNT INI MAI MAS MIC NWH NWJ NWY OHI PEN VER. Hemicr. or rhizome geophyte.
**Carex muricata* var. *cephaloidea* Dewey, Amer. J. Bot. 11: 308 (1826). *Carex sparganioides* var. *cephaloidea* (Dewey) Carey in A.Gray, Manual: 513 (1848).
Carex sparganioides var. *minor* Boott, Ill. Carex 3: 123 (1862).

Carex cephalophora Muhl. ex Willd., Sp. Pl. 4: 220 (1805). *Vignea cephalophora* (Muhl. ex Willd.) Rchb. in J.C.Mössler, Handb. Gewächsk. ed. 2, 3: 1607 (1830). *Diemisa cephalophora* (Muhl. ex Willd.) Raf., Good Book: 27 (1840).
SE. Canada to C. & E. U.S.A. 72 ONT QUE 74 ILL IOW KAN MIN MSO NEB OKL WIS 75 CNT INI MAI MAS MIC NWH NWJ NWY OHI PEN RHO VER WVA 77 TEX 78 ALA ARK DEL FLA GEO KTY LOU MRY NCA SCA TEN VRG WDC. Hemicr. or rhizome geophyte.
Carex cephalophora var. *maxima* Dewey, Amer. J. Sci. Arts 43: 92 (1842).
Carex cephalophora var. *bracteosa* Farw., Rep. (Annual) Michigan Acad. Sci. 22: 180 (1921).
Carex cephalophora var. *anomala* Farw., Pap. Michigan Acad. Sci. 1: 90 (1923).

Carex cephalotes F.Muell., Trans. Philos. Soc. Victoria 1: 110 (1855). *Carex pyrenaica* var. *cephalotes* (F.Muell.) Kük. in H.G.A.Engler (ed.), Pflanzenr., IV, 20(38): 106 (1909).
SE. Australia, New Zealand. 50 NSW VIC 51 NZN NZS. Hemicr. or rhizome geophyte.

Carex cercidascus C.B.Clarke, J. Linn. Soc., Bot. 36: 279 (1903).
SE. China. 36 CHS. Hemicr. or rhizome geophyte.

Carex cespitosa L., Sp. Pl.: 978 (1753). *Vignea cespitosa* (L.) Rchb. in J.C.Mössler, Handb. Gewächsk. ed. 2, 3: 1625 (1830). *Neskiza cespitosa* (L.) Raf., Good Book: 27 (1840). *Limivasculum cespitosum* (L.) Fedde & J.Schust., Just's Bot. Jahresber. 41(2): 11 (1918).
Temp. Eurasia. 10 DEN FIN NOR SWE 11 AUT CZE GER HUN NET POL SWI? 12 FRA 13 ITA ROM YUG 14 BLR BLT RUC RUE RUN RUS RUW UKR 30 KRA WSB YAK 31 AMU PRM SAK 32 KAZ TKM 33 NCS TCS 34 TUR 36 CHI CHM CHQ CHX 37 MON 38 JAP. Hemicr. or rhizome geophyte.

var. ***cespitosa***
Temp. Eurasia. 10 DEN FIN NOR SWE 11 AUT CZE GER HUN NET POL SWI? 12 FRA 13 ITA ROM YUG 14 BLR BLT RUC RUE RUN RUS RUW UKR 30 KRA WSB YAK 31 AMU PRM SAK 32 KAZ TKM 33 NCS TCS 34 TUR 36 CHI CHM CHQ CHX 37 MON 38 JAP. Hemicr. or rhizome geophyte.
Carex alpestris Lam., Encycl. 3: 389 (1792), nom. illeg.
Carex erecta Gilib., Excerc. Phyt. 2: 549 (1792).
Carex ferax Raeusch., Nomencl. Bot., ed. 3: 270 (1797).
Carex inaequalis Raeusch., Nomencl. Bot., ed. 3: 270 (1797).
Carex nutans Raeusch., Nomencl. Bot., ed. 3: 270 (1797).
Carex quinata Raeusch., Nomencl. Bot., ed. 3: 270 (1797).
Carex humilis Willd. ex Kunth, Enum. Pl. 2: 411 (1837), nom. illeg.
Carex obesa Schleich. ex Kunth, Enum. Pl. 2: 411 (1837), nom. illeg.
Carex rigida Clairv. ex Steud., Nomencl. Bot., ed. 2, 1: 295 (1840), nom. illeg.
Carex pacifica Drejer, Rev. Car. Bor.: 292 (1841). *Vignea pacifica* (Drejer) Rchb. ex Nyman, Consp. Fl. Eur.: 777 (1882).
Carex drejeri Lange, Flora 25: 548 (1842). *Vignantha drejeri* (Lange) Schur, Enum. Pl. Transsilv.: 706 (1866). *Vignea drejeri* (Lange) Fuss, Fl. Transsilv.: 680 (1866).
Carex neglecta var. *major* Peterm., Flora 25: 331 (1842). *Carex cespitosa* f. *major* (Peterm.) Kük. in H.G.A.Engler (ed.), Pflanzenr., IV, 20(38): 329 (1909).
Carex cespitosa var. *retorta* Fr., Bot. Not. 1843: 10 (1843). *Carex retorta* (Fr.) V.I.Krecz. in V.L. Komarov (ed.), Fl. URSS 3: 598 (1935).
Carex neglecta Peterm., Flora 27: 331 (1844), nom. illeg.
Carex cespitosa f. *polystachya* Peterm., Anal. Pfl.-Schlüss.: 499 (1846).
Carex fridrichsthaliana Steud., Syn. Pl. Glumac. 2: 211 (1855).
Carex spreta Steud., Syn. Pl. Glumac. 2: 212 (1855). *Carex cespitosa* var. *spreta* (Steud.) Nyman, Consp. Fl. Eur.: 777 (1882).
Carex cespitosa f. *fuliginosa* Döll, Fl. Baden: 260 (1857).
Carex fascicularis Siev. ex Trevir., Bull. Soc. Imp. Naturalistes Moscou 36(2): 543 (1863).
Vignantha drejeri var. *alpina* Schur, Enum. Pl. Transsilv.: 706 (1866). *Carex cespitosa* var. *alpina* (Schur) Kük. in H.G.A.Engler (ed.), Pflanzenr., IV, 20(38): 329 (1909), nom. illeg.
Carex oligophylla Freyn, Oesterr. Bot. Z. 40: 305 (1890).
Carex usta Franch., Bull. Soc. Philom. Paris, VIII, 7: 41 (1895), nom. illeg.
Carex cespitosa f. *acutiuscula* Lackow., Fl. Berl., ed. 12: 27 (1901).
Carex cespitosa var. *waisbeckeri* Kük., Allg. Bot. Z. Syst. 7: 53 (1901).
Carex cespitosa var. *latifolia* Uechtr. ex Asch. & Graebn., Syn. Mitteleur. Fl. 2(2): 87 (1902).
Carex cespitosa var. *multifibrillosa* Kük. in H.G.A.Engler (ed.), Pflanzenr., IV, 20(38): 329 (1909).
Carex cespitosa var. *strictissima* Kük. in H.G.A.Engler (ed.), Pflanzenr., IV, 20(38): 329 (1909).
Carex rubra H.Lév. & Vaniot, Bull. Acad. Int. Géogr. Bot. 19: 33 (1909).
Carex glabella A.Pabrez ex C.Regel, Scripta Horti Bot. Univ. Vytauti Magni 2: 32, 64 (1932).
Carex inumbrata V.I.Krecz. in V.L.Komarov (ed.), Fl. URSS 3: 218 (1935).

var. ***minuta*** (Franch.) Kük. in H.G.A.Engler (ed.), Pflanzenr., IV, 20(38): 329 (1909).
Sakhalin, N. Japan. 31 SAK 38 JAP. Hemicr. or rhizome geophyte.
**Carex minuta* Franch., Bull. Soc. Philom. Paris, VIII, 7: 41 (1895).

Carex ceylanica Boeck., Linnaea 40: 341 (1876). *Carex filicina* subsp. *ceylanica* (Boeck.) T.Koyama in M.D.Daddanayake & al. (eds.), Revised Handb. Fl. Ceylon 5: 377 (1985).
Sri Lanka. 40 SRL. Hemicr. or rhizome geophyte.

Carex chaffanjonii E.G.Camus, Notul. Syst. (Paris) 2: 205 (1912).
Mongolia. 37 MON. Hemicr. or rhizome geophyte.

Carex chalciolepis T.Holm, Amer. J. Sci. 166: 28 (1903). *Carex atrata* var. *chalciolepis* (T.Holm) Kük. in H.G.A.Engler (ed.), Pflanzenr., IV, 20(38): 399 (1909). *Carex heteroneura* var. *chalciolepis* (T.Holm) F.J.Herm., Rhodora 70: 421 (1968).
WC. U.S.A. 73 COL MNT 76 ARI NEV UTA 77 NWM. Hemicr. or rhizome geophyte.

Carex chamissonis Meinsh., Trudy Imp. S.-Peterburgsk. Bot. Sada 18: 361 (1900).
Russian Far East to Alaska. 31+ 70 ASK. Hemicr. or rhizome geophyte.

Carex changmuensis Tang & F.T.Wang ex Y.C.Yang, in Fl. Xizangica 5: 410 (1987).
Tibet. 36 CHT. Hemicr. or rhizome geophyte.

Carex chaofangii C.Z.Zheng & X.F.Jin, Acta Phytotax. Sin. 42: 548 (2004).
China (Zhejiang). 36 CHS. Hemicr. or rhizome geophyte.

Carex chapmanii Steud., Syn. Pl. Glumac. 2: 222 (1855).
SE. U.S.A. 78 FLA GEO NCA SCA. Hemicr. or rhizome geophyte.

Carex charkeviczii A.E.Kozhevn., Byull. Moskovsk. Obshch. Isp. Prir., Otd. Biol. 92: 135 (1987).
Russian Far East. 31+. Hemicr. or rhizome geophyte.

Carex chathamica Petrie, Trans. & Proc. New Zealand Inst. 47: 55 (1914 publ. 1915).
Chatham Is. 51 CTM. Hemicr. or rhizome geophyte.

Carex cheniana Tang & F.T.Wang ex S.Y.Liang, Acta Phytotax. Sin. 36: 532 (1998).
SE. China. 36 CHS. Hemicr. or rhizome geophyte.

Carex cherokeensis Schwein., Ann. Lyceum Nat. Hist. New York 1: 369 (1824).
C. & SE. U.S.A. 74 MSO OKL 77 TEX 78 ALA ARK FLA GEO KTY LOU MSI NCA SCA TEN VRG. Hemicr. or rhizome geophyte.
Carex recurva Muhl., Descr. Gram.: 55 (1817), nom. illeg. *Edritria recurva* Raf., Good Book: 26 (1840).
Carex christyana Boott, Boston J. Nat. Hist. 5: 114 (1847).
Carex baazasana Steud., Syn. Pl. Glumac. 2: 236 (1855).
Carex brazatana Boeck., Flora 39: 321 (1856).

Carex chiapensis F.J.Herm., Brittonia 19: 68 (1967).
S. Mexico. 79 MXS MXT. Hemicr. or rhizome geophyte.

Carex chihuahuensis Mack., Bull. Torrey Bot. Club 35: 265 (1908).
Arizona to Mexico (Chihuahua). 76 ARI 79 MXE. Hemicr. or rhizome geophyte.

Carex chikungana L.H.Bailey, Gentes Herb. 1920: 13 (1920).
C. China. 36 CHC CHS. Hemicr. or rhizome geophyte.

Carex chillanensis Phil., Anales Univ. Chile 21(2): 384 (1862).
Chile, Argentina. 85 AGS AGW CLC. Hemicr. or rhizome geophyte.

Carex chinensis Retz., Observ. Bot. 3: 42 (1783).
China, Taiwan, Nansei-shoto. 36 CHC CHN CHS 38 NNS TAI. Hemicr. or rhizome geophyte.
Carex sinensis J.F.Gmel., Syst. Nat. 2: 144 (1791).
Carex retzii Nees in R.Wight, Contr. Bot. India: 128 (1834).
Carex longkiensis Franch., Bull. Soc. Philom. Paris, VIII, 7: 86 (1895). *Carex chinensis* var. *longkiensis* (Franch.) Kük. in H.G.A.Engler (ed.), Pflanzenr., IV, 20(38): 625 (1909).

Carex chinoi Ohwi ex T.Koyama, J. Jap. Bot. 32: 147 (1957).
Japan (Honshu). 38 JAP. Hemicr. or rhizome geophyte.

Carex chiovendae Pamp., Nuovo Giorn. Bot. Ital., n.s., 22: 256 (1915).
China (Hubei). 36 CHC. Hemicr. or rhizome geophyte.

Carex chiwuana F.T.Wang & Tang ex P.C.Li, Acta Bot. Yunnan. 12: 143 (1990).
SC. China. 36 CHC. Hemicr. or rhizome geophyte.

Carex chlorantha R.Br., Prodr.: 242 (1810).
SE. Australia. 50 NSW TAS VIC. Hemicr. or rhizome geophyte.

Carex chlorocephalula F.T.Wang & Tang ex P.C.Li, Acta Bot. Yunnan. 12: 143 (1990).
SC. China. 36 CHC. Hemicr. or rhizome geophyte.

Carex chloroleuca Meinsh., Trudy Imp. S.-Peterburgsk. Bot. Sada 18: 398 (1900). *Carex amgunensis* var. *chloroleuca* (Meinsh.) Kük. in H.G.A.Engler (ed.), Pflanzenr., IV, 20(38): 447 (1909).
Siberia to N. China. 30+ 31+ 36 CHI CHM. Hemicr. or rhizome geophyte.
Carex chinganensis Litv., Spisok Rast. Gerb. Russk. Fl. Bot. Muz. Imp. Akad. Nauk 6: 135 (1908).

Carex chlorosaccus C.B.Clarke, J. Linn. Soc., Bot. 34: 298 (1899). *Carex echinochloe* var. *chlorosaccus* (C.B.Clarke) Kük. in H.G.A.Engler (ed.), Pflanzenr., IV, 20(38): 271 (1909).
Bioko, SW. Cameroon, Ethiopia to E. Zaïre. 23 CMN GGI RWA ZAI 24 ETH SUD 25 KEN TAN UGA. Hemicr. or rhizome geophyte.

Carex chorda H.Lév. & Vaniot, Bull. Soc. Bot. France 53: 316 (1906).
China (Guizhou). 36 CHC. Hemicr. or rhizome geophyte.

Carex chordorrhiza L.f., Suppl. Pl.: 414 (1782). *Vignea chordorrhiza* (L.f.) Rchb. in J.C.Mössler, Handb. Gewächsk. ed. 2, 3: 1608 (1830). *Carex chordorrhiza* var. *genuina* Trautv., Trudy Imp. S.-Peterburgsk. Bot. Sada 5: 123 (1877), nom. inval. *Caricina chordorrhiza* (L.f.) St.-Lag. in A.Cariot, Étude Fl., ed. 8, 2: 874 (1889).
Temp. Northern Hemisphere. 10 DEN FIN GRB ICE NOR SWE 11 AUT CZE GER POL SWI 12 FRA 13 ROM 14 BLR BLT RUC RUE RUN RUS RUW UKR 30 ALT KRA WSB YAK 31 KAM KHA MAG 32 KAZ 33 NCS 38 KOR 70 ASK GNL NUN NWT YUK 71 ABT BRC MAN SAS 72 NBR NFL ONT PEI QUE 73 MNT ore 74 ILL IOW MIN WIS 75 INI MAI MIC NWH NWY PEN VER. Hemicr. or rhizome geophyte.
Carex funiformis Clairv., Man. Herbor. Suisse: 287 (1811).
Carex fulvicoma Dewey, Amer. J. Sci. Arts 29: 249 (1836).
Carex fischeriana J.Gay, Ann. Sci. Nat., Bot., II, 10: 286 (1838).
Carex teretiuscula Willk., Bot. Zeitung (Berlin) 6: 414 (1848).
Carex chordorrhiza var. *sphagnicola* Laest. ex Th.Fr., Bot. Not. 1857: 208 (1857).
Carex chordorrhiza var. *aestivalis* Asch. & Graebn., Syn. Mitteleur. Fl. 2(2): 23 (1902).

Carex chosenica Ohwi, Acta Phytotax. Geobot. 1: 73 (1932). *Carex dahurica* subsp. *chosenica* (Ohwi) T.V.Egorova, Novosti Sist. Vyssh. Rast. 22: 61 (1985).
Russian Far East, Korea. 31+ 38 KOR. Hemicr. or rhizome geophyte.

Carex chrysolepis Franch. & Sav., Enum. Pl. Jap. 2: 571 (1878).
Japan, Taiwan. 38 JAP TAI. Hemicr. or rhizome geophyte.

var. ***chrysolepis***
S. Japan, Taiwan. 38 JAP TAI. Hemicr. or rhizome geophyte.
Carex odontostoma Kük., Bull. Herb. Boissier, II, 5: 1162 (1905). *Carex chrysolepis* var. *odontostoma* (Kük.) Ohwi, Mem. Coll. Sci. Kyoto Imp. Univ., Ser. B, Biol. 6: 250 (1931).
Carex odontostoma var. *variegata* Kük. in H.G.A.Engler (ed.), Pflanzenr., IV, 20(38): 571 (1909).
Carex gokwanensis Hayata, Icon. Pl. Formosan. 10: 65 (1921).
Carex kiusiuana Ohwi, Mem. Coll. Sci. Kyoto Imp. Univ., Ser. B, Biol. 5(3): 268 (1930).

var. ***glabrior*** (Ohwi) Ohwi, Mem. Coll. Sci. Kyoto Imp. Univ., Ser. B, Biol. 6: 250 (1931).
Japan (Honshu). 38 JAP. Hemicr. or rhizome geophyte.

Carex odontostoma var. *glabrior* Ohwi, Mem. Coll. Sci. Kyoto Imp. Univ., Ser. B, Biol. 5: 253 (1930). *Carex glabrior* (Ohwi) Akiyama, Caric. Far East. Reg. As.: 129 (1955).

Carex chuiana F.T.Wang & Tang ex P.C.Li, Acta Phytotax. Sin. 37: 167 (1999).
China (Sichuan). 36 CHC. Hemicr. or rhizome geophyte.

Carex chuii Nelmes, Kew Bull. 3: 108 (1948).
China (Sichuan). 36 CHC. Hemicr. or rhizome geophyte.

Carex chungii Z.P.Wang, J. Nanjing Univ. (Biology) 1962(2): 45 (1962).
S. China. 36 CHC CHS. Hemicr.

var. ***chungii***
S. China. 36 CHC CHS. Hemicr.
Carex anhuiensis S.W.Su & S.M.Xu, Bull. Bot. Res., Harbin 9(1): 47 (1989).
Carex truncatirostris S.W.Su & S.M.Xu, Bull. Bot. Res., Harbin 9(1): 45 (1989).
Carex truncatirostris f. *erostris* S.W.Su & S.M.Xu, Bull. Bot. Res., Harbin 9(1): 47 (1989).
Carex xuanchengensis S.W.Su & S.M.Xu, Bull. Bot. Res., Harbin 9(1): 48 (1989).

var. ***rigida*** Y.C.Tang & S.Yun Liang, in Fl. Reipubl. Popul. Sin. 12: 522 (2000).
SE. China. 36 CHS. Hemicr.

Carex cilicica Boiss., Diagn. Pl. Orient., II, 4: 124 (1859).
Turkey to Transcaucasus. 33 TCS 34 IRQ TUR. Hemicr. or rhizome geophyte.

subsp. ***cilicica***
S. & E. Turkey, N. Iraq, Transcaucasus. 33 TCS 34 IRQ TUR. Hemicr. or rhizome geophyte.

subsp. ***muglaica*** Ö.Nilsson in P.H.Davis (ed.), Fl. Turkey 9: 622 (1985).
S. Turkey. 34 TUR. Hemicr. or rhizome geophyte.

subsp. ***muratica*** Ö.Nilsson in P.H.Davis (ed.), Fl. Turkey 9: 622 (1985).
C. Turkey. 34 TUR. Hemicr. or rhizome geophyte.

Carex cinerascens Kük., Bull. Herb. Boissier, II, 2: 1017 (1902).
Siberia to C. Japan. 30+ 31+ 36 CHI CHM CHN CHS 38 JAP KOR. Hemicr. or rhizome geophyte.
Carex micrantha Kük., Bull. Herb. Boissier, II, 2: 1018 (1902).
Carex ouensanensis Ohwi, Mem. Coll. Sci. Kyoto Imp. Univ., Ser. B, Biol. 6: 262 (1931).
Carex cheilungkiangnica A.I.Baranov & Skvortsov, Quart. J. Taiwan Mus. 18: 225 (1965).

Carex circinnata C.A.Mey., Mém. Sav. Étr. Acad. St. Pétersbourg 1: 209 (1831).
Komandor Is. to Washington. 31 KAM 70 ALU ASK 72 BRC 73 WAS. Hemicr. or rhizome geophyte.

Carex cirrhosa Berggr., Minneskr. Fisiog. Sallks. Lund 8: 29 (1877).
New Zealand. 51 NZN NZS. Hemicr. or rhizome geophyte.
Carex cirrhosa var. *lutescens* Kük. in T.F.Cheeseman, Man. New Zealand Fl.: 825 (1906).

Carex cirrhulosa Nees, Hooker's J. Bot. Kew Gard. Misc. 6: 29 (1854). *Carex fuirenoides* var. *cirrhulosa* (Nees) Kük. in H.G.A.Engler (ed.), Pflanzenr., IV, 20(38): 287 (1909).
Philippines (Cebu). 42 PHI. Hemicr. or rhizome geophyte.
Carex densiflora C.Presl, Reliq. Haenk. 1: 204 (1828), provisional synonym.
Carex fibrata S.Vidal, Phan. Cuming. Philipp.: 156 (1885).

Carex ×clausa E.Holmb., Bot. Not. 1929: 12 (1929). *C. canescens* × *C. lapponica*.
N. Europe to Siberia. 10 SWE 14 RUN 30+. Hemicr. or rhizome geophyte.

Carex clavata Thunb., Prodr. Pl. Cap.: 14 (1794). *Anithista clavata* (Thunb.) Raf., Good Book: 26 (1840).
Cape Prov. 27 CPP. Hemicr. or rhizome geophyte.
Carex clavata var. *campylostachya* Nees, Linnaea 10: 204 (1836).
Carex lutensis Kunth, Enum. Pl. 2: 487 (1837). *Carex clavata* f. *lutensis* (Kunth) Kük. in H.G.A.Engler (ed.), Pflanzenr., IV, 20(38): 737 (1909).
Carex triticea Wahlenb. ex Steud., Nomencl. Bot., ed. 2, 1: 299 (1840).
Carex macrocystis Boeck., Beitr. Cyper. 1: 50 (1888).

Carex clivorum Ohwi, Repert. Spec. Nov. Regni Veg. 36: 42 (1934).
Japan (Honshu). 38 JAP. Hemicr. or rhizome geophyte.

Carex cochinchinensis Raymond, Mém. Jard. Bot. Montréal 53: 32 (1959).
Vietnam. 41 VIE. Hemicr. or rhizome geophyte.

Carex cochranei Reznicek, in Fl. Novo-Galiciana 13: 249 (1993).
C. & SW. Mexico. 79 MXC MXS. Hemicr. or rhizome geophyte.

Carex cockayniana Kük. in T.F.Cheeseman, Man. New Zealand Fl.: 836 (1906). *Carex forsteri* var. *cockayniana* (Kük.) Kük. in H.G.A.Engler (ed.), Pflanzenr., IV, 20(38): 695 (1909).
New Zealand North I. 51 NZN. Hemicr. or rhizome geophyte.

Carex cognata Kunth, Enum. Pl. 2: 502 (1837). *Carex pseudocyperus* var. *cognata* (Kunth) Boott, Ill. Carex 4: 141 (1867).
Ethiopia to Cape Prov. 23 RWA ZAI 24 ETH 25 TAN UGA 27 BOT CPP LES NAT OFS SWZ TVL. Hemicr.
**Carex retrorsa* Nees, Linnaea 10: 204 (1836), nom. illeg.

var. ***abyssinica*** (Chiov.) Lye, Nordic J. Bot. 3: 244 (1983).
Ethiopia. 24 ETH. Hemicr. or rhizome geophyte.
**Carex abyssinica* Chiov., Ann. Bot. (Rome) 9: 150 (1911).

var. ***cognata***
S. Tanzania to Cape Prov. 25 TAN 27 BOT CPP LES NAT OFS SWZ TVL. Hemicr.

var. ***congolensis*** (Turrill) Lye, Nordic J. Bot. 3: 244 (1983).
Zaïre to Uganda. 23 RWA ZAI 25 UGA. Hemicr. or rhizome geophyte.
**Carex congolensis* Turrill, Bull. Misc. Inform. Kew 1912: 240 (1912).

var. ***drakensbergensis*** (C.B.Clarke) Kük. in H.G.A. Engler (ed.), Pflanzenr., IV, 20(38): 699 (1909).

S. Africa. 27 CPP LES NAT OFS TVL. Hemicr. or rhizome geophyte.
**Carex drakensbergensis* C.B.Clarke in W.H.Harvey & auct. suc. (eds.), Fl. Cap. 7: 309 (1898).

Carex colchica J.Gay, Ann. Sci. Nat., Bot., II, 10: 303 (1838). *Carex arenaria* subsp. *colchica* (J.Gay) Nyman, Consp. Fl. Eur.: 782 (1882). *Carex arenaria* var. *colchica* (J.Gay) H.Christ, Bull. Soc. Roy. Bot. Belgique 24(2): 17 (1885). *Vignea colchica* (J.Gay) Soják, Cas. Nár. Mus., Odd. Prír. 148: 195 (1979 publ. 1980).
Europe to C. Asia. 10 DEN SWE 11 GER NET POL 12 FRA 13 BUL ROM TUE 14 BLT KRY RUC RUE RUS UKR 30 WSB 32 KAZ TKM 33 TCS 34 TUR. Hemicr. or rhizome geophyte.
Carex schoenoides Schrank, Baier. Fl. 1: 274 (1789), provisional synonym.
Carex ligerica J.Gay, Ann. Sci. Nat., Bot., II, 10: 360 (1838). *Carex schreberi* subsp. *ligerica* (J.Gay) Hartm., Handb. Skand. Fl., ed. 11: 475 (1879). *Carex colchica* subsp. *ligerica* (J.Gay) T.V.Egorova, Novosti Sist. Vyssh. Rast. 10: 104 (1973). *Vignea ligerica* (J.Gay) Soják, Cas. Nár. Mus., Odd. Prír. 148: 196 (1979 publ. 1980).
Carex pseudoarenaria Rchb., Icon. Fl. Germ. Helv. 8: 8 (1846). *Vignea pseudoarenaria* Rchb., Icon. Fl. Germ. Helv. 8: t. 208 (1846).
Carex ligerina Guépin ex Boreau, Mém. Soc. Acad. Maine Loire 6: 171 (1859).
Carex arenaria var. *castanea* Boott, Ill. Carex 4: 178 (1867).
Carex ligerica f. *capitata* Lackow. & Paul, Doubl.-Verz. Berl. Bot. Tauschv. 36: 25 (1904).

Carex colensoi Boott in J.D.Hooker, Fl. Nov.-Zel.: 281 (1853).
New Zealand North I. 51 NZN. Hemicr. or rhizome geophyte.

Carex collifera Ohwi, Acta Phytotax. Geobot. 2: 26 (1933).
Nansei-shoto. 38 NNS. Hemicr. or rhizome geophyte.

Carex collimitanea V.I.Krecz., Bot. Mater. Gerb. Bot. Inst. Komarova Akad. Nauk S.S.S.R. 9: 195 (1946).
China (Hubei). 36 CHC. Hemicr. or rhizome geophyte.

Carex collinsii Nutt., Gen. N. Amer. Pl. 2: 205 (1818). *Olotrema collinsii* (Nutt.) Raf., Good Book: 25 (1840).
E. U.S.A. 75 CNT NWJ NWY PEN RHO 78 DEL GEO MRY NCA SCA VRG. Hemicr. or rhizome geophyte.
Carex subulata Michx., Fl. Bor.-Amer. 2: 173 (1803), nom. illeg. *Carex michauxii* Dewey, Amer. J. Sci. Arts 10: 273 (1826), nom. illeg.

Carex collumanthus (Steyerm.) L.E.Mora, Acta Biol. Colomb. 1: 40 (1982).
W. South America to W. Venezuela. 82 VEN 83 BOL CLM PER. Hemicr. or rhizome geophyte.
**Vesicarex collumanthus* Steyerm., Fieldiana, Bot. 28(1): 63 (1951).

Carex comans Berggr., Minneskr. Fisiog. Sallks. Lund 8: 28 (1877).
New Zealand. 51 NZN NZS. Hemicr. or rhizome geophyte.
Carex pulchella Berggr., Minneskr. Fisiog. Sallks. Lund 8: 29 (1877), nom. illeg. *Carex comans* var. *pulchella* Kük. in H.G.A.Engler (ed.), Pflanzenr., IV, 20(38): 689 (1909).
Carex cheesemanii Petrie, Trans. & Proc. New Zealand Inst. 15: 358 (1883). *Carex comans* var. *cheesemanii* (Petrie) Kük. in H.G.A.Engler (ed.), Pflanzenr., IV, 20(38): 689 (1909).
Carex comans f. *stricta* Cheeseman, Trans. & Proc. New Zealand Inst. 24: 415 (1892).
Carex vilmorinii Mottet, Rev. Hort. 1897: 79 (1897).

Carex commixta Steud., Syn. Pl. Glumac. 2: 207 (1855).
Hainan, Indo-China, W. Malesia. 36 CHH 41 MYA THA VIE 42 JAW SUM. Hemicr. or rhizome geophyte.
Carex spatiosa Boott, Ill. Carex 2: 86 (1860).
Carex spatiosa var. *bogorensis* C.B.Clarke, J. Linn. Soc., Bot. 37: 12 (1904).
Carex blepharolepis Nelmes, Kew Bull. 1: 18 (1946).
Carex chuniana F.T.Wang & Tang, Contr. Inst. Bot. Natl. Acad. Peiping 6: 56 (1949).
Carex humbertii F.T.Wang & Tang, Contr. Inst. Bot. Natl. Acad. Peiping 6: 57 (1949), nom. illeg.
Carex hypolytrifolia T.Koyama, Bot. Mag. (Tokyo) 69: 209 (1956).
Carex hainanensis Merr. ex Chun & F.C.How, Acta Phytotax. Sin. 7: 80 (1958).
Carex smitinandii Raymond, Dansk Bot. Ark. 23: 255 (1965).

Carex communis L.H.Bailey, Mem. Torrey Bot. Club 1: 41 (1889).
E. Canada to NC. & E. U.S.A. 72 LAB NBR NFL NSC ONT QUE 74 ILL MIN MSO WIS 75 CNT INI MAI MAS MIC NWH NWJ NWY OHI PEN RHO VER WVA 78 ARK DEL GEO KTY MRY NCA SCA TEN VRG. Hemicr. or rhizome geophyte.
Carex communis var. *gynandra* Farw., Amer. Midl. Naturalist 12: 52 (1930). *Carex communis* f. *gynandra* (Farw.) E.G.Voss, Rhodora 68: 451 (1966).

var. ***amplisquama*** (F.J.Herm.) Rettig, Novon 6: 117 (1996).
N. Georgia, W. South Carolina. 78 GEO SCA. Hemicr. or rhizome geophyte.
**Carex amplisquama* F.J.Herm., Rhodora 57: 158 (1955).

var. ***communis***
E. Canada to NC. & E. U.S.A. 72 LAB NBR NFL NSC ONT QUE 74 ILL MIN MSO WIS 75 CNT INI MAI MAS MIC NWH NWJ NWY OHI PEN RHO VER WVA 78 ARK DEL KTY MRY NCA TEN VRG. Hemicr. or rhizome geophyte.

Carex comosa Boott, Trans. Linn. Soc. London 20: 117 (1846). *Carex pseudocyperus* var. *comosa* (Boott) Boott, Ill. Carex 4: 141 (1867).
E. Canada to Mexico. 72 ONT QUE 73 IDA 74 ILL IOW MIN WIS 75 CNT INI MAI MAS MIC NWJ NWY OHI PEN VER 76 CAL 77 TEX 78 ARK MRY MSI NCA SCA TEN VRG 79 MXC MXE MXS. Hemicr. or rhizome geophyte.
Carex furcata Elliott, Sketch Bot. S. Carolina 2: 552 (1824), nom. illeg.

Carex complanata Torr. & Hook., Ann. Lyceum Nat. Hist. New York 3: 408 (1836). *Facolos complanata* (Torr. & Hook.) Raf., Good Book: 26 (1840).
C. & E. U.S.A. to Guatemala. 74 IOW MSO 75 NWJ 77 TEX 78 ALA ARK DEL FLA GEO LOU MRY MSI NCA SCA TEN VRG WDC 79 MXS MXT 80 GUA. Hemicr. or rhizome geophyte.

Carex complanata var. *robusta* Burnham, Torreya 19: 134 (1919).

subsp. ***complanata***
C. & E. U.S.A. 74 IOW MSO 75 NWJ 77 TEX 78 ALA ARK DEL FLA GEO LOU MRY MSI NCA SCA TEN VRG WDC. Hemicr. or rhizome geophyte.
Carex bolliana Boeck., Flora 61: 40 (1878).

subsp. ***tropicalis*** Reznicek & S.González, Contr. Univ. Michigan Herb. 23: 344 (2001).
S. Mexico to Guatemala. 79 MXS MXT 80 GUA. Hemicr.

Carex composita Boott, Ill. Carex 1: 3 (1858).
S. Bhutan to N. Thailand. 36 CHC 40 ASS EHM NEP? 41 THA. Hemicr. or rhizome geophyte.
Carex composita f. *simplicior* Kük. in H.G.A.Engler (ed.), Pflanzenr., IV, 20(38): 260 (1909).

Carex concinna R.Br. in J.Richardson, Bot. App.: 751 (1823). *Edritria concinna* (R.Br.) Raf., Good Book: 26 (1840).
Alaska to N. U.S.A. 70 ASK NUN NWT YUK 71 ABT BRC MAN SAS 72 LAB NBR NFL ONT QUE 73 COL MNT ORE WYO 74 WIS 75 MIC. Hemicr. or rhizome geophyte.

Carex concinnoides Mack., Bull. Torrey Bot. Club 33: 440 (1906).
W. Canada to NW. U.S.A. 71 ABT BRC 73 IDA MNT ORE WAS 76 CAL. Hemicr. or rhizome geophyte.

Carex condensata Nees in R.Wight, Contr. Bot. India: 123 (1834).
E. Himalaya to Thailand. 40 EHM 41 THA. Hemicr. or rhizome geophyte.

Carex conferta Hochst. ex A.Rich., Tent. Fl. Abyss. 2: 512 (1850).
Ethiopia to E. Trop. Africa. 23 BUR CMN? RWA ZAI 24 ETH 25 KEN TAN UGA. Hemicr. or rhizome geophyte.

var. ***conferta***
Ethiopia to E. Trop. Africa. 23 ZAI 24 ETH 25 KEN UGA. Hemicr. or rhizome geophyte.

var. ***leptosaccus*** (C.B.Clarke) Kük., Nordic J. Bot. 3: 244 (1983).
Ethiopia to Rwanda. 23 BUR RWA ZAI 24 ETH 25 KEN TAN UGA. Hemicr. or rhizome geophyte.
Carex erythrorrhiza Steud., Syn. Pl. Glumac. 2: 194 (1855), nom. inval.
Carex koestlinii Hochst. ex Steud., Syn. Pl. Glumac. 2: 193 (1855).
**Carex leptosaccus* C.B.Clarke in D.Oliver & auct. suc. (eds.), Fl. Trop. Afr. 8: 516 (1902). *Carex tolmiei* var. *leptosaccus* (C.B.Clarke) Kük. in H.G.A.Engler (ed.), Pflanzenr., IV, 20(38): 411 (1909).

var. ***lycurus*** (K.Schum. ex Engl.) Lye, Nordic J. Bot. 3: 244 (1983).
E. Trop. Africa (Mts.). 23 CMN? 25 KEN TAN. Hemicr. or rhizome geophyte.
**Carex lycurus* K.Schum. ex Engl., Abh. Königl. Ges. Wiss. Göttingen, Math.-Phys. Kl. 39: 58 (1894).

Carex confertospicata Boeck., Bot. Jahrb. Syst. 8: 206 (1887).
Panama to N. Ecuador. 80 PAN 83 CLM ECU. Hemicr. or rhizome geophyte.

Carex congdonii L.H.Bailey, Bot. Gaz. 21: 6 (1896).
California. 76 CAL. Hemicr. or rhizome geophyte.

Carex congestiflora Reznicek & S.González, Contr. Univ. Michigan Herb. 22: 125 (1999).
Mexico (Chiapas). 79 MXT. Hemicr. or rhizome geophyte.

Carex conica Boott in M.C.Perry, Jap. Exped. 2: 325 (1856).
S. Korea, Japan. 38 JAP KOR. Hemicr. or rhizome geophyte.
Carex excisa Boott in M.C.Perry, Jap. Exped. 2: 326 (1856).
Carex conica var. *leucolepis* Franch. & Sav., Enum. Pl. Jap. 2: 143 (1877).
Carex discolor Boeck., Bot. Jahrb. Syst. 5: 519 (1884).
Carex naumanniana Boeck., Bot. Jahrb. Syst. 5: 518 (1884).
Carex conica f. *longirostrata* Kük. in H.G.A.Engler (ed.), Pflanzenr., IV, 20(38): 479 (1909).
Carex conica f. *mollis* Kük. in H.G.A.Engler (ed.), Pflanzenr., IV, 20(38): 479 (1909).
Carex conica f. *rubens* Kük. in H.G.A.Engler (ed.), Pflanzenr., IV, 20(38): 479 (1909).
Carex digama Nakai, Bot. Mag. (Tokyo) 28: 326 (1914).
Carex doiana Akiyama, J. Fac. Sci. Hokkaido Univ., Ser. 5, Bot. 1: 59 (1931).
Carex yoshinoi Ohwi, Mem. Coll. Sci. Kyoto Imp. Univ., Ser. B, Biol. 6: 265 (1931).
Carex okushirensis Akiyama, Trans. Sapporo Nat. Hist. Soc. 14: 108 (1935).

Carex conjuncta Boott, Ill. Carex 3: 122 (1862). *Vignea conjuncta* (Boott) Soják, Cas. Nár. Mus., Odd. Prír. 148: 195 (1979 publ. 1980).
NC. & E. U.S.A. 74 ILL IOW KAN MSO NEB 75 INI MIC NWJ NWY OHI WVA 78 ALA ARK MRY TEN VRG. Hemicr. or rhizome geophyte.
Carex vulpina Carey ex Dewey, Amer. J. Sci. Arts, II, 8: 348 (1849), nom. illeg.

Carex* ×*connectens E.Holmb., Bot. Not. 1929: 226 (1929). *C. limosa* × *C. magellanica* subsp. *irrigua*.
N. Europe to Russian Far East, E. Canada. 10+ 14+ 31+ 72 QUE. Hemicr. or rhizome geophyte.
Carex paupercula var. *brevisquama* Fernald, Rhodora 20: 152 (1918). *Carex magellanica* f. *brevisquama* (Fernald) B.Boivin, Provancheria 25: 170 (1992).

Carex conoidea Willd., Sp. Pl. 4: 280 (1805).
C. & E. Canada, NC. & NE. U.S.A. to North Carolina. 71 MAN 72 LAB NBR NFL NSC ONT QUE 74 ILL IOW MIN MSO WIS 75 CNT INI MAI MAS MIC NWH NWJ NWY OHI PEN RHO VER 78 ARK DEL NCA VRG. Hemicr. or rhizome geophyte.
Carex granularioides Schwein., Ann. Lyceum Nat. Hist. New York 1: 69 (1824).
Carex tetanica Schwein. & Torr., Ann. Lyceum Nat. Hist. New York 1: 347 (1824), nom. illeg.
Carex illinoensis Dewey, Amer. J. Sci. Arts, II, 4: 245 (1847).
Carex katahdinensis Fernald, Rhodora 3: 171 (1901). *Carex conoidea* f. *katahdinensis* (Fernald) B.Boivin, Naturaliste Canad. 94: 522 (1967).

Carex conoides Kük., Repert. Spec. Nov. Regni Veg. Beih. 12: 311 (1922).
China (Sichuan). 36 CHC. Hemicr. or rhizome geophyte.

Carex conspecta Mack. in N.L.Britton & al. (eds.), N. Amer. Fl. 18: 294 (1935).
Mexico. 79 MXC MXG. Hemicr. or rhizome geophyte.

Carex conspissata V.I.Krecz. in V.L.Komarov (ed.), Fl. URSS 3: 610 (1935).
Siberia, Mongolia, Russian Far East. 30+ 31+ 37 MON. Hemicr. or rhizome geophyte.

Carex constanceana Stacey, Leafl. W. Bot. 2: 123 (1938).
Washington. 73 WAS. Hemicr. or rhizome geophyte.

Carex continua C.B.Clarke in J.D.Hooker, Fl. Brit. India 6: 717 (1894).
Nepal to SC. China. 36 CHC 40 EHM NEP 41 MYA THA VIE. Hemicr. or rhizome geophyte.
Carex plebeia C.B.Clarke in J.D.Hooker, Fl. Brit. India 6: 718 (1894). *Carex continua* var. *plebeia* (C.B. Clarke) Kük. in H.G.A.Engler (ed.), Pflanzenr., IV, 20(38): 282 (1909).

Carex contracta F.Muell., Fragm. 8: 258 (1874). *Carex gaudichaudiana* var. *contracta* (F.Muell.) Kük. in H.G.A.Engler (ed.), Pflanzenr., IV, 20(38): 313 (1909).
New South Wales. 50 NSW. Hemicr. or rhizome geophyte.

Carex cordillerana Saarela & B.A.Ford, Syst. Bot. 26: 715 (2001).
W. Canada to WC. U.S.A. 71 ABT BRC 73 IDA MNT ORE WAS WYO 76 UTA. Hemicr. or rhizome geophyte.

Carex cordouei H.Lév., Bull. Acad. Int. Géogr. Bot. 19: 34 (1909).
Sakhalin. 31 SAK. Hemicr. or rhizome geophyte.

Carex coriacea Hamlin, Trans. & Proc. Roy. Soc. New Zealand 82: 63 (1954).
New Zealand North I. 51 NZN. Hemicr. or rhizome geophyte.
**Carex ternaria* var. *pallida* Cheeseman, Trans. & Proc. New Zealand Inst. 16: 431 (1884). *Carex ternaria* f. *pallida* (Cheeseman) Kük. in H.G.A. Engler (ed.), Pflanzenr., IV, 20(38): 369 (1909).

Carex corrugata Fernald, Rhodora 44: 76 (1942).
C. & E. U.S.A. 74 ILL KAN MSO OKL 75 INI 77 TEX 78 ALA ARK FLA GEO KTY LOU MRY MSI NCA SCA TEN VRG. Hemicr. or rhizome geophyte.
**Carex rugata* Fernald, Rhodora 43: 545 (1941), nom. illeg.

Carex cortesii Liebm., Mexic. Halvgr.: 268 (1850).
Mexico to C. America. 79 MXC MXE MXG MXS MXT 80 COS GUA. Hemicr.
Carex galeottiana C.A.Mey., Bull. Acad. Roy. Sci. Bruxelles 9(2): 248 (1842), nom. nud.

Carex coulteri Boott ex Hemsl., Biol. Cent.-Amer., Bot. 3: 473 (1885). *Carex longicaulis* var. *coulteri* (Boott ex Hemsl.) Kük. in H.G.A.Engler (ed.), Pflanzenr., IV, 20(38): 580 (1909).
Mexico. 79 MXC MXE MXG. Hemicr. or rhizome geophyte.

Carex coxiana Petrie, Trans. & Proc. New Zealand Inst. 56: 6 (1926).
Chatham Is. 51 CTM. Hemicr. or rhizome geophyte.

Carex cranaocarpa Nelmes, Bull. Misc. Inform. Kew 1939: 184 (1939).
China (Shanxi, Hebei). 36 CHN. Hemicr. or rhizome geophyte.

Carex crassibasis H.Lév. & Vaniot, Repert. Spec. Nov. Regni Veg. 5: 194 (1908).
Korea. 38 KOR. Hemicr. or rhizome geophyte.

Carex crassiflora Kük. in H.G.A.Engler (ed.), Pflanzenr., IV, 20(38): 403 (1909).
Argentina (Salta, Jujuy, Tucumán). 85 AGW. Hemicr. or rhizome geophyte.

Carex crawei Dewey ex Torr., Fl. New York 2: 408 (1843).
Canada to N., C. & E. U.S.A. 71 ABT BRC MAN SAS 72 NBR NFL ONT QUE 73 IDA MNT WAS WYO 74 ILL IOW KAN MIN MSO NDA NEB OKL SDA WIS 75 CNT INI MAI MIC NWJ NWY OHI 76 UTA 78 ALA ARK GEO KTY MSI TEN VRG. Hemicr. or rhizome geophyte.
Carex crawei var. *heterostachya* Dewey, Amer. J. Sci. Arts, II, 42: 4 (1866).

Carex crawfordii Fernald, Proc. Amer. Acad. Arts 37: 469 (1902).
Alaska to N. U.S.A. (11) bgm 70 ASK NWT YUK 71 ABT BRC MAN SAS 72 LAB NBR NFL NSC ONT PEI QUE 73 IDA MNT ORE WAS 74 MIN MSO WIS 75 CNT MAI MAS MIC NWH NWJ NWY PEN VER. Hemicr. or rhizome geophyte.
Carex scoparia var. *minor* Boott, Ill. Carex 3: 116 (1862).
Carex crawfordii var. *vigens* Fernald, Proc. Amer. Acad. Arts 37: 470 (1902). *Carex crawfordii* f. *vigens* (Fernald) Kük. in H.G.A.Engler (ed.), Pflanzenr., IV, 20(38): 203 (1909).

Carex crebra V.I.Krecz., Bot. Mater. Gerb. Bot. Inst. Komarova Akad. Nauk S.S.S.R. 9: 190 (1946).
Tibet to C. China. 36 CHC CHN CHQ CHT. Hemicr. or rhizome geophyte.

Carex crebriflora Wiegand, Rhodora 24: 197 (1922).
SE. U.S.A. to Texas. 77 TEX 78 ALA ARK FLA GEO LOU MSI NCA SCA TEN VRG. Hemicr. or rhizome geophyte.

Carex cretica Gradst. & Kern, Acta Bot. Neerl. 17: 242 (1968).
W. Kriti, E. Aegean Is. (Ikaria). 13 KRI 34 EAI. Hemicr. or rhizome geophyte.

Carex crinita Lam., Encycl. 3: 393 (1792). *Diemisa crinita* (Lam.) Raf., Good Book: 27 (1840). *Neskiza crinita* (Lam.) Raf., Good Book: 27 (1840).
C. & E. Canada to C. & E. U.S.A. 71 MAN 72 NBR NSC ONT PEI QUE 74 ILL MIN MSO NEB OKL WIS 75 CNT INI MAI MAS MIC NWH NWJ NWY OHI PEN RHO VER WVA 77 TEX 78 ALA ARK DEL GEO KTY LOU MRY MSI NCA SCA TEN VRG WDC. Hemicr. or rhizome geophyte.

var. ***brevicrinis*** Fernald, Rhodora 48: 54 (1946).
C. & E. U.S.A. 74 MSO NEB OKL 75 CNT MAS NWJ NWY OHI PEN RHO WVA 77 TEX 78 ALA ARK DEL GEO KTY LOU MRY MSI NCA SCA TEN VRG WDC. Hemicr. or rhizome geophyte.

var. ***crinita***
C. & E. Canada to C. & E. U.S.A. 71 MAN 72 NBR NSC ONT PEI QUE 74 ILL MIN MSO

WIS 75 CNT INI MAI MAS MIC NWH NWJ NWY OHI PEN RHO VER WVA 77 TEX 78 ALA ARK DEL GEO KTY LOU MRY MSI NCA SCA TEN VRG WDC. Hemicr. or rhizome geophyte.
Carex leonura Wahlenb., Kongl. Vetensk. Acad. Nya Handl. 1803: 161 (1803).
Carex crinita var. *minor* Boott, Ill. Carex 1: 18 (1858).
Carex crinita f. *minor* Boott, Ill. Carex 1: 18 (1858).

var. ***porteri*** (Olney) Fernald, Proc. Portland Soc. Nat. Hist. 2: 135 (1897).
NE. U.S.A. 75 MAI. Hemicr. or rhizome geophyte.
**Carex porteri* Olney, Car. Bor. Am.: 12 (1871). *Carex gynandra* var. *porteri* (Olney) Britton in N.L.Britton & A.Brown, Ill. Fl. N. U.S. 1: 315 (1896).

Carex* ×*crinitoides Lepage, Naturaliste Canad. 83: 122 (1956). *C. aquatilis* × *C. crinita*.
Québec. 72 QUE. Hemicr. or rhizome geophyte.

Carex cristatella Britton & A.Br., Ill. Fl. N. U.S. 1: 357 (1896).
Canada, NC. & E. U.S.A. 71 MAN SAS 72 ONT QUE 74 ILL IOW KAN MIN MSO MSO NDA NEB SDA WIS 75 CNT INI MAS MIC NWH NWJ NWY OHI PEN VER WVA 78 DEL KTY TEN VRG WDC. Hemicr. or rhizome geophyte.
**Carex cristata* Schwein., Ann. Lyceum Nat. Hist. New York 1: 66 (1824), nom. illeg. *Carex straminea* var. *cristata* Tuck., Enum. Meth. Caric.: 18 (1843). *Carex lagopodioides* var. *cristata* (Tuck.) Carey in A.Gray, Manual: 545 (1848). *Carex tribuloides* var. *cristata* (Tuck.) L.H.Bailey, Proc. Amer. Acad. Arts 22: 148 (1886). *Carex tribuloides* var. *cristata* (Tuck.) Macoun, Cat. Canad. Pl. 4: 130 (1888).
Carex cristata f. *ellipsoidalis* Farw., Rep. (Annual) Michigan Acad. Sci. 22: 180 (1921).
Carex cristata var. *catelliformis* Farw., Pap. Michigan Acad. Sci. 2: 17 (1923 publ. 1924). *Carex cristatella* f. *catelliformis* (Farw.) Fernald, Rhodora 44: 284 (1942).

Carex cruciata Wahlenb., Kongl. Vetensk. Acad. Nya Handl. 24: 149 (1803).
Madagascar, Nepal to Japan and Malesia. 29 MDG 36 CHC CHH CHS CHT 38 JAP NNS TAI 40 EHM IND NEP 41 MYA NCB THA VIE 42 BOR JAW LSI MLY MOL PHI SUL SUM 43 BIS NWG 50 QLD. Hemicr. or rhizome geophyte.

var. ***cruciata***
Madagascar, Nepal to Japan and W. Malesia. 29 MDG 36 CHC CHH CHS CHT 38 JAP NNS TAI 40 EHM IND NEP 41 MYA THA VIE 42 BOR MLY SUM. Hemicr. or rhizome geophyte.
Carex bengalensis Roxb., Fl. Ind. ed. 1832, 3: 572 (1832).
Carex cruciata Nees in R.Wight, Contr. Bot. India: 123 (1834), nom. illeg.
Carex valida Nees in R.Wight, Contr. Bot. India: 123 (1834).
Carex bengalensis f. *glauca* Boeck. ex Kuntze, Revis. Gen. Pl. 2: 748 (1891).
Carex cruciata var. *argocarpus* C.B.Clarke in J.D.Hooker, Fl. Brit. India 6: 716 (1894).
Carex cruciata var. *nagporensis* C.B.Clarke in J.D.Hooker, Fl. Brit. India 6: 716 (1894).
Carex hakkuensis Hayata, Icon. Pl. Formosan. 6: 122 (1916).
Carex cruciata var. *rubrobrunnea* Ohwi, Mem. Coll. Sci. Kyoto Imp. Univ., Ser. B, Biol. 11: 461 (1936). *Carex cruciata* subsp. *rubrobrunnea* (Ohwi) T.Koyama, in Fl. Taiwan 5: 312 (1978).
Carex galactolepis Nelmes, Kew Bull. 1: 20 (1946).
Carex spongocrepis Nelmes, Mém. Mus. Natl. Hist. Nat., Sér. B, Bot. 4: 128 (1955).

var. ***rafflesiana*** (Boott) Noot., in Fl. Males. 9(1): 122 (1979).
Nicobar Is. to Queensland. 38 TAI 41 NCB THA? 42 BOR JAW LSI MLY MOL PHI SUL SUM 43 BIS NWG 50 QLD. Hemicr. or rhizome geophyte.
**Carex rafflesiana* Boott, Proc. Linn. Soc. London 1: 285 (1846).
Carex macrothyrsa Miq., Fl. Ned. Ind. 3: 351 (1856). *Carex rafflesiana* var. *macrithyrsa* (Miq.) Kük. in H.G.A.Engler (ed.), Pflanzenr., IV, 20(38): 282 (1909).
Carex virgata Miq., Fl. Ned. Ind. 3: 351 (1856), nom. illeg. *Carex bengalensis* var. *virgata* Boeck., Linnaea 40: 347 (1876).
Carex pentacarpa Boeck., Flora 58: 265 (1875).
Carex bengalensis var. *scaberrima* Boeck., Linnaea 40: 347 (1876). *Carex scaberrima* (Boeck.) C.B.Clarke, Kew Bull., Addit. Ser. 8: 72 (1908), nom. illeg. *Carex rafflesiana* var. *scaberrima* (Boeck.) Kük. in H.G.A.Engler (ed.), Pflanzenr., IV, 20(38): 283 (1909). *Carex rafflesiana* subsp. *scaberrima* (Boeck.) T.Koyama, in Fl. Taiwan 5: 310 (1978).
Carex clarkeana Kük., Bull. Herb. Boissier, II, 4: 52 (1904).
Carex gembolensis var. *timorensis* C.B.Clarke, J. Linn. Soc., Bot. 37: 10 (1904). *Carex timorensis* (C.B.Clarke) Nelmes, Kew Bull. 1: 24 (1946).
Carex rafflesiana var. *tenuior* C.B.Clarke, J. Linn. Soc., Bot. 37: 10 (1904).
Carex scaberima C.B.Clarke, J. Linn. Soc., Bot. 37: 10 (1904), nom. illeg.
Carex pycnothyrsos Kük., Philipp. J. Sci., C 6: 60 (1911).
Carex semiglomerata Kük., Bull. Jard. Bot. Buitenzorg, III, 16: 315 (1940).
Carex spongoneura Nelmes, Kew Bull. 1: 18 (1946).
Carex xestogyne Nelmes, Kew Bull. 1: 16 (1946).
Carex buennemeijeri Nelmes, Kew Bull. 5: 191 (1950).
Carex ceramica Nelmes, Kew Bull. 5: 193 (1950).
Carex oblonga Nelmes, Kew Bull. 5: 192 (1950), nom. illeg.

Carex cruenta Nees in R.Wight, Contr. Bot. India: 128 (1834).
NE. Pakistan to SC. China. 36 CHC CHT CHT 40 EHM NEP PAK WHM. Hemicr. or rhizome geophyte.
Carex heterolepis Boeck., Beitr. Cyper. 1: 48 (1888), nom. illeg.

Carex crus-corvi Shuttlew. ex Kunze in C.Schkuhr, Beschr. Riedgräs., Suppl.: 128 (1844).
SE. Canada to C. & E. U.S.A. 72 ONT 74 ILL IOW KAN MIN MSO NEB OKL 75 INI MIC OHI WVA 77 TEX 78 ALA ARK FLA GEO KTY LOU NCA SCA TEN VRG. Hemicr. or rhizome geophyte.
Carex crus-corvi f. *orthoclados* Kunze, Suppl. Riedgräs.: 166 (1844).
Carex siciformis Boott, Boston J. Nat. Hist. 5: 113 (1845).

Carex halei Dewey, Amer. J. Sci. Arts, II, 2: 248 (1846).
Carex hystrix A.Gray ex Boott, Ill. Carex 1: 26 (1858).
Carex ornithorhyncha Boott, Ill. Carex 1: 26 (1858).
Carex crus-corvi var. *virginiana* Fernald, Rhodora 39: 393 (1937). *Carex virginiana* (Fernald) Fernald, Rhodora 43: 542 (1941), nom. illeg. *Carex bayardii* Fernald, Rhodora 44: 71 (1942).

Carex* ×*cryptochlaena T.Holm, Amer. J. Sci. 159: 305 (1900). *C. lyngbyei* × *C. ramenskii.*
Alaska. 70 ASK. Hemicr. or rhizome geophyte.

Carex cryptolepis Mack., Torreya 14: 157 (1914).
E. Canada to NC. & NE. U.S.A. 72 LAB NBR NFL NSC ONT QUE 74 MIN WIS 75 CNT INI MAI MAS MIC NWH NWJ NWY OHI RHO VER. Hemicr. or rhizome geophyte.
Carex cryptolepis var. *prolifera* (H.B.Lord) House, Bull. New York State Mus. Nat. Hist. 266: 26 (1925).

Carex cryptostachys Brongn. in L.I.Duperrey, Voy. Monde, Phan.: 152 (1833).
Thailand to Queensland. 36 CHC CHH CHS 38 TAI 41 THA VIE 42 BOR JAW MLY PHI SUM 43 NWG 50 QLD. Hemicr. or rhizome geophyte.

Carex* ×*csomadensis Simonk., Enum. Fl. Transsilv.: 556 (1887). *C. riparia* × *C. vesicaria.*
E. & SE. Europe. 13 ROM 14+. Hemicr. or rhizome geophyte.

Carex cubensis Kük., Repert. Spec. Nov. Regni Veg. 23: 220 (1926).
E. Cuba to Haiti. 81 CUB HAI. Hemicr. or rhizome geophyte.
Carex cubensis var. *flaccida* Kük., Repert. Spec. Nov. Regni Veg. 23: 221 (1926).

Carex cuchumatanensis Standl. & Steyerm., Ceiba 4: 62 (1953).
Guatemala. 80 GUA. Hemicr. or rhizome geophyte.

Carex culmenicola Steyerm., Fieldiana, Bot. 28: 65 (1951).
Venezuela. 82 VEN. Hemicr. or rhizome geophyte.

Carex cumberlandensis Naczi, Kral & Bryson, Sida 19: 994 (2001).
E. U.S.A. 75 INI OHI PEN WVA 78 ALA ARK GEO KTY MSI NCA SCA TEN VRG. Hemicr. or rhizome geophyte.

Carex cumulata (L.H.Bailey) Mack., Bull. Torrey Bot. Club 49: 366 (1922 publ. 1923).
C. & E. Canada to NC. & NE. U.S.A. 71 MAN 72 LAB NBR NFL NSC ONT PEI QUE 74 ILL WIS 75 CNT INI MAI MAS MIC NWH NWJ NWY PEN RHO VER. Hemicr. or rhizome geophyte.
**Carex straminea* var. *cumulata* L.H.Bailey, Mem. Torrey Bot. Club 1: 23 (1889). *Carex albolutescens* var. *cumulata* (L.H.Bailey) L.H.Bailey, Bull. Torrey Bot. Club 20: 422 (1893).
Carex cumulata f. *soluta* Fernald, Rhodora 44: 285 (1942).

Carex cuprina (I.Sándor ex Heuff.) Nendtv. ex A.Kern., Verh. K. K. Zool.-Bot. Ges. Wien 13: 566 (1863).
Macaronesia, Europe, Medit. to C. Asia. 10 DEN FIN GRB IRE NOR SWE 11 AUT BGM CZE GER HUN NET POL SWI 12 BAL COR FRA POR SAR SPA 13 ALB BUL GRC ITA KRI ROM SIC TUE YUG 14 BLR BLT KRY RUC RUN RUS RUW UKR 20 ALG MOR TUN 21 AZO CNY 32 KGZ TKM TZK 34 AFG CYP IRN IRQ LBS PAL TUR 36 CHX 40 PAK (51) nzn. Hemicr. or rhizome geophyte.
Carex vulpina f. *bracteata* G.Mey., Chloris Han.: 578 (1836). *Carex otrubae* f. *bracteata* (G.Mey.) Soó, Acta Bot. Acad. Sci. Hung. 16: 369 (1970 publ. 1971).
Carex vulpina f. *interrupta* Peterm., Anal. Pfl.-Schlüss.: 490 (1846). *Carex otrubae* f. *interrupta* (Peterm.) Soó, Acta Bot. Acad. Sci. Hung. 16: 369 (1970 publ. 1971).
**Carex nemorosa* var. *cuprina* I.Sándor ex Heuff., Linnaea 31: 662 (1862). *Vignea cuprina* (I.Sándor ex Heuff.) Soják, Cas. Nár. Mus., Odd. Prír. 150: 140 (1981 publ. 1982). *Carex vulpina* subsp. *cuprina* (I.Sándor ex Heuff.) O.Bolòs & Vigo, Fl. Man. Paisos Catalans 4: 251 (2001).
Carex vulpina var. *subcontigua* Kük. in H.G.A.Engler (ed.), Pflanzenr., IV, 20(38): 170 (1909). *Carex cuprina* var. *subcontigua* (Kük.) De Langhe & Lambinon, Nouv. Fl. Belg., ed. 3: 922 (1983).
Carex otrubae Podp., Spisy Prír. Fak. Masarykovy Univ. 12: 15 (1922). *Vignea otrubae* (Podp.) Soják, Cas. Nár. Mus., Odd. Prír. 148: 196 (1979 publ. 1980).
Carex subvulpina Senay, Bull. Mus. Natl. Hist. Nat., II, 17: 339, 444 (1945).

Carex curaica Kunth, Enum. Pl. 2: 375 (1837). *Vignea curaica* (Kunth) Soják, Cas. Nár. Mus., Odd. Prír. 148: 195 (1979 publ. 1980).
Siberia to Mongolia. 30 ALT WSB 32 KAZ KGZ TZK 36 CHX 37 MON 40 PAK. Hemicr. or rhizome geophyte.
Carex commutata Turcz., Bull. Soc. Imp. Naturalistes Moscou 27(2): 205 (1854), nom. illeg.
Carex curaica var. *angustifolia* Turcz., Fl. Baical.-Dahur. 2(1): 261 (1856).
Carex curaica var. *latifolia* Turcz., Fl. Baical.-Dahur. 2(1): 261 (1856).
Carex borotalicola Regel, Trudy Imp. S.-Peterburgsk. Bot. Sada 7: 566 (1880).
Carex curaica var. *rigida* Meinsh., Trudy Imp. S.-Peterburgsk. Bot. Sada 18: 317 (1901).

Carex curatorum Stacey, Leafl. W. Bot. 2: 13 (1937). *Carex scirpoidea* var. *curatorum* (Stacey) Cronquist in A.Cronquist & al, Intermountain Fl. 6: 113 (1977).
Utah to Arizona. 76 ARI UTA. Hemicr. or rhizome geophyte.
Carex haysii S.L.Welsh, Mem. New York Bot. Gard. 64: 124 (1990).

Carex curvata All., Fl. Pedem. 2: 264 (1785).
Mts. of Europe. 11 AUT BGM? SWI 12 FRA SPA 13 BUL ITA ROM YUG 14 UKR. Hemicr. or rhizome geophyte.

Carex curvicollis Franch. & Sav., Enum. Pl. Jap. 2: 579 (1878).
Japan. 38 JAP. Hemicr. or rhizome geophyte.
Carex viridula Franch. & Sav., Enum. Pl. Jap. 2: 579 (1878), nom. illeg.
Carex savatieri Franch., Nouv. Arch. Mus. Hist. Nat., III, 10: 71 (1898).

Carex curviculmis Reznicek, Contr. Univ. Michigan Herb. 19: 107 (1993).
C. & SE. Mexico. 79 MXC MXT. Hemicr. or rhizome geophyte.

Carex curvula All., Fl. Pedem. 2: 264 (1785). *Cryptoglochin curvulus* (All.) Heuff., Flora 27: 528 (1844). *Caricina curvula* (All.) St.-Lag. in A.Cariot, Étude Fl., ed. 8, 2: 876 (1889). *Kobresia curvula* (All.) K.V.Ivanova, Bot. Zhurn. S.S.S.R. 24: 489 (1939).
Mts. of Europe. 11 AUT SWI 12 FRA SPA 13 ALB BUL ITA ROM YUG 14 UKR. Hemicr. or rhizome geophyte.

subsp. ***curvula***
Mts. of Europe. 11 SWI 12 FRA SPA 13 ALB BUL ITA ROM YUG 14 UKR. Hemicr. or rhizome geophyte.
Carex curvula subsp. *mirabilis* Arcang. in ?
Carex lobata Bellardi, App. Fl. Pedem.: 42 (1793), nom. illeg.
Carex mirabilis Schleich. ex Gaudin, Fl. Helv. 6: 63 (1830), nom. illeg. *Carex curvula* var. *mirabilis* Nyman, Consp. Fl. Eur.: 783 (1882).
Carex ovata C.A.Mey. in C.F.von Ledebour, Fl. Altaic. 4: 207 (1833), nom. illeg.
Carex orbelica Velen., Fl. Bulg.: 575 (1891). *Carex curvula* f. *orbelica* (Velen.) Kük. in H.G.A. Engler (ed.), Pflanzenr., IV, 20(38): 186 (1909).
Carex curvula f. *pallida* Asch. & Graebn., Syn. Mitteleur. Fl. 2(2): 51 (1902).
Carex curvula f. *pygmaea* Holler ex Asch. & Graebn., Syn. Mitteleur. Fl. 2(2): 51 (1902).
Carex curvula f. *rodnensis* Porcius ex Asch. & Graebn., Syn. Mitteleur. Fl. 2(2): 51 (1902).
Carex curvula f. *elongata* Husn., Cypéracées 1: 12 (1905).
Carex rodnensis (Porcius ex Asch. & Graebn.) Rouy in G.Rouy & J.Foucaud, Fl. France 13: 404 (1912).

subsp. ***rosae*** Gilomen, Mitth. Naturf. Ges. Bern 1937: xxvii (1937). *Carex rosae* (Gilomen) Holub, Folia Geobot. Phytotax. 12: 429 (1977).
Pyrenees, Alps. 11 AUT SWI 12 FRA SPA 13 ITA. Hemicr. or rhizome geophyte.

Carex cusickii Mack. in C.V.Piper & R.K.Beattie, Fl. N.W. Coast: 72 (1915).
W. Canada to W. U.S.A. 71 BRC 73 IDA MNT ORE WAS WYO 76 CAL. Hemicr. or rhizome geophyte.
Carex obovoidea Cronquist, Madroño 7: 78 (1943).

Carex cuspidosa Dunn, J. Linn. Soc., Bot. 38: 373 (1908).
China. 36+. Hemicr. or rhizome geophyte.

Carex cylindrostachya Franch., Bull. Soc. Philom. Paris, VIII, 7: 32 (1895).
SC. China to Vietnam. 36 CHC 41 VIE. Hemicr. or rhizome geophyte.

Carex cyrtosaccus C.B.Clarke in D.Oliver & auct. suc. (eds.), Fl. Trop. Afr. 8: 524 (1902).
Tanzania to Malawi. 25 TAN 26 MLW. Hemicr. or rhizome geophyte.

Carex dabieensis S.W.Su, Acta Phytotax. Sin. 20: 221 (1982).
China (Anhui). 36 CHS. Hemicr. or rhizome geophyte.

Carex dahurica Kük., Repert. Spec. Nov. Regni Veg. 8: 326 (1910).
Siberia to NE. China. 30+ 31+ 36 CHM 37 MON. Hemicr. or rhizome geophyte.

Carex dailingensis Y.L.Chou, Bull. Bot. Lab. N. E. Forest. Inst., Harbin 4(1): 42 (1979).
NE. China. 36 CHM. Hemicr. or rhizome geophyte.

Carex daisenensis Nakai, Bot. Mag. (Tokyo) 33: 41 (1919).
Japan (Honshu, N. Kyushu). 38 JAP. Hemicr. or rhizome geophyte.

Carex dallii Kirk, Trans. & Proc. New Zealand Inst. 26: 261 (1893 publ. 1894).
New Zealand South I. 51 NZS. Hemicr. or rhizome geophyte.
Carex gibbsii Petrie, Trans. & Proc. New Zealand Inst. 46: 35 (1913 publ. 1914).

Carex daltonii Boott, Ill. Carex 1: 5 (1858).
C. & E. Himalaya. 40 EHM NEP. Hemicr. or rhizome geophyte.
Carex crassipes Boeck., Linnaea 40: 329 (1876).

Carex darwinii Boott, Proc. Linn. Soc. London 1: 261 (1845).
S. Chile to S. Argentina. 85 AGS CLS. Hemicr. or rhizome geophyte.

var. ***darwinii***
S. Argentina, S. Chile. 85 AGS CLS. Hemicr. or rhizome geophyte.
Carex nigricans Hombr. ex Boott, Ill. Carex 4: 156 (1867), nom. illeg.
Carex incompta Franch., Miss. Sci. Cap Horn 5: 376 (1889).

var. ***serranoi*** (Phil.) Kük. in H.G.A.Engler (ed.), Pflanzenr., IV, 20(38): 367 (1909).
S. Chile. 85 AGS? CLS. Hemicr. or rhizome geophyte.
**Carex serranoi* Phil., Anales Univ. Chile 93: 501 (1896).
Carex darwinii var. *robustior* Kük., Bot. Jahrb. Syst. 27: 528 (1899).

var. ***urolepis*** (Franch.) Kük. in H.G.A.Engler (ed.), Pflanzenr., IV, 20(38): 367 (1909).
S. Chile to S. Argentina. 85 AGS CLS. Hemicr. or rhizome geophyte.
**Carex urolepis* Franch., Miss. Sci. Cap Horn 5: 376 (1889).
Carex darwinii var. *aristata* C.B.Clarke ex Kük., Bot. Jahrb. Syst. 27: 529 (1899).

Carex dasycarpa Muhl., Descr. Gram.: 236 (1817). *Edritria dasycarpa* (Muhl.) Raf., Good Book: 26 (1840).
SE. U.S.A. 78 ALA FLA GEO MSI SCA. Hemicr. or rhizome geophyte.

Carex davalliana Sm., Trans. Linn. Soc. London 5: 266 (1800). *Carex dioica* var. *davalliana* (Sm.) Wahlenb., Kongl. Vetensk. Acad. Nya Handl. 24: 138 (1803). *Vignea davalliana* (Sm.) Rchb. in J.C.Mössler, Handb. Gewächsk. ed. 2, 3: 1605 (1830). *Psyllophora davalliana* (Sm.) Schur, Verh. Mitth. Siebenbürg. Vereins Naturwiss. Hermannstadt 4: 78 (1853).
Europe to Turkey. 10 DEN† GRB† 11 AUT CZE GER HUN POL SWI 12 COR FRA SPA 13 GRC ITA ROM YUG 14 BLT RUW UKR 34 TUR. Hemicr. or rhizome geophyte.
Carex scabra Hoppe, Bot. Taschenb. 1800: 242 (1800). *Maukschia scabra* (Hoppe) Heuff., Flora 27: 528 (1844). *Caricinella scabra* (Hoppe) St.-Lag. in A.Cariot, Étude Fl., ed. 8, 2: 880 (1889).
Carex reflexa Gaudin, Etrennes Fl.: 53 (1804).
Carex recurvirostra Hallier f. ex Steud., Nomencl. Bot. 1: 156 (1821).
Carex sieberiana Opiz, Naturalientausch 11: 413

(1826). *Carex davalliana* f. *sieberiana* (Opiz) Kunth, Enum. Pl. 2: 369 (1837). *Psyllophora sieberiana* (Opiz) Opiz, Seznam: 80 (1852). *Carex davalliana* var. *sieberiana* (Opiz) Nyman, Consp. Fl. Eur.: 784 (1882).
Carex curvula Willd. ex Kunth, Enum. Pl. 2: 269 (1837), nom. inval.
Carex davalliana var. *androgyna* Döll, Fl. Baden: 244 (1857).
Carex leucorhizia Dulac, Fl. Hautes -Pyrénées: 452 (1867).
Carex villosa Franch. & Sav., Enum. Pl. Jap. 2: 142 (1877), nom. illeg.
Carex dioicotrigona St.-Lag., Ann. Soc. Bot. Lyon 7: 121 (1880).

Carex david-smithii Reznicek, Novon 2: 433 (1992).
Peru. 83 PER. Hemicr. or rhizome geophyte.

Carex davidii Franch., Pl. David. 1: 319 (1884).
China. 36 CHC CHN CHS. Hemicr. or rhizome geophyte.
Carex kengiana Z.P.Wang, J. Nanjing Univ. (Biology) 1962(2): 46 (1962).

Carex davisii Schwein. & Torr., Ann. Lyceum Nat. Hist. New York 1: 326 (1824). *Loxotrema davisii* (Schwein. & Torr.) Raf., Good Book: 25 (1840).
SE. Canada to C. & E. U.S.A. 72 ONT 74 ILL IOW KAN MIN MSO NEB OKL WIS 75 CNT INI MAS MIC NWJ NWY OHI PEN VER WVA 77 TEX 78 ARK DEL KTY MRY TEN. Hemicr. or rhizome geophyte.
Carex aristata Dewey, Amer. J. Sci. Arts 7: 277 (1824), nom. illeg.
Carex torreyana Dewey, Amer. J. Sci. Arts 10: 47 (1826).

Carex davyi Mack., Bull. Torrey Bot. Club 43: 606 (1916 publ. 1917).
California. 76 CAL. Hemicr. or rhizome geophyte.

Carex dayuongensis Z.P.Wang, Acta Bot. Yunnan. 11: 166 (1989).
China. 36+. Hemicr. or rhizome geophyte.

Carex* ×*deamii F.J.Herm., Rhodora 40: 81 (1938). *C. shortiana* × *C. typhina*.
EC. U.S.A. 74 ILL MSO 75 INI 78 KTY. Hemicr. or rhizome geophyte.

Carex debeauxii H.Lév. & Vaniot, Repert. Spec. Nov. Regni Veg. 6: 264 (1909).
China. 36+. Hemicr. or rhizome geophyte.

Carex debilis Michx., Fl. Bor.-Amer. 2: 172 (1803). *Carex debilis* var. *typica* Fernald, Rhodora 44: 310 (1942), nom. inval.
Azores (Faial), E. Canada to C. & E. U.S.A. 21 AZO 72 LAB NBR NFL NSC ONT PEI QUE 74 ILL MIN MSO OKL WIS 75 CNT INI MAI MAS MIC NWH NWJ NWY OHI PEN WVA 77 TEX 78 ALA ARK DEL FLA GEO KTY LOU MRY MSI NCA SCA TEN VRG WDC. Hemicr. or rhizome geophyte.

var. ***debilis***
Azores (Faial), C. & E. U.S.A. 21 AZO 74 ILL MSO OKL 75 CNT INI MAS MIC NWJ NWY OHI PEN WVA 77 TEX 78 ALA ARK DEL FLA GEO KTY LOU MRY MSI NCA SCA TEN VRG WDC. Hemicr. or rhizome geophyte.
Carex debilis var. *pubera* A.Gray, Manual, ed. 5: 593 (1867).
Carex debilis var. *prolixa* L.H.Bailey, Proc. Amer. Acad. Arts 22: 105 (1886).
Carex debilis var. *intercursa* Fernald, Rhodora 44: 307 (1942).

var. ***rudgei*** L.H.Bailey, Mem. Torrey Bot. Club 1: 34 (1889). *Carex debilis* subsp. *rudgei* (L.H.Bailey) Á.Löve & D.Löve, Univ. Colorado Stud. Ser. Biol. 17: 16 (1965).
E. Canada to NC. & E. U.S.A. 72 LAB NBR NFL NSC ONT PEI QUE 74 ILL MIN WIS 75 CNT INI MAI MAS MIC NWH NWY PEN RHO VER WVA 78 DEL GEO KTY MRY NCA SCA TEN VRG WDC. Hemicr. or rhizome geophyte.
Carex tenuis Rudge, Trans. Linn. Soc. London 7: 97 (1804).
Carex flexuosa Muhl. ex Willd., Sp. Pl. 4: 297 (1805).
Carex debilis var. *strictior* L.H.Bailey, Mem. Torrey Bot. Club 1: 34 (1889).
Carex debilis var. *interjecta* L.H.Bailey, Bull. Torrey Bot. Club 20: 418 (1893). *Carex tenuis* var. *interjecta* (L.H.Bailey) Britton in N.L.Britton & A.Brown, Ill. Fl. N. U.S. 1: 320 (1896).
Carex tenuis var. *erectior* Britton in N.L.Britton & A.Brown, Ill. Fl. N. U.S. 1: 321 (1896).
Carex allegheniensis Mack. in N.L.Britton & al. (eds.), N. Amer. Fl. 18: 291 (1935).

Carex decaulescens V.I.Krecz. in V.L.Komarov (ed.), Fl. URSS 3: 599 (1935).
C. Asia to Iran. 32 KAZ KGZ UZB 34 IRN 40 PAK. Hemicr. or rhizome geophyte.

subsp. ***alsia*** (Raymond) Kukkonen, Ann. Bot. Fenn. 21: 387 (1984).
N. Pakistan. 40 PAK. Hemicr. or rhizome geophyte.
**Carex alsia* Raymond, Biol. Skr. 14(4): 23 (1965).

subsp. ***brunneola*** Kukkonen, Ann. Bot. Fenn. 21: 388 (1984).
NC. Iran. 34 IRN. Hemicr. or rhizome geophyte.

subsp. ***decaulescens***.
C. Asia. 32 KAZ KGZ UZB. Hemicr. or rhizome geophyte.

Carex decidua Boott, Proc. Linn. Soc. London 1: 255 (1845).
Bolivia to Falkland Is. 83 BOL 85 AGE AGS CLS 90 FAL. Hemicr. or rhizome geophyte.
Carex brehmeri Boeck., Allg. Bot. Z. Syst. 2: 190 (1896). *Carex decidua* var. *brehmeri* (Boeck.) Kük. in H.G.A.Engler (ed.), Pflanzenr., IV, 20(38): 307 (1909).
Carex pedicellata Phil., Anales Univ. Chile 93: 500 (1896), nom. illeg.
Carex rahmeri Phil., Anales Univ. Chile 93: 502 (1896).
Carex decidua var. *minor* Kük., Bot. Jahrb. Syst. 27: 526 (1899).
Carex vulgaris var. *antucensis* Kük., Bot. Jahrb. Syst. 27: 525 (1899).

Carex deciduisquama F.T.Wang & Tang ex P.C.Li, Acta Phytotax. Sin. 37: 160 (1999).
China (Yunnan). 36 CHC. Hemicr. or rhizome geophyte.

Carex declinata Boott, Ill. Carex 4: 171 (1867).
E. Australia. 50 NSW QLD. Hemicr. or rhizome geophyte.

Carex decomposita Muhl., Descr. Gram.: 264 (1817). *Carex paniculata* var. *decomposita* (Muhl.) Dewey, Amer. J. Sci. Arts 10: 276 (1826).
C. & E. U.S.A. 74 ILL MSO OKL 75 INI MIC NWY OHI 77 TEX 78 ALA ARK FLA GEO KTY LOU MRY MSI NCA SCA TEN VRG WDC. Hemicr. or rhizome geophyte.

Carex decora Boott, Ill. Carex 1: 5 (1858).
C. & E. Himalaya. 40 EHM NEP. Hemicr. or rhizome geophyte.

Carex decurtata Cheeseman, Trans. & Proc. New Zealand Inst. 24: 414 (1891 publ. 1892).
New Zealand South I. 51 NZS. Hemicr. or rhizome geophyte.
Carex cryptocarpa Cheeseman, Trans. & Proc. New Zealand Inst. 16: 412 (1884), nom. illeg.

Carex deflexa Hornem. in G.C.Oeder & al. (eds.), Fl. Dan.: t. 3051 (1883). *Carex pilulifera* var. *deflexa* (Hornem.) Drejer, Rev. Car. Bor.: 54 (1841). *Carex novae-angliae* var. *deflexa* (Hornem.) L.H.Bailey, Proc. Amer. Acad. Arts 22: 124 (1886).
Subarctic America to N. & W. U.S.A. 70 ASK GNL NUN NWT YUK 71 ABT BRC MAN SAS 72 LAB NBR NFL NSC ONT QUE 73 COL COL COL COL IDA IDA IDA IDA MNT MNT MNT MNT ORE ORE ORE ORE WAS WAS WAS WAS WYO WYO WYO WYO 74 WIS 75 CNT MAI MAS MIC NWH NWY PEN VER WVA 76 CAL UTA. Hemicr. or rhizome geophyte.

var. ***boottii*** L.H.Bailey, Mem. Torrey Bot. Club 1: 43 (1889).
W. Canada to W. U.S.A. 71 ABT BRC 73 COL IDA MNT ORE WAS WYO 76 CAL UTA. Hemicr. or rhizome geophyte.

var. ***deflexa***
Subarctic America to N. U.S.A. 70 ASK GNL NUN NWT YUK 71 ABT BRC MAN SAS 72 LAB NBR NFL NSC ONT QUE 73 COL IDA MNT ORE WAS WYO 74 WIS 75 CNT MAI MAS MIC NWH NWY PEN VER WVA. Hemicr. or rhizome geophyte.
Carex deflexa var. *deanei* L.H.Bailey, Mem. Torrey Bot. Club 1: 42 (1889).

Carex delavayi Franch., Bull. Soc. Philom. Paris, VIII, 7: 29 (1895).
SC. China. 36 CHC. Hemicr. or rhizome geophyte.

Carex delicata C.B.Clarke, Kew Bull., Addit. Ser. 8: 79 (1908).
Siberia to Mongolia. 30 WSB 37 MON. Hemicr. or rhizome geophyte.

Carex densa (L.H.Bailey) L.H.Bailey, Mem. Torrey Bot. Club 1: 50 (1889).
Washington to Mexico. 73 ORE WAS 76 CAL 79 MXC MXE MXS? MXT. Hemicr. or rhizome geophyte.
Carex brongniartii L.H.Bailey, Proc. Amer. Acad. Arts 22: 137 (1886), nom. illeg.
**Carex brongniartii* var. *densa* L.H.Bailey, Proc. Amer. Acad. Arts 22: 137 (1886).
Carex vicaria L.H.Bailey, Mem. Torrey Bot. Club 1: 49 (1889). *Carex vulpinoidea* var. *vicaria* (L.H.Bailey) Kük. in H.G.A.Engler (ed.), Pflanzenr., IV, 20(38): 148 (1909).
Carex vicaria var. *costata* L.H.Bailey, Mem. Torrey Bot. Club 1: 49 (1889). *Carex vulpinoidea* f. *costata* (L.H.Bailey) Kük. in H.G.A.Engler (ed.), Pflanzenr., IV, 20(38): 148 (1909).
Carex chrysoleuca T.Holm, Amer. J. Sci. 166: 454 (1903).
Carex breviligulata Mack., Erythea 8: 92 (1922).
Carex dudleyi Mack., Erythea 8: 30 (1922).

Carex densenervosa Chiov., Ann. Bot. (Rome) 9: 149 (1911).
Ethiopia. 24 ETH. Hemicr. or rhizome geophyte.

Carex densicaespitosa L.K.Dai, Acta Phytotax. Sin. 32: 173 (1994).
China (Guangxi). 36 CHS. Hemicr. or rhizome geophyte.

Carex densifimbriata Tang & F.T.Wang, Acta Phytotax. Sin. 24: 240 (1986).
S. China. 36 CHC CHS. Hemicr. or rhizome geophyte.
Carex densifimbriata var. *hirsuta* P.C.Li, Acta Phytotax. Sin. 37: 162 (1999).

Carex densipilosa C.Z.Zheng & X.F.Jin, Acta Phytotax. Sin. 42: 544 (2004).
China (Zhejiang). 36 CHS. Hemicr. or rhizome geophyte.

Carex depauperata Curtis ex With, Bot. Arr. Brit. Pl. ed. 2, 2: 1049 (1787). *Trasus depauperatus* (Curtis ex With) Gray, Nat. Arr. Brit. Pl. 2: 63 (1821).
Europe to C. Asia. 10 GRB IRE 11 BGM GER SWI 12 COR FRA SPA 13 ALB BUL GRC ITA ROM SIC YUG 14 KRY 32 KGZ TZK 33 NCS TCS 34 TUR. Hemicr. or rhizome geophyte.
Carex ventricosa Curtis, Fl. Londin. 5: t. 68 (1788).
Carex monilifera Thuill., Fl. Env. Paris: 490 (1790).
Carex triflora Schkuhr, Beschr. Riedgräs. 1: 94 (1801).
Loxotrema triflora (Schkuhr) Raf., Good Book: 25 (1840).

Carex depressa Link, J. Bot. (Schrader) 2: 309 (1799). *Carex halleriana* subsp. *depressa* (Link) Nyman, Consp. Fl. Eur.: 773 (1882).
Carpathians, Medit. to Iran. 11 CZE POL 12 BAL? COR FRA POR 13 ITA ROM TUE 14 KRY UKR 20 ALG MOR 33 NCS TCS 34 IRN TUR. Hemicr. or rhizome geophyte.

subsp. ***depressa***
W. & C. Medit. 12 BAL? COR FRA POR 13 ITA 20 ALG MOR. Hemicr. or rhizome geophyte.
Carex basilaris Jord., Observ. Pl. Nouv. 3: 426 (1846). *Carex depressa* var. *basilaris* (Jord.) Asch. & Graebn., Syn. Mitteleur. Fl. 1: 148 (1896). *Carex depressa* subsp. *basilaris* (Jord.) Kerguélen, Lejeunia 120: 62 (1987).
Carex pilulifera var. *vaginata* Boeck., Flora 61: 41 (1878).
Carex laxiflora Merino, Contr. Fl. Galicia: 48 (1897), nom. illeg. *Carex merinoi* Gand., Nov. Consp. Fl. Eur.: 475 (1910).
Carex depressa f. *tenella* Kük. in H.G.A.Engler (ed.), Pflanzenr., IV, 20(38): 463 (1909).

subsp. ***transsilvanica*** (Schur) T.V.Egorova, Novosti Sist. Vyssh. Rast. 9: 80 (1972).
Carpathians, Krym, SE. Europe to Iran. 11 CZE POL 13 ROM TUE 14 KRY UKR 33 NCS TCS 34 IRN TUR. Hemicr. or rhizome geophyte.
**Carex transsilvanica* Schur, Verh. Mitth. Siebenbürg. Vereins Naturwiss. Hermannstadt 4: 80 (1853). *Carex depressa* var. *transsilvanica*

(Schur) H.Christ, Bull. Soc. Roy. Bot. Belgique 24(2): 14 (1885).
Carex debilis Boiss. & Balansa in P.E.Boissier, Fl. Orient. 5: 410 (1882).
Carex baritiana Porcius, Analele Acad. Române, II, 14: 285 (1892).
Carex euxina (Woronow ex Marcow.) V.I.Krecz., Allg. Bot. Z. Syst. 11: 33 (1906).

Carex deqinensis L.K.Dai, Acta Phytotax. Sin. 32: 182 (1994).
China (Yunnan). 36 CHC. Hemicr. or rhizome geophyte.

Carex* ×*descendens Kük., Finska Vetensk. Soc. Forh. 45(8): 6 (1903). *C. appendiculatra* × *C. juncella*.
Russian Far East. 31+. Hemicr. or rhizome geophyte.

Carex desponsa Boott, Ill. Carex 2: 82 (1860).
Bhutan to Assam. 40 ASS EHM. Hemicr. or rhizome geophyte.

Carex devia Cheeseman, Trans. & Proc. New Zealand Inst. 15: 301 (1883).
New Zealand South I. 51 NZS. Hemicr. or rhizome geophyte.

Carex deweyana Schwein., Ann. Lyceum Nat. Hist. New York 1: 65 (1824). *Vignea deweyana* (Schwein.) Raf., Good Book: 27 (1840).
C. Japan, Canada to N. & WC. U.S.A. 38 JAP 70 ASK NWT YUK 71 ABT BRC MAN SAS 72 LAB NBR NFL NSC ONT PEI QUE 73 COL MNT WAS WYO 74 ILL IOW MIN NDA SDA WIS 75 CNT MAI MAS MIC NWH NWJ NWY OHI PEN VER 77 NWM. Hemicr. or rhizome geophyte.

var. ***collectanea*** Fernald, Rhodora 15: 93 (1913).
E. Canada. 72 QUE. Hemicr. or rhizome geophyte.

var. ***deweyana***
Canada to N. & WC. U.S.A. 70 ASK NWT YUK 71 ABT BRC MAN SAS 72 LAB NBR NFL NSC ONT PEI QUE 73 COL MNT WAS WYO 74 ILL IOW MIN NDA SDA WIS 75 CNT MAI MAS MIC NWH NWJ NWY OHI PEN VER 77 NWM. Hemicr. or rhizome geophyte.
Carex deweyi Eaton, Man. Bot., ed. 5: 150 (1829).
Carex deweyana var. *stricta* Farw., Rep. (Annual) Michigan Acad. Sci. 20: 169 (1918 publ. 1919).

var. ***senanensis*** (Ohwi) T.Koyama in ?
Japan (C. Honshu). 38 JAP. Hemicr. or rhizome geophyte.
Carex hondae Akiyama, Bot. Mag. (Tokyo) 45: 474 (1931).
**Carex senanensis* Ohwi, Mem. Coll. Sci. Kyoto Imp. Univ., Ser. B, Biol. 6: 248 (1931).

Carex dianae Steud., Syn. Pl. Glumac. 2: 238 (1855).
St. Helena. 28 STH. Hemicr.
**Carex pedunculata* Willd. ex Roxb. in A.Beatson, Tracts St. Helena: 302 (1816), nom. illeg.
Carex aequabilis Boott, Ill. Carex 1: 66 (1858). *Carex dianae* var. *aequablis* (Boott) Kük. in H.G.A.Engler (ed.), Pflanzenr., IV, 20(38): 650 (1909).
Carex praealta Boott, Ill. Carex 1: 66 (1858).
Carex longifolia Burch. ex Hemsl., Rep. Challenger, Bot. 1(2): 87 (1884), nom. illeg.

Carex diandra Schrank, Acta Acad. Elect. Mogunt. Sci. Util. Erfordiae: 49 (1782). *Carex paniculata* var. *diandra* (Schrank) Fiori, Nuov. Fl. Italia 1: 185 (1923). *Vignea diandra* (Schrank) Soják, Cas. Nár. Mus., Odd. Přír. 148: 195 (1979 publ. 1980).
Temp. Northern Hemisphere, New Zealand. 10 DEN FIN GRB ICE IRE NOR SWE 11 AUT BGM CZE GER HUN NET POL SWI 12 FRA 13 ITA ROM YUG 14 BLR BLT RUC RUE RUN RUW UKR 30 ALT KRA WSB YAK 31 KAM KUR SAK 32 KAZ KGZ TKM UZB 33 NCS TCS 34 TUR 36 CHI 37 MON 38 JAP 40 EHM 51 NZN 70 ASK NUN NWT YUK 71 ABT BRC MAN SAS 72 LAB NBR NFL NSC ONT PEI QUE 73 COL MNT ORE WAS WYO 74 ILL IOW MIN NDA NEB WIS 75 CNT INI MAI MAS MIC NWH NWJ NWY OHI PEN RHO VER 76 CAL UTA 78 MRY. Hemicr. or rhizome geophyte.
Carex paniculata var. *minor* Retz., Fl. Scand. Prodr.: 1152 (1779).
Carex fulva Thuill., Fl. Env. Paris: 483 (1790).
Carex teretiuscula Gooden., Trans. Linn. Soc. London 2: 163 (1794). *Vignea teretiuscula* (Gooden.) Rchb. in J.C.Mössler, Handb. Gewächsk. ed. 2, 3: 1621 (1830). *Physiglochis teretiuscula* (Gooden.) Raf., Good Book: 27 (1840). *Caricina teretiuscula* (Gooden.) St.-Lag. in A.Cariot, Étude Fl., ed. 8, 2: 875 (1889).
Carex teretiuscula var. *major* W.D.J.Koch, Syn. Fl. Germ. Helv.: 751 (1837). *Carex diandra* f. *major* (W.D.J.Koch) Kük. in H.G.A.Engler (ed.), Pflanzenr., IV, 20(38): 177 (1909).
Carex teretiuscula var. *crassior* Hartm., Handb. Skand. Fl., ed. 4: (1843).
Carex pseudoparadoxa S.Gibson, Phytologist 1: 778 (1844). *Carex paradoxa* var. *pseudoparadoxa* (S.Gibson) Nyman, Consp. Fl. Eur.: 781 (1882).
Carex ehrhartiana Hoppe ex Boott, Ill. Carex 4: 145 (1867). *Carex teretiuscula* var. *ehrhartiana* (Hoppe ex Boott) Nyman, Consp. Fl. Eur.: 782 (1882).
Carex teretiuscula var. *ramosa* Boott, Ill. Carex 4: 145 (1867). *Carex diandra* var. *ramosa* (Boott) Fernald, Rhodora 10: 48 (1908).
Carex teretiuscula f. *simplicior* Sanio, Bot. Centralbl. 6: 430 (1881).
Carex teretiuscula var. *ampla* L.H.Bailey, Mem. Torrey Bot. Club 1: 53 (1889). *Carex diandra* var. *ampla* (L.H.Bailey) Kük. in H.G.A.Engler (ed.), Pflanzenr., IV, 20(38): 177 (1909).
Carex teretiuscula var. *tenella* Beckm., Flor. Bassum.: 508 (1889). *Carex diandra* f. *tenella* (Beckm.) Kük. in H.G.A.Engler (ed.), Pflanzenr., IV, 20(38): 177 (1909).
Carex bernardina Parish, Bull. S. Calif. Acad. Sci. 5: 24 (1906).
Carex diandra f. *longibracteata* Chwastowski, Fragm. Florist. Geobot. 24: 641 (1978 publ. 1979).

Carex diastena V.I.Krecz. in V.L.Komarov (ed.), Fl. URSS 3: 594 (1935).
Kamchatka. 31 KAM. Hemicr. or rhizome geophyte.

Carex dickinsii Franch. & Sav., Enum. Pl. Jap. 2: 581 (1878).
SE. China, Korea, Japan. 36 CHS 38 JAP KOR. Hemicr. or rhizome geophyte.
Carex coreana L.H.Bailey, Mem. Torrey Bot. Club 1: 8 (1889).

Carex diclina Phil., Linnaea 33: 271 (1865).
Peru to S. South America. 83 BOL PER 85 AGE AGS AGW CLC CLN CLS. Hemicr. or rhizome geophyte.
Carex gayana Desv. in C.Gay, Fl. Chil. 6: 205 (1854), nom. illeg.

Carex schedonautos Steud., Syn. Pl. Glumac. 2: 189 (1855). *Carex gayana* var. *schedonautos* (Steud.) Kük. in H.G.A.Engler (ed.), Pflanzenr., IV, 20(38): 123 (1909).
Carex taurina Phil., Anales Univ. Chile 93: 490 (1896). *Carex gayana* var. *taurina* (Phil.) Kük. in H.G.A. Engler (ed.), Pflanzenr., IV, 20(38): 123 (1909).
Carex gayana var. *densa* Kük., Bot. Jahrb. Syst. 27: 501 (1899).
Carex gayana var. *rufa* Kük., Bot. Jahrb. Syst. 27: 501 (1899).

Carex dielsiana Kük., Notes Roy. Bot. Gard. Edinburgh 8: 10 (1913).
SC. China. 36 CHC. Hemicr. or rhizome geophyte.

Carex digitalis Willd., Sp. Pl. 4: 298 (1805). *Edritria digitalis* (Willd.) Raf., Good Book: 26 (1840).
SE. Canada to C. & E. U.S.A. 72 NSC ONT QUE 74 ILL MSO OKL WIS 75 CNT INI MAI MAS MIC NWH NWJ NWY OHI PEN RHO VER WVA 77 TEX 78 ALA ARK DEL FLA GEO KTY LOU MRY NCA SCA TEN VRG WDC. Hemicr. or rhizome geophyte.

var. ***digitalis***
SE. Canada to C. & E. U.S.A. 72 NSC ONT QUE 74 ILL MSO OKL WIS 75 CNT INI MAI MAS MIC NWH NWJ NWY OHI PEN RHO VER WVA 77 TEX 78 ALA ARK DEL GEO KTY MRY NCA SCA TEN VRG WDC. Hemicr. or rhizome geophyte.
Carex oligocarpa var. *van-vleckii* Schwein. ex Dewey, Amer. J. Sci. Arts 10: 281 (1826).
Carex digitalis var. *van-weckii* Dewey, Class-book Bot.: 423 (1845).
Carex podostachys Steud., Syn. Pl. Glumac. 2: 232 (1855). *Carex digitalis* f. *podostachys* (Steud.) Kük. in H.G.A.Engler (ed.), Pflanzenr., IV, 20(38): 528 (1909).
Carex oligocarpa Muhl. ex Boott, Ill. Carex 1: 40 (1858), nom. illeg.

var. ***floridana*** (L.H.Bailey) Naczi & Bryson, Novon 12: 531 (2002).
SE. U.S.A. to Texas. 77 TEX 78 ALA FLA GEO MSI SCA VRG. Hemicr. or rhizome geophyte.
**Carex laxiculmis* var. *floridana* L.H.Bailey, Mem. Torrey Bot. Club 1: 47 (1889). *Carex retrocurva* var. *floridana* (L.H.Bailey) Kük. in H.G.A. Engler (ed.), Pflanzenr., IV, 20(38): 529 (1909).
Carex digitalis var. *assymetrica* Fernald, Rhodora 43: 544 (1941).

var. ***macropoda*** Fernald, Rhodora 40: 400 (1938).
C. & E. U.S.A. 74 ILL MSO OKL 75 INI PEN 77 TEX 78 ALA ARK FLA GEO KTY LOU MRY MSI NCA SCA TEN VRG. Hemicr. or rhizome geophyte.

Carex digitata L., Sp. Pl.: 975 (1753). *Trasus digitatus* (L.) Gray, Nat. Arr. Brit. Pl. 2: 57 (1821). *Bitteria digitata* (L.) Fedde & J.Schust., Just's Bot. Jahresber. 41(2): 7 (1913 publ. 1918).
Europe to Iran. 10 DEN FIN GRB NOR SWE 11 AUT BGM CZE GER HUN NET POL SWI 12 COR FRA SPA 13 BUL ITA ROM YUG 14 BLR BLT KRY RUC RUE RUN RUW UKR 30 WSB 33 NCS TCS 34 IRN TUR. Hemicr. or rhizome geophyte.
Carex digitata subsp. *piroskana* (Nyár.) Serb. in ?
Carex digitata var. *intermedia* Crép., Notes Pl. Rar. Belgique 1(1): 26 (1859).
Carex digitata f. *brevifolia* Asch., Fl. Brandenburg 1: 786 (1864).
Carex digitata var. *bulgarica* Velen., Abh. Böhm. Ges. Wiss. 7: 67 (1890). *Carex ornithopoda* subsp. *bulgarica* (Velen.) Stoeva & E.D.Popova, Folia Geobot. Phytotax. 26: 392 (1991).
Carex digitata var. *hungarica* Borbás, Exsicc. (Herb. Europ.) 29: 2 (1896).
Carex digitata var. *albida* Freyn & Sint., Bull. Herb. Boissier, II, 2: 419 (1902).
Carex digitata f. *speirostchys* Neuman, Bot. Not. 1902: 190 (1902).
Carex digitata f. *compactior* Kük. in K.W. von Dalla Torre & L.von Sarntheim, Fl. Tirol 6(1): 354 (1906).
Carex digitata f. *glabrescens* Kük. in H.G.A.Engler (ed.), Pflanzenr., IV, 20(38): 497 (1909).
Carex digitata f. *robustior* Kük. in H.G.A.Engler (ed.), Pflanzenr., IV, 20(38): 497 (1909).

Carex diluta M.Bieb., Fl. Taur.-Caucas. 2: 388 (1808). *Carex hornschuchiana* var. *diluta* (M.Bieb.) Nyman, Consp. Fl. Eur.: 770 (1882).
Morocco, E. Europe to C. Nepal. 14 RUC RUS UKR 20 MOR 30+ 32 KAZ KGZ TKM TZK UZB 33 NCS TCS 34 AFG IRN IRQ TUR 37 MON 40 NEP PAK WHM (73) mnt. Hemicr. or rhizome geophyte.

subsp. ***diluta***
E. Europe to C. Nepal. 14 RUC RUS UKR 30+ 32 KAZ KGZ TKM TZK UZB 33 NCS TCS 34 AFG IRN IRQ TUR 37 MON 40 NEP PAK WHM (73) mnt. Hemicr. or rhizome geophyte.
Carex axillaris Link, Enum. Hort. Berol. Alt. 2: 329 (1822), nom. illeg.
Carex extensa C.A.Mey., Verz. Pfl. Casp. Meer.: 31 (1831), nom. illeg.
Carex microlepis Boeck., Linnaea 41: 269 (1877). *Carex diluta* var. *microlepis* (Boeck.) Kük. in H.G.A.Engler (ed.), Pflanzenr., IV, 20(38): 660 (1909).
Carex aitchisonii Boeck., Flora 63: 456 (1880).
Carex chorgosica Meinsh., Trudy Imp. S.-Peterburgsk. Bot. Sada 18: 381 (1900). *Carex diluta* var. *chorgosica* (Meinsh.) Kük. in H.G.A. Engler (ed.), Pflanzenr., IV, 20(38): 660 (1909).
Carex karelinii Meinsh., Trudy Imp. S.-Peterburgsk. Bot. Sada 18: 380 (1900). *Carex diluta* var. *karelinii* (Meinsh.) Kük. in H.G.A.Engler (ed.), Pflanzenr., IV, 20(38): 660 (1909).
Carex diluta f. *subdistans* Kük. in H.G.A.Engler (ed.), Pflanzenr., IV, 20(38): 660 (1909).
Carex czarwakensis Litv., Trudy Bot. Muz. Imp. Akad. Nauk 7: 97 (1910).
Carex aspratilis V.I.Krecz. in V.L.Komarov (ed.), Fl. URSS 3: 618 (1935). *Carex distans* subsp. *aspratilis* (V.I.Krecz.) T.V.Egorova, in Fl. Evropeiskoi Chasti SSSR 2: 177 (1976).

subsp. ***fissirostris*** (Ball) Maire & Weiller, in Fl. Afr. Nord 4: 158 (1957).
Morocco. 20 MOR. Hemicr. or rhizome geophyte.
**Carex fissirostris* Ball, J. Bot. 13: 206 (1875). *Carex diluta* var. *fissirostris* (Ball) Kük. in H.G.A.Engler (ed.), Pflanzenr., IV, 20(38): 660 (1909).

Carex diminuta Boeck., Beitr. Cyper. 2: 36 (1890).
China. 36+. Hemicr. or rhizome geophyte.

Carex dimorpholepis Steud., Syn. Pl. Glumac. 2: 214 (1855).

Pakistan to Japan. 36 CHC CHM CHN CHS 38 JAP KOR NNS 40 NEP PAK 41 MYA THA VIE. Hemicr. or rhizome geophyte.
Carex cernua Boott, Ill. Carex 4: 171 (1867), nom. illeg.
Carex cernua var. *minor* Boott, Ill. Carex 4: 171 (1867).
Carex rubescens Boeck., Flora 65: 60 (1882).
Carex schkuhriana H.Lév. & Vaniot, Bull. Acad. Int. Géogr. Bot. 11: 59 (1902).

Carex dimorphotheca Stschegl., Bull. Soc. Imp. Naturalistes Moscou 1854: 206 (1854).
Caucasus, C. Asia. 32 KAZ 33 TCS. Hemicr. or rhizome geophyte.

Carex dioica L., Sp. Pl.: 972 (1753). *Vignea dioica* (L.) Rchb. in J.C.Mössler, Handb. Gewächsk. ed. 2, 3: 1604 (1830). *Physiglochis dioica* (L.) Raf., Good Book: 27 (1840). *Psyllophora dioica* (L.) Schur, Verh. Mitth. Siebenbürg. Vereins Naturwiss. Hermannstadt 4: 78 (1853).
Europe to Siberia. 10 DEN FIN FOR GRB ICE IRE NOR SVA SWE 11 AUT BGM CZE GER NET POL SWI 12 FRA SPA 13 ITA ROM YUG 14 BLR BLT RUC RUE RUN RUS RUW UKR 30 WSB. Hemicr. or rhizome geophyte.
Carex laevis Hoppe, Bot. Taschenb. 1800: 243 (1800), nom. illeg. *Maukschia laevis* (Hoppe) Heuff., Flora 27: 527 (1844). *Caricinella laevis* (Hoppe) St.-Lag. in A.Cariot, Étude Fl., ed. 8, 2: 880 (1889).
Carex linnaeana Host, Icon. Descr. Gram. Austriac. 3: 77 (1805).
Carex linnaei Degl. in J.L.A.Loiseleur-Deslongchamps, Fl. Gall.: 627 (1807).
Carex nigricans Dewey, Amer. J. Sci. Arts 29: 249 (1836), nom. illeg.
Carex custoriana Heer ex Hegetschw., Fl. Schweiz: 906 (1840). *Carex dioica* var. *custoriana* (Heer ex Hegetschw.) Nyman, Consp. Fl. Eur.: 784 (1882).
Carex dioica f. *isogyna* Fr., Novit. Fl. Suec. Mant. 3: 135 (1842).
Carex dioica var. *scabrella* Fr., Summa Veg. Scand.: 73 (1845).
Carex dioica var. *subcaespitosa* O.Lang, Linnaea 24: 495 (1851).
Carex dioica f. *metteniana* Asch., Fl. Brandenburg 1: 760 (1864). *Carex metteniana* (Asch.) Lehm. ex Schur, Enum. Pl. Transsilv.: 697 (1866). *Carex dioica* var. *metteniana* (Asch.) Nyman, Consp. Fl. Eur.: 784 (1882).
Carex casteriana Heer ex Boott, Ill. Carex 4: 143 (1867).
Carex dioiscostrongyla St.-Lag., Ann. Soc. Bot. Lyon 7: 121 (1880).
Carex dioica f. *laxa* Junge, Verh. Naturwiss. Vereins Hamburg 14: 94 (1906).

Carex diplodon Nelmes, Bull. Misc. Inform. Kew 1939: 184 (1939).
NC. China. 36 CHN. Hemicr. or rhizome geophyte.

Carex dipsacea Berggr., Minneskr. Fisiog. Sallks. Lund 8: 28 (1877).
New Zealand. 51 NZN NZS. Hemicr. or rhizome geophyte.
Carex tahoata Hamlin, Rec. Domin. Mus. 6: 109 (1968).

Carex discessa V.I.Krecz. ex Grubov, Bot. Mater. Gerb. Bot. Inst. Komarova Akad. Nauk S.S.S.R. 17: 7 (1955).
Mongolia. 37 MON. Hemicr. or rhizome geophyte.

Carex dispalata Boott in M.C.Perry, Jap. Exped. 2: 325 (1856).
Nepal, Russian Far East, China, Korea, Japan. 31 KUR PRM SAK 36 CHI CHM CHN CHS 38 JAP KOR 40 NEP. Hemicr. or rhizome geophyte.

subsp. ***dispalata***
Russian Far East, China, Korea, Japan. 31 KUR PRM SAK 36 CHI CHM CHN CHS 38 JAP KOR. Hemicr. or rhizome geophyte.
Carex dispalata var. *niigatensis* Franch. & Sav., Enum. Pl. Jap. 2: 580 (1878).
Carex subanceps Boeck., Bot. Jahrb. Syst. 5: 520 (1884).
Carex pollens C.B.Clarke, J. Linn. Soc., Bot. 36: 305 (1904).
Carex pollens var. *angustior* C.B.Clarke, J. Linn. Soc., Bot. 36: 305 (1904).
Carex dispalata var. *costata* Kük. in H.G.A.Engler (ed.), Pflanzenr., IV, 20(38): 617 (1909).
Carex coronata H.Lév., Repert. Spec. Nov. Regni Veg. 11: 66 (1912).
Carex persistens Ohwi, Mem. Coll. Sci. Kyoto Imp. Univ., Ser. B, Biol. 5(3): 249 (1930).
Carex persistens var. *takeuchii* Ohwi, Mem. Coll. Sci. Kyoto Imp. Univ., Ser. B, Biol. 5: 249 (1930).

subsp. ***laxiflorens*** T.Koyama, Bot. Mag. (Tokyo) 86: 275 (1973).
Nepal. 40 NEP. Hemicr. or rhizome geophyte.

Carex disperma Dewey, Amer. J. Sci. Arts 8: 266 (1824). *Vignea disperma* (Dewey) Raf., Good Book: 27 (1840).
Temp. & Subarctic Northern Hemisphere. 10 FIN NOR SWE 11 POL 14 BLR BLT RUC RUE RUN RUS RUW 30 WSB YAK 31 AMU KAM PRM SAK 36 CHI CHM 38 JAP KOR 70 ASK GNL NUN NWT YUK 71 ABT BRC MAN SAS 72 LAB NBR NFL NSC ONT PEI QUE 73 COL IDA MNT ORE WAS WYO 74 MIN NDA SDA WIS 75 CNT INI MAI MAS MIC NWH NWJ NWY OHI PEN VER 76 CAL NEV UTA 77 NWM. Hemicr. or rhizome geophyte.
Carex tenella Schkuhr, Beschr. Riedgräs. 1: 23 (1801), nom. illeg.
Carex blyttii F.Nyl., Spic. Fl. Fenn. 2: 35 (1844).
Carex inops Kunze in C.Schkuhr, Beschr. Riedgräs., Suppl.: 131 (1844), pro syn.
Carex gracilis A.Gray, Amer. J. Sci. Arts, II, 4: 19 (1847), nom. illeg.
Carex misera Franch., Bull. Soc. Philom. Paris, VIII, 7: 51 (1895), nom. illeg. *Carex tenella* var. *misera* Franch., Nouv. Arch. Mus. Hist. Nat., III, 8: 224 (1896).
Carex dominii H.Lév. & Vaniot, Bull. Acad. Int. Géogr. Bot. 19: 35 (1909).
Carex nakaii H.Lév., Bull. Acad. Int. Géogr. Bot. 19: 33 (1909).

Carex dissita Sol. ex Boott in J.D.Hooker, Fl. Nov.-Zel. 1: 284 (1853).
New Zealand. 51 NZN NZS. Hemicr. or rhizome geophyte.
Carex quadrangulata Colenso, Trans. & Proc. New Zealand Inst. 17: 254 (1885).
Carex longiacuminata Colenso, Trans. & Proc. New Zealand Inst. 21: 104 (1888 publ. 1889).
Carex polyneura Colenso, Trans. & Proc. New Zealand Inst. 21: 184 (1888 publ. 1889).

Carex dissita var. *monticola* Kük. in T.F. Cheeseman, Man. New Zealand Fl.: 831 (1906).

Carex dissitiflora Franch., Bull. Soc. Philom. Paris, VIII, 7: 32 (1895).
S. Kuril Is., Japan, Taiwan. 31 KUR 38 JAP TAI. Hemicr. or rhizome geophyte.

subsp. ***dissitiflora***
S. Kuril Is., Japan. 31 KUR 38 JAP. Hemicr. or rhizome geophyte.
Carex subdissitiflora Kük. in J.Matsumura, Index Pl. Jap. 2(1): 135 (1905).

subsp. ***taiwanensis*** (Ohwi) T.Koyama, in Fl. Taiwan 5: 349 (1978).
Taiwan. 38 TAI. Hemicr. or rhizome geophyte.
**Carex dissitiflora* var. *taiwanensis* Ohwi, Jap. J. Bot. 7: 193 (1934). *Carex taiwanensis* (Ohwi) Akiyama, Caric. Far East. Reg. As.: 156 (1955).

Carex dissitispicula Ohwi, Acta Phytotax. Geobot. 7: 35 (1938).
Japan (Honshu: Karuisawa). 38 JAP. Hemicr. or rhizome geophyte.
Carex kunioi T.Koyama, J. Jap. Bot. 30: 131 (1955).

Carex distachya Desf., Fl. Atlant. 2: 336 (1799).
Canary Is. (Gran Canaria), Medit. 12 BAL COR FRA POR SAR SPA 13 ALB? BUL GRC ITA KRI SIC TUE YUG 20 MOR 21 CNY 34 EAI TUR. Hemicr. or rhizome geophyte.

var. ***distachya***
Canary Is. (Gran Canaria), Medit. 12 BAL COR FRA POR SAR SPA 13 ALB? BUL GRC ITA KRI SIC TUE YUG 20 MOR 21 CNY 34 EAI TUR. Hemicr. or rhizome geophyte.
Carex longiseta var. *latisquama* Cuatrec. in ?. *Carex latisquama* (Cuatrec.) Pau ex Cuatrec., Trab. Mus. Ci. Nat. Barcelona 12: 242 (1929).
Carex bina Schkuhr, Beschr. Riedgräs. 1: 36 (1801). *Carex nigra* f. *bina* (Schkuhr) Asch. & Graebn., Syn. Mitteleur. Fl. 2(2): 110 (1902).
Carex dimorpha Brot., Fl. Lusit. 1: 64 (1804).
Carex longiseta Brot., Fl. Lusit. 1: 63 (1804).
Carex linkii Willd., Sp. Pl. 4: 223 (1805). *Edritria linkii* (Willd.) Raf., Good Book: 26 (1840).
Carex gynomane Bertol., Rar. Lig. [Ital.] Pl. 2: 43 (1806).
Carex tuberosa Degl. in J.L.A.Loiseleur-Deslongchamps, Fl. Gall. 2: 629 (1807). *Loxotrema tuberosa* (Degl.) Raf., Good Book: 25 (1840).
Carex sarda Link, Hort. Berol. 2: 328 (1833). *Carex distachya* f. *sarda* (Link) Boeck., Linnaea 40: 387 (1876). *Carex linkii* var. *sarda* (Link) Nyman, Consp. Fl. Eur.: 783 (1882).
Carex gynandra Sieber ex Kunth, Enum. Pl. 2: 478 (1837), nom. illeg.
Carex muelleri Steud., Syn. Pl. Glumac. 2: 204 (1855), pro syn.
Carex olyssiponensis Steud., Syn. Pl. Glumac. 2: 204 (1855). *Carex linkii* var. *olyssiponensis* (Steud.) Nyman, Consp. Fl. Eur.: 783 (1882).
Carex olbiensis var. *angustifolia* Rohlena, Sitzungsber. Königl. Böhm. Ges. Wiss. Prag, Math.-Naturwiss. Cl. 1904: 97 (1904).

var. ***phyllostachioidea*** Ö.Nilsson in P.H.Davis (ed.), Fl. Turkey 9: 622 (1985).
E. Aegean Is., W. Turkey. 34 EAI TUR. Hemicr. or rhizome geophyte.

Carex distans L., Syst. Nat. ed. 10, 2: 1263 (1759). *Trasus distans* (L.) Gray, Nat. Arr. Brit. Pl. 2: 62 (1821).
Macaronesia, Europe, Medit. to Japan. 10 DEN FIN GRB ICE IRE NOR SWE 11 AUT BGM CZE GER HUN NET POL SWI 12 BAL COR FRA POR SAR SPA 13 ALB BUL GRC ITA KRI ROM SIC TUE YUG 14 BLR BLT KRY RUC RUE RUN RUS RUW UKR 20 ALG EGY LBY MOR TUN 21 AZO 33 NCS TCS 34 CYP EAI IRN IRQ LBS PAL SIN TUR 35 SAU YEM 36 CHN 38 JAP KOR (75) pen (78) mry. Hemicr. or rhizome geophyte.
Carex bessarabica (Savul. & Rayss) Zahar. in ?
Carex distans subsp. *adriatica* Degen in ?
Carex distans var. *crassa* C.B.Clarke in ?
Carex fulva P.Gaertn., B.Mey. & Scherb., Oekon. Fl. Wetterau 3: 330 (1801), nom. illeg.
Carex neglecta Degl. in J.L.A.Loiseleur-Deslongchamps, Fl. Gall. 2: 298 (1807). *Carex distans* var. *neglecta* (Degl.) Nyman, Consp. Fl. Eur.: 770 (1882).
Carex flavescens Host, Icon. Descr. Gram. Austriac. 4: 53 (1809).
Carex sicula C.Presl, Cyper. Gramin. Sicul.: 9 (1820), nom. illeg.
Carex flavescens f. *minor* Gaudin, Fl. Helv. 1: 134 (1828).
Carex nervosa Willd. ex Kunth, Enum. Pl. 2: 449 (1837), nom. illeg.
Carex baetica Auersw. ex Willk., Bot. Zeitung (Berlin) 6: 414 (1848). *Carex distans* var. *baetica* (Auersw. ex Willk.) Nyman, Consp. Fl. Eur.: 770 (1882).
Carex sinai Boott, Trans. Linn. Soc. London 20: 146 (1851). *Carex distans* f. *sinai* (Boott) Boeck., Linnaea 41: 269 (1877).
Carex sinaica Nees ex Steud., Syn. Pl. Glumac. 2: 223 (1855), nom. inval.
Carex pseudoflava Schur, Enum. Pl. Transsilv.: 712 (1866). *Carex distans* var. *pseudoflava* (Schur) Asch. & Graebn., Syn. Mitteleur. Fl. 1: 193 (1897).
Carex sicula C.Presl ex Boott, Ill. Carex 4: 185 (1867), nom. illeg.
Carex portae K.Richt., Pl. Eur. 1: 165 (1890).
Carex distans var. *oranensis* Trab. in J.A.Battandier & al., Fl. Algérie 1(2): 111 (1895).
Carex oranensis Trab. in J.A.Battandier & al., Fl. Algérie 1(2): 111 (1895).
Carex distans f. *minor* Post, Fl. Syria: 838 (1896).
Carex distans f. *major* Kneuck., Allg. Bot. Z. Syst. 5: 127 (1899).
Carex distans f. *pendula* Lackow., Verh. Bot. Vereins Prov. Brandenburg 44: 165 (1903).
Carex diamantina H.Lév. & Vaniot, Repert. Spec. Nov. Regni Veg. 4: 226 (1907).
Carex vikingensis C.B.Clarke, Kew Bull., Addit. Ser. 8: 84 (1908).
Carex distans f. *appropinquata* Kük. in H.G.A.Engler (ed.), Pflanzenr., IV, 20(38): 664 (1909).
Carex forficula var. *scabida* Kük. in H.G.A.Engler (ed.), Pflanzenr., IV, 20(38): 342 (1909).
Carex mariae-victorinii Sennen, Butl. Inst. Catalana Hist. Nat. 32: 105 (1932).
Carex forficula Sennen, Diagn. Nouv.: 78 (1936), nom. illeg.

Carex distentiformis F.J.Herm., Brittonia 23: 146 (1971).
Mexico (Chiapas). 79 MXT. Hemicr. or rhizome geophyte.

Carex disticha Huds., Fl. Angl.: 347 (1762). *Vignea disticha* (Huds.) Peterm., Fl. Bienitz: 16 (1841). *Caricina disticha* (Huds.) St.-Lag. in A.Cariot, Étude Fl., ed. 8, 2: 880 (1889).

Temp. Eurasia, SE. Australia. 10 DEN FIN GRB IRE NOR SWE 11 AUT BGM CZE GER HUN NET POL SWI 12 FRA SPA 13 BUL GRC? ITA ROM YUG 14 BLR BLT RUC RUE RUS RUW UKR 20 ALG LBY MOR TUN 30+ 32 KAZ 33 NCS TCS 34 TUR 36 CHC CHM 37 MON 50 VIC (72) ont que. Hemicr. or rhizome geophyte.

Carex arenaria Leers, Fl. Herborn.: 195 (1775).
Carex multiformis Thuill., Fl. Env. Paris: 479 (1790).
Carex spicata Lam., Encycl. 3: 381 (1792), nom. illeg.
Carex intermedia Gooden., Trans. Linn. Soc. London 2: 154 (1794), nom. illeg. *Vignea intermedia* Rchb. in J.C.Mössler, Handb. Gewächsk. ed. 2, 3: 1618 (1830).
Carex schoenoides DC. in J.B.A.M.de Lamarck & A.P.de Candolle, Fl. Franç., ed. 3, 3: 104 (1805), nom. illeg.
Carex pseudoarenaria Pers., Syn. Pl. 2: 536 (1807).
Carex disticha var. *longibracteata* Schleich., Cat. Pl. Helv., ed. 4: 11 (1821). *Carex intermedia* f. *longibracteata* (Schleich.) Kük. in H.G.A.Engler (ed.), Pflanzenr., IV, 20(38): 136 (1909).
Carex longibracteata Schleich., Cat. Pl. Helv., ed. 4: 11 (1821), nom. illeg.
Carex intermedia f. *abbreviata* Klett & Richt., Fl. Leipzig: 744 (1830).
Carex modesta J.Gay, Ann. Sci. Nat., Bot., II, 10: 304 (1838). *Carex teretiuscula* subsp. *modesta* (J.Gay) Nyman, Consp. Fl. Eur.: 782 (1882).
Vignea disticha f. *floribunda* Peterm. in H.G.L.Reichenbach, Fl. Saxon.: 43 (1842). *Carex disticha* f. *floribunda* (Peterm.) Peterm., Anal. Pfl.-Schlüss.: 492 (1846). *Carex intermedia* f. *floribunda* (Peterm.) Kük. in H.G.A.Engler (ed.), Pflanzenr., IV, 20(38): 136 (1909).
Carex disticha var. *minor* Peterm., Flora 27: 327 (1844). *Carex intermedia* f. *minor* (Peterm.) Kük. in H.G.A.Engler (ed.), Pflanzenr., IV, 20(38): 136 (1909).
Carex disticha var. *gracilis* Metsch, Fl. Henneberg.: 355 (1845), nom. illeg.
Carex intermedia var. *repens* Rchb., Icon. Fl. Germ. Helv. 8: 8 (1846).
Carex disticha var. *capitata* Hampe ex Láng, Linnaea 24: 512 (1851).
Carex disticha var. *distachya* O.Lang, Linnaea 24: 512 (1851). *Carex intermedia* f. *distachya* (O.Lang) Kük. in H.G.A.Engler (ed.), Pflanzenr., IV, 20(38): 136 (1909).
Carex disticha var. *elatior* Boeck., Linnaea 39: 140 (1875). *Carex intermedia* f. *elatior* (Boeck.) Kük. in H.G.A.Engler (ed.), Pflanzenr., IV, 20(38): 136 (1909).
Carex disticha var. *luxurians* Beck, Fl. Nieder-Österreich: 130 (1890).
Carex disticha f. *pangyna* Beck, Fl. Nieder-Österreich: 130 (1890). *Carex intermedia* f. *pangyna* (Beck) Kük. in H.G.A.Engler (ed.), Pflanzenr., IV, 20(38): 136 (1909).
Carex disticha f. *permixta* Beck, Fl. Nieder-Österreich: 130 (1890). *Carex intermedia* f. *permixta* (Beck) Kük. in H.G.A.Engler (ed.), Pflanzenr., IV, 20(38): 136 (1909).
Carex teretiuscula var. *modesta* K.Richt., Pl. Eur. 1: 150 (1890).
Carex disticha var. *compactior* Kük., Bot. Centralbl. 75: 108 (1898).
Carex disticha var. *decipiens* De Wild. & T.Durand, Prodr. Fl. Belg. 3: 60 (1899).
Carex grossheimii V.I.Krecz. in V.L.Komarov (ed.), Fl. URSS 3: 589 (1935). *Carex disticha* subsp. *grossheimii* (V.I.Krecz.) T.V.Egorova, Novosti Sist. Vyssh. Rast. 22: 54 (1985).
Carex disticha var. *pedunculata* Pamp., Arch. Bot. (Forlì) 12: 21 (1936).

Carex distracta C.B.Clarke in J.D.Hooker, Fl. Brit. India 6: 715 (1894).

Assam. 40 ASS. Hemicr. or rhizome geophyte.

Carex divisa Huds., Fl. Angl.: 348 (1762). *Vignea divisa* (Huds.) Rchb. in J.C.Mössler, Handb. Gewächsk. ed. 2, 3: 1615 (1830). *Caricina divisa* (Huds.) St.-Lag. in A.Cariot, Étude Fl., ed. 8, 2: 874 (1889).

Europe to C. Asia, introduced elsewere. 10 GRB IRE† 11 AUT BGM CZE NET 12 BAL COR FRA POR SAR SPA 13 ALB BUL GRC ITA KRI ROM SIC TUE YUG 14 KRY UKR 20 ALG EGY LBY MOR TUN (21) mdr (27) cpp 32 KAZ KGZ TKM TZK UZB 33 TCS 34 AFG CYP IRN IRQ LBS PAL SIN TUR 35 SAU 36 CHT CHX 40 PAK WHM (50) qld tas vic wau (51) nzn (78) mry nca vrg. Hemicr. or rhizome geophyte.

Carex marginata Gorter, Fl. Belg.: 247 (1767).
Carex tripartita Lam., Encycl. 3: 339 (1789), nom. illeg.
Carex schoenoides Thuill., Fl. Env. Paris: 480 (1790), nom. illeg.
Carex hybrida Lam., Encycl. 3: 382 (1792).
Carex lobata Lam., Encycl. 3: 379 (1792), nom. illeg.
Carex schoenoides Desf., Fl. Atlant. 2: 336 (1799), nom. illeg.
Carex rivularis Schkuhr, Beschr. Riedgräs. 1: 30 (1801). *Vignea rivularis* (Schkuhr) Schur, Verh. Mitth. Siebenbürg. Vereins Naturwiss. Hermannstadt 4: 78 (1853). *Carex divisa* var. *rivularis* (Schkuhr) Nyman, Consp. Fl. Eur.: 782 (1882). *Carex divisa* f. *rivularis* (Schkuhr) Kük. in P.F.A.Ascherson & K.O.R. Graebner, Syn. Mitteleur. Fl. 2(2): 26 (1902).
Carex cuspidata Bertol., Mém. Soc. Méd. Emul. Gen. 2: 145 (1803), nom. illeg. *Carex divisa* var. *cuspidata* Nyman, Consp. Fl. Eur.: 782 (1882).
Carex ammophila Willd., Sp. Pl. 4: 226 (1805). *Vignea ammophila* (Willd.) Rchb. in J.C.Mössler, Handb. Gewächsk. ed. 2, 3: 1615 (1830). *Carex divisa* var. *ammophila* (Willd.) Nyman, Consp. Fl. Eur.: 782 (1882).
Carex austriaca Schkuhr, Beschr. Riedgräs. 2: 10 (1806). *Carex divisa* var. *austriaca* (Schkuhr) Nyman, Consp. Fl. Eur., Suppl. 2: 325 (1890).
Carex bertolonii Schkuhr, Beschr. Riedgräs. 2: 5 (1806). *Vignea bertolonii* (Schkuhr) Rchb. in J.C. Mössler, Handb. Gewächsk. ed. 2, 3: 1608 (1830).
Carex splendens Thuill. ex Pers., Syn. Pl. 2: 536 (1807). *Diemisa splendens* (Thuill. ex Pers.) Raf., Good Book: 27 (1840). *Carex divisa* var. *splendens* (Thuill. ex Pers.) Nyman, Consp. Fl. Eur.: 782 (1882).
Carex incurva Sm. in J.Sibthorp & J.E.Smith, Fl. Graec. Prodr. 1: 227 (1809), nom. illeg.
Carex fontanesii Poir. in J.B.A.M.de Lamarck, Encycl., Suppl. 3: 257 (1813).
Carex paradoxa Benth., Cat. Pl. Pyrénées: 67 (1826), nom. illeg.
Carex consanguinea Kunth, Enum. Pl. 2: 374 (1837).

Carex fuscescens Willd. ex Kunth, Enum. Pl. 2: 373 (1837).
Carex hostii Hoppe ex Kunth, Enum. Pl. 2: 372 (1837).
Carex leporina Pall. ex Kunth, Enum. Pl. 2: 393 (1837).
Carex spicata Brot. ex Kunth, Enum. Pl. 2: 373 (1837), nom. inval.
Carex coacta Boott, Proc. Linn. Soc. London 1: 285 (1846). *Carex curaica* var. *coacta* (Boott) Boott, Ill. Carex 4: 204 (1867).
Carex setifolia Godr., Mém. Soc. Émul. Doubs, II, 5: 14 (1854), nom. illeg. *Vignea setifolia* Fourr., Ann. Soc. Linn. Lyon, n.s., 17: 175 (1869). *Caricina setifolia* (Fourr.) St.-Lag. in A.Cariot, Étude Fl., ed. 8, 2: 874 (1889).
Carex chaetophylla Steud., Syn. Pl. Glumac. 2: 187 (1855). *Carex divisa* subsp. *chaetophylla* (Steud.) Nyman, Consp. Fl. Eur.: 782 (1882). *Carex divisa* var. *chaetophylla* (Steud.) Daveau, Cyper. Portug.: 47 (1892).
Carex hymelayensis Steud., Syn. Pl. Glumac. 2: 204 (1855).
Carex divisa var. *longiculmis* Willk. in M.Willkomm & J.M.C.Lange, Prodr. Fl. Hispan. 1: 119 (1861).
Carex rivalis Willk. & Lange, Prodr. Fl. Hispan. 1: 130 (1861), nom. illeg.
Carex coarctata Kit., Linnaea 32: 319 (1863).
Carex moniziana Lowe ex Boott, Ill. Carex 4: 174 (1867). *Carex divisa* var. *moniziana* (Lowe ex Boott) Kük. in H.G.A.Engler (ed.), Pflanzenr., IV, 20(38): 126 (1909).
Carex schreberi Ten. ex Boott, Ill. Carex 4: 186 (1867), nom. illeg.
Carex divisa f. *pallidiflora* Kük. in H.G.A.Engler (ed.), Pflanzenr., IV, 20(38): 126 (1909).
Carex algarbiensis Samp., Ann. Sci. Acad. Polytecn. Porto 14: 144 (1921).
Carex divisa var. *platyphylla* Braun-Blanq., Maire & Trabut, Mém. Soc. Sci. Nat. Maroc 8: 170 (1925).
Carex austroafghanica Raymond, Biol. Skr. 14(4): 24 (1965). *Vignea austroafghanica* (Raymond) Soják, Cas. Nár. Mus., Odd. Prír. 148: 194 (1979 publ. 1980).

Carex divulsa Stokes in W.Withering, Bot. Arr. Brit. Pl. ed. 2, 2: 1035 (1787). *Carex muricata* subsp. *divulsa* (Stokes) Husn. in ?. *Carex muricata* var. *divulsa* (Stokes) Wahlenb., Kongl. Vetensk. Acad. Nya Handl. 24: 153 (1803). *Carex viridis* var. *divulsa* (Stokes) Spenn., Fl. Friburg. 1: 48 (1825). *Vignea divulsa* (Stokes) Rchb. in J.C.Mössler, Handb. Gewächsk. ed. 2, 3: 1619 (1830). *Caricina divulsa* (Stokes) St.-Lag. in A.Cariot, Étude Fl., ed. 8, 2: 875 (1889).
Macaronesia, Europe, Medit. to W. Himalaya, introduced elsewere. 10 DEN GRB IRE NOR SWE 11 AUT BGM CZE GER HUN NET POL SWI 12 BAL COR FRA POR SPA 13 ALB BUL GRC ITA KRI ROM SIC TUE YUG 14 BLT KRY RUE RUS UKR 20 ALG MOR TUN 21 AZO CNY MDR 30 ALT 32 KAZ KGZ TKM UZB 33 NCS TCS 34 CYP EAI IRN IRQ LBS PAL TUR 40 WHM (50) tas vic (72) ont (75) pen (78) wdc (85) age. Hemicr. or rhizome geophyte.

subsp. ***coriogyne*** (Nelmes) Ö.Nilsson in P.H.Davis (ed.), Fl. Turkey 9: 97 (1985).
Turkey. 34 TUR. Hemicr. or rhizome geophyte.
**Carex coriogyne* Nelmes, Kew Bull. 2: 74 (1947).

subsp. ***divulsa***
Macaronesia, Europe, Medit. to C. Asia, introduced elsewere. 10 GRB IRE 11 AUT BGM CZE GER HUN NET POL SWI 12 BAL COR FRA POR SPA 13 ALB BUL GRC ITA KRI ROM SIC TUE YUG 14 BLT KRY UKR 20 ALG MOR TUN 21 AZO CNY MDR 32 TKM 33 NCS TCS 34 EAI IRN LBS TUR (50) tas vic (72) ont (75) pen (78) wdc (85) age. Hemicr. or rhizome geophyte.
Carex virens Lam., Encycl. 3: 384 (1792), nom. illeg. *Carex divulsa* var. *virens* Durieu, Bull. Soc. Bot. France 6: 633 (1859). *Carex divulsa* subsp. *virens* (Durieu) Nyman, Consp. Fl. Eur.: 781 (1882).
Carex muricata Desf., Fl. Atlant. 2: 335 (1799), nom. illeg.
Vignea guestphalica Rchb., Fl. Germ. Excurs.: 140(7) (1831). *Carex guestphalica* (Rchb.) Boenn. ex W.D.J.Koch, Syn. Fl. Germ. Helv., ed. 2: 867 (1844). *Carex divulsa* var. *guestphalica* (Rchb.) F.W.Schultz, Étude Carex: 8 (1868). *Carex virens* var. *guestphalica* (Rchb.) Garcke, Fl. N. Mitt.-Deutschland, ed. 11: 421 (1873).
Carex muricata var. *subramosa* Neilr., Fl. Wien: 67 (1846).
Carex virens var. *duriaei* F.W.Schultz, Phytost. Pfalz: 161 (1863).
Carex divulsa var. *intermedia* Lange, Haandb. Danske Fl., ed. 3: 678 (1864).
Carex divulsa f. *misera* Kük. ex Vollm., Denkschr. Königl. Bot. Ges. Regensburg 8(2): 82 (1903).
Carex divulsa f. *polycarpa* Vollm., Denkschr. Königl. Bot. Ges. Regensburg 8(2): 81 (1903).
Carex leersiana Rouy in G.Rouy & J.Foucaud, Fl. France 13: 413 (1912).
Carex muricata f. *angustifolia* Podp., Spisy Prír. Fak. Masarykovy Univ. 12: 12 (1922). *Carex divulsa* f. *angustifolia* (Podp.) Soó, Acta Bot. Acad. Sci. Hung. 16: 370 (1970 publ. 1971).
Carex persica Nelmes, Bull. Misc. Inform. Kew 1939: 185 (1939). *Vignea persica* (Nelmes) Soják, Cas. Nár. Mus., Odd. Prír. 148: 196 (1979 publ. 1980).

subsp. ***leersii*** (Kneuck.) W.Koch, Mitt. Bad. Landesvereins Naturk. Naturschuz, n.s., 11: 259 (1923).
Europe to W. Himalaya. 10 DEN GRB IRE NOR SWE 11 AUT BGM CZE GER HUN NET POL SWI 12 FRA SPA 13 ALB BUL GRC ITA KRI ROM SIC TUE YUG 14 BLT KRY RUE RUS UKR 30 ALT 32 KAZ KGZ TKM UZB 33 NCS TCS 34 CYP IRN IRQ LBS PAL TUR 40 WHM (51) nzn. Hemicr. or rhizome geophyte.
Carex divulsa subsp. *leersiana* Rauschert in ?. *Vignea divulsa* subsp. *leersiana* (Rauschert) Dostál, Folia Mus. Rerum Nat. Bohemiae Occid., Bot. 21: 16 (1984).
Carex polyphylla Kar. & Kir., Bull. Soc. Imp. Naturalistes Moscou 14: 859 (1841). *Vignea polyphylla* (Kar. & Kir.) Soják, Cas. Nár. Mus., Odd. Prír. 148: 197 (1979 publ. 1980).
Carex leersii F.W.Schultz, Flora 53: 455 (1870), nom. illeg. **Carex muricata* var. *leersii* Kneuck. in M.A.Seubert, Excurs.-Fl. Baden: 52 (1891). *Carex muricata* subsp. *leersii* (Kneuck.) Asch. & Graebn., Syn. Mitteleur. Fl. 2(2): 40 (1902). *Carex echinata* var. *leersii* (Kneuck.) Kük. in H.G.A.Engler (ed.), Pflanzenr., IV, 20(38): 161 (1909). *Carex pairae* var. *leersii* (Kneuck.) Kük. in H.G.A.Engler (ed.), Pflanzenr., IV, 20(38): 161 (1909). *Carex leersiana* Rauschert, Feddes Repert. 83: 660 (1972 publ. 1973).

Carex chabertii F.W.Schultz, Flora 54: 21 (1871). *Carex divulsa* var. *chabertii* (F.W.Schultz) Nyman, Consp. Fl. Eur.: 781 (1882). *Carex divulsa* var. *chabertii* (F.W.Schultz) Kneuck. in M.A.Seubert, Excurs.-Fl. Baden: 52 (1891).

Carex doenitzii Boeck., Flora 65: 61 (1882).
Russian Far East, Japan. 31+ 38 JAP. Hemicr. or rhizome geophyte.

var. ***doenitzii***
Russian Far East, Japan. 31+ 38 JAP. Hemicr. or rhizome geophyte.
Carex plocamostyla Maxim., Bull. Acad. Imp. Sci. Saint-Pétersbourg 31: 117 (1887).
Carex dicuspis Franch., Bull. Soc. Philom. Paris, VIII, 7: 42 (1895).
Carex chrysolepis var. *modesta* H.Lév. & Vaniot, Bull. Acad. Int. Géogr. Bot. 10: 279 (1901).
Carex dicraea C.B.Clarke, Kew Bull., Addit. Ser. 8: 71 (1908).
Carex nagatadakensis Masam., Mem. Fac. Sci. Taihoku Imp. Univ. 11(4): 526 (1934).
Carex pseudodoenitzii Akiyama, J. Jap. Bot. 11: 288 (1935).

var. ***okuboi*** (Franch.) Kük. in H.G.A.Engler (ed.), Pflanzenr., IV, 20(38): 537 (1909).
Japan (Izu-shoto). 38 JAP. Hemicr. or rhizome geophyte.
**Carex okuboi* Franch., Bull. Soc. Philom. Paris, VIII, 7: 43 (1895).

Carex doisutepensis T.Koyama, Naturaliste Canad. 82: 196 (1955).
China (Yunnan) to N. Thailand. 36 CHC 41 THA. Hemicr. or rhizome geophyte.
Carex doisutepensis var. *flaccidior* T.Koyama, Naturaliste Canad. 82: 197 (1955).

Carex dolichogyne T.Koyama, Jap. J. Bot. 15: 164 (1956).
China (Hunan). 36 CHS. Hemicr. or rhizome geophyte.

Carex dolichostachya Hayata, Icon. Pl. Formosan. 10: 61 (1921).
S. China, Temp. E. Asia, Philippines. 36 CHC CHS 38 JAP NNS TAI 42 PHI. Hemicr. or rhizome geophyte.

subsp. ***dolichostachya***
S. China, Temp. E. Asia, Philippines. 36 CHC CHS 38 JAP NNS TAI 42 PHI. Hemicr. or rhizome geophyte.
Carex rankanensis Hayata, Icon. Pl. Formosan. 10: 64 (1921).
Diplocarex matsudae Hayata, Icon. Pl. Formosan. 10: 70 (1921). *Carex matsudae* (Hayata) Hayata ex Makino & Nemoto, Fl. Japan: 1353 (1925).
Carex multifolia Ohwi, Mem. Coll. Sci. Kyoto Imp. Univ., Ser. B, Biol. 5(3): 264 (1930).
Carex qimenensis S.W.Su & S.M.Xu, Bull. Bot. Res., Harbin 9(1): 49 (1989).

var. ***imbecillis*** Ohwi in ?
Japan (Kyushu). 38 JAP. Hemicr. or rhizome geophyte.

subsp. ***trichosperma*** (Ohwi) T.Koyama, in Fl. Taiwan 5: 341 (1978).
Taiwan (Mt. Alishan). 38 TAI. Hemicr. or rhizome geophyte.
Carex transalpina Hayata, J. Coll. Sci. Imp. Univ. Tokyo 30(1): 398 (1911).
Carex daibuensis Hayata, Icon. Pl. Formosan. 10: 61 (1921).
**Carex trichosperma* Ohwi, Jap. J. Bot. 7: 204 (1934).

Carex dolomitica Heenan & de Lange, New Zealand J. Bot. 35: 423 (1997).
New Zealand (NW. Nelson). 51 NZS. Hemicr. or rhizome geophyte.

Carex donnell-smithii L.H.Bailey, Mem. Torrey Bot. Club 1: 56 (1889).
S. Mexico to C. America. 79 MXG MXS MXT 80 COS GUA HON NIC PAN. Hemicr. or rhizome geophyte.
Carex pittieri Boeck., Allg. Bot. Z. Syst. 2: 190 (1896).
Carex jovis C.B.Clarke, Contr. U. S. Natl. Herb. 10: 470 (1908).

Carex douglasii Boott in W.J.Hooker, Fl. Bor.-Amer. 2: 213 (1839).
W. Canada to W. & C. U.S.A., Mexico (Baja California). 71 ABT BRC SAS 73 COL IDA MNT ORE WAS WYO 74 MSO NDA NEB SDA 76 ARI CAL NEV UTA 77 NWM 79 MXN. Hemicr. or rhizome geophyte.
Carex nuttallii Dewey, Amer. J. Sci. Arts 43: 93 (1842).
Carex meekii Dewey, Amer. J. Sci. Arts, II, 24: 48 (1857).
Carex douglasii var. *densispicata* Dewey, Amer. J. Sci. Arts, II, 32: 41 (1861).
Carex douglasii f. *brunnea* Olney in S.Watson, Botany [Fortieth Parallel]: 363 (1871).
Carex douglasii var. *minor* Olney in S.Watson, Botany [Fortieth Parallel]: 363 (1871).
Carex fendleriana Boeck., Linnaea 39: 135 (1875).
Carex douglasii f. *laxiflora* L.H.Bailey, Mem. Torrey Bot. Club 1: 20 (1889).
Carex usta L.H.Bailey, Mem. Torrey Bot. Club 1: 20 (1889).
Carex irrasa L.H.Bailey, Bot. Gaz. 25: 271 (1898).
Carex douglasii var. *williamsii* Rydb., Mem. New York Bot. Gard. 1: 82 (1900).

Carex drepanorhyncha Franch., Nouv. Arch. Mus. Hist. Nat., II, 10: 183 (1887).
China (Sichuan, Yunnan). 36 CHC. Hemicr. or rhizome geophyte.

Carex druceana Hamlin, Rec. Domin. Mus. 6: 101 (1968).
New Zealand North I. 51 NZN. Hemicr. or rhizome geophyte.

var. ***astonii*** (Hamlin) Edgar in Allan, Fl. N. Zeal. 2: 252 (1970).
New Zealand North I. 51 NZN. Hemicr. or rhizome geophyte.
**Carex astonii* Hamlin, Rec. Domin. Mus. 6: 102 (1968).

var. ***druceana***
New Zealand North I. 51 NZN. Hemicr. or rhizome geophyte.

Carex drymophila Turcz., Bull. Soc. Imp. Naturalistes Moscou 1838: 104 (1838). *Carex orthostachys* var. *drymophila* (Turcz.) Maxim., Prim. Fl. Amur.: 316 (1859). *Carex amurensis* var. *drymophila* (Turcz.) Kük., Bot. Centralbl. 77: 96 (1899).
Siberia to Vietnam. 30+ 31 PRM 36 CHI CHM 38 KOR 41 VIE. Hemicr. or rhizome geophyte.
Carex bongardiana var. *gracilis* Trautv. & C.A.Mey., Fl. Ochot.: 101 (1856).

Carex burejana Meinsh., Trudy Imp. S.-Peterburgsk. Bot. Sada 18: 368 (1900).
Carex pseudohirta Meinsh., Trudy Imp. S.-Peterburgsk. Bot. Sada 18: 371 (1900).
Carex turczaninowiana var. *beaurepairoei* H.Lév. & Vaniot, Bull. Soc. Agric. Sarthe 60: 79 (1905).
Carex drymophila var. *reducta* Kük. in H.G.A.Engler (ed.), Pflanzenr., IV, 20(38): 755 (1909).

Carex* ×*duereriana Kük. in P.F.A.Ascherson & K.O.R. Graebner, Syn. Mitteleur. Fl. 2: 231 (1904). *C. horschuschiana* × *C. panicea*.
WC. Europe. 11 GER. Hemicr. or rhizome geophyte.

Carex* ×*dufftii Hausskn., Mitt. Geogr. Ges. (Thüringen) Jena 6: 10 (1888). *C. digitata* × *C. ornithopoda*.
WC. Europe. 11 GER. Hemicr. or rhizome geophyte.

Carex* ×*dumanii Lepage, Naturaliste Canad. 83: 143 (1956). *C. paleacea* × *C. subspathacea*. *Carex* × *saxenii* var. *dumanii* (Lepage) Boivin in ?
E. Canada. 72 LAB ONT QUE. Hemicr. or rhizome geophyte.

Carex dunniana H.Lév., Repert. Spec. Nov. Regni Veg. 7: 231 (1909).
China. 36+. Hemicr. or rhizome geophyte.

Carex durangensis Reznicek & S.González, Contr. Univ. Michigan Herb. 20: 220 (1995).
Mexico (Durango). 79 MXE. Hemicr. or rhizome geophyte.

Carex durieui Steud., Nomencl. Bot., ed. 2, 1: 289 (1840).
NW. Spain to NW. Portugal. 12 POR SPA. Hemicr. or rhizome geophyte.
**Carex filifolia* J.Gay, Ann. Sci. Nat., Bot., II, 6: 125 (1836), nom. illeg.

Carex duriuscula C.A.Mey., Mém. Sav. Étr. Acad. St. Pétersbourg 1: 214 (1831). *Carex stenophylla* var. *duriuscula* (C.A.Mey.) Trautv., Trudy Imp. S.-Peterburgsk. Bot. Sada 10: 537 (1888). *Vignea duriuscula* (C.A.Mey.) Soják, Cas. Nár. Mus., Odd. Prír. 148: 195 (1979 publ. 1980).
Temp. Asia, New Guinea, N. America. 30 ALT WSB YAK 31 AMU MAG PRM 32 KAZ 36 CHI CHM CHN CHQ CHS 37 MON 38 KOR 43 NWG 70 ASK NWT YUK 71 ABT BRC MAN SAS 73 COL IDA MNT ORE WAS WYO 74 ILL IOW KAN MIN MSO NDA NEB SDA 75 MIC 76 ARI CAL NEV UTA 77 NWM. Hemicr. or rhizome geophyte.

subsp. ***duriuscula***
Temp. Asia, New Guinea, N. America. 30 ALT WSB YAK 31 AMU MAG PRM 32 KAZ 36 CHI CHM CHN 37 MON 38 KOR 43 NWG 70 ASK NWT YUK 71 ABT BRC MAN SAS 73 COL IDA MNT ORE WAS WYO 74 ILL IOW KAN MIN MSO NDA NEB SDA 75 MIC 76 ARI CAL NEV UTA 77 NWM. Hemicr. or rhizome geophyte.
Carex eleocharis L.H.Bailey, Mem. Torrey Bot. Club 1: 6 (1889). *Carex stenophylla* subsp. *eleocharis* (L.H.Bailey) Hultén, Acta Univ. Lund., n.s., 38(1): 314 (1942). *Carex stenophylla* var. *eleocharis* (L.H.Bailey) Breitung, Canad. Field-Naturalist 71: 48 (1957).
Carex duriuscula var. *tenuispica* X.Y.Yuan, Bull. Bot. Res., Harbin 5(4): 155 (1985).

subsp. ***rigescens*** (Franch.) S.Yun Liang & Y.C.Tang, Acta Phytotax. Sin. 28: 155 (1990).
China. 36 CHI CHM CHN CHQ CHS. Hemicr. or rhizome geophyte.
**Carex stenophylla* var. *rigescens* Franch., Pl. David. 1: 318 (1884). *Carex rigescens* (Franch.) V.I.Krecz. in V.L.Komarov (ed.), Fl. URSS 3: 592 (1935).

Carex dusenii Kük. ex Dusén, Ark. Bot. 7(2): 13 (1908).
Argentina (Santa Cruz). 85 AGS. Hemicr. or rhizome geophyte.

Carex duvaliana Franch. & Sav., Enum. Pl. Jap. 2: 568 (1878). *Carex tenuissima* var. *duvaliana* (Franch. & Sav.) Kük. in H.G.A.Engler (ed.), Pflanzenr., IV, 20(38): 476 (1909).
SE. China, C. & S. Japan. 36 CHS 38 JAP. Hemicr. or rhizome geophyte.
Carex hilgendorfiana Boeck., Bot. Jahrb. Syst. 5: 18 (1884).
Carex hololasius H.Lév. & Vaniot, Bull. Acad. Int. Géogr. Bot. 10: 280 (1901).

Carex earistata F.T.Wang & Y.L.Chang ex S.Yun Liang, Acta Phytotax. Sin. 28: 153 (1990).
China (Gansu). 36 CHN. Hemicr. or rhizome geophyte.

Carex ebenea Rydb., Bull. Torrey Bot. Club 28: 266 (1901). *Carex festiva* var. *ebenea* (Rydb.) J.K.Henry, Fl. S. Brit. Columbia: 58 (1915). *Carex haydeniana* var. *ebenea* (Rydb.) Kelso, Biol. Leafl. 66: 23 (1953).
WC. U.S.A. 73 COL WYO 76 ARI UTA 77 NWM. Hemicr. or rhizome geophyte.

Carex eburnea Boott in W.J.Hooker, Fl. Bor.-Amer. 2: 226 (1839).
Subarctic America to N. & E. U.S.A., Mexico. 70 ASK NWT YUK 71 ABT BRC MAN SAS 72 LAB NBR NFL NSC ONT QUE 73 MNT WYO 74 ILL IOW MIN MSO NDA NEB SDA WIS 75 CNT INI MAI MAS MIC NWH NWJ NWY OHI PEN VER WVA 77 TEX 78 ALA ARK KTY NCA SCA TEN VRG 79 MXC. Hemicr. or rhizome geophyte.
Carex alba Dewey, Amer. J. Sci. Arts 7: 266 (1824), nom. illeg.
Carex alba var. *setifolia* Dewey, Amer. J. Sci. Arts 11: 316 (1826). *Carex setifolia* (Dewey) Britton & A.Br., Ill. Fl. N. U.S. 1: 332 (1896), nom. illeg.

Carex echinata Murray, Prod. Stirp. Gott.: 76 (1770). *Vignea echinata* (Murray) Fourr., Ann. Soc. Linn. Lyon, n.s., 17: 176 (1869).
Temp. Northern Hemisphere, Sumatera, New Guinea, SE. Australia, New Zealand. 10 DEN FIN GRB ICE IRE NOR NOR SWE 11 AUT BGM CZE GER HUN NET POL SWI 12 COR FRA POR SPA 13 ALB BUL GRC ITA KRI ROM SIC TUE YUG 14 BLR BLT RUC RUE RUN RUS RUW UKR 20 MOR 21 AZO 31 KUR 33 TCS 34 IRN LBS PAL TUR 36 CHM CHT 38 JAP KOR 40 EHM 42 SUM 43 NWG 50 NSW VIC 51 NZN 63 HAW 70 ALU ASK NUN NWT YUK 71 BRC SAS 72 NBR NFL NSC ONT PEI QUE 73 COL IDA MNT ORE WAS WYO 74 ILL IOW MIN MSO NDA WIS 75 CNT INI MAI MAS MIC NWH NWJ NWY OHI PEN RHO VER WVA 76 CAL NEV UTA 78 DEL MRY NCA TEN VRG WDC 79 MXC MXT 80 GUA. Hemicr. or rhizome geophyte.

subsp. ***echinata***
Temp. Northern Hemisphere, Sumatera, New Guinea, SE. Australia, New Zealand. 10 DEN

FIN GRB ICE IRE NOR NOR SWE 11 AUT BGM CZE GER HUN NET POL SWI 12 COR FRA POR SPA 13 ALB BUL GRC ITA KRI ROM SIC TUE YUG 14 BLR BLT RUC RUE RUN RUS RUW UKR 20 MOR 21 AZO 31 KUR 33 TCS 34 IRN LBS TUR 36 CHM CHT 38 JAP KOR 40 EHM 42 SUM 43 NWG 50 NSW VIC 51 NZN 63 HAW 70 ALU ASK NUN NWT YUK 71 BRC SAS 72 NBR NFL NSC ONT PEI QUE 73 COL IDA MNT ORE WAS WYO 74 ILL IOW MIN MSO NDA WIS 75 CNT INI MAI MAS MIC NWH NWJ NWY OHI PEN RHO VER WVA 76 CAL NEV UTA 78 DEL MRY NCA TEN VRG WDC 79 MXC MXT 80 GUA. Hemicr. or rhizome geophyte.

Carex muricata Huds., Fl. Angl.: 406 (1762), nom. illeg.

Carex muricata Leers, Fl. Herborn.: t. 13, f. 8 (1775), nom. illeg. *Carex leersii* Willd., Fl. Berol. Prodr.: 28 (1787).

Carex echinata Roth, Tent. Fl. Germ. 1: 395 (1788), nom. illeg.

Carex stellulata Gooden., Trans. Linn. Soc. London 2: 144 (1794). *Vignea stellulata* (Gooden.) Rchb. in J.C.Mössler, Handb. Gewächsk. ed. 2, 3: 1610 (1830). *Caricina stellulata* (Gooden.) St.-Lag. in A.Cariot, Étude Fl., ed. 8, 2: 870 (1889).

Carex fasciculata Link ex Schkuhr, Beschr. Riedgräs. 1: 119 (1801).

Carex fasciculata Willd., Sp. Pl. 4: 248 (1805), nom. illeg.

Carex grypos Schkuhr, Beschr. Riedgräs. 2: 18 (1806). *Vignea grypos* (Schkuhr) Rchb. in J.C.Mössler, Handb. Gewächsk. ed. 2, 3: 1608 (1830). *Carex stellulata* var. *grypos* (Schkuhr) W.D.J.Koch, Syn. Fl. Germ. Helv., ed. 2: 869 (1844). *Carex echinata* var. *grypos* (Schkuhr) Nyman, Consp. Fl. Eur.: 780 (1882).

Carex muricata var. *loliacea* Schkuhr, Beschr. Riedgräs. 2: 12 (1806).

Carex retusa Degl. in J.L.A.Loiseleur-Deslongchamps, Fl. Gall.: 307 (1806).

Carex provincialis Degl. in J.L.A.Loiseleur-Deslongchamps, Fl. Gall. 2: 307 (1807).

Carex echinata var. *loliacea* Lej., Rev. Fl. Spa: 193 (1824).

Carex hydrophila Dumort., Fl. Belg.: 146 (1827). *Vignea hydrophila* (Dumort.) Rchb., Fl. Germ. Excurs.: 140(7) (1831). *Carex echinata* var. *hydrophila* (Dumort.) Nyman, Consp. Fl. Eur.: 780 (1882). *Carex stellulata* f. *hydrophila* (Dumort.) Asch. & Graebn., Syn. Mitteleur. Fl. 2(2): 56 (1902).

Carex soleirolii DC. & Duby, Bot. Gall. 1: 498 (1828). *Carex hispida* var. *soleirolii* (DC. & Duby) Nyman, Consp. Fl. Eur.: 774 (1882). *Carex hispida* f. *soleirolii* (DC. & Duby) Asch., Oesterr. Bot. Z. 35: 355 (1885).

Carex serrulata Mutel, Fl. Franç.: 395 (1836), nom. illeg.

Carex subramosa Willd. ex Kunth, Enum. Pl. 2: 385 (1837).

Carex stellulata var. *scirpina* Tuck., Enum. Meth. Caric.: 9 (1843).

Carex stellulata f. *pseudodivulsa* F.W.Schultz, Fl. Pfalz: 499 (1845).

Carex muricata var. *virens* Rchb., Icon. Fl. Germ. Helv. 8: 9 (1846).

Carex stellulata var. *angustata* Carey in A.Gray, Manual: 544 (1848). *Carex echinata* var. *angustata* (Carey) L.H.Bailey, Mem. Torrey Bot. Club 1: 59 (1889). *Carex leersii* var. *angustata* (Carey) Burnham, Torreya 19: 131 (1919). *Carex phyllomanica* var. *angustata* (Carey) B.Boivin, Phytologia 43: 51 (1979).

Carex muricata var. *virens* Andersson, Cyper. Scand.: 65 (1849), nom. illeg.

Carex leptophylla Heuff., Linnaea 31: 728 (1862).

Carex convexa Kit., Linnaea 32: 317 (1863).

Carex muricata var. *monostachya* Asch., Fl. Brandenburg 1: 766 (1864).

Carex obtusangula Salzm. ex Boott, Ill. Carex 4: 206 (1867).

Carex muricata var. *depauperata* Hampe, Prodr. Fl. Hercyn.: 290 (1873).

Carex echinata var. *microstachys* Boeck., Linnaea 39: 125 (1875).

Carex virens var. *pairaei* Garcke, Fl. N. Mitt.-Deutschland, ed. 13: 418 (1878). *Carex muricata* var. *pairaei* (Garcke) Èelak., Prodr. Fl. Böhmen: 731 (1881).

Carex caflischii Brügger, Jahresber. Naturf. Ges. Graubündens, II, 23–24: 119 (1880).

Carex stellulata f. *oligantha* Callmé, Deutsche Bot. Monatsschr. 6: 49 (1888).

Carex echinata var. *cephalantha* L.H.Bailey, Mem. Torrey Bot. Club 1: 58 (1889). *Carex sterilis* var. *cephalantha* (L.H.Bailey) L.H.Bailey, Bull. Torrey Bot. Club 20: 42 (1893). *Carex stellulata* var. *cephalantha* (L.H.Bailey) Fernald, Rhodora 4: 222 (1902). *Carex cephalantha* (L.H.Bailey) E.P.Bicknell, Bull. Torrey Bot. Club 35: 493 (1908). *Carex leersii* var. *cephalantha* (L.H.Bailey) J.K.Henry, Fl. S. Brit. Columbia: 60 (1915). *Carex muricata* var. *cephalantha* (L.H.Bailey) Wiegand & Eames, Cornell Univ. Agric. Exp. Sta. Mem. 92: 120 (1926). *Carex muricata* subsp. *cephalantha* (L.H.Bailey) R.T.Clausen, Cornell Univ. Agric. Exp. Sta. Mem. 291: 10 (1949).

Carex sterilis var. *excelsior* L.H.Bailey, Bull. Torrey Bot. Club 20: 424 (1893). *Carex echinata* var. *excelsior* (L.H.Bailey) Fernald, Proc. Amer. Acad. Arts 37: 484 (1902). *Carex stellulata* f. *excelsior* (L.H.Bailey) Kük. in H.G.A.Engler (ed.), Pflanzenr., IV, 20(38): 231 (1909).

Carex sterilis var. *excelsior* L.H.Bailey, Bull. Torrey Bot. Club 20: 424 (1893).

Carex muricata subsp. *microcarpa* Neuman, Bot. Not. 1894: 107 (1894).

Carex echinata var. *ormantha* Fernald, Proc. Amer. Acad. Arts 37: 483 (1902). *Carex stellulata* var. *ormantha* (Fernald) Fernald, Rhodora 4: 222 (1902). *Carex ormantha* (Fernald) Mack., Erythea 8: 35 (1922).

Carex stellulata f. *hylogiton* Asch. & Graebn., Syn. Mitteleur. Fl. 2(2): 56 (1902).

Carex stellulata var. *subalpina* Asch. & Graebn., Syn. Mitteleur. Fl. 2(2): 56 (1902).

Carex leersii var. *angustifolia* Vollm., Denkschr. Königl. Bot. Ges. Regensburg 8(2): 80 (1903).

Carex leersii subvar. *subramosa* Vollm., Denkschr. Königl. Bot. Ges. Regensburg 8(2): 80 (1903).

Carex interior var. *josselynii* Fernald, Rhodora 8: 115 (1906). *Carex scirpoides* var. *josselynii* (Fernald) Fernald, Rhodora 10: 48 (1908). *Carex josselynii* (Fernald) Mack. ex Pease, Proc. Boston Soc. Nat. Hist. 37: 188 (1924).

Carex echinata var. *tenuior* Kük. in H.G.A.Engler (ed.), Pflanzenr., IV, 20(38): 161 (1909).
Carex stellulata var. *australis* Kük. in H.G.A.Engler (ed.), Pflanzenr., IV, 20(38): 230 (1909).
Carex angustior Mack. in P.A.Rydberg, Fl. Rocky Mts.: 124 (1917). *Vignea angustior* (Mack.) Soják, Cas. Nár. Mus., Odd. Prír. 148: 194 (1979 publ. 1980).
Carex echinata f. *brevispicata* Podp., Práce Morav. Prír. Spolecn. 5(F 46): 209 (1929). *Carex pairae* f. *brevispicata* (Podp.) Soó, Acta Bot. Acad. Sci. Hung. 16: 370 (1970 publ. 1971).
Carex echinata f. *chlorocarpa* Podp., Práce Morav. Prír. Spolecn. 5(F 46): 209 (1929). *Carex pairae* f. *chlorocarpa* (Podp.) Soó, Acta Bot. Acad. Sci. Hung. 16: 370 (1970 publ. 1971).
Carex echinata f. *remotiuscula* Podp., Práce Morav. Prír. Spolecn. 5(F 46): 209 (1929). *Carex pairae* f. *remotiuscula* (Podp.) Soó, Acta Bot. Acad. Sci. Hung. 16: 370 (1970 publ. 1971).
Carex laricina Mack. ex Bright, Trillia 9: 4, 19 (1930). *Carex muricata* var. *laricina* (Mack. ex Bright) Gleason, Phytologia 4: 22 (1952).
Carex angustior var. *gracilenta* Clausen & Wahl, Rhodora 41: 30 (1939).
Carex basilata Ohwi, Acta Phytotax. Geobot. 11: 258 (1942).
Carex svenonis Skottsb., Acta Horti Gothob. 15: 329 (1944).
Carex perileia S.T.Blake, J. Arnold Arbor. 28: 102 (1947). *Carex echinata* var. *perileia* (S.T.Blake) P.Royen, Alp. Fl. New Guinea 2: 1029 (1979). *Vignea perileia* (S.T.Blake) Soják, Cas. Nár. Mus., Odd. Prír. 148: 196 (1979 publ. 1980).
Carex gajonum Nelmes, Kew Bull. 7: 84 (1952).
Carex minganinsularum Raymond, Bull. Soc. Bot. France 99: 197 (1952).
Carex hawaiiensis H.St.John, Pacific Sci. 37: 25 (1983).

subsp. ***phyllomanica*** (Boott) Reznicek, Contr. Univ. Michigan Herb. 14: 195 (1980).
S. Alaska to N. California. 70 ASK 71 BRC 73 ORE WAS 76 CAL. Hemicr. or rhizome geophyte.
**Carex phyllomanica* Boott in S.Watson & al., Bot. California 2: 233 (1880). *Carex echinata* var. *phyllomanica* (Boott) B.Boivin, Provancheria 25: 102 (1992).

Carex echinochloe Kunze in C.Schkuhr, Beschr. Riedgräs., Suppl.: 47 (1841).
Bioko, W. Cameroon, Ethiopia to Zimbabwe. 23 BUR CMN GGI RWA ZAI 24 ETH SUD 25 KEN TAN UGA 26 MLW ZIM. Hemicr. or rhizome geophyte.

subsp. ***echinochloe***
Bioko, W. Cameroon, Ethiopia to Tanzania. 23 BUR CMN GGI RWA ZAI 24 ETH SUD 25 KEN TAN UGA. Hemicr. or rhizome geophyte.

subsp. ***nyasensis*** (C.B.Clarke) Lye, Nordic J. Bot. 3: 244 (1983).
Kenya to Zimbabwe. 25 KEN TAN 26 MLW ZIM. Hemicr. or rhizome geophyte.
**Carex nyasensis* C.B.Clarke in D.Oliver & auct. suc. (eds.), Fl. Trop. Afr. 8: 519 (1902). *Carex echinochloe* var. *nyasensis* (C.B.Clarke) Kük. in H.G.A.Engler (ed.), Pflanzenr., IV, 20(38): 271 (1909).

Carex echinochloiformis Y.L.Chang ex Y.C.Yang, in Fl. Xizangica 5: 436 (1987).
Tibet to China (Yunnan). 36 CHC CHT. Hemicr. or rhizome geophyte.

Carex echinus Ohwi, Acta Phytotax. Geobot. 5: 183 (1936).
China (Sichuan). 36 CHC. Hemicr. or rhizome geophyte.

Carex ecklonii Nees, Linnaea 10: 203 (1836). *Carex extensa* var. *ecklonii* (Nees) Kük. in H.G.A.Engler (ed.), Pflanzenr., IV, 20(38): 667 (1909).
Cape Prov. 27 CPP. Hemicr. or rhizome geophyte.

Carex ecostata C.B.Clarke in J.D.Hooker, Fl. Brit. India 6: 720 (1894).
Assam (Nagaland). 40 ASS. Hemicr. or rhizome geophyte.

Carex ecuadorica Kük., Bot. Jahrb. Syst. 34(78): 7 (1904).
Ecuador to Argentina (San Juan). 83 BOL ECU PER 85 AGW. Hemicr. or rhizome geophyte.

Carex edgariae Hamlin, Rec. Domin. Mus. 6: 103 (1968).
New Zealand South I. 51 NZS. Hemicr. or rhizome geophyte.

Carex edwardsiana E.L.Bridges & Orzell, Phytologia 67: 148 (1989).
C. Texas. 77 TEX. Hemicr. or rhizome geophyte.

Carex egena H.Lév. & Vaniot, Repert. Spec. Nov. Regni Veg. 4: 227 (1907).
Russian Far East, Manchuria, Korea. 31 PRM 36 CHM 38 KOR. Hemicr. or rhizome geophyte.
**Carex oligostachys* Meinsh. ex Maxim., Bull. Acad. Imp. Sci. Saint-Pétersbourg 31: 117 (1887), nom. illeg. *Carex filipes* var. *oligostachys* Kük. in H.G.A. Engler (ed.), Pflanzenr., IV, 20(38): 641 (1909).

Carex egglestonii Mack., Bull. Torrey Bot. Club 42: 614 (1915).
WC. U.S.A. 73 COL WYO 76 UTA. Hemicr. or rhizome geophyte.
Carex egglestonii var. *festivelliformis* F.J.Herm., Brittonia 12: 78 (1960).

Carex ekmanii Kük., Repert. Spec. Nov. Regni Veg. 23: 221 (1926).
Haiti. 81 HAI. Hemicr. or rhizome geophyte.
Carex ekmanii var. *hottensis* Kük. & Ekman, Ark. Bot. 22A(17): 9 (1929).

Carex elata All., Fl. Pedem. 2: 272 (1785). *Carex cespitosa* var. *elata* (All.) Fiori, Nuov. Fl. Italia 1: 188 (1923).
Europe to C. Asia. 10 DEN FIN GRB IRE NOR SWE 11 AUT BGM CZE GER HUN NET POL SWI 12 COR FRA POR SAR SPA 13 GRC ITA ROM SIC YUG 14 BLR BLT RUC RUE RUN RUS RUW UKR 20 ALG 30 ALT 32 KAZ TKM 33 NCS TCS 34 IRN TUR. Hemicr. or rhizome geophyte.

subsp. ***elata***
Europe to Iran. 10 DEN FIN GRB IRE NOR SWE 11 AUT BGM CZE GER HUN NET POL SWI 12 COR FRA POR SAR SPA 13 GRC ITA ROM SIC YUG 14 BLR BLT RUS RUW UKR 20 ALG 33 NCS TCS 34 IRN TUR. Hemicr. or rhizome geophyte.
Carex melanochloros Thuill., Fl. Env. Paris: 488 (1790).
Carex mixta J.F.Gmel., Syst. Nat. 2: 146 (1791).
Carex stricta Gooden., Trans. Linn. Soc. London 2: 196 (1794), nom. illeg.

Carex compressa Gaudin, Etrennes Fl.: 111 (1804), nom. illeg.
Carex homalocarpa Peterm., Flora 27: 333 (1844). *Carex stricta* var. *homalocarpa* (Peterm.) Rchb., Icon. Fl. Germ. Helv. 8: 14 (1846). *Carex hudsonii* var. *homalocarpa* (Peterm.) Kük. in H.G.A.Engler (ed.), Pflanzenr., IV, 20(38): 332 (1909).
Carex reticulosa Peterm., Flora 27: 333 (1844). *Carex hudsonii* f. *reticulosa* (Peterm.) Kük. in H.G.A.Engler (ed.), Pflanzenr., IV, 20(38): 332 (1909).
Carex stricta f. *humilis* Fr., Summa Veg. Scand.: 227 (1845). *Carex hudsonii* f. *humilis* (Fr.) Kük. in H.G.A.Engler (ed.), Pflanzenr., IV, 20(38): 332 (1909).
Carex stricta var. *helvetica* Rion ex Rchb., Icon. Fl. Germ. Helv. 8: 14 (1846).
Carex macra Steud., Syn. Pl. Glumac. 2: 212 (1855). *Carex stricta* var. *macra* (Steud.) Nyman, Consp. Fl. Eur.: 777 (1882).
Carex subrotunda M.Serres, Bull. Soc. Bot. France 4: 440 (1857).
Carex stricta var. *gracilis* Asch., Fl. Brandenburg 1: 773 (1864).
Carex stricta f. *pendula* Lange, Haandb. Danske Fl., ed. 4: 131 (1886). *Carex hudsonii* f. *pendula* (Lange) Kük. in H.G.A.Engler (ed.), Pflanzenr., IV, 20(38): 332 (1909). *Carex elata* f. *pendula* (Lange) Soó, Acta Bot. Acad. Sci. Hung. 16: 370 (1970 publ. 1971).
Carex stricta f. *nigrans* Beck, Fl. Nieder-Österreich: 136 (1890). *Carex hudsonii* f. *nigrans* (Beck) Kük. in H.G.A.Engler (ed.), Pflanzenr., IV, 20(38): 332 (1909).
Carex stricta f. *riartiana* Legrand, Fl. Anal. Berry, ed. 2: 338 (1894). *Carex elata* f. *riartiana* (Legrand) Soó, Acta Bot. Acad. Sci. Hung. 16: 370 (1970 publ. 1971).
Carex hudsonii A.Benn. in H.C.Watson, London Cat. Brit. Pl., ed. 9: 41 (1895).
Carex stricta f. *polystachya* Asch. & Graebn., Syn. Mitteleur. Fl. 2(2): 85 (1902). *Carex hudsonii* var. *polystachya* (Asch. & Graebn.) Kük. in H.G.A. Engler (ed.), Pflanzenr., IV, 20(38): 332 (1909).
Carex hudsonii f. *stenocarpa* Junge, Jahrb. Hamburg. Wiss. Anst. 25(Beih. 3): 178 (1905). *Carex elata* f. *stenocarpa* (Junge) Soó, Acta Bot. Acad. Sci. Hung. 16: 370 (1970 publ. 1971), without basionym date.
Carex hudsonii f. *glabra* Junge, Repert. Spec. Nov. Regni Veg. 13: 6 (1913). *Carex elata* f. *glabra* (Junge) Soó, Acta Bot. Acad. Sci. Hung. 16: 370 (1970 publ. 1971).
Carex hudsonii f. *brachystachys* Podp., Práce Morav. Prír. Spolecn. 5(5): 261 (1929). *Carex elata* f. *brachystachys* (Podp.) Soó, Acta Bot. Acad. Sci. Hung. 16: 370 (1970 publ. 1971).
Carex broteroana Samp., Anais Fac. Ci. Univ. Porto 19: 68 (1934). *Carex gracilis* subsp. *broteroana* (Samp.) M.Laínz, Bol. Inst. Estud. Adsturianos, Supl. Ci. 16: 200 (1973). *Carex nigra* subsp. *broteroana* (Samp.) Rivas Mart., Anales Inst. Bot. Cavanilles 34: 547 (1977 publ. 1978). *Carex acuta* subsp. *broteroana* (Samp.) Rivas Mart. in S.Rivas-Martinez & al., Veg. Picos de Europa: 259 (1984).

subsp. ***omskiana*** (Meinsh.) Jalas, Ann. Bot. Fenn. 1: 49 (1964).
EC. & E. Europe to C. Asia. 11 CZE POL 13 ROM 14 BLR RUC RUE RUN RUS RUW UKR 30 ALT 32 KAZ TKM 33 NCS TCS 34 TUR. Hemicr. or rhizome geophyte.
**Carex omskiana* Meinsh., Trudy Imp. S.-Peterburgsk. Bot. Sada 18: 340 (1900).

Carex elatior Boeck., Abh. Naturwiss. Vereine Bremen 7: 41 (1880).
C. Madagascar. 29 MDG. Hemicr. or rhizome geophyte.

Carex eleusinoides Turcz. ex Kunth, Enum. Pl. 2: 407 (1837).
Siberia to N. Montana. 30 BRY IRK KRA YAK 31 AMU KAM KHA KUR MAG PRM 36 CHM 37 MON 38 JAP KOR 70 ASK YUK 71 BRC. Hemicr. or rhizome geophyte.
Carex kokrinensis A.E.Porsild, Rhodora 41: 206 (1939).

Carex elgonensis Nelmes, Bull. Misc. Inform. Kew 1938: 245 (1938).
E. Trop. Africa (Mt. Elgon). 25 KEN UGA. Hemicr. or rhizome geophyte.
Carex mildbraediana var. *friesiorum* Kük. in H.G.A. Engler (ed.), Pflanzenr., IV, 20(38): 767 (1909).

Carex elingamita Hamlin, Trans. Roy. Soc. New Zealand 85: 393 (1958).
New Zealand (Three Kings Is.). 51 NZN. Hemicr. or rhizome geophyte.

Carex elliottii Schwein. & Torr., Ann. Lyceum Nat. Hist. New York 1: 357 (1824). *Olamblis elliottii* (Schwein. & Torr.) Raf., Good Book: 26 (1840).
SE. U.S.A. 78 ALA FLA GEO NCA SCA. Hemicr. or rhizome geophyte.
Carex fulva Muhl., Descr. Gram.: 246 (1817), nom. illeg.
**Carex castanea* Elliott, Sketch Bot. S. Carolina 2: 546 (1824), nom. illeg.
Carex baldwinia Dewey, Amer. J. Sci. Arts 26: 107 (1834).
Carex baldwiniana Steud., Nomencl. Bot., ed. 2, 1: 286 (1840).

Carex elongata L., Sp. Pl.: 974 (1753). *Vignea elongata* (L.) Rchb. in J.C.Mössler, Handb. Gewächsk. ed. 2, 3: 1617 (1830). *Caricina elongata* (L.) St.-Lag. in A.Cariot, Étude Fl., ed. 8, 2: 878 (1889). *Leptovignea elongata* (L.) Fedde & J.Schust., Just's Bot. Jahresber. 41(2): 11 (1913 publ. 1918).
Europe to Caucasus. 10 DEN FIN GRB IRE NOR SWE 11 AUT BGM CZE GER HUN NET POL SWI 12 FRA 13 GRC? ITA ROM YUG 14 BLR BLT RUC RUE RUN RUS RUW UKR 30 WSB 32 KAZ 33 NCS. Hemicr. or rhizome geophyte.
Carex davalliana subsp. *cyrnea* (Briq.) Cif. & Giacom. in ?
Carex multiculmis Ehrh., Beitr. Naturk. 6: 83 (1791).
Carex divergens Thuill., Fl. Env. Paris, ed. 2: 481 (1799).
Carex multiceps Gaudin, Etrennes Fl.: 78 (1804).
Carex gebhardii Willd., Sp. Pl. 4: 240 (1805). *Vignea gebhardii* (Willd.) Rchb. in J.C.Mössler, Handb. Gewächsk. ed. 2, 3: 1613 (1830). *Olotrema gebhardii* (Willd.) Raf., Good Book: 25 (1840). *Carex elongata* var. *gebhardii* (Willd.) Asch., Fl. Brandenburg 1: 769 (1864).
Carex pinnata Moench ex Rchb., Fl. Germ. Excurs.: 59 (1830).

Carex elongata var. *simplicior* Andersson, Cyper. Scand.: 60 (1849).
Carex elongata var. *alpestris* Trevir. in C.F.von Ledebour, Fl. Ross. 4: 278 (1852).
Carex elongata f. *heterostachya* Wimm., Fl. Schles., ed. 3: 86 (1857).
Carex elongata var. *brachystachya* A.Blytt, Norges Fl. 1: 199 (1861).
Carex elongata var. *minor* Boott, Ill. Carex 4: 187 (1867).
Carex elongata var. *pauciflora* Meinsh., Fl. Ingr.: 399 (1878).
Carex elongata f. *umbrosa* Kneuck. in M.A.Seubert, Excurs.-Fl. Baden, ed. 5: 55 (1891).

Carex eluta Nelmes, Bull. Misc. Inform. Kew 1940: 271 (1940).
S. India. 40 IND. Hemicr. or rhizome geophyte.

Carex elynoides T.Holm, Amer. J. Sci., IV, 9: 356 (1900).
WC. U.S.A. 73 COL IDA MNT WYO 76 NEV UTA 77 NWM. Hemicr. or rhizome geophyte.
Carex filifolia var. *miser* L.H.Bailey, Proc. Amer. Acad. Arts 22: 122 (1886).

Carex* ×*elytroides Fr., Bot. Not. 1846: 106 (1846). *C. acuta* × *C. nigra*. *Carex vulgaris* var. *elytroides* (Fr.) Boott, Ill. Carex 4: 167 (1867). *Carex vulgaris* subsp. *elytroides* (Fr.) Nyman, Consp. Fl. Eur.: 778 (1882).
Europe. 14 RUN. Hemicr.
Carex tricostata var. *praticola* Fr., Summa Veg. Scand.: 72 (1845).
Carex acuta var. *elegans* Wimm., Flora 33: 618 (1850).
Carex acuta var. *sparganioides* Èelak., Prodr. Fl. Böhmen: 734 (1881).

Carex emirnensis Baker, J. Bot. 21: 129 (1883).
Madagascar. 29 MDG. Hemicr. or rhizome geophyte.

Carex* ×*emmae L.Gross, Mitt. Bad. Bot. Vereins 1906: 74 (1906). *C. divulsa* × *C. remota*. *Vignea* ×*emmae* (L.Gross) Soják, Cas. Nár. Mus., Odd. Prír. 148: 195 (1979 publ. 1980).
WC. Europe. 11 GER. Hemicr. or rhizome geophyte.

Carex* ×*emmanuelis Luceño, Anales Jard. Bot. Madrid 44: 442 (1987). *C. caryophyllea* × *C. depressa*.
SW. Europe. 12 SPA. Hemicr. or rhizome geophyte.

Carex emoryi Dewey in W.H.Emory, Rep. U.S. Mex. Bound. 2(1): 230 (1858). *Carex stricta* var. *emoryi* (Dewey) L.H.Bailey, Proc. Amer. Acad. Arts 22: 85 (1886). *Carex stricta* f. *emoryi* (Dewey) Kük. in H.G.A.Engler (ed.), Pflanzenr., IV, 20(38): 330 (1909).
S. Canada, C. & E. U.S.A. to Mexico (Coahuila). 71 MAN 72 ONT 73 COL WYO 74 ILL IOW KAN MIN MSO NDA NEB OKL SDA WIS 75 CNT INI MIC NWJ NWY OHI PEN WVA 77 NWM TEX 78 ARK DEL MRY VRG WDC 79 MXE. Hemicr. or rhizome geophyte.
Carex virginiana var. *elongata* Boeck., Linnaea 40: 432 (1876). *Carex stricta* var. *elongata* (Boeck.) Gleason, Phytologia 4: 22 (1952).
Carex variabilis var. *elatior* L.H.Bailey, Mem. Torrey Bot. Club 1: 19 (1889).
Carex variabilis var. *altior* Rydb., Mem. New York Bot. Gard. 1: 76 (1900). *Carex aquatilis* var. *altior* (Rydb.) Fernald, Rhodora 44: 295 (1942). *Carex aquatilis* subsp. *altior* (Rydb.) Hultén, Kongl. Svenska Vetensk. Acad. Handl. 8(5): 74 (1962).
Carex millegrana T.Holm, Amer. J. Sci. 166: 457 (1903).

Carex endlichii Kük., Repert. Spec. Nov. Regni Veg. 8: 291 (1910).
S. Arizona to Guatemala. 76 ARI 79 MXC MXE MXG MXS MXT 80 GUA. Hemicr. or rhizome geophyte.

Carex enervis C.A.Mey. in C.F.von Ledebour, Fl. Altaic. 4: 209 (1833). *Carex stenophylla* var. *enervis* (C.A.Mey.) Kük. in H.G.A.Engler (ed.), Pflanzenr., IV, 20(38): 122 (1909). *Vignea enervis* (C.A.Mey.) Soják, Cas. Nár. Mus., Odd. Prír. 148: 195 (1979 publ. 1980).
Siberia to China. 30 ALT 32 KAZ KGZ UZB 36 CHC CHI CHM CHN CHQ CHT CHX 37 MON. Hemicr. or rhizome geophyte.

subsp. ***chuanxibeiensis*** S.Yun Liang & Y.C.Tang, Acta Phytotax. Sin. 28: 155 (1990).
China (Sichuan). 36 CHC. Hemicr. or rhizome geophyte.

subsp. ***enervis***
Siberia to China. 30 ALT 32 KAZ KGZ UZB 36 CHC CHI CHM CHN CHQ CHT CHX 37 MON. Hemicr. or rhizome geophyte.
Carex carlantoniana J.Gay, Ann. Sci. Nat., Bot., II, 10: 293 (1838).
Carex similigena V.I.Krecz. in V.L.Komarov (ed.), Fl. URSS 3: 596 (1935).

Carex engelmannii L.H.Bailey, Proc. Amer. Acad. Arts 22: 132 (1887).
SW. Canada to W. U.S.A. 71 BRC 73 COL IDA MNT WAS WYO 76 NEV UTA. Hemicr. or rhizome geophyte.
Carex paddoensis Suksd., Allg. Bot. Z. Syst. 12: 43 (1906). *Carex engelmannii* var. *paddoensis* (Suksd.) Kneuck., Allg. Bot. Z. Syst. 17: 90 (1911). *Carex breweri* var. *paddoensis* (Suksd.) Cronquist, Vasc. Pl. Pacific Northw. 1: 249 (1969).

Carex enneastachya C.B.Clarke, Kew Bull., Addit. Ser. 8: 70 (1908).
Colombia. 83 CLM. Hemicr. or rhizome geophyte.

Carex enysii Petrie, Trans. & Proc. New Zealand Inst. 54: 570 (1923).
New Zealand. 51 NZN NZS. Hemicr. or rhizome geophyte.

Carex erawinensis Korotky, Repert. Spec. Nov. Regni Veg. 13: 293 (1914).
Siberia. 30+. Hemicr. or rhizome geophyte.

Carex ereica Tang & F.T.Wang ex L.K.Dai, Acta Phytotax. Sin. 32: 177 (1994).
China (Sichuan). 36 CHC. Hemicr. or rhizome geophyte.

Carex eremitica Paine, Palestine Explor. Soc. 3: 126 (1875).
Israel. 34 PAL. Hemicr. or rhizome geophyte.

Carex eremopyroides V.I.Krecz. in V.L.Komarov (ed.), Fl. URSS 3: 617 (1935).
Siberia to NE. China. 30+ 36 CHI CHM 37 MON. Hemicr. or rhizome geophyte.

Carex eremostachya S.T.Blake, J. Arnold Arbor. 28: 99 (1947).
W. New Guinea. 43 NWG. Hemicr. or rhizome geophyte.

Carex ericetorum Pollich, Hist. Pl. Palat. 2: 580 (1777).
Europe to Mongolia. 10 DEN FIN GRB NOR SWE

11 AUT BGM CZE GER HUN NET POL SWI 12 FRA SPA 13 BUL ITA ROM YUG 14 BLR BLT RUC RUE RUN RUS UKR 30 ALT WSB 32 KAZ 33 NCS 37 MON. Hemicr. or rhizome geophyte.

Carex approximata All., Fl. Pedem. 2: 267 (1785). *Carex ericetorum* var. *approximata* (All.) Nyman, Consp. Fl. Eur.: 773 (1882). *Caricina approximata* (All.) St.-Lag. in A.Cariot, Étude Fl., ed. 8, 2: 879 (1889).

Carex ciliata Willd., Mém. Acad. Roy. Sci. Hist. (Berlin) 1794: 47 (1797).

Carex membranacea Hoppe in J.Sturm, Deutschl. Fl. 61: t. 71 (1835), nom. illeg. *Carex ericetorum* var. *membranacea* W.D.J.Koch, Syn. Fl. Germ. Helv., ed. 2: 876 (1844).

Carex brachyphylla Turcz., Bull. Soc. Imp. Naturalistes Moscou 1838: 104 (1838).

Carex paradoxa f. *interrupta* Peterm., Anal. Pfl.-Schlüss. 1846: 491 (1846). *Carex approximata* f. *interrupta* (Peterm.) Soó, Acta Bot. Acad. Sci. Hung. 16: 369 (1970 publ. 1971).

Carex ericetorum var. *alpicola* Schur, Enum. Pl. Transsilv.: 716 (1866).

Carex ericetorum f. *pallescens* Kneuck. in M.A.Seubert, Excurs.-Fl. Baden, ed. 5: 58 (1891).

Carex ericetorum f. *tenella* Grutt., Deutsche Bot. Monatsschr. 10: 69 (1892).

Carex ericetorum f. *nigra* Hülsen ex Asch. & Graebn., Syn. Mitteleur. Fl. 1: 122 (1896).

Carex paradoxa f. *ramosa* Asch. & Graebn., Syn. Mitteleur. Fl. 2(2): 45 (1902). *Carex approximata* f. *ramosa* (Asch. & Graebn.) Soó, Acta Bot. Acad. Sci. Hung. 16: 369 (1970 publ. 1971).

Carex ericetorum f. *elongata* Kük. in H.G.A.Engler (ed.), Pflanzenr., IV, 20(38): 440 (1909).

Carex hostiana var. *froedinii* Fagerstr., Acta Soc. Fauna Fl. Fenn. 79(3): 2 (1967).

Carex eriocarpa Hausskn. & Kük., Bull. Herb. Boissier, II, 4: 57 (1904).
E. Turkey. 34 TUR. Hemicr. or rhizome geophyte.

Carex erythrobasis H.Lév. & Vaniot, Repert. Spec. Nov. Regni Veg. 5: 240 (1908).
Russian Far East, Korea. 31+ 36 CHM 38 KOR. Hemicr. or rhizome geophyte.

Carex hallaisanensis H.Lév. & Vaniot, Repert. Spec. Nov. Regni Veg. 5: 240 (1908).

Carex erythrorrhiza Boeck., Linnaea 39: 103 (1875).
Ethiopia to E. Zaïre. 23 ZAI 24 ETH 25 KEN UGA. Hemicr. or rhizome geophyte.

Carex esquiroliana H.Lév., Repert. Spec. Nov. Regni Veg. 12: 506 (1913).
China (Guizhou). 36 CHC. Hemicr. or rhizome geophyte.

Carex esquirolii H.Lév. & Vaniot, Bull. Soc. Bot. France 53: 315 (1906).
China. 36+. Hemicr. or rhizome geophyte.

Carex euprepes Nelmes, Bull. Misc. Inform. Kew 1939: 305 (1939).
Indo-China. 41 LAO VIE. Hemicr. or rhizome geophyte.

Carex euryphylla Cherm., Bull. Soc. Bot. France 70: 411 (1923).
C. Madagascar. 29 MDG. Hemicr. or rhizome geophyte.

Carex evadens S.González & Reznicek, Acta Bot. Mex. 40: 19 (1997).
Mexico (Querétaro). 79 MXE. Hemicr. or rhizome geophyte.

Carex ×evoluta Hartm., Kongl. Vetensk. Acad. Handl. 1818: 156 (1818). *C. lasiocarpa* × *C. riparia*.
Europe. 11 GER 12 FRA. Hemicr. or rhizome geophyte.

Carex ×*kochiana* Schübl., Fl. Würtemberg: 601 (1834), nom. illeg.

Carex ×*acuminata* Rchb., Icon. Fl. Germ. Helv. 8: 28 (1846), nom. illeg.

Carex excelsa Poepp. ex Kunth, Enum. Pl. 2: 502 (1837).
S. South America. 85 AGE AGS AGW CLC CLS. Hemicr. or rhizome geophyte.

Carex lechleri Steud., Syn. Pl. Glumac. 2: 234 (1855).

Carex psilostachys Steud., Syn. Pl. Glumac. 2: 215 (1855).

Carex exilis Dewey, Amer. J. Sci. Arts 14: 351 (1828).
Vignea exilis (Dewey) Raf., Good Book: 27 (1840).
E. Canada, NC. & E. U.S.A. 72 NBR NFL NSC ONT QUE 74 MIN WIS 75 MAI MAS MIC NWH NWJ NWY RHO VER 78 ALA DEL MRY MSI NCA. Hemicr. or rhizome geophyte.

Carex exilis var. *squamacea* Dewey, Amer. J. Sci. Arts 14: 351 (1828).

Carex exilis var. *androgyna* Dewey ex Alph.Wood, Class-book Bot.: 750 (1861).

Carex ×exsalina Lepage, Naturaliste Canad. 83: 133 (1956). *C. bigelowii* × *C. paleacea*.
Québec. 72 QUE. Hemicr. or rhizome geophyte.

Carex exsiccata L.H.Bailey, Mem. Torrey Bot. Club 1: 6 (1889).
W. Canada to W. U.S.A. 71 BRC 73 IDA MNT ORE WAS 76 CAL. Hemicr. or rhizome geophyte.

Carex vesicaria var. *major* Boott in W.J.Hooker, Fl. Bor.-Amer. 2: 221 (1839).

Carex extensa Gooden., Trans. Linn. Soc. London 2: 175 (1794). *Trasus extensus* (Gooden.) Gray, Nat. Arr. Brit. Pl. 2: 61 (1821).
Macaronesia, Europe, Medit. to N. Iran. 10 DEN FIN GRB IRE NOR SWE 11 BGM CZE GER NET POL 12 BAL COR FRA POR SAR SPA 13 ALB BUL GRC ITA KRI ROM SIC YUG 14 BLT KRY UKR 20 ALG EGY LBY MOR TUN 21 AZO MDR 33 TCS 34 CYP IRN PAL SIN TUR (75) nwy (78) mry vrg. Hemicr. or rhizome geophyte.

Carex nervosa Desf., Fl. Atlant. 2: 337 (1799). *Carex extensa* var. *nervosa* (Desf.) Nyman, Consp. Fl. Eur.: 771 (1882).

Carex arcuata Wahlenb., Kongl. Vetensk. Acad. Nya Handl. 24: 151 (1803).

Carex balbisii Schkuhr ex Balbis, Cat. Pl.: 21 (1813).

Carex balbisii Schkuhr ex Spreng., Pl. Min. Cogn. Pug. 2: 86 (1815). *Carex extensa* f. *balbisii* (Schkuhr ex Spreng.) Rchb., Icon. Fl. Germ. Helv. 8: 30 (1846). *Carex extensa* var. *balbisii* (Schkuhr ex Spreng.) Nyman, Consp. Fl. Eur.: 771 (1882).

Carex extensa f. *tenuifolia* DC. in J.B.A.M.de Lamarck & A.P.de Candolle, Fl. Franç., ed. 3, 6: 292 (1815).

Carex extensa f. *pumila* Andersson, Cyper. Scand.: 26 (1849).

Carex peruviana Dombey ex Boott, Ill. Carex 4: 138 (1867), nom. illeg.

Carex extensa var. *graeca* Hausskn., Mitth. Thüring. Bot. Vereins, n.s., 13–14: 36 (1899).

Carex extensa subsp. *viestina* Fen., Webbia 29: 276 (1974).

Carex faberiana Loes., Beih. Bot. Centralbl. 37(2): 94 (1919).
NC. China. 36 CHN. Hemicr. or rhizome geophyte.

Carex fastigiata Franch., Bull. Soc. Philom. Paris, VIII, 7: 34 (1895).
Himalaya to SC. China. 36 CHC 40 EHM NEP. Hemicr. or rhizome geophyte.

Carex feanii F.Br., Bernice P. Bishop Mus. Bull. 84: 115 (1931).
Marquesas. 61 MRQ. Hemicr. or rhizome geophyte.

Carex fecunda Steud., Syn. Pl. Glumac. 2: 194 (1855).
Peru to NW. Argentina. 83 BOL PER 84 BZC 85 AGW. Hemicr. or rhizome geophyte.

var. ***atropurpurea*** (Boeck.) J.F.Macbr., Publ. Field Mus. Nat. Hist., Bot. Ser. 11: 42 (1931).
Peru. 83 PER. Hemicr. or rhizome geophyte.
**Carex atropurpurea* Boeck., Linnaea 39: 150 (1875), nom. illeg.

var. ***fecunda***
Peru to NW. Argentina. 83 BOL PER 84 BZC 85 AGW. Hemicr. or rhizome geophyte.
Carex seditiosa Steud. in W.Lechler, Berberid. Amer. Austral.: 56 (1857).

Carex feddeana H.Pfeiff., Repert. Spec. Nov. Regni Veg. 17: 30 (1921).
Brazil (Paraná) to N. Argentina. 84 BZS 85 AGE AGW PAR URU. Hemicr. or rhizome geophyte.
Carex bonariensis var. *achalensis* Kurtz ex Kük., Bot. Jahrb. Syst. 27: 513 (1899).

Carex fedia Nees in R.Wight, Contr. Bot. India: 129 (1834).
C. Asia, Afghanistan to Indo-China. 32 TZK 34 AFG 36 CHC 40 NEP PAK 41 THA VIE. Hemicr. or rhizome geophyte.
Carex wallichiana Prescott ex Nees in R.Wight, Contr. Bot. India: 129 (1834), nom. illeg.
Carex chaetogyne Nelmes, Kew Bull. 1: 9 (1946).

Carex fenghuangshanica F.T.Wang & Tang ex P.C.Li, Acta Phytotax. Sin. 37: 168 (1999).
China (Guangxi). 36 CHS. Hemicr. or rhizome geophyte.

Carex ferruginea Scop., Fl. Carniol., ed. 2, 2: 225 (1772). *Edritria ferruginea* (Scop.) Raf., Good Book: 26 (1840).
NW. Spain, Alps to NW. Balkan Pen, S. Carpathians to SW. Bulgaria. 11 AUT GER SWI 12 FRA SPA 13 BUL ITA ROM YUG 14 UKR (75) nwj. Hemicr. or rhizome geophyte.

subsp. ***australpina*** (Bech.) W.Dietr., Feddes Repert. 75: 18 (1967).
S. Alps. 13 ITA. Hemicr. or rhizome geophyte.
Carex tenax Reut., Compt.-Rend. Trav. Soc. Hallér 1854–1856: 130 (1855), nom. illeg. *Carex sempervirens* var. *tenax* H.Christ, Pflanzenleben Schweiz: 50 (1879). *Carex ferruginea* var. *tenax* (H.Christ) H.Christ, Bull. Soc. Roy. Bot. Belgique 24(2): 14 (1885).
**Carex australpina* Bech., Candollea 8: 12 (1939).

subsp. ***caudata*** (Kük.) Pereda & Laínz, Bol. Inst. Estud. Adsturianos, Supl. Ci. 3: 183 (1961).
NW. Spain. 12 SPA. Hemicr. or rhizome geophyte.
**Carex ferruginea* var. *caudata* Kük. in H.G.A.Engler (ed.), Pflanzenr., IV, 20(38): 573 (1909). *Carex caudata* (Kük.) Pereda & Laínz, Feddes Repert. 81: 483 (1970).

subsp. ***ferruginea***
Alps to NW. Balkan Pen, S. Carpathians to SW. Bulgaria. 11 AUT GER SWI 12 FRA 13 BUL ITA ROM YUG 14 UKR (75) nwj. Hemicr. or rhizome geophyte.
Carex frigida Vill., Hist. Pl. Dauphiné 2: 215 (1787), nom. illeg.
Carex rigida Schrank, Baier. Fl. 1: 290 (1789).
Carex wazmannii Schrank, Baier. Fl. 1: 297 (1789).
Carex spadicea J.F.Gmel., Syst. Nat. 2: 144 (1791).
Carex nana Lam., Encycl. 3: 389 (1792).
Carex scopoliana Willd., Sp. Pl. 4: 292 (1805).
Carex mielichhoferi Schkuhr, Beschr. Riedgräs. 2: 66 (1806).
Carex brevifolia Host, Icon. Descr. Gram. Austriac. 4: 50 (1809). *Carex tenuis* var. *brevifolia* (Host) Nyman, Consp. Fl. Eur.: 772 (1882).
Carex spadicea Host, Icon. Descr. Gram. Austriac. 4: 51 (1809), nom. illeg. *Carex ferruginea* f. *spadicea* W.D.J.Koch, Syn. Fl. Germ. Helv.: 763 (1837).
Carex cristata Clairv., Man. Herbor. Suisse: 291 (1811).
Carex scopolii Gaudin, Agrost. Helv. 2: 168 (1811).
Carex gracillima Steud. & Hochst. ex Steud., Nomencl. Bot., ed. 2, 1: 290 (1840).
Carex brachyrhyncha Gsaller, Oesterr. Bot. Z. 20: 199 (1870).
Carex kerneri Kohts, Oesterr. Bot. Z. 20: 164 (1870). *Carex ferruginea* var. *kerneri* (Kohts) K.Richt., Pl. Eur. 1: 162 (1890).
Carex alpigena A.Kern. ex Dalla Torre in A.Hartinger & K.W.von Dalla Torre, Atlas Alpenfl.: 224 (1882). *Carex ferruginea* var. *alpigena* (A.Kern. ex Dalla Torre) Nyman, Consp. Fl. Eur.: 772 (1882).
Carex ferruginea f. *angustifolia* Beck, Fl. Nieder-Österreich: 142 (1890).
Carex ferruginea f. *latifolia* Beck, Fl. Nieder-Österreich: 142 (1890).
Carex ferruginea f. *leiocarpa* Beck, Fl. Nieder-Österreich: 142 (1890).
Carex ferruginea f. *capillarioides* Murr, Oesterr. Bot. Z. 41: 124 (1891).
Carex tenerrima Murr & Appel, Oesterr. Bot. Z. 41: 124 (1891). *Carex ferruginea* f. *tenerrima* (Murr & Appel) Kük., Allg. Bot. Z. Syst. 6: 8 (1900).
Carex ferruginea f. *transiens* Kük., Allg. Bot. Z. Syst. 6: 8 (1900).
Carex ferruginea f. *crassispica* Kük. in H.G.A.Engler (ed.), Pflanzenr., IV, 20(38): 573 (1909).

subsp. ***macrostachys*** (Bertol.) Arcang., Comp. Fl. Ital.: 745 (1882).
Italy (Alpi Apuane). 13 ITA. Hemicr. or rhizome geophyte.
**Carex macrostachys* Bertol., Amoen. Ital.: 421 (1819). *Carex ferruginea* var. *macrostachys* (Bertol.) Nyman, Consp. Fl. Eur.: 772 (1882).
Carex manostachys Spreng., Syst. Veg. 3: 824 (1826), sphalm.

subsp. ***tenax*** (H.Christ) K.Richt., Pl. Eur.: s.p. (1890).
SW. Alps. 12 FRA 13 ITA. Hemicr. or rhizome geophyte.

subsp. ***tendae*** W.Dietr., Feddes Repert. 75: 20 (1967). *Carex tendae* (W.Dietr.) Pawl. in ?
SW. Alps. 12 FRA 13 ITA. Hemicr. or rhizome geophyte.
Carex refracta Willd., Sp. Pl. 4: 297 (1805). *Carex brachystachys* var. *refracta* (Willd.) Fiori, Nuov. Fl. Italia 1: 197 (1923).
Carex refracta var. *subferruginea* Kük. in H.G.A. Engler (ed.), Pflanzenr., IV, 20(38): 574 (1909).

Carex festivelloides Reznicek, Contr. Univ. Michigan Herb. 19: 109 (1993).
Mexico (Chihuahua). 79 MXE. Hemicr. or rhizome geophyte.

Carex festucacea Schkuhr ex Willd., Sp. Pl. 4: 242 (1805). *Vignea festucacea* (Schkuhr ex Willd.) Rchb. in J.C.Mössler, Handb. Gewächsk. ed. 2, 3: 1614 (1830). *Carex straminea* var. *festucacea* (Schkuhr ex Willd.) Tuck., Enum. Meth. Caric.: 18 (1843).
SE. Canada to C. & E. U.S.A. 72 ONT 74 ILL IOW KAN MIN MSO OKL WIS 75 CNT INI MAS MIC NWJ NWY OHI PEN VER WVA 77 TEX 78 ALA ARK DEL GEO KTY LOU MRY MSI NCA SCA TEN VRG WDC. Hemicr. or rhizome geophyte.

Carex feta L.H.Bailey, Bull. Torrey Bot. Club 20: 417 (1893).
British Columbia to California. 71 BRC 73 ORE WAS 76 CAL. Hemicr. or rhizome geophyte.
**Carex straminea* var. *mixta* L.H.Bailey, Proc. Amer. Acad. Arts 22: 151 (1886).
Carex feta var. *multa* L.H.Bailey, Bot. Gaz. 21: 8 (1896).

Carex ×figertii Asch. & Graebn., Syn. Mitteleur. Fl. 2(2): 11 (1902). *C. davalliana* × *C. dioica*. *Vignea* × *figertii* (Asch. & Graebn.) Soják, Cas. Nár. Mus., Odd. Přír. 148: 195 (1979 publ. 1980).
N. Europe. 10 FIN. Hemicr. or rhizome geophyte.

Carex filamentosa Petrie, Trans. & Proc. New Zealand Inst. 48: 191 (1915 publ. 1916).
New Zealand (Stewart I.). 51 NZS. Hemicr. or rhizome geophyte.

Carex filicina Nees in R.Wight, Contr. Bot. India: 123 (1834), nom. cons.
China to Trop. Asia. 36 CHC CHH CHS CHT 38 TAI 40 ASS EHM IND NEP PAK SRL WHM 41 LAO MYA THA VIE 42 BOR CKI JAW LSI MLY MOL PHI SUL SUM XMS. Hemicr. or rhizome geophyte.
Carex meiogyna Nees in R.Wight, Contr. Bot. India: 123 (1834). *Carex filicina* var. *meiogyna* (Nees) Strachey, Cat. Pl. Kumaon: 73 (1854).
Carex nilagirica Hochst. ex Steud., Syn. Pl. Glumac. 2: 207 (1855).
Carex glaucina Boeck., Linnaea 40: 353 (1876). *Carex filicina* var. *laucina* (Boeck.) Kük. in H.G.A.Engler (ed.), Pflanzenr., IV, 20(38): 274 (1909). *Carex caricina* var. *glaucina* (Boeck.) Ghildyal & U.C.Bhattach., J. Econ. Taxon. Bot. 7: 602 (1985 publ. 1986).
Carex filicina var. *ciliata* Kuntze, Revis. Gen. Pl. 2: 748 (1891).
Carex filicina var. *hirta* Kuntze, Revis. Gen. Pl. 2: 748 (1891).
Carex filicina var. *laevis* Kuntze, Revis. Gen. Pl. 2: 748 (1891).
Carex filicina var. *microgyna* C.B.Clarke in J.D.Hooker, Fl. Brit. India 6: 718 (1894). *Carex caricina* var. *microgyna* (C.B.Clarke) Karth., in Fl. Ind. Enum., Monocot.: 35 (1989).
Carex leptocarpa C.B.Clarke in J.D.Hooker, Fl. Brit. India 6: 719 (1894). *Carex filicina* var. *leptocarpa* (C.B.Clarke) Kük. in H.G.A.Engler (ed.), Pflanzenr., IV, 20(38): 274 (1909). *Carex caricina* var. *leptocarpa* (C.B.Clarke) Ghildyal & U.C.Bhattach., J. Econ. Taxon. Bot. 7: 602 (1985 publ. 1986).
Carex saturata C.B.Clarke, J. Linn. Soc., Bot. 37: 12 (1904). *Carex filicina* var. *saturata* (C.B.Clarke) Kük. in H.G.A.Engler (ed.), Pflanzenr., IV, 20(38): 275 (1909).
Carex neoguineensis C.B.Clarke, Kew Bull., Addit. Ser. 8: 73 (1908).
Carex filicina f. *depauperata* Kük. in H.G.A.Engler (ed.), Pflanzenr., IV, 20(38): 275 (1909).
Carex pseudofilicina Hayata, J. Coll. Sci. Imp. Univ. Tokyo 30(1): 391 (1911). *Carex filicina* subsp. *pseudofilicina* (Hayata) T.Koyama, in Fl. Taiwan 5: 309 (1978).
Carex sclerioides Ridl., Trans. Linn. Soc. London, Bot. 9: 247 (1916).
Carex subfilicina Ohwi, Bot. Mag. (Tokyo) 56: 213 (1942).
Carex caricina var. *minor* Ghildyal & U.C.Bhattach., J. Econ. Taxon. Bot. 7: 602 (1985 publ. 1986).

Carex filifolia Nutt., Gen. N. Amer. Pl. 2: 204 (1818). *Olotrema filifolia* (Nutt.) Raf., Good Book: 25 (1840). *Uncinia filifolia* (Nutt.) Nees in M.A.P.Wied-Neuwied, Reise Nord-Amer. 2: 450 (1841).
Alaska to W. & C. U.S.A. 70 ASK NWT YUK 71 ABT BRC MAN SAS 73 COL IDA MNT ORE WAS WYO 74 KAN MIN NDA NEB SDA 76 ARI CAL NEV UTA 77 NWM. Hemicr. or rhizome geophyte.

var. ***erostrata*** Kük. in H.G.A.Engler (ed.), Pflanzenr., IV, 20(38): 86 (1909).
W. U.S.A. 73 ORE 76 CAL NEV. Hemicr. or rhizome geophyte.
Carex exserta Mack., Bull. Torrey Bot. Club 42: 620 (1915).

var. ***filifolia***
Alaska to W. & C. U.S.A. 70 ASK NWT YUK 71 ABT BRC MAN SAS 73 COL IDA MNT ORE WAS WYO 74 KAN MIN NDA NEB SDA 76 ARI CAL NEV UTA 77 NWM. Hemicr. or rhizome geophyte.
Kobresia globularis Dewey, Amer. J. Sci. Arts 29: 253 (1836).
Uncinia breviseta Torr., Ann. Lyceum Nat. Hist. New York 3: 428 (1836). *Itheta breviseta* (Torr.) Raf., Good Book: 28 (1840).

Carex filiformis L., Sp. Pl.: 976 (1753). *Trasus filiformis* (L.) Gray, Nat. Arr. Brit. Pl. 2: 59 (1821). *Diemisa filiformis* (L.) Raf., Good Book: 27 (1840).
Europe to C. Asia. 10 GRB 11 AUT BGM CZE GER HUN NET POL SWI 12 FRA SPA 13 ALB? BUL GRC ITA ROM TUE YUG 14 BLR BLT KRY RUC RUE RUS UKR 30 WSB 32 KAZ 33 NCS TCS 34 IRN TUR. Hemicr. or rhizome geophyte.
Carex tomentosa L., Mant. Pl. 1: 123 (1767).
Carex sphaerocarpa Ehrh., Beitr. Naturk. 6: 83 (1791).
Carex subvillosa M.Bieb., Fl. Taur.-Caucas. 2: 386 (1808). *Carex tomentosa* subsp. *subvillosa* (M.Bieb.) Nyman, Consp. Fl. Eur.: 774 (1882).

Trasus tomentosus Gray, Nat. Arr. Brit. Pl. 2: 57 (1821).
Carex caesia Griseb., Spic. Fl. Rumel. 2: 412 (1846). *Carex tomentosa* var. *caesia* (Griseb.) Nyman, Consp. Fl. Eur.: 774 (1882).
Carex tomentosa f. *gracilis* O.Lang, Linnaea 24: 586 (1851).
Carex tomentosa f. *grasmanniana* O.Lang, Linnaea 24: 587 (1851). *Carex grasmanniana* (O.Lang) Rabenh. ex Steud., Syn. Pl. Glumac. 2: 218 (1855). *Carex tomentosa* var. *grasmanniana* (O.Lang) Nyman, Consp. Fl. Eur.: 774 (1882).
Carex tomentosa f. *globifera* Schur, Enum. Pl. Transsilv.: 715 (1866).
Carex tomentosa var. *gracillima* Schur, Enum. Pl. Transsilv.: 715 (1866).
Carex nordmannii A.Kern. ex Palla, Verh. K. K. Zool.-Bot. Ges. Wien 38: 69 (1888). *Carex tomentosa* f. *nordmannii* (A.Kern. ex Palla) K.Richt., Pl. Eur. 1: 159 (1890).
Carex brueckneri Kük., Deutsche Bot. Monatsschr. 8: 107 (1890).
Carex tomentosa f. *luxuriosa* Waisb., Oesterr. Bot. Z. 47: 4 (1897).
Carex tomentosa f. *pallida* Waisb., Magyar Bot. Lapok 4: 73 (1905).
Carex tomentosa var. *lucana* A.Terracc., Nuovo Giorn. Bot. Ital. 14: 206 (1907).
Carex tomentosa var. *decalvescens* Kük. in H.G.A.Engler (ed.), Pflanzenr., IV, 20(38): 436 (1909).
Carex tomentosa f. *nigrescens* Kük. & Litv. in H.G.A.Engler (ed.), Pflanzenr., IV, 20(38): 436 (1909).
Carex tomentosa f. *gynandra* A.T.Szabó, Index Seminum (CLA) 1970: 68 (1971).

Carex filipedunculata S.W.Su, J. Anhui Univ. Nat. Sci. 20(3): 105 (1996).
China (Anhui). 36 CHS. Hemicr.

Carex filipes Franch. & Sav., Enum. Pl. Jap. 2: 576 (1878).
China to Japan. 36 CHC CHS 38 JAP. Hemicr. or rhizome geophyte.

var. ***arakiana*** (Ohwi) Ohwi, Mem. Coll. Sci. Kyoto Imp. Univ., Ser. B, Biol. 11: 424 (1936).
Japan (Honshu). 38 JAP. Hemicr. or rhizome geophyte.
**Carex rouyana* var. *arakiana* Ohwi, Acta Phytotax. Geobot. 1: 999 (1932). *Carex arakiana* (Ohwi) Ohwi, Acta Phytotax. Geobot. 7: 34 (1938).

var. ***filipes***
Japan (Honshu, Shikoku). 38 JAP. Hemicr. or rhizome geophyte.

var. ***sparsinux*** (C.B.Clarke ex Franch.) Kük. in H.G.A. Engler (ed.), Pflanzenr., IV, 20(38): 639 (1909).
SC. & E. China, Japan (Honshu). 36 CHC CHS 38 JAP. Hemicr. or rhizome geophyte.
Carex rouyana Franch., Bull. Soc. Philom. Paris, VIII, 7: 51 (1895). *Carex filipes* var. *rouyana* (Franch.) Kük. in H.G.A.Engler (ed.), Pflanzenr., IV, 20(38): 640 (1909).
**Carex sparsinux* C.B.Clarke ex Franch., Nouv. Arch. Mus. Hist. Nat., III, 10: 66 (1898).

var. ***tremula*** (Ohwi) Ohwi, Mem. Coll. Sci. Kyoto Imp. Univ., Ser. B, Biol. 11: 423 (1936).
S. Japan. 38 JAP. Hemicr. or rhizome geophyte.
**Carex arisanensis* var. *tremula* Ohwi, Mem. Coll. Sci. Kyoto Imp. Univ., Ser. B, Biol. 5(3): 255 (1930). *Carex tremula* (Ohwi) Ohwi, Acta Phytotax. Geobot. 2: 28 (1933).

Carex fimbriata Schkuhr, Beschr. Riedgräs. 2: 61 (1806). *Carex ferruginea* var. *fimbriata* (Schkuhr) Fiori, Nuov. Fl. Italia 1: 197 (1923).
W. Alps. 11 SWI 12 FRA ITA. Hemicr. or rhizome geophyte.
Carex fuliginosa Host, Icon. Descr. Gram. Austriac. 4: 52 (1809), nom. illeg.
Carex hispida Gaudin, Agrost. Helv. 2: 136 (1811), nom. illeg.
Carex hispidula Gaudin, Agrost. Helv. 2: 136 (1811).

Carex finitima Boott, Ill. Carex 1: 44 (1858). *Carex fusiformis* subsp. *finitima* (Boott) Noltie, Edinburgh J. Bot. 50: 200 (1993).
Himalaya to Taiwan, Malesia. 36 CHC 38 TAI 40 EHM NEP 42 SUM 43 NWG. Hemicr. or rhizome geophyte.
Carex finitima var. *attenuata* C.B.Clarke in J.D.Hooker, Fl. Brit. India 6: 736 (1894).
Carex remotiflora Hayata, Icon. Pl. Formosan. 10: 68 (1921).
Carex atjehensis Kük., Bull. Jard. Bot. Buitenzorg, III, 16: 314 (1940).

Carex firma Host, Syn. Pl.: 509 (1797). *Carex ferruginea* var. *firma* (Host) Fiori, Nuov. Fl. Italia 1: 197 (1923).
European Mts. 11 AUT CZE GER POL SWI 12 FRA 13 ITA ROM YUG 14 UKR. Hemicr. or rhizome geophyte.
Carex strigosa Suter ex Boott, Ill. Carex 4: 208 (1867), nom. illeg.
Carex firma f. *longipedunculata* Hausskn., Mitth. Thüring. Bot. Vereins, n.s., 7: 17 (1895).

Carex firmicaulis Kalela, Ann. Acad. Sci. Fenn., Ser. A 54(5): 53 (1940).
Argentina. 85 AGE AGS AGW. Hemicr. or rhizome geophyte.
Carex pycnostachys var. *major* Kük., Bot. Jahrb. Syst. 27: 503 (1899). *Carex nebularum* f. *major* (Kük.) Kük. in H.G.A.Engler (ed.), Pflanzenr., IV, 20(38): 118 (1909).

Carex* ×*firmior (Norman) E.Holmb., Bot. Not. 1929: 227 (1929). *C. limosa* × *C. rariflora*.
E. Canada, N. Europe. 10+ 72 NFL QUE. Hemicr. or rhizome geophyte.
**Carex rariflora* var. *firmior* Norman, Forh. Vidensk.-Selsk. Kristiana 1893: 51 (1893).

Carex fischeri K.Schum. in H.G.A.Engler, Pflanzenw. Ost-Afrikas, C: 130 (1895).
Kenya. 25 KEN. Hemicr. or rhizome geophyte.

Carex fissa Mack. in N.L.Britton & al. (eds.), N. Amer. Fl. 18: 64 (1931).
C. & SE. U.S.A. 74 KAN MSO OKL 77 TEX 78 FLA GEO MSI. Hemicr. or rhizome geophyte.

var. ***aristata*** F.J.Herm., Rhodora 67: 198 (1965).
SE. U.S.A. 78 FLA GEO MSI. Hemicr. or rhizome geophyte.

var. ***fissa***
C. U.S.A. 74 KAN MSO OKL 77 TEX. Hemicr. or rhizome geophyte.
Carex bonariensis Schltdl., Linnaea 10: 116 (1836), nom. illeg.

Carex fissuricola Mack., Muhlenbergia 5: 53 (1909).
W. U.S.A. 73 IDA MNT 76 CAL NEV UTA. Hemicr. or rhizome geophyte.
Carex ablata var. *luzuliformis* L.H.Bailey, Bot. Gaz. 25: 272 (1898). *Carex luzulifolia* var. *luzuliformis* (L.H.Bailey) Kük. in H.G.A.Engler (ed.), Pflanzenr., IV, 20(38): 558 (1909).

Carex flabellata H.Lév. & Vaniot, Bull. Acad. Int. Géogr. Bot. 11: 111 (1902).
Japan (Honshu, Yezo). 38 JAP. Hemicr. or rhizome geophyte.

Carex flacca Schreb., Spic. Fl. Lips.: 669 (1771).
Europe, Medit to Pakistan. 10 DEN FIN FOR GRB ICE IRE NOR SWE 11 AUT BGM CZE GER HUN NET POL SWI 12 BAL COR FRA POR SAR SPA 13 ALB BUL GRC ITA KRI ROM SIC TUE YUG 14 BLR BLT KRY RUC RUE RUS RUW UKR 20 ALG EGY LBY MOR TUN 33 NCS TCS 34 CYP EAI IRN IRQ LBS PAL TUR 40 PAK (50) tas (51) nzn (72) nsc ont que (75) mic nwy (85) ags (90) fal. Hemicr. or rhizome geophyte.

subsp. ***erythrostachys*** (Hoppe) Holub, Folia Geobot. Phytotax. 23: 413 (1988).
Medit. to Pakistan. 12 FRA POR SPA 13 ALB BUL GRC ITA ROM TUE YUG 20 ALG EGY MOR TUN 33 NCS TCS 34 CYP EAI IRN IRQ LBS PAL TUR 40 PAK. Hemicr. or rhizome geophyte.
Carex diversicolor Crantz, Inst. Rei Herb. 1: 405 (1766).
Carex cuspidata Host, Icon. Descr. Gram. Austriac. 1: 71 (1801). *Loxotrema cuspidata* (Host) Raf., Good Book: 25 (1840). *Carex glauca* var. *cuspidata* (Host) Asch. & Graebn., Syn. Mitteleur. Fl. 2(2): 138 (1902). *Carex diversicolor* var. *cuspidata* (Host) Fiori, Nuov. Fl. Italia 1: 195 (1923).
Carex serrulata Biv. ex Spreng., Syst. Veg. 3: 827 (1826). *Carex glauca* subsp. *serrulata* (Biv. ex Spreng.) Arcang., Comp. Fl. Ital., ed. 2: 92 (1894). *Carex glauca* f. *serrulata* (Biv. ex Spreng.) Asch. & Graebn., Syn. Mitteleur. Fl. 2(2): 138 (1902). *Carex flacca* subsp. *serrulata* (Biv. ex Spreng.) Greuter, Boissiera 13: 167 (1967).
**Carex erythrostachys* Hoppe in J.Sturm, Deutschl. Fl.: 69 (1835). *Carex glauca* var. *erythrostachys* (Hoppe) Andersson, Cyper. Scand.: 31 (1849). *Carex recurva* subsp. *erythrostachys* (Hoppe) Arcang., Comp. Fl. Ital.: 744 (1882). *Carex glauca* f. *erythrostachys* (Hoppe) Kük. in H.G.A.Engler (ed.), Pflanzenr., IV, 20(38): 418 (1909).

subsp. ***flacca***
Europe, Medit. 10 DEN FIN FOR GRB ICE IRE NOR SWE 11 AUT BGM CZE GER HUN NET POL SWI 12 BAL COR FRA POR SAR SPA 13 ALB BUL GRC ITA KRI ROM SIC TUE YUG 14 BLR BLT KRY RUC RUE RUS RUW UKR 20 ALG LBY MOR TUN 34 TUR (50) tas (51) nzn (72) nsc ont que (75) mic nwy (85) ags (90) fal. Hemicr. or rhizome geophyte.
Carex diversicolor var. *claviformis* Hoppe in ?
Carex pendula Schreb., Spic. Fl. Lips.: 62 (1771), nom. illeg.
Carex glauca Scop., Fl. Carniol., ed. 2, 2: 223 (1772). *Trasus glaucus* (Scop.) Gray, Nat. Arr. Brit. Pl. 2: 67 (1821). *Carex glauca* var. *typica* Regel, Trudy Imp. S.-Peterburgsk. Bot. Sada 7: 572 (1880), nom. inval.
Carex recurva Huds., Fl. Angl., ed. 2: 413 (1778).
Carex verna Lam., Fl. Franç. 2: 185 (1779).
Carex aspera Willd., Fl. Berol. Prodr.: 32 (1787). *Carex glauca* var. *aspera* (Willd.) Poir. in J.B. A.M.de Lamarck, Encycl., Suppl. 3: 277 (1813).
Carex trachycarpos Link, J. Bot. (Schrader) 2: 310 (1799).
Carex micheliana Sm., Trans. Linn. Soc. London 5: 270 (1800). *Carex glauca* var. *micheliana* (Sm.) Nyman, Consp. Fl. Eur.: 774 (1882). *Carex glauca* f. *micheliana* (Sm.) Asch. & Graebn., Syn. Mitteleur. Fl. 2(2): 136 (1902).
Carex claviformis Hoppe in J.Sturm, Deutschl. Fl. 13: 68 (1803). *Carex glauca* var. *claviformis* (Hoppe) Boeck., Linnaea 41: 188 (1877).
Carex nigrolutea Gaudin, Etrennes Fl.: 196 (1804).
Carex acuminata Willd., Sp. Pl. 4: 300 (1805). *Olotrema acuminata* (Willd.) Raf., Good Book: 25 (1840). *Carex glauca* var. *acuminata* (Willd.) Nyman, Consp. Fl. Eur., Suppl. 2: 322 (1890). *Carex glauca* f. *acuminata* (Willd.) Asch. & Graebn., Syn. Mitteleur. Fl. 2(2): 138 (1902).
Carex ambleocarpa Willd., Sp. Pl. 4: 307 (1805). *Carex glauca* var. *ambleocarpa* (Willd.) Schur, Enum. Pl. Transsilv.: 713 (1866).
Carex thuringiaca Willd., Sp. Pl. 4: 250 (1805). *Carex glauca* var. *thuringiaca* (Willd.) Nyman, Consp. Fl. Eur.: 774 (1882).
Carex nigrolutea Gaudin, Agrost. Helv. 2: 196 (1811), nom. illeg.
Carex genuensis DC., Cat. Pl. Horti Monsp.: 87 (1813). *Carex glauca* var. *genuensis* (DC.) Nyman, Consp. Fl. Eur.: 774 (1882).
Carex stictocarpa Sm., Engl. Bot.: t. 2772 (1835). *Carex glauca* var. *stictocarpa* (Sm.) Nyman, Consp. Fl. Eur.: 774 (1882).
Carex obtusangula Kunth, Enum. Pl. 2: 459 (1837).
Carex truncata Fenzl in E.von Friederichthal, Reise Griechenl.: 278 (1838).
Carex glauca var. *arecta* Drejer, Symb. Caric.: 20 (1844).
Carex glauca f. *bulbosa* Drejer, Symb. Caric.: 20 (1844).
Carex glauca f. *scabra* Peterm., Flora 27: 337 (1844).
Carex praetutiana Parl., Fl. Ital. 2: 182 (1852). *Carex glauca* subsp. *praetutiana* (Parl.) Nyman, Consp. Fl. Eur.: 774 (1882). *Carex recurva* subsp. *praetutiana* (Parl.) Arcang., Comp. Fl. Ital.: 744 (1882). *Carex flacca* subsp. *praetutiana* (Parl.) Holub, Folia Geobot. Phytotax. 8: 176 (1973).
Carex glauca var. *leiocarpa* Willk. in M.Willkomm & J.M.C.Lange, Prodr. Fl. Hispan. 1: 123 (1861). *Carex glauca* f. *leiocarpa* (Willk.) Kük. in H.G.A. Engler (ed.), Pflanzenr., IV, 20(38): 419 (1909).
Carex glauca var. *sphaerostachys* Lange in G.C.Oeder & al. (eds.), Fl. Dan.: t. 2666 (1861).
Carex flacca var. *divaricata* Heuff., Linnaea 31: 710 (1863). *Carex glauca* f. *dinarica* (Heuff.) Asch. & Graebn., Syn. Mitteleur. Fl. 2(2): 137 (1902).
Carex glauca var. *alpina* Heuff., Linnaea 31: 710 (1863).
Carex glauca f. *tenuicula* Martrin-Donos, Fl. Tarn.: 756 (1864). *Carex flacca* f. *tenuicula* (Martrin-Donos) Soó, Acta Bot. Acad. Sci. Hung. 16: 370 (1970 publ. 1971).

Carex glauca var. *arenosa* Schur, Enum. Pl. Transsilv.: 713 (1866).
Carex glauca f. *laxiflora* Schur, Enum. Pl. Transsilv.: 713 (1866).
Carex glauca f. *leptostachys* Schur, Enum. Pl. Transsilv.: 713 (1866).
Carex glauca var. *attenuata* Ball, J. Linn. Soc., Bot. 16: 706 (1877).
Carex flacca f. *melanostachya* Uechtr. in E.Fiek, Fl. Schlesien: 486 (1881). *Carex glauca* f. *melanostachya* (Uechtr.) Kük. in H.G.A.Engler (ed.), Pflanzenr., IV, 20(38): 417 (1909).
Carex recurva subsp. *parlatoreana* Ces. ex Arcang., Comp. Fl. Ital.: 744 (1882).
Carex flacca var. *oogyna* Murr, Deutsche Bot. Monatsschr. 4: 166 (1886).
Carex dinarica Simonk., Enum. Fl. Transsilv.: 533 (1887). *Carex glauca* var. *dinarica* (Simonk.) Nyman, Consp. Fl. Eur., Suppl. 2: 322 (1890).
Carex glauca var. *rotundata* Callmé, Deutsche Bot. Monatsschr. 6: 1 (1888).
Carex flacca f. *pallida* Beck, Fl. Nieder-Österreich: 141 (1890). *Carex glauca* f. *pallida* (Beck) Kük. in H.G.A.Engler (ed.), Pflanzenr., IV, 20(38): 417 (1909).
Carex glauca f. *subrotunda* Kük., Deutsche Bot. Monatsschr. 8: 107 (1890).
Carex glauca var. *melanocarpa* Murr, Oesterr. Bot. Z. 41: 90 (1891).
Carex glauca var. *praetutiana* Arcang., Comp. Fl. Ital., ed. 2: 92 (1894).
Carex serrulata var. *corcyrensis* Kük., Exsicc. (Herb. Europ.) 30: 7 (1897). *Carex glauca* f. *corcyrensis* (Kük.) Kük. in H.G.A.Engler (ed.), Pflanzenr., IV, 20(38): 419 (1909).
Carex glauca f. *lemaniana* H.Christ, Annuaire Conserv. Jard. Bot. Genève 6: 154 (1902).
Carex glauca f. *silvatica* Asch. & Graebn., Syn. Mitteleur. Fl. 2(2): 135 (1902).
Carex glauca var. *subustulata* H.Christ, Annuaire Conserv. Jard. Bot. Genève 6: 155 (1902). *Carex glauca* f. *subustulata* (H.Christ) Kük. in H.G.A.Engler (ed.), Pflanzenr., IV, 20(38): 418 (1909).
Carex flacca f. *chlorocarpa* R.Keller, Bull. Herb. Boissier, II, 3: 377 (1903). *Carex glauca* f. *chlorocarpa* (R.Keller) Kük. in H.G.A.Engler (ed.), Pflanzenr., IV, 20(38): 417 (1909).
Carex glauca f. *pubicarpa* Rohlena, Sitzungsber. Königl. Böhm. Ges. Wiss. Prag, Math.-Naturwiss. Cl. 1904: 95 (1904). *Carex glauca* var. *pubicarpa* (Rohlena) Rohlena, Magyar Bot. Lapok 3: 320 (1905). *Carex pubicarpa* (Rohlena) A.W.Hill, Index Kew., Suppl. 6: 37 (1926).
Carex glauca f. *aristolepis* Kük. ex L.Gross, Mitt. Bad. Bot. Vereins 1906: 75 (1906).
Carex glauca var. *murensis* A.Terracc., Nuovo Giorn. Bot. Ital. 14: 206 (1907).
Carex glauca f. *abbreviata* Kük. in H.G.A.Engler (ed.), Pflanzenr., IV, 20(38): 418 (1909).
Carex glauca f. *flavescens* Kük. in H.G.A.Engler (ed.), Pflanzenr., IV, 20(38): 417 (1909).
Carex flacca f. *castriferrei* Soó, Acta Bot. Acad. Sci. Hung. 16: 370 (1970 publ. 1971).

Carex flaccosperma Dewey, Amer. J. Sci. Arts, II, 2: 245 (1846).
C. & SE. U.S.A. 74 KAN MSO OKL 77 TEX 78 ALA ARK FLA GEO KTY LOU MSI NCA SCA TEN VRG. Hemicr. or rhizome geophyte.
Carex laxiflora var. *mutica* Torr., Ann. Lyceum Nat. Hist. New York 3: 414 (1836). *Carex grisea* var. *mutica* (Torr.) Carey in A.Gray, Manual: 552 (1848).
Carex microsperma Steud., Syn. Pl. Glumac. 2: 231 (1855).
Carex xanthosperma Dewey, Amer. J. Sci. Arts, II, 42: 334 (1866).

Carex flagellifera Colenso, Trans. & Proc. New Zealand Inst. 16: 342 (1884).
Tasmania, New Zealand, Kermadec Is. 50 TAS 51 KER NZN NZS. Hemicr. or rhizome geophyte.
Carex lucida Boott in J.D.Hooker, Fl. Nov.-Zel. 1: 283 (1853), nom. illeg.
Carex pulla Sol. ex Boott, Ill. Carex 1: 64 (1858), nom. illeg.

Carex flava L., Sp. Pl.: 975 (1753). *Trasus flavus* (L.) Gray, Nat. Arr. Brit. Pl. 2: 60 (1821). *Anithista flava* (L.) Raf., Good Book: 26 (1840). *Proteocarpus flavus* (L.) Fedde & J.Schust., Just's Bot. Jahresber. 41(2): 11 (1913 publ. 1918).
Europe to NW. Iran, Canada, NC. & NE. U.S.A. 10 DEN FIN FOR GRB ICE IRE NOR SWE 11 AUT BGM CZE GER HUN NET POL SWI 12 COR FRA POR SPA 13 ALB BUL GRC ITA ROM YUG 14 KRY RUN RUS 20 ALG MOR 30 YAK 33 NCS TCS 34 IRN TUR 70 ASK NUN 71 ABT BRC MAN 72 LAB NBR NFL NSC ONT PEI QUE 73 IDA MNT 74 MIN WIS 75 CNT INI MAI MAS MIC MIN NWH NWJ NWY OHI PEN RHO VER. Hemicr. or rhizome geophyte.

subsp. ***flava***
Europe to NW. Iran, Canada, NC. & NE. U.S.A. 10 DEN FIN FOR GRB ICE IRE NOR SWE 11 AUT BGM CZE GER HUN NET POL SWI 12 COR FRA POR SPA 13 ALB BUL GRC ITA ROM YUG 14 KRY RUN RUS 20 ALG MOR 30 YAK 33 NCS TCS 34 IRN TUR 70 ASK NUN 71 ABT BRC MAN 72 LAB NBR NFL NSC ONT PEI QUE 73 IDA MNT 74 MIN WIS 75 CNT INI MAI MAS MIC MIN NWH NWJ NWY OHI PEN RHO VER. Hemicr. or rhizome geophyte.
Carex flava var. *rectirostra* Gaudin in ?
Carex divisa Oeder in G.C.Oeder & al. (eds.), Fl. Dan.: t. 371 (1768), nom. illeg.
Carex echinata Lam., Fl. Franç. 2: 177 (1779), nom. illeg.
Carex foliosa All., Fl. Pedem. 2: 270 (1785), nom. illeg.
Carex viridis Honck., Syn. Pl. Germ. 1: 372 (1792).
Carex patula Host, Icon. Descr. Gram. Austriac. 1: 2, 221 (1801), nom. illeg. *Carex flava* var. *patula* Coss. & Germ., Fl. Descr. Anal. Paris, ed. 2: 752 (1861). *Carex flava* f. *patula* (Coss. & Germ.) Schur, Enum. Pl. Transsilv.: 711 (1866).
Carex uetliaca Suter, Helvet. Fl. 1: 251 (1802). *Carex flava* var. *uetliaca* (Suter) Nyman, Consp. Fl. Eur.: 771 (1882). *Carex flava* f. *uetliaca* (Suter) Asch. & Graebn., Syn. Mitteleur. Fl. 2(2): 200 (1903).
Carex flava f. *densa* Gaudin, Fl. Helv. 6: 97 (1830).
Carex flava var. *polystachya* Gaudin, Fl. Helv. 6: 97 (1830).
Carex lipsiensis Peterm., Fl. Lips. Excurs.: 58 (1838).
Carex flava var. *vulgaris* Döll, Rhein. Fl.: 147 (1843).
Carex flava f. *deficiens* Peterm., Flora 27: 339 (1844).

Carex flava f. *rectirostris* Peterm., Flora 27: 338 (1844).
Carex flava var. *pygmaea* Andersson, Cyper. Scand.: 25 (1849).
Carex flavofulva Beurl., Bot. Not. 1853: 37 (1853).
Carex pyriformis Tausch ex F.W.Schultz, Arch. Fl. France Allemagne 2: 285 (1854).
Carex flava f. *remotiuscula* Schur, Enum. Pl. Transsilv.: 711 (1866).
Carex flava var. *macrorrhyncha* Èelak., Prodr. Fl. Böhmen: 71 (1867).
Carex flava var. *graminis* L.H.Bailey, Mem. Torrey Bot. Club 1: 30 (1889). *Carex oederi* f. *graminis* (L.H.Bailey) Kük. in H.G.A.Engler (ed.), Pflanzenr., IV, 20(38): 674 (1909). *Carex flava* f. *graminis* (L.H.Bailey) Scoggan, Fl. Canada 1: 51 (1978).
Carex flava var. *fertilis* Peck, Rep. (Annual) Regents Univ. State New York State Mus. 48: 197 (1896).
Carex flava f. *umbrosa* Kneuck., Allg. Bot. Z. Syst. 5: 8 (1899).
Carex flava var. *congesta* Neuman, Sver. Fl.: 696 (1901).
Carex flava var. *dispersa* Neuman, Sver. Fl.: 696 (1901).
Carex flava var. *brevirostris* Asch. & Graebn., Syn. Mitteleur. Fl. 2(2): 201 (1903).
Carex flava f. *brevirostris* Junge, Schriften Naturwiss. Vereins Schleswig-Holstein 13(2): 4 (1906).
Carex flava var. *gaspensis* Fernald, Rhodora 8: 200 (1906).
Carex lepidocarpa var. *laxior* Kük. in H.G.A.Engler (ed.), Pflanzenr., IV, 20(38): 673 (1909). *Carex laxior* (Kük.) Mack. in N.L.Britton & al. (eds.), N. Amer. Fl. 18: 306 (1935). *Carex flava* var. *laxior* (Kük.) Gleason, Phytologia 4: 22 (1952).
Carex flavella V.I.Krecz. in P.F.Majevski, Fl. Centr. Russ., ed. 6: 185 (1933).
Carex oederi var. *microcarpa* Font Quer, Bol. Real Soc. Esp. Hist. Nat. 35: 140 (1935).
Carex flava var. *barrerae* O.Bolòs, Masalles & Vigo, Collect. Bot. (Barcelona) 17: 95 (1987 publ. 1988).

var. ***nevadensis*** (Boiss. & Reut.) Briq., Prodr. Fl. Corse 1: 219 (1910).
Spain (Sierra Nevada), Morocco. 12 SPA 20 MOR. Hemicr. or rhizome geophyte.
**Carex nevadensis* Boiss. & Reut., Pugill. Pl. Afr. Bor. Hispan.: 118 (1852). *Carex oederi* subsp. *nevadensis* (Boiss. & Reut.) Nyman, Consp. Fl. Eur.: 771 (1882). *Carex oederi* var. *nevadensis* (Boiss. & Reut.) K.Richt., Pl. Eur. 1: 164 (1890). *Carex lepidocarpa* var. *nevadensis* (Boiss. & Reut.) Kük. in H.G.A.Engler (ed.), Pflanzenr., IV, 20(38): 673 (1909). *Carex mairii* subsp. *nevadensis* (Boiss. & Reut.) Malag., Subesp. Variac. Geogr.: 27 (1973), without basionym date. *Carex viridula* subsp. *nevadensis* (Boiss. & Reut.) B.Schmid, Watsonia 14: 317 (1983). *Carex viridula* var. *nevadensis* (Boiss. & Reut.) Crins, Canad. J. Bot. 67: 1060 (1989). *Carex lepidocarpa* subsp. *nevadensis* (Boiss. & Reut.) Luceño, Anales Jard. Bot. Madrid 57: 176 (1999).
Carex oederi Willk., Flora 35: 521 (1852), nom. illeg.
Carex flava var. *alpina* Kneuck., Allg. Bot. Z. Syst. 5: 8 (1899). *Carex flava* subsp. *alpina* (Kneuck.) O.Bolòs, Masalles & Vigo, Collect. Bot. (Barcelona) 17: 95 (1987 publ. 1988). *Carex nevadensis* subsp. *alpina* (Kneuck.) Podlech, Bull. Soc. Échange Pl. Vasc. Eur. Occid. Bassin Médit. 28: 113 (2000).
Carex flava f. *minuta* Foucaud & Rotges, Bull. Soc. Bot. France 47: 98 (1900).

subsp. ***scotica*** (E.W.Davies) P.D.Sell in P.D.Sell & G.Murrell, Fl. Great Britain Ireland 5: 363 (1996).
Great Britain (Scotland). 10 GRB. Hemicr. or rhizome geophyte.
**Carex lepidocarpa* subsp. *scotica* E.W.Davies, Watsonia 3: 71 (1953). *Carex viridula* var. *scotica* (E.W.Davies) B.Schmid, Watsonia 14: 318 (1983).

Carex* ×*flavicans F.Nyl., Spic. Fl. Fenn. 3: 15 (1846). *C. aquatilis* × *C. subspathacea*. *Carex halophila* var. *flavicans* (F.Nyl.) Boivin in ?. *Carex salina* var. *flavicans* (F.Nyl.) Almq., Bot. Not. 1891: 127 (1891).
Subarctic to E. Canada. 10 FIN NOR 70 GNL NUN 72 ONT QUE. Hemicr. or rhizome geophyte.
Carex ×*substans* Lepage, Naturaliste Canad. 83: 152 (1956).

Carex flaviformis Nelmes, Kew Bull. 10: 84 (1955).
New Zealand. 51 NZN NZS. Hemicr. or rhizome geophyte.

Carex flavocuspis Franch. & Sav., Enum. Pl. Jap. 2: 574 (1878). *Carex macrochaeta* subsp. *flavocuspis* (Franch. & Sav.) Kük. in H.G.A.Engler (ed.), Pflanzenr., IV, 20(38): 414 (1909).
Russian Far East to N. China, Japan. 31 KAM KUR SAK 36 CHN 38 JAP. Hemicr. or rhizome geophyte.
Carex gansuensis Franch., Bull. Soc. Philom. Paris, VIII, 7: 39 (1895).
Carex flavocuspis var. *caudatifrons* Akiyama, Consp. Caric. Jap.: 123 (1932). *Carex caudatifrons* (Akiyama) Akiyama, J. Fac. Sci. Hokkaido Univ., Ser. 5, Bot. 3: 262 (1935).

Carex flexirostris Reznicek, Syst. Bot. 11: 75 (1986).
Mexico (Oaxaca). 79 MXS. Hemicr. or rhizome geophyte.

Carex floridana Schwein., Ann. Lyceum Nat. Hist. New York 1: 66 (1824). *Vignea floridana* (Schwein.) Raf., Good Book: 27 (1840). *Carex lucorum* var. *floridana* (Schwein.) Chapm., Fl. South. U.S.: 539 (1860). *Carex nigromarginata* var. *floridana* (Schwein.) Kük. in H.G.A.Engler (ed.), Pflanzenr., IV, 20(38): 444 (1909).
SE. U.S.A. to Texas. 77 TEX 78 ALA FLA GEO LOU MSI NCA SCA. Hemicr. or rhizome geophyte.

Carex fluviatilis Boott, Ill. Carex 4: 172 (1867). *Vignea fluviatilis* (Boott) Soják, Cas. Nár. Mus., Odd. Prír. 148: 195 (1979 publ. 1980).
N. Myanmar to SC. China. 36 CHC CHT 41 MYA. Hemicr. or rhizome geophyte.
Carex pterolepta Franch., Nouv. Arch. Mus. Hist. Nat., III, 8: 215 (1896).
Carex yungningensis Hand.-Mazz. & Kük. in H.F.von Handel-Mazzetti, Symb. Sin. 5: 1260 (1936).

Carex foenea Willd., Enum. Pl. 2: 957 (1809). *Vignea foenea* (Willd.) Rchb. in J.C.Mössler, Handb. Gewächsk. ed. 2, 3: 1614 (1830).
Subarctic America to N. U.S.A. 70 ASK NUN NWT YUK 71 ABT BRC MAN SAS 72 LAB NBR NFL NSC ONT PEI QUE 73 MNT 74 MIN NDA SDA

WIS 75 CNT MAI MAS MIC NWH NWY PEN RHO VER. Hemicr. or rhizome geophyte.

Carex aenea Fernald, Proc. Amer. Acad. Arts 37: 480 (1902). *Carex foenea* var. *aenea* (Fernald) Kük. in H.G.A.Engler (ed.), Pflanzenr., IV, 20(38): 205 (1909). *Carex argyrantha* var. *aenea* (Fernald) B.Boivin, Phytologia 43: 60 (1979).

Carex aenea f. *extrapolata* Butters & Abbe, Rhodora 55: 130 (1953).

Carex aenea f. *flumini-regalis* Butters & Abbe, Rhodora 55: 130 (1953).

Carex foetida All., Fl. Pedem. 2: 265 (1785). *Caricina foetida* (All.) St.-Lag. in A.Cariot, Étude Fl., ed. 8, 2: 878 (1889).

C. & E. Pyrenees, Alps to NW. Balkan Pen. 10 AUT 12 FRA SPA 13 ITA YUG. Hemicr. or rhizome geophyte.

Carex baldensis Vill., Hist. Pl. Dauphiné 2: 196 (1787).

Carex lobata Vill., Hist. Pl. Dauphiné 2: 197 (1787). *Carex foetida* var. *lobata* (Vill.) Nyman, Consp. Fl. Eur.: 782 (1882).

Carex villarsii Willd., Sp. Pl. 4: 220 (1805).

Carex foetida f. *elongata* H.Christ, Bull. Soc. Roy. Bot. Belgique 24(2): 19 (1885).

Carex foetida f. *pallescens* Kneuck., Allg. Bot. Z. Syst. 2: 25 (1896).

Carex fokienensis Dunn, J. Linn. Soc., Bot. 38: 371 (1908).

SE. China. 36 CHS. Hemicr. or rhizome geophyte.

Carex foliabunda A.E.Kozhevn., Byull. Moskovsk. Obshch. Isp. Prir., Otd. Biol. 92: 130 (1987).

Russian Far East. 31+. Hemicr. or rhizome geophyte

Carex foliosa D.Don ex Tilloch & Taylor, Philos. Mag. J. 62: 455 (Dec. 1823). *Carex muricata* var. *foliosa* (D.Don ex Tilloch & Taylor) C.B.Clarke in J.D.Hooker, Fl. Brit. India 6: 703 (1894).

Pakistan to Himalaya. 40 ASS EHM IND NEP PAK WHM. Hemicr. or rhizome geophyte.

Carex wallichiana Spreng., Syst. Veg. 3: 812 (1826). *Vignea wallichiana* (Spreng.) Soják, Cas. Nár. Mus., Odd. Přír. 148: 197 (1979 publ. 1980).

Carex notoleia Nees in R.Wight, Contr. Bot. India: 121 (1834).

Carex muricata var. *indica* Boott, Ill. Carex 4: 193 (1867).

Carex foliosissima F.Schmidt, Reis. Amur-Land., Bot.: 195 (1868).

Sakhalin, Japan. 31 SAK 38 JAP. Hemicr. or rhizome geophyte.

var. ***foliosissima***

Sakhalin, Japan. 31 SAK 38 JAP. Hemicr. or rhizome geophyte.

Carex yesinensis Franch., Bull. Soc. Philom. Paris, VIII, 7: 88 (1895). *Carex morrowii* var. *yesinensis* (Franch.) Kük. in H.G.A.Engler (ed.), Pflanzenr., IV, 20(38): 626 (1909).

Carex crassicaulis Franch., Nouv. Arch. Mus. Hist. Nat., III, 10: 41 (1898).

Carex niigatensis Koidz., Fl. Symb. Orient.-Asiat.: 46 (1930).

var. ***latissima*** (Ohwi) Akiyama, Consp. Caric. Jap.: 203 (1932).

Japan (Honshu). 38 JAP. Hemicr. or rhizome geophyte.

**Carex morrowii* var. *latissima* Ohwi, Mem. Coll. Sci. Kyoto Imp. Univ., Ser. B, Biol. 6: 251 (1931).

var. ***pallidivaginata*** J.Oda & Nagam., Acta Phytotax. Geobot. 54: 29 (2003).

Japan. 38 JAP. Hemicr. or rhizome geophyte.

Carex folliculata L., Sp. Pl.: 978 (1753). *Carex folliculacea* Crantz, Inst. Rei Herb. 1: 405 (1766), nom. illeg.

E. Canada to E. U.S.A. 72 LAB NBR NFL NSC ONT PEI QUE 74 WIS 75 CNT INI MAI MAS MIC NWH NWJ NWY OHI PEN RHO VER WVA 78 DEL GEO MRY NCA SCA TEN VRG WDC. Hemicr. or rhizome geophyte.

Carex xanthophysa Wahlenb., Kongl. Vetensk. Acad. Nya Handl. 1803: 152 (1803). *Carex folliculata* var. *xanthophysa* (Wahlenb.) Muhl., Descr. Gram.: 244 (1817).

Carex folliculata var. *bispicata* Peck, Rep. (Annual) Regents Univ. State New York State Mus. 49: 42 (1896).

Carex foraminata C.B.Clarke, J. Linn. Soc., Bot. 36: 285 (1903).

S. China. 36 CHC CHS. Hemicr. or rhizome geophyte.

Carex foraminatiformis Y.C.Tang & S.Yun Liang, in Fl. Reipubl. Popul. Sin. 12: 521 (2000).

SC. China. 36 CHC. Hemicr.

Carex forficula Franch. & Sav., Enum. Pl. Jap. 2: 557 (1878).

Russian Far East to E. China, Japan. 31+ 36 CHM CHN CHS 38 JAP KOR. Hemicr. or rhizome geophyte.

Carex formosa Dewey, Amer. J. Sci. Arts 8: 98 (1824). *Edritria formosa* (Dewey) Raf., Good Book: 26 (1840).

E. Canada to NC. & NE. U.S.A. 72 ONT QUE 74 ILL IOW MIN NDA WIS 75 MAS MIC NWY OHI VER. Hemicr. or rhizome geophyte.

Carex forrestii Kük., Notes Roy. Bot. Gard. Edinburgh 8: 10 (1913).

Tibet to China (Yunnan). 36 CHC CHT. Hemicr. or rhizome geophyte.

Carex forsteri Wahlenb., Kongl. Vetensk. Acad. Nya Handl. 24: 154 (1803).

New Zealand. 51 NZN NZS. Hemicr. or rhizome geophyte.

**Carex debilis* G.Forst., Fl. Ins. Austr.: 92 (1786), nom. nud.

Carex hamata J.R.Forst. ex Steud., Nomencl. Bot., ed. 2, 1: 290 (1840), nom. inval.

Carex cinnamomea Cheeseman, Trans. & Proc. New Zealand Inst. 14: 301 (1882), nom. illeg.

Carex sexspicata Colenso, Trans. & Proc. New Zealand Inst. 16: 842 (1884).

Carex semiforsteri C.B.Clarke ex Cheeseman, Man. New Zealand Fl.: 836 (1906).

Carex semiforsteri C.B.Clarke, Kew Bull., Addit. Ser. 8: 78 (1908).

Carex fossa G.A.Wheeler, Darwiniana 40: 205 (2002).

S. Brazil to S. South America. 84 BZS 85 AGE AGS AGW PAR URU. Hemicr. or rhizome geophyte.

Carex involucrata Boott, Ill. Carex 2: 76 (1860), nom. illeg. *Carex sororia* var. *involucrata* Kük. in H.G.A. Engler (ed.), Pflanzenr., IV, 20(38): 150 (1909).

Carex fracta Mack., Erythea 8: 38 (1922).

W. U.S.A. 73 ORE WAS 76 CAL. Hemicr. or rhizome geophyte.

Carex fragilis Boott, Ill. Carex 1: 7 (1858).
Sikkim. 40 EHM. Hemicr. or rhizome geophyte.

Carex frankii Kunth, Enum. Pl. 2: 498 (1837).
S. Ontario to Mexico (Coahuila). 72 ONT 74 ILL IOW KAN MSO NEB OKL 75 INI MIC NWJ NWY OHI PEN WVA 78 ALA ARK DEL GEO KTY MRY NCA SCA TEN VRG WDC 79 MXE. Hemicr. or rhizome geophyte.
Carex stenolepis Torr., Amer. J. Sci. Arts 30: 59 (1836), nom. illeg. *Carex shortii* Steud., Syn. Pl. Glumac. 2: 202 (1855), nom. illeg.

Carex fretalis Hamlin, Rec. Domin. Mus. 6: 107 (1968).
New Zealand South I. (incl. Stewart I.). 51 NZS. Hemicr. or rhizome geophyte.

Carex ×fridtzii E.Holmb., Bot. Not. 1929: 21 (1929). *C. brunnescens* × *C. parallela*.
N. Europe. 10 NOR. Hemicr. or rhizome geophyte.

Carex ×friesii Blytt in ?. *C. rostrata* × *C. utriculata*. *Carex vesicaria* subsp. *friesii* (Blytt) Nyman, Consp. Fl. Eur.: 768 (1882). *Carex laevirostris* var. *friesii* (Blytt) Kük. in H.G.A.Engler (ed.), Pflanzenr., IV, 20(38): 724 (1909).
N. Europe. 14 RUN. Hemicr. or rhizome geophyte.

Carex frigida All., Fl. Pedem. 2: 270 (1785).
Mts. of SC. & S. Europe. 11 AUT GER SWI 12 COR FRA SPA 13 ITA YUG. Hemicr. or rhizome geophyte.
Carex helvetica Honck., Syn. Pl. Germ. 1: 375 (1792).
Carex spadicea Schkuhr, Beschr. Riedgräs. 1: 90 (1801), nom. illeg.
Carex geniculata Host, Icon. Descr. Gram. Austriac. 4: 52 (1809). *Carex frigida* var. *geniculata* (Host) Nyman, Consp. Fl. Eur.: 771 (1882).
Carex frigida f. *pyrenaica* H.Christ, Bull. Soc. Roy. Bot. Belgique 24(2): 14 (1885).
Carex frigida f. *flavescens* H.Christ, Bull. Soc. Roy. Bot. Belgique 27(2): 165 (1888).
Carex comari H.Lév. & Vaniot, Bull. Acad. Int. Géogr. Bot. 11: 16 (1902). *Carex frigida* var. *comari* (H.Lév. & Vaniot) Kük. in H.G.A.Engler (ed.), Pflanzenr., IV, 20(38): 556 (1909).
Carex pseudofrigida C.B.Clarke, Kew Bull., Addit. Ser. 8: 81 (1908).
Carex frigida var. *brevirostris* Kük. in H.G.A.Engler (ed.), Pflanzenr., IV, 20(38): 556 (1909).

Carex fritschii Waisb., Verh. K. K. Zool.-Bot. Ges. Wien 44: 51 (1895). *Carex montana* subsp. *fritschii* (Waisb.) O.Schwarz in ?
C. Europe to France and NW. Balkan Pen. 11 AUT CZE GER HUN SWI 12 FRA 13 ITA YUG. Hemicr. or rhizome geophyte.

Carex fructus Reznicek, Contr. Univ. Michigan Herb. 17: 286 (1990).
Mexico (Tamaulipas). 79 MXE. Hemicr. or rhizome geophyte.

Carex fucata Boott ex C.B.Clarke in J.D.Hooker, Fl. Brit. India 6: 710 (1894).
EC. Himalaya. 40 EHM NEP. Hemicr. or rhizome geophyte.
Carex sikkimensis C.B.Clarke in J.D.Hooker, Fl. Brit. India 6: 708 (1894). *Carex fucata* var. *sikkimensis* (C.B.Clarke) Kük. in H.G.A.Engler (ed.), Pflanzenr., IV, 20(38): 344 (1909).

Carex fuliginosa Schkuhr, Beschr. Riedgräs. 1: 91 (1801). *Carex frigida* var. *fuliginosa* (Schkuhr) Kunth in ?
Subarctic and Subalpine Northern Hemisphere. 10 FIN NOR SVA SWE 11 AUT CZE GER 13 BUL ITA ROM YUG 14 RUN UKR 30 KRA WSB YAK 31 KAM KUR MAG 70 ASK GNL NUN NWT YUK 71 ABT BRC MAN 72 QUE 73 COL MNT WYO 76 UTA. Hemicr. or rhizome geophyte.
Carex frigida Wahlenb., Kongl. Vetensk. Acad. Nya Handl. 24: 154 (1803), nom. illeg.
Carex misandra R.Br., Chlor. Melvill.: 283 (1823). *Carex fuliginosa* var. *misandra* (R.Br.) O.Lang, Linnaea 24: 597 (1851). *Carex fuliginosa* subsp. *misandra* (R.Br.) Nyman, Consp. Fl. Eur.: 771 (1882).
Carex frigida var. *huebneri* Hising., Anteckn. 5: 20 (1837).
Carex fuliginosa f. *ochrostachys* Schur, Enum. Pl. Transsilv.: 709 (1866).
Carex misandra var. *elatior* Lange, Consp. Fl. Groenland.: 130 (1881).
Carex immanis C.B.Clarke, J. Linn. Soc., Bot. 36: 290 (1903).
Carex misandra f. *ochrolochin* Ostenf., Meddel. Grønland 43: 10 (1910).
Carex misandra f. *flavida* Fernald, Rhodora 36: 91 (1934).

Carex fulta Franch., Bull. Soc. Philom. Paris, VIII, 7: 28 (1895).
Japan (Honshu). 38 JAP. Hemicr. or rhizome geophyte.

Carex fulvorubescens Hayata, J. Coll. Sci. Imp. Univ. Tokyo 30(1): 383 (1911).
Taiwan. 38 TAI. Hemicr. or rhizome geophyte.

subsp. ***fulvorubescens***
Taiwan. 38 TAI. Hemicr. or rhizome geophyte.

subsp. ***longistipes*** (Hayata) T.Koyama, in Fl. Taiwan 5: 361 (1978).
S. Taiwan. 38 TAI. Hemicr. or rhizome geophyte.
**Carex longistipes* Hayata, Icon. Pl. Formosan. 10: 66 (1921).
Carex tagawana T.Koyama, Acta Phytotax. Geobot. 16: 5 (1955).

Carex funhuangshanica F.T.Wang & Tang ex P.C.Li, Acta Phytotax. Sin. 37: 168 (1999).
China (Guangxi). 36 CHS. Hemicr. or rhizome geophyte.

Carex funingensis Tang & F.T.Wang ex S.Y.Liang, Acta Phytotax. Sin. 36: 534 (1998).
China (Yunnan). 36 CHC. Hemicr. or rhizome geophyte.

Carex fusanensis Ohwi, Acta Phytotax. Geobot. 1: 74 (1932).
Korea. 38 KOR. Hemicr. or rhizome geophyte.

Carex fuscolutea Boeck., Bot. Jahrb. Syst. 7: 278 (1886).
Mexico (San Luis Potosí). 79 MXE. Hemicr. or rhizome geophyte.

Carex fuscula d'Urv., Mém. Soc. Linn. Paris 4: 599 (1826). *Carex inconspicua* var. *fuscula* (d'Urv.) Kük., Bot. Jahrb. Syst. 27: 543 (1899).
Brazil to Falkland Is. (75) pen 84 BZL BZS 85 AGE AGS AGW CLC CLS URU 90 FAL. Hemicr. or rhizome geophyte.

var. ***distenta*** (Kunze ex Kunth) Kük. in H.G.A.Engler (ed.), Pflanzenr., IV, 20(38): 669 (1909).

Brazil to S. South America. (75) pen 84 BZS 85 AGE AGS AGW CLC URU. Hemicr. or rhizome geophyte.
Carex distenta Kunze ex Kunth, Enum. Pl. 2: 449 (1837).
Carex huitensis Steud., Syn. Pl. Glumac. 2: 223 (1855).
Carex inconspicua Steud., Syn. Pl. Glumac. 2: 221 (1855).
Carex pseudopunctata Boeck., Allg. Bot. Z. Syst. 2: 191 (1896).

var. ***fuscula***
SE. Brazil to Falkland Is. 84 BZL BZS 85 AGS AGW CLC CLS 90 FAL. Hemicr. or rhizome geophyte.
Carex fuscula var. *ciliata* Boeck. in ?
Carex indecora Kunth, Enum. Pl. 2: 448 (1837). *Carex inconspicua* f. *indecora* (Kunth) Kük., Bot. Jahrb. Syst. 27: 544 (1899). *Carex fuscula* f. *indecora* (Kunth) Kük. in H.G.A.Engler (ed.), Pflanzenr., IV, 20(38): 669 (1909).
Carex indecora var. *humilis* Boott, Fl. Antarct. 2: 367 (1846).
Carex valdiviana Phil., Linnaea 29: 84 (1857).
Carex brachycephala Poepp. ex Boott, Ill. Carex 1: 67 (1858).
Carex araucana Phil., Anales Univ. Chile 93: 495 (1896).
Carex calbucana Phil., Anales Univ. Chile 93: 499 (1896).
Carex catharinensis Boeck., Allg. Bot. Z. Syst. 2: 191 (1896). *Carex fuscula* subsp. *catharinensis* (Boeck.) Luceño & Alves, Anales Jard. Bot. Madrid 57: 173 (1999).
Carex sagei Phil., Anales Univ. Chile 93: 494 (1896).
Carex inconspicua var. *pseudoextensa* Kük., Bot. Jahrb. Syst. 27: 544 (1899). *Carex fuscula* var. *pseudextensa* (Kük.) Kük. in H.G.A.Engler (ed.), Pflanzenr., IV, 20(38): 669 (1909).
Carex fuscula f. *elatior* Kurtz, Bol. Acad. Nac. Ci. 16: 252 (1900).
Carex incurva var. *humilis* Macloskie, Rep. Princeton Univ. Exped. Patagonia, Botany 8: 283 (1904).
Carex decidua var. *caespitosa* H.Pfeiff., Repert. Spec. Nov. Regni Veg. 17: 31 (1921).

Carex fusiformis Nees in R.Wight, Contr. Bot. India: 128 (1834).
Himalaya to China. 36 CHN CHS 40 EHM NEP WHM. Hemicr. or rhizome geophyte.

var. ***fusiformis***
Himalaya to S. China. 36 CHS 40 EHM NEP WHM. Hemicr. or rhizome geophyte.
Carex borneensis C.B.Clarke, Kew Bull., Addit. Ser. 8: 76 (1908), nom. illeg. *Carex fusiformis* var. *borneensis* Kük. in H.G.A.Engler (ed.), Pflanzenr., IV, 20(38): 598 (1909).
Carex fusiformis var. *austrosinensis* Kük. in H.G.A. Engler (ed.), Pflanzenr., IV, 20(38): 598 (1909).

var. ***maijishanica*** Y.C.Yang, Acta Biol. Plateau Sin. 3: 90 (1984).
China (Gansu). 36 CHN. Hemicr. or rhizome geophyte.

Carex ×fussii Simonk., Enum. Fl. Transsilv.: 548 (1887). *C. elongata* × *C. paniculata*. *Vignea ×fussii* (Simonk.) Soják, Cas. Nár. Mus., Odd. Prír. 148: 195 (1979 publ. 1980).
SE. & E. Europe. 13 ROM 14 UKR. Hemicr. or rhizome geophyte.

Carex garberi Fernald, Rhodora 37: 253 (1935).
Subarctic America to N. & WC. U.S.A. 70 ASK NUN NWT YUK 71 ABT BRC MAN SAS 72 NBR ONT QUE 73 ORE WAS WYO 74 MIN NDA WIS 75 INI MAI MIC NWH NWY OHI PEN VER 76 CAL UTA. Hemicr. or rhizome geophyte.
**Carex aurea* var. *androgyna* Olney in S.Watson, Botany [Fortieth Parallel] 5: 371 (1871).
Carex garberi var. *bifaria* Fernald, Rhodora 37: 255 (1935). *Carex garberi* subsp. *bifaria* (Fernald) Hultén, Ark. Bot., n.s., 7(1): 27 (1968).

Carex gaudichaudiana Kunth, Enum. Pl. 2: 417 (1837). *Carex vulgaris* var. *gaudichaudiana* (Kunth) Boott, Ill. Carex 4: 169 (1867). *Carex gaudichaudiana* var. *typica* Domin, Biblioth. Bot. 89: 494 (1915), nom. inval.
Australia to New Zealand. 50 NSW QLD SOA TAS VIC 51 NZN NZS. Hemicr. or rhizome geophyte.
Carex semiplena var. *tenuinervis* Kük., Bull. Herb. Boissier, II, 2: 1018 (1902). *Carex micrantha* var. *tenuinervis* (Kük.) Kük. in H.G.A.Engler (ed.), Pflanzenr., IV, 20(38): 341 (1909).
Carex lacerans Kük. in H.G.A.Engler (ed.), Pflanzenr., IV, 20(38): 326 (1909).
Carex vulpi-caudata Akiyama, J. Fac. Sci. Hokkaido Univ., Ser. 5, Bot. 1: 63 (1931).

Carex ×gaudiniana Guthnick, Flora 15: 241 (1832). *C. dioica* × *C. echinata*. *Carex dioica* var. *gaudiniana* (Guthnick) Sendtn., Flora 34: 737 (1851). *Carex ×microstachya* var. *gaudiniana* (Guthnick) Trevir. in C.F.von Ledebour, Fl. Ross. 4: 285 (1852). *Vignea ×gaudiniana* (Guthnick) Pritz., Icon. Bot. Index: t. 1161 (1855). *Carex ×microstachya* subsp. *gaudiniana* (Guthnick) Nyman, Consp. Fl. Eur., Suppl. 2: 324 (1890). *Vignea ×gaudiniana* (Guthnick) Soják, Cas. Nár. Mus., Odd. Prír. 148: 195 (1979 publ. 1980).
C. Europe. 11 GER SWI. Hemicr. or rhizome geophyte.

Carex gayana Steud., Nomencl. Bot., ed. 2, 1: 290 (1840).
NW. Spain to N. Portugal. 12 POR SPA. Hemicr. or rhizome geophyte.
**Carex leiocarpa* J.Gay, Ann. Sci. Nat., Bot., II, 6: 351 (1836), nom. illeg.
Carex asturica Boiss., Pugill. Pl. Afr. Bor. Hispan.: 117 (1852).
Carex gallaecica H.Lév. & Vaniot, Bull. Acad. Int. Géogr. Bot. 12: 96 (1903).

Carex gemella Hochst. ex Steud., Syn. Pl. Glumac. 2: 214 (1855).
India (Nilgiri Hills). 40 IND. Hemicr. or rhizome geophyte.

Carex geminata Schkuhr, Beschr. Riedgräs. 1: 65 (1801). *Olotrema geminata* (Schkuhr) Raf., Good Book: 25 (1840).
New Zealand. 51 NZN NZS. Hemicr. or rhizome geophyte.

Carex genkaiensis Ohwi, Acta Phytotax. Geobot. 7: 34 (1938).
S. Korea, Japan (Kyushu). 38 JAP KOR. Hemicr. or rhizome geophyte.
Carex mitrata var. *vigens* Kük. in H.G.A.Engler (ed.), Pflanzenr., IV, 20(38): 471 (1909).

Carex gentilis Franch., Bull. Soc. Philom. Paris, VIII, 7: 84 (1895).
Tibet to S. China. 36 CHC CHS CHT. Hemicr. or rhizome geophyte.
Carex tonnerrei C.B.Clarke, J. Linn. Soc., Bot. 36: 314 (1904). *Carex gentilis* var. *tonnerrei* (C.B.Clarke) Kük. in H.G.A.Engler (ed.), Pflanzenr., IV, 20(38): 603 (1909).
Carex gentilis var. *intermedia* F.T.Wang & Tang ex Y.C.Yang, in Fl. Xizangica 5: 408 (1987).
Carex gentilis var. *intermedia* Tang & F.T.Wang ex L.K.Dai, Acta Phytotax. Sin. 37: 187 (1999), nom. illeg.
Carex gentilis var. *macrocarpa* Tang & F.T.Wang ex L.K.Dai, Acta Phytotax. Sin. 37: 188 (1999).

Carex geophila Mack., Bull. Torrey Bot. Club 40: 546 (1913).
Colorado to Guatemala. 73 COL 76 ARI CAL 77 NWM TEX 79 MXC MXG MXS 80 GUA. Hemicr. or rhizome geophyte.

Carex* ×*gerhaldtii Figert, Deutsche Bot. Monatsschr. 4: 153 (1886). *C. echinata* × *C. remota*. *Vignea* ×*gerhaldtii* (Figert) Soják, Cas. Nár. Mus., Odd. Prír. 148: 195 (1979 publ. 1980).
Europe. 11 GER 14+. Hemicr. or rhizome geophyte.
Carex × *vierhapperi* Beck in ?

Carex* ×*germanica K.Richt., Pl. Eur. 1: 169 (1890). *C. diandra* × *C. paniculata*. *Vignea* ×*germanica* (K. Richt.) Soják, Cas. Nár. Mus., Odd. Prír. 148: 195 (1979 publ. 1980).
C. Europe. 11 GER. Hemicr. or rhizome geophyte.

Carex geyeri Boott, Trans. Linn. Soc. London 20: 118 (1846).
SW. Canada to W. U.S.A. 71 ABT BRC 73 COL IDA MNT ORE WAS WYO (75) pen 76 NEV UTA. Hemicr. or rhizome geophyte.

Carex gholsonii Naczi & Cochrane, Novon 12: 524 (2002).
SE. U.S.A. 78 ALA FLA GEO NCA SCA. Hemicr. or rhizome geophyte.

Carex gibba Wahlenb., Kongl. Vetensk. Acad. Nya Handl. 24: 148 (1803).
Korea to Vietnam, Japan. 36 CHC CHM CHN CHS 38 JAP KOR 41 VIE. Hemicr. or rhizome geophyte.
Carex pteroloma Kunze ex Steud., Syn. Pl. Glumac. 2: 242 (1855).
Carex anomala Boott in M.C.Perry, Jap. Exped. 2: 327 (1856), nom. illeg.
Carex leucocarpa Boeck., Flora 39: 227 (1856).
Carex alta var. *brevior* H.Lév. & Vaniot, Bull. Acad. Int. Géogr. Bot. 10: 126 (1901).

Carex gibbsiae Rendle, J. Linn. Soc., Bot. 39: 180 (1909).
Fiji. 60 FIJ. Hemicr. or rhizome geophyte.
Carex vitiensis H.St.John, Pacific Sci. 1: 116 (1947).

Carex gibertii G.A.Wheeler, Sida 17: 475 (1996).
S. Brazil to Argentina (Buenos Aires). 84 BZS 85 AGE URU. Hemicr. or rhizome geophyte.
**Carex phalaroides* var. *crassiflora* Kük., Verh. Bot. Vereins Prov. Brandenburg 47: 209 (1905 publ. 1906).
Carex phalaroides subsp. *crassiflora* (Kük.) Luceño & Alves, Anales Jard. Bot. Madrid 57: 174 (1999).

Carex gifuensis Franch., Bull. Soc. Philom. Paris, VIII, 7: 47 (1895).
Japan (Honshu). 38 JAP. Hemicr. or rhizome geophyte.
Carex argyrostachys H.Lév. & Vaniot, Bull. Acad. Int. Géogr. Bot. 11: 27 (1902). *Carex gifuensis* f. *argyrostachys* (H.Lév. & Vaniot) Kük. in H.G.A. Engler (ed.), Pflanzenr., IV, 20(38): 443 (1909).

Carex gigantea Rudge, Trans. Linn. Soc. London 7: 99 (1804). *Carex lacustris* var. *gigantea* (Rudge) Pursh, Fl. Amer. Sept. 1: 45 (1813). *Carex lupulina* var. *gigantea* (Rudge) Britton, Mem. Torrey Bot. Club 5: 84 (1894).
Illinois to SE. U.S.A. 74 ILL MSO OKL 75 INI 77 TEX 78 ALA ARK DEL FLA GEO KTY LOU MRY MSI NCA SCA TEN VRG. Hemicr. or rhizome geophyte.
Carex grandis L.H.Bailey, Mem. Torrey Bot. Club 1: 13 (1889). *Carex gigantea* var. *grandis* (L.H.Bailey) Farw., Rhodora 23: 87 (1921).
Carex grandis var. *helleri* L.H.Bailey, Bot. Gaz. 21: 7 (1896).
Carex gigantea f. *minor* Farw., Rhodora 23: 87 (1921).
Carex gigantea f. *pedunculata* Farw., Rhodora 23: 87 (1921).

Carex* ×*ginsiensis Waisb., Magyar Bot. Lapok 4: 62, 74 (1905). *C. montana* × *C. pilulifera*.
C. Europe. 11 HUN. Hemicr. or rhizome geophyte.
Carex fritschii var. *oxystachya* Waisb., Oesterr. Bot. Z. 47: 5 (1897).

Carex giraldiana Kük., Bot. Jahrb. Syst. 36(82): 10 (1905).
NC. China. 36 CHN. Hemicr. or rhizome geophyte.

Carex giraudiasii H.Lév., Repert. Spec. Nov. Regni Veg. 12: 288 (1913).
China (Yunnan). 36 CHC. Hemicr. or rhizome geophyte.

Carex glabrescens (Kük.) Ohwi, Mem. Coll. Sci. Kyoto Imp. Univ., Ser. B, Biol. 6: 245 (1931).
Russian Far East to Korea. 31+ 36 CHM 38 KOR. Hemicr. or rhizome geophyte.
Carex drymophila var. *pilifera* Kük. in H.G.A.Engler (ed.), Pflanzenr., IV, 20(38): 756 (1909).
**Carex wallichiana* f. *glabrescens* Kük. in H.G.A. Engler (ed.), Pflanzenr., IV, 20(38): 749 (1909).

Carex glacialis Mack., Bull. Torrey Bot. Club 37: 244 (1910).
N. Europe to Russian Far East, Alaska to Canada. 10 FIN ICE NOR SWE 12 FRA 14 RUE RUN 30 KRA WSB YAK 31 MAG 70 ASK GNL NUN NWT YUK 71 ABT BRC MAN SAS 72 LAB NFL ONT QUE 73 MNT. Hemicr. or rhizome geophyte.
Carex pedata Wahlenb., Fl. Lapp.: 239 (1812), nom. illeg.
Carex terrae-novae Fernald, Rhodora 44: 290 (1942).
Carex glacialis var. *terrae-novae* (Fernald) Boivin, Naturaliste Canad. 94: 522 (1967).

Carex glareosa Schkuhr ex Wahlenb., Kongl. Vetensk. Acad. Nya Handl. 24: 146 (1803). *Vignea glareosa* (Schkuhr ex Wahlenb.) Rchb. in J.C.Mössler, Handb. Gewächsk. ed. 2, 3: 1626 (1830). *Neskiza glareosa* (Schkuhr ex Wahlenb.) Raf., Good Book: 27 (1840). *Carex lachenalii* var. *glareosa* (Schkuhr ex Wahlenb.) Gelting, Meddel. Grønland 101(2): 162 (1934). *Carex bipartita* var. *glareosa* (Schkuhr ex Wahlenb.) Polunin, Bull. Natl. Mus. Canada 92: 114 (1940).
N. Europe to Sakhalin, Alaska to Canada. 10 FIN ICE NOR SVA SWE 14 BLT RUN RUW 30 WSB YAK

31 KAM MAG SAK 70 ALU ASK GNL NUN NWT YUK 71 BRC MAN 72 NBR NFL ONT PEI QUE. Hemicr. or rhizome geophyte.

subsp. ***glareosa***

N. Europe to Sakhalin, Alaska to Canada. 10 FIN ICE NOR SVA SWE 14 BLT RUN RUW 30 WSB YAK 31 KAM MAG SAK 70 ALU ASK GNL NUN NWT YUK 71 BRC MAN 72 NBR NFL ONT PEI QUE. Hemicr. or rhizome geophyte.

Carex heleonastes Lilj., Kongl. Vetensk. Acad. Nya Handl. 1793: 73 (1793), nom. illeg.

Carex cryptantha T.Holm, Amer. J. Sci., IV, 10: 267 (1900).

Carex glareosa var. *amphigena* Fernald, Rhodora 8: 47 (1906). *Carex amphigena* (Fernald) Mack., Bull. Torrey Bot. Club 37: 246 (1910). *Carex bipartita* var. *amphigena* (Fernald) Polunin, Bull. Natl. Mus. Canada 92: 115 (1940). *Carex glareosa* subsp. *amphigena* (Fernald) Á.Löve, Taxon 19: 300 (1970).

Carex soriofkensis H.Lév. & Vaniot, Bull. Acad. Int. Géogr. Bot. 19: 35 (1909). *Carex glareosa* subsp. *soriofkensis* (H.Lév. & Vaniot) McKean, Notes Roy. Bot. Gard. Edinburgh 46: 116 (1989).

Carex chishimana Ohwi, J. Jap. Bot. 11: 406 (1935).

Carex ushishirensis Ohwi, Acta Phytotax. Geobot. 4: 63 (1935).

subsp. ***pribylovensis*** (Macoun) G.Halliday & Chater, Feddes Repert. 80: 100 (1969).

Aleutian Is. to Alaska (Pribilov Is.). 31 MAG? 70 ALU ASK. Hemicr. or rhizome geophyte.

**Carex pribylovensis* Macoun in D.J.Jordan & al., Fur Seals Isl. North Pacif. Ocean 3: 572 (1899). *Carex lagopina* var. *pribylovensis* (Macoun) Kük. in H.G.A.Engler (ed.), Pflanzenr., IV, 20(38): 214 (1909). *Vignea pribylovensis* (Macoun) Soják, Cas. Nár. Mus., Odd. Prír. 148: 197 (1979 publ. 1980).

Carex glaucescens Elliott, Sketch Bot. S. Carolina 2: 553 (1824). *Edritria glaucescens* (Elliott) Raf., Good Book: 26 (1840).

SE. U.S.A. to Texas. 77 TEX 78 ALA FLA GEO LOU MSI NCA SCA TEN VRG. Hemicr. or rhizome geophyte.

Carex sempervirens Schwein., Ann. Lyceum Nat. Hist. New York 1: 70 (1824), nom. illeg.

Carex inundata Willd. ex Kunth, Enum. Pl. 2: 492 (1837).

Carex rufidula Steud., Syn. Pl. Glumac. 2: 220 (1855).

Carex glandulosa Elliott ex Boott, Ill. Carex 2: 89 (1860).

Carex glauca Bosc ex Boott, Ill. Carex 2: 89 (1860).

Carex glaucescens var. *polystachya* Boott, Ill. Carex 2: 89 (1860).

Carex brasiliensis var. *gracilis* Boeck., Linnaea 41: 292 (1877).

Carex glauciformis Meinsh., Trudy Imp. S.-Peterburgsk. Bot. Sada 18: 389 (1900).

Siberia to Korea. 30+ 31+ 36 CHI CHM 38 KOR. Hemicr. or rhizome geophyte.

Carex glaucodea Tuck. ex Olney, Proc. Amer. Acad. Arts 7: 395 (1868). *Carex flaccosperma* var. *glaucodea* (Tuck. ex Olney) Kük. in H.G.A.Engler (ed.), Pflanzenr., IV, 20(38): 518 (1909).

SE. Canada to C. & E. U.S.A. 72 ONT 74 ILL MSO OKL 75 CNT INI MAS NWH NWJ NWY OHI PEN VER WVA 77 TEX 78 ALA ARK DEL KTY LOU MRY MSI NCA SCA TEN VRG WDC. Hemicr. or rhizome geophyte.

Carex globistylosa P.C.Li, Acta Bot. Yunnan. 12: 144 (1990).

China (Sichuan). 36 CHC. Hemicr. or rhizome geophyte.

Carex globosa Boott, Proc. Linn. Soc. London 1: 259 (1845). *Carex umbellata* var. *globosa* (Boott) Kük. in H.G.A.Engler (ed.), Pflanzenr., IV, 20(38): 453 (1909).

California to Mexico (Baja California). 76 CAL 79 MXN. Hemicr. or rhizome geophyte.

Carex globularis L., Sp. Pl.: 976 (1753). *Onkerma globularis* (L.) Raf., Good Book: 27 (1840).

N. Europe to Russian Far East. 10 FIN NOR SWE 11 POL 14 BLR BLT RUC RUE RUN RUW 30 IRK KRA WSB YAK 31 KAM KHA MAG SAK 36 CHI CHM 37 MON 38 KOR. Hemicr. or rhizome geophyte.

Carex tomentosa Hornem. in G.C.Oeder & al. (eds.), Fl. Dan.: t. 1404 (1810), nom. illeg.

Carex oligogyna Less. ex Kunth, Enum. Pl. 2: 442 (1837).

Carex mitsuriokensis H.Lév. & Vaniot, Bull. Acad. Int. Géogr. Bot. 19: 33 (1909).

Carex glomerabilis V.I.Krecz., Bot. Mater. Gerb. Bot. Inst. Komarova Akad. Nauk S.S.S.R. 7: 34 (1937).

S. Africa. 27 CPP LES NAT OFS TVL. Hemicr. or rhizome geophyte.

**Carex glomerata* Thunb., Prodr. Pl. Cap.: 14 (1794), nom. illeg.

Carex glossostigma Hand.-Mazz., Anz. Akad. Wiss. Wien, Math.-Naturwiss. Kl. 59: 140 (1922).

SE. China. 36 CHS. Hemicr. or rhizome geophyte.

Carex exerta C.H.Chu, Contr. Biol. Lab. Chin. Assoc. Advancem. Sci., Sect. Bot. 10: 213 (1938), nom. illeg. *Carex mandarinorum* Raymond, Mém. Jard. Bot. Montréal 53: 69 (1959).

Carex ferruspiculata C.H.Chu, Contr. Biol. Lab. Chin. Assoc. Advancem. Sci., Sect. Bot. 10: 215 (1938).

Carex dayunshanensis L.K.Ling & Y.Z.Huang, Fl. Fujianica 6: 689 (1995).

Carex gmelinii Hook. & Arn., Bot. Beechey Voy.: 118 (1832).

Russian Far East to Japan, Alaska to W. Canada. 31 KAM KHA KUR MAG PRM SAK 36 CHM 38 JAP KOR 70 ALU ASK 71 BRC. Hemicr. or rhizome geophyte.

Carex acrolepis Ledeb., Denkschr. Königl.-Baier. Bot. Ges. Regensburg 3: 56 (1841).

Carex laticuspis Franch., Bull. Soc. Philom. Paris, VIII, 7: 58 (1895).

Carex binnensis Kneuck., Allg. Bot. Z. Syst. 5: 195 (1899).

Carex godfreyi Naczi, Contr. Univ. Michigan Herb. 19: 200 (1993).

SE. U.S.A. 78 ALA FLA GEO NCA SCA. Hemicr. or rhizome geophyte.

Carex goligongshanensis P.C.Li, Acta Bot. Yunnan. 12: 140 (1990).

China (Yunnan). 36 CHC. Hemicr. or rhizome geophyte.

Carex gonggaensis P.C.Li, Acta Bot. Yunnan. 12: 147 (1990).
China (Sichuan). 36 CHC. Hemicr. or rhizome geophyte.

Carex gongshanensis Tang & F.T.Wang ex Y.C.Yang, in Fl. Xizangica 5: 404 (1987).
Tibet to China (Yunnan). 36 CHC CHT. Hemicr. or rhizome geophyte.

Carex gonochorica Cherm., Bull. Soc. Bot. France 72: 618 (1925).
WC. Madagascar. 29 MDG. Hemicr. or rhizome geophyte.

Carex gotoi Ohwi, Mem. Coll. Sci. Kyoto Imp. Univ., Ser. B, Biol. 5(3): 248 (1930).
Siberia to Korea. 30+ 31+ 36 CHI CHM CHN 37 MON 38 KOR. Hemicr. or rhizome geophyte.
Carex sukaczovii V.I.Krecz., Fl. Transbaic. 2: 136 (1931).

Carex goyenii Petrie, Trans. & Proc. New Zealand Inst. 14: 363 (1882).
New Zealand South I. 51 NZS. Hemicr. or rhizome geophyte.

Carex gracilenta Boott ex Boeck., Trudy Bot. Muz. Imp. Akad. Nauk 7: 95 (1910). *Carex alpina* var. *gracilenta* (Boott ex Boeck.) C.B.Clarke in J.D.Hooker, Fl. Brit. India 6: 730 (1894). *Carex infuscata* var. *gracilenta* (Boott ex Boeck.) P.C.Li, in Vasc. Pl. Hengduan Mt. 2: 2373 (1994).
E. Himalaya. 40 EHM. Hemicr. or rhizome geophyte.

Carex graciliflora Dunn, J. Linn. Soc., Bot. 38: 372 (1908).
China. 36+. Hemicr. or rhizome geophyte.

Carex gracilior Mack., Bull. Torrey Bot. Club 43: 614 (1916 publ. 1917).
California. 76 CAL. Hemicr. or rhizome geophyte.

Carex gracilispica Hayata, Icon. Pl. Formosan. 10: 62 (1921).
Taiwan to Pen. Malaysia. 38 TAI 41 VIE 42 MLY. Hemicr. or rhizome geophyte.
Carex malayana Nelmes, Kew Bull. 5: 209 (1950).

Carex gracillima Schwein., Ann. Lyceum Nat. Hist. New York 1: 66 (1824). *Loxanisa gracillima* (Schwein.) Raf., Good Book: 26 (1840).
C. & E. Canada, NC. & E. U.S.A. 71 MAN 72 LAB NBR NFL NSC ONT PEI QUE 74 ILL IOW MIN MSO NDA WIS 75 CNT INI MAI MAS MIC NWH NWJ NWY OHI PEN RHO VER WVA 78 ALA ARK DEL GEO KTY NCA SCA TEN VRG. Hemicr. or rhizome geophyte.
Carex digitalis Schwein. & Torr., Ann. Lyceum Nat. Hist. New York 1: 324 (1824), nom. illeg.
Carex gracillima f. *humilis* L.H.Bailey, Mem. Torrey Bot. Club 1: 71 (1889). *Carex gracillima* var. *humilis* (L.H.Bailey) L.H.Bailey, Mem. Torrey Bot. Club 1: 71 (1889).
Carex gracillima var. *macerrima* Fernald & Wiegand, Rhodora 15: 133 (1913).

Carex graeffeana Boeck., Flora 58: 123 (1875).
Malesia to S. Pacific. 42 BOR JAW LSI PHI 43 BIS NWG 60 FIJ SAM 61 COO. Hemicr. or rhizome geophyte.
Carex rechingeri Palla, Oesterr. Bot. Z. 57: 424 (1907).
Carex exploratorum Nelmes, Bull. Misc. Inform. Kew 1938: 108 (1938).
Carex graeffeana var. *samoensis* Nelmes, Bull. Misc. Inform. Kew 1938: 110 (1938).
Carex philippinensis Nelmes, Bull. Misc. Inform. Kew 1938: 109 (1938).
Carex pandanus Ohwi, Bot. Mag. (Tokyo) 56: 214 (1942).
Carex euphlebia S.T.Blake, J. Arnold Arbor. 28: 108 (1947).

Carex grallatoria Maxim., Bull. Acad. Imp. Sci. Saint-Pétersbourg 31: 113 (1887).
C. & S. Japan, Taiwan. 38 JAP TAI. Hemicr. or rhizome geophyte.

var. ***grallatoria***
C. & S. Japan. 38 JAP. Hemicr. or rhizome geophyte.

var. ***heteroclita*** (Franch.) Kük. in H.G.A.Engler (ed.), Pflanzenr., IV, 20(38): 89 (1909).
C. & S. Japan, Taiwan. 38 JAP TAI. Hemicr. or rhizome geophyte.
**Carex heteroclita* Franch., Bull. Soc. Philom. Paris, VIII, 7: 28 (1895). *Carex grallatoria* subsp. *heteroclita* (Franch.) T.Koyama, in Fl. Taiwan 5: 369 (1978).
Carex terniflora Maxim. ex Matsum., Index Pl. Jap. 2: 136 (1905).

Carex graminiculmis T.Koyama, Jap. J. Bot. 15: 170 (1956).
China (Shanxi). 36 CHN. Hemicr. or rhizome geophyte.

Carex graminifolia Cherm., Bull. Soc. Bot. France 70: 409 (1923).
Madagascar. 29 MDG. Hemicr. or rhizome geophyte.

Carex grandiligulata Kük., Bot. Jahrb. Syst. 36(82): 9 (1905).
C. & S. China. 36 CHC CHN CHS. Hemicr. or rhizome geophyte.

Carex granifera Dunn, J. Linn. Soc., Bot. 38: 372 (1908).
China. 36+. Hemicr. or rhizome geophyte.

Carex ×grantii A.Benn., J. Bot. 35: 250 (1897). *C. aquatilis* × *C. recta*. *Carex halophila* var. *grantii* (A.Benn.) B.Boivin, Provancheria 25: 178 (1992).
N. Europe, E. Canada. 10 GRB 72 LAB ONT QUE. Hemicr. or rhizome geophyte.

Carex granularis Muhl. ex Willd., Sp. Pl. 4: 279 (1805). *Deweya granularis* (Muhl. ex Willd.) Raf., Good Book: 26 (1840).
C. & E. Canada to C. & E. U.S.A. 71 MAN SAS 72 NBR NSC ONT QUE 73 WYO 74 ILL IOW KAN MIN MSO NDA NEB OKL SDA WIS 75 CNT INI MAI MAS MIC NWH NWJ NWY OHI PEN VER WVA 77 TEX 78 ALA ARK DEL FLA GEO KTY LOU MRY MSI NCA SCA TEN VRG WDC. Hemicr. or rhizome geophyte.
Carex chalaros Steud., Syn. Pl. Glumac. 2: 231 (1855).
Carex granularis var. *recta* Dewey ex Alph.Wood, Class-book Bot.: 763 (1860).
Carex haleana Olney, Car. Bor. Am.: 6 (1871). *Carex granularis* var. *haleana* (Olney) Porter, Proc. Acad. Nat. Sci. Philadelphia 1887: 74 (1888). *Carex granularis* var. *shriveri* Britton in N.L.Britton & A.Brown, Ill. Fl. N. U.S. 1: 322 (1896).
Carex shriveri (Britton) Britton, Man. Fl. N. States: 208 (1901).
Carex rectior Mack. in N.L.Britton & al. (eds.), N. Amer. Fl. 18: 261 (1935).

Carex gravida L.H.Bailey, Mem. Torrey Bot. Club 1: 5 (1889).
S. Canada to N. U.S.A. 71 MAN SAS 72 ONT 73 COL MNT WYO 74 ILL IOW KAN MIN MSO NDA NEB OKL SDA WIS 75 INI MIC NWY OHI PEN 77 NWM TEX 78 ARK KTY nca TEN VRG. Hemicr. or rhizome geophyte.
Carex gravida var. *laxifolia* L.H.Bailey, Mem. Torrey Bot. Club 1: 6 (1889).
Carex gravida f. *laxifolia* (L.H.Bailey) Kük. in H.G.A. Engler (ed.), Pflanzenr., IV, 20(38): 157 (1909).
Carex lunelliana Mack., Bull. Torrey Bot. Club 42: 615 (1915). *Carex gravida* var. *lunelliana* (Mack.) F.J. Herm., Amer. Midl. Naturalist 17: 855 (1936).

Carex grayi Carey in A.Gray, Manual: 563 (1848).
E. Canada to NC. & E. U.S.A. 72 ONT QUE 74 ILL IOW KAN MIN MSO OKL WIS 75 CNT INI MAS MIC NWY OHI PEN VER 78 ALA ARK DEL FLA GEO KTY MRY MSI NCA SCA TEN VRG WDC. Hemicr. or rhizome geophyte.
**Carex intumescens* var. *globularis* A.Gray, Ann. Lyceum Nat. Hist. New York 3: 236 (1835).
Carex grayi var. *hispidula* A.Gray ex L.H.Bailey, Mem. Torrey Bot. Club 1: 54 (1889). *Carex asa-grayi* var. *hispidula* (A.Gray ex L.H.Bailey) L.H.Bailey, Bull. Torrey Bot. Club 20: 427 (1893).
Carex asa-grayi L.H.Bailey, Bull. Torrey Bot. Club 20: 427 (1893).
Carex grayi var. *rariflora* Farw., Rep. (Annual) Michigan Acad. Sci. 22: 181 (1921).

Carex griersonii Noltie, Edinburgh J. Bot. 50: 195 (1993).
Himalaya. 40 EHM NEP WHM. Hemicr. or rhizome geophyte.

Carex grioletii Roem. ex Schkuhr, Beschr. Riedgräs. 2: 76 (1806). *Carex virescens* var. *grioletii* (Roem. ex Schkuhr) Gay, Bull. Soc. Bot. France 4: 165 (1857). *Carex tomentosa* var. *grioletii* (Roem. ex Schkuhr) Boeck., Linnaea 41: 205 (1877).
Spain to Iran. 12 COR fra SPA 13 ITA SIC YUG 14 KRY? 33 NCS TCS 34 IRN TUR. Hemicr. or rhizome geophyte.
**Carex grisea* Viv., Ann. di Bot. 1(2): 186 (1804), nom. illeg.
Carex anceps Willd. ex Kunth, Enum. Pl. 2: 457 (1837).

Carex grisea Wahlenb., Kongl. Vetensk. Acad. Nya Handl. 24: 154 (1803). *Manochlaenia grisea* (Wahlenb.) Fedde & J.Schust., Just's Bot. Jahresber. 41(2): 11 (1913 publ. 1918).
E. Canada to C. & E. U.S.A. 72 NBR ONT QUE 74 ILL IOW KAN MIN MSO NEB OKL SDA WIS 75 CNT MAI MAS MIC NWH NWJ NWY OHI PEN VER WVA 77 TEX 78 ARK DEL KTY MRY MSI TEN VRG WDC. Hemicr. or rhizome geophyte.
Carex amphibola var. *turgida* Fernald, Rhodora 44: 311 (1942). *Carex turgida* (Fernald) J.W.Moore, Rhodora 59: 8 (1957).

Carex* × *groenlandica Lange, Consp. Fl. Groenland., ed. 2: 144 (1890). *C. bigelowii* × *C. nigra.*
Greenland. 70 GNL. Hemicr. or rhizome geophyte.

Carex* × *grossii Fiek in P.F.A.Ascherson & K.O.R. Graebner, Syn. Mitteleur. Fl. 1: 228 (1897). *C. hirta* × *C. vesicaria.*
Europe. 11+. Hemicr. or rhizome geophyte.

Carex guatemalensis F.J.Herm., Brittonia 23: 145 (1971).
Guatemala, Costa Rica. 80 COS GUA. Hemicr. or rhizome geophyte.

Carex guffroyi H.Lév. & Perrier, Monde Pl. 19: 15 (1918).
Madagascar. 29 MDG. Hemicr. or rhizome geophyte.

Carex gunniana Boott, Proc. Linn. Soc. London 1: 258 (1845).
E. & SE. Australia. 50 QLD SOA TAS VIC. Hemicr. or rhizome geophyte.
Carex gunniana var. *brevior* Kük. in H.G.A.Engler (ed.), Pflanzenr., IV, 20(38): 663 (1909).

Carex gynaecandra H.Pfeiff., Repert. Spec. Nov. Regni Veg. 17: 31 (1921).
S. Brazil. 84 BZS. Hemicr. or rhizome geophyte.

Carex gynandra Schwein., Ann. Lyceum Nat. Hist. New York 1: 70 (1824). *Carex crinita* var. *gynandra* (Schwein.) Schwein. & Torr., Ann. Lyceum Nat. Hist. New York 1: 360 (1826).
E. Canada to NC. & E. U.S.A. 70 NUN 72 NBR NSC ONT PEI QUE 74 MIN WIS 75 CNT MAI MAS MIC NWH NWJ NWY OHI PEN RHO VER WVA 78 GEO KTY MRY NCA SCA TEN VRG. Hemicr. or rhizome geophyte.
Carex crinita var. *simulans* Fernald, Rhodora 4: 219 (1902). *Carex gynandra* var. *simulans* (Fernald) Roland, Proc. & Trans. Nova Scotian Inst. Sci. 21: 234 (1947).

Carex gynocrates Wormsk. in S.T.N.Drejer, Rev. Car. Bot.: 16 (1841). *Carex dioica* subsp. *gynocrates* (Wormsk.) Hultén in ?. *Carex chordorrhiza* subsp. *gynocrates* (Wormsk.) Nyman, Consp. Fl. Eur.: 782 (1882). *Carex dioica* var. *gynocrates* (Wormsk.) Ostenf., Fl. Arct.: 60 (1902). *Vignea gynocrates* (Wormsk.) Soják, Cas. Nár. Mus., Odd. Prír. 148: 195 (1979 publ. 1980).
Siberia to Japan, Alaska to N. & WC. U.S.A. 30 KRA YAK 31 KAM KUR MAG SAK 36 CHM 38 JAP 70 ALU ASK GNL NUN NWT YUK 71 ABT BRC MAN SAS 72 LAB NBR NFL NSC ONT QUE 73 COL IDA MNT ORE WAS WYO 74 MIN NDA WIS 75 MAI MIC NWY PEN 76 NEV UTA. Hemicr. or rhizome geophyte.
Carex alascana Boeck., Bot. Jahrb. Syst. 7: 277 (1885).
Carex gynocrates var. *substaminata* Peck, Rep. (Annual) Regents Univ. State New York State Mus. 48: 148 (1896).
Carex cobresiiformis A.I.Baranov & Skvortsov, Quart. J. Taiwan Mus. 18: 224 (1965).

Carex gynodynama Olney, Proc. Amer. Acad. Arts 7: 394 (1868).
Oregon to California. 73 ORE 76 CAL. Hemicr. or rhizome geophyte.
Carex blankinshipii Fernald, Erythea 7: 121 (1899).

Carex hachijoensis Akiyama, J. Jap. Bot. 13: 645 (1937).
Japan (Hachijo I.). 38 JAP. Hemicr. or rhizome geophyte.

Carex* × *haematolepis Drejer in G.C.Oeder & al. (eds.), Fl. Dan.: t. 2370 (1843). *C. bigelowii* × *C. lyngbyei. Carex cryptocarpa* var. *haematolepis* (Drejer) L.H. Bailey, Mem. Torrey Bot. Club 1: 27 (1889). *Carex salina* var. *haematolepis* (Drejer) Almq., Bot. Not. 1891: 127 (1891).
Greenland, N. Europe. 10 FOR ICE 70 GNL. Hemicr. or rhizome geophyte.

Carex haematorrhyncha Ohwi & T.Koyama, Bull. Natl. Sci. Mus. Tokyo, n.s., 3(38): 21 (1956).
SE. China. 36 CHS. Hemicr. or rhizome geophyte.

Carex haematosaccus C.B.Clarke, J. Linn. Soc., Bot. 34: 297 (1899). *Carex renschiana* var. *haematosaccus* (C.B.Clarke) Kük. in H.G.A.Engler (ed.), Pflanzenr., IV, 20(38): 279 (1909).
Madagascar. 29 MDG. Hemicr. or rhizome geophyte.

Carex haematostoma Nees in R.Wight, Contr. Bot. India: 125 (1834).
Himalaya to China. 32 KAZ KGZ TZK UZB 34 AFG 36 CHC CHQ CHT CHX 40 EHM NEP PAK WHM. Hemicr. or rhizome geophyte.
Carex haematostoma var. *submacrogyna* Kük. in H.G.A. Engler (ed.), Pflanzenr., IV, 20(38): 561 (1909).
Carex alexeenkoana Litv., Trudy Bot. Muz. Imp. Akad. Nauk 7: 98 (1910).
Carex kenkolensis Litv., Trudy Bot. Muz. Imp. Akad. Nauk 7: 88 (1910).
Carex nakaoana T.Koyama, Acta Phytotax. Geobot. 15: 112 (1954).
Carex bhutanica T.Koyama, Naturaliste Canad. 88: 252 (1961).

Carex hakkodensis Franch., Bull. Soc. Philom. Paris, VIII, 7: 28 (1895).
E. China, Kuril Is., N. & C. Japan. 31 KUR 36 CHS 38 JAP. Hemicr. or rhizome geophyte.
Carex rhizopoda Maxim., Bull. Acad. Imp. Sci. Saint-Pétersbourg 31: 114 (1887).
Carex rhizopoda f. *longior* Maxim., Mélanges Biol. Bull. Phys.-Math. Acad. Imp. Sci. Saint-Pétersbourg 12: 562 (1887).
Carex krebsiana Boeck., Beitr. Cyper. 2: 32 (1890).

Carex hakonemontana Katsuy., Bull. Kanagawa Pref. Mus., Nat. Sci. 20: 3 (1991).
Japan. 38 JAP. Hemicr. or rhizome geophyte.

Carex hakonensis Franch. & Sav., Enum. Pl. Jap. 2: 550 (1878). *Carex onoei* f. *hakonensis* (Franch. & Sav.) Kük. in H.G.A.Engler (ed.), Pflanzenr., IV, 20(38): 102 (1909).
Korea, Japan. 38 JAP KOR. Hemicr. or rhizome geophyte.
Carex krameri Franch. & Sav., Enum. Pl. Jap. 1: 551 (1875). *Carex onoei* var. *krameri* (Franch. & Sav.) Kük. in H.G.A.Engler (ed.), Pflanzenr., IV, 20(38): 101 (1909).

Carex halleriana Asso, Syn. Stirp. Aragon. 1: 372 (1779).
Europe, Medit. to Afghanistan. 11 AUT CZE GER HUN SWI 12 BAL COR FRA POR SAR SOC 13 ALB BUK GRC ITA KRI ROM SIC TUE YUG 14 KRY 20 ALG MOR TUN 33 NCS TCS 34 AFG CYP EAI IRN PAL TUR 40 PAK? Hemicr. or rhizome geophyte.
Carex alpestris All., Fl. Pedem. 2: 270 (1785).
Carex gynobasis Vill., Hist. Pl. Dauphiné 2: 206 (1787). *Olotrema gynobasis* (Vill.) Raf., Good Book: 25 (1840).
Sparganium tenuifolium Poir., Voy. Barbarie 2: 254 (1789), nom. inval.
Carex rhizantha J.F.Gmel., Syst. Nat. 2: 144 (1791).
Carex ptychocarpa Link, J. Bot. (Schrader) 2: 309 (1799).
Carex diversiflora Host, Icon. Descr. Gram. Austriac. 1: 53 (1801).
Carex halleriana f. *corsica* Mabille, Rech. Pl. Corse 1: 32 (1867). *Carex mabilliana* Rouy in G.Rouy & J.Foucaud, Fl. France 13: 440 (1912).
Carex halleriana f. *lerinensis* H.Christ, Bull. Soc. Roy. Bot. Belgique 24(2): 14 (1885).
Carex halleriana f. *occulata* Kük. in H.G.A.Engler (ed.), Pflanzenr., IV, 20(38): 488 (1909).
Carex halleriana f. *pedunculata* Kük. in H.G.A.Engler (ed.), Pflanzenr., IV, 20(38): 488 (1909).
Carex allorgei Sennen, Treb. Inst. Catalana Hist. Nat. 15(1): 37 (1931).

Carex halliana L.H.Bailey, Bot. Gaz. 9: 117 (1884).
W. U.S.A. 73 ORE WAS 76 CAL. Hemicr. or rhizome geophyte.
Carex oregonensis Olney ex A.Gray, Proc. Amer. Acad. Arts 8: 407 (1872), nom. nud.

Carex hallii Olney in F.V.Hayden, Rep.: 496 (1871). *Carex parryana* var. *unica* L.H.Bailey, Mem. Torrey Bot. Club 1: 54 (1889). *Carex parryana* var. *hallii* (Olney) Kük. in H.G.A.Engler (ed.), Pflanzenr., IV, 20(38): 388 (1909). *Carex parryana* subsp. *hallii* (Olney) D.F.Murray, Brittonia 21: 71 (1969).
SC. Canada to WC. U.S.A. 71 MAN 73 COL WYO 74 MIN NDA NEB. Hemicr. or rhizome geophyte.
Carex elrodii M.E.Jones, Bull. Montana State Univ., Biol. Ser. 15: 70 (1910).

Carex hanamninhensis K.K.Nguyen, Bot. Zhurn. (Moscow & Leningrad) 64: 860 (1979).
Vietnam. 41 VIE. Hemicr. or rhizome geophyte.

Carex hancockiana Maxim., Bull. Soc. Imp. Naturalistes Moscou 54(1): 66 (1879).
Siberia to Korea. 30 WSB 36 CHI CHM CHN CHQ 37 MON 38 KOR. Hemicr. or rhizome geophyte.

Carex hanensis Dunn, Kew Bull., Addit. Ser. 10: 305 (1912).
China (Guangdong). 36 CHS. Hemicr. or rhizome geophyte.

Carex hangtongensis H.Lév. & Vaniot, Bull. Soc. Bot. France 53: 316 (1906).
China (Guizhou). 36 CHC. Hemicr. or rhizome geophyte.

Carex hangzhouensis C.Z.Zheng, X.F.Jin & B.Y.Ding, Novon 15: 157 (2005).
China (Zhejiang). 36 CHS. Hemicr. or rhizome geophyte.

Carex harfordii Mack., Bull. Torrey Bot. Club 43: 615 (1916 publ. 1917).
California. 76 CAL. Hemicr. or rhizome geophyte.
Carex montereyensis Mack., Erythea 8: 92 (1922).

Carex harlandii Boott, Ill. Carex 2: 87 (1860).
SE. China to Jawa. 36 CHH? CHS 41 MYA THA VIE 42 JAW SUM. Hemicr. or rhizome geophyte.
Carex chlorocystis Boeck., Bot. Jahrb. Syst. 5: 520 (1884).
Carex harlandii f. *longibracteata* R.Gross, Notizbl. Bot. Gart. Berlin-Dahlem 14: 193 (1938).
Carex manca var. *contigua* R.Gross, Notizbl. Bot. Gart. Berlin-Dahlem 14: 193 (1938).
Carex anomocarya Nelmes, Kew Bull. 5: 202 (1950).
Carex harlandii var. *xiuningensis* S.W.Su, J. Anhui Univ. Nat. Sci. 1985: 78 (1985).
Carex harlandii var. *liuguensis* S.Yun Liang & D.Y.Liu, Bull. Bot. Res., Harbin 6(1): 181 (1986).

Carex harrysmithii Kük., Acta Horti Gothob. 5: 40 (1930).
China (Sichuan). 36 CHC. Hemicr. or rhizome geophyte.

Carex* ×*hartii Dewey, Amer. J. Sci. Arts, II, 41: 226 (1866). *C. lurida* × *C. retrorsa*. *Carex retrorsa* var. *hartii* (Dewey) A.Gray, Manual, ed. 5: 600 (1867).
E. Canada to NE. U.S.A. 72 ONT QUE 75 MAI MAS NWY VER. Hemicr. or rhizome geophyte.
Carex ×*hartii* var. *bradleyi* Dewey, Amer. J. Sci. Arts, II, 41: 226 (1866). *Carex retrorsa* var. *bradleyi* (Dewey) Farw., Rhodora 23: 87 (1921).

Carex hartmanii Cojander, Ann. Bot. Soc. Zool.-Bot. Fenn. "Vanamo" 5: 23 (1935). *Carex buxbaumii* subsp. *hartmanii* (Cojander) Domin in ?
Europe to C. Asia. 10 DEN FIN NOR SWE 11 AUT CZE GER HUN NET POL SWI 12 FRA? 13 ITA YUG 14 BLR BLT RUC RUE RUS RUW UKR 30 WSB 32 KAZ 33 NCS TCS 34 TUR. Hemicr. or rhizome geophyte.
Carex buxbaumii f. *macrostachya* Hartm., Handb. Skand. Fl., ed. 5: 268 (1849).
Carex emasculata V.I.Krecz. in V.L.Komarov (ed.), Fl. URSS 3: 604 (1935).

Carex hashimotoi Ohwi, Acta Phytotax. Geobot. 2: 102 (1933).
Japan (Honshu). 38 JAP. Hemicr. or rhizome geophyte.

Carex hassei L.H.Bailey, Bot. Gaz. 21: 5 (1896).
SW. Canada to Mexico (Baja California). 71 BRC 73 ORE WAS 76 ARI CAL NEV UTA 79 MXN. Hemicr. or rhizome geophyte.
Carex aurea var. *celsa* L.H.Bailey, Mem. Torrey Bot. Club 1: 75 (1889). *Carex celsa* (L.H.Bailey) Piper in C.V.Piper & R.K.Beattie, Fl. N.W. Coast: 79 (1915), nom. illeg.
Carex saliniformis Mack., Bull. Torrey Bot. Club 36: 477 (1909).

Carex hassiana Loes., Beih. Bot. Centralbl. 37(2): 95 (1919).
China (Shandong). 36 CHN. Hemicr. or rhizome geophyte.

Carex hastata Kük., Repert. Spec. Nov. Regni Veg. 27: 110 (1929).
China (Hunan, Zhejiang). 36 CHS. Hemicr. or rhizome geophyte.

Carex hatuyenensis K.K.Nguyen, Bot. Zhurn. (Moscow & Leningrad) 64: 858 (1979).
Vietnam. 41 VIE. Hemicr. or rhizome geophyte.

Carex haydeniana Olney in S.Watson, Botany [Fortieth Parallel]: 366 (1871). *Carex festiva* var. *haydeniana* (Olney) Boott in S.Watson & al., Bot. California 2: 234 (1880). *Carex macloviana* var. *haydeniana* (Olney) Kük. in H.G.A.Engler (ed.), Pflanzenr., IV, 20(38): 196 (1909). *Carex macloviana* subsp. *haydeniana* (Olney) R.L.Taylor & MacBryde, Canad. J. Bot. 56: 190 (1978).
W. Canada to W. U.S.A. 71 ABT BRC 73 COL IDA MNT ORE WAS WYO 76 CAL UTA. Hemicr. or rhizome geophyte.
Carex festiva var. *decumbens* T.Holm, Amer. J. Sci., IV, 16: 20 (1903).
Carex nubicola Mack., Bull. Torrey Bot. Club 36: 480 (1909).

Carex haydenii Dewey, Amer. J. Sci. Arts, II, 18: 103 (1854). *Carex stricta* var. *haydenii* (Dewey) Kük. in H.G.A.Engler (ed.), Pflanzenr., IV, 20(38): 330 (1909).
E. Canada to NC. & NE. U.S.A. 72 NBR ONT QUE 74 ILL IOW MIN MSO NDA NEB SDA WIS 75 INI MAI MAS MIC NWH NWY OHI PEN VER. Hemicr. or rhizome geophyte.
Carex stricta var. *decora* L.H.Bailey, Bot. Gaz. 13: 85 (1888).
Carex rousseaui Raymond, Naturaliste Canad. 82: 31 (1955).

Carex hebecarpa C.A.Mey., Mém. Sav. Étr. Acad. St. Pétersbourg 1: 223 (1831).
India to New Guinea. 36 CHC CHN CHS 38 JAP KOR TAI 40 ASS EHM IND NEP SRL 41 VIE 42 BOR JAW SUM 43 NWG. Hemicr. or rhizome geophyte.

subsp. ***hebecarpa***
India to New Guinea. 36 CHS 38 TAI 40 EHM IND NEP 41 VIE 42 BOR JAW SUM 43 NWG. Hemicr. or rhizome geophyte.
Carex lachnosperma Wall. ex Kunth, Enum. Pl. 2: 505 (1837).
Carex breviculmis Thwaites, Enum. Pl. Zeyl.: 356 (1864), nom. illeg.
Carex thwaitesii Hance, J. Bot. 5: 235 (1867).
Carex hebecarpa f. *latifolia* Makino, Bot. Mag. (Tokyo) 26: 294 (1912).

subsp. ***ligulata*** (Nees) T.Koyama, in Fl. E. Himal., 3rd. Rep.: 123 (1975).
India to Japan. 36 CHC CHN CHS 38 JAP KOR TAI 40 IND NEP SRL. Hemicr. or rhizome geophyte.
**Carex ligulata* Nees in R.Wight, Contr. Bot. India: 127 (1834). *Carex hebecarpa* var. *ligulata* (Nees) Kük. in H.G.A.Engler (ed.), Pflanzenr., IV, 20(38): 745 (1909).
Carex keiskei Miq., Ann. Mus. Bot. Lugduno-Batavi 2: 148 (1866).
Carex bakanensis H.Lév. & Vaniot, Bull. Soc. Bot. France 51: 203 (1904).
Carex ligulata var. *glabriutriculata* Q.S.Wang, J. Wuhan Bot. Res. 5: 344 (1987).

var. ***maubertiana*** (Boott) Franch., Nouv. Arch. Mus. Hist. Nat., III, 10: 70 (1898).
Himalaya to Vietnam. 36 CHC CHS 40 ASS EHM NEP 41 VIE. Hemicr. or rhizome geophyte.
**Carex maubertiana* Boott, Ill. Carex 1: 45 (1858).

Carex hebes Nelmes, Bull. Misc. Inform. Kew 1939: 310 (1939).
SE. Australia. 50 NSW SOA? TAS VIC. Hemicr. or rhizome geophyte.

Carex hebetata Boott, Ill. Carex 4: 172 (1867).
Peru. 83 PER. Hemicr. or rhizome geophyte.

Carex hectori Petrie, Trans. & Proc. New Zealand Inst. 27: 405 (1894 publ. 1895).
New Zealand. 51 NZN NZS. Hemicr. or rhizome geophyte.

Carex heleonastes Ehrh. ex L.f., Suppl. Pl.: 414 (1782). *Vignea heleonastes* (Ehrh. ex L.f.) Rchb. in J.C.Mössler, Handb. Gewächsk. ed. 2, 3: 1609 (1830).
Europe to Russian Far East, W. & C. Canada. 10 FIN ICE NOR SWE 11 AUT GER POL SWI 12 FRA 13 ITA ROM 14 BLR BLT RUC RUE RUN RUS RUW UKR 30 ALT WSB YAK 31 KAM 33 NCS TCS 34 TUR? 70 ASK NWT YUK 71 ABT BRC MAN SAS 72 LAB NFL ONT QUE 75 MIC. Hemicr. or rhizome geophyte.

subsp. ***heleonastes***
Europe to Russian Far East, W. & C. Canada. 10 FIN ICE NOR SWE 11 AUT GER POL SWI 12 FRA 13 ITA ROM 14 BLR BLT RUC RUE RUN

RUS RUW UKR 30 ALT WSB YAK 31 KAM 33 NCS TCS 34 TUR? 70 ASK NWT YUK 71 ABT BRC MAN SAS 72 LAB NFL ONT QUE 75 MIC. Hemicr. or rhizome geophyte.

Carex carltoniana Dewey, Amer. J. Sci. Arts 28: 238 (1835).

Carex leporina Schkuhr ex Ledeb., Fl. Ross. 4: 279 (1852).

Carex sendtneriana Brügger, Jahresber. Naturf. Ges. Graubündens, II, 23–24: 119 (1880). *Carex heleonastes* f. *sendtneriana* (Brügger) Kük. in H.G.A.Engler (ed.), Pflanzenr., IV, 20(38): 215 (1909).

Carex canescens var. *maxima* Kük. in H.G.A.Engler (ed.), Pflanzenr., IV, 20(38): 218 (1909). *Carex curta* var. *maxima* (Kük.) R.R.Stewart, Annot. Cat. Vasc. Pl. W. Pakistan & Kashmir: 75 (1972).

Carex aa Kom., Fl. Kamtschatka 1: 234 (1927).

subsp. ***neurochlaena*** (T.Holm) Böcher, Acta Arctica 5: 10 (1952).

W. Subarctic America. 70 ASK NWT YUK. Hemicr. or rhizome geophyte.

**Carex neurochlaena* T.Holm, Amer. J. Sci. 166: 453 (1903).

Carex helferi Boeck., Linnaea 40: 365 (1876).

Indo-China, W. Jawa. 41 MYA THA 42 JAW. Hemicr. or rhizome geophyte.

Carex mapaniifolia Ridl., J. Fed. Malay States Mus. 10: 124 (1920). *Carex helferi* subsp. *mapaniifolia* (Ridl.) T.Koyama, Bot. Mag. (Tokyo) 92: 224 (1979).

Carex helleri Mack., Erythea 8: 70 (1922).

California to Nevada. 76 CAL NEV. Hemicr. or rhizome geophyte.

Carex ×helvola Blytt, Bot. Not. 1849: 58 (1849). *C. canescens* × *C. tripartita.*

N. Europe. 10 NOR. Hemicr. or rhizome geophyte.

Carex hemineuros T.Koyama, Bot. Mag. (Tokyo) 86: 276 (1973).

Nepal, Tibet. 36 CHT 40 NEP. Hemicr. or rhizome geophyte.

Carex hendersonii L.H.Bailey, Proc. Amer. Acad. Arts 22: 115 (1887).

SW. Canada to California. 71 BRC 73 IDA ORE WAS 76 CAL. Hemicr. or rhizome geophyte.

Carex laxiflora var. *plantaginea* Olney, Proc. Amer. Acad. Arts 8: 407 (1872).

Carex hermannii Cochrane, Brittonia 33: 228 (1981).

C. Mexico. 79 MXC. Hemicr. or rhizome geophyte.

Carex herteri G.A.Wheeler, Sida 17: 471 (1996).

Uruguay. 85 URU. Hemicr. or rhizome geophyte.

Carex heshuonensis S.Yun Liang, in Fl. Reipubl. Popul. Sin. 12: 525 (2000).

Xinjiang. 36 CHX. Hemicr.

Carex heterodoxa Cherm., Bull. Soc. Bot. France 70: 413 (1923).

C. Madagascar. 29 MDG. Hemicr. or rhizome geophyte.

Carex heterolepis Bunge, Enum. Pl. China Bor.: 69 (1833).

Temp. Asia. 30+ 31+ 36 CHI CHM CHN CHS 38 JAP KOR. Hemicr. or rhizome geophyte.

Carex trappistarum Franch., J. Bot. (Morot) 4: 320 (1890).

Carex marginaria Franch., Bull. Soc. Philom. Paris, VIII, 7: 40 (1895).

Carex latinervia H.Lév. & Vaniot, Bull. Acad. Int. Géogr. Bot. 10: 219 (1901).

Carex trappistarum var. *obtegens* Kük., Bot. Jahrb. Syst. 36(82): 8 (1905). *Carex heterolepis* var. *obtegens* (Kük.) Kük. in H.G.A.Engler (ed.), Pflanzenr., IV, 20(38): 342 (1909).

Carex periculosa Honda, Bot. Mag. (Tokyo) 44: 408 (1930).

Carex tobae Honda, Bot. Mag. (Tokyo) 44: 409 (1930).

Carex chamarensis T.V.Egorova, Novosti Sist. Vyssh. Rast. 2: 84 (1965).

Carex heteroneura S.Watson in S.Watson & al., Bot. California 2: 240 (1880).

SW. Canada to W. U.S.A. 71 ABT BRC 73 COL IDA MNT ORE WAS WYO 76 CAL NEV UTA. Hemicr. or rhizome geophyte.

var. ***brevisquama*** F.J.Herm., Rhodora 70: 421 (1968).

NW. U.S.A. to Nevada. 73 COL IDA MNT WYO 76 NEV. Hemicr. or rhizome geophyte.

var. ***epapillosa*** (Mack.) F.J.Herm., Rhodora 70: 421 (1968).

SW. Canada to W. U.S.A. 71 ABT BRC 73 COL IDA MNT ORE WAS WYO 76 CAL UTA. Hemicr. or rhizome geophyte.

**Carex epapillosa* Mack. in P.A.Rydberg, Fl. Rocky Mts.: 138 (1917).

Carex hagiana Kelso, Biol. Leafl. 30: 2 (1945).

var. ***heteroneura***

W. U.S.A. 73 ORE WYO 76 CAL NEV. Hemicr. or rhizome geophyte.

Carex atrata var. *erecta* Boott in S.Watson & al., Bot. California 2: 239 (1880).

Carex quadrifida L.H.Bailey, Proc. Calif. Acad. Sci., II, 3: 104 (1891).

Carex quadrifida var. *lenis* L.H.Bailey, Proc. Calif. Acad. Sci., II, 3: 104 (1891). *Carex atrata* f. *lenis* (L.H.Bailey) Kük. in H.G.A.Engler (ed.), Pflanzenr., IV, 20(38): 399 (1909).

Carex quadrifida var. *caeca* L.H.Bailey, Bot. Gaz. 21: 8 (1896). *Carex atrata* f. *caeca* (L.H.Bailey) Kük. in H.G.A.Engler (ed.), Pflanzenr., IV, 20(38): 400 (1909).

Carex ×heterophyta E.Holmb., Bot. Not. 1929: 16 (1929). *C. canescens* × *C. heleonastes.*

Canada, N. Europe. 10 SWE 71 BRC. Hemicr. or rhizome geophyte.

Carex ×calderi B.Boivin, Naturaliste Canad. 94: 522 (1967).

Carex heterostachya Bunge, Enum. Pl. China Bor.: 69 (1833).

Temp. Asia to Himalaya. 30 ALT 32 KAZ KGZ TKM TZK UZB 33 TCS 34 AFG IRN TUR 36 CHM CHN CHX 37 MON 40 PAK WHM (75) ill. Hemicr. or rhizome geophyte.

Carex songorica f. *minor* Boott in ?. *Carex heterostachya* f. *minor* (Boott) Kük. in H.G.A. Engler (ed.), Pflanzenr., IV, 20(38): 741 (1909).

Carex gebleri Prescott ex Boott, Proc. Linn. Soc. London 1: 28 (1839).

Carex songorica Kar. & Kir., Bull. Soc. Imp. Naturalistes Moscou 15: 525 (1842).

Carex nutans var. *microcarpa* Boeck., Linnaea 41: 297 (1877).

Carex bungeana Debeaux, Actes Soc. Linn. Bordeaux 33: 68 (1879).

Carex orientalis Meinsh., Trudy Imp. S.-Peterburgsk. Bot. Sada 18: 393 (1900).
Carex haematostachys H.Lév. & Vaniot, Bull. Acad. Int. Géogr. Bot. 40: 305 (1902).
Carex fedtschenkoana Kük., Repert. Spec. Nov. Regni Veg. 8: 327 (1910).
Carex elbursensis Gilli, Repert. Spec. Nov. Regni Veg. 43: 296 (1938).
Carex fulleri H.E.Ahles, Rhodora 58: 320 (1956).

Carex heudesii H.Lév. & Vaniot, Bull. Acad. Int. Géogr. Bot. 12: 12 (1903).
C. China. 36 CHC CHN. Hemicr. or rhizome geophyte.

Carex hexinensis S.Yun Liang & Y.Z.Huang, Fl. Fujianica 6: 690 (1995).
China (Fujian). 36 CHS. Hemicr. or rhizome geophyte.

Carex hezhouensis H.Wang & S.N.Wang, Guihaia 25: 105 (2005).
China (Guangxi). 36 CHS. Hemicr. or rhizome geophyte.

Carex hieronymi Boeck., Beitr. Cyper. 1: 46 (1888). *Carex inconspicua* var. *hieronymi* (Boeck.) Kük., Bot. Jahrb. Syst. 27: 544 (1899). *Carex fuscula* var. *hieronymi* (Boeck.) Kük. in H.G.A.Engler (ed.), Pflanzenr., IV, 20(38): 669 (1909).
Brazil to Chile. 84 BZS 85 AGE AGW CLC. Hemicr. or rhizome geophyte.

Carex hilairei Boott, Ill. Carex 4: 146 (1867).
Brazil to Argentina (Misiones). 84 BZL 85 AGE. Hemicr. or rhizome geophyte.

var. ***glazioviana*** (Boeck.) Kük., Verh. Bot. Vereins Prov. Brandenburg 47: 205 (1905 publ. 1906).
Brazil, Argentina (Misiones). 84 BZL 85 AGE. Hemicr. or rhizome geophyte.
**Carex glazioviana* Boeck., Flora 65: 59 (1882).

var. ***hilairei***
SE. Brazil. 84 BZL. Hemicr. or rhizome geophyte.

Carex hilaireioides C.B.Clarke ex Kük., Verh. Bot. Vereins Prov. Brandenburg 47: 205 (1905 publ. 1906).
S. Brazil. 84 BZS. Hemicr. or rhizome geophyte.

Carex hildebrandtiana Boeck., Bot. Jahrb. Syst. 5: 516 (1884).
C. Madagascar. 29 MDG. Hemicr. or rhizome geophyte.

Carex himalaica T.Koyama, Bot. Mag. (Tokyo) 86: 278 (1973).
Nepal. 40 NEP. Hemicr. or rhizome geophyte.

Carex hinnulea C.B.Clarke in I.Urban, Symb. Antill. 2: 159 (1900).
Jamaica. 81 JAM. Hemicr. or rhizome geophyte.

Carex hirsutella Mack., Bull. Torrey Bot. Club 50: 349 (1923). *Carex complanata* subsp. *hirsutella* (Mack.) R.T.Clausen, Torreya 39: 3 (1939).
SE. Canada to C. & E. U.S.A. 72 ONT QUE 74 ILL IOW KAN MSO OKL 75 CNT INI MAI MAS MIC NWJ NWY OHI PEN VER WVA 77 TEX 78 ALA ARK DEL GEO KTY MRY MSI NCA SCA TEN VRG WDC. Hemicr. or rhizome geophyte.
Carex triceps Michx., Fl. Bor.-Amer. 2: 170 (1803), nom. illeg.
**Carex hirsuta* Willd., Sp. Pl. 4: 252 (1805), nom. illeg. *Olamblis hirsuta* Raf., Good Book: 26 (1840). *Carex triceps* var. *hirsuta* (Raf.) L.H.Bailey, Mem. Torrey Bot. Club 1: 35 (1889). *Carex complanata* var. *hirsuta* (Raf.) Gleason, Phytologia 4: 22 (1952).
Carex viridula Schwein. & Torr., Ann. Lyceum Nat. Hist. New York 1: 320 (1824), nom. illeg.
Carex capillata Schkuhr ex Kunth, Enum. Pl. 2: 430 (1837), pro syn.
Carex triceps var. *pedunculata* (Schwein. & Torr.) Torr., Fl. New York 2: 384 (1843).
Carex triceps var. *smithii* L.H.Bailey, Bot. Gaz. 13: 88 (1888).
Carex triceps var. *longicuspis* Kük. in H.G.A.Engler (ed.), Pflanzenr., IV, 20(38): 431 (1909).

Carex hirta L., Sp. Pl.: 975 (1753). *Trasus hirtus* (L.) Gray, Nat. Arr. Brit. Pl. 2: 58 (1821).
Europe to Iran, N. Africa. 10 DEN FIN GRB IRE NOR SWE 11 AUT BGM CZE GER HUN NET POL SWI 12 COR FRA POR SPA 13 ALB BUL GRC ITA ROM YUG 14 BLR BLT KRY RUC RUE RUN RUS RUW UKR 20 ALG MOR 33 TCS 34 IRN TUR (51) nzs (72) nsc ont pei que (74) wis (75) cnt mai mas mic nwj nwy pen (78) mry. Hemicr. or rhizome geophyte.
Carex hirtiformis Pers., Syn. Pl. 2: 547 (1807). *Carex hirta* var. *hirtiformis* (Pers.) Lej., Fl. Spa 2: 230 (1813). *Carex hirta* f. *hirtiformis* (Pers.) Kunth, Enum. Pl. 2: 485 (1837).
Carex villosa Stokes, Bot. Mat. Med. 4: 346 (1812).
Carex hirta var. *sublaevis* Hornem. in G.C.Oeder & al. (eds.), Fl. Dan.: t. 1711 (1821).
Carex hirta var. *glabra* Lej. in A.L.S.Lejeune & R.J.Courtois, Comp. Fl. Belg. 3: 237 (1836).
Carex hirta f. *humilis* Peterm., Fl. Lips. Excurs.: 62 (1838).
Carex hirta f. *villosa* Peterm., Fl. Lips. Excurs.: 62 (1838).
Carex hirta f. *glabrata* Peterm., Anal. Pfl.-Schlüss.: 516 (1846).
Carex hirta f. *major* Peterm., Anal. Pfl.-Schlüss.: 516 (1846).
Carex aristata Siegert ex Wimm., Fl. Schles., ed. 3: 72 (1857), nom. illeg.
Carex hirta var. *vera* Neilr., Fl. Nied.-Oesterr.: 122 (1857).
Carex hirta f. *pseudohirta* Schur, Enum. Pl. Transsilv.: 711 (1866).
Carex hirta var. *pilosa* Èelak., Prodr. Fl. Böhmen: 74 (1867).
Carex hirta var. *subglabra* Èelak., Prodr. Fl. Böhmen: 74 (1867).
Carex hirta f. *spinosa* H.Mort., Bot. Tidsskr., II, 1: 94 (1872).
Carex orthostachys Trevir. ex Nyman, Consp. Fl. Eur.: 769 (1882), nom. illeg.
Carex hirta var. *glabrescens* Cariot, Étude Fl., ed. 8, 2: 871 (1889).
Carex hirta f. *subhirtiformis* Kneuck., Allg. Bot. Z. Syst. 4: 165 (1898).
Carex hirta f. *paludosa* A.Winkl. ex Asch. & Graebn., Syn. Mitteleur. Fl. 2(2): 223 (1903).
Carex hirta var. *aquatica* Waisb., Magyar Bot. Lapok 4: 76 (1905).
Carex hirta f. *latifolia* Waisb. ex Kük. in H.G.A.Engler (ed.), Pflanzenr., IV, 20(38): 751 (1909).

Carex hirtelloides (Kük.) F.T.Wang & Tang ex P.C.Li, in Vasc. Pl. Hengduan Mt. 2: 2379 (1994).
China (Sichuan, Yunnan). 36 CHC. Hemicr. or rhizome geophyte.
**Carex haematostoma* var. *hirtelloides* Kük., Repert. Spec. Nov. Regni Veg. Beih. 12: 311 (1922).

Carex hirticaulis P.C.Li, Acta Bot. Yunnan. 12: 149 (1990).
China (Yunnan). 36 CHC. Hemicr. or rhizome geophyte.

Carex hirtifolia Mack., Bull. Torrey Bot. Club 37: 244 (1910).
E. Canada, NC. & E. U.S.A. 72 NBR NSC ONT QUE 74 ILL IOW KAN MIN MSO WIS 75 CNT INI MAI MAS MIC NWH NWJ OHI PEN RHO VER WVA 78 DEL KTY MRY TEN VRG WDC. Hemicr. or rhizome geophyte.
**Carex pubescens* Muhl. ex Willd., Sp. Pl. 4: 281 (1805), nom. illeg. *Diemisa pubescens* Raf., Good Book: 27 (1840). *Edritria pubescens* Raf., Good Book: 26 (1840).

Carex hirtigluma C.B.Clarke, Kew Bull., Addit. Ser. 8: 73 (1908).
C. Madagascar. 29 MDG. Hemicr.
Carex hirtiglumis C.B.Clarke ex Kük. in H.G.A.Engler (ed.), Pflanzenr., IV, 20(38): 276 (1909).
Carex perrieri H.Lév., Bull. Acad. Int. Géogr. Bot. 27: 5 (1917).
Carex hirtiglumis var. *arcuata* Cherm., Bull. Soc. Bot. France 72: 21 (1925).

Carex hirtissima Boott in S.Watson & al., Bot. California 2: 247 (1880).
California. 76 CAL. Hemicr. or rhizome geophyte.

Carex hirtiutriculata L.K.Dai, Acta Phytotax. Sin. 32: 181 (1994).
China (Yunnan). 36 CHC. Hemicr. or rhizome geophyte.

Carex hispida Willd. ex Schkuhr, Beschr. Riedgräs. 1: 63 (1801). *Carex hispida* var. *genuina* Gren. & Godr., Fl. France 3: 412 (1855), nom. inval.
Medit. 12 BAL COR FRA POR SAR SPA 13 ALG GRC ITA KRI LBY MOR SIC TUN 34 CYP PAL TUR. Hemicr. or rhizome geophyte.
Carex hispida Link, Hort. Berol. 2: 331 (1833), nom. illeg.
Carex lasiochlaena Kunth, Enum. Pl. 2: 485 (1837).
Carex hispida var. *anacantha* Gren. & Godr., Fl. France 3: 412 (1855).
Carex obtusangula Salzm. ex Nyman, Consp. Fl. Eur.: 774 (1882).
Carex hispida f. *marina* Heldr. ex Kük. in H.G.A. Engler (ed.), Pflanzenr., IV, 20(38): 422 (1909).

Carex hitchcockiana Dewey, Amer. J. Sci. Arts 10: 274 (1826). *Loxotrema hitchcockiana* (Dewey) Raf., Good Book: 25 (1840). *Carex oligocarpa* var. *hitchcockiana* (Dewey) Boeck., Linnaea 41: 149 (1877).
SE. Canada to NC. & E. U.S.A. 72 ONT QUE 74 ILL IOW KAN MIN MSO OKL WIS 75 CNT INI MAS MIC NWJ NWY OHI PEN VER WVA 78 ALA ARK KTY MRY NCA TEN VRG WDC. Hemicr. or rhizome geophyte.
Carex hitchcockii Eaton, Man. Bot., ed. 5: 158 (1829).
Carex hitchcockiana var. *triflora* Peck, Rep. (Annual) Regents Univ. State New York State Mus. 48: 164 (1896).

Carex hoatiensis H.Lév., Repert. Spec. Nov. Regni Veg. 11: 66 (1912).
Korea. 38 KOR. Hemicr. or rhizome geophyte.

Carex hochstetteriana J.Gay ex Seub., Fl. Azor.: 22 (1840).
Azores. 21 AZO. Hemicr. or rhizome geophyte.
Carex rigidifolia Hochst. ex Seub., Fl. Azor.: 21 (1840). *Carex hochstetteriana* var. *rigidifolia* (Hochst. ex Seub.) Kük. in H.G.A.Engler (ed.), Pflanzenr., IV, 20(38): 649 (1909).

Carex holostoma Drejer in G.C.Oeder & al. (eds.), Fl. Dan.: t. 2428 (1845). *Carex alpina* var. *holostoma* (Drejer) L.H.Bailey, Proc. Amer. Acad. Arts 22: 76 (1886).
Subarctic. 10 FIN ICE? NOR SWE 14 RUN 30 WSB 31 MAG 70 ASK GNL GNL NUN NWT YUK 72 LAB NFL QUE. Hemicr. or rhizome geophyte.

Carex holotricha Ohwi, Acta Phytotax. Geobot. 7: 131 (1938).
Russian Far East to Korea. 31+ 38 KOR. Hemicr. or rhizome geophyte.

Carex hondoensis Ohwi, Mem. Coll. Sci. Kyoto Imp. Univ., Ser. B, Biol. 3: 252 (1930).
Japan (C. Honshu). 38 JAP. Hemicr. or rhizome geophyte.

Carex hongnoensis H.Lév., Repert. Spec. Nov. Regni Veg. 8: 426 (1910).
Korea. 38 KOR. Hemicr. or rhizome geophyte.

Carex hongyuanensis Y.C.Tang & S.Yun Liang, in Fl. Reipubl. Popul. Sin. 12: 520 (2000).
China (Sichuan). 36 CHC. Hemicr.

Carex hoodii Boott in W.J.Hooker, Fl. Bor.-Amer. 2: 211 (1839).
Alaska to W. & WC. U.S.A. 70 ASK YUK 71 ABT BRC SAS 73 COL IDA MNT ORE WAS WYO 74 SDA 76 CAL NEV UTA. Hemicr. or rhizome geophyte.
Carex congesta C.A.Mey. ex Boott, Ill. Carex 1: 17 (1858).
Carex muricata var. *confixa* L.H.Bailey, Bot. Gaz. 10: 203 (1885).
Carex hoodii var. *nervosa* L.H.Bailey, Mem. Torrey Bot. Club 1: 14 (1889).
Carex hoodii var. *neurocarpa* Piper, Contr. U. S. Natl. Herb. 11: 167 (1906).

Carex hookeri Kunth, Enum. Pl. 2: 490 (1837).
Ecuador to Chile. 83 ECU PER 85 CLC. Hemicr.
**Carex hebecarpa* Hook. & Arn., Bot. Beechey Voy.: 50 (1832), nom. illeg.
Carex beechyana Boott, Ill. Carex 1: 67 (1858).

Carex hookeriana Dewey, Amer. J. Sci. Arts 29: 248 (1836).
S. Canada to NC. U.S.A. 71 ABT MAN SAS 72 ONT 74 NDA SDA. Hemicr. or rhizome geophyte.
Carex muricata var. *gracilis* Boott, Ill. Carex 4: 193 (1867).
Carex muricata var. *americana* L.H.Bailey, Proc. Amer. Acad. Arts 22: 140 (1886).

Carex hoozanensis Hayata, Icon. Pl. Formosan. 10: 67 (1921).
Vietnam, Taiwan. 38 TAI 41 VIE. Hemicr. or rhizome geophyte.

Carex hopeiensis F.T.Wang & Tang, Acta Phytotax. Sin. 1: 133 (1951).
China (Hebei). 36 CHN. Hemicr. or rhizome geophyte.

Carex hordeistichos Vill., Prosp. Hist. Pl. Dauphiné: 18 (1779).
Europe to Iran, NW. Africa. 11 AUT CZE GER HUN 12 COR? FRA SPA 13 BUL ROM YUG 14 KRY RUC RUS UKR 20 ALG MOR TUN 32 KAZ 33

TCS 34 IRN IRQ TUR. Hemicr. or rhizome geophyte.
Carex hordeiformis Wahlenb., Kongl. Vetensk. Acad. Nya Handl. 24: 152 (1803).
Carex lerchenfeldiana Schur, Enum. Pl. Transsilv.: 711 (1866).

Carex hormathodes Fernald, Rhodora 8: 165 (1906).
Coastal E. Canada to E. U.S.A. 72 NBR NFL NSC PEI QUE 75 CNT MAI MAS NWH NWJ NWY RHO 78 DEL MRY NCA VRG. Hemicr. or rhizome geophyte.
**Carex straminea* var. *aperta* Boott, Ill. Carex 3: 120 (1862).
Carex straminea var. *indivisa* Boott, Bot. Gaz. 9: 86 (1884). *Carex straminea* f. *indivisa* (Boott) Britton in N.L.Britton & A.Brown, Ill. Fl. N. U.S. 1: 358 (1896). *Carex tenera* var. *invisa* (Boott) Britton in N.L.Britton & A.Brown, Ill. Fl. N. U.S. 1: 358 (1896).
Carex hormathodes var. *invisa* (W.Boott) Fernald, Rhodora 8: 166 (1906).
Carex hormathodes f. *invisa* (W.Boott) Fernald, Rhodora 44: 285 (1942).

Carex horsfieldii Boott, Proc. Linn. Soc. London 1: 257 (1845).
Indo-China to Queensland. 41 MYA THA VIE 42 JAW MOL 43 NWG 50 QLD. Hemicr. or rhizome geophyte.
Carex fleckeri Nelmes, Bull. Misc. Inform. Kew 1939: 313 (1939).
Carex plagiostoma Nelmes, Kew Bull. 1: 19 (1946).

Carex hostiana DC., Cat. Pl. Horti Monsp.: 88 (1813).
Trasus hostianus (DC.) Gray, Nat. Arr. Brit. Pl. 2: 61 (1821). *Carex distans* var. *hostiana* (DC.) Fiori, Nuov. Fl. Italia 1: 199 (1923).
Europe to Caucasus, E. Canada, NE. U.S.A. 10 DEN FIN FOR GRB IRE NOR SWE 11 AUT BGM CZE GER HUN NET POL SWI 12 FRA SPA 13 BUL GRC ITA ROM YUG 14 BLT RUW UKR 33 TCS 72 LAB NFL QUE 75 MAS†. Hemicr. or rhizome geophyte.
Carex hornschuchiana subsp. *eckeroeensis* H.Lindb. & Palmgr. in ?
Carex trigona All., Fl. Pedem. 2: 269 (1785), provisional synonym.
**Carex fulva* Host, Icon. Descr. Gram. Austriac. 4: 53 (1809), nom. illeg. *Trasus fulvus* Gray, Nat. Arr. Brit. Pl. 2: 62 (1821). *Carex hornschuchiana* var. *fulva* Nyman, Consp. Fl. Eur.: 770 (1882).
Carex hornschuchiana Hoppe, Flora 7: 599 (1824).
Carex speirostachya Sw. ex Sm., Engl. Fl. 4: 98 (1828).
Carex hornschuchiana f. *angustifolia* Gaudin, Fl. Helv. 6: 100 (1830).
Carex greeniana Dewey, Amer. J. Sci. Arts 30: 61 (1836).
Carex biformis F.W.Schultz, Flora 24: 55 (1841).
Carex biformis var. *fertilis* F.W.Schultz, Flora 24: 55 (1841).
Carex hornschuchiana f. *longibracteata* Peterm., Anal. Pfl.-Schlüss.: 511 (1846).
Carex hornschuchiana f. *remota* Peterm., Anal. Pfl.-Schlüss.: 511 (1846).
Carex halophila Heuff., Linnaea 31: 695 (1862), nom. illeg.
Carex subsalsa Nyman, Syll. Fl. Eur., Suppl.: 69 (1865). *Carex* ×*fulva* var. *subsalsa* (Nyman) Nyman, Consp. Fl. Eur.: 770 (1882).
Carex armena Boiss., Fl. Orient. 5: 426 (1882).
Carex hornschuchiana f. *approximata* Murr, Oesterr. Bot. Z. 38: 239 (1888).
Carex hornschuchiana f. *atrata* Buchenau, Fl. Nordwestdeut. Tiefebene: 125 (1894).
Carex hornschuchiana f. *subdistans* Neuman, Sver. Fl.: 695 (1901).
Carex hornschuchiana f. *discolor* Vollm., Ber. Bayer. Bot. Ges. 1904: 57 (1904).
Carex fulvescens Mack., Bull. Torrey Bot. Club 37: 239 (1910).
Carex hornschuchiana var. *laurentiana* Fernald & Wiegand, Rhodora 13: 130 (1911). *Carex hostiana* var. *laurentiana* (Fernald & Wiegand) Fernald & Wiegand, Rhodora 26: 122 (1924).

Carex hotaizanensis Akiyama, J. Jap. Bot. 16: 101 (1940).
Korea. 38 KOR. Hemicr. or rhizome geophyte.

Carex houghtoniana Torr. ex Dewey, Amer. J. Sci. Arts 30: 63 (1836).
C. & E. Canada, NC. & NE. U.S.A. 71 ABT MAN SAS 72 LAB NBR NFL NSC ONT QUE 74 ILL MIN WIS 75 MAI MIC NWH NWY VER. Hemicr. or rhizome geophyte.
Carex houghtonii Torr., Ann. Lyceum Nat. Hist. New York 3: 413 (1836), orth. var.

Carex hovarum Cherm., Bull. Soc. Bot. France 70: 414 (1923).
C. Madagascar. 29 MDG. Hemicr. or rhizome geophyte.

Carex huashanica Tang & F.T.Wang ex L.K.Dai, Acta Phytotax. Sin. 32: 186 (1994).
China (Shaanxi). 36 CHN. Hemicr. or rhizome geophyte.

Carex hubbardii Nelmes, Proc. Linn. Soc. London 1944: 155, 279 (1944).
E. Australia. 50 NSW QLD. Hemicr. or rhizome geophyte.

Carex huehueteca Standl. & Steyerm., Publ. Field Mus. Nat. Hist., Bot. Ser. 23: 195 (1947).
Mexico (Chiapas) to Guatemala. 79 MXT 80 GUA. Hemicr. or rhizome geophyte.

Carex hultenii Aspl., Svensk Bot. Tidskr. 48: 95 (1954).
Mexico (Hidalgo, Puebla, Chiapas). 79 MXC MXE MXT. Hemicr. or rhizome geophyte.

Carex humahuacaensis G.A.Wheeler, Syst. Bot. 14: 37 (1989).
Argentina (Jujuy). 85 AGW. Hemicr. or rhizome geophyte.

Carex humbertiana Ohwi, Acta Phytotax. Geobot. 2: 102 (1933).
Korea. 38 KOR. Hemicr. or rhizome geophyte.

Carex humbertii Cherm., Bull. Soc. Bot. France 73: 554 (1926 publ. 1927).
C. Madagascar. 29 MDG. Hemicr. or rhizome geophyte.

Carex humboldtiana Steud., Syn. Pl. Glumac. 2: 208 (1855).
C. Mexico to Brazil. 79 MXC MXG MXS MXT 80 COS GUA HON NIC PAN 83 CLM 84. Hemicr. or rhizome geophyte.
Carex macrosperma Mack., Bull. Torrey Bot. Club 36: 477 (1909).

Carex humida Y.L.Chang & Y.L.Yang, in Fl. Plant. Herb. Chinae Bor.-Or. 11: 204 (1976).
N. China. 36 CHI CHM. Hemicr. or rhizome geophyte.

Carex humilis Leyss., Fl. Halens.: 175 (1761).
Europe to Japan. 10 GRB 11 AUT BGM CZE GER HUN POL SWI 12 COR FRA SPA 13 ALB BUL ITA ROM YUG 14 KRY RUC RUS UKR 30 ALT WSB 31+ 33 NCS TCS 34 TUR? 36 CHC CHM CHN CHS 38 JAP KOR. Hemicr. or rhizome geophyte.

var. ***humilis***
Europe to Siberia. 10 GRB 11 AUT BGM CZE GER HUN POL SWI 12 COR FRA SPA 13 ALB BUL ITA ROM YUG 14 KRY RUC RUS UKR 30 ALT WSB 33 NCS TCS 34 TUR? Hemicr. or rhizome geophyte.
Carex prostrata All., Fl. Pedem. 2: 267 (1785).
Carex argentea Vill., Hist. Pl. Dauphiné 2: 206 (1787).
Carex scariosa Lam., Encycl. 3: 388 (1792).
Carex clandestina Gooden., Trans. Linn. Soc. London 2: 167 (1794). *Trasus clandestinus* (Gooden.) Gray, Nat. Arr. Brit. Pl. 2: 56 (1821).
Carex gersneri Suter, Helvet. Fl. 2: 248 (1802).
Carex buschiorum V.I.Krecz. ex Kolak., Trudy Inst. Abkhaz. Kult. 13: 218 (1938).
Carex humilis var. *longifolia* Stoeva & E.D.Popova, Folia Geobot. Phytotax. 26: 392 (1991).

var. ***nana*** (H.Lév. & Vaniot) Ohwi, Cyper. Jap. 1: 399 (1936).
Siberia to Japan. 30+ 31+ 36 CHM CHS 38 JAP KOR. Hemicr. or rhizome geophyte.
**Carex lanceolata* var. *nana* H.Lév. & Vaniot, Bull. Acad. Int. Géogr. Bot. 11: 269 (1901). *Carex callitrichos* var. *nana* (H.Lév. & Vaniot) Ohwi, Cyper. Jap. 1: 399 (1936), nom. illeg.
Carex nanella Ohwi, Mem. Coll. Sci. Kyoto Imp. Univ., Ser. B, Biol. 5(3): 263 (1930).
Carex callitrichos var. *austrohinganica* Y.L.Chang & Y.L.Yang, in Fl. Plant. Herb. Chinae Bor.-Or. 11: 205 (1976).

var. ***scirrobasis*** (Kitag.) Y.L.Chang & Y.L.Yang, in Fl. Plant. Herb. Chinae Bor.-Or. 11: 121 (1976).
Russian Far East to C. China. 31+ 36 CHC CHM CHN. Hemicr. or rhizome geophyte.
**Carex scirrobasis* Kitag., Rep. Inst. Sci. Res. Manchoukuo 2: 285 (1938).

Carex humpatensis H.E.Hess, Ber. Schweiz. Bot. Ges. 63: 353 (1953).
Angola. 26 ANG. Hemicr. or rhizome geophyte.

Carex huolushanensis P.C.Li, Acta Bot. Yunnan. 12: 145 (1990).
China (Sichuan). 36 CHC. Hemicr. or rhizome geophyte.

Carex husnotiana H.Lév., Repert. Spec. Nov. Regni Veg. 9: 444 (1911).
Korea. 38 KOR. Hemicr. or rhizome geophyte.

Carex hwangii Matsuda, Bot. Mag. (Tokyo) 28: 297 (1914).
China (Jiangsu). 36 CHS. Hemicr. or rhizome geophyte.

Carex hyalina Boott, Boston J. Nat. Hist. 5: 112 (1847). *Carex straminea* var. *hyalina* (Boott) Boott, Ill. Carex 3: 11 (1862).
C. & SE. U.S.A. 74 OKL 77 TEX 78 ARK LOU MSI TEN. Hemicr. or rhizome geophyte.

Carex hyalinolepis Steud., Syn. Pl. Glumac. 2: 235 (1855).
S. Ontario to C. & E. U.S.A. 72 ONT 74 ILL IOW KAN MSO NEB OKL 75 INI MIC OHI PEN 77 TEX 78 ARK FLA GEO KTY LOU MRY MSI NCA SCA TEN VRG. Hemicr. or rhizome geophyte.
Carex riparia var. *impressa* S.H.Wright in ?. *Carex impressa* (S.H.Wright) Mack., Bull. Torrey Bot. Club 37: 236 (1910).
Carex lacustris var. *laxiflora* Dewey, Amer. J. Sci. Arts, II, 35: 60 (1863).

Carex hymenodon Ohwi, Acta Phytotax. Geobot. 1: 298 (1932).
Japan (Honshu: Shimotsuke Prov.). 38 JAP. Hemicr. or rhizome geophyte.
Carex aequibilirostris Sutô & Suzuki, Bull. Coll. Agric. Utsunomiya Univ. 8: (1933).

Carex hymenolepis Nees in R.Wight, Contr. Bot. India: 126 (1834).
Pakistan to Sikkim. 40 EHM NEP PAK WHM. Hemicr. or rhizome geophyte.

Carex hypandra F.Muell., Fragm. 8: 259 (1874).
SE. Australia. 50 NSW TAS. Hemicr. or rhizome geophyte.

Carex hypoblephara Ohwi & Ryu, J. Jap. Bot. 12: 654 (1936).
China (Jiangxi). 36 CHS. Hemicr. or rhizome geophyte.

Carex hypolytroides Ridl., J. Fed. Malay States Mus. 8(4): 124 (1917).
Indo-China, C. Sumatera, N. Borneo. 41 THA VIE 42 BOR SUM. Hemicr. or rhizome geophyte.
Carex hypolytroides R.Gross & Mattf., Notizbl. Bot. Gart. Berlin-Dahlem 14: 190 (1938), nom. illeg. *Carex hypolytropsis* R.Gross, Repert. Spec. Nov. Regni Veg. 50: 213 (1941).

Carex hypsipedos C.B.Clarke, Bot. Jahrb. Syst. 37: 518 (1906).
Peru to NW. Argentina. 83 BOL PER 85 AGW. Hemicr. or rhizome geophyte.
Carex umbellata var. *depressa* Kük. in H.G.A.Engler (ed.), Pflanzenr., IV, 20(38): 453 (1909).

Carex hypsobates Nelmes, Bull. Misc. Inform. Kew 1940: 272 (1940).
Kashmir. 40 WHM. Hemicr. or rhizome geophyte.

Carex hystericina Muhl. ex Willd., Sp. Pl. 4: 282 (1805).
Canada to Mexico, Jamaica. 71 BRC MAN SAS 72 LAB NFL NSC ONT QUE 73 MNT WYO 74 ILL IOW KAN MIN MSO NEB SDA WIS 75 CNT INI MAS MIC NWJ NWY OHI PEN WVA 76 UTA 77 NWM TEX 78 ARK TEN VRG 79 MXC MXE MXG 81 DOM JAM 84. Hemicr. or rhizome geophyte.
Carex erinacea Muhl. ex Steud., Nomencl. Bot., ed. 2, 1: 288 (1840).
Carex cooleyi Dewey, Amer. J. Sci. Arts 48: 144 (1845). *Carex hystericina* var. *cooleyi* (Dewey) Dewey in A.Wood, Class-book Bot.: 766 (1861).
Carex georgiana Dewey ex Steud., Syn. Pl. Glumac. 2: 225 (1855).
Carex hystericina var. *dudleyi* L.H.Bailey, Mem. Torrey Bot. Club 1: 54 (1889). *Carex hystericina* f. *dudleyi* (L.H.Bailey) Wiegand, Rhodora 26: 2 (1924).

Carex ichangensis C.B.Clarke, J. Linn. Soc., Bot. 36: 290 (1903).
China. 36+. Hemicr. or rhizome geophyte.

Carex idaea Greuter, Matthäs & Risse, Willdenowia 15: 25 (1985).
Kriti. 13 KRI. Hemicr. or rhizome geophyte.

Carex idahoa L.H.Bailey, Bot. Gaz. 21: 5 (1896). *Carex parryana* subsp. *idahoa* (L.H.Bailey) D.F.Murray, Brittonia 21: 73 (1969).
W. U.S.A. 73 IDA MNT ORE 76 CAL UTA. Hemicr. or rhizome geophyte.

Carex idzuroei Franch. & Sav., Enum. Pl. Jap. 2: 583 (1878).
E. China, C. & S. Japan. 36 CHS 38 JAP. Hemicr. or rhizome geophyte.
Carex pseudovesicaria H.Lév. & Vaniot, Bull. Acad. Int. Géogr. Bot. 11: 180 (1902).

Carex ikonnikovii T.V.Egorova, Novosti Sist. Vyssh. Rast. 4: 31 (1968).
C. Asia. 32+. Hemicr. or rhizome geophyte.

Carex iljinii V.I.Krecz. in V.L.Komarov (ed.), Fl. URSS 3: 590 (1935).
Siberia to Mongolia. 30 WSB 37 MON. Hemicr. or rhizome geophyte.

Carex illegitima Ces. in E.von Friedrichsthal, Reise: 271 (1838).
Sicilia (Pantelleria), W. Balkan Pen., W. Turkey, E. Aegean Is., Cyprus. 13 GRC SIC YUG 34 CYP EAI TUR. Hemicr. or rhizome geophyte.
Carex naufragii Hochst. & Steud., Nomencl. Bot., ed. 2, 1: 293 (1840).
Carex pharensis Vis., Fl. Dalmat. 3: 346 (1851).
Carex oliveriana J.Gay ex Boiss., Fl. Orient. 5: 408 (1882).
Coleachyron oliverianum J.Gay in P.E.Boissier, Fl. Orient. 5: 408 (1882).

Carex illota L.H.Bailey, Mem. Torrey Bot. Club 1: 15 (1889).
SW. Canada to W. U.S.A. 71 ABT BRC 73 COL IDA MNT ORE WAS WYO 76 CAL NEV UTA. Hemicr. or rhizome geophyte.
Carex bonplandii var. *minor* Boott, Proc. Acad. Nat. Sci. Philadelphia 15: 77 (1863).
Carex bonplandii var. *angustifolia* W.Boott in S.Watson & al., Bot. California 2: 233 (1880).
Carex dieckii Boeck., Beitr. Cyper. 2: 33 (1890).

Carex impexa K.A.Ford, New Zealand J. Bot. 36: 587 (1998).
New Zealand South I. 51 NZS. Hemicr. or rhizome geophyte.

Carex impressinervia Bryson, Kral & Manhart, Rhodora 89: 357 (1987).
SE. U.S.A. 78 ALA MSI NCA SCA. Hemicr. or rhizome geophyte.

Carex impura Ohwi, Mem. Coll. Sci. Kyoto Imp. Univ., Ser. B, Biol. 6: 251 (1931).
Japan (Honshu: Mt. Hiruzen). 38 JAP. Hemicr. or rhizome geophyte.

Carex inamii Ohwi, Acta Phytotax. Geobot. 11: 256 (1942).
NE. China. 36 CHM. Hemicr. or rhizome geophyte.

Carex inanis Kunth, Enum. Pl. 2: 522 (1837).
C. Asia to Himalaya. 32+ 36 CHC CHT 40 EHM NEP WHM. Hemicr. or rhizome geophyte.
Carex setigera var. *humilis* Nees in R.Wight, Contr. Bot. India: 126 (1834).
Carex setigera var. *fascicularis* Strachey, Cat. Pl. Kumaon: 73 (1854).
Carex setigera var. *minor* Strachey, Cat. Pl. Kumaon: 73 (1854).

Carex incisa Boott in M.C.Perry, Jap. Exped. 2: 327 (1856).
Russian Far East, N. & C. Japan. 31+ 38 JAP. Hemicr. or rhizome geophyte.
Carex textori Miq., Ann. Mus. Bot. Lugduno-Batavi 2: 151 (1866).

Carex inclinis Boott ex C.B.Clarke in J.D.Hooker, Fl. Brit. India 6: 728 (1894).
Nepal to Bhutan. 40 EHM NEP. Hemicr. or rhizome geophyte.

Carex incomitata K.R.Thiele, Muelleria 6: 201 (1986).
SE. Australia. 50 NSW TAS VIC. Hemicr. or rhizome geophyte.

Carex incurviformis Mack. in P.A.Rydberg, Fl. Rocky Mts.: 120 (1917). *Carex maritima* var. *incurviformis* (Mack.) Boivin, Naturaliste Canad. 94: 523 (1967).
Subarctic to WC. U.S.A. 10 FOR ICE NOR SVA SWE 14 RUN 30 WSB 31 70 ASK 71 ABT BRC 73 COL MNT WYO 76 CAL. Hemicr. or rhizome geophyte.
Carex danaensis Stacey, Leafl. W. Bot. 2: 166 (1939).
Carex incurviformis var. *danaensis* (Stacey) F.J.Herm., Leafl. W. Bot. 7: 288 (1955).

Carex indica L., Mant. Pl. 2: 574 (1771).
India to W. Pacific. 36 CHC CHS 40 BAN EHM IND SRL 41 LAO MYA THA VIE 42 BOR JAW MLY PHI SUL 43 NWG SOL 50 QLD 60 FIJ NWC VAN 62 CRL MRN. Hemicr. or rhizome geophyte.

var. ***indica***
India to W. Pacific. 36 CHC CHS 40 BAN EHM IND SRL 41 LAO MYA THA VIE 42 BOR JAW MLY PHI SUL 43 NWG SOL 50 QLD 60 FIJ NWC VAN 62 CRL MRN. Hemicr. or rhizome geophyte.
Carex fuirenioides Gaudich., Voy. Uranie: 412 (1829).
Carex moritzii Steud. in H.Zollinger, Syst. Verz. 2: 60 (1854).
Carex moritzii var. *laxior* Steud. in H.Zollinger, Syst. Verz.: 60 (1854).
Carex fissilis Boott, Ill. Carex 2: 86 (1860). *Carex indica* var. *fissilis* (Boott) Kük. in H.G.A.Engler (ed.), Pflanzenr., IV, 20(38): 264 (1909).
Carex bengalensis Thwaites, Enum. Pl. Zeyl.: 355 (1864).
Carex dietrichiae Boeck., Flora 58: 122 (1875).
Carex longiaristata Boott ex Kurz, J. Asiat. Soc. Bengal, Pt. 2, Nat. Hist. 45(2): 350 (1876).
Carex indica F.Muell. ex Benth., Fl. Austral. 7: 441 (1878), nom. illeg.
Carex indica var. *laetebrunnea* C.B.Clarke in J.D.Hooker, Fl. Brit. India 6: 715 (1894). *Carex indica* subsp. *laetebrunnea* (C.B.Clarke) T.Koyama in M.D.Daddanayake & al. (eds.), Revised Handb. Fl. Ceylon 5: 374 (1985).
Carex indica var. *milnei* C.B.Clarke in J.D.Hooker, Fl. Brit. India 6: 715 (1894).
Carex milnei Boott ex C.B.Clarke in J.D.Hooker, Fl. Brit. India 6: 715 (1894).
Carex fuirenioides var. *gracilis* Hosok., Trans. Nat. Hist. Soc. Taiwan 25: 264 (1935).

var. ***microcarpa*** T.Koyama, Naturaliste Canad. 82: 194 (1955).
N. Thailand. 41 THA. Hemicr. or rhizome geophyte.

Carex indiciformis F.T.Wang & Tang ex P.C.Li, Acta Phytotax. Sin. 37: 156 (1999).
S. China to Hainan. 36 CHC CHH CHS. Hemicr. or rhizome geophyte.

Carex indistincta H.Lév. & Vaniot, Repert. Spec. Nov. Regni Veg. 5: 194 (1908).
Korea. 38 KOR. Hemicr. or rhizome geophyte.

Carex indosinica Raymond, Naturaliste Canad. 82: 154 (1955).
Vietnam. 41 VIE. Hemicr. or rhizome geophyte.

Carex infirminervia Naczi, Novon 12: 528 (2002).
W. Canada to W. U.S.A. 71 ABT BRC 73 IDA MNT ORE VOL WAS WYO 76 CAL NEV UTA. Hemicr. or rhizome geophyte.

Carex infossa Z.P.Wang, J. Nanjing Univ. (Biology) 1962(2): 52 (1962).
SE. China. 36 CHS. Hemicr.

var. ***extensa*** S.W.Su, J. Anhui Univ. Nat. Sci. 20(3): 108 (1996).
China (Anhui). 36 CHS. Hemicr.

var. ***infossa***
E. China. 36 CHS. Hemicr.

Carex infuscata Nees in R.Wight, Contr. Bot. India: 125 (1834).
Afghanistan to Sikkim. 34 AFG 36 CHC 40 EHM NEP PAK WHM. Hemicr. or rhizome geophyte.
Carex atrofurfur T.Koyama, Acta Phytotax. Geobot. 16: 166 (1956).

Carex inopinata V.J.Cook, Trans. & Proc. Roy. Soc. New Zealand 81: 162 (1953).
New Zealand South I. 51 NZS. Hemicr. or rhizome geophyte.

Carex inops L.H.Bailey, Proc. Amer. Acad. Arts 22: 126 (1887).
Canada to W. & C. U.S.A. 71 ABT BRC MAN SAS 72 ONT 73 COL MNT ORE WAS WYO 74 ILL IOW KAN MIN MSO NDA NEB SDA WIS 75 INI MIC 76 CAL 77 NWM. Hemicr. or rhizome geophyte.

subsp. ***heliophila*** (Mack.) Crins, Canad. J. Bot. 61: 1709 (1983).
Canada to WC. & C. U.S.A. 71 ABT BRC MAN SAS 72 ONT 73 COL MNT WYO 74 ILL IOW KAN MIN MSO NDA NEB SDA WIS 75 INI MIC 77 NWM. Hemicr. or rhizome geophyte.
Carex pensylvanica var. *digyna* Boeck., Linnaea 41: 220 (1877).
**Carex heliophila* Mack., Torreya 13: 15 (1913). *Carex pensylvanica* subsp. *heliophila* (Mack.) W.A.Weber, Brittonia 33: 325 (1981).
Carex erxlebeniana Kelso, Biol. Leafl. 51: 1 (1950).

subsp. ***inops***
SW. Canada to W. U.S.A. 71 BRC 73 ORE WAS 76 CAL. Hemicr. or rhizome geophyte.
Carex pensylvanica var. *vespertina* L.H.Bailey, Mem. Torrey Bot. Club 1: 71 (1889). *Carex vespertina* (L.H.Bailey) Howell, Fl. N. W. Amer. 1: 705 (1903).
Carex verecunda T.Holm, Amer. J. Sci. 166: 457 (1903).

Carex insaniae Koidz., Fl. Symb. Orient.-Asiat.: 44 (1930).
S. Kuril Is. to Japan. 31 KUR 38 JAP. Hemicr. or rhizome geophyte.

var. ***insaniae***
S. Kuril Is., N. & C. Japan. 31 KUR 38 JAP. Hemicr. or rhizome geophyte.
Carex fauriei Franch., Bull. Soc. Philom. Paris, VII, 10: 105 (1886), nom. illeg.

var. ***papillaticulmis*** (Ohwi) Ohwi, Acta Phytotax. Geobot. 1: 70 (1932).
C. & S. Japan. 38 JAP. Hemicr. or rhizome geophyte.
**Carex papillaticulmis* Ohwi, Mem. Coll. Sci. Kyoto Imp. Univ., Ser. B, Biol. 5(3): 257 (1930).

var. ***subdita*** (Ohwi) Ohwi, Mem. Coll. Sci. Kyoto Imp. Univ., Ser. B, Biol. 11: 387 (1936).
Japan (Honshu, Shikoku). 38 JAP. Hemicr. or rhizome geophyte.
Carex tosana Makino, Fl. Japan: 1373 (1925).
Carex nankaiensis Honda, Bot. Mag. (Tokyo) 44: 409 (1930).
**Carex subdita* Ohwi, Mem. Coll. Sci. Kyoto Imp. Univ., Ser. B, Biol. 3: 259 (1930).
Carex kiyozumiensis Akiyama, Bot. Mag. (Tokyo) 45: 472 (1931).

Carex insignis Boott, Ill. Carex 1: 5 (1858).
C. & E. Himalaya to Vietnam. 36 CHC 40 ASS EHM NEP 41 VIE. Hemicr. or rhizome geophyte.

Carex insularis Carmich., Trans. Linn. Soc. London 12: 508 (1818).
Tristan da Cunda. 90 TDC. Hemicr. or rhizome geophyte.

Carex integra Mack., Bull. Torrey Bot. Club 43: 608 (1916 publ. 1917).
Oregon to California. 73 ORE 76 CAL. Hemicr. or rhizome geophyte.

Carex interior L.H.Bailey, Bull. Torrey Bot. Club 20: 426 (1873).
Subarctic America to Mexico (Chihuahua). 70 ASK NUN NWT YUK 71 ABT BRC MAN SAS 72 NBR NFL NSC ONT PEI QUE 73 COL IDA MNT ORE WAS WYO 74 ILL IOW KAN MIN MSO NDA NEB SDA WIS 75 CNT INI MAI MAS MIC NWH NWJ NWY OHI PEN VER WVA 76 ARI CAL NEV UTA 78 ARK VRG 79 MXE. Hemicr. or rhizome geophyte.
Carex scirpoides Willd., Sp. Pl. 4: 237 (1805), nom. illeg. *Vignea scirpoides* Rchb. in J.C.Mössler, Handb. Gewächsk. ed. 2, 3: 1611 (1830). *Carex stellulata* var. *scirpoides* Carey in A.Gray, Manual: 544 (1848).
Carex interior subsp. *charlestonensis* Clokey, Bull. S. Calif. Acad. Sci. 38: 1 (1939).
Carex interior var. *keweenawensis* F.J.Herm., Amer. Midl. Naturalist 25: 19 (1941). *Carex interior* f. *keweenawensis* (F.J.Herm.) Fernald, Rhodora 44: 284 (1942).

Carex interjecta Reznicek, Contr. Univ. Michigan Herb. 19: 110 (1993).
C. Mexico. 79 MXC. Hemicr. or rhizome geophyte.

Carex interrupta Boeck., Linnaea 40: 432 (1876).
SW. Canada to NW. U.S.A. 71 BRC 73 ORE WAS. Hemicr. or rhizome geophyte.
Carex angustata var. *verticillata* Boott in W.J.Hooker, Fl. Bor.-Amer. 2: 218 (1839).
Carex verticillata Boott, Ill. Carex 1: 67 (1858), nom. illeg.
Carex interrupta var. *distenta* Kük., Repert. Spec. Nov. Regni Veg. 26: 254 (1929).

Carex intumescens Rudge, Trans. Linn. Soc. London 7: 97 (1804).
Canada to C. & E. U.S.A. 71 MAN 72 LAB NBR NFL NSC ONT PEI QUE 74 ILL IOW MIN MSO OKL SDA WIS 75 CNT INI MAI MAS MIC NWH NWJ NWY OHI PEN RHO VER WVA 77 TEX 78 ALA ARK DEL FLA GEO KTY LOU MRY MSI NCA SCA TEN VRG WDC. Hemicr. or rhizome geophyte.
Carex folliculata Wahlenb., Kongl. Vetensk. Acad. Nya Handl. 24: 152 (1803), nom. illeg.
Carex folliculata var. *major* Pursh, Fl. Amer. Sept. 1: 42 (1813).
Carex intumescens var. *fernaldii* L.H.Bailey, Bull. Torrey Bot. Club 20: 418 (1893).
Carex intumescens f. *ventriosa* Fernald, Rhodora 44: 321 (1942).

Carex inversa R.Br., Prodr.: 242 (1810). *Carex inversa* f. *parvula* Kük. in H.G.A.Engler (ed.), Pflanzenr., IV, 20(38): 189 (1909), nom. inval. *Vignea inversa* (R.Br.) Soják, Cas. Nár. Mus., Odd. Prír. 148: 195 (1979 publ. 1980).
Australia, New Zealand. 50 NFK NSW QLD SOA TAS VIC WAU 51 NZN. Hemicr. or rhizome geophyte.
Carex inversa var. *major* Boott, Ill. Carex 4: 151 (1867). *Carex inversa* f. *major* (Boott) Kük. in H.G.A.Engler (ed.), Pflanzenr., IV, 20(38): 189 (1909).
Carex inversa var. *minor* Boott, Ill. Carex 4: 151 (1867).
Carex inversa var. *perigynium* Boott, Ill. Carex 4: 151 (1867).
Carex inversa var. *leichardtii* Boeck., Linnaea 39: 70 (1875).
Carex smaragdina Colenso, Trans. & Proc. New Zealand Inst. 27: 398 (1894 publ. 1895).
Carex inversa var. *costata* Kük. in H.G.A.Engler (ed.), Pflanzenr., IV, 20(38): 189 (1909).

Carex inversonervosa Nelmes, Kew Bull. 1: 11 (1946).
New Caledonia. 60 NWC. Hemicr. or rhizome geophyte.

Carex ×involuta (Bab.) Syme in J.E.Smith, Engl. Bot., ed. 3, 10: 169 (1870). *C. rostrata* × *C. vesicaria*.
Canada, Europe. 11 GER 72 QUE. Hemicr. or rhizome geophyte.
**Carex vesicaria* var. *involuta* Bab., Man. Brit. Bot., ed. 8: 411 (1881).
Carex ×pannewitziana Figert, Deutsche Bot. Monatsschr. 5: 97 (1887).
Carex vesicaria var. *rostratiformis* Waisb., Magyar Bot. Lapok 4: 76 (1905).

Carex iraqensis S.S.Hooper & Kukkonen, Kew Bull. 38: 673 (1984).
E. Turkey to N. Iraq. 34 IRQ TUR. Hemicr. or rhizome geophyte.

Carex ischnogyne Gilli, Ann. Naturhist. Mus. Wien, B 84: 7 (1980 publ. 1983).
Papua New Guinea. 43 NWG. Hemicr. or rhizome geophyte.

Carex ischnostachya Steud., Syn. Pl. Glumac. 2: 222 (1855).
E. China, Temp. E. Asia. 36 CHC CHS 38 JAP KOR NNS. Hemicr. or rhizome geophyte.
Carex ringgoldiana Boott, Mem. Amer. Acad. Arts, n.s., 6: 419 (1859).
Carex ringgoldiana var. *stenandra* Franch. & Sav., Enum. Pl. Jap. 2: 577 (1878).
Carex nagasakensis C.B.Clarke, Kew Bull., Addit. Ser. 8: 88 (1908).
Carex ischnostachya var. *subtumida* Kük., Repert. Spec. Nov. Regni Veg. 27: 109 (1929). *Carex subtumida* (Kük.) Ohwi, Acta Phytotax. Geobot. 1: 75 (1932).

Carex ivanoviae T.V.Egorova, Novosti Sist. Vyssh. Rast. 3: 34 (1966).
C. Asia to Tibet. 32+ 36 CHQ CHT. Hemicr. or rhizome geophyte.
Carex ivanoviae f. *subglacialis* Yu.E.Alexeev, Byull. Moskovsk. Obshch. Isp. Prir., Otd. Biol. 96: 115 (1991).

Carex ixtapalucensis Reznicek, Syst. Bot. 11: 84 (1986).
Mexico (México State). 79 MXC. Hemicr. or rhizome geophyte.

Carex iynx Nelmes, Proc. Linn. Soc. London 1944: 155, 279 (1944).
SE. Australia. 50 NSW SOA TAS VIC (51) nzn. Hemicr. or rhizome geophyte.

Carex jacens C.B.Clarke, Kew Bull., Addit. Ser. 8: 80 (1908).
S. Kuril Is., N. & C. Japan. 31 KUR 38 JAP. Hemicr. or rhizome geophyte.
Carex geantha Ohwi, Mem. Coll. Sci. Kyoto Imp. Univ., Ser. B, Biol. 5(3): 265 (1930).

Carex jackiana Boott, Proc. Linn. Soc. London 1: 260 (1845).
Assam to Sakhalin and Jawa, SE. Australia. 31 KUR SAK 36 CHC 38 JAP KOR 40 ASS EHM SRL 42 JAW SUM 50 NSW VIC. Hemicr. or rhizome geophyte.

subsp. ***jackiana***
Assam to SC. China, Sri Lanka to Jawa, SE. Australia. 36 CHC 40 ASS EHM SRL 42 JAW SUM 50 NSW VIC. Hemicr. or rhizome geophyte.
Carex jackiana f. *minor* C.B.Clarke in J.D.Hooker, Fl. Brit. India 6: 735 (1894).
Carex ochrolepis Franch., Nouv. Arch. Mus. Hist. Nat., III, 10: 48 (1898). *Carex jackiana* f. *ochrolepis* (Franch.) Kük. in H.G.A.Engler (ed.), Pflanzenr., IV, 20(38): 638 (1909).
Carex oxyphylla Franch., Nouv. Arch. Mus. Hist. Nat., III, 10: 57 (1898). *Carex jackiana* f. *oxyphylla* (Franch.) Kük. in H.G.A.Engler (ed.), Pflanzenr., IV, 20(38): 638 (1909).
Carex pleiorhiza H.Lév. & Vaniot, Bull. Soc. Bot. France 51: 205 (1904).

var. ***macroglossa*** (Franch. & Sav.) Kük. in H.G.A. Engler (ed.), Pflanzenr., IV, 20(38): 638 (1909).
Korea, Japan. 38 JAP KOR. Hemicr. or rhizome geophyte.
Carex cryptandra Franch. & Sav., Enum. Pl. Jap. 2: 148 (1877).
**Carex macroglossa* Franch. & Sav., Enum. Pl. Jap. 2: 576 (1878).

subsp. ***parciflora*** (Boott) Kük. in H.G.A.Engler (ed.), Pflanzenr., IV, 20(38): 638 (1909).
S. Sakhalin, S. Korea, Kuril Is., Japan. 31 KUR SAK 38 JAP KOR. Hemicr. or rhizome geophyte.
**Carex parciflora* Boott, Mem. Amer. Acad. Arts, n.s., 6: 418 (1859).

Carex glehnii F.Schmidt, Reis. Amur-Land., Bot.: 194 (1868).
Carex vaniotii H.Lév., Bull. Acad. Int. Géogr. Bot. 11: 108 (1902). *Carex filipes* var. *vaniotii* (H.Lév.) Kük. in H.G.A.Engler (ed.), Pflanzenr., IV, 20(38): 641 (1909).
Carex kamikochiana Nakai, Rep. Veg. Kamik.: 34, 42 (1928).
Carex gagaensis Akiyama, J. Fac. Sci. Hokkaido Univ., Ser. 5, Bot. 1: 61 (1931).

Carex jacutica V.I.Krecz. in V.L.Komarov (ed.), Fl. URSS 3: 621 (1935).
Siberia to Russian Far East. 30+ 31+. Hemicr. or rhizome geophyte.

Carex jaluensis Kom., Trudy Imp. S.-Peterburgsk. Bot. Sada 20: 369 (1901).
Russian Far East, Korea, China. 31+ 36 CHM CHN 38 KOR. Hemicr. or rhizome geophyte.
Carex dineuros C.B.Clarke, J. Linn. Soc., Bot. 36: 283 (1903).

Carex jamesii Schwein., Ann. Lyceum Nat. Hist. New York 1: 67 (1824).
S. Ontario to C. & E. U.S.A. 72 ONT 74 ILL IOW KAN MIN MSO NEB OKL WIS 75 INI MIC NWJ NWY OHI PEN WVA 77 TEX 78 ALA ARK KTY MRY MSI NCA SCA TEN VRG WDC. Hemicr. or rhizome geophyte.
Carex disperma Kunth, Enum. Pl. 2: 480 (1837), nom. illeg.
Carex steudelii Kunth, Enum. Pl. 2: 480 (1837). *Dapedostachys steudelii* (Kunth) Fedde & J.Schust., Just's Bot. Jahresber. 41(2): 9 (1913 publ. 1918).
Carex austromontana Parish, Bull. S. Calif. Acad. Sci. 4: 108 (1905). *Carex jamesii* var. *austromontana* (Parish) Kük. in H.G.A.Engler (ed.), Pflanzenr., IV, 20(38): 318 (1909).
Carex jacintoensis Parish, Bull. S. Calif. Acad. Sci. 4: 110 (1905).

Carex jamesonii Boott, Proc. Linn. Soc. London 1: 258 (1845).
Mexico to Andes. 79 MXC MXE MXG MXS MXT 80 COS GUA HON NIC PAN 82 VEN 83 BOL CLM ECU PER. Hemicr. or rhizome geophyte.
Carex jamesonii var. *subfulva* Kük., Repert. Spec. Nov. Regni Veg. 8: 7 (1910).

var. ***gracilis*** L.H.Bailey, Bot. Gaz. 13: 88 (1888).
Mexico to Andes. 79 MXC MXE MXG MXS MXT 80 COS GUA 83 CLM ECU PER. Hemicr. or rhizome geophyte.
Carex chordalis Liebm., Mexic. Halvgr.: 269 (1850). *Carex jamesonii* var. *chordalis* (Liebm.) Kük. in H.G.A.Engler (ed.), Pflanzenr., IV, 20(38): 409 (1909).

var. ***jamesonii***
Mexico to Andes. 79 MXG MXS MXT 80 COS GUA HON NIC PAN 82 VEN 83 BOL CLM ECU PER. Hemicr. or rhizome geophyte.
Carex jamesonii var. *minor* Boeck., Bot. Jahrb. Syst. 8: 207 (1887).
Carex jamesonii var. *brevifructus* Kük. in H.G.A.Engler (ed.), Pflanzenr., IV, 20(38): 409 (1909).

var. ***melanosperma*** (Liebm.) Kük. in H.G.A.Engler (ed.), Pflanzenr., IV, 20(38): 409 (1909).
Mexico to Guatemala. 79 MXC MXG MXS MXT 80 GUA. Hemicr. or rhizome geophyte.
**Carex melanosperma* Liebm., Mexic. Halvgr.: 270 (1850).

Carex jankowskii Gorodkov, Trudy Bot. Muz. 20: 216 (1927). *Carex biwensis* subsp. *jankowskii* (Gorodkov) T.V.Egorova, Novosti Sist. Vyssh. Rast. 22: 56 (1985).
Russian Far East. 31+. Hemicr. or rhizome geophyte.

Carex japonica Thunb. in J.A.Murray, Syst. Veg. ed. 14: 845 (1784).
Temp. Asia. 31+ 36 CHC CHI CHM CHN CHS 38 JAP KOR. Hemicr. or rhizome geophyte.
Carex japonica var. *minor* Boott, Ill. Carex 2: 88 (1860).
Carex motoskei Miq., Ann. Mus. Bot. Lugduno-Batavi 2: 148 (1866).
Carex trichostyles Franch. & Sav., Enum. Pl. Jap. 2: 581 (1878).
Carex tokioensis Boeck., Flora 65: 63 (1882).
Carex patens Franch., Bull. Soc. Philom. Paris, VIII, 7: 90 (1895).
Carex baviensis Franch., Nouv. Arch. Mus. Hist. Nat., III, 10: 77 (1898).
Carex japonica var. *gracilis* Franch., Nouv. Arch. Mus. Hist. Nat., III, 10: 78 (1898).
Carex tricarinata H.Lév. & Vaniot, Bull. Soc. Bot. France 53: 317 (1906).
Carex japoniciformis Nakai, Rep. Veg. Kamik.: 34, 42 (1928).

Carex jeanpertii E.G.Camus, Notul. Syst. (Paris) 1: 295 (1910).
Vietnam. 41 VIE. Hemicr. or rhizome geophyte.

Carex jiaodongensis Y.M.Zhang & X.D.Chen, Acta Phytotax. Sin. 31: 381 (1993).
China (Shandong). 36 CHN. Hemicr. or rhizome geophyte.

Carex jinfoshanensis Tang & F.T.Wang ex S.Y.Liang, Acta Phytotax. Sin. 37: 191 (1999).
China (Sichuan). 36 CHC. Hemicr. or rhizome geophyte.

Carex jisaburo-ohwiana T.Koyama, Jap. J. Bot. 15: 176 (1956). *Carex caucasica* subsp. *jisaburo-ohwiana* (T.Koyama) T.Koyama, J. Fac. Sci. Univ. Tokyo, Sect. 3, Bot. 8(4): 197 (1962).
Taiwan. 38 TAI. Hemicr. or rhizome geophyte.

Carex jiuxianshanensis L.K.Dai & Y.Z.Huang, Fl. Fujianica 6: 690 (1995).
China (Zhejiang, Fujian). 36 CHS. Hemicr. or rhizome geophyte.

Carex jizhuangensis S.Yun Liang, in Fl. Reipubl. Popul. Sin. 12: 522 (2000).
China (Guangdong). 36 CHS. Hemicr.

Carex johnstonii Boeck., Bot. Jahrb. Syst. 7: 278 (1886).
Uganda to Malawi. 23 RWA ZAI 25 KEN TAN UGA 26 MLW. Hemicr. or rhizome geophyte.

Carex jonesii L.H.Bailey, Mem. Torrey Bot. Club 1: 16 (1889). *Carex nervina* var. *jonesii* (L.H.Bailey) Kük. in H.G.A.Engler (ed.), Pflanzenr., IV, 20(38): 167 (1909).
W. U.S.A. 73 COL IDA MNT ORE WAS WYO 76 CAL NEV. Hemicr. or rhizome geophyte.

Carex joorii L.H.Bailey, Proc. Amer. Acad. Arts 22: 72 (1887).
SE. U.S.A. to Texas. 74 MSO 77 TEX 78 ALA ARK FLA GEO KTY LOU MRY MSI NCA SCA TEN VRG. Hemicr. or rhizome geophyte.

Carex* ×*josephi-schmittii Raymond, Canad. J. Bot. 28: 428 (1950). *C. intumescens* var. *fernaldii* × *C. retrorsa*.
Québec. 72 QUE. Hemicr. or rhizome geophyte.

Carex jubozanensis J.Oda & A.Tanaka, Acta Phytotax. Geobot. 54: 128 (2003).
Japan (Honshu). 38 JAP. Hemicr.

Carex juniperorum Catling, Reznicek & Crins, Syst. Bot. 18: 497 (1993).
Ontario to Kentucky. 72 ONT 75 OHI 78 KTY VRG. Hemicr. or rhizome geophyte.

Carex juvenilis C.B.Clarke ex E.G.Camus, Notul. Syst. (Paris) 2: 207 (1912).
N. & NE. Thailand. 41 THA. Hemicr. or rhizome geophyte.
Carex sychnostachya Nelmes, Bull. Misc. Inform. Kew 1939: 658 (1940).

Carex kabanovii V.I.Krecz. in V.L.Komarov (ed.), Fl. URSS 3: 593 (1935).
Sakhalin, N. Japan. 31 SAK 38 JAP. Hemicr. or rhizome geophyte.
Carex yezo-alpina Akiyama, Biogeogrophica (Tokyo) 1: 53 (1935).

Carex kaloides Petrie, Trans. & Proc. New Zealand Inst. 13: 332 (1881).
New Zealand South I. 51 NZS. Hemicr. or rhizome geophyte.

Carex kansuensis Nelmes, Bull. Misc. Inform. Kew 1939: 201 (1939).
Tibet to C. China. 36 CHC CHN CHQ CHT. Hemicr. or rhizome geophyte.

Carex kaoi Tang & F.T.Wang ex S.Y.Liang, Acta Phytotax. Sin. 36: 530 (1998).
SE. China. 36 CHS. Hemicr. or rhizome geophyte.

Carex karisimbiensis Cherm., Bull. Soc. Bot. France 82: 344 (1935).
Zaïre. 23 ZAI. Hemicr. or rhizome geophyte.

Carex karlongensis Kük., Acta Horti Gothob. 5: 47 (1930).
SC. China. 36 CHC. Hemicr. or rhizome geophyte.

subsp. ***handelii*** (Kük.) P.C.Li, in Vasc. Pl. Hengduan Mt. 2: 2380 (1994).
China (Sichuan, Yunnan). 36 CHC. Hemicr. or rhizome geophyte.
**Carex handelii* Kük. in H.F.von Handel-Mazzetti, Symb. Sin. 7: 1268 (1936).

subsp. ***karlongensis***
China (Sichuan). 36 CHC. Hemicr. or rhizome geophyte.

Carex karoi (Freyn) Freyn, Oesterr. Bot. Z. 40: 303 (1890).
Afghanistan to N. China. 32 KAZ KGZ TZK UZB 34 AFG 36 CHI CHM CHN 40 PAK WHM. Hemicr. or rhizome geophyte.
**Carex capillaris* subsp. *karoi* Freyn, Oesterr. Bot. Z. 40: 303 (1890).
Carex capillaris var. *parvirostris* Kük., Acta Horti Gothob. 5: 46 (1929).
Carex selengensis K.V.Ivanova, Bot. Mater. Gerb. Bot. Inst. Komarova Akad. Nauk S.S.S.R. 7: 276 (1937). *Carex karoi* subsp. *selengensis* (K.V.Ivanova) T.V. Egorova, Novosti Sist. Vyssh. Rast. 10: 103 (1973).
Carex licentii Nelmes, Bull. Misc. Inform. Kew 1939: 185 (1940).
Carex takenakai Nakai, J. Jap. Bot. 19: 318 (1943).

Carex kashmirensis C.B.Clarke in J.D.Hooker, Fl. Brit. India 6: 743 (1894).
Kashmir to C. Asia. 32 KGZ TZK 40 PAK WHM. Hemicr. or rhizome geophyte.

Carex* ×*kattaeana Kük., Bull. Herb. Boissier, II, 5: 1161 (1905). *C. doenitzii* × *C. splendens*.
Japan. 38 JAP. Hemicr. or rhizome geophyte.
Carex ×*levicuspis* Ohwi, Mem. Coll. Sci. Kyoto Imp. Univ., Ser. B, Biol. 3: 253 (1930).

Carex kauaiensis R.W.Krauss, Pacific Sci. 4: 279 (1950).
Hawaiian Is. (Kauai). 63 HAW. Hemicr. or rhizome geophyte.

Carex* ×*kenaica Lepage, Naturaliste Canad. 84: 95 (1957). *C. ramenskii* × *C. subspathacea*.
Alaska. 70 ASK. Hemicr. or rhizome geophyte.

Carex kermadecensis Petrie, Trans. & Proc. New Zealand Inst. 47: 56 (1914 publ. 1915).
Kermadec Is. 51 KER. Hemicr. or rhizome geophyte.

Carex* ×*ketonensis Akiyama, J. Fac. Sci. Hokkaido Univ., Ser. 5, Bot. 3: 262 (1935). *C. lyngbyei* × *C. tuminensis*.
Sakhalin. 31 SAK. Hemicr. or rhizome geophyte.

Carex khoii T.V.Egorova & Aver., Novosti Sist. Vyssh. Rast. 26: 13 (1989).
Vietnam. 41 VIE. Hemicr. or rhizome geophyte.

Carex kiangsuensis Kük., Repert. Spec. Nov. Regni Veg. 27: 109 (1929).
China (Anhui, Jiangsu). 36 CHS. Hemicr. or rhizome geophyte.
Carex conica var. *densa* Kük., Repert. Spec. Nov. Regni Veg. Beih. 12: 311 (1922).

Carex killickii Nelmes, Kew Bull. 10: 89 (1955).
S. Africa. 27 LES. Hemicr. or rhizome geophyte.

Carex kimurae Ohwi & T.Koyama, Sci. Rep. Tohoku Imp. Univ., Ser. 4, Biol. 29: 143 (1963).
Nansei-shoto. 38 NNS. Hemicr. or rhizome geophyte.

Carex kingii (R.Br. ex Boott) Reznicek, Canad. J. Bot. 68: 1419 (1990).
S. Chile to S. Argentina. 85 AGS CLS. Hemicr. or rhizome geophyte.
**Uncinia kingii* R.Br. ex Boott in J.D.Hooker, Fl. Antarct. 2: 370 (1846).

Carex kiotensis Franch. & Sav., Enum. Pl. Jap. 2: 556 (1878). *Carex prescottiana* var. *kiotensis* (Franch. & Sav.) Kük. in H.G.A.Engler (ed.), Pflanzenr., IV, 20(38): 356 (1909).
Japan, Taiwan. 38 JAP TAI. Hemicr. or rhizome geophyte.
Carex fuscescens Boeck., Bot. Jahrb. Syst. 5: 517 (1884). *Carex prescottiana* var. *fuscescens* (Boeck.) Kük. in H.G.A.Engler (ed.), Pflanzenr., IV, 20(38): 356 (1909).
Carex jizogatakensis H.Lév. & Vaniot, Bull. Soc. Bot. France 51: 203 (1904).
Carex longispicata Hayata, J. Coll. Sci. Imp. Univ. Tokyo 30(1): 386 (1911).
Carex phaeopoda Ohwi, Acta Phytotax. Geobot. 2: 159 (1933).

Carex kirganica Kom., Repert. Spec. Nov. Regni Veg. 13: 164 (1914).

Siberia to Korea. 30+ 31 KAM 36 CHI CHM 38 KOR. Hemicr. or rhizome geophyte.
Carex graciliculmis Ohwi, Mem. Coll. Sci. Kyoto Imp. Univ., Ser. B, Biol. 6: 264 (1931).
Carex mukdenensis Kitag., Bot. Mag. (Tokyo) 48: 29 (1934).

Carex kirinensis W.Wang & Y.L.Chang, in Fl. Plant. Herb. Chinae Bor.-Or. 11: 207 (1976).
NE. China. 36 CHM. Hemicr. or rhizome geophyte.

Carex kirkii Petrie, Trans. & Proc. New Zealand Inst. 18: 297 (1885 publ. 1886).
New Zealand South I. 51 NZS. Hemicr. or rhizome geophyte.

var. ***elatior*** Kük. in T.F.Cheeseman, Man. New Zealand Fl.: 811 (1906).
New Zealand South I. 51 NZS. Hemicr. or rhizome geophyte.

var. ***kirkii***
New Zealand South I. 51 NZS. Hemicr. or rhizome geophyte.
Carex kirkii var. *membranacea* Kük. in T.F. Cheeseman, Man. New Zealand Fl.: 811 (1906).

Carex kitaibeliana Degen ex Bech., Ber. Schweiz. Bot. Ges. 70: 178 (1960).
SE. Europe to Turkey. 13 ALB BUL GRC ITA YUG 34 TUR. Hemicr. or rhizome geophyte.

var. ***capillata*** (Acht.) Stoeva, E.D.Popova & Uzunova in S.I.Kozhukhorov & B.A.Kuzmanov (eds.), Evolyutsiya Tsvet. Rast. Florogen. 2: 152 (1991).
Balkan Pen. 13 GRC. Hemicr. or rhizome geophyte.
**Carex laevis* f. *capillata* Acht., Gen. Carex Bulgar.: 123 (1957). *Carex kitaibeliana* subsp. *capillata* (Acht.) Hartvig in P.A.K.Strid & Kit Tan (eds.), Mount. Fl. Greece 2: 860 (1991).

var. ***kitaibeliana***
SE. Europe, Turkey. 13 ALB BUL GRC ITA YUG 34 TUR. Hemicr. or rhizome geophyte.
Carex laevis f. *balcana* Degen & Nejceff in ?. *Carex kitaibeliana* f. *balcana* (Degen & Nejceff) Stoeva, E.D.Popova & Uzunova in S.I. Kozhukhorov & B.A.Kuzmanov (eds.), Evolyutsiya Tsvet. Rast. Florogen. 2: 152 (1991).
Carex laevis Kit. ex Willd., Sp. Pl. 4: 292 (1805), nom. illeg.
Carex ferruginea var. *laevis* Fiori, Nuov. Fl. Italia 1: 197 (1923).

Carex klaphakei K.L.Wilson, Telopea 6: 569 (1996).
New South Wales (C. Tablelands). 50 NSW. Hemicr. or rhizome geophyte.

Carex × knieskernii Dewey, Amer. J. Sci. Arts 32: 247 (1846). *C. arctata* × *C. castanea*. *Carex castanea* var. *knieskernii* (Dewey) Mack. in N.L.Britton & al. (eds.), N. Amer. Fl. 18: 287 (1935).
E. Canada to NC. & NE. U.S.A. 72 NBR ONT QUE 74 MIN WIS 75 MAI MAS MIC NWH NWY VER. Hemicr. or rhizome geophyte.

Carex knorringiae Kük. ex Krecz. in V.L.Komarov (ed.), Fl. URSS 3: 597 (1935).
C. Asia. 32 KGZ UZB. Hemicr. or rhizome geophyte.

Carex kobomugi Ohwi, Mem. Coll. Sci. Kyoto Imp. Univ., Ser. B, Biol. 3: 281 (1930). *Vignea kobomugi* (Ohwi) Soják, Cas. Nár. Mus., Odd. Prír. 148: 195 (1979 publ. 1980).
Temp. Asia. 31+ 36 CHM CHN CHQ CHS 38 JAP KOR TAI (73) ore (75) nwj nwy rho (78) del mry nca vrg. Hemicr. or rhizome geophyte.
Carex macrocephala var. *longibracteata* Oliv., J. Linn. Soc., Bot. 9: 170 (1866).

Carex korkischkoae A.E.Kozhevn., Byull. Moskovsk. Obshch. Isp. Prir., Otd. Biol. 92: 128 (1987).
Russian Far East. 31+. Hemicr. or rhizome geophyte.

Carex korshinskyi Kom., Trudy Imp. S.-Peterburgsk. Bot. Sada 20: 394 (1901). *Carex supina* var. *korshinskyi* (Kom.) Kük. in H.G.A.Engler (ed.), Pflanzenr., IV, 20(38): 457 (1909). *Carex supina* subsp. *korshinskyi* (Kom.) Malyschev, in Fl. Sibir., Cyperac. 3: 133 (1990).
Siberia to China. 30+ 31+ 36 CHI CHM CHN CHX 37 MON. Hemicr. or rhizome geophyte.
Carex costata Turcz. ex Ledeb., Fl. Ross. 4: 305 (1852), nom. illeg. *Carex supina* var. *costata* Meinsh., Trudy Imp. S.-Peterburgsk. Bot. Sada 18: 392 (1900).

Carex koshewnikowii Litv., Trudy Bot. Muz. Imp. Akad. Nauk 7: 92 (1910).
Afghanistan to W. Himalaya. 32 KAZ KGZ TZK UZB 34 AFG 40 PAK WHM. Hemicr. or rhizome geophyte.

subsp. ***chitralensis*** (Nelmes) Dickore, Stapfia 39: 48 (1995).
Afghanistan to W. Himalaya. 34 AFG 40 PAK WHM. Hemicr. or rhizome geophyte.
**Carex chitralensis* Nelmes, Nytt Mag. Bot. 1: 22 (1952).

subsp. ***koshewnikowii***
C. Asia. 32 KAZ KGZ TZK UZB. Hemicr. or rhizome geophyte.

Carex kraliana Naczi & Bryson, Novon 12: 520 (2002).
E. U.S.A. to Texas. 75 INI OHI 77 TEX 78 ALA ARK FLA GEO KTY LOU MRY MSI NCA SCA TEN VRG. Hemicr. or rhizome geophyte.

Carex krascheninnikovii Kom. ex V.I.Krecz., Bot. Mater. Gerb. Bot. Inst. Komarova Akad. Nauk S.S.S.R. 9: 22 (1941). *Carex flavocuspis* subsp. *krascheninnikovii* (Kom. ex V.I.Krecz.) T.V.Egorova, Novosti Sist. Vyssh. Rast. 16: 58 (1979).
Russian Far East. 31+. Hemicr. or rhizome geophyte.

Carex krausei Boeck., Bot. Jahrb. Syst. 7: 279 (1886). *Carex capillaris* var. *krausei* (Boeck.) Kranz, Cat. Canad. Pl. 2: 163 (1888). *Carex capillaris* f. *krausei* (Boeck.) Kük. in H.G.A.Engler (ed.), Pflanzenr., IV, 20(38): 591 (1909). *Carex capillaris* subsp. *krausei* (Boeck.) Böcher, Meddel. Grønland: 147 (1952).
Subarctic to N. Québec. 10 ICE 14 RUN 30 ALT KRA WSB YAK 31+ 70 ASK GNL NUN NWT YUK 72 QUE. Hemicr. or rhizome geophyte.

subsp. ***krausei***
Subarctic. 14 RUN 30 ALT KRA WSB YAK 31+ 70 ASK NWT YUK. Hemicr. or rhizome geophyte.

subsp. ***porsildiana*** (Polunin) Á.Löve, D.Löve & Raymond, Acta Horti Gothob. 20: 175 (1956).
Subarctic to N. Québec. 10 ICE 70 ASK GNL NUN 72 QUE. Hemicr. or rhizome geophyte.
**Carex capillaris* var. *porsildiana* Polunin, J. Linn. Soc., Bot. 52: 373 (1943). *Carex capillaris* f. *porsildiana* (Polunin) Raymond, Naturaliste Canad. 77: 59 (1950). *Carex capillaris* subsp. *porsildiana* (Polunin) Böcher, Meddel. Grønland: 147 (1952).

Carex kreczetoviczii T.V.Egorova, Bot. Mater. Gerb. Bot. Inst. Komarova Akad. Nauk S.S.S.R. 19: 62 (1959).
Siberia to Russian Far East. 30 YAK 31 KAM MAG. Hemicr. or rhizome geophyte.

Carex kuchunensis Tang & F.T.Wang ex S.Y.Liang, Acta Phytotax. Sin. 36: 536 (1998).
China (Guangxi). 36 CHS. Hemicr. or rhizome geophyte.

Carex kucyniakii Raymond, Naturaliste Canad. 82: 155 (1955).
China (Yunnan) to Vietnam. 36 CHC 41 VIE. Hemicr. or rhizome geophyte.

Carex kujuzana Ohwi, Acta Phytotax. Geobot. 1: 298 (1932).
Japan (Kyushu: Mt. Kuju). 38 JAP. Hemicr. or rhizome geophyte.

Carex kukkonenii Ö.Nilsson in P.H.Davis (ed.), Fl. Turkey 9: 622 (1985).
E. Turkey to N. Iran. 34 IRN IRQ TUR. Hemicr. or rhizome geophyte.

Carex kulingana L.H.Bailey, Gentes Herb. 1920: 13 (1920).
China (Jiangxi). 36 CHS. Hemicr. or rhizome geophyte.

Carex kumaonensis Kük. in H.G.A.Engler (ed.), Pflanzenr., IV, 20(38): 544 (1909).
Himalaya to SC. China. 36 CHC 40 NEP WHM. Hemicr. or rhizome geophyte.

Carex kurdica Kük. ex Hand.-Mazz., Ann. K. K. Naturhist. Hofmus. 28: 23 (1914).
E. Turkey to NW. Afghanistan. 34 AFG IRN IRQ TUR. Hemicr. or rhizome geophyte.

Carex* ×*kurilensis Ohwi, Acta Phytotax. Geobot. 2: 27 (1933). *C. cespitosa* × *C. cryptocarpa*.
Kuril Is. 31 KUR. Hemicr. or rhizome geophyte.

Carex kurtziana Kük., Bot. Jahrb. Syst. 27: 503 (1899). *Carex nebularum* var. *kurtziana* (Kük.) Kük. in H.G.A.Engler (ed.), Pflanzenr., IV, 20(38): 118 (1909).
WC. & SW. Argentina, C. & S. Chile. 85 AGS CLS. Hemicr. or rhizome geophyte.

Carex kwangsiensis F.T.Wang & Tang ex P.C.Li, Acta Phytotax. Sin. 37: 164 (1999).
China (Guangxi). 36 CHS. Hemicr. or rhizome geophyte.

Carex kwangtoushanica K.T.Fu, in Fl. Tsinlingensis 1(1): 447 (1976).
China (Shaanxi). 36 CHN. Hemicr. or rhizome geophyte.

Carex lachenalii Schkuhr, Beschr. Riedgräs. 1: 51 (1801), nom. cons. *Carex heleonastes* var. *lachenalii* (Schkuhr) Fiori, Nuov. Fl. Italia 1: 186 (1923). *Vignea lachenalii* (Schkuhr) Soják, Cas. Nár. Mus., Odd. Prír. 148: 195 (1979 publ. 1980).
Temp. & Subarctic Northern Hemisphere, New Zealand. 10 FIN GRB ICE NOR SVA SWE 11 AUT CZE POL SWI 12 FRA POR SPA 13 ITA ROM YUG? 14 BLR? RUE RUN UKR 30 ALT KRA WSB YAK 31 KAM KHA KUR MAG 32 KAZ UZB 36 CHM 37 MON 38 JAP KOR 51 NZS 70 ASK GNL NUN NWT YUK 71 ABT BRC MAN 72 LAB NBR NFL QUE 73 COL IDA MNT WAS WYO 76. Hemicr. or rhizome geophyte.

Carex lagopina var. *subtenuiflora* Almq., Öfvers. Kongl. Vetensk.-Akad. Förh. 41(7): 72 (1884).

subsp. ***baetica*** (Gay) Luceño & Muñoz Garm., Fontqueria 11: 3 (1986).
Iberian Pen. 12 POR SPA. Hemicr. or rhizome geophyte.

Carex furva Webb, Iter Hispan.: 5 (1838). *Carex lagopina* var. *furva* (Webb) Nyman, Consp. Fl. Eur.: 780 (1882).

**Carex lagopina* var. *baetica* Gay, Ann. Sci. Nat., Bot., II, 11: 181 (1839). *Carex bipartita* subsp. *baetica* (Gay) Luceño & Muñoz Garm., Anales Jard. Bot. Madrid 44: 439 (1987).

subsp. ***lachenalii***
Temp. & Subarctic Northern Hemisphere, New Zealand. 10 FIN GRB ICE NOR SVA SWE 11 AUT CZE POL SWI 12 FRA POR SPA 13 ITA ROM YUG? 14 BLR? RUE RUN UKR 30 ALT KRA WSB YAK 31 KAM KHA KUR MAG 36 CHM 37 MON 38 JAP KOR 51 NZS 70 ASK GNL NUN NWT YUK 71 ABT BRC MAN 72 LAB NBR NFL QUE 73 COL IDA MNT WAS WYO 76. Hemicr. or rhizome geophyte.

Carex tripartita All., Fl. Pedem. 2: 265 (1785). *Vignea tripartita* (All.) Rchb. in J.C.Mössler, Handb. Gewächsk. ed. 2, 3: 1609 (1830). *Carex curvula* var. *tripartita* (All.) Nyman, Consp. Fl. Eur.: 783 (1882).

Carex tripartita Suter, Helvet. Fl. 2: 241 (1802), nom. illeg.

Carex lagopina Wahlenb., Kongl. Vetensk. Acad. Nya Handl. 1803: 145 (1803). *Vignea lagopina* (Wahlenb.) Rchb. in J.C.Mössler, Handb. Gewächsk. ed. 2, 3: 1609 (1830).

Carex parviflora Gaudin, Etrennes Fl.: 84 (1804), nom. illeg.

Carex lagopina var. *pleiostachya* Drejer, Rev. Car. Bor.: 25 (1841).

Carex lagopina var. *gracilescens* Th.Fr., Bot. Not. 1857: 208 (1857).

Carex parkeri Petrie, Trans. & Proc. New Zealand Inst. 13: 332 (1881). *Carex lachenalii* subsp. *parkeri* (Petrie) Toivonen, Ann. Bot. Fenn. 16: 154 (1979).

Carex sewellii A.Benn. & C.B.Clarke, Trans. Bot. Soc. Edinburgh 17: 456 (1889).

Carex lagopina var. *debilis* Lange, Consp. Fl. Groenland., ed. 2, 1: 135 (1890).

Carex lagopina var. *major* Lange, Consp. Fl. Groenland., ed. 2, 1: 135 (1890).

Carex lagopina f. *angustifolia* Norman, Forh. Vidensk.-Selsk. Kristiania 1893: 47 (1893).

Carex lagopina var. *laxior* Norman, Forh. Vidensk.-Selsk. Kristiania 1893: 47 (1893).

Carex lagopina f. *pauciflora* Norman, Forh. Vidensk.-Selsk. Kristiania 1893: 47 (1893).

Carex cooptanda C.B.Clarke in J.D.Hooker, Fl. Brit. India 6: 707 (1894).

Carex lagopina var. *longisquama* Kük. ex J.M.Macoun, The Fur Seals Is. 3: 572 (1899).

Carex bipartita var. *austromontana* F.J.Herm., Leafl. W. Bot. 10: 16 (1963).

Carex lacistoma R.Br., Prodr.: 243 (1810).
New South Wales. 50 NSW. Hemicr. or rhizome geophyte.

Carex lacustris Willd., Sp. Pl. 4: 306 (1805). *Anithista lacustris* (Willd.) Raf., Good Book: 26 (1840). *Carex*

riparia var. *lacustris* (Willd.) Kük. in H.G.A.Engler (ed.), Pflanzenr., IV, 20(38): 736 (1909).
Canada to N. & C. U.S.A. 71 ABT MAN SAS 72 NBR NFL NSC ONT QUE 73 IDA MNT 74 ILL IOW MIN MSO NDA NEB SDA WIS 75 CNT INI MAI MAS MIC NWH NWJ NWY OHI PEN RHO VER WVA 78 DEL MRY VRG WDC. Hemicr. or rhizome geophyte.
Carex riparia Muhl., Descr. Gram.: 259 (1817), nom. illeg. *Anithista riparia* Raf., Good Book: 26 (1840).

Carex laeta Boott, Ill. Carex 1: 69 (1858).
Nepal to SC. China. 36 CHC CHT 40 EHM NEP. Hemicr. or rhizome geophyte.

subsp. ***gelongii*** Noltie, Edinburgh J. Bot. 50: 198 (1993).
Bhutan, Tibet. 36 CHT 40 EHM. Hemicr. or rhizome geophyte.

subsp. ***laeta***
Nepal to SC. China. 36 CHC CHT 40 EHM NEP. Hemicr. or rhizome geophyte.

Carex laeviconica Dewey, Amer. J. Sci. Arts 24: 47 (1857). *Carex aristata* var. *laeviconica* (Dewey) Kük. in H.G.A.Engler (ed.), Pflanzenr., IV, 20(38): 754 (1909).
C. Canada to NW. & NC. U.S.A. 71 MAN SAS 72 ONT 73 MNT 74 ILL IOW KAN MIN MSO NDA NEB SDA WIS. Hemicr. or rhizome geophyte.

Carex laeviculmis Meinsh., Bot. Centralbl. 55: 195 (1893).
Alaska to California. 70 ASK 71 BRC 73 IDA MNT ORE WAS WYO 76 CAL. Hemicr. or rhizome geophyte.
Carex bolanderi var. *sparsiflora* Olney, Proc. Amer. Acad. Arts 7: 407 (1872). *Carex deweyana* var. *sparsiflora* (Olney) L.H.Bailey, Bot. Gaz. 13: 87 (1888).

Carex laevigata Sm., Trans. Linn. Soc. London 5: 272 (1799). *Trasus laevigatus* (Sm.) Gray, Nat. Arr. Brit. Pl. 2: 66 (1821). *Edritria laevigata* (Sm.) Raf., Good Book: 26 (1840).
W. & SW. Europe to Scotland, S. Bulgaria, NW. Morocco. 10 GRB IRE 11 BGM GER NET 12 COR FRA POR SPA 13 BUL SIC 20 MOR. Hemicr. or rhizome geophyte.
Carex helodes Link, J. Bot. (Schrader) 2: 309 (1799). *Carex distans* var. *helodes* (Link) Fiori, Nuov. Fl. Italia 1: 199 (1923).
Carex patula Schkuhr, Beschr. Riedgräs. 1: 115 (1801), nom. illeg.
Carex schraderi Willd., Sp. Pl. 4: 286 (1805).
Carex longirostris Krock., Fl. Siles. 3: 169 (1814).
Carex biligularis DC. in J.B.A.M.de Lamarck & A.P.de Candolle, Fl. Franç., ed. 3, 5: 296 (1815). *Carex laevigata* var. *biligularis* (DC.) Nyman, Consp. Fl. Eur.: 769 (1882). *Carex laevigata* f. *biligularis* (DC.) Asch. & Graebn., Syn. Mitteleur. Fl. 2(2): 182 (1903). *Carex helodes* f. *biligularis* (DC.) Kük. in H.G.A.Engler (ed.), Pflanzenr., IV, 20(38): 656 (1909).
Carex binervis Dewey, Amer. J. Sci. Arts 30: 61 (1836).
Carex flacciformis Hoffmanns. ex Kunth, Enum. Pl. 2: 493 (1837).
Carex welwitschii Boiss. ex Steud., Syn. Pl. Glumac. 2: 230 (1855). *Carex laevigata* var. *welwitschii* (Boiss. ex Steud.) Boott, Ill. Carex 4: 163 (1867).
Carex helodes var. *maurusia* Font Quer & Maire in É.Jahandiez & al., Cat. Pl. Maroc: 1154 (1931).
Carex helodes subsp. *andalusica* H.Lindb., Itin. Medit.: 28 (1932).

Carex laevissima Nakai, Repert. Spec. Nov. Regni Veg. 13: 245 (1914).
Russian Far East to Korea, C. Japan. 31 PRM 36 CHI CHM 38 JAP KOR. Hemicr. or rhizome geophyte.

Carex laevivaginata (Kük.) Mack. in N.L.Britton & A.Brown, Ill. Fl. N. U.S., ed. 2, 1: 371 (1913).
S. Ontario to NC. & E. U.S.A. 72 ONT 74 ILL IOW MIN MSO WIS 75 CNT INI MAS MIC NWJ NWY OHI PEN RHO VER WVA 78 ALA ARK DEL FLA GEO KTY MRY MSI NCA SCA TEN VRG WDC. Hemicr. or rhizome geophyte.
**Carex stipata* var. *laevivaginata* Kük. in H.G.A. Engler (ed.), Pflanzenr., IV, 20(38): 172 (1909).

Carex lageniformis Nelmes, Bull. Misc. Inform. Kew 1939: 658 (1940).
Thailand. 41 THA. Hemicr. or rhizome geophyte.

Carex lagunensis M.E.Jones, Contr. W. Bot. 18: 26 (1933).
N. & C. Mexico. 79 MXC MXE MXN. Hemicr. or rhizome geophyte.
Carex diehlii M.E.Jones ex F.J.Herm., Amer. Midl. Naturalist 51: 268 (1954).

Carex lambertiana Boott in J.D.Hooker, Fl. Nov.-Zel. 1: 284 (1853). *Carex dissita* var. *lambertiana* (Boott) Cheeseman, Trans. & Proc. New Zealand Inst. 16: 437 (1884).
New Zealand. 51 NZN NZS. Hemicr. or rhizome geophyte.

Carex lamprocarpa Phil., Anales Univ. Chile 93: 497 (1896).
S. Chile. 85 CLS. Hemicr. or rhizome geophyte.

Carex lamprochlamys S.T.Blake, J. Arnold Arbor. 28: 28 (1947).
New Guinea. 43 NWG. Hemicr. or rhizome geophyte.
Carex papuana Nelmes, Kew Bull. 4: 379 (1949).
Carex tyttholepis Nelmes, Kew Bull. 4: 388 (1949).

Carex lancangensis S.Yun Liang, Acta Phytotax. Sin. 37: 187 (1999).
China (Yunnan). 36 CHC. Hemicr. or rhizome geophyte.

Carex lanceolata Boott in M.C.Perry, Jap. Exped. 2: 326 (1856).
Siberia to Japan. 30+ 31 PRM 36 CHC CHI CHM CHN CHS 37 MON 38 JAP KOR. Hemicr. or rhizome geophyte.

var. ***lanceolata***
Siberia to Japan. 30+ 31 PRM 36 CHC CHI CHM CHN 37 MON 38 JAP KOR. Hemicr. or rhizome geophyte.
Carex pediformis var. *floribunda* Korsh., Trudy Imp. S.-Peterburgsk. Bot. Sada 12: 409 (1892).
Carex floribunda Meinsh., Trudy Imp. S.-Peterburgsk. Bot. Sada 18: 402 (1900).
Carex longisquamata Meinsh. ex Kom., Trudy Imp. S.-Peterburgsk. Bot. Sada 20: 399 (1901).
Carex delicatula C.B.Clarke, Kew Bull., Addit. Ser. 8: 79 (1908).
Carex yesoensis Koidz., Bot. Mag. (Tokyo) 32: 55 (1918).
Carex lanceolata var. *albomediana* Makino, J. Jap. Bot. 6: 38 (1929).

Carex lanceolata var. *alashanica* T.V.Egorova in V.I.Grubov, Pl. Asia Cent. 3: 74 (1967).

var. ***laxa*** Ohwi, Cyper. Jap. 1: 402 (1936). *Carex karafutoana* Ohwi, Mem. Coll. Sci. Kyoto Imp. Univ., Ser. B, Biol. 18(1): 170 (1944).
Russian Far East to N. China, Japan. 31+ 36 CHI CHM 38 JAP. Hemicr. or rhizome geophyte.

var. ***subpediformis*** Kük. in H.G.A.Engler (ed.), Pflanzenr., IV, 20(38): 493 (1909). *Carex subpediformis* (Kük.) Sutô & Suzuki, Bull. Coll. Agric. Utsunomiya Univ. 8: 11 (1933).
China. 36 CHC CHI CHM CHN. Hemicr. or rhizome geophyte.
Carex prevernalis Kitag., Bot. Mag. (Tokyo) 48: 15 (1934).
Carex pseudolanceolata V.I.Krecz. in V.L.Komarov (ed.), Fl. URSS 3: 613 (1935). *Carex lanceolata* var. *pseudolanceolata* (V.I.Krecz.) Vorosch. in A.K.Skvortsov (ed.), Florist. Issl. Razn. Raionakh SSSR: 155 (1985).

Carex lancibracteata A.E.Kozhevn., Byull. Moskovsk. Obshch. Isp. Prir., Otd. Biol. 92: 134 (1987).
Russian Far East. 31+. Hemicr. or rhizome geophyte.

Carex lancifolia C.B.Clarke, J. Linn. Soc., Bot. 36: 293 (1903).
C. China. 36 CHC CHN. Hemicr. or rhizome geophyte.
Carex glandulifolia Kük., Bull. Herb. Boissier, II, 4: 57 (1904).

Carex lancisquamata L.K.Dai, Acta Phytotax. Sin. 37: 181 (1999).
China (Yunnan). 36 CHC. Hemicr. or rhizome geophyte.

Carex ×langeana Fernald, Rhodora 35: 217 (1933). *C. gynocrates* × *C. maritima*.
Subarctic America to N. Canada, N. Europe. 10 SVA 70 GNL NUN NWT 71 MAN 72 NFL ONT. Hemicr. or rhizome geophyte.
Carex ×dutillyi O'Neill & Duman, Rhodora 43: 413 (1941).

Carex lankana T.Koyama, Bot. Mag. (Tokyo) 83: 184 (1970).
Sri Lanka. 40 SRL. Hemicr. or rhizome geophyte.

Carex laosensis Nelmes, Bull. Misc. Inform. Kew 1939: 305 (1940).
Laos. 41 LAO. Hemicr. or rhizome geophyte.

Carex lapazensis C.B.Clarke, Kew Bull., Addit. Ser. 8: 76 (1908).
Bolivia. 83 BOL. Hemicr. or rhizome geophyte.

Carex lapponica Láng, Linnaea 24: 539 (1851). *Carex canescens* var. *lapponica* (Láng) Nyman, Consp. Fl. Eur.: 780 (1882).
Subarctic to Canada. 10 FIN NOR SWE 14 RUN 30 ALT WSB YAK 31 KAM MAG SAK 70 ASK NWT YUK 71 ABT BRC MAN SAS 72 QUE. Hemicr. or rhizome geophyte.
Carex canescens var. *subloliacea* Laest. in C.J.Hartmann, Handb. Skand. Fl., ed. 4: 299 (1843). *Carex subloliacea* (Laest.) E.P.Bicknell, Bull. Torrey Bot. Club 35: 496 (1908).
Carex lapponica f. *major* Holmboe, Bot. Not. 1929: 12 (1929).

Carex larensis Steyerm., Fieldiana, Bot. 28: 66 (1951).
Venezuela. 82 VEN. Hemicr. or rhizome geophyte.

Carex laricetorum Y.L.Chou, Bull. Bot. Lab. N. E. Forest. Inst., Harbin 1: 19 (1959).
NE. China. 36 CHM. Hemicr. or rhizome geophyte.

Carex lasiocarpa Ehrh., Hannover. Mag. 9: 132 (1784).
Subarctic & Temp. Northern Hemisphere. 10 DEN FIN GRB IRE NOR SWE 11 AUT BGM CZE GER HUN NET POL SWI 12 FRA SPA 13 ITA ROM YUG 14 BLR BLT RUC RUE RUN RUW UKR 30 WSB 31 AMU KAM KUR PRM SAK 32 KAZ 33 NCS 36 CHI CHM 38 JAP KOR 70 ASK NWT 71 BRC MAN SAS 72 LAB NFL NSC ONT PEI QUE 73 IDA MNT WAS 74 ILL IOW MIN WIS 75 CNT INI MAI MAS MIC NWH NWJ NWY OHI PEN RHO VER 76 CAL 78 VRG 79 MXN. Hemicr. or rhizome geophyte.

var. ***americana*** Fernald, Rhodora 44: 304 (1942). *Carex lasiocarpa* subsp. *americana* (Fernald) Á.Löve & J.-P.Bernard, Svensk Bot. Tidskr. 53: 381 (1959). *Carex lanuginosa* var. *americana* (Fernald) Boivin, Naturaliste Canad. 94: 523 (1967).
Subarctic America to N. & WC. U.S.A., Mexico (Baja California). 70 ASK NWT 71 BRC MAN SAS 72 LAB NFL NSC ONT PEI QUE 73 IDA MNT WAS 74 ILL IOW MIN WIS 75 CNT INI MAI MAS MIC NWH NWJ NWY OHI PEN RHO VER 76 CAL 78 VRG 79 MXN. Hemicr. or rhizome geophyte.
Carex lanuginosa Michx., Fl. Bor.-Amer. 2: 175 (1803). *Carex filiformis* var. *lanuginosa* (Michx.) Britton, Sterns & Poggenb., Prelim. Cat.: 63 (1888). *Carex lasiocarpa* var. *lanuginosa* (Michx.) Kük. in H.G.A.Engler (ed.), Pflanzenr., IV, 20(38): 748 (1909). *Carex lasiocarpa* subsp. *lanuginosa* (Michx.) R.T.Clausen & Wahl, Rhodora 41: 31 (1939).
Carex lanuginosa var. *kansana* Britton in N.L.Britton & A.Brown, Ill. Fl. N. U.S. 1: 305 (1896).
Carex lasiocarpa f. *kansana* (Britton) Kük. in H.G.A. Engler (ed.), Pflanzenr., IV, 20(38): 748 (1909).
Carex lanuginosa var. *oriens* Raymond, Naturaliste Canad. 77: 59 (1950).

var. ***lasiocarpa***
Temp. Eurasia. 10 DEN FIN GRB IRE NOR SWE 11 AUT BGM CZE GER HUN NET POL SWI 12 FRA SPA 13 ITA ROM YUG 14 BLR BLT RUC RUE RUN RUW UKR 30 WSB 31 AMU KAM PRM 32 KAZ 33 NCS 36 CHI CHM 70 ASK NWT 71 BRC MAN SAS 72 LAB NFL NSC ONT PEI QUE 73 IDA MNT WAS 74 ILL IOW MIN WIS 75 CNT INI MAI MAS MIC NWH NWJ NWY OHI PEN RHO VER 76 CAL 78 VRG. Hemicr. or rhizome geophyte.
Carex splendida Willd., Fl. Berol. Prodr.: 33 (1787).
Carex filiformis var. *latifolia* Boeck., Linnaea 41: 309 (1877).
Carex filiformis var. *australis* L.H.Bailey, Mem. Torrey Bot. Club 1: 56 (1889).
Carex lasiocarpa f. *meduaensis* H.Lév. & Vaniot, Bull. Acad. Int. Géogr. Bot. 10: 35 (1901).
Carex lasiocarpa f. *robusta* Junge, Verh. Naturwiss. Vereins Hamburg, III, 12: 21 (1904).
Carex lasiocarpa f. *stricta* Junge, Verh. Naturwiss. Vereins Hamburg, III, 12: 22 (1904).
Carex lasiocarpa var. *fuscata* Ohwi, Mem. Coll. Sci. Kyoto Imp. Univ., Ser. B, Biol. 6: 242 (1931).

var. ***occultans*** (Franch.) Kük. in H.G.A.Engler (ed.), Pflanzenr., IV, 20(38): 748 (1909).
Russian Far East, Korea, Japan. 31 KUR SAK 38 JAP KOR. Hemicr. or rhizome geophyte.
**Carex filiformis* var. *occultans* Franch., Nouv. Arch. Mus. Hist. Nat., III, 10: 89 (1898). *Carex occultans* (Franch.) V.I.Krecz. in V.L.Komarov (ed.), Fl. URSS 3: 620 (1935).
Carex koidzumii Honda, Bot. Mag. (Tokyo) 45: 3 (1931).

Carex lasiolepis Franch., Bull. Soc. Philom. Paris, VIII, 7: 46 (1895).
Japan. 38 JAP. Hemicr. or rhizome geophyte.
Carex adumana Makino, Bot. Mag. (Tokyo) 9: 258 (1895).

Carex latebracteata Waterf., Rhodora 56: 23 (1954).
Oklahoma to Arkansas (Ouachita Mts.). 74 OKL 78 ARK. Hemicr. or rhizome geophyte.

Carex lateralis Kük. in H.G.A.Engler (ed.), Pflanzenr., IV, 20(38): 639 (1909).
India to Philippines. 40 IND SRL 42 JAW LSI PHI SUL SUM. Hemicr. or rhizome geophyte.
Carex elmeri Kük., Repert. Spec. Nov. Regni Veg. 8: 326 (1910).

Carex lateriflora Phil., Anales Univ. Chile 93: 488 (1896). *Carex trichodes* var. *lateriflora* (Phil.) Kük. in H.G.A.Engler (ed.), Pflanzenr., IV, 20(38): 99 (1909).
SC. & S. Chile, S. Argentina. 85 AGS CLC CLS. Hemicr. or rhizome geophyte.
Carex reicheana Boeck., Allg. Bot. Z. Syst. 2: 173 (1896).
Carex trichodes var. *major* Kük., Bot. Jahrb. Syst. 27: 496 (1899).

Carex laticeps C.B.Clarke ex Franch., Nouv. Arch. Mus. Hist. Nat., III, 9: 178 (1897).
China, Korea, C. Japan. 36 CHN CHS 38 JAP KOR. Hemicr. or rhizome geophyte.
Carex hancei C.B.Clarke, J. Linn. Soc., Bot. 36: 288 (1902).
Carex cavaleriei H.Lév. & Vaniot, Bull. Soc. Agric. Sarthe 1905: 79 (1905).

Carex latisquamea Kom., Trudy Imp. S.-Peterburgsk. Bot. Sada 18: 447 (1900). *Carex villosa* var. *latisquamea* (Kom.) Kük. in H.G.A.Engler (ed.), Pflanzenr., IV, 20(38): 641 (1909).
Russian Far East, NE. China, Korea, Japan. 31 PRM 36 CHM 38 JAP KOR. Hemicr. or rhizome geophyte.
Carex villosa Boott in M.C.Perry, Jap. Exped. 2: 327 (1856), nom. illeg.

Carex lativena S.D.Jones & G.D.Jones, Sida 15: 510 (1993).
Arizona to Texas. 76 ARI 77 NWM TEX. Hemicr. or rhizome geophyte.

Carex laxa Wahlenb., Kongl. Vetensk. Acad. Nya Handl. 1803: 156 (1803). *Olamblis laxa* (Wahlenb.) Raf., Good Book: 26 (1840).
N. Europe to Japan, W. Subarctic America. 10 FIN NOR SWE 14 RUN 30 WSB 31 KUR 36 CHI CHM 38 JAP KOR 70 ASK YUK. Hemicr. or rhizome geophyte.
Carex macrochlamys Franch., Bull. Soc. Philom. Paris, VIII, 7: 49 (1895).

Carex laxiculmis Schwein., Ann. Lyceum Nat. Hist. New York 1: 70 (1824).
SE. Canada to NC. & E. U.S.A. 72 ONT 74 ILL IOW MSO WIS 75 CNT INI MAI MAS MIC NWH NWJ NWY OHI PEN RHO VER WVA 78 ALA ARK DEL GEO KTY MRY NCA SCA TEN VRG WDC. Hemicr. or rhizome geophyte.

var. ***copulata*** (L.H.Bailey) Fernald, Rhodora 8: 183 (1906).
EC. & E. U.S.A. 72 ONT 74 ILL IOW MSO WIS 75 INI MIC OHI WVA 78 ALA ARK KTY NCA TEN VRG. Hemicr. or rhizome geophyte.
Carex retrocurva var. *copulata* L.H.Bailey, Prelim. Syn. N. Amer. Carices: 161 (1886).
Carex digitalis var. *copulata* (L.H.Bailey) L.H.Bailey, Mem. Torrey Bot. Club 1: 47 (1889). *Carex copulata* (L.H.Bailey) Mack. in P.A.Rydberg, Fl. Plains N. Amer.: 181 (1932).

var. ***laxiculmis***
SE. Canada to NC. & E. U.S.A. 72 ONT 74 ILL IOW MSO WIS 75 CNT INI MAI MAS MIC NWH NWJ NWY OHI PEN RHO VER WVA 78 ALA ARK DEL GEO KTY MRY NCA SCA TEN VRG WDC. Hemicr. or rhizome geophyte.
Carex retrocurva Dewey, Amer. J. Sci. Arts, II, 42: 243 (1866).
Carex laxiculmis f. *serotina* Howe, Rep. (Annual) Regents Univ. State New York State Mus. 48: 169 (1896).
Carex absconditiformis Fernald, Rhodora 44: 387 (1942).

Carex laxiflora Lam., Encycl. 3: 378 (1792).
E. Canada to C. & E. U.S.A. 72 NSC ONT QUE 74 ILL MSO OKL WIS 75 CNT INI MAI MAS MIC NWH NWJ NWY OHI PEN RHO VER WVA 77 TEX 78 ALA ARK DEL GEO KTY MRY MSI NCA SCA TEN VRG WDC. Hemicr. or rhizome geophyte.

var. ***gracillima*** Boott, Ill. Carex 1: 38 (1858).
SE. Canada to NC. & E. U.S.A. 72 ONT 74 ILL MSO OKL WIS 75 CNT INI MAS MIC NWH NWJ NWY OHI PEN WVA 78 ALA ARK DEL FLA GEO KTY MRY MSI NCA SCA TEN VRG WDC. Hemicr. or rhizome geophyte.
Carex gracilescens Steud., Syn. Pl. Glumac. 2: 226 (1855), nom. illeg.

var. ***laxiflora***
E. Canada to C. & E. U.S.A. 72 NSC ONT QUE 74 ILL MSO OKL WIS 75 CNT INI MAI MAS MIC NWH NWJ NWY OHI PEN RHO VER WVA 77 TEX 78 ALA ARK DEL GEO KTY MRY MSI NCA SCA TEN VRG. Hemicr. or rhizome geophyte.
Carex heterosperma Wahlenb., Kongl. Vetensk. Acad. Nya Handl. 24: 151 (1803).
Carex anceps Muhl. ex Willd., Sp. Pl. 4: 278 (1805). *Deweya anceps* (Muhl. ex Willd.) Raf., Good Book: 26 (1840).
Carex conoidea Muhl., Descr. Gram.: 248 (1817), nom. illeg.
Carex plantaginea Muhl., Descr. Gram.: 235 (1817), nom. illeg.
Carex nematosperma Willd. ex Kunth, Enum. Pl. 2: 456 (1837).
Carex anceps var. *angustifolia* Dewey ex Alph.Wood, Class-book Bot.: 423 (1845).
Carex anceps var. *patulifolia* Dewey ex Alph.Wood, Class-book Bot.: 423 (1845). *Carex laxiflora* var. *patulifolia* (Dewey ex Alph.Wood) Kük. in

H.G.A.Engler (ed.), Pflanzenr., IV, 20(38): 526 (1909).
Carex anceps var. *striatula* Carey in A.Gray, Manual: 54 (1848).
Carex bulbosa Boeck., Flora 38: 597 (1855).
Carex protracta Steud., Syn. Pl. Glumac. 2: 234 (1855).
Carex fusiformis Chapm. ex Boott, Ill. Carex 1: 37 (1858), nom. illeg. *Carex ignota* var. *fusiformis* Dewey in A.Wood, Class-book Bot.: 765 (1861). *Carex styloflexa* var. *fusiformis* Wiegand, Rhodora 24: 200 (1922).
Carex laxiflora var. *intermedia* Boott, Ill. Carex 1: 37 (1858).
Carex laxiflora var. *major* Boott, Ill. Carex 1: 37 (1858).
Carex laxiflora var. *minor* Boott, Ill. Carex 1: 38 (1858).
Carex laxiflora var. *plantaginea* Boott, Ill. Carex 1: 37 (1858).
Carex truncata Boeck., Flora 41: 649 (1858).
Carex laxiflora var. *heterosperma* Britton, Sterns & Poggenb., Prelim. Cat.: 63 (1888).
Carex laxiflora var. *divaricata* L.H.Bailey, Mem. Torrey Bot. Club 1: 33 (1889).
Carex laxiflora var. *serrulata* F.J.Herm., Rhodora 40: 80 (1938).

Carex leavenworthii Dewey, Amer. J. Sci. Arts, II, 2: 246 (1846). *Carex cephalophora* var. *leavenworthii* (Dewey) Kük. in H.G.A.Engler (ed.), Pflanzenr., IV, 20(38): 158 (1909).
S. Ontario to C. & E. U.S.A. (38) jap 72 ONT 74 ILL IOW KAN MSO NEB OKL wis 75 INI MIC NWJ NWY OHI PEN WVA (76) cal 77 TEX 78 ALA ARK FLA GEO KTY LOU MRY MSI NCA SCA TEN VRG WDC. Hemicr. or rhizome geophyte.
Carex cephalophora var. *angustifolia* Boott, Ill. Carex 3: 123 (1862).

Carex lebrunii H.Lév., Monde Pl. 17: 15 (1915).
China (Yunnan). 36 CHC. Hemicr. or rhizome geophyte.

Carex ledebouriana C.A.Mey. ex Trevir., Bull. Soc. Imp. Naturalistes Moscou 36(1): 540 (1863). *Carex capillaris* var. *ledebouriana* (C.A.Mey. ex Trevir.) F.Schmidt, Mém. Acad. Imp. Sci. Saint Pétersbourg, Sér. 7, VII, 12: 69 (1868). *Carex capillaris* subsp. *ledebouriana* (C.A.Mey. ex Trevir.) Vorosch. in A.K.Skvortsov (ed.), Florist. Issl. Razn. Raionakh SSSR: 154 (1985).
N. European Russia to Russia Far East. 14 RUN 30 ALT BRY CTA KRA WSB YAK 31 MAG 32 KAZ 36 CHX 37 MON. Hemicr. or rhizome geophyte.

subsp. ***ledebouriana***
N. European Russia to Russia Far East and Mongolia. 14 RUN 30 KRA WSB YAK 31 MAG 32+ 36 CHX 37 MON. Hemicr. or rhizome geophyte.
Carex lenaensis Kük., Finska Vetensk. Soc. Forh. 45(8): 10 (1903).
Carex seiskoensis Freyn, Oesterr. Bot. Z. 53: 27 (1903).

subsp. ***substepposa*** Malyschev, in Fl. Sibir. 3: 117 (1990).
SW. Siberia. 30 ALT. Hemicr. or rhizome geophyte.

subsp. ***transbaicalensis*** Malyschev, in Fl. Sibir. 3: 118 (1990).
Siberia to Russian Far East. 30 BRY CTA 31+. Hemicr. or rhizome geophyte.

Carex lehmanii Drejer, Symb. Caric.: 13 (1844). *Carex alpina* var. *lehmanii* (Drejer) R.C.Srivast., Novon 8: 203 (1998).
Himalaya to Japan. 36 CHC CHN CHQ 38 JAP KOR 40 EHM NEP. Hemicr. or rhizome geophyte.
Carex infuscata var. *microcarpa* Nees in R.Wight, Contr. Bot. India: 125 (1834).
Carex hidewoi Ohwi, Mem. Coll. Sci. Kyoto Imp. Univ., Ser. B, Biol. 3: 276 (1930).

Carex leiorhyncha C.A.Mey., Mém. Sav. Étr. Acad. St. Pétersbourg 1: 217 (1831). *Vignea leiorhyncha* (C.A.Mey.) Soják, Cas. Nár. Mus., Odd. Prír. 148: 196 (1979 publ. 1980).
Siberia to N. China. 30+ 31+ 36 CHI CHM CHN CHS. Hemicr. or rhizome geophyte.
Carex nemarosa Kunth, Enum. Pl. 2: 388 (1837).
Carex setariiformis Turcz. ex Boott, Ill. Carex 4: 189 (1867).

Carex lemanniana Boott, Proc. Linn. Soc. London 1: 256 (1845).
Costa Rica to Peru. 80 COS PAN 83 CLM ECU PER. Hemicr. or rhizome geophyte.
Carex lehmanniana Boott ex Walp., Ann. Bot. Syst. 1: 910 (1849).
Carex lemanniana var. *simplex* Kük. in H.G.A.Engler (ed.), Pflanzenr., IV, 20(38): 405 (1909).

Carex lemmonii W.Boott, Bot. Gaz. 9: 93 (1884).
California. 76 CAL. Hemicr. or rhizome geophyte.
Carex abramsii Mack., Bull. Torrey Bot. Club 36: 482 (1909).

Carex lenta D.Don, Trans. Linn. Soc. London 14: 327 (1824).
Trop. & Subtrop. Asia to Lord Howe I. 36 CHC CHN CHS 38 JAP KOR 40 EHM IND NEP PAK SRL 41 VIE 42 BOR JAW LSI MOL PHI SUL SUM 43 NWG 50 NFK QLD. Hemicr. or rhizome geophyte.

var. ***lenta***
Trop. & Subtrop. Asia to Lord Howe I. 36+ 38 JAP KOR 40 EHM IND NEP PAK SRL 41 VIE 42 BOR JAW LSI MOL PHI SUL SUM 43 NWG 50 NFK QLD. Hemicr. or rhizome geophyte.
Carex gracilis R.Br., Prodr.: 242 (1810), nom. illeg.
Carex flexilis D.Don, Tilloch Phil. Mag., Dec. 1823: 455 (1823), nom. illeg.

var. ***sendaica*** (Franch.) T.Koyama, J. Fac. Sci. Univ. Tokyo, Sect. 3, Bot. 8(4): 166 (1962).
C. China to C. & S. Japan. 36 CHC CHN CHS 38 JAP KOR. Hemicr. or rhizome geophyte.
**Carex sendaica* Franch., Bull. Soc. Philom. Paris, VIII, 7: 42 (1895). *Carex brunnea* var. *sendaica* (Franch.) Kük. in H.G.A.Engler (ed.), Pflanzenr., IV, 20(38): 601 (1909).
Carex longistolon C.B.Clarke ex Franch., Nouv. Arch. Mus. Hist. Nat., III, 8: 243 (1896).
Carex franchetiana H.Lév. & Vaniot, Bull. Acad. Int. Géogr. Bot. 10: 268 (1901).
Carex sendaica var. *pseudosendaica* T.Koyama, J. Jap. Bot. 29: 46 (1954).

Carex lenticularis Michx., Fl. Bor.-Amer. 2: 172 (1803).
Subarctic America to U.S.A. 70 ALU ASK NUN NWT YUK 71 ABT BRC MAN SAS 72 LAB NBR NFL NSC ONT PEI QUE 73 COL IDA MNT ORE WAS WYO 74 MIN WIS 75 MAI MAS MIC

NWH NWY VER 76 ARI CAL UTA 77 NWM. Hemicr. or rhizome geophyte.

var. ***dolia*** (M.E.Jones) L.A.Standl., Syst. Bot. Monogr. 7: 62 (1985).
Alaska to Montana. 70 ASK YUK 71 ABT BRC 73 MNT. Hemicr. or rhizome geophyte.
**Carex goodenowii* var. *dolia* M.E.Jones, Bull. Montana State Univ., Biol. Ser. 15: 72 (1910).
Carex enanderi Hultén, Acta Univ. Lund., n.s., 38: 355 (1942).
Carex eurystachya F.J.Herm., Leafl. W. Bot. 8: 109 (1957).
Carex plectocarpa F.J.Herm., Leafl. W. Bot. 10: 66 (1964).

var. ***impressa*** (L.H.Bailey) L.A.Standl., Syst. Bot. Monogr. 7: 65 (1985).
W. U.S.A. 73 IDA ORE WAS 76 CAL. Hemicr. or rhizome geophyte.
**Carex interrupta* var. *impressa* L.H.Bailey, Mem. Torrey Bot. Club 1: 18 (1889).
Carex limnaea T.Holm, Amer. J. Sci. 170: 301 (1905).
Carex paucicostata Mack., Erythea 8: 74 (1922).

var. ***lenticularis***
Subarctic America to NC. & NE. U.S.A. 70 NUN NWT 71 ABT MAN SAS 72 LAB NBR NFL NSC ONT PEI QUE 74 MIN WIS 75 MAI MAS MIC NWH NWY VER. Hemicr. or rhizome geophyte.
Carex lenticularis var. *albimontana* Dewey in A.Wood, Class-book Bot.: 755 (1861).
Carex lenticularis var. *blakei* Dewey in A.Wood, Class-book Bot.: 755 (1861).
Carex lenticularis var. *merens* Howe, Rep. (Annual) Regents Univ. State New York New York State Mus. 47: 149 (1895).
Carex lenticularis var. *eucycla* Fernald, Rhodora 15: 134 (1913).
Carex aleutica Akiyama, Bot. Mag. (Tokyo) 47: 69 (1933).

var. ***limnophila*** (T.Holm) Cronquist, Vasc. Pl. Pacific Northw. 1: 279 (1969).
Aleutian Is. to W. California. 70 ALU ASK 71 BRC 73 ORE WAS 76 CAL. Hemicr. or rhizome geophyte.
**Carex vulgaris* var. *limnophila* T.Holm, Amer. J. Sci., IV, 17: 307 (1904).
Carex hindsii C.B.Clarke, Kew Bull., Addit. Ser. 8: 70 (1908).

var. ***lipocarpa*** (T.Holm) L.A.Standl., Syst. Bot. Monogr. 7: 67 (1985).
Alaska to New Mexico. 70 ALU ASK 71 ABT BRC 73 COL IDA MNT ORE WAS WYO 76 ARI CAL UTA 77 NWM. Hemicr. or rhizome geophyte.
Carex acuta var. *pallida* Boott, Ill. Carex 4: 166 (1867). *Carex lenticularis* var. *pallida* (Boott) Dorn, Vasc. Pl. Wyoming: 296 (1988).
Carex vulgaris var. *juncella* Olney in S.Watson, Botany [Fortieth Parallel]: 368 (1871).
Carex kelloggii W.Boott in S.Watson & al., Bot. California 2: 240 (1880).
**Carex vulgaris* var. *lipocarpa* T.Holm, Amer. J. Sci., IV, 17: 308 (1904).
Carex hindsii var. *brevigluma* Kük. in H.G.A.Engler (ed.), Pflanzenr., IV, 20(38): 307 (1909).
Carex lenticularis var. *paullifructus* Kük. in H.G.A. Engler (ed.), Pflanzenr., IV, 20(38): 308 (1909).

Carex lepida Boott, Ill. Carex 4: 211 (1867).
Ecuador (Pichincha). 83 ECU. Hemicr. or rhizome geophyte.

Carex leporina L., Sp. Pl.: 973 (1753). *Carex nuda* Lam., Fl. Franç. 2: 172 (1779), nom. illeg. *Carex mollis* Gilib., Excerc. Phyt. 2: 546 (1792), nom. illeg. *Vignea leporina* (L.) Rchb. in J.C.Mössler, Handb. Gewächsk. ed. 2, 3: 1613 (1830). *Olotrema leporina* (L.) Raf., Good Book: 25 (1840).
Azores, Europe to Caucasus, W. Canada to W. U.S.A. 10 DEN FIN FOR GRB IRE NOR SWE 11 AUT BGM CZE GER HUN NET POL SWI 12 COR FRA POR SAR SPA 13 ALB BUL GRC ITA ROM SIC YUG 14 BLR BLT RUC RUE RUN RUS RUW UKR 20 ALG MOR 21 AZO 30 IRK WSB (31) 33 TCS 34 TUR 37 MON (50) tas (51) nzn 71 BRC (72) nbr nfl nsc pei que 73 ORE WAS (74) wis (75) cnt mai nwh nwy pen 76 CAL NEV (78) nca ten. Hemicr. or rhizome geophyte.
Carex ovalis Gooden., Trans. Linn. Soc. London 2: 148 (1794). *Caricina ovalis* (Gooden.) St.-Lag. in A.Cariot, Étude Fl., ed. 8, 2: 878 (1889). *Vignea ovalis* (Gooden.) Soják, Cas. Nár. Mus., Odd. Prír. 148: 196 (1979 publ. 1980).
Carex brizoides Geners., Elench.: 857 (1798).
Carex sicula Tineo, Pl. Rar. Sicil.: 30 (1817). *Carex leporina* subsp. *sicula* (Tineo) Nyman, Consp. Fl. Eur.: 779 (1882).
Carex argyroglochin Hornem., Fors. Oecon. Pantel., ed. 3, 1: 927 (1821). *Vignea argyroglochin* (Hornem.) Rchb. in J.C.Mössler, Handb. Gewächsk. ed. 2, 3: 1612 (1830). *Carex leporina* f. *argyroglochin* (Hornem.) W.D.J.Koch, Syn. Fl. Germ. Helv.: 753 (1837). *Carex leporina* var. *argyroglochin* (Hornem.) Rchb., Icon. Fl. Germ. Helv. 8: 8 (1846). *Carex ovalis* var. *argyroglochin* (Hornem.) De Langhe & Lambinon, Nouv. Fl. Belg.: 759 (1973).
Carex malvernensis S.Gibson, Phytologist 1: 715 (1843).
Carex leporina var. *argyrolepis* Peterm., Anal. Pfl.-Schlüss.: 493 (1846).
Carex leporina f. *longibracteata* Peterm., Anal. Pfl.-Schlüss.: 493 (1846).
Carex leporina f. *monostachya* Peterm., Anal. Pfl.-Schlüss.: 493 (1846).
Carex leporina f. *capitata* Sond., Fl. Hamburg.: 489 (1850).
Carex leporina var. *nemoralis* Trevir. in C.F.von Ledebour, Fl. Ross. 4: 279 (1852).
Carex leporina var. *pallescens* Gren. & Godr., Fl. France 3: 397 (1855).
Carex leporina var. *subfestiva* Lange, Pug. Pl. Hispan.: 62 (1860). *Carex ovalis* var. *subfestiva* (Lange) O.Bolòs & Vigo, Fl. Man. Paisos Catalans 4: 255 (2001).
Carex leporina f. *minor* Kuntze, Taschen-Fl. Leipzig: 16 (1867).
Carex leporina var. *atrofusca* H.Christ, Bull. Soc. Roy. Bot. Belgique 27(2): 166 (1888).
Carex leporina f. *robusta* Fiek, Jahresber. Schles. Ges. Vaterl. Cult. 1889: 203 (1889).
Carex leporina var. *alpina* Asch. & Graebn., Syn. Mitteleur. Fl. 2(2): 52 (1902). *Vignea leporina* var. *alpina* (Asch. & Graebn.) Dostál, Folia Mus. Rerum Nat. Bohemiae Occid., Bot. 21: 16 (1984).
Carex leporina f. *nana* Asch. & Graebn., Syn. Mitteleur. Fl. 2(2): 52 (1902).
Carex leporina var. *radicans* Zabel ex Asch. & Graebn., Syn. Mitteleur. Fl. 2(2): 53 (1902).

Carex cousturieri Gand., Bull. Soc. Bot. France 60: 556 (1913 publ. 1914).
Carex tracyi Mack., Erythea 8: 41 (1922).

Carex leporinella Mack., Bull. Torrey Bot. Club 43: 605 (1916 publ. 1917).
W. U.S.A. 73 IDA MNT ORE WAS WYO 76 CAL NEV UTA. Hemicr. or rhizome geophyte.

Carex leptalea Wahlenb., Kongl. Vetensk. Acad. Nya Handl. 24: 139 (1803).
W. Subarctic America to Mexico, Hispaniola. 70 ASK NUN NWT YUK 71 ABT BRC MAN SAS 72 LAB NBR NFL NSC ONT PEI QUE 73 COL IDA MNT ORE WAS WYO 74 ILL IOW MIN MSO NDA OKL SDA WIS 75 CNT INI MAI MAS MIC NWH NWJ NWY OHI PEN RHO VER WVA 76 CAL UTA 77 NWM TEX 78 ALA ARK DEL FLA KTY LOU MRY NCA SCA TEN VRG WDC 79+ 81 DOM. Hemicr. or rhizome geophyte.

subsp. ***harperi*** (Fernald) W.Stone, Canad. J. Bot. 43: 1391 (1963).
C. & E. U.S.A. 74 MSO 75 INI NWJ PEN 77 TEX 78 ALA FLA GEO KTY LOU SCA TEN VRG. Hemicr. or rhizome geophyte.
**Carex harperi* Fernald, Rhodora 8: 181 (1906).
Carex leptalea var. *harperi* (Fernald) W.Stone, Pl. S. New Jersey: 305 (1912).

subsp. ***leptalea***
W. Subarctic America to Mexico, Hispaniola. 70 ASK NUN NWT YUK 71 ABT BRC MAN SAS 72 LAB NBR NFL NSC ONT PEI QUE 73 COL IDA MNT ORE WAS WYO 74 ILL IOW MIN MSO NDA OKL SDA WIS 75 CNT INI MAI MAS MIC NWH NWJ NWY OHI PEN RHO VER WVA 76 CAL UTA 77 NWM TEX 78 ALA ARK ARK DEL DEL FLA KTY KTY LOU MRY MRY NCA NCA SCA TEN TEN VRG WDC WDC 79+ + 81 DOM DOM. Hemicr. or rhizome geophyte.
Carex microstachya Michx., Fl. Bor.-Amer. 2: 169 (1803), nom. illeg.
Carex polytrichoides Muhl. ex Willd., Sp. Pl. 4: 213 (1805).
Carex hochstetteri Frank ex Steud., Nomencl. Bot., ed. 2, 1: 291 (1840).
Carex leptalea var. *tayloris* B.Boivin, Phytologia 43: 67 (1979).

subsp. ***pacifica*** Calder & R.L.Taylor, Canad. J. Bot. 43: 1391 (1963). *Carex jimcalderi* B.Boivin, Phytologia 43: 67 (1979).
SE. Alaska to Washington. 70 ASK 71 BRC 73 WAS. Hemicr. or rhizome geophyte.

Carex × leptoblasta E.Holmb., Bot. Not. 1929: 19 (1929).
C. brunnescens × *C. disperma.*
N. Europe. 10+. Hemicr. or rhizome geophyte.

Carex leptocladus C.B.Clarke in D.Oliver & auct. suc. (eds.), Fl. Trop. Afr. 8: 519 (1902).
São Tomé. 23 GGI. Hemicr. or rhizome geophyte.

Carex leptonervia (Fernald) Fernald, Rhodora 16: 214 (1914).
E. Canada to NC. & E. U.S.A. 72 LAB NBR NFL NSC ONT PEI QUE 74 MIN WIS 75 CNT INI MAI MAS MIC NWH NWJ NWY OHI PEN RHO VER WVA 78 DEL KTY MRY NCA SCA TEN VRG. Hemicr. or rhizome geophyte.
Carex laxiflora var. *varians* L.H.Bailey, Mem. Torrey Bot. Club 1: 32 (1889).
Carex varians (L.H.Bailey) Hay, Palmer & C.T.White, Proc. Biol. Soc. Wash. 14: 196 (1901), nom. inval.
**Carex laxiflora* var. *leptonervia* Fernald, Rhodora 8: 184 (1906).
Carex laxiflora f. *varians* (L.H.Bailey) Kük. in H.G.A.Engler (ed.), Pflanzenr., IV, 20(38): 526 (1909).
Carex blanda var. *varians* (L.H.Bailey) Burnham, Torreya 19: 133 (1919).

Carex leptopoda Mack. in P.A.Rydberg, Fl. Rocky Mts.: 124 (1917). *Carex deweyana* subsp. *leptopoda* (Mack.) Calder & R.L.Taylor, Canad. J. Bot. 43: 1389 (1963). *Carex deweyana* var. *leptopoda* (Mack.) Boivin, Naturaliste Canad. 94: 522 (1967).
SW. Canada to W. U.S.A., Mexico (Chihuahua, Durango). 71 BRC 73 IDA MNT ORE WAS 76 CAL 79 MXE. Hemicr. or rhizome geophyte.

Carex leribensis Nelmes, Bull. Misc. Inform. Kew 1940: 269 (1940).
S. Africa. 27 CPP LES. Hemicr. or rhizome geophyte.

Carex lessoniana Steud., Nomencl. Bot., ed. 2, 1: 292 (1840).
New Zealand. 51 NZN NZS. Hemicr. or rhizome geophyte.
**Carex polystachya* R.Lesson & A.Rich. in J.S.C.Dumont d'Urville, Voy. Astrolabe: 118 (1832), nom. illeg.

Carex leucantha Arn. ex Boott, Proc. Linn. Soc. London 1: 257 (1845).
India, Sri Lanka. 40 IND SRL. Hemicr. or rhizome geophyte.

Carex lianchengensis S.Yun Liang & Y.Z.Huang, Fl. Fujianica 6: 690 (1995).
China (Fujian). 36 CHS. Hemicr. or rhizome geophyte.

Carex libera (Kük.) Hamlin, Rec. Domin. Mus. 6: 103 (1968).
New Zealand South I. 51 NZS. Hemicr. or rhizome geophyte.
**Carex uncifolia* var. *libera* Kük. in H.G.A.Engler (ed.), Pflanzenr., IV, 20(38): 685 (1909).

Carex ligata Boott in G.Bentham, Fl. Hongk.: 402 (1861).
Japan to Philippines. 36 CHC CHH CHS 38 JAP KOR NNS TAI 42 PHI. Hemicr. or rhizome geophyte.

subsp. ***formosensis*** (H.Lév. & Vaniot) T.Koyama, in Fl. Taiwan 5: 333 (1978).
C. & S. Japan to Philippines. 36 CHC CHH CHS 38 JAP KOR NNS TAI 42 PHI. Hemicr. or rhizome geophyte.
Carex ascocentra C.B.Clarke ex Franch., Nouv. Arch. Mus. Hist. Nat., III, 9: 182 (1897). *Carex davidii* var. *ascocentra* (C.B.Clarke ex Franch.) Kük. in H.G.A.Engler (ed.), Pflanzenr., IV, 20(38): 479 (1909).
**Carex formosensis* H.Lév. & Vaniot, Mém. Soc. Sci. Nat. Cherbourg 35: 216 (1905). *Carex ligata* var. *formosensis* (H.Lév. & Vaniot) Kük. in H.G.A.Engler (ed.), Pflanzenr., IV, 20(38): 474 (1909).
Carex kelungensis Hayata, Icon. Pl. Formosan. 10: 63 (1921).
Carex shimotsukensis Honda, Bot. Mag. (Tokyo) 44: 667 (1930).

subsp. ***ligata***
SE. China. 36 CHS. Hemicr. or rhizome geophyte.
Carex nanca Boott in G.Bentham, Fl. Hongk.: 402 (1861).
Carex nexa Boott in G.Bentham, Fl. Hongk.: 402 (1861). *Carex ligata* var. *nexa* (Boott) Kük. in H.G.A.Engler (ed.), Pflanzenr., IV, 20(38): 474 (1909).
Carex hongkongensis Franch., Nouv. Arch. Mus. Hist. Nat., III, 10: 93 (1898).
Carex ligata var. *cucullata* Kük., J. Coll. Sci. Imp. Univ. Tokyo 22: 495 (1906). *Carex cucullata* (Kük.) Ohwi, Mem. Coll. Sci. Kyoto Imp. Univ., Ser. B, Biol. 5(3): 272 (1930).

Carex limosa L., Sp. Pl.: 977 (1753). *Trasus limosus* (L.) Gray, Nat. Arr. Brit. Pl. 2: 67 (1821). *Facolos limosa* (L.) Raf., Good Book: 26 (1840).
Temp. Northern Hemisphere. 10 DEN FIN GRB ICE IRE NOR SWE 11 AUT BGM CZE GER HUN† NET POL SWI 12 FRA SPA 13 BUL ITA ROM YUG 14 BLR BLT RUC RUE RUN RUS RUW UKR 30 KRA WSB YAK 31 KAM KHA KOR MAG SAK 32 KAZ 33 NCS TCS 34 TUR 36 CHM 37 MON 38 JAP KOR 70 ASK NUN NWT YUK 71 ABT BRC MAN SAS 72 LAB NBR NFL NSC ONT PEI QUE 73 COL MNT ORE WAS 74 IOW MIN NDA NEB WIS 75 CNT INI MAI MAS MIC NWH NWJ NWY OHI PEN VER 76 CAL NEV UTA 78 DEL. Hemicr. or rhizome geophyte.
Carex elegans Willd., Fl. Berol. Prodr.: 34 (1787).
Carex limosa var. *oblonga* Dewey, Amer. J. Sci. Arts 10: 42 (1826).
Carex laxa Dewey, Amer. J. Sci. Arts 26: 376 (1834).
Carex limosa var. *painei* Dewey, Amer. J. Sci. Arts 29: 71 (1836).
Carex limosa f. *pauciflora* Asch., Fl. Brandenburg 1: 783 (1864).
Carex limosa f. *stans* Bolle, Verh. Bot. Vereins Prov. Brandenburg 7: 27 (1865).
Carex limosa f. *laxiformis* Lackow., Verh. Bot. Vereins Prov. Brandenburg 44: 164 (1903).
Carex limosa f. *robustior* Lackow., Verh. Bot. Vereins Prov. Brandenburg 44: 164 (1903).
Carex limosa var. *fuscocuprea* Kük. in H.G.A.Engler (ed.), Pflanzenr., IV, 20(38): 505 (1909). *Carex fuscocuprea* (Kük.) V.I.Krecz. in V.L.Komarov (ed.), Fl. URSS 3: 599 (1935).

Carex × limosoides J.Cay., Naturaliste Canad. 117: 62 (1990 publ. 1991). *C. limosa* × *C. paleacea.*
Québec. 72 QUE. Hemicr. or rhizome geophyte.

Carex limprichtiana Kük., Repert. Spec. Nov. Regni Veg. Beih. 12: 311 (1922).
China (Sichuan). 36 CHC CHT?. Hemicr. or rhizome geophyte.

Carex × limula Fr., Summa Veg. Scand.: 229 (1845). *C. aquatilis* × *C. bigelowii.*
NE. U.S.A. to N. Europe. 10 NOR 70 GNL 72 LAB QUE 75 MAI. Hemicr. or rhizome geophyte.
Carex acuta f. *ripensis* Laest., Nova Acta Regiae Soc. Sci. Upsal. 11: 286 (1839).
Carex × decolorans Wimm., Flora 33: 622 (1850). *Carex × limula* var. *decolorans* (Wimm.) Nyman, Consp. Fl. Eur.: 778 (1882).
Carex acuta var. *minor* Boeck., Linnaea 40: 419 (1876), nom. illeg.
Carex × nearctica Raymond, Bull. Soc. Bot. France 99: 195 (1952), nom. nud.

Carex lindleyana Nees in R.Wight, Contr. Bot. India: 121 (1834).
S. India, Sri Lanka, Indo-China. 40 IND SRL 41 VIE. Hemicr. or rhizome geophyte.
Carex thyrsiflora Boott, Ill. Carex 1: 12 (1858).

Carex lingii F.T.Wang & Tang, Contr. Inst. Bot. Natl. Acad. Peiping 6: 61 (1949).
China (Fujian, Zhejinag). 36 CHS. Hemicr. or rhizome geophyte.
Carex sinocrispa Raymond, Naturaliste Canad. 84: 69 (1957).

Carex liouana F.T.Wang & Tang, Contr. Inst. Bot. Natl. Acad. Peiping 6: 62 (1949).
SE. China. 36 CHS. Hemicr. or rhizome geophyte.

Carex liparocarpos Gaudin, Etrennes Fl.: 153 (1804).
Europe to Iran, Morocco. 11 AUT CZE HUN SWI 12 FRA SPA 13 BUL GRC ITA ROM SIC? TUE YUG 14 KRY RUS UKR 20 MOR 33 TCS 34 IRN TUR. Hemicr. or rhizome geophyte.

subsp. ***bordzilowskii*** (V.I.Krecz.) T.V.Egorova, Novosti Sist. Vyssh. Rast. 9: 84 (1972).
S. Ukraine, Mts. of Krym, Caucasus, Turkey, N. Iran. 14 KRY UKR 33 TCS 34 IRN TUR. Hemicr. or rhizome geophyte.
Carex schkuhrii Willd., Sp. Pl. 4: 264 (1805). *Edritria schkuhrii* (Willd.) Raf., Good Book: 26 (1840).
**Carex bordzilowskii* V.I.Krecz. in V.L.Komarov (ed.), Fl. URSS 3: 615 (1935).

subsp. ***liparocarpos***
Europe, Morocco. 11 AUT CZE HUN SWI 12 FRA SPA 13 BUL GRC ITA ROM SIC? TUE YUG 14 RUS UKR 20 MOR. Hemicr. or rhizome geophyte.
Carex nitida Host, Icon. Descr. Gram. Austriac. 1: 71 (1801), nom. illeg. *Lamprochlaenia nitida* Fedde & J.Schust., Just's Bot. Jahresber. 41(1): 11 (1913 publ. 1918).
Carex lamarckii Wood in A.Rees, Cycl. 6: 103 (1806).
Carex lucida Clairv., Man. Herbor. Suisse: 292 (1811).
Carex brevirostrata Poir. in J.B.A.M.de Lamarck, Encycl., Suppl. 3: 273 (1813).
Carex nitida var. *supiniformis* Kük. in H.G.A. Engler (ed.), Pflanzenr., IV, 20(38): 458 (1909).
Carex palentina Losa & P.Monts., Anales Inst. Bot. Cavanilles 10: 452 (1952).

Carex liqingii Tang & F.T.Wang ex S.Y.Liang, Acta Phytotax. Sin. 33: 489 (1995).
China (Guangxi). 36 CHS. Hemicr. or rhizome geophyte.

Carex lithophila Turcz., Bull. Soc. Imp. Naturalistes Moscou 1838: 104 (1838). *Vignea lithophila* (Turcz.) Soják, Cas. Nár. Mus., Odd. Prír. 148: 196 (1979 publ. 1980).
Siberia to N. Japan. 30+ 31 PRM 36 CHI CHM CHN 37 MON 38 JAP KOR. Hemicr. or rhizome geophyte.
Carex distichoidea H.Lév. & Vaniot, Bull. Acad. Int. Géogr. Bot. 10: 108 (1901).

Carex litorhyncha Franch., Bull. Soc. Philom. Paris, VIII, 7: 84 (1895).
China (Yunnan). 36 CHC. Hemicr. or rhizome geophyte.

Carex litorosa L.H.Bailey, Mem. Torrey Bot. Club 1: 72 (1889).
New Zealand. 51 NZN NZS. Hemicr. or rhizome geophyte.
**Carex littoralis* Petrie, Trans. & Proc. New Zealand Inst. 15: 358 (1883), nom. illeg.
Carex australis Kirk, Trans. & Proc. New Zealand Inst. 26: 262 (1893 publ. 1894).

Carex litvinovii Kük., Bull. Herb. Boissier, II, 4: 59 (1904).
Altay to C. Asia. 30 ALT 32 KAZ KGZ UZB. Hemicr. or rhizome geophyte.

Carex liuii T.Koyama & T.I.Chuang, Quart. J. Taiwan Mus. 13: 224 (1960).
China (Zhejiang), Taiwan. 36 CHS 38 TAI. Hemicr. or rhizome geophyte.

Carex livida (Wahlenb.) Willd., Sp. Pl. 4: 285 (1805).
N. Europe to Kuril Is., Subarctic America to N. U.S.A. 10 FIN ICE NOR SWE 14 RUN RUW 30 WSB 31 KAM KUR 38 JAP KOR 70 ALU ASK NUN NWT YUK 71 ABT BRC MAN SAS 72 LAB NBR NFL NSC ONT PEI QUE 73 MNT ORE WAS WYO 74 MIN WIS 75 INI MAI MAS MIC NWH NWJ NWY VER 76 CAL 80 PAN. Hemicr. or rhizome geophyte.
**Carex limosa* var. *livida* Wahlenb., Kongl. Vetensk. Acad. Nya Handl. 24: 162 (1803). *Edritria livida* (Wahlenb.) Raf., Good Book: 26 (1840). *Carex livida* var. *typica* Fernald, Rhodora 28: 8 (1926), nom. inval.
Carex grayana Dewey, Amer. J. Sci. Arts 25: 141 (1834). *Carex livida* var. *grayana* (Dewey) Fernald, Rhodora 28: 8 (1926).
Carex livida var. *radicalis* Paine, Rep. (Annual) Regents Univ. State New York State Mus.: 159 (1865).
Carex livida var. *radicalis* Paine ex Dewey, Amer. J. Sci. Arts, II, 41: 329 (1866).
Carex fujitae Kudô, J. Coll. Agric. Hokkaido Imp. Univ. 11: 82 (1922).
Carex livida var. *rufiniformis* Fernald, Rhodora 28: 8 (1926).
Carex lividulla Nakai, Bull. Natl. Sci. Mus. Tokyo 31: 130 (1952).
Carex fujimakii M.Kikuchi, Rep. (Annual) Gakugei Fac. Iwate Univ. 13(2): 47 (1958).

Carex lobolepis F.Muell., Fragm. 8: 258 (1874). *Carex cernua* var. *lobolepis* (F.Muell.) Kük. in H.G.A.Engler (ed.), Pflanzenr., IV, 20(38): 354 (1909).
E. Australia. 50 NSW QLD. Hemicr. or rhizome geophyte.

Carex lobulirostris Drejer, Symb. Caric.: 27 (1844).
Sri Lanka. 40 SRL. Hemicr. or rhizome geophyte.
Carex neesiana Arn. ex Boott, Proc. Linn. Soc. London 1: 260 (1845).
Carex arnottiana Boott, Ann. Mag. Nat. Hist. 17: 56 (1846).

Carex loheri C.B.Clarke, J. Linn. Soc., Bot. 37: 14 (1904).
Philippines. 42 PHI. Hemicr. or rhizome geophyte.

Carex loliacea L., Sp. Pl.: 974 (1753). *Vignea loliacea* (L.) Rchb. in J.C.Mössler, Handb. Gewächsk. ed. 2, 3: 1610 (1830). *Neskiza loliacea* (L.) Raf., Good Book: 27 (1840). *Leptovignea loliacea* (L.) Fedde & J.Schust., Just's Bot. Jahresber. 41(2): 11 (1913 publ. 1918).
N. & EC. Europe to Japan, Subarctic America to Canada. 10 FIN NOR SWE 11 POL 13 ROM 14 BLR BLT RUC RUE RUN RUS RUW UKR? 30 ALT WSB 31 AMU KAM KHA PRM SAK 32 KAZ 36 CHM 37 MON 38 JAP KOR 70 ASK NUN NWT YUK 71 ABT BRC MAN SAS 72 ONT. Hemicr. or rhizome geophyte.
Carex quaternaria Spreng., Syst. Veg. 3: 809 (1826).
Carex sibirica Willd. ex Kunth, Enum. Pl. 2: 406 (1837).
Carex tenuiflora Hartm. ex Kunth, Enum. Pl. 2: 405 (1837), nom. illeg.

Carex lonchocarpa Willd. ex Spreng., Syst. Veg. 3: 817 (1826).
SE. U.S.A. to Texas. 77 TEX 78 ALA DEL FLA GEO LOU MRY MSI NCA SCA TEN VRG. Hemicr. or rhizome geophyte.
Carex folliculata var. *australis* L.H.Bailey, Proc. Amer. Acad. Arts 22: 62 (1886).
Carex smalliana Mack., Bull. Torrey Bot. Club 36: 483 (1909).

Carex longhiensis Franch., Bull. Soc. Philom. Paris, VIII, 7: 85 (1895).
China (Yunnan). 36 CHC. Hemicr. or rhizome geophyte.

Carex longibrachiata Boeck., Linnaea 41: 282 (1877).
E. & SE. Australia. 50 NSW QLD TAS VIC (51) nzn. Hemicr. or rhizome geophyte.
Carex longifolia R.Br., Prodr.: 242 (1810), nom. illeg. *Carex brownei* Steud., Syn. Pl. Glumac. 2: 209 (1855), non C. brownii Torr.
Carex longifolia var. *minor* Boott in J.D.Hooker, Fl. Tasman. 2: 101 (1858).

Carex longicaulis Boeck., Flora 65: 62 (1882).
Mexico. 79 MXC MXE MXG MXS. Hemicr. or rhizome geophyte.
Carex ciliaris Fernald, Proc. Amer. Acad. Arts 43: 61 (1907).

Carex longicruris Nees in R.Wight, Contr. Bot. India: 124 (1834).
Himalaya, India (Nilgiri Hills), Sri Lanka. 36 CHC CHS 40 EHM IND NEP SRL WHM. Hemicr. or rhizome geophyte.
Carex henryi C.B.Clarke ex Franch., Nouv. Arch. Mus. Hist. Nat., III, 8: 243 (1896). *Carex longicruris* var. *henryi* (C.B.Clarke ex Franch.) C.B.Clarke, J. Linn. Soc., Bot. 36: 295 (1903).

Carex longiculmis Petrie, Trans. & Proc. New Zealand Inst. 14: 363 (1882).
New Zealand (Stewart I.). 51 NZS. Hemicr. or rhizome geophyte.

Carex longicuspis Boeck., Beitr. Cyper. 1: 49 (1888).
Tibet. 36 CHT. Hemicr. or rhizome geophyte.

Carex longii Mack., Bull. Torrey Bot. Club 49: 373 (1922).
SE. Canada to Venezuela, NE. Argentina to Uruguay. (51) nzn (63) haw 72 NSC ONT 74 ILL MSO OKL WIS 75 CNT INI MAI MAS MIC NWJ NWY OHI PEN RHO VER WVA 77 TEX 78 ALA ARK DEL FLA GEO KTY LOU MRY MSI NCA SCA TEN VRG WDC 79 MXC MXE MXG MXS MXT 80 COS GUA HON 81 BER HAI pue 82 VEN 83 CLM ECU 84 BZS 85 AGE URU. Hemicr. or rhizome geophyte.

subsp. ***longii***
SE. Canada to Venezuela. (51) nzn (63) haw 72

NSC ONT 74 ILL MSO OKL WIS 75 CNT INI MAI MAS MIC NWJ NWY OHI PEN RHO VER WVA 77 TEX 78 ALA ARK DEL FLA GEO KTY LOU MRY MSI NCA SCA TEN VRG WDC 79 MXC MXE MXG MXS MXT 80 COS GUA HON 81 BER HAI pue 82 VEN 83 CLM ECU. Hemicr. or rhizome geophyte.
Carex leporina var. *bracteata* Liebm., Mexic. Halvgr.: 264 (1850).

subsp. ***meridionalis*** (Kük.) Luceño & Alves, Anales Jard. Bot. Madrid 57: 173 (1999).
NE. Argentina to Uruguay. 84 BZS 85 AGE URU. Hemicr. or rhizome geophyte.
**Carex albolutescens* var. *meridionalis* Kük., Verh. Bot. Vereins Prov. Brandenburg 47: 208 (1905 publ. 1906). *Carex meridionalis* (Kük.) Herter, Revista Sudamer. Bot. 9: 162 (1953). *Carex longii* var. *meridionalis* (Kük.) G.A.Wheeler, Aliso 11: 536 (1987).

Carex longiligula Reznicek & S.González, Contr. Univ. Michigan Herb. 23: 333 (2001).
Mexico (Chiapas). 79 MXT. Hemicr.

Carex longipes D.Don ex Tilloch & Taylor, Philos. Mag. J. 62: 455 (Dec. 1823).
India (Nilgiri Hills), Himalaya to New Guinea. 36 CHC 40 EHM IND NEP SRL WHM 41 VIE 42 JAW SUL 43 NWG. Hemicr. or rhizome geophyte.
Carex macrolepis D.Don, Tilloch Phil. Mag., Dec. 1823: 455 (1823), nom. illeg.
Carex nepalensis Spreng., Syst. Veg. 3: 811 (1826). *Forexeta nepalensis* (Spreng.) Raf., Good Book: 28 (1840). *Carex longipes* var. *nepalensis* (Spreng.) Boott, Ill. Carex 4: 191 (1867).
Carex longipes var. *lissitiflora* C.B.Clarke in J.D. Hooker, Fl. Brit. India 6: 705 (1894).
Carex longipes var. *sessilis* Tang & F.T.Wang ex L.K.Dai, Acta Phytotax. Sin. 32: 179 (1994).

Carex longirostrata C.A.Mey., Mém. Sav. Étr. Acad. St. Pétersbourg 1: 220 (1831). *Carex michelii* var. *ultra* L.H.Bailey, Proc. Amer. Acad. Arts 22: 102 (1886).
Russian Far East, China to Japan. 31 KUR PRM SAK 36 CHM CHN CHS 38 JAP KOR. Hemicr. or rhizome geophyte.

var. ***exaristata*** X.F.Jin & C.Z.Zheng, Acta Phytotax. Sin. 42: 548 (2004).
China (Zhejiang). 36 CHS. Hemicr. or rhizome geophyte.

var. ***hoi*** K.L.Chu ex S.Y.Liang, Acta Phytotax. Sin. 36: 537 (1998).
China (Zhejiang). 36 CHS. Hemicr. or rhizome geophyte.

var. ***longirostrata***
Russian Far East, China to Japan. 31 KUR PRM SAK 36 CHM CHN 38 JAP KOR. Hemicr. or rhizome geophyte.
Carex bispicata Hook. & Arn., Bot. Beechey Voy.: 118 (1832).
Carex camtschatcensis Kunth, Enum. Pl. 2: 477 (1837).
Carex longirostrata var. *recurvifolia* Kük. in H.G.A. Engler (ed.), Pflanzenr., IV, 20(38): 636 (1909).
Carex tenuistachya Nakai, Bot. Mag. (Tokyo) 36: 127 (1922).
Carex fuscofibrosa Ohwi, Mem. Coll. Sci. Kyoto Imp. Univ., Ser. B, Biol. 6: 246 (1931).
Carex tenuistachya var. *pallida* Kitag., Bot. Mag. (Tokyo) 48: 25 (1934). *Carex longirostrata* var. *pallida* (Kitag.) Ohwi, Acta Phytotax. Geobot. 4: 43 (1935).
Carex longirostrata var. *tsinlingensis* K.T.Fu, in Fl. Tsinlingensis 1(1): 449 (1976).

Carex longispiculata Y.C.Yang, Acta Biol. Plateau Sin. 3: 91 (1984).
China (Sichuan, Gansu). 36 CHC CHN. Hemicr. or rhizome geophyte.

Carex longissima M.E.Jones, Contr. W. Bot. 18: 26 (1933).
Mexico (Baja California Sur). 79 MXN. Hemicr. or rhizome geophyte.

Carex longpanlaensis S.Yun Liang, Acta Phytotax. Sin. 37: 191 (1999).
China (Yunnan). 36 CHC. Hemicr. or rhizome geophyte.

Carex longshengensis Y.C.Tang & S.Yun Liang, in Fl. Reipubl. Popul. Sin. 12: 522 (2000).
China (Guangxi). 36 CHS. Hemicr.

Carex longxishanensis S.Yun Liang in Z.Y.Li (ed.), Pl. Longqui Mt.: 603 (1994).
China (Fujian). 36 CHS. Hemicr. or rhizome geophyte.

Carex lophocarpa C.B.Clarke, Kew Bull., Addit. Ser. 8: 69 (1908).
E. Australia. 50 NSW QLD. Hemicr. or rhizome geophyte.

Carex louisianica L.H.Bailey, Bull. Torrey Bot. Club 20: 428 (1893).
C. & E. U.S.A. 74 ILL MSO OKL 75 INI NWJ OHI 77 TEX 78 ALA ARK DEL FLA GEO KTY LOU MRY MSI NCA SCA TEN VRG. Hemicr. or rhizome geophyte.
Carex turgescens Dewey, Amer. J. Sci. Arts, II, 3: 356 (1847), nom. illeg.
**Carex halei* J.Carey in A.W.Chapman, Fl. South. U.S.: 543 (1860), nom. illeg.
Carex halei var. *minor* Boott, Ill. Carex 2: 94 (1860).
Carex eggertii L.H.Bailey, Bot. Gaz. 21: 6 (1896).

Carex lowei Bech., Candollea 8: 15 (1939).
N. & C. Madeira. 21 MDR. Hemicr. or rhizome geophyte.
**Carex elata* A.Lowe, Trans. Cambridge Philos. Soc. 4: 11 (1831), nom. illeg.
Carex elata f. *simplex* Bornm., Bot. Jahrb. Syst. 33: 406 (1903).
Carex malato-belizii Raymond, Contr. Inst. Bot. Univ. Montréal 70: 73 (1957).

Carex lucorum Willd., Enum. Pl., Suppl.: 63 (1814). *Vignea lucorum* (Willd.) Rchb. in J.C.Mössler, Handb. Gewächsk. ed. 2, 3: 1614 (1830). *Carex pensylvanica* var. *lucorum* (Willd.) Boeck., Linnaea 41: 220 (1877).
E. Canada to NC. & E. U.S.A. 72 NBR NSC NSC ONT 74 MIN WIS 75 CNT MAI MAS MIC NWH NWJ NWY OHI PEN RHO VER 78 GEO KTY NCA SCA TEN VRG. Hemicr. or rhizome geophyte.

var. ***austrolucorum*** Rettig, Sida 13: 449 (1989).
SE. U.S.A. 78 GEO NCA SCA TEN VRG. Hemicr. or rhizome geophyte.

var. ***lucorum***
E. Canada to NC. & NE. U.S.A. 72 NBR NSC NSC

ONT 74 MIN WIS 75 CNT MAI MAS MIC NWH NWJ NWY OHI PEN RHO VER 78 KTY. Hemicr. or rhizome geophyte.

Carex michiganensis Dewey ex Alph.Wood, Class-book Bot.: 765 (1861).

Carex pensylvanica var. *distans* Peck, Rep. (Annual) Regents Univ. State New York State Mus. 46: 131 (1893).

Carex pensylvanica var. *separans* Peck, Rep. (Annual) Regents Univ. State New York New York State Mus. 47: 174 (1895). *Carex pennsylvanica* f. *separans* (Peck) Kük. in H.G.A.Engler (ed.), Pflanzenr., IV, 20(38): 446 (1909).

Carex luctuosa Franch., Nouv. Arch. Mus. Hist. Nat., III, 9: 130 (1897).

China (Sichuan to Gansu). 36 CHC CHN. Hemicr. or rhizome geophyte.

Carex luctuosa f. *brevisquama* K.T.Fu, in Fl. Tsinlingensis 1(1): 444 (1976).

Carex luctuosa f. *mucronata* K.T.Fu, in Fl. Tsinlingensis 1(1): 444 (1976).

Carex × ludibunda J.Gay, Ann. Sci. Nat., Bot., II, 10: 357 (1838). *C. canescens* × *C. paniculata*. *Carex schreberi* subsp. *ludibunda* (J.Gay) Nyman, Consp. Fl. Eur.: 779 (1882). *Carex paniculata* var. *ludibunda* (J.Gay) K.Richt., Pl. Eur. 1: 149 (1890). *Vignea × ludibunda* (J.Gay) Soják, Cas. Nár. Mus., Odd. Prír. 148: 196 (1979 publ. 1980).

SW. & WC. Europe. 11 GER 12 FRA. Hemicr. or rhizome geophyte.

Carex × silesiaca Figert, Deutsche Bot. Monatsschr. 6: 146 (1888).

Carex lupuliformis Sartwell ex Dewey, Amer. J. Sci. Arts, II, 9: 29 (1850).

SE. Canada to C. & E. U.S.A. 72 ONT QUE 74 ILL IOW MIN MSO OKL WIS 75 CNT INI MAS MIC NWJ NWY OHI PEN VER WVA 77 TEX 78 ARK DEL FLA GEO LOU MRY MSI NCA SCA VRG. Hemicr. or rhizome geophyte.

Carex lupulina Muhl. ex Willd., Sp. Pl. 4: 266 (1805). *Carex gigantea* var. *lupulina* (Muhl. ex Willd.) Farw., Rep. (Annual) Commiss. Parks Boulevards Detroit 11: 50 (1900).

E. Canada to C. & E. U.S.A., NE. & NC. Mexico. 72 NBR NSC ONT QUE 74 ILL IOW KAN MIN MSO NEB OKL WIS 75 CNT INI MAI MAS MIC NWH NWJ NWY OHI PEN RHO VER WVA 77 TEX 78 ALA ARK DEL FLA GEO KTY LOU MRY MSI NCA TEN VRG WDC 79 MXE MXG. Hemicr. or rhizome geophyte.

Carex lupulina var. *pedunculata* A.Gray in ?

Carex bella-villa Dewey, Amer. J. Sci. Arts, II, 41: 229 (1866). *Carex lupulina* var. *bella-villa* (Dewey) L.H.Bailey, Mem. Torrey Bot. Club 1: 12 (1889). *Carex gigantea* f. *bella-villa* (Dewey) Farw., Rhodora 23: 87 (1921).

Carex canadensis Dewey, Amer. J. Sci. Arts, II, 41: 229 (1866).

Carex lupulina var. *longipedunculata* Dudley, Cayuga Fl.: 119 (1886).

Carex lurida L.H.Bailey, Proc. Amer. Acad. Arts 22: 63 (1886), nom. illeg.

Carex lupulina var. *albomarginata* Sherff, Bull. Torrey Bot. Club 38: 482 (1911).

Carex lurida Wahlenb., Kongl. Vetensk. Acad. Nya Handl. 24: 153 (1803).

E. Canada to Ecuador. (51) nzn 72 LAB NBR NFL NSC ONT QUE 74 ILL MSO OKL WIS 75 CNT INI MAI MAS MIC NWH NWJ NWY OHI PEN RHO VER WVA 77 TEX 78 ALA ARK FLA GEO KTY LOU MRY MSI NCA SCA TEN VRG WDC 79 MXE MXG MXT 82 VEN 83 ECU. Hemicr. or rhizome geophyte.

Carex lurida var. *exudans* L.H.Bailey ex Britton & A.Br. in ?

Carex rostrata Muhl. ex Willd., Sp. Pl. 4: 282 (1805), nom. illeg. *Carex purshii* Olney, Car. Bor. Am.: 2 (1871).

Carex tentaculata var. *rostrata* Pursh, Fl. Amer. Sept. 1: 41 (1813).

Carex lupulina var. *polystachya* Schwein. & Torr., Ann. Lyceum Nat. Hist. New York 1: 337 (1825). *Carex lurida* var. *polystachya* (Schwein. & Torr.) L.H.Bailey, Proc. Amer. Acad. Arts 22: 63 (1886).

Carex tentaculata var. *altior* Boott, Ill. Carex 2: 94 (1860).

Carex tentaculata var. *gracilis* Boott, Ill. Carex 2: 94 (1860). *Carex lurida* var. *gracilis* (Boott) Kük. in H.G.A.Engler (ed.), Pflanzenr., IV, 20(38): 714 (1909).

Carex tentaculata var. *parvula* Paine, Cat. Pl. Oneida Co. 1865: 105 (1865). *Carex lurida* var. *parvula* (Paine) Kük. in H.G.A.Engler (ed.), Pflanzenr., IV, 20(38): 714 (1909). *Carex lurida* f. *parvula* (Paine) D.S.Carp., Fl. Vermont, ed. 3: 64 (1937).

Carex beyrichiana Boeck., Linnaea 41: 239 (1877).

Carex lurida var. *flaccida* L.H.Bailey, Mem. Torrey Bot. Club 1: 73 (1889). *Carex tentaculata* var. *flaccida* (L.H.Bailey) Howe, Rep. (Annual) Regents Univ. State New York State Mus. 48: 190 (1896).

Carex luridiformis Mack. ex Reznicek & S.González, Contr. Univ. Michigan Herb. 20: 225 (1995).

Colombia. 83 CLM. Hemicr. or rhizome geophyte.

Carex lushanensis Kük., Acta Horti Gothob. 5: 47 (1930).

China (Sichuan). 36 CHC. Hemicr. or rhizome geophyte.

Carex lutea Leblond, Sida 16: 155 (1994).

North Carolina. 78 NCA. Hemicr. or rhizome geophyte.

Carex luzulifolia Boott in S.Watson & al., Bot. California 2: 250 (1880).

Montana to California. 73 IDA MNT 76 CAL. Hemicr. or rhizome geophyte.

Carex luzulifolia var. *strobilantha* T.Holm, Amer. J. Sci., IV, 20: 305 (1905).

Carex pseudojaponica C.B.Clarke, Kew Bull., Addit. Ser. 8: 81 (1908).

Carex luzulina Olney, Proc. Amer. Acad. Arts 7: 395 (1868).

W. Canada to W. U.S.A. 71 BRC 73 IDA MNT ORE WAS WYO 76 CAL. Hemicr. or rhizome geophyte.

var. ***ablata*** (L.H.Bailey) F.J.Herm., Rhodora 70: 420 (1968).

SW. Canada to W. U.S.A. 71 BRC 73 IDA MNT ORE WAS WYO 76 CAL NEV UTA. Hemicr. or rhizome geophyte.

**Carex ablata* L.H.Bailey, Bot. Gaz. 13: 82 (1888). *Carex luzulifolia* var. *ablta* (L.H.Bailey) Kük. in H.G.A.Engler (ed.), Pflanzenr., IV, 20(38): 558 (1909).

Carex owyheensis A.Nelson, Bot. Gaz. 53: 219 (1912).

var. ***atropurpurea*** Dorn, Vasc. Pl. Wyoming: 296 (1988).
NW. U.S.A. 73 IDA MNT WYO. Hemicr. or rhizome geophyte.

var. ***luzulina***
W. Canada to W. U.S.A. 71 BRC 73 IDA MNT ORE WAS WYO 76 CAL. Hemicr. or rhizome geophyte.

Carex lyi H.Lév. & Vaniot, Repert. Spec. Nov. Regni Veg. 3: 349 (1907).
China. 36+. Hemicr. or rhizome geophyte.

Carex lyngbyei Hornem. in G.C.Oeder & al. (eds.), Fl. Dan.: t. 1888 (1827). *Carex maritima* subsp. *lyngbyei* (Hornem.) Nyman, Consp. Fl. Eur.: 776 (1882).
Greenland, N. Europe, Russian Far East to W. U.S.A. 10 FOR ICE 31 KAM KHA KUR MAG SAK 36? 38 JAP 70 ALU ASK GNL 71 BRC 73 ORE WAS 76 CAL. Hemicr. or rhizome geophyte.
Carex cryptocarpa C.A.Mey., Mém. Sav. Étr. Acad. St. Pétersbourg 1: 226 (1831). *Carex halophila* subsp. *cryptocarpa* (C.A.Mey.) Nyman, Consp. Fl. Eur.: 776 (1882). *Carex lyngbyei* subsp. *cryptocarpa* (C.A.Mey.) Hultén, Kongl. Svenska Vetensk. Acad. Handl., III, 5(1): 188 (1927).
Carex scouleri Torr., Ann. Lyceum Nat. Hist. New York 3: 398 (1836). *Temnemis scouleri* (Torr.) Raf., Good Book: 27 (1840).
Carex salina Boott in W.J.Hooker, Fl. Bor.-Amer.: 219 (1839), nom. illeg.
Carex steenstrupii Liebm., Forhand. Scand. Naturf. 1840: 321 (1840).
Carex filipendula Drejer, Rev. Car. Bor.: 46 (1841). *Carex salina* f. *filipendula* (Drejer) Almq. in A.H.Hjelt, Consp. Fl. Fennic. 3: 281 (1895).
Carex filipendula var. *concolor* Drejer, Rev. Car. Bor.: 46 (1841). *Carex salina* var. *concolor* (Drejer) Almq., Bot. Not. 1891: 127 (1891).
Carex filipendula var. *littoralis* Drejer, Rev. Car. Bor.: 46 (1841).
Carex filipendula var. *variegata* Drejer, Rev. Car. Bor.: 46 (1841). *Carex cryptocarpa* var. *variegata* (Drejer) Britton, Mem. Torrey Bot. Club 5: 76 (1894). *Carex lyngbyei* var. *variegata* (Drejer) Kük. in H.G.A.Engler (ed.), Pflanzenr., IV, 20(38): 364 (1909).
Carex romanzowiana Cham. ex Steud., Syn. Pl. Glumac. 2: 216 (1855).
Carex capillipes Drejer in G.C.Oeder & al. (eds.), Fl. Dan.: t. 2844 (1871). *Carex lyngbyei* var. *capillipes* (Drejer) Kük. in H.G.A.Engler (ed.), Pflanzenr., IV, 20(38): 364 (1909).
Carex macounii A.Benn. in J.Macoun, Cat. Canad. Pl. 4: 147 (1888), nom. illeg.
Carex salina var. *robusta* L.H.Bailey, Bot. Gaz. 13: 87 (1888). *Carex lyngbyei* var. *robusta* (L.H.Bailey) Cronquist, Vasc. Pl. Pacific Northw. 1: 287 (1969).
Carex cryptocarpa var. *pumila* L.H.Bailey, Bull. Torrey Bot. Club 1: 27 (1889).
Carex qualicumensis L.H.Bailey, Bull. Torrey Bot. Club 20: 428 (1893).
Carex prionocarpa Franch., Bull. Soc. Philom. Paris, VIII, 7: 87 (1895). *Carex lyngbyei* var. *prionocarpa* (Franch.) Kük. in H.G.A.Engler (ed.), Pflanzenr., IV, 20(38): 364 (1909).
Carex pedunculifera Kom., Repert. Spec. Nov. Regni Veg. 13: 163 (1914).
Carex riabushinskii Kom., Repert. Spec. Nov. Regni Veg. 13: 163 (1914).
Carex behringensis Gand., Bull. Soc. Bot. France 66: 295 (1919 publ. 1920), nom. illeg.
Carex lyngbyei var. *gigas* Hultén, Fl. Aleutian Isl.: 113 (1937).

Carex maackii Maxim., Prim. Fl. Amur.: 308 (1859).
Russian Far East, E. China to Japan. 31 AMU PRM 36 CHM CHS 38 JAP KOR. Hemicr. or rhizome geophyte.
Carex nipponica Franch., Bull. Soc. Bot. France 26: 89 (1879).
Carex calcitrapa Franch., Bull. Soc. Philom. Paris, VIII, 7: 30 (1895).

Carex × macilenta Fr., Summa Veg. Scand.: 224 (1845). *C. brunnescens* × *C. loliacea*. *Carex tenuiflora* var. *macilenta* (Fr.) O.Lang, Linnaea 24: 541 (1851). *Carex canescens* var. *macilenta* (Fr.) H.Christ, Bull. Soc. Roy. Bot. Belgique 24(2): 18 (1885).
N. Europe. 10 NOR 14+. Hemicr. or rhizome geophyte.

Carex mackenziana Weath., Contr. Gray Herb. 114: 36 (1936).
Mexico (Nuevo León). 79 MXE. Hemicr. or rhizome geophyte.

Carex mackenziei V.I.Krecz. in V.L.Komarov (ed.), Fl. URSS 3: 183 (1935).
N. Europe to Japan, Subarctic America to NE. U.S.A. 10 FIN ICE NOR SWE 14 BLT RUN RUW 30+ 31 KAM MAG SAK 38 JAP 70 ASK GNL NUN NWT YUK 71 BRC MAN SAS 72 LAB NBR NFL NSC ONT PEI QUE 75 MAI. Hemicr. or rhizome geophyte.
Carex norvegica Willd. ex Schkuhr, Beschr. Riedgräs. 1: 50 (1801), nom. illeg.

Carex macloviana d'Urv., Mém. Soc. Linn. Paris 5: 599 (1826). *Vignea macloviana* (d'Urv.) Soják, Cas. Nár. Mus., Odd. Prír. 148: 196 (1979 publ. 1980).
N. Europe, Russian Far East, Subarctic America to Falkland Is. 10 FIN ICE NOR SWE 31 KAM 63 HAW 70 ALU ASK GNL NWT YUK 71 ABT BRC SAS 72 LAB QUE 73 COL IDA MNT ORE WAS WYO 76 ARI CAL NEV UTA 77 NWM 79 MXG MXN 83 BOL PER 85 AGS AGW CLC CLN CLS 90 FAL. Hemicr. or rhizome geophyte.

subsp. ***macloviana***
N. Europe, Russian Far East, Subarctic America to WC. U.S.A., Peru to Falkland Is. 10 FIN ICE NOR SWE 31 KAM 70 ALU ASK GNL NWT YUK 71 ABT BRC 72 LAB QUE 73 COL MNT WYO 79 MXG 83 BOL PER 85 AGS AGW CLC CLN CLS 90 FAL. Hemicr. or rhizome geophyte.
Carex ovalis var. *minor* Brongn. in L.I.Duperrey, Voy. Monde, Phan.: 149 (1833).
Carex festiva Dewey, Amer. J. Sci. Arts 29: 246 (1835).
Carex leporina var. *ovata* Laest. in C.J.Hartmann, Handb. Skand. Fl., ed. 3: 216 (1838).
Carex oreades C.A.Mey., Bull. Acad. Roy. Sci. Bruxelles 9(2): 248 (1842).
Carex incisodentata Lechl. ex Boott, Ill. Carex 1: 26 (1848).
Carex allomacros Steud., Syn. Pl. Glumac. 2: 189 (1855).
Carex incisodentata Steud., Syn. Pl. Glumac. 2: 189 (1855).
Carex pratensis Phil., Linnaea 29: 81 (1857), nom. illeg.

Carex tenuirostris Olney, Amer. Naturalist 8: 214 (1874).
Carex festiva var. *stricta* L.H.Bailey, Mem. Torrey Bot. Club 1: 51 (1889).
Carex festiva var. *viridis* L.H.Bailey, Mem. Torrey Bot. Club 1: 51 (1889).
Carex multimoda L.H.Bailey, Bot. Gaz. 21: 5 (1896).
Carex macloviana f. *decumbens* T.Holm, Amer. J. Sci., IV, 16: 20 (1903).
Carex festiva var. *horneri* Piper, Contr. U. S. Natl. Herb. 11: 164 (1906).
Carex macloviana f. *involucrata* Kük. in H.G.A. Engler (ed.), Pflanzenr., IV, 20(38): 197 (1909).
Carex macloviana var. *stricta* (L.H.Bailey) Kük. in H.G.A.Engler (ed.), Pflanzenr., IV, 20(38): 196 (1909).
Carex macloviana f. *viridis* (L.H.Bailey) Kük. in H.G.A.Engler (ed.), Pflanzenr., IV, 20(38): 197 (1909).
Carex pyrophila Gand., Bull. Soc. Bot. France 60: 420 (1913).
Carex olympica Mack., Bull. Torrey Bot. Club 43: 610 (1916 publ. 1917).
Carex soperi Raup, Sargentia 6: 129 (1947).
Carex incondita F.J.Herm., Leafl. W. Bot. 8: 112 (1957).

subsp. ***pachystachya*** (Cham. ex Steud.) Hultén, Acta Univ. Lund., n.s., 38(1): 318 (1942).
Russian Far East to W. U.S.A. 31 70 ALU ASK YUK 71 ABT BRC SAS 73 COL IDA MNT ORE WAS WYO 76 CAL NEV UTA. Hemicr. or rhizome geophyte.
**Carex pachystachya* Cham. ex Steud., Syn. Pl. Glumac. 2: 197 (1855). *Carex festiva* var. *pachystachya* (Cham. ex Steud.) L.H.Bailey, Mem. Torrey Bot. Club 1: 51 (1889). *Carex macloviana* var. *pachystachya* (Cham. ex Steud.) Kük. in H.G.A.Engler (ed.), Pflanzenr., IV, 20(38): 197 (1909).
Carex pachystachya var. *monds-coulteri* Kelso, Biol. Leafl. 64: 2 (1953). *Carex pachystachya* f. *monds-coulteri* (Kelso) F.J.Herm., Leafl. W. Bot. 9: 16 (1959).

var. ***pseudoleporina*** Kük., Bot. Jahrb. Syst. 27: 506 (1899).
WC. & SW. Argentina, C. & S. Chile. 85 AGS AGW CLC CLS. Hemicr. or rhizome geophyte.
Carex leporina var. *elatior* Boeck., Linnaea 39: 120 (1875).

subsp. ***subfusca*** (W.Boott) T.Koyama, Bishop Mus. Occas. Pap. 29: 118 (1989).
Hawaiian Is., W. U.S.A. to Mexico (Baja California). 63 HAW 73 IDA ORE WAS 76 ARI CAL NEV UTA 77 NWM 79 MXN. Hemicr. or rhizome geophyte.
Carex propinqua Nees & Meyen ex Kunth, Enum. Pl. 2: 396 (1837).
Carex festiva var. *gracilis* Olney, Proc. Amer. Acad. Arts 8: 407 (1872). *Carex macloviana* var. *gracilis* (Olney) Kük. in H.G.A.Engler (ed.), Pflanzenr., IV, 20(38): 197 (1909). *Carex pachystachya* var. *gracilis* (Olney) Mack. in N.L.Britton & al. (eds.), N. Amer. Fl. 18: 136 (1931).
**Carex subfusca* W.Boott in S.Watson & al., Bot. California 2: 234 (1880). *Carex macloviana* var. *subfusca* (W.Boott) Kük. in H.G.A.Engler (ed.), Pflanzenr., IV, 20(38): 197 (1909).
Carex macloviana var. *bracteata* Kük., Bot. Jahrb. Syst. 27: 506 (1899).
Carex flaviceps Kük. in H.G.A.Engler (ed.), Pflanzenr., IV, 20(38): 133 (1909).
Carex macloviana f. *bracteata* (Kük.) Kük. in H.G.A.Engler (ed.), Pflanzenr., IV, 20(38): 196 (1909).
Carex teneriformis Mack., Bull. Torrey Bot. Club 43: 609 (1916 publ. 1917).
Carex stenoptera Mack., Erythea 8: 28 (1922).

var. ***thermarum*** (Phil.) Kük. in H.G.A.Engler (ed.), Pflanzenr., IV, 20(38): 196 (1909).
WC. & SW. Argentina, C. & S. Chile. 85 AGS AGW CLC CLS. Hemicr. or rhizome geophyte.
**Carex thermarum* Phil., Anales Univ. Chile 21(2): 385 (1862).
Carex macloviana var. *incrassata* Kük., Bot. Jahrb. Syst. 27: 506 (1899).

Carex* × *macounii Dewey, Amer. J. Sci. Arts, II, 41: 228 (1866). *C. lupuliformis* × *C. retrorsa*. *Carex retrorsa* var. *macounii* (Dewey) Fernald, Rhodora 3: 55 (1901).
E. Canada. 72 ONT. Hemicr. or rhizome geophyte.
Carex lurida var. *divergens* L.H.Bailey, Proc. Amer. Acad. Arts 22: 63 (1886).

Carex macrandrolepis H.Lév. & Vaniot, Repert. Spec. Nov. Regni Veg. 5: 241 (1908).
Korea, Japan (Honshu, Kyushu), Taiwan, Vietnam. 38 JAP KOR TAI 41 VIE. Hemicr. or rhizome geophyte.
Carex distantiflora Nakai, Repert. Spec. Nov. Regni Veg. 13: 245 (1914).
Carex tatsutakensis Hayata, Icon. Pl. Formosan. 6: 133 (1916).
Carex sharyotensis Hayata, Icon. Pl. Formosan. 10: 69 (1921).

Carex macrocephala Willd. ex Spreng., Syst. Veg. 3: 808 (1826). *Vignea macrocephala* (Willd. ex Spreng.) Soják, Cas. Nár. Mus., Odd. Prír. 148: 196 (1979 publ. 1980).
Russian Far East, N. Japan, Alaska to Oregon. 31 KAM KUR SAK 38 JAP 70 ASK 71 BRC 73 ORE WAS. Hemicr. or rhizome geophyte.
Carex anthericoides C.Presl, Reliq. Haenk. 1: 204 (1828). *Carex macrocephala* subsp. *anthericoides* (C.Presl) Hultén, Acta Univ. Lund. 3(38; 1): 317 (1942).
Carex menziesiana Sm. ex Boott, Ill. Carex 1: 27 (1858).
Carex macrocephala var. *bracteata* T.Holm, Amer. J. Sci. 160: 266 (1900). *Carex macrocephala* f. *bracteata* (T.Holm) Kük. in H.G.A.Engler (ed.), Pflanzenr., IV, 20(38): 187 (1909).

Carex macrochaeta C.A.Mey., Mém. Sav. Étr. Acad. St. Pétersbourg 1: 224 (1831).
Siberia to Oregon. 30 ALT 31 KAM KUR MAG 70 ALU ASK YUK 71 BRC 73 ORE WAS. Hemicr. or rhizome geophyte.
Carex macrochaeta var. *subrigida* Kük. ex J.M.Macoun, Fur Seals Isl. North Pacif. Ocean 3: 573 (1899).
Carex macrochaeta var. *emarginata* T.Holm, Amer. J. Sci., IV, 17: 314 (1904).
Carex macrochaeta var. *macrochlaena* T.Holm, Amer. J. Sci., IV, 17: 315 (1904).
Carex macrochaeta var. *denticulata* Kük. in H.G.A. Engler (ed.), Pflanzenr., IV, 20(38): 414 (1909).
Carex macrochaeta var. *platycarpa* Kük. in H.G.A. Engler (ed.), Pflanzenr., IV, 20(38): 414 (1909).

Carex kuehleweinii Gand., Bull. Soc. Bot. France 66: 295 (1919 publ. 1920).

Carex macrolepis DC., Cat. Pl. Horti Monsp.: 89 (1813).
Italy (Appennini), Albania, W. & S. Greece, S. Turkey. 13 ALB GRC ITA 34 TUR. Hemicr. or rhizome geophyte.
Carex platystachys Ten., Fl. Napol. 4: 133 (1830).
Carex tenorii Kunth, Enum. Pl. 2: 437 (1837).

Carex macrophyllidion Nelmes, Bull. Misc. Inform. Kew 1940: 161 (1940).
Tanzania to S. Trop. Africa. 25 TAN 26 ANG ZAM. Hemicr. or rhizome geophyte.

Carex macrorrhiza Boeck., Beitr. Cyper. 1: 43 (1888).
Bolivia to S. South America. 83 BOL 85 AGE AGS AGW CLC CLN URU. Hemicr. or rhizome geophyte.

var. ***macrorrhiza***
Bolivia to S. South America. 83 BOL 85 AGE AGS AGW CLC CLN URU. Hemicr. or rhizome geophyte.
Carex curvifolia Boeck., Beitr. Cyper. 1: 41 (1888).

var. ***simplex*** Kük. in H.G.A.Engler (ed.), Pflanzenr., IV, 20(38): 127 (1909).
C. Argentina, C. Chile. 85 AGE AGS AGW CLC. Hemicr. or rhizome geophyte.

Carex macrosandra (Franch.) V.I.Krecz., Bot. Mater. Gerb. Bot. Inst. Komarova Akad. Nauk S.S.S.R. 9: 189 (1946).
SC. China. 36 CHC. Hemicr. or rhizome geophyte.
**Carex lanceolata* var. *macrosandra* Franch., Nouv. Arch. Mus. Hist. Nat., III, 9: 169 (1897). *Carex pediformis* var. *macrosandra* (Franch.) C.B.Clarke, J. Linn. Soc., Bot. 36: 303 (1904).

Carex macrosolen Steud., Syn. Pl. Glumac. 2: 210 (1855).
Magellan Reg. 85 AGS CLS. Hemicr. or rhizome geophyte.
Carex macrocarpa Phil., Linnaea 29: 86 (1857).

Carex macrostigmatica Kük., Finska Vetensk. Soc. Forh. 45(8): 5 (1903).
Siberia. 30+. Hemicr. or rhizome geophyte.

Carex macrostylon Lapeyr., Hist. Pl. Pyrénées: 562 (1813). *Vignea macrostylon* (Lapeyr.) Soják, Cas. Nár. Mus., Odd. Prír. 148: 196 (1979 publ. 1980).
Pyrenees to NW. Spain. 12 FRA SPA. Hemicr. or rhizome geophyte.

Carex maculata Boott, Trans. Linn. Soc. London 20: 128 (1851).
Trop. & Subtrop. Asia, E. Australia, SW. Pacific. 36 CHC CHH CHS 38 JAP KOR NNS TAI 40 IND SRL 41 THA 42 JAW LSI MLY SUL SUM 43 NWG 50 NSW QLD 60 NWC SAM. Hemicr. or rhizome geophyte.

var. ***maculata***
Trop. & Subtrop. Asia, E. Australia, SW. Pacific. 36 CHC CHH CHS 38 JAP KOR NNS TAI 40 IND SRL 41 THA 42 JAW LSI MLY SUL SUM 43 NWG 50 NSW QLD 60 NWC SAM. Hemicr. or rhizome geophyte.
Carex micans Boott, Mem. Amer. Acad. Arts, n.s., 6: 419 (1859).
Carex sieboldii Miq., Ann. Mus. Bot. Lugduno-Batavi 2: 150 (1866).
Carex neurochlamys F.Muell., Fragm. 8: 258 (1874). *Carex maculata* var. *neurochlamys* (F.Muell.) Kük. in H.G.A.Engler (ed.), Pflanzenr., IV, 20(38): 428 (1909).
Carex samoensis Boeck., Bot. Jahrb. Syst. 25: 588 (1898).
Carex acrogyna H.Lév. & Vaniot, Bull. Acad. Int. Géogr. Bot. 10: 201 (1901).
Carex malacocarpa H.Lév. & Vaniot, Bull. Acad. Int. Géogr. Bot. 10: 200 (1901).
Carex chilosacca C.B.Clarke, Kew Bull., Addit. Ser. 8: 82 (1908).
Carex latitans H.Lév. & Vaniot, Repert. Spec. Nov. Regni Veg. 5: 240 (1908).
Carex elibates Nelmes, Bull. Misc. Inform. Kew 1937: 353 (1937).

var. ***tetsuoi*** (Ohwi) Hatus. ex T.Koyama, Bot. Mag. (Tokyo) 72: 307 (1959).
Nansei-shoto. 38 NNS. Hemicr. or rhizome geophyte.
**Carex tetsuoi* Ohwi, Misc. Rep. Natl. Sci. Mus. 5: 2 (1952).

Carex madagascariensis Boeck., Bot. Jahrb. Syst. 5: 517 (1884).
Madagascar. 29 MDG. Hemicr. or rhizome geophyte.

Carex madrensis L.H.Bailey, Bot. Gaz. 25: 270 (1898). *Carex polystachya* var. *madrensis* (L.H.Bailey) Kük. in H.G.A.Engler (ed.), Pflanzenr., IV, 20(38): 267 (1909).
Mexico. 79 MXC MXE MXG MXS. Hemicr. or rhizome geophyte.

Carex magellanica Lam., Encycl. 3: 385 (1792). *Carex atrata* var. *magellanica* (Lam.) Wahlenb., Kongl. Vetensk. Acad. Nya Handl. 24: 160 (1803). *Carex limosa* var. *magellanica* (Lam.) Fiori, Nuov. Fl. Italia 1: 194 (1923).
Temp. & Subarctic Northern Hemisphere, Chile to Falkland Is. 10 FIN GRB ICE IRE NOR SWE 11 AUT CZE GER POL SWI 12 FRA 13 ITA ROM 14 BLR BLT RUC RUE RUN RUS RUW UKR? 30 ALT KRA WSB 31 KAM KUR 32 KAZ 33 NCS TCS 34 TUR 37 MON 38 JAP 70 ASK GNL NUN NWT YUK 71 ABT BRC MAN SAS 72 LAB NBR NFL NSC ONT PEI QUE 73 COL IDA WAS 74 MIN WIS 75 CNT MAI MIC NWH NWY OHI PEN VER 76 UTA 85 AGS CHC CHS 90 FAL. Hemicr. or rhizome geophyte.

subsp. ***irrigua*** (Wahlenb.) Hultén, Kuomen Kasvio: 161 (1933).
Temp. & Subarctic Northern Hemisphere. 10 FIN GRB ICE IRE NOR SWE 11 AUT CZE GER POL SWI 12 FRA 13 ITA ROM 14 BLR BLT RUC RUE RUN RUS RUW UKR? 30 ALT KRA WSB 31 KAM KUR 32 KAZ 33 NCS TCS 34 TUR 37 MON 38 JAP 70 ASK GNL NUN NWT YUK 71 ABT BRC MAN SAS 72 LAB NBR NFL NSC ONT PEI QUE 73 COL IDA WAS 74 MIN WIS 75 CNT MAI MIC NWH NWY OHI PEN VER 76 UTA. Hemicr. or rhizome geophyte.
Carex majae A.P.Khokhr. in ?
**Carex limosa* var. *irrigua* Wahlenb., Kongl. Vetensk. Acad. Nya Handl. 24: 162 (1803). *Carex irrigua* (Wahlenb.) Sm. ex Hoppe, Caricolog. Germ.: t. 92 (1826). *Carex magellanica* var. *irrigua* (Wahlenb.) Britton, Sterns & Poggenb.,

Prelim. Cat.: 63 (1888). *Carex paupercula* var. *irrigua* (Wahlenb.) Fernald, Rhodora 8: 76 (1906).
Carex paupercula Michx., Fl. Bor.-Amer. 2: 172 (1803).
Carex planifolia Kohts, Oesterr. Bot. Z. 20: 141 (1870).
Carex subalpina Brügger, Jahresber. Naturf. Ges. Graubündens 1886: 121 (1886).
Carex subalpina Brügger ex Appel, Ber. Bayer. Bot. Ges. 1: 76 (1891), nom. illeg.
Carex gentiliana H.Lév., Bull. Acad. Int. Géogr. Bot. 10: 221 (1901), nom. nud.
Carex magellanica var. *planitiei* Asch. & Graebn., Syn. Mitteleur. Fl. 2(2): 132 (1902).
Carex ontakensis H.Lév., Mém. Soc. Sci. Nat. Cherbourg 35: 218 (1905), nom. illeg.
Carex paupercula var. *pallens* Fernald, Rhodora 8: 77 (1906).

subsp. ***magellanica***
S. Argentina, C. & S. Chile, Falkland Is. 85 AGS CHC CHS 90 FAL. Hemicr. or rhizome geophyte.
Carex lenticularis Dewey, Amer. J. Sci. Arts 7: 263 (1824), nom. illeg.
Carex abjiciens Steud., Syn. Pl. Glumac. 2: 209 (1855).
Carex cernua Phil., Linnaea 29: 83 (1857), nom. illeg.
Carex limosa var. *humilior* St.-Lag. in A.Cariot, Étude Fl., ed. 8, 2: 121 (1886).

Carex magnoutriculata Tang & F.T.Wang ex L.K.Dai, Acta Phytotax. Sin. 32: 179 (1994).
China (Sichuan, Yunnan). 36 CHC. Hemicr. or rhizome geophyte.

Carex mairii Coss. & Germ., Obs. Pl. Env. Paris: 18 (1840).
Spain, France, NW. Italy, NW. Africa. 12 FRA SPA 13 ITA 20 ALG MOR. Hemicr. or rhizome geophyte.
Carex loscosii Lange, Vidensk. Meddel. Naturhist. Foren. Kjøbenhavn 1877–1878: 223 (1878). *Carex mairii* f. *loscosii* (Lange) Kük., Allg. Bot. Z. Syst. 7: 29 (1901).
Carex winkleri Lange, Vidensk. Meddel. Naturhist. Foren. Kjøbenhavn 1877–1878: 224 (1878).
Carex rodriguesii Merino, Fl. Galicia 3: 186 (1909).

Carex makinoensis Franch., Bull. Soc. Philom. Paris, VIII, 7: 47 (1895).
S. Japan to Taiwan. 38 JAP NNS TAI. Hemicr. or rhizome geophyte.
Carex shikokiana Makino, Bot. Mag. (Tokyo) 6: 47 (1892), nom. illeg.
Carex warburgiana Kük., Bull. Herb. Boissier, II, 5: 1162 (1905).
Carex shimadai Hayata, J. Coll. Sci. Imp. Univ. Tokyo 30(1): 396 (1911).

Carex makuensis P.C.Li, Acta Bot. Yunnan. 12: 139 (1990).
China (Yunnan). 36 CHC. Hemicr. or rhizome geophyte.

Carex malaccensis C.B.Clarke in J.D.Hooker, Fl. Brit. India 6: 722 (1894).
Pen. Malaysia (Lanykawi I.). 42 MLY. Hemicr. or rhizome geophyte.

Carex malmei Kalela, Ann. Acad. Sci. Fenn., Ser. A 54(5): 65 (1940).
N. Chile to NW. Argentina. 85 AGW CLN. Hemicr. or rhizome geophyte.

Carex malyschevii T.V.Egorova, Novosti Sist. Vyssh. Rast. 4: 32 (1968).
Siberia. 30+ 31?. Hemicr. or rhizome geophyte.

Carex manca Boott in G.Bentham, Fl. Hongk.: 402 (1861).
S. China to Taiwan. 36 CHC CHS 38 TAI. Hemicr. or rhizome geophyte.

subsp. ***jiuhuaensis*** (S.W.Su) S.Y.Liang, Acta Phytotax. Sin. 36: 537 (1998).
SE. China. 36 CHS. Hemicr. or rhizome geophyte.
**Carex jiuhuaensis* S.W.Su, Bull. Bot. Res., Harbin 10(1): 5 (1990).

subsp. ***manca***
SE. China. 36 CHC CHS. Hemicr. or rhizome geophyte.

subsp. ***takasagoana*** (Akiyama) T.Koyama, in Fl. Taiwan 5: 328 (1978).
Taiwan. 38 TAI. Hemicr. or rhizome geophyte.
**Carex takasagoana* Akiyama, Caric. Far East. Reg. As.: 214 (1955).

subsp. ***wichurae*** (Boeck.) S.Y.Liang, Acta Phytotax. Sin. 36: 537 (1998).
China (Macao). 36 CHS. Hemicr. or rhizome geophyte.
**Carex wichurae* Boeck., Bot. Jahrb. Syst. 5: 519 (1884). *Carex manca* var. *wichurae* (Boeck.) Kük. in H.G.A.Engler (ed.), Pflanzenr., IV, 20(38): 629 (1909).

Carex manciformis C.B.Clarke ex Franch., Nouv. Arch. Mus. Hist. Nat., III, 10: 62 (1898).
SC. China. 36 CHC. Hemicr. or rhizome geophyte.

Carex mandoniana Boeck., Allg. Bot. Z. Syst. 2: 174 (1896).
Ecuador to NW. Argentina. 83 BOL ECU PER 85 AGW. Hemicr. or rhizome geophyte.

Carex mandshurica Meinsh., Bot. Centralbl. 55: 197 (1893).
Russian Far East to Korea. 31 PRM 36 CHM 38 KOR. Hemicr. or rhizome geophyte.

Carex manginii E.G.Camus, Notul. Syst. (Paris) 2: 206 (1912).
NE. China. 36 CHM. Hemicr. or rhizome geophyte.

Carex manhartii Bryson, Castanea 50: 15 (1985).
West Virginia to SE. U.S.A. 75 WVA 78 GEO NCA SCA TEN VRG. Hemicr. or rhizome geophyte.

Carex mannii E.A.Bruce, Bull. Misc. Inform. Kew 1933: 150 (1933).
Bioko, SW. Cameroon, NE. Zaïre to Uganda. 23 CMN GGI RWA ZAI 25 UGA. Hemicr. or rhizome geophyte.
Carex boryana var. *simplicissima* Kük. in H.G.A. Engler (ed.), Pflanzenr., IV, 20(38): 651 (1909).

Carex manongarivensis Cherm., Bull. Soc. Bot. France 70: 412 (1923).
C. Madagascar. 29 MDG. Hemicr. or rhizome geophyte.

Carex maorica Hamlin, Trans. Roy. Soc. New Zealand 84: 684 (1957).
New Zealand. 51 NZN NZS. Hemicr. or rhizome geophyte.

Carex maorshanica Y.L.Chou, Bull. Bot. Lab. N. E. Forest. Inst., Harbin 4(1): 40 (1979).
NE. China. 36 CHM. Hemicr. or rhizome geophyte.

Carex maquensis Y.C.Yang, Acta Biol. Plateau Sin. 3: 85 (1984).
China (Gansu). 36 CHN. Hemicr. or rhizome geophyte.

Carex marahuacana Reznicek, Novon 6: 419 (1996).
S. Venezuela (Cerro Marahuaca). 82 VEN. Hemicr. or rhizome geophyte.

Carex marianensis Stacey, Leafl. W. Bot. 2: 31 (1937).
Mexico. 79 MXC MXE MXN MXS MXT. Hemicr. or rhizome geophyte.
**Carex autumnalis* Mack. in N.L.Britton & al. (eds.), N. Amer. Fl. 18: 66 (1931), nom. illeg.

Carex marina Dewey, Amer. J. Sci. Arts 29: 247 (1836).
Subarctic to Canada. 10 SVA 14 RUN 30 ALT KRA WSB YAK 31 MAG 70 ASK GNL NUN NWT YUK 71 MAN 72 LAB NFL QUE. Hemicr. or rhizome geophyte.

subsp. ***marina***
Siberia, Subarctic America to Canada. 30 ALT KRA WSB YAK 31 MAG 70 ASK GNL NUN NWT YUK 71 MAN 72 LAB NFL QUE. Hemicr. or rhizome geophyte.
Carex amblyorhyncha V.I.Krecz. in V.L.Komarov (ed.), Fl. URSS 3: 595 (1935).
Carex amblyorhyncha var. *fusca* Böcher, Acta Arctica 5: 25 (1952).
Carex amblyorhyncha var. *tundrarum* Böcher, Acta Arctica 5: 25 (1952).
Carex marina var. *fusca* Böcher, Feddes Repert. 80: 105 (1969).
Carex marina var. *tundrarum* Böcher, Feddes Repert. 80: 105 (1969).

subsp. ***pseudolagopina*** (T.J.Sørensen) Böcher, Feddes Repert. 80: 106 (1969).
Greenland to N. Russia. 10 SVA 14 RUN 70 GNL. Hemicr. or rhizome geophyte.
**Carex pseudolagopina* T.J.Sørensen, Meddel. Grønland 101(4): 167 (1937). *Carex amblyorhyncha* subsp. *pseudolagopina* (T.J. Sørensen) Böcher, Acta Arctica 5: 12 (1952).
Carex pseudolagopina var. *solida* T.J.Sørensen, Meddel. Grønland 101(4): 167 (1937). *Carex marina* var. *solida* (T.J.Sørensen) Böcher, Feddes Repert. 80: 106 (1969).

Carex mariposana L.H.Bailey ex Mack., Bull. Torrey Bot. Club 43: 619 (1916 publ. 1917).
California to Nevada. 73 IDA? 76 CAL NEV. Hemicr. or rhizome geophyte.
Carex paucifructus Mack., Bull. Torrey Bot. Club 43: 615 (1916 publ. 1917).

Carex maritima Gunnerus, Fl. Norveg. 2: 131 (1772). *Vignea maritima* (Gunnerus) Rchb. in J.C.Mössler, Handb. Gewächsk. ed. 2, 3: 1623 (1830).
Europe to Himalaya, E. Canada, NE. U.S.A., Ecuador to Argentina. 10 DEN GRB ICE NOR SVA SWE 11 AUT SWI 12 FRA SPA? 13 ITA 14 BLT RUN 20 MOR 30 KRA WSB YAK 31 MAG 40 NEP 70 ASK GNL NUN NWT YUK (71) abt man SAS 72 LAB NFL ONT QUE 83 ECU 85 AGS AGW CLC CLN CLS. Hemicr. or rhizome geophyte.
Carex incurva Lightf., Fl. Scot. 2: 544 (1777). *Vignea incurva* (Lightf.) Rchb. in J.C.Mössler, Handb. Gewächsk. ed. 2, 3: 1607 (1830). *Caricina incurva* (Lightf.) St.-Lag. in A.Cariot, Étude Fl., ed. 8, 2: 873 (1889). *Rhaptocalymna incurva* (Lightf.) Fedde & J.Schust., Just's Bot. Jahresber. 41(2): 11 (1913 publ. 1918).
Carex juncifolia All., Fl. Pedem. 2: 264 (1785). *Olotrema juncifolia* (All.) Raf., Good Book: 25 (1840).
Carex banata Sm. in A.Rees, Cycl. 39: s.p. (1818).
Carex amphilogos K.Koch, Linnaea 21: 615 (1848).
Carex incurva f. *erecta* O.Lang, Linnaea 24: 507 (1851).
Carex incurva var. *pratensis* Hartm., Handb. Skand. Fl., ed. 6: 502 (1854).
Carex melanocystis Desv. in C.Gay, Fl. Chil. 6: 203 (1854). *Carex incurva* var. *melanocystis* (Desv.) Kük. in H.G.A.Engler (ed.), Pflanzenr., IV, 20(38): 114 (1909).
Carex psammogaea Steud., Syn. Pl. Glumac. 2: 187 (1855).
Carex misera Phil., Fl. Atacam.: 53 (1860), nom. illeg. *Carex incurva* var. *misera* (Phil.) Kük., Bot. Jahrb. Syst. 27: 499 (1899). *Carex melanocystis* var. *misera* Kük., Bot. Jahrb. Syst. 27: 499 (1899). *Carex maritima* var. *misera* Fernald, Rhodora 35: 397 (1933).
Carex stenophylla L.Thienem. ex Boott, Ill. Carex 4: 210 (1867), nom. illeg.
Carex incurva var. *setina* H.Christ, Kongl. Vetensk. Acad. Handl. 22: 174 (1887). *Carex setina* (H.Christ) V.I.Krecz., Izv. Bot. Sada Akad. Nauk S.S.S.R. 30: 123 (1931). *Carex maritima* var. *setina* (H.Christ) Fernald, Rhodora 35: 397 (1933). *Carex maritima* subsp. *setina* (H.Christ) T.V.Egorova, Novosti Sist. Vyssh. Rast. 1965: 73 (1965). *Vignea maritima* subsp. *setina* (H.Christ) Soják, Cas. Nár. Mus., Odd. Prír. 148: 196 (1979 publ. 1980).
Carex oligantha Phil., Anales Mus. Nac. Santiago de Chile 2: 80 (1892), nom. illeg.
Carex incurva var. *inflata* Simmons, Vasc. Pl. Ellesmereland: 146 (1906). *Carex maritima* f. *inflata* (Simmons) Polunin, Bull. Natl. Mus. Canada 92: 110 (1940).
Carex incurva var. *chartacea* Kük. in H.G.A.Engler (ed.), Pflanzenr., IV, 20(38): 114 (1909).
Carex incurva f. *pallens* Kük. in H.G.A.Engler (ed.), Pflanzenr., IV, 20(38): 113 (1909).
Carex bucculenta V.I.Krecz., Izv. Bot. Sada Akad. Nauk S.S.S.R. 30: 122 (1931).
Carex jucunda V.I.Krecz., Izv. Bot. Sada Akad. Nauk S.S.S.R. 30: 124 (1931).
Carex orthocaula V.I.Krecz., Izv. Bot. Sada Akad. Nauk S.S.S.R. 30: 123 (1931).
Carex transmarina Krecz., Izv. Bot. Sada Akad. Nauk S.S.S.R. 30: 122 (1931).
Carex camptotropa V.I.Krecz., Izv. Bot. Sada Akad. Nauk S.S.S.R. 30: 124 (1931 publ. 1932).
Carex psychroluta V.I.Krecz., Izv. Bot. Sada Akad. Nauk S.S.S.R. 30: 138 (1931 publ. 1932).
Carex maritima var. *melanocystis* (É.Desv.) Fernald, Rhodora 35: 297 (1933).
Carex maritima subsp. *yukonensis* A.E.Porsild, Bull. Natl. Mus. Canada 216: 19 (1966).

Carex markgrafii Kük., Ber. Deutsch. Bot. Ges. 44: 420 (1926).
C. Albania (Shën Ndue). 13 ALB. Hemicr. or rhizome geophyte.

Carex × marshallii A.Benn., Trans. & Proc. Bot. Soc. Edinburgh 29: 128 (1925). *C. flava* × *C. saxatilis*.
Europe. 1+. Hemicr. or rhizome geophyte.

Carex martinii H.Lév. & Vaniot, Bull. Acad. Int. Géogr. Bot. 11: 57 (1902).
China (Guizhou). 36 CHC. Hemicr. or rhizome geophyte.

Carex martynenkoi Zolot., Novosti Sist. Vyssh. Rast. 21: 226 (1984).
Siberia. 30+. Hemicr. or rhizome geophyte.

Carex matsumurae Franch., Bull. Soc. Philom. Paris, VIII, 7: 50 (1895).
S. Korea, C. & S. Japan. 38 JAP KOR. Hemicr. or rhizome geophyte.
Carex taquetii H.Lév., Repert. Spec. Nov. Regni Veg. 8: 426 (1910).
Carex viridissima Nakai, Bot. Mag. (Tokyo) 31: 281 (1917).
Carex matsumurae var. *variegata* Makino, J. Jap. Bot. 6: 38 (1929).

Carex mayebarana Ohwi, Mem. Coll. Sci. Kyoto Imp. Univ., Ser. B, Biol. 3: 256 (1930).
S. Japan. 38 JAP. Hemicr. or rhizome geophyte.

Carex mckittrickensis P.W.Ball, Novon 8: 221 (1998).
Texas (Guadalupe Mts.). 77 TEX. Hemicr. or rhizome geophyte.

Carex mcvaughii Reznicek, Syst. Bot. 7: 342 (1982).
Mexico (Jalisco). 79 MXS. Hemicr. or rhizome geophyte.

Carex meadii Dewey, Amer. J. Sci. Arts 43: 90 (1842). *Carex tetanica* var. *meadii* (Dewey) L.H.Bailey, Mem. Torrey Bot. Club 1: 118 (1889). *Carex douglasii* f. *meadii* (Dewey) Kük. in H.G.A.Engler (ed.), Pflanzenr., IV, 20(38): 122 (1909).
SC. & SE. Canada to C. & E. U.S.A., Mexico (Chihuahua). 71 MAN SAS 72 ONT 74 ILL IOW KAN MIN MSO NDA NEB OKL SDA WIS 75 INI MIC NWJ OHI PEN RHO 76 ARI 77 TEX 78 ALA ARK GEO KTY MSI NCA TEN VRG 79 MXE. Hemicr. or rhizome geophyte.
Carex panicea Carey in A.Gray, Manual: 521 (1848), nom. illeg.
Carex meadii var. *bebbii* Arthur, Proc. Davenport Acad. Nat. Sci. 4: 70 (1884).

Carex media R.Br. in J.Richardson, Bot. App.: 750 (1823).
Subarctic to N. U.S.A. 10 FIN 14 RUN 30 ALT KRA WSB YAK 31 MAG 32 KAZ KGZ 36 CHI CHM 37 MON 38 KOR 70 ASK NUN NWT YUK 71 ABT BRC MAN SAS 72 LAB ONT QUE 73 IOW MNT ORE WAS 74 WIS 75 MAI MIC. Hemicr. or rhizome geophyte.
Carex alpina var. *inferalpina* Wahlenb., Fl. Lapp.: 241 (1812). *Carex vahlii* var. *inferalpina* (Wahlenb.) Fernald, Rhodora 35: 398 (1933). *Carex norvegica* subsp. *inferalpina* (Wahlenb.) Hultén, Acta Univ. Lund., n.s., 38(1): 348 (1942). *Carex norvegica* var. *inferalpina* (Wahlenb.) Boivin, Naturaliste Canad. 94: 523 (1967).
Carex brachylepis Turcz. ex Steud., Nomencl. Bot., ed. 2, 1: 286 (1840).
Carex angarae Steud., Syn. Pl. Glumac. 2: 190 (1855).
Carex komaroviana A.I.Baranov & Skvortsov, Quart. J. Taiwan Mus. 18: 229 (1965).

Carex mediterranea C.B.Clarke ex Post, Fl. Syria: 837 (1896).
Syria. 34 LBS. Hemicr. or rhizome geophyte.

Carex medwedewii Leskov, Zhurn. Russk. Bot. Obshch. 16(1): 70 (1931). *Carex aterrima* subsp. *medwedewii* (Leskov) T.V.Egorova, Novosti Sist. Vyssh. Rast. 22: 54 (1985).
Caucasus. 33+. Hemicr. or rhizome geophyte.

Carex meeboldiana Kük., Repert. Spec. Nov. Regni Veg. 16: 434 (1920).
Myanmar to SW. Thailand. 41 MYA THA. Hemicr. or rhizome geophyte.
Carex hypotracheia Nelmes, Kew Bull. 4: 39 (1949).

Carex meihsienica K.T.Fu, in Fl. Tsinlingensis 1(1): 445 (1976).
China (Shaanxi). 36 CHN. Hemicr. or rhizome geophyte.

Carex meiocarpa H.Lév. & Vaniot, Repert. Spec. Nov. Regni Veg. 4: 226 (1907).
Korea. 38 KOR. Hemicr. or rhizome geophyte.
Carex harmsiana Kük. ex H.Lév., Bull. Acad. Int. Géogr. Bot. 20: 47 (1910).

Carex melanantha C.A.Mey. in C.F.von Ledebour, Fl. Ross. 4: 216 (1852). *Carex melanantha* var. *genuina* Turcz., Fl. Baical.-Dahur. 2(1): 270 (1856), nom. inval. *Carex nigra* var. *melanantha* (C.A.Mey.) Regel, Trudy Imp. S.-Peterburgsk. Bot. Sada 7: 568 (1880).
Siberia to Afghanistan and C. China. 30+ 32 KAZ KGZ TZK UZB 34 AFG IRN 36 CHC CHQ CHT CHX 37 MON 40 EHM NEP PAK WHM. Hemicr. or rhizome geophyte.

var. ***melanantha***
Siberia to Afghanistan and Himalaya. 30+ 32 KAZ KGZ TZK UZB 34 AFG IRN 36 CHX 37 MON 40 EHM NEP PAK WHM. Hemicr. or rhizome geophyte.
Carex melanantha var. *baicalensis* Turcz., Fl. Baical.-Dahur. 2(1): 270 (1856).
Carex nigra var. *orientalis* Regel, Trudy Imp. S.-Peterburgsk. Bot. Sada 7: 568 (1880).
Carex macrantha Boeck., Beitr. Cyper. 1: 49 (1888).
Carex prattii Franch., Bull. Soc. Philom. Paris, VIII, 7: 32 (1895).
Carex melanantha f. *gracilescens* Kük. in H.G.A. Engler (ed.), Pflanzenr., IV, 20(38): 391 (1909).

var. ***moorcroftii*** (Falc. ex Boott) Kük. in H.G.A. Engler (ed.), Pflanzenr., IV, 20(38): 391 (1909).
W. Himalaya to C. China. 36 CHC CHQ CHT 40 WHM. Hemicr. or rhizome geophyte.
**Carex moorcroftii* Falc. ex Boott, Proc. Linn. Soc. London 1: 288 (1846).

Carex melananthiformis Litv., Trudy Bot. Muz. Imp. Akad. Nauk 7: 90 (1910).
Caucasus, C. Asia, Siberia. 30+ 32 KAZ KGZ 33+ 37 MON. Hemicr. or rhizome geophyte.

Carex melanocarpa Cham. ex Trautv. in A.T.von Middendorff, Reise Sibir. 1(2): 21 (1856). *Carex ericetorum* subsp. *melanocarpa* (Cham. ex Trautv.) Kük. in H.G.A.Engler (ed.), Pflanzenr., IV, 20(38): 440 (1909).
N. European Russia to N. Japan. 14 RUN 30 KRA WSB YAK 31 MAG SAK 38 JAP. Hemicr. or rhizome geophyte.

Carex melanocephala Turcz. ex Kunth, Enum. Pl. 2: 437 (1837).
Siberia to Mongolia. 30 ALT 32 KAZ KGZ 36 CHX 37 MON. Hemicr. or rhizome geophyte.

Carex melanorrhyncha Nelmes, Notes Roy. Bot. Gard. Edinburgh 22: 84 (1956).
NE. Turkey (Rize). 34 TUR. Hemicr. or rhizome geophyte.

Carex melanostachya M.Bieb. ex Willd., Sp. Pl. 4: 299 (1805).
Europe to NW. China. 11 AUT CZE GER HUN 12 FRA 13 BUL GRC? ITA ROM YUG 14 KRY RUC RUS UKR 30 ALT 32 KAZ KGZ TKM TZK UZB 33 NCS TCS 34 AFG IRN IRQ TUR 36 CHX (74) kan neb okl. Hemicr. or rhizome geophyte.
Carex nutans Host, Icon. Descr. Gram. Austriac. 1: 61 (1801), nom. illeg.
Carex juncoides J.Presl & C.Presl, Fl. Cech.: 190 (1819). *Vignea juncoides* (J.Presl & C.Presl) Opiz, Seznam: 104 (1852).
Uncinia digyna Hochst. ex Steud., Syn. Pl. Glumac. 2: 244 (1855).
Carex sulcata Schur, Verh. Mitth. Siebenbürg. Vereins Naturwiss. Hermannstadt 4: 80 (1858). *Carex nutans* f. *sulcata* (Schur) Kük. in H.G.A.Engler (ed.), Pflanzenr., IV, 20(38): 740 (1909).
Carex ledebourii Boiss. & Buhse, Nouv. Mém. Soc. Imp. Naturalistes Moscou 12: 222 (1860).
Carex nutans var. *japonica* Franch. & Sav., Enum. Pl. Jap. 2: 154 (1877).
Carex nutans var. *major* Boeck., Linnaea 41: 296 (1877).
Carex bornmulleri Kük., Oesterr. Bot. Z. 47: 136 (1897).
Carex ripariiformis Litv., Trudy Bot. Muz. Imp. Akad. Nauk 7: 96 (1910).
Carex bicuspidata Regel ex Krecz. in V.L.Komarov (ed.), Fl. URSS 3: 412 (1935), pro syn.

Carex melinacra Franch., Nouv. Arch. Mus. Hist. Nat., III, 9: 135 (1897). *Carex forficula* var. *melinacra* (Franch.) Kük. in H.G.A.Engler (ed.), Pflanzenr., IV, 20(38): 342 (1909).
SC. China. 36 CHC. Hemicr. or rhizome geophyte.
Carex melinacra var. *changningensis* S.Yun Liang, Acta Phytotax. Sin. 37: 193 (1999).

Carex membranacea Hook. in G.F.Lyon, Brief Narrative Attempt Reach Repulse Bay: 195 (1824 publ. 1825).
Russian Far East, Subarctic America to Canada. 31 KAM MAG 70 ASK NUN NWT YUK 71 BRC MAN 72 LAB NFL QUE. Hemicr. or rhizome geophyte.
Carex compacta R.Br., Voy. Explor. Baffin's Bay: 163 (1819), nom. illeg. *Carex vesicaria* var. *compacta* Dewey, Amer. J. Sci. Arts 11: 310 (1826). *Carex rotundata* var. *compacta* (Dewey) Boivin, Naturaliste Canad. 94: 523 (1967).
Carex membranopacta L.H.Bailey, Bull. Torrey Bot. Club 20: 428 (1893).
Carex physochlaena T.Holm, Amer. J. Sci. 166: 464 (1903).

Carex* × *mendica Lepage, Naturaliste Canad. 83: 147 (1956). *C. recta* × *C. salina*.
E. Canada. 72 LAB ONT QUE. Hemicr. or rhizome geophyte.

Carex mendocinensis Olney ex Boott in S.Watson & al., Bot. California 2: 249 (1880).
Oregon to California. 73 ORE 76 CAL. Hemicr. or rhizome geophyte.
Carex cinnamomea Olney, Proc. Amer. Acad. Arts 7: 896 (1868), nom. illeg.
Carex debiliformis Mack., Bull. Torrey Bot. Club 37: 244 (1910).

Carex merritt-fernaldii Mack., Bull. Torrey Bot. Club 49: 370 (1922 publ. 1923).
S. Canada to NC. & NE. U.S.A. 71 MAN 72 NBR ONT QUE 74 MIN WIS 75 CNT MAI MAS MIC NWH NWY OHI VER. Hemicr. or rhizome geophyte.
Carex brevior var. *pseudofestucacea* Farw., Pap. Michigan Acad. Sci. 2: 18 (1923 publ. 1924).

Carex mertensii Prescott ex Bong., Veg. Sitcha: 51 (1833).
Kuril Is., N. & C. Japan, Alaska to W. U.S.A. 31 KUR 38 JAP 70 ASK 71 ABT BRC 72 IDA MNT ORE WAS 76 CAL. Hemicr. or rhizome geophyte.

var. ***mertensii***
Alaska to W. U.S.A. 70 ASK 71 ABT BRC 72 IDA MNT ORE WAS 76 CAL. Hemicr. or rhizome geophyte.
Carex columbiana Dewey, Amer. J. Sci. Arts 30: 62 (1836).

var. ***urostachys*** (Franch.) Kük. in H.G.A.Engler (ed.), Pflanzenr., IV, 20(38): 401 (1909).
Kuril Is., N. & C. Japan. 31 KUR 38 JAP. Hemicr. or rhizome geophyte.
**Carex urostachys* Franch., Bull. Soc. Philom. Paris, VIII, 7: 35 (1895). *Carex mertensii* subsp. *urostachys* (Franch.) Calder & T.Koyama, Canad. J. Bot. 43: 1392 (1965).

Carex merxmuelleri Podlech, Mitt. Bot. Staatssamml. München 4: 121 (1961).
S. Africa. 27 NAT TVL. Hemicr. or rhizome geophyte.

Carex mesochorea Mack., Bull. Torrey Bot. Club 37: 246 (1910). *Carex cephalophora* var. *mesochorea* (Mack.) Gleason, Phytologia 4: 22 (1952).
SE. Canada to C. & E. U.S.A. 72 ONT 74 ILL KAN MSO NEB 75 CNT INI MAS MIC NWJ NWY OHI PEN WVA (76) cal 77 TEX 78 ALA GEO KTY MRY NCA SCA TEN VRG WDC. Hemicr. or rhizome geophyte.
**Carex mediterranea* Mack., Bull. Torrey Bot. Club 33: 441 (1906), nom. illeg.

Carex metallica H.Lév., Repert. Spec. Nov. Regni Veg. 5: 239 (1908).
C. & S. Japan to Taiwan, China (Fujian). 36 CHS 38 JAP KOR NNS TAI. Hemicr. or rhizome geophyte.
Carex pachinensis Hayata, Icon. Pl. Formosan. 10: 58 (1921).

Carex meyenii Nees, Nova Acta Acad. Caes. Leop.-Carol. German. Nat. Cur. 19(Suppl. 1): 123 (1843). *Carex brunnea* var. *meyenii* (Nees) Kük. in H.G.A. Engler (ed.), Pflanzenr., IV, 20(38): 601 (1909).
N. Pacific. 62 CRL 63 HAW. Hemicr.
Carex remyi Boeck., Flora 58: 264 (1875).
Carex kanehirae Ohwi, Acta Phytotax. Geobot. 8: 67 (1939).

Carex meyeriana Kunth, Enum. Pl. 2: 438 (1837).
Siberia to N. & C. Japan. 30+ 31+ 36 CHC CHI CHM 37 MON 38 JAP KOR. Hemicr. or rhizome geophyte.
Carex asperula Turcz., Bull. Soc. Imp. Naturalistes Moscou 1838: 104 (1838).
Carex crassinervia Franch., Bull. Soc. Philom. Paris, VIII, 7: 38 (1895).
Carex funicularis Franch., Bull. Soc. Philom. Paris, VIII, 7: 57 (1895).
Carex meyeriana var. *scabra* Meinsh., Trudy Imp. S.-Peterburgsk. Bot. Sada 18: 349 (1901).

Carex michauxiana Boeck., Linnaea 41: 336 (1877).
Russian Far East, Japan, China, Papua New Guinea, C. Canada to NC. & NE. U.S.A. 31 KAM KUR 36 CHS 38 JAP 43 NWG 71 MAN SAS 72 LAB NBR NFL NSC ONT QUE 74 MIN WIS 75 MAI MAS MIC NWH NWY VER 78 MRY. Hemicr. or rhizome geophyte.

subsp. ***asiatica*** Hultén, Kongl. Svenska Vetensk. Acad. Handl. 5: 207 (1927). *Carex michauxiana* var. *asiatica* (Hultén) Ohwi, Mem. Coll. Sci. Kyoto Imp. Univ., Ser. B, Biol. 11(5): 491 (1936).
Russian Far East, Japan, China, Papua New Guinea. 31 KAM KUR 36 CHS 38 JAP 43 NWG. Hemicr. or rhizome geophyte.
Carex dolichocarpa C.A.Mey. ex Kom., Fl. Kamtschatka 1: 264 (1927), pro syn.

subsp. ***michauxiana***
C. Canada to NC. & NE. U.S.A. 71 MAN SAS 72 LAB NBR NFL NSC ONT QUE 74 MIN WIS 75 MAI MAS MIC NWH NWY VER 78 MRY. Hemicr. or rhizome geophyte.
Carex rostrata Michx., Fl. Bor.-Amer. 2: 173 (1803), nom. illeg.
Carex xanthophysa var. *nana* Dewey, Amer. J. Sci. Arts 14: 353 (1828).
Carex abacta L.H.Bailey, Bull. Torrey Bot. Club 20: 427 (1893).

Carex michelii Host, Syn. Pl.: 507 (1797).
Europe to Caucasus. 11 AUT CZE HUN POL 12 FRA 13 BUL ITA ROM YUG 14 KRY RUC RUS UKR 33 NCS TCS 34 TUR. Hemicr. or rhizome geophyte.
Carex rostrata Hoppe ex Schkuhr, Beschr. Riedgräs. 1: 99 (1801).
Carex ampullacea Wulfen ex Wahlenb., Kongl. Vetensk. Acad. Nya Handl. 24: 155 (1803), nom. illeg.
Carex beckeri C.A.Mey. ex Claus, Beitr. Pflanzenk. Russ. Reiches 8: 55 (1851). *Carex michelii* var. *beckeri* (C.A.Mey. ex Claus) Nyman, Consp. Fl. Eur., Suppl. 2: 321 (1890).
Carex vexilis Wormsk. ex Boott, Ill. Carex 4: 191 (1867).
Carex michelii f. *puberula* Beck, Fl. Nieder-Österreich: 145 (1890).

Carex microcarpa Bertol. ex Moris, Stirp. Sard. Elench. 1: 48 (1827).
Corse, Sardegne, C. Italy. 12 COR SAR 13 ITA. Hemicr. or rhizome geophyte.
Carex corsica Degl. in J.L.A.Loiseleur-Deslongchamps, Fl. Gall., ed. 2, 2: 307 (1828).
Carex aquatilis Ten., Fl. Napol. 4: 133 (1830), nom. illeg.

Carex microchaeta T.Holm, Amer. J. Sci. 166: 457 (1903).
Russian Far East, Subarctic America to W. Canada. 31 KAM 70 ALU ASK NWT YUK 71 BRC. Hemicr. or rhizome geophyte.

subsp. ***microchaeta***
Subarctic America to W. Canada. 70 ASK NWT YUK 71 BRC. Hemicr. or rhizome geophyte.

subsp. ***nesophila*** (T.Holm) D.F.Murray, Canad. J. Bot. 48: 323 (1970).
Russian Far East, Subarctic America. 31 KAM 70 ALU ASK. Hemicr. or rhizome geophyte.
**Carex nesophila* T.Holm, Amer. J. Sci. 166: 458 (1903).
Carex arakamensis C.B.Clarke, Kew Bull., Addit. Ser. 8: 82 (1908).
Carex melanostoma V.I.Krecz. in V.L.Komarov (ed.), Fl. URSS 3: 602 (1935).

Carex microdonta Torr., Ann. Lyceum Nat. Hist. New York 3: 423 (1836).
C. & S. U.S.A. to NE. Mexico. 74 KAN MSO OKL 76 ARI 77 NWM TEX 78 ALA ARK FLA LOU MSI 79 MXE. Hemicr. or rhizome geophyte.
Carex roemeriana Scheele, Linnaea 22: 346 (1849).
Carex wrightii Dewey ex Torr. in W.H.Emory, Rep. U.S. Mex. Bound. 2(1): 232 (1858).
Carex microdonta var. *controversa* L.H.Bailey, Proc. Amer. Acad. Arts 22: 110 (1886).
Carex microdonta var. *latifolia* L.H.Bailey, Proc. Amer. Acad. Arts 22: 110 (1886).

Carex microglochin Wahlenb., Kongl. Vetensk. Acad. Nya Handl. 1803: 140 (1803). *Carex pauciflora* var. *microglochin* (Wahlenb.) Poir. in J.B.A.M.de Lamarck, Encycl., Suppl. 3: 247 (1813). *Uncinia microglochin* (Wahlenb.) Spreng., Syst. Veg. 3: 830 (1826). *Caricinella microglochin* (Wahlenb.) St.-Lag. in A.Cariot, Étude Fl., ed. 8, 2: 880 (1889). *Kobresia microglochin* (Wahlenb.) Tang & W.T.Wang, Rep. Invest. Fl. Fauna Ah Li: 101 (1979), without basionym ref.
Europe to Himalaya, Subarctic America to WC. U.S.A., Colombia to S. South America, Falkland Is. 10 FIN GRB ICE NOR SWE 11 AUT GER† POL SWI 12 FRA 13 ITA 14 BLT? RUN 30 ALT WSB 32 KAZ KGZ TZK UZB 33 NCS TCS 34 AFG IRN TUR 36 CHC CHQ CHT CHX 37 MON 40 EHM NEP WHM 70 ASK GNL NUN NWT YUK 71 ABT BRC MAN 72 LAB NFL ONT QUE 73 COL MNT? WYO 76 UTA 83 BOL CLM ECU PER 85 AGS AGW CLC CLS 90 FAL. Hemicr. or rhizome geophyte.

subsp. ***fuegina*** Kük., Bot. Jahrb. Syst. 27: 546 (1899).
Colombia to S. South America, Falkland Is. 83 CLM ECU 85 CLS 90 FAL. Hemicr. or rhizome geophyte.
Carex parviflora C.A.Mey. ex Kunth, Enum. Pl. 2: 436 (1837), nom. illeg.
Carex oligantha Boott, Ill. Carex 4: 174 (1867), nom. illeg. *Carex microglochin* var. *oligantha* Kük. in H.G.A.Engler (ed.), Pflanzenr., IV, 20(38): 109 (1909). *Carex camptoglochin* V.I. Krecz., Bot. Mater. Gerb. Bot. Inst. Komarova Akad. Nauk S.S.S.R. 7: 34 (1937).

subsp. ***microglochin***
Europe to Himalaya, Subarctic & Subalpine America. 10 FIN GRB ICE NOR SWE 11 AUT GER† POL SWI 12 FRA 13 ITA 14 BLT? RUN 30 ALT WSB 32 KAZ KGZ TZK UZB 33 NCS TCS 34 AFG IRN TUR 36 CHC CHQ CHT CHX 37 MON 40 EHM NEP WHM 70 ASK GNL NUN NWT YUK 71 ABT BRC MAN 72 LAB NFL ONT QUE 73 COL MNT? WYO 76 UTA 83 BOL ECU PER 85 AGS AGW CLC CLS. Hemicr. or rhizome geophyte.
Uncinia europaea J.Gay, Flora 10: 26 (1827).
Carex lyonii Boott in W.J.Hooker, Fl. Bor.-Amer. 2: 209 (1839).
Carex pauciflora Balb. ex Boott, Ill. Carex 4: 174 (1867), nom. illeg.

Carex micropoda C.A.Mey., Mém. Sav. Étr. Acad. St. Pétersbourg 1: 210 (1831). *Carex pyrenaica* subsp. *micropoda* (C.A.Mey.) Hultén in ?
Japan to W. U.S.A. 31 KAM KUR MAG SAK 38 JAP

70 ALU ASK YUK 71 ABT BRC 73 COL IDA MNT ORE WAS WYO 76 CAL NEV UTA. Hemicr. or rhizome geophyte.
Carex nivalis Cham. ex Steud., Syn. Pl. Glumac. 2: 184 (1855), nom. inval.
Carex crandallii Gand., Bull. Soc. Bot. France 66: 295 (1919 publ. 1920).
Carex jacobi-peteri Hultén, Acta Univ. Lund., n.s., 38(1): 300 (1942).

Carex micropterа Mack., Muhlenbergia 5: 56 (1909). *Carex macloviana* var. *microptera* (Mack.) Boivin, Naturaliste Canad. 94: 523 (1967).
Yukon to W. & WC. U.S.A., Mexico. 70 NWT YUK 71 ABT BRC MAN SAS 73 COL IDA MNT ORE WAS WYO 74 SDA 76 ARI CAL NEV UTA 77 NWM 79 MXC MXE. Hemicr. or rhizome geophyte.
Carex festivella Mack., Bull. Torrey Bot. Club 42: 609 (1915). *Carex macloviana* subsp. *festivella* (Mack.) Á.Löve & D.Löve, Taxon 30: 847 (1981).
Carex limnophila F.J.Herm., Leafl. W. Bot. 8: 28 (1956). *Carex microptera* var. *limnophila* (F.J.Herm.) Dorn, Vasc. Pl. Wyoming: 296 (1988).
Carex microptera var. *crassinervia* F.J.Herm., Rhodora 70: 420 (1968).

Carex microrhyncha Mack., Bull. Torrey Bot. Club 40: 548 (1913).
C. U.S.A. to Mexico (Tamaulipas). 74 KAN MSO OKL 77 TEX 79 MXE. Hemicr. or rhizome geophyte.

Carex ×microstachya Ehrh., Hannover. Mag. 9: 132 (1784). *C. canescens* × *C. dioica*. *Vignea ×microstachya* (Ehrh.) Rchb. in J.C.Mössler, Handb. Gewächsk. ed. 2, 3: 1613 (1830).
Europe. 10 FIN NOR 11 GER 14 BLT. Hemicr. or rhizome geophyte.

Carex ×microstachyoides Mantell in ?. *C. dioica* × *C. heleonastes*.
European Russia to Siberia. 14+ 30+. Hemicr. or rhizome geophyte.

Carex ×microstyla J.Gay ex Gaudin, Fl. Helv. 6: 37 (1830). *C. brunnescens* × *C. foetida*. *Vignea ×microstylis* (J.Gay ex Gaudin) Rchb., Fl. Germ. Excurs.: 140(6) (1831). *Caricina ×microstyla* (J.Gay ex Gaudin) St.-Lag. in A.Cariot, Étude Fl., ed. 8, 2: 873 (1889).
Europe. 12 FRA SPA. Hemicr. or rhizome geophyte.

Carex middendorffii F.Schmidt, Reis. Amur-Land., Bot.: 70 (1868).
Russian Far East, Japan. 31 KAM KUR PRM SAK 36+ 38 JAP. Hemicr. or rhizome geophyte.

var. ***kirigaminensis*** (Ohwi) Ohwi, J. Jap. Bot. 11: 407 (1935).
Japan. 38 JAP. Hemicr. or rhizome geophyte.
Carex leiogona Franch., Bull. Soc. Philom. Paris, VIII, 7: 88 (1865).
Carex caulorrhiza H.Lév. & Vaniot, Bull. Acad. Int. Géogr. Bot. 11: 179 (1902).
Carex semiplena Kük., Bull. Herb. Boissier, II, 2: 1018 (1902).
**Carex caulorrhiza* var. *kirigaminensis* Ohwi, Mem. Coll. Sci. Kyoto Imp. Univ., Ser. B, Biol. 6(5): 241 (1931).

var. ***middendorffii***
Russian Far East, Japan. 31 KAM KUR PRM SAK 36+ 38 JAP. Hemicr. or rhizome geophyte.
Carex levicaulis Franch., Bull. Soc. Philom. Paris, VIII, 7: 37 (1895).
Carex grandilimosa Akiyama, J. Fac. Sci. Hokkaido Univ., Ser. 5, Bot. 1: 57 (1931).

Carex mildbraediana Kük. in H.G.A.Engler (ed.), Pflanzenr., IV, 20(38): 767 (1909).
E. Zaïre to Tanzania. 23 BUR RWA ZAI 25 TAN UGA. Hemicr. or rhizome geophyte.
Carex ramosipes Cherm., Bull. Soc. Bot. France 82: 343 (1935).

Carex millsii Dunn, Bull. Misc. Inform. Kew 1912: 109 (1912).
Korea. 38 KOR. Hemicr. or rhizome geophyte.

Carex mingrelica Kük. in H.G.A.Engler (ed.), Pflanzenr., IV, 20(38): 574 (1909).
Caucasus. 33 TCS. Hemicr. or rhizome geophyte.

Carex minquinensis Z.P.Wang, Acta Bot. Yunnan. 11: 169 (1989).
China. 36+. Hemicr. or rhizome geophyte.

Carex minuticulmis S.W.Su & S.M.Xu, Bull. Bot. Res., Harbin 8(3): 79 (1988).
China. 36+. Hemicr. or rhizome geophyte.

Carex minutiscabra Kük. ex Krecz. in V.L.Komarov (ed.), Fl. URSS 3: 616 (1935).
C. Asia. 32 KAZ. Hemicr. or rhizome geophyte.

Carex minutissima Barros, Darwiniana 12: 137 (1960).
S. Chile to SW. Argentina. 85 AGS CLS. Hemicr. or rhizome geophyte.

Carex minxianensis S.Yun Liang, in Fl. Reipubl. Popul. Sin. 12: 520 (2000).
China (Gansu). 36 CHN. Hemicr.

Carex minxianica Y.C.Yang, Acta Biol. Plateau Sin. 3: 86 (1984).
China (Gansu). 36 CHN. Hemicr. or rhizome geophyte.

Carex mira Kük., Bull. Herb. Boissier, II, 5: 1161 (1905).
Korea, Japan (Honshu). 38 JAP KOR. Hemicr. or rhizome geophyte.

Carex misera Buckley, Amer. J. Sci. Arts 45: 173 (1843).
EC. U.S.A. 78 GEO NCA TEN. Hemicr. or rhizome geophyte.
Carex juncea Willd., Enum. Pl., Suppl.: 63 (1814), nom. illeg.
Carex rugeliana Kunze, Linnaea 20: 8 (1847).

Carex missouriensis P.Rothr. & Reznicek, Novon 11: 220 (2001).
NC. U.S.A. to Indiana. 74 ILL IOW KAN MSO NEB 75 INI. Hemicr.
Carex straminea var. *meadii* Boott, Ill. Carex 3: 121 (1862).

Carex mitchelliana M.A.Curtis, Amer. J. Sci. Arts 44: 84 (1843). *Carex crinita* var. *mitchelliana* (M.A.Curtis) Gleason, Phytologia 4: 23 (1952).
E. U.S.A. 75 MAS NWJ NWY PEN RHO 78 ALA DEL FLA GEO KTY MRY MSI NCA SCA TEN VRG. Hemicr. or rhizome geophyte.

Carex ×mithala Callier, Deutsche Bot. Monatsschr. 6: 40 (1888). *C. canescens* × *C. loliacea*. *Vignea ×mithala* (Callier) Soják, Cas. Nár. Mus., Odd. Prír. 148: 196 (1979 publ. 1980).
N. Europe to Siberia. 10+ 30+. Hemicr. or rhizome geophyte.
Carex ×albidula E.Holmb., Bot. Not. 1910: 88 (1910).

Carex mitrata Franch., Bull. Soc. Philom. Paris, VIII, 7: 88 (1895).
S. China, Temp. E. Asia. 36 CHC CHS 38 JAP KOR TAI 41 VIE?. Hemicr. or rhizome geophyte.

subsp. ***aristata*** (Ohwi) T.Koyama, in Fl. Taiwan 5: 334 (1978).
S. China, C. & S. Japan, Taiwan. 36 CHC CHS 38 JAP TAI. Hemicr. or rhizome geophyte.
**Carex mitrata* var. *aristata* Ohwi, Mem. Coll. Sci. Kyoto Imp. Univ., Ser. B, Biol. 5: 269 (1930).

subsp. ***mitrata***
SE. China S. Korea, C. & S. Japan. 36 CHS 38 JAP KOR 41 VIE?. Hemicr. or rhizome geophyte.
Carex kingiana H.Lév. & Vaniot, Bull. Acad. Int. Géogr. Bot. 10: 272 (1901). *Carex breviculmis* var. *kingiana* (H.Lév. & Vaniot) Kük. in H.G.A. Engler (ed.), Pflanzenr., IV, 20(38): 470 (1909).

Carex miyabei Franch., Bull. Soc. Philom. Paris, VIII, 7: 52 (1895). *Carex wallichiana* var. *miyabei* (Franch.) Kük. in H.G.A.Engler (ed.), Pflanzenr., IV, 20(38): 749 (1909).
SE. China, Japan. 36 CHS 38 JAP. Hemicr. or rhizome geophyte.

var. ***maopengensis*** S.W.Su, J. Anhui Univ. Nat. Sci. 20(3): 107 (1996).
China (Anhui). 36 CHS. Hemicr.

var. ***miyabei***
Japan. 38 JAP. Hemicr. or rhizome geophyte.
Carex saruensis Franch., Bull. Soc. Philom. Paris, VIII, 7: 85 (1895).
Carex kinugawaensis Sutô & Suzuki, Bull. Coll. Agric. Utsunomiya Univ. 8: (1933).

Carex mokkwaensis Akiyama, Caric. Far East. Reg. As.: 101 (1955).
Taiwan. 38 TAI. Hemicr. or rhizome geophyte.

Carex molesta Mack. in N.L.Britton & al. (eds.), N. Amer. Fl. 18: 151 (1931). *Carex brevior* var. *molesta* (Mack.) F.C.Gates, Trans. Kansas Acad. Sci. 42: 135 (1940).
SE. Canada to WC. & E. U.S.A. 72 ONT QUE 74 ILL IOW KAN MIN MSO NDA NEB OKL WIS 75 CNT INI MAS MIC NWH NWY OHI PEN VER WVA (76) cal 78 ALA ARK KTY MRY MSI TEN VRG. Hemicr. or rhizome geophyte.

Carex molestiformis Reznicek & Rothrock, Contr. Univ. Michigan Herb. 21: 300 (1997).
EC. U.S.A. 74 MSO OKL 78 ARK KTY TEN VRG. Hemicr. or rhizome geophyte.

Carex molinae Phil., Anales Univ. Chile 93: 486 (1896).
C. Chile. 85 CLC. Hemicr. or rhizome geophyte.

Carex mollicula Boott, Ill. Carex 4: 192 (1867).
S. Kuril Is., Japan, S. Korea, China (Zhejiang), Taiwan. 31 KUR 36 CHS 38 JAP KOR TAI. Hemicr. or rhizome geophyte.
Carex arcuata Franch., Bull. Soc. Philom. Paris, VII, 10: 106 (1886).
Carex mollicula f. *lutescens* Kük. in H.G.A.Engler (ed.), Pflanzenr., IV, 20(38): 622 (1909).

Carex mollissima H.Christ ex Scheutz, Kongl. Svenska Vetensk. Acad. Handl., n.s., 22(10): 181 (1887).
NE. Europe to Russian Far East and South China Sea (Spratly Is.). 14 RUE RUN 30 WSB 31+ 36 CHI CHM 41 SCS. Hemicr. or rhizome geophyte.
Carex vesicaria var. *reflexa* Meinsh., Trudy Imp. S.-Peterburgsk. Bot. Sada 18: 373 (1900). *Carex mollissima* var. *reflexa* (Meinsh.) Kük. in H.G.A. Engler (ed.), Pflanzenr., IV, 20(38): 720 (1909).
Carex divaricata Kük., Forh. Finska Vetensk. Soc. 45: 12 (1902).
Carex yingkiliensis A.I.Baranov & Skvortsov, Quart. J. Taiwan Mus. 18: 226 (1965).

Carex mongolica A.I.Baranov & Skvortsov, Quart. J. Taiwan Mus. 18: 224 (1965).
NE. China. 36 CHM. Hemicr. or rhizome geophyte.

Carex monodynama (Griseb.) G.A.Wheeler, Syst. Bot. 15: 656 (1990).
Argentina (Mendoza, La Rioja, Córdoba). 85 AGE AGW. Hemicr. or rhizome geophyte.
**Carex atropicta* f. *monodynama* Griseb., Symb. Fl. Argent.: 314 (1879).
Carex atropicta f. *pallescens* Kurtz ex Kük., Bot. Jahrb. Syst. 27: 520 (1899).

Carex monostachya A.Rich., Tent. Fl. Abyss. 2: 512 (1850).
N. Ethiopia to Tanzania (Mt. Kilimanjaro). 24 ETH 25 KEN TAN. Rhizome geophyte.
Carex parasitica Kunze in C.Schkuhr, Beschr. Riedgräs., Suppl.: 83 (1842), nom. nud.
Carex triquetrifolia Boeck., Bot. Jahrb. Syst. 7: 279 (1886). *Carex monostachya* var. *triquetrifolia* (Boeck.) Kük. in H.G.A.Engler (ed.), Pflanzenr., IV, 20(38): 74 (1909).

Carex monotropa Nelmes, Kew Bull. 10: 86 (1955).
S. Africa. 27 LES. Hemicr. or rhizome geophyte.

Carex montana L., Sp. Pl.: 975 (1753).
Europe to Siberia. 10 DEN FIN† GRB SWE 11 AUT BGM CZE GER HUN POL SWI 12 COR FRA SPA 13 BUL ITA ROM YUG 14 BLR BLT RUC RUE RUS RUW UKR 30 WSB 33 NCS?. Hemicr. or rhizome geophyte.
Carex conglobata All., Fl. Pedem. 2: 268 (1785). *Carex montana* var. *conglobata* (All.) Nyman, Consp. Fl. Eur.: 773 (1882).
Carex caryophyllaea J.F.Gmel., Syst. Nat. 2: 142 (1791).
Carex pubescens Gilib., Excerc. Phyt. 2: 547 (1792).
Carex pumila Honck., Syn. Pl. Germ. 1: 373 (1792), nom. illeg.
Carex gracilis Moench, Methodus: 323 (1794), nom. illeg.
Carex collina Willd., Sp. Pl. 4: 260 (1805).
Carex emarginata Willd., Sp. Pl. 4: 262 (1805). *Carex montana* var. *emarginata* (Willd.) Nyman, Consp. Fl. Eur.: 773 (1882).
Carex montana f. *procerior* Gaudin, Fl. Helv. 6: 76 (1830).
Carex caryophyllata Kunth, Enum. Pl. 2: 437 (1837).
Carex perennis Krock. ex Steud., Nomencl. Bot., ed. 2, 1: 294 (1840).
Carex praecox Pollich ex Steud., Nomencl. Bot., ed. 2, 1: 294 (1840), nom. inval.
Carex montana f. *pallescens* Döll, Fl. Baden: 271 (1857).
Carex czetzii Janka, Oesterr. Bot. Z. 8: 93 (1858). *Carex montana* subsp. *csetzii* (Janka) Nyman, Consp. Fl. Eur.: 773 (1882).
Carex montana var. *luxurians* Èelak., Prodr. Fl. Böhmen: 736 (1867).
Carex montana var. *flavida* Waisb., Oesterr. Bot. Z. 45: 109 (1895).

Carex montana f. *albescens* Bornm. & Kük., Mitth. Thüring. Bot. Vereins, n.s., 10: 39 (1897).
Carex montana f. *bulboides* Waisb., Oesterr. Bot. Z. 47: 431 (1897).
Carex montana f. *fimbriata* Waisb., Oesterr. Bot. Z. 47: 4 (1897).
Carex montana f. *gracilior* Waisb., Oesterr. Bot. Z. 47: 432 (1897).
Carex montana f. *marginata* Waisb., Oesterr. Bot. Z. 47: 4 (1897).
Carex montana f. *remotiflora* Waisb., Oesterr. Bot. Z. 47: 432 (1897).
Carex montana f. *truncata* Waisb., Oesterr. Bot. Z. 47: 432 (1897).
Carex montana var. *verniformis* Waisb., Oesterr. Bot. Z. 47: 4 (1897).
Carex montana f. *pseudopallescens* Kneuck., Allg. Bot. Z. Syst. 5(Beih. 1): 61 (1899).
Carex montana var. *rigida* Waisb., Magyar Bot. Lapok 4: 73 (1905).

Carex montanoaltaica Zolot., Novosti Sist. Vyssh. Rast. 21: 228 (1984).
Siberia. 30 ALT. Hemicr. or rhizome geophyte.

Carex montis-eeka Hillebr., Fl. Hawaiian Isl.: 486 (1888).
Hawaiian Is. (Kauai, Maui, Molokai). 63 HAW. Hemicr. or rhizome geophyte.
Carex montis-eeka var. *viridens* Kük., Repert. Spec. Nov. Regni Veg. 16: 435 (1920).
Carex montis-eeka var. *filifolia* Skottsb., Acta Horti Gothob. 15: 333 (1944).

Carex montis-everestii Kük., Bull. Misc. Inform. Kew 1934: 261 (1934).
Tibet to Sikkim. 36 CHT 40 EHM NEP. Hemicr. or rhizome geophyte.

Carex montis-wutaii T.Koyama, Jap. J. Bot. 15: 172 (1956).
NC. China (Shanxi). 36 CHN. Hemicr. or rhizome geophyte.

Carex moorei G.A.Wheeler, Syst. Bot. 13: 202 (1988).
S. Chile to S. Argentina. 85 AGS CLS. Hemicr. or rhizome geophyte.

Carex morii Hayata, Icon. Pl. Formosan. 6: 135 (1916).
Taiwan. 38 TAI. Hemicr. or rhizome geophyte.

Carex* ×*moriyoshiensis Fujiw. & Y.Matsuda, J. Phytogeogr. Taxon. 39: 11 (1991). *C. curvicollis* × *C. sadoensis*.
Japan. 38 JAP. Hemicr. or rhizome geophyte.

Carex morrowii Boott in M.C.Perry, Jap. Exped. 2: 326 (1856).
C. & S. Japan. 38 JAP. Hemicr. or rhizome geophyte.

var. ***laxa*** Ohwi, Mem. Coll. Sci. Kyoto Imp. Univ., Ser. B, Biol. 5: 260 (1930).
Japan (Yakushima). 38 JAP. Hemicr. or rhizome geophyte.

var. ***morrowii***
C. & S. Japan. 38 JAP. Hemicr. or rhizome geophyte.
Carex morrowii f. *niigatensis* Franch. & Sav., Enum. Pl. Jap. 2: 145 (1877).
Carex morrowii f. *virginea* Franch. & Sav., Enum. Pl. Jap. 2: 145 (1877).
Carex morrowii var. *crassicaulis* Franch., Nouv. Arch. Mus. Hist. Nat., III, 10: 41 (1898).
Carex morrowii var. *albomarginata* Makino, J. Jap. Bot. 6: 38 (1929).

var. ***temnolepis*** (Franch.) Ohwi, Acta Phytotax. Geobot. 2: 134 (1933).
Japan (Honshu). 38 JAP. Hemicr. or rhizome geophyte.
**Carex temnolepis* Franch., Nouv. Arch. Mus. Hist. Nat., III, 10: 61 (1898). *Carex foliosissima* var. *temnolepis* (Franch.) Kük. in H.G.A.Engler (ed.), Pflanzenr., IV, 20(38): 478 (1909).
Carex kinpokusanensis Akiyama, J. Fac. Sci. Hokkaido Univ., Ser. 5, Bot. 3: 265 (1935).

Carex mossii Nelmes, Bull. Misc. Inform. Kew 1940: 137 (1940).
S. Africa. 27 CPP NAT TVL. Hemicr. or rhizome geophyte.

Carex motuoensis Y.C.Yang, in Fl. Xizangica 5: 443 (1987).
Tibet. 36 CHT. Hemicr. or rhizome geophyte.

Carex moupinensis Franch., Nouv. Arch. Mus. Hist. Nat., II, 10: 102 (1887).
SC. China. 36 CHC. Hemicr. or rhizome geophyte.
Homalostachys sinensis Boeck., Beitr. Cyper. 1: 38 (1888). *Scleria sinensis* (Boeck.) H.Pfeiff., Repert. Spec. Nov. Regni Veg. 26: 263 (1929).

Carex mucronata All., Fl. Pedem. 2: 268 (1785). *Vignea mucronata* (All.) Rchb., Fl. Germ. Excurs.: 140(7) (1831). *Edritria mucronata* (All.) Raf., Good Book: 26 (1840). *Vignantha mucronata* (All.) Schur, Enum. Pl. Transsilv.: 705 (1866).
C. & S. Europe to Caucasus. 11 AUT GER SWI 12 COR? FRA 13 ITA ROM? YUG 33+. Hemicr. or rhizome geophyte.
Carex juncifolia J.F.Gmel., Syst. Nat. 2: 142 (1791), nom. illeg.
Carex gracilis Honck., Syn. Pl. Germ. 1: 368 (1792), nom. illeg.
Carex nitida Hoppe, Bot. Taschenb. 1800: 198 (1800).
Carex bracteata Suter, Helvet. Fl. 2: 250 (1802).
Carex setacea Froel. ex Kunth, Enum. Pl. 2: 409 (1837), nom. illeg.
Carex illustranda Steud., Syn. Pl. Glumac. 2: 190 (1855).
Carex mucronata f. *androgyna* Camperio, Jahreskat. Wiener Bot. Tauschanst. 1904: 222 (1904).
Carex abnormis Rouy in G.Rouy & J.Foucaud, Fl. France 13: 506 (1912).

Carex mucronatiformis Tang & F.T.Wang ex S.Yun Liang, in Fl. Reipubl. Popul. Sin. 12: 523 (2000).
Qinghai. 36 CHQ. Hemicr.

Carex* ×*mucronulata E.Holmb., Bot. Not. 1929: 20 (1929). *C. heleonastes* × *C. tenuiflora*.
N. Europe. 10+ 14+. Hemicr. or rhizome geophyte.

Carex muehlenbergii Willd., Sp. Pl. 4: 231 (1805). *Vignea muhlenbergii* (Willd.) Raf., Good Book: 27 (1840).
E. Canada to Guatemala. 72 ONT QUE 74 ILL IOW KAN MIN MSO NEB OKL WIS 75 CNT INI MAI MAS MIC NWH NWJ NWY OHI PEN RHO VER WVA 77 TEX 78 ALA ARK GEO KTY LOU MRY MSI NCA SCA TEN VRG WDC 79 MXC MXE MXG MXS MXT 80 GUA. Hemicr. or rhizome geophyte.

var. ***enervis*** Boott, Ill. Carex 3: 124 (1862).
S. Ontario, C. & E. U.S.A. to Mexico. 72 ONT 74 ILL KAN MIN MSO NEB OKL 75 CNT INI MAS MIC NWH NWJ NWY OHI PEN RHO

VER WVA 77 TEX 78 ALA ARK GEO KTY LOU MRY MSI NCA SCA TEN VRG WDC 79 MXE MXG MXS. Hemicr. or rhizome geophyte.
Carex onusta Mack., Bull. Torrey Bot. Club 42: 618 (1915).
Carex plana Mack., Bull. Torrey Bot. Club 50: 350 (1923).

var. ***muehlenbergii***
E. Canada to C. & E. U.S.A. 72 ONT QUE 74 ILL IOW KAN MIN MSO OKL WIS 75 CNT INI MAI MAS MIC NWH NWJ NWY OHI PEN RHO VER WVA 77 TEX 78 ALA ARK DEL FLA GEO KTY LOU MRY MSI NCA SCA VRG WDC. Hemicr. or rhizome geophyte.
Carex pinetorum Willd. ex Kunth, Enum. Pl. 2: 381 (1837).
Carex piniaria Bosc ex Boott, Ill. Carex 3: 124 (1862).

var. ***xalapensis*** (Kunth) Britton, Mem. Torrey Bot. Club 5: 86 (1894).
Mexico to Guatemala. 79 MXC MXE MXG MXS MXT 80 GUA. Hemicr. or rhizome geophyte.
**Carex xalapensis* Kunth, Enum. Pl. 2: 380 (1837).
Carex involucratella Mack. in N.L.Britton & al. (eds.), N. Amer. Fl. 18: 50 (1931).

Carex muelleri Petrie, Trans. & Proc. New Zealand Inst. 18: 298 (1885 publ. 1886).
New Zealand South I. 51 NZS. Hemicr. or rhizome geophyte.
Carex viridis Petrie, Trans. & Proc. New Zealand Inst. 13: 332 (1881), nom. illeg.

Carex muliensis Hand.-Mazz., Symb. Sin. 7: 1262 (1936).
China (Sichuan, Yunnan). 36 CHC. Hemicr. or rhizome geophyte.

Carex multicaulis L.H.Bailey, Bot. Gaz. 9: 118 (1884).
Oregon to California. 73 ORE 76 CAL. Hemicr. or rhizome geophyte.

Carex multicostata Mack., Bull. Torrey Bot. Club 43: 604 (1916 publ. 1917).
W. U.S.A. 73 IDA MNT ORE WAS 76 CAL NEV UTA. Hemicr. or rhizome geophyte.
Carex pachycarpa Mack., Bull. Torrey Bot. Club 43: 616 (1916 publ. 1917).

Carex munda Boott, Ill. Carex 1: 7 (1858).
C. & E. Himalaya to Tibet. 36 CHT 40 EHM NEP. Hemicr. or rhizome geophyte.
Carex munda var. *mundiformis* T.Koyama, Acta Phytotax. Geobot. 15: 113 (1954).

Carex munipoorensis C.B.Clarke in J.D.Hooker, Fl. Brit. India 6: 729 (1894).
N. Myanmar. 41 MYA. Hemicr. or rhizome geophyte.

Carex munroi Boott ex C.B.Clarke in J.D.Hooker, Fl. Brit. India 6: 738 (1894).
W. Himalaya. 40 WHM. Hemicr. or rhizome geophyte.
Carex chinensis Munro ex Hook.f., Fl. Brit. India 6: 738 (1894).

Carex muricata L., Sp. Pl.: 974 (1753). *Vignea muricata* (L.) Rchb. in J.C.Mössler, Handb. Gewächsk. ed. 2, 3: 1611 (1830). *Caricina muricata* (L.) St.-Lag. in A.Cariot, Étude Fl., ed. 8, 2: 875 (1889).
Europe to Caucasus. 10 DEN FIN NOR SWE 11 CZE POL 14 BLR BLT RUC RUE RUS RUW UKR 20 MOR 30 WSB 33 TCS 34 TUR (51) nzn (75) mic pen (83) ecu. Hemicr. or rhizome geophyte.
Carex pairae subsp. *borealis* Hyl. in ?
Carex intermedia Retz., Fl. Scand. Prodr.: 178 (1779).
Carex nemorosa Lumn. ex Honck., Syn. Pl. Germ. 1: 355 (1792), nom. illeg. *Carex muricata* var. *nemorosa* Nyman, Consp. Fl. Eur.: 781 (1882). *Carex muricata* race *lumnitzeri* Rouy in G.Rouy & J.Foucaud, Fl. France 13: 412 (1912). *Carex lumnitzeri* (Rouy) V.I.Krecz. in V.L.Komarov (ed.), Fl. URSS 3: 154 (1935). *Carex spicata* subsp. *lumnitzeri* (Rouy) Soó, Feddes Repert. 85: 438 (1974).
Carex stellulata M.Bieb., Fl. Taur.-Caucas. 2: 383 (1808), nom. illeg.
Carex divulsa Gaudin, Fl. Helv. 6: 47 (1830), nom. illeg.
Carex serotina Ten., Fl. Napol. 4: 133 (1830), nom. illeg.
Vignea virens Rchb. in J.C.Mössler, Handb. Gewächsk. ed. 2, 3: 1611 (1830).
Carex vulpina Hohen., Bull. Soc. Imp. Naturalistes Moscou 6: 259 (1833), nom. illeg.
Carex orsiniana Ten., Fl. Napol. 5: 244 (1835). *Carex muricata* subsp. *orsiniana* (Ten.) Nyman, Consp. Fl. Eur.: 781 (1882).
Carex astracanica Willd. ex Kunth, Enum. Pl. 2: 385 (1837).
Vignea altissima Schur, Verh. Mitth. Siebenbürg. Vereins Naturwiss. Hermannstadt 4: 79 (1858).
Carex tenuissima Schur, Enum. Pl. Transsilv.: 704 (1866), nom. illeg. *Vignea tenuissima* (Schur) Schur, Enum. Pl. Transsilv.: 979 (1866).
Carex tergestina Hoppe ex Boott, Ill. Carex 4: 198 (1867).
Carex muricata f. *arcuata* P.Silva, Bol. Soc. Brot., II, 54: 240 (1980–1981 publ. 1981).

Carex muriculata F.J.Herm., Field & Lab. 17: 132 (1949).
SC. U.S.A. to NE. Mexico. 77 NWM TEX 79 MXE. Hemicr. or rhizome geophyte.

Carex × musashiensis Ohwi, Mem. Coll. Sci. Kyoto Imp. Univ., Ser. B, Biol. 11(5): 512 (1936). *C. curvicollis* × *C. gaudichaudiana* var. *thunbergii*.
Japan. 38 JAP. Hemicr. or rhizome geophyte.
**Carex × sharensis* Akiyama, Trans. Sapporo Nat. Hist. Soc. 14: 109 (1935), nom. illeg.

Carex musei Steud., Syn. Pl. Glumac. 2: 205 (1855).
Réunion. 29 REU. Hemicr. or rhizome geophyte.

Carex muskingumensis Schwein., Ann. Lyceum Nat. Hist. New York 1: 66 (1824). *Carex scoparia* var. *muskingumensis* (Schwein.) Tuck., Enum. Meth. Caric.: 8, 17 (1843). *Thysanocarex muskingumensis* (Schwein.) Fedde & J.Schust., Just's Bot. Jahresber. 41(2): 12 (1918). *Vignea muskingumensis* (Schwein.) Soják, Cas. Nár. Mus., Odd. Přír. 148: 196 (1979 publ. 1980).
E. Canada to EC. U.S.A. 72 ONT 74 ILL IOW KAN MIN MSO OKL WIS 75 INI MIC OHI 78 ARK KTY TEN. Hemicr. or rhizome geophyte.
Carex arida Schwein. & Torr., Ann. Lyceum Nat. Hist. New York 1: 312 (1824).
Carex scoparia Torr., Ann. Lyceum Nat. Hist. New York 3: 394 (1836), nom. illeg.

Carex myosurus Nees in R.Wight, Contr. Bot. India: 122 (1834).
S. China to Trop. Asia. 36 CHC CHH CHT 40 EHM IND NEP 41 MYA VIE 42 JAW LSI PHI SUL SUM. Hemicr. or rhizome geophyte.

Carex eminens Nees in R.Wight, Contr. Bot. India: 122 (1834). *Carex myosurus* var. *eminens* (Nees) Boeck., Linnaea 40: 335 (1876).
Carex myosurus subsp. *spiculata* Boott, Proc. Linn. Soc. London 1: 288 (1846).
Carex spiculata Boott, Trans. Linn. Soc. London 20: 139 (1851).
Carex longibracteata Steud., Syn. Pl. Glumac. 2: 205 (1855), nom. illeg.
Carex macrophylla Hochst. ex Steud., Syn. Pl. Glumac. 2: 207 (1855).
Carex nobilis Boott, Ill. Carex 1: 4 (1858). *Carex myosurus* var. *nobilis* (Boott) C.B.Clarke in J.D.Hooker, Fl. Brit. India 6: 724 (1894).
Carex aequata Nees ex Boott, Ill. Carex 2: 83 (1860).
Carex floribunda Boeck., Linnaea 40: 335 (1876). *Carex myosurus* var. *floribunda* (Boeck.) Kük. in H.G.A.Engler (ed.), Pflanzenr., IV, 20(38): 259 (1909).
Carex kuntzeana Boeck., Beitr. Cyper. 2: 51 (1890).
Carex praestans C.B.Clarke in J.D.Hooker, Fl. Brit. India 6: 723 (1894). *Carex myosurus* var. *praestans* (C.B.Clarke) Kük. in H.G.A.Engler (ed.), Pflanzenr., IV, 20(38): 259 (1909).
Carex longibracteata f. *angustifolia* Kük. in H.G.A. Engler (ed.), Pflanzenr., IV, 20(38): 260 (1909).
Carex longibracteata f. *distans* Kük. in H.G.A.Engler (ed.), Pflanzenr., IV, 20(38): 260 (1909).
Carex myosurus var. *gracilior* Kük. in H.G.A.Engler (ed.), Pflanzenr., IV, 20(38): 259 (1909).
Carex peiana F.T.Wang & Tang, Contr. Inst. Bot. Natl. Acad. Peiping 6: 55 (1949).

Carex nachiana Ohwi, Acta Phytotax. Geobot. 2: 103 (1933).
Japan (Honshu, Kyushu) to SE. China. 36 CHS 38 JAP TAI. Hemicr. or rhizome geophyte.

Carex nairii Ghildyal & U.C.Bhattach., J. Econ. Taxon. Bot. 8: 479 (1986).
Kashmir. 40 WHM. Hemicr. or rhizome geophyte.

Carex nakasimae Ohwi, Mem. Coll. Sci. Kyoto Imp. Univ., Ser. B, Biol. 18(1): 171 (1944).
Korea. 38 KOR. Hemicr. or rhizome geophyte.

Carex nanchuanensis K.L.Chu ex S.Y.Liang, Acta Phytotax. Sin. 36: 532 (1998).
China (Sichuan). 36 CHC. Hemicr. or rhizome geophyte.

Carex nandadeviensis Ghildyal, U.C.Bhattach. & Hajra, Indian J. Forest. 9: 90 (1986).
Uttaranchal. 40 WHM. Hemicr. or rhizome geophyte.

Carex nanelliformis A.E.Kozhevn., Byull. Moskovsk. Obshch. Isp. Prir., Otd. Biol. 92: 132 (1987).
Russian Far East. 31+. Hemicr. or rhizome geophyte.

Carex nangtciangensis Pamp., Nuovo Giorn. Bot. Ital., n.s., 22: 260 (1915).
China (Hubei). 36 CHC. Hemicr. or rhizome geophyte.

Carex nardina Fr., Novit. Fl. Suec. Mant. 2: 55 (1839). *Vignea nardina* (Fr.) Soják, Cas. Nár. Mus., Odd. Prír. 148: 196 (1979 publ. 1980).
Subarctic to W. U.S.A. 10 ICE NOR SVA SWE 31 MAG 70 ASK GNL NUN NWT YUK 71 ABT BRC 72 LAB NFL ONT QUE 73 COL IDA MNT WAS WYO 76 NEV UTA. Hemicr. or rhizome geophyte.

var. ***hepburnii*** (Boott) Kük. in H.G.A.Engler (ed.), Pflanzenr., IV, 20(38): 70 (1909).
Russian Far East, Alaska to W. U.S.A. 31 MAG 70 ASK NWT 71 ABT BRC 73 COL IDA MNT WAS WYO 76 NEV UTA. Hemicr. or rhizome geophyte.
**Carex hepburnii* Boott in W.J.Hooker, Fl. Bor.-Amer. 2: 209 (1839). *Kobresia hepburnii* (Boott) K.V.Ivanova, Bot. Zhurn. S.S.S.R. 24: 490 (1939). *Carex nardina* subsp. *hepburnii* (Boott) Á.Löve, D.Löve & S.L.Kapoor, Arctic Alpine Res. 3: 145 (1971). *Vignea hepburnii* (Boott) Soják, Cas. Nár. Mus., Odd. Prír. 148: 195 (1979 publ. 1980).
Carex stantonensis M.E.Jones, Bull. Montana State Univ., Biol. Ser. 61: 20 (1910).

var. ***nardina***
N. Europe, Subarctic America to Canada. 10 ICE NOR SVA SWE 70 ASK GNL NUN NWT YUK 71 ABT BRC 72 LAB NFL ONT QUE. Hemicr. or rhizome geophyte.
Carex nardina var. *atriceps* Kük., Repert. Spec. Nov. Regni Veg. 8: 7 (1910).
Kobresia arctica K.V.Ivanova, Bot. Zhurn. S.S.S.R. 24: 490 (1939).
Carex elyniformis A.E.Porsild, Sargentia 4: 17 (1943).

Carex neblinensis Reznicek, Novon 6: 421 (1996).
Venezuela/Brazil border (Cerro de la Neblina). 82 VEN 84 BZN. Hemicr. or rhizome geophyte.

Carex nebraskensis Dewey, Amer. J. Sci. Arts, II, 18: 102 (1854). *Carex jamesii* var. *nebraskensis* (Dewey) L.H.Bailey, Proc. Amer. Acad. Arts 22: 84 (1886).
S. Canada to W. & NC. U.S.A. 71 ABT SAS 73 COL IDA MNT ORE WAS WYO (74) ill KAN mso NDA NEB SDA 76 ARI CAL NEV UTA 77 NWM. Hemicr. or rhizome geophyte.
Carex jamesii Torr., Ann. Lyceum Nat. Hist. New York 3: 398 (1836), nom. illeg.
Carex nebraskensis var. *praevia* L.H.Bailey, Mem. Torrey Bot. Club 1: 49 (1889).
Carex nebraskensis var. *ultriformis* L.H.Bailey, Bot. Gaz. 21: 8 (1896). *Carex jamesii* var. *ultriformis* (L.H.Bailey) Kük. in H.G.A.Engler (ed.), Pflanzenr., IV, 20(38): 318 (1909).
Carex nebraskensis var. *eruciformis* Suksd., Werdenda 1: 5 (1923).

Carex nebularum Phil., Anales Univ. Chile 93: 492 (1896).
Chile, Argentina. 85 AGE AGS AGW CLC CLS. Hemicr. or rhizome geophyte.
Carex pycnostachya Desv. in C.Gay, Fl. Chil. 6: 204 (1854), nom. illeg.

Carex neesiana Endl., Prodr. Fl. Norfolk.: 24 (1833). *Carex dissita* var. *neesiana* (Endl.) Kük. in H.G.A.Engler (ed.), Pflanzenr., IV, 20(38): 691 (1909).
Norfolk I. 50 NFK. Hemicr.

Carex negrii Chiov., Ann. Bot. (Rome) 10: 406 (1912).
Ethiopia to Somalia, Yemen. 24 ETH SOM 35 YEM. Hemicr. or rhizome geophyte.

Carex nelmesiana Barros, Bol. Soc. Argent. Bot. 6: 209 (1957).
S. Argentina. 85 AGS. Hemicr. or rhizome geophyte.

Carex nelmesii H.E.Hess, Ber. Schweiz. Bot. Ges. 63: 350 (1953).
Angola. 26 ANG. Hemicr. or rhizome geophyte.

Carex nelsonii Mack. in P.A.Rydberg, Fl. Rocky Mts.: 137 (1917).
WC. U.S.A. 73 COL MNT WYO 76 UTA. Hemicr. or rhizome geophyte.
Carex estesiana Kelso, Biol. Leafl. 30: 3 (1945).

Carex nemostachys Steud., Flora 29: 23 (1846).
Indo-China to C. & S. Japan. 36 CHC CHS 38 JAP NNS TAI 41 MYA THA VIE. Hemicr. or rhizome geophyte.
Carex excurva Boott, Ill. Carex 1: 57 (1858).

Carex nemurensis Franch., Bull. Soc. Philom. Paris, VIII, 7: 30 (1895). *Vignea nemurensis* (Franch.) Soják, Cas. Nár. Mus., Odd. Prír. 148: 196 (1979 publ. 1980).
Russian Far East, N. & C. Japan. 31 KUR SAK 38 JAP. Hemicr. or rhizome geophyte.

Carex* × *neobigelowii Lepage, Naturaliste Canad. 91: 166 (1964). *C. bigelowii* × *C. lenticularis*.
E. Canada. 72 QUE. Hemicr. or rhizome geophyte.

Carex neochevalieri Kük. ex A.Chev., Bull. Mus. Natl. Hist. Nat., II, 3: 467 (1931).
W. & WC. Trop. Africa. 22 GHA GUI MLI NGA SIE 23 CAF CMN. Hemicr. or rhizome geophyte.

Carex neodigyna P.C.Li, Acta Bot. Yunnan. 12: 146 (1990).
China (Sichuan). 36 CHC. Hemicr. or rhizome geophyte.

Carex* × *neofilipendula Lepage, Naturaliste Canad. 83: 123 (1956). *C. aquatilis* × *C. paleacea*.
E. Canada. 72 NFL QUE. Hemicr. or rhizome geophyte.

Carex neofilipes Nakai, Repert. Spec. Nov. Regni Veg. 13: 243 (1914).
Korea. 38 KOR. Hemicr. or rhizome geophyte.

Carex neoguinensis C.B.Clarke, J. Linn. Soc., Bot. 37: 12 (1904).
Papua New Guinea. 43 NWG. Hemicr. or rhizome geophyte.

Carex neohebridensis Guillaumin & Kük., Bull. Mus. Natl. Hist. Nat., II, 9: 304 (1937).
Vanuatu. 60 VAN. Hemicr. or rhizome geophyte.

Carex neokukenthaliana H.Lév. & Vaniot, Repert. Spec. Nov. Regni Veg. 3: 350 (1907).
China. 36+. Hemicr. or rhizome geophyte.

Carex* × *neomiliaris Lepage, Naturaliste Canad. 84: 89 (1957). *C. aquatilis* × *C. saxatilis* var. *militaris*.
E. Canada. 72 QUE. Hemicr. or rhizome geophyte.

Carex neopetelotii Raymond, Naturaliste Canad. 82: 159 (1955).
Vietnam. 41 VIE. Hemicr. or rhizome geophyte.

Carex neopolycephala Tang & F.T.Wang ex L.K.Dai, Acta Phytotax. Sin. 32: 175 (1994).
China (Yunnan). 36 CHC. Hemicr. or rhizome geophyte.

var. ***neopolycephala***.
China (Yunnan). 36 CHC. Hemicr. or rhizome geophyte.

var. ***simplex*** Tang & F.T.Wang ex L.K.Dai, Acta Phytotax. Sin. 32: 175 (1994).
China (Yunnan). 36 CHC. Hemicr. or rhizome geophyte.

Carex* × *neorigida Lepage, Naturaliste Canad. 84: 46 (1957). *C. bigelowii* × *C. salina*.
E. Canada. 72 LAB. Hemicr. or rhizome geophyte.

Carex neosachalinensis A.E.Kozhevn., Byull. Moskovsk. Obshch. Isp. Prir., Otd. Biol. 92: 131 (1987).
Sakhalin. 31 SAK. Hemicr. or rhizome geophyte.

Carex nervata Franch. & Sav., Enum. Pl. Jap. 2: 566 (1878).
Russian Far East to S. Korea, Japan. 31+ 36 CHI CHM 38 JAP KOR. Hemicr. or rhizome geophyte.
Carex vidalii Franch. & Sav. in ?
Carex homoiolepis Franch. & Sav., Enum. Pl. Jap. 2: 567 (1878).

Carex nervina L.H.Bailey, Bot. Gaz. 10: 203 (1885).
Oregon to Nevada. 73 ORE 76 CAL NEV. Hemicr. or rhizome geophyte.

Carex neurocarpa Maxim., Prim. Fl. Amur.: 306 (1859).
Amur to C. Japan. 31 AMU 36 CHI CHM CHN CHS 38 JAP KOR. Hemicr. or rhizome geophyte.

Carex neurophora Mack. in L.Abrams, Ill. Fl. Pacific States 1: 298 (1923).
NW. U.S.A. 73 COL IDA MNT ORE WAS WYO. Hemicr. or rhizome geophyte.

Carex* × *nicoloff Pamp., Ann. Bot. (Rome) 1: 135 (1903). *C. hudsonii* × *C. riparia*.
Italy. 13 ITA. Hemicr. or rhizome geophyte.

Carex niederleiniana Boeck., Beitr. Cyper. 1: 50 (1888). *Carex acutata* var. *niederleiniana* (Boeck.) Kük. in H.G.A.Engler (ed.), Pflanzenr., IV, 20(38): 702 (1909).
W. Argentina. 85 AGS AGW. Hemicr. or rhizome geophyte.
Carex acutata var. *hirtisquama* Kük., Bot. Jahrb. Syst. 27: 549 (1899).

Carex nigerrima Nelmes, Bull. Misc. Inform. Kew 1939: 200 (1939).
N. Pakistan to Kashmir. 40 PAK WHM. Hemicr. or rhizome geophyte.

Carex nigra (L.) Reichard, Fl. Moeno-Francof. 2: 96 (1778).
Temp. Northern Hemisphere. 10 DEN FIN FOR GRB ICE IRE NOR SWE 11 AUT BGM CZE GER HUN NET POL SWI 12 COR FRA POR SAR SPA 13 ALB BUL GRC ITA ROM SIC YUG 14 BLR BLT KRY RUC RUE RUN RUS RUW UKR 20 MOR 30 ALT BRY CTA IRK KRA WSB YAK 32 KAZ 33 NCS TCS 34 TUR 37 MON 40 EHM 70 GNL 72 LAB NBR NFL NSC PEI QUE 75 CNT MAI MAS MIC NWH NWY VER (90) sge. Hemicr. or rhizome geophyte.
**Carex acuta* var. *nigra* L., Sp. Pl.: 978 (1753).

subsp. ***alpina*** (Gaudin) Lemke, Feddes Repert. Spec. Nov. Regni Veg. 67: 3 (1963).
C. & SW. European Mts., NW. Turkey (Ulu Da.). 11 AUT SWI 12 FRA SPA 13 ITA 34 TUR. Hemicr. or rhizome geophyte.
Carex obesa All., Fl. Pedem. 2: 270 (1785).
Carex stolonifera Hoppe in J.Sturm, Deutschl. Fl. 7: 6 (1835), nom. illeg. *Carex goodenowii* var. *stolonifera* (Hoppe) Asch., Fl. Brandenburg 1: 777 (1864). *Carex goodenowii* subsp. *stolonifera* (Hoppe) Asch. & Graebn., Syn. Mitteleur. Fl. 2(2): 98 (1904).
**Carex cespitosa* var. *alpina* Gaudin, Syn. Fl. Helv.: 782 (1836).
Carex chlorocarpa Wimm., Flora 33: 620 (1849).

subsp. ***dacica*** (Heuff.) Soó, Acta Bot. Acad. Sci. Hung. 16: 370 (1970 publ. 1971).
Carpathians, Balkan Pen. to Caucasus. 11 CZE 13 BUL GRC ROM YUG 14 UKR 33 NCS TCS 34 TUR. Hemicr. or rhizome geophyte.
Carex bigelowii subsp. *nardeticola* Holub in ?
**Carex dacica* Heuff., Flora 18: 247 (1835). *Vignantha dacica* (Heuff.) Schur, Enum. Pl. Transsilv.: 705 (1866). *Vignea dacica* (Heuff.) Fuss, Fl. Transsilv.: 681 (1866). *Carex cespitosa* var. *dacica* (Heuff.) Nyman, Consp. Fl. Eur.: 777 (1882). *Carex rigida* var. *dacica* (Heuff.) Kük. in H.G.A.Engler (ed.), Pflanzenr., IV, 20(38): 302 (1909). *Carex bigelowii* subsp. *dacica* (Heuff.) T.V.Egorova, in Fl. Evropeiskoi Chasti SSSR 2: 202 (1976).

subsp. ***drukyulensis*** Noltie, Edinburgh J. Bot. 50: 189 (1993).
Bhutan. 40 EHM. Hemicr. or rhizome geophyte.

subsp. ***intricata*** (Tineo) Maire & Wieller, in Fl. Afr. Nord 4: 126 (1957).
Spain (Sierra Nevada), Morocco (Atlas Mts.), Corse, Sicilia. 12 COR SPA 13 SIC 20 MOR. Hemicr. or rhizome geophyte.
**Carex intricata* Tineo in G.Gussone, Fl. Sicul. Syn. 2: 574 (1844). *Carex rigida* subsp. *intricata* (Tineo) Nyman, Consp. Fl. Eur.: 778 (1882). *Carex cespitosa* var. *intricata* (Tineo) Fiori, Nuov. Fl. Italia 1: 188 (1923). *Carex fusca* f. *intricata* (Tineo) Gamisans, Candollea 29: 43 (1974).
Carex navasii Merino, Bol. Soc. Aragonesa Ci. Nat. 1: 65 (1902).

subsp. ***juncella*** (Fr.) Lemke, Feddes Repert. Spec. Nov. Regni Veg. 67: 4 (1963).
N. & NC. Europe to Russian Far East. 10 DEN FIN NOR SWE 11 GER POL 14 BLR RUC RUE RUN RUS RUW UKR 30 ALT BRY CTA IRK KRA WSB YAK 31+ 32 KAZ 37 MON. Hemicr. or rhizome geophyte.
Carex vulgaris var. *juncea* Hartm., Handb. Skand. Fl., ed. 4: 303 (1843). *Carex vulgaris* subsp. *juncea* (Hartm.) Fr., Novit. Fl. Suec. Mant. 3: 154 (1845). *Carex rigida* var. *juncea* (Hartm.) L.H.Bailey, J. Bot. 28: 172 (1890). *Carex goodenowii* var. *juncea* (Hartm.) Kük. in H.G.A.Engler (ed.), Pflanzenr., IV, 20(38): 317 (1909). *Carex nigra* var. *juncea* (Hartm.) Hyl., Nordisk Karlvaxtfl. 2: 115 (1966), no basionym page ref. *Carex nigra* subsp. *juncea* (Hartm.) Soó, Feddes Repert. 83: 148 (1972).
**Carex vulgaris* subsp. *juncella* Fr., Bot. Not. 1843: 105 (1843). *Carex juncella* (Fr.) Th.Fr., Bot. Not. 1857: 207 (1857). *Carex goodenowii* var. *juncella* (Fr.) Crép., Man. Fl. Belgique, ed. 2: 321 (1866). *Carex vulgaris* f. *juncella* (Fr.) Kük., Allg. Bot. Z. Syst. 4: 3 (1898).
Carex wiluica Meinsh. ex Maack, Vilyuisk. Okruge Fakutsk. Obl. 2: 308 (1886). *Carex juncella* subsp. *wiluica* (Meinsh. ex Maack) T.V.Egorova, Novosti Sist. Vyssh. Rast. 9: 92 (1972). *Carex nigra* subsp. *wiluica* (Meinsh. ex Maack) Á.Löve & D.Löve, Bot. Not. 128: 505 (1975 publ. 1976).

subsp. ***nigra***
Europe to Russian Far East, E. Canada to NE. U.S.A. 10 DEN FIN FOR GRB ICE IRE NOR SWE 11 AUT BGM CZE GER HUN NET POL SWI 12 COR FRA POR SAR SPA 13 ALB BUL GRC ITA ROM SIC YUG 14 BLR BLT KRY RUC RUE RUN RUS RUW UKR 20 MOR 30 WSB 34 TUR 70 GNL 72 LAB NBR NFL NSC PEI QUE 75 CNT MAI MAS MIC NWH NWY VER (90) sge. Hemicr. or rhizome geophyte.
Carex rufa Lam., Fl. Franç. 2: 184 (1779).
Carex fusca All., Fl. Pedem. 2: 269 (1785).
Carex alpina Honck., Syn. Pl. Germ. 1: 374 (1792), nom. illeg.
Carex polyandra Schkuhr, Beschr. Riedgräs. 1: 59 (1801). *Carex goodenowii* f. *polyandra* (Schkuhr) Kük. in H.G.A.Engler (ed.), Pflanzenr., IV, 20(38): 314 (1909).
Carex acuta var. *minor* Sw., Svensk Bot. Tidskr. 6: t. 408, f. B (1809).
Carex compacta Krock. ex Hoppe, Caricolog. Germ.: 37 (1826), nom. illeg.
Carex angustifolia Sm., Engl. Fl. 4: 127 (1828).
Vignea compacta Rchb. in J.C.Mössler, Handb. Gewächsk. ed. 2, 3: 1623 (1830).
Vignea cerina Rchb., Fl. Germ. Excurs.: 140(7) (1831).
Carex cespitosa var. *curvata* J.Fleisch., Riedgr. Wurtemb.: 15 (1832).
Carex cespitosa var. *recta* J.Fleisch., Riedgr. Wurtemb.: 15 (1832). *Carex goodenowii* var. *recta* (J.Fleisch.) Kük. in H.G.A.Engler (ed.), Pflanzenr., IV, 20(38): 316 (1909).
Carex alboatra Willd. ex Kunth, Enum. Pl. 2: 432 (1837).
Carex antucensis Kunze ex Kunth, Enum. Pl. 2: 412 (1837).
Carex cespitosa var. *polymorpha* Laest., Nova Acta Regiae Soc. Sci. Upsal. 11: 282 (1839).
Carex goodenowii J.Gay, Ann. Sci. Nat., Bot., II, 11: 191 (1839). *Carex rigida* var. *goodenowii* (J.Gay) L.H.Bailey, J. Bot. 28: 173 (1890). *Carex cespitosa* var. *goodenowii* (J.Gay) Fiori, Nuov. Fl. Italia 1: 188 (1923). *Carex nigra* subsp. *goodenowii* (J.Gay) Maire & Weiller, in Fl. Afr. Nord 4: 126 (1957).
Carex cespitosa f. *fuliginosa* A.Br. in J.C.Döll, Rhein. Fl.: 260 (1843). *Carex goodenowii* f. *fuliginosa* (A.Br.) Kük. in H.G.A.Engler (ed.), Pflanzenr., IV, 20(38): 315 (1909).
Carex gibsonii Bab., Ann. Mag. Nat. Hist. 11: 168 (1843).
Carex vulgaris f. *zonata* F.Nyl., Spic. Fl. Fenn. 2: 19 (1844). *Carex vulgaris* var. *zonata* (F.Nyl.) Nyman, Consp. Fl. Eur., Suppl. 2: 324 (1890). *Carex goodenowii* f. *zonata* (F.Nyl.) Kük. in H.G.A. Engler (ed.), Pflanzenr., IV, 20(38): 315 (1909).
Carex xonata F.Nyl., Spic. Fl. Fenn. 2: 19 (1844).
Carex vulgaris R.E.Fr., Novit. Fl. Suec. Mant. 3: 153 (1845). *Vignantha vulgaris* (R.E.Fr.) Schur, Enum. Pl. Transsilv.: 706 (1866). *Vignea vulgaris* (R.E.Fr.) Fuss, Fl. Transsilv.: 680 (1866). *Carex vulgaris* var. *communis* Meinsh., Trudy Imp. S.-Peterburgsk. Bot. Sada 18: 333 (1900), nom. inval.
Carex vulgaris var. *tornata* Fr., Novit. Fl. Suec. Mant. 3: 154 (1845). *Carex goodenowii* var. *tornata* (Fr.) Kük. in H.G.A.Engler (ed.), Pflanzenr., IV, 20(38): 316 (1909).
Carex vulgaris f. *abbreviata* Peterm., Anal. Pfl.-Schlüss.: 500 (1846). *Carex nigra* f. *abbreviata* (Peterm.) Soó, Acta Bot. Acad. Sci. Hung. 16: 370 (1970 publ. 1971).

Carex vulgaris f. *androgyna* Peterm., Anal. Pfl.-Schlüss.: 500 (1846). *Carex nigra* lus *androgyna* (Peterm.) Soó, Acta Bot. Acad. Sci. Hung. 16: 370 (1970 publ. 1971), no basionym stated.
Carex vulgaris var. *atra* Peterm., Anal. Pfl.-Schlüss.: 500 (1846).
Carex vulgaris f. *basigyna* Rchb., Icon. Fl. Germ. Helv. 8: t. 227 (1846). *Carex goodenowii* f. *basigyna* (Rchb.) Kük. in H.G.A.Engler (ed.), Pflanzenr., IV, 20(38): 314 (1909).
Carex vulgaris f. *chlorostachya* Rchb., Icon. Fl. Germ. Helv. 8: t. 226 (1846). *Carex goodenowii* f. *chlorostachya* (Rchb.) Kük. in H.G.A.Engler (ed.), Pflanzenr., IV, 20(38): 315 (1909).
Carex vulgaris var. *polygama* Rchb., Icon. Fl. Germ. Helv. 8: t. 227 (1846).
Carex vulgaris f. *ramosa* Peterm., Anal. Pfl.-Schlüss.: 500 (1846). *Carex nigra* f. *ramosa* (Peterm.) Soó, Acta Bot. Acad. Sci. Hung. 16: 370 (1970 publ. 1971).
Carex vulgaris f. *tenuis* Peterm., Anal. Pfl.-Schlüss.: 500 (1846). *Carex nigra* f. *tenuis* (Peterm.) Soó, Acta Bot. Acad. Sci. Hung. 16: 370 (1970 publ. 1971).
Carex vulgaris var. *humilior* Maly, Enum. Pl. Phan. Austriac.: 34 (1848).
Carex vulgaris var. *bructeri* G.Mey., Fl. Hanov. Excurs.: 598 (1849).
Carex melaena Wimm., Flora 33: 619 (1850). *Carex vulgaris* var. *melaena* (Wimm.) Nyman, Consp. Fl. Eur.: 777 (1882).
Carex tumida Beilschm., Arch. Skand. Beitr. Naturgesch. 2: 214 (1850).
Carex bolina Láng, Linnaea 24: 551 (1851). *Carex vulgaris* var. *bolina* (Láng) K.Richt., Pl. Eur. 1: 156 (1890).
Carex malazena Steud., Syn. Pl. Glumac. 2: 211 (1855).
Vignea chlorocarpa Schur, Verh. Mitth. Siebenbürg. Vereins Naturwiss. Hermannstadt 10: 193 (1859).
Carex vulgaris var. *rigida* A.Blytt, Norges Fl. 1: 213 (1861).
Carex intermedia Miégev., Bull. Soc. Bot. France 10: 83 (1863), nom. illeg. *Carex vulgaris* var. *intermedia* Nyman, Consp. Fl. Eur.: 777 (1882).
Vignantha vulgaris var. *subfastigiata* Schur, Enum. Pl. Transsilv.: 706 (1866). *Carex goodenowii* f. *subfastigiata* (Schur) Kük. in H.G.A.Engler (ed.), Pflanzenr., IV, 20(38): 314 (1909). *Carex nigra* f. *subfastigiata* (Schur) Soó, Acta Bot. Acad. Sci. Hung. 16: 370 (1970 publ. 1971).
Vignea crassinervis Schur, Enum. Pl. Transsilv.: 704 (1866).
Vignea rigida Fuss, Fl. Transsilv.: 681 (1866).
Carex acuta var. *angustifolia* Èelak., Prodr. Fl. Böhmen: 63 (1867).
Carex vulgaris var. *intricata* Boott, Ill. Carex 4: 169 (1867).
Carex vulgaris var. *teres* Boott, Ill. Carex 4: 168 (1867).
Carex vulgaris var. *angustifolia* A.Blytt, Veg. Sognefjord.: 83 (1869).
Carex goodenowii var. *nigra* Cogn., Herb. Glum. Belg. 3: 176 (1870).
Carex vulgaris var. *subramosa* Meinsh., Fl. Ingr.: 409 (1878). *Carex goodenowii* f. *subramosa* (Meinsh.) Kük. in H.G.A.Engler (ed.), Pflanzenr., IV, 20(38): 314 (1909). *Carex nigra* f. *subramosa* (Meinsh.) Soó, Acta Bot. Acad. Sci. Hung. 16: 370 (1970 publ. 1971).
Carex acuta f. *badia* Sanio, Bot. Centralbl. 6: 434 (1881). *Carex goodenowii* f. *badia* (Sanio) Kük. in H.G.A.Engler (ed.), Pflanzenr., IV, 20(38): 314 (1909). *Carex nigra* f. *badia* (Sanio) Soó, Acta Bot. Acad. Sci. Hung. 16: 370 (1970 publ. 1971).
Carex acuta f. *diluta* Sanio, Bot. Centralbl. 6: 437 (1881).
Carex acuta var. *macrocarpa* Èelak., Prodr. Fl. Böhmen 4: 754 (1881).
Carex acuta f. *melanolepis* Sanio, Bot. Centralbl. 6: 437 (1881).
Carex acuta var. *oxylepis* Sanio, Bot. Centralbl. 6: 437 (1881). *Carex goodenowii* f. *oxylepis* (Sanio) Kük. in H.G.A.Engler (ed.), Pflanzenr., IV, 20(38): 315 (1909).
Carex acuta f. *pseudosparganioides* Èelak., Prodr. Fl. Böhmen 4: 754 (1881).
Carex acuta f. *stygia* Sanio, Bot. Centralbl. 6: 437 (1881).
Carex acuta var. *turfosa* Sanio, Bot. Centralbl. 6: 432 (1881).
Carex goodenowii f. *grata* Sanio, Bot. Centralbl. 6: 435 (1881).
Carex vulgaris var. *tenuis* Hartm. ex Lange, Haandb. Danske Fl., ed. 4: 132 (1886).
Carex vulgaris var. *strictiformis* L.H.Bailey, Mem. Torrey Bot. Club 1: 74 (1889). *Carex goodenowii* var. *strictiformis* (L.H.Bailey) Kük. in H.G.A.Engler (ed.), Pflanzenr., IV, 20(38): 316 (1909). *Carex nigra* var. *strictiformis* (L.H.Bailey) Fernald, Rhodora 44: 299 (1942).
Carex nigra var. *alpicola* Beck, Fl. Nieder-Österreich: 136 (1890).
Carex rigida var. *strictiformis* L.H.Bailey, J. Bot. 28: 172 (1890).
Carex vulgaris f. *densicarpa* Kneuck. in M.A.Seubert, Excurs.-Fl. Baden, ed. 5: 62 (1891). *Carex goodenowii* f. *densicarpa* (Kneuck.) Kük. in H.G.A.Engler (ed.), Pflanzenr., IV, 20(38): 316 (1909). *Carex nigra* f. *densicarpa* (Kneuck.) Soó, Acta Bot. Acad. Sci. Hung. 16: 370 (1970 publ. 1971).
Carex goodenowii f. *microlepis* Norman, Forh. Vidensk.-Selsk. Kristiania, extr.: 49 (1893).
Carex vulgaris var. *salinoides* Kük., Mitth. Thüring. Bot. Vereins, n.s., 5: 17 (1893). *Carex goodenowii* var. *salinoides* (Kük.) Kük. in H.G.A.Engler (ed.), Pflanzenr., IV, 20(38): 317 (1909).
Carex vulgaris f. *androgyna* Russow ex E.Lehm., Fl. Poln.-Livl.: 162 (1895). *Carex goodenowii* f. *androgyna* (Russow ex E.Lehm.) Kük. in H.G.A.Engler (ed.), Pflanzenr., IV, 20(38): 314 (1909).
Carex vulgaris f. *silvatica* Russow ex E.Lehm., Fl. Poln.-Livl.: 162 (1895). *Carex goodenowii* f. *silvatica* (Russow ex E.Lehm.) Kük. in H.G.A.Engler (ed.), Pflanzenr., IV, 20(38): 316 (1909). *Carex nigra* f. *silvatica* (Russow ex E.Lehm.) Soó, Acta Bot. Acad. Sci. Hung. 16: 370 (1970 publ. 1971).
Carex melanolepis Phil., Anales Univ. Chile 93: 499 (1896).
Carex vulgaris f. *aquatiliformis* Kük., Oesterr. Bot. Z. 46: 163 (1896). *Carex goodenowii* f. *aquatiliformis* (Kük.) Kük. in H.G.A.Engler (ed.), Pflanzenr., IV, 20(38): 317 (1909).

Carex vulgaris f. *angustifolia* Kük., Allg. Bot. Z. Syst. 4: 3 (1898).
Carex vulgaris f. *latifolia* Kük., Allg. Bot. Z. Syst. 4: 3 (1898).
Carex vulgaris f. *pumila* Kük., Allg. Bot. Z. Syst. 4: 2 (1898). *Carex goodenowii* f. *pumila* (Kük.) Kük. in H.G.A.Engler (ed.), Pflanzenr., IV, 20(38): 314 (1909).
Carex goodenoughii Asch. & Graebn., Syn. Mitteleur. Fl. 2(1): 94 (1899).
Carex vulgaris f. *leucolepis* Meinsh., Trudy Imp. S.-Peterburgsk. Bot. Sada 18: 334 (1900). *Carex goodenowii* f. *leucolepis* (Meinsh.) Kük. in H.G.A. Engler (ed.), Pflanzenr., IV, 20(38): 315 (1909). *Carex nigra* f. *leucolepis* (Meinsh.) Soó, Acta Bot. Acad. Sci. Hung. 16: 370 (1970 publ. 1971).
Carex vulgaris var. *sabulosa* Meinsh., Trudy Imp. S.-Peterburgsk. Bot. Sada 18: 333 (1900).
Carex goodenowii f. *roussioei* H.Lév., Bull. Acad. Int. Géogr. Bot. 10: 37 (1901). *Carex nigra* f. *roussioei* (H.Lév.) Soó, Acta Bot. Acad. Sci. Hung. 16: 370 (1970 publ. 1971).
Carex vulgaris f. *subovalis* Kneuck., Allg. Bot. Z. Syst. 7: 56 (1901). *Carex goodenowii* f. *subovalis* (Kneuck.) Kük. in H.G.A.Engler (ed.), Pflanzenr., IV, 20(38): 317 (1909). *Carex nigra* f. *subovalis* (Kneuck.) Soó, Acta Bot. Acad. Sci. Hung. 16: 370 (1970 publ. 1971).
Carex goodenowii f. *stenostachya* Uechtr. ex Asch. & Graebn., Syn. Mitteleur. Fl. 2(2): 96 (1902). *Carex nigra* f. *stenostachya* (Uechtr. ex Asch. & Graebn.) Soó, Acta Bot. Acad. Sci. Hung. 16: 370 (1970 publ. 1971).
Carex goodenowii f. *microstachys* Zapal., Consp. Fl. Gallic. Crit. 1: 97 (1904). *Carex nigra* f. *microstachys* (Zapal.) Soó, Acta Bot. Acad. Sci. Hung. 16: 370 (1970 publ. 1971).
Carex vulgaris var. *elatior* T.Holm, Amer. J. Sci., IV, 17: 308 (1904), nom. illeg.
Carex vulgaris var. *hydrophila* T.Holm, Amer. J. Sci., IV, 17: 308 (1904).
Carex goodenowii var. *subrigida* Kük. in H.Brockmann-Jerosch, Fl. Puschlav.: 91 (1907).
Carex goodenowii f. *pleiandra* Kük. in H.G.A.Engler (ed.), Pflanzenr., IV, 20(38): 316 (1909). *Carex nigra* f. *pleiandra* (Kük.) Soó, Acta Bot. Acad. Sci. Hung. 16: 370 (1970 publ. 1971), without basionym date.
Carex goodenowii var. *stenocarpa* Kük. in H.G.A.Engler (ed.), Pflanzenr., IV, 20(38): 317 (1909). *Carex nigra* var. *stenocarpa* (Kük.) Soó, Acta Bot. Acad. Sci. Hung. 16: 370 (1970 publ. 1971), without basionym date.
Carex goodenowii var. *subcaespitosa* Kük. in H.G.A.Engler (ed.), Pflanzenr., IV, 20(38): 316 (1909).
Carex goodenowii f. *subsetacea* Kük. in H.G.A.Engler (ed.), Pflanzenr., IV, 20(38): 314 (1909). *Carex nigra* f. *subsetacea* (Kük.) Soó, Acta Bot. Acad. Sci. Hung. 16: 370 (1970 publ. 1971), without basionym date.
Carex goodenowii var. *hydrophila* M.E.Jones, Bull. Montana State Univ., Biol. Ser. 15: 72 (1910).
Carex goodenowii var. *limnophila* M.E.Jones, Bull. Montana State Univ., Biol. Ser. 15: 72 (1910).
Carex goodenowii var. *lipocarpa* M.E.Jones, Bull. Montana State Univ., Biol. Ser. 15: 72 (1910).
Carex aquanigra B.Boivin, Naturaliste Canad. 94: 522 (1967).
Carex nigra subsp. *iberica* Rivas Mart., Fontqueria 24: 14 (1989).

Carex nigricans C.A.Mey., Mém. Sav. Étr. Acad. St. Pétersbourg 1: 210 (1831).
Alaska to W. U.S.A. 31? 70 ALU ASK 71 ABT BRC 73 COL MNT ORE WAS WYO 76 CAL UTA. Hemicr. or rhizome geophyte.
Carex nigricans f. *feminea* Kneuck., Allg. Bot. Z. Syst. 17: 90 (1911).

Carex nigromarginata Schwein., Ann. Lyceum Nat. Hist. New York 1: 68 (1824). *Carex lucorum* var. *nigromarginata* (Schwein.) Chapm., Fl. South. U.S.: 539 (1860).
SE. Canada to C. & E. U.S.A., Mexico (Puebla). 72 ONT 74 ILL MSO 75 CNT INI MAS NWJ NWY OHI PEN RHO WVA 77 TEX 78 ALA ARK DEL GEO KTY MSI NCA SCA TEN VRG WDC 79 MXC. Hemicr. or rhizome geophyte.
Carex lucorum Sartwell ex Boott, Ill. Carex 2: 99 (1860).
Carex nigromarginata var. *subdigyna* Boeck., Linnaea 41: 220 (1877).

Carex nikolskensis Kom., Izv. Imp. Bot. Sada Petra Velikago 16: 154 (1916).
Russian Far East. 31+. Hemicr. or rhizome geophyte.

Carex nitidiutriculata L.K.Dai, Acta Phytotax. Sin. 37: 186 (1999).
China (Yunnan). 36 CHC. Hemicr. or rhizome geophyte.

Carex nivalis Boott, Proc. Linn. Soc. London 1: 256 (1845).
Afghanistan to SC. China. 32 KGZ UZB 34 AFG 36 CHC CHT 40 NEP PAK WHM. Hemicr. or rhizome geophyte.
Carex cinnamomea Boott, Proc. Linn. Soc. London 1: 257 (1845). *Carex nivalis* f. *cinnamomea* (Boott) Kük. in H.G.A.Engler (ed.), Pflanzenr., IV, 20(38): 553 (1909).
Carex griffithii Boott, Proc. Linn. Soc. London 1: 286 (1846). *Carex nivalis* f. *griffithii* (Boott) Kük. in H.G.A.Engler (ed.), Pflanzenr., IV, 20(38): 553 (1909).
Carex oliveri Boeck., Flora 63: 455 (1880).
Carex ustulata var. *aterrima* Regel, Trudy Imp. S.-Peterburgsk. Bot. Sada 7: 571 (1880).
Carex ustulata var. *coriophora* Regel, Trudy Imp. S.-Peterburgsk. Bot. Sada 7: 571 (1880).
Carex nivalis f. *luteobrunnea* Kük. in H.G.A.Engler (ed.), Pflanzenr., IV, 20(38): 553 (1909).
Carex gilesii Nelmes, Bull. Misc. Inform. Kew 1939: 306 (1939).
Carex luteobrunnea Nelmes, Bull. Misc. Inform. Kew 1939: 308 (1939).
Carex lepus-aestatis T.Koyama, Acta Phytotax. Geobot. 16: 167 (1956).

Carex nodaeana A.I.Baranov & Skvortsov, Quart. J. Taiwan Mus. 18: 225 (1965).
N. China. 36 CHM CHN. Hemicr. or rhizome geophyte.
Carex pseudolongirostrata Y.L.Chang & Y.L.Yang, in Fl. Plant. Herb. Chinae Bor.-Or. 11: 206 (1976).

Carex nodiflora Boeck., Bot. Jahrb. Syst. 5: 516 (1884).
Philippines. 42 PHI. Hemicr. or rhizome geophyte.
Carex vulcanica Elmer, Leafl. Philipp. Bot. 10: 3526 (1938), no latin descr.

Carex normalis Mack., Bull. Torrey Bot. Club 37: 244 (1919).
E. Canada, NC. & E. U.S.A. 72 ONT QUE 74 ILL IOW KAN MIN MSO SDA WIS 75 CNT INI MAI MAS MIC NWH NWJ NWY OHI PEN RHO VER WVA 78 ARK DEL GEO KTY MRY NCA SCA TEN VRG WDC. Hemicr. or rhizome geophyte.
Carex mirabilis var. *perlonga* Fernald, Proc. Amer. Acad. Arts 37: 473 (1902). *Carex normalis* var. *perlonga* (Fernald) Burnham, Torreya 19: 131 (1919). *Carex normalis* f. *perlonga* (Fernald) Fernald, Rhodora 44: 285 (1942).

Carex norvegica Retz., Fl. Scand. Prodr.: 179 (1779). *Vignea norvegica* (Retz.) Rchb. in J.C.Mössler, Handb. Gewächsk. ed. 2, 3: 1610 (1830).
Europe to Russian Far East, Afghanistan to Himlaya, Subarctic America to E. Canada. 10 FIN GRB ICE NOR SWE 11 AUT SWI 12 FRA? 13 ITA YUG 14 RUC RUN 30 ALT KRA WSB YAK 31 MAG 34 AFG 37 MON 40 EHM NEP WHM 70 GNL NUN 71 MAN 72 LAB NFL ONT QUE. Hemicr. or rhizome geophyte.

subsp. ***norvegica***
Europe to Russian Far East, Afghanistan to Himlaya, Subarctic America to E. Canada. 10 FIN GRB ICE NOR SWE 11 AUT SWI 12 FRA? 13 ITA 14 RUC RUN 30 ALT KRA WSB YAK 31 MAG 34 AFG 37 MON 40 EHM NEP WHM 70 GNL NUN 71 MAN 72 LAB NFL ONT QUE. Hemicr. or rhizome geophyte.
Carex halleri Gunnerus, Fl. Norveg. 2: 106 (1772), provisional synonym.
Carex ovata Honck., Syn. Pl. Germ. 1: 341 (1792), nom. illeg.
Carex vahlii Schkuhr, Beschr. Riedgräs. 1: 87 (1801).
Carex alpina Sw. ex Wahlenb., Kongl. Vetensk. Acad. Nya Handl. 1803: 160 (1803), nom. illeg.
Carex hirsuta Suter, Helvet. Fl., ed. 2, 2: 324 (1822).
Carex alpina subsp. *infuscata* Nees in R.Wight, Contr. Bot. India: 125 (1834). *Carex alpina* var. *infuscata* (Nees) Boott, Ill. Carex 3: 113 (1862).
Carex alpina f. *nigrescens* Andersson, Cyper. Scand.: 39 (1849).
Carex alpina var. *eriostrata* Boott, Ill. Carex 1: 71 (1858).
Carex dimorpha Boeck., Beitr. Cyper. 1: 45 (1888), nom. illeg.
Carex trispiculata Boeck., Beitr. Cyper. 1: 45 (1888).
Carex norvegica var. *isostachya* Norman, Forh. Vidensk.-Selsk. Kristiania 1893: 48 (1893).
Carex norvegica subsp. *glareoides* Jörg., Nyt Mag. Naturvidensk. 34: 99 (1895).
Carex alpina f. *robustior* Kük. in H.G.A.Engler (ed.), Pflanzenr., IV, 20(38): 386 (1909).
Carex gracilenta Litv., Trudy Bot. Muz. Imp. Akad. Nauk 7: 95 (1910).
Carex mimula V.I.Krecz. in V.L.Komarov (ed.), Fl. URSS 3: 603 (1935).
Carex norvegica subsp. *conicorostrata* Kalela, Ann. Bot. Soc. Zool.-Bot. Fenn. "Vanamo" 19(3): 30 (1944).
Carex norvegica subsp. *inserrulata* Kalela, Ann. Bot. Soc. Zool.-Bot. Fenn. "Vanamo" 19(3): 25 (1944).

subsp. ***pusteriana*** (Kalela) Chater, Bot. J. Linn. Soc. 76: 365 (1978).
E. Alps. 11 AUT 13 ITA YUG. Hemicr. or rhizome geophyte.
**Carex angarae* subsp. *pusteriana* Kalela, Ann. Bot. Soc. Zool.-Bot. Fenn. "Vanamo" 19(3): 4 (1944). *Carex media* subsp. *pusteriana* (Kalela) W.Schultze-Motel in G.Hegi, Ill. Fl. Mitt.-Eur., ed. 3, 2(1): 168 (1969).

Carex notha Kunth, Enum. Pl. 2: 421 (1837).
Himalaya. 40 EHM NEP WHM. Hemicr. or rhizome geophyte.
Carex punctata Nees in R.Wight, Contr. Bot. India: 127 (1834), nom. illeg.
Carex victorialis Nees in R.Wight, Contr. Bot. India: 126 (1834), provisional synonym.
Carex benthamiana Boott, Ill. Carex 1: 8 (1858).
Carex rostellata Gay ex Boott, Ill. Carex 1: 8 (1858).

Carex nova L.H.Bailey, J. Bot. 26: 322 (1888).
WC. U.S.A. 73 COL WYO 76 NEV UTA 77 NWM. Hemicr. or rhizome geophyte.
Carex atrata var. *nigra* Olney in S.Watson, Botany [Fortieth Parallel] 5: 371 (1871).
Carex violacea C.B.Clarke, Kew Bull., Addit. Ser. 8: 87 (1908).
Carex elbertiana Kelso, Biol. Leafl. 31: 3 (1945).
Carex nova var. *erxlebenii* Kelso, Biol. Leafl. 66: 31 (1953).

Carex novae-angliae Schwein., Ann. Lyceum Nat. Hist. New York 1: 67 (1824). *Carex varia* var. *novae-angliae* (Schwein.) Kük. in H.G.A.Engler (ed.), Pflanzenr., IV, 20(38): 449 (1909).
E. Canada to NE. U.S.A. 72 LAB NBR NFL NSC ONT QUE 74 WIS 75 CNT MAI MAS MIC NWH NWY PEN VER WVA 78 SCA. Hemicr. or rhizome geophyte.

Carex novogaliciana Reznicek, Contr. Univ. Michigan Herb. 17: 284 (1990).
Mexico (Jalisco). 79 MXS. Hemicr.

Carex ×nubens Lepage, Naturaliste Canad. 84: 40 (1957). *C. recta* × *C. saxatilis* var. *militaris*.
E. Canada. 72 QUE. Hemicr. or rhizome geophyte.

Carex nubigena D.Don ex Tilloch & Taylor, Philos. Mag. J. 62: 455 (Dec. 1823). *Vignea nubigena* (D.Don ex Tilloch & Taylor) Soják, Cas. Nár. Mus., Odd. Prír. 148: 196 (1979 publ. 1980).
Afghanistan to Japan, India, Sri Lanka, C. & E. Jawa. 34 AFG? 36 CHC CHN CHT 38 JAP TAI 40 EHM IND NEP PAK? SRL 41 VIE 42 JAW. Hemicr. or rhizome geophyte.

var. ***franchetiana*** Ohwi, Mem. Coll. Sci. Kyoto Imp. Univ., Ser. B, Biol. 5: 283 (1930).
Japan (S. Honshu, Kyushu). 38 JAP. Hemicr. or rhizome geophyte.
Carex phaeoleuca Ohwi, Acta Phytotax. Geobot. 1: 68 (1932). *Vignea phaeoleuca* (Ohwi) Soják, Cas. Nár. Mus., Odd. Prír. 148: 196 (1979 publ. 1980).

subsp. ***nubigena***
Afghanistan to SC. China, India, Sri Lanka, C. & E. Jawa. 34 AFG? 36 CHC CHN CHT 40 EHM IND NEP PAK? SRL 41 VIE 42 JAW. Hemicr. or rhizome geophyte.
Carex fallax Steud., Syn. Pl. Glumac. 2: 189 (1855). *Carex nubigena* var. *fallax* (Steud.) C.B.Clarke, J. Linn. Soc., Bot. 36: 301 (1904).

Carex argyrolepis Maxim. ex Franch. & Sav., Enum. Pl. Jap. 1: 126 (1873).
Carex nubigena f. *laxiuscula* Kük. in H.G.A.Engler (ed.), Pflanzenr., IV, 20(38): 146 (1909).
Carex pleistogyna V.I.Krecz., Bot. Mater. Gerb. Bot. Inst. Komarova Akad. Nauk S.S.S.R. 9: 194 (1946). *Vignea pleistogyna* (V.I.Krecz.) Soják, Cas. Nár. Mus., Odd. Prír. 148: 196 (1979 publ. 1980).

subsp. ***pseudoarenicola*** (Hayata) T.Koyama, in Fl. Taiwan 5: 367 (1978).
Taiwan. 38 TAI. Hemicr. or rhizome geophyte.
**Carex pseudoarenicola* Hayata, Icon. Pl. Formosan. 6: 118 (1916).

Carex nudata Boott in S.Watson & al., Bot. California 2: 241 (1880).
W. U.S.A. 73 ORE WAS 76 CAL. Hemicr. or rhizome geophyte.
Carex aperta var. *angustifolia* Boott in W.J.Hooker, Fl. Bor.-Amer. 2: 218 (1839). *Carex nudata* var. *angustifolia* (Boott) L.H.Bailey, Mem. Torrey Bot. Club 1: 16 (1889).
Carex hallii L.H.Bailey, Proc. Amer. Acad. Arts 22: 82 (1887), nom. illeg. *Carex bishallii* C.B.Clarke, Kew Bull., Addit. Ser. 8: 70 (1908).
Carex acutina L.H.Bailey, Mem. Torrey Bot. Club 1: 52 (1889).
Carex nudata var. *anomala* L.H.Bailey, Mem. Torrey Bot. Club 1: 17 (1889).
Carex nudata var. *versuta* Piper, Contr. U. S. Natl. Herb. 11: 171 (1906).
Carex nudata f. *firmior* Kük. in H.G.A.Engler (ed.), Pflanzenr., IV, 20(38): 337 (1909).
Carex nudata f. *sessiliflora* Kük. in H.G.A.Engler (ed.), Pflanzenr., IV, 20(38): 337 (1909).
Carex suborbiculata Mack. in L.Abrams, Ill. Fl. Pacific States 1: 338 (1923).
Carex tenacissima Suksd., Werdenda 1: 5 (1923).

Carex* ×*oberrodensis B.Walln., Linzer Biol. Beitr. 25: 734 (1993). *C. elata* × *C. randalpina*.
C. Europe. 11 AUT 13 ITA. Hemicr. or rhizome geophyte.

Carex obispoensis Stacey, Leafl. W. Bot. 1: 240 (1936).
California (San Luis Obispo Co.). 76 CAL. Hemicr. or rhizome geophyte.

Carex oblanceolata T.Koyama, Willdenowia 5: 489 (1969).
China (Guangdong). 36 CHS. Hemicr. or rhizome geophyte.

Carex obliquicarpa X.F.Jin, C.Z.Zheng & B.Y.Ding, Ann. Bot. Fenn. 42: 223 (2005).
China (Guangxi). 36 CHS. Hemicr. or rhizome geophyte.

Carex obnupta L.H.Bailey, Proc. Calif. Acad. Sci., II, 3: 104 (1891).
British Columbia to California. 71 BRC 73 ORE WAS 76 CAL. Hemicr. or rhizome geophyte.
Carex magnifica Dewey ex C.B.Clarke, Kew Bull., Addit. Ser. 8: 71 (1908).

Carex obovatosquamata F.T.Wang & Y.L.Chang ex P.C.Li, Acta Phytotax. Sin. 37: 175 (1999).
Tibet to China (Yunnan). 36 CHC CHT. Hemicr. or rhizome geophyte.

Carex obscura Nees in R.Wight, Contr. Bot. India: 126 (1834).
N. Pakistan to SC. China. 36 CHC CHT 40 EHM NEP PAK WHM. Hemicr. or rhizome geophyte.

var. ***brachycarpa*** C.B.Clarke in J.D.Hooker, Fl. Brit. India 6: 731 (1894).
Himalaya to SC. China. 36 CHC CHT 40 EHM NEP WHM. Hemicr. or rhizome geophyte.
Carex souliei Franch., Bull. Soc. Philom. Paris, VIII, 7: 35 (1895).

var. ***obscura***
N. Pakistan to Himalaya. 40 NEP PAK WHM. Hemicr. or rhizome geophyte.

Carex obtusata Lilj., Kongl. Vetensk. Acad. Nya Handl. 1793: 69 (1793). *Genersichia obtusata* (Lilj.) Heuff., Flora 27(2): 528 (1844). *Psyllophora obtusata* (Lilj.) Schur, Verh. Mitth. Siebenbürg. Vereins Naturwiss. Hermannstadt 4: 78 (1853).
NE. Germany to Czech Rep., S. Sweden to Russian Far East, Canada to WC. U.S.A. 10 SWE 11 CZE GER 14 RUC RUN RUS UKR 30 ALT KRA TVA WSB YAK 31 AMU KHA MAG PRM 32 KAZ 33 NCS 36 CHM CHX 37 MON 70 ASK NWT YUK 71 ABT BRC MAN SAS 72 ONT 73 COL MNT WAS WYO 74 MIN NDA SDA 76 UTA 77 NWM. Hemicr. or rhizome geophyte.
Carex spicata Schkuhr, Beschr. Riedgräs. 1: 11 (1801), nom. illeg. *Carex obtusata* f. *spicata* Kük. in H.G.A.Engler (ed.), Pflanzenr., IV, 20(38): 88 (1909).
Carex leucoglochin Ehrh. ex Wahlenb., Kongl. Vetensk. Acad. Nya Handl. 1803: 139 (1803).
Carex affinis R.Br. in J.Richardson, Bot. App.: 750 (1823).
Carex microcephala C.A.Mey. in C.F.von Ledebour, Fl. Altaic. 2: 205 (1830).
Carex backiana Dewey, Amer. J. Sci. Arts 29: 250 (1836).
Carex decipiens Turcz., Bull. Soc. Imp. Naturalistes Moscou 1838: 102 (1838), nom. illeg.
Carex obtusata subf. *bracteata* Peterm., Anal. Pfl.-Schlüss.: 486 (1846).
Carex obtusata subf. *capitata* Peterm., Anal. Pfl.-Schlüss.: 486 (1846).
Carex obtusata subf. *cuspidata* Peterm., Anal. Pfl.-Schlüss.: 486 (1846).
Carex obtusata subf. *oblonga* Peterm., Anal. Pfl.-Schlüss.: 486 (1846).
Carex obesa var. *monostachya* Boeck., Linnaea 41: 185 (1877).
Carex obtusata f. *bracteata* Neuman, Bot. Not. 1887: 21 (1887).

Carex occidentalis L.H.Bailey, Mem. Torrey Bot. Club 1: 14 (1889). *Carex hookeriana* var. *occidentalis* (L.H.Bailey) Kük. in H.G.A.Engler (ed.), Pflanzenr., IV, 20(38): 162 (1909).
SW. & WC. U.S.A. to NE. Mexico. 73 COL MNT WYO 74 NDA SDA 76 ARI CAL NEV UTA 77 NWM TEX 79 MXE. Hemicr. or rhizome geophyte.
Carex neomexicana Mack., Bull. Torrey Bot. Club 34: 153 (1907).

Carex ochrochlamys Ohwi, Mem. Coll. Sci. Kyoto Imp. Univ., Ser. B, Biol. 6: 240 (1931). *Vignea ochrochlamys* (Ohwi) Soják, Cas. Nár. Mus., Odd. Prír. 148: 196 (1979 publ. 1980).
Korea. 38 KOR. Hemicr. or rhizome geophyte.

Carex ochrosaccus (C.B.Clarke) Hamlin, Rec. Domin. Mus. 6: 99 (1968).

New Zealand North I. 51 NZN. Hemicr. or rhizome geophyte.
Carex dissita var. *ochrosaccus* C.B.Clarke in T.F.Cheeseman, Man. New Zealand Fl.: 831 (1906).

Carex oedipostyla Duval-Jouve, Mém. Sect. Sci. Acad. Sci. Montpellier 7: 442 (1869).
Canary Is., SW. Europe, NW. Morocco. 12 BAL FRA POR SAR SPA 20 MOR 21 CNY. Hemicr. or rhizome geophyte.
Carex ambigua Link, J. Bot. (Schrader) 2: 308 (1799), nom. illeg. *Carex linkii* subsp. *ambigua* Nyman, Consp. Fl. Eur.: 783 (1882).
Carex valesiaca Wahlenb., Kongl. Vetensk. Acad. Nya Handl. 1803: 155 (1803), nom. illeg.

Carex oedorrhampha Nelmes, Bull. Misc. Inform. Kew 1939: 659 (1940).
Bhutan to New Guinea. 36 CHC CHS 40 ASS EHM 41 THA VIE 42 JAW SUM 43 NWG. Hemicr. or rhizome geophyte.
**Carex tumida* Boott, Ill. Carex 1: 66 (1858), nom. illeg.

Carex × oenensis A.Neumann ex B.Walln., Linzer Biol. Beitr. 24: 832 (1992). *C. gracilis* × *C. randalpina.*
C. Europe. 11 AUT 13 ITA. Hemicr. or rhizome geophyte.

Carex okamotoi Ohwi, Acta Phytotax. Geobot. 5: 53 (1936).
Korea. 38 KOR. Hemicr. or rhizome geophyte.

Carex oklahomensis Mack., Torreya 14: 126 (1914). *Carex stipata* var. *oklahomensis* (Mack.) Gleason, Phytologia 4: 22 (1952).
C. & SE. U.S.A. 74 ILL KAN MSO OKL 75 INI 77 TEX 78 ARK DEL KTY MRY MSI NCA TEN VRG. Hemicr. or rhizome geophyte.

Carex olbiensis Jord., Observ. Pl. Nouv. 3: 241 (1846).
S. Europe, NW. Africa. 12 COR FRA SPA 13 ITA SIC YUG 20 ALG TUN. Hemicr. or rhizome geophyte.
Carex ardoiniana De Not., Ann. Sci. Nat., Bot., III, 9: 324 (1848).
Carex andoiniana De Not. ex Steud., Syn. Pl. Glumac. 2: 233 (1855).

Carex oligantha Steud., Syn. Pl. Glumac. 2: 203 (1855).
NE. Turkey to NW. Iran. 33 TCS 34 IRN TUR. Hemicr. or rhizome geophyte.

Carex oligocarpa Willd., Sp. Pl. 4: 279 (1805). *Deweya oligocarpa* (Willd.) Raf., Good Book: 26 (1840). *Olotrema oligocarpa* (Willd.) Raf., Good Book: 25 (1840).
SE. Canada to NC. & E. U.S.A. 72 ONT QUE 74 ILL IOW KAN MIN MSO NEB OKL WIS 75 CNT INI MAS MIC NWJ NWY OHI PEN VER WVA 78 ALA ARK DEL GEO KTY MRY MSI NCA TEN VRG WDC. Hemicr. or rhizome geophyte.
Carex vleckii Schwein., Ann. Lyceum Nat. Hist. New York 1: 69 (1824).
Carex oligocarpa var. *major* Torr., Fl. New York 2: 406 (1843).
Carex oligocarpa var. *sartwelliana* Dewey, Amer. J. Sci. Arts, II, 5: 176 (1848).
Carex subuniflora Steud., Syn. Pl. Glumac. 2: 234 (1855). *Carex oligocarpa* f. *subuniflora* (Steud.) Kük. in H.G.A.Engler (ed.), Pflanzenr., IV, 20(38): 521 (1909).
Carex oligocarpa var. *minor* Boott, Ill. Carex 1: 40 (1858).
Carex oligocarpa f. *triflora* Peck, Rep. (Annual) Regents Univ. State New York New York State Mus. 47: 164 (1895).

Carex oligocarya C.B.Clarke in J.D.Hooker, Fl. Brit. India 6: 746 (1894). *Carex annulata* Kük. in H.G.A.Engler (ed.), Pflanzenr., IV, 20(38): 494 (1909), nom. illeg.
N. Pakistan, W. Himalaya. 40 PAK WHM. Hemicr. or rhizome geophyte.

Carex oligosperma Michx., Fl. Bor.-Amer. 2: 174 (1803).
S. Kuril Is., N. & C. Japan, Subarctic America to NC. & NE. U.S.A. 31 KUR 38 JAP 70 NUN YUK 71 MAN SAS 72 LAB NBR NFL NSC ONT QUE 74 MIN WIS 75 INI MAI MIC NWH NWY OHI PEN VER 78 NCA. Hemicr. or rhizome geophyte.
Carex okesiana Dewey, Amer. J. Sci. Arts 14: 351 (1828).
Carex okesii Eaton, Man. Bot., ed. 5: 157 (1829).
Carex oakesiana Torr., Ann. Lyceum Nat. Hist. New York 3: 421 (1836). *Olamblis oakesiana* (Torr.) Raf., Good Book: 26 (1840).
Carex despreauxii Steud., Syn. Pl. Glumac. 2: 237 (1855).
Carex tsuishikarensis Koidz. & Ohwi, J. Fac. Agric. Hokkaido Univ. 26: 273 (1931). *Carex oligosperma* var. *tsuishikarensis* (Koidz. & Ohwi) B.Boivin, Phytologia 43: 104 (1979).
Carex oligosperma var. *churchilliana* Raymond, Canad. J. Bot. 34: 292 (1956).

Carex oligostachya Nees, Hooker's J. Bot. Kew Gard. Misc. 6: 29 (1854).
Assam to Solomon Is. 36 CHC CHS 40 ASS EHM 41 MYA VIE 42 LSI MOL? PHI SUL SUM 43 NWG SOL. Hemicr. or rhizome geophyte.
Carex cumingiana Steud., Syn. Pl. Glumac. 2: 206 (1855).
Carex rhizomatosa Steud., Syn. Pl. Glumac. 2: 206 (1855).
Carex impunctata Boott, Ill. Carex 3: 107 (1862). *Carex rhizomatosa* var. *impunctata* (Boott) Kük. in H.G.A.Engler (ed.), Pflanzenr., IV, 20(38): 291 (1909).
Carex capitulata Boott ex C.B.Clarke in J.D.Hooker, Fl. Brit. India 6: 721 (1894), pro syn.
Carex bukaensis Palla, Denkschr. Kaiserl. Akad. Wiss., Wien. Math.-Naturwiss. Kl. 89: 500 (1913).
Carex breviceps Kük., Bot. Jahrb. Syst. 69: 263 (1938).

Carex olivacea Boott, Proc. Linn. Soc. London 1: 286 (1846).
E. Himalaya to Japan, W. Sumatera, W. Jawa. 36 CHC CHS 38 JAP 40 ASS EHM 42 JAW SUM. Hemicr. or rhizome geophyte.

subsp. ***confertiflora*** (Boott) T.Koyama, Bot. Mag. (Tokyo) 72: 307 (1959).
China (Yunnan to Hunan), Japan. 36 CHC CHS 38 JAP. Hemicr. or rhizome geophyte.
**Carex confertiflora* Boott, Mem. Amer. Acad. Arts, n.s., 6: 418 (1859).
Carex olivacea var. *angustior* Kük. in H.G.A.Engler (ed.), Pflanzenr., IV, 20(38): 618 (1909).

subsp. ***olivacea***
E. Himalaya to SC. China, W. Sumatera, W. Jawa. 36 CHC 40 ASS EHM 42 JAW SUM. Hemicr. or rhizome geophyte.

subsp. ***recurvisaccus*** (T.Koyama) T.Koyama, Bot. Mag. (Tokyo) 72: 307 (1959).
China (Yunnan, Guangdong). 36 CHC CHS. Hemicr. or rhizome geophyte.
**Carex recurvisaccus* T.Koyama, Jap. J. Bot. 15: 166 (1956).

Carex olivieri H.Lév., Rev. Sci. Bourbonnais Centr. France 17: 166 (1904).
China. 36+. Hemicr. or rhizome geophyte.

Carex × olneyi Boott, Ill. Carex 1: 15 (1858). *C. bullata* × *C. rostrata*. *Carex bullata* var. *olneui* (Boott) Fernald, Rhodora 3: 52 (1901).
E. U.S.A. 75 CNT MAS RHO 78 DEL. Hemicr. or rhizome geophyte.

Carex omeiensis Tang, Bull. Fan Mem. Inst. Biol. 3: 361 (1932).
C. China. 36 CHC CHN CHS. Hemicr. or rhizome geophyte.
Carex omeiensis var. *multifascula* Q.S.Wang, J. Wuhan Bot. Res. 5: 343 (1987).

Carex omiana Franch. & Sav., Enum. Pl. Jap. 2: 554 (1878). *Carex stellulata* var. *omiana* (Franch. & Sav.) Kük. in H.G.A.Engler (ed.), Pflanzenr., IV, 20(38): 230 (1909). *Vignea omiana* (Franch. & Sav.) Soják, Cas. Nár. Mus., Odd. Prír. 148: 196 (1979 publ. 1980).
NE. China, S. Kuril Is., Japan. 31 KUR 36 CHM 38 JAP. Hemicr. or rhizome geophyte.

var. ***monticola*** Ohwi, Acta Phytotax. Geobot. 1: 71 (1932).
N. & C. Japan. 38 JAP. Hemicr. or rhizome geophyte.

var. ***omiana***
NE. China, S. Kuril Is., Japan. 31 KUR 36 CHM 38 JAP. Hemicr. or rhizome geophyte.

var. ***yakushimana*** Ohwi, Acta Phytotax. Geobot. 1: 71 (1932).
Japan (Yakushima). 38 JAP. Hemicr. or rhizome geophyte.
Carex ohwii Masam., Mem. Fac. Sci. Taihoku Imp. Univ. 11(4): 526 (1934).

Carex omurae T.Koyama, J. Jap. Bot. 30: 313 (1955).
Japan (Honshu: Tokaido Distr.). 38 JAP. Hemicr. or rhizome geophyte.

Carex × oneillii Lepage, Naturaliste Canad. 83: 115 (1956). *C. hindsii* × *C. rostrata*.
W. Alaska. 70 ASK. Hemicr. or rhizome geophyte.

Carex onoei Franch. & Sav., Enum. Pl. Jap. 1: 551 (1875).
Russian Far East to N. & C. Japan. 31 PRM 36 CHC CHM CHN CHS 38 JAP KOR. Hemicr. or rhizome geophyte.
Carex capituliformis Meinsh. ex Maxim., Bull. Acad. Imp. Sci. Saint-Pétersbourg 31: 115 (1887). *Carex onoei* var. *capituliformis* (Meinsh. ex Maxim.) Kitag., Neo-Lineam. Fl. Manshur.: 133 (1979). *Carex onoei* subsp. *capituliformis* (Meinsh. ex Maxim.) T.V.Egorova, Novosti Sist. Vyssh. Rast. 22: 60 (1985).
Carex heleochariformis H.Lév. & Vaniot, Bull. Soc. Bot. France 51: 202 (1904).
Carex onoei f. *latifolia* Kük. in H.G.A.Engler (ed.), Pflanzenr., IV, 20(38): 101 (1909).
Carex onoei var. *macrogyna* Tang, Bull. Fan Mem. Inst. Biol. 3: 137 (1932).
Carex perpusilla Meinsh. ex Krecz. in V.L.Komarov (ed.), Fl. URSS 3: 300 (1935), pro syn.

Carex opaca (F.J.Herm.) P.Rothr. & Reznicek, Novon 11: 223 (2001).
C. & EC. U.S.A. 74 ILL KAN MSO OKL 78 ARK MSI. Hemicr.
**Carex bicknellii* var. *opaca* F.J.Herm., Sida 5: 49 (1972).

Carex ophiolithica Heenan & de Lange, New Zealand J. Bot. 35: 429 (1997).
NW. New Zealand North I. 51 NZN. Hemicr. or rhizome geophyte.

Carex ophiopogon H.Lév., Bull. Acad. Int. Géogr. Bot. 24: 283 (1914).
China (Yunnan). 36 CHC. Hemicr. or rhizome geophyte.

Carex orbicularinucis L.K.Dai, Acta Phytotax. Sin. 37: 183 (1999).
China (Yunnan). 36 CHC. Hemicr. or rhizome geophyte.

Carex orbicularis Boott, Proc. Linn. Soc. London 1: 254 (1845).
Temp. Asia to Himalaya. 30 ALT 32 KAZ KGZ TZK UZB 33 TCS 34 AFG IRN IRQ TUR 36 CHN CHQ CHT CHX 37 MON 38 KOR 40 EHM NEP PAK WHM. Hemicr. or rhizome geophyte.

subsp. ***altaica*** (Gorodkov) T.V.Egorova, Novosti Sist. Vyssh. Rast. 22: 54 (1985).
Siberia to NW. China. 30 ALT 36 CHX. Hemicr. or rhizome geophyte.
**Carex rigida* subsp. *altaica* Gorodkov, Zhurn. Russk. Bot. Obshch. 15: 183 (1930). *Carex altaica* (Gorodkov) V.I.Krecz. in V.L.Komarov (ed.), Fl. URSS 3: 227 (1935). *Carex bigelowii* subsp. *altaica* (Gorodkov) Malyschev, in Fl. Tsentral'noi Sibiri 1: 199 (1979).

var. ***caucasica*** Ö.Nilsson in P.H.Davis (ed.), Fl. Turkey 9: 623 (1985).
NE. Turkey to Transcaucasus. 33 TCS 34 TUR. Hemicr. or rhizome geophyte.

subsp. ***kotschyana*** (Boiss. & Hohen.) Kukkonen, Ann. Bot. Fenn. 21: 389 (1984).
SE. Turkey to Iran. 33 TCS 34 IRN IRQ TUR. Hemicr. or rhizome geophyte.
**Carex kotschyana* Boiss. & Hohen. in P.E.Boissier, Diagn. Pl. Orient. 13: 38 (1854).

subsp. ***orbicularis***
Temp. Asia to Himalaya. 30 ALT 32 KGZ TZK UZB 34 AFG IRN 36 CHN CHQ CHT CHX 37 MON 38 KOR 40 EHM NEP PAK WHM. Hemicr. or rhizome geophyte.
Carex cespitosa var. *microsachys* Regel, Trudy Imp. S.-Peterburgsk. Bot. Sada 7: 574 (1880).
Carex glauca var. *brachylepis* Regel, Trudy Imp. S.-Peterburgsk. Bot. Sada 7: 572 (1880). *Carex orbicularis* var. *brachylepis* (Regel) Kük. in H.G.A.Engler (ed.), Pflanzenr., IV, 20(38): 304 (1909).
Carex cyclocystis Boeck., Beitr. Cyper. 1: 47 (1888).
Carex melanolepis Boeck., Beitr. Cyper. 1: 47 (1888).
Carex erostrata Boott ex C.B.Clarke in J.D.Hooker, Fl. Brit. India 6: 711 (1894).
Carex vulgaris var. *distracta* C.B.Clarke in J.D.Hooker, Fl. Brit. India 6: 711 (1894).
Carex glauca var. *kotschyi* Kük. & Bornm., Oesterr. Bot. Z. 47: 135 (1897).

Carex arcatica Meinsh., Trudy Imp. S.-Peterburgsk. Bot. Sada 18: 336 (1900).
Carex arcatica var. *pedunculata* Meinsh., Trudy Imp. S.-Peterburgsk. Bot. Sada 18: 337 (1900).
Carex taldycola Meinsh., Trudy Imp. S.-Peterburgsk. Bot. Sada 18: 339 (1900). *Carex orbicularis* var. *taldycola* (Meinsh.) Kük. in H.G.A. Engler (ed.), Pflanzenr., IV, 20(38): 304 (1909).
Carex petunnikowii Litv., Trudy Bot. Muz. Imp. Akad. Nauk 7: 87 (1910).
Carex satakeana T.Koyama, Acta Phytotax. Geobot. 15: 113 (1954).
Carex pakistanica T.Koyama, Acta Phytotax. Geobot. 17: 100 (1958).
Carex transalaica Tzvelev, Bot. Mater. Gerb. Bot. Inst. Komarova Akad. Nauk S.S.S.R. 20: 430 (1960).

Carex oreocharis T.Holm, Amer. J. Sci., IV, 9: 358 (1900).
WC. U.S.A. to Mexico (Coahuila). 73 COL WYO 79 MXE. Hemicr. or rhizome geophyte.
Carex filifolia var. *valida* Olney ex L.H.Bailey in J.M.Coulter, Man. Bot. Rocky Mt.: 374 (1885).
Carex oreocharis var. *enantiomorpha* Kelso, Biol. Leafl. 64: 2 (1953).

Carex oreophila C.A.Mey., Verz. Pfl. Casp. Meer.: 29 (1831). *Vignea oreophila* (C.A.Mey.) Soják, Cas. Nár. Mus., Odd. Prír. 148: 196 (1979 publ. 1980).
Turkey to Iran. 33 TCS 34 IRN IRQ TUR. Hemicr. or rhizome geophyte.
Carex oreophila f. *minor* Kük. in H.G.A.Engler (ed.), Pflanzenr., IV, 20(38): 72 (1909).

Carex orizabae Liebm., Mexic. Halvgr.: 263 (1850).
Mexico to Guatemala. 79 MXC MXG MXS 80 GUA. Hemicr. or rhizome geophyte.
Carex volcanica F.J.Herm., Brittonia 23: 144 (1971).

Carex ormostachya Wiegand, Rhodora 24: 196 (1922). *Carex laxiflora* var. *ormostachya* (Wiegand) Gleason, Phytologia 4: 22 (1952).
E. Canada to Virginia. 72 NBR ONT QUE 74 MIN WIS 75 CNT MAI MAS MIC NWH NWY PEN VER 78 VRG. Hemicr. or rhizome geophyte.

Carex ornithopoda Willd., Sp. Pl. 4: 255 (1805). *Carex digitata* var. *ornithopoda* (Willd.) G.Mey., Fl. Hanov. Excurs.: 607 (1849).
Europe to Turkey. 10 FIN GRB NOR SWE 11 AUT BGM CZE GER POL SWI 12 COR FRA SPA 13 ALB GRC ITA ROM YUG 14 BLT RUC RUN RUW UKR 34 TUR. Hemicr. or rhizome geophyte.

subsp. ***ornithopoda***
Europe to Turkey. 10 FIN GRB NOR SWE 11 AUT BGM CZE GER POL SWI 12 COR FRA SPA 13 ALB ITA ROM YUG 14 BLT RUC RUN RUW UKR 34 TUR. Hemicr. or rhizome geophyte.
Carex pedata L., Sp. Pl. ed. 2: 1384 (1763), nom. rejic. prop.
Carex pedata All., Fl. Pedem. 2: 168 (1785), nom. illeg.
Carex ornithopoda var. *alpina* Gaudin, Fl. Helv. 6: 87 (1830).
Carex ornithopoda var. *hausmannii* Döll, Fl. Baden 1: 227 (1857).
Carex subnivalis Arv.-Touv., Ber. Deutsch. Bot. Ges. 7: 169 (1890).
Carex ornithopoda var. *castanea* Murb., Beitr. Fl. Südbosnien: 30 (1891).
Carex ornithopoda var. *alpina* Kük., Allg. Bot. Z. Syst. 5: 128 (1899), nom. illeg.
Carex ornithopoda f. *major* Bornm. ex Kük. in H.G.A.Engler (ed.), Pflanzenr., IV, 20(38): 498 (1909).

subsp. ***ornithopodioides*** (Hausm.) Nyman, Consp. Fl. Eur.: 772 (1882).
C. & S. Europe. 11 AUT CZE GER SWI 12 FRA SPA 13 GRC ITA ROM YUG. Hemicr. or rhizome geophyte.
**Carex ornithopodioides* Hausm., Flora 36: 225 (1853). *Carex ornithopoda* var. *ornithopodioides* (Hausm.) Garcke, Fl. N. Mitt.-Deutschland, ed. 15: 446 (1885). *Carex digitata* var. *ornithopodioides* (Hausm.) Fiori, Nuov. Fl. Italia 1: 190 (1923).
Carex reclinata Facchini, Ambr. Fir. Mer. 1: 333, 828 (1854).
Carex ornithopodioides f. *elongata* Leyb., Flora 38: 348 (1855). *Carex ornithopoda* var. *elongata* (Leyb.) Asch. & Graebn., Syn. Mitteleur. Fl. 1: 162 (1897).
Carex ornithopodioides f. *supina* Leyb., Flora 38: 348 (1855).

Carex oronensis Fernald, Proc. Amer. Acad. Arts 37: 471 (1902).
Maine. 75 MAI. Hemicr. or rhizome geophyte.

Carex orthostemon Hayata, J. Coll. Sci. Imp. Univ. Tokyo 30(1): 389 (1911).
Taiwan. 38 TAI. Hemicr. or rhizome geophyte.

Carex oshimensis Nakai, Bot. Mag. (Tokyo) 28: 326 (1914).
Japan (Izu-shoto). 38 JAP. Hemicr. or rhizome geophyte.
Carex oshimensis f. *variegata* Hid.Takah., J. Geobot. 21: 20 (1973).

Carex otaruensis Franch., Bull. Soc. Philom. Paris, VIII, 7: 34 (1895).
E. China, Japan. 36 CHS 38 JAP. Hemicr. or rhizome geophyte.
Carex cardioglochis H.Lév. & Vaniot, Bull. Acad. Int. Géogr. Bot. 11: 31 (1902).
Carex peniculacea H.Lév. & Vaniot, Bull. Acad. Int. Géogr. Bot. 11: 110 (1902).
Carex mitoensis Ohwi, Mem. Coll. Sci. Kyoto Imp. Univ., Ser. B, Biol. 5(3): 279 (1930).

Carex* ×*otayai Ohwi, Mem. Coll. Sci. Kyoto Imp. Univ., Ser. B, Biol. 6: 265 (1931). *C. aphyllopus* × *C. podogyna.*
Japan. 38 JAP. Hemicr. or rhizome geophyte.

Carex ouachitana Kral, Manhart & Bryson, Ann. Missouri Bot. Gard. 74: 440 (1987).
Oklahoma to Tennessee. 74 OKL 78 ARK KTY TEN. Hemicr. or rhizome geophyte.

Carex ovatispiculata F.T.Wang & Y.L.Chang ex S.Yun Liang, Acta Phytotax. Sin. 28: 155 (1990).
Tibet to C. China. 36 CHC CHN CHT. Hemicr. or rhizome geophyte.

Carex ovoidoconica Ohwi, Acta Phytotax. Geobot. 11: 261 (1942).
NC. China. 36 CHN. Hemicr. or rhizome geophyte.

Carex ownbeyi G.A.Wheeler, Darwiniana 40: 200 (2002).
Bolivia. 83 BOL. Hemicr. or rhizome geophyte.

Carex oxyandra (Franch. & Sav.) Kudô, Rep. Veg. N. Saghalien: 72 (1923).
Russian Far East, Japan, Taiwan. 31 KAM KUR SAK 38 JAP TAI. Hemicr. or rhizome geophyte.
**Carex montana* var. *oxyandra* Franch. & Sav., Enum. Pl. Jap. 2: 141 (1877).

var. ***oxyandra***
Russian Far East, Japan, Taiwan. 31 KUR SAK 38 JAP TAI. Hemicr. or rhizome geophyte.
Carex wrightii Franch., Bull. Soc. Philom. Paris, VIII, 7: 47 (1895), nom. illeg.
Carex stolonifera H.Lév. & Vaniot, Bull. Acad. Int. Géogr. Bot. 11: 27 (1902), nom. illeg.
Carex wrightii var. *lanceata* Kük. in H.G.A.Engler (ed.), Pflanzenr., IV, 20(38): 443 (1909).
Carex oxyandra var. *globosa* Akiyama, Bot. Mag. (Tokyo) 45: 473 (1931).

var. ***pauzhetica*** (A.E.Kozhevn.) A.E.Kozhevn., Sosud. Rast. Sovet. Dal'nego Vostoka 3: 288 (1988).
Russian Far East. 31 KAM KUR. Hemicr. or rhizome geophyte.
**Carex pauzhetica* A.E.Kozhevn., Komarovskie Chteniya (Moscow & Leningrad) 30: 46 (1983).

Carex oxylepis Torr. & Hook., Ann. Lyceum Nat. Hist. New York 3: 409 (1836).
C. & SE. U.S.A. 74 ILL MSO OKL 77 TEX 78 ALA ARK FLA GEO KTY LOU MSI NCA SCA TEN VRG. Hemicr. or rhizome geophyte.

var. ***oxylepis***
C. & SE. U.S.A. 74 ILL MSO OKL 77 TEX 78 ALA ARK FLA GEO KTY LOU MSI NCA SCA TEN VRG. Hemicr. or rhizome geophyte.
Carex familiaris Steud., Syn. Pl. Glumac. 2: 226 (1855).
Carex oxylepis f. *glabra* Boeck., Linnaea 40: 408 (1876).

var. ***pubescens*** J.K.Underw., Amer. Midl. Naturalist 33: 635 (1945).
EC. U.S.A. 78 ARK KTY TEN. Hemicr. or rhizome geophyte.

Carex ozarkana P.Rothr. & Reznicek, Brittonia 48: 104 (1996).
Oklahoma to Louisiana. 74 OKL 77 TEX 78 ARK LOU. Hemicr. or rhizome geophyte.

Carex pachygyna Franch. & Sav., Enum. Pl. Jap. 2: 560 (1878).
Japan (Honshu, Shikoku). 38 JAP. Hemicr. or rhizome geophyte.

Carex pachyneura Kitag., Rep. Inst. Sci. Res. Manchoukuo 4: 78 (1940).
Inner Mongolia to NE. China. 36 CHI CHM. Hemicr. or rhizome geophyte.

Carex pachystylis J.Gay, Ann. Sci. Nat., Bot., II, 10: 301 (1838). *Carex stenophylla* f. *pachystylis* (J.Gay) Asch. & Graebn., Syn. Mitteleur. Fl. 2(2): 25 (1902). *Vignea pachystylis* (J.Gay) Soják, Cas. Nár. Mus., Odd. Prír. 148: 196 (1979 publ. 1980).
E. Medit. to C. Asia. 20 EGY 32 KAZ KGZ TKM TZK UZB 33 TCS 34 AFG IRN IRQ PAL SIN TUR. Hemicr. or rhizome geophyte.
Carex stenophylla var. *planifolia* Boiss., Fl. Orient. 5: 400 (1882).
Carex desertorum Litv., Trudy Bot. Muz. Imp. Akad. Nauk 1: 19 (1902). *Carex stenophylla* var. *desertorum* (Litv.) Litv., Trudy Imp. S.-Peterburgsk. Bot. Sada 21: 19 (1902).

Carex paeninsulae Naczi, E.L.Bridges & Orzell, Novon 12: 514 (2002).
Florida. 78 FLA. Hemicr. or rhizome geophyte.
Carex grisea var. *angustifolia* Boott, Ill. Carex 1: 34 (1858).

Carex pairae F.W.Schultz, Flora 51: 303 (1868). *Carex divulsa* var. *pairae* (F.W.Schultz) Nyman, Consp. Fl. Eur.: 781 (1882). *Carex muricata* subsp. *pairae* (F.W.Schultz) Asch. & Graebn., Syn. Mitteleur. Fl. 2(2): 40 (1902). *Carex continua* subsp. *pairae* (F.W.Schultz) Degen in H.G.A.Engler (ed.), Pflanzenr., IV, 20(38): 282 (1909).
Macaronesia, S. & W. Europe. 11 BGM CZE 12 FRA SPA 13 GRC ITA YUG 20 ALG MOR 21 AZO CNY MDR (72) nbr ont (75) mic pen. Hemicr. or rhizome geophyte.
Carex muricata subsp. *lamprocarpa* Èelak., Anal. Kvet. Ceska: 88 (1879). *Vignea muricata* subsp. *lamprocarpa* (Èelak.) Soják, Cas. Nár. Mus., Odd. Prír. 148: 196 (1979 publ. 1980).
Carex pairae var. *capitata* H.Christ, Bull. Herb. Boissier, II, 2: 780 (1902). *Carex echinata* f. *capitata* (H.Christ) Kük. in H.G.A.Engler (ed.), Pflanzenr., IV, 20(38): 161 (1909).
Carex pairae var. *monostachya* Asch. & Graebn., Syn. Mitteleur. Fl. 2(2): 40 (1902). *Carex echinata* f. *monostachya* (Asch. & Graebn.) Kük. in H.G.A. Engler (ed.), Pflanzenr., IV, 20(38): 161 (1909).
Carex bullockiana Nelmes, Bot. Mater. Gerb. Bot. Inst. Komarova Akad. Nauk S.S.S.R. 19: 77 (1959).

Carex paishanensis Nakai, Bot. Mag. (Tokyo) 28: 301 (1914). *Carex bigelowii* subsp. *paishanensis* (Nakai) Vorosch. in A.K.Skvortsov (ed.), Florist. Issl. Razn. Raionakh SSSR: 153 (1985).
Russian Far East, Korea. 31+ 38 KOR. Hemicr. or rhizome geophyte.

Carex palawanensis Kük., Leafl. Philipp. Bot. 4: 1169 (1911).
N. Borneo, Philippines (Palawan). 42 BOR PHI. Hemicr. or rhizome geophyte.

Carex paleacea Schreb. ex Wahlenb., Kongl. Vetensk. Acad. Nya Handl. 24: 164 (1803). *Carex crinita* var. *paleacea* (Schreb. ex Wahlenb.) Dewey, Amer. J. Sci. Arts 10: 270 (1826).
N. & NE. Europe, C. & E. Canada to NE. U.S.A. 10 FIN NOR SWE 14 RUN 70 NUN 71 MAN 72 LAB NBR NFL NSC ONT PEI QUE 75 MAI MAS NWH. Hemicr. or rhizome geophyte.
Carex maritima O.F.Müll. in G.C.Oeder & al. (eds.), Fl. Dan.: t. 703 (1777), nom. illeg.
Carex maritima var. *brunnescens* F.Nyl., Spic. Fl. Fenn. 2: 20 (1844).
Carex maritima var. *erectiuscula* Fernald, Rhodora 2: 170 (1900). *Carex paleacea* f. *erectiuscula* (Fernald) Fernald, Rhodora 44: 293 (1942).
Carex paleacea var. *transatlantica* Fernald, Rhodora 35: 397 (1933).
Carex paralia V.I.Krecz. in V.L.Komarov (ed.), Fl. URSS 3: 237 (1935).
Carex neopaleacea Lepage, Naturaliste Canad. 83: 137 (1956). *Carex paleacea* f. *neopaleacea* (Lepage) B.Boivin, Provancheria 25: 182 (1992).
Carex sublimosa Lepage, Naturaliste Canad. 83: 139 (1956).

Carex gardneri Lepage, Naturaliste Canad. 84: 42 (1957).
Carex paleaceoides Lepage, Naturaliste Canad. 94: 273 (1967). *Carex paleacea* var. *lepagei* B.Boivin, Provancheria 25: 182 (1992).
Carex gauthieri Lepage, Naturaliste Canad. 103: 387 (1976).

Carex pallescens L., Sp. Pl.: 977 (1753). *Carex pallida* Salisb., Prodr. Stirp. Chap. Allerton: 29 (1796), nom. illeg. *Trasus pallescens* (L.) Gray, Nat. Arr. Brit. Pl. 2: 69 (1821).
Europe to C. Asia, E. Canada to NC. & NE. U.S.A. 10 DEN FIN GRB ICE IRE NOR SWE 11 AUT BGM CZE GER HUN NET POL SWI 12 COR FRA SPA 13 ALB BUL GRC ITA ROM SIC TUE YUG 14 BLR BLT KRY RUC RUE RUN RUS RUW UKR 30+ 32 KAZ 33 NCS TCS 34 IRN LBS TUR 72 LAB NBR NFL NSC ONT QUE 74 MIN 75 CNT MAI MAS MIC NWH NWJ NWY OHI PEN RHO VER. Hemicr. or rhizome geophyte.

var. ***chalcodeta*** (V.I.Krecz.) Ö.Nilsson in P.H.Davis (ed.), Fl. Turkey 9: 134 (1985).
Turkey to Caucasus. 33 NCS TCS 34 TUR. Hemicr. or rhizome geophyte.
**Carex chalcodeta* V.I.Krecz. in V.L.Komarov (ed.), Fl. URSS 3: 604 (1935).

var. ***pallescens***
Europe to C. Asia, E. Canada to NC. & NE. U.S.A. 10 DEN FIN GRB ICE IRE NOR SWE 11 AUT BGM CZE GER HUN NET POL SWI 12 COR FRA SPA 13 ALB BUL GRC ITA ROM SIC TUE YUG 14 BLR BLT KRY RUC RUE RUN RUS RUW UKR 30+ 32 KAZ 33 TCS 34 IRN LBS TUR 72 LAB NBR NFL NSC ONT QUE 74 MIN 75 CNT MAI MAS MIC NWH NWJ NWY OHI PEN RHO VER. Hemicr. or rhizome geophyte.
Carex undulata Kunze in C.Schkuhr, Beschr. Riedgräs., Suppl.: 23 (1840).
Carex pallescens f. *cylindrica* Peterm., Anal. Pfl.-Schlüss.: 506 (1846).
Carex pallescens f. *glaberrima* K.Koch, Linnaea 21: 612 (1848).
Carex leucantha Schur, Enum. Pl. Transsilv.: 714 (1866). *Carex pallescens* var. *leucantha* (Schur) Asch. & Graebn., Syn. Mitteleur. Fl. 1: 146 (1896).
Carex pallescens f. *alpestris* Schur, Enum. Pl. Transsilv.: 714 (1866).
Carex undulata Schur, Enum. Pl. Transsilv.: 714 (1866), nom. illeg.
Carex pallescens var. *alpestris* Kohts, Oesterr. Bot. Z. 20: 166 (1870).
Carex pallescens var. *subglabra* Beck, Fl. Nieder-Österreich: 140 (1890).
Carex microstoma Franch., Bull. Soc. Philom. Paris, VIII, 7: 48 (1895).
Carex pallescens f. *elatior* Asch. & Graebn., Syn. Mitteleur. Fl. 1: 145 (1896).
Carex tymphaea Formánek, Verh. Naturf. Vereins Brünn 34: 279 (1896).
Carex pallescens f. *brevibracteata* Neuman, Sver. Fl.: 702 (1901).
Carex pallescens f. *pygmaea* Lackow., Verh. Bot. Vereins Prov. Brandenburg 44: 165 (1902).
Carex pallescens var. *luxuriosa* Kük. in H.G.A.Engler (ed.), Pflanzenr., IV, 20(38): 433 (1909).
Carex pallescens var. *subsilvatica* Kük. in H.G.A.Engler (ed.), Pflanzenr., IV, 20(38): 433 (1909).
Carex pallescens var. *neogaea* Fernald, Rhodora 44: 306 (1942).

Carex pallidula Harmaja, Ann. Bot. Fenn. 42: 221 (2005).
N. & NE. Europe. 10 FIN SWE 14 RUW. Hemicr. or rhizome geophyte.
**Carex digitata* var. *pallens* Fristedt, Vaxtgeogr. Skildr. Sodrav Angermanl.: 36 (1857). *Carex pallens* (Fristedt) Harmaja, Ann. Bot. Fenn. 23: 148 (1986), nom. illeg. *Carex digitata* subsp. *pallens* (Fristedt) Tzvelev, Novosti Sist. Vyssh. Rast. 32: 183 (2000).

Carex* ×*paludivagans D.G.Drury, Rhodora 58: 66 (1956). *C. rostrata* × *C. rotundifolia*.
Alaska. 70 ASK. Hemicr. or rhizome geophyte.

Carex pamirica (O.Fedtsch.) O.Fedtsch. & B.Fedtsch., Bot. Zhurn. (St. Petersburg) 1(6): 191 (1907).
Afghanistan to W. Himalaya and Mongolia. 14+ 30 IRK 32 KAZ KGZ TZK UZB 33 NCS 34 AFG 36 CHC CHI 37 MON 40 PAK WHM. Hemicr. or rhizome geophyte.
**Carex vesicaria* var. *pamirica* O.Fedtsch., Fl. Pamir: 200 (1903). *Carex obscuriceps* var. *pamirica* (O.Fedtsch.) Kük. in H.G.A.Engler (ed.), Pflanzenr., IV, 20(38): 724 (1909).

subsp. ***dichroa*** (Malyschev) T.V.Egorova, Sedges Russia & Adj. States: 183 (1999).
E. Europe to Mongolia. 14+ 30 IRK 32 KAZ TZK 33 NCS 36 CHI 37 MON. Hemicr.
Carex pulla subsp. *dichroa* Freyn, Oesterr. Bot. Z. 40: 304 (1890), nom. illeg. *Carex dichroa* V.I. Krecz. in V.L.Komarov (ed.), Fl. URSS 3: 447 (1935), nom. illeg. **Carex pamirensis* subsp. *dichroa* Malyschev, in Fl. Sibir., Cyperac. 3: 95 (1990).
Carex kryloviana Schischk. & Serg., Sist. Zametki Mater. Gerb. Krylova Tomsk. Gosud. Univ. Kuybysheva 6: 1 (1928).

subsp. ***pamirica***
Afghanistan to W. Himalaya and C. Asia. 32 KGZ TZK UZB 34 AFG 36 CHC 40 PAK WHM. Hemicr. or rhizome geophyte.
Carex lurida C.B.Clarke in J.D.Hooker, Fl. Brit. India 6: 742 (1894), nom. illeg. *Carex obscuriceps* Kük. in H.G.A.Engler (ed.), Pflanzenr., IV, 20(38): 723 (1909).
Carex pamirensis C.B.Clarke, Kew Bull., Addit. Ser. 8: 87 (1908). *Carex dichroa* subsp. *pamirensis* (C.B.Clarke) T.V.Egorova, Novosti Sist. Vyssh. Rast. 10: 103 (1973).
Carex pamirensis var. *angustispicata* Y.C.Yang, Acta Biol. Plateau Sin. 3: 89 (1984).

Carex pandanophylla C.B.Clarke in J.D.Hooker, Fl. Brit. India 6: 714 (1894).
Myanmar. 41 MYA. Hemicr. or rhizome geophyte.

Carex panicea L., Sp. Pl.: 977 (1753). *Trasus paniceus* (L.) Gray, Nat. Arr. Brit. Pl. 2: 63 (1821).
Greenland, Azores, Europe to C. Asia. 10 DEN FIN FOR GRB ICE IRE NOR SWE 11 AUT BGM CZE GER HUN NET POL SWI 12 COR FRA POR SPA 13 ALB BUL GRC ITA ROM YUG 14 BLR BLT RUC RUE RUN RUS RUW UKR 20 MOR 21 AZO 30 WSB 32 KAZ KGZ TZK UZB 33 NCS TCS 34 IRN IRQ TUR 70 GNL (72) lab

nbr nfl nsc ont (75) cnt mai mas nwh nwy rho. Hemicr. or rhizome geophyte.
Carex panicea subsp. *dalmatica* Degen & Lengyel in ?
Carex panicea f. *tumidula* Laest., Loca Parall.: 283 (1831).
Carex algida Turcz. ex V.I.Krecz., Bull. Soc. Imp. Naturalistes Moscou 1838: 104 (1838).
Carex panicea f. *microcarpa* Sond. in W.D.J.Koch, Syn. Fl. Germ. Helv., ed. 2: 897 (1844).
Carex panicea f. *refracta* Peterm., Anal. Pfl.-Schlüss.: 507 (1846).
Carex panicea var. *sublivida* Hartm. ex Andersson, Cyper. Scand.: 33 (1849).
Carex pelia O.Lang, Linnaea 24: 575 (1851). *Carex panicea* subsp. *pelia* (O.Lang) Nyman, Consp. Fl. Eur.: 775 (1882). *Carex panicea* var. *pelia* (O.Lang) K.Richt., Pl. Eur. 1: 159 (1890).
Carex efflata Laest. ex Ledeb., Fl. Ross. 4: 291 (1852).
Carex panicea f. *pallida* A.Blytt, Norges Fl. 1: 235 (1861).
Carex panicea f. *nigra* Kuntze, Taschen-Fl. Leipzig: 19 (1867).
Carex panicea f. *gracilis* Lange, Haandb. Danske Fl., ed. 4: 139 (1886).
Carex panicea f. *robusta* Warnst., Verh. Bot. Vereins Prov. Brandenburg 38: 60 (1896).
Carex panicea f. *latifolia* Waisb., Oesterr. Bot. Z. 47: 433 (1897).
Carex panicea f. *melanocarpa* Waisb., Oesterr. Bot. Z. 47: 433 (1897).
Carex panicea var. *praestabilis* Waisb., Oesterr. Bot. Z. 47: 433 (1897).
Carex panicea f. *humilis* Meinsh., Trudy Imp. S.-Peterburgsk. Bot. Sada 18: 390 (1900).
Carex panicea var. *intermedia* Meinsh., Trudy Imp. S.-Peterburgsk. Bot. Sada 18: 390 (1900).
Carex panicea f. *conferta* N.H.Nilsson ex Neuman, Sver. Fl.: 701 (1901).
Carex panicea f. *ferruginea* Neuman, Sver. Fl.: 701 (1901).
Carex panicea f. *longipedunculata* Asch. & Graebn., Syn. Mitteleur. Fl. 2: 142 (1902).
Carex panicea f. *binata* Lackow. ex Kük. in H.G.A. Engler (ed.), Pflanzenr., IV, 20(38): 511 (1909).

Carex paniculata L., Cent. Pl. I: 31 (1755). *Vignea paniculata* (L.) Rchb. in J.C.Mössler, Handb. Gewächsk. ed. 2, 3: 1622 (1830). *Physiglochis paniculata* (L.) Raf., Good Book: 27 (1840). *Caricina paniculata* (L.) St.-Lag. in A.Cariot, Étude Fl., ed. 8, 2: 876 (1889). *Rhynchopera paniculata* (L.) Fedde & J.Schust., Just's Bot. Jahresber. 41(2): 11 (1913 publ. 1918).
Macaronesia, Europe to Caucasus. 10 DEN FIN GRB IRE NOR SWE 11 AUT BGM CZE GER HUN NET POL SWI 12 FRA POR SPA 13 ALB BUL GRC ITA ROM SIC YUG 14 BLR BLT KRY RUC RUN RUW UKR 20 MOR 21 CNY CVI 33 NCS TCS 34 TUR (51) nzn. Hemicr. or rhizome geophyte.

subsp. ***calderae*** (A.Hansen) Lewej. & Lobin, Senckenberg. Biol. 67: 438 (1987).
Canary Is. (Tenerife). 21 CNY. Hemicr. or rhizome geophyte.
Carex calderae A.Hansen, Cuad. Bot. Canaria 14–15: 61 (1972).

subsp. ***hansenii*** Lewej. & Lobin, Senckenberg. Biol. 67: 440 (1987).
Cape Verde. 21 CVI. Hemicr. or rhizome geophyte.

subsp. ***lusitanica*** (Willd.) Maire, Bull. Soc. Hist. Nat. Afrique N. 20: 205 (1929).
W. Medit. 12 FRA? POR SPA 20 MOR. Hemicr. or rhizome geophyte.
Carex lusitanica Willd., Sp. Pl. 4: 245 (1805). *Loxotrema lusitanica* (Willd.) Raf., Good Book: 25 (1840). *Carex paniculata* var. *lusitanica* (Willd.) Nyman, Consp. Fl. Eur.: 781 (1882). *Vignea paniculata* subsp. *lusitanica* (Willd.) Soják, Cas. Nár. Mus., Odd. Prír. 148: 196 (1979 publ. 1980).

subsp. ***paniculata***
Europe to Caucasus. 10 DEN FIN GRB IRE NOR SWE 11 AUT BGM CZE GER HUN NET POL SWI 12 FRA POR SPA 13 ALB BUL GRC ITA ROM SIC YUG 14 BLR BLT RUC RUN RUW UKR 20 MOR 33 NCS TCS 34 TUR (51) nzn. Hemicr. or rhizome geophyte.
Carex punctulata Leers, Fl. Herborn.: 713 (1775).
Carex diandra Roth, Tent. Fl. Germ. 2(2): 437 (1793), nom. illeg.
Carex paniculata f. *simplex* Peterm., Anal. Pfl.-Schlüss.: 490 (1846).
Carex rotae De Not., Index Seminum (G) 1848: 24 (1848). *Carex paniculata* var. *rotae* (De Not.) Nyman, Consp. Fl. Eur.: 781 (1882). *Vignea rotae* (De Not.) Soják, Cas. Nár. Mus., Odd. Prír. 148: 197 (1979 publ. 1980).
Carex paniculata var. *simplicior* Andersson, Cyper. Scand.: 67 (1849).
Carex ×*boenninghauseniana* var. *subpaniculata* Lange, Haandb. Danske Fl., ed. 3: 683 (1864).
Carex paniculata f. *elongata* Èelak., Prodr. Fl. Böhmen: 60 (1867).
Carex paniculata var. *squarrosa* Èelak., Prodr. Fl. Böhmen: 60 (1867).
Carex paniculata f. *pallida* Lange in G.C.Oeder & al. (eds.), Fl. Dan.: t. 2793 (1869).
Carex paniculata var. *tenuior* Grantzow, Fl. Uckermarck: 309 (1880).
Carex paniculata f. *longibracteata* Zapal., Spraw. Komis. Fizjogr.: 24 (1889).
Carex paniculata var. *rigida* A.Blytt ex Benner, J. Bot. 32: 365 (1894).
Carex paniculata f. *brevis* Asch. & Graebn., Syn. Mitteleur. Fl. 2(2): 46 (1902).
Carex paniculata f. *gracilis* Asch. & Graebn., Syn. Mitteleur. Fl. 2(2): 46 (1902).
Carex paniculata f. *rigidior* Kük. in H.G.A.Engler (ed.), Pflanzenr., IV, 20(38): 182 (1909).

subsp. ***szovitsii*** (V.I.Krecz.) Ö.Nilsson in P.H.Davis (ed.), Fl. Turkey 9: 91 (1985).
Krym, NE. Turkey, W. Caucasus. 14 KRY 33 NCS TCS 34 TUR. Hemicr. or rhizome geophyte.
Carex favratii H.Christ, Ber. Deutsch. Bot. Ges. 8: 173 (1890). *Carex paniculata* var. *favratii* (H.Christ) Kük. in H.G.A.Engler (ed.), Pflanzenr., IV, 20(38): 182 (1909).
Carex szovitsii V.I.Krecz. in V.L.Komarov (ed.), Fl. URSS 3: 593 (1935).

Carex panormitana Guss., Fl. Sicul. Syn. 2: 575 (1844). *Carex acuta* var. *panormitana* (Guss.) Nyman, Consp. Fl. Eur.: 777 (1882). *Carex cespitosa* var. *panormitana* (Guss.) Fiori, Nuov. Fl. Italia 1: 188 (1923).

Sardegna, Sicilia. 12 SAR 13 SIC. Hemicr. or rhizome geophyte.

Carex pansa L.H.Bailey, Bot. Gaz. 13: 82 (1888). *Carex arenicola* subsp. *pansa* (L.H.Bailey) T.Koyama & Calder, Canad. J. Bot. 43: 1389 (1965).
NE. China to Japan, W. Canada to W. U.S.A. 31 KUR PRM SAK 36 CHM 38 JAP KOR 71 BRC 73 ORE WAS 76 CAL. Hemicr. or rhizome geophyte.
Carex arenicola F.Schmidt, Reis. Amur-Land., Bot.: 195 (1868), nom. illeg.
Carex curaica Franch. & Sav., Enum. Pl. Jap. 2: 252 (1877), nom. illeg.
Carex chaetorhiza Franch. & Sav., Enum. Pl. Jap. 2: 552 (1878).
Carex chaetorrhiza var. *stenostachys* Franch. & Sav., Enum. Pl. Jap. 2: 552 (1878).
Carex yedoensis Boeck., Bot. Jahrb. Syst. 5: 515 (1884).

Carex papillosissima Nelmes, Bull. Misc. Inform. Kew 1939: 158 (1939).
Tanzania (Sao Hill). 25 TAN. Hemicr.

Carex papulosa Boott, Mem. Amer. Acad. Arts, n.s., 6: 418 (1859).
Russian Far East, Korea, Japan. 31 PRM 37 MON 38 JAP KOR. Hemicr. or rhizome geophyte.
Carex flectens Boott, Ill. Carex 4: 171 (1867).
Carex grandisquama Franch., Bull. Soc. Philom. Paris, VIII, 7: 50 (1895).
Carex papulosa f. *laxiflora* Kük. in H.G.A.Engler (ed.), Pflanzenr., IV, 20(38): 638 (1909).
Carex papulosa f. *paupercula* Kük. in H.G.A.Engler (ed.), Pflanzenr., IV, 20(38): 638 (1909).
Carex sekimotoi Honda, Bot. Mag. (Tokyo) 43: 543 (1929).

Carex paracuraica F.T.Wang & Y.L.Chang ex S.Yun Liang, in Fl. Reipubl. Popul. Sin. 12: 527 (2000).
China (Shaanxi). 36 CHN. Hemicr.

Carex parallela (Laest.) Sommerf., Suppl. Fl. Lapp.: 39 (1826).
Greenland, N. Europe to Siberia. 10 FIN NOR SVA SWE 14 RUE RUN 30 ALT IRK KRA WSC YAK 32 KAZ 70 GNL. Hemicr. or rhizome geophyte.
**Carex dioica* var. *parallela* Laest., Kongl. Vetensk. Acad. Handl. 1822: 338 (1822). *Vignea parallela* (Laest.) Soják, Cas. Nár. Mus., Odd. Prír. 148: 196 (1979 publ. 1980).

subsp. ***parallela***
Greenland, N. Europe to Siberia. 10 FIN NOR SVA SWE 14 RUN 30 WSB 70 GNL. Hemicr. or rhizome geophyte.
Carex parallela f. *androgyna* O.Lang, Linnaea 24: 497 (1851).
Carex parallela var. *pauciflora* O.Lang, Linnaea 24: 497 (1851).
Carex parallela var. *subcaespitosa* O.Lang, Linnaea 24: 496 (1851).
Carex dioica var. *subtriangularis* Boeck., Linnaea 39: 18 (1875).

subsp. ***redowskiana*** (C.A.Mey.) T.V.Egorova, Novosti Sist. Vyssh. Rast. 10: 104 (1973).
NE. Europe to Siberia. 14 RUE RUN 30 ALT IRK KRA WSB YAK 32 KAZ. Hemicr. or rhizome geophyte.
**Carex redowskiana* C.A.Mey., Mém. Sav. Étr. Acad. St. Pétersbourg 1: 207 (1831). *Carex dioica* var. *redowskiana* (C.A.Mey.) Ostenf., Fl. Arct.: 60 (1902). *Vignea redowskiana* (C.A.Mey.) Soják, Cas. Nár. Mus., Odd. Prír. 148: 197 (1979 publ. 1980).
Carex davalliana var. *pallida* H.Christ ex Scheutz, Kongl. Svenska Vetensk. Acad. Handl. 22: 173 (1887).

Carex parryana Dewey, Amer. J. Sci. Arts 27: 239 (1835).
W. Subarctic America to WC. U.S.A. 70 ASK YUK 71 ABT BRC MAN SAS 72 ONT 73 COL IDA MNT WYO 76 NEV UTA. Hemicr. or rhizome geophyte.

var. ***brevisquama*** F.J.Herm., Rhodora 70: 420 (1968).
WC. U.S.A. 73 IDA 76 UTA. Hemicr. or rhizome geophyte.

var. ***parryana***
W. Subarctic America to WC. U.S.A. 70 ASK YUK 71 ABT BRC MAN SAS 72 ONT 73 COL IDA MNT WYO 76 NEV UTA. Hemicr. or rhizome geophyte.
Carex arctica Dewey, Amer. J. Sci. Arts 27: 239 (1835).
Carex parryana var. *statoni* M.E.Jones, Bull. Montana State Univ., Biol. Ser. 15: 72 (1910).

Carex parva Nees in R.Wight, Contr. Bot. India: 120 (1834).
Afghanistan to C. China. 32 KAZ KGZ TKM UZB 34 AFG 36 CHC CHN CHQ CHT 37 MON 40 EHM NEP PAK WHM. Hemicr. or rhizome geophyte.
Carex macrorhyncha Kar. & Kir., Bull. Soc. Imp. Naturalistes Moscou 15: 521 (1842).
Kobresia lolonum Hand.-Mazz., Anz. Akad. Wiss. Wien, Math.-Naturwiss. Kl. 57: 389 (1920).
Carex unifoliata Kük. & Hand.-Mazz. in H.F.von Handel-Mazzetti, Symb. Sin. 7: 1259 (1936).

Carex parviflora Host, Icon. Descr. Gram. Austriac. 1: 64 (1801). *Edritria parviflora* (Host) Raf., Good Book: 26 (1840). *Carex alpina* var. *parviflora* (Host) Boott, Ill. Carex 1: 113 (1858). *Carex atrata* var. *parviflora* (Host) Beck, Fl. Nieder-Österreich: 137 (1890).
WC. & S. Europe. 11 AUT CZE GER POL SWI 12 FRA SPA 13 BUL ITA ROM. Hemicr. or rhizome geophyte.
Carex nigra All., Fl. Pedem. 2: 267 (1785), nom. illeg.
Carex atrata var. *nigra* Gaudin, Fl. Helv. 6: 66 (1830). *Carex atrata* subsp. *nigra* (Gaudin) Hartm., Sv. Norsk Exc.-Fl.: 131 (1846).
Carex binata Poir. in J.B.A.M.de Lamarck, Encycl., Suppl. 3: 257 (1813).
Carex atrata var. *conglomerata* Neilr., Fl. Nied.-Oesterr.: 107 (1857).
Carex pirinensis Acht., Gen. Carex Bulgar.: 65 (1957).

Carex parvigluma C.B.Clarke in J.D.Hooker, Fl. Brit. India 6: 716 (1894).
Assam. 40 ASS. Hemicr. or rhizome geophyte.

Carex patagonica Speg., Revista Fac. Agron. Univ. Nac. La Plata 1897: 626 (1897).
WC. & SW. Argentina, C. & S. Chile. 85 AGS AGW CLC CLS. Hemicr. or rhizome geophyte.

Carex ×patuensis Lepage, Naturaliste Canad. 89: 113 (1962). *C. atratiformis* × *C. saxatilis* var. *militaris*.
E. Canada. 72 QUE. Hemicr. or rhizome geophyte.

Carex pauciflora Lightf., Fl. Scot. 2: 543 (1777). *Trasus pauciflorus* (Lightf.) Gray, Nat. Arr. Brit. Pl. 2: 56 (1821) (1821). *Leucoglochin pauciflorus* (Lightf.) Heuff., Flora 27: 528 (1844). *Psyllophora pauciflora* (Lightf.) Schur, Verh. Mitth. Siebenbürg. Vereins Naturwiss. Hermannstadt 10: 201 (1859). *Caricinella pauciflora* (Lightf.) St.-Lag. in A.Cariot, Étude Fl., ed. 8, 2: 881 (1889).
Temp. Northern Hemisphere. 10 DEN FIN GRB IRE NOR SWE 11 AUT BGM CZE GER POL SWI 12 FRA 13 ITA ROM YUG 14 BLR BLT RUC RUE RUN RUS RUW UKR 30 WSB 31 KAM KUR SAK 32 KAZ 38 JAP KOR 70 ASK YUK 71 ABT BRC MAN 72 LAB NBR NFL NSC ONT QUE 73 WAS 74 MIN WIS 75 CNT MAI MAS MIC NWH NWY PEN VER WVA. Hemicr. or rhizome geophyte.
Carex patula Huds., Fl. Angl., ed. 2, 2: 402 (1778), nom. illeg. *Anithista patula* Raf., Good Book: 26 (1840).
Carex leucoglochin L.f., Suppl. Pl.: 413 (1782).
Carex phyllophora Kunth, Enum. Pl. 2: 424 (1837).
Carex pauciflora f. *elatior* Kük. & J.Schust., Mitt. Bayer. Bot. Ges. 1905: 449 (1905).

Carex paucimascula H.Lév. & Vaniot, Bull. Soc. Bot. France 53: 315 (1906).
China. 36+. Hemicr. or rhizome geophyte.

Carex paulo-vargasii Luceño & Marín, Anales Jard. Bot. Madrid 59: 348 (2001 publ. 2002).
Morocco. 20 MOR. Hemicr. or rhizome geophyte.

Carex paupera Nelmes, Proc. Linn. Soc. London 1944: 155, 282 (1944).
Victoria. 50 VIC. Hemicr. or rhizome geophyte.

Carex paxii Kük. in H.G.A.Engler (ed.), Pflanzenr., IV, 20(38): 765 (1909).
SE. China, Korea, C. Japan. 36 CHS 38 JAP KOR. Hemicr. or rhizome geophyte.
Carex succedanea Nakai, Repert. Spec. Nov. Regni Veg. 13: 244 (1914).
Carex kengii Kük., Repert. Spec. Nov. Regni Veg. 27: 108 (1929).

Carex paysonis Clokey, Amer. J. Sci., V, 3: 89 (1922). *Carex podocarpa* var. *paysonis* (Clokey) B.Boivin, Phytologia 43: 88 (1979).
SW. Canada to NW. U.S.A. 71 ABR BRC 73 IDA MNT ORE WYO 76 UTA. Hemicr. or rhizome geophyte.

Carex peckii Howe, Rep. (Annual) New York State Mus. Nat. Hist. 47: 166 (1893 publ. 1894).
Alaska to N. U.S.A. 70 ASK YUK 71 ABT BRC MAN SAS 72 NBR NSC ONT QUE 73 WYO 74 IOW MIN NDA NEB SDA WIS 75 MAI MAS MIC NWH NWY VER. Hemicr. or rhizome geophyte.
Carex emmonsii var. *ellpitica* Boott, Ill. Carex 2: 97 (1860). *Carex nigromarginata* var. *elliptica* (Boott) Gleason, Phytologia 4: 22 (1952).
Carex clivicola Fernald & Weath., Rhodora 33: 233 (1931).
Carex peckii f. *elongata* Lepage, Naturaliste Canad. 78: 346 (1951).

Carex pediformis C.A.Mey., Mém. Sav. Étr. Acad. St. Pétersbourg 1: 219 (1831). *Carex pediformis* var. *genuina* Maxim., Prim. Fl. Amur.: 309 (1859), nom. inval. *Bitteria pediformis* (C.A.Mey.) Fedde & J.Schust., Just's Bot. Jahresber. 41(2): 7 (1913 publ. 1918).
Europe to Russian Far East. 10 FIN NOR SWE 11 CZE POL 13 ROM 14 BLR BLT RUC RUE RUN RUS RUW UKR 30 ALT WSB YAK 31 AMU KHA PRM 32 KAZ KGZ 36 CHI CHM CHN CHX 37 MON. Hemicr. or rhizome geophyte.

subsp. ***pediformis***
European Russia to Russian Far East. 14 RUC RUE RUN RUS UKR 30 ALT WSB YAK 31 AMU KHA PRM 32 KAZ 36 CHI CHM CHN CHX 37 MON. Hemicr. or rhizome geophyte.
Carex obliqua Turcz., Bull. Soc. Imp. Naturalistes Moscou 1838: 104 (1838).
Carex obovata Goldb. ex Ledeb., Fl. Ross. 4: 290 (1852).
Carex kirilovii Turcz., Bull. Soc. Imp. Naturalistes Moscou 28(1): 340 (1855). *Carex macroura* subsp. *kirilovii* (Turcz.) Malyschev, in Fl. Sibir., Cyperac. 3: 123 (1990).
Carex pediformis var. *pedunculata* Maxim., Prim. Fl. Amur.: 309 (1859).
Carex pellucida Turcz. ex Boott, Ill. Carex 4: 196 (1867).
Carex chamissoi Boeck., Linnaea 41: 145 (1877).
Carex pediformis subsp. *pododactyla* Norman, Fl. Arct. Norv. Sp.: 50 (1893).
Carex macroura Meinsh., Trudy Imp. S.-Peterburgsk. Bot. Sada 18: 404 (1900). *Carex pediformis* var. *macroura* (Meinsh.) Kük., Finska Vetensk. Soc. Forh. 45(8): 10 (1903). *Carex pediformis* subsp. *macroura* (Meinsh.) Podp., Spisy Prír. Fak. Masarykovy Univ. 101: 21 (1982).
Carex rhizodes var. *abbreviata* Meinsh., Trudy Imp. S.-Peterburgsk. Bot. Sada 18: 404 (1900).
Carex pediformis f. *leopolitana* Kük. in H.G.A.Engler (ed.), Pflanzenr., IV, 20(38): 491 (1909).
Carex sutchanensis Kom., Izv. Imp. Bot. Sada Petra Velikago 16: 155 (1916).
Carex aneurocarpa V.I.Krecz. in V.L.Komarov (ed.), Fl. URSS 3: 614 (1935).
Carex hankaensis Kitag., J. Jap. Bot. 17: 236 (1941).
Carex supermascula V.I.Krecz., Bot. Mater. Gerb. Bot. Inst. Komarova Akad. Nauk S.S.S.R. 9: 188 (1946).
Carex ayouensis X.Y.Mao & Y.C.Yang, in Fl. Intramongolica 8: 347 (1985).

subsp. ***rhizodes*** (Blytt ex Boott) H.Lindb., Exsicc. (Sched. Pl. Finl.) 1–8: 44 (1907).
Europe. 10 FIN NOR SWE 11 CZE POL 13 ROM 14 BLR BLT RUC RUE RUN RUS RUW UKR. Hemicr. or rhizome geophyte.
Carex rhizina Blytt ex Lindblom, Bot. Not. 1839(8): 98 (1839). *Carex pediformis* var. *rhizina* (Blytt ex Lindblom) Kük. in H.G.A.Engler (ed.), Pflanzenr., IV, 20(38): 491 (1909). *Carex digitata* var. *rhizina* (Blytt ex Lindblom) Fiori, Nuov. Fl. Italia 1: 190 (1923).
**Carex rhizodes* Blytt ex Boott, Ill. Carex 4: 196 (1867).

Carex pedunculata Muhl. ex Willd., Sp. Pl. 4: 222 (1805).
Canada to NC. & E. U.S.A. 71 ABT BRC MAN SAS 72 LAB NBR NFL NSC ONT QUE 74 ILL IOW MIN SDA WIS 75 CNT INI MAI MAS MIC NWH NWJ NWY OHI PEN RHO VER WVA 78 ALA DEL GEO KTY MRY NCA TEN VRG. Hemicr. or rhizome geophyte.

Carex peiktusanii Kom., Trudy Imp. S.-Peterburgsk. Bot. Sada 18: 445 (1900). *Carex hancockiana* var. *peiktusanii* (Kom.) Kük. in H.G.A.Engler (ed.), Pflanzenr., IV, 20(38): 395 (1909).
Russian Far East, China, Korea, C. Japan. 31+ 36 CHM CHN 38 JAP KOR. Hemicr. or rhizome geophyte.

Carex peliosanthifolia F.T.Wang & Tang ex P.C.Li, Acta Phytotax. Sin. 37: 162 (1999).
China (Guangxi). 36 CHC. Hemicr. or rhizome geophyte.

Carex pellita Muhl. ex Willd., Sp. Pl. 4: 302 (1805). *Diemisa pellita* (Muhl. ex Willd.) Raf., Good Book: 27 (1840).
N. America. 70 ASK YUK 71 ABT BRC MAN SAS 72 NBR NFL NSC ONT QUE 73 COL IDA MNT ORE WAS WYO 74 ILL IOW KAN MIN MSO NDA NEB OKL SDA WIS 75 CNT INI MAI MAS MIC NWH NWJ NWY OHI PEN RHO VER WVA 76 ARI CAL NEV UTA 77 NWM TEX 78 ARK DEL KTY MRY VRG 79 MXE. Hemicr. or rhizome geophyte.

Carex pelocarpa F.J.Herm., Rhodora 39: 492 (1937). *Carex nova* var. *pelocarpa* (F.J.Herm.) Dorn, Vasc. Pl. Wyoming: 296 (1988).
WC. U.S.A. 73 COL IDA MNT ORE WYO 76 NEV UTA. Hemicr. or rhizome geophyte.

Carex pendula Huds., Fl. Angl.: 352 (1762). *Trasus pendulus* (Huds.) Gray, Nat. Arr. Brit. Pl. 2: 65 (1821). *Manochlaenia pendula* (Huds.) Fedde & J.Schust., Just's Bot. Jahresber. 41(2): 11 (1913 publ. 1918).
Macaronesia, Europe, Medit. to Afghanistan. 10 DEN GRB IRE SWE 11 AUT BGM CZE GER HUN NET POL SWI 12 COR FRA POR SAR SPA 13 ALB BUL GRC ITA KRI ROM SIC TUE YUG 14 KRY UKR 20 ALG MOR TUN 21 AZO MDR 33 NCS TCS 34 AFG CYP IRN LBS PAL TUR (51) nzs (73) was (75) vrg. Hemicr. or rhizome geophyte.
Carex maxima Scop., Fl. Carniol., ed. 2, 2: 229 (1772).
Carex agastachys L.f., Suppl. Pl.: 414 (1782).
Carex mutabilis Willd., Fl. Berol. Prodr.: 37 (1787).
Carex myosuroides Lowe, Trans. Cambridge Philos. Soc. 4: 10 (1833), nom. illeg. *Carex pendula* var. *myosuroides* Boott, Ill. Carex 4: 197 (1867).
Carex maxima var. *angustifolia* Albov, Prodr. Fl. Colchic.: 250 (1895). *Carex pendula* var. *angustifolia* (Albov) Kük. in H.G.A.Engler (ed.), Pflanzenr., IV, 20(38): 426 (1909).
Carex pendula f. *pallida* Asch. & Graebn., Syn. Mitteleur. Fl. 2(2): 155 (1903).

Carex penduliformis Cherm., Bull. Soc. Bot. France 70: 414 (1923).
C. Madagascar. 29 MDG. Hemicr. or rhizome geophyte.

Carex pensylvanica Lam., Encycl. 3: 388 (1792).
C. & E. Canada to NC. & E. U.S.A. 71 MAN 72 ONT QUE 74 ILL IOW MIN MSO SDA WIS 75 CNT INI MAI MAS MIC NWH NWJ NWY OHI PEN RHO VER WVA 77 NWM 78 ARK DEL GEO KTY MRY NCA TEN VRG. Hemicr. or rhizome geophyte.
Carex marginata Muhl. ex Willd., Sp. Pl. 4: 261 (1805), nom. illeg.
Carex subbiflora Steud., Syn. Pl. Glumac. 2: 234 (1855).
Carex pensylvanica var. *angustifolia* Peck, Rep. (Annual) Regents Univ. State New York State Mus. 46: 131 (1893).
Carex pensylvanica f. *bracteata* Peck, Rep. (Annual) Regents Univ. State New York State Mus. 46: 131 (1893).
Carex pensylvanica f. *paleacea* Peck, Rep. (Annual) Regents Univ. State New York State Mus. 46: 131 (1893).
Carex pensylvanica var. *glumabunda* Peck, Rep. (Annual) Regents Univ. State New York New York State Mus. 47: 174 (1895). *Carex pensylvanica* f. *glumabunda* (Peck) Kük. in H.G.A.Engler (ed.), Pflanzenr., IV, 20(38): 446 (1909).
Carex pensylvanica var. *gracilifolia* Peck, Rep. (Annual) Regents Univ. State New York New York State Mus. 47: 174 (1895). *Carex pensylvanica* f. *gracilifolia* (Peck) Kük. in H.G.A.Engler (ed.), Pflanzenr., IV, 20(38): 445 (1909).
Carex pensylvanica f. *latifolia* Kük. in H.G.A.Engler (ed.), Pflanzenr., IV, 20(38): 446 (1909).
Carex pennsylvanica var. *pinicola* M.E.Jones, Bull. Montana State Univ., Biol. Ser. 15: 21 (1910).
Carex pensylvanica f. *androgyna* C.F.Wheeler ex F.J.Herm., Amer. Midl. Naturalist 25: 33 (1941).
Carex pensylvanica var. *livoniensis* Farw., Pap. Michigan Acad. Sci. 26(1): 8 (1941).

Carex perakensis C.B.Clarke in J.D.Hooker, Fl. Brit. India 6: 720 (1894). *Carex wightiana* var. *perakensis* (C.B.Clarke) Kük. in H.G.A.Engler (ed.), Pflanzenr., IV, 20(38): 288 (1909).
Assam to Taiwan and C. Malesia. 36 CHC CHH CHS 38 TAI 40 ASS 41 LAO MYA THA VIE 42 BOR MLY SUL SUM. Hemicr. or rhizome geophyte.

var. ***borneensis*** (C.B.Clarke) Noot., in Fl. Males. 9(1): 141 (1979).
Vietnam to C. Malesia. 41 VIE 42 BOR MLY SUL. Hemicr. or rhizome geophyte.
**Carex borneensis* C.B.Clarke, J. Linn. Soc., Bot. 37: 14 (1904).
Carex kinabaluensis Stapf, J. Linn. Soc., Bot. 42: 183 (1914).
Carex rivulorum Ridl., J. Fed. Malay States Mus. 6: 195 (1915), nom. illeg. *Carex pseudorivulorum* Kük., Bull. Jard. Bot. Buitenzorg, III, 16: 319 (1940).
Carex breviglumis Ridl., J. Straits Branch Roy. Asiat. Soc. 82: 202 (1920).
Carex breviglumis Ridl., Fl. Malay Penins. 5: 183 (1925), nom. illeg.
Carex clemensii Kük., Repert. Spec. Nov. Regni Veg. 29: 202 (1931).
Carex eymae Nelmes, Kew Bull. 5: 199 (1950).

var. ***perakensis***
Assam to Taiwan and C. Malesia. 36 CHC CHH CHS 38 TAI 40 ASS 41 MYA THA VIE 42 BOR MLY SUL SUM. Hemicr. or rhizome geophyte.
Carex arridens C.B.Clarke in J.D.Hooker, Fl. Brit. India 6: 726 (1894).
Carex tonkinensis Franch., Bull. Soc. Philom. Paris, VIII, 7: 33 (1895).
Carex prainii C.B.Clarke, J. Linn. Soc., Bot. 36: 305 (1904).
Carex dunnii Hayata, J. Coll. Sci. Imp. Univ. Tokyo 30(1): 382 (1911).
Carex leucostachys Ridl., Bull. Misc. Inform. Kew 1928: 77 (1928).
Carex tatewakiana Ohwi, Acta Phytotax. Geobot. 1: 299 (1932).
Carex setulifolia Nelmes, Mém. Mus. Natl. Hist. Nat., Sér. B, Bot. 4: 116 (1955).

var. ***vansteenisii*** (Kük.) Noot., in Fl. Males. 9(1): 142 (1979).
Indo-China, N. Sumatera. 41 LAO VIE 42 SUM. Hemicr. or rhizome geophyte.
**Carex vansteenisii* Kük., Bull. Jard. Bot. Buitenzorg, III, 16: 320 (1940).

Carex percostata F.J.Herm., J. Wash. Acad. Sci. 40: 282 (1950).
Mexico (Chihuahua, Durango). 79 MXE. Hemicr. or rhizome geophyte.

Carex perdentata S.D.Jones, Sida 16: 342 (1994).
Oklahoma to Texas. 74 OKL 77 TEX. Hemicr. or rhizome geophyte.

Carex peregrina Link, Hort. Berol. 1: 334 (1827). *Carex macrostylon* var. *peregrina* (Link) L.H.Bailey ex Trel., Rep. (Annual) Missouri Bot. Gard. 8: 159 (1897).
Azores, N. Madeira, Ethiopia (Bale Mts.) to Tanzania (Mt. Meru). 21 AZO MDR 24 ETH 25 KEN TAN. Hemicr. or rhizome geophyte.
Carex sagittifera Lowe, Trans. Cambridge Philos. Soc. 6: 531 (1838).

Carex perglobosa Mack., Bull. Torrey Bot. Club 34: 606 (1907 publ. 1908).
WC. U.S.A. 73 COL 76 UTA. Hemicr. or rhizome geophyte.

Carex pergracilis Nelmes, Kew Bull. 3: 107 (1948).
SC. China. 36 CHC. Hemicr. or rhizome geophyte.

Carex perlonga Fernald, Proc. Amer. Acad. Arts 43: 61 (1907).
Mexico (Veracruz, Hidalgo, Oaxaca). 79 MXE MXG MXS. Hemicr. or rhizome geophyte.
Carex viridis Schltdl. & Cham., Linnaea 6: 30 (1831), nom. illeg. *Carex viridiflora* Mack. in N.L.Britton & al. (eds.), N. Amer. Fl. 18: 275 (1935).
Carex longa Mack. in N.L.Britton & al. (eds.), N. Amer. Fl. 18: 277 (1935).

Carex perprava C.B.Clarke, Kew Bull., Addit. Ser. 8: 76 (1908).
Bolivia. 83 BOL. Hemicr. or rhizome geophyte.

Carex perraudieriana (Kük. ex Bornm.) Gay ex Kük. in H.G.A.Engler (ed.), Pflanzenr., IV, 20(38): 649 (1909).
Canary Is. (Tenerife). 21 CNY. Hemicr. or rhizome geophyte.
**Carex elata* var. *perraudieriana* Kük. ex Bornm., Bot. Jahrb. Syst. 33: 406 (1903).

Carex* ×*persalina Lepage, Naturaliste Canad. 83: 148 (1956). *C. salina* × *C. subspathacea*.
E. Canada. 72 ONT QUE. Hemicr. or rhizome geophyte.

Carex pertenuis L.H.Bailey, Mem. Torrey Bot. Club 1: 9 (1889).
Mexico to Guatemala. 79 MXG MXT 80 GUA. Hemicr. or rhizome geophyte.
Carex venosivaginata Standl. & Steyerm., Ceiba 4: 67 (1953).

Carex peruviana J.Presl & C.Presl, Reliq. Haenk. 1: 205 (1828).
Colombia to Peru. 83 CLM ECU PER. Hemicr. or rhizome geophyte.
Carex crinalis Boott, Proc. Linn. Soc. London 1: 256 (1845).

Carex peruvida G.A.Wheeler, Darwiniana 40: 199 (2002).
Bolivia. 83 BOL. Hemicr. or rhizome geophyte.

Carex petasata Dewey, Amer. J. Sci. Arts 29: 246 (1836).
Alaska to W. U.S.A. 70 ASK NWT YUK 71 ABT BRC SAS 73 COL IDA MNT ORE WAS WYO 76 ARI CAL NEV UTA. Hemicr. or rhizome geophyte.
Carex adusta var. *minor* Boott in W.J.Hooker, Fl. Bor.-Amer. 2: 215 (1839). *Carex petasata* var. *minor* (Boott) B.Boivin, Phytologia 43: 57 (1979).
Carex liddonii Boott in W.J.Hooker, Fl. Bor.-Amer. 2: 214 (1839).
Carex leporina var. *americana* Olney, Proc. Amer. Acad. Arts 8: 407 (1872).
Carex liddonii var. *incerta* L.H.Bailey, Bot. Gaz. 13: 88 (1888).
Carex rufovariegata Boeck., Beitr. Cyper. 1: 33 (1888).
Carex petasata var. *pleiostachya* Kük. in H.G.A.Engler (ed.), Pflanzenr., IV, 20(38): 210 (1909).

Carex petelotii R.Gross, Notizbl. Bot. Gart. Berlin-Dahlem 14: 190 (1938).
Vietnam. 41 VIE. Hemicr. or rhizome geophyte.

Carex petitiana A.Rich., Tent. Fl. Abyss. 2: 513 (1850).
Nigeria to Cameroon, Ethiopia to Tanzania. 22 NGA 23 CMN RWA ZAI 24 ETH 25 KEN TAN UGA. Hemicr. or rhizome geophyte.
Carex anomala Steud., Syn. Pl. Glumac. 2: 230 (1855).
Carex longipedunculata K.Schum. in H.G.A.Engler, Pflanzenw. Ost-Afrikas, C: 130 (1895).
Carex preussii K.Schum., Bot. Jahrb. Syst. 24: 340 (1897). *Carex longipedunculata* var. *preussii* (K.Schum.) Kük. in H.G.A.Engler (ed.), Pflanzenr., IV, 20(38): 652 (1909).
Carex longipedunculata var. *ninagongensis* Kük. in H.G.A.Engler (ed.), Pflanzenr., IV, 20(38): 767 (1909). *Carex ninagongensis* (Kük.) Nelmes ex Robyns & Tournay in W.Robyns, Fl. Spermatophyt. Parc Nat. Albert 3: 292 (1955).
Carex longipedunculata var. *cuprea* Kük., Notizbl. Bot. Gart. Berlin-Dahlem 9: 315 (1925). *Carex cuprea* (Kük.) Nelmes, Bull. Misc. Inform. Kew 1938: 247 (1938).
Carex preussii var. *camerunensis* Nelmes, Bull. Misc. Inform. Kew 1938: 247 (1938).

Carex petricosa Dewey, Amer. J. Sci. Arts 29: 246 (1836).
Siberia to Russian Far East, Alaska to Montana and Canada. 30 KRA YAK 31 MAG 32 KAZ KGZ 37 MON 70 ASK NWT YUK 71 ABT BRC 72 LAB NFL QUE 73 MNT. Hemicr. or rhizome geophyte.

var. ***misandroides*** (Fernald) B.Boivin, Phytologia 43: 83 (1979).
E. Canada. 72 LAB NFL QUE. Hemicr. or rhizome geophyte.
**Carex misandroides* Fernald, Rhodora 17: 158 (1915). *Carex franklinii* var. *misandroides* (Fernald) Raymond, Canad. Field-Naturalist 66: 102 (1952).

var. ***petricosa***
Siberia to Russian Far East, Alaska to Montana. 30 KRA YAK 31 MAG 32 KAZ KGZ 37 MON 70 ASK NWT YUK 71 ABT BRC 73 MNT. Hemicr. or rhizome geophyte.
Carex franklinii Boott in W.J.Hooker, Fl. Bor.-Amer. 2: 217 (1839). *Carex petricosa* var. *franklinii* (Boott) Boivin, Naturaliste Canad. 94: 523 (1967).
Carex macrogyna Turcz. ex Steud., Syn. Pl. Glumac. 2: 236 (1855). *Carex ustulata* var. *macrogyna* (Turcz. ex Steud.) Regel, Trudy Imp. S.-Peterburgsk. Bot. Sada 7: 571 (1880).

Carex distichiflora B.Boivin, Naturaliste Canad. 75: 206 (1948). *Carex petricosa* var. *distichiflora* (B.Boivin) B.Boivin, Naturaliste Canad. 94: 523 (1967).
Carex franklinii var. *nicholsonis* B.Boivin, Naturaliste Canad. 75: 207 (1948).
Carex petricosa var. *edwardsii* B.Boivin, Naturaliste Canad. 75: 207 (1948).
Carex lepageana Raymond, Canad. Field-Naturalist 66: 101 (1952).
Carex magnursina Raymond, Canad. Field-Naturalist 66: 100 (1952).

Carex petriei Cheeseman, Trans. & Proc. New Zealand Inst. 16: 413 (1884).
New Zealand. 51 NZN NZS. Hemicr. or rhizome geophyte.

Carex peucophila T.Holm, Contr. U. S. Natl. Herb. 8: 290 (1905).
Mexico to Guatemala. 79 MXC MXE MXG MXS MXT 80 GUA. Hemicr. or rhizome geophyte.
**Carex pinetorum* Liebm., Mexic. Halvgr.: 263 (1850), nom. illeg.
Carex straminea var. *australis* L.H.Bailey, Mem. Torrey Bot. Club 1: 23 (1889).
Carex pinetorum var. *elatior* Kük. in H.G.A.Engler (ed.), Pflanzenr., IV, 20(38): 195 (1909).

Carex phacota Spreng., Syst. Veg. 3: 826 (1826).
Trop. & Subtrop. Asia. 36 CHC CHN CHS 38 JAP KOR NNS TAI 40 EHM IND NEP SRL 41 THA VIE 42 JAW LSI PHI SUL SUM 43 NWG. Hemicr. or rhizome geophyte.
Carex biconvexis D.Don, Trans. Linn. Soc. London 14: 331 (1824), nom. illeg.
Carex platycarpa Hochst. ex Steud., Syn. Pl. Glumac. 2: 214 (1855).
Carex gracilipes Miq., Ann. Mus. Bot. Lugduno-Batavi 2: 151 (1866).
Carex fauriae Franch., Bull. Soc. Bot. France 26: 89 (1879).
Carex pruinosa var. *aristata* Kuntze, Revis. Gen. Pl. 2: 748 (1891).
Carex cincta Franch., Bull. Soc. Philom. Paris, VIII, 7: 35 (1895).
Carex lepidopristis H.Lév. & Vaniot, Bull. Acad. Int. Géogr. Bot. 10: 198 (1901).
Carex cincta var. *subphacota* Kük. in H.G.A.Engler (ed.), Pflanzenr., IV, 20(38): 353 (1909). *Carex subphacota* (Kük.) Nakai, Veg. Ins. Quelp.: 24 (1914).
Carex phacota var. *gracilispica* Kük. in H.G.A.Engler (ed.), Pflanzenr., IV, 20(38): 352 (1909).
Carex shichiseitensis Hayata, Icon. Pl. Formosan. 10: 58 (1921).
Carex lenticularis D.Don ex Tilloch & Taylor, Philos. Mag. J. 62: 455 (Dec. 1823).

Carex phaeocephala Piper, Contr. U. S. Natl. Herb. 11: 172 (1906).
Alaska to W. U.S.A. 70 ASK 71 ABT BRC 73 COL IDA MNT ORE WAS WYO 76 CAL NEV UTA. Hemicr. or rhizome geophyte.

Carex phaeodon T.Koyama, J. Jap. Bot. 30: 315 (1955).
Japan (Honshu: Suruga Prov.). 38 JAP. Hemicr. or rhizome geophyte.

Carex phaeothrix Ohwi, Mem. Coll. Sci. Kyoto Imp. Univ., Ser. B, Biol. 6: 239 (1931).
Korea. 38 KOR. Hemicr. or rhizome geophyte.

Carex phalaroides Kunth, Enum. Pl. 2: 482 (1837). *Uncinia phalaroides* (Kunth) Nees in C.F.P.von Martius & auct. suc. (eds.), Fl. Bras. 2(1): 202 (1842).
Ecuador to Uruguay. 83 BOL ECU 84 BZL BZS 85 AGE AGW CLC CLN PAR URU. Hemicr. or rhizome geophyte.

var. ***elongata*** Kük., Bot. Jahrb. Syst. 27: 537 (1899).
Bolivia to NW. Argentina and Brazil. 83 BOL 84 BZS 85 AGW PAR URU. Hemicr. or rhizome geophyte.
Carex moesta Kunth, Enum. Pl. 2: 482 (1837). *Carex phalaroides* var. *moesta* (Kunth) Kük., Verh. Bot. Vereins Prov. Brandenburg 47: 208 (1905 publ. 1906). *Carex phalaroides* subsp. *moesta* (Kunth) Luceño & Alves, Anales Jard. Bot. Madrid 57: 174 (1999).

var. ***macella*** (Kunth) Kük., Verh. Bot. Vereins Prov. Brandenburg 47: 208 (1905 publ. 1906).
Brazil to N. Argentina. 84 BZS 85 AGE AGW PAR URU. Hemicr. or rhizome geophyte.
**Carex macella* Kunth, Enum. Pl. 2: 482 (1837).

var. ***paraguayensis*** (Maury ex Micheli) Kük., Verh. Bot. Vereins Prov. Brandenburg 47: 208 (1905 publ. 1906).
Brazil to NE. Argentina and Bolivia. 83 BOL 84 BZS 85 AGE PAR URU. Hemicr. or rhizome geophyte.
**Carex paraguayensis* Maury ex Micheli, Mém. Soc. Phys. Genève 31(1): 151 (1890). *Carex macella* var. *paraguayensis* (Maury ex Micheli) Kük., Bot. Jahrb. Syst. 27: 538 (1899). *Carex phalaroides* subsp. *paraguayensis* (Maury ex Micheli) Luceño & Alves, Anales Jard. Bot. Madrid 57: 174 (1999).

var. ***phalaroides***
Ecuador to Uruguay. 83 BOL ECU 84 BZL BZS 85 AGE AGW CLC CLN PAR URU. Hemicr. or rhizome geophyte.
Carex chlorolepis Steud., Syn. Pl. Glumac. 2: 204 (1855).
Carex schenckiana Boeck., Beitr. Cyper. 2: 35 (1890).
Carex capitellata Boeck., Allg. Bot. Z. Syst. 2: 174 (1896). *Carex phalaroides* var. *capitellata* Kük., Verh. Bot. Vereins Prov. Brandenburg 47: 209 (1905 publ. 1906).
Carex phalaroides var. *humilis* Boeck., Allg. Bot. Z. Syst. 2: 190 (1896).
Carex phalaroides var. *parvula* R.Gross, Repert. Spec. Nov. Regni Veg. 50: 211 (1941).

Carex phankei K.K.Nguyen, Bot. Zhurn. (Moscow & Leningrad) 64: 857 (1979).
Vietnam. 41 VIE. Hemicr. or rhizome geophyte.

Carex phoenicis Dunn, Kew Bull., Addit. Ser. 10: 305 (1912).
China (Guangdong). 36 CHS. Hemicr. or rhizome geophyte.

Carex phoenocarpa Franch., Bull. Soc. Philom. Paris, VIII, 7: 90 (1895).
China (Yunnan). 36 CHC. Hemicr. or rhizome geophyte.

Carex phragmitoides Kük., Repert. Spec. Nov. Regni Veg. 21: 329 (1925).
Kenya (Aberdare Range), Tanzania (Ossirwa Crater). 25 KEN TAN. Hemicr. or rhizome geophyte.
Carex taylorii Nelmes, Bull. Misc. Inform. Kew 1937: 472 (1937).

Carex phyllocaula Nelmes, Kew Bull. 1: 21 (1946).
N. Thailand. 41 THA. Hemicr. or rhizome geophyte.

Carex phyllocephala T.Koyama, Acta Phytotax. Geobot. 16: 40 (1955).
China (Fujian). 36 CHS (38) jap. Hemicr. or rhizome geophyte.

Carex phyllostachys C.A.Mey., Verz. Pfl. Casp. Meer.: 30 (1831). *Forexeta phyllostachys* (C.A.Mey.) Raf., Good Book: 28 (1840).
W. Macedonia, Turkey to N. Iran. 13 YUG 33 TCS 34 IRN TUR. Hemicr. or rhizome geophyte.

Carex ×physocarpoides Lepage, Naturaliste Canad. 83: 113 (1956). *C. saxatilis* × *C. utriculata*.
Alaska to W. Canada. 70 ASK 72 BRC 73 COL IDA MNT ORE WAS WYO 76 NEV UTA. Hemicr. or rhizome geophyte.

Carex physodes M.Bieb., Mém. Soc. Imp. Naturalistes Moscou 2: 104 (1809).
S. European Russia to C. Asia. 14 RUS 32 KAZ KGZ TKM UZB 33 NCS 34 AFG IRN 36 CHX 40 PAK. Hemicr. or rhizome geophyte.

Carex pichinchensis Kunth in F.W.H.von Humboldt, A.J.A.Bonpland & C.S.Kunth, Nov. Gen. Sp. 1: 233 (1817).
Colombia to Bolivia. 83 BOL CLM ECU PER. Hemicr. or rhizome geophyte.
Carex atrorufa Willd. ex Kunth, Enum. Pl. 2: 392 (1837).
Carex atrofusca Willd. ex Steud., Nomencl. Bot., ed. 2, 1: 286 (1840), nom. inval.
Carex dura Boott, Proc. Linn. Soc. London 1: 255 (1845).
Carex pichinchensis var. *dura* Boott, Trans. Linn. Soc. London 20: 121 (1846).
Carex sachapata Steud. in W.Lechler, Berberid. Amer. Austral.: 56 (1857).
Carex tabina Steud. in W.Lechler, Berberid. Amer. Austral.: 56 (1857).
Carex ruiziana Boeck., Linnaea 39: 30 (1875).
Carex fuscoatia Boeck., Linnaea 40: 378 (1876).
Carex pichinchensis var. *obtusisquama* R.Gross, Notizbl. Bot. Gart. Berlin-Dahlem 14: 193 (1938).
Carex pichinchensis var. *simplex* R.Gross, Repert. Spec. Nov. Regni Veg. 50: 211 (1941).

Carex picta Steud., Syn. Pl. Glumac. 2: 184 (1855).
Indiana to Louisiana. 75 INI 78 ALA KTY LOU MSI TEN. Hemicr. or rhizome geophyte.
Carex boottiana Benth. ex Boott, Boston J. Nat. Hist. 5: 112 (1845), nom. illeg.

Carex ×pieperiana Junge, Verh. Naturwiss. Vereins Hamburg, III, 12: 118 (1904). *C. flava* × *C. lepidocarpa*.
C. Europe. 11 GER 72 NFL QUE. Hemicr. or rhizome geophyte.

Carex pigra Naczi, Novon 7: 67 (1997).
SE. U.S.A. 78 ALA FLA GEO MSI NCA SCA TEN VRG. Hemicr. or rhizome geophyte.

Carex pilosa Scop., Fl. Carniol., ed. 2, 2: 226 (1772). *Loxotrema pilosa* (Scop.) Raf., Good Book: 25 (1840).
C. France to E. & SE. Europe, Russian Far East, Korea, Japan. 11 AUT CZE GER HUN POL SWI 12 FRA 13 BUL ITA ROM TUE YUG 14 BLR BLT RUC RUE RUW UKR 31 SAK 36 CHM 38 JAP KOR. Hemicr. or rhizome geophyte.
Carex nemorensis J.F.Gmel., Syst. Nat. 2: 143 (1791).
Carex prostrata J.F.Gmel., Syst. Nat. 2: 142 (1791).
Carex nemorosa J.F.Gmel., Syst. Nat. 2: 1520 (1792), nom. illeg.
Carex pilosa var. *beckeri* C.A.Mey. ex Claus, Beitr. Pflanzenk. Russ. Reiches 8: 162 (1851).
Carex pilosa var. *densiflora* Schur, Enum. Pl. Transsilv.: 713 (1866).
Carex auriculata Franch., Bull. Soc. Philom. Paris, VII, 10: 106 (1886). *Carex pilosa* var. *auriculata* (Franch.) Kük. in H.G.A.Engler (ed.), Pflanzenr., IV, 20(38): 637 (1909).
Carex foliata F.Schmidt ex Meinsh., Trudy Imp. S.-Peterburgsk. Bot. Sada 18: 385 (1900).
Carex hakodatensis H.Lév. & Vaniot, Bull. Soc. Bot. France 51: 206 (1904).

Carex ×pilosiuscula Gobi, Bot. Jahresber. (Just) 4: 1078 (1876). *C. hirta* × *C. rhynchophysa*.
Europe. 1+. Hemicr. or rhizome geophyte.

Carex pilulifera L., Sp. Pl.: 976 (1753). *Trasus piluliferus* (L.) Gray, Nat. Arr. Brit. Pl. 2: 58 (1821). *Bitteria pilulifera* (L.) Fedde & J.Schust., Just's Bot. Jahresber. 41(2): 7 (1913 publ. 1918).
Macaronesia, Morocco, Europe. 10 DEN FIN GRB ICE IRE NOR SWE 11 AUT BGM CZE GER HUN NET POL SWI 12 COR FRA POR SPA 13 ITA ROM YUG 14 BLR BLT RUC RUN RUW UKR 20 MOR 21 AZO MDR. Hemicr. or rhizome geophyte.

subsp. ***azorica*** (J.Gay) Franco & Rocha Afonso, Bot. J. Linn. Soc. 76: 366 (1978).
Azores. 21 AZO. Hemicr. or rhizome geophyte.
**Carex azorica* J.Gay, Ann. Sci. Nat., Bot., II, 11: 185 (1839). *Carex pilulifera* var. *azorica* (J.Gay) H.Christ, Bull. Soc. Roy. Bot. Belgique 24(2): 14 (1885).

subsp. ***pilulifera***
Madeira, Morocco, Europe. 10 DEN FIN GRB ICE IRE NOR SWE 11 AUT BGM CZE GER HUN NET POL SWI 12 COR FRA POR SPA 13 ITA ROM YUG 14 BLR BLT RUC RUN RUW UKR 20 MOR 21 MDR. Hemicr. or rhizome geophyte.
Carex oederi Retz., Fl. Scand. Prodr.: 179 (1779). *Carex flava* var. *oederi* (Retz.) DC. in J.B.A.M.de Lamarck & A.P.de Candolle, Fl. Franç., ed. 3, 3: 121 (1805). *Trasus oederi* (Retz.) Gray, Nat. Arr. Brit. Pl. 2: 60 (1821). *Onkerma oederi* (Retz.) Raf., Good Book: 27 (1840). *Carex flava* subsp. *oederi* (Retz.) Syme, Engl. Bot., ed. 10, 3: 157 (1870).
Carex decumbens Ehrh., Beitr. Naturk. 6: 82 (1791).
Carex alba Bastard, Essai Fl. Maine et Loire: 338 (1809), nom. illeg.
Carex flava var. *lutescens* Wahlenb., Fl. Lapp.: 234 (1812).
Carex bastardiana DC. in J.B.A.M.de Lamarck & A.P.de Candolle, Fl. Franç., ed. 3, 6: 293 (1815). *Carex pilulifera* var. *bastardiana* (DC.) Nyman, Consp. Fl. Eur.: 774 (1882).
Carex flava var. *minor* Wahlenb., Fl. Suec. 2: 596 (1826).
Carex aederi Desv., Fl. Anjou: 74 (1827), sphalm.
Carex pilulifera var. *bertolonii* Pi.Savi, Fl. Gorgonica (1844).
Carex pilulifera var. *mucronata* Peterm., Anal. Pfl.-Schlüss.: 505 (1846).

Carex pilulifera f. *pallida* Peterm., Anal. Pfl.-Schlüss.: 505 (1846).
Carex oederi f. *pygmaea* Andersson, Cyper. Scand.: 25 (1849).
Carex oedocarpa Andersson, Cyper. Scand.: 26 (1849). *Carex oederi* var. *oedocarpa* (Andersson) Nyman, Consp. Fl. Eur.: 771 (1882).
Carex flava subvar. *elongata* Coss. & Germ., Fl. Descr. Anal. Paris, ed. 2: 753 (1861).
Carex oederi var. *minor* Boeck., Verh. Bot. Vereins Prov. Brandenburg 3(4): 287 (1862).
Carex pilulifera f. *longibracteata* Lange, Haandb. Danske Fl., ed. 3: 694 (1864).
Carex flava var. *orthorhyncha* Èelak., Prodr. Fl. Böhmen: 71 (1867).
Carex pullulans Dulac, Fl. Hautes -Pyrénées: 57 (1867).
Carex pilulifera f. *major* Wirtg. ex Herrenkohl, Verh. Naturhist. Vereines Preuss. Rheinl. Westphalens 28: 213 (1871).
Carex pilulifera var. *digyna* Boeck., Linnaea 41: 217 (1877).
Carex flava var. *argillacea* Townson, J. Bot. 19: 263 (1881). *Carex oederi* f. *argillacea* (Townson) Kük. in H.G.A.Engler (ed.), Pflanzenr., IV, 20(38): 674 (1909). *Carex serotina* f. *argillacea* (Townson) Soó, Acta Bot. Acad. Sci. Hung. 16: 371 (1970 publ. 1971).
Carex flava var. *minor* Townson, J. Bot. 19: 263 (1881), nom. illeg.
Carex oederi var. *fallax* Heimerl, Verh. K. K. Zool.-Bot. Ges. Wien 31: 132 (1881).
Carex pilulifera var. *leesii* Ridl., J. Bot. 19: 98 (1884).
Carex oederi var. *glomerata* Callmé, Deutsche Bot. Monatsschr. 5: 19 (1887).
Carex oederi f. *tularia* Callmé, Deutsche Bot. Monatsschr. 5: 18 (1887).
Carex oederi f. *virescens* Callmé, Deutsche Bot. Monatsschr. 5: 19 (1887).
Carex oederi f. *pumila* Zahn, Oesterr. Bot. Z. 40: 363 (1890).
Carex pilulifera var. *pallescens* Beckh., Fl. Westfalen: 917 (1893).
Carex pilulifera f. *fallax* Vatke ex Asch. & Graebn., Syn. Mitteleur. Fl. 1: 115 (1896).
Carex pilulifera f. *golenzii* Asch. & Graebn., Syn. Mitteleur. Fl. 1: 115 (1896).
Carex pilulifera f. *fuscidula* Waisb., Oesterr. Bot. Z. 47: 5 (1897).
Carex flava subsp. *brevirostrata* Kük., Bot. Jahrb. Syst. 27: 545 (1899).
Carex oederi var. *congesta* Neuman, Sver. Fl.: 696 (1901).
Carex oederi f. *pinguis* Neuman, Sver. Fl.: 697 (1901).
Carex oederi var. *brevirostris* Asch. & Graebn., Syn. Mitteleur. Fl. 2(2): 201 (1903).
Carex pilulifera f. *laxa* Waisb., Magyar Bot. Lapok 4: 73 (1905).
Carex pilulifera f. *pedunculata* Waisb., Magyar Bot. Lapok 4: 73 (1905).
Carex pilulifera f. *refracta* Waisb., Magyar Bot. Lapok 4: 73 (1905).
Carex bergrothii Palmgr., Commentat. Biol. 20(3): 4 (1958). *Carex viridula* var. *bergrothii* (Palmgr.) B.Schmid, Watsonia 14: 315 (1983). *Carex flava* subsp. *bergrothii* (Palmgr.) P.D.Sell in P.D.Sell & G.Murrell, Fl. Great Britain Ireland 5: 363 (1996). *Carex viridula* subsp. *bergrothii* (Palmgr.) Tzvelev, Novosti Sist. Vyssh. Rast. 32: 183 (2000). *Carex oederi* var. *bergrothii* (Palmgr.) Hedrén & Lassen, Nordic J. Bot. 22: 262 (2002 publ. 2003).

Carex pinophila Reznicek & S.González, Contr. Univ. Michigan Herb. 21: 294 (1997).
Mexico (Querétaro). 79 MXE. Hemicr. or rhizome geophyte.

Carex pisanensis T.Koyama, Acta Phytotax. Geobot. 15: 112 (1954).
Himalaya. 40 NEP. Hemicr.

Carex pisanoi G.A.Wheeler, Syst. Bot. 12: 579 (1987).
S. Chile. 85 CLS. Hemicr. or rhizome geophyte.

Carex pisiformis Boott in M.C.Perry, Jap. Exped. 2: 324 (1856).
Japan (Honshu). 38 JAP. Hemicr. or rhizome geophyte.
Carex amphora Franch. & Sav., Enum. Pl. Jap. 2: 566 (1878).

Carex pityophila Mack., Bull. Torrey Bot. Club 40: 545 (1913).
Colorado to New Mexico. 73 COL 76 UTA 77 NWM. Hemicr. or rhizome geophyte.

Carex planata Franch. & Sav., Enum. Pl. Jap. 2: 555 (1878). *Vignea planata* (Franch. & Sav.) Soják, Cas. Nár. Mus., Odd. Prír. 148: 196 (1979 publ. 1980).
Japan. 38 JAP. Hemicr. or rhizome geophyte.
Carex ponmoshiriensis Miyabe & Kudô, Trans. Sapporo Nat. Hist. Soc. 6: 165 (1917).

Carex planiculmis Kom., Trudy Imp. S.-Peterburgsk. Bot. Sada 17: 448 (1899).
Russian Far East, China to N. & C. Japan. 31 PRM SAK 36 CHM CHN 38 JAP KOR. Hemicr. or rhizome geophyte.
Carex dispalatha H.Lév. & Vaniot, Bull. Acad. Int. Géogr. Bot. 11: 177 (1902), nom. illeg.
Carex paniculigera Nakai, Repert. Spec. Nov. Regni Veg. 13: 244 (1914).
Carex macromollicula Nakai, Rep. Veg. Kamik.: 34, 41 (1928).

Carex planiscapa Chun & F.C.How, Acta Phytotax. Sin. 7: 82 (1958).
Hainan. 36 CHH. Hemicr. or rhizome geophyte.

Carex planispicata Naczi, J. Kentucky Acad. Sci. 60: 37 (1999).
E. U.S.A. to Texas. 74 ILL MSO OKL 75 INI NWJ OHI PEN WVA 77 TEX 78 ALA ARK DEL GEO KTY LOU MRY MSI NCA SCA TEN VRG WDC. Hemicr. or rhizome geophyte.
Carex grisea var. *rigida* L.H.Bailey, Mem. Torrey Bot. Club 1: 56 (1889). *Carex amphibola* var. *rigida* (L.H.Bailey) Fernald, Rhodora 44: 311 (1942).

Carex planostachys Kunze in C.Schkuhr, Beschr. Riedgräs., Suppl.: 138 (1844).
Oklahoma to Guatemala. 74 OKL 77 TEX 78 ARK 79 MXC MXE MXG MXN MXS MXT 80 GUA. Hemicr. or rhizome geophyte.

Carex plantaginea Lam., Encycl. 3: 392 (1792). *Deweya plantaginea* (Lam.) Raf., Good Book: 26 (1840).
E. Canada to NC. & E. U.S.A. 72 NBR ONT QUE 74 IOW MIN WIS 75 CNT INI MAI MAS MIC NWH NWJ NWY OHI PEN VER WVA 78 GEO KTY

MRY NCA SCA TEN VRG. Hemicr. or rhizome geophyte.
Carex latifolia Moench, Methodus: 324 (1794), nom. illeg. *Anithista latifolia* (Moench) Raf., Good Book: 26 (1840).

Carex platyphylla Carey, Amer. J. Sci. Arts, II, 4: 23 (1847).
SE. Canada, NC. & E. U.S.A. 72 NBR ONT QUE 74 WIS 75 CNT INI MAI MAS MIC NWH NWJ NWY OHI PEN RHO VER WVA 78 DEL KTY MRY NCA TEN VRG WDC. Hemicr. or rhizome geophyte.
Carex platyphylla var. *longipedunculata* Kük. in H.G.A.Engler (ed.), Pflanzenr., IV, 20(38): 529 (1909).

Carex platysperma Y.L.Chang & Y.L.Yang, in Fl. Plant. Herb. Chinae Bor.-Or. 11: 206 (1976).
NE. China. 36 CHM. Hemicr. or rhizome geophyte.
Carex platysperma var. *sungareensis* Y.L.Chang & Y.L.Yang, in Fl. Plant. Herb. Chinae Bor.-Or. 11: 206 (1976).

Carex plectobasis V.I.Krecz., Bot. Mater. Gerb. Bot. Inst. Komarova Akad. Nauk S.S.S.R. 9: 192 (1946).
Afghanistan to SC. China. 34 AFG 36 CHC CHT 40 NEP WHM. Hemicr. or rhizome geophyte.
Carex tenuis Nees in R.Wight, Contr. Bot. India: 128 (1834), nom. illeg.
**Carex hirtella* Drejer, Symb. Caric.: 21 (1844), nom. illeg.

Carex pleiandra Ohwi, Acta Phytotax. Geobot. 1: 73 (1932).
NE. China. 36 CHM. Hemicr. or rhizome geophyte.

Carex pleioneura G.A.Wheeler, Aliso 11: 533 (1987).
Chile, Argentina (San Juan). 85 AGW CLN. Hemicr. or rhizome geophyte.

Carex pleiostachys C.B.Clarke in T.F.Cheeseman, Man. New Zealand Fl.: 829 (1906).
New Zealand South I. 51 NZS. Hemicr. or rhizome geophyte.

Carex pleurocaula Nelmes, Kew Bull. 1: 7 (1946).
Indo-China. 41 CBD THA VIE. Hemicr. or rhizome geophyte.
Carex ligata var. *indochinensis* R.Gross, Notizbl. Bot. Gart. Berlin-Dahlem 14: 161 (1938).

Carex ×ploettneriana Beyer, Verh. Bot. Vereins Prov. Brandenburg 47: 194 (1905 publ. 1906). *C. elongata* × *C. remota*. *Vignea ×ploettneriana* (Beyer) Soják, Cas. Nár. Mus., Odd. Prír. 148: 196 (1979 publ. 1980).
Europe. 11 GER. Hemicr. or rhizome geophyte.

Carex pluriflora Hultén, Acta Univ. Lund., n.s., 38(1): 367 (1942). *Carex rariflora* var. *pluriflora* (Hultén) Boivin, Naturaliste Canad. 94: 523 (1967). *Carex rariflora* subsp. *pluriflora* (Hultén) T.V.Egorova, Novosti Sist. Vyssh. Rast. 22: 54 (1985).
Russian Far East, Alaska to NW. U.S.A. 31+ 70 ALU ASK 71 BRC 73 WAS. Hemicr. or rhizome geophyte.

Carex poculisquama Kük., Repert. Spec. Nov. Regni Veg. 27: 111 (1929).
E. China, Japan (Honshu: Mt. Iwafune, Nagato). 36 CHS 38 JAP. Hemicr. or rhizome geophyte.

Carex podocarpa R.Br. in J.Richardson, Bot. App.: 751 (1823).
Siberia, Russian Far East, Alaska to NW. U.S.A. 30 BRY YAK 31 KAM KHA KUR MAG 32 KGZ 70 ASK NWT YUK 71 ABT BRC 73 IDA MNT ORE. Hemicr. or rhizome geophyte.
Carex excurrens Cham. ex Steud., Syn. Pl. Glumac. 2: 228 (1855).
Carex trichostachya Fisch. ex Boott, Ill. Carex 4: 198 (1867).
Carex montanensis L.H.Bailey, Bot. Gaz. 17: 152 (1892).
Carex venustula T.Holm, Amer. J. Sci. 166: 457 (1903).
Carex behringensis C.B.Clarke, Kew Bull., Addit. Ser. 8: 81 (1908).
Carex musartiana Kük. ex Krecz. in V.L.Komarov (ed.), Fl. URSS 3: 602 (1935).
Carex pauxilla V.I.Krecz. in V.L.Komarov (ed.), Fl. URSS 3: 601 (1935).
Carex podocarpa f. *pallidior* Lepage, Amer. Midl. Naturalist 46: 756 (1952).

Carex podogyna Franch. & Sav., Enum. Pl. Jap. 2: 557 (1878).
N. & C. Japan. 38 JAP. Hemicr. or rhizome geophyte.
Carex trichopoda Franch., Bull. Soc. Philom. Paris, VIII, 7: 43 (1895). *Carex doenitzii* var. *trichopoda* (Franch.) Kük. in H.G.A.Engler (ed.), Pflanzenr., IV, 20(38): 537 (1909).
Carex podogyna f. *aristodentata* Kük. in H.G.A.Engler (ed.), Pflanzenr., IV, 20(38): 538 (1909).

Carex poeppigii C.B.Clarke ex G.A.Wheeler, Aliso 12: 97 (1988).
S. Chile. 85 CLS. Hemicr. or rhizome geophyte.
**Carex lamprocarpa* var. *rotundata* Kük., Bot. Jahrb. Syst. 27: 541 (1899).

Carex poilanei Raymond, Mém. Jard. Bot. Montréal 53: 37 (1959).
Laos. 41 LAO. Hemicr. or rhizome geophyte.

Carex polyantha F.Muell., Trans. Philos. Soc. Victoria 1: 110 (1855).
E. & SE. Australia. 50 NSW QLD VIC. Hemicr. or rhizome geophyte.

Carex polycephala Boott, Ill. Carex 1: 4 (1858).
C. & E. Himalaya. 40 ASS EHM NEP. Hemicr. or rhizome geophyte.

Carex polymascula P.C.Li, Acta Bot. Yunnan. 12: 140 (1990).
China (Sichuan). 36 CHC. Hemicr. or rhizome geophyte.

Carex polymorpha Muhl., Descr. Gram.: 239 (1817).
NE. U.S.A. to Virginia. 75 CNT MAI MAS NWH NWJ NWY PEN RHO WVA 78 DEL MRY VRG. Hemicr. or rhizome geophyte.
Carex halseyana Dewey, Amer. J. Sci. Arts 11: 313 (1826). *Loxotrema halseyana* (Dewey) Raf., Good Book: 25 (1840).
Carex halseyi Eaton, Man. Bot., ed. 5: 157 (1829).
Carex striata Torr. ex Boott, Ill. Carex 1: 22 (1858), nom. illeg.

Carex polyschoenoides K.T.Fu, in Fl. Tsinlingensis 1(1): 444 (1976).
China. 36 CHN CHS. Hemicr. or rhizome geophyte.
Carex polyschoenoides var. *fractiflexa* Y.C.Yang, Acta Biol. Plateau Sin. 3: 91 (1984).

Carex polystachya Sw. ex Wahlenb., Kongl. Vetensk. Acad. Nya Handl. 24: 149 (1803). *Carex cladostachya* f. *polystachya* (Sw. ex Wahlenb.) C.B.Clarke ex

Lindm., Bih. Kongl. Svenska Vetensk.-Akad. Handl. 26(3)9: 36 (1900).
Trop. & Subtrop. America. 77 TEX 79 MXC MXE MXG MXS MXT 80 BLZ COS ELS GUA HON NIC PAN 81 CUB JAM LEE PUE WIN 82 GUY VEN 83 BOL CLM ECU PER 85 AGW. Hemicr. or rhizome geophyte.

var. ***bartlettii*** (O'Neill) Standl. & Steyerm., Ceiba 4: 68 (1953).
SE. Mexico to C. America. 79 MXT 80 BLZ COS GUA HON PAN. Hemicr. or rhizome geophyte.
**Carex bartlettii* O'Neill, Publ. Carnegie Inst. Wash. 522: 255 (1940).

var. ***polystachya***
Trop. & Subtrop. America. 77 TEX 79 MXC MXE MXG MXS MXT 80 BLZ COS ELS GUA HON NIC PAN 81 CUB JAM LEE PUE WIN 82 GUY VEN 83 BOL CLM ECU PER 85 AGW. Hemicr. or rhizome geophyte.
Carex cladostachya Wahlenb., Kongl. Vetensk. Acad. Nya Handl. 24: 149 (1803).
Carex mexicana C.Presl, Reliq. Haenk. 1: 204 (1828). *Carex cladostachya* var. *mexicana* (C.Presl) Kük. in H.G.A.Engler (ed.), Pflanzenr., IV, 20(38): 268 (1909).
Carex flaccida Sw. ex Kunth, Enum. Pl. 2: 150 (1837).
Carex hartwegii Boott in G.Bentham, Pl. Hartw.: 96 (1842).
Carex acrolepis Liebm., Mexic. Halvgr.: 271 (1850).
Carex oerstedii Liebm., Mexic. Halvgr.: 272 (1850). *Carex cladostachya* var. *oerstedii* (Liebm.) Kük. in H.G.A.Engler (ed.), Pflanzenr., IV, 20(38): 268 (1909).
Carex regnelliana Boeck., Beitr. Cyper. 1: 34 (1888).
Carex dussiana Boeck., Beitr. Cyper. 2: 42 (1890).
Carex pseudocyperus subsp. *platygluma* C.B.Clarke ex Kük., Bot. Jahrb. Syst. 27: 550 (1899).
Carex cladostachya var. *maxima* Kük. in H.G.A. Engler (ed.), Pflanzenr., IV, 20(38): 268 (1909).

Carex polysticha Boeck., Vidensk. Meddel. Naturhist. Foren. Kjøbenhavn 1869: 157 (1869). *Carex pseudocyperus* var. *polysticha* (Boeck.) Kük., Verh. Bot. Vereins Prov. Brandenburg 47: 210 (1905 publ. 1906).
Mexico (Veracruz) to Argentina, Jamaica. 79 MXG 80 COS GUA PAN 81 JAM 83 CLM ECU PER 84 BZL BZS 85 AGE AGS AGW PAR URU. Hemicr. or rhizome geophyte.
Carex underwoodii Britton, Torreya 5: 10 (1905). *Carex hystericina* var. *underwoodii* (Britton) Kük. in H.G.A.Engler (ed.), Pflanzenr., IV, 20(38): 700 (1909).

Carex pomiensis Y.C.Yang, in Fl. Xizangica 5: 408 (1987).
Tibet. 36 CHT. Hemicr. or rhizome geophyte.

Carex pontica Albov, Bull. Herb. Boissier 3: 93 (1895).
Turkey to Tanscaucasus. 33 TCS 34 TUR. Hemicr. or rhizome geophyte.
Carex tristis var. *lazica* Boiss. & Balansa in P.E. Boissier, Fl. Orient. 5: 424 (1882). *Carex lazica* (Boiss. & Balansa) Boiss. & Balansa ex B.D.Jacks., Index Kew. 1: 433 (1893), nom. inval. *Carex lazica* (Boiss. & Balansa) Kük. in H.G.A.Engler (ed.), Pflanzenr., IV, 20(38): 543 (1909).
Carex lazica var. *clavata* Kük. in H.G.A.Engler (ed.), Pflanzenr., IV, 20(38): 543 (1909).

Carex popovii V.I.Krecz., Byull. Sredne-Aziatsk. Gosud. Univ. 21: 181 (1935).
C. Asia. 32 UZB. Hemicr. or rhizome geophyte.

Carex porrecta Reznicek & Camelb., Novon 6: 423 (1996).
Costa Rica to Ecuador. 80 COS 82 VEN 83 CLM ECU. Hemicr. or rhizome geophyte.

Carex potosina Hemsl., Biol. Cent.-Amer., Bot. 3: 474 (1885).
NE. Mexico. 79 MXE. Hemicr. or rhizome geophyte.
**Carex schaffneri* W.Boott ex S.Watson, Proc. Amer. Acad. Arts 18: 172 (1882), nom. illeg.

Carex praeceptorium Mack. in N.L.Britton & al. (eds.), N. Amer. Fl. 18: 95 (1931).
W. Canada to W. U.S.A. 71 BRC 73 COL IDA MNT ORE WAS WYO 76 CAL NEV UTA. Hemicr. or rhizome geophyte.
Carex canescens var. *dubia* L.H.Bailey, Bot. Gaz. 9: 119 (1884). *Carex heleonastes* var. *dubia* (L.H.Bailey) B.Boivin, Phytologia 43: 48 (1979).

Carex praeclara Nelmes, Hooker's Icon. Pl. 35: t. 3403 (1940).
E. Himalaya to China (Yunnan). 36 CHC CHT 40 EHM. Hemicr. or rhizome geophyte.

Carex praecox Schreb., Spic. Fl. Lips.: 63 (1771). *Trasus praecox* (Schreb.) Gray, Nat. Arr. Brit. Pl. 2: 57 (1821). *Carex brizoides* var. *praecox* (Schreb.) Fiori, Nuov. Fl. Italia 1: 186 (1923). *Vignea praecox* (Schreb.) Soják, Cas. Nár. Mus., Odd. Prír. 148: 197 (1979 publ. 1980).
Europe to C. Asia. (10) fin 11 AUT BGM CZE GER HUN NET POL SWI 12 COR? FRA SAR? SPA 13 ALB BUL ITA ROM TUE? YUG 14 BLR BLT KRY RUC RUE RUN RUS RUW UKR 30 ALT 32 KAZ TKM 33 TCS 34 TUR. Hemicr. or rhizome geophyte.
Carex montana Pollich ex Lightf., Fl. Scot. 2: 551 (1777), nom. illeg.
Carex schreberi Schrank, Baier. Fl. 1: 278 (1789). *Vignea schreberi* (Schrank) Rchb. in J.C.Mössler, Handb. Gewächsk. ed. 2, 3: 1614 (1830). *Caricina multicaulis* (Schrank) St.-Lag. in A.Cariot, Étude Fl., ed. 8, 2: 877 (1889).
Carex tenella Thuill., Fl. Env. Paris: 429 (1790).
Carex stolonifera Ehrh., Beitr. Naturk. 6: 83 (1791).
Carex aristata Honck, Syn. Pl. Germ. 1: 361 (1792).
Carex brisoides Lam., Encycl. 3: 382 (1792).
Carex curvula Lam., Encycl. 3: 380 (1792), nom. illeg. *Vignea curvula* Opiz, Seznam: 103 (1852).
Carex schreberi Willd., Mém. Acad. Roy. Sci. Hist. (Berlin) 1794: 38 (1797), nom. illeg.
Carex reflexa Hoppe, Caricolog. Germ.: 65 (1826), nom. illeg. *Carex umbrosa* var. *reflexa* Nyman, Consp. Fl. Eur.: 773 (1882).
Carex praeceps Borkh. ex Rchb., Fl. Germ. Excurs.: 64 (1830).
Carex princeps Kunth, Enum. Pl. 2: 439 (1837).
Carex verna Torr. ex Kunth, Enum. Pl. 2: 439 (1837), nom. illeg.
Carex weiheana Boeck. ex Kunth, Enum. Pl. 2: 439 (1837).
Carex arenaria Dubois ex Steud., Nomencl. Bot., ed. 2, 1: 286 (1840).
Carex heterophylla Krock. ex Steud., Nomencl. Bot., ed. 2, 1: 290 (1840).
Carex brizoides var. *campestris* Wimm., Fl. Schles., ed. 2, 1: 401 (1844).

Carex praecox f. *pallida* Peterm., Flora 27: 331 (1844). *Carex schreberi* var. *pallida* Peterm., Flora 27: 330 (1844). *Carex praecox* f. *petermanii* Soó, Acta Bot. Acad. Sci. Hung. 16: 370 (1970 publ. 1971).
Carex curvata Knaf, Flora 30: 184 (1847), nom. illeg. *Carex praecox* var. *curvata* Asch., Fl. Brandenburg 1: 764 (1864). *Carex brizoides* subsp. *curvata* (Asch.) Èelak., Prodr. Fl. Böhmen, Nachtr.: 733 (1881). *Carex schreberi* var. *curvata* (Asch.) Nyman, Consp. Fl. Eur.: 779 (1882). *Carex brizoides* var. *curvata* (Asch.) Beck, Fl. Nieder-Österreich: 135 (1890). *Carex praecox* subsp. *curvata* (Asch.) Knaf in H.G.A.Engler (ed.), Pflanzenr., IV, 20(38): 131 (1909). *Vignea curvata* (Asch.) Soják, Cas. Nár. Mus., Odd. Prír. 148: 195 (1979 publ. 1980). *Vignea praecox* subsp. *curvata* (Asch.) Dostál, Folia Mus. Rerum Nat. Bohemiae Occid., Bot. 21: 16 (1984).
Carex sicyocarpa Lebel, Rech. Pl. Manche: 18 (1848). *Carex praecox* var. *sicyocarpa* (Lebel) Nyman, Consp. Fl. Eur.: 773 (1882).
Carex schreberi var. *gracillima* Asch., Fl. Brandenburg 2: 170 (1859). *Carex praecox* f. *gracillima* (Asch.) Soó, Acta Bot. Acad. Sci. Hung. 16: 370 (1970 publ. 1971).
Carex pseudopraecox Schur, Enum. Pl. Transsilv.: 716 (1866).
Carex brizoides var. *brunnea* Èelak., Prodr. Fl. Böhmen: 62 (1867).
Carex brizoides var. *intermedia* Èelak., Prodr. Fl. Böhmen: 62 (1867). *Carex praecox* subsp. *intermedia* (Èelak.) W.Schultze-Motel in G.Hegi, Ill. Fl. Mitt.-Eur. 2/1(2): 119 (1968).
Carex pilulifera Geners. ex Boott, Ill. Carex 4: 216 (1867), nom. illeg.
Neilreichia umbrosa Kotula, Spraw. Komis. Fizjogr. 17: 136 (1883).
Carex praecox f. *rumelica* Velen., Fl. Bulg., Suppl. 1: 283 (1898).
Carex praecox f. *monostachys* Lackow., Fl. Berl., ed. 12: 22 (1901).
Carex brizoides var. *adrianopolitana* Podp., Verh. K. K. Zool.-Bot. Ges. Wien 52: 678 (1902).
Carex schreberi f. *nana* Zapal., Consp. Fl. Gallic. Crit. 1: 89 (1904). *Carex praecox* f. *nana* (Zapal.) Soó, Acta Bot. Acad. Sci. Hung. 16: 370 (1970 publ. 1971), without basionym date.
Carex schreberi f. *podolica* Zapal., Consp. Fl. Gallic. Crit. 1: 89 (1904). *Carex praecox* f. *podolica* (Zapal.) Soó, Acta Bot. Acad. Sci. Hung. 16: 370 (1970 publ. 1971), without basionym date.
Carex praecox f. *laxa* Kük. in H.G.A.Engler (ed.), Pflanzenr., IV, 20(38): 131 (1909).
Carex velenovskyi Domin, Beih. Bot. Centralbl. 26(2): 281 (1910).
Carex praecox var. *alata* Zlinska, Biologia Bot. Czech. 48: 37 (1993).

Carex praegracilis W.Boott, Bot. Gaz. 9: 87 (1884).
Alaska to Mexico, W. South America. 70 ASK YUK 71 ABT BRC MAN SAS 72 ONT QUE 73 COL IDA MNT ORE WAS WYO 74 ILL IOW KAN MIN MSO NDA NEB SDA WIS 75 INI MAI MIC NWY OHI PEN VER 76 ARI CAL NEV UTA 77 NWM 78 KTY VRG 79 MXC MXE MXN MXS 80 GUA 83 PER 85 AGE. Hemicr. or rhizome geophyte.
Carex marcida Boott in W.J.Hooker, Fl. Bor.-Amer. 2: 212 (1839), nom. illeg.
Carex gayana var. *hyalina* L.H.Bailey, Proc. Amer. Acad. Arts 22: 135 (1886). *Carex latebrosa* Mack., Bull. Torrey Bot. Club 34: 603 (1907 publ. 1908).
Carex marcida var. *debilis* L.H.Bailey, Proc. Amer. Acad. Arts 22: 136 (1886). *Carex marcida* f. *debilis* (L.H.Bailey) Kük. in H.G.A.Engler (ed.), Pflanzenr., IV, 20(38): 124 (1909).
Carex marcida var. *alterna* L.H.Bailey in J.Macoun, Cat. Canad. Pl. 5: 374 (1890). *Carex marcida* f. *alterna* (L.H.Bailey) Kük. in H.G.A.Engler (ed.), Pflanzenr., IV, 20(38): 124 (1909).
Carex sartwellii var. *occidentalis* L.H.Bailey in J.Macoun, Cat. Canad. Pl. 5: 374 (1890).
Carex camporum Mack., Bull. Torrey Bot. Club 37: 244 (1910).

Carex prairea Dewey ex Alph.Wood, Class-book Bot.: 750 (1861). *Carex teretiuscula* var. *prairea* (Dewey ex Alph.Wood) Britton in N.L.Britton & A.Brown, Ill. Fl. N. U.S. 1: 344 (1896). *Carex diandra* var. *prairea* (Dewey ex Alph.Wood) Kük. in H.G.A.Engler (ed.), Pflanzenr., IV, 20(38): 177 (1909).
Canada to N. & CE. U.S.A. 70 NWT YUK 71 ABT BRC MAN SAS 72 NBR NSC ONT QUE 73 MNT 74 ILL IOW MIN NDA NEB SDA WIS 75 CNT INI MAI MAS MIC NWJ NWY OHI PEN VER WVA 78 VRG. Hemicr. or rhizome geophyte.

Carex prasina Wahlenb., Kongl. Vetensk. Acad. Nya Handl. 1803: 161 (1803).
E. Canada to NC. & E. U.S.A. 72 ONT QUE 74 ILL MSO WIS 75 CNT INI MAI MAS MIC NWH NWJ NWY OHI PEN RHO VER WVA 78 ALA ARK DEL GEO KTY MRY MSI NCA SCA TEN VRG WDC. Hemicr. or rhizome geophyte.
Carex miliacea Muhl. ex Willd., Sp. Pl. 4: 290 (1805), nom. illeg. *Olamblis miliacea* Raf., Good Book: 26 (1840).
Carex subcompressa Steud., Syn. Pl. Glumac. 2: 221 (1855).

Carex praticola Rydb., Mem. New York Bot. Gard. 1: 84 (1900).
Subarctic America to N. & WC. U.S.A. 70 ASK GNL NUN NWT YUK 71 ABT BRC MAN SAS 72 LAB NFL ONT QUE 73 COL IDA MNT ORE WAS WYO (74) ill MIN NDA SDA 75 MAI MIC 76 CAL NEV UTA. Hemicr. or rhizome geophyte.
**Carex pratensis* Drejer, Rev. Car. Bor.: 24 (1841), nom. illeg.
Carex pratensis var. *furva* L.H.Bailey in J.Macoun, Cat. Canad. Pl. 5: 377 (1890). *Carex pratericola* var. *furva* (L.H.Bailey) A.Heller, Cat. N. Amer. Pl., ed. 2: 3 (1900). *Carex furva* (L.H.Bailey) Piper, Contr. U. S. Natl. Herb. 11: 166 (1906), nom. illeg. *Carex pratensis* f. *furva* (L.H.Bailey) Kük. in H.G.A.Engler (ed.), Pflanzenr., IV, 20(38): 198 (1909). *Carex piperi* Mack. in C.V.Piper & R.K.Beattie, Fl. N.W. Coast: 75 (1915).
Carex platylepis Mack. in N.L.Britton & al. (eds.), N. Amer. Fl. 18: 142 (1931).
Carex praticola var. *subcoriacea* F.J.Herm., Leafl. W. Bot. 8: 113 (1957).

Carex preissii Nees in J.G.C.Lehmann, Pl. Preiss. 2: 94 (1846).
Western Australia. 50 WAU. Hemicr. or rhizome geophyte.
Carex thecata Boott, Trans. Linn. Soc. London 20: 143 (1851).

Carex prescottiana Boott, Proc. Linn. Soc. London 1: 286 (1846).
Nepal to EC. China. 36 CHC CHS 40 ASS NEP. Hemicr. or rhizome geophyte.

Carex polyantha Franch. & Sav., Enum. Pl. Jap. 2: 556 (1878), nom. illeg.
Carex cremostachys Franch., Bull. Soc. Philom. Paris, VIII, 7: 34 (1895). *Carex prescottiana* var. *cremostachys* (Franch.) Kük. in H.G.A.Engler (ed.), Pflanzenr., IV, 20(38): 356 (1909).
Carex fargesii Franch., Bull. Soc. Philom. Paris, VIII, 7: 34 (1895). *Carex prescottiana* var. *fargesii* (Franch.) Kük. in H.G.A.Engler (ed.), Pflanzenr., IV, 20(38): 356 (1909).
Carex guffroyana H.Lév. & Vaniot, Bull. Acad. Int. Géogr. Bot. 10: 203 (1901).
Carex kinashii H.Lév. & Vaniot, Bull. Soc. Bot. France 51: 204 (1904).

Carex preslii Steud., Syn. Pl. Glumac. 2: 242 (1855).
SE. Alaska to W. U.S.A. 70 ASK YUK 71 ABT BRC 73 IDA MNT ORE WAS 76 CAL. Hemicr. or rhizome geophyte.

Carex pringlei L.H.Bailey, Bot. Gaz. 17: 151 (1892).
NE. Mexico. 79 MXE. Hemicr. or rhizome geophyte.

Carex projecta Mack., Bull. Torrey Bot. Club 35: 264 (1908).
C. & E. Canada to NC. & E. U.S.A. 71 MAN SAS 72 LAB NBR NFL ONT PEI QUE 74 ILL IOW MIN WIS 75 CNT INI MAI MAS MIC NWH NWJ NWY OHI PEN RHO VER WVA 78 MRY NCA VRG. Hemicr. or rhizome geophyte.

Carex prolongata Kük. in H.F.von Handel-Mazzetti, Symb. Sin. 7: 1262 (1936).
China (Yunnan). 36 CHC. Hemicr. or rhizome geophyte.

Carex proposita Mack. in N.L.Britton & al. (eds.), N. Amer. Fl. 18: 126 (1931).
W. U.S.A. 73 IDA WAS 76 CAL. Hemicr. or rhizome geophyte.

Carex provotii Franch., Bull. Soc. Philom. Paris, VIII, 7: 52 (1895).
China. 36+. Hemicr. or rhizome geophyte.

Carex proxima Cherm., Bull. Soc. Bot. France 70: 411 (1923).
E. Madagascar. 29 MDG. Hemicr. or rhizome geophyte.

Carex pruinosa Boott, Proc. Linn. Soc. London 1: 255 (1845).
Kuril Is. to Jawa. 31 KUR 36 CHC CHM CHN CHS 38 JAP KOR NNS 40 ASS EHM 41 THA VIE 42 JAW. Hemicr. or rhizome geophyte.

subsp. ***maximowiczii*** (Miq.) Kük. in H.G.A.Engler (ed.), Pflanzenr., IV, 20(38): 253 (1909).
NE. China to Temp. E. Asia, Kuril Is. 31 KUR 36 CHM CHN 38 JAP KOR NNS. Hemicr. or rhizome geophyte.
**Carex maximowiczii* Miq., Ann. Mus. Bot. Lugduno-Batavi 2: 180 (1866).
Carex sylvatica Maxim. ex Boeck., Linnaea 41: 238 (1877), nom. illeg.
Carex maximowiczii var. *levisaccus* Ohwi, Mem. Coll. Sci. Kyoto Imp. Univ., Ser. B, Biol. 5: 277 (1930). *Carex pruinosa* var. *levisaccus* (Ohwi) Makino & Nemoto, Fl. Japan, ed. 2: 1446 (1931).

subsp. ***pruinosa***
Bhutan to Jawa. 36 CHC CHN CHS 40 ASS EHM 41 THA VIE 42 JAW. Hemicr. or rhizome geophyte.
Carex pruinosa f. *submutica* Kuntze, Revis. Gen. Pl. 2: 748 (1891).

Carex przewalskii T.V.Egorova, Trudy Bot. Inst. Akad. Nauk S.S.S.R., Rast. Tsentral. Azii 3: 80 (1967).
Qinghai to C. China. 36 CHC CHN CHQ. Hemicr. or rhizome geophyte.

var. ***przewalskii***
Qinghai to C. China. 36 CHC CHN CHQ. Hemicr. or rhizome geophyte.
Carex haematostoma var. *digyae* Kük., Acta Horti Gothob. 5: 45 (1929).

var. ***ramosa*** Y.C.Yang, Acta Biol. Plateau Sin. 3: 89 (1984).
China (Gansu). 36 CHN. Hemicr. or rhizome geophyte.

Carex pseudoaperta Boeck. ex Kük. in H.G.A.Engler (ed.), Pflanzenr., IV, 20(38): 325 (1909).
S. India. 40 IND. Hemicr. or rhizome geophyte.

Carex ×pseudoaxillaris K.Richt., Pl. Eur. 1: 168 (1890).
C. otrubae × *C. remota*. *Vignea* ×*pseudoaxillaris* (K. Richt.) Dostál, Folia Mus. Rerum Nat. Bohemiae Occid., Bot. 21: 16 (1984).
C. Europe. 11 GER. Hemicr. or rhizome geophyte.
Carex ×*kneuckeriana* Zahn, Oesterr. Bot. Z. 40: 412 (1890). *Vignea* ×*kneuckeriana* (Zahn) Soják, Cas. Nár. Mus., Odd. Prír. 148: 195 (1979 publ. 1980).

Carex pseudobicolor Boeck., Beitr. Cyper. 1: 44 (1888).
E. Afghanistan to Tibet. 34 AFG 36 CHT 40 PAK WHM. Hemicr. or rhizome geophyte.

Carex pseudobrizoides Chabaud, Bull. Trav. Soc. Pharm. Bordeaux 13: 156 (1873). *Vignea pseudobrizoides* (Chabaud) Schur, Enum. Pl. Transsilv.: 700 (1866). *Carex arenaria* var. *pseudobrizoides* (Chabaud) Nyman, Consp. Fl. Eur., Suppl. 2: 325 (1890).
W. Europe to Belarus. 11 BGM CZE GER NET POL 12 FRA 14 BLR. Hemicr. or rhizome geophyte. Close to C. reichenbachii.
Carex brizoides var. *stricta* Peterm., Flora 27: 330 (1844).
Carex arenaria var. *umbrosa* Coss. & Germ., Fl. Descr. Anal. Paris, ed. 2: 742 (1861).
Carex reichenbachii Bonnet, Petite Fl. Paris: 420 (1883). *Vignea reichenbachii* (Bonnet) Soják, Cas. Nár. Mus., Odd. Prír. 148: 197 (1979 publ. 1980).
Carex reichenbachiana H.Lév. & Vaniot, Bull. Acad. Int. Géogr. Bot. 11: 60 (1902).
Carex clavaudiana Rouy in G.Rouy & J.Foucaud, Fl. France 13: 420 (1912).

Carex pseudochinensis H.Lév. & Vaniot, Bull. Acad. Int. Géogr. Bot. 11: 306 (1902).
Korea. 38 KOR. Hemicr. or rhizome geophyte.

Carex pseudocuraica F.Schmidt, Reis. Amur-Land., Bot.: 67 (1868). *Carex chordorrhiza* var. *pseudocuraica* (F.Schmidt) Trautv., Trudy Imp. S.-Peterburgsk. Bot. Sada 5: 123 (1877).
Siberia to Japan. 30+ 31 KUR PRM SAK 36 CHI CHM 38 JAP KOR. Hemicr. or rhizome geophyte.
Carex curaica var. *major* Boeck. in ?
Carex chordorrhiza var. *major* Boeck., Linnaea 39: 54 (1875).

Carex pseudocyperus L., Sp. Pl.: 978 (1753).
Temp. & Subtrop. 10 DEN FIN GRB IRE NOR SWE 11 AUT BGM CZE GER HUN NET POL SWI 12 COR FRA POR SPA 13 ALB BUL ITA ROM SIC

YUG 14 BLR BLT KRY RUC RUE RUN RUS RUW UKR 20 ALG 30 ALT TVA 31+ 32 KAZ KGZ TZK 33 NCS TCS 34 LBS PAL TUR 36 CHN 38 JAP 43 NWG 50 NSW NTA QLD SOA TAS VIC WAU 51 NZN 71 MAN 72 NBR NSC ONT QUE 74 MIN WIS 75 CNT MAI MAS MIC NWY OHI PEN VER 85 AGS CLC CLN CLS. Hemicr. or rhizome geophyte.

var. ***fascicularis*** (Sol. ex Hook.f.) Boott, Ill. Carex 4: 141 (1867).
New Guinea, Australia, New Zealand. 43 NWG 50 NSW NTA QLD SOA TAS VIC WAU 51 NZN. Hemicr. or rhizome geophyte.
**Carex fascicularis* Sol. ex Hook.f., Fl. Nov.-Zel. 1: 283 (1853).

var. ***haenkeana*** (C.Presl) Kük., Bot. Jahrb. Syst. 27: 550 (1899).
Chile, Argentina. 85 AGS CLC CLN CLS. Hemicr. or rhizome geophyte.
**Carex haenkeana* C.Presl, Reliq. Haenk. 1: 203 (1828). *Carex pseudocyperus* subsp. *haenkeana* (C.Presl) Kük., Bot. Jahrb. Syst. 27: 550 (1899).

var. ***pseudocyperus***
Temp. & Subtrop. Northern Hemisphere, New Zealand. 10 DEN FIN GRB IRE NOR SWE 11 AUT BGM CZE GER HUN NET POL SWI 12 COR FRA POR SPA 13 ALB BUL ITA ROM SIC YUG 14 BLR BLT KRY RUC RUE RUN RUS RUW UKR 20 ALG 30 TVA 31+ 32 KAZ KGZ TZK 33 NCS TCS 34 LBS PAL TUR 36 CHN 38 JAP 51 NZS 71 MAN 72 NBR NSC ONT QUE 74 MIN WIS 75 CNT MAI MAS MIC NWY OHI PEN VER 85 AGS CLC CLN CLS. Hemicr. or rhizome geophyte.
Carex reversa Gilib., Excerc. Phyt. 2: 549 (1792).
Carex recurva Schkuhr, Beschr. Riedgräs. 1: 120 (1801), nom. illeg.
Trasus chlorostachyos Gray, Nat. Arr. Brit. Pl. 2: 68 (1821).
Carex punctulata R.Lesson ex A.Rich. in J.S.C.Dumont d'Urville, Voy. Astrolabe: 119 (1832).
Carex fascicularis var. *minor* Boott, Ill. Carex 1: 53 (1858).
Carex latifolia Sol. ex Boott, Ill. Carex 1: 52 (1858), nom. illeg.
Carex forsteri var. *fascicularis* Hook.f., Handb. N. Zeal. Fl. 1: 315 (1867).
Carex forsteri var. *minor* Hook.f., Handb. N. Zeal. Fl. 1: 315 (1867).
Carex longibracteata Dulac, Fl. Hautes -Pyrénées: 57 (1867), nom. illeg.
Carex pseudocyperus f. *minor* Hampe, Prodr. Fl. Hercyn.: 299 (1873).
Carex novae-selandiae Boeck., Flora 61: 169 (1878). *Carex pseudocyperus* f. *novae-selandiae* (Boeck.) Kük. in H.G.A.Engler (ed.), Pflanzenr., IV, 20(38): 696 (1909).
Carex longibracteata St.-Lag. in A.Cariot, Étude Fl., ed. 8, 2: 863 (1889), nom. illeg.
Carex pseudocyperus var. *americana* Hochst. ex L.H.Bailey, Mem. Torrey Bot. Club 1: 54 (1889).
Carex pseudocyperus var. *furcata* Kük., Allg. Bot. Z. Syst. 7: 195 (1901).
Carex pseudocyperus f. *pallidior* Kük. in H.G.A. Engler (ed.), Pflanzenr., IV, 20(38): 697 (1909).
Carex pseudocyperus f. *multispicula* Lepage, Naturaliste Canad. 79: 177 (1952).

Carex pseudodahurica A.P.Khokhr. in ?
Russian Far East. 31+. Hemicr. or rhizome geophyte.

Carex pseudodispalata K.T.Fu, in Fl. Tsinlingensis 1(1): 448 (1976).
China (Shaanxi). 36 CHN. Hemicr. or rhizome geophyte.

Carex pseudofoetida Kük. ex Ostenf., Bot. Tidsskr. 28: 226 (1907).
NE. Turkey to Mongolia. 30 ALT 32 KGZ TZK UZB 33 TCS 34 AFG IRN TUR 36 CHQ CHT CHX 37 MON 40 EHM NEP PAK WHM. Hemicr. or rhizome geophyte.

subsp. ***acrifolia*** (V.I.Krecz.) Kukkonen, Ann. Bot. Fenn. 21: 384 (1984).
NE. Turkey to NW. & N. Iran. 33 TCS 34 IRN TUR. Hemicr. or rhizome geophyte.
Carex foetida var. *minor* Kük., Oesterr. Bot. Z. 47: 134 (1897).
Carex curaica var. *armena* Schischk., Izv. Imp. Tomsk. Univ. 81: 317 (1901).
**Carex acrifolia* V.I.Krecz. in V.L.Komarov (ed.), Fl. URSS 3: 591 (1935). *Carex enervis* subsp. *acrifolia* (V.I.Krecz.) T.V.Egorova, Novosti Sist. Vyssh. Rast. 22: 54 (1985).

subsp. ***afghanica*** Kukkonen, Ann. Bot. Fenn. 21: 384 (1984).
Afghanistan to Himalaya. 34 AFG 40 EHM NEP PAK WHM. Hemicr. or rhizome geophyte.

subsp. ***pseudofoetida***
Siberia to Mongolia. 30 ALT 32 KGZ TZK UZB 36 CHQ CHT CHX 37 MON. Hemicr. or rhizome geophyte.
Carex slobodovii V.I.Krecz. in V.L.Komarov (ed.), Fl. URSS 3: 595 (1935).
Carex roborowskii V.I.Krecz., Bot. Mater. Gerb. Bot. Inst. Komarova Akad. Nauk S.S.S.R. 7: 36 (1937). *Vignea roborowskii* (V.I.Krecz.) Soják, Cas. Nár. Mus., Odd. Prír. 148: 197 (1979 publ. 1980).

Carex ×pseudofulva Fernald, Rhodora 35: 231 (1933). *C. hostiana* var. *laurentiana* × *C. lepidocarpa*. *Carex ×leutzii* var. *pseudofulva* (Fernald) Boivin in ?
E. Canada. 72 NFL QUE. Hemicr. or rhizome geophyte.

Carex ×pseudohelvola Kihlm., Meddeland. Soc. Fauna Fl. Fenn. 1888: 71 (1888). *C. canescens* × *C. mackenziei*. *Carex ×helvola* subsp. *pseudohelvola* (Kihlm.) Nyman, Consp. Fl. Eur., Suppl. 2, 2: 324 (1890).
Subarctic to NE. U.S.A. 10 FIN 14+ 70+ 72 NBR NFL QUE 75 MAI. Hemicr. or rhizome geophyte.

Carex pseudohumilis F.T.Wang & Y.L.Chang ex P.C.Li, Acta Phytotax. Sin. 37: 172 (1999).
China (Yunnan). 36 CHC. Hemicr. or rhizome geophyte.

Carex pseudohypochlora Y.L.Chang & Y.L.Yang, in Fl. Plant. Herb. Chinae Bor.-Or. 11: 205 (1976).
China. 36+. Hemicr. or rhizome geophyte.
Carex pseudohypochlora f. *denticulata* Y.L.Chang & Y.L.Yang, in Fl. Plant. Herb. Chinae Bor.-Or. 11: 205 (1976).

Carex pseudolaticeps Tang & F.T.Wang ex S.Y.Liang, Acta Phytotax. Sin. 36: 536 (1998).
China (Hong Kong). 36 CHS. Hemicr. or rhizome geophyte.

Carex pseudoligulata L.K.Dai, Acta Phytotax. Sin. 32: 187 (1994).
China (Hunan). 36 CHS. Hemicr. or rhizome geophyte.

Carex pseudololiacea F.Schmidt, Reis. Amur-Land., Bot.: 193 (1868).
Russian Far East, N. & C. Japan. 31 KUR SAK 38 JAP. Hemicr. or rhizome geophyte.
Carex pseudololiacea f. *sejuncta* Kük. in H.G.A. Engler (ed.), Pflanzenr., IV, 20(38): 225 (1909).

Carex pseudomacloviana G.A.Wheeler, Hickenia 2: 193 (1996).
Argentina (Salta, Tucumán). 85 AGW. Hemicr. or rhizome geophyte.

Carex pseudophyllocephala L.K.Dai, Acta Phytotax. Sin. 32: 187 (1994).
China (Hunan). 36 CHS. Hemicr. or rhizome geophyte.

Carex pseudosabynensis (T.V.Egorova) A.E.Kozhevn., Sosud. Rast. Sovet. Dal'nego Vostoka 3: 266 (1988).
Russian Far East. 31+. Hemicr. or rhizome geophyte.
**Carex umbrosa* subsp. *pseudosabynensis* T.V. Egorova, Novosti Sist. Vyssh. Rast. 17: 104 (1980).

Carex pseudosadoensis Akiyama, J. Fac. Sci. Hokkaido Univ., Ser. 5, Bot. 2: 24, 119 (1932). *C. angustisquama* × *C. sadoensis*.
Japan. 38 JAP. Hemicr. or rhizome geophyte.

Carex pseudospachiana H.Lév. & Vaniot, Repert. Spec. Nov. Regni Veg. 3: 349 (1907).
China. 36+. Hemicr. or rhizome geophyte.

Carex pseudosphaemgyna Nelmes, Bull. Misc. Inform. Kew 1937: 473 (1937).
EC. Trop. Africa. 23 BUR. Hemicr.

Carex pseudosupina Y.C.Tang ex L.K.Dai, Acta Phytotax. Sin. 37: 177 (1999).
China (Sichuan). 36 CHC. Hemicr. or rhizome geophyte.

Carex pseudotristachya X.F.Jin & C.Z.Zheng, Acta Phytotax. Sin. 42: 543 (2004).
China (Zhejiang). 36 CHS. Hemicr. or rhizome geophyte.

Carex psilocarpa Steud., Syn. Pl. Glumac. 2: 195 (1855).
Mexico to El Salvador. 79 MXC MXE MXG MXS MXT 80 ELS HON. Hemicr. or rhizome geophyte.
**Carex maculata* Liebm., Mexic. Halvgr.: 268 (1850), nom. illeg.
Carex consors C.B.Clarke, Kew Bull., Addit. Ser. 8: 76 (1908).

Carex psychrophila Nees in R.Wight, Contr. Bot. India: 127 (1834).
E. Afghanistan to SC. China. 34 AFG 36 CHC 40 EHM NEP PAK WHM. Hemicr. or rhizome geophyte.
Carex asperula Nees in R.Wight, Contr. Bot. India: 124 (1834).
Carex parvibracteata Nees in R.Wight, Contr. Bot. India: 125 (1834).
Carex celsa Boott, Ill. Carex 3: 108 (1862).

Carex pterocarpa Petrie, Trans. & Proc. New Zealand Inst. 31: 353 (1898).
New Zealand South I. 51 NZS. Hemicr. or rhizome geophyte.
Carex thomsonii Petrie, Trans. & Proc. New Zealand Inst. 18: 298 (1885 publ. 1886), nom. illeg.

Carex pterocaulos Nelmes, Hooker's Icon. Pl. 35: t. 3467 (1947).
N. Myanmar to N. Thailand. 36 CHC 41 MYA THA. Hemicr. or rhizome geophyte.

Carex pubigluma Reznicek, Contr. Univ. Michigan Herb. 17: 282 (1990).
Mexico (Chiapas). 79 MXT. Hemicr. or rhizome geophyte.

Carex pudica Honda, Bot. Mag. (Tokyo) 43: 543 (1929).
Japan (Honshu). 38 JAP. Hemicr. or rhizome geophyte.
Carex iseana Akiyama, J. Fac. Sci. Hokkaido Univ., Ser. 5, Bot. 3: 261 (1935).

Carex pulchra Boott, Ill. Carex 1: 4 (1858).
Nepal, Sikkim. 40 EHM NEP. Hemicr. or rhizome geophyte.

Carex pulchrifolia A.E.Kozhevn., Byull. Moskovsk. Obshch. Isp. Prir., Otd. Biol. 92: 133 (1987).
Russian Far East. 31+. Hemicr. or rhizome geophyte.

Carex pulicaris L., Sp. Pl.: 972 (1753). *Vignea pulicaris* (L.) Rchb. in J.C.Mössler, Handb. Gewächsk. ed. 2, 3: 1606 (1830). *Psyllophora pulicaris* (L.) Schur, Verh. Mitth. Siebenbürg. Vereins Naturwiss. Hermannstadt 4: 78 (1853). *Caricinella pulicaris* (L.) St.-Lag. in A.Cariot, Étude Fl., ed. 8, 2: 881 (1889).
Europe. 10 DEN FIN GRB ICE IRE NOR SWE 11 AUT BGM CZE GER HUN NET POL SWI 12 FRA SPA 13 ITA YUG 14 BLT. Hemicr. or rhizome geophyte.
Carex psyllophora L.f., Suppl. Pl.: 413 (1782). *Psyllophora vulgaris* Heuff., Flora 27: 528 (1844).
Carex decipiens J.Gay, Ann. Sci. Nat. (Paris) 26: 209 (1832). *Carex macrostylon* var. *decipiens* (J.Gay) Nyman, Consp. Fl. Eur.: 784 (1882).
Carex guthnickiana J.Gay, Ann. Sci. Nat., Bot., II, 10: 289 (1838).
Psyllophora puliciformis Montandon in F.Friche-Joset, Syn. Fl. Jura, ed. 2: 341 (1868).
Carex pulicaris var. *caespitosa* Vollm., Ber. Bayer. Bot. Ges. 9: 54 (1904).

Carex pumila Thunb. in J.A.Murray, Syst. Veg. ed. 14: 846 (1784). *Carex nutans* var. *pumila* (Thunb.) Boeck., Linnaea 41: 298 (1877).
Temp. Asia, Australia, New Zealand. 31 KUR SAK 36 CHM CHN CHS 38 JAP KOR NNS TAI 50 NFK NSW QLD VIC 51 CTM NZN NZS (73) ore (78) nca (85) clc. Hemicr. or rhizome geophyte.
Carex littorea Labill., Nov. Holl. Pl. 2: 69 (1806). *Anithista littorea* (Labill.) Raf., Good Book: 26 (1840). *Carex pumila* subsp. *littorea* (Labill.) Kük., Bot. Jahrb. Syst. 27: 551 (1899).
Carex urvillei Brongn. in L.I.Duperrey, Voy. Monde, Phan.: 157 (1833). *Carex pumila* var. *urvillei* (Brongn.) Kük., Bot. Jahrb. Syst. 27: 552 (1899).
Carex fusca Sol. ex Boott, Ill. Carex 4: 217 (1867), nom. illeg.
Carex platyrhyncha Franch. & Sav., Enum. Pl. Jap. 2: 582 (1878). *Carex nutans* var. *platyrhyncha* (Franch. & Sav.) Kük. in H.G.A.Engler (ed.), Pflanzenr., IV, 20(38): 740 (1909).
Carex krullii Boeck., Flora 65: 59 (1882).
Carex sepulta Phil., Anales Univ. Chile 93: 494 (1896).
Carex forbesii C.B.Clarke, J. Linn. Soc., Bot. 36: 286 (1903). *Carex pumila* var. *forbesii* (C.B.Clarke) Kük. in H.G.A.Engler (ed.), Pflanzenr., IV, 20(38): 740 (1909).

Carex punctata Gaudin, Agrost. Helv. 2: 152 (1811).
Macaronesia, Europe to Turkey, NW. Africa. 10 GRB IRE NOR SWE 11 AUT GER NET POL SWI 12 COR FRA POR SAR SPA 13 BUL GRC ITA KRI SIC TUE YUG 20 ALG MOR TUN 21 AZO MDR 34 TUR. Hemicr. or rhizome geophyte.

var. ***laevicaulis*** (Hochst. ex Seub.) Boott, Ill. Carex 4: 155 (1867).
Macaronesia. 21 AZO MDR. Hemicr. or rhizome geophyte.
**Carex laevicaulis* Hochst. ex Seub., Fl. Azor.: 21 (1840).

var. ***punctata***
Europe to Turkey, NW. Africa. 10 GRB IRE NOR SWE 11 AUT GER NET POL SWI 12 COR FRA POR SAR SPA 13 BUL GRC ITA KRI SIC TUE YUG 20 ALG MOR TUN 34 TUR. Hemicr. or rhizome geophyte.
Carex corsicana Link, Enum. Hort. Berol. Alt. 1: 358 (1821).
Carex pallidior Degl. in J.L.A.Loiseleur-Deslongchamps, Fl. Gall., ed. 2: 299 (1828).
Carex helvetica Schleich. ex Kunth, Enum. Pl. 2: 448 (1837).
Carex lainzii Luceño, E.Rico & T.Romero, Anales Jard. Bot. Madrid 44: 429 (1987). *Carex punctata* subsp. *lainzii* (Luceño, E.Rico & T.Romero) Rivas Mart., Itinera Geobot. 15: 699 (2002).

Carex purdiei Boott, Ill. Carex 1: 26 (1858).
C. America to Ecuador. 80 COS GUA PAN 83 CLM ECU. Hemicr. or rhizome geophyte.

Carex purplevaginalis Q.S.Wang, J. Wuhan Bot. Res. 5: 343 (1987).
China. 36+. Hemicr. or rhizome geophyte.

Carex purpureosquamata L.K.Dai, Acta Phytotax. Sin. 37: 183 (1999).
China (Yunnan). 36 CHC. Hemicr. or rhizome geophyte.

Carex purpureotincta Ohwi, Acta Phytotax. Geobot. 2: 159 (1933).
Taiwan. 38 TAI. Hemicr. or rhizome geophyte.

Carex purpureovagina F.T.Wang & Y.L.Chang ex S.Yun Liang, in Fl. Reipubl. Popul. Sin. 12: 626 (2000).
China (Guangxi). 36 CHS. Hemicr.

Carex purpureovaginata Boeck., Vidensk. Meddel. Naturhist. Foren. Kjøbenhavn 1879–1880: 30 (1879).
SE. Brazil. 84 BZL. Hemicr. or rhizome geophyte.
Carex sonderiana Boott ex Lindm., Bih. Kongl. Svenska Vetensk.-Akad. Handl. 26(9): 38 (1900).
Carex sphaerogyna var. *brasiliensis* H.Pfeiff., Repert. Spec. Nov. Regni Veg. 17: 32 (1921).

Carex purpurifera Mack. in J.K.Small, Man. S.E. Fl.: 216 (1933). *Carex laxiflora* var. *purpurifera* (Mack.) Gleason, Phytologia 4: 22 (1952).
Ohio to SE. U.S.A. 75 OHI 78 ALA GEO KTY NCA TEN VRG. Hemicr. or rhizome geophyte.

Carex* ×*putjatini Kom., Izv. Imp. Bot. Sada Petra Velikago 16: 154 (1916). *C. meyeriana* × *C. schmidtii*.
Temp. Asia. 3+. Hemicr. or rhizome geophyte.

Carex putuoensis S.Yun Liang, Acta Phytotax. Sin. 33: 486 (1995).
China (Zhejiang). 36 CHS. Hemicr. or rhizome geophyte.

Carex pycnostachys Kar. & Kir., Bull. Soc. Imp. Naturalistes Moscou 15: 522 (1842). *Carex curaica* var. *pycnostachys* (Kar. & Kir.) Kük. in H.G.A.Engler (ed.), Pflanzenr., IV, 20(38): 124 (1909). *Carex curaica* subsp. *pycnostachys* (Kar. & Kir.) T.V.Egorova, Novosti Sist. Vyssh. Rast. 10: 104 (1973).
Siberia to Afghanistan. 30+ 32 KAZ KGZ TZK 34 AFG 36 CHX. Hemicr. or rhizome geophyte.

Carex pygmaea Boeck., Linnaea 40: 379 (1876).
Costa Rica to Ecuador. 80 COS PAN 83 CLM ECU. Hemicr. or rhizome geophyte.

Carex pyramidalis Kük., Bull. Herb. Boissier, II, 4: 52 (1904).
Comoros, Madagsascar. 29 COM MDG. Hemicr. or rhizome geophyte.
Carex crinigera var. *minor* Boott, Ill. Carex 2: 102 (1860).

Carex pyrenaica Wahlenb., Kongl. Vetensk. Acad. Nya Handl. 1803: 139 (1803). *Callistachys pyrenaica* (Wahlenb.) Heuff., Flora 27: 528 (1844). *Heuffelia pyrenaica* (Wahlenb.) Opiz, Ost & West, Beibl. 98: 392 (1845). *Psyllophora pyrenaica* (Wahlenb.) Schur, Verh. Mitth. Siebenbürg. Vereins Naturwiss. Hermannstadt 4: 78 (1853).
S. Europe to Caucasus. 12 FRA SPA 13 BUL ROM 33 TCS 34 TUR. Hemicr. or rhizome geophyte.
Carex fontanesiana DC. in J.B.A.M.de Lamarck & A.P.de Candolle, Fl. Franç., ed. 3, 3: 102 (1805).
Carex pulicarioides Ramond ex DC. in J.B.A.M.de Lamarck & A.P.de Candolle, Fl. Franç., ed. 3, 3: 101 (1805).
Carex ramondiana DC. in J.B.A.M.de Lamarck & A.P.de Candolle, Fl. Franç., ed. 3, 3: 101 (1805).
Carex acutissima Degl. in J.L.A.Loiseleur-Deslongchamps, Fl. Gall.: 628 (1807).
Carex denudata Lapeyr. ex Steud., Nomencl. Bot. 1: 160 (1821).
Carex marchandiana Lapeyr. ex Spreng., Syst. Veg. 3: 806 (1826).
Carex pyrenaica f. *longibracteata* Schur, Enum. Pl. Transsilv.: 698 (1866).
Dornera alpina Heuff. ex Schur, Enum. Pl. Transsilv.: 697 (1866).
Carex grosseckii Heuff. ex Boott, Ill. Carex 4: 148 (1867). *Carex pyrenaica* var. *grosseckii* (Heuff. ex Boott) Nyman, Consp. Fl. Eur.: 783 (1882).
Carex pyrenaica var. *altior* Kük. in H.G.A.Engler (ed.), Pflanzenr., IV, 20(38): 106 (1909).
Carex pyrenaica var. *articulata* Kük. in H.G.A.Engler (ed.), Pflanzenr., IV, 20(38): 106 (1909).
Carex micropodioides V.I.Krecz. in V.L.Komarov (ed.), Fl. URSS 3: 608 (1935). *Carex pyrenaica* subsp. *micropodioides* (V.I.Krecz.) Chandjian, Fl. Armenii 10: 526 (2001).

Carex qingdaoensis F.Z.Li & S.J.Fan, Bull. Bot. Res., Harbin 13: 71 (1993).
China (Shandong). 36 CHN. Hemicr. or rhizome geophyte.

Carex qinghaiensis Y.C.Yang, Acta Phytotax. Sin. 18: 364 (1980).
Qinghai. 36 CHQ. Hemicr. or rhizome geophyte.

Carex qingyangensis S.W.Su & S.M.Xu, Bull. Bot. Res., Harbin 8(3): 77 (1988).
China (Anhui). 36 CHS. Hemicr. or rhizome geophyte.

Carex qiyunensis S.W.Su & S.M.Xu, Bull. Bot. Res., Harbin 9(1): 50 (1989).
China (Anhui). 36 CHS. Hemicr. or rhizome geophyte.

Carex quadriflora (Kük.) Ohwi, Acta Phytotax. Geobot. 1: 74 (1932).
Primorye to Korea, N. Japan. 31 PRM 36 CHC CHM 38 JAP KOR. Hemicr. or rhizome geophyte.
Carex digitata var. *pallida* Meinsh., Trudy Imp. S.-Peterburgsk. Bot. Sada 18: 401 (1901).
**Carex digitata* subsp. *quadriflora* Kük. in H.G.A. Engler (ed.), Pflanzenr., IV, 20(38): 497 (1909).

Carex quasivaginata C.B.Clarke, Kew Bull., Addit. Ser. 8: 79 (1908). *Carex vaginata* subsp. *quasivaginata* (C.B.Clarke) Malyschev in ?
NE. Europe to Russian Far East. 10 FIN 14 RUE RUN 30 WSB 31+. Hemicr. or rhizome geophyte.

Carex* × *quebecensis Lepage, Naturaliste Canad. 91: 168 (1964). *C. bigelowii* × *C. saxatilis* var. *rhomalea*.
E. Canada. 72 QUE. Hemicr. or rhizome geophyte.

Carex queretarensis Reznicek & S.González, Contr. Univ. Michigan Herb. 21: 291 (1997).
Mexico (Querétaro). 79 MXE. Hemicr. or rhizome geophyte.

Carex quichensis F.J.Herm., J. Wash. Acad. Sci. 40: 284 (1950).
Guatemala. 80 GUA. Hemicr. or rhizome geophyte.

Carex* × *quirponensis Fernald, Rhodora 28: 164 (1926). *C. atratiformis* × *C. norvegica*.
E. Canada. 72 NFL QUE. Hemicr. or rhizome geophyte.

Carex raddei Kük., Bot. Centralbl. 77: 97 (1899). *Carex aristata* subsp. *raddei* (Kük.) Kük. in H.G.A.Engler (ed.), Pflanzenr., IV, 20(38): 755 (1909).
Russian Far East to N. China. 31+ 36 CHI CHM CHN. Hemicr. or rhizome geophyte.
Carex aristata var. *eriophylla* Kük. in H.G.A.Engler (ed.), Pflanzenr., IV, 20(38): 755 (1909). *Carex eriophylla* (Kük.) Kom., Mal. Opr. Rast. Dal'nevost. Kr.: 135 (1925).

Carex radfordii Gaddy, Novon 5: 259 (1995).
SE. U.S.A. 78 NCA SCA. Hemicr. or rhizome geophyte.

Carex radiata (Wahlenb.) Small, Fl. S.E. U.S.: 218 (1903).
S. Canada to NC. & E. U.S.A. 71 MAN 72 NBR NSC ONT QUE 74 ILL IOW KAN MIN MSO NDA NEB OKL SDA WIS 75 CNT INI MAI MAS MIC NWH NWJ NWY OHI PEN VER WVA 78 ALA ARK DEL KTY LOU MRY MSI NCA SCA TEN VRG WDC. Hemicr. or rhizome geophyte.
Carex radiata var. *staminea* Beck ex Howe in ?
**Carex stellulata* var. *radiata* Wahlenb., Kongl. Vetensk. Acad. Nya Handl. 24: 147 (1803). *Carex rosea* var. *radiata* (Wahlenb.) Dewey, Amer. J. Sci. Arts 10: 276 (1826).
Carex rosea var. *minor* Boott, Ill. Carex 2: 81 (1860).
Carex sylvicola J.M.Webber & P.W.Ball, Taxon 28: 613 (1979).

Carex radicalis Boott, Ill. Carex 1: 56 (1858).
Sikkim. 40 EHM. Hemicr. or rhizome geophyte.

Carex radiciflora Dunn, J. Linn. Soc., Bot. 38: 371 (1908).
SE. China. 36 CHS. Hemicr. or rhizome geophyte.

Carex radicina Z.P.Wang, J. Nanjing Univ. (Biology) 1962(2): 53 (1962).
China (Jiangsu). 36 CHS. Hemicr.

Carex raleighii Nelmes, Bull. Misc. Inform. Kew 1939: 310 (1939).
SE. Australia. 50 NSW TAS VIC. Hemicr. or rhizome geophyte.

Carex ramenskii Kom., Repert. Spec. Nov. Regni Veg. 13: 164 (1914). *Carex salina* subsp. *ramenskii* (Kom.) T.V.Egorova, Novosti Sist. Vyssh. Rast. 22: 54 (1985). *Carex subspathacea* subsp. *ramenskii* (Kom.) T.V.Egorova, Bot. Zhurn. (Moscow & Leningrad) 75: 864 (1990).
Russian Far East to Alaska, N. Japan. 31 KAM SAK 38 JAP 70 ALU ASK YUK. Hemicr. or rhizome geophyte.
Carex salina var. *tristigmatica* Kük. in H.G.A.Engler (ed.), Pflanzenr., IV, 20(38): 362 (1909).
Carex ramenskii var. *caudata* Hultén, Acta Univ. Lund., n.s., 38(1): 342 (1942).

Carex ramentaceofructus K.T.Fu, in Fl. Tsinlingensis 1(1): 447 (1976).
China. 36 CHC. Hemicr. or rhizome geophyte.

Carex ramosa Willd., Sp. Pl. 4: 247 (1805).
Comoros, Mascarenes. 29 COM MAU REU. Hemicr. or rhizome geophyte.
Carex rhynchosporoides Steud., Syn. Pl. Glumac. 2: 195 (1855).
Carex crinigera Boott, Ill. Carex 2: 102 (1860).

Carex ramosii Kük., Repert. Spec. Nov. Regni Veg. 8: 8 (1910).
Philippines. 42 PHI. Hemicr. or rhizome geophyte.

Carex randalpina B.Walln., Linzer Biol. Beitr. 25: 720 (1993).
Alps. 11 AUT GER 13 ITA YUG. Hemicr. or rhizome geophyte.

Carex raoulii Boott in J.D.Hooker, Fl. Nov.-Zel.: 283 (1853).
New Zealand. 51 NZN NZS. Hemicr. or rhizome geophyte.
Carex haastiana Boeck., Flora 61: 168 (1878). *Carex raoulii* f. *haastiana* (Boeck.) Kük. in H.G.A.Engler (ed.), Pflanzenr., IV, 20(38): 689 (1909).
Carex mcmahonii Petrie, Trans. & Proc. New Zealand Inst. 56: 6 (1926).

Carex raphidocarpa Nees in R.Wight, Contr. Bot. India: 122 (1834).
India. 40 IND. Hemicr. or rhizome geophyte.
Carex rhaphidocarpa Kunth, Enum. Pl. 2: 512 (1837).
Carex mercarensis Hochst. ex Steud., Syn. Pl. Glumac. 2: 194 (1855). *Carex raphidocarpa* var. *mercarensis* (Hochst. ex Steud.) Kük. in H.G.A.Engler (ed.), Pflanzenr., IV, 20(38): 277 (1909).
Carex mercarensis var. *major* Steud., Syn. Pl. Glumac. 2: 194 (1855). *Carex raphidocarpa* var. *major* (Steud.) Kük. in H.G.A.Engler (ed.), Pflanzenr., IV, 20(38): 277 (1909).
Carex amoena Boott, Ill. Carex 3: 106 (1862).

Carex rara Boott, Trans. Linn. Soc. London 20: 139 (1846). *Carex rara* var. *typica* Domin, Biblioth. Bot. 89: 491 (1915), nom. inval.
Himalaya to Japan, Sri Lanka. 31 PRM 36 CHC CHM CHS CHT 38 JAP KOR 40 ASS EHM NEP SRL. Hemicr. or rhizome geophyte.

subsp. ***patanicola*** T.Koyama, Bull. Natl. Sci. Mus. Tokyo, n.s., 17: 65 (1974).
Sri Lanka. 40 SRL. Hemicr. or rhizome geophyte.

subsp. ***rara***
Himalaya to Japan. 31 PRM 36 CHC CHM CHS CHT 38 JAP KOR 40 ASS EHM NEP. Hemicr. or rhizome geophyte.
Carex biwensis Franch., Bull. Soc. Philom. Paris, VIII, 7: 28 (1895). *Carex rara* var. *biwensis* (Franch.) Kük. in H.G.A.Engler (ed.), Pflanzenr., IV, 20(38): 102 (1909).
Carex rara f. *macrolepis* Franch., Nouv. Arch. Mus. Hist. Nat., III, 8: 198 (1896).
Carex rara f. *yunnanensis* Franch., Nouv. Arch. Mus. Hist. Nat., III, 8: 198 (1896).
Carex oldhamii C.B.Clarke, Kew Bull., Addit. Ser. 8: 71 (1908).
Carex pseudobiwensis Kitag., Rep. Inst. Sci. Res. Manchoukuo 6(4): 119 (1942).

Carex rariflora (Wahlenb.) Sm., Engl. Fl. 4: 100 (1828).
N. & E. Europe to Russian Far East, Subarctic America to Maine. 10 FIN GRB ICE IRE NOR SWE 14 RUN 30 KRA WSB YAK 31 KAM KUR MAG SAK 70 ASK GNL NUN NWT YUK 71 MAN 72 LAB NFL ONT QUE 75 MAI. Hemicr. or rhizome geophyte.
**Carex limosa* var. *rariflora* Wahlenb., Kongl. Vetensk. Acad. Nya Handl. 24: 162 (1803). *Trasus rariflorus* (Wahlenb.) Gray, Nat. Arr. Brit. Pl. 2: 68 (1821).

var. ***androgyna*** Porsild, Sargentia 4: 20 (1943).
W. Subarctic America. 70 NWT. Hemicr. or rhizome geophyte.

var. ***rariflora***
N. & E. Europe to Russian Far East, Subarctic America to Maine. 10 FIN GRB ICE IRE NOR SWE 14 RUN 30 KRA WSB YAK 31 KAM KUR MAG SAK 70 ASK GNL NUN NWT YUK 71 MAN 72 LAB NFL ONT QUE 75 MAI. Hemicr. or rhizome geophyte.
Carex limosa var. *pauciflora* Vahl ex Schkuhr, Beschr. Riedgräs. 1: 105 (1801).
Carex rariflora f. *pallidior* A.Blytt, Norges Fl. 1: 240 (1861).
Carex rariflora f. *baeostachya* Norman, Forh. Vidensk.-Selsk. Kristiana 1893: 52 (1893).
Carex rariflora f. *expallida* Norman, Forh. Vidensk.-Selsk. Kristiana 1893: 51 (1893).
Carex rariflora f. *rufescens* Norman, Forh. Vidensk.-Selsk. Kristiana 1893: 51 (1893).
Carex rariflora f. *brevipedumculata* Meinsh., Trudy Imp. S.-Peterburgsk. Bot. Sada 18: 351 (1900).

Carex ratongensis (C.B.Clarke) C.B.Clarke, Kew Bull., Addit. Ser. 8: 75 (1908).
India. 40 IND. Hemicr. or rhizome geophyte.
**Carex myosurus* var. *ratongensis* C.B.Clarke in J.D.Hooker, Fl. Brit. India 6: 723 (1894).

Carex raynoldsii Dewey, Amer. J. Sci. Arts, II, 32: 39 (1861).
W. Canada to W. U.S.A. 71 ABT BRC SAS 73 COL IDA MNT ORE WAS WYO 76 CAL NEV UTA. Hemicr. or rhizome geophyte.
Carex lyallii Boott, Ill. Carex 4: 150 (1867).

Carex recta Boott in W.J.Hooker, Fl. Bor.-Amer. 2: 220 (1839).
N. & NE. Europe, E. Canada to NE. U.S.A. 10 FIN GRB NOR SWE 14 RUN 71 MAN 72 LAB NFL NSC ONT PEI QUE 75 MAI MAS NWH. Hemicr. or rhizome geophyte.
Carex halophila F.Nyl., Spic. Fl. Fenn. 2: 50 (1844). *Carex salina* subsp. *halophila* (F.Nyl.) Almq. in C.J. Hartmann, Handb. Skand. Fl., ed. 11: 466 (1879).
Carex salina f. *elata* A.Blytt, Norges Fl. 1: 219 (1861).
Carex salina subsp. *fuliginea* A.Blytt, Norges Fl. 1: 219 (1861).
Carex salina f. *pallida* A.Blytt, Norges Fl. 1: 219 (1861).
Carex salina var. *caespitosa* A.Blytt, Veg. Sognefjord: 84 (1869). *Carex salina* f. *caespitosa* (A.Blytt) Kük. in H.G.A.Engler (ed.), Pflanzenr., IV, 20(38): 363 (1909).
Carex salina subsp. *obtusa* A.Blytt, Veg. Sognefjord.: 84 (1869). *Carex salina* f. *obtusa* (A.Blytt) Kük. in H.G.A.Engler (ed.), Pflanzenr., IV, 20(38): 363 (1909).
Carex salina f. *ostrobottnica* Almq. in A.H.Hjlet, Consp. Fl. Fennic. 3: 280 (1895). *Carex ostrobottnica* (Almq.) Saxén, Acta Bot. Fenn. 22: 10, 18 (1938).
Carex ×*saxenii* var. *ferruginea* Lepage, Naturaliste Canad. 83: 142 (1956).

Carex* ×*reducta Drejer, Rev. Car. Bor.: 36 (1841). *Carex subspathacea* var. *reducta* (Drejer) Fr., Summa Veg. Scand.: 232 (1845). *Carex salina* var. *reducta* (Drejer) Nyman, Consp. Fl. Eur.: 776 (1882). *Carex salina* f. *reducta* (Drejer) Almq., Bot. Not. 1891: 127 (1891).
Greenland. 70 GNL. Hemicr. or rhizome geophyte.

Carex regeliana (Kük.) Litv., Trudy Bot. Muz. Imp. Akad. Nauk 7: 94 (1910).
C. Asia. 32 KGZ UZB. Hemicr. or rhizome geophyte.
**Carex capillaris* var. *regeliana* Kük. in H.G.A.Engler (ed.), Pflanzenr., IV, 20(38): 591 (1909).

Carex reichei Kük., Bot. Jahrb. Syst. 27: 504 (1899).
C. & S. Chile, S. Argentina. 85 AGS CLC CLS. Hemicr. or rhizome geophyte.

Carex reinii Franch. & Sav., Enum. Pl. Jap. 2: 559 (1878).
C. & S. Japan. 38 JAP. Hemicr. or rhizome geophyte.
Carex nambuensis Franch., Bull. Soc. Philom. Paris, VIII, 7: 44 (1895).
Carex ogawae Akiyama, J. Fac. Sci. Hokkaido Univ., Ser. 5, Bot. 2: 36, 172 (1932).

Carex relaxa V.I.Krecz. in V.L.Komarov (ed.), Fl. URSS 3: 616 (1935).
Siberia to Mongolia. 30+ 37 MON. Hemicr. or rhizome geophyte.

Carex remota L., Fl. Angl.: 24 (1754). *Vignea remota* (L.) Rchb. in J.C.Mössler, Handb. Gewächsk. ed. 2, 3: 1616 (1830). *Diemisa remota* (L.) Raf., Good Book: 27 (1840). *Caricina remota* (L.) St.-Lag. in A.Cariot, Étude Fl., ed. 8, 2: 879 (1889).
N. Africa, Europe to Iran and Lebanon. 10 DEN FIN GRB IRE NOR SWE 11 AUT BGM CZE GER HUN NET POL SWI 12 COR FRA POR SAR SPA 13 ALB BUL GRC ITA KRI ROM SIC TUE YUG 14 BLR BLT KRY RUC RUE RUS RUW UKR 20 ALG MOR TUN (31) 33 NCS TCS 34 AFG CYP? IRN LBS TUR 36 CHC CHN CHS CHT 38 JAP TAI 40 ASS EHM IND NEP PAK SRL WHM 41 THA VIE 42 JAW SUM. Hemicr. or rhizome geophyte.

subsp. ***alta*** (Boott) Kük. in H.G.A.Engler (ed.), Pflanzenr., IV, 20(38): 234 (1909).
S. China to Jawa. 36 CHC CHS CHT 40 ASS 41 THA VIE 42 JAW. Hemicr. or rhizome geophyte.

**Carex alta* Boott, Proc. Linn. Soc. London 1: 254 (1845). *Vignea alta* (Boott) Soják, Cas. Nár. Mus., Odd. Prír. 148: 194 (1979 publ. 1980).
Carex brizopyrum Kunze in C.Schkuhr, Beschr. Riedgräs., Suppl.: 169 (1851). *Carex remota* var. *brizopyrum* (Kunze) Boeck., Linnaea 39: 126 (1875).
Carex alta var. *gymnocarpa* H.Lév. & Vaniot, Bull. Acad. Int. Géogr. Bot. 10: 125 (1901).
Carex imbricata Kük. in H.F.von Handel-Mazzetti, Symb. Sin. 7: 1260 (1936).
Carex craspedotricha Nelmes, Bull. Misc. Inform. Kew 1939: 657 (1940). *Vignea craspedotricha* (Nelmes) Soják, Cas. Nár. Mus., Odd. Prír. 148: 195 (1979 publ. 1980).
Carex squamata V.I.Krecz., Bot. Mater. Gerb. Bot. Inst. Komarova Akad. Nauk S.S.S.R. 9: 196 (1946).

subsp. ***remota***
N. Africa, Europe to Iran and Lebanon. 10 DEN FIN GRB IRE NOR SWE 11 AUT BGM CZE GER HUN NET POL SWI 12 COR FRA POR SAR SPA 13 ALB BUL GRC ITA KRI ROM SIC TUE YUG 14 BLR BLT KRY RUC RUE RUS RUW UKR 20 ALG MOR TUN 33 NCS TCS 34 CYP? IRN LBS TUR 36 CHC CHN CHS CHT 38 JAP TAI 40 ASS EHM IND NEP PAK SRL WHM 41 THA VIE 42 JAW SUM. Hemicr. or rhizome geophyte.
Carex axillaris L., Sp. Pl. ed. 2: 1382 (1763). *Vignea axillaris* (L.) Rchb. in J.C.Mössler, Handb. Gewächsk. ed. 2, 3: 1617 (1830). *Carex remota* var. *axillaris* (L.) Döll, Fl. Baden: 258 (1857).
Carex hailstonii S.Gibson, Phytologist 1: 870 (1841).
Carex remota var. *repens* Brittinger ex Rchb., Icon. Fl. Germ. Helv. 8: 9 (1846).
Carex remota f. *stricta* Asch. in P.F.A.Ascherson, Fl. Brandenburg 1: 770 (1864).
Vignea remota var. *subloliacea* Schur, Enum. Pl. Transsilv.: 703 (1866). *Carex remota* var. *subloliacea* (Schur) Kük. in H.G.A.Engler (ed.), Pflanzenr., IV, 20(38): 234 (1909).
Carex remota f. *rigida* Waisb., Oesterr. Bot. Z. 49: 67 (1899).
Carex remota f. *doelliana* Asch. & Graebn., Syn. Mitteleur. Fl. 2(2): 67 (1902).

var. ***remotispicula*** (Hayata) Ohwi, Mem. Coll. Sci. Kyoto Imp. Univ., Ser. B, Biol. 6: 258 (1931).
China, Taiwan. 36 CHC CHN CHS 38 TAI. Hemicr. or rhizome geophyte.
**Carex remotispicula* Hayata, Icon. Pl. Formosan. 10: 57 (1921). *Carex rochebrunii* subsp. *remotispicula* (Hayata) T.Koyama, in Fl. Taiwan 5: 369 (1978).

var. ***reptans*** Franch., Nouv. Arch. Mus. Hist. Nat., III, 8: 235 (1896). *Carex rochebrunii* subsp. *reptans* (Franch.) S.Yun Liang & Y.C.Tang, Acta Phytotax. Sin. 28: 158 (1990).
C. China. 36 CHC CHN. Hemicr. or rhizome geophyte.

subsp. ***rochebrunii*** (Franch. & Sav.) Kük. in H.G.A. Engler (ed.), Pflanzenr., IV, 20(38): 234 (1909).
Nepal to Jawa and C. & S. Japan. (31) 36 CHS 38 JAP TAI 40 EHM IND NEP SRL 42 JAW SUM. Hemicr. or rhizome geophyte.
**Carex rochebrunii* Franch. & Sav., Enum. Pl. Jap. 2: 555 (1878). *Carex remota* var. *rochebrunii* (Franch. & Sav.) C.B.Clarke in J.D.Hooker, Fl. Brit. India 6: 706 (1894). *Vignea rochebrunii* (Franch. & Sav.) Soják, Cas. Nár. Mus., Odd. Prír. 148: 197 (1979 publ. 1980).
Carex monopleura V.I.Krecz., Bot. Mater. Gerb. Bot. Inst. Komarova Akad. Nauk S.S.S.R. 7: 35 (1937). *Vignea monopleura* (V.I.Krecz.) Soják, Cas. Nár. Mus., Odd. Prír. 148: 196 (1979 publ. 1980).

subsp. ***stewartii*** Kukkonen, Ann. Bot. Fenn. 21: 385 (1984).
E. Afghanistan to Himalaya. 34 AFG 40 EHM PAK WHM. Hemicr. or rhizome geophyte.

Carex remotiuscula Wahlenb., Kongl. Vetensk. Acad. Nya Handl. 1803: 147 (1803). *Vignea remotiuscula* (Wahlenb.) Soják, Cas. Nár. Mus., Odd. Prír. 148: 197 (1979 publ. 1980).
Russian Far East, Korea, N. & C. Japan. 31 PRM SAK 36 CHC CHM CHN CHS 38 JAP KOR. Hemicr. or rhizome geophyte.
Carex remotiformis Kom., Trudy Imp. S.-Peterburgsk. Bot. Sada 18: 444 (1900). *Carex remota* var. *remotiformis* (Kom.) Kük. in H.G.A.Engler (ed.), Pflanzenr., IV, 20(38): 235 (1909).
Carex remota var. *enervulosa* Kük., Bot. Jahrb. Syst. 36(82): 7 (1905). *Carex remotiuscula* var. *enervulosa* (Kük.) K.T.Fu, in Fl. Tsinlingensis 1(1): 241 (1976).

Carex renauldii H.Lév., Cat. Pl. Yunnan: 289 (1917).
C. Madagascar. 29 MDG. Hemicr.
**Carex lebrunei* H.Lév., Bull. Acad. Int. Géogr. Bot. 27: 4 (1917), nom. illeg.

Carex reniformis (L.H.Bailey) Small, Fl. S.E. U.S.: 220 (1903).
C. & E. U.S.A. 74 ILL MSO OKL 75 INI 77 TEX 78 ALA ARK FLA GEO KTY LOU MSI NCA SCA TEN VRG. Hemicr. or rhizome geophyte.
**Carex straminea* var. *reniformis* L.H.Bailey, Mem. Torrey Bot. Club 1: 73 (1889). *Carex straminea* f. *reniformis* (L.H.Bailey) Kük. in H.G.A.Engler (ed.), Pflanzenr., IV, 20(38): 207 (1909).

Carex renschiana Boeck., Bot. Jahrb. Syst. 5: 515 (1884).
C. Madagascar. 29 MDG. Hemicr. or rhizome geophyte.
Carex renschiana var. *minor* Boeck., Bot. Jahrb. Syst. 5: 516 (1884).
Carex renschiana var. *laxissima* Cherm., Bull. Soc. Bot. France 70: 412 (1923).
Carex renschiana var. *brachystachya* Cherm., Bull. Soc. Bot. France 73: 554 (1926 publ. 1927).

Carex repanda C.B.Clarke in J.D.Hooker, Fl. Brit. India 6: 720 (1894). *Carex wightiana* var. *repanda* (C.B. Clarke) Kük. in H.G.A.Engler (ed.), Pflanzenr., IV, 20(38): 288 (1909).
E. Himalaya to SC. China. 36 CHC 40 ASS. Hemicr. or rhizome geophyte.

Carex repens Bellardi, App. Fl. Pedem.: 42 (1793). *Carex arenaria* var. *repens* (Bellardi) Wahlenb., Kongl. Vetensk. Acad. Nya Handl. 24: 144 (1803). *Vignea repens* (Bellardi) Rchb. in J.C.Mössler, Handb. Gewächsk. ed. 2, 3: 1618 (1830). *Carex disticha* var. *repens* (Bellardi) Asch., Fl. Brandenburg 1: 763 (1864). *Vignea repens* (Bellardi) Schur, Enum. Pl. Transsilv.: 701 (1866).

C. Europe, N. Italy, Romania. 11 AUT GER HUN POL 12 FRA? 13 ITA ROM. Hemicr. or rhizome geophyte.

Carex laggeri Wimm., Flora 37: 161 (1854). *Carex repens* var. *laggeri* (Wimm.) Nyman, Consp. Fl. Eur.: 782 (1882).

Carex disticha var. *gracilis* Boeck., Linnaea 39: 140 (1875).

Carex ligerica var. *umbrosa* Sprib., Verz. Inowrazlaw Gefässpfl. 2: 8 (1889). *Carex arenaria* var. *umbrosa* (Sprib.) Sprib., Deutsche Ges. Kunst Posen Z. Naturwiss. Abt. 1895: 45 (1895).

Carex posnaniensis Sprib., Allg. Bot. Z. Syst. 2: 184 (1896). *Carex arenaria* var. *posnaniensis* (Sprib.) Asch. & Graebn., Fl. Nordostdeut. Flachl.: 146 (1898).

Carex reptabunda (Trautv.) V.I.Krecz., Izv. Bot. Sada Akad. Nauk S.S.S.R. 30: 134 (1931 publ. 1932).

Siberia to Mongolia. 30+ 36 CHI CHM CHN 37 MON. Hemicr. or rhizome geophyte.

**Carex stenophylla* var. *reptabunda* Trautv., Trudy Imp. S.-Peterburgsk. Bot. Sada 1(2): 194 (1871). *Vignea reptabunda* (Trautv.) Soják, Cas. Nár. Mus., Odd. Prír. 148: 197 (1979 publ. 1980).

Carex decurticaulis Ohwi, Acta Phytotax. Geobot. 11: 260 (1942).

Carex heilongjiangensis Y.L.Chou, Bull. Bot. Lab. N. E. Forest. Inst., Harbin 1980(9): 85 (1980).

Carex resectans Cheeseman, Trans. & Proc. New Zealand Inst. 24: 413 (1891 publ. 1892).

New Zealand North I. 51 NZN. Hemicr. or rhizome geophyte.

Carex inversa var. *radicata* Cheeseman, Trans. & Proc. New Zealand Inst. 16: 425 (1884).

Carex retroflexa Muhl. ex Willd., Sp. Pl. 4: 235 (1805). *Vignea retroflexa* (Muhl. ex Willd.) Rchb. in J.C.Mössler, Handb. Gewächsk. ed. 2, 3: 1610 (1830). *Carex rosea* var. *retroflexa* (Muhl. ex Willd.) Torr., Ann. Lyceum Nat. Hist. New York 3: 389 (1836). *Diemisa retroflexa* (Muhl. ex Willd.) Raf., Good Book: 27 (1840).

S. Ontario to C. & E. U.S.A. 72 ONT 74 ILL KAN MSO OKL 75 CNT INI MAS MIC NWJ NWY OHI PEN RHO WVA 77 TEX 78 ARK DEL FLA GEO KTY LOU MRY MSI NCA SCA TEN VRG WDC. Hemicr. or rhizome geophyte.

Carex retrofracta Kük., Repert. Spec. Nov. Regni Veg. 27: 110 (1929).

China (Zhejiang). 36 CHS. Hemicr. or rhizome geophyte.

Carex retrorsa Schwein., Ann. Lyceum Nat. Hist. New York 1: 71 (1824).

Canada to N. & WC. U.S.A. 71 BRC MAN SAS 72 NBR ONT QUE 73 IDA MNT ORE WAS WYO 74 ILL IOW MIN WIS 75 CNT INI MAI MAS MIC NWY PEN VER 76 UTA. Hemicr. or rhizome geophyte.

Carex reversa Spreng., Syst. Veg. 3: 827 (1826).

Carex lupulina var. *gigantoidea* Dewey, Amer. J. Sci. Arts, II, 41: 328 (1866).

Carex retrorsa var. *robinsonii* Fernald, Rhodora 8: 201 (1906).

Carex retrorsa var. *gigantoides* Farw., Rhodora 23: 87 (1921).

Carex retrorsa var. *multispicula* Lepage, Naturaliste Canad. 86: 68 (1959).

Carex reuteriana Boiss., Pugill. Pl. Afr. Bor. Hispan.: 116 (1852). *Carex fusca* var. *reuteriana* (Boiss.) Vicioso in ?. *Carex goodenowii* var. *reuteriana* (Boiss.) Daveau, Cyper. Portug.: 56 (1892). *Carex gracilis* f. *reuteriana* (Boiss.) Kük., Allg. Bot. Z. Syst. 3: 171 (1897). *Carex acuta* subsp. *reuteriana* (Boiss.) Rivas Mart. & J.M.Pizarro, Fontqueria 24: 13 (1989). *Carex elata* subsp. *reuteriana* (Boiss.) Luceño & Aedo, Anales Jard. Bot. Madrid 47: 143 (1989 publ. 1990).

SW. Europe. 12 FRA POR SPA. Hemicr. or rhizome geophyte.

subsp. ***reuteriana***

SW. Europe. 12 FRA SPA. Hemicr. or rhizome geophyte.

subsp. ***tartessiana*** (Luceño & Aedo) Rivas Mart., Itinera Geobot. 15: 699 (2002).

Iberian Pen. 12 POR SPA. Hemicr. or rhizome geophyte.

**Carex elata* subsp. *tartessiana* Luceño & Aedo, Bot. J. Linn. Soc. 114: 205 (1994).

Carex reventa V.I.Krecz. in V.L.Komarov (ed.), Fl. URSS 3: 614 (1935). *Carex pediformis* var. *reventa* (V.I.Krecz.) Vorosch. in A.K.Skvortsov (ed.), Florist. Issl. Razn. Raionakh SSSR: 155 (1985). *Carex rhizina* subsp. *reventa* (V.I.Krecz.) T.V.Egorova, Bot. Zhurn. (Moscow & Leningrad) 70: 1552 (1985). *Carex pediformis* subsp. *reventa* (V.I.Krecz.) Malyschev, in Fl. Sibir., Cyperac. 3: 127 (1990).

Russian Far East. 31 KAM. Hemicr. or rhizome geophyte.

Carex rhodesiaca Nelmes, Bull. Misc. Inform. Kew 1939: 159 (1939).

Zambia. 26 ZAM. Hemicr. or rhizome geophyte.

Carex rhombifructus Ohwi, J. Jap. Bot. 44: 60 (1969).

Nepal. 40 NEP. Hemicr. or rhizome geophyte.

Carex rhynchoperigynium S.D.Jones & Reznicek, Sida 15: 215 (1992).

Mexico (Hidalgo). 79 MXE. Hemicr. or rhizome geophyte.

Carex rhyncophora Franch., Bull. Soc. Philom. Paris, VIII, 7: 90 (1895).

S. China. 36 CHC CHS. Hemicr. or rhizome geophyte.

Carex rhytidocarpa Nelmes, Proc. Linn. Soc. London 1944: 155, 282 (1944).

Queensland. 50 QLD. Hemicr. or rhizome geophyte.

Carex richardsonii R.Br. in J.Richardson, Bot. App.: 751, 35 (1823).

Subarctic America to N. U.S.A. 70 NWT 71 ABT BRC MAN SAS 72 ONT QUE 73 WYO 74 ILL IOW MIN NDA SDA WIS 75 INI MIC NWY OHI PEN VER 78 MRY. Hemicr. or rhizome geophyte.

Carex richardsonii var. *exserta* Fernald, Rhodora 44: 290 (1942).

Carex ridongensis P.C.Li, Acta Bot. Yunnan. 12: 148 (1990).

Tibet. 36 CHT. Hemicr. or rhizome geophyte.

Carex rigidioides (Gorodkov) V.I.Krecz. in V.L. Komarov (ed.), Fl. URSS 3: 228 (1935).

Siberia, Russian Far East. 30 YAK 31 PRM 37 MON. Hemicr. or rhizome geophyte.

**Carex rigida* subsp. *rigidioides* Gorodkov, Zhurn. Russk. Bot. Obshch. 15: 182 (1930). *Carex bigelowii* subsp. *rigidioides* (Gorodkov) T.V.Egorova, Novosti Sist. Vyssh. Rast. 10: 103 (1973).

Carex* ×*rikuchiuensis Akiyama, J. Fac. Sci. Hokkaido Univ., Ser. 5, Bot. 3: 268 (1935). *C. heterolepis* × *C. sadoensis*.
Japan. 38 JAP. Hemicr. or rhizome geophyte.

Carex riparia Curtis, Fl. Londin. 4: t. 60 (1783). *Trasus riparius* (Curtis) Gray, Nat. Arr. Brit. Pl. 2: 64 (1821).
Europe to C. Asia, S. South America. 10 DEN FIN GRB IRE NOR SWE 11 AUT BGM CZE GER HUN NET POL SWI 12 COR FRA POR SAR SPA 13 BLR BLT RUC RUE RUN RUS RUW UKR 14 BLR BLT KRY RUC RUE RUN RUS RUW UKR 20 ALG MOR 30 ALT WSB 32 KAZ KGZ TZK UZB 33 NCS TCS 34 IRN LBS TUR 36 CHX (51) nzn 85 AGE AGS AGW CLC CLS URU. Hemicr. or rhizome geophyte.

subsp. ***chilensis*** (Brongn.) Kük., Bot. Jahrb. Syst. 27: 553 (1899).
Uruguay, Argentina, Chile. 85 AGE AGS AGW CLC CLS URU. Hemicr. or rhizome geophyte.
**Carex chilensis* Brongn. in L.I.Duperrey, Voy. Monde, Phan.: 156 (1833). *Carex riparia* var. *chilensis* (Brongn.) Kük. in H.G.A.Engler (ed.), Pflanzenr., IV, 20(38): 736 (1909).

subsp. ***riparia***
Europe to C. Asia. 10 DEN FIN GRB IRE NOR SWE 11 AUT BGM CZE GER HUN NET POL SWI 12 COR FRA POR SAR SPA 13 BLR BLT RUC RUE RUN RUS RUW UKR 14 BLR BLT KRY RUC RUE RUN RUS RUW UKR 20 ALG MOR 30 ALT WSB 32 KAZ KGZ TZK UZB 33 NCS TCS 34 IRN LBS TUR 36 CHX (51) nzn. Hemicr. or rhizome geophyte.
Carex crassa Ehrh., Beitr. Naturk. 4: 43 (1789).
Carex striata Gilib., Excerc. Phyt. 2: 550 (1792), nom. inval.
Carex bifurca Moench, Methodus: 326 (1794).
Carex vesicaria var. *plumbea* Wahlenb., Kongl. Vetensk. Acad. Nya Handl. 24: 163 (1803). *Carex plumbea* (Wahlenb.) Willd., Sp. Pl. 4: 308 (1805).
Carex ruffa Brot., Fl. Lusit. 1: 66 (1804).
Carex secalina Lapeyr., Hist. Pl. Pyrénées 2: 576 (1813), nom. illeg.
Carex incrassata Schltdl., Linnaea 10: 119 (1836). *Carex riparia* f. *incrassata* (Schltdl.) Kük. in H.G.A.Engler (ed.), Pflanzenr., IV, 20(38): 736 (1909).
Carex exaltata Peterm., Flora 27: 340 (1844). *Carex riparia* var. *exaltata* (Peterm.) Nyman, Consp. Fl. Eur.: 767 (1882).
Carex riparia f. *laevigata* Peterm., Flora 27: 341 (1844).
Carex gracilescens Hartm. ex Andersson, Cyper. Scand.: 16 (1849). *Carex riparia* subsp. *gracilescens* (Hartm. ex Andersson) Nyman, Consp. Fl. Eur.: 767 (1882).
Carex vacillans Steud., Syn. Pl. Glumac. 2: 235 (1855), nom. illeg.
Carex riparia var. *pseudopaludosa* Schur, Enum. Pl. Transsilv.: 720 (1866).
Carex watsonii Olney in S.Watson, Botany [Fortieth Parallel]: 370 (1871).
Carex riparia var. *humilis* Uechtr. in E.Fiek, Fl. Schlesien: 492 (1881).
Carex riparia var. *inferior* Peck, Rep. (Annual) Regents Univ. State New York State Mus. 49: 43 (1896).
Carex riparia f. *reticulosa* Torges, Katal. Wiener Bot. Tauschenstalt 1897–1898: 84 (1898).
Carex riparia f. *brevipalea* Neuman, Sver. Fl.: 690 (1901).
Carex riparia f. *aristata* Asch. & Graebn., Syn. Mitteleur. Fl. 2(2): 216 (1903).
Carex riparia var. *claviformis* J.A.Schmidt, Jahrb. Hamburg. Wiss. Anst. 22: 71 (1904).
Carex riparia f. *flagelliformis* Waisb., Magyar Bot. Lapok 3: 19 (1904).
Carex riparia var. *subgracilescens* Kük. in H.G.A. Engler (ed.), Pflanzenr., IV, 20(38): 735 (1909).

Carex rivulorum Dunn, J. Linn. Soc., Bot. 38: 372 (1908).
China. 36+. Hemicr. or rhizome geophyte.

Carex roanensis F.J.Herm., Castanea 12: 113 (1947).
SE. U.S.A. 78 TEN VRG?. Hemicr. or rhizome geophyte.

Carex robinsonii Podlech, Mitt. Bot. Staatssamml. München 4: 122 (1961).
Zambia. 26 ZAM. Hemicr. or rhizome geophyte.

Carex* ×*rollandii Lepage, Naturaliste Canad. 84: 92 (1957). *C. aquatilis* × *C. nigra*.
E. Canada. 72 QUE. Hemicr. or rhizome geophyte.

Carex roraimensis Steyerm., Fieldiana, Bot. 28: 67 (1951).
Brazil (Mt. Roraima), Guyana (Mt. Roraima), Venezuela (Mt. Rorayma, Macizo del Chimantá). 82 GUY VEN 84 BZN. Hemicr. or rhizome geophyte.

Carex rorulenta Porta, Nuovo Giorn. Bot. Ital. 19: 321 (1887). *Carex halleriana* subsp. *rorulenta* (Porta) Malag., Subesp. Variac. Geogr.: 26 (1973).
Baleares. 12 BAL. Hemicr. or rhizome geophyte.

Carex rosea Willd., Sp. Pl. 4: 237 (1805). *Vignea rosea* (Willd.) Rchb. in J.C.Mössler, Handb. Gewächsk. ed. 2, 3: 1611 (1830). *Diemisa rosea* (Willd.) Raf., Good Book: 27 (1840).
Canada to C. & E. U.S.A. 71 MAN 72 NBR NSC ONT QUE 73 WYO 74 ILL IOW KAN MIN MSO NEB OKL SDA WIS 75 CNT INI MAI MAS MIC NWH NWJ NWY OHI PEN RHO VER WVA 77 TEX 78 ALA ARK DEL FLA GEO KTY MRY NCA SCA TEN VRG WDC. Hemicr. or rhizome geophyte.
Carex neglecta Tuck., Enum. Meth. Caric.: 19 (1843), nom. illeg.
Carex disperma Kunze ex Carey in A.Gray, Manual: 542 (1848), nom. illeg.
Carex flaccidula Steud., Syn. Pl. Glumac. 2: 199 (1855).
Carex trisperma A.Gray ex Boott, Ill. Carex 2: 81 (1860), nom. illeg.
Carex rosea var. *pusilla* Peck, Rep. (Annual) Regents Univ. State New York New York State Mus. 47: 132 (1895).
Carex rosea var. *staminata* Peck, Rep. (Annual) Regents Univ. State New York New York State Mus. 47: 132 (1895).
Carex convoluta Mack., Bull. Torrey Bot. Club 43: 428 (1916).
Carex rosea f. *debilis* Farw., Pap. Michigan Acad. Sci. 2: 19 (1923).

Carex rossii Boott in W.J.Hooker, Fl. Bor.-Amer. 2: 222 (1839). *Carex novae-angliae* var. *rossii* (Boott) L.H.Bailey, Bot. Gaz. 10: 207 (1885). *Carex deflexa*

var. *rossii* (Boott) L.H.Bailey, Mem. Torrey Bot. Club 1: 43 (1889).
Alaska to W. & N. U.S.A. 70 ASK NWT YUK 71 ABT BRC MAN SAS 72 ONT 73 COL IDA MNT ORE WAS WYO 74 MIN NEB SDA 75 MIC 76 ARI CAL NEV UTA 77 NWM. Hemicr. or rhizome geophyte.
Carex brevipes S.Watson in S.Watson & al., Bot. California 2: 246 (1880), nom. inval.
Carex globosa var. *brevipes* Boott in S.Watson & al., Bot. California 2: 485 (1880). *Carex rossii* var. *brevipes* (Boott) Kük. in H.G.A.Engler (ed.), Pflanzenr., IV, 20(38): 452 (1909).
Carex deflexa var. *media* L.H.Bailey, Mem. Torrey Bot. Club 1: 43 (1889).
Carex pilulifera var. *novae-angliae* Kurtz, Bot. Jahrb. Syst. 19: 419 (1894).
Carex deflexa var. *farwellii* Britton in N.L.Britton & A.Brown, Ill. Fl. N. U.S. 1: 334 (1896).
Carex farwellii Mack., Bull. Torrey Bot. Club 37: 244 (1910).
Carex diversistylis A.Roach, Madroño 11: 277 (1952).

Carex rostellifera Y.L.Chang & Y.L.Yang, in Fl. Plant. Herb. Chinae Bor.-Or. 11: 205 (1976).
China. 36+. Hemicr. or rhizome geophyte.

Carex rostrata Stokes in W.Withering, Bot. Arr. Brit. Pl. ed. 2, 2: 1059 (1787), nom. cons.
Europe to W. Himalaya, Subarctic America to N. U.S.A. 10 DEN FIN FOR GRB ICE IRE NOR SWE 11 AUT BGM CZE GER HUN NET POL SWI 12 COR? FRA SPA 13 ALB BUL GRC ITA ROM YUG 14 BLR BLT RUC RUE RUN RUS RUW UKR 30 ALT KRA WSB YAK 31 KAM KHA MAG 32 KAZ 33 NCS TCS 34 IRN TUR 36 CHI CHM 37 MON 38 KOR 40 WHM 70 ASK GNL NUN NWT YUK 71 ABT BRC MAN SAS 72 LAB NBR NFL NSC ONT QUE 73 MNT WAS 74 MIN WIS 75 MAI MIC NWH. Hemicr. or rhizome geophyte.
Carex inflata Huds., Fl. Angl.: 354 (1762), nom. rejic.

var. ***ambigens*** Fernald, Rhodora 3: 51 (1901).
E. Canada. 72 NBR QUE. Hemicr. or rhizome geophyte.

var. ***rostrata***
Europe to W. Himalaya, Subarctic America to N. U.S.A. 10 DEN FIN FOR GRB ICE IRE NOR SWE 11 AUT BGM CZE GER HUN NET POL SWI 12 COR? FRA SPA 13 ALB BUL GRC ITA ROM YUG 14 BLR BLT RUC RUE RUN RUS RUW UKR 30 ALT KRA WSB YAK 31 KAM KHA MAG 32 KAZ 33 NCS TCS 34 IRN TUR 36 CHI CHM 37 MON 38 KOR 40 WHM 70 ASK GNL NUN NWT YUK 71 ABT BRC MAN SAS 72 LAB NBR NFL NSC ONT QUE 73 MNT WAS 74 MIN WIS 75 MAI MIC NWH. Hemicr. or rhizome geophyte.
Carex bifurca Schrank, Baier. Fl. 1: 304 (1789).
Carex longifolia Thuill., Fl. Env. Paris: 490 (1790).
Carex obtusangula Ehrh., Beitr. Naturk. 6: 82 (1791).
Carex torfacea J.F.Gmel., Syst. Nat. 2: 145 (1791).
Carex ampullacea Gooden., Trans. Linn. Soc. London 2: 207 (1794). *Trasus ampullaceus* (Gooden.) Gray, Nat. Arr. Brit. Pl. 2: 59 (1821).
Carex inflata Suter, Helvet. Fl. 2: 265 (1802), nom. illeg.
Carex ampullacea var. *borealis* Hartm., Handb. Skand. Fl.: 39 (1820). *Carex rostrata* var. *borealis* (Hartm.) Kük. in H.G.A.Engler (ed.), Pflanzenr., IV, 20(38): 723 (1909).
Carex vesicaria var. *inalpina* Laest., Loca Parall.: 88 (1831).
Carex ampullacea var. *altissima* Andersson, Cyper. Scand.: 20 (1849).
Carex ampullacea var. *brunnescens* Andersson, Cyper. Scand.: 20 (1849). *Carex rostrata* var. *brunnescens* (Andersson) Kük. in H.G.A.Engler (ed.), Pflanzenr., IV, 20(38): 722 (1909).
Carex ampullacea var. *maxima* Andersson, Bot. Not. 1894: (1849).
Carex ampullacea var. *robusta* Sond., Fl. Hamburg.: 505 (1850).
Carex ampullacea f. *elatior* A.Blytt, Norges Fl. 1: 254 (1861). *Carex rostrata* f. *elatior* (A.Blytt) Kük. in H.G.A.Engler (ed.), Pflanzenr., IV, 20(38): 722 (1909), nom. illeg.
Carex ampullacea f. *pendulina* A.Blytt, Norges Fl. 1: 254 (1861). *Carex rostrata* f. *pendulina* (A.Blytt) Kük. in H.G.A.Engler (ed.), Pflanzenr., IV, 20(38): 722 (1909).
Carex rostrata var. *latifolia* Asch., Fl. Brandenburg 1: 792 (1864).
Carex hymenocarpa Drejer in G.C.Oeder & al. (eds.), Fl. Dan.: t. 2849 (1871). *Carex ampullacea* subsp. *hymenocarpa* (Drejer) Nyman, Consp. Fl. Eur.: 768 (1882). *Carex rostrata* var. *hymenocarpa* (Drejer) H.Christ, Bull. Soc. Roy. Bot. Belgique 24(2): 11 (1885).
Carex rostrata f. *elatior* Lange, Consp. Fl. Groenland., ed. 2: 152 (1890).
Carex ampullacea f. *planifolia* Norman, Forh. Vidensk.-Selsk. Kristiania, Extr.: 53 (1893).
Carex rostrata f. *laeta* Norman, Fl. Arct. Norv. Sp.: 52 (1893).
Carex rostrata f. *plumosa* Norman, Forh. Vidensk.-Selsk. Kristiania 1893: 53 (1893).
Carex rostrata f. *sommieri* H.Christ in S.Sommier, Fl. Ob Inf.: 103 (1896).
Carex rostrata f. *sparangiformis* Murr, Deutsche Bot. Monatsschr. 16: 147 (1898).
Carex rotundata var. *major* Meinsh., Trudy Imp. S.-Peterburgsk. Bot. Sada 18: 376 (1900).
Carex rostrata f. *longipalea* Neuman, Sver. Fl.: 691 (1901).
Carex rostrata var. *cliftonii* Farw., Rep. (Annual) Michigan Acad. Sci. 6: 204 (1904).
Carex inflata var. *ambigens* Fernald, Rhodora 44: 330 (1942).

Carex rotundata Wahlenb., Kongl. Vetensk. Acad. Nya Handl. 1803: 153 (1803). *Onkerma rotundata* (Wahlenb.) Raf., Good Book: 27 (1840). *Carex rostrata* subsp. *rotundata* (Wahlenb.) Kük. in H.G.A.Engler (ed.), Pflanzenr., IV, 20(38): 723 (1909). *Carex saxatilis* var. *rotundata* (Wahlenb.) Gelting, Meddel. Grønland 101(2): 173 (1934).
N. Europe to Russian Far East, Subarctic America to Canada. 10 BOR FIN SWE 14 RUN 30 KRA WSB YAK 31 AMU KAM KHA KUR MAG SAK 70 ASK NUN NWT YUK 71 MAN 72 LAB NFL QUE. Hemicr. or rhizome geophyte.
Carex ruesanensis Kudô, J. Coll. Agric. Hokkaido Imp. Univ. 11: 83 (1922).
Carex melozitnensis A.E.Porsild, Rhodora 41: 209 (1939).

Carex royleana Nees in R.Wight, Contr. Bot. India: 127 (1834).

E. Afghanistan to Himalaya. 34 AFG 40 NEP WHM. Hemicr. or rhizome geophyte.

Carex rubicunda Petrie, Trans. & Proc. New Zealand Inst. 31: 353 (1898). *Carex petriei* var. *rubicunda* (Petrie) Kük. in H.G.A.Engler (ed.), Pflanzenr., IV, 20(38): 690 (1909).
New Zealand. 51 NZN NZS. Hemicr. or rhizome geophyte.
Carex novae-zelandiae Petrie, Trans. & Proc. New Zealand Inst. 25: 273 (1893), nom. illeg.

Carex rubrobrunnea C.B.Clarke in J.D.Hooker, Fl. Brit. India 6: 710 (1894).
Bhutan to China and Vietnam. 36 CHC CHS CHT 40 ASS EHM 41 VIE. Hemicr. or rhizome geophyte.
Carex stachydesma Franch., Bull. Soc. Philom. Paris, VIII, 7: 85 (1895).
Carex taliensis Franch., Bull. Soc. Philom. Paris, VIII, 7: 34 (1895). *Carex rubrobrunnea* var. *taliensis* (Franch.) Kük. in H.G.A.Engler (ed.), Pflanzenr., IV, 20(38): 344 (1909).
Carex gracilirostris C.B.Clarke ex Franch., Nouv. Arch. Mus. Hist. Nat., III, 9: 138 (1897).

Carex rufina Drejer, Rev. Car. Bor.: 28 (1841).
Subarctic America to WC. Canada, N. Europe. 10 ICE NOR SWE 70 GNL NUN 71 MAN 72 QUE. Hemicr. or rhizome geophyte.

Carex rufulistolon T.Koyama, Acta Phytotax. Geobot. 15: 111 (1954).
C. Nepal. 40 NEP. Hemicr. or rhizome geophyte.

Carex rugata Ohwi, Acta Phytotax. Geobot. 1: 76 (1932).
SE. China, Japan. 36 CHS 38 JAP. Hemicr. or rhizome geophyte.

Carex rugulosa Kük., Bull. Herb. Boissier, II, 4: 58 (1904). *Carex riparia* var. *rugulosa* (Kük.) Kük. in H.G.A.Engler (ed.), Pflanzenr., IV, 20(38): 736 (1909).
Siberia to Japan. 30+ 31 PRM 36 CHI CHM CHN 38 JAP KOR. Hemicr. or rhizome geophyte.
Carex smirnovii V.I.Krecz. in V.L.Komarov (ed.), Fl. URSS 3: 619 (1935).

Carex runssoroensis K.Schum. in H.G.A.Engler, Pflanzenw. Ost-Afrikas, C: 129 (1895). *Uncinia runssoroensis* (K.Schum.) Chiov. in F.de Filippi, Ruwenzori: 466 (1909).
NE. Zaïre to Kenya (Aberdare, Mt. Kenya). 23 RWA ZAI 25 KEN UGA. Hemicr. or rhizome geophyte.
Carex runssoroensis var. *aberdarensis* Kük., Notizbl. Bot. Gart. Berlin-Dahlem 9: 311 (1925).

Carex rupestris All., Fl. Pedem. 2: 264 (1785). *Edritria rupestris* (All.) Raf., Good Book: 26 (1840). *Caricinella rupestris* (All.) St.-Lag. in A.Cariot, Étude Fl., ed. 8, 2: 882 (1889).
Temp. & Subarctic Northern Hemisphere. 10 FIN GRB ICE NOR SVA SWE 11 AUT CZE GER SWI 12 COR? FRA SPA 13 BUL ITA ROM YUG 14 BLR RUE RUN RUS UKR 30 ALT KRA WSB YAK 31 AMU KHA MAG PRM SAK 32 KAZ 33 NCS 37 MON 38 KOR 70 ASK GNL NUN NWT YUK 71 ABT BRC MAN 72 LAB NFL ONT QUE 73 COL MNT WYO 74 SDA 76 UTA. Hemicr. or rhizome geophyte.

subsp. ***altimontana*** T.V.Ebel, Turczaninowia 5(4): 5 (2002).
Altay. 30 ALT. Hemicr. or rhizome geophyte.

subsp. ***rupestris***
Temp. & Subarctic Northern Hemisphere. 10 FIN GRB ICE NOR SVA SWE 11 AUT CZE GER SWI 12 COR? FRA SPA 13 BUL ITA ROM YUG 14 BLR RUE RUN RUS UKR 30 ALT KRA WSB YAK 31 AMU KHA MAG PRM SAK 32 KAZ 33 NCS 37 MON 38 KOR 70 ASK GNL NUN NWT YUK 71 ABT BRC MAN 72 LAB NFL ONT QUE 73 COL MNT WYO 74 SDA 76 UTA. Hemicr. or rhizome geophyte.
Carex petraea Wahlenb., Kongl. Vetensk. Acad. Nya Handl. 1803: 139 (1803). *Edritria petraea* (Wahlenb.) Raf., Good Book: 26 (1840).
Carex dufourei Lapeyr., Hist. Pl. Pyrénées, Suppl.: 140 (1818).
Carex attenuata R.Br. in J.Richardson, Bot. App.: 750 (1823).
Carex drummondiana Dewey, Amer. J. Sci. Arts 29: 251 (1836). *Carex rupestris* var. *drummondiana* (Dewey) L.H.Bailey, Carex Cat.: 1 (1884).

Carex rupicola (Pedersen) G.A.Wheeler, Hickenia 2: 180 (1996).
Argentina (Buenos Aires), Uruguay. 85 AGE URU. Hemicr. or rhizome geophyte.
**Carex bonariensis* var. *rupicola* Pedersen, Colecç. Ci. Inst. Nac. Tecnol. Agropecu 4(1): 332 (1969).

Carex rutenbergiana Boeck., Abh. Naturwiss. Vereine Bremen 7: 40 (1880).
C. Madagascar. 29 MDG. Hemicr. or rhizome geophyte.

Carex ruthii Mack. in N.L.Britton & al. (eds.), N. Amer. Fl. 18: 113 (1931). *Carex muricata* var. *ruthii* (Mack.) Gleason, Phytologia 4: 22 (1952).
SE. U.S.A. 78 GEO NCA SCA TEN VRG. Hemicr. or rhizome geophyte.

Carex ruthsatziae G.A.Wheeler, Darwiniana 40: 191 (2002).
Bolivia. 83 BOL. Hemicr. or rhizome geophyte.

Carex rzedowskii Reznicek & S.González, Contr. Univ. Michigan Herb. 20: 218 (1995).
Mexico (Michoacán). 79 MXS. Hemicr. or rhizome geophyte.

Carex sabulosa Turcz. ex Kunth, Enum. Pl. 2: 432 (1837). *Carex melanantha* var. *sabulosa* (Turcz. ex Kunth) Kük. in H.G.A.Engler (ed.), Pflanzenr., IV, 20(38): 392 (1909).
C. Asia, Siberia, Subarctic America. 30 WSB 32 KAZ 37 MON 70 ASK YUK. Hemicr. or rhizome geophyte.
Carex leiophylla Mack. in N.L.Britton & al. (eds.), N. Amer. Fl. 18: 365 (1935). *Carex sabulosa* subsp. *leiophylla* (Mack.) Porsild, Bull. Natl. Mus. Canada 216: 20 (1966).

Carex sacerdotis Nelmes, Mém. Mus. Natl. Hist. Nat., Sér. B, Bot. 4: 172 (1955).
Vietnam. 41 VIE. Hemicr. or rhizome geophyte.

Carex sachalinensis F.Schmidt, Reis. Amur-Land., Bot.: 194 (1868). *Carex pisiformis* var. *sachalinensis* (F.Schmidt) Kük. in H.G.A.Engler (ed.), Pflanzenr., IV, 20(38): 477 (1909).
Russian Far East, China to Vietnam, Temp. E. Asia. 31 KUR SAK 36 CHM CHN CHS 38 JAP KOR TAI 41 VIE. Hemicr. or rhizome geophyte.

var. ***alterniflora*** (Franch.) Ohwi, Bull. Natl. Sci. Mus. Tokyo 33: 67 (1953).

Japan, Taiwan. 38 JAP TAI. Hemicr. or rhizome geophyte.
Carex alterniflora Franch., Bull. Soc. Philom. Paris, VIII, 7: 51 (1895). *Carex sachalinensis* subsp. *alterniflora* (Franch.) T.Koyama, in Fl. Taiwan 5: 336 (1978).
Carex pseudostrigosa H.Lév. & Vaniot, Bull. Acad. Int. Géogr. Bot. 11: 109 (1902).
Carex scabroaristata Akiyama, J. Fac. Sci. Hokkaido Univ., Ser. 5, Bot. 1: 58 (1931).

var. ***arimaensis*** (Ohwi) Ohwi in ?
Japan (Honshu: Kinki Distr.). 38 JAP. Hemicr. or rhizome geophyte.
**Carex alterniflora* var. *arimaensis* Ohwi, Mem. Coll. Sci. Kyoto Imp. Univ., Ser. B, Biol. 6: 259 (1931).

var. ***aureobrunnea*** (Ohwi) Ohwi in ?
S. Japan. 38 JAP. Hemicr. or rhizome geophyte.
**Carex alterniflora* var. *aureobrunnea* Ohwi, Mem. Coll. Sci. Kyoto Imp. Univ., Ser. B, Biol. 6: 259 (1931).

var. ***conicoides*** (Honda) Ohwi, Bull. Natl. Sci. Mus. Tokyo 33: 68 (1953).
Japan (Kyushu). 38 JAP. Hemicr. or rhizome geophyte.
**Carex conicoides* Honda, Bot. Mag. (Tokyo) 42: 506 (1928).

var. ***elongatula*** (Ohwi) Ohwi in ?
Japan (Kyushu: Mt. Kuju). 38 JAP. Hemicr. or rhizome geophyte.
**Carex alterniflora* var. *elongatula* Ohwi, Mem. Coll. Sci. Kyoto Imp. Univ., Ser. B, Biol. 6: 260 (1931).

var. ***fernaldiana*** (H.Lév. & Vaniot) T.Koyama, Bull. Arts Sci. Div. Ryukyu Univ. 3: 72 (1959).
Japan, Taiwan. 38 JAP TAI. Hemicr. or rhizome geophyte.
**Carex fernaldiana* H.Lév. & Vaniot, Bull. Acad. Int. Géogr. Bot. 10: 276 (1901). *Carex sachalinensis* subsp. *fernaldiana* (H.Lév. & Vaniot) T.Koyama, in Fl. Taiwan 5: 337 (1978).
Carex ischne C.B.Clarke, Kew Bull., Addit. Ser. 8: 80 (1908).
Carex mariesii C.B.Clarke, Kew Bull., Addit. Ser. 8: 80 (1908).

var. ***fulva*** (Ohwi) Ohwi in ?
Japan (Honshu). 38 JAP. Hemicr. or rhizome geophyte.
Carex artinux C.B.Clarke, Kew Bull., Addit. Ser. 8: 80 (1908).
**Carex alterniflora* var. *fulva* Ohwi, Mem. Coll. Sci. Kyoto Imp. Univ., Ser. B, Biol. 11: 367 (1936).

var. ***iwakiana*** Ohwi, Mem. Coll. Sci. Kyoto Imp. Univ., Ser. B, Biol. 11: 368 (1936).
Japan (Honshu: Iwaki Prov.). 38 JAP. Hemicr. or rhizome geophyte.

var. ***longiuscula*** Ohwi, Mem. Coll. Sci. Kyoto Imp. Univ., Ser. B, Biol. 11: 368 (1936).
Japan (C. Honshu). 38 JAP. Hemicr. or rhizome geophyte.
Carex niko-montana Akiyama, Bot. Mag. (Tokyo) 45: 472 (1931).

var. ***pineticola*** (Ohwi) Ohwi in ?
Japan (Honshu: Shimosa Prov.). 38 JAP. Hemicr. or rhizome geophyte.
**Carex pineticola* Ohwi, Bull. Natl. Sci. Mus. Tokyo 26: 5 (1949).

var. ***sachalinensis***
Russian Far East, China, N. & NC. Japan. 31 KUR SAK 36 CHM CHN CHS 38 JAP KOR 41 VIE. Hemicr. or rhizome geophyte.
Carex tenuissima Boott, Proc. Linn. Soc. London 1: 288 (1846), provisional synonym.
Carex pseudoconica Franch. & Sav., Enum. Pl. Jap. 2: 570 (1878).
Carex albomas C.B.Clarke, J. Linn. Soc., Bot. 36: 270 (1903).
Carex polyschoena H.Lév. & Vaniot, Bull. Acad. Int. Géogr. Bot. 12: 9 (1903). *Carex pisiformis* f. *polyschoena* (H.Lév. & Vaniot) Kük. in H.G.A.Engler (ed.), Pflanzenr., IV, 20(38): 477 (1909).
Carex korsakoviensis H.Lév., Bull. Acad. Int. Géogr. Bot. 19: 34 (1909).
Carex umbrosa var. *koreana* Nakai, Bot. Mag. (Tokyo) 28: 327 (1914).
Carex bauhwaensis Z.P.Wang, J. Nanjing Univ. (Biology) 1962(2): 47 (1962).

var. ***sikokiana*** (Franch. & Sav.) Ohwi, Bull. Natl. Sci. Mus. Tokyo 33: 67 (1953).
S. Korea, C. & S. Japan, Vietnam. 38 JAP KOR 41 VIE. Hemicr. or rhizome geophyte.
**Carex sikokiana* Franch. & Sav., Enum. Pl. Jap. 2: 573 (1878). *Carex tenuissima* var. *sikokiana* (Franch. & Sav.) Kük. in H.G.A.Engler (ed.), Pflanzenr., IV, 20(38): 475 (1909).

Carex sacrosancta Honda, Bot. Mag. (Tokyo) 43: 190 (1929).
Temp. E. Asia. 38 JAP NNS TAI. Hemicr. or rhizome geophyte.

var. ***sacrosancta***
C. & S. Japan. Taiwan. 38 JAP TAI. Hemicr. or rhizome geophyte.
Carex kasugayamensis Akiyama, J. Fac. Sci. Hokkaido Univ., Ser. 5, Bot. 1: 57 (1931).

var. ***tamakii*** (T.Koyama) T.Koyama, J. Fac. Sci. Univ. Tokyo, Sect. 3, Bot. 8(4): 164 (1962).
Nansei-shoto. 38 NNS. Hemicr. or rhizome geophyte.
**Carex tamakii* T.Koyama, Bull. Arts Sci. Div. Ryukyu Univ. 3: 75 (1959).

Carex sadoensis Franch., Bull. Soc. Philom. Paris, VIII, 7: 41 (1895).
Russian Far East to E. China, N. & C. Japan. 31 KUR SAK 36 CHS 38 JAP. Hemicr. or rhizome geophyte.

Carex sagaensis Y.C.Yang, in Fl. Xizangica 5: 439 (1987).
Tibet. 36 CHT. Hemicr. or rhizome geophyte.

Carex sahnii Ghildyal & U.C.Bhattach., J. Econ. Taxon. Bot. 7: 447 (1985 publ. 1986).
Sikkim. 40 EHM. Hemicr. or rhizome geophyte.
**Carex kingiana* C.B.Clarke, Kew Bull., Addit. Ser. 8: 75 (1908), nom. illeg.

Carex sajanensis V.I.Krecz., Izv. Bot. Sada Akad. Nauk S.S.S.R. 30: 124 (1931 publ. 1932).
Siberia to Mongolia. 30 IRK 37 MON. Hemicr. or rhizome geophyte.

Carex × sakaguchii Ohwi, Mem. Coll. Sci. Kyoto Imp.

Univ., Ser. B, Biol. 11(5): 512 (1936). *C. breviculmis* × *C. nervata*.
Japan. 38 JAP. Hemicr. or rhizome geophyte.

Carex sakonis T.Koyama, Bot. Mag. (Tokyo) 70: 351 (1957).
Nansei-shoto. 38 NNS. Hemicr. or rhizome geophyte.

Carex salina Wahlenb., Kongl. Vetensk. Acad. Nya Handl. 24: 165 (1803). *Vignea salina* (Wahlenb.) Rchb. in J.C.Mössler, Handb. Gewächsk. ed. 2, 3: 1623 (1830). *Neskiza salina* (Wahlenb.) Raf., Good Book: 27 (1840).
N. & NE. Europe Subarctic America to E. Canada. 10 FIN? ICE NOR 14 RUN 30 WSB 72 LAB NBR NFL ONT QUE. Hemicr. or rhizome geophyte.
Carex salina f. *corylorum* B.Boivin, Provancheria 25: 183 (1992).

var. ***pseudofilipendula*** Kük. in H.G.A.Engler (ed.), Pflanzenr., IV, 20(38): 363 (1909).
E. Canada, N. Europe. 10 NOR 70 LAB NFL QUE. Hemicr. or rhizome geophyte.

var. ***salina***
N. & NE. Europe Subarctic America to E. Canada. 10 FIN? ICE NOR 14 RUN 30 WSB 72 LAB NBR NFL ONT QUE. Hemicr. or rhizome geophyte.
Carex cuspidata Wahlenb., Kongl. Vetensk. Acad. Nya Handl. 24: 164 (1803), nom. illeg. *Carex salina* var. *cuspidata* Wahlenb., Fl. Lapp.: 246 (1812).
Carex cuspidata Degl. in J.L.A.Loiseleur-Deslongchamps, Fl. Gall.: 294 (1806), nom. illeg.
Carex lanceata Dewey, Amer. J. Sci. Arts 29: 249 (1836). *Carex salina* var. *lanceata* (Dewey) Kük. in H.G.A.Engler (ed.), Pflanzenr., IV, 20(38): 362 (1909).
Carex epigeios Fr., Summa Veg. Scand.: 223 (1845). *Carex aquatilis* subsp. *epigeios* (Fr.) Nyman, Consp. Fl. Eur. (1878). *Carex salina* var. *epigeios* (Fr.) Nyman, Consp. Fl. Eur.: 776 (1882).
Carex discolor F.Nyl., Spic. Fl. Fenn. 3: 12 (1846). *Carex salina* f. *discolor* (F.Nyl.) Almq. in A.H.Hjlet, Consp. Fl. Fennic. 3: 281 (1895).
Carex salinoides Beurl., Bot. Not. 1853: 35 (1853).
Carex salina var. *filipendula* A.Blytt, Norges Fl. 1: 219 (1861), nom. illeg.
Carex bracteata Giseke ex Boott, Ill. Carex 4: 160 (1867), nom. illeg.
Carex salina var. *borealis* Almq. in C.J.Hartmann, Handb. Skand. Fl., ed. 11: 465 (1879).
Carex salina f. *thulensis* Kük. in J.M.Macoun, The Fur Seals Is. 3: 573 (1899).

Carex saltaensis R.Gross, Repert. Spec. Nov. Regni Veg. 50: 211 (1941).
Argentina (Salta). 85 AGW. Hemicr. or rhizome geophyte.

Carex sambiranensis (H.Lév.) Cherm., Bull. Soc. Bot. France 70: 412 (1923).
N. Madagascar. 29 MDG. Hemicr. or rhizome geophyte.
**Carex spicatopaniculata* var. *sambiranensis* H.Lév., Bull. Acad. Int. Géogr. Bot. 27: 5 (1917).

Carex sampsonii Hance, J. Bot. 6: 89 (1868).
China. 36+. Hemicr. or rhizome geophyte.

Carex sanctae-marthae L.E.Mora & J.O.Rangel, Revista Acad. Colomb. Ci. Exact. 58: 13 (1983).
Colombia. 83 CLM. Hemicr. or rhizome geophyte.

Carex sarawaketensis Kük., Bot. Jahrb. Syst. 69: 262 (1938).
New Guinea. 43 NWG. Hemicr. or rhizome geophyte.
Carex melanophora S.T.Blake, J. Arnold Arbor. 28: 106 (1947).

Carex* × *sardloqensis Å.E.Dahl, Nyt Mag. Naturvidensk. 82: 101 (1941).
Greenland. 70 GNL. Hemicr. or rhizome geophyte.

Carex sartwelliana Olney, Proc. Amer. Acad. Arts 7: 396 (1868).
California. 76 CAL. Hemicr. or rhizome geophyte.
Carex yosemitana L.H.Bailey, Mem. Torrey Bot. Club 1: 8 (1889).

Carex sartwellii Dewey, Amer. J. Sci. Arts 43: 90 (1842).
Carex disticha var. *sartwellii* (Dewey) Dewey, Amer. J. Sci. Arts, II, 41: 330 (1866).
Canada to N. U.S.A. 70 NWT 71 ABT BRC MAN SAS 72 ONT QUE 73 COL IDA MNT WYO 74 ILL IOW MIN MSO NDA NEB SDA WIS 75 INI MIC NWY OHI PEN 76 UTA 78 ARK MRY. Hemicr. or rhizome geophyte.

var. ***sartwellii***
Canada to N. U.S.A. 70 NWT 71 ABT BRC MAN SAS 72 ONT QUE 73 COL IDA MNT WYO 74 ILL IOW MIN MSO NDA NEB SDA WIS 75 INI MIC NWY OHI PEN 76 UTA 78 ARK MRY. Hemicr. or rhizome geophyte.
Carex disticha A.Gray, Manual, ed. 5: 574 (1867), nom. illeg.

var. ***stenorrhyncha*** F.J.Herm., Rhodora 40: 78 (1938).
Indiana. 75 INI. Hemicr. or rhizome geophyte.

Carex satsumensis Franch. & Sav, Enum. Pl. Jap. 2: 558 (1878).
Japan to Vietnam and Philippines. 38 JAP TAI 41 VIE 42 PHI. Hemicr. or rhizome geophyte.
Carex nikoensis Franch. & Sav., Enum. Pl. Jap. 2: 558 (1878). *Vignidula nikoensis* (Franch. & Sav.) Fedde & J.Schust., Just's Bot. Jahresber. 41(2): 12 (1913 publ. 1918).
Carex contracta Boeck., Beitr. Cyper. 2: 34 (1890).

Carex savaiiensis Kük., Bernice P. Bishop Mus. Bull. 128: 24 (1935).
Samoa. 60 SAM. Hemicr. or rhizome geophyte.

Carex saxatilis L., Sp. Pl.: 976 (1753). *Vignea saxatilis* (L.) Rchb. in J.C.Mössler, Handb. Gewächsk. ed. 2, 3: 1624 (1830).
N. Europe to Russian Far East, Subarctic America to WC. U.S.A. 10 FIN FOR GRB IRE NOR SVA SWE 14 RUE RUN 30 KRA WSB YAK 31 KAM KUR MAG 37 MON 70 ALU ASK GNL NUN NWT YUK 71 ABT BRC MAN SAS 72 LAB NBR NFL NSC ONT QUE 73 COL MNT WAS WYO 75 MAI 76 UTA. Hemicr. or rhizome geophyte.
Carex saxatilis var. *major* Olney in ?
Carex pulla Gooden., Trans. Linn. Soc. London 3: 78 (1797). *Vignea pulla* (Gooden.) Rchb. in J.C. Mössler, Handb. Gewächsk. ed. 2, 3: 1624 (1830). *Carex vesicaria* subsp. *pulla* (Gooden.) Andersson, Cyper. Scand.: 19 (1849).
Carex fusca Schkuhr, Beschr. Riedgräs. 1: 64 (1801), nom. illeg.

Carex miliaris Michx., Fl. Bor.-Amer. 2: 174 (1803). *Carex pulla* var. *miliaris* (Michx.) A.Gray, Manual, ed. 5: 602 (1867). *Carex saxatilis* var. *miliaris* (Michx.) L.H.Bailey, Bot. Gaz. 9: 120 (1884).
Carex physocarpa C.Presl, Reliq. Haenk. 1: 202 (1828). *Carex vesicaria* var. *physocarpa* (C.Presl) Kük. in H.G.A.Engler (ed.), Pflanzenr., IV, 20(38): 728 (1909).
Carex ensifolia Turcz. ex Ledeb., Fl. Ross. 4: 309 (1852).
Carex vesicaria var. *distenta* Fr., Exsicc. (Herb. Norm. Suec.) 15: 84 (1853).
Carex incerta Steud., Syn. Pl. Glumac. 2: 210 (1855).
Carex jungenda Steud., Syn. Pl. Glumac. 2: 210 (1855).
Carex ochroleuca Cham. ex Steud., Syn. Pl. Glumac. 2: 216 (1855).
Carex ambusta Boott, Ill. Carex 1: 64 (1858).
Carex pulla f. *laxa* Trautv., Trudy Imp. S.-Peterburgsk. Bot. Sada 5: 130 (1877). *Carex vesicaria* f. *laxa* (Trautv.) Kük. in H.G.A.Engler (ed.), Pflanzenr., IV, 20(38): 728 (1909). *Carex saxatilis* var. *laxa* (Trautv.) Ohwi, J. Jap. Bot. 11: 408 (1935). *Carex saxatilis* subsp. *laxa* (Trautv.) Kalela, Ann. Bot. Soc. Zool.-Bot. Fenn. "Vanamo" 14(2): 15 (1940).
Carex saxatilis var. *grahamii* L.H.Bailey, Proc. Amer. Acad. Arts 22: 66 (1886), nom. illeg.
Carex miliaris var. *aurea* L.H.Bailey, Mem. Torrey Bot. Club 1: 37 (1889).
Carex miliaris f. *major* L.H.Bailey, Mem. Torrey Bot. Club 1: 36 (1889). *Carex saxatilis* var. *rhomalea* Fernald, Rhodora 3: 50 (1901). *Carex rhomalea* (Fernald) Mack., Bull. Torrey Bot. Club 37: 246 (1910).
Carex miliaris var. *obtusa* L.H.Bailey, Mem. Torrey Bot. Club 1: 35 (1889).
Carex pulla var. *vesicarioides* Kurtz, Bot. Jahrb. Syst. 19: 417 (1894).
Carex vesicaria var. *jejuna* Fernald, Rhodora 3: 53 (1901).
Carex miliaris f. *major* (L.H.Bailey) Kük. in H.G.A. Engler (ed.), Pflanzenr., IV, 20(38): 719 (1909).
Carex pulla f. *debilis* Porsild, Meddel. Grønland 50: 370 (1912).
Carex vesicaria var. *laurentiana* Fernald, Rhodora 35: 232 (1933).
Carex procerula V.I.Krecz. in V.L.Komarov (ed.), Fl. URSS 3: 622 (1935).
Carex miliaris var. *ungavensis* Raymond, Bull. Soc. Bot. France 99: 196 (1952).
Carex miliaris f. *longipedunculata* Lepage, Naturaliste Canad. 82: 189 (1955). *Carex saxatilis* f. *longipedunculata* (Lepage) Lepage, Naturaliste Canad. 84: 93 (1957).

Carex* ×*saxenii Raymond, Canad. J. Bot. 28: 421 (1950). *C. paleacea* × *C. recta*.
E. Canada. 72 LAB NSC QUE. Hemicr. or rhizome geophyte.

Carex saxicola Tang & F.T.Wang, in Fl. Hainan. 4: 535 (1977).
Hainan. 36 CHH. Hemicr. or rhizome geophyte.

Carex saximontana Mack., Bull. Torrey Bot. Club 33: 439 (1906). *Carex backii* var. *saximontana* (Mack.) Boivin, Naturaliste Canad. 94: 522 (1967).
S. Canada to WC. U.S.A. 71 MAN SAS 72 ONT 73 COL MNT WYO 74 MIN NDA NEB SDA. Hemicr. or rhizome geophyte.
Carex durifolia var. *subrostrata* Bates, Univ. Nebraska, Sci. Rep. 14: 161 (1914). *Carex backii* var. *subrostrata* (Bates) Dorn, Vasc. Pl. Wyoming: 296 (1988).

Carex scabrata Schwein., Ann. Lyceum Nat. Hist. New York 1: 69 (1824).
E. Canada to EC. & E. U.S.A. 72 NBR NSC ONT PEI QUE 74 MSO OKL WIC 75 CNT INI MAI MAS MIC NWH NWJ NWY OHI PEN RHO VER WVA 78 DEL GEO KTY MRY NCA SCA TEN VRG WDC. Hemicr. or rhizome geophyte.

Carex scabrella Wahlenb., Kongl. Vetensk. Acad. Nya Handl. 24: 149 (1803).
Caribbean. 81 CUB DOM JAM. Hemicr. or rhizome geophyte.
Carex laxa Sw. ex Kunth, Enum. Pl. 2: 513 (1837).

Carex scabrifolia Steud., Syn. Pl. Glumac. 2: 237 (1855).
Russian Far East, China, Temp. E. Asia. 31 PRM 36 CHM CHN CHS 38 JAP KOR NNS TAI. Hemicr. or rhizome geophyte.
Carex pierotii Miq., Ann. Mus. Bot. Lugduno-Batavi 2: 148 (1866).
Carex suberea Boott, Ill. Carex 4: 165 (1867).
Carex malinvaldii H.Lév. & Vaniot, Bull. Soc. Bot. France 51: 204 (1904).
Carex yabei H.Lév. & Vaniot, Bull. Soc. Agric. Sarthe 1905: 79 (1905).

Carex scabripes Cherm., Bull. Soc. Bot. France 70: 410 (1923).
Madagascar. 29 MDG. Hemicr. or rhizome geophyte.

Carex scabrirostris Kük., Bot. Jahrb. Syst. 36(82): 9 (1905).
Tibet to C. China. 36 CHC CHN CHQ CHT. Hemicr. or rhizome geophyte.

Carex scabrisacca Ohwi & Ryu, J. Jap. Bot. 12: 655 (1936).
China (Jiangxi). 36 CHS. Hemicr. or rhizome geophyte.

Carex scabriuscula Mack., Bull. Torrey Bot. Club 35: 268 (1908).
Oregon to N. California. 73 ORE 76 CAL. Hemicr. or rhizome geophyte.
Carex scirpoidea var. *gigas* T.Holm, Amer. J. Sci. 167: 20 (1904). *Carex gigas* (T.Holm) Mack., Bull. Torrey Bot. Club 35: 268 (1908).

Carex scaposa C.B.Clarke, Bot. Mag. 113: t. 6940 (1887).
S. China to Vietnam. 36 CHC CHS 38 TAI 41 VIE. Hemicr. or rhizome geophyte.

var. ***dolichostachya*** F.T.Wang & Tang, Contr. Inst. Bot. Natl. Acad. Peiping 6(2): 58 (1949).
SE. China. 36 CHS. Hemicr. or rhizome geophyte.

var. ***hirsuta*** P.C.Li, Acta Phytotax. Sin. 37: 162 (1999).
China (Guangdong, Fujian). 36 CHS. Hemicr. or rhizome geophyte.

var. ***scaposa***
S. China to Vietnam. 36 CHC CHS 38 TAI 41 VIE. Hemicr. or rhizome geophyte.
Carex scaposa var. *marantacea* Raymond, Mém. Jard. Bot. Montréal 53: 26 (1959).

Carex schaffneri Boeck., Flora 61: 39 (1878).
C. Mexico. 79 MXC. Hemicr. or rhizome geophyte.

Carex schenkiana Boeck., Beitr. Cyper. 1: 36 (1888).
Brazil. 84+. Hemicr. or rhizome geophyte.

Carex schiedeana Kunze in C.Schkuhr, Beschr. Riedgräs., Suppl.: 119 (1842).

Mexico. 79 MXC MXE MXS. Hemicr. or rhizome geophyte.
Carex perstricta Mack., Bull. Torrey Bot. Club 36: 479 (1909). *Carex schiedeana* var. *perstricta* (Mack.) F.J.Herm., U.S.D.A. Agric. Handb. 467: 96 (1974).
Carex stellata Mack. in N.L.Britton & al. (eds.), N. Amer. Fl. 18: 226 (1935). *Carex schiedeana* var. *stellata* (Mack.) F.J.Herm., U.S.D.A. Agric. Handb. 467: 98 (1974).

Carex schlechteri Nelmes, Bull. Misc. Inform. Kew 1940: 269 (1940).
S. Africa. 27 CPP OFS TVL. Hemicr. or rhizome geophyte.

Carex schliebenii Podlech, Mitt. Bot. Staatssamml. München 4: 123 (1961).
Tanzania. 25 TAN. Hemicr. or rhizome geophyte.

Carex schmidtii Meinsh. in F.Schmidt, Beitr. Kenntn. Russ. Reich. 26: 224 (1871).
Siberia to N. Japan. 30 BRY CTA IRK KRA YAK 31 AMU KAM KHA KUR MAG PRM SAK 36 CHI CHM 37 MON 38 JAP KOR. Hemicr. or rhizome geophyte.
**Carex maximowiczii* F.Schmidt, Reis. Amur-Land., Bot.: 71 (1868), nom. illeg.
Carex lineolata Cham. ex Meinsh., Trudy Imp. S.-Peterburgsk. Bot. Sada 18: 338 (1901).
Carex subvaginata Meinsh. ex C.B.Clarke, Kew Bull., Addit. Ser. 8: 70 (1908).
Carex vladimiroviensis H.Lév., Bull. Acad. Int. Géogr. Bot. 19: 34 (1909).

Carex schneideri Nelmes, Bull. Misc. Inform. Kew 1939: 201 (1939).
Tibet to SC. China. 36 CHC CHT. Hemicr. or rhizome geophyte.

Carex schottii Dewey in W.H.Emory, Rep. U.S. Mex. Bound. 2(1): 231 (1858).
California to Mexico (Baja California). 76 CAL 79 MXN. Hemicr. or rhizome geophyte.
Carex barbarae Parish, Bull. S. Calif. Acad. Sci. 4: 108 (1905), nom. illeg.

Carex schwackeana Boeck., Allg. Bot. Z. Syst. 2: 191 (1896).
Brazil. 84+. Hemicr. or rhizome geophyte.

Carex schweinitzii Dewey ex Schwein., Ann. Lyceum Nat. Hist. New York 1: 71 (1824).
S. Ontario to NC. & E. U.S.A. 72 ONT 74 MSO WIS 75 CNT MAS MIC NWY PEN VER 78 VRG. Hemicr. or rhizome geophyte.
Carex schweinitzia Eaton, Man. Bot.: 159 (1829), orth. var.

Carex scirpoidea Michx., Fl. Bor.-Amer. 2: 171 (1803).
Carex scirpoidea var. *typica* B.Boivin, Naturaliste Canad. 75: 203 (1948), nom. inval.
Norway (Solvagtind), Siberia, Russian Far East, Subarctic America to W. & N. U.S.A. 10 NOR 30 YAK 31 KAM MAG 70 ASK GNL NUN NWT YUK 71 ABT BRC MAN SAS 72 LAB NBR NFL NSC ONT QUE 73 COL IDA MNT ORE WAS WYO 74 MIN NDA 75 MAI MIC NWH NWY VER 76 CAL NEV UTA. Hemicr. or rhizome geophyte.

var. ***convoluta*** Kük. in H.G.A.Engler (ed.), Pflanzenr., IV, 20(38): 81 (1909). *Carex scirpoidea* subsp. *convoluta* (Kük.) Dunlop, Novon 7: 355 (1997).
S. Ontario to Michigan. 72 ONT 75 MIC. Hemicr. or rhizome geophyte.

var. ***pseudoscirpoidea*** (Rydb.) Cronquist, Univ. Wash. Publ. Biol. 17: 325 (1969).
W. Canada to W. U.S.A. 71 BRC 73 COL IDA MNT ORE WAS WYO 76 CAL UTA. Hemicr. or rhizome geophyte.
**Carex pseudoscirpoidea* Rydb., Mem. New York Bot. Gard. 1: 78 (1900). *Carex scirpoidea* subsp. *pseudoscirpoidea* (Rydb.) Dunlop, Novon 7: 355 (1997).

var. ***scirpoidea***
Norway (Solvagtind), Siberia, Russian Far East, Subarctic America to W. & N. U.S.A. 10 NOR 30 YAK 31 KAM MAG 70 ASK GNL NUN NWT YUK 71 ABT BRC MAN SAS 72 LAB NBR NFL NSC ONT QUE 73 COL IDA MNT ORE WAS WYO 74 MIN NDA 75 MAI MIC NWH NWY VER 76 NEV UTA. Hemicr. or rhizome geophyte.
Carex wormskioldiana Hornem. in G.C.Oeder & al. (eds.), Fl. Dan.: t. 1528 (1816).
Carex michauxii Schwein., Ann. Lyceum Nat. Hist. New York 1: 64 (1824).
Physiglochis wormskioldii Raf., Good Book: 27 (1840). *Carex wormskioldii* (Raf.) Drejer, Rev. Car. Bor.: 18 (1841).
Carex scirpina Tuck. ex Boott, Ill. Carex 4: 150 (1867).
Carex scirpoidea f. *basigyna* Lange, Consp. Fl. Groenland., ed. 2: 132 (1890).
Carex scirpoides Sartwell, Bull. Torrey Bot. Club 20: 426 (1893), nom. illeg.
Carex scirpiformis Mack., Bull. Torrey Bot. Club 35: 270 (1908). *Carex scirpoidea* var. *scirpiformis* (Mack.) O'Neill & Duman, Rhodora 43: 417 (1941).
Carex scirpoidea var. *europaea* Kük. in H.G.A. Engler (ed.), Pflanzenr., IV, 20(38): 81 (1909).
Carex athabascensis F.J.Herm., Leafl. W. Bot. 8: 111 (1957).

var. ***stenochlaena*** T.Holm, Amer. J. Sci., IV, 18: 20 (1904). *Carex stenochlaena* (T.Holm) Mack., Bull. Torrey Bot. Club 35: 269 (1908). *Carex scirpoidea* subsp. *stenochlaena* (T.Holm) Á.Löve & D.Löve, Taxon 13: 202 (1964).
Alaska to NW. U.S.A. 70 ASK YUK 71 BRC 73 MNT WAS. Hemicr. or rhizome geophyte.

Carex scita Maxim., Bull. Acad. Imp. Sci. Saint-Pétersbourg 31: 115 (1887).
Russian Far East, Japan. 31 KAM KUR SAK 38 JAP. Hemicr. or rhizome geophyte.

var. ***parvisquama*** T.Koyama in ?
Japan (Honshu: Mt. Daisen). 38 JAP. Hemicr. or rhizome geophyte.

var. ***riishirensis*** (Franch.) Kük. in H.G.A.Engler (ed.), Pflanzenr., IV, 20(38): 414 (1909).
Russian Far East, N. Japan. 31 KAM KUR SAK 38 JAP. Hemicr. or rhizome geophyte.
Carex ciliolata Franch., Bull. Soc. Philom. Paris, VIII, 7: 38 (1895).
**Carex riishirensis* Franch., Bull. Soc. Philom. Paris, VIII, 7: 81 (1895).
Carex koraginensis Meinsh., Trudy Imp. S.-Peterburgsk. Bot. Sada 18: 351 (1901). *Carex scita* var. *korginensis* (Meinsh.) Kük. in H.G.A.

Engler (ed.), Pflanzenr., IV, 20(38): 415 (1909). *Carex podocarpa* var. *koraginensis* (Meinsh.) Vorosch. in A.K.Skvortsov (ed.), Florist. Issl. Razn. Raionakh SSSR: 155 (1985).

var. ***scabrinervia*** (Franch.) Kük. in H.G.A.Engler (ed.), Pflanzenr., IV, 20(38): 415 (1909).
Russian Far East, N. Japan. 31 KUR SAK 38 JAP. Hemicr. or rhizome geophyte.
Carex scabrinervia Franch., Bull. Soc. Philom. Paris, VIII, 7: 37 (1895).
Carex xanthoathera Franch., Bull. Soc. Philom. Paris, VIII, 7: 88 (1895).

var. ***scita***
Japan (Honshu). 38 JAP. Hemicr. or rhizome geophyte.
Carex tenuiseta var. *brevisquama* Koidz. in ?. *Carex scita* var. *brevisquama* (Koidz.) Ohwi, Mem. Coll. Sci. Kyoto Imp. Univ., Ser. B, Biol. 11: 322 (1936), nom. illeg.

Carex scitiformis Kük., Bull. Herb. Boissier, II, 5: 1162 (1905).
Japan (Honshu). 38 JAP. Hemicr. or rhizome geophyte.

Carex scitula Boott, Ill. Carex 4: 177 (1867).
Arunachal Pradesh (Mishmee Hills). 40 EHM. Hemicr. or rhizome geophyte.

Carex sclerocarpa Franch., Bull. Soc. Philom. Paris, VIII, 7: 91 (1895).
China (Sichuan to Anhui), Taiwan. 36 CHC CHS 38 TAI. Hemicr. or rhizome geophyte.
Carex taiheiensis Hayata, Icon. Pl. Formosan. 10: 59 (1921).

Carex scolopendriformis F.T.Wang & Tang ex P.C.Li, Acta Bot. Yunnan. 12: 142 (1990).
Tibet to SC. China. 36 CHC CHT. Hemicr. or rhizome geophyte.

Carex scoparia Willd., Sp. Pl. 4: 230 (1805). *Vignea scoparia* (Willd.) Rchb. in J.C.Mössler, Handb. Gewächsk. ed. 2, 3: 1613 (1830). *Loncoperis scoparia* (Willd.) Raf., Good Book: 27 (1840).
Canada to U.S.A. (51) nzn 71 ABT BRC MAN 72 NBR NFL NSC ONT PEI QUE 73 COL IDA MNT ORE WAS 74 ILL IOW KAN MIN MSO NEB OKL SDA WIS 75 CNT INI MAI MAS MIC NWH NWJ NWY OHI PEN RHO VER WVA 76 CAL 78 ARK DEL GEO KTY MRY MSI NCA SCA TEN VRG WDC. Hemicr. or rhizome geophyte.

var. ***scoparia***
Canada to U.S.A. (51) nzn 71 ABT BRC MAN 72 NBR NFL NSC ONT PEI QUE 73 COL IDA MNT ORE WAS 74 ILL IOW KAN MIN MSO NEB OKL SDA WIS 75 CNT INI MAI MAS MIC NWH NWJ NWY OHI PEN RHO VER WVA 76 CAL 78 ARK DEL GEO KTY MRY MSI NCA SCA TEN VRG WDC. Hemicr. or rhizome geophyte.
Carex leporina Michx., Fl. Bor.-Amer. 2: 170 (1803), nom. illeg.
Carex ovalis Willd. ex Kunth, Enum. Pl. 2: 397 (1837), nom. illeg.
Carex scoparia var. *moniliformis* Tuck., Enum. Meth. Caric.: 17 (1843). *Carex scoparia* f. *moniliformis* (Tuck.) Kük. in H.G.A.Engler (ed.), Pflanzenr., IV, 20(38): 203 (1909).
Carex scoparia var. *vera* Tuck., Enum. Meth. Caric.: 17 (1843).
Carex scoparia var. *aggregata* Dewey, Class-book Bot.: 416 (1845).
Carex lagopodioides var. *scoparia* Boeck., Linnaea 39: 114 (1875).
Carex scoparia f. *elatior* Peck, Rep. (Annual) Regents Univ. State New York State Mus. 48: 141 (1896).
Carex tribuloides var. *moniliformis* (Tuck.) Britton, Ill. Fl. N. U.S. 1: 356 (1896).
Carex scoparia var. *condensa* Fernald, Proc. Amer. Acad. Arts 37: 468 (1902). *Carex scoparia* f. *condensa* (Fernald) Kük. in H.G.A.Engler (ed.), Pflanzenr., IV, 20(38): 203 (1909), nom. illeg.
Carex scoparia f. *tenerrima* Kük. in H.G.A.Engler (ed.), Pflanzenr., IV, 20(38): 203 (1909).
Carex scoparia var. *subturbinata* Fernald & Wiegand, Rhodora 14: 116 (1912). *Carex scoparia* f. *subturbinata* (Fernald & Wiegand) Fernald, Rhodora 44: 284 (1942).
Carex scoparia f. *peracuta* Fernald, Rhodora 23: 234 (1922).

var. ***tessellata*** Fernald & Wiegand, Rhodora 12: 135 (1910).
E. Canada to Maine. 72 NBR 75 MAI. Hemicr. or rhizome geophyte.

Carex scopulorum T.Holm, Amer. J. Sci. 164: 422 (1902).
W. Canada to W. U.S.A. 70 YUK 71 ABT BRC 73 COL IDA MNT ORE WAS WYO 76 CAL UTA. Hemicr. or rhizome geophyte.

var. ***bracteosa*** (L.H.Bailey) F.J.Herm., Leafl. W. Bot. 9: 16 (1959).
W. Canada to W. U.S.A. 70 YUK 71 ABT BRC 73 COL IDA MNT ORE WAS WYO 76 CAL UTA. Hemicr. or rhizome geophyte.
Carex vulgaris var. *bracteosa* L.H.Bailey, Proc. Amer. Acad. Arts 22: 81 (1886). *Carex rigida* var. *bracteosa* (L.H.Bailey) L.H.Bailey, J. Bot. 28: 173 (1890).
Carex tolmiei var. *angusta* L.H.Bailey, Mem. Torrey Bot. Club 1: 47 (1889).
Carex gymnoclada T.Holm, Amer. J. Sci. 164: 424 (1902).
Carex brachypoda T.Holm, Amer. J. Sci. 170: 302 (1905).
Carex campylocarpa T.Holm, Amer. J. Sci. 170: 304 (1905).
Carex campylocarpa subsp. *affinis* Maguire & A.H.Holmgren, Leafl. W. Bot. 4: 262 (1946).

var. ***prionophylla*** (T.Holm) L.A.Standl., Syst. Bot. Monogr. 7: 86 (1985).
SW. Canada to NW. U.S.A. 71 BRC 73 IDA MNT WAS. Hemicr. or rhizome geophyte.
Carex prionophylla T.Holm, Amer. J. Sci. 164: 423 (1902).
Carex miserabilis Mack. in N.L.Britton & al. (eds.), N. Amer. Fl. 18: 385 (1935).

var. ***scopulorum***
WC. U.S.A. 73 COL WYO. Hemicr. or rhizome geophyte.
Carex tolmiei var. *subsessilis* L.H.Bailey, Mem. Torrey Bot. Club 1: 47 (1889).
Carex chimaphila T.Holm, Amer. J. Sci. 166: 33 (1903). *Carex scopulorum* var. *chimaphila* (T.Holm) Kük. in H.G.A.Engler (ed.), Pflanzenr., IV, 20(38): 303 (1909).

Carex secalina Willd. ex Wahlenb., Kongl. Vetensk. Acad. Nya Handl. 1803: 151 (1803). *Carex hordeistichos* var. *secalina* (Willd. ex Wahlenb.) Boott, Ill. Carex 4: 133 (1867). *Kuekenthalia secalina* (Willd. ex Wahlenb.) Fedde & J.Schust., Just's Bot. Jahresber. 41(2): 10 (1913 publ. 1918).
C. Europe to C. Asia. 11 AUT CZE GER HUN POL 13 ROM 14 RUS UKR 30+ (31) 32 KAZ 33 TCS 34 AFG IRN. Hemicr. or rhizome geophyte.

var. ***alpina*** Kük. & Bornm., Oesterr. Bot. Z. 47: 137 (1897).
Iran. 34 IRN. Hemicr. or rhizome geophyte.

var. ***secalina***
C. Europe to C. Asia. 11 AUT CZE GER HUN POL 13 ROM 14 RUS UKR 30+ (31) 32 KAZ 33 TCS 34 AFG IRN. Hemicr. or rhizome geophyte.
Carex secale Ledeb., Fl. Ross. 4: 300 (1852).
Carex hordeistichos var. *microstachys* Schur, Enum. Pl. Transsilv.: 710 (1866).
Carex hordeistichos var. *microcarpa* Boeck., Linnaea 41: 280 (1877).

Carex secta Boott in J.D.Hooker, Fl. Nov.-Zel. 1: 281 (1853). *Carex virgata* var. *secta* (Boott) Hook.f., Handb. N. Zeal. Fl. 1: 313 (1867). *Carex paniculata* var. *secta* (Boott) Cheeseman, Trans. & Proc. New Zealand Inst. 16: 428 (1884). *Carex appressa* var. *secta* (Boott) Kük. in H.G.A.Engler (ed.), Pflanzenr., IV, 20(38): 179 (1909).
New Zealand (incl. Stewart I.). 51 NZN NZS. Hemicr. or rhizome geophyte.

Carex sectoides (Kük.) Edgar in Allan, Fl. N. Zeal. 2: 276 (1970).
Snares I., Antipodean Is., Chatham Is. 51 ATP CTM NZS. Hemicr. or rhizome geophyte.
**Carex appressa* var. *sectoides* Kük. in H.G.A.Engler (ed.), Pflanzenr., IV, 20(38): 179 (1909).

Carex sedakowii C.A.Mey. ex Meinsh., Trudy Imp. S.-Peterburgsk. Bot. Sada 18: 360 (1902).
Siberia to Korea. 30+ 31+ 36 CHI CHN 37 MON 38 KOR. Hemicr. or rhizome geophyte.

Carex sellowiana Schltdl., Linnaea 10: 117 (1836). *Uncinia sellowiana* (Schltdl.) Nees in C.F.P.von Martius & auct. suc. (eds.), Fl. Bras. 2(1): 201 (1842).
S. Brazil, S. South America. 84 BZS 85 AGE AGW PAR URU. Hemicr. or rhizome geophyte.
Carex sellowiana var. *tenuis* Boeck., Beitr. Cyper. 2: 33 (1890).
Carex sellowiana var. *subrepens* Kük., Bot. Jahrb. Syst. 27: 536 (1899).
Carex sellowiana var. *prorepens* Kük., Verh. Bot. Vereins Prov. Brandenburg 47: 204 (1905 publ. 1906).

Carex semihyalofructa Tak.Shimizu, Acta Phytotax. Geobot. 56: 33 (2005).
Japan (Honshu). 38 JAP. Hemicr. or rhizome geophyte.

Carex sempervirens Vill., Hist. Pl. Dauphiné 2: 214 (1787). *Trasus erectus* (Vill.) Gray, Nat. Arr. Brit. Pl. 2: 67 (1821). *Carex ferruginea* var. *sempervirens* (Vill.) Fiori, Nuov. Fl. Italia 1: 197 (1923).
European Mts. 11 AUT CZE GER POL SWI 12 FRA SPA 13 ALB BUL GRC? ITA ROM YUG 14 UKR. Hemicr. or rhizome geophyte.
Carex sempervirens var. *pseudotristis* Domin in ?. *Carex sempervirens* subsp. *pseudotristis* (Domin) Pawl. in ?
Carex tristis var. *tatrorum* Zapal. in ?. *Carex tatrorum* (Zapal.) Racib. in W.Szafer, Fl. Polska 1: 215 (1919).
Carex alpina Schrank, Baier. Fl. 1: 298 (1789).
Carex variegata Lam., Encycl. 3: 389 (1792).
Carex ferruginea Schkuhr, Beschr. Riedgräs. 1: 92 (1801), nom. illeg.
Carex varia Host, Icon. Descr. Gram. Austriac. 1: 59 (1801).
Carex alpestris Gaudin, Etrennes Fl.: 173 (1804), nom. illeg.
Carex erecta DC. in J.B.A.M.de Lamarck & A.P.de Candolle, Fl. Franç., ed. 3, 3: 120 (1805), nom. illeg.
Carex sempervirens subsp. *laevis* Kit. ex Willd., Sp. Pl. 4: 292 (1805). *Carex sempervirens* var. *laevis* (Kit. ex Willd.) Asch. & Kantiz, Cat. Cormophyt. Anthophyt. Serbiae: 15 (1877).
Carex firma var. *subalpina* Wahlenb., Veg. Clim. Helv. Sept.: 160 (1813).
Carex arida Schleich. ex Kunth, Enum. Pl. 2: 464 (1837).
Carex sempervirens var. *setifolia* Kumm. & Sendtn., Flora 32: 759 (1849).
Carex sempervirens f. *laxiflora* Schur, Enum. Pl. Transsilv.: 709 (1866).
Carex sempervirens f. *pumila* Schur, Enum. Pl. Transsilv.: 10 (1866).
Carex sempervirens f. *rigida* Schur, Enum. Pl. Transsilv.: 709 (1866).
Carex sempervirens f. *stenophylla* Schur, Enum. Pl. Transsilv.: 710 (1866).
Carex sempervirens var. *trichocarpa* Schur, Enum. Pl. Transsilv.: 709 (1866).
Carex trichocarpa Schur, Enum. Pl. Transsilv.: 709 (1866), nom. illeg.
Carex sempervirens f. *schkuhriana* Bonnet & J.A.Richt., Bull. Soc. Bot. France 29: lxvi (1882).
Carex sempervirens f. *villarsiana* Bonnet & J.A.Richt., Bull. Soc. Bot. France 29: lxv (1882).
Carex sempervirens var. *aurigerana* Marcailhou, Rev. Bot. Bull. Mens., Sept. 1894: 9 (1894).
Carex frigida var. *densa* Meinsh., Trudy Imp. S.-Peterburgsk. Bot. Sada 18: 359 (1900).
Carex sempervirens f. *coarctata* Huter ex Kneuck., Allg. Bot. Z. Syst. 6: 43 (1900).
Carex sempervirens f. *semipallescens* Kneuck., Allg. Bot. Z. Syst. 6: 44 (1900).
Carex sempervirens var. *pendulina* Kük., Bull. Herb. Boissier, II, 4: 58 (1904).
Carex sempervirens f. *albicans* Lackow. ex Kük. in H.G.A.Engler (ed.), Pflanzenr., IV, 20(38): 569 (1909).
Carex sempervirens f. *fastigiata* Lackow. ex Kük. in H.G.A.Engler (ed.), Pflanzenr., IV, 20(38): 569 (1909).
Carex granitica Braun-Blanq., Commun. Stat. Int. Géobot. Médit. Montpellier 87: 224 (1945).

Carex senta Boott, Ill. Carex 4: 174 (1867).
California to Mexico (Baja California). 76 ARI CAL 79 MXN. Hemicr. or rhizome geophyte.

Carex seorsa Howe, Rep. (Annual) Regents Univ. State New York New York State Mus. 48: 138 (1894).
SE. Canada to E. U.S.A. 72 ONT 75 CNT INI MAS MIC NWH NWJ NWY OHI PEN RHO 78 ALA ARK DEL FLA GEO KTY MRY MSI NCA SCA TEN VRG WDC. Hemicr. or rhizome geophyte.

Carex seposita C.B.Clarke, Kew Bull., Addit. Ser. 8: 87 (1908).
India. 40 IND. Hemicr. or rhizome geophyte.

Carex serpenticola Zika, Madroño 45: 261 (1998 publ. 1999).
Oregon to California. 73 ORE 76 CAL. Hemicr. or rhizome geophyte.

Carex serratodens S.Watson in S.Watson & al., Bot. California 2: 245 (1880).
Oregon to Arizona. 73 ORE 76 ARI CAL. Hemicr. or rhizome geophyte.
Carex bifida Boott, Proc. Amer. Acad. Arts 7: 394 (1868), nom. illeg.
Carex aequa C.B.Clarke, Kew Bull., Addit. Ser. 8: 86 (1908).

Carex serreana Hand.-Mazz., Oesterr. Bot. Z. 85: 225 (1936).
Qinghai to C. China. 36 CHC CHN CHQ. Hemicr. or rhizome geophyte.
Carex wenchenii F.T.Wang & Tang, Acta Phytotax. Sin. 1: 134 (1951).

Carex seticulmis Boeck., Vidensk. Meddel. Naturhist. Foren. Kjøbenhavn 1869: 156 (1869).
SE. & S. Brazil. 84 BZL BZS. Hemicr. or rhizome geophyte.
Uncinia phyllostachya Nees in C.F.P.von Martius & auct. suc. (eds.), Fl. Bras. 2(1): 201 (1842).
Carex uleana Boeck., Beitr. Cyper. 2: 42 (1890).

Carex setifolia Kunze, Syn. Pl. Crypt.: 26 (1834). *Carex berteroana* var. *setifolia* (Kunze) Kük., Bot. Jahrb. Syst. 27: 534 (1899).
Peru to Chile. 83 PER 85 CLC CLN. Hemicr. or rhizome geophyte.

var. ***pungens*** (Boeck.) Kük. in H.G.A.Engler (ed.), Pflanzenr., IV, 20(38): 84 (1909).
Chile. 85 CLC CLN. Hemicr. or rhizome geophyte.
**Carex pungens* Boeck., Linnaea 39: 46 (1875).

var. ***setifolia***
Peru to Chile. 83 PER 85 CLC CLN. Hemicr. or rhizome geophyte.
Carex piptolepis Desv. in C.Gay, Fl. Chil. 6: 201 (1854).
Isolepis heterolepis Steud., Syn. Pl. Glumac. 2: 93 (1855).
Isolepis variegata Steud., Syn. Pl. Glumac. 2: 93 (1855).
Carex pissisi Phil., Linnaea 33: 272 (1865).
Carex berteroana Hohen. ex Boott, Ill. Carex 4: 149 (1867).
Carex setifolia var. *minor* Boott, Ill. Carex 4: 149 (1867).
Scirpus cumingii Boeck., Linnaea 36: 491 (1870).
Carex colchaguensis Phil., Anales Univ. Chile 93: 484 (1896). *Carex setifolia* var. *colchaguensis* (Phil.) Kük. in H.G.A.Engler (ed.), Pflanzenr., IV, 20(38): 84 (1909).
Carex moelleri Phil., Anales Univ. Chile 93: 487 (1896).

Carex setigera D.Don, Prodr. Fl. Nepal.: 43 (1825).
Himalaya to N. Thailand. 36 CHC CHT 40 EHM NEP WHM 41 THA. Hemicr. or rhizome geophyte.

var. ***schlagintweitiana*** (Boeck.) Kük. in H.G.A.Engler (ed.), Pflanzenr., IV, 20(38): 419 (1909).
Himalaya to SC. China. 36 CHC CHT 40 EHM NEP WHM. Hemicr. or rhizome geophyte.
**Carex schlagintweitiana* Boeck., Beitr. Cyper. 1: 48 (1888).
Carex tsangensis Franch., Bull. Soc. Philom. Paris, VIII, 7: 44 (1895).
Carex schlagintweitiana subsp. *deformis* Noltie, Edinburgh J. Bot. 50: 194 (1993).

var. ***setigera***
Himalaya to N. Thailand. 36 CHC CHT 40 EHM NEP WHM 41 THA. Hemicr. or rhizome geophyte.
Carex setigera var. *ustilaginosa* Nees in R.Wight, Contr. Bot. India: 126 (1834).
Carex minutiflora Boeck., Beitr. Cyper. 1: 51 (1888).

Carex setigluma Reznicek & S.González, Contr. Univ. Michigan Herb. 20: 222 (1995).
Colombia to Ecuador. 83 CLM ECU. Hemicr. or rhizome geophyte.

Carex setispica C.B.Clarke, Kew Bull., Addit. Ser. 8: 77 (1908).
Brazil. 84+. Hemicr. or rhizome geophyte.

Carex setosa Boott, Ill. Carex 3: 108 (1862).
Himalaya to China. 36 CHC CHN CHQ CHS CHT 40 EHM NEP WHM. Hemicr. or rhizome geophyte.

var. ***mianxianica*** S.Yun Liang, in Fl. Reipubl. Popul. Sin. 12: 525 (2000).
China (Shaanxi). 36 CHN. Hemicr.

var. ***punctata*** S.Yun Liang, in Fl. Reipubl. Popul. Sin. 12: 525 (2000).
China (Sichuan, Guangxi). 36 CHC CHS. Hemicr.

var. ***setosa***
Himalaya to C. China. 36 CHC CHN CHQ CHT 40 EHM NEP WHM. Hemicr. or rhizome geophyte.
Carex pachyrrhiza Franch., Bull. Soc. Philom. Paris, VIII, 7: 45 (1895).

Carex shaanxiensis F.T.Wang & Tang ex P.C.Li, Acta Phytotax. Sin. 37: 174 (1999).
China (Shaanxi, Gansu). 36 CHN. Hemicr. or rhizome geophyte.

Carex* × *shakushizawaensis Akiyama, Bot. Mag. (Tokyo) 45: 473 (1931). *C. doenitzii* × *C. forficula*.
Japan. 38 JAP. Hemicr. or rhizome geophyte.

Carex shandanica Y.C.Yang, Acta Biol. Plateau Sin. 3: 86 (1984).
China (Gansu). 36 CHN. Hemicr. or rhizome geophyte.

Carex shanensis C.B.Clarke, Kew Bull., Addit. Ser. 8: 72 (1908).
Myanmar. 41 MYA. Hemicr. or rhizome geophyte.

Carex shangchengensis S.Yun Liang, Acta Phytotax. Sin. 24: 242 (1986).
China (Henan). 36 CHS. Hemicr. or rhizome geophyte.

Carex shanghaiensis S.X.Qian & Y.Q.Liu, Acta Phytotax. Sin. 37: 103 (1999).
China (Jiangsu, Shanghai). 36 CHS. Hemicr. or rhizome geophyte.

Carex shanghangensis S.Yun Liang, Acta Phytotax. Sin. 34: 94 (1996).
China (Fujian). 36 CHS. Hemicr. or rhizome geophyte.

Carex sheldonii Mack., Bull. Torrey Bot. Club 42: 618 (1915).
W. U.S.A. 73 IDA ORE 76 CAL NEV. Hemicr. or rhizome geophyte.

Carex shimidzensis Franch., Bull. Soc. Philom. Paris, VIII, 7: 33 (1895).
Korea (Kangwon, Ullung Do), Kuril Is., Japan. 31 KUR 38 JAP KOR. Hemicr. or rhizome geophyte.
Carex nervulosa Franch., Nouv. Arch. Mus. Hist. Nat., III, 9: 120 (1897). *Carex shimidzensis* var. *nervulosa* (Franch.) Kük. in H.G.A.Engler (ed.), Pflanzenr., IV, 20(38): 357 (1909).
Carex sorachensis H.Lév. & Vaniot, Bull. Soc. Bot. France 51: 205 (1904).
Carex takeshimensis Nakai, Bot. Mag. (Tokyo) 31: 281 (1917).

Carex ×shinanoana Nakai ex Aliyama, Consp. Caric. Jap.: 107 (1932). *C. heterolepis* × *C. otaruensis*.
Japan. 38 JAP. Hemicr. or rhizome geophyte.

Carex shinnersii P.Rothr. & Reznicek, Novon 11: 225 (2001).
Kansas to Texas. 74 KAN OKL 77 TEX 78 ARK. Hemicr.

Carex shiriyajirensis Akiyama ex Tatew., Bull. Biogeogr. Soc. Japan 4(4): 257–334 (1934).
Kuril Is. 31 KUR. Hemicr. or rhizome geophyte.

Carex shortiana Dewey & Torr., Amer. J. Sci. Arts 30: 60 (1836).
S. Ontario to NC. & E. U.S.A. 72 ONT 74 ILL IOW KAN MSO OKL 75 INI OHI PEN WVA 78 ARL KTY MRY TEN VRG WDC. Hemicr. or rhizome geophyte.
Carex shortii Torr., Ann. Lyceum Nat. Hist. New York 3: 407 (1836).
Carex ohiotica Hochst. & Steud., Nomencl. Bot., ed. 2, 1: 293 (1840).
Carex aporandra Kunze ex Boott, Ill. Carex 2: 92 (1860).

Carex shuangbaiensis L.K.Dai, Acta Phytotax. Sin. 37: 187 (1999).
China (Yunnan). 36 CHC. Hemicr. or rhizome geophyte.

Carex shuchengensis S.W.Su & Q.Zhang, Acta Phytotax. Sin. 27: 465 (1989).
China (Anhui). 36 CHS. Hemicr. or rhizome geophyte.

Carex siccata Dewey, Amer. J. Sci. Arts 10: 27 (1826). *Vignea siccata* (Dewey) Raf., Good Book: 27 (1840).
N. America. 70 NWT YUK 71 ABT BRC MAN SAS 72 ONT QUE 73 COL WAS WYO 74 ILL MIN SDA WIS 75 CNT INI MAI MAS MIC NWH NWJ NWY OHI PEN RHO VER 76 ARI UTA 77 NWM. Hemicr. or rhizome geophyte.
Carex liddonii Carey in A.Gray, Manual: 545 (1848), nom. illeg.
Carex adusta var. *argyrantha* L.H.Bailey, Carex Cat.: 2 (1884). *Carex albolutescens* var. *argyrantha* (L.H. Bailey) Olney ex L.H.Bailey, Proc. Amer. Acad. Arts 22: 149 (1886).
Carex adusta var. *sparsiflora* L.H.Bailey, Carex Cat.: 2 (1884).
Carex foenea var. *sparsiflora* Howe, Rep. (Annual) Regents Univ. State New York New York State Mus. 47: 142 (1895).
Carex siccata var. *obscurior* Kük. in H.G.A.Engler (ed.), Pflanzenr., IV, 20(38): 133 (1909).
Carex foenea var. *tuberculata* F.J.Herm., Rhodora 70: 419 (1968).
Carex foenea var. *enervis* D.K.Evans & Mohlenbr., Trans. Illinois State Acad. Sci. 64: 270 (1971 publ. 1972).

Carex sichouensis P.C.Li, Acta Phytotax. Sin. 37: 173 (1999).
China (Yunnan). 36 CHC. Hemicr. or rhizome geophyte.

Carex siderosticta Hance, J. Linn. Soc., Bot. 13: 89 (1873).
Russian Far East, China, Korea, Japan. 31 PRM 36 CHM CHN CHS 38 JAP KOR. Hemicr. or rhizome geophyte.

var. ***pilosa*** H.Lév. ex T.Koyama, Repert. Spec. Nov. Regni Veg. 13: 244 (1914).
E. China to C. & S. Japan. 36 CHM CHS 38 JAP KOR. Hemicr. or rhizome geophyte.
**Carex ciliatomarginata* Nakai, Repert. Spec. Nov. Regni Veg. 13: 244 (1914).

var. ***siderosticta***
Russian Far East, China, Korea, Japan. 31 PRM 36 CHM CHN CHS 38 JAP KOR. Hemicr. or rhizome geophyte.
Pseudocarex plantaginea Miq., Ann. Mus. Bot. Lugduno-Batavi 2: 146 (1865).
Carex platyphylla Franch., Bull. Soc. Philom. Paris, VIII, 7: 50 (1895), nom. illeg.
Carex siderosticta var. *bracteosa* Franch., Nouv. Arch. Mus. Hist. Nat., III, 9: 196 (1897).
Carex siderosticta f. *stenophylla* Kitag., J. Jap. Bot. 44: 278 (1969). *Carex siderosticta* var. *stenophylla* (Kitag.) Kitag., J. Jap. Bot. 50: 88 (1975).

Carex silicea Olney, Proc. Amer. Acad. Arts 7: 393 (1868). *Carex straminea* var. *silicea* (Olney) L.H. Bailey, Carex Cat.: 4 (1884).
E. Canada to Virginia. 72 NBR NFL NSC PEI QUE 75 CNT MAI MAS NWH NWJ NWY RHO 78 DEL MRY VRG. Hemicr. or rhizome geophyte.
Carex straminea var. *moniliformis* Tuck., Enum. Meth. Caric.: 7 (1843).
Carex adusta Carey ex Boott, Ill. Carex 3: 118 (1862).
Carex foenea var. *sabulonum* A.Gray, Manual, ed. 5: 580 (1867).
Carex moniliformis Britton, Geol. Surv. New Jersey 2: 278 (1879).

Carex silvestrii Pamp., Nuovo Giorn. Bot. Ital. 17: 235 (1910).
China (Hubei). 36 CHC. Hemicr. or rhizome geophyte.

Carex simensis Hochst. ex A.Rich., Tent. Fl. Abyss. 2: 514 (1850).
Ethiopia to Kenya. 24 ETH 25 KEN UGA. Hemicr. or rhizome geophyte.
Carex aethiopica var. *stolonifera* Boeck., Linnaea 41: 286 (1877). *Carex simensis* var. *stolonifera* (Boeck.) Kük. in H.G.A.Engler (ed.), Pflanzenr., IV, 20(38): 654 (1909).
Carex simensis var. *mauensis* Kük. in H.G.A.Engler (ed.), Pflanzenr., IV, 20(38): 654 (1909).
Carex simensis var. *longistipitata* Kük., Notizbl. Bot. Gart. Berlin-Dahlem 9: 316 (1925).
Carex simensis var. *lanuriensis* De Wild., Pl. Bequaert. 4: 247 (1927).

Carex simulans C.B.Clarke, J. Linn. Soc., Bot. 36: 310 (1904).
China (Sichuan to Zhejiang). 36 CHC CHS. Hemicr. or rhizome geophyte.
Carex schistorhyncha H.Lév. & Vaniot, Bull. Soc. Bot. France 53: 317 (1906).
Carex simulans var. *densiflora* Tang & F.T.Wang ex S.Y.Liang, Acta Phytotax. Sin. 36: 537 (1998).

Carex simulata Mack., Bull. Torrey Bot. Club 34: 604 (1907 publ. 1908). *Carex praegracilis* var. *simulata* (Mack.) B.Boivin, Phytologia 43: 41 (1979).
W. Canada to W. & WC. U.S.A. 71 ABT SAS 73 COL IDA MNT ORE WAS WYO 74 NDA 76 ARI CAL UTA 77 NWM. Hemicr. or rhizome geophyte.

Carex sinclairii Boott ex Cheeseman, Man. New Zealand Fl.: 820 (1906).
New Zealand. 51 NZN NZS. Hemicr. or rhizome geophyte.

Carex sinoaristata Tang & F.T.Wang ex L.K.Dai, Acta Phytotax. Sin. 32: 184 (1994).
SC. China (Sichuan). 36 CHC. Hemicr. or rhizome geophyte.

Carex sinodissitiflora Tang & F.T.Wang ex L.K.Dai, Acta Phytotax. Sin. 32: 177 (1994).
China (Yunnan). 36 CHC. Hemicr. or rhizome geophyte.

Carex sinomairei H.Lév., Repert. Spec. Nov. Regni Veg. 11: 550 (1913).
China (Yunnan). 36 CHC. Hemicr. or rhizome geophyte.

Carex siroumensis Koidz., Bot. Mag. (Tokyo) 32: 54 (1918).
NE. China to N. Korea, C. Japan. 38 JAP. Hemicr. or rhizome geophyte.
Carex siroumensis Koidz., Acta Phytotax. Geobot. 3: 199 (1934), nom. illeg.

Carex skottsbergiana Kük., Repert. Spec. Nov. Regni Veg. 8: 7 (1910).
Chile. 85+. Hemicr. or rhizome geophyte.

Carex socialis Mohlenbr. & Schwegman, Brittonia 21: 77 (1969).
EC. & SE. U.S.A. 74 ILL MSO OKL 75 INI 77 TEX 78 ALA ARK GEO KTY MSI NCA SCA TEN. Hemicr. or rhizome geophyte.

Carex sociata Boott, Mem. Amer. Acad. Arts, n.s., 6: 420 (1859).
S. Japan, Nansei-shoto, Taiwan. 38 JAP NNS TAI. Hemicr. or rhizome geophyte.
Carex legendrei H.Lév. & Vaniot, Mém. Soc. Sci. Nat. Cherbourg 35: 216 (1905).
Carex nexa var. *strictior* Kük., J. Coll. Sci. Imp. Univ. Tokyo 22: 496 (1906). *Carex ligata* var. *strictior* (Kük.) Kük. in H.G.A.Engler (ed.), Pflanzenr., IV, 20(38): 474 (1909).
Carex atronucula Hayata, J. Coll. Sci. Imp. Univ. Tokyo 30(1): 379 (1911).
Carex uraiensis Hayata, Icon. Pl. Formosan. 10: 60 (1921).
Carex uber Ohwi, Mem. Coll. Sci. Kyoto Imp. Univ., Ser. B, Biol. 3: 261 (1930).

Carex soczavaeana Gorodkov, Zhurn. Russk. Bot. Obshch. 15: 185 (1930). *Carex kamtschatica* (Gorodkov) V.I.Krecz. in ?. *Carex ensifolia* subsp. *soczavaeana* (Gorodkov) Vorosch. in A.K.Skvortsov (ed.), Florist. Issl. Razn. Raionakh SSSR: 154 (1985).
Russian Far East. 31 KAM MAG. Hemicr. or rhizome geophyte.

Carex sodiroi Kük., Bot. Jahrb. Syst. 34(78): 7 (1904).
Ecuador (Pichincha). 83 ECU. Hemicr. or rhizome geophyte.

Carex* × *soerensenii Lepage, Naturaliste Canad. 83: 145 (1956). *C. rariflora* × *C. subspathacea*.
Greenland. 70 GNL. Hemicr. or rhizome geophyte.

Carex sohachii Ohwi, Acta Phytotax. Geobot. 11: 254 (1942).
China (Shanxi). 36 CHN. Hemicr. or rhizome geophyte.

Carex solanderi Boott in J.D.Hooker, Fl. Nov.-Zel. 1: 284 (1853). *Carex dissita* var. *solanderi* (Boott) Kük. in H.G.A.Engler (ed.), Pflanzenr., IV, 20(38): 692 (1909).
New Zealand. 51 NZN NZS. Hemicr. or rhizome geophyte.

Carex sordida Van Heurck & Müll.Arg., Observ. Bot. Descript. Pl. Nov. Herb. Van Heurckiani 1: 33 (1870).
Siberia to Japan. 30 BRY 31 AMU KAM MAG PRM SAK 36 CHI CHM 37 MON 38 JAP KOR. Hemicr. or rhizome geophyte.
Carex akanensis Franch., Bull. Soc. Philom. Paris, VIII, 7: 51 (1895). *Carex drymophila* var. *akanensis* (Franch.) Kük. in H.G.A.Engler (ed.), Pflanzenr., IV, 20(38): 756 (1909).
Carex amurensis Kük., Bot. Centralbl. 77: 94 (1899).
Carex amurensis var. *abbreviata* Kük., Bot. Centralbl. 77: 94 (1899). *Carex drymophila* var. *abbreviata* (Kük.) Ohwi, Acta Phytotax. Geobot. 12: 107 (1943).

Carex sorianoi Barros, Darwiniana 9: 166 (1949).
S. Chile to S. Argentina. 85 AGS CLS. Hemicr. or rhizome geophyte.

Carex sororia Kunth, Enum. Pl. 2: 379 (1837).
S. Ecuador to S. South America. 83 ECU 84 BZS 85 AGE AGS AGW PAR URU. Hemicr. or rhizome geophyte.
Carex sororia f. *paupera* Osten, Anales Mus. Hist. Nat. Montevideo, II, 3: 236 (1931).

var. ***pseudobracteosa*** (Kük.) Kük. in H.G.A.Engler (ed.), Pflanzenr., IV, 20(38): 150 (1909).
NE. Argentina, Uruguay. 85 AGE URU. Hemicr. or rhizome geophyte.
**Carex involucrata* var. *pseudobracteosa* Kük., Bot. Jahrb. Syst. 25: 510 (1899).

var. ***sororia***
S. Ecuador to S. South America. 83 ECU 84 BZS 85 AGE AGS AGW PAR URU. Hemicr. or rhizome geophyte.
Carex involucrata var. *submuricata* C.B.Clarke ex Kük., Bot. Jahrb. Syst. 27: 510 (1899).

Carex sozusensis Ohwi, Mem. Coll. Sci. Kyoto Imp. Univ., Ser. B, Biol. 6: 243 (1931).
Korea. 38 KOR. Hemicr. or rhizome geophyte.

Carex spachiana Boott, Ill. Carex 1: 43 (1858).
China (Guangdong, Macao). 36 CHS. Hemicr. or rhizome geophyte.

Carex sparganioides Muhl. ex Willd., Sp. Pl. 4: 237 (1805).
SE. Canada to NC. & E. U.S.A. 72 ONT QUE 74 ILL IOW KAN MIN MSO NEB SDA WIS 75 CNT MAI MAS MIC NWH NWJ NWY OHI PEN VER WVA 78 ARK DEL GEO KTY MRY TEN VRG WDC. Hemicr. or rhizome geophyte.
Carex sparganioides var. *ramea* Dewey, Class-book Bot.: 415 (1845).
Carex sparganioides var. *lutea* Farw., Pap. Michigan Acad. Sci. 1: 91 (1923).

Carex specifica L.H.Bailey, Mem. Torrey Bot. Club 1: 21 (1889).
SW. U.S.A. 76 CAL NEV. Hemicr. or rhizome geophyte.
Carex adusta var. *congesta* Boott in S.Watson & al., Bot. California 2: 238 (1880).
Carex loddonii var. *incerta* L.H.Bailey, Bot. Gaz. 13: 88 (1888).
Carex specifica f. *brevifructus* Kük. in H.G.A.Engler (ed.), Pflanzenr., IV, 20(38): 199 (1909).
Carex lancifructus Mack., Bull. Torrey Bot. Club 43: 607 (1916 publ. 1917).

Carex speciosa Kunth, Enum. Pl. 2: 504 (1837).
Himalaya to SC. China and W. Malesia. 36 CHC 40 ASS EHM IND NEP 41 THA VIE 42 BOR JAW MLY SUM. Hemicr. or rhizome geophyte.

subsp. ***dilatata*** Noltie, Edinburgh J. Bot. 50: 197 (1993).
NE. India to China (Yunnan). 36 CHC 40 ASS EHM IND NEP. Hemicr. or rhizome geophyte.

subsp. ***latifolia*** T.Koyama, Bot. Mag. (Tokyo) 92: 230 (1979).
Thailand. 41 THA. Hemicr. or rhizome geophyte.

subsp. ***pinetorum*** Noltie, Edinburgh J. Bot. 50: 197 (1993).
Bhutan. 40 EHM. Hemicr. or rhizome geophyte.

subsp. ***platyrhina*** (Ohwi) T.Koyama, Bot. Mag. (Tokyo) 92: 228 (1979).
Thailand. 41 THA. Hemicr. or rhizome geophyte.
**Carex platyrhina* Ohwi, Acta Phytotax. Geobot. 22: 139 (1967).

subsp. ***speciosa***
Himalaya to SC. China and W. Malesia. 36 CHC 40 EHM IND NEP 41 THA VIE 42 BOR JAW MLY SUM. Hemicr. or rhizome geophyte.
Carex concolor Nees in R.Wight, Contr. Bot. India: 125 (1834), nom. illeg.
Carex courtallensis Nees ex Boott, Ill. Carex 1: 52 (1858). *Carex speciosa* var. *courtallensis* (Nees ex Boott) Kük. in H.G.A.Engler (ed.), Pflanzenr., IV, 20(38): 481 (1909).
Carex peduncularis Wall. ex Boott, Ill. Carex 1: 53 (1858).
Carex speciosa var. *angustifolia* Boott, Ill. Carex 1: 52 (1858).
Carex speciosa var. *minor* Boeck., Linnaea 40: 388 (1876).
Carex longispica Boeck., Allg. Bot. Z. Syst. 2: 174 (1896).
Carex speciosa var. *abscondita* Kük., Bull. Jard. Bot. Buitenzorg, III, 16: 318 (1940).
Carex stenura Nelmes, Kew Bull. 5: 202 (1950).

Carex spectabilis Dewey, Amer. J. Sci. Arts 29: 248 (1836).
Russian Far East, Subarctic America to W. U.S.A. 31 KAM MAG 70 ALU ASK YUK 71 ABT BRC 73 IDA MNT ORE WAS WYO 76 CAL UTA. Hemicr. or rhizome geophyte.
Carex nigella Boott in W.J.Hooker, Fl. Bor.-Amer. 2: 225 (1839). *Carex tolmiei* var. *nigella* (Boott) Kük. in H.G.A.Engler (ed.), Pflanzenr., IV, 20(38): 411 (1909).
Carex tolmiei Boott in W.J.Hooker, Fl. Bor.-Amer. 2: 224 (1839).
Carex melastoma Fisch. ex Boott, Ill. Carex 2: 100 (1860).
Carex invisa L.H.Bailey, Proc. Amer. Acad. Arts 22: 82 (1887). *Carex tolmiei* var. *invisa* (L.H.Bailey) Kük. in H.G.A.Engler (ed.), Pflanzenr., IV, 20(38): 412 (1909).
Carex tolmiei var. *longiuscula* Hultén, Kongl. Svenska Vetensk. Acad. Handl., III, 5(1): 195 (1927).

Carex specuicola J.T.Howell, Leafl. W. Bot. 5: 148 (1949).
Utah to Arizona. 76 ARI UTA. Hemicr. or rhizome geophyte.

Carex sphaerogyna Baker, J. Bot. 21: 129 (1883).
C. Madagascar. 29 MDG. Hemicr. or rhizome geophyte.

Carex spicata Huds., Fl. Angl.: 349 (1762). *Vignea spicata* (Huds.) Soják, Cas. Nár. Mus., Odd. Prír. 148: 197 (1979 publ. 1980).
Europe to NW. Iran. 10 DEN FIN GRB IRE NOR SWE 11 AUT BGM CZE GER HUN NET POL SWI 12 FRA SPA 13 ALB? BUL GRC ITA KRI ROM TUE YUG 14 BLR BLT KRY RUC RUE RUN RUS RUW UKR 33 NCS TCS 34 IRN TUR (51) nzs (72) nsc ont (74) ill (75) cnt ini mai mas mic nwh nwj nwy ohi pen rho (78) del mry wdc. Hemicr. or rhizome geophyte.
Carex muricata subsp. *macrocarpa* Neuman in ?
Carex contigua Hoppe in J.Sturm, Deutschl. Fl.: 61 (1833). *Carex divulsa* var. *virens* (Hoppe) Nyman, Consp. Fl. Eur.: 781 (1882). *Desmiograstis contigua* (Hoppe) Fedde & J.Schust., Just's Bot. Jahresber. 41(2): 9 (1913 publ. 1918). *Carex muricata* var. *contigua* (Hoppe) Fiori, Nuov. Fl. Italia 1: 185 (1923).
Vignea spicata f. *humilis* Peterm., Fl. Lips. Excurs.: 52 (1838). *Carex spicata* f. *humilis* (Peterm.) Soó, Acta Bot. Acad. Sci. Hung. 16: 369 (1970 publ. 1971).
Carex furculata Peterm., Flora 27: 329 (1844). *Carex muricata* var. *furculata* (Peterm.) Nyman, Consp. Fl. Eur.: 781 (1882). *Carex spicata* f. *furculata* (Peterm.) Soó, Acta Bot. Acad. Sci. Hung. 16: 370 (1970 publ. 1971).
Carex fraterna Rupr., Symb. Pl. Ross.: 208 (1846).
Carex muricata f. *argyrolepis* Peterm., Anal. Pfl.-Schlüss.: 488 (1846). *Carex spicata* f. *argyrolepis* (Peterm.) Soó, Acta Bot. Acad. Sci. Hung. 16: 369 (1970 publ. 1971).
Carex muricata var. *incrassata* Crép., Notes Pl. Rar. Belgique 1: 23 (1859).
Carex muricata var. *fumosa* Gren., Fl. Jurass. 2: 835 (1869). *Carex contigua* f. *fumosa* (Gren.) Kük. in H.G.A.Engler (ed.), Pflanzenr., IV, 20(38): 156 (1909). *Carex fumosa* (Gren.) Rouy in G.Rouy & J.Foucaud, Fl. France 13: 412 (1912). *Carex spicata* f. *fumosa* (Gren.) Soó, Acta Bot. Acad. Sci. Hung. 16: 370 (1970 publ. 1971).
Carex contigua f. *remota* F.W.Schultz, Flora 53: 459 (1870).
Carex contigua f. *pallida* Appel ex Woerl., Phan.-Fl. Münch. Thalebene: 164 (1893).
Carex contigua var. *mediterranea* Asch. & Graebn., Syn. Mitteleur. Fl. 2(2): 39 (1902).
Carex contigua f. *pseudoguestphalica* Asch. & Graebn., Syn. Mitteleur. Fl. 2(2): 39 (1902).
Carex contigua f. *submonostachya* Asch. & Graebn., Syn. Mitteleur. Fl. 2(2): 39 (1902).
Carex muricata f. *mediterranea* Asch. & Graebn., Syn. Mitteleur. Fl. 2(2): 39 (1902). *Carex spicata* f. *mediterranea* (Asch. & Graebn.) Soó, Acta Bot. Acad. Sci. Hung. 16: 370 (1970 publ. 1971).

Carex muricata f. *pseudoguestephalica* Asch. & Graebn., Syn. Mitteleur. Fl. 2(2): 39 (1902). *Carex spicata* f. *pseudoguestephalica* (Asch. & Graebn.) Soó, Acta Bot. Acad. Sci. Hung. 16: 370 (1970 publ. 1971).
Carex contigua var. *longissima* Tauscher ex Vollm., Denkschr. Königl. Bot. Ges. Regensburg 8(2): 27 (1903).

Carex spicatopaniculata Boeck. ex C.B.Clarke in W.H.Harvey & auct. suc. (eds.), Fl. Cap. 7: 304 (1898).
Ethiopia to S. Africa. 24 ETH 25 TAN 26 MOZ 27 CPP LES NAT OFS SWZ TVL. Hemicr. or rhizome geophyte.

Carex spicigera Nees in R.Wight, Contr. Bot. India: 121 (1834).
Sri Lanka. 40 SRL. Hemicr. or rhizome geophyte.
Carex spicigera var. *minor* Boott, Ill. Carex 1: 11 (1858). *Carex spicigera* subsp. *minor* (Boott) T.Koyama in M.D.Dassanayake & al. (eds.), Revised Handb. Fl. Ceylon 5: 385 (1985).
Carex rubella Boott, Ill. Carex 4: 176 (1867). *Carex spicigera* var. *rubella* (Boott) Boeck., Linnaea 40: 369 (1876). *Carex spicigera* f. *rubella* (Boott) T.Koyama in M.D.Daddanayake & al. (eds.), Revised Handb. Fl. Ceylon 5: 385 (1985).
Carex spicigera var. *rostrata* Boeck., Linnaea 40: 369 (1876). *Carex spicigera* f. *rostrata* (Boeck.) T.Koyama in M.D.Daddanayake & al. (eds.), Revised Handb. Fl. Ceylon 5: 384 (1985).
Carex exigua Boeck., Bot. Jahrb. Syst. 5: 514 (1884).

Carex* ×*spiculosa F.Nyl., Spic. Fl. Fenn. 2: 49 (1844). *C. nigra* × *C. salina.*
E. Canada, N. & NE. Europe. 10 NOR 14 RUN 72 NSC. Hemicr. or rhizome geophyte.

Carex spinirostris Colenso, Trans. & Proc. New Zealand Inst. 15: 335 (1883).
New Zealand. 51 NZN NZS. Hemicr. or rhizome geophyte.
Carex vacillans Sol. ex Boott in J.D.Hooker, Fl. Nov.-Zel. 1: 285 (1853), nom. illeg.

Carex spissa L.H.Bailey ex Hemsl., Biol. Cent.-Amer., Bot. 4: 94 (1886).
California to New Mexico and Mexico. 76 ARI CAL 77 NWM 79 MXC MXE MXN MXS. Hemicr. or rhizome geophyte.

var. ***seatoniana*** (L.H.Bailey) Kük. in H.G.A.Engler (ed.), Pflanzenr., IV, 20(38): 422 (1909).
SW. Mexico (to Hidalgo). 79 MXE MXS. Hemicr. or rhizome geophyte.
**Carex seatoniana* L.H.Bailey, Bot. Gaz. 25: 270 (1898).

var. ***spissa***
California to New Mexico and NW. & C. Mexico. 76 ARI CAL 77 NWM 79 MXC MXN. Hemicr. or rhizome geophyte.
Carex ultra L.H.Bailey, Proc. Amer. Acad. Arts 22: 83 (1887). *Carex spissa* var. *ultra* (L.H.Bailey) Kük. in H.G.A.Engler (ed.), Pflanzenr., IV, 20(38): 422 (1909).

Carex spongiosa Ohwi, Mem. Coll. Sci. Kyoto Imp. Univ., Ser. B, Biol. 5(3): 284 (1930).
Korea. 38 KOR. Hemicr. or rhizome geophyte.

Carex sprengelii Dewey ex Spreng., Syst. Veg. 3: 837 (1826). *Uncinia sprengelii* (Dewey ex Spreng.) Nees, Linnaea 10: 205 (1836).
Canada to N. U.S.A. 71 ABT BRC MAN SAS 72 NBR ONT QUE 73 COL MNT WYO 74 ILL IOW MIN MSO NDA NEB SDA WIS 75 CNT INI MAI MAS MIC NWH NWJ NWY PEN VER 78 DEL. Hemicr. or rhizome geophyte.
Carex longirostris Torr. ex Schwein., Ann. Lyceum Nat. Hist. New York 1: 71 (1824), nom. illeg.
Carex longirostris var. *minor* Boott, Proc. Acad. Nat. Sci. Philadelphia 15: 78 (1863).
Carex longirostris var. *microcystis* Boeck., Linnaea 41: 241 (1877).

Carex* ×*squamigera V.I.Krecz. & Luchnik, Trudy Dal'nevost. Fil. Akad. Nauk S.S.S.R., Ser. Bot. 2: 890 (1937). *C. appendiculata* × *C. schmidtii.*
Siberia. 30+. Hemicr. or rhizome geophyte.

Carex squarrosa L., Sp. Pl.: 973 (1753). *Vignea squarrosa* (L.) Rchb. in J.C.Mössler, Handb. Gewächsk. ed. 2, 3: 1606 (1830).
SE. Canada to NC. & E. U.S.A. 72 ONT 74 ILL IOW KAN MIN MSO NEB OKL 75 CNT INI MAS MIC NWJ NWY OHI PEN WVA 78 ALA ARK DEL GEO KTY LOU MRY MSI NCA SCA TEN VRG WDC. Hemicr. or rhizome geophyte.
Carex typhinoides Schwein., Ann. Lyceum Nat. Hist. New York 1: 66 (1824). *Carex squarrosa* var. *typhinoides* (Schwein.) Dewey, Amer. J. Sci. Arts 11: 316 (1826).
Carex squarrosa f. *robusta* Peck, Rep. (Annual) Regents Univ. State New York State Mus. 48: 198 (1896).

Carex standleyana Steyerm., Ceiba 3: 23 (1952).
Mexico to C. America. 79 MXG MXS MXT 80 COS GUA HON PAN. Hemicr. or rhizome geophyte.

Carex stenandra Kük. in H.G.A.Engler (ed.), Pflanzenr., IV, 20(38): 272 (1909).
Madagascar. 29 MDG. Hemicr. or rhizome geophyte.

Carex stenantha Franch. & Sav., Enum. Pl. Jap. 2: 573 (1878).
Russian Far East, Japan. 31 KAM KHA KUR MAG SAK 38 JAP. Hemicr. or rhizome geophyte.

var. ***stenantha***
Japan (Honshu). 38 JAP. Hemicr. or rhizome geophyte.
Carex schmidtii Boeck., Flora 65: 63 (1882), nom. illeg.

var. ***taisetsuensis*** Akiyama, J. Fac. Sci. Hokkaido Univ., Ser. 5, Bot. 1: 60 (1931).
Russian Far East, N. Japan. 31 KAM KHA KUR MAG SAK 38 JAP. Hemicr. or rhizome geophyte.
Carex ktausipalii Meinsh., Trudy Imp. S.-Peterburgsk. Bot. Sada 18: 359 (1900).
Carex gorodkovii V.I.Krecz. in V.L.Komarov (ed.), Fl. URSS 3: 606 (1935).

Carex stenocarpa Turcz. ex V.I.Krecz. in V.L.Komarov (ed.), Fl. URSS 3: 607 (1935). *Carex tristis* subsp. *stenocarpa* (Turcz. ex V.I.Krecz.) T.V.Egorova, Sedges Russia & Adj. States: 147 (1999).
Siberia to W. Himalaya. 30 TVA 32 KAZ KGZ TZK UZB 34 AFG 36 CHX 37 MON 40 NEP PAK WHM. Hemicr. or rhizome geophyte.
Carex lineariterminalis T.Koyama, Acta Phytotax. Geobot. 17: 99 (1958).

Carex* ×*stenolepis Less., Reise Norwegen: 301 (1831). *C. saxatilis* × *C. vesicaria.*

Greenland, N. & NE. Europe. 10 FIN NOR SWE 14 RUN 70 GNL 72 LAB NBR NFL ONT QUE 75 MAI. Hemicr. or rhizome geophyte.

Carex ×*grahamii* Boott, Proc. Linn. Soc. London 1: 180 (1844). *Carex saxatilis* var. *grahamii* (Boott) Hook. & Arn., Brit. Fl., ed. 8: 510 (1860). *Carex saxatilis* subsp. *grahamii* (Boott) Nyman, Consp. Fl. Eur.: 768 (1882). *Carex vesicaria* var. *grahamii* (Boott) Kük. in H.G.A.Engler (ed.), Pflanzenr., IV, 20(38): 727 (1909).

Carex stenophylla Wahlenb., Kongl. Vetensk. Acad. Nya Handl. 24: 142 (1803). *Vignea stenophylla* (Wahlenb.) Rchb. in J.C.Mössler, Handb. Gewächsk. ed. 2, 3: 1607 (1830). *Carex divisa* var. *stenophylla* (Wahlenb.) Fiori, Nuov. Fl. Italia 1: 184 (1923).

Temp. Eurasia, W. & C. Canada, NW. & NC. U.S.A. 11 AUT CZE HUN 12 SPA? 13 ITA ROM YUG 14 KRY RUC RUS UKR 30 ALT WSB 32 KAZ KGZ TKM TZK UZB 33 NCS TCS 34 AFG IRN TUR 35 YEM 36 CHI CHN CHT CHX 37 MON 40 PAK WHM. Hemicr. or rhizome geophyte.

subsp. ***stenophylla***

Europe to C. Asia, W. & C. Canada, NW. & NC. U.S.A. 11 AUT CZE HUN 12 SPA? 13 ITA ROM YUG 14 KRY RUC RUS UKR 30 ALT WSB 32 KAZ 35 YEM 36 CHX. Hemicr. or rhizome geophyte.

Carex juncifolia Host, Syn. Pl.: 504 (1797), nom. illeg.

Carex glomerata Host, Icon. Descr. Gram. Austriac. 1: t. 44 (1801), nom. illeg.

Carex hostii Schkuhr, Beschr. Riedgräs. 2: 7 (1806).

Carex deinbolliana J.Gay, Ann. Sci. Nat., Bot., II, 11: 183 (1839). *Carex incurva* subsp. *deinbolliana* (J.Gay) Nyman, Consp. Fl. Eur.: 782 (1882). *Vignea deinbolliana* (J.Gay) Soják, Cas. Nár. Mus., Odd. Prír. 148: 195 (1979 publ. 1980).

Carex arctica Deinboll ex Fr., Novit. Fl. Suec. Mant. 3: 135 (1842), nom. illeg. *Carex incurva* var. *arctica* (Deinboll ex Fr.) Hartm., Handb. Skand. Fl., ed. 4: 273 (1843).

Vignea stenophylla f. *elata* Schur, Enum. Pl. Transsilv.: 699 (1866). *Carex stenophylla* f. *elata* (Schur) Kük. in H.G.A.Engler (ed.), Pflanzenr., IV, 20(38): 121 (1909).

Vignea stenophylla var. *exigua* Schur, Enum. Pl. Transsilv.: 698 (1866). *Carex stenophylla* var. *exigua* (Schur) Kük. in H.G.A.Engler (ed.), Pflanzenr., IV, 20(38): 121 (1909).

Carex stenophylla var. *curaica* Litv., Allg. Bot. Z. Syst. 4: 125 (1898).

Carex curaica var. *rigida* Meinsh., Trudy Imp. S.-Peterburgsk. Bot. Sada 18: 317 (1901).

Carex stenophylla var. *humilis* Meinsh., Trudy Imp. S.-Peterburgsk. Bot. Sada 18: 316 (1901).

Carex stenophylla f. *elatior* Kük., Finska Vetensk. Soc. Forh. 45(8): 2 (1903).

Carex uralensis C.B.Clarke, Kew Bull., Addit. Ser. 8: 76 (1908).

subsp. ***stenophylloides*** (V.I.Krecz.) T.V.Egorova in A.L.Takhtajan & A.A.Fedorov, Fl. Erevana: 339 (1972).

Turkey to Mongolia and W. Himalaya. 32 KAZ KGZ TKM TZK UZB 33 NCS TCS 34 AFG IRN TUR 36 CHI CHN CHT CHX 37 MON 40 PAK WHM. Hemicr. or rhizome geophyte.

Carex longipedicellata Boeck., Beitr. Cyper. 1: 41 (1888). *Carex stenophylla* var. *longipedicellata* (Boeck.) Kük. in H.G.A.Engler (ed.), Pflanzenr., IV, 20(38): 121 (1909).

**Carex stenophylloides* V.I.Krecz. in B.A. Fedtschenko & al., Fl. Turkm. 1: 230 (1932). *Carex duriuscula* subsp. *stenophylloides* (V.I.Krecz.) S.Yun Liang & Y.C.Tang, Acta Phytotax. Sin. 28: 155 (1990).

Carex duriusculiformis V.I.Krecz. in V.L.Komarov (ed.), Fl. URSS 3: 142 (1935).

Carex stenoptila F.J.Herm., Leafl. W. Bot. 4: 194 (1945).

Montana to Colorado. 73 COL MNT WYO 76 UTA. Hemicr. or rhizome geophyte.

Carex stenostachys Franch. & Sav., Enum. Pl. Jap. 2: 569 (1878).

Japan (Honshu). 38 JAP. Hemicr. or rhizome geophyte.

var. ***cuneata*** (Ohwi) Ohwi & T.Koyama in ?

Japan (Honshu). 38 JAP. Hemicr. or rhizome geophyte.

**Carex cuneata* Ohwi, Mem. Coll. Sci. Kyoto Imp. Univ., Ser. B, Biol. 6: 256 (1931).

var. ***ikegamii*** T.Koyama in ?

Japan (WC. Honshu). 38 JAP. Hemicr. or rhizome geophyte.

var. ***stenostachys***

Japan (Honshu). 38 JAP. Hemicr. or rhizome geophyte.

Carex sterilis Willd., Sp. Pl. 4: 208 (1805). *Carex stellulata* var. *sterilis* (Willd.) Carey in A.Gray, Manual: 544 (1848). *Carex muricata* var. *sterilis* (Willd.) Gleason, Phytologia 4: 22 (1952).

Canada to NC. & E. U.S.A. 70 NUN 71 MAN SAS 72 NFL ONT QUE 74 ILL IOW MIN MSO WIS 75 CNT INI MAI MAS MIC NWJ NWY OHI PEN WVA 78 TEN VRG. Hemicr. or rhizome geophyte.

Carex sterilis var. *angustata* L.H.Bailey, Bull. Torrey Bot. Club 20: 425 (1893).

Carex sterilis var. *aequidistans* Peck, Rep. (Annual) Regents Univ. State New York New York State Mus. 47: 137 (1895).

Carex elachycarpa Fernald, Proc. Amer. Acad. Arts 37: 492 (1902). *Kobresia elachycarpa* (Fernald) Fernald, Rhodora 5: 251 (1903).

Carex sterilis f. *flexibilis* Farw., Rep. (Annual) Michigan Acad. Sci. 21: 361 (1920).

Carex sterilis var. *rousseauiana* Raymond, Naturaliste Canad. 77: 62 (1950).

Carex steudneri Boeck., Linnaea 40: 364 (1876).

Ethiopia to Tanzania. 24 ETH SUD 25 KEN TAN. Hemicr. or rhizome geophyte.

Carex wahlenbergiana var. *schimperi* Boott, Ill. Carex 2: 101 (1860).

Carex stevenii (T.Holm) Kalela, Ann. Bot. Soc. Zool.-Bot. Fenn. "Vanamo" 19(3): 11 (1944).

WC. U.S.A. 73 COL IDA MNT WYO 76 UTA 77 NWM. Hemicr. or rhizome geophyte.

**Carex alpina* var. *stevenii* T.Holm, Amer. J. Sci., IV, 16: 21 (1903). *Carex vahlii* var. *stevenii* (T.Holm) Fernald, Rhodora 35: 398 (1933). *Carex angarae* var. *stevenii* (T.Holm) A.E.Porsild, Rhodora 41: 204 (1939). *Carex media* var. *stevenii* (T.Holm) Fernald, Rhodora 44: 303 (1942). *Carex norvegica* subsp.

stevenii (T.Holm) D.F.Murray, Brittonia 21: 74 (1969). *Carex norvegica* var. *stevenii* (T.Holm) Dorn, Vasc. Pl. Wyoming: 296 (1988).

Carex steyermarkii Standl., Publ. Field Mus. Nat. Hist., Bot. Ser. 23: 196 (1947).
Mexico (Chiapas) to Guatemala. 79 MXT 80 GUA. Hemicr. or rhizome geophyte.

Carex stipata Muhl. ex Willd., Sp. Pl. 4: 233 (1805). *Vignea stipata* (Muhl. ex Willd.) Rchb. in J.C. Mössler, Handb. Gewächsk. ed. 2, 3: 1620 (1830). *Loncoperis stipata* (Muhl. ex Willd.) Raf., Good Book: 27 (1840).
Russian Far East to C. China, N. & C. Japan, Canada, U.S.A. 31+ 36 CHC CHM 38 JAP 70 ASK 71 ABT BRC MAN SAS 72 LAB NBR NFL NSC ONT PEI QUE 73 COL IDA MNT ORE WAS WYO 74 ILL IOW KAN MIN MSO NDA NEB OKL SDA WIS 75 CNT INI MAI MAS MIC NWH NWJ NWY OHI PEN RHO VER WVA 76 ARI CAL UTA 77 NWM TEX 78 ALA DEL FLA GEO KTY MRY NCA SCA TEN VRG WDC. Hemicr. or rhizome geophyte.

var. ***maxima*** Chapm. ex Boott, Ill. Carex 3: 121 (1862).
E. U.S.A. 75 OHI WVA 78 ALA FLA GEO MRY NCA SCA VRG. Hemicr. or rhizome geophyte.
Carex stipata var. *uberior* C.Mohr, Contr. U. S. Natl. Herb. 6: 419 (1901). *Carex uberior* (C.Mohr) Mack. in N.L.Britton & al. (eds.), N. Amer. Fl. 18: 79 (1931).

var. ***stipata***
Russian Far East to C. China, N. & C. Japan, Canada, U.S.A. 31+ 36 CHC CHM 38 JAP 70 ASK 71 ABT BRC MAN SAS 72 LAB NBR NFL NSC ONT PEI QUE 73 COL IDA MNT ORE WAS WYO 74 ILL IOW KAN MIN MSO NDA NEB OKL SDA WIS 75 CNT INI MAI MAS MIC NWH NWJ NWY OHI PEN RHO VER WVA 76 ARI CAL UTA 77 NWM TEX 78 DEL KTY MRY NCA SCA TEN VRG WDC. Hemicr. or rhizome geophyte.
Carex setacea Frank ex Kunth, Enum. Pl. 2: 382 (1837), nom. illeg.
Carex sparganioides Leconte ex Kunth, Enum. Pl. 2: 382 (1837), nom. illeg.
Carex stipitata Boeck., Linnaea 39: 85 (1875).
Carex stipata var. *crassicurta* Peck, Rep. (Annual) Regents Univ. State New York New York State Mus. 47: 128 (1895).
Carex stipata var. *subsecuta* Peck, Rep. (Annual) Regents Univ. State New York New York State Mus. 47: 128 (1895).

Carex stiphrogyne Gilli, Ann. Naturhist. Mus. Wien, B 84: 8 (1980 publ. 1983).
Papua New Guinea. 43 NWG. Hemicr. or rhizome geophyte.

Carex stipitiutriculata P.C.Li, Acta Bot. Yunnan. 12: 137 (1990).
China (Yunnan). 36 CHC. Hemicr. or rhizome geophyte.

Carex stokesii F.Br., Bernice P. Bishop Mus. Bull. 84: 116 (1931).
Tubuai Is. (Rapa-Iti). 61 TUB. Hemicr. or rhizome geophyte.

Carex stracheyi Boott ex C.B.Clarke in J.D.Hooker, Fl. Brit. India 6: 727 (1894).
Himalaya. 40 NEP WHM. Hemicr. or rhizome geophyte.

Carex stramentitia Boott ex Boeck., Linnaea 40: 351 (1876).
Indian Subcontinent, SC. China, Indo-China, W. Jawa. 36 CHC CHS 40 ASS EHM IND NEP 41 LAO MYA THA VIE 42 JAW. Hemicr. or rhizome geophyte.
Carex condensata var. *flava* Nees in R.Wight, Contr. Bot. India: 123 (1834).
Carex filicina Boeck. ex C.B.Clarke in J.D.Hooker, Fl. Brit. India 6: 717 (1894), pro syn.

Carex straminea Willd. ex Schkuhr, Beschr. Riedgräs. 1: 49 (1801). *Vignea straminea* (Willd. ex Schkuhr) Raf., Good Book: 27 (1840). *Carex straminea* var. *typica* Macoun, Cat. Canad. Pl. 2: 131 (1888), nom. inval.
NC. & E. U.S.A. 74 ILL MSO WIS 75 CNT INI MAS MIC NWJ NWY OHI PEN RHO WVA 78 DEL KTY MRY NCA VRG WDC. Hemicr. or rhizome geophyte.
Carex straminea var. *minor* Dewey, Amer. J. Sci. Arts 11: 318 (1826).
Carex mirabilis Dewey, Amer. J. Sci. Arts 30: 63 (1836), nom. illeg.
Carex straminea var. *schkuhrii* Gay, Ann. Sci. Nat., Bot., II, 10: 363 (1838).
Carex festucacea var. *tenera* Carey in A.Gray, Manual: 545 (1848). *Carex straminea* var. *tenera* (Carey) Boott, Ill. Carex 3: 120 (1862).
Carex cristata var. *mirabilis* Boott in S.Watson & al., Bot. California 2: 237 (1880), nom. illeg.
Carex scoparia var. *fulva* Boott in S.Watson & al., Bot. California 2: 237 (1880).
Carex straminea var. *brevior* L.H.Bailey, Mem. Torrey Bot. Club 1: 22 (1889), nom. illeg.
Carex straminea f. *multa* L.H.Bailey, Bot. Gaz. 21: 8 (1896).
Carex straminea f. *perlonga* Fernald, Proc. Amer. Acad. Arts 37: 473 (1902).
Carex straminea f. *tincta* Fernald, Proc. Amer. Acad. Arts 37: 473 (1902).
Carex tenera var. *richii* Fernald, Proc. Amer. Acad. Arts 37: 475 (1902). *Carex hormathodes* var. *richii* (Fernald) Fernald, Rhodora 8: 166 (1906).
Carex straminea var. *prorepens* Kük. in H.G.A.Engler (ed.), Pflanzenr., IV, 20(38): 208 (1909).
Carex straminea f. *richii* (Fernald) Kük. in H.G.A.Engler (ed.), Pflanzenr., IV, 20(38): 208 (1909).
Carex richii (Fernald) Mack., Bull. Torrey Bot. Club 49: 362 (1922 publ. 1923).

Carex straminiformis L.H.Bailey, Mem. Torrey Bot. Club 1: 24 (1889).
W. U.S.A. 73 IDA MNT ORE WAS 76 CAL NEV UTA. Hemicr. or rhizome geophyte.
Carex straminea var. *congesta* Olney, Proc. Amer. Acad. Arts 7: 393 (1868).

Carex streptorrhampha Nelmes, Bull. Misc. Inform. Kew 1939: 253 (1939).
Tibet to Myanmar. 36 CHT 41 MYA. Hemicr. or rhizome geophyte.

Carex striata Michx., Fl. Bor.-Amer. 2: 174 (1803).
E. U.S.A. 75 MAS NWH NWJ NWY 78 ALA DEL FLA GEO MRY MSI NCA SCA VRG. Hemicr. or rhizome geophyte.

Carex bullata Elliott, Sketch Bot. S. Carolina 2: 556 (1824), nom. illeg.
Carex polymorpha Torr., Ann. Lyceum Nat. Hist. New York 3: 413 (1836), nom. illeg.
Carex walteriana L.H.Bailey, Bull. Torrey Bot. Club 20: 420 (1893).
Carex striata var. *brevis* L.H.Bailey, Mem. Torrey Bot. Club 1: 34 (1899).

Carex striatula Michx., Fl. Bor.-Amer. 2: 173 (1803). *Carex laxiflora* var. *striatula* (Michx.) Chapm., Fl. South. U.S.: 540 (1860). *Carex laxiflora* f. *striatula* (Michx.) Kük. in H.G.A.Engler (ed.), Pflanzenr., IV, 20(38): 526 (1909).
E. U.S.A. to Texas. 75 NWJ NWY PEN WVA 77 TEX 78 ALA ARK DEL FLA GEO KTY LOU MRY MSI NCA SCA TEN VRG WDC. Hemicr. or rhizome geophyte.
Carex laxiflora var. *angustifolia* Dewey in ?
Carex ignota Dewey, Amer. J. Sci. Arts, II, 8: 348 (1849).
Carex laxiflora var. *michauxii* L.H.Bailey, Mem. Torrey Bot. Club 1: 32 (1889).

Carex stricta Lam., Encycl. 3: 387 (1792). *Vignea stricta* (Lam.) Rchb. in J.C.Mössler, Handb. Gewächsk. ed. 2, 3: 1624 (1830). *Diemisa stricta* (Lam.) Raf., Good Book: 27 (1840). *Vignantha stricta* (Lam.) Schur, Enum. Pl. Transsilv.: 707 (1866).
E. Canada to C. & E. U.S.A. 72 NBR NSC ONT PEI QUE 74 ILL IOW KAN MIN MSO NDA NEB SDA WIS 75 CNT INI MAI MAS MIC NWH NWJ NWY OHI RHO VER WVA 77 TEX 78 DEL KTY MRY MSI NCA TEN VRG WDC. Hemicr. or rhizome geophyte.
Carex verrucosa Schwein. ex Torr., Ann. Lyceum Nat. Hist. New York 3: 400 (1836), nom. illeg.
Carex sempervirens Elliott ex Kunth, Enum. Pl. 2: 460 (1837), nom. illeg.
Carex commutata J.Gay, Ann. Sci. Nat., Bot., II, 11: 198 (1839).
Carex littoralis Krock. ex Steud., Nomencl. Bot., ed. 2, 1: 292 (1840).
Carex strictior Dewey ex Alph.Wood, Class-book Bot.: 755 (1845). *Carex stricta* var. *strictior* (Dewey ex Alph.Wood) Carey in A.Gray, Manual: 548 (1848).
Carex kelvingtoniana Steud., Syn. Pl. Glumac. 2: 215 (1855).
Carex watsoniana Steud., Syn. Pl. Glumac. 2: 215 (1855).
Carex tristicha Boeck., Flora 41: 651 (1858).
Carex prolixa Schur, Enum. Pl. Transsilv.: 707 (1866), nom. illeg.
Carex xerocarpa S.H.Wright, Amer. J. Sci. Arts, II, 42: 334 (1866). *Carex stricta* var. *xerocarpa* (S.H. Wright) Britton, Bull. Torrey Bot. Club 22: 222 (1895). *Carex stricta* f. *xerocarpa* (S.H.Wright) Kük. in H.G.A.Engler (ed.), Pflanzenr., IV, 20(38): 330 (1909).
Carex stricta var. *curtissima* Peck, Rep. (Annual) Regents Univ. State New York State Mus. 48: 151 (1895).
Carex stricta f. *curtissima* (Peck) Kük. in H.G.A. Engler (ed.), Pflanzenr., IV, 20(38): 330 (1909).
Carex stricta f. *brevior* House, Bull. New York State Mus. Nat. Hist. 243–244: 61 (1923).
Carex stricta f. *pedicellaris* House, Bull. New York State Mus. Nat. Hist. 243–244: 61 (1923).

Carex* ×*stricticulmis E.Holmb., Bot. Not. 1929: 22 (1929). *C. dioica* × *C. glareosa*.
N. Europe. 10 NOR. Hemicr. or rhizome geophyte.

Carex strigosa Huds., Fl. Angl.: 411 (1762). *Trasus strigosus* (Huds.) Gray, Nat. Arr. Brit. Pl. 2: 65 (1821).
Europe to N. Iran. 10 DEN GRB IRE 11 AUT BGM CZE GER HUN NET POL SWI 12 FRA SPA 13 BUL ITA ROM YUG 14 UKR 33 TCS 34 IRN. Hemicr. or rhizome geophyte.
Carex leptostachys Ehrh. ex L.f., Suppl. Pl.: 414 (1782). *Loxotrema leptostachys* (Ehrh. ex L.f.) Raf., Good Book: 25 (1840).
Carex godefrinii Willemet, Phytogr. Encycl. 3: 1114 (1805).
Carex leptostachya Boiss., Fl. Orient. 5: 418 (1882), nom. illeg.

Carex* ×*stygia Fr., Novit. Fl. Suec. Mant. 3: 141 (1845). *C. magellanica* subsp. *irrigua* × *C. rariflora*. *Carex rariflora* subsp. *stygia* (Fr.) Andersson, Cyper. Scand.: 35 (1849). *Carex limosa* var. *stygia* (Fr.) L.H.Bailey, Proc. Amer. Acad. Arts 22: 95 (1886). *Carex rariflora* var. *stygia* (Fr.) Kük. in H.G.A.Engler (ed.), Pflanzenr., IV, 20(38): 504 (1909).
Subarctic. 10 NOR 14+ 70 ASK YUK 71 BRC 73 WAS. Hemicr. or rhizome geophyte.

Carex styloflexa Buckley, Amer. J. Sci. Arts 45: 174 (1843). *Carex laxiflora* var. *styloflexa* (Buckley) Boott, Ill. Carex 1: 37 (1858).
E. U.S.A. to Texas. 75 CNT NWJ NWY OHI PEN WVA 77 TEX 78 ALA DEL FLA GEO KTY LOU MRY MSI NCA SCA TEN VRG WDC. Hemicr. or rhizome geophyte.
Carex styloflexa var. *remotiflora* Wiegand, Rhodora 24: 200 (1922).

Carex stylosa C.A.Mey., Mém. Sav. Étr. Acad. St. Pétersbourg 1: 222 (1831).
Subarctic Northern Hemisphere. 10 NOR 30+ 31 KAM MAG 70 ALU ASK GNL YUK 71 BRC 72 LAB NFL QUE 73 WAS. Hemicr. or rhizome geophyte.
Carex nigritella Drejer in G.C.Oeder & al. (eds.), Fl. Dan.: t. 2369 (1843). *Carex stylosa* var. *nigritella* (Drejer) Fernald, Rhodora 44: 302 (1942).
Carex rigida Tuck., Enum. Meth. Caric.: 11 (1843), nom. illeg.
Carex beringiana Cham. ex Steud., Syn. Pl. Glumac. 2: 229 (1855).
Carex spreta L.H.Bailey, Mem. Torrey Bot. Club 1: 6 (1889), nom. illeg.

Carex subandrogyna G.A.Wheeler & Guagl., Hickenia 3: 163 (2003).
NW. Argentina. 85 AGW. Hemicr.

Carex subantarctica Speg., Anales Mus. Nac. Buenos Aires 7: 180 (1907).
WC. & S. Argentina. 85 AGS AGW. Hemicr. or rhizome geophyte.
Carex chubutensis C.B.Clarke, Kew Bull., Addit. Ser. 8: 68 (1908).
Carex pampae Kalela, Ann. Acad. Sci. Fenn., Ser. A 54(5): 67 (1940), nom. illeg.

Carex subbracteata Mack., Bull. Torrey Bot. Club 43: 612 (1916 publ. 1917).
W. & WC. California. 76 CAL. Hemicr. or rhizome geophyte.

Carex subcapitata X.F.Jin, C.Z.Zheng & B.Y.Ding, Acta Phytotax. Sin. 41: 566 (2003).
China (Zhejiang). 36 CHS. Hemicr.

Carex subcernua Ohwi, Mem. Coll. Sci. Kyoto Imp. Univ., Ser. B, Biol. 3: 278 (1930).
China (Zhejiang), Japan (Kyushu). 36 CHS 38 JAP. Hemicr. or rhizome geophyte.

Carex* ×*subcostata E.Holmb., Bot. Not. 1929: 26 (1929). *C. dioica* × *C. loliacea.*
N. Europe. 10 FIN. Hemicr. or rhizome geophyte.

Carex subdivulsa (Kük.) G.A.Wheeler, Darwiniana 40: 203 (2002).
Argentina (La Rioja). 85 AGW. Hemicr. or rhizome geophyte.
**Carex involucrata* var. *subdivulsa* Kük., Bot. Jahrb. Syst. 27: 511 (1899). *Carex sororia* f. *subdivulsa* (Kük.) Kük. in H.G.A.Engler (ed.), Pflanzenr., IV, 20(38): 150 (1909).

Carex subdola Boott, Proc. Linn. Soc. London 1: 225 (1845).
New Zealand. 51 NZN NZS. Hemicr. or rhizome geophyte.

Carex subebracteata (Kük.) Ohwi, Mem. Coll. Sci. Kyoto Imp. Univ., Ser. B, Biol. 6: 252 (1931).
Siberia to NE. China. 30+ 31+ 36 CHI CHM. Hemicr. or rhizome geophyte.
**Carex pisiformis* var. *subebracteata* Kük. in H.G.A. Engler (ed.), Pflanzenr., IV, 20(38): 477 (1909).

Carex suberecta (Olney) Britton, Man. Fl. N. States, ed. 2: 1057 (1905).
SE. Canada, EC. U.S.A. to Virginia. 72 ONT 74 ILL IOW MIN MSO WIS 75 INI MIC OHI WVA 77 TEX 78 ARK VRG. Hemicr. or rhizome geophyte.
**Carex tenera* var. *suberecta* Olney, Car. Bor. Am.: 3 (1871).

Carex subfilicinoides Kük., Acta Horti Gothob. 5: 42 (1930).
China (Sichuan), Taiwan. 36 CHC 38 TAI. Hemicr. or rhizome geophyte.

Carex* ×*subimpressa Clokey, Rhodora 21: 84 (1919). *C. hyalinolepis* × *C. lanuginosa.*
NC. & NE. U.S.A. 74 ILL KAN MSO 75 INI MIC OHI. Hemicr. or rhizome geophyte.

Carex subinclinata T.Koyama, Bot. Mag. (Tokyo) 92: 217 (1979).
NE. Thailand. 41 THA. Hemicr. or rhizome geophyte.

Carex subinflata Nelmes, Bull. Misc. Inform. Kew 1940: 270 (1940).
KwaZulu-Natal, Lesotho. 27 LES NAT. Hemicr. or rhizome geophyte.

Carex sublateralis T.Koyama, Jap. J. Bot. 15: 180 (1956).
China (Jiangsu). 36 CHS. Hemicr. or rhizome geophyte.

Carex submollicula Tang & F.T.Wang ex L.K.Dai, Acta Phytotax. Sin. 32: 186 (1994).
SE. China. 36 CHS. Hemicr. or rhizome geophyte.

Carex subnigricans Stacey, Leafl. W. Bot. 2: 167 (1939).
W. U.S.A. 73 IDA ORE WYO 76 NEV UTA. Hemicr. or rhizome geophyte.
Carex rachillis Maguire, Brittonia 5: 199 (1944).

Carex* ×*subpaleacea J.Cay., Naturaliste Canad. 117: 66 (1990 publ. 1991). *C. paleacea* × *C. salina.*
E. Canada. 72 LAB ONT QUE. Hemicr. or rhizome geophyte.

Carex* ×*subpatula E.Holmb., Bot. Not. 1929: 14 (1929). *C. brunnescens* × *C. lapponica.*
N. & NE. Europe. 10 FIN SWE 14 RUN. Hemicr. or rhizome geophyte.

Carex subperakensis L.K.Ling & Y.Z.Huang, Fl. Fujianica 6: 689 (1995).
China (Fujian). 36 CHS. Hemicr. or rhizome geophyte.

Carex subphysodes Popov ex V.I.Krecz., Trudy Sredne-Aziatsk. Gosud. Univ., Ser. 8b, Bot. 20: 23 (1935). *Carex physodes* subsp. *subphysodes* (Popov ex V.I. Krecz.) Kukkonen, Ann. Bot. Fenn. 21: 383 (1984).
C. Asia to Afghanistan. 32 KAZ TKM TZK UZB 34 AFG IRN. Hemicr. or rhizome geophyte.

Carex subpumila Tang & F.T.Wang ex L.K.Dai, Acta Phytotax. Sin. 37: 182 (1999).
China (Hebei). 36 CHN. Hemicr. or rhizome geophyte.

Carex* ×*subrecta J.Cay., Naturaliste Canad. 117: 69 (1990 publ. 1991). *C. recta* × *C. subspathacea.*
E. Canada. 72 QUE. Hemicr. or rhizome geophyte.

Carex* ×*subreducta Lepage, Naturaliste Canad. 83: 135 (1956). *C. bigelowii* × *S. subspathacea.*
E. Canada, Greenland. 70 GNL 72 ONT QUE. Hemicr. or rhizome geophyte.

Carex* ×*subsalina Lepage, Naturaliste Canad. 83: 127 (1956). *C. aquatilis* × *C. salina.*
Subarctic America, Canada. 70 NWT 71 MAN 72 NFL ONT QUE. Hemicr. or rhizome geophyte.

nothovar. ***stansalina*** (Lepage) Govaerts in R.H.A.Govaerts & D.A.Simpson, World Checklist of Cyperaceae: 186 (2007). *C. aquatilis* var. *stans* × *C. salina.*
Subarctic America, Canada. 70 NWT 71 MAN 72 ONT QUE. Hemicr. or rhizome geophyte.
**Carex* ×*stansalina* Lepage, Naturaliste Canad. 83: 150 (1956).

nothovar. ***subsalina***. *C. aquatilis* × *C. salina.*
E. Canada. 72 NFL ONT QUE. Hemicr. or rhizome geophyte.

Carex subscabrella Kük., Ark. Bot. 22A(17): 8 (1929).
Dominican Rep. 81 DOM. Hemicr. or rhizome geophyte.

Carex subspathacea Wormsk. ex Hornem. in G.C.Oeder & al. (eds.), Fl. Dan. 9: 4 (1813). *Carex salina* var. *subspathacea* (Wormsk. ex Hornem.) Tuck., Enum. Meth. Caric.: 12 (1843). *Carex salina* subsp. *subspathacea* (Wormsk. ex Hornem.) Nyman, Consp. Fl. Eur.: 776 (1882).
Subarctic Northern Hemisphere to Japan. 10 ICE NOR SVA 14 RUN 30 KRA WSB YAK 31 KAM KUR MAG SAK 38 JAP KOR 70 ASK GNL NUN NWT YUK 71 MAN 72 LAB NFL ONT QUE. Hemicr. or rhizome geophyte.
Carex aquatilis var. *nardifolia* Wahlenb., Kongl. Vetensk. Acad. Nya Handl. 24: 165 (1803). *Carex subspathacea* f. *nardifolia* (Wahlenb.) Fr., Novit. Fl. Suec. Mant. 3: 149 (1845).
Carex hoppneri Boott in W.J.Hooker, Fl. Bor.-Amer. 2: 219 (1839).

Carex subspathacea f. *curvata* Drejer, Rev. Car. Bor.: 34 (1841).
Carex subspathacea f. *stricta* Drejer, Rev. Car. Bor.: 34 (1841). *Carex salina* f. *stricta* (Drejer) Almq., Bot. Not. 1891: 127 (1891).
Carex subspathacea var. *planifolia* Fr., Novit. Fl. Suec. Mant. 3: 149 (1845).
Carex salina var. *pumila* A.Blytt, Norges Fl. 1: 218 (1861).
Carex salina var. *minor* Boott, Ill. Carex 4: 160 (1867).
Carex salina f. *nana* Trautv., Consp. Fl. Now.-Semlja: 82 (1871).

Carex subumbellata Meinsh., Trudy Imp. S.-Peterburgsk. Bot. Sada 18: 395 (1900).
Sakhalin, Japan. 31 SAK 38 JAP. Hemicr. or rhizome geophyte.

var. ***subumbellata***
Sakhalin, N. Japan (Mt. Yubaridake). 31 SAK 38 JAP. Hemicr. or rhizome geophyte.
Carex multifiliculmis H.Lév. & Vaniot, Bull. Acad. Int. Géogr. Bot. 19: 34 (1909).

var. ***verecunda*** Ohwi, Mem. Coll. Sci. Kyoto Imp. Univ., Ser. B, Biol. 6: 260 (1931).
C. Japan. 38 JAP. Hemicr. or rhizome geophyte.

Carex suifunensis Kom., Trudy Imp. S.-Peterburgsk. Bot. Sada 18: 445 (1900). *Carex pruinosa* var. *suifunensis* (Kom.) Kük. in H.G.A.Engler (ed.), Pflanzenr., IV, 20(38): 353 (1909). *Carex maximowiczii* subsp. *suifunensis* (Kom.) Vorosch. in A.K.Skvortsov (ed.), Florist. Issl. Razn. Raionakh SSSR: 155 (1985).
Russian Far East. 31+. Hemicr. or rhizome geophyte.

Carex* ×*sullivantii Boott, Amer. J. Sci. Arts 42: 29 (1842). *C. gracillima* × *C. hirtifolia*.
E. U.S.A. 75 MAS NWY OHI 78 DEL VRG. Hemicr. or rhizome geophyte.

Carex* ×*sumikawaensis Fujiw. & Y.Matsuda, J. Phytogeogr. Taxon. 39: 65 (1991). *C. lyngbyi* × *C. middendorffii*.
Japan. 38 JAP. Hemicr. or rhizome geophyte.

Carex superata Naczi, Reznicek & B.A.Ford, Amer. J. Bot. 85: 443 (1998).
SE. U.S.A. 78 ALA FLA GEO KTY MSI SCA TEN VRG. Hemicr. or rhizome geophyte.
Carex willdenowii var. *megarrhyncha* F.J.Herm., Amer. Midl. Naturalist 51: 277 (1954).

Carex supina Willd. ex Wahlenb., Kongl. Vetensk. Acad. Nya Handl. 1808: 158 (1808). *Edritria supina* (Willd. ex Wahlenb.) Raf., Good Book: 26 (1840). *Carex obtusata* var. *supina* (Willd. ex Wahlenb.) Garcke, Fl. N. Mitt.-Deutschland, ed. 16: 469 (1890). *Carex nitida* var. *supina* (Willd. ex Wahlenb.) Fiori, Nuov. Fl. Italia 1: 194 (1923).
Temp. Eurasia, Subarctic America to NC. U.S.A. 11 AUT CZE GER HUN POL 13 ITA ROM YUG 14 KRY RUC RUS 30 ALT KRA WSB 32 KAZ KGZ TKM 33 NCS TCS 34 IRN TUR 36 CHM 38 KOR 40 NEP WHM 41 AMU 70 ASK GNL NUN NWT YUK 71 BRC MAN SAS 72 ONT 74 MIN. Hemicr. or rhizome geophyte.

var. ***spaniocarpa*** (Steud.) Boivin, Naturaliste Canad. 94: 523 (1967).
Siberia to Korea, Subarctic America to NC. U.S.A. 30 KRA 31 AMU 36 CHM 38 KOR 70 ASK GNL NUN NWT YUK 71 BRC MAN SAS 72 ONT 74 MIN. Hemicr. or rhizome geophyte.
**Carex spaniocarpa* Steud., Syn. Pl. Glumac. 2: 225 (1855). *Carex supina* subsp. *spaniocarpa* (Steud.) Hultén, Acta Univ. Lund., n.s., 38(1): 365 (1942).

var. ***supina***
Europe to W. Nepal. 11 AUT CZE GER HUN POL 13 ITA ROM YUG 14 KRY RUC RUS 30 ALT WSB 32 KAZ KGZ TKM 33 NCS TCS 34 IRN TUR 40 NEP WHM. Hemicr. or rhizome geophyte.
Carex wohllebii Hoppe in ?
Carex verna Schkuhr, Beschr. Riedgräs. 1: 89 (1801), nom. illeg.
Carex conglobata Kit. ex Willd., Sp. Pl. 4: 281 (1805), nom. illeg. *Carex nitida* var. *conglobata* Nyman, Consp. Fl. Eur.: 775 (1882). *Carex liparocarpos* f. *conglobata* (Nyman) Soó, Acta Bot. Acad. Sci. Hung. 16: 370 (1970 publ. 1971).
Carex sphaerocarpa Willd., Sp. Pl. 4: 265 (1805), nom. illeg. *Onkerma sphaerocarpa* Raf., Good Book: 27 (1840).
Carex campestris Host, Icon. Descr. Gram. Austriac. 4: t. 38 (1809).
Carex costata J.Presl & C.Presl, Fl. Cech.: 191 (1819).
Carex aprica Turcz. ex Besser, Flora 17(1): 27 (1834).
Carex campestris Boott, Ill. Carex 1: 161 (1858), nom. illeg.
Carex obtusata var. *spicata* Asch., Fl. Brandenburg 1: 779 (1864).
Carex oligocarpa Hornem. ex Boott, Ill. Carex 4: 161 (1867), nom. illeg.
Carex turczaninofii Steud. ex Boott, Ill. Carex 4: 161 (1867).
Carex obesa var. *minor* Boeck., Linnaea 41: 184 (1877).
Carex supina f. *pallida* Bubela in A.Oborn?, Fl. Mähren: 185 (1882).
Carex supina f. *elatior* Behrendsen, Allg. Bot. Z. Syst. 2: 89 (1896).
Carex supina f. *humilior* Behrendsen, Allg. Bot. Z. Syst. 2: 89 (1896).
Carex supina f. *pseudomonostachys* Asch., Verh. Bot. Vereins Prov. Brandenburg 40: 41 (1897).

Carex sutchuensis Franch., Bull. Soc. Philom. Paris, VIII, 7: 45 (1895).
China (Sichuan). 36 CHC. Hemicr. or rhizome geophyte.

Carex swanii (Fernald) Mack., Bull. Torrey Bot. Club 37: 246 (1910).
E. Canada to NC. & E. U.S.A. 72 NBR NSC ONT QUE 74 ILL KAN MSO WIS 75 CNT INI MAI MAS MIC NWH NWJ NWY OHI PEN RHO VER WVA 78 ARK DEL GEO KTY MRY MSI NCA TEN VRG WDC. Hemicr. or rhizome geophyte.
**Carex virescens* var. *swanii* Fernald, Rhodora 8: 183 (1906).

Carex sychnocephala Carey, Amer. J. Sci. Arts, II, 10: 24 (1847).
Subarctic America to N. U.S.A. 70 ASK NWT YUK 71 ABT BRC MAN SAS 72 ONT QUE 73 COL MNT WAS 74 IOW MIN MSO NDA SDA WIS 75 MIC NWY. Hemicr. or rhizome geophyte.

Carex sylvatica Huds., Fl. Angl.: 353 (1762). *Trasus sylvaticus* (Huds.) Gray, Nat. Arr. Brit. Pl. 2: 65 (1821). *Edritria sylvatica* (Huds.) Raf., Good Book: 26 (1840). *Proteocarpus sylvaticus* (Huds.) Fedde & J.Schust., Just's Bot. Jahresber. 41(2): 11 (1913 publ. 1918).
Europe to Iran, NW. Africa. 10 DEN GRB IRE NOR SWE 11 AUT BGM CZE GER HUN NET POL SWI 12 COR FRA SPA 13 ALB BUL GRC ITA ROM SIC TUE YUG 14 BLR BLT KRY RUC RUE RUS RUW UKR 20 ALG MOR TUN 30 ALT WSB 33 NCS TCS 34 CYP IRN LBS TUR (51) nzn (72) ont (75) nwy (78) nca. Hemicr. or rhizome geophyte.

subsp. ***latifrons*** (V.I.Krecz.) Ö.Nilsson in P.H.Davis (ed.), Fl. Turkey 9: 119 (1985).
Turkey to W. Caucasus. 33 NCS TCS 34 TUR. Hemicr. or rhizome geophyte.
Carex latifolia Boiss. & Balansa in P.E.Boissier, Fl. Orient. 5: 421 (1882), nom. illeg.
**Carex latifrons* V.I.Krecz. in V.L.Komarov (ed.), Fl. URSS 3: 425 (1935).

subsp. ***paui*** (Sennen) A.Bolòs & O.Bolòs, Veg. Com. Barcelon.: 246 (1950).
NE. Spain, NW. Africa. 12 SPA 20 MOR. Hemicr. or rhizome geophyte.
Carex paui Sennen, Exsicc. (Pl. Esp.) 1925: 5435 (1925).

subsp. ***sylvatica***
Europe to Iran, NW. Africa. 10 DEN GRB IRE NOR SWE 11 AUT BGM CZE GER HUN NET POL SWI 12 COR FRA SPA 13 ALB BUL GRC ITA ROM SIC TUE YUG 14 BLR BLT KRY RUC RUE RUS RUW UKR 20 ALG MOR TUN 30 ALT WSB 33 NCS TCS 34 CYP IRN LBS TUR (51) nzn (72) ont (75) nwy (78) nca. Hemicr. or rhizome geophyte.
Carex patula Scop., Fl. Carniol., ed. 2, 2: 226 (1772).
Carex drymeia L.f., Suppl. Pl.: 414 (1782).
Carex emarcida Suter, Helvet. Fl. 2: 263 (1802).
Carex psilostachya Kit. ex Willd., Sp. Pl. 4: 289 (1805).
Carex strigosa Willd. ex Kunth, Enum. Pl. 2: 470 (1837), nom. illeg.
Carex sylvatica var. *tommasinii* Rchb., Icon. Fl. Germ. Helv. 8: 19 (1846).
Carex laxula Tineo ex Boott, Ill. Carex 4: 202 (1867). *Carex sylvatica* var. *laxula* (Tineo ex Boott) Nyman, Consp. Fl. Eur.: 769 (1882). *Carex sylvatica* f. *laxula* (Tineo ex Boott) Kük. in H.G.A.Engler (ed.), Pflanzenr., IV, 20(38): 607 (1909).
Carex sylvatica f. *pumila* Fiek ex Uechtr., Jahresber. Schles. Ges. Vaterl. Cult. 57: 332 (1880).
Carex sylvatica f. *gracilis* Èelak., Abh. Böhm. Ges. Wiss. 1887: 131 (1887).
Carex sismanii Velen., Sitzungsber. Königl. Böhm. Ges. Wiss. Prag, Math.-Naturwiss. Cl. 1892: 67 (1893).
Carex loncholepis Less. ex Krecz., Trudy Imp. S.-Peterburgsk. Bot. Sada 18: 338 (1901).
Carex sylvatica f. *latifolia* Kneuck., Allg. Bot. Z. Syst. 7: 31 (1901).
Carex hypaneura V.I.Krecz. in V.L.Komarov (ed.), Fl. URSS 3: 621 (1935).
Carex algeriensis Nelmes, Bull. Misc. Inform. Kew 1939: 99 (1939). *Carex sylvatica* subsp. *algeriensis* (Nelmes) Maire & Weiller, in Fl. Afr. Nord 4: 154 (1957).

Carex* × *sylvenii E.Holmb., Bot. Not. 1929: 26 (1929). *C. dioica* × *C. mackenziei*.
N. Europe. 10 NOR. Hemicr. or rhizome geophyte.

Carex tachirensis Steyerm., Fieldiana, Bot. 28: 68 (1951).
Venezuela. 82 VEN. Hemicr. or rhizome geophyte.

Carex tahitensis F.Br., Occas. Pap. Bernice Pauahi Bishop Mus. 9(4): 8 (1930).
Society Is. (Tahiti). 61 SCI. Hemicr.

Carex tahoensis Smiley, Univ. Calif. Publ. Bot. 9: 119 (1921).
Yukon to W. U.S.A. 70 YUK 71 ABT BRC 73 COL MNT ORE WAS WYO 76 CAL UTA. Hemicr. or rhizome geophyte.
Carex eastwoodiana Stacey, Leafl. W. Bot. 2: 121 (1938). *Carex phaeocephala* f. *eastwoodiana* (Stacey) F.J.Herm., Leafl. W. Bot. 9: 15 (1959).

Carex taihokuensis Hayata, Icon. Pl. Formosan. 10: 70 (1921).
Nansei-shoto to Taiwan, S. China to Indo-China. 36 CHC 38 NNS TAI 41 VIE. Hemicr. or rhizome geophyte.

Carex taihuensis S.W.Su & S.M.Xu, Bull. Bot. Res., Harbin 9(1): 51 (1989).
SE. China. 36 CHS. Hemicr. or rhizome geophyte.

Carex taipaishanica K.T.Fu, in Fl. Tsinlingensis 1(1): 449 (1976).
C. China. 36 CHC CHN. Hemicr. or rhizome geophyte.

Carex* × *takoensis Y.Endo & Yashiro, J. Jap. Bot. 70: 276 (1995). *C. lasiocarpa* var. *occultans* × *C. rugulosa*.
Japan. 38 JAP. Hemicr. or rhizome geophyte.

Carex tamana Steyerm., Fieldiana, Bot. 28: 70 (1951).
S. Venezuela to Ecuador. 82 VEN 83 CLM ECU. Hemicr. or rhizome geophyte.
Carex tenuispica Boeck., Flora 39: 225 (1856), nom. illeg.
Carex tristicha Spruce ex Boott, Ill. Carex 4: 153 (1867), nom. illeg.

Carex tangiana Ohwi, J. Jap. Bot. 12: 656 (1936).
China (Hebei). 36 CHN. Hemicr. or rhizome geophyte.

Carex tangii Kük., Bull. Fan Mem. Inst. Biol. 3: 144 (1932).
China (Hebei). 36 CHN. Hemicr. or rhizome geophyte.

Carex tangulashanensis Y.C.Yang, Acta Phytotax. Sin. 18: 362 (1980).
Qinghai to Tibet. 36 CHQ CHT CHX. Hemicr. or rhizome geophyte.

Carex tapintzensis Franch., Bull. Soc. Philom. Paris, VIII, 7: 44 (1895).
Tibet to SC. China. 36 CHC CHT. Hemicr. or rhizome geophyte.
Carex lamprosandra Franch., Bull. Soc. Philom. Paris, VIII, 7: 45 (1895). *Carex tapintzensis* var. *lamprosandra* (Franch.) Kük. in H.G.A.Engler (ed.), Pflanzenr., IV, 20(38): 490 (1909).

Carex taprobanensis T.Koyama, Bot. Mag. (Tokyo) 93: 342 (1980).
Sri Lanka. 40 SRL. Hemicr. or rhizome geophyte.

Carex tarumensis Franch., Bull. Soc. Philom. Paris, VIII, 7: 86 (1895).
NE. China. 36 CHM. Hemicr. or rhizome geophyte.

Carex tashiroana Ohwi, Mem. Coll. Sci. Kyoto Imp. Univ., Ser. B, Biol. 5(3): 267 (1930).
Japan (Honshu: Aki Prov.). 38 JAP. Hemicr. or rhizome geophyte. Perhaps identical with C. tenuinervis.

Carex tasmanica Kük., Bull. Herb. Boissier, II, 4: 59 (1904). *Echinochlaenia tasmanica* (Kük.) Fedde & J.Schust., Just's Bot. Jahresber. 41(2): 10 (1913 publ. 1918).
SE. Australia. 50 TAS VIC. Hemicr. or rhizome geophyte.

Carex tatjanae Malyschev, Bot. Mater. Gerb. Bot. Inst. Komarova Akad. Nauk S.S.S.R. 21: 461 (1961).
Siberia. 30+. Hemicr. or rhizome geophyte.

Carex tatsiensis (Franch.) Kük. in H.G.A.Engler (ed.), Pflanzenr., IV, 20(38): 563 (1909).
China (Sichuan). 36 CHC. Hemicr. or rhizome geophyte.
**Carex ferruginea* var. *tatsiensis* Franch., Nouv. Arch. Mus. Hist. Nat., III, 10: 55 (1898).

Carex tavoyensis Nelmes, Kew Bull. 3: 67 (1948).
Myanmar. 41 MYA. Hemicr. or rhizome geophyte.

Carex tegulata H.Lév. & Vaniot, Bull. Acad. Int. Géogr. Bot. 11: 306 (1902).
Russian Far East to Korea. 31+ 38 KOR. Hemicr. or rhizome geophyte.

Carex teinogyna Boott, Ill. Carex 1: 60 (1858).
Assam to Sumatera and Japan. 36 CHC CHS 38 JAP KOR 40 ASS 41 MYA THA VIE 42 SUM. Hemicr. or rhizome geophyte.
Carex teinogyna var. *scabriculmis* Kük. in H.G.A. Engler (ed.), Pflanzenr., IV, 20(38): 602 (1909). *Carex scabriculmis* (Kük.) Ohwi, Acta Phytotax. Geobot. 2: 27 (1933).

Carex tenax Chapm. ex Dewey, Amer. J. Sci. Arts, II, 19: 254 (1855). *Carex chapmanii* Sartwell ex L.H.Bailey, Proc. Amer. Acad. Arts 22: 127 (1886), nom. illeg. *Carex dasycarpa* var. *tenax* (Chapm. ex Dewey) Kük. in H.G.A.Engler (ed.), Pflanzenr., IV, 20(38): 437 (1909).
SE. U.S.A. to Texas. 77 TEX 78 ALA FLA GEO LOU MSI NCA SCA. Hemicr. or rhizome geophyte.
Carex chapmanii Sartwell ex Dewey, Amer. J. Sci. Arts 19: 254 (1855), nom. illeg.

Carex ×tenebricans E.Holmb., Bot. Not. 1929: 21 (1929). *C. brunnescens* × *C. dioica*.
N. & NE. Europe. 10 FIN 14 RUN. Hemicr. or rhizome geophyte.

Carex tenebrosa Boott, Ill. Carex 1: 88 (1858). *Carex limprichtii* Kük., Repert. Spec. Nov. Regni Veg. Beih. 12: 312 (1922).
China (Guangdong, Hong Kong). 36 CHS. Hemicr. or rhizome geophyte.

Carex tenejapensis Reznicek & S.González, Contr. Univ. Michigan Herb. 22: 121 (1999).
Mexico (Chiapas). 79 MXT. Hemicr. or rhizome geophyte.

Carex ×tenelliformis E.Holmb., Bot. Not. 1929: 18 (1929). *C. canescens* × *C. disperma*.
N. Europe. 10 FIN. Hemicr. or rhizome geophyte.

Carex tenera Dewey, Amer. J. Sci. Arts 8: 97 (1824). *Diemisa tenera* (Dewey) Raf., Good Book: 27 (1840). *Carex mirabilis* var. *tenera* (Dewey) Prov., Fl. Canad. 2: 648 (1863). *Carex straminea* var. *tenera* (Dewey) L.H.Bailey, Bot. Gaz. 10: 381 (1885).
Canada to N. & E. U.S.A. 71 ABT BRC MAN SAS 72 NBR NSC ONT QUE 73 MNT ORE WYO 74 ILL IOW KAN MIN MSO NDA NEB SDA WIS 75 CNT INI MAI MAS MIC NWH NWJ NWY OHI PEN RHO VER WVA 77 NWM 78 KTY MRY NCA VRG WDC. Hemicr. or rhizome geophyte.

var. ***echinodes*** (Fernald) Wiegand, Rhodora 26: 2 (1924).
C. Canada to NC. & NE. U.S.A. 71 MAN 72 ONT 74 ILL IOW MIN MSO NDA NEB SDA WIS 75 INI MIC OHI. Hemicr. or rhizome geophyte.
**Carex straminea* var. *echinodes* Fernald, Proc. Amer. Acad. Arts 37: 474 (1902). *Carex straminea* f. *echinodes* (Fernald) Kük. in H.G.A. Engler (ed.), Pflanzenr., IV, 20(38): 206 (1909). *Carex festucacea* var. *echinodes* (Fernald) Farw., Pap. Michigan Acad. Sci. 2: 17 (1923).

var. ***tenera***
Canada to N. & E. U.S.A. 71 ABT BRC MAN SAS 72 NBR NSC ONT QUE 73 MNT ORE WYO 74 ILL IOW KAN MIN MSO NDA NEB SDA WIS 75 CNT INI MAI MAS MIC NWH NWJ NWY OHI PEN RHO VER WVA 77 NWM 78 KTY MRY NCA VRG WDC. Hemicr. or rhizome geophyte.
Carex tenera var. *major* Olney, Car. Bor. Am.: 3 (1871).

Carex tenuiculmis (Petrie) Heenan & de Lange, New Zealand J. Bot. 35: 159 (1997).
New Zealand South I. (incl. Stewart I.), Chatham Is. (Pitt I.). 51 CTM NZS. Hemicr. or rhizome geophyte.
**Carex secta* var. *tenuiculmis* Petrie, Trans. & Proc. New Zealand Inst. 52: 19 (1920).

Carex tenuiflora Wahlenb., Kongl. Vetensk. Acad. Nya Handl. 24: 147 (1803). *Vignea tenuiflora* (Wahlenb.) Soják, Cas. Nár. Mus., Odd. Prír. 148: 197 (1979 publ. 1980).
N. Europe to Japan, Subarctic America to N. U.S.A. 10 FIN NOR SWE 14 RUC RUE RUN RUS RUW 30 ALT KRA WSB YAK 31 AMU KAM KHA KUR PRM SAK 33 NCS 36 CHI CHM 37 MON 38 JAP KOR 70 ASK NUN NWT YUK 71 ABT MAN SAS 72 LAB NBR NFL ONT PEI QUE 73 COL MNT 74 MIN WIS 75 MAI MAS MIC NWY OHI VER. Hemicr. or rhizome geophyte.
Carex leucolepis Turcz. ex Steud., Nomencl. Bot., ed. 2, 1: 292 (1840).
Carex arrhyncha Franch., Bull. Soc. Philom. Paris, VIII, 7: 30 (1895). *Carex tenuiflora* var. *arrhyncha* (Franch.) Kük. in H.G.A.Engler (ed.), Pflanzenr., IV, 20(38): 224 (1909).
Carex tenuiflora var. *setacea* Kük. in H.G.A.Engler (ed.), Pflanzenr., IV, 20(38): 224 (1909).

Carex tenuiformis H.Lév. & Vaniot, Bull. Acad. Int. Géogr. Bot. 11: 104 (1902). *Carex ledebouriana* subsp. *tenuiformis* (H.Lév. & Vaniot) T.V.Egorova, Novosti Sist. Vyssh. Rast. 22: 54 (1985).
Russian Far East, Korea, N. & C. Japan. 31 KUR PRM SAK 36 CHI CHM 38 JAP KOR. Hemicr. or rhizome geophyte.
Carex koreana Kom., Trudy Imp. S.-Peterburgsk. Bot. Sada 18: 446 (1900), non *C. coreana*.

Carex tenuiformis f. *paupera* Kük. in H.G.A.Engler (ed.), Pflanzenr., IV, 20(38): 571 (1909).
Carex tasorum Kom., Izv. Imp. Bot. Sada Petra Velikago 16: 153 (1916).
Carex komarovii Koidz., Bot. Mag. (Tokyo) 33: 208 (1919).
Carex fraudulans Printz, Veg. Siber.-Mongol. Front.: 160, 456 (1921).

Carex tenuinervis Ohwi, Mem. Coll. Sci. Kyoto Imp. Univ., Ser. B, Biol. 5(3): 266 (1930).
Japan (Kyushu). 38 JAP. Hemicr. or rhizome geophyte.

Carex tenuior T.Koyama & T.I.Chuang, Quart. J. Taiwan Mus. 13: 227 (1960).
Nansei-shoto. 38 NNS. Hemicr. or rhizome geophyte.

Carex tenuipaniculata P.C.Li, Acta Phytotax. Sin. 37: 157 (1999).
China (Yunnan). 36 CHC. Hemicr. or rhizome geophyte.

Carex tenuiseta Franch., Bull. Soc. Philom. Paris, VIII, 7: 43 (1895).
Russian Far East, Japan. 31+ 38 JAP. Hemicr. or rhizome geophyte.

Carex tenuispicula Tang ex S.Y.Liang, Acta Phytotax. Sin. 34: 92 (1996).
SE. China. 36 CHS. Hemicr. or rhizome geophyte.

Carex teres Boott, Ill. Carex 1: 62 (1858).
Nepal to Jawa. 36 CHC CHT 40 ASS EHM NEP 41 LAO 42 JAW SUM. Hemicr. or rhizome geophyte.
Carex phacota var. *minor* C.B.Clarke, J. Linn. Soc., Bot. 28: 141 (1890).
Carex praelonga C.B.Clarke in J.D.Hooker, Fl. Brit. India 6: 707 (1894).
Carex praelonga var. *angustior* C.B.Clarke in J.D. Hooker, Fl. Brit. India 6: 707 (1894).
Carex kerrii Nelmes, Bull. Misc. Inform. Kew 1939: 304 (1939).
Carex teres var. *spathulata* Kük., Bull. Jard. Bot. Buitenzorg, III, 16: 316 (1940). *Carex spathulata* (Kük.) Nelmes, Kew Bull. 5: 207 (1950), nom. illeg.
Carex kemiriensis Nelmes, Kew Bull. 5: 206 (1950).
Carex petecticalis Nelmes, Kew Bull. 5: 205 (1950).

Carex tereticaulis F.Muell., Fragm. 8: 256 (1874). *Carex tereticaulis* var. *typica* Domin, Biblioth. Bot. 89: 492 (1915), nom. inval. *Vignea tereticaulis* (F.Muell.) Soják, Cas. Nár. Mus., Odd. Prír. 148: 197 (1979 publ. 1980).
S. & SE. Australia. 50 NSW SOA TAS VIC WAU. Hemicr. or rhizome geophyte.
Carex appressa var. *elatior* Nees in J.G.C.Lehmann, Pl. Preiss. 2: 94 (1846).
Carex halmaturina Boeck., Linnaea 39: 100 (1875).
Carex tereticaulis var. *septentrionalis* Domin, Biblioth. Bot. 85: 492 (1915).

Carex ternaria G.Forst., Fl. Ins. Austr.: 92 (1786).
Antipodean Is., Chatham Is. 51 ATP CTM. Hemicr. or rhizome geophyte.
Carex polystachya A.Rich. in J.S.C.Dumont d'Urville, Voy. Astrolabe: 111 (1832), nom. illeg.
Carex ternaria var. *minor* Boott, Ill. Carex 4: 176 (1867).
Carex ternaria var. *gracilis* Cheeseman, Trans. & Proc. New Zealand Inst. 16: 431 (1884). *Carex confusa* Hamlin, Trans. & Proc. Roy. Soc. New Zealand 82: 58 (1954).
Carex ternaria f. *nigrescens* Kük. in H.G.A.Engler (ed.), Pflanzenr., IV, 20(38): 369 (1909).
Carex martinii Petrie, Trans. & Proc. New Zealand Inst. 56: 7 (1926), nom. illeg.

Carex tessellata Spruce ex C.B.Clarke, Kew Bull., Addit. Ser. 8: 86 (1908).
C. Ecuador. 83 ECU. Hemicr. or rhizome geophyte.

Carex testacea Sol. ex Boott in J.D.Hooker, Fl. Nov.-Zel. 1: 282 (1853).
New Zealand. 51 NZN NZS. Hemicr. or rhizome geophyte.

Carex tetanica Schkuhr, Beschr. Riedgräs. 2: 68 (1806). *Olotrema tetanica* (Schkuhr) Raf., Good Book: 25 (1840).
S. Canada to NC. & E. U.S.A. 71 MAN SAS 72 ONT 74 ILL IOW MIN MSO NDA SDA WIS 75 CNT INI MAS MIC NWJ NWY OHI PEN WVA 78 MRY NCA VRG. Hemicr. or rhizome geophyte.
Carex multinervis Boeck., Flora 38: 596 (1855).
Carex alveata Boott, Ill. Carex 1: 21 (1858).
Carex tetanica var. *canbyi* Porter, Proc. Acad. Nat. Sci. Philadelphia 1887: 76 (1887). *Carex tetanica* f. *canbyi* (Porter) Kük. in H.G.A.Engler (ed.), Pflanzenr., IV, 20(38): 515 (1909).
Carex tetanica var. *carteri* Porter, Proc. Acad. Nat. Sci. Philadelphia 1887: 76 (1887).
Carex meadii Porter, Bull. Torrey Bot. Club 21: 122 (1894), nom. illeg.

Carex tetrastachya Scheele, Linnaea 22: 347 (1849).
Oklahoma to Texas. 74 OKL 77 TEX 78 LOU. Hemicr. or rhizome geophyte.
Carex straminea var. *maxima* L.H.Bailey, Mem. Torrey Bot. Club 1: 22 (1889).
Carex brittoniana J.M.Coult., Contr. U. S. Natl. Herb. 2: 484 (1894).

Carex texensis (Torr. ex L.H.Bailey) L.H.Bailey, Mem. Torrey Bot. Club 5: 97 (1894).
C. & SE. U.S.A. 74 KAN MSO NEB (75) nwj nwy ohi (76) cal 77 TEX 78 ALA ARK FLA GEO KTY LOU MRY MSI NCA SCA TEN VRG. Hemicr. or rhizome geophyte.
**Carex rosea* var. *texensis* Torr. ex L.H.Bailey, Mem. Torrey Bot. Club 1: 57 (1889). *Carex retroflexa* var. *texensis* (Torr. ex L.H.Bailey) Kük. in H.G.A.Engler (ed.), Pflanzenr., IV, 20(38): 158 (1909).

Carex thailandica T.Koyama, Bot. Mag. (Tokyo) 92: 220 (1979).
Thailand. 41 THA. Hemicr. or rhizome geophyte.

Carex thanikaimoniana Govind., J. Econ. Taxon. Bot. 20: 305 (1996).
India (Tamil Nadu). 40 IND. Hemicr. or rhizome geophyte.

Carex thibetica Franch., Nouv. Arch. Mus. Hist. Nat., II, 10: 103 (1887).
China to Vietnam. 36 CHC CHN CHS 41 VIE. Hemicr. or rhizome geophyte.

var. ***pauciflora*** Tang & F.T.Wang ex S.Yun Liang, in Fl. Reipubl. Popul. Sin. 12: 526 (2000).
China (Yunnan). 36 CHC. Hemicr.

var. ***thibetica***
China to Vietnam. 36 CHC CHN CHS 41 VIE. Hemicr. or rhizome geophyte.
Carex dichroa Franch., Bull. Soc. Philom. Paris, VIII, 7: 86 (1895).

Carex thibetica var. *minor* Kük., Bot. Jahrb. Syst. 36(82): 10 (1905).
Carex tchenkeouensis E.G.Camus, Notul. Syst. (Paris) 1: 294 (1910).

Carex thomasii Nelmes, Bull. Misc. Inform. Kew 1938: 245 (1938).
Sudan (Imatong Mts.) to Uganda. 24 SUD 25 UGA. Hemicr. or rhizome geophyte.

Carex thompsonii Franch., Nouv. Arch. Mus. Hist. Nat., III, 8: 212 (1896).
SC. China. 36 CHC. Hemicr. or rhizome geophyte.

Carex thomsonii Boott, Ill. Carex 1: 1 (1858).
Himalaya to Indo-China. 36 CHC CHS 40 ASS EHM NEP WHM 41 THA VIE. Hemicr. or rhizome geophyte.

Carex thornei Naczi, Novon 12: 516 (2002).
SE. U.S.A. 78 ALA FLA GEO. Hemicr. or rhizome geophyte.

Carex thouarsii Carmich., Trans. Linn. Soc. London 12: 508 (1818).
Tristan da Cunha. 90 TDC. Hemicr. or rhizome geophyte.

Carex thunbergii Steud., Flora 29: 23 (1846). *Carex gaudichaudiana* var. *thunbergii* (Steud.) Kük. in H.G.A.Engler (ed.), Pflanzenr., IV, 20(38): 313 (1909).
Siberia to Hawaiian Is. 30 ALT BRY CTA IRK KRA WSB 31 AMU KAM KHA KUR MAG PRM SAK 36 CHI CHM CHN CHS 37 MON 38 JAP KOR 43 NWG 63 HAW. Hemicr. or rhizome geophyte.
Carex acuta var. *appendiculata* Trautv. & C.A.Mey., Fl. Ochot.: 100 (1856). *Carex appendiculata* (Trautv. & C.A.Mey.) Kük., Bull. Herb. Boissier, II, 4: 54 (1904). *Carex thunbergii* var. *appendiculata* (Trautv. & C.A.Mey.) Ohwi, J. Jap. Bot. 11: 409 (1935). *Carex gaudichaudiana* subsp. *appendiculata* (Trautv. & C.A.Mey.) Á.Löve & D.Löve, Bot. Not. 128: 504 (1975 publ. 1976).
Carex thunbergii var. *platycarpa* Franch. & Sav., Enum. Pl. Jap. 2: 129 (1877).
Carex thunbergii var. *quinquenervis* Franch. & Sav., Enum. Pl. Jap. 2: 129 (1877).
Carex brachysandra Franch., Bull. Soc. Philom. Paris, VIII, 7: 40 (1894). *Carex gaudichaudiana* f. *brachysandra* (Franch.) Kük. in H.G.A.Engler (ed.), Pflanzenr., IV, 20(38): 313 (1909).
Carex gaudichaudiana var. *humilior* Kük. in H.G.A. Engler (ed.), Pflanzenr., IV, 20(38): 313 (1909).
Carex koidzumiana Ohwi, Mem. Coll. Sci. Kyoto Imp. Univ., Ser. B, Biol. 6: 263 (1931).
Carex nealiae R.W.Krauss, Pacific Sci. 4: 267 (1950).
Carex appendiculata var. *sacculiformis* Y.L.Chang & Y.L.Yang, in Fl. Plant. Herb. Chinae Bor.-Or. 11: 206 (1976).
Carex spongiifolia A.E.Kozhevn., Bot. Zhurn. (Moscow & Leningrad) 72: 1660 (1987).

Carex thurberi Dewey ex Torr. in W.H.Emory, Rep. U.S. Mex. Bound. 2(1): 232 (1858).
S. Arizona to Guatemala. 76 ARI 79 MXE MXG MXN MXS MXT 80 GUA 81 DOM?. Hemicr. or rhizome geophyte.
Carex arizonensis Kük., Bot. Jahrb. Syst. 27: 549 (1899).

Carex tianmushanica C.Z.Zheng & X.F.Jin, Acta Phytotax. Sin. 42: 541 (2004).
China (Zhejiang). 36 CHS. Hemicr. or rhizome geophyte.

Carex tianschanica T.V.Egorova, Opred. Rast. Sred. Azii 5: 245 (1976).
C. Asia. 32 KGZ. Hemicr. or rhizome geophyte.

Carex timida Naczi & B.A.Ford, Sida 19: 879 (2001).
EC. & E. U.S.A. 74 MSO OKL 75 INI OHI 78 ALA ARK KTY TEN. Hemicr. or rhizome geophyte.

Carex ×timmiana Junge, Jahrb. Hamburg. Wiss. Anst. 22(Beih.): 70 (1904). *C. nigra* × *C. trinervis*.
Coasts of NW. Europe. 10 DEN 11 GER. Hemicr. or rhizome geophyte.

Carex tincta (Fernald) Fernald, Rhodora 15: 186 (1913).
E. Canada to NE. U.S.A. 72 NBR NSC ONT PEI QUE 74 ILL WIS 75 MAI MAS MIC NWH VER. Hemicr. or rhizome geophyte.
**Carex mirabilis* var. *tincta* Fernald, Proc. Amer. Acad. Arts 37: 473 (1902).

Carex titovii V.I.Krecz., Trudy Sredne-Aziatsk. Gosud. Univ., Ser. 8b, Bot. 20: 15, 28 (1935).
C. Asia. 32 KAZ KGZ. Hemicr. or rhizome geophyte.

Carex ×toezensis Simonk., Oesterr. Bot. Z. 41: 426 (1891). *C. melanostachya* × *C. riparia*.
EC. Europe. 11 HUN. Hemicr. or rhizome geophyte.
Carex ×*fleischeri* Podp., Oesterr. Bot. Z. 50: 213 (1900).

Carex tojquianensis Standl. & Steyerm., Ceiba 4: 64 (1953).
Guatemala. 80 GUA. Hemicr. or rhizome geophyte.

Carex tokarensis T.Koyama, Acta Phytotax. Geobot. 16: 7 (1955).
Nansei-shoto. 38 NNS. Hemicr. or rhizome geophyte.

Carex tolucensis (F.J.Herm.) Reznicek, Contr. Univ. Michigan Herb. 19: 128 (1993).
Mexico (México State). 79 MXC. Hemicr. or rhizome geophyte.
**Carex subbracteata* var. *tolucensis* F.J.Herm., Brittonia 23: 145 (1971).

Carex tompkinsii J.T.Howell, Leafl. W. Bot. 9: 185 (1961).
California (Fresno Co.). 76 CAL. Hemicr. or rhizome geophyte.

Carex tonsa (Fernald) E.P.Bicknell, Bull. Torrey Bot. Club 35: 492 (1908).
Canada to NC. & E. U.S.A. 71 ABT BRC MAN SAS 72 LAB NBR NSC ONT PEI QUE 74 ILL MIN WIS 75 CNT INI MAI MAS MIC NWH NWJ NWY OHI PEN VER WVA 78 GEO MRY NCA SCA WDC. Hemicr. or rhizome geophyte.
**Carex umbellata* var. *tonsa* Fernald, Proc. Amer. Acad. Arts 37: 507 (1902). *Carex umbellata* f. *tonsa* (Fernald) Kük. in H.G.A.Engler (ed.), Pflanzenr., IV, 20(38): 453 (1909). *Carex rugosperma* var. *tonsa* (Fernald) E.G.Voss, Rhodora 68: 453 (1966).

var. ***rugosperma*** (Mack.) Crins, Novon 6: 118 (1996).
S. Canada to NC. & E. U.S.A. 71 MAN SAS 72 ONT QUE 74 ILL MIN WIS 75 CNT INI MAI MAS MIC NWH NWJ NWY OHI PEN VER WVA 78 GEO MRY NCA SCA. Hemicr. or rhizome geophyte.
**Carex rugosperma* Mack., Bull. Torrey Bot. Club 42: 621 (1915).

var. ***tonsa***
Canada to NC. & E. U.S.A. 71 ABT BRC SAS 72 LAB NBR NSC ONT PEI QUE 74 ILL WIS 75 INI MAI MAS MIC NWH NWJ NWY OHI PEN 78 NCA WDC. Hemicr. or rhizome geophyte.

Carex toreadora Steyerm., Phytologia 9: 338 (1963).
Ecuador (Azuay). 83 ECU. Hemicr. or rhizome geophyte.

Carex toroensis G.A.Wheeler, Syst. Bot. 14: 182 (1989).
Chile (Magellanes). 85 CLS. Hemicr. or rhizome geophyte.

Carex torreyi Tuck., Enum. Meth. Caric.: 21 (1843).
Canada to NW. & NC. U.S.A. 71 ABT MAN SAS 73 COL MNT WYO 74 MIN NDA SDA WIS. Hemicr. or rhizome geophyte.
Carex abbreviata Prescott ex Boott, Ann. Mag. Nat. Hist. 17: 57 (1846).

Carex torta Boott ex Tuck., Enum. Meth. Caric.: 11 (1843).
E. Canada to NC. & E. U.S.A. 72 NBR NSC ONT QUE 74 ILL IOW MSO OKL 75 CNT INI MAI MAS NWH NWJ NWY OHI PEN VER WVA 78 ARK KTY MRY NCA SCA TEN VRG. Hemicr. or rhizome geophyte.
Carex torta var. *staminata* Peck, Rep. (Annual) Regents Univ. State New York State Mus. 46: 131 (1893).

Carex tovarensis Reznicek & G.A.Wheeler, Contr. Univ. Michigan Herb. 19: 137 (1993).
Venezuela. 82 VEN. Hemicr. or rhizome geophyte.

Carex townsendii Mack. in N.L.Britton & al. (eds.), N. Amer. Fl. 18: 111 (1931). *Carex echinata* subsp. *townsendii* (Mack.) Reznicek, Contr. Univ. Michigan Herb. 16: 207 (1987).
Mexico to Guatemala. 79 MXC MXE MXG MXS MXT 80 GUA. Hemicr. or rhizome geophyte.

Carex toyoshimae Tuyama, Bot. Mag. (Tokyo) 49: 367 (1935).
Ogasawara-shoto (Chichi-jima, Haha-jima). 38 OGA. Hemicr. or rhizome geophyte.

Carex trachycarpa Cheeseman, Trans. & Proc. New Zealand Inst. 24: 413 (1891 publ. 1892).
New Zealand South I. 51 NZS. Hemicr. or rhizome geophyte.

Carex trachycystis Griseb., Abh. Königl. Ges. Wiss. Göttingen 24: 314 (1879). *Carex bonariensis* var. *trachycystis* (Griseb.) Kük., Bot. Jahrb. Syst. 27: 513 (1899).
Chile, Argentina, Uruguay. 85 AGE AGS AGW CLC URU. Hemicr. or rhizome geophyte.
Carex bonariensis var. *remota* Kük., Bot. Jahrb. Syst. 27: 513 (1899). *Carex bonariensis* f. *remota* (Kük.) Kük. in H.G.A.Engler (ed.), Pflanzenr., IV, 20(38): 151 (1909).

Carex traiziscana F.Schmidt, Reis. Amur-Land., Bot.: 192 (1868).
Sakhalin, N. & C. Japan. 31 SAK 38 JAP. Hemicr. or rhizome geophyte.

Carex transandina G.A.Wheeler, Brittonia 38: 317 (1986).
S. Chile (Magallanes). 85 CLS. Hemicr. or rhizome geophyte.

Carex transcaucasica T.V.Egorova, Novosti Sist. Vyssh. Rast. 26: 11 (1989).
Transcaucasus. 33 TCS. Hemicr. or rhizome geophyte.

Carex trautvetteriana Kom., Trudy Imp. S.-Peterburgsk. Bot. Sada 20: 393 (1901).
Siberia, Russian Far East. 30 YAK 31+. Hemicr. or rhizome geophyte.
**Carex ebracteata* Trautv., Trudy Imp. S.-Peterburgsk. Bot. Sada 5: 125 (1877), nom. illeg. *Carex pisiformis* var. *ebracteata* (Trautv.) Kük. in H.G.A.Engler (ed.), Pflanzenr., IV, 20(38): 477 (1909).

Carex traversii Kirk, Trans. & Proc. New Zealand Inst. 26: 262 (1893 publ. 1894).
New Zealand South I. 51 NZS. Hemicr. or rhizome geophyte.

Carex triangularis Boeck., Flora 39: 226 (1856). *Carex vulpinoidea* var. *triangularis* (Boeck.) Kük. in H.G.A. Engler (ed.), Pflanzenr., IV, 20(38): 148 (1909).
C. & SE. U.S.A. 74 KAN MSO OKL 77 TEX 78 ARK LOU MSI TEN. Hemicr. or rhizome geophyte.
Carex vulpinoidea var. *platycarpa* Olney in E.Hall, Pl. Tex.: 25 (1873).
Carex vulpinoidea var. *drummondiana* Boeck., Linnaea 39: 96 (1875).

Carex tribuloides Wahlenb., Kongl. Vetensk. Acad. Nya Handl. 1803: 145 (1803).
E. Canada to EC. & E. U.S.A., Mexico (Veracruz). 72 NBR NSC ONT QUE 74 ILL ILL IOW KAN KAN MIN MSO MSO NEB OKL WIS 75 CNT INI MAI MAS MIC NWH NWJ NWY OHI PEN RHO VER WVA 77 TEX 78 ALA ARK GEO KTY LOU MRY MSI NCA SCA TEN VRG WDC 79 MXG. Hemicr. or rhizome geophyte.

var. ***sangamonensis*** Clokey, Rhodora 21: 84 (1919).
EC. & E. U.S.A., Mexico (Veracruz). 74 ILL KAN MSO OKL 75 INI OHI 77 TEX 78 ALA ARK GEO KTY LOU MSI NCA SCA TEN 79 MXG. Hemicr. or rhizome geophyte.

var. ***tribuloides***
E. Canada to NC. & E. U.S.A. 72 NBR NSC ONT QUE 74 ILL IOW KAN MIN MSO NEB WIS 75 CNT INI MAI MAS MIC NWH NWJ NWY OHI PEN RHO VER WVA 78 DEL FLA GEO KTY MRY NCA SCA TEN VRG WDC. Hemicr. or rhizome geophyte.
Carex richardii Michx., Fl. Bor.-Amer. 2: 170 (1803), nom. illeg.
Carex lagopodioides Willd., Sp. Pl. 4: 230 (1805). *Vignea lagopodioides* (Willd.) Rchb. in J.C. Mössler, Handb. Gewächsk. ed. 2, 3: 1614 (1830). *Carex scoparia* var. *lagopodioides* (Willd.) Torr., Ann. Lyceum Nat. Hist. New York 3: 394 (1836).
Carex lagopodioides var. *moniliformis* Olney ex L.H.Bailey, Bot. Gaz. 10: 380 (1885).
Carex tribuloides var. *reducta* L.H.Bailey, Proc. Amer. Acad. Arts 22: 148 (1886). *Carex cristata* var. *reducta* (L.H.Bailey) Kük. in H.G.A.Engler (ed.), Pflanzenr., IV, 20(38): 201 (1909).
Carex tribuloides var. *turbata* L.H.Bailey, Mem. Torrey Bot. Club 1: 55 (1889).
Carex tribuloides f. *aggregata* Peck, Rep. (Annual) Regents Univ. State New York State Mus. 46: 132 (1893).

Carex tricephala Boeck., Flora 58: 263 (1875).
SC. China, Indo-China, NW. Jawa (Madura I.). 36

CHC 41 MYA THA VIE 42 JAW. Hemicr. or rhizome geophyte.
Carex madoerensis C.B.Clarke, J. Linn. Soc., Bot. 37: 15 (1904).
Carex plesiocephala Turrill, Bull. Misc. Inform. Kew 1910: 385 (1910).
Carex thorelii E.G.Camus, Notul. Syst. (Paris) 1: 295 (1910).
Schoenoxiphium leucocephalum Thorel ex E.G.Camus, Notul. Syst. (Paris) 1: 295 (1910).
Carex hispidangula T.Koyama, Naturaliste Canad. 82: 200 (1955).

Carex ×trichina Fernald, Rhodora 35: 219 (1933). *C. tenuiflora* × *C. trisperma*.
E. Canada to NE. U.S.A. 72 LAB NFL 75 MAI MIC. Hemicr. or rhizome geophyte.

Carex trichocarpa Muhl. ex Willd., Sp. Pl. 4: 302 (1805).
SE. Canada to NC. & E. U.S.A. 72 ONT QUE 73 ORE 74 ILL IOW MIN MSO WIS 75 CNT INI MAS MIC NWJ NWY OHI PEN VER WVA 78 DEL MRY NCA TEN VRG. Hemicr. or rhizome geophyte.

Carex trichodes Steud., Linnaea 39: 29 (1875).
S. Chile. 85 CLS. Hemicr. or rhizome geophyte.

Carex tricholepis Nelmes, Bull. Misc. Inform. Kew 1940: 160 (1940).
Zambia. 26 ZAM. Hemicr. or rhizome geophyte.

Carex trichophylla Nelmes, Mém. Mus. Natl. Hist. Nat., Sér. B, Bot. 4: 106 (1955).
Vietnam. 41 VIE. Hemicr. or rhizome geophyte.

Carex tricolor Velen., Sitzungsber. Königl. Böhm. Ges. Wiss. Prag, Math.-Naturwiss. Cl. 1889(2): 58 (1890).
W. Bulgaria. 13 BUL. Hemicr. or rhizome geophyte.

Carex trifida Cav., Icon. 5: 41 (1799).
New Zealand, S. Chile, Subantarctic Is. 51 ATP CTM NZS 85 CLS 90 FAL MAQ. Hemicr. or rhizome geophyte.
Carex incrassata Sol. ex Boott, Ill. Carex 4: 138 (1867).
Carex rekohu Petrie, Trans. & Proc. New Zealand Inst. 56: 8 (1926).

Carex trigonosperma Ohwi, Acta Phytotax. Geobot. 5: 54 (1936).
China (Sichuan). 36 CHC. Hemicr. or rhizome geophyte.

Carex trinervis Degl. in J.L.A.Loiseleur-Deslongchamps, Fl. Gall. 2: 731 (1807). *Vignea trinervis* (Degl.) Rchb. in J.C.Mössler, Handb. Gewächsk. ed. 2, 3: 1625 (1830). *Osculisa trinervis* (Degl.) Raf., Good Book: 27 (1840). *Carex acuta* var. *trinervis* (Degl.) Jess., Deut. Excurs.-Fl.: 543 (1879). *Carex glauca* subsp. *trinervis* (Degl.) Asch. & Graebn., Syn. Mitteleur. Fl. 2(2): 138 (1903).
Coasts of Denmark to C. Portugal. 10 DEN GRB NOR? 11 BGM GER NET 12 FRA POR SPA. Hemicr. or rhizome geophyte.
Carex cerina Dumort., Fl. Belg.: 147 (1827), nom. illeg.
Carex frisica H.Koch, Flora 29: 273 (1846).
Carex acuta var. *nervosa* G.Mey., Fl. Hanov. Excurs.: 98 (1849).
Carex trinervis var. *laxa* Lange, Bot. Tidsskr. 2(1): 274 (1872).
Carex glauca var. *digyna* Boeck., Linnaea 41: 189 (1877).

Carex triquetra Boott, Trans. Linn. Soc. London 20: 126 (1851).
California to Mexico (Baja California). 76 CAL 79 MXN. Hemicr. or rhizome geophyte.
Carex monticola Dewey ex Torr. in W.H.Emory, Rep. U.S. Mex. Bound. 2(1): 229 (1858).

Carex trisperma Dewey, Amer. J. Sci. Arts 9: 63 (1825). *Neskiza trisperma* (Dewey) Raf., Good Book: 27 (1840).
Greenland to NC. & NE. U.S.A. 70 GNL NUN NWT 71 ABT BRC MAN SAS 72 LAB NBR NFL NSC ONT PEI QUE 74 ILL MIN WIS 75 CNT INI MAI MAS MIC NWH NWJ NWY OHI PEN RHO VER WVA 78 DEL MRY. Hemicr. or rhizome geophyte.

var. ***billingsii*** O.W.Knight, Rhodora 8: 185 (1906). *Carex trisperma* f. *billingsii* (O.W.Knight) B.Boivin, Naturaliste Canad. 94: 523 (1967).
E. Canada to NE. U.S.A. 72 NBR NFL NSC ONT PEI QUE 75 MAI MAS MIC NWJ NWY PEN RHO. Hemicr. or rhizome geophyte.

var. ***trisperma***
Greenland to NC. & NE. U.S.A. 70 GNL NUN NWT 71 ABT BRC MAN SAS 72 LAB NBR NFL NSC ONT PEI QUE 74 ILL MIN WIS 75 CNT INI MAI MAS MIC NWH NWJ NWY OHI PEN RHO VER WVA 78 DEL MRY. Hemicr. or rhizome geophyte.

Carex tristachya Thunb. in J.A.Murray, Syst. Veg. ed. 14: 845 (1784).
Temp. E. Asia, Malesia to New Guinea. 36 CHC CHS 38 JAP KOR NNS TAI 42 BOR PHI 43 NWG. Hemicr. or rhizome geophyte.

var. ***pocilliformis*** (Boott) Kük. in H.G.A.Engler (ed.), Pflanzenr., IV, 20(38): 473 (1909).
Temp. E. Asia, Malesia to New Guinea. 36 CHC CHS 38 JAP KOR NNS TAI 42 BOR PHI 43 NWG. Hemicr. or rhizome geophyte.
**Carex pocilliformis* Boott, Ill. Carex 4: 175 (1867).
Carex tristachya subsp. *pocilliformis* (Boott) T.Koyama, in Fl. Taiwan 5: 337 (1978).

var. ***tristachya***
China, Korea, Japan. 36 CHH CHS 38 JAP KOR. Hemicr. or rhizome geophyte.
Carex monadelpha Boott in M.C.Perry, Jap. Exped. 2: 323 (1856).

Carex tristis M.Bieb., Fl. Taur.-Caucas. 3: 615 (1819). *Carex ferruginea* subsp. *tristis* (M.Bieb.) Nyman, Consp. Fl. Eur.: 772 (1882). *Carex sempervirens* subsp. *tristis* (M.Bieb.) Kük. in H.G.A.Engler (ed.), Pflanzenr., IV, 20(38): 569 (1909).
NE. Turkey to NW. Iran. 33 NCS TCS 34 IRN TUR. Hemicr. or rhizome geophyte.
Carex ferruginea Steven, Mém. Soc. Imp. Naturalistes Moscou 4: 68 (1813), nom. illeg.
Carex schottii Boiss., Diagn. Pl. Orient., II, 4: 123 (1859), nom. illeg.
Carex meinshauseniana V.I.Krecz. in V.L.Komarov (ed.), Fl. URSS 3: 606 (1935).

Carex trongii K.K.Nguyen, Bot. Zhurn. (Moscow & Leningrad) 64: 856 (1979).
Vietnam. 41 VIE. Hemicr. or rhizome geophyte.

Carex troodi Turrill, Bull. Misc. Inform. Kew 1930: 125 (1930).
W. Cyprus. 13 KRI? 34 CYP. Hemicr. or rhizome geophyte.

Carex truncatigluma C.B.Clarke, J. Linn. Soc., Bot. 36: 315 (1904).
S. China to Philippines. 36 CHC CHH CHS 38 TAI 41 VIE 42 PHI. Hemicr. or rhizome geophyte.

subsp. ***rhynchachaenium*** (C.B.Clarke) Y.C.Tang & S.Yun Liang, in Fl. Reipubl. Popul. Sin. 12: 146 (2000).
Vietnam, Taiwan, Philippines. 36 CHT 41 VIE 42 PHI. Hemicr.
**Carex rhynchachaenium* C.B.Clarke, Publ. Bur. Sci. Gov. Lab. 35: 5 (1905 publ. 1906).
Carex hatusimana Ohwi, Jap. J. Bot. 7: 196 (1934).

subsp. ***truncatigluma***
S. China to Taiwan. 36 CHC CHH CHS 38 TAI. Hemicr. or rhizome geophyte.

Carex tsaiana F.T.Wang & Tang ex P.C.Li, Acta Phytotax. Sin. 37: 169 (1999).
China (Yunnan). 36 CHC. Hemicr. or rhizome geophyte.

Carex tsaratananensis Cherm., Bull. Soc. Bot. France 72: 21 (1925).
C. Madagascar. 29 MDG. Hemicr. or rhizome geophyte.
Carex tsaratananensis var. *laevis* Cherm., Mém. Acad. Malagache 10: 46 (1931).

Carex tsiangii F.T.Wang & Tang, Contr. Inst. Bot. Natl. Acad. Peiping 6: 59 (1949).
China (Guangdong). 36 CHS. Hemicr. or rhizome geophyte.
Carex surculosa Raymond, Mém. Jard. Bot. Montréal 53: 22 (1959).

Carex tsoi Merr. & Chun, Sunyatsenia 2: 207 (1935).
Hainan. 36 CHH. Hemicr. or rhizome geophyte.

Carex tsushimensis (Ohwi) Ohwi, Acta Phytotax. Geobot. 1: 67 (1932).
Japan (Kyushu). 38 JAP. Hemicr. or rhizome geophyte.
**Carex chinensis* var. *tsushimensis* Ohwi, Mem. Coll. Sci. Kyoto Imp. Univ., Ser. B, Biol. 5(3): 263 (1930).

Carex tuberculata Liebm., Mexic. Halvgr.: 270 (1850).
C. Mexico (to Hidalgo). 79 MXC MXE. Hemicr. or rhizome geophyte.
Carex ehrenbergiana Boeck., Linnaea 41: 175 (1877). *Carex tuberculata* var. *ehrenbergiana* (Boeck.) Kük. in H.G.A.Engler (ed.), Pflanzenr., IV, 20(38): 408 (1909).

Carex tubulosa Pamp., Nuovo Giorn. Bot. Ital., n.s., 22: 258 (1915).
China (Hubei). 36 CHC. Hemicr. or rhizome geophyte.

Carex tuckermanii Boott, London J. Bot. 5: 73 (1846).
E. Canada to NC. & NE. U.S.A. 72 NBR ONT QUE 74 MIN WIS 75 CNT INI MAI MIC NWY OHI PEN VER 78 MRY. Hemicr. or rhizome geophyte.
Carex bullata Dewey, Amer. J. Sci. Arts 9: 71 (1825), nom. illeg.
Carex tuckermanii var. *cylindracea* Dewey, Amer. J. Sci. Arts 49: 48 (1845).
Carex tuckermanii var. *niagarensis* C.P.Sm., Rhodora 17: 57 (1915).

Carex tucumanensis G.A.Wheeler, Hickenia 2: 190 (1996).
Argentina (Tucumán). 85 AGW. Hemicr. or rhizome geophyte.

Carex tumidula Ohwi, Mem. Coll. Sci. Kyoto Imp. Univ., Ser. B, Biol. 6: 244 (1931).
Japan (Shikoku: Iyo Prov.). 38 JAP. Hemicr. or rhizome geophyte.

Carex tuminensis Kom., Trudy Imp. S.-Peterburgsk. Bot. Sada 18: 444 (1900).
Russian Far East to Korea. 31+ 36 CHM 38 KOR. Hemicr. or rhizome geophyte.

Carex tumulicola Mack., Bull. Torrey Bot. Club 34: 154 (1907).
W. Canada to California. 71 BRC 73 ORE WAS 76 CAL. Hemicr. or rhizome geophyte.

Carex tungfangensis L.K.Dai & S.M.Huang, in Fl. Hainan. 4: 536 (1977).
Hainan. 36 CHH. Hemicr. or rhizome geophyte.

Carex tunimanensis Standl. & Steyerm., Ceiba 4: 65 (1953).
Mexico (Chiapas) to Guatemala. 79 MXT 80 GUA. Hemicr. or rhizome geophyte.

Carex turbinata Liebm., Mexic. Halvgr.: 265 (1850).
Arizona to Mexico. 76 ARI 79 MXC MXE MXN MXS. Hemicr. or rhizome geophyte.
Carex rigens L.H.Bailey, Bot. Gaz. 9: 117 (1884), nom. illeg. *Carex leucodonta* T.Holm, Amer. J. Sci. 166: 461 (1903). *Carex turbinata* var. *rigens* Kük. in H.G.A.Engler (ed.), Pflanzenr., IV, 20(38): 445 (1909).

Carex ×turfosa Fr., Summa Veg. Scand.: 228 (1845). *C. elata* × *C. nigra*. *Carex goodenowii* var. *turfosa* (Fr.) Asch., Fl. Brandenburg 1: 777 (1864). *Carex vulgaris* var. *turfosa* (Fr.) Boott, Ill. Carex 4: 169 (1867). *Carex stricta* var. *turfosa* (Fr.) Almq. in C.J.Hartmann, Handb. Skand. Fl., ed. 11: 468 (1879). *Carex nigra* subsp. *turfosa* (Fr.) Podlech in E.Janchen, Fl. Wien Niederost. & Nordburg., ed. 2: 649 (1977), without basionym ref.
Europe. 10 SWE 11 GER NET SWI 12 FRA 13 ITA. Hemicr. or rhizome geophyte.
Carex stricta var. *minor* Gaudin, Fl. Helv. 6: 71 (1830).
Carex ×dematranea Lagger, Flora 38: 207 (1855).
Carex stricta var. *fallax* T.Marsson, Fl. Neu-Vorpommern: 530 (1869). *Carex elata* subsp. *fallax* (T.Marsson) Suess. in G.Hegi, Ill. Fl. Mitt.-Eur., ed. 2, 2: 101 (1939).
Carex acuta f. *minor* Boeck., Linnaea 40: 418 (1876), nom. illeg.

Carex turgescens Torr., Ann. Lyceum Nat. Hist. New York 3: 419 (1836).
SE. U.S.A. 78 ALA FLA GEO LOU MSI NCA SCA. Hemicr. or rhizome geophyte.
Carex folliculata Elliott, Sketch Bot. S. Carolina 2: 545 (1824), nom. illeg.

Carex turkestanica Regel, Trudy Imp. S.-Peterburgsk. Bot. Sada 7: 570 (1880).
Siberia to Pakistan. 30+ 32 KAZ KGZ TZK UZB 34 AFG 36 CHN CHX 40 PAK WHM. Hemicr. or rhizome geophyte.

subsp. ***beleensis*** Zolot., Novosti Sist. Vyssh. Rast. 21: 228 (1984).
Siberia. 30+. Hemicr. or rhizome geophyte.

subsp. ***turkestanica***
Siberia to Pakistan. 30+ 32 KAZ KGZ TZK UZB 34 AFG 36 CHN CHX 40 PAK WHM. Hemicr. or rhizome geophyte.

Carex obesa var. *aspera* Boeck., Linnaea 41: 185 (1877). *Carex nitida* var. *aspera* (Boeck.) Kük. in H.G.A.Engler (ed.), Pflanzenr., IV, 20(38): 458 (1909).
Carex verae Ovcz. & Czukav., in Fl. Tadzhikskoi SSR 2: 423 (1963).

Carex turrita C.B.Clarke, J. Linn. Soc., Bot. 37: 13 (1904). *Carex walkeri* var. *turrita* (C.B.Clarke) Kük. in H.G.A.Engler (ed.), Pflanzenr., IV, 20(38): 546 (1909).
Philippines, New Guinea. 42 PHI 43 NWG. Hemicr. or rhizome geophyte.

var. ***merrillii*** (Kük.) Noot., in Fl. Males. 9(1): 143 (1979).
Philippines, New Guinea. 42 PHI 43 NWG. Hemicr. or rhizome geophyte.
**Carex merrillii* Kük., Repert. Spec. Nov. Regni Veg. 8: 7 (1910).
Carex pullei Nelmes, Kew Bull. 5: 198 (1950).

var. ***turrita***
Philippines, New Guinea. 42 PHI 43 NWG. Hemicr. or rhizome geophyte.
Carex atrosanguinea Nelmes, Kew Bull. 5: 197 (1950).

Carex turumiquirensis Steyerm., Fieldiana, Bot. 28: 70 (1951).
Venezuela. 82 VEN. Hemicr. or rhizome geophyte.

Carex tweediana Nees, J. Bot. (Hooker) 2: 398 (1840).
S. Brazil to NE. Argentina. 84 BZS 85 AGE PAR URU. Hemicr. or rhizome geophyte.

Carex typhina Michx., Fl. Bor.-Amer. 2: 169 (1803). *Carex squarrosa* var. *typhina* (Michx.) Nutt. in ?
E. Canada to C. & E. U.S.A. 72 ONT QUE 74 ILL IOW MIN MSO OKL WIS 75 CNT INI MAI MAS MIC NWJ NWY OHI PEN VER WVA 77 TEX 78 ARK DEL GEO KTY LOU MRY MSI NCA SCA TEN VRG WDC. Hemicr. or rhizome geophyte.

Carex uda Maxim., Prim. Fl. Amur.: 303 (1859).
Russian Far East to N. & C. Japan. 31 AMU PRM SAK 36 CHM 38 JAP KOR. Hemicr. or rhizome geophyte.

Carex ulobasis V.I.Krecz. in V.L.Komarov (ed.), Fl. URSS 3: 608 (1935).
Inner Mongolia to Russian Far East. 31+ 36 CHI CHM. Hemicr. or rhizome geophyte.
Carex montana var. *manshuriensis* Kom., Fl. Manshur. 1: 396 (1901).

Carex umbellata Willd., Sp. Pl. 4: 290 (1805). *Olotrema umbellata* (Willd.) Raf., Good Book: 25 (1840).
Canada to N. & E. U.S.A. 70 GNL 71 BRC SAS 72 LAB NBR NFL NSC ONT PEI QUE 74 ILL KAN MIN MSO NEB OKL WIS 75 INI MAI MAS MIC NWJ NWY OHI PEN RHO WVA 78 ALA ARK GEO KTY MRY MSI NCA. Hemicr. or rhizome geophyte.
Carex umbellata var. *vicina* Dewey, Amer. J. Sci. Arts 10: t. 13 (1826). *Carex umbellata* f. *vicina* (Dewey) Wiegand, Rhodora 26: 2 (1924).
Carex umbellata var. *brevirostris* Boott, Ill. Carex 2: 99 (1860).
Carex umbellata var. *brachyrhina* Piper, Contr. U. S. Natl. Herb. 11: 175 (1906).
Carex abdita E.P.Bicknell, Bull. Torrey Bot. Club 35: 492 (1908). *Carex umbellata* subsp. *abdita* (E.P. Bicknell) W.Stone, Pl. S. New Jersey: 305 (1912).

Carex umbrosa Host, Icon. Descr. Gram. Austriac. 1: 52 (1801). *Carex caryophyllea* var. *umbrosa* (Host) Fiori, Nuov. Fl. Italia 1: 192 (1923).
Europe to Japan. 11 AUT BGM CZE GER HUN POL 12 FRA SAR SPA 13 BUL ITA ROM YUG 14 BLR RUW UKR 30 ALT KRA WSB 31 KUR PRM SAK 33 NCS TCS 34 TUR 36 CHM 37 MON 38 JAP KOR. Hemicr. or rhizome geophyte.

subsp. ***huetiana*** (Boiss.) Soó, Acta Bot. Acad. Sci. Hung. 16: 371 (1970 publ. 1971).
Pyrenees, Balkan Pen., Turkey, Caucasus. 12 FRA SPA 13 BUL YUG 33 NCS TCS 34 TUR. Hemicr. or rhizome geophyte.
**Carex huetiana* Boiss., Diagn. Pl. Orient., II, 4: 123 (1859). *Carex umbrosa* var. *huetiana* (Boiss.) Kük., Allg. Bot. Z. Syst. 7: 193 (1901).
Carex mixta Miégev., Bull. Soc. Bot. France 12: 343 (1865), nom. illeg.
Carex reuteriana Boiss. & Reut. in P.E.Boissier, Fl. Orient. 5: 411 (1882), nom. illeg.

var. ***rostrata*** (Maxim.) Govaerts, World Checklist Seed Pl. 3(1): 13 (1999).
Russian Far East, Korea, Japan. 31+ 38 JAP KOR. Hemicr. or rhizome geophyte.
**Carex pediformis* var. *rostrata* Maxim., Prim. Fl. Amur.: 310 (1859). *Carex sabynensis* var. *rostrata* (Maxim.) Ohwi, Mem. Coll. Sci. Kyoto Imp. Univ., Ser. B, Biol. 11: 353 (1936).
Carex lucidula Franch., Bull. Soc. Philom. Paris, VIII, 7: 48 (1895).
Carex umbrosa var. *stolonifera* Kük. in H.G.A. Engler (ed.), Pflanzenr., IV, 20(38): 468 (1909).
Carex praestabilis Ohwi, Mem. Coll. Sci. Kyoto Imp. Univ., Ser. B, Biol. 6: 262 (1931).

subsp. ***sabynensis*** (Less. ex Kunth) Kük. in H.G.A. Engler (ed.), Pflanzenr., IV, 20(38): 468 (1909).
European Russia (Ural'skiy Chrebet) to Japan, Caucasus. 14 RUE RUN 30 ALT KRA WSB 31 KUR PRM SAK 33 NCS TCS 36 CHM 37 MON 38 JAP KOR. Hemicr. or rhizome geophyte.
**Carex sabynensis* Less. ex Kunth, Enum. Pl. 2: 440 (1837).
Carex pediformis var. *caespitosa* F.Schmidt, Mém. Acad. Imp. Sci. Saint Pétersbourg, Sér. 7, VII, 17: 127 (1872).
Carex kamikawensis Franch., Bull. Soc. Philom. Paris, VIII, 7: 48 (1895).
Carex recticulmis Franch., Bull. Soc. Philom. Paris, VIII, 7: 48 (1895).
Carex eriandrolepis H.Lév., Bull. Acad. Int. Géogr. Bot. 19: 34 (1909).

subsp. ***umbrosa***
Europe. 11 AUT BGM CZE GER HUN POL 12 FRA SAR SPA 13 BUL ITA ROM YUG 14 BLR RUW UKR. Hemicr. or rhizome geophyte.
Carex longifolia Host, Icon. Descr. Gram. Austriac. 4: t. 85 (1809), nom. illeg. *Carex praecox* var. *longifolia* Wahlenb., Fl. Carpat. Princ.: 300 (1814). *Carex umbrosa* var. *longifolia* (Wahlenb.) Nyman, Consp. Fl. Eur.: 773 (1882).
Carex praecox var. *procerior* Gaudin, Agrost. Helv. 2: 131 (1811).
Carex polyrrhiza Wallr., Sched. Crit.: 492 (1822).
Carex stolonifera var. *procera* Klett & Richt., Fl. Leibzig: 756 (1830).

Carex polyrrhiza f. *breviscapa* Peterm., Anal. Pfl.-Schlüss.: 504 (1846). *Carex umbrosa* f. *breviscapa* (Peterm.) Soó, Acta Bot. Acad. Sci. Hung. 16: 370 (1970 publ. 1971).
Carex polyrrhiza var. *decalvata* Hausm. ex Döll, Fl. Baden 1: 274 (1857). *Carex umbrosa* var. *decalvata* (Hausm. ex Döll) Kük. in H.G.A.Engler (ed.), Pflanzenr., IV, 20(38): 468 (1909).
Carex praecox var. *polyrrhiza* Boott, Ill. Carex 4: 216 (1867).
Carex longifolia f. *gracilis* Kohts, Oesterr. Bot. Z. 20: 140 (1870). *Carex umbrosa* f. *gracilis* (Kohts) Kük. in H.G.A.Engler (ed.), Pflanzenr., IV, 20(38): 468 (1909).
Carex sabynensis var. *laxior* Trautv., Trudy Imp. S.-Peterburgsk. Bot. Sada 5: 128 (1877).
Carex brenneri H.Christ in N.J.W.Scheutz, Pl. Vasc. Jenis.: 178 (1888).
Carex polyrrhiza f. *subintegra* Kük., Mitth. Thüring. Bot. Vereins, n.s., 10: 36 (1897). *Carex umbrosa* f. *subintegra* (Kük.) Kük. in H.G.A.Engler (ed.), Pflanzenr., IV, 20(38): 466 (1909).
Carex bremurii H.Christ ex Meinsh., Trudy Imp. S.-Peterburgsk. Bot. Sada 18: 411 (1900).
Carex polyrrhiza f. *brevifolia* Waisb., Oesterr. Bot. Z. 51: 128 (1901). *Carex umbrosa* f. *brevifolia* (Waisb.) Soó, Acta Bot. Acad. Sci. Hung. 16: 371 (1970 publ. 1971).
Carex umbrosa f. *diandrostachya* Waisb., Oesterr. Bot. Z. 51: 128 (1901).
Carex umbrosa f. *longipedunculata* Torges ex Kük. in H.G.A.Engler (ed.), Pflanzenr., IV, 20(38): 468 (1909).
Carex mixta var. *gynobasis* Rouy in G.Rouy & J.Foucaud, Fl. France 13: 444 (1912). *Carex umbrosa* f. *gynobasis* (Rouy) O.Bolòs & Vigo, Fl. Man. Paisos Catalans 4: 282 (2001).

Carex umbrosiformis H.Lév., Repert. Spec. Nov. Regni Veg. 11: 33 (1912).
Korea. 38 KOR. Hemicr. or rhizome geophyte.

Carex uncifolia Cheeseman, Trans. & Proc. New Zealand Inst. 16: 412 (1884).
New Zealand. 51 NZN NZS. Hemicr. or rhizome geophyte.
Carex cardrona Hamlin, Rec. Domin. Mus. 2: 12 (1956).

Carex ×ungavensis Lepage, Naturaliste Canad. 83: 132 (1956). *C. paleacea* × *C. stylosa* var. *nigritella*.
E. Canada. 72 QUE. Hemicr. or rhizome geophyte.

Carex ungurensis Litv., Trudy Bot. Muz. Imp. Akad. Nauk 7: 93 (1910).
C. Asia. 32 KGZ. Hemicr. or rhizome geophyte.

Carex unilateralis Mack., Erythea 8: 43 (1922). *Carex athrostachya* var. *unilateralis* (Mack.) B.Boivin, Phytologia 43: 55 (1979).
SW. Canada to W. U.S.A. 71 BRC 73 ORE WAS 76 CAL. Hemicr. or rhizome geophyte.

Carex unisexualis C.B.Clarke, J. Linn. Soc., Bot. 36: 316 (1904). *Carex fluviatilis* var. *unisexualis* (C.B.Clarke) Kük. in H.G.A.Engler (ed.), Pflanzenr., IV, 20(38): 144 (1909).
S. China. 36 CHC CHS. Hemicr. or rhizome geophyte.

Carex urelytra Ohwi, Jap. J. Bot. 7: 204 (1934).
Taiwan. 38 TAI. Hemicr. or rhizome geophyte.

Carex ursina Dewey, Amer. J. Sci. Arts 27: 240 (1835). *Carex glareosa* var. *ursina* (Dewey) Britton, Mem. Torrey Bot. Club 5: 80 (1894). *Vignea ursina* (Dewey) Soják, Cas. Nár. Mus., Odd. Prír. 148: 197 (1979 publ. 1980).
Subarctic. 10 SVA 14 RUN 30 KRA WSB YAK 31 MAG 70 ASK GNL NUN NWT YUK 71 MAN 72 QUE. Hemicr. or rhizome geophyte.
Carex glareosa var. *caespitosa* Boeck., Linnaea 39: 76 (1875).

Carex uruguensis Boeck., Bot. Jahrb. Syst. 7: 277 (1886). *Carex sororia* subsp. *uruguensis* (Boeck.) Luceño & Alves, Anales Jard. Bot. Madrid 57: 173 (1999).
SE. Ecuador to N. Argentina. 83 ECU 84 BZS 85 AGE AGW URU. Hemicr. or rhizome geophyte.

var. ***angustata*** (Kük.) Kük. in H.G.A.Engler (ed.), Pflanzenr., IV, 20(38): 149 (1909).
Brazil to NE. Argentina. 84 BZS 85 AGE URU. Hemicr. or rhizome geophyte.
**Carex involucrata* var. *angustata* Kük., Bot. Jahrb. Syst. 27: 511 (1899).

var. ***pseudoechinata*** (Boeck.) Kük., Bot. Jahrb. Syst. 27: 512 (1899).
NE. Argentina. 85 AGE. Hemicr. or rhizome geophyte.
**Carex pseudoechinata* Boeck., Beitr. Cyper. 1: 34 (1888).

var. ***uruguensis***
SE. Ecuador to N. Argentina. 83 ECU 84 BZS 85 AGE AGW URU. Hemicr. or rhizome geophyte.

Carex ussuriensis Kom., Trudy Imp. S.-Peterburgsk. Bot. Sada 18: 443 (1901).
Russian Far East to China. 31 PRM 36 CHC CHI CHM CHN. Hemicr. or rhizome geophyte.

f. ***latifolia*** K.T.Fu, in Fl. Tsinlingensis 1(1): 446 (1976).
China. 36 CHC. Hemicr. or rhizome geophyte.

f. ***ussuriensis***
Russian Far East to N. China. 31 PRM 36 CHI CHM CHN. Hemicr. or rhizome geophyte.

Carex utriculata Boott in W.J.Hooker, Fl. Bor.-Amer. 2: 221 (1839). *Carex vesicaria* var. *utriculata* (Boott) Dewey, Class-book Bot.: 428 (1845). *Carex ampullacea* var. *utriculata* (Boott) Carey in A.Gray, Manual: 566 (1848). *Carex rostrata* var. *utriculata* (Boott) L.H.Bailey, Proc. Amer. Acad. Arts 22: 67 (1886). *Carex inflata* var. *utriculata* (Boott) Druce, Bot. Soc. Exch. Club Brit. Isles 9(1): 141 (1930). *Carex rostrata* subsp. *utriculata* (Boott) Soják, Cas. Nár. Mus., Odd. Prír. 148: 77 (1979 publ. 1980).
Temp. Northern Hemisphere. 10 FIN NOR SWE 11 POL 13 ROM? 14 BLR BLT RUC RUE RUN RUW 30 ALT KRA WSB YAK 31 KAM KHA KUR MAG PRM SAK 36 CHM CHX 37 MON 38 JAP KOR 70 ASK NUN NWT YUK 71 ABT BRC MAN SAS 72 LAB NBR NFL NSC ONT PEI QUE 73 COL IDA MNT ORE WAS WYO 74 ILL MIN SDA WIS 75 CNT INI MAI MAS MIC NWH NWY OHI PEN VER WVA 76 ARI CAL NEV UTA 77 NWM 78 MRY TEN VRG 79 MXE. Hemicr. or rhizome geophyte.
Carex bullata var. *laevirostris* Blytt ex Fr., Novit. Fl. Suec. Mant. 2: 59 (1839). *Carex laevirostris* (Blytt ex Fr.) Andersson, Cyper. Scand.: 17 (1849).

Carex utriculata var. *minor* Boott in W.J.Hooker, Fl. Bor.-Amer. 2: 221 (1839). *Carex rostrata* f. *minor* (Boott) Kük. in H.G.A.Engler (ed.), Pflanzenr., IV, 20(38): 722 (1909).
Carex rhynchophysa Fisch., C.A.Mey. & Avé-Lall., Index Seminum (LE) 9(Suppl.): 9 (1843).
Carex robusta F.Nyl., Spic. Fl. Fenn.: 33 (1843).
Carex vesicaria subsp. *lacustris* Th.Fr., Bot. Not. 1857: 200 (1857).
Carex utriculata var. *sparsiflora* Dewey in A.Wood, Class-book Bot.: 768 (1861).
Carex utriculata var. *globosa* Olney in S.Watson, Botany [Fortieth Parallel]: 374 (1871).
Carex ventricosa Franch., Bull. Soc. Philom. Paris, VIII, 7: 91 (1895), nom. illeg.
Carex laevirostris f. *gracilior* Kük. in H.G.A.Engler (ed.), Pflanzenr., IV, 20(38): 724 (1909).

Carex* × *uzenensis Koidz., Bot. Mag. (Tokyo) 30: 332 (1916). *C. forficula* × *C. podogyna*.
Japan. 38 JAP. Hemicr. or rhizome geophyte.

Carex vacillans Drejer in C.J.Hartmann, Fl. Excurs. Hafn.: 3 (1838). *Carex* ×*haematolepis* var. *vacillans* (Drejer) Nyman, Consp. Fl. Eur.: 776 (1882).
N. Europe, E. Canada to NE. U.S.A. 10 NOR SWE 72 LAB NBR NFL NSC PEI QUE 75 MAI MAS. Hemicr. or rhizome geophyte.
Carex salina var. *kattegatensis* Almq. in C.J.Hartmann, Handb. Skand. Fl., ed. 11: 466 (1879). *Carex salina* subsp. *kattegatensis* (Almq.) Nyman, Consp. Fl. Eur., Suppl. 2: 323 (1890). *Carex kattegatensis* (Almq.) Fr. ex Krechet. in V.L.Komarov (ed.), Fl. URSS 3: 237 (1935).
Carex ×*haematolepis* subsp. *kattegatensis* Nyman, Consp. Fl. Eur.: 776 (1882).
Carex supergoodenoughii Lepage, Naturaliste Canad. 83: 146 (1956).
Carex subnigra Lepage, Naturaliste Canad. 84: 46 (1957).

Carex vaginata Tausch, Flora 4: 557 (1821). *Carex panicea* subsp. *vaginata* (Tausch) Rouy in ?
Temp. Northern Hemisphere. 10 FIN GRB ICE NOR SWE 11 AUT CZE GER POL SWI 12 FRA SPA 14 BLR BLT RUC RUE RUN RUW UKR 30 KRA WSB YAK 31 KAM KUR MAG PRM 32 KAZ 33 NCS TCS 36 CHM 37 MON 38 JAP KOR 70 ASK GNL NUN NWT YUK 71 ABT BRC MAN SAS 72 LAB NBR NFL QUE 73 MNT 74 MIN WIS 75 MAI MIC NWH NWY VER. Hemicr. or rhizome geophyte.

var. ***petersii*** (C.A.Mey. ex Schmidt) Akiyama, J. Jap. Bot. 11: 499 (1935).
Siberia to N. & C. Japan. 30 YAK 31 KAM KUR PRM SAK 36 CHM 37 MON 38 JAP KOR. Hemicr. or rhizome geophyte.
Carex falcata Turcz., Bull. Soc. Imp. Naturalistes Moscou 1838: 104 (1838).
**Carex petersii* C.A.Mey. ex Schmidt, Reis. Amur-Land., Bot.: 194 (1868). *Carex sparsiflora* var. *petersii* (C.A.Mey. ex Schmidt) Kük. in H.G.A. Engler (ed.), Pflanzenr., IV, 20(38): 513 (1909).

var. ***vaginata***
Greenland, Europe to Siberia, Canada to N. U.S.A. 10 FIN GRB ICE NOR SWE 11 AUT CZE GER POL SWI 12 FRA SPA 14 BLR BLT RUC RUE RUN RUW UKR 30 KRA WSB YAK 31 MAG 32 KAZ 33 NCS TCS 36 CHM 37 MON 70 ASK GNL NUN NWT YUK 71 ABT BRC MAN SAS 72 LAB NBR NFL QUE 73 MNT 74 MIN WIS 75 MAI MIC NWH NWY VER. Hemicr. or rhizome geophyte.
Carex depauperata Hornem., Fors. Oecon. Plantel., ed. 2: 829 (1806), nom. illeg.
Carex mielichhoferi Sm., Engl. Bot.: t. 2293 (1811), nom. illeg.
Carex panicea var. *sparsiflora* Wahlenb., Fl. Lapp.: 236 (1812). *Carex sparsiflora* (Wahlenb.) Steud., Nomencl. Bot., ed. 2, 1: 296 (1840).
Carex smithii Tausch, Flora 4: 559 (1821).
Carex panicea var. *pauciflora* Wahlenb., Fl. Suec. 2: 602 (1826). *Carex sparsiflora* var. *pauciflora* (Wahlenb.) Andersson, Cyper. Scand.: 34 (1849).
Carex scotica Spreng., Syst. Veg. 3: 819 (1826).
Carex phaeostachya Sm., Engl. Fl. 4: 99 (1828).
Carex tetanica Rchb., Fl. Germ. Excurs.: 63 (1830), nom. illeg.
Carex curvirostra Hartm., Handb. Skand. Fl., ed. 2: 221 (1832). *Carex vaginata* var. *curvirostra* (Hartm.) Nyman, Consp. Fl. Eur.: 775 (1882).
Carex phaecostachya Sm., Engl. Bot., Suppl.: t. 2731 (1835).
Carex sudetica C.Presl ex Kunth, Enum. Pl. 2: 451 (1837), nom. illeg.
Carex triflora Vahl ex Kunth, Enum. Pl. 2: 460 (1837), nom. illeg.
Carex sparsiflora var. *borealis* Andersson, Cyper. Scand.: 34 (1849).
Carex vaginata var. *altocaulis* Dewey, Amer. J. Sci. Arts, II, 41: 227 (1866). *Carex sparsiflora* subsp. *altocaulis* (Dewey) Britton in N.L.Britton & A.Brown, Ill. Fl. N. U.S. 1: 326 (1896).
Carex nivicola G.Don ex Boott, Ill. Carex 4: 149 (1867).
Carex saltuensis L.H.Bailey, Mem. Torrey Bot. Club 1: 7 (1889).
Carex sparsiflora var. *distracta* Norman, Fl. Arct. Norv. Sp.: 51 (1893).
Carex altocaulis Britton in N.L.Britton & A.Brown, Ill. Fl. N. U.S. 1: 326 (1896).
Carex sparsiflora var. *grutteri* Asch. & Graebn., Syn. Mitteleur. Fl. 2: 144 (1902).

Carex valbrayi H.Lév., Bull. Acad. Int. Géogr. Bot. 27: 6 (1917).
E. Madagascar. 29 MDG. Hemicr. or rhizome geophyte.

Carex vallicola Dewey, Amer. J. Sci. Arts, II, 32: 40 (1861).
W. Canada, W. U.S.A. to South Dakota, C. Mexico (to Hidalgo). 71 BRC 73 COL IDA MNT ORE WAS WYO 74 SDA 76 ARI CAL NEV UTA 77 NWM TEX 79 MXC MXE. Hemicr. or rhizome geophyte.
Carex phaeolepis T.Holm, Amer. J. Sci. 166: 454 (1903).
Carex vagans T.Holm, Amer. J. Sci. 166: 454 (1903).
Carex brevisquama Mack., Bull. Torrey Bot. Club 34: 152 (1907).
Carex rusbyi Mack., Smithsonian Misc. Collect. 65(7): 2 (1915). *Carex vallicola* var. *rusbyi* (Mack.) F.J.Herm., Rhodora 70: 419 (1968).
Carex vallicola var. *hidalgensis* F.J.Herm., Brittonia 12: 78 (1960).

Carex vallis-pulchrae Phil., Anales Univ. Chile 93: 487 (1896).
Chile to Falkland Is. 85 AGS AGW CLC CLS 90 FAL. Hemicr. or rhizome geophyte.

var. ***barrosiana*** G.A.Wheeler, Syst. Bot. 14: 178 (1989).
S. Chile to Falkland Is. 85 AGS CLS 90 FAL. Hemicr. or rhizome geophyte.

var. ***vallis-pulchrae***
C. Chile to W. Argentina. 85 AGS AGW CLC. Hemicr. or rhizome geophyte.
Carex castellanosii Barros, Lilloa 23: 415 (1950).

Carex vallis-rosetto K.Schum. in H.G.A.Engler (ed.), Pflanzenr., IV, 20(38): 130 (1909).
E. Trop. Africa (Mts.). 25 KEN TAN. Hemicr. or rhizome geophyte.
Carex vallis-rosetto var. *purpurea* Kük. in H.G.A. Engler (ed.), Pflanzenr., IV, 20(38): 647 (1909).
Carex greenwayi Nelmes, Bull. Misc. Inform. Kew 1938: 244 (1938).

Carex van-heurckii Müll.Arg., Observ. Bot. 1: 30 (1870).
Siberia to N. & C. Japan. 30 BRY CTA IRK YAK 31 AMU KAM KHA KUR MAG PRM SAK 36 CHI CHM? 37 MON 38 JAP. Hemicr. or rhizome geophyte.

subsp. ***crassispiculata*** (Malyschev) Malyschev, in Fl. Tsentral'noi Sibiri 1: 186 (1979).
Siberia. 30+. Hemicr. or rhizome geophyte.
**Carex van-heurckii* var. *crassispiculata* Malyschev in ?

subsp. ***van-heurckii***
Siberia to N. & C. Japan. 30 BRY CTA IRK YAK 31 AMU KAM KHA KUR MAG PRM SAK 36 CHI CHM? 37 MON 38 JAP. Hemicr. or rhizome geophyte.
Carex amblyolepis Trautv. & C.A.Mey., Fl. Ochot.: 99 (1856), provisional synonym. *Carex pensylvanica* var. *amblyolepis* (Trautv. & C.A. Mey.) Kük., Finska Vetensk. Soc. Forh. 45(8): 8 (1903).
Carex pseudowrightii Honda, Bot. Mag. (Tokyo) 43: 291 (1929).

Carex ventosa C.B.Clarke in T.F.Cheeseman, Man. New Zealand Fl.: 832 (1906).
Chatham Is. 51 CTM. Hemicr. or rhizome geophyte.
Carex mcclurgii Petrie, Trans. & Proc. New Zealand Inst. 56: 8 (1926).

Carex venusta Dewey, Amer. J. Sci. Arts 26: 107 (1834).
E. U.S.A. 75 NWJ NWY 78 ALA DEL FLA GEO LOU MRY MSI NCA SCA TEN VRG WDC. Hemicr. or rhizome geophyte.
Carex oblita Steud., Syn. Pl. Glumac. 2: 231 (1855). *Carex venusta* var. *oblita* (Steud.) Kük. in H.G.A. Engler (ed.), Pflanzenr., IV, 20(38): 597 (1909).
Carex glabra Boott, Ill. Carex 2: 93 (1860).
Carex venusta var. *minor* Boeck., Linnaea 41: 255 (1877).
Carex venusta var. *glabra* L.H.Bailey, Proc. Amer. Acad. Arts 22: 105 (1886).

Carex vernacula L.H.Bailey, Bull. Torrey Bot. Club 20: 417 (1893). *Carex foetida* var. *vernacula* (L.H.Bailey) Kük. in H.G.A.Engler (ed.), Pflanzenr., IV, 20(38): 115 (1909).
NW. U.S.A. to California. 73 COL IDA ORE WAS WYO 76 CAL. Hemicr. or rhizome geophyte.
Carex vernacula var. *hobsonii* Maguire, Brittonia 5: 199 (1944).

Carex verrucosa Muhl., Descr. Gram.: 261 (1817).
SE. U.S.A. to Texas. 77 TEX 78 ALA FLA GEO LOU NCA SCA. Hemicr. or rhizome geophyte.
Carex macrokolea Steud., Syn. Pl. Glumac. 2: 223 (1855). *Carex glaucescens* f. *macrokolea* (Steud.) Kük. in H.G.A.Engler (ed.), Pflanzenr., IV, 20(38): 733 (1909).
Carex glaucescens var. *androgyna* Boott, Ill. Carex 2: 89 (1860).

Carex verticillata Zoll. & Moritzi in H.Zollinger, Syst. Verz.: 98 (1854). *Carex hypsophila* var. *verticillata* (Zoll. & Moritzi) Kük. in H.G.A.Engler (ed.), Pflanzenr., IV, 20(38): 547 (1909), nom. illeg.
Indo-China, W. Malesia. 41 LAO VIE 42 BOR JAW MLY SUM. Hemicr. or rhizome geophyte.
Carex hypsophila Miq., Fl. Ned. Ind. 3: 354 (1857).
Carex tartarea Ridl., J. Bot. 21: 262 (1883).
Carex havilandii C.B.Clarke, J. Linn. Soc., Bot. 37: 13 (1904). *Carex hypsophila* var. *havilandii* (C.B. Clarke) Kük. in H.G.A.Engler (ed.), Pflanzenr., IV, 20(38): 546 (1909).
Carex sumatrensis C.B.Clarke, J. Linn. Soc., Bot. 37: 13 (1904).
Carex havilandii C.B.Clarke, Kew Bull., Addit. Ser. 8: 74 (1908), nom. illeg.
Carex phacelostachys Nelmes, Kew Bull. 5: 195 (1950).

Carex vesca C.B.Clarke ex Kük., Verh. Bot. Vereins Prov. Brandenburg 47: 206 (1905 publ. 1906).
Brazil (Rio de Janeiro). 84 BZL. Hemicr. or rhizome geophyte.

Carex vesicaria L., Sp. Pl.: 979 (1753). *Trasus vesicarius* (L.) Gray, Nat. Arr. Brit. Pl. 2: 59 (1821).
NW. Africa, Europe to Japan, Canada to W. U.S.A. 10 DEN FIN GRB IRE NOR SWE 11 AUT BGM CZE GER HUN NET POL SWI 12 COR FRA SPA 13 BUL ITA ROM SIC YUG 14 BLR BLT KRY RUC RUE RUN RUS RUW UKR 20 ALG 30 ALT WSB YAK 32 KAZ KGZ TZK 33 NCS TCS 34 IRN TUR 36 CHI CHM 37 MON 38 JAP KOR 71 BRC MAN 72 LAB NBR NFL NSC ONT QUE 73 MNT ORE WAS WYO 74 MIN WIS 75 CNT INI MAI MAS MIC NWH NWY OHI PEN RHO VER WVA 76 CAL UTA 78 KTY VRG. Hemicr. or rhizome geophyte.
Carex vesicaria var. *brachystachys* Uechtr. ex Asch. & Graebn. in ?
Carex vesicaria subsp. *saxatilis* L., Sp. Pl.: 976 (1753).
Carex pendula Moench, Methodus: 324 (1794), nom. illeg.
Carex vesicaria var. *calcifoenum* Laest., Loca Parall.: 88 (1831).
Carex vesicaria var. *subalpina* Laest., Loca Parall.: 88 (1831).
Carex suilla J.Fellm., Bull. Soc. Imp. Naturalistes Moscou 8: 285 (1835).
Carex vesicaria var. *alpigena* Fr., Novit. Fl. Suec. Mant. 3: 142 (1845).
Carex monile Tuck. ex Boott, London J. Bot. 5: 69 (1846). *Carex vesicaria* var. *monile* (Tuck. ex Boott) Tuck. ex Boott, Linnaea 41: 320 (1877).
Carex vesicaria var. *dichroa* Andersson, Cyper. Scand.: 18 (1849).
Carex vesicaria f. *elatior* Andersson, Bot. Not. 1894: (1849).
Carex vesicaria var. *robusta* Sond., Fl. Hamburg.: 506 (1850).

Carex raeana Boott in J.Richardson, Arct. Search. Exped. 2: 344 (1851). *Carex monile* var. *raeana* (Boott) L.H.Bailey, Mem. Torrey Bot. Club 1: 39 (1889). *Carex vesicaria* var. *raeana* (Boott) Fernald, Rhodora 3: 54 (1901). *Carex miliaris* var. *raeana* (Boott) Kük. in H.G.A.Engler (ed.), Pflanzenr., IV, 20(38): 719 (1909).
Carex schrenkiana C.A.Mey. ex Ledeb., Fl. Ross. 4: 317 (1852).
Carex vesicaria f. *brachystachys* Lindeb., Bot. Not. 1855: 12 (1855).
Carex bongardiana C.A.Mey., Fl. Ochot.: 101 (1856).
Carex vaseyi Dewey, Amer. J. Sci. Arts, II, 29: 347 (1860).
Carex pulla f. *dichroa* A.Blytt, Norges Fl. 1: 212 (1861).
Carex pulla f. *polystachya* A.Blytt, Norges Fl. 1: 212 (1861). *Carex vesicaria* f. *polystachya* (A.Blytt) Kük. in H.G.A.Engler (ed.), Pflanzenr., IV, 20(38): 728 (1909).
Carex vesicaria var. *latifolia* A.Blytt, Norges Fl. 1: 252 (1861).
Carex vesicaria f. *pendulina* A.Blytt, Norges Fl. 1: 262 (1861).
Carex vesicaria var. *globosa* Olney, Proc. Amer. Acad. Arts 8: 408 (1872). *Carex exsiccata* var. *globosa* (Olney) L.H.Bailey, Mem. Torrey Bot. Club 1: 7 (1889). *Carex vesicaria* f. *globosa* (Olney) Kük. in H.G.A.Engler (ed.), Pflanzenr., IV, 20(38): 727 (1909).
Carex vesicaria var. *lanceolata* Olney, Proc. Amer. Acad. Arts 8: 407 (1872).
Carex pulla f. *pedunculata* Kjellm., Vega Exped. Vet. Jaktag. 1: 560 (1882).
Carex salina var. *ambusta* L.H.Bailey, Carex Cat.: 1 (1884).
Carex monile var. *obtusisquamis* (L.H.Bailey) L.H.Bailey, Proc. Amer. Acad. Arts 22: 68 (1886).
Carex vesicaria f. *aurea* Lange, Haandb. Danske Fl., ed. 4: 149 (1886).
Carex pulla var. *sibirica* H.Christ, Kongl. Vetensk. Acad. Handl. 22: 181 (1887).
Carex exsiccata var. *pungens* L.H.Bailey, Mem. Torrey Bot. Club 1: 7 (1889). *Carex vesicaria* f. *pungens* (L.H.Bailey) Kük. in H.G.A.Engler (ed.), Pflanzenr., IV, 20(38): 727 (1909).
Carex monile var. *colorata* L.H.Bailey, Mem. Torrey Bot. Club 1: 39 (1889). *Carex vesicaria* var. *colorata* (L.H.Bailey) Kük. in H.G.A.Engler (ed.), Pflanzenr., IV, 20(38): 726 (1909).
Carex monile var. *monstrosa* L.H.Bailey, Mem. Torrey Bot. Club 1: 40 (1889). *Carex vesicaria* f. *monstrosa* (L.H.Bailey) Kük. in H.G.A.Engler (ed.), Pflanzenr., IV, 20(38): 726 (1909).
Carex monile var. *monstrosa* L.H.Bailey, Mem. Torrey Bot. Club 1: 40 (1889). *Carex vesicaria* var. *monstrosa* (L.H.Bailey) Farw., Rep. (Annual) Michigan Acad. Sci. 6: 204 (1904).
Carex monile var. *pacifica* L.H.Bailey, Proc. Calif. Acad. Sci., II, 3: 105 (1891). *Carex vesicaria* var. *pacifica* (L.H.Bailey) Kük. in H.G.A.Engler (ed.), Pflanzenr., IV, 20(38): 726 (1909).
Carex vesicaria var. *tenuistachya* Kük., Bot. Centralbl. 77: 58 (1899).
Carex udensis Meinsh., Trudy Imp. S.-Peterburgsk. Bot. Sada 18: 366 (1900). *Carex drymophila* var. *udensis* (Meinsh.) Kük. in H.G.A.Engler (ed.), Pflanzenr., IV, 20(38): 755 (1909).
Carex vesicata Meinsh., Trudy Imp. S.-Peterburgsk. Bot. Sada 18: 367 (1900). *Carex vesicaria* subsp. *vesicata* (Meinsh.) T.V.Egorova, Novosti Sist. Vyssh. Rast. 10: 103 (1973).
Carex vesicaria f. *hexasticha* Hoschede, Bull. Acad. Int. Géogr. Bot. 12: 205 (1903).
Carex vesicaria f. *gracillima* Kük. in H.G.A.Engler (ed.), Pflanzenr., IV, 20(38): 727 (1909).
Carex vesicaria f. *umbrosa* Kük. in H.G.A.Engler (ed.), Pflanzenr., IV, 20(38): 725 (1909).
Carex vesicaria f. *fluitans* Raymond, Naturaliste Canad. 77: 65 (1950).

Carex vesiculosa Boott, Ill. Carex 3: 107 (1862).
Nepal to Lesser Sunda Is. 40 ASS EHM NEP 41 THA 42 JAW LSI. Hemicr. or rhizome geophyte.
Carex diffusa Boott ex C.B.Clarke in J.D.Hooker, Fl. Brit. India 6: 717 (1894).
Carex vesiculosa var. *paniculata* C.B.Clarke in J.D.Hooker, Fl. Brit. India 6: 717 (1894).
Carex gembolensis C.B.Clarke, J. Linn. Soc., Bot. 37: 10 (1904).
Carex gembolensis C.B.Clarke, Kew Bull., Addit. Ser. 8: 72 (1908).
Carex vesiculosa var. *congesta* Kük. in H.G.A.Engler (ed.), Pflanzenr., IV, 20(38): 183 (1909).
Carex vesiculosa f. *pallida* Kük. in H.G.A.Engler (ed.), Pflanzenr., IV, 20(38): 283 (1909). *Carex vesiculosa* var. *pallida* (Kük.) R.C.Srivast., Novon 8: 203 (1998).

Carex vestita Willd., Sp. Pl. 4: 268 (1805). *Loxotrema vestita* (Willd.) Raf., Good Book: 25 (1840).
E. U.S.A. 75 CNT MAI MAS NWH NWJ NWY PEN RHO 78 ALA DEL MRY NCA TEN VRG. Hemicr. or rhizome geophyte.
Carex vestita var. *kennedyi* Fernald, Rhodora 2: 170 (1900).

Carex vexans F.J.Herm., Rhodora 57: 156 (1955).
Florida. 78 FLA. Hemicr. or rhizome geophyte.

Carex vicinalis Boott, Ill. Carex 4: 133 (1867).
S. India. 40 IND. Hemicr. or rhizome geophyte.

Carex vietnamica Raymond, Mém. Jard. Bot. Montréal 53: 32 (1959).
Vietnam. 41 VIE. Hemicr. or rhizome geophyte.

Carex virescens Muhl. ex Willd., Sp. Pl. 4: 251 (1805). *Olamblis virescens* (Muhl. ex Willd.) Raf., Good Book: 26 (1840).
E. Canada, NC. & E. U.S.A. 72 ONT QUE 74 MSO 75 CNT INI MAI MAS MIC NWH NWJ NWY OHI PEN RHO WVA 78 ARK DEL GEO KTY MRY MSI NCA SCA TEN VRG WDC. Hemicr. or rhizome geophyte.
Carex costata Schwein., Ann. Lyceum Nat. Hist. New York 1: 67 (1824), nom. illeg. *Carex virescens* var. *costata* Dewey, Amer. J. Sci. Arts 9: 260 (1825).
Carex virescens var. *enormis* Farw., Amer. Midl. Naturalist 12: 51 (1930).

Carex viridimarginata Kük., Acta Horti Gothob. 5: 113 (1930).
China (Shanxi). 36 CHN. Hemicr. or rhizome geophyte.

Carex viridula Michx., Fl. Bor.-Amer. 2: 170 (1803). *Edritria viridula* (Michx.) Raf., Good Book: 26 (1840). *Carex flava* var. *viridula* (Michx.) L.H.Bailey, Mem. Torrey Bot. Club 1: 31 (1889). *Carex oederi* var. *viridula* (Michx.) Kük. in H.G.A.Engler (ed.), Pflanzenr., IV, 20(38): 674 (1909). *Carex oederi* subsp. *viridula* (Michx.) Hultén, Ark. Bot., n.s., 7: 30

(1968). *Carex flava* subsp. *viridula* (Michx.) O.Bolòs & Vigo, Fl. Man. Paisos Catalans 4: 305 (2001).

Temp. & Subarctic Northern Hemisphere. 10 DEN FIN FOR GRB ICE IRE NOR SWE 11 AUT BGM CZE GER HUN NET POL SWI 12 COR FRA POR SAR SPA 13 ALB BUL GRC ITA ROM SIC YUG 14 BLR BLT RUC RUN RUW UKR 21 AZO MDR 30 WSB 31 KAM KUR SAK 32 KAZ KGZ UZB 33 NCS 34 AFG IRN 38 JAP 40 EHM NEP WHM (50) tas (51) nzn 70 ASK GNL NUN NWT YUK 71 ABT BRC MAN SAS 72 LAB NBR NFL NSC ONT PEI QUE 73 COL IDA MNT ORE WAS WYO 74 ILL MIN NDA WIS 75 CNT INI MAI MAS MIC NWH NWJ NWY OHI PEN RHO VER 76 CAL NEV UTA 77 NWM. Hemicr. or rhizome geophyte.

subsp. ***brachyrhyncha*** (Èelak.) B.Schmid, Watsonia 14: 317 (1983).

Europe. 11 CZE SWI 12 FRA. Hemicr. or rhizome geophyte.

**Carex flava* var. *brachyrhyncha* Èelak., Prodr. Fl. Böhmen: 71 (1867).

subsp. ***cedercreutzii*** (Fagerstr.) B.Schmid, Watsonia 14: 317 (1983).

Azores. 21 AZO. Hemicr. or rhizome geophyte.

**Carex tumidicarpa* subsp. *cedercreutzii* Fagerstr., Acta Soc. Fauna Fl. Fenn. 79(3): 3 (1967).

var. ***elatior*** (Schltr.) Crins, Canad. J. Bot. 67: 1058 (1989).

E. Canada to Europe. 10 DEN FIN GRB? IRE? NOR SWE 11 CZE GER 12 FRA 13 ALB BUL GRC ITA KRI ROM SIC TUE YUG 14 BLR BLT RUC RUN UKR 72 LAB NBR NFL NSC QUE. Hemicr. or rhizome geophyte.

**Carex flava* var. *elatior* Schltr., Fl. Berol. 1: 477 (1823). *Carex oederi* f. *elatior* (Schltr.) Andersson, Cyper. Scand.: 25 (1849).

Carex lepidocarpa Tausch, Flora 17: 179 (1834). *Carex flava* var. *lepidocarpa* (Tausch) Godr., Fl. Lorraine 3: 118 (1844). *Carex flava* subsp. *lepidocarpa* (Tausch) Nyman, Consp. Fl. Eur.: 771 (1882). *Carex viridula* var. *lepidocarpa* (Tausch) B.Schmid, Watsonia 14: 317 (1983).

Carex flava f. *laevigata* Peterm., Flora 27: 338 (1844).

Carex flava var. *minor* Peterm., Flora 27: 338 (1844), nom. illeg.

Carex lepidocarpa f. *laevigata* Peterm., Flora 27: 338 (1844).

Carex flava var. *intermedia* Coss. & Germ., Fl. Descr. Anal. Paris: 602 (1845). *Carex lepidocarpa* f. *intermedia* (Coss. & Germ.) Asch. & Graebn., Syn. Mitteleur. Fl. 2(2): 200 (1903).

Carex marssonii Auersw., Bot. Zeitung (Berlin) 10: 409 (1852). *Carex flava* var. *marssonii* (Auersw.) Nyman, Consp. Fl. Eur.: 771 (1882).

Carex oederi var. *major* Boeck., Verh. Bot. Vereins Prov. Brandenburg 3(4): 285 (1862).

Carex lepidocarpa f. *laeviculmis* Kneuck., Allg. Bot. Z. Syst. 5: 10 (1899).

Carex lepidocarpa f. *major* Kneuck., Allg. Bot. Z. Syst. 5: 10 (1899).

Carex lepidocarpa var. *pseudolepidocarpa* Kneuck., Allg. Bot. Z. Syst. 5: 9 (1899).

Carex lepidocarpa f. *rectirostris* Kneuck., Allg. Bot. Z. Syst. 5: 9 (1899).

Carex lepidocarpa var. *cylindrostachys* Neuman, Sver. Fl.: 696 (1901).

Carex lepidocarpa var. *sphaerostachys* Neuman, Sver. Fl.: 696 (1901).

Carex lepidocarpa var. *nelmesiana* Raymond, Bull. Soc. Bot. France 99: 194 (1952). *Carex flava* var. *nelmesiana* (Raymond) Boivin, Naturaliste Canad. 94: 522 (1967).

var. ***jemtlandica*** (Palmgr.) Blackst. & P.A.Ashton, Watsonia 23: 515 (2001).

N. & NE. Europe. 10 FIN GRB? IRE? NOR SWE 11 SWI? 14 BLT RUN. Hemicr.

**Carex lepidocarpa* subsp. *jemtlandica* Palmgr. in C.A.M.Lindman, Sv. Fanerogamfl., ed. 2: 153 (1926). *Carex jemtlandica* (Palmgr.) Palmgr., Memoranda Soc. Fauna Fl. Fenn. 13: 126 (1937). *Carex flava* subsp. *jemtlandica* (Palmgr.) P.D.Sell in P.D.Sell & G.Murrell, Fl. Great Britain Ireland 5: 363 (1996).

subsp. ***oedocarpa*** (Andersson) B.Schmid, Watsonia 14: 316 (1983).

Macaronesia, Europe to C. Asia. 10 DEN FIN FOR GRB ICE IRE NOR SWE 11 AUT BGM CZE GER HUN NET POL SWI 12 COR FRA POR SPA 13 ALB? BUL GRC ITA ROM SIC YUG 14 BLR BLT RUC RUN RUW UKR 20 MOR 21 AZO MDR 30 WSB 32+ 33 NCS 34 AFG IRN 40 WHM (50) tas (51) nzn (72) lab nfl nsw que (75) cnt nwj. Hemicr. or rhizome geophyte.

Carex demissa Hornem. in G.C.Oeder & al. (eds.), Fl. Dan.: t. 1342 (1808). *Carex flava* f. *demissa* (Hornem.) Kük. in H.G.A.Engler (ed.), Pflanzenr., IV, 20(38): 672 (1909). *Carex flava* subsp. *demissa* (Hornem.) O.Bolòs, Masalles & Vigo, Collect. Bot. (Barcelona) 17: 95 (1987 publ. 1988).

**Carex oederi* subsp. *oedocarpa* Andersson, Cyper. Scand.: 25 (1849). *Carex oederi* f. *oedocarpa* (Andersson) Kük. in H.G.A.Engler (ed.), Pflanzenr., IV, 20(38): 674 (1909). *Carex flava* subsp. *oedocarpa* (Andersson) P.D.Sell in P.D.Sell & G.Murrell, Fl. Great Britain Ireland 5: 363 (1996).

Carex tumidicarpa Andersson, Bot. Not. 1849: s.p. (1849).

Carex demissa subsp. *iranica* Kukkonen, Ann. Bot. Fenn. 21: 385 (1984).

var. ***saxilittoralis*** (A.Robertson) Crins, Canad. J. Bot. 67: 1061 (1989).

E. Canada. 72 LAB NFL NSC. Hemicr. or rhizome geophyte.

**Carex saxilittoralis* A.Robertson, Rhodora 82: 370 (1980).

subsp. ***viridula***

Temp. & Subarctic Northern Hemisphere. 10 DEN FIN FOR GRB ICE IRE NOR SWE 11 AUT BGM CZE GER HUN NET POL SWI 12 COR FRA POR SAR SPA 13 ALB BUL GRC ITA ROM SIC YUG 14 BLR BLT RUC RUN RUW UKR 21 AZO 30 WSB 31 KAM KUR SAK 32 KAZ KGZ UZB 38 JAP 40 EHM NEP (50) tas (51) nzn 70 ASK GNL NUN NWT YUK 71 ABT BRC MAN SAS 72 LAB NBR NFL NSC ONT PEI QUE 73 COL IDA MNT ORE WAS WYO 74 ILL MIN NDA WIS 75 CNT INI MAI MAS MIC NWH NWJ NWY OHI PEN RHO VER 76 CAL NEV UTA 77 NWM. Hemicr. or rhizome geophyte.

Carex oederi Ehrh., Beitr. Naturk. 6: 83 (1791), nom. illeg.

Carex serotina Mérat, Nouv. Fl. Env. Paris, ed. 2, 2: 54 (1821). *Carex flava* var. *serotina* (Mérat) Lej., Rev. Fl. Spa: 194 (1824). *Carex oederi* var. *serotina* (Mérat) Nyman, Consp. Fl. Eur.: 771 (1882). *Carex viridula* subsp. *serotina* (Mérat) Malyschev, in Fl. Sibir., Cyperac. 3: 130 (1990). *Carex flava* subsp. *serotina* (Mérat) P.D.Sell in P.D.Sell & G.Murrell, Fl. Great Britain Ireland 5: 363 (1996).

Carex irregularis Schwein., Ann. Lyceum Nat. Hist. New York 1: 66 (1824).

Carex subglobosa Miel., Flora 22: 256 (1839). *Carex oederi* f. *subglobosa* (Miel.) Storch, Skizz. Naturh. Herzogth. Salzburg: 76 (1857). *Carex oederi* var. *subglobosa* (Miel.) Nyman, Consp. Fl. Eur.: 771 (1882). *Carex serotina* f. *subglobosa* (Miel.) Soó, Acta Bot. Acad. Sci. Hung. 16: 371 (1970 publ. 1971).

Carex oederi subsp. *pulchella* Lönnr., Obs. Crit. Pl. Suec.: 24 (1854). *Carex serotina* subsp. *pulchella* (Lönnr.) Ooststr. in H.Heukels & W.H.Wachter, Bekn. Schoolfl. Nederl., ed. 7: 319 (1949). *Carex viridula* var. *pulchella* (Lönnr.) B.Schmid, Watsonia 14: 316 (1983). *Carex viridula* subsp. *pulchella* (Lönnr.) Malyschev, in Fl. Sibir., Cyperac. 3: 130 (1990). *Carex flava* subsp. *pulchella* (Lönnr.) P.D.Sell in P.D.Sell & G.Murrell, Fl. Great Britain Ireland 5: 363 (1996). *Carex oederi* var. *pulchella* (Lönnr.) Hedrén & Lassen, Nordic J. Bot. 22: 262 (2002 publ. 2003).

Carex oederi var. *exigua* Schur, Enum. Pl. Transsilv.: 711 (1866). *Carex serotina* f. *exigua* (Schur) Soó, Acta Bot. Acad. Sci. Hung. 16: 371 (1970 publ. 1971).

Carex oederi f. *cyperoides* T.Marsson, Fl. Neu-Vorpommern: 537 (1869). *Carex serotina* f. *cyperoides* (T.Marsson) Soó, Acta Bot. Acad. Sci. Hung. 16: 371 (1970 publ. 1971).

Carex urbanii Boeck., Bot. Jahrb. Syst. 7: 280 (1885).

Carex oederi f. *intermedia* Dudley, Cornell Univ. Sci. Bull. 2: 117 (1886). *Carex viridula* f. *intermedia* (Dudley) F.J.Herm., Fl. Indiana: 256 (1940).

Carex flava var. *recterostrata* L.H.Bailey, Bot. Gaz. 13: 84 (1888). *Carex oederi* f. *recterostrata* (L.H.Bailey) Kük. in H.G.A.Engler (ed.), Pflanzenr., IV, 20(38): 674 (1909). *Carex oederi* var. *recterostrata* (L.H.Bailey) Dorn, Vasc. Pl. Wyoming: 296 (1988).

Carex oederi f. *canaliculata* Callmé, Deutsche Bot. Monatsschr. 6: 3 (1888). *Carex serotina* f. *canaliculata* (Callmé) Soó, Acta Bot. Acad. Sci. Hung. 16: 371 (1970 publ. 1971).

Carex oederi f. *alpestris* Asch. & Graebn., Syn. Mitteleur. Fl. 2(2): 201 (1903). *Carex serotina* f. *alpestris* (Asch. & Graebn.) Soó, Acta Bot. Acad. Sci. Hung. 16: 371 (1970 publ. 1971).

Carex oederi f. *thalassica* Asch. & Graebn., Syn. Mitteleur. Fl. 2(2): 202 (1903). *Carex serotina* f. *thalassica* (Asch. & Graebn.) Soó, Acta Bot. Acad. Sci. Hung. 16: 371 (1970 publ. 1971).

Carex oederi var. *pumila* (Coss. & Germ.) Fernald, Rhodora 8: 201 (1906).

Carex oederi f. *tenuis* Zapal., Consp. Fl. Gallic. Crit. 1: 114 (1906). *Carex serotina* f. *tenuis* (Zapal.) Soó, Acta Bot. Acad. Sci. Hung. 16: 371 (1970 publ. 1971).

Carex oederi var. *rousseauiana* Vict., Contr. Lab. Bot. Univ. Montréal 15: 262 (1929). *Carex viridula* f. *rousseauiana* (Vict.) Raymond, Naturaliste Canad. 77: 63 (1950).

Carex philocrena V.I.Krecz., Trudy Sredne-Aziatsk. Gosud. Univ., Ser. 8b, Bot. 17: 75 (1934). *Carex serotina* subsp. *philocrena* (V.I.Krecz.) Kukkonen, Ann. Bot. Fenn. 21: 387 (1984).

Carex chlorophila Mack. in N.L.Britton & al. (eds.), N. Amer. Fl. 18: 304 (1935).

Carex oederoides Tatew. & Akiyama, Trans. Sapporo Nat. Hist. Soc. 14: 107 (1935).

Carex kotilainii Palmgr., Memoranda Soc. Fauna Fl. Fenn. 19: 89 (1944).

Carex viridula f. *clandestina* Raymond, Naturaliste Canad. 77: 63 (1950).

Carex scandinavica E.W.Davies, Watsonia 3: 66 (1953).

Carex oederi subsp. *fennica* Palmgr., Commentat. Biol. 20(3): 8 (1958). *Carex serotina* var. *fennica* (Palmgr.) T.V.Egorova, Sedges Russia & Adj. States: 278 (1999).

Carex viridula f. *pygmaea* Lepage, Naturaliste Canad. 89: 115 (1962).

Carex viridula var. *czatkalensis* Yu.E.Alexeev, Byull. Moskovsk. Obshch. Isp. Prir., Otd. Biol. 98: 126 (1992).

Carex vixdentata (Kük. ex Barros) G.A.Wheeler, Aliso 12: 98 (1988).
NE. & SC. Argentina, Uruguay. 85 AGE AGS URU. Hemicr. or rhizome geophyte.
Carex extensa var. *vixdentata* Kük. ex Barros, Physis (Buenos Aires) 9: 392 (1939).

Carex vorobjevii A.E.Kozhevn., Byull. Moskovsk. Obshch. Isp. Prir., Otd. Biol. 92: 127 (1987).
Russian Far East. 31+. Hemicr. or rhizome geophyte.

Carex vulcani Hochst. ex Seub., Fl. Azor.: 22 (1840).
Azores. 21 AZO. Hemicr. or rhizome geophyte.
Carex floresiana Hochst. ex Seub., Fl. Azor.: 22 (1840). *Carex vulcani* f. *floresiana* (Hochst. ex Seub.) Kük. in H.G.A.Engler (ed.), Pflanzenr., IV, 20(38): 604 (1909).
Carex vulcani var. *minor* Boott, Ill. Carex 4: 203 (1867).
Carex watsonii Boott, Ill. Carex 4: 203 (1867), nom. illeg.

Carex vulpina L., Sp. Pl.: 973 (1753). *Vignea vulpina* (L.) Rchb. in J.C.Mössler, Handb. Gewächsk. ed. 2, 3: 1620 (1830). *Edritria vulpina* (L.) Raf., Good Book: 26 (1840). *Carex vulpina* var. *vulgaris* Èelak., Prodr. Fl. Böhmen: 731 (1881), nom. inval.
Europe to NW. China. 10 DEN FIN GRB NOR SWE 11 AUT BGM CZE GER HUN NET POL SWI 12 COR FRA SPA 13 ALB BUL GRC ITA KRI? ROM TUE? YUG 14 BLR BLT KRY RUC RUE RUN RUS RUW UKR 30 WSB 32 KAZ UZB 33 NCS 36 CHX. Hemicr. or rhizome geophyte.
Carex compacta Lam., Fl. Franç. 2: 172 (1779). *Carex vulpina* var. *compacta* (Lam.) Velen., Fl. Bulg.: 576 (1891).
Carex spicata Thuill., Fl. Env. Paris: 480 (1790), nom. illeg.
Carex glomerata Gilib., Excerc. Phyt. 2: 545 (1792).
Carex nemorosa Rebent., Prodr. Fl. Neomarch.: 21 (1804), nom. illeg. *Vignea nemorosa* Rchb. in J.C.Mössler, Handb. Gewächsk. ed. 2, 3: 1620 (1830). *Carex vulpina* var. *nemorosa* Lej. & Courtois, Comp. Fl. Belg. 3: 223 (1836).

Carex muricata Jungh., Linnaea 6: 29 (1831), nom. illeg.
Carex mertensis Weihe ex Kunth, Enum. Pl. 2: 383 (1837).
Carex reflexa D.Dietr. ex Kunth, Enum. Pl. 2: 384 (1837), nom. illeg.
Vignea vulpina f. *capitulata* Peterm., Fl. Lips. Excurs.: 53 (1838). *Carex vulpina* f. *capitulata* (Peterm.) Soó, Acta Bot. Acad. Sci. Hung. 16: 369 (1970 publ. 1971).
Carex vulpina f. *minor* Peterm., Anal. Pfl.-Schlüss.: 490 (1846).
Carex vulpina f. *elongata* Andersson, Cyper. Scand.: 67 (1849).
Carex vulpina var. *tenuior* Ledeb., Fl. Ross. 4: 275 (1852).
Vignea vulpina var. *crassinervis* Schur, Enum. Pl. Transsilv.: 704 (1866). *Carex vulpina* var. *crassinervis* (Schur) Kük. in H.G.A.Engler (ed.), Pflanzenr., IV, 20(38): 171 (1909).
Carex vulpina f. *aristata* Asch., Verh. K. K. Zool.-Bot. Ges. Wien 17: 570 (1867).
Carex vulpina var. *remotiflora* Lange, Bot. Tidsskr. 2: 41 (1868).
Carex vulpina var. *divulsa* Èelak., Prodr. Fl. Böhmen: 731 (1881).
Carex vulpina var. *longibracteata* Beck, Ann. Wiener Mus. Naturgesch. 2: 61 (1887).
Carex vulpina var. *stribnyi* Velen., Fl. Bulg., Suppl. 1: 283 (1898).
Carex vulpina var. *pallidior* Meinsh., Trudy Imp. S.-Peterburgsk. Bot. Sada 18: 323 (1901).
Carex vulpina f. *laeviuscula* Sanio ex Asch. & Graebn., Syn. Mitteleur. Fl. 2(2): 37 (1902).
Carex vulpina var. *litoralis* Nolte in P.F.A.Ascherson & K.O.R.Graebner, Syn. Mitteleur. Fl. 2(2): 36 (1902).

Carex vulpinaris Nees in R.Wight, Contr. Bot. India: 121 (1834). *Vignea vulpinaris* (Nees) Soják, Cas. Nár. Mus., Odd. Prír. 148: 197 (1979 publ. 1980).
Afghanistan to W. Himalaya. 34 AFG 40 PAK WHM. Hemicr. or rhizome geophyte.
Carex henningsiana Boeck., Beitr. Cyper. 1: 42 (1888).
Carex vulpinaris f. *angustifolia* Kük. in H.G.A.Engler (ed.), Pflanzenr., IV, 20(38): 117 (1909).
Carex wendelboi Nelmes, Nytt Mag. Bot. 1: 20 (1952). *Vignea wendelboi* (Nelmes) Soják, Cas. Nár. Mus., Odd. Prír. 148: 197 (1979 publ. 1980).

Carex vulpinoidea Michx., Fl. Bor.-Amer. 2: 169 (1803). *Vignea vulpinoidea* (Michx.) Soják, Cas. Nár. Mus., Odd. Prír. 148: 197 (1979 publ. 1980).
C. & E. Canada to U.S.A., Mexico, introduced elsewere. (10) grb (11) cze net pol swi (12) fra (14) (34) tur (51) nzn 71 ABT BRC MAN SAS 72 LAB NBR NFL NSC ONT PEI QUE 73 COL IDA MNT ORE WAS WYO 74 ILL IOW KAN MIN MSO NDA NEB OKL SDA WIS 75 CNT INI MAI MAS MIC NWH NWJ NWY OHI PEN RHO VER WVA 76 ARI CAL 77 NWM TEX 78 ALA ARK DEL FLA GEO KTY LOU MRY MSI NCA SCA TEN VRG WDC 79 MXG MXN MXS. Hemicr. or rhizome geophyte.
Carex microsperma Wahlenb., Kongl. Vetensk. Acad. Nya Handl. 1803: 144 (1803). *Carex multiflora* var. *microsperma* (Wahlenb.) Dewey, Amer. J. Sci. Arts 11: 317 (1826). *Carex vulpinoidea* var. *microsperma* (Wahlenb.) Dewey, Class-book Bot.: 414 (1845).
Carex multiflora Muhl. ex Willd., Sp. Pl. 4: 243 (1805). *Vignea multiflora* (Muhl. ex Willd.) Rchb. in J.C.Mössler, Handb. Gewächsk. ed. 2, 3: 1621 (1830).
Carex bracteosa Schwein., Ann. Lyceum Nat. Hist. New York 1: 65 (1824).
Carex setacea Dewey, Amer. J. Sci. Arts 9: 61 (1825). *Vignea setacea* (Dewey) Raf., Good Book: 27 (1840). *Carex vulpinoidea* var. *setacea* (Dewey) Kük. in H.G.A.Engler (ed.), Pflanzenr., IV, 20(38): 148 (1909).
Carex vulpinoidea var. *glomerata* Barratt, N. Amer. Caric.: 61 (1840).
Carex vulpiniformis Tuck., Enum. Meth. Caric.: 9 (1843).
Carex moniezii Timb.-Lagr., Bull. Soc. Bot. France 4: 164 (1857).
Carex muhlenbergii Kunth ex Boott, Ill. Carex 3: 125 (1862).
Carex polymorpha Schwein. ex Boott, Ill. Carex 3: 125 (1862), nom. illeg.
Carex scabrior Sartwell ex Boott, Ill. Carex 3: 125 (1862).
Carex vulpinoidea var. *segregata* Farw., Rep. (Annual) Michigan Acad. Sci. 22: 180 (1921). *Carex vulpinoidea* f. *segregata* (Farw.) Raymond, Naturaliste Canad. 70: 278 (1943).
Carex vulpinoidea var. *pycnocephala* F.J.Herm., Rhodora 38: 363 (1936).

Carex wahlenbergiana Boott, Ill. Carex 2: 101 (1860).
Mascarenes. 29 MAU REU. Hemicr. or rhizome geophyte.
Carex indica Wahlenb., Kongl. Vetensk. Acad. Nya Handl. 24: 149 (1803), nom. illeg.
Carex personii Schkuhr ex Kunth, Enum. Pl. 2: 509 (1837), pro syn.
Carex wahlenbergiana var. *hirtisaccus* C.B.Clarke ex Kük. in H.G.A.Engler (ed.), Pflanzenr., IV, 20(38): 276 (1909).

Carex wahuensis C.A.Mey., Mém. Sav. Étr. Acad. St. Pétersbourg 1: 218 (1831).
Hawaiian Is. 63 HAW. Hemicr. or rhizome geophyte.

subsp. ***herbstii*** T.Koyama, Bishop Mus. Occas. Pap. 29: 118 (1989).
Hawaiian Is. (Oahu). 63 HAW. Hemicr. or rhizome geophyte.

subsp. ***rubiginosa*** (R.W.Krauss) T.Koyama, Bishop Mus. Occas. Pap. 29: 119 (1989).
Hawaiian Is. 63 HAW. Hemicr. or rhizome geophyte.
**Carex wahuensis* var. *rubiginosa* R.W.Krauss, Pacific Sci. 4: 257 (1950).

subsp. ***wahuensis***
Hawaiian Is. 63 HAW. Hemicr. or rhizome geophyte.
Carex nuptialis Boott, Ill. Carex 4: 175 (1867).
Carex oahuensis var. *bongardii* Franch. & Sav., Enum. Pl. Jap. 2: 563 (1878).
Carex oahuensis var. *robusta* Franch. & Sav., Enum. Pl. Jap. 2: 563 (1878).
Carex wahuensis var. *meyeri* Franch. & Sav., Enum. Pl. Jap. 2: 563 (1878).
Carex oahuensis var. *meyeri* Franch. & Sav., Enum. Pl. Jap. 2: 563 (1879).
Carex oahuensis Hillebr., Fl. Hawaiian Isl.: 487 (1888).

Carex oahuensis f. *bracteosa* C.B.Clarke, J. Linn. Soc., Bot. 36: 275 (1903).
Carex erythrolepis Kük. in H.G.A.Engler (ed.), Pflanzenr., IV, 20(38): 628 (1909).
Carex oahuensis var. *angustior* Kük. in H.G.A. Engler (ed.), Pflanzenr., IV, 20(38): 632 (1909).

Carex wakatipu Petrie, Trans. & Proc. New Zealand Inst. 14: 363 (1882).
New Zealand South I. 51 NZS. Hemicr. or rhizome geophyte.

Carex ×walasii Ceynowa, Fragm. Florist. Geobot. 38: 47 (1993). *C. atherodes* × *C. hirta*.
Poland. 11 POL. Hemicr. or rhizome geophyte.

Carex walkeri Arn. ex Boott, Proc. Linn. Soc. London 1: 257 (1845).
India, Sri Lanka. 40 IND SRL. Hemicr. or rhizome geophyte.

Carex wawuensis W.M.Chu ex S.Yun Liang, Acta Phytotax. Sin. 37: 192 (1999).
China (Sichuan). 36 CHC. Hemicr. or rhizome geophyte.

Carex wenshanensis L.K.Dai, Acta Phytotax. Sin. 37: 179 (1999).
China (Yunnan). 36 CHC. Hemicr. or rhizome geophyte.

Carex werdermannii L.Gross, Notizbl. Bot. Gart. Berlin-Dahlem 10: 763 (1929).
Chile. 85+. Hemicr. or rhizome geophyte.

Carex whitneyi Olney, Proc. Amer. Acad. Arts 7: 394 (1868).
Oregon to Nevada. 73 ORE 76 CAL NEV. Hemicr. or rhizome geophyte.
Carex pilosiuscula Boeck., Flora 65: 61 (1882).
Carex flaccifolia Mack., Erythea 8: 92 (1922).
Carex jepsonii J.T.Howell, Leafl. W. Bot. 8: 222 (1958).

Carex wiegandii Mack. in N.L.Britton & al. (eds.), N. Amer. Fl. 18: 108 (1931).
E. Canada to NE. U.S.A. 72 LAB NFL NSC ONT PEI QUE 75 MAI MAS MIC NWH NWY PEN VER. Hemicr. or rhizome geophyte.

Carex wightiana Nees in R.Wight, Contr. Bot. India: 122 (1834).
India. 40 IND. Hemicr. or rhizome geophyte.

Carex willdenowii Willd., Sp. Pl. 4: 211 (1805).
SE. Canada to C. & E. U.S.A. 72 ONT 74 ILL MSO 75 CNT INI MAS MIC NWJ NWY OHI PEN VER WVA 77 TEX 78 ALA ARK DEL KTY MRY NCA SCA TEN VRG WDC. Hemicr. or rhizome geophyte.

Carex williamsii Britton, Bull. New York Bot. Gard. 2: 159 (1901). *Carex capillaris* var. *williamsii* (Britton) Boivin, Naturaliste Canad. 94: 522 (1967).
European Russia to Russian Far East, Subarctic America to Canada. 14 RUN 30 WSB 31 KAM MAG 37 MON 70 ASK NUN NWT YUK 71 MAN 72 LAB NFL ONT QUE. Hemicr. or rhizome geophyte.
Carex novograblenovii Kom., Izv. Bot. Sada Akad. Nauk S.S.S.R. 30: 199 (1931 publ. 1932).
Carex peshemskyi Malyschev, Bot. Mater. Gerb. Bot. Inst. Komarova Akad. Nauk S.S.S.R. 21: 462 (1961).

Carex winterbottomii C.B.Clarke in J.D.Hooker, Fl. Brit. India 6: 727 (1894).
Himalaya. 40 NEP WHM. Hemicr. or rhizome geophyte.
Carex setigera var. *fasciculata* Strachey, Cat. Pl. Kumaon: 73 (1854).

Carex ×wolteri L.Gross, Allg. Bot. Z. Syst. 11: 23 (1905). *C. pseudocyperus* × *C. vesicaria*.
Europe. 11 AUT. Hemicr. or rhizome geophyte.

Carex woodii Dewey, Amer. J. Sci. Arts, II, 2: 249 (1846). *Carex tetanica* var. *woodii* (Dewey) L.H.Bailey, Mem. Torrey Bot. Club 1: 53 (1889). *Carex tetanica* f. *woodii* (Dewey) Kük. in H.G.A. Engler (ed.), Pflanzenr., IV, 20(38): 515 (1909).
SE. Canada to E. U.S.A. 72 ONT 74 ILL IOW MIN WIS 75 INI MIC NWY OHI PEN WVA 78 GEO KTY MRY NCA SCA VRG WDC. Hemicr. or rhizome geophyte.
Carex colorata Mack., Bull. Torrey Bot. Club 37: 232 (1910).

Carex wootonii Mack., Smithsonian Misc. Collect. 65(7): 1 (1915).
Arizona to New Mexico and NE. Mexico. 76 ARI 77 NWM 79 MXE. Hemicr. or rhizome geophyte.

Carex woroschilovii A.E.Kozhevn., Byull. Moskovsk. Obshch. Isp. Prir., Otd. Biol. 92: 128 (1987).
Russian Far East. 31+. Hemicr. or rhizome geophyte.

Carex wui W.M.Chu ex L.K.Dai, Acta Phytotax. Sin. 37: 179 (1999).
China (Sichuan, Guizhou). 36 CHC. Hemicr. or rhizome geophyte.

Carex wushanensis S.Yun Liang, Acta Phytotax. Sin. 33: 484 (1995).
China (Gansu). 36 CHN. Hemicr. or rhizome geophyte.

Carex wutuensis K.T.Fu, in Fl. Tsinlingensis 1(1): 448 (1976).
China (Gansu). 36 CHN. Hemicr. or rhizome geophyte.

Carex wuyishanensis S.Yun Liang, Acta Phytotax. Sin. 34: 94 (1996).
China (Fujian). 36 CHS. Hemicr. or rhizome geophyte.

Carex ×xanthocarpa Degl. in J.L.A.Loiseleur-Deslongchamps, Fl. Gall. 2: 299 (1807). *C. flava* × *C. hostiana*.
E. Canada, Europe. 11 GER 12 FRA 72 QUE. Hemicr. or rhizome geophyte.
Carex ×juncea Scop., Fl. Carniol., ed. 2, 2: 227 (1772), provisional synonym.

Carex xerantica L.H.Bailey, Bot. Gaz. 17: 151 (1892). *Carex foenea* var. *xerantica* (L.H.Bailey) Kük. in H.G.A.Engler (ed.), Pflanzenr., IV, 20(38): 205 (1909).
W. & C. Canada to NW. & NC. U.S.A. 71 ABT BRC MAN SAS 72 ONT 73 MNT WYO 74 MIN NDA NEB SDA. Hemicr. or rhizome geophyte.

Carex xiangxiensis Z.P.Wang, Acta Bot. Yunnan. 11: 168 (1989).
China. 36+. Hemicr. or rhizome geophyte.

Carex xiphium Kom., Trudy Imp. S.-Peterburgsk. Bot. Sada 18: 446 (1900).
Russian Far East to NE. China. 31+ 36 CHM. Hemicr. or rhizome geophyte.

Carex yajiangensis Tang & F.T.Wang ex S.Yun Liang, in Fl. Reipubl. Popul. Sin. 12: 524 (2000).
China (Sichuan). 36 CHC. Hemicr.

Carex yamatsutana Ohwi, Acta Phytotax. Geobot. 1: 72 (1932).
Siberia to NE. China. 30+ 31+ 36 CHI CHM. Hemicr. or rhizome geophyte.
Carex diplasiocarpa V.I.Krecz. in V.L.Komarov (ed.), Fl. URSS 3: 590 (1935).

Carex yangshuoensis Tang & F.T.Wang ex S.Y.Liang, Acta Phytotax. Sin. 33: 488 (1995).
China (Guangxi). 36 CHS. Hemicr. or rhizome geophyte.

Carex yonganensis L.K.Dai & Y.Z.Huang, Fl. Fujianica 6: 689 (1995).
China (Fujian). 36 CHS. Hemicr. or rhizome geophyte.

Carex ypsilandrifolia F.T.Wang & Tang, Contr. Inst. Bot. Natl. Acad. Peiping 6: 58 (1949).
SE. China. 36 CHS. Hemicr. or rhizome geophyte.

Carex yuexiensis S.W.Su & S.M.Xu, Bull. Bot. Res., Harbin 8(3): 78 (1988).
China (Anhui). 36 CHS. Hemicr. or rhizome geophyte.

Carex yulungshanensis P.C.Li, Acta Bot. Yunnan. 12: 138 (1990).
China (Yunnan). 36 CHC. Hemicr. or rhizome geophyte.

Carex yunlingensis P.C.Li, Acta Bot. Yunnan. 12: 149 (1990).
China (Yunnan). 36 CHC. Hemicr. or rhizome geophyte.

Carex yunnanensis Franch., Bull. Soc. Philom. Paris, VIII, 7: 31 (1895).
China (Sichuan, Yunnan). 36 CHC. Hemicr. or rhizome geophyte.

Carex yushuensis Y.C.Yang, Acta Phytotax. Sin. 18: 364 (1980).
China. 36+. Hemicr. or rhizome geophyte.

Carex* × *zahnii Kneuck., Deutsche Bot. Monatsschr. 9: 60 (1891). *C. brunnescens* × *C. lachenalii*.
Europe. 11 SWI. Hemicr. or rhizome geophyte.

Carex zekogensis Y.C.Yang, Acta Phytotax. Sin. 18: 362 (1980).
Qinghai. 36 CHQ. Hemicr. or rhizome geophyte.

Carex zhenkangensis Tang & F.T.Wang ex S.Yun Liang, in Fl. Reipubl. Popul. Sin. 12: 523 (2000).
China (Yunnan, Guangxi). 36 CHC CHS. Hemicr.

Carex zhonghaiensis S.Yun Liang, in Fl. Reipubl. Popul. Sin. 12: 526 (2000).
Xinjiang. 36 CHX. Hemicr.

Carex zizaniifolia Raymond, Mém. Jard. Bot. Montréal 53: 36 (1959).
China (Yunnan). 36 CHC. Hemicr. or rhizome geophyte.

Carex zuluensis C.B.Clarke, Kew Bull., Addit. Ser. 8: 74 (1908).
SC. Trop. & S. Africa. 26 MLW 27 CPP LES NAT SWZ TVL. Hemicr. or rhizome geophyte.
Carex huttoniana Kük. in H.G.A.Engler (ed.), Pflanzenr., IV, 20(38): 271 (1909).

Carex zunyiensis Tang & F.T.Wang, Acta Phytotax. Sin. 24: 241 (1986).
S. China. 36 CHC CHS. Hemicr. or rhizome geophyte.

Synonyms:

Carex aa Kom. = ***Carex heleonastes*** Ehrh. ex L.f. subsp. ***heleonastes***
Carex abacta L.H.Bailey = ***Carex michauxiana*** Boeck. subsp. ***michauxiana***
Carex abbreviata Prescott ex Boott = ***Carex torreyi*** Tuck.
Carex abdita E.P.Bicknell = ***Carex umbellata*** Willd.
Carex abjiciens Steud. = ***Carex magellanica*** Lam. subsp. ***magellanica***
Carex ablata L.H.Bailey = ***Carex luzulina*** var. ***ablata*** (L.H.Bailey) F.J.Herm.
Carex ablata var. *luzuliformis* L.H.Bailey = ***Carex fissuricola*** Mack.
Carex abnormis Rouy = ***Carex mucronata*** All.
Carex abramsii Mack. = ***Carex lemmonii*** W.Boott
Carex abscondita var. *glauca* (Chapm.) Fernald = ***Carex abscondita*** Mack.
Carex abscondita var. *rostellata* Fernald = ***Carex abscondita*** Mack.
Carex absconditiformis Fernald = ***Carex laxiculmis*** Schwein. var. ***laxiculmis***
Carex abyssinica Chiov. = ***Carex cognata*** var. ***abyssinica*** (Chiov.) Lye
Carex accedens T.Holm = ***Carex aperta*** Boott
Carex acrifolia V.I.Krecz. = ***Carex pseudofoetida*** subsp. ***acrifolia*** (V.I.Krecz.) Kukkonen
Carex acroandra Schur = ?
Carex acrogyna H.Lév. & Vaniot = ***Carex maculata*** Boott var. ***maculata***
Carex acrolepis Ledeb. = ***Carex gmelinii*** Hook. & Arn.
Carex acrolepis Liebm. = ***Carex polystachya*** Sw. ex Wahlenb. var. ***polystachya***
Carex acrophila S.T.Blake = ***Carex bilateralis*** Hayata
Carex × *acuminata* Rchb. = ***Carex* × *evoluta*** Hartm.
Carex acuminata Willd. = ***Carex flacca*** Schreb. subsp. ***flacca***
Carex acuta amblylepis Peterm. = ***Carex acuta*** L.
Carex acuta var. *amblylepis* (Peterm.) Nyman = ***Carex acuta*** L.
Carex acuta var. *angustifolia* Èelak. = ***Carex nigra*** (L.) Reichard subsp. ***nigra***
Carex acuta var. *appendiculata* Trautv. & C.A.Mey. = ***Carex thunbergii*** Steud.
Carex acuta f. *badia* Sanio = ***Carex nigra*** (L.) Reichard subsp. ***nigra***
Carex acuta subsp. *broteroana* (Samp.) Rivas Mart. = ***Carex elata*** All. subsp. ***elata***
Carex acuta var. *chlorotica* Èelak. = ***Carex brunnea*** Thunb. subsp. ***brunnea***
Carex acuta var. *colorata* F.Nyl. = ***Carex brunnea*** Thunb. subsp. ***brunnea***
Carex acuta f. *diluta* Sanio = ***Carex nigra*** (L.) Reichard subsp. ***nigra***
Carex acuta dioica Peterm. = ***Carex acuta*** L.
Carex acuta var. *elegans* Wimm. = ***Carex* × *elytroides*** Fr.
Carex acuta var. *erecta* Dewey = ***Carex angustata*** Boott
Carex acuta var. *fluviatilis* Hartm. = ***Carex brunnea*** Thunb. subsp. ***brunnea***
Carex acuta f. *graciliflora* Legrand = ***Carex acuta*** L.
Carex acuta f. *gracilis* (Curtis) Sanio = ***Carex acuta*** L.
Carex acuta var. *intermedia* Èelak. = ***Carex acuta*** L.
Carex acuta var. *longiglumis* Cariot = ***Carex brunnea*** Thunb. subsp. ***brunnea***
Carex acuta var. *macrocarpa* Èelak. = ***Carex nigra*** (L.) Reichard subsp. ***nigra***

Carex acuta f. *melanolepis* Sanio = ***Carex nigra*** (L.) Reichard subsp. ***nigra***
Carex acuta var. *minor* Trevir. = ***Carex brunnea*** Thunb. subsp. ***brunnea***
Carex acuta var. *minor* Boeck. = ***Carex* × *limula*** Fr.
Carex acuta var. *minor* Sw. = ***Carex nigra*** (L.) Reichard subsp. ***nigra***
Carex acuta f. *minor* Boeck. = ***Carex* × *turfosa*** Fr.
Carex acuta var. *moenchiana* (Wender.) Nyman = ***Carex acuta*** L.
Carex acuta var. *nervosa* G.Mey. = ***Carex trinervis*** Degl.
Carex acuta var. *nigra* L. = ***Carex nigra*** (L.) Reichard
Carex acuta var. *nutans* Èelak. = ***Carex brunnea*** Thunb. subsp. ***brunnea***
Carex acuta var. *oxylepis* Sanio = ***Carex nigra*** (L.) Reichard subsp. ***nigra***
Carex acuta var. *pallida* Boott = ***Carex lenticularis*** var. ***lipocarpa*** (T.Holm) L.A.Standl.
Carex acuta var. *panormitana* (Guss.) Nyman = ***Carex panormitana*** Guss.
Carex acuta f. *pedicellata* Peterm. = ***Carex acuta*** L.
Carex acuta var. *personata* Fr. = ***Carex brunnea*** Thunb. subsp. ***brunnea***
Carex acuta subsp. *prolixa* (Fr.) Nyman = ***Carex acuta*** L.
Carex acuta var. *prolixa* (Fr.) Hartm. = ***Carex acuta*** L.
Carex acuta var. *pseudaquatilis* Appel = ***Carex brunnea*** Thunb. subsp. ***brunnea***
Carex acuta f. *pseudosparganioides* Èelak. = ***Carex nigra*** (L.) Reichard subsp. ***nigra***
Carex acuta subsp. *reuteriana* (Boiss.) Rivas Mart. & J.M.Pizarro = ***Carex reuteriana*** Boiss.
Carex acuta f. *rigida* Sanio = ***Carex brunnea*** Thunb. subsp. ***brunnea***
Carex acuta f. *ripensis* Laest. = ***Carex* × *limula*** Fr.
Carex acuta var. *rufa* L. = ***Carex brunnea*** Thunb. subsp. ***brunnea***
Carex acuta var. *sparganioides* Èelak. = ***Carex* × *elytroides*** Fr.
Carex acuta var. *sparsiflora* Dewey = ***Carex acuta*** L.
Carex acuta var. *sphaerocarpa* Uechtr. = ***Carex brunnea*** Thunb. subsp. ***brunnea***
Carex acuta var. *staminifera* Boott = ***Carex brunnea*** Thunb. subsp. ***brunnea***
Carex acuta f. *stygia* Sanio = ***Carex nigra*** (L.) Reichard subsp. ***nigra***
Carex acuta f. *tourangiana* (Boreau) Sanio = ***Carex acuta*** L.
Carex acuta var. *tricostata* (Fr.) Crép. = ***Carex acuta*** L.
Carex acuta var. *trinervis* (Degl.) Jess. = ***Carex trinervis*** Degl.
Carex acuta var. *turfosa* Sanio = ***Carex nigra*** (L.) Reichard subsp. ***nigra***
Carex acuta f. *vaginata* Peterm. = ***Carex acuta*** L.
Carex acuta f. *zygostachya* Rchb. = ***Carex brunnea*** Thunb. subsp. ***brunnea***
Carex acutata var. *ciliata* Kük. = ***Carex acutata*** Boott
Carex acutata var. *hirtisquama* Kük. = ***Carex niederleiniana*** Boeck.
Carex acutata var. *multispicata* (Kunze) Kük. = ***Carex acutata*** Boott
Carex acutata var. *niederleiniana* (Boeck.) Kük. = ***Carex niederleiniana*** Boeck.
Carex acutatiformis H.E.Hess = ***Carex acutiformis*** Ehrh.
Carex acutiformis Brot. = ?
Carex acutiformis var. *abbreviata* Beck = ***Carex acutiformis*** Ehrh.
Carex acutiformis f. *depauperata* (Lange) Kük. = ***Carex acutiformis*** Ehrh.
Carex acutiformis f. *gracilior* Kük. = ***Carex acutiformis*** Ehrh.
Carex acutiformis f. *maxima* (Urb.) Kük. = ***Carex acutiformis*** Ehrh.
Carex acutiformis f. *minor* Peterm. = ***Carex acutiformis*** Ehrh.
Carex acutiformis f. *spadicea* (Nyman) Asch. & Graebn. = ***Carex acutiformis*** Ehrh.
Carex acutina L.H.Bailey = ***Carex nudata*** Boott
Carex acutina var. *petrophila* T.Holm = ***Carex aperta*** Boott
Carex acutina var. *tenuior* L.H.Bailey = ***Carex aperta*** Boott
Carex acutinella Mack. = ***Carex aquatilis*** Wahlenb. var. ***aquatilis***
Carex acutinella var. *congesta* Boott = ***Carex aquatilis*** Wahlenb. var. ***aquatilis***
Carex acutissima Degl. = ***Carex pyrenaica*** Wahlenb.
Carex adumana Makino = ***Carex lasiolepis*** Franch.
Carex adusta Carey ex Boott = ***Carex silicea*** Olney
Carex adusta var. *argyrantha* L.H.Bailey = ***Carex siccata*** Dewey
Carex adusta var. *congesta* Boott = ***Carex specifica*** L.H.Bailey
Carex adusta var. *glomerata* L.H.Bailey = ***Carex adusta*** Boott
Carex adusta var. *minor* Boott = ***Carex petasata*** Dewey
Carex adusta var. *sparsiflora* L.H.Bailey = ***Carex siccata*** Dewey
Carex aederi Desv. = ***Carex pilulifera*** L. subsp. ***pilulifera***
Carex aemulans Liebm. & Drejer = ***Carex brunnea*** Thunb. subsp. ***brunnea***
Carex aenea Fernald = ***Carex foenea*** Willd.
Carex aenea f. *extrapolata* Butters & Abbe = ***Carex foenea*** Willd.
Carex aenea f. *flumini-regalis* Butters & Abbe = ***Carex foenea*** Willd.
Carex aequa C.B.Clarke = ***Carex serratodens*** S.Watson
Carex aequabilis Boott = ***Carex dianae*** Steud.
Carex aequalis Notø = ?
Carex aequata Nees ex Boott = ***Carex myosurus*** Nees
Carex aequibilirostris Sutô & Suzuki = ***Carex hymenodon*** Ohwi
Carex aethiopica var. *iridifolia* (Kunth) C.B.Clarke = ***Carex aethiopica*** Schkuhr
Carex aethiopica var. *latispica* C.B.Clarke = ***Carex aethiopica*** Schkuhr
Carex aethiopica var. *stolonifera* Boeck. = ***Carex simensis*** Hochst. ex A.Rich.
Carex aethostachya Schkuhr = ***Carex atrata*** L. subsp. ***atrata***
Carex affinis R.Br. = ***Carex obtusata*** Lilj.
Carex agastachys L.f. = ***Carex pendula*** Huds.
Carex agglomerata Mack. = ***Carex aggregata*** Mack.
Carex agrostoides Mack. = ***Carex alma*** L.H.Bailey
Carex aitchisonii Boeck. = ***Carex diluta*** M.Bieb. subsp. ***diluta***
Carex ajanensis Vorosch. = ***Carex alba*** Scop.
Carex akanensis Franch. = ***Carex sordida*** Van Heurck & Müll.Arg.
Carex akiensis K.Okamoto = [38 JAP]
Carex alascana Boeck. = ***Carex gynocrates*** Wormsk.
Carex alata var. *ferruginea* (A.Gray) Fernald = ***Carex albolutescens*** Schwein.
Carex alba Dewey = ***Carex eburnea*** Boott
Carex alba Bastard = ***Carex pilulifera*** L. subsp. ***pilulifera***

Carex alba f. *aprica* Kük. = ***Carex alba*** Scop.
Carex alba var. *setifolia* Dewey = ***Carex eburnea*** Boott
Carex alba subsp. *ussuriensis* Kom. = ***Carex alba*** Scop.
Carex ×*albertii* H.Lév. = ?
Carex albescens F.Nyl. = ***Carex brunnea*** Thunb. subsp. ***brunnea***
Carex ×*albidula* E.Holmb. = ***Carex* × *mithala*** Callier
Carex albirostris C.B.Clarke = ?
Carex alboatra Willd. ex Kunth = ***Carex nigra*** (L.) Reichard subsp. ***nigra***
Carex albolutescens var. *alata* (Torr.) Kük. = ***Carex alata*** Torr.
Carex albolutescens var. *argyrantha* (L.H.Bailey) Olney ex L.H.Bailey = ***Carex siccata*** Dewey
Carex albolutescens var. *cumulata* (L.H.Bailey) L.H. Bailey = ***Carex cumulata*** (L.H.Bailey) Mack.
Carex albolutescens var. *ferruginea* (A.Gray) Kük. = ***Carex albolutescens*** Schwein.
Carex albolutescens var. *glomerata* Olney ex L.H.Bailey = ***Carex adusta*** Boott
Carex albolutescens var. *meridionalis* Kük. = ***Carex longii*** subsp. ***meridionalis*** (Kük.) Luceño & Alves
Carex albomas C.B.Clarke = ***Carex sachalinensis*** F.Schmidt var. ***sachalinensis***
Carex aleutica Akiyama = ***Carex lenticularis*** Michx. var. ***lenticularis***
Carex alexeenkoana Litv. = ***Carex haematostoma*** Nees
Carex algarbiensis Samp. = ***Carex divisa*** Huds.
Carex algeriensis Nelmes = ***Carex sylvatica*** Huds. subsp. ***sylvatica***
Carex algida Turcz. ex V.I.Krecz. = ***Carex panicea*** L.
Carex allegheniensis Mack. = ***Carex debilis*** var. ***rudgei*** L.H.Bailey
Carex ×*allensteiniensis* R.Gross = ?
Carex alligata var. *degeneri* R.W.Krauss = ***Carex alligata*** Boott
Carex allomacros Steud. = ***Carex macloviana*** d'Urv. subsp. ***macloviana***
Carex allorgei Sennen = ***Carex halleriana*** Asso
Carex ×*alluvialis* Figert = ?
Carex ×*alluvialis* Schube = ?
Carex alopecoidea f. *sparsispicata* Dewey = ***Carex alopecoidea*** Tuck.
Carex alopecuroides var. *chlorostachys* C.B.Clarke = ***Carex alopecuroides*** D.Don ex Tilloch & Taylor
Carex alopecuroides subsp. *subtransversa* (C.B.Clarke) T.Koyama = ***Carex alopecuroides*** D.Don ex Tilloch & Taylor
Carex alopecurus Lapeyr. = ***Eriophorum latifolium*** Hoppe
Carex alpestris Dewey = ***Carex albicans*** Willd. ex Spreng. var. ***albicans***
Carex alpestris Lam. = ***Carex cespitosa*** L. var. ***cespitosa***
Carex alpestris All. = ***Carex halleriana*** Asso
Carex alpestris Gaudin = ***Carex sempervirens*** Vill.
Carex alpestris var. *tripla* Dewey = ***Carex albicans*** Willd. ex Spreng. var. ***albicans***
Carex alpigena A.Kern. ex Dalla Torre = ***Carex ferruginea*** Scop. subsp. ***ferruginea***
Carex alpina Honck. = ***Carex nigra*** (L.) Reichard subsp. ***nigra***
Carex alpina Sw. ex Wahlenb. = ***Carex norvegica*** Retz. subsp. ***norvegica***
Carex alpina Schrank = ***Carex sempervirens*** Vill.
Carex alpina var. *eriostrata* Boott = ***Carex norvegica*** Retz. subsp. ***norvegica***
Carex alpina var. *gracilenta* (Boott ex Boeck.) C.B. Clarke = ***Carex gracilenta*** Boott ex Boeck.
Carex alpina var. *holostoma* (Drejer) L.H.Bailey = ***Carex holostoma*** Drejer
Carex alpina var. *inferalpina* Wahlenb. = ***Carex media*** R.Br.
Carex alpina subsp. *infuscata* Nees = ***Carex norvegica*** Retz. subsp. ***norvegica***
Carex alpina var. *infuscata* (Nees) Boott = ***Carex norvegica*** Retz. subsp. ***norvegica***
Carex alpina var. *lehmanii* (Drejer) R.C.Srivast. = ***Carex lehmanii*** Drejer
Carex alpina f. *nigrescens* Andersson = ***Carex norvegica*** Retz. subsp. ***norvegica***
Carex alpina var. *parviflora* (Host) Boott = ***Carex parviflora*** Host
Carex alpina f. *robustior* Kük. = ***Carex norvegica*** Retz. subsp. ***norvegica***
Carex alpina var. *stevenii* T.Holm = ***Carex stevenii*** (T.Holm) Kalela
Carex alsia Raymond = ***Carex decaulescens*** subsp. ***alsia*** (Raymond) Kukkonen
Carex alta Boott = ***Carex remota*** subsp. ***alta*** (Boott) Kük.
Carex alta var. *brevior* H.Lév. & Vaniot = ***Carex gibba*** Wahlenb.
Carex alta var. *gymnocarpa* H.Lév. & Vaniot = ***Carex remota*** subsp. ***alta*** (Boott) Kük.
Carex altaica (Gorodkov) V.I.Krecz. = ***Carex orbicularis*** subsp. ***altaica*** (Gorodkov) T.V.Egorova
Carex alterna C.B.Clarke = ?
Carex alterniflora Franch. = ***Carex sachalinensis*** var. ***alterniflora*** (Franch.) Ohwi
Carex alterniflora var. *arimaensis* Ohwi = ***Carex sachalinensis*** var. ***arimaensis*** (Ohwi) Ohwi
Carex alterniflora var. *aureobrunnea* Ohwi = ***Carex sachalinensis*** var. ***aureobrunnea*** (Ohwi) Ohwi
Carex alterniflora var. *elongatula* Ohwi = ***Carex sachalinensis*** var. ***elongatula*** (Ohwi) Ohwi
Carex alterniflora var. *fulva* Ohwi = ***Carex sachalinensis*** var. ***fulva*** (Ohwi) Ohwi
Carex alticola Popl. ex Sukaczev = ***Carex argunensis*** subsp. ***alticola*** (Popl. ex Sukaczev) Malyschev
Carex altocaulis Britton = ***Carex vaginata*** Tausch var. ***vaginata***
Carex alveata Boott = ***Carex tetanica*** Schkuhr
Carex amami-oshimensis Akiyama = ***Carex brunnea*** Thunb. subsp. ***brunnea***
Carex ambigua Moench = ***Carex acuta*** L.
Carex ambigua Link = ***Carex oedipostyla*** Duval-Jouve
Carex ambleocarpa Willd. = ***Carex flacca*** Schreb. subsp. ***flacca***
Carex amblyolepis Trautv. & C.A.Mey. = ***Carex van-heurckii*** Müll.Arg. subsp. ***van-heurckii***
Carex amblyorhyncha V.I.Krecz. = ***Carex marina*** Dewey subsp. ***marina***
Carex amblyorhyncha var. *fusca* Böcher = ***Carex marina*** Dewey subsp. ***marina***
Carex amblyorhyncha subsp. *pseudolagopina* (T.J. Sørensen) Böcher = ***Carex marina*** subsp. ***pseudolagopina*** (T.J.Sørensen) Böcher
Carex amblyorhyncha var. *tundrarum* Böcher = ***Carex marina*** Dewey subsp. ***marina***
Carex ambusta Boott = ***Carex saxatilis*** L.
Carex amgunensis var. *chloroleuca* (Meinsh.) Kük. = ***Carex chloroleuca*** Meinsh.
Carex ammophila Willd. = ***Carex divisa*** Huds.
Carex amoena Boott = ***Carex raphidocarpa*** Nees
Carex amphibola var. *globosa* (L.H.Bailey) L.H.Bailey = ***Carex bulbostylis*** Mack.
Carex amphibola var. *rigida* (L.H.Bailey) Fernald = ***Carex planispicata*** Naczi
Carex amphibola var. *turgida* Fernald = ***Carex grisea*** Wahlenb.

Carex amphigena (Fernald) Mack. = ***Carex glareosa*** Schkuhr ex Wahlenb. subsp. ***glareosa***
Carex amphilogos K.Koch = ***Carex maritima*** Gunnerus
Carex amphora Franch. & Sav. = ***Carex pisiformis*** Boott
Carex amplisquama F.J.Herm. = ***Carex communis*** var. ***amplisquama*** (F.J.Herm.) Rettig
Carex ampullacea Wulfen ex Wahlenb. = ***Carex michelii*** Host
Carex ampullacea Gooden. = ***Carex rostrata*** Stokes var. ***rostrata***
Carex ampullacea var. *altissima* Andersson = ***Carex rostrata*** Stokes var. ***rostrata***
Carex ampullacea var. *borealis* Hartm. = ***Carex rostrata*** Stokes var. ***rostrata***
Carex ampullacea var. *brunnescens* Andersson = ***Carex rostrata*** Stokes var. ***rostrata***
Carex ampullacea f. *elatior* A.Blytt = ***Carex rostrata*** Stokes var. ***rostrata***
Carex ampullacea subsp. *hymenocarpa* (Drejer) Nyman = ***Carex rostrata*** Stokes var. ***rostrata***
Carex ampullacea var. *maxima* Andersson = ***Carex rostrata*** Stokes var. ***rostrata***
Carex ampullacea f. *pendulina* A.Blytt = ***Carex rostrata*** Stokes var. ***rostrata***
Carex ampullacea f. *planifolia* Norman = ***Carex rostrata*** Stokes var. ***rostrata***
Carex ampullacea var. *robusta* Sond. = ***Carex rostrata*** Stokes var. ***rostrata***
Carex ampullacea var. *utriculata* (Boott) Carey = ***Carex utriculata*** Boott
Carex amurensis Kük. = ***Carex sordida*** Van Heurck & Müll.Arg.
Carex amurensis var. *abbreviata* Kük. = ***Carex sordida*** Van Heurck & Müll.Arg.
Carex amurensis var. *drymophila* (Turcz.) Kük. = ***Carex drymophila*** Turcz.
Carex amurensis var. *mandschurica* Kük. = ***Carex atherodes*** Spreng.
Carex anceps Willd. ex Kunth = ***Carex grioletii*** Roem. ex Schkuhr
Carex anceps Muhl. ex Willd. = ***Carex laxiflora*** Lam. var. ***laxiflora***
Carex anceps var. *angustifolia* Dewey ex Alph.Wood = ***Carex laxiflora*** Lam. var. ***laxiflora***
Carex anceps var. *blanda* (Dewey) Boott = ***Carex blanda*** Dewey
Carex anceps var. *patulifolia* Dewey ex Alph.Wood = ***Carex laxiflora*** Lam. var. ***laxiflora***
Carex anceps var. *striatula* Carey = ***Carex laxiflora*** Lam. var. ***laxiflora***
Carex andalusica H.Lindb. = [12 SPA]
Carex andina var. *subabscondita* Kük. = ***Carex andina*** Phil.
Carex andoiniana De Not. ex Steud. = ***Carex olbiensis*** Jord.
Carex androgyna L.H.Bailey = ***Carex anisostachys*** Liebm.
Carex androgyna Balb. = ***Carex bicolor*** Bellardi ex All.
Carex aneurocarpa V.I.Krecz. = ***Carex pediformis*** C.A.Mey. subsp. ***pediformis***
Carex angarae Steud. = ***Carex media*** R.Br.
Carex angarae subsp. *pusteriana* Kalela = ***Carex norvegica*** subsp. ***pusteriana*** (Kalela) Chater
Carex angarae var. *stevenii* (T.Holm) A.E.Porsild = ***Carex stevenii*** (T.Holm) Kalela
Carex anguillata Drejer = ***Carex bigelowii*** Torr. ex Schwein. subsp. ***bigelowii***
Carex angustata var. *strictior* Dudley = ***Carex angustata*** Boott
Carex angustata var. *verticillata* Boott = ***Carex interrupta*** Boeck.
Carex angustifolia Sm. = ***Carex nigra*** (L.) Reichard subsp. ***nigra***
Carex angustior Mack. = ***Carex echinata*** Murray subsp. ***echinata***
Carex angustior var. *gracilenta* Clausen & Wahl = ***Carex echinata*** Murray subsp. ***echinata***
Carex anhuiensis S.W.Su & S.M.Xu = ***Carex chungii*** Z.P.Wang var. ***chungii***
Carex anisostachys var. *chlorocarpos* (Liebm.) L.H. Bailey = ***Carex anisostachys*** Liebm.
Carex anisostachys var. *monticola* Kük. = ***Carex anisostachys*** Liebm.
Carex annectens var. *ambigua* (Barratt) Gleason = ***Carex annectens*** (E.P.Bicknell) E.P.Bicknell
Carex annectens var. *xanthocarpa* (Kük.) Wiegand = ***Carex annectens*** (E.P.Bicknell) E.P.Bicknell
Carex annulata Kük. = ***Carex oligocarya*** C.B.Clarke
Carex anomala Janka = ***Carex caryophyllea*** Latourr. var. ***caryophyllea***
Carex anomala Boott = ***Carex gibba*** Wahlenb.
Carex anomala Steud. = ***Carex petitiana*** A.Rich.
Carex anomocarya Nelmes = ***Carex harlandii*** Boott
Carex ×*anselmii* Sennen = ?
Carex antarctogena Roiv. = ***Carex capitata*** L.
Carex anthericoides C.Presl = ***Carex macrocephala*** Willd. ex Spreng.
Carex anthoxanthea var. *leiocarpa* (C.A.Mey.) Kük. = ***Carex anthoxanthea*** C.Presl
Carex ×*anticostensis* nothovar. *longidens* Lepage = ***Carex* × *anticostensis*** nothovar. ***inflatior*** Lepage
Carex ×*anticostensis* nothovar. *minor* Lepage = ***Carex* × *anticostensis*** (Fernald) Lepage nothovar. ***anticostensis***
Carex antucensis Kunze ex Kunth = ***Carex nigra*** (L.) Reichard subsp. ***nigra***
Carex anwandteri Phil. = ***Carex acutata*** Boott
Carex aomorensis Franch. = ***Carex capillacea*** var. ***sachalinensis*** (F.Schmidt) Ohwi
Carex aperta var. *angustifolia* Boott = ***Carex nudata*** Boott
Carex aperta f. *concinnula* T.Holm = ***Carex aperta*** Boott
Carex aperta var. *divaricata* L.H.Bailey = ***Carex aquatilis*** Wahlenb. var. ***aquatilis***
Carex aperta f. *hydroessa* T.Holm = ***Carex aperta*** Boott
Carex aperta f. *mimetica* T.Holm = ***Carex aperta*** Boott
Carex aperta var. *umbrosa* Kük. = ***Carex aperta*** Boott
Carex aperta var. *viridans* Kük. = ***Carex aperta*** Boott
Carex aphanandra Franch. & Sav. = ***Carex breviculmis*** var. ***discoidea*** (Boott) Boott
Carex apoda Clokey = ***Carex atrosquama*** Mack.
Carex apodostachya Ohwi = ***Carex atrata*** subsp. ***apodostachya*** (Ohwi) T.Koyama
Carex aporandra Kunze ex Boott = ***Carex shortiana*** Dewey & Torr.
Carex ×*appeliana* Zahn = ?
Carex appendiculata (Trautv. & C.A.Mey.) Kük. = ***Carex thunbergii*** Steud.
Carex appendiculata var. *sacculiformis* Y.L.Chang & Y.L.Yang = ***Carex thunbergii*** Steud.
Carex appressa var. *diaphana* (Boott) Kük. = ***Carex appressa*** R.Br. var. ***appressa***
Carex appressa var. *elatior* Nees = ***Carex tereticaulis*** F.Muell.
Carex appressa f. *minor* Kük. = ***Carex appressa*** R.Br. var. ***appressa***
Carex appressa var. *secta* (Boott) Kük. = ***Carex secta*** Boott

Carex appressa var. *sectoides* Kük. = ***Carex sectoides*** (Kük.) Edgar
Carex appressa var. *typica* Domin = ***Carex appressa*** R.Br.
Carex approximata All. = ***Carex ericetorum*** Pollich
Carex approximata f. *interrupta* (Peterm.) Soó = ***Carex ericetorum*** Pollich
Carex approximata f. *ramosa* (Asch. & Graebn.) Soó = ***Carex ericetorum*** Pollich
Carex aprica Turcz. ex Besser = ***Carex supina*** Willd. ex Wahlenb. var. ***supina***
Carex aquanigra B.Boivin = ***Carex nigra*** (L.) Reichard subsp. ***nigra***
Carex aquatilis Ten. = ***Carex microcarpa*** Bertol. ex Moris
Carex aquatilis var. *altior* (Rydb.) Fernald = ***Carex emoryi*** Dewey
Carex aquatilis subsp. *altior* (Rydb.) Hultén = ***Carex emoryi*** Dewey
Carex aquatilis var. *anguillata* (Drejer) Nyman = ***Carex bigelowii*** Torr. ex Schwein. subsp. ***bigelowii***
Carex aquatilis f. *angustata* Kük. = ***Carex aquatilis*** var. ***dives*** (T.Holm) Kük.
Carex aquatilis f. *cuspidata* Laest. ex Andersson = ***Carex aquatilis*** Wahlenb. var. ***aquatilis***
Carex aquatilis var. *elatior* Bab. = ***Carex aquatilis*** Wahlenb. var. ***aquatilis***
Carex aquatilis f. *epigaea* Kurtz = ***Carex aquatilis*** Wahlenb. var. ***aquatilis***
Carex aquatilis subsp. *epigeios* (Fr.) Nyman = ***Carex salina*** Wahlenb. var. ***salina***
Carex aquatilis var. *genuina* Syme = ***Carex aquatilis*** Wahlenb.
Carex aquatilis f. *laxa* Kük. = ***Carex aquatilis*** Wahlenb. var. ***aquatilis***
Carex aquatilis var. *nardifolia* Wahlenb. = ***Carex subspathacea*** Wormsk. ex Hornem.
Carex aquatilis var. *pachystoma* (T.Holm) Kük. = ***Carex aquatilis*** var. ***dives*** (T.Holm) Kük.
Carex aquatilis var. *planifolia* F.Nyl. = ***Carex aquatilis*** Wahlenb. var. ***aquatilis***
Carex aquatilis var. *plumbeophila* Kelso = ***Carex aquatilis*** Wahlenb. var. ***aquatilis***
Carex aquatilis var. *rhomboidea* (T.Holm) Kelso = ***Carex aquatilis*** Wahlenb. var. ***aquatilis***
Carex aquatilis f. *sciaphila* (T.Holm) Kük. = ***Carex aquatilis*** Wahlenb. var. ***aquatilis***
Carex aquatilis var. *sciaphila* (T.Holm) Kelso = ***Carex aquatilis*** Wahlenb. var. ***aquatilis***
Carex aquatilis var. *sitchensis* (J.D.Prescott ex Bong.) Kelso = ***Carex aquatilis*** var. ***dives*** (T.Holm) Kük.
Carex aquatilis var. *sphagnophila* Fr. = ***Carex aquatilis*** Wahlenb. var. ***aquatilis***
Carex aquatilis var. *stans* (Drejer) Boott = ***Carex aquatilis*** var. ***minor*** Boott
Carex aquatilis subsp. *stans* (Drejer) Hultén = ***Carex aquatilis*** var. ***minor*** Boott
Carex aquatilis f. *virescens* Andersson = ***Carex aquatilis*** Wahlenb. var. ***aquatilis***
Carex aquatilis var. *watsonii* Syme = ***Carex aquatilis*** Wahlenb. var. ***aquatilis***
Carex arakamensis C.B.Clarke = ***Carex microchaeta*** subsp. ***nesophila*** (T.Holm) D.F.Murray
Carex arakiana (Ohwi) Ohwi = ***Carex filipes*** var. ***arakiana*** (Ohwi) Ohwi
Carex araucana Phil. = ***Carex fuscula*** d'Urv. var. ***fuscula***
Carex arcatica Meinsh. = ***Carex orbicularis*** Boott subsp. ***orbicularis***
Carex arcatica var. *pedunculata* Meinsh. = ***Carex orbicularis*** Boott subsp. ***orbicularis***
Carex arcta var. *oregana* (L.H.Bailey) B.Boivin = ***Carex arcta*** Boott
Carex arctata var. *faxonii* L.H.Bailey = ***Carex arctata*** Boott
Carex arctica Dewey = ***Carex parryana*** Dewey var. ***parryana***
Carex arctica Deinboll ex Fr. = ***Carex stenophylla*** Wahlenb. subsp. ***stenophylla***
Carex arctisibirica (Jurtzev) Czern. = ***Carex bigelowii*** subsp. ***arctisibirica*** (Jurtzev) Á.Löve & D.Löve
Carex arctogena Harry Sm. = ***Carex capitata*** L.
Carex arctophila Nyl. ex Fr. = ***Carex aquatilis*** Wahlenb. var. ***aquatilis***
Carex arcuata Wahlenb. = ***Carex extensa*** Gooden.
Carex arcuata Franch. = ***Carex mollicula*** Boott
Carex arcuata C.Laest. = ?
Carex ardoiniana De Not. = ***Carex olbiensis*** Jord.
Carex arenaria Leers = ***Carex disticha*** Huds.
Carex arenaria Dubois ex Steud. = ***Carex praecox*** Schreb.
Carex arenaria var. *castanea* Boott = ***Carex colchica*** J.Gay
Carex arenaria subsp. *colchica* (J.Gay) Nyman = ***Carex colchica*** J.Gay
Carex arenaria var. *colchica* (J.Gay) H.Christ = ***Carex colchica*** J.Gay
Carex arenaria f. *pallida* Lackow. = ***Carex arenaria*** L.
Carex arenaria var. *posnaniensis* (Sprib.) Asch. & Graebn. = ***Carex repens*** Bellardi
Carex arenaria var. *prostrata* De Langhe = ***Carex arenaria*** L.
Carex arenaria var. *pseudoarenaria* T.Marsson = ***Carex arenaria*** L.
Carex arenaria var. *pseudobrizoides* (Chabaud) Nyman = ***Carex pseudobrizoides*** Chabaud
Carex arenaria f. *pumila* Lackow. = ***Carex arenaria*** L.
Carex arenaria f. *remota* T.Marsson = ***Carex arenaria*** L.
Carex arenaria var. *repens* (Bellardi) Wahlenb. = ***Carex repens*** Bellardi
Carex arenaria var. *spiralis* Asch. & Graebn. = ***Carex arenaria*** L.
Carex arenaria var. *tenuis* Asch. & Graebn. = ***Carex arenaria*** L.
Carex arenaria var. *umbrosa* Coss. & Germ. = ***Carex pseudobrizoides*** Chabaud
Carex arenaria var. *umbrosa* (Sprib.) Sprib. = ***Carex repens*** Bellardi
Carex arenicola F.Schmidt = ***Carex pansa*** L.H.Bailey
Carex arenicola subsp. *pansa* (L.H.Bailey) T.Koyama & Calder = ***Carex pansa*** L.H.Bailey
Carex argentea J.F.Gmel. = ***Carex alba*** Scop.
Carex argentea Vill. = ***Carex humilis*** Leyss. var. ***humilis***
Carex argyrantha var. *aenea* (Fernald) B.Boivin = ***Carex foenea*** Willd.
Carex argyroglochin Hornem. = ***Carex leporina*** L.
Carex argyrolepis Maxim. ex Franch. & Sav. = ***Carex nubigena*** D.Don ex Tilloch & Taylor subsp. ***nubigena***
Carex argyrostachys H.Lév. & Vaniot = ***Carex gifuensis*** Franch.
Carex arida Schwein. & Torr. = ***Carex muskingumensis*** Schwein.
Carex arida Schleich. ex Kunth = ***Carex sempervirens*** Vill.
Carex arisanensis var. *tremula* Ohwi = ***Carex filipes*** var. ***tremula*** (Ohwi) Ohwi
Carex aristata R.Br. = ***Carex atherodes*** Spreng.
Carex aristata Dewey = ***Carex davisii*** Schwein. & Torr.

Carex aristata Siegert ex Wimm. = ***Carex hirta*** L.
Carex aristata Honck = ***Carex praecox*** Schreb.
Carex aristata D.Don ex Tilloch & Taylor = ?
Carex aristata var. *cujavica* Asch. & Sprib. = ***Carex atherodes*** Spreng.
Carex aristata var. *eriophylla* Kük. = ***Carex raddei*** Kük.
Carex aristata var. *glabra* Uechtr. = ***Carex atherodes*** Spreng.
Carex aristata var. *hirtiformis* H.Lév. & Vaniot = ***Carex atherodes*** Spreng.
Carex aristata var. *imberbis* A.Gray = ***Carex atherodes*** Spreng.
Carex aristata var. *kirschsteiniana* Asch., Graebn. & Kük. = ***Carex atherodes*** Spreng.
Carex aristata var. *kirschsteinii* Asch. & Graebn. = ***Carex atherodes*** Spreng.
Carex aristata var. *laeviconica* (Dewey) Kük. = ***Carex laeviconica*** Dewey
Carex aristata var. *lanceisquama* Hand.-Mazz. = ***Carex atherodes*** Spreng.
Carex aristata var. *longilanceolata* Dewey = ***Carex atherodes*** Spreng.
Carex aristata var. *maxima* (Kük.) Kük. = ***Carex atherodes*** Spreng.
Carex aristata subsp. *orostachys* C.A.Mey. = ***Carex atherodes*** Spreng.
Carex aristata var. *orostachys* (C.A.Mey.) C.B.Clarke = ***Carex atherodes*** Spreng.
Carex aristata subsp. *raddei* (Kük.) Kük. = ***Carex raddei*** Kük.
Carex aristata var. *subaristata* Kük. = ***Carex atherodes*** Spreng.
Carex aristata var. *vix-vaginans* Kük. = ***Carex atherodes*** Spreng.
Carex arizonensis C.B.Clarke = ***Carex alma*** L.H.Bailey
Carex arizonensis Kük. = ***Carex thurberi*** Dewey ex Torr.
Carex armena Boiss. = ***Carex hostiana*** DC.
Carex arnottiana Boott = ***Carex lobulirostris*** Drejer
Carex arrhyncha Franch. = ***Carex tenuiflora*** Wahlenb.
Carex arridens C.B.Clarke = ***Carex perakensis*** C.B.Clarke var. ***perakensis***
Carex artinux C.B.Clarke = ***Carex sachalinensis*** var. ***fulva*** (Ohwi) Ohwi
Carex artitecta Mack. = ***Carex albicans*** Willd. ex Spreng. var. ***albicans***
Carex artitecta f. *colorata* (L.H.Bailey) D.S.Carp. = ***Carex albicans*** Willd. ex Spreng. var. ***albicans***
Carex artitecta var. *subtilirostris* F.J.Herm. = ***Carex albicans*** Willd. ex Spreng. var. ***albicans***
Carex asa-grayi L.H.Bailey = ***Carex grayi*** Carey
Carex asa-grayi var. *hispidula* (A.Gray ex L.H.Bailey) L.H.Bailey = ***Carex grayi*** Carey
Carex × *aschersonii* H.Lév. = ?
Carex ascocentra C.B.Clarke ex Franch. = ***Carex ligata*** subsp. ***formosensis*** (H.Lév. & Vaniot) T.Koyama
Carex aspera Willd. = ***Carex flacca*** Schreb. subsp. ***flacca***
Carex asperifolia Steud. = ***Carex brunnea*** Thunb. subsp. ***brunnea***
Carex asperinervis T.Koyama = ***Carex bilateralis*** Hayata
Carex asperula Turcz. = ***Carex meyeriana*** Kunth
Carex asperula Nees = ***Carex psychrophila*** Nees
Carex aspratilis V.I.Krecz. = ***Carex diluta*** M.Bieb. subsp. ***diluta***
Carex assimilis Steud. = ***Carex brunnea*** Thunb. subsp. ***brunnea***
Carex assiniboinensis f. *ambulans* J.-P.Bernard = ***Carex assiniboinensis*** Boott
Carex astonii Hamlin = ***Carex druceana*** var. ***astonii*** (Hamlin) Edgar
Carex astracanica Willd. ex Kunth = ***Carex muricata*** L.
Carex asturica Boiss. = ***Carex gayana*** Steud.
Carex aterrima Hoppe = ***Carex atrata*** subsp. ***aterrima*** (Hoppe) Èelak.
Carex aterrima W.D.J.Koch = ***Carex atrata*** L. subsp. ***atrata***
Carex aterrima subsp. *medwedewii* (Leskov) T.V. Egorova = ***Carex medwedewii*** Leskov
Carex aterrima var. *wolfii* Kneuck. = ***Carex atrata*** L. subsp. ***atrata***
Carex athabascensis F.J.Herm. = ***Carex scirpoidea*** Michx. var. ***scirpoidea***
Carex atherodes var. *cujavica* (Asch. & Sprib.) Rauschert = ***Carex atherodes*** Spreng.
Carex atherodes f. *glabra* (Uechtr.) Lepage = ***Carex atherodes*** Spreng.
Carex atherodes var. *glabra* (Uechtr. ex Garcke) Rauschert = ***Carex atherodes*** Spreng.
Carex atherodes f. *imberbis* (A.Gray) B.Boivin = ***Carex atherodes*** Spreng.
Carex atherodes var. *kirschsteinii* (Asch., Graebn. & Kük.) Rauschert = ***Carex atherodes*** Spreng.
Carex atherodes var. *longilanceolata* (Dewey) Gilly = ***Carex atherodes*** Spreng.
Carex atherodes var. *maxima* (Kük.) A.E.Kozhevn. = ***Carex atherodes*** Spreng.
Carex atherodes var. *orthostachys* (C.A.Mey.) A.E. Kozhevn. = ***Carex atherodes*** Spreng.
Carex atherodes var. *siegertiana* (Uechtr. ex Garcke) Rauschert = ***Carex atherodes*** Spreng.
Carex atherodes var. *vix-vaginans* (Kük.) A.E.Kozhevn. = ***Carex atherodes*** Spreng.
Carex athrostachya var. *unilateralis* (Mack.) B.Boivin = ***Carex unilateralis*** Mack.
Carex atjehensis Kük. = ***Carex finitima*** Boott
Carex atlantica var. *capillacea* (L.H.Bailey) Cronquist = ***Carex atlantica*** subsp. ***capillacea*** (L.H.Bailey) Reznicek
Carex atlantica var. *incomperta* (E.P.Bicknell) F.J.Herm. = ***Carex atlantica*** L.H.Bailey subsp. ***atlantica***
Carex atrata Banks ex Boott = ***Carex banksii*** Boott var. ***banksii***
Carex atrata var. *aterrima* (Hoppe) Hartm. = ***Carex atrata*** subsp. ***aterrima*** (Hoppe) Èelak.
Carex atrata subsp. *atratiformis* (Britton) Kük. = ***Carex atratiformis*** Britton
Carex atrata subsp. *atrosquama* (Mack.) Hultén = ***Carex atrosquama*** Mack.
Carex atrata var. *atrosquama* (Mack.) Cronquist = ***Carex atrosquama*** Mack.
Carex atrata var. *bicolor* Èelak. = ***Carex atrata*** L. subsp. ***atrata***
Carex atrata var. *brunnescens* Andersson = ***Carex atrata*** L. subsp. ***atrata***
Carex atrata f. *caeca* (L.H.Bailey) Kük. = ***Carex heteroneura*** S.Watson var. ***heteroneura***
Carex atrata var. *castanea* Nyman = ***Carex atrata*** L. subsp. ***atrata***
Carex atrata f. *castanea* (Nyman) K.Richt. = ***Carex atrata*** L. subsp. ***atrata***
Carex atrata var. *caucasica* (Steven) Boott = ***Carex atrata*** subsp. ***caucasica*** (Steven) Kük.
Carex atrata var. *chalciolepis* (T.Holm) Kük. = ***Carex chalciolepis*** T.Holm
Carex atrata var. *conglomerata* Neilr. = ***Carex parviflora*** Host
Carex atrata f. *decolorans* Neuman = ***Carex atrata*** L. subsp. ***atrata***
Carex atrata var. *discolor* L.H.Bailey = ***Carex bella*** L.H.Bailey

Carex atrata var. *dubia* Gaudin = ***Carex atrata*** L. subsp. ***atrata***
Carex atrata var. *erecta* Boott = ***Carex heteroneura*** S.Watson var. ***heteroneura***
Carex atrata f. *gelida* Schur = ***Carex atrata*** L. subsp. ***atrata***
Carex atrata var. *glacialis* Boott = ***Carex atrata*** subsp. ***pullata*** (Boott) Kük.
Carex atrata var. *heterostachya* Almq. = ***Carex atrata*** L. subsp. ***atrata***
Carex atrata var. *japonoalpina* T.Koyama = ***Carex atrata*** L. subsp. ***atrata***
Carex atrata var. *larimerana* (Kelso) Kelso = ***Carex chalciolepis*** T.Holm
Carex atrata var. *laxa* Neilr. = ***Carex atrata*** L. subsp. ***atrata***
Carex atrata f. *lenis* (L.H.Bailey) Kük. = ***Carex heteroneura*** S.Watson var. ***heteroneura***
Carex atrata var. *longistolonifera* Kük. = ***Carex atrata*** L. subsp. ***atrata***
Carex atrata subsp. *longistolonifera* (Kük.) S.Yun Liang = ***Carex atrata*** L. subsp. ***atrata***
Carex atrata var. *magellanica* (Lam.) Wahlenb. = ***Carex magellanica*** Lam.
Carex atrata var. *nigra* Olney = ***Carex nova*** L.H.Bailey
Carex atrata var. *nigra* Gaudin = ***Carex parviflora*** Host
Carex atrata subsp. *nigra* (Gaudin) Hartm. = ***Carex parviflora*** Host
Carex atrata var. *ovata* Boott = ***Carex atrata*** L. subsp. ***atrata***
Carex atrata var. *parviflora* (Host) Beck = ***Carex parviflora*** Host
Carex atrata subsp. *perfusca* (V.I.Krecz.) T.Koyama = ***Carex atrata*** subsp. ***aterrima*** (Hoppe) Èelak.
Carex atrata var. *pseudocilicica* Kük. = ***Carex atrata*** L. subsp. ***atrata***
Carex atrata var. *pullata* Boott = ***Carex atrata*** subsp. ***pullata*** (Boott) Kük.
Carex atrata var. *rectiuscula* Hartm. = ***Carex atrata*** L. subsp. ***atrata***
Carex atrata var. *sinensis* Kük. = ***Carex atrata*** subsp. ***pullata*** (Boott) Kük.
Carex atrata var. *spadicea* Beurl. = ***Carex atrata*** L. subsp. ***atrata***
Carex atrata var. *spiculosior* Norman = ***Carex atrata*** L. subsp. ***atrata***
Carex atrata var. *typica* Beck = ***Carex atrata*** L.
Carex atrata var. *varia* Gaudin = ***Carex atrata*** L. subsp. ***atrata***
Carex atrata f. *wolfii* (Kneuck.) Kük. = ***Carex atrata*** L. subsp. ***atrata***
Carex atratiformis subsp. *raymondii* (Calder) Scoggan = ***Carex atratiformis*** Britton
Carex atrofurfur T.Koyama = ***Carex infuscata*** Nees
Carex atrofusca Steven = ***Carex atrata*** L. subsp. ***atrata***
Carex atrofusca Willd. ex Steud. = ***Carex pichinchensis*** Kunth
Carex atrofusca var. *angustifructus* Kük. = ***Carex angustifructus*** (Kük.) Nelmes
Carex atrofusca var. *coriophora* (Fisch. & C.A.Mey. ex Kunth) Kük. = ***Carex atrofusca*** var. ***major*** (Boeck.) Raymond
Carex atrofusca var. *decolorata* A.E.Porsild = ***Carex atrofusca*** Schkuhr subsp. ***atrofusca***
Carex atrofusca f. *decolorata* (A.E.Porsild) B.Boivin = ***Carex atrofusca*** Schkuhr subsp. ***atrofusca***
Carex atrofusca f. *flavescens* Kük. = ***Carex atrofusca*** Schkuhr subsp. ***atrofusca***
Carex atrofusca var. *minor* (Boott) Kük. = ***Carex atrofusca*** subsp. ***minor*** (Boott) T.Koyama
Carex atrofusca var. *nortoniana* B.Boivin = ***Carex atrofusca*** Schkuhr subsp. ***atrofusca***
Carex atronucula Hayata = ***Carex sociata*** Boott
Carex atropicta f. *monodynama* Griseb. = ***Carex monodynama*** (Griseb.) G.A.Wheeler
Carex atropicta f. *pallescens* Kurtz ex Kük. = ***Carex monodynama*** (Griseb.) G.A.Wheeler
Carex atropicta var. *platycarpa* Phil. ex Kük. = ***Carex atropicta*** Steud.
Carex atropurpurea Fisch. ex Trevir. = ***Carex atrata*** L. subsp. ***atrata***
Carex atropurpurea Boeck. = ***Carex fecunda*** var. ***atropurpurea*** (Boeck.) J.F.Macbr.
Carex atrorufa Willd. ex Kunth = ***Carex pichinchensis*** Kunth
Carex atrosanguinea Nelmes = ***Carex turrita*** C.B.Clarke var. ***turrita***
Carex atrovirens Boeck. = ***Carex aureolensis*** Steud.
Carex attenuata R.Br. = ***Carex rupestris*** All. subsp. ***rupestris***
Carex augustinowiczii subsp. *soyaeensis* (Kük.) T.V. Egorova = ***Carex augustinowiczii*** Meinsh. var. ***augustinowiczii***
Carex aurea var. *androgyna* Olney = ***Carex garberi*** Fernald
Carex aurea var. *celsa* L.H.Bailey = ***Carex hassei*** L.H.Bailey
Carex aurea f. *colorata* Farw. = ***Carex aurea*** Nutt.
Carex auriculata L.H.Bailey = ***Carex angustata*** Boott
Carex auriculata Franch. = ***Carex pilosa*** Scop.
Carex ×*auroniensis* L.C.Lamb. = ?
Carex australis Kirk = ***Carex litorosa*** L.H.Bailey
Carex australpina Bech. = ***Carex ferruginea*** subsp. ***australpina*** (Bech.) W.Dietr.
Carex austriaca Schkuhr = ***Carex divisa*** Huds.
Carex austroafghanica Raymond = ***Carex divisa*** Huds.
Carex austromontana Parish = ***Carex jamesii*** Schwein.
Carex autumnalis Mack. = ***Carex marianensis*** Stacey
Carex axillaris Link = ***Carex diluta*** M.Bieb. subsp. ***diluta***
Carex axillaris L. = ***Carex remota*** L. subsp. ***remota***
Carex axillaris subsp. *boenninghausiana* (Weihe) Nyman = ***Carex ×boenninghausiana*** Weihe
Carex ayouensis X.Y.Mao & Y.C.Yang = ***Carex pediformis*** C.A.Mey. subsp. ***pediformis***
Carex azorica J.Gay = ***Carex pilulifera*** subsp. ***azorica*** (J.Gay) Franco & Rocha Afonso
Carex baazasana Steud. = ***Carex cherokeensis*** Schwein.
Carex baccans f. *aurantiaca* Kuntze = ***Carex baccans*** Nees
Carex baccans f. *nigra* Kuntze = ***Carex baccans*** Nees
Carex baccans f. *normalis* Kuntze = ***Carex baccans*** Nees
Carex baccans f. *recurvirostris* Kuntze = ***Carex baccans*** Nees
Carex baccans var. *siccifructus* C.B.Clarke = ***Carex baccans*** Nees
Carex backiana Dewey = ***Carex obtusata*** Lilj.
Carex backii var. *saximontana* (Mack.) Boivin = ***Carex saximontana*** Mack.
Carex backii var. *subrostrata* (Bates) Dorn = ***Carex saximontana*** Mack.
Carex badia Pers. = ***Carex canescens*** L. subsp. ***canescens***
Carex baetica Auersw. ex Willk. = ***Carex distans*** L.
Carex bakanensis H.Lév. & Vaniot = ***Carex hebecarpa*** subsp. ***ligulata*** (Nees) T.Koyama

Carex balbisii Schkuhr ex Balbis = ***Carex extensa*** Gooden.
Carex balbisii Schkuhr ex Spreng. = ***Carex extensa*** Gooden.
Carex baldensis Vill. = ***Carex foetida*** All.
Carex baldwinia Dewey = ***Carex elliottii*** Schwein. & Torr.
Carex baldwiniana Steud. = ***Carex elliottii*** Schwein. & Torr.
Carex banata Sm. = ***Carex maritima*** Gunnerus
Carex banatica Heuff. = ***Carex buekii*** Wimm.
Carex banksii f. *brevifolia* Kük. = ***Carex banksii*** Boott var. ***banksii***
Carex banksii var. *cilliesii* (Phil.) Kük. = ***Carex banksii*** Boott var. ***banksii***
Carex banksii var. *distans* Kük. = ***Carex banksii*** Boott var. ***banksii***
Carex banksii var. *gilliesii* (Phil.) Kük. = ***Carex banksii*** Boott
Carex banksii f. *ochrostachys* (Phil.) Kük. = ***Carex banksii*** Boott var. ***banksii***
Carex banksii f. *pallescens* Kük. = ***Carex banksii*** Boott var. ***banksii***
Carex barbarae Parish = ***Carex schottii*** Dewey
Carex baritiana Porcius = ***Carex depressa*** subsp. ***transsilvanica*** (Schur) T.V.Egorova
Carex ×*barrei* H.Lév. = ?
Carex bartlettii O'Neill = ***Carex polystachya*** var. ***bartlettii*** (O'Neill) Standl. & Steyerm.
Carex basiflora C.B.Clarke = ***Carex brevicuspis*** var. ***basiflora*** (C.B.Clarke) Kük.
Carex basilaris Jord. = ***Carex depressa*** Link subsp. ***depressa***
Carex basilata Ohwi = ***Carex echinata*** Murray subsp. ***echinata***
Carex bastardiana DC. = ***Carex pilulifera*** L. subsp. ***pilulifera***
Carex bauhwaensis Z.P.Wang = ***Carex sachalinensis*** F.Schmidt var. ***sachalinensis***
Carex baviensis Franch. = ***Carex japonica*** Thunb.
Carex bayardii Fernald = ***Carex crus-corvi*** Shuttlew. ex Kunze
Carex beckeri C.A.Mey. ex Claus = ***Carex michelii*** Host
Carex beechyana Boott = ***Carex hookeri*** Kunth
Carex behringensis Gand. = ***Carex lyngbyei*** Hornem.
Carex behringensis C.B.Clarke = ***Carex podocarpa*** R.Br.
Carex bella-villa Dewey = ***Carex lupulina*** Muhl. ex Willd.
Carex bellardii All. = ***Kobresia myosuroides*** (Vill.) Fiori & Paol.
Carex bengalensis Roxb. = ***Carex cruciata*** Wahlenb. var. ***cruciata***
Carex bengalensis Thwaites = ***Carex indica*** L. var. ***indica***
Carex bengalensis f. *glauca* Boeck. ex Kuntze = ***Carex cruciata*** Wahlenb. var. ***cruciata***
Carex bengalensis var. *scaberrima* Boeck. = ***Carex cruciata*** var. ***rafflesiana*** (Boott) Noot.
Carex bengalensis var. *virgata* Boeck. = ***Carex cruciata*** var. ***rafflesiana*** (Boott) Noot.
Carex ×*bengyana* H.Lév. & L.C.Lamb. = ?
Carex benthamiana Boott = ***Carex notha*** Kunth
Carex bergrothii Palmgr. = ***Carex pilulifera*** L. subsp. ***pilulifera***
Carex beringiana Cham. ex Steud. = ***Carex stylosa*** C.A.Mey.
Carex bernardina Parish = ***Carex diandra*** Schrank
Carex berteroana Desv. = ***Carex aphylla*** Kunth
Carex berteroana Hohen. ex Boott = ***Carex setifolia*** Kunze var. ***setifolia***
Carex berteroana var. *colchaguensis* (Phil.) Kük. = ***Carex setifolia*** Kunze
Carex berteroana var. *setifolia* (Kunze) Kük. = ***Carex setifolia*** Kunze
Carex bertolonii Schkuhr = ***Carex divisa*** Huds.
Carex bessarabica (Savul. & Rayss) Zahar. = ***Carex distans*** L.
Carex beyrichiana Boeck. = ***Carex lurida*** Wahlenb.
Carex bhutanica T.Koyama = ***Carex haematostoma*** Nees
Carex bicknellii E.G.Camus = ***Carex annectens*** (E.P. Bicknell) E.P.Bicknell
Carex bicknellii var. *opaca* F.J.Herm. = ***Carex opaca*** (F.J.Herm.) P.Rothr. & Reznicek
Carex bicolor var. *minima* (Boullu) Nyman = ***Carex bicolor*** Bellardi ex All.
Carex biconvexis D.Don = ***Carex phacota*** Spreng.
Carex biconvexis Michx. = ?
Carex bicostata Olney = ?
Carex bicuspidata Regel ex Krecz. = ***Carex melanostachya*** M.Bieb. ex Willd.
Carex bidentula Franch. = ***Carex augustinowiczii*** Meinsh. var. ***augustinowiczii***
Carex bifida Boott = ***Carex serratodens*** S.Watson
Carex bifida Roth ex Steud. = ?
Carex biformis F.W.Schultz = ***Carex hostiana*** DC.
Carex biformis var. *fertilis* F.W.Schultz = ***Carex hostiana*** DC.
Carex biformis var. *sterilis* F.W.Schultz = ***Carex annectens*** (E.P.Bicknell) E.P.Bicknell
Carex bifurca Moench = ***Carex riparia*** Curtis subsp. ***riparia***
Carex bifurca Schrank = ***Carex rostrata*** Stokes var. ***rostrata***
Carex bigelowii subsp. *altaica* (Gorodkov) Malyschev = ***Carex orbicularis*** subsp. ***altaica*** (Gorodkov) T.V. Egorova
Carex bigelowii f. *anguillata* (Drejer) Fernald = ***Carex bigelowii*** Torr. ex Schwein. subsp. ***bigelowii***
Carex bigelowii var. *arctisibirica* (Jurtzev) Malyschev = ***Carex bigelowii*** subsp. ***arctisibirica*** (Jurtzev) Á.Löve & D.Löve
Carex bigelowii subsp. *dacica* (Heuff.) T.V.Egorova = ***Carex nigra*** subsp. ***dacica*** (Heuff.) Soó
Carex bigelowii f. *glacialis* (Fr.) Raymond = ***Carex bigelowii*** Torr. ex Schwein. subsp. ***bigelowii***
Carex bigelowii subsp. *hyperborea* (Drejer) Böcher = ***Carex bigelowii*** Torr. ex Schwein. subsp. ***bigelowii***
Carex bigelowii subsp. *nardeticola* Holub = ***Carex nigra*** subsp. ***dacica*** (Heuff.) Soó
Carex bigelowii subsp. *paishanensis* (Nakai) Vorosch. = ***Carex paishanensis*** Nakai
Carex bigelowii subsp. *rigida* (Raf.) W.Schultze-Motel = ***Carex bigelowii*** Torr. ex Schwein. subsp. ***bigelowii***
Carex bigelowii subsp. *rigidioides* (Gorodkov) T.V. Egorova = ***Carex rigidioides*** (Gorodkov) V.I.Krecz.
Carex biligularis DC. = ***Carex laevigata*** Sm.
Carex bina Schkuhr = ***Carex distachya*** Desf. var. ***distachya***
Carex binata Poir. = ***Carex parviflora*** Host
Carex ×*binderi* Podp. = ?
Carex binensis Franch. = ?
Carex binerviformis Holmboe = ?
Carex binervis Dewey = ***Carex laevigata*** Sm.
Carex binervis var. *alpina* Drejer = ***Carex binervis*** Sm.
Carex binervis f. *nigrescens* Druce = ***Carex binervis*** Sm.
Carex binnensis Kneuck. = ***Carex gmelinii*** Hook. & Arn.
Carex bipartita F.Dietr. =
Carex bipartita All. = ***Kobresia simpliciuscula*** (Wahlenb.) Mack. subsp. ***simpliciuscula***

Carex bipartita var. *amphigena* (Fernald) Polunin = ***Carex glareosa*** Schkuhr ex Wahlenb. subsp. ***glareosa***
Carex bipartita var. *austromontana* F.J.Herm. = ***Carex lachenalii*** Schkuhr subsp. ***lachenalii***
Carex bipartita subsp. *baetica* (Gay) Luceño & Muñoz Garm. = ***Carex lachenalii*** subsp. ***baetica*** (Gay) Luceño & Muñoz Garm.
Carex bipartita var. *glareosa* (Schkuhr ex Wahlenb.) Polunin = ***Carex glareosa*** Schkuhr ex Wahlenb.
Carex bisexualis C.B.Clarke = ***Schoenoxiphium ecklonii*** Nees
Carex bishallii C.B.Clarke = ***Carex nudata*** Boott
Carex bispicata Hook. & Arn. = ***Carex longirostrata*** C.A.Mey. var. ***longirostrata***
Carex biwensis Franch. = ***Carex rara*** Boott subsp. ***rara***
Carex biwensis subsp. *jankowskii* (Gorodkov) T.V.Egorova = ***Carex jankowskii*** Gorodkov
Carex blanda var. *varians* (L.H.Bailey) Burnham = ***Carex leptonervia*** (Fernald) Fernald
Carex blankinshipii Fernald = ***Carex gynodynama*** Olney
Carex blepharicarpa f. *duneensis* (Meinsh.) T.Koyama = ***Carex blepharicarpa*** var. ***duneensis*** (Meinsh.) Akiyama
Carex blepharicarpa var. *stenocarpa* Ohwi = ***Carex blepharicarpa*** var. ***hirtifructus*** (Kük.) Ohwi
Carex blepharolepis Nelmes = ***Carex commixta*** Steud.
Carex blepharophora A.Gray = ***Carex castanea*** Wahlenb.
Carex blyttii F.Nyl. = ***Carex disperma*** Dewey
Carex ×*boenninghauseniana* Kunth = ***Carex ×boenninghausiana*** Weihe
Carex ×*boenninghauseniana* var. *subpaniculata* Lange = ***Carex paniculata*** L. subsp. ***paniculata***
Carex ×*bogdanensis* Zapal. = ?
Carex bohemica f. *aggregata* (Domin) Soó = ***Carex bohemica*** Schreb.
Carex boissieri Steud. = ***Carex brunnea*** Thunb. subsp. ***brunnea***
Carex bolanderi Gand. = ?
Carex bolanderi var. *artemisiana* Kelso = ***Carex bolanderi*** Olney
Carex bolanderi var. *sparsiflora* Olney = ***Carex laeviculmis*** Meinsh.
Carex bolina Láng = ***Carex nigra*** (L.) Reichard subsp. ***nigra***
Carex bolliana Boeck. = ***Carex complanata*** Torr. & Hook. subsp. ***complanata***
Carex bolusii C.B.Clarke = ***Schoenoxiphium sparteum*** (Wahlenb.) C.B.Clarke
Carex bonariensis Schltdl. = ***Carex fissa*** Mack. var. ***fissa***
Carex bonariensis var. *achalensis* Kurtz ex Kük. = ***Carex feddeana*** H.Pfeiff.
Carex bonariensis f. *pumila* Osten = ***Carex bonariensis*** Desf. ex Poir.
Carex bonariensis var. *remota* Kük. = ***Carex trachycystis*** Griseb.
Carex bonariensis f. *remota* (Kük.) Kük. = ***Carex trachycystis*** Griseb.
Carex bonariensis var. *rupicola* Pedersen = ***Carex rupicola*** (Pedersen) G.A.Wheeler
Carex bonariensis var. *tolimensis* Maury = ***Carex bonariensis*** Desf. ex Poir.
Carex bonariensis var. *trachycystis* (Griseb.) Kük. = ***Carex trachycystis*** Griseb.
Carex bonatiana (Kük.) K.V.Ivanova = ***Kobresia fragilis*** C.B.Clarke
Carex bongardiana C.A.Mey. = ***Carex vesicaria*** L.
Carex bongardiana var. *gracilis* Trautv. & C.A.Mey. = ***Carex drymophila*** Turcz.
Carex bongardii Boott = ***Carex boottiana*** Hook. & Arn.
Carex bongardii var. *robusta* Franch. & Sav. = ***Carex boottiana*** Hook. & Arn.
Carex boniana Boeck. = ***Carex breviculmis*** R.Br. var. ***breviculmis***
Carex boninensis Koidz. = ***Carex boottiana*** Hook. & Arn.
Carex bonplandii var. *angustifolia* Boott = ***Carex bonplandii*** Kunth
Carex bonplandii var. *angustifolia* W.Boott = ***Carex illota*** L.H.Bailey
Carex bonplandii var. *humilior* Kük. = ***Carex bonplandii*** Kunth
Carex bonplandii var. *minor* Boott = ***Carex illota*** L.H. Bailey
Carex bonplandii var. *robustior* Kük. = ***Carex bonplandii*** Kunth
Carex boottiana Benth. ex Boott = ***Carex picta*** Steud.
Carex borbonica var. *pallidior* Kük. = ***Carex borbonica*** Lam.
Carex bordabensis Litv. = ?
Carex bordzilowskii V.I.Krecz. = ***Carex liparocarpos*** subsp. ***bordzilowskii*** (V.I.Krecz.) T.V.Egorova
Carex borealis Láng = ***Carex aquatilis*** Wahlenb. var. ***aquatilis***
Carex borii f. *lutea* R.R.Stewart = ***Carex borii*** Nelmes
Carex borneensis C.B.Clarke = ***Carex fusiformis*** Nees var. ***fusiformis***
Carex borneensis C.B.Clarke = ***Carex perakensis*** var. ***borneensis*** (C.B.Clarke) Noot.
Carex bornmulleri Kük. = ***Carex melanostachya*** M.Bieb. ex Willd.
Carex borotalicola Regel = ***Carex curaica*** Kunth
Carex boryana var. *minor* Boott = ***Carex boryana*** Schkuhr
Carex boryana var. *rigidifolia* Boeck. = ***Carex borbonica*** Lam.
Carex boryana var. *simplicissima* Kük. = ***Carex mannii*** E.A.Bruce
Carex boscii Willd. ex Spreng. = ?
Carex ×*botasgdensis* Kük. = ?
Carex bovina Howell = ***Carex aperta*** Boott
Carex brachycephala Poepp. ex Boott = ***Carex fuscula*** d'Urv. var. ***fuscula***
Carex brachyglossa Mack. = ***Carex annectens*** (E.P. Bicknell) E.P.Bicknell
Carex brachylepis Turcz. ex Steud. = ***Carex media*** R.Br.
Carex brachyphylla Turcz. = ***Carex ericetorum*** Pollich
Carex brachypoda T.Holm = ***Carex scopulorum*** var. ***bracteosa*** (L.H.Bailey) F.J.Herm.
Carex brachyrhyncha Gsaller = ***Carex ferruginea*** Scop. subsp. ***ferruginea***
Carex brachysandra Franch. = ***Carex thunbergii*** Steud.
Carex brachystachya (Regel & Maack) Akiyama = ***Carex capricornis*** Meinsh. ex Maxim.
Carex brachystachys var. *refracta* (Willd.) Fiori = ***Carex ferruginea*** subsp. ***tendae*** W.Dietr.
Carex bracteata Suter = ***Carex mucronata*** All.
Carex bracteata Giseke ex Boott = ***Carex salina*** Wahlenb. var. ***salina***
Carex bracteosa Schwein. = ***Carex vulpinoidea*** Michx.
Carex brasiliensis var. *gracilis* Boeck. = ***Carex glaucescens*** Elliott
Carex brasiliensis f. *robusta* C.B.Clarke ex Kük. = ***Carex brasiliensis*** A.St.-Hil.
Carex brazatana Boeck. = ***Carex cherokeensis*** Schwein.
Carex brehmeri Boeck. = ***Carex decidua*** Boott

Carex bremurii H.Christ ex Meinsh. = ***Carex umbrosa*** Host subsp. ***umbrosa***
Carex brenneri H.Christ = ***Carex umbrosa*** Host subsp. ***umbrosa***
Carex breutelii Schnizl. ex Lange = ***Carex canescens*** L. subsp. ***canescens***
Carex brevicaulis Mack. = [71 BRC 73 ORE WAS]
Carex brevicaulis Thouars = ***Uncinia brevicaulis*** (Thouars) Kunth
Carex breviceps Kük. = ***Carex oligostachya*** Nees
Carex breviculmis Thwaites = ***Carex hebecarpa*** C.A. Mey. subsp. ***hebecarpa***
Carex breviculmis f. *aphanandra* (Franch. & Sav.) Kük. = ***Carex breviculmis*** var. ***discoidea*** (Boott) Boott
Carex breviculmis f. *discoidea* (Boott) Kük. = ***Carex breviculmis*** var. ***discoidea*** (Boott) Boott
Carex breviculmis f. *fibrillosa* (Franch. & Sav.) Kük. = ***Carex breviculmis*** var. ***pluricostata*** Kük.
Carex breviculmis subsp. *fibrillosa* (Franch. & Sav.) T.Koyama = ***Carex breviculmis*** var. ***pluricostata*** Kük.
Carex breviculmis f. *filiculmis* (Franch. & Sav.) Kük. = ***Carex breviculmis*** R.Br. var. ***breviculmis***
Carex breviculmis var. *kingiana* (H.Lév. & Vaniot) Kük. = ***Carex mitrata*** Franch. subsp. ***mitrata***
Carex breviculmis f. *longiaristata* Kük. = ***Carex breviculmis*** R.Br. var. ***breviculmis***
Carex breviculmis subsp. *royleana* Nees = ***Carex breviculmis*** R.Br. var. ***breviculmis***
Carex breviculmis var. *stipitata* Kük. = ***Carex breviculmis*** R.Br. var. ***breviculmis***
Carex breviculmis var. *typica* Domin = ***Carex breviculmis*** R.Br.
Carex breviculmis f. *wardiana* (H.Lév. & Vaniot) Kük. = ***Carex breviculmis*** R.Br. var. ***breviculmis***
Carex brevifolia Host = ***Carex ferruginea*** Scop. subsp. ***ferruginea***
Carex breviglumis Ridl. = ***Carex perakensis*** var. ***borneensis*** (C.B.Clarke) Noot.
Carex breviglumis Ridl. = ***Carex perakensis*** var. ***borneensis*** (C.B.Clarke) Noot.
Carex breviglumis Le Gall = ?
Carex breviligulata Mack. = ***Carex densa*** (L.H.Bailey) L.H.Bailey
Carex brevior var. *crawei* (Boott) B.Boivin = ***Carex bicknellii*** Britton & A.Br.
Carex brevior var. *molesta* (Mack.) F.C.Gates = ***Carex molesta*** Mack.
Carex brevior var. *pseudofestucacea* Farw. = ***Carex merritt-fernaldii*** Mack.
Carex brevipes S.Watson = ***Carex rossii*** Boott
Carex brevirostrata Poir. = ***Carex liparocarpos*** Gaudin subsp. ***liparocarpos***
Carex × *brevirostris* Cederstr. = ?
Carex brevirostris Blytt = ?
Carex brevis S.T.Blake = ***Carex breviculmis*** var. ***perciliata*** Kük.
Carex brevisquama Mack. = ***Carex vallicola*** Dewey
Carex breweri var. *paddoensis* (Suksd.) Cronquist = ***Carex engelmannii*** L.H.Bailey
Carex brisoides Lam. = ***Carex praecox*** Schreb.
Carex brittoniana J.M.Coult. = ***Carex tetrastachya*** Scheele
Carex brizoides Huds. = ***Carex canescens*** L. subsp. ***canescens***
Carex brizoides Geners. = ***Carex leporina*** L.
Carex brizoides var. *adrianopolitana* Podp. = ***Carex praecox*** Schreb.
Carex brizoides var. *brunnea* Èelak. = ***Carex praecox*** Schreb.
Carex brizoides f. *brunnescens* Kük. = ***Carex brizoides*** L.
Carex brizoides var. *campestris* Wimm. = ***Carex praecox*** Schreb.
Carex brizoides subsp. *curvata* (Asch.) Èelak. = ***Carex praecox*** Schreb.
Carex brizoides var. *curvata* (Asch.) Beck = ***Carex praecox*** Schreb.
Carex brizoides var. *genuina* Èelak. = ***Carex brizoides*** L.
Carex brizoides var. *intermedia* Èelak. = ***Carex praecox*** Schreb.
Carex brizoides var. *nemoralis* Wimm. = ***Carex brizoides*** L.
Carex brizoides subsp. *ohmuelleriana* (O.Lang) Nyman = ***Carex arenaria*** L.
Carex brizoides var. *praecox* (Schreb.) Fiori = ***Carex praecox*** Schreb.
Carex brizoides f. *remota* Peterm. = ***Carex brizoides*** L.
Carex brizoides var. *stricta* Peterm. = ***Carex pseudobrizoides*** Chabaud
Carex brizoides f. *subheterostachya* Asch. & Graebn. = ***Carex brizoides*** L.
Carex brizopyrum Kunze = ***Carex remota*** subsp. ***alta*** (Boott) Kük.
Carex brongniartii L.H.Bailey = ***Carex densa*** (L.H. Bailey) L.H.Bailey
Carex brongniartii var. *densa* L.H.Bailey = ***Carex densa*** (L.H.Bailey) L.H.Bailey
Carex broteroana Samp. = ***Carex elata*** All. subsp. ***elata***
Carex brotherorum H.Christ = ?
Carex brownei Steud. = ***Carex longibrachiata*** Boeck.
Carex brownii var. *transversa* (Boott) Kük. = ***Carex brownii*** subsp. ***transversa*** (Boott) J.Kern
Carex brueckneri Kük. = ***Carex filiformis*** L.
Carex × *brueggeri* K.Richt. = ?
Carex brunnea var. *meyenii* (Nees) Kük. = ***Carex meyenii*** Nees
Carex brunnea var. *minor* (Boott) Kük. = ***Carex brunnea*** Thunb. subsp. ***brunnea***
Carex brunnea var. *scabrifolia* Boeck. = ***Carex brunnea*** Thunb. subsp. ***brunnea***
Carex brunnea var. *sendaica* (Franch.) Kük. = ***Carex lenta*** var. ***sendaica*** (Franch.) T.Koyama
Carex brunnea f. *simplex* Kük. = ***Carex brunnea*** Thunb. subsp. ***brunnea***
Carex brunnescens subsp. *alaskana* Kalela = ***Carex brunnescens*** (Pers.) Poir. subsp. ***brunnescens***
Carex brunnescens var. *gracilior* Britton = ***Carex brunnescens*** subsp. ***sphaerostachya*** (Tuck.) Kalela
Carex brunnescens subsp. *pacifica* Kalela = ***Carex brunnescens*** (Pers.) Poir. subsp. ***brunnescens***
Carex brunnescens var. *sphaerostachya* (Tuck.) Kük. = ***Carex brunnescens*** subsp. ***sphaerostachya*** (Tuck.) Kalela
Carex brunnescens subsp. *vitilis* (Fr.) Kalela = ***Carex brunnescens*** subsp. ***sphaerostachya*** (Tuck.) Kalela
Carex bucculenta V.I.Krecz. = ***Carex maritima*** Gunnerus
Carex buchananii C.B.Clarke = ***Schoenoxiphium rufum*** Nees
Carex buckleyi Dewey = ***Carex brunnescens*** subsp. ***sphaerostachya*** (Tuck.) Kalela
Carex buekii var. *melanostachya* Uechtr. =
Carex buennemeijeri Nelmes = ***Carex cruciata*** var. ***rafflesiana*** (Boott) Noot.
Carex bufonia Gueldenst. ex Ledeb. = ?
Carex bukaensis Palla = ***Carex oligostachya*** Nees
Carex bulbosa Boeck. = ***Carex laxiflora*** Lam. var. ***laxiflora***
Carex bulbostylis Kük. = ***Carex breviculmis*** var. ***montivaga*** (S.T.Blake) Noot.
Carex bullata Elliott = ***Carex striata*** Michx.

Carex bullata Dewey = ***Carex tuckermanii*** Boott
Carex bullata var. *greenii* (Boeck.) Fernald = ***Carex bullata*** Willd.
Carex bullata var. *laevirostris* Blytt ex Fr. = ***Carex utriculata*** Boott
Carex bullata var. *olneui* (Boott) Fernald = ***Carex × olneyi*** Boott
Carex bullata var. *olneyi* Fernald = ***Carex bullata*** Willd.
Carex bullockiana Nelmes = ***Carex pairae*** F.W.Schultz
Carex bungeana Debeaux = ***Carex heterostachya*** Bunge
Carex burejana Meinsh. = ***Carex drymophila*** Turcz.
Carex burjatorum Korotky = ?
Carex buruensis Nelmes = ***Carex brunnea*** Thunb. subsp. ***brunnea***
Carex buschiorum V.I.Krecz. ex Kolak. = ***Carex humilis*** Leyss. var. ***humilis***
Carex buxbaumii var. *alpicola* Hartm. = ***Carex adelostoma*** V.I.Krecz.
Carex buxbaumii var. *anticostensis* Raymond = ***Carex buxbaumii*** Wahlenb.
Carex buxbaumii var. *australis* Andersson = ***Carex buxbaumii*** Wahlenb.
Carex buxbaumii f. *dilutior* Kük. = ***Carex buxbaumii*** Wahlenb.
Carex buxbaumii subsp. *hartmanii* (Cojander) Domin = ***Carex hartmanii*** Cojander
Carex buxbaumii f. *heterostachya* Andersson = ***Carex buxbaumii*** Wahlenb.
Carex buxbaumii f. *macrostachya* Hartm. = ***Carex hartmanii*** Cojander
Carex buxbaumii f. *mitis* Norman ex Christian = ***Carex buxbaumii*** Wahlenb.
Carex buxbaumii var. *mutica* Hartm. = ***Carex adelostoma*** V.I.Krecz.
Carex buxbaumii subsp. *mutica* (Hartm.) Isov. = ***Carex adelostoma*** V.I.Krecz.
Carex buxbaumii f. *oenipontana* Grembl. ex Appel = ***Carex buxbaumii*** Wahlenb.
Carex buxbaumii f. *pedunculata* Raymond = ***Carex buxbaumii*** Wahlenb.
Carex buxbaumii subsp. *subulata* (Schumach.) Liro = ***Carex buxbaumii*** Wahlenb.
Carex buxbaumii f. *virescens* Norman ex Christian = ***Carex buxbaumii*** Wahlenb.
Carex caduca var. *ortegae* (Phil.) Kük. = ***Carex caduca*** Boott
Carex caesia Griseb. = ***Carex filiformis*** L.
Carex caflischii Brügger = ***Carex echinata*** Murray subsp. ***echinata***
Carex cajanderi Kük. = ***Carex bonanzensis*** Britton
Carex calbucana Phil. = ***Carex fuscula*** d'Urv. var. ***fuscula***
Carex calcitrapa Franch. = ***Carex maackii*** Maxim.
Carex calderae A.Hansen = ***Carex paniculata*** subsp. ***calderae*** (A.Hansen) Lewej. & Lobin
Carex ×*calderi* B.Boivin = ***Carex × heterophyta*** E.Holmb.
Carex callitrichos var. *austrohinganica* Y.L.Chang & Y.L.Yang = ***Carex humilis*** var. ***nana*** (H.Lév. & Vaniot) Ohwi
Carex callitrichos var. *nana* (H.Lév. & Vaniot) Ohwi = ***Carex humilis*** var. ***nana*** (H.Lév. & Vaniot) Ohwi
Carex campestris Host = ***Carex supina*** Willd. ex Wahlenb. var. ***supina***
Carex campestris Boott = ***Carex supina*** Willd. ex Wahlenb. var. ***supina***
Carex camporum Mack. = ***Carex praegracilis*** W.Boott
Carex camptoglochin V.I.Krecz. = ***Carex microglochin*** subsp. ***fuegina*** Kük.
Carex camptotropa V.I.Krecz. = ***Carex maritima*** Gunnerus
Carex campylocarpa T.Holm = ***Carex scopulorum*** var. ***bracteosa*** (L.H.Bailey) F.J.Herm.
Carex campylocarpa subsp. *affinis* Maguire & A.H. Holmgren = ***Carex scopulorum*** var. ***bracteosa*** (L.H.Bailey) F.J.Herm.
Carex campyloxys Steud. = ***Carex aematorrhyncha*** Desv. var. ***aematorrhyncha***
Carex camtschatcensis Kunth = ***Carex longirostrata*** C.A.Mey. var. ***longirostrata***
Carex canadensis Dewey = ***Carex lupulina*** Muhl. ex Willd.
Carex candolleana H.Lév. & Vaniot = ***Carex breviculmis*** R.Br. var. ***breviculmis***
Carex canescens var. *alpicola* Wahlenb. = ***Carex brunnescens*** (Pers.) Poir. subsp. ***brunnescens***
Carex canescens var. *alpicola* Kurtz = ***Carex canescens*** var. ***robustior*** Blytt ex Andersson
Carex canescens subsp. *arctiformis* (Mack.) Calder & R.L.Taylor = ***Carex arctiformis*** Mack.
Carex canescens f. *bracteata* Klett & Richt. = ***Carex canescens*** L. subsp. ***canescens***
Carex canescens var. *brunnea* Macoun = ***Carex canescens*** L. subsp. ***canescens***
Carex canescens var. *brunnescens* (Pers.) W.D.J.Koch = ***Carex brunnescens*** (Pers.) Poir.
Carex canescens subsp. *brunnescens* (Pers.) Asch. & Graebn. = ***Carex brunnescens*** (Pers.) Poir.
Carex canescens var. *congesta* Domin = ***Carex canescens*** L. subsp. ***canescens***
Carex canescens var. *disjuncta* Fernald = ***Carex canescens*** subsp. ***disjuncta*** (Fernald) Toivonen
Carex canescens var. *dubia* Kük. = ***Carex canescens*** var. ***robustior*** Blytt ex Andersson
Carex canescens var. *dubia* L.H.Bailey = ***Carex praeceptorium*** Mack.
Carex canescens var. *fallax* Kurtz ex Kük. = ***Carex canescens*** var. ***robustior*** Blytt ex Andersson
Carex canescens var. *gracillima* Schur = ***Carex canescens*** L. subsp. ***canescens***
Carex canescens var. *heterostachya* Farw. = ***Carex canescens*** L. subsp. ***canescens***
Carex canescens var. *laetevirens* Asch. = ***Carex canescens*** L. subsp. ***canescens***
Carex canescens var. *lapponica* (Láng) Nyman = ***Carex lapponica*** Láng
Carex canescens var. *longibracteata* Porcius = ***Carex canescens*** L. subsp. ***canescens***
Carex canescens var. *macilenta* (Fr.) H.Christ = ***Carex × macilenta*** Fr.
Carex canescens var. *maxima* Kük. = ***Carex heleonastes*** Ehrh. ex L.f. subsp. ***heleonastes***
Carex canescens var. *oregana* L.H.Bailey = ***Carex arcta*** Boott
Carex canescens var. *persoonii* H.Christ = ***Carex brunnescens*** (Pers.) Poir. subsp. ***brunnescens***
Carex canescens var. *polystachya* Boott = ***Carex arcta*** Boott
Carex canescens var. *robusta* A.Blytt = ***Carex canescens*** L. subsp. ***canescens***
Carex canescens var. *robustina* Macoun = ***Carex canescens*** L. subsp. ***canescens***
Carex canescens var. *sphaerostachya* Tuck. = ***Carex brunnescens*** subsp. ***sphaerostachya*** (Tuck.) Kalela
Carex canescens var. *subloliacea* Laest. = ***Carex lapponica*** Láng
Carex canescens f. *subsabulosa* (Norman) Toivonen = ***Carex canescens*** L. subsp. ***canescens***
Carex canescens var. *subtenella* Kük. = ***Carex canescens*** L. subsp. ***canescens***

Carex canescens var. *subvitilis* Neuman = ***Carex canescens*** L. subsp. ***canescens***
Carex canescens var. *tenuis* Láng = ***Carex canescens*** L. subsp. ***canescens***
Carex canescens var. *vitilis* (Fr.) J.Carey = ***Carex brunnescens*** subsp. ***sphaerostachya*** (Tuck.) Kalela
Carex canescens var. *vulgaris* L.H.Bailey = ***Carex brunnescens*** subsp. ***sphaerostachya*** (Tuck.) Kalela
Carex cantabrorum Gand. = ?
Carex capensis Schkuhr = ***Schoenoxiphium ecklonii*** Nees
Carex capensis Thunb. = ***Schoenoxiphium thunbergii*** Nees
Carex capillacea subsp. *aomorensis* (Franch.) T.V. Egorova = ***Carex capillacea*** var. ***sachalinensis*** (F.Schmidt) Ohwi
Carex capillacea var. *linzensis* Y.C.Yang = ***Carex capillacea*** Boott var. ***capillacea***
Carex capillacea var. *yunnanensis* Franch. = ***Carex capillacea*** Boott var. ***capillacea***
Carex capillaris var. *castanea* Nyman = ***Carex atrata*** L. subsp. ***atrata***
Carex capillaris subsp. *chlorostachys* (Steven) Á.Löve, D.Löve & Raymond = ***Carex capillaris*** L. subsp. ***capillaris***
Carex capillaris var. *elongata* Olney ex Fernald = ***Carex capillaris*** L. subsp. ***capillaris***
Carex capillaris subsp. *karoi* Freyn = ***Carex karoi*** (Freyn) Freyn
Carex capillaris var. *krausei* (Boeck.) Kranz = ***Carex krausei*** Boeck.
Carex capillaris f. *krausei* (Boeck.) Kük. = ***Carex krausei*** Boeck.
Carex capillaris subsp. *krausei* (Boeck.) Böcher = ***Carex krausei*** Boeck.
Carex capillaris var. *ledebouriana* (C.A.Mey. ex Trevir.) F.Schmidt = ***Carex ledebouriana*** C.A.Mey. ex Trevir.
Carex capillaris subsp. *ledebouriana* (C.A.Mey. ex Trevir.) Vorosch. = ***Carex ledebouriana*** C.A.Mey. ex Trevir.
Carex capillaris f. *major* Drejer = ***Carex capillaris*** L. subsp. ***capillaris***
Carex capillaris f. *minima* Beck = ***Carex capillaris*** L. subsp. ***capillaris***
Carex capillaris var. *nana* (Cham. ex Steud.) Kük. = ***Carex capillaris*** L. subsp. ***capillaris***
Carex capillaris var. *paludosa* F.Schmidt = ***Carex capillaris*** L. subsp. ***capillaris***
Carex capillaris var. *parvirostris* Kük. = ***Carex karoi*** (Freyn) Freyn
Carex capillaris var. *porsildiana* Polunin = ***Carex krausei*** subsp. ***porsildiana*** (Polunin) Á.Löve, D.Löve & Raymond
Carex capillaris f. *porsildiana* (Polunin) Raymond = ***Carex krausei*** subsp. ***porsildiana*** (Polunin) Á.Löve, D.Löve & Raymond
Carex capillaris subsp. *porsildiana* (Polunin) Böcher = ***Carex krausei*** subsp. ***porsildiana*** (Polunin) Á.Löve, D.Löve & Raymond
Carex capillaris var. *regeliana* Kük. = ***Carex regeliana*** (Kük.) Litv.
Carex capillaris subsp. *robustior* Drejer ex Böcher = ***Carex boecheriana*** Á.Löve, D.Löve & Raymond
Carex capillaris f. *saskatschewana* (Boeck.) L.H.Bailey = ***Carex capillaris*** L. subsp. ***capillaris***
Carex capillaris var. *tenuior* Drejer = ***Carex capillaris*** L. subsp. ***capillaris***
Carex capillaris f. *typica* B.Boivin = ***Carex capillaris*** L.
Carex capillaris var. *williamsii* (Britton) Boivin = ***Carex williamsii*** Britton
Carex capillata Schkuhr ex Kunth = ***Carex hirsutella*** Mack.
Carex capilliformis var. *major* Kük. = ***Carex capilliformis*** Franch.
Carex capillipes Drejer = ***Carex lyngbyei*** Hornem.
Carex capitata f. *alpicola* Andersson = ***Carex capitata*** L.
Carex capitata var. *arctogena* (Harry Sm.) Hultén = ***Carex capitata*** L.
Carex capitata f. *arctogena* (Harry Sm.) Raymond = ***Carex capitata*** L.
Carex capitata subsp. *arctogena* (Harry Sm.) Böcher ex Á.Löve, D.Löve & B.M.Kapoor = ***Carex capitata*** L.
Carex capitellata Boeck. = ***Carex phalaroides*** Kunth var. ***phalaroides***
Carex capitulata Boott ex C.B.Clarke = ***Carex oligostachya*** Nees
Carex capituliformis Meinsh. ex Maxim. = ***Carex onoei*** Franch. & Sav.
Carex cardioglochis H.Lév. & Vaniot = ***Carex otaruensis*** Franch.
Carex cardrona Hamlin = ***Carex uncifolia*** Cheeseman
Carex caricina (D.Don) Ghildyal & U.C.Bhattach. = ***Carex filicina*** Nees
Carex caricina var. *glaucina* (Boeck.) Ghildyal & U.C.Bhattach. = ***Carex filicina*** Nees
Carex caricina var. *leptocarpa* (C.B.Clarke) Ghildyal & U.C.Bhattach. = ***Carex filicina*** Nees
Carex caricina var. *microgyna* (C.B.Clarke) Karth. = ***Carex filicina*** Nees
Carex caricina var. *minor* Ghildyal & U.C.Bhattach. = ***Carex filicina*** Nees
Carex carlantoniana J.Gay = ***Carex enervis*** C.A.Mey. subsp. ***enervis***
Carex carltoniana Dewey = ***Carex heleonastes*** Ehrh. ex L.f. subsp. ***heleonastes***
Carex caroliniana Buckley = ***Carex austrocaroliniana*** L.H.Bailey
Carex caroliniana var. *cuspidata* (Dewey) Shinners = ***Carex bushii*** Mack.
Carex caryophyllaea J.F.Gmel. = ***Carex montana*** L.
Carex caryophyllata Kunth = ***Carex montana*** L.
Carex caryophyllea var. *angustifolia* Asch. & Graebn. = ***Carex caryophyllea*** Latourr. var. ***caryophyllea***
Carex caryophyllea f. *argyrata* (Waisb.) Kük. = ***Carex caryophyllea*** Latourr. var. ***caryophyllea***
Carex caryophyllea var. *caespitosa* (J.Fleisch.) Kük. = ***Carex caryophyllea*** Latourr. var. ***caryophyllea***
Carex caryophyllea f. *chlorantha* (Waisb.) Kük. = ***Carex caryophyllea*** Latourr. var. ***caryophyllea***
Carex caryophyllea f. *dissita* Franch. ex Kük. = ***Carex caryophyllea*** Latourr. var. ***caryophyllea***
Carex caryophyllea f. *globuliflora* (Waisb.) Kük. = ***Carex caryophyllea*** Latourr. var. ***caryophyllea***
Carex caryophyllea f. *glomerata* (Waisb.) Kük. = ***Carex caryophyllea*** Latourr. var. ***caryophyllea***
Carex caryophyllea subsp. *insularis* (H.Christ) Arrigoni = ***Carex caryophyllea*** Latourr. var. ***caryophyllea***
Carex caryophyllea f. *laxiflora* (Waisb.) Kük. = ***Carex caryophyllea*** Latourr. var. ***caryophyllea***
Carex caryophyllea f. *leiocarpa* Franch. ex Kük. = ***Carex caryophyllea*** Latourr. var. ***caryophyllea***
Carex caryophyllea f. *longibracteata* (Beck) Kük. = ***Carex caryophyllea*** Latourr. var. ***caryophyllea***
Carex caryophyllea f. *longivaginata* (Kük.) Kük. = ***Carex caryophyllea*** Latourr. var. ***caryophyllea***
Carex caryophyllea f. *luxurians* (Kük.) Kük. = ***Carex caryophyllea*** Latourr. var. ***caryophyllea***
Carex caryophyllea var. *mollis* Asch. & Graebn. = ***Carex caryophyllea*** Latourr. var. ***caryophyllea***

Carex caryophyllea subsp. *nervata* Franch. & Sav. = ***Carex caryophyllea*** Latourr. var. ***caryophyllea***
Carex caryophyllea var. *oxycarpa* (Waisb.) Kük. = ***Carex caryophyllea*** Latourr. var. ***caryophyllea***
Carex caryophyllea f. *pallescens* Kneuck. = ***Carex caryophyllea*** Latourr. var. ***caryophyllea***
Carex caryophyllea f. *pedunculata* (Beck) Kük. = ***Carex caryophyllea*** Latourr. var. ***caryophyllea***
Carex caryophyllea f. *pygmaea* (J.Fleisch.) Kük. = ***Carex caryophyllea*** Latourr. var. ***caryophyllea***
Carex caryophyllea f. *refracta* (Roth) Kük. = ***Carex caryophyllea*** Latourr. var. ***caryophyllea***
Carex caryophyllea f. *subpolyrrhiza* (Kük.) Kük. = ***Carex caryophyllea*** Latourr. var. ***caryophyllea***
Carex caryophyllea subsp. *trachyantha* (Dorner ex Heuff.) Degen = ***Carex caryophyllea*** Latourr. var. ***caryophyllea***
Carex caryophyllea var. *umbrosa* (Host) Fiori = ***Carex umbrosa*** Host
Carex caryophyllea f. *umbrosiformis* (J.Fleisch.) Kük. = ***Carex caryophyllea*** Latourr. var. ***caryophyllea***
Carex caryophyllea f. *vidalii* (Franch. & Sav.) Kük. = ***Carex caryophyllea*** Latourr. var. ***caryophyllea***
Carex castanea Miel. = ***Carex atrata*** L. subsp. ***atrata***
Carex castanea Elliott = ***Carex elliottii*** Schwein. & Torr.
Carex castanea var. *knieskernii* (Dewey) Mack. = ***Carex* × *knieskernii*** Dewey
Carex castellanosii Barros = ***Carex vallis-pulchrae*** Phil. var. ***vallis-pulchrae***
Carex casteriana Heer ex Boott = ***Carex dioica*** L.
Carex ×*castriferrei* Waisb. = ?
Carex ×*catalaunica* Sennen = ?
Carex catharinensis Boeck. = ***Carex fuscula*** d'Urv. var. ***fuscula***
Carex ×*catteyensis* A.Benn. = [10 GRB]
Carex caucasica Steven = ***Carex atrata*** subsp. ***caucasica*** (Steven) Kük.
Carex caucasica subsp. *jisaburo-ohwiana* (T.Koyama) T.Koyama = ***Carex jisaburo-ohwiana*** T.Koyama
Carex caudata (Kük.) Pereda & Laínz = ***Carex ferruginea*** subsp. ***caudata*** (Kük.) Pereda & Laínz
Carex caudatifrons (Akiyama) Akiyama = ***Carex flavocuspis*** Franch. & Sav.
Carex caulorrhiza H.Lév. & Vaniot = ***Carex middendorffii*** var. ***kirigaminensis*** (Ohwi) Ohwi
Carex caulorrhiza var. *kirigaminensis* Ohwi = ***Carex middendorffii*** var. ***kirigaminensis*** (Ohwi) Ohwi
Carex cavaleriei H.Lév. & Vaniot = ***Carex laticeps*** C.B.Clarke ex Franch.
Carex celsa (L.H.Bailey) Piper = ***Carex hassei*** L.H.Bailey
Carex celsa Boott = ***Carex psychrophila*** Nees
Carex cenisia Balb. = ***Carex bicolor*** Bellardi ex All.
Carex cephalantha (L.H.Bailey) E.P.Bicknell = ***Carex echinata*** Murray subsp. ***echinata***
Carex cephalophora var. *angustifolia* Boott = ***Carex leavenworthii*** Dewey
Carex cephalophora var. *anomala* Farw. = ***Carex cephalophora*** Muhl. ex Willd.
Carex cephalophora var. *bracteosa* Farw. = ***Carex cephalophora*** Muhl. ex Willd.
Carex cephalophora var. *leavenworthii* (Dewey) Kük. = ***Carex leavenworthii*** Dewey
Carex cephalophora var. *maxima* Dewey = ***Carex cephalophora*** Muhl. ex Willd.
Carex cephalophora var. *mesochorea* (Mack.) Gleason = ***Carex mesochorea*** Mack.
Carex ceramica Nelmes = ***Carex cruciata*** var. ***rafflesiana*** (Boott) Noot.
Carex cercostachys Franch. = ***Kobresia cercostachys*** (Franch.) C.B.Clarke
Carex cerina Dumort. = ***Carex trinervis*** Degl.
Carex cernua Boott = ***Carex dimorpholepis*** Steud.
Carex cernua Phil. = ***Carex magellanica*** Lam. subsp. ***magellanica***
Carex cernua Raeusch. = ?
Carex cernua var. *austroafricana* Kük. = ***Carex austroafricana*** (Kük.) Raymond
Carex cernua var. *lobolepis* (F.Muell.) Kük. = ***Carex lobolepis*** F.Muell.
Carex cernua var. *minor* Boott = ***Carex dimorpholepis*** Steud.
Carex cespitosa f. *acutiuscula* Lackow. = ***Carex cespitosa*** L. var. ***cespitosa***
Carex cespitosa var. *alpina* (Schur) Kük. = ***Carex cespitosa*** L. var. ***cespitosa***
Carex cespitosa var. *alpina* Gaudin = ***Carex nigra*** subsp. ***alpina*** (Gaudin) Lemke
Carex cespitosa var. *buekii* (Wimm.) Fiori = ***Carex buekii*** Wimm.
Carex cespitosa var. *curvata* J.Fleisch. = ***Carex nigra*** (L.) Reichard subsp. ***nigra***
Carex cespitosa var. *dacica* (Heuff.) Nyman = ***Carex nigra*** subsp. ***dacica*** (Heuff.) Soó
Carex cespitosa var. *elata* (All.) Fiori = ***Carex elata*** All.
Carex cespitosa var. *elliptica* Drejer =
Carex cespitosa f. *fuliginosa* Döll = ***Carex cespitosa*** L. var. ***cespitosa***
Carex cespitosa f. *fuliginosa* A.Br. = ***Carex nigra*** (L.) Reichard subsp. ***nigra***
Carex cespitosa var. *goodenowii* (J.Gay) Fiori = ***Carex nigra*** (L.) Reichard subsp. ***nigra***
Carex cespitosa var. *gracilis* (Curtis) Fiori = ***Carex acuta*** L.
Carex cespitosa var. *intricata* (Tineo) Fiori = ***Carex nigra*** subsp. ***intricata*** (Tineo) Maire & Wieller
Carex cespitosa var. *latifolia* Uechtr. ex Asch. & Graebn. = ***Carex cespitosa*** L. var. ***cespitosa***
Carex cespitosa f. *major* (Peterm.) Kük. = ***Carex cespitosa*** L. var. ***cespitosa***
Carex cespitosa var. *microsachys* Regel = ***Carex orbicularis*** Boott subsp. ***orbicularis***
Carex cespitosa var. *multifibrillosa* Kük. = ***Carex cespitosa*** L. var. ***cespitosa***
Carex cespitosa var. *panormitana* (Guss.) Fiori = ***Carex panormitana*** Guss.
Carex cespitosa var. *polymorpha* Laest. = ***Carex nigra*** (L.) Reichard subsp. ***nigra***
Carex cespitosa f. *polystachya* Peterm. = ***Carex cespitosa*** L. var. ***cespitosa***
Carex cespitosa var. *ramosa* Dewey = [75 MAS]
Carex cespitosa var. *recta* J.Fleisch. = ***Carex nigra*** (L.) Reichard subsp. ***nigra***
Carex cespitosa var. *retorta* Fr. = ***Carex cespitosa*** L. var. ***cespitosa***
Carex cespitosa var. *rigida* (Raf.) Benth. & Hook.f. = ***Carex bigelowii*** Torr. ex Schwein. subsp. ***bigelowii***
Carex cespitosa var. *spreta* (Steud.) Nyman = ***Carex cespitosa*** L. var. ***cespitosa***
Carex cespitosa var. *strictissima* Kük. = ***Carex cespitosa*** L. var. ***cespitosa***
Carex cespitosa var. *waisbeckeri* Kük. = ***Carex cespitosa*** L. var. ***cespitosa***
Carex ×*cetica* Rech. = ?
Carex chabertii F.W.Schultz = ***Carex divulsa*** subsp. ***leersii*** (Kneuck.) W.Koch
Carex chaetogyne Nelmes = ***Carex fedia*** Nees
Carex chaetophylla Steud. = ***Carex divisa*** Huds.

Carex chaetorhiza Franch. & Sav. = ***Carex pansa*** L.H.Bailey
Carex chaetorrhiza var. *stenostachys* Franch. & Sav. = ***Carex pansa*** L.H.Bailey
Carex chalaros Steud. = ***Carex granularis*** Muhl. ex Willd.
Carex chalciolepis var. *larimerana* Kelso = ***Carex atrata*** L. subsp. ***atrata***
Carex chalcodeta V.I.Krecz. = ***Carex pallescens*** var. ***chalcodeta*** (V.I.Krecz.) Ö.Nilsson
Carex chamarensis T.V.Egorova = ***Carex heterolepis*** Bunge
Carex chamissoi Boeck. = ***Carex pediformis*** C.A.Mey. subsp. ***pediformis***
Carex chapmanii Torr. ex Boott = ***Carex baltzellii*** Chapm.
Carex chapmanii Sartwell ex Dewey = ***Carex tenax*** Chapm. ex Dewey
Carex chapmanii Sartwell ex L.H.Bailey = ***Carex tenax*** Chapm. ex Dewey
Carex cheesemanii Petrie = ***Carex comans*** Berggr.
Carex cheilungkiangnica A.I.Baranov & Skvortsov = ***Carex cinerascens*** Kük.
Carex ×*chevalieri* Corb. = ***Carex ×xanthocarpa*** nothovar. ***leutzii*** (Kneuck.) Govaerts
Carex chienii F.T.Wang & Tang = ***Carex adrienii*** E.G.Camus
Carex chilensis Brongn. = ***Carex riparia*** subsp. ***chilensis*** (Brongn.) Kük.
Carex chilosacca C.B.Clarke = ***Carex maculata*** Boott var. ***maculata***
Carex chimaphila T.Holm = ***Carex scopulorum*** T.Holm var. ***scopulorum***
Carex chinensis Munro ex Hook.f. = ***Carex munroi*** Boott ex C.B.Clarke
Carex chinensis var. *longkiensis* (Franch.) Kük. = ***Carex chinensis*** Retz.
Carex chinensis var. *tsushimensis* Ohwi = ***Carex tsushimensis*** (Ohwi) Ohwi
Carex chinganensis Litv. = ***Carex chloroleuca*** Meinsh.
Carex chionophila T.Holm = ***Carex aquatilis*** Wahlenb. var. ***aquatilis***
Carex chishimana Ohwi = ***Carex glareosa*** Schkuhr ex Wahlenb. subsp. ***glareosa***
Carex chitralensis Nelmes = ***Carex koshewnikowii*** subsp. ***chitralensis*** (Nelmes) Dickore
Carex chlamydea Norman = ?
Carex chlorantha var. *composita* F.Muell. = ***Carex appressa*** R.Br. var. ***appressa***
Carex chlorocarpa Wimm. = ***Carex nigra*** subsp. ***alpina*** (Gaudin) Lemke
Carex chlorocarpos Liebm. = ***Carex anisostachys*** Liebm.
Carex chlorocystis Boeck. = ***Carex harlandii*** Boott
Carex chlorolepis Steud. = ***Carex phalaroides*** Kunth var. ***phalaroides***
Carex chlorophila Mack. = ***Carex viridula*** Michx. subsp. ***viridula***
Carex chlorostachys D.Don ex Tilloch & Taylor = ***Carex alopecuroides*** D.Don ex Tilloch & Taylor
Carex chlorostachys Steven = ***Carex capillaris*** L. subsp. ***capillaris***
Carex chlorostachys var. *conferta* Tang & F.T.Wang ex L.K.Dai = ***Carex capillaris*** L. subsp. ***capillaris***
Carex chordalis Liebm. = ***Carex jamesonii*** var. ***gracilis*** L.H.Bailey
Carex chordorrhiza var. *aestivalis* Asch. & Graebn. = ***Carex chordorrhiza*** L.f.
Carex chordorrhiza var. *genuina* Trautv. = ***Carex chordorrhiza*** L.f.
Carex chordorrhiza subsp. *gynocrates* (Wormsk.) Nyman = ***Carex gynocrates*** Wormsk.
Carex chordorrhiza var. *major* Boeck. = ***Carex pseudocuraica*** F.Schmidt
Carex chordorrhiza var. *pseudocuraica* (F.Schmidt) Trautv. = ***Carex pseudocuraica*** F.Schmidt
Carex chordorrhiza var. *sphagnicola* Laest. ex Th.Fr. = ***Carex chordorrhiza*** L.f.
Carex chorgosica Meinsh. = ***Carex diluta*** M.Bieb. subsp. ***diluta***
Carex ×*christii* Boeck. = ?
Carex christyana Boott = ***Carex cherokeensis*** Schwein.
Carex chrysolepis var. *modesta* H.Lév. & Vaniot = ***Carex doenitzii*** Boeck. var. ***doenitzii***
Carex chrysolepis var. *odontostoma* (Kük.) Ohwi = ***Carex chrysolepis*** Franch. & Sav. var. ***chrysolepis***
Carex chrysoleuca T.Holm = ***Carex densa*** (L.H.Bailey) L.H.Bailey
Carex chubutensis C.B.Clarke = ***Carex subantarctica*** Speg.
Carex chuniana F.T.Wang & Tang = ***Carex commixta*** Steud.
Carex ciliaris Fernald = ***Carex longicaulis*** Boeck.
Carex ciliata Willd. = ***Carex ericetorum*** Pollich
Carex ×*ciliatifructus* Ohwi = ?
Carex ciliatomarginata Nakai = ***Carex siderosticta*** var. ***pilosa*** H.Lév. ex T.Koyama
Carex ciliolata Franch. = ***Carex scita*** var. ***riishirensis*** (Franch.) Kük.
Carex cincta Franch. = ***Carex phacota*** Spreng.
Carex cincta var. *subphacota* Kük. = ***Carex phacota*** Spreng.
Carex cinerea Pollich = ***Carex canescens*** L. subsp. ***canescens***
Carex cinerea var. *robustior* (Kük.) B.Boivin = ***Carex arctiformis*** Mack.
Carex cinnamomea Cheeseman = ***Carex forsteri*** Wahlenb.
Carex cinnamomea Olney = ***Carex mendocinensis*** Olney ex Boott
Carex cinnamomea Boott = ***Carex nivalis*** Boott
Carex cirrhosa var. *lutescens* Kük. = ***Carex cirrhosa*** Berggr.
Carex cladostachya Wahlenb. = ***Carex polystachya*** Sw. ex Wahlenb. var. ***polystachya***
Carex cladostachya var. *maxima* Kük. = ***Carex polystachya*** Sw. ex Wahlenb. var. ***polystachya***
Carex cladostachya var. *mexicana* (C.Presl) Kük. = ***Carex polystachya*** Sw. ex Wahlenb. var. ***polystachya***
Carex cladostachya var. *oerstedii* (Liebm.) Kük. = ***Carex polystachya*** Sw. ex Wahlenb. var. ***polystachya***
Carex cladostachya f. *polystachya* (Sw. ex Wahlenb.) C.B.Clarke ex Lindm. = ***Carex polystachya*** Sw. ex Wahlenb.
Carex clandestina Gooden. = ***Carex humilis*** Leyss. var. ***humilis***
Carex clarkeana Kük. = ***Carex cruciata*** var. ***rafflesiana*** (Boott) Noot.
Carex clavata var. *campylostachya* Nees = ***Carex clavata*** Thunb.
Carex clavata f. *lutensis* (Kunth) Kük. = ***Carex clavata*** Thunb.
Carex clavaudiana Rouy = ***Carex pseudobrizoides*** Chabaud
Carex claviformis Hoppe = ***Carex flacca*** Schreb. subsp. ***flacca***
Carex clemensii Kük. = ***Carex perakensis*** var. ***borneensis*** (C.B.Clarke) Noot.
Carex clivicola Fernald & Weath. = ***Carex peckii*** Howe
Carex coacta Boott = ***Carex divisa*** Huds.
Carex coarctata Kit. = ***Carex divisa*** Huds.

Carex cobresiiformis A.I.Baranov & Skvortsov = ***Carex gynocrates*** Wormsk.
Carex ×*cokae* Podp. = [11 CZE]
Carex colchaguensis Phil. = ***Carex setifolia*** Kunze var. ***setifolia***
Carex colchica subsp. *ligerica* (J.Gay) T.V.Egorova = ***Carex colchica*** J.Gay
Carex collata Boott = ***Carex appressa*** R.Br. var. ***appressa***
Carex collecta Dewey = ***Carex albicans*** Willd. ex Spreng. var. ***albicans***
Carex collina Willd. = ***Carex montana*** L.
Carex colorata Mack. = ***Carex woodii*** Dewey
Carex columbiana Dewey = ***Carex mertensii*** Prescott ex Bong. var. ***mertensii***
Carex comans var. *cheesemanii* (Petrie) Kük. = ***Carex comans*** Berggr.
Carex comans var. *pulchella* Kük. = ***Carex comans*** Berggr.
Carex comans f. *stricta* Cheeseman = ***Carex comans*** Berggr.
Carex comari H.Lév. & Vaniot = ***Carex frigida*** All.
Carex commersoniana Sieber ex Kunth = ***Carex brunnea*** Thunb. subsp. ***brunnea***
Carex communis var. *gynandra* Farw. = ***Carex communis*** L.H.Bailey
Carex communis f. *gynandra* (Farw.) E.G.Voss = ***Carex communis*** L.H.Bailey
Carex communis var. *wheeleri* L.H.Bailey = ***Carex austrina*** Mack.
Carex commutata Turcz. = ***Carex curaica*** Kunth
Carex commutata J.Gay = ***Carex stricta*** Lam.
Carex compacta Wimm. = ***Carex brunnea*** Thunb. subsp. ***brunnea***
Carex compacta R.Br. = ***Carex membranacea*** Hook.
Carex compacta Krock. ex Hoppe = ***Carex nigra*** (L.) Reichard subsp. ***nigra***
Carex compacta Lam. = ***Carex vulpina*** L.
Carex compacta (R.Br.) Poir. = ***Uncinia compacta*** R.Br.
Carex complanata var. *hirsuta* (Raf.) Gleason = ***Carex hirsutella*** Mack.
Carex complanata subsp. *hirsutella* (Mack.) R.T.Clausen = ***Carex hirsutella*** Mack.
Carex complanata var. *robusta* Burnham = ***Carex complanata*** Torr. & Hook.
Carex composita f. *simplicior* Kük. = ***Carex composita*** Boott
Carex compressa Vahl ex Kunth = ***Carex bonariensis*** Desf. ex Poir. var. ***bonariensis***
Carex compressa Kit. ex Willd. = ***Carex brachystachys*** Schrank
Carex compressa Hosé = ***Carex canescens*** L. subsp. ***canescens***
Carex compressa Gaudin = ***Carex elata*** All. subsp. ***elata***
Carex concolor R.Br. = ***Carex aquatilis*** var. ***minor*** Boott
Carex concolor Nees = ***Carex speciosa*** Kunth subsp. ***speciosa***
Carex concolor var. *minuscula* Kuvaev = ***Carex aquatilis*** var. ***minor*** Boott
Carex condensata var. *flava* Nees = ***Carex stramentitia*** Boott ex Boeck.
Carex confertiflora Boott = ***Carex olivacea*** subsp. ***confertiflora*** (Boott) T.Koyama
Carex confusa Hamlin = ***Carex ternaria*** G.Forst.
Carex congesta C.A.Mey. ex Boott = ***Carex hoodii*** Boott
Carex conglobata All. = ***Carex montana*** L.
Carex conglobata Kit. ex Willd. = ***Carex supina*** Willd. ex Wahlenb. var. ***supina***
Carex congolensis Turrill = ***Carex cognata*** var. ***congolensis*** (Turrill) Lye
Carex conica var. *densa* Kük. = ***Carex kiangsuensis*** Kük.
Carex conica var. *leucolepis* Franch. & Sav. = ***Carex conica*** Boott
Carex conica f. *longirostrata* Kük. = ***Carex conica*** Boott
Carex conica f. *mollis* Kük. = ***Carex conica*** Boott
Carex conica f. *rubens* Kük. = ***Carex conica*** Boott
Carex conicoides Honda = ***Carex sachalinensis*** var. ***conicoides*** (Honda) Ohwi
Carex conjugata Willd. ex Kunth = ***Carex brunnea*** Thunb. subsp. ***brunnea***
Carex conoidea Muhl. = ***Carex laxiflora*** Lam. var. ***laxiflora***
Carex conoidea f. *katahdinensis* (Fernald) B.Boivin = ***Carex conoidea*** Willd.
Carex conorrhyncha Nelmes = ***Carex breviculmis*** R.Br. var. ***breviculmis***
Carex consanguinea Kunth = ***Carex divisa*** Huds.
Carex consimilis T.Holm = ***Carex bigelowii*** subsp. ***lugens*** (T.Holm) T.V.Egorova
Carex consocialis Steud. = ***Carex alopecuroides*** D.Don ex Tilloch & Taylor
Carex consors C.B.Clarke = ***Carex psilocarpa*** Steud.
Carex conspicua Boott ex C.B.Clarke = ?
Carex constricta S.T.Blake = ***Carex celebica*** Kük.
Carex contigua Hoppe = ***Carex spicata*** Huds.
Carex contigua f. *fumosa* (Gren.) Kük. = ***Carex spicata*** Huds.
Carex contigua subsp. *leersii* (Kneuck.) Degen = ***Carex divulsa*** subsp. ***leersii*** (Kneuck.) W.Koch
Carex contigua var. *longissima* Tauscher ex Vollm. = ***Carex spicata*** Huds.
Carex contigua var. *mediterranea* Asch. & Graebn. = ***Carex spicata*** Huds.
Carex contigua f. *pallida* Appel ex Woerl. = ***Carex spicata*** Huds.
Carex contigua f. *pseudoguestphalica* Asch. & Graebn. = ***Carex spicata*** Huds.
Carex contigua f. *remota* F.W.Schultz = ***Carex spicata*** Huds.
Carex contigua f. *submonostachya* Asch. & Graebn. = ***Carex spicata*** Huds.
Carex continua subsp. *pairae* (F.W.Schultz) Degen = ***Carex pairae*** F.W.Schultz
Carex continua var. *plebeia* (C.B.Clarke) Kük. = ***Carex continua*** C.B.Clarke
Carex contracta Boeck. = ***Carex satsumensis*** Franch. & Sav
Carex convexa Kit. = ***Carex echinata*** Murray subsp. ***echinata***
Carex convoluta Mack. = ***Carex rosea*** Willd.
Carex cooleyi Dewey = ***Carex hystericina*** Muhl. ex Willd.
Carex cooptanda C.B.Clarke = ***Carex lachenalii*** Schkuhr subsp. ***lachenalii***
Carex copulata (L.H.Bailey) Mack. = ***Carex laxiculmis*** var. ***copulata*** (L.H.Bailey) Fernald
Carex ×*corcoutica* Domin = [11 CZE]
Carex coreana L.H.Bailey = ***Carex dickinsii*** Franch. & Sav.
Carex coriogyne Nelmes = ***Carex divulsa*** subsp. ***coriogyne*** (Nelmes) Ö.Nilsson
Carex coriophora Fisch. & C.A.Mey. ex Kunth = ***Carex atrofusca*** var. ***major*** (Boeck.) Raymond
Carex coriophora subsp. *langtaodianensis* S.Yun Liang = [36 CHN]
Carex coronata H.Lév. = ***Carex dispalata*** Boott subsp. ***dispalata***
Carex corralensis Phil. = ***Carex aematorrhyncha*** var. ***corralensis*** (Phil.) Kük.

Carex corsica Degl. = ***Carex microcarpa*** Bertol. ex Moris
Carex corsicana Link = ***Carex punctata*** Gaudin var. ***punctata***
Carex ×*corstorphinei* Druce = ?
Carex corynophora Peterm. = ***Carex acuta*** L.
Carex costata Turcz. ex Ledeb. = ***Carex korshinskyi*** Kom.
Carex costata J.Presl & C.Presl = ***Carex supina*** Willd. ex Wahlenb. var. ***supina***
Carex costata Schwein. = ***Carex virescens*** Muhl. ex Willd.
Carex ×*costei* Rouy = ?
Carex courtallensis Nees ex Boott = ***Carex speciosa*** Kunth subsp. ***speciosa***
Carex cousturieri Gand. = ***Carex leporina*** L.
Carex crandallii Gand. = ***Carex micropoda*** C.A.Mey.
Carex craspedotricha Nelmes = ***Carex remota*** subsp. ***alta*** (Boott) Kük.
Carex crassa Ehrh. = ***Carex riparia*** Curtis subsp. ***riparia***
Carex crassicaulis Franch. = ***Carex foliosissima*** F.Schmidt var. ***foliosissima***
Carex crassinervia Franch. = ***Carex meyeriana*** Kunth
Carex crassipes Boeck. = ***Carex daltonii*** Boott
Carex crawei var. *heterostachya* Dewey = ***Carex crawei*** Dewey ex Torr.
Carex crawfordii var. *vigens* Fernald = ***Carex crawfordii*** Fernald
Carex crawfordii f. *vigens* (Fernald) Kük. = ***Carex crawfordii*** Fernald
Carex cremostachys Franch. = ***Carex prescottiana*** Boott
Carex crinalis Boott = ***Carex peruviana*** J.Presl & C.Presl
Carex crinigera Boott = ***Carex ramosa*** Willd.
Carex crinigera var. *minor* Boott = ***Carex pyramidalis*** Kük.
Carex crinita var. *gynandra* (Schwein.) Schwein. & Torr. = ***Carex gynandra*** Schwein.
Carex crinita var. *minor* Boott = ***Carex crinita*** Lam. var. ***crinita***
Carex crinita f. *minor* Boott = ***Carex crinita*** Lam. var. ***crinita***
Carex crinita var. *mitchelliana* (M.A.Curtis) Gleason = ***Carex mitchelliana*** M.A.Curtis
Carex crinita var. *paleacea* (Schreb. ex Wahlenb.) Dewey = ***Carex paleacea*** Schreb. ex Wahlenb.
Carex crinita var. *simulans* Fernald = ***Carex gynandra*** Schwein.
Carex cristata Schwein. = ***Carex cristatella*** Britton & A.Br.
Carex cristata Clairv. = ***Carex ferruginea*** Scop. subsp. ***ferruginea***
Carex cristata var. *catelliformis* Farw. = ***Carex cristatella*** Britton & A.Br.
Carex cristata f. *ellipsoidalis* Farw. = ***Carex cristatella*** Britton & A.Br.
Carex cristata var. *mirabilis* (Tuck.) A.Gray = ***Carex normalis*** Mack.
Carex cristata var. *mirabilis* Boott = ***Carex straminea*** Willd. ex Schkuhr
Carex cristata var. *reducta* (L.H.Bailey) Kük. = ***Carex tribuloides*** Wahlenb. var. ***tribuloides***
Carex cristatella f. *catelliformis* (Farw.) Fernald = ***Carex cristatella*** Britton & A.Br.
Carex cruciata Nees = ***Carex cruciata*** Wahlenb. var. ***cruciata***
Carex cruciata var. *argocarpus* C.B.Clarke = ***Carex cruciata*** Wahlenb. var. ***cruciata***
Carex cruciata var. *nagporensis* C.B.Clarke = ***Carex cruciata*** Wahlenb. var. ***cruciata***
Carex cruciata var. *rubrobrunnea* Ohwi = ***Carex cruciata*** Wahlenb. var. ***cruciata***
Carex cruciata subsp. *rubrobrunnea* (Ohwi) T.Koyama = ***Carex cruciata*** Wahlenb. var. ***cruciata***
Carex crus-corvi f. *orthoclados* Kunze = ***Carex crus-corvi*** Shuttlew. ex Kunze
Carex crus-corvi var. *virginiana* Fernald = ***Carex crus-corvi*** Shuttlew. ex Kunze
Carex crustacea Nelmes = ?
Carex cryptandra Franch. & Sav. = ***Carex jackiana*** var. ***macroglossa*** (Franch. & Sav.) Kük.
Carex cryptantha T.Holm = ***Carex glareosa*** Schkuhr ex Wahlenb. subsp. ***glareosa***
Carex cryptocarpa Cheeseman = ***Carex decurtata*** Cheeseman
Carex cryptocarpa C.A.Mey. = ***Carex lyngbyei*** Hornem.
Carex cryptocarpa var. *haematolepis* (Drejer) L.H. Bailey = ***Carex* × *haematolepis*** Drejer
Carex cryptocarpa var. *pumila* L.H.Bailey = ***Carex lyngbyei*** Hornem.
Carex cryptocarpa var. *variegata* (Drejer) Britton = ***Carex lyngbyei*** Hornem.
Carex cryptolepis var. *prolifera* (H.B.Lord) House = ***Carex cryptolepis*** Mack.
Carex ×*cuatrecasasii* Sennen = [12 SPA]
Carex cubensis var. *flaccida* Kük. = ***Carex cubensis*** Kük.
Carex cucullata (Kük.) Ohwi = ***Carex ligata*** Boott subsp. ***ligata***
Carex cumingiana Steud. = ***Carex oligostachya*** Nees
Carex cumulata f. *soluta* Fernald = ***Carex cumulata*** (L.H.Bailey) Mack.
Carex cuneata Ohwi = ***Carex stenostachys*** var. ***cuneata*** (Ohwi) Ohwi & T.Koyama
Carex cunninghamii Boott = ***Carex acuta*** L.
Carex cuprea (Kük.) Nelmes = ***Carex petitiana*** A.Rich.
Carex cuprina subsp. *polyphylla* (Kar. & Kir.) = ***Carex divulsa*** subsp. ***leersii*** (Kneuck.) W.Koch
Carex cuprina var. *subcontigua* (Kük.) De Langhe & Lambinon = ***Carex cuprina*** (I.Sándor ex Heuff.) Nendtv. ex A.Kern.
Carex curaica Franch. & Sav. = ***Carex pansa*** L.H.Bailey
Carex curaica var. *angustifolia* Turcz. = ***Carex curaica*** Kunth
Carex curaica var. *armena* Schischk. = ***Carex pseudofoetida*** subsp. ***acrifolia*** (V.I.Krecz.) Kukkonen
Carex curaica var. *coacta* (Boott) Boott = ***Carex divisa*** Huds.
Carex curaica var. *latifolia* Turcz. = ***Carex curaica*** Kunth
Carex curaica var. *major* Boeck. = ***Carex pseudocuraica*** F.Schmidt
Carex curaica var. *pycnostachys* (Kar. & Kir.) Kük. = ***Carex pycnostachys*** Kar. & Kir.
Carex curaica subsp. *pycnostachys* (Kar. & Kir.) T.V.Egorova = ***Carex pycnostachys*** Kar. & Kir.
Carex curaica var. *rigida* Meinsh. = ***Carex curaica*** Kunth
Carex curaica var. *rigida* Meinsh. = ***Carex stenophylla*** Wahlenb. subsp. ***stenophylla***
Carex curta Gooden. = ***Carex canescens*** L. subsp. ***canescens***
Carex curta var. *brunnescens* Pers. = ***Carex brunnescens*** (Pers.) Poir.
Carex curta var. *maxima* (Kük.) R.R.Stewart = ***Carex heleonastes*** Ehrh. ex L.f. subsp. ***heleonastes***
Carex curta var. *robustior* (Blytt ex Andersson) D.M. Moore & Chater = ***Carex canescens*** var. ***robustior*** Blytt ex Andersson
Carex curticeps C.B.Clarke = ***Kobresia curticeps*** (C.B.Clarke) Kük.
Carex curtiglumis C.B.Clarke = ***Carex augustinowiczii*** Meinsh. var. ***augustinowiczii***
Carex curtisii Ridl. = ***Carex breviscapa*** C.B.Clarke

Carex curvata Knaf = ***Carex praecox*** Schreb.
Carex curvifolia Boeck. = ***Carex macrorrhiza*** Boeck. var. ***macrorrhiza***
Carex curvirostra Hartm. = ***Carex vaginata*** Tausch var. ***vaginata***
Carex curvirostris Kunze = ***Carex baccans*** Nees
Carex curvula Willd. ex Kunth = ***Carex davalliana*** Sm.
Carex curvula Lam. = ***Carex praecox*** Schreb.
Carex curvula f. *elongata* Husn. = ***Carex curvula*** All. subsp. ***curvula***
Carex curvula subsp. *mirabilis* Arcang. = ***Carex curvula*** All. subsp. ***curvula***
Carex curvula var. *mirabilis* Nyman = ***Carex curvula*** All. subsp. ***curvula***
Carex curvula f. *orbelica* (Velen.) Kük. = ***Carex curvula*** All. subsp. ***curvula***
Carex curvula f. *pallida* Asch. & Graebn. = ***Carex curvula*** All. subsp. ***curvula***
Carex curvula f. *pygmaea* Holler ex Asch. & Graebn. = ***Carex curvula*** All. subsp. ***curvula***
Carex curvula f. *rodnensis* Porcius ex Asch. & Graebn. = ***Carex curvula*** All. subsp. ***curvula***
Carex curvula var. *tripartita* (All.) Nyman = ***Carex lachenalii*** Schkuhr subsp. ***lachenalii***
Carex cuspidata Bertol. = ***Carex divisa*** Huds.
Carex cuspidata Host = ***Carex flacca*** subsp. ***erythrostachys*** (Hoppe) Holub
Carex cuspidata Wahlenb. = ***Carex salina*** Wahlenb. var. ***salina***
Carex cuspidata Degl. = ***Carex salina*** Wahlenb. var. ***salina***
Carex custoriana Heer ex Hegetschw. = ***Carex dioica*** L.
Carex cyclocarpa T.Holm = ***Carex bigelowii*** subsp. ***lugens*** (T.Holm) T.V.Egorova
Carex cyclocystis Boeck. = ***Carex orbicularis*** Boott subsp. ***orbicularis***
Carex cylindrica Miel. ex Kunth = ***Carex atrata*** L. subsp. ***atrata***
Carex cylindrica Schwein. = ***Carex bullata*** Willd.
Carex cyperoidea Houtt. = ***Carex bohemica*** Schreb.
Carex cyperoides L. = ***Carex bohemica*** Schreb.
Carex cyperoides f. *aggregata* Domin = ***Carex bohemica*** Schreb.
Carex czarwakensis Litv. = ***Carex diluta*** M.Bieb. subsp. ***diluta***
Carex czetzii Janka = ***Carex montana*** L.
Carex dacica Heuff. = ***Carex nigra*** subsp. ***dacica*** (Heuff.) Soó
Carex dahurica subsp. *chosenica* (Ohwi) T.V.Egorova = ***Carex chosenica*** Ohwi
Carex daibuensis Hayata = ***Carex dolichostachya*** subsp. ***trichosperma*** (Ohwi) T.Koyama
Carex danaensis Stacey = ***Carex incurviformis*** Mack.
Carex ×*dandoensis* T.Koyama = ?
Carex ×*danielis* H.Lév. = ?
Carex darlingtonii Schwein. ex M.A.Curtis = ***Carex aestivalis*** M.A.Curtis ex A.Gray
Carex darwinii var. *aristata* C.B.Clarke ex Kük. = ***Carex darwinii*** var. ***urolepis*** (Franch.) Kük.
Carex darwinii var. *robustior* Kük. = ***Carex darwinii*** var. ***serranoi*** (Phil.) Kük.
Carex dasycarpa var. *tenax* (Chapm. ex Dewey) Kük. = ***Carex tenax*** Chapm. ex Dewey
Carex davalliana var. *androgyna* Döll = ***Carex davalliana*** Sm.
Carex davalliana subsp. *cyrnea* (Briq.) Cif. & Giacom. = ***Carex elongata*** L.
Carex davalliana var. *pallida* H.Christ ex Scheutz = ***Carex parallela*** subsp. ***redowskiana*** (C.A.Mey.) T.V.Egorova
Carex davalliana f. *sieberiana* (Opiz) Kunth = ***Carex davalliana*** Sm.
Carex davalliana var. *sieberiana* (Opiz) Nyman = ***Carex davalliana*** Sm.
Carex davidii var. *ascocentra* (C.B.Clarke ex Franch.) Kük. = ***Carex ligata*** subsp. ***formosensis*** (H.Lév. & Vaniot) T.Koyama
Carex davisii Dewey = ***Carex albicans*** var. ***emmonsii*** (Dewey ex Torr.) Rettig
Carex davisii f. *glabrescens* Kük. = ***Carex albicans*** Willd. ex Spreng. var. ***albicans***
Carex dayunshanensis L.K.Ling & Y.Z.Huang = ***Carex glossostigma*** Hand.-Mazz.
Carex de-la-costa (Steud.) Kuntze = ***Uncinia macloviana*** Gaudich.
Carex debiliformis Mack. = ***Carex mendocinensis*** Olney ex Boott
Carex debilis Boiss. & Balansa = ***Carex depressa*** subsp. ***transsilvanica*** (Schur) T.V.Egorova
Carex debilis G.Forst. = ***Carex forsteri*** Wahlenb.
Carex debilis var. *intercursa* Fernald = ***Carex debilis*** Michx. var. ***debilis***
Carex debilis var. *interjecta* L.H.Bailey = ***Carex debilis*** var. ***rudgei*** L.H.Bailey
Carex debilis var. *prolixa* L.H.Bailey = ***Carex debilis*** Michx. var. ***debilis***
Carex debilis subsp. *pubera* (A.Gray) Á.Löve & D.Löve = ***Carex debilis*** Michx.
Carex debilis var. *pubera* A.Gray = ***Carex debilis*** Michx. var. ***debilis***
Carex debilis subsp. *rudgei* (L.H.Bailey) Á.Löve & D.Löve = ***Carex debilis*** var. ***rudgei*** L.H.Bailey
Carex debilis var. *strictior* L.H.Bailey = ***Carex debilis*** var. ***rudgei*** L.H.Bailey
Carex debilis var. *typica* Fernald = ***Carex debilis*** Michx.
Carex decidua var. *brehmeri* (Boeck.) Kük. = ***Carex decidua*** Boott
Carex decidua var. *caespitosa* H.Pfeiff. = ***Carex fuscula*** d'Urv. var. ***fuscula***
Carex decidua var. *minor* Kük. = ***Carex decidua*** Boott
Carex decipiens Turcz. = ***Carex obtusata*** Lilj.
Carex decipiens J.Gay = ***Carex pulicaris*** L.
Carex ×*decolorans* Wimm. = ***Carex* × *limula*** Fr.
Carex decumbens Ehrh. = ***Carex pilulifera*** L. subsp. ***pilulifera***
Carex decurticaulis Ohwi = ***Carex reptabunda*** (Trautv.) V.I.Krecz.
Carex deflexa var. *brevicaulis* (Mack.) Boivin =
Carex deflexa var. *deanei* L.H.Bailey = ***Carex deflexa*** Hornem. var. ***deflexa***
Carex deflexa var. *farwellii* Britton = ***Carex rossii*** Boott
Carex deflexa var. *media* L.H.Bailey = ***Carex rossii*** Boott
Carex deflexa var. *rossii* (Boott) L.H.Bailey = ***Carex rossii*** Boott
Carex deinbolliana J.Gay = ***Carex stenophylla*** Wahlenb. subsp. ***stenophylla***
Carex delicatula E.P.Bicknell = ***Carex atlantica*** subsp. ***capillacea*** (L.H.Bailey) Reznicek
Carex delicatula C.B.Clarke = ***Carex lanceolata*** Boott var. ***lanceolata***
Carex delicatula Lojac. =
Carex ×*dematranea* Lagger = ***Carex* × *turfosa*** Fr.
Carex demissa Hornem. = ***Carex viridula*** subsp. ***oedocarpa*** (Andersson) B.Schmid
Carex demissa subsp. *iranica* Kukkonen = ***Carex viridula*** subsp. ***oedocarpa*** (Andersson) B.Schmid
Carex densifimbriata var. *hirsuta* P.C.Li = ***Carex densifimbriata*** Tang & F.T.Wang
Carex densiflora C.Presl = ***Carex cirrhulosa*** Nees

Carex denudata Lapeyr. ex Steud. = ***Carex pyrenaica*** Wahlenb.
Carex depauperata Hornem. = ***Carex vaginata*** Tausch var. ***vaginata***
Carex depressa var. *basilaris* (Jord.) Asch. & Graebn. = ***Carex depressa*** Link subsp. ***depressa***
Carex depressa subsp. *basilaris* (Jord.) Kerguélen = ***Carex depressa*** Link subsp. ***depressa***
Carex depressa f. *tenella* Kük. = ***Carex depressa*** Link subsp. ***depressa***
Carex depressa var. *transsilvanica* (Schur) H.Christ = ***Carex depressa*** subsp. ***transsilvanica*** (Schur) T.V. Egorova
Carex deserta Merino = ?
Carex desertorum Litv. = ***Carex pachystylis*** J.Gay
Carex despreauxii Steud. = ***Carex oligosperma*** Michx.
Carex dessaueri Phil. = ***Carex banksii*** Boott var. ***banksii***
Carex deweyana var. *bolanderi* (Olney) Boott = ***Carex bolanderi*** Olney
Carex deweyana subsp. *leptopoda* (Mack.) Calder & R.L.Taylor = ***Carex leptopoda*** Mack.
Carex deweyana var. *leptopoda* (Mack.) Boivin = ***Carex leptopoda*** Mack.
Carex deweyana var. *sparsiflora* (Olney) L.H.Bailey = ***Carex laeviculmis*** Meinsh.
Carex deweyana var. *stricta* Farw. = ***Carex deweyana*** Schwein. var. ***deweyana***
Carex deweyi Eaton = ***Carex deweyana*** Schwein. var. ***deweyana***
Carex diamantina H.Lév. & Vaniot = ***Carex distans*** L.
Carex dianae var. *aequablis* (Boott) Kük. = ***Carex dianae*** Steud.
Carex diandra Roth = ***Carex paniculata*** L. subsp. ***paniculata***
Carex diandra var. *ampla* (L.H.Bailey) Kük. = ***Carex diandra*** Schrank
Carex diandra f. *longibracteata* Chwastowski = ***Carex diandra*** Schrank
Carex diandra f. *major* (W.D.J.Koch) Kük. = ***Carex diandra*** Schrank
Carex diandra var. *prairea* (Dewey ex Alph.Wood) Kük. = ***Carex prairea*** Dewey ex Alph.Wood
Carex diandra var. *ramosa* (Boott) Fernald = ***Carex diandra*** Schrank
Carex diandra f. *tenella* (Beckm.) Kük. = ***Carex diandra*** Schrank
Carex diaphana Boott = ***Carex appressa*** R.Br. var. ***appressa***
Carex dichroa V.I.Krecz. = ***Carex pamirica*** subsp. ***dichroa*** (Malyschev) T.V.Egorova
Carex dichroa Franch. = ***Carex thibetica*** Franch. var. ***thibetica***
Carex dichroa subsp. *pamirensis* (C.B.Clarke) T.V. Egorova = ***Carex pamirica*** (O.Fedtsch.) O.Fedtsch. & B.Fedtsch. subsp. ***pamirica***
Carex dichroandra V.I.Krecz. = ***Carex acuta*** L.
Carex dicraea C.B.Clarke = ***Carex doenitzii*** Boeck. var. ***doenitzii***
Carex dicuspis Franch. = ***Carex doenitzii*** Boeck. var. ***doenitzii***
Carex dieckii Boeck. = ***Carex illota*** L.H.Bailey
Carex diehlii M.E.Jones ex F.J.Herm. = ***Carex lagunensis*** M.E.Jones
Carex dietrichiae Boeck. = ***Carex indica*** L. var. ***indica***
Carex diffusa Boott ex C.B.Clarke = ***Carex vesiculosa*** Boott
Carex digama Nakai = ***Carex conica*** Boott
Carex digitalis Schwein. & Torr. = ***Carex gracillima*** Schwein.
Carex digitalis var. *assymetrica* Fernald = ***Carex digitalis*** var. ***floridana*** (L.H.Bailey) Naczi & Bryson
Carex digitalis var. *copulata* (L.H.Bailey) L.H.Bailey = ***Carex laxiculmis*** var. ***copulata*** (L.H.Bailey) Fernald
Carex digitalis var. *glauca* Chapm. = ***Carex abscondita*** Mack.
Carex digitalis f. *podostachys* (Steud.) Kük. = ***Carex digitalis*** Willd. var. ***digitalis***
Carex digitalis var. *van-weckii* Dewey = ***Carex digitalis*** Willd. var. ***digitalis***
Carex digitata var. *albida* Freyn & Sint. = ***Carex digitata*** L.
Carex digitata f. *brevifolia* Asch. = ***Carex digitata*** L.
Carex digitata subsp. *bulgarica* (Velen.) Nyman = ***Carex digitata*** L.
Carex digitata var. *bulgarica* Velen. = ***Carex digitata*** L.
Carex digitata f. *compactior* Kük. = ***Carex digitata*** L.
Carex digitata f. *glabrescens* Kük. = ***Carex digitata*** L.
Carex digitata var. *hungarica* Borbás = ***Carex digitata*** L.
Carex digitata var. *intermedia* Crép. = ***Carex digitata*** L.
Carex digitata var. *ornithopoda* (Willd.) G.Mey. = ***Carex ornithopoda*** Willd.
Carex digitata var. *ornithopodioides* (Hausm.) Fiori = ***Carex ornithopoda*** subsp. ***ornithopodioides*** (Hausm.) Nyman
Carex digitata var. *pallens* Fristedt = ***Carex pallidula*** Harmaja
Carex digitata subsp. *pallens* (Fristedt) Tzvelev = ***Carex pallidula*** Harmaja
Carex digitata var. *pallida* Meinsh. = ***Carex quadriflora*** (Kük.) Ohwi
Carex digitata subsp. *piroskana* (Nyár.) Serb. = ***Carex digitata*** L.
Carex digitata subsp. *quadriflora* Kük. = ***Carex quadriflora*** (Kük.) Ohwi
Carex digitata var. *rhizina* (Blytt ex Lindblom) Fiori = ***Carex pediformis*** subsp. ***rhizodes*** (Blytt ex Boott) H.Lindb.
Carex digitata f. *robustior* Kük. = ***Carex digitata*** L.
Carex digitata f. *speirostchys* Neuman = ***Carex digitata*** L.
Carex diluta var. *chorgosica* (Meinsh.) Kük. = ***Carex diluta*** M.Bieb. subsp. ***diluta***
Carex diluta var. *fissirostris* (Ball) Kük. = ***Carex diluta*** subsp. ***fissirostris*** (Ball) Maire & Weiller
Carex diluta var. *karelinii* (Meinsh.) Kük. = ***Carex diluta*** M.Bieb. subsp. ***diluta***
Carex diluta var. *microlepis* (Boeck.) Kük. = ***Carex diluta*** M.Bieb. subsp. ***diluta***
Carex diluta f. *subdistans* Kük. = ***Carex diluta*** M.Bieb. subsp. ***diluta***
Carex dimorpha Brot. = ***Carex distachya*** Desf. var. ***distachya***
Carex dimorpha Boeck. = ***Carex norvegica*** Retz. subsp. ***norvegica***
Carex dinarica Simonk. = ***Carex flacca*** Schreb. subsp. ***flacca***
Carex dineuros C.B.Clarke = ***Carex jaluensis*** Kom.
Carex dioica var. *custoriana* (Heer ex Hegetschw.) Nyman = ***Carex dioica*** L.
Carex dioica var. *davalliana* (Sm.) Wahlenb. = ***Carex davalliana*** Sm.
Carex dioica var. *gaudiniana* (Guthnick) Sendtn. = ***Carex × gaudiniana*** Guthnick
Carex dioica subsp. *gynocrates* (Wormsk.) Hultén = ***Carex gynocrates*** Wormsk.
Carex dioica var. *gynocrates* (Wormsk.) Ostenf. = ***Carex gynocrates*** Wormsk.
Carex dioica f. *isogyna* Fr. = ***Carex dioica*** L.
Carex dioica f. *laxa* Junge = ***Carex dioica*** L.

Carex dioica f. *metteniana* Asch. = ***Carex dioica*** L.
Carex dioica var. *metteniana* (Asch.) Nyman = ***Carex dioica*** L.
Carex dioica var. *parallela* Laest. = ***Carex parallela*** (Laest.) Sommerf.
Carex dioica var. *redowskiana* (C.A.Mey.) Ostenf. = ***Carex parallela*** subsp. ***redowskiana*** (C.A.Mey.) T.V.Egorova
Carex dioica var. *scabrella* Fr. = ***Carex dioica*** L.
Carex dioica var. *subcaespitosa* O.Lang = ***Carex dioica*** L.
Carex dioica var. *subtriangularis* Boeck. = ***Carex parallela*** (Laest.) Sommerf. subsp. ***parallela***
Carex dioicotrigona St.-Lag. = ***Carex davalliana*** Sm.
Carex dioiscostrongyla St.-Lag. = ***Carex dioica*** L.
Carex diplasiocarpa V.I.Krecz. = ***Carex yamatsutana*** Ohwi
Carex discoidea Boott = ***Carex breviculmis*** var. ***discoidea*** (Boott) Boott
Carex discolor Reinw. ex de Vriese = ***Carex appressa*** var. ***virgata*** (Sol. ex Boott) Kük.
Carex discolor Boeck. = ***Carex conica*** Boott
Carex discolor F.Nyl. = ***Carex salina*** Wahlenb. var. ***salina***
Carex disjuncta E.P.Bicknell = ***Carex canescens*** L. subsp. ***canescens***
Carex dispalata var. *costata* Kük. = ***Carex dispalata*** Boott subsp. ***dispalata***
Carex dispalata var. *niigatensis* Franch. & Sav. = ***Carex dispalata*** Boott subsp. ***dispalata***
Carex dispalatha H.Lév. & Vaniot = ***Carex planiculmis*** Kom.
Carex disperma Kunth = ***Carex jamesii*** Schwein.
Carex disperma Kunze ex Carey = ***Carex rosea*** Willd.
Carex displodens Printz = ?
Carex dissimilis J.F.Gmel. = ?
Carex dissita var. *lambertiana* (Boott) Cheeseman = ***Carex lambertiana*** Boott
Carex dissita var. *monticola* Kük. = ***Carex dissita*** Sol. ex Boott
Carex dissita var. *neesiana* (Endl.) Kük. = ***Carex neesiana*** Endl.
Carex dissita var. *ochrosaccus* C.B.Clarke = ***Carex ochrosaccus*** (C.B.Clarke) Hamlin
Carex dissita var. *solanderi* (Boott) Kük. = ***Carex solanderi*** Boott
Carex dissitiflora var. *taiwanensis* Ohwi = ***Carex dissitiflora*** subsp. ***taiwanensis*** (Ohwi) T.Koyama
Carex distachya Willd. = ***Carex atrata*** L. subsp. ***atrata***
Carex distachya f. *sarda* (Link) Boeck. = ***Carex distachya*** Desf. var. ***distachya***
Carex distans Lightf. = ***Carex binervis*** Sm.
Carex distans subsp. *adriatica* Degen = ***Carex distans*** L.
Carex distans f. *appropinquata* Kük. = ***Carex distans*** L.
Carex distans subsp. *aspratilis* (V.I.Krecz.) T.V.Egorova = ***Carex diluta*** M.Bieb. subsp. ***diluta***
Carex distans var. *baetica* (Auersw. ex Willk.) Nyman = ***Carex distans*** L.
Carex distans var. *crassa* C.B.Clarke = ***Carex distans*** L.
Carex distans var. *helodes* (Link) Fiori = ***Carex laevigata*** Sm.
Carex distans var. *hostiana* (DC.) Fiori = ***Carex hostiana*** DC.
Carex distans f. *major* Kneuck. = ***Carex distans*** L.
Carex distans f. *minor* Post = ***Carex distans*** L.
Carex distans var. *neglecta* (Degl.) Nyman = ***Carex distans*** L.
Carex distans var. *oranensis* Trab. = ***Carex distans*** L.
Carex distans f. *pendula* Lackow. = ***Carex distans*** L.
Carex distans var. *pseudoflava* (Schur) Asch. & Graebn. = ***Carex distans*** L.
Carex distans f. *sinai* (Boott) Boeck. = ***Carex distans*** L.
Carex distantiflora Nakai = ***Carex macrandrolepis*** H.Lév. & Vaniot
Carex distenta Kunze ex Kunth = ***Carex fuscula*** var. ***distenta*** (Kunze ex Kunth) Kük.
Carex disticha A.Gray = ***Carex sartwellii*** Dewey var. ***sartwellii***
Carex disticha var. *capitata* Hampe ex Láng = ***Carex disticha*** Huds.
Carex disticha var. *compactior* Kük. = ***Carex disticha*** Huds.
Carex disticha var. *decipiens* De Wild. & T.Durand = ***Carex disticha*** Huds.
Carex disticha var. *distachya* O.Lang = ***Carex disticha*** Huds.
Carex disticha var. *elatior* Boeck. = ***Carex disticha*** Huds.
Carex disticha f. *floribunda* (Peterm.) Peterm. = ***Carex disticha*** Huds.
Carex disticha var. *gracilis* Metsch = ***Carex disticha*** Huds.
Carex disticha var. *gracilis* Boeck. = ***Carex repens*** Bellardi
Carex disticha subsp. *grossheimii* (V.I.Krecz.) T.V. Egorova = ***Carex disticha*** Huds.
Carex disticha var. *longibracteata* Schleich. = ***Carex disticha*** Huds.
Carex disticha var. *luxurians* Beck = ***Carex disticha*** Huds.
Carex disticha var. *minor* Peterm. = ***Carex disticha*** Huds.
Carex disticha f. *pangyna* Beck = ***Carex disticha*** Huds.
Carex disticha var. *pedunculata* Pamp. = ***Carex disticha*** Huds.
Carex disticha f. *permixta* Beck = ***Carex disticha*** Huds.
Carex disticha var. *repens* (Bellardi) Asch. = ***Carex repens*** Bellardi
Carex disticha var. *sartwellii* (Dewey) Dewey = ***Carex sartwellii*** Dewey
Carex distichiflora B.Boivin = ***Carex petricosa*** Dewey var. ***petricosa***
Carex distichoidea H.Lév. & Vaniot = ***Carex lithophila*** Turcz.
Carex divaricata Kük. = ***Carex mollissima*** H.Christ ex Scheutz
Carex divergens Thuill. = ***Carex elongata*** L.
Carex diversicolor Crantz = ***Carex flacca*** subsp. ***erythrostachys*** (Hoppe) Holub
Carex diversicolor var. *claviformis* Hoppe = ***Carex flacca*** Schreb. subsp. ***flacca***
Carex diversicolor var. *cuspidata* (Host) Fiori = ***Carex flacca*** subsp. ***erythrostachys*** (Hoppe) Holub
Carex diversiflora Host = ***Carex halleriana*** Asso
Carex diversistylis A.Roach = ***Carex rossii*** Boott
Carex dives T.Holm = ***Carex aquatilis*** var. ***dives*** (T.Holm) Kük.
Carex divisa Oeder = ***Carex flava*** L. subsp. ***flava***
Carex divisa subsp. *ammophila* (Willd.) Vicioso = ***Carex divisa*** Huds.
Carex divisa var. *ammophila* (Willd.) Nyman = ***Carex divisa*** Huds.
Carex divisa var. *austriaca* (Schkuhr) Nyman = ***Carex divisa*** Huds.
Carex divisa subsp. *chaetophylla* (Steud.) Nyman = ***Carex divisa*** Huds.
Carex divisa var. *chaetophylla* (Steud.) Daveau = ***Carex divisa*** Huds.
Carex divisa var. *cuspidata* Nyman = ***Carex divisa*** Huds.
Carex divisa var. *longiculmis* Willk. = ***Carex divisa*** Huds.
Carex divisa var. *moniziana* (Lowe ex Boott) Kük. = ***Carex divisa*** Huds.
Carex divisa f. *pallidiflora* Kük. = ***Carex divisa*** Huds.

Carex divisa var. *platyphylla* Braun-Blanq., Maire & Trabut = ***Carex divisa*** Huds.
Carex divisa var. *rivularis* (Schkuhr) Nyman = ***Carex divisa*** Huds.
Carex divisa f. *rivularis* (Schkuhr) Kük. = ***Carex divisa*** Huds.
Carex divisa var. *splendens* (Thuill. ex Pers.) Nyman = ***Carex divisa*** Huds.
Carex divisa var. *stenophylla* (Wahlenb.) Fiori = ***Carex stenophylla*** Wahlenb.
Carex divulsa Gaudin = ***Carex muricata*** L.
Carex divulsa f. *angustifolia* (Podp.) Soó = ***Carex divulsa*** Stokes subsp. ***divulsa***
Carex divulsa var. *chabertii* (F.W.Schultz) Nyman = ***Carex divulsa*** subsp. ***leersii*** (Kneuck.) W.Koch
Carex divulsa var. *chabertii* (F.W.Schultz) Kneuck. = ***Carex divulsa*** subsp. ***leersii*** (Kneuck.) W.Koch
Carex divulsa var. *guestphalica* (Rchb.) F.W.Schultz = ***Carex divulsa*** Stokes subsp. ***divulsa***
Carex divulsa var. *intermedia* Lange = ***Carex divulsa*** Stokes subsp. ***divulsa***
Carex divulsa subsp. *leersiana* Rauschert = ***Carex divulsa*** subsp. ***leersii*** (Kneuck.) W.Koch
Carex divulsa f. *misera* Kük. ex Vollm. = ***Carex divulsa*** Stokes subsp. ***divulsa***
Carex divulsa var. *pairae* (F.W.Schultz) Nyman = ***Carex pairae*** F.W.Schultz
Carex divulsa f. *polycarpa* Vollm. = ***Carex divulsa*** Stokes subsp. ***divulsa***
Carex divulsa var. *virens* Durieu = ***Carex divulsa*** Stokes subsp. ***divulsa***
Carex divulsa subsp. *virens* (Durieu) Nyman = ***Carex divulsa*** Stokes subsp. ***divulsa***
Carex divulsa var. *virens* (Hoppe) Nyman = ***Carex spicata*** Huds.
Carex doenitzii var. *trichopoda* (Franch.) Kük. = ***Carex podogyna*** Franch. & Sav.
Carex doiana Akiyama = ***Carex conica*** Boott
Carex doisutepensis var. *flaccidior* T.Koyama = ***Carex doisutepensis*** T.Koyama
Carex dolichocarpa C.A.Mey. ex Kom. = ***Carex michauxiana*** subsp. ***asiatica*** Hultén
Carex dominii H.Lév. & Vaniot = ***Carex disperma*** Dewey
Carex doniana Spreng. = ***Carex alopecuroides*** D.Don ex Tilloch & Taylor
Carex douglasii f. *brunnea* Olney = ***Carex douglasii*** Boott
Carex douglasii var. *densispicata* Dewey = ***Carex douglasii*** Boott
Carex douglasii f. *laxiflora* L.H.Bailey = ***Carex douglasii*** Boott
Carex douglasii f. *meadii* (Dewey) Kük. = ***Carex meadii*** Dewey
Carex douglasii var. *minor* Olney = ***Carex douglasii*** Boott
Carex douglasii var. *williamsii* Rydb. = ***Carex douglasii*** Boott
Carex drakensbergensis C.B.Clarke = ***Carex cognata*** var. ***drakensbergensis*** (C.B.Clarke) Kük.
Carex dregeana Kunth = ***Schoenoxiphium sparteum*** (Wahlenb.) C.B.Clarke
Carex drejeri Lange = ***Carex cespitosa*** L. var. ***cespitosa***
Carex drejeriana Lange = ***Carex bigelowii*** Torr. ex Schwein. subsp. ***bigelowii***
Carex drejeriana f. *cuspidata* Rosenv. = ***Carex bigelowii*** Torr. ex Schwein. subsp. ***bigelowii***
Carex drummondiana Dewey = ***Carex rupestris*** All. subsp. ***rupestris***
Carex drymeia L.f. = ***Carex sylvatica*** Huds. subsp. ***sylvatica***
Carex drymophila var. *abbreviata* (Kük.) Ohwi = ***Carex sordida*** Van Heurck & Müll.Arg.
Carex drymophila var. *akanensis* (Franch.) Kük. = ***Carex sordida*** Van Heurck & Müll.Arg.
Carex drymophila var. *pilifera* Kük. = ***Carex glabrescens*** (Kük.) Ohwi
Carex drymophila var. *reducta* Kük. = ***Carex drymophila*** Turcz.
Carex drymophila var. *udensis* (Meinsh.) Kük. = ***Carex vesicaria*** L.
Carex dubia Hosé = ***Carex acutiformis*** Ehrh.
Carex dubia Gaudin = ***Carex atrata*** L. subsp. ***atrata***
Carex dubia Missbach & E.H.L.Krause = ?
Carex dubia Sibth. = (Poaceae)
Carex dubitata Dewey ex Boott = ***Carex bigelowii*** Torr. ex Schwein. subsp. ***bigelowii***
Carex ×*ducellieri* Beauverd = ?
Carex dudleyi Mack. = ***Carex densa*** (L.H.Bailey) L.H. Bailey
Carex dueensis Meinsh. = ***Carex blepharicarpa*** var. ***duneensis*** (Meinsh.) Akiyama
Carex ×*duernbergeri* Ritzb. = ?
Carex dufourei Lapeyr. = ***Carex rupestris*** All. subsp. ***rupestris***
Carex duneensis Meinsh. = ***Carex blepharicarpa*** var. ***duneensis*** (Meinsh.) Akiyama
Carex duneensis var. *rostrata* Meinsh. = ***Carex blepharicarpa*** var. ***duneensis*** (Meinsh.) Akiyama
Carex dunnii Hayata = ***Carex perakensis*** C.B.Clarke var. ***perakensis***
Carex dura Boott = ***Carex pichinchensis*** Kunth
Carex durandii Boeck. = ***Carex bonplandii*** Kunth
Carex durifolia L.H.Bailey = ***Carex backii*** Boott
Carex durifolia var. *subrostrata* Bates = ***Carex saximontana*** Mack.
Carex duriuscula subsp. *stenophylloides* (V.I.Krecz.) S.Yun Liang & Y.C.Tang = ***Carex stenophylla*** subsp. ***stenophylloides*** (V.I.Krecz.) T.V.Egorova
Carex duriuscula var. *tenuispica* X.Y.Yuan = ***Carex duriuscula*** C.A.Mey. subsp. ***duriuscula***
Carex duriusculiformis V.I.Krecz. = ***Carex stenophylla*** subsp. ***stenophylloides*** (V.I.Krecz.) T.V.Egorova
Carex dussiana Boeck. = ***Carex polystachya*** Sw. ex Wahlenb. var. ***polystachya***
Carex duthiei C.B.Clarke = ***Carex atrata*** subsp. ***pullata*** (Boott) Kük.
Carex ×*dutillyi* O'Neill & Duman = ***Carex ×langeana*** Fernald
Carex dyrrachiensis Nelmes = ?
Carex eastwoodiana Stacey = ***Carex tahoensis*** Smiley
Carex eboracensis Nelmes = ?
Carex ebracteata Phil. = ***Carex aphylla*** Kunth
Carex ebracteata Trautv. = ***Carex trautvetteriana*** Kom.
Carex echinata Roth = ***Carex echinata*** Murray subsp. ***echinata***
Carex echinata Lam. = ***Carex flava*** L. subsp. ***flava***
Carex echinata var. *angustata* (Carey) L.H.Bailey = ***Carex echinata*** Murray subsp. ***echinata***
Carex echinata f. *brevispicata* Podp. = ***Carex echinata*** Murray subsp. ***echinata***
Carex echinata f. *capitata* (H.Christ) Kük. = ***Carex pairae*** F.W.Schultz
Carex echinata var. *cephalantha* L.H.Bailey = ***Carex echinata*** Murray subsp. ***echinata***
Carex echinata f. *chlorocarpa* Podp. = ***Carex echinata*** Murray subsp. ***echinata***

Carex echinata var. *conferta* (Chapm.) L.H.Bailey = ***Carex atlantica*** L.H.Bailey subsp. ***atlantica***
Carex echinata var. *excelsior* (L.H.Bailey) Fernald = ***Carex echinata*** Murray subsp. ***echinata***
Carex echinata subsp. *grypos* (Schkuhr) Arcang. = ***Carex echinata*** Murray subsp. ***echinata***
Carex echinata var. *grypos* (Schkuhr) Nyman = ***Carex echinata*** Murray subsp. ***echinata***
Carex echinata var. *hydrophila* (Dumort.) Nyman = ***Carex echinata*** Murray subsp. ***echinata***
Carex echinata var. *leersii* (Kneuck.) Kük. = ***Carex divulsa*** subsp. ***leersii*** (Kneuck.) W.Koch
Carex echinata var. *loliacea* Lej. = ***Carex echinata*** Murray subsp. ***echinata***
Carex echinata var. *microstachys* Boeck. = ***Carex echinata*** Murray subsp. ***echinata***
Carex echinata f. *monostachya* (Asch. & Graebn.) Kük. = ***Carex pairae*** F.W.Schultz
Carex echinata var. *ormantha* Fernald = ***Carex echinata*** Murray subsp. ***echinata***
Carex echinata var. *perileia* (S.T.Blake) P.Royen = ***Carex echinata*** Murray subsp. ***echinata***
Carex echinata var. *phyllomanica* (Boott) B.Boivin = ***Carex echinata*** subsp. ***phyllomanica*** (Boott) Reznicek
Carex echinata f. *remotiuscula* Podp. = ***Carex echinata*** Murray subsp. ***echinata***
Carex echinata var. *tenuior* Kük. = ***Carex echinata*** Murray subsp. ***echinata***
Carex echinata subsp. *townsendii* (Mack.) Reznicek = ***Carex townsendii*** Mack.
Carex echinochloe var. *chlorosaccus* (C.B.Clarke) Kük. = ***Carex chlorosaccus*** C.B.Clarke
Carex echinochloe var. *nyasensis* (C.B.Clarke) Kük. = ***Carex echinochloe*** subsp. ***nyasensis*** (C.B.Clarke) Lye
Carex efflata Laest. ex Ledeb. = ***Carex panicea*** L.
Carex eggertii L.H.Bailey = ***Carex louisianica*** L.H.Bailey
Carex egglestonii var. *festivelliformis* F.J.Herm. = ***Carex egglestonii*** Mack.
Carex eggytera Steud. = ***Carex breviculmis*** R.Br. var. ***breviculmis***
Carex egregia Mack. = ***Carex angustata*** Boott
Carex ehrenbergiana Boeck. = ***Carex tuberculata*** Liebm.
Carex ehrhartiana Hoppe ex Boott = ***Carex diandra*** Schrank
Carex ekmanii var. *hottensis* Kük. & Ekman = ***Carex ekmanii*** Kük.
Carex elachycarpa Fernald = ***Carex sterilis*** Willd.
Carex ×*elanescens* Cif. & Giacom. = ?
Carex elata A.Lowe = ***Carex lowei*** Bech.
Carex elata f. *brachystachys* (Podp.) Soó = ***Carex elata*** All. subsp. ***elata***
Carex elata subsp. *fallax* (T.Marsson) Suess. = ***Carex ×turfosa*** Fr.
Carex elata f. *glabra* (Junge) Soó = ***Carex elata*** All. subsp. ***elata***
Carex elata subsp. *mauritanica* (Boiss. & Reut.) A.Galán = ***Carex acuta*** L.
Carex elata f. *pendula* (Lange) Soó = ***Carex elata*** All. subsp. ***elata***
Carex elata var. *perraudieriana* Kük. ex Bornm. = ***Carex perraudieriana*** (Kük. ex Bornm.) Gay ex Kük.
Carex elata subsp. *reuteriana* (Boiss.) Luceño & Aedo = ***Carex reuteriana*** Boiss.
Carex elata f. *riartiana* (Legrand) Soó = ***Carex elata*** All. subsp. ***elata***
Carex elata f. *simplex* Bornm. = ***Carex lowei*** Bech.
Carex elata f. *stenocarpa* (Junge) Soó = ***Carex elata*** All. subsp. ***elata***
Carex elata subsp. *tartessiana* Luceño & Aedo = ***Carex reuteriana*** subsp. ***tartessiana*** (Luceño & Aedo) Rivas Mart.
Carex elbertiana Kelso = ***Carex nova*** L.H.Bailey
Carex elbursensis Gilli = ***Carex heterostachya*** Bunge
Carex elegans Willd. = ***Carex limosa*** L.
Carex eleocharis L.H.Bailey = ***Carex duriuscula*** C.A. Mey. subsp. ***duriuscula***
Carex eleopsammodes Steud. = ***Carex andersonii*** Boott
Carex eleusinoides var. *flaccidior* F.Schmidt = ***Carex augustinowiczii*** Meinsh. var. ***augustinowiczii***
Carex elibates Nelmes = ***Carex maculata*** Boott var. ***maculata***
Carex ×*elisabeth* J.Andrés & al. = ?
Carex elmeri Kük. = ***Carex lateralis*** Kük.
Carex elongata var. *alpestris* Trevir. = ***Carex elongata*** L.
Carex elongata var. *brachystachya* A.Blytt = ***Carex elongata*** L.
Carex elongata var. *gebhardii* (Willd.) Asch. = ***Carex elongata*** L.
Carex elongata f. *heterostachya* Wimm. = ***Carex elongata*** L.
Carex elongata var. *minor* Boott = ***Carex elongata*** L.
Carex elongata var. *pauciflora* Meinsh. = ***Carex elongata*** L.
Carex elongata var. *simplicior* Andersson = ***Carex elongata*** L.
Carex elongata f. *umbrosa* Kneuck. = ***Carex elongata*** L.
Carex eloyi C.B.Clarke = ?
Carex elrodii M.E.Jones = ***Carex hallii*** Olney
Carex elyniformis A.E.Porsild = ***Carex nardina*** Fr. var. ***nardina***
Carex emarcida Suter = ***Carex sylvatica*** Huds. subsp. ***sylvatica***
Carex emarginata Willd. = ***Carex montana*** L.
Carex emasculata V.I.Krecz. = ***Carex hartmanii*** Cojander
Carex eminens Nees = ***Carex myosurus*** Nees
Carex emmonsii Dewey ex Torr. = ***Carex albicans*** var. ***emmonsii*** (Dewey ex Torr.) Rettig
Carex emmonsii var. *australis* (L.H.Bailey) Rettig = ***Carex albicans*** var. ***australis*** (L.H.Bailey) Rettig
Carex emmonsii var. *colorata* Howe = ***Carex albicans*** var. ***emmonsii*** (Dewey ex Torr.) Rettig
Carex emmonsii var. *distincta* Howe = ***Carex albicans*** var. ***emmonsii*** (Dewey ex Torr.) Rettig
Carex emmonsii var. *ellpitica* Boott = ***Carex peckii*** Howe
Carex emmonsii var. *muehlenbergii* (A.Gray ex Torr.) Rettig = ***Carex albicans*** Willd. ex Spreng. var. ***albicans***
Carex emodorum Spreng. = ***Carex alopecuroides*** D.Don ex Tilloch & Taylor
Carex enanderi Hultén = ***Carex lenticularis*** var. ***dolia*** (M.E.Jones) L.A.Standl.
Carex enervis Dulac = ***Carex brizoides*** L.
Carex enervis subsp. *acrifolia* (V.I.Krecz.) T.V.Egorova = ***Carex pseudofoetida*** subsp. ***acrifolia*** (V.I.Krecz.) Kukkonen
Carex engelmannii var. *paddoensis* (Suksd.) Kneuck. = ***Carex engelmannii*** L.H.Bailey
Carex engleriana H.Lév. & Vaniot = ?
Carex ×*enshuensis* T.Koyama = ?
Carex ensifolia (Turcz. ex Gorodkov) V.I.Krecz. = ***Carex bigelowii*** subsp. ***ensifolia*** (Turcz. ex Gorodkov) Holub
Carex ensifolia Turcz. ex Ledeb. = ***Carex saxatilis*** L.
Carex ensifolia subsp. *arctisibirica* Jurtzev = ***Carex bigelowii*** subsp. ***arctisibirica*** (Jurtzev) Á.Löve & D.Löve
Carex ensifolia subsp. *lugens* (T.Holm) Vorosch. = ***Carex bigelowii*** subsp. ***lugens*** (T.Holm) T.V.Egorova

Carex ensifolia subsp. *soczavaeana* (Gorodkov) Vorosch. = ***Carex soczavaeana*** Gorodkov
Carex epapillosa Mack. = ***Carex heteroneura*** var. ***epapillosa*** (Mack.) F.J.Herm.
Carex epigeios Fr. = ***Carex salina*** Wahlenb. var. ***salina***
Carex erecta Gilib. = ***Carex cespitosa*** L. var. ***cespitosa***
Carex erecta DC. = ***Carex sempervirens*** Vill.
Carex eriandrolepis H.Lév. = ***Carex umbrosa*** subsp. ***sabynensis*** (Less. ex Kunth) Kük.
Carex ericetorum var. *alpicola* Schur = ***Carex ericetorum*** Pollich
Carex ericetorum var. *approximata* (All.) Nyman = ***Carex ericetorum*** Pollich
Carex ericetorum f. *elongata* Kük. = ***Carex ericetorum*** Pollich
Carex ericetorum subsp. *melanocarpa* (Cham. ex Trautv.) Kük. = ***Carex melanocarpa*** Cham. ex Trautv.
Carex ericetorum var. *membranacea* W.D.J.Koch = ***Carex ericetorum*** Pollich
Carex ericetorum f. *nigra* Hülsen ex Asch. & Graebn. = ***Carex ericetorum*** Pollich
Carex ericetorum f. *pallescens* Kneuck. = ***Carex ericetorum*** Pollich
Carex ericetorum f. *tenella* Grutt. = ***Carex ericetorum*** Pollich
Carex erinacea Muhl. ex Steud. = ***Carex hystericina*** Muhl. ex Willd.
Carex erinacea Cav. = ***Uncinia erinacea*** (Cav.) Pers.
Carex eriophylla (Kük.) Kom. = ***Carex raddei*** Kük.
Carex erostrata Boott ex C.B.Clarke = ***Carex orbicularis*** Boott subsp. ***orbicularis***
Carex erxlebeniana Kelso = ***Carex inops*** subsp. ***heliophila*** (Mack.) Crins
Carex erythraea Schrad. = ?
Carex erythrolepis Kük. = ***Carex wahuensis*** C.A.Mey. subsp. ***wahuensis***
Carex erythrorrhiza Steud. = ***Carex conferta*** var. ***leptosaccus*** (C.B.Clarke) Kük.
Carex erythrostachys Hoppe = ***Carex flacca*** subsp. ***erythrostachys*** (Hoppe) Holub
Carex esenbeckiana Boeck. = ***Schoenoxiphium sparteum*** (Wahlenb.) C.B.Clarke
Carex esenbeckii Kunth = ***Kobresia esenbeckii*** (Kunth) Noltie
Carex essenbeckii Boott = ***Kobresia nepalensis*** (Nees) Kük. var. ***nepalensis***
Carex estesiana Kelso = ***Carex nelsonii*** Mack.
Carex euphlebia S.T.Blake = ***Carex graeffeana*** Boeck.
Carex eurycarpa T.Holm = ***Carex angustata*** Boott
Carex eurycarpa var. *attenuata* Kük. = ***Carex angustata*** Boott
Carex eurycarpa var. *oxycarpa* (T.Holm) Kük. = ***Carex angustata*** Boott
Carex eurystachya F.J.Herm. = ***Carex lenticularis*** var. ***dolia*** (M.E.Jones) L.A.Standl.
Carex euxina (Woronow ex Marcow.) V.I.Krecz. = ***Carex depressa*** subsp. ***transsilvanica*** (Schur) T.V.Egorova
Carex ×*ewingii* E.S.Marshall = ?
Carex exaltata Peterm. = ***Carex riparia*** Curtis subsp. ***riparia***
Carex exaltata Kit. = ?
Carex excisa Boott = ***Carex conica*** Boott
Carex excurrens Cham. ex Steud. = ***Carex podocarpa*** R.Br.
Carex excurva Boott = ***Carex nemostachys*** Steud.
Carex exerta C.H.Chu = ***Carex glossostigma*** Hand.-Mazz.
Carex exigua Boeck. = ***Carex spicigera*** Nees
Carex exilis var. *androgyna* Dewey ex Alph.Wood = ***Carex exilis*** Dewey
Carex exilis var. *squamacea* Dewey = ***Carex exilis*** Dewey
Carex explens Kük. = ***Carex bostrychostigma*** Maxim.
Carex exploratorum Nelmes = ***Carex graeffeana*** Boeck.
Carex exserta Mack. = ***Carex filifolia*** var. ***erostrata*** Kük.
Carex exsiccata var. *globosa* (Olney) L.H.Bailey = ***Carex vesicaria*** L.
Carex exsiccata var. *pungens* L.H.Bailey = ***Carex vesicaria*** L.
Carex extensa C.A.Mey. = ***Carex diluta*** M.Bieb. subsp. ***diluta***
Carex extensa f. *balbisii* (Schkuhr ex Spreng.) Rchb. = ***Carex extensa*** Gooden.
Carex extensa var. *balbisii* (Schkuhr ex Spreng.) Nyman = ***Carex extensa*** Gooden.
Carex extensa var. *ecklonii* (Nees) Kük. = ***Carex ecklonii*** Nees
Carex extensa var. *graeca* Hausskn. = ***Carex extensa*** Gooden.
Carex extensa var. *nervosa* (Desf.) Nyman = ***Carex extensa*** Gooden.
Carex extensa f. *pumila* Andersson = ***Carex extensa*** Gooden.
Carex extensa f. *tenuifolia* DC. = ***Carex extensa*** Gooden.
Carex extensa subsp. *viestina* Fen. = ***Carex extensa*** Gooden.
Carex extensa var. *vixdentata* Kük. ex Barros = ***Carex vixdentata*** (Kük. ex Barros) G.A.Wheeler
Carex eymae Nelmes = ***Carex perakensis*** var. ***borneensis*** (C.B.Clarke) Noot.
Carex fabri Hance = ***Carex brunnea*** Thunb. subsp. ***brunnea***
Carex falcata Turcz. = ***Carex vaginata*** var. ***petersii*** (C.A.Mey. ex Schmidt) Akiyama
Carex fallax Steud. = ***Carex nubigena*** D.Don ex Tilloch & Taylor subsp. ***nubigena***
Carex familiaris Steud. = ***Carex oxylepis*** Torr. & Hook. var. ***oxylepis***
Carex fargesii Franch. = ***Carex prescottiana*** Boott
Carex farwellii Mack. = ***Carex rossii*** Boott
Carex fascicularis Siev. ex Trevir. = ***Carex cespitosa*** L. var. ***cespitosa***
Carex fascicularis Sol. ex Hook.f. = ***Carex pseudocyperus*** var. ***fascicularis*** (Sol. ex Hook.f.) Boott
Carex fascicularis var. *minor* Boott = ***Carex pseudocyperus*** L. var. ***pseudocyperus***
Carex fasciculata Link ex Schkuhr = ***Carex echinata*** Murray subsp. ***echinata***
Carex fasciculata Willd. = ***Carex echinata*** Murray subsp. ***echinata***
Carex fauriae Franch. = ***Carex phacota*** Spreng.
Carex fauriei Franch. = ***Carex insaniae*** Koidz. var. ***insaniae***
Carex favratii H.Christ = ***Carex paniculata*** subsp. ***szovitsii*** (V.I.Krecz.) Ö.Nilsson
Carex fedtschenkoana Kük. = ***Carex heterostachya*** Bunge
Carex felipensis C.B.Clarke = ***Carex anisostachys*** Liebm.
Carex ×*felixii* L.C.Lamb. = ?
Carex feminea Steud. ex Boott = ***Carex acutata*** Boott
Carex fendleriana Boeck. = ***Carex douglasii*** Boott
Carex ferax Raeusch. = ***Carex cespitosa*** L. var. ***cespitosa***
Carex fernaldiana H.Lév. & Vaniot = ***Carex sachalinensis*** var. ***fernaldiana*** (H.Lév. & Vaniot) T.Koyama
Carex ferruginea Schkuhr = ***Carex sempervirens*** Vill.
Carex ferruginea Steven = ***Carex tristis*** M.Bieb.
Carex ferruginea var. *alpigena* (A.Kern. ex Dalla Torre) Nyman = ***Carex ferruginea*** Scop. subsp. ***ferruginea***

Carex ferruginea f. *angustifolia* Beck = ***Carex ferruginea*** Scop. subsp. ***ferruginea***
Carex ferruginea f. *capillarioides* Murr = ***Carex ferruginea*** Scop. subsp. ***ferruginea***
Carex ferruginea var. *caudata* Kük. = ***Carex ferruginea*** subsp. ***caudata*** (Kük.) Pereda & Laínz
Carex ferruginea f. *crassispica* Kük. = ***Carex ferruginea*** Scop. subsp. ***ferruginea***
Carex ferruginea var. *fimbriata* (Schkuhr) Fiori = ***Carex fimbriata*** Schkuhr
Carex ferruginea var. *firma* (Host) Fiori = ***Carex firma*** Host
Carex ferruginea subsp. *kerneri* (Kohts) Schinz & R.Keller = ***Carex ferruginea*** Scop. subsp. ***ferruginea***
Carex ferruginea var. *kerneri* (Kohts) K.Richt. = ***Carex ferruginea*** Scop. subsp. ***ferruginea***
Carex ferruginea var. *laevis* Fiori = ***Carex kitaibeliana*** Degen ex Bech. var. ***kitaibeliana***
Carex ferruginea f. *latifolia* Beck = ***Carex ferruginea*** Scop. subsp. ***ferruginea***
Carex ferruginea f. *leiocarpa* Beck = ***Carex ferruginea*** Scop. subsp. ***ferruginea***
Carex ferruginea var. *macrostachys* (Bertol.) Nyman = ***Carex ferruginea*** subsp. ***macrostachys*** (Bertol.) Arcang.
Carex ferruginea var. *sempervirens* (Vill.) Fiori = ***Carex sempervirens*** Vill.
Carex ferruginea f. *spadicea* W.D.J.Koch = ***Carex ferruginea*** Scop. subsp. ***ferruginea***
Carex ferruginea var. *tatsiensis* Franch. = ***Carex tatsiensis*** (Franch.) Kük.
Carex ferruginea var. *tenax* (H.Christ) H.Christ = ***Carex ferruginea*** subsp. ***australpina*** (Bech.) W.Dietr.
Carex ferruginea f. *tenerrima* (Murr & Appel) Kük. = ***Carex ferruginea*** Scop. subsp. ***ferruginea***
Carex ferruginea f. *transiens* Kük. = ***Carex ferruginea*** Scop. subsp. ***ferruginea***
Carex ferruginea subsp. *tristis* (M.Bieb.) Nyman = ***Carex tristis*** M.Bieb.
Carex ferruspiculata C.H.Chu = ***Carex glossostigma*** Hand.-Mazz.
Carex festiva Dewey = ***Carex macloviana*** d'Urv. subsp. ***macloviana***
Carex festiva var. *decumbens* T.Holm = ***Carex haydeniana*** Olney
Carex festiva var. *ebenea* (Rydb.) J.K.Henry = ***Carex ebenea*** Rydb.
Carex festiva var. *gracilis* Olney = ***Carex macloviana*** subsp. ***subfusca*** (W.Boott) T.Koyama
Carex festiva var. *haydeniana* (Olney) Boott = ***Carex haydeniana*** Olney
Carex festiva var. *horneri* Piper = ***Carex macloviana*** d'Urv. subsp. ***macloviana***
Carex festiva var. *pachystachya* (Cham. ex Steud.) L.H. Bailey = ***Carex macloviana*** subsp. ***pachystachya*** (Cham. ex Steud.) Hultén
Carex festiva var. *stricta* L.H.Bailey = ***Carex macloviana*** d'Urv. subsp. ***macloviana***
Carex festiva var. *viridis* L.H.Bailey = ***Carex macloviana*** d'Urv. subsp. ***macloviana***
Carex festivella Mack. = ***Carex microptera*** Mack.
Carex festucacea var. *brevior* (Dewey) Fernald = ***Carex brevior*** (Dewey) Mack. ex Lunell
Carex festucacea var. *echinodes* (Fernald) Farw. = ***Carex tenera*** var. ***echinodes*** (Fernald) Wiegand
Carex festucacea var. *ferruginea* (A.Gray) Farw. = ***Carex albolutescens*** Schwein.
Carex festucacea var. *mirabilis* (Tuck.) Carey = ***Carex normalis*** Mack.
Carex festucacea var. *tenera* Carey = ***Carex straminea*** Willd. ex Schkuhr
Carex feta var. *multa* L.H.Bailey = ***Carex feta*** L.H.Bailey
Carex fibrata S.Vidal = ***Carex cirrhulosa*** Nees
Carex fibrillosa Franch. & Sav. = ***Carex breviculmis*** var. ***pluricostata*** Kük.
Carex filamentosa K.T.Fu = ***Carex capilliculmis*** S.R. Zhang
Carex filicina Boeck. ex C.B.Clarke = ***Carex stramentitia*** Boott ex Boeck.
Carex filicina subsp. *ceylanica* (Boeck.) T.Koyama = ***Carex ceylanica*** Boeck.
Carex filicina var. *ciliata* Kuntze = ***Carex filicina*** Nees
Carex filicina f. *depauperata* Kük. = ***Carex filicina*** Nees
Carex filicina var. *hirta* Kuntze = ***Carex filicina*** Nees
Carex filicina var. *laevis* Kuntze = ***Carex filicina*** Nees
Carex filicina var. *laucina* (Boeck.) Kük. = ***Carex filicina*** Nees
Carex filicina var. *leptocarpa* (C.B.Clarke) Kük. = ***Carex filicina*** Nees
Carex filicina var. *meiogyna* (Nees) Strachey = ***Carex filicina*** Nees
Carex filicina var. *microgyna* C.B.Clarke = ***Carex filicina*** Nees
Carex filicina subsp. *pseudofilicina* (Hayata) T.Koyama = ***Carex filicina*** Nees
Carex filicina var. *saturata* (C.B.Clarke) Kük. = ***Carex filicina*** Nees
Carex filiculmis Franch. & Sav. = ***Carex breviculmis*** R.Br. var. ***breviculmis***
Carex filifolia J.Gay = ***Carex durieui*** Steud.
Carex filifolia var. *miser* L.H.Bailey = ***Carex elynoides*** T.Holm
Carex filifolia var. *valida* Olney ex L.H.Bailey = ***Carex oreocharis*** T.Holm
Carex filiformis var. *aematorrhyncha* (Desv.) W.Boott = ***Carex aematorrhyncha*** Desv.
Carex filiformis subsp. *aematorrhyncha* (Desv.) Kük. = ***Carex aematorrhyncha*** Desv.
Carex filiformis var. *australis* L.H.Bailey = ***Carex lasiocarpa*** Ehrh. var. ***lasiocarpa***
Carex filiformis var. *gracilis* Kurtz ex Kük. = ***Carex aematorrhyncha*** Desv. var. ***aematorrhyncha***
Carex filiformis var. *lanuginosa* (Michx.) Britton, Sterns & Poggenb. = ***Carex lasiocarpa*** var. ***americana*** Fernald
Carex filiformis var. *latifolia* Boeck. = ***Carex lasiocarpa*** Ehrh. var. ***lasiocarpa***
Carex filiformis var. *occultans* Franch. = ***Carex lasiocarpa*** var. ***occultans*** (Franch.) Kük.
Carex filipendula Drejer = ***Carex lyngbyei*** Hornem.
Carex filipendula var. *concolor* Drejer = ***Carex lyngbyei*** Hornem.
Carex filipendula var. *littoralis* Drejer = ***Carex lyngbyei*** Hornem.
Carex filipendula var. *variegata* Drejer = ***Carex lyngbyei*** Hornem.
Carex filipes var. *oligostachys* Kük. = ***Carex egena*** H.Lév. & Vaniot
Carex filipes var. *rouyana* (Franch.) Kük. = ***Carex filipes*** var. ***sparsinux*** (C.B.Clarke ex Franch.) Kük.
Carex filipes var. *vaniotii* (H.Lév.) Kük. = ***Carex jackiana*** subsp. ***parciflora*** (Boott) Kük.
Carex ×*filkukae* Podp. = ?
Carex finitima var. *attenuata* C.B.Clarke = ***Carex finitima*** Boott
Carex firma f. *longipedunculata* Hausskn. = ***Carex firma*** Host

Carex firma var. *subalpina* Wahlenb. = ***Carex sempervirens*** Vill.
Carex fischeriana J.Gay = ***Carex chordorrhiza*** L.f.
Carex fissilis Boott = ***Carex indica*** L. var. ***indica***
Carex fissirostris Ball = ***Carex diluta*** subsp. ***fissirostris*** (Ball) Maire & Weiller
Carex flacca f. *castriferrei* Soó = ***Carex flacca*** Schreb. subsp. ***flacca***
Carex flacca f. *chlorocarpa* R.Keller = ***Carex flacca*** Schreb. subsp. ***flacca***
Carex flacca subsp. *claviformis* (Hoppe) Degen = ***Carex flacca*** Schreb. subsp. ***flacca***
Carex flacca subsp. *cuspidata* (Host) Degen = ***Carex flacca*** subsp. ***erythrostachys*** (Hoppe) Holub
Carex flacca var. *divaricata* Heuff. = ***Carex flacca*** Schreb. subsp. ***flacca***
Carex flacca f. *melanostachya* Uechtr. = ***Carex flacca*** Schreb. subsp. ***flacca***
Carex flacca var. *oogyna* Murr = ***Carex flacca*** Schreb. subsp. ***flacca***
Carex flacca f. *pallida* Beck = ***Carex flacca*** Schreb. subsp. ***flacca***
Carex flacca subsp. *praetutiana* (Parl.) Holub = ***Carex flacca*** Schreb. subsp. ***flacca***
Carex flacca subsp. *serrulata* (Biv. ex Spreng.) Greuter = ***Carex flacca*** subsp. ***erythrostachys*** (Hoppe) Holub
Carex flacca f. *tenuicula* (Martrin-Donos) Soó = ***Carex flacca*** Schreb. subsp. ***flacca***
Carex flaccida Sw. ex Kunth = ***Carex polystachya*** Sw. ex Wahlenb. var. ***polystachya***
Carex flaccidior (F.Schmidt) Miyabe & Kudô = ***Carex augustinowiczii*** Meinsh. var. ***augustinowiczii***
Carex flaccidula Steud. = ***Carex rosea*** Willd.
Carex flaccifolia Mack. = ***Carex whitneyi*** Olney
Carex flacciformis Hoffmanns. ex Kunth = ***Carex laevigata*** Sm.
Carex flaccosperma var. *glaucodea* (Tuck. ex Olney) Kük. = ***Carex glaucodea*** Tuck. ex Olney
Carex flagellum J.F.Gmel. = ***Scleria secans*** (L.) Urb.
Carex flava var. *alpina* Kneuck. = ***Carex flava*** var. ***nevadensis*** (Boiss. & Reut.) Briq.
Carex flava subsp. *alpina* (Kneuck.) O.Bolòs, Masalles & Vigo = ***Carex flava*** var. ***nevadensis*** (Boiss. & Reut.) Briq.
Carex flava var. *argillacea* Townson = ***Carex pilulifera*** L. subsp. ***pilulifera***
Carex flava var. *barrerae* O.Bolòs, Masalles & Vigo = ***Carex flava*** L. subsp. ***flava***
Carex flava subsp. *bergrothii* (Palmgr.) P.D.Sell = ***Carex pilulifera*** L. subsp. ***pilulifera***
Carex flava var. *brachyrhyncha* Èelak. = ***Carex viridula*** subsp. ***brachyrhyncha*** (Èelak.) B.Schmid
Carex flava subsp. *brevirostrata* Kük. = ***Carex pilulifera*** L. subsp. ***pilulifera***
Carex flava var. *brevirostris* Asch. & Graebn. = ***Carex flava*** L. subsp. ***flava***
Carex flava f. *brevirostris* Junge = ***Carex flava*** L. subsp. ***flava***
Carex flava var. *congesta* Neuman = ***Carex flava*** L. subsp. ***flava***
Carex flava f. *deficiens* Peterm. = ***Carex flava*** L. subsp. ***flava***
Carex flava f. *demissa* (Hornem.) Kük. = ***Carex viridula*** subsp. ***oedocarpa*** (Andersson) B.Schmid
Carex flava subsp. *demissa* (Hornem.) O.Bolòs, Masalles & Vigo = ***Carex viridula*** subsp. ***oedocarpa*** (Andersson) B.Schmid
Carex flava f. *densa* Gaudin = ***Carex flava*** L. subsp. ***flava***
Carex flava var. *dispersa* Neuman = ***Carex flava*** L. subsp. ***flava***
Carex flava var. *elatior* Schltr. = ***Carex viridula*** var. ***elatior*** (Schltr.) Crins
Carex flava subvar. *elongata* Coss. & Germ. = ***Carex pilulifera*** L. subsp. ***pilulifera***
Carex flava var. *fertilis* Peck = ***Carex flava*** L. subsp. ***flava***
Carex flava var. *gaspensis* Fernald = ***Carex flava*** L. subsp. ***flava***
Carex flava var. *graminis* L.H.Bailey = ***Carex flava*** L. subsp. ***flava***
Carex flava f. *graminis* (L.H.Bailey) Scoggan = ***Carex flava*** L. subsp. ***flava***
Carex flava var. *intermedia* Coss. & Germ. = ***Carex viridula*** var. ***elatior*** (Schltr.) Crins
Carex flava subsp. *jemtlandica* (Palmgr.) P.D.Sell = ***Carex viridula*** var. ***jemtlandica*** (Palmgr.) Blackst. & P.A.Ashton
Carex flava f. *laevigata* Peterm. = ***Carex viridula*** var. ***elatior*** (Schltr.) Crins
Carex flava var. *laxior* (Kük.) Gleason = ***Carex flava*** L. subsp. ***flava***
Carex flava var. *lepidocarpa* (Tausch) Godr. = ***Carex viridula*** var. ***elatior*** (Schltr.) Crins
Carex flava subsp. *lepidocarpa* (Tausch) Nyman = ***Carex viridula*** var. ***elatior*** (Schltr.) Crins
Carex flava var. *lutescens* Wahlenb. = ***Carex pilulifera*** L. subsp. ***pilulifera***
Carex flava var. *macrorrhyncha* Èelak. = ***Carex flava*** L. subsp. ***flava***
Carex flava var. *marssonii* (Auersw.) Nyman = ***Carex viridula*** var. ***elatior*** (Schltr.) Crins
Carex flava var. *minor* Wahlenb. = ***Carex pilulifera*** L. subsp. ***pilulifera***
Carex flava var. *minor* Townson = ***Carex pilulifera*** L. subsp. ***pilulifera***
Carex flava var. *minor* Peterm. = ***Carex viridula*** var. ***elatior*** (Schltr.) Crins
Carex flava f. *minuta* Foucaud & Rotges = ***Carex flava*** var. ***nevadensis*** (Boiss. & Reut.) Briq.
Carex flava var. *nelmesiana* (Raymond) Boivin = ***Carex viridula*** var. ***elatior*** (Schltr.) Crins
Carex flava var. *oederi* (Retz.) DC. = ***Carex pilulifera*** L. subsp. ***pilulifera***
Carex flava subsp. *oederi* (Retz.) Syme = ***Carex pilulifera*** L. subsp. ***pilulifera***
Carex flava subsp. *oedocarpa* (Andersson) P.D.Sell = ***Carex viridula*** subsp. ***oedocarpa*** (Andersson) B.Schmid
Carex flava var. *orthorhyncha* Èelak. = ***Carex pilulifera*** L. subsp. ***pilulifera***
Carex flava var. *patula* Coss. & Germ. = ***Carex flava*** L. subsp. ***flava***
Carex flava f. *patula* (Coss. & Germ.) Schur = ***Carex flava*** L. subsp. ***flava***
Carex flava var. *polystachya* Gaudin = ***Carex flava*** L. subsp. ***flava***
Carex flava subsp. *pulchella* (Lönnr.) P.D.Sell = ***Carex viridula*** Michx. subsp. ***viridula***
Carex flava var. *pygmaea* Andersson = ***Carex flava*** L. subsp. ***flava***
Carex flava var. *recterostrata* L.H.Bailey = ***Carex viridula*** Michx. subsp. ***viridula***
Carex flava var. *rectirostra* Gaudin = ***Carex flava*** L. subsp. ***flava***
Carex flava f. *rectirostris* Peterm. = ***Carex flava*** L. subsp. ***flava***
Carex flava f. *remotiuscula* Schur = ***Carex flava*** L. subsp. ***flava***

Carex flava var. *serotina* (Mérat) Lej. = ***Carex viridula*** Michx. subsp. ***viridula***
Carex flava subsp. *serotina* (Mérat) P.D.Sell = ***Carex viridula*** Michx. subsp. ***viridula***
Carex flava var. *uetliaca* (Suter) Nyman = ***Carex flava*** L. subsp. ***flava***
Carex flava f. *uetliaca* (Suter) Asch. & Graebn. = ***Carex flava*** L. subsp. ***flava***
Carex flava f. *umbrosa* Kneuck. = ***Carex flava*** L. subsp. ***flava***
Carex flava var. *viridula* (Michx.) L.H.Bailey = ***Carex viridula*** Michx.
Carex flava subsp. *viridula* (Michx.) O.Bolòs & Vigo = ***Carex viridula*** Michx.
Carex flava var. *vulgaris* Döll = ***Carex flava*** L. subsp. ***flava***
Carex flavella V.I.Krecz. = ***Carex flava*** L. subsp. ***flava***
Carex flavescens Host = ***Carex distans*** L.
Carex flavescens f. *minor* Gaudin = ***Carex distans*** L.
Carex flaviceps Kük. = ***Carex macloviana*** subsp. ***subfusca*** (W.Boott) T.Koyama
Carex flavocuspis var. *caudatifrons* Akiyama = ***Carex flavocuspis*** Franch. & Sav.
Carex flavocuspis subsp. *krascheninnikovii* (Kom. ex V.I.Krecz.) T.V.Egorova = ***Carex krascheninnikovii*** Kom. ex V.I.Krecz.
Carex flavofulva Beurl. = ***Carex flava*** L. subsp. ***flava***
Carex fleckeri Nelmes = ***Carex horsfieldii*** Boott
Carex flectens Boott = ***Carex papulosa*** Boott
Carex ×*fleischeri* Podp. = ***Carex* × *toezensis*** Simonk.
Carex flexilis Rudge = ***Carex castanea*** Wahlenb.
Carex flexilis D.Don = ***Carex lenta*** D.Don var. ***lenta***
Carex flexuosa Muhl. ex Willd. = ***Carex debilis*** var. ***rudgei*** L.H.Bailey
Carex floresiana Hochst. ex Seub. = ***Carex vulcani*** Hochst. ex Seub.
Carex floribunda Meinsh. = ***Carex lanceolata*** Boott var. ***lanceolata***
Carex floribunda Boeck. = ***Carex myosurus*** Nees
Carex fluviatilis var. *unisexualis* (C.B.Clarke) Kük. = ***Carex unisexualis*** C.B.Clarke
Carex foeminea Steud. = ***Carex acutata*** Boott
Carex foenea Muhl. ex Torr. = ***Carex albolutescens*** Schwein.
Carex foenea var. *aenea* (Fernald) Kük. = ***Carex foenea*** Willd.
Carex foenea var. *enervis* D.K.Evans & Mohlenbr. = ***Carex siccata*** Dewey
Carex foenea var. *ferruginea* A.Gray = ***Carex albolutescens*** Schwein.
Carex foenea var. *perplexa* L.H.Bailey = ***Carex argyrantha*** Tuck. ex Boott
Carex foenea var. *sabulonum* A.Gray = ***Carex silicea*** Olney
Carex foenea var. *sparsiflora* Howe = ***Carex siccata*** Dewey
Carex foenea var. *tuberculata* F.J.Herm. = ***Carex siccata*** Dewey
Carex foenea var. *xerantica* (L.H.Bailey) Kük. = ***Carex xerantica*** L.H.Bailey
Carex foetida f. *elongata* H.Christ = ***Carex foetida*** All.
Carex foetida var. *lobata* (Vill.) Nyman = ***Carex foetida*** All.
Carex foetida var. *minor* Kük. = ***Carex pseudofoetida*** subsp. ***acrifolia*** (V.I.Krecz.) Kukkonen
Carex foetida f. *pallescens* Kneuck. = ***Carex foetida*** All.
Carex foetida var. *vernacula* (L.H.Bailey) Kük. = ***Carex vernacula*** L.H.Bailey
Carex foliata F.Schmidt ex Meinsh. = ***Carex pilosa*** Scop.
Carex foliosa All. = ***Carex flava*** L. subsp. ***flava***
Carex foliosissima var. *temnolepis* (Franch.) Kük. = ***Carex morrowii*** var. ***temnolepis*** (Franch.) Ohwi
Carex folliculacea Crantz = ***Carex folliculata*** L.
Carex folliculata Wahlenb. = ***Carex intumescens*** Rudge
Carex folliculata Elliott = ***Carex turgescens*** Torr.
Carex folliculata var. *australis* L.H.Bailey = ***Carex lonchocarpa*** Willd. ex Spreng.
Carex folliculata var. *bispicata* Peck = ***Carex folliculata*** L.
Carex folliculata var. *major* Pursh = ***Carex intumescens*** Rudge
Carex folliculata var. *xanthophysa* (Wahlenb.) Muhl. = ***Carex folliculata*** L.
Carex fonkii Phil. = ***Carex banksii*** var. ***fonkii*** (Phil.) Kük.
Carex fontanesiana DC. = ***Carex pyrenaica*** Wahlenb.
Carex fontanesii Poir. = ***Carex divisa*** Huds.
Carex ×*fontis-sancti* Podp. = ?
Carex forbesii C.B.Clarke = ***Carex pumila*** Thunb.
Carex forficula Sennen = ***Carex distans*** L.
Carex forficula var. *melinacra* (Franch.) Kük. = ***Carex melinacra*** Franch.
Carex forficula var. *scabida* Kük. = ***Carex distans*** L.
Carex formosensis H.Lév. & Vaniot = ***Carex ligata*** subsp. ***formosensis*** (H.Lév. & Vaniot) T.Koyama
Carex forsteri var. *cockayniana* (Kük.) Kük. = ***Carex cockayniana*** Kük.
Carex forsteri var. *fascicularis* Hook.f. = ***Carex pseudocyperus*** L. var. ***pseudocyperus***
Carex forsteri var. *minor* Hook.f. = ***Carex pseudocyperus*** L. var. ***pseudocyperus***
Carex ×*fragosoana* Pau = ?
Carex franchetiana H.Lév. & Vaniot = ***Carex lenta*** var. ***sendaica*** (Franch.) T.Koyama
Carex frankii Steud. = ***Carex atherodes*** Spreng.
Carex franklinii Boott = ***Carex petricosa*** Dewey var. ***petricosa***
Carex franklinii var. *misandroides* (Fernald) Raymond = ***Carex petricosa*** var. ***misandroides*** (Fernald) B.Boivin
Carex franklinii var. *nicholsonis* B.Boivin = ***Carex petricosa*** Dewey var. ***petricosa***
Carex fraseri Andrews = ***Cymophyllus fraserianus*** (Ker Gawl.) Kartesz & Gandhi
Carex fraseriana Ker Gawl. = ***Cymophyllus fraserianus*** (Ker Gawl.) Kartesz & Gandhi
Carex fraterna Rupr. = ***Carex spicata*** Huds.
Carex fraudulans Printz = ***Carex tenuiformis*** H.Lév. & Vaniot
Carex fridrichsthaliana Steud. = ***Carex cespitosa*** L. var. ***cespitosa***
Carex frigida Vill. = ***Carex ferruginea*** Scop. subsp. ***ferruginea***
Carex frigida Wahlenb. = ***Carex fuliginosa*** Schkuhr
Carex frigida var. *brevirostris* Kük. = ***Carex frigida*** All.
Carex frigida var. *comari* (H.Lév. & Vaniot) Kük. = ***Carex frigida*** All.
Carex frigida var. *densa* Meinsh. = ***Carex sempervirens*** Vill.
Carex frigida f. *flavescens* H.Christ = ***Carex frigida*** All.
Carex frigida var. *fuliginosa* (Schkuhr) Kunth = ***Carex fuliginosa*** Schkuhr
Carex frigida var. *geniculata* (Host) Nyman = ***Carex frigida*** All.
Carex frigida var. *huebneri* Hising. = ***Carex fuliginosa*** Schkuhr
Carex frigida f. *pyrenaica* H.Christ = ***Carex frigida*** All.
Carex frisica H.Koch = ***Carex trinervis*** Degl.
Carex fritschii var. *oxystachya* Waisb. = ***Carex* × *ginsiensis*** Waisb.
Carex fucata var. *sikkimensis* (C.B.Clarke) Kük. = ***Carex fucata*** Boott ex C.B.Clarke

Carex fuegiana Gand. = ?
Carex fuegina Phil. = ***Carex atropicta*** Steud.
Carex fuirenioides Gaudich. = ***Carex indica*** L. var. ***indica***
Carex fuirenioides var. *gracilis* Hosok. = ***Carex indica*** L. var. ***indica***
Carex fuirenoides var. *cirrhulosa* (Nees) Kük. = ***Carex cirrhulosa*** Nees
Carex fujimakii M.Kikuchi = ***Carex livida*** (Wahlenb.) Willd.
Carex fujitae Kudô = ***Carex livida*** (Wahlenb.) Willd.
Carex fuliginosa Host = ***Carex fimbriata*** Schkuhr
Carex fuliginosa var. *misandra* (R.Br.) O.Lang = ***Carex fuliginosa*** Schkuhr
Carex fuliginosa subsp. *misandra* (R.Br.) Nyman = ***Carex fuliginosa*** Schkuhr
Carex fuliginosa f. *ochrostachys* Schur = ***Carex fuliginosa*** Schkuhr
Carex fulleri H.E.Ahles = ***Carex heterostachya*** Bunge
Carex fulva Hornem. = ***Carex binervis*** Sm.
Carex fulva Thuill. = ***Carex diandra*** Schrank
Carex fulva P.Gaertn., B.Mey. & Scherb. = ***Carex distans*** L.
Carex fulva Muhl. = ***Carex elliottii*** Schwein. & Torr.
Carex fulva Host = ***Carex hostiana*** DC.
Carex ×fulva Gooden. = ?
Carex ×fulva var. *subsalsa* (Nyman) Nyman = ***Carex hostiana*** DC.
Carex fulvescens Mack. = ***Carex hostiana*** DC.
Carex fulvicoma Dewey = ***Carex chordorrhiza*** L.f.
Carex fumosa (Gren.) Rouy = ***Carex spicata*** Huds.
Carex funicularis Franch. = ***Carex meyeriana*** Kunth
Carex funiformis Clairv. = ***Carex chordorrhiza*** L.f.
Carex furcata Elliott = ***Carex comosa*** Boott
Carex furcata Lapeyr. = ?
Carex furculata Peterm. = ***Carex spicata*** Huds.
Carex ×furusei T.Koyama = [38 JAP] *C. brownii* × *C. brownii* subsp. *transversa*
Carex furva Webb = ***Carex lachenalii*** subsp. ***baetica*** (Gay) Luceño & Muñoz Garm.
Carex furva (L.H.Bailey) Piper = ***Carex praticola*** Rydb.
Carex fusca All. = ***Carex nigra*** (L.) Reichard subsp. ***nigra***
Carex fusca Sol. ex Boott = ***Carex pumila*** Thunb.
Carex fusca Schkuhr = ***Carex saxatilis*** L.
Carex fusca var. *broteroana* (Samp.) Vicioso = ***Carex elata*** All. subsp. ***elata***
Carex fusca subsp. *dacica* (Heuff.) Serb. & Nyár. = ***Carex nigra*** subsp. ***dacica*** (Heuff.) Soó
Carex fusca var. *intricata* (Tineo) Vicioso = ***Carex nigra*** subsp. ***intricata*** (Tineo) Maire & Wieller
Carex fusca f. *intricata* (Tineo) Gamisans = ***Carex nigra*** subsp. ***intricata*** (Tineo) Maire & Wieller
Carex fusca var. *reuteriana* (Boiss.) Vicioso = ***Carex reuteriana*** Boiss.
Carex fusca var. *stolonifera* (Hoppe) Vicioso = ***Carex nigra*** subsp. ***alpina*** (Gaudin) Lemke
Carex fuscescens Willd. ex Kunth = ***Carex divisa*** Huds.
Carex fuscescens Boeck. = ***Carex kiotensis*** Franch. & Sav.
Carex fuscidula V.I.Krecz. ex Egorova = ***Carex capillaris*** subsp. ***fuscidula*** (V.I.Krecz. ex Egorova) Á.Löve & D.Löve
Carex fuscifructus C.B.Clarke = ***Carex atherodes*** Spreng.
Carex fuscoatia Boeck. = ***Carex pichinchensis*** Kunth
Carex fuscocuprea (Kük.) V.I.Krecz. = ***Carex limosa*** L.
Carex fuscofibrosa Ohwi = ***Carex longirostrata*** C.A.Mey. var. ***longirostrata***
Carex fuscotincta Mack. = ***Carex aztecica*** Mack.
Carex fuscotincta Merino = ***Carex caryophyllea*** Latourr. var. ***caryophyllea***
Carex fuscotincta Merino = ?
Carex fuscovaginata Kük. = ***Carex acuta*** L.
Carex fuscula subsp. *catharinensis* (Boeck.) Luceño & Alves = ***Carex fuscula*** d'Urv. var. ***fuscula***
Carex fuscula var. *ciliata* Boeck. = ***Carex fuscula*** d'Urv. var. ***fuscula***
Carex fuscula f. *elatior* Kurtz = ***Carex fuscula*** d'Urv. var. ***fuscula***
Carex fuscula var. *hieronymi* (Boeck.) Kük. = ***Carex hieronymi*** Boeck.
Carex fuscula f. *indecora* (Kunth) Kük. = ***Carex fuscula*** d'Urv. var. ***fuscula***
Carex fuscula var. *pseudextensa* (Kük.) Kük. = ***Carex fuscula*** d'Urv. var. ***fuscula***
Carex fusiformis Chapm. ex Boott = ***Carex laxiflora*** Lam. var. ***laxiflora***
Carex fusiformis var. *austrosinensis* Kük. = ***Carex fusiformis*** Nees var. ***fusiformis***
Carex fusiformis var. *borneensis* Kük. = ***Carex fusiformis*** Nees var. ***fusiformis***
Carex fusiformis subsp. *finitima* (Boott) Noltie = ***Carex finitima*** Boott
Carex fyllae T.Holm = ***Carex bigelowii*** Torr. ex Schwein. subsp. ***bigelowii***
Carex gagaensis Akiyama = ***Carex jackiana*** subsp. ***parciflora*** (Boott) Kük.
Carex gaimardioides É.Desv. = ***Zameioscirpus gaimardioides*** (É.Desv.) Dhooge & Goetgh.
Carex gajonum Nelmes = ***Carex echinata*** Murray subsp. ***echinata***
Carex galactolepis Nelmes = ***Carex cruciata*** Wahlenb. var. ***cruciata***
Carex galbana L.H.Bailey = ***Carex boliviensis*** Van Heurck & Müll.Arg. subsp. ***boliviensis***
Carex galeottiana C.A.Mey. = ***Carex cortesii*** Liebm.
Carex gallaecica H.Lév. & Vaniot = ***Carex gayana*** Steud.
Carex gandogeri H.Lév. & Vaniot = ?
Carex gandogeri H.Lév. ex Gand. = ?
Carex gansuensis Franch. = ***Carex flavocuspis*** Franch. & Sav.
Carex garberi var. *bifaria* Fernald = ***Carex garberi*** Fernald
Carex garberi subsp. *bifaria* (Fernald) Hultén = ***Carex garberi*** Fernald
Carex gardneri Lepage = ***Carex paleacea*** Schreb. ex Wahlenb.
Carex gasparrinii Parl. = ?
Carex gaudichaudiana subsp. *appendiculata* (Trautv. & C.A.Mey.) Á.Löve & D.Löve = ***Carex thunbergii*** Steud.
Carex gaudichaudiana f. *brachysandra* (Franch.) Kük. = ***Carex thunbergii*** Steud.
Carex gaudichaudiana var. *contracta* (F.Muell.) Kük. = ***Carex contracta*** F.Muell.
Carex gaudichaudiana var. *humilior* Kük. = ***Carex thunbergii*** Steud.
Carex gaudichaudiana var. *thunbergii* (Steud.) Kük. = ***Carex thunbergii*** Steud.
Carex gaudichaudiana var. *typica* Domin = ***Carex gaudichaudiana*** Kunth
Carex gauthieri Lepage = ***Carex paleacea*** Schreb. ex Wahlenb.
Carex gayana Desv. = ***Carex diclina*** Phil.
Carex gayana var. *densa* Kük. = ***Carex diclina*** Phil.
Carex gayana var. *hyalina* L.H.Bailey = ***Carex praegracilis*** W.Boott

Carex gayana var. *rufa* Kük. = ***Carex diclina*** Phil.
Carex gayana var. *schedonautos* (Steud.) Kük. = ***Carex diclina*** Phil.
Carex gayana var. *taurina* (Phil.) Kük. = ***Carex diclina*** Phil.
Carex geantha Ohwi = ***Carex jacens*** C.B.Clarke
Carex gebhardii Hoppe = ***Carex brunnescens*** (Pers.) Poir. subsp. ***brunnescens***
Carex gebhardii Willd. = ***Carex elongata*** L.
Carex gebleri Prescott ex Boott = ***Carex heterostachya*** Bunge
Carex geihokuensis K.Okamoto = [38 JAP]
Carex gembolensis C.B.Clarke = ***Carex vesiculosa*** Boott
Carex gembolensis C.B.Clarke = ***Carex vesiculosa*** Boott
Carex gembolensis var. *timorensis* C.B.Clarke = ***Carex cruciata*** var. ***rafflesiana*** (Boott) Noot.
Carex geniculata Host = ***Carex frigida*** All.
Carex gentiliana H.Lév. = ***Carex magellanica*** subsp. ***irrigua*** (Wahlenb.) Hultén
Carex gentilis var. *intermedia* F.T.Wang & Tang ex Y.C.Yang = ***Carex gentilis*** Franch.
Carex gentilis var. *intermedia* Tang & F.T.Wang ex L.K.Dai = ***Carex gentilis*** Franch.
Carex gentilis var. *macrocarpa* Tang & F.T.Wang ex L.K.Dai = ***Carex gentilis*** Franch.
Carex gentilis subsp. *nakaharae* (Hayata) T.Koyama = ***Carex brunnea*** Thunb. subsp. ***brunnea***
Carex gentilis var. *oshimensis* Kük. = ***Carex brunnea*** Thunb. subsp. ***brunnea***
Carex gentilis var. *tonnerrei* (C.B.Clarke) Kük. = ***Carex gentilis*** Franch.
Carex genuensis DC. = ***Carex flacca*** Schreb. subsp. ***flacca***
Carex georgiana Dewey ex Steud. = ***Carex hystericina*** Muhl. ex Willd.
Carex germana Boott = ***Carex banksii*** Boott var. ***banksii***
Carex gersneri Suter = ***Carex humilis*** Leyss. var. ***humilis***
Carex × *ghristii* Boeck. = ?
Carex gibbsii Petrie = ***Carex dallii*** Kirk
Carex gibsonii Bab. = ***Carex nigra*** (L.) Reichard subsp. ***nigra***
Carex gifuensis f. *argyrostachys* (H.Lév. & Vaniot) Kük. = ***Carex gifuensis*** Franch.
Carex gigantea f. *bella-villa* (Dewey) Farw. = ***Carex lupulina*** Muhl. ex Willd.
Carex gigantea var. *grandis* (L.H.Bailey) Farw. = ***Carex gigantea*** Rudge
Carex gigantea var. *lupulina* (Muhl. ex Willd.) Farw. = ***Carex lupulina*** Muhl. ex Willd.
Carex gigantea f. *minor* Farw. = ***Carex gigantea*** Rudge
Carex gigantea f. *pedunculata* Farw. = ***Carex gigantea*** Rudge
Carex gigas (T.Holm) Mack. = ***Carex scabriuscula*** Mack.
Carex gilesii Nelmes = ***Carex nivalis*** Boott
Carex gilliesii Phil. = ***Carex banksii*** Boott var. ***banksii***
Carex giraudiana Steud. = ***Carex borbonica*** Lam.
Carex glabella A.Pabrez ex C.Regel = ***Carex cespitosa*** L. var. ***cespitosa***
Carex glaberrima Meinsh. = ***Carex atherodes*** Spreng.
Carex glabra Boott = ***Carex venusta*** Dewey
Carex glabrior (Ohwi) Akiyama = ***Carex chrysolepis*** var. ***glabrior*** (Ohwi) Ohwi
Carex glacialis var. *terrae-novae* (Fernald) Boivin = ***Carex glacialis*** Mack.
Carex glandulifolia Kük. = ***Carex lancifolia*** C.B.Clarke
Carex glandulosa Elliott ex Boott = ***Carex glaucescens*** Elliott
Carex glareosa var. *amphigena* Fernald = ***Carex glareosa*** Schkuhr ex Wahlenb. subsp. ***glareosa***
Carex glareosa subsp. *amphigena* (Fernald) Á.Löve = ***Carex glareosa*** Schkuhr ex Wahlenb. subsp. ***glareosa***
Carex glareosa var. *caespitosa* Boeck. = ***Carex ursina*** Dewey
Carex glareosa subsp. *soriofkensis* (H.Lév. & Vaniot) McKean = ***Carex glareosa*** Schkuhr ex Wahlenb. subsp. ***glareosa***
Carex glareosa var. *ursina* (Dewey) Britton = ***Carex ursina*** Dewey
Carex glauca Scop. = ***Carex flacca*** Schreb. subsp. ***flacca***
Carex glauca Bosc ex Boott = ***Carex glaucescens*** Elliott
Carex glauca f. *abbreviata* Kük. = ***Carex flacca*** Schreb. subsp. ***flacca***
Carex glauca var. *acuminata* (Willd.) Nyman = ***Carex flacca*** Schreb. subsp. ***flacca***
Carex glauca f. *acuminata* (Willd.) Asch. & Graebn. = ***Carex flacca*** Schreb. subsp. ***flacca***
Carex glauca var. *alpina* Heuff. = ***Carex flacca*** Schreb. subsp. ***flacca***
Carex glauca var. *ambleocarpa* (Willd.) Schur = ***Carex flacca*** Schreb. subsp. ***flacca***
Carex glauca var. *arecta* Drejer = ***Carex flacca*** Schreb. subsp. ***flacca***
Carex glauca var. *arenosa* Schur = ***Carex flacca*** Schreb. subsp. ***flacca***
Carex glauca f. *aristolepis* Kük. ex L.Gross = ***Carex flacca*** Schreb. subsp. ***flacca***
Carex glauca var. *aspera* (Willd.) Poir. = ***Carex flacca*** Schreb. subsp. ***flacca***
Carex glauca var. *attenuata* Ball = ***Carex flacca*** Schreb. subsp. ***flacca***
Carex glauca var. *brachylepis* Regel = ***Carex orbicularis*** Boott subsp. ***orbicularis***
Carex glauca f. *bulbosa* Drejer = ***Carex flacca*** Schreb. subsp. ***flacca***
Carex glauca f. *chlorocarpa* (R.Keller) Kük. = ***Carex flacca*** Schreb. subsp. ***flacca***
Carex glauca subsp. *claviformis* (Hoppe) Schinz & R. Keller = ***Carex flacca*** Schreb. subsp. ***flacca***
Carex glauca var. *claviformis* (Hoppe) Boeck. = ***Carex flacca*** Schreb. subsp. ***flacca***
Carex glauca f. *corcyrensis* (Kük.) Kük. = ***Carex flacca*** Schreb. subsp. ***flacca***
Carex glauca subsp. *cuspidata* (Host) Suess. = ***Carex flacca*** subsp. ***erythrostachys*** (Hoppe) Holub
Carex glauca var. *cuspidata* (Host) Asch. & Graebn. = ***Carex flacca*** subsp. ***erythrostachys*** (Hoppe) Holub
Carex glauca var. *digyna* Boeck. = ***Carex trinervis*** Degl.
Carex glauca var. *dinarica* (Simonk.) Nyman = ***Carex flacca*** Schreb. subsp. ***flacca***
Carex glauca f. *dinarica* (Heuff.) Asch. & Graebn. = ***Carex flacca*** Schreb. subsp. ***flacca***
Carex glauca subsp. *diversicolor* (Crantz) Moravec = ***Carex flacca*** subsp. ***erythrostachys*** (Hoppe) Holub
Carex glauca var. *erythrostachys* (Hoppe) Andersson = ***Carex flacca*** subsp. ***erythrostachys*** (Hoppe) Holub
Carex glauca f. *erythrostachys* (Hoppe) Kük. = ***Carex flacca*** subsp. ***erythrostachys*** (Hoppe) Holub
Carex glauca f. *flavescens* Kük. = ***Carex flacca*** Schreb. subsp. ***flacca***
Carex glauca var. *genuensis* (DC.) Nyman = ***Carex flacca*** Schreb. subsp. ***flacca***
Carex glauca var. *kotschyi* Kük. & Bornm. = ***Carex orbicularis*** Boott subsp. ***orbicularis***
Carex glauca f. *laxiflora* Schur = ***Carex flacca*** Schreb. subsp. ***flacca***
Carex glauca var. *leiocarpa* Willk. = ***Carex flacca*** Schreb. subsp. ***flacca***

Carex glauca f. *leiocarpa* (Willk.) Kük. = ***Carex flacca*** Schreb. subsp. ***flacca***
Carex glauca f. *lemaniana* H.Christ = ***Carex flacca*** Schreb. subsp. ***flacca***
Carex glauca f. *leptostachys* Schur = ***Carex flacca*** Schreb. subsp. ***flacca***
Carex glauca var. *melanocarpa* Murr = ***Carex flacca*** Schreb. subsp. ***flacca***
Carex glauca f. *melanostachya* (Uechtr.) Kük. = ***Carex flacca*** Schreb. subsp. ***flacca***
Carex glauca var. *micheliana* (Sm.) Nyman = ***Carex flacca*** Schreb. subsp. ***flacca***
Carex glauca f. *micheliana* (Sm.) Asch. & Graebn. = ***Carex flacca*** Schreb. subsp. ***flacca***
Carex glauca var. *murensis* A.Terracc. = ***Carex flacca*** Schreb. subsp. ***flacca***
Carex glauca f. *pallida* (Beck) Kük. = ***Carex flacca*** Schreb. subsp. ***flacca***
Carex glauca subsp. *praetutiana* (Parl.) Nyman = ***Carex flacca*** Schreb. subsp. ***flacca***
Carex glauca var. *praetutiana* Arcang. = ***Carex flacca*** Schreb. subsp. ***flacca***
Carex glauca f. *pubicarpa* Rohlena = ***Carex flacca*** Schreb. subsp. ***flacca***
Carex glauca var. *pubicarpa* (Rohlena) Rohlena = ***Carex flacca*** Schreb. subsp. ***flacca***
Carex glauca var. *rotundata* Callmé = ***Carex flacca*** Schreb. subsp. ***flacca***
Carex glauca f. *scabra* Peterm. = ***Carex flacca*** Schreb. subsp. ***flacca***
Carex glauca subsp. *serrulata* (Biv. ex Spreng.) Arcang. = ***Carex flacca*** subsp. ***erythrostachys*** (Hoppe) Holub
Carex glauca f. *serrulata* (Biv. ex Spreng.) Asch. & Graebn. = ***Carex flacca*** subsp. ***erythrostachys*** (Hoppe) Holub
Carex glauca f. *silvatica* Asch. & Graebn. = ***Carex flacca*** Schreb. subsp. ***flacca***
Carex glauca var. *sphaerostachys* Lange = ***Carex flacca*** Schreb. subsp. ***flacca***
Carex glauca var. *stictocarpa* (Sm.) Nyman = ***Carex flacca*** Schreb. subsp. ***flacca***
Carex glauca f. *subrotunda* Kük. = ***Carex flacca*** Schreb. subsp. ***flacca***
Carex glauca var. *subustulata* H.Christ = ***Carex flacca*** Schreb. subsp. ***flacca***
Carex glauca f. *subustulata* (H.Christ) Kük. = ***Carex flacca*** Schreb. subsp. ***flacca***
Carex glauca f. *tenuicula* Martrin-Donos = ***Carex flacca*** Schreb. subsp. ***flacca***
Carex glauca f. *thuringiaca* (Willd.) Kunth = ***Carex flacca*** Schreb. subsp. ***flacca***
Carex glauca var. *thuringiaca* (Willd.) Nyman = ***Carex flacca*** Schreb. subsp. ***flacca***
Carex glauca subsp. *trinervis* (Degl.) Asch. & Graebn. = ***Carex trinervis*** Degl.
Carex glauca var. *typica* Regel = ***Carex orbicularis*** Boott subsp. ***orbicularis***
Carex glaucescens var. *androgyna* Boott = ***Carex verrucosa*** Muhl.
Carex glaucescens f. *macrokolea* (Steud.) Kük. = ***Carex verrucosa*** Muhl.
Carex glaucescens var. *polystachya* Boott = ***Carex glaucescens*** Elliott
Carex glaucina Boeck. = ***Carex filicina*** Nees
Carex glazioviana Boeck. = ***Carex hilairei*** var. ***glazioviana*** (Boeck.) Kük.
Carex glehnii F.Schmidt = ***Carex jackiana*** subsp. ***parciflora*** (Boott) Kük.
Carex globosa var. *brevipes* Boott = ***Carex rossii*** Boott
Carex glomerata Thunb. = ***Carex glomerabilis*** V.I.Krecz.
Carex glomerata Host = ***Carex stenophylla*** Wahlenb. subsp. ***stenophylla***
Carex glomerata Gilib. = ***Carex vulpina*** L.
Carex godefrinii Willemet = ***Carex strigosa*** Huds.
Carex × gogelana Podp. = ?
Carex gokwanensis Hayata = ***Carex chrysolepis*** Franch. & Sav. var. ***chrysolepis***
Carex goodenoughii Asch. & Graebn. = ***Carex nigra*** (L.) Reichard subsp. ***nigra***
Carex goodenowii J.Gay = ***Carex nigra*** (L.) Reichard subsp. ***nigra***
Carex goodenowii f. *androgyna* (Russow ex E.Lehm.) Kük. = ***Carex nigra*** (L.) Reichard subsp. ***nigra***
Carex goodenowii f. *aquatiliformis* (Kük.) Kük. = ***Carex nigra*** (L.) Reichard subsp. ***nigra***
Carex goodenowii f. *badia* (Sanio) Kük. = ***Carex nigra*** (L.) Reichard subsp. ***nigra***
Carex goodenowii f. *basigyna* (Rchb.) Kük. = ***Carex nigra*** (L.) Reichard subsp. ***nigra***
Carex goodenowii f. *chlorostachya* (Rchb.) Kük. = ***Carex nigra*** (L.) Reichard subsp. ***nigra***
Carex goodenowii f. *densicarpa* (Kneuck.) Kük. = ***Carex nigra*** (L.) Reichard subsp. ***nigra***
Carex goodenowii var. *dolia* M.E.Jones = ***Carex lenticularis*** var. ***dolia*** (M.E.Jones) L.A.Standl.
Carex goodenowii f. *fuliginosa* (A.Br.) Kük. = ***Carex nigra*** (L.) Reichard subsp. ***nigra***
Carex goodenowii f. *grata* Sanio = ***Carex nigra*** (L.) Reichard subsp. ***nigra***
Carex goodenowii var. *hydrophila* M.E.Jones = ***Carex nigra*** (L.) Reichard subsp. ***nigra***
Carex goodenowii var. *intricata* (Tineo) H.Lindb. = ***Carex nigra*** subsp. ***intricata*** (Tineo) Maire & Wieller
Carex goodenowii var. *juncea* (Hartm.) Kük. = ***Carex nigra*** subsp. ***juncella*** (Fr.) Lemke
Carex goodenowii var. *juncella* (Fr.) Crép. = ***Carex nigra*** subsp. ***juncella*** (Fr.) Lemke
Carex goodenowii f. *leucolepis* (Meinsh.) Kük. = ***Carex nigra*** (L.) Reichard subsp. ***nigra***
Carex goodenowii var. *limnophila* M.E.Jones = ***Carex nigra*** (L.) Reichard subsp. ***nigra***
Carex goodenowii var. *lipocarpa* M.E.Jones = ***Carex nigra*** (L.) Reichard subsp. ***nigra***
Carex goodenowii f. *microlepis* Norman = ***Carex nigra*** (L.) Reichard subsp. ***nigra***
Carex goodenowii f. *microstachys* Zapal. = ***Carex nigra*** (L.) Reichard subsp. ***nigra***
Carex goodenowii var. *nigra* Cogn. = ***Carex nigra*** (L.) Reichard subsp. ***nigra***
Carex goodenowii f. *oxylepis* (Sanio) Kük. = ***Carex nigra*** (L.) Reichard subsp. ***nigra***
Carex goodenowii f. *pleiandra* Kük. = ***Carex nigra*** (L.) Reichard subsp. ***nigra***
Carex goodenowii f. *polyandra* (Schkuhr) Kük. = ***Carex nigra*** (L.) Reichard subsp. ***nigra***
Carex goodenowii f. *pumila* (Kük.) Kük. = ***Carex nigra*** (L.) Reichard subsp. ***nigra***
Carex goodenowii var. *recta* (J.Fleisch.) Kük. = ***Carex nigra*** (L.) Reichard subsp. ***nigra***
Carex goodenowii var. *reuteriana* (Boiss.) Daveau = ***Carex reuteriana*** Boiss.
Carex goodenowii f. *roussioei* H.Lév. = ***Carex nigra*** (L.) Reichard subsp. ***nigra***
Carex goodenowii var. *salinoides* (Kük.) Kük. = ***Carex nigra*** (L.) Reichard subsp. ***nigra***
Carex goodenowii f. *silvatica* (Russow ex E.Lehm.) Kük. = ***Carex nigra*** (L.) Reichard subsp. ***nigra***

Carex goodenowii var. *stenocarpa* Kük. = ***Carex nigra*** (L.) Reichard subsp. ***nigra***
Carex goodenowii f. *stenostachya* Uechtr. ex Asch. & Graebn. = ***Carex nigra*** (L.) Reichard subsp. ***nigra***
Carex goodenowii var. *stolonifera* (Hoppe) Asch. = ***Carex nigra*** subsp. ***alpina*** (Gaudin) Lemke
Carex goodenowii subsp. *stolonifera* (Hoppe) Asch. & Graebn. = ***Carex nigra*** subsp. ***alpina*** (Gaudin) Lemke
Carex goodenowii var. *strictiformis* (L.H.Bailey) Kük. = ***Carex nigra*** (L.) Reichard subsp. ***nigra***
Carex goodenowii var. *subcaespitosa* Kük. = ***Carex nigra*** (L.) Reichard subsp. ***nigra***
Carex goodenowii f. *subfastigiata* (Schur) Kük. = ***Carex nigra*** (L.) Reichard subsp. ***nigra***
Carex goodenowii f. *subovalis* (Kneuck.) Kük. = ***Carex nigra*** (L.) Reichard subsp. ***nigra***
Carex goodenowii f. *subramosa* (Meinsh.) Kük. = ***Carex nigra*** (L.) Reichard subsp. ***nigra***
Carex goodenowii var. *subrigida* Kük. = ***Carex nigra*** (L.) Reichard subsp. ***nigra***
Carex goodenowii f. *subsetacea* Kük. = ***Carex nigra*** (L.) Reichard subsp. ***nigra***
Carex goodenowii var. *tornata* (Fr.) Kük. = ***Carex nigra*** (L.) Reichard subsp. ***nigra***
Carex goodenowii var. *turfosa* (Fr.) Asch. = ***Carex ×turfosa*** Fr.
Carex goodenowii f. *zonata* (F.Nyl.) Kük. = ***Carex nigra*** (L.) Reichard subsp. ***nigra***
Carex gorodkovii V.I.Krecz. = ***Carex stenantha*** var. ***taisetsuensis*** Akiyama
Carex ×goroi T.Koyama = ?
Carex ×gotlandica Englund = ?
Carex ×goyozanensis M.Kikuchi & Mochizuki = ?
Carex gracilenta Litv. = ***Carex norvegica*** Retz. subsp. ***norvegica***
Carex gracilescens Steud. = ***Carex laxiflora*** var. ***gracillima*** Boott
Carex gracilescens Hartm. ex Andersson = ***Carex riparia*** Curtis subsp. ***riparia***
Carex graciliculmis Ohwi = ***Carex kirganica*** Kom.
Carex graciliformis V.I.Krecz. = ***Carex acuta*** L.
Carex gracilipes Miq. = ***Carex phacota*** Spreng.
Carex gracilirostris C.B.Clarke ex Franch. = ***Carex rubrobrunnea*** C.B.Clarke
Carex gracilis Curtis = ***Carex acuta*** L.
Carex gracilis A.Gray = ***Carex disperma*** Dewey
Carex gracilis R.Br. = ***Carex lenta*** D.Don var. ***lenta***
Carex gracilis Moench = ***Carex montana*** L.
Carex gracilis Honck. = ***Carex mucronata*** All.
Carex gracilis f. *amblylepis* (Peterm.) Asch. & Graebn. = ***Carex acuta*** L.
Carex gracilis var. *angustifolia* Kük. = ***Carex acuta*** L.
Carex gracilis var. *approximata* Kük. ex Junge = ***Carex acuta*** L.
Carex gracilis f. *brachystachya* Kük. = ***Carex acuta*** L.
Carex gracilis subsp. *broteroana* (Samp.) M.Laínz = ***Carex elata*** All. subsp. ***elata***
Carex gracilis var. *chlorotica* (Èelak.) Kük. = ***Carex brunnea*** Thunb. subsp. ***brunnea***
Carex gracilis var. *compacta* (St.-Lag.) Kük. = ***Carex brunnea*** Thunb. subsp. ***brunnea***
Carex gracilis subsp. *corynophora* Asch. & Graebn. = ***Carex acuta*** L.
Carex gracilis f. *corynophora* (Peterm.) Kük. = ***Carex acuta*** L.
Carex gracilis lus. *dioica* (Peterm.) Soó = ***Carex acuta*** L.
Carex gracilis subsp. *erecta* Kük. = ***Carex acuta*** L.
Carex gracilis subsp. *eugracilis* Kük. = ***Carex acuta*** L.
Carex gracilis var. *fluviatilis* (Hartm.) Kük. = ***Carex brunnea*** Thunb. subsp. ***brunnea***
Carex gracilis f. *graciliflora* (Legrand) Soó = ***Carex acuta*** L.
Carex gracilis var. *humilis* Kük. = ***Carex acuta*** L.
Carex gracilis subsp. *intermedia* (Èelak.) Soó = ***Carex acuta*** L.
Carex gracilis var. *libanotica* Kük. = ***Carex acuta*** L.
Carex gracilis var. *minor* Boott = ***Carex brunnea*** Thunb. subsp. ***brunnea***
Carex gracilis f. *pedicellata* (Peterm.) Soó = ***Carex acuta*** L.
Carex gracilis var. *personata* (Fr.) Kük. = ***Carex brunnea*** Thunb. subsp. ***brunnea***
Carex gracilis f. *podperae* Soó = ***Carex acuta*** L.
Carex gracilis f. *protensa* (Fr.) Kük. = ***Carex brunnea*** Thunb. subsp. ***brunnea***
Carex gracilis f. *reuteriana* (Boiss.) Kük. = ***Carex reuteriana*** Boiss.
Carex gracilis f. *rudis* (Wimm.) Kük. = ***Carex acuta*** L.
Carex gracilis var. *rudis* (Wimm.) Kük. = ***Carex acuta*** L.
Carex gracilis f. *seminuda* (Beck) Kük. = ***Carex acuta*** L.
Carex gracilis var. *sphaerocarpa* (Uechtr.) Kük. = ***Carex brunnea*** Thunb. subsp. ***brunnea***
Carex gracilis var. *strictifolia* (Opiz ex Nyman) Asch. = ***Carex acuta*** L.
Carex gracilis var. *tricostata* (Fr.) Asch. = ***Carex acuta*** L.
Carex gracilis f. *vaginata* (Peterm.) Soó = ***Carex acuta*** L.
Carex gracilis f. *zygostachya* (Rchb.) Kük. = ***Carex brunnea*** Thunb. subsp. ***brunnea***
Carex gracillima Steud. & Hochst. ex Steud. = ***Carex ferruginea*** Scop. subsp. ***ferruginea***
Carex gracillima f. *humilis* L.H.Bailey = ***Carex gracillima*** Schwein.
Carex gracillima var. *humilis* (L.H.Bailey) L.H.Bailey = ***Carex gracillima*** Schwein.
Carex gracillima var. *macerrima* Fernald & Wiegand = ***Carex gracillima*** Schwein.
Carex graeffeana var. *samoensis* Nelmes = ***Carex graeffeana*** Boeck.
Carex ×grahamii Boott = ***Carex ×stenolepis*** Less. nothovar. ***stenolepis***
Carex grallatoria subsp. *heteroclita* (Franch.) T.Koyama = ***Carex grallatoria*** var. ***heteroclita*** (Franch.) Kük.
Carex grandilimosa Akiyama = ***Carex middendorffii*** F.Schmidt var. ***middendorffii***
Carex grandis L.H.Bailey = ***Carex gigantea*** Rudge
Carex grandis var. *helleri* L.H.Bailey = ***Carex gigantea*** Rudge
Carex grandisquama Franch. = ***Carex papulosa*** Boott
Carex granitica Braun-Blanq. = ***Carex sempervirens*** Vill.
Carex granularioides Schwein. = ***Carex conoidea*** Willd.
Carex granularis var. *haleana* (Olney) Porter = ***Carex granularis*** Muhl. ex Willd.
Carex granularis var. *recta* Dewey ex Alph.Wood = ***Carex granularis*** Muhl. ex Willd.
Carex granularis var. *shriveri* Britton = ***Carex granularis*** Muhl. ex Willd.
Carex grasmanniana (O.Lang) Rabenh. ex Steud. = ***Carex filiformis*** L.
Carex gravida var. *laxifolia* L.H.Bailey = ***Carex gravida*** L.H.Bailey
Carex gravida f. *laxifolia* (L.H.Bailey) Kük. = ***Carex gravida*** L.H.Bailey
Carex gravida var. *lunelliana* (Mack.) F.J.Herm. = ***Carex gravida*** L.H.Bailey
Carex grayana Dewey = ***Carex livida*** (Wahlenb.) Willd.
Carex grayi var. *hispidula* A.Gray ex L.H.Bailey = ***Carex grayi*** Carey

Carex grayi var. *rariflora* Farw. = ***Carex grayi*** Carey
Carex greeniana Dewey = ***Carex hostiana*** DC.
Carex greenii Boeck. = ***Carex bullata*** Willd.
Carex greenwayi Nelmes = ***Carex vallis-rosetto*** K.Schum.
Carex griffithii Boott = ***Carex nivalis*** Boott
Carex grisea Viv. = ***Carex grioletii*** Roem. ex Schkuhr
Carex grisea var. *amphibola* (Steud.) Kük. = ***Carex amphibola*** Steud.
Carex grisea var. *angustifolia* Boott = ***Carex paeninsulae*** Naczi, E.L.Bridges & Orzell
Carex grisea var. *globosa* L.H.Bailey = ***Carex bulbostylis*** Mack.
Carex grisea var. *minor* Olney = ***Carex amphibola*** Steud.
Carex grisea var. *mutica* (Torr.) Carey = ***Carex flaccosperma*** Dewey
Carex grisea var. *rigida* L.H.Bailey = ***Carex planispicata*** Naczi
Carex grosseckii Heuff. ex Boott = ***Carex pyrenaica*** Wahlenb.
Carex grossheimii V.I.Krecz. = ***Carex disticha*** Huds.
Carex grypos Schkuhr = ***Carex echinata*** Murray subsp. ***echinata***
Carex guestphalica (Rchb.) Boenn. ex W.D.J.Koch = ***Carex divulsa*** Stokes subsp. ***divulsa***
Carex guffroyana H.Lév. & Vaniot = ***Carex prescottiana*** Boott
Carex gunniana var. *barbata* (Boott) Kük. = ***Carex barbata*** Boott
Carex gunniana var. *brevior* Kük. = ***Carex gunniana*** Boott
Carex guthnickiana J.Gay = ***Carex pulicaris*** L.
Carex gymnoclada T.Holm = ***Carex scopulorum*** var. ***bracteosa*** (L.H.Bailey) F.J.Herm.
Carex gynandra Sieber ex Kunth = ***Carex distachya*** Desf. var. ***distachya***
Carex gynandra var. *porteri* (Olney) Britton = ***Carex crinita*** var. ***porteri*** (Olney) Fernald
Carex gynandra var. *simulans* (Fernald) Roland = ***Carex gynandra*** Schwein.
Carex gynobasis Vill. = ***Carex halleriana*** Asso
Carex gynocrates var. *substaminata* Peck = ***Carex gynocrates*** Wormsk.
Carex gynomane Bertol. = ***Carex distachya*** Desf. var. ***distachya***
Carex haastiana Boeck. = ***Carex raoulii*** Boott
Carex ×*haematolepis* subsp. *kattegatensis* Nyman = ***Carex vacillans*** Drejer
Carex ×*haematolepis* var. *vacillans* (Drejer) Nyman = ***Carex vacillans*** Drejer
Carex haematostachys H.Lév. & Vaniot = ***Carex heterostachya*** Bunge
Carex haematostoma var. *digyae* Kük. = ***Carex przewalskii*** T.V.Egorova var. ***przewalskii***
Carex haematostoma var. *hirtelloides* Kük. = ***Carex hirtelloides*** (Kük.) F.T.Wang & Tang ex P.C.Li
Carex haematostoma var. *submacrogyna* Kük. = ***Carex haematostoma*** Nees
Carex haenkeana C.Presl = ***Carex pseudocyperus*** var. ***haenkeana*** (C.Presl) Kük.
Carex ×*hageri* E.Baumann = ?
Carex hagiana Kelso = ***Carex heteroneura*** var. ***epapillosa*** (Mack.) F.J.Herm.
Carex hailstonii S.Gibson = ***Carex remota*** L. subsp. ***remota***
Carex hainanensis Merr. ex Chun & F.C.How = ***Carex commixta*** Steud.
Carex hakkuensis Hayata = ***Carex cruciata*** Wahlenb. var. ***cruciata***
Carex hakodatensis H.Lév. & Vaniot = ***Carex pilosa*** Scop.
Carex haleana Olney = ***Carex granularis*** Muhl. ex Willd.
Carex halei Dewey = ***Carex crus-corvi*** Shuttlew. ex Kunze
Carex halei J.Carey = ***Carex louisianica*** L.H.Bailey
Carex halei L.H.Bailey ex C.B.Clarke = ?
Carex halei var. *minor* Boott = ***Carex louisianica*** L.H. Bailey
Carex hallaisanensis H.Lév. & Vaniot = ***Carex erythrobasis*** H.Lév. & Vaniot
Carex halleri Gunnerus = ***Carex norvegica*** Retz. subsp. ***norvegica***
Carex halleri Vest ex Steud. =
Carex halleriana subsp. *corsica* (Mabille) Cif. & Giacom. = ***Carex halleriana*** Asso
Carex halleriana f. *corsica* Mabille = ***Carex halleriana*** Asso
Carex halleriana subsp. *depressa* (Link) Nyman = ***Carex depressa*** Link
Carex halleriana f. *lerinensis* H.Christ = ***Carex halleriana*** Asso
Carex halleriana f. *occulata* Kük. = ***Carex halleriana*** Asso
Carex halleriana f. *pedunculata* Kük. = ***Carex halleriana*** Asso
Carex halleriana subsp. *planostachys* (Kunze) Kük. = ***Carex halleriana*** Asso
Carex halleriana subsp. *rorulenta* (Porta) Malag. = ***Carex rorulenta*** Porta
Carex hallii L.H.Bailey = ***Carex nudata*** Boott
Carex halmaturina Boeck. = ***Carex tereticaulis*** F.Muell.
Carex halophila Heuff. = ***Carex hostiana*** DC.
Carex halophila F.Nyl. = ***Carex recta*** Boott
Carex halophila subsp. *cryptocarpa* (C.A.Mey.) Nyman = ***Carex lyngbyei*** Hornem.
Carex halophila var. *flavicans* (F.Nyl.) Boivin = ***Carex ×flavicans*** F.Nyl.
Carex halophila var. *grantii* (A.Benn.) B.Boivin = ***Carex ×grantii*** A.Benn.
Carex halseyana Dewey = ***Carex polymorpha*** Muhl.
Carex halseyi Eaton = ***Carex polymorpha*** Muhl.
Carex hamata J.R.Forst. ex Steud. = ***Carex forsteri*** Wahlenb.
Carex hamata Sw. = ***Uncinia hamata*** (Sw.) Urb.
Carex hamosa Thouars = ***Uncinia uncinata*** (L.f.) Kük.
Carex ×*hanasakensis* T.Koyama = ?
Carex hancei C.B.Clarke = ***Carex laticeps*** C.B.Clarke ex Franch.
Carex hancockiana var. *peiktusanii* (Kom.) Kük. = ***Carex peiktusanii*** Kom.
Carex handelii Kük. = ***Carex karlongensis*** subsp. ***handelii*** (Kük.) P.C.Li
Carex hankaensis Kitag. = ***Carex pediformis*** C.A.Mey. subsp. ***pediformis***
Carex ×*hanseniana* Junge = ?
Carex harlandii var. *liuguensis* S.Yun Liang & D.Y.Liu = ***Carex harlandii*** Boott
Carex harlandii f. *longibracteata* R.Gross = ***Carex harlandii*** Boott
Carex harlandii var. *xiuningensis* S.W.Su = ***Carex harlandii*** Boott
Carex harmsiana Kük. ex H.Lév. = ***Carex meiocarpa*** H.Lév. & Vaniot
Carex harperi Fernald = ***Carex leptalea*** subsp. ***harperi*** (Fernald) W.Stone
Carex ×*hartii* var. *bradleyi* Dewey = ***Carex ×hartii*** Dewey
Carex hartwegii Boott = ***Carex polystachya*** Sw. ex Wahlenb. var. ***polystachya***
Carex hartzii Gand. = ***Carex bigelowii*** Torr. ex Schwein. subsp. ***bigelowii***

Carex hattoriana Nakai ex Tuyama = ***Carex brunnea*** Thunb. subsp. ***brunnea***
Carex hatusimana Ohwi = ***Carex truncatigluma*** subsp. ***rhynchachaenium*** (C.B.Clarke) Y.C.Tang & S.Yun Liang
Carex hausmannii Tapp. ex Kalela = ?
Carex × *haussknechtii* Senay =
Carex havilandii C.B.Clarke = ***Carex verticillata*** Zoll. & Moritzi
Carex havilandii C.B.Clarke = ***Carex verticillata*** Zoll. & Moritzi
Carex hawaiiensis H.St.John = ***Carex echinata*** Murray subsp. ***echinata***
Carex hayatae H.Lév. = ***Carex blepharicarpa*** Franch. var. ***blepharicarpa***
Carex hayatana Honda = ***Carex alopecuroides*** D.Don ex Tilloch & Taylor
Carex haydeniana var. *ebenea* (Rydb.) Kelso = ***Carex ebenea*** Rydb.
Carex haysii S.L.Welsh = ***Carex curatorum*** Stacey
Carex hebecarpa Hook. & Arn. = ***Carex hookeri*** Kunth
Carex hebecarpa f. *latifolia* Makino = ***Carex hebecarpa*** C.A.Mey. subsp. ***hebecarpa***
Carex hebecarpa var. *ligulata* (Nees) Kük. = ***Carex hebecarpa*** subsp. ***ligulata*** (Nees) T.Koyama
Carex heilongjiangensis Y.L.Chou = ***Carex reptabunda*** (Trautv.) V.I.Krecz.
Carex heleochariformis H.Lév. & Vaniot = ***Carex onoei*** Franch. & Sav.
Carex heleonastes Lilj. = ***Carex glareosa*** Schkuhr ex Wahlenb. subsp. ***glareosa***
Carex heleonastes var. *dubia* (L.H.Bailey) B.Boivin = ***Carex praeceptorium*** Mack.
Carex heleonastes var. *lachenalii* (Schkuhr) Fiori = ***Carex lachenalii*** Schkuhr
Carex heleonastes var. *robustior* Kük. = ***Carex arctiformis*** Mack.
Carex heleonastes var. *scabriuscula* Kük. = ***Carex arcta*** Boott
Carex heleonastes f. *sendtneriana* (Brügger) Kük. = ***Carex heleonastes*** Ehrh. ex L.f. subsp. ***heleonastes***
Carex helferi subsp. *mapaniifolia* (Ridl.) T.Koyama = ***Carex helferi*** Boeck.
Carex heliophila Mack. = ***Carex inops*** subsp. ***heliophila*** (Mack.) Crins
Carex helodes Link = ***Carex laevigata*** Sm.
Carex helodes subsp. *andalusica* H.Lindb. = ***Carex laevigata*** Sm.
Carex helodes f. *biligularis* (DC.) Kük. = ***Carex laevigata*** Sm.
Carex helodes var. *maurusia* Font Quer & Maire = ***Carex laevigata*** Sm.
Carex helodes subsp. *pseudohelvola* (Kihlm.) Nyman = ***Carex ×pseudohelvola*** Kihlm.
Carex helvetica Honck. = ***Carex frigida*** All.
Carex helvetica Schleich. ex Kunth = ***Carex punctata*** Gaudin var. ***punctata***
Carex × *helvola* subsp. *pseudohelvola* (Kihlm.) Nyman = ***Carex ×pseudohelvola*** Kihlm.
Carex henningsiana Boeck. = ***Carex vulpinaris*** Nees
Carex henryi C.B.Clarke ex Franch. = ***Carex longicruris*** Nees
Carex hepburnii Boott = ***Carex nardina*** var. ***hepburnii*** (Boott) Kük.
Carex heptastachya Boeck. = ***Carex bonplandii*** Kunth
Carex herbariorum L.H.Bailey = ?
Carex heribaudiana H.Lév. & Vaniot = ***Carex caryophyllea*** Latourr. var. ***caryophyllea***
Carex hermaphrodita Jacq. = ***Cyperus hermaphroditus*** (Jacq.) Standl.
Carex hermaphrodita J.F.Gmel. = ***Kobresia myosuroides*** (Vill.) Fiori & Paol.
Carex heteroclita Franch. = ***Carex grallatoria*** var. ***heteroclita*** (Franch.) Kük.
Carex heterolepis Boeck. = ***Carex cruenta*** Nees
Carex heterolepis var. *obtegens* (Kük.) Kük. = ***Carex heterolepis*** Bunge
Carex heteroneura var. *chalciolepis* (T.Holm) F.J.Herm. = ***Carex chalciolepis*** T.Holm
Carex heterophylla Krock. ex Steud. = ***Carex praecox*** Schreb.
Carex heterosperma Wahlenb. = ***Carex laxiflora*** Lam. var. ***laxiflora***
Carex heterostachya Torr. = ***Carex careyana*** Torr. ex Dewey
Carex heterostachya f. *minor* (Boott) Kük. = ***Carex heterostachya*** Bunge
Carex hians Steud. = ?
Carex × *hibernica* A.Benn. = ?
Carex hidewoi Ohwi = ***Carex lehmanii*** Drejer
Carex hilgendorfiana Boeck. = ***Carex duvaliana*** Franch. & Sav.
Carex hindsii C.B.Clarke = ***Carex lenticularis*** var. ***limnophila*** (T.Holm) Cronquist
Carex hindsii var. *brevigluma* Kük. = ***Carex lenticularis*** var. ***lipocarpa*** (T.Holm) L.A.Standl.
Carex hirsuta Willd. = ***Carex hirsutella*** Mack.
Carex hirsuta Suter = ***Carex norvegica*** Retz. subsp. ***norvegica***
Carex hirsuta var. *cuspidata* Dewey = ***Carex bushii*** Mack.
Carex hirta var. *aquatica* Waisb. = ***Carex hirta*** L.
Carex hirta var. *glabra* Lej. = ***Carex hirta*** L.
Carex hirta f. *glabrata* Peterm. = ***Carex hirta*** L.
Carex hirta var. *glabrescens* Cariot = ***Carex hirta*** L.
Carex hirta var. *hirtiformis* (Pers.) Lej. = ***Carex hirta*** L.
Carex hirta f. *hirtiformis* (Pers.) Kunth = ***Carex hirta*** L.
Carex hirta f. *humilis* Peterm. = ***Carex hirta*** L.
Carex hirta f. *latifolia* Waisb. ex Kük. = ***Carex hirta*** L.
Carex hirta f. *major* Peterm. = ***Carex hirta*** L.
Carex hirta f. *paludosa* A.Winkl. ex Asch. & Graebn. = ***Carex hirta*** L.
Carex hirta var. *pilosa* Èelak. = ***Carex hirta*** L.
Carex hirta f. *pseudohirta* Schur = ***Carex hirta*** L.
Carex hirta subsp. *siegertiana* (Uechtr. ex Garcke) Nyman = ***Carex atherodes*** Spreng.
Carex hirta f. *spinosa* H.Mort. = ***Carex hirta*** L.
Carex hirta var. *subglabra* Èelak. = ***Carex hirta*** L.
Carex hirta f. *subhirtiformis* Kneuck. = ***Carex hirta*** L.
Carex hirta var. *sublaevis* Hornem. = ***Carex hirta*** L.
Carex hirta var. *vera* Neilr. = ***Carex hirta*** L.
Carex hirta f. *villosa* Peterm. = ***Carex hirta*** L.
Carex hirtella Drejer = ***Carex plectobasis*** V.I.Krecz.
Carex hirtella (Sw.) J.F.Gmel. = ***Scleria hirtella*** Sw.
Carex hirtiformis Pers. = ***Carex hirta*** L.
Carex hirtiformis Dalla Torre & Sarnth. = ?
Carex hirtifructus Kük. = ***Carex blepharicarpa*** var. ***hirtifructus*** (Kük.) Ohwi
Carex hirtiglumis C.B.Clarke ex Kük. = ***Carex hirtigluma*** C.B.Clarke
Carex hirtiglumis var. *arcuata* Cherm. = ***Carex hirtigluma*** C.B.Clarke
Carex hispida Gaudin = ***Carex fimbriata*** Schkuhr
Carex hispida Link = ***Carex hispida*** Willd. ex Schkuhr
Carex hispida var. *anacantha* Gren. & Godr. = ***Carex hispida*** Willd. ex Schkuhr
Carex hispida var. *genuina* Gren. & Godr. = ***Carex hispida*** Willd. ex Schkuhr
Carex hispida f. *marina* Heldr. ex Kük. = ***Carex hispida*** Willd. ex Schkuhr

Carex hispida subsp. *retusa* (Degl.) Arcang. = ***Carex echinata*** Murray subsp. ***echinata***
Carex hispida var. *soleirolii* (DC. & Duby) Nyman = ***Carex echinata*** Murray subsp. ***echinata***
Carex hispida f. *soleirolii* (DC. & Duby) Asch. = ***Carex echinata*** Murray subsp. ***echinata***
Carex hispidangula T.Koyama = ***Carex tricephala*** Boeck.
Carex hispidula Gaudin = ***Carex fimbriata*** Schkuhr
Carex hitchcockiana var. *triflora* Peck = ***Carex hitchcockiana*** Dewey
Carex hitchcockii Eaton = ***Carex hitchcockiana*** Dewey
Carex hobkodensis Franch. = ?
Carex hochstetteri Frank ex Steud. = ***Carex leptalea*** Wahlenb. subsp. ***leptalea***
Carex hochstetteriana var. *rigidifolia* (Hochst. ex Seub.) Kük. = ***Carex hochstetteriana*** J.Gay ex Seub.
Carex × *hokariensis* T.Koyama = ?
Carex holmiana Mack. = ***Carex buxbaumii*** Wahlenb.
Carex hololasius H.Lév. & Vaniot = ***Carex duvaliana*** Franch. & Sav.
Carex homalocarpa Peterm. = ***Carex elata*** All. subsp. ***elata***
Carex homoiolepis Franch. & Sav. = ***Carex nervata*** Franch. & Sav.
Carex hondae Akiyama = ***Carex deweyana*** var. ***senanensis*** (Ohwi) T.Koyama
Carex hongkongensis Franch. = ***Carex ligata*** Boott subsp. ***ligata***
Carex hoodii var. *nervosa* L.H.Bailey = ***Carex hoodii*** Boott
Carex hoodii var. *neurocarpa* Piper = ***Carex hoodii*** Boott
Carex hooker-grayi C.B.Clarke = ?
Carex hookeriana var. *occidentalis* (L.H.Bailey) Kük. = ***Carex occidentalis*** L.H.Bailey
Carex hoppneri Boott = ***Carex subspathacea*** Wormsk. ex Hornem.
Carex hordeiformis Wahlenb. = ***Carex hordeistichos*** Vill.
Carex hordeistichos var. *microcarpa* Boeck. = ***Carex secalina*** Willd. ex Wahlenb. var. ***secalina***
Carex hordeistichos var. *microstachys* Schur = ***Carex secalina*** Willd. ex Wahlenb. var. ***secalina***
Carex hordeistichos var. *secalina* (Willd. ex Wahlenb.) Boott = ***Carex secalina*** Willd. ex Wahlenb.
Carex horikawae K.Okamoto = ***Carex breviculmis*** R.Br. var. ***breviculmis***
Carex hormathodes var. *invisa* (W.Boott) Fernald = ***Carex hormathodes*** Fernald
Carex hormathodes f. *invisa* (W.Boott) Fernald = ***Carex hormathodes*** Fernald
Carex hormathodes var. *richii* (Fernald) Fernald = ***Carex straminea*** Willd. ex Schkuhr
Carex hornschuchiana Hoppe = ***Carex hostiana*** DC.
Carex hornschuchiana f. *angustifolia* Gaudin = ***Carex hostiana*** DC.
Carex hornschuchiana f. *approximata* Murr = ***Carex hostiana*** DC.
Carex hornschuchiana f. *atrata* Buchenau = ***Carex hostiana*** DC.
Carex hornschuchiana var. *diluta* (M.Bieb.) Nyman = ***Carex diluta*** M.Bieb.
Carex hornschuchiana f. *discolor* Vollm. = ***Carex hostiana*** DC.
Carex hornschuchiana subsp. *eckeroeensis* H.Lindb. & Palmgr. = ***Carex hostiana*** DC.
Carex hornschuchiana var. *fulva* Nyman = ***Carex hostiana*** DC.
Carex hornschuchiana var. *laurentiana* Fernald & Wiegand = ***Carex hostiana*** DC.
Carex hornschuchiana f. *longibracteata* Peterm. = ***Carex hostiana*** DC.
Carex hornschuchiana f. *remota* Peterm. = ***Carex hostiana*** DC.
Carex hornschuchiana f. *subdistans* Neuman = ***Carex hostiana*** DC.
Carex × *hosoii* T.Koyama = ?
Carex hostiana var. *froedinii* Fagerstr. = ***Carex ericetorum*** Pollich
Carex hostiana var. *laurentiana* (Fernald & Wiegand) Fernald & Wiegand = ***Carex hostiana*** DC.
Carex hostii Hoppe ex Kunth = ***Carex divisa*** Huds.
Carex hostii Schkuhr = ***Carex stenophylla*** Wahlenb. subsp. ***stenophylla***
Carex houghtonii Torr. = ***Carex houghtoniana*** Torr. ex Dewey
Carex howei Mack. = ***Carex atlantica*** subsp. ***capillacea*** (L.H.Bailey) Reznicek
Carex howei var. *capillacea* (L.H.Bailey) Fernald = ***Carex atlantica*** subsp. ***capillacea*** (L.H.Bailey) Reznicek
Carex howellii L.H.Bailey = ***Carex aquatilis*** var. ***dives*** (T.Holm) Kük.
Carex hudsonii A.Benn. = ***Carex elata*** All. subsp. ***elata***
Carex hudsonii f. *brachystachys* Podp. = ***Carex elata*** All. subsp. ***elata***
Carex hudsonii f. *glabra* Junge = ***Carex elata*** All. subsp. ***elata***
Carex hudsonii var. *homalocarpa* (Peterm.) Kük. = ***Carex elata*** All. subsp. ***elata***
Carex hudsonii f. *humilis* (Fr.) Kük. = ***Carex elata*** All. subsp. ***elata***
Carex hudsonii f. *nigrans* (Beck) Kük. = ***Carex elata*** All. subsp. ***elata***
Carex hudsonii f. *pendula* (Lange) Kük. = ***Carex elata*** All. subsp. ***elata***
Carex hudsonii var. *polystachya* (Asch. & Graebn.) Kük. = ***Carex elata*** All. subsp. ***elata***
Carex hudsonii f. *reticulosa* (Peterm.) Kük. = ***Carex elata*** All. subsp. ***elata***
Carex hudsonii f. *stenocarpa* Junge = ***Carex elata*** All. subsp. ***elata***
Carex huetiana Boiss. = ***Carex umbrosa*** subsp. ***huetiana*** (Boiss.) Soó
Carex huitensis Steud. = ***Carex fuscula*** var. ***distenta*** (Kunze ex Kunth) Kük.
Carex humbertii F.T.Wang & Tang = ***Carex commixta*** Steud.
Carex humilis Willd. ex Kunth = ***Carex cespitosa*** L. var. ***cespitosa***
Carex humilis var. *longifolia* Stoeva & E.D.Popova = ***Carex humilis*** Leyss. var. ***humilis***
Carex huttoniana Kük. = ***Carex zuluensis*** C.B.Clarke
Carex hybrida Lam. = ***Carex divisa*** Huds.
Carex hybrida Schkuhr ex Kunth = ***Kobresia simpliciuscula*** (Wahlenb.) Mack. subsp. ***simpliciuscula***
Carex hydrophila Dumort. = ***Carex echinata*** Murray subsp. ***echinata***
Carex hylaea V.I.Krecz. = ***Carex canescens*** L. subsp. ***canescens***
Carex hymelayensis Steud. = ***Carex divisa*** Huds.
Carex hymenina Nees ex Boott = ?
Carex hymenocarpa Drejer = ***Carex rostrata*** Stokes var. ***rostrata***
Carex hypaneura V.I.Krecz. = ***Carex sylvatica*** Huds. subsp. ***sylvatica***
Carex hyperborea Drejer = ***Carex bigelowii*** Torr. ex Schwein. subsp. ***bigelowii***
Carex hyperborea f. *cuspidata* Rosenv. = ***Carex bigelowii*** Torr. ex Schwein. subsp. ***bigelowii***

Carex hyperborea var. *latifolia* Andersson = ***Carex bigelowii*** Torr. ex Schwein. subsp. ***bigelowii***
Carex hyperborea var. *paradoxa* Drejer = ***Carex bigelowii*** Torr. ex Schwein. subsp. ***bigelowii***
Carex hyperborea f. *pumila* Lange = ***Carex bigelowii*** Torr. ex Schwein. subsp. ***bigelowii***
Carex hypochlora Freyn = ***Carex breviculmis*** R.Br. var. ***breviculmis***
Carex hypolytrifolia T.Koyama = ***Carex commixta*** Steud.
Carex hypolytroides R.Gross & Mattf. = ***Carex hypolytroides*** Ridl.
Carex hypolytropsis R.Gross = ***Carex hypolytroides*** Ridl.
Carex hypotracheia Nelmes = ***Carex meeboldiana*** Kük.
Carex hypoxanthos Steud. = ***Carex brongniartii*** Kunth
Carex hypsophila Miq. = ***Carex verticillata*** Zoll. & Moritzi
Carex hypsophila var. *havilandii* (C.B.Clarke) Kük. = ***Carex verticillata*** Zoll. & Moritzi
Carex hypsophila var. *verticillata* (Zoll. & Moritzi) Kük. = ***Carex verticillata*** Zoll. & Moritzi
Carex hystericina var. *angustior* L.H.Bailey = [76 ARI]
Carex hystericina var. *cooleyi* (Dewey) Dewey = ***Carex hystericina*** Muhl. ex Willd.
Carex hystericina var. *dudleyi* L.H.Bailey = ***Carex hystericina*** Muhl. ex Willd.
Carex hystericina f. *dudleyi* (L.H.Bailey) Wiegand = ***Carex hystericina*** Muhl. ex Willd.
Carex hystericina var. *underwoodii* (Britton) Kük. = ***Carex polysticha*** Boeck.
Carex hystrix A.Gray ex Boott = ***Carex crus-corvi*** Shuttlew. ex Kunze
Carex ignota Dewey = ***Carex striatula*** Michx.
Carex ignota var. *fusiformis* Dewey = ***Carex laxiflora*** Lam. var. ***laxiflora***
Carex illinoensis Dewey = ***Carex conoidea*** Willd.
Carex illustranda Steud. = ***Carex mucronata*** All.
Carex × *ilseana* Ruhmer = ?
Carex imbricata Kük. = ***Carex remota*** subsp. ***alta*** (Boott) Kük.
Carex imbricata Drobow = [32]
Carex immanis C.B.Clarke = ***Carex fuliginosa*** Schkuhr
Carex impressa (S.H.Wright) Mack. = ***Carex hyalinolepis*** Steud.
Carex impunctata Boott = ***Carex oligostachya*** Nees
Carex × *inaensis* T.Koyama = ?
Carex inaequabilis V.I.Krecz. = ***Carex anomoea*** Hand.-Mazz.
Carex inaequalis Boott ex C.B.Clarke = ***Carex anomoea*** Hand.-Mazz.
Carex inaequalis Raeusch. = ***Carex cespitosa*** L. var. ***cespitosa***
Carex inaequalis J.F.Gmel. = ?
Carex inandrensis Kihlm. ex Hjelt = ?
Carex incerta Steud. = ***Carex saxatilis*** L.
Carex incisodentata Lechl. ex Boott = ***Carex macloviana*** d'Urv. subsp. ***macloviana***
Carex incisodentata Steud. = ***Carex macloviana*** d'Urv. subsp. ***macloviana***
Carex inclusa Turcz. ex Boott = ***Carex alba*** Scop.
Carex incomperta E.P.Bicknell = ***Carex atlantica*** L.H.Bailey subsp. ***atlantica***
Carex incompta Franch. = ***Carex darwinii*** Boott var. ***darwinii***
Carex incondita F.J.Herm. = ***Carex macloviana*** d'Urv. subsp. ***macloviana***
Carex inconspicua Colenso = ***Carex acicularis*** Boott
Carex inconspicua Steud. = ***Carex fuscula*** var. ***distenta*** (Kunze ex Kunth) Kük.
Carex inconspicua var. *fuscula* (d'Urv.) Kük. = ***Carex fuscula*** d'Urv.
Carex inconspicua var. *hieronymi* (Boeck.) Kük. = ***Carex hieronymi*** Boeck.
Carex inconspicua f. *indecora* (Kunth) Kük. = ***Carex fuscula*** d'Urv. var. ***fuscula***
Carex inconspicua var. *pseudoextensa* Kük. = ***Carex fuscula*** d'Urv. var. ***fuscula***
Carex incrassata Schltdl. = ***Carex riparia*** Curtis subsp. ***riparia***
Carex incrassata Sol. ex Boott = ***Carex trifida*** Cav.
Carex incurva Sm. = ***Carex divisa*** Huds.
Carex incurva Lightf. = ***Carex maritima*** Gunnerus
Carex incurva var. *arctica* (Deinboll ex Fr.) Hartm. = ***Carex stenophylla*** Wahlenb. subsp. ***stenophylla***
Carex incurva var. *brevirostris* (Cederstr.) Nyman =
Carex incurva var. *chartacea* Kük. = ***Carex maritima*** Gunnerus
Carex incurva subsp. *deinbolliana* (J.Gay) Nyman = ***Carex stenophylla*** Wahlenb. subsp. ***stenophylla***
Carex incurva f. *erecta* O.Lang = ***Carex maritima*** Gunnerus
Carex incurva var. *humilis* Macloskie = ***Carex fuscula*** d'Urv. var. ***fuscula***
Carex incurva var. *inflata* Simmons = ***Carex maritima*** Gunnerus
Carex incurva var. *melanocystis* (Desv.) Kük. = ***Carex maritima*** Gunnerus
Carex incurva var. *misera* (Phil.) Kük. = ***Carex maritima*** Gunnerus
Carex incurva f. *pallens* Kük. = ***Carex maritima*** Gunnerus
Carex incurva var. *pratensis* Hartm. = ***Carex maritima*** Gunnerus
Carex incurva var. *setina* H.Christ = ***Carex maritima*** Gunnerus
Carex incurviformis var. *danaensis* (Stacey) F.J.Herm. = ***Carex incurviformis*** Mack.
Carex × *indagata* Torges & Hausskn. = ?
Carex indecora Kunth = ***Carex fuscula*** d'Urv. var. ***fuscula***
Carex indecora var. *humilis* Boott = ***Carex fuscula*** d'Urv. var. ***fuscula***
Carex indica F.Muell. ex Benth. = ***Carex indica*** L. var. ***indica***
Carex indica Wahlenb. = ***Carex wahlenbergiana*** Boott
Carex indica Schkuhr = ***Schoenoxiphium sparteum*** (Wahlenb.) C.B.Clarke
Carex indica J.König ex Willd. = ***Scleria tessellata*** Willd.
Carex indica var. *fissilis* (Boott) Kük. = ***Carex indica*** L. var. ***indica***
Carex indica var. *laetebrunnea* C.B.Clarke = ***Carex indica*** L. var. ***indica***
Carex indica subsp. *laetebrunnea* (C.B.Clarke) T.Koyama = ***Carex indica*** L. var. ***indica***
Carex indica var. *milnei* C.B.Clarke = ***Carex indica*** L. var. ***indica***
Carex infirma C.B.Clarke = ***Carex augustinowiczii*** Meinsh. var. ***augustinowiczii***
Carex inflata Huds. = ***Carex rostrata*** Stokes
Carex inflata Suter = ***Carex rostrata*** Stokes var. ***rostrata***
Carex inflata var. *ambigens* Fernald = ***Carex rostrata*** Stokes var. ***rostrata***
Carex inflata var. *anticostensis* Fernald = ***Carex* × *anticostensis*** (Fernald) Lepage
Carex inflata var. *utriculata* (Boott) Druce = ***Carex utriculata*** Boott
Carex infuscata var. *gracilenta* (Boott ex Boeck.) P.C.Li = ***Carex gracilenta*** Boott ex Boeck.

Carex infuscata var. *microcarpa* Nees = ***Carex lehmanii*** Drejer
Carex × *innocua* Wallr. = ?
Carex inops Kunze = ***Carex disperma*** Dewey
Carex × *insolita* T.Koyama = ?
Carex insulana (H.Christ) Rouy = ***Carex caryophyllea*** Latourr. var. ***caryophyllea***
Carex intacta Samp. = ?
Carex interimus Maguire = ***Carex aquatilis*** Wahlenb. var. ***aquatilis***
Carex interior var. *capillacea* L.H.Bailey = ***Carex atlantica*** subsp. ***capillacea*** (L.H.Bailey) Reznicek
Carex interior subsp. *charlestonensis* Clokey = ***Carex interior*** L.H.Bailey
Carex interior var. *josselynii* Fernald = ***Carex echinata*** Murray subsp. ***echinata***
Carex interior var. *keweenawensis* F.J.Herm. = ***Carex interior*** L.H.Bailey
Carex interior f. *keweenawensis* (F.J.Herm.) Fernald = ***Carex interior*** L.H.Bailey
Carex intermedia Suter = ***Carex acutiformis*** Ehrh.
Carex intermedia d'Urv. = ***Carex arenaria*** L.
Carex intermedia Gooden. = ***Carex disticha*** Huds.
Carex intermedia Retz. = ***Carex muricata*** L.
Carex intermedia Miégev. = ***Carex nigra*** (L.) Reichard subsp. ***nigra***
Carex intermedia f. *abbreviata* Klett & Richt. = ***Carex disticha*** Huds.
Carex intermedia f. *distachya* (O.Lang) Kük. = ***Carex disticha*** Huds.
Carex intermedia f. *elatior* (Boeck.) Kük. = ***Carex disticha*** Huds.
Carex intermedia f. *floribunda* (Peterm.) Kük. = ***Carex disticha*** Huds.
Carex intermedia f. *longibracteata* (Schleich.) Kük. = ***Carex disticha*** Huds.
Carex intermedia f. *minor* (Peterm.) Kük. = ***Carex disticha*** Huds.
Carex intermedia f. *pangyna* (Beck) Kük. = ***Carex disticha*** Huds.
Carex intermedia f. *permixta* (Beck) Kük. = ***Carex disticha*** Huds.
Carex intermedia var. *repens* Rchb. = ***Carex disticha*** Huds.
Carex interrupta var. *distenta* Kük. = ***Carex interrupta*** Boeck.
Carex interrupta var. *impressa* L.H.Bailey = ***Carex lenticularis*** var. ***impressa*** (L.H.Bailey) L.A.Standl.
Carex intricata Tineo = ***Carex nigra*** subsp. ***intricata*** (Tineo) Maire & Wieller
Carex intumescens var. *fernaldii* L.H.Bailey = ***Carex intumescens*** Rudge
Carex intumescens var. *globularis* A.Gray = ***Carex grayi*** Carey
Carex intumescens f. *ventriosa* Fernald = ***Carex intumescens*** Rudge
Carex inumbrata V.I.Krecz. = ***Carex cespitosa*** L. var. ***cespitosa***
Carex inundata Willd. ex Kunth = ***Carex glaucescens*** Elliott
Carex inversa var. *costata* Kük. = ***Carex inversa*** R.Br.
Carex inversa var. *leichardtii* Boeck. = ***Carex inversa*** R.Br.
Carex inversa var. *major* Boott = ***Carex inversa*** R.Br.
Carex inversa f. *major* (Boott) Kük. = ***Carex inversa*** R.Br.
Carex inversa var. *minor* Boott = ***Carex inversa*** R.Br.
Carex inversa f. *parvula* Kük. = ***Carex inversa*** R.Br.
Carex inversa var. *perigynium* Boott = ***Carex inversa*** R.Br.
Carex inversa var. *radicata* Cheeseman = ***Carex resectans*** Cheeseman
Carex invisa L.H.Bailey = ***Carex spectabilis*** Dewey
Carex involucrata Boeck. = ***Carex aureolensis*** Steud.
Carex involucrata Boott = ***Carex fossa*** G.A.Wheeler
Carex involucrata var. *angustata* Kük. = ***Carex uruguensis*** var. ***angustata*** (Kük.) Kük.
Carex involucrata var. *pseudobracteosa* Kük. = ***Carex sororia*** var. ***pseudobracteosa*** (Kük.) Kük.
Carex involucrata var. *subdivulsa* Kük. = ***Carex subdivulsa*** (Kük.) G.A.Wheeler
Carex involucrata var. *submuricata* C.B.Clarke ex Kük. = ***Carex sororia*** Kunth var. ***sororia***
Carex involucratella Mack. = ***Carex muehlenbergii*** var. ***xalapensis*** (Kunth) Britton
Carex iridifolia Kunth = ***Carex aethiopica*** Schkuhr
Carex irrasa L.H.Bailey = ***Carex douglasii*** Boott
Carex irregularis Schwein. = ***Carex viridula*** Michx. subsp. ***viridula***
Carex irrigua (Wahlenb.) Sm. ex Hoppe = ***Carex magellanica*** subsp. ***irrigua*** (Wahlenb.) Hultén
Carex ischnantha Steud. = ***Carex brunnea*** Thunb. subsp. ***brunnea***
Carex ischne C.B.Clarke = ***Carex sachalinensis*** var. ***fernaldiana*** (H.Lév. & Vaniot) T.Koyama
Carex ischnostachya var. *subtumida* Kük. = ***Carex ischnostachya*** Steud.
Carex iseana Akiyama = ***Carex pudica*** Honda
Carex isomadensis Simonk. ex K.Richt. = ?
Carex ivanoviae f. *subglacialis* Yu.E.Alexeev = ***Carex ivanoviae*** T.V.Egorova
Carex jacca Carey ex Boott = ***Carex barrattii*** Torr. ex Schwein.
Carex jacintoensis Parish = ***Carex jamesii*** Schwein.
Carex jackiana var. *breviculmis* Thwaites & Hook.f. = ***Carex breviscapa*** C.B.Clarke
Carex jackiana f. *minor* C.B.Clarke = ***Carex jackiana*** Boott subsp. ***jackiana***
Carex jackiana f. *ochrolepis* (Franch.) Kük. = ***Carex jackiana*** Boott subsp. ***jackiana***
Carex jackiana f. *oxyphylla* (Franch.) Kük. = ***Carex jackiana*** Boott subsp. ***jackiana***
Carex jacobi-peteri Hultén = ***Carex micropoda*** C.A.Mey.
Carex jamaicensis Poir. = ***Uncinia hamata*** (Sw.) Urb.
Carex jamesii Torr. = ***Carex nebraskensis*** Dewey
Carex jamesii var. *austromontana* (Parish) Kük. = ***Carex jamesii*** Schwein.
Carex jamesii var. *nebraskensis* (Dewey) L.H.Bailey = ***Carex nebraskensis*** Dewey
Carex jamesii var. *ultriformis* (L.H.Bailey) Kük. = ***Carex nebraskensis*** Dewey
Carex jamesonii var. *brevifructus* Kük. = ***Carex jamesonii*** Boott var. ***jamesonii***
Carex jamesonii var. *chordalis* (Liebm.) Kük. = ***Carex jamesonii*** var. ***gracilis*** L.H.Bailey
Carex jamesonii var. *minor* Boeck. = ***Carex jamesonii*** Boott var. ***jamesonii***
Carex jamesonii var. *subfulva* Kük. = ***Carex jamesonii*** Boott
Carex jankae Gand. = ?
Carex japonica var. *alopecuroides* (D.Don ex Tilloch & Taylor) C.B.Clarke = ***Carex alopecuroides*** D.Don ex Tilloch & Taylor
Carex japonica var. *aphanolepis* (Franch. & Sav.) Kük. = ***Carex aphanolepis*** Franch. & Sav.
Carex japonica var. *chlorostachys* (C.B.Clarke) Kük. = ***Carex alopecuroides*** D.Don ex Tilloch & Taylor
Carex japonica subsp. *chlorostachys* (C.B.Clarke) T.Koyama = ***Carex alopecuroides*** D.Don ex Tilloch & Taylor
Carex japonica var. *gracilis* Franch. = ***Carex japonica*** Thunb.

Carex japonica var. *humilis* Franch. = ***Carex aphanolepis*** Franch. & Sav.
Carex japonica var. *minor* Boott = ***Carex japonica*** Thunb.
Carex japonica var. *phaenocarpa* (Franch.) Kük. = ***Carex alopecuroides*** D.Don ex Tilloch & Taylor
Carex japoniciformis Nakai = ***Carex japonica*** Thunb.
Carex japonoalpina (T.Koyama) T.Koyama = ***Carex atrata*** L. subsp. ***atrata***
Carex javanica Boeck. = ***Carex baccans*** Nees
Carex jemtlandica (Palmgr.) Palmgr. = ***Carex viridula*** var. ***jemtlandica*** (Palmgr.) Blackst. & P.A.Ashton
Carex jepsonii J.T.Howell = ***Carex whitneyi*** Olney
Carex jimcalderi B.Boivin = ***Carex leptalea*** subsp. ***pacifica*** Calder & R.L.Taylor
Carex jiuhuaensis S.W.Su = ***Carex manca*** subsp. ***jiuhuaensis*** (S.W.Su) S.Y.Liang
Carex jizogatakensis H.Lév. & Vaniot = ***Carex kiotensis*** Franch. & Sav.
Carex josselynii (Fernald) Mack. ex Pease = ***Carex echinata*** Murray subsp. ***echinata***
Carex jovis C.B.Clarke = ***Carex donnell-smithii*** L.H. Bailey
Carex jucunda V.I.Krecz. = ***Carex maritima*** Gunnerus
Carex juncea Willd. = ***Carex misera*** Buckley
Carex ×*juncea* Scop. = ***Carex ×xanthocarpa*** Degl. nothovar. ***xanthocarpa***
Carex juncella (Fr.) Th.Fr. = ***Carex nigra*** subsp. ***juncella*** (Fr.) Lemke
Carex juncella subsp. *wiluica* (Meinsh. ex Maack) T.V. Egorova = ***Carex nigra*** subsp. ***juncella*** (Fr.) Lemke
Carex juncifolia All. = ***Carex maritima*** Gunnerus
Carex juncifolia J.F.Gmel. = ***Carex mucronata*** All.
Carex juncifolia Host = ***Carex stenophylla*** Wahlenb. subsp. ***stenophylla***
Carex juncoides J.Presl & C.Presl = ***Carex melanostachya*** M.Bieb. ex Willd.
Carex jungenda Steud. = ***Carex saxatilis*** L.
Carex kamagariensis K.Okamoto = [38 JAP]
Carex kamikawensis Franch. = ***Carex umbrosa*** subsp. ***sabynensis*** (Less. ex Kunth) Kük.
Carex kamikochiana Nakai = ***Carex jackiana*** subsp. ***parciflora*** (Boott) Kük.
Carex kamtschatica (Gorodkov) V.I.Krecz. = ***Carex soczavaeana*** Gorodkov
Carex kanehirae Ohwi = ***Carex meyenii*** Nees
Carex kanitzii Porcius = ***Carex canescens*** L. subsp. ***canescens***
Carex karacolica Polozhij = ***Carex atrata*** subsp. ***caucasica*** (Steven) Kük.
Carex karafutoana Ohwi = ***Carex lanceolata*** var. ***laxa*** Ohwi
Carex karashidaniensis Akiyama = ?
Carex karelinii Meinsh. = ***Carex diluta*** M.Bieb. subsp. ***diluta***
Carex karoi subsp. *selengensis* (K.V.Ivanova) T.V. Egorova = ***Carex karoi*** (Freyn) Freyn
Carex kasugayamensis Akiyama = ***Carex sacrosancta*** Honda var. ***sacrosancta***
Carex katahdinensis Fernald = ***Carex conoidea*** Willd.
Carex kattegatensis (Almq.) Fr. ex Krechet. = ***Carex vacillans*** Drejer
Carex kawakamii Hayata = ***Carex alopecuroides*** D.Don ex Tilloch & Taylor
Carex keiskei Miq. = ***Carex hebecarpa*** subsp. ***ligulata*** (Nees) T.Koyama
Carex kelloggii W.Boott = ***Carex lenticularis*** var. ***lipocarpa*** (T.Holm) L.A.Standl.
Carex kelungensis Hayata = ***Carex ligata*** subsp. ***formosensis*** (H.Lév. & Vaniot) T.Koyama
Carex kelvingtoniana Steud. = ***Carex stricta*** Lam.
Carex kemiriensis Nelmes = ***Carex teres*** Boott
Carex kengiana Z.P.Wang = ***Carex davidii*** Franch.
Carex kengii Kük. = ***Carex paxii*** Kük.
Carex kenkolensis Litv. = ***Carex haematostoma*** Nees
Carex kerneri Kohts = ***Carex ferruginea*** Scop. subsp. ***ferruginea***
Carex kerrii Nelmes = ***Carex teres*** Boott
Carex kinabaluensis Stapf = ***Carex perakensis*** var. ***borneensis*** (C.B.Clarke) Noot.
Carex kinashii H.Lév. & Vaniot = ***Carex prescottiana*** Boott
Carex kingiana H.Lév. & Vaniot = ***Carex mitrata*** Franch. subsp. ***mitrata***
Carex kingiana C.B.Clarke = ***Carex sahnii*** Ghildyal & U.C.Bhattach.
Carex kinpokusanensis Akiyama = ***Carex morrowii*** var. ***temnolepis*** (Franch.) Ohwi
Carex kinugawaensis Sutô & Suzuki = ***Carex miyabei*** Franch. var. ***miyabei***
Carex kirilovii Turcz. = ***Carex pediformis*** C.A.Mey. subsp. ***pediformis***
Carex kirkii var. *membranacea* Kük. = ***Carex kirkii*** Petrie var. ***kirkii***
Carex kitaibeliana f. *balcana* (Degen & Nejceff) Stoeva, E.D.Popova & Uzunova = ***Carex kitaibeliana*** Degen ex Bech. var. ***kitaibeliana***
Carex kitaibeliana subsp. *capillata* (Acht.) Hartvig = ***Carex kitaibeliana*** var. ***capillata*** (Acht.) Stoeva, E.D. Popova & Uzunova
Carex kiusiuana Ohwi = ***Carex chrysolepis*** Franch. & Sav. var. ***chrysolepis***
Carex kiyozumiensis Akiyama = ***Carex insaniae*** var. ***subdita*** (Ohwi) Ohwi
Carex ×*kneuckeriana* Zahn = ***Carex ×pseudoaxillaris*** K.Richt.
Carex kobresia Missbach & E.H.L.Krause = ***Kobresia simpliciuscula*** (Wahlenb.) Mack. subsp. ***simpliciuscula***
Carex kochiana DC. = ***Carex acutiformis*** Ehrh.
Carex ×*kochiana* Schübl. = ***Carex ×evoluta*** Hartm.
Carex koestlinii Hochst. ex Steud. = ***Carex conferta*** var. ***leptosaccus*** (C.B.Clarke) Kük.
Carex ×*kohtsii* K.Richt. = ?
Carex koidzumiana Ohwi = ***Carex thunbergii*** Steud.
Carex koidzumii Honda = ***Carex lasiocarpa*** var. ***occultans*** (Franch.) Kük.
Carex kokrinensis A.E.Porsild = ***Carex eleusinoides*** Turcz. ex Kunth
Carex komaroviana A.I.Baranov & Skvortsov = ***Carex media*** R.Br.
Carex komarovii Koidz. = ***Carex tenuiformis*** H.Lév. & Vaniot
Carex komarovii V.I.Krecz. = ?
Carex koraginensis Meinsh. = ***Carex scita*** var. ***riishirensis*** (Franch.) Kük.
Carex koreana Kom. = ***Carex tenuiformis*** H.Lév. & Vaniot
Carex korsakoviensis H.Lév. = ***Carex sachalinensis*** F.Schmidt var. ***sachalinensis***
Carex kotilainii Palmgr. = ***Carex viridula*** Michx. subsp. ***viridula***
Carex kotoensis Hayata ex Ohwi = ***Carex bilateralis*** Hayata
Carex kotschyana Boiss. & Hohen. = ***Carex orbicularis*** subsp. ***kotschyana*** (Boiss. & Hohen.) Kukkonen
Carex ×*krajinae* Domin = ?
Carex krameri Franch. & Sav. = ***Carex hakonensis*** Franch. & Sav.
Carex ×*krausei* P.Fourn. = ?
Carex krebsiana Boeck. = ***Carex hakkodensis*** Franch.

Carex krullii Boeck. = ***Carex pumila*** Thunb.
Carex kryloviana Schischk. & Serg. = ***Carex pamirica*** subsp. ***dichroa*** (Malyschev) T.V.Egorova
Carex ktausipalii Meinsh. = ***Carex stenantha*** var. ***taisetsuensis*** Akiyama
Carex kuehleweinii Gand. = ***Carex macrochaeta*** C.A.Mey.
Carex ×*kuekenthaliana* Appel & Brückn. = ?
Carex kuekenthalii K.Schum. ex C.B.Clarke = ?
Carex ×*kukenthalii* Dörfl. = ?
Carex kunioi T.Koyama = ***Carex dissitispicula*** Ohwi
Carex kuntzeana Boeck. = ***Carex myosurus*** Nees
Carex kunzei Olney ex A.Gray = ***Carex arcta*** Boott
Carex kwangtungensis F.T.Wang & Tang = ***Carex adrienii*** E.G.Camus
Carex ×*kyyhkynenii* Hiitonen = *C. canescens* × *C. tenuiflora*
Carex lacerans Kük. = ***Carex gaudichaudiana*** Kunth
Carex lachenalii var. *glareosa* (Schkuhr ex Wahlenb.) Gelting = ***Carex glareosa*** Schkuhr ex Wahlenb.
Carex lachenalii subsp. *parkeri* (Petrie) Toivonen = ***Carex lachenalii*** Schkuhr subsp. ***lachenalii***
Carex lachnosperma Wall. ex Kunth = ***Carex hebecarpa*** C.A.Mey. subsp. ***hebecarpa***
Carex laciniata Boott = ***Carex barbarae*** Dewey
Carex ×*lackowitziana* Paul = ?
Carex lacunarum T.Holm = ***Carex barbarae*** Dewey
Carex lacustris var. *gigantea* (Rudge) Pursh = ***Carex gigantea*** Rudge
Carex lacustris var. *laxiflora* Dewey = ***Carex hyalinolepis*** Steud.
Carex ×*laestadii* E.Holmb. = ?
Carex laevicaulis Hochst. ex Seub. = ***Carex punctata*** var. ***laevicaulis*** (Hochst. ex Seub.) Boott
Carex laevigata Wahlenb. = ***Carex aethiopica*** Schkuhr
Carex laevigata Boiss. = ***Carex camposii*** Boiss. & Reut.
Carex laevigata var. *biligularis* (DC.) Nyman = ***Carex laevigata*** Sm.
Carex laevigata f. *biligularis* (DC.) Asch. & Graebn. = ***Carex laevigata*** Sm.
Carex laevigata var. *welwitschii* (Boiss. ex Steud.) Boott = ***Carex laevigata*** Sm.
Carex laevirostris (Blytt ex Fr.) Andersson = ***Carex utriculata*** Boott
Carex laevirostris var. *friesii* (Blytt) Kük. = ***Carex*** ***×friesii*** Blytt
Carex laevirostris f. *gracilior* Kük. = ***Carex utriculata*** Boott
Carex laevis Hoppe = ***Carex dioica*** L.
Carex laevis Kit. ex Willd. = ***Carex kitaibeliana*** Degen ex Bech. var. ***kitaibeliana***
Carex laevis (Retz.) J.F.Gmel. = ***Scleria levis*** Retz.
Carex laevis f. *balcana* Degen & Nejceff = ***Carex kitaibeliana*** Degen ex Bech. var. ***kitaibeliana***
Carex laevis f. *capillata* Acht. = ***Carex kitaibeliana*** var. ***capillata*** (Acht.) Stoeva, E.D.Popova & Uzunova
Carex laggeri Wimm. = ***Carex repens*** Bellardi
Carex lagopina Wahlenb. = ***Carex lachenalii*** Schkuhr subsp. ***lachenalii***
Carex lagopina f. *angustifolia* Norman = ***Carex lachenalii*** Schkuhr subsp. ***lachenalii***
Carex lagopina var. *baetica* Gay = ***Carex lachenalii*** subsp. ***baetica*** (Gay) Luceño & Muñoz Garm.
Carex lagopina var. *debilis* Lange = ***Carex lachenalii*** Schkuhr subsp. ***lachenalii***
Carex lagopina var. *furva* (Webb) Nyman = ***Carex lachenalii*** subsp. ***baetica*** (Gay) Luceño & Muñoz Garm.
Carex lagopina var. *gracilescens* Th.Fr. = ***Carex lachenalii*** Schkuhr subsp. ***lachenalii***
Carex lagopina var. *laxior* Norman = ***Carex lachenalii*** Schkuhr subsp. ***lachenalii***
Carex lagopina var. *longisquama* Kük. ex J.M.Macoun = ***Carex lachenalii*** Schkuhr subsp. ***lachenalii***
Carex lagopina var. *major* Lange = ***Carex lachenalii*** Schkuhr subsp. ***lachenalii***
Carex lagopina f. *pauciflora* Norman = ***Carex lachenalii*** Schkuhr subsp. ***lachenalii***
Carex lagopina var. *pleiostachya* Drejer = ***Carex lachenalii*** Schkuhr subsp. ***lachenalii***
Carex lagopina var. *pribylovensis* (Macoun) Kük. = ***Carex glareosa*** subsp. ***pribylovensis*** (Macoun) G.Halliday & Chater
Carex lagopina var. *subtenuiflora* Almq. = ***Carex lachenalii*** Schkuhr
Carex lagopodioides Willd. = ***Carex tribuloides*** Wahlenb. var. ***tribuloides***
Carex lagopodioides var. *cristata* (Tuck.) Carey = ***Carex tribuloides*** Wahlenb. var. ***tribuloides***
Carex lagopodioides var. *moniliformis* Olney ex L.H. Bailey = ***Carex tribuloides*** Wahlenb. var. ***tribuloides***
Carex lagopodioides var. *scoparia* Boeck. = ***Carex scoparia*** Willd. var. ***scoparia***
Carex lagopus Muhl. = ***Cymophyllus fraserianus*** (Ker Gawl.) Kartesz & Gandhi
Carex lainzii Luceño, E.Rico & T.Romero = ***Carex punctata*** Gaudin var. ***punctata***
Carex lamarckii Wood = ***Carex liparocarpos*** Gaudin subsp. ***liparocarpos***
Carex ×*lambertiana* H.Lév. = ?
Carex lamprocarpa var. *rotundata* Kük. = ***Carex poeppigii*** C.B.Clarke ex G.A.Wheeler
Carex lamprophysa Sam. ex Nordh. = ***Carex cuprina*** (I.Sándor ex Heuff.) Nendtv. ex A.Kern.
Carex lamprosandra Franch. = ***Carex tapintzensis*** Franch.
Carex lancea (Thunb.) Baill. = ***Schoenoxiphium lanceum*** (Thunb.) Kük.
Carex lanceata Dewey = ***Carex salina*** Wahlenb. var. ***salina***
Carex lanceisquama (Hand.-Mazz.) V.I.Krecz. = ***Carex atherodes*** Spreng.
Carex lanceolata var. *alashanica* T.V.Egorova = ***Carex lanceolata*** Boott var. ***lanceolata***
Carex lanceolata var. *albomediana* Makino = ***Carex lanceolata*** Boott var. ***lanceolata***
Carex lanceolata var. *macrosandra* Franch. = ***Carex macrosandra*** (Franch.) V.I.Krecz.
Carex lanceolata var. *nana* H.Lév. & Vaniot = ***Carex humilis*** var. ***nana*** (H.Lév. & Vaniot) Ohwi
Carex lanceolata var. *pseudolanceolata* (V.I.Krecz.) Vorosch. = ***Carex lanceolata*** var. ***subpediformis*** Kük.
Carex lancifructus Mack. = ***Carex specifica*** L.H.Bailey
Carex lanescens J.Fellm. = ?
Carex ×*langii* Steud. = ?
Carex langsdorffii Boott = ***Carex breviculmis*** R.Br. var. ***breviculmis***
Carex lanosa Gueldenst. ex Ledeb. = ?
Carex lanuginosa Michx. = ***Carex lasiocarpa*** var. ***americana*** Fernald
Carex lanuginosa var. *americana* (Fernald) Boivin = ***Carex lasiocarpa*** var. ***americana*** Fernald
Carex lanuginosa var. *kansana* Britton = ***Carex lasiocarpa*** var. ***americana*** Fernald
Carex lanuginosa var. *oriens* Raymond = ***Carex lasiocarpa*** var. ***americana*** Fernald
Carex lapponica f. *major* Holmboe = ***Carex lapponica*** Láng
Carex laricina Mack. ex Bright = ***Carex echinata*** Murray subsp. ***echinata***

Carex × *larminatii* H.Lév. = ?
Carex lasiocarpa subsp. *americana* (Fernald) Á.Löve & J.-P.Bernard = ***Carex lasiocarpa*** var. ***americana*** Fernald
Carex lasiocarpa var. *fuscata* Ohwi = ***Carex lasiocarpa*** Ehrh. var. ***lasiocarpa***
Carex lasiocarpa f. *kansana* (Britton) Kük. = ***Carex lasiocarpa*** var. ***americana*** Fernald
Carex lasiocarpa var. *lanuginosa* (Michx.) Kük. = ***Carex pellita*** Muhl. ex Willd.
Carex lasiocarpa subsp. *lanuginosa* (Michx.) R.T.Clausen & Wahl = ***Carex pellita*** Muhl. ex Willd.
Carex lasiocarpa var. *latifolia* (Boott) Gilly = ***Carex albursina*** E.Sheld.
Carex lasiocarpa f. *meduaensis* H.Lév. & Vaniot = ***Carex lasiocarpa*** Ehrh. var. ***lasiocarpa***
Carex lasiocarpa f. *robusta* Junge = ***Carex lasiocarpa*** Ehrh. var. ***lasiocarpa***
Carex lasiocarpa f. *stricta* Junge = ***Carex lasiocarpa*** Ehrh. var. ***lasiocarpa***
Carex lasiochlaena Kunth = ***Carex hispida*** Willd. ex Schkuhr
Carex latebrosa Mack. = ***Carex praegracilis*** W.Boott
Carex latibracteolata Kük. = ***Carex catamarcensis*** C.B.Clarke ex Kük.
Carex laticuspis Franch. = ***Carex gmelinii*** Hook. & Arn.
Carex latifolia Moench = ***Carex plantaginea*** Lam.
Carex latifolia Sol. ex Boott = ***Carex pseudocyperus*** L. var. ***pseudocyperus***
Carex latifolia Boiss. & Balansa = ***Carex sylvatica*** subsp. ***latifrons*** (V.I.Krecz.) Ö.Nilsson
Carex latifolia (Sw.) J.F.Gmel. = ***Scleria latifolia*** Sw.
Carex latifrons V.I.Krecz. = ***Carex sylvatica*** subsp. ***latifrons*** (V.I.Krecz.) Ö.Nilsson
Carex latinervia H.Lév. & Vaniot = ***Carex heterolepis*** Bunge
Carex latisquama (Cuatrec.) Pau ex Cuatrec. = ***Carex distachya*** Desf. var. ***distachya***
Carex latitans H.Lév. & Vaniot = ***Carex maculata*** Boott var. ***maculata***
Carex × *lausii* Podp. = ?
Carex laxa Dewey = ***Carex limosa*** L.
Carex laxa Sw. ex Kunth = ***Carex scabrella*** Wahlenb.
Carex laxiculmis var. *floridana* L.H.Bailey = ***Carex digitalis*** var. ***floridana*** (L.H.Bailey) Naczi & Bryson
Carex laxiculmis f. *serotina* Howe = ***Carex laxiculmis*** Schwein. var. ***laxiculmis***
Carex laxiflora Elliott = ***Carex amphibola*** Steud.
Carex laxiflora Merino = ***Carex depressa*** Link subsp. ***depressa***
Carex laxiflora var. *angustifolia* Dewey = ***Carex striatula*** Michx.
Carex laxiflora var. *blanda* (Dewey) Boott = ***Carex blanda*** Dewey
Carex laxiflora var. *divaricata* L.H.Bailey = ***Carex laxiflora*** Lam. var. ***laxiflora***
Carex laxiflora var. *heterosperma* Britton, Sterns & Poggenb. = ***Carex laxiflora*** Lam. var. ***laxiflora***
Carex laxiflora var. *intermedia* Boott = ***Carex laxiflora*** Lam. var. ***laxiflora***
Carex laxiflora var. *latifolia* Boott = ***Carex albursina*** E.Sheld.
Carex laxiflora var. *leptonervia* Fernald = ***Carex leptonervia*** (Fernald) Fernald
Carex laxiflora var. *major* Boott = ***Carex laxiflora*** Lam. var. ***laxiflora***
Carex laxiflora var. *michauxii* L.H.Bailey = ***Carex striatula*** Michx.
Carex laxiflora var. *minor* Boott = ***Carex laxiflora*** Lam. var. ***laxiflora***
Carex laxiflora var. *mutica* Torr. = ***Carex flaccosperma*** Dewey
Carex laxiflora var. *ormostachya* (Wiegand) Gleason = ***Carex ormostachya*** Wiegand
Carex laxiflora var. *patulifolia* (Dewey ex Alph.Wood) Kük. = ***Carex laxiflora*** Lam. var. ***laxiflora***
Carex laxiflora var. *plantaginea* Olney = ***Carex hendersonii*** L.H.Bailey
Carex laxiflora var. *plantaginea* Boott = ***Carex laxiflora*** Lam. var. ***laxiflora***
Carex laxiflora var. *purpurifera* (Mack.) Gleason = ***Carex purpurifera*** Mack.
Carex laxiflora var. *serrulata* F.J.Herm. = ***Carex laxiflora*** Lam. var. ***laxiflora***
Carex laxiflora var. *striatula* (Michx.) Chapm. = ***Carex striatula*** Michx.
Carex laxiflora f. *striatula* (Michx.) Kük. = ***Carex striatula*** Michx.
Carex laxiflora var. *styloflexa* (Buckley) Boott = ***Carex styloflexa*** Buckley
Carex laxiflora var. *varians* L.H.Bailey = ***Carex leptonervia*** (Fernald) Fernald
Carex laxiflora f. *varians* (L.H.Bailey) Kük. = ***Carex leptonervia*** (Fernald) Fernald
Carex laxior (Kük.) Mack. = ***Carex flava*** L. subsp. ***flava***
Carex laxisquamata Tang & F.T.Wang = ***Carex breviscapa*** C.B.Clarke
Carex laxula Tineo ex Boott = ***Carex sylvatica*** Huds. subsp. ***sylvatica***
Carex lazica (Boiss. & Balansa) Boiss. & Balansa ex B.D.Jacks. = ***Carex pontica*** Albov
Carex lazica (Boiss. & Balansa) Kük. = ***Carex pontica*** Albov
Carex lazica var. *clavata* Kük. = ***Carex pontica*** Albov
Carex lebrunei H.Lév. = ***Carex renauldii*** H.Lév.
Carex lechleri Phil. = ***Carex atropicta*** Steud.
Carex lechleri Steud. = ***Carex excelsa*** Poepp. ex Kunth
Carex ledebouriana subsp. *tenuiformis* (H.Lév. & Vaniot) T.V.Egorova = ***Carex tenuiformis*** H.Lév. & Vaniot
Carex ledebourii Boiss. & Buhse = ***Carex melanostachya*** M.Bieb. ex Willd.
Carex leersiana Rouy = ***Carex divulsa*** Stokes subsp. ***divulsa***
Carex leersiana Rauschert = ***Carex divulsa*** subsp. ***leersii*** (Kneuck.) W.Koch
Carex leersii F.W.Schultz = ***Carex divulsa*** subsp. ***leersii*** (Kneuck.) W.Koch
Carex leersii Willd. = ***Carex echinata*** Murray subsp. ***echinata***
Carex leersii var. *angustata* (Carey) Burnham = ***Carex echinata*** Murray subsp. ***echinata***
Carex leersii var. *angustifolia* Vollm. = ***Carex echinata*** Murray subsp. ***echinata***
Carex leersii var. *cephalantha* (L.H.Bailey) J.K.Henry = ***Carex echinata*** Murray subsp. ***echinata***
Carex leersii subvar. *subramosa* Vollm. = ***Carex echinata*** Murray subsp. ***echinata***
Carex legendrei H.Lév. & Vaniot = ***Carex sociata*** Boott
Carex lehmanniana Boott ex Walp. = ***Carex lemanniana*** Boott
Carex leiocarpa C.A.Mey. = ***Carex anthoxanthea*** C.Presl
Carex leiocarpa J.Gay = ***Carex gayana*** Steud.
Carex leiogona Franch. = ***Carex middendorffii*** var. ***kirigaminensis*** (Ohwi) Ohwi
Carex leiophylla Mack. = ***Carex sabulosa*** Turcz. ex Kunth
Carex lemanniana var. *simplex* Kük. = ***Carex lemanniana*** Boott
Carex lenaensis Kük. = ***Carex ledebouriana*** C.A.Mey. ex Trevir. subsp. ***ledebouriana***

Carex lenticularis Dewey = ***Carex magellanica*** Lam. subsp. ***magellanica***
Carex lenticularis D.Don ex Tilloch & Taylor = ***Carex phacota*** Spreng.
Carex lenticularis var. *albimontana* Dewey = ***Carex lenticularis*** Michx. var. ***lenticularis***
Carex lenticularis var. *blakei* Dewey = ***Carex lenticularis*** Michx. var. ***lenticularis***
Carex lenticularis var. *eucycla* Fernald = ***Carex lenticularis*** Michx. var. ***lenticularis***
Carex lenticularis var. *merens* Howe = ***Carex lenticularis*** Michx. var. ***lenticularis***
Carex lenticularis var. *pallida* (Boott) Dorn = ***Carex lenticularis*** var. ***lipocarpa*** (T.Holm) L.A.Standl.
Carex lenticularis var. *paullifructus* Kük. = ***Carex lenticularis*** var. ***lipocarpa*** (T.Holm) L.A.Standl.
Carex leonura Wahlenb. = ***Carex crinita*** Lam. var. ***crinita***
Carex lepageana Raymond = ***Carex petricosa*** Dewey var. ***petricosa***
Carex lepidocarpa Tausch = ***Carex viridula*** var. ***elatior*** (Schltr.) Crins
Carex lepidocarpa var. *cylindrostachys* Neuman = ***Carex viridula*** var. ***elatior*** (Schltr.) Crins
Carex lepidocarpa f. *intermedia* (Coss. & Germ.) Asch. & Graebn. = ***Carex viridula*** var. ***elatior*** (Schltr.) Crins
Carex lepidocarpa subsp. *jemtlandica* Palmgr. = ***Carex viridula*** var. ***jemtlandica*** (Palmgr.) Blackst. & P.A.Ashton
Carex lepidocarpa f. *laeviculmis* Kneuck. = ***Carex viridula*** var. ***elatior*** (Schltr.) Crins
Carex lepidocarpa f. *laevigata* Peterm. = ***Carex viridula*** var. ***elatior*** (Schltr.) Crins
Carex lepidocarpa var. *laxior* Kük. = ***Carex flava*** L. subsp. ***flava***
Carex lepidocarpa f. *major* Kneuck. = ***Carex viridula*** var. ***elatior*** (Schltr.) Crins
Carex lepidocarpa var. *nelmesiana* Raymond = ***Carex viridula*** var. ***elatior*** (Schltr.) Crins
Carex lepidocarpa var. *nevadensis* (Boiss. & Reut.) Kük. = ***Carex flava*** var. ***nevadensis*** (Boiss. & Reut.) Briq.
Carex lepidocarpa subsp. *nevadensis* (Boiss. & Reut.) Luceño = ***Carex flava*** var. ***nevadensis*** (Boiss. & Reut.) Briq.
Carex lepidocarpa var. *pseudolepidocarpa* Kneuck. = ***Carex viridula*** var. ***elatior*** (Schltr.) Crins
Carex lepidocarpa f. *rectirostris* Kneuck. = ***Carex viridula*** var. ***elatior*** (Schltr.) Crins
Carex lepidocarpa subsp. *scotica* E.W.Davies = ***Carex flava*** subsp. ***scotica*** (E.W.Davies) P.D.Sell
Carex lepidocarpa var. *sphaerostachys* Neuman = ***Carex viridula*** var. ***elatior*** (Schltr.) Crins
Carex lepidopristis H.Lév. & Vaniot = ***Carex phacota*** Spreng.
Carex leporina Pall. ex Kunth = ***Carex divisa*** Huds.
Carex leporina Schkuhr ex Ledeb. = ***Carex heleonastes*** Ehrh. ex L.f. subsp. ***heleonastes***
Carex leporina Michx. = ***Carex scoparia*** Willd. var. ***scoparia***
Carex leporina var. *alpina* Asch. & Graebn. = ***Carex leporina*** L.
Carex leporina var. *americana* Olney = ***Carex petasata*** Dewey
Carex leporina f. *argyroglochin* (Hornem.) W.D.J.Koch = ***Carex leporina*** L.
Carex leporina var. *argyroglochin* (Hornem.) Rchb. = ***Carex leporina*** L.
Carex leporina var. *argyrolepis* Peterm. = ***Carex leporina*** L.
Carex leporina subsp. *atlasica* H.Lindb. = ***Carex atlasica*** (H.Lindb.) Tattou
Carex leporina var. *atrofusca* H.Christ = ***Carex leporina*** L.
Carex leporina var. *bracteata* Liebm. = ***Carex longii*** Mack. subsp. ***longii***
Carex leporina f. *capitata* Sond. = ***Carex leporina*** L.
Carex leporina var. *elatior* Boeck. = ***Carex macloviana*** var. ***pseudoleporina*** Kük.
Carex leporina f. *longibracteata* Peterm. = ***Carex leporina*** L.
Carex leporina f. *minor* Kuntze = ***Carex leporina*** L.
Carex leporina f. *monostachya* Peterm. = ***Carex leporina*** L.
Carex leporina f. *nana* Asch. & Graebn. = ***Carex leporina*** L.
Carex leporina var. *nemoralis* Trevir. = ***Carex leporina*** L.
Carex leporina var. *ovata* Laest. = ***Carex macloviana*** d'Urv. subsp. ***macloviana***
Carex leporina var. *pallescens* Gren. & Godr. = ***Carex leporina*** L.
Carex leporina var. *radicans* Zabel ex Asch. & Graebn. = ***Carex leporina*** L.
Carex leporina f. *robusta* Fiek = ***Carex leporina*** L.
Carex leporina subsp. *sicula* (Tineo) Nyman = ***Carex leporina*** L.
Carex leporina var. *subfestiva* Lange = ***Carex leporina*** L.
Carex leptalea var. *harperi* (Fernald) W.Stone = ***Carex leptalea*** subsp. ***harperi*** (Fernald) W.Stone
Carex leptalea var. *tayloris* B.Boivin = ***Carex leptalea*** Wahlenb. subsp. ***leptalea***
Carex leptocarpa C.B.Clarke = ***Carex filicina*** Nees
Carex leptophylla Heuff. = ***Carex echinata*** Murray subsp. ***echinata***
Carex leptosaccus C.B.Clarke = ***Carex conferta*** var. ***leptosaccus*** (C.B.Clarke) Kük.
Carex leptostachya Boiss. = ***Carex strigosa*** Huds.
Carex leptostachys Ehrh. ex L.f. = ***Carex strigosa*** Huds.
Carex lepus-aestatis T.Koyama = ***Carex nivalis*** Boott
Carex lerchenfeldiana Schur = ***Carex hordeistichos*** Vill.
Carex leucantha Schur = ***Carex pallescens*** L. var. ***pallescens***
Carex leucocarpa Phil. = ***Carex banksii*** Boott var. ***banksii***
Carex leucocarpa Boeck. = ***Carex gibba*** Wahlenb.
Carex leucochlora Bunge = ***Carex breviculmis*** R.Br. var. ***breviculmis***
Carex leucochlora f. *fibrillosa* (Franch. & Sav.) K.T.Fu = ***Carex breviculmis*** var. ***pluricostata*** Kük.
Carex leucochlora var. *horikawae* (K.Okamoto) Katsuy. = ***Carex breviculmis*** R.Br. var. ***breviculmis***
Carex leucochlora var. *morrisonicola* (Hayata) Katsuy. = ***Carex breviculmis*** R.Br. var. ***breviculmis***
Carex leucochlora var. *petrogena* Kitag. = ***Carex breviculmis*** R.Br. var. ***breviculmis***
Carex leucodonta T.Holm = ***Carex turbinata*** Liebm.
Carex leucoglochin Ehrh. ex Wahlenb. = ***Carex obtusata*** Lilj.
Carex leucoglochin L.f. = ***Carex pauciflora*** Lightf.
Carex leucolepis Turcz. ex Steud. = ***Carex tenuiflora*** Wahlenb.
Carex leucorhizia Dulac = ***Carex davalliana*** Sm.
Carex leucostachys Ridl. = ***Carex perakensis*** C.B.Clarke var. ***perakensis***
Carex ×*leutzii* Kneuck. = ***Carex ×xanthocarpa*** nothovar. ***leutzii*** (Kneuck.) Govaerts
Carex ×*leutzii* var. *pseudofulva* (Fernald) Boivin = ***Carex ×pseudofulva*** Fernald
Carex ×*leutzii* var. *xanthina* (Fernald) Boivin = ***Carex ×xanthocarpa*** nothovar. ***leutzii*** (Kneuck.) Govaerts
Carex leveillei Husn. = [12 FRA]
Carex levicaulis Franch. = ***Carex middendorffii*** F.Schmidt var. ***middendorffii***

Carex ×*levicuspis* Ohwi = ***Carex* × *kattaeana*** Kük.
Carex lhotzkiana (Nees) Steud. = ***Trilepis lhotzkiana*** Nees
Carex licentii Nelmes = ***Carex karoi*** (Freyn) Freyn
Carex liddonii Boott = ***Carex petasata*** Dewey
Carex liddonii Carey = ***Carex siccata*** Dewey
Carex liddonii var. *incerta* L.H.Bailey = ***Carex petasata*** Dewey
Carex ×*lidii* Hadac = ?
Carex ×*lidii* Flatberg = *C. canescens* × *C. chordorrhiza*
Carex liebmannii Walp. = ***Carex anisostachys*** Liebm.
Carex ligata var. *cucullata* Kük. = ***Carex ligata*** Boott subsp. ***ligata***
Carex ligata var. *formosensis* (H.Lév. & Vaniot) Kük. = ***Carex ligata*** subsp. ***formosensis*** (H.Lév. & Vaniot) T.Koyama
Carex ligata var. *indochinensis* R.Gross = ***Carex pleurocaula*** Nelmes
Carex ligata var. *nexa* (Boott) Kük. = ***Carex ligata*** Boott subsp. ***ligata***
Carex ligata var. *strictior* (Kük.) Kük. = ***Carex sociata*** Boott
Carex ligerica J.Gay = ***Carex colchica*** J.Gay
Carex ligerica f. *capitata* Lackow. & Paul = ***Carex colchica*** J.Gay
Carex ligerica var. *umbrosa* Sprib. = ***Carex repens*** Bellardi
Carex ligerina Guépin ex Boreau = ***Carex colchica*** J.Gay
Carex ×*ligniciensis* Figert = ?
Carex ligulata Nees = ***Carex hebecarpa*** subsp. ***ligulata*** (Nees) T.Koyama
Carex ligulata var. *glabriutriculata* Q.S.Wang = ***Carex hebecarpa*** subsp. ***ligulata*** (Nees) T.Koyama
Carex lilibaea Lojac. = ***Carex acuta*** L.
Carex limnaea T.Holm = ***Carex lenticularis*** var. ***impressa*** (L.H.Bailey) L.A.Standl.
Carex limnocharis T.Holm = ***Carex aquatilis*** Wahlenb. var. ***aquatilis***
Carex ×*limnogena* Appel ex Kneuck. = ?
Carex limnophila F.J.Herm. = ***Carex microptera*** Mack.
Carex limosa var. *fuscocuprea* Kük. = ***Carex limosa*** L.
Carex limosa var. *humilior* St.-Lag. = ***Carex magellanica*** Lam. subsp. ***magellanica***
Carex limosa var. *irrigua* Wahlenb. = ***Carex magellanica*** subsp. ***irrigua*** (Wahlenb.) Hultén
Carex limosa f. *laxiformis* Lackow. = ***Carex limosa*** L.
Carex limosa var. *livida* Wahlenb. = ***Carex livida*** (Wahlenb.) Willd.
Carex limosa var. *magellanica* (Lam.) Fiori = ***Carex magellanica*** Lam.
Carex limosa var. *oblonga* Dewey = ***Carex limosa*** L.
Carex limosa var. *painei* Dewey = ***Carex limosa*** L.
Carex limosa f. *pauciflora* Asch. = ***Carex limosa*** L.
Carex limosa var. *pauciflora* Vahl ex Schkuhr = ***Carex rariflora*** (Wahlenb.) Sm. var. ***rariflora***
Carex limosa var. *rariflora* Wahlenb. = ***Carex rariflora*** (Wahlenb.) Sm.
Carex limosa f. *robustior* Lackow. = ***Carex limosa*** L.
Carex limosa f. *stans* Bolle = ***Carex limosa*** L.
Carex limosa var. *stygia* (Fr.) L.H.Bailey = ***Carex* × *stygia*** Fr.
Carex limprichtii Kük. = ***Carex tenebrosa*** Boott
Carex ×*limula* var. *decolorans* (Wimm.) Nyman = ***Carex* × *limula*** Fr.
Carex linearis Clairv. ex Boott = ***Carex brachystachys*** Schrank
Carex linearis Boott = ***Kobresia nepalensis*** (Nees) Kük. var. ***nepalensis***
Carex linearis var. *elachista* C.B.Clarke = ***Kobresia nepalensis*** (Nees) Kük. var. ***nepalensis***
Carex lineariterminalis T.Koyama = ***Carex stenocarpa*** Turcz. ex V.I.Krecz.
Carex lineola C.A.Mey. ex Fr. = [10 SWE]
Carex lineolata Cham. ex Meinsh. = ***Carex schmidtii*** Meinsh.
Carex linkii Willd. = ***Carex distachya*** Desf. var. ***distachya***
Carex linkii subsp. *ambigua* Nyman = ***Carex oedipostyla*** Duval-Jouve
Carex linkii var. *olyssiponensis* (Steud.) Nyman = ***Carex distachya*** Desf. var. ***distachya***
Carex linkii var. *sarda* (Link) Nyman = ***Carex distachya*** Desf. var. ***distachya***
Carex linnaeana Host = ***Carex dioica*** L.
Carex linnaei Degl. = ***Carex dioica*** L.
Carex liparocarpos f. *conglobata* (Nyman) Soó = ***Carex supina*** Willd. ex Wahlenb. var. ***supina***
Carex lipsiensis Peterm. = ***Carex flava*** L. subsp. ***flava***
Carex lithosperma (L.) L. = ***Scleria flagellum-nigrorum*** P.J.Bergius
Carex lithosperma G.Forst. = ***Scleria polycarpa*** Boeck.
Carex lithosperma Walter = ***Scleria reticularis*** Michx.
Carex littoralis Schwein. = ***Carex barrattii*** Torr. ex Schwein.
Carex littoralis Petrie = ***Carex litorosa*** L.H.Bailey
Carex littoralis Krock. ex Steud. = ***Carex stricta*** Lam.
Carex littorea Labill. = ***Carex pumila*** Thunb.
Carex livida var. *grayana* (Dewey) Fernald = ***Carex livida*** (Wahlenb.) Willd.
Carex livida var. *radicalis* Paine = ***Carex livida*** (Wahlenb.) Willd.
Carex livida var. *radicalis* Paine ex Dewey = ***Carex livida*** (Wahlenb.) Willd.
Carex livida var. *rufiniformis* Fernald = ***Carex livida*** (Wahlenb.) Willd.
Carex livida var. *typica* Fernald = ***Carex livida*** (Wahlenb.) Willd.
Carex lividulla Nakai = ***Carex livida*** (Wahlenb.) Willd.
Carex lobata Bellardi = ***Carex curvula*** All. subsp. ***curvula***
Carex lobata Lam. = ***Carex divisa*** Huds.
Carex lobata Vill. = ***Carex foetida*** All.
Carex loddonii var. *incerta* L.H.Bailey = ***Carex specifica*** L.H.Bailey
Carex loncholepis Less. ex Krecz. = ***Carex sylvatica*** Huds. subsp. ***sylvatica***
Carex lonchophora Ohwi = ***Carex breviculmis*** R.Br. var. ***breviculmis***
Carex lonchosacca C.B.Clarke = ?
Carex longa Mack. = ***Carex perlonga*** Fernald
Carex longiacuminata Colenso = ***Carex dissita*** Sol. ex Boott
Carex longiaristata Boott ex Kurz = ***Carex indica*** L. var. ***indica***
Carex longiaristata Biv. = ?
Carex longibracteata Schleich. = ***Carex disticha*** Huds.
Carex longibracteata Steud. = ***Carex myosurus*** Nees
Carex longibracteata Dulac = ***Carex pseudocyperus*** L. var. ***pseudocyperus***
Carex longibracteata St.-Lag. = ***Carex pseudocyperus*** L. var. ***pseudocyperus***
Carex longibracteata Hornem. = ?
Carex longibracteata f. *angustifolia* Kük. = ***Carex myosurus*** Nees
Carex longibracteata f. *distans* Kük. = ***Carex myosurus*** Nees
Carex longicaulis var. *coulteri* (Boott ex Hemsl.) Kük. = ***Carex coulteri*** Boott ex Hemsl.
Carex longicruris var. *henryi* (C.B.Clarke ex Franch.) C.B.Clarke = ***Carex longicruris*** Nees
Carex longifolia Burch. ex Hemsl. = ***Carex dianae*** Steud.

Carex longifolia R.Br. = ***Carex longibrachiata*** Boeck.
Carex longifolia Thuill. = ***Carex rostrata*** Stokes var. ***rostrata***
Carex longifolia Host = ***Carex umbrosa*** Host subsp. ***umbrosa***
Carex longifolia f. *gracilis* Kohts = ***Carex umbrosa*** Host subsp. ***umbrosa***
Carex longifolia var. *minor* Boott = ***Carex longibrachiata*** Boeck.
Carex longii var. *meridionalis* (Kük.) G.A.Wheeler = ***Carex longii*** subsp. ***meridionalis*** (Kük.) Luceño & Alves
Carex longipedicellata Boeck. = ***Carex stenophylla*** subsp. ***stenophylloides*** (V.I.Krecz.) T.V.Egorova
Carex longipedunculata K.Schum. = ***Carex petitiana*** A.Rich.
Carex longipedunculata var. *cuprea* Kük. = ***Carex petitiana*** A.Rich.
Carex longipedunculata var. *ninagongensis* Kük. = ***Carex petitiana*** A.Rich.
Carex longipedunculata var. *preussii* (K.Schum.) Kük. = ***Carex petitiana*** A.Rich.
Carex longipes var. *lissitiflora* C.B.Clarke = ***Carex longipes*** D.Don ex Tilloch & Taylor
Carex longipes var. *nepalensis* (Spreng.) Boott = ***Carex longipes*** D.Don ex Tilloch & Taylor
Carex longipes var. *sessilis* Tang & F.T.Wang ex L.K.Dai = ***Carex longipes*** D.Don ex Tilloch & Taylor
Carex longirostrata var. *pallida* (Kitag.) Ohwi = ***Carex longirostrata*** C.A.Mey. var. ***longirostrata***
Carex longirostrata var. *recurvifolia* Kük. = ***Carex longirostrata*** C.A.Mey. var. ***longirostrata***
Carex longirostrata var. *tsinlingensis* K.T.Fu = ***Carex longirostrata*** C.A.Mey. var. ***longirostrata***
Carex longirostris Krock. = ***Carex laevigata*** Sm.
Carex longirostris Torr. ex Schwein. = ***Carex sprengelii*** Dewey ex Spreng.
Carex longirostris var. *microcystis* Boeck. = ***Carex sprengelii*** Dewey ex Spreng.
Carex longirostris var. *minor* Boott = ***Carex sprengelii*** Dewey ex Spreng.
Carex longiseta Brot. = ***Carex distachya*** Desf. var. ***distachya***
Carex longiseta var. *latisquama* Cuatrec. = ***Carex distachya*** Desf. var. ***distachya***
Carex longispica Boeck. = ***Carex speciosa*** Kunth subsp. ***speciosa***
Carex longispicata Hayata = ***Carex kiotensis*** Franch. & Sav.
Carex longisquamata Meinsh. ex Kom. = ***Carex lanceolata*** Boott var. ***lanceolata***
Carex longistipes Hayata = ***Carex fulvorubescens*** subsp. ***longistipes*** (Hayata) T.Koyama
Carex longistolon C.B.Clarke ex Franch. = ***Carex lenta*** var. ***sendaica*** (Franch.) T.Koyama
Carex longkiensis Franch. = ***Carex chinensis*** Retz.
Carex lorentziana Griseb. = ***Carex boliviensis*** Van Heurck & Müll.Arg. subsp. ***boliviensis***
Carex ×*loretii* Rouy = ?
Carex loscosii Lange = ***Carex mairii*** Coss. & Germ.
Carex lucida Boott = ***Carex flagellifera*** Colenso
Carex lucida Clairv. = ***Carex liparocarpos*** Gaudin subsp. ***liparocarpos***
Carex lucida var. *buchananii* (Berggr.) Kük. = ***Carex buchananii*** Berggr.
Carex lucidula Franch. = ***Carex umbrosa*** var. ***rostrata*** (Maxim.) Govaerts
Carex lucorum Sartwell ex Boott = ***Carex nigromarginata*** Schwein.
Carex lucorum var. *emmonsii* (Dewey ex Torr.) Chapm. = ***Carex albicans*** var. ***emmonsii*** (Dewey ex Torr.) Rettig
Carex lucorum var. *floridana* (Schwein.) Chapm. = ***Carex floridana*** Schwein.
Carex lucorum var. *nigromarginata* (Schwein.) Chapm. = ***Carex nigromarginata*** Schwein.
Carex luctuosa f. *brevisquama* K.T.Fu = ***Carex luctuosa*** Franch.
Carex luctuosa f. *mucronata* K.T.Fu = ***Carex luctuosa*** Franch.
Carex lugens T.Holm = ***Carex bigelowii*** subsp. ***lugens*** (T.Holm) T.V.Egorova
Carex lumnitzeri (Rouy) V.I.Krecz. = ***Carex spicata*** Huds.
Carex lunelliana Mack. = ***Carex gravida*** L.H.Bailey
Carex lupulina var. *albomarginata* Sherff = ***Carex lupulina*** Muhl. ex Willd.
Carex lupulina var. *bella-villa* (Dewey) L.H.Bailey = ***Carex lupulina*** Muhl. ex Willd.
Carex lupulina var. *gigantea* (Rudge) Britton = ***Carex gigantea*** Rudge
Carex lupulina var. *gigantoidea* Dewey = ***Carex retrorsa*** Schwein.
Carex lupulina var. *longipedunculata* Dudley = ***Carex lupulina*** Muhl. ex Willd.
Carex lupulina var. *pedunculata* A.Gray = ***Carex lupulina*** Muhl. ex Willd.
Carex lupulina var. *polystachya* Schwein. & Torr. = ***Carex lurida*** Wahlenb.
Carex lurida L.H.Bailey = ***Carex lupulina*** Muhl. ex Willd.
Carex lurida C.B.Clarke = ***Carex pamirica*** (O.Fedtsch.) O.Fedtsch. & B.Fedtsch. subsp. ***pamirica***
Carex lurida var. *divergens* L.H.Bailey = ***Carex* ×*macounii*** Dewey
Carex lurida var. *exudans* L.H.Bailey ex Britton & A.Br. = ***Carex lurida*** Wahlenb.
Carex lurida var. *flaccida* L.H.Bailey = ***Carex lurida*** Wahlenb.
Carex lurida var. *gracilis* (Boott) Kük. = ***Carex lurida*** Wahlenb.
Carex lurida var. *parvula* (Paine) Kük. = ***Carex lurida*** Wahlenb.
Carex lurida f. *parvula* (Paine) D.S.Carp. = ***Carex lurida*** Wahlenb.
Carex lurida var. *polystachya* (Schwein. & Torr.) L.H. Bailey = ***Carex lurida*** Wahlenb.
Carex lusitanica Willd. = ***Carex paniculata*** subsp. ***lusitanica*** (Willd.) Maire
Carex lutchuensis Ohwi = ***Carex breviscapa*** C.B.Clarke
Carex lutensis Kunth = ***Carex clavata*** Thunb.
Carex luteobrunnea Nelmes = ***Carex nivalis*** Boott
Carex luteola Nees ex Boott = ***Carex bracteosa*** Kunze
Carex luzulifolia var. *ablta* (L.H.Bailey) Kük. = ***Carex luzulina*** var. ***ablata*** (L.H.Bailey) F.J.Herm.
Carex luzulifolia f. *albida* (L.H.Bailey) Kük. = ***Carex albida*** L.H.Bailey
Carex luzulifolia var. *luzuliformis* (L.H.Bailey) Kük. = ***Carex fissuricola*** Mack.
Carex luzulifolia var. *strobilantha* T.Holm = ***Carex luzulifolia*** Boott
Carex lyallii Boott = ***Carex raynoldsii*** Dewey
Carex lycurus K.Schum. ex Engl. = ***Carex conferta*** var. ***lycurus*** (K.Schum. ex Engl.) Lye
Carex lyngbyei var. *capillipes* (Drejer) Kük. = ***Carex lyngbyei*** Hornem.
Carex lyngbyei subsp. *cryptocarpa* (C.A.Mey.) Hultén = ***Carex lyngbyei*** Hornem.
Carex lyngbyei var. *gigas* Hultén = ***Carex lyngbyei*** Hornem.

Carex lyngbyei var. *prionocarpa* (Franch.) Kük. = ***Carex lyngbyei*** Hornem.
Carex lyngbyei var. *robusta* (L.H.Bailey) Cronquist = ***Carex lyngbyei*** Hornem.
Carex lyngbyei var. *variegata* (Drejer) Kük. = ***Carex lyngbyei*** Hornem.
Carex lyonii Boott = ***Carex microglochin*** Wahlenb. subsp. ***microglochin***
Carex lyonii Kurtz = ?
Carex mabilliana Rouy = ***Carex halleriana*** Asso
Carex macella Kunth = ***Carex phalaroides*** var. ***macella*** (Kunth) Kük.
Carex macella var. *paraguayensis* (Maury ex Micheli) Kük. = ***Carex phalaroides*** var. ***paraguayensis*** (Maury ex Micheli) Kük.
Carex macloviana var. *bracteata* Kük. = ***Carex macloviana*** subsp. ***subfusca*** (W.Boott) T.Koyama
Carex macloviana f. *bracteata* (Kük.) Kük. = ***Carex macloviana*** subsp. ***subfusca*** (W.Boott) T.Koyama
Carex macloviana f. *decumbens* T.Holm = ***Carex macloviana*** d'Urv. subsp. ***macloviana***
Carex macloviana subsp. *festivella* (Mack.) Á.Löve & D.Löve = ***Carex microptera*** Mack.
Carex macloviana var. *gracilis* (Olney) Kük. = ***Carex macloviana*** subsp. ***subfusca*** (W.Boott) T.Koyama
Carex macloviana var. *haydeniana* (Olney) Kük. = ***Carex haydeniana*** Olney
Carex macloviana subsp. *haydeniana* (Olney) R.L.Taylor & MacBryde = ***Carex haydeniana*** Olney
Carex macloviana var. *incrassata* Kük. = ***Carex macloviana*** var. ***thermarum*** (Phil.) Kük.
Carex macloviana f. *involucrata* Kük. = ***Carex macloviana*** d'Urv. subsp. ***macloviana***
Carex macloviana var. *microptera* (Mack.) Boivin = ***Carex microptera*** Mack.
Carex macloviana var. *pachystachya* (Cham. ex Steud.) Kük. = ***Carex macloviana*** subsp. ***pachystachya*** (Cham. ex Steud.) Hultén
Carex macloviana var. *stricta* (L.H.Bailey) Kük. = ***Carex macloviana*** d'Urv. subsp. ***macloviana***
Carex macloviana var. *subfusca* (W.Boott) Kük. = ***Carex macloviana*** subsp. ***subfusca*** (W.Boott) T.Koyama
Carex macloviana f. *viridis* (L.H.Bailey) Kük. = ***Carex macloviana*** d'Urv. subsp. ***macloviana***
Carex macounii A.Benn. = ***Carex lyngbyei*** Hornem.
Carex macra Steud. = ***Carex elata*** All. subsp. ***elata***
Carex macrantha Boeck. = ***Carex melanantha*** C.A.Mey. var. ***melanantha***
Carex macrocarpa Phil. = ***Carex macrosolen*** Steud.
Carex macrocephala subsp. *anthericoides* (C.Presl) Hultén = ***Carex macrocephala*** Willd. ex Spreng.
Carex macrocephala var. *bracteata* T.Holm = ***Carex macrocephala*** Willd. ex Spreng.
Carex macrocephala f. *bracteata* (T.Holm) Kük. = ***Carex macrocephala*** Willd. ex Spreng.
Carex macrocephala var. *longibracteata* Oliv. = ***Carex kobomugi*** Ohwi
Carex macrochaeta var. *denticulata* Kük. = ***Carex macrochaeta*** C.A.Mey.
Carex macrochaeta var. *emarginata* T.Holm = ***Carex macrochaeta*** C.A.Mey.
Carex macrochaeta subsp. *flavocuspis* (Franch. & Sav.) Kük. = ***Carex flavocuspis*** Franch. & Sav.
Carex macrochaeta var. *macrochlaena* T.Holm = ***Carex macrochaeta*** C.A.Mey.
Carex macrochaeta var. *platycarpa* Kük. = ***Carex macrochaeta*** C.A.Mey.
Carex macrochaeta var. *subrigida* Kük. ex J.M.Macoun = ***Carex macrochaeta*** C.A.Mey.
Carex macrochlamys Franch. = ***Carex laxa*** Wahlenb.
Carex macrocystis Boeck. = ***Carex clavata*** Thunb.
Carex macroglossa Franch. & Sav. = ***Carex jackiana*** var. ***macroglossa*** (Franch. & Sav.) Kük.
Carex macrogyna Turcz. ex Steud. = ***Carex petricosa*** Dewey var. ***petricosa***
Carex macrokolea Steud. = ***Carex verrucosa*** Muhl.
Carex macrolepis D.Don = ***Carex longipes*** D.Don ex Tilloch & Taylor
Carex macromollicula Nakai = ***Carex planiculmis*** Kom.
Carex macrophylla Hochst. ex Steud. = ***Carex myosurus*** Nees
Carex macrophylla K.Schum. = ?
Carex macrorhyncha Kar. & Kir. = ***Carex parva*** Nees
Carex macrosperma Mack. = ***Carex humboldtiana*** Steud.
Carex macrostachyos Link = ?
Carex macrostachys Bertol. = ***Carex ferruginea*** subsp. ***macrostachys*** (Bertol.) Arcang.
Carex macrostylon var. *decipiens* (J.Gay) Nyman = ***Carex pulicaris*** L.
Carex macrostylon var. *guthnickiana* (J.Gay) C.B.Clarke = ***Carex peregrina*** Link
Carex macrostylon var. *peregrina* (Link) L.H.Bailey ex Trel. = ***Carex peregrina*** Link
Carex macrothyrsa Miq. = ***Carex cruciata*** var. ***rafflesiana*** (Boott) Noot.
Carex macroura Meinsh. = ***Carex pediformis*** C.A.Mey. subsp. ***pediformis***
Carex macroura subsp. *kirilovii* (Turcz.) Malyschev = ***Carex pediformis*** C.A.Mey. subsp. ***pediformis***
Carex maculata Turcz. ex Ledeb. = ***Carex atrofusca*** Schkuhr subsp. ***atrofusca***
Carex maculata Turcz. ex Steud. = ***Carex atrofusca*** var. ***major*** (Boeck.) Raymond
Carex maculata Liebm. = ***Carex psilocarpa*** Steud.
Carex maculata var. *neurochlamys* (F.Muell.) Kük. = ***Carex maculata*** Boott var. ***maculata***
Carex madoerensis C.B.Clarke = ***Carex tricephala*** Boeck.
Carex magellanica Banks ex Boott = ***Carex banksii*** Boott var. ***banksii***
Carex magellanica f. *brevisquama* (Fernald) B.Boivin = ***Carex × connectens*** E.Holmb.
Carex magellanica var. *irrigua* (Wahlenb.) Britton, Sterns & Poggenb. = ***Carex magellanica*** subsp. ***irrigua*** (Wahlenb.) Hultén
Carex magellanica subsp. *planitiei* (Asch. & Graebn.) W.Schultze-Motel = ***Carex magellanica*** subsp. ***irrigua*** (Wahlenb.) Hultén
Carex magellanica var. *planitiei* Asch. & Graebn. = ***Carex magellanica*** subsp. ***irrigua*** (Wahlenb.) Hultén
Carex magnifica Dewey ex C.B.Clarke = ***Carex obnupta*** L.H.Bailey
Carex magnifica var. *lacunarum* (T.Holm) Kük. = ***Carex obnupta*** L.H.Bailey
Carex magnifolia Mack. = ***Carex abscondita*** Mack.
Carex magnursina Raymond = ***Carex petricosa*** Dewey var. ***petricosa***
Carex maidenii Gand. = ?
Carex ×mainensis Porter = ***Carex × stenolepis*** nothovar. ***mainensis*** (Porter) Govaerts
Carex mairii f. *loscosii* (Lange) Kük. = ***Carex mairii*** Coss. & Germ.
Carex mairii subsp. *nevadensis* (Boiss. & Reut.) Malag. = ***Carex flava*** var. ***nevadensis*** (Boiss. & Reut.) Briq.
Carex majae A.P.Khokhr. = ***Carex magellanica*** subsp. ***irrigua*** (Wahlenb.) Hultén
Carex malacocarpa H.Lév. & Vaniot = ***Carex maculata*** Boott var. ***maculata***
Carex malato-belizii Raymond = ***Carex lowei*** Bech.

Carex malayana Nelmes = ***Carex gracilispica*** Hayata
Carex malazena Steud. = ***Carex nigra*** (L.) Reichard subsp. ***nigra***
Carex malinvaldii H.Lév. & Vaniot = ***Carex scabrifolia*** Steud.
Carex malvernensis S.Gibson = ***Carex leporina*** L.
Carex manca var. *contigua* R.Gross = ***Carex harlandii*** Boott
Carex manca var. *wichurae* (Boeck.) Kük. = ***Carex manca*** subsp. ***wichurae*** (Boeck.) S.Y.Liang
Carex mandarinorum Raymond = ***Carex glossostigma*** Hand.-Mazz.
Carex manostachys Spreng. = ***Carex ferruginea*** subsp. ***macrostachys*** (Bertol.) Arcang.
Carex mapaniifolia Ridl. = ***Carex helferi*** Boeck.
Carex marchandiana Lapeyr. ex Spreng. = ***Carex pyrenaica*** Wahlenb.
Carex marcida Boott = ***Carex praegracilis*** W.Boott
Carex marcida Raeusch. = ?
Carex marcida var. *alterna* L.H.Bailey = ***Carex praegracilis*** W.Boott
Carex marcida f. *alterna* (L.H.Bailey) Kük. = ***Carex praegracilis*** W.Boott
Carex marcida var. *debilis* L.H.Bailey = ***Carex praegracilis*** W.Boott
Carex marcida f. *debilis* (L.H.Bailey) Kük. = ***Carex praegracilis*** W.Boott
Carex margaritifera Steud. = ***Scleria secans*** (L.) Urb.
Carex marginaria Franch. = ***Carex heterolepis*** Bunge
Carex marginata Gorter = ***Carex divisa*** Huds.
Carex marginata Muhl. ex Willd. = ***Carex pensylvanica*** Lam.
Carex mariae-victorinii Sennen = ***Carex distans*** L.
Carex mariesii C.B.Clarke = ***Carex sachalinensis*** var. ***fernaldiana*** (H.Lév. & Vaniot) T.Koyama
Carex marina var. *fusca* Böcher = ***Carex marina*** Dewey subsp. ***marina***
Carex marina var. *solida* (T.J.Sørensen) Böcher = ***Carex marina*** subsp. ***pseudolagopina*** (T.J.Sørensen) Böcher
Carex marina var. *tundrarum* Böcher = ***Carex marina*** Dewey subsp. ***marina***
Carex maritima O.F.Müll. = ***Carex paleacea*** Schreb. ex Wahlenb.
Carex maritima var. *brunnescens* F.Nyl. = ***Carex paleacea*** Schreb. ex Wahlenb.
Carex maritima var. *erectiuscula* Fernald = ***Carex paleacea*** Schreb. ex Wahlenb.
Carex maritima var. *incurviformis* (Mack.) Boivin = ***Carex incurviformis*** Mack.
Carex maritima f. *inflata* (Simmons) Polunin = ***Carex maritima*** Gunnerus
Carex maritima subsp. *lyngbyei* (Hornem.) Nyman = ***Carex lyngbyei*** Hornem.
Carex maritima var. *melanocystis* (É.Desv.) Fernald = ***Carex maritima*** Gunnerus
Carex maritima var. *misera* Fernald = ***Carex maritima*** Gunnerus
Carex maritima var. *setina* (H.Christ) Fernald = ***Carex maritima*** Gunnerus
Carex maritima subsp. *setina* (H.Christ) T.V.Egorova = ***Carex maritima*** Gunnerus
Carex maritima subsp. *yukonensis* A.E.Porsild = ***Carex maritima*** Gunnerus
Carex marssonii Auersw. = ***Carex viridula*** var. ***elatior*** (Schltr.) Crins
Carex martinii Petrie = ***Carex ternaria*** G.Forst.
Carex masoalensis Cherm. = ***Carex brunnea*** Thunb. subsp. ***brunnea***
Carex ×massonii Cay. & Lepage = ?
Carex ×mastnyana Otruba & Podp. = ?
Carex matsudae (Hayata) Hayata ex Makino & Nemoto = ***Carex dolichostachya*** Hayata subsp. ***dolichostachya***
Carex matsumurae var. *variegata* Makino = ***Carex matsumurae*** Franch.
Carex maubertiana Boott = ***Carex hebecarpa*** var. ***maubertiana*** (Boott) Franch.
Carex mauritanica Boiss. & Reut. = ***Carex acuta*** L.
Carex mauritanica Steud. = ***Carex brunnea*** Thunb. subsp. ***brunnea***
Carex maxima Scop. = ***Carex pendula*** Huds.
Carex maxima var. *angustifolia* Albov = ***Carex pendula*** Huds.
Carex maximowiczii Boeck. = ***Carex arnellii*** H.Christ
Carex maximowiczii Miq. = ***Carex pruinosa*** subsp. ***maximowiczii*** (Miq.) Kük.
Carex maximowiczii F.Schmidt = ***Carex schmidtii*** Meinsh.
Carex maximowiczii var. *levisaccus* Ohwi = ***Carex pruinosa*** subsp. ***maximowiczii*** (Miq.) Kük.
Carex maximowiczii subsp. *suifunensis* (Kom.) Vorosch. = ***Carex suifunensis*** Kom.
Carex mcclurgii Petrie = ***Carex ventosa*** C.B.Clarke
Carex mcmahonii Petrie = ***Carex raoulii*** Boott
Carex meadii Porter = ***Carex tetanica*** Schkuhr
Carex meadii var. *bebbii* Arthur = ***Carex meadii*** Dewey
Carex media subsp. *pusteriana* (Kalela) W.Schultze-Motel = ***Carex norvegica*** subsp. ***pusteriana*** (Kalela) Chater
Carex media var. *stevenii* (T.Holm) Fernald = ***Carex stevenii*** (T.Holm) Kalela
Carex mediterranea Mack. = ***Carex mesochorea*** Mack.
Carex meekii Dewey = ***Carex douglasii*** Boott
Carex megacarpa T.Koyama = ***Carex brunnea*** Thunb. subsp. ***brunnea***
Carex megalandra Gand. = ?
Carex meinshauseniana V.I.Krecz. = ***Carex tristis*** M.Bieb.
Carex meiogyna Nees = ***Carex filicina*** Nees
Carex melaena Wimm. = ***Carex nigra*** (L.) Reichard subsp. ***nigra***
Carex melanantha var. *baicalensis* Turcz. = ***Carex melanantha*** C.A.Mey. var. ***melanantha***
Carex melanantha var. *genuina* Turcz. = ***Carex melanantha*** C.A.Mey.
Carex melanantha f. *gracilescens* Kük. = ***Carex melanantha*** C.A.Mey. var. ***melanantha***
Carex melanantha var. *sabulosa* (Turcz. ex Kunth) Kük. = ***Carex sabulosa*** Turcz. ex Kunth
Carex melanocephala Kar. & Kir. = ?
Carex melanochloros Thuill. = ***Carex elata*** All. subsp. ***elata***
Carex melanocystis Desv. = ***Carex maritima*** Gunnerus
Carex melanocystis var. *misera* Kük. = ***Carex maritima*** Gunnerus
Carex melanolepis Phil. = ***Carex nigra*** (L.) Reichard subsp. ***nigra***
Carex melanolepis Boeck. = ***Carex orbicularis*** Boott subsp. ***orbicularis***
Carex melanophora S.T.Blake = ***Carex sarawaketensis*** Kük.
Carex melanosperma Liebm. = ***Carex jamesonii*** var. ***melanosperma*** (Liebm.) Kük.
Carex melanostoma V.I.Krecz. = ***Carex microchaeta*** subsp. ***nesophila*** (T.Holm) D.F.Murray
Carex melanostylis Desv. = ?
Carex melastoma Fisch. ex Boott = ***Carex spectabilis*** Dewey
Carex melinacra var. *changningensis* S.Yun Liang = ***Carex melinacra*** Franch.
Carex melozitnensis A.E.Porsild = ***Carex rotundata*** Wahlenb.
Carex membranacea Hoppe = ***Carex ericetorum*** Pollich

Carex membranopacta L.H.Bailey = ***Carex membranacea*** Hook.
Carex menziesiana Sm. ex Boott = ***Carex macrocephala*** Willd. ex Spreng.
Carex mercarensis Hochst. ex Steud. = ***Carex raphidocarpa*** Nees
Carex mercarensis var. *major* Steud. = ***Carex raphidocarpa*** Nees
Carex meridiana Akiyama = ?
Carex meridionalis (Kük.) Herter = ***Carex longii*** subsp. ***meridionalis*** (Kük.) Luceño & Alves
Carex merinoi Gand. = ***Carex depressa*** Link subsp. ***depressa***
Carex merrillii Kük. = ***Carex turrita*** var. ***merrillii*** (Kük.) Noot.
Carex mertensii subsp. *urostachys* (Franch.) Calder & T.Koyama = ***Carex mertensii*** var. ***urostachys*** (Franch.) Kük.
Carex mertensis Weihe ex Kunth = ***Carex vulpina*** L.
Carex mesomorpha Gand. = ?
Carex metteniana (Asch.) Lehm. ex Schur = ***Carex dioica*** L.
Carex mexicana C.Presl = ***Carex polystachya*** Sw. ex Wahlenb. var. ***polystachya***
Carex meyeriana var. *scabra* Meinsh. = ***Carex meyeriana*** Kunth
Carex micans Boott = ***Carex maculata*** Boott var. ***maculata***
Carex michauxiana var. *asiatica* (Hultén) Ohwi = ***Carex michauxiana*** subsp. ***asiatica*** Hultén
Carex michauxii Dewey = ***Carex collinsii*** Nutt.
Carex michauxii Schwein. = ***Carex scirpoidea*** Michx. var. ***scirpoidea***
Carex micheliana Sm. = ***Carex flacca*** Schreb. subsp. ***flacca***
Carex michelii var. *beckeri* (C.A.Mey. ex Claus) Nyman = ***Carex michelii*** Host
Carex michelii f. *puberula* Beck = ***Carex michelii*** Host
Carex michelii var. *ultra* L.H.Bailey = ***Carex longirostrata*** C.A.Mey.
Carex michiganensis Dewey ex Alph.Wood = ***Carex lucorum*** Willd. var. ***lucorum***
Carex micrantha Kük. = ***Carex cinerascens*** Kük.
Carex micrantha var. *tenuinervis* (Kük.) Kük. = ***Carex gaudichaudiana*** Kunth
Carex microcephala C.A.Mey. = ***Carex obtusata*** Lilj.
Carex microdonta var. *controversa* L.H.Bailey = ***Carex microdonta*** Torr.
Carex microdonta var. *latifolia* L.H.Bailey = ***Carex microdonta*** Torr.
Carex microglochin var. *oligantha* Kük. = ***Carex microglochin*** subsp. ***fuegina*** Kük.
Carex microlepis Boeck. = ***Carex diluta*** M.Bieb. subsp. ***diluta***
Carex micropodioides V.I.Krecz. = ***Carex pyrenaica*** Wahlenb.
Carex microptera var. *crassinervia* F.J.Herm. = ***Carex microptera*** Mack.
Carex microptera var. *limnophila* (F.J.Herm.) Dorn = ***Carex microptera*** Mack.
Carex microsacca C.B.Clarke = [38 JAP]
Carex microsperma Steud. = ***Carex flaccosperma*** Dewey
Carex microsperma Wahlenb. = ***Carex vulpinoidea*** Michx.
Carex microstachya Michx. = ***Carex leptalea*** Wahlenb. subsp. ***leptalea***
Carex ×*microstachya* var. *gaudiniana* (Guthnick) Trevir. = ***Carex ×gaudiniana*** Guthnick
Carex ×*microstachya* subsp. *gaudiniana* (Guthnick) Nyman = ***Carex ×gaudiniana*** Guthnick
Carex microstoma Franch. = ***Carex pallescens*** L. var. ***pallescens***
Carex microtricha Franch. = ***Carex caryophyllea*** var. ***microtricha*** (Franch.) Kük.
Carex mielichhoferi Willd. ex Kunth = ***Carex brachystachys*** Schrank
Carex mielichhoferi Schkuhr = ***Carex ferruginea*** Scop. subsp. ***ferruginea***
Carex mielichhoferi Sm. = ***Carex vaginata*** Tausch var. ***vaginata***
Carex ×*mihashii* T.Koyama = ?
Carex mildbraediana var. *friesiorum* Kük. = ***Carex elgonensis*** Nelmes
Carex miliacea Muhl. ex Willd. = ***Carex prasina*** Wahlenb.
Carex miliacea Schrank = ?
Carex miliaris Michx. = ***Carex saxatilis*** L.
Carex miliaris var. *aurea* L.H.Bailey = ***Carex saxatilis*** L.
Carex miliaris f. *longipedunculata* Lepage = ***Carex saxatilis*** L.
Carex miliaris f. *major* L.H.Bailey = ***Carex saxatilis*** L.
Carex miliaris f. *major* (L.H.Bailey) Kük. = ***Carex saxatilis*** L.
Carex miliaris var. *obtusa* L.H.Bailey = ***Carex saxatilis*** L.
Carex miliaris var. *raeana* (Boott) Kük. = ***Carex vesicaria*** L.
Carex miliaris var. *ungavensis* Raymond = ***Carex saxatilis*** L.
Carex millegrana T.Holm = ***Carex emoryi*** Dewey
Carex milnei Boott ex C.B.Clarke = ***Carex indica*** L. var. ***indica***
Carex mimula V.I.Krecz. = ***Carex norvegica*** Retz. subsp. ***norvegica***
Carex minganinsularum Raymond = ***Carex echinata*** Murray subsp. ***echinata***
Carex minima Boullu = ***Carex bicolor*** Bellardi ex All.
Carex minuscula (Kuvaev) Rebrist. = ***Carex aquatilis*** var. ***minor*** Boott
Carex minuta Franch. = ***Carex cespitosa*** var. ***minuta*** (Franch.) Kük.
Carex minutiflora Boeck. = ***Carex setigera*** D.Don var. ***setigera***
Carex mirabilis Schleich. ex Gaudin = ***Carex curvula*** All. subsp. ***curvula***
Carex mirabilis Dewey = ***Carex straminea*** Willd. ex Schkuhr
Carex mirabilis Host = ***Kobresia simpliciuscula*** (Wahlenb.) Mack. subsp. ***simpliciuscula***
Carex mirabilis var. *adusta* (Boott) Prov. = ***Carex adusta*** Boott
Carex mirabilis var. *perlonga* Fernald = ***Carex normalis*** Mack.
Carex mirabilis var. *tenera* (Dewey) Prov. = ***Carex tenera*** Dewey
Carex mirabilis var. *tincta* Fernald = ***Carex tincta*** (Fernald) Fernald
Carex mirata Dewey = ***Carex atherodes*** Spreng.
Carex mirata var. *minor* Dewey = ***Carex atherodes*** Spreng.
Carex misandra R.Br. = ***Carex fuliginosa*** Schkuhr
Carex misandra var. *elatior* Lange = ***Carex fuliginosa*** Schkuhr
Carex misandra f. *flavida* Fernald = ***Carex fuliginosa*** Schkuhr
Carex misandra f. *ochrolochin* Ostenf. = ***Carex fuliginosa*** Schkuhr
Carex misandroides Fernald = ***Carex petricosa*** var. ***misandroides*** (Fernald) B.Boivin

Carex misera Franch. = ***Carex disperma*** Dewey
Carex misera Phil. = ***Carex maritima*** Gunnerus
Carex miserabilis Mack. = ***Carex scopulorum*** var. ***prionophylla*** (T.Holm) L.A.Standl.
Carex mitis (P.J.Bergius) J.F.Gmel. = ***Scleria mitis*** P.J.Bergius
Carex mitoensis Ohwi = ***Carex otaruensis*** Franch.
Carex mitrata var. *aristata* Ohwi = ***Carex mitrata*** subsp. ***aristata*** (Ohwi) T.Koyama
Carex mitrata var. *vigens* Kük. = ***Carex genkaiensis*** Ohwi
Carex mitsuriokensis H.Lév. & Vaniot = ***Carex globularis*** L.
Carex × *mixta* Corb. = ***Carex* × *alsatica*** Zahn nothosubsp. ***alsatica***
Carex mixta J.F.Gmel. = ***Carex elata*** All. subsp. ***elata***
Carex mixta Miégev. = ***Carex umbrosa*** subsp. ***huetiana*** (Boiss.) Soó
Carex mixta var. *gynobasis* Rouy = ***Carex umbrosa*** Host subsp. ***umbrosa***
Carex modesta J.Gay = ***Carex disticha*** Huds.
Carex moelleri Phil. = ***Carex setifolia*** Kunze var. ***setifolia***
Carex moenchiana Wender. = ***Carex acuta*** L.
Carex moesta Kunth = ***Carex phalaroides*** var. ***elongata*** Kük.
Carex mohriana Mack. = ***Carex atlantica*** subsp. ***capillacea*** (L.H.Bailey) Reznicek
Carex mollicula f. *lutescens* Kük. = ***Carex mollicula*** Boott
Carex mollis Host = ***Carex caryophyllea*** Latourr. var. ***caryophyllea***
Carex mollis Gilib. = ***Carex leporina*** L.
Carex mollissima var. *reflexa* (Meinsh.) Kük. = ***Carex mollissima*** H.Christ ex Scheutz
Carex monadelpha Boott = ***Carex tristachya*** Thunb. var. ***tristachya***
Carex moniezii Timb.-Lagr. = ***Carex vulpinoidea*** Michx.
Carex monile Tuck. ex Boott = ***Carex vesicaria*** L.
Carex monile var. *colorata* L.H.Bailey = ***Carex vesicaria*** L.
Carex monile var. *monstrosa* L.H.Bailey = ***Carex vesicaria*** L.
Carex monile var. *monstrosa* L.H.Bailey = ***Carex vesicaria*** L.
Carex monile var. *obtusisquamis* (L.H.Bailey) L.H. Bailey = ***Carex vesicaria*** L.
Carex monile var. *pacifica* L.H.Bailey = ***Carex vesicaria*** L.
Carex monile var. *raeana* (Boott) L.H.Bailey = ***Carex vesicaria*** L.
Carex monilifera Thuill. = ***Carex depauperata*** Curtis ex With
Carex moniliformis Britton = ***Carex silicea*** Olney
Carex moniziana Lowe ex Boott = ***Carex divisa*** Huds.
Carex monopleura V.I.Krecz. = ***Carex remota*** subsp. ***rochebrunii*** (Franch. & Sav.) Kük.
Carex monostachya var. *triquetrifolia* (Boeck.) Kük. = ***Carex monostachya*** A.Rich.
Carex montana Pollich ex Lightf. = ***Carex praecox*** Schreb.
Carex montana f. *albescens* Bornm. & Kük. = ***Carex montana*** L.
Carex montana f. *bulboides* Waisb. = ***Carex montana*** L.
Carex montana var. *conglobata* (All.) Nyman = ***Carex montana*** L.
Carex montana subsp. *csetzii* (Janka) Nyman = ***Carex montana*** L.
Carex montana var. *emarginata* (Willd.) Nyman = ***Carex montana*** L.
Carex montana f. *fimbriata* Waisb. = ***Carex montana*** L.
Carex montana var. *flavida* Waisb. = ***Carex montana*** L.
Carex montana subsp. *fritschii* (Waisb.) O.Schwarz = ***Carex fritschii*** Waisb.
Carex montana f. *gracilior* Waisb. = ***Carex montana*** L.
Carex montana var. *luxurians* Èelak. = ***Carex montana*** L.
Carex montana var. *manshuriensis* Kom. = ***Carex ulobasis*** V.I.Krecz.
Carex montana f. *marginata* Waisb. = ***Carex montana*** L.
Carex montana var. *oxyandra* Franch. & Sav. = ***Carex oxyandra*** (Franch. & Sav.) Kudô
Carex montana f. *pallescens* Döll = ***Carex montana*** L.
Carex montana f. *procerior* Gaudin = ***Carex montana*** L.
Carex montana f. *pseudopallescens* Kneuck. = ***Carex montana*** L.
Carex montana f. *remotiflora* Waisb. = ***Carex montana*** L.
Carex montana var. *rigida* Waisb. = ***Carex montana*** L.
Carex montana f. *truncata* Waisb. = ***Carex montana*** L.
Carex montana var. *verniformis* Waisb. = ***Carex montana*** L.
Carex montanensis L.H.Bailey = ***Carex podocarpa*** R.Br.
Carex montereyensis Mack. = ***Carex harfordii*** Mack.
Carex monticola Boeck. = ***Carex anisostachys*** Liebm.
Carex monticola Dewey ex Torr. = ***Carex triquetra*** Boott
Carex montis-eeka var. *filifolia* Skottsb. = ***Carex montis-eeka*** Hillebr.
Carex montis-eeka var. *viridens* Kük. = ***Carex montis-eeka*** Hillebr.
Carex montivaga S.T.Blake = ***Carex breviculmis*** var. ***montivaga*** (S.T.Blake) Noot.
Carex moorcroftii Falc. ex Boott = ***Carex melanantha*** var. ***moorcroftii*** (Falc. ex Boott) Kük.
Carex moritzii Steud. = ***Carex indica*** L. var. ***indica***
Carex moritzii var. *laxior* Steud. = ***Carex indica*** L. var. ***indica***
Carex morrisonicola Hayata = ***Carex breviculmis*** R.Br. var. ***breviculmis***
Carex morrisseyi A.E.Porsild = ***Carex adelostoma*** V.I.Krecz.
Carex morrowii var. *albomarginata* Makino = ***Carex morrowii*** Boott var. ***morrowii***
Carex morrowii var. *crassicaulis* Franch. = ***Carex morrowii*** Boott var. ***morrowii***
Carex morrowii var. *latissima* Ohwi = ***Carex foliosissima*** var. ***latissima*** (Ohwi) Akiyama
Carex morrowii f. *niigatensis* Franch. & Sav. = ***Carex morrowii*** Boott var. ***morrowii***
Carex morrowii f. *virginea* Franch. & Sav. = ***Carex morrowii*** Boott var. ***morrowii***
Carex morrowii var. *yesinensis* (Franch.) Kük. = ***Carex foliosissima*** F.Schmidt var. ***foliosissima***
Carex moscowensis C.B.Clarke ex B.Fedtsch. = [14]
Carex mosoynensis Franch. = ***Carex brunnea*** Thunb. subsp. ***brunnea***
Carex motoskei Miq. = ***Carex japonica*** Thunb.
Carex mucronata f. *androgyna* Camperio = ***Carex mucronata*** All.
Carex muehlenbergii Brongn. = ***Carex brongniartii*** Kunth
Carex muehlenbergii var. *australis* Olney = ***Carex austrina*** Mack.
Carex muehlenbergii var. *austrina* (Mack.) Small = ***Carex austrina*** Mack.
Carex muelleri Steud. = ***Carex distachya*** Desf. var. ***distachya***
Carex muhlenbergii Kunth ex Boott = ***Carex vulpinoidea*** Michx.
Carex mukdenensis Kitag. = ***Carex kirganica*** Kom.
Carex multiceps Gaudin = ***Carex elongata*** L.

Carex multiculmis Ehrh. = ***Carex elongata*** L.
Carex multifiliculmis H.Lév. & Vaniot = ***Carex subumbellata*** Meinsh. var. ***subumbellata***
Carex multiflora Muhl. ex Willd. = ***Carex vulpinoidea*** Michx.
Carex multiflora var. *microsperma* (Wahlenb.) Dewey = ***Carex vulpinoidea*** Michx.
Carex multifolia Ohwi = ***Carex dolichostachya*** Hayata subsp. ***dolichostachya***
Carex multiformis Thuill. = ***Carex disticha*** Huds.
Carex multimoda L.H.Bailey = ***Carex macloviana*** d'Urv. subsp. ***macloviana***
Carex multinervis Krock. ex Steud. = ***Carex binervis*** Sm.
Carex multinervis Boeck. = ***Carex tetanica*** Schkuhr
Carex multipes H.Lév. & Vaniot = ***Carex caryophyllea*** Latourr. var. ***caryophyllea***
Carex multispicata Kunze = ***Carex acutata*** Boott
Carex munda var. *mundiformis* T.Koyama = ***Carex munda*** Boott
Carex muricata Schltdl. = ***Carex brongniartii*** Kunth
Carex muricata Desf. = ***Carex divulsa*** Stokes subsp. ***divulsa***
Carex muricata Huds. = ***Carex echinata*** Murray subsp. ***echinata***
Carex muricata Leers = ***Carex echinata*** Murray subsp. ***echinata***
Carex muricata Jungh. = ***Carex vulpina*** L.
Carex muricata var. *americana* L.H.Bailey = ***Carex hookeriana*** Dewey
Carex muricata var. *angustata* (Carey) Carey ex Gleason = ***Carex echinata*** Murray subsp. ***echinata***
Carex muricata f. *angustifolia* Podp. = ***Carex divulsa*** Stokes subsp. ***divulsa***
Carex muricata f. *arcuata* P.Silva = ***Carex muricata*** L.
Carex muricata f. *argyrolepis* Peterm. = ***Carex spicata*** Huds.
Carex muricata var. *cephalantha* (L.H.Bailey) Wiegand & Eames = ***Carex echinata*** Murray subsp. ***echinata***
Carex muricata subsp. *cephalantha* (L.H.Bailey) R.T. Clausen = ***Carex echinata*** Murray subsp. ***echinata***
Carex muricata var. *cephaloidea* Dewey = ***Carex cephaloidea*** (Dewey) Dewey ex Boott
Carex muricata var. *confixa* L.H.Bailey = ***Carex hoodii*** Boott
Carex muricata subsp. *contigua* (Hoppe) Moravec = ***Carex spicata*** Huds.
Carex muricata var. *contigua* (Hoppe) Fiori = ***Carex spicata*** Huds.
Carex muricata var. *depauperata* Hampe = ***Carex echinata*** Murray subsp. ***echinata***
Carex muricata subsp. *divulsa* (Stokes) Husn. = ***Carex divulsa*** Stokes
Carex muricata var. *divulsa* (Stokes) Wahlenb. = ***Carex divulsa*** Stokes
Carex muricata var. *foliosa* (D.Don ex Tilloch & Taylor) C.B.Clarke = ***Carex foliosa*** D.Don ex Tilloch & Taylor
Carex muricata var. *fumosa* Gren. = ***Carex spicata*** Huds.
Carex muricata var. *furculata* (Peterm.) Nyman = ***Carex spicata*** Huds.
Carex muricata var. *gracilis* Boott = ***Carex hookeriana*** Dewey
Carex muricata var. *incrassata* Crép. = ***Carex spicata*** Huds.
Carex muricata var. *indica* Boott = ***Carex foliosa*** D.Don ex Tilloch & Taylor
Carex muricata subsp. *lamprocarpa* Èelak. = ***Carex pairae*** F.W.Schultz
Carex muricata var. *laricina* (Mack. ex Bright) Gleason = ***Carex echinata*** Murray subsp. ***echinata***
Carex muricata var. *leersii* Kneuck. = ***Carex divulsa*** subsp. ***leersii*** (Kneuck.) W.Koch
Carex muricata subsp. *leersii* (Kneuck.) Asch. & Graebn. = ***Carex divulsa*** subsp. ***leersii*** (Kneuck.) W.Koch
Carex muricata var. *loliacea* Schkuhr = ***Carex echinata*** Murray subsp. ***echinata***
Carex muricata subsp. *lumnitzeri* (Rouy) Soó = ***Carex spicata*** Huds.
Carex muricata race *lumnitzeri* Rouy = ***Carex spicata*** Huds.
Carex muricata subsp. *macrocarpa* Neuman = ***Carex spicata*** Huds.
Carex muricata f. *mediterranea* Asch. & Graebn. = ***Carex spicata*** Huds.
Carex muricata subsp. *microcarpa* Neuman = ***Carex echinata*** Murray subsp. ***echinata***
Carex muricata var. *monostachya* Asch. = ***Carex echinata*** Murray subsp. ***echinata***
Carex muricata var. *nemorosa* Nyman = ***Carex muricata*** L.
Carex muricata subsp. *orsiniana* (Ten.) Nyman = ***Carex muricata*** L.
Carex muricata subsp. *pairae* (F.W.Schultz) Asch. & Graebn. = ***Carex pairae*** F.W.Schultz
Carex muricata var. *pairaei* (Garcke) Èelak. = ***Carex echinata*** Murray subsp. ***echinata***
Carex muricata f. *pseudoguestephalica* Asch. & Graebn. = ***Carex spicata*** Huds.
Carex muricata var. *ruthii* (Mack.) Gleason = ***Carex ruthii*** Mack.
Carex muricata var. *sterilis* (Willd.) Gleason = ***Carex sterilis*** Willd.
Carex muricata var. *subramosa* Neilr. = ***Carex divulsa*** Stokes subsp. ***divulsa***
Carex muricata var. *virens* Rchb. = ***Carex echinata*** Murray subsp. ***echinata***
Carex muricata var. *virens* Andersson = ***Carex echinata*** Murray subsp. ***echinata***
Carex ×*murrii* Appel = ?
Carex musartiana Kük. ex Krecz. = ***Carex podocarpa*** R.Br.
Carex mutabilis Willd. = ***Carex pendula*** Huds.
Carex mutans Boott ex C.B.Clarke = ***Kobresia esenbeckii*** (Kunth) Noltie var. ***esenbeckii***
Carex mutica R.Br. = ***Carex aurea*** Nutt.
Carex myosuroides Lowe = ***Carex pendula*** Huds.
Carex myosuroides Vill. = ***Kobresia myosuroides*** (Vill.) Fiori & Paol.
Carex myosurus var. *eminens* (Nees) Boeck. = ***Carex myosurus*** Nees
Carex myosurus var. *floribunda* (Boeck.) Kük. = ***Carex myosurus*** Nees
Carex myosurus var. *gracilior* Kük. = ***Carex myosurus*** Nees
Carex myosurus var. *nobilis* (Boott) C.B.Clarke = ***Carex myosurus*** Nees
Carex myosurus var. *praestans* (C.B.Clarke) Kük. = ***Carex myosurus*** Nees
Carex myosurus var. *ratongensis* C.B.Clarke = ***Carex ratongensis*** (C.B.Clarke) C.B.Clarke
Carex myosurus subsp. *spiculata* Boott = ***Carex myosurus*** Nees
Carex ×*naegelii* Brügger ex Sennholz = ?
Carex nagasakensis C.B.Clarke = ***Carex ischnostachya*** Steud.
Carex nagatadakensis Masam. = ***Carex doenitzii*** Boeck. var. ***doenitzii***
Carex nakaharae Hayata = ***Carex brunnea*** Thunb. subsp. ***brunnea***
Carex nakaii H.Lév. = ***Carex disperma*** Dewey

Carex nakaoana T.Koyama = ***Carex haematostoma*** Nees
Carex nakiri Ohwi = ***Carex brunnea*** Thunb. subsp. ***brunnea***
Carex nambuensis Franch. = ***Carex reinii*** Franch. & Sav.
Carex nana Boott = ***Carex capillacea*** var. ***sachalinensis*** (F.Schmidt) Ohwi
Carex nana Cham. ex Steud. = ***Carex capillaris*** L. subsp. ***capillaris***
Carex nana Lam. = ***Carex ferruginea*** Scop. subsp. ***ferruginea***
Carex nanca Boott = ***Carex ligata*** Boott subsp. ***ligata***
Carex nanella Ohwi = ***Carex humilis*** var. ***nana*** (H.Lév. & Vaniot) Ohwi
Carex nanella var. *callitrichos* (V.I.Krecz.) Vorosch. = ***Carex callitrichos*** V.I.Krecz.
Carex nankaiensis Honda = ***Carex insaniae*** var. ***subdita*** (Ohwi) Ohwi
Carex nardifolia Wahlenb. = ***Carex aquatilis*** Wahlenb. var. ***aquatilis***
Carex nardina var. *atriceps* Kük. = ***Carex nardina*** Fr. var. ***nardina***
Carex nardina subsp. *hepburnii* (Boott) Á.Löve, D.Löve & S.L.Kapoor = ***Carex nardina*** var. ***hepburnii*** (Boott) Kük.
Carex naufragii Hochst. & Steud. = ***Carex illegitima*** Ces.
Carex naumanniana Boeck. = ***Carex conica*** Boott
Carex navasii Merino = ***Carex nigra*** subsp. ***intricata*** (Tineo) Maire & Wieller
Carex nealiae R.W.Krauss = ***Carex thunbergii*** Steud.
Carex ×*nearctica* Raymond = ***Carex* × *limula*** Fr.
Carex nebraskensis var. *eruciformis* Suksd. = ***Carex nebraskensis*** Dewey
Carex nebraskensis var. *praevia* L.H.Bailey = ***Carex nebraskensis*** Dewey
Carex nebraskensis var. *ultriformis* L.H.Bailey = ***Carex nebraskensis*** Dewey
Carex nebrodensis Lojac. = ?
Carex nebularum var. *kurtziana* (Kük.) Kük. = ***Carex kurtziana*** Kük.
Carex nebularum f. *major* (Kük.) Kük. = ***Carex firmicaulis*** Kalela
Carex neesiana Arn. ex Boott = ***Carex lobulirostris*** Drejer
Carex neglecta Peterm. = ***Carex cespitosa*** L. var. ***cespitosa***
Carex neglecta Degl. = ***Carex distans*** L.
Carex neglecta Tuck. = ***Carex rosea*** Willd.
Carex neglecta var. *major* Peterm. = ***Carex cespitosa*** L. var. ***cespitosa***
Carex nemarosa Kunth = ***Carex leiorhyncha*** C.A.Mey.
Carex nematosperma Willd. ex Kunth = ***Carex laxiflora*** Lam. var. ***laxiflora***
Carex nemorensis J.F.Gmel. = ***Carex pilosa*** Scop.
Carex nemorosa Schrank = ***Carex alba*** Scop.
Carex nemorosa Lumn. ex Honck. = ***Carex muricata*** L.
Carex nemorosa J.F.Gmel. = ***Carex pilosa*** Scop.
Carex nemorosa Rebent. = ***Carex vulpina*** L.
Carex nemorosa var. *cuprina* I.Sándor ex Heuff. = ***Carex cuprina*** (I.Sándor ex Heuff.) Nendtv. ex A.Kern.
Carex nenningiana Opiz = ?
Carex neoguineensis C.B.Clarke = ***Carex filicina*** Nees
Carex neomexicana Mack. = ***Carex occidentalis*** L.H.Bailey
Carex neonelmesii Raymond = ?
Carex neopaleacea Lepage = ***Carex paleacea*** Schreb. ex Wahlenb.
Carex nepalensis Spreng. = ***Carex longipes*** D.Don ex Tilloch & Taylor
Carex nervina var. *jonesii* (L.H.Bailey) Kük. = ***Carex jonesii*** L.H.Bailey
Carex nervosa Willd. ex Kunth = ***Carex distans*** L.
Carex nervosa Desf. = ***Carex extensa*** Gooden.
Carex nervulosa Franch. = ***Carex shimidzensis*** Franch.
Carex nesophila T.Holm = ***Carex microchaeta*** subsp. ***nesophila*** (T.Holm) D.F.Murray
Carex neurochlaena T.Holm = ***Carex heleonastes*** subsp. ***neurochlaena*** (T.Holm) Böcher
Carex neurochlamys F.Muell. = ***Carex maculata*** Boott var. ***maculata***
Carex nevadensis Boiss. & Reut. = ***Carex flava*** var. ***nevadensis*** (Boiss. & Reut.) Briq.
Carex nevadensis subsp. *alpina* (Kneuck.) Podlech = ***Carex flava*** var. ***nevadensis*** (Boiss. & Reut.) Briq.
Carex nevadensis subsp. *flavella* (V.I.Krecz.) Janch. = ***Carex flava*** L. subsp. ***flava***
Carex nexa Boott = ***Carex ligata*** Boott subsp. ***ligata***
Carex nexa var. *strictior* Kük. = ***Carex sociata*** Boott
Carex nigella Boott = ***Carex spectabilis*** Dewey
Carex nigra Schkuhr = ***Carex atrata*** subsp. ***aterrima*** (Hoppe) Èelak.
Carex nigra All. = ***Carex parviflora*** Host
Carex nigra f. *abbreviata* (Peterm.) Soó = ***Carex nigra*** (L.) Reichard subsp. ***nigra***
Carex nigra var. *alpicola* Beck = ***Carex nigra*** (L.) Reichard subsp. ***nigra***
Carex nigra lus *androgyna* (Peterm.) Soó = ***Carex nigra*** (L.) Reichard subsp. ***nigra***
Carex nigra f. *badia* (Sanio) Soó = ***Carex nigra*** (L.) Reichard subsp. ***nigra***
Carex nigra f. *bina* (Schkuhr) Asch. & Graebn. = ***Carex distachya*** Desf. var. ***distachya***
Carex nigra subsp. *broteroana* (Samp.) Rivas Mart. = ***Carex elata*** All. subsp. ***elata***
Carex nigra f. *densicarpa* (Kneuck.) Soó = ***Carex nigra*** (L.) Reichard subsp. ***nigra***
Carex nigra var. *distachya* Nyman = ***Carex atrata*** L. subsp. ***atrata***
Carex nigra subsp. *goodenowii* (J.Gay) Maire & Weiller = ***Carex nigra*** (L.) Reichard subsp. ***nigra***
Carex nigra subsp. *iberica* Rivas Mart. = ***Carex nigra*** (L.) Reichard subsp. ***nigra***
Carex nigra var. *juncea* (Hartm.) Hyl. = ***Carex nigra*** subsp. ***juncella*** (Fr.) Lemke
Carex nigra subsp. *juncea* (Hartm.) Soó = ***Carex nigra*** subsp. ***juncella*** (Fr.) Lemke
Carex nigra f. *leucolepis* (Meinsh.) Soó = ***Carex nigra*** (L.) Reichard subsp. ***nigra***
Carex nigra var. *melanantha* (C.A.Mey.) Regel = ***Carex melanantha*** C.A.Mey.
Carex nigra f. *microstachys* (Zapal.) Soó = ***Carex nigra*** (L.) Reichard subsp. ***nigra***
Carex nigra subsp. *obesa* (All.) H.P.Fuchs ex Janch. = ***Carex nigra*** subsp. ***alpina*** (Gaudin) Lemke
Carex nigra var. *orientalis* Regel = ***Carex melanantha*** C.A.Mey. var. ***melanantha***
Carex nigra f. *pleiandra* (Kük.) Soó = ***Carex nigra*** (L.) Reichard subsp. ***nigra***
Carex nigra f. *ramosa* (Peterm.) Soó = ***Carex nigra*** (L.) Reichard subsp. ***nigra***
Carex nigra subsp. *recta* (J.Fleisch.) Rothm. = ***Carex nigra*** (L.) Reichard subsp. ***nigra***
Carex nigra f. *roussioei* (H.Lév.) Soó = ***Carex nigra*** (L.) Reichard subsp. ***nigra***
Carex nigra f. *silvatica* (Russow ex E.Lehm.) Soó = ***Carex nigra*** (L.) Reichard subsp. ***nigra***
Carex nigra var. *stenocarpa* (Kük.) Soó = ***Carex nigra*** (L.) Reichard subsp. ***nigra***
Carex nigra f. *stenostachya* (Uechtr. ex Asch. & Graebn.) Soó = ***Carex nigra*** (L.) Reichard subsp. ***nigra***
Carex nigra var. *strictiformis* (L.H.Bailey) Fernald = ***Carex nigra*** (L.) Reichard subsp. ***nigra***

Carex nigra f. *subfastigiata* (Schur) Soó = ***Carex nigra*** (L.) Reichard subsp. ***nigra***
Carex nigra f. *subovalis* (Kneuck.) Soó = ***Carex nigra*** (L.) Reichard subsp. ***nigra***
Carex nigra f. *subramosa* (Meinsh.) Soó = ***Carex nigra*** (L.) Reichard subsp. ***nigra***
Carex nigra f. *subsetacea* (Kük.) Soó = ***Carex nigra*** (L.) Reichard subsp. ***nigra***
Carex nigra f. *tenuis* (Peterm.) Soó = ***Carex nigra*** (L.) Reichard subsp. ***nigra***
Carex nigra subsp. *tornata* (Fr.) Rothm. = ***Carex nigra*** (L.) Reichard subsp. ***nigra***
Carex nigra subsp. *turfosa* (Fr.) Podlech = ***Carex* × *turfosa*** Fr.
Carex nigra subsp. *wiluica* (Meinsh. ex Maack) Á.Löve & D.Löve = ***Carex nigra*** subsp. ***juncella*** (Fr.) Lemke
Carex nigricans Hombr. ex Boott = ***Carex darwinii*** Boott var. ***darwinii***
Carex nigricans Dewey = ***Carex dioica*** L.
Carex nigricans Kit. = ?
Carex nigricans f. *feminea* Kneuck. = ***Carex nigricans*** C.A.Mey.
Carex nigritella Drejer = ***Carex stylosa*** C.A.Mey.
Carex nigrolutea Gaudin = ***Carex flacca*** Schreb. subsp. ***flacca***
Carex nigrolutea Gaudin = ***Carex flacca*** Schreb. subsp. ***flacca***
Carex nigromarginata var. *elliptica* (Boott) Gleason = ***Carex peckii*** Howe
Carex nigromarginata var. *floridana* (Schwein.) Kük. = ***Carex floridana*** Schwein.
Carex nigromarginata f. *minor* Boeck. = ***Carex albicans*** var. ***australis*** (L.H.Bailey) Rettig
Carex nigromarginata var. *minor* (Boott) Gleason = ***Carex albicans*** var. ***emmonsii*** (Dewey ex Torr.) Rettig
Carex nigromarginata var. *muehlenbergii* (A.Gray ex Torr.) Gleason = ***Carex albicans*** Willd. ex Spreng. var. ***albicans***
Carex nigromarginata var. *subdigyna* Boeck. = ***Carex nigromarginata*** Schwein.
Carex niigatensis Koidz. = ***Carex foliosissima*** F.Schmidt var. ***foliosissima***
Carex × *nikaii* T.Koyama = ?
Carex niko-montana Akiyama = ***Carex sachalinensis*** var. ***longiuscula*** Ohwi
Carex nikoensis Franch. & Sav. = ***Carex satsumensis*** Franch. & Sav
Carex nilagirica Hochst. ex Steud. = ***Carex filicina*** Nees
Carex ninagongensis (Kük.) Nelmes ex Robyns & Tournay = ***Carex petitiana*** A.Rich.
Carex nipponica Franch. = ***Carex maackii*** Maxim.
Carex nipposinica Ohwi = ***Carex brownii*** Tuck. subsp. ***brownii***
Carex nitens Phil. = ?
Carex nitida Host = ***Carex liparocarpos*** Gaudin subsp. ***liparocarpos***
Carex nitida Hoppe = ***Carex mucronata*** All.
Carex nitida var. *aspera* (Boeck.) Kük. = ***Carex turkestanica*** Regel subsp. ***turkestanica***
Carex nitida var. *conglobata* Nyman = ***Carex supina*** Willd. ex Wahlenb. var. ***supina***
Carex nitida var. *supina* (Willd. ex Wahlenb.) Fiori = ***Carex supina*** Willd. ex Wahlenb.
Carex nitida var. *supiniformis* Kük. = ***Carex liparocarpos*** Gaudin subsp. ***liparocarpos***
Carex nivalis Cham. ex Steud. = ***Carex micropoda*** C.A.Mey.
Carex nivalis f. *cinnamomea* (Boott) Kük. = ***Carex nivalis*** Boott
Carex nivalis f. *griffithii* (Boott) Kük. = ***Carex nivalis*** Boott
Carex nivalis f. *luteobrunnea* Kük. = ***Carex nivalis*** Boott
Carex nivicola G.Don ex Boott = ***Carex vaginata*** Tausch var. ***vaginata***
Carex nobilis Boott = ***Carex myosurus*** Nees
Carex nordmannii A.Kern. ex Palla = ***Carex filiformis*** L.
Carex normalis var. *perlonga* (Fernald) Burnham = ***Carex normalis*** Mack.
Carex normalis f. *perlonga* (Fernald) Fernald = ***Carex normalis*** Mack.
Carex normalis var. *tincta* (Fernald) House = ***Carex normalis*** Mack.
Carex norvegica Willd. ex Schkuhr = ***Carex mackenziei*** V.I.Krecz.
Carex norvegica subsp. *conicorostrata* Kalela = ***Carex norvegica*** Retz. subsp. ***norvegica***
Carex norvegica subsp. *glareoides* Jörg. = ***Carex norvegica*** Retz. subsp. ***norvegica***
Carex norvegica subsp. *inferalpina* (Wahlenb.) Hultén = ***Carex media*** R.Br.
Carex norvegica var. *inferalpina* (Wahlenb.) Boivin = ***Carex media*** R.Br.
Carex norvegica var. *inserrulata* (Kalela) Raymond = ***Carex norvegica*** Retz.
Carex norvegica subsp. *inserrulata* Kalela = ***Carex norvegica*** Retz. subsp. ***norvegica***
Carex norvegica var. *isostachya* Norman = ***Carex norvegica*** Retz. subsp. ***norvegica***
Carex norvegica subsp. *stevenii* (T.Holm) D.F.Murray = ***Carex stevenii*** (T.Holm) Kalela
Carex norvegica var. *stevenii* (T.Holm) Dorn = ***Carex stevenii*** (T.Holm) Kalela
Carex notoleia Nees = ***Carex foliosa*** D.Don ex Tilloch & Taylor
Carex nova var. *erxlebenii* Kelso = ***Carex nova*** L.H.Bailey
Carex nova var. *pelocarpa* (F.J.Herm.) Dorn = ***Carex pelocarpa*** F.J.Herm.
Carex novae-angliae var. *collecta* (Dewey) Dewey = ***Carex albicans*** Willd. ex Spreng. var. ***albicans***
Carex novae-angliae var. *deflexa* (Hornem.) L.H.Bailey = ***Carex deflexa*** Hornem.
Carex novae-angliae var. *emmonsii* (Dewey ex Torr.) Carey = ***Carex albicans*** var. ***emmonsii*** (Dewey ex Torr.) Rettig
Carex novae-angliae var. *rossii* (Boott) L.H.Bailey = ***Carex rossii*** Boott
Carex novae-selandiae Boeck. = ***Carex pseudocyperus*** L. var. ***pseudocyperus***
Carex novae-zelandiae Petrie = ***Carex rubicunda*** Petrie
Carex novograblenovii Kom. = ***Carex williamsii*** Britton
Carex nubicola Mack. = ***Carex haydeniana*** Olney
Carex nubigena var. *albata* (Boott ex Franch.) Kük. = ***Carex albata*** Boott ex Franch.
Carex nubigena subsp. *albata* (Boott ex Franch.) T.Koyama = ***Carex albata*** Boott ex Franch.
Carex nubigena var. *fallax* (Steud.) C.B.Clarke = ***Carex nubigena*** D.Don ex Tilloch & Taylor subsp. ***nubigena***
Carex nubigena f. *laxiuscula* Kük. = ***Carex nubigena*** D.Don ex Tilloch & Taylor subsp. ***nubigena***
Carex nuda Lam. = ***Carex leporina*** L.
Carex nudata var. *angustifolia* (Boott) L.H.Bailey = ***Carex nudata*** Boott
Carex nudata var. *anomala* L.H.Bailey = ***Carex nudata*** Boott
Carex nudata f. *firmior* Kük. = ***Carex nudata*** Boott
Carex nudata f. *sessiliflora* Kük. = ***Carex nudata*** Boott
Carex nudata var. *versuta* Piper = ***Carex nudata*** Boott

Carex nuptialis Boott = ***Carex wahuensis*** C.A.Mey. subsp. ***wahuensis***
Carex nutans Raeusch. = ***Carex cespitosa*** L. var. ***cespitosa***
Carex nutans Host = ***Carex melanostachya*** M.Bieb. ex Willd.
Carex nutans var. *japonica* Franch. & Sav. = ***Carex melanostachya*** M.Bieb. ex Willd.
Carex nutans var. *major* Boeck. = ***Carex melanostachya*** M.Bieb. ex Willd.
Carex nutans var. *microcarpa* Boeck. = ***Carex heterostachya*** Bunge
Carex nutans var. *platyrhyncha* (Franch. & Sav.) Kük. = ***Carex pumila*** Thunb.
Carex nutans var. *pumila* (Thunb.) Boeck. = ***Carex pumila*** Thunb.
Carex nutans f. *sulcata* (Schur) Kük. = ***Carex melanostachya*** M.Bieb. ex Willd.
Carex nuttallii Dewey = ***Carex douglasii*** Boott
Carex nyasensis C.B.Clarke = ***Carex echinochloe*** subsp. ***nyasensis*** (C.B.Clarke) Lye
Carex oahuensis Hillebr. = ***Carex wahuensis*** C.A.Mey. subsp. ***wahuensis***
Carex oahuensis var. *angustior* Kük. = ***Carex wahuensis*** C.A.Mey. subsp. ***wahuensis***
Carex oahuensis var. *bongardii* Franch. & Sav. = ***Carex wahuensis*** C.A.Mey. subsp. ***wahuensis***
Carex oahuensis var. *boottiana* (Hook. & Arn.) Kük. = ***Carex boottiana*** Hook. & Arn.
Carex oahuensis f. *bracteosa* C.B.Clarke = ***Carex wahuensis*** C.A.Mey. subsp. ***wahuensis***
Carex oahuensis var. *meyeri* Franch. & Sav. = ***Carex wahuensis*** C.A.Mey. subsp. ***wahuensis***
Carex oahuensis var. *robusta* Franch. & Sav. = ***Carex wahuensis*** C.A.Mey. subsp. ***wahuensis***
Carex oakesiana Torr. = ***Carex oligosperma*** Michx.
Carex oaxacana L.H.Bailey = ***Carex anisostachys*** Liebm.
Carex obesa Schleich. ex Kunth = ***Carex cespitosa*** L. var. ***cespitosa***
Carex obesa All. = ***Carex nigra*** subsp. ***alpina*** (Gaudin) Lemke
Carex obesa var. *aspera* Boeck. = ***Carex turkestanica*** Regel subsp. ***turkestanica***
Carex obesa var. *minor* Boeck. = ***Carex supina*** Willd. ex Wahlenb. var. ***supina***
Carex obesa var. *monostachya* Boeck. = ***Carex obtusata*** Lilj.
Carex obliqua Turcz. = ***Carex pediformis*** C.A.Mey. subsp. ***pediformis***
Carex oblita Steud. = ***Carex venusta*** Dewey
Carex oblonga Nelmes = ***Carex cruciata*** var. ***rafflesiana*** (Boott) Noot.
Carex oblonga J.F.Gmel. = ?
Carex obovata Goldb. ex Ledeb. = ***Carex pediformis*** C.A.Mey. subsp. ***pediformis***
Carex obovoidea Cronquist = ***Carex cusickii*** Mack.
Carex obscuriceps Kük. = ***Carex pamirica*** (O.Fedtsch.) O.Fedtsch. & B.Fedtsch. subsp. ***pamirica***
Carex obscuriceps var. *pamirica* (O.Fedtsch.) Kük. = ***Carex pamirica*** (O.Fedtsch.) O.Fedtsch. & B.Fedtsch.
Carex obtusa Kit. = ?
Carex obtusangula Salzm. ex Boott = ***Carex echinata*** Murray subsp. ***echinata***
Carex obtusangula Kunth = ***Carex flacca*** Schreb. subsp. ***flacca***
Carex obtusangula Salzm. ex Nyman = ***Carex hispida*** Willd. ex Schkuhr
Carex obtusangula Ehrh. = ***Carex rostrata*** Stokes var. ***rostrata***
Carex obtusata subf. *bracteata* Peterm. = ***Carex obtusata*** Lilj.
Carex obtusata f. *bracteata* Neuman = ***Carex obtusata*** Lilj.
Carex obtusata subf. *capitata* Peterm. = ***Carex obtusata*** Lilj.
Carex obtusata subf. *cuspidata* Peterm. = ***Carex obtusata*** Lilj.
Carex obtusata subf. *oblonga* Peterm. = ***Carex obtusata*** Lilj.
Carex obtusata f. *spicata* Kük. = ***Carex obtusata*** Lilj.
Carex obtusata var. *spicata* Asch. = ***Carex supina*** Willd. ex Wahlenb. var. ***supina***
Carex obtusata var. *supina* (Willd. ex Wahlenb.) Garcke = ***Carex supina*** Willd. ex Wahlenb.
Carex occultans (Franch.) V.I.Krecz. = ***Carex lasiocarpa*** var. ***occultans*** (Franch.) Kük.
Carex ochrolepis Franch. = ***Carex jackiana*** Boott subsp. ***jackiana***
Carex ochroleuca Cham. ex Steud. = ***Carex saxatilis*** L.
Carex ochrostachya Phil. = ***Carex banksii*** Boott var. ***banksii***
Carex odontolepis Phil. = ***Carex banksii*** var. ***odontolepis*** (Phil.) Kük.
Carex odontostoma Kük. = ***Carex chrysolepis*** Franch. & Sav. var. ***chrysolepis***
Carex odontostoma var. *glabrior* Ohwi = ***Carex chrysolepis*** var. ***glabrior*** (Ohwi) Ohwi
Carex odontostoma var. *variegata* Kük. = ***Carex chrysolepis*** Franch. & Sav. var. ***chrysolepis***
Carex oederi Willk. = ***Carex flava*** var. ***nevadensis*** (Boiss. & Reut.) Briq.
Carex oederi Retz. = ***Carex pilulifera*** L. subsp. ***pilulifera***
Carex oederi Ehrh. = ***Carex viridula*** Michx. subsp. ***viridula***
Carex oederi f. *alpestris* Asch. & Graebn. = ***Carex viridula*** Michx. subsp. ***viridula***
Carex oederi f. *argillacea* (Townson) Kük. = ***Carex pilulifera*** L. subsp. ***pilulifera***
Carex oederi var. *bergrothii* (Palmgr.) Hedrén & Lassen = ***Carex pilulifera*** L. subsp. ***pilulifera***
Carex oederi var. *brevirostris* Asch. & Graebn. = ***Carex pilulifera*** L. subsp. ***pilulifera***
Carex oederi f. *canaliculata* Callmé = ***Carex viridula*** Michx. subsp. ***viridula***
Carex oederi var. *cataractae* (R.Br.) Kük. = ***Carex cataractae*** R.Br.
Carex oederi var. *congesta* Neuman = ***Carex pilulifera*** L. subsp. ***pilulifera***
Carex oederi f. *cyperoides* T.Marsson = ***Carex viridula*** Michx. subsp. ***viridula***
Carex oederi subsp. *demissa* (Hornem.) Vicioso = ***Carex viridula*** subsp. ***oedocarpa*** (Andersson) B.Schmid
Carex oederi f. *elatior* (Schltr.) Andersson = ***Carex viridula*** var. ***elatior*** (Schltr.) Crins
Carex oederi var. *exigua* Schur = ***Carex viridula*** Michx. subsp. ***viridula***
Carex oederi var. *fallax* Heimerl = ***Carex pilulifera*** L. subsp. ***pilulifera***
Carex oederi subsp. *fennica* Palmgr. = ***Carex viridula*** Michx. subsp. ***viridula***
Carex oederi var. *glomerata* Callmé = ***Carex pilulifera*** L. subsp. ***pilulifera***
Carex oederi f. *graminis* (L.H.Bailey) Kük. = ***Carex flava*** L. subsp. ***flava***
Carex oederi f. *intermedia* Dudley = ***Carex viridula*** Michx. subsp. ***viridula***
Carex oederi var. *major* Boeck. = ***Carex viridula*** var. ***elatior*** (Schltr.) Crins
Carex oederi var. *microcarpa* Font Quer = ***Carex flava*** L. subsp. ***flava***

Carex oederi var. *minor* Boeck. = ***Carex pilulifera*** L. subsp. ***pilulifera***
Carex oederi subsp. *nevadensis* (Boiss. & Reut.) Nyman = ***Carex flava*** var. ***nevadensis*** (Boiss. & Reut.) Briq.
Carex oederi var. *nevadensis* (Boiss. & Reut.) K.Richt. = ***Carex flava*** var. ***nevadensis*** (Boiss. & Reut.) Briq.
Carex oederi var. *oedocarpa* (Andersson) Nyman = ***Carex pilulifera*** L. subsp. ***pilulifera***
Carex oederi subsp. *oedocarpa* Andersson = ***Carex viridula*** subsp. ***oedocarpa*** (Andersson) B.Schmid
Carex oederi f. *oedocarpa* (Andersson) Kük. = ***Carex viridula*** subsp. ***oedocarpa*** (Andersson) B.Schmid
Carex oederi f. *pinguis* Neuman = ***Carex pilulifera*** L. subsp. ***pilulifera***
Carex oederi subsp. *pulchella* Lönnr. = ***Carex viridula*** Michx. subsp. ***viridula***
Carex oederi var. *pulchella* (Lönnr.) Hedrén & Lassen = ***Carex viridula*** Michx. subsp. ***viridula***
Carex oederi f. *pumila* Zahn = ***Carex pilulifera*** L. subsp. ***pilulifera***
Carex oederi var. *pumila* (Coss. & Germ.) Fernald = ***Carex viridula*** Michx. subsp. ***viridula***
Carex oederi f. *pygmaea* Andersson = ***Carex pilulifera*** L. subsp. ***pilulifera***
Carex oederi f. *recterostrata* (L.H.Bailey) Kük. = ***Carex viridula*** Michx. subsp. ***viridula***
Carex oederi var. *recterostrata* (L.H.Bailey) Dorn = ***Carex viridula*** Michx. subsp. ***viridula***
Carex oederi var. *rousseauiana* Vict. = ***Carex viridula*** Michx. subsp. ***viridula***
Carex oederi var. *serotina* (Mérat) Nyman = ***Carex viridula*** Michx. subsp. ***viridula***
Carex oederi f. *subglobosa* (Miel.) Storch = ***Carex viridula*** Michx. subsp. ***viridula***
Carex oederi var. *subglobosa* (Miel.) Nyman = ***Carex viridula*** Michx. subsp. ***viridula***
Carex oederi f. *tenuis* Zapal. = ***Carex viridula*** Michx. subsp. ***viridula***
Carex oederi f. *thalassica* Asch. & Graebn. = ***Carex viridula*** Michx. subsp. ***viridula***
Carex oederi f. *tularia* Callmé = ***Carex pilulifera*** L. subsp. ***pilulifera***
Carex oederi f. *virescens* Callmé = ***Carex pilulifera*** L. subsp. ***pilulifera***
Carex oederi var. *viridula* (Michx.) Kük. = ***Carex viridula*** Michx.
Carex oederi subsp. *viridula* (Michx.) Hultén = ***Carex viridula*** Michx.
Carex oederoides Tatew. & Akiyama = ***Carex viridula*** Michx. subsp. ***viridula***
Carex oedocarpa Andersson = ***Carex pilulifera*** L. subsp. ***pilulifera***
Carex oerstedii Liebm. = ***Carex polystachya*** Sw. ex Wahlenb. var. ***polystachya***
Carex ogawae Akiyama = ***Carex reinii*** Franch. & Sav.
Carex ohiotica Hochst. & Steud. = ***Carex shortiana*** Dewey & Torr.
Carex ohmuelleriana O.Lang = ***Carex arenaria*** L.
Carex ohwii Masam. = ***Carex omiana*** var. ***yakushimana*** Ohwi
Carex okesiana Dewey = ***Carex oligosperma*** Michx.
Carex okesii Eaton = ***Carex oligosperma*** Michx.
Carex okuboi Franch. = ***Carex doenitzii*** var. ***okuboi*** (Franch.) Kük.
Carex okushirensis Akiyama = ***Carex conica*** Boott
Carex olbiensis var. *angustifolia* Rohlena = ***Carex distachya*** Desf. var. ***distachya***
Carex oldhamii C.B.Clarke = ***Carex rara*** Boott subsp. ***rara***
Carex olgae Regel = ***Carex acutiformis*** Ehrh.
Carex oligandra F.Muell. ex Boott = ***Carex buxbaumii*** Wahlenb.
Carex oligantha Phil. = ***Carex maritima*** Gunnerus
Carex oligantha Boott = ***Carex microglochin*** subsp. ***fuegina*** Kük.
Carex oligocarpa Muhl. ex Boott = ***Carex digitalis*** Willd. var. ***digitalis***
Carex oligocarpa Hornem. ex Boott = ***Carex supina*** Willd. ex Wahlenb. var. ***supina***
Carex oligocarpa var. *hitchcockiana* (Dewey) Boeck. = ***Carex hitchcockiana*** Dewey
Carex oligocarpa var. *major* Torr. = ***Carex oligocarpa*** Willd.
Carex oligocarpa var. *minor* Boott = ***Carex oligocarpa*** Willd.
Carex oligocarpa var. *sartwelliana* Dewey = ***Carex oligocarpa*** Willd.
Carex oligocarpa f. *subuniflora* (Steud.) Kük. = ***Carex oligocarpa*** Willd.
Carex oligocarpa f. *triflora* Peck = ***Carex oligocarpa*** Willd.
Carex oligocarpa var. *van-vleckii* Schwein. ex Dewey = ***Carex digitalis*** Willd. var. ***digitalis***
Carex oligogyna Less. ex Kunth = ***Carex globularis*** L.
Carex oligophylla Freyn = ***Carex cespitosa*** L. var. ***cespitosa***
Carex oligosperma var. *churchilliana* Raymond = ***Carex oligosperma*** Michx.
Carex oligosperma var. *tsuishikarensis* (Koidz. & Ohwi) B.Boivin = ***Carex oligosperma*** Michx.
Carex oligostachys Meinsh. ex Maxim. = ***Carex egena*** H.Lév. & Vaniot
Carex olivacea Liebm. = ***Carex anisostachys*** Liebm.
Carex olivacea var. *angustior* Kük. = ***Carex olivacea*** subsp. ***confertiflora*** (Boott) T.Koyama
Carex oliveri Boeck. = ***Carex nivalis*** Boott
Carex oliveriana J.Gay ex Boiss. = ***Carex illegitima*** Ces.
Carex olympica Mack. = ***Carex macloviana*** d'Urv. subsp. ***macloviana***
Carex olyssiponensis Steud. = ***Carex distachya*** Desf. var. ***distachya***
Carex omeiensis var. *multifascula* Q.S.Wang = ***Carex omeiensis*** Tang
Carex omskiana Meinsh. = ***Carex elata*** subsp. ***omskiana*** (Meinsh.) Jalas
Carex onoei var. *capituliformis* (Meinsh. ex Maxim.) Kitag. = ***Carex onoei*** Franch. & Sav.
Carex onoei subsp. *capituliformis* (Meinsh. ex Maxim.) T.V.Egorova = ***Carex onoei*** Franch. & Sav.
Carex onoei f. *hakonensis* (Franch. & Sav.) Kük. = ***Carex hakonensis*** Franch. & Sav.
Carex onoei var. *krameri* (Franch. & Sav.) Kük. = ***Carex hakonensis*** Franch. & Sav.
Carex onoei f. *latifolia* Kük. = ***Carex onoei*** Franch. & Sav.
Carex onoei var. *macrogyna* Tang = ***Carex onoei*** Franch. & Sav.
Carex ontakensis Franch. & Sav. = ***Carex capillacea*** Boott var. ***capillacea***
Carex ontakensis H.Lév. = ***Carex magellanica*** subsp. ***irrigua*** (Wahlenb.) Hultén
Carex onusta Mack. = ***Carex muehlenbergii*** var. ***enervis*** Boott
Carex oranensis Trab. = ***Carex distans*** L.
Carex orbelica Velen. = ***Carex curvula*** All. subsp. ***curvula***
Carex orbicularis var. *brachylepis* (Regel) Kük. = ***Carex orbicularis*** Boott subsp. ***orbicularis***
Carex orbicularis var. *taldycola* (Meinsh.) Kük. = ***Carex orbicularis*** Boott subsp. ***orbicularis***

Carex oreades C.A.Mey. = ***Carex macloviana*** d'Urv. subsp. ***macloviana***
Carex oregonensis Olney ex A.Gray = ***Carex halliana*** L.H.Bailey
Carex oreocharis var. *enantiomorpha* Kelso = ***Carex oreocharis*** T.Holm
Carex oreophila f. *minor* Kük. = ***Carex oreophila*** C.A.Mey.
Carex orientalis Meinsh. = ***Carex heterostachya*** Bunge
Carex ormantha (Fernald) Mack. = ***Carex echinata*** Murray subsp. ***echinata***
Carex ornithopoda var. *alpina* Gaudin = ***Carex ornithopoda*** Willd. subsp. ***ornithopoda***
Carex ornithopoda var. *alpina* Kük. = ***Carex ornithopoda*** Willd. subsp. ***ornithopoda***
Carex ornithopoda subsp. *bulgarica* (Velen.) Stoeva & E.D.Popova = ***Carex digitata*** L.
Carex ornithopoda var. *castanea* Murb. = ***Carex ornithopoda*** Willd. subsp. ***ornithopoda***
Carex ornithopoda subsp. *elongata* (Leyb.) Vierh. = ***Carex ornithopoda*** subsp. ***ornithopodioides*** (Hausm.) Nyman
Carex ornithopoda var. *elongata* (Leyb.) Asch. & Graebn. = ***Carex ornithopoda*** subsp. ***ornithopodioides*** (Hausm.) Nyman
Carex ornithopoda var. *hausmannii* Döll = ***Carex ornithopoda*** Willd. subsp. ***ornithopoda***
Carex ornithopoda f. *major* Bornm. ex Kük. = ***Carex ornithopoda*** Willd. subsp. ***ornithopoda***
Carex ornithopoda var. *ornithopodioides* (Hausm.) Garcke = ***Carex ornithopoda*** subsp. ***ornithopodioides*** (Hausm.) Nyman
Carex ornithopodioides Hausm. = ***Carex ornithopoda*** subsp. ***ornithopodioides*** (Hausm.) Nyman
Carex ornithopodioides f. *elongata* Leyb. = ***Carex ornithopoda*** subsp. ***ornithopodioides*** (Hausm.) Nyman
Carex ornithopodioides f. *supina* Leyb. = ***Carex ornithopoda*** subsp. ***ornithopodioides*** (Hausm.) Nyman
Carex ornithorhyncha Boott = ***Carex crus-corvi*** Shuttlew. ex Kunze
Carex orsiniana Ten. = ***Carex muricata*** L.
Carex ortegae Phil. = ***Carex caduca*** Boott
Carex orthocaula V.I.Krecz. = ***Carex maritima*** Gunnerus
Carex orthostachys C.A.Mey. = ***Carex atherodes*** Spreng.
Carex orthostachys Trevir. ex Nyman = ***Carex hirta*** L.
Carex orthostachys var. *drymophila* (Turcz.) Maxim. = ***Carex drymophila*** Turcz.
Carex orthostachys var. *spuria* Y.L.Chang & Y.L.Yang = ***Carex atherodes*** Spreng.
Carex orthostemon var. *cupulifera* Hayata = ***Carex breviculmis*** var. ***cupulifera*** (Hayata) Y.C.Tang & S.Yun Liang
Carex oshimensis f. *variegata* Hid.Takah. = ***Carex oshimensis*** Nakai
Carex ostrobottnica (Almq.) Saxén = ***Carex recta*** Boott
Carex otrubae Podp. = ***Carex cuprina*** (I.Sándor ex Heuff.) Nendtv. ex A.Kern.
Carex otrubae f. *bracteata* (G.Mey.) Soó = ***Carex cuprina*** (I.Sándor ex Heuff.) Nendtv. ex A.Kern.
Carex otrubae f. *interrupta* (Peterm.) Soó = ***Carex cuprina*** (I.Sándor ex Heuff.) Nendtv. ex A.Kern.
Carex ouensanensis Ohwi = ***Carex cinerascens*** Kük.
Carex ovalis Gooden. = ***Carex leporina*** L.
Carex ovalis Willd. ex Kunth = ***Carex scoparia*** Willd. var. ***scoparia***
Carex ovalis var. *argyroglochin* (Hornem.) De Langhe & Lambinon = ***Carex leporina*** L.
Carex ovalis var. *minor* Brongn. = ***Carex macloviana*** d'Urv. subsp. ***macloviana***
Carex ovalis var. *subfestiva* (Lange) O.Bolòs & Vigo = ***Carex leporina*** L.
Carex ovata Rudge = ***Carex atratiformis*** Britton
Carex ovata C.A.Mey. = ***Carex curvula*** All. subsp. ***curvula***
Carex ovata Honck. = ***Carex norvegica*** Retz. subsp. ***norvegica***
Carex ovata Burm.f. = ***Fimbristylis ovata*** (Burm.f.) J.Kern
Carex owyheensis A.Nelson = ***Carex luzulina*** var. ***ablata*** (L.H.Bailey) F.J.Herm.
Carex oxyandra var. *globosa* Akiyama = ***Carex oxyandra*** (Franch. & Sav.) Kudô var. ***oxyandra***
Carex oxycarpa T.Holm = ***Carex angustata*** Boott
Carex oxylepis f. *glabra* Boeck. = ***Carex oxylepis*** Torr. & Hook. var. ***oxylepis***
Carex oxyleuca V.I.Krecz. = ***Carex atrofusca*** subsp. ***minor*** (Boott) T.Koyama
Carex oxyphylla Franch. = ***Carex jackiana*** Boott subsp. ***jackiana***
Carex pachinensis Hayata = ***Carex metallica*** H.Lév.
Carex pachyandra C.B.Clarke = ?
Carex pachycarpa Mack. = ***Carex multicostata*** Mack.
Carex pachyrrhiza Franch. = ***Carex setosa*** Boott var. ***setosa***
Carex pachystachya Cham. ex Steud. = ***Carex macloviana*** subsp. ***pachystachya*** (Cham. ex Steud.) Hultén
Carex pachystachya var. *gracilis* (Olney) Mack. = ***Carex macloviana*** subsp. ***subfusca*** (W.Boott) T.Koyama
Carex pachystachya var. *monds-coulteri* Kelso = ***Carex macloviana*** subsp. ***pachystachya*** (Cham. ex Steud.) Hultén
Carex pachystachya f. *monds-coulteri* (Kelso) F.J.Herm. = ***Carex macloviana*** subsp. ***pachystachya*** (Cham. ex Steud.) Hultén
Carex pachystoma T.Holm = ***Carex aquatilis*** var. ***dives*** (T.Holm) Kük.
Carex pacifica Drejer = ***Carex cespitosa*** L. var. ***cespitosa***
Carex ×*paczoskii* Zapal. = ?
Carex paddoensis Suksd. = ***Carex engelmannii*** L.H.Bailey
Carex pagesii H.Lév. = ?
Carex pairae subsp. *borealis* Hyl. = ***Carex muricata*** L.
Carex pairae f. *brevispicata* (Podp.) Soó = ***Carex echinata*** Murray subsp. ***echinata***
Carex pairae var. *capitata* H.Christ = ***Carex pairae*** F.W.Schultz
Carex pairae f. *chlorocarpa* (Podp.) Soó = ***Carex echinata*** Murray subsp. ***echinata***
Carex pairae var. *javanica* Nelmes = ?
Carex pairae subsp. *leersii* (Kneuck.) Soó & Jáv. = ***Carex divulsa*** subsp. ***leersii*** (Kneuck.) W.Koch
Carex pairae var. *leersii* (Kneuck.) Kük. = ***Carex divulsa*** subsp. ***leersii*** (Kneuck.) W.Koch
Carex pairae var. *monostachya* Asch. & Graebn. = ***Carex pairae*** F.W.Schultz
Carex pairae f. *remotiuscula* (Podp.) Soó = ***Carex echinata*** Murray subsp. ***echinata***
Carex pakistanica T.Koyama = ***Carex orbicularis*** Boott subsp. ***orbicularis***
Carex paleacea f. *erectiuscula* (Fernald) Fernald = ***Carex paleacea*** Schreb. ex Wahlenb.
Carex paleacea var. *lepagei* B.Boivin = ***Carex paleacea*** Schreb. ex Wahlenb.
Carex paleacea f. *neopaleacea* (Lepage) B.Boivin = ***Carex paleacea*** Schreb. ex Wahlenb.

Carex paleacea var. *transatlantica* Fernald = ***Carex paleacea*** Schreb. ex Wahlenb.
Carex paleaceoides Lepage = ***Carex paleacea*** Schreb. ex Wahlenb.
Carex paleata Boott = ***Carex berteroniana*** Steud.
Carex palentina Losa & P.Monts. = ***Carex liparocarpos*** Gaudin subsp. ***liparocarpos***
Carex ×*palezieuxii* Kneuck. = ?
Carex pallens C.P.Wang = ***Carex breviculmis*** var. ***pluricostata*** Kük.
Carex pallens (Fristedt) Harmaja = ***Carex pallidula*** Harmaja
Carex pallescens f. *alpestris* Schur = ***Carex pallescens*** L. var. ***pallescens***
Carex pallescens var. *alpestris* Kohts = ***Carex pallescens*** L. var. ***pallescens***
Carex pallescens f. *brevibracteata* Neuman = ***Carex pallescens*** L. var. ***pallescens***
Carex pallescens f. *cylindrica* Peterm. = ***Carex pallescens*** L. var. ***pallescens***
Carex pallescens f. *elatior* Asch. & Graebn. = ***Carex pallescens*** L. var. ***pallescens***
Carex pallescens f. *glaberrima* K.Koch = ***Carex pallescens*** L. var. ***pallescens***
Carex pallescens var. *leucantha* (Schur) Asch. & Graebn. = ***Carex pallescens*** L. var. ***pallescens***
Carex pallescens var. *luxuriosa* Kük. = ***Carex pallescens*** L. var. ***pallescens***
Carex pallescens var. *neogaea* Fernald = ***Carex pallescens*** L. var. ***pallescens***
Carex pallescens f. *pygmaea* Lackow. = ***Carex pallescens*** L. var. ***pallescens***
Carex pallescens var. *subglabra* Beck = ***Carex pallescens*** L. var. ***pallescens***
Carex pallescens var. *subsilvatica* Kük. = ***Carex pallescens*** L. var. ***pallescens***
Carex pallescens var. *undulata* (Kunze) J.Carey = ***Carex pallescens*** L.
Carex pallida C.A.Mey. = ***Carex accrescens*** Ohwi
Carex pallida Salisb. = ***Carex pallescens*** L.
Carex pallida var. *angustifolia* Y.L.Chang = ***Carex accrescens*** Ohwi
Carex pallida f. *angustifolia* (Y.L.Chang) Y.L.Chou = ***Carex accrescens*** Ohwi
Carex pallida var. *lithophiloides* A.E.Kozhevn. = ***Carex accrescens*** Ohwi
Carex pallida var. *papillosa* Kük. = ***Carex accrescens*** Ohwi
Carex pallidior Degl. = ***Carex punctata*** Gaudin var. ***punctata***
Carex paludicola (Ohwi) Ohwi = ***Carex aphyllopus*** Kük.
Carex paludosa Gooden. = ***Carex acutiformis*** Ehrh.
Carex paludosa var. *aemulans* (Liebm. & Drejer) Nyman = ***Carex brunnea*** Thunb. subsp. ***brunnea***
Carex paludosa f. *depauperata* Lange = ***Carex acutiformis*** Ehrh.
Carex paludosa subsp. *kochiana* (DC.) Arcang. = ***Carex acutiformis*** Ehrh.
Carex paludosa var. *longiglumis* St.-Lag. = ***Carex acutiformis*** Ehrh.
Carex paludosa var. *spadicea* Nyman = ***Carex acutiformis*** Ehrh.
Carex palustris Suter = ***Carex acutiformis*** Ehrh.
Carex pamirensis C.B.Clarke = ***Carex pamirica*** (O. Fedtsch.) O.Fedtsch. & B.Fedtsch. subsp. ***pamirica***
Carex pamirensis var. *angustispicata* Y.C.Yang = ***Carex pamirica*** (O.Fedtsch.) O.Fedtsch. & B.Fedtsch. subsp. ***pamirica***
Carex pamirensis subsp. *dichroa* Malyschev = ***Carex pamirica*** subsp. ***dichroa*** (Malyschev) T.V.Egorova
Carex pampae Kalela = ***Carex subantarctica*** Speg.
Carex panda C.B.Clarke = ***Carex aquatilis*** var. ***dives*** (T.Holm) Kük.
Carex pandanus Ohwi = ***Carex graeffeana*** Boeck.
Carex panicea Carey = ***Carex meadii*** Dewey
Carex panicea f. *binata* Lackow. ex Kük. = ***Carex panicea*** L.
Carex panicea f. *conferta* N.H.Nilsson ex Neuman = ***Carex panicea*** L.
Carex panicea subsp. *dalmatica* Degen & Lengyel = ***Carex panicea*** L.
Carex panicea f. *ferruginea* Neuman = ***Carex panicea*** L.
Carex panicea f. *gracilis* Lange = ***Carex panicea*** L.
Carex panicea f. *humilis* Meinsh. = ***Carex panicea*** L.
Carex panicea var. *intermedia* Meinsh. = ***Carex panicea*** L.
Carex panicea f. *latifolia* Waisb. = ***Carex panicea*** L.
Carex panicea f. *longipedunculata* Asch. & Graebn. = ***Carex panicea*** L.
Carex panicea f. *melanocarpa* Waisb. = ***Carex panicea*** L.
Carex panicea f. *microcarpa* Sond. = ***Carex panicea*** L.
Carex panicea f. *nigra* Kuntze = ***Carex panicea*** L.
Carex panicea f. *pallida* A.Blytt = ***Carex panicea*** L.
Carex panicea var. *pauciflora* Wahlenb. = ***Carex vaginata*** Tausch var. ***vaginata***
Carex panicea subsp. *pelia* (O.Lang) Nyman = ***Carex panicea*** L.
Carex panicea var. *pelia* (O.Lang) K.Richt. = ***Carex panicea*** L.
Carex panicea var. *praestabilis* Waisb. = ***Carex panicea*** L.
Carex panicea f. *refracta* Peterm. = ***Carex panicea*** L.
Carex panicea f. *robusta* Warnst. = ***Carex panicea*** L.
Carex panicea var. *sparsiflora* Wahlenb. = ***Carex vaginata*** Tausch var. ***vaginata***
Carex panicea var. *sublivida* Hartm. ex Andersson = ***Carex panicea*** L.
Carex panicea f. *tumidula* Laest. = ***Carex panicea*** L.
Carex panicea subsp. *vaginata* (Tausch) Rouy = ***Carex vaginata*** Tausch
Carex paniculata var. *appressa* (R.Br.) Cheeseman = ***Carex appressa*** R.Br.
Carex paniculata f. *brevis* Asch. & Graebn. = ***Carex paniculata*** L. subsp. ***paniculata***
Carex paniculata var. *decomposita* (Muhl.) Dewey = ***Carex decomposita*** Muhl.
Carex paniculata var. *diandra* (Schrank) Fiori = ***Carex diandra*** Schrank
Carex paniculata f. *elongata* Èelak. = ***Carex paniculata*** L. subsp. ***paniculata***
Carex paniculata var. *favratii* (H.Christ) Kük. = ***Carex paniculata*** subsp. ***szovitsii*** (V.I.Krecz.) Ö.Nilsson
Carex paniculata f. *gracilis* Asch. & Graebn. = ***Carex paniculata*** L. subsp. ***paniculata***
Carex paniculata f. *longibracteata* Zapal. = ***Carex paniculata*** L. subsp. ***paniculata***
Carex paniculata var. *ludibunda* (J.Gay) K.Richt. = ***Carex* × *ludibunda*** J.Gay
Carex paniculata var. *lusitanica* (Willd.) Nyman = ***Carex paniculata*** subsp. ***lusitanica*** (Willd.) Maire
Carex paniculata var. *minor* Retz. = ***Carex diandra*** Schrank
Carex paniculata f. *pallida* Lange = ***Carex paniculata*** L. subsp. ***paniculata***
Carex paniculata var. *paradoxa* (Rchb.) Fiori = ***Carex appropinquata*** Schumach.
Carex paniculata var. *rigida* A.Blytt ex Benner = ***Carex paniculata*** L. subsp. ***paniculata***
Carex paniculata f. *rigidior* Kük. = ***Carex paniculata*** L. subsp. ***paniculata***
Carex paniculata var. *rotae* (De Not.) Nyman = ***Carex paniculata*** L. subsp. ***paniculata***

Carex paniculata var. *secta* (Boott) Cheeseman = ***Carex secta*** Boott
Carex paniculata f. *simplex* Peterm. = ***Carex paniculata*** L. subsp. ***paniculata***
Carex paniculata var. *simplicior* Andersson = ***Carex paniculata*** L. subsp. ***paniculata***
Carex paniculata var. *squarrosa* Èelak. = ***Carex paniculata*** L. subsp. ***paniculata***
Carex paniculata var. *subdiaphana* F.Muell. = ***Carex appressa*** R.Br. var. ***appressa***
Carex paniculata var. *tenuior* Grantzow = ***Carex paniculata*** L. subsp. ***paniculata***
Carex paniculata var. *virgata* (Sol. ex Boott) Cheeseman = ***Carex appressa*** var. ***virgata*** (Sol. ex Boott) Kük.
Carex paniculigera Nakai = ***Carex planiculmis*** Kom.
Carex ×*pannewitziana* Figert = ***Carex* × *involuta*** (Bab.) Syme
Carex papillaticulmis Ohwi = ***Carex insaniae*** var. ***papillaticulmis*** (Ohwi) Ohwi
Carex papillosa Nees = ***Carex bonariensis*** Desf. ex Poir. var. ***bonariensis***
Carex ×*paponii* Muret ex T.Durand & Pittier = ?
Carex papuana Nelmes = ***Carex lamprochlamys*** S.T. Blake
Carex papulosa f. *laxiflora* Kük. = ***Carex papulosa*** Boott
Carex papulosa f. *paupercula* Kük. = ***Carex papulosa*** Boott
Carex paradoxa Willd. = ***Carex appropinquata*** Schumach.
Carex paradoxa Benth. = ***Carex divisa*** Huds.
Carex paradoxa J.F.Gmel. = ?
Carex paradoxa f. *brachystachya* Schatz = ***Carex appropinquata*** Schumach.
Carex paradoxa var. *gracilis* Asch. & Graebn. = ***Carex appropinquata*** Schumach.
Carex paradoxa f. *interrupta* Peterm. = ***Carex ericetorum*** Pollich
Carex paradoxa var. *pseudoparadoxa* (S.Gibson) Nyman = ***Carex diandra*** Schrank
Carex paradoxa f. *ramosa* Asch. & Graebn. = ***Carex ericetorum*** Pollich
Carex paradoxa f. *sparsiflora* Lange = ***Carex appropinquata*** Schumach.
Carex paraguayensis Maury ex Micheli = ***Carex phalaroides*** var. ***paraguayensis*** (Maury ex Micheli) Kük.
Carex paralia V.I.Krecz. = ***Carex paleacea*** Schreb. ex Wahlenb.
Carex parallela f. *androgyna* O.Lang = ***Carex parallela*** (Laest.) Sommerf. subsp. ***parallela***
Carex parallela var. *pauciflora* O.Lang = ***Carex parallela*** (Laest.) Sommerf. subsp. ***parallela***
Carex parallela var. *subcaespitosa* O.Lang = ***Carex parallela*** (Laest.) Sommerf. subsp. ***parallela***
Carex parasitica Kunze = ***Carex monostachya*** A.Rich.
Carex parciflora Boott = ***Carex jackiana*** subsp. ***parciflora*** (Boott) Kük.
Carex parkeri Petrie = ***Carex lachenalii*** Schkuhr subsp. ***lachenalii***
Carex parryana var. *hallii* (Olney) Kük. = ***Carex hallii*** Olney
Carex parryana subsp. *hallii* (Olney) D.F.Murray = ***Carex hallii*** Olney
Carex parryana subsp. *idahoa* (L.H.Bailey) D.F.Murray = ***Carex idahoa*** L.H.Bailey
Carex parryana var. *statoni* M.E.Jones = ***Carex parryana*** Dewey var. ***parryana***
Carex parryana var. *unica* L.H.Bailey = ***Carex hallii*** Olney
Carex parvibracteata Nees = ***Carex psychrophila*** Nees
Carex parviflora Gaudin = ***Carex lachenalii*** Schkuhr subsp. ***lachenalii***
Carex parviflora C.A.Mey. ex Kunth = ***Carex microglochin*** subsp. ***fuegina*** Kük.
Carex patens Franch. = ***Carex japonica*** Thunb.
Carex patula Host = ***Carex flava*** L. subsp. ***flava***
Carex patula Schkuhr = ***Carex laevigata*** Sm.
Carex patula Huds. = ***Carex pauciflora*** Lightf.
Carex patula Scop. = ***Carex sylvatica*** Huds. subsp. ***sylvatica***
Carex paucicostata Mack. = ***Carex lenticularis*** var. ***impressa*** (L.H.Bailey) L.A.Standl.
Carex pauciflora Balb. ex Boott = ***Carex microglochin*** Wahlenb. subsp. ***microglochin***
Carex pauciflora f. *elatior* Kük. & J.Schust. = ***Carex pauciflora*** Lightf.
Carex pauciflora var. *microglochin* (Wahlenb.) Poir. = ***Carex microglochin*** Wahlenb.
Carex paucifructus Mack. = ***Carex mariposana*** L.H.Bailey ex Mack.
Carex paui Sennen = ***Carex sylvatica*** subsp. ***paui*** (Sennen) A.Bolòs & O.Bolòs
Carex paupercula Michx. = ***Carex magellanica*** subsp. ***irrigua*** (Wahlenb.) Hultén
Carex paupercula var. *brevisquama* Fernald = ***Carex* × *connectens*** E.Holmb.
Carex paupercula subsp. *irrigua* (Wahlenb.) Á.Löve & D.Löve = ***Carex magellanica*** subsp. ***irrigua*** (Wahlenb.) Hultén
Carex paupercula var. *irrigua* (Wahlenb.) Fernald = ***Carex magellanica*** subsp. ***irrigua*** (Wahlenb.) Hultén
Carex paupercula var. *pallens* Fernald = ***Carex magellanica*** subsp. ***irrigua*** (Wahlenb.) Hultén
Carex pauxilla V.I.Krecz. = ***Carex podocarpa*** R.Br.
Carex pauzhetica A.E.Kozhevn. = ***Carex oxyandra*** var. ***pauzhetica*** (A.E.Kozhevn.) A.E.Kozhevn.
Carex peckii f. *elongata* Lepage = ***Carex peckii*** Howe
Carex pedata Wahlenb. = ***Carex glacialis*** Mack.
Carex pedata L. = ***Carex ornithopoda*** Willd. subsp. ***ornithopoda***
Carex pedata All. = ***Carex ornithopoda*** Willd. subsp. ***ornithopoda***
Carex pedicellata (Dewey) Britton = ***Carex austrina*** Mack.
Carex pedicellata Phil. = ***Carex decidua*** Boott
Carex pedicellata var. *wheeleri* (L.H.Bailey) Britton = ***Carex austrina*** Mack.
Carex pediformis var. *caespitosa* F.Schmidt = ***Carex umbrosa*** subsp. ***sabynensis*** (Less. ex Kunth) Kük.
Carex pediformis var. *floribunda* Korsh. = ***Carex lanceolata*** Boott var. ***lanceolata***
Carex pediformis var. *genuina* Maxim. = ***Carex pediformis*** C.A.Mey.
Carex pediformis f. *leopolitana* Kük. = ***Carex pediformis*** C.A.Mey. subsp. ***pediformis***
Carex pediformis var. *macrosandra* (Franch.) C.B.Clarke = ***Carex macrosandra*** (Franch.) V.I.Krecz.
Carex pediformis var. *macroura* (Meinsh.) Kük. = ***Carex pediformis*** C.A.Mey. subsp. ***pediformis***
Carex pediformis subsp. *macroura* (Meinsh.) Podp. = ***Carex pediformis*** C.A.Mey. subsp. ***pediformis***
Carex pediformis var. *pedunculata* Maxim. = ***Carex pediformis*** C.A.Mey. subsp. ***pediformis***
Carex pediformis subsp. *pododactyla* Norman = ***Carex pediformis*** C.A.Mey. subsp. ***pediformis***
Carex pediformis var. *reventa* (V.I.Krecz.) Vorosch. = ***Carex reventa*** V.I.Krecz.
Carex pediformis subsp. *reventa* (V.I.Krecz.) Malyschev = ***Carex reventa*** V.I.Krecz.

Carex pediformis var. *rhizina* (Blytt ex Lindblom) Kük. = ***Carex pediformis*** subsp. ***rhizodes*** (Blytt ex Boott) H.Lindb.
Carex pediformis var. *rostrata* F.Schmidt = ***Carex caryophyllea*** Latourr. var. ***caryophyllea***
Carex pediformis var. *rostrata* Maxim. = ***Carex umbrosa*** var. ***rostrata*** (Maxim.) Govaerts
Carex peduncularis Wall. ex Boott = ***Carex speciosa*** Kunth subsp. ***speciosa***
Carex pedunculata Willd. ex Roxb. = ***Carex dianae*** Steud.
Carex pedunculifera Kom. = ***Carex lyngbyei*** Hornem.
Carex peiana F.T.Wang & Tang = ***Carex myosurus*** Nees
Carex pelia O.Lang = ***Carex panicea*** L.
Carex pellucida Turcz. ex Boott = ***Carex pediformis*** C.A.Mey. subsp. ***pediformis***
Carex pendula Geners. = ***Carex capillaris*** L. subsp. ***capillaris***
Carex pendula Schreb. = ***Carex flacca*** Schreb. subsp. ***flacca***
Carex pendula Moench = ***Carex vesicaria*** L.
Carex pendula var. *angustifolia* (Albov) Kük. = ***Carex pendula*** Huds.
Carex pendula var. *myosuroides* Boott = ***Carex pendula*** Huds.
Carex pendula f. *pallida* Asch. & Graebn. = ***Carex pendula*** Huds.
Carex pendulina Lojac. = [13 SIC]
Carex peniculacea H.Lév. & Vaniot = ***Carex otaruensis*** Franch.
Carex pennsylvanica var. *pinicola* M.E.Jones = ***Carex pensylvanica*** Lam.
Carex pennsylvanica f. *separans* (Peck) Kük. = ***Carex pensylvanica*** Lam.
Carex pensylvanica Torr. = ***Carex albicans*** Willd. ex Spreng. var. ***albicans***
Carex pensylvanica var. *amblyolepis* (Trautv. & C.A. Mey.) Kük. = ***Carex van-heurckii*** Müll.Arg. subsp. ***van-heurckii***
Carex pensylvanica f. *androgyna* C.F.Wheeler ex F.J.Herm. = ***Carex pensylvanica*** Lam.
Carex pensylvanica var. *angustifolia* Peck = ***Carex pensylvanica*** Lam.
Carex pensylvanica f. *bracteata* Peck = ***Carex pensylvanica*** Lam.
Carex pensylvanica var. *digyna* Boeck. = ***Carex inops*** subsp. ***heliophila*** (Mack.) Crins
Carex pensylvanica var. *distans* Peck = ***Carex lucorum*** Willd. var. ***lucorum***
Carex pensylvanica var. *glumabunda* Peck = ***Carex pensylvanica*** Lam.
Carex pensylvanica f. *glumabunda* (Peck) Kük. = ***Carex pensylvanica*** Lam.
Carex pensylvanica var. *gracilifolia* Peck = ***Carex pensylvanica*** Lam.
Carex pensylvanica f. *gracilifolia* (Peck) Kük. = ***Carex pensylvanica*** Lam.
Carex pensylvanica subsp. *heliophila* (Mack.) W.A. Weber = ***Carex inops*** subsp. ***heliophila*** (Mack.) Crins
Carex pensylvanica f. *latifolia* Kük. = ***Carex pensylvanica*** Lam.
Carex pensylvanica var. *livoniensis* Farw. = ***Carex pensylvanica*** Lam.
Carex pensylvanica var. *lucorum* (Willd.) Boeck. = ***Carex lucorum*** Willd.
Carex pensylvanica var. *marginata* Dewey = ***Carex pensylvanica*** Lam.
Carex pensylvanica var. *muehlenbergii* A.Gray ex Torr. = ***Carex albicans*** Willd. ex Spreng. var. ***albicans***
Carex pensylvanica f. *paleacea* Peck = ***Carex pensylvanica*** Lam.
Carex pensylvanica var. *separans* Peck = ***Carex lucorum*** Willd. var. ***lucorum***
Carex pensylvanica var. *vespertina* L.H.Bailey = ***Carex inops*** L.H.Bailey subsp. ***inops***
Carex pentacarpa Boeck. = ***Carex cruciata*** var. ***rafflesiana*** (Boott) Noot.
Carex ×*peraffinis* Appel = ?
Carex peraltae Phil. = ***Carex atropicta*** Steud.
Carex perangusta Ohwi = ***Carex breviculmis*** var. ***discoidea*** (Boott) Boott
Carex perciliata (Kük.) Nelmes = ***Carex breviculmis*** var. ***perciliata*** Kük.
Carex perennis Krock. ex Steud. = ***Carex montana*** L.
Carex perfusca V.I.Krecz. = ***Carex atrata*** subsp. ***aterrima*** (Hoppe) Èelak.
Carex perfusca var. *japonoalpina* (T.Koyama) Kitag. = ***Carex atrata*** L. subsp. ***atrata***
Carex pergrandis V.I.Krecz. & Luchnik = ***Carex atherodes*** Spreng.
Carex periculosa Honda = ***Carex heterolepis*** Bunge
Carex perileia S.T.Blake = ***Carex echinata*** Murray subsp. ***echinata***
Carex perpusilla Meinsh. ex Krecz. = ***Carex onoei*** Franch. & Sav.
Carex perrieri H.Lév. = ***Carex hirtigluma*** C.B.Clarke
Carex persica Nelmes = ***Carex divulsa*** Stokes subsp. ***divulsa***
Carex persistens Ohwi = ***Carex dispalata*** Boott subsp. ***dispalata***
Carex persistens var. *takeuchii* Ohwi = ***Carex dispalata*** Boott subsp. ***dispalata***
Carex personata C.B.Clarke = ?
Carex personii Schkuhr ex Kunth = ***Carex wahlenbergiana*** Boott
Carex persoonii Sieber ex W.D.J.Koch = ***Carex brunnescens*** (Pers.) Poir. subsp. ***brunnescens***
Carex persoonii Láng = ***Carex canescens*** L. subsp. ***canescens***
Carex persoonii subsp. *vitilis* (Fr.) Nyman = ***Carex brunnescens*** subsp. ***sphaerostachya*** (Tuck.) Kalela
Carex perstricta Mack. = ***Carex schiedeana*** Kunze
Carex peruviana Dombey ex Boott = ***Carex extensa*** Gooden.
Carex peshemskyi Malyschev = ***Carex williamsii*** Britton
Carex petasata var. *minor* (Boott) B.Boivin = ***Carex petasata*** Dewey
Carex petasata var. *pleiostachya* Kük. = ***Carex petasata*** Dewey
Carex petecticalis Nelmes = ***Carex teres*** Boott
Carex petersii C.A.Mey. ex Schmidt = ***Carex vaginata*** var. ***petersii*** (C.A.Mey. ex Schmidt) Akiyama
Carex ×*petrae-fu* F.Murr = ?
Carex petraea Wahlenb. = ***Carex rupestris*** All. subsp. ***rupestris***
Carex petricosa var. *distichiflora* (B.Boivin) B.Boivin = ***Carex petricosa*** Dewey var. ***petricosa***
Carex petricosa var. *edwardsii* B.Boivin = ***Carex petricosa*** Dewey var. ***petricosa***
Carex petricosa var. *franklinii* (Boott) Boivin = ***Carex petricosa*** Dewey var. ***petricosa***
Carex petriei var. *rubicunda* (Petrie) Kük. = ***Carex rubicunda*** Petrie
Carex petunnikowii Litv. = ***Carex orbicularis*** Boott subsp. ***orbicularis***
Carex phacelostachys Nelmes = ***Carex verticillata*** Zoll. & Moritzi
Carex phacota var. *gracilispica* Kük. = ***Carex phacota*** Spreng.
Carex phacota var. *minor* C.B.Clarke = ***Carex teres*** Boott

Carex phaecostachya Sm. = ***Carex vaginata*** Tausch var. ***vaginata***
Carex phaenocarpa Franch. = ***Carex alopecuroides*** D.Don ex Tilloch & Taylor
Carex phaeocephala f. *eastwoodiana* (Stacey) F.J.Herm. = ***Carex tahoensis*** Smiley
Carex phaeolepis T.Holm = ***Carex vallicola*** Dewey
Carex phaeoleuca Ohwi = ***Carex nubigena*** var. ***franchetiana*** Ohwi
Carex phaeopoda Ohwi = ***Carex kiotensis*** Franch. & Sav.
Carex phaeostachya Sm. = ***Carex vaginata*** Tausch var. ***vaginata***
Carex phalaroides var. *capitellata* Kük. = ***Carex phalaroides*** Kunth var. ***phalaroides***
Carex phalaroides var. *crassiflora* Kük. = ***Carex gibertii*** G.A.Wheeler
Carex phalaroides subsp. *crassiflora* (Kük.) Luceño & Alves = ***Carex gibertii*** G.A.Wheeler
Carex phalaroides var. *humilis* Boeck. = ***Carex phalaroides*** Kunth var. ***phalaroides***
Carex phalaroides var. *moesta* (Kunth) Kük. = ***Carex phalaroides*** var. ***elongata*** Kük.
Carex phalaroides subsp. *moesta* (Kunth) Luceño & Alves = ***Carex phalaroides*** var. ***elongata*** Kük.
Carex phalaroides subsp. *paraguayensis* (Maury ex Micheli) Luceño & Alves = ***Carex phalaroides*** var. ***paraguayensis*** (Maury ex Micheli) Kük.
Carex phalaroides var. *parvula* R.Gross = ***Carex phalaroides*** Kunth var. ***phalaroides***
Carex pharensis Vis. = ***Carex illegitima*** Ces.
Carex philippinensis Nelmes = ***Carex graeffeana*** Boeck.
Carex philocrena V.I.Krecz. = ***Carex viridula*** Michx. subsp. ***viridula***
Carex phleoides Cav. = ***Uncinia phleoides*** (Cav.) Pers.
Carex phyllomanica Boott = ***Carex echinata*** subsp. ***phyllomanica*** (Boott) Reznicek
Carex phyllomanica var. *angustata* (Carey) B.Boivin = ***Carex echinata*** Murray subsp. ***echinata***
Carex phyllomanica var. *ormantha* (Fernald) B.Boivin = ***Carex echinata*** Murray subsp. ***echinata***
Carex phyllophora Kunth = ***Carex pauciflora*** Lightf.
Carex physema Dewey = ***Carex bullata*** Willd.
Carex physocarpa Nees ex Boott = ***Carex acutata*** Boott
Carex physocarpa C.Presl = ***Carex saxatilis*** L.
Carex physochlaena T.Holm = ***Carex membranacea*** Hook.
Carex physodes subsp. *subphysodes* (Popov ex V.I.Krecz.) Kukkonen = ***Carex subphysodes*** Popov ex V.I.Krecz.
Carex physorhyncha Liebm. ex Steud. = ***Carex albicans*** var. ***australis*** (L.H.Bailey) Rettig
Carex picea Franch. = ***Carex buxbaumii*** Wahlenb.
Carex pichinchensis var. *dura* Boott = ***Carex pichinchensis*** Kunth
Carex pichinchensis var. *obtusisquama* R.Gross = ***Carex pichinchensis*** Kunth
Carex pichinchensis var. *simplex* R.Gross = ***Carex pichinchensis*** Kunth
Carex pierotii Miq. = ***Carex scabrifolia*** Steud.
Carex pilosa var. *auriculata* (Franch.) Kük. = ***Carex pilosa*** Scop.
Carex pilosa var. *beckeri* C.A.Mey. ex Claus = ***Carex pilosa*** Scop.
Carex pilosa var. *densiflora* Schur = ***Carex pilosa*** Scop.
Carex pilosa f. *gracilior* Meinsh. = ***Carex campylorhina*** V.I.Krecz.
Carex pilosiuscula Boeck. = ***Carex whitneyi*** Olney
Carex pilulifera Geners. ex Boott = ***Carex praecox*** Schreb.
Carex pilulifera f. *americana* Boeck. = ***Carex albicans*** Willd. ex Spreng. var. ***albicans***
Carex pilulifera var. *azorica* (J.Gay) H.Christ = ***Carex pilulifera*** subsp. ***azorica*** (J.Gay) Franco & Rocha Afonso
Carex pilulifera var. *bastardiana* (DC.) Nyman = ***Carex pilulifera*** L. subsp. ***pilulifera***
Carex pilulifera var. *bertolonii* Pi.Savi = ***Carex pilulifera*** L. subsp. ***pilulifera***
Carex pilulifera var. *deflexa* (Hornem.) Drejer = ***Carex deflexa*** Hornem.
Carex pilulifera var. *digyna* Boeck. = ***Carex pilulifera*** L. subsp. ***pilulifera***
Carex pilulifera f. *fallax* Vatke ex Asch. & Graebn. = ***Carex pilulifera*** L. subsp. ***pilulifera***
Carex pilulifera f. *fuscidula* Waisb. = ***Carex pilulifera*** L. subsp. ***pilulifera***
Carex pilulifera f. *golenzii* Asch. & Graebn. = ***Carex pilulifera*** L. subsp. ***pilulifera***
Carex pilulifera f. *laxa* Waisb. = ***Carex pilulifera*** L. subsp. ***pilulifera***
Carex pilulifera var. *leesii* Ridl. = ***Carex pilulifera*** L. subsp. ***pilulifera***
Carex pilulifera f. *longibracteata* Lange = ***Carex pilulifera*** L. subsp. ***pilulifera***
Carex pilulifera f. *major* Wirtg. ex Herrenkohl = ***Carex pilulifera*** L. subsp. ***pilulifera***
Carex pilulifera var. *mucronata* Peterm. = ***Carex pilulifera*** L. subsp. ***pilulifera***
Carex pilulifera var. *novae-angliae* Kurtz = ***Carex rossii*** Boott
Carex pilulifera var. *pallescens* Beckh. = ***Carex pilulifera*** L. subsp. ***pilulifera***
Carex pilulifera f. *pallida* Peterm. = ***Carex pilulifera*** L. subsp. ***pilulifera***
Carex pilulifera f. *pedunculata* Waisb. = ***Carex pilulifera*** L. subsp. ***pilulifera***
Carex pilulifera f. *refracta* Waisb. = ***Carex pilulifera*** L. subsp. ***pilulifera***
Carex pilulifera var. *vaginata* Boeck. = ***Carex depressa*** Link subsp. ***depressa***
Carex pineticola Ohwi = ***Carex sachalinensis*** var. ***pineticola*** (Ohwi) Ohwi
Carex pinetorum Willd. ex Kunth = ***Carex muehlenbergii*** Willd. var. ***muehlenbergii***
Carex pinetorum Liebm. = ***Carex peucophila*** T.Holm
Carex pinetorum var. *elatior* Kük. = ***Carex peucophila*** T.Holm
Carex pinguipes C.B.Clarke = ?
Carex pinguis L.H.Bailey = ***Carex adusta*** Boott
Carex piniaria Bosc ex Boott = ***Carex muehlenbergii*** Willd. var. ***muehlenbergii***
Carex pinnata Moench ex Rchb. = ***Carex elongata*** L.
Carex piperi Mack. = ***Carex praticola*** Rydb.
Carex piptolepis Desv. = ***Carex setifolia*** Kunze var. ***setifolia***
Carex pirinensis Acht. = ***Carex parviflora*** Host
Carex pisiformis var. *ebracteata* (Trautv.) Kük. = ***Carex trautvetteriana*** Kom.
Carex pisiformis f. *polyschoena* (H.Lév. & Vaniot) Kük. = ***Carex sachalinensis*** F.Schmidt var. ***sachalinensis***
Carex pisiformis var. *sachalinensis* (F.Schmidt) Kük. = ***Carex sachalinensis*** F.Schmidt
Carex pisiformis var. *subebracteata* Kük. = ***Carex subebracteata*** (Kük.) Ohwi
Carex pissisi Phil. = ***Carex setifolia*** Kunze var. ***setifolia***
Carex pittieri Boeck. = ***Carex donnell-smithii*** L.H.Bailey
Carex plagiostoma Nelmes = ***Carex horsfieldii*** Boott
Carex plana Mack. = ***Carex muehlenbergii*** var. ***enervis*** Boott

Carex planifolia Kohts = ***Carex magellanica*** subsp. ***irrigua*** (Wahlenb.) Hultén
Carex plantaginea Muhl. = ***Carex laxiflora*** Lam. var. ***laxiflora***
Carex platycarpa Hochst. ex Steud. = ***Carex phacota*** Spreng.
Carex platycarpos Rchb. = ?
Carex platylepis Mack. = ***Carex praticola*** Rydb.
Carex platyphylla Franch. = ***Carex siderosticta*** Hance var. ***siderosticta***
Carex platyphylla var. *longipedunculata* Kük. = ***Carex platyphylla*** Carey
Carex platyrhina Ohwi = ***Carex speciosa*** subsp. ***platyrhina*** (Ohwi) T.Koyama
Carex platyrhyncha Franch. & Sav. = ***Carex pumila*** Thunb.
Carex platysperma var. *sungareensis* Y.L.Chang & Y.L. Yang = ***Carex platysperma*** Y.L.Chang & Y.L.Yang
Carex platystachys Ten. = ***Carex macrolepis*** DC.
Carex plebeia C.B.Clarke = ***Carex continua*** C.B.Clarke
Carex plectocarpa F.J.Herm. = ***Carex lenticularis*** var. ***dolia*** (M.E.Jones) L.A.Standl.
Carex pleiorhiza H.Lév. & Vaniot = ***Carex jackiana*** Boott subsp. ***jackiana***
Carex pleistogyna V.I.Krecz. = ***Carex nubigena*** D.Don ex Tilloch & Taylor subsp. ***nubigena***
Carex plena Clairv. = ***Carex capillaris*** L. subsp. ***capillaris***
Carex plesiocephala Turrill = ***Carex tricephala*** Boeck.
Carex plocamostyla Maxim. = ***Carex doenitzii*** Boeck. var. ***doenitzii***
Carex plumbea (Wahlenb.) Willd. = ***Carex riparia*** Curtis subsp. ***riparia***
Carex pluvia R.W.Krauss = ***Carex alligata*** Boott
Carex pluvia var. *koolauensis* R.W.Krauss = ***Carex alligata*** Boott
Carex pocilliformis Boott = ***Carex tristachya*** var. ***pocilliformis*** (Boott) Kük.
Carex podocarpa var. *koraginensis* (Meinsh.) Vorosch. = ***Carex scita*** var. ***riishirensis*** (Franch.) Kük.
Carex podocarpa f. *pallidior* Lepage = ***Carex podocarpa*** R.Br.
Carex podocarpa var. *paysonis* (Clokey) B.Boivin = ***Carex paysonis*** Clokey
Carex podogyna f. *aristodentata* Kük. = ***Carex podogyna*** Franch. & Sav.
Carex podostachys Steud. = ***Carex digitalis*** Willd. var. ***digitalis***
Carex podperae Otruba = ***Carex acuta*** L.
Carex poiformis (Retz.) J.F.Gmel. = ***Scleria poiformis*** Retz.
Carex poiretii J.F.Gmel. = ***Fuirena pubescens*** (Poir.) Kunth var. ***pubescens***
Carex pollens C.B.Clarke = ***Carex dispalata*** Boott subsp. ***dispalata***
Carex pollens var. *angustior* C.B.Clarke = ***Carex dispalata*** Boott subsp. ***dispalata***
Carex polyandra Schkuhr = ***Carex nigra*** (L.) Reichard subsp. ***nigra***
Carex polyantha Franch. & Sav. = ***Carex prescottiana*** Boott
Carex polygama Schkuhr = ***Carex buxbaumii*** Wahlenb.
Carex polygama J.F.Gmel. = ?
Carex polygama var. *heterostachya* (Andersson) Farw. = ***Carex buxbaumii*** Wahlenb.
Carex polygyna Boeck. = ***Kobresia esenbeckii*** (Kunth) Noltie var. ***esenbeckii***
Carex polymorpha Torr. = ***Carex striata*** Michx.
Carex polymorpha Schwein. ex Boott = ***Carex vulpinoidea*** Michx.
Carex polymorpha var. *californica* (L.H.Bailey) Kük. = ***Carex californica*** L.H.Bailey
Carex polyneura Colenso = ***Carex dissita*** Sol. ex Boott
Carex polyphylla Kar. & Kir. = ***Carex divulsa*** subsp. ***leersii*** (Kneuck.) W.Koch
Carex polyrrhiza Wallr. = ***Carex umbrosa*** Host subsp. ***umbrosa***
Carex polyrrhiza f. *brevifolia* Waisb. = ***Carex umbrosa*** Host subsp. ***umbrosa***
Carex polyrrhiza f. *breviscapa* Peterm. = ***Carex umbrosa*** Host subsp. ***umbrosa***
Carex polyrrhiza var. *decalvata* Hausm. ex Döll = ***Carex umbrosa*** Host subsp. ***umbrosa***
Carex polyrrhiza f. *subintegra* Kük. = ***Carex umbrosa*** Host subsp. ***umbrosa***
Carex polyschoena H.Lév. & Vaniot = ***Carex sachalinensis*** F.Schmidt var. ***sachalinensis***
Carex polyschoenoides var. *fractiflexa* Y.C.Yang = ***Carex polyschoenoides*** K.T.Fu
Carex polystachya R.Lesson & A.Rich. = ***Carex lessoniana*** Steud.
Carex polystachya A.Rich. = ***Carex ternaria*** G.Forst.
Carex polystachya var. *madrensis* (L.H.Bailey) Kük. = ***Carex madrensis*** L.H.Bailey
Carex polytrichoides Muhl. ex Willd. = ***Carex leptalea*** Wahlenb. subsp. ***leptalea***
Carex ponmoshiriensis Miyabe & Kudô = ***Carex planata*** Franch. & Sav.
Carex portae K.Richt. = ***Carex distans*** L.
Carex porteri Olney = ***Carex crinita*** var. ***porteri*** (Olney) Fernald
Carex posnaniensis Sprib. = ***Carex repens*** Bellardi
Carex potamostauros Steud. = ***Carex aematorrhyncha*** Desv. var. ***aematorrhyncha***
Carex praealta Boott = ***Carex dianae*** Steud.
Carex praeceps Borkh. ex Rchb. = ***Carex praecox*** Schreb.
Carex praecox J.Jundz. = ***Carex brizoides*** L.
Carex praecox Jacq. = ***Carex caryophyllea*** Latourr. var. ***caryophyllea***
Carex praecox Pollich ex Steud. = ***Carex montana*** L.
Carex praecox var. *alata* Zlinska = ***Carex praecox*** Schreb.
Carex praecox var. *caespitosa* J.Fleisch. = ***Carex caryophyllea*** Latourr. var. ***caryophyllea***
Carex praecox var. *curvata* Asch. = ***Carex praecox*** Schreb.
Carex praecox subsp. *curvata* (Asch.) Knaf = ***Carex praecox*** Schreb.
Carex praecox var. *diastachya* Callmé = ***Carex caryophyllea*** Latourr. var. ***caryophyllea***
Carex praecox var. *dumetorum* A.Blytt = ***Carex caryophyllea*** Latourr. var. ***caryophyllea***
Carex praecox var. *elatior* Bogenh. = ***Carex caryophyllea*** Latourr. var. ***caryophyllea***
Carex praecox f. *fissa* Kük. = ***Carex caryophyllea*** Latourr. var. ***caryophyllea***
Carex praecox f. *gracillima* (Asch.) Soó = ***Carex praecox*** Schreb.
Carex praecox var. *insularis* H.Christ = ***Carex caryophyllea*** Latourr. var. ***caryophyllea***
Carex praecox subsp. *intermedia* (Èelak.) W.Schultze-Motel = ***Carex praecox*** Schreb.
Carex praecox f. *laxa* Kük. = ***Carex praecox*** Schreb.
Carex praecox var. *longifolia* Wahlenb. = ***Carex umbrosa*** Host subsp. ***umbrosa***
Carex praecox var. *longifrons* Waisb. & Kük. = ***Carex caryophyllea*** Latourr. var. ***caryophyllea***
Carex praecox f. *longivaginata* Kük. = ***Carex caryophyllea*** Latourr. var. ***caryophyllea***
Carex praecox f. *luxurians* Kük. = ***Carex caryophyllea*** Latourr. var. ***caryophyllea***

Carex praecox var. *magna* Schur = ***Carex caryophyllea*** Latourr. var. ***caryophyllea***
Carex praecox f. *monostachys* Lackow. = ***Carex praecox*** Schreb.
Carex praecox f. *nana* (Zapal.) Soó = ***Carex praecox*** Schreb.
Carex praecox f. *pallida* Peterm. = ***Carex praecox*** Schreb.
Carex praecox f. *petermanii* Soó = ***Carex praecox*** Schreb.
Carex praecox f. *podolica* (Zapal.) Soó = ***Carex praecox*** Schreb.
Carex praecox var. *polyrrhiza* Boott = ***Carex umbrosa*** Host subsp. ***umbrosa***
Carex praecox var. *procerior* Gaudin = ***Carex umbrosa*** Host subsp. ***umbrosa***
Carex praecox f. *pygmaea* J.Fleisch. = ***Carex caryophyllea*** Latourr. var. ***caryophyllea***
Carex praecox var. *reflexa* (Roth) Rchb. = ***Carex caryophyllea*** Latourr. var. ***caryophyllea***
Carex praecox f. *rumelica* Velen. = ***Carex praecox*** Schreb.
Carex praecox var. *sicyocarpa* (Lebel) Nyman = ***Carex praecox*** Schreb.
Carex praecox f. *subpolyrrhiza* Kük. = ***Carex caryophyllea*** Latourr. var. ***caryophyllea***
Carex praecox var. *trachyantha* (Dorner ex Heuff.) Nyman = ***Carex caryophyllea*** Latourr. var. ***caryophyllea***
Carex praecox var. *umbrosa* De Wild. & T.Durand = ***Carex caryophyllea*** Latourr. var. ***caryophyllea***
Carex praecox f. *umbrosiformis* J.Fleisch. = ***Carex caryophyllea*** Latourr. var. ***caryophyllea***
Carex praecox subsp. *velenovskyi* (Domin) Jílek = ***Carex praecox*** Schreb.
Carex praecox var. *vidalii* Franch. & Sav. = ***Carex caryophyllea*** Latourr. var. ***caryophyllea***
Carex praegracilis var. *simulata* (Mack.) B.Boivin = ***Carex simulata*** Mack.
Carex praelonga C.B.Clarke = ***Carex teres*** Boott
Carex praelonga var. *angustior* C.B.Clarke = ***Carex teres*** Boott
Carex praestabilis Ohwi = ***Carex umbrosa*** var. ***rostrata*** (Maxim.) Govaerts
Carex praestans C.B.Clarke = ***Carex myosurus*** Nees
Carex praetutiana Parl. = ***Carex flacca*** Schreb. subsp. ***flacca***
Carex ×*prahliana* Junge = ?
Carex ×*prahliana* Junge = ?
Carex prainii C.B.Clarke = ***Carex perakensis*** C.B.Clarke var. ***perakensis***
Carex prainii Kük. = ***Kobresia sikkimensis*** Kük.
Carex pratensis Hosé = ***Carex acuta*** L.
Carex pratensis Phil. = ***Carex macloviana*** d'Urv. subsp. ***macloviana***
Carex pratensis Drejer = ***Carex praticola*** Rydb.
Carex pratensis var. *furva* L.H.Bailey = ***Carex praticola*** Rydb.
Carex pratensis f. *furva* (L.H.Bailey) Kük. = ***Carex praticola*** Rydb.
Carex pratericola var. *furva* (L.H.Bailey) A.Heller = ***Carex praticola*** Rydb.
Carex praticola var. *subcoriacea* F.J.Herm. = ***Carex praticola*** Rydb.
Carex prattii Franch. = ***Carex melanantha*** C.A.Mey. var. ***melanantha***
Carex prescottiana var. *cremostachys* (Franch.) Kük. = ***Carex prescottiana*** Boott
Carex prescottiana var. *fargesii* (Franch.) Kük. = ***Carex prescottiana*** Boott
Carex prescottiana var. *fuscescens* (Boeck.) Kük. = ***Carex kiotensis*** Franch. & Sav.
Carex prescottiana var. *kiotensis* (Franch. & Sav.) Kük. = ***Carex kiotensis*** Franch. & Sav.
Carex preussii K.Schum. = ***Carex petitiana*** A.Rich.
Carex preussii var. *camerunensis* Nelmes = ***Carex petitiana*** A.Rich.
Carex prevernalis Kitag. = ***Carex lanceolata*** var. ***subpediformis*** Kük.
Carex pribylovensis Macoun = ***Carex glareosa*** subsp. ***pribylovensis*** (Macoun) G.Halliday & Chater
Carex princeps Kunth = ***Carex praecox*** Schreb.
Carex prionocarpa Franch. = ***Carex lyngbyei*** Hornem.
Carex prionophylla T.Holm = ***Carex scopulorum*** var. ***prionophylla*** (T.Holm) L.A.Standl.
Carex procera Kunth = ***Carex brasiliensis*** A.St.-Hil.
Carex procera var. *brasiliensis* (A.St.-Hil.) Boott = ***Carex brasiliensis*** A.St.-Hil.
Carex procerula V.I.Krecz. = ***Carex saxatilis*** L.
Carex prolifera Kit. = ?
Carex prolixa Fr. = ***Carex acuta*** L.
Carex prolixa Schur = ***Carex stricta*** Lam.
Carex prolixa var. *protensa* Fr. = ***Carex brunnea*** Thunb. subsp. ***brunnea***
Carex promaucana Phil. = ***Carex banksii*** Boott var. ***banksii***
Carex propinqua Nees & Meyen ex Kunth = ***Carex macloviana*** subsp. ***subfusca*** (W.Boott) T.Koyama
Carex prostrata All. = ***Carex humilis*** Leyss. var. ***humilis***
Carex prostrata J.F.Gmel. = ***Carex pilosa*** Scop.
Carex protracta Steud. = ***Carex laxiflora*** Lam. var. ***laxiflora***
Carex provincialis Degl. = ***Carex echinata*** Murray subsp. ***echinata***
Carex pruinosa var. *aristata* Kuntze = ***Carex phacota*** Spreng.
Carex pruinosa var. *levisaccus* (Ohwi) Makino & Nemoto = ***Carex pruinosa*** subsp. ***maximowiczii*** (Miq.) Kük.
Carex pruinosa f. *submutica* Kuntze = ***Carex pruinosa*** Boott subsp. ***pruinosa***
Carex pruinosa var. *suifunensis* (Kom.) Kük. = ***Carex suifunensis*** Kom.
Carex psammogaea Steud. = ***Carex maritima*** Gunnerus
Carex psammophila Schur = ***Carex brizoides*** L.
Carex ×*pseudoaphanolepis* Ohwi = ?
Carex pseudoarenaria Rchb. = ***Carex colchica*** J.Gay
Carex pseudoarenaria Pers. = ***Carex disticha*** Huds.
Carex pseudoarenicola Hayata = ***Carex nubigena*** subsp. ***pseudoarenicola*** (Hayata) T.Koyama
Carex pseudoatrata Meinsh. = ***Carex atrata*** L. subsp. ***atrata***
Carex pseudobiwensis Kitag. = ***Carex rara*** Boott subsp. ***rara***
Carex pseudobuxbaumii M.Winkl. = ***Carex buxbaumii*** Wahlenb.
Carex pseudocilicica Hausskn. = ***Carex atrata*** L. subsp. ***atrata***
Carex pseudoconica Franch. & Sav. = ***Carex sachalinensis*** F.Schmidt var. ***sachalinensis***
Carex pseudocyperus var. *americana* Hochst. ex L.H. Bailey = ***Carex pseudocyperus*** L. var. ***pseudocyperus***
Carex pseudocyperus var. *brachystachya* Regel & Maack = ***Carex capricornis*** Meinsh. ex Maxim.
Carex pseudocyperus var. *cognata* (Kunth) Boott = ***Carex cognata*** Kunth
Carex pseudocyperus var. *comosa* (Boott) Boott = ***Carex comosa*** Boott
Carex pseudocyperus var. *furcata* Kük. = ***Carex pseudocyperus*** L. var. ***pseudocyperus***

Carex pseudocyperus subsp. *haenkeana* (C.Presl) Kük. = ***Carex pseudocyperus*** var. ***haenkeana*** (C.Presl) Kük.
Carex pseudocyperus f. *minor* Hampe = ***Carex pseudocyperus*** L. var. ***pseudocyperus***
Carex pseudocyperus f. *multispicula* Lepage = ***Carex pseudocyperus*** L.
Carex pseudocyperus f. *novae-selandiae* (Boeck.) Kük. = ***Carex pseudocyperus*** L. var. ***pseudocyperus***
Carex pseudocyperus f. *pallidior* Kük. = ***Carex pseudocyperus*** L. var. ***pseudocyperus***
Carex pseudocyperus subsp. *platygluma* C.B.Clarke ex Kük. = ***Carex polystachya*** Sw. ex Wahlenb. var. ***polystachya***
Carex pseudocyperus var. *polysticha* (Boeck.) Kük. = ***Carex polysticha*** Boeck.
Carex pseudodiandra Waisb. = ?
Carex pseudodoenitzii Akiyama = ***Carex doenitzii*** Boeck. var. ***doenitzii***
Carex pseudoechinata Boeck. = ***Carex uruguensis*** var. ***pseudoechinata*** (Boeck.) Kük.
Carex pseudofilicina Hayata = ***Carex filicina*** Nees
Carex pseudoflava Schur = ***Carex distans*** L.
Carex pseudofrigida C.B.Clarke = ***Carex frigida*** All.
Carex ×*pseudoheleonastes* Hand.-Mazz. = ?
Carex pseudohirta Meinsh. = ***Carex drymophila*** Turcz.
Carex pseudohypochlora f. *denticulata* Y.L.Chang & Y.L.Yang = ***Carex pseudohypochlora*** Y.L.Chang & Y.L.Yang
Carex pseudojaponica Hayata = ***Carex alopecuroides*** D.Don ex Tilloch & Taylor
Carex pseudojaponica C.B.Clarke = ***Carex luzulifolia*** Boott
Carex pseudolagopina T.J.Sørensen = ***Carex marina*** subsp. ***pseudolagopina*** (T.J.Sørensen) Böcher
Carex pseudolagopina var. *solida* T.J.Sørensen = ***Carex marina*** subsp. ***pseudolagopina*** (T.J.Sørensen) Böcher
Carex pseudolanceolata V.I.Krecz. = ***Carex lanceolata*** var. ***subpediformis*** Kük.
Carex pseudololiacea f. *sejuncta* Kük. = ***Carex pseudololiacea*** F.Schmidt
Carex pseudolongirostrata Y.L.Chang & Y.L.Yang = ***Carex nodaeana*** A.I.Baranov & Skvortsov
Carex pseudomacrogyna Kük. = ?
Carex ×*pseudomairii* Camus = ?
Carex pseudonutans Boreau ex É.Martin = ?
Carex pseudoparadoxa S.Gibson = ***Carex diandra*** Schrank
Carex pseudopraecox Schur = ***Carex praecox*** Schreb.
Carex pseudopunctata Boeck. = ***Carex fuscula*** var. ***distenta*** (Kunze ex Kunth) Kük.
Carex pseudorivulorum Kük. = ***Carex perakensis*** var. ***borneensis*** (C.B.Clarke) Noot.
Carex pseudoscirpoidea Rydb. = ***Carex scirpoidea*** var. ***pseudoscirpoidea*** (Rydb.) Cronquist
Carex pseudostrigosa H.Lév. & Vaniot = ***Carex sachalinensis*** var. ***alterniflora*** (Franch.) Ohwi
Carex pseudothunbergii Akiyama = ?
Carex pseudoturfosa Dalla Torre & Sarnth. = ?
Carex ×*pseudoturfosa* H.Lév. & L.C.Lamb. = ?
Carex pseudovesicaria H.Lév. & Vaniot = ***Carex idzuroei*** Franch. & Sav.
Carex × *pseudovulpina* K.Richt. = ?
Carex pseudowrightii Honda = ***Carex van-heurckii*** Müll.Arg. subsp. ***van-heurckii***
Carex psilochloa Wall. = ?
Carex psilostachya Kit. ex Willd. = ***Carex sylvatica*** Huds. subsp. ***sylvatica***
Carex psilostachys Steud. = ***Carex excelsa*** Poepp. ex Kunth
Carex psychroluta V.I.Krecz. = ***Carex maritima*** Gunnerus
Carex psyllophora L.f. = ***Carex pulicaris*** L.
Carex pterolepta Franch. = ***Carex fluviatilis*** Boott
Carex pteroloma Kunze ex Steud. = ***Carex gibba*** Wahlenb.
Carex pterygocarpa Steud. = ***Carex banksii*** Boott var. ***banksii***
Carex ptychocarpa Steud. = ***Carex abscondita*** Mack.
Carex ptychocarpa Link = ***Carex halleriana*** Asso
Carex ptychocarpa var. *macrophylla* L.H.Bailey = ***Carex abscondita*** Mack.
Carex puberula Boott = ***Carex breviculmis*** var. ***puberula*** (Boott) T.Koyama
Carex pubescens Muhl. ex Willd. = ***Carex hirtifolia*** Mack.
Carex pubescens Gilib. = ***Carex montana*** L.
Carex pubescens Poir. = ***Fuirena pubescens*** (Poir.) Kunth
Carex pubicarpa (Rohlena) A.W.Hill = ***Carex flacca*** Schreb. subsp. ***flacca***
Carex pulchella Nyl. ex Trevir = ***Carex acuta*** L.
Carex pulchella Berggr. = ***Carex comans*** Berggr.
Carex pulchella T.Holm = ?
Carex pulicarioides Ramond ex DC. = ***Carex pyrenaica*** Wahlenb.
Carex pulicaris var. *caespitosa* Vollm. = ***Carex pulicaris*** L.
Carex pulla Sol. ex Boott = ***Carex flagellifera*** Colenso
Carex pulla Gooden. = ***Carex saxatilis*** L.
Carex pulla f. *debilis* Porsild = ***Carex saxatilis*** L.
Carex pulla subsp. *dichroa* Freyn = ***Carex pamirica*** subsp. ***dichroa*** (Malyschev) T.V.Egorova
Carex pulla f. *dichroa* A.Blytt = ***Carex vesicaria*** L.
Carex pulla f. *laxa* Trautv. = ***Carex saxatilis*** L.
Carex pulla var. *miliaris* (Michx.) A.Gray = ***Carex saxatilis*** L.
Carex pulla f. *pedunculata* Kjellm. = ***Carex vesicaria*** L.
Carex pulla f. *polystachya* A.Blytt = ***Carex vesicaria*** L.
Carex pulla var. *sibirica* H.Christ = ***Carex vesicaria*** L.
Carex pulla var. *vesicarioides* Kurtz = ***Carex saxatilis*** L.
Carex pullei Nelmes = ***Carex turrita*** var. ***merrillii*** (Kük.) Noot.
Carex pullulans Dulac = ***Carex pilulifera*** L. subsp. ***pilulifera***
Carex pumila Honck. = ***Carex montana*** L.
Carex pumila var. *bichenoviana* (Boott) Kük. = ***Carex bichenoviana*** Boott
Carex pumila var. *forbesii* (C.B.Clarke) Kük. = ***Carex pumila*** Thunb.
Carex pumila subsp. *littorea* (Labill.) Kük. = ***Carex pumila*** Thunb.
Carex pumila var. *urvillei* (Brongn.) Kük. = ***Carex pumila*** Thunb.
Carex punctata Nees = ***Carex notha*** Kunth
Carex punctata subsp. *lainzii* (Luceño, E.Rico & T.Romero) Rivas Mart. = ***Carex punctata*** Gaudin var. ***punctata***
Carex punctulata Leers = ***Carex paniculata*** L. subsp. ***paniculata***
Carex punctulata R.Lesson ex A.Rich. = ***Carex pseudocyperus*** L. var. ***pseudocyperus***
Carex pungens Boeck. = ***Carex setifolia*** var. ***pungens*** (Boeck.) Kük.
Carex purpurascens Kük. = ***Carex alliiformis*** C.B.Clarke
Carex purpurea Boott = ***Carex acutata*** Boott
Carex purshii Olney = ***Carex lurida*** Wahlenb.
Carex pusilla Pers. = ***Carex bicolor*** Bellardi ex All.
Carex pusilla Valbusa = ?
Carex pycnostachya Desv. = ***Carex nebularum*** Phil.
Carex pycnostachys var. *major* Kük. = ***Carex firmicaulis*** Kalela

Carex pycnothyrsos Kük. = ***Carex cruciata*** var. ***rafflesiana*** (Boott) Noot.
Carex pyrenaica F.Muell. = ***Carex acicularis*** Boott
Carex pyrenaica var. *altior* Kük. = ***Carex pyrenaica*** Wahlenb.
Carex pyrenaica var. *articulata* Kük. = ***Carex pyrenaica*** Wahlenb.
Carex pyrenaica var. *cephalotes* (F.Muell.) Kük. = ***Carex cephalotes*** F.Muell.
Carex pyrenaica var. *grosseckii* (Heuff. ex Boott) Nyman = ***Carex pyrenaica*** Wahlenb.
Carex pyrenaica f. *longibracteata* Schur = ***Carex pyrenaica*** Wahlenb.
Carex pyrenaica subsp. *micropoda* (C.A.Mey.) Hultén = ***Carex micropoda*** C.A.Mey.
Carex pyrenaica subsp. *micropodioides* (V.I.Krecz.) Chandjian = ***Carex pyrenaica*** Wahlenb.
Carex pyriformis F.W.Schultz = ***Carex aurea*** Nutt.
Carex pyriformis Tausch ex F.W.Schultz = ***Carex flava*** L. subsp. ***flava***
Carex pyrophila Gand. = ***Carex macloviana*** d'Urv. subsp. ***macloviana***
Carex qimenensis S.W.Su & S.M.Xu = ***Carex dolichostachya*** Hayata subsp. ***dolichostachya***
Carex quadrangulata Colenso = ***Carex dissita*** Sol. ex Boott
Carex quadrifida L.H.Bailey = ***Carex heteroneura*** S.Watson var. ***heteroneura***
Carex quadrifida var. *caeca* L.H.Bailey = ***Carex heteroneura*** S.Watson var. ***heteroneura***
Carex quadrifida var. *lenis* L.H.Bailey = ***Carex heteroneura*** S.Watson var. ***heteroneura***
Carex qualicumensis L.H.Bailey = ***Carex lyngbyei*** Hornem.
Carex quaternaria Spreng. = ***Carex loliacea*** L.
Carex quinata Raeusch. = ***Carex cespitosa*** L. var. ***cespitosa***
Carex rachillis Maguire = ***Carex subnigricans*** Stacey
Carex ×*raciborskii* Zapal. = ?
Carex radiata var. *staminea* Beck ex Howe = ***Carex radiata*** (Wahlenb.) Small
Carex raeana Boott = ***Carex vesicaria*** L.
Carex rafflesiana Boott = ***Carex cruciata*** var. ***rafflesiana*** (Boott) Noot.
Carex rafflesiana var. *macrithyrsa* (Miq.) Kük. = ***Carex cruciata*** var. ***rafflesiana*** (Boott) Noot.
Carex rafflesiana var. *scaberrima* (Boeck.) Kük. = ***Carex cruciata*** var. ***rafflesiana*** (Boott) Noot.
Carex rafflesiana subsp. *scaberrima* (Boeck.) T.Koyama = ***Carex cruciata*** var. ***rafflesiana*** (Boott) Noot.
Carex rafflesiana var. *tenuior* C.B.Clarke = ***Carex cruciata*** var. ***rafflesiana*** (Boott) Noot.
Carex rahmeri Phil. = ***Carex decidua*** Boott
Carex ramenskii var. *caudata* Hultén = ***Carex ramenskii*** Kom.
Carex ramondiana DC. = ***Carex pyrenaica*** Wahlenb.
Carex ramosa Eckl. ex Kunth = ***Schoenoxiphium lanceum*** (Thunb.) Kük.
Carex ramosipes Cherm. = ***Carex mildbraediana*** Kük.
Carex rankanensis Hayata = ***Carex dolichostachya*** Hayata subsp. ***dolichostachya***
Carex raoulii f. *haastiana* (Boeck.) Kük. = ***Carex raoulii*** Boott
Carex raphidocarpa var. *major* (Steud.) Kük. = ***Carex raphidocarpa*** Nees
Carex raphidocarpa var. *mercarensis* (Hochst. ex Steud.) Kük. = ***Carex raphidocarpa*** Nees
Carex rara var. *biwensis* (Franch.) Kük. = ***Carex rara*** Boott subsp. ***rara***
Carex rara subsp. *capillacea* (Boott) Kük. = ***Carex capillacea*** Boott
Carex rara f. *macrolepis* Franch. = ***Carex rara*** Boott subsp. ***rara***
Carex rara var. *nana* Kük. = ***Carex capillacea*** var. ***sachalinensis*** (F.Schmidt) Ohwi
Carex rara var. *typica* Domin = ***Carex rara*** Boott
Carex rara f. *yunnanensis* Franch. = ***Carex rara*** Boott subsp. ***rara***
Carex rariflora f. *baeostachya* Norman = ***Carex rariflora*** (Wahlenb.) Sm. var. ***rariflora***
Carex rariflora f. *brevipedumculata* Meinsh. = ***Carex rariflora*** (Wahlenb.) Sm. var. ***rariflora***
Carex rariflora f. *expallida* Norman = ***Carex rariflora*** (Wahlenb.) Sm. var. ***rariflora***
Carex rariflora var. *firmior* Norman = ***Carex ×firmior*** (Norman) E.Holmb.
Carex rariflora f. *pallidior* A.Blytt = ***Carex rariflora*** (Wahlenb.) Sm. var. ***rariflora***
Carex rariflora var. *pluriflora* (Hultén) Boivin = ***Carex pluriflora*** Hultén
Carex rariflora subsp. *pluriflora* (Hultén) T.V.Egorova = ***Carex pluriflora*** Hultén
Carex rariflora f. *rufescens* Norman = ***Carex rariflora*** (Wahlenb.) Sm. var. ***rariflora***
Carex rariflora subsp. *stygia* (Fr.) Andersson = ***Carex ×stygia*** Fr.
Carex rariflora var. *stygia* (Fr.) Kük. = ***Carex ×stygia*** Fr.
Carex raymondii Calder = ***Carex atratiformis*** Britton
Carex rechingeri Palla = ***Carex graeffeana*** Boeck.
Carex reclinata Facchini = ***Carex ornithopoda*** subsp. ***ornithopodioides*** (Hausm.) Nyman
Carex recticulmis Franch. = ***Carex umbrosa*** subsp. ***sabynensis*** (Less. ex Kunth) Kük.
Carex rectior Mack. = ***Carex granularis*** Muhl. ex Willd.
Carex recurva Willd. ex Kunth = ***Carex acuta*** L.
Carex recurva Muhl. = ***Carex cherokeensis*** Schwein.
Carex recurva Huds. = ***Carex flacca*** Schreb. subsp. ***flacca***
Carex recurva Schkuhr = ***Carex pseudocyperus*** L. var. ***pseudocyperus***
Carex recurva subsp. *erythrostachys* (Hoppe) Arcang. = ***Carex flacca*** subsp. ***erythrostachys*** (Hoppe) Holub
Carex recurva subsp. *parlatoreana* Ces. ex Arcang. = ***Carex flacca*** Schreb. subsp. ***flacca***
Carex recurva subsp. *praetutiana* (Parl.) Arcang. = ***Carex flacca*** Schreb. subsp. ***flacca***
Carex recurvirostra Kunze = ***Carex baccans*** Nees
Carex recurvirostra Hallier f. ex Steud. = ***Carex davalliana*** Sm.
Carex recurvisaccus T.Koyama = ***Carex olivacea*** subsp. ***recurvisaccus*** (T.Koyama) T.Koyama
Carex redowskiana C.A.Mey. = ***Carex parallela*** subsp. ***redowskiana*** (C.A.Mey.) T.V.Egorova
Carex reflexa Gaudin = ***Carex davalliana*** Sm.
Carex reflexa Hoppe = ***Carex praecox*** Schreb.
Carex reflexa D.Dietr. ex Kunth = ***Carex vulpina*** L.
Carex reflexistyla Hayata = ***Carex boottiana*** Hook. & Arn.
Carex refracta Roth = ***Carex caryophyllea*** Latourr. var. ***caryophyllea***
Carex refracta Willd. = ***Carex ferruginea*** subsp. ***tendae*** W.Dietr.
Carex refracta var. *subferruginea* Kük. = ***Carex ferruginea*** subsp. ***tendae*** W.Dietr.
Carex regelii C.B.Clarke ex O.Fedtsch. = [32]
Carex regnelliana Boeck. = ***Carex polystachya*** Sw. ex Wahlenb. var. ***polystachya***
Carex reicheana Boeck. = ***Carex lateriflora*** Phil.

Carex reichenbachiana H.Lév. & Vaniot = ***Carex pseudobrizoides*** Chabaud
Carex reichenbachii Bonnet = ***Carex pseudobrizoides*** Chabaud
Carex rekohu Petrie = ***Carex trifida*** Cav.
Carex remota var. *axillaris* (L.) Döll = ***Carex remota*** L. subsp. ***remota***
Carex remota var. *brizopyrum* (Kunze) Boeck. = ***Carex remota*** subsp. ***alta*** (Boott) Kük.
Carex remota f. *doelliana* Asch. & Graebn. = ***Carex remota*** L. subsp. ***remota***
Carex remota var. *enervulosa* Kük. = ***Carex remotiuscula*** Wahlenb.
Carex remota var. *remotiformis* (Kom.) Kük. = ***Carex remotiuscula*** Wahlenb.
Carex remota var. *repens* Brittinger ex Rchb. = ***Carex remota*** L. subsp. ***remota***
Carex remota f. *rigida* Waisb. = ***Carex remota*** L. subsp. ***remota***
Carex remota var. *rochebrunii* (Franch. & Sav.) C.B. Clarke = ***Carex remota*** subsp. ***rochebrunii*** (Franch. & Sav.) Kük.
Carex remota f. *stricta* Asch. = ***Carex remota*** L. subsp. ***remota***
Carex remota var. *subloliacea* (Schur) Kük. = ***Carex remota*** L. subsp. ***remota***
Carex remotiflora Hayata = ***Carex finitima*** Boott
Carex remotiformis Kom. = ***Carex remotiuscula*** Wahlenb.
Carex remotispicula Hayata = ***Carex remota*** var. ***remotispicula*** (Hayata) Ohwi
Carex remotiuscula var. *enervulosa* (Kük.) K.T.Fu = ***Carex remotiuscula*** Wahlenb.
Carex remyi Boeck. = ***Carex meyenii*** Nees
Carex renschiana var. *brachystachya* Cherm. = ***Carex renschiana*** Boeck.
Carex renschiana var. *haematosaccus* (C.B.Clarke) Kük. = ***Carex haematosaccus*** C.B.Clarke
Carex renschiana var. *laxissima* Cherm. = ***Carex renschiana*** Boeck.
Carex renschiana var. *minor* Boeck. = ***Carex renschiana*** Boeck.
Carex repens var. *laggeri* (Wimm.) Nyman = ***Carex repens*** Bellardi
Carex reticulosa Peterm. = ***Carex elata*** All. subsp. ***elata***
Carex retorta (Fr.) V.I.Krecz. = ***Carex cespitosa*** L. var. ***cespitosa***
Carex retrocurva Dewey = ***Carex laxiculmis*** Schwein. var. ***laxiculmis***
Carex retrocurva var. *copulata* L.H.Bailey = ***Carex laxiculmis*** var. ***copulata*** (L.H.Bailey) Fernald
Carex retrocurva var. *floridana* (L.H.Bailey) Kük. = ***Carex digitalis*** var. ***floridana*** (L.H.Bailey) Naczi & Bryson
Carex retroflexa var. *texensis* (Torr. ex L.H.Bailey) Kük. = ***Carex texensis*** (Torr. ex L.H.Bailey) L.H.Bailey
Carex retrorsa Nees = ***Carex cognata*** Kunth
Carex retrorsa var. *bradleyi* (Dewey) Farw. = ***Carex* × *hartii*** Dewey
Carex retrorsa var. *gigantoides* Farw. = ***Carex retrorsa*** Schwein.
Carex retrorsa var. *hartii* (Dewey) A.Gray = ***Carex* × *hartii*** Dewey
Carex retrorsa var. *macounii* (Dewey) Fernald = ***Carex* × *macounii*** Dewey
Carex retrorsa var. *multispicula* Lepage = ***Carex retrorsa*** Schwein.
Carex retrorsa var. *robinsonii* Fernald = ***Carex retrorsa*** Schwein.
Carex retusa Degl. = ***Carex echinata*** Murray subsp. ***echinata***
Carex retzii Nees = ***Carex chinensis*** Retz.
Carex reuteriana Boiss. & Reut. = ***Carex umbrosa*** subsp. ***huetiana*** (Boiss.) Soó
Carex reversa Gilib. = ***Carex pseudocyperus*** L. var. ***pseudocyperus***
Carex reversa Spreng. = ***Carex retrorsa*** Schwein.
Carex rhaphidocarpa Kunth = ***Carex raphidocarpa*** Nees
Carex rhizantha J.F.Gmel. = ***Carex halleriana*** Asso
Carex rhizina Blytt ex Lindblom = ***Carex pediformis*** subsp. ***rhizodes*** (Blytt ex Boott) H.Lindb.
Carex rhizina subsp. *reventa* (V.I.Krecz.) T.V.Egorova = ***Carex reventa*** V.I.Krecz.
Carex rhizodes Blytt ex Boott = ***Carex pediformis*** subsp. ***rhizodes*** (Blytt ex Boott) H.Lindb.
Carex rhizodes var. *abbreviata* Meinsh. = ***Carex pediformis*** C.A.Mey. subsp. ***pediformis***
Carex rhizomatosa Steud. = ***Carex oligostachya*** Nees
Carex rhizomatosa var. *impunctata* (Boott) Kük. = ***Carex oligostachya*** Nees
Carex rhizopoda Maxim. = ***Carex hakkodensis*** Franch.
Carex rhizopoda f. *longior* Maxim. = ***Carex hakkodensis*** Franch.
Carex rhomalea (Fernald) Mack. = ***Carex saxatilis*** L.
Carex rhomboidea T.Holm = ***Carex aquatilis*** Wahlenb. var. ***aquatilis***
Carex rhynchachaenium C.B.Clarke = ***Carex truncatigluma*** subsp. ***rhynchachaenium*** (C.B. Clarke) Y.C.Tang & S.Yun Liang
Carex rhynchocarpa Heuff. = ***Carex brevicollis*** DC.
Carex rhynchophysa Liebm. = ***Carex albicans*** var. ***australis*** (L.H.Bailey) Rettig
Carex rhynchophysa Fisch., C.A.Mey. & Avé-Lall. = ***Carex utriculata*** Boott
Carex rhynchosporoides Steud. = ***Carex ramosa*** Willd.
Carex riabushinskii Kom. = ***Carex lyngbyei*** Hornem.
Carex richardii Thuill. = ***Carex canescens*** L. subsp. ***canescens***
Carex richardii Michx. = ***Carex tribuloides*** Wahlenb. var. ***tribuloides***
Carex richardsonii var. *exserta* Fernald = ***Carex richardsonii*** R.Br.
Carex richii (Fernald) Mack. = ***Carex straminea*** Willd. ex Schkuhr
Carex richmondii Boott ex C.B.Clarke = ?
Carex ×*rieseana* Figert = *C. paradoxa* × *C. remota*
Carex rigens Thuill. = ***Carex acutiformis*** Ehrh.
Carex rigens Boott = ***Carex brownii*** Tuck. subsp. ***brownii***
Carex rigens L.H.Bailey = ***Carex turbinata*** Liebm.
Carex rigescens (Franch.) V.I.Krecz. = ***Carex duriuscula*** subsp. ***rigescens*** (Franch.) S.Yun Liang & Y.C.Tang
Carex rigida Gooden. = ***Carex bigelowii*** Torr. ex Schwein. subsp. ***bigelowii***
Carex rigida Clairv. ex Steud. = ***Carex cespitosa*** L. var. ***cespitosa***
Carex rigida Schrank = ***Carex ferruginea*** Scop. subsp. ***ferruginea***
Carex rigida Tuck. = ***Carex stylosa*** C.A.Mey.
Carex rigida f. *aggregata* Almq. ex Neuman = ***Carex bigelowii*** Torr. ex Schwein. subsp. ***bigelowii***
Carex rigida subsp. *altaica* Gorodkov = ***Carex orbicularis*** subsp. ***altaica*** (Gorodkov) T.V.Egorova
Carex rigida f. *anguillata* (Drejer) Kük. = ***Carex bigelowii*** Torr. ex Schwein. subsp. ***bigelowii***
Carex rigida var. *bigelowii* (Torr. ex Schwein.) Tuck. = ***Carex bigelowii*** Torr. ex Schwein.
Carex rigida var. *bracteosa* (L.H.Bailey) L.H.Bailey = ***Carex scopulorum*** var. ***bracteosa*** (L.H.Bailey) F.J.Herm.

Carex rigida var. *concolor* (R.Br.) Kük. = ***Carex aquatilis*** var. ***minor*** Boott
Carex rigida f. *cuspidata* (Rosenv.) Kük. = ***Carex bigelowii*** Torr. ex Schwein. subsp. ***bigelowii***
Carex rigida var. *dacica* (Heuff.) Kük. = ***Carex nigra*** subsp. ***dacica*** (Heuff.) Soó
Carex rigida f. *drejeriana* (Lange) Kük. = ***Carex bigelowii*** Torr. ex Schwein. subsp. ***bigelowii***
Carex rigida var. *ensifolia* Turcz. ex Gorodkov = ***Carex bigelowii*** subsp. ***ensifolia*** (Turcz. ex Gorodkov) Holub.
Carex rigida f. *glacialis* Fr. = ***Carex bigelowii*** Torr. ex Schwein. subsp. ***bigelowii***
Carex rigida var. *goodenowii* (J.Gay) L.H.Bailey = ***Carex nigra*** (L.) Reichard subsp. ***nigra***
Carex rigida var. *hesperia* Piper = ***Carex bigelowii*** Torr. ex Schwein. subsp. ***bigelowii***
Carex rigida var. *inferalpina* Laest. = ***Carex bigelowii*** Torr. ex Schwein. subsp. ***bigelowii***
Carex rigida f. *infuscata* (Drejer) Kük. = ***Carex bigelowii*** Torr. ex Schwein. subsp. ***bigelowii***
Carex rigida var. *intricata* (Tineo) Briq. = ***Carex nigra*** subsp. ***intricata*** (Tineo) Maire & Wieller
Carex rigida subsp. *intricata* (Tineo) Nyman = ***Carex nigra*** subsp. ***intricata*** (Tineo) Maire & Wieller
Carex rigida var. *juncea* (Hartm.) L.H.Bailey = ***Carex nigra*** subsp. ***juncella*** (Fr.) Lemke
Carex rigida f. *juncelliformis* Almq. ex Neuman = ***Carex bigelowii*** Torr. ex Schwein. subsp. ***bigelowii***
Carex rigida f. *latifolia* (Andersson) Kük. = ***Carex bigelowii*** Torr. ex Schwein. subsp. ***bigelowii***
Carex rigida f. *luteosa* (Drejer) Kük. = ***Carex bigelowii*** Torr. ex Schwein. subsp. ***bigelowii***
Carex rigida f. *paradoxa* (Drejer) Kük. = ***Carex bigelowii*** Torr. ex Schwein. subsp. ***bigelowii***
Carex rigida f. *pudica* (Drejer) Kük. = ***Carex bigelowii*** Torr. ex Schwein. subsp. ***bigelowii***
Carex rigida subsp. *rigidioides* Gorodkov = ***Carex rigidioides*** (Gorodkov) V.I.Krecz.
Carex rigida var. *ripensis* Almq. =
Carex rigida var. *saxatilis* Fr. = ***Carex bigelowii*** Torr. ex Schwein. subsp. ***bigelowii***
Carex rigida var. *strictiformis* L.H.Bailey = ***Carex nigra*** (L.) Reichard subsp. ***nigra***
Carex rigida var. *subconcoloroides* (Kelso) Kelso =
Carex rigidifolia Hochst. ex Seub. = ***Carex hochstetteriana*** J.Gay ex Seub.
Carex riishirensis Franch. = ***Carex scita*** var. ***riishirensis*** (Franch.) Kük.
Carex riloensis Stoeva & E.D.Popova = [13 BUL]
Carex ringgoldiana Boott = ***Carex ischnostachya*** Steud.
Carex ringgoldiana var. *stenandra* Franch. & Sav. = ***Carex ischnostachya*** Steud.
Carex riparia Moench ex Steud. = ***Carex acuta*** L.
Carex riparia Geners. = ***Carex acutiformis*** Ehrh.
Carex riparia Muhl. = ***Carex lacustris*** Willd.
Carex riparia (R.Br.) Poir. = ***Uncinia riparia*** R.Br.
Carex riparia var. *acutiformis* (Ehrh.) Fiori = ***Carex acutiformis*** Ehrh.
Carex riparia f. *aristata* Asch. & Graebn. = ***Carex riparia*** Curtis subsp. ***riparia***
Carex riparia f. *brevipalea* Neuman = ***Carex riparia*** Curtis subsp. ***riparia***
Carex riparia var. *chilensis* (Brongn.) Kük. = ***Carex riparia*** subsp. ***chilensis*** (Brongn.) Kük.
Carex riparia var. *claviformis* J.A.Schmidt = ***Carex riparia*** Curtis subsp. ***riparia***
Carex riparia var. *exaltata* (Peterm.) Nyman = ***Carex riparia*** Curtis subsp. ***riparia***
Carex riparia f. *flagelliformis* Waisb. = ***Carex riparia*** Curtis subsp. ***riparia***
Carex riparia subsp. *gracilescens* (Hartm. ex Andersson) Nyman = ***Carex riparia*** Curtis subsp. ***riparia***
Carex riparia var. *humilis* Uechtr. = ***Carex riparia*** Curtis subsp. ***riparia***
Carex riparia var. *impressa* S.H.Wright = ***Carex hyalinolepis*** Steud.
Carex riparia f. *incrassata* (Schltdl.) Kük. = ***Carex riparia*** Curtis subsp. ***riparia***
Carex riparia var. *inferior* Peck = ***Carex riparia*** Curtis subsp. ***riparia***
Carex riparia var. *lacustris* (Willd.) Kük. = ***Carex lacustris*** Willd.
Carex riparia f. *laevigata* Peterm. = ***Carex riparia*** Curtis subsp. ***riparia***
Carex riparia var. *pseudopaludosa* Schur = ***Carex riparia*** Curtis subsp. ***riparia***
Carex riparia f. *reticulosa* Torges = ***Carex riparia*** Curtis subsp. ***riparia***
Carex riparia var. *rugulosa* (Kük.) Kük. = ***Carex rugulosa*** Kük.
Carex riparia var. *subgracilescens* Kük. = ***Carex riparia*** Curtis subsp. ***riparia***
Carex ripariiformis Litv. = ***Carex melanostachya*** M.Bieb. ex Willd.
Carex rivalis Willk. & Lange = ***Carex divisa*** Huds.
Carex rivularis Schkuhr = ***Carex divisa*** Huds.
Carex rivulorum Ridl. = ***Carex perakensis*** var. ***borneensis*** (C.B.Clarke) Noot.
Carex roborowskii V.I.Krecz. = ***Carex pseudofoetida*** Kük. ex Ostenf. subsp. ***pseudofoetida***
Carex robusta F.Nyl. = ***Carex utriculata*** Boott
Carex rochebrunii Franch. & Sav. = ***Carex remota*** subsp. ***rochebrunii*** (Franch. & Sav.) Kük.
Carex rochebrunii subsp. *remotispicula* (Hayata) T.Koyama = ***Carex remota*** var. ***remotispicula*** (Hayata) Ohwi
Carex rochebrunii subsp. *reptans* (Franch.) S.Yun Liang & Y.C.Tang = ***Carex remota*** var. ***reptans*** Franch.
Carex rodnensis (Porcius ex Asch. & Graebn.) Rouy = ***Carex curvula*** All. subsp. ***curvula***
Carex rodriguesii Merino = ***Carex mairii*** Coss. & Germ.
Carex roemeriana Scheele = ***Carex microdonta*** Torr.
Carex romanzowiana Cham. ex Steud. = ***Carex lyngbyei*** Hornem.
Carex rosae (Gilomen) Holub = ***Carex curvula*** subsp. ***rosae*** Gilomen
Carex rosea var. *arkansana* L.H.Bailey = ***Carex arkansana*** (L.H.Bailey) L.H.Bailey
Carex rosea f. *debilis* Farw. = ***Carex rosea*** Willd.
Carex rosea var. *minor* Boott = ***Carex radiata*** (Wahlenb.) Small
Carex rosea var. *pusilla* Peck = ***Carex rosea*** Willd.
Carex rosea var. *radiata* (Wahlenb.) Dewey = ***Carex radiata*** (Wahlenb.) Small
Carex rosea var. *retroflexa* (Muhl. ex Willd.) Torr. = ***Carex retroflexa*** Muhl. ex Willd.
Carex rosea var. *staminata* Peck = ***Carex rosea*** Willd.
Carex rosea var. *texensis* Torr. ex L.H.Bailey = ***Carex texensis*** (Torr. ex L.H.Bailey) L.H.Bailey
Carex ×*rossiana* Degen = [13 YUG]
Carex rossii var. *brevipes* (Boott) Kük. = ***Carex rossii*** Boott
Carex rostellata Gay ex Boott = ***Carex notha*** Kunth
Carex rostrata Muhl. ex Willd. = ***Carex lurida*** Wahlenb.
Carex rostrata Michx. = ***Carex michauxiana*** Boeck. subsp. ***michauxiana***
Carex rostrata Hoppe ex Schkuhr = ***Carex michelii*** Host

Carex rostrata var. *anticostensis* (Fernald) Fernald = ***Carex* × *anticostensis*** (Fernald) Lepage
Carex rostrata var. *borealis* (Hartm.) Kük. = ***Carex rostrata*** Stokes var. ***rostrata***
Carex rostrata var. *brunnescens* (Andersson) Kük. = ***Carex rostrata*** Stokes var. ***rostrata***
Carex rostrata var. *cliftonii* Farw. = ***Carex rostrata*** Stokes var. ***rostrata***
Carex rostrata f. *elatior* Lange = ***Carex rostrata*** Stokes var. ***rostrata***
Carex rostrata f. *elatior* (A.Blytt) Kük. = ***Carex rostrata*** Stokes var. ***rostrata***
Carex rostrata var. *hymenocarpa* (Drejer) H.Christ = ***Carex rostrata*** Stokes var. ***rostrata***
Carex rostrata f. *laeta* Norman = ***Carex rostrata*** Stokes var. ***rostrata***
Carex rostrata var. *latifolia* Asch. = ***Carex rostrata*** Stokes var. ***rostrata***
Carex rostrata f. *longipalea* Neuman = ***Carex rostrata*** Stokes var. ***rostrata***
Carex rostrata f. *minor* (Boott) Kük. = ***Carex utriculata*** Boott
Carex rostrata f. *pendulina* (A.Blytt) Kük. = ***Carex rostrata*** Stokes var. ***rostrata***
Carex rostrata f. *plumosa* Norman = ***Carex rostrata*** Stokes var. ***rostrata***
Carex rostrata subsp. *rotundata* (Wahlenb.) Kük. = ***Carex rotundata*** Wahlenb.
Carex rostrata f. *sommieri* H.Christ = ***Carex rostrata*** Stokes var. ***rostrata***
Carex rostrata f. *sparangiformis* Murr = ***Carex rostrata*** Stokes var. ***rostrata***
Carex rostrata var. *utriculata* (Boott) L.H.Bailey = ***Carex utriculata*** Boott
Carex rostrata subsp. *utriculata* (Boott) Soják = ***Carex utriculata*** Boott
Carex rotae De Not. = ***Carex paniculata*** L. subsp. ***paniculata***
Carex rotoensis Petrie = [51 CTM]
Carex rotundata var. *compacta* (Dewey) Boivin = ***Carex membranacea*** Hook.
Carex rotundata var. *major* Meinsh. = ***Carex rostrata*** Stokes var. ***rostrata***
Carex rousseaui Raymond = ***Carex haydenii*** Dewey
Carex rouyana Franch. = ***Carex filipes*** var. ***sparsinux*** (C.B.Clarke ex Franch.) Kük.
Carex rouyana var. *arakiana* Ohwi = ***Carex filipes*** var. ***arakiana*** (Ohwi) Ohwi
Carex rubella Boott = ***Carex spicigera*** Nees
Carex rubescens Boeck. = ***Carex dimorpholepis*** Steud.
Carex rubra H.Lév. & Vaniot = ***Carex cespitosa*** L. var. ***cespitosa***
Carex rubrobrunnea var. *taliensis* (Franch.) Kük. = ***Carex rubrobrunnea*** C.B.Clarke
Carex rudis Wimm. = ***Carex acuta*** L.
Carex ruesanensis Kudô = ***Carex rotundata*** Wahlenb.
Carex rufa Schrank = ***Carex acutiformis*** Ehrh.
Carex rufa Lam. = ***Carex nigra*** (L.) Reichard subsp. ***nigra***
Carex rufa (Nees) Baill. = ***Schoenoxiphium rufum*** Nees
Carex rufa var. *seminuda* Beck = ***Carex acuta*** L.
Carex ruffa Brot. = ***Carex riparia*** Curtis subsp. ***riparia***
Carex rufidula Steud. = ***Carex glaucescens*** Elliott
Carex rufovariegata Boeck. = ***Carex petasata*** Dewey
Carex rugata Fernald = ***Carex corrugata*** Fernald
Carex rugeliana Kunze = ***Carex misera*** Buckley
Carex rugosperma Mack. = ***Carex tonsa*** var. ***rugosperma*** (Mack.) Crins
Carex rugosperma var. *tonsa* (Fernald) E.G.Voss = ***Carex tonsa*** (Fernald) E.P.Bicknell
Carex ruiziana Boeck. = ***Carex pichinchensis*** Kunth
Carex runssoroensis var. *aberdarensis* Kük. = ***Carex runssoroensis*** K.Schum.
Carex rupestris Turcz. ex Ledeb. = ***Carex argunensis*** Turcz. ex Ledeb. subsp. ***argunensis***
Carex rupestris subsp. *alticola* (Popl. ex Sukaczev) Vorosch. = ***Carex argunensis*** subsp. ***alticola*** (Popl. ex Sukaczev) Malyschev
Carex rupestris subsp. *argunensis* (Turcz. ex Ledeb.) Vorosch. = ***Carex argunensis*** Turcz. ex Ledeb.
Carex rupestris subsp. *drummondiana* (Dewey) Holub = ***Carex rupestris*** All. subsp. ***rupestris***
Carex rupestris var. *drummondiana* (Dewey) L.H.Bailey = ***Carex rupestris*** All. subsp. ***rupestris***
Carex rusbyi Mack. = ***Carex vallicola*** Dewey
Carex ruthenica V.I.Krecz. = ***Carex caryophyllea*** Latourr. var. ***caryophyllea***
Carex rutilus Humn. = ?
Carex sabuletorum M.Bieb. ex Boott = ***Carex arenaria*** L.
Carex sabulosa subsp. *leiophylla* (Mack.) Porsild = ***Carex sabulosa*** Turcz. ex Kunth
Carex sabynensis Less. ex Kunth = ***Carex umbrosa*** subsp. ***sabynensis*** (Less. ex Kunth) Kük.
Carex sabynensis var. *laxior* Trautv. = ***Carex umbrosa*** Host subsp. ***umbrosa***
Carex sabynensis var. *rostrata* (Maxim.) Ohwi = ***Carex umbrosa*** var. ***rostrata*** (Maxim.) Govaerts
Carex sachalinensis subsp. *alterniflora* (Franch.) T.Koyama = ***Carex sachalinensis*** var. ***alterniflora*** (Franch.) Ohwi
Carex sachalinensis subsp. *fernaldiana* (H.Lév. & Vaniot) T.Koyama = ***Carex sachalinensis*** var. ***fernaldiana*** (H.Lév. & Vaniot) T.Koyama
Carex sachapata Steud. = ***Carex pichinchensis*** Kunth
Carex sadae H.Lév. & Vaniot = ***Carex blepharicarpa*** Franch. var. ***blepharicarpa***
Carex sadleri E.F.Linton = ***Carex binervis*** Sm.
Carex sagei Phil. = ***Carex fuscula*** d'Urv. var. ***fuscula***
Carex sagittifera Lowe = ***Carex peregrina*** Link
Carex salina Bong. ex Ledeb. = ***Carex canescens*** L. subsp. ***canescens***
Carex salina Boott = ***Carex lyngbyei*** Hornem.
Carex salina var. *ambusta* L.H.Bailey = ***Carex vesicaria*** L.
Carex salina var. *borealis* Almq. = ***Carex salina*** Wahlenb. var. ***salina***
Carex salina var. *caespitosa* A.Blytt = ***Carex recta*** Boott
Carex salina f. *caespitosa* (A.Blytt) Kük. = ***Carex recta*** Boott
Carex salina var. *concolor* (Drejer) Almq. = ***Carex lyngbyei*** Hornem.
Carex salina f. *corylorum* B.Boivin = ***Carex salina*** Wahlenb.
Carex salina var. *cuspidata* Wahlenb. = ***Carex salina*** Wahlenb. var. ***salina***
Carex salina f. *discolor* (F.Nyl.) Almq. = ***Carex salina*** Wahlenb. var. ***salina***
Carex salina f. *elata* A.Blytt = ***Carex recta*** Boott
Carex salina var. *epigeios* (Fr.) Nyman = ***Carex salina*** Wahlenb. var. ***salina***
Carex salina f. *filipendula* (Drejer) Almq. = ***Carex lyngbyei*** Hornem.
Carex salina var. *filipendula* A.Blytt = ***Carex salina*** Wahlenb. var. ***salina***
Carex salina var. *flavicans* (F.Nyl.) Almq. = ***Carex* × *flavicans*** F.Nyl.
Carex salina subsp. *fuliginea* A.Blytt = ***Carex recta*** Boott
Carex salina var. *haematolepis* (Drejer) Almq. = ***Carex* × *haematolepis*** Drejer

Carex salina subsp. *halophila* (F.Nyl.) Almq. = ***Carex recta*** Boott
Carex salina var. *hyperborea* (Drejer) Nyman = ***Carex bigelowii*** Torr. ex Schwein. subsp. ***bigelowii***
Carex salina var. *kattegatensis* Almq. = ***Carex vacillans*** Drejer
Carex salina subsp. *kattegatensis* (Almq.) Nyman = ***Carex vacillans*** Drejer
Carex salina var. *lanceata* (Dewey) Kük. = ***Carex salina*** Wahlenb. var. ***salina***
Carex salina var. *minor* Boott = ***Carex subspathacea*** Wormsk. ex Hornem.
Carex salina var. *mutica* Wahlenb. = *C. aquatilis* × *C. flavicans*
Carex salina f. *nana* Trautv. = ***Carex subspathacea*** Wormsk. ex Hornem.
Carex salina subsp. *obtusa* A.Blytt = ***Carex recta*** Boott
Carex salina f. *obtusa* (A.Blytt) Kük. = ***Carex recta*** Boott
Carex salina f. *ostrobottnica* Almq. = ***Carex recta*** Boott
Carex salina f. *pallida* A.Blytt = ***Carex recta*** Boott
Carex salina var. *pumila* A.Blytt = ***Carex subspathacea*** Wormsk. ex Hornem.
Carex salina subsp. *ramenskii* (Kom.) T.V.Egorova = ***Carex ramenskii*** Kom.
Carex salina var. *reducta* (Drejer) Nyman = ***Carex ×reducta*** Drejer
Carex salina f. *reducta* (Drejer) Almq. = ***Carex ×reducta*** Drejer
Carex salina var. *robusta* L.H.Bailey = ***Carex lyngbyei*** Hornem.
Carex salina f. *stricta* (Drejer) Almq. = ***Carex subspathacea*** Wormsk. ex Hornem.
Carex salina var. *subspathacea* (Wormsk. ex Hornem.) Tuck. = ***Carex subspathacea*** Wormsk. ex Hornem.
Carex salina subsp. *subspathacea* (Wormsk. ex Hornem.) Nyman = ***Carex subspathacea*** Wormsk. ex Hornem.
Carex salina f. *thulensis* Kük. = ***Carex salina*** Wahlenb. var. ***salina***
Carex salina var. *tristigmatica* Kük. = ***Carex ramenskii*** Kom.
Carex saliniformis Mack. = ***Carex hassei*** L.H.Bailey
Carex salinoides Beurl. = ***Carex salina*** Wahlenb. var. ***salina***
Carex ×salisiana Brügger = ?
Carex saltuensis L.H.Bailey = ***Carex vaginata*** Tausch var. ***vaginata***
Carex samoensis Boeck. = ***Carex maculata*** Boott var. ***maculata***
Carex sandwicensis Boeck. = [63 HAW]
Carex sanguinea Boott = ***Kobresia sanguinea*** (Boott) Raymond
Carex ×sanionis K.Richt. = ?
Carex sarda Link = ***Carex distachya*** Desf. var. ***distachya***
Carex sareptana V.I.Krecz. = ***Carex acuta*** L.
Carex ×sarntheinii Vetter = ?
Carex sartwellii var. *occidentalis* L.H.Bailey = ***Carex praegracilis*** W.Boott
Carex saruensis Franch. = ***Carex miyabei*** Franch. var. ***miyabei***
Carex sasakii Hayata = ***Carex alopecuroides*** D.Don ex Tilloch & Taylor
Carex saskatschewana Boeck. = ***Carex capillaris*** L. subsp. ***capillaris***
Carex satakeana T.Koyama = ***Carex orbicularis*** Boott subsp. ***orbicularis***
Carex saturata C.B.Clarke = ***Carex filicina*** Nees
Carex savatieri Franch. = ***Carex curvicollis*** Franch. & Sav.
Carex saxatilis var. *bigelowii* (Torr. ex Schwein.) Torr. = ***Carex bigelowii*** Torr. ex Schwein.
Carex saxatilis var. *grahamii* L.H.Bailey = ***Carex saxatilis*** L.
Carex saxatilis var. *grahamii* (Boott) Hook. & Arn. = ***Carex ×stenolepis*** Less. nothovar. ***stenolepis***
Carex saxatilis subsp. *grahamii* (Boott) Nyman = ***Carex ×stenolepis*** Less. nothovar. ***stenolepis***
Carex saxatilis f. *infuscata* Drejer = ***Carex bigelowii*** Torr. ex Schwein. subsp. ***bigelowii***
Carex saxatilis var. *laxa* (Trautv.) Ohwi = ***Carex saxatilis*** L.
Carex saxatilis subsp. *laxa* (Trautv.) Kalela = ***Carex saxatilis*** L.
Carex saxatilis f. *longipedunculata* (Lepage) Lepage = ***Carex saxatilis*** L.
Carex saxatilis f. *luteosa* Drejer = ***Carex bigelowii*** Torr. ex Schwein. subsp. ***bigelowii***
Carex saxatilis var. *major* Olney = ***Carex saxatilis*** L.
Carex saxatilis var. *miliaris* (Michx.) L.H.Bailey = ***Carex saxatilis*** L.
Carex saxatilis var. *pallidior* A.Blytt = ***Carex bigelowii*** Torr. ex Schwein. subsp. ***bigelowii***
Carex saxatilis f. *pudica* Drejer = ***Carex bigelowii*** Torr. ex Schwein. subsp. ***bigelowii***
Carex saxatilis var. *rhomalea* Fernald = ***Carex saxatilis*** L.
Carex saxatilis var. *rotundata* (Wahlenb.) Gelting = ***Carex rotundata*** Wahlenb.
Carex ×saxenii var. *dumanii* (Lepage) Boivin = ***Carex ×dumanii*** Lepage
Carex ×saxenii var. *ferruginea* Lepage = ***Carex recta*** Boott
Carex saxilittoralis A.Robertson = ***Carex viridula*** var. ***saxilittoralis*** (A.Robertson) Crins
Carex scaberima C.B.Clarke = ***Carex cruciata*** var. ***rafflesiana*** (Boott) Noot.
Carex scaberrima (Boeck.) C.B.Clarke = ***Carex cruciata*** var. ***rafflesiana*** (Boott) Noot.
Carex scaberrima Scheele = ?
Carex scabra Hoppe = ***Carex davalliana*** Sm.
Carex scabriculmis (Kük.) Ohwi = ***Carex teinogyna*** Boott
Carex scabricuspis V.I.Krecz. = ***Carex caryophyllea*** Latourr. var. ***caryophyllea***
Carex scabrinervia Franch. = ***Carex scita*** var. ***scabrinervia*** (Franch.) Kük.
Carex scabrior Sartwell ex Boott = ***Carex vulpinoidea*** Michx.
Carex scabroaristata Akiyama = ***Carex sachalinensis*** var. ***alterniflora*** (Franch.) Ohwi
Carex scandinavica E.W.Davies = ***Carex viridula*** Michx. subsp. ***viridula***
Carex scaposa var. *baviensis* Franch. = ***Carex adrienii*** E.G.Camus
Carex scaposa var. *marantacea* Raymond = ***Carex scaposa*** C.B.Clarke var. ***scaposa***
Carex scariosa Lam. = ***Carex humilis*** Leyss. var. ***humilis***
Carex schaffneri W.Boott ex S.Watson = ***Carex potosina*** Hemsl.
Carex ×schalkertii Murr = ?
Carex schedonautos Steud. = ***Carex diclina*** Phil.
Carex schenckiana Boeck. = ***Carex phalaroides*** Kunth var. ***phalaroides***
Carex scheuchzeri J.F.Gmel. = ***Carex acutiformis*** Ehrh.
Carex scheuchzeri Gaudin ex Boott = ***Carex brachystachys*** Schrank
Carex schiedeana var. *perstricta* (Mack.) F.J.Herm. = ***Carex schiedeana*** Kunze
Carex schiedeana var. *stellata* (Mack.) F.J.Herm. = ***Carex schiedeana*** Kunze
Carex schimperiana Boeck. = ***Schoenoxiphium lanceum*** (Thunb.) Kük.

Carex schistorhyncha H.Lév. & Vaniot = ***Carex simulans*** C.B.Clarke
Carex schkuhriana H.Lév. & Vaniot = ***Carex dimorpholepis*** Steud.
Carex schkuhrii Willd. = ***Carex liparocarpos*** subsp. ***bordzilowskii*** (V.I.Krecz.) T.V.Egorova
Carex schlagintweitiana Boeck. = ***Carex setigera*** var. ***schlagintweitiana*** (Boeck.) Kük.
Carex schlagintweitiana subsp. *deformis* Noltie = ***Carex setigera*** var. ***schlagintweitiana*** (Boeck.) Kük.
Carex schmidtii Boeck. = ***Carex stenantha*** Franch. & Sav. var. ***stenantha***
Carex schoenoides Schrank = ***Carex colchica*** J.Gay
Carex schoenoides DC. = ***Carex disticha*** Huds.
Carex schoenoides Thuill. = ***Carex divisa*** Huds.
Carex schoenoides Desf. = ***Carex divisa*** Huds.
Carex schottii Boiss. = ***Carex tristis*** M.Bieb.
Carex schraderi Willd. = ***Carex laevigata*** Sm.
Carex schreberi Desv. = ***Carex arenaria*** L.
Carex schreberi Ten. ex Boott = ***Carex divisa*** Huds.
Carex schreberi Schrank = ***Carex praecox*** Schreb.
Carex schreberi Willd. = ***Carex praecox*** Schreb.
Carex schreberi subsp. *curvata* (Asch.) P.Fourn. = ***Carex praecox*** Schreb.
Carex schreberi var. *curvata* (Asch.) Nyman = ***Carex praecox*** Schreb.
Carex schreberi var. *gracillima* Asch. = ***Carex praecox*** Schreb.
Carex schreberi subsp. *ligerica* (J.Gay) Hartm. = ***Carex colchica*** J.Gay
Carex schreberi subsp. *ludibunda* (J.Gay) Nyman = ***Carex × ludibunda*** J.Gay
Carex schreberi f. *nana* Zapal. = ***Carex praecox*** Schreb.
Carex schreberi subsp. *ohmuelleriana* (O.Lang) Nyman = ***Carex arenaria*** L.
Carex schreberi var. *pallida* Peterm. = ***Carex praecox*** Schreb.
Carex schreberi f. *podolica* Zapal. = ***Carex praecox*** Schreb.
Carex schrenkiana C.A.Mey. ex Ledeb. = ***Carex vesicaria*** L.
Carex × schuetzeana Figert = ?
Carex schummelii Siegert = ***Carex acuta*** L.
Carex schweinitzia Eaton = ***Carex schweinitzii*** Dewey ex Schwein.
Carex scirpiformis Mack. = ***Carex scirpoidea*** Michx. var. ***scirpoidea***
Carex scirpina Tuck. ex Boott = ***Carex scirpoidea*** Michx. var. ***scirpoidea***
Carex scirpina (Willd.) Missbach & E.H.L.Krause = ***Kobresia myosuroides*** (Vill.) Fiori & Paol.
Carex scirpoidea f. *basigyna* Lange = ***Carex scirpoidea*** Michx. var. ***scirpoidea***
Carex scirpoidea subsp. *convoluta* (Kük.) Dunlop = ***Carex scirpoidea*** var. ***convoluta*** Kük.
Carex scirpoidea var. *curatorum* (Stacey) Cronquist = ***Carex curatorum*** Stacey
Carex scirpoidea var. *europaea* Kük. = ***Carex scirpoidea*** Michx. var. ***scirpoidea***
Carex scirpoidea var. *gigas* T.Holm = ***Carex scabriuscula*** Mack.
Carex scirpoidea subsp. *pseudoscirpoidea* (Rydb.) Dunlop = ***Carex scirpoidea*** var. ***pseudoscirpoidea*** (Rydb.) Cronquist
Carex scirpoidea var. *scirpiformis* (Mack.) O'Neill & Duman = ***Carex scirpoidea*** Michx. var. ***scirpoidea***
Carex scirpoidea subsp. *stenochlaena* (T.Holm) Á.Löve & D.Löve = ***Carex scirpoidea*** var. ***stenochlaena*** T.Holm
Carex scirpoidea var. *typica* B.Boivin = ***Carex scirpoidea*** Michx.
Carex scirpoides Willd. = ***Carex interior*** L.H.Bailey
Carex scirpoides Sartwell = ***Carex scirpoidea*** Michx. var. ***scirpoidea***
Carex scirpoides var. *capillacea* (L.H.Bailey) Fernald = ***Carex atlantica*** subsp. ***capillacea*** (L.H.Bailey) Reznicek
Carex scirpoides var. *josselynii* (Fernald) Fernald = ***Carex echinata*** Murray subsp. ***echinata***
Carex scirrobasis Kitag. = ***Carex humilis*** var. ***scirrobasis*** (Kitag.) Y.L.Chang & Y.L.Yang
Carex scita var. *brevisquama* (Koidz.) Ohwi = ***Carex scita*** Maxim. var. ***scita***
Carex scita var. *korginensis* (Meinsh.) Kük. = ***Carex scita*** var. ***riishirensis*** (Franch.) Kük.
Carex sclerioides Ridl. = ***Carex filicina*** Nees
Carex scoparia Torr. = ***Carex muskingumensis*** Schwein.
Carex scoparia var. *aggregata* Dewey = ***Carex scoparia*** Willd. var. ***scoparia***
Carex scoparia var. *condensa* Fernald = ***Carex scoparia*** Willd. var. ***scoparia***
Carex scoparia f. *condensa* (Fernald) Kük. = ***Carex scoparia*** Willd. var. ***scoparia***
Carex scoparia f. *elatior* Peck = ***Carex scoparia*** Willd. var. ***scoparia***
Carex scoparia var. *fulva* Boott = ***Carex straminea*** Willd. ex Schkuhr
Carex scoparia var. *lagopodioides* (Willd.) Torr. = ***Carex tribuloides*** Wahlenb. var. ***tribuloides***
Carex scoparia var. *minor* Boott = ***Carex crawfordii*** Fernald
Carex scoparia var. *moniliformis* Tuck. = ***Carex scoparia*** Willd. var. ***scoparia***
Carex scoparia f. *moniliformis* (Tuck.) Kük. = ***Carex scoparia*** Willd. var. ***scoparia***
Carex scoparia var. *muskingumensis* (Schwein.) Tuck. = ***Carex muskingumensis*** Schwein.
Carex scoparia f. *peracuta* Fernald = ***Carex scoparia*** Willd. var. ***scoparia***
Carex scoparia var. *subturbinata* Fernald & Wiegand = ***Carex scoparia*** Willd. var. ***scoparia***
Carex scoparia f. *subturbinata* (Fernald & Wiegand) Fernald = ***Carex scoparia*** Willd. var. ***scoparia***
Carex scoparia f. *tenerrima* Kük. = ***Carex scoparia*** Willd. var. ***scoparia***
Carex scoparia var. *vera* Tuck. = ***Carex scoparia*** Willd. var. ***scoparia***
Carex scopoliana Willd. = ***Carex ferruginea*** Scop. subsp. ***ferruginea***
Carex scopolii Gaudin = ***Carex ferruginea*** Scop. subsp. ***ferruginea***
Carex scopulorum var. *chimaphila* (T.Holm) Kük. = ***Carex scopulorum*** T.Holm var. ***scopulorum***
Carex scopulorum var. *subconcoloroides* Kelso =
Carex scordiae Lojac. = [13 SIC]
Carex scotica Spreng. = ***Carex vaginata*** Tausch var. ***vaginata***
Carex scouleri Torr. = ***Carex lyngbyei*** Hornem.
Carex seatoniana L.H.Bailey = ***Carex spissa*** var. ***seatoniana*** (L.H.Bailey) Kük.
Carex secale Ledeb. = ***Carex secalina*** Willd. ex Wahlenb. var. ***secalina***
Carex secalina Lapeyr. = ***Carex riparia*** Curtis subsp. ***riparia***
Carex seckeri C.B.Clarke = ?
Carex secta var. *tenuiculmis* Petrie = ***Carex tenuiculmis*** (Petrie) Heenan & de Lange
Carex seditiosa Steud. = ***Carex fecunda*** Steud. var. ***fecunda***

Carex seiskoensis Freyn = ***Carex ledebouriana*** C.A.Mey. ex Trevir. subsp. ***ledebouriana***
Carex sekimotoi Honda = ***Carex papulosa*** Boott
Carex selengensis K.V.Ivanova = ***Carex karoi*** (Freyn) Freyn
Carex sellowiana var. *prorepens* Kük. = ***Carex sellowiana*** Schltdl.
Carex sellowiana var. *subrepens* Kük. = ***Carex sellowiana*** Schltdl.
Carex sellowiana var. *tenuis* Boeck. = ***Carex sellowiana*** Schltdl.
Carex semicrinita C.B.Clarke = ?
Carex semicylindrica Kit. = ?
Carex semiforsteri C.B.Clarke ex Cheeseman = ***Carex forsteri*** Wahlenb.
Carex semiforsteri C.B.Clarke = ***Carex forsteri*** Wahlenb.
Carex semiglomerata Kük. = ***Carex cruciata*** var. ***rafflesiana*** (Boott) Noot.
Carex semiplena Kük. = ***Carex middendorffii*** var. ***kirigaminensis*** (Ohwi) Ohwi
Carex semiplena var. *tenuinervis* Kük. = ***Carex gaudichaudiana*** Kunth
Carex sempervirens Schwein. = ***Carex glaucescens*** Elliott
Carex sempervirens Elliott ex Kunth = ***Carex stricta*** Lam.
Carex sempervirens f. *albicans* Lackow. ex Kük. = ***Carex sempervirens*** Vill.
Carex sempervirens var. *aurigerana* Marcailhou = ***Carex sempervirens*** Vill.
Carex sempervirens f. *coarctata* Huter ex Kneuck. = ***Carex sempervirens*** Vill.
Carex sempervirens f. *fastigiata* Lackow. ex Kük. = ***Carex sempervirens*** Vill.
Carex sempervirens subsp. *granitica* (Braun-Blanq.) Vicioso = ***Carex sempervirens*** Vill.
Carex sempervirens subsp. *laevis* Kit. ex Willd. = ***Carex sempervirens*** Vill.
Carex sempervirens var. *laevis* (Kit. ex Willd.) Asch. & Kantiz = ***Carex sempervirens*** Vill.
Carex sempervirens f. *laxiflora* Schur = ***Carex sempervirens*** Vill.
Carex sempervirens var. *pendulina* Kük. = ***Carex sempervirens*** Vill.
Carex sempervirens var. *pseudotristis* Domin = ***Carex sempervirens*** Vill.
Carex sempervirens subsp. *pseudotristis* (Domin) Pawl. = ***Carex sempervirens*** Vill.
Carex sempervirens f. *pumila* Schur = ***Carex sempervirens*** Vill.
Carex sempervirens f. *rigida* Schur = ***Carex sempervirens*** Vill.
Carex sempervirens f. *schkuhriana* Bonnet & J.A.Richt. = ***Carex sempervirens*** Vill.
Carex sempervirens f. *semipallescens* Kneuck. = ***Carex sempervirens*** Vill.
Carex sempervirens var. *setifolia* Kumm. & Sendtn. = ***Carex sempervirens*** Vill.
Carex sempervirens f. *stenophylla* Schur = ***Carex sempervirens*** Vill.
Carex sempervirens var. *tenax* H.Christ = ***Carex ferruginea*** subsp. ***australpina*** (Bech.) W.Dietr.
Carex sempervirens var. *trichocarpa* Schur = ***Carex sempervirens*** Vill.
Carex sempervirens subsp. *tristis* (M.Bieb.) Kük. = ***Carex tristis*** M.Bieb.
Carex sempervirens f. *villarsiana* Bonnet & J.A.Richt. = ***Carex sempervirens*** Vill.
Carex senanensis Ohwi = ***Carex deweyana*** var. ***senanensis*** (Ohwi) T.Koyama
Carex × *senayana* Soó = ?
Carex sendaica Franch. = ***Carex lenta*** var. ***sendaica*** (Franch.) T.Koyama
Carex sendaica var. *pseudosendaica* T.Koyama = ***Carex lenta*** var. ***sendaica*** (Franch.) T.Koyama
Carex sendtneriana Brügger = ***Carex heleonastes*** Ehrh. ex L.f. subsp. ***heleonastes***
Carex sepulta Phil. = ***Carex pumila*** Thunb.
Carex serotina Ten. = ***Carex muricata*** L.
Carex serotina Mérat = ***Carex viridula*** Michx. subsp. ***viridula***
Carex serotina f. *alpestris* (Asch. & Graebn.) Soó = ***Carex viridula*** Michx. subsp. ***viridula***
Carex serotina f. *argillacea* (Townson) Soó = ***Carex pilulifera*** L. subsp. ***pilulifera***
Carex serotina f. *canaliculata* (Callmé) Soó = ***Carex viridula*** Michx. subsp. ***viridula***
Carex serotina f. *cyperoides* (T.Marsson) Soó = ***Carex viridula*** Michx. subsp. ***viridula***
Carex serotina f. *exigua* (Schur) Soó = ***Carex viridula*** Michx. subsp. ***viridula***
Carex serotina var. *fennica* (Palmgr.) T.V.Egorova = ***Carex viridula*** Michx. subsp. ***viridula***
Carex serotina subsp. *philocrena* (V.I.Krecz.) Kukkonen = ***Carex viridula*** Michx. subsp. ***viridula***
Carex serotina subsp. *pulchella* (Lönnr.) Ooststr. = ***Carex viridula*** Michx. subsp. ***viridula***
Carex serotina f. *subglobosa* (Miel.) Soó = ***Carex viridula*** Michx. subsp. ***viridula***
Carex serotina f. *tenuis* (Zapal.) Soó = ***Carex viridula*** Michx. subsp. ***viridula***
Carex serotina f. *thalassica* (Asch. & Graebn.) Soó = ***Carex viridula*** Michx. subsp. ***viridula***
Carex serranoi Phil. = ***Carex darwinii*** var. ***serranoi*** (Phil.) Kük.
Carex × *serravalensis* Beauverd = ?
Carex serrulata Mutel = ***Carex echinata*** Murray subsp. ***echinata***
Carex serrulata Biv. ex Spreng. = ***Carex flacca*** subsp. ***erythrostachys*** (Hoppe) Holub
Carex serrulata var. *corcyrensis* Kük. = ***Carex flacca*** Schreb. subsp. ***flacca***
Carex setacea Froel. ex Kunth = ***Carex mucronata*** All.
Carex setacea Frank ex Kunth = ***Carex stipata*** Muhl. ex Willd. var. ***stipata***
Carex setacea Dewey = ***Carex vulpinoidea*** Michx.
Carex setacea var. *ambigua* (Barratt) Fernald = ***Carex annectens*** (E.P.Bicknell) E.P.Bicknell
Carex setariiformis Turcz. ex Boott = ***Carex leiorhyncha*** C.A.Mey.
Carex setifolia Godr. = ***Carex divisa*** Huds.
Carex setifolia (Dewey) Britton & A.Br. = ***Carex eburnea*** Boott
Carex setifolia Heer = ?
Carex setifolia var. *colchaguensis* (Phil.) Kük. = ***Carex setifolia*** Kunze var. ***setifolia***
Carex setifolia var. *minor* Boott = ***Carex setifolia*** Kunze var. ***setifolia***
Carex setigera var. *fascicularis* Strachey = ***Carex inanis*** Kunth
Carex setigera var. *fasciculata* Strachey = ***Carex winterbottomii*** C.B.Clarke
Carex setigera var. *humilis* Nees = ***Carex inanis*** Kunth
Carex setigera var. *minor* Strachey = ***Carex inanis*** Kunth
Carex setigera var. *ustilaginosa* Nees = ***Carex setigera*** D.Don var. ***setigera***
Carex setina (H.Christ) V.I.Krecz. = ***Carex maritima*** Gunnerus
Carex setosa Franch. & Sav. = ***Carex caryophyllea*** Latourr. var. ***caryophyllea***

Carex setulifolia Nelmes = ***Carex perakensis*** C.B.Clarke var. ***perakensis***
Carex sewellii A.Benn. & C.B.Clarke = ***Carex lachenalii*** Schkuhr subsp. ***lachenalii***
Carex sexspicata Colenso = ***Carex forsteri*** Wahlenb.
Carex sharensis Franch. = ***Carex augustinowiczii*** var. ***sharensis*** (Franch.) Ohwi
Carex ×*sharensis* Akiyama = ***Carex ×musashiensis*** Ohwi
Carex sharyotensis Hayata = ***Carex macrandrolepis*** H.Lév. & Vaniot
Carex shichiseitensis Hayata = ***Carex phacota*** Spreng.
Carex shikokiana Makino = ***Carex makinoensis*** Franch.
Carex shimadai Hayata = ***Carex makinoensis*** Franch.
Carex shimidzensis var. *nervulosa* (Franch.) Kük. = ***Carex shimidzensis*** Franch.
Carex shimotsukensis Honda = ***Carex ligata*** subsp. ***formosensis*** (H.Lév. & Vaniot) T.Koyama
Carex shortii Steud. = ***Carex frankii*** Kunth
Carex shortii Torr. = ***Carex shortiana*** Dewey & Torr.
Carex shriveri (Britton) Britton = ***Carex granularis*** Muhl. ex Willd.
Carex sibirica Willd. ex Kunth = ***Carex loliacea*** L.
Carex siccata var. *obscurior* Kük. = ***Carex siccata*** Dewey
Carex siciformis Boott = ***Carex crus-corvi*** Shuttlew. ex Kunze
Carex sicula C.Presl = ***Carex distans*** L.
Carex sicula C.Presl ex Boott = ***Carex distans*** L.
Carex sicula Tineo = ***Carex leporina*** L.
Carex sicyocarpa Lebel = ***Carex praecox*** Schreb.
Carex siderosticta var. *bracteosa* Franch. = ***Carex siderosticta*** Hance var. ***siderosticta***
Carex siderosticta f. *stenophylla* Kitag. = ***Carex siderosticta*** Hance var. ***siderosticta***
Carex siderosticta var. *stenophylla* (Kitag.) Kitag. = ***Carex siderosticta*** Hance var. ***siderosticta***
Carex sieberi Nees ex Kunth = ***Cyperus aggregatus*** (Willd.) Endl. var. ***aggregatus***
Carex sieberiana Opiz = ***Carex davalliana*** Sm.
Carex sieboldii Miq. = ***Carex maculata*** Boott var. ***maculata***
Carex siegertiana Uechtr. ex Garcke = ***Carex atherodes*** Spreng.
Carex siegertiana var. *glabra* Uechtr. ex Garcke = ***Carex atherodes*** Spreng.
Carex sikkimensis C.B.Clarke = ***Carex fucata*** Boott ex C.B.Clarke
Carex sikokiana Franch. & Sav. = ***Carex sachalinensis*** var. ***sikokiana*** (Franch. & Sav.) Ohwi
Carex ×*silesiaca* Figert = ***Carex ×ludibunda*** J.Gay
Carex simensis var. *lanuriensis* De Wild. = ***Carex simensis*** Hochst. ex A.Rich.
Carex simensis var. *longistipitata* Kük. = ***Carex simensis*** Hochst. ex A.Rich.
Carex simensis var. *mauensis* Kük. = ***Carex simensis*** Hochst. ex A.Rich.
Carex simensis var. *stolonifera* (Boeck.) Kük. = ***Carex simensis*** Hochst. ex A.Rich.
Carex similigena V.I.Krecz. = ***Carex enervis*** C.A.Mey. subsp. ***enervis***
Carex similis d'Urv. = ***Carex canescens*** var. ***robustior*** Blytt ex Andersson
Carex simplicissima F.Muell. = ***Carex capillacea*** Boott var. ***capillacea***
Carex simpliciuscula Wahlenb. = ***Kobresia simpliciuscula*** (Wahlenb.) Mack.
Carex simulans var. *densiflora* Tang & F.T.Wang ex S.Y. Liang = ***Carex simulans*** C.B.Clarke
Carex sinai Boott = ***Carex distans*** L.
Carex sinaica Nees ex Steud. = ***Carex distans*** L.
Carex sinensis J.F.Gmel. = ***Carex chinensis*** Retz.
Carex sinocrispa Raymond = ***Carex lingii*** F.T.Wang & Tang
Carex siroumensis Koidz. = ***Carex siroumensis*** Koidz.
Carex sismanii Velen. = ***Carex sylvatica*** Huds. subsp. ***sylvatica***
Carex sisukensis Akiyama = ***Carex atrofusca*** Schkuhr subsp. ***atrofusca***
Carex sitchensis Prescott ex Bong. = ***Carex aquatilis*** var. ***dives*** (T.Holm) Kük.
Carex skorpili Velen. = ?
Carex skottsbergii Gand. = ***Carex canescens*** var. ***robustior*** Blytt ex Andersson
Carex slobodovii V.I.Krecz. = ***Carex pseudofoetida*** Kük. ex Ostenf. subsp. ***pseudofoetida***
Carex smalliana Mack. = ***Carex lonchocarpa*** Willd. ex Spreng.
Carex smaragdina Colenso = ***Carex inversa*** R.Br.
Carex smirnovii V.I.Krecz. = ***Carex rugulosa*** Kük.
Carex smithii Tausch = ***Carex vaginata*** Tausch var. ***vaginata***
Carex smitinandii Raymond = ***Carex commixta*** Steud.
Carex socia Boott = ***Carex arnottiana*** Nees ex Drejer
Carex soleirolii DC. & Duby = ***Carex echinata*** Murray subsp. ***echinata***
Carex ×*solstitialis* Figert = ?
Carex sonderiana Boott ex Lindm. = ***Carex purpureovaginata*** Boeck.
Carex sonderiana Boott ex C.B.Clarke = ?
Carex songorica Kar. & Kir. = ***Carex heterostachya*** Bunge
Carex songorica f. *minor* Boott = ***Carex heterostachya*** Bunge
Carex sonomensis Stacey = ***Carex albida*** L.H.Bailey
Carex ×*sooi* Jákucs = ?
Carex soperi Raup = ***Carex macloviana*** d'Urv. subsp. ***macloviana***
Carex sorachensis H.Lév. & Vaniot = ***Carex shimidzensis*** Franch.
Carex soriofkensis H.Lév. & Vaniot = ***Carex glareosa*** Schkuhr ex Wahlenb. subsp. ***glareosa***
Carex ×*sororia* Meinsh. =
Carex sororia var. *involucrata* Kük. = ***Carex fossa*** G.A. Wheeler
Carex sororia f. *paupera* Osten = ***Carex sororia*** Kunth
Carex sororia f. *subdivulsa* (Kük.) Kük. = ***Carex subdivulsa*** (Kük.) G.A.Wheeler
Carex sororia subsp. *uruguensis* (Boeck.) Luceño & Alves = ***Carex uruguensis*** Boeck.
Carex souliei Franch. = ***Carex obscura*** var. ***brachycarpa*** C.B.Clarke
Carex soyaeensis Kük. = ***Carex augustinowiczii*** Meinsh. var. ***augustinowiczii***
Carex spadicea Roth = ***Carex acutiformis*** Ehrh.
Carex spadicea Gilib. = ***Carex arenaria*** L.
Carex spadicea J.F.Gmel. = ***Carex ferruginea*** Scop. subsp. ***ferruginea***
Carex spadicea Host = ***Carex ferruginea*** Scop. subsp. ***ferruginea***
Carex spadicea Schkuhr = ***Carex frigida*** All.
Carex spadicea var. *kochiana* (DC.) Crép. = ***Carex acutiformis*** Ehrh.
Carex spadicea f. *maxima* Urb. = ***Carex acutiformis*** Ehrh.
Carex spadiceovaginata Ohwi = ***Carex brunnea*** Thunb. subsp. ***brunnea***
Carex spaniocarpa Steud. = ***Carex supina*** var. ***spaniocarpa*** (Steud.) Boivin

Carex sparganioides Leconte ex Kunth = ***Carex stipata*** Muhl. ex Willd. var. ***stipata***
Carex sparganioides var. *aggregata* (Mack.) Gleason = ***Carex aggregata*** Mack.
Carex sparganioides var. *cephaloidea* (Dewey) Carey = ***Carex cephaloidea*** (Dewey) Dewey ex Boott
Carex sparganioides var. *lutea* Farw. = ***Carex sparganioides*** Muhl. ex Willd.
Carex sparganioides var. *minor* Boott = ***Carex cephaloidea*** (Dewey) Dewey ex Boott
Carex sparganioides var. *ramea* Dewey = ***Carex sparganioides*** Muhl. ex Willd.
Carex sparsiflora (Wahlenb.) Steud. = ***Carex vaginata*** Tausch var. ***vaginata***
Carex sparsiflora subsp. *altocaulis* (Dewey) Britton = ***Carex vaginata*** Tausch var. ***vaginata***
Carex sparsiflora var. *borealis* Andersson = ***Carex vaginata*** Tausch var. ***vaginata***
Carex sparsiflora var. *distracta* Norman = ***Carex vaginata*** Tausch var. ***vaginata***
Carex sparsiflora var. *grutteri* Asch. & Graebn. = ***Carex vaginata*** Tausch var. ***vaginata***
Carex sparsiflora var. *pauciflora* (Wahlenb.) Andersson = ***Carex vaginata*** Tausch var. ***vaginata***
Carex sparsiflora var. *petersii* (C.A.Mey. ex Schmidt) Kük. = ***Carex vaginata*** var. ***petersii*** (C.A.Mey. ex Schmidt) Akiyama
Carex sparsinux C.B.Clarke ex Franch. = ***Carex filipes*** var. ***sparsinux*** (C.B.Clarke ex Franch.) Kük.
Carex spartea Wahlenb. = ***Schoenoxiphium sparteum*** (Wahlenb.) C.B.Clarke
Carex spathaceobracteata Kük. = ***Carex bilateralis*** Hayata
Carex spathulata (Kük.) Nelmes = ***Carex teres*** Boott
Carex spathulata Becc. = ?
Carex spatiosa Boott = ***Carex commixta*** Steud.
Carex spatiosa var. *bogorensis* C.B.Clarke = ***Carex commixta*** Steud.
Carex specifica f. *brevifructus* Kük. = ***Carex specifica*** L.H.Bailey
Carex speciosa var. *abscondita* Kük. = ***Carex speciosa*** Kunth subsp. ***speciosa***
Carex speciosa var. *angustifolia* Boott = ***Carex speciosa*** Kunth subsp. ***speciosa***
Carex speciosa var. *courtallensis* (Nees ex Boott) Kük. = ***Carex speciosa*** Kunth subsp. ***speciosa***
Carex speciosa var. *minor* Boeck. = ***Carex speciosa*** Kunth subsp. ***speciosa***
Carex speirostachya Sw. ex Sm. = ***Carex hostiana*** DC.
Carex sphacelata T.Holm = ***Carex aquatilis*** Wahlenb. var. ***aquatilis***
Carex sphaerica Lapeyr. = ?
Carex sphaerocarpa Ehrh. = ***Carex filiformis*** L.
Carex sphaerocarpa Willd. = ***Carex supina*** Willd. ex Wahlenb. var. ***supina***
Carex sphaerogyna var. *brasiliensis* H.Pfeiff. = ***Carex purpureovaginata*** Boeck.
Carex sphaerostachya (Tuck.) Dewey = ***Carex brunnescens*** subsp. ***sphaerostachya*** (Tuck.) Kalela
Carex spicata Lam. = ***Carex disticha*** Huds.
Carex spicata Brot. ex Kunth = ***Carex divisa*** Huds.
Carex spicata Schkuhr = ***Carex obtusata*** Lilj.
Carex spicata Thuill. = ***Carex vulpina*** L.
Carex spicata f. *argyrolepis* (Peterm.) Soó = ***Carex spicata*** Huds.
Carex spicata f. *fumosa* (Gren.) Soó = ***Carex spicata*** Huds.
Carex spicata f. *furculata* (Peterm.) Soó = ***Carex spicata*** Huds.
Carex spicata f. *humilis* (Peterm.) Soó = ***Carex spicata*** Huds.
Carex spicata subsp. *lumnitzeri* (Rouy) Soó = ***Carex spicata*** Huds.
Carex spicata f. *mediterranea* (Asch. & Graebn.) Soó = ***Carex spicata*** Huds.
Carex spicata f. *pseudoguestephalica* (Asch. & Graebn.) Soó = ***Carex spicata*** Huds.
Carex spicatopaniculata var. *sambiranensis* H.Lév. = ***Carex sambiranensis*** (H.Lév.) Cherm.
Carex spicigera var. *minor* Boott = ***Carex spicigera*** Nees
Carex spicigera subsp. *minor* (Boott) T.Koyama = ***Carex spicigera*** Nees
Carex spicigera var. *rostrata* Boeck. = ***Carex spicigera*** Nees
Carex spicigera f. *rostrata* (Boeck.) T.Koyama = ***Carex spicigera*** Nees
Carex spicigera var. *rubella* (Boott) Boeck. = ***Carex spicigera*** Nees
Carex spicigera f. *rubella* (Boott) T.Koyama = ***Carex spicigera*** Nees
Carex spiculata Boott = ***Carex myosurus*** Nees
Carex spiralis Ewing = ***Carex bigelowii*** Torr. ex Schwein. subsp. ***bigelowii***
Carex spissa var. *ultra* (L.H.Bailey) Kük. = ***Carex spissa*** L.H.Bailey ex Hemsl. var. ***spissa***
Carex splendens Thuill. ex Pers. = ***Carex divisa*** Huds.
Carex splendida Willd. = ***Carex lasiocarpa*** Ehrh. var. ***lasiocarpa***
Carex spongiifolia A.E.Kozhevn. = ***Carex thunbergii*** Steud.
Carex spongocrepis Nelmes = ***Carex cruciata*** Wahlenb. var. ***cruciata***
Carex spongoneura Nelmes = ***Carex cruciata*** var. ***rafflesiana*** (Boott) Noot.
Carex sprengelii Boeck. = ***Schoenoxiphium sparteum*** (Wahlenb.) C.B.Clarke
Carex spreta Steud. = ***Carex cespitosa*** L. var. ***cespitosa***
Carex spreta L.H.Bailey = ***Carex stylosa*** C.A.Mey.
Carex squamata V.I.Krecz. = ***Carex remota*** subsp. ***alta*** (Boott) Kük.
Carex squamoidea Akiyama = ***Carex caryophyllea*** var. ***microtricha*** (Franch.) Kük.
Carex squarrosa f. *robusta* Peck = ***Carex squarrosa*** L.
Carex squarrosa var. *typhina* (Michx.) Nutt. = ***Carex typhina*** Michx.
Carex squarrosa var. *typhinoides* (Schwein.) Dewey = ***Carex squarrosa*** L.
Carex stachydesma Franch. = ***Carex rubrobrunnea*** C.B.Clarke
Carex × *stagnalis* L.C.Lamb. = [12 FRA]
Carex stans Drejer = ***Carex aquatilis*** var. ***minor*** Boott
Carex × *stansalina* Lepage = ***Carex* × *subsalina*** nothosubsp. ***stansalina*** (Lepage) Govaerts
Carex stantonensis M.E.Jones = ***Carex nardina*** var. ***hepburnii*** (Boott) Kük.
Carex steenstrupii Liebm. = ***Carex lyngbyei*** Hornem.
Carex stellata Mack. = ***Carex schiedeana*** Kunze
Carex stellulata Gooden. = ***Carex echinata*** Murray subsp. ***echinata***
Carex stellulata M.Bieb. = ***Carex muricata*** L.
Carex stellulata var. *angustata* Carey = ***Carex echinata*** Murray subsp. ***echinata***
Carex stellulata var. *australis* Kük. = ***Carex echinata*** Murray subsp. ***echinata***
Carex stellulata f. *capillacea* (L.H.Bailey) Kük. = ***Carex atlantica*** subsp. ***capillacea*** (L.H.Bailey) Reznicek
Carex stellulata var. *cephalantha* (L.H.Bailey) Fernald = ***Carex echinata*** Murray subsp. ***echinata***
Carex stellulata var. *conferta* Chapm. = ***Carex atlantica*** L.H.Bailey subsp. ***atlantica***

Carex stellulata f. *excelsior* (L.H.Bailey) Kük. = ***Carex echinata*** Murray subsp. ***echinata***
Carex stellulata var. *excelsior* (L.H.Bailey) Fernald =
Carex stellulata var. *grypos* (Schkuhr) W.D.J.Koch = ***Carex echinata*** Murray subsp. ***echinata***
Carex stellulata f. *hydrophila* (Dumort.) Asch. & Graebn. = ***Carex echinata*** Murray subsp. ***echinata***
Carex stellulata f. *hylogiton* Asch. & Graebn. = ***Carex echinata*** Murray subsp. ***echinata***
Carex stellulata f. *oligantha* Callmé = ***Carex echinata*** Murray subsp. ***echinata***
Carex stellulata var. *omiana* (Franch. & Sav.) Kük. = ***Carex omiana*** Franch. & Sav.
Carex stellulata var. *ormantha* (Fernald) Fernald = ***Carex echinata*** Murray subsp. ***echinata***
Carex stellulata f. *pseudodivulsa* F.W.Schultz = ***Carex echinata*** Murray subsp. ***echinata***
Carex stellulata var. *radiata* Wahlenb. = ***Carex radiata*** (Wahlenb.) Small
Carex stellulata var. *scirpina* Tuck. = ***Carex echinata*** Murray subsp. ***echinata***
Carex stellulata var. *scirpoides* Carey = ***Carex interior*** L.H.Bailey
Carex stellulata var. *sterilis* (Willd.) Carey = ***Carex sterilis*** Willd.
Carex stellulata var. *subalpina* Asch. & Graebn. = ***Carex echinata*** Murray subsp. ***echinata***
Carex stenochlaena (T.Holm) Mack. = ***Carex scirpoidea*** var. ***stenochlaena*** T.Holm
Carex stenolepis Torr. = ***Carex frankii*** Kunth
Carex × *stenolepis* f. *gracilior* Kük. = ?
Carex stenophylla L.Thienem. ex Boott = ***Carex maritima*** Gunnerus
Carex stenophylla var. *curaica* Litv. = ***Carex stenophylla*** Wahlenb. subsp. ***stenophylla***
Carex stenophylla var. *desertorum* (Litv.) Litv. = ***Carex pachystylis*** J.Gay
Carex stenophylla var. *duriuscula* (C.A.Mey.) Trautv. = ***Carex duriuscula*** C.A.Mey.
Carex stenophylla f. *elata* (Schur) Kük. = ***Carex stenophylla*** Wahlenb. subsp. ***stenophylla***
Carex stenophylla f. *elatior* Kük. = ***Carex stenophylla*** Wahlenb. subsp. ***stenophylla***
Carex stenophylla subsp. *eleocharis* (L.H.Bailey) Hultén = ***Carex duriuscula*** C.A.Mey. subsp. ***duriuscula***
Carex stenophylla var. *eleocharis* (L.H.Bailey) Breitung = ***Carex duriuscula*** C.A.Mey. subsp. ***duriuscula***
Carex stenophylla var. *enervis* (C.A.Mey.) Kük. = ***Carex enervis*** C.A.Mey.
Carex stenophylla var. *exigua* (Schur) Kük. = ***Carex stenophylla*** Wahlenb. subsp. ***stenophylla***
Carex stenophylla var. *humilis* Meinsh. = ***Carex stenophylla*** Wahlenb. subsp. ***stenophylla***
Carex stenophylla var. *longipedicellata* (Boeck.) Kük. = ***Carex stenophylla*** subsp. ***stenophylloides*** (V.I. Krecz.) T.V.Egorova
Carex stenophylla f. *pachystylis* (J.Gay) Asch. & Graebn. = ***Carex pachystylis*** J.Gay
Carex stenophylla var. *planifolia* Boiss. = ***Carex pachystylis*** J.Gay
Carex stenophylla var. *reptabunda* Trautv. = ***Carex reptabunda*** (Trautv.) V.I.Krecz.
Carex stenophylla var. *rigescens* Franch. = ***Carex duriuscula*** subsp. ***rigescens*** (Franch.) S.Yun Liang & Y.C.Tang
Carex stenophylloides V.I.Krecz. = ***Carex stenophylla*** subsp. ***stenophylloides*** (V.I.Krecz.) T.V.Egorova
Carex stenoptera Mack. = ***Carex macloviana*** subsp. ***subfusca*** (W.Boott) T.Koyama
Carex stenura Nelmes = ***Carex speciosa*** Kunth subsp. ***speciosa***
Carex sterilis var. *aequidistans* Peck = ***Carex sterilis*** Willd.
Carex sterilis var. *angustata* L.H.Bailey = ***Carex sterilis*** Willd.
Carex sterilis var. *cephalantha* (L.H.Bailey) L.H.Bailey = ***Carex echinata*** Murray subsp. ***echinata***
Carex sterilis var. *excelsior* L.H.Bailey = ***Carex echinata*** Murray subsp. ***echinata***
Carex sterilis var. *excelsior* L.H.Bailey = ***Carex echinata*** Murray subsp. ***echinata***
Carex sterilis f. *flexibilis* Farw. = ***Carex sterilis*** Willd.
Carex sterilis var. *rousseauiana* Raymond = ***Carex sterilis*** Willd.
Carex steudelii Kunth = ***Carex jamesii*** Schwein.
Carex stictocarpa Sm. = ***Carex flacca*** Schreb. subsp. ***flacca***
Carex stilbophaea V.I.Krecz. = ***Carex atrofusca*** Schkuhr subsp. ***atrofusca***
Carex stipata var. *crassicurta* Peck = ***Carex stipata*** Muhl. ex Willd. var. ***stipata***
Carex stipata var. *laevivaginata* Kük. = ***Carex laevivaginata*** (Kük.) Mack.
Carex stipata var. *oklahomensis* (Mack.) Gleason = ***Carex oklahomensis*** Mack.
Carex stipata var. *subsecuta* Peck = ***Carex stipata*** Muhl. ex Willd. var. ***stipata***
Carex stipata var. *uberior* C.Mohr = ***Carex stipata*** var. ***maxima*** Chapm. ex Boott
Carex stipitata Boeck. = ***Carex stipata*** Muhl. ex Willd. var. ***stipata***
Carex stipitinux C.B.Clarke ex Franch. = ***Carex brunnea*** var. ***stipitinux*** (C.B.Clarke ex Franch.) Kük.
Carex stolonifera Hoppe = ***Carex nigra*** subsp. ***alpina*** (Gaudin) Lemke
Carex stolonifera H.Lév. & Vaniot = ***Carex oxyandra*** (Franch. & Sav.) Kudô var. ***oxyandra***
Carex stolonifera Ehrh. = ***Carex praecox*** Schreb.
Carex stolonifera var. *procera* Klett & Richt. = ***Carex umbrosa*** Host subsp. ***umbrosa***
Carex straminea var. *alata* (Torr.) Boeck. = ***Carex alata*** Torr.
Carex straminea var. *aperta* Boott = ***Carex hormathodes*** Fernald
Carex straminea var. *australis* L.H.Bailey = ***Carex peucophila*** T.Holm
Carex straminea var. *brevior* Dewey = ***Carex brevior*** (Dewey) Mack. ex Lunell
Carex straminea var. *brevior* L.H.Bailey = ***Carex straminea*** Willd. ex Schkuhr
Carex straminea var. *chlorostachys* Boeck. = ***Carex albolutescens*** Schwein.
Carex straminea var. *congesta* Olney = ***Carex straminiformis*** L.H.Bailey
Carex straminea var. *crawei* Boott = ***Carex bicknellii*** Britton & A.Br.
Carex straminea var. *cristata* Tuck. = ***Carex tribuloides*** Wahlenb. var. ***tribuloides***
Carex straminea var. *cumulata* L.H.Bailey = ***Carex cumulata*** (L.H.Bailey) Mack.
Carex straminea var. *echinodes* Fernald = ***Carex tenera*** var. ***echinodes*** (Fernald) Wiegand
Carex straminea f. *echinodes* (Fernald) Kük. = ***Carex tenera*** var. ***echinodes*** (Fernald) Wiegand
Carex straminea var. *ferruginea* (A.Gray) L.H.Bailey = ***Carex albolutescens*** Schwein.
Carex straminea var. *festucacea* (Schkuhr ex Willd.) Tuck. = ***Carex festucacea*** Schkuhr ex Willd.

Carex straminea var. *foenea* Torr. = ***Carex albolutescens*** Schwein.
Carex straminea var. *hyalina* (Boott) Boott = ***Carex hyalina*** Boott
Carex straminea var. *indivisa* Boott = ***Carex hormathodes*** Fernald
Carex straminea f. *indivisa* (Boott) Britton = ***Carex hormathodes*** Fernald
Carex straminea var. *intermedia* Gay = ***Carex albolutescens*** Schwein.
Carex straminea var. *maxima* L.H.Bailey = ***Carex tetrastachya*** Scheele
Carex straminea var. *meadii* Boott = ***Carex missouriensis*** P.Rothr. & Reznicek
Carex straminea var. *minor* Dewey = ***Carex straminea*** Willd. ex Schkuhr
Carex straminea var. *mirabilis* Tuck. = ***Carex normalis*** Mack.
Carex straminea var. *mixta* L.H.Bailey = ***Carex feta*** L.H.Bailey
Carex straminea var. *moniliformis* Tuck. = ***Carex silicea*** Olney
Carex straminea f. *multa* L.H.Bailey = ***Carex straminea*** Willd. ex Schkuhr
Carex straminea f. *perlonga* Fernald = ***Carex straminea*** Willd. ex Schkuhr
Carex straminea var. *prorepens* Kük. = ***Carex straminea*** Willd. ex Schkuhr
Carex straminea var. *reniformis* L.H.Bailey = ***Carex reniformis*** (L.H.Bailey) Small
Carex straminea f. *reniformis* (L.H.Bailey) Kük. = ***Carex reniformis*** (L.H.Bailey) Small
Carex straminea f. *richii* (Fernald) Kük. = ***Carex straminea*** Willd. ex Schkuhr
Carex straminea var. *schkuhrii* Gay = ***Carex straminea*** Willd. ex Schkuhr
Carex straminea var. *silicea* (Olney) L.H.Bailey = ***Carex silicea*** Olney
Carex straminea var. *tenera* (Carey) Boott = ***Carex straminea*** Willd. ex Schkuhr
Carex straminea var. *tenera* (Dewey) L.H.Bailey = ***Carex tenera*** Dewey
Carex straminea f. *tincta* Fernald = ***Carex straminea*** Willd. ex Schkuhr
Carex straminea var. *typica* Macoun = ***Carex straminea*** Willd. ex Schkuhr
Carex striata R.Br. = ***Carex brownii*** Tuck. subsp. ***brownii***
Carex striata Torr. ex Boott = ***Carex polymorpha*** Muhl.
Carex striata Gilib. = ***Carex riparia*** Curtis subsp. ***riparia***
Carex striata var. *brevis* L.H.Bailey = ***Carex striata*** Michx.
Carex striata var. *japonica* Koidz. = [38 JAP]
Carex stricta Gooden. = ***Carex elata*** All. subsp. ***elata***
Carex stricta var. *angustata* (Boott) L.H.Bailey = ***Carex angustata*** Boott
Carex stricta f. *angustata* (Boott) Kük. = ***Carex angustata*** Boott
Carex stricta f. *brevior* House = ***Carex stricta*** Lam.
Carex stricta var. *curtissima* Peck = ***Carex stricta*** Lam.
Carex stricta f. *curtissima* (Peck) Kük. = ***Carex stricta*** Lam.
Carex stricta var. *decora* L.H.Bailey = ***Carex haydenii*** Dewey
Carex stricta var. *elongata* (Boeck.) Gleason = ***Carex emoryi*** Dewey
Carex stricta var. *emoryi* (Dewey) L.H.Bailey = ***Carex emoryi*** Dewey
Carex stricta f. *emoryi* (Dewey) Kük. = ***Carex emoryi*** Dewey
Carex stricta var. *erecta* (Dewey) C.F.Reed = ***Carex angustata*** Boott
Carex stricta var. *fallax* T.Marsson = ***Carex × turfosa*** Fr.
Carex stricta var. *gracilis* Asch. = ***Carex elata*** All. subsp. ***elata***
Carex stricta var. *haydenii* (Dewey) Kük. = ***Carex haydenii*** Dewey
Carex stricta var. *helvetica* Rion ex Rchb. = ***Carex elata*** All. subsp. ***elata***
Carex stricta var. *homalocarpa* (Peterm.) Rchb. = ***Carex elata*** All. subsp. ***elata***
Carex stricta f. *humilis* Fr. = ***Carex elata*** All. subsp. ***elata***
Carex stricta var. *macra* (Steud.) Nyman = ***Carex elata*** All. subsp. ***elata***
Carex stricta var. *minor* Gaudin = ***Carex × turfosa*** Fr.
Carex stricta f. *nigrans* Beck = ***Carex elata*** All. subsp. ***elata***
Carex stricta f. *pedicellaris* House = ***Carex stricta*** Lam.
Carex stricta f. *pendula* Lange = ***Carex elata*** All. subsp. ***elata***
Carex stricta f. *polystachya* Asch. & Graebn. = ***Carex elata*** All. subsp. ***elata***
Carex stricta f. *riartiana* Legrand = ***Carex elata*** All. subsp. ***elata***
Carex stricta var. *strictior* (Dewey ex Alph.Wood) Carey = ***Carex stricta*** Lam.
Carex stricta var. *turfosa* (Fr.) Almq. = ***Carex × turfosa*** Fr.
Carex stricta var. *xerocarpa* (S.H.Wright) Britton = ***Carex stricta*** Lam.
Carex stricta f. *xerocarpa* (S.H.Wright) Kük. = ***Carex stricta*** Lam.
Carex strictifolia Opiz ex Nyman = ***Carex acuta*** L.
Carex strictior Dewey ex Alph.Wood = ***Carex stricta*** Lam.
Carex strictior f. *angustata* (Boott) House = ***Carex angustata*** Boott
Carex strictior f. *curtissima* (Peck) House =
Carex strigosa All. = ***Carex brachystachys*** Schrank
Carex strigosa Suter ex Boott = ***Carex firma*** Host
Carex strigosa Willd. ex Kunth = ***Carex sylvatica*** Huds. subsp. ***sylvatica***
Carex × strigosula Chatenier = [12 FRA]
Carex stupenda H.Lév. & Vaniot = ***Carex boottiana*** Hook. & Arn.
Carex styloflexa var. *fusiformis* Wiegand = ***Carex laxiflora*** Lam. var. ***laxiflora***
Carex styloflexa var. *remotiflora* Wiegand = ***Carex styloflexa*** Buckley
Carex stylosa var. *nigritella* (Drejer) Fernald = ***Carex stylosa*** C.A.Mey.
Carex stylosa var. *virens* L.H.Bailey = ***Carex aperta*** Boott
Carex subalpina Brügger = ***Carex magellanica*** subsp. ***irrigua*** (Wahlenb.) Hultén
Carex subalpina Brügger ex Appel = ***Carex magellanica*** subsp. ***irrigua*** (Wahlenb.) Hultén
Carex subanceps Boeck. = ***Carex dispalata*** Boott subsp. ***dispalata***
Carex subbiflora Steud. = ***Carex pensylvanica*** Lam.
Carex subbracteata var. *tolucensis* F.J.Herm. = ***Carex tolucensis*** (F.J.Herm.) Reznicek
Carex subcaespitosa (Kük.) Wiinst. = ***Carex nigra*** (L.) Reichard subsp. ***nigra***
Carex subcompressa Steud. = ***Carex prasina*** Wahlenb.
Carex subconcolor Kitag. = ***Carex arnellii*** H.Christ
Carex × subdecolorans Nyár. = ?
Carex subdissitiflora Kük. = ***Carex dissitiflora*** Franch. subsp. ***dissitiflora***

Carex subdita Ohwi = ***Carex insaniae*** var. ***subdita*** (Ohwi) Ohwi
Carex suberea Boott = ***Carex scabrifolia*** Steud.
Carex subfilicina Ohwi = ***Carex filicina*** Nees
Carex subfusca W.Boott = ***Carex macloviana*** subsp. ***subfusca*** (W.Boott) T.Koyama
Carex subglobosa Miel. = ***Carex viridula*** Michx. subsp. ***viridula***
Carex sublimosa Lepage = ***Carex paleacea*** Schreb. ex Wahlenb.
Carex sublivida Norrl. = ?
Carex subloliacea (Laest.) E.P.Bicknell = ***Carex lapponica*** Láng
Carex subnigra Lepage = ***Carex vacillans*** Drejer
Carex subnivalis Arv.-Touv. = ***Carex ornithopoda*** Willd. subsp. ***ornithopoda***
Carex suborbiculata Mack. = ***Carex nudata*** Boott
Carex subpediformis (Kük.) Sutô & Suzuki = ***Carex lanceolata*** var. ***subpediformis*** Kük.
Carex subphacota (Kük.) Nakai = ***Carex phacota*** Spreng.
Carex subramosa Willd. ex Kunth = ***Carex echinata*** Murray subsp. ***echinata***
Carex subrotunda M.Serres = ***Carex elata*** All. subsp. ***elata***
Carex subsabulosa Norman = ***Carex canescens*** L. subsp. ***canescens***
Carex subsabulosa Norman = ?
Carex subsalsa Nyman = ***Carex hostiana*** DC.
Carex subspathacea f. *curvata* Drejer = ***Carex subspathacea*** Wormsk. ex Hornem.
Carex subspathacea f. *nardifolia* (Wahlenb.) Fr. = ***Carex subspathacea*** Wormsk. ex Hornem.
Carex subspathacea var. *planifolia* Fr. = ***Carex subspathacea*** Wormsk. ex Hornem.
Carex subspathacea subsp. *ramenskii* (Kom.) T.V. Egorova = ***Carex ramenskii*** Kom.
Carex subspathacea var. *reducta* (Drejer) Fr. = ***Carex ×reducta*** Drejer
Carex subspathacea f. *stricta* Drejer = ***Carex subspathacea*** Wormsk. ex Hornem.
Carex ×substans Lepage = ***Carex ×flavicans*** F.Nyl.
Carex substricta (Kük.) Mack. = ***Carex aquatilis*** var. ***substricta*** Kük.
Carex subteinogyna Ohwi = ***Carex bilateralis*** Hayata
Carex subtransversa C.B.Clarke = ***Carex alopecuroides*** D.Don ex Tilloch & Taylor
Carex subtumida (Kük.) Ohwi = ***Carex ischnostachya*** Steud.
Carex subulata Wahlenb. = ***Carex boryana*** Schkuhr
Carex subulata Schumach. = ***Carex buxbaumii*** Wahlenb.
Carex subulata Michx. = ***Carex collinsii*** Nutt.
Carex subulata J.F.Gmel. = ***Scleria lithosperma*** (L.) Sw. var. ***lithosperma***
Carex subuniflora Steud. = ***Carex oligocarpa*** Willd.
Carex subvaginata Meinsh. ex C.B.Clarke = ***Carex schmidtii*** Meinsh.
Carex subvillosa M.Bieb. = ***Carex filiformis*** L.
Carex ×subviridula Fernald = ***Carex ×alsatica*** nothosubsp. ***subviridula*** (Fernald) Govaerts
Carex subvulpina Senay = ***Carex cuprina*** (I.Sándor ex Heuff.) Nendtv. ex A.Kern.
Carex succedanea Nakai = ***Carex paxii*** Kük.
Carex sudetica Opiz = ***Carex atrata*** L. subsp. ***atrata***
Carex sudetica C.Presl ex Kunth = ***Carex vaginata*** Tausch var. ***vaginata***
Carex suilla J.Fellm. = ***Carex vesicaria*** L.
Carex sukaczovii V.I.Krecz. = ***Carex gotoi*** Ohwi
Carex suksdorfii Kük. = ***Carex aquatilis*** Wahlenb. var. ***aquatilis***
Carex suksdorfii var. *ovalis* Kük. = ***Carex aquatilis*** Wahlenb. var. ***aquatilis***
Carex sulcata Schur = ***Carex melanostachya*** M.Bieb. ex Willd.
Carex sumatrensis C.B.Clarke = ***Carex verticillata*** Zoll. & Moritzi
Carex supergoodenoughii Lepage = ***Carex vacillans*** Drejer
Carex supermascula V.I.Krecz. = ***Carex pediformis*** C.A.Mey. subsp. ***pediformis***
Carex ×superriparia Hübl = ?
Carex supina var. *costata* Meinsh. = ***Carex korshinskyi*** Kom.
Carex supina f. *elatior* Behrendsen = ***Carex supina*** Willd. ex Wahlenb. var. ***supina***
Carex supina f. *humilior* Behrendsen = ***Carex supina*** Willd. ex Wahlenb. var. ***supina***
Carex supina var. *korshinskyi* (Kom.) Kük. = ***Carex korshinskyi*** Kom.
Carex supina subsp. *korshinskyi* (Kom.) Malyschev = ***Carex korshinskyi*** Kom.
Carex supina f. *pallida* Bubela = ***Carex supina*** Willd. ex Wahlenb. var. ***supina***
Carex supina f. *pseudomonostachys* Asch. = ***Carex supina*** Willd. ex Wahlenb. var. ***supina***
Carex supina subsp. *spaniocarpa* (Steud.) Hultén = ***Carex supina*** var. ***spaniocarpa*** (Steud.) Boivin
Carex surculosa Raymond = ***Carex tsiangii*** F.T.Wang & Tang
Carex sutchanensis Kom. = ***Carex pediformis*** C.A.Mey. subsp. ***pediformis***
Carex suteri Kunth = ?
Carex ×suziella Podp. = ?
Carex svenonis Skottsb. = ***Carex echinata*** Murray subsp. ***echinata***
Carex sychnostachya Nelmes = ***Carex juvenilis*** C.B.Clarke ex E.G.Camus
Carex sylvatica Dewey = ***Carex arctata*** Boott
Carex sylvatica Maxim. ex Boeck. = ***Carex pruinosa*** subsp. ***maximowiczii*** (Miq.) Kük.
Carex sylvatica subsp. *algeriensis* (Nelmes) Maire & Weiller = ***Carex sylvatica*** Huds. subsp. ***sylvatica***
Carex sylvatica f. *gracilis* Èelak. = ***Carex sylvatica*** Huds. subsp. ***sylvatica***
Carex sylvatica f. *latifolia* Kneuck. = ***Carex sylvatica*** Huds. subsp. ***sylvatica***
Carex sylvatica var. *laxula* (Tineo ex Boott) Nyman = ***Carex sylvatica*** Huds. subsp. ***sylvatica***
Carex sylvatica f. *laxula* (Tineo ex Boott) Kük. = ***Carex sylvatica*** Huds. subsp. ***sylvatica***
Carex sylvatica f. *pumila* Fiek ex Uechtr. = ***Carex sylvatica*** Huds. subsp. ***sylvatica***
Carex sylvatica var. *tommasinii* Rchb. = ***Carex sylvatica*** Huds. subsp. ***sylvatica***
Carex sylvicola J.M.Webber & P.W.Ball = ***Carex radiata*** (Wahlenb.) Small
Carex szovitsii V.I.Krecz. = ***Carex paniculata*** subsp. ***szovitsii*** (V.I.Krecz.) Ö.Nilsson
Carex tabina Steud. = ***Carex pichinchensis*** Kunth
Carex tabularia M.A.Curtis ex Boott = ***Carex aestivalis*** M.A.Curtis ex A.Gray
Carex tagawana T.Koyama = ***Carex fulvorubescens*** subsp. ***longistipes*** (Hayata) T.Koyama
Carex tahoata Hamlin = ***Carex dipsacea*** Berggr.
Carex taiheiensis Hayata = ***Carex sclerocarpa*** Franch.
Carex taiwanensis (Ohwi) Akiyama = ***Carex dissitiflora*** subsp. ***taiwanensis*** (Ohwi) T.Koyama
Carex takasagoana Akiyama = ***Carex manca*** subsp. ***takasagoana*** (Akiyama) T.Koyama

Carex takenakai Nakai = ***Carex karoi*** (Freyn) Freyn
Carex takeshimensis Nakai = ***Carex shimidzensis*** Franch.
Carex taldycola Meinsh. = ***Carex orbicularis*** Boott subsp. ***orbicularis***
Carex taliensis Franch. = ***Carex rubrobrunnea*** C.B.Clarke
Carex tamakii T.Koyama = ***Carex sacrosancta*** var. ***tamakii*** (T.Koyama) T.Koyama
Carex tapintzensis var. *lamprosandra* (Franch.) Kük. = ***Carex tapintzensis*** Franch.
Carex taquetii H.Lév. = ***Carex matsumurae*** Franch.
Carex taraspensis Brügger & Killias = ?
Carex tartarea Ridl. = ***Carex verticillata*** Zoll. & Moritzi
Carex taskanensis A.P.Khokhr. = ***Carex atrofusca*** Schkuhr subsp. ***atrofusca***
Carex tasorum Kom. = ***Carex tenuiformis*** H.Lév. & Vaniot
Carex tatewakiana Ohwi = ***Carex perakensis*** C.B.Clarke var. ***perakensis***
Carex tatrorum (Zapal.) Racib. = ***Carex sempervirens*** Vill.
Carex tatsutakensis Hayata = ***Carex macrandrolepis*** H.Lév. & Vaniot
Carex tauricola Boiss. = ***Carex atrata*** L. subsp. ***atrata***
Carex taurina Phil. = ***Carex diclina*** Phil.
Carex taylorii Nelmes = ***Carex phragmitoides*** Kük.
Carex tchenkeouensis E.G.Camus = ***Carex thibetica*** Franch. var. ***thibetica***
Carex teinogyna var. *scabriculmis* Kük. = ***Carex teinogyna*** Boott
Carex temnolepis Franch. = ***Carex morrowii*** var. ***temnolepis*** (Franch.) Ohwi
Carex tenacissima Suksd. = ***Carex nudata*** Boott
Carex tenax Reut. = ***Carex ferruginea*** subsp. ***australpina*** (Bech.) W.Dietr.
Carex tendae (W.Dietr.) Pawl. = ***Carex ferruginea*** subsp. ***tendae*** W.Dietr.
Carex tenella Ehrh. = ***Carex canescens*** L. subsp. ***canescens***
Carex tenella Schkuhr = ***Carex disperma*** Dewey
Carex tenella Thuill. = ***Carex praecox*** Schreb.
Carex tenella Poir. = ***Uncinia tenella*** R.Br.
Carex tenella var. *misera* Franch. = ***Carex disperma*** Dewey
Carex tenera var. *invisa* (Boott) Britton = ***Carex hormathodes*** Fernald
Carex tenera var. *major* Olney = ***Carex tenera*** Dewey var. ***tenera***
Carex tenera var. *richii* Fernald = ***Carex straminea*** Willd. ex Schkuhr
Carex tenera var. *suberecta* Olney = ***Carex suberecta*** (Olney) Britton
Carex teneriformis Mack. = ***Carex macloviana*** subsp. ***subfusca*** (W.Boott) T.Koyama
Carex tenerrima Murr & Appel = ***Carex ferruginea*** Scop. subsp. ***ferruginea***
Carex tenorii Kunth = ***Carex macrolepis*** DC.
Carex tentaculata Muhl. ex Willd. = ***Carex lurida*** Wahlenb.
Carex tentaculata var. *altior* Boott = ***Carex lurida*** Wahlenb.
Carex tentaculata var. *flaccida* (L.H.Bailey) Howe = ***Carex lurida*** Wahlenb.
Carex tentaculata var. *gracilis* Boott = ***Carex lurida*** Wahlenb.
Carex tentaculata var. *parvula* Paine = ***Carex lurida*** Wahlenb.
Carex tentaculata var. *rostrata* Pursh = ***Carex lurida*** Wahlenb.
Carex tenuiflora Hartm. ex Kunth = ***Carex loliacea*** L.
Carex tenuiflora var. *arrhyncha* (Franch.) Kük. = ***Carex tenuiflora*** Wahlenb.
Carex tenuiflora var. *macilenta* (Fr.) O.Lang = ***Carex × macilenta*** Fr.
Carex tenuiflora var. *setacea* Kük. = ***Carex tenuiflora*** Wahlenb.
Carex tenuifolia Poir. = ***Fuirena pubescens*** (Poir.) Kunth var. ***pubescens***
Carex tenuiformis f. *paupera* Kük. = ***Carex tenuiformis*** H.Lév. & Vaniot
Carex tenuirostris Olney = ***Carex macloviana*** d'Urv. subsp. ***macloviana***
Carex tenuirostris Boeck. = ?
Carex tenuis Host = ***Carex brachystachys*** Schrank
Carex tenuis Rudge = ***Carex debilis*** var. ***rudgei*** L.H.Bailey
Carex tenuis Nees = ***Carex plectobasis*** V.I.Krecz.
Carex tenuis (Retz.) J.F.Gmel. = ***Scleria lithosperma*** (L.) Sw. var. ***lithosperma***
Carex tenuis var. *brevifolia* (Host) Nyman = ***Carex ferruginea*** Scop. subsp. ***ferruginea***
Carex tenuis var. *erectior* Britton = ***Carex debilis*** var. ***rudgei*** L.H.Bailey
Carex tenuis var. *interjecta* (L.H.Bailey) Britton = ***Carex debilis*** var. ***rudgei*** L.H.Bailey
Carex tenuiseta var. *brevisquama* Koidz. = ***Carex scita*** Maxim. var. ***scita***
Carex tenuispica Steud. = ***Carex brunnea*** Thunb. subsp. ***brunnea***
Carex tenuispica Boeck. = ***Carex tamana*** Steyerm.
Carex tenuissima Schur = ***Carex muricata*** L.
Carex tenuissima Boott = ***Carex sachalinensis*** F.Schmidt var. ***sachalinensis***
Carex tenuissima var. *duvaliana* (Franch. & Sav.) Kük. = ***Carex duvaliana*** Franch. & Sav.
Carex tenuissima var. *sikokiana* (Franch. & Sav.) Kük. = ***Carex sachalinensis*** var. ***sikokiana*** (Franch. & Sav.) Ohwi
Carex tenuistachya Nakai = ***Carex longirostrata*** C.A.Mey. var. ***longirostrata***
Carex tenuistachya var. *pallida* Kitag. = ***Carex longirostrata*** C.A.Mey. var. ***longirostrata***
Carex × *teramotoi* T.Koyama = ?
Carex teres var. *spathulata* Kük. = ***Carex teres*** Boott
Carex tereticaulis var. *septentrionalis* Domin = ***Carex tereticaulis*** F.Muell.
Carex tereticaulis var. *typica* Domin = ***Carex tereticaulis*** F.Muell.
Carex teretiuscula Willk. = ***Carex chordorrhiza*** L.f.
Carex teretiuscula Gooden. = ***Carex diandra*** Schrank
Carex teretiuscula var. *ampla* L.H.Bailey = ***Carex diandra*** Schrank
Carex teretiuscula var. *crassior* Hartm. = ***Carex diandra*** Schrank
Carex teretiuscula var. *ehrhartiana* (Hoppe ex Boott) Nyman = ***Carex diandra*** Schrank
Carex teretiuscula var. *major* W.D.J.Koch = ***Carex diandra*** Schrank
Carex teretiuscula subsp. *modesta* (J.Gay) Nyman = ***Carex disticha*** Huds.
Carex teretiuscula var. *modesta* K.Richt. = ***Carex disticha*** Huds.
Carex teretiuscula var. *prairea* (Dewey ex Alph.Wood) Britton = ***Carex prairea*** Dewey ex Alph.Wood
Carex teretiuscula var. *ramosa* Boott = ***Carex diandra*** Schrank
Carex teretiuscula f. *simplicior* Sanio = ***Carex diandra*** Schrank
Carex teretiuscula var. *tenella* Beckm. = ***Carex diandra*** Schrank

Carex tergestina Hoppe ex Boott = ***Carex muricata*** L.
Carex ternaria var. *gracilis* Cheeseman = ***Carex ternaria*** G.Forst.
Carex ternaria var. *minor* Boott = ***Carex ternaria*** G.Forst.
Carex ternaria f. *nigrescens* Kük. = ***Carex ternaria*** G.Forst.
Carex ternaria var. *pallida* Cheeseman = ***Carex coriacea*** Hamlin
Carex ternaria f. *pallida* (Cheeseman) Kük. = ***Carex coriacea*** Hamlin
Carex terniflora Maxim. ex Matsum. = ***Carex grallatoria*** var. ***heteroclita*** (Franch.) Kük.
Carex terrae-novae Fernald = ***Carex glacialis*** Mack.
Carex tetanica Schwein. & Torr. = ***Carex conoidea*** Willd.
Carex tetanica Rchb. = ***Carex vaginata*** Tausch var. ***vaginata***
Carex tetanica var. *canbyi* Porter = ***Carex tetanica*** Schkuhr
Carex tetanica f. *canbyi* (Porter) Kük. = ***Carex tetanica*** Schkuhr
Carex tetanica var. *carteri* Porter = ***Carex tetanica*** Schkuhr
Carex tetanica var. *meadii* (Dewey) L.H.Bailey = ***Carex meadii*** Dewey
Carex tetanica var. *woodii* (Dewey) L.H.Bailey = ***Carex woodii*** Dewey
Carex tetanica f. *woodii* (Dewey) Kük. = ***Carex woodii*** Dewey
Carex ×*tetrastachya* Traunst. = ***Carex* × *biharica*** Simonk.
Carex tetrasticha Boeck. = ***Carex brasiliensis*** A.St.-Hil.
Carex tetsuoi Ohwi = ***Carex maculata*** var. ***tetsuoi*** (Ohwi) Hatus. ex T.Koyama
Carex textori Miq. = ***Carex incisa*** Boott
Carex thecata Boott = ***Carex preissii*** Nees
Carex thermarum Phil. = ***Carex macloviana*** var. ***thermarum*** (Phil.) Kük.
Carex thibetica var. *minor* Kük. = ***Carex thibetica*** Franch. var. ***thibetica***
Carex thomsonii Petrie = ***Carex pterocarpa*** Petrie
Carex thomsonii Gand. = ?
Carex thorelii E.G.Camus = ***Carex tricephala*** Boeck.
Carex thunbergii var. *appendiculata* (Trautv. & C.A.Mey.) Ohwi = ***Carex thunbergii*** Steud.
Carex thunbergii var. *platycarpa* Franch. & Sav. = ***Carex thunbergii*** Steud.
Carex thunbergii var. *quinquenervis* Franch. & Sav. = ***Carex thunbergii*** Steud.
Carex thuringiaca Willd. = ***Carex flacca*** Schreb. subsp. ***flacca***
Carex thwaitesii Hance = ***Carex hebecarpa*** C.A.Mey. subsp. ***hebecarpa***
Carex thyrsiflora Boott = ***Carex lindleyana*** Nees
Carex timorensis (C.B.Clarke) Nelmes = ***Carex cruciata*** var. ***rafflesiana*** (Boott) Noot.
Carex tiogana D.W.Taylor & J.D.Mastrog. = ***Carex capillaris*** L. subsp. ***capillaris***
Carex tobae Honda = ***Carex heterolepis*** Bunge
Carex tokioensis Boeck. = ***Carex japonica*** Thunb.
Carex tolmiei Boott = ***Carex spectabilis*** Dewey
Carex tolmiei var. *angusta* L.H.Bailey = ***Carex scopulorum*** var. ***bracteosa*** (L.H.Bailey) F.J.Herm.
Carex tolmiei var. *invisa* (L.H.Bailey) Kük. = ***Carex spectabilis*** Dewey
Carex tolmiei var. *leptosaccus* (C.B.Clarke) Kük. = ***Carex conferta*** var. ***leptosaccus*** (C.B.Clarke) Kük.
Carex tolmiei var. *longiuscula* Hultén = ***Carex spectabilis*** Dewey
Carex tolmiei var. *nigella* (Boott) Kük. = ***Carex spectabilis*** Dewey
Carex tolmiei var. *subsessilis* L.H.Bailey = ***Carex scopulorum*** T.Holm var. ***scopulorum***
Carex tomentosa L. = ***Carex filiformis*** L.
Carex tomentosa Hornem. = ***Carex globularis*** L.
Carex tomentosa var. *caesia* (Griseb.) Nyman = ***Carex filiformis*** L.
Carex tomentosa var. *decalvescens* Kük. = ***Carex filiformis*** L.
Carex tomentosa f. *globifera* Schur = ***Carex filiformis*** L.
Carex tomentosa f. *gracilis* O.Lang = ***Carex filiformis*** L.
Carex tomentosa var. *gracillima* Schur = ***Carex filiformis*** L.
Carex tomentosa subsp. *grasmanniana* (O.Lang) Asch. = ***Carex filiformis*** L.
Carex tomentosa f. *grasmanniana* O.Lang = ***Carex filiformis*** L.
Carex tomentosa var. *grasmanniana* (O.Lang) Nyman = ***Carex filiformis*** L.
Carex tomentosa var. *grioletii* (Roem. ex Schkuhr) Boeck. = ***Carex grioletii*** Roem. ex Schkuhr
Carex tomentosa f. *gynandra* A.T.Szabó = ***Carex filiformis*** L.
Carex tomentosa var. *lucana* A.Terracc. = ***Carex filiformis*** L.
Carex tomentosa f. *luxuriosa* Waisb. = ***Carex filiformis*** L.
Carex tomentosa f. *nigrescens* Kük. & Litv. = ***Carex filiformis*** L.
Carex tomentosa f. *nordmannii* (A.Kern. ex Palla) K.Richt. = ***Carex filiformis*** L.
Carex tomentosa f. *pallida* Waisb. = ***Carex filiformis*** L.
Carex tomentosa subsp. *subvillosa* (M.Bieb.) Nyman = ***Carex filiformis*** L.
Carex tonkinensis Franch. = ***Carex perakensis*** C.B.Clarke var. ***perakensis***
Carex tonnerrei C.B.Clarke = ***Carex gentilis*** Franch.
Carex torfacea J.F.Gmel. = ***Carex rostrata*** Stokes var. ***rostrata***
Carex ×*torgesiana* Kük. = ?
Carex ×*tornabenii* Chiov. = ?
Carex torreyana Dewey = ***Carex davisii*** Schwein. & Torr.
Carex torta var. *staminata* Peck = ***Carex torta*** Boott ex Tuck.
Carex tosaensis Akiyama = ***Carex breviculmis*** var. ***pluricostata*** Kük.
Carex tosana Makino = ***Carex insaniae*** var. ***subdita*** (Ohwi) Ohwi
Carex ×*toshironis* T.Koyama = ?
Carex touranginiana Boreau = ***Carex acuta*** L.
Carex ×*tourletii* Gillies ex Tourlet = ?
Carex trachyantha Dorner ex Heuff. = ***Carex caryophyllea*** Latourr. var. ***caryophyllea***
Carex trachycarpos Link = ***Carex flacca*** Schreb. subsp. ***flacca***
Carex tracyi Mack. = ***Carex leporina*** L.
Carex transalaica Tzvelev = ***Carex orbicularis*** Boott subsp. ***orbicularis***
Carex transalpina Hayata = ***Carex dolichostachya*** subsp. ***trichosperma*** (Ohwi) T.Koyama
Carex transmarina Krecz. = ***Carex maritima*** Gunnerus
Carex transsilvanica Schur = ***Carex depressa*** subsp. ***transsilvanica*** (Schur) T.V.Egorova
Carex transversa Miq. = ***Carex brownii*** Tuck. subsp. ***brownii***
Carex transversa Boott = ***Carex brownii*** subsp. ***transversa*** (Boott) J.Kern
Carex trappistarum Franch. = ***Carex heterolepis*** Bunge

Carex trappistarum var. *obtegens* Kük. = ***Carex heterolepis*** Bunge
Carex tremula (Ohwi) Ohwi = ***Carex filipes*** var. ***tremula*** (Ohwi) Ohwi
Carex tribuloides f. *aggregata* Peck = ***Carex tribuloides*** Wahlenb. var. ***tribuloides***
Carex tribuloides var. *bebbii* (Olney ex Britton) L.H.Bailey = ***Carex bebbii*** Olney ex Britton
Carex tribuloides var. *cristata* (Tuck.) L.H.Bailey = ***Carex tribuloides*** Wahlenb. var. ***tribuloides***
Carex tribuloides var. *cristata* (Tuck.) Macoun = ***Carex tribuloides*** Wahlenb. var. ***tribuloides***
Carex tribuloides var. *moniliformis* (Tuck.) Britton = ***Carex scoparia*** Willd. var. ***scoparia***
Carex tribuloides var. *reducta* L.H.Bailey = ***Carex tribuloides*** Wahlenb. var. ***tribuloides***
Carex tribuloides var. *turbata* L.H.Bailey = ***Carex tribuloides*** Wahlenb. var. ***tribuloides***
Carex tricarinata H.Lév. & Vaniot = ***Carex japonica*** Thunb.
Carex triceps Michx. = ***Carex hirsutella*** Mack.
Carex triceps Schrank = ?
Carex triceps var. *hirsuta* (Raf.) L.H.Bailey = ***Carex hirsutella*** Mack.
Carex triceps var. *longicuspis* Kük. = ***Carex hirsutella*** Mack.
Carex triceps var. *pedunculata* (Schwein. & Torr.) Torr. = ***Carex hirsutella*** Mack.
Carex triceps var. *smithii* L.H.Bailey = ***Carex hirsutella*** Mack.
Carex trichocarpa Schur = ***Carex sempervirens*** Vill.
Carex trichocarpa var. *aristata* L.H.Bailey = ***Carex atherodes*** Spreng.
Carex trichocarpa var. *deweyi* L.H.Bailey = ***Carex atherodes*** Spreng.
Carex trichocarpa var. *imberbis* A.Gray = ***Carex atherodes*** Spreng.
Carex trichocarpa var. *laeviconica* Hitchc. = ***Carex atherodes*** Spreng.
Carex trichocarpa var. *maxima* Kük. = ***Carex atherodes*** Spreng.
Carex trichocarpa var. *orostachys* (C.A.Mey.) Kük. = ***Carex atherodes*** Spreng.
Carex trichocarpa var. *turbinata* Dewey = ***Carex atherodes*** Spreng.
Carex trichodes var. *lateriflora* (Phil.) Kük. = ***Carex lateriflora*** Phil.
Carex trichodes var. *major* Kük. = ***Carex lateriflora*** Phil.
Carex tricholoma S.T.Blake = ***Carex breviculmis*** var. ***perciliata*** Kük.
Carex trichopoda Franch. = ***Carex podogyna*** Franch. & Sav.
Carex trichosperma Ohwi = ***Carex dolichostachya*** subsp. ***trichosperma*** (Ohwi) T.Koyama
Carex trichostachya Fisch. ex Boott = ***Carex podocarpa*** R.Br.
Carex trichostyles Franch. & Sav. = ***Carex japonica*** Thunb.
Carex tricostata Fr. = ***Carex acuta*** L.
Carex tricostata var. *praticola* Fr. = ***Carex × elytroides*** Fr.
Carex tricostata var. *schummelii* (Siegert) Nyman = ***Carex acuta*** L.
Carex tricuspidata Kük. = ***Carex brachyanthera*** Ohwi
Carex trifida var. *franchetii* Kurtz = ***Carex banksii*** Boott var. ***banksii***
Carex triflora Schkuhr = ***Carex depauperata*** Curtis ex With
Carex triflora Vahl ex Kunth = ***Carex vaginata*** Tausch var. ***vaginata***
Carex trigona All. = ***Carex hostiana*** DC.
Carex trinervis Nees = ***Kobresia esenbeckii*** (Kunth) Noltie var. ***esenbeckii***
Carex trinervis var. *laxa* Lange = ***Carex trinervis*** Degl.
Carex tripartita Lam. = ***Carex divisa*** Huds.
Carex tripartita All. = ***Carex lachenalii*** Schkuhr subsp. ***lachenalii***
Carex tripartita Suter = ***Carex lachenalii*** Schkuhr subsp. ***lachenalii***
Carex triquetrifolia Boeck. = ***Carex monostachya*** A.Rich.
Carex trisperma A.Gray ex Boott = ***Carex rosea*** Willd.
Carex trisperma f. *billingsii* (O.W.Knight) B.Boivin = ***Carex trisperma*** var. ***billingsii*** O.W.Knight
Carex trispiculata Boeck. = ***Carex norvegica*** Retz. subsp. ***norvegica***
Carex tristachya subsp. *pocilliformis* (Boott) T.Koyama = ***Carex tristachya*** var. ***pocilliformis*** (Boott) Kük.
Carex tristicha Boeck. = ***Carex stricta*** Lam.
Carex tristicha Spruce ex Boott = ***Carex tamana*** Steyerm.
Carex tristis var. *lazica* Boiss. & Balansa = ***Carex pontica*** Albov
Carex tristis subsp. *stenocarpa* (Turcz. ex V.I.Krecz.) T.V.Egorova = ***Carex stenocarpa*** Turcz. ex V.I.Krecz.
Carex tristis var. *tatrorum* Zapal. = ***Carex sempervirens*** Vill.
Carex triticea Wahlenb. ex Steud. = ***Carex clavata*** Thunb.
Carex truncata Fenzl = ***Carex flacca*** Schreb. subsp. ***flacca***
Carex truncata Boeck. = ***Carex laxiflora*** Lam. var. ***laxiflora***
Carex truncatirostris S.W.Su & S.M.Xu = ***Carex chungii*** Z.P.Wang var. ***chungii***
Carex truncatirostris f. *erostris* S.W.Su & S.M.Xu = ***Carex chungii*** Z.P.Wang var. ***chungii***
Carex truncatula Kit. = ?
Carex tsangensis Franch. = ***Carex setigera*** var. ***schlagintweitiana*** (Boeck.) Kük.
Carex tsaratananensis var. *laevis* Cherm. = ***Carex tsaratananensis*** Cherm.
Carex tschonoskii V.I.Krecz. = ?
Carex tsuishikarensis Koidz. & Ohwi = ***Carex oligosperma*** Michx.
Carex tuberculata Nees ex Boott = ***Carex bonariensis*** Desf. ex Poir. var. ***bonariensis***
Carex tuberculata var. *ehrenbergiana* (Boeck.) Kük. = ***Carex tuberculata*** Liebm.
Carex tuberosa Degl. = ***Carex distachya*** Desf. var. ***distachya***
Carex tuberosa Blanco = ***Eleocharis dulcis*** (Burm.f.) Trin. ex Hensch.
Carex tubulata K.Schum. ex Boott = ***Carex buxbaumii*** Wahlenb.
Carex tuckermanii var. *cylindracea* Dewey = ***Carex tuckermanii*** Boott
Carex tuckermanii var. *niagarensis* C.P.Sm. = ***Carex tuckermanii*** Boott
Carex tumida Beilschm. = ***Carex nigra*** (L.) Reichard subsp. ***nigra***
Carex tumida Boott = ***Carex oedorrhampha*** Nelmes
Carex tumidicarpa Andersson = ***Carex viridula*** subsp. ***oedocarpa*** (Andersson) B.Schmid
Carex tumidicarpa subsp. *cedercreutzii* Fagerstr. = ***Carex viridula*** subsp. ***cedercreutzii*** (Fagerstr.) B.Schmid
Carex turbinata var. *rigens* Kük. = ***Carex turbinata*** Liebm.
Carex turczaninofii Steud. ex Boott = ***Carex supina*** Willd. ex Wahlenb. var. ***supina***
Carex turczaninowiana Meinsh. ex Korsh. = ***Carex arnellii*** H.Christ
Carex turczaninowiana var. *beaurepairoei* H.Lév. & Vaniot = ***Carex drymophila*** Turcz.

Carex turgescens Dewey = ***Carex louisianica*** L.H.Bailey
Carex turgida (Fernald) J.W.Moore = ***Carex grisea*** Wahlenb.
Carex turgidula L.H.Bailey = ***Carex aperta*** Boott
Carex turuli Simonk. = ?
Carex tymphaea Formánek = ***Carex pallescens*** L. var. ***pallescens***
Carex typhinoides Schwein. = ***Carex squarrosa*** L.
Carex typhoides Bory = ***Carex borbonica*** Lam.
Carex tyttholepis Nelmes = ***Carex lamprochlamys*** S.T.Blake
Carex uber Ohwi = ***Carex sociata*** Boott
Carex uberior (C.Mohr) Mack. = ***Carex stipata*** var. ***maxima*** Chapm. ex Boott
Carex uda var. *sachalinensis* F.Schmidt = ***Carex capillacea*** var. ***sachalinensis*** (F.Schmidt) Ohwi
Carex udensis Meinsh. = ***Carex vesicaria*** L.
Carex ×*uechtritziana* K.Richt. = ?
Carex uetliaca Suter = ***Carex flava*** L. subsp. ***flava***
Carex uhligii K.Schum. ex C.B.Clarke = ***Schoenoxiphium lehmannii*** (Nees) Kunth ex Steud.
Carex ulaviformis Hoppe = ?
Carex uleana Boeck. = ***Carex seticulmis*** Boeck.
Carex uliginosa L. = ***Blysmus compressus*** (L.) Panz. ex Link subsp. ***compressus***
Carex ultra L.H.Bailey = ***Carex spissa*** L.H.Bailey ex Hemsl. var. ***spissa***
Carex umbellata subsp. *abdita* (E.P.Bicknell) W.Stone = ***Carex umbellata*** Willd.
Carex umbellata var. *brachyrhina* Piper = ***Carex umbellata*** Willd.
Carex umbellata var. *brevirostris* Boott = ***Carex umbellata*** Willd.
Carex umbellata var. *depressa* Kük. = ***Carex hypsipedos*** C.B.Clarke
Carex umbellata var. *globosa* (Boott) Kük. = ***Carex globosa*** Boott
Carex umbellata var. *tonsa* Fernald = ***Carex tonsa*** (Fernald) E.P.Bicknell
Carex umbellata f. *tonsa* (Fernald) Kük. = ***Carex tonsa*** (Fernald) E.P.Bicknell
Carex umbellata var. *vicina* Dewey = ***Carex umbellata*** Willd.
Carex umbellata f. *vicina* (Dewey) Wiegand = ***Carex umbellata*** Willd.
Carex umbrosa f. *brevifolia* (Waisb.) Soó = ***Carex umbrosa*** Host subsp. ***umbrosa***
Carex umbrosa f. *breviscapa* (Peterm.) Soó = ***Carex umbrosa*** Host subsp. ***umbrosa***
Carex umbrosa var. *decalvata* (Hausm. ex Döll) Kük. = ***Carex umbrosa*** Host subsp. ***umbrosa***
Carex umbrosa f. *diandrostachya* Waisb. = ***Carex umbrosa*** Host subsp. ***umbrosa***
Carex umbrosa f. *gracilis* (Kohts) Kük. = ***Carex umbrosa*** Host subsp. ***umbrosa***
Carex umbrosa f. *gynobasis* (Rouy) O.Bolòs & Vigo = ***Carex umbrosa*** Host subsp. ***umbrosa***
Carex umbrosa var. *huetiana* (Boiss.) Kük. = ***Carex umbrosa*** subsp. ***huetiana*** (Boiss.) Soó
Carex umbrosa subsp. *hypochlora* (Freyn) T.V.Egorova = ***Carex breviculmis*** R.Br. var. ***breviculmis***
Carex umbrosa var. *koreana* Nakai = ***Carex sachalinensis*** F.Schmidt var. ***sachalinensis***
Carex umbrosa var. *longifolia* (Wahlenb.) Nyman = ***Carex umbrosa*** Host subsp. ***umbrosa***
Carex umbrosa f. *longipedunculata* Torges ex Kük. = ***Carex umbrosa*** Host subsp. ***umbrosa***
Carex umbrosa subsp. *pseudosabynensis* T.V.Egorova = ***Carex pseudosabynensis*** (T.V.Egorova) A.E.Kozhevn.
Carex umbrosa var. *reflexa* Nyman = ***Carex praecox*** Schreb.
Carex umbrosa var. *stolonifera* Kük. = ***Carex umbrosa*** var. ***rostrata*** (Maxim.) Govaerts
Carex umbrosa f. *subintegra* (Kük.) Kük. = ***Carex umbrosa*** Host subsp. ***umbrosa***
Carex uncifolia var. *libera* Kük. = ***Carex libera*** (Kük.) Hamlin
Carex uncinata Schkuhr ex Steud. = ***Uncinia hamata*** (Sw.) Urb.
Carex uncinata L.f. = ***Uncinia uncinata*** (L.f.) Kük.
Carex unciniiformis Boeck. = ***Kobresia nepalensis*** (Nees) Kük. var. ***nepalensis***
Carex uncinioides Boott = ***Kobresia uncinioides*** (Boott) C.B.Clarke
Carex uncompahgre Kelso = ***Carex bella*** L.H.Bailey
Carex underwoodii Britton = ***Carex polysticha*** Boeck.
Carex undulata Kunze = ***Carex pallescens*** L. var. ***pallescens***
Carex undulata Schur = ***Carex pallescens*** L. var. ***pallescens***
Carex unifoliata Kük. & Hand.-Mazz. = ***Carex parva*** Nees
Carex uraiensis Hayata = ***Carex sociata*** Boott
Carex uralensis C.B.Clarke = ***Carex stenophylla*** Wahlenb. subsp. ***stenophylla***
Carex urbanii Boeck. = ***Carex viridula*** Michx. subsp. ***viridula***
Carex urbis-malorum Popov = ***Carex atrata*** subsp. ***caucasica*** (Steven) Kük.
Carex urolepis Franch. = ***Carex darwinii*** var. ***urolepis*** (Franch.) Kük.
Carex urostachys Franch. = ***Carex mertensii*** var. ***urostachys*** (Franch.) Kük.
Carex urvillei Brongn. = ***Carex pumila*** Thunb.
Carex ushishirensis Ohwi = ***Carex glareosa*** Schkuhr ex Wahlenb. subsp. ***glareosa***
Carex usta Franch. = ***Carex cespitosa*** L. var. ***cespitosa***
Carex usta L.H.Bailey = ***Carex douglasii*** Boott
Carex ustulata Wahlenb. = ***Carex atrofusca*** Schkuhr subsp. ***atrofusca***
Carex ustulata var. *aterrima* Regel = ***Carex nivalis*** Boott
Carex ustulata var. *coriophora* Regel = ***Carex nivalis*** Boott
Carex ustulata var. *macrogyna* (Turcz. ex Steud.) Regel = ***Carex petricosa*** Dewey var. ***petricosa***
Carex ustulata var. *major* Boeck. = ***Carex atrofusca*** var. ***major*** (Boeck.) Raymond
Carex ustulata var. *minor* Boott = ***Carex atrofusca*** subsp. ***minor*** (Boott) T.Koyama
Carex ustulata var. *pallisa* Meinsh. = ***Carex atrofusca*** Schkuhr subsp. ***atrofusca***
Carex utriculata var. *globosa* Olney = ***Carex utriculata*** Boott
Carex utriculata var. *minor* Boott = ***Carex utriculata*** Boott
Carex utriculata var. *sparsiflora* Dewey = ***Carex utriculata*** Boott
Carex uzoni Kom. = ***Carex aquatilis*** var. ***minor*** Boott
Carex vacillans Steud. = ***Carex riparia*** Curtis subsp. ***riparia***
Carex vacillans Sol. ex Boott = ***Carex spinirostris*** Colenso
Carex vagans T.Holm = ***Carex vallicola*** Dewey
Carex vaginata var. *altocaulis* Dewey = ***Carex vaginata*** Tausch var. ***vaginata***
Carex vaginata var. *curvirostra* (Hartm.) Nyman = ***Carex vaginata*** Tausch var. ***vaginata***
Carex vaginata subsp. *quasivaginata* (C.B.Clarke) Malyschev = ***Carex quasivaginata*** C.B.Clarke

Carex vahlii Schkuhr = ***Carex norvegica*** Retz. subsp. ***norvegica***
Carex vahlii var. *inferalpina* (Wahlenb.) Fernald = ***Carex media*** R.Br.
Carex vahlii var. *stevenii* (T.Holm) Fernald = ***Carex stevenii*** (T.Holm) Kalela
Carex valdiviana Phil. = ***Carex fuscula*** d'Urv. var. ***fuscula***
Carex valesiaca Suter = ***Carex brachystachys*** Schrank
Carex valesiaca Wahlenb. = ***Carex oedipostyla*** Duval-Jouve
Carex valida Nees = ***Carex cruciata*** Wahlenb. var. ***cruciata***
Carex vallicola var. *hidalgensis* F.J.Herm. = ***Carex vallicola*** Dewey
Carex vallicola var. *rusbyi* (Mack.) F.J.Herm. = ***Carex vallicola*** Dewey
Carex vallis-rosetto var. *purpurea* Kük. = ***Carex vallis-rosetto*** K.Schum.
Carex van-heurckii var. *crassispiculata* Malyschev = ***Carex van-heurckii*** subsp. ***crassispiculata*** (Malyschev) Malyschev
Carex vaniotii H.Lév. = ***Carex jackiana*** subsp. ***parciflora*** (Boott) Kük.
Carex vansteenisii Kük. = ***Carex perakensis*** var. ***vansteenisii*** (Kük.) Noot.
Carex varia Muhl. ex Wahlenb. = ***Carex albicans*** Willd. ex Spreng. var. ***albicans***
Carex varia Host = ***Carex sempervirens*** Vill.
Carex varia var. *australis* L.H.Bailey = ***Carex albicans*** var. ***australis*** (L.H.Bailey) Rettig
Carex varia f. *australis* (L.H.Bailey) Kük. = ***Carex albicans*** var. ***australis*** (L.H.Bailey) Rettig
Carex varia var. *colorata* L.H.Bailey = ***Carex albicans*** Willd. ex Spreng. var. ***albicans***
Carex varia f. *colorata* (L.H.Bailey) Kük. = ***Carex albicans*** Willd. ex Spreng. var. ***albicans***
Carex varia f. *distincta* Howe = ***Carex albicans*** Willd. ex Spreng. var. ***albicans***
Carex varia var. *minor* Boott = ***Carex albicans*** var. ***emmonsii*** (Dewey ex Torr.) Rettig
Carex varia var. *novae-angliae* (Schwein.) Kük. = ***Carex novae-angliae*** Schwein.
Carex varia var. *pedicellata* Dewey = ***Carex austrina*** Mack.
Carex variabilis L.H.Bailey = ***Carex aquatilis*** Wahlenb. var. ***aquatilis***
Carex variabilis var. *altior* Rydb. = ***Carex emoryi*** Dewey
Carex variabilis var. *elatior* L.H.Bailey = ***Carex emoryi*** Dewey
Carex variabilis var. *sciaphila* T.Holm = ***Carex aquatilis*** Wahlenb. var. ***aquatilis***
Carex varians (L.H.Bailey) Hay, Palmer & C.T.White = ***Carex leptonervia*** (Fernald) Fernald
Carex varica C.B.Clarke = ?
Carex variegata Scheele = ***Carex barrattii*** Torr. ex Schwein.
Carex variegata Lam. = ***Carex sempervirens*** Vill.
Carex vaseyi Dewey = ***Carex vesicaria*** L.
Carex velenovskyi Domin = ***Carex praecox*** Schreb.
Carex venosivaginata Standl. & Steyerm. = ***Carex pertenuis*** L.H.Bailey
Carex ventricosa Curtis = ***Carex depauperata*** Curtis ex With
Carex ventricosa Franch. = ***Carex utriculata*** Boott
Carex venusta var. *glabra* L.H.Bailey = ***Carex venusta*** Dewey
Carex venusta var. *minor* Boeck. = ***Carex venusta*** Dewey
Carex venusta var. *oblita* (Steud.) Kük. = ***Carex venusta*** Dewey
Carex venustula T.Holm = ***Carex podocarpa*** R.Br.
Carex verae Ovcz. & Czukav. = ***Carex turkestanica*** Regel subsp. ***turkestanica***
Carex verecunda T.Holm = ***Carex inops*** L.H.Bailey subsp. ***inops***
Carex verna Chaix & Vill. = ***Carex caryophyllea*** Latourr. var. ***caryophyllea***
Carex verna Lam. = ***Carex flacca*** Schreb. subsp. ***flacca***
Carex verna Torr. ex Kunth = ***Carex praecox*** Schreb.
Carex verna Schkuhr = ***Carex supina*** Willd. ex Wahlenb. var. ***supina***
Carex verna f. *argyrata* Waisb. = ***Carex caryophyllea*** Latourr. var. ***caryophyllea***
Carex verna var. *caespitiformis* Waisb. = ***Carex caryophyllea*** Latourr. var. ***caryophyllea***
Carex verna f. *chlorantha* Waisb. = ***Carex caryophyllea*** Latourr. var. ***caryophyllea***
Carex verna f. *globuliflora* Waisb. = ***Carex caryophyllea*** Latourr. var. ***caryophyllea***
Carex verna f. *glomerata* Waisb. = ***Carex caryophyllea*** Latourr. var. ***caryophyllea***
Carex verna f. *laxiflora* Waisb. = ***Carex caryophyllea*** Latourr. var. ***caryophyllea***
Carex verna f. *longibracteata* Beck = ***Carex caryophyllea*** Latourr. var. ***caryophyllea***
Carex verna var. *minor* Beck = ***Carex caryophyllea*** Latourr. var. ***caryophyllea***
Carex verna var. *oxycarpa* Waisb. = ***Carex caryophyllea*** Latourr. var. ***caryophyllea***
Carex verna f. *pedunculata* Beck = ***Carex caryophyllea*** Latourr. var. ***caryophyllea***
Carex vernacula var. *hobsonii* Maguire = ***Carex vernacula*** L.H.Bailey
Carex vernicosa Franch. ex C.B.Clarke = ***Carex aphanolepis*** Franch. & Sav.
Carex verrucosa Schwein. ex Torr. = ***Carex stricta*** Lam.
Carex versicolor Dumort. = ***Carex binervis*** Sm.
Carex verticillata Boott = ***Carex interrupta*** Boeck.
Carex vesicaria var. *alpigena* Fr. = ***Carex vesicaria*** L.
Carex vesicaria f. *aurea* Lange = ***Carex vesicaria*** L.
Carex vesicaria var. *brachystachys* Uechtr. ex Asch. & Graebn. = ***Carex vesicaria*** L.
Carex vesicaria f. *brachystachys* Lindeb. = ***Carex vesicaria*** L.
Carex vesicaria var. *calcifoenum* Laest. = ***Carex vesicaria*** L.
Carex vesicaria var. *colorata* (L.H.Bailey) Kük. = ***Carex vesicaria*** L.
Carex vesicaria var. *compacta* Dewey = ***Carex membranacea*** Hook.
Carex vesicaria f. *dichroa* (A.Blytt) Kük. = ***Carex saxatilis*** L.
Carex vesicaria var. *dichroa* Andersson = ***Carex vesicaria*** L.
Carex vesicaria var. *distenta* Fr. = ***Carex saxatilis*** L.
Carex vesicaria f. *elatior* Andersson = ***Carex vesicaria*** L.
Carex vesicaria f. *fluitans* Raymond = ***Carex vesicaria*** L.
Carex vesicaria subsp. *friesii* (Blytt) Nyman = ***Carex ×friesii*** Blytt
Carex vesicaria var. *globosa* Olney = ***Carex vesicaria*** L.
Carex vesicaria f. *globosa* (Olney) Kük. = ***Carex vesicaria*** L.
Carex vesicaria f. *gracillima* Kük. = ***Carex vesicaria*** L.
Carex vesicaria var. *grahamii* (Boott) Kük. = ***Carex ×stenolepis*** Less. nothovar. ***stenolepis***
Carex vesicaria f. *hexasticha* Hoschede = ***Carex vesicaria*** L.

Carex vesicaria var. *inalpina* Laest. = ***Carex rostrata*** Stokes var. ***rostrata***
Carex vesicaria var. *involuta* Bab. = ***Carex ×involuta*** (Bab.) Syme
Carex vesicaria var. *jejuna* Fernald = ***Carex saxatilis*** L.
Carex vesicaria subsp. *lacustris* Th.Fr. = ***Carex utriculata*** Boott
Carex vesicaria var. *lanceolata* Olney = ***Carex vesicaria*** L.
Carex vesicaria var. *latifolia* A.Blytt = ***Carex vesicaria*** L.
Carex vesicaria var. *laurentiana* Fernald = ***Carex saxatilis*** L.
Carex vesicaria f. *laxa* (Trautv.) Kük. = ***Carex saxatilis*** L.
Carex vesicaria var. *major* Boott = ***Carex exsiccata*** L.H.Bailey
Carex vesicaria var. *monile* (Tuck. ex Boott) Tuck. ex Boott = ***Carex vesicaria*** L.
Carex vesicaria var. *monstrosa* (L.H.Bailey) Farw. = ***Carex vesicaria*** L.
Carex vesicaria f. *monstrosa* (L.H.Bailey) Kük. = ***Carex vesicaria*** L.
Carex vesicaria var. *pacifica* (L.H.Bailey) Kük. = ***Carex vesicaria*** L.
Carex vesicaria var. *pamirica* O.Fedtsch. = ***Carex pamirica*** (O.Fedtsch.) O.Fedtsch. & B.Fedtsch.
Carex vesicaria f. *pendulina* A.Blytt = ***Carex vesicaria*** L.
Carex vesicaria var. *physocarpa* (C.Presl) Kük. = ***Carex saxatilis*** L.
Carex vesicaria var. *plumbea* Wahlenb. = ***Carex riparia*** Curtis subsp. ***riparia***
Carex vesicaria f. *polystachya* (A.Blytt) Kük. = ***Carex vesicaria*** L.
Carex vesicaria subsp. *pulla* (Gooden.) Andersson = ***Carex saxatilis*** L.
Carex vesicaria f. *pungens* (L.H.Bailey) Kük. = ***Carex vesicaria*** L.
Carex vesicaria var. *raeana* (Boott) Fernald = ***Carex vesicaria*** L.
Carex vesicaria var. *reflexa* Meinsh. = ***Carex mollissima*** H.Christ ex Scheutz
Carex vesicaria var. *robusta* Sond. = ***Carex vesicaria*** L.
Carex vesicaria var. *rostratiformis* Waisb. = ***Carex ×involuta*** (Bab.) Syme
Carex vesicaria subsp. *saxatilis* L. = ***Carex vesicaria*** L.
Carex vesicaria var. *subalpina* Laest. = ***Carex vesicaria*** L.
Carex vesicaria var. *tenuistachya* Kük. = ***Carex vesicaria*** L.
Carex vesicaria f. *umbrosa* Kük. = ***Carex vesicaria*** L.
Carex vesicaria var. *utriculata* (Boott) Dewey = ***Carex utriculata*** Boott
Carex vesicaria subsp. *vesicata* (Meinsh.) T.V.Egorova = ***Carex vesicaria*** L.
Carex vesicata Meinsh. = ***Carex vesicaria*** L.
Carex vesiciformis Peterm. = ?
Carex vesiculosa var. *congesta* Kük. = ***Carex vesiculosa*** Boott
Carex vesiculosa f. *pallida* Kük. = ***Carex vesiculosa*** Boott
Carex vesiculosa var. *pallida* (Kük.) R.C.Srivast. = ***Carex vesiculosa*** Boott
Carex vesiculosa var. *paniculata* C.B.Clarke = ***Carex vesiculosa*** Boott
Carex vespertina (L.H.Bailey) Howell = ***Carex inops*** L.H.Bailey subsp. ***inops***
Carex vestita var. *kennedyi* Fernald = ***Carex vestita*** Willd.
Carex vexilis Wormsk. ex Boott = ***Carex michelii*** Host
Carex ×viadrina Figert = ?
Carex ×viadrina Schube = ?
Carex vicaria L.H.Bailey = ***Carex densa*** (L.H.Bailey) L.H.Bailey
Carex vicaria var. *costata* L.H.Bailey = ***Carex densa*** (L.H.Bailey) L.H.Bailey
Carex victorialis Nees = ***Carex notha*** Kunth
Carex vidalii Franch. & Sav. = ***Carex nervata*** Franch. & Sav.
Carex vidua Boott ex C.B.Clarke = ***Kobresia vidua*** (Boott ex C.B.Clarke) Kük.
Carex ×vierhapperi Beck = ***Carex ×gerhaldtii*** Figert
Carex vikingensis C.B.Clarke = ***Carex distans*** L.
Carex ×villacensis Kük. = ?
Carex villarsii Willd. = ***Carex foetida*** All.
Carex villosa Franch. & Sav. = ***Carex davalliana*** Sm.
Carex villosa Stokes = ***Carex hirta*** L.
Carex villosa Boott = ***Carex latisquamea*** Kom.
Carex villosa var. *latisquamea* (Kom.) Kük. = ***Carex latisquamea*** Kom.
Carex vilmorinii Mottet = ***Carex comans*** Berggr.
Carex ×vimariensis Hausskn. ex Berthold = ?
Carex violacea C.B.Clarke = ***Carex nova*** L.H.Bailey
Carex virens Thuill. = ***Carex acuta*** L.
Carex virens Lam. = ***Carex divulsa*** Stokes subsp. ***divulsa***
Carex virens var. *duriaei* F.W.Schultz = ***Carex divulsa*** Stokes subsp. ***divulsa***
Carex virens var. *guestphalica* (Rchb.) Garcke = ***Carex divulsa*** Stokes subsp. ***divulsa***
Carex virens var. *pairaei* Garcke = ***Carex echinata*** Murray subsp. ***echinata***
Carex virescens var. *costata* Dewey = ***Carex virescens*** Muhl. ex Willd.
Carex virescens var. *enormis* Farw. = ***Carex virescens*** Muhl. ex Willd.
Carex virescens var. *grioletii* (Roem. ex Schkuhr) Gay = ***Carex grioletii*** Roem. ex Schkuhr
Carex virescens var. *swanii* Fernald = ***Carex swanii*** (Fernald) Mack.
Carex virgata Sol. ex Boott = ***Carex appressa*** var. ***virgata*** (Sol. ex Boott) Kük.
Carex virgata Miq. = ***Carex cruciata*** var. ***rafflesiana*** (Boott) Noot.
Carex virgata var. *abbreviata* Boeck. = ***Carex appressa*** R.Br. var. ***appressa***
Carex virgata var. *secta* (Boott) Hook.f. = ***Carex secta*** Boott
Carex virginiana Woods = ***Carex acuta*** L.
Carex virginiana (Fernald) Fernald = ***Carex crus-corvi*** Shuttlew. ex Kunze
Carex virginiana var. *elongata* Boeck. = ***Carex emoryi*** Dewey
Carex virginica Steud. = ***Carex acuta*** L.
Carex viridiflora Mack. = ***Carex perlonga*** Fernald
Carex viridior Mack. = ***Carex atrosquama*** Mack.
Carex viridis Honck. = ***Carex flava*** L. subsp. ***flava***
Carex viridis Petrie = ***Carex muelleri*** Petrie
Carex viridis Schltdl. & Cham. = ***Carex perlonga*** Fernald
Carex viridis Spenn. = ?
Carex viridis var. *divulsa* (Stokes) Spenn. = ***Carex divulsa*** Stokes
Carex viridissima Nakai = ***Carex matsumurae*** Franch.
Carex viridula Franch. & Sav. = ***Carex curvicollis*** Franch. & Sav.
Carex viridula Schwein. & Torr. = ***Carex hirsutella*** Mack.
Carex viridula var. *bergrothii* (Palmgr.) B.Schmid = ***Carex pilulifera*** L. subsp. ***pilulifera***
Carex viridula subsp. *bergrothii* (Palmgr.) Tzvelev = ***Carex pilulifera*** L. subsp. ***pilulifera***
Carex viridula f. *clandestina* Raymond = ***Carex viridula*** Michx. subsp. ***viridula***

Carex viridula var. *czatkalensis* Yu.E.Alexeev = ***Carex viridula*** Michx. subsp. ***viridula***
Carex viridula f. *intermedia* (Dudley) F.J.Herm. = ***Carex viridula*** Michx. subsp. ***viridula***
Carex viridula var. *lepidocarpa* (Tausch) B.Schmid = ***Carex viridula*** var. ***elatior*** (Schltr.) Crins
Carex viridula subsp. *nevadensis* (Boiss. & Reut.) B.Schmid = ***Carex flava*** var. ***nevadensis*** (Boiss. & Reut.) Briq.
Carex viridula var. *nevadensis* (Boiss. & Reut.) Crins = ***Carex flava*** var. ***nevadensis*** (Boiss. & Reut.) Briq.
Carex viridula var. *pulchella* (Lönnr.) B.Schmid = ***Carex viridula*** Michx. subsp. ***viridula***
Carex viridula subsp. *pulchella* (Lönnr.) Malyschev = ***Carex viridula*** Michx. subsp. ***viridula***
Carex viridula f. *pygmaea* Lepage = ***Carex viridula*** Michx. subsp. ***viridula***
Carex viridula f. *rousseauiana* (Vict.) Raymond = ***Carex viridula*** Michx. subsp. ***viridula***
Carex viridula var. *scotica* (E.W.Davies) B.Schmid = ***Carex flava*** subsp. ***scotica*** (E.W.Davies) P.D.Sell
Carex viridula subsp. *serotina* (Mérat) Malyschev = ***Carex viridula*** Michx. subsp. ***viridula***
Carex vitiensis H.St.John = ***Carex gibbsiae*** Rendle
Carex vitilis Fr. = ***Carex brunnescens*** subsp. ***sphaerostachya*** (Tuck.) Kalela
Carex vitilis var. *sylvatica* Meinsh. = ***Carex brunnescens*** (Pers.) Poir. subsp. ***brunnescens***
Carex vitrea T.Holm = ***Carex alma*** L.H.Bailey
Carex vladimiroviensis H.Lév. = ***Carex schmidtii*** Meinsh.
Carex vleckii Schwein. = ***Carex oligocarpa*** Willd.
Carex volcanica F.J.Herm. = ***Carex orizabae*** Liebm.
Carex volkensii K.Schum. = [25]
Carex ×*vratislaviensis* Figert = ?
Carex vulcani f. *floresiana* (Hochst. ex Seub.) Kük. = ***Carex vulcani*** Hochst. ex Seub.
Carex vulcani var. *minor* Boott = ***Carex vulcani*** Hochst. ex Seub.
Carex vulcanica Elmer = ***Carex nodiflora*** Boeck.
Carex vulcanicola Nakai = ***Kobresia myosuroides*** (Vill.) Fiori & Paol.
Carex vulgaris R.E.Fr. = ***Carex nigra*** (L.) Reichard subsp. ***nigra***
Carex vulgaris f. *abbreviata* Peterm. = ***Carex nigra*** (L.) Reichard subsp. ***nigra***
Carex vulgaris var. *alpina* Boott = ***Carex bigelowii*** Torr. ex Schwein. subsp. ***bigelowii***
Carex vulgaris subsp. *alpina* (Gaudin) Arcang. = ***Carex nigra*** subsp. ***alpina*** (Gaudin) Lemke
Carex vulgaris f. *androgyna* Peterm. = ***Carex nigra*** (L.) Reichard subsp. ***nigra***
Carex vulgaris f. *androgyna* Russow ex E.Lehm. = ***Carex nigra*** (L.) Reichard subsp. ***nigra***
Carex vulgaris var. *angustifolia* A.Blytt = ***Carex nigra*** (L.) Reichard subsp. ***nigra***
Carex vulgaris f. *angustifolia* Kük. = ***Carex nigra*** (L.) Reichard subsp. ***nigra***
Carex vulgaris var. *antucensis* Kük. = ***Carex decidua*** Boott
Carex vulgaris f. *aquatiliformis* Kük. = ***Carex nigra*** (L.) Reichard subsp. ***nigra***
Carex vulgaris var. *atra* Peterm. = ***Carex nigra*** (L.) Reichard subsp. ***nigra***
Carex vulgaris f. *basigyna* Rchb. = ***Carex nigra*** (L.) Reichard subsp. ***nigra***
Carex vulgaris var. *bolina* (Láng) K.Richt. = ***Carex nigra*** (L.) Reichard subsp. ***nigra***
Carex vulgaris var. *bracteosa* L.H.Bailey = ***Carex scopulorum*** var. ***bracteosa*** (L.H.Bailey) F.J.Herm.
Carex vulgaris var. *bructeri* G.Mey. = ***Carex nigra*** (L.) Reichard subsp. ***nigra***
Carex vulgaris f. *chlorostachya* Rchb. = ***Carex nigra*** (L.) Reichard subsp. ***nigra***
Carex vulgaris var. *communis* Meinsh. = ***Carex nigra*** (L.) Reichard subsp. ***nigra***
Carex vulgaris f. *densicarpa* Kneuck. = ***Carex nigra*** (L.) Reichard subsp. ***nigra***
Carex vulgaris var. *distracta* C.B.Clarke = ***Carex orbicularis*** Boott subsp. ***orbicularis***
Carex vulgaris var. *elatior* T.Holm = ***Carex nigra*** (L.) Reichard subsp. ***nigra***
Carex vulgaris var. *elytroides* (Fr.) Boott = ***Carex ×elytroides*** Fr.
Carex vulgaris subsp. *elytroides* (Fr.) Nyman = ***Carex ×elytroides*** Fr.
Carex vulgaris var. *gaudichaudiana* (Kunth) Boott = ***Carex gaudichaudiana*** Kunth
Carex vulgaris var. *humilior* Maly = ***Carex nigra*** (L.) Reichard subsp. ***nigra***
Carex vulgaris var. *hydrophila* T.Holm = ***Carex nigra*** (L.) Reichard subsp. ***nigra***
Carex vulgaris var. *hyperborea* Boott = ***Carex bigelowii*** Torr. ex Schwein. subsp. ***bigelowii***
Carex vulgaris var. *intermedia* Nyman = ***Carex nigra*** (L.) Reichard subsp. ***nigra***
Carex vulgaris var. *intricata* Boott = ***Carex nigra*** (L.) Reichard subsp. ***nigra***
Carex vulgaris var. *juncea* Hartm. = ***Carex nigra*** subsp. ***juncella*** (Fr.) Lemke
Carex vulgaris subsp. *juncea* (Hartm.) Fr. = ***Carex nigra*** subsp. ***juncella*** (Fr.) Lemke
Carex vulgaris var. *juncella* Olney = ***Carex lenticularis*** var. ***lipocarpa*** (T.Holm) L.A.Standl.
Carex vulgaris subsp. *juncella* Fr. = ***Carex nigra*** subsp. ***juncella*** (Fr.) Lemke
Carex vulgaris f. *juncella* (Fr.) Kük. = ***Carex nigra*** subsp. ***juncella*** (Fr.) Lemke
Carex vulgaris f. *latifolia* Kük. = ***Carex nigra*** (L.) Reichard subsp. ***nigra***
Carex vulgaris f. *leucolepis* Meinsh. = ***Carex nigra*** (L.) Reichard subsp. ***nigra***
Carex vulgaris var. *limnophila* T.Holm = ***Carex lenticularis*** var. ***limnophila*** (T.Holm) Cronquist
Carex vulgaris var. *lipocarpa* T.Holm = ***Carex lenticularis*** var. ***lipocarpa*** (T.Holm) L.A.Standl.
Carex vulgaris var. *melaena* (Wimm.) Nyman = ***Carex nigra*** (L.) Reichard subsp. ***nigra***
Carex vulgaris var. *polygama* Rchb. = ***Carex nigra*** (L.) Reichard subsp. ***nigra***
Carex vulgaris var. *pseudohyperborea* Sagorski ex Kük. =
Carex vulgaris f. *pumila* Kük. = ***Carex nigra*** (L.) Reichard subsp. ***nigra***
Carex vulgaris f. *ramosa* Peterm. = ***Carex nigra*** (L.) Reichard subsp. ***nigra***
Carex vulgaris var. *rigida* A.Blytt = ***Carex nigra*** (L.) Reichard subsp. ***nigra***
Carex vulgaris var. *sabulosa* Meinsh. = ***Carex nigra*** (L.) Reichard subsp. ***nigra***
Carex vulgaris var. *salinoides* Kük. = ***Carex nigra*** (L.) Reichard subsp. ***nigra***
Carex vulgaris f. *silvatica* Russow ex E.Lehm. = ***Carex nigra*** (L.) Reichard subsp. ***nigra***
Carex vulgaris var. *strictiformis* L.H.Bailey = ***Carex nigra*** (L.) Reichard subsp. ***nigra***
Carex vulgaris f. *subovalis* Kneuck. = ***Carex nigra*** (L.) Reichard subsp. ***nigra***
Carex vulgaris var. *subramosa* Meinsh. = ***Carex nigra*** (L.) Reichard subsp. ***nigra***

Carex vulgaris f. *tenuis* Peterm. = ***Carex nigra*** (L.) Reichard subsp. ***nigra***
Carex vulgaris var. *tenuis* Hartm. ex Lange = ***Carex nigra*** (L.) Reichard subsp. ***nigra***
Carex vulgaris var. *teres* Boott = ***Carex nigra*** (L.) Reichard subsp. ***nigra***
Carex vulgaris var. *tornata* Fr. = ***Carex nigra*** (L.) Reichard subsp. ***nigra***
Carex vulgaris var. *tricostata* Boott = ***Carex brunnea*** Thunb. subsp. ***brunnea***
Carex vulgaris var. *turfosa* (Fr.) Boott = ***Carex × turfosa*** Fr.
Carex vulgaris f. *zonata* F.Nyl. = ***Carex nigra*** (L.) Reichard subsp. ***nigra***
Carex vulgaris var. *zonata* (F.Nyl.) Nyman = ***Carex nigra*** (L.) Reichard subsp. ***nigra***
Carex vulpi-caudata Akiyama = ***Carex gaudichaudiana*** Kunth
Carex vulpina Carey ex Dewey = ***Carex conjuncta*** Boott
Carex vulpina Hohen. = ***Carex muricata*** L.
Carex vulpina f. *aristata* Asch. = ***Carex vulpina*** L.
Carex vulpina f. *bracteata* G.Mey. = ***Carex cuprina*** (I.Sándor ex Heuff.) Nendtv. ex A.Kern.
Carex vulpina f. *capitulata* (Peterm.) Soó = ***Carex vulpina*** L.
Carex vulpina var. *compacta* (Lam.) Velen. = ***Carex vulpina*** L.
Carex vulpina var. *crassinervis* (Schur) Kük. = ***Carex vulpina*** L.
Carex vulpina subsp. *cuprina* (I.Sándor ex Heuff.) O.Bolòs & Vigo = ***Carex cuprina*** (I.Sándor ex Heuff.) Nendtv. ex A.Kern.
Carex vulpina var. *divulsa* Èelak. = ***Carex vulpina*** L.
Carex vulpina f. *elongata* Andersson = ***Carex vulpina*** L.
Carex vulpina f. *interrupta* Peterm. = ***Carex cuprina*** (I.Sándor ex Heuff.) Nendtv. ex A.Kern.
Carex vulpina f. *laeviuscula* Sanio ex Asch. & Graebn. = ***Carex vulpina*** L.
Carex vulpina var. *litoralis* Nolte = ***Carex vulpina*** L.
Carex vulpina var. *longibracteata* Beck = ***Carex vulpina*** L.
Carex vulpina f. *minor* Peterm. = ***Carex vulpina*** L.
Carex vulpina subsp. *nemorosa* (Lej. & Courtois) Schinz & R.Keller = ***Carex cuprina*** (I.Sándor ex Heuff.) Nendtv. ex A.Kern.
Carex vulpina f. *nemorosa* (Lej. & Courtois) W.D.J.Koch = ***Carex cuprina*** (I.Sándor ex Heuff.) Nendtv. ex A.Kern.
Carex vulpina var. *nemorosa* Lej. & Courtois = ***Carex vulpina*** L.
Carex vulpina var. *pallidior* Meinsh. = ***Carex vulpina*** L.
Carex vulpina var. *remotiflora* Lange = ***Carex vulpina*** L.
Carex vulpina var. *stribnyi* Velen. = ***Carex vulpina*** L.
Carex vulpina var. *subcontigua* Kük. = ***Carex cuprina*** (I.Sándor ex Heuff.) Nendtv. ex A.Kern.
Carex vulpina var. *tenuior* Ledeb. = ***Carex vulpina*** L.
Carex vulpina var. *vulgaris* Èelak. = ***Carex vulpina*** L.
Carex vulpinaris f. *angustifolia* Kük. = ***Carex vulpinaris*** Nees
Carex vulpiniformis Tuck. = ***Carex vulpinoidea*** Michx.
Carex vulpinoidea var. *ambigua* Barratt = ***Carex annectens*** (E.P.Bicknell) E.P.Bicknell
Carex vulpinoidea f. *annectens* (E.P.Bicknell) Kük. = ***Carex annectens*** (E.P.Bicknell) E.P.Bicknell
Carex vulpinoidea var. *annectens* (E.P.Bicknell) Farw. = ***Carex annectens*** (E.P.Bicknell) E.P.Bicknell
Carex vulpinoidea f. *costata* (L.H.Bailey) Kük. = ***Carex densa*** (L.H.Bailey) L.H.Bailey
Carex vulpinoidea var. *drummondiana* Boeck. = ***Carex triangularis*** Boeck.
Carex vulpinoidea var. *glomerata* Barratt = ***Carex vulpinoidea*** Michx.
Carex vulpinoidea var. *microsperma* (Wahlenb.) Dewey = ***Carex vulpinoidea*** Michx.
Carex vulpinoidea var. *platycarpa* Olney = ***Carex triangularis*** Boeck.
Carex vulpinoidea var. *pycnocephala* F.J.Herm. = ***Carex vulpinoidea*** Michx.
Carex vulpinoidea var. *segregata* Farw. = ***Carex vulpinoidea*** Michx.
Carex vulpinoidea f. *segregata* (Farw.) Raymond = ***Carex vulpinoidea*** Michx.
Carex vulpinoidea var. *setacea* (Dewey) Kük. = ***Carex vulpinoidea*** Michx.
Carex vulpinoidea var. *triangularis* (Boeck.) Kük. = ***Carex triangularis*** Boeck.
Carex vulpinoidea var. *vicaria* (L.H.Bailey) Kük. = ***Carex densa*** (L.H.Bailey) L.H.Bailey
Carex vulpinoidea var. *xanthocarpa* Kük. = ***Carex annectens*** (E.P.Bicknell) E.P.Bicknell
Carex wahlenbergiana var. *hirtisaccus* C.B.Clarke ex Kük. = ***Carex wahlenbergiana*** Boott
Carex wahlenbergiana var. *schimperi* Boott = ***Carex steudneri*** Boeck.
Carex wahuensis var. *meyeri* Franch. & Sav. = ***Carex wahuensis*** C.A.Mey. subsp. ***wahuensis***
Carex wahuensis subsp. *robusta* (Franch. & Sav.) T.Koyama = ***Carex boottiana*** Hook. & Arn.
Carex wahuensis var. *rubiginosa* R.W.Krauss = ***Carex wahuensis*** subsp. ***rubiginosa*** (R.W.Krauss) T.Koyama
Carex walkeri var. *turrita* (C.B.Clarke) Kük. = ***Carex turrita*** C.B.Clarke
Carex wallichiana Prescott ex Nees = ***Carex fedia*** Nees
Carex wallichiana Spreng. = ***Carex foliosa*** D.Don ex Tilloch & Taylor
Carex wallichiana f. *glabrescens* Kük. = ***Carex glabrescens*** (Kük.) Ohwi
Carex wallichiana var. *miyabei* (Franch.) Kük. = ***Carex miyabei*** Franch.
Carex wallii Petrie = ***Carex capillacea*** Boott var. ***capillacea***
Carex walteriana L.H.Bailey = ***Carex striata*** Michx.
Carex walteriana var. *brevis* (L.H.Bailey) L.H.Bailey = ***Carex striata*** Michx.
Carex warburgiana Kük. = ***Carex makinoensis*** Franch.
Carex wardiana H.Lév. & Vaniot = ***Carex breviculmis*** R.Br. var. ***breviculmis***
Carex warmingii Holm ex Lange = ***Carex bigelowii*** Torr. ex Schwein. subsp. ***bigelowii***
Carex washingtonia Eaton = ***Carex aquatilis*** var. ***minor*** Boott
Carex washingtoniana Dewey = ***Carex bigelowii*** Torr. ex Schwein. subsp. ***bigelowii***
Carex watsoniana Steud. = ***Carex stricta*** Lam.
Carex watsonii Olney = ***Carex riparia*** Curtis subsp. ***riparia***
Carex watsonii Boott = ***Carex vulcani*** Hochst. ex Seub.
Carex wazmannii Schrank = ***Carex ferruginea*** Scop. subsp. ***ferruginea***
Carex weiheana Boeck. ex Kunth = ***Carex praecox*** Schreb.
Carex welwitschii Boiss. ex Steud. = ***Carex laevigata*** Sm.
Carex wenchenii F.T.Wang & Tang = ***Carex serreana*** Hand.-Mazz.
Carex wendelboi Nelmes = ***Carex vulpinaris*** Nees
Carex wichurae Boeck. = ***Carex manca*** subsp. ***wichurae*** (Boeck.) S.Y.Liang
Carex wightiana var. *perakensis* (C.B.Clarke) Kük. = ***Carex perakensis*** C.B.Clarke

Carex wightiana var. *repanda* (C.B.Clarke) Kük. = ***Carex repanda*** C.B.Clarke
Carex wilfordii C.B.Clarke = ***Carex boottiana*** Hook. & Arn.
Carex wilkesii Olney = ***Carex aquatilis*** var. ***dives*** (T.Holm) Kük.
Carex willdenowii var. *megarrhyncha* F.J.Herm. = ***Carex superata*** Naczi, Reznicek & B.A.Ford
Carex willdenowii var. *pauciflora* Olney ex L.H.Bailey = ***Carex basiantha*** Steud.
Carex wiluica Meinsh. ex Maack = ***Carex nigra*** subsp. ***juncella*** (Fr.) Lemke
Carex wimmeri Steud. = ***Carex acuta*** L.
Carex winkleri Lange = ***Carex mairii*** Coss. & Germ.
Carex witheringii Gray = ***Carex arenaria*** L.
Carex wohllebii Hoppe = ***Carex supina*** Willd. ex Wahlenb. var. ***supina***
Carex wormskioldiana Hornem. = ***Carex scirpoidea*** Michx. var. ***scirpoidea***
Carex wormskioldii (Raf.) Drejer = ***Carex scirpoidea*** Michx. var. ***scirpoidea***
Carex wrightii Dewey ex Torr. = ***Carex microdonta*** Torr.
Carex wrightii Franch. = ***Carex oxyandra*** (Franch. & Sav.) Kudô var. ***oxyandra***
Carex wrightii var. *lanceata* Kük. = ***Carex oxyandra*** (Franch. & Sav.) Kudô var. ***oxyandra***
Carex xalapensis Kunth = ***Carex muehlenbergii*** var. ***xalapensis*** (Kunth) Britton
Carex ×*xanthina* Fernald = ***Carex ×xanthocarpa*** nothovar. ***leutzii*** (Kneuck.) Govaerts
Carex xanthoathera Franch. = ***Carex scita*** var. ***scabrinervia*** (Franch.) Kük.
Carex xanthocarpa E.P.Bicknell = ***Carex annectens*** (E.P.Bicknell) E.P.Bicknell
Carex xanthocarpa var. *annectens* E.P.Bicknell = ***Carex annectens*** (E.P.Bicknell) E.P.Bicknell
Carex xanthophysa Wahlenb. = ***Carex folliculata*** L.
Carex xanthophysa var. *nana* Dewey = ***Carex michauxiana*** Boeck. subsp. ***michauxiana***
Carex xanthosperma Dewey = ***Carex flaccosperma*** Dewey
Carex ×*xenostachya* T.Koyama = ?
Carex xerocarpa S.H.Wright = ***Carex stricta*** Lam.
Carex xestogyne Nelmes = ***Carex cruciata*** var. ***rafflesiana*** (Boott) Noot.
Carex xonata F.Nyl. = ***Carex nigra*** (L.) Reichard subsp. ***nigra***
Carex xuanchengensis S.W.Su & S.M.Xu = ***Carex chungii*** Z.P.Wang var. ***chungii***
Carex yabei H.Lév. & Vaniot = ***Carex scabrifolia*** Steud.
Carex yakusimensis Masam. = ***Carex atroviridis*** Ohwi
Carex yedoensis Boeck. = ***Carex pansa*** L.H.Bailey
Carex yesinensis Franch. = ***Carex foliosissima*** F.Schmidt var. ***foliosissima***
Carex yesoensis Koidz. = ***Carex lanceolata*** Boott var. ***lanceolata***
Carex yezo-alpina Akiyama = ***Carex kabanovii*** V.I.Krecz.
Carex yezo-montana Akiyama = ***Carex blepharicarpa*** Franch. var. ***blepharicarpa***
Carex yingkiliensis A.I.Baranov & Skvortsov = ***Carex mollissima*** H.Christ ex Scheutz
Carex yodoensis Boeck. = ?
Carex yosemitana L.H.Bailey = ***Carex sartwelliana*** Olney
Carex yoshinoi Ohwi = ***Carex conica*** Boott
Carex yukonensis Britton = ***Carex bigelowii*** subsp. ***lugens*** (T.Holm) T.V.Egorova
Carex yungningensis Hand.-Mazz. & Kük. = ***Carex fluviatilis*** Boott
Carex zallichiana Prescott ex Nees = ?
Carex zeyheri C.B.Clarke = ***Schoenoxiphium ecklonii*** Nees
Carex zollingeri Kunze ex Steud. = ***Carex alopecuroides*** D.Don ex Tilloch & Taylor

***Unplaced Names*:**

Carex acroandra Schur, Enum. Pl. Transsilv.: 699 (1866). = ?
Carex acutiformis Brot., Fl. Lusit. 1: 66 (1804), nom. illeg. = ?
Carex aequalis Notø, Nyt Mag. Naturvidensk. 61: 223 (1924). = ?
Carex akiensis K.Okamoto, Bull. Okayama Coll. Sci. 1: 53 (1965). = [38 JAP]
Carex ×*albertii* H.Lév., Bull. Acad. Int. Géogr. Bot. 22: 48 (1912). = ?
Carex albirostris C.B.Clarke, Kew Bull., Addit. Ser. 8: 78 (1908). = ?
Carex ×*allensteiniensis* R.Gross, Jahrb. Preuss. Bot. Ver. 1927–9: 9 (1930). = ?
Carex ×*alluvialis* Figert, Allg. Bot. Z. Syst. 13: 4 (1907). = ?
Carex ×*alluvialis* Schube, Jahresber. Schles. Ges. Vaterl. Cult. 86(2): 51 (1909), nom. illeg. = ?
Carex alterna C.B.Clarke, Kew Bull., Addit. Ser. 8: 69 (1908). = ?
Carex andalusica H.Lindb., Acta Soc. Sci. Fenn., Ser. B, Opera Biol. 1(2): 28 (1932). = [12 SPA]
Carex ×*anselmii* Sennen, Treb. Inst. Catalana Hist. Nat. 15(1): 37 (1931). = ?
Carex ×*appeliana* Zahn, Oesterr. Bot. Z. 40: 364 (1890). = ?
Carex arcuata C.Laest., Bidr. Växtligh.: 26 (1860). = ?
Carex aristata D.Don ex Tilloch & Taylor, Philos. Mag. J. 62: 455 (Dec. 1823). = ?
Carex ×*aschersonii* H.Lév., Bull. Acad. Int. Géogr. Bot. 22: 48 (1912). = ?
Carex ×*auroniensis* L.C.Lamb., Bull. Acad. Int. Géogr. Bot. 18: 339 (1908). = ?
Carex ×*barrei* H.Lév., Bull. Acad. Int. Géogr. Bot. 22: 48 (1912). = ?
Carex ×*bengyana* H.Lév. & L.C.Lamb., Bull. Acad. Int. Géogr. Bot. 21: 267 (1911). = ?
Carex biconvexis Michx. in ? = ?
Carex bicostata Olney, Car. Bar. Am.: 9 (1871). = ?
Carex bifida Roth ex Steud., Nomencl. Bot. 1: 155 (1821). = ?
Carex ×*binderi* Podp., Oesterr. Bot. Z. 1: 212 (1900). = ?
Carex binensis Franch., Bull. Soc. Philom. Paris, VIII, 7: 28 (1895). = ?
Carex binerviformis Holmboe, Stud. Veg. Cypr.: 42 (1914). = ?
Carex ×*bogdanensis* Zapal., Consp. Fl. Gallic. Crit. 3: 233 (1911). = ?
Carex bolanderi Gand., Bull. Soc. Bot. France 66: 296 (1919 publ. 1920), nom. illeg. = ?
Carex bordabensis Litv., Trudy Bot. Muz. Imp. Akad. Nauk 7: 100 (1910). = ?
Carex boscii Willd. ex Spreng., Syst. Veg. 3: 812 (1826). = ?
Carex ×*botasgdensis* Kük., Mitth. Thüring. Bot. Vereins, n.s., 5: 19 (1893). = ?
Carex brevicaulis Mack., Bull. Torrey Bot. Club 40: 547 (1913), nom. illeg. = [71 BRC 73 ORE WAS]
Carex breviglumis Le Gall, Fl. Morbihan: 658 (1852). = ?
Carex ×*brevirostris* Cederstr., Kongl. Vetensk. Acad. Handl. 14: 199 (1857). = ?
Carex brevirostris Blytt, Bot. Not. 1857: 133 (1857). = ?

Carex brotherorum H.Christ, Trudy Imp. S.-Peterburgsk. Bot. Sada 16: 434 (1900). = ?
Carex ×*brueggeri* K.Richt., Pl. Eur. 1: 170 (1890). = ?
Carex bufonia Gueldenst. ex Ledeb., Fl. Ross. 4: 320 (1852). = ?
Carex burjatorum Korotky, Repert. Spec. Nov. Regni Veg. 13: 293 (1914). = ?
Carex cantabrorum Gand., Bull. Soc. Bot. France 56: 134 (1909). = ?
Carex ×*castriferrei* Waisb., Magyar Bot. Lapok 4: 75 (1905). = ?
Carex ×*catalaunica* Sennen, Treb. Inst. Catalana Hist. Nat. 1917: 236, 182 (1917). = ?
Carex ×*catteyensis* A.Benn., Ann. Scott. Nat. Hist. 1911: 49 (1911). = [10 GRB]
Carex cernua Raeusch., Nomencl. Bot., ed. 3: 270 (1797). = ?
Carex cespitosa var. *ramosa* Dewey, Amer. J. Sci. Arts 12: 297 (1827). = [75 MAS]
Carex ×*cetica* Rech., Allg. Bot. Z. Syst. 20: 21 (1914). = ?
Carex chlamydea Norman, Forh. Vidensk.-Selsk. Kristiana 1893: 45 (1893). = ?
Carex ×*christii* Boeck., Bot. Centralbl. 42: 134 (1890). = ?
Carex ×*ciliatifructus* Ohwi, Acta Phytotax. Geobot. 1: 297 (1932). = ?
Carex ×*cokae* Podp., Vestn. Klubu Prír. v Prostejove 10: 7 (1907 publ. 1908). = [11 CZE]
Carex conspicua Boott ex C.B.Clarke, Kew Bull., Addit. Ser. 8: 82 (1908). = ?
Carex ×*corcoutica* Domin in K.Domin & J.Podpera, Klíc Kvet. Ceskoslov.: 961 (1928). = [11 CZE]
Carex coriophora subsp. *langtaodianensis* S.Yun Liang, in Fl. Reipubl. Popul. Sin. 12: 524 (2000). = [36 CHN]
Carex ×*corstorphinei* Druce, Rep. Bot. Exch. Club Brit. Isles 4: 216 (1915 publ. 1916). = ?
Carex ×*costei* Rouy, Ill. Pl. Eur.: t. 149 (1897). = ?
Carex crustacea Nelmes, Kew Bull. 1: 11 (1946). = ?
Carex ×*cuatrecasasii* Sennen, Treb. Inst. Catalana Hist. Nat. 15(1): 37 (1931). = [12 SPA]
Carex ×*dandoensis* T.Koyama, J. Jap. Bot. 31: 287 (1956). = ?
Carex ×*danielis* H.Lév., Bull. Acad. Int. Géogr. Bot. 22: 48 (1912). = ?
Carex deserta Merino, Fl. Galicia 3: 189 (1909). = ?
Carex displodens Printz, Veg. Siber.-Mongol. Front.: 153 (1921). = ?
Carex dissimilis J.F.Gmel., Syst. Nat. 2: 144 (1791). = ?
Carex dubia Missbach & E.H.L.Krause in J.Sturm, Deutschl. Fl., ed. 2, 2: 142 (1900), nom. illeg. = ?
Carex ×*ducellieri* Beauverd, Bull. Soc. Bot. Genève, II, 11: 6 (1919). = ?
Carex ×*duerrnbergeri* Ritzb., Oesterr. Bot. Z. 41: 246 (1891). = ?
Carex dyrrachiensis Nelmes, Kew Bull. 7: 67 (1952). = ?
Carex eboracensis Nelmes, J. Bot. 77: 114 (1939). = ?
Carex ×*elanescens* Cif. & Giacom., Nomencl. Fl. Ital. 1: 65 (1950). = ?
Carex ×*elisabeth* J.Andrés & al., Stud. Bot. Univ. Salamanca 7: 243 (1988 publ. 1989), contrary to Art. H.5.2 ICBN (1988). = ?
Carex eloyi C.B.Clarke, Kew Bull., Addit. Ser. 8: 88 (1908). = ?
Carex engleriana H.Lév. & Vaniot, Bull. Acad. Int. Géogr. Bot. 10: 274 (1901). = ?
Carex ×*enshuensis* T.Koyama, J. Jap. Bot. 30: 134 (1955). = ?
Carex erythraea Schrad., Linnaea 11: 87 (1837). = ?
Carex ×*ewingii* E.S.Marshall, J. Bot. 49: 197 (1911). = ?
Carex exaltata Kit., Linnaea 32: 322 (1863), nom. illeg. = ?
Carex ×*felixii* L.C.Lamb., Bull. Acad. Int. Géogr. Bot. 16(202): vii (1906). = ?
Carex ×*filkukae* Podp., Práce Morav. Prír. Spolecn. 6(3): 199 (1928). = ?
Carex ×*fontis-sancti* Podp., Spisy Prír. Fak. Masarykovy Univ. 12: 15 (1922). = ?
Carex × *fragosoana* Pau, Bol. Soc. Ibér. Ci. Nat. 25: 33 (1926). = ?
Carex fuegiana Gand., Bull. Soc. Bot. France 66: 296 (1919 publ. 1920). = ?
Carex ×*fulva* Gooden., Trans. Linn. Soc. London 2: 177 (1794), nom. illeg. = ?
Carex furcata Lapeyr., Hist. Pl. Pyrénées: 568 (1813). = ?
Carex ×*furusei* T.Koyama, J. Jap. Bot. 30: 135 (1955). = [38 JAP] *C. brownii* × *C. brownii* subsp. *transversa*
Carex fuscotincta Merino, Mem. Real Soc. Esp. Hist. Nat. 2: 467 (1904). = ?
Carex gandogeri H.Lév. & Vaniot, Bull. Acad. Int. Géogr. Bot. 14: 184 (1905). = ?
Carex gandogeri H.Lév. ex Gand., Bull. Soc. Bot. France 55: 223 (1908). = ?
Carex gasparrinii Parl., Fl. Ital. 2: 149 (1852). = ?
Carex geihokuensis K.Okamoto, Bull. Okayama Coll. Sci. 1: 51 (1965). = [38 JAP]
Carex ×*ghristii* Boeck., Bot. Centralbl. 42: 134 (1890). = ?
Carex ×*gogelana* Podp., Práce Morav. Prír. Spolecn. 6(3): 375 (1928). = ?
Carex ×*goroi* T.Koyama, J. Jap. Bot. 30: 133 (1955). = ?
Carex ×*gotlandica* Englund, Memoranda Soc. Fauna Fl. Fenn. 6: 104 (1930). = ?
Carex ×*goyozanensis* M.Kikuchi & Mochizuki, J. Geobot. 18: 73 (1970). = ?
Carex ×*hageri* E.Baumann, Ber. Schweiz. Bot. Ges. 26–29: 180 (1920). = ?
Carex halei L.H.Bailey ex C.B.Clarke, Kew Bull., Addit. Ser. 8: 87 (1908), nom. illeg. = ?
Carex ×*hanasakensis* T.Koyama, J. Jap. Bot. 30: 133 (1955). = ?
Carex ×*hanseniana* Junge, Jahrb. Hamburg. Wiss. Anst. 25: 157, 232 (1908). = ?
Carex hausmannii Tapp. ex Kalela, Ann. Bot. Soc. Zool.-Bot. Fenn. "Vanamo" 19(3): 57 (1944). = ?
Carex herbariorum L.H.Bailey, Bot. Gaz. 17: 150 (1892). = ?
Carex hians Steud., Syn. Pl. Glumac. 2: 236 (1855). = ?
Carex ×*hibernica* A.Benn., J. Bot. 35: 250 (1897). = ?
Carex hirtiformis Dalla Torre & Sarnth., Fl. Tirol 6(2): 380 (1906). = ?
Carex hobkodensis Franch., Bull. Soc. Philom. Paris, VIII, 7: 28 (1895). = ?
Carex ×*hokariensis* T.Koyama, J. Jap. Bot. 31: 288 (1956). = ?
Carex hooker-grayi C.B.Clarke, Kew Bull., Addit. Ser. 8: 85 (1908). = ?
Carex ×*hosoii* T.Koyama, J. Jap. Bot. 31: 287 (1956). = ?
Carex hymenina Nees ex Boott, Ill. Carex 1: 67 (1858). = ?
Carex hystericina var. *angustior* L.H.Bailey, Contr. U. S. Natl. Herb. 1: 126 (1891). = [76 ARI]
Carex ×*ilseana* Ruhmer, Jahrb. Königl. Bot. Gart. Berlin 1: 258 (1881). = ?
Carex imbricata Drobow in M.G.Popov (ed.), Key Pl. Envir. Tashkent: 51 (1923). = [32]
Carex ×*inaensis* T.Koyama, J. Jap. Bot. 30: 133 (1955). = ?

Carex inaequalis J.F.Gmel., Syst. Nat. 2: 146 (1791), nom. illeg. = ?
Carex inandrensis Kihlm. ex Hjelt, Acta Soc. Sci. Fenn. 5: 298 (1895). = ?
Carex ×*indagata* Torges & Hausskn., Mitt. Geogr. Ges. (Thüringen) Jena 9: 26 (1891). = ?
Carex ×*innocua* Wallr., Erst. Beitr. Fl. Hercyn.: 299 (1840). = ?
Carex ×*insolita* T.Koyama, J. Jap. Bot. 30: 135 (1955). = ?
Carex intacta Samp., Bol. Soc. Brot. 24: 118 (1908). = ?
Carex isomadensis Simonk. ex K.Richt., Pl. Eur. 1: (1890). = ?
Carex jankae Gand., Contr. Fl. Terr. Slav. Merid. 1: 30 (1883). = ?
Carex kamagariensis K.Okamoto, Bull. Okayama Coll. Sci. 1: 66 (1965). = [38 JAP]
Carex karashidaniensis Akiyama, J. Jap. Bot. 13: 652 (1937). = ?
Carex ×*kohtsii* K.Richt., Pl. Eur. 1: 170 (1890). = ?
Carex komarovii V.I.Krecz., Bot. Mater. Gerb. Bot. Inst. Komarova Akad. Nauk S.S.S.R. 7: 33 (1937), nom. illeg. = ?
Carex ×*krajinae* Domin, Preslia 13–15: 30 (1935). = ?
Carex ×*krausei* P.Fourn., Fl. Compl. Plaine Franç.: 558 (1928), nom. illeg. = ?
Carex ×*kuekenthaliana* Appel & Brückn., Ber. Bayer. Bot. Ges. 1: 75 (1891). = ?
Carex kuekenthalii K.Schum. ex C.B.Clarke, Bot. Jahrb. Syst. 38: 136 (1906), nom. illeg. = ?
Carex ×*kukenthalii* Dörfl., Jahreskat. Wiener Bot. Tauschanst. 1900: 130 (1900). = ?
Carex ×*kyyhkynenii* Hiitonen in ? = *C. canescens* × *C. tenuiflora*
Carex ×*lackowitziana* Paul, Allg. Bot. Z. Syst. 12: 160 (1906). = ?
Carex ×*laestadii* E.Holmb., Bot. Not. 1910: 87 (1910). = ?
Carex ×*lambertiana* H.Lév., Bull. Acad. Int. Géogr. Bot. 21: 268 (1911), nom. illeg. = ?
Carex lanescens J.Fellm., Bull. Soc. Imp. Naturalistes Moscou 8: 283 (1835). = ?
Carex ×*langii* Steud., Syn. Pl. Glumac. 2: 227 (1855). = ?
Carex lanosa Gueldenst. ex Ledeb., Fl. Ross. 4: 320 (1852). = ?
Carex ×*larminatii* H.Lév., Bull. Acad. Int. Géogr. Bot. 23: 286 (1914). = ?
Carex ×*lausii* Podp., Vestn. Klubu Prír. v Prostejove 10: 7 (1907 publ. 1908). = ?
Carex leveillei Husn., Cypéracées: 12 (1905). = [12 FRA]
Carex ×*lidii* Hadac, Stud. Bot. Cech. 5: 2 (1942). = ?
Carex ×*lidii* Flatberg, Norweg. J. Bot. 19: 93 (1972), nom. illeg. = *C. canescens* × *C. chordorrhiza*
Carex ×*ligniciensis* Figert, Allg. Bot. Z. Syst. 6: 38 (1900). = ?
Carex ×*limnogena* Appel ex Kneuck., Deutsche Bot. Monatsschr. 10: 168 (1892). = ?
Carex lineola C.A.Mey. ex Fr., Novit. Fl. Suec. Mant. 3: 145 (1842). = [10 SWE]
Carex lonchosacca C.B.Clarke, Kew Bull., Addit. Ser. 8: 77 (1908). = ?
Carex longiaristata Biv., Stirp. Rar. Sicilia 4: 8 (1818). = ?
Carex longibracteata Hornem., Hort. Bot. Hafn. 2: 881 (1815). = ?
Carex ×*loretii* Rouy in G.Rouy & J.Foucaud, Fl. France 13: 486 (1912). = ?
Carex lyonii Kurtz, Bot. Jahrb. Syst. 19: 419 (1894). = ?
Carex macrophylla K.Schum., Bot. Jahrb. Syst. 30: 381 (1901), nom. illeg. = ?
Carex macrostachyos Link, J. Bot. (Schrader) 2: 811 (1799). = ?
Carex maidenii Gand., Bull. Soc. Bot. France 46: 392 (1900). = ?
Carex marcida Raeusch., Nomencl. Bot., ed. 3: 270 (1797). = ?
Carex ×*massonii* Cay. & Lepage, Naturaliste Canad. 104: 567 (1977). = ?
Carex ×*mastnyana* Otruba & Podp., Sched. Fl. Exsicc. Reip. Bohem. Slov. Cent. 6: 21 (1931). = ?
Carex megalandra Gand., Bull. Soc. Bot. France 66: 295 (1919 publ. 1920). = ?
Carex melanocephala Kar. & Kir., Bull. Soc. Imp. Naturalistes Moscou 14: 436 (1841), nom. illeg. = ?
Carex melanostylis Desv. in C.Gay, Fl. Chil. 6: t. 73 (1854). = ?
Carex meridiana Akiyama, J. Jap. Bot. 13: 652 (1937). = ?
Carex mesomorpha Gand., Contr. Fl. Terr. Slav. Merid. 1: 30 (1883). = ?
Carex microsacca C.B.Clarke, Kew Bull., Addit. Ser. 8: 83 (1908). = [38 JAP]
Carex ×*mihashii* T.Koyama, J. Jap. Bot. 31: 288 (1956). = ?
Carex miliacea Schrank, Baier. Fl. 1: 300 (1789). = ?
Carex moscowensis C.B.Clarke ex B.Fedtsch., Bot. Zhurn. (St. Petersburg) 1: 191 (1906). = [14]
Carex ×*murrii* Appel, Oesterr. Bot. Z. 41: 124 (1891). = ?
Carex ×*naegelii* Brügger ex Sennholz, Verh. K. K. Zool.-Bot. Ges. Wien 37: 72 (1887). = ?
Carex nebrodensis Lojac., Fl. Sicul. 3: 231 (1909). = ?
Carex nenningiana Opiz, Seznam: 103 (1852). = ?
Carex neonelmesii Raymond, Naturaliste Canad. 82: 158 (1955). = ?
Carex nigricans Kit., Linnaea 32: 320 (1863), nom. illeg. = ?
Carex ×*nikaii* T.Koyama, J. Jap. Bot. 30: 132 (1955). = ?
Carex nitens Phil., Anales Univ. Chile 1873: 557 (1873). = ?
Carex oblonga J.F.Gmel., Syst. Nat. 2: 146 (1791). = ?
Carex obtusa Kit., Linnaea 32: 319 (1863). = ?
Carex pachyandra C.B.Clarke, Kew Bull., Addit. Ser. 8: 82 (1908). = ?
Carex ×*paczoskii* Zapal., Consp. Fl. Gallic. Crit. 3: 234 (1911). = ?
Carex pagesii H.Lév., Bull. Acad. Int. Géogr. Bot. 20: 208 (1910). = ?
Carex pairae var. *javanica* Nelmes, Kew Bull. 5: 208 (1950). = ?
Carex ×*palezieuxii* Kneuck., Allg. Bot. Z. Syst. 19: 1 (1913). = ?
Carex ×*paponii* Muret ex T.Durand & Pittier, Bull. Soc. Roy. Bot. Belgique 25: 336 (1886). = ?
Carex paradoxa J.F.Gmel., Syst. Nat. 2: 145 (1791). = ?
Carex pendulina Lojac., Fl. Sicul. 3: 224 (1909). = [13 SIC]
Carex ×*peraffinis* Appel, Ber. Deutsch. Bot. Ges. 10: 77 (1892). = ?
Carex personata C.B.Clarke, Kew Bull., Addit. Ser. 8: 71 (1908). = ?
Carex ×*petrae-fu* F.Murr, Progr. K. K. Oberrealschule Insbr.: 57 (1891). = ?
Carex pinguipes C.B.Clarke, Kew Bull., Addit. Ser. 8: 85 (1908). = ?
Carex platycarpos Rchb. in J.C.Mössler, Handb. Gewächsk. ed. 2, 3: 1656 (1830). = ?
Carex polygama J.F.Gmel., Syst. Nat. 2: 145 (1791). = ?
Carex ×*prahliana* Junge, Jahrb. Hamburg. Wiss. Anst. 22(Beih.): 71 (1904). = ?

Carex ×*prahliana* Junge, Verh. Vereins Naturwiss. Heimatf. Hamburg, III, 14: 118 (1906). = ?
Carex prolifera Kit., Linnaea 32: 320 (1863). = ?
Carex ×*pseudoaphanolepis* Ohwi, Mem. Coll. Sci. Kyoto Imp. Univ., Ser. B, Biol. 11(5): 512 (1936). = ?
Carex pseudodiandra Waisb., Magyar Bot. Lapok 4: 60, 72 (1905). = ?
Carex ×*pseudoheleonastes* Hand.-Mazz., Verh. K. K. Zool.-Bot. Ges. Wien 58: 102 (1908). = ?
Carex pseudomacrogyna Kük., Izv. Glavn. Bot. Sada S.S.S.R. 28: 220 (1929). = ?
Carex ×*pseudomairii* Camus, Bull. Soc. Bot. France 33: 479 (1886). = ?
Carex pseudonutans Boreau ex É.Martin, Cat. Pl. Romorantin: 309 (1875). = ?
Carex pseudothunbergii Akiyama, J. Jap. Bot. 16: 100 (1940). = ?
Carex pseudoturfosa Dalla Torre & Sarnth., Fl. Tirol 6(1): 330 (1906). = ?
Carex ×*pseudoturfosa* H.Lév. & L.C.Lamb., Bull. Acad. Int. Géogr. Bot. 21: 267 (1911), nom. illeg. = ?
Carex ×*pseudovulpina* K.Richt., Pl. Eur. 1: 168 (1890). = ?
Carex psilochloa Wall., Numer. List: 3534 (1831), nom. inval. = ?
Carex pulchella T.Holm, Amer. J. Sci., IV, 16: 457 (1903), nom. illeg. = ?
Carex pusilla Valbusa, Nuovo Giorn. Bot. Ital., n.s., 4: 174 (1897). = ?
Carex ×*raciborskii* Zapal., Consp. Fl. Gallic. Crit. 3: 233 (1911). = ?
Carex regelii C.B.Clarke ex O.Fedtsch., Fl. Pamir: 71 (1903). = [32]
Carex richmondii Boott ex C.B.Clarke, Kew Bull., Addit. Ser. 8: 83 (1908). = ?
Carex ×*rieseana* Figert, Allg. Bot. Z. Syst. 9: 112 (1903). = *C. paradoxa* × *C. remota*
Carex riloensis Stoeva & E.D.Popova, Fragm. Florist. Geobot. 38: 39 (1993). = [13 BUL]
Carex ×*rossiana* Degen, Fl. Veleb. 1: 585 (1936). = [13 YUG]
Carex rotoensis Petrie, Trans. & Proc. New Zealand Inst. 56: 9 (1926). = [51 CTM]
Carex rutilus Humn., Cat. Pl. Luxeuil: 25 (1876). = ?
Carex salina var. *mutica* Wahlenb., Fl. Lapp.: 246 (1812). = *C. aquatilis* × *C. flavicans*
Carex ×*salisiana* Brügger, Jahresber. Naturf. Ges. Graubündens, II, 23–24: 120 (1880). = ?
Carex sandwicensis Boeck., Flora 18: 265 (1835). = [63 HAW]
Carex ×*sanionis* K.Richt., Pl. Eur. 1: 170 (1890). = ?
Carex ×*sarntheinii* Vetter, Verh. K. K. Zool.-Bot. Ges. Wien 57: 236 (1907). = ?
Carex scaberrima Scheele, Linnaea 22: 345 (1849). = ?
Carex ×*schalkertii* Murr, Neue Übers. Bl.-Pfl. Voralberg: 57 (1923). = ?
Carex ×*schuetzeana* Figert. = ?
Carex scordiae Lojac., Fl. Sicul. 3: 230 (1909). = [13 SIC]
Carex seckeri C.B.Clarke, Kew Bull., Addit. Ser. 8: 88 (1908). = ?
Carex semicrinita C.B.Clarke, Kew Bull., Addit. Ser. 8: 70 (1908). = ?
Carex semicylindrica Kit., Linnaea 32: 319 (1863). = ?
Carex ×*senayana* Soó, Acta Bot. Acad. Sci. Hung. 16: 371 (1970 publ. 1971). = ?
Carex ×*serravalensis* Beauverd, Bull. Soc. Bot. Genève, II, 14: 39 (1922 publ. 1923). = ?
Carex setifolia Heer in J.Hegetschweiler-Bodmer, Fl. Schweiz: 929 (1840), nom. illeg. = ?
Carex skorpili Velen., Reliq. Mrkvic.: 31 (1922). = ?
Carex ×*solstitialis* Figert, Deutsche Bot. Monatsschr. 7: 86 (1889). = ?
Carex sonderiana Boott ex C.B.Clarke, Kew Bull., Addit. Ser. 8: 86 (1908), nom. illeg. = ?
Carex ×*sooi* Jákucs, Ann. Hist.-Nat. Mus. Natl. Hung., n.s., 3: 90 (1953). = ?
Carex spathulata Becc. in J.D.Hooker, Fl. Brit. India 6: 559 (1893). = ?
Carex sphaerica Lapeyr., Hist. Pl. Pyrénées: 570 (1813). = ?
Carex ×*stagnalis* L.C.Lamb., Bull. Acad. Int. Géogr. Bot. 21: 10 (1911). = [12 FRA]
Carex ×*stenolepis* f. *gracilior* Kük. in H.G.A.Engler (ed.), Pflanzenr., IV, 20(38): 715 (1909). = ?
Carex striata var. *japonica* Koidz., Bot. Mag. (Tokyo) 29: 388 (1915). = [38 JAP]
Carex ×*strigosula* Chatenier, Bull. Soc. Bot. France 58: 347 (1911). = [12 FRA]
Carex ×*subdecolorans* Nyár. in T.Savulescu, Fl. Reipubl. Popul. Roman. 11: 842 (1966). = ?
Carex sublivida Norrl., Not. Sällsk. Fauna Fl. Fenn. Förh. 10: 51 (1871). = ?
Carex subsabulosa Norman, Forh. Vidensk.-Selsk. Kristiania 1893: 48 (1893). = ?
Carex ×*superriparia* Hübl, Verh. Zool.-Bot. Ges. Wien 82: 22, 56 (1932). = ?
Carex suteri Kunth, Enum. Pl. 2: 523 (1837). = ?
Carex ×*suziella* Podp., Práce Morav. Prír. Spolecn. 6(3): 191 (1928). = ?
Carex taraspensis Brügger & Killias, Jahresber. Naturf. Ges. Graubündens, II, 31(Beibl.): 193 (1888). = ?
Carex tenuirostris Boeck., Flora 63: 456 (1880). = ?
Carex ×*teramotoi* T.Koyama, J. Jap. Bot. 30: 134 (1955). = ?
Carex thomsonii Gand., Bull. Soc. Bot. France 60: 27 (1913), nom. illeg. = ?
Carex ×*torgesiana* Kük., Mitth. Thüring. Bot. Vereins, n.s., 28: 62 (1911). = ?
Carex ×*tornabenii* Chiov., Ann. Bot. (Rome) 17: 83 (1927). = ?
Carex ×*toshironis* T.Koyama, J. Jap. Bot. 31: 287 (1956). = ?
Carex ×*tourletii* Gillies ex Tourlet, Bull. Soc. Bot. France 1: 313 (1854). = ?
Carex triceps Schrank, Baier. Fl. 1: 287 (1789). = ?
Carex truncatula Kit., Linnaea 32: 318 (1863). = ?
Carex tschonoskii V.I.Krecz., Bot. Mater. Gerb. Bot. Inst. Komarova Akad. Nauk S.S.S.R. 7: 32 (1937). = ?
Carex turuli Simonk., Magyar Bot. Lapok 3: 82 (1904). = ?
Carex ×*uechtritziana* K.Richt., Pl. Eur. 1: 170 (1896). = ?
Carex ulaviformis Hoppe in J.Sturm, Deutschl. Fl.: 61 (1833). = ?
Carex varica C.B.Clarke, Kew Bull., Addit. Ser. 8: 69 (1908). = ?
Carex vesiciformis Peterm., Flora 27: 339 (1844). = ?
Carex ×*viadrina* Figert, Allg. Bot. Z. Syst. 13: 3 (1907). = ?
Carex ×*viadrina* Schube, Jahresber. Schles. Ges. Vaterl. Cult. 86(2): 51 (1909). = ?
Carex ×*villacensis* Kük., Mitth. Thüring. Bot. Vereins, n.s., 38: 35 (1929). = ?
Carex ×*vimariensis* Hausskn. ex Berthold, Ber. Deutsch. Bot. Ges. 10: 131 (1890). = ?
Carex viridis Spenn., Fl. Friburg. 1: 48 (1825), nom. illeg. = ?
Carex volkensii K.Schum. in H.G.A.Engler, Pflanzenw. Ost-Afrikas, C: 130 (1895). = [25]
Carex ×*vratislaviensis* Figert, Allg. Bot. Z. Syst. 6: 39 (1900). = ?

Carex ×*xenostachya* T.Koyama, J. Jap. Bot. 31: 286 (1956). = ?
Carex yodoensis Boeck., Bot. Jahrb. Syst. 5: 515 (1884). = ?
Carex zallichiana Prescott ex Nees in R.Wight, Contr. Bot. India: 129 (1834). = ?

Caricella

Caricella Ehrh. = ***Carex*** L.

Caricina

Caricina St.-Lag. = ***Carex*** L.
Caricina approximata (All.) St.-Lag. = ***Carex ericetorum*** Pollich
Caricina brizoidea (L.) St.-Lag. = ***Carex brizoides*** L.
Caricina canescens (L.) St.-Lag. = ***Carex canescens*** L.
Caricina chordorrhiza (L.f.) St.-Lag. = ***Carex chordorrhiza*** L.f.
Caricina compacta St.-Lag. = ***Carex brunnea*** Thunb. subsp. ***brunnea***
Caricina curvula (All.) St.-Lag. = ***Carex curvula*** All.
Caricina cyperoides (L.) St.-Lag. = ***Carex bohemica*** Schreb.
Caricina disticha (Huds.) St.-Lag. = ***Carex disticha*** Huds.
Caricina divisa (Huds.) St.-Lag. = ***Carex divisa*** Huds.
Caricina divulsa (Stokes) St.-Lag. = ***Carex divulsa*** Stokes
Caricina elongata (L.) St.-Lag. = ***Carex elongata*** L.
Caricina foetida (All.) St.-Lag. = ***Carex foetida*** All.
Caricina incurva (Lightf.) St.-Lag. = ***Carex maritima*** Gunnerus
Caricina ×*microstyla* (J.Gay ex Gaudin) St.-Lag. = ***Carex* × *microstyla*** J.Gay ex Gaudin
Caricina multicaulis (Schrank) St.-Lag. = ***Carex praecox*** Schreb.
Caricina muricata (L.) St.-Lag. = ***Carex muricata*** L.
Caricina ovalis (Gooden.) St.-Lag. = ***Carex leporina*** L.
Caricina paniculata (L.) St.-Lag. = ***Carex paniculata*** L.
Caricina paradoxa (Rchb.) St.-Lag. = ***Carex appropinquata*** Schumach.
Caricina remota (L.) St.-Lag. = ***Carex remota*** L.
Caricina setifolia (Fourr.) St.-Lag. = ***Carex divisa*** Huds.
Caricina stellulata (Gooden.) St.-Lag. = ***Carex echinata*** Murray subsp. ***echinata***
Caricina teretiuscula (Gooden.) St.-Lag. = ***Carex diandra*** Schrank

Caricinella

Caricinella St.-Lag. = ***Carex*** L.
Caricinella laevis (Hoppe) St.-Lag. = ***Carex dioica*** L.
Caricinella microglochin (Wahlenb.) St.-Lag. = ***Carex microglochin*** Wahlenb.
Caricinella pauciflora (Lightf.) St.-Lag. = ***Carex pauciflora*** Lightf.
Caricinella pulicaris (L.) St.-Lag. = ***Carex pulicaris*** L.
Caricinella rupestris (All.) St.-Lag. = ***Carex rupestris*** All.
Caricinella scabra (Hoppe) St.-Lag. = ***Carex davalliana*** Sm.

Carpha

Carpha Banks & Sol. ex R.Br., Prodr.: 230 (1810).
Uganda to S. Africa, W. Indian Ocean, S. Japan, New Guinea to New Zealand, S. South America. 23 RWA ZAI 25 TAN UGA 27 CPP LES NAT OFS TVL 29 MAU MDG REU 38 JAP 43 NWG 50 NSW TAS VIC 51 ATP NZN NZS 85 AGS CLS.
16 Species
Asterochaete Nees, Linnaea 9: 300 (1834).
Oreograstis K.Schum. in H.G.A.Engler, Pflanzenw. Ost-Afrikas, C: 127 (1895).

Carpha alpina R.Br., Prodr.: 230 (1810). *Rhynchospora alpina* (R.Br.) Spreng., Syst. Veg. 1: 195 (1824). *Chaetospora alpina* (R.Br.) F.Muell., Fragm. 9: 39 (1875).
New Guinea, SE. Australia, New Zealand, S. South America. 43 NWG 50 NSW TAS VIC 51 ATP NZN NZS 85 AGS CLS. Hemicr.

var. ***alpina***
New Guinea, SE. Australia, New Zealand. 43 NWG 50 NSW TAS VIC 51 ATP NZN NZS. Hemicr.
Carpha novae-zelandiae Gand., Bull. Soc. Bot. France 66: 297 (1919 publ. 1920).
Carpha tasmanica Gand., Bull. Soc. Bot. France 66: 297 (1919 publ. 1920).

var. ***schoenoides*** (Banks & Sol. ex Hook.f.) Kük., Repert. Spec. Nov. Regni Veg. 47: 113 (1939).
S. Chile to S. Argentina. 85 AGS CLS. Hemicr.
**Carpha schoenoides* Banks & Sol. ex Hook.f., Fl. Antarct.: 362 (1846).

Carpha angustissima Cherm., Bull. Soc. Bot. France 82: 341 (1935). *Carpha eminii* var. *angustissima* (Cherm.) Kük., Repert. Spec. Nov. Regni Veg. 47: 210 (1939).
SW. Uganda to E. Zaïre. 23 RWA ZAI 25 UGA. Hemicr. or rhizome geophyte.

Carpha aristata Kük., Repert. Spec. Nov. Regni Veg. 47: 118 (1939).
Japan (Kyushu). 38 JAP. Hemicr.

Carpha borbonica (Steud.) C.B.Clarke in T.A.Durand & H.Schinz, Consp. Fl. Afric. 5: 655 (1894).
Réunion. 29 REU. Hemicr.
**Elynanthus borbonicus* Steud., Syn. Pl. Glumac. 2: 155 (1855).

Carpha bracteosa C.B.Clarke in W.H.Harvey & auct. suc. (eds.), Fl. Cap. 7: 270 (1898). *Carpha capitellata* var. *bracteosa* (C.B.Clarke) Kük., Repert. Spec. Nov. Regni Veg. 47: 117 (1939).
Cape Prov. 27 CPP. Hemicr.

Carpha capitellata (Nees) Boeck., Linnaea 38: 266 (1874).
Cape Prov., Northern Prov. 27 CPP TVL. Hemicr.
**Asterochaete capitellata* Nees, Linnaea 10: 194 (1836).
Asterochaete ludwigii Hochst., Flora 28: 759 (1845).

Carpha costularioides C.B.Clarke, Kew Bull., Addit. Ser. 8: 43 (1908).
Mauritius, Réunion. 29 MAU REU. Hemicr.

Carpha curvata W.M.Curtis, Brunonia 7: 303 (1985).
Tasmania. 50 TAS. Hemicr.

Carpha eminii (K.Schum.) C.B.Clarke in D.Oliver & auct. suc. (eds.), Fl. Trop. Afr. 8: 483 (1902).
Uganda (Ruwenzori). 25 UGA. Hemicr. or rhizome geophyte.
**Oreograstis eminii* K.Schum. in H.G.A.Engler, Pflanzenw. Ost-Afrikas, C: 127 (1895).

Carpha filifolia C.Reid & T.H.Arnold, Bothalia 15: 139 (1984).
S. Africa. 27 LES NAT OFS TVL. Hemicr.

Carpha glomerata Nees, Linnaea 7: 529 (1832). *Asterochaete glomerata* (Nees) Nees, Linnaea 9: 300 (1834).

Tanzania (Mt. Uluguru), Cape Prov., KwaZulu-Natal. 25 TAN 27 CPP NAT. Hemicr.
**Schoenus glomeratus* Thunb., Prodr. Pl. Cap.: 17 (1794), nom. illeg.
Schoenus dactyloides Vahl, Enum. Pl. 2: 224 (1805). *Chaetospora dactyloides* (Vahl) A.Dietr., Sp. Pl. 2: 32 (1832).
Asterochaete angustifolia Nees, Linnaea 9: 300 (1834), nom. inval.
Asterochaete tenuis Kunth, Enum. Pl. 2: 312 (1837).

Carpha nitens (Kunth) Kük., Repert. Spec. Nov. Regni Veg. 47: 118 (1939).
Mauritius. 29 MAU. Hemicr.
**Asterochaete nitens* Kunth, Enum. Pl. 2: 313 (1837). *Cyclocampe nitens* (Kunth) Hook.f., Hooker's Icon. Pl. 15: t. 1434 (1883).
Schoenus nitens Willd. ex Kunth, Enum. Pl. 2: 313 (1837), pro syn.

Carpha nivicola F.Muell., Trans. Philos. Soc. Victoria 1: 111 (1855).
SE. Australia. 50 NSW VIC. Hemicr.
Carpha alpina var. *subacaulis* Kük., Repert. Spec. Nov. Regni Veg. 47: 113 (1939).

Carpha perrieri Cherm., Bull. Soc. Bot. France 69: 721 (1922 publ. 1923).
C. Madagascar. 29 MDG. Hemicr.

Carpha rodwayi W.M.Curtis, Brunonia 7: 302 (1985).
Tasmania. 50 TAS. Hemicr.

Carpha schlechteri C.B.Clarke in É.A.J.de Wildeman, Pl. Nov. Hort. Then. 1: 37 (1904).
Cape Prov., KwaZulu-Natal. 27 CPP NAT. Hemicr.

Synonyms:
Carpha alpina var. *subacaulis* Kük. = ***Carpha nivicola*** F.Muell.
Carpha andina Phil. = ***Schoenus andinus*** (Phil.) H.Pfeiff.
Carpha antarctica (Hook.f.) C.B.Clarke = ***Schoenus antarcticus*** (Hook.f.) Dusén
Carpha arundinacea (Sol. ex Vahl) Brongn. = ***Costularia arundinacea*** (Sol. ex Vahl) Kük.
Carpha aubertii Nees = ***Costularia elongata*** (Kunth) Kük.
Carpha avenacea R.Br. = ***Cyathochaeta avenacea*** (R.Br.) Benth.
Carpha capensis (Steud.) M.Martens ex Steud. = ***Trianoptiles capensis*** (Steud.) Harv.
Carpha capensis (Steud.) H.Pfeiff. = ***Trianoptiles capensis*** (Steud.) Harv.
Carpha capitellata var. *bracteosa* (C.B.Clarke) Kük. = ***Carpha bracteosa*** C.B.Clarke
Carpha clandestina R.Br. = ***Cyathochaeta clandestina*** (R.Br.) Benth.
Carpha deusta R.Br. = ***Ptilothrix deusta*** (R.Br.) K.L.Wilson
Carpha diandra R.Br. = ***Cyathochaeta diandra*** (R.Br.) Nees
Carpha elongata (Kunth) Boeck. = ***Costularia elongata*** (Kunth) Kük.
Carpha eminii var. *angustissima* (Cherm.) Kük. = ***Carpha angustissima*** Cherm.
Carpha graciliceps C.B.Clarke = ***Mesomelaena graciliceps*** (C.B.Clarke) K.L.Wilson
Carpha hexandra Nees = ***Cyathocoma hexandra*** (Nees) Browning
Carpha junciformis Boeck. = ***Tricostularia undulata*** (Thwaites) J.Kern
Carpha laxa (Hook.f.) H.Pfieff. = ***Schoenus rhynchosporoides*** (Steud.) Kük.
Carpha novae-zelandiae Gand. = ***Carpha alpina*** R.Br. var. ***alpina***
Carpha paniculata Phil. = ***Schoenus rhynchosporoides*** (Steud.) Kük.
Carpha schoenoides Banks & Sol. ex Hook.f. = ***Carpha alpina*** var. ***schoenoides*** (Banks & Sol. ex Hook.f.) Kük.
Carpha schweinfurthiana Boeck. = ***Coleochloa schweinfurthiana*** (Boeck.) Nelmes
Carpha solitaria (C.B.Clarke) H.Pfeiff. = ***Trianoptiles solitaria*** (C.B.Clarke) Levyns
Carpha tasmanica Gand. = ***Carpha alpina*** R.Br. var. ***alpina***
Carpha urvilleana Gaudich. ex Boeck. = ***Costularia pilisepala*** (Steud.) F.Kern

Catagyna

Catagyna Hutch. & Dalziel = ***Afrotrilepis*** (Gilg) J.Raynal
Catagyna P.Beauv. ex Lestib. = ***Scleria*** P.J.Bergius
Catagyna pilosa (Boeck.) Hutch. = ***Afrotrilepis pilosa*** (Boeck.) J.Raynal

Caustis

Caustis R.Br., Prodr.: 239 (1810).
SW. & E. Australia. 50 NSW QLD SOA TAS VIC WAU.
5 Species

Caustis blakei Kük. ex S.T.Blake, Proc. Roy. Soc. Queensland 54: 71 (1943).
SE. Queensland to NE. New South Wales. 50 NSW QLD. Hemicr. or rhizome geophyte.

subsp. ***blakei***
SE. Queensland to NE. New South Wales. 50 NSW QLD. Hemicr. or rhizome geophyte.

subsp. ***macrantha*** M.E.Johnst. & J.H.Webber, Austrobaileya 4: 615 (1997).
SE. Queensland. 50 QLD. Hemicr. or rhizome geophyte.

Caustis dioica R.Br., Prodr.: 239 (1810).
W. & SW. Western Australia. 50 WAU. Hemicr. or rhizome geophyte.
Caustis hexandra Nees in J.G.C.Lehmann, Pl. Preiss. 2: 88 (1846).
Caustis dioica f. *multiandra* H.Pfeiff., Repert. Spec. Nov. Regni Veg. 21: 365 (1925).
Caustis dioica f. *multiradiata* H.Pfeiff., Repert. Spec. Nov. Regni Veg. 21: 365 (1925).
Caustis dioica var. *eciliata* Kük., Feddes Repert. Spec. Nov. Regni Veg. 53: 99 (1944).

Caustis flexuosa R.Br., Prodr.: 239 (1810).
Queensland to Victoria. 50 NSW QLD VIC. Hemicr. or rhizome geophyte.
Caustis restiacea F.Muell. ex Benth., Fl. Austral. 7: 421 (1878).
Caustis tenuirama Gand., Bull. Soc. Bot. France 66: 297 (1919 publ. 1920). *Caustis flexuosa* var. *tenuirama* (Gand.) Kük., Feddes Repert. Spec. Nov. Regni Veg. 53: 96 (1944).
Caustis flexuosa var. *rectiramulosa* Kük., Feddes Repert. Spec. Nov. Regni Veg. 53: 96 (1944).

Caustis pentandra R.Br., Prodr.: 239 (1810).
E. & SE. Australia. 50 NSW QLD SOA TAS VIC. Hemicr. or rhizome geophyte.

Caustis squamellata C.B.Clarke, Kew Bull., Addit. Ser. 8: 48 (1908).

Caustis recurvata Spreng., Syst. Veg. 4(2): 26 (1827).
E. Australia. 50 NSW QLD. Hemicr. or rhizome geophyte.
Caustis recurvata var. *foliosa* Kük., Feddes Repert. Spec. Nov. Regni Veg. 53: 98 (1944).
Caustis recurvata var. *hirsuta* Kük., Feddes Repert. Spec. Nov. Regni Veg. 53: 98 (1944).

Synonyms:
Caustis dioica var. *eciliata* Kük. = ***Caustis dioica*** R.Br.
Caustis dioica f. *multiandra* H.Pfeiff. = ***Caustis dioica*** R.Br.
Caustis dioica f. *multiradiata* H.Pfeiff. = ***Caustis dioica*** R.Br.
Caustis filifolia (C.Presl) H.Pfeiff. = ***Gahnia filifolia*** (C.Presl) Kük. ex Benl
Caustis flexuosa var. *rectiramulosa* Kük. = ***Caustis flexuosa*** R.Br.
Caustis flexuosa var. *tenuirama* (Gand.) Kük. = ***Caustis flexuosa*** R.Br.
Caustis hexandra Nees = ***Caustis dioica*** R.Br.
Caustis pilisepala (Steud.) H.Pfieff. = ***Costularia pilisepala*** (Steud.) F.Kern
Caustis recurvata var. *foliosa* Kük. = ***Caustis recurvata*** Spreng.
Caustis recurvata var. *hirsuta* Kük. = ***Caustis recurvata*** Spreng.
Caustis restiacea F.Muell. ex Benth. = ***Caustis flexuosa*** R.Br.
Caustis sieberi (Schrad. ex Schult. & Schult.f.) Kunth = ***Gahnia filifolia*** (C.Presl) Kük. ex Benl
Caustis squamellata C.B.Clarke = ***Caustis pentandra*** R.Br.
Caustis tenuirama Gand. = ***Caustis flexuosa*** R.Br.

Cephalocarpus

Cephalocarpus Nees in C.F.P.von Martius & auct. suc. (eds.), Fl. Bras. 2(1): 162 (1842).
S. Trop. America. 82 GUY SUR VEN 83 CLM 84 BZN.
4 Species

Cephalocarpus confertus Gilly, Bull. Torrey Bot. Club 69: 293 (1942).
S. Venezuela. 82 VEN. Hemicr. or cham.

Cephalocarpus dracaenula Nees in C.F.P.von Martius & auct. suc. (eds.), Fl. Bras. 2(1): 162 (1842). *Cryptangium dracaenula* (Nees) Boeck., Linnaea 38: 415 (1874). *Lagenocarpus dracaenula* (Nees) H.Pfeiff., Repert. Spec. Nov. Regni Veg. 18: 92 (1922).
Colombia (Caquetá, Vaupés), Venezuela (Ayán-tepui). 82 VEN 83 CLM. Cham.

Cephalocarpus obovoideus T.Koyama, Mem. New York Bot. Gard. 18(2): 22 (1969).
Venezuela (Sierra de la Neblina), Brazil (Serra da Neblina). 82 VEN 84 BZN. Cham.

Cephalocarpus rigidus Gilly ex Gleason & Killip, Brittonia 3: 152 (1939). *Cephalocarpus rigidus* var. *typicus* Gilly, Bull. Torrey Bot. Club 69: 296 (1942).
Guayana Highland to Brazil (Roraima). 82 GUY SUR VEN 83 CLM 84 BZN. Cham.
Cephalocarpus longibracteatus Gilly ex Gleason & Killip, Brittonia 3: 153 (1939).
Cephalocarpus lineariifolius Gilly, Bull. Torrey Bot. Club 69: 295 (1942).
Cephalocarpus rigidus var. *mucronatus* Gilly, Bull. Torrey Bot. Club 69: 296 (1942).
Cephalocarpus maguirei Gilly, Bull. Torrey Bot. Club 76: 292 (1949).
Cephalocarpus steyermarkii Gilly, Fieldiana, Bot. 28: 54 (1951).

Synonyms:
Cephalocarpus clarkei H.Pfeiff. = ***Lagenocarpus clarkei*** H.Pfeiff.
Cephalocarpus comatus (Boeck.) H.Pfeiff. = ***Lagenocarpus comatus*** (Boeck.) H.Pfeiff.
Cephalocarpus humilis (Nees) H.Pfeiff. = ***Lagenocarpus humilis*** (Nees) Kuntze
Cephalocarpus lineariifolius Gilly = ***Cephalocarpus rigidus*** Gilly ex Gleason & Killip
Cephalocarpus longibracteatus Gilly ex Gleason & Killip = ***Cephalocarpus rigidus*** Gilly ex Gleason & Killip
Cephalocarpus maguirei Gilly = ***Cephalocarpus rigidus*** Gilly ex Gleason & Killip
Cephalocarpus polyphyllus (Nees) H.Pfeiff. = ***Lagenocarpus polyphyllus*** (Nees) Kuntze
Cephalocarpus rigidus var. *mucronatus* Gilly = ***Cephalocarpus rigidus*** Gilly ex Gleason & Killip
Cephalocarpus rigidus var. *typicus* Gilly = ***Cephalocarpus rigidus*** Gilly ex Gleason & Killip
Cephalocarpus steyermarkii Gilly = ***Cephalocarpus rigidus*** Gilly ex Gleason & Killip

Cephaloschoenus

Cephaloschoenus Nees = ***Rhynchospora*** Vahl
Cephaloschoenus armeria Nees = ***Rhynchospora globosa*** (Kunth) Roem. & Schult. var. ***globosa***
Cephaloschoenus articulatus (Roxb.) Nees = ***Rhynchospora corymbosa*** (L.) Britton
Cephaloschoenus ceylanicus Nees = ***Rhynchospora triflora*** Vahl
Cephaloschoenus divergens Nees = ***Rhynchospora corymbosa*** (L.) Britton
Cephaloschoenus globosus (Kunth) Nees = ***Rhynchospora globosa*** (Kunth) Roem. & Schult.
Cephaloschoenus globosus var. *armeria* Nees = ***Rhynchospora globosa*** (Kunth) Roem. & Schult. var. ***globosa***
Cephaloschoenus globosus var. *minor* Nees = ***Rhynchospora globosa*** (Kunth) Roem. & Schult. var. ***globosa***
Cephaloschoenus longirostris (Nees) Nees = ***Rhynchospora amazonica*** subsp. ***guianensis*** (Kük.) T.Koyama
Cephaloschoenus longisetis (R.Br.) Nees = ***Rhynchospora longisetis*** R.Br.
Cephaloschoenus marginatus Liebm. = ***Rhynchospora globosa*** (Kunth) Roem. & Schult. var. ***globosa***
Cephaloschoenus oligocephalus Hochst. = ***Rhynchospora holoschoenoides*** (Rich.) Herter
Cephaloschoenus parvus Nees = ***Rhynchospora rubra*** (Lour.) Makino subsp. ***rubra***
Cephaloschoenus pohlianus Nees = ***Rhynchospora globosa*** (Kunth) Roem. & Schult. var. ***globosa***
Cephaloschoenus polycephalus Nees = ***Rhynchospora holoschoenoides*** (Rich.) Herter
Cephaloschoenus pseudomariscus Nees = ***Rhynchospora gigantea*** Link
Cephaloschoenus tenuirostris Nees = ***Rhynchospora holoschoenoides*** (Rich.) Herter

Cephaloscirpus

Cephaloscirpus Kurz = ***Mapania*** Aubl.

Cephaloscirpus macrocephalus (Gaudich.) Kurz = ***Mapania macrocephala*** (Gaudich.) K.Schum.

Ceratoschoenus

Ceratoschoenus Nees = ***Rhynchospora*** Vahl
Ceratoschoenus capitatus Chapm. = ***Rhynchospora tracyi*** Britton
Ceratoschoenus corniculatus (Lam.) Nees = ***Rhynchospora corniculata*** (Lam.) A.Gray
Ceratoschoenus inundatus (Oakes) House = ***Rhynchospora inundata*** (Oakes) Fernald
Ceratoschoenus longirostris (Michx.) A.Gray = ***Rhynchospora corniculata*** (Lam.) A.Gray
Ceratoschoenus macrophyllus Tuck. = ***Rhynchospora macrostachya*** Torr. ex A.Gray
Ceratoschoenus macrostachys (Torr. ex A.Gray) A.Gray = ***Rhynchospora macrostachya*** Torr. ex A.Gray
Ceratoschoenus macrostachys var. *inundatus* Oakes = ***Rhynchospora inundata*** (Oakes) Fernald
Ceratoschoenus macrostachys var. *patulus* Chapm. = ***Rhynchospora careyana*** Fernald

Chaetocyperus

Chaetocyperus Nees = ***Eleocharis*** R.Br.
Chaetocyperus acicularis (L.) Nees = ***Eleocharis acicularis*** (L.) Roem. & Schult.
Chaetocyperus albibracteatus (Nees & Meyen ex Kunth) Nees & Meyen = ***Eleocharis albibracteata*** Nees & Meyen ex Kunth
Chaetocyperus arenicola Steud. = ***Eleocharis retroflexa*** subsp. ***chaetaria*** (Roem. & Schult.) T.Koyama
Chaetocyperus baldwinii Torr. = ***Eleocharis baldwinii*** (Torr.) Chapm.
Chaetocyperus bonariensis (Nees) Nees = ***Eleocharis bonariensis*** Nees
Chaetocyperus bonariensis var. *fluitans* Nees = ***Eleocharis bonariensis*** Nees
Chaetocyperus capillaceus (Kunth) Nees = ***Eleocharis capillacea*** Kunth
Chaetocyperus costulatus (Nees & Meyen ex Kunth) Nees = ***Eleocharis radicans*** Kunth
Chaetocyperus depauperatus Steud. = ***Eleocharis retroflexa*** subsp. ***chaetaria*** (Roem. & Schult.) T.Koyama
Chaetocyperus emarginatus Nees = ***Eleocharis emarginata*** (Nees) Klotzsch ex Boeck.
Chaetocyperus jamesonii Steud. = ***Eleocharis minima*** Kunth var. ***minima***
Chaetocyperus limnocharis Nees = ***Eleocharis retroflexa*** subsp. ***chaetaria*** (Roem. & Schult.) T.Koyama
Chaetocyperus membranaceus Buckley = ***Eleocharis parvula*** (Roem. & Schult.) Palla
Chaetocyperus niveus Liebm. = ***Eleocharis retroflexa*** (Poir.) Urb. subsp. ***retroflexa***
Chaetocyperus obtusatus Nees = ***Eleocharis bonariensis*** Nees
Chaetocyperus polymorphus Nees & Lindl. = ***Eleocharis nana*** Kunth
Chaetocyperus polymorphus var. *capillaceus* (Michx.) Nees = ***Eleocharis nana*** Kunth
Chaetocyperus polymorphus var. *depauperatus* (Vahl) Nees = ***Eleocharis retroflexa*** (Poir.) Urb. subsp. ***retroflexa***
Chaetocyperus polymorphus var. *sphagnicola* Nees = ***Eleocharis nana*** Kunth
Chaetocyperus punctatus Nees = ***Eleocharis nana*** Kunth
Chaetocyperus pusillus Steud. = ***Eleocharis parvula*** (Roem. & Schult.) Palla
Chaetocyperus pygmaeus (Torr.) Torr. ex A.Gray = ***Eleocharis parvula*** (Roem. & Schult.) Palla
Chaetocyperus radicans (Poir.) Steud. = ***Eleocharis radicans*** Kunth
Chaetocyperus reptans Boeck. = ***Eleocharis minima*** Kunth var. ***minima***
Chaetocyperus rugulosus Nees = ***Eleocharis retroflexa*** (Poir.) Urb. subsp. ***retroflexa***
Chaetocyperus setaceus (R.Br.) Nees = ***Eleocharis retroflexa*** subsp. ***chaetaria*** (Roem. & Schult.) T.Koyama
Chaetocyperus stoloniferus Nees = ***Eleocharis dombeyana*** Kunth
Chaetocyperus subarticulatus Nees = ***Eleocharis subarticulata*** (Nees) Boeck.
Chaetocyperus tenuiculus (Schrad. ex Schult.) Nees = ***Eleocharis filiculmis*** Kunth
Chaetocyperus trichodes (Muhl.) Steud. = ***Eleocharis acicularis*** (L.) Roem. & Schult. subsp. ***acicularis***
Chaetocyperus urceolatus Liebm. = ***Eleocharis urceolata*** (Liebm.) Svenson
Chaetocyperus viviparus Nees = ***Eleocharis minima*** Kunth var. ***minima***

Chaetospora

Chaetospora Kunth = ***Rhynchospora*** Vahl
Chaetospora R.Br. = ***Schoenus*** L.
Chaetospora albescens Franch. & Sav. = ***Schoenus apogon*** Roem. & Schult. var. ***apogon***
Chaetospora alpina (R.Br.) F.Muell. = ***Carpha alpina*** R.Br.
Chaetospora anceps R.Br. = ***Gymnoschoenus anceps*** (R.Br.) C.B.Clarke
Chaetospora antarctica Hook.f. = ***Schoenus antarcticus*** (Hook.f.) Dusén
Chaetospora arundinacea (Sol. ex Vahl) A.Dietr. = ***Costularia arundinacea*** (Sol. ex Vahl) Kük.
Chaetospora aurata Nees = ***Schoenus curvifolius*** (R.Br.) Roem. & Schult.
Chaetospora aurea Cham. & Schltdl. = ***Rhynchospora aristata*** Boeck. var. ***aristata***
Chaetospora aurea (Vahl) Kunth = ***Rhynchospora corymbosa*** (L.) Britton
Chaetospora avenacea (R.Br.) F.Muell. = ***Cyathochaeta avenacea*** (R.Br.) Benth.
Chaetospora axillaris R.Br. = ***Schoenus maschalinus*** Roem. & Schult.
Chaetospora axillaris f. *terrestris* Boeck. = ***Schoenus maschalinus*** Roem. & Schult.
Chaetospora brevisetis R.Br. = ***Schoenus brevisetis*** (R.Br.) Roem. & Schult.
Chaetospora brevisetis F.Muell. = ***Schoenus subbulbosus*** Benth.
Chaetospora burmanii (Vahl) Schrad. = ***Tetraria burmanii*** (Vahl) C.B.Clarke
Chaetospora calostachys R.Br. = ***Schoenus calostachyus*** (R.Br.) Poir.
Chaetospora capillacea (Thunb.) Nees = ***Tetraria capillacea*** (Thunb.) C.B.Clarke
Chaetospora capillacea Hook.f. = ***Tetraria capillaris*** (F.Muell.) J.M.Black
Chaetospora capillaris F.Muell. = ***Tetraria capillaris*** (F.Muell.) J.M.Black
Chaetospora capitata Kunth = ***Rhynchospora capitata*** (Kunth) Roem. & Schult.
Chaetospora capitellum (Thunb.) A.Dietr. = ***Ficinia capitella*** (Thunb.) Nees

Chaetospora ciliata Schrad. ex Nees = ***Rhynchospora elatior*** Kunth
Chaetospora circinalis Schrad. = ***Tetraria microstachys*** (Vahl) H.Pfeiff.
Chaetospora clandestina (R.Br.) F.Muell. = ***Cyathochaeta clandestina*** (R.Br.) Benth.
Chaetospora compressa (L.) Gray = ***Blysmus compressus*** (L.) Panz. ex Link
Chaetospora compressa Nees = ***Schoenus benthamii*** F.Muell.
Chaetospora concava Nees = ***Lepidosperma concavum*** R.Br.
Chaetospora concinna Hook.f. = ***Schoenus nitens*** var. ***concinnus*** (Hook.f.) Cheeseman
Chaetospora cruenta Nees = ***Schoenus cruentus*** (Nees) Benth.
Chaetospora curvifolia R.Br. = ***Schoenus curvifolius*** (R.Br.) Roem. & Schult.
Chaetospora cuspidata (Rottb.) Nees = ***Tetraria cuspidata*** (Rottb.) C.B.Clarke
Chaetospora cygnea Nees = ***Schoenus cygneus*** (Nees) Nees
Chaetospora dactyloides (Vahl) A.Dietr. = ***Carpha glomerata*** Nees
Chaetospora deformis R.Br. = ***Schoenus deformis*** (R.Br.) Roem. & Schult.
Chaetospora deusta (R.Br.) F.Muell. = ***Ptilothrix deusta*** (R.Br.) K.L.Wilson
Chaetospora diandra (R.Br.) F.Muell. = ***Cyathochaeta diandra*** (R.Br.) Nees
Chaetospora diodon Schrad. ex Nees = ***Rhynchospora diodon*** (Nees) Griseb.
Chaetospora dispar (Spreng.) A.Dietr. = ***Ficinia secunda*** (Vahl) Kunth
Chaetospora distachya Nees = ***Bulbostylis hispidula*** (Vahl) R.W.Haines subsp. ***hispidula***
Chaetospora distans F.Muell. = ***Schoenus rigens*** S.T.Blake
Chaetospora effusa Schrad. ex Kunth = ***Pleurostachys sparsiflora*** Kunth
Chaetospora elongata Nees = ***Gymnoschoenus anceps*** (R.Br.) C.B.Clarke
Chaetospora ferruginea Cham. & Schltdl. = ***Rhynchospora eximia*** (Nees) Boeck.
Chaetospora ferruginea Kunth = ***Rhynchospora rugosa*** subsp. ***brownii*** (Roem. & Schult.) T.Koyama
Chaetospora ferruginea (L.) Rchb. = ***Schoenus ferrugineus*** L.
Chaetospora fimbriolata Nees = ***Tetraria fimbriolata*** (Nees) C.B.Clarke
Chaetospora fimbristyloides F.Muell. = ***Tricostularia undulata*** (Thwaites) J.Kern
Chaetospora flexuosa (Thunb.) Schrad. = ***Tetraria flexuosa*** (Thunb.) C.B.Clarke
Chaetospora globosa Kunth = ***Rhynchospora globosa*** (Kunth) Roem. & Schult.
Chaetospora hexandra Boeck. = ***Tetraria macowaniana*** B.L.Burtt
Chaetospora imberbis R.Br. = ***Schoenus apogon*** Roem. & Schult.
Chaetospora ×*intermedia* (Brügger) Beck = ***Schoenus* ×*scheuchzeri*** Brügger
Chaetospora japonica Franch. & Sav. = ***Schoenus apogon*** Roem. & Schult. var. ***apogon***
Chaetospora lanata (Labill.) R.Br. = ***Schoenus lanatus*** Labill.
Chaetospora laxa Hook.f. = ***Schoenus rhynchosporoides*** (Steud.) Kük.
Chaetospora lepidosperma F.Muell. = ***Schoenus lepidosperma*** (F.Muell.) K.L.Wilson
Chaetospora lucens Poir. = ***Schoenus nitens*** (R.Br.) Roem. & Schult. var. ***nitens***
Chaetospora madagascariensis Steud. = ***Rhynchospora rubra*** subsp. ***africana*** J.Raynal
Chaetospora mauritii Steud. = ***Rhynchospora rugosa*** subsp. ***brownii*** (Roem. & Schult.) T.Koyama
Chaetospora microstachya Nees = ***Schoenus submicrostachyus*** Kük.
Chaetospora microstachys (Vahl) Schrad. = ***Tetraria microstachys*** (Vahl) H.Pfeiff.
Chaetospora nana Nees = ***Schoenus nanus*** (Nees) Benth.
Chaetospora natans F.Muell. = ***Schoenus natans*** (F.Muell.) Benth.
Chaetospora neesii (Nees) Boeck. = ***Tricostularia neesii*** Nees
Chaetospora nigricans (L.) Kunth = ***Schoenus nigricans*** L.
Chaetospora nitens R.Br. = ***Schoenus nitens*** (R.Br.) Roem. & Schult.
Chaetospora oligostachya F.Muell. = ***Schoenus bifidus*** (Nees) Boeck.
Chaetospora oligostachya Boeck. = ***Schoenus nigricans*** L.
Chaetospora paludosa R.Br. = ***Schoenus paludosus*** (R.Br.) Roem. & Schult.
Chaetospora paniculata Steud. = [29 REU]
Chaetospora pauciflora Hook.f. = ***Schoenus pauciflorus*** (Hook.f.) Hook.f.
Chaetospora pedicellata R.Br. = ***Schoenus pedicellatus*** (R.Br.) Roem. & Schult.
Chaetospora pterocarpa Kunth = ***Rhynchospora barbata*** (Vahl) Kunth
Chaetospora pterosperma Kunth = ***Rhynchospora barbata*** (Vahl) Kunth
Chaetospora pterosperma Nees = ***Rhynchospora barbata*** (Vahl) Kunth
Chaetospora punctoria (Vahl) A.Dietr. = ***Neesenbeckia punctoria*** (Vahl) Levyns
Chaetospora rhynchospermoides F.Phil. = ***Schoenus rhynchosporoides*** (Steud.) Kük.
Chaetospora rhynchosporoides Steud. = ***Schoenus rhynchosporoides*** (Steud.) Kük.
Chaetospora robusta Kunth = ***Tetraria robusta*** (Kunth) C.B.Clarke
Chaetospora rufa (Huds.) Gray = ***Blysmus rufus*** (Huds.) Link
Chaetospora schiedeana Schltdl. = ***Rhynchospora schiedeana*** (Schltdl.) Kunth
Chaetospora sphaerocephala R.Br. = ***Gymnoschoenus sphaerocephalus*** (R.Br.) Hook.f.
Chaetospora spicata Boeck. = ***Tricostularia compressa*** Nees
Chaetospora striata (Thunb.) A.Dietr. = ***Ficinia indica*** (Lam.) H.Pfeiff.
Chaetospora stygia R.Br. = ***Mesomelaena stygia*** (R.Br.) Nees
Chaetospora tenax Hook.f. = ***Schoenus brevifolius*** R.Br.
Chaetospora tendo Hook.f. = ***Schoenus tendo*** (Hook.f.) Banks & Sol. ex Hook.f.
Chaetospora tenella Rupr. = [32]
Chaetospora tenera (Spreng.) A.Dietr. =
Chaetospora tenuissima Steud. = ***Schoenus apogon*** Roem. & Schult. var. ***apogon***
Chaetospora tenuissima Hook.f. = ***Schoenus lepidosperma*** (F.Muell.) K.L.Wilson subsp. ***lepidosperma***
Chaetospora tetragona R.Br. = ***Mesomelaena tetragona*** (R.Br.) Benth.
Chaetospora triceps Cham. & Schltdl. = ***Rhynchospora exaltata*** Kunth

Chaetospora turbinata R.Br. = ***Schoenus turbinatus*** (R.Br.) Roem. & Schult.
Chaetospora umbellulifera Boeck. = ***Schoenus apogon*** Roem. & Schult. var. ***apogon***
Chaetospora vicozensis Schrad. ex Schult. = ***Rhynchospora rugosa*** subsp. ***brownii*** (Roem. & Schult.) T.Koyama
Chaetospora villosa (R.Br.) Nees ex Steud. = ***Schoenus villosus*** R.Br.

***Unplaced Names*:**
Chaetospora paniculata Steud., Syn. Pl. Glumac. 2: 161 (1855). = [29 REU]
Chaetospora tenella Rupr. in F.von der Osten-Saken, Sert. Tianschan.: 34 (1869). = [32]

Chamaegyne

Chamaegyne Suess. = ***Eleocharis*** R.Br.
Chamaegyne pygmaea Suess. = ***Eleocharis chamaegyne*** L.T.Eiten

Chamaeschoenus

Chamaeschoenus Ehrh. = ***Scirpus*** Tourn. ex L.

Chamaexiphium

Chamaexiphium Hochst. ex Steud. = ***Ficinia*** Schrad.
Chamaexiphium clandestinum (Steud.) Hochst. = ***Ficinia clandestina*** (Steud.) Boeck.
Chamaexiphium dregeanum Steud. = ***Ficinia pygmaea*** Boeck.

Chapelliera

Chapelliera Nees = ***Machaerina*** Vahl
Chapelliera arthrophylla Nees = ***Machaerina arthrophylla*** (Nees) T.Koyama
Chapelliera articulata (R.Br.) Nees = ***Machaerina articulata*** (R.Br.) T.Koyama
Chapelliera glomerata (R.Br.) Nees = ***Machaerina rubiginosa*** (Spreng.) T.Koyama
Chapelliera iridifolia Nees = ***Machaerina flexuosa*** (Boeck.) J.Kern subsp. ***flexuosa***
Chapelliera juncea (R.Br.) Nees = ***Machaerina juncea*** (R.Br.) T.Koyama
Chapelliera laxa Nees = ***Machaerina laxa*** (Nees) T.Koyama
Chapelliera pauciflora (R.Br.) Nees = ***Tricostularia pauciflora*** (R.Br.) Benth.
Chapelliera riparia Nees = ***Machaerina riparia*** (Nees) T.Koyama
Chapelliera teretifolia (R.Br.) Nees = ***Machaerina teretifolia*** (R.Br.) T.Koyama

Chillania

Chillania Roiv., Ann. Bot. Soc. Zool.-Bot. Fenn. "Vanamo" 4(7): 2 (1933).
Chile. 85 CLC.
1 Species

Chillania pusilla Roiv., Ann. Bot. Soc. Zool.-Bot. Fenn. "Vanamo" 4(7): 2 (1933). *Eleocharis uniflora* O.Seberg, Syst. Bot. 10: 240 (1985).
Chile (Biobío). 85 CLC. Hemicr.

Chionanthula

Chionanthula Börner = ***Carex*** L.
Chionanthula baldensis (L.) Fedde & J.Schust. = ***Carex baldensis*** L.

Chlorocharis

Chlorocharis Rikli = ***Eleocharis*** R.Br.
Chlorocharis balansaeana (Boeck.) Rikli = ***Eleocharis filiculmis*** Kunth
Chlorocharis capitata (L.) Rikli = ***Eleocharis geniculata*** (L.) Roem. & Schult.
Chlorocharis emarginata (Nees) Rikli = ***Eleocharis emarginata*** (Nees) Klotzsch ex Boeck.
Chlorocharis geniculata (L.) Rikli = ***Eleocharis geniculata*** (L.) Roem. & Schult.
Chlorocharis palustris (L.) Rikli = ***Eleocharis palustris*** (L.) Roem. & Schult.
Chlorocharis subprolifera Rikli = ***Eleocharis pellucida*** C.Presl
Chlorocharis tuberculosa (Michx.) Rikll = ***Eleocharis tuberculosa*** (Michx.) Roem. & Schult.
Chlorocharis vivipara (Link) Rikli = ***Eleocharis vivipara*** Link

Chlorocyperus

Chlorocyperus Rikli = ***Cyperus*** L.
Chlorocyperus abyssinicus (Hochst. ex A.Rich.) Rikli = ***Pycreus flavescens*** (L.) P.Beauv. ex Rchb. subsp. ***flavescens***
Chlorocyperus aegyptiacus (Gloxin) Rikli = ***Cyperus capitatus*** Vand.
Chlorocyperus alopecuroides (Rottb.) Grossh. = ***Cyperus alopecuroides*** Rottb.
Chlorocyperus amuricus (Maxim.) Palla ex Kük. = ***Cyperus amuricus*** Maxim.
Chlorocyperus arsenii Palla = ***Cyperus ischnos*** Schltdl.
Chlorocyperus articulatus (L.) Rikli = ***Cyperus articulatus*** L.
Chlorocyperus aureus Palla ex Kneuck. = ***Cyperus esculentus*** L.
Chlorocyperus badius (Desf.) Palla = ***Cyperus longus*** subsp. ***badius*** (Desf.) Murb.
Chlorocyperus balansae (Maury) Palla = ***Cyperus impolitus*** Kunth
Chlorocyperus barteri (Boeck.) Rikli = ***Cyperus pustulatus*** Vahl
Chlorocyperus cimicinus (J.Presl & C.Presl) Rikli = ***Pycreus niger*** (Ruiz & Pav.) Cufod.
Chlorocyperus confertus (Sw.) Rikll = ***Cyperus confertus*** Sw.
Chlorocyperus congestus (Vahl) Palla = ***Cyperus congestus*** Vahl
Chlorocyperus cordobensis Palla = ***Cyperus articulatus*** L.
Chlorocyperus eragrostis (Lam.) Rikli = ***Cyperus eragrostis*** Lam.
Chlorocyperus erythrorrhizus (Muhl.) Palla = ***Cyperus erythrorrhizos*** Muhl.
Chlorocyperus fendlerianus (Boeck.) Palla = ***Cyperus fendlerianus*** Boeck.
Chlorocyperus filiculmis (Vahl) Falla = ***Cyperus filiculmis*** Vahl
Chlorocyperus flavescens (L.) Rikli = ***Pycreus flavescens*** (L.) P.Beauv. ex Rchb.
Chlorocyperus franchetii Palla = ***Cyperus amuricus*** Maxim.
Chlorocyperus giganteus (Vahl) Palla = ***Cyperus giganteus*** Vahl
Chlorocyperus glaber (L.) Palla = ***Cyperus glaber*** L.
Chlorocyperus glaber f. *fasciculariforme* (Lojac.) Soó = ***Cyperus glaber*** L.
Chlorocyperus globosus (Rchb.) Palla = ***Pycreus flavidus*** (Retz.) T.Koyama

Chlorocyperus glomeratus (L.) Palla = ***Cyperus glomeratus*** L.
Chlorocyperus glomeratus f. *stenostachys* (Borbás) Soó = ***Cyperus glomeratus*** L.
Chlorocyperus grammicus (Kunze ex Kunth) Rikli = ***Pycreus grammicus*** (Kunze ex Kunth) C.B.Clarke
Chlorocyperus inflexus (Muhl.) Palla = ***Cyperus squarrosus*** L.
Chlorocyperus intermedius Rikli = ***Pycreus flavescens*** subsp. ***intermedius*** (Rikli) Lye
Chlorocyperus iria (L.) Rikli = ***Cyperus iria*** L.
Chlorocyperus jacquinii (Schrad.) Rikli = ***Cyperus nutans*** Vahl var. ***nutans***
Chlorocyperus junciformis (Desf.) Rikli = ***Cyperus laevigatus*** L. subsp. ***laevigatus***
Chlorocyperus laevigatus (L.) Palla = ***Cyperus laevigatus*** L.
Chlorocyperus latispicatus (Boeck.) Rikli = ***Pycreus diaphanus*** (Schrad. ex Roem. & Schult.) S.S.Hooper & T.Koyama
Chlorocyperus longus (L.) Palla = ***Cyperus longus*** L.
Chlorocyperus longus subsp. *badius* (Desf.) Soó = ***Cyperus longus*** subsp. ***badius*** (Desf.) Murb.
Chlorocyperus longus f. *pallidior* (Kük.) Soó = ***Cyperus longus*** L. subsp. ***longus***
Chlorocyperus malaccensis (Lam.) Palla = ***Cyperus malaccensis*** Lam.
Chlorocyperus megapotamicus (A.Dietr.) Rikli = ***Pycreus megapotamicus*** (A.Dietr.) Nees
Chlorocyperus melanostachyus (Kunth) Rikli = ***Pycreus niger*** (Ruiz & Pav.) Cufod.
Chlorocyperus meridionalis Palla = ***Cyperus pohlii*** (Nees) Steud. subsp. ***pohlii***
Chlorocyperus mexicanus Palla = ***Cyperus fendlerianus*** Boeck.
Chlorocyperus michoacanensis Palla = ***Cyperus sanguineoater*** Boeck.
Chlorocyperus mundtii (Nees) Rikli = ***Pycreus mundtii*** Nees
Chlorocyperus nilagiricus (Hochst. ex Steud.) Rikli = ***Pycreus flavidus*** (Retz.) T.Koyama
Chlorocyperus oostachyus (Nees) Palla = ***Cyperus rigens*** J.Presl & C.Presl subsp. ***rigens***
Chlorocyperus pannonicus (Jacq.) Rikli = ***Cyperus pannonicus*** Jacq.
Chlorocyperus papyrus (L.) Rikli = ***Cyperus papyrus*** L.
Chlorocyperus paranaguensis Palla = ***Cyperus meyenianus*** Kunth
Chlorocyperus patuliflorus (Boeck.) Rikli = ***Pycreus macrostachyos*** (Lam.) J.Raynal subsp. ***macrostachyos***
Chlorocyperus pauper (Hochst. ex A.Rich.) Rikli = ***Pycreus pauper*** (Hochst. ex A.Rich.) C.B.Clarke
Chlorocyperus phymatodes (Muhl.) Palla = ***Cyperus esculentus*** L.
Chlorocyperus polystachyus (Rottb.) Rikli = ***Pycreus polystachyos*** (Rottb.) P.Beauv.
Chlorocyperus prolixus (Kunth) Palla = ***Cyperus prolixus*** Kunth
Chlorocyperus riograndensis Palla = ***Cyperus pohlii*** (Nees) Steud. subsp. ***pohlii***
Chlorocyperus rivularis (Kunth) Rikli = ***Pycreus bipartitus*** (Torr.) C.B.Clarke
Chlorocyperus rotundus (L.) Palla = ***Cyperus rotundus*** L.
Chlorocyperus salaamensis Palla = ***Cyperus tuberosus*** Rottb.
Chlorocyperus schweinitzii (Torr.) Palla = ***Cyperus schweinitzii*** Torr.
Chlorocyperus serotinus Palla = ***Cyperus serotinus*** Rottb.
Chlorocyperus serrae (Boeck.) Palla = ***Cyperus rigens*** subsp. ***serrae*** (Boeck.) Pedersen

Chondrachne

Chondrachne R.Br. = ***Lepironia*** Rich. ex Pers.
Chondrachne articulata (Retz.) R.Br. = ***Lepironia articulata*** (Retz.) Domin

Chondrolomia

Chondrolomia Nees = ***Scleria*** P.J.Bergius
Chondrolomia sellowiana (Kunth) Nees = ***Scleria sellowiana*** Kunth

Chordorrhiza

Chordorrhiza Ehrh. = ***Carex*** L.

Choricarpha

Choricarpha Boeck. = ***Lepironia*** Rich. ex Pers.
Choricarpha aphylla Boeck. = ***Lepironia articulata*** (Retz.) Domin

Chorizandra

Chorizandra R.Br., Prodr.: 221 (1810).
E. & S. Australia. 50 NSW QLD SOA TAS VIC WAU.
5 Species

Chorizandra australis K.L.Wilson, Telopea 5: 594 (1994).
SE. South Australia, Victoria, Tasmania. 50 SOA TAS VIC. Hemicr.

Chorizandra cymbaria R.Br., Prodr.: 221 (1810). *Lepironia cymbaria* (R.Br.) T.Koyama, J. Fac. Sci. Univ. Tokyo, Sect. 3, Bot. 8: 69 (1961).
S. & SE. Australia. 50 NSW QLD TAS VIC WAU. Hemicr. or rhizome geophyte.

Chorizandra enodis Nees in J.G.C.Lehmann, Pl. Preiss. 2: 73 (1846). *Lepironia neesii* T.Koyama, J. Fac. Sci. Univ. Tokyo, Sect. 3, Bot. 8: 69 (1961).
S. Australia. 50 SOA TAS VIC WAU. Hemicr. or rhizome geophyte.

Chorizandra multiarticulata Nees, Ann. Mag. Nat. Hist. 6: 48 (1841). *Lepironia multiarticulata* (Nees) T.Koyama, J. Fac. Sci. Univ. Tokyo, Sect. 3, Bot. 8: 69 (1961).
SW. Australia. 50 WAU. Hemicr. or rhizome geophyte.

Chorizandra sphaerocephala R.Br., Prodr.: 221 (1810). *Lepironia sphaerocephala* (R.Br.) T.Koyama, J. Fac. Sci. Univ. Tokyo, Sect. 3, Bot. 8: 69 (1961).
SE. Queensland to E. New South Wales. 50 NSW QLD. Hemicr. or rhizome geophyte.

Synonyms:
Chorizandra involucrata (J.V.Suringar) Ridl. = ***Capitularina involucrata*** (J.V.Suringar) J.Kern

Chrysitrix

Chrysitrix L., Mant. Pl. 2: 165 (1771).
Cape Prov., SW. Australia. 27 CPP 50 WAU.
4 Species

Chrysitrix capensis L., Mant. Pl. 2: 304 (1771).
Cape Prov. 27 CPP. Hemicr.

var. ***capensis***
Cape Prov. 27 CPP. Hemicr.
Willdenowia compressa Steud., Flora 12: 134 (1829), nom. illeg.

var. ***subteres*** C.B.Clarke in W.H.Harvey & auct. suc. (eds.), Fl. Cap. 7: 293 (1898).
SW. Cape Prov. 27 CPP. Hemicr.
Chrysitrix subteres H.Pfieff., Bot. Arch. 12: 448 (1925).

Chrysitrix distigmatosa C.B.Clarke ex Diels & Pritz., Bot. Jahrb. Syst. 35: 82 (1904).
SW. Australia. 50 WAU. Hemicr.

Chrysitrix dodii C.B.Clarke in W.H.Harvey & auct. suc. (eds.), Fl. Cap. 7: 760 (1900).
SW. Cape Prov. 27 CPP. Hemicr.

Chrysitrix junciformis Nees, Linnaea 10: 144 (1836).
Cape Prov. (Riversdale Div.). 27 CPP. Hemicr.

Synonyms:
Chrysitrix subteres H.Pfieff. = ***Chrysitrix capensis*** var. ***subteres*** C.B.Clarke

Cladium

Cladium P.Browne, Civ. Nat. Hist. Jamaica: 114 (1756).
Cosmopolitan. 10 DEN FIN GRB IRE NOR SWE 11 AUT BGM CZE GER HUN NET POL SWI 12 BAL COR FRA POR SAR SPA 13 ALB BUL GRC ITA KRI ROM SIC TUE YUG 14 BLR BLT KRY RUC RUE RUS RUW UKR 20 ALG EGY LBY MOR 21 CVI 22 SEN 23 BUR RWA ZAI 24 SOC 25 KEN TAN UGA 26 ANG 27 BOT CPP NAM NAT TVL 29 MAU MDG 32 KAZ KGZ TKM UZB 33 NCS TCS 34 CYP IRN IRQ LBS PAL TUR 35 SAU 36 CHC CHS 38 JAP KOR NNS OGA TAI 40 IND NEP WHM 42 LSI MOL SUM 43 NWG SOL 50 NSW QLD SOA VIC WAU 60 NWC 61 SCI 63 HAW 71 MAN SAS 72 LAB NBR NFL NSC ONT QUE 74 ILL MIN WIS 75 CNT INI MAI MAS MIC NWH NWJ NWY OHI PEN RHO VER 76 ARI CAL NEV UTA 77 TEX 78 ALA ARK DEL FLA GEO LOU MRY MSI NCA SCA TEN VRG 79 MXE MXG MXN MXS MXT 80 BLZ COS GUA HON NIC PAN 81 BAH BER CAY CUB DOM HAI JAM LEE PUE WIN 82 GUY SUR? VEN 83 BOL PER 84 BZE BZL BZS 85 AGE AGW PAR URU.
3 Species
Mariscus Scop., Meth. Pl.: 22 (1754).
Trasis P.Beauv. in T.G.Lestiboudois, Essai Cypér.: 32 (1819).

Cladium costatum Steyerm., Fieldiana, Bot. 28: 50 (1951).
S. Venezuela to Guyana (Mt. Roraima). 82 GUY VEN. Hemicr.

Cladium mariscoides (Muhl.) Torr., Ann. Lyceum Nat. Hist. New York 3: 372 (1836).
S. & E. Canada, EC. & E. U.S.A. 71 MAN SAS 72 LAB NBR NFL NSC ONT QUE 74 ILL MIN WIS 75 CNT INI MAI MAS MIC NWH NWJ NWY OHI PEN RHO VER 77 TEX 78 DEL FLA MRY NCA SCA TEN. Hemicr.
**Schoenus mariscoides* Muhl., Descr. Gram.: 4 (1817). *Mariscus mariscoides* (Muhl.) Kuntze, Revis. Gen. Pl. 2: 755 (1891).
Cladium mariscoides f. *congestum* Fernald, Rhodora 23: 234 (1922). *Mariscus mariscoides* f. *congestus* (Fernald) Fernald, Rhodora 25: 53 (1923).

Cladium mariscus (L.) Pohl, Tent. Fl. Bohem. 1: 32 (1809).
Cosmopolitan. 10 DEN FIN GRB IRE NOR SWE 11 AUT BGM CZE GER HUN NET POL SWI 12 BAL COR FRA POR SAR SPA 13 ALB BUL GRC ITA KRI ROM SIC YUG 14 BLR BLT KRY RUC RUE RUS RUW UKR 20 ALG EGY LBY MOR 21 CVI 22 SEN 23 BUR RWA ZAI 24 SOC 25 KEN TAN UGA 26 ANG 27 BOT CPP NAM NAT TVL 29 MAU MDG 32 KAZ KGZ UZB 33 NCS TCS 34 CYP PAL 36 CHC CHS 38 JAP KOR NNS OGA TAI 40 IND NEP WHM 42 LSI MOL SUM 43 NWG SOL 60 NWC 61 SCI 63 HAW 76 ARI CAL NEV UTA 77 TEX 78 ALA ARK FLA GEO LOU MSI NCA SCA VRG 79 MXE MXG MXN MXS MXT 80 BLZ COS GUA HON NIC PAN 81 BAH BER CAY CUB DOM HAI JAM LEE PUE WIN 82 SUR? 83 BOL PER 84 BZE BZL BZS 85 AGE AGW PAR URU. Hemicr.
**Schoenus mariscus* L., Sp. Pl.: 42 (1753). *Mariscus mariscus* (L.) Borbás, Balaton Fl.: 321 (1900), nom. inval.

subsp. ***californicum*** (S.Watson) Govaerts, World Checklist Seed Pl. 3(1): 15 (1999).
SW. U.S.A. to Mexico (Sonora, Coahuila). 76 ARI CAL NEV UTA 77 TEX 79 MXE MXN. Hemicr.
**Cladium mariscus* var. *californicum* S.Watson in S.Watson & al., Bot. California 2: 224 (1880). *Mariscus californicus* (S.Watson) Fernald, Rhodora 25: 52 (1923). *Cladium californicum* (S.Watson) O'Neill in I.Tidestrom & T.Kittell, Fl. Ariz. New Mex.: 773 (1941).

subsp. ***chinense*** (Nees ex Hook. & Arn.) T.Koyama, Enum. Fl. Pl. Nepal 1: 105 (1978).
Nepal to Temp. E. Asia. 36 CHC CHS 38 JAP KOR NNS OGA TAI 40 IND NEP. Hemicr.
**Cladium chinense* Nees ex Hook. & Arn., Bot. Beechey Voy.: 277 (1838). *Mariscus chiuensis* (Nees ex Hook. & Arn.) Fernald, Rhodora 25: 52 (1923). *Cladium jamaicense* subsp. *chinense* (Nees ex Hook. & Arn.) T.Koyama, Enum. Fl. Pl. Nepal 1: 105 (1978).
Cladium japonicum Steud., Syn. Pl. Glumac. 2: 152 (1855).

subsp. ***intermedium*** Kük., Repert. Spec. Nov. Regni Veg. 51: 188 (1942).
Australia, New Caledonia. 50 NSW QLD SOA VIC WAU 60 NWC. Hemicr.
Cladium procerum S.T.Blake, Trans. Roy. Soc. South Australia 67: 57 (1943). *Machaerina procera* (S.T.Blake) T.Koyama, Bot. Mag. (Tokyo) 69: 65 (1956).

subsp. ***jamaicense*** (Crantz) Kük., Repert. Spec. Nov. Regni Veg. Beih. 40(1): 523 (1938).
Trop. & Subtrop. 21 CVI 22 SEN 23 BUR RWA ZAI 24 SOC 25 KEN TAN UGA 26 ANG 27 BOT CPP NAM NAT TVL 29 MAU MDG 42 LSI MOL SUM 43 NWG SOL 60 NWC 61 SCI 63 HAW 77 TEX 78 ALA ARK FLA GEO LOU MSI NCA SCA VRG 79 MXE MXG MXS MXT 80 BLZ COS GUA HON NIC PAN 81 BAH BER CAY CUB DOM HAI JAM LEE PUE WIN 82 SUR? VEN 83 BOL PER 84 BZE BZL BZS 85 AGE AGW PAR URU. Hemicr. or rhizome geophyte.
**Cladium jamaicense* Crantz, Inst. Rei Herb. 1: 362 (1766). *Mariscus jamaicensis* (Crantz) Britton in J.K.Small, Fl. Miami: 31 (1913).
Cladium occidentale Schrad., Fl. Germ. 1: 76 (1806).
Cladium bahiense Nees, Linnaea 9: 301 (1834), nom. inval.

Cladium leptostachyum Nees in F.J.F.Meyen, Observ. Bot. 2: 119 (1835). *Mariscus leptostachyus* (Nees) Kuntze, Revis. Gen. Pl. 2: 755 (1891).
Cladium effusum Torr., Ann. Lyceum Nat. Hist. New York 3: 374 (1836).
Hypolytrum kuntzeanum Boeck., Beitr. Cyper. 1: 23 (1888).
Mariscus elevatus J.W.Moore, Bernice P. Bishop Mus. Bull. 102: 20 (1933).
Cladium mariscus var. *ferruginescens* Kük., Repert. Spec. Nov. Regni Veg. 51: 192 (1942).

subsp. ***mariscus***
Europe, Medit. to Caucasus. 10 DEN FIN GRB IRE NOR SWE 11 AUT BGM CZE GER HUN NET POL SWI 12 BAL COR FRA POR SAR SPA 13 ALB BUL GRC ITA KRI ROM SIC YUG 14 BLR BLT KRY RUC RUE RUS RUW UKR 20 ALG EGY LBY MOR 33 NCS TCS 34 CYP PAL. Hemicr.
Schoenus elevatus Sol. ex G.Forst., Fl. Ins. Austr.: 89 (1786).
Schoenus cladium Sw., Prodr.: 19 (1788). *Mariscus cladium* (Sw.) Kuntze, Revis. Gen. Pl. 2: 755 (1891).
Schoenus effusus Sw., Prodr.: 19 (1788).
Mariscus serratus Gilib., Excerc. Phyt. 2: 512 (1792).
Scirpus excelsus Salisb., Prodr. Stirp. Chap. Allerton: 31 (1796).
Cladium germanicum Schrad., Fl. Germ. 1: 75 (1806).
Cladium palustre Poir. in G.-F.Cuvier, Dict. Sci. Nat., ed. 2, 9: 344 (1817).
Scirpus martii Dufour ex Roem. & Schult., Syst. Veg. 2: 117 (1817). *Mariscus martii* (Dufour ex Roem. & Schult.) Fernald, Rhodora 25: 54 (1923).
Schoenus congestus Willd. ex Link, Jahrb. Gewächsk. 1(3): 76 (1820).
Cladium floribundum C.Presl, Isis (Oken) 21: 270 (1828).
Schoenus floribundus Willd. ex C.Presl, Isis (Oken) 21: 270 (1828).
Schoenus castaneus Willd. ex Kunth, Enum. Pl. 2: 303 (1837).
Schoenus altissimus Salzm. ex Steud., Syn. Pl. Glumac. 2: 152 (1855).
Schoenus medwedewii Meinsh., Trudy Imp. S.-Peterburgsk. Bot. Sada 18: 272 (1901).

subsp. ***martii*** (Roem. & Schult.) T.V.Egorova, in Fl. Evropeiskoi Chasti SSSR 2: 129 (1976).
S. Europe to W. Himalaya. 12 BAL COR POR SPA 13 GRC ITA TUE 32 KAZ KGZ TKM UZB 33 TCS 34 IRN IRQ LBS PAL TUR 35 SAU 40 WHM. Hemicr.
**Isolepis martii* Roem. & Schult., Syst. Veg. 2: 117 (1817).
Cladium giganteum Willk. in M.Willkomm & J.M.C.Lange, Prodr. Fl. Hispan. 1: 136 (1861).
Cladium lucense Merino, Bol. Soc. Esp. Hist. Nat. 2: 65 (1901).
Cladium grossheimii Pobed., Bot. Mater. Gerb. Bot. Inst. Komarova Akad. Nauk S.S.S.R. 12: 13 (1950).

Synonyms:
Cladium acutum (Labill.) Poir. = ***Machaerina acuta*** (Labill.) J.Kern
Cladium anceps Hook.f. = ***Machaerina sinclairii*** (Hook.f.) T.Koyama
Cladium ancistrophyllum F.Muell. ex Benth. = ***Gahnia ancistrophylla*** Benth.
Cladium angustifolium (Gaudich.) Benth. & Hook.f. ex Drake = ***Machaerina angustifolia*** (Gaudich.) T.Koyama
Cladium arfakense Rendle = ***Machaerina teretifolia*** (R.Br.) T.Koyama
Cladium aristatum F.Muell. ex Benth. = ***Gahnia aristata*** Benth.
Cladium aromaticum Merr. = ***Schoenus punctatus*** R.Br.
Cladium arthrophyllum (Nees) F.Muell. = ***Machaerina arthrophylla*** (Nees) T.Koyama
Cladium arthrophyllum var. *riparium* (Nees) Kük. = ***Machaerina riparia*** (Nees) T.Koyama
Cladium articulatum R.Br. = ***Machaerina articulata*** (R.Br.) T.Koyama
Cladium aspericaule Kük. = ***Machaerina aspericaulis*** (Kük.) T.Koyama
Cladium asperum (R.Br.) F.Muell. = ***Gahnia aspera*** (R.Br.) Spreng.
Cladium australe (Nees) H.Pfeiff. = ***Gahnia australis*** (Nees) K.L.Wilson
Cladium australe (A.Rich.) Druce = ***Lepidosperma australe*** (A.Rich.) Hook.f.
Cladium bahiense Nees = ***Cladium mariscus*** subsp. ***jamaicense*** (Crantz) Kük.
Cladium bidwellii (Stapf ex Setch.) Kük. = ***Machaerina bidwellii*** (Stapf ex Setch.) T.Koyama
Cladium boninsimae Nakai = ***Machaerina glomerata*** (Gaudich.) T.Koyama
Cladium borneense C.B.Clarke ex Stapf = ***Lepidosperma chinense*** Nees & Meyen ex Kunth
Cladium brevipaniculatum (Kük.) Kük. = ***Machaerina gunnii*** (Hook.f.) J.Kern
Cladium brevistigma Nakai ex Tuyama = ***Machaerina rubiginosa*** (Spreng.) T.Koyama
Cladium californicum (S.Watson) O'Neill = ***Cladium mariscus*** subsp. ***californicum*** (S.Watson) Govaerts
Cladium capillaceum (Benth.) C.B.Clarke = ***Tetraria capillaris*** (F.Muell.) J.M.Black
Cladium capillaceum var. *laminatum* Kük. = ***Tetraria capillaris*** (F.Muell.) J.M.Black
Cladium chinense Nees ex Hook. & Arn. = ***Cladium mariscus*** subsp. ***chinense*** (Nees ex Hook. & Arn.) T.Koyama
Cladium colpodes Lauterb. = ***Machaerina glomerata*** (Gaudich.) T.Koyama
Cladium complanatum Berggr. = ***Machaerina complanata*** (Berggr.) T.Koyama
Cladium crassum (Thwaites) Kük. = ***Machaerina rubiginosa*** (Spreng.) T.Koyama
Cladium crinitum (Stapf) Merr. = ***Machaerina sinclairii*** (Hook.f.) T.Koyama
Cladium cubense Kük. = ***Machaerina cubensis*** (Kük.) T.Koyama
Cladium cyperoides Merr. = ***Fimbristylis vanoverberghii*** Kük.
Cladium decompositum R.Br. = ***Gahnia decomposita*** (R.Br.) Benth.
Cladium deplanchei (Boeck.) C.B.Clarke = ***Machaerina deplanchei*** (Boeck.) T.Koyama
Cladium deustum R.Br. = ***Gahnia deusta*** (R.Br.) Benth.
Cladium distichum C.B.Clarke = ***Machaerina disticha*** (C.B.Clarke) T.Koyama
Cladium drummondii C.B.Clarke = ***Machaerina preissii*** (Nees) L.A.S.Johnson & Koyama
Cladium dubium Spreng. = ***Machaerina rubiginosa*** (Spreng.) T.Koyama
Cladium durandoi Chabert = ?

Cladium effusum Torr. = ***Cladium mariscus*** subsp. ***jamaicense*** (Crantz) Kük.
Cladium ekmanii Kük. = ***Machaerina ekmanii*** (Kük.) T.Koyama
Cladium elynanthoides F.Muell. = ***Gahnia australis*** (Nees) K.L.Wilson
Cladium ensifolium (Boeck.) Benth. & Hook.f. ex B.D.Jacks. = ***Machaerina iridifolia*** (Bory) T.Koyama
Cladium ensigerum Hance = ***Machaerina ensigera*** (Hance) T.Koyama
Cladium erythrocarpum (R.Br.) F.Muell. = ***Gahnia erythrocarpa*** R.Br.
Cladium falcatum (Nees) C.B.Clarke = ***Machaerina falcata*** (Nees) T.Koyama
Cladium falcatum var. *celebicum* Kük. = ***Machaerina falcata*** (Nees) T.Koyama
Cladium falcatum var. *sumatrense* C.B.Clarke = ***Machaerina falcata*** (Nees) T.Koyama
Cladium ficticium Hemsl. = ***Machaerina ficticia*** (Hemsl.) T.Koyama
Cladium ficticium var. *macrophyllum* (Boeck.) Kük. = ***Machaerina macrophylla*** (Boeck.) T.Koyama
Cladium filifolium (Griseb.) Benth. & Hook.f. = ***Machaerina filifolia*** Griseb.
Cladium filiforme Merr. = ***Schoenus delicatulus*** (Fernald) J.Kern
Cladium filum (Labill.) R.Br. = ***Gahnia filum*** (Labill.) F.Muell.
Cladium fimbristyloides Baker = ***Costularia leucocarpa*** (Ridl.) H.Pfeiff.
Cladium flexuosum (Boeck.) C.B.Clarke = ***Machaerina flexuosa*** (Boeck.) J.Kern
Cladium flexuosum var. *polyanthemum* Kük. = ***Machaerina flexuosa*** subsp. ***polyanthemum*** (Kük.) Lye
Cladium floribundum C.Presl = ***Cladium mariscus*** (L.) Pohl subsp. ***mariscus***
Cladium gahnoides Colenso = ***Machaerina sinclairii*** (Hook.f.) T.Koyama
Cladium gaudichaudii W.Wight = ***Machaerina glomerata*** (Gaudich.) T.Koyama
Cladium germanicum Schrad. = ***Cladium mariscus*** (L.) Pohl subsp. ***mariscus***
Cladium giganteum Willk. = ***Cladium mariscus*** subsp. ***martii*** (Roem. & Schult.) T.V.Egorova
Cladium globiceps C.B.Clarke = ***Machaerina glomerata*** (Gaudich.) T.Koyama
Cladium glomeratum (Gaudich.) H.Pfieff. = ***Machaerina glomerata*** (Gaudich.) T.Koyama
Cladium glomeratum R.Br. = ***Machaerina rubiginosa*** (Spreng.) T.Koyama
Cladium glomeratum f. *major* Kük. ex Carse = ***Machaerina rubiginosa*** (Spreng.) T.Koyama
Cladium glomeratum var. *subseptatum* Benth. = ***Machaerina rubiginosa*** (Spreng.) T.Koyama
Cladium gracile J.M.Black = ***Machaerina laxa*** (Nees) T.Koyama
Cladium grossheimii Pobed. = ***Cladium mariscus*** subsp. ***martii*** (Roem. & Schult.) T.V.Egorova
Cladium gunnii Hook.f. = ***Machaerina gunnii*** (Hook.f.) J.Kern
Cladium gunnii var. *brevipaniculatum* Kük. = ***Machaerina gunnii*** (Hook.f.) J.Kern
Cladium hirtum (Boeck.) C.B.Clarke = ***Machaerina hirta*** Boeck.
Cladium huttonii Kirk = ***Machaerina huttonii*** (Kirk) T.Koyama
Cladium insulare Benth. = ***Machaerina insularis*** (Benth.) T.Koyama
Cladium iridifolium (Bory) Baker = ***Machaerina iridifolia*** (Bory) T.Koyama
Cladium iris Ohwi = ***Machaerina falcata*** (Nees) T.Koyama
Cladium jamaicense Crantz = ***Cladium mariscus*** subsp. ***jamaicense*** (Crantz) Kük.
Cladium jamaicense subsp. *chinense* (Nees ex Hook. & Arn.) T.Koyama = ***Cladium mariscus*** subsp. ***chinense*** (Nees ex Hook. & Arn.) T.Koyama
Cladium japonicum Steud. = ***Cladium mariscus*** subsp. ***chinense*** (Nees ex Hook. & Arn.) T.Koyama
Cladium junceum R.Br. = ***Machaerina juncea*** (R.Br.) T.Koyama
Cladium junceum var. *pauciflorum* (R.Br.) Kük. = ***Tricostularia pauciflora*** (R.Br.) Benth.
Cladium juncoides Elmer = ***Machaerina glomerata*** (Gaudich.) T.Koyama
Cladium lamii Kük. = ***Machaerina lamii*** (Kük.) J.Kern
Cladium lanigerum R.Br. = ***Gahnia lanigera*** (R.Br.) Benth.
Cladium latifolium Merr. = ***Machaerina falcata*** (Nees) T.Koyama
Cladium latifolium (Kunth) Benth. & Hook.f. ex Drake = ***Machaerina iridifolia*** (Bory) T.Koyama
Cladium latissimum F.Muell. = ***Machaerina preissii*** (Nees) L.A.S.Johnson & Koyama
Cladium lavarum (Poir.) Cordem. = ***Machaerina lavarum*** (Poir.) Bojer
Cladium laxiflorum Hook.f. = ***Machaerina gunnii*** (Hook.f.) J.Kern
Cladium laxum (Nees) Benth. = ***Machaerina laxa*** (Nees) T.Koyama
Cladium leptostachyum Nees = ***Cladium mariscus*** subsp. ***jamaicense*** (Crantz) Kük.
Cladium lucense Merino = ***Cladium mariscus*** subsp. ***martii*** (Roem. & Schult.) T.V.Egorova
Cladium maingayi C.B.Clarke = ***Machaerina maingayi*** (C.B.Clarke) T.Koyama
Cladium maingayi var. *subsetosum* Kük. = ***Machaerina sinclairii*** (Hook.f.) T.Koyama
Cladium mariscoides (Gaudich.) Fern.-Vill. = ***Machaerina mariscoides*** (Gaudich.) J.Kern
Cladium mariscoides f. *congestum* Fernald = ***Cladium mariscoides*** (Muhl.) Torr.
Cladium mariscus var. *californicum* S.Watson = ***Cladium mariscus*** subsp. ***californicum*** (S.Watson) Govaerts
Cladium mariscus var. *ferruginescens* Kük. = ***Cladium mariscus*** subsp. ***jamaicense*** (Crantz) Kük.
Cladium medium R.Br. = ***Gahnia lanigera*** (R.Br.) Benth.
Cladium melanocarpum (R.Br.) F.Muell. = ***Gahnia melanocarpa*** R.Br.
Cladium melleri Baker = ***Costularia melleri*** (Baker) C.B.Clarke
Cladium meyenii (Kunth) Benth. & Hook.f. ex Drake = ***Machaerina mariscoides*** (Gaudich.) J.Kern
Cladium micranthes C.B.Clarke = ***Machaerina disticha*** (C.B.Clarke) T.Koyama
Cladium microstachyum (Benth.) F.Muell. = ***Gahnia microstachya*** Benth.
Cladium milnei C.B.Clarke = ***Machaerina milnei*** (C.B.Clarke) T.Koyama
Cladium monocarpum J.M.Black = ***Schoenus carsei*** Cheeseman
Cladium muelleri C.B.Clarke = ***Machaerina muelleri*** (C.B.Clarke) T.Koyama
Cladium myrianthum Chun & F.C.How = ***Machaerina myriantha*** (Chun & F.C.How) Y.C.Tang

Cladium nipponense Ohwi = ***Machaerina rubiginosa*** (Spreng.) T.Koyama
Cladium nudum (Steud.) Boeck. = ***Machaerina nuda*** (Steud.) J.Kern
Cladium nudum var. *rigidulum* Kük. = ***Machaerina nuda*** (Steud.) J.Kern
Cladium nukuhivense F.Br. = ***Machaerina nukuhivensis*** (F.Br.) T.Koyama
Cladium occidentale Schrad. = ***Cladium mariscus*** subsp. ***jamaicense*** (Crantz) Kük.
Cladium ouveanum (Däniker) Guillaumin = ***Machaerina juncea*** (R.Br.) T.Koyama
Cladium palustre Poir. = ***Cladium mariscus*** (L.) Pohl subsp. ***mariscus***
Cladium pantopodum Baker = ***Costularia pantopoda*** (Baker) C.B.Clarke
Cladium pauciflorum R.Br. = ***Tricostularia pauciflora*** (R.Br.) Benth.
Cladium philippinense Merr. = ***Machaerina disticha*** (C.B.Clarke) T.Koyama
Cladium polyphyllum (Benth.) F.Muell. = ***Gahnia drummondii*** (Steud.) K.L.Wilson
Cladium ponapense Ohwi = ***Machaerina falcata*** (Nees) T.Koyama
Cladium preissii (Nees) F.Muell. = ***Gahnia decomposita*** (R.Br.) Benth.
Cladium preissii var. *laxum* (Nees) Kük. = ***Machaerina laxa*** (Nees) T.Koyama
Cladium procerum S.T.Blake = ***Cladium mariscus*** subsp. ***intermedium*** Kük.
Cladium psittacorum (Labill.) F.Muell. = ***Gahnia psittacorum*** Labill.
Cladium pulchrum Ridl. = ***Tricostularia undulata*** (Thwaites) J.Kern
Cladium punctatum (R.Br.) H.Pfieff. = ***Schoenus punctatus*** R.Br.
Cladium quadrangulare Nees = ***Morelotia gahniiformis*** Gaudich.
Cladium radula R.Br. = ***Gahnia radula*** (R.Br.) Benth.
Cladium restioides (Sw.) Benth. ex C.B.Clarke = ***Machaerina restioides*** (Sw.) Vahl
Cladium restioides var. *effusa* (Griseb.) C.B.Clarke = ***Machaerina restioides*** (Sw.) Vahl
Cladium rigidifolium C.Presl = ***Machaerina teretifolia*** (R.Br.) T.Koyama
Cladium riparium (Nees) Benth. = ***Machaerina riparia*** (Nees) T.Koyama
Cladium rubiginosum (Spreng.) Domin = ***Machaerina rubiginosa*** (Spreng.) T.Koyama
Cladium rubiginosum var. *subriparium* Kük. = ***Machaerina rubiginosa*** (Spreng.) T.Koyama
Cladium rubiginosum var. *subseptatum* (Benth.) Domin = ***Machaerina rubiginosa*** (Spreng.) T.Koyama
Cladium samoense C.B.Clarke ex Stapf = ***Machaerina falcata*** (Nees) T.Koyama
Cladium samoense var. *uniseta* C.B.Clarke ex Stapf = ***Machaerina falcata*** (Nees) T.Koyama
Cladium schoenoides R.Br. = ***Machaerina acuta*** (Labill.) J.Kern
Cladium schoenoides var. *elongatum* Benth. = ***Machaerina muelleri*** (C.B.Clarke) T.Koyama
Cladium scirpoideum (Steud.) Benth. & Hook.f. ex Hemsl. = ***Machaerina scirpoidea*** (Steud.) T.Koyama
Cladium sclerioides F.Muell. = ***Exocarya sclerioides*** (F.Muell.) Benth.
Cladium sieberi (Schrad. ex Schult. & Schult.f.) F.Muell. = ***Gahnia filifolia*** (C.Presl) Kük. ex Benl
Cladium sinclairii Hook.f. = ***Machaerina sinclairii*** (Hook.f.) T.Koyama
Cladium sinuatum Ridl. = ***Machaerina glomerata*** (Gaudich.) T.Koyama
Cladium steyermarkii T.Koyama = ***Rhynchocladium steyermarkii*** (T.Koyama) T.Koyama
Cladium stradbrokense Domin = ***Trachystylis stradbrokensis*** (Domin) Kük.
Cladium tenax (Hook.f.) Druce = ***Machaerina tenax*** (Hook.f.) T.Koyama
Cladium teretifolium R.Br. = ***Machaerina teretifolia*** (R.Br.) T.Koyama
Cladium teretifolium var. *asperrimum* Kük. = ***Machaerina tetragona*** (Labill.) T.Koyama
Cladium teretifolium var. *planifolium* (Benth.) Domin =
Cladium teretifolium f. *planifolium* (Benth.) Kük. =
Cladium teretifolium var. *tetraquetrum* (Hook.f.) Domin = ***Machaerina tetragona*** (Labill.) T.Koyama
Cladium teretifolium var. *typicum* Domin = ***Machaerina teretifolia*** (R.Br.) T.Koyama
Cladium tetragonocarpum (Boeck.) F.Muell. = ***Gahnia sieberiana*** Kunth
Cladium tetragonum (Labill.) J.M.Black = ***Machaerina tetragona*** (Labill.) T.Koyama
Cladium tetraquetrum Hook.f. = ***Machaerina tetragona*** (Labill.) T.Koyama
Cladium tetraquetrum var. *planifolium* Benth. = [50 NSW QLD VIC]
Cladium trifidum (Labill.) F.Muell. = ***Gahnia trifida*** Labill.
Cladium triglomeratum (Michx.) Nees = ***Scleria triglomerata*** Michx.
Cladium undulatum Thwaites = ***Tricostularia undulata*** (Thwaites) J.Kern
Cladium undulatum var. *fimbristyloides* (F.Muell.) Domin = ***Tricostularia undulata*** (Thwaites) J.Kern
Cladium vaginale Benth. = ***Machaerina vaginalis*** (Benth.) T.Koyama
Cladium vauthiera C.B.Clarke = ***Lepidosperma australe*** (A.Rich.) Hook.f.
Cladium vincentia C.B.Clarke = ***Machaerina angustifolia*** (Gaudich.) T.Koyama
Cladium xanthocarpum (Hook.f.) F.Muell. = ***Gahnia xanthocarpa*** (Hook.f.) Hook.f.
Cladium xipholepis Baker = ***Costularia elongata*** (Kunth) Kük.

***Unplaced Names*:**
Cladium durandoi Chabert, Bull. Soc. Bot. France 36: 321 (1889). = ?
Cladium tetraquetrum var. *planifolium* Benth., Fl. Austral. 7: 407 (1878). = [50 NSW QLD VIC]

Cladotheca

Cladotheca Steud. = ***Lagenocarpus*** Nees
Cladotheca vaginans Steud. = ***Lagenocarpus verticillatus*** (Spreng.) T.Koyama & Maguire

Clavula

Clavula Dumort. = ***Eleocharis*** R.Br.
Clavula acicularis (L.) Dumort. = ***Eleocharis acicularis*** (L.) Roem. & Schult.
Clavula boeotryon (L.f.) Dumort. = ***Eleocharis quinqueflora*** (Hartmann) O.Schwarz
Clavula cespitosa (L.) Dumort. = ***Trichophorum cespitosum*** (L.) Hartm.
Clavula comosa Dumort. = ***Eleocharis parvula*** (Roem. & Schult.) Palla
Clavula fluitans (L.) Dumort. = ***Isolepis fluitans*** (L.) R.Br.

Clavula multicaulis (Sm.) Dumort. = ***Eleocharis multicaulis*** (Sm.) Desv.
Clavula multicaulis var. *vivipara* Dumort. = ***Eleocharis multicaulis*** (Sm.) Desv.
Clavula ovata (Roth) Dumort. = ***Eleocharis ovata*** (Roth) Roem. & Schult.
Clavula palustris (L.) Dumort. = ***Eleocharis palustris*** (L.) Roem. & Schult.
Clavula uniglumis (Link) Dumort. = ***Eleocharis uniglumis*** (Link) Schult.

Cleistocalyx

Cleistocalyx Steud. = ***Rhynchospora*** Vahl
Cleistocalyx junciformis (Steud.) Steud. = ***Rhynchospora oligantha*** A.Gray var. ***oligantha***

Coleachyron

Coleachyron J.Gay ex Boiss. = ***Carex*** L.
Coleachyron oliverianum J.Gay = ***Carex illegitima*** Ces.

Coleochloa

Coleochloa Gilly, Brittonia 5: 12 (1943).
Trop. & S. Africa, Madagascar. 22 NGA 23 CMN 24 ERI ETH SUD 25 KEN TAN UGA 26 ANG MLW MOZ ZAM ZIM 27 NAT SWZ TVL 29 MDG.
7 Species
**Eriospora* Hochst. ex A.Rich., Tent. Fl. Abyss. 2: 508 (1850).

Coleochloa abyssinica (Hochst. ex A.Rich.) Gilly, Brittonia 5: 14 (1943).
Trop. Africa. 22 NGA 23 CMN 24 ERI ETH SUD 25 TAN UGA 26 ANG. Hemicr.
**Eriospora abyssinica* Hochst. ex A.Rich., Tent. Fl. Abyss. 2: 508 (1850). *Trilepis abyssinica* (Hochst. ex A.Rich.) Boeck., Linnaea 39: 9 (1875).

var. ***abyssinica***
NE. Trop. Africa. 24 ERI ETH. Hemicr.

var. ***castanea*** (C.B.Clarke) Pic.Serm., Webbia 7: 347 (1951).
Nigeria to Ethiopia and Angola. 22 NGA 23 CMN 24 ETH SUD 25 TAN UGA 26 ANG. Hemicr.
**Eriospora abyssinica* var. *castanea* C.B.Clarke in D.Oliver & auct. suc. (eds.), Fl. Trop. Afr. 8: 513 (1902).

Coleochloa glabra Nelmes, Kew Bull. 8: 380 (1953).
Sudan (Jebel Oda). 24 SUD. Hemicr.

Coleochloa microcephala Nelmes, Kew Bull. 8: 377 (1953).
Tanzania. 25 TAN. Hemicr.

Coleochloa pallidior Nelmes, Kew Bull. 8: 378 (1953).
Northern Prov. 27 TVL. Hemicr.

Coleochloa schweinfurthiana (Boeck.) Nelmes, Kew Bull. 8: 375 (1953).
Sudan (Jebel Bangenze). 24 SUD. Hemicr.
**Carpha schweinfurthiana* Boeck., Flora 62: 569 (1879). *Eriospora schweinfurthiana* (Boeck.) Benth. ex C.B.Clarke in T.A.Durand & H.Schinz, Consp. Fl. Afric. 5: 676 (1894).

Coleochloa setifera (Ridl.) Gilly, Brittonia 5: 14 (1943).
Kenya to S. Africa, Madagascar. 25 KEN TAN 26 MLW MOZ ZAM ZIM 27 NAT SWZ TVL 29 MDG. Hemicr. or rhizome geophyte.
**Fintelmannia setifera* Ridl., J. Linn. Soc., Bot. 20: 837 (1883). *Trilepis setifera* (Ridl.) Kuntze, Revis. Gen. Pl. 2: 758 (1891). *Eriospora setifera* (Ridl.) C.B.Clarke in T.A.Durand & H.Schinz, Consp. Fl. Afric. 5: 676 (1894).
Trilepis oliveri Boeck., Beitr. Cyper. 1: 18 (1888). *Eriospora oliveri* (Boeck.) C.B.Clarke in T.A. Durand & H.Schinz, Consp. Fl. Afric. 5: 676 (1894). *Coleochloa oliveri* (Boeck.) Gilly, Brittonia 5: 14 (1943).
Eriospora villosula C.B.Clarke, Trans. Linn. Soc. London, Bot. 4: 54 (1894). *Coleochloa villosula* (C.B.Clarke) Gilly, Brittonia 5: 14 (1943).
Trilepis rehmanniana Boeck. ex C.B.Clarke in T.A.Durand & H.Schinz, Consp. Fl. Afric. 5: 676 (1894), nom. inval.
Eriospora rehmanniana C.B.Clarke in W.H.Harvey & auct. suc. (eds.), Fl. Cap. 7: 297 (1898). *Coleochloa rehmanniana* (C.B.Clarke) Gilly, Brittonia 5: 14 (1943).

Coleochloa virgata (K.Schum.) Nelmes, Kew Bull. 8: 381 (1953).
Tanzania (Mt. Meru), Malawi (Mt. Mlanje). 25 TAN 26 MLW. Hemicr.
**Eriospora virgata* K.Schum. in H.G.A.Engler, Pflanzenw. Ost-Afrikas, C: 128 (1895).

Synonyms:
Coleochloa oliveri (Boeck.) Gilly = ***Coleochloa setifera*** (Ridl.) Gilly
Coleochloa rehmanniana (C.B.Clarke) Gilly = ***Coleochloa setifera*** (Ridl.) Gilly
Coleochloa villosula (C.B.Clarke) Gilly = ***Coleochloa setifera*** (Ridl.) Gilly

Comostemum

Comostemum Nees = ***Androtrichum*** (Brongn.) Brongn.
Comostemum globuliferum (J.Presl & C.Presl) Nees = ***Androtrichum trigynum*** (Spreng.) H.Pfeiff.
Comostemum impolitum (Kunth) Nees = ***Cyperus impolitus*** Kunth
Comostemum laetum (J.Presl & C.Presl) Nees = ***Cyperus laetus*** J.Presl & C.Presl
Comostemum montevidense (Link) Nees = ***Androtrichum trigynum*** (Spreng.) H.Pfeiff.
Comostemum prolixum (Kunth) Nees = ***Cyperus prolixus*** Kunth
Comostemum schottii Nees = ***Cyperus prolixus*** Kunth

Costularia

Costularia C.B.Clarke in W.H.Harvey & auct. suc. (eds.), Fl. Cap. 7: 274 (1898).
W. Indian Ocean, Borneo to New Caledonia. 29 MAU MDG REU SEY 42 BOR MOL? 43 NWG 60 NWC.
24 Species
Cyclocampe Benth. & Hook.f., Gen. Pl. 3: 1063 (1883), nom. illeg.
Lophoschoenus Stapf, J. Linn. Soc., Bot. 42: 179 (1914).

Costularia arundinacea (Sol. ex Vahl) Kük., Repert. Spec. Nov. Regni Veg. 44: 185 (1938).
New Caledonia. 60 NWC. Hemicr. or rhizome geophyte.
**Schoenus arundinaceus* Sol. ex Vahl, Enum. Pl. 2: 220 (1805). *Chaetospora arundinacea* (Sol. ex Vahl) A.Dietr., Sp. Pl. 2: 31 (1832). *Carpha arundinacea* (Sol. ex Vahl) Brongn. in L.I.Duperrey, Voy. Monde, Phan.: 169 (1833). *Asterochaete arundinacea* (Sol. ex Vahl) Kunth, Enum. Pl. 2: 312

(1837). *Cyclocampe arundinacea* (Sol. ex Vahl) Benth. & Hook.f., Gen. Pl. 3: 1063 (1883). *Lophoschoenus arundinaceus* (Sol. ex Vahl) Stapf, J. Linn. Soc., Bot. 42: 180 (1914). *Tetraria arundinacea* (Sol. ex Vahl) T.Koyama, J. Fac. Sci. Univ. Tokyo, Sect. 3, Bot. 8: 74 (1961).
Schoenus radiatus J.R.Forst. ex Vahl, Enum. Pl. 2: 220 (1805).
Costularia daenikeri Kük., Repert. Spec. Nov. Regni Veg. 46: 31 (1939). *Tetraria daenikeri* (Kük.) T.Koyama, J. Fac. Sci. Univ. Tokyo, Sect. 3, Bot. 8: 74 (1961).

Costularia brevifolia Cherm., Bull. Soc. Bot. France 69: 723 (1922 publ. 1923). *Tetraria brevifolia* (Cherm.) T.Koyama, J. Fac. Sci. Univ. Tokyo, Sect. 3, Bot. 8: 74 (1961).
EC. & E. Madagascar. 29 MDG. Cham.

Costularia breviseta J.Raynal, Adansonia, n.s., 14: 363 (1974).
E. New Caledonia. 60 NWC. Hemicr. or rhizome geophyte.

Costularia chamaedendron (Guillaumin) Kük., Repert. Spec. Nov. Regni Veg. 46: 24 (1939).
EC. New Caledonia. 60 NWC. Cham.
**Lophoschoenus chamaedendron* Guillaumin, Bull. Mus. Natl. Hist. Nat., II, 6: 199 (1934). *Tetraria chamaedendron* (Guillaumin) T.Koyama, J. Fac. Sci. Univ. Tokyo, Sect. 3, Bot. 8: 74 (1961).

Costularia comosa (C.B.Clarke) Kük., Repert. Spec. Nov. Regni Veg. 44: 186 (1938).
New Caledonia. 60 NWC. Hemicr. or rhizome geophyte.
**Schoenus comosus* C.B.Clarke, Kew Bull., Addit. Ser. 8: 44 (1908). *Lophoschoenus conosus* (C.B.Clarke) Stapf, J. Linn. Soc., Bot. 42: 180 (1914). *Tetraria comosa* (C.B.Clarke) T.Koyama, J. Fac. Sci. Univ. Tokyo, Sect. 3, Bot. 8: 74 (1961).

Costularia elongata (Kunth) Kük., Repert. Spec. Nov. Regni Veg. 44: 187 (1938).
Mascarenes. 29 MAU REU. Hemicr.
Carpha aubertii Nees, Linnaea 9: 300 (1834), nom. inval. *Cyclocampe aubertii* Pax in H.G.A.Engler & K.A.E.Prantl, Nat. Pflanzenfam. 2(2): 115 (1887).
**Asterochaete elongata* Kunth, Enum. Pl. 2: 312 (1837). *Carpha elongata* (Kunth) Boeck., Linnaea 38: 273 (1874). *Cyclocampe elongata* (Kunth) Benth. & Hook.f., Gen. Pl. 3: 1063 (1883). *Lophoschoenus elongatus* (Kunth) H.Pfeiff., Beih. Bot. Centralbl. 44(1): 133 (1927). *Tetraria elongata* (Kunth) T.Koyama, J. Fac. Sci. Univ. Tokyo, Sect. 3, Bot. 8: 74 (1961).
Schoenus elongatus Willd. ex Kunth, Enum. Pl. 2: 312 (1837), nom. inval.
Cladium xipholepis Baker, Fl. Mauritius: 424 (1877). *Schoenus xipholepis* (Baker) Summerh., Bull. Misc. Inform. Kew 1928: 394 (1928).

Costularia fragilis (Däniker) Kük., Repert. Spec. Nov. Regni Veg. 46: 23 (1939).
SE. New Caledonia. 60 NWC. Cham.
**Lophoschoenus fragilis* Däniker, Vierteljahrsschr. Naturf. Ges. Zürich 77(19): 67 (1932). *Tetraria fragilis* (Däniker) T.Koyama, J. Fac. Sci. Univ. Tokyo, Sect. 3, Bot. 8: 75 (1961).

Costularia hornei (C.B.Clarke) Kük., Repert. Spec. Nov. Regni Veg. 44: 189 (1938).
Seychelles. 29 SEY. Hemicr.
**Schoenus hornei* C.B.Clarke in T.A.Durand & H.Schinz, Consp. Fl. Afric. 5: 657 (1894). *Lophoschoenus hornei* (C.B.Clarke) Stapf, J. Linn. Soc., Bot. 42: 181 (1914). *Tetraria hornei* (C.B.Clarke) T.Koyama, J. Fac. Sci. Univ. Tokyo, Sect. 3, Bot. 8: 75 (1961).

Costularia humbertii Bosser, Naturaliste Malgache 7: 121 (1955).
Madagascar. 29 MDG. Hemicr. or rhizome geophyte.

Costularia laxa Cherm., Bull. Soc. Bot. France 69: 723 (1922 publ. 1923). *Tetraria laxa* (Cherm.) T.Koyama, J. Fac. Sci. Univ. Tokyo, Sect. 3, Bot. 8: 75 (1961).
Madagascar. 29 MDG. Hemicr. or rhizome geophyte.
Costularia laxa var. *macrantha* Cherm., Bull. Soc. Bot. France 72: 21 (1925).

Costularia leucocarpa (Ridl.) H.Pfeiff., Repert. Spec. Nov. Regni Veg. 23: 346 (1927).
Madagascar. 29 MDG. Hemicr. or rhizome geophyte.
**Rhynchospora leucocarpa* Ridl., J. Linn. Soc., Bot. 20: 335 (1883). *Costularia recurva* C.B.Clarke in T.A.Durand & H.Schinz, Consp. Fl. Afric. 5: 658 (1894). *Tetraria leucocarpa* (Ridl.) T.Koyama, J. Fac. Sci. Univ. Tokyo, Sect. 3, Bot. 8: 75 (1961).
Cladium fimbristyloides Baker, J. Linn. Soc., Bot. 22: 531 (1887).

Costularia melicoides (Poir.) C.B.Clarke, Kew Bull., Addit. Ser. 8: 48 (1908).
Mascarenes. 29 MAU REU. Hemicr. or rhizome geophyte.
**Cyperus melicoides* Poir. in J.B.A.M.de Lamarck, Encycl. 7: 273 (1806). *Machaerina melicoides* (Poir.) Bojer, Hortus Maurit.: 386 (1837).

Costularia melleri (Baker) C.B.Clarke in T.A.Durand & H.Schinz, Consp. Fl. Afric. 5: 658 (1894).
C. Madagascar. 29 MDG. Hemicr. or rhizome geophyte.
**Cladium melleri* Baker, J. Linn. Soc., Bot. 21: 451 (1885). *Machaerina melleri* (Baker) T.Koyama, Bot. Mag. (Tokyo) 69: 64 (1956). *Tetraria melleri* (Baker) T.Koyama, J. Fac. Sci. Univ. Tokyo, Sect. 3, Bot. 8: 75 (1961).

Costularia microcarpa (Cherm.) Kük., Repert. Spec. Nov. Regni Veg. 46: 69 (1939).
C. Madagascar (Isalo). 29 MDG. Hemicr. or rhizome geophyte.
**Costularia baronii* var. *microcarpa* Cherm., Bull. Soc. Bot. France 72: 617 (1925).

Costularia neocaledonica (C.B.Clarke) Rendle, J. Linn. Soc., Bot. 45: 261 (1921).
SE. New Caledonia. 60 NWC. Hemicr. or rhizome geophyte.
Schoenus juvenis C.B.Clarke, Bot. Jahrb. Syst. 39: 25 (1906).
**Schoenus neocaledonicus* C.B.Clarke, Bot. Jahrb. Syst. 39: 25 (1906). *Lophoschoenus neocaledonicus* (C.B.Clarke) H.Pfeiff., Beih. Bot. Centralbl. 44(1): 133 (1927).
Lophoschoenus montis-fontium Guillaumin, Mém. Mus. Natl. Hist. Nat., Sér. B, Bot. 8: 31 (1957).

Costularia nervosa J.Raynal, Adansonia, n.s., 14: 366 (1974).
New Caledonia. 60 NWC. Hemicr. or rhizome geophyte.

Costularia pantopoda (Baker) C.B.Clarke in T.A.Durand & H.Schinz, Consp. Fl. Afric. 5: 658 (1894).
Madagascar. 29 MDG. Hemicr.
**Cladium pantopodum* Baker, J. Linn. Soc., Bot. 21: 451 (1885). *Mariscus pantopodus* (Baker) Fernald, Rhodora 25: 54 (1923). *Machaerina pantopoda* (Baker) T.Koyama, Bot. Mag. (Tokyo) 69: 65 (1956). *Tetraria pantopoda* (Baker) T.Koyama, J. Fac. Sci. Univ. Tokyo, Sect. 3, Bot. 8: 75 (1961).

var. ***baronii*** (C.B.Clarke) Kük., Repert. Spec. Nov. Regni Veg. 41: 67 (1939).
WC. & C. Madagascar. 29 MDG. Hemicr. or rhizome geophyte.
**Costularia baronii* C.B.Clarke in W.H.Harvey & auct. suc. (eds.), Fl. Cap. 7: 274 (1898).

var. ***gracilescens*** Kük., Repert. Spec. Nov. Regni Veg. 41: 67 (1939).
Madagascar. 29 MDG. Hemicr. or rhizome geophyte.

subsp. ***pantopoda***
C. Madagascar. 29 MDG. Hemicr.

var. ***robusta*** (Cherm.) Kük., Repert. Spec. Nov. Regni Veg. 41: 68 (1939).
C. Madagascar. 29 MDG. Hemicr. or rhizome geophyte.
**Costularia baronii* var. *robusta* Cherm., Bull. Soc. Bot. France 67: 723 (1922).

Costularia pilisepala (Steud.) F.Kern, in Fl. Males. 5: 420 (1957).
N. Borneo, W. New Guinea (incl. Waigiou I.). 42 BOR MOL? 43 NWG. Hemicr. or rhizome geophyte.
**Restio pilisepalus* Steud., Syn. Pl. Glumac. 2: 256 (1855). *Caustis pilisepala* (Steud.) H.Pfieff., Repert. Spec. Nov. Regni Veg. 21: 366 (1925).
Carpha urvilleana Gaudich. ex Boeck., Linnaea 38: 272 (1874). *Lophoschoenus urvilleanus* (Gaudich. ex Boeck.) Stapf, J. Linn. Soc., Bot. 42: 180 (1914). *Costularia urvilleana* (Gaudich. ex Boeck.) Kük., Repert. Spec. Nov. Regni Veg. 46: 28 (1939). *Tetraria urvilleana* (Gaudich. ex Boeck.) T.Koyama, J. Fac. Sci. Univ. Tokyo, Sect. 3, Bot. 8: 75 (1961).

Costularia pubescens J.Raynal, Adansonia, n.s., 14: 363 (1974).
New Caledonia. 60 NWC. Hemicr. or rhizome geophyte.
Costularia arundinacea var. *crassicaudex* Kük., Repert. Spec. Nov. Regni Veg. 46: 65 (1939).

Costularia purpurea Cherm., Bull. Soc. Bot. France 69: 722 (1922 publ. 1923). *Tetraria purpurea* (Cherm.) T.Koyama, J. Fac. Sci. Univ. Tokyo, Sect. 3, Bot. 8: 75 (1961).
C. Madagascar. 29 MDG. Hemicr. or rhizome geophyte.

Costularia setacea J.Raynal, Adansonia, n.s., 14: 366 (1974).
SE. New Caledonia. 60 NWC. Hemicr. or rhizome geophyte.

Costularia stagnalis (Däniker) Kük., Repert. Spec. Nov. Regni Veg. 46: 30 (1939).
SE. New Caledonia. 60 NWC. Hemicr. or rhizome geophyte.
**Lophoschoenus stagnalis* Däniker, Vierteljahrsschr. Naturf. Ges. Zürich 77(19): 79 (1932). *Tetraria stagnalis* (Däniker) T.Koyama, J. Fac. Sci. Univ. Tokyo, Sect. 3, Bot. 8: 75 (1961).

Costularia sylvestris J.Raynal, Adansonia, n.s., 14: 361 (1974).
New Caledonia. 60 NWC. Hemicr. or rhizome geophyte.
Costularia arundinacea var. *perlaxiflora* Kük., Repert. Spec. Nov. Regni Veg. 46: 65 (1939).

Costularia xyridioides (Däniker) Kük., Repert. Spec. Nov. Regni Veg. 4: 193 (1938).
SE. New Caledonia. 60 NWC. Hemicr. or rhizome geophyte.
Lophoschoenus falcifolius Däniker, Vierteljahrsschr. Naturf. Ges. Zürich 77(19): 77 (1932). *Costularia falcifolia* (Däniker) Kük., Repert. Spec. Nov. Regni Veg. 46: 26 (1939). *Tetraria falcifolia* (Däniker) T.Koyama, J. Fac. Sci. Univ. Tokyo, Sect. 3, Bot. 8: 75 (1961).
**Schoenus xyridioides* Däniker, Vierteljahrsschr. Naturf. Ges. Zürich 77(19): 81 (1932). *Lophoschoenus xyridioides* (Däniker) Guillaumin, Bull. Mus. Natl. Hist. Nat., II, 6: 200 (1934). *Tetraria xyridioides* (Däniker) T.Koyama, J. Fac. Sci. Univ. Tokyo, Sect. 3, Bot. 8: 75 (1961).

***Synonyms*:**
Costularia arundinacea var. *crassicaudex* Kük. = ***Costularia pubescens*** J.Raynal
Costularia arundinacea var. *perlaxiflora* Kük. = ***Costularia sylvestris*** J.Raynal
Costularia baronii C.B.Clarke = ***Costularia pantopoda*** var. ***baronii*** (C.B.Clarke) Kük.
Costularia baronii var. *microcarpa* Cherm. = ***Costularia microcarpa*** (Cherm.) Kük.
Costularia baronii var. *robusta* Cherm. = ***Costularia pantopoda*** var. ***robusta*** (Cherm.) Kük.
Costularia brevicaulis C.B.Clarke = ***Capeobolus brevicaulis*** (C.B.Clarke) Browning
Costularia daenikeri Kük. = ***Costularia arundinacea*** (Sol. ex Vahl) Kük.
Costularia falcifolia (Däniker) Kük. = ***Costularia xyridioides*** (Däniker) Kük.
Costularia laxa var. *macrantha* Cherm. = ***Costularia laxa*** Cherm.
Costularia natalensis C.B.Clarke = ***Tetraria natalensis*** (C.B.Clarke) T.Koyama
Costularia paludosa (R.Br.) C.B.Clarke = ***Schoenus paludosus*** (R.Br.) Roem. & Schult.
Costularia recurva C.B.Clarke = ***Costularia leucocarpa*** (Ridl.) H.Pfeiff.
Costularia urvilleana (Gaudich. ex Boeck.) Kük. = ***Costularia pilisepala*** (Steud.) F.Kern
Costularia yarrabensis (Domin) H.Pfeiff. = ***Schoenus yarrabensis*** Domin

Courtoisia

Courtoisia Nees = ***Courtoisina*** Soják
Courtoisia assimilis (Steud.) C.B.Clarke = ***Courtoisina assimilis*** (Steud.) Maquet
Courtoisia cyperoides (Roxb.) Nees = ***Courtoisina cyperoides*** (Roxb.) Soják
Courtoisia cyperoides var. *africana* C.B.Clarke = ***Courtoisina cyperoides*** (Roxb.) Soják
Courtoisia olivacea Boeck. = [82 SUR] Perhaps identical with Oxycaryum cubense.

***Unplaced Names*:**
Courtoisia olivacea Boeck., Flora 44: 331 (1861). = [82 SUR] Perhaps identical with Oxycaryum cubense.

Courtoisina

Courtoisina Soják, Cas. Nár. Mus., Odd. Prír. 148: 193 (1979 publ. 1980).
Ethiopia to S. Africa and Indo-China. 23 RWA ZAI 24 CHA ETH 25 KEN TAN UGA 26 MLW ZIM 27 BOT CPP NAM NAT OFS SWZ TVL 29 MDG 36 CHC 40 ASS EHM IND NEP 41 MYA THA VIE.
2 Species
**Courtoisia* Nees, Linnaea 9: 286 (1834), nom. illeg. *Indocourtoisia* Bennet & Raizada, Indian Forester 107: 432 (1981), nom. illeg. *Pseudomariscus* Rauschert, Taxon 31: 559 (1982), nom. illeg.

Courtoisina assimilis (Steud.) Maquet, Bull. Jard. Bot. Natl. Belg. 58: 265 (1988).
Ethiopia to S. Africa, Madagascar. 23 RWA ZAI 24 ETH 25 KEN TAN UGA 26 MLW ZIM 27 BOT CPP NAM OFS TVL 29 MDG. Ther.
**Cyperus assimilis* Steud., Flora 25: 584 (1842). *Courtoisia assimilis* (Steud.) C.B.Clarke in T.A.Durand & H.Schinz, Consp. Fl. Afric. 5: 596 (1894). *Mariscus assimilis* (Steud.) Podlech, Mitt. Bot. Staatssamml. München 3: 523 (1960). *Indocourtoisia assimilis* (Steud.) Bennet & Raizada, Indian Forester 107: 432 (1981).

Courtoisina cyperoides (Roxb.) Soják, Cas. Nár. Mus., Odd. Prír. 148: 193 (1979 publ. 1980).
Chad to S. Africa, Madagascar, Himalaya to Indo-China. 24 CHA 25 KEN TAN 26 MLW ZIM 27 BOT NAM NAT SWZ TVL 29 MDG 36 CHC 40 ASS EHM IND NEP 41 MYA THA VIE. Ther.
**Kyllinga cyperoides* Roxb., Fl. Ind. 1: 182 (1820). *Mariscus cyperoides* (Roxb.) A.Dietr., Sp. Pl. 2: 348 (1832). *Courtoisia cyperoides* (Roxb.) Nees, Linnaea 9: 286 (1834). *Cyperus pseudokyllingioides* Kük. in H.G.A.Engler (ed.), Pflanzenr., IV, 20(101): 14 (1935). *Indocourtoisia cyperoides* (Roxb.) Bennet & Raizada, Indian Forester 107: 432 (1981). *Pseudomariscus cyperoides* (Roxb.) Rauschert, Taxon 31: 559 (1982).
Cyperus kleinianus Hochst. ex Steud., Syn. Pl. Glumac. 2: 71 (1854).
Courtoisia cyperoides var. *africana* C.B.Clarke in T.A.Durand & H.Schinz, Consp. Fl. Afric. 5: 596 (1894), nom. inval. *Cyperus pseudokyllingioides* var. *africanus* (C.B.Clarke) Kük. in H.G.A.Engler (ed.), Pflanzenr., IV, 20(101): 501 (1936).

Crepidocarpus

Crepidocarpus Klotzsch ex Boeck = ***Oxycaryum*** Nees
Crepidocarpus cubensis (Poepp. & Kunth) Klotzsch ex Boeck. = ***Oxycaryum cubense*** (Poepp. & Kunth) Palla
Crepidocarpus schinzii Klotzsch ex Boeck. = ***Oxycaryum cubense*** (Poepp. & Kunth) Palla

Crosslandia

Crosslandia W.Fitzg., J. Proc. Roy. Soc. Western Australia 3: 122 (1918).
N. Australia. 50 NTA WAU.
1 Species

Crosslandia setifolia W.Fitzg., J. Proc. Roy. Soc. Western Australia 3: 122 (1918).
N. Western Australia to N. Northern Territory. 50 NTA WAU. Hemicr.

Cryptangium

Cryptangium Schrad. ex Nees = ***Lagenocarpus*** Nees
Cryptangium arundinaceum Boeck. = ***Lagenocarpus rigidus*** (Kunth) Nees subsp. ***rigidus***
Cryptangium brevifolium Boeck. = ***Lagenocarpus rigidus*** subsp. ***tenuifolius*** (Boeck.) T.Koyama & Maguire
Cryptangium campestre (Nees) Boeck. = ***Lagenocarpus minarum*** (Nees) Kuntze
Cryptangium ciliatum Boeck. = ***Lagenocarpus verticillatus*** (Spreng.) T.Koyama & Maguire
Cryptangium claussenii C.B.Clarke = ***Lagenocarpus minarum*** (Nees) Kuntze
Cryptangium comatum Boeck. = ***Lagenocarpus comatus*** (Boeck.) H.Pfeiff.
Cryptangium densifolium (Nees) Boeck. = ***Lagenocarpus minarum*** (Nees) Kuntze
Cryptangium dioicum Boeck. = ***Lagenocarpus rigidus*** (Kunth) Nees subsp. ***rigidus***
Cryptangium distichophyllum Boeck. = ***Lagenocarpus distichophyllus*** (Boeck.) H.Pfeiff.
Cryptangium dracaenula (Nees) Boeck. = ***Cephalocarpus dracaenula*** Nees
Cryptangium glaziovii Boeck. = ***Lagenocarpus minarum*** (Nees) Kuntze
Cryptangium griseum Boeck. = ***Lagenocarpus griseus*** (Boeck.) H.Pfeiff.
Cryptangium humile (Nees) Boeck. = ***Lagenocarpus humilis*** (Nees) Kuntze
Cryptangium insigne Boeck. = ***Lagenocarpus alboniger*** (A.St.-Hil.) C.B.Clarke
Cryptangium junciforme (Kunth) Boeck. = ***Lagenocarpus junciformis*** (Kunth) Kuntze
Cryptangium kuntzeanum Boeck. = ***Lagenocarpus rigidus*** (Kunth) Nees subsp. ***rigidus***
Cryptangium leptocladium (Poepp. ex Kunth) Boeck. = ***Lagenocarpus verticillatus*** (Spreng.) T.Koyama & Maguire
Cryptangium leptocladium var. *longifolium* Boeck. = ***Lagenocarpus verticillatus*** (Spreng.) T.Koyama & Maguire
Cryptangium melanocarpum Boeck. = ***Lagenocarpus minarum*** (Nees) Kuntze
Cryptangium minarum (Nees) Boeck. = ***Lagenocarpus minarum*** (Nees) Kuntze
Cryptangium nudipes C.B.Clarke = ***Lagenocarpus junciformis*** (Kunth) Kuntze
Cryptangium parvulum C.B.Clarke = ***Lagenocarpus parvulus*** (C.B.Clarke) H.Pfeiff.
Cryptangium pauciflorum Boeck. = ***Lagenocarpus junciformis*** (Kunth) Kuntze
Cryptangium paucifolium Boeck. = ***Lagenocarpus minarum*** (Nees) Kuntze
Cryptangium polyphyllum (Nees) Boeck. = ***Lagenocarpus polyphyllus*** (Nees) Kuntze
Cryptangium schenckianum Boeck. = ***Lagenocarpus triqueter*** (Boeck.) Kuntze
Cryptangium stellatum Boeck. = ***Lagenocarpus stellatus*** (Boeck.) Kuntze
Cryptangium strictum C.B.Clarke = ***Lagenocarpus clarkei*** H.Pfeiff.
Cryptangium strictum (Kunth) H.Pfeiff. = ***Lagenocarpus verticillatus*** (Spreng.) T.Koyama & Maguire
Cryptangium tenuifolium Boeck. = ***Lagenocarpus rigidus*** subsp. ***tenuifolius*** (Boeck.) T.Koyama & Maguire
Cryptangium tremulum (Nees) Boeck. = ***Lagenocarpus rigidus*** (Kunth) Nees subsp. ***rigidus***

Cryptangium triquetrum Boeck. = ***Lagenocarpus triqueter*** (Boeck.) Kuntze
Cryptangium uliginosum Schrad. ex Nees = ***Lagenocarpus verticillatus*** (Spreng.) T.Koyama & Maguire
Cryptangium uliginosum f. *humilis* C.B.Clarke = ***Lagenocarpus verticillatus*** (Spreng.) T.Koyama & Maguire

Cryptoglochin

Cryptoglochin Heuff. = ***Carex*** L.
Cryptoglochin curvulus (All.) Heuff. = ***Carex curvula*** All.

Cryptopodium

Cryptopodium Schrad. ex Nees = ***Scleria*** P.J.Bergius

Cyathochaeta

Cyathochaeta Nees in J.G.C.Lehmann, Pl. Preiss. 2: 86 (1846).
SW. & E. Australia. 50 NSW QLD VIC WAU.
6 Species
Tetralepis Steud., Syn. Pl. Glumac. 2: 159 (1855).

Cyathochaeta aequitans K.L.Wilson, Nuytsia 11: 269 (1997).
SW. Australia. 50 WAU. Hemicr.

Cyathochaeta avenacea (R.Br.) Benth., Fl. Austral. 7: 351 (1878).
SW. Australia. 50 WAU. Hemicr. or rhizome geophyte.
**Carpha avenacea* R.Br., Prodr.: 230 (1810). *Rhynchospora avenacea* (R.Br.) Spreng., Syst. Veg. 1: 197 (1824). *Chaetospora avenacea* (R.Br.) F.Muell., Fragm. 9: 40 (1875).
Tetralepis australis Steud., Syn. Pl. Glumac. 2: 159 (1855). *Cyathochaeta australis* (Steud.) C.A. Gardner, Enum. Pl. Austr. Occ.: 14 (1930).

Cyathochaeta clandestina (R.Br.) Benth., Fl. Austral. 7: 351 (1878).
SW. Australia. 50 WAU. Hemicr. or rhizome geophyte.
**Carpha clandestina* R.Br., Prodr.: 231 (1810). *Rhynchospora clandestina* (R.Br.) Spreng., Syst. Veg. 1: 194 (1824). *Chaetospora clandestina* (R.Br.) F.Muell., Fragm. 9: 40 (1875).

Cyathochaeta diandra (R.Br.) Nees in J.G.C.Lehmann, Pl. Preiss. 2: 86 (1846).
E. & SE. Australia. 50 NSW QLD VIC. Hemicr. or rhizome geophyte.
**Carpha diandra* R.Br., Prodr.: 231 (1810). *Rhynchospora diandra* (R.Br.) Spreng., Syst. Veg. 1: 197 (1824). *Chaetospora diandra* (R.Br.) F.Muell., Fragm. 9: 39 (1875).
Cyathochaeta teretifolia W.Fitzg., Proc. Linn. Soc. New South Wales 28: 109 (1903).

Cyathochaeta equitans K.L.Wilson, Nuytsia 11: 269 (1997).
SW. Australia. 50 WAU. Hemicr. or rhizome geophyte.

Cyathochaeta stipoides K.L.Wilson, Nuytsia 11: 271 (1997).
SW. Australia. 50 WAU. Hemicr.

Synonyms:
Cyathochaeta australis (Steud.) C.A.Gardner = ***Cyathochaeta avenacea*** (R.Br.) Benth.
Cyathochaeta teretifolia W.Fitzg. = ***Cyathochaeta diandra*** (R.Br.) Nees

Cyathocoma

Cyathocoma Nees, Linnaea 9: 300 (1834).
S. Africa. 27 CPP NAT.
3 Species

Cyathocoma bachmannii (Kük.) C.Archer, S. African J. Bot. 63: 170 (1997).
SE. Cape Prov. to KwaZulu-Natal. 27 CPP NAT. Hemicr. or rhizome geophyte.
Tetraria thuarii var. *gracilior* C.B.Clarke in W.H.Harvey & auct. suc. (eds.), Fl. Cap. 7: 289 (1898).

Cyathocoma ecklonii Nees, Linnaea 10: 195 (1836).
Macrochaetium ecklonii (Nees) Levyns, J. S. African Bot. 13: 77 (1947).
W. Cape Prov. 27 CPP. Hemicr. or rhizome geophyte.

Cyathocoma hexandra (Nees) Browning, S. African J. Bot. 62: 250 (1996).
SW. & S. Cape Prov. 27 CPP. Hemicr. or rhizome geophyte.
**Carpha hexandra* Nees, Linnaea 10: 193 (1836). *Ideleria neesii* Kunth, Enum. Pl. 2: 311 (1837), nom. illeg. *Macrochaetium neesii* (Kunth) Kük., Repert. Spec. Nov. Regni Veg. 29: 194 (1931). *Cyathocoma neesii* Boeck., Linnaea 38: 312 (1874), nom. illeg. *Ideleria hexandra* (Nees) H.Pfeiff., Beih. Bot. Centralbl. 44(1): 142 (1927). *Macrochaetium hexandrum* (Nees) H.Pfeiff., Repert. Spec. Nov. Regni Veg. 29: 186 (1931). *Tetraria hexandra* (Nees) Kük., Repert. Spec. Nov. Regni Veg. 47: 211 (1939).
Ideleria capensis Kunth, Enum. Pl. 2: 311 (1837).
Elynanthus kraussii Hochst., Flora 28: 761 (1845).
Macrochaetium dregei Steud., Syn. Pl. Glumac. 2: 159 (1855).

Synonyms:
Cyathocoma neesii Boeck. = ***Cyathocoma hexandra*** (Nees) Browning
Cyathocoma nigrovaginata Nees = ***Tetraria nigrovaginata*** (Nees) C.B.Clarke

Cyclocampe

Cyclocampe Benth. & Hook.f. = ***Costularia*** C.B.Clarke
Cyclocampe Steud. = ***Schoenus*** L.
Cyclocampe arundinacea (Sol. ex Vahl) Benth. & Hook.f. = ***Costularia arundinacea*** (Sol. ex Vahl) Kük.
Cyclocampe aubertii Pax = ***Costularia elongata*** (Kunth) Kük.
Cyclocampe elongata (Kunth) Benth. & Hook.f. = ***Costularia elongata*** (Kunth) Kük.
Cyclocampe nitens (Kunth) Hook.f. = ***Carpha nitens*** (Kunth) Kük.
Cyclocampe waigiouensis Steud. = ***Schoenus calostachyus*** (R.Br.) Poir.

Cylindrolepis

Cylindrolepis Boeck. = ***Cyperus*** L.
Cylindrolepis rehmanniana Boeck. = ***Cyperus vestitus*** Hochst. ex C.Krauss

Cylindropus

Cylindropus Nees = ***Scleria*** P.J.Bergius
Cylindropus junciformis Nees = ***Scleria pilosa*** Boeck.

Cymophyllus

Cymophyllus Mack. ex Britton & A.Br., Ill. Fl. N. U.S., ed. 2, 1: 441 (1913).
EC. U.S.A. 75 PEN WVA 78 GEO KTY MRY NCA SCA TEN VRG.
1 Species

Cymophyllus fraserianus (Ker Gawl.) Kartesz & Gandhi, Rhodora 93: 138 (1991).
EC. U.S.A. (Appalachian Mts.). 75 PEN WVA 78 GEO KTY MRY NCA SCA TEN VRG. Hemicr.
Carex fraseri Andrews, Bot. Repos. 10: t. 639 (1811). *Olamblis fraseri* (Andrews) Raf., Good Book: 26 (1840). *Cymophyllus fraseri* (Andrews) Mack. in N.L.Britton & A.Brown, Ill. Fl. N. U.S., ed. 2, 1: 441 (1913).
**Carex fraseriana* Ker Gawl., Bot. Mag. 33: t. 1391 (1811).
Mapania sylvatica Pursh, Fl. Amer. Sept. 1: 47 (1813), nom. illeg.
Carex lagopus Muhl., Descr. Gram.: 265 (1817).

Synonyms:
Cymophyllus fraseri (Andrews) Mack. = ***Cymophyllus fraserianus*** (Ker Gawl.) Kartesz & Gandhi

Cyperoides

Cyperoides Ség. = ***Carex*** L.

Cyperus

Cyperus L., Sp. Pl.: 44 (1753).
Cosmopolitan. 10 DEN GRB SWE 11 AUT BGM CZE GER HUN NET POL SWI 12 BAL COR FRA POR SAR SPA 13 ALB BUL GRC ITA KRI ROM SIC TUE YUG 14 BLR BLT KRY RUC RUE RUN RUS RUW UKR 20 ALG EGY LBY MOR TUN WSA 21 AZO CNY CVI MDR SEL 22 BEN BKN GAM GHA GNB GUI IVO LBR MLI MTN NGA NGR SEN SIE TOG 23 BUR CAB CAF CMN CON EQG GAB GGI RWA ZAI 24 CHA DJI ERI ETH SOC SOM SUD 25 KEN TAN TEN UGA 26 ANG MLW MOZ ZAM ZIM 27 BOT CPP CPV LES MLW NAM NAT OFS SWZ TVL 28 ASC STH 29 ALD CGS COM MAU MCI MDG REU ROD SEY 30 ALT WSB 31 AMU PRM 32 KAZ KGZ TKM TZK UZB 33 AFG NCS TCS 34 AFG CYP EAI IRN IRQ LBS PAL SIN TUR 35 GST KUW OMA SAU YEM 36 CHC CHH CHI CHM CHN CHS 37 MON 38 JAP KOR KZN NNS OGA TAI 40 ASS BAN EHM IND NEP PAK SRL WHM 41 AND CBD LAO MYA NCB SCS THA VIE 42 BOR CKI JAW LSI MLY MOL PHI SUL SUM XMS 43 BIS NWG SOL 50 NFK NSW NTA QLD SOA TAS VIC WAU 51 KER NZN 60 FIJ GIL NUE NWC SAM TON VAN 61 COO EAS MRQ PIT SCI TUA TUB 62 CRL MRN MRS WAK 63 HAW 71 ABT BRC MAN SAS 72 NBR NSC ONT QUE 73 COL IDA MNT ORE WAS WYO 74 ILL IOW KAN MIN MSO NDA NEB OKL SDA WIS 75 CNT INI MAI MAS MIC NWH NWJ NWY OHI PEN RHO VER WVA 76 ARI CAL NEV UTA 77 NWM TEX 78 ALA ARK DEL FLA GEO KTY LOU MRY MSI NCA SCA TEN VRG WDC 79 MXC MXE MXG MXI MXN MXS MXT 80 BLZ COS CPI ELS GUA HON NIC PAN 81 ARU BAH BER CAY CUB DOM HAI JAM LEE NLA PUE TRT WIN 82 FRG GUY SUR VEN 83 BOL CLM ECU GAL PER 84 AGW BZC BZE BZL BZN BZS PAR 85 AGE AGS AGW CLC CLN JNF PAR URU (90) tdc.
694 Species
Mariscus Gaertn., Fruct. Sem. Pl. 1: 11 (1788).
Opetiola Gaertn., Fruct. Sem. Pl. 1: 14 (1788).
Adupla Bosc in J.H.Jaume Saint-Hilaire, Expos. Fam. Nat. 1: 65 (1805).
Mariscus Vahl, Enum. Pl. 2: 372 (1805), nom. cons.
Papyrus Willd., Abh. Königl. Akad. Wiss. Berlin 1812–1813: 70 (1812).
Epiphystis Trin., Fund. Agrost.: 97 (1820).
Torulinium Desv. in W.Hamilton, Prodr. Pl. Ind. Occid.: 15 (1825).
Anosporum Nees, Linnaea 9: 287 (1834).
Diclidium Schrad. ex Nees in C.F.P.von Martius & auct. suc. (eds.), Fl. Bras. 2(1): 51 (1842).
Galilea Parl., Fl. Palerm. 1: 297 (1845).
Hydroschoenus Zoll. & Moritzi, Syst. Verz.: 95 (1846).
Atomostylis Steud., Flora 33: 229 (1850).
Pterocyperus Opiz, Seznam: 80 (1852).
Sorostachys Steud., Syn. Pl. Glumac. 2: 71 (1854).
Borabora Steud., Syn. Pl. Glumac. 2: 71 (1855).
Trentepohlia Boeck., Bot. Zeitung (Berlin) 16: 249 (1858).
Ungeria Nees ex C.B.Clarke, J. Linn. Soc., Bot. 21: 34 (1884).
Didymia Phil., Bot. Jahrb. Syst. 8: 57 (1887).
Cylindrolepis Boeck., Bot. Centralbl. 39: 75 (1889).
Iria Kuntze, Revis. Gen. Pl. 2: 751 (1891).
Juncellus C.B.Clarke in J.D.Hooker, Fl. Brit. India 7: 594 (1893).
Chlorocyperus Rikli, Jahrb. Wiss. Bot. 27: 563 (1895).
Eucyperus Rikli, Jahrb. Wiss. Bot. 27: 568 (1895).
Acorellus Palla ex Kneuck., Allg. Bot. Z. Syst. 9: 68 (1903).
Duval-jouvea Palla in W.D.J.Koch, Syn. Deut. Schweiz. Fl., ed. 3, 3: 2555 (1905).
Sphaeromariscus E.G.Camus, Notul. Syst. (Paris) 1: 239 (1910).
Ascopholis C.E.C.Fisch., Bull. Misc. Inform. Kew 1931: 104 (1931).

Cyperus acuminatus Torr. & Hook., Ann. Lyceum Nat. Hist. New York 3: 435 (1836).
U.S.A. to NE. Mexico. 73 COL MNT ORE WAS WYO 74 ILL IOW KAN MIN MSO NDA NEB OKL SDA 75 INI MIC nwh nwy OHI pen 76 ARI CAL UTA 77 TEX 78 ALA ARK KTY LOU NCA TEN VRG 79 MXE. Ther.
Cyperus cyrtolepis Torr. & Hook., Ann. Lyceum Nat. Hist. New York 3: 436 (1836). *Cyperus acuminatus* var. *cyrtolepis* (Torr. & Hook.) Kük. in H.G.A. Engler (ed.), Pflanzenr., IV, 20(101): 178 (1936).
Cyperus cyrtolepis var. *caepitosus* Boeck., Linnaea 35: 558 (1868).
Cyperus cyrtolepis var. *glaucescens* Boeck., Linnaea 35: 557 (1868).
Cyperus rufescens var. *denticarinatus* Britton, Bull. Torrey Bot. Club 11: 85 (1884). *Cyperus cyrtolepis* var. *denticarinatus* (Britton) Britton, Bull. Torrey Bot. Club 13: 209 (1886).

Cyperus afroalpinus Lye, Nordic J. Bot. 3: 226 (1983).
Zaïre to Kenya. 23 ZAI 25 KEN UGA. Hemicr. or rhizome geophyte.

Cyperus afrodunensis Lye, Nordic J. Bot. 3: 222 (1983).
S. Somalia to Coastal Kenya (incl. Kiunga Arch.). 24 SOM 25 KEN. Hemicr. or bulb geophyte.

Cyperus afroechinatus Lye, Nordic J. Bot. 3: 231 (1983).
W. Trop. Africa. 22 GHA. Hemicr. or rhizome geophyte.

Cyperus afromontanus Lye, Nordic J. Bot. 3: 225 (1983).
E. Trop. Africa. 25 TAN UGA. Hemicr. or rhizome geophyte.

Cyperus afrovaricus Lye, Nordic J. Bot. 3: 222 (1983).
Tanzania. 25 TAN. Hemicr. or rhizome geophyte.

Cyperus aggregatus (Willd.) Endl., Cat. Horti Vindob. 1: 93 (1842).
Trop. & Subtrop. America. (50) (73) ore (75) nwj pen 76 ARI 77 NWM TEX 78 LOU MSI 79 MXC MXE MXG MXN MXS MXT 80 BLZ COS ELS GUA HON NIC PAN 81 CUB DOM HAI JAM LEE PUE TRT WIN 82 FRG GUY SUR VEN 83 BOL CLM ECU PER 84 BZC BZE BZL BZN BZS 85 AGE AGW CLN PAR URU. Hemicr. or rhizome geophyte.
**Mariscus aggregatus* Willd., Enum. Pl.: 70 (1809). *Cyperus flavus* var. *aggregatus* (Willd.) Kük. in H.G.A.Engler (ed.), Pflanzenr., IV, 20(101): 532 (1936).

var. ***aggregatus***
Trop. & Subtrop. America. (50) (73) ore (75) nwj pen 76 ARI 77 NWM TEX 78 LOU MSI 79 MXC MXE MXG MXN MXS MXT 80 BLZ COS ELS GUA HON NIC PAN 81 CUB DOM HAI JAM LEE PUE TRT WIN 82 FRG GUY SUR VEN 83 BOL CLM ECU PER 84 BZC BZE BZL BZN BZS 85 AGE AGW CLN PAR URU. Hemicr. or rhizome geophyte.
Kyllinga cayennensis Lam., Tabl. Encycl. 1: 149 (1791). *Mariscus cayennensis* (Lam.) Urb., Symb. Antill. 2: 165 (1900). *Cyperus cayennensis* (Lam.) Britton, Bull. Dept. Agric. Jamaica 5(Suppl. 1): 8 (1907), nom. illeg.
Mariscus flavus Vahl, Enum. Pl. 2: 374 (1805). *Cyperus flavus* (Vahl) Nees, Linnaea 19: 698 (1847), nom. illeg.
Mariscus confertus Kunth in F.W.H.von Humboldt, A.J.A.Bonpland & C.S.Kunth, Nov. Gen. Sp. 1: 213 (1816).
Mariscus laevis Kunth in F.W.H.von Humboldt, A.J.A.Bonpland & C.S.Kunth, Nov. Gen. Sp. 1: 214 (1816). *Cyperus flavus* var. *laevis* (Kunth) Kük. in H.G.A.Engler (ed.), Pflanzenr., IV, 20(101): 531 (1936).
Mariscus sylvestris Kunth in F.W.H.von Humboldt, A.J.A.Bonpland & C.S.Kunth, Nov. Gen. Sp. 1: 377 (1816).
Mariscus laevigatus Roem. & Schult., Syst. Veg. 2: 243 (1817).
Mariscus littoralis G.Mey., Prim. Fl. Esseq.: 37 (1818).
Mariscus mexicanus Willd. ex Link, Jahrb. Gewächsk. 1(3): 90 (1820).
Kyllinga squarrosa Baldwin, Trans. Amer. Philos. Soc., n.s., 2: 170 (1825).
Mariscus rigidus Spreng., Syst. Veg. 4(2): 29 (1827).
Carex sieberi Nees ex Kunth, Enum. Pl. 2: 118 (1837).
Mariscus elatus Kunth, Enum. Pl. 2: 116 (1837).
Cyperus obesus Liebm., Mexic. Halvgr.: 211 (1850). *Cyperus flavus* var. *obesus* (Liebm.) Kük. in H.G.A.Engler (ed.), Pflanzenr., IV, 20(101): 532 (1936).
Cyperus bracteiferus Steud., Syn. Pl. Glumac. 2: 28 (1854).
Mariscus confertospicatus Steud., Syn. Pl. Glumac. 2: 64 (1854).
Mariscus toluccensis Steud., Syn. Pl. Glumac. 2: 65 (1854).
Cyperus flavomariscus Griseb., Fl. Brit. W. I.: 567 (1864).
Cyperus productus C.Wright, Anales Acad. Ci. Méd. Habana 8: 77 (1871).
Cyperus argentinus Boeck., Bot. Jahrb. Syst. 5: 502 (1884). *Cyperus flavus* var. *argentinus* (Boeck.) Kük. ex Osten, Anales Mus. Hist. Nat. Montevideo 3(2): 151 (1931).
Didymia cyperomorpha Phil., Bot. Jahrb. Syst. 8: t. 1 (1887).
Mariscus latibracteatus Palla, Oesterr. Bot. Z. 63: 404 (1913).
Cyperus flavus var. *angustatus* Kük., Repert. Spec. Nov. Regni Veg. 23: 186 (1926).
Cyperus flavus var. *australis* Osten, Anales Mus. Hist. Nat. Montevideo 3(2): 152 (1931).

var. ***gigas*** (Lindm.) Guagl., Hickenia 2(31): 137 (1995).
Trop. America. 81 CUB 83 BOL 84 BZC 85 AGE AGW PAR URU. Hemicr. or rhizome geophyte.
Mariscus cylindricus var. *australis* Lindm., Bih. Kongl. Svenska Vetensk.-Akad. Handl. 26(3; 9): 10 (1900). *Cyperus retrorsus* var. *australis* (Lindm.) Kük. in H.G.A.Engler (ed.), Pflanzenr., IV, 20(101): 512 (1936).
**Mariscus flavus* var. *gigas* Lindm., Bih. Kongl. Svenska Vetensk.-Akad. Handl. 26(3; 9): 10 (1900). *Cyperus flavus* var. *gigas* (Lindm.) Kük. in H.G.A.Engler (ed.), Pflanzenr., IV, 20(101): 532 (1936). *Cyperus cayennensis* var. *gigas* (Lindm.) Barros, Anales Mus. Argent. Ci. Nat. "Bernardino Rivadavia" 39: 355 (1938).

Cyperus ajax C.B.Clarke in D.Oliver & auct. suc. (eds.), Fl. Trop. Afr. 8: 343 (1901).
WC. Trop. Africa to Malawi. 23 BUR GGI RWA ZAI 25 KEN TAN UGA 26 MLW. Hemicr. or rhizome geophyte.

Cyperus albopilosus (C.B.Clarke) Kük., Bot. Not. 1934: 69 (1934).
Ethiopia to Zimbabwe. 24 ETH 25 KEN TAN 26 MLW ZAM ZIM. Hemicr. or rhizome geophyte.
**Mariscus albopilosus* C.B.Clarke in D.Oliver & auct. suc. (eds.), Fl. Trop. Afr. 8: 394 (1902).

Cyperus albopurpureus Cherm., Bull. Soc. Bot. France 66: 34 (1919 publ. 1920).
E. Madagascar. 29 MDG. Hemicr. or rhizome geophyte.

Cyperus albosanguineus Kük. in H.G.A.Engler (ed.), Pflanzenr., IV, 20(101): 555 (1936). *Mariscus albosanguineus* (Kük.) Napper, J. E. Africa Nat. Hist. Soc. Natl. Mus. 28: 16 (1971).
E. Trop. Africa. 25 KEN TAN UGA. Hemicr. or tuber geophyte.

Cyperus albostriatus Schrad., Anal. Fl. Cap. 1: 7 (1832).
S. Trop. & S. Africa. 26 ZIM 27 BOT CPP LES NAM SWZ TVL (50) nfk. Hemicr. or rhizome geophyte.
Cyperus mariscus Nees, Linnaea 7: 515 (1832).
Cyperus braunii Vatke, Index Seminum (B) 1872 (App.): 5 (1872).
Cyperus mossii Turrill, Bull. Misc. Inform. Kew 1925: 67 (1925). *Cyperus albostriatus* var. *mossii* (Turrill) Kük., Bot. Not. 1934: 66 (1934).

Cyperus albus J.Presl & C.Presl, Reliq. Haenk. 1: 175 (1828).

S. Africa. 27 BOT CPP LES NAM NAT OFS SWZ TVL. Hemicr. or rhizome geophyte.

Cyperus algeriensis Väre & Kukkonen, Ann. Bot. Fenn. 42: 481 (2005).
Algeria. 20 ALG. Hemicr. or rhizome geophyte

Cyperus almensis D.A.Simpson, Kew Bull. 48: 699 (1993).
Brazil (Bahia). 84 BZE. Hemicr. or rhizome geophyte.

Cyperus alopecuroides Rottb., Descr. Icon. Rar. Pl.: 38 (1773). *Juncellus alopecuroides* (Rottb.) C.B.Clarke in J.D.Hooker, Fl. Brit. India 6: 595 (1893). *Chlorocyperus alopecuroides* (Rottb.) Grossh., Fl. Kavkaza 1: 150 (1928).
Africa, SW. Syria to Trop. Asia, N. Australia. 20 EGY 21 CNY CVI 22 BEN BKN GHA IVO MLI MTN NGA NGR SEN TOG 24 CHA ERI ETH SOM 25 KEN TAN 26 ANG MLW 27 BOT NAT SWZ TVL 29 COM MDG SEY 34 LBS PAL 35 SAU YEM 40 IND PAK SRL 41 VIE 42 JAW LSI 50 NTA QLD WAU (78) fla (81) lee win. Hemicr. or rhizome geophyte.
Cyperus fastigiatus Forssk., Fl. Aegypt.-Arab.: 14 (1775), nom. illeg.
Cyperus glomeratus Willd., Sp. Pl. 1: 277 (1797), nom. illeg.
Cyperus bidentatus Poir. in J.B.A.M.de Lamarck, Encycl. 7: 250 (1806), nom. illeg.
Cyperus compositus R.Br., Prodr.: 217 (1810).
Cyperus poiretii Roem. & Schult., Syst. Veg. 2: 223 (1817).
Cyperus glaucus Willd. ex Kunth, Enum. Pl. 2: 20 (1837).
Cyperus alopecuroides var. *digynus* Boeck., Linnaea 36: 322 (1870). *Cyperus exaltatus* var. *digynus* (Boeck.) F.N.Williams, Bull. Herb. Boissier, II, 4: 224 (1904).
Juncellus pallidiflorus Peter, Abh. Preuss. Akad. Wiss., Phys.-Math. Kl., n.f., 13(2): 48 (1928). *Cyperus alopecuroides* f. *pallidiflorus* (Peter) Kük. in H.G.A. Engler (ed.), Pflanzenr., IV, 20(101): 72 (1935).
Cyperus alopecuroides f. *latifolius* A.Peter ex Kük. in H.G.A.Engler (ed.), Pflanzenr., IV, 20(101): 73 (1935).

Cyperus alterniflorus R.Br., Prodr.: 216 (1810). *Mariscus alterniflorus* (R.Br.) C.B.Clarke, Kew Bull., Addit. Ser. 8: 18 (1908).
Australia. 50 QLD SOA WAU. Hemicr. or rhizome geophyte.
Cyperus alternifolius Steud., Syn. Pl. Glumac. 2: 54 (1854), orth. var.
Cyperus pictus Steud., Syn. Pl. Glumac. 2: 43 (1854), nom. illeg.

Cyperus alternifolius L., Mant. Pl. 1: 28 (1767). *Eucyperus alternifolius* (L.) Rikli, Jahrb. Wiss. Bot. 27: 568 (1895).
Ethiopia to Mozambique, W. Indian Ocean, Arabian Pen. (20) egy mor (21) azo cny cvi (22) ben 23 BUR RWA ZAI 24 ERI ETH SOM SUD 25 KEN TAN UGA 26 ANG MLW MOZ ZAM ZIM 29 MAU MDG REU 35 SAU YEM (36) chs (38) nns oga tai (40) ehm pak srl (41) tha (42) jaw phi sum (50) nfk (51) nzn (60) fij (61) sci (63) haw (76) cal (77) tex (78) fla lou (79) mxc mxe mxn mxs mxt (80) cos (81) jam lee pue (85) cln. Hemicr. or rhizome geophyte.

subsp. ***alternifolius***
E. Madagascar. 29 MDG. Hemicr. or rhizome geophyte.
Cyperus frondosus Salisb., Prodr. Stirp. Chap. Allerton: 30 (1796).
Cyperus onustus Steud., Syn. Pl. Glumac. 2: 20 (1854).

subsp. ***flabelliformis*** Kük. in H.G.A.Engler (ed.), Pflanzenr., IV, 20(101): 193 (1936).
Ethiopia to Mozambique, W. Indian Ocean, Arabian Pen. (20) egy mor (21) azo cny cvi (22) ben 23 BUR RWA ZAI 24 ERI ETH SOM SUD 25 KEN TAN UGA 26 ANG MLW MOZ ZAM ZIM 29 MAU MDG REU 35 SAU YEM (36) chs (38) nns oga tai (40) ehm pak srl (41) tha (42) jaw phi sum (50) nfk (51) nzn (60) fij (61) sci (63) haw (76) cal (77) tex (78) fla lou (79) mxc mxe mxn mxs mxt (80) cos (81) jam lee pue (85) cln. Hemicr. or rhizome geophyte. Widely cultivated as an ornamental.
Cyperus involucratus Rottb., Descr. Pl. Rar.: 22 (1772). **Cyperus flabelliformis* Rottb., Descr. Icon. Rar. Pl.: 42 (1773), nom. illeg.
Cyperus gradatus Forssk., Fl. Aegypt.-Arab.: 13 (1775).
Cyperus flagellatus Hochst., Flora 24(1): 21 (1841).
Cyperus proximus Steud., Syn. Pl. Glumac. 2: 19 (1854).
Cyperus alternifolius var. *albovariegatus* auct., J. Gén. Hort. 14: 233 (1861).
Cyperus petersianus Boeck. in W.C.H.Peters, Naturw. Reise Mossambique: 537 (1864). *Cyperus alternifolius* var. *petersianus* (Boeck.) Kük. in H.G.A.Engler (ed.), Pflanzenr., IV, 20(101): 194 (1936).

Cyperus altochrysocephalus Lye, Candollea 43: 508 (1988).
Zambia. 26 ZAM. Hemicr. or rhizome geophyte.

Cyperus altomicroglumis Lye, Bull. Jard. Bot. Natl. Belg. 65: 249 (1997).
S. Somalia. 24 SOM. Hemicr. or rhizome geophyte.

Cyperus altsonii Kük., Bull. Misc. Inform. Kew 1932: 322 (1932).
Guyana. 82 GUY. Hemicr. or rhizome geophyte.

Cyperus alulatus J.Kern, Reinwardtia 1: 463 (1952).
Afghanistan to Nepal and C. India. 34 AFG 40 IND NEP PAK WHM. Ther.

Cyperus amabilis Vahl, Enum. Pl. 2: 318 (1805).
Trop. & Subtrop. 21 CVI 22 BEN BKN GAM GHA GUI IVO MLI NGA NGR SEN SIE TOG 23 BUR CMN CON GAB ZAI 24 CHA ETH SOC SOM SUD 25 KEN TAN UGA 26 ANG MLW MOZ 27 BOT NAM SWZ TVL 29 COM MDG 40 IND WHM 41 VIE 76 ARI 79 MXC MXE MXG MXN MXS MXT 80 COS ELS GUA HON NIC PAN 81 ARU CUB DOM NLA 82 GUY SUR VEN 83 CLM ECU PER 84 BZC BZN. Ther.
Cyperus microstachyus Vahl, Enum. Pl. 2: 318 (1805).
Cyperus aurantiacus Kunth in F.W.H.von Humboldt, A.J.A.Bonpland & C.S.Kunth, Nov. Gen. Sp. 1: 205 (1817). *Cyperus aureus* var. *aurantiacus* (Kunth) Boeck., Linnaea 35: 495 (1868).
Cyperus aureus Kunth in F.W.H.von Humboldt, A.J.A.Bonpland & C.S.Kunth, Nov. Gen. Sp. 1: 205 (1817), nom. illeg.
Cyperus oligostachyus Kunth in F.W.H.von Humboldt, A.J.A.Bonpland & C.S.Kunth, Nov. Gen. Sp. 1: 204 (1817). *Cyperus aureus* var. *oligostachyus* (Kunth) Boeck., Linnaea 35: 495 (1868).

Cyperus amabilis var. *oligostachyus* (Kunth) Kük., Bot. Not. 1934: 67 (1934).
Cyperus brachyphyllus Willd. ex Link, Jahrb. Gewächsk. 1(3): 82 (1818).
Cyperus quitensis Spreng., Syst. Veg. 1: 224 (1824).
Cyperus anisostachyos Willd. ex Kunth, Enum. Pl. 2: 20 (1837).
Cyperus glareosus Liebm., Mexic. Halvgr.: 204 (1850).
Cyperus lepidus Hochst. ex Steud., Syn. Pl. Glumac. 2: 14 (1854).
Cyperus muelleri Boeck., Flora 42: 434 bis (1859).
Cyperus aureus var. *macrostachyus* Boeck., Linnaea 35: 494 (1868). *Cyperus amabilis* var. *macrostachyus* (Boeck.) Kük. in H.G.A.Engler (ed.), Pflanzenr., IV, 20(101): 266 (1936).
Cyperus amabilis var. *macra* C.B.Clarke in D.Oliver & auct. suc. (eds.), Fl. Trop. Afr. 8: 328 (1901).
Cyperus guadalajaranus M.E.Jones, Contr. W. Bot. 18: 25 (1933).
Cyperus amabilis var. *subacaulis* Kük. in H.G.A. Engler (ed.), Pflanzenr., IV, 20(101): 266 (1936).

Cyperus amauropus Steud., Syn. Pl. Glumac. 2: 33 (1854). *Mariscus amauropus* (Steud.) Cufod., Bull. Jard. Bot. Natl. Belg. 40(Suppl.): 1448 (1970). *Cyperus cruentus* subsp. *amauropus* (Steud.) Lye, Lidia 3: 131 (1994).
Ethiopia to Zambia, Arabian Pen. 23 RWA 24 ERI ETH SOM SUD 25 KEN TAN UGA 26 ZAM 35 SAU YEM. Hemicr. or rhizome geophyte.
Cyperus leptophyllus Hochst. ex Steud., Syn. Pl. Glumac. 2: 33 (1854). *Mariscus leptophyllus* (Hochst. ex Steud.) C.B.Clarke in T.A.Durand & H.Schinz, Consp. Fl. Afric. 5: 589 (1894).
Cyperus ibeensis C.B.Clarke in D.Oliver & auct. suc. (eds.), Fl. Trop. Afr. 8: 369 (1901). *Cyperus leptophyllus* var. *ibeensis* (C.B.Clarke) Kük. in H.G.A.Engler (ed.), Pflanzenr., IV, 20(101): 550 (1936).
Cyperus friesii Kük. in R.E.Fries, Wiss. Erg. Schwed. Rhod.-Kongo Exped. 1: 4 (1921). *Cyperus leptophyllus* var. *friesii* (Kük.) Kük. in H.G.A. Engler (ed.), Pflanzenr., IV, 20(101): 549 (1936).
Cyperus leptophyllus f. *conglobata* Kük. in H.G.A. Engler (ed.), Pflanzenr., IV, 20(101): 549 (1936).
Cyperus leptophyllus var. *deliciosus* Kük. in H.G.A. Engler (ed.), Pflanzenr., IV, 20(101): 549 (1936).

Cyperus amomodorus K.Schum. in H.G.A.Engler, Pflanzenw. Ost-Afrikas, C: 122 (1895). *Cyperus mollipes* var. *amomodorus* (K.Schum.) Kük. in H.G.A.Engler (ed.), Pflanzenr., IV, 20(101): 557 (1936). *Mariscus amomodorus* (K.Schum.) Cufod., Bull. Jard. Bot. Natl. Belg. 40(Suppl.): 1449 (1970).
E. Zaïre to E. Trop. Africa. 23 RWA ZAI 25 KEN TAN UGA. Hemicr. or tuber geophyte.
Mariscus oblonginux C.B.Clarke in D.Oliver & auct. suc. (eds.), Fl. Trop. Afr. 8: 526 (1902).
Mariscus boeckeleri C.B.Clarke, Kew Bull., Addit. Ser. 8: 13 (1908).

Cyperus amuricus Maxim., Prim. Fl. Amur.: 296 (1859). *Chlorocyperus amuricus* (Maxim.) Palla ex Kük. in H.G.A.Engler (ed.), Pflanzenr., IV, 20(101): 152 (1935), pro syn.
Temp. Asia. 31 AMU PRM 36 CHC CHM CHN CHS 38 JAP KOR NNS TAI (75) mas nwj nwy ohi pen (78) del kty. Ther.
Cyperus amuricus var. *japonicus* Miq., Prolus. Fl. Jap.: 73 (1866).
Cyperus krameri Franch. & Sav., Enum. Pl. Jap. 2: 104, 539 (1877).
Cyperus aristatus var. *maingayi* C.B.Clarke, J. Linn. Soc., Bot. 21: 94 (1884).
Cyperus japonicus Makino, Bot. Mag. (Tokyo) 18: 53 (1904), nom. illeg.
Cyperus pterygorrachis H.Lév., Bull. Acad. Int. Géogr. Bot. 14: 202 (1904). *Cyperus amuricus* var. *pterygorrachis* (H.Lév.) Ohwi, Bot. Mag. (Tokyo) 45: 184 (1931).
Chlorocyperus franchetii Palla, Oesterr. Bot. Z. 59: 193 (1909).
Cyperus amuricus var. *subiroides* Kük., Repert. Spec. Nov. Regni Veg. 27: 107 (1929).
Cyperus amuricus f. *nanus* Kük. in H.G.A.Engler (ed.), Pflanzenr., IV, 20(101): 153 (1935).
Cyperus amuricus f. *perdebilis* Kük. in H.G.A.Engler (ed.), Pflanzenr., IV, 20(101): 153 (1935).

Cyperus anderssonii Boeck., Linnaea 36: 388 (1870).
Galápagos. 83 GAL. Hemicr. or rhizome geophyte.
Mariscus brachystachys Hook.f., Trans. Linn. Soc. London 20: 179 (1847). *Cyperus brachystachys* (Hook.f.) Andersson, Galapagos Veg.: 53 (1854), nom. illeg.

Cyperus androhibensis D.A.Simpson, Kew Bull. 47: 745 (1992).
Madagascar. 29 MDG. Hemicr. or rhizome geophyte.

Cyperus angolensis Boeck., Flora 63: 435 (1880).
Trop. & S. Africa. 22 GHA NGA SIE 23 CMN CON GAB RWA ZAI 25 TAN UGA 26 ANG MLW ZAM ZIM 27 TVL. Hemicr. or tuber geophyte.
Cyperus ochrocephalus C.B.Clarke, Trans. Linn. Soc. London, Bot. 4: 53 (1894), nom. illeg.
Cyperus angolensis var. *ampibulbus* A.Peter & Kük. in H.G.A.Engler (ed.), Pflanzenr., IV, 20(101): 282 (1936).

Cyperus angustatus R.Br., Prodr.: 214 (1810). *Mariscus angustatus* (R.Br.) C.B.Clarke, Kew Bull., Addit. Ser. 8: 19 (1908).
New Guinea to N. & E. Australia. 43 NWG 50 NSW NTA QLD WAU. Hemicr. or rhizome geophyte.

Cyperus anisitsii Kük. in H.G.A.Engler (ed.), Pflanzenr., IV, 20(101): 221 (1936).
Paraguay. 85 PAR. Hemicr. or rhizome geophyte.

Cyperus ankaizinensis Cherm., Bull. Soc. Bot. France 72: 19 (1925).
NC. Madagascar. 29 MDG. Hemicr. or rhizome geophyte.

Cyperus ankaratrensis Cherm., Bull. Soc. Bot. France 67: 329 (1920 publ. 1921).
C. Madagascar. 29 MDG. Hemicr. or rhizome geophyte.

Cyperus antillanus (Kük.) O'Neill, Contr. Ocas. Mus. Hist. Nat. Colegio "De La Salle" 6: 9 (1946).
WC. Cuba. 81 CUB. Hemicr. or rhizome geophyte.
**Cyperus refractus* var. *antillanus* Kük., Repert. Spec. Nov. Regni Veg. 23: 222 (1926).

Cyperus appendiculatus (Brongn.) Kunth, Enum. Pl. 2: 81 (1837).
Brazil (Trindade, Fernando de Noronha), Ascension. 28 ASC 84 BZE BZL. Hemicr. or rhizome geophyte.
**Mariscus appendiculatus* Brongn. in L.I.Duperrey, Voy. Monde, Phan.: 178 (1834).

var. ***appendiculatus***
Ascension. 28 ASC. Hemicr. or rhizome geophyte.
Mariscus polycephalus A.Rich. in J.S.C.Dumont d'Urville, Voy. Astrolabe 2: lii (1834).
Mariscus polystachyus Brongn. ex A.Rich. in J.S.C.Dumont d'Urville, Voy. Astrolabe 2: liv (1834).
Cyperus capito Steud., Syn. Pl. Glumac. 2: 28 (1854).
Cyperus ligularis Hemsl., Rep. Challenger, Bot. 1(2): 23 (1884), nom. illeg.
Cyperus umbellatus Hemsl., Rep. Challenger, Bot. 1(2): 23 (1884), nom. illeg.
Cyperus appendiculatus var. *gordonii* Kük. in H.G.A.Engler (ed.), Pflanzenr., IV, 20(101): 436 (1936).

var. ***atlanticus*** (Hemsl.) Kük. in H.G.A.Engler (ed.), Pflanzenr., IV, 20(101): 436 (1936).
Brazil (Trindade). 84 BZL. Hemicr. or rhizome geophyte.
**Cyperus atlanticus* Hemsl., Rep. Challenger, Bot. 1(2): 130 (1884).

var. ***noronhae*** (Ridl.) Kük. in H.G.A.Engler (ed.), Pflanzenr., IV, 20(101): 436 (1936).
Brazil (Fernando de Noronha). 84 BZE. Hemicr. or rhizome geophyte.
**Cyperus noronhae* Ridl., J. Linn. Soc., Bot. 27: 66 (1890).

Cyperus aquatilis R.Br., Prodr.: 213 (1810). *Cyperus trinervis* var. *aquatilis* (R.Br.) Kük. in H.G.A.Engler (ed.), Pflanzenr., IV, 20(101): 294 (1936).
New Guinea to N. & E. Australia. 43 NWG 50 NSW NTA QLD WAU. Hemicr. or rhizome geophyte.

Cyperus arenarius Retz., Observ. Bot. 4: 9 (1786).
Arabian Pen., S. Iran to Indo-China. (27) nat 34 IRN 35 GST SAU 40 IND PAK SRL 41 VIE (50). Hemicr. or rhizome geophyte.
Scirpus glomeratus B.Heyne ex Wall., Numer. List: 3460 (1831), nom. illeg.
Cyperus persicus Boiss. & Hohen. in P.E.Boissier, Diagn. Pl. Orient. 13: 39 (1854).

Cyperus armstrongii Benth., Fl. Austral. 7: 289 (1878). *Mariscus armstrongii* (Benth.) C.B.Clarke, Kew Bull., Addit. Ser. 8: 17 (1908). *Cyperus pennatus* var. *armstrongii* (Benth.) Kük. in H.G.A.Engler (ed.), Pflanzenr., IV, 20(101): 479 (1936).
Queensland. 50 QLD. Hemicr. or rhizome geophyte.

Cyperus arsenei O'Neill & Ben.Ayers, Leafl. W. Bot. 4: 33 (1944).
Mexico. 79 MXC MXN MXS. Hemicr. or rhizome geophyte.

Cyperus articulatus L., Sp. Pl.: 44 (1753). *Chlorocyperus articulatus* (L.) Rikli, Jahrb. Wiss. Bot. 27: 563 (1895).
Trop. & Subtrop. 20 ALG EGY 21 CVI 22 BEN BKN GAM GHA GNB GUI IVO LBR NGA NGR SEN SIE TOG 23 BUR CAB CAF CMN CON EQG GAB GGI RWA ZAI 24 CHA ERI ETH SOM 25 KEN TAN UGA 26 ANG MLW MOZ 27 BOT CPP NAM NAT SWZ TVL 29 MAU MDG REU SEY 34 IRN IRQ LBS PAL SIN 35 SAU YEM (38) nns 40 ASS BAN IND NEP PAK SRL WHM 41 MYA THA VIE 50 QLD wau 77 TEX 78 ALA FLA GEO LOU MSI SCA 79 MXC MXE MXG MXN MXS MXT 80 BLZ COS ELS GUA HON NIC PAN 81 CUB DOM HAI JAM LEE NLA PUE TRT WIN 82 FRG GUY SUR VEN 83 BOL CLM ECU PER 84 BZL BZS 85 AGE AGW PAR URU. Hemicr. or rhizome geophyte.
Cyperus niloticus Forssk., Fl. Aegypt.-Arab.: 13 (1775).
Cyperus nodosus Humb. & Bonpl. ex Willd., Enum. Pl.: 72 (1809). *Cyperus articulatus* var. *nodosus* (Humb. & Bonpl. ex Willd.) Kük. in H.G.A.Engler (ed.), Pflanzenr., IV, 20(101): 79 (1935).
Cyperus autumnalis Pursh, Fl. Amer. Sept. 1: 51 (1813), nom. illeg.
Cyperus pertenuis Roxb., Fl. Ind. 1: 202 (1820).
Cyperus gymnos Schult. in J.J.Roemer & J.A.Schultes, Mant. 2: 97 (1824).
Cyperus subarticulatus Nees & Meyen, Linnaea 9: 285 (1834), nom. inval.
Papyrus pangorei Nees in R.Wight, Contr. Bot. India: 88 (1834).
Cyperus subnodosus Nees & Meyen, Nova Acta Acad. Caes. Leop.-Carol. German. Nat. Cur. 19(Suppl. 1): 59 (1843). *Cyperus nodosus* var. *subnodosus* (Nees & Meyen) Boeck., Linnaea 36: 276 (1870). *Cyperus corymbosus* var. *subnodosus* (Nees & Meyen) Kük., Repert. Spec. Nov. Regni Veg. 23: 185 (1926).
Cyperus articulatus Benth., Pl. Hartw.: 124 (1844), nom. illeg.
Cyperus borbonicus Steud., Syn. Pl. Glumac. 2: 31 (1854).
Cyperus interceptus Steud., Syn. Pl. Glumac. 2: 31 (1854).
Cyperus fistulosus Ehrenb. ex Boeck., Linnaea 36: 275 (1870).
Cyperus nodosus var. *aphyllus* Boeck., Linnaea 36: 275 (1870).
Cyperus articulatus var. *conglomeratus* Britton, Bull. Torrey Bot. Club 13: 210 (1886).
Chlorocyperus cordobensis Palla, Allg. Bot. Z. Syst. 9: 69 (1903). *Cyperus cordobensis* (Palla) Hicken, Darwiniana 1: 112 (1924).
Cyperus articulatus var. *erythrostachys* Graebn., Repert. Spec. Nov. Regni Veg. 16: 25 (1919).
Cyperus articulatus f. *longispiculosus* Kük., Repert. Spec. Nov. Regni Veg. 26: 250 (1929).
Cyperus articulatus var. *fistulosus* Kük. in H.G.A. Engler (ed.), Pflanzenr., IV, 20(101): 79 (1935).
Cyperus articulatus var. *multiflorus* Kük. in H.G.A. Engler (ed.), Pflanzenr., IV, 20(101): 79 (1935).

Cyperus astartodes K.L.Wilson, Telopea 4: 421 (1991).
N. Australia. 50 NTA WAU. Hemicr. or rhizome geophyte.

Cyperus aster (C.B.Clarke ex Cherm.) Kük. in H.G.A. Engler (ed.), Pflanzenr., IV, 20(101): 551 (1936).
E. Trop. Africa, Madagascar. 25 TAN 29 MDG. Hemicr. or rhizome geophyte.
**Mariscus aster* C.B.Clarke ex Cherm., Bull. Mus. Natl. Hist. Nat. 25: 302 (1919).

var. ***aster***
C. Madagascar. 29 MDG. Hemicr. or rhizome geophyte.
Mariscus andringitrensis Cherm., Bull. Mus. Natl. Hist. Nat. 27: 552 (1921). *Cyperus aster* var. *andringitrensis* (Cherm.) A.Peter & Kük. in H.G.A.Engler (ed.), Pflanzenr., IV, 20(101): 551 (1936).

var. ***biflorus*** A.Peter & Kük. in H.G.A.Engler (ed.), Pflanzenr., IV, 20(101): 551 (1936).
E. Trop. Africa. 25 TAN. Hemicr. or rhizome geophyte.

Cyperus aterrimus Hochst. ex Steud., Syn. Pl. Glumac. 2: 31 (1854).
Bioko to Ethiopia and Malawi. 23 BUR CMN GGI RWA ZAI 24 ETH 25 KEN TEN UGA 26 MLW. Hemicr. or rhizome geophyte.
Cyperus atroviridis C.B.Clarke in D.Oliver & auct. suc. (eds.), Fl. Trop. Afr. 8: 359 (1901). *Cyperus aterrimus* var. *atroviridis* (C.B.Clarke) Kük. in H.G.A.Engler (ed.), Pflanzenr., IV, 20(101): 142 (1935).
Cyperus aterrimus var. *agglomeratus* Kük., Notizbl. Bot. Gart. Berlin-Dahlem 9: 304 (1925).

Cyperus atkinsonii C.B.Clarke, J. Linn. Soc., Bot. 21: 109 (1884).
S. Iran to NW. India. 34 IRN 40 IND PAK WHM. Ther. or hemicr.

Cyperus atractocarpus Ridl., Trans. Linn. Soc. London, Bot. 2: 141 (1884).
Zambia to Angola. 26 ANG ZAM. Hemicr.

Cyperus aureoalatus Lye, Lidia 3: 171 (1995).
Ethiopia to Uganda. 24 ETH SOM 25 UGA. Hemicr.

Cyperus aureobrunneus C.B.Clarke in D.Oliver & auct. suc. (eds.), Fl. Trop. Afr. 8: 346 (1901). *Cyperus denudatus* var. *aureobrunneus* (C.B.Clarke) Kük. in H.G.A.Engler (ed.), Pflanzenr., IV, 20(101): 256 (1936).
S. Trop. Africa. 26 MLW. Hemicr. or rhizome geophyte.

Cyperus auriculatus Nees & Meyen ex Kunth, Enum. Pl. 2: 83 (1837). *Cyperus ferax* var. *auriculatus* (Nees & Meyen ex Kunth) Kük. in H.G.A.Engler (ed.), Pflanzenr., IV, 20(101): 618 (1936). *Torulinium odoratum* subsp. *auriculatum* (Nees & Meyen ex Kunth) T.Koyama, Bishop Mus. Occas. Pap. 29: 125 (1989).
Hawaiian Is. 63 HAW. Ther. or hemicr. Close to C. odoratus.

Cyperus aurifer Cherm., Bull. Soc. Bot. France 66: 351 (1919 publ. 1920).
NW. Madagascar. 29 MDG. Hemicr. or rhizome geophyte.
Cyperus aurifer var. *brachyanthus* Kük. in H.G.A. Engler (ed.), Pflanzenr., IV, 20(101): 83 (1935).

Cyperus austrochrysanthus Lye, Candollea 43: 505 (1988).
Angola. 26 ANG. Hemicr. or rhizome geophyte.

Cyperus babakan Steud., Syn. Pl. Glumac. 2: 6 (1854).
Assam to New Guinea. 40 ASS 41 THA VIE 42 BOR JAW MLY PHI SUL 43 NWG. Hemicr. or rhizome geophyte.
Cyperus babakensis Steud. ex Miq., Fl. Ned. Ind. 3: 257 (1856). *Cyperus pilosus* var. *babakensis* (Steud. ex Miq.) C.B.Clarke, J. Linn. Soc., Bot. 21: 151 (1884). *Duval-jouvea babakensis* (Steud. ex Miq.) H.Pfeiff., Mitt. Inst. Allg. Bot. Hamburg 7: 167 (1928).

Cyperus balfourii C.B.Clarke, J. Linn. Soc., Bot. 20: 289 (1883).
W. Indian Ocean. 29 COM MDG REU. Hemicr. or rhizome geophyte.
Cyperus confusus var. *laxissimus* Cherm., Bull. Soc. Bot. France 66: 345 (1919). *Cyperus balfourii* var. *laxissimus* (Cherm.) Kük. in H.G.A.Engler (ed.), Pflanzenr., IV, 20(101): 204 (1936).
Cyperus confusus var. *scabricaulis* Cherm., Bull. Soc. Bot. France 66: 345 (1919). *Cyperus balfourii* var. *scabricaulis* (Cherm.) Kük. in H.G.A.Engler (ed.), Pflanzenr., IV, 20(101): 204 (1936).
Cyperus confusus Cherm., Bull. Soc. Bot. France 66: 345 (1919 publ. 1920).
Cyperus confusus var. *tsiafajavonensis* Cherm., Ann. Inst. Bot.-Géol. Colon. Marseille, III, 10: 25 (1922). *Cyperus balfourii* var. *tsiafajavonensis* (Cherm.) Kük. in H.G.A.Engler (ed.), Pflanzenr., IV, 20(101): 204 (1936).
Cyperus balfourii var. *rubiginosus* Kük. in H.G.A. Engler (ed.), Pflanzenr., IV, 20(101): 204 (1936).

Cyperus baobab Lye, Nordic J. Bot. 16: 371 (1996).
Coastal S. Somalia. 24 SOM. Hemicr.

Cyperus baoulensis Kük. in H.G.A.Engler (ed.), Pflanzenr., IV, 20(101): 467 (1936).
W. & WC. Trop. Africa. 22 GHA IVO TOG 23 GAB. Hemicr. or rhizome geophyte.
Mariscus baoulensis Hutch. in J.Hutchinson & J.M.Dalziel, Fl. W. Trop. Afr. 2: 485 (1936), no latin descr.

Cyperus baronii C.B.Clarke, J. Linn. Soc., Bot. 20: 289 (1883).
Trop. Africa, W. Indian Ocean. 22 LBR NGA SIE 23 CMN GGI 25 KEN TAN 26 MLW 29 COM MDG REU. Hemicr. or rhizome geophyte.
Cyperus mannii C.B.Clarke in D.Oliver & auct. suc. (eds.), Fl. Trop. Afr. 8: 341 (1901). *Cyperus baronii* var. *mannii* (C.B.Clarke) Kük. in H.G.A.Engler (ed.), Pflanzenr., IV, 20(101): 202 (1936).
Cyperus baronii var. *densus* C.B.Clarke ex Cherm., Bull. Soc. Bot. France 66: 344 (1919).
Cyperus baronii var. *interpositus* Kük. in H.G.A. Engler (ed.), Pflanzenr., IV, 20(101): 202 (1936).
Cyperus baronii var. *squarrosus* C.B.Clarke ex Kük. in H.G.A.Engler (ed.), Pflanzenr., IV, 20(101): 202 (1936).

Cyperus bellus Kunth, Enum. Pl. 2: 52 (1837).
S. Africa. 27 BOT CPP NAM OFS TVL. Hemicr. or rhizome geophyte.
Cyperus remotiflorus Kük., Repert. Spec. Nov. Regni Veg. 18: 345 (1922).
Cyperus bellus f. *auratus* Kük. in H.G.A.Engler (ed.), Pflanzenr., IV, 20(101): 303 (1936).

Cyperus benadirensis Chiov., Fl. Somala 2: 434 (1932).
C. & S. Somalia. 24 SOM. Hemicr. or rhizome geophyte.
Cyperus poecilus var. *evolutus* Kük. in H.G.A.Engler (ed.), Pflanzenr., IV, 20(101): 283 (1936).

Cyperus bernieri Cherm., Bull. Soc. Bot. France 66: 317 (1919 publ. 1920).
E. Madagascar. 29 MDG. Hemicr. or rhizome geophyte.

Cyperus berroi (C.B.Clarke) Barros, Darwiniana 6: 600 (1944).
NE. Argentina to Uruguay. 85 AGE URU. Hemicr. or rhizome geophyte.
**Hoppia berroi* C.B.Clarke, Kew Bull., Addit. Ser. 8: 62 (1908). *Bisboeckelera berroi* (C.B.Clarke) Standl., Contr. U. S. Natl. Herb. 17: 429 (1914).

Cyperus betafensis Cherm., Ann. Inst. Bot.-Géol. Colon. Marseille, III, 10(1): 21 (1922).
C. Madagascar. 29 MDG. Hemicr. or rhizome geophyte.
Cyperus cuspidatus Baker, J. Linn. Soc., Bot. 22: 532 (1887), nom. illeg.

Cyperus betchei (Kük.) S.T.Blake, Proc. Roy. Soc. Queensland 51: 43 (1939 publ. 1940).
Australia. 50 NSW NTA QLD WAU. Hemicr. or rhizome geophyte.
**Cyperus angustatus* var. *betchei* Kük. in H.G.A.Engler (ed.), Pflanzenr., IV, 20(101): 452 (1936).

subsp. ***betchei***
E. Australia. 50 NSW QLD. Hemicr. or rhizome geophyte.

subsp. ***commiscens*** K.L.Wilson, Telopea 4: 457 (1991).
N. Australia. 50 NTA QLD WAU. Hemicr. or rhizome geophyte.

Cyperus beyrichii (Schrad. ex Nees) Steud., Syn. Pl. Glumac. 2: 47 (1854).
Brazil (Rio de Janeiro). 84 BZL. Hemicr. or rhizome geophyte.
**Mariscus beyrichii* Schrad. ex Nees in C.F.P.von Martius & auct. suc. (eds.), Fl. Bras. 2(1): 48 (1842).

Cyperus bifolius Lye, Lidia 3: 206 (1996).
Ethiopia. 24 ETH. Hemicr. or rhizome geophyte.

Cyperus blakeanus K.L.Wilson, Telopea 4: 441 (1991).
N. Australia. 50 NTA QLD WAU. Hemicr. or rhizome geophyte.

Cyperus blysmoides Hochst. ex C.B.Clarke in D.Oliver & auct. suc. (eds.), Fl. Trop. Afr. 8: 354 (1902).
E. Trop. Africa, Arabian Pen. 25 KEN TAN 35 SAU YEM. Bulb geophyte.
**Hemichlaena bulbosa* Hochst. ex A.Rich., Tent. Fl. Abyss. 2: 509 (1850).
Cyperus bulbosus var. *spicatus* Boeck., Linnaea 36: 301 (1870).

Cyperus boreobellus Lye, Nordic J. Bot. 3: 220 (1983).
Kenya. 25 KEN. Hemicr.

Cyperus boreochrysocephalus Lye, Nordic J. Bot. 3: 216 (1983).
E. Trop. Africa. 25 KEN TAN UGA. Hemicr. or tuber geophyte.

Cyperus boreohemisphaericus Lye, Lidia 3: 77 (1993).
Ethiopia. 24 ETH. Hemicr. or rhizome geophyte.

Cyperus bowmanii F.Muell. ex Benth., Fl. Austral. 7: 287 (1878). *Mariscus bowmanii* (F.Muell. ex Benth.) C.B.Clarke, Kew Bull., Addit. Ser. 8: 20 (1908).
Queensland to NE. New South Wales. 50 NSW QLD. Hemicr. or rhizome geophyte.
Mariscus dietrichiae subsp. *bowmannii* (F.Muell. ex Benth.) Domin, Biblioth. Bot. 85: 436 (1915).

Cyperus breedlovei G.C.Tucker, Rhodora 88: 505 (1986).
Mexico (Chiapas). 79 MXT. Hemicr. or rhizome geophyte.

Cyperus breviculmis R.Br., Prodr.: 214 (1810). *Cyperus trinervis* var. *breviculmis* (R.Br.) Kük. in H.G.A.Engler (ed.), Pflanzenr., IV, 20(101): 294 (1936).
N. Australia. 50 NTA QLD WAU. Hemicr. or rhizome geophyte.

Cyperus brumadoi D.A.Simpson, Kew Bull. 48: 701 (1993).
Brazil (Bahia). 84 BZE. Hemicr. or rhizome geophyte.

Cyperus brunneofibrosus Lye, Candollea 51: 423 (1996).
C. Somalia. 24 SOM. Hemicr. or rhizome geophyte.

Cyperus brunnescens Boeck., Beitr. Cyper. 2: 3 (1890).
E. India, Vietnam. 40 IND 41 VIE. Hemicr. or rhizome geophyte.

Cyperus bulamensis Steud., Syn. Pl. Glumac. 2: 19 (1854).
W. Trop. Africa. 22 GNB. Hemicr. or rhizome geophyte.

Cyperus bulbosus Vahl, Enum. Pl. 2: 342 (1805).
Africa to Australia. 20 ALG 21 CVI 22 GHA GUI NGA NGR SEN 24 CHA ERI ETH SOM SUD 25 KEN TAN 26 MOZ 29 SEY 34 IRN 35 SAU YEM 40 IND PAK SRL 41 VIE 42 JAW LSI 50 NSW NTA QLD SOA WAU. Bulb geophyte.
Cyperus jeminicus Retz., Observ. Bot. 4: 11 (1786), nom. illeg.
Cyperus polyphyllus Vahl, Enum. Pl. 2: 317 (1805).
Cyperus geminatus J.König ex Ainslie, Mat. Med. Hind.: 250 (1813).
Cyperus laxus R.Br. in H.Salt, Voy. Abyss., App.: lxii (1814), nom. illeg.
Cyperus yemenicus Spreng., Syst. Veg. 1: 221 (1824).
Cyperus bulbiferus A.Dietr., Sp. Pl. 2: 324 (1832).
Cyperus oleraceus Roxb. ex Nees in R.Wight, Contr. Bot. India: 81 (1834).
Cyperus rotundus var. *pendulus* Nees in R.Wight, Contr. Bot. India: 82 (1834).
Cyperus stolonifer Willd. ex Kunth, Enum. Pl. 2: 59 (1837).
Cyperus subulatus var. *confertus* Benth., Fl. Austral. 7: 281 (1878).
Cyperus bulbosus var. *elatior* Kük. in H.G.A.Engler (ed.), Pflanzenr., IV, 20(101): 126 (1935).
Cyperus bulbosus var. *melanolepis* Kük. in H.G.A. Engler (ed.), Pflanzenr., IV, 20(101): 127 (1935).

Cyperus burkartii Guagl., Darwiniana 30: 233 (1990 publ. 1991).
Argentina (Misiones). 85 AGE. Hemicr. or rhizome geophyte.

Cyperus cadamostii Bolle ex E.H.L.Krause, Bot. Jahrb. Syst. 14: 400 (1892).
Cape Verde, N. Africa. 20 TUN 21 CVI. Hemicr. or rhizome geophyte.
Cyperus olivetorum Murb., Acta Univ. Lund. 35(2; 3): 25 (1899).

Cyperus caesius Boeck., Linnaea 36: 331 (1870). *Mariscus caesius* (Boeck.) C.B.Clarke, Kew Bull., Addit. Ser. 8: 19 (1908).
Venezuela. 82 VEN. Hemicr. or rhizome geophyte.

Cyperus calderoniae S.González, Phytologia 57: 384 (1985).
WC. & C. Mexico. 79 MXC MXE MXS. Hemicr. or rhizome geophyte.

Cyperus camphoratus Liebm., Mexic. Halvgr.: 216 (1850). *Mariscus camphoratus* (Liebm.) T.Koyama, Phytologia 29: 73 (1974).
S. Mexico to Paraguay. 79 MXG MXS MXT 80 COS HON NIC 82 GUY VEN 83 CLM 84 BZN 85 PAR. (Tuber) hemicr.
Cyperus camphoratus var. *boliviensis* Kük. in H.G.A. Engler (ed.), Pflanzenr., IV, 20(101): 422 (1936).
Cyperus herndoniae G.C.Tucker, Syst. Bot. 7: 345 (1982).

Cyperus cancrorum Cherm., Bull. Soc. Bot. France 66: 352 (1919 publ. 1920).
C. Madagascar. 29 MDG. Hemicr. or rhizome geophyte.

Cyperus canus J.Presl & C.Presl, Reliq. Haenk. 1: 170 (1828).
Mexico to Colombia. 79 MXC MXE MXG MXN MXS MXT 80 COS GUA HON NIC 83 CLM. Hemicr. or rhizome geophyte.
Cyperus flabelliformis var. *obtusangulum* Boeck., Linnaea 35: 566 (1868).

Cyperus capensis (Steud.) Endl., Cat. Horti Vindob. 1: 94 (1842).
Mozambique to S. Africa. 26 MOZ 27 CPP NAT OFS SWZ TVL. Hemicr. or rhizome geophyte.
**Kyllinga capensis* Steud., Flora 12: 153 (1829). *Mariscus capensis* (Steud.) Schrad., Anal. Fl. Cap. 1: 13 (1832).
Mariscus uitenhagensis Steud., Syn. Pl. Glumac. 2: 317 (1855).
Mariscus marlothii var. *globospica* C.B.Clarke in W.H.Harvey & auct. suc. (eds.), Fl. Cap. 7: 187 (1897). *Cyperus capensis* f. *globospica* (C.B. Clarke) Kük. in H.G.A.Engler (ed.), Pflanzenr., IV, 20(101): 540 (1936).
Cyperus capensis var. *pseudomarlothii* Kük. in H.G.A. Engler (ed.), Pflanzenr., IV, 20(101): 540 (1936).

Cyperus capitatus Vand., Fasc. Pl.: 5 (1771).
Macaronesia, Medit. to Caucasus. 12 BAL COR FRA POR SAR SPA 13 ALB BUL GRC ITA KRI SIC TUE YUG 20 ALG EGY LBY MOR TUN WSA 21 CNY CVI 33 TCS 34 CYP EAI LBS PAL SIN TUR. Hemicr. or rhizome geophyte.
Schoenus mucronatus L., Sp. Pl.: 42 (1753). *Mariscus mucronatus* (L.) Gaertn., Fruct. Sem. Pl. 1: 11 (1788). *Galilea mucronata* (L.) Parl., Fl. Palerm. 1: 299 (1845). *Cyperus mucronatus* (L.) Mabille, Rech. Pl. Corse 1: 27 (1867), nom. illeg.
Scirpus kalli Forssk., Fl. Aegypt.-Arab.: 15 (1775). *Cyperus kalli* (Forssk.) Murb., Contr. Fl. Nord-Ouest Afrique 3: 24 (1899).
Schoenus maritimus Lam., Fl. Franç. 3: 543 (1778). *Cyperus maritimus* (Lam.) P.Silva, Agron. Lusit. 10: 92 (1948), nom. illeg.
Cyperus aegyptiacus Gloxin, Observ. Bot.: 20 (1785). *Chlorocyperus aegyptiacus* (Gloxin) Rikli, Jahrb. Wiss. Bot. 27: 564 (1895).
Mariscus mucronatus C.Presl, Cyper. Gramin. Sicul.: 12 (1820).
Cyperus schoenoides Griseb., Spic. Fl. Rumel. 2: 421 (1846), nom. illeg.
Cyperus macrorrhizus Nees ex Boiss., Fl. Orient. 5: 369 (1882), nom. illeg.

Cyperus carinatus R.Br., Prodr.: 216 (1810). *Mariscus carinatus* (R.Br.) C.B.Clarke, Kew Bull., Addit. Ser. 8: 19 (1908).
N. Australia. 50 NTA QLD WAU. Hemicr. or rhizome geophyte.

Cyperus castaneobellus Lye, Nordic J. Bot. 3: 227 (1983).
Tanzania. 25 TAN. Hemicr.

Cyperus castaneus Willd., Sp. Pl. 1: 278 (1797). *Dichostylis castanea* (Willd.) Palla, Bot. Jahrb. Syst. 10: 296 (1888).
S. Africa, Trop. Asia, N. & E. Australia. 27 NAM TVL 40 EHM IND NEP SRL 41 MYA THA VIE 42 BOR MLY SUM 50 NSW NTA QLD WAU. Ther.
Cyperus castaneus var. *brevimucronatus* Kük. in H.G.A.Engler (ed.), Pflanzenr., IV, 20(101): 264 (1936).

Cyperus cearaensis R.Gross ex Kük. in H.G.A.Engler (ed.), Pflanzenr., IV, 20(101): 466 (1936). *Mariscus cearaensis* (R.Gross ex Kük.) T.Koyama, Phytologia 29: 73 (1974).
NE. Brazil. 84 BZE. Hemicr. or rhizome geophyte.

Cyperus celans Kukkonen, Ann. Bot. Fenn. 32: 159 (1995).
S. Iran. 34 IRN. Hemicr. or rhizome geophyte.

Cyperus cellulosoreticulatus Boeck., Allg. Bot. Z. Syst. 1: 202 (1895).
S. Brazil. 84 BZS. Hemicr. or rhizome geophyte.

Cyperus centralis K.L.Wilson, Telopea 4: 470 (1991).
C. Australia. 50 NTA SOA WAU. Hemicr. or rhizome geophyte.

Cyperus cephalanthus Torr. & Hook., Ann. Lyceum Nat. Hist. New York 3: 431 (1836). *Mariscus cephalanthus* (Torr. & Hook.) C.B.Clarke, Kew Bull., Addit. Ser. 8: 19 (1908). *Cyperus laetus* var. *cephalanthus* (Torr. & Hook.) Kük. in H.G.A.Engler (ed.), Pflanzenr., IV, 20(101): 442 (1936). *Cyperus rigens* subsp. *cephalanthus* (Torr. & Hook.) Pedersen, Darwiniana 17: 539 (1972).
Texas to Louisiana, Brazil to N. Argentina. 77 TEX 78 LOU 84 BZS 85 AGE PAR URU. Hemicr. or rhizome geophyte.

Cyperus cephalotes Vahl, Enum. Pl. 2: 311 (1805). *Anosporum cephalotes* (Vahl) Kurz, J. Asiat. Soc. Bengal, Pt. 2, Nat. Hist. 39(2): 84 (1870).
Trop. Asia to NE. Australia. 40 ASS IND SRL 41 MYA NCB THA VIE 42 BOR JAW 43 NWG 50 QLD. Hemicr. or rhizome geophyte.
Cyperus monocephalus Roxb., Fl. Ind. 1: 193 (1820). *Anosporum monocephalum* (Roxb.) Nees, Linnaea 9: 287 (1834). *Ungeria monocephala* (Roxb.) Nees ex C.B.Clarke, J. Linn. Soc., Bot. 21: 34 (1884).
Cyperus monocephaloides Roxb. ex Nees in R.Wight, Contr. Bot. India: 92 (1834).
Hydroschoenus kyllingioides Zoll. & Moritzi, Syst. Verz.: 95 (1846).
Trentepohlia bifoliata Boeck., Bot. Zeitung (Berlin) 16: 249 (1858).
Cyperus hookerianus Thwaites, Enum. Pl. Zeyl.: 342 (1864).
Cyperus monogynus Boeck., Linnaea 35: 565 (1868).
Cyperus natans Buch.-Ham. ex C.B.Clarke, J. Linn. Soc., Bot. 21: 34 (1884).
Ficinia foliaceobracteata H.Pfeiff., Revis. Ficinia: 35 (1921).

Cyperus chaetophyllus (Chiov.) Kük. in H.G.A.Engler (ed.), Pflanzenr., IV, 20(101): 551 (1936).
S. Somalia. 24 SOM. Hemicr. or rhizome geophyte.
**Mariscus chaetophyllus* Chiov., Result. Sci. Miss. Stefan.-Paoli Somal. Ital. 1: 180 (1916).

Cyperus chalaranthus J.Presl & C.Presl, Reliq. Haenk. 1: 177 (1830). *Cyperus diffusus* var. *chalaranthus* (J. Presl & C.Presl) C.B.Clarke, Bull. Herb. Boissier, II, 3: 1008 (1903). *Cyperus diffusus* subsp. *chalaranthus* (J.Presl & C.Presl) Kük. in H.G.A.Engler (ed.), Pflanzenr., IV, 20(101): 210 (1936).
W. South America to Paraguay. 83 BOL CLM ECU PER 85 PAR. Hemicr. or rhizome geophyte.

Cyperus chamaecephalus Cherm., Bull. Soc. Bot. France 72: 20 (1925).
E. Madagascar. 29 MDG. Hemicr. or rhizome geophyte.

Cyperus chermezonianus Robyns & Tournay in W.Robyns, Fl. Spermatophyt. Parc Nat. Albert 3: 246 (1955).
WC. Trop. Africa to Tanzania. 23 CMN ZAI 25 TAN. Hemicr. or rhizome geophyte.
Cyperus foliosus K.Schum. in H.G.A.Engler, Pflanzenw. Ost-Afrikas, C: 122 (1895), nom. illeg.

Cyperus chersinus (N.E.Br.) Kük. in H.G.A.Engler (ed.), Pflanzenr., IV, 20(101): 525 (1936).
S. Trop. & S. Africa. 26 ZIM 27 BOT CPP NAM NAT TVL. Hemicr. or rhizome geophyte.
**Mariscus chersinus* N.E.Br., Bull. Misc. Inform. Kew 1921: 300 (1921).
Mariscus angularis Turrill, Bull. Misc. Inform. Kew 1925: 69 (1925). *Cyperus chersinus* var. *angularis* (Turrill) Kük. in H.G.A.Engler (ed.), Pflanzenr., IV, 20(101): 525 (1936).
Cyperus bullatus Kük., Repert. Spec. Nov. Regni Veg. 29: 198 (1931). *Mariscus bullatus* (Kük.) Podlech, Mitt. Bot. Staatssamml. München 3: 523 (1960).
Cyperus capensis var. *polyanthemus* Kük. in H.G.A. Engler (ed.), Pflanzenr., IV, 20(101): 540 (1936).

Cyperus chevalieri Kük. in H.G.A.Engler (ed.), Pflanzenr., IV, 20(101): 557 (1936).
Congo to Chad. 23 CON 24 CHA. Hemicr. or rhizome geophyte.
**Mariscus stolonifer* C.B.Clarke, Bull. Soc. Bot. France 54(8): 27 (1907).

Cyperus chinsalensis Podlech, Mitt. Bot. Staatssamml. München 4: 107 (1961).
S. Tanzania to Zambia. 25 TAN 26 ZAM. Hemicr. or rhizome geophyte.

Cyperus chionocephalus (Chiov.) Chiov. ex Chiarugi, Webbia 8: 52 (1951).
Somalia. 24 SOM. Hemicr. or rhizome geophyte.
**Mariscus chionocephalus* Chiov., Miss. Biol. Borana, Racc. Bot., Angiosp.-Gymnosp.: 292 (1939).

Cyperus chlorocephalus (C.B.Clarke) Kük. in H.G.A. Engler (ed.), Pflanzenr., IV, 20(101): 527 (1936).
Mexico to Peru. 79 MXS 80 BLZ GUA 82 FRG GUY VEN 83 ECU PER. Hemicr. or rhizome geophyte.
**Mariscus chlorocephalus* C.B.Clarke, Kew Bull., Addit. Ser. 8: 14 (1908).

Cyperus chordorrhizus Chiov., Agric. Colon. 20: 105 (1926).
Coastal Somalia to E. Kenya. 24 SOM 25 KEN. Hemicr. or rhizome geophyte.

Cyperus chorisanthos C.B.Clarke, Contr. U. S. Natl. Herb. 10: 449 (1908). *Cyperus diffusus* var. *chorisanthos* (C.B.Clarke) Kük. in H.G.A.Engler (ed.), Pflanzenr., IV, 20(101): 210 (1936).
Mexico to C. America. 79 MXT 80 COS NIC PAN. Hemicr. or rhizome geophyte.

Cyperus chrysocephalus (K.Schum.) Kük. in R.E.Fries, Wiss. Erg. Schwed. Rhod.-Kongo Exped. 1: 5 (1921).
Tanzania to S. Trop. Africa. 23 BUR ZAI 25 TAN 26 ANG ZAM. Hemicr. or tuber geophyte.
**Mariscus chrysocephalus* K.Schum. in O.Warburg (ed.), Kunene-Sambesi Exped.: 178 (1903).

Cyperus ciliatus Jungh., Linnaea 6: 25 (1831). *Mariscus ciliatus* (Jungh.) T.Koyama, Phytologia 29: 73 (1974).
S. Mexico (to Veracruz). 79 MXG MXS MXT. Hemicr. or rhizome geophyte.

Cyperus cinereobrunneus Kük., Mitth. Thüring. Bot. Vereins, n.s., 50: 3 (1943).
Papua New Guinea. 43 NWG. Hemicr. or rhizome geophyte.

Cyperus circumclusus (C.B.Clarke) Schweinf., Bull. Herb. Boissier 2(App. 2): 103 (1894).
Trop. Africa. 23 RWA 24 ETH 25 KEN TAN UGA. Hemicr.
Cyperus macropus Boeck., Flora 62: 550 (1879), nom. illeg. *Mariscus macropus* C.B.Clarke in T.A.Durand & H.Schinz, Consp. Fl. Afric. 5: 590 (1894). *Cyperus submacropus* Kük. in H.G.A.Engler (ed.), Pflanzenr., IV, 20(101): 561 (1936).
**Mariscus circumclusus* C.B.Clarke in D.Oliver & auct. suc. (eds.), Fl. Trop. Afr. 8: 387 (1902).
Cyperus submacropus var. *abbreviatus* Kük. in H.G.A. Engler (ed.), Pflanzenr., IV, 20(101): 561 (1936).
Cyperus submacropus var. *calocephalus* A.Peter ex Kük. in H.G.A.Engler (ed.), Pflanzenr., IV, 20(101): 562 (1936).
Cyperus submacropus var. *fuscofibrosus* A.Peter ex Kük. in H.G.A.Engler (ed.), Pflanzenr., IV, 20(101): 562 (1936).

Cyperus clarkei T.Cooke, Fl. Bombay 2: 873 (1908). *Mariscus clarkei* (T.Cooke) T.Koyama, J. Jap. Bot. 51: 316 (1976), nom. illeg.
S. India, Sri Lanka, Vietnam. 40 IND SRL 41 VIE. Hemicr. or rhizome geophyte.
**Mariscus bulbosus* C.B.Clarke in J.D.Hooker, Fl. Brit. India 6: 620 (1893), nom. illeg.

Cyperus clarus S.T.Blake, Proc. Roy. Soc. Queensland 51: 44 (1939 publ. 1940).
E. Australia. 50 NSW QLD. Hemicr. or rhizome geophyte.

Cyperus clavinux C.B.Clarke in D.Oliver & auct. suc. (eds.), Fl. Trop. Afr. 8: 319 (1901).
Nigeria, Chad to S. Africa. 22 NGA 24 CHA 25 TAN 26 MLW ZAM ZIM 27 BOT. Hemicr. or tuber geophyte.
Cyperus monostigma C.B.Clarke, Bull. Soc. Bot. France 58: 26 (1907).

Cyperus clelandii J.M.Black, Trans. & Proc. Roy. Soc. South Australia 48: 253 (1924).
South Australia. 50 SOA. Hemicr. or rhizome geophyte. Provisionally accepted.

Cyperus columbiensis Palla, Oesterr. Bot. Z. 58: 389 (1908).
Colombia. 83 CLM. Hemicr. or rhizome geophyte.

Cyperus colymbetes Kotschy & Peyr., Pl. Tinn.: 49 (1867). *Anosporum colymbetes* (Kotschy & Peyr.) Boeck., Bot. Zeitung (Berlin) 27: 26 (1869).
Sudan to Mozambique, Madagascar. 24 SOM SUD 25 KEN TAN UGA 26 MOZ 29 MDG. Hemicr. or rhizome geophyte.

Cyperus commixtus Kük., Repert. Spec. Nov. Regni Veg. 29: 195 (1931).
N. Somalia. 24 SOM. Hemicr. or rhizome geophyte.

Cyperus compactus Retz., Observ. Bot. 5: 10 (1788). *Mariscus compactus* (Retz.) Bold., Zakfl. Java: 77 (1916).
Madagascar, Trop. & Subtrop. Asia, N. Australia. (29) mau MDG reu 36 CHS 38 TAI 40 ASS EHM IND

NEP PAK SRL 41 AND CBD MYA NCB THA VIE 42 BOR JAW LSI MLY MOL PHI SUL SUM 43 NWG 50 NTA QLD WAU (62) mrn. Hemicr. or rhizome geophyte.

Cyperus congestus Poir. in J.B.A.M.de Lamarck, Encycl. 7: 239 (1806), nom. illeg.

Cyperus spinulosus Roxb., Fl. Ind. 1: 207 (1820).

Cyperus luzonensis J.Presl & C.Presl, Reliq. Haenk. 1: 174 (1828).

Mariscus microcephalus J.Presl & C.Presl, Reliq. Haenk. 1: 182 (1828). *Sphaeromariscus microcephalus* (J.Presl & C.Presl) E.G.Camus, Notul. Syst. (Paris) 1: 239 (1910).

Mariscus dilutus Nees in R.Wight, Contr. Bot. India: 90 (1834).

Cyperus haenkeanus Kunth, Enum. Pl. 2: 93 (1837).

Cyperus rufus Nees ex Kunth, Enum. Pl. 2: 93 (1837), nom. illeg.

Cyperus cuadriflorus Llanos, Fragm. Pl. Filip.: 18 (1851).

Cyperus septatus Steud., Syn. Pl. Glumac. 2: 46 (1854).

Mariscus giganteus Boeck., Flora 42: 443 (1859).

Cyperus dilutus var. *macrostachys* Boeck., Linnaea 36: 355 (1870). *Cyperus compactus* var. *macrostachys* (Boeck.) Kük. in H.G.A.Engler (ed.), Pflanzenr., IV, 20(101): 424 (1936). *Mariscus compactus* var. *macrostachys* (Boeck.) H.B.Naithani & S.Biswas in H.B.Naithani, Flow. Pl. India, Nepal & Bhutan: 471 (1990).

Cyperus grabowskianus Boeck., Bot. Jahrb. Syst. 5: 502 (1884). *Mariscus grabowskianus* (Boeck.) Merr., J. Straits Branch Roy. Asiat. Soc., Spec. No.: 57 (1921).

Cyperus compactus f. *decolorans* Kük., Bot. Jahrb. Syst. 59: 45 (1924).

Mariscus microcephalus var. *pauciflorus* H.Pfeiff., Mitt. Inst. Allg. Bot. Hamburg 7(2): 167 (1928). *Cyperus compactus* var. *pauciflorus* (H.Pfeiff.) Kük. in H.G.A.Engler (ed.), Pflanzenr., IV, 20(101): 424 (1936).

Cyperus compressus L., Sp. Pl.: 46 (1753).

Trop. & Subtrop. 20 EGY 21 CVI 22 BEN BKN GHA GNB GUI IVO NGA SEN SIE TOG 23 CMN EQG GAB GGI 24 ETH SOC SOM 25 KEN TAN UGA 26 ANG MLW MOZ ZAM ZIM 27 BOT NAM NAT SWZ TVL 29 COM MAU MDG REU SEY 34 AFG 36 CHC CHH CHS 38 JAP KZN NNS TAI 40 ASS BAN EHM IND NEP PAK SRL WHM 41 AND LAO MYA NCB THA VIE 42 BOR JAW LSI MOL PHI SUL SUM xms 43 NWG 50 NSW NTA QLD (60) fij GIL nue SAM TON (61) sci 62 CRL MRN MRS (63) haw 74 ILL MSO 75 NWY OHI PEN 77 TEX 78 ALA ARK DEL FLA GEO LOU MRY MSI NCA SCA VRG 79 MXG MXN MXS MXT 80 BLZ COS ELS GUA HON NIC PAN 81 BAH CUB DOM HAI JAM LEE PUE TRT 82 FRG GUY SUR VEN 83 BOL CLM ECU PER 84 BZE BZN. Ther.

Cyperus dilutus Vahl, Enum. Pl. 2: 357 (1805). *Duval-jouvea diluta* (Vahl) Palla, Allg. Bot. Z. Syst. 17(Beil.): 8 (1911).

Cyperus brachiatus Poir. in J.B.A.M.de Lamarck, Encycl. 7: 259 (1806). *Cyperus compressus* var. *brachiatus* (Poir.) Nees in R.Wight, Contr. Bot. India: 77 (1834).

Cyperus pectinatus Roxb., Fl. Ind. 1: 195 (1820), nom. illeg. *Cyperus pectiniformis* Schult. in J.J.Roemer & J.A.Schultes, Mant. 2: 128 (1824). *Cyperus compressus* var. *pectiniformis* (Schult.) C.B.Clarke, J. Linn. Soc., Bot. 21: 97 (1884).

Cyperus compressus var. *compositus* C.Presl, Reliq. Haenk. 1: 177 (1828).

Cyperus compressus var. *simplex* C.Presl, Reliq. Haenk. 1: 177 (1828).

Cyperus myyenii Nees & Arn., Nova Acta Acad. Caes. Leop.-Carol. German. Nat. Cur. 19(Suppl. 1): 57 (1843).

Cyperus caffer G.Bertol., Mem. Reale Accad. Sci. Ist. Bologna 5: 464 (1854).

Cyperus compressus var. *capillaceus* C.B.Clarke in I.Urban, Symb. Antill. 2: 32 (1900).

Cyperus compressus var. *floribundus* E.G.Camus, Notul. Syst. (Paris) 1: 243 (1910).

Cyperus compressus var. *laxus* E.G.Camus, Notul. Syst. (Paris) 1: 242 (1910).

Cyperus compressus f. *depauperatus* Domin, Biblioth. Bot. 85: 430 (1915).

Cyperus compressus subsp. *micranthus* T.Koyama, Gard. Bull. Singapore 30: 139 (1977).

Cyperus concinniformis Kük. in H.G.A.Engler (ed.), Pflanzenr., IV, 20(101): 548 (1936).

E. Trop. Africa. 25 KEN TAN. Hemicr. or rhizome geophyte.

**Mariscus concinnus* C.B.Clarke, J. Bot. 34: 224 (1896).

Cyperus concinnus R.Br., Prodr.: 214 (1810).

N. & E. Australia. 50 NSW NTA QLD VIC WAU. Hemicr. or rhizome geophyte.

Cyperus lepidus F.Muell. ex C.B.Clarke, J. Linn. Soc., Bot. 21: 124 (1884), nom. inval.

Cyperus concinnus var. *parvifructus* Kük. in H.G.A. Engler (ed.), Pflanzenr., IV, 20(101): 252 (1936).

Cyperus confertus Sw., Prodr.: 20 (1788). *Chlorocyperus confertus* (Sw.) Rikll, Jahrb. Wiss. Bot. 27: 564 (1895). *Mariscus confertus* (Sw.) C.B.Clarke in I.Urban, Symb. Antill. 2: 50 (1900), nom. illeg.

Caribbean, Venezuela, Galápagos. 81 ARU CUB DOM HAI JAM LEE NLA PUE TRT WIN 82 VEN 83 GAL. Ther. or hemicr.

Cyperus bidentatus Vahl, Enum. Pl. 2: 330 (1805). *Cyperus confertus* var. *bidentatus* (Vahl) Kük. in H.G.A.Engler (ed.), Pflanzenr., IV, 20(101): 498 (1936). *Mariscus bidentatus* (Vahl) J.Raynal, Adansonia, n.s., 17: 273 (1978).

Cyperus jamaicensis Poir. in J.B.A.M.de Lamarck, Encycl., Suppl. 5: 166 (1817).

Cyperus biuncialis Andersson, Kongl. Vetensk. Acad. Handl. 1853: 156 (1855).

Cyperus dissitiflorus Andersson, Kongl. Vetensk. Acad. Handl. 1853: 153 (1855), nom. illeg.

Cyperus confertus var. *biuncialis* Boeck., Linnaea 35: 515 (1868).

Cyperus confertus f. *debilior* (Kük.) Kük., Repert. Spec. Nov. Regni Veg. 23: 188 (1926).

Cyperus confertus var. *debilior* Kük. ex Ekman, Ark. Bot. 22A(9): 14 (1929).

Cyperus confertus f. *gracillimus* Kük. in H.G.A.Engler (ed.), Pflanzenr., IV, 20(101): 498 (1936).

Cyperus congensis C.B.Clarke in T.A.Durand & H.Schinz, Études Fl. Congo: 285 (1896).

W. Trop. Africa to Tanzania. 22 BEN IVO LBR SEN SIE TOG 23 GAB ZAI 25 TAN. Hemicr. or bulb geophyte.

Cyperus eleusinoides var. *dinklageanus* Kük. in H.G.A. Engler (ed.), Pflanzenr., IV, 20(101): 145 (1935).

Cyperus congestus Vahl, Enum. Pl. 2: 358 (1805). *Mariscus congestus* (Vahl) C.B.Clarke, J. Bot. 35: 72

(1897). *Chlorocyperus congestus* (Vahl) Palla, Allg. Bot. Z. Syst. 8: 66 (1902). *Pycreus congestus* (Vahl) Hayek, Repert. Spec. Nov. Regni Veg. Beih. 30(3): 146 (1932).

S. Africa, introduced elsewere. (11) swi 27 CPP LES NAM NAT OFS TVL (28) sth (33) tcs (50) nfk nsw qld wau (63) haw (90) tdc. Hemicr. or rhizome geophyte.

Cyperus paramatta Mart., Pl. Hort. Erlang.: 24 (1814).

Cyperus bulbosus Lag., Gen. Sp. Pl.: 2 (1816), nom. illeg.

Cyperus subbulbosus Roem. & Schult., Syst. Veg. 2: 186 (1817).

Cyperus polycephalus Link, Enum. Hort. Berol. Alt. 1: 46 (1821), nom. illeg.

Cyperus linkianus Schult. in J.J.Roemer & J.A.Schultes, Mant. 2: 118 (1824).

Cyperus martianus Schult. in J.J.Roemer & J.A.Schultes, Mant. 2: 108 (1824).

Cyperus incarnatus Link, Hort. Berol. 1: 306 (1827).

Cyperus multiceps Link, Hort. Berol. 1: 311 (1827).

Cyperus spiculosus Rchb., Index Seminum (DR) 1828: s.p. (1828).

Cyperus badius Steud., Flora 12: 153 (1829), nom. illeg.

Cyperus schottianus auct., Index Seminum (WU) 1836: s.p. (1836).

Cyperus paramatta Balb. ex Kunth, Enum. Pl. 2: 81 (1837), nom. illeg.

Cyperus carinatus Nees in J.G.C.Lehmann, Pl. Preiss. 2: 72 (1846).

Mariscus congestus var. *glanduliferus* C.B.Clarke in W.H.Harvey & auct. suc. (eds.), Fl. Cap. 7: 192 (1897). *Cyperus congestus* var. *glanduliferus* (C.B.Clarke) Kük. in H.G.A.Engler (ed.), Pflanzenr., IV, 20(101): 446 (1936).

Mariscus cooperi C.B.Clarke in W.H.Harvey & auct. suc. (eds.), Fl. Cap. 7: 197 (1898). *Cyperus cooperi* (C.B.Clarke) K.Schum., Just's Bot. Jahresber. 26(1): 328 (1900).

Cyperus congestus var. *grandiceps* Kük. in H.G.A. Engler (ed.), Pflanzenr., IV, 20(101): 446 (1936).

Cyperus congestus var. *pseudonatalensis* Kük. in H.G.A.Engler (ed.), Pflanzenr., IV, 20(101): 446 (1936).

Cyperus conglomeratus Rottb., Descr. Icon. Rar. Pl.: 21 (1773).

W. Trop. Africa to Sri Lanka. 20 ALG EGY LBY TUN 22 BKN MLI MTN NGR SEN 23 CAF GAB 24 CHA ERI ETH SOC SOM SUD 25 KEN 29 ALD MAU MDG SEY 34 AFG IRN IRQ SIN 35 GST KUW OMA SAU YEM 40 IND PAK SRL. Hemicr. or rhizome geophyte.

subsp. ***conglomeratus***

W. Trop. Africa to S. Iran. 20 ALG EGY LBY TUN 22 BKN MLI MTN NGR SEN 23 CAF GAB 24 CHA ERI ETH SOM SUD 25 KEN 29 ALD MAU MDG SEY 34 AFG IRN IRQ SIN 35 GST KUW OMA SAU YEM 40 IND PAK. Hemicr. or rhizome geophyte.

Cyperus complanatus Forssk., Fl. Aegypt.-Arab.: 14 (1775).

Cyperus globosus Forssk., Fl. Aegypt.-Arab.: 13 (1775).

Cyperus involutus R.Br. in H.Salt, Voy. Abyss., App.: lxii (1814).

Cyperus aucheri Jaub. & Spach, Ill. Pl. Orient. 2: 1, t. 101 (1844). *Cyperus conglomeratus* var. *aucheri* (Jaub. & Spach) C.B.Clarke in T.A. Durand & H.Schinz, Consp. Fl. Afric. 5: 554 (1894).

Cyperus acutiflorus Steud., Syn. Pl. Glumac. 2: 33 (1854).

Cyperus proteinolepis Steud., Syn. Pl. Glumac. 2: 15 (1854).

Cyperus arcuatus Boeck., Linnaea 35: 542 (1868).

Cyperus conglomeratus var. *major* Boeck., Linnaea 35: 544 (1868). *Cyperus conglomeratus* f. *major* (Boeck.) Kük. in H.G.A.Engler (ed.), Pflanzenr., IV, 20(101): 275 (1936).

Cyperus ensifolius Nees & Ehrenb. ex Boeck., Linnaea 35: 545 (1868). *Cyperus conglomeratus* var. *ensifolius* (Nees & Ehrenb. ex Boeck.) Kük. in H.G.A.Engler (ed.), Pflanzenr., IV, 20(101): 276 (1936).

Cyperus falcatus Ehrenb. ex Boeck., Linnaea 35: 547 (1868).

Cyperus proteinolepis var. *pumilus* Boeck., Linnaea 35: 523 (1868). *Cyperus conglomeratus* f. *pumilus* (Boeck.) Kük. in H.G.A.Engler (ed.), Pflanzenr., IV, 20(101): 274 (1936).

Cyperus pungens Boeck., Linnaea 35: 587 (1868), nom. illeg.

Cyperus squarrosus F.Muell., Fragm. 8: 262 (1874), nom. illeg.

Cyperus squarrosus var. *stenocarpus* F.Muell., Fragm. 8: 262 (1874).

Cyperus pungens Boeck. ex Aitch. & Hemsl., Trans. Linn. Soc. London, Bot. 3: 121 (1886), nom. illeg.

Cyperus senegalensis C.B.Clarke in T.A.Durand & H.Schinz, Consp. Fl. Afric. 5: 577 (1894), nom. nud.

Schoenus glaucus Sieber ex C.B.Clarke in T.A. Durand & H.Schinz, Consp. Fl. Afric. 5: 554 (1894), nom. inval.

Schoenus macrorrhizus Rich. ex C.B.Clarke in T.A.Durand & H.Schinz, Consp. Fl. Afric. 5: 554 (1894), nom. inval.

Cyperus adansonii C.B.Clarke in D.Oliver & auct. suc. (eds.), Fl. Trop. Afr. 8: 335 (1901).

Cyperus nubianus Gand., Bull. Soc. Bot. France 66: 297 (1919 publ. 1920).

Cyperus pappii Gand., Bull. Soc. Bot. France 66: 297 (1919 publ. 1920).

Cyperus conglomeratus f. *oligostachys* Kük. in H.G.A.Engler (ed.), Pflanzenr., IV, 20(101): 274 (1936).

Cyperus conglomeratus f. *strictus* Kük. in H.G.A. Engler (ed.), Pflanzenr., IV, 20(101): 274 (1936).

subsp. ***curvulus*** (Boeck.) Kukkonen, Ann. Bot. Fenn. 32: 158 (1995).

Arabian Pen. to Pakistan. 24 SOC (29) mdg 34 IRN 35 SAU YEM 40 PAK. Hemicr. or rhizome geophyte.

Cyperus effusus Rottb., Descr. Icon. Rar. Pl.: 22 (1773). *Cyperus conglomeratus* var. *effusus* (Rottb.) Boiss., Fl. Orient. 5: 369 (1882).

**Cyperus curvulus* Boeck., Linnaea 35: 541 (1868). *Cyperus conglomeratus* f. *curvulus* (Boeck.) Kük. in H.G.A.Engler (ed.), Pflanzenr., IV, 20(101): 275 (1936).

Cyperus conglomeratus var. *socotrensis* C.B.Clarke, J. Linn. Soc., Bot. 21: 113 (1884).

subsp. ***pachyrrhizus*** (Nees) T.Koyama, Bot. Mag. (Tokyo) 93: 344 (1980).

S. India, Sri Lanka. 40 IND SRL. Hemicr. or rhizome geophyte.
**Cyperus pachyrrhizus* Nees, Linnaea 35: 545 (1868). *Cyperus conglomeratus* f. *pachyrrhizus* (Nees) Kük. in H.G.A.Engler (ed.), Pflanzenr., IV, 20(101): 275 (1936).

Cyperus conicus (R.Br.) Boeck., Linnaea 38: 371 (1874).
N. & E. Australia. 50 NSW NTA QLD WAU. Hemicr. or rhizome geophyte.
**Mariscus conicus* R.Br., Prodr.: 218 (1810).
Cyperus glaucinus Boeck., Flora 58: 89 (1875). *Cyperus conicus* var. *glaucinus* (Boeck.) Kük. in H.G.A.Engler (ed.), Pflanzenr., IV, 20(101): 481 (1936).

Cyperus consors C.B.Clarke, Kew Bull., Addit. Ser. 8: 6 (1908).
SE. & S. Brazil. 84 BZL BZS. Hemicr. or rhizome geophyte.
Cyperus riedelianus C.B.Clarke, Kew Bull., Addit. Ser. 8: 6 (1908). *Cyperus consors* var. *riedelianus* (C.B. Clarke) Kük. in H.G.A.Engler (ed.), Pflanzenr., IV, 20(101): 231 (1936).
Cyperus consors var. *pernudus* Kük. in H.G.A.Engler (ed.), Pflanzenr., IV, 20(101): 231 (1936).

Cyperus constanzae Urb., Symb. Antill. 7: 168 (1912).
Caribbean. 81 CUB DOM HAI JAM. Hemicr. or rhizome geophyte.
Cyperus ignotus Britton, Bull. Torrey Bot. Club 43: 442 (1916).

Cyperus coriifolius Boeck., Linnaea 36: 391 (1870).
Trop. America. 80 PAN 81 PUE WIN 84 BZL. Hemicr. or rhizome geophyte.
Mariscus elatus Vahl, Enum. Pl. 2: 377 (1805).
Mariscus alatus Schult. in J.J.Roemer & J.A.Schultes, Mant. 2: 144 (1824).
Mariscus latifolius Schrad. ex Nees in C.F.P.von Martius & auct. suc. (eds.), Fl. Bras. 2(1): 50 (1842).
Mariscus procerus Schrad. ex Nees in C.F.P.von Martius & auct. suc. (eds.), Fl. Bras. 2(1): 46 (1842).
Cyperus coriifolius var. *abbreviatus* Boeck., Linnaea 36: 392 (1870).
Mariscus incompletus Urb., Symb. Antill. 2: 165 (1900).

Cyperus cornelii-ostenii Kük., Repert. Spec. Nov. Regni Veg. 29: 198 (1931). *Mariscus cornelii-ostenii* (Kük.) T.Koyama, Phytologia 29: 73 (1974).
S. Trop. America. 82 VEN 83 BOL 84+ 85 AGE AGW PAR. Hemicr. or rhizome geophyte.

var. ***cornelii-ostenii***
S. Trop. America. 82 VEN 83 BOL 84 85 AGE AGW PAR. Hemicr. or rhizome geophyte.
Cyperus laetus var. *obtusiflorus* Boeck., Allg. Bot. Z. Syst. 2: 2 (1896).

var. ***pilcomayensis*** Kük., Repert. Spec. Nov. Regni Veg. 29: 199 (1931).
Paraguay to N. Argentina. 85 AGE AGW PAR. Hemicr. or rhizome geophyte.

Cyperus coronarius (Vahl) Kunth, Enum. Pl. 2: 44 (1837).
Bangladesh. 40 BAN. Hemicr. or rhizome geophyte. Provisionally accepted.
**Scirpus coronarius* Vahl, Enum. Pl. 2: 261 (1805).

Cyperus correllii (T.Koyama) G.C.Tucker, Sida 10: 305 (1984).
Bahamas. 81 BAH. Hemicr. or rhizome geophyte.
**Torulinium correllii* T.Koyama, Brittonia 28: 252 (1976).

Cyperus corymbosus Rottb., Descr. Icon. Rar. Pl.: 42 (1773).
Africa to Indo-China, Trop. America. 20 ALG 22 IVO TOG 25 TAN 26 ANG MOZ 27 NAT TVL 29 MDG 34 IRQ PAL 40 ASS BAN IND NEP PAK SRL WHM 41 MYA THA VIE (50) wau 81 CUB 82 GUY SUR VEN 84 BZL. Hel. or rhizome geophyte.
Cyperus diphyllus Retz., Observ. Bot. 5: 11 (1788).
Cyperus pallescens Desf., Fl. Atlant. 1: 45 (1798).
Cyperus koenigii Vahl, Enum. Pl. 2: 302 (1805).
Cyperus seminudus Roxb., Fl. Ind. 1: 191 (1820).
Cyperus bengalensis Spreng., Neue Entdeck. Pflanzenk. 3: 101 (1822).
Cyperus gula-metthi Schult. in J.J.Roemer & J.A.Schultes, Mant. 2: 125 (1824).
Cyperus nudus Roxb., Fl. Ind. ed. 1832, 1: 209 (1832), nom. illeg.
Cyperus tegetiformis Roxb. ex Arn. in R.Wight, Contr. Bot. India: 89 (1834).
Cyperus roestelii Kunth, Enum. Pl. 2: 58 (1837).
Cyperus tenuicomus Miq., Linnaea 19: 223 (1847).
Cyperus corymbosus var. *macrostachyus* Boeck., Linnaea 36: 277 (1870).
Cyperus diphyllus var. *triangularis* Boeck., Linnaea 36: 273 (1870).
Cyperus enodis Boeck., Linnaea 36: 271 (1870).
Cyperus diphyllus var. *elatior* Benth., Fl. Austral. 7: 279 (1878).
Cyperus corymbosus var. *pangorei* C.B.Clarke, J. Linn. Soc., Bot. 21: 159 (1884).
Cyperus corymbosus var. *brevispiculosus* Kuntze, Revis. Gen. Pl. 2: 749 (1891).
Cyperus enodis var. *longispiculatus* Kuntze, Revis. Gen. Pl. 2: 749 (1891). *Cyperus corymbosus* var. *longispiculatus* (Kuntze) Kük. in H.G.A.Engler (ed.), Pflanzenr., IV, 20(101): 82 (1935).

Cyperus costaricensis Gómez-Laur., Phytologia 52: 228 (1982).
Costa Rica to Panama. 80 COS PAN. Hemicr. or rhizome geophyte.
Cyperus amplus G.C.Tucker, Syst. Bot. Monogr. 2: 23 (1983).

Cyperus cracens K.L.Wilson, Telopea 4: 435 (1991).
N. Australia. 50 NTA WAU. Hemicr. or rhizome geophyte.

Cyperus crassipes Vahl, Enum. Pl. 2: 299 (1805). *Cyperus maritimus* var. *crassipes* (Vahl) C.B.Clarke in T.A.Durand & H.Schinz, Consp. Fl. Afric. 5: 569 (1894).
Africa. 21 CVI 22 BEN GAM GHA GNB GUI IVO LBR NGA SEN SIE TOG 23 EQG GAB ZAI 24 SOM 25 KEN TAN 26 ANG MOZ 27 CPP NAT 29 MDG. Hemicr. or rhizome geophyte.
Cyperus scirpoides Vahl, Enum. Pl. 2: 311 (1805).
Cyperus maritimus Poir. in J.B.A.M.de Lamarck, Encycl. 7: 240 (1806).
Cyperus jardinei Steud., Syn. Pl. Glumac. 2: 28 (1854).
Cyperus cephalostachyus Steud., Syn. Pl. Glumac. 2: 315 (1855).

Cyperus heterophyllus Boeck., Flora 42: 441 (1859).
Cyperus brevis Boeck., Linnaea 36: 341 (1870). *Cyperus congestus* var. *brevis* (Boeck.) Kük. in H.G.A.Engler (ed.), Pflanzenr., IV, 20(101): 445 (1936).
Cyperus maritimus var. *gracilescens* Kük. in H.G.A. Engler (ed.), Pflanzenr., IV, 20(101): 270 (1936).

Cyperus crassivaginatus Lye, Nordic J. Bot. 3: 217 (1983).
Uganda to W. Kenya. 25 KEN UGA. Hemicr. or tuber geophyte.

Cyperus cremeomariscus Lye, Nordic J. Bot. 3: 218 (1983).
Tanzania (Sumbawanga Distr.). 25 TAN. Hemicr. or tuber geophyte.

Cyperus crispulus K.L.Wilson, Telopea 4: 430 (1991).
N. Australia. 50 NTA WAU. Hemicr. or rhizome geophyte.

Cyperus cristulatus S.T.Blake, Proc. Roy. Soc. Queensland 51: 35 (1939 publ. 1940).
N. Australia. 50 NTA QLD WAU. Hemicr. or rhizome geophyte.

Cyperus croceus Vahl, Enum. Pl. 2: 357 (1805).
C. & E. U.S.A. to Trop. America. 74 MSO OKL 75 NWJ PEN WVA 77 TEX 78 ALA ARK DEL FLA GEO LOU MRY MSI NCA SCA TEN VRG 79 MXT 80 COS GUA HON NIC PAN 81 BAH BER CUB HAI JAM PUE 82 FRG GUY SUR VEN 83 PER. Hemicr. or rhizome geophyte.
Cyperus baldwinii Torr., Ann. Lyceum Nat. Hist. New York 3: 270 (1836). *Dichostylis baldwinii* (Torr.) Palla, Bot. Jahrb. Syst. 10: 299 (1888).
Cyperus cyclostachyus Griseb., Fl. Brit. W. I.: 567 (1864).
Cyperus echinatus var. *multiflorus* Britton, Bull. Torrey Bot. Club 15: 99 (1888). *Cyperus multiflorus* (Britton) Small, Man. S.E. Fl.: 153 (1933).
Cyperus globulosus var. *multiflorus* Kük. & Ekman, Ark. Bot. 22A(17): 4 (1929).
Cyperus globulosus var. *pseudofiliculmis* Kük. in H.G.A.Engler (ed.), Pflanzenr., IV, 20(101): 511 (1936).

Cyperus cruentus Rottb., Descr. Icon. Rar. Pl.: 21 (1773).
N. & NE. Trop. Africa, Arabian Pen. 20 24 ETH SOM 35 SAU YEM. Hemicr. or rhizome geophyte.
Mariscus schimperi Hochst. ex A.Rich., Tent. Fl. Abyss. 2: 491 (1850). *Cyperus neoschimperi* Kük. in H.G.A.Engler (ed.), Pflanzenr., IV, 20(101): 552 (1936).
Mariscus viridis Hochst. ex Schweinf., Beitr. Fl. Aethiop.: 217 (1867). *Cyperus neoschimperi* var. *viridis* (Hochst. ex Schweinf.) Kük. in H.G.A.Engler (ed.), Pflanzenr., IV, 20(101): 552 (1936).
Cyperus variegatus Boeck., Linnaea 36: 337 (1870), nom. illeg.
Cyperus neoschimperi var. *subvirescens* A.Peter & Kük. in H.G.A.Engler (ed.), Pflanzenr., IV, 20(101): 552 (1936).

Cyperus crypsoides J.Kern, Reinwardtia 2: 128 (1952).
Sulawesi. 42 SUL. Hemicr. or rhizome geophyte. Provisionally accepted.

Cyperus cunduduensis Chiov., Atti Reale Accad. Ital., Mem. Cl. Sci. Fis. 11: 60 (1940).
E. Ethiopia to N. Somalia. 24 ETH SOM. Hemicr.

Cyperus cunninghamii (C.B.Clarke) C.A.Gardner, Enum. Pl. Austr. Occ.: 12 (1930).
Australia. 50 NTA QLD SOA WAU. Hemicr. or rhizome geophyte.
**Mariscus cunninghamii* C.B.Clarke, Kew Bull., Addit. Ser. 8: 18 (1908).

subsp. ***cheradicus*** K.L.Wilson, Telopea 4: 417 (1991).
N. Australia. 50 NTA QLD WAU. Hemicr. or rhizome geophyte.

subsp. ***cunninghamii***
Australia. 50 NTA QLD SOA WAU. Hemicr. or rhizome geophyte.
Cyperus umbellatus var. *fasciculigerus* F.Muell., Sci. Exp. Centr. Austr. 3: 181 (1896). *Mariscus fasciculigerus* (F.Muell.) Domin, Biblioth. Bot. 85: 443 (1915).

subsp. ***uniflorus*** K.L.Wilson, Telopea 4: 419 (1991).
N. Australia. 50 NTA WAU. Hemicr. or rhizome geophyte.

Cyperus curvistylis J.Kern, Blumea 13: 123 (1965).
Queensland. 50 QLD. Hemicr. or rhizome geophyte.

Cyperus cuspidatus Kunth in F.W.H.von Humboldt, A.J.A.Bonpland & C.S.Kunth, Nov. Gen. Sp. 1: 204 (1817). *Dichostylis cuspidata* (Kunth) Palla, Bot. Jahrb. Syst. 10: 296 (1888).
Trop. & S. Africa to N. Australia, Trop. America. 22 BEN BKN GHA GNB GUI IVO LBR MLI NGA NGR SEN SIE TOG 23 BUR CAF GAB GGI ZAI 24 CHA ETH SOM SUD 25 KEN TAN UGA 26 ANG MLW MOZ ZAM ZIM 27 BOT NAM NAT SWZ TVL 29 MAU MDG 36 CHC CHS 38 TAI 40 ASS BAN EHM IND NEP PAK SRL WHM 41 THA VIE 42 BOR JAW LSI MLY MOL PHI SUL SUM 50 NTA QLD WAU 78 ALA FLA GEO SCA 79 MXC MXG MXN MXS MXT 80 ELS HON 81 CUB 82 FRG GUY SUR VEN 83 CLM PER 84 BZE BZN. Ther.
Cyperus capitatus Retz., Observ. Bot. 4: 9 (1786), nom. illeg.
Cyperus recurvus Vahl, Enum. Pl. 2: 310 (1805), provisional synonym.
Cyperus uncinatus Poir. in J.B.A.M.de Lamarck, Encycl. 7: 247 (1806).
Cyperus setifolius D.Don, Prodr. Fl. Nepal.: 38 (1825).
Cyperus angustifolius Schumach. & Thonn. in C.F.Schumacher, Beskr. Guin. Pl.: 38 (1827). *Cyperus cuspidatus* f. *angustifolius* (Schumach. & Thonn.) Kük. in H.G.A.Engler (ed.), Pflanzenr., IV, 20(101): 264 (1936).
Cyperus exiguus Nees, Linnaea 9: 284 (1834), nom. inval. *Cyperus cuspidatus* f. *exiguus* (Nees) Kük. in H.G.A.Engler (ed.), Pflanzenr., IV, 20(101): 264 (1936).
Cyperus hamulatus Kunth, Enum. Pl. 2: 22 (1837).
Cyperus recurvatus Willd. ex Kunth, Enum. Pl. 2: 22 (1837).
Cyperus hamatus Schrad. ex Nees in C.F.P.von Martius & auct. suc. (eds.), Fl. Bras. 2(1): 23 (1842).
Cyperus contractus Steud., Syn. Pl. Glumac. 2: 13 (1854).
Cyperus salzmannii Steud., Syn. Pl. Glumac. 2: 15 (1854).
Cyperus solutus Steud., Syn. Pl. Glumac. 2: 14 (1854).
Cyperus cuspidatus var. *burchellii* C.B.Clarke, J. Linn. Soc., Bot. 21: 90 (1884).
Cyperus gratus C.B.Clarke, Bot. Jahrb. Syst. 38: 133 (1906). *Cyperus uncinatus* var. *gratus* (C.B.Clarke)

Kük. in H.G.A.Engler (ed.), Pflanzenr., IV, 20(101): 261 (1936).

Cyperus waterloti Cherm., Bull. Soc. Bot. France 66: 339 (1919 publ. 1920).

Cyperus cylindrostachyus Boeck., Linnaea 36: 383 (1870).

Africa, Caribbean to N. South America. 22 BEN BKN GAM GHA GNB IVO LBR MLI NGA SEN SIE TOG 23 CMN GAB GGI ZAI 24 CHA ERI ETH SUD 25 KEN TAN UGA 26 ANG MLW 27 NAM TVL 28 STH 29 COM MDG 81 CUB TRT WIN 82 SUR. Hemicr. or rhizome geophyte.

Kyllinga umbellata Rottb., Descr. Icon. Rar. Pl.: 15 (1773). *Mariscus umbellatus* (Rottb.) Vahl, Enum. Pl. 2: 376 (1805). *Cyperus umbellatus* (Rottb.) Benth., Fl. Hongk.: 386 (1861), nom. illeg. *Cyperus subumbellatus* Kük. in H.G.A.Engler (ed.), Pflanzenr., IV, 20(101): 523 (1936), nom. illeg.

Cyperus umbellatus var. *laxiflorus* Benth., Fl. Austral. 7: 289 (1878).

Mariscus thomensis C.B.Clarke in D.Oliver & auct. suc. (eds.), Fl. Trop. Afr. 8: 395 (1902). *Cyperus subumbellatus* var. *thomensis* (C.B.Clarke) Kük. in H.G.A.Engler (ed.), Pflanzenr., IV, 20(101): 525 (1936).

Cyperus subumbellatus var. *sessilispicatus* Kük. in H.G.A.Engler (ed.), Pflanzenr., IV, 20(101): 525 (1936).

Cyperus subumbellatus var. *subglobosus* Kük. in H.G.A. Engler (ed.), Pflanzenr., IV, 20(101): 524 (1936).

Cyperus cymulosus Willemet, Ann. Bot. (Usteri) 18: 7 (1796).

Mauritius. 29 MAU. Hemicr. or rhizome geophyte.

Cyperus cyperinus (Retz.) Suringar, Cyperus Mal. Arch.: 154 (1898).

Trop. & Subtrop. Asia to Pacific. 29 SEY 35 YEM 36 CHC CHH CHS 38 NNS OGA TAI 40 ASS BAN EHM IND NEP SRL 41 MYA THA VIE 42 BOR CKI JAW LSI MLY MOL PHI SUL SUM XMS 43 NWG 50 QLD 60 FIJ NWC SAM TON VAN 61 COO EAS MRQ SCI TUB 62 CRL MRN 63 HAW. Hemicr. or rhizome geophyte.

Cyperus venustus G.Forst., Fl. Ins. Austr.: 89 (1786), nom. nud.

Opetiola myosuroides Gaertn., Fruct. Sem. Pl. 1: 14 (1788), provisional synonym.

**Kyllinga cyperina* Retz., Observ. Bot. 6: 21 (1791). *Mariscus cyperinus* (Retz.) Vahl, Enum. Pl. 2: 377 (1805).

Kyllinga bulbosa J.König ex Vahl, Enum. Pl. 2: 376 (1805), nom. illeg.

Kyllinga rigida Link, Hort. Berol. 2: 323 (1827), nom. illeg.

Mariscus pictus Nees in R.Wight, Contr. Bot. India: 90 (1834). *Cyperus cyperinus* var. *pictus* (Nees) Kük. in H.G.A.Engler (ed.), Pflanzenr., IV, 20(101): 519 (1936). *Mariscus cyperinus* var. *pictus* (Nees) Karthik. in S.Karthikeyan & al., Fl. Ind. Enumerat. – Monocot.: 63 (1989).

Mariscus tenuifolius Schrad. ex Nees in C.F.P.von Martius & auct. suc. (eds.), Fl. Bras. 2(1): 46 (1842).

Cyperus ischnostachys Steud., Syn. Pl. Glumac. 2: 53 (1854).

Mariscus sundaicus Miq., Fl. Ned. Ind. 3: 289 (1859).

Cyperus graeffei Boeck., Flora 58: 84 (1875).

Cyperus andersonianus Boeck., Bot. Jahrb. Syst. 5: 502 (1884).

Cyperus nanilensis Boeck., Bot. Jahrb. Syst. 5: 501 (1884).

Cyperus umbellatus var. *laxatus* C.B.Clarke, J. Linn. Soc., Bot. 21: 201 (1884). *Cyperus cyperinus* var. *laxatus* (C.B.Clarke) Kük. in H.G.A.Engler (ed.), Pflanzenr., IV, 20(101): 519 (1936). *Mariscus cyperinus* subsp. *laxatus* (C.B.Clarke) T.Koyama, Gard. Bull. Singapore 30: 157 (1977).

Mariscus cyperinus var. *bengalensis* C.B.Clarke in J.D.Hooker, Fl. Brit. India 6: 621 (1893). *Cyperus cyperinus* var. *bengalensis* (C.B.Clarke) Kük. in H.G.A.Engler (ed.), Pflanzenr., IV, 20(101): 520 (1936).

Cyperus samoensis Boeck., Bot. Jahrb. Syst. 25: 586 (1898). *Torulinium samoense* (Boeck.) Palla, Denkschr. Kaiserl. Akad. Wiss., Wien. Math.-Naturwiss. Kl. 84: 452 (1909).

Mariscus cyperinus var. *maximus* C.B.Clarke, J. Linn. Soc., Bot. 34: 43 (1898). *Cyperus cyperinus* var. *maximus* (C.B.Clarke) Kük. in H.G.A.Engler (ed.), Pflanzenr., IV, 20(101): 520 (1936).

Cyperus cyperinus var. *scabriculmis* Kük. in H.G.A. Engler (ed.), Pflanzenr., IV, 20(101): 520 (1936).

Cyperus cyperoides (L.) Kuntze, Revis. Gen. Pl. 3(2): 333 (1898).

Africa to Pacific. 21 CVI 22 BEN BKN NGA TOG 23 BUR CMN EQG GGI RWA ZAI 24 CHA 25 KEN TAN UGA 26 ANG MLW MOZ 27 BOT NAM NAT SWZ TVL 28 ASC STH 29 COM MAU MDG 36 CHC CHH CHS 38 JAP KOR KZN NNS OGA TAI 40 ASS EHM IND NEP PAK SRL WHM 41 MYA THA VIE 42 BOR JAW LSI MLY MOL PHI SUL SUM xms 43 BIS NWG SOL 50 NSW QLD (60) fij NUE SAM ton 61 COO EAS SCI 62 MRN (81) jam trt. Hemicr. or rhizome geophyte.

**Scirpus cyperoides* L., Mant. Pl. 2: 181 (1771). *Mariscus cyperoides* (L.) Urb., Symb. Antill. 2: 164 (1900), nom. illeg.

subsp. ***cyperoides***

Africa to Pacific. 21 CVI 22 BEN BKN NGA TOG 23 BUR CMN EQG GAB GGI RWA ZAI 24 CHA 25 KEN TAN UGA 26 ANG MLW MOZ 27 BOT NAM NAT SWZ TVL 28 ASC STH 29 COM MAU MDG 36 CHC CHH CHS 38 JAP KOR KZN NNS OGA TAI 40 ASS EHM IND NEP PAK SRL WHM 41 MYA THA VIE 42 BOR JAW LSI MLY MOL PHI SUL SUM xms 43 BIS NWG SOL 50 NSW QLD (60) fij NUE SAM ton 61 COO EAS SCI 62 MRN (81) jam trt. Hemicr. or rhizome geophyte.

Kyllinga sumatrensis Retz., Observ. Bot. 4: 13 (1786). *Mariscus sumatrensis* (Retz.) J.Raynal, Adansonia, n.s., 15: 110 (1975).

Mariscus biglumis Gaertn., Fruct. Sem. Pl. 1: 12 (1788). *Kyllinga biglumis* (Gaertn.) Steud., Nomencl. Bot., ed. 2, 1: 851 (1840).

Mariscus alternifolius Vahl, Enum. Pl. 2: 376 (1805).

Mariscus umbellatus J.Presl & C.Presl, Reliq. Haenk. 1: 181 (1828), nom. illeg.

Mariscus concinnus Schrad. ex Nees in C.F.P.von Martius & auct. suc. (eds.), Fl. Bras. 2(1): 47 (1842).

Mariscus polyphyllus Steud., Flora 25: 596 (1842), nom. illeg. *Cyperus cyperoides* var. *polyphyllus* Kük. in H.G.A.Engler (ed.), Pflanzenr., IV, 20(101): 517 (1936).

Mariscus cylindristachyus Steud., Syn. Pl. Glumac.

2: 65 (1854). *Cyperus cyperoides* subsp. *flavus* Lye, Nordic J. Bot. 3: 231 (1983).
Mariscus nossibeensis Steud., Syn. Pl. Glumac. 2: 63 (1854). *Mariscus sieberianus* var. *nossibeensis* (Steud.) Cherm., Ann. Inst. Bot.-Géol. Colon. Marseille 27: 55 (1919). *Cyperus cyperoides* var. *nossibeensis* (Steud.) Kük. in H.G.A.Engler (ed.), Pflanzenr., IV, 20(101): 517 (1936).
Mariscus pachystachyus Steud., Syn. Pl. Glumac. 2: 65 (1854).
Mariscus philippensis Steud., Syn. Pl. Glumac. 2: 66 (1854).
Mariscus corymbosus Boeck., Flora 41: 409 (1858).
Cyperus steudelianus Boeck., Linnaea 36: 382 (1870). *Mariscus steudelianus* (Boeck.) Cufod., Bull. Jard. Bot. Natl. Belg. 40(Suppl.): 1457 (1970).
Cyperus biglumis C.B.Clarke, J. Linn. Soc., Bot. 21: 199 (1884).
Cyperus manilensis Boeck. ex Warb., Bot. Jahrb. Syst. 13: 264 (1891).
Mariscus sieberianus Nees ex C.B.Clarke in J.D.Hooker, Fl. Brit. India 6: 622 (1893). *Cyperus sieberianus* (Nees ex C.B.Clarke) K.Schum., Notizbl. Königl. Bot. Gart. Berlin 2: 96 (1898), nom. illeg.
Mariscus sieberianus var. *evolutior* C.B.Clarke in J.D.Hooker, Fl. Brit. India 6: 622 (1893). *Cyperus cyperoides* var. *evolutior* (C.B.Clarke) Kük., Sinensia 3: 80 (1932). *Mariscus sumatrensis* var. *evolutior* (C.B.Clarke) Karthik. in S.Karthikeyan & al., Fl. Ind. Enumerat. - Monocot.: 64 (1989). *Mariscus concinnus* var. *evolutior* (C.B.Clarke) Panigrahi in S.K.Murti & G.Panigrahi, Fl. Bilaspur 2: 699 (1999).
Mariscus sieberianus var. *khasianus* C.B.Clarke in J.D.Hooker, Fl. Brit. India 6: 623 (1893). *Cyperus cyperoides* var. *khasianus* (C.B.Clarke) Kük. in H.G.A.Engler (ed.), Pflanzenr., IV, 20(101): 518 (1936). *Mariscus sumatrensis* var. *khasianus* (C.B.Clarke) Karthik. in S.Karthikeyan & al., Fl. Ind. Enumerat. - Monocot.: 64 (1989). *Mariscus concinnus* var. *khasianus* (C.B.Clarke) Panigrahi in S.K.Murti & G.Panigrahi, Fl. Bilaspur 2: 699 (1999).
Mariscus sieberianus var. *subcompositus* C.B.Clarke in J.D.Hooker, Fl. Brit. India 6: 622 (1893). *Cyperus cyperoides* var. *subcompositus* (C.B.Clarke) Kük. in H.G.A.Engler (ed.), Pflanzenr., IV, 20(101): 516 (1936). *Mariscus sumatrensis* var. *subcompositus* (C.B.Clarke) Karthik. in S.Karthikeyan & al., Fl. Ind. Enumerat. - Monocot.: 64 (1989). *Mariscus concinnus* var. *subcompositus* (C.B.Clarke) Panigrahi in S.K.Murti & G.Panigrahi, Fl. Bilaspur 2: 699 (1999).
Mariscus sublimis C.B.Clarke in T.A.Durand & H.Schinz, Consp. Fl. Afric. 5: 594 (1894). *Cyperus subumbellatus* var. *sublimis* (C.B. Clarke) Kük. in H.G.A.Engler (ed.), Pflanzenr., IV, 20(101): 525 (1936). *Cyperus sublimis* (C.B.Clarke) Dandy in A.W.Exell, Cat. Vasc. Pl. S. Tome: 362 (1944).
Cyperus flexifolius Boeck. ex Reinecke, Bot. Jahrb. Syst. 25: 586 (1898), nom. illeg.
Cyperus konkanensis T.Cooke, Fl. Bombay 2: 874 (1908).
Mariscus clarkei Turrill ex Craib, Bull. Misc. Inform. Kew 1912: 422 (1912). *Cyperus cyperoides* var. *clarkei* (Turrill ex Craib) Kük. in H.G.A.Engler (ed.), Pflanzenr., IV, 20(101): 518 (1936).
Mariscus cyperoides var. *capitatus* Domin, Biblioth. Bot. 85: 438 (1915).
Mariscus quarrei Cherm., Rev. Zool. Bot. Africaines 24: 294 (1934).
Cyperus cyperoides var. *brevispicatus* Kük. in H.G.A.Engler (ed.), Pflanzenr., IV, 20(101): 517 (1936).
Cyperus cyperoides var. *microstachyys* Kük. in H.G.A.Engler (ed.), Pflanzenr., IV, 20(101): 518 (1936).
Cyperus cyperoides var. *ovalis* Kük. in H.G.A. Engler (ed.), Pflanzenr., IV, 20(101): 518 (1936).
Cyperus cyperoides var. *repens* Kük. in H.G.A. Engler (ed.), Pflanzenr., IV, 20(101): 518 (1936).

subsp. ***pseudoflavus*** (C.B.Clarke) Lye, Nordic J. Bot. 3: 231 (1983).
E. Trop. Africa. 25 KEN TAN UGA. Hemicr. or rhizome geophyte.
Mariscus macer Kunth, Enum. Pl. 2: 121 (1837).
**Mariscus pseudoflavus* C.B.Clarke, Ill. Cyper.: t. 23, f. 8–9 (1909). *Cyperus macrocarpus* var. *pseudoflavus* (C.B.Clarke) Kük. in H.G.A. Engler (ed.), Pflanzenr., IV, 20(101): 529 (1936).

Cyperus cyprius Post, Mém. Herb. Boissier 18: 102 (1900).
W. Cyprus. 34 CYP. Hemicr. or rhizome geophyte.

Cyperus dactyliformis Boeck., Linnaea 36: 329 (1870). *Mariscus dactyliformis* (Boeck.) C.B.Clarke, Trans. Linn. Soc. London, Bot. 4: 53 (1894).
S. Trop. & S. Africa. 26 ZIM 27 CPP NAT TVL. Hemicr. or rhizome geophyte.
Cyperus tabularis var. *major* Nees, Linnaea 10: 137 (1836).
Cyperus solidus var. *elatior* Kunth, Enum. Pl. 2: 76 (1837). *Cyperus dactyliformis* var. *elatior* (Kunth) Kük., Bot. Not. 1934: 70 (1934).
Cyperus elatior Boeck., Linnaea 36: 327 (1870). *Mariscus elatior* (Boeck.) C.B.Clarke in T.A.Durand & H.Schinz, Consp. Fl. Afric. 5: 586 (1894).
Cyperus toisensis Kuntze, Revis. Gen. Pl. 3(2): 334 (1898).
Mariscus gueinzii C.B.Clarke in W.H.Harvey & auct. suc. (eds.), Fl. Cap. 7: 195 (1898). *Cyperus dactyliformis* var. *gueinzii* (C.B.Clarke) Kük., Bot. Not. 1934: 70 (1934).
Mariscus involutus C.B.Clarke in W.H.Harvey & auct. suc. (eds.), Fl. Cap. 7: 194 (1898). *Cyperus dactyliformis* var. *involutus* (C.B.Clarke) K.Schum., Just's Bot. Jahresber. 26: 328 (1900). *Cyperus involutus* (C.B.Clarke) K.Schum., Just's Bot. Jahresber. 26(1): 328 (1900).

Cyperus dactylotes Benth., Fl. Austral. 7: 273 (1878).
Australia. 50 NSW NTA QLD SOA WAU. Hemicr. or rhizome geophyte.

Cyperus × deamii O'Neill, Rhodora 44: 81 (1942). *C. esculentus* × *C. strigosus*.
E. U.S.A. 75 INI NWJ 78 DEL. Hemicr. or rhizome geophyte.
**Cyperus strigosus* var. *multiflorus* Geise, Amer. Midl. Naturalist 15: 253 (1934).

Cyperus debilis R.Br., Prodr.: 213 (1810).
E. Australia. 50 NSW QLD. Hemicr. or rhizome geophyte.

Cyperus debilissimus Baker, J. Linn. Soc., Bot. 22: 532 (1887).
C. Madagascar. 29 MDG. Hemicr. or rhizome geophyte.
Cyperus calochrous Cherm., Bull. Soc. Bot. France 66: 312 (1919 publ. 1920). *Cyperus debilissimus* var. *calochrous* (Cherm.) Kük. in H.G.A.Engler (ed.), Pflanzenr., IV, 20(101): 188 (1936).
Cyperus debilissimus var. *triqueter* Cherm., Bull. Mus. Natl. Hist. Nat. 27: 553 (1921).

Cyperus deciduus Boeck., Flora 62: 547 (1879). *Mariscus deciduus* (Boeck.) C.B.Clarke in W.H. Harvey & auct. suc. (eds.), Fl. Cap. 7: 191 (1897).
Zaïre to S. Africa. 23 ZAI 25 TAN 26 ANG ZIM 27 BOT NAM SWZ TVL. Hemicr. or rhizome geophyte.
Cyperus wittei Cherm., Rev. Zool. Bot. Africaines 22: 64 (1932).

Cyperus decompositus (R.Br.) F.Muell., Fragm. 8: 267 (1874).
N. Australia. 50 NTA QLD. Hemicr. or rhizome geophyte.
**Mariscus decompositus* R.Br., Prodr.: 218 (1810).

Cyperus densibulbosus Lye, Nordic J. Bot. 16: 369 (1996).
N. Somalia. 24 SOM. Hemicr. or rhizome geophyte.

Cyperus dentatus Torr., Fl. N. Middle United States 1: 61 (1823).
SE. Canada to E. U.S.A. 72 NBR NSC ONT QUE 75 CNT INI MAI MAS NWH NWJ NWY PEN RHO VER 78 DEL GEO MRY NCA SCA TEN VRG WDC. Hemicr. or rhizome geophyte.
Cyperus parviflorus Muhl., Descr. Gram.: 19 (1817), nom. illeg.
Cyperus micranthus Schult. in J.J.Roemer & J.A. Schultes, Mant. 2: 121 (1824).
Cyperus watsonianus Boeck., Flora 42: 435 bis (1859).
Cyperus dentatus var. *ctenostachys* Fernald, Rhodora 8: 126 (1906).
Cyperus dentatus f. *ctenostachys* (Fernald) Fernald, Rhodora 44: 151 (1940).

Cyperus dentoniae G.C.Tucker, Syst. Bot. Monogr. 2: 56 (1983).
S. Arizona to C. America. 76 ARI 79 MXE MXG MXN MXS MXT 80 COS ELS GUA HON NIC PAN. Hemicr. or rhizome geophyte.
**Mariscus pubescens* C.Presl, Reliq. Haenk. 1: 181 (1828).
Cyperus flavomariscus var. *peduncularis* Britton, Bull. Torrey Bot. Club 13: 215 (1886). *Cyperus flavus* var. *peduncularis* (Britton) Kük. in H.G.A.Engler (ed.), Pflanzenr., IV, 20(101): 532 (1936). *Mariscus hermaphroditus* var. *peduncularis* (Britton) Stehlé, Notul. Syst. (Paris) 13: 87 (1947).

Cyperus denudatus L.f., Suppl. Pl.: 98 (1782).
Trop. & S. Africa to N. Australia. 22 GHA IVO MLI NGA NGR SEN SIE 23 BUR CMN GAB RWA ZAI 24 ETH 25 KEN TAN UGA 26 MLW MOZ ZAM 27 BOT CPP LES NAM OFS SWZ TVL 29 MDG 40 IND 41 VIE 50 QLD. Hemicr. or rhizome geophyte.
Fimbristylis androgyna R.Br., Prodr.: 226 (1810). *Scirpus androgyna* (R.Br.) Poir. in J.B.A.M.de Lamarck, Encycl., Suppl. 5: 98 (1817), nom. inval. *Iria androgyna* (R.Br.) Kuntze, Revis. Gen. Pl. 2: 751 (1891).
Cyperus corymbosus Steud., Flora 12: 153 (1829), nom. illeg.
Cyperus flavissimus Steud., Flora 12: 152 (1829), nom. illeg.
Cyperus tristis Kunth, Enum. Pl. 2: 35 (1837).
Cyperus amphibolus Steud., Syn. Pl. Glumac. 2: 22 (1854).
Cyperus spretus Steud., Syn. Pl. Glumac. 2: 21 (1854).
Cyperus huillensis Ridl., Trans. Linn. Soc. London, Bot. 2: 139 (1884).
Cyperus haspanoides C.B.Clarke ex K.Schum. in H.G.A.Engler, Pflanzenw. Ost-Afrikas, C: 119 (1895).
Cyperus phaeorhizus K.Schum. in H.G.A.Engler, Pflanzenw. Ost-Afrikas, C: 119 (1895).
Cyperus princeae C.B.Clarke, Bot. Jahrb. Syst. 38: 133 (1906). *Cyperus phaeorhizus* var. *princeae* (C.B.Clarke) Kük. in H.G.A.Engler (ed.), Pflanzenr., IV, 20(101): 253 (1936).
Cyperus verrucinux C.B.Clarke, Bot. Jahrb. Syst. 38: 132 (1906).
Cyperus sphaerospermoides Cherm., Bull. Soc. Bot. France 66: 340 (1919 publ. 1920). *Cyperus denudatus* var. *sphaerospermoides* (Cherm.) Kük. in H.G.A.Engler (ed.), Pflanzenr., IV, 20(101): 256 (1936).
Cyperus denudatus f. *longispiculosus* Kük. in H.G.A. Engler (ed.), Pflanzenr., IV, 20(101): 255 (1936).
Cyperus denudatus f. *singularis* Kük. in H.G.A.Engler (ed.), Pflanzenr., IV, 20(101): 255 (1936).

Cyperus derreilema Steud., Flora 25: 585 (1842).
Ethiopia to Malawi. 23 RWA ZAI 24 ETH 25 KEN TAN UGA 26 MLW. Hemicr. or rhizome geophyte.
Cyperus deckenii Boeck., Linnaea 38: 361 (1874). *Cyperus derreilema* subsp. *deckenii* (Boeck.) Kük. in H.G.A.Engler (ed.), Pflanzenr., IV, 20(101): 199 (1936).
Cyperus tsaratananensis Cherm., Bull. Soc. Bot. France 66: 343 (1919 publ. 1920). *Cyperus derreilema* var. *tsaratananensis* (Cherm.) Kük. in H.G.A.Engler (ed.), Pflanzenr., IV, 20(101): 199 (1936).
Cyperus tschinsendensis Turrill, Bull. Misc. Inform. Kew 1925: 68 (1925). *Cyperus pustulatus* var. *tschinsendensis* (Turrill) Kük. in H.G.A.Engler (ed.), Pflanzenr., IV, 20(101): 161 (1936).
Cyperus derreilema var. *brevispiculosus* Kük. in H.G.A. Engler (ed.), Pflanzenr., IV, 20(101): 199 (1936).

Cyperus diamantinus (D.A.Simpson) Govaerts & D.A.Simpson, World Checklist of Cyperaceae: 316 (2007).
SE. Brazil. 84 BZL. Hemicr. or rhizome geophyte.
**Mariscus diamantinus* D.A.Simpson, Kew Bull. 44: 283 (1989).

Cyperus dianae Steud., Syn. Pl. Glumac. 2: 28 (1854).
St. Helena. 28 STH. Hemicr. or rhizome geophyte.

Cyperus dichromeniformis Kunth, Enum. Pl. 2: 26 (1837).
SE. Brazil. 84 BZL. Hemicr. or rhizome geophyte.
Cyperus tephrodes Schrad. ex Nees in C.F.P.von Martius & auct. suc. (eds.), Fl. Bras. 2(1): 30 (1842).

Cyperus dichromus C.B.Clarke, Bot. Jahrb. Syst. 38: 133 (1906).
S. Somalia to N. Kenya. 24 SOM 25 KEN. Hemicr. or rhizome geophyte.

Cyperus dichrostachyus Hochst. ex A.Rich., Tent. Fl. Abyss. 2: 481 (1850).

Trop. & S. Africa, Madagascar, India to SC. China. 23 BUR CMN RWA ZAI 24 ERI ETH 25 KEN TAN UGA 26 MLW 27 TVL 29 MDG 36 CHC 40. Hemicr. or rhizome geophyte.
Cyperus andschoa A.Rich., Tent. Fl. Abyss. 2: 481 (1850).

Cyperus dietrichiae Boeck., Flora 58: 87 (1875). *Mariscus dietrichiae* (Boeck.) C.B.Clarke, Kew Bull., Addit. Ser. 8: 20 (1908).
Bismarck Arch. to Queensland. 43 BIS 50 QLD. Hemicr. or rhizome geophyte.

var. ***brevibracteatus*** (Domin) Kük. in H.G.A.Engler (ed.), Pflanzenr., IV, 20(101): 426 (1936).
Queensland. 50 QLD. Hemicr. or rhizome geophyte.
Mariscus dietrichiae subsp. *brevibracteatus* Domin, Biblioth. Bot. 85: 436 (1915).
**Mariscus dietrichiae* var. *brevibracteatus* Domin, Biblioth. Bot. 85: 436 (1915).

var. ***dietrichiae***
Bismarck Arch. to Queensland. 43 BIS 50 QLD. Hemicr. or rhizome geophyte.
Cyperus trichostachys Benth., Fl. Austral. 7: 287 (1878).
Cyperus dietrichiae var. *fractuosus* Kük. in H.G.A. Engler (ed.), Pflanzenr., IV, 20(101): 424 (1936).

Cyperus difformis L., Cent. Pl. II: 6 (1756).
Trop. & Subtrop. Old World. 12 COR FRA POR SPA 13 ALB BUL GRC ITA KRI ROM SIC TUE YUG 14 UKR 20 EGY LBY MOR 21 AZO mdr 22 BEN BKN GAM GHA GNB GUI IVO LBR NGA NGR SEN SIE TOG 23 BUR CMN EQG GAB RWA ZAI 24 CHA ETH SOM 25 KEN TAN UGA 26 MLW 27 BOT CPP CPV LES NAM NAT OFS SWZ TVL (28) sth 29 COM MAU MDG REU SEY 31 32 KAZ KGZ TZK UZB 33 TCS 34 AFG IRN IRQ PAL TUR 35 YEM 36 CHC 38 JAP KOR KZN NNS OGA TAI 40 EHM IND NEP PAK SRL 41 MYA THA VIE 42 JAW LSI MLY PHI SUL SUM 43 50 NFK NSW NTA QLD SOA VIC WAU (60) fij 62 MRN (63) haw (73) (74) neb okl (75) pen (76) ari cal (77) nwm tex (78) ala fla geo kty lou msi nca ten vrg (79) mxe mxn mxs (80) cos pan (81) jam pue (82) guy ven (83) bol ecu per. Ther.
Cyperus protractus Link, Enum. Hort. Berol. Alt. 1: 45 (1821), nom. illeg.
Cyperus holoschoenoides Jan ex Schult. in J.J.Roemer & J.A.Schultes, Mant. 2: 111 (1824).
Cyperus viridis Willd. ex Kunth, Enum. Pl. 2: 38 (1837), nom. illeg.
Cyperus subrotundus Llanos, Fragm. Pl. Filip.: 14 (1851).
Cyperus goeringii Steud., Syn. Pl. Glumac. 2: 24 (1854).
Cyperus oryzetorum Steud., Syn. Pl. Glumac. 2: 24 (1854).
Cyperus lateriflorus Torr. in W.H.Emory, Rep. U.S. Mex. Bound. 2(1): 226 (1858), nom. illeg.
Cyperus difformis f. *humilis* Debeaux, Actes Soc. Linn. Bordeaux 1: 144 (1876).
Cyperus difformis f. *maximus* C.B.Clarke, J. Linn. Soc., Bot. 21: 135 (1884).
Cyperus difformis var. *breviglobosus* Kük. in H.G.A. Engler (ed.), Pflanzenr., IV, 20(101): 240 (1936).
Cyperus difformis var. *subdecompositus* Kük. in H.G.A.Engler (ed.), Pflanzenr., IV, 20(101): 240 (1936).

Cyperus diffusus Vahl, Enum. Pl. 2: 321 (1805). *Eucyperus diffusus* (Vahl) Rikli, Jahrb. Wiss. Bot. 27: 568 (1895).
Trop. & Subtrop. Asia to SW. Pacific. 36 CHH CHS 38 TAI 40 ASS EHM IND NEP SRL 41 AND MYA NCB THA VIE 42 BOR JAW MLY MOL SUL SUM 43 BIS NWG SOL 50 QLD. Hemicr. or rhizome geophyte.
Cyperus parciflorus Link, Jahrb. Gewächsk. 1(3): 85 (1818).
Cyperus sarzogonensis J.Presl & C.Presl, Reliq. Haenk. 1: 351 (1830).
Cyperus scirpoides J.Presl & C.Presl, Reliq. Haenk. 1: 178 (1830), nom. illeg.
Cyperus iridifolius A.Dietr., Sp. Pl. 2: 315 (1832).
Cyperus moestus Kunth, Enum. Pl. 2: 31 (1837).
Cyperus calacaryensis Steud., Syn. Pl. Glumac. 2: 34 (1854).
Cyperus lagorensis Steud., Syn. Pl. Glumac. 2: 36 (1854).
Cyperus pubisquama Steud., Syn. Pl. Glumac. 2: 20 (1854).
Cyperus holophyllus var. *celebicus* Miq., Fl. Ned. Ind. 3: 283 (1856). *Cyperus diffusus* var. *celebicus* (Miq.) Kük., Bot. Jahrb. Syst. 59: 43 (1924).
Cyperus holophyllus Miq. ex de Vriese, Pl. Ind. Bat. Orient.: 139 (1857).
Cyperus nigroviridis Thwaites, Enum. Pl. Zeyl.: 344 (1864).
Cyperus diffusus var. *macrostachyus* Boeck., Linnaea 35: 534 (1868). *Cyperus diffusus* subsp. *macrostachyus* (Boeck.) T.Koyama, Gard. Bull. Singapore 30: 139 (1977). *Cyperus laxus* var. *macrostachyus* (Boeck.) Karth., in Fl. Ind. Enum., Monocot.: 46 (1989). *Cyperus laxus* subsp. *macrostachyus* (Boeck.) V.P.Prasad & N.P.Singh, Phytotaxonomy 1: 64 (2001).
Cyperus buettneri Boeck., Beitr. Cyper. 1: 3 (1888).
Cyperus bancanus var. *brevibracteatus* E.G.Camus, Notul. Syst. (Paris) 1: 243 (1910). *Cyperus diffusus* var. *brevibracteatus* (E.G.Camus) Kük. in H.G.A. Engler (ed.), Pflanzenr., IV, 20(101): 209 (1936).
Cyperus diffusus var. *multispicatus* S.M.Huang, in Fl. Hainan. 4: 536 (1977).

Cyperus digitatus Roxb., Fl. Ind. 1: 209 (1820).
Trop. & Subtrop. (12) spa 20 EGY 22 BKN GAM GHA GUI IVO MLI MTN NGA NGR SEN 23 CMN GAB RWA ZAI 24 CHA ETH SUD 25 KEN TAN UGA 26 ANG ZAM ZIM 27 BOT NAM NAT TVL 36 CHC CHH CHS 38 TAI 40 ASS BAN EHM IND NEP PAK SRL 41 MYA THA VIE 42 BOR JAW LSI MLY MOL PHI SUL SUM 43 NWG 50 NTA QLD 62 CRL 77 TEX 78 FLA 79 MXE MXG MXS MXT 80 BLZ COS GUA PAN 81 CUB DOM HAI JAM PUE TRT 82 GUY SUR VEN 83 BOL CLM ECU PER 84 BZE BZN 85 AGE AGW PAR. Hemicr. or rhizome geophyte.

subsp. ***auricomus*** (Sieber ex Spreng.) Kük., Bot. Not. 1934: 65 (1934).
Africa. (12) spa 20 EGY 22 BKN GAM GHA GUI IVO MLI MTN NGA NGR SEN 23 CMN GAB RWA ZAI 24 CHA ETH SUD 25 KEN TAN UGA 26 ANG ZAM ZIM 27 BOT NAM NAT TVL. Hemicr. or rhizome geophyte.
**Cyperus auricomus* Sieber ex Spreng., Syst. Veg. 1: 230 (1824). *Papyrus auricomus* (Sieber ex Spreng.) Schrad., Index Seminum (GOET) 1835: 5 (1935).

Papyrus elatus Nees in R.Wight, Contr. Bot. India: 88 (1834).
Cyperus subalatus Boeck., Linnaea 36: 291 (1870), nom. illeg.
Cyperus aureorufus Boeck., Linnaea 38: 369 (1874). *Cyperus digitatus* var. *aureorufus* (Boeck.) Kük. in H.G.A.Engler (ed.), Pflanzenr., IV, 20(101): 58 (1935).
Cyperus auricomus var. *khasiana* C.B.Clarke, J. Linn. Soc., Bot. 21: 189 (1884).
Cyperus auricomus var. *microstachyus* Boeck. in K.M.Schumann & U.M.Hollrung, Fl. Kais. Wilh. Land: 23 (1889).
Cyperus auricomus var. *minor* C.B.Clarke in D.Oliver & auct. suc. (eds.), Fl. Trop. Afr. 8: 373 (1901).
Cyperus digitatus var. *abbreviatus* Kük. in H.G.A. Engler (ed.), Pflanzenr., IV, 20(101): 58 (1935).
Cyperus digitatus var. *bruntii* S.S.Hooper, Kew Bull. 26: 577 (1972).

subsp. ***digitatus***
E. & Trop. Asia, N. Australia, Trop. & Subtrop. America. 36 CHC CHH CHS 38 TAI 40 ASS BAN EHM IND NEP PAK SRL 41 MYA THA VIE 42 BOR JAW LSI MLY MOL PHI SUL SUM 43 NWG 50 NTA QLD 62 CRL 77 TEX 78 FLA 79 MXE MXG MXS MXT 80 BLZ COS GUA PAN 81 CUB DOM HAI JAM PUE TRT 82 GUY SUR VEN 83 BOL CLM ECU PER 84 BZE BZN 85 AGE AGW PAR. Hemicr. or rhizome geophyte.
Cyperus ornithopodioides Delile, Descr. Egypte, Hist. Nat. 3: 50 (1813), nom. nud.
Cyperus xanthocomus Link, Hort. Berol. 1: 316 (1827).
Cyperus glomeratus Roxb. ex Arn. in R.Wight, Contr. Bot. India: 85 (1834), nom. illeg.
Papyrus tuberiferus Schrad., Index Seminum (GOET) 1835: s.p. (1835).
Cyperus neesii Kunth, Enum. Pl. 2: 101 (1837).
Cyperus rhaphiostachys Humb. & Bonpl. ex Kunth, Enum. Pl. 2: 68 (1837).
Cyperus tuberiferus Steud., Nomencl. Bot., ed. 2, 1: 474 (1840).
Cyperus mexicanus Liebm., Mexic. Halvgr.: 227 (1850). *Cyperus digitatus* var. *mexicanus* (Liebm.) Kük. in H.G.A.Engler (ed.), Pflanzenr., IV, 20(101): 57 (1935).
Cyperus wiestii Steud., Syn. Pl. Glumac. 2: 35 (1854).
Cyperus giganteus Griseb., Cat. Pl. Cub.: 238 (1866), nom. illeg.
Cyperus brachystachys Nees ex Boeck., Linnaea 36: 310 (1870), nom. illeg.
Cyperus hookeri Boeck., Linnaea 36: 308 (1870). *Cyperus digitatus* var. *hookeri* (Boeck.) C.B.Clarke in J.D.Hooker, Fl. Brit. India 6: 618 (1893).
Cyperus venustus Boeck., Linnaea 36: 316 (1870), nom. illeg.
Cyperus exaltatus var. *oatesii* C.B.Clarke, J. Linn. Soc., Bot. 21: 188 (1884). *Cyperus digitatus* var. *oatesi* (C.B.Clarke) Kük. in H.G.A.Engler (ed.), Pflanzenr., IV, 20(101): 56 (1935).
Cyperus spongiosovaginatus Boeck., Beitr. Cyper. 1: 5 (1888).
Cyperus trinidadensis Boeck., Beitr. Cyper. 2: 7 (1890).
Cyperus oatesii C.B.Clarke in J.D.Hooker, Fl. Brit. India 6: 618 (1893).
Cyperus digitatus f. *contractior* Kük. in H.G.A. Engler (ed.), Pflanzenr., IV, 20(101): 57 (1935).
Cyperus digitatus var. *obtusifructus* Kük. in H.G.A. Engler (ed.), Pflanzenr., IV, 20(101): 56 (1935).

Cyperus dilatatus Schumach., Beskr. Guin. Pl.: 38 (1827).
W. Trop. Africa to Tanzania. 22 BEN BKN GAM GHA GUI IVO LBR MLI NGA SEN SIE TOG 23 GGI ZAI 24 CHA SOM 25 TAN. Hemicr.
Cyperus tenuiculmis Boeck., Flora 62: 554 (1879), nom. illeg.
Cyperus gracilinux C.B.Clarke, J. Linn. Soc., Bot. 21: 162 (1884).
Cyperus pseudosphacelatus Chiov., Ann. Bot. (Rome) 13: 374 (1915), nom. illeg. *Cyperus esphacelatus* Kük. in H.G.A.Engler (ed.), Pflanzenr., IV, 20(101): 131 (1935).

Cyperus dioicus I.M.Johnst., Proc. Calif. Acad. Sci., IV, 12: 988 (1924).
Mexico (Baja California Sur). 79 MXN. Hemicr. or rhizome geophyte.

Cyperus dipsaceus Liebm., Mexic. Halvgr.: 219 (1850). *Mariscus dipsaceus* (Liebm.) C.B.Clarke, Kew Bull., Addit. Ser. 8: 16 (1908).
Arizona to Mexico. 76 ARI 77 NWM 79 MXE MXG MXN MXS. Hemicr. or rhizome geophyte.
Cyperus wrightii Britton, Bull. Torrey Bot. Club 13: 215 (1886). *Mariscus wrightii* (Britton) T.Koyama, Phytologia 29: 74 (1974).

Cyperus disjunctus C.B.Clarke, Kew Bull., Addit. Ser. 8: 8 (1908).
E. Australia. 50 NSW QLD. Hemicr. or rhizome geophyte.
Cyperus tetraphyllus var. *monocephalus* Benth., Fl. Austral. 7: 269 (1878).

Cyperus dissolutus Kunth, Jahrb. Gewächsk. 1(3): 87 (1820). *Mariscus dissolutus* (Kunth) T.Koyama, Phytologia 29: 73 (1974).
S. America (?). 8+. Hemicr. or rhizome geophyte.
Cyperus ubertus Willd. ex Link, Jahrb. Gewächsk. 1(3): 87 (1818).

Cyperus distans L.f., Suppl. Pl.: 103 (1782).
Trop. & Subtrop. 22 BEN GHA GUI IVO LBR MLI NGA SEN SIE TOG 23 BUR CMN EQG GAB GGI RWA ZAI 24 ETH SOM 25 KEN TAN UGA 26 MLW ZIM 27 BOT NAT SWZ TVL 28 ASC STH 29 COM MAU MDG REU SEY 36 CHH CHS 38 NNS TAI 40 ASS EHM IND NEP SRL WHM 41 AND CBD MYA THA VIE 42 BOR JAW LSI MLY MOL PHI SUL SUM 43 NWG 50 QLD WAU (60) fij 62 CRL 78 FLA NCA 79 MXS MXT 80 NIC 81 DOM JAM LEE PUE TRT WIN 82 VEN 83 CLM ECU GAL PER 84 BZE BZL BZS 85 AGW. Hemicr. or rhizome geophyte.
**Cyperus elatus* Rottb., Descr. Icon. Rar. Pl.: 37 (1773), nom. illeg.
Cyperus laxus Vahl, Enum. Pl. 2: 362 (1805), nom. illeg.
Cyperus nutans Sieber ex C.Presl, Isis (Oken) 21: 271 (1828), nom. illeg.
Cyperus graminicolus Steud., Syn. Pl. Glumac. 2: 49 (1854).
Cyperus kurrii Steud., Syn. Pl. Glumac. 2: 38 (1854).
Cyperus psilostachys Steud., Syn. Pl. Glumac. 2: 47 (1854).
Cyperus schraderianus Mart. ex Steud., Syn. Pl. Glumac. 2: 38 (1854).

Cyperus squamulatus Steud., Syn. Pl. Glumac. 2: 49 (1854).
Cyperus densiflorus Hemsl., Rep. Challenger, Bot. 1(2): 23 (1885), nom. illeg. *Cyperus distans* var. *densiflorus* Kük. in H.G.A.Engler (ed.), Pflanzenr., IV, 20(101): 140 (1935).
Cyperus purpureovaginatus Boeck., Beitr. Cyper. 2: 8 (1890).
Cyperus distans f. *capillarioides* Kük. in H.G.A.Engler (ed.), Pflanzenr., IV, 20(101): 140 (1935).
Cyperus distans var. *crassispiculosus* R.Gross & Kük. in H.G.A.Engler (ed.), Pflanzenr., IV, 20(101): 141 (1935).
Cyperus distans f. *pachyanthos* Kük. in H.G.A.Engler (ed.), Pflanzenr., IV, 20(101): 140 (1935).
Cyperus distans f. *pallidior* Kük. in H.G.A.Engler (ed.), Pflanzenr., IV, 20(101): 140 (1935).
Cyperus distans var. *pseudonutans* Kük. in H.G.A. Engler (ed.), Pflanzenr., IV, 20(101): 140 (1935).

Cyperus distinctus Steud., Syn. Pl. Glumac. 2: 24 (1854).
SE. U.S.A., Bahamas. 78 FLA GEO LOU SCA 81 BAH. Hemicr. or rhizome geophyte.
Cyperus virens var. *brittonii* C.B.Clarke in I.Urban, Symb. Antill. 2: 26 (1900).

Cyperus diurensis Boeck., Flora 62: 556 (1879). *Mariscus diurensis* (Boeck.) C.B.Clarke in T.A.Durand & H.Schinz, Consp. Fl. Afric. 5: 586 (1894).
E. Zaïre to NE. & E. Trop. Africa. 23 RWA ZAI 24 ETH SUD 25 KEN TAN UGA. Hemicr. or rhizome geophyte.
Cyperus gondanus Boeck., Beitr. Cyper. 1: 3 (1888). *Cyperus diurensis* var. *gondanus* (Boeck.) Kük. in H.G.A.Engler (ed.), Pflanzenr., IV, 20(101): 560 (1936).
Cyperus diurensis var. *longistolon* Kük. in R.E.Fries, Wiss. Erg. Schwed. Rhod.-Kongo Exped. 1: 4 (1921).
Cyperus diurensis var. *acuminatosquamatus* Kük. in H.G.A.Engler (ed.), Pflanzenr., IV, 20(101): 561 (1936).
Cyperus diurensis var. *laetevirens* A.Peter & Kük. in H.G.A.Engler (ed.), Pflanzenr., IV, 20(101): 560 (1936).

Cyperus dives Delile, Descr. Egypte, Hist. Nat.: 149 (1813). *Cyperus alopecuroides* var. *dives* (Delile) Boeck., Linnaea 36: 321 (1870).
SW. Syria to Africa, India to Vietnam. 20 EGY 21 CVI 22 IVO NGR SEN 23 BUR GAB RWA ZAI 24 ERI ETH SOM SUD 25 KEN TAN UGA 26 MLW MOZ ZAM 27 BOT 29 MDG 34 LBS PAL 40 ASS IND PAK 41 VIE. Hemicr. or rhizome geophyte.
Cyperus canariensis Steud., Syn. Pl. Glumac. 2: 42 (1854).
Cyperus alopecuroides var. *microstachys* Boeck., Linnaea 36: 322 (1870).
Cyperus immensus C.B.Clarke, J. Linn. Soc., Bot. 20: 294 (1883).
Cyperus immensus var. *taylorii* C.B.Clarke in D.Oliver & auct. suc. (eds.), Fl. Trop. Afr. 8: 372 (1901).
Cyperus petherickii C.B.Clarke in D.Oliver & auct. suc. (eds.), Fl. Trop. Afr. 8: 371 (1901). *Cyperus immensus* var. *petherickii* (C.B.Clarke) Kük. in H.G.A.Engler (ed.), Pflanzenr., IV, 20(101): 67 (1935).

Cyperus drakensbergensis (Vorster) Govaerts in R.H.A.Govaerts & D.A.Simpson, World Checklist of Cyperaceae: 319 (2007).
KwaZulu-Natal. 27 NAT. Hemicr. or rhizome geophyte.
**Mariscus drakensbergensis* Vorster, S. African J. Bot. 52: 181 (1986).

Cyperus dregeanus Kunth, Enum. Pl. 2: 31 (1837).
S. Africa. 27 BOT CPP NAM NAT SWZ TVL. Hemicr. or rhizome geophyte.

Cyperus drummondii Torr. & Hook., Ann. Lyceum Nat. Hist. New York 3: 437 (1836). *Cyperus virens* var. *drummondii* (Torr. & Hook.) Kük. in H.G.A.Engler (ed.), Pflanzenr., IV, 20(101): 181 (1936). *Cyperus virens* subsp. *drummondii* (Torr. & Hook.) T.Koyama, Madroño 20: 254 (1970).
SE. U.S.A. to Trop. America. 77 TEX 78 ALA FLA GEO LOU MSI SCA 80 COS NIC 81 JAM 82 SUR 83 GAL 84 BZL. Hemicr. or rhizome geophyte.
Cyperus robustus Kunth, Enum. Pl. 2: 41 (1837). *Cyperus virens* var. *robustus* (Kunth) Kük. in H.G.A. Engler (ed.), Pflanzenr., IV, 20(101): 181 (1936).

Cyperus dubius Rottb., Descr. Icon. Rar. Pl.: 20 (1773). *Mariscus dubius* (Rottb.) Kük. ex G.E.C.Fischer in J.S.Gamble, Fl. Madras: 1644 (1931).
Trop. & S. Africa to Trop. Asia. 22 BEN GHA GUI IVO LBR NGA TOG 23 BUR CMN EQG GAB GGI RWA ZAI 24 ERI ETH SOM 25 KEN TAN UGA 26 MLW MOZ ZAM 27 BOT CPP NAM NAT SWZ TVL 29 ALD CGS COM MAU MCI MDG REU ROD SEY 35 YEM 40 ASS IND SRL 41 SCS THA VIE 42 BOR JAW PHI SUM (50). Hemicr.

subsp. ***dubius***
Trop. & S. Africa to Trop. Asia. 22 BEN GHA GUI IVO LBR NGA TOG 23 BUR CMN EQG GAB GGI RWA ZAI 24 ERI ETH SOM 25 KEN TAN UGA 26 MLW MOZ ZAM 27 BOT CPP NAM NAT SWZ TVL 29 ALD CGS COM MAU MCI MDG REU ROD SEY 35 YEM 40 ASS IND SRL 41 SCS THA VIE 42 BOR JAW PHI SUM (50). Hemicr.
Scirpus glomeratus L., Sp. Pl.: 52 (1753).
Schoenus bobartiae Retz., Observ. Bot. 4: 9 (1786).
Cyperus coloratus Vahl, Enum. Pl. 2: 312 (1805). *Mariscus coloratus* (Vahl) Nees, Linnaea 9: 286 (1834). *Cyperus dubius* var. *coloratus* (Vahl) Kük. in H.G.A.Engler (ed.), Pflanzenr., IV, 20(101): 565 (1936). *Cyperus dubius* subsp. *coloratus* (Vahl) Lye, Nordic J. Bot. 3: 231 (1983).
Cyperus kyllingioides Vahl, Enum. Pl. 2: 312 (1805).
Cyperus capitatus Poir. in J.B.A.M.de Lamarck, Encycl. 7: 246 (1806), nom. illeg. *Cyperus capitellatus* Schult. in J.J.Roemer & J.A.Schultes, Mant. 2: 101 (1824). *Mariscus kraussii* var. *capitatus* Cherm., Ann. Inst. Bot.-Géol. Colon. Marseille 27: 48 (1919). *Cyperus dubius* var. *capitatus* (Cherm.) Kük. in H.G.A. Engler (ed.), Pflanzenr., IV, 20(101): 564 (1936).
Cyperus mollis Poir. in J.B.A.M.de Lamarck, Encycl. 7: 247 (1806).
Cyperus thouarsianus C.Presl, Isis (Oken) 21: 271 (1828).
Mariscus dregeanus Kunth, Enum. Pl. 2: 120 (1837).
Mariscus kraussii Hochst., Flora 28: 756 (1845).
Kyllinga mariae Steud., Syn. Pl. Glumac. 2: 69 (1854).

Mariscus irroratus Nees, Hooker's J. Bot. Kew Gard. Misc. 6: 28 (1854).
Kyllinga multinervia Steud., Syn. Pl. Glumac. 2: 317 (1855).
Mariscus kyllingiiformis Boeck., Flora 42: 443 (1859).
Mariscus maritimus Miq., Fl. Ned. Ind., Eerste Bijv.: 600 (1861). *Cyperus maritimus* (Miq.) Suringar, Gesl. Cyp. Mal. Arch.: 160 (1898), nom. illeg.
Cyperus dubius Boeck., Linnaea 36: 282 (1870), nom. illeg.
Isolepis boeckeleri Oliv., Trans. Linn. Soc. London 29: 167 (1875).
Cyperus daphaenus Ridl., J. Bot. 22: 16 (1884).
Mariscus dregeanus var. *buchananii* C.B.Clarke in W.H.Harvey & auct. suc. (eds.), Fl. Cap. 7: 188 (1897). *Cyperus dubius* var. *buchananii* (C.B.Clarke) Kük. in H.G.A.Engler (ed.), Pflanzenr., IV, 20(101): 564 (1936).
Mariscus merrillii C.B.Clarke, Philipp. J. Sci., C 2: 87 (1907). *Cyperus merrillii* (C.B.Clarke) Kük. in H.G.A.Engler (ed.), Pflanzenr., IV, 20(101): 562 (1936).
Mariscus detersus C.B.Clarke ex Cherm., Bull. Mus. Natl. Hist. Nat. 25: 300 (1919). *Cyperus dubius* var. *detersus* (C.B.Clarke ex Cherm.) Kük. in H.G.A.Engler (ed.), Pflanzenr., IV, 20(101): 565 (1936).
Mariscus decaryi Cherm., Bull. Soc. Bot. France 79: 719 (1922 publ. 1923). *Cyperus dubius* var. *decaryi* (Cherm.) Kük. in H.G.A.Engler (ed.), Pflanzenr., IV, 20(101): 566 (1936).
Mariscus decaryi var. *pauciflorus* Cherm., Bull. Soc. Bot. France 72: 614 (1925).
Cyperus dubius var. *polyactis* A.Peter ex Kük. in H.G.A.Engler (ed.), Pflanzenr., IV, 20(101): 565 (1936).
Cyperus dubius var. *stenactis* A.Peter ex Kük. in H.G.A.Engler (ed.), Pflanzenr., IV, 20(101): 565 (1936).

subsp. ***macrocephalus*** (C.B.Clarke) Lye, Nordic J. Bot. 3: 231 (1983).
Trop. Africa. 23 RWA ZAI 24 ETH 25 KEN TAN UGA. Hemicr.
**Mariscus coloratus* var. *macrocephalus* C.B.Clarke in D.Oliver & auct. suc. (eds.), Fl. Trop. Afr. 8: 381 (1901). *Cyperus dubius* var. *macrocephalus* (C.B.Clarke) Kük. in H.G.A. Engler (ed.), Pflanzenr., IV, 20(101): 564 (1936). *Mariscus dubius* subsp. *macrocephalus* (C.B.Clarke) J.-P.Lebrun & Stork, Énum. Pl. Fl. Afr. Trop. 3: 197 (1995).

Cyperus duclouxii Camus, Notul. Syst. (Paris) 1: 244 (1910).
China (Yunnan). 36 CHC. Hemicr. or rhizome geophyte.

Cyperus dunensis (Cherm.) Kük. in H.G.A.Engler (ed.), Pflanzenr., IV, 20(101): 420 (1936).
W. Madagascar. 29 MDG. Hemicr. or rhizome geophyte.
**Mariscus dunensis* Cherm., Bull. Soc. Bot. France 72: 18 (1925).

Cyperus duripes I.M.Johnst., Proc. Calif. Acad. Sci., IV, 20: 54 (1931). *Mariscus duripes* (I.M.Johnst.) Conz., Fl. Taxon. Mexic. 2: 236 (1943).
Revillagigedo Is. (I. Clarión). 79 MXI. Hemicr. or rhizome geophyte.

Cyperus durus Kunth, Enum. Pl. 2: 76 (1837). *Mariscus durus* (Kunth) C.B.Clarke in T.A.Durand & H.Schinz, Consp. Fl. Afric. 5: 586 (1894).
S. Angola, Cape Prov. 26 ANG 27 CPP. Hemicr. or rhizome geophyte.

Cyperus echinatus (L.) Alph.Wood, Class-book Bot.: 734 (1861).
C. & E. U.S.A. to Venezuela. (21) azo 74 ILL KAN MSO OKL WIS 75 INI NWJ NWY OHI PEN RHO WVA 77 NWM TEX 78 ALA ARK DEL FLA GEO KTY LOU MRY MSI NCA SCA TEN VRG WDC 79 MXT 81 CUB HAI JAM PUE 82 VEN. Hemicr. or rhizome geophyte.
**Scirpus echinatus* L., Sp. Pl.: 50 (1753). *Mariscus echinatus* (L.) Elliott, Sketch Bot. S. Carolina 1: 75 (1816).
Schoenus umbellatus Jacq., Collectanea 1: 113 (1787).
Kyllinga ovularis Michx., Fl. Bor.-Amer. 1: 29 (1803). *Mariscus ovularis* (Michx.) Vahl, Enum. Pl. 2: 374 (1805). *Cyperus ovularis* (Michx.) Torr., Ann. Lyceum Nat. Hist. New York 3: 278 (1836). *Cyperus ovularis* var. *americanus* Boeck., Linnaea 36: 377 (1862), nom. inval.
Mariscus ovularis var. *tenellus* Torr., Fl. N. Middle United States 1: 58 (1824).
Mariscus ovularis Kunth, Enum. Pl. 2: 121 (1837), nom. illeg.
Mariscus retroflexus Willd. ex Kunth, Enum. Pl. 2: 121 (1837).
Mariscus sphaericus Willd. ex Kunth, Enum. Pl. 2: 121 (1837).
Mariscus rivularis Walp., Ann. Bot. Syst. 3: 674 (1852).
Mariscus bracheilema Steud., Syn. Pl. Glumac. 2: 317 (1855).
Mariscus drummondii Steud., Syn. Pl. Glumac. 2: 316 (1855).
Mariscus sphaerocephalus Steud., Syn. Pl. Glumac. 2: 316 (1855).
Cyperus ovularis var. *sphaericus* Boeck., Linnaea 36: 378 (1870).
Cyperus wolfii Alph.Wood, Bot. Gaz. 1: 38 (1876). *Cyperus ovularis* var. *wolfii* (Alph.Wood) Kük. in H.G.A.Engler (ed.), Pflanzenr., IV, 20(101): 512 (1936).
Mariscus globulosus Urb., Symb. Antill. 2: 165 (1900).

Cyperus eglobosus K.L.Wilson, Telopea 5: 596 (1994).
E. Australia. 50 NSW QLD. Hemicr. or rhizome geophyte.

Cyperus ekmanii Kük., Repert. Spec. Nov. Regni Veg. 23: 186 (1926). *Mariscus ekmanii* (Kük.) T.Koyama, Phytologia 29: 73 (1974).
C. & E. Cuba. 81 CUB. Hemicr. or rhizome geophyte.

Cyperus elatus L., Cent. Pl. II: 5 (1756).
Trop. Asia. 40 ASS BAN IND 41 CBD THA VIE 42 BOR JAW LSI MLY PHI SUL SUM. Hemicr. or rhizome geophyte.
Cyperus racemosus Retz., Observ. Bot. 6: 20 (1791).
Cyperus bispicatus Steud., Syn. Pl. Glumac. 2: 49 (1854).
Cyperus elatus var. *macronux* C.B.Clarke, J. Linn. Soc., Bot. 21: 190 (1884).
Cyperus elatus var. *laxus* E.G.Camus in H.Lecomte, Fl. Indo-Chine 7: 72 (1912).

Cyperus elegans L., Sp. Pl.: 45 (1753). *Papyrus elegans* (L.) G.Don ex Loudon, Hort. Brit.: 22 (1830).
S. U.S.A. to Peru, Caribbean. (61) sci 77 NWM TEX 78 FLA LOU MSI 79 MXC MXE MXG MXN

MXS MXT 80 BLZ COS NIC 81 ARU CAY CUB DOM HAI JAM LEE NLA PUE TRT WIN 83 CLM ECU GAL PER. Ther. or hemicr.

subsp. ***elegans***

S. U.S.A. to Peru, Caribbean. (61) sci 77 NWM TEX 78 FLA LOU MSI 79 MXC MXE MXG MXN MXS MXT 80 BLZ COS NIC 81 ARU CAY CUB DOM HAI JAM LEE NLA PUE TRT WIN 83 ECU PER. Ther. or hemicr.

Cyperus viscosus Sw., Prodr.: 20 (1788).

Scirpus viscosus Lam., Tabl. Encycl. 1: 142 (1791).

Cyperus viviparus Steud., Syn. Pl. Glumac. 2: 316 (1855).

Cyperus trachynotus Torr. in W.H.Emory, Rep. U.S. Mex. Bound. 2(1): 227 (1858).

Cyperus elegans var. *breviradiata* Maury, J. Bot. (Morot) 3: 196 (1889).

Scirpus potosiensis C.B.Clarke, Kew Bull., Addit. Ser. 8: 89 (1908), nom. inval.

Cyperus elegans var. *major* Kük. in H.G.A.Engler (ed.), Pflanzenr., IV, 20(101): 165 (1936).

subsp. ***rubiginosus*** (Hook.f.) Eliasson, Svensk Bot. Tidskr. 59: 475 (1965).

Galápagos. 83 GAL. Hemicr. or rhizome geophyte.

Mariscus mutisii Hook.f., Trans. Linn. Soc. London 20: 178 (1847), nom. illeg.

**Cyperus rubiginosus* Hook.f., Trans. Linn. Soc. London 20: 178 (1851).

Cyperus cornutus Andersson, Galapagos Veg.: 53 (1854). *Cyperus rubiginosus* var. *cornutus* (Andersson) Kük. in H.G.A.Engler (ed.), Pflanzenr., IV, 20(101): 484 (1936).

Mariscus cornutus Andersson, Kongl. Vetensk. Acad. Handl. 1853: 151 (1855).

Cyperus elephantinus (C.B.Clarke) C.B.Clarke, Ill. Cyper.: t. XX, f. 1–2 (1909).

S. Africa. 27 NAT TVL. Hemicr. or rhizome geophyte.

**Mariscus elephantinus* C.B.Clarke in W.H.Harvey & auct. suc. (eds.), Fl. Cap. 7: 195 (1898).

Cyperus endlichii Kük., Repert. Spec. Nov. Regni Veg. 21: 327 (1925).

Tanzania (Kilimanjaro, Pare). 25 TAN. Hemicr. or rhizome geophyte.

Cyperus enervis R.Br., Prodr.: 213 (1810). *Cyperus gracilis* var. *enervis* (R.Br.) Kük. in H.G.A.Engler (ed.), Pflanzenr., IV, 20(101): 297 (1936).

Queensland, NE. New South Wales. 50 NSW QLD. Hemicr. or rhizome geophyte.

Cyperus enervis var. *fallax* Domin, Biblioth. Bot. 89: 421 (1915).

Cyperus entrerianus Boeck., Flora 61: 139 (1878). *Cyperus luzulae* var. *entrerianus* (Boeck.) M.Barros, Anales Mus. Argent. Ci. Nat. "Bernardino Rivadavia" 39: 309 (1938).

Mexico to Argentina. (77) tex (78) ala fla geo lou msi 79 MXE MXG MXN MXS MXT 81 CUB JAM 83 BOL CLM 84 BZS 85 AGE AGW PAR URU. Hemicr. or rhizome geophyte.

Cyperus tucumanensis Boeck., Bot. Jahrb. Syst. 7: 274 (1886). *Cyperus luzulae* var. *tucumanensis* (Boeck.) C.B.Clarke ex Chodat, Bull. Herb. Boissier, II, 3: 1007 (1903).

Cyperus entrerianus f. *abbreviatus* Kük. in H.G.A. Engler (ed.), Pflanzenr., IV, 20(101): 170 (1936).

Cyperus entrerianus var. *parvicapitulatus* Kük. in H.G.A.Engler (ed.), Pflanzenr., IV, 20(101): 170 (1936).

Cyperus surinamensis var. *strictus* Kük. in H.G.A. Engler (ed.), Pflanzenr., IV, 20(101): 175 (1936).

Cyperus ephemerus Kukkonen & Väre, Ann. Bot. Fenn. 42: 475 (2005).

Iran. 34 IRN. Hemicr. or rhizome geophyte

Cyperus eragrostis Lam., Tabl. Encycl. 1: 196 (1791). *Chlorocyperus eragrostis* (Lam.) Rikli, Jahrb. Wiss. Bot. 27: 563 (1895).

Easter I., America. (11) cze (12) fra por spa (21) azo cny mdr (27) cpp nat ofs tvl (38) nns (50) nfk 61 EAS 71 BRC 73 ORE WAS 75 NWJ PEN 76 CAL 77 TEX 78 ALA LOU MSI SCA 82 SUR 83 BOL PER 84 BZS 85 AGE AGS AGW CLC JNF URU. Hemicr. or rhizome geophyte.

Cyperus declinatus Moench, Methodus: 317 (1794).

Cyperus monandrus Roth, Catal. Bot. 1: 3 (1797).

Cyperus vegetus Willd., Sp. Pl. 1: 283 (1797).

Cyperus ochrocephalus Steud., Flora 25: 601 (1842).

Cyperus prionotropis Steud., Flora 25: 601 (1842).

Cyperus vegetus var. *compactus* Desv. in C.Gay, Fl. Chil. 7: 167 (1853). *Cyperus eragrostis* var. *compactus* (Desv.) Kük. in H.G.A.Engler (ed.), Pflanzenr., IV, 20(101): 179 (1936).

Cyperus luzulae Hochst. ex Steud., Syn. Pl. Glumac. 2: 25 (1854), nom. illeg.

Cyperus vegetus var. *triangularis* Boeck., Flora 42: 436 (1859).

Cyperus gracilis Buchanan, Trans. & Proc. New Zealand Inst. 3: 210 (1871), nom. illeg.

Cyperus serrulatus S.Watson, Proc. Amer. Acad. Arts 17: 382 (1882), nom. illeg.

Cyperus lutescens Phil., Anales Univ. Chile 93: 345 (1896), nom. illeg.

Cyperus vegetus var. *obtusangulus* Kuntze, Revis. Gen. Pl. 3(2): 334 (1898).

Cyperus vegetus var. *trigonus* Kuntze, Revis. Gen. Pl. 3(2): 334 (1898).

Cyperus vegetus var. *limbata* C.B.Clarke, Bot. Jahrb. Syst. 30(68): 12 (1901).

Cyperus declinatus var. *limbatus* Druce, Rep. Bot. Exch. Club Brit. Isles 5: 311 (1919).

Cyperus eragrostis f. *latifrons* Kük. in H.G.A.Engler (ed.), Pflanzenr., IV, 20(101): 179 (1936).

Cyperus eragrostis f. *tener* Kük. in H.G.A.Engler (ed.), Pflanzenr., IV, 20(101): 179 (1936).

Cyperus eremicus Kukkonen, Ann. Bot. Fenn. 32: 159 (1995).

Arabian Pen. to Afghanistan. 34 AFG IRN 35 SAU. Hemicr. or rhizome geophyte.

Cyperus erythrorrhizos Muhl., Descr. Gram.: 20 (1817). *Chlorocyperus erythrorrhizus* (Muhl.) Palla, Allg. Bot. Z. Syst. 9: 69 (1903).

S. Ontario to U.S.A., NW. Mexico. 71 MAN 72 ONT 73 ORE WAS 74 ILL IOW KAN MIN MSO NDA NEB OKL SDA WIS 75 CNT INI MAI MAS MIC NWH NWJ NWY OHI PEN RHO WVA 76 ARI CAL NEV UTA 77 NWM TEX 78 ALA ARK DEL FLA GEO KTY LOU MRY MSI NCA SCA TEN VRG WDC 79 MXN. Ther.

Cyperus tenuiflorus Elliott, Sketch Bot. S. Carolina 1: 70 (1816), nom. illeg.

Cyperus cupreus J.Presl & C.Presl, Reliq. Haenk. 1: 172 (1828). *Cyperus erythrorrhizos* var. *cupreus* (J.Presl & C.Presl) Kük. in H.G.A.Engler (ed.), Pflanzenr., IV, 20(101): 59 (1935).

Cyperus occidentalis Torr., Ann. Lyceum Nat. Hist. New York 3: 259 (1836).
Cyperus chrysokerkos Steud., Syn. Pl. Glumac. 2: 316 (1855).
Cyperus infirmus Boeck., Flora 42: 438 bis (1859). *Cyperus erythrorrhizos* var. *infirmus* (Boeck.) Kük. in H.G.A.Engler (ed.), Pflanzenr., IV, 20(101): 59 (1935).
Cyperus halei Torr. ex Britton, Bull. Torrey Bot. Club 13: 213 (1886). *Cyperus erythrorrhizos* var. *halei* (Torr. ex Britton) Kük. in H.G.A.Engler (ed.), Pflanzenr., IV, 20(101): 59 (1935).
Cyperus spiculatus Alph.Wood in B.B.Smyth, Checkl. Pl. Kansas: 26 (1892).
Cyperus washingtonensis Gand., Bull. Soc. Bot. France 66: 296 (1919 publ. 1920).

Cyperus esculentus L., Sp. Pl.: 45 (1753). *Pterocyperus esculentus* (L.) Opiz, Seznam: 80 (1852). *Cyperus aureus* subsp. (L.) Nyman, Consp. Fl. Eur.: 759 (1882), nom. superfl. *Pycreus esculentus* (L.) Hayek, Repert. Spec. Nov. Regni Veg. Beih. 30(3): 147 (1932).
Trop. & Subtrop. 12 COR FRA POR 13 ALB BUL GRC ITA KRI SIC (14) ukr 20 ALG EGY LBY 21 AZO CNY CVI MDR 22 BEN BKN GAM GHA GNB GUI IVO LBR NGA NGR SEN SIE TOG 23 BUR CAB CAF CMN CON EQG GAB GGI RWA ZAI 24 CHA ERI ETH SOM 25 KEN TAN UGA 26 ANG 27 BOT CPP CPV LES NAM NAT OFS SWZ TVL (28) sth 29 COM MAU MDG 33 TCS 34 AFG IRN IRQ TUR 35 YEM 40 IND NEP PAK WHM 41 VIE 42 JAW (50) (63) haw 71 BRC 72 NBR ONT QUE 73 COL IDA ORE WAS 74 ILL IOW KAN MIN MSO NDA NEB OKL SDA WIS 75 CNT INI MAI MAS MIC NWH NWJ NWY OHI PEN RHO VER WVA 76 ARI CAL NEV UTA 77 NWM TEX 78 ALA ARK DEL FLA GEO KTY LOU MRY MSI NCA SCA TEN VRG WDC 79 MXC MXE MXG MXN MXS MXT 80 BLZ COS CPI PAN 81 ARU BER CUB JAM LEE NLA PUE TRT WIN 82 VEN 83 BOL CLM ECU GAL PER 84 BZS 85 AGE AGW PAR URU. Tuber geophyte. Widely cultivated for its edible tubers.
Cyperus pallidus Savi, Cat. Piant. Egiz.: 23, 31, nom. illeg.
Cyperus melanorhizus Delile, Descr. Egypte, Hist. Nat. 3: 50 (1813).
Cyperus tuberosus Pursh, Fl. Amer. Sept. 1: 52 (1813), nom. illeg.
Cyperus scirpoides R.Br. in H.Salt, Voy. Abyss., App.: lxii (1814), nom. illeg. *Cyperus fresenii* Steud., Nomencl. Bot., ed. 2, 1: 469 (1840).
Cyperus repens Elliott, Sketch Bot. S. Carolina 1: 68 (1816).
Cyperus hydra Kunth in F.W.H.von Humboldt, A.J.A.Bonpland & C.S.Kunth, Nov. Gen. Sp. 1: 205 (1817), nom. illeg.
Cyperus phymatodes Muhl., Descr. Gram.: 23 (1817). *Chlorocyperus phymatodes* (Muhl.) Palla, Allg. Bot. Z. Syst. 9: 69 (1903). *Cyperus esculentus* var. *phymatodes* (Muhl.) Kük., Repert. Spec. Nov. Regni Veg. 23: 185 (1926).
Cyperus gracilis Link, Jahrb. Gewächsk. 1(3): 84 (1818), nom. illeg.
Cyperus nervosus Bertol., Opusc. Sci. 3: 407 (1819).
Cyperus aureus Ten., Fl. Napol. 3: 45 (1824), nom. illeg. *Chlorocyperus aureus* Palla ex Kneuck., Allg. Bot. Z. Syst. 9: 69 (1903).
Cyperus gracilescens Schult. in J.J.Roemer & J.A.Schultes, Mant. 2: 100 (1824).
Cyperus tenorianus Schult. in J.J.Roemer & J.A.Schultes, Mant. 2: 544 (1824).
Cyperus tenorei C.Presl, Fl. Sicul.: xliii (1826).
Cyperus sieberianus Link, Hort. Berol. 1: 313 (1827).
Cyperus officinalis T.Nees, Pl. Med., Abbild.: t. 25, f. B (1830).
Cyperus damiettensis A.Dietr., Sp. Pl. 2: 269 (1832).
Cyperus lutescens Torr. & Hook., Ann. Lyceum Nat. Hist. New York 3: 433 (1836). *Cyperus esculentus* var. *macrostachyus* Boeck., Linnaea 36: 291 (1870). *Cyperus esculentus* f. *macrostachyus* (Boeck.) Fernald, Rhodora 44: 151 (1942). *Cyperus esculentus* var. *lutescens* (Torr. & Hook.) Kük. ex Osten, Anales Mus. Hist. Nat. Montevideo, II, 3: 146 (1931).
Cyperus helodes Schrad. ex Nees in C.F.P.von Martius & auct. suc. (eds.), Fl. Bras. 2(1): 37 (1842). *Cyperus esculentus* var. *helodes* (Schrad. ex Nees) C.B.Clarke, J. Linn. Soc., Bot. 21: 181 (1884).
Cyperus fulvescens Liebm., Mexic. Halvgr.: 210 (1850).
Cyperus bahiensis Steud., Syn. Pl. Glumac. 2: 35 (1854).
Cyperus cubensis Steud., Syn. Pl. Glumac. 2: 42 (1854).
Cyperus variabilis Salzm. ex Steud., Syn. Pl. Glumac. 2: 46 (1854).
Cyperus chrysostachys Boeck., Flora 42: 438 bis (1859).
Cyperus heermannii Buckley, Proc. Acad. Nat. Sci. Philadelphia 15: 10 (1863). *Cyperus phymatodes* var. *heermannii* (Buckley) S.Watson in S.Watson & al., Bot. California 2: 215 (1880). *Cyperus esculentus* var. *heermannii* (Buckley) Britton, Bull. Torrey Bot. Club 23: 214 (1886).
Cyperus ruficomus Buckley, Proc. Acad. Nat. Sci. Philadelphia 15: 9 (1863).
Cyperus esculentus var. *leptostachyus* Boeck., Linnaea 36: 290 (1870).
Cyperus esculentus var. *sativus* Boeck., Linnaea 36: 290 (1870).
Cyperus callistus Ridl., Trans. Linn. Soc. London, Bot. 2: 143 (1884).
Cyperus esculentus f. *evolutus* C.B.Clarke, J. Linn. Soc., Bot. 21: 180 (1884).
Cyperus esculentus f. *princeps* C.B.Clarke, J. Linn. Soc., Bot. 21: 179 (1884).
Cyperus esculentus var. *sprucei* C.B.Clarke, J. Linn. Soc., Bot. 21: 181 (1884).
Cyperus esculentus var. *angustispicatus* Britton, Bull. Torrey Bot. Club 23: 211 (1886). *Cyperus esculentus* f. *angustispicatus* (Britton) Fernald, Rhodora 44: 151 (1942).
Cyperus buchananii Boeck., Beitr. Cyper. 1: 4 (1888).
Cyperus esculentus var. *aureus* K.Richt., Pl. Eur. 1: 135 (1890).
Cyperus esculentus var. *cyclolepis* Boeck. ex Kük. in H.G.A.Engler (ed.), Pflanzenr., IV, 20(101): 119 (1935).

Cyperus exaltatus Retz., Observ. Bot. 5: 11 (1788).
Trop. Africa, Trop. & Subtrop. Asia, Australia, Argentina. 22 BEN BKN GHA IVO MLI NGA SEN TOG 23 CMN GGI 24 CHA ETH SOM SUD 25 KEN TAN UGA 26 ANG 29 SEY 36 CHH CHS 38 JAP KOR TAI 40 ASS BAN IND NEP PAK SRL WHM 41 THA VIE 42 JAW LSI MLY 43 NWG 50 NSW NTA QLD SOA VIC 85 AGE. Hemicr. or rhizome geophyte.
Cyperus odoratus Burm.f., Fl. Indica: 20 (1768), nom. illeg.
Cyperus canaliculatus Retz., Observ. Bot. 6: 20 (1791).
Cyperus venustus R.Br., Prodr.: 217 (1810). *Papyrus venustus* (R.Br.) Nees, Linnaea 10: 138 (1836).

Cyperus festivus Link, Jahrb. Gewächsk. 1(3): 88 (1818).
Cyperus alopecuroides J.König ex Roxb., Fl. Ind. 1: 208 (1820), nom. illeg.
Cyperus umbellatus Roxb., Fl. Ind. 1: 205 (1820), nom. illeg.
Cyperus altus Nees in R.Wight, Contr. Bot. India: 84 (1834).
Cyperus oryzeticola Steud., Syn. Pl. Glumac. 2: 37 (1854).
Cyperus racemosus B.Heyne ex Boeck., Flora 62: 555 (1879), nom. illeg.
Cyperus iwasakii Makino, Bot. Mag. (Tokyo) 6: 47 (1892). *Cyperus exaltatus* var. *iwasakii* (Makino) T.Koyama, Acta Phytotax. Geobot. 16: 11 (1955).
Cyperus tokiensis C.B.Clarke, Kew Bull., Addit. Ser. 8: 9 (1908).
Cyperus exaltatus var. *minor* J.M.Black, Fl. S. Austral. 1: 90 (1922).
Cyperus exaltatus var. *megalanthus* Kük., Repert. Spec. Nov. Regni Veg. 27: 107 (1929).
Cyperus exaltatus var. *divergens* Kük. in H.G.A.Engler (ed.), Pflanzenr., IV, 20(101): 66 (1935).
Cyperus exaltatus var. *serpens* Kük. in H.G.A.Engler (ed.), Pflanzenr., IV, 20(101): 67 (1935).

Cyperus exilis Willd. ex Kunth, Enum. Pl. 2: 50 (1837).
W. Indian Ocean, India. 29 MAU MDG REU 40 IND. Hemicr. or rhizome geophyte.
Cyperus pentadactylus Willd. ex Kunth, Enum. Pl. 2: 50 (1837).

Cyperus expansus Poir. in J.B.A.M.de Lamarck, Encycl. 7: 269 (1806).
W. Indian Ocean. 29 MDG REU. Hemicr. or rhizome geophyte.
Cyperus dupetitii C.B.Clarke, Kew Bull., Addit. Ser. 8: 10 (1908).

Cyperus familiaris Steud., Syn. Pl. Glumac. 2: 46 (1854). *Mariscus familiaris* (Steud.) T.Koyama, Phytologia 29: 73 (1974).
Cuba. 81 CUB. Hemicr. or rhizome geophyte.

Cyperus fastigiatus Rottb., Descr. Icon. Rar. Pl.: 31 (1773).
S. Africa. 27 CPP LES NAT OFS SWZ TVL. Hemicr. or rhizome geophyte.
Cyperus fastigiosus J.F.Gmel., Syst. Nat. 2: 134 (1791).
Cyperus flabellaris Schrad. ex Nees, Linnaea 7: 519 (1832).
Cyperus egregius Kunth, Enum. Pl. 2: 69 (1837).
Cyperus semiangulatus Boeck., Linnaea 38: 367 (1874).

Cyperus fauriei Kük., Repert. Spec. Nov. Regni Veg. 16: 431 (1920). *Mariscus fauriei* (Kük.) T.Koyama, Bishop Mus. Occas. Pap. 29: 122 (1989).
Hawaiian Is. 63 HAW. Hemicr. or rhizome geophyte.

Cyperus feani F.Br., Bernice P. Bishop Mus. Bull. 84: 94 (1931).
Marquesas. 61 MRQ. Hemicr. or rhizome geophyte.

Cyperus felipponei Kük. in H.G.A.Engler (ed.), Pflanzenr., IV, 20(101): 213 (1936).
S. Venezuela to NE. Argentina. 82 VEN 84 BZN 85 AGE URU. Hemicr. or rhizome geophyte.

Cyperus fendlerianus Boeck., Linnaea 35: 520 (1868). *Chlorocyperus fendlerianus* (Boeck.) Palla, Oesterr. Bot. Z. 58: 392 (1908). *Mariscus fendlerianus* (Boeck.) T.Koyama, Phytologia 29: 73 (1974).
WC. U.S.A. to Mexico. 73 COL WYO 76 ARI 77 NWM TEX 79 MXC MXE MXN MXT. Hemicr. or rhizome geophyte.
Cyperus rusbyi Britton, Contr. U. S. Natl. Herb. 2: 461 (1894), nom. illeg.
Chlorocyperus mexicanus Palla, Oesterr. Bot. Z. 58: 391 (1908).
Cyperus fendlerianus var. *major* Kük. in H.G.A.Engler (ed.), Pflanzenr., IV, 20(101): 466 (1936).

Cyperus ferrugineoviridis (C.B.Clarke) Kük. in H.G.A.Engler (ed.), Pflanzenr., IV, 20(101): 412 (1936).
EC. & E. Trop. Africa. 23 BUR RWA ZAI 24 SOM 25 KEN TAN UGA. Hemicr. or rhizome geophyte.
**Cyperus maranguensis* var. *ferrugineoviridis* C.B. Clarke in D.Oliver & auct. suc. (eds.), Fl. Trop. Afr. 8: 359 (1901). *Mariscus ferrugineoviridis* (C.B. Clarke) Cherm., Bull. Jard. Bot. État 14: 330 (1937).
Cyperus ferrugineoviridis var. *distantiformis* Kük. in H.G.A.Engler (ed.), Pflanzenr., IV, 20(101): 413 (1936).
Cyperus ferrugineoviridis var. *luteiformis* Kük. in H.G.A.Engler (ed.), Pflanzenr., IV, 20(101): 412 (1936).
Mariscus bequaertii Cherm., Bull. Jard. Bot. État 14: 329 (1937). *Cyperus bequaertii* (Cherm.) Robyns & Tournay in W.Robyns, Fl. Spermatophyt. Parc Nat. Albert 3: 246 (1955).

Cyperus fertilis Boeck., Bot. Jahrb. Syst. 5: 90 (1884).
W. Trop. Africa to Angola. 22 GHA IVO LBR NGA 23 CAF CMN EQG GAB ZAI 25 UGA 26 ANG. Hemicr. or rhizome geophyte.
Cyperus lanceola Ridl., Trans. Linn. Soc. London, Bot. 2: 134 (1884).

Cyperus filiculmis Vahl, Enum. Pl. 2: 328 (1805). *Chlorocyperus filiculmis* (Vahl) Palla, Allg. Bot. Z. Syst. 8: 66 (1902). *Mariscus filiculmis* (Vahl) T.Koyama, Phytologia 29: 73 (1974).
SE. U.S.A. to Texas. 77 TEX 78 ALA FLA GEO LOU MRY MSI NCA SCA VRG. Hemicr. or rhizome geophyte.
Cyperus kyllingioides Pursh, Fl. Amer. Sept. 1: 80 (1813), nom. illeg.
Cyperus annuus Spreng. & Link, Jahrb. Gewächsk. 1(3): 83 (1818).
Cyperus mariscoides Nutt., Gen. N. Amer. Pl. 1: 35 (1818), nom. illeg.
Cyperus etesios Schult. in J.J.Roemer & J.A.Schultes, Mant. 2: 114 (1824).
Cyperus martindalei Britton, Bull. Torrey Bot. Club 15: 98 (1888).
Cyperus filiculmis var. *oblitus* Fernald, Rhodora 37: 153 (1935).

Cyperus filifolius Willd. ex Kunth, Enum. Pl. 2: 85 (1837).
Venezuela, Guyana, N. Brazil. 82 GUY VEN 84 BZN. Hemicr. or rhizome geophyte.
**Mariscus filiformis* Kunth in F.W.H.von Humboldt, A.J.A.Bonpland & C.S.Kunth, Nov. Gen. Sp. 1: 213 (1816), nom. illeg.
Cyperus filiculmis Schrad. ex Nees in C.F.P.von Martius & auct. suc. (eds.), Fl. Bras. 2(1): 40 (1842), nom. illeg.
Cyperus nemacaulos Steud., Syn. Pl. Glumac. 2: 43 (1854).

Cyperus filiformis Sw., Prodr.: 20 (1788). *Mariscus filiformis* (Sw.) Roem. & Schult., Syst. Veg. 2: 242

(1817), nom. illeg. *Torulinium filiforme* (Sw.) C.B.Clarke in I.Urban, Symb. Antill. 2: 57 (1900).
S. Florida, Caribbean, Venezuela. 78 FLA 81 ARU BAH CAY CUB DOM HAI JAM LEE NLA 82 VEN. Hemicr.
Cyperus ornithopus Pers., Syn. Pl. 1: 61 (1805).
Mariscus havanensis Kunth in F.W.H.von Humboldt, A.J.A.Bonpland & C.S.Kunth, Nov. Gen. Sp. 1: 212 (1816).
Epiphystis ophiuroides Trin., Fund. Agrost.: 97 (1820).
Cyperus havanensis Kunth, Enum. Pl. 2: 84 (1837).
Cyperus filiformis var. *havanensis* (Kunth) Kük., Repert. Spec. Nov. Regni Veg. 23: 190 (1926).
Cyperus underwoodii Britton, Mem. Torrey Bot. Club 16: 59 (1920).
Cyperus filiformis f. *filifolius* Kük., Repert. Spec. Nov. Regni Veg. 23: 190 (1926).
Cyperus filiformis f. *robustior* Kük., Repert. Spec. Nov. Regni Veg. 23: 190 (1926).
Cyperus filiformis f. *subfilifolius* Kük., Repert. Spec. Nov. Regni Veg. 23: 191 (1926).

Cyperus filipes Benth., Fl. Austral. 7: 271 (1878).
New South Wales. 50 NSW. Hemicr. or rhizome geophyte.

Cyperus firmipes (C.B.Clarke) Kük. in H.G.A.Engler (ed.), Pflanzenr., IV, 20(101): 562 (1936).
Tanzania, Malawi. 25 TAN 26 MLW. Hemicr. or rhizome geophyte.
**Mariscus firmipes* C.B.Clarke in D.Oliver & auct. suc. (eds.), Fl. Trop. Afr. 8: 382 (1901).

Cyperus fischerianus Schimp. ex A.Rich., Tent. Fl. Abyss. 2: 488 (1850).
Ethiopia to Malawi. 23 ZAI 24 ERI ETH 25 KEN TAN UGA 26 MLW. Hemicr. or rhizome geophyte.
Cyperus fischerianus var. *ugandensis* Lye, Nordic J. Bot. 3: 230 (1983).

Cyperus fissus Steud., Flora 25: 593 (1842).
Ethiopia, Yemen. 24 ETH 35 YEM. Hemicr. or rhizome geophyte. Provisionally accepted.

Cyperus flaccidus R.Br., Prodr.: 213 (1810). *Cyperus trinervis* var. *flaccidus* (R.Br.) Kük. in H.G.A.Engler (ed.), Pflanzenr., IV, 20(101): 294 (1936).
N. & E. Australia. 50 NSW NTA QLD VIC. Hemicr. or rhizome geophyte.

Cyperus flavoculmis Lye, Nordic J. Bot. 3: 223 (1983).
Kenya. 25 KEN. Hemicr. or rhizome geophyte.

Cyperus flexuosus Vahl, Enum. Pl. 2: 359 (1805). *Diclidium vahlii* Schrad. ex Nees in C.F.P.von Martius & auct. suc. (eds.), Fl. Bras. 2(1): 53 (1842), nom. superfl. *Cyperus vahlii* (Schrad. ex Nees) Steud., Syn. Pl. Glumac. 2: 48 (1854), nom. superfl. *Torulinium vahlii* (Schrad. ex Nees) C.B.Clarke in I.Urban, Symb. Antill. 2: 56 (1900), nom. superfl. *Torulinium flexuosum* (Vahl) T.Koyama, Phytologia 29: 74 (1974).
Mexico, Caribbean, SE. Brazil. 79 MXC 81 CUB DOM HAI JAM LEE 84 BZL. Hemicr. or rhizome geophyte.
Cyperus ehrenbergii Kunth, Enum. Pl. 2: 89 (1837).
Cyperus insignis Kunth, Enum. Pl. 2: 92 (1837). *Cyperus flexuosus* var. *insignis* (Kunth) Kük. in H.G.A. Engler (ed.), Pflanzenr., IV, 20(101): 622 (1936).

Cyperus floribundus (Kük.) R.Carter & S.D.Jones, Rhodora 99: 330 (1997).
S. Texas to NE. Mexico. 77 TEX 79 MXE. Hemicr. or rhizome geophyte.
**Cyperus uniflorus* var. *floribundus* Kük. in H.G.A. Engler (ed.), Pflanzenr., IV, 20(101): 521 (1936).

Cyperus floridanus Britton in J.K.Small, Fl. S.E. U.S.: 170 (1903).
Florida, Caribbean. 78 FLA 81 BAH CAY CUB JAM PUE. Hemicr. or rhizome geophyte.
Mariscus floridensis C.B.Clarke, Kew Bull., Addit. Ser. 8: 20 (1908).
Cyperus filiformis f. *coloratus* Kük., Repert. Spec. Nov. Regni Veg. 23: 190 (1926).
Cyperus filiformis var. *densiceps* Kük., Repert. Spec. Nov. Regni Veg. 23: 190 (1926).
Cyperus kingsii C.D.Adams ex Proctor, Sloanea 1: 1 (1977).

Cyperus foliaceus C.B.Clarke, Bot. Jahrb. Syst. 38: 134 (1906).
Trop. Africa. 22 TOG 24 ETH 25 KEN TAN UGA. Ther.

Cyperus forskalianus Väre & Kukkonen, Ann. Bot. Fenn. 42: 477 (2005).
Yemen. 35 YEM.

Cyperus frerei C.B.Clarke in D.Oliver & auct. suc. (eds.), Fl. Trop. Afr. 8: 327 (1901).
Coastal S. Somalia to E. Kenya. 24 SOM 25 KEN. Hemicr. or rhizome geophyte.
Cyperus chisimajensis Chiov., Agric. Colon. 20: 104 (1926).

Cyperus friburgensis Boeck., Beitr. Cyper. 2: 2 (1890).
S. Trop. America. 82 VEN 83 BOL ECU PER 84 BZC BZL 85 AGE AGW. Hemicr. or rhizome geophyte.
Cyperus bulbosus Schrank, Denkschr. Königl.-Baier. Bot. Ges. Regensburg 2: 24 (1822), nom. illeg.
Cyperus martianus Schrad. ex Nees in C.F.P.von Martius & auct. suc. (eds.), Fl. Bras. 2(1): 32 (1842), nom. illeg. *Eucyperus martianus* Rikli, Jahrb. Wiss. Bot. 27: 568 (1895). *Cyperus schraderi* J.F.Macbr., Publ. Field Columbian Mus., Bot. Ser. 4: 167 (1929).
Cyperus fuscoferrugineus Boeck., Allg. Bot. Z. Syst. 1: 226 (1895).

Cyperus fucosus K.L.Wilson, Telopea 4: 461 (1991).
N. Australia. 50 NTA QLD. Hemicr. or rhizome geophyte.

Cyperus fulgens C.B.Clarke in D.Oliver & auct. suc. (eds.), Fl. Trop. Afr. 8: 355 (1901).
C. Tanzania to S. Africa. 25 TAN 26 MLW 27 BOT NAM TVL. Hemicr. or tuber geophyte.

var. ***contractus*** Kük. in H.G.A.Engler (ed.), Pflanzenr., IV, 20(101): 122 (1935).
S. Africa. 27 BOT NAM TVL. Hemicr. or rhizome geophyte.

var. ***fulgens***
C. Tanzania to S. Africa. 25 TAN 26 MLW 27 TVL. Hemicr. or tuber geophyte.

Cyperus fuligineus Chapm., Fl. South. U.S.: 511 (1860). *Mariscus fuligineus* (Chapm.) C.B.Clarke in I.Urban, Symb. Antill. 5: 290 (1907).
S. Florida, Caribbean. 78 FLA 81 BAH CUB HAI LEE NLA PUE. Hemicr.

Cyperus fulvoalbescens T.Koyama, Naturaliste Canad. 82: 211 (1955).
S. Vietnam. 41 VIE. Hemicr. or rhizome geophyte.

Cyperus fulvus R.Br., Prodr.: 215 (1810). *Mariscus fulvus* (R.Br.) C.B.Clarke, Kew Bull., Addit. Ser. 8: 18 (1908).

New Guinea to N. & E. Australia. 43 NWG 50 NSW NTA QLD. Hemicr. or rhizome geophyte.
Cyperus sieberi Kunth, Enum. Pl. 2: 96 (1837).
Mariscus fulvus var. *canescens* C.B.Clarke ex Domin, Biblioth. Bot. 85: 444 (1915). *Cyperus fulvus* var. *canescens* (C.B.Clarke ex Domin) Kük. in H.G.A. Engler (ed.), Pflanzenr., IV, 20(101): 456 (1936).
Mariscus fulvus var. *confusus* C.B.Clarke ex Domin, Biblioth. Bot. 85: 444 (1915). *Cyperus fulvus* var. *confusus* (C.B.Clarke ex Domin) Kük. in H.G.A. Engler (ed.), Pflanzenr., IV, 20(101): 456 (1936).
Mariscus fulvus var. *densispiculatus* Domin, Biblioth. Bot. 85: 444 (1915). *Cyperus fulvus* var. *densispiculatus* (Domin) Kük. in H.G.A.Engler (ed.), Pflanzenr., IV, 20(101): 457 (1936).
Mariscus fulvus var. *jucundus* Domin, Biblioth. Bot. 85: 444 (1915).
Cyperus fulvus var. *viscidus* Kük. in H.G.A.Engler (ed.), Pflanzenr., IV, 20(101): 456 (1936).

Cyperus fuscescens Willd. ex Link, Jahrb. Gewächsk. 1(3): 83 (1818).
Cape Prov. 27 CPP. Hemicr. or rhizome geophyte.

Cyperus fuscovaginatus Kük. in R.E.Fries, Wiss. Erg. Schwed. Rhod.-Kongo Exped. 1: 4 (1921).
Zambia. 26 ZAM. Hemicr. or rhizome geophyte.

Cyperus fuscus L., Sp. Pl.: 46 (1753). *Eucyperus fuscus* (L.) Rikli, Jahrb. Wiss. Bot. 27: 568 (1895).
Macaronesia, Europe, Medit. to China. 10 DEN GRB SWE 11 AUT BGM CZE GER HUN NET POL SWI 12 BAL COR FRA POR SAR SPA 13 ALB BUL GRC ITA KRI ROM SIC TUE YUG 14 BLR BLT KRY RUC RUE RUN RUS RUW UKR 20 ALG EGY MOR TUN 21 CNY MDR 30 ALT WSB 32 KAZ KGZ TKM TZK UZB 33 NCS TCS 34 AFG CYP EAI IRN IRQ LBS PAL TUR 35 YEM 36 CHN 37 MON 40 IND PAK WHM 41 LAO THA VIE (72) ont que (74) mso neb sda (75) cnt mas nwj pen (76) cal nev (78) mry vrg (81) lee. Ther.
Cyperus compressus Krock., Fl. Siles. 1: 64 (1787), nom. illeg.
Cyperus viridis Krock., Fl. Siles. 1: 64 (1787).
Cyperus virescens Hoffm., Deutschl. Fl., ed. 2, 1(1): 21 (1791). *Cyperus fuscus* f. *virescens* (Hoffm.) Vahl, Enum. Pl. 2: 336 (1805). *Cyperus fuscus* var. *virescens* (Hoffm.) Dumort., Fl. Belg.: 145 (1827).
Cyperus protractus Delile, Descr. Egypte, Hist. Nat. 3: 50 (1813).
Cyperus haworthii Gray, Nat. Arr. Brit. Pl. 2: 730 (1821).
Cyperus forsskalii A.Dietr., Sp. Pl. 2: 251 (1832).
Cyperus fuscus var. *nigricans* Neilr., Fl. Nied.-Oesterr.: 135 (1859).
Cyperus calidus A.Kern., Oesterr. Bot. Z. 14: 84 (1864). *Cyperus fuscus* f. *calidus* (A.Kern.) Asch. & Graebn., Syn. Mitteleur. Fl. 2(2): 276 (1904).
Cyperus sabaudus Billet ex Gave, Liste Contrib. Fl. Savoie: 14 (1906).
Cyperus fuscus var. *condensatus* Kük. in H.G.A.Engler (ed.), Pflanzenr., IV, 20(101): 237 (1936).
Cyperus fuscus var. *subjuncellus* Kük. in H.G.A.Engler (ed.), Pflanzenr., IV, 20(101): 237 (1936).

Cyperus gardneri Nees in C.F.P.von Martius & auct. suc. (eds.), Fl. Bras. 2(1): 34 (1842).
Cuba, SE. Mexico to NE. Argentina. 79 MXT 80 ELS 81 CUB 82 VEN 83 ECU 84 BZE 85 AGE. Hemicr. or rhizome geophyte.
Cyperus corticatus C.Wright, Anales Acad. Ci. Méd. Habana 8: 45 (1871).
Cyperus gardneri var. *vegetior* Kük. in H.G.A.Engler (ed.), Pflanzenr., IV, 20(101): 164 (1936).

Cyperus gayi (C.B.Clarke) Kük. in H.G.A.Engler (ed.), Pflanzenr., IV, 20(101): 484 (1936).
French Guiana. 82 FRG. Ther.
**Mariscus gayi* C.B.Clarke, Kew Bull., Addit. Ser. 8: 16 (1908).

Cyperus giganteus Vahl, Enum. Pl. 2: 364 (1805). *Papyrus giganteus* (Vahl) Schrad. ex Nees in C.F.P.von Martius & auct. suc. (eds.), Fl. Bras. 2(1): 58 (1842). *Chlorocyperus giganteus* (Vahl) Palla, Denkschr. Kaiserl. Akad. Wiss., Wien. Math.-Naturwiss. Kl. 79: 181 (1908).
Trop. & Subtrop. America. (77) tex (78) lou 79 MXG MXS MXT 80 BLZ COS ELS HON NIC PAN 81 CUB DOM HAI JAM PUE TRT 82 FRG GUY SUR VEN 83 BOL CLM ECU PER 84 BZC BZE BZL 85 AGE AGW PAR URU. Hemicr. or rhizome geophyte.
Papyrus comosus Kunth, Nov. Gen. Sp. Pl. 1: 218 (1815). *Cyperus comosus* (Kunth) Poir. in J.B.A.M.de Lamarck, Encycl., Suppl. 5: 185 (1817), nom. illeg. *Cyperus giganteus* var. *comosus* (Kunth) Kük., Repert. Spec. Nov. Regni Veg. 32: 74 (1933).
Cyperus affinis Roem. & Schult., Syst. Veg. 2: 213 (1817).
Cyperus berteroi Kunth, Enum. Pl. 2: 67 (1837).
Cyperus densiflorus Rchb. ex Kunth, Enum. Pl. 2: 65 (1837), nom. illeg.
Cyperus princeps Kunth, Enum. Pl. 2: 66 (1837).
Papyrus elegans Schrad. ex Nees in C.F.P.von Martius & auct. suc. (eds.), Fl. Bras. 2(1): 59 (1842).
Papyrus junciformis Schrad. ex Nees in C.F.P.von Martius & auct. suc. (eds.), Fl. Bras. 2(1): 57 (1842).
Papyrus odoratus Willd. ex Nees in C.F.P.von Martius & auct. suc. (eds.), Fl. Bras. 2(1): 58 (1842).
Papyrus radiatus Schrad. ex Nees in C.F.P.von Martius & auct. suc. (eds.), Fl. Bras. 2(1): 57 (1842).
Papyrus spectabilis Schrad. ex Nees in C.F.P.von Martius & auct. suc. (eds.), Fl. Bras. 2(1): 58 (1842).
Cyperus conspicuus Steud., Syn. Pl. Glumac. 2: 39 (1854).
Cyperus pseudogiganteus Steud., Syn. Pl. Glumac. 2: 39 (1854).
Cyperus elatus Griseb., Fl. Brit. W. I.: 566 (1864), nom. illeg.

Cyperus gigantobulbes Lye, Nordic J. Bot. 3: 219 (1983).
Tanzania (Uzaramo Distr.). 25 TAN. Hemicr. or tuber geophyte.

Cyperus gilesii Benth., Fl. Austral. 7: 274 (1878).
Australia. 50 NSW NTA QLD SOA. Hemicr. or rhizome geophyte.

Cyperus glaber L., Mant. Pl. 2: 179 (1771). *Chlorocyperus glaber* (L.) Palla, Allg. Bot. Z. Syst. 6: 201 (1900). *Dichostylis glabra* (Palla) Palla, Vestn. Tiflissk. Bot. Sada 21: 24 (1912). *Pycreus glaber* (L.) Hayek, Repert. Spec. Nov. Regni Veg. Beih. 30(3): 146 (1932).
C. Europe to Pakistan. 11 AUT 13 BUL GRC ITA SIC TUE YUG 14 KRY RUS UKR (27) cpp 32 KAZ KGZ TZK UZB 33 TCS 34 AFG CYP EAI IRN IRQ LBS PAL TUR 40 PAK. Ther.
Cyperus patulus Kit. ex Host, Icon. Descr. Gram. Austriac. 3: 49 (1805).

Cyperus caucasicus Link, Jahrb. Gewächsk. 1(3): 89 (1818).
Cyperus erubescens Link, Enum. Hort. Berol. Alt. 1: 44 (1821).
Cyperus pictus Ten., Fl. Napol. 3: 47 (1824).
Cyperus viridis Sieber ex C.Presl, Isis (Oken) 21: 270 (1828), nom. illeg.
Cyperus fascicularis Phil. ex Kunth, Enum. Pl. 2: 25 (1837), nom. illeg.
Cyperus banaticus Kit. ex Nyman, Consp. Fl. Eur.: 759 (1882).
Cyperus blancheanus Desv. ex Boiss., Fl. Orient. 5: 371 (1882).
Cyperus glaber var. *blancheanus* Boiss., Fl. Orient. 5: 371 (1882).
Cyperus subchoristachys C.B.Clarke in W.H.Harvey & auct. suc. (eds.), Fl. Cap. 7: 177 (1897). *Cyperus glaber* var. *subchoristachys* (C.B.Clarke) Kük., Repert. Spec. Nov. Regni Veg. 26: 249 (1929).
Cyperus glaber f. *contractus* Asch. & Graebn., Syn. Mitteleur. Fl. 2(2): 275 (1904).
Cyperus glaber f. *fasciculariformis* Lojac., Fl. Sicul. 3: 201 (1909). *Chlorocyperus glaber* f. *fasciculariforme* (Lojac.) Soó, Acta Bot. Acad. Sci. Hung. 16: 269 (1970 publ. 1971), without basionym date.

Cyperus glaucophyllus Boeck., Beitr. Cyper. 1: 4 (1888).
Kenya to Malawi. 23 RWA ZAI 25 KEN TAN 27 MLW. Hemicr. or rhizome geophyte.
Cyperus glaucophyllus var. *longispiculosus* Kük. in H.G.A.Engler (ed.), Pflanzenr., IV, 20(101): 203 (1936).

Cyperus globifer (C.B.Clarke) Lye, Nordic J. Bot. 3: 232 (1983).
E. Trop. Africa. 25 KEN TAN. Hemicr. or tuber geophyte.
**Mariscus globifer* C.B.Clarke, J. Bot. 34: 225 (1896). *Cyperus mollipes* var. *globifer* (C.B.Clarke) Kük. in H.G.A.Engler (ed.), Pflanzenr., IV, 20(101): 558 (1936).

Cyperus glomeratus L., Cent. Pl. II: 5 (1756). *Chlorocyperus glomeratus* (L.) Palla, Allg. Bot. Z. Syst. 6: 201 (1900). *Pycreus glomeratus* (L.) Hayek, Repert. Spec. Nov. Regni Veg. Beih. 30(3): 145 (1932).
Temp. Eurasia. 11 AUT CZE HUN SWI 12 FRA 13 BUL GRC ITA ROM YUG 14 RUS UKR 31 AMU PRM 32 KAZ TKM UZB 33 TCS 34 IRN TUR 36 CHM CHN 38 JAP KOR 40 WHM. Ther. or hemicr.
Cyperus cinnamomeus Retz., Observ. Bot. 4: 10 (1786).
Cyperus aureus Georgi, Beschr. Nation. Russ. Reich, Nachtr.: 257 (1802).
Cyperus australis Schrad., Fl. Germ. 1: 116 (1806).
Cyperus multumbelliferus Roem. & Schult., Syst. Veg. 2: 200 (1817).
Cyperus sparsiflorus Baldwin, Trans. Amer. Philos. Soc., n.s., 2: 168 (1825).
Cyperus granulatus Bosc ex Kunth, Enum. Pl. 2: 95 (1837).
Cyperus semiradiciflorus Boeck., Flora 42: 434 bis (1859). *Cyperus glomeratus* var. *semiradiciflorus* (Boeck.) Kük. in H.G.A.Engler (ed.), Pflanzenr., IV, 20(101): 148 (1935).
Cyperus glomeratus f. *stenostachys* Borbás, Term. Füz. 8: 56 (1884). *Chlorocyperus glomeratus* f. *stenostachys* (Borbás) Soó, Acta Bot. Acad. Sci. Hung. 16: 369 (1970 publ. 1971).

Cyperus graciliculmis Lye, Nordic J. Bot. 3: 224 (1983).
Tanzania (Mt. Mnyera). 23 RWA? 25 TAN. Hemicr. or rhizome geophyte.

Cyperus gracilis R.Br., Prodr.: 213 (1810). *Eucyperus gracilis* (R.Br.) Rikli, Jahrb. Wiss. Bot. 27: 568 (1895).
New Caledonia, E. Australia. (50) nfk NSW QLD 60 NWC (63) haw (76) cal. Ther. or hemicr.
Cyperus gracilis f. *viviparus* Domin, Biblioth. Bot. 89: 421 (1915).

Cyperus granatensis C.B.Clarke, Kew Bull., Addit. Ser. 8: 11 (1908).
Colombia. 83 CLM. Hemicr. or rhizome geophyte.

Cyperus grandibulbosus C.B.Clarke in D.Oliver & auct. suc. (eds.), Fl. Trop. Afr. 8: 353 (1901).
Ethiopia to N. Tanzania. 24 ETH SOM 25 KEN TAN. Hemicr. or bulb geophyte.
Cyperus giolii Chiov., Ann. Bot. (Rome) 13: 375 (1915).
Mariscus nogalensis Chiov., Pl. Nov. Aethiop.: 14 (1928). *Cyperus giolii* var. *nogalensis* (Chiov.) Kük. in H.G.A.Engler (ed.), Pflanzenr., IV, 20(101): 127 (1935).
Cyperus grandibulbosus var. *amplus* Kük. in H.G.A. Engler (ed.), Pflanzenr., IV, 20(101): 125 (1935).

Cyperus grandifolius J.G.Anderson, Kongl. Vetensk. Acad. Handl. 1853: 157 (1855).
Galápagos. 83 GAL. Hemicr. or rhizome geophyte.
Cyperus galapagensis Caruel., Atti Reale Accad. Lincei, Rendiconti Cl. Sci. Fis. 5: 621 (1889).

Cyperus grandis C.B.Clarke in D.Oliver & auct. suc. (eds.), Fl. Trop. Afr. 8: 372 (1901).
E. Trop. Africa, W. Madagascar. 25 KEN TAN 29 MDG 41 VIE. Hemicr. or rhizome geophyte.

Cyperus grandisimplex C.B.Clarke, Kew Bull., Addit. Ser. 8: 7 (1908).
S. Venezuela to Paraguay. 82 VEN 83 PER 84 BZE 85 PAR. Hemicr. or rhizome geophyte.

Cyperus granitophilus McVaugh, Castanea 2: 103 (1937).
SE. U.S.A. 78 ALA GEO NCA SCA TEN VRG. Ther.

Cyperus grantii Boeck., Flora 58: 260 (1875).
S. Africa. 27 CPP NAT. Hemicr. or rhizome geophyte.

Cyperus grayi Torr., Ann. Lyceum Nat. Hist. New York 3: 260 (1836). *Cyperus filiculmis* var. *grayi* (Torr.) Boeck., Linnaea 35: 609 (1868). *Mariscus grayi* (Torr.) C.B.Clarke, Kew Bull., Addit. Ser. 8: 89 (1908).
E. U.S.A. 75 CNT MAS NWH NWJ NWY PEN RHO 78 DEL GEO MRY NCA SCA VRG. Hemicr. or rhizome geophyte.
Cyperus setifolius Torr. ex C.B.Clarke, J. Linn. Soc., Bot. 21: 198 (1884), nom. inval.

Cyperus grayioides Mohlenbr., Brittonia 11: 255 (1959).
Illinois to Texas. 74 ILL MSO 77 TEX 78 ARK LOU. Hemicr. or rhizome geophyte.

Cyperus grossianus Pedersen, Darwiniana 17: 543 (1972).
Brazil. 84 BZE. Hemicr. or rhizome geophyte.
**Cyperus laetus* var. *tenuissimus* Grosser & Kük. in H.G.A.Engler (ed.), Pflanzenr., IV, 20(101): 441 (1936).

Cyperus gubanii Väre & Kukkonen, Ann. Bot. Fenn. 42: 479 (2005).
Somalia. 24 SOM.

Cyperus guanipensis Schnee, Bol. Soc. Venez. Ci. Nat. 9: 23 (1944).
S. Caribbean to Venezuela. 81 ARU NLA 82 VEN. Ther.

Cyperus guaricensis Schnee, Bol. Soc. Venez. Ci. Nat. 9: 17 (1944).
Venezuela. 82 VEN. Hemicr. or rhizome geophyte.

Cyperus guianensis (C.B.Clarke) Kük. in H.G.A.Engler (ed.), Pflanzenr., IV, 20(101): 483 (1936).
Guianas, N. Brazil. 82 GUY SUR 84 BZN. Hemicr. or rhizome geophyte.
**Mariscus guianensis* C.B.Clarke, Kew Bull., Addit. Ser. 8: 16 (1908).

Cyperus gunnii Hook.f., Fl. Tasman. 2: 80 (1858). *Cyperus lucidus* var. *gunnii* (Hook.f.) C.Moore & Betche, Handb. Fl. N.S.W.: 449 (1893). *Mariscus gunnii* (Hook.f.) C.B.Clarke, Kew Bull., Addit. Ser. 8: 18 (1908). *Mariscus gunnii* var. *typicus* Domin, Biblioth. Bot. 85: 442 (1915), nom. inval.
Australia. 50 NSW NTA QLD SOA TAS VIC. Hemicr. or rhizome geophyte.

var. ***gunnii***
Australia. 50 NSW NTA QLD SOA TAS VIC. Hemicr. or rhizome geophyte.
Cyperus compositus Boeck., Linnaea 36: 333 (1870), nom. illeg.
Cyperus nodulosus F.Muell. ex Benth., Fl. Austral. 7: 283 (1878).

var. ***novae-hollandiae*** (Boeck.) Kük. in H.G.A.Engler (ed.), Pflanzenr., IV, 20(101): 450 (1936).
Queensland. 50 QLD. Hemicr. or rhizome geophyte.
**Cyperus novae-hollandiae* Boeck., Linnaea 36: 344 (1870). *Mariscus gunnii* var. *novae-hollandiae* (Boeck.) Domin, Biblioth. Bot. 85: 443 (1915). *Cyperus gunnii* subsp. *novae-hollandiae* (Boeck.) K.L.Wilson, Telopea 4: 465 (1991).
Cyperus inornatus Boeck., Flora 58: 86 (1875).

Cyperus gymnocaulos Steud., Syn. Pl. Glumac. 2: 12 (1854). *Cyperus vaginatus* subsp. *gymnocaulos* (Steud.) Kük. in H.G.A.Engler (ed.), Pflanzenr., IV, 20(101): 190 (1936).
Australia. 50 NSW NTA QLD SOA VIC WAU. Hemicr. or rhizome geophyte.
Cyperus vaginatus var. *gymnocaulos* (Steud.) Probst, Wolladventivfl. Mitteleur.: 49 (1949).

Cyperus gypsophilus Lye, Nordic J. Bot. 16: 374 (1996).
N. Somalia. 24 SOM. Hemicr. or rhizome geophyte.

Cyperus gyraudinii Steud., Syn. Pl. Glumac. 2: 28 (1854).
Réunion. 29 REU. Hemicr. or rhizome geophyte. Provisionally accepted.

Cyperus haematocephalus Boeck. ex C.B.Clarke in W.H.Harvey & auct. suc. (eds.), Fl. Cap. 7: 171 (1897).
S. Africa. 27 CPP LES NAT OFS TVL. Hemicr. or rhizome geophyte.

Cyperus haematodes Endl., Prodr. Fl. Norfolk.: 22 (1833). *Mariscus haematodes* (Endl.) C.B.Clarke, Kew Bull., Addit. Ser. 8: 18 (1908).
Queensland, Lord Howe I., Norfolk I. 50 NFK QLD. Hemicr. or rhizome geophyte.
Cyperus congestus f. *gigantea* F.Muell., Fragm. 8: 18 (1872).
Cyperus ferax Benth., Fl. Austral. 7: 286 (1878), nom. illeg.

Cyperus hakonensis Franch. & Sav., Enum. Pl. Jap. 2: 104, 538 (1877).
China, Korea, Japan. 36 CHN 38 JAP KOR. Ther.

Cyperus hamulosus M.Bieb., Fl. Taur.-Caucas. 1: 35 (1808). *Scirpus hamulosus* (M.Bieb.) Steven, Mém. Soc. Imp. Naturalistes Moscou 4: 356 (1814). *Dichostylis hamulosa* (M.Bieb.) Nees, Linnaea 9: 289 (1834). *Isolepis hamulosa* (M.Bieb.) Kunth, Enum. Pl. 2: 204 (1837). *Mariscus hamulosus* (M.Bieb.) S.S.Hooper, Kew Bull. 26: 578 (1972).
Bulgaria to C. Asia, Morocco to W. Trop. & S. Africa. 13 BUL ROM 14 RUC RUS UKR 20 MOR 22 MLI NGA NGR SEN 27 BOT NAM 30 32 KAZ TKM UZB 33 TCS (50). Ther.
Cyperus pygmaeus Cav., Anales Ci. Nat. 3: 7 (1801), nom. illeg.
Scirpus lugardii C.B.Clarke in D.Oliver & auct. suc. (eds.), Fl. Trop. Afr. 8: 458 (1902).
Scirpus pitardii Trab. ex Pit., Contr. Fl. Maroc.: 37 (1918).

Cyperus harrisii Kük., Repert. Spec. Nov. Regni Veg. 23: 191 (1926). *Torulinium harrisii* (Kük.) T.Koyama, Phytologia 29: 74 (1974).
Jamaica. 81 JAM. Hemicr. or rhizome geophyte.

Cyperus haspan L., Sp. Pl.: 45 (1753).
Trop. & Subtrop. 22 BEN BKN GAM GHA GNB GUI IVO LBR NGA NGR SEN SIE TOG 23 BUR CMN EQG GAB ZAI 24 CHA ETH SOM SUD 25 KEN TAN UGA 26 ANG MLW MOZ ZAM ZIM 27 TVL 29 COM MDG SEY 34 IRN 36 CHS 38 JAP NNS TAI 40 ASS EHM IND NEP PAK SRL 41 AND CBD LAO MYA NCB THA VIE 42 BOR JAW LSI MLY MOL PHI SUL SUM 43 BIS NWG SOL 50 NSW NTA QLD WAU (60) fij 62 CRL (63) haw 77 TEX 78 ALA FLA GEO LOU MSI NCA SCA VRG 79 MXG MXS MXT 80 BLZ COS GUA HON NIC PAN 81 CUB DOM PUE TRT 82 FRG GUY SUR VEN 83 BOL CLM ECU PER 84 BZE BZL BZS 85 PAR. Ther. or hemicr.
Scirpus autumnalis Rottb., Descr. Icon. Rar. Pl.: 58 (1773), nom. illeg. *Cyperus autumnalis* Vahl, Enum. Pl. 2: 318 (1805).
Cyperus juncoides Lam., Tabl. Encycl. 1: 147 (1791). *Cyperus haspan* subsp. *juncoides* (Lam.) Kük. in H.G.A.Engler (ed.), Pflanzenr., IV, 20(101): 249 (1936).
Cyperus aphyllus Vahl, Eclog. Amer. 2: 6 (1798).
Cyperus graminifolius Poir. in J.B.A.M.de Lamarck, Encycl. 7: 267 (1806).
Cyperus platyculmis R.Br., Prodr.: 214 (1810).
Cyperus gracilis Muhl., Descr. Gram.: 18 (1817), nom. illeg.
Cyperus gramineus Roem. & Schult., Syst. Veg. 2: 205 (1817).
Cyperus nudus Kunth in F.W.H.von Humboldt, A.J.A.Bonpland & C.S.Kunth, Nov. Gen. Sp. 1: 203 (1817).
Cyperus cayennensis Link, Jahrb. Gewächsk. 1(3): 84 (1818).
Cyperus junceus Link, Jahrb. Gewächsk. 1(3): 85 (1818).
Cyperus stellatus Link, Jahrb. Gewächsk. 1(3): 84 (1818), nom. illeg.
Cyperus vaginatus Link, Jahrb. Gewächsk. 1(3): 85 (1818), nom. illeg.
Cyperus laticulmis Spreng., Syst. Veg. 1: 228 (1824).
Cyperus leptos Schult. in J.J.Roemer & J.A.Schultes, Mant. 2: 105 (1824).

Cyperus tunicatus Roem. & Schult., Mant. 2: 115 (1824).
Cyperus muehlenbergii Spreng., Syst. Veg. 5: 221 (1828).
Cyperus nudus var. *firmior* C.Presl, Reliq. Haenk. 1: 166 (1828).
Cyperus nudus var. *flaccidior* C.Presl, Reliq. Haenk. 1: 166 (1828).
Cyperus adenophorus Schrad. ex Nees in C.F.P.von Martius & auct. suc. (eds.), Fl. Bras. 2(1): 28 (1842). *Cyperus haspan* var. *adenophorus* (Schrad. ex Nees) Kük. in H.G.A.Engler (ed.), Pflanzenr., IV, 20(101): 250 (1936).
Cyperus haspan var. *coarctatus* Nees in C.F.P.von Martius & auct. suc. (eds.), Fl. Bras. 2(1): 25 (1842). *Cyperus haspan* f. *coarctatus* (Nees) Osten, Anales Mus. Hist. Nat. Montevideo, II, 3: 135 (1931).
Cyperus riparius Schrad. ex Nees in C.F.P.von Martius & auct. suc. (eds.), Fl. Bras. 2(1): 28 (1842), nom. illeg. *Cyperus haspan* var. *riparius* Kük. in H.G.A. Engler (ed.), Pflanzenr., IV, 20(101): 250 (1936).
Cyperus pedatus Nees in F.J.F.Meyen, Observ. Bot. 1: 431 (1843).
Cyperus caespitosus Llanos, Fragm. Pl. Filip.: 14 (1851), nom. illeg.
Cyperus planiculmis Steud., Syn. Pl. Glumac. 2: 22 (1854).
Cyperus adenophorus var. *aphyllus* Boeck., Linnaea 35: 576 (1868).
Cyperus haspan var. *americanus* Boeck., Linnaea 35: 575 (1868). *Cyperus americanus* (Boeck.) Palla, Denkschr. Kaiserl. Akad. Wiss., Wien. Math.-Naturwiss. Kl. 79: 173 (1908), nom. illeg.
Cyperus triqueter Boeck., Flora 62: 548 (1879).
Cyperus microcarpus Boeck., Abh. Naturwiss. Vereine Bremen 7: 37 (1880).
Cyperus cancellatus Ridl., Trans. Linn. Soc. London, Bot. 2: 131 (1884). *Cyperus haspan* var. *cancellatus* (Ridl.) Kük. in H.G.A.Engler (ed.), Pflanzenr., IV, 20(101): 249 (1936).
Cyperus hyemalis Pursh ex C.B.Clarke, J. Linn. Soc., Bot. 21: 120 (1884).
Cyperus efoliatus Boeck., Allg. Bot. Z. Syst. 1: 226 (1895).
Cyperus haspan f. *major* C.B.Clarke, Bull. Herb. Boissier, II, 3: 940 (1903).
Cyperus haspan var. *dietrichiae* Domin, Biblioth. Bot. 89: 426 (1915).
Cyperus haspan var. *bulboides* Kük. in R.E.Fries, Wiss. Erg. Schwed. Rhod.-Kongo Exped. 1: 3 (1921).
Cyperus haspan var. *amplissimus* Kük., Repert. Spec. Nov. Regni Veg. 23: 184 (1926).
Cyperus haspan var. *firmicaulis* Kük., Repert. Spec. Nov. Regni Veg. 23: 184 (1926).
Cyperus haspan var. *flaccidissimus* Kük. in H.G.A. Engler (ed.), Pflanzenr., IV, 20(101): 248 (1936).

Cyperus hayesii (C.B.Clarke) Standl., J. Wash. Acad. Sci. 15: 457 (1925).
C. America. 80 COS GUA HON PAN. Hemicr. or rhizome geophyte.
**Torulinium hayesii* C.B.Clarke, Kew Bull., Addit. Ser. 8: 20 (1908). *Cyperus ferax* var. *hayesii* (C.B.Clarke) Kük. in H.G.A.Engler (ed.), Pflanzenr., IV, 20(101): 619 (1936).

Cyperus helferi Boeck., Linnaea 38: 360 (1874).
Indo-China. 41 AND CBD MYA THA. Hemicr. or rhizome geophyte.

Cyperus hemisphaericus Boeck., Flora 42: 436 (1859). *Mariscus hemisphaericus* (Boeck.) C.B.Clarke in T.A. Durand & H.Schinz, Consp. Fl. Afric. 5: 589 (1894).
S. Somalia to Mozambique, Madagascar. 24 SOM 25 KEN TAN 26 MLW MOZ ZIM 29 MDG. Hemicr. or rhizome geophyte.
Cyperus hildebrandtii Boeck., Flora 63: 436 (1880).
Mariscus gregorii C.B.Clarke, J. Bot. 34: 225 (1896). *Cyperus hemisphaericus* var. *gregorii* (C.B.Clarke) Kük. in H.G.A.Engler (ed.), Pflanzenr., IV, 20(101): 407 (1936).
Cyperus hemisphaericus var. *longibracteus* A.Peter ex Kük. in H.G.A.Engler (ed.), Pflanzenr., IV, 20(101): 407 (1936).

Cyperus hensii C.B.Clarke in T.A.Durand & H.Schinz, Études Fl. Congo: 289 (1896).
WC. Trop. Africa to Angola. 23 CON ZAI 26 ANG. Hemicr. or rhizome geophyte.
Cyperus kuebensis Kük., Repert. Spec. Nov. Regni Veg. 18: 346 (1922).

Cyperus hermaphroditus (Jacq.) Standl., Contr. U. S. Natl. Herb. 18: 88 (1916).
W. Texas to Trop. America. 76 ARI 77 TEX 79 MXC MXE MXG MXN MXS MXT 80 BLZ COS ELS GUA HON NIC PAN 81 CUB DOM JAM LEE WIN 82 FRG? GUY VEN 83 BOL CLM ECU PER 84 BZS 85 AGE AGW. Hemicr. or rhizome geophyte.
**Carex hermaphrodita* Jacq., Collectanea 4: 174 (1791). *Mariscus hermaphroditus* (Jacq.) Urb., Symb. Antill. 2: 165 (1900).
Kyllinga incompleta Jacq., Collectanea 4: 101 (1791). *Cyperus incompletus* (Jacq.) Link, Hort. Berol. 1: 319 (1827).
Kobresia cyperina Willd., Sp. Pl. 4: 206 (1805).
Mariscus jacquinii Kunth in F.W.H.von Humboldt, A.J.A.Bonpland & C.S.Kunth, Nov. Gen. Sp. 1: 216 (1816).
Mariscus tovarii Kunth in F.W.H.von Humboldt, A.J.A.Bonpland & C.S.Kunth, Nov. Gen. Sp. 1: 215 (1816).
Eleocharis cyperina Roem. & Schult., Syst. Veg. 2: 155 (1817).
Mariscus conflexus Link, Enum. Hort. Berol. Alt. 1: 47 (1821).
Mariscus patulus Schrad., Index Seminum (GOET) 1835: 4 (1835).
Cyperus alpinus Liebm., Mexic. Halvgr.: 215 (1850). *Mariscus alpinus* (Liebm.) C.B.Clarke, Kew Bull., Addit. Ser. 8: 15 (1908).
Cyperus sertularinus Liebm., Mexic. Halvgr.: 213 (1850). *Mariscus sertularinus* (Liebm.) T.Koyama, Phytologia 29: 74 (1974).
Mariscus ambiguus Liebm., Mexic. Neldeagt. Pl., V, 2: 231 (1851). *Cyperus subambiguus* Kük. in H.G.A. Engler (ed.), Pflanzenr., IV, 20(101): 527 (1936).
Mariscus longiradiatus Liebm., Mexic. Neldeagt. Pl., V, 2: 232 (1851). *Cyperus hermaphroditus* var. *longiradiatus* (Liebm.) Kük. in H.G.A.Engler (ed.), Pflanzenr., IV, 20(101): 489 (1936).
Cyperus incompletus Boeck., Linnaea 36: 361 (1870), nom. illeg.
Cyperus ruizianus Boeck., Linnaea 36: 360 (1870).
Cyperus tetragonus Hemsl., Biol. Cent.-Amer., Bot. 3: 452 (1885), nom. illeg.
Cyperus longispicatus Boeck., Allg. Bot. Z. Syst. 1: 229 (1895). *Cyperus hermaphroditus* var. *longispicatus* (Boeck.) Kük. in H.G.A.Engler (ed.), Pflanzenr., IV, 20(101): 489 (1936).

Mariscus jacquinii var. *angustior* C.B.Clarke in I.Urban, Symb. Antill. 2: 46 (1900). *Cyperus hermaphroditus* var. *angustior* (C.B.Clarke) Kük. in H.G.A.Engler (ed.), Pflanzenr., IV, 20(101): 489 (1936).
Mariscus jacquinii subsp. *angustior* C.B.Clarke, Contr. U. S. Natl. Herb. 10: 453 (1908).
Cyperus hermaphroditus var. *condensatus* Kük., Repert. Spec. Nov. Regni Veg. 26: 251 (1929).

Cyperus hesperius K.L.Wilson, Telopea 4: 422 (1991).
Western Australia. 50 WAU. Hemicr. or rhizome geophyte.

Cyperus heterocladus Baker, J. Linn. Soc., Bot. 20: 292 (1883).
C. Madagascar. 29 MDG. Hemicr. or rhizome geophyte.

Cyperus hieronymi Boeck., Beitr. Cyper. 1: 7 (1888).
Paraguay to N. Argentina. 85 AGE AGW PAR. Hemicr. or rhizome geophyte.
Cyperus fiebrigii Kük., Repert. Spec. Nov. Regni Veg. 18: 346 (1922). *Cyperus hieronymi* var. *fiebrigii* (Kük.) Kük. in H.G.A.Engler (ed.), Pflanzenr., IV, 20(101): 184 (1936).

Cyperus hilairenus Steud., Syn. Pl. Glumac. 2: 29 (1854).
SE. Brazil. 84 BZL. Hemicr. or rhizome geophyte.
**Cyperus caespitosus* A.St.-Hil., Voy. Distr. Diam. 2: 390 (1833), nom. illeg.

Cyperus hillebrandii Boeck., Flora 63: 436 (1880). *Mariscus hillebrandii* (Boeck.) T.Koyama, Bishop Mus. Occas. Pap. 29: 122 (1989).
Hawaiian Is. 63 HAW. Hemicr. or rhizome geophyte.

var. ***decipiens*** (Hillebr.) Kük. in H.G.A.Engler (ed.), Pflanzenr., IV, 20(101): 484 (1936).
Hawaiian Is. 63 HAW. Hemicr. or rhizome geophyte.
**Cyperus decipiens* Hillebr., Fl. Hawaiian Isl.: 467 (1888). *Mariscus decipiens* (Hillebr.) C.B. Clarke, Kew Bull., Addit. Ser. 8: 16 (1908). *Mariscus hillebrandii* subsp. *decipiens* (Hillebr.) T.Koyama, Bishop Mus. Occas. Pap. 29: 122 (1989).

var. ***hillebrandii***
Hawaiian Is. 63 HAW. Hemicr. or rhizome geophyte.
Cyperus kunthianus H.Mann, Proc. Amer. Acad. Arts 7: 208 (1867), nom. illeg.
Cyperus hawaicensis Hillebr., Fl. Hawaiian Isl.: 470 (1888).
Cyperus mauiensis Hillebr., Fl. Hawaiian Isl.: 469 (1888). *Cyperus hillebrandii* var. *mauiensis* (Hillebr.) Kük. in H.G.A.Engler (ed.), Pflanzenr., IV, 20(101): 484 (1936).
Cyperus hillebrandii var. *helleri* Kük. in H.G.A. Engler (ed.), Pflanzenr., IV, 20(101): 484 (1936).

Cyperus hirtellus (Chiov.) Kük. in H.G.A.Engler (ed.), Pflanzenr., IV, 20(101): 7 (1935).
Uganda to S. Zaïre. 23 ZAI 25 TAN UGA 26 ZIM. Hemicr. or rhizome geophyte.
**Mariscus hirtellus* Chiov., Nuovo Giorn. Bot. Ital., n.s., 26: 71 (1919).
Mariscus rhodesicus Podlech, Mitt. Bot. Staatssamml. München 4: 115 (1961).

Cyperus holoschoenus R.Br., Prodr.: 215 (1810). *Mariscus holoschoenus* (R.Br.) C.B.Clarke, Kew Bull., Addit. Ser. 8: 19 (1908).
New Guinea to N. Australia. 43 NWG 50 NTA QLD WAU. Hemicr. or rhizome geophyte.
Cyperus holoschoenus var. *viscidus* W.Fitzg., J. Proc. Roy. Soc. Western Australia 3: 117 (1918).
Cyperus holoschoenus var. *fuscisquamatus* Kük., Bull. Jard. Bot. Buitenzorg, III, 16: 301 (1940).

Cyperus holostigma C.B.Clarke ex Schweinf., Bull. Herb. Boissier 2(App. 2): 103 (1894).
NE. Trop. Africa. 24 ERI ETH. Hemicr. or rhizome geophyte.

Cyperus holstii Kük., Repert. Spec. Nov. Regni Veg. 21: 328 (1925).
S. Zaïre, Tanzania. 23 ZAI 25 KEN TAN. Hemicr. or rhizome geophyte.

Cyperus hoppiifolius Uittien, Recueil Trav. Bot. Néerl. 30: 186 (1933).
S. Trop. America. 82 FRG SUR VEN 83 PER. Hemicr. or rhizome geophyte.

Cyperus houghtonii Torr., Ann. Lyceum Nat. Hist. New York 3: 277 (1836). *Mariscus houghtonii* (Torr.) T.Koyama, Phytologia 29: 73 (1974).
C. & E. Canada to NC. & NE. U.S.A. 71 MAN 72 ONT QUE 73 COL IDA ORE WAS 74 ILL MIN WIS 75 INI MAI MAS MIC NWH NWJ NWY OHI PEN VER WVA 78 MRY VRG. Hemicr. or rhizome geophyte.
Cyperus houghtonii var. *uberior* Kük. in H.G.A.Engler (ed.), Pflanzenr., IV, 20(101): 469 (1936).

Cyperus humilis Kunth, Enum. Pl. 2: 23 (1837).
Mexico to Trop. America. 79 MXE MXG MXS MXT 80 BLZ COS ELS GUA HON NIC 81 CUB DOM HAI JAM WIN 83 ECU PER. Ther.
Mariscus cephalotes Poepp. ex Kunth, Enum. Pl. 2: 23 (1837).
Cyperus humilis var. *elatior* Britton, Bull. Torrey Bot. Club 15: 99 (1888).

Cyperus hylophilus Cherm., Bull. Soc. Bot. France 72: 616 (1925).
E. Madagascar. 29 MDG. Hemicr. or rhizome geophyte.

Cyperus hypochlorus Hillebr., Fl. Hawaiian Isl.: 468 (1888). *Mariscus hypochlorus* (Hillebr.) C.B.Clarke, Kew Bull., Addit. Ser. 8: 19 (1908).
Hawaiian Is. 63 HAW. Hemicr. or rhizome geophyte.

var. ***brevior*** Kük. in H.G.A.Engler (ed.), Pflanzenr., IV, 20(101): 409 (1936). *Mariscus hypochlorus* subsp. *brevior* (Kük.) T.Koyama, Bishop Mus. Occas. Pap. 29: 122 (1989).
Hawaiian Is. 63 HAW. Hemicr. or rhizome geophyte.

var. ***hypochlorus***
Hawaiian Is. 63 HAW. Hemicr. or rhizome geophyte.
Cyperus hypochlorus var. *kauaiensis* Kük. in H.G.A.Engler (ed.), Pflanzenr., IV, 20(101): 409 (1936).

Cyperus hypopitys G.C.Tucker, Syst. Bot. Monogr. 43: 129 (1994).
Arizona to N. Mexico. 76 ARI 79 MXE MXN. Hemicr. or rhizome geophyte.

Cyperus hystricinus Fernald, Rhodora 8: 127 (1906). *Cyperus retrofractus* var. *hystricinus* (Fernald) Kük. in H.G.A.Engler (ed.), Pflanzenr., IV, 20(101): 509 (1936).

New Jersey to Texas. 74 MSO 75 NWJ 77 TEX 78 ALA ARK DEL GEO LOU MRY MSI NCA VRG WDC. Hemicr. or rhizome geophyte.

Cyperus imbecillis R.Br., Prodr.: 213 (1810).
NE. New South Wales. 50 NSW. Hemicr. or rhizome geophyte.

Cyperus imbricatus Retz., Observ. Bot. 5: 12 (1788).
Trop. & Subtrop. 20 EGY 22 BEN BKN GHA GUI IVO LBR MLI NGA NGR SEN TOG 23 CMN CON GAB ZAI 24 CHA ETH SUD 25 TAN UGA 27 BOT NAM NAT SWZ TVL 29 MDG SEY 34 AFG IRN 36 CHS 38 TAI 40 ASS BAN IND NEP 41 MYA THA VIE 42 BOR JAW LSI MLY MOL PHI SUL SUM 43 NWG (50) 79 MXG MXS MXT 80 BLZ COS ELS GUA HON NIC PAN 81 CUB DOM HAI JAM PUE TRT 82 GUY VEN 83 BOL CLM ECU PER 84 BZE BZN. Hemicr. or rhizome geophyte.
Cyperus biceps Vahl, Enum. Pl. 2: 368 (1805).
Cyperus radiatus Vahl, Enum. Pl. 2: 369 (1805), nom. illeg. *Dichostylis radiata* (Vahl) Palla, Allg. Bot. Z. Syst. 17(Beil.): 8 (1911).
Cyperus involucratus Poir. in J.B.A.M.de Lamarck, Encycl. 7: 253 (1806), nom. illeg.
Cyperus densiflorus Link, Jahrb. Gewächsk. 1(3): 88 (1818).
Cyperus verticillatus Roxb., Fl. Ind. 1: 209 (1820).
Cyperus confertiflorus Schult. in J.J.Roemer & J.A.Schultes, Mant. 2: 131 (1824).
Cyperus spicatus J.Presl & C.Presl, Reliq. Haenk. 1: 173 (1828).
Cyperus digitatus Nees in R.Wight, Contr. Bot. India: 85 (1834), nom. illeg.
Cyperus roylei Arn. in R.Wight, Contr. Bot. India: 85 (1834).
Cyperus campestris Schrad. ex Nees in C.F.P.von Martius & auct. suc. (eds.), Fl. Bras. 2(1): 35 (1842).
Cyperus anabaptistus Steud., Syn. Pl. Glumac. 2: 37 (1854).
Cyperus macrosciadion Steud., Syn. Pl. Glumac. 2: 37 (1854).
Cyperus obscurus Nees, Hooker's J. Bot. Kew Gard. Misc. 6: 27 (1854).
Cyperus semidives Steud., Syn. Pl. Glumac. 2: 36 (1854).
Mariscus leptochloides Steud., Syn. Pl. Glumac. 2: 317 (1855).
Cyperus radiatus var. *capitatus* Boeck., Linnaea 36: 319 (1870). *Cyperus imbricatus* var. *capitatus* (Boeck.) Kük. in H.G.A.Engler (ed.), Pflanzenr., IV, 20(101): 70 (1935).
Cyperus radiatus var. *elongatus* Boeck., Linnaea 36: 319 (1870). *Cyperus imbricatus* var. *elongatus* (Boeck.) Kük. in H.G.A.Engler (ed.), Pflanzenr., IV, 20(101): 71 (1935). *Cyperus imbricatus* subsp. *elongatus* (Boeck.) T.Koyama, in Fl. Taiwan 5: 260 (1978).
Cyperus radiatus var. *minor* Boeck., Linnaea 36: 319 (1870). *Cyperus imbricatus* var. *minor* (Boeck.) Kük. in H.G.A.Engler (ed.), Pflanzenr., IV, 20(101): 70 (1935).
Cyperus flexifolius Boeck., Flora 62: 549 (1879).
Cyperus viridispicatus Boeck., Beitr. Cyper. 2: 8 (1890). *Cyperus imbricatus* var. *viridispicatus* (Boeck.) Kük. in H.G.A.Engler (ed.), Pflanzenr., IV, 20(101): 71 (1935).
Cyperus densespicatus Hayata, Icon. Pl. Formosan. 6: 105 (1916).
Cyperus mediorubescens Hayata, Icon. Pl. Formosan. 6: 106 (1916).
Cyperus densispicatus Cherm., Bull. Mus. Natl. Hist. Nat. 27: 317 (1921).
Cyperus imbricatus var. *multiflorus* Kük. in H.G.A. Engler (ed.), Pflanzenr., IV, 20(101): 71 (1935).

Cyperus impolitus Kunth, Enum. Pl. 2: 78 (1837).
Comostemum impolitum (Kunth) Nees, J. Bot. (Hooker) 2: 397 (1840). *Cyperus laetus* var. *impolitus* (Kunth) Kük., Repert. Spec. Nov. Regni Veg. 26: 253 (1929). *Mariscus impolitus* (Kunth) J.Raynal, Adansonia, n.s., 17: 274 (1978).
Brazil to Chile. 84 BZL BZS 85 AGE CLC PAR URU. Hemicr. or rhizome geophyte.
Cyperus laetus var. *capitatus* Boeck., Linnaea 36: 296 (1870).
Cyperus balansae Maury, Mém. Soc. Phys. Genève 31: 130 (1890). *Mariscus balansae* (Maury) C.B.Clarke, Bull. Herb. Boissier, II, 3: 1011 (1903). *Chlorocyperus balansae* (Maury) Palla, Denkschr. Kaiserl. Akad. Wiss., Wien. Math.-Naturwiss. Kl. 79: 180 (1908). *Cyperus laetus* var. *balansae* (Maury) Kük., Repert. Spec. Nov. Regni Veg. 26: 253 (1929).
Mariscus hassleri C.B.Clarke, Bull. Herb. Boissier, II, 3: 1011 (1903). *Cyperus laetus* var. *hassleri* (C.B. Clarke) Kük., Repert. Spec. Nov. Regni Veg. 26: 253 (1929).
Cyperus wacketii Kük. in H.G.A.Engler (ed.), Pflanzenr., IV, 20(101): 443 (1936). *Mariscus wacketii* (Kük.) T.Koyama, Phytologia 29: 74 (1974).

Cyperus impubes Steud., Syn. Pl. Glumac. 2: 45 (1854).
Mariscus impubes (Steud.) Napper, J. E. Africa Nat. Hist. Soc. Natl. Mus. 28: 12 (1971).
NE. & E. Trop. Africa, W. Indian Ocean. 24 ERI ETH SOC SOM 25 KEN TAN UGA 29 MAU MDG. Hemicr. or rhizome geophyte.

var. ***fallax*** (Cherm.) Kük. in H.G.A.Engler (ed.), Pflanzenr., IV, 20(101): 492 (1936).
Madagascar. 29 MDG. Hemicr. or rhizome geophyte.
**Mariscus fallax* Cherm., Bull. Mus. Natl. Hist. Nat. 25: 405 (1919).
Mariscus fallax var. *major* Cherm., Bull. Soc. Bot. France 72: 19 (1925). *Cyperus impubes* f. *major* (Cherm.) Kük. in H.G.A.Engler (ed.), Pflanzenr., IV, 20(101): 493 (1936).

var. ***impubes***
NE. & E. Trop. Africa. 24 ERI ETH SOC SOM 25 KEN TAN UGA. Hemicr. or rhizome geophyte.
Mariscus procerus A.Rich., Tent. Fl. Abyss. 2: 489 (1850).
Mariscus richardii Steud., Syn. Pl. Glumac. 2: 66 (1854).
Cyperus quadriflorus Boeck., Linnaea 36: 367 (1870).
Mariscus cupreus Hochst. ex Boeck., Linnaea 36: 367 (1870).
Cyperus impubes var. *brevispiculosus* Kük., Notizbl. Bot. Gart. Berlin-Dahlem 9: 306 (1925).
Mariscus moniliferus Chiov., Miss. Biol. Borana, Racc. Bot., Angiosp.-Gymnosp.: 296 (1939).

var. ***viguieri*** (Cherm.) Kük. in H.G.A.Engler (ed.), Pflanzenr., IV, 20(101): 493 (1936).
Mascarenes, Madagascar. 29 MAU MDG. Hemicr. or rhizome geophyte.
**Mariscus viguieri* Cherm., Bull. Mus. Natl. Hist. Nat. 25: 303 (1919).

Mariscus viguieri var. *contractus* Cherm., Bull. Soc. Bot. France 69: 719 (1922).

Cyperus inaequalis Willemet, Ann. Bot. (Usteri) 18: 5 (1796).
Mauritius. 29 MAU. Hemicr. or rhizome geophyte. Provisionally accepted.

Cyperus incompressus C.B.Clarke in D.Oliver & auct. suc. (eds.), Fl. Trop. Afr. 8: 318 (1901).
W. & WC. Trop. Africa, N. Madagascar. 22 GUI IVO SEN SIE TOG 23 CON 29 MDG. Ther.

Cyperus incomtus Kunth, Enum. Pl. 2: 39 (1837).
Peru to N. Argentina. 83 BOL PER 84 BZS 85 AGE AGW PAR URU. Hemicr. or rhizome geophyte.
Cyperus curvifolius Boeck., Flora 61: 138 (1878).
Cyperus cinereus Maury ex Micheli, Mém. Soc. Phys. Genève 31(1): 127 (1890).
Cyperus incomtus var. *dissolutior* Kük. in H.G.A. Engler (ed.), Pflanzenr., IV, 20(101): 184 (1936).
Cyperus incomtus var. *miguelii* Kük. in H.G.A.Engler (ed.), Pflanzenr., IV, 20(101): 183 (1936).
Cyperus venturii Kük. in H.G.A.Engler (ed.), Pflanzenr., IV, 20(101): 182 (1936). *Cyperus incomtus* var. *venturii* (Kük.) Barros in H.R.Descole (ed.), Gen. Sp. Pl. Argent. 4(1): 56 (1947).

Cyperus indecorus Kunth, Enum. Pl. 2: 85 (1837). *Mariscus indecorus* (Kunth) Podlech, Mitt. Bot. Staatssamml. München 3: 524 (1960).
S. Trop. & S. Africa. 26 MLW MOZ 27 CPP NAT TVL. Hemicr. or rhizome geophyte.
Cyperus luzuliformis Boeck., Linnaea 36: 356 (1870). *Mariscus luzuliformis* (Boeck.) C.B.Clarke in T.A. Durand & H.Schinz, Consp. Fl. Afric. 5: 589 (1894).

Cyperus infucatus Kunth, Enum. Pl. 2: 86 (1837). *Mariscus infucatus* (Kunth) T.Koyama, Phytologia 29: 73 (1974).
E. Brazil to N. Argentina. 84 AGW BZE BZL PAR. Hemicr. or rhizome geophyte.
Mariscus longus Schrad. ex Nees in C.F.P.von Martius & auct. suc. (eds.), Fl. Bras. 2(1): 44 (1842).
Cyperus brevispicatus Boeck., Flora 42: 442 (1859).

Cyperus inops C.B.Clarke, Kew Bull., Addit. Ser. 8: 5 (1908).
S. Brazil. 84 BZS. Hemicr. or rhizome geophyte.

Cyperus × insidiosus Cherm., Bull. Mus. Natl. Hist. Nat. 27: 318 (1921). *C. grandis* × *C. mangorensis*.
Madagascar. 29 MDG. Hemicr. or rhizome geophyte.

Cyperus insularis Heenan & de Lange, New Zealand J. Bot. 43: 353 (2005).
New Zealand North I. 51 NZN.

Cyperus intricatus Schrad. ex Schult. in J.J.Roemer & J.A.Schultes, Mant. 2: 98 (1824). *Cyperus reflexus* var. *intricatus* (Schrad. ex Schult.) Kük. in H.G.A.Engler (ed.), Pflanzenr., IV, 20(101): 174 (1936). *Cyperus schraderi* Herter, Revista Sudamer. Bot. 9: 143 (1953), nom. illeg.
Costa Rica to N. Argentina. 80 COS 84 BZL 85 AGE AGW PAR URU. Hemicr. or rhizome geophyte.
Cyperus consanguineus Kunth, Enum. Pl. 2: 42 (1837).
Cyperus chamissoi Schrad. ex Nees in C.F.P.von Martius & auct. suc. (eds.), Fl. Bras. 2(1): 33 (1842). *Cyperus consanguineus* var. *chamissoi* (Schrad. ex Nees) Kük. in H.G.A.Engler (ed.), Pflanzenr., IV, 20(101): 171 (1936).
Cyperus widgrenii Boeck., Bot. Jahrb. Syst. 5: 499 (1884).
Cyperus ciliolatus Boeck., Beitr. Cyper. 2: 5 (1890).
Cyperus schenckianus Boeck., Beitr. Cyper. 2: 4 (1890).
Cyperus virens Boeck., Beitr. Cyper. 2: 6 (1890), nom. illeg.
Cyperus longicaulis Boeck., Allg. Bot. Z. Syst. 1: 202 (1895).
Cyperus varius Boeck., Allg. Bot. Z. Syst. 1: 202 (1895). *Cyperus consanguineus* var. *varius* (Boeck.) Kük. in H.G.A.Engler (ed.), Pflanzenr., IV, 20(101): 171 (1936).
Cyperus usterii Palla, Oesterr. Bot. Z. 57: 257 (1907). *Cyperus consanguineus* var. *usterii* (Palla) Kük. in H.G.A.Engler (ed.), Pflanzenr., IV, 20(101): 172 (1936).
Cyperus pauloensis Palla, Denkschr. Kaiserl. Akad. Wiss., Wien. Math.-Naturwiss. Kl. 79: 175 (1908).

Cyperus iria L., Sp. Pl.: 45 (1753). *Chlorocyperus iria* (L.) Rikli, Jahrb. Wiss. Bot. 27: 564 (1895). *Cyperus iria* var. *typicus* Domin, Biblioth. Bot. 89: 428 (1915), nom. inval.
Trop. & Subtrop. Old World to C. Asia. 22 BEN BKN GAM GHA IVO MLI NGA NGR SEN 24 CHA ERI ETH SOM SUD 25 KEN TAN UGA 27 BOT NAM SWZ TVL 29 MAU MDG 32 TKM UZB 34 AFG IRN IRQ 36 CHC CHM CHS 38 JAP KOR KZN NNS OGA TAI 40 ASS BAN EHM IND NEP PAK SRL WHM 41 MYA THA VIE 42 BOR JAW LSI MLY MOL PHI SUL SUM 43 NWG 50 NSW NTA QLD SOA WAU (60) fij (61) sci 62 CRL MRN (73) ore (74) ill mso okl (75) cnt nwy wva (76) cal (77) tex (78) ala ark del fla geo kty lou mry msi nca sca ten vrg wdc (79) mxg mxn mxs mxt (80) blz cos pan (81) cub pue trt win (82) ven (83) bol ecu per (84) bzc bzn (85) age agw. Ther.
Cyperus santonici Rottb., Descr. Icon. Rar. Pl.: 41 (1773). *Cyperus iria* var. *santonici* (Rottb.) Fernald & Griscom, Rhodora 37: 147 (1935).
Cyperus panicoides Lam., Tabl. Encycl. 1: 145 (1791).
Cyperus chrysomelinus Link, Hort. Berol. 2: 305 (1833). *Cyperus iria* f. *chrysomelinus* (Link) Kük. in H.G.A.Engler (ed.), Pflanzenr., IV, 20(101): 151 (1935).
Cyperus diaphaniria Steud., Syn. Pl. Glumac. 2: 23 (1854). *Cyperus iria* f. *diaphaniria* (Steud.) Miq., Fl. Ned. Ind. 3: 270 (1856).
Cyperus resinosus Hochst. ex Steud., Syn. Pl. Glumac. 2: 23 (1854).
Cyperus iria var. *flavescens* Benth., Fl. Austral. 7: 276 (1878). *Cyperus iria* f. *flavescens* (Benth.) Domin, Biblioth. Bot. 85: 429 (1915).
Cyperus paniciformis Franch. & Sav., Enum. Pl. Jap. 2: 537 (1878).
Cyperus iria var. *paniciformis* C.B.Clarke in J.D.Hooker, Fl. Brit. India 6: 607 (1893).
Cyperus iria f. *multiflora* Domin, Biblioth. Bot. 89: 428 (1915).
Cyperus nangtciangensis Pamp., Nuovo Giorn. Bot. Ital., n.s., 22: 260 (1915).

Cyperus isabellinus K.L.Wilson, Telopea 4: 475 (1991).
Queensland. 50 QLD. Hemicr. or rhizome geophyte.

Cyperus ischnos Schltdl., Bot. Zeitung (Berlin) 7: 99 (1849). *Mariscus ischnos* (Schltdl.) C.B.Clarke in J.D.Hooker, Fl. Brit. India 6: 623 (1893).
Mexico to Bolivia. 79 MXC MXE MXG MXN MXS MXT 80 BLZ COS ELS GUA HON NIC PAN 82 VEN 83 BOL CLM. Hemicr. or rhizome geophyte.
Cyperus botteri Boeck., Allg. Bot. Z. Syst. 1: 227 (1895).

Chlorocyperus arsenii Palla, Oesterr. Bot. Z. 63: 403 (1913). *Cyperus ischnos* var. *arsenii* (Palla) Kük. in H.G.A.Engler (ed.), Pflanzenr., IV, 20(101): 462 (1936).

Cyperus ivohibensis (Cherm.) Kük. in H.G.A.Engler (ed.), Pflanzenr., IV, 20(101): 438 (1936).
C. & E. Madagascar. 29 MDG. Hemicr. or rhizome geophyte.
**Mariscus ivohibensis* Cherm., Bull. Soc. Bot. France 74: 603 (1927 publ. 1928).

Cyperus ixiocarpus F.Muell., Trans. Roy. Soc. Victoria 23: 55 (1887).
N. Australia. 50 NTA QLD WAU. Hemicr. or rhizome geophyte.

Cyperus javanicus Houtt., Nat. Hist.: t. 88, f. 1 (1782). *Mariscus javanicus* (Houtt.) Merr. & F.P.Metcalf, Lingnan Sci. J. 21: 4 (1945), nom. illeg.
W. Indian Ocean, Trop. & Subtrop. Asia to Pacific. 29 MDG SEY 36 CHS 38 KZN NNS OGA TAI 40 IND SRL 41 THA VIE 42 BOR CKI JAW LSI MLY MOL PHI SUL SUM XMS 43 BIS NWG 50 NTA QLD WAU 60 FIJ GIL nue TON 61 COO PIT SCI TUA 62 CRL MRN MRS 63 HAW. Hemicr. or rhizome geophyte.
Cyperus stuppeus G.Forst., Fl. Ins. Austr.: 89 (1786). *Mariscus stuppeus* (G.Forst.) Merr., Philipp. J. Sci., C 3: 398 (1909).
Cyperus pennatus Lam., Tabl. Encycl. 1: 144 (1791). *Duval-jouvea pennata* (Lam.) Palla, Denkschr. Kaiserl. Akad. Wiss., Wien. Math.-Naturwiss. Kl. 84: 453 (1909). *Mariscus pennatus* (Lam.) Domin, Biblioth. Bot. 85: 440 (1915).
Cyperus canescens Vahl, Enum. Pl. 2: 355 (1805).
Cyperus parviflorus Vahl, Enum. Pl. 2: 352 (1805).
Cyperus firmus J.Presl & C.Presl, Reliq. Haenk. 1: 171 (1828).
Cyperus holciflorus J.Presl & C.Presl, Reliq. Haenk. 1: 173 (1828).
Mariscus albescens Gaudich., Voy. Uranie: 415 (1829).
Cyperus caricifolius Hook. & Arn., Bot. Beechey Voy.: 99 (1832).
Cyperus pinnatus A.Rich. in J.S.C.Dumont d'Urville, Voy. Astrolabe 2: vii (1834).
Cyperus owahuensis Nees in F.J.F.Meyen, Observ. Bot. 2: 119 (1843).
Cyperus ovatus Llanos, Fragm. Pl. Filip.: 15 (1851), nom. illeg.
Cyperus anomalus Steud., Syn. Pl. Glumac. 2: 37 (1854).
Cyperus stigmatosus Steud., Syn. Pl. Glumac. 2: 38 (1854).
Cyperus taitensis Steud., Syn. Pl. Glumac. 2: 37 (1854).
Cyperus valdiviae Steud., Syn. Pl. Glumac. 2: 45 (1854).
Cyperus macreilema Steud. ex Jard., Mém. Soc. Sci. Nat. Cherbourg 4: 10 (1856).
Cyperus spongiosocoriaceus Boeck., Flora 42: 441 (1859).
Cyperus tongatabuensis Boeck., Flora 58: 85 (1875).
Cyperus fucatus Boeck., Flora 65: 13 (1882).

Cyperus jeminicus Rottb., Descr. Icon. Rar. Pl.: 25 (1773). *Cyperus conglomeratus* subsp. *jeminicus* (Rottb.) Lye, Lidia 3: 131 (1994).
W. Trop. Africa to Arabian Pen., Madagascar, India. 20 EGY 22 BKN MLI MTN NGR SEN 24 CHA ERI ETH SOM 29 MDG 34 PAL 35 SAU YEM 40 IND. Hemicr. or rhizome geophyte.
Cyperus excisus Boeck., Linnaea 35: 523 (1868). *Cyperus conglomeratus* f. *excisus* (Boeck.) Kük. in H.G.A.Engler (ed.), Pflanzenr., IV, 20(101): 275 (1936).
Cyperus pungens var. *multiculmis* Boeck., Linnaea 35: 538 (1868). *Cyperus conglomeratus* var. *multiculmis* (Boeck.) Kük. in H.G.A.Engler (ed.), Pflanzenr., IV, 20(101): 274 (1936).

Cyperus juncelliformis Peter & Kük. in H.G.A.Engler (ed.), Pflanzenr., IV, 20(101): 352 (1936).
Tanzania. 25 TAN. Hemicr. or rhizome geophyte.

Cyperus kabarensis Cherm., Bull. Jard. Bot. État 14: 310 (1937).
Zaïre. 23 ZAI. Hemicr. or rhizome geophyte.

Cyperus kaessneri C.B.Clarke, Bot. Jahrb. Syst. 38: 133 (1906).
Eritrea to Tanzania. 24 ERI ETH 25 KEN TAN. Ther.

Cyperus kappleri Hochst. ex Steud., Syn. Pl. Glumac. 2: 38 (1854).
N. South America. 82 FRG GUY SUR VEN. Hemicr. or rhizome geophyte.

Cyperus karisimbiensis (Cherm.) Kük. in H.G.A.Engler (ed.), Pflanzenr., IV, 20(101): 559 (1936).
EC. & E. Trop. African Mts. 23 RWA ZAI 25 KEN TAN UGA. Hemicr. or tuber geophyte.
**Mariscus karisimbiensis* Cherm., Bull. Soc. Bot. France 82: 335 (1935).

var. ***karisimbiensis***
EC. Trop. Africa. 23 RWA ZAI 25 KEN UGA. Hemicr. or tuber geophyte.
Mariscus maritimus C.B.Clarke, J. Bot. 34: 226 (1896).

var. ***longinux*** (Kük.) Kük. in H.G.A.Engler (ed.), Pflanzenr., IV, 20(101): 560 (1936).
E. Trop. Africa. 25 KEN TAN. Hemicr. or rhizome geophyte.
**Cyperus coloratus* var. *longinux* Kük., Notizbl. Bot. Gart. Berlin-Dahlem 9: 305 (1925).

Cyperus karlschumannii C.B.Clarke, Bot. Jahrb. Syst. 38: 134 (1906). *Cyperus margaritaceus* var. *karlschumannii* (C.B.Clarke) Kük. in H.G.A.Engler (ed.), Pflanzenr., IV, 20(101): 285 (1936).
W. Trop. Africa. 22 BKN GHA IVO MLI NGA TOG. Hemicr. or rhizome geophyte.

Cyperus kasamensis Podlech, Mitt. Bot. Staatssamml. München 4: 108 (1961).
Zambia. 26 ZAM. Hemicr. or rhizome geophyte.

Cyperus keniensis Kük., Notizbl. Bot. Gart. Berlin-Dahlem 9: 306 (1925). *Mariscus keniensis* (Kük.) S.S.Hooper, Kew Bull. 26: 579 (1972). *Mariscus longibracteatus* var. *keniensis* (Kük.) Maquet, Bull. Jard. Bot. Natl. Belg. 58: 265 (1988).
Trop. & S. Africa, Madagascar. 22 NGA 23 CMN 25 KEN 27 NAT SWZ TVL 29 MDG. Hemicr. or rhizome geophyte.

Cyperus kerstenii Boeck., Linnaea 36: 373 (1870). *Mariscus kerstenii* (Boeck.) C.B.Clarke in D.Oliver & auct. suc. (eds.), Fl. Trop. Afr. 8: 392 (1902).
E. Trop. Africa. 25 KEN TAN UGA. Hemicr. or rhizome geophyte.
Cyperus vaginatissimus K.Schum. in H.G.A.Engler, Pflanzenw. Ost-Afrikas, C: 121 (1895).
Cyperus kerstenii var. *irregularis* Kük. in H.G.A. Engler (ed.), Pflanzenr., IV, 20(101): 555 (1936).

Cyperus kibweanus J.Duvign., Bull. Soc. Roy. Bot. Belgique 96: 157 (1963).
Zaïre. 23 ZAI. Hemicr. or rhizome geophyte.

Cyperus kilimandscharicus Kük., Repert. Spec. Nov. Regni Veg. 21: 326 (1925).
Ethiopia to Tanzania. 24 ETH 25 TAN. Hemicr. or rhizome geophyte.
Cyperus kilimandscharicus var. *chlorilepis* Peter & Kük. in H.G.A.Engler (ed.), Pflanzenr., IV, 20(101): 106 (1935).

Cyperus kipasensis Cherm., Rev. Zool. Bot. Africaines 24: 297 (1934). *Cyperus platycaulis* var. *kipasensis* (Cherm.) Peter & Kük. in H.G.A.Engler (ed.), Pflanzenr., IV, 20(101): 636 (1936).
Zaïre to W. Tanzania. 23 ZAI 25 TAN. Hemicr. or rhizome geophyte.

Cyperus kirkii C.B.Clarke in D.Oliver & auct. suc. (eds.), Fl. Trop. Afr. 8: 318 (1901).
Tanzania to S. Trop. Africa. 25 TAN 26 MLW MOZ ZIM. Hemicr.
Cyperus ducis Buscal. & Muschl., Bot. Jahrb. Syst. 49: 460 (1913).

Cyperus kituiensis Muasya, Kew Bull. 59: 247 (2004).
Kenya. 25 KEN. Rhizome geophyte.

Cyperus koyaliensis Cherm., Arch. Bot. Mém. 7(3): 7 (1936).
W. & WC. Trop. Africa. 22 GUI IVO LBR NGA SIE 23 CAF CMN GGI. Hemicr.
Cyperus immanis Nelmes, Kew Bull. 6: 164 (1951).

Cyperus kurzii C.B.Clarke, J. Linn. Soc., Bot. 21: 129 (1884).
Andaman Is. 41 AND. Hemicr. or rhizome geophyte.

Cyperus kwaleensis Lye, Nordic J. Bot. 3: 221 (1983).
Kenya (Kwale Distr.). 25 KEN. Hemicr. or rhizome geophyte.

Cyperus kyllingiformis Lye, Nordic J. Bot. 3: 218 (1983).
Kenya (Kitale reg.). 25 KEN. Hemicr. or tuber geophyte.

Cyperus lacunosus Griseb., Cat. Pl. Cub.: 237 (1866).
W. Cuba. 81 CUB. Hemicr. or rhizome geophyte.

Cyperus laeteflorens (C.B.Clarke) Kük., Bernice P. Bishop Mus. Bull. 81: 23 (1931).
New Caledonia. 60 NWC. Hemicr. or rhizome geophyte.
**Mariscus laeteflorens* C.B.Clarke, Kew Bull., Addit. Ser. 8: 17 (1908).

Cyperus laetus J.Presl & C.Presl, Reliq. Haenk. 1: 172 (1828). *Comostemum laetum* (J.Presl & C.Presl) Nees, J. Bot. (Hooker) 2: 397 (1840). *Mariscus laetus* (J.Presl & C.Presl) Lindm., Bih. Kongl. Svenska Vetensk.-Akad. Handl. 26(9): 11 (1900). *Cyperus laetus* subsp. *eu-laetus* Kük., Repert. Spec. Nov. Regni Veg. 26: 252 (1929), nom. inval.
S. Trop. America. 83 PER 84 BZC BZL BZN BZS. Hemicr. or rhizome geophyte.
Cyperus laetus var. *elatior* Boeck., Linnaea 36: 296 (1870).
Cyperus arechavaletae Boeck., Beitr. Cyper. 1: 9 (1888). *Cyperus laetus* var. *arechavaletae* (Boeck.) Kük. in H.G.A.Engler (ed.), Pflanzenr., IV, 20(101): 441 (1936). *Cyperus rigens* subsp. *arechavaletae* (Boeck.) Pedersen, Darwiniana 17: 539 (1972).
Cyperus laetus var. *obtusiflora* Boeck., Allg. Bot. Z. Syst. 2: 2 (1896).
Cyperus nocha Phil., Anales Univ. Chile 93: 346 (1896).
Cyperus paposanus Phil., Anales Univ. Chile 93: 343 (1896).
Cyperus rancoanus Phil., Anales Univ. Chile 93: 345 (1896).
Cyperus laetus var. *laeviculmis* Kük., Repert. Spec. Nov. Regni Veg. 26: 253 (1929).
Cyperus laetus var. *maximus* Kük., Repert. Spec. Nov. Regni Veg. 26: 252 (1929). *Cyperus rigens* var. *maximus* (Kük.) Pedersen, Darwiniana 17: 538 (1972).
Cyperus laetus f. *contracta* Osten, Anales Mus. Hist. Nat. Montevideo, II, 3: 157 (1931).

Cyperus laevigatus L., Mant. Pl. 2: 179 (1771). *Pycreus laevigatus* (L.) Nees, Linnaea 10: 130 (1836). *Juncellus laevigatus* (L.) C.B.Clarke in J.D.Hooker, Fl. Brit. India 6: 596 (1893). *Chlorocyperus laevigatus* (L.) Palla, Allg. Bot. Z. Syst. 6: 221 (1900). *Acorellus laevigatus* (L.) Palla in W.D.J.Koch, Syn. Deut. Schweiz. Fl., ed. 3, 2: 2558 (1905).
Trop. & Subtrop. 12 BAL FRA POR SAR SPA 13 GRC ITA KRI SIC 20 ALG EGY LBY MOR TUN 21 CNY CVI MDR 22 MLI MTN NGA NGR SEN 23 BUR RWA ZAI 24 CHA ERI ETH SOC SOM SUD 25 KEN TAN UGA 26 ANG MLW 27 BOT CPP CPV LES NAM NAT OFS SWZ TVL 29 MAU MDG 32 TKM TZK UZB 34 AFG CYP EAI IRN IRQ LBS PAL SIN TUR 35 GST OMA SAU YEM 40 IND PAK 41 VIE 50 NSW QLD SOA WAU 60 GIL 63 HAW 76 ARI CAL 77 TEX 78 NCA 79 MXC MXE MXG MXN MXS MXT 80 COS GUA NIC 81 CUB DOM JAM LEE PUE 83 BOL ECU GAL PER 85 AGE AGW CLC cln. Hemicr. or rhizome geophyte.

subsp. ***distachyos*** (All.) Maire & Weiller, in Fl. Afr. Nord 4: 35 (1957).
Medit. to C. Asia. 12 BAL FRA POR SAR SPA 13 GRC ITA KRI SIC 20 ALG EGY LBY MOR TUN 21 CNY 32 TKM TZK UZB 34 AFG CYP IRN IRQ LBS PAL SIN TUR 35 SAU (50) soa. Hemicr. or rhizome geophyte.
Cyperus mucronatus Rottb., Descr. Pl. Rar.: 17 (1772). *Pycreus mucronatus* (Rottb.) Nees, Linnaea 9: 283 (1834).
**Cyperus distachyos* All., Auct. Fl. Pedem.: 48 (1789). *Cyperus laevigatus* var. *distachyos* (All.) Coss. & Durieu, Expl. Sci. Algérie 2: 251 (1868). *Acorellus distachyus* (All.) Palla, Allg. Bot. Z. Syst. 9: 68 (1903). *Juncellus distachyos* (All.) Turrill, Bull. Misc. Inform. Kew 1926: 375 (1926). *Juncellus laevigatus* subsp. *distachyos* (All.) P.H.Davis, Fl. Turkey 9: 42 (1985). *Acorellus laevigatus* subsp. *distachyos* (All.) Holub, Folia Geobot. Phytotax. 23: 413 (1988).
Cyperus subaphyllus Boeck. ex Schinz, Verh. Bot. Vereins Prov. Brandenburg 30: 139 (1888). *Cyperus laevigatus* var. *subaphyllus* (Boeck. ex Schinz) Kük. in R.E.Fries, Wiss. Erg. Schwed. Rhod.-Kongo Exped. 1: 278 (1921).
Juncellus calanthus Peter, Abh. Preuss. Akad. Wiss., Phys.-Math. Kl., n.f., 13: 48 (1928).

subsp. ***laevigatus***
Trop. & Subtrop. 13 SIC 20 ALG EGY LBY MOR 21 CNY CVI MDR 22 MLI MTN NGA NGR SEN 23 BUR RWA ZAI 24 CHA ERI ETH SOC SOM SUD 25 KEN TAN UGA 26 ANG MLW

27 BOT CPP CPV LES NAM NAT OFS SWZ TVL 29 MAU MDG 32 UZB 34 AFG EAI IRN IRQ LBS PAL SIN 35 GST OMA SAU YEM 40 IND PAK 41 VIE 50 NSW QLD SOA WAU 60 GIL 63 HAW 76 ARI CAL 77 TEX 78 NCA 79 MXC MXE MXG MXN MXS MXT 80 COS GUA NIC 81 CUB DOM JAM LEE PUE 83 BOL ECU GAL PER 85 AGE AGW CLC cln. Hemicr. or rhizome geophyte.

Cyperus lateralis Forssk., Fl. Aegypt.-Arab.: 13 (1775). *Pycreus lateralis* (Forssk.) Nees, Linnaea 9: 283 (1834).

Cyperus junciformis Desf., Fl. Atlant. 1: 42 (1798). *Chlorocyperus junciformis* (Desf.) Rikli, Jahrb. Wiss. Bot. 27: 563 (1895).

Cyperus laevigatus var. *albidus* Vahl, Enum. Pl. 2: 306 (1805).

Cyperus leucostachys Link, Jahrb. Gewächsk. 1(3): 81 (1818).

Cyperus uliginosus Thouars ex Link, Jahrb. Gewächsk. 1(3): 82 (1818).

Cyperus maritimus Gouan ex Spreng., Syst. Veg. 1: 217 (1824), nom. illeg.

Cyperus monostachyos Link in C.L.von Buch, Phys. Beschr. Canar. Ins.: 138 (1828), nom. illeg.

Cyperus roxburghianus C.Presl, Isis (Oken) 21: 272 (1828).

Cyperus roxburghii A.Dietr., Sp. Pl. 2: 214 (1832).

Cyperus pleuranthus Nees in R.Wight, Contr. Bot. India: 73 (1834).

Cyperus incurvus Spreng. ex Kunth, Enum. Pl. 2: 18 (1837).

Cyperus cossyrensis Tineo ex Guss., Fl. Sicul. Syn. 2: 779 (1844).

Cyperus teretifolius A.Rich., Tent. Fl. Abyss. 2: 477 (1850).

Cyperus rivularis Steud. in W.Lechler, Berberid. Amer. Austral.: 56 (1857), nom. illeg.

Cyperus laevigatus var. *pictus* Boeck., Linnaea 35: 487 (1868).

Cyperus reptans Boeck., Linnaea 35: 485 (1868). *Cyperus laevigatus* f. *reptans* (Boeck.) C.B.Clarke, J. Linn. Soc., Bot. 21: 78 (1884). *Cyperus laevigatus* var. *reptans* (Boeck.) Kük. in H.G.A.Engler (ed.), Pflanzenr., IV, 20(101): 324 (1936).

Cyperus viridulus Boeck., Linnaea 35: 485 (1868). *Cyperus laevigatus* var. *viridulus* (Boeck.) C.B.Clarke, J. Linn. Soc., Bot. 21: 78 (1884).

Cyperus submonastichyus Steud. & Jard., Bull. Soc. Linn. Normandie, II, 9: 278 (1875).

Cyperus laevigatus var. *caespitosus* C.B.Clarke, J. Linn. Soc., Bot. 20: 282 (1883).

Cyperus laevigatus f. *macra* C.B.Clarke, J. Linn. Soc., Bot. 21: 78 (1884).

Cyperus laevigatus var. *ramlehensis* Sickenb., Contr. Fl. Egypte: 299 (1901).

Acorellus pallae Kneuck., Allg. Bot. Z. Syst. 13: 31 (1907). *Cyperus laevigatus* var. *pallae* (Kneuck.) Kük. in H.G.A.Engler (ed.), Pflanzenr., IV, 20(101): 323 (1936).

Cyperus juncellus Dinter, Deut. Südw. Afrik.: 41 (1909).

Cyperus careyi Britton in J.K.Small, Fl. S.E. U.S., ed. 2: 166 (1913).

Cyperus laevigatus f. *atratus* Peter ex Kük. in H.G.A.Engler (ed.), Pflanzenr., IV, 20(101): 325 (1936).

Cyperus laevis R.Br., Prodr.: 213 (1810).
E. Australia. 50 NSW QLD. Hemicr. or rhizome geophyte.

Cyperus lancastriensis Porter in A.Gray, Manual, ed. 5: 555 (1867). *Mariscus lancastriensis* (Porter) C.B. Clarke, Kew Bull., Addit. Ser. 8: 20 (1908). *Cyperus refractus* var. *lancastriensis* (Porter) Kük. in H.G.A. Engler (ed.), Pflanzenr., IV, 20(101): 510 (1936).
C. & E. U.S.A. 74 MSO 75 NWJ OHI PEN WVA 77 TEX 78 ALA ARK DEL FLA GEO KTY MRY MSI NCA SCA TEN VRG WDC. Hemicr. or rhizome geophyte.

Cyperus lateriticus J.Raynal, Adansonia, n.s., 6: 308 (1966).
Senegal. 22 SEN. Hemicr. or rhizome geophyte.

Cyperus latifolius Poir. in J.B.A.M.de Lamarck, Encycl. 7: 268 (1806).
Trop. & S. Africa, W. Indian Ocean. 22 BEN 23 BUR CMN CON RWA 24 ETH 25 KEN TAN UGA 26 ANG MLW MOZ 27 BOT CPP NAT SWZ TVL 29 COM MAU MDG REU 34 PAL. Hemicr. or rhizome geophyte.

Cyperus latifolius var. *angustifolius* C.Krauss, Flora 28: 754 (1845).

Cyperus herana Cherm., Bull. Soc. Bot. France 66: 319 (1919 publ. 1920). *Cyperus latifolius* var. *herana* (Cherm.) Cherm., Bull. Soc. Bot. France 74: 606 (1927).

Cyperus latifolius var. *austroafricanus* Kük. in H.G.A. Engler (ed.), Pflanzenr., IV, 20(101): 88 (1935).

Cyperus latifolius var. *fimbristyloides* Kük. in H.G.A. Engler (ed.), Pflanzenr., IV, 20(101): 88 (1935).

Cyperus latzii K.L.Wilson, Telopea 4: 487 (1991).
N. Western Australia. 50 WAU. Hemicr. or rhizome geophyte.

Cyperus laxespicatus Kük. in R.E.Fries, Wiss. Erg. Schwed. Rhod.-Kongo Exped. 1: 3 (1921).
E. & S. Trop. Africa. 25 TAN 26 ANG ZAM ZIM. Hemicr. or rhizome geophyte.

Cyperus laxespicatus var. *brunneotinctus* Kük., Repert. Spec. Nov. Regni Veg. 21: 325 (1925).

Cyperus laxespicatus var. *percrassus* Kük., Repert. Spec. Nov. Regni Veg. 21: 325 (1925).

Cyperus laxiflorus Poir. in J.B.A.M.de Lamarck, Encycl. 7: 270 (1806).
W. Indian Ocean. 29 MAU MDG. Hemicr. or rhizome geophyte.

Cyperus laxus Lam., Tabl. Encycl. 1: 146 (1791).
Trop. Africa, Mexico to Trop. America. 22 BEN GHA GUI IVO LBR NGA TOG 23 CAF CMN CON GAB GGI ZAI 25 TAN UGA 26 ANG 79 MXC MXG MXS MXT 80 BLZ COS ELS GUA HON NIC PAN 81 CUB JAM LEE TRT WIN 82 FRG GUY SUR VEN 83 BOL CLM ECU PER 84 BZC BZE BZL BZN 85 AGE AGW PAR. Hemicr. or rhizome geophyte.

subsp. ***buchholzii*** (Boeck.) Lye, Nordic J. Bot. 3: 232 (1983).
Trop. Africa. 22 BEN GHA GUI IVO LBR NGA TOG 23 CAF CMN CON EQG GAB GGI ZAI 25 TAN UGA 26 ANG. Hemicr. or rhizome geophyte.

Cyperus sylvestris Ridl., Trans. Linn. Soc. London, Bot. 2: 134 (1884). *Cyperus diffusus* subsp. *sylvestris* (Ridl.) Kük. in H.G.A.Engler (ed.),

Pflanzenr., IV, 20(101): 209 (1936). *Cyperus laxus* subsp. *sylvestris* (Ridl.) Lye, Nordic J. Bot. 3: 232 (1983).

**Cyperus buchholzii* Boeck., Beitr. Cyper. 1: 3 (1888). *Cyperus diffusus* subsp. *buchholzii* (Boeck.) Kük. in H.G.A.Engler (ed.), Pflanzenr., IV, 20(101): 210 (1936).

Torulinium angolense Turrill, Bull. Misc. Inform. Kew 1915: 18 (1915).

subsp. ***laxus***

Mexico to Trop. America. 22 GHA GUI IVO LBR NGA TOG 23 CAF CMN CON GAB GGI ZAI 25 TAN UGA 26 ANG 79 MXC MXG MXS MXT 80 BLZ COS ELS GUA HON NIC PAN 81 CUB JAM LEE TRT WIN 82 FRG GUY SUR VEN 83 BOL CLM ECU PER 84 BZC BZE BZL BZN 85 AGE AGW PAR. Hemicr. or rhizome geophyte.

Cyperus tolucensis Kunth in F.W.H.von Humboldt, A.J.A.Bonpland & C.S.Kunth, Nov. Gen. Sp. 1: 206 (1817). *Cyperus diffusus* var. *tolucensis* (Kunth) Kük. in H.G.A.Engler (ed.), Pflanzenr., IV, 20(101): 210 (1936). *Cyperus diffusus* subsp. *tolucensis* (Kunth) Borhidi, Acta Bot. Acad. Sci. Hung. 25: 1 (1979).

Cyperus umbrosus Lindl. & Nees in C.F.P.von Martius & auct. suc. (eds.), Fl. Bras. 2(1): 31 (1842). *Cyperus diffusus* var. *umbrosus* (Lindl. & Nees) Kük. in H.G.A.Engler (ed.), Pflanzenr., IV, 20(101): 210 (1936).

Cyperus homoiiria Steud., Syn. Pl. Glumac. 2: 23 (1854).

Cyperus octophyllus Hochst. ex Steud., Syn. Pl. Glumac. 2: 20 (1854).

Cyperus laxus Griseb., Fl. Brit. W. I.: 563 (1864), nom. illeg.

Cyperus macrostachyus Ten. ex Boeck., Linnaea 36: 284 (1870).

Cyperus limbatus Maury ex Micheli, Mém. Soc. Phys. Genève 31: 129 (1890).

Cyperus lecontei Torr. ex Steud., Syn. Pl. Glumac. 2: 21 (1854).

SE. U.S.A. 78 ALA FLA LOU MSI NCA SCA. Hemicr. or rhizome geophyte.

Cyperus dentatus var. *multiradiatus* Torr., Ann. Lyceum Nat. Hist. New York 3: 272 (1836). *Cyperus multiradiatus* (Torr.) C.Mohr, Contr. U. S. Natl. Herb. 6: 392 (1901).

Cyperus leiocaulon Benth., Fl. Austral. 7: 287 (1878).

E. Australia. 50 NSW QLD. Hemicr. or rhizome geophyte.

**Mariscus laevis* R.Br., Prodr.: 218 (1810).

Cyperus lentiginosus Millsp. & Chase, Publ. Field Columbian Mus., Bot. Ser. 3: 74 (1903). *Cyperus tenuis* var. *lentiginosus* (Millsp. & Chase) Kük. in H.G.A.Engler (ed.), Pflanzenr., IV, 20(101): 418 (1936). *Mariscus lentiginosus* (Millsp. & Chase) Conz., Fl. Taxon. Mexic. 2: 239 (1943).

Texas to C. America. 77 TEX 78 FLA 79 MXE MXG MXS MXT 80 BLZ GUA HON. Hemicr. or rhizome geophyte.

Cyperus strigosus var. *gracilis* Britton, Bull. Torrey Bot. Club 13: 212 (1886).

Cyperus leptocladus Kunth, Enum. Pl. 2: 32 (1837).

S. Africa. 27 CPP NAT SWZ TVL. Hemicr. or rhizome geophyte.

Cyperus boucheanus Regel, Trudy Imp. S.-Peterburgsk. Bot. Sada 2: 316 (1873).

Cyperus smithii McClean, Bull. Misc. Inform. Kew 1927: 77 (1927), nom. illeg.

Cyperus leptolepis Peter ex Kük. in H.G.A.Engler (ed.), Pflanzenr., IV, 20(101): 550 (1936).

Tanzania. 25 TAN. Hemicr. or rhizome geophyte.

Cyperus leucocephalus Retz., Observ. Bot. 5: 11 (1788). *Sorostachys leucocephalus* (Retz.) Lye, Nordic J. Bot. 1: 190 (1981).

Indian Subcontinent, Indo-China. 40 ASS BAN IND 41 MYA THA VIE. Hemicr. or rhizome geophyte.

Isolepis coronaria Roem. & Schult., Syst. Veg. 2: 113 (1817).

Sorostachys kyllingioides Steud., Syn. Pl. Glumac. 2: 71 (1854).

Kyllinga pierreana E.G.Camus, Notul. Syst. (Paris) 1: 290 (1910).

Cyperus lhotskyanus Beck, Bot. Jahrb. Syst. 5: 498 (1884).

SE. Australia. 50 NSW SOA VIC. Hemicr. or rhizome geophyte.

Mariscus rutilans C.B.Clarke, Kew Bull., Addit. Ser. 8: 18 (1908). *Cyperus rutilans* (C.B.Clarke) Maiden & Betche, Census N.S.W. Pl.: 28 (1916).

Cyperus ligularis L., Amoen. Acad. 5: 31 (1760). *Mariscus ligularis* (L.) Urb., Symb. Antill. 2: 165 (1900).

Trop. & Subtrop. America, Cape Verde to W. & WC. Trop. Africa, Aldabra. 21 CVI 22 BEN GAM GHA GNB GNB GUI IVO LBR NGA SEN SIE TOG 23 CMN EQG GAB GGI ZAI 29 ALD SEY (62) mrn (76) cal 78 ALA FLA FLA LOU 79 MXE MXG MXN MXS MXT 80 BLZ COS ELS GUA HON NIC PAN 81 ARU BAH CAY CUB DOM HAI JAM LEE NLA PUE TRT WIN 82 SUR VEN 83 BOL CLM ECU GAL PER 84 BZC BZN 85 PAR. Hemicr.

Mariscus rufus Kunth in F.W.H.von Humboldt, A.J.A. Bonpland & C.S.Kunth, Nov. Gen. Sp. 1: 216 (1816).

Cyperus rufus Kunth in F.W.H.von Humboldt, A.J.A. Bonpland & C.S.Kunth, Nov. Gen. Sp. 1: 210 (1817).

Cyperus callophorus G.Mey., Prim. Fl. Esseq.: 32 (1818).

Mariscus coriaceus G.Mey., Prim. Fl. Esseq.: 38 (1818). *Cyperus coriaceus* (G.Mey.) Schrad. ex Nees in C.F.P.von Martius & auct. suc. (eds.), Fl. Bras. 2(1): 42 (1842).

Mariscus glandulosus Bojer, Hortus Maurit.: 382 (1837). *Cyperus glandulosus* (Bojer) Rolfe, J. Bot. 20: 362 (1882).

Cyperus vulnerans Salzm. ex Schltdl., Bot. Zeitung (Berlin) 7: 98 (1849).

Cyperus punctatifolius Steud., Syn. Pl. Glumac. 2: 27 (1854).

Cyperus spicatocapitatus Steud., Syn. Pl. Glumac. 2: 27 (1854). *Cyperus ligularis* var. *spicatocapitatus* (Steud.) Kük. in H.G.A.Engler (ed.), Pflanzenr., IV, 20(101): 476 (1936).

Cyperus thyrsiflorus Boeck., Linnaea 36: 5 (1869), nom. illeg.

Cyperus rionensis Boeck., Beitr. Cyper. 1: 12 (1888).

Cyperus sintenisii Boeck., Beitr. Cyper. 1: 12 (1888).

Cyperus trigonus Boeck., Beitr. Cyper. 1: 11 (1888).

Cyperus glaucoviridis Boeck., Allg. Bot. Z. Syst. 1: 230 (1895).

Cyperus limosus Maxim., Prim. Fl. Amur.: 294 (1859). *Juncellus limosus* (Maxim.) C.B.Clarke, Kew Bull., Addit. Ser. 8: 3 (1908). *Pycreus limosus* (Maxim.) Schischk. in V.L.Komarov (ed.), Fl. URSS 3: 8 (1935). *Dichostylis limosa* (Maxim.) A.E.Kozhevn., Sosud. Rast. Sovet. Dal'nego Vostoka 3: 221 (1988).
Russian Far East, China to Vietnam. 31 AMU PRM 36 CHM 41 VIE. Hemicr. or rhizome geophyte.

Cyperus linearispiculatus T.L.Dai, in Fl. Reipubl. Popul. Sin. 11: 230 (1961).
China. 36+. Hemicr. or rhizome geophyte.

Cyperus locuples C.B.Clarke in D.Oliver & auct. suc. (eds.), Fl. Trop. Afr. 8: 362 (1901).
Nigeria. 22 NGA. Hemicr. or rhizome geophyte.

Cyperus longibracteatus (Cherm.) Kük., Repert. Spec. Nov. Regni Veg. 26: 250 (1929).
Trop. Africa, Madagascar. 22 BEN GHA GNB GUI IVO LBR MLI NGA SEN SIE TOG 23 CMN CON GAB GGI RWA ZAI 24 ETH 25 KEN TAN UGA 26 ANG ZIM 29 MDG. Hemicr. or rhizome geophyte.
Mariscus bojeri C.B.Clarke in T.A.Durand & H.Schinz, Consp. Fl. Afric. 5: 584 (1894), nom. inval.
Cyperus distans var. *niger* C.B.Clarke in D.Oliver & auct. suc. (eds.), Fl. Trop. Afr. 8: 350 (1901). *Cyperus longibracteatus* var. *niger* (C.B.Clarke) Lye, Lidia 3: 131 (1994).
**Mariscus longibracteatus* Cherm., Bull. Mus. Natl. Hist. Nat. 25: 407 (1919). *Cyperus distans* subsp. *longibracteatus* (Cherm.) Lye, Nordic J. Bot. 3: 231 (1983).
Mariscus rubrotinctus Cherm., Bull. Mus. Natl. Hist. Nat. 25: 407 (1919). *Cyperus longibracteatus* var. *rubrotinctus* (Cherm.) Kük., Repert. Spec. Nov. Regni Veg. 26: 251 (1929). *Cyperus rubrotinctus* (Cherm.) Kük., Repert. Spec. Nov. Regni Veg. 26: 251 (1929). *Cyperus distans* var. *rubrotinctus* (Cherm.) Lye, Nordic J. Bot. 3: 231 (1983).

Cyperus longifolius Poir. in J.B.A.M.de Lamarck, Encycl. 7: 270 (1806). *Eucyperus longifolius* (Poir.) Rikli, Jahrb. Wiss. Bot. 27: 568 (1895).
W. Indian Ocean. 29 MAU MDG REU. Hemicr. or rhizome geophyte.
Cyperus longifolius var. *condensatus* Cherm., Bull. Soc. Bot. France 66: 345 (1919).

Cyperus longiinvolucratus Lye, Nordic J. Bot. 3: 220 (1983).
Tanzania. 25 TAN. Hemicr. or rhizome geophyte.

Cyperus longispicula Muasya & D.A.Simpson, Kew Bull. 59: 595 (2004).
Zimbabwe to Mozambique. 26 MOZ ZIM.

Cyperus longistylus Kük., Repert. Spec. Nov. Regni Veg. 26: 250 (1929).
Solomon Is. 43 SOL. Hemicr. or rhizome geophyte.

Cyperus longus L., Sp. Pl.: 45 (1753). *Eucyperus longus* (L.) Rikli, Jahrb. Wiss. Bot. 27: 568 (1895). *Chlorocyperus longus* (L.) Palla, Allg. Bot. Z. Syst. 6: 201 (1900). *Pycreus longus* (L.) Hayek, Repert. Spec. Nov. Regni Veg. Beih. 30(3): 146 (1932).
Macaronesia, Europe, Medit. to Himalaya. 10 GRB 11 AUT GER† HUN SWI 12 BAL COR FRA POR SAR SPA 13 ALB BUL GRC ITA KRI ROM SIC TUE YUG 14 KRY RUS 20 ALG EGY MOR 21 AZO CNY CVI MDR 22 BEN BKN MLI NGA NGR SEN TOG 23 CMN 24 CHA ETH SOM 25 KEN TAN UGA 27 BOT CPP CPV LES NAM NAT OFS SWZ TVL 32 KAZ KGZ TKM UZB 33 TCS 34 AFG CYP IRN IRQ LBS PAL TUR 35 SAU YEM 40 IND NEP PAK. Hemicr. or rhizome geophyte.

subsp. ***badius*** (Desf.) Murb., Contr. Fl. Tun. 3: 27 (1899).
Macaronesia to India and Madagascar. 12 COR FRA POR SAR SPA 13 ALB GRC ITA SIC YUG 20 ALG EGY LBY MOR TUN 21 AZO CNY MDR 24 ETH 25 KEN 27 CPP NAT TVL 29 MDG 32 KAZ KGZ TZK 33 TCS 34 AFG LBS TUR 40 IND PAK. Hemicr. or rhizome geophyte.
**Cyperus badius* Desf., Fl. Atlant. 1: 45 (1798). *Cyperus longus* var. *badius* (Desf.) Cambess., Mém. Mus. Hist. Nat. 14: 323 (1827). *Chlorocyperus badius* (Desf.) Palla, Allg. Bot. Z. Syst. 6: 61 (1900). *Pycreus badius* (Desf.) Hayek, Repert. Spec. Nov. Regni Veg. Beih. 30(3): 147 (1932). *Chlorocyperus longus* subsp. *badius* (Desf.) Soó, Acta Bot. Acad. Sci. Hung. 16: 369 (1970 publ. 1971).

subsp. ***longus***
Macaronesia, Europe, Medit. to Himalaya. 10 GRB 11 AUT GER† HUN SWI 12 BAL COR FRA POR SAR SPA 13 ALB BUL GRC ITA KRI ROM SIC TUE YUG 14 KRY RUS 20 ALG MOR 21 AZO CNY CVI MDR 24 ETH SOM 25 KEN TAN UGA 27 BOT CPP CPV LES NAM NAT OFS SWZ TVL 32 KAZ KGZ TKM UZB 33 TCS 34 AFG CYP IRN IRQ LBS PAL TUR 35 YEM 40 IND IND NEP NEP PAK PAK. Hemicr. or rhizome geophyte.
Cyperus racemosus Ten., Fl. Napol.: viii. (1811), nom. illeg.
Cyperus brachystachys C.Presl, Cyper. Gramin. Sicul.: 15 (1820). *Cyperus badius* var. *brachystachys* (C.Presl) Nyman, Consp. Fl. Eur.: 759 (1882).
Cyperus myriostachyus Ten., Fl. Napol. 3: 47 (1824). *Cyperus longus* f. *myriostachyus* (Ten.) Kük. in H.G.A.Engler (ed.), Pflanzenr., IV, 20(101): 102 (1935).
Cyperus thermalis Dumort., Fl. Belg.: 145 (1827).
Cyperus lateriflorus Steud., Flora 12: 152 (1829).
Cyperus emarginatus Schrad., Anal. Fl. Cap. 1: 5 (1832).
Cyperus intermedius Guss., Fl. Sicul. Prodr., Suppl.: 10 (1832). *Cyperus preslii* subsp. *intermedius* (Guss.) Nyman, Consp. Fl. Eur.: 759 (1882). *Cyperus longus* f. *intermedius* (Guss.) Kük. in H.G.A.Engler (ed.), Pflanzenr., IV, 20(101): 101 (1935).
Cyperus elongatus Lej. ex Nees, Linnaea 9: 285 (1834), nom. inval.
Cyperus amoenus Kunth, Enum. Pl. 2: 58 (1837).
Cyperus fastigiatus Willd. ex Kunth, Enum. Pl. 2: 60 (1837), nom. illeg.
Cyperus neglectus Parl., Giorn. Bot. Ital. 2(1): 131 (1846). *Cyperus badius* var. *neglectus* (Parl.) Nyman, Consp. Fl. Eur.: 759 (1882).
Cyperus kunthii Parl., Fl. Ital. 2: 41 (1852).
Cyperus preslii Parl., Fl. Ital. 2: 40 (1852), nom. illeg.
Cyperus heldreichianus Boiss., Diagn. Pl. Orient. 13: 39 (1854). *Cyperus longus* var. *heldreichianus* (Boiss.) Boiss., Fl. Orient. 5: 375 (1882).

Cyperus italicus Tod., Index Seminum (PAL) 1861: 32 (1861), nom. illeg.
Cyperus longus var. *auratus* Boeck., Linnaea 36: 279 (1870).
Cyperus longus var. *elongatus* Boeck., Linnaea 36: 279 (1870).
Cyperus oxystachys Nees ex Boeck., Linnaea 36: 280 (1870).
Cyperus longus f. *cypricus* C.B.Clarke, J. Linn. Soc., Bot. 21: 165 (1884).
Cyperus longus var. *elongatus* C.B.Clarke, J. Linn. Soc., Bot. 21: 166 (1884).
Cyperus longus var. *pallescens* C.B.Clarke, J. Linn. Soc., Bot. 21: 164 (1884).
Cyperus longus f. *longispiculatus* Reynolds, Bull. Assoc. Pyrén. Échange Pl. 17: 16 (1907).
Cyperus longus f. *lucanus* A.Terracc., Nuovo Giorn. Bot. Ital. 14: 205 (1907).
Cyperus guthnickii Tineo ex Lojac., Fl. Sicul. 3: 205 (1909). *Cyperus longus* var. *guthnickii* (Tineo ex Lojac.) Kük. in H.G.A.Engler (ed.), Pflanzenr., IV, 20(101): 100 (1935).
Cyperus parlatoris Lojac., Fl. Sicul. 3: 204 (1909). *Cyperus longus* f. *parlatoris* (Lojac.) Kük. in H.G.A.Engler (ed.), Pflanzenr., IV, 20(101): 102 (1935).
Cyperus parlatoris var. *difformis* Lojac., Fl. Sicul. 3: 204 (1909). *Cyperus longus* f. *difformis* (Lojac.) Kük. in H.G.A.Engler (ed.), Pflanzenr., IV, 20(101): 102 (1935).
Cyperus speciosus Lojac., Fl. Sicul. 3: 206 (1909), nom. illeg. *Cyperus longus* var. *speciosus* Kük. in H.G.A.Engler (ed.), Pflanzenr., IV, 20(101): 100 (1935).
Cyperus longus var. *pallidior* Kük. in H.G.A.Engler (ed.), Pflanzenr., IV, 20(101): 100 (1935). *Chlorocyperus longus* f. *pallidior* (Kük.) Soó, Acta Bot. Acad. Sci. Hung. 16: 369 (1970 publ. 1971).

var. ***pallidus*** Boeck., Linnaea 36: 280 (1870).
N. & Trop. Africa , Arabian Pen., S. India. 20 ALG EGY 22 BEN BKN MLI NGA NGR SEN TOG 23 CMN 24 CHA SOM 25 TAN UGA 35 SAU 40 IND. Hemicr. or rhizome geophyte.
Cyperus fenzelianus Steud., Syn. Pl. Glumac. 2: 33 (1854). *Cyperus rotundus* var. *fenzelianus* (Steud.) El-Hadidy in L.Boulos, Fl. Egypt Checklist: 213 (1995).
Cyperus fenzelianus var. *badiiformis* Chiov., Ann. Bot. (Rome) 13: 376 (1915). *Cyperus longus* f. *badiiformis* (Chiov.) Kük. in H.G.A.Engler (ed.), Pflanzenr., IV, 20(101): 101 (1935).

Cyperus lucidus R.Br., Prodr.: 218 (1810). *Mariscus lucidus* (R.Br.) C.B.Clarke, Kew Bull., Addit. Ser. 8: 18 (1908).
New Guinea to E. & SE. Australia. 43 NWG 50 NSW QLD SOA TAS VIC. Hemicr. or rhizome geophyte.
Cyperus acutus R.Br., Prodr.: 217 (1810). *Mariscus acutus* (R.Br.) C.B.Clarke, Kew Bull., Addit. Ser. 8: 19 (1908). *Cyperus lucidus* var. *acutus* (R.Br.) Kük. in H.G.A.Engler (ed.), Pflanzenr., IV, 20(101): 449 (1936).
Cyperus sanguineofuscus Nees, Ann. Mag. Nat. Hist. 6: 46 (1841). *Mariscus sanguineofuscus* (Nees) C.B. Clarke, Kew Bull., Addit. Ser. 8: 18 (1908). *Cyperus lucidus* var. *sanguineofuscus* (Nees) Kük. in H.G.A. Engler (ed.), Pflanzenr., IV, 20(101): 449 (1936).
Cyperus lucidus var. *grandiusculus* Kük. in H.G.A. Engler (ed.), Pflanzenr., IV, 20(101): 449 (1936).

Cyperus luerssenii Boeck., Flora 58: 86 (1875).
Queensland. 50 QLD. Hemicr. or rhizome geophyte.

Cyperus lundellii O'Neill, Publ. Carnegie Inst. Wash. 522: 276 (1940).
Mexico to Guatemala. 79 MXE MXT 80 GUA. Hemicr. or rhizome geophyte.

Cyperus lupulinus (Spreng.) Marcks, Trans. Wisconsin Acad. Sci. 62: 271 (1974).
E. Canada to U.S.A. 72 NBR NSC ONT QUE 73 COL IDA ORE 74 ILL IOW KAN MIN MSO NEB WIS 75 CNT INI MAS MIC NWJ NWY OHI PEN RHO VER WVA 77 TEX 78 ARK DEL GEO KTY MRY NCA TEN VRG WDC. Hemicr. or rhizome geophyte.
**Scirpus lupulinus* Spreng., Mant. Prim. Fl. Hal.: 30 (1807).

subsp. ***lupulinus***
E. Canada to U.S.A. 72 ONT 73 COL IDA ORE 74 ILL KAN MIN MSO WIS 75 CNT INI MAS MIC NWJ NWY OHI PEN RHO VER WVA 77 TEX 78 ARK DEL KTY NCA TEN VRG. Hemicr. or rhizome geophyte.

subsp. ***macilentus*** (Fernald) Marcks, Trans. Wisconsin Acad. Sci 62: 275 (1974).
E. Canada to NC. & E. U.S.A. 72 NBR NSC ONT QUE 74 ILL IOW MIN NEB WIS 75 CNT INI MAI MAS MIC NWH NWJ NWY OHI PEN RHO VER WVA 78 DEL GEO MRY NCA VRG WDC. Hemicr. or rhizome geophyte.
Mariscus glomeratus W.Barton, Fl. Philadelph. Prodr.: 18 (1815).
Scirpus cyperiformis Muhl., Descr. Gram.: 41 (1817). *Mariscus cyperiformis* (Muhl.) Torr., Cat. Pl. New York: 14 (1819).
Cyperus bushii Britton, Man. Fl. N. States: 1044 (1901). *Cyperus houghtonii* var. *bushii* (Britton) Kük. in H.G.A.Engler (ed.), Pflanzenr., IV, 20(101): 469 (1936).
**Cyperus filiculmis* var. *macilentus* Fernald, Rhodora 8: 128 (1906). *Cyperus macilentus* (Fernald) E.P.Bicknell, Bull. Torrey Bot. Club 35: 478 (1908).

Cyperus luteus Boeck., Linnaea 38: 370 (1874). *Mariscus luteus* (Boeck.) C.B.Clarke in T.A.Durand & H.Schinz, Consp. Fl. Afric. 5: 589 (1894).
SW. Cameroon, Uganda to Malawi, Comoros, Madagascar. 23 CMN RWA ZAI 25 KEN TAN UGA 26 MLW 29 COM MDG. Hemicr. or rhizome geophyte.
Mariscus foliosus C.B.Clarke in D.Oliver & auct. suc. (eds.), Fl. Trop. Afr. 8: 399 (1902).
Mariscus manongarivensis Cherm., Bull. Mus. Natl. Hist. Nat. 25: 406 (1919). *Cyperus luteus* var. *manongarivensis* (Cherm.) Kük. in H.G.A.Engler (ed.), Pflanzenr., IV, 20(101): 414 (1936).

Cyperus luzulae (L.) Retz., Observ. Bot. 4: 11 (1786).
Mexico to Trop. America. 79 MXE MXG MXS MXT 80 BLZ COS GUA HON NIC PAN 81 BAH BER CUB DOM HAI JAM LEE PUE TRT WIN 82 FRG GUY SUR VEN 83 BOL CLM ECU PER 84 BZC BZE BZN BZS 85 AGE AGW PAR URU. Hemicr. or rhizome geophyte.
**Scirpus luzulae* L., Syst. Nat. ed. 10: 868 (1759). *Eucyperus luzulae* (L.) Rikli, Jahrb. Wiss. Bot. 27: 568 (1895).
Cyperus globulosus Aubl., Hist. Pl. Guiane 1: 47 (1775).

Cyperus polycephalus Lam., Tabl. Encycl. 1: 147 (1791). *Kyllinga polycephala* (Lam.) Link, Enum. Hort. Berol. Alt. 1: 47 (1821). *Mariscus polycephalus* (Lam.) Link, Hort. Berol. 1: 324 (1827).
Cyperus conoideus Rich., Actes Soc. Hist. Nat. Paris 1: 106 (1792). *Cyperus luzulae* f. *conoideus* (Rich.) Nees in C.F.P.von Martius & auct. suc. (eds.), Fl. Bras. 2(1): 20 (1842).
Scirpus elegans Salisb., Prodr. Stirp. Chap. Allerton: 31 (1796).
Cyperus globuliferus Link, Jahrb. Gewächsk. 1(3): 89 (1818).
Cyperus sphaerostachys Link, Jahrb. Gewächsk. 1(3): 89 (1818).
Cyperus glomuliferus Schrad. ex Schult. in J.J.Roemer & J.A.Schultes, Mant. 2: 108 (1824).
Mariscus parviflorus Nees, Flora 11: 329 (1828).
Kyllinga scirpina Rchb. ex Steud., Nomencl. Bot., ed. 2, 1: 470 (1840).
Cyperus hostmannii Steud., Syn. Pl. Glumac. 2: 25 (1854).
Cyperus trinitatis Steud., Syn. Pl. Glumac. 2: 26 (1854).
Cyperus luzulae var. *minor* Boeck., Linnaea 35: 561 (1868).
Cyperus luzulae var. *umbellulatus* Britton, Bull. Torrey Bot. Club 13: 208 (1886).
Cyperus pseudosurinamensis Boeck., Allg. Bot. Z. Syst. 1: 201 (1895).
Cyperus bangianus Gand., Bull. Soc. Bot. France 66: 297 (1919).
Cyperus guatemalensis Gand., Bull. Soc. Bot. France 66: 297 (1919 publ. 1920), nom. illeg.
Cyperus luzulae f. *pallidiflorens* Kük. in H.G.A.Engler (ed.), Pflanzenr., IV, 20(101): 171 (1936).

Cyperus macer C.B.Clarke, J. Linn. Soc., Bot. 21: 160 (1884).
Assam to Pen. Malaysia. 40 ASS BAN 42 MLY. Hemicr. or rhizome geophyte.

Cyperus macrocarpus (Kunth) Boeck., Linnaea 36: 380 (1870).
Trop. & S. Africa, Madagascar. 22 BEN NGA 23 BUR CON GAB RWA ZAI 24 SUD 25 KEN TAN UGA 26 ANG ZIM 27 NAT TVL 29 MDG. Hemicr. or rhizome geophyte.
**Mariscus macrocarpus* Kunth, Enum. Pl. 2: 120 (1837). *Cyperus cyperoides* subsp. *macrocarpus* (Kunth) Lye, Nordic J. Bot. 3: 231 (1983). *Mariscus sumatrensis* var. *macrocarpus* (Kunth) Maquet, in Fl. Rwanda 4: 630 (1988).
Mariscus radiatus Hochst., Flora 28: 757 (1845).
Cyperus kraussii Boeck., Linnaea 36: 379 (1870). *Cyperus macrocarpus* var. *kraussii* (Boeck.) Kük. in H.G.A.Engler (ed.), Pflanzenr., IV, 20(101): 528 (1936).
Mariscus humbertii Cherm., Bull. Mus. Natl. Hist. Nat. 25: 303 (1919). *Cyperus macrocarpus* var. *humbertii* (Cherm.) Kük. in H.G.A.Engler (ed.), Pflanzenr., IV, 20(101): 529 (1936).
Cyperus macrocarpus var. *excelsior* Kük., Bot. Not. 1934: 70 (1934).
Cyperus cyperoides var. *aureus* A.Peter ex Kük. in H.G.A.Engler (ed.), Pflanzenr., IV, 20(101): 517 (1936).
Cyperus macrocarpus var. *submacrocarpus* Kük. in H.G.A.Engler (ed.), Pflanzenr., IV, 20(101): 528 (1936).

Cyperus macropachycephalus Goetgh., Blumea 34: 161 (1989).
Papua New Guinea. 43 NWG. Hemicr. or rhizome geophyte.

Cyperus macrophyllus (Brongn.) Boeck., Linnaea 36: 376 (1870).
SC. Pacific. 61 COO PIT SCI. Hemicr. or rhizome geophyte.
**Mariscus macrophyllus* Brongn. in L.I.Duperrey, Voy. Monde, Phan.: 179 (1834).
Borabora cyperoidea Steud., Syn. Pl. Glumac. 2: 71 (1854).

Cyperus macrorrhizus Nees in R.Wight, Contr. Bot. India: 73 (1834).
NE. Africa to S. Iran. 20 EGY 24 SOM 34 IRN PAL 35+. Hemicr. or rhizome geophyte.
Cyperus pungens var. *elatus* Boeck., Linnaea 35: 538 (1868).

Cyperus maculatus Boeck. in W.C.H.Peters, Naturw. Reise Mossambique: 539 (1864). *Cyperus longus* var. *maculatus* (Boeck.) Boeck., Linnaea 36: 282 (1870).
Africa. 20 EGY 22 BEN BKN GHA GUI IVO MLI NGA NGR SEN SIE 23 BUR CMN CON RWA ZAI 24 CHA ETH SOM SUD 25 KEN TAN UGA 26 MLW 27 BOT NAM NAT TVL 29 MAU MDG SEY. Hemicr. or rhizome geophyte.

subsp. ***maculatus***
Africa. 20 EGY 22 BEN BKN GHA GUI IVO MLI NGA NGR SEN SIE 23 BUR CMN CON RWA ZAI 24 CHA ETH SOM SUD 25 KEN TAN UGA 26 MLW 27 BOT NAM NAT TVL 29 MAU MDG SEY. Hemicr. or rhizome geophyte.
Cyperus vulgaris Hochst. ex Boeck., Linnaea 35: 490 (1868), nom. illeg.
Cyperus naumannianus Boeck., Flora 62: 552 (1879). *Cyperus maculatus* var. *naumannianus* (Boeck.) Kük. in H.G.A.Engler (ed.), Pflanzenr., IV, 20(101): 104 (1935).
Cyperus galegensis C.B.Clarke, J. Linn. Soc., Bot. 20: 285 (1883).
Cyperus kamphoeveneri Boeck., Beitr. Cyper. 1: 5 (1888).
Cyperus heudelotii C.B.Clarke in D.Oliver & auct. suc. (eds.), Fl. Trop. Afr. 8: 364 (1901).

subsp. ***ogadensis*** Lye, Lidia 4: 206 (1996).
Ethiopia. 24 ETH. Hemicr. or rhizome geophyte.

Cyperus maderaspatanus Willd., Sp. Pl. 1: 278 (1797). *Mariscus maderaspatanus* (Willd.) Napper, J. E. Africa Nat. Hist. Soc. Natl. Mus. 28: 10 (1971).
Trop. Old World. 25 KEN TAN UGA 40 IND. Ther.

Cyperus majungensis Cherm., Bull. Soc. Bot. France 67: 329 (1920 publ. 1921).
Mozambique, Madagascar. 26 MOZ 29 MDG. Hemicr. or rhizome geophyte.

Cyperus malaccensis Lam., Tabl. Encycl. 1: 146 (1791). *Chlorocyperus malaccensis* (Lam.) Palla, Allg. Bot. Z. Syst. 17(Beil.): 6 (1911).
Trop. & Subtrop. Asia, N. Australia. 34 IRN IRQ 36 CHS 38 JAP NNS oga TAI 40 IND NEP PAK 41 MYA THA VIE 42 BOR JAW LSI MLY MOL PHI SUL SUM 43 NWG 50 NTA. Hemicr. or rhizome geophyte.

var. ***brevifolius*** Boeck., Linnaea 35: 604 (1868).
Japan to Indo-China. 36 CHS 38 JAP NNS oga TAI 41 VIE. Hemicr. or rhizome geophyte.

Cyperus monophyllus Vahl, Enum. Pl. 2: 352 (1805). *Cyperus malaccensis* subsp. *monophyllus* (Vahl) T.Koyama, in Fl. Taiwan 5: 266 (1978).

var. ***malaccensis***
Trop. & Subtrop. Asia, N. Australia. 34 IRN IRQ 36 CHS 38 TAI 40 IND NEP PAK 41 MYA THA VIE 42 BOR JAW LSI MLY MOL PHI SUL SUM 43 NWG 50 NTA. Hemicr. or rhizome geophyte.
Cyperus incurvatus Roxb., Fl. Ind. 1: 200 (1820).
Cyperus wallichii Nees in R.Wight, Contr. Bot. India: 83 (1834).
Cyperus neesii D.Dietr., Syn. Pl. 1: 223 (1839), nom. illeg.
Cyperus fortunei Steud., Syn. Pl. Glumac. 2: 21 (1854).
Cyperus spaniophyllus Steud., Syn. Pl. Glumac. 2: 21 (1854).
Cyperus tegetiformis Benth., Fl. Austral. 7: 278 (1878), nom. illeg.

Cyperus mangorensis Cherm., Bull. Soc. Bot. France 66: 352 (1919 publ. 1920).
Madagascar. 29 MDG. Hemicr. or rhizome geophyte.
Cyperus volodioides Cherm., Bull. Mus. Natl. Hist. Nat. 27: 316 (1921). *Cyperus mangorensis* var. *volodioides* (Cherm.) Kük. in H.G.A.Engler (ed.), Pflanzenr., IV, 20(101): 53 (1936).

Cyperus manimae Kunth in F.W.H.von Humboldt, A.J.A.Bonpland & C.S.Kunth, Nov. Gen. Sp. 1: 209 (1817). *Mariscus manimae* (Kunth) C.B.Clarke, Contr. U. S. Natl. Herb. 10: 452 (1908).
Arizona to N. Argentina. 76 ARI 79 MXC MXE MXG MXN MXS MXT 80 COS GUA HON NIC PAN 82 VEN 83 BOL CLM ECU PER 85 AGW. Hemicr. or rhizome geophyte.

var. ***apiculatus*** (Liebm.) C.D.Adams, Ann. Missouri Bot. Gard. 78: 254 (1991).
Mexico to C. America. 79 MXC MXE MXG MXS MXT 80 GUA. Hemicr. or rhizome geophyte.
**Cyperus apiculatus* Liebm., Mexic. Halvgr.: 220 (1850). *Mariscus manimae* var. *apiculatus* (Liebm.) C.B.Clarke, Contr. U. S. Natl. Herb. 10: 452 (1908). *Mariscus apiculatus* (Liebm.) T.Koyama, Phytologia 29: 73 (1974).
Cyperus apiculatus var. *latifolius* Kük. in H.G.A. Engler (ed.), Pflanzenr., IV, 20(101): 463 (1936).

var. ***asperrimus*** (Liebm.) Kük. in H.G.A.Engler (ed.), Pflanzenr., IV, 20(101): 463 (1936).
Arizona to Ecuador. 76 ARI 79 MXC MXE MXG MXN MXS MXT 80 COS GUA HON PAN 83 CLM ECU. Hemicr. or rhizome geophyte.
Cyperus triceps Nees, Linnaea 19: 697 (1847), nom. illeg.
**Cyperus asperrimus* Liebm., Mexic. Halvgr.: 218 (1850).
Cyperus asperrimus var. *multiflorus* Liebm., Mexic. Halvgr.: 31 (1850).
Mariscus guatemalensis C.B.Clarke, Kew Bull., Addit. Ser. 8: 16 (1908). *Cyperus bernoullii* Kük. in H.G.A.Engler (ed.), Pflanzenr., IV, 20(101): 464 (1936).

var. ***manimae***
Mexico to N. Argentina. 79 MXC MXE MXG MXN MXS MXT 80 COS GUA HON NIC PAN 82 VEN 83 BOL ECU PER 85 AGW. Hemicr. or rhizome geophyte.
Cyperus divergens Kunth in F.W.H.von Humboldt, A.J.A.Bonpland & C.S.Kunth, Nov. Gen. Sp. 1: 208 (1817). *Mariscus manimae* var. *divergens* (Kunth) C.B.Clarke, Contr. U. S. Natl. Herb. 10: 452 (1908). *Cyperus manimae* var. *divergens* (Kunth) Kük. in H.G.A.Engler (ed.), Pflanzenr., IV, 20(101): 463 (1936).
Cyperus inaequalis Link, Jahrb. Gewächsk. 1(3): 88 (1818), nom. illeg.
Cyperus anisos Roem. & Schult. in J.J.Roemer & J.A.Schultes, Mant. 2: 131 (1824).
Cyperus triceps var. *multiflorus* Nees, Linnaea 19: 697 (1847).
Cyperus spectabilis var. *coarctatus* Boeck., Linnaea 35: 605 (1868).
Cyperus phaeocephalus Griseb., Abh. Königl. Ges. Wiss. Göttingen 19: 264 (1874). *Mariscus phaeocephalus* (Griseb.) C.B.Clarke, Kew Bull., Addit. Ser. 8: 16 (1908). *Cyperus manimae* var. *phaeocephalus* (Griseb.) O'Neill & Ben.Ayers, Leafl. W. Bot. 4: 36 (1944).
Mariscus manimae var. *hartwegianus* C.B.Clarke, Contr. U. S. Natl. Herb. 10: 452 (1908).
Cyperus manimae var. *pseudospectabilis* Kük. in H.G.A.Engler (ed.), Pflanzenr., IV, 20(101): 463 (1936).
Cyperus phaeocephalus var. *major* Kük. in H.G.A. Engler (ed.), Pflanzenr., IV, 20(101): 464 (1936).

Cyperus mapanioides C.B.Clarke in D.Oliver & auct. suc. (eds.), Fl. Trop. Afr. 8: 340 (1901).
Trop. Africa. 22 GUI IVO NGA SIE TOG 23 CMN CON GAB ZAI 24 CHA 25 KEN TAN UGA 26 ANG. Hemicr. or rhizome geophyte.
Cyperus dichromeniformis var. *major* Boeck., Flora 62: 549 (1879). *Cyperus major* (Boeck.) Cherm., Ann. Inst. Bot.-Géol. Colon. Marseille, III, 10: 29 (1922). *Cyperus mapanioides* var. *major* (Boeck.) Kük. in H.G.A.Engler (ed.), Pflanzenr., IV, 20(101): 230 (1936).
Cyperus major var. *micranthus* Cherm., Bull. Jard. Bot. État 13: 282 (1935). *Cyperus mapanioides* f. *micranthus* (Cherm.) Kük. in H.G.A.Engler (ed.), Pflanzenr., IV, 20(101): 231 (1936).

Cyperus maranguensis K.Schum. in H.G.A.Engler, Pflanzenw. Ost-Afrikas, C: 120 (1895).
Ethiopia to Tanzania. 24 ETH 25 KEN TAN UGA. Hemicr. or rhizome geophyte.

Cyperus margaritaceus Vahl, Enum. Pl. 2: 307 (1805).
Trop. & S. Africa. 22 BEN BKN GAM GHA GNB GUI IVO NGA SEN TOG 23 BUR CMN CON GAB ZAI 24 CHA 25 KEN TAN UGA 26 ANG MLW MOZ ZAM ZIM 27 BOT CPP CPV LES NAM NAT OFS SWZ TVL. Hemicr. or tuber geophyte.
Cyperus eburneus Thonn. ex Kunth, Enum. Pl.: 46 (1835).
Cyperus pseudoniveus Boeck., Verh. Bot. Vereins Prov. Brandenburg 29: 45 (1888). *Cyperus margaritaceus* var. *pseudoniveus* (Boeck.) C.B.Clarke in D.Oliver & auct. suc. (eds.), Fl. Trop. Afr. 8: 322 (1901).
Cyperus margaritaceus var. *prorepens* Kük. in H.G.A. Engler (ed.), Pflanzenr., IV, 20(101): 285 (1936).

Cyperus marginatus Thunb., Prodr. Pl. Cap.: 18 (1794).
Kenya to S. Africa. 25 KEN 27 CPP LES NAM NAT OFS SWZ TVL. Hemicr. or rhizome geophyte.
Cyperus blandus Kunth, Enum. Pl. 2: 36 (1837). *Cyperus marginatus* var. *blandus* (Kunth) Kük. in H.G.A. Engler (ed.), Pflanzenr., IV, 20(101): 189 (1936).

Cyperus fonticola Kunth, Enum. Pl. 2: 36 (1837).
Cyperus prionodes Steud., Syn. Pl. Glumac. 2: 22 (1854).
Cyperus brunneovaginatus Boeck., Bot. Jahrb. Syst. 11: 407 (1890). *Eucyperus bruneovaginatus* (Boeck.) Rikli, Jahrb. Wiss. Bot. 27: 568 (1895).

Cyperus marlothii Boeck., Bot. Jahrb. Syst. 11: 407 (1890). *Mariscus marlothii* (Boeck.) C.B.Clarke in T.A. Durand & H.Schinz, Consp. Fl. Afric. 5: 590 (1894).
S. Africa. 27 BOT CPP NAM OFS TVL. Hemicr. or rhizome geophyte.
Cyperus marlothii var. *seineri* Kük. in H.G.A.Engler (ed.), Pflanzenr., IV, 20(101): 539 (1936).

Cyperus marojejyensis Bosser, Naturaliste Malgache 7: 120 (1955).
Madagascar. 29 MDG. Hemicr. or rhizome geophyte.

Cyperus marquisensis F.Br., Bernice P. Bishop Mus. Bull. 84: 98 (1931).
Marquesas. 61 MRQ. Hemicr. or rhizome geophyte.

Cyperus matagoroensis Muasya & D.A.Simpson, Kew Bull. 59: 593 (2004).
Tanzania to Zambia. 25 TAN 26 ZAM.

Cyperus matudae G.C.Tucker, Rhodora 88: 503 (1986).
SE. Mexico (Chiapas). 79 MXT. Hemicr. or rhizome geophyte.

Cyperus mauretaniensis Väre & Kukkonen, Ann. Bot. Fenn. 42: 480 (2005).
Mauritania. 22 MTN.

Cyperus medusaeus Chiov., Pl. Nov. Aethiop.: 13 (1928).
Coastal C. & S. Somalia. 24 SOM. Hemicr. or rhizome geophyte.

Cyperus meeboldii Kük., Repert. Spec. Nov. Regni Veg. 18: 345 (1922).
Trop. Africa, S. India. 22 NGR SEN 23 GAB 24 CHA ETH SOM 25 KEN TAN UGA 40 IND. Hemicr.
Cyperus sphaeranthelus Chiov., Miss. Biol. Borana, Racc. Bot., Angiosp.-Gymnosp.: 290 (1939).
Cyperus adamii Raymond, Bull. Soc. Bot. France 106: 145 (1959).

Cyperus megalanthus (Kük.) G.C.Tucker, Syst. Bot. Monogr. 43: 39 (1994).
Mexico to C. America. 79 MXE MXG MXT 80 GUA. Hemicr. or rhizome geophyte.
**Cyperus pseudovegetus* var. *megalanthus* Kük. in H.G.A.Engler (ed.), Pflanzenr., IV, 20(101): 176 (1936).

Cyperus meistostylus S.T.Blake, J. Arnold Arbor. 28: 217 (1947).
New Guinea. 43 NWG. Hemicr. or rhizome geophyte.

Cyperus meridionalis Barros, Anales Mus. Argent. Ci. Nat. "Bernardino Rivadavia" 39: 357 (1938).
S. Venezuela to NE. Argentina. 82 VEN 83 BOL CLM 84 BZS 85 AGE URU. Hemicr. or rhizome geophyte.

Cyperus meyenianus Kunth, Enum. Pl. 2: 88 (1837). *Mariscus meyenianus* (Kunth) Nees in C.F.P.von Martius & auct. suc. (eds.), Fl. Bras. 2(1): 49 (1842).
Windward Is. to N. Argentina. (63) haw 81 WIN 82 VEN 83 ECU PER 84 BZC BZE BZL BZS 85 AGE AGW PAR URU. Hemicr. or rhizome geophyte.
Mariscus palustris Schrad. in J.J.Roemer & J.A. Schultes, Mant. 2: 141 (1824). *Cyperus palustris* (Schrad.) Kük. in H.G.A.Engler (ed.), Pflanzenr., IV, 20(101): 421 (1936), nom. illeg.
**Cyperus subulatus* Meyen, Observ. Bot. 1: 108 (1834), nom. illeg. *Mariscus subulatus* Nees, Linnaea 9: 286 (1834).
Mariscus oligostachys Schrad. ex Nees in C.F.P.von Martius & auct. suc. (eds.), Fl. Bras. 2(1): 48 (1842). *Cyperus meyenianus* var. *oligostachys* (Schrad. ex Nees) Kük. in H.G.A.Engler (ed.), Pflanzenr., IV, 20(101): 491 (1936).
Mariscus uliginosus Schrad. ex Nees in C.F.P.von Martius & auct. suc. (eds.), Fl. Bras. 2(1): 46 (1842).
Cyperus kunthianus Nees, Nova Acta Acad. Caes. Leop. -Carol. German. Nat. Cur. 19(Suppl. 1): 69 (1843).
Cyperus trispicatus Boeck., Flora 65: 14 (1882).
Cyperus meyenianus var. *breviradiatus* Maury, J. Bot. (Morot) 2: 392 (1888).
Cyperus prionensis Boeck., Beitr. Cyper. 2: 10 (1890).
Chlorocyperus paranaguensis Palla, Denkschr. Kaiserl. Akad. Wiss., Wien. Math.-Naturwiss. Kl. 79: 181 (1908). *Cyperus meyenianus* var. *paranaguensis* (Palla) Kük. in H.G.A.Engler (ed.), Pflanzenr., IV, 20(101): 491 (1936).
Cyperus meyenianus var. *patentissimus* Kük. in H.G.A. Engler (ed.), Pflanzenr., IV, 20(101): 491 (1936).

Cyperus meyerianus Kunth, Enum. Pl. 2: 49 (1837).
Cape Prov. 27 CPP. Hemicr. or rhizome geophyte.

Cyperus michelianus (L.) Delile, Descr. Egypte, Hist. Nat. 3: 50 (1813).
Old World. 11 AUT GER HUN POL SWI† 12 COR FRA POR 13 BUL GRC ITA KRI ROM YUG 14 BLR RUC RUS UKR 20 ALG EGY MOR 21 CNY 22 GHA MLI NGR SEN 23 GAB 24 ETH SOM SUD 25 KEN TAN 27 NAM TVL 29 MAU MDG 30 ALT 31 AMU 32 KAZ 33 AFG TCS 34 AFG IRQ LBS PAL TUR 35 SAU 36 CHM CHN CHS 38 JAP KOR TAI 40 ASS EHM IND NEP PAK SRL WHM 41 MYA THA VIE 42 JAW PHI SUL 43 NWG 50 NSW NTA QLD SOA VIC WAU. Ther.
**Scirpus michelianus* L., Sp. Pl.: 52 (1753). *Isolepis micheliana* (L.) Roem. & Schult., Syst. Veg. 2: 114 (1817). *Fimbristylis micheliana* (L.) Rchb., Fl. Germ. Excurs.: 75 (1830). *Dichostylis micheliana* (L.) Nees in R.Wight, Contr. Bot. India: 94 (1834). *Cyperus pygmaeus* var. *michelianus* (L.) Boeck., Linnaea 35: 494 (1868). *Juncellus michelianus* (L.) Blatt. & McCann, J. Bombay Nat. Hist. Soc. 37: 33 (1934).

subsp. ***michelianus***
Europe to Himalaya, Africa. 11 AUT GER HUN POL SWI† 12 COR FRA POR 13 BUL GRC ITA KRI ROM YUG 14 BLR RUC RUS UKR 20 ALG EGY MOR 21 CNY 22 GHA MLI SEN 23 GAB 24 ETH 27 TVL 30 ALT 31 AMU 32 KAZ 33 AFG TCS 34 IRQ TUR 36 CHM CHN CHS 38 JAP KOR 40 ASS EHM IND NEP PAK WHM 41 MYA MYA THA VIE VIE 42 JAW PHI SUL 43 NWG 50 NSW NTA QLD SOA VIC WAU. Ther.
Cyperus italicus Rottb., Descr. Pl. Rar.: 17 (1772).
Cyperus minutulus Vahl, Enum. Pl. 2: 314 (1805).
Cyperus diffusus Roxb., Fl. Ind. 1: 193 (1820), nom. illeg.
Cyperus mulen-pullu Schult. in J.J.Roemer & J.A.Schultes, Mant. 2: 99 (1824).
Cyperus musarius Buch.-Ham. ex Wall., Numer. List: 3325 (1831), nom. inval.
Cyperus pugioniformis A.Dietr., Sp. Pl. 2: 230 (1832).
Schoenus capitatus Dubois, Méth. Éprouv., ed. 2: 247 (1833), nom. illeg.

Cyperus hyalinolepis Steud., Syn. Pl. Glumac. 2: 316 (1855).
Cyperus hilgendorfianus Boeck., Bot. Jahrb. Syst. 5: 501 (1884). *Cyperus michelianus* f. *hilgendorfianus* (Boeck.) Kük. in H.G.A.Engler (ed.), Pflanzenr., IV, 20(101): 315 (1936).
Scirpus michelianus f. *conglomeratus* Beck, Fl. Nieder-Österreich: 124 (1890). *Dichostylis micheliana* f. *conglomerata* (Beck) Soó, Acta Bot. Acad. Sci. Hung. 16: 369 (1970 publ. 1971).
Scirpus michelii Bubani, Fl. Pyren. 4: 204 (1902).
Cyperus michelianus f. *elatus* Kük. in H.G.A. Engler (ed.), Pflanzenr., IV, 20(101): 312 (1936). *Dichostylis micheliana* f. *elata* (Kük.) Soó, Acta Bot. Acad. Sci. Hung. 16: 369 (1970 publ. 1971).
Cyperus michelianus f. *exiguus* Kük. in H.G.A. Engler (ed.), Pflanzenr., IV, 20(101): 315 (1936).
Cyperus extremiorientalis Ohwi, Mem. Coll. Sci. Kyoto Imp. Univ., Ser. B, Biol. 18(1): 150 (1944).
Dichostylis wolgensis A.O.Tarassov, Pochvy Rast. Yugo-Vostoka 1970: 93 (1970).

subsp. ***pygmaeus*** (Rottb.) Asch. & Graebn., Syn. Mitteleur. Fl. 2(2): 273 (1904).
Trop. & Subtrop. Old World. 13 GRC 20 ALG EGY 22 GHA NGR 24 ETH SOM SUD 25 KEN TAN 27 NAM 29 MAU MDG 31 AMU 33 TCS 34 AFG IRQ LBS TUR 35 SAU 36 CHM CHS 38 JAP KOR TAI 40 IND PAK SRL WHM 41 MYA THA VIE 42 JAW PHI SUL 43 NWG 50 NSW NTA QLD SOA VIC WAU. Ther.
**Cyperus pygmaeus* Rottb., Descr. Icon. Rar. Pl.: 20 (1773). *Dichostylis pygmaea* (Rottb.) Nees, Linnaea 9: 289 (1834). *Pycreus pygmaeus* (Rottb.) Nees, Linnaea 9: 283 (1834). *Juncellus pygmaeus* (Rottb.) C.B.Clarke in J.D.Hooker, Fl. Brit. India 6: 596 (1893).

Cyperus michoacanensis Britton ex C.B.Clarke, Kew Bull., Addit. Ser. 8: 5 (1908).
SW. Mexico (Sierra de Manantlán). 79 MXS. Hemicr. or rhizome geophyte.

Cyperus micrantherus Cherm., Bull. Mus. Natl. Hist. Nat. 27: 554 (1921).
C. Madagascar. 29 MDG. Hemicr. or rhizome geophyte.

Cyperus microbolbos C.B.Clarke in D.Oliver & auct. suc. (eds.), Fl. Trop. Afr. 8: 354 (1901). *Mariscus microbolbos* (C.B.Clarke) Vorster, Kew Bull. 42: 200 (1987).
NE. Sudan to Eritrea. 24 ERI SUD. Hemicr. or rhizome geophyte.

Cyperus microbrunneus G.C.Tucker, Bull. Torrey Bot. Club 110: 163 (1983).
S. Mexico to C. America. 79 MXS MXT 80 COS GUA HON. Hemicr. or rhizome geophyte.

Cyperus microcephalus R.Br., Prodr.: 215 (1810). *Cyperus sporobolus* var. *microcephalus* (R.Br.) Benth., Fl. Austral. 7: 282 (1878).
N. Australia. 50 NTA QLD WAU. Hemicr. or rhizome geophyte.

subsp. ***chersophilus*** K.L.Wilson, Telopea 4: 427 (1991).
N. Australia. 50 NTA WAU. Hemicr. or rhizome geophyte.

subsp. ***microcephalus***
N. Australia. 50 NTA QLD WAU. Hemicr. or rhizome geophyte.

subsp. ***saxicola*** K.L.Wilson, Telopea 4: 429 (1991).
N. Australia. 50 NTA QLD WAU. Hemicr. or rhizome geophyte.

Cyperus microglumis D.A.Simpson, Kew Bull. 45: 492 (1990).
C. Somalia. 24 SOM. Hemicr. or tuber geophyte.

Cyperus microiria Steud., Syn. Pl. Glumac. 2: 23 (1854).
Japan to Himalaya. 36 CHC CHM CHN CHS 38 JAP KOR 40 ASS IND 41 THA VIE (75) cnt mas nwj nwy ohi pen (78) kty. Hemicr. or rhizome geophyte.
Cyperus parviflorus Nees in R.Wight, Contr. Bot. India: 87 (1834), nom. illeg. *Cyperus iria* var. *parviflorus* (Nees) Miq., Fl. Ned. Ind. 3: 270 (1856).
Cyperus textori Miq., Ann. Mus. Bot. Lugduno-Batavi 2: 141 (1866). *Cyperus amuricus* var. *textori* (Miq.) Kük., Sinensia 3(3): 80 (1933).
Cyperus iria var. *rectangularis* Kük. in H.G.A.Engler (ed.), Pflanzenr., IV, 20(101): 152 (1935). *Cyperus rectangularis* (Kük.) Bennet, Indian Forester 95: 692 (196Ï9).

Cyperus micromariscus Lye, Nordic J. Bot. 3: 216 (1983).
NE. Tanzania. 25 TAN. Ther.

Cyperus micromedusaeus Lye, Nordic J. Bot. 16: 373 (1996).
N. Somalia. 24 SOM. Ther. or hemicr.

Cyperus micropelophilus Lye, Willdenowia 26: 233 (1996).
S. Somalia. 24 SOM. Ther.

Cyperus microumbellatus Lye, Nordic J. Bot. 3: 223 (1983).
Kenya (Shimba Hills). 25 KEN. Hemicr. or rhizome geophyte.

Cyperus miliifolius Poepp. & Kunth, Enum. Pl. 2: 29 (1837).
C. & S. Trop. America. 80 COS PAN 82 FRG GUY SUR VEN 83 BOL CLM ECU PER 84 BZE BZN. Hemicr. or rhizome geophyte.
Cyperus lacustris Schrad. ex Nees in C.F.P.von Martius & auct. suc. (eds.), Fl. Bras. 2(1): 1 (1842).
Cyperus tenuispicatus Boeck., Linnaea 35: 525 (1868).
Cyperus macrolepis Boeck., Allg. Bot. Z. Syst. 1: 226 (1895).
Mariscus saturatus Donn.Sm., Contr. U. S. Natl. Herb. 10: 453 (1905). *Cyperus saturatus* (Donn.Sm.) C.B.Clarke, Bot. Jahrb. Syst. 37: 517 (1906). *Cyperus miliifolius* var. *saturatus* (Donn.Sm.) Kük. in H.G.A. Engler (ed.), Pflanzenr., IV, 20(101): 221 (1936).
Cyperus miliifolius var. *rotheryi* A.Gross, Arq. Mus. Nac. Rio de Janeiro 34: 52 (1932).

Cyperus mirus C.B.Clarke, Kew Bull., Addit. Ser. 8: 4 (1908).
E. Australia. 50 NSW QLD. Hemicr. or rhizome geophyte.

Cyperus mitis Steud., Syn. Pl. Glumac. 2: 316 (1855).
W. Indian Ocean, India to Indo-China. 29 MAU MDG 40 IND SRL 41 MYA THA. Hemicr. or rhizome geophyte.
Cyperus tenuiflorus Roxb., Fl. Ind. 1: 199 (1820), nom. illeg.
Cyperus longus var. *gracilis* Boeck., Linnaea 36: 282 (1870).
Cyperus stenostachyus var. *indica* C.B.Clarke in J.D.Hooker, Fl. Brit. India 6: 614 (1893).
Cyperus rotundus var. *pseudotuberosus* Cherm., Bull. Soc. Bot. France 74: 607 (1927).

Cyperus mogadoxensis Chiov., Atti Reale Accad. Ital., Mem. Cl. Sci. Fis. 11: 60 (1940).
Coastal C. & S. Somalia. 24 SOM. Hemicr. or tuber geophyte.

Cyperus molliglumis Cherm., Bull. Soc. Bot. France 72: 615 (1925).
C. Madagascar. 29 MDG. Hemicr. or rhizome geophyte.

Cyperus mollipes (C.B.Clarke) K.Schum. in H.G.A.Engler, Pflanzenw. Ost-Afrikas, C: 122 (1895).
Trop. Africa, India to Thailand. 23 ZAI 24 ETH SOM SUD 25 TAN 26 MLW ZAM 40 IND 41 MYA THA. Hemicr. or rhizome geophyte.
Kyllinga bulbocaulis Boeck., Flora 58: 258 (1875). *Rhynchospora bulbocaulis* (Boeck.) Boeck., Flora 62: 567 (1879). **Mariscus mollipes* C.B.Clarke in T.A.Durand & H.Schinz, Consp. Fl. Afric. 5: 590 (1894). *Cyperus mollipes* var. *bulbocaulis* (Boeck.) Kük. in H.G.A.Engler (ed.), Pflanzenr., IV, 20(101): 557 (1936).
Ascopholis gamblei C.E.C.Fisch., Bull. Misc. Inform. Kew 1931: 105 (1931).

Cyperus moutona F.Br., Bernice P. Bishop Mus. Bull. 84: 99 (1931).
Marquesas. 61 MRQ. Hemicr. or rhizome geophyte.

Cyperus mudugensis D.A.Simpson, Kew Bull. 49: 365 (1994).
EC. Somalia. 24 SOM. Hemicr. or rhizome geophyte.

Cyperus multifolius Kunth, Enum. Pl. 2: 91 (1837). *Mariscus multifolius* (Kunth) T.Koyama, Phytologia 29: 74 (1974), nom. illeg.
Panama to Ecuador. 80 PAN 83 ECU. Hemicr. or rhizome geophyte.
**Mariscus polyphyllus* Kunth in F.W.H.von Humboldt, A.J.A.Bonpland & C.S.Kunth, Nov. Gen. Sp. 1: 217 (1816).

Cyperus multinervatus Bosser, Naturaliste Malgache 7: 119 (1955).
Madagascar. 29 MDG. Hemicr. or rhizome geophyte.

Cyperus multispicatus Boeck., Linnaea 38: 362 (1874).
Assam to W. Jawa. 40 ASS 41 MYA VIE 42 JAW. Ther.

Cyperus mundulus Kunth, Enum. Pl. 2: 74 (1837).
SE. & S. Brazil. 84 BZL BZS. Hemicr. or rhizome geophyte.
Cyperus glaziovianus Boeck., Vidensk. Meddel. Naturhist. Foren. Kjøbenhavn 1869: 128 (1869).

Cyperus mutisii (Kunth) Andersson, Galapagos Veg.: 53 (1854).
S. Arizona to Trop. America. 76 ARI 79 MXC MXE MXG MXN MXS MXT 80 BLZ COS GUA HON NIC PAN 81 DOM HAI JAM PUE TRT WIN 82 FRG GUY VEN 83 BOL CLM ECU PER. Hemicr. or rhizome geophyte.
**Mariscus mutisii* Kunth in F.W.H.von Humboldt, A.J.A.Bonpland & C.S.Kunth, Nov. Gen. Sp. 1: 216 (1816).
Mariscus karwinskianus Kunth, Enum. Pl. 2: 123 (1837).
Mariscus poeppigianus Kunth, Enum. Pl. 2: 123 (1837).
Mariscus micans Nees in C.F.P.von Martius & auct. suc. (eds.), Fl. Bras. 2(1): 50 (1842).
Mariscus asper Liebm., Mexic. Neldeagt. Pl., V, 2: 229 (1851). *Cyperus mutisii* var. *asper* (Liebm.) Kük. in H.G.A.Engler (ed.), Pflanzenr., IV, 20(101): 486 (1936). *Cyperus asper* (Liebm.) O'Neill in I.Tidestrom & T.Kittell, Fl. Ariz. New Mex.: 767 (1941).
Mariscus mexicanus Steud., Syn. Pl. Glumac. 2: 64 (1854).
Mariscus oaxacensis Steud., Syn. Pl. Glumac. 2: 65 (1854).
Cyperus ehrenbergianus Boeck., Linnaea 36: 391 (1870). *Mariscus ehrenbergianus* (Boeck.) C.B. Clarke in I.Urban, Symb. Antill. 5: 290 (1907). *Cyperus mutisii* var. *ehrenbergianus* (Boeck.) Kük. in H.G.A.Engler (ed.), Pflanzenr., IV, 20(101): 485 (1936).
Cyperus ochreatus Boeck., Linnaea 36: 386 (1870).
Cyperus semitribrachiatus Boeck., Linnaea 36: 386 (1870). *Cyperus mutisii* var. *semitribrachiatus* (Boeck.) Kük. in H.G.A.Engler (ed.), Pflanzenr., IV, 20(101): 485 (1936).
Cyperus compressotriqueter Boeck., Beitr. Cyper. 1: 10 (1888).
Cyperus martinicensis Boeck., Beitr. Cyper. 2: 38 (1890). *Cyperus mutisii* var. *martinicensis* (Boeck.) Kük. in H.G.A.Engler (ed.), Pflanzenr., IV, 20(101): 485 (1936).
Cyperus mutisii var. *contractus* Kük. in H.G.A.Engler (ed.), Pflanzenr., IV, 20(101): 486 (1936).
Cyperus mutisii var. *fuscisquamatus* Kük. in H.G.A. Engler (ed.), Pflanzenr., IV, 20(101): 486 (1936).

Cyperus mwinilungensis Podlech, Mitt. Bot. Staatssamml. München 4: 109 (1961).
Zambia to Botswana. 26 ZAM 27 BOT. Hemicr. or rhizome geophyte.

Cyperus myrmecias Ridl., Trans. Linn. Soc. London, Bot. 2: 144 (1884). *Mariscus myrmecias* (Ridl.) C.B.Clarke in T.A.Durand & H.Schinz, Consp. Fl. Afric. 5: 590 (1894).
Angola. 26 ANG. Hemicr. or rhizome geophyte.

Cyperus nanellus Tang & F.T.Wang, in Fl. Reipubl. Popul. Sin. 11: 232 (1961).
China. 36+. Hemicr. or rhizome geophyte.

Cyperus nanus Willd., Sp. Pl. 1: 272 (1797). *Mariscus nanus* (Willd.) T.Koyama, Phytologia 29: 74 (1974).
Mexico, Caribbean. 79 MXG 81 ARU? CUB DOM HAI JAM LEE NLA PUE. Hemicr.

var. **liebmannii** (Steud.) Kük. in H.G.A.Engler (ed.), Pflanzenr., IV, 20(101): 536 (1936).
Mexico. 79 MXG. Hemicr. or rhizome geophyte.
**Kyllinga liebmannii* Steud., Syn. Pl. Glumac. 2: 67 (1854).

var. **nanus**
Caribbean. 81 ARU? CUB DOM HAI JAM LEE NLA PUE. Hemicr.
Schoenus capillaris Sw., Prodr.: 20 (1788). *Mariscus capillaris* (Sw.) Vahl, Enum. Pl. 2: 372 (1805). *Kyllinga filiformis* var. *capillaris* (Sw.) Griseb., Fl. Brit. W. I.: 568 (1864). *Cyperus tenuis* var. *capillaris* (Sw.) Kük., Repert. Spec. Nov. Regni Veg. 23: 188 (1926).
Cyperus nanus var. *rigidulus* Ekman & Kük. in H.G.A.Engler (ed.), Pflanzenr., IV, 20(101): 536 (1936).

Cyperus natalensis Hochst. ex C.Krauss, Flora 28: 755 (1845).
KwaZulu-Natal to Mozambique. 26 MOZ 27 NAT. Hemicr. or rhizome geophyte.

Cyperus nayaritensis G.C.Tucker, Bull. Torrey Bot. Club 110: 161 (1983).
SW. Mexico. 79 MXS. Hemicr. or rhizome geophyte.

Cyperus nduru Cherm., Arch. Bot. Mém. 4(7): 18 (1931). *Cyperus margaritaceus* var. *nduru* (Cherm.) Kük., Bot. Not. 1934: 67 (1934).
Trop. Africa. 22 GHA GUI NGA SIE 23 CMN CON ZAI 25 TAN 26 MLW MOZ ZAM ZIM. Hemicr. or rhizome geophyte.

Cyperus nemoralis Cherm., Bull. Mus. Natl. Hist. Nat. 27: 553 (1921).
C. Madagascar. 29 MDG. Hemicr. or rhizome geophyte.
Cyperus onivensis Cherm., Bull. Soc. Bot. France 72: 614 (1925). *Cyperus nemoralis* var. *onivensis* (Cherm.) Kük. in H.G.A.Engler (ed.), Pflanzenr., IV, 20(101): 200 (1936).

Cyperus neoguinensis Kük., Bot. Jahrb. Syst. 59: 43 (1924).
New Guinea. 43 NWG. Hemicr. or rhizome geophyte.
**Cyperus platyphyllos* Ridl., Trans. Linn. Soc. London, Bot. 9: 242 (1916), nom. illeg.
Cyperus rigidulus Ridl., Trans. Linn. Soc. London, Bot. 9: 211 (1916), nom. illeg. *Cyperus subrigidulus* Kük. in H.G.A.Engler (ed.), Pflanzenr., IV, 20(101): 230 (1936).

Cyperus neokunthianus Kük. in H.G.A.Engler (ed.), Pflanzenr., IV, 20(101): 427 (1936).
Hawaiian Is. (W. Maui). 63 HAW. Hemicr. or rhizome geophyte.
**Mariscus kunthianus* Gaudich., Voy. Uranie: 415 (1829). *Cyperus kunthianus* (Gaudich.) Hillebr., Fl. Hawaiian Isl.: 468 (1888), nom. illeg.

Cyperus nervosostriatus Turrill, Bull. Misc. Inform. Kew 1925: 68 (1925). *Cyperus esculentus* var. *nervosostriatus* (Turrill) Kük. in H.G.A.Engler (ed.), Pflanzenr., IV, 20(101): 119 (1935).
Northern Prov. 27 TVL. Hemicr. or rhizome geophyte.

Cyperus nigrofuscus T.L.Dai, in Fl. Reipubl. Popul. Sin. 11: 232 (1961).
China. 36+. Hemicr. or rhizome geophyte.

Cyperus niigatensis Ohwi, Repert. Spec. Nov. Regni Veg. 36: 43 (1934).
Japan (Honshu). 38 JAP. Ther.

Cyperus nipponicus Franch. & Sav., Enum. Pl. Jap. 2: 537 (1878). *Juncellus nipponicus* (Franch. & Sav.) C.B.Clarke, Kew Bull., Addit. Ser. 8: 3 (1908). *Dichostylis nipponica* (Franch. & Sav.) Palla, Monde Pl. 12: 40 (1910). *Cyperus michelianus* var. *nipponicus* (Franch. & Sav.) Kük. in H.G.A.Engler (ed.), Pflanzenr., IV, 20(101): 314 (1936).
Temp. Asia. 31 PRM 36 CHM CHN 38 JAP KOR. Ther.
Scirpus stauntonii C.B.Clarke, J. Linn. Soc., Bot. 36: 253 (1903). *Cyperus stauntonii* (C.B.Clarke) Ohwi, Repert. Spec. Nov. Regni Veg. 36: 43 (1934).

Cyperus niveoides C.B.Clarke in D.Oliver & auct. suc. (eds.), Fl. Trop. Afr. 8: 319 (1901).
WC. Trop. Africa. 23 CMN ZAI. Hemicr. or rhizome geophyte.

Cyperus niveus Retz., Observ. Bot. 5: 12 (1788).
Trop. & S. Africa to SC. China. 22 BEN BKN GHA IVO NGA SEN TOG 23 BUR CAF CMN GAB RWA ZAI 24 ETH SOM 25 KEN TAN UGA 27 BOT CPP LES NAM NAT OFS SWZ TVL 29 ALD 34 AFG IRN 35 SAU YEM 36 CHC 40 ASS EHM NEP PAK WHM 41 MYA THA VIE. Hemicr. or rhizome geophyte.

var. ***flavissimus*** (Schrad.) Lye in R.W.Haines & K.A.Lye, Sedges & Rushes E. Afr.: 257 (1983).
Somalia to S. Africa. 24 SOM 25 TAN 27+. Hemicr. or rhizome geophyte.
**Cyperus flavissimus* Schrad., Gött. Gel. Anz. 3: 2067 (1821). *Cyperus obtusiflorus* var. *flavissimus* (Schrad.) Boeck., Linnaea 35: 529 (1868).

var. ***ledermannii*** (Kük.) Lye, Nordic J. Bot. 3: 231 (1983).
W. Trop. Africa to Tanzania. 22 BEN NGA 23 CMN 25 TAN. Hemicr. or rhizome geophyte.
**Cyperus obtusiflorus* var. *ledermannii* Kük. in H.G.A.Engler (ed.), Pflanzenr., IV, 20(101): 287 (1936). *Cyperus ledermannii* (Kük.) S.S.Hooper, Kew Bull. 26: 578 (1972).

var. ***leucocephalus*** (Kunth) Fosberg, Kew Bull. 31: 835 (1977).
Trop. & S. Africa, Aldabra, Arabian Pen. 23 BUR GAB RWA ZAI 24 ETH SOM 25 KEN TAN UGA 27 BOT CPP LES NAM NAT OFS SWZ TVL 29 ALD 35 SAU YEM. Hemicr. or rhizome geophyte.
Cyperus compactus Lam., Tabl. Encycl. 1: 144 (1791), nom. illeg.
Cyperus obtusiflorus Vahl, Enum. Pl. 2: 308 (1805).
Cyperus rigidus Vahl, Enum. Pl. 2: 309 (1805). *Cyperus obtusiflorus* var. *rigidus* (Vahl) Kük. in H.G.A.Engler (ed.), Pflanzenr., IV, 20(101): 286 (1936).
Cyperus serrulatus Vahl, Enum. Pl. 2: 300 (1805).
Cyperus sphaerocephalus Vahl, Enum. Pl. 2: 310 (1805).
Cyperus striatulus Vahl, Enum. Pl. 2: 311 (1805).
**Cyperus sphaerocephalus* var. *leucocephalus* Kunth, Enum. Pl. 2: 45 (1837).
Cyperus lacteus Steud., Syn. Pl. Glumac. 2: 27 (1854).
Cyperus ambongensis Boeck., Linnaea 38: 361 (1874).
Cyperus argenteus Ridl., Trans. Linn. Soc. London, Bot. 2: 133 (1884).
Cyperus compactus var. *tenerior* C.B.Clarke in D.Oliver & auct. suc. (eds.), Fl. Trop. Afr. 8: 321 (1901).
Cyperus schlechteri C.B.Clarke, Kew Bull., Addit. Ser. 8: 3 (1908), nom. illeg.
Cyperus somalicus Gand., Bull. Soc. Bot. France 69: 350 (1922).
Cyperus wilmsii Gand., Bull. Soc. Bot. France 69: 350 (1922).
Cyperus obtusiflorus var. *membranaceus* Kük. in H.G.A.Engler (ed.), Pflanzenr., IV, 20(101): 287 (1936).
Cyperus obtusiflorus var. *peramoenus* Kük. in H.G.A. Engler (ed.), Pflanzenr., IV, 20(101): 287 (1936).

var. ***niveus***
S. Iran to SC. China. 25? 34 AFG IRN 35 SAU 36 CHC 40 ASS EHM NEP PAK WHM 41 MYA THA VIE. Hemicr. or rhizome geophyte.

var. ***tisserantii*** (Cherm.) Lye, Nordic J. Bot. 3: 231 (1983).

Trop. Africa. 22 BEN BKN GHA IVO NGA SEN TOG 23 CAF CMN GAB 24 ETH. Hemicr. or rhizome geophyte.
**Cyperus tisserantii* Cherm., Arch. Bot. Mém. 4(7): 18 (1931). *Cyperus margaritaceus* var. *tisserantii* (Cherm.) Kük. in H.G.A.Engler (ed.), Pflanzenr., IV, 20(101): 285 (1936).

Cyperus noeanus Boiss., Fl. Orient. 5: 371 (1882).
Turkey. 34 TUR. Hemicr. or rhizome geophyte.

Cyperus nutans Vahl, Enum. Pl. 2: 363 (1805).
Trop. Africa, Asia to N. Australia. 22 NGR 24 ERI ETH SUD 25 UGA 26 MOZ 34 AFG PAL 35 YEM 36 CHC CHS 38 NNS TAI 40 ASS EHM IND NEP PAK SRL WHM 41 CBD THA VIE 42 JAW LSI PHI SUL SUM 43 NWG 50 NTA QLD WAU. Hemicr. or rhizome geophyte.

var. ***eleusinoides*** (Kunth) Haines, Bot. Bihar Orissa 5: 898 (1924).
Trop. Africa to Australia. 22 NGR 24 ERI ETH SUD 25 UGA 26 MOZ 34 AFG PAL 35 YEM 36 CHC CHS 38 NNS 40 ASS BAN EHM IND NEP PAK SRL WHM 41 CBD THA VIE 42 PHI? 50 NTA QLD WAU. Hemicr. or rhizome geophyte.
Cyperus mysurensis B.Heyne ex Wall., Numer. List: 3346 (1831), nom. inval.
**Cyperus eleusinoides* Kunth, Enum. Pl. 2: 39 (1837). *Cyperus nutans* subsp. *eleusinoides* (Kunth) T.Koyama, Gard. Bull. Singapore 30: 136 (1977).
Cyperus xanthopus Steud., Flora 25: 595 (1842).
Cyperus kotschyanus Fenzl ex Steud., Syn. Pl. Glumac. 2: 36 (1854).
Cyperus infra-apicalis Nees ex Aitch., Cat. Pl. Punjab Sindh: 155 (1869).
Cyperus indicus Boeck., Beitr. Cyper. 1: 6 (1888), nom. illeg.

var. ***nutans***
Trop. & Subtrop. Asia to N. Australia. 36 CHS 40 ASS EHM IND NEP SRL 41 THA VIE 42 JAW LSI PHI SUL SUM 43 NWG 50 NTA. Hemicr. or rhizome geophyte.
Cyperus jacquinii Schrad., Linnaea 11(Litt.): 87 (1837). *Chlorocyperus jacquinii* (Schrad.) Rikli, Jahrb. Wiss. Bot. 27: 563 (1895).
Cyperus calopterus Miq., Fl. Ned. Ind. 3: 282 (1856).

var. ***subprolixus*** (Kük.) Karth., in Fl. Ind. Enum., Monocot.: 46 (1989).
Himalaya, Temp. E. Asia, Malesia. 38 NNS TAI 40 WHM 42 LSI PHI 43 NWG. Hemicr. or rhizome geophyte.
**Cyperus eleusinoides* var. *subprolixus* Kük. in H.G.A.Engler (ed.), Pflanzenr., IV, 20(101): 145 (1935). *Cyperus nutans* subsp. *subprolixus* (Kük.) T.Koyama, in Fl. Taiwan 5: 260 (1978).

Cyperus nyassensis (Podlech) Lye in R.W.Haines & K.A.Lye, Sedges & Rushes E. Afr.: 211 (1983).
Tanzania to Malawi. 25 TAN 26 MLW. Hemicr.
**Mariscus nyassensis* Podlech, Mitt. Bot. Staatssamml. München 4: 114 (1961).

Cyperus nyererei Lye, Nordic J. Bot. 3: 225 (1983).
S. Tanzania. 25 TAN. Hemicr.

Cyperus obbiadensis Chiov., Pl. Nov. Aethiop.: 12 (1928).
C. Somalia. 24 SOM. Hemicr.

Cyperus oblongoincrassatus Kük. in H.G.A.Engler (ed.), Pflanzenr., IV, 20(101): 550 (1936).
E. Trop. Africa. 25 KEN TAN. Hemicr. or rhizome geophyte.
**Mariscus taylorii* C.B.Clarke in D.Oliver & auct. suc. (eds.), Fl. Trop. Afr. 8: 384 (1901).
Cyperus oblongoincrassatus var. *clarior* Kük. in H.G.A.Engler (ed.), Pflanzenr., IV, 20(101): 550 (1936).
Cyperus oblongoincrassatus var. *groteanus* Kük. in H.G.A.Engler (ed.), Pflanzenr., IV, 20(101): 550 (1936). *Mariscus taylorii* var. *groteanus* (Kük.) Napper, J. E. Africa Nat. Hist. Soc. Natl. Mus. 28: 14 (1971).
Cyperus oblongoincrassatus var. *udigensis* Kük. in H.G.A.Engler (ed.), Pflanzenr., IV, 20(101): 550 (1936). *Mariscus taylorii* var. *udigensis* (Kük.) Napper, J. E. Africa Nat. Hist. Soc. Natl. Mus. 28: 14 (1971).

Cyperus obsoletenervosus Peter & Kük. in H.G.A.Engler (ed.), Pflanzenr., IV, 20(101): 548 (1936). *Mariscus obsoletenervosus* (Peter & Kük.) Greenway, J. E. Africa Nat. Hist. Soc. Natl. Mus. 27: 203 (1969).
Tanzania. 25 TAN. Hemicr. or rhizome geophyte.

Cyperus obtusus Willemet, Ann. Bot. (Usteri) 18: 5 (1796).
Mauritius. 29 MAU. Hemicr. or rhizome geophyte.

Cyperus ochraceus Vahl, Enum. Pl. 2: 325 (1805).
Trop. & Subtrop. America. (75) pen 77 TEX 78 FLA GEO LOU MSI 79 MXC MXE MXG MXN MXS MXT 80 BLZ COS GUA HON NIC PAN 81 BAH CUB DOM HAI JAM LEE PUE WIN 82 VEN 83 BOL CLM ECU PER 84 BZE BZL 85 AGE AGW PAR. Hemicr. or rhizome geophyte.
Cyperus bakeri C.B.Clarke, J. Linn. Soc., Bot. 20: 290 (1883).
Cyperus ochraceus var. *excelsior* Kük. in H.G.A.Engler (ed.), Pflanzenr., IV, 20(101): 182 (1936).
Cyperus ochraceus var. *minor* Kük. in H.G.A.Engler (ed.), Pflanzenr., IV, 20(101): 182 (1936).

Cyperus ochroleucus Boeck., Flora 58: 85 (1875).
N. Australia. 50 NTA QLD. Hemicr. or rhizome geophyte.

Cyperus odoratus L., Sp. Pl.: 46 (1753). *Diclidium odoratum* (L.) Schrad. ex Nees in C.F.P.von Martius & auct. suc. (eds.), Fl. Bras. 2(1): 54 (1842). *Torulinium odoratum* (L.) S.S.Hooper, Kew Bull. 26: 579 (1972).
Trop. & Subtrop. to Caucasus. 22 BEN GHA IVO NGA SEN 23 EQG GGI ZAI 26 ANG 29 MDG 33 TCS 34 IRN IRQ 38 JAP KOR KZN NNS OGA TAI 41 MYA THA VIE 42 BOR CKI JAW LSI MLY MOL PHI SUL SUM XMS 43 NWG 50 NSW QLD 60 FIJ GIL (61) sci TUA 62 CRL MRN MRS 72 ONT QUE 73 ORE WAS 74 ILL IOW KAN MIN MSO NDA NEB OKL SDA WIS 75 CNT INI MAI MAS MIC NWH NWJ NWY OHI PEN RHO VER WVA 76 ARI CAL 77 NWM TEX 78 ALA ARK DEL FLA GEO KTY LOU MRY MSI NCA SCA TEN VRG WDC 79 MXC MXE MXG MXN MXS MXT 80 BLZ COS CPI ELS GUA HON NIC PAN 81 ARU BAH BER CAY CUB DOM HAI JAM LEE NLA PUE TRT WIN 82 FRG GUY SUR VEN 83 BOL CLM ECU GAL PER 84 BZC BZE BZL BZN BZS 85 AGE AGW CLC PAR URU. Ther. or hemicr.

subsp. ***odoratus***

Trop. & Subtrop. 22 BEN GHA IVO NGA SEN 23 EQG GGI ZAI 26 ANG 29 MDG 38 JAP KOR KZN NNS OGA TAI 41 MYA THA VIE 42 BOR CKI JAW LSI MLY MOL PHI SUL SUM XMS 43 NWG 50 NSW QLD 60 FIJ GIL (61) sci TUA 62 CRL MRN MRS 72 ONT QUE 73 ORE WAS 74 ILL IOW KAN MIN MSO NDA NEB OKL SDA WIS 75 CNT INI MAI MAS MIC NWH NWJ NWY OHI PEN RHO VER WVA 76 ARI CAL 77 NWM TEX 78 ALA ARK DEL FLA GEO KTY LOU MRY MSI NCA SCA TEN VRG WDC 79 MXC MXE MXG MXN MXS MXT 80 BLZ COS CPI ELS GUA HON NIC PAN 81 ARU BAH BER CAY CUB DOM HAI JAM LEE NLA PUE TRT WIN 82 FRG GUY SUR VEN 83 BOL CLM ECU GAL PER 84 BZC BZE BZL BZN BZS 85 AGE AGW CLC PAR URU. Ther. or hemicr.

Cyperus ferax Rich., Actes Soc. Hist. Nat. Paris 1: 106 (1792). *Torulinium ferax* (Rich.) Ham. in W.Hamilton, Prodr. Pl. Ind. Occid.: 15 (1825). *Diclidium ferox* (Rich.) Schrad. ex Nees in C.F.P.von Martius & auct. suc. (eds.), Fl. Bras. 2(1): 54 (1842), orth. var. *Diclidium ferax* (Rich.) Steud., Syn. Pl. Glumac. 2: 332 (1855). *Mariscus ferax* (Rich.) C.B.Clarke in J.D.Hooker, Fl. Brit. India 6: 624 (1893).

Cyperus ferox Vahl, Enum. Pl. 2: 357 (1805).

Cyperus jubiflorus Rudge, Pl. Guian.: 17 (1805).

Cyperus speciosus Vahl, Enum. Pl. 2: 364 (1805). *Cyperus ferax* subsp. *speciosus* (Vahl) Kük. in H.G.A.Engler (ed.), Pflanzenr., IV, 20(101): 619 (1936).

Cyperus stellatus Rudge, Pl. Guian.: 17 (1805).

Cyperus flavicomus Pursh, Fl. Amer. Sept. 1: 53 (1813), nom. illeg.

Mariscus huarmensis Kunth in F.W.H.von Humboldt, A.J.A.Bonpland & C.S.Kunth, Nov. Gen. Sp. 1: 214 (1816). *Cyperus huarmensis* (Kunth) M.C. Johnst., S. W. Naturalist 11: 124 (1966).

Cyperus conglobatus Link, Jahrb. Gewächsk. 1(3): 87 (1818). *Diclidium conglobatum* (Link) Nees ex B.D.Jacks., Index Kew. 1: 751 (1893), nom. inval. *Cyperus ferax* var. *conglobatus* (Link) Kük. in H.G.A.Engler (ed.), Pflanzenr., IV, 20(101): 617 (1936).

Cyperus densiflorus G.Mey., Prim. Fl. Esseq.: 34 (1818). *Cyperus ferax* var. *densiflorus* (G.Mey.) Kük. in H.G.A.Engler (ed.), Pflanzenr., IV, 20(101): 618 (1936).

Cyperus distans G.Mey., Prim. Fl. Esseq.: 35 (1818), nom. illeg.

Cyperus raphiostachys Link, Jahrb. Gewächsk. 1(3): 87 (1818).

Cyperus erythrorhizus Torr., Fl. N. Middle United States 1: 61 (1823).

Cyperus michauxianus Schult. in J.J.Roemer & J.A.Schultes, Mant. 2: 123 (1824).

Cyperus fastuosus Desv. ex Ham., Prodr. Pl. Ind. Occid.: 12 (1825).

Cyperus poeoides Desv. ex Ham., Prodr. Pl. Ind. Occid.: 12 (1825).

Torulinium confertum Desv. ex Ham., Prodr. Pl. Ind. Occid.: 15 (1825).

Cyperus cephalophorus J.Presl & C.Presl, Reliq. Haenk. 1: 170 (1828).

Cyperus haenkei J.Presl & C.Presl, Reliq. Haenk. 1: 172 (1828).

Cyperus multiceps Hook. & Arn., Bot. Beechey Voy.: 100 (1832), nom. illeg.

Cyperus prescottianus Hook. & Arn., Bot. Beechey Voy.: 100 (1832).

Cyperus hamiltonii Kunth, Enum. Pl. 2: 90 (1837).

Cyperus lomentaceus Nees & Meyen ex Kunth, Enum. Pl. 2: 83 (1837).

Cyperus lucidus Kunth, Enum. Pl. 2: 89 (1837), nom. illeg.

Diclidium aciculare Schrad. ex Nees in C.F.P.von Martius & auct. suc. (eds.), Fl. Bras. 2(1): 55 (1842). *Cyperus acicularis* (Schrad. ex Nees) Steud., Syn. Pl. Glumac. 2: 45 (1854), nom. illeg. *Cyperus ferax* var. *acicularis* (Schrad. ex Nees) Kük. in H.G.A.Engler (ed.), Pflanzenr., IV, 20(101): 619 (1936). *Cyperus odoratus* var. *acicularis* (Schrad. ex Nees) O'Neill in Léon, Fl. Cuba 1: 196 (1946).

Diclidium auriculatum Nees in C.F.P.von Martius & auct. suc. (eds.), Fl. Bras. 2(1): 51, 217 (1842).

Diclidium fuscatum Schrad. ex Nees in C.F.P.von Martius & auct. suc. (eds.), Fl. Bras. 2(1): 52 (1842).

Diclidium lenticulare Schrad. ex Nees in C.F.P.von Martius & auct. suc. (eds.), Fl. Bras. 2(1): 53 (1842).

Diclidium lomentaceum Nees in C.F.P.von Martius & auct. suc. (eds.), Fl. Bras. 2(1): 55 (1842).

Diclidium maximiliani Schrad. ex Nees in C.F.P.von Martius & auct. suc. (eds.), Fl. Bras. 2(1): 52 (1842). *Cyperus ferax* var. *maximiliani* (Schrad. ex Nees) Boeck., Linnaea 36: 400 (1870). *Cyperus maximiliani* (Schrad. ex Nees) Griseb., Abh. Königl. Ges. Wiss. Göttingen 24: 311 (1879).

Diclidium reinwardtii Schrad. ex Nees in C.F.P.von Martius & auct. suc. (eds.), Fl. Bras. 2(1): 51 (1842).

Diclidium uliginosum Schrad. ex Nees in C.F.P.von Martius & auct. suc. (eds.), Fl. Bras. 2(1): 56 (1842).

Mariscus pohlianus Nees in C.F.P.von Martius & auct. suc. (eds.), Fl. Bras. 2(1): 50 (1842). *Cyperus pohlianus* (Nees) Kuntze, Revis. Gen. Pl. 3(2): 334 (1898).

Cyperus cubanus Liebm., Mexic. Halvgr.: 222 (1850).

Cyperus fossarum Liebm., Mexic. Halvgr.: 225 (1850).

Cyperus fragilis Liebm., Mexic. Halvgr.: 226 (1850).

Cyperus granadinus Liebm., Mexic. Halvgr.: 224 (1850).

Cyperus macrocephalus Liebm., Mexic. Halvgr.: 221 (1850). *Torulinium macrocephalum* (Liebm.) C.B.Clarke, Kew Bull., Addit. Ser. 8: 20 (1908).

Cyperus oerstedii Liebm., Mexic. Halvgr.: 224 (1850).

Cyperus rufinus Liebm., Mexic. Halvgr.: 222 (1850).

Cyperus sanctae-crucis Liebm., Mexic. Halvgr.: 35 (1850).

Mariscus rhaphiostachys Liebm., Mexic. Neldeagt. Pl., V, 2: 232 (1851).

Cyperus bracteolatus Steud., Syn. Pl. Glumac. 2: 49 (1854).

Cyperus engelmannii Steud., Syn. Pl. Glumac. 2: 47 (1854). *Cyperus ferax* var. *engelmannii* (Steud.) Kük. ex Osten, Anales Mus. Hist. Nat. Montevideo, II, 3: 159 (1931). *Cyperus ferax*

subsp. *engelmannii* (Steud.) Kük. in H.G.A.Engler (ed.), Pflanzenr., IV, 20(101): 620 (1936). *Cyperus odoratus* var. *engelmannii* (Steud.) R.Carter, S.D.Jones & Wipff, Phytologia 80: 289 (1996).
Cyperus horizontalis Steud., Syn. Pl. Glumac. 2: 49 (1854), nom. illeg.
Cyperus lenticularis Steud., Syn. Pl. Glumac. 2: 48 (1854).
Cyperus longiradiatus Steud., Syn. Pl. Glumac. 2: 37 (1854).
Cyperus pseudostrigosus Steud., Syn. Pl. Glumac. 2: 46 (1854).
Cyperus virginicus Jacq. ex Steud., Syn. Pl. Glumac. 2: 47 (1854).
Cyperus tenuior Engelm. in A.Gray, Manual, ed. 2: 492 (1856).
Cyperus speciosus Torr. in W.H.Emory, Rep. U.S. Mex. Bound. 2(1): 226 (1858), nom. illeg.
Cyperus flexuosus Griseb., Fl. Brit. W. I.: 566 (1864), nom. illeg.
Cyperus phleoides Seem., Fl. Vit.: 319 (1868), nom. illeg.
Cyperus ferruginescens Boeck., Linnaea 36: 396 (1870). *Cyperus speciosus* var. *ferruginescens* (Boeck.) Britton, Mem. Torrey Bot. Club 5: 61 (1894).
Cyperus ignoratus Boeck., Linnaea 36: 395 (1870).
Cyperus nitidulus Boeck., Linnaea 36: 364 (1870).
Cyperus parvus Boeck., Linnaea 36: 397 (1870), nom. illeg. *Cyperus speciosus* var. *parvus* Britton, Bull. Torrey Bot. Club 13: 214 (1886). *Cyperus ferax* f. *parvus* (Boeck.) Kük. in H.G.A. Engler (ed.), Pflanzenr., IV, 20(101): 620 (1936).
Cyperus pennatus Boeck., Linnaea 36: 404 (1870), nom. illeg.
Cyperus vahlii Boeck., Linnaea 36: 398 (1870), nom. illeg.
Cyperus multibracteatus Boeck., Flora 58: 107 (1875).
Cyperus californicus S.Watson in S.Watson & al., Bot. California 2: 216 (1880).
Cyperus novae-hannoverae Boeck., Bot. Jahrb. Syst. 5: 91 (1884). *Cyperus ferax* var. *novae-hannoverae* (Boeck.) Kük., Bot. Jahrb. Syst. 59: 4, 46 (1924). *Cyperus odoratus* var. *novae-hannoverae* (Boeck.) Fosberg & Sachet, Micronesica 20: 168 (1987).
Cyperus oxycarioides Britton, Bull. Torrey Bot. Club 11: 86 (1884).
Cyperus speciosus var. *squarrosus* Britton, Bull. Torrey Bot. Club 13: 214 (1886). *Cyperus ferax* var. *squarrosus* (Britton) Kük. in H.G.A.Engler (ed.), Pflanzenr., IV, 20(101): 620 (1936). *Cyperus odoratus* var. *squarrosus* (Britton) S.D.Jones, Wipff & R.Carter, Phytologia 80: 289 (1996).
Cyperus eggersii Boeck., Beitr. Cyper. 1: 53 (1888). *Torulinium eggersii* (Boeck.) C.B.Clarke in I.Urban, Symb. Antill. 2: 56 (1900). *Torulinium macrocephalum* var. *eggersii* (Boeck.) C.D.Adams, Ann. Missouri Bot. Gard. 78: 254 (1991). *Cyperus macrocephalus* var. *eggersii* (Boeck.) S.D.Jones, Wipff & R.Carter, Phytologia 80: 289 (1996).
Cyperus parvispiculatus Boeck., Beitr. Cyper. 1: 7 (1888).
Cyperus carruthii Alph.Wood in B.B.Smyth, Checkl. Pl. Kansas: 26 (1892).
Cyperus oxycarioides Britton ex Coult., Contr. U. S. Natl. Herb. 2: 465 (1894), nom. illeg.
Cyperus reineckei Boeck., Bot. Jahrb. Syst. 25: 585 (1898).
Torulinium michauxianum C.B.Clarke in I.Urban, Symb. Antill. 2: 56 (1900).
Cyperus longispicatus Norton, Trans. Acad. Sci. St. Louis 12: 37 (1902), nom. illeg.
Cyperus nortonii A.Heller, Muhlenbergia 6: 12 (1910).
Torulinium caucasicum Palla, Vestn. Tiflissk. Bot. Sada 30: 27 (1913).
Cyperus raiateensis J.W.Moore, Bernice P. Bishop Mus. Bull. 102: 19 (1933).
Cyperus eggersii var. *laticeps* Kük. in H.G.A.Engler (ed.), Pflanzenr., IV, 20(101): 623 (1936).
Cyperus ferax var. *bulbiferus* Barros, Anales Mus. Argent. Ci. Nat. "Bernardino Rivadavia" 39: 371 (1938).
Cyperus odoratus var. *attenuatus* Fosberg & Sachet, Micronesica 20: 167 (1987).
Cyperus odoratus var. *curtispiculus* Fosberg & Sachet, Micronesica 20: 167 (1987).
Torulinium macrocephalum var. *davidsei* C.D. Adams, Phytologia 63: 300 (1987).

subsp. ***transcaucasicus*** (Kük.) Kukkonen, Ann. Bot. Fenn. 32: 162 (1995).
SE. Transcaucasus to NW. Iran. 33 TCS 34 IRN IRQ. Hemicr. or rhizome geophyte.
**Cyperus ferax* subsp. *transcaucasicus* Kük. in H.G.A.Engler (ed.), Pflanzenr., IV, 20(101): 620 (1936).

Cyperus ohwii Kük., Repert. Spec. Nov. Regni Veg. 29: 197 (1931).
Trop. & Subtrop. Asia, Queensland. 38 JAP 40 ASS 41 THA VIE 42 JAW 50 QLD. Hemicr. or rhizome geophyte.

Cyperus onerosus M.C.Johnst., S. W. Naturalist 9: 308 (1964).
Texas (Andrews Co., Winkler Co.). 77 TEX. Hemicr. or rhizome geophyte.

Cyperus orgadophilus K.L.Wilson, Telopea 4: 439 (1991).
N. Australia. 50 NTA QLD WAU. Hemicr. or rhizome geophyte.

Cyperus ornatus R.Br., Prodr.: 217 (1810).
N. Australia. 50 NTA. Hemicr. or rhizome geophyte.

Cyperus orthostachyus Franch. & Sav., Enum. Pl. Jap. 2: 539 (1878). *Cyperus truncatus* var. *orthostachyus* (Franch. & Sav.) C.B.Clarke, J. Linn. Soc., Bot. 36: 218 (1903), nom. illeg.
Temp. Asia to Vietnam. 30+ 31 AMU PRM 36 CHC CHN 38 JAP KOR 41 VIE. Hemicr. or rhizome geophyte.
Cyperus truncatus C.A.Mey. ex Turcz. in C.F.von Ledebour, Fl. Ross. 4: 241 (1852), nom. illeg.

Cyperus ossicaulis Lye, Kew Bull. 51: 205 (1996).
C. Somalia. 24 SOM. Hemicr.

Cyperus ovatus Baldwin, Trans. Amer. Philos. Soc., n.s., 2: 168 (1825).
SE. U.S.A., Cuba. 78 ALA FLA GEO MSI NCA SCA 81 CUB. Hemicr. or rhizome geophyte.
Cyperus cylindricus Chapm., Bot. Gaz. 3: 18 (1877), nom. illeg.
Cyperus pollardii Britton ex Small, Fl. S.E. U.S.: 1321 (1903).

Mariscus curtisii C.B.Clarke, Kew Bull., Addit. Ser. 8: 15 (1908). *Cyperus retrorsus* var. *curtisii* (C.B. Clarke) Kük. in H.G.A.Engler (ed.), Pflanzenr., IV, 20(101): 512 (1936).

Mariscus litoreus C.B.Clarke, Kew Bull., Addit. Ser. 8: 19 (1908). *Cyperus litoreus* (C.B.Clarke) Britton ex Small, Fl. Florida Keys: 22 (1913).

Cyperus deeringianus Britton & Small in J.K.Small, Man. S.E. Fl.: 151 (1933). *Cyperus retrorsus* var. *deeringianus* (Britton & Small) Fernald & Griscom, Rhodora 37: 152 (1935).

Cyperus winkleri Britton ex Small, Man. S.E. Fl.: 152, 1503 (1933).

Cyperus owanii Boeck., Flora 61: 29 (1878). *Mariscus owanii* (Boeck.) C.B.Clarke in T.A.Durand & H.Schinz, Consp. Fl. Afric. 5: 590 (1894).

Mozambique to Cape Prov. 26 MOZ 27 BOT CPP NAT OFS TVL (76) cal. Hemicr. or rhizome geophyte.

Mariscus umbilensis C.B.Clarke ex W.Watson, Gard. Chron. 1891(2): 190 (1891). *Cyperus umbilensis* (C.B.Clarke ex W.Watson) Boeck. ex Kuntze, Revis. Gen. Pl. 3(2): 334 (1898).

Mariscus bolusii C.B.Clarke in T.A.Durand & H. Schinz, Consp. Fl. Afric. 5: 585 (1894), nom. inval.

Cyperus owanii var. *rogersii* Kük. in H.G.A.Engler (ed.), Pflanzenr., IV, 20(101): 432 (1936).

Cyperus oxycarpus S.T.Blake, Proc. Roy. Soc. Queensland 51: 46 (1939 publ. 1940).

N. Australia. 50 NTA QLD WAU. Hemicr. or rhizome geophyte.

Cyperus oxylepis Nees ex Steud., Syn. Pl. Glumac. 2: 25 (1853).

Caribbean to S. Trop. America. (77) tex (78) ark fla geo lou msi sca 79 MXN MXS MXT (80) gua nic 81 ARU JAM LEE NLA 82 GUY 83 CLM ECU PER 85 AGE AGW PAR. Hemicr.

Cyperus jenmanii C.B.Clarke ex Morong, Ann. New York Acad. Sci 7: 252 (1893).

Cyperus oxylepis var. *punensis* Kük. in H.G.A.Engler (ed.), Pflanzenr., IV, 20(101): 167 (1936).

Cyperus pachycephalus J.Kern, Reinwardtia 2: 119 (1952).

New Guinea. 43 NWG. Ther.

Cyperus pacificus (Ohwi) Ohwi, Mem. Coll. Sci. Kyoto Imp. Univ., Ser. B, Biol. 18(1): 150 (1944).

Korea, N. & C. Japan. 38 JAP KOR. Ther.

**Cyperus michelianus* var. *pacificus* Ohwi, Acta Phytotax. Geobot. 7: 137 (1933). *Dichostylis pacificus* (Ohwi) Nakai, Bull. Natl. Sci. Mus. Tokyo 31: 132 (1952).

Cyperus palianparaiensis Govind., Proc. Indian Acad. Sci., Pl. Sci. 100: 409 (1990).

India. 40 IND. Hemicr. or rhizome geophyte.

Cyperus pallidicolor (Kük.) G.C.Tucker, in Fl. Novo-Galiciana 13: 324 (1993).

Arizona to Peru. 76 ARI 77 NWM 79 MXC MXE MXG MXN MXS MXT 80 GUA 83 PER. Hemicr. or rhizome geophyte.

Mariscus schaffnerianus Schltdl., Linnaea 26: 379 (1853), provisional synonym.

**Cyperus subambiguus* var. *pallidicolor* Kük. in H.G.A. Engler (ed.), Pflanzenr., IV, 20(101): 528 (1936).

Cyperus panamensis (C.B.Clarke) Britton ex Standl., J. Wash. Acad. Sci. 15: 457 (1925).

S. Mexico to Paraguay. 79 MXN MXS MXT 80 COS ELS GUA HON NIC PAN 83 ECU PER 85 PAR. Hemicr. or rhizome geophyte.

**Mariscus panamensis* C.B.Clarke, Kew Bull., Addit. Ser. 8: 15 (1908).

Mariscus panamensis var. *deformis* C.B.Clarke, Kew Bull., Addit. Ser. 8: 15 (1908). *Cyperus panamensis* var. *deformis* (C.B.Clarke) Kük. in H.G.A.Engler (ed.), Pflanzenr., IV, 20(101): 498 (1936).

Cyperus panamensis var. *ramigerus* Kük. in H.G.A. Engler (ed.), Pflanzenr., IV, 20(101): 498 (1936).

Cyperus pandanophyllum C.B.Clarke, Kew Bull., Addit. Ser. 8: 8 (1908).

E. Madagascar. 29 MDG. Hemicr. or rhizome geophyte.

Cyperus pangorei Rottb., Descr. Icon. Rar. Pl.: 31 (1773).

Pakistan to China. (29) mau 36 CHC 40 ASS EHM NEP PAK SRL 41 MYA (42) sul. Hemicr. or rhizome geophyte.

Cyperus tegetum Roxb., Fl. Ind. 1: 211 (1820).

Papyrus corymbosus Nees in R.Wight, Contr. Bot. India: 89 (1834).

Papyrus dehiscens Nees in R.Wight, Contr. Bot. India: 89 (1834).

Cyperus dehiscens Kunth, Enum. Pl. 2: 56 (1837).

Cyperus pangorei var. *ambiguus* C.B.Clarke, J. Linn. Soc., Bot. 21: 162 (1884).

Cyperus paniceus (Rottb.) Boeck., Linnaea 36: 381 (1870).

Seychelles, India to Vietnam, Jawa. 29 SEY 40 ASS BAN EHM IND NEP SRL 41 THA VIE 42 JAW. Hemicr. or rhizome geophyte.

**Kyllinga panicea* Rottb., Descr. Icon. Rar. Pl.: 15 (1773). *Mariscus paniceus* (Rottb.) Vahl, Enum. Pl. 2: 373 (1805).

Kyllinga punicea J.F.Gmel., Syst. Nat. 2: 122 (1791).

Mariscus wallichianus Kunth, Enum. Pl. 2: 117 (1837).

Mariscus pullu Steud., Syn. Pl. Glumac. 2: 66 (1854).

Cyperus umbellatus Thwaites, Enum. Pl. Zeyl.: 345 (1864), nom. illeg.

Cyperus cariciformis F.Muell., Fragm. 5: 92 (1865), nom. illeg.

Mariscus cariciformis F.Muell., Fragm. 5: 92 (1865).

Mariscus paniceus var. *roxburghianus* C.B.Clarke in J.D.Hooker, Fl. Brit. India 6: 621 (1893). *Mariscus roxburghianus* (C.B.Clarke) Alston in H.Trimen, Handb. Fl. Ceylon 6(Suppl.): 309 (1931). *Cyperus paniceus* var. *roxburghianus* (C.B.Clarke) Kük. in H.G.A.Engler (ed.), Pflanzenr., IV, 20(101): 526 (1936).

Mariscus paniceus var. *queenslandicus* Domin, Biblioth. Bot. 85: 435 (1915). *Cyperus paniceus* var. *queenslandicus* (Domin) Kük. in H.G.A.Engler (ed.), Pflanzenr., IV, 20(101): 526 (1936).

Cyperus pannonicus Jacq., Fl. Austriac. 5: 29 (1778). *Pycreus pannonicus* (Jacq.) P.Beauv. ex Rchb., Fl. Germ. Excurs.: 72 (1830). *Chlorocyperus pannonicus* (Jacq.) Rikli, Jahrb. Wiss. Bot. 27: 563 (1895). *Acorellus pannonicus* (Jacq.) Palla in W.D.J.Koch, Syn. Deut. Schweiz. Fl., ed. 3, 2: 2557 (1905). *Juncellus pannonicus* (Jacq.) C.B.Clarke, Kew Bull., Addit. Ser. 8: 3 (1908).

C. Europe to China. 11 AUT CZE HUN 13 ALB BUL GRC ROM YUG 14 KRY RUS UKR 30 ALT 32 KAZ KGZ TKM UZB 33 TCS 34 TUR 36 CHN 37

MON. Ther. or hemicr.
Cyperus tataricus Less., Linnaea 9: 211 (1834).

Cyperus paolii Chiov., Ann. Bot. (Rome) 13: 373 (1915). *Cyperus mollipes* var. *paolii* (Chiov.) Kük. in H.G.A. Engler (ed.), Pflanzenr., IV, 20(101): 558 (1936).
S. Somalia. 24 SOM. Hemicr.

Cyperus papyrus L., Sp. Pl.: 47 (1753). *Chlorocyperus papyrus* (L.) Rikli, Jahrb. Wiss. Bot. 27: 563 (1895).
Africa. (13) sic 20 EGY 22 BEN GUI IVO LBR NGA SEN 23 BUR CON GAB RWA ZAI 24 CHA ETH SUD 25 KEN TAN UGA 26 ANG MLW MOZ ZAM 27 BOT CPP NAM NAT TVL 29 MAU MDG REU (34) pal (41) tha (63) haw (78) fla (80) cos (81) pue (85) jnf. Hemicr. or rhizome geophyte.

subsp. ***madagascariensis*** (Willd.) Kunth, Enum. Pl. 2: 64 (1837).
E. Trop. Africa, W. Indian Ocean. 25 TAN 29 MAU MDG REU. Hemicr. or rhizome geophyte.
**Papyrus madagascariensis* Willd., Abh. Königl. Akad. Wiss. Berlin 1812–1813: 72 (1812). *Cyperus madagascariensis* (Willd.) Roem. & Schult., Syst. Veg. 2: 876 (1817).
Cyperus imerinensis Boeck., Bot. Jahrb. Syst. 5: 500 (1884).

subsp. ***nyassicus*** Chiov., Mem. Reale Ist. Bot. Modena 1: 72 (1931).
Malawi, N. Northern Prov. 26 MLW 27 TVL. Hemicr. or rhizome geophyte.

subsp. ***papyrus***
Egypt, Trop. & S. Africa. (13) sic 20 EGY 22 BEN GUI IVO LBR NGA SEN 23 BUR GAB RWA ZAI 24 CHA ETH SUD 25 KEN TAN UGA 26 MLW MOZ ZAM 27 BOT CPP NAM NAT TVL (34) pal (41) tha (63) haw (78) fla (80) cos (81) pue (85) jnf. Hemicr. or rhizome geophyte.
Cyperus papyrus subsp. *niliacus* Tournay in ?
Cyperus papyraceus Crantz, Inst. Rei Herb. 1: 411 (1766).
Papyrus antiquorum Willd., Abh. Königl. Akad. Wiss. Berlin 1812–1813: 70 (1812). *Cyperus papyrus* var. *antiquorum* (Willd.) C.B.Clarke in T.A.Durand & H.Schinz, Consp. Fl. Afric. 5: 571 (1894). *Cyperus antiquorum* (Willd.) Chiov., Popiro Ital.: 72 (1931). *Cyperus papyrus* subsp. *antiquorum* (Willd.) Chiov., Mem. Reale Ist. Bot. Modena 1: 73 (1931).
Papyrus domesticus Poir. in G.-F.Cuvier, Dict. Sci. Nat. 37: 447 (1826).
Papyrus siculus Parl., Hooker's J. Bot. Kew Gard. Misc. 3: 189 (1851). *Cyperus papyrus* subsp. *siculus* (Parl.) Chiov., Index Seminum (Catinensis) 1927: s.p. (1927).
Cyperus syriacus Parl., Mém. Divers Savants Acad. Roy. Sci. Inst. Roy. France, Sci. Math. 12: 498 (1854).
Papyrus mossambicensis Parl. in W.C.H.Peters, Naturw. Reise Mossambique: 582, t. 59 (1864).
Cyperus elapsus Chiov., Papiro Ital.: 73 (1931).
Cyperus panormitanus Chiov., Papiro Ital.: 73 (1931).
Cyperus papyrus var. *palaestinae* Chiov., Mem. Reale Ist. Bot. Modena 1: 119 (1931).
Cyperus papyrus subsp. *ugandensis* Chiov., Mem. Reale Ist. Bot. Modena 1: 73 (1931).
Cyperus papyrus subsp. *hadidii* Chrtek & Slavíková, Preslia 49: 184 (1977).

subsp. ***zairensis*** Chiov., Mem. Reale Ist. Bot. Modena 1: 73 (1931).
Congo to Angola. 23 CON ZAI 26 ANG. Hemicr. or rhizome geophyte.

Cyperus parishii Britton, Bull. S. Calif. Acad. Sci. 3: 52 (1904). *Cyperus congestus* var. *parishii* (Britton) Kük. in H.G.A.Engler (ed.), Pflanzenr., IV, 20(101): 446 (1936).
SW. U.S.A. to New Mexico and Mexico (Sonora). 76 ARI CAL 77 NWM 79 MXN. Ther.

Cyperus paucispiculatus Cherm., Bull. Soc. Bot. France 66: 342 (1919 publ. 1920).
C. Madagascar. 29 MDG. Hemicr. or rhizome geophyte.

Cyperus pearcei C.B.Clarke, Kew Bull., Addit. Ser. 8: 7 (1908).
Peru to Bolivia. 83 BOL PER. Hemicr. or rhizome geophyte.

Cyperus pectinatus Vahl, Enum. Pl. 2: 298 (1805). *Anosporum pectinatum* (Vahl) Lye, Nordic J. Bot. 1: 188 (1981).
Trop. & S. Africa, Madagascar. 22 BEN GHA GUI IVO LBR MLI NGA NGR SEN SIE TOG 23 CMN CON GAB RWA ZAI 24 CHA ETH SUD 25 KEN TAN UGA 26 ANG MLW ZAM 27 BOT NAT SWZ 29 MDG (81) win. Hemicr. or rhizome geophyte.
Cyperus nudicaulis Poir. in J.B.A.M.de Lamarck, Encycl. 7: 240 (1806). *Anosporum nudicaule* (Poir.) Boeck., Bot. Zeitung (Berlin) 27: 26 (1869).
Atomostylis flavescens Steud., Flora 33: 229 (1850).
Atomostylis cyperiformis Steud., Syn. Pl. Glumac. 2: 315 (1855).

Cyperus pedunculosus F.Muell., Fragm. 8: 266 (1874).
New Guinea to N. Australia. 43 NWG 50 NTA QLD. Hemicr. or rhizome geophyte.
Cyperus montis-sellae K.Schum. ex Warb., Bot. Jahrb. Syst. 18: 186 (1894).
Cyperus pedunculosus var. *longibracteatus* Domin, Biblioth. Bot. 85: 425 (1915).
Cyperus papuanus Ridl., Trans. Linn. Soc. London, Bot. 9: 241 (1916).
Cyperus pedunculosus var. *floribundus* Kük., Bot. Jahrb. Syst. 59: 44 (1924).

Cyperus pendulus Cherm., Bull. Soc. Bot. France 66: 340 (1919 publ. 1920).
Madagascar. 29 MDG. Hemicr. or rhizome geophyte.

Cyperus penicillatus Conz., Fl. Taxon. Mexic., ed. 2, 1: 334 (1946).
Mexico (Oaxaca). 79 MXS. Hemicr. or rhizome geophyte.

Cyperus pennatiformis Kük., Bernice P. Bishop Mus. Bull. 81: 23 (1931). *Mariscus pennatiformis* (Kük.) T.Koyama, Bishop Mus. Occas. Pap. 29: 123 (1989).
Hawaiian Is. 63 HAW. Hemicr. or rhizome geophyte.

var. ***bryanii*** Kük., Bernice P. Bishop Mus. Bull. 81: 23 (1931). *Mariscus pennatiformis* subsp. *bryanii* (Kük.) T.Koyama, Bishop Mus. Occas. Pap. 29: 123 (1989).
Hawaiian Is. (SE. Laysan). 63 HAW. Hemicr. or rhizome geophyte.

var. ***pennatiformis***
Hawaiian Is. 63 HAW. Hemicr. or rhizome geophyte.

Cyperus pennellii O'Neill & Ben.Ayers, Leafl. W. Bot. 4: 34 (1944).
Mexico (Durango, México State). 79 MXC MXE. Hemicr. or rhizome geophyte.

Cyperus pentabracteatus Govind. & Hemadri, Proc. Indian Acad. Sci., B 82: 207 (1975).
India. 40 IND. Hemicr. or rhizome geophyte.

Cyperus penzoanus Pic.Serm., Miss. Stud. Lago Tana 7(1): 183 (1951).
Ethiopia to WC. & E. Trop. Africa. 23 EQG 24 ETH 25 KEN TAN UGA. Hemicr. or rhizome geophyte.
Cyperus morandinii Pic.Serm., Miss. Stud. Lago Tana 7(1): 185 (1951).

Cyperus perangustus S.T.Blake, Proc. Roy. Soc. Queensland 51: 43 (1939 publ. 1940).
Queensland. 50 QLD. Hemicr. or rhizome geophyte.

Cyperus perennis (M.E.Jones) O'Neill, Contr. U. S. Natl. Herb. 29: 93 (1945).
NW. Mexico. 79 MXN. Hemicr. or rhizome geophyte.
**Cyperus aristatus* var. *perennis* M.E.Jones, Contr. W. Bot. 18: 25 (1933).

Cyperus permacer C.B.Clarke, Bull. Soc. Bot. France 54(8): 27 (1907). *Cyperus zollingeri* var. *permacer* (C.B.Clarke) Kük. in H.G.A.Engler (ed.), Pflanzenr., IV, 20(101): 134 (1935).
W. Trop. Africa. 22 GHA GUI NGA SIE. Hemicr. or rhizome geophyte.

Cyperus pernambucensis Steud., Syn. Pl. Glumac. 2: 42 (1854).
NE. Brazil. 84 BZE. Hemicr. or rhizome geophyte.

Cyperus phaeolepis Cherm., Bull. Soc. Bot. France 66: 341 (1919 publ. 1920).
Madagascar. 29 MDG. Hemicr. or rhizome geophyte.
Cyperus albiflorus Cherm., Bull. Soc. Bot. France 66: 311 (1919 publ. 1920). *Cyperus phaeolepis* var. *albiflorus* (Cherm.) Kük. in H.G.A.Engler (ed.), Pflanzenr., IV, 20(101): 190 (1936).

Cyperus phillipsiae (C.B.Clarke) Kük. in H.G.A.Engler (ed.), Pflanzenr., IV, 20(101): 497 (1936).
Somalia to E. Kenya. 24 SOM 25 KEN. Hemicr.
**Mariscus phillipsiae* C.B.Clarke in D.Oliver & auct. suc. (eds.), Fl. Trop. Afr. 8: 391 (1902).

Cyperus phleoides (Nees ex Kunth) H.Mann, Proc. Amer. Acad. Arts 7: 208 (1867).
Tuamotu, Hawaiian Is. 61 TUA 63 HAW. Hemicr. or rhizome geophyte.
**Mariscus phleoides* Nees ex Kunth, Enum. Pl. 2: 127 (1837).

var. ***hawaiiensis*** (H.Mann) Kük. in H.G.A.Engler (ed.), Pflanzenr., IV, 20(101): 428 (1936).
Hawaiian Is. 63 HAW. Hemicr. or rhizome geophyte.
**Cyperus hawaiiensis* H.Mann, Proc. Amer. Acad. Arts 7: 208 (1867). *Mariscus phleoides* subsp. *hawaiiensis* (H.Mann) T.Koyama, Bishop Mus. Occas. Pap. 29: 124 (1989).

var. ***phleoides***
Tuamotu, Hawaiian Is. 61 TUA 63 HAW. Hemicr. or rhizome geophyte.

Cyperus picardae Boeck., Allg. Bot. Z. Syst. 1: 229 (1895). *Mariscus picardae* (Boeck.) T.Koyama, Phytologia 29: 74 (1974).
Hispaniola. 81 DOM HAI. Hemicr. or rhizome geophyte.
Cyperus picardae var. *brevinux* Kük., Repert. Spec. Nov. Regni Veg. 23: 187 (1926).

Cyperus pilosulus (C.B.Clarke) K.Schum. ex Kük. in H.G.A.Engler (ed.), Pflanzenr., IV, 20(101): 551 (1936).
Tanzania (Zanzibar). 25 TAN. Hemicr. or rhizome geophyte.
**Mariscus pilosulus* C.B.Clarke in D.Oliver & auct. suc. (eds.), Fl. Trop. Afr. 8: 384 (1901).

Cyperus pilosus Vahl, Enum. Pl. 2: 354 (1805). *Duval-jouvea pilosa* (Vahl) Palla in W.D.J.Koch, Syn. Deut. Schweiz. Fl., ed. 3, 3: 2555 (1905).
Trop. & Subtrop. Asia to NE. Australia. 36 CHC 38 JAP NNS TAI 40 ASS BAN EHM IND NEP SRL 41 MYA THA VIE 42 BOR JAW LSI MLY MOL PHI SUL SUM 43 NWG 50 NSW QLD (60) fij 62 CRL (63) haw (78) fla lou msi sca. Ther. or hemicr.
Cyperus paniculatus D.Don, Prodr. Fl. Nepal.: 39 (1825), nom. illeg.
Cyperus donianus A.Dietr., Sp. Pl. 2: 290 (1832).
Cyperus fimbriatus Nees in R.Wight, Contr. Bot. India: 86 (1834). *Cyperus pilosus* var. *fimbriatus* (Nees) Kük. in H.G.A.Engler (ed.), Pflanzenr., IV, 20(101): 94 (1935).
Cyperus marginellus Nees in R.Wight, Contr. Bot. India: 83 (1834).
Cyperus obliquus Nees in R.Wight, Contr. Bot. India: 86 (1834). *Cyperus pilosus* var. *obliquus* (Nees) C.B.Clarke, J. Linn. Soc., Bot. 21: 151 (1884).
Cyperus honestus Kunth, Enum. Pl. 2: 74 (1837).
Cyperus pauciflorus Steud., Syn. Pl. Glumac. 2: 34 (1854).
Cyperus piptolepis Steud., Syn. Pl. Glumac. 2: 40 (1854).
Cyperus quinqueflorus Hochst. ex Steud., Syn. Pl. Glumac. 2: 37 (1854).
Cyperus subalatus Steud., Syn. Pl. Glumac. 2: 31 (1854).
Cyperus hebes Steud., Syn. Pl. Glumac. 2: 315 (1855).
Cyperus piptolepis var. *contractus* Miq., Fl. Ned. Ind. 3: 280 (1856). *Cyperus pilosus* var. *contractus* (Miq.) Kük. in H.G.A.Engler (ed.), Pflanzenr., IV, 20(101): 94 (1935).
Cyperus piptolepis var. *pyrophytus* Miq., Fl. Ned. Ind. 3: 279 (1856).
Cyperus pyrophilus Reinw. ex de Vriese, Pl. Ind. Bat. Orient.: 139 (1857).
Cyperus heyneanus Boeck., Flora 42: 440 bis (1859).
Cyperus pilosus var. *muticus* Boeck., Linnaea 35: 599 (1868).
Cyperus pilosus var. *polyanthus* C.B.Clarke, J. Linn. Soc., Bot. 21: 151 (1884).
Cyperus pilosus f. *badius* Kuntze, Revis. Gen. Pl. 2: 750 (1891).
Cyperus pilosus f. *pallidus* Kuntze, Revis. Gen. Pl. 2: 750 (1891).
Cyperus pilosus var. *pseudobabakensis* Kük., Repert. Spec. Nov. Regni Veg. 27: 107 (1929).
Cyperus pilosus var. *laevirhachis* Kük. in H.G.A. Engler (ed.), Pflanzenr., IV, 20(101): 94 (1935).

Cyperus pinetorum Britton, Bull. Torrey Bot. Club 43: 462 (1916).
Cuba (I. de la Juventud). 81 CUB. Hemicr. or rhizome geophyte.

Cyperus planifolius Rich., Actes Soc. Hist. Nat. Paris 1: 106 (1792). *Mariscus planifolius* (Rich.) Urb., Symb. Antill. 2: 165 (1900).
SE. U.S.A. to NE. Brazil. 78 FLA GEO 79 MXT 80

BLZ HON 81 ARU BAH BER CAY CUB DOM HAI JAM LEE NLA PUE TRT WIN 84 BZE. Hemicr.

Cyperus brizaeus Rich., Actes Soc. Hist. Nat. Paris 1: 106 (1792).

Cyperus brunneus Sw., Fl. Ind. Occid. 1: 116 (1797). *Mariscus brunneus* (Sw.) C.B.Clarke in I.Urban, Symb. Antill. 2: 52 (1900). *Cyperus planifolius* var. *brunneus* (Sw.) Kük., Repert. Spec. Nov. Regni Veg. 23: 189 (1926).

Cyperus brizaeus Vahl, Enum. Pl. 2: 316 (1805), nom. illeg.

Cyperus purpurascens Vahl, Enum. Pl. 2: 359 (1805). *Mariscus purpurascens* (Vahl) C.B.Clarke, Bull. Misc. Inform. Kew 1893: 283 (1893).

Cyperus sanguineus Balb. ex Steud., Nomencl. Bot., ed. 2, 1: 473 (1840).

Cyperus glaucus Steud., Syn. Pl. Glumac. 2: 42 (1854), nom. illeg.

Cyperus brunneus Griseb., Fl. Brit. W. I.: 565 (1864), nom. illeg.

Cyperus evaginatus Boeck., Linnaea 36: 351 (1870).

Cyperus glaucus var. *gracilis* Boeck., Linnaea 36: 339 (1870). *Cyperus planifolius* f. *gracilis* (Boeck.) Kük. in H.G.A.Engler (ed.), Pflanzenr., IV, 20(101): 448 (1936).

Cyperus ottonis Boeck., Linnaea 36: 350 (1870). *Cyperus planifolius* var. *ottonis* (Boeck.) Kük., Repert. Spec. Nov. Regni Veg. 23: 188 (1926).

Cyperus ottonis var. *humilior* Boeck., Linnaea 36: 351 (1870). *Cyperus planifolius* f. *humilior* (Boeck.) Kük. in H.G.A.Engler (ed.), Pflanzenr., IV, 20(101): 448 (1936).

Cyperus discolor Boeck., Beitr. Cyper. 1: 7 (1888).

Cyperus krugii Boeck., Beitr. Cyper. 1: 8 (1888).

Cyperus solmsii Boeck., Allg. Bot. Z. Syst. 1: 228 (1895).

Mariscus brizaeus C.B.Clarke in I.Urban, Symb. Antill. 2: 52 (1900).

Cyperus planifolius var. *navassanus* Kük. & Ekman, Ark. Bot. 22A(17): 6 (1929).

Cyperus planifolius var. *pickelii* Grosser ex Kük. in H.G.A.Engler (ed.), Pflanzenr., IV, 20(101): 447 (1936).

Cyperus plantaginifolius Cherm., Bull. Soc. Bot. France 66: 316 (1920).

Madagascar. 29 MDG. Hemicr. or rhizome geophyte.

Cyperus plantaginifolius var. *minor* Cherm., Bull. Soc. Bot. France 72: 615 (1925).

Cyperus plateilema (Steud.) Kük. in H.G.A.Engler (ed.), Pflanzenr., IV, 20(101): 558 (1936).

Ethiopia to Tanzania, Yemen (J. Ta'kar). 24 ETH 25 KEN TAN UGA 35 YEM. Hemicr. or tuber geophyte.

**Mariscus plateilema* Steud., Flora 25: 596 (1842).

Mariscus atrosanguineus Hochst. ex A.Rich., Tent. Fl. Abyss. 2: 490 (1850).

Mariscus bulbocaulis A.Rich., Tent. Fl. Abyss. 2: 490 (1850). *Cyperus bulbocaulis* (A.Rich.) Boeck., Linnaea 36: 372 (1870).

Mariscus hochstetteri Walp., Ann. Bot. Syst. 3: 678 (1852).

Kyllinga bulbosa Steud., Syn. Pl. Glumac. 2: 62 (1854), nom. illeg.

Mariscus bulbosus Steud., Syn. Pl. Glumac. 2: 62 (1854).

Cyperus variegatus var. *atrosanguineus* Boeck., Linnaea 36: 337 (1870). *Cyperus bulbocaulis* var. *atrosanguineus* (Boeck.) Kük. in H.G.A.Engler (ed.), Pflanzenr., IV, 20(101): 559 (1936).

Cyperus platycaulis Baker, J. Linn. Soc., Bot. 22: 532 (1887).

Chad to KwaZulu-Natal, Madagascar. 23 BUR RWA ZAI 24 CHA 25 KEN TAN UGA 26 ZIM 27 NAT TVL 29 MDG. Hemicr. or rhizome geophyte.

Cyperus lucentinigricans K.Schum., Abh. Königl. Ges. Wiss. Göttingen, Math.-Phys. Kl. 39: 59 (1894). *Cyperus denudatus* var. *lucentinigricans* (K.Schum.) Kük., Notizbl. Bot. Gart. Berlin-Dahlem 9: 303 (1925). *Cyperus platycaulis* var. *lucentinigricans* (K.Schum.) Kük. in H.G.A.Engler (ed.), Pflanzenr., IV, 20(101): 254 (1936).

Cyperus denudatus var. *delicatulus* C.B.Clarke in D.Oliver & auct. suc. (eds.), Fl. Trop. Afr. 8: 338 (1901).

Cyperus serpens Cherm., Rev. Zool. Bot. Africaines 24: 297 (1934).

Cyperus kivuensis Cherm., Bull. Soc. Bot. France 82: 337 (1935).

Cyperus platycaulis var. *recedens* Peter & Kük. in H.G.A.Engler (ed.), Pflanzenr., IV, 20(101): 254 (1936).

Cyperus platyphyllus Roem. & Schult., Syst. Veg. 2: 876 (1817).

S. India, Sri Lanka. 40 IND SRL. Hemicr. or rhizome geophyte.

Cyperus fastigiatus Vahl, Enum. Pl. 2: 367 (1805), nom. illeg.

Papyrus latifolius Willd., Abh. Königl. Akad. Wiss. Berlin 1812–1813: 74 (1812).

Cyperus elatus Roxb., Fl. Ind. 1: 204 (1820), nom. illeg.

Cyperus maximus Roxb. ex Nees in R.Wight, Contr. Bot. India: 85 (1834).

Cyperus roxburghii Nees in R.Wight, Contr. Bot. India: 84 (1834), nom. illeg.

Cyperus eminens Klein ex Kunth, Enum. Pl. 2: 70 (1837).

Cyperus platystylis R.Br., Prodr.: 214 (1810).

Trop. & Subtrop. Asia, Australia. 38 TAI 40 ASS BAN IND NEP SRL 41 MYA THA VIE 42 BOR JAW LSI MLY MOL SUL SUM 43 NWG 50 NSW NTA QLD. Hemicr. or rhizome geophyte.

Cyperus pallidus Nees in R.Wight, Contr. Bot. India: 79 (1834), nom. illeg. *Anosporum pallidum* Boeck., Linnaea 36: 412 (1870).

Cyperus fluitans Buch.-Ham. ex C.B.Clarke, J. Linn. Soc., Bot. 21: 118 (1884), pro syn.

Cyperus plukenetii Fernald, Rhodora 47: 110 (1945).

C. & E. U.S.A. 74 MSO OKL 75 NWJ 77 TEX 78 ALA ARK DEL FLA GEO KTY LOU MRY MSI NCA SCA TEN VRG. Hemicr. or rhizome geophyte.

Cyperus pluribracteatus (Kük.) Govaerts in R.H.A. Govaerts & D.A.Simpson, World Checklist of Cyperaceae: 350 (2007).

E. & S. Trop. Africa. 23 BUR RWA 25 KEN TAN UGA 26 ZIM. Hemicr. or rhizome geophyte.

Mariscus psilostachys C.B.Clarke, J. Bot. 34: 225 (1896). *Cyperus psilostachys* (C.B.Clarke) Kük., Bot. Not. 1934: 69 (1934), nom. illeg.

Cyperus psilostachys f. *glabrescens* Kük. in H.G.A. Engler (ed.), Pflanzenr., IV, 20(101): 546 (1936).

**Cyperus psilostachys* var. *pluribracteatus* Kük. in H.G.A.Engler (ed.), Pflanzenr., IV, 20(101): 546 (1936).

Cyperus psilostachys var. *subrufus* Kük. in H.G.A. Engler (ed.), Pflanzenr., IV, 20(101): 546 (1936).

Cyperus pluricephalus Lye, Nordic J. Bot. 16: 371 (1996).
C. Somalia. 24 SOM. Hemicr. or rhizome geophyte.

Cyperus plurinervosus Bodard, Bull. Soc. Bot. France 99: 61 (1952).
Central African Rep. 23 CAF. Hemicr. or rhizome geophyte.

Cyperus podocarpus Boeck., Flora 62: 551 (1879).
W. Trop. Africa to Sudan. 22 BEN BKN GHA GUI IVO MLI NGA NGR SEN SIE TOG 24 CHA SUD. Ther.

Cyperus poecilus C.B.Clarke in D.Oliver & auct. suc. (eds.), Fl. Trop. Afr. 8: 323 (1901).
Eritrea, N. Somalia. 24 ERI SOM. Hemicr. or rhizome geophyte.

Cyperus poeppigii Kunth, Enum. Pl. 2: 90 (1837). *Torulinium poeppigii* (Kunth) C.B.Clarke, Bot. Jahrb. Syst. 30(68): 17 (1902).
S. Texas to Mexico. 77 TEX 79 MXE. Hemicr. or rhizome geophyte.
Mariscus castaneus Kunze ex Kunth, Enum. Pl. 2: 91 (1837).
Cyperus acaulis Steud., Flora 25: 599 (1842).

Cyperus pohlii (Nees) Steud., Syn. Pl. Glumac. 2: 40 (1854).
S. Trop. America. 83 ECU 84 BZC BZE BZL BZS 85 AGE AGW PAR. Hemicr. or rhizome geophyte.
**Papyrus pohlii* Nees in C.F.P.von Martius & auct. suc. (eds.), Fl. Bras. 2(1): 59 (1842). *Mariscus pohlii* (Nees) D.A.Simpson, Kew Bull. 44: 285 (1989).

subsp. ***bahiensis*** (D.A.Simpson) D.A.Simpson, Kew Bull. 48: 701 (1993).
NE. Brazil. 84 BZE. Hemicr. or rhizome geophyte.
**Mariscus pohlii* subsp. *bahiensis* D.A.Simpson, Kew Bull. 44: 286 (1989).

subsp. ***pohlii***
S. Trop. America. 83 ECU 84 BZC BZL BZS 85 AGE AGW PAR. Hemicr. or rhizome geophyte.
Mariscus bracteosus Schltdl., Linnaea 25: 164 (1852).
Cyperus warmingii Boeck., Vidensk. Meddel. Naturhist. Foren. Kjøbenhavn 1869: 129 (1869). *Cyperus pohlii* var. *warmingii* (Boeck.) Kük. in H.G.A.Engler (ed.), Pflanzenr., IV, 20(101): 412 (1936).
Cyperus mendoncae Boeck., Beitr. Cyper. 2: 9 (1890). *Cyperus pohlii* var. *mendoncae* (Boeck.) Kük. in H.G.A.Engler (ed.), Pflanzenr., IV, 20(101): 412 (1936).
Cyperus validus Boeck., Allg. Bot. Z. Syst. 1: 229 (1895).
Mariscus setiglumis C.B.Clarke ex S.Moore, Trans. Linn. Soc. London, Bot. 4: 512 (1895).
Chlorocyperus meridionalis Palla, Denkschr. Kaiserl. Akad. Wiss., Wien. Math.-Naturwiss. Kl. 79: 179 (1908).
Chlorocyperus riograndensis Palla, Denkschr. Kaiserl. Akad. Wiss., Wien. Math.-Naturwiss. Kl. 79: 179 (1908).
Cyperus pohlii var. *obscuricoloratus* Kük. in H.G.A.Engler (ed.), Pflanzenr., IV, 20(101): 412 (1936).

Cyperus polyanthelus Govind., Proc. Indian Acad. Sci., B 88: 234 (1979).
India. 40 IND. Hemicr. or rhizome geophyte.

Cyperus portae-tartari K.L.Wilson, Telopea 1: 460 (1980).
Northern Territory. 50 NTA. Hemicr. or rhizome geophyte.

Cyperus praemorsus Boeck., Flora 42: 436 bis (1859).
Jamaica. 81 JAM. Hemicr. or rhizome geophyte.

Cyperus pratensis Boeck., Linnaea 38: 364 (1874). *Mariscus pratensis* (Boeck.) Cufod., Bull. Jard. Bot. Natl. Belg. 40(Suppl.): 1454 (1970).
Ethiopia. 24 ETH. Hemicr. or rhizome geophyte.

Cyperus procerus Rottb., Descr. Icon. Rar. Pl.: 29 (1773). *Duval-jouvea procera* (Rottb.) H.Pfeiff., Mitt. Inst. Allg. Bot. Hamburg 7: 167 (1928).
Trop. & Subtrop. Old World. 20 EGY 22 BKN GHA IVO MLI NGA NGR TOG 23 CMN 24 CHA ETH SOM SUD 25 KEN TAN 26 ZAM 27 NAM TVL 29 MDG 38 TAI 40 ASS BAN IND NEP SRL 41 THA VIE 42 BOR JAW MLY PHI 50 NSW NTA QLD. Hemicr. or rhizome geophyte.
Cyperus carnosus B.Heyne in N.Wallich, Numer. List: 3355 A (1831), nom. inval.
Cyperus anceps B.Heyne ex Steud., Nomencl. Bot., ed. 2, 1: 466 (1840), nom. nud.
Cyperus griffithianus Boeck., Linnaea 35: 601 (1868). *Cyperus procerus* var. *griffithianus* (Boeck.) Kük. in H.G.A.Engler (ed.), Pflanzenr., IV, 20(101): 91 (1935).
Cyperus heynei Boeck., Linnaea 35: 600 (1868).
Cyperus procerus var. *lasiorrhachis* C.B.Clarke in J.D.Hooker, Fl. Brit. India 6: 610 (1893).
Cyperus straminicolor Cherm., Bull. Soc. Bot. France 66: 349 (1919 publ. 1920). *Cyperus procerus* var. *straminicolor* (Cherm.) Kük. in H.G.A.Engler (ed.), Pflanzenr., IV, 20(101): 91 (1935).
Cyperus trachycladus Cherm., Bull. Soc. Bot. France 66: 348 (1919 publ. 1920). *Cyperus procerus* var. *trachycladus* (Cherm.) Kük. in H.G.A.Engler (ed.), Pflanzenr., IV, 20(101): 92 (1935).
Cyperus straminicolor var. *betsileensis* Cherm., Bull. Mus. Natl. Hist. Nat. 27: 554 (1921).
Mariscus thwaitesii Livera, Ann. Roy. Bot. Gard. (Peradeniya) 11: 103 (1928).
Cyperus procerus var. *stenanthus* Kük. in H.G.A. Engler (ed.), Pflanzenr., IV, 20(101): 92 (1935).
Cyperus procerus var. *vanderystii* Kük. in H.G.A. Engler (ed.), Pflanzenr., IV, 20(101): 92 (1935).

Cyperus prolifer Lam., Tabl. Encycl. 1: 147 (1791).
Somalia to S. Africa, W. Indian Ocean. 24 SOM 25 KEN TAN 26 MOZ 27 CPP NAT 29 MAU MDG REU (50) (63) haw (76) cal (78) fla. Hemicr. or rhizome geophyte.
Cyperus aequalis Vahl, Enum. Pl. 2: 320 (1805). *Papyrus aequalis* (Vahl) Bojer, Hortus Maurit.: 381 (1837).
Cyperus papyroides Poir. in J.B.A.M.de Lamarck, Encycl. 7: 272 (1806).
Papyrus laxiflorus Spreng., Syst. Veg. 1: 232 (1824).
Cyperus isocladus Kunth, Enum. Pl. 2: 37 (1837). *Cyperus prolifer* var. *isocladus* (Kunth) Kük. in H.G.A.Engler (ed.), Pflanzenr., IV, 20(101): 257 (1936).
Cyperus jocladus E.Mey. in J.F.Drège, Zwei Pflanzengeogr. Dokum.: 177 (1843).
Cyperus aequalis C.Krauss, Flora 28: 754 (1845), nom. illeg.

Cyperus prolixus Kunth in F.W.H.von Humboldt, A.J.A.Bonpland & C.S.Kunth, Nov. Gen. Sp. 1: 206 (1817). *Comostemum prolixum* (Kunth) Nees, J. Bot. (Hooker) 2: 397 (1840). *Chlorocyperus prolixus* (Kunth) Palla, Denkschr. Kaiserl. Akad. Wiss., Wien. Math.-Naturwiss. Kl. 79: 180 (1908).
Trop. & Subtrop. America. 78 LOU 79 MXC MXE MXG MXS MXT 80 BLZ COS GUA HON PAN 82 FRG GUY SUR VEN 83 BOL CLM ECU PER 84 BZC BZE BZL BZN BZS 85 AGE AGW PAR URU. Hemicr. or rhizome geophyte.
Comostemum schottii Nees, Linnaea 9: 283 (1834). *Cyperus schottii* (Nees) D.Dietr., Syn. Pl. 1: 220 (1839).
Cyperus bisumbellatus Steud., Syn. Pl. Glumac. 2: 46 (1854).
Cyperus amplissimus Steud., Syn. Pl. Glumac. 2: 316 (1855).
Cyperus vialis Ridl., J. Linn. Soc., Bot. 2 7: 65 (1890). *Cyperus prolixus* var. *vialis* (Ridl.) Kük. in H.G.A. Engler (ed.), Pflanzenr., IV, 20(101): 147 (1935).

Cyperus psammophilus Cherm., Bull. Soc. Bot. France 66: 338 (1919 publ. 1920).
Madagascar. 29 MDG. Hemicr. or rhizome geophyte.
Cyperus psammophilus var. *gracillimus* Cherm., Bull. Soc. Bot. France 66: 339 (1919 publ. 1920).

Cyperus pseuderemicus Kukkonen & Väre, Ann. Bot. Fenn. 42: 474 (2005).
Saudi Arabia. 35 SAU.

Cyperus pseudobrunneus (C.B.Clarke ex Cherm.) Kük. in H.G.A.Engler (ed.), Pflanzenr., IV, 20(101): 435 (1936).
E. Trop. Africa, W. Indian Ocean. 25 TAN 29 COM MDG. Hemicr. or rhizome geophyte.
Mariscus hilsenbergii C.B.Clarke in T.A.Durand & H.Schinz, Consp. Fl. Afric. 5: 589 (1894), nom. nud. *Cyperus pseudobrunneus* var. *hilsenbergii* Kük. in H.G.A.Engler (ed.), Pflanzenr., IV, 20(101): 436 (1936).
Cyperus pseudobrunneus var. *continentalis* Kük. in H.G.A.Engler (ed.), Pflanzenr., IV, 20(101): 436 (1936).
Cyperus pseudobrunneus var. *pseudobojeri* Kük. in H.G.A.Engler (ed.), Pflanzenr., IV, 20(101): 436 (1936).
**Mariscus pseudobrunneus* C.B.Clarke ex Cherm., Notul. Syst. (Paris) 5: 271 (1936).

Cyperus pseudodistans Uittien, Recueil Trav. Bot. Néerl. 22: 334 (1925).
Suriname. 82 SUR. Hemicr. or rhizome geophyte.

Cyperus pseudoleptocladus Kük., Repert. Spec. Nov. Regni Veg. 29: 196 (1931).
Trop. & S. Africa. 23 BUR CMN RWA ZAI 25 KEN TAN UGA 26 MLW 27 SWZ TVL. Hemicr. or rhizome geophyte.
Cyperus pseudoleptocladus var. *polycarpus* Kük., Repert. Spec. Nov. Regni Veg. 29: 196 (1931).

Cyperus pseudopilosus (T.Durand & De Wild.) Govaerts in R.H.A.Govaerts & D.A.Simpson, World Checklist of Cyperaceae: 352 (2007).
Trop. Africa. 22 GUI IVO LBR 23 CMN CON EQG GAB ZAI 25 UGA 26 ANG. Hemicr. or rhizome geophyte.
Mariscus trinervis C.B.Clarke in T.A.Durand & H.Schinz, Consp. Fl. Afric. 5: 577 (1894), nom. inval.
**Mariscus pseudopilosus* T.Durand & De Wild., Bull. Soc. Roy. Bot. Belgique, Compt. Rend. 36: 89 (1897).
Cyperus socialis C.B.Clarke in D.Oliver & auct. suc. (eds.), Fl. Trop. Afr. 8: 351 (1901). *Mariscus socialis* (C.B.Clarke) S.S.Hooper, Kew Bull. 26: 578 (1972).

Cyperus pseudosomaliensis Kük. in H.G.A.Engler (ed.), Pflanzenr., IV, 20(101): 552 (1936).
N. Somalia. 24 SOM. Hemicr.

Cyperus pseudothyrsiflorus (Kük.) R.Carter & S.D.Jones, Rhodora 99: 330 (1997).
Texas to NE. Mexico. 77 TEX 79 MXE. Hemicr. or rhizome geophyte.
**Cyperus uniflorus* var. *pseudothyrsiflorus* Kük. in H.G.A.Engler (ed.), Pflanzenr., IV, 20(101): 521 (1936).

Cyperus pseudovegetus Steud., Syn. Pl. Glumac. 2: 24 (1854).
C. & E. U.S.A. 74 ILL KAN MSO OKL 75 INI mas NWJ 77 TEX 78 ALA ARK DEL FLA GEO KTY LOU MRY MSI? NCA SCA TEN VRG WDC. Hemicr. or rhizome geophyte.
Cyperus calcaratus S.Watson in A.Gray, Manual, ed. 6: 570 (1890).

Cyperus pseudovestitus (C.B.Clarke) Kük. in H.G.A.Engler (ed.), Pflanzenr., IV, 20(101): 547 (1936).
Uganda to S. Africa, Madagascar. 25 KEN TAN UGA 26 MOZ ZIM 27 CPP NAM NAT SWZ TVL 29 MDG. Hemicr. or rhizome geophyte.
**Mariscus pseudovestitus* C.B.Clarke, Bull. Herb. Boissier, II, 6: 709 (1906).
Mariscus goniobolbus Cherm., Bull. Mus. Natl. Hist. Nat. 25: 301 (1919).
Mariscus perrieri Cherm., Bull. Mus. Natl. Hist. Nat. 25: 301 (1919). *Mariscus goniobolbus* var. *perrieri* (Cherm.) Cherm., Bull. Soc. Bot. France 72: 169 (1925). *Cyperus pseudovestitus* var. *perrieri* (Cherm.) Kük. in H.G.A.Engler (ed.), Pflanzenr., IV, 20(101): 547 (1936).
Mariscus goniobolbus var. *angustifolius* Cherm., Bull. Soc. Bot. France 72: 614 (1925). *Cyperus pseudovestitus* f. *angustifolius* (Cherm.) Kük. in H.G.A.Engler (ed.), Pflanzenr., IV, 20(101): 548 (1936).
Kyllinga hymenopoda Peter, Abh. Preuss. Akad. Wiss., Phys.-Math. Kl., n.f., 13(2): 47 (1928).
Cyperus pseudovestitus var. *polycarpus* Kük. in H.G.A. Engler (ed.), Pflanzenr., IV, 20(101): 548 (1936).

Cyperus pubens Kük., Repert. Spec. Nov. Regni Veg. 29: 200 (1931). *Mariscus pubens* (Kük.) Podlech, Mitt. Bot. Staatssamml. München 4: 115 (1961).
S. Tanzania to Zimbabwe. 25 TAN 26 ZIM. Hemicr. or rhizome geophyte.

Cyperus pulchellus R.Br., Prodr.: 213 (1810). *Sorostachys pulchellus* (R.Br.) Lye, Nordic J. Bot. 1: 189 (1981).
Trop. Old World. 22 BKN GAM GNB GUI MLI NGA NGR SEN TOG 24 CHA ETH SOM 25 KEN TAN UGA 42 PHI 50 NTA QLD WAU. Hemicr. or rhizome geophyte.
Cyperus sorostachyus Boeck., Linnaea 35: 588 (1868).
Cyperus zambesiensis C.B.Clarke, Trans. Linn. Soc. London, Bot. 4: 53 (1894). *Cyperus glaucophyllus* var. *zambesiensis* (C.B.Clarke) Kük. in H.G.A. Engler (ed.), Pflanzenr., IV, 20(101): 203 (1936).

Cyperus pulcher Thunb., Prodr. Pl. Cap.: 18 (1794).
Cape Prov., KwaZulu-Natal. 27 CPP NAT. Hemicr. or rhizome geophyte.
Cyperus ingratus Kunth, Enum. Pl. 2: 31 (1837).
Cyperus prasinus Kunth, Enum. Pl. 2: 31 (1837). *Cyperus pulcher* var. *prasinus* (Kunth) Kük. in H.G.A.Engler (ed.), Pflanzenr., IV, 20(101): 235 (1936).

Cyperus pulcherrimus Willd. ex Kunth, Enum. Pl. 2: 35 (1837).
Trop. Asia. 40 ASS BAN IND PAK? SRL 41 THA VIE 42 BOR JAW MLY PHI SUL SUM 43 NWG. Hemicr. or rhizome geophyte.
Cyperus eumorphus Steud., Syn. Pl. Glumac. 2: 22 (1854).
Cyperus silletensis Thwaites, Enum. Pl. Zeyl.: 343 (1864), nom. illeg.
Cyperus pulcherrimus var. *rectiglumis* Suringar, Cyperus Mal. Arch.: 95 (1898).

Cyperus pulguerensis M.T.Strong, Contr. U. S. Natl. Herb. 52: 284 (2005).
Puerto Rico. 81 PUE.

Cyperus pulicaris Kük., Repert. Spec. Nov. Regni Veg. 23: 187 (1926).
Haiti. 81 HAI. Hemicr. or rhizome geophyte.
Cyperus pulicaris var. *explicatus* Kük., Ark. Bot. 22A(17): 5 (1929).

Cyperus purpureoviridis Lye, Nordic J. Bot. 3: 224 (1983).
Tanzania (Nguru Mts.). 25 TAN. Hemicr. or rhizome geophyte.

Cyperus pustulatus Vahl, Enum. Pl. 2: 341 (1805). *Juncellus pustulatus* (Vahl) C.B.Clarke in T.A.Durand & H.Schinz, Consp. Fl. Afric. 5: 546 (1894).
Trop. Africa. 22 BEN BKN GAM GHA GNB GUI IVO MLI NGA SEN TOG 23 CAF CMN GAB RWA ZAI 24 CHA ETH 25 TAN UGA 26 ANG. Ther.
Cyperus barteri Boeck., Linnaea 35: 460 (1868). *Chlorocyperus barteri* (Boeck.) Rikli, Jahrb. Wiss. Bot. 27: 563 (1895).
Pycreus djalonis A.Chev., Explor. Bot. Afrique Occ. Franç. 1: 696 (1920), nom. nud. *Cyperus pustulatus* var. *djalonis* Kük. in H.G.A.Engler (ed.), Pflanzenr., IV, 20(101): 161 (1936).
Cyperus pustulatus var. *debilis* Kük. in H.G.A.Engler (ed.), Pflanzenr., IV, 20(101): 161 (1936).

Cyperus pycnostachyus (Kunth) Kunth, Enum. Pl. 2: 77 (1837).
Mexico. 79 MXC MXI MXN MXS. Hemicr. or rhizome geophyte.
**Mariscus pycnostachyus* Kunth in F.W.H.von Humboldt, A.J.A.Bonpland & C.S.Kunth, Nov. Gen. Sp. 1: 215 (1816).
Cyperus schmitzianus Boeck., Allg. Bot. Z. Syst. 1: 226 (1895).

Cyperus radians Nees & Meyen ex Kunth, Enum. Pl. 2: 95 (1837). *Mariscus radians* (Nees & Meyen ex Kunth) Tang & F.T.Wang, in Fl. Reipubl. Popul. Sin. 11: 177 (1961).
SE. China to W. Malesia. 36 CHS 38 TAI 40 SRL 41 MYA THA VIE 42 BOR JAW MLY. Hemicr. or rhizome geophyte.
Cyperus griffithii Steud., Syn. Pl. Glumac. 2: 316 (1855). *Cyperus radians* var. *griffithii* (Steud.) Kük. in H.G.A.Engler (ed.), Pflanzenr., IV, 20(101): 215 (1936).
Cyperus macropus Miq., Fl. Ned. Ind., Eerste Bijv.: 599 (1861).
Cyperus sinensis Debeaux, Actes Soc. Linn. Bordeaux 31: t. 2 (1877).
Cyperus sinensis var. *floribundus* E.G.Camus, Notul. Syst. (Paris) 1: 243 (1910). *Cyperus radians* var. *floribundus* (E.G.Camus) Kük. in H.G.A.Engler (ed.), Pflanzenr., IV, 20(101): 215 (1936). *Mariscus radians* var. *floribundus* (E.G.Camus) S.M.Huang, in Fl. Hainan. 4: 537 (1977).
Cyperus shimadai Ohwi, Acta Phytotax. Geobot. 1: 300 (1932).
Cyperus radians var. *compositus* Kük. in H.G.A.Engler (ed.), Pflanzenr., IV, 20(101): 215 (1936).

Cyperus ramosus (Benth.) Kük. in H.G.A.Engler (ed.), Pflanzenr., IV, 20(101): 486 (1936).
N. Australia. 50 NTA QLD. Hemicr. or rhizome geophyte.
**Cyperus conicus* var. *ramosus* Benth., Fl. Austral. 7: 290 (1878). *Mariscus ramosus* (Benth.) Domin, Biblioth. Bot. 85: 441 (1915).

Cyperus rapensis F.Br., Bernice P. Bishop Mus. Bull. 84: 98 (1931).
Tubuai Is. (Rapa-Iti). 61 TUB. Hemicr. or rhizome geophyte.

Cyperus recurvispicatus Lye, Nordic J. Bot. 16: 376 (1996).
C. Somalia. 24 SOM. Hemicr. or rhizome geophyte.

Cyperus redolens Maury ex Micheli, Mém. Soc. Phys. Genève 31: 126 (1890). *Cyperus cayennensis* var. *redolens* (Maury ex Micheli) R.Gross ex J.F.Macbr., Publ. Field Mus. Nat. Hist., Bot. Ser. 11: 42 (1931). *Cyperus flavus* subsp. *redolens* (Maury ex Micheli) Osten, Anales Mus. Hist. Nat. Montevideo 3(2): 152 (1931).
S. America. 82 VEN 83 BOL PER 84 BZE BZS 85 AGE AGW PAR URU. Hemicr. or rhizome geophyte.
Mariscus flavus var. *humilis* Benth. ex C.B.Clarke, Consp. Fl. Afric. 5: 588 (1894). *Mariscus cayennensis* var. *humilis* (Benth. ex C.B.Clarke) Stehlé, Notul. Syst. (Paris) 13: 86 (1947).

Cyperus reduncus Hochst. ex Boeck., Linnaea 35: 580 (1868).
W. Trop. Africa to Uganda. 22 BEN BKN GHA GNB GUI IVO MLI NGA SEN TOG 23 CON ZAI 24 CHA ETH SUD 25 UGA. Ther.
Cyperus aristatus Hook.f. & Thomson ex C.B.Clarke, J. Linn. Soc., Bot. 21: 90 (1884), nom. illeg.

Cyperus reflexus Vahl, Enum. Pl. 2: 42 (1805). *Cyperus reflexus* var. *genuinus* Osten, Anales Mus. Hist. Nat. Montevideo, II, 3: 136 (1931), nom. inval.
Oklahoma to Mexico, Costa Rica to Argentina. (12) fra (50) 74 OKL 77 TEX 78 FLA LOU 79 MXC MXE MXG MXS 80 COS 82 VEN 83 BOL ECU PER 84 BZL BZS 85 AGE AGS AGW CLC CLN JNF PAR URU. Hemicr. or rhizome geophyte.
Cyperus sellowii Link, Hort. Berol. 1: 307 (1827).
Cyperus fernandezianus Colla, Mem. Reale Accad. Sci. Torino 39: 21 (1836).
Cyperus rufescens Torr. & Hook., Ann. Lyceum Nat. Hist. New York 3: 436 (1836).
Cyperus fraternus Kunth, Enum. Pl. 2: 42 (1837). *Cyperus reflexus* var. *fraternus* (Kunth) Kuntze, Revis. Gen. Pl. 3(2): 334 (1898).
Cyperus sparteus Schrad. ex Nees in C.F.P.von Martius & auct. suc. (eds.), Fl. Bras. 2(1): 19 (1842).

Cyperus arenicola Steud., Syn. Pl. Glumac. 2: 51 (1854). *Cyperus cyrtolepis* var. *arenicola* (Steud.) Boeck., Linnaea 35: 558 (1868). *Cyperus pseudovegetus* var. *arenicola* (Steud.) Kük. in H.G.A.Engler (ed.), Pflanzenr., IV, 20(101): 176 (1936). *Cyperus virens* var. *arenicola* (Steud.) Shinners, Field & Lab. 22: 30 (1954).
Cyperus haematostachys Steud., Syn. Pl. Glumac. 2: 41 (1854).
Cyperus baazas Steud., Syn. Pl. Glumac. 2: 316 (1855).
Cyperus juncoides Regel, Index Seminum (LE) 1866: 77 (1866), nom. illeg.
Cyperus brazas Boeck., Linnaea 35: 559 (1868).
Cyperus reflexus var. *macrostachys* Boeck., Linnaea 35: 559 (1868).
Cyperus reflexus f. *anterna* C.B.Clarke, Bull. Herb. Boissier, II, 1: 438 (1901).
Pycreus fernandezianus C.B.Clarke, Bot. Jahrb. Syst. 30(68): 5 (1902).
Cyperus herteri Kük. ex Osten, Anales Mus. Hist. Nat. Montevideo 3: 137 (1931), pro syn.
Cyperus reflexus f. *capitatus* Osten, Anales Mus. Hist. Nat. Montevideo, II, 3: 136 (1931).
Cyperus reflexus f. *evolutus* Osten, Anales Mus. Hist. Nat. Montevideo, II, 3: 136 (1931).
Cyperus reflexus subf. *flaccidus* Osten, Anales Mus. Hist. Nat. Montevideo, II, 3: 136 (1931).
Cyperus reflexus subf. *glaucovirens* Osten, Anales Mus. Hist. Nat. Montevideo, II, 3: 137 (1931).

Cyperus refractus Engelm. ex Boeck., Linnaea 36: 369 (1870).
C. & E. U.S.A. 74 MSO 75 PEN 77 TEX 78 ALA ARK GEO MRY NCA SCA TEN VRG WDC. Hemicr. or rhizome geophyte.
Cyperus retrofractus Engelm. ex Steud., Syn. Pl. Glumac. 2: 47 (1854), nom. illeg.
Cyperus refractus f. *grandis* Kük. in H.G.A.Engler (ed.), Pflanzenr., IV, 20(101): 510 (1936).

Cyperus regiomontanus Britton, Contr. U. S. Natl. Herb. 1: 362 (1895).
Mexico to C. America. 79 MXN MXS MXT 80 COS 83 ECU? Hemicr. or rhizome geophyte.
**Mariscus haenkei* J.Presl & C.Presl, Reliq. Haenk. 1: 181 (1828).

Cyperus rehmii Merxm., Mitt. Bot. Staatssamml. München 3: 84 (1951).
Namibia. 27 NAM. Hemicr. or rhizome geophyte.

Cyperus remotispicatus S.S.Hooper, Kew Bull. 26: 577 (1972).
W. & WC. Trop. Africa. 22 BKN GHA IVO NGA SEN 23 GAB. Ther.

Cyperus remotus (C.B.Clarke) Kük. in H.G.A.Engler (ed.), Pflanzenr., IV, 20(101): 561 (1936).
Zaïre. 23 ZAI. Hemicr. or rhizome geophyte.
**Mariscus remotus* C.B.Clarke in D.Oliver & auct. suc. (eds.), Fl. Trop. Afr. 8: 382 (1901).

Cyperus renschii Boeck., Flora 65: 11 (1882).
Trop. Africa, Comoros. 22 GHA GUI IVO LBR NGA SIE 23 CAF CMN EQG GAB GGI RWA ZAI 25 KEN TAN UGA 26 ANG 29 COM. Hemicr. or rhizome geophyte.
Cyperus hylaeus Ridl., Trans. Linn. Soc. London, Bot. 2: 134 (1884).
Cyperus sylvicola Ridl., Bol. Soc. Brot. 5: 208 (1887).
Cyperus deremensis K.Schum. ex Engl., Abh. Königl. Ges. Wiss. Göttingen, Math.-Phys. Kl. 39: 50 (1894).
Cyperus ochrocarpus K.Schum. in H.G.A.Engler, Pflanzenw. Ost-Afrikas, C: 122 (1895).
Cyperus renschii var. *scabridus* Lye, Nordic J. Bot. 3: 229 (1983).

Cyperus retroflexus Buckley, Proc. Acad. Nat. Sci. Philadelphia 15: 9 (1863). *Cyperus uniflorus* var. *retroflexus* (Buckley) Kük. in H.G.A.Engler (ed.), Pflanzenr., IV, 20(101): 521 (1936).
C. & SE. U.S.A. to N. Mexico. 74 MSO OKL 76 ARI 77 NWM TEX 78 ALA ARK MSI 79 MXE MXG MXN. Hemicr. or rhizome geophyte.

var. ***pumilus*** (Britton) R.Carter & S.D.Jones, Rhodora 99: 329 (1997).
Texas to NE. Mexico. 77 TEX 79 MXE. Hemicr. or rhizome geophyte.
**Cyperus uniflorus* var. *pumilus* Britton, Bull. Torrey Bot. Club 11: 87 (1884). *Cyperus subuniflorus* Britton in J.K.Small, Fl. S.E. U.S.: 173 (1903). *Mariscus subuniflorus* (Britton) T.Koyama, Phytologia 29: 74 (1974).

var. ***retroflexus***
C. & SE. U.S.A. to N. Mexico. 74 MSO OKL 76 ARI 77 NWM TEX 78 ALA ARK MSI 79 MXE MXG MXN. Hemicr. or rhizome geophyte.
Mariscus retrofractus Gaertn., Fruct. Sem. Pl. 1: 12 (1788).
Cyperus uniflorus Torr. & Hook., Ann. Lyceum Nat. Hist. New York 3: 431 (1836), nom. illeg. *Mariscus uniflorus* Steud., Syn. Pl. Glumac. 2: 64 (1854).
Cyperus uniflorus var. *dicarpus* Boeck., Flora 61: 30 (1878).

Cyperus retrofractus (L.) Torr., Fl. New York 2: 344 (1843).
E. U.S.A. to Missouri. 74 MSO 75 NWJ OHI PEN 78 ALA ARK DEL GEO KTY LOU MRY NCA SCA TEN VRG WDC. Hemicr. or rhizome geophyte.
**Scirpus retrofractus* L., Sp. Pl.: 50 (1753). *Mariscus retrofractus* (L.) Vahl, Enum. Pl. 2: 375 (1805).
Cyperus dipsaciformis Fernald, Rhodora 8: 127 (1906). *Cyperus retrofractus* var. *dipsaciformis* (Fernald) Kük. in H.G.A.Engler (ed.), Pflanzenr., IV, 20(101): 509 (1936).

Cyperus retrorsus Chapm., Bot. Gaz. 3: 17 (1878). *Mariscus retrorsus* (Chapm.) C.B.Clarke, Kew Bull., Addit. Ser. 8: 15 (1908). *Cyperus retrorsus* var. *typicus* Fernald & Griscom, Rhodora 37: 152 (1935), nom. inval.
C. & E. U.S.A. to Colombia. (21) azo 74 MSO OKL 75 NWJ NWY 77 TEX 78 ALA ARK DEL FLA GEO LOU MRY MSI NCA SCA TEN VRG WDC 79 MXE MXS 80 PAN 83 CLM. Hemicr. or rhizome geophyte.
Mariscus umbellatus Pursh, Fl. Amer. Sept. 1: 59 (1813), nom. illeg.
Mariscus cylindricus Elliott, Sketch Bot. S. Carolina 1: 74 (1816). *Cyperus ovularis* var. *cylindricus* (Elliott) Torr., Ann. Lyceum Nat. Hist. New York 3: 279 (1836). *Cyperus cylindricus* (Elliott) Britton, Bull. Torrey Bot. Club 7: 48 (1880), nom. illeg. *Cyperus retrorsus* var. *cylindricus* (Elliott) Fernald & Griscom, Rhodora 37: 152 (1935).
Mariscus ovularis Torr., Fl. N. Middle United States 1: 58 (1823), nom. illeg.
Mariscus neglectus Schult. in J.J.Roemer & J.A.Schultes, Mant. 2: 144 (1824).

Cyperus ovularis var. *tenellus* Torr., Ann. Lyceum Nat. Hist. New York 3: 279 (1836).
Mariscus aureus Bertol., Misc. Bot. 8: 29 (1849).
Cyperus ovularis var. *robustus* Boeck., Linnaea 36: 378 (1870). *Cyperus retrorsus* var. *robustus* (Boeck.) Kük. in H.G.A.Engler (ed.), Pflanzenr., IV, 20(101): 513 (1936). *Cyperus globulosus* var. *robustus* (Boeck.) Shinners, Field & Lab. 20: 34 (1952).
Cyperus retroversus Chapm., Fl. South. U.S., ed. 2: 659 (1883), orth. var.
Cyperus torreyi Britton, Bull. Torrey Bot. Club 13: 215 (1886).
Cyperus blodgettii Britton, Bull. Torrey Bot. Club 17: 316 (1890). *Mariscus blodgettii* (Britton) T.Koyama, Phytologia 29: 73 (1974).
Mariscus avenicola C.B.Clarke ex Britton, Bull. Torrey Bot. Club 17: 316 (1890).
Cyperus nashii Britton in J.K.Small, Fl. S.E. U.S.: 1321 (1903). *Cyperus retrorsus* var. *nashii* (Britton) Fernald & Griscom, Rhodora 37: 152 (1935).
Cyperus plankii Britton ex Small, Fl. S.E. U.S.: 172 (1903).
Mariscus brittonii C.B.Clarke, Kew Bull., Addit. Ser. 8: 14 (1908).
Cyperus retrorsus var. *multiflorens* Kük. in H.G.A. Engler (ed.), Pflanzenr., IV, 20(101): 513 (1936).

Cyperus rhynchosporoides Kük. in H.G.A.Engler (ed.), Pflanzenr., IV, 20(101): 537 (1936).
S. Zaïre to Angola. 23 ZAI 26 ANG ZAM. Hemicr. or rhizome geophyte.
**Rhynchospora ochrocephala* Boeck., Flora 62: 568 (1879).

Cyperus rigens J.Presl & C.Presl, Reliq. Haenk. 1: 170 (1828). *Mariscus rigens* (J.Presl & C.Presl) T.Koyama, Phytologia 29: 74 (1974).
Peru to Uruguay. (50) 83 BOL PER 84 BZS 85 AGE AGW CLN PAR URU. Hemicr. or rhizome geophyte.

subsp. ***rigens***
Peru to Uruguay. (50) 83 BOL PER 84 BZS 85 AGE AGW CLN PAR URU. Hemicr. or rhizome geophyte.
Cyperus rigens var. *capitatus* C.Presl, Reliq. Haenk. 1: 170 (1828).
Cyperus rigens var. *umbellatus* C.Presl, Reliq. Haenk. 1: 170 (1828).
Cyperus oostachyus Nees in C.F.P.von Martius & auct. suc. (eds.), Fl. Bras. 2(1): 39 (1842). *Chlorocyperus oostachyus* (Nees) Palla, Denkschr. Kaiserl. Akad. Wiss., Wien. Math.-Naturwiss. Kl. 79: 180 (1908). *Cyperus laetus* subsp. *oostachyus* (Nees) Kük., Repert. Spec. Nov. Regni Veg. 26: 253 (1929).
Cyperus conceptionis Steud., Syn. Pl. Glumac. 2: 42 (1854). *Mariscus conceptionis* (Steud.) C.B.Clarke, Bot. Jahrb. Syst. 30(68): 16 (1902). *Cyperus laetus* var. *conceptionis* (Steud.) Kük., Repert. Spec. Nov. Regni Veg. 26: 253 (1929).
Cyperus rufidulus Steud., Syn. Pl. Glumac. 2: 316 (1855). *Cyperus rigens* var. *rufidulus* (Steud.) Kük., Repert. Spec. Nov. Regni Veg. 26: 252 (1929).
Cyperus lacustris Phil., Linnaea 33: 270 (1865), nom. illeg.
Cyperus chilensis Boeck., Linnaea 36: 323 (1870).
Cyperus aurescens Boeck., Beitr. Cyper. 1: 8 (1888).
Cyperus quinquespicatus Boeck., Allg. Bot. Z. Syst. 1: 226 (1895).
Cyperus laetus var. *parciflorus* Kük., Repert. Spec. Nov. Regni Veg. 26: 252 (1929).
Cyperus rigens var. *latifolius* Kük., Repert. Spec. Nov. Regni Veg. 26: 252 (1929).

var. ***selmirae*** Pedersen, Darwiniana 17: 538 (1972).
Paraguay to NE. Argentina. 85 AGE PAR. Hemicr. or rhizome geophyte.

subsp. ***serrae*** (Boeck.) Pedersen, Darwiniana 17: 540 (1972).
Brazil to Argentina. 84 BZS 85 AGE URU. Hemicr. or rhizome geophyte.
**Cyperus serrae* Boeck., Allg. Bot. Z. Syst. 1: 227 (1895). *Chlorocyperus serrae* (Boeck.) Palla, Denkschr. Kaiserl. Akad. Wiss., Wien. Math.-Naturwiss. Kl. 79: 180 (1908).
Cyperus serrae var. *elongata* Boeck., Allg. Bot. Z. Syst. 2: 3 (1896).
Cyperus serrae var. *minor* Boeck., Allg. Bot. Z. Syst. 2: 3 (1896).

Cyperus rigidellus (Benth.) J.M.Black, Fl. S. Austral.: 676 (1929).
Australia. 50 NSW NTA QLD SOA VIC WAU. Hemicr. or rhizome geophyte.
**Cyperus gracilis* var. *rigidellus* Benth., Fl. Austral. 7: 266 (1878). *Mariscus rigidellus* (Benth.) C.B.Clarke, Kew Bull., Addit. Ser. 8: 18 (1908).

Cyperus rigidifolius Steud., Flora 25: 593 (1842).
Ethiopia to S. Africa, Arabian Pen. 23 BUR RWA ZAI 24 ETH 25 KEN TAN UGA 27 CPP LES NAT OFS SWZ TVL 35 SAU YEM. Hemicr. or rhizome geophyte.
Cyperus adoensis Hochst. ex A.Rich., Tent. Fl. Abyss. 2: 484 (1850). *Cyperus longus* var. *adoensis* (Hochst. ex A.Rich.) Boeck., Linnaea 36: 281 (1870).
Cyperus atrosanguineus Hochst. ex Steud., Syn. Pl. Glumac. 2: 30 (1854).
Cyperus rigidifolius var. *intercedens* Kük., Notizbl. Bot. Gart. Berlin-Dahlem 9: 304 (1925).

Cyperus robinsonii Podlech, Mitt. Bot. Staatssamml. München 4: 111 (1961).
Zambia. 26 ZAM. Hemicr. or rhizome geophyte.

Cyperus rockii Kük., Repert. Spec. Nov. Regni Veg. 16: 431 (1920). *Mariscus rockii* (Kük.) T.Koyama, Bishop Mus. Occas. Pap. 29: 124 (1989).
Hawaiian Is. (Kauai: Wai'alae valley). 63 HAW. Hemicr. or rhizome geophyte.

Cyperus rohlfsii Boeck., Flora 65: 13 (1882). *Mariscus rohlfsii* (Boeck.) C.B.Clarke in T.A.Durand & H.Schinz, Consp. Fl. Afric. 5: 592 (1894). *Cyperus impubes* var. *rohlfsii* (Boeck.) Kük. in H.G.A.Engler (ed.), Pflanzenr., IV, 20(101): 492 (1936).
NE. & E. Trop. Africa. 24 ERI ETH SOM 25 KEN TAN. Hemicr. or rhizome geophyte.

Cyperus rotundus L., Sp. Pl.: 45 (1753). *Chlorocyperus rotundus* (L.) Palla, Allg. Bot. Z. Syst. 6: 61 (1900). *Pycreus rotundus* (L.) Hayek, Repert. Spec. Nov. Regni Veg. Beih. 30(3): 147 (1932).
Cosmopolitan. 11 AUT SWI 12 BAL COR FRA POR SAR SPA 13 ALB BUL GRC ITA KRI ROM SIC TUE YUG 20 ALG EGY LBY MOR TUN WSA 21 AZO CNY CVI MDR SEL 22 BEN BKN GHA GUI IVO MLI MTN NGA NGR SEN SIE TOG 23

BUR EQG GAB RWA ZAI 24 CHA DJI ERI ETH SOC SOM SUD 25 KEN TAN UGA 26 ANG MLW MOZ ZAM ZIM 27 BOT CPP CPV LES NAM NAT OFS SWZ TVL 28 STH 29 ALD CGS COM MAU MCI MDG REU ROD SEY 32 KAZ KGZ TKM UZB 33 NCS TCS 34 AFG CYP IRN IRQ LBS PAL SIN TUR 35 GST SAU YEM 36 CHC CHS 38 JAP KOR KZN NNS OGA TAI 40 EHM IND NEP PAK SRL 41 AND MYA THA VIE 42 BOR CKI JAW LSI MLY MOL PHI SUL SUM XMS (50) nfk NSW NTA QLD SOA WAU (60) fij nue ton (61) coo sci tua 62 CRL MRN MRS WAK 63 HAW 74 MSO (76) ari cal 77 NWM TEX 78 ALA ARK FLA GEO LOU MSI NCA SCA TEN VRG 79 MXC MXE MXG MXN MXS MXT 80 BLZ COS CPI ELS GUA HON NIC PAN 81 LEE PUE 82 FRG GUY SUR VEN 83 BOL CLM ECU PER 84 BZE BZL BZS 85 AGE AGS AGW clc. Tuber geophyte.

subsp. ***divaricatus*** Lye, Nordic J. Bot. 16: 367 (1996).
N. Somalia. 24 SOM. Hemicr. or rhizome geophyte.

subsp. ***merkeri*** (C.B.Clarke) Kük. in H.G.A.Engler (ed.), Pflanzenr., IV, 20(101): 115 (1935).
Trop. Africa. 23 GAB RWA 24 SOM 25 KEN TAN UGA. Hemicr.
**Cyperus merkeri* C.B.Clarke, Bot. Jahrb. Syst. 38: 134 (1906).

subsp. ***retzii*** Kük. in H.G.A.Engler (ed.), Pflanzenr., IV, 20(101): 114 (1935).
Trop. 22 GHA NGA 23 GGI ZAI 24 SOM SUD 25 TAN UGA 26 ANG 35 YEM 40 IND SRL 41 THA 42 LSI 43 BIS NWG 50 NSW NTA QLD SOA WAU 83 ECU 84 BZE BZL. Tuber geophyte.
**Cyperus retzii* Nees in R.Wight, Contr. Bot. India: 82 (1834), nom. illeg.
Cyperus rotundus var. *pallidus* Benth., Fl. Austral. 7: 280 (1878).
Cyperus nubicus C.B.Clarke in D.Oliver & auct. suc. (eds.), Fl. Trop. Afr. 8: 360 (1901). *Cyperus rotundus* var. *nubicus* (C.B.Clarke) Kük. in H.G.A.Engler (ed.), Pflanzenr., IV, 20(101): 115 (1935).
Cyperus bifax C.B.Clarke, Kew Bull., Addit. Ser. 8: 13 (1908).
Cyperus disruptus C.B.Clarke, Kew Bull., Addit. Ser. 8: 12 (1908). *Cyperus rotundus* var. *disruptus* (C.B.Clarke) Kük. in H.G.A.Engler (ed.), Pflanzenr., IV, 20(101): 115 (1935).
Cyperus weinlandii Kük. in H.G.A.Engler (ed.), Pflanzenr., IV, 20(101): 131 (1935).

subsp. ***rotundus***
Cosmopolitan. 11 AUT SWI 12 BAL COR FRA POR SAR SPA 13 ALB BUL GRC ITA KRI ROM SIC TUE YUG 20 ALG EGY LBY MOR TUN WSA 21 AZO CNY CVI MDR SEL 22 BEN BKN GHA GUI IVO MLI MTN NGA NGR SEN SIE TOG 23 BUR EQG GAB RWA ZAI 24 CHA DJI ERI ETH SOC SOM SUD 25 KEN TAN UGA 26 ANG MLW MOZ ZAM ZIM 27 BOT CPP CPV LES NAM NAT OFS SWZ TVL 28 STH 29 ALD CGS COM MAU MCI MDG REU ROD SEY 32 KAZ KGZ TKM UZB 33 NCS TCS 34 AFG CYP IRN IRQ LBS PAL SIN TUR 35 GST SAU YEM 36 CHC CHS 38 JAP KOR KZN NNS OGA TAI 40 EHM IND NEP PAK SRL 41 AND MYA THA VIE 42 BOR CKI JAW LSI MLY MOL PHI SUL SUM XMS (50) nfk NSW NTA QLD SOA WAU (60) fij nue ton (61) coo sci tua 62 CRL MRN MRS WAK 63 HAW 74 MSO (76) ari cal 77 NWM TEX 78 ALA ARK FLA GEO LOU MSI NCA SCA TEN VRG 79 MXC MXE MXG MXN MXS MXT 80 BLZ COS CPI ELS GUA HON NIC PAN 81 LEE PUE 82 FRG GUY SUR VEN 83 BOL CLM ECU PER 84 BZE BZL BZS 85 AGE AGS AGW clc. Tuber geophyte.
Schoenus tuberosus Burm.f., Fl. Indica: 190 (1768).
Cyperus hexastachyos Rottb., Descr. Icon. Rar. Pl.: 28 (1773).
Cyperus tetrastachyos Desf., Fl. Atlant. 1: 45 (1798).
Cyperus hydra Michx., Fl. Bor.-Amer. 1: 27 (1803).
Cyperus rotundus var. *hydra* (Michx.) A.Gray, Manual, ed. 2: 493 (1856).
Cyperus bicolor Vahl, Enum. Pl. 2: 340 (1805).
Cyperus comosus Sm. in J.Sibthorp & J.E.Smith, Fl. Graec. 1: 31 (1806). *Cyperus rotundus* var. *comosus* (Sm.) Nyman, Consp. Fl. Eur.: 759 (1882). *Cyperus rotundus* f. *comosus* (Sm.) Kük. in H.G.A.Engler (ed.), Pflanzenr., IV, 20(101): 112 (1935).
Cyperus hildra Poir. in J.B.A.M.de Lamarck, Encycl. 7: 251 (1806).
Cyperus pallescens Poir. in J.B.A.M.de Lamarck, Encycl. 7: 260 (1806), nom. illeg.
Cyperus radicosus Sm. in J.Sibthorp & J.E.Smith, Fl. Graec. Prodr. 1: 30 (1806).
Cyperus olivaris O.Targ.Tozz., Mem. Mat. Fis. Soc. Ital. Sci. 13(2): 338 (1807).
Cyperus agrestis Willd. ex Spreng. & Link, Jahrb. Gewächsk. 1(3): 86 (1818).
Cyperus rubicundus Willd. ex Link, Jahrb. Gewächsk. 1(3): 87 (1818), nom. illeg.
Cyperus patulus M.Bieb., Fl. Taur.-Caucas. 3: 47 (1819), nom. illeg.
Cyperus procerulus Nees in R.Wight, Contr. Bot. India: 82 (1834).
Cyperus elongatus Sieber ex Kunth, Enum. Pl. 2: 59 (1837), nom. illeg.
Cyperus leptostachyus Griff., Itin. Pl. Khasyah Mts.: 321 (1848).
Cyperus tenuifolius Walp., Ann. Bot. Syst. 1: 898 (1849), sphalm.
Cyperus inconspicuus Gennari, Linnaea 24: 200 (1851), nom. illeg. *Cyperus rotundus* var. *inconspicuus* Nyman, Consp. Fl. Eur.: 759 (1882). *Cyperus badius* var. *inconspicuus* Nyman, Consp. Fl. Eur., Suppl. 2: 317 (1890). *Cyperus rotundus* f. *inconspicuus* (Gennari) K.Richt., Pl. Eur. 1: 135 (1890).
Cyperus rotundus var. *major* Parl., Fl. Ital. 2: 37 (1852).
Cyperus laevissimus Steud., Syn. Pl. Glumac. 2: 32 (1854).
Cyperus micreilema Steud., Syn. Pl. Glumac. 2: 32 (1854).
Cyperus ochreoides Steud., Syn. Pl. Glumac. 2: 34 (1854).
Cyperus officinalis Nees ex Gren. & Godr., Fl. France 3: 359 (1855), nom. illeg.
Cyperus bulbosostoloniferus Miq., Fl. Ned. Ind., Eerste Bijv.: 260, 559 (1861).
Cyperus stoloniferus var. *pallidus* Boeck., Linnaea 35: 490 (1868).
Cyperus rotundus var. *acutus* Boeck., Linnaea 36: 284 (1870).

Cyperus rotundus var. *elongatus* Boeck., Linnaea 36: 285 (1870).
Cyperus rotundus var. *spadiceus* Boeck., Linnaea 36: 284 (1870).
Cyperus herbicavus Melliss, St. Helena: 343 (1875).
Cyperus rotundus var. *carinalis* Benth., Fl. Austral. 7: 280 (1878).
Cyperus proteinolepis Boeck., Flora 62: 549 (1879), nom. illeg.
Cyperus pseudovariegatus Boeck., Beitr. Cyper. 2: 37 (1880).
Cyperus longus Boeck., J. Linn. Soc., Bot. 18: 104 (1881), nom. illeg.
Cyperus pallescens Boiss., Fl. Orient. 5: 375 (1882), nom. illeg.
Cyperus rotundus var. *macrostachyus* Boiss., Fl. Orient. 5: 377 (1882).
Cyperus rudioi Boeck., Flora 65: 12 (1882).
Cyperus rudioi var. *minor* Boeck., Flora 65: 12 (1882).
Cyperus rotundus var. *amaliae* C.B.Clarke, J. Linn. Soc., Bot. 21: 172 (1884).
Cyperus rotundus var. *centiflorus* C.B.Clarke, J. Linn. Soc., Bot. 21: 171 (1884).
Cyperus rotundus var. *procerula* C.B.Clarke, J. Linn. Soc., Bot. 21: 172 (1884).
Cyperus rotundus var. *salsolus* C.B.Clarke, J. Linn. Soc., Bot. 21: 171 (1884).
Cyperus viridis Roxb. ex C.B.Clarke, J. Linn. Soc., Bot. 21: 172 (1884), nom. inval.
Cyperus purpureovariegatus Boeck., Beitr. Cyper. 2: 37 (1890).
Cyperus subcapitatus C.B.Clarke in J.D.Hooker, Fl. Brit. India 6: 616 (1893). *Cyperus rotundus* var. *subcapitatus* (C.B.Clarke) Kük. in H.G.A.Engler (ed.), Pflanzenr., IV, 20(101): 113 (1935).
Cyperus rotundus var. *platystachys* Bojer ex C.B.Clarke in W.H.Harvey & auct. suc. (eds.), Fl. Cap. 7: 182 (1897).
Cyperus olivaris var. *brevibracteatus* Legrand, Bull. Assoc. Franç. Bot. 4: 61 (1901). *Cyperus rotundus* subsp. *brevibracteatus* (Legrand) M.Laínz, Aport. Conocim. Fl. Gallega 7: 34 (1971).
Cyperus taylorii C.B.Clarke in D.Oliver & auct. suc. (eds.), Fl. Trop. Afr. 8: 367 (1901). *Cyperus rotundus* var. *taylorii* (C.B.Clarke) Kük. in H.G.A.Engler (ed.), Pflanzenr., IV, 20(101): 114 (1935).
Cyperus rotundus var. *carinatus* F.M.Bailey, Queensl. Fl. 6: 1745 (1902), orth. var.
Cyperus platystachys Cherm., Ann. Inst. Bot.-Géol. Colon. Marseille, III, 10: 48 (1922).
Cyperus rotundus f. *contractus* Kük. in H.G.A. Engler (ed.), Pflanzenr., IV, 20(101): 112 (1935).
Cyperus rotundus f. *depallescens* Ekman & Kük. in H.G.A.Engler (ed.), Pflanzenr., IV, 20(101): 112 (1935).
Cyperus rotundus f. *latifolius* Kük. in H.G.A.Engler (ed.), Pflanzenr., IV, 20(101): 112 (1935).
Cyperus rotundus f. *latimarginatus* Kük. in H.G.A. Engler (ed.), Pflanzenr., IV, 20(101): 112 (1935).
Cyperus yoshinagae Ohwi, J. Jap. Bot. 13: 332 (1937).

Cyperus rubicundus Vahl, Enum. Pl. 2: 308 (1805).
Africa to N. Australia. 21 CNY 23 BUR RWA ZAI 24 ERI ETH SOM 25 KEN TAN UGA 26 ANG 27 CPP NAM NAT SWZ TVL 29 MDG 35 GST SAU YEM 40 IND 42 LSI 50 QLD. Ther. or hemicr.
Cyperus teneriffae Poir. in J.B.A.M.de Lamarck, Encycl. 7: 245 (1806).
Cyperus tersus Link, Jahrb. Gewächsk. 1(3): 82 (1818).
Cyperus coromandelinus Spreng., Syst. Veg. 1: 217 (1824).
Cyperus wightii Nees in R.Wight, Contr. Bot. India: 78 (1834).
Cyperus delesserianus Webb & Berthel., Hist. Nat. Iles Canaries 3: 363 (1848).
Cyperus serra A.Rich., Tent. Fl. Abyss. 2: 479 (1850).
Cyperus petraeus Hochst. ex Steud., Syn. Pl. Glumac. 2: 29 (1854). *Cyperus teneriffae* f. *petraeus* (Hochst. ex Steud.) Kük. in H.G.A.Engler (ed.), Pflanzenr., IV, 20(101): 307 (1936).
Cyperus calcicola Domin, Biblioth. Bot. 85: 422 (1915), nom. illeg.
Cyperus teneriffae var. *longimucronatus* Kük., Notizbl. Bot. Gart. Berlin-Dahlem 9: 302 (1925).

Cyperus rufostriatus C.B.Clarke ex Cherm., Bull. Soc. Bot. France 66: 347 (1919).
E. Madagascar. 29 MDG. Hemicr. or rhizome geophyte.

Cyperus rupestris Kunth, Enum. Pl. 2: 52 (1837).
S. Trop. & S. Africa, Indo-China. 26 ANG MLW 27 BOT CPP LES NAM NAT OFS SWZ TVL 41 VIE. Hemicr. or rhizome geophyte.
Cyperus amnicola Kunth, Enum. Pl. 2: 52 (1837). *Cyperus rupestris* var. *amnicola* (Kunth) Kük. in H.G.A.Engler (ed.), Pflanzenr., IV, 20(101): 304 (1936).
Cyperus cognatus Kunth, Enum. Pl. 2: 52 (1837).
Cyperus parvinux C.B.Clarke in W.H.Harvey & auct. suc. (eds.), Fl. Cap. 7: 167 (1897). *Cyperus rupestris* var. *parvinux* (C.B.Clarke) Kük. in H.G.A. Engler (ed.), Pflanzenr., IV, 20(101): 304 (1936).

Cyperus rupicola S.T.Blake, Proc. Roy. Soc. Queensland 51: 32 (1939 publ. 1940).
Queensland to NE. New South Wales. 50 NSW QLD. Hemicr. or rhizome geophyte.

Cyperus sahelii Väre & Kukkonen, Ann. Bot. Fenn. 42: 478 (2005).
Niger. 22 NGR.

Cyperus sandwicensis Kük., Repert. Spec. Nov. Regni Veg. 16: 430 (1920). *Mariscus sandwicensis* (Kük.) T.Koyama, Bishop Mus. Occas. Pap. 29: 124 (1989).
Hawaiian Is. 63 HAW. Hemicr. or rhizome geophyte.
Cyperus sandwicensis var. *pseudoprescottianus* Kük. in H.G.A.Engler (ed.), Pflanzenr., IV, 20(101): 406 (1936).

Cyperus sanguineoater Boeck., Bot. Jahrb. Syst. 1: 363 (1881).
Mexico to Guatemala. 79 MXC MXE MXN MXS 80 GUA. Hemicr. or rhizome geophyte.
**Mariscus atrosanguineus* Liebm., Mexic. Neldeagt. Pl., V, 2: 228 (1851).
Cyperus atrosanguineus Hemsl., Biol. Cent.-Amer., Bot. 3: 445 (1885), nom. illeg.
Chlorocyperus michoacanensis Palla, Oesterr. Bot. Z. 63: 403 (1913).
Cyperus sanguineoater var. *floribundus* Kük. in H.G.A. Engler (ed.), Pflanzenr., IV, 20(101): 464 (1936).

Cyperus sartorii Kük. in H.G.A.Engler (ed.), Pflanzenr., IV, 20(101): 494 (1936).
Cape Prov. 27 CPP. Hemicr. or rhizome geophyte.

Cyperus scaber (R.Br.) Boeck., Flora 58: 107 (1875).
N. & NE. Australia. 50 NSW NTA QLD. Hemicr. or rhizome geophyte.
**Mariscus scaber* R.Br., Prodr.: 218 (1810).

Cyperus scabricaulis Lye, Nordic J. Bot. 16: 369 (1996).
Coastal S. Somalia. 24 SOM. Hemicr. or rhizome geophyte.

Cyperus scariosus R.Br., Prodr.: 216 (1810). *Cyperus corymbosus* var. *scariosus* (R.Br.) Kük. in H.G.A. Engler (ed.), Pflanzenr., IV, 20(101): 83 (1935).
Bangladesh to N. Australia. 40 BAN 41 MYA 42+ 43 NWG 50 NTA QLD WAU. Hemicr. or rhizome geophyte.

Cyperus schaffneri Boeck., Flora 61: 30 (1878).
Mexico (México D.F.). 79 MXC.

Cyperus schimperianus Steud., Syn. Pl. Glumac. 2: 34 (1854).
Egypt to Cameroon and Tanzania, Arabian Pen. 20 EGY 23 CMN ZAI 24 ETH SUD 25 KEN TAN UGA 35 SAU. Hemicr. or rhizome geophyte.
Cyperus truncatus A.Rich., Tent. Fl. Abyss. 2: 487 (1850).
Cyperus amblyleptos Steud., Syn. Pl. Glumac. 2: 30 (1854).
Cyperus schimperianus var. *minor* Boeck., Flora 62: 551 (1879).
Cyperus elongatus Hochst. ex C.B.Clarke, J. Linn. Soc., Bot. 21: 160 (1884), nom. illeg.
Cyperus nudiculmis Sieber ex C.B.Clarke, J. Linn. Soc., Bot. 21: 161 (1884).
Cyperus tegetum C.B.Clarke, J. Linn. Soc., Bot. 21: 160 (1884), nom. illeg.
Cyperus tegetum var. *protracta* C.B.Clarke, J. Linn. Soc., Bot. 21: 162 (1884).

Cyperus schinzii Boeck., Verh. Bot. Vereins Prov. Brandenburg 29: 45 (1888).
S. Africa. 27 BOT NAM TVL. Hemicr. or rhizome geophyte.
Cyperus purpureus Boeck., Verh. Bot. Vereins Prov. Brandenburg 29: 45 (1888).

Cyperus schlechteri C.B.Clarke, Bull. Herb. Boissier, II, 3: 663 (1903).
S. Africa. 27 CPP LES NAT TVL. Hemicr. or rhizome geophyte.

Cyperus schoenomorphus Steud., Syn. Pl. Glumac. 2: 39 (1854).
Brazil. 84 BZL BZN. Hemicr. or rhizome geophyte.
Cyperus spicigerus Vahl, Enum. Pl. 2: 300 (1805), provisional synonym.

Cyperus schomburgkianus Nees, J. Bot. (Hooker) 2: 393 (1840).
N. South America to Brazil. 82 GUY VEN 84 BZC BZE BZL. Hemicr. or rhizome geophyte.
Cyperus leucanthus Schrad. ex Nees in C.F.P.von Martius & auct. suc. (eds.), Fl. Bras. 2(1): 18 (1842). *Cyperus schomburgkianus* var. *leucanthus* (Schrad. ex Nees) Kük. in H.G.A.Engler (ed.), Pflanzenr., IV, 20(101): 277 (1936).
Cyperus schomburgkianus var. *trilobatus* Kük. in H.G.A. Engler (ed.), Pflanzenr., IV, 20(101): 277 (1936).

Cyperus schweinfurthii (Chiov.) Kük. in H.G.A.Engler (ed.), Pflanzenr., IV, 20(101): 553 (1936).
Eritrea to Ethiopia. 24 ERI ETH. Hemicr. or rhizome geophyte.
**Mariscus schweinfurthii* Chiov., Annuario Reale Ist. Bot. Roma 8: 78 (1903).

Cyperus schweinitzii Torr., Ann. Lyceum Nat. Hist. New York 3: 276 (1836). *Chlorocyperus schweinitzii* (Torr.) Palla, Oesterr. Bot. Z. 58: 392 (1908). *Mariscus schweinitzii* (Torr.) T.Koyama, Phytologia 29: 74 (1974).
S. Canada to NE. Mexico. 71 ABT MAN SAS 72 ONT QUE 73 COL MNT WAS WYO 74 ILL IOW MIN MSO OKL SDA WIS 75 MAS MIC NWJ NWY OHI PEN 76 UTA 77 NWM TEX 78 KTY 79 MXE. Hemicr. or rhizome geophyte.
**Cyperus alterniflorus* Schwein. in W.H.Keating, Narr. Exp. St. Peter's River 2: 381 (1824), nom. illeg.
Cyperus mesochorus Geise, Amer. Midl. Naturalist 15: 249 (1934).

Cyperus sciaphilus Cherm., Bull. Soc. Bot. France 66: 346 (1919 publ. 1920).
E. Madagascar. 29 MDG. Hemicr. or rhizome geophyte.

Cyperus scleropodus Chiov., Pl. Nov. Aethiop.: 11 (1928). *Mariscus scleropodus* (Chiov.) Cufod., Bull. Jard. Bot. Natl. Belg. 40(Suppl.): 1456 (1970).
N. Somalia. 24 SOM. Hemicr.

Cyperus sculptus S.T.Blake, Proc. Roy. Soc. Queensland 51: 34 (1939 publ. 1940).
Queensland to NE. New South Wales. 50 NSW QLD. Hemicr. or rhizome geophyte.

Cyperus secubans K.L.Wilson, Telopea 4: 474 (1991).
New South Wales. 50 NSW. Hemicr. or rhizome geophyte.

Cyperus seemannianus Boeck., Linnaea 36: 390 (1870). *Mariscus seemannianus* (Boeck.) Palla, Denkschr. Kaiserl. Akad. Wiss., Wien. Math.-Naturwiss. Kl. 84: 452 (1909).
S. Pacific. 60 FIJ NUE SAM TON 61 SCI. Hemicr. or rhizome geophyte.
Mariscus laevigatus Hook. & Arn., Bot. Beechey Voy.: 72 (1832).
**Mariscus flavus* Seem., Fl. Vit.: 319 (1868), nom. illeg.
Cyperus monostachys Boeck., Linnaea 36: 389 (1870). *Cyperus seemannianus* var. *monostachys* (Boeck.) Kük., Bernice P. Bishop Mus. Bull. 128: 18 (1934).
Cyperus upoluensis Boeck., Flora 58: 88 (1875).
Cyperus seemannianus var. *altus* Kük., Bernice P. Bishop Mus. Bull. 128: 17 (1934).
Cyperus seemannianus var. *perdensus* Kük., Bernice P. Bishop Mus. Bull. 128: 19 (1934).

Cyperus semifertilis S.T.Blake, Proc. Roy. Soc. Queensland 51: 33 (1939 publ. 1940).
Queensland. 50 QLD. Hemicr. or rhizome geophyte.

Cyperus semiochraceus Boeck., Flora 61: 29 (1878).
Mexico to C. America. 79 MXC MXE MXG MXN MXS MXT 80 BLZ GUA NIC. Hemicr. or rhizome geophyte.
Cyperus bourgaei C.B.Clarke ex Lundell, Publ. Carnegie Inst. Wash. 436: 287 (1934), nom. subnud.

Cyperus semitrifidus Schrad., Anal. Fl. Cap. 1: 6 (1832).
S. Africa. 27 CPP LES NAT OFS SWZ TVL. Hemicr. or rhizome geophyte.
Cyperus crinitus Spreng., Flora 12(Beibl. 1): 5 (1829), nom. illeg.
Cyperus usitatus var. *sanguinolentus* Nees, Linnaea 10: 136 (1836). *Cyperus semitrifidus* var. *sanguinolentus*

(Nees) Kük. in H.G.A.Engler (ed.), Pflanzenr., IV, 20(101): 301 (1936).
Cyperus herbivagus Kunth, Enum. Pl. 2: 53 (1837).
Cyperus apricus Ridl., Trans. Linn. Soc. London, Bot. 2: 141 (1884). *Cyperus semitrifidus* var. *apricus* (Ridl.) Kük. in H.G.A.Engler (ed.), Pflanzenr., IV, 20(101): 301 (1936).
Cyperus multiglumis Turrill, Bull. Misc. Inform. Kew 1925: 67 (1925). *Cyperus semitrifidus* var. *multiglumis* (Turrill) Kük. in H.G.A.Engler (ed.), Pflanzenr., IV, 20(101): 301 (1936).

Cyperus sensilis Baijnath, J. S. African Bot. 42: 369 (1976).
KwaZulu-Natal. 27 NAT. Hemicr. or rhizome geophyte.

Cyperus serotinus Rottb., Descr. Icon. Rar. Pl.: 31 (1773). *Juncellus serotinus* (Rottb.) C.B.Clarke in J.D.Hooker, Fl. Brit. India 6: 594 (1893). *Duval-jouvea serotina* (Rottb.) Palla in W.D.J.Koch, Syn. Deut. Schweiz. Fl., ed. 3, 3: 2556 (1905). *Pycreus serotinus* (Rottb.) Hayek, Repert. Spec. Nov. Regni Veg. Beih. 30(3): 145 (1932).
Temp. Eurasia to Vietnam. 11 AUT HUN SWI† 12 COR FRA POR SPA 13 ALB BUL ITA ROM TUE YUG 14 RUC RUS 31+ 32 KAZ KGZ TKM UZB 33 TCS 34 AFG EAI IRN TUR 36 CHC CHI CHM CHS 38 JAP KOR KZN NNS TAI 40 ASS IND PAK WHM 41 VIE (50) nta (75) nwj pen (78) del. Hemicr. or rhizome geophyte.
Cyperus monti L.f., Suppl. Pl.: 102 (1782). *Pycreus monti* (L.f.) P.Beauv. ex Rchb., Fl. Germ. Excurs.: 72 (1830).
Cyperus glaber Vill., Hist. Pl. Dauphiné 2: 182 (1787), nom. illeg.
Cyperus inundatus Roxb., Fl. Ind. 1: 205 (1820), nom. illeg. *Juncellus inundatus* C.B.Clarke in J.D.Hooker, Fl. Brit. India 6: 595 (1893). *Cyperus serotinus* var. *inundatus* (C.B.Clarke) Kük. in H.G.A.Engler (ed.), Pflanzenr., IV, 20(101): 318 (1936).
Cyperus diluvialis Schult. in J.J.Roemer & J.A. Schultes, Mant. 2: 124 (1824).
Cyperus colchicus K.Koch, Linnaea 21: 623 (1848). *Pycreus colchicus* (K.Koch) Schischk. ex Grossh., Fl. Kavkaza 1: 152 (1928). *Cyperus tremulus* var. *colchicus* (K.Koch) Kük. in H.G.A.Engler (ed.), Pflanzenr., IV, 20(101): 362 (1936).
Cyperus amoyensis Hance, Ann. Sci. Nat., Bot., V, 5: 249 (1866).
Cyperus japonicus Miq., Ann. Mus. Bot. Lugduno-Batavi 2: 140 (1866).
Cyperus puncticulatus Aitch., Cat. Pl. Punjab Sindh: 189 (1869), nom. illeg.
Cyperus monti var. *stylosa* C.B.Clarke, J. Linn. Soc., Bot. 21: 73 (1885). *Juncellus stylosus* (C.B.Clarke) C.B.Clarke in J.D.Hooker, Fl. Brit. India 6: 595 (1893).
Cyperus krebsii Boeck., Beitr. Cyper. 2: 2 (1890).
Chlorocyperus serotinus Palla, Allg. Bot. Z. Syst. 6: 221 (1900).
Cyperus litvinovii Meinsh., Trudy Imp. S.-Peterburgsk. Bot. Sada 18: 233 (1900).
Cyperus serotinus f. *contortus* Ces. ex Asch. & Graebn., Syn. Mitteleur. Fl. 2(2): 270 (1904). *Juncellus serotinus* f. *contortus* (Ces. ex Asch. & Graebn.) Soó, Acta Bot. Acad. Sci. Hung. 16: 369 (1970 publ. 1971).
Cyperus scleropoae Lojac., Fl. Sicul. 3: 199 (1909).
Cyperus makinoi Nakai, Bot. Mag. (Tokyo) 26: 187 (1912).
Cyperus serotinus f. *depauperatus* Kük. in H.G.A. Engler (ed.), Pflanzenr., IV, 20(101): 318 (1936).
Cyperus serotinus f. *grandis* Kük. in H.G.A.Engler (ed.), Pflanzenr., IV, 20(101): 318 (1936). *Juncellus serotinus* f. *grandis* (Kük.) Soó, Acta Bot. Acad. Sci. Hung. 16: 369 (1970 publ. 1971).
Juncellus serotinus var. *capitatus* D.Z.Ma, Fl. Ningxiaensis 2: 522 (1988 publ. 1990).

Cyperus seslerioides Kunth in F.W.H.von Humboldt, A.J.A.Bonpland & C.S.Kunth, Nov. Gen. Sp. 1: 209 (1817).
S. U.S.A. to Peru. 76 ARI 77 TEX 79 MXC MXE MXG MXN MXS MXT 80 COS GUA HON NIC 82 VEN 83 BOL PER 85 AGE AGW. Hemicr. or rhizome geophyte.
Cyperus divergens Benth., Pl. Hartw.: 28 (1840), nom. illeg.
Cyperus perpusillus Boeck., Linnaea 35: 499 (1868). *Cyperus seslerioides* f. *minor* C.B.Clarke ex Donn. Sm., Enum. Pl. Guatem. 3: 87 (1893).
Cyperus seslerioides var. *minor* C.B.Clarke in J.D. Smith, Enum. Pl. Guatem. 3: 87 (1893).
Scirpus hahnii C.B.Clarke, Kew Bull., Addit. Ser. 8: 30 (1908).
Cyperus andinus Palla ex Kük. in H.G.A.Engler (ed.), Pflanzenr., IV, 20(101): 310 (1936).

Cyperus setigerus Torr. & Hook., Ann. Lyceum Nat. Hist. New York 3: 434 (1836).
C. U.S.A. 74 KAN MSO OKL 77 NWM TEX. Hemicr. or rhizome geophyte.
Cyperus lutescens var. *fulvescens* C.B.Clarke, J. Linn. Soc., Bot. 21: 182 (1884). *Cyperus setigerus* var. *fulvescens* (C.B.Clarke) Kük. in H.G.A.Engler (ed.), Pflanzenr., IV, 20(101): 97 (1935).
Cyperus hallii Britton, Bull. Torrey Bot. Club 13: 211 (1886). *Cyperus setigerus* var. *hallii* (Britton) Kük. in H.G.A.Engler (ed.), Pflanzenr., IV, 20(101): 97 (1935).
Cyperus berlandieri C.B.Clarke, Kew Bull., Addit. Ser. 8: 10 (1908).

Cyperus sexangularis Nees, Linnaea 10: 135 (1836).
S. Trop. & S. Africa. 26 ZAM 27 BOT CPP NAM SWZ TVL. Hemicr. or rhizome geophyte.
Cyperus webbianus Steud., Syn. Pl. Glumac. 2: 20 (1854).

Cyperus sexflorus R.Br., Prodr.: 215 (1810). *Cyperus sporobolus* var. *sexflorus* (R.Br.) Benth., Fl. Austral. 7: 282 (1878).
N. Australia. 50 NTA WAU. Hemicr. or rhizome geophyte.

Cyperus shandongense F.Z.Li, Acta Phytotax. Sin. 22: 151 (1984).
China. 36+. Hemicr. or rhizome geophyte.

Cyperus sharonensis Danin & Kukkonen, Israel J. Pl. Sci. 43: 77 (1995).
W. Israel. 34 PAL. Hemicr. or rhizome geophyte.

Cyperus sieberianus Spreng., Syst. Veg. 4(2): 29 (1827).
Cape Verde (Santo Antão). 21 CVI. Hemicr. or rhizome geophyte.

Cyperus sikkimensis Kük. in H.G.A.Engler (ed.), Pflanzenr., IV, 20(101): 467 (1936).
Sikkim. 40 EHM. Hemicr. or rhizome geophyte.
**Mariscus hookerianus* C.B.Clarke in J.D.Hooker, Fl. Brit. India 6: 623 (1893).

Cyperus silletensis Nees in R.Wight, Contr. Bot. India: 79 (1834).
Assam to Vietnam. 40 ASS BAN EHM 41 MYA VIE. Hemicr. or rhizome geophyte.

Cyperus simplex Kunth in F.W.H.von Humboldt, A.J.A. Bonpland & C.S.Kunth, Nov. Gen. Sp. 1: 207 (1817).
S. Mexico to Trop. America. 79 MXS MXT 80 COS ELS GUA HON NIC PAN 81 TRT 82 FRG GUY SUR VEN 83 BOL CLM ECU PER 84 BZC BZE BZN. Hemicr. or rhizome geophyte.

Cyperus solidifolius Boeck., Bot. Jahrb. Syst. 5: 499 (1884).
Madagascar. 29 MDG. Hemicr. or rhizome geophyte.
Cyperus herana var. *angustifolius* Cherm., Ann. Inst. Bot.-Géol. Colon. Marseille, III, 10: 37 (1922).

Cyperus solidus Kunth, Enum. Pl. 2: 76 (1837). *Mariscus solidus* (Kunth) Vorster, S. African J. Bot. 52: 265 (1986).
S. Africa. 27 CPP NAT SWZ TVL. Hemicr. or rhizome geophyte.

Cyperus somalidunensis Lye, Nordic J. Bot. 16: 374 (1996).
S. Somalia. 24 SOM. Hemicr. or rhizome geophyte.

Cyperus somaliensis C.B.Clarke, Bull. Misc. Inform. Kew 1895: 229 (1895).
N. Somalia. 24 SOM. Hemicr.
Mariscus somaliensis C.B.Clarke, Bull. Misc. Inform. Kew 1895: 229 (1895).

Cyperus soongoricus Kar. & Kir., Bull. Soc. Imp. Naturalistes Moscou 14: 859 (1841).
E. Kazakhstan. 32 KAZ. Hemicr. or rhizome geophyte.

Cyperus sordidus J.Presl & C.Presl, Reliq. Haenk. 1: 171 (1828).
W. Mexico. 79 MXN MXS. Hemicr. or rhizome geophyte.
Cyperus planifolius f. *decolorans* Kük. in H.G.A. Engler (ed.), Pflanzenr., IV, 20(101): 448 (1936).
Cyperus howellii O'Neill & Ben.Ayers, Leafl. W. Bot. 4: 35 (1944).

Cyperus soyauxii Boeck., Bot. Jahrb. Syst. 5: 501 (1884). *Mariscus soyauxii* (Boeck.) C.B.Clarke in T.A.Durand & H.Schinz, Consp. Fl. Afric. 5: 593 (1894).
W. Trop. Africa to Kenya. 22 BEN GHA IVO NGA SEN TOG 23 GAB 25 KEN. Hemicr. or rhizome geophyte.

subsp. ***pallescens*** Lye, Nordic J. Bot. 3: 227 (1983).
Kenya. 25 KEN. Hemicr. or rhizome geophyte.

subsp. ***soyauxii***
W. & WC. Trop. Africa. 22 BEN GHA IVO NGA SEN TOG 23 GAB. Hemicr. or rhizome geophyte.

Cyperus spectabilis Link, Hort. Berol. 1: 318 (1827). *Mariscus spectabilis* (Link) T.Koyama, Phytologia 29: 74 (1974).
S. U.S.A. to NW. Argentina. 76 ARI 77 TEX 79 MXC MXE MXG MXN MXS 80 COS GUA HON 83 PER 85 AGW. Hemicr. or rhizome geophyte.

var. ***jujuyensis*** Kük. in H.G.A.Engler (ed.), Pflanzenr., IV, 20(101): 461 (1936).
NW. Argentina. 85 AGW. Hemicr. or rhizome geophyte.

var. ***spectabilis***
S. U.S.A. to Peru. 76 ARI 77 TEX 79 MXC MXE MXN MXS 80 COS GUA HON 83 PER. Hemicr. or rhizome geophyte.
Cyperus scaberrimus Nees, Linnaea 19: 697 (1847). *Cyperus spectabilis* var. *scaberrimus* (Nees) Boeck., Linnaea 35: 606 (1868).
Cyperus spectabilis var. *filiformis* Boeck., Linnaea 35: 607 (1868).
Cyperus buckleyi Britton ex Coult., Contr. U. S. Natl. Herb. 2: 461 (1894).
Cyperus mohrii Britton ex C.B.Clarke, Kew Bull., Addit. Ser. 8: 10 (1908). *Mariscus mohrii* (Britton ex C.B.Clarke) T.Koyama, Phytologia 29: 74 (1974).
Cyperus parryi Britton ex C.B.Clarke, Kew Bull., Addit. Ser. 8: 9 (1908). *Cyperus spectabilis* var. *parryi* (Britton ex C.B.Clarke) Kük. in H.G.A. Engler (ed.), Pflanzenr., IV, 20(101): 461 (1936).

Cyperus sphacelatus Rottb., Descr. Icon. Rar. Pl.: 26 (1773).
Trop. Africa, Madagascar, Trop. America, introduced elsewere. 22 BEN BKN GHA GUI IVO LBR MLI MTN NGA SEN SIE TOG 23 BUR CMN EQG GAB GGI ZAI 25 TAN UGA UGA 26 ANG MOZ 29 MDG (40) ind srl (41) tha vie (42) bor mly sum (43) bis nwg (50) qld (61) sci (78) fla 80 COS HON NIC PAN 81 CUB DOM HAI JAM LEE PUE TRT WIN 82 FRG GUY SUR VEN 83 BOL CLM ECU PER 84 BZC BZE BZN. Ther. or hemicr.
Cyperus hexastachyos Sw., Fl. Ind. Occid. 1: 117 (1797), nom. illeg.
Cyperus nemorosus G.Mey., Prim. Fl. Esseq.: 31 (1818).
Cyperus balbisii Kunth, Enum. Pl. 2: 63 (1837).
Cyperus chromatolepis Steud., Syn. Pl. Glumac. 2: 32 (1854).
Cyperus duchaisingii Steud., Syn. Pl. Glumac. 2: 45 (1854).

Cyperus sphaeroideus L.A.S.Johnson & O.D.Evans, Contr. New South Wales Natl. Herb. 4: 378 (1973).
Queensland to Victoria. 50 NSW QLD VIC. Hemicr. or rhizome geophyte.

Cyperus sphaerolepis Boeck., Linnaea 35: 609 (1868).
S. U.S.A. to C. America. 76 ARI 77 NWM TEX 79 MXC MXE MXN 80 GUA. Hemicr. or rhizome geophyte.
Cyperus leucolepis Boeck., Linnaea 35: 610 (1868). *Cyperus fendlerianus* var. *leucolepis* (Boeck.) Kük. in H.G.A.Engler (ed.), Pflanzenr., IV, 20(101): 466 (1936).
Cyperus rusbyi Britton, Bull. Torrey Bot. Club 11: 29 (1884).
Cyperus schweinitzii var. *debilis* Britton, Bull. Torrey Bot. Club 13: 208 (1886). *Cyperus fendlerianus* var. *debilis* (Britton) Kük. in H.G.A.Engler (ed.), Pflanzenr., IV, 20(101): 465 (1936).

Cyperus sphaerospermus Schrad., Anal. Fl. Cap. 1: 8 (1832). *Cyperus denudatus* var. *sphaerospermus* (Schrad.) Kük. in H.G.A.Engler (ed.), Pflanzenr., IV, 20(101): 255 (1936).
Mozambique, S. Africa. 26 MOZ 27 BOT CPP LES NAM NAT OFS SWZ TVL. Hemicr. or rhizome geophyte.

Cyperus splendens (Cherm.) Kük. in H.G.A.Engler (ed.), Pflanzenr., IV, 20(101): 414 (1936).
E. Madagascar. 29 MDG. Hemicr. or rhizome geophyte.

**Mariscus splendens* Cherm., Bull. Mus. Natl. Hist. Nat. 25: 405 (1919).

Cyperus sporobolus R.Br., Prodr.: 215 (1810).
N. Australia. 50 NTA QLD. Hemicr. or rhizome geophyte.
Cyperus spherobolos Poir. in J.B.A.M.de Lamarck, Encycl., Suppl. 5: 187 (1817).

Cyperus squarrosus L., Cent. Pl. II: 6 (1756). *Pycreus squarrosus* (L.) Nees, Linnaea 9: 283 (1834). *Dichostylis squarrosa* (L.) Palla, Bot. Jahrb. Syst. 10: 296 (1888). *Mariscus squarrosus* (L.) C.B.Clarke in J.D.Hooker, Fl. Brit. India 6: 623 (1893).
N. America to Trop. & Subtrop. 21 CVI 22 BEN BKN GAM GHA GUI IVO MLI MTN NGA NGR SEN TOG 23 BUR CAF RWA ZAI 24 CHA SOC SOM 25 KEN TAN UGA 26 ANG MLW MOZ 27 BOT CPP NAM NAT OFS SWZ TVL 29 MDG 34 AFG 35 SAU YEM 36 CHC 40 ASS BAN EHM IND NEP PAK SRL WHM 41 MYA THA VIE 42 LSI 50 NSW NTA QLD SOA VIC WAU 71 ABT BRC MAN 72 NBR ONT QUE 73 COL IDA MNT ORE WAS WYO 74 ILL IOW KAN MIN MSO NDA NEB OKL SDA WIS 75 CNT INI MAI MAS MIC NWH NWJ NWY OHI PEN RHO VER WVA 76 ARI CAL NEV UTA 77 NWM TEX 78 ALA ARK DEL FLA GEO KTY LOU MRY MSI NCA SCA TEN VRG WDC 79 MXC MXE MXG MXN MXS MXT 80 81 ARU CUB LEE NLA 83 BOL CLM ECU GAL PER 84 BZN 85 AGE AGS AGW CLC. Ther.
Scirpus intricatus L., Mant. Pl. 2: 182 (1771). *Cyperus aristatus* Rottb., Descr. Icon. Rar. Pl.: 22 (1773), nom. superfl. *Dichostylis aristata* (Rottb.) Palla, Bot. Jahrb. Syst. 10: 299 (1888). *Mariscus aristatus* (Rottb.) Cherm., Bull. Soc. Bot. France 85: 366 (1938), nom. illeg. *Mariscus intricatus* (L.) Cufod., Bull. Jard. Bot. Natl. Belg. 40(Suppl.): 1453 (1970).
Scirpus lappaceus Lam., Tabl. Encycl. 1: 139 (1791).
Cyperus uncinatus R.Br., Prodr.: 215 (1810), nom. illeg.
Cyperus brownii Roem. & Schult., Syst. Veg. 2: 228 (1817).
Cyperus inflexus Muhl., Descr. Gram.: 16 (1817). *Cyperus aristatus* f. *inflexus* (Muhl.) C.B.Clarke, J. Linn. Soc., Bot. 21: 92 (1884). *Chlorocyperus inflexus* (Muhl.) Palla, Allg. Bot. Z. Syst. 17(Beil.): 6 (1911). *Cyperus aristatus* var. *inflexus* (Muhl.) Boeck. ex Kük. in H.G.A.Engler (ed.), Pflanzenr., IV, 20(101): 504 (1936).
Cyperus purshii Roem. & Schult., Syst. Veg. 2: 177 (1817).
Cyperus aureus J.Presl & C.Presl, Reliq. Haenk. 1: 168 (1828), nom. illeg.
Cyperus versicolor Nees in R.Wight, Contr. Bot. India: 78 (1834). *Cyperus aristatus* var. *versicolor* (Nees) Boeck., Linnaea 35: 500 (1868).
Cyperus pygmaeus Nutt., Trans. Amer. Philos. Soc., n.s., 5: 142 (1837), nom. illeg.
Isolepis echinulata Kunth, Enum. Pl. 2: 205 (1837).
Cyperus inflexus var. *acaulis* Hook.f., Trans. Linn. Soc. London 20: 178 (1847).
Cyperus inflexus var. *elongatus* Hook.f., Trans. Linn. Soc. London 20: 178 (1847).
Cyperus falciculosus Liebm., Mexic. Halvgr.: 204 (1850). *Cyperus aristatus* f. *falciculosus* (Liebm.) Kük. in H.G.A.Engler (ed.), Pflanzenr., IV, 20(101): 505 (1936).
Cyperus squarrosus var. *congestus* Benth., Fl. Austral. 7: 268 (1878).
Cyperus squarrosus var. *cylindraceus* Benth., Fl. Austral. 7: 268 (1878).
Cyperus arsitatus var. *perennis* M.E.Jones, Contr. W. Bot. 18: 25 (1933).
Cyperus aristatus var. *semiglobosus* Kük. in H.G.A. Engler (ed.), Pflanzenr., IV, 20(101): 504 (1936).
Cyperus aristatus var. *runyonii* O'Neill, Rhodora 44: 56 (1942). *Cyperus squarrosus* var. *runyonii* (O'Neill) S.D.Jones & Wipff, Phytologia 77: 462 (1994 publ. 1995).

Cyperus steadii Kük., Repert. Spec. Nov. Regni Veg. 29: 199 (1931).
S. Iran to Pakistan. 34 IRN 40 PAK. Ther.

Cyperus stenophyllus J.V.Suringar, Nova Guinea 8: 701 (1912). *Cyperus ornans* var. *stenophyllus* (J.V.Suringar) Kük. in H.G.A.Engler (ed.), Pflanzenr., IV, 20(101): 419 (1936).
Malesia to Solomon Is. 41 THA 42 JAW LSI MOL PHI 43 NWG SOL. Hemicr. or rhizome geophyte.
Cyperus ornans J.V.Suringar, Nova Guinea 8: 700 (1912). *Cyperus stenophyllus* var. *ornans* (J.V. Suringar) Kük., Bot. Jahrb. Syst. 59: 47 (1924).
Mariscus rechingeri Palla, Denkschr. Kaiserl. Akad. Wiss., Wien. Math.-Naturwiss. Kl. 89: 499 (1913 publ. 1914). *Cyperus ornans* var. *rechingeri* (Palla) Kük. in H.G.A.Engler (ed.), Pflanzenr., IV, 20(101): 419 (1936).

Cyperus stoloniferus Retz., Observ. Bot. 4: 10 (1786).
W. Indian Ocean to SW. Pacific. 29 MAU MDG 36 CHS 38 NNS TAI 40 IND PAK SRL 41 NCB THA VIE 42 CKI MLY PHI SUL SUM 50 NTA QLD 60 SAM. Hemicr. or rhizome geophyte.
Cyperus spadiceus Lam., Tabl. Encycl. 1: 147 (1791).
Cyperus littoralis R.Br., Prodr.: 216 (1810).
Cyperus bulbosostoloniferus Steud., Syn. Pl. Glumac. 2: 18 (1854).
Cyperus conjunctus Steud., Syn. Pl. Glumac. 2: 18 (1854).
Cyperus lamprocarpus Nees, Hooker's J. Bot. Kew Gard. Misc. 6: 27 (1854).
Cyperus arenarius Hance ex C.B.Clarke, J. Linn. Soc., Bot. 21: 173 (1884), nom. illeg.
Cyperus bulbosus var. *elatus* E.G.Camus, Notul. Syst. (Paris) 1: 244 (1910).
Cyperus mayeri Kük., Repert. Spec. Nov. Regni Veg. 29: 194 (1931).

Cyperus stradbrokensis Domin, Biblioth. Bot. 85: 422 (1915).
Queensland, NE. New South Wales. 50 NSW QLD. Hemicr. or rhizome geophyte.

Cyperus stramineoferrugineus Kük. in H.G.A.Engler (ed.), Pflanzenr., IV, 20(101): 555 (1936). *Mariscus stramineoferrugineus* (Kük.) Napper, J. E. Africa Nat. Hist. Soc. Natl. Mus. 28: 16 (1971).
Tanzania (Usaramo reg.). 25 TAN. Hemicr. or tuber geophyte.

Cyperus strigosus L., Sp. Pl.: 47 (1753). *Mariscus strigosus* (L.) C.B.Clarke, J. Linn. Soc., Bot. 30: 308 (1894).
C. & E. Canada to U.S.A. (13) ita 71 MAN SAS 72 ONT QUE 73 COL IDA ORE WAS 74 ILL IOW KAN MIN MSO NEB OKL SDA WIS 75 CNT INI MAI MAS MIC NWH NWJ NWY OHI PEN RHO VER WVA 76 ARI CAL NEV UTA 77 NWM TEX 78 ALA ARK DEL FLA GEO KTY LOU MRY MSI NCA SCA TEN VRG WDC 81

CUB. Hemicr. or rhizome geophyte.
Cyperus distans Pursh, Fl. Amer. Sept. 1: 53 (1813), nom. illeg.
Cyperus enslenii Pursh, Fl. Amer. Sept. 1: 53 (1813).
Cyperus michauxianus var. *elongatus* Torr., Ann. Lyceum Nat. Hist. New York 3: 432 (1836). *Cyperus strigosus* var. *elongatus* (Torr.) Britton, Bull. Torrey Bot. Club 13: 212 (1886).
Cyperus stenolepis Torr., Ann. Lyceum Nat. Hist. New York 3: 263 (1836). *Mariscus stenolepis* (Torr.) C.B.Clarke, Kew Bull., Addit. Ser. 8: 19 (1908). *Cyperus strigosus* var. *stenolepis* (Torr.) Kük., Repert. Spec. Nov. Regni Veg. 23: 189 (1926).
Cyperus strigosus f. *robustior* Kunth, Enum. Pl. 2: 88 (1837).
Cyperus praelongatus Steud., Syn. Pl. Glumac. 2: 46 (1854). *Mariscus praelongatus* (Steud.) C.B. Clarke, Kew Bull., Addit. Ser. 8: 20 (1908). *Cyperus strigosus* f. *praelongatus* (Steud.) Kük. in H.G.A.Engler (ed.), Pflanzenr., IV, 20(101): 408 (1936).
Cyperus strigosus var. *capitatus* Boeck., Linnaea 36: 347 (1870). *Cyperus strigosus* f. *capitatus* (Boeck.) Kük. in H.G.A.Engler (ed.), Pflanzenr., IV, 20(101): 408 (1936).
Cyperus strigosus f. *compositus* Britton, Bull. Torrey Bot. Club 13: 212 (1886).
Cyperus strigosus var. *compositus* Britton, Bull. Torrey Bot. Club 13: 212 (1886).
Cyperus strigosus var. *pygmaeus* Gatt., Tennessee Fl.: 89 (1887).
Cyperus hansenii Britton in L.Abrams, Ill. Fl. Pacific States 1: 260 (1923). *Cyperus strigosus* var. *hansenii* (Britton) Kük. in H.G.A.Engler (ed.), Pflanzenr., IV, 20(101): 408 (1936).
Cyperus naguensis Britton, Bull. Torrey Bot. Club 51: 1 (1924). *Cyperus strigosus* var. *naguensis* (Britton) O'Neill, Contr. Ocas. Mus. Hist. Nat. Colegio "De La Salle" 6: 14 (1946).

Cyperus subaequalis Baker, J. Linn. Soc., Bot. 22: 531 (1887).
Madagascar. 29 MDG. Hemicr. or rhizome geophyte.

Cyperus subbadius Kük. in H.G.A.Engler (ed.), Pflanzenr., IV, 20(101): 510 (1936).
Madagascar. 29 MDG. Hemicr. or rhizome geophyte.
**Mariscus badius* Willd. ex Kunth, Enum. Pl. 2: 123 (1837). *Cyperus badius* (Willd. ex Kunth) Boeck., Linnaea 36: 375 (1870), nom. illeg.

Cyperus subcaracasanus Kük., Ark. Bot. 22A(17): 5 (1929). *Mariscus subcaracasanus* (Kük.) T.Koyama, Phytologia 29: 74 (1974).
Haiti (Navassa I.). 81 HAI. Hemicr. or rhizome geophyte.

Cyperus subcastaneus D.A.Simpson, Kew Bull. 42: 897 (1987).
Brazil (Bahia to Minas Gerais). 84 BZE BZL. Hemicr. or rhizome geophyte.

subsp. ***graomogolii*** D.A.Simpson, Kew Bull. 44: 279 (1989).
Brazil (Minas Gerais). 84 BZL. Hemicr. or rhizome geophyte.

subsp. ***subcastaneus***
Brazil (Bahia to Minas Gerais). 84 BZE BZL. Hemicr. or rhizome geophyte.

Cyperus subfuscus Debeaux, Actes Soc. Linn. Bordeaux 21: t. 3 (1877).
NC. China. 36 CHN. Hemicr. or rhizome geophyte.

Cyperus submicrolepis Kük. in H.G.A.Engler (ed.), Pflanzenr., IV, 20(101): 241 (1936).
Trop. Africa. 22 BEN BKN GHA GUI IVO MLI NGA NGR SEN 23 CAF 24 CHA SUD 25 UGA 26 ANG. Ther.
**Cyperus microlepis* Boeck., Flora 62: 551 (1879), nom. illeg.

Cyperus subpapuanus Kük., Repert. Spec. Nov. Regni Veg. 29: 196 (1931).
Papua New Guinea. 43 NWG. Hemicr. or rhizome geophyte.

Cyperus subpinnatus Kük., Repert. Spec. Nov. Regni Veg. 29: 199 (1931).
New South Wales. 50 NSW. Hemicr. or rhizome geophyte.
Cyperus subpinnatus var. *subrigidellus* Kük., Repert. Spec. Nov. Regni Veg. 29: 200 (1931).

Cyperus subtenax Kük., Bull. Misc. Inform. Kew 1932: 322 (1932).
Angola. 26 ANG. Hemicr. or rhizome geophyte.

Cyperus subtenuis (Kük.) M.T.Strong, Contr. U. S. Natl. Herb. 52: 288 (2005).
**Cyperus nanus* var. *subtenuis* Kük.
Caribbean. 81 CUB DOM HAI JAM LEE PUE. Hemicr. or rhizome geophyte

Cyperus subtilis (Kük.) Väre & Kukkonen, Ann. Bot. Fenn. 42: 482 (2005).
**Cyperus maritimus* var. *subtilis* Kük.
Senegal. 22 SEN. Hemicr. or rhizome geophyte

Cyperus subulatus R.Br., Prodr.: 217 (1810).
E. & SE. Australia. 50 NSW QLD VIC. Hemicr. or rhizome geophyte.
Cyperus subulatus var. *confertus* Benth., Fl. Austral. 7: 281 (1878).
Cyperus andrewsii C.B.Clarke, Kew Bull., Addit. Ser. 8: 12 (1908).

Cyperus surinamensis Rottb., Descr. Icon. Rar. Pl.: 35 (1773).
Trop. & Subtrop. America. (22) ivo 74 KAN OKL 77 TEX 78 ALA ARK FLA GEO LOU MSI SCA TEN 79 MXC MXE MXG MXN MXS MXT 80 BLZ COS ELS GUA HON NIC PAN 81 CUB DOM JAM LEE PUE TRT WIN 82 FRG GUY SUR VEN 83 BOL CLM ECU PER 84 BZC BZE BZL BZN BZS 85 AGE AGW PAR URU. Ther. or hemicr.

var. ***lutescens*** Boeck., Linnaea 35: 555 (1868).
Mexico to Argentina. 79+ 80 BLZ 81 CUB DOM LEE PUE TRT 82 GUY VEN 83 BOL ECU 84 BZL BZS 85 AGE PAR. Hemicr. or rhizome geophyte.

var. ***surinamensis***
Trop. & Subtrop. America. (22) ivo 74 KAN OKL 77 TEX 78 ALA ARK FLA GEO LOU MSI SCA TEN 79 MXC MXE MXG MXN MXS MXT 80 BLZ COS ELS GUA HON NIC PAN 81 CUB JAM LEE PUE TRT WIN 82 FRG GUY SUR VEN 83 BOL CLM ECU PER 84 BZC BZE BZL BZN BZS 85 AGE AGW PAR URU. Ther. or hemicr.
Cyperus compressus Jacq., Hort. Bot. Vindob. 3: 10 (1777), nom. illeg.

Cyperus denticulatus Schrad. ex Schult. in J.J.Roemer & J.A.Schultes, Mant. 2: 104 (1824).
Cyperus tenuicapitatus Steud., Nomencl. Bot., ed. 2, 1: 470 (1840).
Cyperus subenervius Steud., Syn. Pl. Glumac. 2: 27 (1854).
Cyperus bipontinii Boeck., Flora 40: 33 (1857).
Cyperus surinamensis var. *viridis* Boeck., Linnaea 35: 555 (1868).
Cyperus barrancae M.E.Jones, Contr. W. Bot. 18: 25 (1933).

Cyperus svensonii G.C.Tucker, Rhodora 88: 510 (1986).
SE. Mexico to Ecuador. 79 MXT 80 GUA HON NIC 83 ECU. Hemicr. or rhizome geophyte.

Cyperus swartzii (A.Dietr.) Boeck. ex Kük., Repert. Spec. Nov. Regni Veg. 23: 186 (1926).
Mexico (San Luis Potosí), Caribbean, Venezuela. 79 MXE 81 CAY CUB DOM HAI JAM LEE PUE 82 VEN. Hemicr. or rhizome geophyte.
Kyllinga filiformis Sw., Prodr.: 20 (1788). *Mariscus filiformis* (Sw.) Spreng., Syst. Veg. 1: 234 (1824). **Mariscus swartzii* A.Dietr., Sp. Pl. 2: 343 (1832).
Mariscus gracilis Vahl, Enum. Pl. 2: 273 (1805).
Kyllinga granularis Desf. ex Link, Jahrb. Gewächsk. 1(3): 91 (1820), provisional synonym from FRG ? *Cyperus granularis* (Desf. ex Link) Britton, Bull. Dept. Agric. Jamaica 5(Suppl. 1): 9 (1907). *Cyperus swartzii* var. *granularis* (Desf. ex Link) Kük., Repert. Spec. Nov. Regni Veg. 23: 186 (1926).
Kyllinga microcephala Liebm., Mexic. Neldeagt. Pl., V, 2: 234 (1851), nom. illeg.
Kyllinga filiformis var. *junciformis* C.Wright ex Griseb., Cat. Pl. Cub.: 238 (1866).
Kyllinga rigidifolia Boeck., Linnaea 35: 432 (1868).
Cyperus caymanensis Millsp., Publ. Field Columbian Mus., Bot. Ser. 2: 120 (1896).
Cyperus swartzii var. *efoliatus* Kük., Repert. Spec. Nov. Regni Veg. 23: 187 (1926).
Cyperus swartzii var. *elongatus* Kük., Repert. Spec. Nov. Regni Veg. 23: 187 (1926).
Cyperus swartzii var. *latifolius* Kük., Repert. Spec. Nov. Regni Veg. 23: 187 (1926).

Cyperus szechuanensis T.Koyama, Jap. J. Bot. 15: 182 (1956).
S. China. 36 CHC CHS. Hemicr. or rhizome geophyte.

Cyperus tabina Steud. ex Boeck., Linnaea 75: 559 (1868). *Mariscus tabina* (Steud. ex Boeck.) C.B.Clarke, Kew Bull., Addit. Ser. 8: 19 (1908).
C. & S. Trop. America. 80 COS PAN 83 BOL ECU PER 84+ 85 AGE AGW PAR. Hemicr. or rhizome geophyte.
Cyperus andreanus Maury, J. Bot. (Morot) 2: 393 (1888).
Cyperus capitinduensis Maury ex Micheli, Mém. Soc. Phys. Genève 31(1): 129 (1890). *Cyperus andreanus* var. *capitinduensis* (Maury ex Micheli) Kük. in H.G.A.Engler (ed.), Pflanzenr., IV, 20(101): 212 (1936).
Cyperus foliosissimus Boeck., Beitr. Cyper. 2: 7 (1890). *Cyperus andreanus* var. *foliosissimus* (Boeck.) Kük. in H.G.A.Engler (ed.), Pflanzenr., IV, 20(101): 212 (1936).
Cyperus nubigenus Britton & Standl., J. Wash. Acad. Sci. 15: 472 (1925).
Cyperus andreanus var. *excelsior* Kük. in H.G.A. Engler (ed.), Pflanzenr., IV, 20(101): 212 (1936).
Cyperus andreanus var. *yguazensis* Kük. in H.G.A. Engler (ed.), Pflanzenr., IV, 20(101): 212 (1936).

Cyperus tabularis Schrad., Anal. Fl. Cap. 1: 10 (1832). *Mariscus tabularis* (Schrad.) C.B.Clarke in T.A.Durand & H.Schinz, Consp. Fl. Afric. 5: 594 (1894).
Cape Prov. 27 CPP. Hemicr. or rhizome geophyte.
Mariscus humilis Zeyh. ex Schrad., Anal. Fl. Cap. 1: 10 (1832).
Cyperus ecklonii Boeck., Linnaea 36: 325 (1870). *Cyperus tabularis* var. *ecklonii* (Boeck.) Kük. in H.G.A.Engler (ed.), Pflanzenr., IV, 20(101): 438 (1936).
Mariscus humilis Nees ex C.B.Clarke. in T.A.Durand & H.Schinz, Consp. Fl. Afric. 5: 594 (1894), nom. inval.
Mariscus tabularis var. *humilis* C.B.Clarke in W.H.Harvey & auct. suc. (eds.), Fl. Cap. 7: 196 (1898). *Cyperus tabularis* var. *humilis* (C.B.Clarke) Kük. in H.G.A.Engler (ed.), Pflanzenr., IV, 20(101): 437 (1936).
Mariscus tabularis var. *zeyheri* C.B.Clarke in W.H.Harvey & auct. suc. (eds.), Fl. Cap. 7: 196 (1898). *Cyperus tabularis* var. *zeyheri* (C.B.Clarke) Kük. in H.G.A.Engler (ed.), Pflanzenr., IV, 20(101): 438 (1936).

Cyperus tacnensis Nees & Meyen, Observ. Bot. 1: 431 (1834). *Mariscus tacnensis* (Nees & Meyen) J.Raynal, Adansonia, n.s., 17: 275 (1978).
Peru to Bolivia. 83 BOL PER. Hemicr. or rhizome geophyte.

var. ***tacnensis***
Peru. 83 PER. Hemicr. or rhizome geophyte.

var. ***tarijensis*** (Kük.) Pedersen, Darwiniana 17: 543 (1972).
Peru to Bolivia. 83 BOL PER. Hemicr. or rhizome geophyte.
**Cyperus rigens* var. *tarijensis* Kük. in H.G.A. Engler (ed.), Pflanzenr., IV, 20(101): 444 (1936).

var. ***weberbaueri*** (Kük.) Pedersen, Darwiniana 17: 543 (1972).
Peru. 83 PER. Hemicr. or rhizome geophyte.
**Cyperus rigens* var. *weberbaueri* Kük., Repert. Spec. Nov. Regni Veg. 26: 251 (1929).

Cyperus tanganyicanus (Kük.) Lye, Nordic J. Bot. 3: 231 (1983).
Tanzania, Arabian Pen. 25 TAN 35 GST. Hemicr.
**Cyperus bellus* var. *tanganyicanus* Kük. in H.G.A. Engler (ed.), Pflanzenr., IV, 20(101): 304 (1936).

Cyperus tanyphyllus Ridl., Trans. Linn. Soc. London, Bot. 2: 143 (1884). *Mariscus tanyphyllus* (Ridl.) C.B.Clarke in T.A.Durand & H.Schinz, Consp. Fl. Afric. 5: 594 (1894).
Angola. 26 ANG. Hemicr. or rhizome geophyte.

Cyperus tatandaensis Muasya & D.A.Simpson, Kew Bull. 59: 595 (2004).
Tanzania. 25 TAN.

Cyperus tempeae G.C.Tucker, Syst. Bot. Monogr. 43: 153 (1994).
Mexico. 79 MXE MXG. Hemicr. or rhizome geophyte.

Cyperus tenax Boeck., Linnaea 35: 504 (1868).
Trop. & S. Africa. 22 BEN IVO LBR NGA SIE 23 EQG GAB ZAI 25 KEN TAN UGA 26 ANG MLW

ZIM 27 CPP NAM NAT OFS TVL. Hemicr. or rhizome geophyte.

Cyperus actinostachys Welw. ex Ridl., Trans. Linn. Soc. London, Bot. 2: 140 (1884). *Cyperus tenax* var. *actinostachys* (Welw. ex Ridl.) Kük. in H.G.A. Engler (ed.), Pflanzenr., IV, 20(101): 260 (1936).

Cyperus andongensis Ridl., Trans. Linn. Soc. London, Bot. 2: 140 (1884). *Cyperus tenax* var. *andongensis* (Ridl.) Kük. in H.G.A.Engler (ed.), Pflanzenr., IV, 20(101): 259 (1936).

Cyperus boehmii Boeck., Bot. Jahrb. Syst. 5: 498 (1884).

Cyperus monroviensis Boeck., Bot. Jahrb. Syst. 5: 90 (1884). *Cyperus tenax* var. *monroviensis* (Boeck.) Kük. in H.G.A.Engler (ed.), Pflanzenr., IV, 20(101): 259 (1936).

Cyperus sabulicolus Ridl., Trans. Linn. Soc. London, Bot. 2: 136 (1884). *Cyperus tenax* var. *sabulicolus* (Ridl.) Kük. in H.G.A.Engler (ed.), Pflanzenr., IV, 20(101): 259 (1936).

Cyperus amabilis var. *pseudocastaneus* Kük. in R.E.Fries, Wiss. Erg. Schwed. Rhod.-Kongo Exped. 1: 2 (1921). *Cyperus tenax* var. *pseudocastaneus* (Kük.) Kük. in H.G.A.Engler (ed.), Pflanzenr., IV, 20(101): 260 (1936).

Cyperus tenax var. *angustissimus* Kük. in H.G.A. Engler (ed.), Pflanzenr., IV, 20(101): 260 (1936).

Cyperus tenellus L.f., Suppl. Pl.: 103 (1782). *Eucyperus tenellus* (L.f.) Palla, Allg. Bot. Z. Syst. 8: 67 (1902).

Cape Prov. 27 CPP (50) nsw qld vic wau (51) nzn (90) tdc. Ther.

Cyperus lateralis L.f., Suppl. Pl.: 102 (1782), nom. illeg.

Cyperus pygmaeus Lam., Tabl. Encycl. 1: 143 (1791), nom. illeg.

Cyperus micromegas Nees, Linnaea 10: 131 (1836). *Cyperus tenellus* var. *micromegas* (Nees) F.Muell., Fragm. 8: 261 (1874).

Cyperus nudiusculus Nees, Linnaea 10: 132 (1836).

Cyperus tenerrimus J.Presl & C.Presl, Reliq. Haenk. 1: 166 (1828).

Mexico to Bolivia. 79 MXC MXG MXN MXS MXT 80 COS CPI GUA HON NIC PAN 82 VEN 83 BOL CLM. Hemicr. or rhizome geophyte.

Cyperus cymbiformis Liebm., Mexic. Halvgr.: 208 (1850).

Cyperus wawrae Boeck., Linnaea 38: 363 (1874).

Cyperus tenuiculmis Boeck., Linnaea 36: 286 (1870).

Trop. Africa to N. Australia. 22 BEN BKN GHA GNB GUI IVO LBR MLI NGA SEN SIE TOG 23 BUR CMN CON EQG GAB GGI RWA ZAI 24 CHA SUD 25 KEN TAN UGA 26 MLW 36 CHC CHS 38 TAI 40 ASS EHM IND NEP SRL 41 CBD MYA THA VIE 42 BOR JAW LSI MLY MOL PHI SUL SUM 43 BIS NWG 50 NTA QLD 62 CRL. Hemicr. or rhizome geophyte.

var. ***guineensis*** (Nelmes) S.S.Hooper, Kew Bull. 26: 583 (1972).

W. & WC. Trop. Africa. 22 BKN GHA GUI IVO NGA SIE 23 CMN. Hemicr. or rhizome geophyte.

**Cyperus guineensis* Nelmes, Kew Bull. 6: 165 (1951).

var. ***schweinfurthianus*** (Boeck.) S.S.Hooper, Kew Bull. 26: 578 (1972).

W. Trop. Africa to Tanzania. 22 BKN GHA GUI IVO MLI NGA SEN SIE TOG 23 CMN CON GAB ZAI 24 CHA SUD 25 TAN UGA. Hemicr. or rhizome geophyte.

**Cyperus schweinfurthianus* Boeck., Flora 62: 553 (1879). *Cyperus zollingeri* var. *schweinfurthianus* (Boeck.) Kük. in H.G.A.Engler (ed.), Pflanzenr., IV, 20(101): 134 (1935).

var. ***tenuiculmis***

Trop. Africa to N. Australia. 22 BEN BKN GHA GNB GUI IVO LBR MLI NGA SEN SIE TOG 23 BUR CMN GAB RWA ZAI 24 CHA 25 KEN TAN UGA 26 MLW 36 CHC CHS 38 TAI 40 ASS EHM IND NEP SRL 41 CBD MYA THA VIE 42 BOR JAW LSI MLY MOL PHI SUL SUM 43 BIS NWG 50 NTA QLD 62 CRL. Hemicr. or rhizome geophyte.

Cyperus hexastachyus Nees in R.Wight, Contr. Bot. India: 82 (1834), nom. illeg.

Cyperus bracteatus Steud., Syn. Pl. Glumac. 2: 19 (1854), provisional synonym.

Cyperus wightii Hance, J. Bot. 16: 14 (1878), nom. illeg.

Cyperus elatior Boeck., Flora 62: 553 (1879), nom. illeg.

Cyperus zollingeri f. *levicaulis* C.B.Clarke in D.Oliver & auct. suc. (eds.), Fl. Trop. Afr. 8: 361 (1901).

Cyperus zollingeri var. *parvus* C.B.Clarke in D.Oliver & auct. suc. (eds.), Fl. Trop. Afr. 8: 361 (1901).

Cyperus zollingeri var. *condensatus* Kük. in H.G.A. Engler (ed.), Pflanzenr., IV, 20(101): 134 (1935).

Cyperus zollingeri var. *livingstonii* Kük. in H.G.A. Engler (ed.), Pflanzenr., IV, 20(101): 135 (1935).

Cyperus zollingeri var. *longiramulosus* Kük. in H.G.A.Engler (ed.), Pflanzenr., IV, 20(101): 135 (1935).

Cyperus tenuiflorus Rottb., Descr. Icon. Rar. Pl.: 30 (1773). *Cyperus longus* var. *tenuiflorus* (Rottb.) Boeck., Linnaea 36: 281 (1870). *Cyperus longus* subsp. *tenuiflorus* (Rottb.) Kük. in H.G.A.Engler (ed.), Pflanzenr., IV, 20(101): 102 (1935).

E. Trop. to S. Africa. 25 KEN TAN UGA 26 ANG MOZ ZAM 27 BOT CPP NAM OFS TVL (50) wau. Hemicr. or rhizome geophyte.

Cyperus torosus Vahl, Enum. Pl. 2: 339 (1805).

Cyperus stenostachyus Benth., Fl. Austral. 7: 280 (1878). *Cyperus longus* var. *stenostachyus* (Benth.) Kük. in H.G.A.Engler (ed.), Pflanzenr., IV, 20(101): 103 (1935).

Cyperus longus var. *discolor* Kük. in H.G.A.Engler (ed.), Pflanzenr., IV, 20(101): 103 (1935).

Cyperus longus var. *leptochlaenus* Kük. in H.G.A. Engler (ed.), Pflanzenr., IV, 20(101): 103 (1935).

Cyperus tenuis Sw., Prodr.: 20 (1788). *Mariscus tenuis* (Sw.) Nelmes, Amer. J. Bot. 39: 385 (1952).

Mexico to Trop. America, W. & WC. Trop. Africa. 22 BEN BKN GHA IVO LBR MLI NGA NGR SEN SIE TOG 23 CMN EQG GAB GGI ZAI 24 CHA 79 MXC MXE MXG MXN MXS MXT 80 BLZ COS CPI ELS GUA HON NIC PAN 81 CUB JAM LEE PUE TRT WIN 82 VEN 83 BOL CLM PER 84 BZS 85 PAR. Hemicr. or rhizome geophyte.

Mariscus flabelliformis Kunth in F.W.H.von Humboldt, A.J.A.Bonpland & C.S.Kunth, Nov. Gen. Sp. 1: 215 (1816). *Cyperus flabelliformis* (Kunth) Spreng., Syst. Veg. 1: 228 (1825), nom. illeg.

Cyperus caracasanus Kunth, Enum. Pl. 2: 86 (1837).

Cyperus breviradiatus Liebm., Mexic. Halvgr.: 214

(1850). *Cyperus camphoratus* var. *breviradiatus* (Liebm.) Kük. in H.G.A.Engler (ed.), Pflanzenr., IV, 20(101): 422 (1936).
Cyperus discigerus Liebm., Mexic. Halvgr.: 212 (1850). *Cyperus sertularinus* var. *discigerus* (Liebm.) Kük. in H.G.A.Engler (ed.), Pflanzenr., IV, 20(101): 411 (1936). *Mariscus dicigerus* (Liebm.) T.Koyama, Phytologia 29: 73 (1974).
Cyperus platystachyus Griseb., Fl. Brit. W. I.: 567 (1864).
Cyperus eurystachys Ridl., Trans. Linn. Soc. London, Bot. 2: 143 (1884). *Mariscus eurystachys* (Ridl.) C.B.Clarke in T.A.Durand & H.Schinz, Consp. Fl. Afric. 5: 587 (1894). *Cyperus tenuis* var. *eurystachys* (Ridl.) Kük. in H.G.A.Engler (ed.), Pflanzenr., IV, 20(101): 418 (1936).
Cyperus flexibilis Maury, J. Bot. (Morot) 2: 392 (1888).
Cyperus hartii Boeck., Beitr. Cyper. 1: 9 (1888).
Cyperus dussii Boeck., Beitr. Cyper. 2: 37 (1890).
Cyperus dussianus Duss, Ann. Inst. Bot.-Géol. Colon. Marseille 3: 541 (1896 publ. 1897).
Mariscus luridus T.Durand & De Wild., Bull. Soc. Roy. Bot. Belgique, Compt. Rend. 36: 89 (1897). *Cyperus tenuis* var. *luridus* (T.Durand & De Wild.) Kük. in H.G.A.Engler (ed.), Pflanzenr., IV, 20(101): 418 (1936).
Mariscus aximensis C.B.Clarke in D.Oliver & auct. suc. (eds.), Fl. Trop. Afr. 8: 398 (1902). *Cyperus tenuis* var. *aximensis* (C.B.Clarke) Kük. in H.G.A.Engler (ed.), Pflanzenr., IV, 20(101): 418 (1936). *Mariscus flabelliformis* var. *aximensis* (C.B.Clarke) S.S. Hooper, Kew Bull. 26: 579 (1972).
Cyperus tenuis var. *brevior* Kük. in H.G.A.Engler (ed.), Pflanzenr., IV, 20(101): 417 (1936).
Cyperus tenuis var. *fulvescens* Kük. in H.G.A.Engler (ed.), Pflanzenr., IV, 20(101): 417 (1936).
Cyperus tenuis var. *grandiceps* Kük. in H.G.A.Engler (ed.), Pflanzenr., IV, 20(101): 417 (1936).

Cyperus tenuispica Steud., Syn. Pl. Glumac. 2: 11 (1854).
Trop. & Subtrop. Old World. 22 BEN GHA GNB GUI IVO LBR MLI MTN NGA NGR SEN SIE TOG 23 CMN RWA ZAI 24 CHA ETH SOM SUD 25 TAN UGA 26 ANG MLW 27 BOT NAM NAT TVL 29 SEY 32 TZK UZB 36 CHS 38 JAP KOR KZN NNS OGA TAI 40 ASS EHM NEP PAK SRL WHM 41 MYA THA VIE 42 JAW LSI MLY PHI SUL SUM 50 NTA QLD WAU (80) cos (82) guy ven (84). Ther.
Cyperus leptostachys Nees, Linnaea 9: 285 (1834), nom. inval.
Cyperus delicatulatus Steud., Syn. Pl. Glumac. 2: 21 (1854).
Cyperus fieldingii Steud., Syn. Pl. Glumac. 2: 2 (1854).
Cyperus pseudohaspan Makino, Bot. Mag. (Tokyo) 6: 47 (1892).

Cyperus tenuispiculatus Boeck., Bot. Jahrb. Syst. 5: 500 (1884).
Madagascar. 29 MDG. Hemicr. or rhizome geophyte.
Cyperus tenuispiculatus var. *pseudatterimus* Kük. in H.G.A.Engler (ed.), Pflanzenr., IV, 20(101): 141 (1935).

Cyperus tetracarpus Boeck., Flora 58: 88 (1875).
Queensland. 50 QLD. Hemicr. or rhizome geophyte.

Cyperus tetragonus Elliott, Sketch Bot. S. Carolina 1: 71 (1816). *Mariscus tetragonus* (Elliott) C.B.Clarke in I.Urban, Symb. Antill. 2: 44 (1900).
S. U.S.A. to Mexico. 76 ARI 77 NWM 78 ALA FLA GEO MSI NCA 79 MXE. Hemicr. or rhizome geophyte.
Cyperus pringlei Britton, Bull. Torrey Bot. Club 12: 7 (1885). *Cyperus tetragonus* var. *pringlei* (Britton) Kük. in H.G.A.Engler (ed.), Pflanzenr., IV, 20(101): 493 (1936).
Cyperus tetragonus var. *intercedens* Kük. in H.G.A. Engler (ed.), Pflanzenr., IV, 20(101): 493 (1936).

Cyperus tetraphyllus R.Br., Prodr.: 214 (1810).
E. Australia. 50 NSW QLD. Hemicr. or rhizome geophyte.

Cyperus textilis Thunb., Prodr. Pl. Cap.: 18 (1794). *Eucyperus textilis* (Thunb.) Rikli, Jahrb. Wiss. Bot. 27: 568 (1895).
S. Africa. (21) azo 27 CPP NAT. Hemicr. or rhizome geophyte.
Cyperus asperifolius Desf., Tabl. École Bot., ed. 3: 387 (1829).
Cyperus burchellii Schrad., Anal. Fl. Cap. 1: 11 (1832).
Cyperus smithii Schrad., Anal. Fl. Cap. 1: 12 (1832).
Eucyperus pungens Rikli, Jahrb. Wiss. Bot. 27: 568 (1895).

Cyperus thomsonii Boeck., Linnaea 36: 294 (1870).
Assam to Vietnam. 40 ASS BAN EHM 41 VIE. Hemicr. or rhizome geophyte.

Cyperus thorelii Camus, Notul. Syst. (Paris) 1: 242 (1910).
N. Vietnam. 41 VIE. Hemicr. or rhizome geophyte.

Cyperus thorncroftii McClean, Bothalia 2: 366 (1927).
S. Africa. 27 SWZ TVL. Hemicr. or rhizome geophyte.

Cyperus thunbergii Vahl, Enum. Pl. 2: 371 (1805). *Mariscus thunbergii* (Vahl) Schrad., Anal. Fl. Cap. 1: 13 (1832).
Cape Prov. 27 CPP. Hemicr. or rhizome geophyte.
**Cyperus alopecuroides* Thunb., Prodr. Pl. Cap.: 18 (1794), nom. illeg.
Mariscus riparius Schrad., Gött. Gel. Anz. 3: 2067 (1821).
Mariscus riparius var. *gillii* C.B.Clarke in W.H.Harvey & auct. suc. (eds.), Fl. Cap. 7: 193 (1898). *Cyperus thunbergii* var. *gillii* (C.B.Clarke) Kük. in H.G.A. Engler (ed.), Pflanzenr., IV, 20(101): 439 (1936).
Mariscus riparius var. *robustior* C.B.Clarke in W.H.Harvey & auct. suc. (eds.), Fl. Cap. 7: 193 (1898). *Cyperus thunbergii* var. *robustior* (C.B.Clarke) Kük. in H.G.A.Engler (ed.), Pflanzenr., IV, 20(101): 439 (1936).
Mariscus riparius var. *trisumbellatus* C.B.Clarke in W.H.Harvey & auct. suc. (eds.), Fl. Cap. 7: 193 (1898). *Cyperus thunbergii* var. *trisumbellatus* (C.B.Clarke) Kük. in H.G.A.Engler (ed.), Pflanzenr., IV, 20(101): 439 (1936).

Cyperus thyrsiflorus Jungh. ex Schltdl., Linnaea 4: 24 (1831). *Mariscus thyrsiflorus* (Jungh. ex Schltdl.) T.Koyama, Phytologia 29: 74 (1974).
Trop. & Subtrop. America. 77 TEX 78 ALA FLA GEO LOU 79 MXE MXG MXS MXT 80 COS ELS GUA HON NIC PAN 81 CUB TRT 82 VEN 83 CLM ECU PER 85 PAR. Hemicr. or rhizome geophyte.
Cyperus dissitiflorus Nees ex Torr., Ann. Lyceum Nat. Hist. New York 3: 266 (1836). *Mariscus dissitiflorus* (Nees ex Torr.) C.B.Clarke in I.Urban, Symb. Antill. 2: 45 (1900).
Cyperus anceps Liebm., Mexic. Halvgr.: 213 (1850). *Cyperus thyrsiflorus* var. *anceps* (Liebm.) Kük. in

H.G.A.Engler (ed.), Pflanzenr., IV, 20(101): 494 (1936).
Mariscus pallens Liebm., Mexic. Neldeagt. Pl., V, 2: 230 (1851). *Cyperus regiomontanus* var. *pallens* (Liebm.) Kük. in H.G.A.Engler (ed.), Pflanzenr., IV, 20(101): 527 (1936). *Cyperus pallens* (Liebm.) Standl. & Steyerm., Fieldiana, Bot. 24: 133 (1958).
Mariscus tribrachiatus Liebm., Mexic. Neldeagt. Pl., V, 2: 230 (1851). *Cyperus tribrachiatus* (Liebm.) Kük. in H.G.A.Engler (ed.), Pflanzenr., IV, 20(101): 489 (1936).
Cyperus amphekes Steud., Syn. Pl. Glumac. 2: 56 (1854).
Cyperus pseudosphacelatus Boeck., Beitr. Cyper. 2: 7 (1890).
Cyperus pittieri Boeck. ex Kneuck., Allg. Bot. Z. Syst. 1: 230 (1895). *Cyperus regiomontanus* var. *pittieri* (Boeck. ex Kneuck.) Kük. in H.G.A.Engler (ed.), Pflanzenr., IV, 20(101): 527 (1936).
Cyperus tonduzianus var. *tenuis* Boeck., Allg. Bot. Z. Syst. 2: 18 (1896). *Cyperus tribrachiatus* var. *tenuis* (Boeck.) Kük. in H.G.A.Engler (ed.), Pflanzenr., IV, 20(101): 490 (1936).
Mariscus dissitiflorus var. *lucidior* C.B.Clarke in I.Urban, Symb. Antill. 2: 45 (1900).
Cyperus thyrsiflorus var. *macrior* F.Aresch., Pl. Itin. Eugeniae: 121 (1910).
Cyperus thyrsiflorus var. *vegetior* F.Aresch., Pl. Itin. Eugeniae: 121 (1910).

Cyperus tomaiophyllus K.Schum. in H.G.A.Engler, Pflanzenw. Ost-Afrikas, C: 122 (1895). *Mariscus tomaiophyllus* (K.Schum.) C.B.Clarke in D.Oliver & auct. suc. (eds.), Fl. Trop. Afr. 8: 392 (1902).
Nigeria to E. Trop. Africa, Madagascar. 22 NGA 23 CMN GGI RWA ZAI 24 ETH 25 KEN TAN UGA 29 MDG. Hemicr. or rhizome geophyte.
Cyperus alpestris K.Schum. in H.G.A.Engler, Pflanzenw. Ost-Afrikas, C: 122 (1895). *Mariscus alpestris* (K.Schum.) C.B.Clarke in D.Oliver & auct. suc. (eds.), Fl. Trop. Afr. 8: 401 (1902). *Cyperus tomaiophyllus* var. *alpestris* (K.Schum.) Kük. in H.G.A.Engler (ed.), Pflanzenr., IV, 20(101): 430 (1936).
Mariscus magnus C.B.Clarke, Bot. Jahrb. Syst. 38: 134 (1906). *Cyperus tomaiophyllus* var. *magnus* (C.B. Clarke) Kük. in H.G.A.Engler (ed.), Pflanzenr., IV, 20(101): 429 (1936).
Cyperus tomaiophyllus var. *crassistolon* Kük. in H.G.A. Engler (ed.), Pflanzenr., IV, 20(101): 429 (1936).
Cyperus tomaiophyllus var. *laxiflorus* Kük. in H.G.A. Engler (ed.), Pflanzenr., IV, 20(101): 430 (1936).

Cyperus tonkinensis C.B.Clarke, Kew Bull., Addit. Ser. 8: 11 (1908).
W. Trop. Africa to Chad, Indo-China. 22 BEN GHA GUI IVO LBR MLI NGA SEN SIE 23 CAF CON 24 CHA 41 LAO THA VIE. Hemicr. or rhizome geophyte.

var. ***baikiei*** (C.B.Clarke ex Kük.) S.S.Hooper, Kew Bull. 26: 577 (1972).
W. Trop. Africa to Chad. 22 BEN GHA GUI IVO LBR MLI NGA SEN SIE 23 CAF CON 24 CHA. Hemicr. or rhizome geophyte.
**Cyperus baikiei* C.B.Clarke ex Kük., Repert. Spec. Nov. Regni Veg. 21: 325 (1925).
Cyperus kottensis Cherm., Arch. Bot. Mém. 4(7): 23 (1931).

var. ***tonkinensis***
Indo-China. 41 LAO THA VIE. Hemicr. or rhizome geophyte.
Cyperus brevicaulis C.B.Clarke, Kew Bull., Addit. Ser. 8: 11 (1908).

Cyperus trachysanthos Hook. & Arn., Bot. Beechey Voy.: 99 (1832).
Hawaiian Is. 63 HAW. Hemicr. or rhizome geophyte.

Cyperus trailii C.B.Clarke, Kew Bull., Addit. Ser. 8: 5 (1908).
N. South America to N. Brazil. 82 GUY SUR VEN 84 BZN. Hemicr. or rhizome geophyte.
Cyperus ruderalis Uittien, Recueil Trav. Bot. Néerl. 30: 185 (1933).

Cyperus trialatus (Boeck.) J.Kern, Reinwardtia 3: 32 (1954).
S. China to W. Malesia. 36 CHS 41 THA VIE 42 BOR JAW MLY SUM. Hemicr. or rhizome geophyte.
**Scirpus trialatus* Boeck., Flora 41: 445 (1859). *Mariscus trialatus* (Boeck.) Tang & F.T.Wang, in Fl. Reipubl. Popul. Sin. 11: 176 (1961).
Cyperus bancanus Miq., Fl. Ned. Ind., Eerste Bijv.: 599 (1861). *Cyperus diffusus* subsp. *bancanus* (Miq.) Kük. in H.G.A.Engler (ed.), Pflanzenr., IV, 20(101): 209 (1936).
Cyperus turgidulus C.B.Clarke, J. Linn. Soc., Bot. 21: 130 (1884).

Cyperus trichodes Griseb., Fl. Brit. W. I.: 564 (1864).
Jamaica. 81 JAM. Hemicr. or rhizome geophyte.

Cyperus trigonellus Suess., Trans. Rhodesia Sci. Assoc. 43: 80 (1951).
Zimbabwe. 26 ZIM. Hemicr. or rhizome geophyte.

Cyperus trinervis R.Br., Prodr.: 213 (1810). *Cyperus trinervis* var. *typicus* Domin, Biblioth. Bot. 85: 423 (1915), nom. inval.
N. & E. Australia. 50 NSW NTA QLD (63) haw. Hemicr. or rhizome geophyte.
Cyperus inundatus R.Br., Prodr.: 214 (1810).
Cyperus lessonianus Kunth, Enum. Pl. 2: 29 (1837).
Cyperus macellus Kunth, Enum. Pl. 2: 33 (1837).
Cyperus enervis var. *laxus* Benth., Fl. Austral. 7: 266 (1878). *Cyperus trinervis* var. *laxus* (Benth.) Kük. in H.G.A.Engler (ed.), Pflanzenr., IV, 20(101): 293 (1936).
Cyperus superatus C.B.Clarke, Kew Bull., Addit. Ser. 8: 4 (1908). *Cyperus trinervis* var. *superatus* (C.B. Clarke) Kük. in H.G.A.Engler (ed.), Pflanzenr., IV, 20(101): 294 (1936).
Cyperus trinervis var. *contractus* Domin, Biblioth. Bot. 85: 423 (1915).

Cyperus trisulcus D.Don, Prodr. Fl. Nepal.: 39 (1825).
C. Nepal. 40 NEP. Hemicr. or rhizome geophyte.

Cyperus tuberosus Rottb., Descr. Icon. Rar. Pl.: 28 (1773). *Cyperus rotundus* subsp. *tuberosus* (Rottb.) Kük. in H.G.A.Engler (ed.), Pflanzenr., IV, 20(101): 113 (1935).
Africa, India, Sri Lanka. 20 ALG LBY 22 GHA GUI IVO MLI NGA SEN TOG 23 CMN GGI ZAI 24 ERI ETH SOM SUD 25 KEN TAN UGA 26 ANG MOZ ZAM 27 NAT 29 COM MDG 35 YEM 38 NNS TAI 40 IND SRL. Hemicr. or rhizome geophyte.
Cyperus arabicus Ehrenb. ex Boeck., Linnaea 36: 284 (1870).
Chlorocyperus salaamensis Palla, Allg. Bot. Z. Syst. 9: 70 (1903).

Cyperus* × *turbatus Baijnath, J. S. African Bot. 42: 372 (1976). *C. prolifer* × *C. sensilis*.
KwaZulu-Natal. 27 NAT. Hemicr. or rhizome geophyte.

Cyperus turrialbanus Gómez-Laur., Brenesia 14–15: 357 (1978).
Costa Rica to Panama. 80 COS PAN. Hemicr. or rhizome geophyte.

Cyperus turrillii Kük., Repert. Spec. Nov. Regni Veg. 29: 199 (1931).
Tanzania to Angola. 23 ZAI? 25 TAN 26 ANG ZAM. Hemicr. or rhizome geophyte.
**Mariscus laxiflorus* Turrill, Bull. Misc. Inform. Kew 1914: 171 (1914).

Cyperus tweediei C.B.Clarke, Kew Bull., Addit. Ser. 8: 4 (1908).
Argentina (Tucumán). 85 AGW. Hemicr. or rhizome geophyte.

Cyperus uncinulatus Schrad. ex Nees in C.F.P.von Martius & auct. suc. (eds.), Fl. Bras. 2(1): 23 (1842).
S. Mexico to Trop. America. 79 MXS MXT 80 GUA 83 BOL 84 BZC BZE BZL 85 AGE AGW PAR. Hemicr. or rhizome geophyte.
Cyperus paraguayensis C.B.Clarke, J. Linn. Soc., Bot. 21: 90 (1884).

Cyperus undulatus Kük., Repert. Spec. Nov. Regni Veg. 21: 328 (1925).
E. Trop. Africa. 25 KEN TAN. Hemicr. or rhizome geophyte.

Cyperus unicolor Boeck., Vidensk. Meddel. Naturhist. Foren. Kjøbenhavn 1879–1880: 24 (1879).
Venezuela to Argentina. 82 VEN 83 BOL 84 BZL 85 AGE AGW PAR URU. Hemicr. or rhizome geophyte.

Cyperus unifolius Boeck., Linnaea 36: 374 (1870). *Mariscus unifolius* (Boeck.) T.Koyama, Phytologia 29: 74 (1974).
Puerto Rico to Virgin Is. 81 LEE PUE. Hemicr.
Cyperus calcicola Britton, Bull. Torrey Bot. Club 42: 387 (1915). *Cyperus fuligineus* var. *calcicola* (Britton) Kük. in H.G.A.Engler (ed.), Pflanzenr., IV, 20(101): 470 (1936).

Cyperus urbanii Boeck., Beitr. Cyper. 1: 10 (1888). *Mariscus urbanii* (Boeck.) C.B.Clarke in I.Urban, Symb. Antill. 2: 47 (1900).
Puerto Rico. 81 PUE. Hemicr. or rhizome geophyte.

Cyperus usitatus Burch. ex Roem. & Schult., Mant. 2: 477 (1824).
NE. Trop. to S. Africa. 24 ETH 25 KEN TAN UGA 27 BOT CPP NAM NAT TVL. Hemicr. or tuber geophyte.

var. ***macrobulbosus*** Kük. in H.G.A.Engler (ed.), Pflanzenr., IV, 20(101): 124 (1935).
C. Tanzania to S. Africa. 25 TAN 27 BOT NAM. Hemicr. or tuber geophyte.

subsp. ***palmatus*** Lye, Nordic J. Bot. 3: 228 (1983).
Tanzania. 25 TAN. Hemicr. or tuber geophyte.

var. ***stuhlmannii*** (C.B.Clarke ex K.Schum.) Lye, Nordic J. Bot. 3: 231 (1983).
E. Trop. Africa. 25 KEN TAN. Hemicr. or tuber geophyte.
**Cyperus stuhlmannii* C.B.Clarke ex K.Schum. in H.G.A.Engler, Pflanzenw. Ost-Afrikas, C: 118 (1895).

subsp. ***usitatus***
NE. Trop. to S. Africa. 24 ETH 25 KEN TAN UGA 27 CPP NAT TVL. Hemicr. or tuber geophyte.
Cyperus bulbifex E.Mey. ex Drège, Linnaea 20: 245 (1847).
Cyperus edulis Dinter, Deut. Südw. Afrik.: 41 (1909).

Cyperus ustulatus A.Rich. in J.S.C.Dumont d'Urville, Voy. Astrolabe 1: 101 (1832). *Mariscus ustulatus* (A.Rich.) C.B.Clarke in T.F.Cheeseman, Man. New Zealand Fl.: 766 (1906).
New Zealand North I., Kermadec Is. 51 KER NZN. Hemicr. or rhizome geophyte.
Cyperus ustulatus f. *grandispiculosus* Kük. in H.G.A.Engler (ed.), Pflanzenr., IV, 20(101): 440 (1936).

Cyperus vaginatus R.Br., Prodr.: 213 (1810).
Australia. 50 NSW NTA QLD SOA VIC WAU. Hemicr. or rhizome geophyte.
Cyperus cruciformis Boeck., Linnaea 35: 572 (1868), nom. illeg.
Cyperus textilis F.Muell., Fragm. 8: 261 (1874).
Cyperus vaginatus var. *densiflorus* Benth., Fl. Austral. 7: 273 (1878).
Cyperus vaginatus var. *contractus* Domin, J. Linn. Soc., Bot. 41: 268 (1912).
Cyperus vaginatus var. *pseudotextilis* Kük. in H.G.A. Engler (ed.), Pflanzenr., IV, 20(101): 190 (1936).

Cyperus varicus (C.B.Clarke ex Cherm.) Kük. in H.G.A. Engler (ed.), Pflanzenr., IV, 20(101): 432 (1936).
Comoros, Madagascar. 29 COM MDG. Hemicr. or rhizome geophyte.
Mariscus arcuatoreflexus Cherm., Bull. Mus. Natl. Hist. Nat. 25: 409 (1919). *Cyperus varicus* var. *arcuatoreflexus* (Cherm.) Kük. in H.G.A.Engler (ed.), Pflanzenr., IV, 20(101): 433 (1936).
**Mariscus varicus* C.B.Clarke ex Cherm., Bull. Mus. Natl. Hist. Nat. 25: 408 (1919).
Cyperus varicus var. *simpliciusculus* Kük. in H.G.A. Engler (ed.), Pflanzenr., IV, 20(101): 433 (1936).

Cyperus ventricosus R.Br., Prodr.: 217 (1810). *Mariscus pennatus* var. *ventricosus* (R.Br.) Domin, Biblioth. Bot. 85: 440 (1915).
Queensland. 50 QLD. Hemicr. or rhizome geophyte.

Cyperus vestitus Hochst. ex C.Krauss, Flora 28: 755 (1845). *Mariscus vestitus* (Hochst. ex C.Krauss) C.B.Clarke in T.A.Durand & H.Schinz, Consp. Fl. Afric. 5: 595 (1894).
Somalia to KwaZulu-Natal. 23 RWA ZAI 24 SOM 25 KEN TAN UGA 26 MLW 27 NAT. Hemicr. or bulb geophyte.
Cylindrolepis rehmanniana Boeck., Bot. Centralbl. 39: 73 (1889). *Mariscus rehmannianus* (Boeck.) C.B.Clarke in T.A.Durand & H.Schinz, Consp. Fl. Afric. 5: 591 (1894).
Cyperus albomarginatus K.Schum. in H.G.A.Engler, Pflanzenw. Ost-Afrikas, C: 122 (1895), nom. illeg.
Mariscus inflatus C.B.Clarke in D.Oliver & auct. suc. (eds.), Fl. Trop. Afr. 8: 384 (1901). *Cyperus indecorus* var. *inflatus* (C.B.Clarke) Kük. in H.G.A. Engler (ed.), Pflanzenr., IV, 20(101): 545 (1936).
Mariscus vestitus var. *decurvatus* C.B.Clarke, Bot. Jahrb. Syst. 38: 134 (1906). *Cyperus indecorus* var. *decurvatus* (C.B.Clarke) Kük. in H.G.A.Engler (ed.), Pflanzenr., IV, 20(101): 545 (1936).
Mariscus binucifer C.B.Clarke, Kew Bull., Addit. Ser. 8: 13 (1908).

Cyperus pseudocallistus Kük., Repert. Spec. Nov. Regni Veg. 21: 329 (1925). *Cyperus vestitus* var. *pseudocallistus* (Kük.) Kük. in H.G.A.Engler (ed.), Pflanzenr., IV, 20(101): 543 (1936).
Cyperus indecorus var. *dinteri* Kük. in H.G.A.Engler (ed.), Pflanzenr., IV, 20(101): 545 (1936).
Cyperus indecorus var. *namaquensis* Kük. in H.G.A. Engler (ed.), Pflanzenr., IV, 20(101): 545 (1936).

Cyperus vexillatus Peter ex Kük. in H.G.A.Engler (ed.), Pflanzenr., IV, 20(101): 547 (1936).
Tanzania. 25 TAN. Hemicr. or rhizome geophyte.

Cyperus victoriensis C.B.Clarke, Kew Bull., Addit. Ser. 8: 12 (1908).
Australia. 50 NSW NTA QLD SOA VIC. Hemicr. or rhizome geophyte.

Cyperus virens Michx., Fl. Bor.-Amer. 1: 28 (1803).
Trop. & Subtrop. America. (63) haw (76) cal 77 TEX 78 ALA FLA GEO LOU MSI NCA SCA 79 MXC MXE MXG MXS MXT 80 COS ELS GUA HON NIC PAN 81 CUB DOM HAI JAM PUE 82 SUR VEN 83 BOL CLM 84 BZL BZS 85 AGE AGW PAR URU. Hemicr. or rhizome geophyte.

var. ***minarum*** (Boeck.) Denton, Contr. Univ. Michigan Herb. 11: 264 (1978).
Mexico to C. America, Hispaniola, Uruguay to NE. Argentina. 79 MXC MXS MXT 80 COS ELS GUA HON NIC PAN 81 DOM 85 AGE URU. Hemicr. or rhizome geophyte.
**Cyperus minarum* Boeck., Beitr. Cyper. 2: 5 (1890). *Cyperus consanguineus* var. *minarum* (Boeck.) Kük. in H.G.A.Engler (ed.), Pflanzenr., IV, 20(101): 172 (1936).

var. ***montanus*** (Boeck.) Deuton, Contr. Univ. Michigan Herb. 11: 265 (1978).
Brazil to S. South America. 84 BZL BZS 85 AGE CLC URU. Hemicr. or rhizome geophyte.
**Cyperus montanus* Boeck., Beitr. Cyper. 2: 4 (1890).
Cyperus uleanus Boeck., Beitr. Cyper. 2: 5 (1890).

var. ***virens***
Trop. & Subtrop. America. (63) haw (76) cal 77 TEX 78 ALA FLA GEO LOU MSI NCA SCA 79 MXC MXE MXG MXS MXT 80 COS ELS GUA HON NIC 81 CUB HAI JAM PUE 82 SUR VEN 83 BOL CLM 84 BZL BZS 85 AGE AGW PAR URU. Hemicr. or rhizome geophyte.
Cyperus glomeratus Walter, Fl. Carol.: 70 (1788), nom. illeg.
Scirpus reticulatus Lam., Tabl. Encycl. 1: 142 (1791).
Cyperus formosus Vahl, Enum. Pl. 2: 327 (1805).
Cyperus acutangulus Boeck., Linnaea 35: 551 (1868). *Cyperus virens* var. *acutangulus* (Boeck.) Kük. in H.G.A.Engler (ed.), Pflanzenr., IV, 20(101): 181 (1936).
Cyperus virens var. *glaucopallidus* Boeck., Linnaea 35: 554 (1868).
Cyperus baenitzii Boeck., Flora 61: 140 (1878).
Cyperus scaberrimus Boeck., Beitr. Cyper. 2: 3 (1890), nom. illeg. *Cyperus virens* var. *scaberrimus* Kük. in H.G.A.Engler (ed.), Pflanzenr., IV, 20(101): 181 (1936).

Cyperus viscidulus K.L.Wilson, Telopea 1: 462 (1980).
N. Australia. 50 NTA WAU. Hemicr. or rhizome geophyte.

Cyperus volckmannii Phil., Linnaea 33: 269 (1865).
NC. Chile. 85 CLC. Ther.
Cyperus araeus Phil., Anales Univ. Chile 93: 346 (1896).

Cyperus volodia Cherm., Bull. Soc. Bot. France 66: 351 (1919 publ. 1920).
W. Madagascar. 29 MDG. Hemicr. or rhizome geophyte.

Cyperus vorsteri K.L.Wilson, Telopea 5: 598 (1994).
S. Africa. 27 NAT (50) nsw wau. Hemicr. or rhizome geophyte.
**Mariscus grantii* C.B.Clarke in W.H.Harvey & auct. suc. (eds.), Fl. Cap. 7: 194 (1898).

Cyperus wallichianus Spreng., Syst. Veg. 4(2): 28 (1827).
C. Nepal. 40 NEP. Hemicr. or rhizome geophyte.
**Cyperus pulcher* D.Don, Prodr. Fl. Nepal.: 38 (1825), nom. illeg.

Cyperus whitmeei (C.B.Clarke) Kük. in H.G.A.Engler (ed.), Pflanzenr., IV, 20(101): 429 (1936).
Samoa. 60 SAM. Hemicr. or rhizome geophyte.
**Mariscus whitmeei* C.B.Clarke, Kew Bull., Addit. Ser. 8: 14 (1908).

Cyperus wilburii G.C.Tucker, Rhodora 88: 508 (1986).
Mexico (Oaxaca). 79 MXS. Ther.

Cyperus wissmannii O.Schwartz, Mitt. Inst. Allg. Bot. Hamburg 10: 341 (1939).
NE. Trop. Africa, Yemen. 24 ETH 35 YEM. Ther.

Cyperus xanthostachyus Steud., Flora 25: 600 (1842).
S. South America. 85 AGS CLC URU. Hemicr. or rhizome geophyte.
Cyperus cumingii Nees, Linnaea 9: 284 (1834), nom. inval.
Cyperus bracteosus Kunze ex Desv. in C.Gay, Fl. Chil. 6: 165 (1854).
Cyperus lechleri Steud., Syn. Pl. Glumac. 2: 27 (1854).
Cyperus lechleri var. *viridiglaucus* Boeck., Beitr. Cyper. 2: 6 (1890).

Cyperus xerophilus Cherm., Bull. Soc. Bot. France 66: 313 (1919 publ. 1920). *Cyperus subxerophilus* Kük. in H.G.A.Engler (ed.), Pflanzenr., IV, 20(101): 200 (1936), nom. illeg.
C. Madagascar. 29 MDG. Hemicr. or rhizome geophyte.
Mariscus aridicola Domin, Biblioth. Bot. 85: 439 (1915). *Cyperus aridicola* (Domin) Kük. in H.G.A. Engler (ed.), Pflanzenr., IV, 20(101): 451 (1936).

Cyperus zairensis Chiov., Papiro Ital.: 73, 78 (1931).
Zaïre. 23 ZAI. Hemicr. or rhizome geophyte.

Cyperus zanzibarensis C.B.Clarke in D.Oliver & auct. suc. (eds.), Fl. Trop. Afr. 8: 323 (1901).
E. Trop. Africa. 25 KEN. Hemicr. or rhizome geophyte.

Cyperus zollingeri Steud., Syn. Pl. Glumac. 2: 17 (1854).
Trop. & S. Africa to W. Pacific. 22 GAM GHA GUI IVO LBR NGA SEN SIE TOG 23 BUR CMN GAB ZAI 24 SUD 25 KEN TAN 26 ANG ZAM 27 BOT NAT TVL 29 MDG 36 CHC CHS 40 ASS EHM IND NEP SRL 41 AND CBD MYA VIE 42 JAW LSI PHI 43 NWG 50 NTA QLD WAU 62 CRL. Ther.
Cyperus lucidulus C.B.Clarke, J. Linn. Soc., Bot. 21: 99 (1884), nom. illeg.

Cyperus sphacelatus var. *tenuior* C.B.Clarke in D.Oliver & auct. suc. (eds.), Fl. Trop. Afr. 8: 347 (1901).
Cyperus rubroviridis Cherm., Bull. Soc. Bot. France 66: 350 (1919 publ. 1920).
Cyperus ramosi Kük., Repert. Spec. Nov. Regni Veg. 21: 326 (1925).
Cyperus rubroviridis var. *unicapitatus* Kük. in H.G.A. Engler (ed.), Pflanzenr., IV, 20(101): 136 (1935).

Cyperus zollingerioides C.B.Clarke, Bot. Jahrb. Syst. 38: 132 (1906).
W. Cape Prov. 27 CPP. Hemicr. or rhizome geophyte.

Cyperus zonatus Kük., Repert. Spec. Nov. Regni Veg. 12: 93 (1913).
WC. Trop. Africa. 23 CON ZAI. Ther.

***Synonyms*:**

Cyperus aberdarensis Kük. = ***Pycreus aethiops*** (Welw. ex Ridl.) C.B.Clarke
Cyperus abnormis Steud. = ?
Cyperus abyssinicus Hochst. ex A.Rich. = ***Pycreus flavescens*** (L.) P.Beauv. ex Rchb. subsp. ***flavescens***
Cyperus acaulis Steud. = ***Cyperus poeppigii*** Kunth
Cyperus acicularis (Schrad. ex Nees) Steud. = ***Cyperus odoratus*** L. subsp. ***odoratus***
Cyperus acicularis (L.) With. = ***Eleocharis acicularis*** (L.) Roem. & Schult.
Cyperus actinostachys Welw. ex Ridl. = ***Cyperus tenax*** Boeck.
Cyperus acuminatus var. *cyrtolepis* (Torr. & Hook.) Kük. = ***Cyperus acuminatus*** Torr. & Hook.
Cyperus acutangulus Boeck. = ***Cyperus virens*** Michx. var. ***virens***
Cyperus acuticarinatus Kük. = ***Pycreus acuticarinatus*** (Kük.) Cherm.
Cyperus acutiflorus Steud. = ***Cyperus conglomeratus*** Rottb. subsp. ***conglomeratus***
Cyperus acutiusculus Lag. = [79]
Cyperus acutus R.Br. = ***Cyperus lucidus*** R.Br.
Cyperus adamii Raymond = ***Cyperus meeboldii*** Kük.
Cyperus adansonii C.B.Clarke = ***Cyperus conglomeratus*** Rottb. subsp. ***conglomeratus***
Cyperus adenophorus Schrad. ex Nees = ***Cyperus haspan*** L.
Cyperus adenophorus var. *aphyllus* Boeck. = ***Cyperus haspan*** L.
Cyperus adoensis Hochst. ex A.Rich. = ***Cyperus rigidifolius*** Steud.
Cyperus adscendens Steud. = ?
Cyperus adustus J.Presl & C.Presl = ***Pycreus niger*** (Ruiz & Pav.) Cufod.
Cyperus aegyptiacus Gloxin = ***Cyperus capitatus*** Vand.
Cyperus aequalis Vahl = ***Cyperus prolifer*** Lam.
Cyperus aequalis C.Krauss = ***Cyperus prolifer*** Lam.
Cyperus aethiops Welw. ex Ridl. = ***Pycreus aethiops*** (Welw. ex Ridl.) C.B.Clarke
Cyperus aethiops var. *aberdarensis* (Kük.) Kük. = ***Pycreus aethiops*** (Welw. ex Ridl.) C.B.Clarke
Cyperus aethiops var. *tessmannii* Kük. = ***Pycreus aethiops*** (Welw. ex Ridl.) C.B.Clarke
Cyperus affinis Roem. & Schult. = ***Cyperus giganteus*** Vahl
Cyperus afropumilus (Lye) Lye = ***Kyllinga afropumila*** Lye
Cyperus afzelii Boeck. = ***Pycreus capillifolius*** (A.Rich.) C.B.Clarke
Cyperus agrestis Willd. ex Spreng. & Link = ***Cyperus rotundus*** L. subsp. ***rotundus***
Cyperus aitonii Spreng. = ***Isolepis prolifera*** (Rottb.) R.Br.
Cyperus alatus (Nees) F.Muell. = ***Kyllinga alba*** Nees subsp. ***alba***
Cyperus alatus subsp. *albus* (Nees) Lye = ***Kyllinga alba*** Nees
Cyperus alatus subsp. *ascolepidioides* (Cherm.) Lye = ***Kyllinga alba*** subsp. ***ascolepidioides*** (Cherm.) Lye
Cyperus alatus subsp. *controversus* (Steud.) Lye = ***Kyllinga controversa*** Steud.
Cyperus alatus var. *serratus* Peter & Kük. = ***Kyllinga alba*** Nees subsp. ***alba***
Cyperus alba-purpureus (Lye) Lye = ***Kyllinga alba-purpurea*** Lye
Cyperus albiceps Ridl. = ***Kyllinga albiceps*** (Ridl.) Rendle
Cyperus albidus Retz. = [36]
Cyperus albidus Lam. = ***Pycreus sanguinolentus*** (Vahl) Nees subsp. ***sanguinolentus***
Cyperus albiflorus Cherm. = ***Cyperus phaeolepis*** Cherm.
Cyperus albogracilis (Lye) Lye = ***Kyllinga albogracilis*** Lye
Cyperus albomarginatus K.Schum. = ***Cyperus vestitus*** Hochst. ex C.Krauss
Cyperus albomarginatus (Mart. & Schrad. ex Nees) Steud. = ***Pycreus macrostachyos*** (Lam.) J.Raynal subsp. ***macrostachyos***
Cyperus albomarginatus var. *pachyanthemos* Kük. = ***Pycreus macrostachyos*** (Lam.) J.Raynal subsp. ***macrostachyos***
Cyperus albomarginatus var. *sabulosus* (Nees) Kük. = ***Pycreus macrostachyos*** (Lam.) J.Raynal subsp. ***macrostachyos***
Cyperus albomarginatus var. *tenuis* (Boeck.) Kük. = ***Pycreus macrostachyos*** (Lam.) J.Raynal subsp. ***macrostachyos***
Cyperus albostriatus var. *mossii* (Turrill) Kük. = ***Cyperus albostriatus*** Schrad.
Cyperus alboviridis C.B.Clarke ex Scott-Elliot = ***Fimbristylis alboviridis*** (C.B.Clarke ex Scott-Elliot) C.B.Clarke
Cyperus alleizettei (Cherm.) Kük. = ***Pycreus alleizettei*** Cherm.
Cyperus alopecuroides J.König ex Roxb. = ***Cyperus exaltatus*** Retz.
Cyperus alopecuroides Thunb. = ***Cyperus thunbergii*** Vahl
Cyperus alopecuroides var. *digynus* Boeck. = ***Cyperus alopecuroides*** Rottb.
Cyperus alopecuroides var. *dives* (Delile) Boeck. = ***Cyperus dives*** Delile
Cyperus alopecuroides f. *latifolius* A.Peter ex Kük. = ***Cyperus alopecuroides*** Rottb.
Cyperus alopecuroides var. *microstachys* Boeck. = ***Cyperus dives*** Delile
Cyperus alopecuroides f. *pallidiflorus* (Peter) Kük. = ***Cyperus alopecuroides*** Rottb.
Cyperus alpestris K.Schum. = ***Cyperus tomaiophyllus*** K.Schum.
Cyperus alpinus Liebm. = ***Cyperus hermaphroditus*** (Jacq.) Standl.
Cyperus alterniflorus Schwein. = ***Cyperus schweinitzii*** Torr.
Cyperus alternifolius Steud. = ***Cyperus alterniflorus*** R.Br.
Cyperus alternifolius var. *albovariegatus* auct. = ***Cyperus alternifolius*** subsp. ***flabelliformis*** Kük.
Cyperus alternifolius var. *obtusangulum* (Boeck.) T.Koyama = ***Cyperus alternifolius*** subsp. ***flabelliformis*** Kük.

Cyperus alternifolius var. *petersianus* (Boeck.) Kük. = ***Cyperus alternifolius*** subsp. ***flabelliformis*** Kük.
Cyperus altus Nees = ***Cyperus exaltatus*** Retz.
Cyperus amabilis var. *macra* C.B.Clarke = ***Cyperus amabilis*** Vahl
Cyperus amabilis var. *macrostachyus* (Boeck.) Kük. = ***Cyperus amabilis*** Vahl
Cyperus amabilis var. *oligostachyus* (Kunth) Kük. = ***Cyperus amabilis*** Vahl
Cyperus amabilis var. *pseudocastaneus* Kük. = ***Cyperus tenax*** Boeck.
Cyperus amabilis var. *subacaulis* Kük. = ***Cyperus amabilis*** Vahl
Cyperus amane Siebold = ***Fimbristylis aestivalis*** Vahl var. ***aestivalis***
Cyperus ambiguus Steud. = ***Isolepis cernua*** (Vahl) Roem. & Schult. var. ***cernua***
Cyperus ambiguus Liebm. = ***Pycreus lanceolatus*** (Poir.) C.B.Clarke
Cyperus amblyleptos Steud. = ***Cyperus schimperianus*** Steud.
Cyperus ambongensis Boeck. = ***Cyperus niveus*** var. ***leucocephalus*** (Kunth) Fosberg
Cyperus amentaceus Rudge = ***Fimbristylis complanata*** (Retz.) Link var. ***complanata***
Cyperus americanus (Boeck.) Palla = ***Cyperus haspan*** L.
Cyperus americanus Garsault = ***Schoenoplectus lacustris*** (L.) Palla subsp. ***lacustris***
Cyperus amnicola Kunth = ***Cyperus rupestris*** Kunth
Cyperus amoenus Kunth = ***Cyperus longus*** L. subsp. ***longus***
Cyperus amoyensis Hance = ***Cyperus serotinus*** Rottb.
Cyperus amphekes Steud. = ***Cyperus thyrsiflorus*** Jungh. ex Schltdl.
Cyperus amphibolus Steud. = ***Cyperus denudatus*** L.f.
Cyperus amplissimus Steud. = ***Cyperus prolixus*** Kunth
Cyperus amplus G.C.Tucker = ***Cyperus costaricensis*** Gómez-Laur.
Cyperus × *amuricocompressus* T.Koyama = [38 JAP]
Cyperus × *amuricoides* T.Koyama = ?
Cyperus amuricus var. *japonicus* Miq. = ***Cyperus amuricus*** Maxim.
Cyperus amuricus f. *nanus* Kük. = ***Cyperus amuricus*** Maxim.
Cyperus amuricus f. *perdebilis* Kük. = ***Cyperus amuricus*** Maxim.
Cyperus amuricus var. *pterygorrachis* (H.Lév.) Ohwi = ***Cyperus amuricus*** Maxim.
Cyperus amuricus var. *subiroides* Kük. = ***Cyperus amuricus*** Maxim.
Cyperus amuricus var. *textori* (Miq.) Kük. = ***Cyperus microiria*** Steud.
Cyperus anabaptistus Steud. = ***Cyperus imbricatus*** Retz.
Cyperus anceps B.Heyne ex Steud. = ***Cyperus procerus*** Rottb.
Cyperus anceps Liebm. = ***Cyperus thyrsiflorus*** Jungh. ex Schltdl.
Cyperus anceps Steud. = ***Fimbristylis complanata*** (Retz.) Link var. ***complanata***
Cyperus andersonianus Boeck. = ***Cyperus cyperinus*** (Retz.) Suringar
Cyperus andinus Palla ex Kük. = ***Cyperus seslerioides*** Kunth
Cyperus andongensis Ridl. = ***Cyperus tenax*** Boeck.
Cyperus andreanus Maury = ***Cyperus tabina*** Steud. ex Boeck.
Cyperus andreanus var. *capitinduensis* (Maury ex Micheli) Kük. = ***Cyperus tabina*** Steud. ex Boeck.
Cyperus andreanus var. *excelsior* Kük. = ***Cyperus tabina*** Steud. ex Boeck.
Cyperus andreanus var. *foliosissimus* (Boeck.) Kük. = ***Cyperus tabina*** Steud. ex Boeck.
Cyperus andreanus var. *yguazensis* Kük. = ***Cyperus tabina*** Steud. ex Boeck.
Cyperus andrewsii C.B.Clarke = ***Cyperus subulatus*** R.Br.
Cyperus andschoa A.Rich. = ***Cyperus dichrostachyus*** Hochst. ex A.Rich.
Cyperus angolensis var. *ampibulbus* A.Peter & Kük. = ***Cyperus angolensis*** Boeck.
Cyperus angulatus W.Watson = ***Pycreus pauper*** (Hochst. ex A.Rich.) C.B.Clarke
Cyperus angulatus Nees = ***Pycreus unioloides*** (R.Br.) Urb.
Cyperus angustatus var. *betchei* Kük. = ***Cyperus betchei*** (Kük.) S.T.Blake
Cyperus angustifolius Schumach. & Thonn. = ***Cyperus cuspidatus*** Kunth
Cyperus angustifolius A.Dietr. = ***Pycreus bipartitus*** (Torr.) C.B.Clarke
Cyperus anisos Roem. & Schult. = ***Cyperus manimae*** Kunth var. ***manimae***
Cyperus anisostachyos Willd. ex Kunth = ***Cyperus amabilis*** Vahl
Cyperus annuus Spreng. & Link = ***Cyperus filiculmis*** Vahl
Cyperus annuus Willd. ex Kunth = ***Fimbristylis dichotoma*** (L.) Vahl subsp. ***dichotoma***
Cyperus anomalus Steud. = ***Cyperus javanicus*** Houtt.
Cyperus antiquorum (Willd.) Chiov. = ***Cyperus papyrus*** L. subsp. ***papyrus***
Cyperus apatris Steud. = ?
Cyperus aphyllus Vahl = ***Cyperus haspan*** L.
Cyperus aphyllus (Vahl) Hassk. = ***Kyllinga tibialis*** Poit. ex Ledeb.
Cyperus aphyllus (Kunth) F.Muell. = ***Kyllinga tibialis*** Poit. ex Ledeb.
Cyperus aphyllus Boeck. = ***Pycreus diaphanus*** (Schrad. ex Roem. & Schult.) S.S.Hooper & T.Koyama
Cyperus apiculatus Liebm. = ***Cyperus manimae*** var. ***apiculatus*** (Liebm.) C.D.Adams
Cyperus apiculatus var. *latifolius* Kük. = ***Cyperus manimae*** var. ***apiculatus*** (Liebm.) C.D.Adams
Cyperus appendiculatus var. *gordonii* Kük. = ***Cyperus appendiculatus*** (Brongn.) Kunth var. ***appendiculatus***
Cyperus apricus Ridl. = ***Cyperus semitrifidus*** Schrad.
Cyperus aquatilis F.Muell. = ***Pycreus polystachyos*** (Rottb.) P.Beauv. subsp. ***polystachyos***
Cyperus arabicus Ehrenb. ex Boeck. = ***Cyperus tuberosus*** Rottb.
Cyperus araeus Phil. = ***Cyperus volckmannii*** Phil.
Cyperus arcuatus Boeck. = ***Cyperus conglomeratus*** Rottb. subsp. ***conglomeratus***
Cyperus arechavaletae Boeck. = ***Cyperus laetus*** J.Presl & C.Presl
Cyperus arenarius Hance ex C.B.Clarke = ***Cyperus stoloniferus*** Retz.
Cyperus arenarius Salzm. ex Steud. = ?
Cyperus arenicola Steud. = ***Cyperus reflexus*** Vahl
Cyperus areolatus R.Br. = ***Pycreus sanguinolentus*** (Vahl) Nees subsp. ***sanguinolentus***
Cyperus argenteus Ridl. = ***Cyperus niveus*** var. ***leucocephalus*** (Kunth) Fosberg
Cyperus argentinus Boeck. = ***Cyperus aggregatus*** (Willd.) Endl. var. ***aggregatus***
Cyperus argentinus C.B.Clarke = ***Pycreus bipartitus*** (Torr.) C.B.Clarke

Cyperus argyraeus Steud. = ***Pycreus macrostachyos*** (Lam.) J.Raynal subsp. ***macrostachyos***
Cyperus aridicola (Domin) Kük. = ***Cyperus xerophilus*** Cherm.
Cyperus aristatus Hook.f. & Thomson ex C.B.Clarke = ***Cyperus reduncus*** Hochst. ex Boeck.
Cyperus aristatus Rottb. = ***Cyperus squarrosus*** L.
Cyperus aristatus Boeck. = ?
Cyperus aristatus f. *falciculosus* (Liebm.) Kük. = ***Cyperus squarrosus*** L.
Cyperus aristatus f. *inflexus* (Muhl.) C.B.Clarke = ***Cyperus squarrosus*** L.
Cyperus aristatus var. *inflexus* (Muhl.) Boeck. ex Kük. = ***Cyperus squarrosus*** L.
Cyperus aristatus var. *maingayi* C.B.Clarke = ***Cyperus amuricus*** Maxim.
Cyperus aristatus var. *perennis* M.E.Jones = ***Cyperus perennis*** (M.E.Jones) O'Neill
Cyperus aristatus var. *runyonii* O'Neill = ***Cyperus squarrosus*** L.
Cyperus aristatus var. *semiglobosus* Kük. = ***Cyperus squarrosus*** L.
Cyperus aristatus var. *versicolor* (Nees) Boeck. = ***Cyperus squarrosus*** L.
Cyperus aromaticus (Ridl.) Mattf. & Kük. = ***Kyllinga polyphylla*** Willd. ex Kunth
Cyperus aromaticus var. *brachyrhizomatosus* Kük. = ***Kyllinga polyphylla*** Willd. ex Kunth
Cyperus aromaticus var. *elatior* (Kunth) Kük. = ***Kyllinga elatior*** Kunth
Cyperus aromaticus var. *elatus* (Steud.) Kük. = ***Kyllinga melanosperma*** subsp. ***elata*** (Steud.) Lye
Cyperus aromaticus var. *repens* Kük. = ***Kyllinga polyphylla*** Willd. ex Kunth
Cyperus aromaticus var. *teres* (C.B.Clarke) Kük. = ***Kyllinga polyphylla*** Willd. ex Kunth
Cyperus arsitatus var. *perennis* M.E.Jones = ***Cyperus squarrosus*** L.
Cyperus articulatus Benth. = ***Cyperus articulatus*** L.
Cyperus articulatus var. *conglomeratus* Britton = ***Cyperus articulatus*** L.
Cyperus articulatus var. *erythrostachys* Graebn. = ***Cyperus articulatus*** L.
Cyperus articulatus var. *fistulosus* Kük. = ***Cyperus articulatus*** L.
Cyperus articulatus f. *longispiculosus* Kük. = ***Cyperus articulatus*** L.
Cyperus articulatus var. *multiflorus* Kük. = ***Cyperus articulatus*** L.
Cyperus articulatus var. *nodosus* (Humb. & Bonpl. ex Willd.) Kük. = ***Cyperus articulatus*** L.
Cyperus arundinaceus L. = ***Dulichium arundinaceum*** (L.) Britton
Cyperus aschenbornianus Boeck. = ***Pycreus aschenbornianus*** (Boeck.) C.B.Clarke
Cyperus ascolepidioides (Cherm.) Kük. = ***Kyllinga alba*** subsp. ***ascolepidioides*** (Cherm.) Lye
Cyperus asper (Liebm.) O'Neill = ***Cyperus mutisii*** (Kunth) Andersson
Cyperus asperifolius Desf. = ***Cyperus textilis*** Thunb.
Cyperus asperrimus Liebm. = ***Cyperus manimae*** var. ***asperrimus*** (Liebm.) Kük.
Cyperus asperrimus var. *multiflorus* Liebm. = ***Cyperus manimae*** var. ***asperrimus*** (Liebm.) Kük.
Cyperus assimilis Steud. = ***Courtoisina assimilis*** (Steud.) Maquet
Cyperus aster var. *andringitrensis* (Cherm.) A.Peter & Kük. = ***Cyperus aster*** (C.B.Clarke ex Cherm.) Kük. var. ***aster***
Cyperus ater Dalzell & A.Gibson = ***Pycreus flavidus*** (Retz.) T.Koyama
Cyperus ater Vahl = ***Pycreus nitidus*** (Lam.) J.Raynal
Cyperus aterrimus var. *agglomeratus* Kük. = ***Cyperus aterrimus*** Hochst. ex Steud.
Cyperus aterrimus var. *atroviridis* (C.B.Clarke) Kük. = ***Cyperus aterrimus*** Hochst. ex Steud.
Cyperus atlanticus Hemsl. = ***Cyperus appendiculatus*** var. ***atlanticus*** (Hemsl.) Kük.
Cyperus atratus Steud. = ***Pycreus sanguinolentus*** (Vahl) Nees subsp. ***sanguinolentus***
Cyperus atribulbus Kük. = ***Pycreus atribulbus*** (Kük.) Napper
Cyperus atrobrunneus Baker = ***Pycreus atrobrunneus*** (Baker) C.B.Clarke
Cyperus atroferrugineus Steud. = ***Pycreus flavidus*** (Retz.) T.Koyama
Cyperus atroglumosa Govind. = ***Pycreus atroglumosus*** (Govind.) P.Singh & V.Singh
Cyperus atronervatus Boeck. = ***Pycreus atronervatus*** (Boeck.) C.B.Clarke
Cyperus atronitens Hochst. = ***Pycreus elegantulus*** (Steud.) C.B.Clarke
Cyperus atropurpureus Pers. = ***Bulbostylis striata*** (Ruiz & Pav.) H.Pfeiff.
Cyperus atropurpureus Boeck. = ***Pycreus atropurpureus*** C.B.Clarke
Cyperus atrorubidus (Nelmes) Raymond = ***Pycreus atrorubidus*** Nelmes
Cyperus atrosanguineus Hochst. ex Steud. = ***Cyperus rigidifolius*** Steud.
Cyperus atrosanguineus Hemsl. = ***Cyperus sanguineoater*** Boeck.
Cyperus atroviridis C.B.Clarke = ***Cyperus aterrimus*** Hochst. ex Steud.
Cyperus aucheri Jaub. & Spach = ***Cyperus conglomeratus*** Rottb. subsp. ***conglomeratus***
Cyperus aurantiacus Kunth = ***Cyperus amabilis*** Vahl
Cyperus aureolus Steud. = [81 WIN]
Cyperus aureorufus Boeck. = ***Cyperus digitatus*** subsp. ***auricomus*** (Sieber ex Spreng.) Kük.
Cyperus aureostramineus Mattf. & Kük. = ***Kyllinga chrysantha*** K.Schum.
Cyperus aureostramineus var. *decolorans* (Kük.) Kük. = ***Kyllinga comosipes*** subsp. ***decolorans*** (Kük.) Lye
Cyperus aureovillosus (Lye) Lye = ***Kyllinga aureovillosa*** Lye
Cyperus aurescens Boeck. = ***Cyperus rigens*** J.Presl & C.Presl subsp. ***rigens***
Cyperus aureus Kunth = ***Cyperus amabilis*** Vahl
Cyperus aureus Ten. = ***Cyperus esculentus*** L.
Cyperus aureus subsp. (L.) Nyman = ***Cyperus esculentus*** L.
Cyperus aureus Georgi = ***Cyperus glomeratus*** L.
Cyperus aureus J.Presl & C.Presl = ***Cyperus squarrosus*** L.
Cyperus aureus var. *aurantiacus* (Kunth) Boeck. = ***Cyperus amabilis*** Vahl
Cyperus aureus var. *macrostachyus* Boeck. = ***Cyperus amabilis*** Vahl
Cyperus aureus var. *oligostachyus* (Kunth) Boeck. = ***Cyperus amabilis*** Vahl
Cyperus auricomus Sieber ex Spreng. = ***Cyperus digitatus*** subsp. ***auricomus*** (Sieber ex Spreng.) Kük.
Cyperus auricomus var. *khasiana* C.B.Clarke = ***Cyperus digitatus*** subsp. ***auricomus*** (Sieber ex Spreng.) Kük.
Cyperus auricomus var. *microstachyus* Boeck. = ***Cyperus digitatus*** subsp. ***auricomus*** (Sieber ex Spreng.) Kük.

Cyperus auricomus var. *minor* C.B.Clarke = ***Cyperus digitatus*** subsp. ***auricomus*** (Sieber ex Spreng.) Kük.
Cyperus aurifer var. *brachyanthus* Kük. = ***Cyperus aurifer*** Cherm.
Cyperus australis Schrad. = ***Cyperus glomeratus*** L.
Cyperus autumnalis Pursh = ***Cyperus articulatus*** L.
Cyperus autumnalis Vahl = ***Cyperus haspan*** L.
Cyperus baazas Steud. = ***Cyperus reflexus*** Vahl
Cyperus × × *babae* T.Koyama = ?
Cyperus babakensis Steud. ex Miq. = ***Cyperus babakan*** Steud.
Cyperus baccha Nees = ***Pycreus puncticulatus*** (Vahl) Nees
Cyperus badius Steud. = ***Cyperus congestus*** Vahl
Cyperus badius Desf. = ***Cyperus longus*** subsp. ***badius*** (Desf.) Murb.
Cyperus badius (Willd. ex Kunth) Boeck. = ***Cyperus subbadius*** Kük.
Cyperus badius var. *brachystachys* (C.Presl) Nyman = ***Cyperus longus*** L. subsp. ***longus***
Cyperus badius var. *inconspicuus* Nyman = ***Cyperus rotundus*** L. subsp. ***rotundus***
Cyperus badius var. *neglectus* (Parl.) Nyman = ***Cyperus longus*** L. subsp. ***longus***
Cyperus baenitzii Boeck. = ***Cyperus virens*** Michx. var. ***virens***
Cyperus bahiensis Steud. = ***Cyperus esculentus*** L.
Cyperus baikiei C.B.Clarke ex Kük. = ***Cyperus tonkinensis*** var. ***baikiei*** (C.B.Clarke ex Kük.) S.S.Hooper
Cyperus bakeri C.B.Clarke = ***Cyperus ochraceus*** Vahl
Cyperus balansae Maury = ***Cyperus impolitus*** Kunth
Cyperus balbisii Kunth = ***Cyperus sphacelatus*** Rottb.
Cyperus baldwinii Torr. = ***Cyperus croceus*** Vahl
Cyperus baldwinii Chapm. = ?
Cyperus balfourii var. *laxissimus* (Cherm.) Kük. = ***Cyperus balfourii*** C.B.Clarke
Cyperus balfourii var. *rubiginosus* Kük. = ***Cyperus balfourii*** C.B.Clarke
Cyperus balfourii var. *scabricaulis* (Cherm.) Kük. = ***Cyperus balfourii*** C.B.Clarke
Cyperus balfourii var. *tsiafajavonensis* (Cherm.) Kük. = ***Cyperus balfourii*** C.B.Clarke
Cyperus banaticus Kit. ex Nyman = ***Cyperus glaber*** L.
Cyperus bancanus Miq. = ***Cyperus trialatus*** (Boeck.) J.Kern
Cyperus bancanus var. *brevibracteatus* E.G.Camus = ***Cyperus diffusus*** Vahl
Cyperus bangianus Gand. = ***Cyperus luzulae*** (L.) Retz.
Cyperus barbata (Rottb.) Poir. = ***Bulbostylis barbata*** (Rottb.) C.B.Clarke
Cyperus baronii var. *densus* C.B.Clarke ex Cherm. = ***Cyperus baronii*** C.B.Clarke
Cyperus baronii var. *interpositus* Kük. = ***Cyperus baronii*** C.B.Clarke
Cyperus baronii var. *mannii* (C.B.Clarke) Kük. = ***Cyperus baronii*** C.B.Clarke
Cyperus baronii var. *squarrosus* C.B.Clarke ex Kük. = ***Cyperus baronii*** C.B.Clarke
Cyperus barrancae M.E.Jones = ***Cyperus surinamensis*** Rottb. var. ***surinamensis***
Cyperus barrosianus Herter = ***Pycreus tener*** C.B.Clarke
Cyperus barteri Boeck. = ***Cyperus pustulatus*** Vahl
Cyperus bellus f. *auratus* Kük. = ***Cyperus bellus*** Kunth
Cyperus bellus var. *tanganyicanus* Kük. = ***Cyperus tanganyicanus*** (Kük.) Lye
Cyperus bengalensis Spreng. = ***Cyperus corymbosus*** Rottb.
Cyperus benghalensis C.B.Clarke = ?
Cyperus bequaertii (Cherm.) Robyns & Tournay = ***Cyperus ferrugineoviridis*** (C.B.Clarke) Kük.
Cyperus berlandieri C.B.Clarke = ***Cyperus setigerus*** Torr. & Hook.
Cyperus bernoullii Kük. = ***Cyperus manimae*** var. ***asperrimus*** (Liebm.) Kük.
Cyperus berteroi Kunth = ***Cyperus giganteus*** Vahl
Cyperus betschuanus Boeck. = ***Pycreus betschuanus*** (Boeck.) C.B.Clarke
Cyperus biceps Vahl = ***Cyperus imbricatus*** Retz.
Cyperus bicolor Vahl = ***Cyperus rotundus*** L. subsp. ***rotundus***
Cyperus bicolor Barton = ***Pycreus bipartitus*** (Torr.) C.B.Clarke
Cyperus bidentatus Poir. = ***Cyperus alopecuroides*** Rottb.
Cyperus bidentatus Vahl = ***Cyperus confertus*** Sw.
Cyperus bifax C.B.Clarke = ***Cyperus rotundus*** subsp. ***retzii*** Kük.
Cyperus bifidus Pourr. = ?
Cyperus bigelowii Steud. = ***Pycreus bipartitus*** (Torr.) C.B.Clarke
Cyperus bigibbosa Fosberg = ***Kyllinga bigibbosa*** (Fosberg) Lye
Cyperus biglumis C.B.Clarke = ***Cyperus cyperoides*** (L.) Kuntze subsp. ***cyperoides***
Cyperus bipartitus Torr. = ***Pycreus bipartitus*** (Torr.) C.B.Clarke
Cyperus bipartitus f. *elongatus* (Boeck.) Mohlenbr. = ***Pycreus bipartitus*** (Torr.) C.B.Clarke
Cyperus bipartitus f. *elutus* (C.B.Clarke) Mohlenbr. = ***Pycreus bipartitus*** (Torr.) C.B.Clarke
Cyperus bipontinii Boeck. = ***Cyperus surinamensis*** Rottb. var. ***surinamensis***
Cyperus biradiatus Boeck. = ***Pycreus niger*** (Ruiz & Pav.) Cufod.
Cyperus bispicatus Steud. = ***Cyperus elatus*** L.
Cyperus bisumbellatus Steud. = ***Cyperus prolixus*** Kunth
Cyperus biuncialis Andersson = ***Cyperus confertus*** Sw.
Cyperus blancheanus Desv. ex Boiss. = ***Cyperus glaber*** L.
Cyperus blandus Kunth = ***Cyperus marginatus*** Thunb.
Cyperus blastophorus (Cherm.) Kük. = ***Pycreus blastophorus*** Cherm.
Cyperus blepharoleptos Steud. = ***Oxycaryum cubense*** (Poepp. & Kunth) Palla
Cyperus blodgettii Britton = ***Cyperus retrorsus*** Chapm.
Cyperus boeckleri Phil. = ?
Cyperus boehmii Boeck. = ***Cyperus tenax*** Boeck.
Cyperus boivinii Boeck. = ***Pycreus polystachyos*** (Rottb.) P.Beauv. subsp. ***polystachyos***
Cyperus bonianus Boeck. = ***Pycreus sulcinux*** (C.B.Clarke) C.B.Clarke
Cyperus boninensis Ohwi = [38 OGA]
Cyperus borbonicus Steud. = ***Cyperus articulatus*** L.
Cyperus ×*borealikiiensis* T.Koyama & Ogawa = ?
Cyperus botteri Boeck. = ***Cyperus ischnos*** Schltdl.
Cyperus boucheanus Regel = ***Cyperus leptocladus*** Kunth
Cyperus bourgaei C.B.Clarke ex Lundell = ***Cyperus semiochraceus*** Boeck.
Cyperus bracheilema (Steud.) Mattf. & Kük. = ***Kyllinga pulchella*** Kunth
Cyperus brachiatus Poir. = ***Cyperus compressus*** L.
Cyperus brachyphyllus Willd. ex Link = ***Cyperus amabilis*** Vahl
Cyperus brachystachys (Hook.f.) Andersson = ***Cyperus anderssonii*** Boeck.
Cyperus brachystachys Nees ex Boeck. = ***Cyperus digitatus*** Roxb. subsp. ***digitatus***

Cyperus brachystachys C.Presl = ***Cyperus longus*** L. subsp. ***longus***
Cyperus bracteatus Steud. = ***Cyperus tenuiculmis*** Boeck. var. ***tenuiculmis***
Cyperus bracteiferus Steud. = ***Cyperus aggregatus*** (Willd.) Endl. var. ***aggregatus***
Cyperus bracteolatus Steud. = ***Cyperus odoratus*** L. subsp. ***odoratus***
Cyperus bracteosus Kunze ex Desv. = ***Cyperus xanthostachyus*** Steud.
Cyperus braunii Vatke = ***Cyperus albostriatus*** Schrad.
Cyperus brazas Boeck. = ***Cyperus reflexus*** Vahl
Cyperus brevicaulis C.B.Clarke = ***Cyperus tonkinensis*** C.B.Clarke var. ***tonkinensis***
Cyperus breviculmis F.Muell. = ***Pycreus pumilus*** (L.) Nees
Cyperus breviflorus A.Dietr. =
Cyperus brevifolioides Thieret & Delahouss. = ***Kyllinga gracillima*** Miq.
Cyperus brevifolius (Rottb.) Hassk. = ***Kyllinga brevifolia*** Rottb.
Cyperus brevifolius f. *firmiculmis* Kük. = ***Kyllinga brevifolia*** Rottb. subsp. ***brevifolia***
Cyperus brevifolius subsp. *intricatus* (Cherm.) Lye = ***Kyllinga brevifolia*** subsp. ***intricata*** (Cherm.) Lye
Cyperus brevifolius subsp. *luridus* (Kük.) Lye = ***Kyllinga brevifolia*** subsp. ***lurida*** (Kük.) Lye
Cyperus brevifolius f. *stellulatus* J.V.Suringar = ***Kyllinga brevifolia*** var. ***stellulata*** (J.V.Suringar) Hooper ex Ohwi
Cyperus brevifolius var. *stellulatus* (J.V.Suringar) Kük. = ***Kyllinga brevifolia*** var. ***stellulata*** (J.V.Suringar) Hooper ex Ohwi
Cyperus brevifolius f. *vaginatus* J.V.Suringar = ***Kyllinga melanosperma*** subsp. ***bifolia*** (Miq.) Karthik.
Cyperus breviradiatus Liebm. = ***Cyperus tenuis*** Sw.
Cyperus brevis Boeck. = ***Cyperus crassipes*** Vahl
Cyperus brevispicatus Boeck. = ***Cyperus infucatus*** Kunth
Cyperus brizaeus Rich. = ***Cyperus planifolius*** Rich.
Cyperus brizaeus Vahl = ***Cyperus planifolius*** Rich.
Cyperus brizaeus Pursh = ***Pycreus filicinus*** (Vahl) T.Koyama
Cyperus brizimorphus Steud. = ?
Cyperus bromoides Britton ex Parish = ?
Cyperus bromoides Link = ***Pycreus unioloides*** (R.Br.) Urb.
Cyperus brownii Roem. & Schult. = ***Cyperus squarrosus*** L.
Cyperus brunneoalbus (Lye) Lye = ***Kyllinga brunneoalba*** Lye
Cyperus brunneoater Boeck. = ***Pycreus nitidus*** (Lam.) J.Raynal
Cyperus brunneovaginatus Boeck. = ***Cyperus marginatus*** Thunb.
Cyperus brunneus Sw. = ***Cyperus planifolius*** Rich.
Cyperus brunneus Griseb. = ***Cyperus planifolius*** Rich.
Cyperus buchananii Boeck. = ***Cyperus esculentus*** L.
Cyperus buchholzii Boeck. = ***Cyperus laxus*** subsp. ***buchholzii*** (Boeck.) Lye
Cyperus buckleyi Britton ex Coult. = ***Cyperus spectabilis*** Link var. ***spectabilis***
Cyperus buettneri Boeck. = ***Cyperus diffusus*** Vahl
Cyperus bufonius (L.) Maack ex Trautv. = (Juncaceae)
Cyperus bulbiferus A.Dietr. = ***Cyperus bulbosus*** Vahl
Cyperus bulbifex E.Mey. ex Drège = ***Cyperus usitatus*** Burch. ex Roem. & Schult. subsp. ***usitatus***
Cyperus bulbipes Mattf. & Kük. = ***Kyllinga crassipes*** Boeck.
Cyperus bulbipes var. *pallescens* Kük. = ***Kyllinga crassipes*** Boeck.
Cyperus bulbocaulis (A.Rich.) Boeck. = ***Cyperus plateilema*** (Steud.) Kük.
Cyperus bulbocaulis var. *atrosanguineus* (Boeck.) Kük. = ***Cyperus plateilema*** (Steud.) Kük.
Cyperus bulbosostoloniferus Miq. = ***Cyperus rotundus*** L. subsp. ***rotundus***
Cyperus bulbosostoloniferus Steud. = ***Cyperus stoloniferus*** Retz.
Cyperus bulbosus Lag. = ***Cyperus congestus*** Vahl
Cyperus bulbosus Schrank = ***Cyperus friburgensis*** Boeck.
Cyperus bulbosus var. *elatior* Kük. = ***Cyperus bulbosus*** Vahl
Cyperus bulbosus var. *elatus* E.G.Camus = ***Cyperus stoloniferus*** Retz.
Cyperus bulbosus var. *melanolepis* Kük. = ***Cyperus bulbosus*** Vahl
Cyperus bulbosus var. *spicatus* Boeck. = ***Cyperus blysmoides*** Hochst. ex C.B.Clarke
Cyperus bullatus Kük. = ***Cyperus chersinus*** (N.E.Br.) Kük.
Cyperus burchellii Schrad. = ***Cyperus textilis*** Thunb.
Cyperus bushii Britton = ***Cyperus lupulinus*** subsp. ***macilentus*** (Fernald) Marcks
Cyperus caducus Steud. = ?
Cyperus caespitosus Llanos = ***Cyperus haspan*** L.
Cyperus caespitosus A.St.-Hil. = ***Cyperus hilairenus*** Steud.
Cyperus caespitosus Poir. = ***Pycreus polystachyos*** (Rottb.) P.Beauv. subsp. ***polystachyos***
Cyperus caffer G.Bertol. = ***Cyperus compressus*** L.
Cyperus calacaryensis Steud. = ***Cyperus diffusus*** Vahl
Cyperus calcaratus S.Watson = ***Cyperus pseudovegetus*** Steud.
Cyperus calcicola Domin = ***Cyperus rubicundus*** Vahl
Cyperus calcicola Britton = ***Cyperus unifolius*** Boeck.
Cyperus calidus A.Kern. = ***Cyperus fuscus*** L.
Cyperus californicus S.Watson = ***Cyperus odoratus*** L. subsp. ***odoratus***
Cyperus callistus Ridl. = ***Cyperus esculentus*** L.
Cyperus callophorus G.Mey. = ***Cyperus ligularis*** L.
Cyperus calochrous Cherm. = ***Cyperus debilissimus*** Baker
Cyperus calopterus Miq. = ***Cyperus nutans*** Vahl var. ***nutans***
Cyperus camagueyensis Britton = ***Pycreus camagueyensis*** (Britton) T.Koyama
Cyperus campestris Schrad. ex Nees = ***Cyperus imbricatus*** Retz.
Cyperus camphoratus var. *boliviensis* Kük. = ***Cyperus camphoratus*** Liebm.
Cyperus camphoratus var. *breviradiatus* (Liebm.) Kük. = ***Cyperus tenuis*** Sw.
Cyperus canaliculatus Retz. = ***Cyperus exaltatus*** Retz.
Cyperus canariensis Steud. = ***Cyperus dives*** Delile
Cyperus cancellatus Ridl. = ***Cyperus haspan*** L.
Cyperus canescens Vahl = ***Cyperus javanicus*** Houtt.
Cyperus capensis Boeck. = ?
Cyperus capensis f. *globospica* (C.B.Clarke) Kük. = ***Cyperus capensis*** (Steud.) Endl.
Cyperus capensis var. *polyanthemus* Kük. = ***Cyperus chersinus*** (N.E.Br.) Kük.
Cyperus capensis var. *pseudomarlothii* Kük. = ***Cyperus capensis*** (Steud.) Endl.
Cyperus capillaris J.König ex Roxb. = ***Pycreus flavidus*** (Retz.) T.Koyama
Cyperus capillaris Hochst. ex Steud. = ***Pycreus stramineus*** C.B.Clarke

Cyperus capillifolius A.Rich. = ***Pycreus capillifolius*** (A.Rich.) C.B.Clarke
Cyperus capillifolius var. *major* (Cherm.) Kük. = ***Pycreus capillifolius*** (A.Rich.) C.B.Clarke
Cyperus capitatus Retz. = ***Cyperus cuspidatus*** Kunth
Cyperus capitatus Poir. = ***Cyperus dubius*** Rottb. subsp. ***dubius***
Cyperus capitatus (L.) Burm.f. ex B.D.Jacks. = ***Eleocharis geniculata*** (L.) Roem. & Schult.
Cyperus capitellatus Schult. = ***Cyperus dubius*** Rottb. subsp. ***dubius***
Cyperus capitinduensis Maury ex Micheli = ***Cyperus tabina*** Steud. ex Boeck.
Cyperus capito Steud. = ***Cyperus appendiculatus*** (Brongn.) Kunth var. ***appendiculatus***
Cyperus caracasanus Kunth = ***Cyperus tenuis*** Sw.
Cyperus careyi Britton = ***Cyperus laevigatus*** L. subsp. ***laevigatus***
Cyperus caribaeus Pers. = ***Fimbristylis ovata*** (Burm.f.) J.Kern
Cyperus caricifolius Hook. & Arn. = ***Cyperus javanicus*** Houtt.
Cyperus cariciformis F.Muell. = ***Cyperus paniceus*** (Rottb.) Boeck.
Cyperus caricinus D.Don = ***Pycreus filicinus*** (Vahl) T.Koyama
Cyperus carinatus Nees = ***Cyperus congestus*** Vahl
Cyperus carnosus B.Heyne = ***Cyperus procerus*** Rottb.
Cyperus carrii Kük. = ?
Cyperus carruthii Alph.Wood = ***Cyperus odoratus*** L. subsp. ***odoratus***
Cyperus cartilagineus (K.Schum.) Mattf. & Kük. = ***Kyllinga cartilaginea*** K.Schum.
Cyperus cartilagineus var. *angustatus* Peter & Kük. = ***Kyllinga comosipes*** (Mattf. & Kük.) Napper subsp. ***comosipes***
Cyperus cartilagineus var. *laevissimus* (Cherm.) Kük. = ***Kyllinga cartilaginea*** K.Schum.
Cyperus cartilagineus var. *serratangulus* Peter & Kük. = ***Kyllinga cartilaginea*** K.Schum.
Cyperus castaneus Bigelow = ***Pycreus bipartitus*** (Torr.) C.B.Clarke
Cyperus castaneus var. *brevimucronatus* Kük. = ***Cyperus castaneus*** Willd.
Cyperus cataractarum (C.B.Clarke) K.Schum. ex Engl. = ***Pycreus cataractarum*** C.B.Clarke
Cyperus catharinensis Boeck. = [84 BZS]
Cyperus caucasicus Link = ***Cyperus glaber*** L.
Cyperus cayennensis (Lam.) Britton = ***Cyperus aggregatus*** (Willd.) Endl. var. ***aggregatus***
Cyperus cayennensis Link = ***Cyperus haspan*** L.
Cyperus cayennensis var. *argentinus* (Boeck.) Kük. ex Barros = ***Cyperus aggregatus*** (Willd.) Endl.
Cyperus cayennensis var. *gigas* (Lindm.) Barros = ***Cyperus aggregatus*** var. ***gigas*** (Lindm.) Guagl.
Cyperus cayennensis var. *redolens* (Maury ex Micheli) R.Gross ex J.F.Macbr. = ***Cyperus redolens*** Maury ex Micheli
Cyperus caymanensis Millsp. = ***Cyperus swartzii*** (A.Dietr.) Boeck. ex Kük.
Cyperus cephalophorus J.Presl & C.Presl = ***Cyperus odoratus*** L. subsp. ***odoratus***
Cyperus cephalostachyus Steud. = ***Cyperus crassipes*** Vahl
Cyperus ceylanicus (Vahl) T.Koyama = ***Lipocarpha gracilis*** (Rich. ex Pers.) Nees
Cyperus chamissoi Schrad. ex Nees = ***Cyperus intricatus*** Schrad. ex Schult.
Cyperus chermezonii Kük. = ***Pycreus atropurpureus*** C.B.Clarke
Cyperus chersinus var. *angularis* (Turrill) Kük. = ***Cyperus chersinus*** (N.E.Br.) Kük.
Cyperus chilensis Boeck. = ***Cyperus rigens*** J.Presl & C.Presl subsp. ***rigens***
Cyperus chisimajensis Chiov. = ***Cyperus frerei*** C.B.Clarke
Cyperus chlorostachys Boeck. = ***Pycreus polystachyos*** (Rottb.) P.Beauv. subsp. ***polystachyos***
Cyperus chlorotropis (Steud.) Mattf. & Kük. = ***Kyllinga chlorotropis*** Steud.
Cyperus chromatolepis Steud. = ***Cyperus sphacelatus*** Rottb.
Cyperus chrysanthus Boeck. = ***Pycreus chrysanthus*** (Boeck.) C.B.Clarke
Cyperus chrysanthus var. *occidentalis* Kük. = ***Pycreus chrysanthus*** (Boeck.) C.B.Clarke
Cyperus chrysokerkos Steud. = ***Cyperus erythrorrhizos*** Muhl.
Cyperus chrysomelinus Link = ***Cyperus iria*** L.
Cyperus chrysostachys Boeck. = ***Cyperus esculentus*** L.
Cyperus ciliatopilosus Mattf. & Kük. = ***Kyllinga platyphylla*** K.Schum.
Cyperus ciliatopilosus var. *longifolius* (Kük.) Kük. = ***Kyllinga platyphylla*** K.Schum.
Cyperus ciliatopilosus var. *rhizomatosus* Kük. = ***Kyllinga platyphylla*** K.Schum.
Cyperus ciliolatus Boeck. = ***Cyperus intricatus*** Schrad. ex Schult.
Cyperus cimicinus J.Presl & C.Presl = ***Pycreus niger*** (Ruiz & Pav.) Cufod.
Cyperus cinereus Maury ex Micheli = ***Cyperus incomtus*** Kunth
Cyperus cinnamomeus Retz. = ***Cyperus glomeratus*** L.
Cyperus circinatus Ridl. = ***Pycreus polystachyos*** (Rottb.) P.Beauv. subsp. ***polystachyos***
Cyperus clandestinus Steud. = ***Ficinia clandestina*** (Steud.) Boeck.
Cyperus clarkeanus K.Schum. = ***Cyperus oblongoincrassatus*** Kük.
Cyperus cleaveri Torr. = ***Pycreus filicinus*** (Vahl) T.Koyama
Cyperus coercens Rchb. ex Kunth = ?
Cyperus cognatus Kunth = ***Cyperus rupestris*** Kunth
Cyperus colchicus K.Koch = ***Cyperus serotinus*** Rottb.
Cyperus collinus Vahl = ***Bulbostylis striata*** (Ruiz & Pav.) H.Pfeiff.
Cyperus coloratus Vahl = ***Cyperus dubius*** Rottb. subsp. ***dubius***
Cyperus coloratus var. *longinux* Kük. = ***Cyperus karisimbiensis*** var. ***longinux*** (Kük.) Kük.
Cyperus commersonii C.B.Clarke = ***Pycreus rhizomatosus*** C.B.Clarke
Cyperus commutatus Steud. = ***Pycreus pumilus*** (L.) Nees
Cyperus comosipes Mattf. & Kük. = ***Kyllinga comosipes*** (Mattf. & Kük.) Napper
Cyperus comosipes subsp. *decolorans* (Kük.) Lye = ***Kyllinga comosipes*** subsp. ***decolorans*** (Kük.) Lye
Cyperus comosus (Kunth) Poir. = ***Cyperus giganteus*** Vahl
Cyperus comosus Sm. = ***Cyperus rotundus*** L. subsp. ***rotundus***
Cyperus compactus Lam. = ***Cyperus niveus*** var. ***leucocephalus*** (Kunth) Fosberg
Cyperus compactus f. *decolorans* Kük. = ***Cyperus compactus*** Retz.
Cyperus compactus var. *macrostachys* (Boeck.) Kük. = ***Cyperus compactus*** Retz.
Cyperus compactus var. *pauciflorus* (H.Pfeiff.) Kük. = ***Cyperus compactus*** Retz.

Cyperus compactus var. *tenerior* C.B.Clarke = ***Cyperus niveus*** var. ***leucocephalus*** (Kunth) Fosberg
Cyperus complanatus Forssk. = ***Cyperus conglomeratus*** Rottb. subsp. ***conglomeratus***
Cyperus complanatus (Retz.) Willd. = ***Fimbristylis complanata*** (Retz.) Link
Cyperus complanatus C.Presl = ***Pycreus flavidus*** (Retz.) T.Koyama
Cyperus compositus R.Br. = ***Cyperus alopecuroides*** Rottb.
Cyperus compositus Boeck. = ***Cyperus gunnii*** Hook.f. var. ***gunnii***
Cyperus compressiformis (Cherm.) Kük. = ***Pycreus compressiformis*** Cherm.
Cyperus compressotriqueter Boeck. = ***Cyperus mutisii*** (Kunth) Andersson
Cyperus compressus Krock. = ***Cyperus fuscus*** L.
Cyperus compressus Jacq. = ***Cyperus surinamensis*** Rottb. var. ***surinamensis***
Cyperus compressus var. *brachiatus* (Poir.) Nees = ***Cyperus compressus*** L.
Cyperus compressus var. *capillaceus* C.B.Clarke = ***Cyperus compressus*** L.
Cyperus compressus var. *compositus* C.Presl = ***Cyperus compressus*** L.
Cyperus compressus f. *depauperatus* Domin = ***Cyperus compressus*** L.
Cyperus compressus var. *floribundus* E.G.Camus = ***Cyperus compressus*** L.
Cyperus compressus var. *laxus* E.G.Camus = ***Cyperus compressus*** L.
Cyperus compressus subsp. *micranthus* T.Koyama = ***Cyperus compressus*** L.
Cyperus compressus var. *pectiniformis* (Schult.) C.B.Clarke = ***Cyperus compressus*** L.
Cyperus compressus var. *simplex* C.Presl = ***Cyperus compressus*** L.
Cyperus conceptionis Steud. = ***Cyperus rigens*** J.Presl & C.Presl subsp. ***rigens***
Cyperus concinnus var. *parvifructus* Kük. = ***Cyperus concinnus*** R.Br.
Cyperus concolor Steud. = ***Pycreus sanguinolentus*** (Vahl) Nees subsp. ***sanguinolentus***
Cyperus ×*condensatus* T.Koyama = ?
Cyperus confertiflorus Schult. = ***Cyperus imbricatus*** Retz.
Cyperus confertus Lam. = ***Pycreus flavidus*** (Retz.) T.Koyama
Cyperus confertus var. *bidentatus* (Vahl) Kük. = ***Cyperus confertus*** Sw.
Cyperus confertus var. *biuncialis* Boeck. = ***Cyperus confertus*** Sw.
Cyperus confertus f. *debilior* (Kük.) Kük. = ***Cyperus confertus*** Sw.
Cyperus confertus var. *debilior* Kük. ex Ekman = ***Cyperus confertus*** Sw.
Cyperus confertus f. *gracillimus* Kük. = ***Cyperus confertus*** Sw.
Cyperus confusus Cherm. = ***Cyperus balfourii*** C.B.Clarke
Cyperus confusus var. *laxissimus* Cherm. = ***Cyperus balfourii*** C.B.Clarke
Cyperus confusus var. *scabricaulis* Cherm. = ***Cyperus balfourii*** C.B.Clarke
Cyperus confusus var. *tsiafajavonensis* Cherm. = ***Cyperus balfourii*** C.B.Clarke
Cyperus congestus Poir. = ***Cyperus compactus*** Retz.
Cyperus congestus var. *brevis* (Boeck.) Kük. = ***Cyperus crassipes*** Vahl
Cyperus congestus f. *gigantea* F.Muell. = ***Cyperus haematodes*** Endl.
Cyperus congestus var. *glanduliferus* (C.B.Clarke) Kük. = ***Cyperus congestus*** Vahl
Cyperus congestus var. *grandiceps* Kük. = ***Cyperus congestus*** Vahl
Cyperus congestus var. *parishii* (Britton) Kük. = ***Cyperus parishii*** Britton
Cyperus congestus var. *pseudonatalensis* Kük. = ***Cyperus congestus*** Vahl
Cyperus conglobatus Link = ***Cyperus odoratus*** L. subsp. ***odoratus***
Cyperus conglomeratus var. *aucheri* (Jaub. & Spach) C.B.Clarke = ***Cyperus conglomeratus*** Rottb. subsp. ***conglomeratus***
Cyperus conglomeratus f. *curvulus* (Boeck.) Kük. = ***Cyperus conglomeratus*** subsp. ***curvulus*** (Boeck.) Kukkonen
Cyperus conglomeratus var. *effusus* (Rottb.) Boiss. = ***Cyperus conglomeratus*** subsp. ***curvulus*** (Boeck.) Kukkonen
Cyperus conglomeratus var. *ensifolius* (Nees & Ehrenb. ex Boeck.) Kük. = ***Cyperus conglomeratus*** Rottb. subsp. ***conglomeratus***
Cyperus conglomeratus f. *excisus* (Boeck.) Kük. = ***Cyperus jeminicus*** Rottb.
Cyperus conglomeratus subsp. *jeminicus* (Rottb.) Lye = ***Cyperus jeminicus*** Rottb.
Cyperus conglomeratus var. *major* Boeck. = ***Cyperus conglomeratus*** Rottb. subsp. ***conglomeratus***
Cyperus conglomeratus f. *major* (Boeck.) Kük. = ***Cyperus conglomeratus*** Rottb. subsp. ***conglomeratus***
Cyperus conglomeratus var. *multiculmis* (Boeck.) Kük. = ***Cyperus jeminicus*** Rottb.
Cyperus conglomeratus f. *oligostachys* Kük. = ***Cyperus conglomeratus*** Rottb. subsp. ***conglomeratus***
Cyperus conglomeratus f. *pachyrrhizus* (Nees) Kük. = ***Cyperus conglomeratus*** subsp. ***pachyrrhizus*** (Nees) T.Koyama
Cyperus conglomeratus f. *pumilus* (Boeck.) Kük. = ***Cyperus conglomeratus*** Rottb. subsp. ***conglomeratus***
Cyperus conglomeratus var. *socotrensis* C.B.Clarke = ***Cyperus conglomeratus*** subsp. ***curvulus*** (Boeck.) Kukkonen
Cyperus conglomeratus f. *strictus* Kük. = ***Cyperus conglomeratus*** Rottb. subsp. ***conglomeratus***
Cyperus conicus var. *glaucinus* (Boeck.) Kük. = ***Cyperus conicus*** (R.Br.) Boeck.
Cyperus conicus var. *ramosus* Benth. = ***Cyperus ramosus*** (Benth.) Kük.
Cyperus conjunctus Steud. = ***Cyperus stoloniferus*** Retz.
Cyperus conoideus Rich. = ***Cyperus luzulae*** (L.) Retz.
Cyperus consanguineus Kunth = ***Cyperus intricatus*** Schrad. ex Schult.
Cyperus consanguineus var. *chamissoi* (Schrad. ex Nees) Kük. = ***Cyperus intricatus*** Schrad. ex Schult.
Cyperus consanguineus var. *minarum* (Boeck.) Kük. = ***Cyperus virens*** var. ***minarum*** (Boeck.) Denton
Cyperus consanguineus var. *usterii* (Palla) Kük. = ***Cyperus intricatus*** Schrad. ex Schult.
Cyperus consanguineus var. *varius* (Boeck.) Kük. = ***Cyperus intricatus*** Schrad. ex Schult.
Cyperus consors var. *pernudus* Kük. = ***Cyperus consors*** C.B.Clarke
Cyperus consors var. *riedelianus* (C.B.Clarke) Kük. = ***Cyperus consors*** C.B.Clarke
Cyperus conspicuus Steud. = ***Cyperus giganteus*** Vahl
Cyperus contractus Steud. = ***Cyperus cuspidatus*** Kunth

Cyperus controversus (Steud.) Mattf. & Kük. = ***Kyllinga controversa*** Steud.
Cyperus controversus var. *subexalatus* (C.B.Clarke) Kük. = ***Kyllinga welwitschii*** Ridl.
Cyperus convallium Steud. = ?
Cyperus cooperi (C.B.Clarke) K.Schum. = ***Cyperus congestus*** Vahl
Cyperus cooperi (C.B.Clarke) Kük. = ***Pycreus cooperi*** C.B.Clarke
Cyperus cordobensis (Palla) Hicken = ***Cyperus articulatus*** L.
Cyperus coriaceus (G.Mey.) Schrad. ex Nees = ***Cyperus ligularis*** L.
Cyperus coriifolius var. *abbreviatus* Boeck. = ***Cyperus coriifolius*** Boeck.
Cyperus cornutus Andersson = ***Cyperus elegans*** subsp. ***rubiginosus*** (Hook.f.) Eliasson
Cyperus coromandelinus Spreng. = ***Cyperus rubicundus*** Vahl
Cyperus coromandelinus Boeck. = ***Pycreus stramineus*** C.B.Clarke
Cyperus correctus Steud. = ***Pycreus lanceolatus*** (Poir.) C.B.Clarke
Cyperus corticatus C.Wright = ***Cyperus gardneri*** Nees
Cyperus corymbosus Steud. = ***Cyperus denudatus*** L.f.
Cyperus corymbosus var. *brevispiculosus* Kuntze = ***Cyperus corymbosus*** Rottb.
Cyperus corymbosus var. *longispiculatus* (Kuntze) Kük. = ***Cyperus corymbosus*** Rottb.
Cyperus corymbosus var. *macrostachyus* Boeck. = ***Cyperus corymbosus*** Rottb.
Cyperus corymbosus var. *pangorei* C.B.Clarke = ***Cyperus corymbosus*** Rottb.
Cyperus corymbosus var. *scariosus* (R.Br.) Kük. = ***Cyperus scariosus*** R.Br.
Cyperus corymbosus var. *subnodosus* (Nees & Meyen) Kük. = ***Cyperus articulatus*** L.
Cyperus cossyrensis Tineo ex Guss. = ***Cyperus laevigatus*** L. subsp. ***laevigatus***
Cyperus costatus Mattf. & Kük. = ***Kyllinga nervosa*** Steud.
Cyperus costatus subsp. *sidamoensis* (Mtot.) Lye = ***Kyllinga nervosa*** subsp. ***sidamoensis*** Mtot.
Cyperus crinitus Spreng. = ***Cyperus semitrifidus*** Schrad.
Cyperus crinitus Poir. = ***Ficinia crinita*** (Poir.) B.L Burtt
Cyperus cristatus (Kunth) Mattf. & Kük. = ***Kyllinga alba*** Nees subsp. ***alba***
Cyperus cristatus subsp. *ascolepidioides* (Cherm.) Lye = ***Kyllinga alba*** subsp. ***ascolepidioides*** (Cherm.) Lye
Cyperus cristatus var. *nigritanus* (C.B.Clarke) Kük. = ***Kyllinga alba*** Nees subsp. ***alba***
Cyperus cruciformis Boeck. = ***Cyperus vaginatus*** R.Br.
Cyperus cruciformis (Schrad. ex Schult.) Endl. = ***Kyllinga brevifolia*** Rottb. subsp. ***brevifolia***
Cyperus cruentus subsp. *amauropus* (Steud.) Lye = ***Cyperus amauropus*** Steud.
Cyperus crustaceus Raymond = ***Pycreus acuticarinatus*** (Kük.) Cherm.
Cyperus cuadriflorus Llanos = ***Cyperus compactus*** Retz.
Cyperus cuanzensis Ridl. = ***Pycreus cuanzensis*** (Ridl.) C.B.Clarke
Cyperus cubanus Liebm. = ***Cyperus odoratus*** L. subsp. ***odoratus***
Cyperus cubensis Steud. = ***Cyperus esculentus*** L.
Cyperus cumingii Nees = ***Cyperus xanthostachyus*** Steud.
Cyperus cupreus J.Presl & C.Presl = ***Cyperus erythrorrhizos*** Muhl.
Cyperus curvatus Vahl = ***Pycreus flavidus*** (Retz.) T.Koyama
Cyperus curvibracteatus Govind. = ***Pycreus curvibracteatus*** (Govind.) P.Singh & V.Singh
Cyperus curvifolius Boeck. = ***Cyperus incomtus*** Kunth
Cyperus curvispiculosus T.Koyama = ***Kyllinga curvispiculosa*** (T.Koyama) P.H.Hô
Cyperus curvulus Boeck. = ***Cyperus conglomeratus*** subsp. ***curvulus*** (Boeck.) Kukkonen
Cyperus cuspidatus Baker = ***Cyperus betafensis*** Cherm.
Cyperus cuspidatus f. *angustifolius* (Schumach. & Thonn.) Kük. = ***Cyperus cuspidatus*** Kunth
Cyperus cuspidatus var. *burchellii* C.B.Clarke = ***Cyperus cuspidatus*** Kunth
Cyperus cuspidatus f. *exiguus* (Nees) Kük. = ***Cyperus cuspidatus*** Kunth
Cyperus cyclostachyus Griseb. = ***Cyperus croceus*** Vahl
Cyperus cylindricus Chapm. = ***Cyperus ovatus*** Baldwin
Cyperus cylindricus (Elliott) Britton = ***Cyperus retrorsus*** Chapm.
Cyperus cylindricus Boeck. = [84]
Cyperus cylindricus var. *elatior* Boeck. =
Cyperus cymbiformis Liebm. = ***Cyperus tenerrimus*** J.Presl & C.Presl
Cyperus cyperinus var. *bengalensis* (C.B.Clarke) Kük. = ***Cyperus cyperinus*** (Retz.) Suringar
Cyperus cyperinus var. *laxatus* (C.B.Clarke) Kük. = ***Cyperus cyperinus*** (Retz.) Suringar
Cyperus cyperinus var. *maximus* (C.B.Clarke) Kük. = ***Cyperus cyperinus*** (Retz.) Suringar
Cyperus cyperinus var. *pictus* (Nees) Kük. = ***Cyperus cyperinus*** (Retz.) Suringar
Cyperus cyperinus var. *scabriculmis* Kük. = ***Cyperus cyperinus*** (Retz.) Suringar
Cyperus cyperoides var. *aureus* A.Peter ex Kük. = ***Cyperus macrocarpus*** (Kunth) Boeck.
Cyperus cyperoides var. *brevispicatus* Kük. = ***Cyperus cyperoides*** (L.) Kuntze subsp. ***cyperoides***
Cyperus cyperoides var. *clarkei* (Turrill ex Craib) Kük. = ***Cyperus cyperoides*** (L.) Kuntze subsp. ***cyperoides***
Cyperus cyperoides var. *evolutior* (C.B.Clarke) Kük. = ***Cyperus cyperoides*** (L.) Kuntze subsp. ***cyperoides***
Cyperus cyperoides subsp. *flavus* Lye = ***Cyperus cyperoides*** (L.) Kuntze subsp. ***cyperoides***
Cyperus cyperoides var. *khasianus* (C.B.Clarke) Kük. = ***Cyperus cyperoides*** (L.) Kuntze subsp. ***cyperoides***
Cyperus cyperoides subsp. *macrocarpus* (Kunth) Lye = ***Cyperus macrocarpus*** (Kunth) Boeck.
Cyperus cyperoides var. *microstachyys* Kük. = ***Cyperus cyperoides*** (L.) Kuntze subsp. ***cyperoides***
Cyperus cyperoides var. *nossibeensis* (Steud.) Kük. = ***Cyperus cyperoides*** (L.) Kuntze subsp. ***cyperoides***
Cyperus cyperoides var. *ovalis* Kük. = ***Cyperus cyperoides*** (L.) Kuntze subsp. ***cyperoides***
Cyperus cyperoides var. *polyphyllus* Kük. = ***Cyperus cyperoides*** (L.) Kuntze subsp. ***cyperoides***
Cyperus cyperoides var. *repens* Kük. = ***Cyperus cyperoides*** (L.) Kuntze subsp. ***cyperoides***
Cyperus cyperoides var. *subcompositus* (C.B.Clarke) Kük. = ***Cyperus cyperoides*** (L.) Kuntze subsp. ***cyperoides***
Cyperus cyrtolepis Torr. & Hook. = ***Cyperus acuminatus*** Torr. & Hook.
Cyperus cyrtolepis var. *arenicola* (Steud.) Boeck. = ***Cyperus reflexus*** Vahl
Cyperus cyrtolepis var. *caepitosus* Boeck. = ***Cyperus acuminatus*** Torr. & Hook.
Cyperus cyrtolepis var. *denticarinatus* (Britton) Britton = ***Cyperus acuminatus*** Torr. & Hook.
Cyperus cyrtolepis var. *glaucescens* Boeck. = ***Cyperus acuminatus*** Torr. & Hook.
Cyperus cyrtostachys C.B.Clarke = ?

Cyperus dactyliformis var. *elatior* (Kunth) Kük. = ***Cyperus dactyliformis*** Boeck.
Cyperus dactyliformis var. *gueinzii* (C.B.Clarke) Kük. = ***Cyperus dactyliformis*** Boeck.
Cyperus dactyliformis var. *involutus* (C.B.Clarke) K.Schum. = ***Cyperus dactyliformis*** Boeck.
Cyperus damiettensis A.Dietr. = ***Cyperus esculentus*** L.
Cyperus daphaenus Ridl. = ***Cyperus dubius*** Rottb. subsp. ***dubius***
Cyperus debilissimus var. *calochrous* (Cherm.) Kük. = ***Cyperus debilissimus*** Baker
Cyperus debilissimus var. *triqueter* Cherm. = ***Cyperus debilissimus*** Baker
Cyperus decipiens Hillebr. = ***Cyperus hillebrandii*** var. ***decipiens*** (Hillebr.) Kük.
Cyperus deckenii Boeck. = ***Cyperus derreilema*** Steud.
Cyperus declinatus Moench = ***Cyperus eragrostis*** Lam.
Cyperus declinatus var. *limbatus* Druce = ***Cyperus eragrostis*** Lam.
Cyperus decumbens Govind. = ***Pycreus govindarajalui*** V.S.Raju
Cyperus deeringianus Britton & Small = ***Cyperus ovatus*** Baldwin
Cyperus dehiscens Kunth = ***Cyperus pangorei*** Rottb.
Cyperus delavayi (C.B.Clarke) Kük. = ***Pycreus delavayi*** C.B.Clarke
Cyperus delesserianus Webb & Berthel. = ***Cyperus rubicundus*** Vahl
Cyperus delicatulatus Steud. = ***Cyperus tenuispica*** Steud.
Cyperus demangei (J.Raynal) Lye = ***Pycreus demangei*** J.Raynal
Cyperus densespicatus Hayata = ***Cyperus imbricatus*** Retz.
Cyperus densicaespitosus Mattf. & Kük. = ***Kyllinga pumila*** Michx.
Cyperus densicaespitosus var. *major* (Nees) Kük. = ***Kyllinga pumila*** Michx.
Cyperus densicaespitosus var. *rigidulus* (Steud.) Kük. = ***Kyllinga pumila*** Michx.
Cyperus densicaespitosus var. *stenophyllus* (K.Schum. ex C.B.Clarke) Kük. = ***Kyllinga stenophylla*** K.Schum. ex C.B.Clarke
Cyperus densiflorus Hemsl. = ***Cyperus distans*** L.f.
Cyperus densiflorus Rchb. ex Kunth = ***Cyperus giganteus*** Vahl
Cyperus densiflorus Link = ***Cyperus imbricatus*** Retz.
Cyperus densiflorus G.Mey. = ***Cyperus odoratus*** L. subsp. ***odoratus***
Cyperus densifolius Kunth = ***Pycreus mundtii*** Nees
Cyperus densifolius Steud. = ***Pycreus mundtii*** Nees
Cyperus densispicatus Cherm. = ***Cyperus imbricatus*** Retz.
Cyperus densus Link = ***Pycreus lanceolatus*** (Poir.) C.B.Clarke
Cyperus dentatus var. *ctenostachys* Fernald = ***Cyperus dentatus*** Torr.
Cyperus dentatus f. *ctenostachys* (Fernald) Fernald = ***Cyperus dentatus*** Torr.
Cyperus dentatus var. *multiradiatus* Torr. = ***Cyperus lecontei*** Torr. ex Steud.
Cyperus denticulatus Schrad. ex Schult. = ***Cyperus surinamensis*** Rottb. var. ***surinamensis***
Cyperus dentifer Kom. = ?
Cyperus denudatus var. *aureobrunneus* (C.B.Clarke) Kük. = ***Cyperus aureobrunneus*** C.B.Clarke
Cyperus denudatus var. *delicatulus* C.B.Clarke = ***Cyperus platycaulis*** Baker
Cyperus denudatus f. *longispiculosus* Kük. = ***Cyperus denudatus*** L.f.
Cyperus denudatus var. *lucentinigricans* (K.Schum.) Kük. = ***Cyperus platycaulis*** Baker
Cyperus denudatus f. *singularis* Kük. = ***Cyperus denudatus*** L.f.
Cyperus denudatus var. *sphaerospermoides* (Cherm.) Kük. = ***Cyperus denudatus*** L.f.
Cyperus denudatus var. *sphaerospermus* (Schrad.) Kük. = ***Cyperus sphaerospermus*** Schrad.
Cyperus depauperatus Steud. = ?
Cyperus depauperatus Vahl = ***Eleocharis retroflexa*** (Poir.) Urb. subsp. ***retroflexa***
Cyperus deremensis K.Schum. ex Engl. = ***Cyperus renschii*** Boeck.
Cyperus derreilema var. *brevispiculosus* Kük. = ***Cyperus derreilema*** Steud.
Cyperus derreilema subsp. *deckenii* (Boeck.) Kük. = ***Cyperus derreilema*** Steud.
Cyperus derreilema var. *tsaratananensis* (Cherm.) Kük. = ***Cyperus derreilema*** Steud.
Cyperus dewildeorum (J.Raynal) Lye = ***Pycreus dewildeorum*** J.Raynal
Cyperus diandrus Torr. = ***Pycreus diander*** (Torr.) C.B.Clarke
Cyperus diandrus var. *bipartitus* (Torr.) Kük. = ***Pycreus bipartitus*** (Torr.) C.B.Clarke
Cyperus diandrus var. *capitatus* Britton = ***Pycreus niger*** (Ruiz & Pav.) Cufod.
Cyperus diandrus var. *castaneus* S.Watson = ***Pycreus niger*** (Ruiz & Pav.) Cufod.
Cyperus diandrus var. *elongatus* Britton = ***Pycreus diander*** (Torr.) C.B.Clarke
Cyperus diandrus f. *elongatus* (Britton) Fernald = ***Pycreus diander*** (Torr.) C.B.Clarke
Cyperus diandrus var. *pauciflorus* Alph.Wood = ***Pycreus diander*** (Torr.) C.B.Clarke
Cyperus diaphaniria Steud. = ***Cyperus iria*** L.
Cyperus diaphanus Schrad. ex Roem. & Schult. = ***Pycreus diaphanus*** (Schrad. ex Roem. & Schult.) S.S.Hooper & T.Koyama
Cyperus diaphanus var. *gracilescens* (Kük.) H.O.Saxena = ***Pycreus diaphanus*** (Schrad. ex Roem. & Schult.) S.S.Hooper & T.Koyama
Cyperus diaphanus var. *latispicatus* (Boeck.) J.Kern = ***Pycreus diaphanus*** (Schrad. ex Roem. & Schult.) S.S.Hooper & T.Koyama
Cyperus diaphanus subsp. *setiformis* (Korsh.) J.Kern = ***Pycreus diaphanus*** (Schrad. ex Roem. & Schult.) S.S.Hooper & T.Koyama
Cyperus dichromeniformis var. *major* Boeck. = ***Cyperus mapanioides*** C.B.Clarke
Cyperus dietrichiae var. *fractuosus* Kük. = ***Cyperus dietrichiae*** Boeck. var. ***dietrichiae***
Cyperus difformis var. *breviglobosus* Kük. = ***Cyperus difformis*** L.
Cyperus difformis f. *humilis* Debeaux = ***Cyperus difformis*** L.
Cyperus difformis f. *maximus* C.B.Clarke = ***Cyperus difformis*** L.
Cyperus difformis var. *subdecompositus* Kük. = ***Cyperus difformis*** L.
Cyperus diffusus Roxb. = ***Cyperus michelianus*** (L.) Delile subsp. ***michelianus***
Cyperus diffusus subsp. *bancanus* (Miq.) Kük. = ***Cyperus trialatus*** (Boeck.) J.Kern
Cyperus diffusus var. *brevibracteatus* (E.G.Camus) Kük. = ***Cyperus diffusus*** Vahl

Cyperus diffusus subsp. *buchholzii* (Boeck.) Kük. = ***Cyperus laxus*** subsp. ***buchholzii*** (Boeck.) Lye
Cyperus diffusus var. *celebicus* (Miq.) Kük. = ***Cyperus diffusus*** Vahl
Cyperus diffusus var. *chalaranthus* (J.Presl & C.Presl) C.B.Clarke = ***Cyperus chalaranthus*** J.Presl & C.Presl
Cyperus diffusus subsp. *chalaranthus* (J.Presl & C.Presl) Kük. = ***Cyperus chalaranthus*** J.Presl & C.Presl
Cyperus diffusus var. *chorisanthos* (C.B.Clarke) Kük. = ***Cyperus chorisanthos*** C.B.Clarke
Cyperus diffusus var. *macrostachyus* Boeck. = ***Cyperus diffusus*** Vahl
Cyperus diffusus subsp. *macrostachyus* (Boeck.) T.Koyama = ***Cyperus diffusus*** Vahl
Cyperus diffusus var. *multispicatus* S.M.Huang = ***Cyperus diffusus*** Vahl
Cyperus diffusus subsp. *sylvestris* (Ridl.) Kük. = ***Cyperus laxus*** subsp. ***buchholzii*** (Boeck.) Lye
Cyperus diffusus var. *tolucensis* (Kunth) Kük. = ***Cyperus laxus*** Lam. subsp. ***laxus***
Cyperus diffusus subsp. *tolucensis* (Kunth) Borhidi = ***Cyperus laxus*** Lam. subsp. ***laxus***
Cyperus diffusus var. *umbrosus* (Lindl. & Nees) Kük. = ***Cyperus laxus*** Lam. subsp. ***laxus***
Cyperus digitatus Nees = ***Cyperus imbricatus*** Retz.
Cyperus digitatus var. *abbreviatus* Kük. = ***Cyperus digitatus*** subsp. ***auricomus*** (Sieber ex Spreng.) Kük.
Cyperus digitatus var. *aureorufus* (Boeck.) Kük. = ***Cyperus digitatus*** subsp. ***auricomus*** (Sieber ex Spreng.) Kük.
Cyperus digitatus var. *bruntii* S.S.Hooper = ***Cyperus digitatus*** subsp. ***auricomus*** (Sieber ex Spreng.) Kük.
Cyperus digitatus f. *contractior* Kük. = ***Cyperus digitatus*** Roxb. subsp. ***digitatus***
Cyperus digitatus var. *hookeri* (Boeck.) C.B.Clarke = ***Cyperus digitatus*** Roxb. subsp. ***digitatus***
Cyperus digitatus var. *mexicanus* (Liebm.) Kük. = ***Cyperus digitatus*** Roxb. subsp. ***digitatus***
Cyperus digitatus var. *oatesi* (C.B.Clarke) Kük. = ***Cyperus digitatus*** Roxb. subsp. ***digitatus***
Cyperus digitatus var. *obtusifructus* Kük. = ***Cyperus digitatus*** Roxb. subsp. ***digitatus***
Cyperus diloloensis (Kük. ex Cherm.) Kük. = ***Pycreus diloloensis*** Kük. ex Cherm.
Cyperus dilutus Vahl = ***Cyperus compressus*** L.
Cyperus dilutus var. *macrostachys* Boeck. = ***Cyperus compactus*** Retz.
Cyperus diluvialis Schult. = ***Cyperus serotinus*** Rottb.
Cyperus diphyllus Retz. = ***Cyperus corymbosus*** Rottb.
Cyperus diphyllus var. *elatior* Benth. = ***Cyperus corymbosus*** Rottb.
Cyperus diphyllus var. *triangularis* Boeck. = ***Cyperus corymbosus*** Rottb.
Cyperus dipsaciformis Fernald = ***Cyperus retrofractus*** (L.) Torr.
Cyperus discigerus Liebm. = ***Cyperus tenuis*** Sw.
Cyperus discolor Boeck. = ***Cyperus planifolius*** Rich.
Cyperus disruptus C.B.Clarke = ***Cyperus rotundus*** subsp. ***retzii*** Kük.
Cyperus dissitiflorus Andersson = ***Cyperus confertus*** Sw.
Cyperus dissitiflorus Nees ex Torr. = ***Cyperus thyrsiflorus*** Jungh. ex Schltdl.
Cyperus distachyos All. = ***Cyperus laevigatus*** subsp. ***distachyos*** (All.) Maire & Weiller
Cyperus distans G.Mey. = ***Cyperus odoratus*** L. subsp. ***odoratus***
Cyperus distans Pursh = ***Cyperus strigosus*** L.
Cyperus distans f. *capillarioides* Kük. = ***Cyperus distans*** L.f.
Cyperus distans var. *crassispiculosus* R.Gross & Kük. = ***Cyperus distans*** L.f.
Cyperus distans var. *densiflorus* Kük. = ***Cyperus distans*** L.f.
Cyperus distans subsp. *longibracteatus* (Cherm.) Lye = ***Cyperus longibracteatus*** (Cherm.) Kük.
Cyperus distans var. *niger* C.B.Clarke = ***Cyperus longibracteatus*** (Cherm.) Kük.
Cyperus distans f. *pachyanthos* Kük. = ***Cyperus distans*** L.f.
Cyperus distans f. *pallidior* Kük. = ***Cyperus distans*** L.f.
Cyperus distans var. *pseudonutans* Kük. = ***Cyperus distans*** L.f.
Cyperus distans var. *rubrotinctus* (Cherm.) Lye = ***Cyperus longibracteatus*** (Cherm.) Kük.
Cyperus distichophyllus Steud. = ***Pycreus mundtii*** Nees
Cyperus diurensis var. *acuminatosquamatus* Kük. = ***Cyperus diurensis*** Boeck.
Cyperus diurensis var. *gondanus* (Boeck.) Kük. = ***Cyperus diurensis*** Boeck.
Cyperus diurensis var. *laetevirens* A.Peter & Kük. = ***Cyperus diurensis*** Boeck.
Cyperus diurensis var. *longistolon* Kük. = ***Cyperus diurensis*** Boeck.
Cyperus divaricatus Lam. = ***Pycreus flavidus*** (Retz.) T.Koyama
Cyperus divergens Kunth = ***Cyperus manimae*** Kunth var. ***manimae***
Cyperus divergens Benth. = ***Cyperus seslerioides*** Kunth
Cyperus divergens Chapm. = ***Pycreus pumilus*** (L.) Nees
Cyperus divulsus Ridl. = ***Pycreus divulsus*** (Ridl.) C.B.Clarke
Cyperus donianus A.Dietr. = ***Cyperus pilosus*** Vahl
Cyperus donianus Boeck. = ?
Cyperus drummondii Nees ex Boeck. = ?
Cyperus dubius Boeck. = ***Cyperus dubius*** Rottb. subsp. ***dubius***
Cyperus dubius Roxb. = ?
Cyperus dubius var. *buchananii* (C.B.Clarke) Kük. = ***Cyperus dubius*** Rottb. subsp. ***dubius***
Cyperus dubius var. *capitatus* (Cherm.) Kük. = ***Cyperus dubius*** Rottb. subsp. ***dubius***
Cyperus dubius var. *coloratus* (Vahl) Kük. = ***Cyperus dubius*** Rottb. subsp. ***dubius***
Cyperus dubius subsp. *coloratus* (Vahl) Lye = ***Cyperus dubius*** Rottb. subsp. ***dubius***
Cyperus dubius var. *decaryi* (Cherm.) Kük. = ***Cyperus dubius*** Rottb. subsp. ***dubius***
Cyperus dubius var. *detersus* (C.B.Clarke ex Cherm.) Kük. = ***Cyperus dubius*** Rottb. subsp. ***dubius***
Cyperus dubius var. *macrocephalus* (C.B.Clarke) Kük. = ***Cyperus dubius*** subsp. ***macrocephalus*** (C.B.Clarke) Lye
Cyperus dubius var. *polyactis* A.Peter ex Kük. = ***Cyperus dubius*** Rottb. subsp. ***dubius***
Cyperus dubius var. *stenactis* A.Peter ex Kük. = ***Cyperus dubius*** Rottb. subsp. ***dubius***
Cyperus duchaisingii Steud. = ***Cyperus sphacelatus*** Rottb.
Cyperus ducis Buscal. & Muschl. = ***Cyperus kirkii*** C.B.Clarke
Cyperus dupetitii C.B.Clarke = ***Cyperus expansus*** Poir.
Cyperus durandii Boeck. = ***Pycreus flavescens*** (L.) P.Beauv. ex Rchb. subsp. ***flavescens***
Cyperus dussianus Duss = ***Cyperus tenuis*** Sw.
Cyperus dussii Boeck. = ***Cyperus tenuis*** Sw.
Cyperus dwarkensis K.C.Sahni & H.B.Naithani = ***Pycreus dwarkensis*** (K.C.Sahni & H.B.Naithani) S.S.Hooper

Cyperus eburneus Thonn. ex Kunth = ***Cyperus margaritaceus*** Vahl
Cyperus echinatus var. *multiflorus* Britton = ***Cyperus croceus*** Vahl
Cyperus echinolepis T.Koyama = ***Lipocarpha albiceps*** Ridl.
Cyperus ecklonii Boeck. = ***Cyperus tabularis*** Schrad.
Cyperus edulis Dinter = ***Cyperus usitatus*** Burch. ex Roem. & Schult. subsp. ***usitatus***
Cyperus effusus Rottb. = ***Cyperus conglomeratus*** subsp. ***curvulus*** (Boeck.) Kukkonen
Cyperus efoliatus Boeck. = ***Cyperus haspan*** L.
Cyperus eggersii Boeck. = ***Cyperus odoratus*** L. subsp. ***odoratus***
Cyperus eggersii var. *laticeps* Kük. = ***Cyperus odoratus*** L. subsp. ***odoratus***
Cyperus egregius Kunth = ***Cyperus fastigiatus*** Rottb.
Cyperus ehrenbergianus Boeck. = ***Cyperus mutisii*** (Kunth) Andersson
Cyperus ehrenbergii Kunth = ***Cyperus flexuosus*** Vahl
Cyperus ehrenbergii f. *fulvescens* (Kük.) Kük. =
Cyperus ehrenbergii f. *pallidiflorus* (Kük.) Kük. =
Cyperus elapsus Chiov. = ***Cyperus papyrus*** L. subsp. ***papyrus***
Cyperus elatior Boeck. = ***Cyperus dactyliformis*** Boeck.
Cyperus elatior Boeck. = ***Cyperus tenuiculmis*** Boeck. var. ***tenuiculmis***
Cyperus elatus Rottb. = ***Cyperus distans*** L.f.
Cyperus elatus Griseb. = ***Cyperus giganteus*** Vahl
Cyperus elatus Roxb. = ***Cyperus platyphyllus*** Roem. & Schult.
Cyperus elatus var. *laxus* E.G.Camus = ***Cyperus elatus*** L.
Cyperus elatus var. *macronux* C.B.Clarke = ***Cyperus elatus*** L.
Cyperus elegans var. *breviradiata* Maury = ***Cyperus elegans*** L. subsp. ***elegans***
Cyperus elegans var. *major* Kük. = ***Cyperus elegans*** L. subsp. ***elegans***
Cyperus elegantulus Steud. = ***Pycreus elegantulus*** (Steud.) C.B.Clarke
Cyperus elegantulus var. *submelanostachyus* Kük. = ***Pycreus elegantulus*** (Steud.) C.B.Clarke
Cyperus eleusinoides Kunth = ***Cyperus nutans*** var. ***eleusinoides*** (Kunth) Haines
Cyperus eleusinoides var. *dinklageanus* Kük. = ***Cyperus congensis*** C.B.Clarke
Cyperus eleusinoides var. *subprolixus* Kük. = ***Cyperus nutans*** var. ***subprolixus*** (Kük.) Karth.
Cyperus elliottianus Schult. = ***Pycreus lanceolatus*** (Poir.) C.B.Clarke
Cyperus elongatus Lej. ex Nees = ***Cyperus longus*** L. subsp. ***longus***
Cyperus elongatus Sieber ex Kunth = ***Cyperus rotundus*** L. subsp. ***rotundus***
Cyperus elongatus Hochst. ex C.B.Clarke = ***Cyperus schimperianus*** Steud.
Cyperus elongatus Steud. = ***Pycreus intactus*** (Vahl) J.Raynal
Cyperus elytropiptos Steud. = ?
Cyperus emarginatus Schrad. = ***Cyperus longus*** L. subsp. ***longus***
Cyperus eminens Klein ex Kunth = ***Cyperus platyphyllus*** Roem. & Schult.
Cyperus enervis var. *fallax* Domin = ***Cyperus enervis*** R.Br.
Cyperus enervis var. *laxus* Benth. = ***Cyperus trinervis*** R.Br.
Cyperus engelmannii Steud. = ***Cyperus odoratus*** L. subsp. ***odoratus***
Cyperus enodis Boeck. = ***Cyperus corymbosus*** Rottb.
Cyperus enodis var. *longispiculatus* Kuntze = ***Cyperus corymbosus*** Rottb.
Cyperus ensifolius Nees & Ehrenb. ex Boeck. = ***Cyperus conglomeratus*** Rottb. subsp. ***conglomeratus***
Cyperus enslenii Pursh = ***Cyperus strigosus*** L.
Cyperus entrerianus f. *abbreviatus* Kük. = ***Cyperus entrerianus*** Boeck.
Cyperus entrerianus var. *parvicapitulatus* Kük. = ***Cyperus entrerianus*** Boeck.
Cyperus eragrostis C.Krauss = ***Pycreus flavescens*** subsp. ***intermedius*** (Rikli) Lye
Cyperus eragrostis Clementi ex Willk. & Lange = ***Pycreus mundtii*** Nees
Cyperus eragrostis Vahl = ***Pycreus sanguinolentus*** (Vahl) Nees subsp. ***sanguinolentus***
Cyperus eragrostis var. *compactus* (Desv.) Kük. = ***Cyperus eragrostis*** Lam.
Cyperus eragrostis var. *cyrtostachys* Miq. = ***Pycreus sanguinolentus*** var. ***cyrtostachys*** (Miq.) L.K.Dai
Cyperus eragrostis var. *flaccidulus* Boeck. = ***Pycreus sanguinolentus*** (Vahl) Nees subsp. ***sanguinolentus***
Cyperus eragrostis var. *humilis* Miq. = ***Pycreus sanguinolentus*** (Vahl) Nees subsp. ***sanguinolentus***
Cyperus eragrostis f. *latifrons* Kük. = ***Cyperus eragrostis*** Lam.
Cyperus eragrostis var. *melanocephalus* Boeck. = ***Pycreus sanguinolentus*** (Vahl) Nees subsp. ***sanguinolentus***
Cyperus eragrostis var. *micronux* C.B.Clarke = ***Pycreus sanguinolentus*** (Vahl) Nees subsp. ***sanguinolentus***
Cyperus eragrostis var. *microstachyus* Boeck. = ***Pycreus sanguinolentus*** (Vahl) Nees subsp. ***sanguinolentus***
Cyperus eragrostis var. *minor* Ewart = ***Pycreus sanguinolentus*** (Vahl) Nees subsp. ***sanguinolentus***
Cyperus eragrostis var. *pauperita* J.M.Black = ***Pycreus sanguinolentus*** (Vahl) Nees subsp. ***sanguinolentus***
Cyperus eragrostis f. *tener* Kük. = ***Cyperus eragrostis*** Lam.
Cyperus erectus (Schumach.) Mattf. & Kük. = ***Kyllinga erecta*** Schumach.
Cyperus erectus subsp. *albescens* (Lye) Lye = ***Kyllinga erecta*** subsp. ***albescens*** Lye
Cyperus erectus var. *auratus* (Nees) Kük. = ***Kyllinga brevifolia*** subsp. ***intricata*** (Cherm.) Lye
Cyperus erectus var. *intercedens* (Kük.) Kük. = ***Kyllinga erecta*** Schumach. subsp. ***erecta***
Cyperus erectus var. *intricatus* (Cherm.) Kük. = ***Kyllinga brevifolia*** subsp. ***intricata*** (Cherm.) Lye
Cyperus erectus subsp. *jubensis* (Chiov.) Lye = ***Kyllinga jubensis*** Chiov.
Cyperus erectus var. *luridus* (Kük.) Kük. = ***Kyllinga brevifolia*** subsp. ***lurida*** (Kük.) Lye
Cyperus erectus f. *minor* Kük. = ***Kyllinga erecta*** Schumach. subsp. ***erecta***
Cyperus erectus f. *pallescens* Kük. = ***Kyllinga erecta*** Schumach. subsp. ***erecta***
Cyperus erectus var. *pleiocarpus* (Kük.) Kük. = ***Kyllinga erecta*** Schumach. subsp. ***erecta***
Cyperus erectus var. *schlechteri* (Kük.) Kük. = ***Kyllinga erecta*** Schumach. subsp. ***erecta***
Cyperus erinaceus (Ridl.) Kük. = ***Sphaerocyperus erinaceus*** (Ridl.) Lye
Cyperus eriophorus Steud. = ***Bulbostylis eriophora*** (Steud.) H.Pfeiff.
Cyperus erubescens Link = ***Cyperus glaber*** L.
Cyperus erythraeus Schrad. ex Schult. = ***Pycreus sanguinolentus*** (Vahl) Nees subsp. ***sanguinolentus***
Cyperus erythrorhizus Torr. = ***Cyperus odoratus*** L. subsp. ***odoratus***

Cyperus erythrorrhizos var. *cupreus* (J.Presl & C.Presl) Kük. = ***Cyperus erythrorrhizos*** Muhl.
Cyperus erythrorrhizos var. *halei* (Torr. ex Britton) Kük. = ***Cyperus erythrorrhizos*** Muhl.
Cyperus erythrorrhizos var. *infirmus* (Boeck.) Kük. = ***Cyperus erythrorrhizos*** Muhl.
Cyperus esculentus var. *angustispicatus* Britton = ***Cyperus esculentus*** L.
Cyperus esculentus f. *angustispicatus* (Britton) Fernald = ***Cyperus esculentus*** L.
Cyperus esculentus var. *aureus* K.Richt. = ***Cyperus esculentus*** L.
Cyperus esculentus var. *cyclolepis* Boeck. ex Kük. = ***Cyperus esculentus*** L.
Cyperus esculentus f. *evolutus* C.B.Clarke = ***Cyperus esculentus*** L.
Cyperus esculentus var. *heermannii* (Buckley) Britton = ***Cyperus esculentus*** L.
Cyperus esculentus var. *helodes* (Schrad. ex Nees) C.B.Clarke = ***Cyperus esculentus*** L.
Cyperus esculentus var. *leptostachyus* Boeck. = ***Cyperus esculentus*** L.
Cyperus esculentus var. *lutescens* (Torr. & Hook.) Kük. ex Osten = ***Cyperus esculentus*** L.
Cyperus esculentus var. *macrostachyus* Boeck. = ***Cyperus esculentus*** L.
Cyperus esculentus f. *macrostachyus* (Boeck.) Fernald = ***Cyperus esculentus*** L.
Cyperus esculentus var. *nervosostriatus* (Turrill) Kük. = ***Cyperus nervosostriatus*** Turrill
Cyperus esculentus var. *phymatodes* (Muhl.) Kük. = ***Cyperus esculentus*** L.
Cyperus esculentus f. *princeps* C.B.Clarke = ***Cyperus esculentus*** L.
Cyperus esculentus var. *sativus* Boeck. = ***Cyperus esculentus*** L.
Cyperus esculentus var. *sprucei* C.B.Clarke = ***Cyperus esculentus*** L.
Cyperus esphacelatus Kük. = ***Cyperus dilatatus*** Schumach.
Cyperus etesios Schult. = ***Cyperus filiculmis*** Vahl
Cyperus eumorphus Steud. = ***Cyperus pulcherrimus*** Willd. ex Kunth
Cyperus eurystachys Ridl. = ***Cyperus tenuis*** Sw.
Cyperus evaginatus Boeck. = ***Cyperus planifolius*** Rich.
Cyperus exaltatus var. *digynus* (Boeck.) F.N.Williams = ***Cyperus alopecuroides*** Rottb.
Cyperus exaltatus var. *divergens* Kük. = ***Cyperus exaltatus*** Retz.
Cyperus exaltatus var. *iwasakii* (Makino) T.Koyama = ***Cyperus exaltatus*** Retz.
Cyperus exaltatus var. *megalanthus* Kük. = ***Cyperus exaltatus*** Retz.
Cyperus exaltatus var. *minor* J.M.Black = ***Cyperus exaltatus*** Retz.
Cyperus exaltatus var. *oatesii* C.B.Clarke = ***Cyperus digitatus*** Roxb. subsp. ***digitatus***
Cyperus exaltatus var. *serpens* Kük. = ***Cyperus exaltatus*** Retz.
Cyperus excisus Boeck. = ***Cyperus jeminicus*** Rottb.
Cyperus exiguus Nees = ***Cyperus cuspidatus*** Kunth
Cyperus eximius (C.B.Clarke) Mattf. & Kük. = ***Kyllinga eximia*** C.B.Clarke
Cyperus eximius var. *kelleri* (C.B.Clarke) Kük. = ***Kyllinga comosipes*** (Mattf. & Kük.) Napper subsp. ***comosipes***
Cyperus extremiorientalis Ohwi = ***Cyperus michelianus*** (L.) Delile subsp. ***michelianus***
Cyperus falcatus Ehrenb. ex Boeck. = ***Cyperus conglomeratus*** Rottb. subsp. ***conglomeratus***
Cyperus falciculosus Liebm. = ***Cyperus squarrosus*** L.
Cyperus fallaciosus (Cherm.) Raymond = ***Pycreus flavescens*** subsp. ***fallaciosus*** (Cherm.) Lye
Cyperus fascicularis Phil. ex Kunth = ***Cyperus glaber*** L.
Cyperus fascicularis DC. = ***Pycreus flavidus*** (Retz.) T.Koyama
Cyperus fascicularis Poir. = ***Pycreus polystachyos*** (Rottb.) P.Beauv. subsp. ***polystachyos***
Cyperus fasciculatus Elliott = ***Pycreus lanceolatus*** (Poir.) C.B.Clarke
Cyperus fastigiatus Forssk. = ***Cyperus alopecuroides*** Rottb.
Cyperus fastigiatus Willd. ex Kunth = ***Cyperus longus*** L. subsp. ***longus***
Cyperus fastigiatus Vahl = ***Cyperus platyphyllus*** Roem. & Schult.
Cyperus fastigiosus J.F.Gmel. = ***Cyperus fastigiatus*** Rottb.
Cyperus fastuosus Desv. ex Ham. = ***Cyperus odoratus*** L. subsp. ***odoratus***
Cyperus fendlerianus var. *debilis* (Britton) Kük. = ***Cyperus sphaerolepis*** Boeck.
Cyperus fendlerianus var. *leucolepis* (Boeck.) Kük. = ***Cyperus sphaerolepis*** Boeck.
Cyperus fendlerianus var. *major* Kük. = ***Cyperus fendlerianus*** Boeck.
Cyperus fenzelianus Steud. = ***Cyperus longus*** var. ***pallidus*** Boeck.
Cyperus fenzelianus var. *badiiformis* Chiov. = ***Cyperus longus*** var. ***pallidus*** Boeck.
Cyperus ferax Benth. = ***Cyperus haematodes*** Endl.
Cyperus ferax Rich. = ***Cyperus odoratus*** L. subsp. ***odoratus***
Cyperus ferax var. *acicularis* (Schrad. ex Nees) Kük. = ***Cyperus odoratus*** L. subsp. ***odoratus***
Cyperus ferax var. *auriculatus* (Nees & Meyen ex Kunth) Kük. = ***Cyperus auriculatus*** Nees & Meyen ex Kunth
Cyperus ferax var. *bulbiferus* Barros = ***Cyperus odoratus*** L. subsp. ***odoratus***
Cyperus ferax var. *conglobatus* (Link) Kük. = ***Cyperus odoratus*** L. subsp. ***odoratus***
Cyperus ferax var. *densiflorus* (G.Mey.) Kük. = ***Cyperus odoratus*** L. subsp. ***odoratus***
Cyperus ferax var. *engelmannii* (Steud.) Kük. ex Osten = ***Cyperus odoratus*** L. subsp. ***odoratus***
Cyperus ferax subsp. *engelmannii* (Steud.) Kük. = ***Cyperus odoratus*** L. subsp. ***odoratus***
Cyperus ferax var. *hayesii* (C.B.Clarke) Kük. = ***Cyperus hayesii*** (C.B.Clarke) Standl.
Cyperus ferax var. *maximiliani* (Schrad. ex Nees) Boeck. = ***Cyperus odoratus*** L. subsp. ***odoratus***
Cyperus ferax var. *novae-hannoverae* (Boeck.) Kük. = ***Cyperus odoratus*** L. subsp. ***odoratus***
Cyperus ferax f. *parvus* (Boeck.) Kük. = ***Cyperus odoratus*** L. subsp. ***odoratus***
Cyperus ferax subsp. *speciosus* (Vahl) Kük. = ***Cyperus odoratus*** L. subsp. ***odoratus***
Cyperus ferax var. *squarrosus* (Britton) Kük. = ***Cyperus odoratus*** L. subsp. ***odoratus***
Cyperus ferax subsp. *transcaucasicus* Kük. = ***Cyperus odoratus*** subsp. ***transcaucasicus*** (Kük.) Kukkonen
Cyperus fernandezianus Colla = ***Cyperus reflexus*** Vahl
Cyperus ferox Vahl = ***Cyperus odoratus*** L. subsp. ***odoratus***
Cyperus ferrugineoviridis var. *distantiformis* Kük. = ***Cyperus ferrugineoviridis*** (C.B.Clarke) Kük.
Cyperus ferrugineoviridis var. *luteiformis* Kük. = ***Cyperus ferrugineoviridis*** (C.B.Clarke) Kük.
Cyperus ferruginescens Boeck. = ***Cyperus odoratus*** L. subsp. ***odoratus***

Cyperus ferrugineus L. = ***Dulichium arundinaceum*** (L.) Britton var. ***arundinaceum***
Cyperus ferrugineus Poir. = ***Pycreus intactus*** (Vahl) J.Raynal
Cyperus ferrugineus var. *baronii* (C.B.Clarke) Kük. = ***Pycreus intactus*** (Vahl) J.Raynal
Cyperus festivus Link = ***Cyperus exaltatus*** Retz.
Cyperus fibrillosus Kük. = ***Pycreus fibrillosus*** (Kük.) Cherm.
Cyperus fibrillosus f. *katangensis* (Cherm.) Kük. = ***Pycreus fibrillosus*** (Kük.) Cherm.
Cyperus fibrillosus var. *scaettae* (Cherm.) Kük. = ***Pycreus fibrillosus*** (Kük.) Cherm.
Cyperus fibrillosus var. *vanderystii* (Cherm.) Kük. = ***Pycreus fibrillosus*** (Kük.) Cherm.
Cyperus fiebrigii Kük. = ***Cyperus hieronymi*** Boeck.
Cyperus fieldingii Steud. = ***Cyperus tenuispica*** Steud.
Cyperus filicinus Vahl = ***Pycreus filicinus*** (Vahl) T.Koyama
Cyperus filicinus f. *cleaveri* (Torr.) Kük. = ***Pycreus filicinus*** (Vahl) T.Koyama
Cyperus filicinus var. *microdontus* (Torr.) Fernald =
Cyperus filicinus f. *splendens* Kük. = ***Pycreus filicinus*** (Vahl) T.Koyama
Cyperus filiculmis Schrad. ex Nees = ***Cyperus filifolius*** Willd. ex Kunth
Cyperus filiculmis A.Rich. = ?
Cyperus filiculmis var. *grayi* (Torr.) Boeck. = ***Cyperus grayi*** Torr.
Cyperus filiculmis var. *macilentus* Fernald = ***Cyperus lupulinus*** subsp. ***macilentus*** (Fernald) Marcks
Cyperus filiculmis var. *oblitus* Fernald = ***Cyperus filiculmis*** Vahl
Cyperus filiformis f. *coloratus* Kük. = ***Cyperus floridanus*** Britton
Cyperus filiformis var. *densiceps* Kük. = ***Cyperus floridanus*** Britton
Cyperus filiformis f. *filifolius* Kük. = ***Cyperus filiformis*** Sw.
Cyperus filiformis var. *havanensis* (Kunth) Kük. = ***Cyperus filiformis*** Sw.
Cyperus filiformis f. *robustior* Kük. = ***Cyperus filiformis*** Sw.
Cyperus filiformis f. *subfilifolius* Kük. = ***Cyperus filiformis*** Sw.
Cyperus fimbriatus Nees = ***Cyperus pilosus*** Vahl
Cyperus fimbristyloides T.Koyama = ***Alinula paradoxa*** (Cherm.) Goetgh. & Vorster
Cyperus firmus J.Presl & C.Presl = ***Cyperus javanicus*** Houtt.
Cyperus fischerianus var. *ugandensis* Lye = ***Cyperus fischerianus*** Schimp. ex A.Rich.
Cyperus fistulosus Ehrenb. ex Boeck. = ***Cyperus articulatus*** L.
Cyperus flabellaris Schrad. ex Nees = ***Cyperus fastigiatus*** Rottb.
Cyperus flabelliformis Rottb. = ***Cyperus alternifolius*** subsp. ***flabelliformis*** Kük.
Cyperus flabelliformis (Kunth) Spreng. = ***Cyperus tenuis*** Sw.
Cyperus flabelliformis var. *obtusangulum* Boeck. = ***Cyperus canus*** J.Presl & C.Presl
Cyperus flagellatus Hochst. = ***Cyperus alternifolius*** subsp. ***flabelliformis*** Kük.
Cyperus flavens Pall. = ***Pycreus flavescens*** (L.) P.Beauv. ex Rchb. subsp. ***flavescens***
Cyperus flavescens L. = ***Pycreus flavescens*** (L.) P.Beauv. ex Rchb.
Cyperus flavescens var. *abyssinicus* (Hochst. ex A.Rich.) C.B.Clarke = ***Pycreus flavescens*** (L.) P.Beauv. ex Rchb. subsp. ***flavescens***
Cyperus flavescens f. *abyssinicus* (Hochst. ex A.Rich.) Kük. = ***Pycreus flavescens*** (L.) P.Beauv. ex Rchb. subsp. ***flavescens***
Cyperus flavescens var. *bicolor* Farw. = ***Pycreus bipartitus*** (Torr.) C.B.Clarke
Cyperus flavescens subvar. *capitatus* (Britton) Farw. = ***Pycreus niger*** (Ruiz & Pav.) Cufod.
Cyperus flavescens var. *castaneus* (Lye) Lye = ***Pycreus flavescens*** var. ***rehmannianus*** (C.B.Clarke) Govaerts
Cyperus flavescens f. *caucasicus* K.Koch = ***Pycreus flavescens*** (L.) P.Beauv. ex Rchb. subsp. ***flavescens***
Cyperus flavescens f. *compactus* Asch. & Graebn. = ***Pycreus flavescens*** (L.) P.Beauv. ex Rchb. subsp. ***flavescens***
Cyperus flavescens subsp. *fallaciosus* (Cherm.) Lye = ***Pycreus flavescens*** subsp. ***fallaciosus*** (Cherm.) Lye
Cyperus flavescens var. *fontanesii* (Kunth) C.B.Clarke = ***Pycreus flavescens*** (L.) P.Beauv. ex Rchb. subsp. ***flavescens***
Cyperus flavescens f. *fontanesii* (Kunth) Kük. = ***Pycreus flavescens*** (L.) P.Beauv. ex Rchb. subsp. ***flavescens***
Cyperus flavescens f. *gracillimus* Asch. & Graebn. = ***Pycreus flavescens*** (L.) P.Beauv. ex Rchb. subsp. ***flavescens***
Cyperus flavescens var. *gussonei* (Gasp. ex Guss.) Nyman = ***Pycreus flavescens*** (L.) P.Beauv. ex Rchb. subsp. ***flavescens***
Cyperus flavescens subsp. *intermedius* (Rikli) Lye = ***Pycreus flavescens*** subsp. ***intermedius*** (Rikli) Lye
Cyperus flavescens f. *major* Lojac. = ***Pycreus flavescens*** (L.) P.Beauv. ex Rchb. subsp. ***flavescens***
Cyperus flavescens subsp. *microglumis* (Lye) Lye = ***Pycreus flavescens*** subsp. ***microglumis*** Lye
Cyperus flavescens f. *minor* Gaudin = ***Pycreus flavescens*** (L.) P.Beauv. ex Rchb. subsp. ***flavescens***
Cyperus flavescens subvar. *mohrii* Farw. = ***Pycreus flavescens*** (L.) P.Beauv. ex Rchb. subsp. ***flavescens***
Cyperus flavescens f. *monostachyus* Sacc. = ***Pycreus flavescens*** (L.) P.Beauv. ex Rchb. subsp. ***flavescens***
Cyperus flavescens var. *paraensis* (Nees) Boeck. = ***Pycreus flavescens*** (L.) P.Beauv. ex Rchb. subsp. ***flavescens***
Cyperus flavescens f. *paraensis* (Nees) Kük. = ***Pycreus flavescens*** (L.) P.Beauv. ex Rchb. subsp. ***flavescens***
Cyperus flavescens var. *piceus* (Liebm.) Fernald = ***Pycreus flavescens*** (L.) P.Beauv. ex Rchb. subsp. ***flavescens***
Cyperus flavescens var. *poiformis* (Pursh) Fernald = ***Pycreus flavescens*** (L.) P.Beauv. ex Rchb. subsp. ***flavescens***
Cyperus flavescens var. *pygmaeus* Goiran = ***Pycreus flavescens*** (L.) P.Beauv. ex Rchb. subsp. ***flavescens***
Cyperus flavescens var. *rehmannianus* (C.B.Clarke) Kük. = ***Pycreus flavescens*** var. ***rehmannianus*** (C.B.Clarke) Govaerts
Cyperus flavescens var. *rubromarginatus* Schrenk = ***Pycreus sanguinolentus*** (Vahl) Nees subsp. ***sanguinolentus***
Cyperus flavescens f. *setaceus* Kük. = ***Pycreus flavescens*** (L.) P.Beauv. ex Rchb. subsp. ***flavescens***
Cyperus flavescens subsp. *tanaensis* (Kük.) Lye = ***Pycreus flavescens*** subsp. ***tanaensis*** (Kük.) Lye
Cyperus flavicomus Pursh = ***Cyperus odoratus*** L. subsp. ***odoratus***
Cyperus flavicomus Schltdl. = ?

Cyperus flavicomus Michx. = ***Pycreus macrostachyos*** (Lam.) J.Raynal subsp. ***macrostachyos***
Cyperus flavicomus Vahl = ***Pycreus macrostachyos*** (Lam.) J.Raynal subsp. ***macrostachyos***
Cyperus flavidus Retz. = ***Pycreus flavidus*** (Retz.) T.Koyama
Cyperus flavissimus Steud. = ***Cyperus denudatus*** L.f.
Cyperus flavissimus Schrad. = ***Cyperus niveus*** var. ***flavissimus*** (Schrad.) Lye
Cyperus flavomariscus Griseb. = ***Cyperus aggregatus*** (Willd.) Endl. var. ***aggregatus***
Cyperus flavomariscus var. *peduncularis* Britton = ***Cyperus dentoniae*** G.C.Tucker
Cyperus flavus (Vahl) Nees = ***Cyperus aggregatus*** (Willd.) Endl. var. ***aggregatus***
Cyperus flavus J.Presl & C.Presl = ***Pycreus lanceolatus*** (Poir.) C.B.Clarke
Cyperus flavus var. *aggregatus* (Willd.) Kük. = ***Cyperus aggregatus*** (Willd.) Endl.
Cyperus flavus var. *angustatus* Kük. = ***Cyperus aggregatus*** (Willd.) Endl. var. ***aggregatus***
Cyperus flavus var. *argentinus* (Boeck.) Kük. ex Osten = ***Cyperus aggregatus*** (Willd.) Endl. var. ***aggregatus***
Cyperus flavus var. *australis* Osten = ***Cyperus aggregatus*** (Willd.) Endl. var. ***aggregatus***
Cyperus flavus var. *genuina* Osten = ***Cyperus aggregatus*** (Willd.) Endl.
Cyperus flavus var. *gigas* (Lindm.) Kük. = ***Cyperus aggregatus*** var. ***gigas*** (Lindm.) Guagl.
Cyperus flavus var. *laevis* (Kunth) Kük. = ***Cyperus aggregatus*** (Willd.) Endl. var. ***aggregatus***
Cyperus flavus var. *obesus* (Liebm.) Kük. = ***Cyperus aggregatus*** (Willd.) Endl. var. ***aggregatus***
Cyperus flavus var. *peduncularis* (Britton) Kük. = ***Cyperus dentoniae*** G.C.Tucker
Cyperus flavus subsp. *redolens* (Maury ex Micheli) Osten = ***Cyperus redolens*** Maury ex Micheli
Cyperus flexibilis Maury = ***Cyperus tenuis*** Sw.
Cyperus flexifolius Boeck. ex Reinecke = ***Cyperus cyperoides*** (L.) Kuntze subsp. ***cyperoides***
Cyperus flexifolius Boeck. = ***Cyperus imbricatus*** Retz.
Cyperus flexuosus Griseb. = ***Cyperus odoratus*** L. subsp. ***odoratus***
Cyperus flexuosus f. *fulvescens* Kük. =
Cyperus flexuosus var. *insignis* (Kunth) Kük. = ***Cyperus flexuosus*** Vahl
Cyperus flexuosus f. *pallidiflorus* Kük. =
Cyperus fluitans Buch.-Ham. ex C.B.Clarke = ***Cyperus platystylis*** R.Br.
Cyperus fluitans (L.) Missbach & E.H.L.Krause = ***Isolepis fluitans*** (L.) R.Br.
Cyperus fluminalis Ridl. = ***Pycreus fluminalis*** (Ridl.) Rendle
Cyperus fluminalis var. *longifolius* Cherm. ex Kük. = ***Pycreus fluminalis*** (Ridl.) Rendle
Cyperus foliosissimus Boeck. = ***Cyperus tabina*** Steud. ex Boeck.
Cyperus foliosus K.Schum. = ***Cyperus chermezonianus*** Robyns & Tournay
Cyperus foliosus Willd. ex Kunth = ***Pycreus intactus*** (Vahl) J.Raynal
Cyperus fontanesii Kunth = ***Pycreus flavescens*** (L.) P.Beauv. ex Rchb. subsp. ***flavescens***
Cyperus fonticola Kunth = ***Cyperus marginatus*** Thunb.
Cyperus fontinalis (Cherm.) Kük. = ***Pycreus fontinalis*** Cherm.
Cyperus formosus Vahl = ***Cyperus virens*** Michx. var. ***virens***
Cyperus forsskalii A.Dietr. = ***Cyperus fuscus*** L.
Cyperus fortunei Steud. = ***Cyperus malaccensis*** Lam. var. ***malaccensis***
Cyperus fossarum Liebm. = ***Cyperus odoratus*** L. subsp. ***odoratus***
Cyperus fossarum Schltdl. = [82 SUR]
Cyperus fragilis Liebm. = ***Cyperus odoratus*** L. subsp. ***odoratus***
Cyperus fraternus Kunth = ***Cyperus reflexus*** Vahl
Cyperus fresenii Steud. = ***Cyperus esculentus*** L.
Cyperus friesii Kük. = ***Cyperus amauropus*** Steud.
Cyperus frionensis Boeck. = ?
Cyperus frondosus Salisb. = ***Cyperus alternifolius*** L. subsp. ***alternifolius***
Cyperus fucatus Boeck. = ***Cyperus javanicus*** Houtt.
Cyperus fugax Liebm. = ***Pycreus fugax*** (Liebm.) C.D.Adams
Cyperus fugax var. *humilis* Boeck. = ***Pycreus fugax*** (Liebm.) C.D.Adams
Cyperus fuligineus var. *calcicola* (Britton) Kük. = ***Cyperus unifolius*** Boeck.
Cyperus fulvescens Liebm. = ***Cyperus esculentus*** L.
Cyperus fulvus Ridl. = ***Pycreus nitidus*** (Lam.) J.Raynal
Cyperus fulvus var. *canescens* (C.B.Clarke ex Domin) Kük. = ***Cyperus fulvus*** R.Br.
Cyperus fulvus var. *confusus* (C.B.Clarke ex Domin) Kük. = ***Cyperus fulvus*** R.Br.
Cyperus fulvus var. *densispiculatus* (Domin) Kük. = ***Cyperus fulvus*** R.Br.
Cyperus fulvus var. *viscidus* Kük. = ***Cyperus fulvus*** R.Br.
Cyperus fusciformis Drobow = ?
Cyperus fusco-ater Meinsh. ex Korsh. = ***Pycreus flavidus*** (Retz.) T.Koyama
Cyperus fuscoferrugineus Boeck. = ***Cyperus friburgensis*** Boeck.
Cyperus fuscus f. *calidus* (A.Kern.) Asch. & Graebn. = ***Cyperus fuscus*** L.
Cyperus fuscus var. *condensatus* Kük. = ***Cyperus fuscus*** L.
Cyperus fuscus var. *nigricans* Neilr. = ***Cyperus fuscus*** L.
Cyperus fuscus var. *subjuncellus* Kük. = ***Cyperus fuscus*** L.
Cyperus fuscus f. *virescens* (Hoffm.) Vahl = ***Cyperus fuscus*** L.
Cyperus fuscus var. *virescens* (Hoffm.) Dumort. = ***Cyperus fuscus*** L.
Cyperus galapagensis Caruel. = ***Cyperus grandifolius*** J.G.Anderson
Cyperus galegensis C.B.Clarke = ***Cyperus maculatus*** Boeck. subsp. ***maculatus***
Cyperus gardneri Baker = ?
Cyperus gardneri var. *vegetior* Kük. = ***Cyperus gardneri*** Nees
Cyperus gatesii Torr. = ***Pycreus polystachyos*** (Rottb.) P.Beauv. subsp. ***polystachyos***
Cyperus geminatus J.König ex Ainslie = ***Cyperus bulbosus*** Vahl
Cyperus geminatus Schrad. ex Schult. = ***Fimbristylis bahiensis*** Steud.
Cyperus geminiflora (Steud.) Wickens = ***Kyllinga bulbosa*** P.Beauv.
Cyperus giganteus Griseb. = ***Cyperus digitatus*** Roxb. subsp. ***digitatus***
Cyperus giganteus var. *comosus* (Kunth) Kük. = ***Cyperus giganteus*** Vahl
Cyperus ginge Welw. = ?
Cyperus giolii Chiov. = ***Cyperus grandibulbosus*** C.B.Clarke
Cyperus giolii var. *nogalensis* (Chiov.) Kük. = ***Cyperus grandibulbosus*** C.B.Clarke
Cyperus glaber Vill. = ***Cyperus serotinus*** Rottb.

Cyperus glaber var. *blancheanus* Boiss. = ***Cyperus glaber*** L.
Cyperus glaber f. *contractus* Asch. & Graebn. = ***Cyperus glaber*** L.
Cyperus glaber f. *fasciculariformis* Lojac. = ***Cyperus glaber*** L.
Cyperus glaber var. *subchoristachys* (C.B.Clarke) Kük. = ***Cyperus glaber*** L.
Cyperus glandulosus (Bojer) Rolfe = ***Cyperus ligularis*** L.
Cyperus glareosus Liebm. = ***Cyperus amabilis*** Vahl
Cyperus glaucinus Boeck. = ***Cyperus conicus*** (R.Br.) Boeck.
Cyperus glaucophyllus var. *longispiculosus* Kük. = ***Cyperus glaucophyllus*** Boeck.
Cyperus glaucophyllus var. *zambesiensis* (C.B.Clarke) Kük. = ***Cyperus pulchellus*** R.Br.
Cyperus glaucoviridis Boeck. = ***Cyperus ligularis*** L.
Cyperus glaucus Willd. ex Kunth = ***Cyperus alopecuroides*** Rottb.
Cyperus glaucus Steud. = ***Cyperus planifolius*** Rich.
Cyperus glaucus var. *gracilis* Boeck. = ***Cyperus planifolius*** Rich.
Cyperus glaziovianus Boeck. = ***Cyperus mundulus*** Kunth
Cyperus globosus Forssk. = ***Cyperus conglomeratus*** Rottb. subsp. ***conglomeratus***
Cyperus globosus All. = ***Pycreus flavidus*** (Retz.) T.Koyama
Cyperus globosus var. *bubanii* Debeaux = ***Pycreus flavidus*** (Retz.) T.Koyama
Cyperus globosus f. *depauperatus* Kük. = ***Pycreus flavidus*** (Retz.) T.Koyama
Cyperus globosus var. *erectus* (C.B.Clarke) Kük. = ***Pycreus flavidus*** (Retz.) T.Koyama
Cyperus globosus f. *fusco-ater* (Meinsh. ex Korsh.) Kük. = ***Pycreus flavidus*** (Retz.) T.Koyama
Cyperus globosus var. *humifusus* (Clemente) Nyman = ***Pycreus flavidus*** (Retz.) T.Koyama
Cyperus globosus var. *khasianus* C.B.Clarke = ***Pycreus flavidus*** (Retz.) T.Koyama
Cyperus globosus f. *minimus* Kük. = ***Pycreus flavidus*** (Retz.) T.Koyama
Cyperus globosus var. *nilagiricus* (Hochst. ex Steud.) C.B.Clarke = ***Pycreus flavidus*** (Retz.) T.Koyama
Cyperus globosus var. *nuerensis* (Boeck.) Kük. = ***Pycreus nuerensis*** (Boeck.) S.S.Hooper
Cyperus globosus f. *pauperior* (Boeck.) Kük. = ***Pycreus flavidus*** (Retz.) T.Koyama
Cyperus globosus var. *strictus* (Lam.) C.B.Clarke = ***Pycreus flavidus*** (Retz.) T.Koyama
Cyperus globosus var. *submonocephalus* Boeck. = ***Pycreus flavidus*** (Retz.) T.Koyama
Cyperus globuliferus Link = ***Cyperus luzulae*** (L.) Retz.
Cyperus globuliferus J.Presl & C.Presl = ***Androtrichum trigynum*** (Spreng.) H.Pfeiff.
Cyperus globulosus Aubl. = ***Cyperus luzulae*** (L.) Retz.
Cyperus globulosus var. *multiflorus* Kük. & Ekman = ***Cyperus croceus*** Vahl
Cyperus globulosus var. *pseudofiliculmis* Kük. = ***Cyperus croceus*** Vahl
Cyperus globulosus var. *robustus* (Boeck.) Shinners = ***Cyperus retrorsus*** Chapm.
Cyperus glomeratus Willd. = ***Cyperus alopecuroides*** Rottb.
Cyperus glomeratus Roxb. ex Arn. = ***Cyperus digitatus*** Roxb. subsp. ***digitatus***
Cyperus glomeratus Walter = ***Cyperus virens*** Michx. var. ***virens***
Cyperus glomeratus var. *semiradiciflorus* (Boeck.) Kük. = ***Cyperus glomeratus*** L.
Cyperus glomeratus f. *stenostachys* Borbás = ***Cyperus glomeratus*** L.
Cyperus glomuliferus Schrad. ex Schult. = ***Cyperus luzulae*** (L.) Retz.
Cyperus goeringii Steud. = ***Cyperus difformis*** L.
Cyperus gondanus Boeck. = ***Cyperus diurensis*** Boeck.
Cyperus gossweileri Kük. = ***Pycreus pubescens*** Turrill
Cyperus grabowskianus Boeck. = ***Cyperus compactus*** Retz.
Cyperus gracilescens Schult. = ***Cyperus esculentus*** L.
Cyperus gracilinux C.B.Clarke = ***Cyperus dilatatus*** Schumach.
Cyperus gracilis Buchanan = ***Cyperus eragrostis*** Lam.
Cyperus gracilis Link = ***Cyperus esculentus*** L.
Cyperus gracilis Muhl. = ***Cyperus haspan*** L.
Cyperus gracilis var. *enervis* (R.Br.) Kük. = ***Cyperus enervis*** R.Br.
Cyperus gracilis var. *rigidellus* Benth. = ***Cyperus rigidellus*** (Benth.) J.M.Black
Cyperus gracilis f. *viviparus* Domin = ***Cyperus gracilis*** R.Br.
Cyperus gracillimus (Chiov.) Kük. = ***Pycreus gracillimus*** Chiov.
Cyperus gradatus Forssk. = ***Cyperus alternifolius*** subsp. ***flabelliformis*** Kük.
Cyperus graeffei Boeck. = ***Cyperus cyperinus*** (Retz.) Suringar
Cyperus gramineus Roem. & Schult. = ***Cyperus haspan*** L.
Cyperus gramineus Steud. = ***Pycreus grammicus*** (Kunze ex Kunth) C.B.Clarke
Cyperus graminicolus Steud. = ***Cyperus distans*** L.f.
Cyperus graminifolius Poir. = ***Cyperus haspan*** L.
Cyperus grammicus Kunze ex Kunth = ***Pycreus grammicus*** (Kunze ex Kunth) C.B.Clarke
Cyperus granadinus Liebm. = ***Cyperus odoratus*** L. subsp. ***odoratus***
Cyperus grandibulbosus var. *amplus* Kük. = ***Cyperus grandibulbosus*** C.B.Clarke
Cyperus granularis (Desf. ex Link) Britton = ***Cyperus swartzii*** (A.Dietr.) Boeck. ex Kük.
Cyperus granulatus Bosc ex Kunth = ***Cyperus glomeratus*** L.
Cyperus gratus C.B.Clarke = ***Cyperus cuspidatus*** Kunth
Cyperus griffithianus Boeck. = ***Cyperus procerus*** Rottb.
Cyperus griffithii Steud. = ***Cyperus radians*** Nees & Meyen ex Kunth
Cyperus guadalajaranus M.E.Jones = ***Cyperus amabilis*** Vahl
Cyperus guatemalensis Gand. = ***Cyperus luzulae*** (L.) Retz.
Cyperus guatemalensis Steud. = ?
Cyperus guineensis Nelmes = ***Cyperus tenuiculmis*** var. ***guineensis*** (Nelmes) S.S.Hooper
Cyperus gula-metthi Schult. = ***Cyperus corymbosus*** Rottb.
Cyperus gunnii subsp. *novae-hollandiae* (Boeck.) K.L.Wilson = ***Cyperus gunnii*** var. ***novae-hollandiae*** (Boeck.) Kük.
Cyperus gussonei Gasp. ex Guss. = ***Pycreus flavescens*** (L.) P.Beauv. ex Rchb. subsp. ***flavescens***
Cyperus guthnickii Tineo ex Lojac. = ***Cyperus longus*** L. subsp. ***longus***
Cyperus gymnoleptus Steud. = ***Pycreus pumilus*** (L.) Nees
Cyperus gymnos Schult. = ***Cyperus articulatus*** L.
Cyperus haematostachys Steud. = ***Cyperus reflexus*** Vahl
Cyperus haenkeanus Kunth = ***Cyperus compactus*** Retz.
Cyperus haenkei J.Presl & C.Presl = ***Cyperus odoratus*** L. subsp. ***odoratus***
Cyperus hahnianus Boeck. = ***Pycreus polystachyos*** (Rottb.) P.Beauv. subsp. ***polystachyos***

Cyperus halei Torr. ex Britton = ***Cyperus erythrorrhizos*** Muhl.
Cyperus hallii Britton = ***Cyperus setigerus*** Torr. & Hook.
Cyperus hamatus Schrad. ex Nees = ***Cyperus cuspidatus*** Kunth
Cyperus hamiltonii Kunth = ***Cyperus odoratus*** L. subsp. ***odoratus***
Cyperus hamulatus Kunth = ***Cyperus cuspidatus*** Kunth
Cyperus hangianus Gand. = ?
Cyperus hansenii Britton = ***Cyperus strigosus*** L.
Cyperus hartii Boeck. = ***Cyperus tenuis*** Sw.
Cyperus haspan var. *adenophorus* (Schrad. ex Nees) Kük. = ***Cyperus haspan*** L.
Cyperus haspan var. *americanus* Boeck. = ***Cyperus haspan*** L.
Cyperus haspan var. *amplissimus* Kük. = ***Cyperus haspan*** L.
Cyperus haspan var. *bulboides* Kük. = ***Cyperus haspan*** L.
Cyperus haspan var. *cancellatus* (Ridl.) Kük. = ***Cyperus haspan*** L.
Cyperus haspan var. *coarctatus* Nees = ***Cyperus haspan*** L.
Cyperus haspan f. *coarctatus* (Nees) Osten = ***Cyperus haspan*** L.
Cyperus haspan var. *dietrichiae* Domin = ***Cyperus haspan*** L.
Cyperus haspan var. *firmicaulis* Kük. = ***Cyperus haspan*** L.
Cyperus haspan var. *flaccidissimus* Kük. = ***Cyperus haspan*** L.
Cyperus haspan subsp. *juncoides* (Lam.) Kük. = ***Cyperus haspan*** L.
Cyperus haspan f. *major* C.B.Clarke = ***Cyperus haspan*** L.
Cyperus haspan var. *riparius* Kük. = ***Cyperus haspan*** L.
Cyperus haspanoides C.B.Clarke ex K.Schum. = ***Cyperus denudatus*** L.f.
Cyperus havanensis Kunth = ***Cyperus filiformis*** Sw.
Cyperus hawaicensis Hillebr. = ***Cyperus hillebrandii*** Boeck. var. ***hillebrandii***
Cyperus hawaiiensis H.Mann = ***Cyperus phleoides*** var. ***hawaiiensis*** (H.Mann) Kük.
Cyperus haworthii Gray = ***Cyperus fuscus*** L.
Cyperus hebes Steud. = ***Cyperus pilosus*** Vahl
Cyperus heermannii Buckley = ***Cyperus esculentus*** L.
Cyperus heldreichianus Boiss. = ***Cyperus longus*** L. subsp. ***longus***
Cyperus helodes Schrad. ex Nees = ***Cyperus esculentus*** L.
Cyperus helvus Liebm. = ***Pycreus lanceolatus*** (Poir.) C.B.Clarke
Cyperus hemisphaericus var. *gregorii* (C.B.Clarke) Kük. = ***Cyperus hemisphaericus*** Boeck.
Cyperus hemisphaericus var. *longibracteus* A.Peter ex Kük. = ***Cyperus hemisphaericus*** Boeck.
Cyperus heptaphyllus Larrañaga = ?
Cyperus herana Cherm. = ***Cyperus latifolius*** Poir.
Cyperus herana var. *angustifolius* Cherm. = ***Cyperus solidifolius*** Boeck.
Cyperus herbicavus Melliss = ***Cyperus rotundus*** L. subsp. ***rotundus***
Cyperus herbivagus Kunth = ***Cyperus semitrifidus*** Schrad.
Cyperus × *heribaudii* Pau = *Pycreus flavescens* × *C. fuscus* ?
Cyperus hermaphroditus var. *angustior* (C.B.Clarke) Kük. = ***Cyperus hermaphroditus*** (Jacq.) Standl.
Cyperus hermaphroditus var. *condensatus* Kük. = ***Cyperus hermaphroditus*** (Jacq.) Standl.
Cyperus hermaphroditus var. *longiradiatus* (Liebm.) Kük. = ***Cyperus hermaphroditus*** (Jacq.) Standl.
Cyperus hermaphroditus var. *longispicatus* (Boeck.) Kük. = ***Cyperus hermaphroditus*** (Jacq.) Standl.
Cyperus herndoniae G.C.Tucker = ***Cyperus camphoratus*** Liebm.
Cyperus herteri Kük. ex Osten = ***Cyperus reflexus*** Vahl
Cyperus heterophyllus Boeck. = ***Cyperus crassipes*** Vahl
Cyperus heudelotii C.B.Clarke = ***Cyperus maculatus*** Boeck. subsp. ***maculatus***
Cyperus hexastachyos Rottb. = ***Cyperus rotundus*** L. subsp. ***rotundus***
Cyperus hexastachyos Sw. = ***Cyperus sphacelatus*** Rottb.
Cyperus hexastachyus Nees = ***Cyperus tenuiculmis*** Boeck. var. ***tenuiculmis***
Cyperus heyneanus Boeck. = ***Cyperus pilosus*** Vahl
Cyperus heynei Boeck. = ***Cyperus procerus*** Rottb.
Cyperus hieronymi var. *fiebrigii* (Kük.) Kük. = ***Cyperus hieronymi*** Boeck.
Cyperus hildebrandtii Boeck. = ***Cyperus hemisphaericus*** Boeck.
Cyperus hildebrandtii (C.B.Clarke) K.Schum. ex T.Durand & B.D.Jacks. = ***Pycreus hildebrandtii*** C.B.Clarke
Cyperus hildra Poir. = ***Cyperus rotundus*** L. subsp. ***rotundus***
Cyperus hilgendorfianus Boeck. = ***Cyperus michelianus*** (L.) Delile subsp. ***michelianus***
Cyperus hillebrandii var. *helleri* Kük. = ***Cyperus hillebrandii*** Boeck. var. ***hillebrandii***
Cyperus hillebrandii var. *mauiensis* (Hillebr.) Kük. = ***Cyperus hillebrandii*** Boeck. var. ***hillebrandii***
Cyperus hirsutus P.J.Bergius = ***Fuirena hirsuta*** (P.J.Bergius) P.L.Forbes
Cyperus hirtus Thunb. = ***Fimbristylis squarrosa*** Vahl var. ***squarrosa***
Cyperus hochstetteri Nees ex Krauss = ***Pycreus macrostachyos*** (Lam.) J.Raynal subsp. ***macrostachyos***
Cyperus hochstetteri var. *tenuis* Boeck. = ***Pycreus macrostachyos*** (Lam.) J.Raynal subsp. ***macrostachyos***
Cyperus holciflorus J.Presl & C.Presl = ***Cyperus javanicus*** Houtt.
Cyperus holophyllus Miq. ex de Vriese = ***Cyperus diffusus*** Vahl
Cyperus holophyllus var. *celebicus* Miq. = ***Cyperus diffusus*** Vahl
Cyperus holoschoenoides Jan ex Schult. = ***Cyperus difformis*** L.
Cyperus holoschoenus (L.) Missbach & E.H.L.Krause = ***Scirpoides holoschoenus*** (L.) Soják
Cyperus holoschoenus var. *fuscisquamatus* Kük. = ***Cyperus holoschoenus*** R.Br.
Cyperus holoschoenus var. *viscidus* W.Fitzg. = ***Cyperus holoschoenus*** R.Br.
Cyperus holosericeus Link = ***Pycreus polystachyos*** var. ***microdontus*** (Torr.) Govaerts
Cyperus homoiiria Steud. = ***Cyperus laxus*** Lam. subsp. ***laxus***
Cyperus honestus Kunth = ***Cyperus pilosus*** Vahl
Cyperus honoluluensis Steud. ex Jard. = ?
Cyperus hookeri Boeck. = ***Cyperus digitatus*** Roxb. subsp. ***digitatus***
Cyperus hookerianus Thwaites = ***Cyperus cephalotes*** Vahl
Cyperus hookerianus Arn. ex C.B.Clarke = ***Pycreus polystachyos*** (Rottb.) P.Beauv. subsp. ***polystachyos***
Cyperus horizontalis Steud. = ***Cyperus odoratus*** L. subsp. ***odoratus***
Cyperus horizontalis Salisb. = ***Blysmus compressus*** (L.) Panz. ex Link subsp. ***compressus***

Cyperus horizontalis Willd. ex Kunth = ***Pycreus flavidus*** (Retz.) T.Koyama
Cyperus hostmannii Steud. = ***Cyperus luzulae*** (L.) Retz.
Cyperus houghtonii var. *bushii* (Britton) Kük. = ***Cyperus lupulinus*** subsp. ***macilentus*** (Fernald) Marcks
Cyperus houghtonii var. *uberior* Kük. = ***Cyperus houghtonii*** Torr.
Cyperus howellii O'Neill & Ben.Ayers = ***Cyperus sordidus*** J.Presl & C.Presl
Cyperus huarmensis (Kunth) M.C.Johnst. = ***Cyperus odoratus*** L. subsp. ***odoratus***
Cyperus huillensis Ridl. = ***Cyperus denudatus*** L.f.
Cyperus humboldtianus Schult. = ***Pycreus lanceolatus*** (Poir.) C.B.Clarke
Cyperus humifusus Clemente = ***Pycreus flavidus*** (Retz.) T.Koyama
Cyperus humilis var. *elatior* Britton = ***Cyperus humilis*** Kunth
Cyperus hyalinolepis Steud. = ***Cyperus michelianus*** (L.) Delile subsp. ***michelianus***
Cyperus hyalinus Vahl = ***Queenslandiella hyalina*** (Vahl) Ballard
Cyperus hyalinus var. *substerilis* (E.G.Camus) Kük. = ***Queenslandiella hyalina*** (Vahl) Ballard
Cyperus ×*hybridus* Hausskn. = *C. longus* × *C. rotundus*
Cyperus hydra Kunth = ***Cyperus esculentus*** L.
Cyperus hydra Michx. = ***Cyperus rotundus*** L. subsp. ***rotundus***
Cyperus hyemalis Pursh ex C.B.Clarke = ***Cyperus haspan*** L.
Cyperus hylaeus Ridl. = ***Cyperus renschii*** Boeck.
Cyperus hypochlorus var. *kauaiensis* Kük. = ***Cyperus hypochlorus*** Hillebr. var. ***hypochlorus***
Cyperus ibeensis C.B.Clarke = ***Cyperus amauropus*** Steud.
Cyperus ignoratus Boeck. = ***Cyperus odoratus*** L. subsp. ***odoratus***
Cyperus ignotus Britton = ***Cyperus constanzae*** Urb.
Cyperus imbricatus var. *capitatus* (Boeck.) Kük. = ***Cyperus imbricatus*** Retz.
Cyperus imbricatus var. *elongatus* (Boeck.) Kük. = ***Cyperus imbricatus*** Retz.
Cyperus imbricatus subsp. *elongatus* (Boeck.) T.Koyama = ***Cyperus imbricatus*** Retz.
Cyperus imbricatus var. *minor* (Boeck.) Kük. = ***Cyperus imbricatus*** Retz.
Cyperus imbricatus var. *multiflorus* Kük. = ***Cyperus imbricatus*** Retz.
Cyperus imbricatus var. *viridispicatus* (Boeck.) Kük. = ***Cyperus imbricatus*** Retz.
Cyperus imerinensis Boeck. = ***Cyperus papyrus*** subsp. ***madagascariensis*** (Willd.) Kunth
Cyperus immanis Nelmes = ***Cyperus koyaliensis*** Cherm.
Cyperus immensus C.B.Clarke = ***Cyperus dives*** Delile
Cyperus immensus var. *petherickii* (C.B.Clarke) Kük. = ***Cyperus dives*** Delile
Cyperus immensus var. *taylorii* C.B.Clarke = ***Cyperus dives*** Delile
Cyperus impubes var. *brevispiculosus* Kük. = ***Cyperus impubes*** Steud. var. ***impubes***
Cyperus impubes f. *major* (Cherm.) Kük. = ***Cyperus impubes*** var. ***fallax*** (Cherm.) Kük.
Cyperus impubes var. *rohlfsii* (Boeck.) Kük. = ***Cyperus rohlfsii*** Boeck.
Cyperus inaequalis Link = ***Cyperus manimae*** Kunth var. ***manimae***
Cyperus inauratus (Nees ex Boeck.) Mattf. & Kük. = ***Kyllinga inaurata*** Nees ex Boeck.
Cyperus inauratus var. *laevicarinatus* Kük. = ***Kyllinga inaurata*** Nees ex Boeck.
Cyperus incarnatus Link = ***Cyperus congestus*** Vahl
Cyperus incompletus (Jacq.) Link = ***Cyperus hermaphroditus*** (Jacq.) Standl.
Cyperus incompletus Boeck. = ***Cyperus hermaphroditus*** (Jacq.) Standl.
Cyperus incomtus var. *dissolutior* Kük. = ***Cyperus incomtus*** Kunth
Cyperus incomtus var. *miguelii* Kük. = ***Cyperus incomtus*** Kunth
Cyperus incomtus var. *venturii* (Kük.) Barros = ***Cyperus incomtus*** Kunth
Cyperus inconspicuus Gennari = ***Cyperus rotundus*** L. subsp. ***rotundus***
Cyperus inconspicuus Liebm. = ***Pycreus polystachyos*** var. ***microdontus*** (Torr.) Govaerts
Cyperus incrassatus Beyr. ex Kunth = ***Pycreus polystachyos*** (Rottb.) P.Beauv. subsp. ***polystachyos***
Cyperus incurvatus Roxb. = ***Cyperus malaccensis*** Lam. var. ***malaccensis***
Cyperus incurvus Spreng. ex Kunth = ***Cyperus laevigatus*** L. subsp. ***laevigatus***
Cyperus indecorus var. *decurvatus* (C.B.Clarke) Kük. = ***Cyperus vestitus*** Hochst. ex C.Krauss
Cyperus indecorus var. *dinteri* Kük. = ***Cyperus vestitus*** Hochst. ex C.Krauss
Cyperus indecorus var. *inflatus* (C.B.Clarke) Kük. = ***Cyperus vestitus*** Hochst. ex C.Krauss
Cyperus indecorus var. *namaquensis* Kük. = ***Cyperus vestitus*** Hochst. ex C.Krauss
Cyperus indicus Boeck. = ***Cyperus nutans*** var. ***eleusinoides*** (Kunth) Haines
Cyperus indicus Rich. ex Pers. = ***Fimbristylis ovata*** (Burm.f.) J.Kern
Cyperus infirmus Boeck. = ***Cyperus erythrorrhizos*** Muhl.
Cyperus inflexus Muhl. = ***Cyperus squarrosus*** L.
Cyperus inflexus var. *acaulis* Hook.f. = ***Cyperus squarrosus*** L.
Cyperus inflexus var. *elongatus* Hook.f. = ***Cyperus squarrosus*** L.
Cyperus infra-apicalis Nees ex Aitch. = ***Cyperus nutans*** var. ***eleusinoides*** (Kunth) Haines
Cyperus ingratus Kunth = ***Cyperus pulcher*** Thunb.
Cyperus inornatus Boeck. = ***Cyperus gunnii*** var. ***novae-hollandiae*** (Boeck.) Kük.
Cyperus insignis Kunth = ***Cyperus flexuosus*** Vahl
Cyperus intactus Vahl = ***Pycreus intactus*** (Vahl) J.Raynal
Cyperus interceptus Steud. = ***Cyperus articulatus*** L.
Cyperus intermedius Guss. = ***Cyperus longus*** L. subsp. ***longus***
Cyperus intermedius Steud. = ***Pycreus flavescens*** subsp. ***intermedius*** (Rikli) Lye
Cyperus intricatoramosus Boeck. = ***Pycreus niger*** (Ruiz & Pav.) Cufod.
Cyperus inundatus Roxb. = ***Cyperus serotinus*** Rottb.
Cyperus inundatus R.Br. = ***Cyperus trinervis*** R.Br.
Cyperus inundatus Nees = ***Pycreus puncticulatus*** (Vahl) Nees
Cyperus involucratus Rottb. = ***Cyperus alternifolius*** subsp. ***flabelliformis*** Kük.
Cyperus involucratus Poir. = ***Cyperus imbricatus*** Retz.
Cyperus involutus R.Br. = ***Cyperus conglomeratus*** Rottb. subsp. ***conglomeratus***
Cyperus involutus (C.B.Clarke) K.Schum. = ***Cyperus dactyliformis*** Boeck.
Cyperus iria f. *chrysomelinus* (Link) Kük. = ***Cyperus iria*** L.
Cyperus iria f. *diaphaniria* (Steud.) Miq. = ***Cyperus iria*** L.
Cyperus iria var. *flavescens* Benth. = ***Cyperus iria*** L.

Cyperus iria f. *flavescens* (Benth.) Domin = ***Cyperus iria*** L.
Cyperus iria f. *multiflora* Domin = ***Cyperus iria*** L.
Cyperus iria var. *paniciformis* C.B.Clarke = ***Cyperus iria*** L.
Cyperus iria var. *parviflorus* (Nees) Miq. = ***Cyperus microiria*** Steud.
Cyperus iria var. *rectangularis* Kük. = ***Cyperus microiria*** Steud.
Cyperus iria var. *santonici* (Rottb.) Fernald & Griscom = ***Cyperus iria*** L.
Cyperus iria var. *typicus* Domin = ***Cyperus iria*** L.
Cyperus iridifolius A.Dietr. = ***Cyperus diffusus*** Vahl
Cyperus ischnocormus Steud. = ***Pycreus elegantulus*** (Steud.) C.B.Clarke
Cyperus ischnos var. *arsenii* (Palla) Kük. = ***Cyperus ischnos*** Schltdl.
Cyperus ischnostachys Steud. = ***Cyperus cyperinus*** (Retz.) Suringar
Cyperus isocladus Kunth = ***Cyperus prolifer*** Lam.
Cyperus italicus Tod. = ***Cyperus longus*** L. subsp. ***longus***
Cyperus italicus Rottb. = ***Cyperus michelianus*** (L.) Delile subsp. ***michelianus***
Cyperus iwasakii Makino = ***Cyperus exaltatus*** Retz.
Cyperus jacquemontii Boeck. = ***Pycreus lanceolatus*** (Poir.) C.B.Clarke
Cyperus jacquinii Schrad. = ***Cyperus nutans*** Vahl var. ***nutans***
Cyperus jaeggii Boeck. = ***Pycreus megapotamicus*** (A.Dietr.) Nees
Cyperus jamaicensis Poir. = ***Cyperus confertus*** Sw.
Cyperus japonicus Makino = ***Cyperus amuricus*** Maxim.
Cyperus japonicus Miq. = ***Cyperus serotinus*** Rottb.
Cyperus jardinei Steud. = ***Cyperus crassipes*** Vahl
Cyperus javanicus Kük. =
Cyperus jeminicus Retz. = ***Cyperus bulbosus*** Vahl
Cyperus jenmanii C.B.Clarke ex Morong = ***Cyperus oxylepis*** Nees ex Steud.
Cyperus jocladus E.Mey. = ***Cyperus prolifer*** Lam.
Cyperus jouannatii Pau = ?
Cyperus jubiflorus Rudge = ***Cyperus odoratus*** L. subsp. ***odoratus***
Cyperus jucundus C.B.Clarke = ***Fimbristylis jucunda*** (C.B.Clarke) J.Kern
Cyperus juncellus Dinter = ***Cyperus laevigatus*** L. subsp. ***laevigatus***
Cyperus junceus Link = ***Cyperus haspan*** L.
Cyperus junciformis Desf. = ***Cyperus laevigatus*** L. subsp. ***laevigatus***
Cyperus juncoides Lam. = ***Cyperus haspan*** L.
Cyperus juncoides Regel = ***Cyperus reflexus*** Vahl
Cyperus jungendus Steud. = ***Pycreus flavidus*** (Retz.) T.Koyama
Cyperus junghuhnii Miq. = ***Pycreus flavidus*** (Retz.) T.Koyama
Cyperus × *kadzusensis* T.Koyama = ?
Cyperus kalli (Forssk.) Murb. = ***Cyperus capitatus*** Vand.
Cyperus kamphoeveneri Boeck. = ***Cyperus maculatus*** Boeck. subsp. ***maculatus***
Cyperus kegelianus Steud. = ***Remirea maritima*** Aubl.
Cyperus kernianus Ohwi & T.Koyama = ***Kyllinga odorata*** Vahl subsp. ***odorata***
Cyperus kerstenii var. *irregularis* Kük. = ***Cyperus kerstenii*** Boeck.
Cyperus kerstingii Engl. = [24 SUD]
Cyperus kilimandscharicus var. *chlorilepis* Peter & Kük. = ***Cyperus kilimandscharicus*** Kük.
Cyperus kingsii C.D.Adams ex Proctor = ***Cyperus floridanus*** Britton
Cyperus kivuensis Cherm. = ***Cyperus platycaulis*** Baker
Cyperus kleinianus Hochst. ex Steud. = ***Courtoisina cyperoides*** (Roxb.) Soják
Cyperus koenigii Vahl = ***Cyperus corymbosus*** Rottb.
Cyperus konkanensis T.Cooke = ***Cyperus cyperoides*** (L.) Kuntze subsp. ***cyperoides***
Cyperus korshinskyi Meinsh. = ***Pycreus sanguinolentus*** (Vahl) Nees subsp. ***sanguinolentus***
Cyperus kotschyanus Fenzl ex Steud. = ***Cyperus nutans*** var. ***eleusinoides*** (Kunth) Haines
Cyperus kottensis Cherm. = ***Cyperus tonkinensis*** var. ***baikiei*** (C.B.Clarke ex Kük.) S.S.Hooper
Cyperus krameri Franch. & Sav. = ***Cyperus amuricus*** Maxim.
Cyperus kraussii Boeck. = ***Cyperus macrocarpus*** (Kunth) Boeck.
Cyperus krebsii Boeck. = ***Cyperus serotinus*** Rottb.
Cyperus krugii Boeck. = ***Cyperus planifolius*** Rich.
Cyperus kuebensis Kük. = ***Cyperus hensii*** C.B.Clarke
Cyperus kuekenthalii Merr. = ?
Cyperus kunthianus H.Mann = ***Cyperus hillebrandii*** Boeck. var. ***hillebrandii***
Cyperus kunthianus Nees = ***Cyperus meyenianus*** Kunth
Cyperus kunthianus (Gaudich.) Hillebr. = ***Cyperus neokunthianus*** Kük.
Cyperus kunthii Parl. = ***Cyperus longus*** L. subsp. ***longus***
Cyperus kuntzei Boeck. = ?
Cyperus kurrii Steud. = ***Cyperus distans*** L.f.
Cyperus kyllingia Endl. = ***Kyllinga nemoralis*** (J.R.Forst. & G.Forst.) Dandy ex Hutch. & Dalziel
Cyperus kyllingia f. *humilis* (Boeck.) Kük. = ***Kyllinga nemoralis*** (J.R.Forst. & G.Forst.) Dandy ex Hutch. & Dalziel
Cyperus kyllingia var. *latifolius* (Boeck.) Kük. = ***Kyllinga planiculmis*** C.B.Clarke ex Cherm.
Cyperus kyllingia f. *subtriceps* (Kunth) Kük. = ***Kyllinga nemoralis*** (J.R.Forst. & G.Forst.) Dandy ex Hutch. & Dalziel
Cyperus kyllingia f. *tenuis* (Boeck.) Kük. = ***Kyllinga nemoralis*** (J.R.Forst. & G.Forst.) Dandy ex Hutch. & Dalziel
Cyperus kyllingioides Vahl = ***Cyperus dubius*** Rottb. subsp. ***dubius***
Cyperus kyllingioides Pursh = ***Cyperus filiculmis*** Vahl
Cyperus lacteus Steud. = ***Cyperus niveus*** var. ***leucocephalus*** (Kunth) Fosberg
Cyperus lacustris Schrad. ex Nees = ***Cyperus miliifolius*** Poepp. & Kunth
Cyperus lacustris Phil. = ***Cyperus rigens*** J.Presl & C.Presl subsp. ***rigens***
Cyperus lacustris (L.) Missbach & E.H.L.Krause = ***Schoenoplectus lacustris*** (L.) Palla
Cyperus laetus C.B.Clarke = [50 QLD]
Cyperus laetus var. *arechavaletae* (Boeck.) Kük. = ***Cyperus laetus*** J.Presl & C.Presl
Cyperus laetus var. *balansae* (Maury) Kük. = ***Cyperus impolitus*** Kunth
Cyperus laetus var. *capitatus* Boeck. = ***Cyperus impolitus*** Kunth
Cyperus laetus var. *cephalanthus* (Torr. & Hook.) Kük. = ***Cyperus cephalanthus*** Torr. & Hook.
Cyperus laetus var. *conceptionis* (Steud.) Kük. = ***Cyperus rigens*** J.Presl & C.Presl subsp. ***rigens***
Cyperus laetus f. *contracta* Osten = ***Cyperus laetus*** J.Presl & C.Presl
Cyperus laetus var. *elatior* Boeck. = ***Cyperus laetus*** J.Presl & C.Presl
Cyperus laetus subsp. *eu-laetus* Kük. = ***Cyperus laetus*** J.Presl & C.Presl

Cyperus laetus var. *hassleri* (C.B.Clarke) Kük. = ***Cyperus impolitus*** Kunth
Cyperus laetus var. *impolitus* (Kunth) Kük. = ***Cyperus impolitus*** Kunth
Cyperus laetus var. *laeviculmis* Kük. = ***Cyperus laetus*** J.Presl & C.Presl
Cyperus laetus var. *maximus* Kük. = ***Cyperus laetus*** J.Presl & C.Presl
Cyperus laetus var. *obtusiflora* Boeck. = ***Cyperus laetus*** J.Presl & C.Presl
Cyperus laetus var. *obtusiflorus* Boeck. = ***Cyperus cornelii-ostenii*** Kük. var. ***cornelii-ostenii***
Cyperus laetus subsp. *oostachyus* (Nees) Kük. = ***Cyperus rigens*** J.Presl & C.Presl subsp. ***rigens***
Cyperus laetus var. *parciflorus* Kük. = ***Cyperus rigens*** J.Presl & C.Presl subsp. ***rigens***
Cyperus laetus var. *tenuissimus* Grosser & Kük. = ***Cyperus grossianus*** Pedersen
Cyperus laevigatus var. *albidus* Vahl = ***Cyperus laevigatus*** L. subsp. ***laevigatus***
Cyperus laevigatus f. *atratus* Peter ex Kük. = ***Cyperus laevigatus*** L. subsp. ***laevigatus***
Cyperus laevigatus var. *caespitosus* C.B.Clarke = ***Cyperus laevigatus*** L. subsp. ***laevigatus***
Cyperus laevigatus var. *distachyos* (All.) Coss. & Durieu = ***Cyperus laevigatus*** subsp. ***distachyos*** (All.) Maire & Weiller
Cyperus laevigatus f. *macra* C.B.Clarke = ***Cyperus laevigatus*** L. subsp. ***laevigatus***
Cyperus laevigatus var. *pallae* (Kneuck.) Kük. = ***Cyperus laevigatus*** L. subsp. ***laevigatus***
Cyperus laevigatus var. *pictus* Boeck. = ***Cyperus laevigatus*** L. subsp. ***laevigatus***
Cyperus laevigatus var. *ramlehensis* Sickenb. = ***Cyperus laevigatus*** L. subsp. ***laevigatus***
Cyperus laevigatus f. *reptans* (Boeck.) C.B.Clarke = ***Cyperus laevigatus*** L. subsp. ***laevigatus***
Cyperus laevigatus var. *reptans* (Boeck.) Kük. = ***Cyperus laevigatus*** L. subsp. ***laevigatus***
Cyperus laevigatus var. *subaphyllus* (Boeck. ex Schinz) Kük. = ***Cyperus laevigatus*** subsp. ***distachyos*** (All.) Maire & Weiller
Cyperus laevigatus var. *viridulus* (Boeck.) C.B.Clarke = ***Cyperus laevigatus*** L. subsp. ***laevigatus***
Cyperus laevissimus Steud. = ***Cyperus rotundus*** L. subsp. ***rotundus***
Cyperus lagorensis Steud. = ***Cyperus diffusus*** Vahl
Cyperus lagunetto Steud. = ***Pycreus bipartitus*** (Torr.) C.B.Clarke
Cyperus lamarckianus Schult. = ***Pycreus flavidus*** (Retz.) T.Koyama
Cyperus lamprocarpus Nees = ***Cyperus stoloniferus*** Retz.
Cyperus lamprocarpus Boeck. = ?
Cyperus lanatus Ruiz ex Boeck. = ***Fimbristylis pilosa*** Vahl
Cyperus lanceola Ridl. = ***Cyperus fertilis*** Boeck.
Cyperus lanceolatus Poir. = ***Pycreus lanceolatus*** (Poir.) C.B.Clarke
Cyperus lanceolatus var. *compositus* C.Presl = ***Pycreus lanceolatus*** (Poir.) C.B.Clarke
Cyperus lanceolatus var. *helvus* (Liebm.) Kük. = ***Pycreus lanceolatus*** (Poir.) C.B.Clarke
Cyperus lanceolatus var. *subunioloides* Kük. = ***Pycreus lanceolatus*** (Poir.) C.B.Clarke
Cyperus lanceolatus var. *tener* (C.B.Clarke) Kük. = ***Pycreus tener*** C.B.Clarke
Cyperus lanceus Thunb. = ***Pycreus nitidus*** (Lam.) J.Raynal
Cyperus lanceus F.Muell. = ***Pycreus unioloides*** (R.Br.) Urb.
Cyperus lanceus var. *angustifolius* Ridl. = ***Pycreus macranthus*** (Boeck.) C.B.Clarke
Cyperus lanceus f. *densior* (Cherm.) Kük. = ***Pycreus nitidus*** (Lam.) J.Raynal
Cyperus lanceus var. *divaricatus* Kük. = ***Pycreus nitidus*** (Lam.) J.Raynal
Cyperus lanceus var. *grantii* C.B.Clarke = ***Pycreus nitidus*** (Lam.) J.Raynal
Cyperus lanceus var. *humilis* Kunth = ***Pycreus nitidus*** (Lam.) J.Raynal
Cyperus lanceus var. *macrostachya* Kunth = ***Pycreus macranthus*** (Boeck.) C.B.Clarke
Cyperus lanceus var. *melanopus* (Boeck.) Kük. = ***Pycreus nitidus*** (Lam.) J.Raynal
Cyperus lanceus var. *mucronatus* Kunth = ***Pycreus macranthus*** (Boeck.) C.B.Clarke
Cyperus lanceus var. *palaestinensis* Kük. = ***Pycreus nitidus*** (Lam.) J.Raynal
Cyperus lanceus var. *ramosus* Kük. = ***Pycreus nitidus*** (Lam.) J.Raynal
Cyperus lanuginosulus Ruiz ex Buchenau. = (Juncaceae)
Cyperus lapidicola Kük. = [27]
Cyperus lateralis Forssk. = ***Cyperus laevigatus*** L. subsp. ***laevigatus***
Cyperus lateralis L.f. = ***Cyperus tenellus*** L.f.
Cyperus lateriflorus Torr. = ***Cyperus difformis*** L.
Cyperus lateriflorus Steud. = ***Cyperus longus*** L. subsp. ***longus***
Cyperus laticulmis Spreng. = ***Cyperus haspan*** L.
Cyperus latifolius var. *angustifolius* C.Krauss = ***Cyperus latifolius*** Poir.
Cyperus latifolius var. *austroafricanus* Kük. = ***Cyperus latifolius*** Poir.
Cyperus latifolius var. *fimbristyloides* Kük. = ***Cyperus latifolius*** Poir.
Cyperus latifolius var. *herana* (Cherm.) Cherm. = ***Cyperus latifolius*** Poir.
Cyperus latispicatus Boeck. = ***Pycreus diaphanus*** (Schrad. ex Roem. & Schult.) S.S.Hooper & T.Koyama
Cyperus latispicatus Boeck. ex C.B.Clarke = ***Pycreus diaphanus*** (Schrad. ex Roem. & Schult.) S.S.Hooper & T.Koyama
Cyperus latispicatus var. *acaulis* C.B.Clarke = ***Pycreus diaphanus*** (Schrad. ex Roem. & Schult.) S.S.Hooper & T.Koyama
Cyperus latispicatus var. *diaphanus* (Schrad. ex Roem. & Schult.) Kük. = ***Pycreus diaphanus*** (Schrad. ex Roem. & Schult.) S.S.Hooper & T.Koyama
Cyperus latispicatus var. *gracilescens* Kük. = ***Pycreus diaphanus*** (Schrad. ex Roem. & Schult.) S.S.Hooper & T.Koyama
Cyperus latispicatus var. *laxus* Kük. = ***Pycreus diaphanus*** (Schrad. ex Roem. & Schult.) S.S.Hooper & T.Koyama
Cyperus latovaginata Govind. = ***Pycreus latovaginatus*** (Govind.) P.Singh & V.Singh
Cyperus laxespicatus var. *brunneotinctus* Kük. = ***Cyperus laxespicatus*** Kük.
Cyperus laxespicatus var. *percrassus* Kük. = ***Cyperus laxespicatus*** Kük.
Cyperus laxespicatus var. *testui* (Cherm.) Kük. = ***Pycreus testui*** Cherm.
Cyperus laxiusculus Steud. = ***Pycreus pumilus*** (L.) Nees
Cyperus laxus R.Br. = ***Cyperus bulbosus*** Vahl
Cyperus laxus Vahl = ***Cyperus distans*** L.f.
Cyperus laxus Griseb. = ***Cyperus laxus*** Lam. subsp. ***laxus***

Cyperus laxus Willemet = [29 MAU]
Cyperus laxus var. *macrostachyus* (Boeck.) Karth. = ***Cyperus diffusus*** Vahl
Cyperus laxus subsp. *macrostachyus* (Boeck.) V.P.Prasad & N.P.Singh = ***Cyperus diffusus*** Vahl
Cyperus laxus subsp. *sylvestris* (Ridl.) Lye = ***Cyperus laxus*** subsp. ***buchholzii*** (Boeck.) Lye
Cyperus lechleri Steud. = ***Cyperus xanthostachyus*** Steud.
Cyperus lechleri var. *viridiglaucus* Boeck. = ***Cyperus xanthostachyus*** Steud.
Cyperus ledermannii (Kük.) S.S.Hooper = ***Cyperus niveus*** var. ***ledermannii*** (Kük.) Lye
Cyperus lehmannii (Nees) F.Muell. = ***Kyllinga lehmannii*** Nees
Cyperus lenticularis Steud. = ***Cyperus odoratus*** L. subsp. ***odoratus***
Cyperus leonensis Steud. = [22 SIE]
Cyperus lepidus Hochst. ex Steud. = ***Cyperus amabilis*** Vahl
Cyperus lepidus F.Muell. ex C.B.Clarke = ***Cyperus concinnus*** R.Br.
Cyperus lepidus Phil. = ***Pycreus lanceolatus*** (Poir.) C.B.Clarke
Cyperus leptanthos Steud. = ?
Cyperus leptophyes C.B.Clarke = ?
Cyperus leptophyllus Hochst. ex Steud. = ***Cyperus amauropus*** Steud.
Cyperus leptophyllus f. *conglobata* Kük. = ***Cyperus amauropus*** Steud.
Cyperus leptophyllus var. *deliciosus* Kük. = ***Cyperus amauropus*** Steud.
Cyperus leptophyllus var. *friesii* (Kük.) Kük. = ***Cyperus amauropus*** Steud.
Cyperus leptophyllus var. *ibeensis* (C.B.Clarke) Kük. = ***Cyperus amauropus*** Steud.
Cyperus leptorhachis Mattf. & Kük. = ***Kyllinga debilis*** C.B.Clarke
Cyperus leptos Schult. = ***Cyperus haspan*** L.
Cyperus leptostachys Nees = ***Cyperus tenuispica*** Steud.
Cyperus leptostachyus Griff. = ***Cyperus rotundus*** L. subsp. ***rotundus***
Cyperus lessonianus Kunth = ***Cyperus trinervis*** R.Br.
Cyperus leucanthus Schrad. ex Nees = ***Cyperus schomburgkianus*** Nees
Cyperus leucocephalus Hassk. = ***Kyllinga nemoralis*** (J.R.Forst. & G.Forst.) Dandy ex Hutch. & Dalziel
Cyperus leucolepis Boeck. = ***Cyperus sphaerolepis*** Boeck.
Cyperus leucolepis Carey ex C.B.Clarke = ***Pycreus pumilus*** (L.) Nees
Cyperus leucoloma Nees = ***Isolepis leucoloma*** (Nees) C.Archer
Cyperus leucostachys Link = ***Cyperus laevigatus*** L. subsp. ***laevigatus***
Cyperus levis F.Muell. ex C.B.Clarke = ***Pycreus sanguinolentus*** (Vahl) Nees subsp. ***sanguinolentus***
Cyperus liebmannii Steud. = ***Pycreus polystachyos*** (Rottb.) P.Beauv. subsp. ***polystachyos***
Cyperus ligularis Hemsl. = ***Cyperus appendiculatus*** (Brongn.) Kunth var. ***appendiculatus***
Cyperus ligularis var. *spicatocapitatus* (Steud.) Kük. = ***Cyperus ligularis*** L.
Cyperus ligusticus Gand. = [13 ITA]
Cyperus limbatus Maury ex Micheli = ***Cyperus laxus*** Lam. subsp. ***laxus***
Cyperus linkianus Schult. = ***Cyperus congestus*** Vahl
Cyperus linnaeanus Kunth = ***Pycreus flavidus*** (Retz.) T.Koyama
Cyperus lipocarpha T.Koyama = ***Lipocarpha chinensis*** (Osbeck) J.Kern
Cyperus lipocarphioides (Kük.) Lye = ***Alinula lipocarphioides*** (Kük.) J.Raynal
Cyperus litoreus (C.B.Clarke) Britton ex Small = ***Cyperus ovatus*** Baldwin
Cyperus litos Schult. = ***Pycreus polystachyos*** (Rottb.) P.Beauv. subsp. ***polystachyos***
Cyperus littoralis R.Br. = ***Cyperus stoloniferus*** Retz.
Cyperus litvinovii Meinsh. = ***Cyperus serotinus*** Rottb.
Cyperus lividus Steud. = ?
Cyperus lomatophyllus Engl. = ?
Cyperus lomentaceus Nees & Meyen ex Kunth = ***Cyperus odoratus*** L. subsp. ***odoratus***
Cyperus longibracteatus var. *niger* (C.B.Clarke) Lye = ***Cyperus longibracteatus*** (Cherm.) Kük.
Cyperus longibracteatus var. *rubrotinctus* (Cherm.) Kük. = ***Cyperus longibracteatus*** (Cherm.) Kük.
Cyperus longicaulis Boeck. = ***Cyperus intricatus*** Schrad. ex Schult.
Cyperus longicruris Boeck. = ***Pycreus niger*** (Ruiz & Pav.) Cufod.
Cyperus longifolius var. *condensatus* Cherm. = ***Cyperus longifolius*** Poir.
Cyperus longiradiatus Steud. = ***Cyperus odoratus*** L. subsp. ***odoratus***
Cyperus longispicatus Boeck. = ***Cyperus hermaphroditus*** (Jacq.) Standl.
Cyperus longispicatus Norton = ***Cyperus odoratus*** L. subsp. ***odoratus***
Cyperus longistolon Peter & Kük. = ***Pycreus longistolon*** (Peter & Kük.) Napper
Cyperus longistolon subsp. *atrofuscus* (Lye) Lye = ***Pycreus longistolon*** subsp. ***atrofuscus*** Lye
Cyperus longivaginans Kük. = ***Pycreus longivaginans*** (Kük.) Cherm.
Cyperus longus Boeck. = ***Cyperus rotundus*** L. subsp. ***rotundus***
Cyperus longus var. *adoensis* (Hochst. ex A.Rich.) Boeck. = ***Cyperus rigidifolius*** Steud.
Cyperus longus var. *auratus* Boeck. = ***Cyperus longus*** L. subsp. ***longus***
Cyperus longus f. *badiiformis* (Chiov.) Kük. = ***Cyperus longus*** var. ***pallidus*** Boeck.
Cyperus longus var. *badius* (Desf.) Cambess. = ***Cyperus longus*** subsp. ***badius*** (Desf.) Murb.
Cyperus longus f. *cypricus* C.B.Clarke = ***Cyperus longus*** L. subsp. ***longus***
Cyperus longus f. *difformis* (Lojac.) Kük. = ***Cyperus longus*** L. subsp. ***longus***
Cyperus longus var. *discolor* Kük. = ***Cyperus tenuiflorus*** Rottb.
Cyperus longus var. *elongatus* Boeck. = ***Cyperus longus*** L. subsp. ***longus***
Cyperus longus var. *elongatus* C.B.Clarke = ***Cyperus longus*** L. subsp. ***longus***
Cyperus longus var. *gracilis* Boeck. = ***Cyperus mitis*** Steud.
Cyperus longus var. *guthnickii* (Tineo ex Lojac.) Kük. = ***Cyperus longus*** L. subsp. ***longus***
Cyperus longus var. *heldreichianus* (Boiss.) Boiss. = ***Cyperus longus*** L. subsp. ***longus***
Cyperus longus f. *intermedius* (Guss.) Kük. = ***Cyperus longus*** L. subsp. ***longus***
Cyperus longus var. *leptochlaenus* Kük. = ***Cyperus tenuiflorus*** Rottb.
Cyperus longus f. *longispiculatus* Reynolds = ***Cyperus longus*** L. subsp. ***longus***
Cyperus longus f. *lucanus* A.Terracc. = ***Cyperus longus*** L. subsp. ***longus***

Cyperus longus var. *maculatus* (Boeck.) Boeck. = ***Cyperus maculatus*** Boeck.
Cyperus longus f. *myriostachyus* (Ten.) Kük. = ***Cyperus longus*** L. subsp. ***longus***
Cyperus longus var. *pallescens* C.B.Clarke = ***Cyperus longus*** L. subsp. ***longus***
Cyperus longus var. *pallidior* Kük. = ***Cyperus longus*** L. subsp. ***longus***
Cyperus longus f. *parlatoris* (Lojac.) Kük. = ***Cyperus longus*** L. subsp. ***longus***
Cyperus longus var. *speciosus* Kük. = ***Cyperus longus*** L. subsp. ***longus***
Cyperus longus var. *stenostachyus* (Benth.) Kük. = ***Cyperus tenuiflorus*** Rottb.
Cyperus longus var. *tenuiflorus* (Rottb.) Boeck. = ***Cyperus tenuiflorus*** Rottb.
Cyperus longus subsp. *tenuiflorus* (Rottb.) Kük. = ***Cyperus tenuiflorus*** Rottb.
Cyperus lorentzianus Boeck. = ***Pycreus niger*** (Ruiz & Pav.) Cufod.
Cyperus lorentzianus var. *compressus* Kük. = ***Pycreus niger*** (Ruiz & Pav.) Cufod.
Cyperus lorentzianus var. *elatus* Kük. = ***Pycreus niger*** (Ruiz & Pav.) Cufod.
Cyperus louisianae Steud. = ***Pycreus filicinus*** (Vahl) T.Koyama
Cyperus louisianensis Thieret = ***Pycreus sanguinolentus*** (Vahl) Nees subsp. ***sanguinolentus***
Cyperus lucentinigricans K.Schum. = ***Cyperus platycaulis*** Baker
Cyperus lucidulus C.B.Clarke = ***Cyperus zollingeri*** Steud.
Cyperus lucidulus Link = ?
Cyperus lucidus Kunth = ***Cyperus odoratus*** L. subsp. ***odoratus***
Cyperus lucidus var. *acutus* (R.Br.) Kük. = ***Cyperus lucidus*** R.Br.
Cyperus lucidus var. *grandiusculus* Kük. = ***Cyperus lucidus*** R.Br.
Cyperus lucidus var. *gunnii* (Hook.f.) C.Moore & Betche = ***Cyperus gunnii*** Hook.f.
Cyperus lucidus var. *sanguineofuscus* (Nees) Kük. = ***Cyperus lucidus*** R.Br.
Cyperus lurida Govind. = ***Pycreus lurida*** (Govind.) P.Singh & V.Singh
Cyperus luteolus Boeck. = ***Pycreus unioloides*** (R.Br.) Urb.
Cyperus luteostramineus Mattf. & Kük. = ***Kyllinga exigua*** Boeck.
Cyperus lutescens Phil. = ***Cyperus eragrostis*** Lam.
Cyperus lutescens Torr. & Hook. = ***Cyperus esculentus*** L.
Cyperus lutescens var. *fulvescens* C.B.Clarke = ***Cyperus setigerus*** Torr. & Hook.
Cyperus luteus var. *manongarivensis* (Cherm.) Kük. = ***Cyperus luteus*** Boeck.
Cyperus luzonensis J.Presl & C.Presl = ***Cyperus compactus*** Retz.
Cyperus luzulae Hochst. ex Steud. = ***Cyperus eragrostis*** Lam.
Cyperus luzulae f. *conoideus* (Rich.) Nees = ***Cyperus luzulae*** (L.) Retz.
Cyperus luzulae var. *entrerianus* (Boeck.) M.Barros = ***Cyperus entrerianus*** Boeck.
Cyperus luzulae var. *minor* Boeck. = ***Cyperus luzulae*** (L.) Retz.
Cyperus luzulae f. *pallidiflorens* Kük. = ***Cyperus luzulae*** (L.) Retz.
Cyperus luzulae var. *tucumanensis* (Boeck.) C.B.Clarke ex Chodat = ***Cyperus entrerianus*** Boeck.
Cyperus luzulae var. *umbellulatus* Britton = ***Cyperus luzulae*** (L.) Retz.
Cyperus luzuliformis Boeck. = ***Cyperus indecorus*** Kunth
Cyperus macellus Kunth = ***Cyperus trinervis*** R.Br.
Cyperus macilentus (Fernald) E.P.Bicknell = ***Cyperus lupulinus*** subsp. ***macilentus*** (Fernald) Marcks
Cyperus macranthus Boeck. = ***Pycreus macranthus*** (Boeck.) C.B.Clarke
Cyperus macranthus f. *acuticarinatus* (Kük.) Kük. = ***Pycreus acuticarinatus*** (Kük.) Cherm.
Cyperus macranthus var. *angustifolius* (Ridl.) Kük. = ***Pycreus macranthus*** (Boeck.) C.B.Clarke
Cyperus macranthus var. *mucronatus* (Kunth) Kük. = ***Pycreus macranthus*** (Boeck.) C.B.Clarke
Cyperus macreilema Steud. ex Jard. = ***Cyperus javanicus*** Houtt.
Cyperus macrocarpus var. *excelsior* Kük. = ***Cyperus macrocarpus*** (Kunth) Boeck.
Cyperus macrocarpus var. *humbertii* (Cherm.) Kük. = ***Cyperus macrocarpus*** (Kunth) Boeck.
Cyperus macrocarpus var. *kraussii* (Boeck.) Kük. = ***Cyperus macrocarpus*** (Kunth) Boeck.
Cyperus macrocarpus var. *pseudoflavus* (C.B.Clarke) Kük. = ***Cyperus cyperoides*** subsp. ***pseudoflavus*** (C.B.Clarke) Lye
Cyperus macrocarpus var. *submacrocarpus* Kük. = ***Cyperus macrocarpus*** (Kunth) Boeck.
Cyperus macrocephalus Liebm. = ***Cyperus odoratus*** L. subsp. ***odoratus***
Cyperus macrocephalus var. *eggersii* (Boeck.) S.D.Jones, Wipff & R.Carter = ***Cyperus odoratus*** L. subsp. ***odoratus***
Cyperus macrolepis Boeck. = ***Cyperus miliifolius*** Poepp. & Kunth
Cyperus macropus Boeck. = ***Cyperus circumclusus*** (C.B.Clarke) Schweinf.
Cyperus macropus Miq. = ***Cyperus radians*** Nees & Meyen ex Kunth
Cyperus macrorrhizus Nees ex Boiss. = ***Cyperus capitatus*** Vand.
Cyperus macrosciadion Steud. = ***Cyperus imbricatus*** Retz.
Cyperus macrostachyos Lam. = ***Pycreus macrostachyos*** (Lam.) J.Raynal
Cyperus macrostachyos subsp. *tremulus* (Poir.) Lye = ***Pycreus macrostachyos*** subsp. ***tremulus*** (Poir.) Lye
Cyperus macrostachys Link = ?
Cyperus macrostachyus Ten. ex Boeck. = ***Cyperus laxus*** Lam. subsp. ***laxus***
Cyperus maculatus var. *naumannianus* (Boeck.) Kük. = ***Cyperus maculatus*** Boeck. subsp. ***maculatus***
Cyperus madagascariensis (Willd.) Roem. & Schult. = ***Cyperus papyrus*** subsp. ***madagascariensis*** (Willd.) Kunth
Cyperus major (Boeck.) Cherm. = ***Cyperus mapanioides*** C.B.Clarke
Cyperus major var. *micranthus* Cherm. = ***Cyperus mapanioides*** C.B.Clarke
Cyperus makinoi Nakai = ***Cyperus serotinus*** Rottb.
Cyperus malabaricus (C.B.Clarke) T.Cooke = ***Pycreus malabaricus*** C.B.Clarke
Cyperus malaccensis subsp. *monophyllus* (Vahl) T.Koyama = ***Cyperus malaccensis*** var. ***brevifolius*** Boeck.
Cyperus malawicus (J.Raynal) Lye = ***Alinula malawica*** (J.Raynal) Goetgh. & Vorster
Cyperus mangorensis var. *volodioides* (Cherm.) Kük. = ***Cyperus mangorensis*** Cherm.
Cyperus manilensis Boeck. ex Warb. = ***Cyperus cyperoides*** (L.) Kuntze subsp. ***cyperoides***

Cyperus manimae var. *divergens* (Kunth) Kük. = ***Cyperus manimae*** Kunth var. ***manimae***
Cyperus manimae var. *phaeocephalus* (Griseb.) O'Neill & Ben.Ayers = ***Cyperus manimae*** Kunth var. ***manimae***
Cyperus manimae var. *pseudospectabilis* Kük. = ***Cyperus manimae*** Kunth var. ***manimae***
Cyperus mannii C.B.Clarke = ***Cyperus baronii*** C.B.Clarke
Cyperus mapanioides var. *major* (Boeck.) Kük. = ***Cyperus mapanioides*** C.B.Clarke
Cyperus mapanioides f. *micranthus* (Cherm.) Kük. = ***Cyperus mapanioides*** C.B.Clarke
Cyperus maranguensis var. *ferrugineoviridis* C.B.Clarke = ***Cyperus ferrugineoviridis*** (C.B.Clarke) Kük.
Cyperus margaritaceus var. *karlschumannii* (C.B. Clarke) Kük. = ***Cyperus karlschumannii*** C.B.Clarke
Cyperus margaritaceus var. *nduru* (Cherm.) Kük. = ***Cyperus nduru*** Cherm.
Cyperus margaritaceus var. *prorepens* Kük. = ***Cyperus margaritaceus*** Vahl
Cyperus margaritaceus var. *pseudoniveus* (Boeck.) C.B.Clarke = ***Cyperus margaritaceus*** Vahl
Cyperus margaritaceus var. *tisserantii* (Cherm.) Kük. = ***Cyperus niveus*** var. ***tisserantii*** (Cherm.) Lye
Cyperus marginatus var. *blandus* (Kunth) Kük. = ***Cyperus marginatus*** Thunb.
Cyperus marginellus Nees = ***Cyperus pilosus*** Vahl
Cyperus mariscoides Nutt. = ***Cyperus filiculmis*** Vahl
Cyperus mariscoides Elliott = ?
Cyperus mariscus Nees = ***Cyperus albostriatus*** Schrad.
Cyperus maritimus (Lam.) P.Silva = ***Cyperus capitatus*** Vand.
Cyperus maritimus Poir. = ***Cyperus crassipes*** Vahl
Cyperus maritimus (Miq.) Suringar = ***Cyperus dubius*** Rottb. subsp. ***dubius***
Cyperus maritimus Gouan ex Spreng. = ***Cyperus laevigatus*** L. subsp. ***laevigatus***
Cyperus maritimus var. *crassipes* (Vahl) C.B.Clarke = ***Cyperus crassipes*** Vahl
Cyperus maritimus var. *gracilescens* Kük. = ***Cyperus crassipes*** Vahl
Cyperus maritimus var. *subtilis* Kük. = **Cyperus subtilis**
Cyperus marlothii var. *seineri* Kük. = ***Cyperus marlothii*** Boeck.
Cyperus martianus Schult. = ***Cyperus congestus*** Vahl
Cyperus martianus Schrad. ex Nees = ***Cyperus friburgensis*** Boeck.
Cyperus martindalei Britton = ***Cyperus filiculmis*** Vahl
Cyperus martinicensis Boeck. = ***Cyperus mutisii*** (Kunth) Andersson
Cyperus mauiensis Hillebr. = ***Cyperus hillebrandii*** Boeck. var. ***hillebrandii***
Cyperus mauritianus Willemet = [29 MAU]
Cyperus mauryi Kuntze = ***Pycreus megapotamicus*** (A.Dietr.) Nees
Cyperus maximiliani (Schrad. ex Nees) Griseb. = ***Cyperus odoratus*** L. subsp. ***odoratus***
Cyperus maximiliani Schrad. ex Nees = ***Pycreus lanceolatus*** (Poir.) C.B.Clarke
Cyperus maximus Roxb. ex Nees = ***Cyperus platyphyllus*** Roem. & Schult.
Cyperus mayeri Kük. = ***Cyperus stoloniferus*** Retz.
Cyperus mediorubescens Hayata = ***Cyperus imbricatus*** Retz.
Cyperus megapotamicus (A.Dietr.) Kunth = ***Pycreus megapotamicus*** (A.Dietr.) Nees
Cyperus megapotamicus var. *genuinus* Osten = ***Pycreus megapotamicus*** (A.Dietr.) Nees
Cyperus megapotamicus var. *jaeggii* (Boeck.) Kük. = ***Pycreus megapotamicus*** (A.Dietr.) Nees
Cyperus megapotamicus var. *nematodes* Kük. ex Osten = ***Pycreus megapotamicus*** (A.Dietr.) Nees
Cyperus melanacme (Nelmes) Raymond = ***Pycreus melanacme*** Nelmes
Cyperus melanocephalus R.Br. = ***Pycreus elegantulus*** (Steud.) C.B.Clarke
Cyperus melanocephalus Miq. = ***Pycreus sanguinolentus*** subsp. ***melanocephalus*** (Miq.) Karthik.
Cyperus melanopus Boeck. = ***Pycreus nitidus*** (Lam.) J.Raynal
Cyperus melanorhizus Delile = ***Cyperus esculentus*** L.
Cyperus melanospermus (Nees) Suringar = ***Kyllinga melanosperma*** Nees
Cyperus melanospermus subsp. *bifolius* (Miq.) J.Kern = ***Kyllinga melanosperma*** subsp. ***bifolia*** (Miq.) Karthik.
Cyperus melanospermus subsp. *elatus* (Steud.) Lye = ***Kyllinga melanosperma*** subsp. ***elata*** (Steud.) Lye
Cyperus melanospermus var. *hexalatus* (Lye) Lye = ***Kyllinga melanosperma*** var. ***hexalata*** Lye
Cyperus melanospermus var. *imerinensis* (Cherm.) Kük. = ***Kyllinga melanosperma*** Nees subsp. ***melanosperma***
Cyperus melanospermus var. *perrieri* (Cherm.) Kük. = ***Kyllinga melanosperma*** Nees subsp. ***melanosperma***
Cyperus melanospermus var. *plurifoliatus* (Kük.) Kük. = ***Kyllinga melanosperma*** var. ***plurifoliata*** Kük.
Cyperus melanostachyus Walp. = ***Pycreus flavescens*** (L.) P.Beauv. ex Rchb. subsp. ***flavescens***
Cyperus melanostachyus Kunth = ***Pycreus niger*** (Ruiz & Pav.) Cufod.
Cyperus melanostachyus var. *caespitosus* Boeck. = ***Pycreus niger*** (Ruiz & Pav.) Cufod.
Cyperus melanostachyus var. *gracilis* Liebm. = ***Pycreus niger*** (Ruiz & Pav.) Cufod.
Cyperus melanostachyus var. *robustus* Liebm. = ***Pycreus niger*** (Ruiz & Pav.) Cufod.
Cyperus melas Ridl. = ***Pycreus melas*** (Ridl.) C.B.Clarke
Cyperus melicoides Poir. = ***Costularia melicoides*** (Poir.) C.B.Clarke
Cyperus membranaceus Vahl = ***Pycreus membranaceus*** (Vahl) Govind.
Cyperus mendoncae Boeck. = ***Cyperus pohlii*** (Nees) Steud. subsp. ***pohlii***
Cyperus merkeri C.B.Clarke = ***Cyperus rotundus*** subsp. ***merkeri*** (C.B.Clarke) Kük.
Cyperus merrillii (C.B.Clarke) Kük. = ***Cyperus dubius*** Rottb. subsp. ***dubius***
Cyperus merxmuelleri (Podlech) Lye = ***Kyllinga albiceps*** (Ridl.) Rendle
Cyperus mesochorus Geise = ***Cyperus schweinitzii*** Torr.
Cyperus metzii (Hochst. ex Steud.) Mattf. & Kük. = ***Kyllinga squamulata*** Vahl
Cyperus mexicanus Liebm. = ***Cyperus digitatus*** Roxb. subsp. ***digitatus***
Cyperus meyenianus var. *breviradiatus* Maury = ***Cyperus meyenianus*** Kunth
Cyperus meyenianus var. *oligostachys* (Schrad. ex Nees) Kük. = ***Cyperus meyenianus*** Kunth
Cyperus meyenianus var. *paranaguensis* (Palla) Kük. = ***Cyperus meyenianus*** Kunth
Cyperus meyenianus var. *patentissimus* Kük. = ***Cyperus meyenianus*** Kunth
Cyperus micans Kunth = ***Pycreus intactus*** (Vahl) J.Raynal
Cyperus michauxianus Schult. = ***Cyperus odoratus*** L. subsp. ***odoratus***
Cyperus michauxianus var. *elongatus* Torr. = ***Cyperus strigosus*** L.
Cyperus michelianus f. *elatus* Kük. = ***Cyperus michelianus*** (L.) Delile subsp. ***michelianus***

Cyperus michelianus f. *exiguus* Kük. = ***Cyperus michelianus*** (L.) Delile subsp. ***michelianus***
Cyperus michelianus f. *hilgendorfianus* (Boeck.) Kük. = ***Cyperus michelianus*** (L.) Delile subsp. ***michelianus***
Cyperus michelianus var. *nipponicus* (Franch. & Sav.) Kük. = ***Cyperus nipponicus*** Franch. & Sav.
Cyperus michelianus var. *pacificus* Ohwi = ***Cyperus pacificus*** (Ohwi) Ohwi
Cyperus micranthus Schult. = ***Cyperus dentatus*** Torr.
Cyperus micranthus J.Presl & C.Presl =
Cyperus micreilema Steud. = ***Cyperus rotundus*** L. subsp. ***rotundus***
Cyperus microaureus Lye = ***Alinula peteri*** (Kük.) Goetgh. & Vorster
Cyperus microbracteatus (Lye) Lye = ***Kyllinga microbracteata*** Lye
Cyperus microbulbosus (Lye) Lye = ***Kyllinga microbulbosa*** Lye
Cyperus microcarpus Boeck. = ***Cyperus haspan*** L.
Cyperus microdontus Torr. = ***Pycreus polystachyos*** var. ***microdontus*** (Torr.) Govaerts
Cyperus microdontus var. *texensis* Torr. = ***Pycreus polystachyos*** var. ***microdontus*** (Torr.) Govaerts
Cyperus microlepis Boeck. = ***Cyperus submicrolepis*** Kük.
Cyperus micromegas Nees = ***Cyperus tenellus*** L.f.
Cyperus micromelas (Lye) Lye = ***Pycreus micromelas*** Lye
Cyperus microstachyus Vahl = ***Cyperus amabilis*** Vahl
Cyperus microstylus (C.B.Clarke) Mattf. & Kük. = ***Kyllinga microstyla*** C.B.Clarke
Cyperus ×*mihashii* T.Koyama = ?
Cyperus miliifolius var. *rotheryi* A.Gross = ***Cyperus miliifolius*** Poepp. & Kunth
Cyperus miliifolius var. *saturatus* (Donn.Sm.) Kük. = ***Cyperus miliifolius*** Poepp. & Kunth
Cyperus minarum Boeck. = ***Cyperus virens*** var. ***minarum*** (Boeck.) Denton
Cyperus minimus Forssk. = ?
Cyperus minimus L. = ***Bulbostylis capillaris*** (L.) Kunth ex C.B.Clarke subsp. ***capillaris***
Cyperus minor Steud. = ***Pycreus intactus*** (Vahl) J.Raynal
Cyperus minutiflorus J.Presl & C.Presl = [42 PHI]
Cyperus minutulus Vahl = ***Cyperus michelianus*** (L.) Delile subsp. ***michelianus***
Cyperus minutulus K.Schum. = ***Pycreus hildebrandtii*** C.B.Clarke
Cyperus minutus J.F.Gmel. = ?
Cyperus minutus (C.B.Clarke) Kük. =
Cyperus minutus Roth = ***Isolepis antarctica*** (L.) Roem. & Schult.
Cyperus modestulus Steud. = ***Isolepis marginata*** (Thunb.) A.Dietr.
Cyperus moestus Kunth = ***Cyperus diffusus*** Vahl
Cyperus mohrii Britton ex C.B.Clarke = ***Cyperus spectabilis*** Link var. ***spectabilis***
Cyperus mollipes var. *amomodorus* (K.Schum.) Kük. = ***Cyperus amomodorus*** K.Schum.
Cyperus mollipes var. *bulbocaulis* (Boeck.) Kük. = ***Cyperus mollipes*** (C.B.Clarke) K.Schum.
Cyperus mollipes var. *globifer* (C.B.Clarke) Kük. = ***Cyperus globifer*** (C.B.Clarke) Lye
Cyperus mollipes var. *paolii* (Chiov.) Kük. = ***Cyperus paolii*** Chiov.
Cyperus mollis Poir. = ***Cyperus dubius*** Rottb. subsp. ***dubius***
Cyperus monandrus Roth = ***Cyperus eragrostis*** Lam.
Cyperus monocephaloides Roxb. ex Nees = ***Cyperus cephalotes*** Vahl
Cyperus monocephalus Roxb. = ***Cyperus cephalotes*** Vahl
Cyperus monocephalus F.Muell. = ***Kyllinga erecta*** Schumach. subsp. ***erecta***
Cyperus monocephalus Baker = ***Pycreus fluminalis*** (Ridl.) Rendle
Cyperus monogynus Boeck. = ***Cyperus cephalotes*** Vahl
Cyperus monophyllus Vahl = ***Cyperus malaccensis*** var. ***brevifolius*** Boeck.
Cyperus monostachyos Link = ***Cyperus laevigatus*** L. subsp. ***laevigatus***
Cyperus monostachyos L. = ***Fimbristylis ovata*** (Burm.f.) J.Kern
Cyperus monostachys Boeck. = ***Cyperus seemannianus*** Boeck.
Cyperus monostigma C.B.Clarke = ***Cyperus clavinux*** C.B.Clarke
Cyperus monroviensis Boeck. = ***Cyperus tenax*** Boeck.
Cyperus montanus Boeck. = ***Cyperus virens*** var. ***montanus*** (Boeck.) Deuton
Cyperus monti L.f. = ***Cyperus serotinus*** Rottb.
Cyperus monti var. *stylosa* C.B.Clarke = ***Cyperus serotinus*** Rottb.
Cyperus montis-sellae K.Schum. ex Warb. = ***Cyperus pedunculosus*** F.Muell.
Cyperus morandinii Pic.Serm. = ***Cyperus penzoanus*** Pic.Serm.
Cyperus mortonii (S.S.Hooper) Lye = ***Pycreus mortonii*** S.S.Hooper
Cyperus mossii Turrill = ***Cyperus albostriatus*** Schrad.
Cyperus mucronatus (L.) Mabille = ***Cyperus capitatus*** Vand.
Cyperus mucronatus Rottb. = ***Cyperus laevigatus*** subsp. ***distachyos*** (All.) Maire & Weiller
Cyperus mucronatus Steud. = ***Pycreus flavidus*** (Retz.) T.Koyama
Cyperus muehlenbergii Spreng. = ***Cyperus haspan*** L.
Cyperus muelleri Boeck. = ***Cyperus amabilis*** Vahl
Cyperus mulen-pullu Schult. = ***Cyperus michelianus*** (L.) Delile subsp. ***michelianus***
Cyperus multibracteatus Boeck. = ***Cyperus odoratus*** L. subsp. ***odoratus***
Cyperus multicaulis (Sm.) Missbach & E.H.L.Krause = ***Eleocharis multicaulis*** (Sm.) Desv.
Cyperus multiceps Link = ***Cyperus congestus*** Vahl
Cyperus multiceps Hook. & Arn. = ***Cyperus odoratus*** L. subsp. ***odoratus***
Cyperus multiflorus (Britton) Small = ***Cyperus croceus*** Vahl
Cyperus multiflorus Kunth =
Cyperus multiglumis Turrill = ***Cyperus semitrifidus*** Schrad.
Cyperus multiradiatus (Torr.) C.Mohr = ***Cyperus lecontei*** Torr. ex Steud.
Cyperus multumbelliferus Roem. & Schult. = ***Cyperus glomeratus*** L.
Cyperus mundtii (Nees) Kunth = ***Pycreus mundtii*** Nees
Cyperus mundtii var. *densispiculosus* Kük. = ***Pycreus mundtii*** Nees
Cyperus mundtii f. *distichophyllus* (Steud.) Kük. = ***Pycreus mundtii*** Nees
Cyperus mundtii var. *distichophyllus* (Steud.) Kük. = ***Pycreus mundtii*** Nees
Cyperus mundtii var. *glaucus* Boeck. = ***Pycreus mundtii*** Nees
Cyperus mundtii var. *uniceps* (C.B.Clarke) Kük. = ***Pycreus mundtii*** Nees
Cyperus muricatus Kük. = ***Pycreus muricatus*** (Kük.) Napper

Cyperus musarius Buch.-Ham. ex Wall. = ***Cyperus michelianus*** (L.) Delile subsp. ***michelianus***
Cyperus mutisii var. *asper* (Liebm.) Kük. = ***Cyperus mutisii*** (Kunth) Andersson
Cyperus mutisii var. *contractus* Kük. = ***Cyperus mutisii*** (Kunth) Andersson
Cyperus mutisii var. *ehrenbergianus* (Boeck.) Kük. = ***Cyperus mutisii*** (Kunth) Andersson
Cyperus mutisii var. *fuscisquamatus* Kük. = ***Cyperus mutisii*** (Kunth) Andersson
Cyperus mutisii var. *martinicensis* (Boeck.) Kük. = ***Cyperus mutisii*** (Kunth) Andersson
Cyperus mutisii var. *semitribrachiatus* (Boeck.) Kük. = ***Cyperus mutisii*** (Kunth) Andersson
Cyperus myriostachyus Ten. = ***Cyperus longus*** L. subsp. ***longus***
Cyperus mysurensis B.Heyne ex Wall. = ***Cyperus nutans*** var. ***eleusinoides*** (Kunth) Haines
Cyperus myyenii Nees & Arn. = ***Cyperus compressus*** L.
Cyperus naguensis Britton = ***Cyperus strigosus*** L.
Cyperus nangtciangensis Pamp. = ***Cyperus iria*** L.
Cyperus nanilensis Boeck. = ***Cyperus cyperinus*** (Retz.) Suringar
Cyperus nanus var. *rigidulus* Ekman & Kük. = ***Cyperus nanus*** Willd. var. ***nanus***
Cyperus nanus var. *subtenuis* Kük. = **Cyperus subtenuis**
Cyperus nashii Britton = ***Cyperus retrorsus*** Chapm.
Cyperus natans Buch.-Ham. ex C.B.Clarke = ***Cyperus cephalotes*** Vahl
Cyperus naumannianus Boeck. = ***Cyperus maculatus*** Boeck. subsp. ***maculatus***
Cyperus navicularis Steud. = ?
Cyperus neesii Kunth = ***Cyperus digitatus*** Roxb. subsp. ***digitatus***
Cyperus neesii D.Dietr. = ***Cyperus malaccensis*** Lam. var. ***malaccensis***
Cyperus neglectus Parl. = ***Cyperus longus*** L. subsp. ***longus***
Cyperus nemacaulos Steud. = ***Cyperus filifolius*** Willd. ex Kunth
Cyperus nematodes Schrad. ex Nees = ***Pycreus megapotamicus*** (A.Dietr.) Nees
Cyperus nemoralis var. *onivensis* (Cherm.) Kük. = ***Cyperus nemoralis*** Cherm.
Cyperus nemorosus G.Mey. = ***Cyperus sphacelatus*** Rottb.
Cyperus neobarteri T.Koyama = ***Lipocarpha barteri*** C.B.Clarke
Cyperus neoschimperi Kük. = ***Cyperus cruentus*** Rottb.
Cyperus neoschimperi var. *subvirescens* A.Peter & Kük. = ***Cyperus cruentus*** Rottb.
Cyperus neoschimperi var. *viridis* (Hochst. ex Schweinf.) Kük. = ***Cyperus cruentus*** Rottb.
Cyperus neotropicalis Alain = ***Lipocarpha maculata*** (Michx.) Torr.
Cyperus neourbanii Kük. = ***Kyllinga urbanii*** Kük.
Cyperus nervosus Bertol. = ***Cyperus esculentus*** L.
Cyperus nervulosus (Kük.) S.T.Blake = ***Pycreus pumilus*** (L.) Nees
Cyperus neurotropis Steud. = ***Pycreus sanguinolentus*** (Vahl) Nees subsp. ***sanguinolentus***
Cyperus niederleinianus Boeck. = ***Pycreus niederleinianus*** (Boeck.) Lindm.
Cyperus nieuwlandii Geise = ***Pycreus flavescens*** (L.) P.Beauv. ex Rchb. subsp. ***flavescens***
Cyperus niger Ruiz & Pav. = ***Pycreus niger*** (Ruiz & Pav.) Cufod.
Cyperus niger var. *capitatus* (Britton) O'Neill = ***Pycreus niger*** (Ruiz & Pav.) Cufod.
Cyperus niger var. *castaneus* (S.Watson) Kük. = ***Pycreus niger*** (Ruiz & Pav.) Cufod.
Cyperus niger subsp. *elegantulus* (Steud.) Lye = ***Pycreus elegantulus*** (Steud.) C.B.Clarke
Cyperus niger var. *intricatoramosus* (Boeck.) Kük. = ***Pycreus niger*** (Ruiz & Pav.) Cufod.
Cyperus niger var. *lorentzianus* (Boeck.) Kük. = ***Pycreus niger*** (Ruiz & Pav.) Cufod.
Cyperus niger var. *pseudoelegantulus* Kük. = ***Pycreus niger*** (Ruiz & Pav.) Cufod.
Cyperus niger var. *rivularis* (Kunth) V.E.Grant = ***Pycreus bipartitus*** (Torr.) C.B.Clarke
Cyperus niger var. *robustus* (Liebm.) Kük. = ***Pycreus niger*** (Ruiz & Pav.) Cufod.
Cyperus nigricans Steud. = ***Pycreus nigricans*** (Steud.) C.B.Clarke
Cyperus nigricans var. *firmior* Kük. = ***Pycreus nigricans*** (Steud.) C.B.Clarke
Cyperus nigricans var. *simulans* (Cherm.) Kük. = ***Pycreus nigricans*** (Steud.) C.B.Clarke
Cyperus nigripes (C.B.Clarke) Kük. = ***Kyllinga nigripes*** C.B.Clarke
Cyperus nigripes var. *grandiceps* Kük. = ***Kyllinga nigripes*** C.B.Clarke
Cyperus nigroviridis Thwaites = ***Cyperus diffusus*** Vahl
Cyperus nilagiricus Hochst. ex Steud. = ***Pycreus flavidus*** (Retz.) T.Koyama
Cyperus nilagiricus var. *pauperior* Boeck. = ***Pycreus flavidus*** (Retz.) T.Koyama
Cyperus niloticus Forssk. = ***Cyperus articulatus*** L.
Cyperus nitens Retz. = ***Pycreus pumilus*** (L.) Nees
Cyperus nitens Vahl = ***Pycreus pumilus*** (L.) Nees
Cyperus nitens var. *patens* (Vahl) Boeck. = ***Pycreus pumilus*** (L.) Nees
Cyperus nitidulus Boeck. = ***Cyperus odoratus*** L. subsp. ***odoratus***
Cyperus nitidus Lam. = ***Pycreus nitidus*** (Lam.) J.Raynal
Cyperus nocha Phil. = ***Cyperus laetus*** J.Presl & C.Presl
Cyperus nodosus Humb. & Bonpl. ex Willd. = ***Cyperus articulatus*** L.
Cyperus nodosus var. *aphyllus* Boeck. = ***Cyperus articulatus*** L.
Cyperus nodosus var. *subnodosus* (Nees & Meyen) Boeck. = ***Cyperus articulatus*** L.
Cyperus nodulosus F.Muell. ex Benth. = ***Cyperus gunnii*** Hook.f. var. ***gunnii***
Cyperus noronhae Ridl. = ***Cyperus appendiculatus*** var. ***noronhae*** (Ridl.) Kük.
Cyperus nortonii A.Heller = ***Cyperus odoratus*** L. subsp. ***odoratus***
Cyperus novae-hannoverae Boeck. = ***Cyperus odoratus*** L. subsp. ***odoratus***
Cyperus novae-hollandiae Boeck. = ***Cyperus gunnii*** var. ***novae-hollandiae*** (Boeck.) Kük.
Cyperus nubianus Gand. = ***Cyperus conglomeratus*** Rottb. subsp. ***conglomeratus***
Cyperus nubicus C.B.Clarke = ***Cyperus rotundus*** subsp. ***retzii*** Kük.
Cyperus nubigenus Britton & Standl. = ***Cyperus tabina*** Steud. ex Boeck.
Cyperus nudicaulis Poir. = ***Cyperus pectinatus*** Vahl
Cyperus nudiceps (C.B.Clarke ex Standl.) O'Neill = ***Kyllinga nudiceps*** C.B.Clarke ex Standl.
Cyperus nudiculmis Sieber ex C.B.Clarke = ***Cyperus schimperianus*** Steud.
Cyperus nudiusculus Nees = ***Cyperus tenellus*** L.f.
Cyperus nudus Roxb. = ***Cyperus corymbosus*** Rottb.
Cyperus nudus Kunth = ***Cyperus haspan*** L.
Cyperus nudus var. *firmior* C.Presl = ***Cyperus haspan*** L.

Cyperus nudus var. *flaccidior* C.Presl = ***Cyperus haspan*** L.
Cyperus nuerensis Boeck. = ***Pycreus nuerensis*** (Boeck.) S.S.Hooper
Cyperus nutans Sieber ex C.Presl = ***Cyperus distans*** L.f.
Cyperus nutans subsp. *eleusinoides* (Kunth) T.Koyama = ***Cyperus nutans*** var. ***eleusinoides*** (Kunth) Haines
Cyperus nutans subsp. *subprolixus* (Kük.) T.Koyama = ***Cyperus nutans*** var. ***subprolixus*** (Kük.) Karth.
Cyperus nuttallii Eddy ex Spreng. = ***Pycreus filicinus*** (Vahl) T.Koyama
Cyperus nuttallii var. *minimus* Alph.Wood = ***Pycreus filicinus*** (Vahl) T.Koyama
Cyperus nyassicus Chiov. = ?
Cyperus oakfortensis Boeck. ex C.B.Clarke = ***Pycreus oakfortensis*** (Boeck. ex C.B.Clarke) C.B.Clarke
Cyperus oatesii C.B.Clarke = ***Cyperus digitatus*** Roxb. subsp. ***digitatus***
Cyperus obesus Liebm. = ***Cyperus aggregatus*** (Willd.) Endl. var. ***aggregatus***
Cyperus obliquus Nees = ***Cyperus pilosus*** Vahl
Cyperus oblongoincrassatus var. *clarior* Kük. = ***Cyperus oblongoincrassatus*** Kük.
Cyperus oblongoincrassatus var. *groteanus* Kük. = ***Cyperus oblongoincrassatus*** Kük.
Cyperus oblongoincrassatus var. *udigensis* Kük. = ***Cyperus oblongoincrassatus*** Kük.
Cyperus oblongus (C.B.Clarke) Kük. = ***Kyllinga nervosa*** subsp. ***oblonga*** (C.B.Clarke) J.-P.Lebrun & Stork
Cyperus oblongus subsp. *flavus* (C.B.Clarke) Lye = ***Kyllinga nervosa*** subsp. ***flava*** (C.B.Clarke) Lye
Cyperus oblongus subsp. *jubensis* (Mtot.) Lye = ***Kyllinga nervosa*** subsp. ***jubensis*** (Mtot.) J.-P.Lebrun & Stork
Cyperus oblongus subsp. *nervosus* (Steud.) Lye = ***Kyllinga nervosa*** Steud.
Cyperus oblongus var. *ruwenzoriensis* (C.B.Clarke) Kük. = ***Kyllinga nervosa*** var. ***ruwenzoriensis*** (C.B.Clarke) Lye
Cyperus obscurus Nees = ***Cyperus imbricatus*** Retz.
Cyperus obstinatus Steud. = ***Pycreus pumilus*** (L.) Nees
Cyperus obtusatus (J.Presl & C.Presl) Mattf. & Kük. = ***Kyllinga vaginata*** Lam.
Cyperus obtusatus var. *africanus* Kük. = ***Kyllinga erecta*** var. ***africana*** (Kük.) S.S.Hooper
Cyperus obtusatus var. *cylindrostachyus* (Boeck.) Kük. = ***Kyllinga vaginata*** Lam.
Cyperus obtusatus var. *tenuis* (Boeck.) Kük. = ***Kyllinga vaginata*** Lam.
Cyperus obtusiflorus Vahl = ***Cyperus niveus*** var. ***leucocephalus*** (Kunth) Fosberg
Cyperus obtusiflorus var. *flavissimus* (Schrad.) Boeck. = ***Cyperus niveus*** var. ***flavissimus*** (Schrad.) Lye
Cyperus obtusiflorus var. *ledermannii* Kük. = ***Cyperus niveus*** var. ***ledermannii*** (Kük.) Lye
Cyperus obtusiflorus var. *membranaceus* Kük. = ***Cyperus niveus*** var. ***leucocephalus*** (Kunth) Fosberg
Cyperus obtusiflorus var. *peramoenus* Kük. = ***Cyperus niveus*** var. ***leucocephalus*** (Kunth) Fosberg
Cyperus obtusiflorus var. *rigidus* (Vahl) Kük. = ***Cyperus niveus*** var. ***leucocephalus*** (Kunth) Fosberg
Cyperus obtusiusculus Gand. = [12 FRA]
Cyperus occidentalis Torr. = ***Cyperus erythrorrhizos*** Muhl.
Cyperus ochraceus var. *excelsior* Kük. = ***Cyperus ochraceus*** Vahl
Cyperus ochraceus var. *minor* Kük. = ***Cyperus ochraceus*** Vahl
Cyperus ochreatus Boeck. = ***Cyperus mutisii*** (Kunth) Andersson
Cyperus ochreoides Steud. = ***Cyperus rotundus*** L. subsp. ***rotundus***
Cyperus ochrocarpus K.Schum. = ***Cyperus renschii*** Boeck.
Cyperus ochrocephalus C.B.Clarke = ***Cyperus angolensis*** Boeck.
Cyperus ochrocephalus Steud. = ***Cyperus eragrostis*** Lam.
Cyperus octophyllus Hochst. ex Steud. = ***Cyperus laxus*** Lam. subsp. ***laxus***
Cyperus odoratus Burm.f. = ***Cyperus exaltatus*** Retz.
Cyperus odoratus var. *acicularis* (Schrad. ex Nees) O'Neill = ***Cyperus odoratus*** L. subsp. ***odoratus***
Cyperus odoratus var. *attenuatus* Fosberg & Sachet = ***Cyperus odoratus*** L. subsp. ***odoratus***
Cyperus odoratus var. *curtispiculus* Fosberg & Sachet = ***Cyperus odoratus*** L. subsp. ***odoratus***
Cyperus odoratus var. *engelmannii* (Steud.) R.Carter, S.D.Jones & Wipff = ***Cyperus odoratus*** L. subsp. ***odoratus***
Cyperus odoratus var. *novae-hannoverae* (Boeck.) Fosberg & Sachet = ***Cyperus odoratus*** L. subsp. ***odoratus***
Cyperus odoratus var. *squarrosus* (Britton) S.D.Jones, Wipff & R.Carter = ***Cyperus odoratus*** L. subsp. ***odoratus***
Cyperus oerstedii Liebm. = ***Cyperus odoratus*** L. subsp. ***odoratus***
Cyperus officinalis T.Nees = ***Cyperus esculentus*** L.
Cyperus officinalis Nees ex Gren. & Godr. = ***Cyperus rotundus*** L. subsp. ***rotundus***
Cyperus ×ogawai T.Koyama = ?
Cyperus oleraceus Roxb. ex Nees = ***Cyperus bulbosus*** Vahl
Cyperus olfersianus Kunth = ***Pycreus lanceolatus*** (Poir.) C.B.Clarke
Cyperus olfersianus var. *elatior* Boeck. = ***Pycreus lanceolatus*** (Poir.) C.B.Clarke
Cyperus olfersianus var. *maximiliani* (Nees) Boeck. = ***Pycreus lanceolatus*** (Poir.) C.B.Clarke
Cyperus olidus Vahl = ***Pycreus polystachyos*** (Rottb.) P.Beauv. subsp. ***polystachyos***
Cyperus oliganthus Gand. = [12 FRA]
Cyperus oligostachyus Kunth = ***Cyperus amabilis*** Vahl
Cyperus olivaceus Vahl = ***Pycreus polystachyos*** (Rottb.) P.Beauv. subsp. ***polystachyos***
Cyperus olivaris O.Targ.Tozz. = ***Cyperus rotundus*** L. subsp. ***rotundus***
Cyperus olivaris var. *brevibracteatus* Legrand = ***Cyperus rotundus*** L. subsp. ***rotundus***
Cyperus olivetorum Murb. = ***Cyperus cadamostii*** Bolle ex E.H.L.Krause
Cyperus onivensis Cherm. = ***Cyperus nemoralis*** Cherm.
Cyperus onustus Steud. = ***Cyperus alternifolius*** L. subsp. ***alternifolius***
Cyperus oostachyus Nees = ***Cyperus rigens*** J.Presl & C.Presl subsp. ***rigens***
Cyperus orbicephalus (Beetle) T.Koyama & McVaugh = ***Karinia mexicana*** (C.B.Clarke ex Britton) Reznicek & McVaugh
Cyperus ornans J.V.Suringar = ***Cyperus stenophyllus*** J.V.Suringar
Cyperus ornans var. *rechingeri* (Palla) Kük. = ***Cyperus stenophyllus*** J.V.Suringar
Cyperus ornans var. *stenophyllus* (J.V.Suringar) Kük. = ***Cyperus stenophyllus*** J.V.Suringar
Cyperus ornatissimus F.Muell. = ***Pycreus macrostachyos*** (Lam.) J.Raynal subsp. ***macrostachyos***
Cyperus ornithopodioides Delile = ***Cyperus digitatus*** Roxb. subsp. ***digitatus***

Cyperus ornithopus Pers. = ***Cyperus filiformis*** Sw.
Cyperus oryzeticola Steud. = ***Cyperus exaltatus*** Retz.
Cyperus oryzetorum Steud. = ***Cyperus difformis*** L.
Cyperus ostenii Kük. ex Osten = ?
Cyperus ottonis Boeck. = ***Cyperus planifolius*** Rich.
Cyperus ottonis var. *humilior* Boeck. = ***Cyperus planifolius*** Rich.
Cyperus ovatus Llanos = ***Cyperus javanicus*** Houtt.
Cyperus ovatus (Roth) Missbach & E.H.L.Krause = ***Eleocharis ovata*** (Roth) Roem. & Schult.
Cyperus overlaetii (Cherm. ex S.S.Hooper & J.Raynal) Lye = ***Pycreus overlaetii*** Cherm. ex S.S.Hooper & J.Raynal
Cyperus ovularis (Michx.) Torr. = ***Cyperus echinatus*** (L.) Alph.Wood
Cyperus ovularis var. *americanus* Boeck. = ***Cyperus retrorsus*** Chapm.
Cyperus ovularis var. *cylindricus* (Elliott) Torr. = ***Cyperus retrorsus*** Chapm.
Cyperus ovularis var. *robustus* Boeck. = ***Cyperus retrorsus*** Chapm.
Cyperus ovularis var. *sphaericus* Boeck. = ***Cyperus echinatus*** (L.) Alph.Wood
Cyperus ovularis var. *tenellus* Torr. = ***Cyperus retrorsus*** Chapm.
Cyperus ovularis var. *wolfii* (Alph.Wood) Kük. = ***Cyperus echinatus*** (L.) Alph.Wood
Cyperus owahuensis Nees = ***Cyperus javanicus*** Houtt.
Cyperus owanii var. *rogersii* Kük. = ***Cyperus owanii*** Boeck.
Cyperus oxycarioides Britton = ***Cyperus odoratus*** L. subsp. ***odoratus***
Cyperus oxycarioides Britton ex Coult. = ***Cyperus odoratus*** L. subsp. ***odoratus***
Cyperus oxylepis var. *punensis* Kük. = ***Cyperus oxylepis*** Nees ex Steud.
Cyperus oxystachys Nees ex Boeck. = ***Cyperus longus*** L. subsp. ***longus***
Cyperus pachyrrhizus Nees = ***Cyperus conglomeratus*** subsp. ***pachyrrhizus*** (Nees) T.Koyama
Cyperus pachystylus (Kük.) Kük. = ***Kyllinga pachystyla*** Kük.
Cyperus pagotii (J.Raynal) Lye = ***Pycreus pagotii*** J.Raynal
Cyperus pallens (Liebm.) Standl. & Steyerm. = ***Cyperus thyrsiflorus*** Jungh. ex Schltdl.
Cyperus pallescens Desf. = ***Cyperus corymbosus*** Rottb.
Cyperus pallescens Poir. = ***Cyperus rotundus*** L. subsp. ***rotundus***
Cyperus pallescens Boiss. = ***Cyperus rotundus*** L. subsp. ***rotundus***
Cyperus pallidus Savi = ***Cyperus esculentus*** L.
Cyperus pallidus Nees = ***Cyperus platystylis*** R.Br.
Cyperus pallidus Link = ?
Cyperus pallidus Kit. ex Kanitz = ***Pycreus flavescens*** (L.) P.Beauv. ex Rchb. subsp. ***flavescens***
Cyperus paluster Missbach & E.H.L.Krause = ***Eleocharis palustris*** (L.) Roem. & Schult. subsp. ***palustris***
Cyperus paluster var. *uniglumoides* E.H.L.Krause = ***Eleocharis palustris*** (L.) Roem. & Schult. subsp. ***palustris***
Cyperus palustris (Schrad.) Kük. = ***Cyperus meyenianus*** Kunth
Cyperus palustris (L.) Sessé & Moç. = ***Eleocharis palustris*** (L.) Roem. & Schult.
Cyperus panamensis var. *deformis* (C.B.Clarke) Kük. = ***Cyperus panamensis*** (C.B.Clarke) Britton ex Standl.
Cyperus panamensis var. *ramigerus* Kük. = ***Cyperus panamensis*** (C.B.Clarke) Britton ex Standl.
Cyperus pangorei var. *ambiguus* C.B.Clarke = ***Cyperus pangorei*** Rottb.
Cyperus paniceus var. *queenslandicus* (Domin) Kük. = ***Cyperus paniceus*** (Rottb.) Boeck.
Cyperus paniceus var. *roxburghianus* (C.B.Clarke) Kük. = ***Cyperus paniceus*** (Rottb.) Boeck.
Cyperus paniciformis Franch. & Sav. = ***Cyperus iria*** L.
Cyperus panicoides Lam. = ***Cyperus iria*** L.
Cyperus paniculatus D.Don = ***Cyperus pilosus*** Vahl
Cyperus paniculatus Hook. & Arn. = [63]
Cyperus paniculatus Blanco = (Poaceae)
Cyperus paniculatus Rottb. = ***Pycreus polystachyos*** (Rottb.) P.Beauv. subsp. ***polystachyos***
Cyperus panormitanus Chiov. = ***Cyperus papyrus*** L. subsp. ***papyrus***
Cyperus paposanus Phil. = ***Cyperus laetus*** J.Presl & C.Presl
Cyperus pappii Gand. = ***Cyperus conglomeratus*** Rottb. subsp. ***conglomeratus***
Cyperus papuanus Ridl. = ***Cyperus pedunculosus*** F.Muell.
Cyperus papyraceus Crantz = ***Cyperus papyrus*** L. subsp. ***papyrus***
Cyperus papyroides Poir. = ***Cyperus prolifer*** Lam.
Cyperus papyrus var. *antiquorum* (Willd.) C.B.Clarke = ***Cyperus papyrus*** L. subsp. ***papyrus***
Cyperus papyrus subsp. *antiquorum* (Willd.) Chiov. = ***Cyperus papyrus*** L. subsp. ***papyrus***
Cyperus papyrus subsp. *hadidii* Chrtek & Slavíková = ***Cyperus papyrus*** L. subsp. ***papyrus***
Cyperus papyrus subsp. *niliacus* Tournay = ***Cyperus papyrus*** L. subsp. ***papyrus***
Cyperus papyrus var. *palaestinae* Chiov. = ***Cyperus papyrus*** L. subsp. ***papyrus***
Cyperus papyrus subsp. *siculus* (Parl.) Chiov. = ***Cyperus papyrus*** L. subsp. ***papyrus***
Cyperus papyrus subsp. *ugandensis* Chiov. = ***Cyperus papyrus*** L. subsp. ***papyrus***
Cyperus paradoxus Steud. = [29 REU]
Cyperus paraensis Mart. & Schrad. ex Nees = ***Pycreus flavescens*** (L.) P.Beauv. ex Rchb. subsp. ***flavescens***
Cyperus paraguayensis C.B.Clarke = ***Cyperus uncinulatus*** Schrad. ex Nees
Cyperus paramatta Mart. = ***Cyperus congestus*** Vahl
Cyperus paramatta Balb. ex Kunth = ***Cyperus congestus*** Vahl
Cyperus paramattensis Link = ?
Cyperus parciflorus Link = ***Cyperus diffusus*** Vahl
Cyperus parlatoris Lojac. = ***Cyperus longus*** L. subsp. ***longus***
Cyperus parlatoris var. *difformis* Lojac. = ***Cyperus longus*** L. subsp. ***longus***
Cyperus parryi Britton ex C.B.Clarke = ***Cyperus spectabilis*** Link var. ***spectabilis***
Cyperus parviflorus Muhl. = ***Cyperus dentatus*** Torr.
Cyperus parviflorus Vahl = ***Cyperus javanicus*** Houtt.
Cyperus parviflorus Nees = ***Cyperus microiria*** Steud.
Cyperus parvinux C.B.Clarke = ***Cyperus rupestris*** Kunth
Cyperus parvispiculatus Boeck. = ***Cyperus odoratus*** L. subsp. ***odoratus***
Cyperus parvulus (Roem. & Schult.) Missbach & E.H.L.Krause = ***Eleocharis parvula*** (Roem. & Schult.) Palla
Cyperus parvulus Steud. = ***Pycreus grammicus*** (Kunze ex Kunth) C.B.Clarke
Cyperus parvus Boeck. = ***Cyperus odoratus*** L. subsp. ***odoratus***
Cyperus parvus Hassk. = ?
Cyperus patens Vahl = ***Pycreus pumilus*** (L.) Nees

Cyperus patuliflorus Boeck. = ***Pycreus macrostachyos*** (Lam.) J.Raynal subsp. ***macrostachyos***
Cyperus patulus Kit. ex Host = ***Cyperus glaber*** L.
Cyperus patulus M.Bieb. = ***Cyperus rotundus*** L. subsp. ***rotundus***
Cyperus patulus Willd. ex Link = ?
Cyperus pauciflorus Steud. = ***Cyperus pilosus*** Vahl
Cyperus pauciflorus (Lightf.) Missbach & E.H.L.Krause = ***Eleocharis quinqueflora*** (Hartmann) O.Schwarz
Cyperus pauloensis Palla = ***Cyperus intricatus*** Schrad. ex Schult.
Cyperus pauper Phil. = ***Pycreus bipartitus*** (Torr.) C.B. Clarke
Cyperus pauper Hochst. ex A.Rich. = ***Pycreus pauper*** (Hochst. ex A.Rich.) C.B.Clarke
Cyperus pectinatus Roxb. = ***Cyperus compressus*** L.
Cyperus pectiniformis Schult. = ***Cyperus compressus*** L.
Cyperus pedatus Nees = ***Cyperus haspan*** L.
Cyperus pedunculatus (R.Br.) J.Kern = ***Remirea maritima*** Aubl.
Cyperus pedunculosus var. *floribundus* Kük. = ***Cyperus pedunculosus*** F.Muell.
Cyperus pedunculosus var. *longibracteatus* Domin = ***Cyperus pedunculosus*** F.Muell.
Cyperus pelophilus Ridl. = ***Pycreus pelophilus*** (Ridl.) C.B.Clarke
Cyperus pelophilus f. *nanus* Kük. = ***Pycreus pelophilus*** (Ridl.) C.B.Clarke
Cyperus pennatus Lam. = ***Cyperus javanicus*** Houtt.
Cyperus pennatus Boeck. = ***Cyperus odoratus*** L. subsp. ***odoratus***
Cyperus pennatus var. *armstrongii* (Benth.) Kük. = ***Cyperus armstrongii*** Benth.
Cyperus pentadactylus Willd. ex Kunth = ***Cyperus exilis*** Willd. ex Kunth
Cyperus pergracilis Boeck. = ***Pycreus niger*** (Ruiz & Pav.) Cufod.
Cyperus permutatus Boeck. = ***Pycreus macranthus*** (Boeck.) C.B.Clarke
Cyperus perpusillus Boeck. = ***Cyperus seslerioides*** Kunth
Cyperus persicus Boiss. & Hohen. = ***Cyperus arenarius*** Retz.
Cyperus persquarrosus T.Koyama = ***Lipocarpha nana*** (A.Rich.) Cherm.
Cyperus pertenuis Roxb. = ***Cyperus articulatus*** L.
Cyperus peruvianus (Lam.) F.N.Williams = ***Kyllinga vaginata*** Lam.
Cyperus peruvianus var. *foliatus* (Kük.) Kük. = ***Kyllinga tibialis*** Poit. ex Ledeb.
Cyperus pervillei Boeck. = ***Pycreus pervillei*** (Boeck.) C.B.Clarke
Cyperus pes-avium Bertol. = ?
Cyperus peteri Kük. = ***Kyllinga peteri*** (Kük.) Lye
Cyperus petersianus Boeck. = ***Cyperus alternifolius*** subsp. ***flabelliformis*** Kük.
Cyperus petherickii C.B.Clarke = ***Cyperus dives*** Delile
Cyperus petraeus Hochst. ex Steud. = ***Cyperus rubicundus*** Vahl
Cyperus phaeocephalus Griseb. = ***Cyperus manimae*** Kunth var. ***manimae***
Cyperus phaeocephalus var. *major* Kük. = ***Cyperus manimae*** Kunth var. ***manimae***
Cyperus phaeolepis var. *albiflorus* (Cherm.) Kük. = ***Cyperus phaeolepis*** Cherm.
Cyperus phaeorhizus K.Schum. = ***Cyperus denudatus*** L.f.
Cyperus phaeorhizus var. *princeae* (C.B.Clarke) Kük. = ***Cyperus denudatus*** L.f.
Cyperus philippensis J.Presl & C.Presl = [42 PHI]
Cyperus phleoides Seem. = ***Cyperus odoratus*** L. subsp. ***odoratus***
Cyperus phymatodes Muhl. = ***Cyperus esculentus*** L.
Cyperus phymatodes var. *heermannii* (Buckley) S.Watson = ***Cyperus esculentus*** L.
Cyperus picardae var. *brevinux* Kük. = ***Cyperus picardae*** Boeck.
Cyperus piceus Liebm. = ***Pycreus flavescens*** (L.) P.Beauv. ex Rchb. subsp. ***flavescens***
Cyperus piceus f. *densior* Kük. = ***Pycreus niger*** (Ruiz & Pav.) Cufod.
Cyperus piceus f. *potosiensis* (C.B.Clarke) Kük. = ***Pycreus flavescens*** (L.) P.Beauv. ex Rchb. subsp. ***flavescens***
Cyperus piceus f. *squalidus* (Liebm.) Kük. = ***Pycreus flavescens*** (L.) P.Beauv. ex Rchb. subsp. ***flavescens***
Cyperus pictus Steud. = ***Cyperus alterniflorus*** R.Br.
Cyperus pictus Ten. = ***Cyperus glaber*** L.
Cyperus pilosus var. *babakensis* (Steud. ex Miq.) C.B.Clarke = ***Cyperus babakan*** Steud.
Cyperus pilosus f. *badius* Kuntze = ***Cyperus pilosus*** Vahl
Cyperus pilosus var. *contractus* (Miq.) Kük. = ***Cyperus pilosus*** Vahl
Cyperus pilosus var. *fimbriatus* (Nees) Kük. = ***Cyperus pilosus*** Vahl
Cyperus pilosus var. *laevirhachis* Kük. = ***Cyperus pilosus*** Vahl
Cyperus pilosus var. *muticus* Boeck. = ***Cyperus pilosus*** Vahl
Cyperus pilosus var. *obliquus* (Nees) C.B.Clarke = ***Cyperus pilosus*** Vahl
Cyperus pilosus f. *pallidus* Kuntze = ***Cyperus pilosus*** Vahl
Cyperus pilosus var. *polyanthus* C.B.Clarke = ***Cyperus pilosus*** Vahl
Cyperus pilosus var. *pseudobabakensis* Kük. = ***Cyperus pilosus*** Vahl
Cyperus pinguis (C.B.Clarke) Mattf. & Kük. = ***Kyllinga elatior*** Kunth
Cyperus pinnatus A.Rich. = ***Cyperus javanicus*** Houtt.
Cyperus piptolepis Steud. = ***Cyperus pilosus*** Vahl
Cyperus piptolepis var. *contractus* Miq. = ***Cyperus pilosus*** Vahl
Cyperus piptolepis var. *pyrophytus* Miq. = ***Cyperus pilosus*** Vahl
Cyperus pittieri Boeck. ex Kneuck. = ***Cyperus thyrsiflorus*** Jungh. ex Schltdl.
Cyperus planiculmis Steud. = ***Cyperus haspan*** L.
Cyperus planifolius var. *brunneus* (Sw.) Kük. = ***Cyperus planifolius*** Rich.
Cyperus planifolius f. *decolorans* Kük. = ***Cyperus sordidus*** J.Presl & C.Presl
Cyperus planifolius f. *gracilis* (Boeck.) Kük. = ***Cyperus planifolius*** Rich.
Cyperus planifolius f. *humilior* (Boeck.) Kük. = ***Cyperus planifolius*** Rich.
Cyperus planifolius var. *navassanus* Kük. & Ekman = ***Cyperus planifolius*** Rich.
Cyperus planifolius var. *ottonis* (Boeck.) Kük. = ***Cyperus planifolius*** Rich.
Cyperus planifolius var. *pickelii* Grosser ex Kük. = ***Cyperus planifolius*** Rich.
Cyperus plankii Britton ex Small = ***Cyperus retrorsus*** Chapm.
Cyperus plantaginifolius var. *minor* Cherm. = ***Cyperus plantaginifolius*** Cherm.
Cyperus platycaulis var. *kipasensis* (Cherm.) Peter & Kük. = ***Cyperus kipasensis*** Cherm.
Cyperus platycaulis var. *lucentinigricans* (K.Schum.) Kük. = ***Cyperus platycaulis*** Baker

Cyperus platycaulis var. *recedens* Peter & Kük. = ***Cyperus platycaulis*** Baker
Cyperus platyculmis R.Br. = ***Cyperus haspan*** L.
Cyperus platyphyllos Ridl. = ***Cyperus neoguinensis*** Kük.
Cyperus platystachys Cherm. = ***Cyperus rotundus*** L. subsp. ***rotundus***
Cyperus platystachyus Griseb. = ***Cyperus tenuis*** Sw.
Cyperus pleuranthus Nees = ***Cyperus laevigatus*** L. subsp. ***laevigatus***
Cyperus plumbeonucea Govind. = ***Pycreus plumbeonucea*** (Govind.) P.Singh & V.Singh
Cyperus plurinodosa Govind. = ***Pycreus plurinodosus*** (Govind.) P.Singh & V.Singh
Cyperus poecilus var. *evolutus* Kük. = ***Cyperus benadirensis*** Chiov.
Cyperus poeoides Desv. ex Ham. = ***Cyperus odoratus*** L. subsp. ***odoratus***
Cyperus pohlianus (Nees) Kuntze = ***Cyperus odoratus*** L. subsp. ***odoratus***
Cyperus pohlii var. *mendoncae* (Boeck.) Kük. = ***Cyperus pohlii*** (Nees) Steud. subsp. ***pohlii***
Cyperus pohlii var. *obscuricoloratus* Kük. = ***Cyperus pohlii*** (Nees) Steud. subsp. ***pohlii***
Cyperus pohlii var. *warmingii* (Boeck.) Kük. = ***Cyperus pohlii*** (Nees) Steud. subsp. ***pohlii***
Cyperus poiformis Pursh = ***Pycreus flavescens*** (L.) P.Beauv. ex Rchb. subsp. ***flavescens***
Cyperus poiretii Roem. & Schult. = ***Cyperus alopecuroides*** Rottb.
Cyperus politus Welw. ex Ridl. = ?
Cyperus pollardii Britton ex Small = ***Cyperus ovatus*** Baldwin
Cyperus polycephalus Link = ***Cyperus congestus*** Vahl
Cyperus polycephalus Lam. = ***Cyperus luzulae*** (L.) Retz.
Cyperus polyphyllus Vahl = ***Cyperus bulbosus*** Vahl
Cyperus polyphyllus Larrañaga = ?
Cyperus polystachyos Rottb. = ***Pycreus polystachyos*** (Rottb.) P.Beauv.
Cyperus polystachyos var. *baronii* C.B.Clarke = ***Pycreus intactus*** (Vahl) J.Raynal
Cyperus polystachyos var. *caespitosus* (Poir.) Kük. = ***Pycreus polystachyos*** (Rottb.) P.Beauv. subsp. ***polystachyos***
Cyperus polystachyos var. *chlorostachys* (Boeck.) Kük. = ***Pycreus polystachyos*** (Rottb.) P.Beauv. subsp. ***polystachyos***
Cyperus polystachyos var. *circinatus* (Ridl.) Kük. = ***Pycreus polystachyos*** (Rottb.) P.Beauv. subsp. ***polystachyos***
Cyperus polystachyos var. *cleaveri* (Torr.) C.B.Clarke = ***Pycreus filicinus*** (Vahl) T.Koyama
Cyperus polystachyos var. *fascicularis* (Poir.) Kuntze = ***Pycreus polystachyos*** (Rottb.) P.Beauv. subsp. ***polystachyos***
Cyperus polystachyos var. *ferrugineus* Boeck. = ***Pycreus intactus*** (Vahl) J.Raynal
Cyperus polystachyos var. *filicinus* (Vahl) C.B.Clarke = ***Pycreus filicinus*** (Vahl) T.Koyama
Cyperus polystachyos f. *fugax* (Liebm.) Kük. = ***Pycreus fugax*** (Liebm.) C.D.Adams
Cyperus polystachyos f. *glaucescens* Boeck. = ***Pycreus polystachyos*** (Rottb.) P.Beauv. subsp. ***polystachyos***
Cyperus polystachyos var. *hahnianus* (Boeck.) Kük. = ***Pycreus polystachyos*** (Rottb.) P.Beauv. subsp. ***polystachyos***
Cyperus polystachyos var. *holosericeus* (Link) C.B.Clarke = ***Pycreus polystachyos*** var. ***microdontus*** (Torr.) Govaerts
Cyperus polystachyos subsp. *holosericeus* (Link) T.Koyama = ***Pycreus polystachyos*** var. ***microdontus*** (Torr.) Govaerts
Cyperus polystachyos f. *inconspicuus* (Liebm.) Kük. = ***Pycreus polystachyos*** var. ***microdontus*** (Torr.) Govaerts
Cyperus polystachyos var. *laxiflorus* Benth. = ***Pycreus polystachyos*** var. ***microdontus*** (Torr.) Govaerts
Cyperus polystachyos subsp. *laxiflorus* (Benth.) Lye = ***Pycreus polystachyos*** var. ***microdontus*** (Torr.) Govaerts
Cyperus polystachyos var. *leptostachyus* Boeck. = ***Pycreus polystachyos*** var. ***microdontus*** (Torr.) Govaerts
Cyperus polystachyos var. *longispiculatus* Suringar = ***Pycreus polystachyos*** (Rottb.) P.Beauv. subsp. ***polystachyos***
Cyperus polystachyos var. *macrostachyos* Boeck. = ***Pycreus polystachyos*** (Rottb.) P.Beauv. subsp. ***polystachyos***
Cyperus polystachyos var. *macrostachyus* Boeck. = ***Pycreus polystachyos*** (Rottb.) P.Beauv. subsp. ***polystachyos***
Cyperus polystachyos var. *micans* (Kunth) C.B.Clarke = ***Pycreus intactus*** (Vahl) J.Raynal
Cyperus polystachyos f. *minor* Cherm. ex Kük. = ***Pycreus polystachyos*** (Rottb.) P.Beauv. subsp. ***polystachyos***
Cyperus polystachyos var. *miser* Kük. = ***Pycreus polystachyos*** (Rottb.) P.Beauv. subsp. ***polystachyos***
Cyperus polystachyos var. *pallidus* Hillebr. = ***Pycreus polystachyos*** (Rottb.) P.Beauv. subsp. ***polystachyos***
Cyperus polystachyos var. *paniculatus* (Rottb.) C.B.Clarke = ***Pycreus polystachyos*** (Rottb.) P.Beauv. subsp. ***polystachyos***
Cyperus polystachyos f. *pernanus* Kük. = ***Pycreus polystachyos*** (Rottb.) P.Beauv. subsp. ***polystachyos***
Cyperus polystachyos var. *sanguineus* Kük. = ***Pycreus polystachyos*** (Rottb.) P.Beauv. subsp. ***polystachyos***
Cyperus polystachyos var. *texensis* (Torr.) Fernald = ***Pycreus polystachyos*** var. ***microdontus*** (Torr.) Govaerts
Cyperus polystachyos var. *thouarsii* (Kunth) Kük. = ***Pycreus polystachyos*** (Rottb.) P.Beauv. subsp. ***polystachyos***
Cyperus polystachyus Jungh. = ***Pycreus flavescens*** (L.) P.Beauv. ex Rchb. subsp. ***flavescens***
Cyperus praealtus Kük. = ***Pycreus altus*** (Turrill) Lye
Cyperus praelongatus Steud. = ***Cyperus strigosus*** L.
Cyperus prasinus Kunth = ***Cyperus pulcher*** Thunb.
Cyperus pratorum Korotky = ***Pycreus pratorum*** (Korotky) Schischk.
Cyperus prescottianus Hook. & Arn. = ***Cyperus odoratus*** L. subsp. ***odoratus***
Cyperus preslii Parl. = ***Cyperus longus*** L. subsp. ***longus***
Cyperus preslii A.Dietr. = ***Androtrichum trigynum*** (Spreng.) H.Pfeiff.
Cyperus preslii subsp. *intermedius* (Guss.) Nyman = ***Cyperus longus*** L. subsp. ***longus***
Cyperus prieurianus (Steud.) T.Koyama = ***Lipocarpha prieuriana*** Steud.
Cyperus princeae C.B.Clarke = ***Cyperus denudatus*** L.f.
Cyperus princeps Kunth = ***Cyperus giganteus*** Vahl
Cyperus pringlei Britton = ***Cyperus tetragonus*** Elliott
Cyperus prionensis Boeck. = ***Cyperus meyenianus*** Kunth
Cyperus prionodes Steud. = ***Cyperus marginatus*** Thunb.
Cyperus prionotropis Steud. = ***Cyperus eragrostis*** Lam.
Cyperus procerulus Nees = ***Cyperus rotundus*** L. subsp. ***rotundus***
Cyperus procerus var. *griffithianus* (Boeck.) Kük. = ***Cyperus procerus*** Rottb.

Cyperus procerus var. *lasiorrhachis* C.B.Clarke = ***Cyperus procerus*** Rottb.
Cyperus procerus var. *stenanthus* Kük. = ***Cyperus procerus*** Rottb.
Cyperus procerus var. *straminicolor* (Cherm.) Kük. = ***Cyperus procerus*** Rottb.
Cyperus procerus var. *trachycladus* (Cherm.) Kük. = ***Cyperus procerus*** Rottb.
Cyperus procerus var. *vanderystii* Kük. = ***Cyperus procerus*** Rottb.
Cyperus productus C.Wright = ***Cyperus aggregatus*** (Willd.) Endl. var. ***aggregatus***
Cyperus prolifer var. *isocladus* (Kunth) Kük. = ***Cyperus prolifer*** Lam.
Cyperus prolixus var. *vialis* (Ridl.) Kük. = ***Cyperus prolixus*** Kunth
Cyperus proteinolepis Steud. = ***Cyperus conglomeratus*** Rottb. subsp. ***conglomeratus***
Cyperus proteinolepis Boeck. = ***Cyperus rotundus*** L. subsp. ***rotundus***
Cyperus proteinolepis var. *pumilus* Boeck. = ***Cyperus conglomeratus*** Rottb. subsp. ***conglomeratus***
Cyperus protractus Link = ***Cyperus difformis*** L.
Cyperus protractus Delile = ***Cyperus fuscus*** L.
Cyperus proximus Steud. = ***Cyperus alternifolius*** subsp. ***flabelliformis*** Kük.
Cyperus psammophilus var. *gracillimus* Cherm. = ***Cyperus psammophilus*** Cherm.
Cyperus pseudobromoides Boeck. = ***Pycreus diaphanus*** (Schrad. ex Roem. & Schult.) S.S.Hooper & T.Koyama
Cyperus pseudobrunneus var. *continentalis* Kük. = ***Cyperus pseudobrunneus*** (C.B.Clarke ex Cherm.) Kük.
Cyperus pseudobrunneus var. *hilsenbergii* Kük. = ***Cyperus pseudobrunneus*** (C.B.Clarke ex Cherm.) Kük.
Cyperus pseudobrunneus var. *pseudobojeri* Kük. = ***Cyperus pseudobrunneus*** (C.B.Clarke ex Cherm.) Kük.
Cyperus pseudocallistus Kük. = ***Cyperus vestitus*** Hochst. ex C.Krauss
Cyperus pseudodiaphanus (S.S.Hooper) Lye = ***Pycreus pseudodiaphanus*** S.S.Hooper
Cyperus pseudogiganteus Steud. = ***Cyperus giganteus*** Vahl
Cyperus pseudohaspan Makino = ***Cyperus tenuispica*** Steud.
Cyperus pseudohildebrandtii Kük. = ***Pycreus hildebrandtii*** C.B.Clarke
Cyperus pseudokyllingioides Kük. = ***Courtoisina cyperoides*** (Roxb.) Soják
Cyperus pseudokyllingioides var. *africanus* (C.B.Clarke) Kük. = ***Courtoisina cyperoides*** (Roxb.) Soják
Cyperus pseudoleptocladus var. *polycarpus* Kük. = ***Cyperus pseudoleptocladus*** Kük.
Cyperus pseudoniveus Boeck. = ***Cyperus margaritaceus*** Vahl
Cyperus pseudosphacelatus Chiov. = ***Cyperus dilatatus*** Schumach.
Cyperus pseudosphacelatus Boeck. = ***Cyperus thyrsiflorus*** Jungh. ex Schltdl.
Cyperus pseudostrigosus Steud. = ***Cyperus odoratus*** L. subsp. ***odoratus***
Cyperus pseudosurinamensis Boeck. = ***Cyperus luzulae*** (L.) Retz.
Cyperus pseudovariegatus Boeck. = ***Cyperus rotundus*** L. subsp. ***rotundus***
Cyperus pseudovegetus var. *arenicola* (Steud.) Kük. = ***Cyperus reflexus*** Vahl
Cyperus pseudovegetus var. *megalanthus* Kük. = ***Cyperus megalanthus*** (Kük.) G.C.Tucker
Cyperus pseudovestitus f. *angustifolius* (Cherm.) Kük. = ***Cyperus pseudovestitus*** (C.B.Clarke) Kük.
Cyperus pseudovestitus var. *perrieri* (Cherm.) Kük. = ***Cyperus pseudovestitus*** (C.B.Clarke) Kük.
Cyperus pseudovestitus var. *polycarpus* Kük. = ***Cyperus pseudovestitus*** (C.B.Clarke) Kük.
Cyperus psilostachys Steud. = ***Cyperus distans*** L.f.
Cyperus psilostachys (C.B.Clarke) Kük. = ***Cyperus pluribracteatus*** (Kük.) Govaerts
Cyperus psilostachys f. *glabrescens* Kük. = ***Cyperus pluribracteatus*** (Kük.) Govaerts
Cyperus psilostachys var. *pluribracteatus* Kük. = ***Cyperus pluribracteatus*** (Kük.) Govaerts
Cyperus psilostachys var. *subrufus* Kük. = ***Cyperus pluribracteatus*** (Kük.) Govaerts
Cyperus pterygorrachis H.Lév. = ***Cyperus amuricus*** Maxim.
Cyperus pubescens J.Presl & C.Presl = ?
Cyperus pubescens Steud. = ***Bulbostylis capillaris*** (L.) Kunth ex C.B.Clarke subsp. ***capillaris***
Cyperus pubifolius Steud. = ***Bulbostylis ciliata*** (C.Presl) C.Presl ex Nees
Cyperus pubisquama Steud. = ***Cyperus diffusus*** Vahl
Cyperus pugioniformis A.Dietr. = ***Cyperus michelianus*** (L.) Delile subsp. ***michelianus***
Cyperus pulcher D.Don = ***Cyperus wallichianus*** Spreng.
Cyperus pulcher var. *prasinus* (Kunth) Kük. = ***Cyperus pulcher*** Thunb.
Cyperus pulcherrimus var. *rectiglumis* Suringar = ***Cyperus pulcherrimus*** Willd. ex Kunth
Cyperus pulicaris var. *explicatus* Kük. = ***Cyperus pulicaris*** Kük.
Cyperus pulvinatus Nees & Meyen = ***Pycreus pumilus*** (L.) Nees
Cyperus pumilio Nees = ***Isolepis cernua*** (Vahl) Roem. & Schult. var. ***cernua***
Cyperus pumilus L. = ***Pycreus pumilus*** (L.) Nees
Cyperus pumilus Rottb. = ***Pycreus sanguinolentus*** (Vahl) Nees subsp. ***sanguinolentus***
Cyperus pumilus f. *gracilis* (E.G.Camus) Kük. = ***Pycreus pumilus*** (L.) Nees
Cyperus pumilus var. *leucolepis* (C.B.Clarke) Kük. = ***Pycreus pumilus*** (L.) Nees
Cyperus pumilus var. *membranaceus* (Vahl) C.B.Clarke = ***Pycreus membranaceus*** (Vahl) Govind.
Cyperus pumilus var. *nervulosus* Kük. = ***Pycreus pumilus*** (L.) Nees
Cyperus pumilus var. *patens* (Vahl) Kük. = ***Pycreus pumilus*** (L.) Nees
Cyperus pumilus var. *prorepens* Kük. = ***Pycreus pumilus*** (L.) Nees
Cyperus pumilus var. *punctatus* C.B.Clarke = ***Pycreus pumilus*** (L.) Nees
Cyperus punctatifolius Steud. = ***Cyperus ligularis*** L.
Cyperus punctatus Lam. = ***Isolepis prolifera*** (Rottb.) R.Br.
Cyperus punctatus Roxb. = ***Pycreus pumilus*** (L.) Nees
Cyperus puncticulatus Aitch. = ***Cyperus serotinus*** Rottb.
Cyperus puncticulatus Vahl = ***Pycreus puncticulatus*** (Vahl) Nees
Cyperus puncticulatus var. *quinquagintiflorus* C.B.Clarke = ***Pycreus puncticulatus*** (Vahl) Nees
Cyperus punctorius Schrad. ex Boeck. = ?
Cyperus pungens Boeck. = ***Cyperus conglomeratus*** Rottb. subsp. ***conglomeratus***
Cyperus pungens Boeck. ex Aitch. & Hemsl. = ***Cyperus conglomeratus*** Rottb. subsp. ***conglomeratus***

Cyperus pungens Kunth = ?
Cyperus pungens (Vahl) Missbach & E.H.L.Krause = ***Schoenoplectus pungens*** (Vahl) Palla
Cyperus pungens var. *elatus* Boeck. = ***Cyperus macrorrhizus*** Nees
Cyperus pungens var. *multiculmis* Boeck. = ***Cyperus jeminicus*** Rottb.
Cyperus purpurascens Vahl = ***Cyperus planifolius*** Rich.
Cyperus purpureoglandulosus Mattf. & Kük. = ***Kyllinga sphaerocephala*** Boeck.
Cyperus purpureovaginatus Boeck. = ***Cyperus distans*** L.f.
Cyperus purpureovariegatus Boeck. = ***Cyperus rotundus*** L. subsp. ***rotundus***
Cyperus purpureus Boeck. = ***Cyperus schinzii*** Boeck.
Cyperus purshii Roem. & Schult. = ***Cyperus squarrosus*** L.
Cyperus pusillus Vahl = ***Pycreus pumilus*** (L.) Nees
Cyperus pustulatus var. *debilis* Kük. = ***Cyperus pustulatus*** Vahl
Cyperus pustulatus var. *djalonis* Kük. = ***Cyperus pustulatus*** Vahl
Cyperus pustulatus var. *tschinsendensis* (Turrill) Kük. = ***Cyperus derreilema*** Steud.
Cyperus pycnocephalus Steud. = ***Pycreus lanceolatus*** (Poir.) C.B.Clarke
Cyperus pygmaeus Cav. = ***Cyperus hamulosus*** M.Bieb.
Cyperus pygmaeus Rottb. = ***Cyperus michelianus*** subsp. ***pygmaeus*** (Rottb.) Asch. & Graebn.
Cyperus pygmaeus Nutt. = ***Cyperus squarrosus*** L.
Cyperus pygmaeus Lam. = ***Cyperus tenellus*** L.f.
Cyperus pygmaeus Retz. = ***Pycreus pumilus*** (L.) Nees
Cyperus pygmaeus var. *michelianus* (L.) Boeck. = ***Cyperus michelianus*** (L.) Delile
Cyperus pyrophilus Reinw. ex de Vriese = ***Cyperus pilosus*** Vahl
Cyperus quadriflorus Boeck. = ***Cyperus impubes*** Steud. var. ***impubes***
Cyperus quinqueflorus Hochst. ex Steud. = ***Cyperus pilosus*** Vahl
Cyperus quinquespicatus Boeck. = ***Cyperus rigens*** J.Presl & C.Presl subsp. ***rigens***
Cyperus quitensis Spreng. = ***Cyperus amabilis*** Vahl
Cyperus racemosus Retz. = ***Cyperus elatus*** L.
Cyperus racemosus B.Heyne ex Boeck. = ***Cyperus exaltatus*** Retz.
Cyperus racemosus Ten. = ***Cyperus longus*** L. subsp. ***longus***
Cyperus radians var. *compositus* Kük. = ***Cyperus radians*** Nees & Meyen ex Kunth
Cyperus radians var. *floribundus* (E.G.Camus) Kük. = ***Cyperus radians*** Nees & Meyen ex Kunth
Cyperus radians var. *griffithii* (Steud.) Kük. = ***Cyperus radians*** Nees & Meyen ex Kunth
Cyperus radiatus Vahl = ***Cyperus imbricatus*** Retz.
Cyperus radiatus var. *capitatus* Boeck. = ***Cyperus imbricatus*** Retz.
Cyperus radiatus var. *elongatus* Boeck. = ***Cyperus imbricatus*** Retz.
Cyperus radiatus var. *minor* Boeck. = ***Cyperus imbricatus*** Retz.
Cyperus radicans Missbach & E.H.L.Krause = ***Scirpus radicans*** Schkuhr
Cyperus radicosus Sm. = ***Cyperus rotundus*** L. subsp. ***rotundus***
Cyperus raiateensis J.W.Moore = ***Cyperus odoratus*** L. subsp. ***odoratus***
Cyperus ramosi Kük. = ***Cyperus zollingeri*** Steud.
Cyperus rancoanus Phil. = ***Cyperus laetus*** J.Presl & C.Presl
Cyperus ranko Steud. = ***Isolepis ranko*** (Steud.) Vegetti
Cyperus raphiostachys Link = ***Cyperus odoratus*** L. subsp. ***odoratus***
Cyperus rarissimus Steud. = ***Bulbostylis rarissima*** (Steud.) C.B.Clarke
Cyperus rectangularis (Kük.) Bennet = ***Cyperus microiria*** Steud.
Cyperus recurvatus Willd. ex Kunth = ***Cyperus cuspidatus*** Kunth
Cyperus recurvus Vahl = ***Cyperus cuspidatus*** Kunth
Cyperus reflexus f. *anterna* C.B.Clarke = ***Cyperus reflexus*** Vahl
Cyperus reflexus f. *capitatus* Osten = ***Cyperus reflexus*** Vahl
Cyperus reflexus f. *evolutus* Osten = ***Cyperus reflexus*** Vahl
Cyperus reflexus subf. *flaccidus* Osten = ***Cyperus reflexus*** Vahl
Cyperus reflexus var. *fraternus* (Kunth) Kuntze = ***Cyperus reflexus*** Vahl
Cyperus reflexus var. *genuinus* Osten = ***Cyperus reflexus*** Vahl
Cyperus reflexus subf. *glaucovirens* Osten = ***Cyperus reflexus*** Vahl
Cyperus reflexus var. *intricatus* (Schrad. ex Schult.) Kük. = ***Cyperus intricatus*** Schrad. ex Schult.
Cyperus reflexus var. *macrostachys* Boeck. = ***Cyperus reflexus*** Vahl
Cyperus refractus var. *antillanus* Kük. = ***Cyperus antillanus*** (Kük.) O'Neill
Cyperus refractus f. *grandis* Kük. = ***Cyperus refractus*** Engelm. ex Boeck.
Cyperus refractus var. *lancastriensis* (Porter) Kük. = ***Cyperus lancastriensis*** Porter
Cyperus regiomontanus var. *pallens* (Liebm.) Kük. = ***Cyperus thyrsiflorus*** Jungh. ex Schltdl.
Cyperus regiomontanus var. *pittieri* (Boeck. ex Kneuck.) Kük. = ***Cyperus thyrsiflorus*** Jungh. ex Schltdl.
Cyperus rehmannianus (C.B.Clarke) Boeck. ex Kuntze = ***Pycreus flavescens*** var. ***rehmannianus*** (C.B.Clarke) Govaerts
Cyperus rehmannianus var. *bathiei* (Cherm.) Kük. = ***Pycreus flavescens*** var. ***rehmannianus*** (C.B.Clarke) Govaerts
Cyperus rehmannianus f. *minor* Kük. = ***Pycreus flavescens*** var. ***rehmannianus*** (C.B.Clarke) Govaerts
Cyperus rehmannianus var. *rigidiculmis* Kük. = ***Pycreus flavescens*** var. ***rehmannianus*** (C.B.Clarke) Govaerts
Cyperus rehmannii Boiss. = ***Pycreus sanguinolentus*** (Vahl) Nees subsp. ***sanguinolentus***
Cyperus reineckei Boeck. = ***Cyperus odoratus*** L. subsp. ***odoratus***
Cyperus remotiflorus Kük. = ***Cyperus bellus*** Kunth
Cyperus renschii var. *scabridus* Lye = ***Cyperus renschii*** Boeck.
Cyperus repens Elliott = ***Cyperus esculentus*** L.
Cyperus reptans Boeck. = ***Cyperus laevigatus*** L. subsp. ***laevigatus***
Cyperus resinosus Hochst. ex Steud. = ***Cyperus iria*** L.
Cyperus retrofractus Engelm. ex Steud. = ***Cyperus refractus*** Engelm. ex Boeck.
Cyperus retrofractus var. *dipsaciformis* (Fernald) Kük. = ***Cyperus retrofractus*** (L.) Torr.
Cyperus retrofractus var. *hystricinus* (Fernald) Kük. = ***Cyperus hystricinus*** Fernald
Cyperus retrorsus var. *australis* (Lindm.) Kük. = ***Cyperus aggregatus*** var. ***gigas*** (Lindm.) Guagl.
Cyperus retrorsus var. *curtisii* (C.B.Clarke) Kük. = ***Cyperus ovatus*** Baldwin

Cyperus retrorsus var. *cylindricus* (Elliott) Fernald & Griscom = ***Cyperus retrorsus*** Chapm.
Cyperus retrorsus var. *deeringianus* (Britton & Small) Fernald & Griscom = ***Cyperus ovatus*** Baldwin
Cyperus retrorsus var. *multiflorens* Kük. = ***Cyperus retrorsus*** Chapm.
Cyperus retrorsus var. *nashii* (Britton) Fernald & Griscom = ***Cyperus retrorsus*** Chapm.
Cyperus retrorsus var. *robustus* (Boeck.) Kük. = ***Cyperus retrorsus*** Chapm.
Cyperus retrorsus var. *typicus* Fernald & Griscom = ***Cyperus retrorsus*** Chapm.
Cyperus retroversus Chapm. = ***Cyperus retrorsus*** Chapm.
Cyperus retusus A.Rich. = ***Pycreus macrostachyos*** (Lam.) J.Raynal subsp. ***macrostachyos***
Cyperus retusus Nees ex Steud. = ***Pycreus macrostachyos*** subsp. ***tremulus*** (Poir.) Lye
Cyperus retzii Nees = ***Cyperus rotundus*** subsp. ***retzii*** Kük.
Cyperus retzii Poir. =
Cyperus rhaphiostachys Humb. & Bonpl. ex Kunth = ***Cyperus digitatus*** Roxb. subsp. ***digitatus***
Cyperus rheedei F.Muell. = ?
Cyperus rhenanus Schrad. ex Kunth = ?
Cyperus rhigiophyllus Hochst. ex C.Krauss = [27 NAT]
Cyperus rhizomafragilis (Lye) Lye = ***Kyllinga rhizomafragilis*** Lye
Cyperus rhizomatosus (C.B.Clarke) Kük. = ***Pycreus rhizomatosus*** C.B.Clarke
Cyperus richardii Steud. = ***Kyllinga bulbosa*** P.Beauv.
Cyperus richardii var. *angustior* (C.B.Clarke) Kük. = ***Kyllinga bulbosa*** P.Beauv.
Cyperus richardii var. *oliganthus* (Cherm.) Kük. = ***Kyllinga bulbosa*** P.Beauv.
Cyperus ridleyi Mattf. & Kük. = ***Kyllinga pauciflora*** Ridl.
Cyperus riedelianus C.B.Clarke = ***Cyperus consors*** C.B.Clarke
Cyperus rigens subsp. *arechavaletae* (Boeck.) Pedersen = ***Cyperus laetus*** J.Presl & C.Presl
Cyperus rigens var. *capitatus* C.Presl = ***Cyperus rigens*** J.Presl & C.Presl subsp. ***rigens***
Cyperus rigens subsp. *cephalanthus* (Torr. & Hook.) Pedersen = ***Cyperus cephalanthus*** Torr. & Hook.
Cyperus rigens var. *latifolius* Kük. = ***Cyperus rigens*** J.Presl & C.Presl subsp. ***rigens***
Cyperus rigens var. *maximus* (Kük.) Pedersen = ***Cyperus laetus*** J.Presl & C.Presl
Cyperus rigens var. *rufidulus* (Steud.) Kük. = ***Cyperus rigens*** J.Presl & C.Presl subsp. ***rigens***
Cyperus rigens var. *tarijensis* Kük. = ***Cyperus tacnensis*** var. ***tarijensis*** (Kük.) Pedersen
Cyperus rigens var. *umbellatus* C.Presl = ***Cyperus rigens*** J.Presl & C.Presl subsp. ***rigens***
Cyperus rigens var. *weberbaueri* Kük. = ***Cyperus tacnensis*** var. ***weberbaueri*** (Kük.) Pedersen
Cyperus rigidifolius var. *intercedens* Kük. = ***Cyperus rigidifolius*** Steud.
Cyperus rigidulus Ridl. = ***Cyperus neoguinensis*** Kük.
Cyperus rigidulus Vahl = ?
Cyperus rigidus Vahl = ***Cyperus niveus*** var. ***leucocephalus*** (Kunth) Fosberg
Cyperus rionensis Boeck. = ***Cyperus ligularis*** L.
Cyperus riparius Schrad. ex Nees = ***Cyperus haspan*** L.
Cyperus riparius Salisb. = ?
Cyperus rivularis Steud. = ***Cyperus laevigatus*** L. subsp. ***laevigatus***
Cyperus rivularis Kunth = ***Pycreus bipartitus*** (Torr.) C.B.Clarke
Cyperus rivularis var. *acutatus* C.B.Clarke = ***Pycreus bipartitus*** (Torr.) C.B.Clarke
Cyperus rivularis f. *acutatus* (C.B.Clarke) Kük. = ***Pycreus bipartitus*** (Torr.) C.B.Clarke
Cyperus rivularis var. *depauperatus* C.B.Clarke = ***Pycreus bipartitus*** (Torr.) C.B.Clarke
Cyperus rivularis f. *depauperatus* (C.B.Clarke) Kük. = ***Pycreus bipartitus*** (Torr.) C.B.Clarke
Cyperus rivularis var. *elongatus* Boeck. = ***Pycreus bipartitus*** (Torr.) C.B.Clarke
Cyperus rivularis var. *elutus* C.B.Clarke = ***Pycreus bipartitus*** (Torr.) C.B.Clarke
Cyperus rivularis f. *elutus* (C.B.Clarke) Kük. = ***Pycreus bipartitus*** (Torr.) C.B.Clarke
Cyperus rivularis subsp. *lagunetto* (Steud.) Kük. = ***Pycreus bipartitus*** (Torr.) C.B.Clarke
Cyperus rivularis var. *lagunetto* (Steud.) O'Neill = ***Pycreus bipartitus*** (Torr.) C.B.Clarke
Cyperus rivularis f. *subacaulis* Kük. = ***Pycreus bipartitus*** (Torr.) C.B.Clarke
Cyperus robustior Gand. = [12 FRA]
Cyperus robustus Kunth = ***Cyperus drummondii*** Torr. & Hook.
Cyperus roestelii Kunth = ***Cyperus corymbosus*** Rottb.
Cyperus rothianus Schult. = ***Isolepis marginata*** (Thunb.) A.Dietr.
Cyperus rotundus var. *acutus* Boeck. = ***Cyperus rotundus*** L. subsp. ***rotundus***
Cyperus rotundus var. *amaliae* C.B.Clarke = ***Cyperus rotundus*** L. subsp. ***rotundus***
Cyperus rotundus subsp. *brevibracteatus* (Legrand) M.Laínz = ***Cyperus rotundus*** L. subsp. ***rotundus***
Cyperus rotundus var. *carinalis* Benth. = ***Cyperus rotundus*** L. subsp. ***rotundus***
Cyperus rotundus var. *carinatus* F.M.Bailey = ***Cyperus rotundus*** L. subsp. ***rotundus***
Cyperus rotundus var. *centiflorus* C.B.Clarke = ***Cyperus rotundus*** L. subsp. ***rotundus***
Cyperus rotundus var. *comosus* (Sm.) Nyman = ***Cyperus rotundus*** L. subsp. ***rotundus***
Cyperus rotundus f. *comosus* (Sm.) Kük. = ***Cyperus rotundus*** L. subsp. ***rotundus***
Cyperus rotundus f. *contractus* Kük. = ***Cyperus rotundus*** L. subsp. ***rotundus***
Cyperus rotundus f. *depallescens* Ekman & Kük. = ***Cyperus rotundus*** L. subsp. ***rotundus***
Cyperus rotundus var. *disruptus* (C.B.Clarke) Kük. = ***Cyperus rotundus*** subsp. ***retzii*** Kük.
Cyperus rotundus var. *elongatus* Boeck. = ***Cyperus rotundus*** L. subsp. ***rotundus***
Cyperus rotundus var. *fenzelianus* (Steud.) El-Hadidy = ***Cyperus longus*** var. ***pallidus*** Boeck.
Cyperus rotundus var. *hydra* (Michx.) A.Gray = ***Cyperus rotundus*** L. subsp. ***rotundus***
Cyperus rotundus var. *inconspicuus* Nyman = ***Cyperus rotundus*** L. subsp. ***rotundus***
Cyperus rotundus f. *inconspicuus* (Gennari) K.Richt. = ***Cyperus rotundus*** L. subsp. ***rotundus***
Cyperus rotundus f. *latifolius* Kük. = ***Cyperus rotundus*** L. subsp. ***rotundus***
Cyperus rotundus f. *latimarginatus* Kük. = ***Cyperus rotundus*** L. subsp. ***rotundus***
Cyperus rotundus var. *macrostachyus* Boiss. = ***Cyperus rotundus*** L. subsp. ***rotundus***
Cyperus rotundus var. *major* Parl. = ***Cyperus rotundus*** L. subsp. ***rotundus***

Cyperus rotundus var. *nubicus* (C.B.Clarke) Kük. = ***Cyperus rotundus*** subsp. ***retzii*** Kük.
Cyperus rotundus var. *pallidus* Benth. = ***Cyperus rotundus*** subsp. ***retzii*** Kük.
Cyperus rotundus var. *pendulus* Nees = ***Cyperus bulbosus*** Vahl
Cyperus rotundus var. *platystachys* Bojer ex C.B.Clarke = ***Cyperus rotundus*** L. subsp. ***rotundus***
Cyperus rotundus var. *procerula* C.B.Clarke = ***Cyperus rotundus*** L. subsp. ***rotundus***
Cyperus rotundus var. *pseudotuberosus* Cherm. = ***Cyperus mitis*** Steud.
Cyperus rotundus var. *salsolus* C.B.Clarke = ***Cyperus rotundus*** L. subsp. ***rotundus***
Cyperus rotundus var. *spadiceus* Boeck. = ***Cyperus rotundus*** L. subsp. ***rotundus***
Cyperus rotundus var. *subcapitatus* (C.B.Clarke) Kük. = ***Cyperus rotundus*** L. subsp. ***rotundus***
Cyperus rotundus var. *taylorii* (C.B.Clarke) Kük. = ***Cyperus rotundus*** L. subsp. ***rotundus***
Cyperus rotundus subsp. *tuberosus* (Rottb.) Kük. = ***Cyperus tuberosus*** Rottb.
Cyperus roxburghianus C.Presl = ***Cyperus laevigatus*** L. subsp. ***laevigatus***
Cyperus roxburghii A.Dietr. = ***Cyperus laevigatus*** L. subsp. ***laevigatus***
Cyperus roxburghii Nees = ***Cyperus platyphyllus*** Roem. & Schult.
Cyperus roylei Arn. = ***Cyperus imbricatus*** Retz.
Cyperus rubicundus Willd. ex Link = ***Cyperus rotundus*** L. subsp. ***rotundus***
Cyperus rubiginosus Hook.f. = ***Cyperus elegans*** subsp. ***rubiginosus*** (Hook.f.) Eliasson
Cyperus rubiginosus var. *cornutus* (Andersson) Kük. = ***Cyperus elegans*** subsp. ***rubiginosus*** (Hook.f.) Eliasson
Cyperus rubriglumosus Govind. = ***Pycreus rubriglumosus*** (Govind.) P.Singh & V.Singh
Cyperus rubromarginatus Meinsh. = ?
Cyperus rubromarginatus (E.G.Camus) Drobow = ***Pycreus sanguinolentus*** (Vahl) Nees subsp. ***sanguinolentus***
Cyperus rubrotinctus (Cherm.) Kük. = ***Cyperus longibracteatus*** (Cherm.) Kük.
Cyperus rubroviridis Cherm. = ***Cyperus zollingeri*** Steud.
Cyperus rubroviridis var. *unicapitatus* Kük. = ***Cyperus zollingeri*** Steud.
Cyperus ruderalis Uittien = ***Cyperus trailii*** C.B.Clarke
Cyperus rudioi Boeck. = ***Cyperus rotundus*** L. subsp. ***rotundus***
Cyperus rudioi var. *minor* Boeck. = ***Cyperus rotundus*** L. subsp. ***rotundus***
Cyperus rufescens Torr. & Hook. = ***Cyperus reflexus*** Vahl
Cyperus rufescens Schrad. ex Steud. = ?
Cyperus rufescens var. *denticarinatus* Britton = ***Cyperus acuminatus*** Torr. & Hook.
Cyperus ruficomus Buckley = ***Cyperus esculentus*** L.
Cyperus rufidulus Steud. = ***Cyperus rigens*** J.Presl & C.Presl subsp. ***rigens***
Cyperus rufinus Liebm. = ***Cyperus odoratus*** L. subsp. ***odoratus***
Cyperus rufus Nees ex Kunth = ***Cyperus compactus*** Retz.
Cyperus rufus Kunth = ***Cyperus ligularis*** L.
Cyperus rufus (Huds.) Missbach & E.H.L.Krause = ***Blysmus rufus*** (Huds.) Link
Cyperus ruizianus Boeck. = ***Cyperus hermaphroditus*** (Jacq.) Standl.
Cyperus rupestris var. *amnicola* (Kunth) Kük. = ***Cyperus rupestris*** Kunth
Cyperus rupestris var. *parvinux* (C.B.Clarke) Kük. = ***Cyperus rupestris*** Kunth
Cyperus rusbyi Britton = ***Cyperus fendlerianus*** Boeck.
Cyperus rusbyi Britton = ***Cyperus sphaerolepis*** Boeck.
Cyperus rutilans (C.B.Clarke) Maiden & Betche = ***Cyperus lhotskyanus*** Beck
Cyperus sabaudus Billet ex Gave = ***Cyperus fuscus*** L.
Cyperus sabulicolus Ridl. = ***Cyperus tenax*** Boeck.
Cyperus sabulosus (Nees) Boeck. = ***Pycreus macrostachyos*** (Lam.) J.Raynal subsp. ***macrostachyos***
Cyperus salzmannii Steud. = ***Cyperus cuspidatus*** Kunth
Cyperus samoensis Boeck. = ***Cyperus cyperinus*** (Retz.) Suringar
Cyperus sanctae-crucis Liebm. = ***Cyperus odoratus*** L. subsp. ***odoratus***
Cyperus sandwicensis var. *pseudoprescottianus* Kük. = ***Cyperus sandwicensis*** Kük.
Cyperus sanguineoater var. *floribundus* Kük. = ***Cyperus sanguineoater*** Boeck.
Cyperus sanguineofuscus Nees = ***Cyperus lucidus*** R.Br.
Cyperus sanguineofuscus Hook.f. = ?
Cyperus sanguineus Balb. ex Steud. = ***Cyperus planifolius*** Rich.
Cyperus sanguinolentus Vahl = ***Pycreus sanguinolentus*** (Vahl) Nees
Cyperus sanguinolentus var. *areolatus* (R.Br.) Kük. = ***Pycreus sanguinolentus*** (Vahl) Nees subsp. ***sanguinolentus***
Cyperus sanguinolentus f. *cyrtostachys* (Miq.) Kük. = ***Pycreus sanguinolentus*** var. ***cyrtostachys*** (Miq.) L.K.Dai
Cyperus sanguinolentus subsp. *cyrtostachys* (Miq.) A.Kern. = ***Pycreus sanguinolentus*** var. ***cyrtostachys*** (Miq.) L.K.Dai
Cyperus sanguinolentus f. *flaccidulus* (Boeck.) Kük. = ***Pycreus sanguinolentus*** (Vahl) Nees subsp. ***sanguinolentus***
Cyperus sanguinolentus f. *humilis* (Miq.) Kük. = ***Pycreus sanguinolentus*** (Vahl) Nees subsp. ***sanguinolentus***
Cyperus sanguinolentus var. *korshinskyi* (Meinsh.) Kük. = ***Pycreus sanguinolentus*** (Vahl) Nees subsp. ***sanguinolentus***
Cyperus sanguinolentus f. *melanocephalus* (Miq.) Kük. = ***Pycreus sanguinolentus*** subsp. ***melanocephalus*** (Miq.) Karthik.
Cyperus sanguinolentus subsp. *melanocephalus* (Miq.) A.Kern. = ***Pycreus sanguinolentus*** subsp. ***melanocephalus*** (Miq.) Karthik.
Cyperus sanguinolentus var. *micronux* (C.B.Clarke) Kük. = ***Pycreus sanguinolentus*** (Vahl) Nees subsp. ***sanguinolentus***
Cyperus sanguinolentus subsp. *nairobiensis* (Lye) Lye = ***Pycreus sanguinolentus*** subsp. ***nairobiensis*** Lye
Cyperus sanguinolentus f. *neurotropis* (Steud.) Kük. = ***Pycreus sanguinolentus*** (Vahl) Nees subsp. ***sanguinolentus***
Cyperus sanguinolentus var. *pratorum* (Korotky) Kük. = ***Pycreus pratorum*** (Korotky) Schischk.
Cyperus sanguinolentus f. *rubromarginatus* (Schrenk) Kük. = ***Pycreus sanguinolentus*** (Vahl) Nees subsp. ***sanguinolentus***
Cyperus sanguinolentus subsp. *sacer* Korlahalli = ***Pycreus sanguinolentus*** (Vahl) Nees subsp. ***sanguinolentus***
Cyperus sanguinolentus f. *setaceus* Kük. = ***Pycreus sanguinolentus*** (Vahl) Nees subsp. ***sanguinolentus***
Cyperus sanguinolentus var. *teysmannii* (Boeck.) Kük. = ***Pycreus sanguinolentus*** (Vahl) Nees subsp. ***sanguinolentus***

Cyperus sanguinolentus subsp. *teysmannii* (Boeck.) A.Kern. = ***Pycreus sanguinolentus*** (Vahl) Nees subsp. ***sanguinolentus***
Cyperus santonici Rottb. = ***Cyperus iria*** L.
Cyperus sarzogonensis J.Presl & C.Presl = ***Cyperus diffusus*** Vahl
Cyperus saturatus (Donn.Sm.) C.B.Clarke = ***Cyperus miliifolius*** Poepp. & Kunth
Cyperus scaberrimus Nees = ***Cyperus spectabilis*** Link var. ***spectabilis***
Cyperus scaberrimus Boeck. = ***Cyperus virens*** Michx. var. ***virens***
Cyperus schaffneri Boeck. = ***Karinia mexicana*** (C.B.Clarke ex Britton) Reznicek & McVaugh
Cyperus schenckianus Boeck. = ***Cyperus intricatus*** Schrad. ex Schult.
Cyperus schimperianus var. *minor* Boeck. = ***Cyperus schimperianus*** Steud.
Cyperus schlechteri C.B.Clarke = ***Cyperus niveus*** var. ***leucocephalus*** (Kunth) Fosberg
Cyperus schmitzianus Boeck. = ***Cyperus pycnostachyus*** (Kunth) Kunth
Cyperus schoenoides Griseb. = ***Cyperus capitatus*** Vand.
Cyperus schoenoides (R.Br.) Poir. = ***Fimbristylis squarrulosa*** F.Muell.
Cyperus schomburgkianus var. *leucanthus* (Schrad. ex Nees) Kük. = ***Cyperus schomburgkianus*** Nees
Cyperus schomburgkianus var. *trilobatus* Kük. = ***Cyperus schomburgkianus*** Nees
Cyperus schottianus auct. = ***Cyperus congestus*** Vahl
Cyperus schottii (Nees) D.Dietr. = ***Cyperus prolixus*** Kunth
Cyperus schraderi J.F.Macbr. = ***Cyperus friburgensis*** Boeck.
Cyperus schraderi Herter = ***Cyperus intricatus*** Schrad. ex Schult.
Cyperus schraderianus Mart. ex Steud. = ***Cyperus distans*** L.f.
Cyperus schweinfurthianus Boeck. = ***Cyperus tenuiculmis*** var. ***schweinfurthianus*** (Boeck.) S.S.Hooper
Cyperus schweinitzii var. *debilis* Britton = ***Cyperus sphaerolepis*** Boeck.
Cyperus scirpinus Salzm. ex Steud. = ***Fimbristylis bahiensis*** Steud.
Cyperus scirpoides Vahl = ***Cyperus crassipes*** Vahl
Cyperus scirpoides J.Presl & C.Presl = ***Cyperus diffusus*** Vahl
Cyperus scirpoides R.Br. = ***Cyperus esculentus*** L.
Cyperus scleropoae Lojac. = ***Cyperus serotinus*** Rottb.
Cyperus scoparius Poir. = [29 MDG] Perhaps identical with Cyperus platyphyllus.
Cyperus scopellatus Rich. = ***Pycreus polystachyos*** (Rottb.) P.Beauv. subsp. ***polystachyos***
Cyperus seemannianus var. *altus* Kük. = ***Cyperus seemannianus*** Boeck.
Cyperus seemannianus var. *monostachys* (Boeck.) Kük. = ***Cyperus seemannianus*** Boeck.
Cyperus seemannianus var. *perdensus* Kük. = ***Cyperus seemannianus*** Boeck.
Cyperus selloanus Boeck. = ***Pycreus unioloides*** (R.Br.) Urb.
Cyperus sellowianus (Kunth) T.Koyama = ***Lipocarpha maculata*** (Michx.) Torr.
Cyperus sellowii Link = ***Cyperus reflexus*** Vahl
Cyperus semiangulatus Boeck. = ***Cyperus fastigiatus*** Rottb.
Cyperus semidives Steud. = ***Cyperus imbricatus*** Retz.
Cyperus seminudus Roxb. = ***Cyperus corymbosus*** Rottb.
Cyperus seminudus Zoll. & Moritzi = ?
Cyperus semiradiciflorus Boeck. = ***Cyperus glomeratus*** L.
Cyperus semiteres B.Heyne ex C.B.Clarke = ***Pycreus flavidus*** (Retz.) T.Koyama
Cyperus semitribrachiatus Boeck. = ***Cyperus mutisii*** (Kunth) Andersson
Cyperus semitrifidus var. *apricus* (Ridl.) Kük. = ***Cyperus semitrifidus*** Schrad.
Cyperus semitrifidus var. *multiglumis* (Turrill) Kük. = ***Cyperus semitrifidus*** Schrad.
Cyperus semitrifidus var. *sanguinolentus* (Nees) Kük. = ***Cyperus semitrifidus*** Schrad.
Cyperus senegalensis C.B.Clarke = ***Cyperus conglomeratus*** Rottb. subsp. ***conglomeratus***
Cyperus senegalensis (C.B.Clarke) Mattf. & Kük. = ***Kyllinga senegalensis*** C.B.Clarke
Cyperus septatus Steud. = ***Cyperus compactus*** Retz.
Cyperus serotinus f. *contortus* Ces. ex Asch. & Graebn. = ***Cyperus serotinus*** Rottb.
Cyperus serotinus f. *depauperatus* Kük. = ***Cyperus serotinus*** Rottb.
Cyperus serotinus f. *grandis* Kük. = ***Cyperus serotinus*** Rottb.
Cyperus serotinus var. *inundatus* (C.B.Clarke) Kük. = ***Cyperus serotinus*** Rottb.
Cyperus serpens Cherm. = ***Cyperus platycaulis*** Baker
Cyperus serra A.Rich. = ***Cyperus rubicundus*** Vahl
Cyperus serrae Boeck. = ***Cyperus rigens*** subsp. ***serrae*** (Boeck.) Pedersen
Cyperus serrae var. *elongata* Boeck. = ***Cyperus rigens*** subsp. ***serrae*** (Boeck.) Pedersen
Cyperus serrae var. *minor* Boeck. = ***Cyperus rigens*** subsp. ***serrae*** (Boeck.) Pedersen
Cyperus serrulatus S.Watson = ***Cyperus eragrostis*** Lam.
Cyperus serrulatus Vahl = ***Cyperus niveus*** var. ***leucocephalus*** (Kunth) Fosberg
Cyperus sertularinus Liebm. = ***Cyperus hermaphroditus*** (Jacq.) Standl.
Cyperus sertularinus var. *discigerus* (Liebm.) Kük. = ***Cyperus tenuis*** Sw.
Cyperus seslerioides var. *minor* C.B.Clarke = ***Cyperus seslerioides*** Kunth
Cyperus seslerioides f. *minor* C.B.Clarke ex Donn.Sm. = ***Cyperus seslerioides*** Kunth
Cyperus sesquiflorus (Torr.) Mattf. & Kük. = ***Kyllinga odorata*** Vahl subsp. ***odorata***
Cyperus sesquiflorus subsp. *appendiculatus* (K.Schum.) Lye = ***Kyllinga odorata*** subsp. ***appendiculata*** (K.Schum.) Lye
Cyperus sesquiflorus f. *bulbiferus* (Kük.) Kük. = ***Kyllinga odorata*** Vahl subsp. ***odorata***
Cyperus sesquiflorus var. *cylindricus* (Nees) Kük. = ***Kyllinga odorata*** subsp. ***cylindrica*** (Nees) T.Koyama
Cyperus sesquiflorus subsp. *cylindricus* (Nees) T.Koyama = ***Kyllinga odorata*** subsp. ***cylindrica*** (Nees) T.Koyama
Cyperus sesquiflorus f. *denticulatus* Kük. = ***Kyllinga odorata*** Vahl subsp. ***odorata***
Cyperus sesquiflorus f. *dunensis* (Osten) Kük. = ***Kyllinga odorata*** Vahl subsp. ***odorata***
Cyperus sesquiflorus f. *elongatus* (Boeck.) Kük. = ***Kyllinga odorata*** Vahl subsp. ***odorata***
Cyperus sesquiflorus var. *fallax* (Kük.) Kük. = ***Kyllinga odorata*** Vahl subsp. ***odorata***
Cyperus sesquiflorus f. *globosus* Kük. = ***Kyllinga odorata*** Vahl subsp. ***odorata***
Cyperus sesquiflorus f. *gracilis* (Boeck.) Kük. = ***Kyllinga odorata*** Vahl subsp. ***odorata***
Cyperus sesquiflorus f. *latifolius* (Boeck.) Kük. = ***Kyllinga planiculmis*** C.B.Clarke ex Cherm.

Cyperus sesquiflorus var. *major* (C.B.Clarke) Kük. = ***Kyllinga odorata*** subsp. ***appendiculata*** (K.Schum.) Lye
Cyperus sesquiflorus var. *minor* (Boeck.) Kük. = ***Kyllinga odorata*** Vahl subsp. ***odorata***
Cyperus sesquiflorus var. *pluriceps* (Kük.) Kük. = ***Kyllinga odorata*** Vahl subsp. ***odorata***
Cyperus sesquiflorus f. *spinulosus* Kük. = ***Kyllinga odorata*** Vahl subsp. ***odorata***
Cyperus sesquiflorus var. *subtriceps* (Nees) T.Koyama = ***Kyllinga odorata*** subsp. ***cylindrica*** (Nees) T.Koyama
Cyperus setaceus Retz. = ***Eleocharis retroflexa*** subsp. ***chaetaria*** (Roem. & Schult.) T.Koyama
Cyperus setaceus (L.) Missbach & E.H.L.Krause = ***Isolepis setacea*** (L.) R.Br.
Cyperus setaceus Wulfen = ***Pycreus flavescens*** (L.) P.Beauv. ex Rchb. subsp. ***flavescens***
Cyperus setifolius D.Don = ***Cyperus cuspidatus*** Kunth
Cyperus setifolius Torr. ex C.B.Clarke = ***Cyperus grayi*** Torr.
Cyperus setiformis Korsh. = ***Pycreus diaphanus*** (Schrad. ex Roem. & Schult.) S.S.Hooper & T.Koyama
Cyperus setigerus var. *fulvescens* (C.B.Clarke) Kük. = ***Cyperus setigerus*** Torr. & Hook.
Cyperus setigerus var. *hallii* (Britton) Kük. = ***Cyperus setigerus*** Torr. & Hook.
Cyperus shimadai Ohwi = ***Cyperus radians*** Nees & Meyen ex Kunth
Cyperus sieberi Kunth = ***Cyperus fulvus*** R.Br.
Cyperus sieberianus (Nees ex C.B.Clarke) K.Schum. = ***Cyperus cyperoides*** (L.) Kuntze subsp. ***cyperoides***
Cyperus sieberianus Link = ***Cyperus esculentus*** L.
Cyperus silletensis Thwaites = ***Cyperus pulcherrimus*** Willd. ex Kunth
Cyperus sinensis Debeaux = ***Cyperus radians*** Nees & Meyen ex Kunth
Cyperus sinensis var. *floribundus* E.G.Camus = ***Cyperus radians*** Nees & Meyen ex Kunth
Cyperus sintenisii Boeck. = ***Cyperus ligularis*** L.
Cyperus smithianus Ridl. = ***Pycreus smithianus*** (Ridl.) C.B.Clarke
Cyperus smithii McClean = ***Cyperus leptocladus*** Kunth
Cyperus smithii Schrad. = ***Cyperus textilis*** Thunb.
Cyperus socialis C.B.Clarke = ***Cyperus pseudopilosus*** (T.Durand & De Wild.) Govaerts
Cyperus solidus var. *elatior* Kunth = ***Cyperus dactyliformis*** Boeck.
Cyperus solmsii Boeck. = ***Cyperus planifolius*** Rich.
Cyperus solutus Steud. = ***Cyperus cuspidatus*** Kunth
Cyperus somalicus Gand. = ***Cyperus niveus*** var. ***leucocephalus*** (Kunth) Fosberg
Cyperus sonderi J.A.Schmidt = ***Pycreus polystachyos*** (Rottb.) P.Beauv. subsp. ***polystachyos***
Cyperus songeensis (Lye) Lye = ***Kyllinga songeensis*** Lye
Cyperus sorostachyus Boeck. = ***Cyperus pulchellus*** R.Br.
Cyperus spadiceus Lam. = ***Cyperus stoloniferus*** Retz.
Cyperus spaniophyllus Steud. = ***Cyperus malaccensis*** Lam. var. ***malaccensis***
Cyperus sparsiflorus Baldwin = ***Cyperus glomeratus*** L.
Cyperus sparteus Schrad. ex Nees = ***Cyperus reflexus*** Vahl
Cyperus spathaceus (L.) L. = ***Dulichium arundinaceum*** (L.) Britton var. ***arundinaceum***
Cyperus speciosus Lojac. = ***Cyperus longus*** L. subsp. ***longus***
Cyperus speciosus Vahl = ***Cyperus odoratus*** L. subsp. ***odoratus***
Cyperus speciosus Torr. = ***Cyperus odoratus*** L. subsp. ***odoratus***
Cyperus speciosus var. *ferruginescens* (Boeck.) Britton = ***Cyperus odoratus*** L. subsp. ***odoratus***
Cyperus speciosus var. *parvus* Britton = ***Cyperus odoratus*** L. subsp. ***odoratus***
Cyperus speciosus var. *squarrosus* Britton = ***Cyperus odoratus*** L. subsp. ***odoratus***
Cyperus spectabilis var. *coarctatus* Boeck. = ***Cyperus manimae*** Kunth var. ***manimae***
Cyperus spectabilis var. *filiformis* Boeck. = ***Cyperus spectabilis*** Link var. ***spectabilis***
Cyperus spectabilis var. *parryi* (Britton ex C.B.Clarke) Kük. = ***Cyperus spectabilis*** Link var. ***spectabilis***
Cyperus spectabilis var. *scaberrimus* (Nees) Boeck. = ***Cyperus spectabilis*** Link var. ***spectabilis***
Cyperus sphacelatus var. *tenuior* C.B.Clarke = ***Cyperus zollingeri*** Steud.
Cyperus sphaeranthelus Chiov. = ***Cyperus meeboldii*** Kük.
Cyperus sphaerocephalus Vahl = ***Cyperus niveus*** var. ***leucocephalus*** (Kunth) Fosberg
Cyperus sphaerocephalus var. *leucocephalus* Kunth = ***Cyperus niveus*** var. ***leucocephalus*** (Kunth) Fosberg
Cyperus sphaerospermoides Cherm. = ***Cyperus denudatus*** L.f.
Cyperus sphaerostachys Link = ***Cyperus luzulae*** (L.) Retz.
Cyperus spherobolos Poir. = ***Cyperus sporobolus*** R.Br.
Cyperus spicatocapitatus Steud. = ***Cyperus ligularis*** L.
Cyperus spicatus J.Presl & C.Presl = ***Cyperus imbricatus*** Retz.
Cyperus spicigerus Vahl = ***Cyperus schoenomorphus*** Steud.
Cyperus spiculatus Alph.Wood = ***Cyperus erythrorrhizos*** Muhl.
Cyperus spiculosus Rchb. = ***Cyperus congestus*** Vahl
Cyperus spiculosus F.N.Williams = ***Pycreus fluminalis*** (Ridl.) Rendle
Cyperus spinuliferus Boeck. = ***Pycreus unioloides*** (R.Br.) Urb.
Cyperus spinulosus Roxb. = ***Cyperus compactus*** Retz.
Cyperus spissiflorus (C.B.Clarke) K.Schum. = ***Pycreus spissiflorus*** C.B.Clarke
Cyperus spongiosocoriaceus Boeck. = ***Cyperus javanicus*** Houtt.
Cyperus spongiosovaginatus Boeck. = ***Cyperus digitatus*** Roxb. subsp. ***digitatus***
Cyperus sporobolus var. *microcephalus* (R.Br.) Benth. = ***Cyperus microcephalus*** R.Br.
Cyperus sporobolus var. *sexflorus* (R.Br.) Benth. = ***Cyperus sexflorus*** R.Br.
Cyperus spretus Steud. = ***Cyperus denudatus*** L.f.
Cyperus squalidus Liebm. = ***Pycreus flavescens*** (L.) P.Beauv. ex Rchb. subsp. ***flavescens***
Cyperus squamulatus Steud. = ***Cyperus distans*** L.f.
Cyperus squarrosulus (Cherm.) Kük. = ***Pycreus squarrosulus*** Cherm.
Cyperus squarrosus F.Muell. = ***Cyperus conglomeratus*** Rottb. subsp. ***conglomeratus***
Cyperus squarrosus var. *congestus* Benth. = ***Cyperus squarrosus*** L.
Cyperus squarrosus var. *cylindraceus* Benth. = ***Cyperus squarrosus*** L.
Cyperus squarrosus var. *runyonii* (O'Neill) S.D.Jones & Wipff = ***Cyperus squarrosus*** L.
Cyperus squarrosus var. *stenocarpus* F.Muell. = ***Cyperus conglomeratus*** Rottb. subsp. ***conglomeratus***
Cyperus st.-crucei Liebm. = ?

Cyperus stachyphorus K.Koch = ?
Cyperus stauntonii (C.B.Clarke) Ohwi = ***Cyperus nipponicus*** Franch. & Sav.
Cyperus stellatus Link = ***Cyperus haspan*** L.
Cyperus stellatus Rudge = ***Cyperus odoratus*** L. subsp. ***odoratus***
Cyperus stenolepis Torr. = ***Cyperus strigosus*** L.
Cyperus stenophyllus var. *ornans* (J.V.Suringar) Kük. = ***Cyperus stenophyllus*** J.V.Suringar
Cyperus stenostachyus Benth. = ***Cyperus tenuiflorus*** Rottb.
Cyperus stenostachyus var. *indica* C.B.Clarke = ***Cyperus mitis*** Steud.
Cyperus steudelianus Boeck. = ***Cyperus cyperoides*** (L.) Kuntze subsp. ***cyperoides***
Cyperus stigmatosus Steud. = ***Cyperus javanicus*** Houtt.
Cyperus stolonifer Willd. ex Kunth = ***Cyperus bulbosus*** Vahl
Cyperus stoloniferus var. *pallidus* Boeck. = ***Cyperus rotundus*** L. subsp. ***rotundus***
Cyperus stramineus Link = ***Pycreus flavescens*** (L.) P.Beauv. ex Rchb. subsp. ***flavescens***
Cyperus stramineus Nees = ***Pycreus stramineus*** C.B.Clarke
Cyperus straminicolor Cherm. = ***Cyperus procerus*** Rottb.
Cyperus straminicolor var. *betsileensis* Cherm. = ***Cyperus procerus*** Rottb.
Cyperus striatulus Vahl = ***Cyperus niveus*** var. ***leucocephalus*** (Kunth) Fosberg
Cyperus striatus Ruiz & Pav. = ***Bulbostylis striata*** (Ruiz & Pav.) H.Pfeiff.
Cyperus stricticulmis Govind. = ***Pycreus stricticulmis*** (Govind.) P.Singh & V.Singh
Cyperus strictus Lam. = ***Pycreus flavidus*** (Retz.) T.Koyama
Cyperus strictus Rottler ex C.B.Clarke = ***Pycreus polystachyos*** (Rottb.) P.Beauv. subsp. ***polystachyos***
Cyperus strictus Wight ex C.B.Clarke = ***Pycreus pumilus*** (L.) Nees
Cyperus strigosus var. *capitatus* Boeck. = ***Cyperus strigosus*** L.
Cyperus strigosus f. *capitatus* (Boeck.) Kük. = ***Cyperus strigosus*** L.
Cyperus strigosus f. *compositus* Britton = ***Cyperus strigosus*** L.
Cyperus strigosus var. *compositus* Britton = ***Cyperus strigosus*** L.
Cyperus strigosus var. *elongatus* (Torr.) Britton = ***Cyperus strigosus*** L.
Cyperus strigosus var. *gracilis* Britton = ***Cyperus lentiginosus*** Millsp. & Chase
Cyperus strigosus var. *hansenii* (Britton) Kük. = ***Cyperus strigosus*** L.
Cyperus strigosus var. *multiflorus* Geise = ***Cyperus × deamii*** O'Neill
Cyperus strigosus var. *naguensis* (Britton) O'Neill = ***Cyperus strigosus*** L.
Cyperus strigosus f. *praelongatus* (Steud.) Kük. = ***Cyperus strigosus*** L.
Cyperus strigosus var. *pygmaeus* Gatt. = ***Cyperus strigosus*** L.
Cyperus strigosus f. *robustior* Kunth = ***Cyperus strigosus*** L.
Cyperus strigosus var. *stenolepis* (Torr.) Kük. = ***Cyperus strigosus*** L.
Cyperus stuhlmannii C.B.Clarke ex K.Schum. = ***Cyperus usitatus*** var. ***stuhlmannii*** (C.B.Clarke ex K.Schum.) Lye
Cyperus stuppeus G.Forst. = ***Cyperus javanicus*** Houtt.
Cyperus suaveolens Boivin ex Cherm. = ***Queenslandiella hyalina*** (Vahl) Ballard
Cyperus subalatus Boeck. = ***Cyperus digitatus*** subsp. ***auricomus*** (Sieber ex Spreng.) Kük.
Cyperus subalatus Steud. = ***Cyperus pilosus*** Vahl
Cyperus subambiguus Kük. = ***Cyperus hermaphroditus*** (Jacq.) Standl.
Cyperus subambiguus var. *pallidicolor* Kük. = ***Cyperus pallidicolor*** (Kük.) G.C.Tucker
Cyperus subanceps Boeck. = [81]
Cyperus subaphyllus Boeck. ex Schinz = ***Cyperus laevigatus*** subsp. ***distachyos*** (All.) Maire & Weiller
Cyperus subarticulatus Nees & Meyen = ***Cyperus articulatus*** L.
Cyperus subbulbosus Roem. & Schult. = ***Cyperus congestus*** Vahl
Cyperus subcapitatus C.B.Clarke = ***Cyperus rotundus*** L. subsp. ***rotundus***
Cyperus subchoristachys C.B.Clarke = ***Cyperus glaber*** L.
Cyperus subenervius Steud. = ***Cyperus surinamensis*** Rottb. var. ***surinamensis***
Cyperus subintermedius Kük. = ***Pycreus flavescens*** subsp. ***intermedius*** (Rikli) Lye
Cyperus subintermedius var. *angustisquamatus* Kük. = ***Pycreus flavescens*** subsp. ***intermedius*** (Rikli) Lye
Cyperus subintermedius var. *vicinus* (Cherm.) Kük. = ***Pycreus flavescens*** subsp. ***intermedius*** (Rikli) Lye
Cyperus sublaevicarinatus Mattf. & Kük. = ***Kyllinga buchananii*** C.B.Clarke
Cyperus sublimis (C.B.Clarke) Dandy = ***Cyperus cyperoides*** (L.) Kuntze subsp. ***cyperoides***
Cyperus submacropus Kük. = ***Cyperus circumclusus*** (C.B.Clarke) Schweinf.
Cyperus submacropus var. *abbreviatus* Kük. = ***Cyperus circumclusus*** (C.B.Clarke) Schweinf.
Cyperus submacropus var. *calocephalus* A.Peter ex Kük. = ***Cyperus circumclusus*** (C.B.Clarke) Schweinf.
Cyperus submacropus var. *fuscofibrosus* A.Peter ex Kük. = ***Cyperus circumclusus*** (C.B.Clarke) Schweinf.
Cyperus submaculatus (Vahl) T.Koyama = ***Lipocarpha chinensis*** (Osbeck) J.Kern
Cyperus submonastichyus Steud. & Jard. = ***Cyperus laevigatus*** L. subsp. ***laevigatus***
Cyperus subnodosus Nees & Meyen = ***Cyperus articulatus*** L.
Cyperus subparadoxus Kük. = ***Alinula paradoxa*** (Cherm.) Goetgh. & Vorster
Cyperus subpinnatus var. *subrigidellus* Kük. = ***Cyperus subpinnatus*** Kük.
Cyperus subrigidulus Kük. = ***Cyperus neoguinensis*** Kük.
Cyperus subrotundus Llanos = ***Cyperus difformis*** L.
Cyperus substramineus Kük. = ***Pycreus stramineus*** C.B.Clarke
Cyperus subtrigonus (C.B.Clarke) Kük. = ***Pycreus subtrigonus*** C.B.Clarke
Cyperus subulatus Meyen = ***Cyperus meyenianus*** Kunth
Cyperus subulatus var. *confertus* Benth. = ***Cyperus bulbosus*** Vahl
Cyperus subulatus var. *confertus* Benth. = ***Cyperus subulatus*** R.Br.
Cyperus subumbellatus Kük. = ***Cyperus cylindrostachyus*** Boeck.
Cyperus subumbellatus var. *sessilispicatus* Kük. = ***Cyperus cylindrostachyus*** Boeck.
Cyperus subumbellatus var. *subglobosus* Kük. = ***Cyperus cylindrostachyus*** Boeck.
Cyperus subumbellatus var. *sublimis* (C.B.Clarke) Kük. = ***Cyperus cyperoides*** (L.) Kuntze subsp. ***cyperoides***

Cyperus subumbellatus var. *thomensis* (C.B.Clarke) Kük. = ***Cyperus cylindrostachyus*** Boeck.
Cyperus subuniflorus Britton = ***Cyperus retroflexus*** var. ***pumilus*** (Britton) R.Carter & S.D.Jones
Cyperus subxerophilus Kük. = ***Cyperus xerophilus*** Cherm.
Cyperus sulcinux C.B.Clarke = ***Pycreus sulcinux*** (C.B.Clarke) C.B.Clarke
Cyperus superatus C.B.Clarke = ***Cyperus trinervis*** R.Br.
Cyperus supinus (L.) Missbach & E.H.L.Krause = ***Schoenoplectiella supina*** (L.) Lye
Cyperus surinamensis var. *formosus* (Vahl) Kük. = ***Cyperus virens*** Michx.
Cyperus surinamensis var. *strictus* Kük. = ***Cyperus entrerianus*** Boeck.
Cyperus surinamensis var. *viridis* Boeck. = ***Cyperus surinamensis*** Rottb. var. ***surinamensis***
Cyperus swartzii var. *efoliatus* Kük. = ***Cyperus swartzii*** (A.Dietr.) Boeck. ex Kük.
Cyperus swartzii var. *elongatus* Kük. = ***Cyperus swartzii*** (A.Dietr.) Boeck. ex Kük.
Cyperus swartzii var. *granularis* (Desf. ex Link) Kük. = ***Cyperus swartzii*** (A.Dietr.) Boeck. ex Kük.
Cyperus swartzii var. *latifolius* Kük. = ***Cyperus swartzii*** (A.Dietr.) Boeck. ex Kük.
Cyperus sylvaticus (L.) Missbach & E.H.L.Krause = ***Scirpus sylvaticus*** L.
Cyperus sylvestris Ridl. = ***Cyperus laxus*** subsp. ***buchholzii*** (Boeck.) Lye
Cyperus sylvicola Ridl. = ***Cyperus renschii*** Boeck.
Cyperus syriacus Parl. = ***Cyperus papyrus*** L. subsp. ***papyrus***
Cyperus tabernaemontani (C.C.Gmel.) Missbach & E.H.L.Krause = ***Schoenoplectus tabernaemontani*** (C.C.Gmel.) Palla
Cyperus tabularis var. *ecklonii* (Boeck.) Kük. = ***Cyperus tabularis*** Schrad.
Cyperus tabularis var. *humilis* (C.B.Clarke) Kük. = ***Cyperus tabularis*** Schrad.
Cyperus tabularis var. *major* Nees = ***Cyperus dactyliformis*** Boeck.
Cyperus tabularis var. *zeyheri* (C.B.Clarke) Kük. = ***Cyperus tabularis*** Schrad.
Cyperus taitensis Steud. = ***Cyperus javanicus*** Houtt.
Cyperus tanaensis Kük. = ***Pycreus flavescens*** subsp. ***tanaensis*** (Kük.) Lye
Cyperus tanzaniae (Lye) Lye = ***Kyllinga tanzaniae*** Lye
Cyperus tataricus Less. = ***Cyperus pannonicus*** Jacq.
Cyperus taubertii Boeck. = ***Pycreus unioloides*** (R.Br.) Urb.
Cyperus taylorii C.B.Clarke = ***Cyperus rotundus*** L. subsp. ***rotundus***
Cyperus tegetiformis Roxb. ex Arn. = ***Cyperus corymbosus*** Rottb.
Cyperus tegetiformis Benth. = ***Cyperus malaccensis*** Lam. var. ***malaccensis***
Cyperus tegetum Roxb. = ***Cyperus pangorei*** Rottb.
Cyperus tegetum C.B.Clarke = ***Cyperus schimperianus*** Steud.
Cyperus tegetum var. *protracta* C.B.Clarke = ***Cyperus schimperianus*** Steud.
Cyperus tenax var. *actinostachys* (Welw. ex Ridl.) Kük. = ***Cyperus tenax*** Boeck.
Cyperus tenax var. *andongensis* (Ridl.) Kük. = ***Cyperus tenax*** Boeck.
Cyperus tenax var. *angustissimus* Kük. = ***Cyperus tenax*** Boeck.
Cyperus tenax var. *monroviensis* (Boeck.) Kük. = ***Cyperus tenax*** Boeck.
Cyperus tenax var. *pseudocastaneus* (Kük.) Kük. = ***Cyperus tenax*** Boeck.
Cyperus tenax var. *sabulicolus* (Ridl.) Kük. = ***Cyperus tenax*** Boeck.
Cyperus tendo Banks & Sol. ex Hook.f. = ***Schoenus tendo*** (Hook.f.) Banks & Sol. ex Hook.f.
Cyperus tenellus Torr. = ***Pycreus filicinus*** (Vahl) T.Koyama
Cyperus tenellus J.Presl & C.Presl = ***Pycreus fugax*** (Liebm.) C.D.Adams
Cyperus tenellus var. *micromegas* (Nees) F.Muell. = ***Cyperus tenellus*** L.f.
Cyperus tener Vahl = ?
Cyperus teneriffae Poir. = ***Cyperus rubicundus*** Vahl
Cyperus teneriffae var. *longimucronatus* Kük. = ***Cyperus rubicundus*** Vahl
Cyperus teneriffae f. *petraeus* (Hochst. ex Steud.) Kük. = ***Cyperus rubicundus*** Vahl
Cyperus teneristolon Mattf. & Kük. = ***Kyllinga pulchella*** Kunth
Cyperus teneristolon var. *robustior* (Kük.) Kük. = ***Kyllinga pulchella*** Kunth
Cyperus tenorei C.Presl = ***Cyperus esculentus*** L.
Cyperus tenorianus Schult. = ***Cyperus esculentus*** L.
Cyperus tenuicapitatus Steud. = ***Cyperus surinamensis*** Rottb. var. ***surinamensis***
Cyperus tenuicomus Miq. = ***Cyperus corymbosus*** Rottb.
Cyperus tenuiculmis Boeck. = ***Cyperus dilatatus*** Schumach.
Cyperus tenuiculus Kunth = ***Pycreus bipartitus*** (Torr.) C.B.Clarke
Cyperus tenuiflorus Elliott = ***Cyperus erythrorrhizos*** Muhl.
Cyperus tenuiflorus Roxb. = ***Cyperus mitis*** Steud.
Cyperus tenuifolius Walp. = ***Cyperus rotundus*** L. subsp. ***rotundus***
Cyperus tenuifolius T.L.Dai = [36]
Cyperus tenuifolius (Steud.) Dandy = ***Kyllinga pumila*** Michx.
Cyperus tenuior Engelm. = ***Cyperus odoratus*** L. subsp. ***odoratus***
Cyperus tenuis Muhl. = ***Pycreus polystachyos*** (Rottb.) P.Beauv. subsp. ***polystachyos***
Cyperus tenuis var. *aximensis* (C.B.Clarke) Kük. = ***Cyperus tenuis*** Sw.
Cyperus tenuis var. *brevior* Kük. = ***Cyperus tenuis*** Sw.
Cyperus tenuis var. *capillaris* (Sw.) Kük. = ***Cyperus nanus*** Willd. var. ***nanus***
Cyperus tenuis var. *eurystachys* (Ridl.) Kük. = ***Cyperus tenuis*** Sw.
Cyperus tenuis var. *fulvescens* Kük. = ***Cyperus tenuis*** Sw.
Cyperus tenuis var. *grandiceps* Kük. = ***Cyperus tenuis*** Sw.
Cyperus tenuis var. *lentiginosus* (Millsp. & Chase) Kük. = ***Cyperus lentiginosus*** Millsp. & Chase
Cyperus tenuis var. *luridus* (T.Durand & De Wild.) Kük. = ***Cyperus tenuis*** Sw.
Cyperus tenuispicatus Boeck. = ***Cyperus miliifolius*** Poepp. & Kunth
Cyperus tenuispiculatus var. *pseudatterimus* Kük. = ***Cyperus tenuispiculatus*** Boeck.
Cyperus tephrodes Schrad. ex Nees = ***Cyperus dichromeniformis*** Kunth
Cyperus teretifolius A.Rich. = ***Cyperus laevigatus*** L. subsp. ***laevigatus***
Cyperus teretifructus Steud. = ***Pycreus polystachyos*** (Rottb.) P.Beauv. subsp. ***polystachyos***
Cyperus terminalis Marsh ex Steud. = ***Pycreus pumilus*** (L.) Nees
Cyperus tersus Link = ***Cyperus rubicundus*** Vahl

Cyperus tetragonus Hemsl. = ***Cyperus hermaphroditus*** (Jacq.) Standl.
Cyperus tetragonus var. *intercedens* Kük. = ***Cyperus tetragonus*** Elliott
Cyperus tetragonus var. *pringlei* (Britton) Kük. = ***Cyperus tetragonus*** Elliott
Cyperus tetraphyllus var. *monocephalus* Benth. = ***Cyperus disjunctus*** C.B.Clarke
Cyperus tetrastachyos Desf. = ***Cyperus rotundus*** L. subsp. ***rotundus***
Cyperus texensis Steud. = ***Pycreus polystachyos*** (Rottb.) P.Beauv. subsp. ***polystachyos***
Cyperus textilis F.Muell. = ***Cyperus vaginatus*** R.Br.
Cyperus textori Miq. = ***Cyperus microiria*** Steud.
Cyperus teysmannii Boeck. = ***Pycreus sanguinolentus*** (Vahl) Nees subsp. ***sanguinolentus***
Cyperus thermalis Dumort. = ***Cyperus longus*** L. subsp. ***longus***
Cyperus thouarsianus C.Presl = ***Cyperus dubius*** Rottb. subsp. ***dubius***
Cyperus thouarsii Kunth = ***Pycreus polystachyos*** (Rottb.) P.Beauv. subsp. ***polystachyos***
Cyperus thunbergii var. *gillii* (C.B.Clarke) Kük. = ***Cyperus thunbergii*** Vahl
Cyperus thunbergii var. *robustior* (C.B.Clarke) Kük. = ***Cyperus thunbergii*** Vahl
Cyperus thunbergii var. *trisumbellatus* (C.B.Clarke) Kük. = ***Cyperus thunbergii*** Vahl
Cyperus thyrsiflorus Boeck. = ***Cyperus ligularis*** L.
Cyperus thyrsiflorus var. *anceps* (Liebm.) Kük. = ***Cyperus thyrsiflorus*** Jungh. ex Schltdl.
Cyperus thyrsiflorus var. *macrior* F.Aresch. = ***Cyperus thyrsiflorus*** Jungh. ex Schltdl.
Cyperus thyrsiflorus var. *vegetior* F.Aresch. = ***Cyperus thyrsiflorus*** Jungh. ex Schltdl.
Cyperus tibialis (Poit. ex Ledeb.) Govaerts = ***Kyllinga tibialis*** Poit. ex Ledeb.
Cyperus tisserantii Cherm. = ***Cyperus niveus*** var. ***tisserantii*** (Cherm.) Lye
Cyperus toisensis Kuntze = ***Cyperus dactyliformis*** Boeck.
Cyperus tokiensis C.B.Clarke = ***Cyperus exaltatus*** Retz.
Cyperus tolucensis Kunth = ***Cyperus laxus*** Lam. subsp. ***laxus***
Cyperus tomaiophyllus var. *alpestris* (K.Schum.) Kük. = ***Cyperus tomaiophyllus*** K.Schum.
Cyperus tomaiophyllus var. *crassistolon* Kük. = ***Cyperus tomaiophyllus*** K.Schum.
Cyperus tomaiophyllus var. *laxiflorus* Kük. = ***Cyperus tomaiophyllus*** K.Schum.
Cyperus tomaiophyllus var. *magnus* (C.B.Clarke) Kük. = ***Cyperus tomaiophyllus*** K.Schum.
Cyperus tonduzianus Boeck. = ***Pycreus flavescens*** (L.) P.Beauv. ex Rchb. subsp. ***flavescens***
Cyperus tonduzianus var. *tenuis* Boeck. = ***Cyperus thyrsiflorus*** Jungh. ex Schltdl.
Cyperus tongatabuensis Boeck. = ***Cyperus javanicus*** Houtt.
Cyperus torosus Vahl = ***Cyperus tenuiflorus*** Rottb.
Cyperus torreyanus Schult. = ***Pycreus filicinus*** (Vahl) T.Koyama
Cyperus torreyi Britton = ***Cyperus retrorsus*** Chapm.
Cyperus tortuosus J.König ex Roxb. = ***Pycreus flavidus*** (Retz.) T.Koyama
Cyperus tosaensis Makino = ***Pycreus unioloides*** (R.Br.) Urb.
Cyperus trachycladus Cherm. = ***Cyperus procerus*** Rottb.
Cyperus trachynotus Torr. = ***Cyperus elegans*** L. subsp. ***elegans***
Cyperus trachyrhachis Steud. = ***Pycreus flavidus*** (Retz.) T.Koyama
Cyperus transitorius Kük. = ***Kyllinga pulchella*** Kunth
Cyperus tremulus Poir. = ***Pycreus macrostachyos*** subsp. ***tremulus*** (Poir.) Lye
Cyperus tremulus var. *colchicus* (K.Koch) Kük. = ***Cyperus serotinus*** Rottb.
Cyperus tremulus var. *minor* Boeck. = ***Pycreus macrostachyos*** subsp. ***tremulus*** (Poir.) Lye
Cyperus tremulus var. *opulentus* Boeck. = ***Pycreus macrostachyos*** subsp. ***tremulus*** (Poir.) Lye
Cyperus tremulus var. *woronowii* (Palla) Kük. = ***Pycreus macrostachyos*** subsp. ***tremulus*** (Poir.) Lye
Cyperus tribrachiatus (Liebm.) Kük. = ***Cyperus thyrsiflorus*** Jungh. ex Schltdl.
Cyperus tribrachiatus var. *tenuis* (Boeck.) Kük. = ***Cyperus thyrsiflorus*** Jungh. ex Schltdl.
Cyperus triceps Nees = ***Cyperus manimae*** var. ***asperrimus*** (Liebm.) Kük.
Cyperus triceps F.N.Williams = ?
Cyperus triceps (Rottb.) Endl. = ***Fimbristylis cymosa*** R.Br. subsp. ***cymosa***
Cyperus triceps var. *angustifolius* Kük. = ***Kyllinga bulbosa*** P.Beauv.
Cyperus triceps var. *ciliatus* (Boeck.) Kük. = ***Kyllinga welwitschii*** Ridl.
Cyperus triceps var. *multiflorus* Nees = ***Cyperus manimae*** Kunth var. ***manimae***
Cyperus triceps var. *obtusiflorus* (Boeck.) Kük. = ***Kyllinga bulbosa*** P.Beauv.
Cyperus trichostachys Benth. = ***Cyperus dietrichiae*** Boeck. var. ***dietrichiae***
Cyperus triflorus L. = ***Fimbristylis triflora*** (L.) K.Schum. ex Engl.
Cyperus trigonus Boeck. = ***Cyperus ligularis*** L.
Cyperus trigynus Spreng. = ***Androtrichum trigynum*** (Spreng.) H.Pfeiff.
Cyperus trinervis var. *aquatilis* (R.Br.) Kük. = ***Cyperus aquatilis*** R.Br.
Cyperus trinervis var. *breviculmis* (R.Br.) Kük. = ***Cyperus breviculmis*** R.Br.
Cyperus trinervis var. *contractus* Domin = ***Cyperus trinervis*** R.Br.
Cyperus trinervis var. *flaccidus* (R.Br.) Kük. = ***Cyperus flaccidus*** R.Br.
Cyperus trinervis var. *laxus* (Benth.) Kük. = ***Cyperus trinervis*** R.Br.
Cyperus trinervis var. *superatus* (C.B.Clarke) Kük. = ***Cyperus trinervis*** R.Br.
Cyperus trinervis var. *typicus* Domin = ***Cyperus trinervis*** R.Br.
Cyperus trinidadensis Boeck. = ***Cyperus digitatus*** Roxb. subsp. ***digitatus***
Cyperus trinitatis Steud. = ***Cyperus luzulae*** (L.) Retz.
Cyperus triphyllus Larrañaga = ?
Cyperus triqueter Boeck. = ***Cyperus haspan*** L.
Cyperus triqueter (L.) Missbach & E.H.L.Krause = ***Schoenoplectus triqueter*** (L.) Palla
Cyperus trispicatus Boeck. = ***Cyperus meyenianus*** Kunth
Cyperus tristachyus Boeck. = ***Pycreus flavescens*** (L.) P.Beauv. ex Rchb. subsp. ***flavescens***
Cyperus tristachyus var. *elongatus* Boeck. = ***Pycreus flavescens*** (L.) P.Beauv. ex Rchb. subsp. ***flavescens***
Cyperus tristachyus var. *minor* Boeck. = ***Pycreus flavescens*** (L.) P.Beauv. ex Rchb. subsp. ***flavescens***
Cyperus tristis Kunth = ***Cyperus denudatus*** L.f.
Cyperus truncatulus Steud. = ***Pycreus pumilus*** (L.) Nees
Cyperus truncatus C.A.Mey. ex Turcz. = ***Cyperus orthostachyus*** Franch. & Sav.

Cyperus ×amuricocompressus T.Koyama, J. Jap. Bot. 33: 256 (1958). = [38 JAP]
Cyperus ×amuricoides T.Koyama, J. Jap. Bot. 30: 130 (1955). = ?
Cyperus apatris Steud., Syn. Pl. Glumac. 2: 47 (1854). = ?
Cyperus arenarius Salzm. ex Steud., Syn. Pl. Glumac. 2: 46 (1854), nom. illeg. = ?
Cyperus aristatus Boeck., Linnaea 35: 501 (1868), nom. illeg. = ?
Cyperus aureolus Steud., Syn. Pl. Glumac. 2: 23 (1854). = [81 WIN]
Cyperus ×babae T.Koyama, Quart. J. Taiwan Mus. 14: 194 (1961). = ?
Cyperus baldwinii Chapm., Fl. South. U.S.: 511 (1860), nom. illeg. = ?
Cyperus benghalensis C.B.Clarke, J. Linn. Soc., Bot. 21: 151 (1884), nom. illeg. = ?
Cyperus bifidus Pourr., Chlor. Hisp.: 174 (1788). = ?
Cyperus boeckleri Phil., Anales Univ. Chile 93: 344 (1896). = ?
Cyperus boninensis Ohwi, Mem. Coll. Sci. Kyoto Imp. Univ., Ser. B, Biol. 18(1): 161 (1944). = [38 OGA]
Cyperus ×borealikiiensis T.Koyama & Ogawa, J. Jap. Bot. 33: 256 (1958). = ?
Cyperus brizimorphus Steud., Syn. Pl. Glumac. 2: 5 (1854). = ?
Cyperus bromoides Britton ex Parish, Bull. S. Calif. Acad. Sci. 3: 49 (1904), nom. illeg. = ?
Cyperus caducus Steud., Syn. Pl. Glumac. 2: 315 (1855). = ?
Cyperus capensis Boeck., Linnaea 36: 378 (1870), nom. illeg. = ?
Cyperus carrii Kük., Bot. Jahrb. Syst. 69: 256 (1938). = ?
Cyperus catharinensis Boeck., Beitr. Cyper. 2: 9 (1890). = [84 BZS]
Cyperus coercens Rchb. ex Kunth, Enum. Pl. 2: 20 (1837). = ?
Cyperus ×condensatus T.Koyama, J. Jap. Bot. 28: 344 (1953). = ?
Cyperus convallium Steud., Syn. Pl. Glumac. 2: 5 (1854). = ?
Cyperus cylindricus Boeck., Flora 42: 437 bis (1859). = [84]
Cyperus cyrtostachys C.B.Clarke, J. Linn. Soc., Bot. 21: 59 (1884). = ?
Cyperus dentifer Kom., Izv. Bot. Sada Akad. Nauk S.S.S.R. 30: 198 (1931 publ. 1932). = ?
Cyperus depauperatus Steud., Syn. Pl. Glumac. 2: 34 (1854), nom. illeg. = ?
Cyperus donianus Boeck., Allg. Bot. Z. Syst. 1: 165 (1895), nom. illeg. = ?
Cyperus drummondii Nees ex Boeck., Linnaea 36: 314 (1870), nom. illeg. = ?
Cyperus dubius Roxb., Fl. Ind. 1: 188 (1820), nom. illeg. = ?
Cyperus elytropiptos Steud., Syn. Pl. Glumac. 2: 27 (1854). = ?
Cyperus filiculmis A.Rich. in R.de la Sagra, Hist. Fis. Cuba, Bot. 11: 285 (1850), nom. illeg. = ?
Cyperus flavicomus Schltdl., Linnaea 6: 24 (1831), nom. illeg. = ?
Cyperus fossarum Schltdl., Linnaea 31: 316 (1862), nom. illeg. = [82 SUR]
Cyperus frionensis Boeck., Beitr. Cyper. 2: 10 (1890). = ?
Cyperus fusciformis Drobow in M.G.Popov (ed.), Key Pl. Envir. Tashkent: 46 (1923). = ?
Cyperus gardneri Baker, Fl. Mauritius: 413 (1877), nom. illeg. = ?
Cyperus ginge Welw., Apont.: 586 (1859). = ?
Cyperus guatemalensis Steud., Syn. Pl. Glumac. 2: 47 (1854). = ?
Cyperus hangianus Gand., Bull. Soc. Bot. France 66: 297 (1919 publ. 1920). = ?
Cyperus heptaphyllus Larrañaga, Escritos D. A. Larrañaga 2: 24 (1923). = ?
Cyperus ×heribaudii Pau, Bol. Soc. Aragonesa Ci. Nat. 4: 328 (1905). = *Pycreus flavescens* × *C. fuscus* ?
Cyperus honoluluensis Steud. ex Jard., Bull. Soc. Linn. Normandie, II, 9: 287 (1875). = ?
Cyperus ×hybridus Hausskn., Mitth. Thüring. Bot. Vereins, n.s., 13–14: 34 (1899). = *C. longus* × *C. rotundus*
Cyperus jouannatii Pau, Bol. Soc. Aragonesa Ci. Nat. 4: 328 (1905). = ?
Cyperus ×kadzusensis T.Koyama, Quart. J. Taiwan Mus. 14: 193 (1961). = ?
Cyperus kerstingii Engl. in H.G.A.Engler & C.G.O. Drude, Veg. Erde 9: 204 (1908). = [24 SUD]
Cyperus kuekenthalii Merr., J. Arnold Arbor. 19: 322 (1938). = ?
Cyperus kuntzei Boeck., Beitr. Cyper. 1: 1 (1888). = ?
Cyperus laetus C.B.Clarke, Kew Bull., Addit. Ser. 8: 9 (1908), nom. illeg. = [50 QLD]
Cyperus lamprocarpus Boeck., Linnaea 35: 490 (1868), nom. illeg. = ?
Cyperus lapidicola Kük. in H.G.A.Engler (ed.), Pflanzenr., IV, 20(101): 17 (1935). = [27]
Cyperus laxus Willemet, Ann. Bot. (Usteri) 18: 6 (1796), nom. illeg. = [29 MAU]
Cyperus leonensis Steud., Nomencl. Bot., ed. 2, 1: 470 (1840). = [22 SIE]
Cyperus leptanthos Steud., Syn. Pl. Glumac. 2: 26 (1854). = ?
Cyperus leptophyes C.B.Clarke, Bot. Jahrb. Syst. 30(68): 14 (1902). = ?
Cyperus ligusticus Gand., Dec. Pl. Nov. 2: 28 (1876). = [13 ITA]
Cyperus lividus Steud., Syn. Pl. Glumac. 2: 35 (1854). = ?
Cyperus lomatophyllus Engl., Pflanzenw. Ost-Afrikas, A: 125 (1895). = ?
Cyperus lucidulus Link, Jahrb. Gewächsk. 1(3): 86 (1818). = ?
Cyperus macrostachys Link, J. Bot. (Schrader) 2: 310 (1799). = ?
Cyperus mariscoides Elliott, Sketch Bot. S. Carolina 1: 67 (1816). = ?
Cyperus mauritianus Willemet, Ann. Bot. (Usteri) 18: 4 (1796). = [29 MAU]
Cyperus ×mihashii T.Koyama, J. Jap. Bot. 28: 346 (1953). = ?
Cyperus minimus Forssk., Fl. Aegypt.-Arab.: 202 (1775), nom. illeg. = ?
Cyperus minutiflorus J.Presl & C.Presl, Reliq. Haenk. 1: 351 (1830). = [42 PHI]
Cyperus minutus J.F.Gmel., Syst. Nat. 2: 130 (1791). = ?
Cyperus navicularis Steud., Syn. Pl. Glumac. 2: 32 (1854). = ?
Cyperus nyassicus Chiov., Papiro Ital.: 72 (1931). = ?
Cyperus obtusiusculus Gand., Dec. Pl. Nov. 2: 29 (1876). = [12 FRA]
Cyperus ×ogawai T.Koyama, J. Jap. Bot. 31: 325 (1956). = ?
Cyperus oliganthus Gand., Dec. Pl. Nov. 2: 37 (1876). = [12 FRA]

Cyperus ostenii Kük. ex Osten, Anales Mus. Hist. Nat. Montevideo 3(2): 135 (1931). = ?
Cyperus pallidus Link, Jahrb. Gewächsk. 1(3): 86 (1818). = ?
Cyperus paniculatus Hook. & Arn., Bot. Beechey Voy.: 99 (1832), nom. illeg. = [63]
Cyperus paradoxus Steud., Syn. Pl. Glumac. 2: 8 (1854). = [29 REU]
Cyperus paramattensis Link, Enum. Pl.: 46 (1809). = ?
Cyperus parvus Hassk., Cat. Hort. Bot. Bogor.: 24 (1844). = ?
Cyperus patulus Willd. ex Link, Enum. Hort. Berol. Alt. 1: 46 (1821), nom. illeg. = ?
Cyperus pes-avium Bertol., Misc. Bot. 8: 30 (1849). = ?
Cyperus philippensis J.Presl & C.Presl, Reliq. Haenk. 1: 174 (1828). = [42 PHI]
Cyperus politus Welw. ex Ridl., Trans. Linn. Soc. London, Bot. 2: 125 (1884). = ?
Cyperus polyphyllus Larrañaga, Escritos D. A. Larrañaga 2: 24 (1923), nom. illeg. = ?
Cyperus pubescens J.Presl & C.Presl, Reliq. Haenk. 1: 181 (1830). = ?
Cyperus punctorius Schrad. ex Boeck., Linnaea 35: 567 (1868). = ?
Cyperus pungens Kunth, Enum. Pl. 2: 32 (1837). = ?
Cyperus rheedei F.Muell., Descr. Notes Papuan Pl. 7: 34 (1886). = ?
Cyperus rhenanus Schrad. ex Kunth, Enum. Pl. 2: 60 (1837). = ?
Cyperus rhigiophyllus Hochst. ex C.Krauss, Flora 28: 756 (1845). = [27 NAT]
Cyperus rigidulus Vahl, Enum. Pl. 2: 342 (1805). = ?
Cyperus riparius Salisb., Prodr. Stirp. Chap. Allerton: 30 (1796). = ?
Cyperus robustior Gand., Dec. Pl. Nov. 2: 37 (1876). = [12 FRA]
Cyperus rubromarginatus Meinsh., Trudy Imp. S.-Peterburgsk. Bot. Sada 18: 236 (1900). = ?
Cyperus rufescens Schrad. ex Steud., Syn. Pl. Glumac. 2: 27 (1854), nom. illeg. = ?
Cyperus sanguineofuscus Hook.f., Fl. Tasman. 2: 80 (1858), nom. illeg. = ?
Cyperus scoparius Poir. in J.B.A.M.de Lamarck, Encycl. 7: 253 (1806). = [29 MDG] Perhaps identical with Cyperus platyphyllus.
Cyperus seminudus Zoll. & Moritzi in H.Zollinger, Syst. Verz.: 96 (1854), nom. illeg. = ?
Cyperus st.-crucei Liebm., Mexic. Halvgr.: 223 (1850). = ?
Cyperus stachyphorus K.Koch, Linnaea 21: 623 (1848). = ?
Cyperus subanceps Boeck., Linnaea 38: 368 (1874). = [81]
Cyperus tener Vahl, Enum. Pl. 2: 299 (1805). = ?
Cyperus tenuifolius T.L.Dai, in Fl. Reipubl. Popul. Sin. 11: 231 (1961), nom. illeg. = [36]
Cyperus triceps F.N.Williams, Bull. Herb. Boissier, II, 7: 90 (1907), nom. illeg. = ?
Cyperus triphyllus Larrañaga, Escritos D. A. Larrañaga 2: 24 (1923). = ?
Cyperus tuberosus Roxb., Fl. Ind. 1: 199 (1820), nom. illeg. = ?
Cyperus ugandensis Chiov., Papiro Ital.: 73 (1931). = ?
Cyperus umbellatus Burm.f., Fl. Indica: 21 (1768). = ?
Cyperus umbellatus Hillebr., Fl. Hawaiian Isl.: 470 (1888), nom. illeg. = ?
Cyperus uncinatus Desv. ex Ham., Prodr. Pl. Ind. Occid.: 12 (1825), nom. illeg. = ?
Cyperus uniflorus Thunb., Fl. Jav.: 8 (1825). = ?
Cyperus venustus Kunth, Enum. Pl. 2: 68 (1837), nom. illeg. = ?
Cyperus ×*weatherbianus* Fernald, Rhodora 20: 190 (1918). = ?
Cyperus ×*yamamotoi* T.Koyama, J. Jap. Bot. 28: 344 (1953). = ?

Cypringlea

Cypringlea M.T.Strong, Novon 13: 123 (2003).
Mexico. 79 MXC MXE MXS MXT.
2 Species

Cypringlea analecta (Beetle) M.T.Strong, Novon 13: 125 (2003).
Mexico. 79 MXC MXE MXS MXT. Hemicr. or rhizome geophyte.
Scirpus potosinus Britton, Trans. New York Acad. Sci. 11: 82 (1892).
**Scirpus analecti* Beetle, Brittonia 5: 148 (1944).
Scirpus analecti var. *evadens* C.D.Adams, Ann. Missouri Bot. Gard. 78: 254 (1991).

Cypringlea coahuilensis (Svenson) M.T.Strong, Novon 13: 129 (2003).
Mexico (W. Coahuila). 79 MXE. Hemicr. or rhizome geophyte.
**Scirpus coahuilensis* Svenson, J. Arnold Arbor. 25: 46 (1944).

Cyprolepis

Cyprolepis Steud. = ***Kyllinga*** Rottb.
Cyprolepis denudata Steud. = ***Kyllinga vaginata*** Lam.

Dapedostachys

Dapedostachys Börner = ***Carex*** L.
Dapedostachys steudelii (Kunth) Fedde & J.Schust. = ***Carex jamesii*** Schwein.

Decalepis

Decalepis Boeck. = ***Tetraria*** P.Beauv.
Decalepis dregeana (Nees) Boeck. = ***Tetraria nigrovaginata*** (Nees) C.B.Clarke

Desmiograstis

Desmiograstis Börner = ***Carex*** L.
Desmiograstis contigua (Hoppe) Fedde & J.Schust. = ***Carex spicata*** Huds.

Desmoschoenus

Desmoschoenus Hook.f., Fl. Nov.-Zel. 1: 271 (1853).
New Zealand. 51 NZN NZS.
1 Species
Anthophyllum Steud., Syn. Pl. Glumac. 2: 160 (1855).

Desmoschoenus spiralis (A.Rich.) Hook.f., Fl. Nov.-Zel. 1: 272 (1853).
New Zealand. 51 NZN NZS. Hemicr.
**Isolepis spiralis* A.Rich. in J.S.C.Dumont d'Urville, Voy. Astrolabe: t. 19 (1832). *Scirpus spiralis* (A.Rich.) Druce, Bot. Soc. Exch. Club Brit. Isles 1916: 646 (1917).
Anthophyllum urvillei Steud., Syn. Pl. Glumac. 2: 160 (1855).
Scirpus frondosus Banks & Sol. ex Boeck., Flora 61: 141 (1878).

Deweya

Deweya Raf. = ***Carex*** L.
Deweya anceps (Muhl. ex Willd.) Raf. = ***Carex laxiflora*** Lam. var. ***laxiflora***
Deweya blanda (Dewey) Raf. = ***Carex blanda*** Dewey
Deweya granularis (Muhl. ex Willd.) Raf. = ***Carex granularis*** Muhl. ex Willd.
Deweya oligocarpa (Willd.) Raf. = ***Carex oligocarpa*** Willd.
Deweya plantaginea (Lam.) Raf. = ***Carex plantaginea*** Lam.

Diaphora

Diaphora Lour. = ***Scleria*** P.J.Bergius
Diaphora cochinchinensis Lour. = ***Scleria terrestris*** (L.) Fassett var. ***terrestris***

Dichismus

Dichismus Raf. = ***Scirpus*** Tourn. ex L.

Dichostylis

Dichostylis P.Beauv. = ***Fimbristylis*** Vahl
Dichostylis aristata (Rottb.) Palla = ***Cyperus squarrosus*** L.
Dichostylis baldwinii (Torr.) Palla = ***Cyperus croceus*** Vahl
Dichostylis castanea (Willd.) Palla = ***Cyperus castaneus*** Willd.
Dichostylis congesta (Torr.) Palla = ***Fimbristylis vahlii*** (Lam.) Link
Dichostylis cuspidata (Kunth) Palla = ***Cyperus cuspidatus*** Kunth
Dichostylis fluitans (L.) Rchb. = ***Isolepis fluitans*** (L.) R.Br.
Dichostylis glabra (Palla) Palla = ***Cyperus glaber*** L.
Dichostylis hamulosa (M.Bieb.) Nees = ***Cyperus hamulosus*** M.Bieb.
Dichostylis limosa (Maxim.) A.E.Kozhevn. = ***Cyperus limosus*** Maxim.
Dichostylis ludwigii (Steud.) Nees ex Kunth = ***Isolepis ludwigii*** (Steud.) Kunth
Dichostylis micheliana (L.) Nees = ***Cyperus michelianus*** (L.) Delile
Dichostylis micheliana f. *conglomerata* (Beck) Soó = ***Cyperus michelianus*** (L.) Delile subsp. ***michelianus***
Dichostylis micheliana f. *elata* (Kük.) Soó = ***Cyperus michelianus*** (L.) Delile subsp. ***michelianus***
Dichostylis nipponica (Franch. & Sav.) Palla = ***Cyperus nipponicus*** Franch. & Sav.
Dichostylis nitens (Retz.) Palla = ***Pycreus pumilus*** (L.) Nees
Dichostylis nitens (Vahl) Rikli = ***Rhynchospora nitens*** (Vahl) A.Gray
Dichostylis pacificus (Ohwi) Nakai = ***Cyperus pacificus*** (Ohwi) Ohwi
Dichostylis patens (Vahl) Palla = ***Pycreus pumilus*** (L.) Nees
Dichostylis patens (Vahl) Rikli = ***Pycreus pumilus*** (L.) Nees
Dichostylis pygmaea (Rottb.) Nees = ***Cyperus michelianus*** subsp. ***pygmaeus*** (Rottb.) Asch. & Graebn.
Dichostylis radiata (Vahl) Palla = ***Cyperus imbricatus*** Retz.
Dichostylis squarrosa (L.) Palla = ***Cyperus squarrosus*** L.
Dichostylis torresiana Gaudich. = ***Fimbristylis umbellaris*** (Lam.) Vahl
Dichostylis wolgensis A.O.Tarassov = ***Cyperus michelianus*** (L.) Delile subsp. ***michelianus***

Dichroma

Dichroma Ham. = ***Rhynchospora*** Vahl
Dichroma ciliatum Ham. = ***Rhynchospora colorata*** (L.) H.Pfeiff.

Dichromena

Dichromena Michx. = ***Rhynchospora*** Vahl
Dichromena aberrans (C.B.Clarke) J.F.Macbr. = ***Rhynchospora aberrans*** C.B.Clarke
Dichromena alba (L.) J.F.Macbr. = ***Rhynchospora alba*** (L.) Vahl
Dichromena albida (Nees) Steud. = ***Rhynchospora albida*** (Nees) Boeck.
Dichromena alta (Boeck.) J.F.Macbr. = ***Rhynchospora globosa*** (Kunth) Roem. & Schult. var. ***globosa***
Dichromena amazonica (Poepp. ex Kunth) J.F.Macbr. = ***Rhynchospora amazonica*** Poepp. ex Kunth
Dichromena atrosanguinea É.Desv. = ***Schoenoplectus californicus*** (C.A.Mey.) Soják
Dichromena barbata (Vahl) J.F.Macbr. = ***Rhynchospora barbata*** (Vahl) Kunth
Dichromena blepharophora J.Presl & C.Presl = ***Rhynchospora blepharophora*** (J.Presl & C.Presl) H.Pfeiff.
Dichromena brevicaulis Suess. = ***Rhynchospora duckei*** R.Gross
Dichromena caespitosa (Muhl.) Spreng. = ***Bulbostylis stenophylla*** (Elliott) C.B.Clarke
Dichromena caespitosum Muhl. = ***Bulbostylis stenophylla*** (Elliott) C.B.Clarke
Dichromena canaliculata Steud. = ***Rhynchospora riparia*** (Nees) Boeck.
Dichromena candida Baker = ?
Dichromena candida (Nees) Ridl. = ***Rhynchospora candida*** (Nees) Boeck.
Dichromena canescens Maury ex Micheli = ***Rhynchospora albiceps*** Kunth
Dichromena capillaris Kunth = ***Rhynchospora tenuis*** subsp. ***tenuis***
Dichromena caracasana Kunth = ***Rhynchospora caracasana*** (Kunth) Boeck.
Dichromena cariciformis (Nees) J.F.Macbr. = ***Rhynchospora cariciformis*** Nees
Dichromena cephalotes Britton = ***Lipocarpha maculata*** (Michx.) Torr.
Dichromena ciliata J.Presl & C.Presl = ***Rhynchospora blepharophora*** (J.Presl & C.Presl) H.Pfeiff.
Dichromena ciliata Pers. = ***Rhynchospora colorata*** (L.) H.Pfeiff.
Dichromena ciliata Vahl = ***Rhynchospora nervosa*** subsp. ***ciliata*** (G.Mey.) T.Koyama
Dichromena ciliata var. *angustifolia* Cherm. = ***Rhynchospora nervosa*** subsp. ***ciliata*** (G.Mey.) T.Koyama
Dichromena ciliata var. *cinnamomea* Kük. = ***Rhynchospora nervosa*** subsp. ***ciliata*** (G.Mey.) T.Koyama
Dichromena ciliata var. *vahliana* C.B.Clarke = ***Rhynchospora nervosa*** (Vahl) Boeck. subsp. ***nervosa***
Dichromena colorata (L.) C.L.Hitchc. = ***Rhynchospora colorata*** (L.) H.Pfeiff.
Dichromena compressa Spreng. = ***Rhynchospora consanguinea*** (Kunth) Boeck.
Dichromena consanguinea Kunth = ***Rhynchospora consanguinea*** (Kunth) Boeck.
Dichromena contracta (Nees) Steud. = ***Rhynchospora contracta*** (Nees) J.Raynal

Dichromena corymbosa (L.) J.F.Macbr. = ***Rhynchospora corymbosa*** (L.) Britton
Dichromena cubensis Poepp. & Kunth = ***Rhynchospora tenuis*** subsp. ***tenuis***
Dichromena cymosa (Willd.) J.F.Macbr. = ***Scirpus polyphyllus*** Vahl
Dichromena cyperoides (Mart.) J.F.Macbr. = ***Rhynchospora holoschoenoides*** (Rich.) Herter
Dichromena distans (Michx.) J.F.Macbr. = ***Rhynchospora fascicularis*** (Michx.) Vahl subsp. ***fascicularis***
Dichromena drummondiana Steud. = ***Rhynchospora colorata*** (L.) H.Pfeiff.
Dichromena ebracteata Standl. = ***Rhynchospora ebracteata*** (Standl.) H.Pfeiff.
Dichromena elatior Kunth = ***Rhynchospora stricta*** Boeck.
Dichromena emaciata (Nees) Steud. = ***Rhynchospora emaciata*** (Nees) Boeck.
Dichromena erostris Nees = ***Rhynchospora pubera*** (Vahl) Boeck. subsp. ***pubera***
Dichromena exaltata (Kunth) J.F.Macbr. = ***Rhynchospora exaltata*** Kunth
Dichromena fasciata Maury = ***Rhynchospora watsonii*** (Britton) Davidse
Dichromena filiformis C.B.Clarke = ***Rhynchospora berteroi*** (Spreng.) C.B.Clarke
Dichromena filiformis Kunth = ***Rhynchospora filiformis*** Vahl
Dichromena floridensis Britton ex Small = ***Rhynchospora floridensis*** (Britton ex Small) H.Pfeiff.
Dichromena foliosa Hochst. ex Boeck. = ***Rhynchospora albiceps*** Kunth
Dichromena fragifera A.Dietr. = ***Calyptrocarya glomerulata*** (Brongn.) Urb.
Dichromena gigantea (Link) J.F.Macbr. = ***Rhynchospora gigantea*** Link
Dichromena glabra Pers. = ***Rhynchospora colorata*** (L.) H.Pfeiff.
Dichromena glauca (Vahl) J.F.Macbr. = ***Rhynchospora rugosa*** subsp. ***brownii*** (Roem. & Schult.) T.Koyama
Dichromena globosa (Kunth) Roem. & Schult. = ***Rhynchospora nervosa*** subsp. ***ciliata*** (G.Mey.) T.Koyama
Dichromena gracilis C.B.Clarke = ***Rhynchospora breviuscula*** H.Pfeiff.
Dichromena gracilis (Spreng.) Kunth = ***Rhynchospora tenuis*** subsp. ***tenuis***
Dichromena grisebachii C.B.Clarke = ***Rhynchospora subsetigera*** H.Pfeiff.
Dichromena heterophylla Boeck. = ***Rhynchospora chapmanii*** M.A.Curtis
Dichromena hirsuta (Vahl) Kunth = ***Rhynchospora hirsuta*** (Vahl) Vahl
Dichromena hispida Schrad. ex Nees = ***Rhynchospora consanguinea*** (Kunth) Boeck.
Dichromena hispidula (Vahl) Kunth = ***Rhynchospora trichodes*** C.B.Clarke
Dichromena humboldtiana Nees = ***Rhynchospora pubera*** (Vahl) Boeck. subsp. ***pubera***
Dichromena inaguensis Britton = ***Rhynchospora floridensis*** (Britton ex Small) H.Pfeiff.
Dichromena junciformis Kunth = ***Rhynchospora junciformis*** (Kunth) Boeck.
Dichromena killipii (R.Gross) J.F.Macbr. = ***Rhynchospora killipii*** R.Gross
Dichromena kuekenthalii (H.Pfeiff.) J.F.Macbr. = ***Rhynchospora kuntzei*** C.B.Clarke ex Kuntze
Dichromena kuntzei (C.B.Clarke ex Kuntze) J.F.Macbr. = ***Rhynchospora kuntzei*** C.B.Clarke ex Kuntze
Dichromena latifolia Baldwin ex Elliott = ***Rhynchospora latifolia*** (Baldwin ex Elliott) W.W.Thomas
Dichromena lechleri (Steud. ex Boeck.) J.F.Macbr. = ***Rhynchospora vulcani*** Boeck.
Dichromena leucocephala Michx. = ***Rhynchospora colorata*** (L.) H.Pfeiff.
Dichromena leucocephalum (Michx.) Pers. = ***Rhynchospora colorata*** (L.) H.Pfeiff.
Dichromena leucophylla Willk. ex Kunth = ***Rhynchospora nervosa*** subsp. ***ciliata*** (G.Mey.) T.Koyama
Dichromena linkii J.F.Macbr. = ***Rhynchospora tenuis*** Link
Dichromena locuples (C.B.Clarke) J.F.Macbr. = ***Rhynchospora locuples*** C.B.Clarke
Dichromena longa Lindm. = ***Rhynchospora longa*** (Lindm.) H.Pfeiff.
Dichromena longifolia (Rudge) Spreng. = ***Bisboeckelera longifolia*** (Rudge) Kuntze
Dichromena macrochaeta (Steud. ex Boeck.) J.F.Macbr. = ***Rhynchospora macrochaeta*** Steud. ex Boeck.
Dichromena mandonii (C.B.Clarke) J.F.Macbr. = ***Rhynchospora mandonii*** C.B.Clarke
Dichromena marisculus (Lindl. & Nees) J.F.Macbr. = ***Rhynchospora marisculus*** Lindl. & Nees
Dichromena micrantha (Vahl) Kunth = ***Rhynchospora rariflora*** (Michx.) Elliott
Dichromena microcephala Bertero ex Spreng. = ***Rhynchospora radicans*** subsp. ***microcephala*** (Bertero ex Spreng.) W.W.Thomas
Dichromena minarum C.B.Clarke = ***Rhynchospora pilosa*** Boeck.
Dichromena monostachya C.B.Clarke = ***Rhynchospora albescens*** (Miq.) Kük.
Dichromena montana Uittien = ***Rhynchospora montana*** (Uittien) H.Pfeiff.
Dichromena nana (Boeck.) Core = ***Rhynchospora pilosa*** Boeck.
Dichromena nardifolia Kunth = ***Rhynchospora nardifolia*** (Kunth) Boeck.
Dichromena nervosa Vahl = ***Rhynchospora nervosa*** (Vahl) Boeck.
Dichromena nitens (Vahl) Kunth = ***Rhynchospora nitens*** (Vahl) A.Gray
Dichromena nivea (Boeck.) Britton = ***Rhynchospora nivea*** Boeck.
Dichromena nivea (Boeck.) Boeck. = ***Rhynchospora nivea*** Boeck.
Dichromena obtusiflora Schrad. ex Nees = ***Rhynchospora nervosa*** subsp. ***ciliata*** (G.Mey.) T.Koyama
Dichromena obtusifolia Steud. = ***Rhynchospora nervosa*** subsp. ***ciliata*** (G.Mey.) T.Koyama
Dichromena pearcei (C.B.Clarke) J.F.Macbr. = ***Pleurostachys pearcei*** C.B.Clarke
Dichromena persooniana Nees = ***Rhynchospora colorata*** (L.) H.Pfeiff.
Dichromena peruviana (C.B.Clarke) J.F.Macbr. = ***Pleurostachys peruviana*** C.B.Clarke
Dichromena pilosa Spreng. = ***Rhynchospora nervosa*** (Vahl) Boeck. subsp. ***nervosa***
Dichromena pilosa Kunth = ***Rhynchospora pilosa*** Boeck.
Dichromena pittieri Standl. = ***Rhynchospora blepharophora*** (J.Presl & C.Presl) H.Pfeiff.
Dichromena pohliana Steud. = ***Rhynchospora emaciata*** (Nees) Boeck.

Dichromena polyphylla (Vahl) J.F.Macbr. = ***Rhynchospora polyphylla*** (Vahl) Vahl
Dichromena polystachys Turrill = ***Rhynchospora polystachys*** (Turrill) H.Pfeiff.
Dichromena preslii A.Dietr. = ***Rhynchospora blepharophora*** (J.Presl & C.Presl) H.Pfeiff.
Dichromena pubera Vahl = ***Rhynchospora pubera*** (Vahl) Boeck.
Dichromena pulchella Poepp. & Kunth = ***Rhynchospora blepharophora*** (J.Presl & C.Presl) H.Pfeiff.
Dichromena pura Nees = ***Rhynchospora nervosa*** subsp. ***ciliata*** (G.Mey.) T.Koyama
Dichromena pusilla (Sw.) Kunth = ***Rhynchospora berteroi*** (Spreng.) C.B.Clarke
Dichromena radicans Schltdl. & Cham. = ***Rhynchospora radicans*** (Schltdl. & Cham.) H.Pfeiff.
Dichromena radicans var. *typica* A.Gross & Brade = ***Rhynchospora radicans*** (Schltdl. & Cham.) H.Pfeiff.
Dichromena repens Vahl = ***Rhynchospora reptans*** (Rich.) Kük.
Dichromena reptans (Rich.) Pers. = ***Rhynchospora reptans*** (Rich.) Kük.
Dichromena reverchonii S.H.Wright = ***Rhynchospora nivea*** Boeck.
Dichromena rigida Kunth = ***Rhynchospora consanguinea*** (Kunth) Boeck.
Dichromena rigidifolia Gilly = ***Rhynchospora rigidifolia*** (Gilly) T.Koyama
Dichromena riparia (Nees) Steud. = ***Rhynchospora riparia*** (Nees) Boeck.
Dichromena robusta Kunth = ***Rhynchospora robusta*** (Kunth) Boeck.
Dichromena rufidula Steud. = ?
Dichromena ruiziana (Boeck.) J.F.Macbr. = ***Rhynchospora ruiziana*** Boeck.
Dichromena ruiziana var. *triceps* J.F.Macbr. = ***Rhynchospora macrochaeta*** var. ***triceps*** Kük.
Dichromena sabulosa Kunth = ***Rhynchospora tenuis*** subsp. ***tenuis***
Dichromena schiedeana Kunth = ***Rhynchospora eximia*** (Nees) Boeck.
Dichromena schoenus J.F.Macbr. = ***Rhynchospora cephalotes*** (L.) Vahl
Dichromena sellowiana Kunth = ***Rhynchospora velutina*** (Kunth) Boeck.
Dichromena semiinvolucrata (J.Presl & C.Presl) J.F.Macbr. = ***Rhynchospora semiinvolucrata*** J.Presl & C.Presl
Dichromena seslerioides (Griseb.) Benth. ex C.B.Clarke = ***Rhynchospora seslerioides*** Griseb.
Dichromena setacea (P.J.Bergius) Kunth = ***Rhynchospora tenerrima*** Nees ex Spreng. subsp. ***tenerrima***
Dichromena setigera Kunth = ***Rhynchospora setigera*** (Kunth) Boeck.
Dichromena speciosa Kunth = ***Rhynchospora consanguinea*** (Kunth) Boeck.
Dichromena stolonifera (Nees) Steud. = ***Rhynchospora stolonifera*** (Nees) Boeck.
Dichromena tenella (Nees) Steud. = ***Rhynchospora tenella*** (Nees) Boeck.
Dichromena tenera Schrad. ex Nees = ***Rhynchospora pubera*** (Vahl) Boeck. subsp. ***pubera***
Dichromena tenuifolia (Kunth) Spreng. = ***Rhynchospora nervosa*** (Vahl) Boeck. subsp. ***nervosa***
Dichromena tenuis Steud. = ***Rhynchospora radicans*** (Schltdl. & Cham.) H.Pfeiff. subsp. ***radicans***
Dichromena trachycarpa Steud. = ?
Dichromena triflora (Vahl) J.F.Macbr. = ***Rhynchospora triflora*** Vahl
Dichromena umbraticola (Poepp. & Kunth) J.F.Macbr. = ***Rhynchospora umbraticola*** Poepp. & Kunth
Dichromena unicolor Salzm. ex Steud. = ***Rhynchospora pubera*** (Vahl) Boeck. subsp. ***pubera***
Dichromena vahlii A.Dietr. = ***Scleria melaleuca*** Rchb. ex Schltdl. & Cham.
Dichromena velutina Kunth = ***Rhynchospora velutina*** (Kunth) Boeck.
Dichromena watsonii Britton = ***Rhynchospora watsonii*** (Britton) Davidse
Dichromena weberbaueri (C.B.Clarke) J.F.Macbr. = ***Rhynchospora marisculus*** Lindl. & Nees

***Unplaced Names*:**
Dichromena candida Baker, J. Linn. Soc., Bot. 21: 450 (1885), nom. illeg. = ?
Dichromena rufidula Steud., Syn. Pl. Glumac. 2: 137 (1855). = ?
Dichromena trachycarpa Steud., Syn. Pl. Glumac. 2: 136 (1855). = ?

Diclidium

Diclidium Schrad. ex Nees = ***Cyperus*** L.
Diclidium aciculare Schrad. ex Nees = ***Cyperus odoratus*** L. subsp. ***odoratus***
Diclidium auriculatum Nees = ***Cyperus odoratus*** L. subsp. ***odoratus***
Diclidium conglobatum (Link) Nees ex B.D.Jacks. = ***Cyperus odoratus*** L. subsp. ***odoratus***
Diclidium ferax (Rich.) Steud. = ***Cyperus odoratus*** L. subsp. ***odoratus***
Diclidium ferox (Rich.) Schrad. ex Nees = ***Cyperus odoratus*** L. subsp. ***odoratus***
Diclidium fuscatum Schrad. ex Nees = ***Cyperus odoratus*** L. subsp. ***odoratus***
Diclidium lenticulare Schrad. ex Nees = ***Cyperus odoratus*** L. subsp. ***odoratus***
Diclidium lomentaceum Nees = ***Cyperus odoratus*** L. subsp. ***odoratus***
Diclidium maximiliani Schrad. ex Nees = ***Cyperus odoratus*** L. subsp. ***odoratus***
Diclidium odoratum (L.) Schrad. ex Nees = ***Cyperus odoratus*** L.
Diclidium reinwardtii Schrad. ex Nees = ***Cyperus odoratus*** L. subsp. ***odoratus***
Diclidium uliginosum Schrad. ex Nees = ***Cyperus odoratus*** L. subsp. ***odoratus***
Diclidium vahlii Schrad. ex Nees = ***Cyperus flexuosus*** Vahl

Didymia

Didymia Phil. = ***Cyperus*** L.
Didymia cyperomorpha Phil. = ***Cyperus aggregatus*** (Willd.) Endl. var. ***aggregatus***

Didymiandrum

Didymiandrum Gilly = ***Lagenocarpus*** Nees
Didymiandrum flexifolium Gilly = ***Everardia flexifolia*** (Gilly) T.Koyama & Maguire
Didymiandrum guaiquinimae Schnee = ***Lagenocarpus stellatus*** (Boeck.) Kuntze
Didymiandrum stellatum (Boeck.) Gilly = ***Lagenocarpus stellatus*** (Boeck.) Kuntze

Didymonema

Didymonema C.Presl = ***Gahnia*** J.R.Forst. & G.Forst.
Didymonema filifolia C.Presl = ***Gahnia filifolia*** (C.Presl) Kük. ex Benl

Diemisa

Diemisa Raf. = ***Carex*** L.
Diemisa capitata (L.) Raf. = ***Carex capitata*** L.
Diemisa cephalophora (Muhl. ex Willd.) Raf. = ***Carex cephalophora*** Muhl. ex Willd.
Diemisa concolor (R.Br.) Raf. = ***Carex aquatilis*** var. ***minor*** Boott
Diemisa crinita (Lam.) Raf. = ***Carex crinita*** Lam.
Diemisa filiformis (L.) Raf. = ***Carex filiformis*** L.
Diemisa pellita (Muhl. ex Willd.) Raf. = ***Carex pellita*** Muhl. ex Willd.
Diemisa pubescens Raf. = ***Carex hirtifolia*** Mack.
Diemisa remota (L.) Raf. = ***Carex remota*** L.
Diemisa retroflexa (Muhl. ex Willd.) Raf. = ***Carex retroflexa*** Muhl. ex Willd.
Diemisa rosea (Willd.) Raf. = ***Carex rosea*** Willd.
Diemisa splendens (Thuill. ex Pers.) Raf. = ***Carex divisa*** Huds.
Diemisa stricta (Lam.) Raf. = ***Carex stricta*** Lam.
Diemisa tenera (Dewey) Raf. = ***Carex tenera*** Dewey

Diplacrum

Diplacrum R.Br., Prodr.: 241 (1810).
Trop. 22 BEN BKN GHA GNB GUI IVO LBR NGA SEN SIE TOG 23 GAB 24 CHA SUD 25 TAN UGA 26 ZAM 29 MDG 36 CHS 38 JAP NNS TAI 40 ASS BAN EHM IND SRL 41 AND MYA THA VIE 42 BOR JAW LSI MLY MOL PHI SUL SUM 43 BIS NWG 50 NTA QLD WAU 62 CRL MRN 80 COS HON NIC PAN 81 JAM LEE TRT 82 FRG GUY SUR VEN 83 BOL CLM ECU PER 84 BZC BZN.
8 Species
Pteroscleria Nees in C.F.P.von Martius & auct. suc. (eds.), Fl. Bras. 2(1): 196 (1842).

Diplacrum africanum (Benth.) C.B.Clarke in T.A.Durand & H.Schinz, Consp. Fl. Afric. 5: 668 (1894).
N. South America, Trop. Africa, Madagascar, S. India. 22 BEN BKN GHA GNB GUI IVO NGA SEN SIE TOG 24 CHA SUD 25 TAN UGA 26 ZAM 29 MDG 40 IND 82 SUR. Ther.
**Scleria africana* Benth. in G.Bentham & J.D.Hooker, Gen. Pl. 3: 1071 (1883).

Diplacrum capitatum (Willd.) Boeck., Linnaea 38: 435 (1874).
Trop. America, W. & WC. Trop. Africa. 22 GHA GUI IVO LBR NGA SIE 23 GAB 80 COS NIC PAN 81 JAM LEE TRT 82 FRG GUY SUR VEN 83 BOL CLM ECU PER 84 BZC BZN. Hemicr.
**Scleria capitata* Willd., Sp. Pl. 4: 319 (1805).
Pteroscleria longifolia Griseb., Fl. Brit. W. I.: 579 (1864). *Diplacrum longifolium* (Griseb.) C.B.Clarke in T.A.Durand & H.Schinz, Consp. Fl. Afric. 5: 669 (1894). *Scleria longifolia* (Griseb.) Roberty, Fl. Ouest-Afr.: 377 (1954), nom. illeg. *Scleria amphigaea* Raymond, Naturaliste Canad. 91: 132 (1964).
Diplacrum longifolium var. *angustifolium* Kük., Bot. Jahrb. Syst. 56(125): 23 (1921). *Scleria amphigaea* var. *angustifolia* (Kük.) Raymond, Naturaliste Canad. 91: 132 (1963).
Diplacrum longifolium f. *graciliores* H.Pfeiff., Repert. Spec. Nov. Regni Veg. 28: 24 (1930).
Bisboeckelera paporiensis Suess., Bot. Jahrb. Syst. 73: 120 (1943).

Diplacrum caricinum R.Br., Prodr.: 241 (1810). *Scleria caricina* (R.Br.) Benth., Fl. Austral. 7: 426 (1878).
Trop. & Subtrop. Asia to W. Pacific. 36 CHS 38 JAP NNS TAI 40 ASS BAN EHM IND SRL 41 AND MYA THA VIE 42 BOR JAW LSI MLY MOL PHI SUL SUM 43 BIS NWG 50 NTA QLD 62 CRL MRN. Ther.
Scleria axillaris Moon, Cat. Pl. Ceylon: 62 (1824).
Diplacrum tridentatum Brongn. in L.I.Duperrey, Voy. Monde, Phan.: 26 (1829).
Diplacrum zeylanicum Nees in R.Wight, Contr. Bot. India: 119 (1834).
Diplacrum caricinum subsp. *sumatranum* Miq., Fl. Ned. Ind., Eerste Bijv.: 602 (1861).

Diplacrum exiguum (J.Kern) T.Koyama, Brittonia 31: 293 (1979).
Vietnam. 41 VIE. Ther.
**Scleria exigua* J.Kern, Adansonia, n.s., 2: 109 (1962).

Diplacrum guianense (Nees) T.Koyama, Mem. New York Bot. Gard. 16: 51 (1967).
N. South America to N. Brazil. 82 FRG GUY SUR VEN 83 CLM 84 BZN. Hemicr. or rhizome geophyte.
**Pteroscleria guianensis* Nees in C.F.P.von Martius & auct. suc. (eds.), Fl. Bras. 2(1): 196 (1842). *Scleria notoptera* Nees ex Steud., Syn. Pl. Glumac. 2: 177 (1855), nom. illeg. *Diplacrum notopterum* C.B.Clarke, Kew Bull., Addit. Ser. 8: 62 (1908).
Diplacrum longifolium var. *laevissimum* Kük., Bot. Jahrb. Syst. 56(125): 23 (1921). *Diplacrum laevissimum* (Kük.) Uittien, Meded. Afd. Handelsmus. Kolon. Inst. 30: 149 (1934).

Diplacrum mitracarpoides (Standl. & L.O.Williams) C.D. Adams, Ann. Missouri Bot. Gard. 78: 254 (1991).
C. America. 80 HON. Ther.
**Scleria mitracarpoides* Standl. & L.O.Williams, Ceiba 3: 36 (1952).

Diplacrum pygmaeopsis (J.Kern) T.Koyama, Mem. New York Bot. Gard. 17(1): 33 (1967).
Lesser Sunda Is. (E. Sumba). 42 LSI. Ther.
**Scleria pygmaeopsis* J.Kern, Blumea 11: 208 (1961).

Diplacrum pygmaeum (R.Br.) Nees ex Boeck., Linnaea 38: 434 (1874).
N. Australia. 50 NTA QLD WAU. Hemicr. or rhizome geophyte.
**Scleria pygmaea* R.Br., Prodr.: 240 (1810). *Hypoporum pygmaeum* (R.Br.) Nees, Linnaea 9: 303 (1834). *Sphaeropus pygmaeus* (R.Br.) Boeck., Flora 56: 90 (1873).

Synonyms:
Diplacrum caricinum subsp. *sumatranum* Miq. = ***Diplacrum caricinum*** R.Br.
Diplacrum laevissimum (Kük.) Uittien = ***Diplacrum guianense*** (Nees) T.Koyama
Diplacrum longifolium (Griseb.) C.B.Clarke = ***Diplacrum capitatum*** (Willd.) Boeck.
Diplacrum longifolium var. *angustifolium* Kük. = ***Diplacrum capitatum*** (Willd.) Boeck.
Diplacrum longifolium f. *graciliores* H.Pfeiff. = ***Diplacrum capitatum*** (Willd.) Boeck.
Diplacrum longifolium var. *laevissimum* Kük. = ***Diplacrum guianense*** (Nees) T.Koyama
Diplacrum monocephalum (Hochst. ex Steud.) H.Pfeiff. = ***Calyptrocarya monocephala*** Hochst. ex Steud.
Diplacrum notopterum C.B.Clarke = ***Diplacrum guianense*** (Nees) T.Koyama

Diplacrum reticulatum Holttum = ***Scleria reticulata*** (Holttum) J.Kern
Diplacrum tridentatum Brongn. = ***Diplacrum caricinum*** R.Br.
Diplacrum zeylanicum Nees = ***Diplacrum caricinum*** R.Br.

Diplarinus

Diplarinus Raf. = ***Scirpus*** Tourn. ex L.

Diplasia

Diplasia Rich. ex Pers., Syn. Pl. 1: 70 (1805).
Trinidad, C. & S. Trop. America. 80 COS PAN 81 TRT 82 FRG GUY SUR VEN 83 BOL CLM ECU PER 84 BZN.
1 Species

Diplasia karatifolia Rich. ex Pers., Syn. Pl. 1: 70 (1805).
Trinidad, C. & S. Trop. America. 80 COS PAN 81 TRT 82 FRG GUY SUR VEN 83 BOL CLM ECU PER 84 BZN. Hemicr.
Scirpus bromeliifolius Rudge, Pl. Guian.: 19 (1805).
Hypolytrum iridifolium Link, Jahrb. Gewächsk. 1(3): 91 (1820).
Fimbristylis bromeliifolia A.Spreng., Tent. Suppl.: 3 (1828).

***Synonyms*:**
Diplasia pycnostachya Benth. = ***Mapania pycnostachya*** (Benth.) T.Koyama
Diplasia tonkinensis E.G.Camus = ***Mapania tonkinensis*** (E.G.Camus) T.Koyama

Diplocarex

Diplocarex Hayata = ***Carex*** L.
Diplocarex matsudae Hayata = ***Carex dolichostachya*** Hayata subsp. ***dolichostachya***

Diplochaete

Diplochaete Nees = ***Rhynchospora*** Vahl
Diplochaete guianensis Nees = ***Rhynchospora barbata*** (Vahl) Kunth

Diploscyphus

Diploscyphus Liebm. = ***Scleria*** P.J.Bergius
Diploscyphus mexicanus Liebm. = ***Scleria sororia*** Kunth

Discopodium

Discopodium Steud. = ***Tricostularia*** Nees
Discopodium drummondii Steud. = ***Tricostularia compressa*** Nees

Distimus

Distimus Raf. = ***Pycreus*** P.Beauv.
Distimus flavescens (L.) Raf. = ***Pycreus flavescens*** (L.) P.Beauv. ex Rchb.
Distimus flavicomus Raf. = ***Pycreus macrostachyos*** (Lam.) J.Raynal subsp. ***macrostachyos***

Dornera

Dornera Heuff. ex Schur = ***Carex*** L.
Dornera alpina Heuff. ex Schur = ***Carex pyrenaica*** Wahlenb.

Drymeia

Drymeia Ehrh. = ***Carex*** L.

Dulichium

Dulichium Pers., Syn. Pl. 1: 65 (1805).
Canada to U.S.A. 71 BRC MAN 72 NBR NFL NSC ONT PEI QUE 73 IDA MNT ORE WAS WYO 74 ILL IOW MIN MSO NEB OKL WIS 75 CNT INI MAI MAS MIC NWH NWJ NWY OHI PEN RHO VER WVA 76 CAL 77 TEX 78 ALA ARK DEL FLA GEO KTY LOU MRY MSI NCA SCA TEN VRG.
1 Species
Pleuranthus Rich. ex Pers., Syn. Pl. 1: 65 (1805).

Dulichium arundinaceum (L.) Britton, Bull. Torrey Bot. Club 21: 29 (1894).
Canada to U.S.A. 71 BRC MAN 72 NBR NFL NSC ONT PEI QUE 73 IDA MNT ORE WAS WYO 74 ILL IOW MIN MSO NEB OKL WIS 75 CNT INI MAI MAS MIC NWH NWJ NWY OHI PEN RHO VER WVA 76 CAL 77 TEX 78 ALA ARK DEL FLA GEO KTY LOU MRY MSI NCA SCA TEN VRG. Hemicr.
**Cyperus arundinaceus* L., Sp. Pl.: 44 (1753).

var. ***arundinaceum***
Canada to U.S.A. 71 BRC MAN 72 NBR NFL NSC ONT PEI QUE 73 IDA MNT ORE WAS WYO 74 ILL IOW MIN MSO NEB OKL WIS 75 CNT INI MAI MAS MIC NWH NWJ NWY OHI PEN RHO VER WVA 76 CAL 77 TEX 78 ALA ARK DEL FLA GEO KTY LOU MRY MSI NCA SCA TEN VRG. Hemicr.
Cyperus ferrugineus L., Sp. Pl.: 44 (1753).
Schoenus spathaceus L., Sp. Pl. ed. 2: 63 (1762).
Cyperus spathaceus (L.) L., Syst. Nat. ed. 12, 2: 735 (1767). *Scirpus spathaceus* (L.) Michx., Fl. Bor.-Amer. 1: 32 (1803). *Dulichium spathaceum* (L.) Rich. ex Pers., Syn. Pl. 1: 65 (1805).
Dulichium canadense Rich. ex Pers., Syn. Pl. 1: 65 (1805).
Schoenus angustifolius Vahl, Enum. Pl. 2: 225 (1805).
Sparganium trifidum Poir. in J.B.A.M.de Lamarck, Encycl., Suppl. 4: 723 (1816).
Scirpus tegetalis Burch., Trav. S. Africa 1: 260 (1822).
Dulichium spathaceum var. *humilius* Hook., Fl. Bor.-Amer. 2: 232 (1839).

var. ***boreale*** Lepage, Naturaliste Canad. 85: 100 (1958).
Québec. 72 QUE. Hemicr.

***Synonyms*:**
Dulichium canadense Rich. ex Pers. = ***Dulichium arundinaceum*** (L.) Britton var. ***arundinaceum***
Dulichium confervoides (Poir.) Alston = ***Websteria confervoides*** (Poir.) S.S.Hooper
Dulichium spathaceum (L.) Rich. ex Pers. = ***Dulichium arundinaceum*** (L.) Britton var. ***arundinaceum***
Dulichium spathaceum var. *humilius* Hook. = ***Dulichium arundinaceum*** (L.) Britton var. ***arundinaceum***

Durandia

Durandia Boeck. = ***Scleria*** P.J.Bergius
Durandia macrophylla (Boeck.) Boeck. = ***Scleria vaginata*** Steud.

Duval-jouvea.

Duval-jouvea Palla = ***Cyperus*** L.

Duval-jouvea babakensis (Steud. ex Miq.) H.Pfeiff. = ***Cyperus babakan*** Steud.
Duval-jouvea diluta (Vahl) Palla = ***Cyperus compressus*** L.
Duval-jouvea maritima (Aubl.) Palla = ***Remirea maritima*** Aubl.
Duval-jouvea pennata (Lam.) Palla = ***Cyperus javanicus*** Houtt.
Duval-jouvea pilosa (Vahl) Palla = ***Cyperus pilosus*** Vahl
Duval-jouvea procera (Rottb.) H.Pfeiff. = ***Cyperus procerus*** Rottb.
Duval-jouvea serotina (Rottb.) Palla = ***Cyperus serotinus*** Rottb.

Echinochlaenia

Echinochlaenia Börner = ***Carex*** L.
Echinochlaenia tasmanica (Kük.) Fedde & J.Schust. = ***Carex tasmanica*** Kük.

Echinolytrum

Echinolytrum Desv. = ***Fimbristylis*** Vahl
Echinolytrum dipsaceum (Rottb.) Desf. = ***Fimbristylis dipsacea*** (Rottb.) C.B.Clarke
Echinolytrum verruciferum (Maxim.) Makino = ***Fimbristylis dipsacea*** (Rottb.) C.B.Clarke

Echinoschoenus

Echinoschoenus Nees & Meyen = ***Rhynchospora*** Vahl
Echinoschoenus beyrichii Nees = ***Rhynchospora exaltata*** Kunth
Echinoschoenus sparganioides Lindl. ex Nees = ***Rhynchospora exaltata*** Kunth
Echinoschoenus triceps (Cham. & Schltdl.) Nees & Meyen = ***Rhynchospora exaltata*** Kunth

Ecklonea

Ecklonea Steud. = ***Trianoptiles*** Fenzl ex Endl.
Ecklonea capensis Steud. = ***Trianoptiles capensis*** (Steud.) Harv.
Ecklonea solitaria C.B.Clarke = ***Trianoptiles solitaria*** (C.B.Clarke) Levyns

Edritria

Edritria Raf. = ***Carex*** L.
Edritria compressa (Hosé) Raf. = ***Carex canescens*** L. subsp. ***canescens***
Edritria concinna (R.Br.) Raf. = ***Carex concinna*** R.Br.
Edritria dasycarpa (Muhl.) Raf. = ***Carex dasycarpa*** Muhl.
Edritria digitalis (Willd.) Raf. = ***Carex digitalis*** Willd.
Edritria ferruginea (Scop.) Raf. = ***Carex ferruginea*** Scop.
Edritria formosa (Dewey) Raf. = ***Carex formosa*** Dewey
Edritria glaucescens (Elliott) Raf. = ***Carex glaucescens*** Elliott
Edritria laevigata (Sm.) Raf. = ***Carex laevigata*** Sm.
Edritria linkii (Willd.) Raf. = ***Carex distachya*** Desf. var. ***distachya***
Edritria livida (Wahlenb.) Raf. = ***Carex livida*** (Wahlenb.) Willd.
Edritria mucronata (All.) Raf. = ***Carex mucronata*** All.
Edritria paludosa (Gooden.) Raf. = ***Carex acutiformis*** Ehrh.
Edritria parviflora (Host) Raf. = ***Carex parviflora*** Host
Edritria petraea (Wahlenb.) Raf. = ***Carex rupestris*** All. subsp. ***rupestris***
Edritria pubescens Raf. = ***Carex hirtifolia*** Mack.
Edritria recurva Raf. = ***Carex cherokeensis*** Schwein.
Edritria rupestris (All.) Raf. = ***Carex rupestris*** All.
Edritria schkuhrii (Willd.) Raf. = ***Carex liparocarpos*** subsp. ***bordzilowskii*** (V.I.Krecz.) T.V.Egorova
Edritria supina (Willd. ex Wahlenb.) Raf. = ***Carex supina*** Willd. ex Wahlenb.
Edritria sylvatica (Huds.) Raf. = ***Carex sylvatica*** Huds.
Edritria viridula (Michx.) Raf. = ***Carex viridula*** Michx.
Edritria vulpina (L.) Raf. = ***Carex vulpina*** L.

Egleria

Egleria L.T.Eiten, Phytologia 9: 481 (1964).
SE. Colombia to N. Brazil. 82 VEN 83 CLM 84 BZN.
1 Species
Egleria fluctuans L.T.Eiten, Phytologia 9: 481 (1964).
SE. Colombia to N. Brazil. 82 VEN 83 CLM 84 BZN. Hydrocham.

Elaeocharis

Elaeocharis Brongn. = ***Eleocharis*** R.Br.

Eleocharis

Eleocharis R.Br., Prodr.: 224 (1810).
Cosmopolitan. 10 DEN FIN FOR GRB ICE IRE NOR SWE 11 AUT BGM CZE DEN FIN FOR GER GRB HUN ICE IRE NET NOR POL SWE SWI 12 BAL COR FRA POR SAR SPA 13 ALB BUL GRC ITA KRI ROM SIC TUE YUG 14 BLR BLT KRY RUC RUE RUN RUS RUW UKR 20 ALG EGY LBY MOR TUN 21 AZO CNY MDR 22 BEN BKN GAM GHA GNB GUI IVO LBR MLI MTN NGA NGR SEN SIE TOG 23 BUR CAF CMN CON EQG GAB GGI RWA ZAI 24 CHA ETH SOC SOM SUD 25 KEN TAN UGA 26? ANG MLW MOZ ZAM ZIM 27? BOT CPP LES NAM NAT OFS SWZ TVL 29 MAU MDG REU SEY 30 ALT BRY CTA IRK KRA TVA WSB YAK 31 KAM KUR PRM SAK 32 KAZ KGZ TKM TZK UZB 33 NCS TCS 34 AFG CYP IRN IRQ LBS PAL SIN TUR 35 GST SAU YEM 36+ CHC CHH CHM CHN CHS CHX 37 MON 38 JAP KOR NNS TAI 40 ASS BAN EHM IND NEP PAK SRL WHM 41 AND CBD MYA THA VIE 42 BOR CKI JAW LSI MLY MOL PHI SUL SUM XMS 43 BIS NWG SOL 50 NFK NSW NTA QLD SOA TAS VIC WAU 51 CTM KER NZN NZS 60 FIJ GIL NWC SAM TON 61 COO SCI 62 CRL MRN 63 HAW 70 ASK GNL NUN NWT YUK 71 ABT BRC MAN SAS 72 ABT BRC LAB MAN NBR NCS NFL NSC ONT PEI QUE 73 COL IDA MNT ORE WAS WYO 74 ILL IOW KAN MIN MSO NDA NEB OKL SDA WIS 75 CNT INI MAI MAS MIC NWH NWJ NWY OHI PEN RHO VER WVA 76 ARI CAL NEV UTA 77 NWM TEX 78 ALA ARK DEL FLA GEO KTY LOU MRY MSI NCA SCA TEN VRG WDC 79 MXC MXE MXG MXI MXN MXS MXT 80 BLZ COS ELS GUA HON NIC PAN 81 ARU BAH BER CAY CUB DOM HAI JAM LEE NLA PUE TRT VNA WIN 82 FRG GUY SUR VEN 83 BOL CLM ECU GAL PER 84 BZC BZE BZL BZN BZS 85 AGE AGS AGW CLC CLN CLS JNF PAR URU 90 FAL.
254 Species
Trichophyllum Ehrh., Beitr. Naturk. 4: 147 (1789), nom. inval.
Bulbostylis Steven, Mém. Soc. Imp. Naturalistes Moscou 5: 355 (1814), nom. illeg.

Heleocharis T.Lestib., Essai Cypér.: 41 (1819).
Limnochloa P.Beauv. ex Lestib., Essai Cypér.: 41 (1819).
Megadenus Raf., Neogenyton: 4 (1825).
Clavula Dumort., Fl. Belg.: 143 (1827).
Heliocharis Lindl., Syn. Brit. Fl.: 281 (1829).
Baeothryon Ehrh. ex A.Dietr., Sp. Pl. 2: 89 (1832).
Chaetocyperus Nees, Linnaea 9: 289 (1834).
Eleogenus Nees, Linnaea 9: 294 (1834).
Scirpidium Nees, Linnaea 9: 293 (1834).
Limnocharis Kunth, Enum. Pl. 2: 153 (1837).
Elaeocharis Brongn., Énum. Pl. Mus. Paris: 12 (1843), orth. var.
Chlorocharis Rikli, Jahrb. Wiss. Bot. 27: 564 (1895).
Chamaegyne Suess., Bot. Jahrb. Syst. 73: 113 (1943).
Helonema Suess., Bot. Jahrb. Syst. 73: 118 (1943).

Eleocharis abnorma Y.D.Chen, Bull. Bot. Res., Harbin 7(2): 124 (1987).
Qinghai. 36CHQ. Hemicr. or rhizome geophyte.

Eleocharis acicularis (L.) Roem. & Schult., Syst. Veg. 2: 154 (1817).
Temp. Northern Hemisphere to W. South America. 10 DEN FIN GRB ICE IRE NOR SWE 11 AUT BGM CZE GER HUN NET POL SWI 12 COR FRA POR SPA 13 ALB? BUL GRC ITA ROM TUE YUG 14 BLR BLT RUC RUE RUN RUS RUW UKR 20 MOR 30 TVA 31 PRM 32 KAZ 33 NCS TCS 34 AFG IRN TUR 36 CHC CHM CHN CHS 37 MON 38 JAP KOR NNS TAI 41 VIE 42 PHI SUM (60) gil (62) crl 70 ASK GNL NUN NWT YUK 71 ABT BRC MAN SAS 72 LAB NBR NFL NSC ONT PEI QUE 73 COL IDA MNT ORE WAS WYO 74 ILL IOW KAN MIN MSO NDA NEB OKL SDA WIS 75 CNT INI MAI MAS MIC NWH NWJ NWY OHI PEN RHO VER WVA 76 ARI CAL NEV UTA 77 NWM TEX 78 ALA ARK DEL FLA GEO KTY LOU MRY MSI NCA SCA TEN VRG WDC 79 MXC MXE MXG MXN MXS MXT 80 COS ELS GUA PAN 83 BOL CLM ECU. Hel. or hemicr.
**Scirpus acicularis* L., Sp. Pl.: 48 (1753). *Mariscus acicularis* (L.) Moench, Methodus: 350 (1794). *Cyperus acicularis* (L.) With., Arr. Brit. Pl. ed. 3, 2: 78 (1796). *Isolepis acicularis* (L.) Schltdl., Fl. Berol. 1: 36 (1823). *Clavula acicularis* (L.) Dumort., Fl. Belg.: 143 (1827). *Limnochloa acicularis* (L.) Rchb. in J.C.Mössler, Handb. Gewächsk. ed. 2, 3: 1808 (1830). *Scirpidium aciculare* (L.) Nees, Linnaea 9: 293 (1834). *Chaetocyperus acicularis* (L.) Nees in C.F.P.von Martius & auct. suc. (eds.), Fl. Bras. 2(1): 94 (1842). *Trichophyllum aciculare* (L.) House, Amer. Midl. Naturalist 6: 204 (1920). *Eleocharis acicularis* var. *typica* Svenson, Rhodora 31: 184 (1929), nom. inval.

subsp. ***acicularis***
Temp. Northern Hemisphere to W. South America. 10 DEN FIN GRB ICE IRE NOR SWE 11 AUT BGM CZE GER HUN NET POL SWI 12 COR FRA POR SPA 13 ALB? BUL GRC ITA ROM TUE YUG 14 BLR BLT RUC RUE RUN RUS RUW UKR 20 MOR 30 TVA 31 32 KAZ 33 NCS TCS 34 AFG IRN TUR 37 MON 38 NNS TAI 41 VIE 42 PHI SUM (60) gil (62) crl 70 ASK GNL NUN NWT YUK 71 ABT BRC MAN SAS 72 LAB NBR NFL NSC ONT PEI QUE 73 COL IDA MNT ORE WAS WYO 74 ILL IOW KAN MIN MSO NDA NEB OKL SDA WIS 75 CNT INI MAI MAS MIC NWH NWJ NWY OHI PEN RHO VER WVA 76 ARI CAL NEV UTA 77 NWM TEX 78 ALA ARK DEL FLA GEO KTY LOU MRY MSI NCA SCA TEN VRG WDC 79 MXC MXE MXG MXN MXS MXT 80 COS ELS GUA PAN 83 BOL CLM ECU. Hel. or hemicr.
Scirpus trichoides Kunth in F.W.H.von Humboldt, A.J.A.Bonpland & C.S.Kunth, Nov. Gen. Sp. 1: 225 (1816). *Baeothryon trichoides* (Kunth) A.Dietr., Sp. Pl. 2: 93 (1832). *Eleocharis trichoides* (Kunth) Kuntze, Enum. Pl. 2: 141 (1837).
Scirpus trichodes Muhl., Descr. Gram.: 30 (1817). *Chaetocyperus trichodes* (Muhl.) Steud., Syn. Pl. Glumac. 2: 73 (1854).
Eleocharis costata J.Presl & C.Presl, Fl. Cech.: 11 (1819).
Scirpus chaeta Schult. in J.J.Roemer & J.A.Schultes, Mant. 2: 72 (1824).
Eleogiton exigua A.Dietr., Sp. Pl. 2: 98 (1832).
Limnochloa costulata Nees ex Kunth, Enum. Pl. 2: 142 (1837), pro syn.
Scirpus pusillus Kunth, Enum. Pl. 2: 141 (1837).
Isolepis exigua Hook. & Arn., Bot. Beechey Voy.: 312 (1838).
Eleocharis acicularis var. *fluitans* Döll, Rhein. Fl.: 160 (1843). *Eleocharis acicularis* f. *fluitans* (Döll) Svenson, Rhodora 31: 186 (1929).
Scirpus acicularis var. *fluitans* Döll, Rhein. Fl.: 160 (1843).
Scirpus acicularis var. *rigidula* Rchb., Icon. Fl. Germ. Helv. 8: 37 (1846). *Eleocharis acicularis* f. *rigidula* (Rchb.) Svenson, Rhodora 31: 187 (1929).
Scirpus uliginosus K.Koch, Linnaea 21: 620 (1848), nom. illeg. *Isolepis uliginosa* Steud., Syn. Pl. Glumac. 2: 91 (1855).
Scirpus polymorphus Salzm. ex Steud., Syn. Pl. Glumac. 2: 74 (1854).
Isolepis longifolia Steud., Syn. Pl. Glumac. 2: 90 (1855).
Eleocharis atricha Benth., Fl. Austral. 7: 295 (1878), nom. illeg.
Eleocharis chaetaria Hance, J. Bot. 16: 112 (1878), nom. illeg.
Scirpus acicularis var. *submersa* Nilsson, Bot. Not. 1888: 144 (1888). *Eleocharis acicularis* f. *submersa* (Nilsson) Norman, Skr. Vidensk.-Selsk. Christiana, Math.-Naturvidensk. Kl. 16: 4388 (1893). *Eleocharis acicularis* var. *submersa* (Nilsson) Svenson, Rhodora 31: 188 (1929).
Eleocharis comosa K.Richt., Pl. Eur. 1: 143 (1890).
Eleocharis acicularis f. *submersa* Druce, Fl. Berkshire: 524 (1898).
Eleocharis triangularis Reinsch, Exsicc. (Herb. Norm): 4384 (1902). *Eleocharis acicularis* f. *triangularis* (Reinsch) Svenson, Rhodora 31: 187 (1929).
Scirpus acicularis f. *submersus* (Osten) Porsild, Meddel. Grønland 50: 370 (1912).
Eleocharis acicularis var. *gracilescens* Svenson, Rhodora 31: 191 (1929).
Eleocharis acicularis f. *inundata* Svenson, Rhodora 31: 186 (1929).
Eleocharis acicularis var. *occidentalis* Svenson, Rhodora 31: 190 (1929). *Eleocharis acicularis* f. *occidentalis* (Svenson) Beetle, Publ. Univ. Wyoming 5: 23 (1938).
Eleocharis nervata Svenson, Rhodora 31: 204 (1929).

var. ***porcata*** S.G.Sm., Novon 11: 255 (2001).
Canada to U.S.A. 71 ABT 73 COL 74 ILL KAN NDA NEB 76 ARI 77 NWM TEX 78 LOU. Hemicr.

subsp. ***yokoscensis*** (Franch. & Sav.) T.V.Egorova, Novosti Sist. Vyssh. Rast. 17: 69 (1980).
E. & S. Asia. 31 PRM 36 CHC CHM CHN CHS 37 MON 38 JAP KOR NNS 41 VIE 42 PHI SUM. Hel. or hemicr.
**Scirpus yokoscensis* Franch. & Sav., Enum. Pl. Jap. 2: 543 (1878). *Eleocharis yokoscensis* (Franch. & Sav.) Tang & F.T.Wang, in Fl. Reipubl. Popul. Sin. 11: 54 (1961).
Eleocharis acicularis var. *longiseta* Svenson, Rhodora 31: 189 (1929).
Eleocharis svensonii Zinserl. in V.L.Komarov (ed.), Fl. URSS 3: 580 (1935).

Eleocharis acuta R.Br., Prodr.: 224 (1810). *Scirpus acutus* (R.Br.) Spreng., Syst. Veg. 1: 203 (1824), nom. illeg. *Eleocharis palustris* var. *mucronulata* Boeck., Flora 61: 35 (1878).
New Guinea, Australia, New Zealand. 43 NWG 50 NFK NSW QLD SOA TAS VIC WAU 51 CTM KER NZN NZS. Hemicr. or rhizome geophyte.
Eleocharis mucronulata Nees, Ann. Mag. Nat. Hist. 6: 46 (1841). *Scirpus mucronulatus* (Nees) Kuntze, Revis. Gen. Pl. 2: 758 (1891).
Eleocharis mucronulata var. *minor* Nees, Ann. Mag. Nat. Hist. 6: 46 (1841).
Eleocharis gracilis Hook.f., Fl. Tasman. 2: 85 (1858), nom. illeg.
Hemicarpha nuda Boeck., Flora 12: 604 (1858).
Eleocharis acuta var. *platylepis* Hook.f., Handb. N. Zeal. Fl. 1: 745 (1867).
Eleocharis ambigua Kirk & Buchanan, Trans. & Proc. New Zealand Inst. 6: 225 (1873).
Eleocharis acuta var. *pallens* Benth., Fl. Austral. 7: 295 (1878).
Eleocharis acuta var. *tenuis* Carse, Trans. & Proc. New Zealand Inst. 57: 89 (1926).

Eleocharis acutangula (Roxb.) Schult. in J.J.Roemer & J.A.Schultes, Mant. 2: 91 (1824).
Trop. & Subtrop. 22 BEN BKN GAM GNB GUI IVO LBR MLI MTN NGA NGR SEN SIE TOG 23 CAF CMN GAB RWA ZAI 24 CHA 25 KEN TAN UGA 26 ZIM 27 BOT CPP NAM NAT SWZ TVL 29 MDG 36 CHH CHS 38 JAP NNS TAI 40 ASS NEP SRL 41 THA VIE 42 BOR JAW LSI MLY PHI SUL SUM 43 NWG 50 QLD 77 TEX 79 MXE MXG MXS MXT 80 BLZ COS ELS GUA HON NIC PAN 81 CUB 82 GUY SUR VEN 83 BOL CLM ECU GAL PER 84 BZC BZN BZS 85 AGE AGW PAR. Hemicr.
Scirpus fistulosus Poir. in J.B.A.M.de Lamarck, Encycl. 6: 749 (1805), nom. illeg. *Eleocharis fistulosa* Link, Jahrb. Gewächsk. 3: 78 (1820). *Baeothryon fistulosum* (Link) A.Dietr., Sp. Pl. 2: 94 (1832). *Limnochloa fistulosa* (Link) Nees, Linnaea 9: 294 (1834). *Scirpus fistulosus* (Link) Kuntze, Revis. Gen. Pl. 2: 757 (1891), nom. illeg.
Scirpus spiraloides Roxb., Hort. Bengal.: 81 (1814), nom. nud.
**Scirpus acutangulus* Roxb., Fl. Ind. 1: 216 (1820). *Limnochloa acutangula* (Roxb.) Nees in R.Wight, Contr. Bot. India: 114 (1834).
Scirpus medius Roxb., Fl. Ind. 1: 216 (1820). *Eleocharis media* (Roxb.) Schult. in J.J.Roemer & J.A.Schultes, Mant. 2: 91 (1824). *Limnochloa media* (Roxb.) Nees in R.Wight, Contr. Bot. India: 114 (1834).
Scirpus angulatus Willd. ex Kunth, Enum. Pl. 2: 155 (1837).
Eleocharis planiculmis Steud., Syn. Pl. Glumac. 2: 80 (1854).

Eleocharis aestuum D.M.Hines ex A.Haines, Novon 11: 45 (2001).
NE. U.S.A. to Delaware. 75 CNT MAI MAS NWJ NWY PEN 78 DEL. Hemicr. or rhizome geophyte.

Eleocharis albibracteata Nees & Meyen ex Kunth, Enum. Pl. 2: 143 (1837). *Chaetocyperus albibracteatus* (Nees & Meyen ex Kunth) Nees & Meyen, Nov. Actorum Acad. Caes. Leop.-Carol. Nat. Cur. 19(Suppl. 1): 95 (1843). *Scirpus albibracteatus* (Nees & Meyen ex Kunth) Kuntze, Revis. Gen. Pl. 2: 757 (1891).
S. Mexico to W. & S. South America. 79 MXS MXT 80 GUA 83 BOL ECU PER 85 AGE AGS AGW CLC CLN CLS JNF. Hemicr.

var. ***albibracteata***
Guatemala, W. & S. South America. 80 GUA 83 BOL PER 85 AGE AGS AGW CLC CLN CLS JNF. Hemicr.
Limnochloa albibracteata Nees ex Kunth, Enum. Pl. 2: 143 (1837), nom. inval.
Eleocharis melanocephala É.Desv. in C.Gay, Fl. Chil. 6: 175 (1854). *Scirpus melanocephalus* (É.Desv.) Griseb., Abh. Königl. Ges. Wiss. Göttingen 24: 311 (1879).

var. ***nubigena*** (C.B.Clarke) Boros in H.R.Descole (ed.), Gen. Sp. Pl. Argent. 4(1): 179 (1949).
Ecuador to NW. Argentina. 83 BOL ECU 85 AGW CLN. Hemicr.
Eleocharis crinalis f. *humilis* Boeck. ex C.B.Clarke, Bot. Jahrb. Syst. 37: 518 (1906).
**Eleocharis nubigena* C.B.Clarke, Bot. Jahrb. Syst. 37: 518 (1906).

Eleocharis albida Torr., Ann. Lyceum Nat. Hist. New York 3: 304 (1836). *Scirpus albidus* (Torr.) Kuntze, Revis. Gen. Pl. 2: 757 (1891). *Trichophyllum albidum* (Torr.) House, Amer. Midl. Naturalist 6: 204 (1920).
Bermuda, SC. & SE. U.S.A. to NE. Mexico. 77 NWM TEX 78 ALA FLA GEO LOU MRY MSI NCA SCA VRG? 79 MXE MXT 81 BER. Hemicr.
Scirpus simplex Elliott, Sketch Bot. S. Carolina 1: 76 (1816). *Eleocharis simplex* (Elliott) A.Dietr., Sp. Pl. 2: 78 (1832). *Trichophyllum simplex* (Elliott) House, Amer. Midl. Naturalist 6: 205 (1920).
Scirpus capitatus Muhl., Descr. Gram.: 80 (1817), nom. illeg.
Isolepis aphylla Boeck., Flora 41: 417 (1858).
Eleocharis berlandieri C.B.Clarke in I.Urban, Symb. Antill. 2: 162 (1900).
Eleocharis bermudiana Britton, J. New York Bot. Gard. 13: 191 (1912).

Eleocharis almensis D.A.Simpson, Kew Bull. 48: 703 (1993).
Brazil (Bahia, Minas Gerais). 84 BZE BZL. Hemicr.

Eleocharis alveolata Svenson, Rhodora 31: 241 (1929).
W. Cuba. 81 CUB. Hemicr.

Eleocharis alveolatoides S.González & Reznicek, Novon 6: 356 (1996).
Venezuela (Bolívar). 82 VEN. Hemicr.

Eleocharis amazonica C.B.Clarke, Kew Bull., Addit. Ser. 8: 22 (1908).
S. Venezuela to N. Brazil. 82 VEN 84 BZN. Hemicr.

Eleocharis andamanensis Govind., Proc. Indian Acad. Sci., Pl. Sci. 100: 71 (1990).
Andaman Is. 41 AND. Hemicr.

Eleocharis angolensis H.E.Hess, Ber. Schweiz. Bot. Ges. 63: 346 (1953).
Angola. 26 ANG. Hemicr.

Eleocharis antunesii H.E.Hess, Ber. Schweiz. Bot. Ges. 63: 334 (1953).
Angola. 26 ANG. Hemicr.

Eleocharis argentina Barros, Lilloa 12: 5 (1947).
Argentina (Mendoza). 85 AGW. Hemicr.

Eleocharis argyrolepis Kierulff in A.A.von Bunge, Beitr. Fl. Russl.: 342 (1852).
Caucasus to NW. China. 32 KGZ TKM TZK UZB 33 NCS 34 AFG IRN 36 CHX 40 PAK? Hemicr.
Eleocharis lehmannii Kierulff in A.A.von Bunge, Beitr. Fl. Russl.: 341 (1852).

Eleocharis arseniflora S.González, Tena & T.Alarcón, Brittonia 57: 150 (2005).
Mexico (Chihuahua). 79 MXE.

Eleocharis atricha R.Br., Prodr.: 225 (1810). *Scirpus atrichus* (R.Br.) Poir. in J.B.A.M.de Lamarck, Encycl., Suppl. 5: 103 (1817).
E. & SE. Australia. 50 NSW QLD SOA VIC. Hemicr.

Eleocharis atropurpurea (Retz.) Kunth, Enum. Pl. 2: 151 (1837).
Trop. & Subtrop. to U.S.A. (11) swi (13) ita 22 BEN BKN GHA GUI MLI MTN NGA NGR SEN TOG 23 GAB RWA ZAI 24 CHA SOM SUD 25 KEN TAN UGA 27 BOT NAM NAT SWZ TVL 29 MAU MDG 32 KAZ TZK UZB 34 AFG IRN TUR 36 CHC CHH CHS 38 JAP TAI 40 EHM IND NEP PAK 41 VIE 42 JAW PHI SUM 43 NWG 50 NTA QLD WAU 71 BRC 73 COL MNT WAS 74 IOW KAN MSO NEB OKL 75 MIC 76 CAL NEV 77 NWM TEX 78 ALA FLA GEO LOU NCA SCA VRG 79 MXN MXS MXT 80 COS NIC 81 ARU CAY NLA 83 BOL CLM ECU GAL PER 84 BZN 85 AGE AGW. Ther.
**Scirpus atropurpureus* Retz., Observ. Bot. 5: 14 (1788). *Isolepis atropurpurea* (Retz.) Roem. & Schult., Syst. Veg. 2: 106 (1817). *Aplostemon atropurpureum* (Retz.) Raf., J. Phys. Chim. Hist. Nat. Arts 89: 105 (1819). *Eleogiton atropurpurea* (Retz.) A.Dietr., Sp. Pl. 2: 97 (1832). *Eleogenus atropurpureus* (Retz.) Nees in R.Wight, Contr. Bot. India: 113 (1834). *Trichophyllum atropurpureum* (Retz.) House, Amer. Midl. Naturalist 6: 204 (1920).
Eleocharis lereschii Shuttlew., Flora 20: 241 (1837). *Eleogenus lereschii* (Shuttlew.) Bluff, Nees & Schauer, Comp. Fl. German. 2: 746 (1838). *Scirpus lereschii* (Shuttlew.) Thomas ex W.D.J.Koch, Syn. Fl. Germ. Helv., ed. 2: 853 (1844).
Eleogenus laetevirens Nees in C.F.P.von Martius & auct. suc. (eds.), Fl. Bras. 2(1): 103 (1842).
Eleogenus laetevirens var. *tenuis* Nees in C.F.P.von Martius & auct. suc. (eds.), Fl. Bras. 2(1): 103 (1842).
Eleogenus laetovirens var. *umbonatus* Nees in C.F.P.von Martius & auct. suc. (eds.), Fl. Bras. 2(1): 103 (1842).
Scirpus laetevirens Schrad. ex Nees in C.F.P.von Martius & auct. suc. (eds.), Fl. Bras. 2(1): 103 (1842). *Eleocharis laetevirens* (Schrad. ex Nees) Steud., Syn. Pl. Glumac. 2: 79 (1854).
Scirpus sulciculmis Schrad. ex Nees in C.F.P.von Martius & auct. suc. (eds.), Fl. Bras. 2(1): 103 (1842).
Scirpus erraticus Rota ex De Not., Ann. Sci. Nat., Bot., III, 5: 366 (1846). *Eleocharis erratica* (Rota ex De Not.) Steud., Syn. Pl. Glumac. 2: 79 (1854).
Eleocharis zanardinii Parl., Fl. Ital. 2: 67 (1852). *Scirpus zanardinii* (Parl.) Ambrosi, Fl. Tirolo Mer. 1: 448 (1854). *Eleocharis atropurpurea* var. *zanardinii* (Parl.) A.Terracc., Malpighia 2: 312 (1889).
Isolepis lereschii Shuttlew. ex Parl., Fl. Ital. 2: 68 (1852).
Scirpus acicularis Zanardini ex Parl., Fl. Ital. 2: 67 (1852), nom. illeg.
Eleocharis monandra Hochst. ex Steud., Syn. Pl. Glumac. 2: 75 (1854).
Isolepis allochroa Steud., Syn. Pl. Glumac. 2: 91 (1855).
Isolepis dichroa Steud., Syn. Pl. Glumac. 2: 91 (1855).
Eleocharis multiflora Chapm., Fl. South. U.S.: 517 (1860).
Eleocharis atropurpurea var. *albivaginata* Boeck., Linnaea 36: 459 (1870).
Eleocharis atropurpurea var. *trigyna* Boeck., Linnaea 36: 459 (1870).
Eleocharis atropurpurea var. *setiformis* Benth., Fl. Austral. 7: 297 (1878).

Eleocharis atrospiculata S.González & Reznicek, Novon 6: 357 (1996).
Venezuela (Cerro Marahuaka). 82 VEN. Hemicr.

Eleocharis attenuata (Franch. & Sav.) Palla, Monde Pl. 12: 40 (1910).
China to E. Asia, New Guinea. 31 PRM 36 CHM CHN CHS 38 JAP KOR NNS 41 VIE 43 NWG. Hemicr.
Eleocharis leviseta Nakai in ?
**Scirpus attenuatus* Franch. & Sav., Enum. Pl. Jap. 2: 543 (1878). *Eleocharis pellucida* f. *attenuata* (Franch. & Sav.) Ohwi, Mem. Coll. Sci. Kyoto Imp. Univ., Ser. B, Biol. 18: 40 (1944).
Eleocharis major Hara, J. Jap. Bot. 11: 820 (1935). *Eleocharis attenuata* var. *major* (Hara) Hara, J. Jap. Bot. 14: 224 (1938). *Eleocharis laeviseta* var. *major* (Hara) Hara, J. Jap. Bot. 14: 521 (1938).
Eleocharis attenuata var. *erhizomatosa* Tang & F.T. Wang, in Fl. Reipubl. Popul. Sin. 11: 226 (1961).

Eleocharis aurea Boeck., Beitr. Cyper. 1: 15 (1888).
NE. Mexico. 79 MXE. Hemicr.

Eleocharis ayacuchensis S.González & Reznicek, Novon 6: 358 (1996).
Venezuela (Amazonas). 82 VEN. Hemicr.

Eleocharis bahamensis Boeck., Beitr. Cyper. 2: 11 (1890).
Bahamas. 81 BAH. Hemicr.
Eleocharis bahamensis f. *minor* Boeck., Beitr. Cyper. 2: 12 (1890).
Eleocharis camptotricha var. *schweinitzii* C.B.Clarke in I.Urban, Symb. Antill. 2: 69 (1900).

Eleocharis bahiensis D.A.Simpson, Kew Bull. 43: 127 (1988).
Brazil (Bahia). 84 BZE. Hemicr.

Eleocharis baldwinii (Torr.) Chapm., Fl. South. U.S.: 519 (1860).

SE. U.S.A. to Texas. 77 TEX 78 ALA ARK FLA GEO LOU MSI NCA SCA. Hemicr.
Chaetocyperus baldwinii Torr., Ann. Lyceum Nat. Hist. New York 3: 295 (1836). *Trichophyllum baldwinii* (Torr.) House, Amer. Midl. Naturalist 6: 204 (1920).
Eleocharis prolifera Torr., Ann. Lyceum Nat. Hist. New York 3: 316 (1836). *Trichophyllum proliferum* (Torr.) House, Amer. Midl. Naturalist 6: 205 (1920).

Eleocharis barrosii Svenson, Rhodora 39: 220 (1937).
Bolivia to S. Brazil. 83 BOL 84 BZS 85 AGE PAR. Hemicr.

Eleocharis bella (Piper) Svenson, Rhodora 31: 201 (1929).
W. U.S.A. to New Mexico, N. Mexico. 73 IDA MNT ORE WAS WYO 76 ARI CAL NEV UTA 77 NWM 79 MXE MXN. Ther.
Eleocharis acicularis var. *minima* Torr., J. New York Microscop. Soc. 5: 104 (1889).
**Eleocharis acicularis* var. *bella* Piper in C.V.Piper & R.K.Beattie, Fl. Palouse Reg.: 35 (1901).

Eleocharis bernardina (Munz & I.M.Johnst.) Munz & I.M.Johnst. in P.A.Munz, Man. S. Calif. Bot.: 597 (1935).
California to Mexico (Baja California). 76 CAL 79 MXN. Hemicr.
**Scirpus bernardinus* Munz & I.M.Johnst., Bull. Torrey Bot. Club 52: 221 (1925). *Eleocharis pauciflora* var. *bernardina* (Munz & I.M.Johnst.) Svenson, Rhodora 31: 174 (1929). *Eleocharis quinqueflora* var. *bernardina* (Munz & I.M.Johnst.) S.González & P.M.Peterson, Sida 17: 414 (1996).

Eleocharis bifida S.G.Sm., Novon 11: 243 (2001).
SE. U.S.A. 78 ALA GEO KTY TEN. Hemicr. or rhizome geophyte.

Eleocharis blakeana L.A.S.Johnson & O.D.Evans, Contr. New South Wales Natl. Herb. 4: 70 (1968).
E. Australia. 50 NSW QLD. Hemicr.

Eleocharis bolanderi A.Gray, Proc. Amer. Acad. Arts 7: 392 (1868). *Eleocharis montevidensis* var. *bolanderi* (A.Gray) V.E.Grant, Aliso 4: 87 (1958).
W. U.S.A. 73 COL IDA ORE 76 CAL NEV UTA. Hemicr.

Eleocharis bonariensis Nees, J. Bot. (Hooker) 2: 398 (1840). *Chaetocyperus bonariensis* (Nees) Nees in C.F.P.von Martius & auct. suc. (eds.), Fl. Bras. 2(1): 96 (1842). *Scirpus bonariensis* (Nees) Kuntze, Revis. Gen. Pl. 2: 757 (1891). *Eleocharis acicularis* subsp. *bonariensis* (Nees) Osten, Anales Mus. Hist. Nat. Montevideo, II, 3: 174 (1931).
Mexico, Ecuador to S. South America. (12) fra 79 MXC MXE MXG MXS 83 BOL ECU 84 BZS 85 AGE AGS AGW CLC CLN CLS PAR URU. Hemicr.
Chaetocyperus bonariensis var. *fluitans* Nees in C.F.P. von Martius & auct. suc. (eds.), Fl. Bras. 2(1): 96 (1842).
Chaetocyperus obtusatus Nees in C.F.P.von Martius & auct. suc. (eds.), Fl. Bras. 2(1): 94 (1842).
Eleocharis striatula É.Desv. in C.Gay, Fl. Chil. 6: 173 (1854). *Scirpus striatulus* (É.Desv.) Griseb., Abh. Königl. Ges. Wiss. Göttingen 19: 266 (1874).
Eleocharis oxyoneura Durand, Bull. Soc. Bot. France 2: 609 (1855).
Eleocharis amphibia Durand, Actes Soc. Linn. Bordeaux 31: 487 (1858). *Scirpus amphibius* (Durand) Bonnier, Fl. Ill. France 11: 80 (1930).

Eleocharis aciculariformis Greenm., Proc. Amer. Acad. Arts 34: 566 (1899).
Eleocharis acicularis var. *lilliputiana* Speg., Anales Mus. Nac. Buenos Aires 7: 179 (1902).
Eleocharis acicularis f. *filiformis* Osten, Anales Mus. Hist. Nat. Montevideo, II, 3: 174 (1931).

Eleocharis brachycarpa Svenson, Rhodora 31: 200 (1929).
Texas to Mexico (Tamaulipas). 77 TEX 79 MXE. Ther.

Eleocharis brainii Svenson, Rhodora 39: 251 (1937).
Trop. Africa. 22 GHA GUI NGA SEN 23 CMN 24 CHA SUD 25 KEN TAN UGA 26 MOZ ZAM. Ther.

Eleocharis brasiliensis Boeck., Beitr. Cyper. 2: 13 (1890).
SE. Brazil. 84 BZE. Hemicr.

Eleocharis brassii S.T.Blake, Proc. Roy. Soc. Queensland 49: 154 (1937 publ. 1938).
N. Australia. 50 NTA QLD WAU. Hemicr.

Eleocharis braunii H.E.Hess, Ber. Schweiz. Bot. Ges. 67: 91 (1957).
N. Brazil. 84 BZN. Hemicr.

Eleocharis brevicollis J.Kern, Blumea 13: 117 (1965).
New Guinea. 43 NWG. Hemicr.

Eleocharis brittonii Svenson ex Small, Man. S.E. Fl.: 164 (1933). *Eleocharis microcarpa* var. *brittonii* (Svenson ex Small) Svenson, Rhodora 39: 230 (1937).
C. & SE. U.S.A. 74 MSO 77 TEX 78 ALA ARK FLA GEO LOU MSI SCA TEN. Ther.

Eleocharis bulbosa Kukkonen, Ann. Bot. Fenn. 27: 112 (1990).
Afghanistan. 34 AFG. Hemicr.

Eleocharis caduca (Delile) Schult. in J.J.Roemer & J.A.Schultes, Mant. 2: 88 (1824).
SW. Sardegna, W. Kriti, Africa to Arabian Pen. 12 SAR 13 KRI 20 ALG EGY LBY 22 NGR 24 CHA ETH SOC SOM SUD 25 TAN 26 MLW ZIM 27 BOT CPP NAT TVL 29 MAU MDG 34 IRN SIN 35 YEM. Ther. or hemicr.
**Scirpus caducus* Delile, Descr. Egypte, Hist. Nat.: 153 (1813).
Eleocharis caduca var. *gracilescens* A.Terracc., Malpighia 2: 305 (1889).
Eleocharis intricata Kük., Repert. Spec. Nov. Regni Veg. 13: 135 (1914).
Eleocharis madagascariensis Cherm., Bull. Soc. Bot. France 75: 284 (1928).
Eleocharis ovata var. *gaetula* Maire, Bull. Soc. Hist. Nat. Afrique N. 22: 70 (1931).
Eleocharis tibestica Quézel, Bull. Soc. Hist. Nat. Afrique N. 48: 85 (1957).
Eleocharis intricata var. *peteri* W.Schultze-Motel, Willdenowia 2: 507 (1960).

Eleocharis caespitosissima Baker, J. Linn. Soc., Bot. 21: 450 (1885).
Madagascar, N. Australia. 29 MDG 50 NTA WAU. Ther. or hemicr.

Eleocharis callensii H.E.Hess, Ber. Schweiz. Bot. Ges. 62: 341 (1953).
Zaïre. 23 ZAI. Hemicr.

Eleocharis callosa Y.D.Chen, Bull. Bot. Res., Harbin 7(2): 123 (1987).
Qinghai. 36 CHQ. Hemicr. or rhizome geophyte.

Eleocharis cancellata S.Watson, Proc. Amer. Acad. Arts 18: 170 (1883).
Texas to C. Mexico. 77 TEX 79 MXC MXE. Ther.

Eleocharis canindeyuensis Mereles & S.González, Candollea 58: 75 (2003).
Paraguay. 85 PAR.

Eleocharis capillacea Kunth, Enum. Pl. 2: 139 (1837). *Chaetocyperus capillaceus* (Kunth) Nees in C.F.P.von Martius & auct. suc. (eds.), Fl. Bras. 2(1): 93 (1842). *Scirpus capillaceus* (Kunth) Griseb., Cat. Pl. Cub.: 239 (1866), nom. illeg.
S. Trop. America. 82 VEN 83 ECU 84 BZC BZL BZN BZS 85 AGE PAR. Hemicr. or hel.

Eleocharis carniolica W.D.J.Koch, Syn. Fl. Germ. Helv., ed. 2: 853 (1844). *Scirpus carniolicus* (W.D.J. Koch) Ambrosi, Fl. Tirolo Mer. 1: 447 (1854).
C. Europe to Ukraine. 11 AUT CZE HUN 13 BUL ITA ROM YUG 14 UKR. Hemicr.
Eleocharis bartoliana De Not., Index Seminum (G) 1848: 24 (1848).
Eleocharis carniolica var. *macrostachya* Boeck., Linnaea 36: 425 (1870).

Eleocharis cellulosa Torr., Ann. Lyceum Nat. Hist. New York 3: 298 (1836). *Scirpus cellulosus* (Torr.) Kuntze, Revis. Gen. Pl. 2: 757 (1891). *Trichophyllum cellulosum* (Torr.) House, Amer. Midl. Naturalist 6: 204 (1920).
SE. U.S.A. to C. America, Caribbean, W. South America. 77 TEX 78 ALA ARK FLA GEO MSI NCA SCA 79 MXE MXG MXT 80 BLZ GUA HON NIC 81 BAH BER CAY CUB JAM LEE PUE 82 VEN 83 ECU. Hemicr.
Scirpus dictyospermus C.Wright, Anales Acad. Ci. Méd. Habana 8: 79 (1871).

Eleocharis chamaegyne L.T.Eiten, Phytologia 22: 423 (1972).
N. Brazil. 84 BZN. Hemicr.
**Chamaegyne pygmaea* Suess., Bot. Jahrb. Syst. 73: 114 (1943). *Eleocharis pygmaea* (Suess.) L.T.Eiten, Phytologia 20: 273 (1970), nom. illeg.

Eleocharis coloradoensis (Britton) Gilly, Amer. Midl. Naturalist 26: 66 (1941).
S. Canada to W. & EC. U.S.A., Mexico. 71 MAN SAS 73 COL IDA ORE 74 IOW KAN MIN MSO NDA OKL SDA 76 CAL NEV UTA 77 NWM TEX 78 ALA ARK LOU MSI 79 MXE MXG MXN. Tuber or rhizome geophyte.
Eleocharis pygmaea var. *anachaeta* Torr., Ann. Lyceum Nat. Hist. New York 3: 441 (1836). *Scirpus nanus* var. *anachaetus* (Torr.) Britton, Trans. New York Acad. Sci. 11: 75 (1892). *Eleocharis parvula* var. *anachaeta* (Torr.) Svenson, Rhodora 36: 386 (1934).
**Scirpus coloradoensis* Britton, Torreya 4: 93 (1904). *Eleocharis leptos* var. *coloradoensis* (Britton) Svenson, Rhodora 31: 177 (1929). *Eleocharis parvula* var. *coloradoensis* (Britton) Beetle, Publ. Univ. Wyoming 5: 22 (1938).

Eleocharis columbiensis L.E.Moro, Caldasia 12: 278 (1978).
Colombia. 83 CLM. Hemicr.

Eleocharis complanata Boeck., Flora 62: 562 (1879).
W. Trop. Africa to Tanzania. 22 BEN GHA GNB IVO MLI NGA NGR SEN SIE TOG 23 CMN 25 TAN. Ther.
Eleocharis anceps Ridl., Trans. Linn. Soc. London, Bot. 2: 148 (1884).

Eleocharis compressa Sull., Amer. J. Sci. Arts 42: 50 (1842). *Eleocharis elliptica* var. *compressa* (Sull.) Drapalik & Mohlenbr., Amer. Midl. Naturalist 64: 503 (1960).
Canada to WC. & E. U.S.A. 71 BRC MAN SAS 72 ONT QUE 73 COL 74 ILL IOW KAN MIN MSO NDA NEB OKL SDA WIS 75 INI MIC NWY OHI PEN WVA 77 NWM TEX 78 ALA ARK KTY MRY MSI TEN VRG WDC. Hemicr.

var. ***acutisquamata*** (Buckley) S.G.Sm., Novon 11: 249 (2001).
S. Canada to WC. & C. U.S.A. 71 BRC MAN SAS 73 COL 74 ILL IOW KAN MIN MSO NDA NEB OKL SDA 77 NWM TEX. Hemicr. or rhizome geophyte.
**Eleocharis acutisquamata* Buckley, Proc. Acad. Nat. Sci. Philadelphia 1862: 10 (1863).

var. ***compressa***
C. & SE. Canada to C. & E. U.S.A. 72 ONT QUE 73 COL 74 ILL IOW KAN MIN MSO NDA NEB SDA WIS 75 INI MIC NWY OHI PEN WVA 77 NWM TEX 78 ALA ARK KTY MRY MSI TEN VRG WDC. Hemicr.
Scirpus acuminatus Muhl., Descr. Gram.: 27 (1817), provisional synonym. *Eleocharis acuminata* (Muhl.) Nees, Linnaea 9: 294 (1834). *Trichophyllum acuminatum* (Muhl.) House, Amer. Midl. Naturalist 6: 204 (1920).
Eleocharis acuminata var. *tenerrima* Boeck., Flora 61: 35 (1878).

Eleocharis congesta D.Don, Prodr. Fl. Nepal.: 41 (1825). *Scirpus congestus* (D.Don) Buch.-Ham. ex D.Don, Prodr. Fl. Nepal.: 41 (1825).
Himalaya to S. China and Malesia. 36 CHC 38 NNS 40 EHM NEP SRL WHM 41 THA VIE 42 BOR JAW MLY PHI SUL SUM 62 CRL. Ther. or hemicr.
Eleocharis purpurascens Boeck., Linnaea 36: 455 (1870). *Scirpus purpurascens* (Boeck.) Kuntze, Revis. Gen. Pl. 2: 758 (1891).
Eleocharis subvivipara Boeck., Linnaea 36: 424 (1870). *Scirpus subviviparus* (Boeck.) Kuntze, Revis. Gen. Pl. 2: 758 (1891). *Eleocharis congesta* subsp. *subvivipara* (Boeck.) T.Koyama, Micronesica 1: 78 (1964).

Eleocharis contracta Maury ex Micheli, Mém. Soc. Phys. Genève 31(1): 139 (1890).
Brazil to S. South America. 84 BZC 85 AGE AGW PAR URU. Hemicr.
Eleocharis posadensis Kük., Feddes Repert. Spec. Nov. Regni Veg. 53: 271 (1944).

Eleocharis coronata Steud., Syn. Pl. Glumac. 2: 77 (1854).
SC. Chile. 85 CLC. Hemicr.

Eleocharis crinalis (Griseb.) C.B.Clarke, Kew Bull., Addit. Ser. 8: 23 (1908).
Ecuador to N. Argentina. 83 BOL ECU 85 AGE AGW PAR. Hemicr.
**Scirpus crinalis* Griseb., Abh. Königl. Ges. Wiss. Göttingen 19: 265 (1874).
Eleocharis brehmeriana Boeck., Allg. Bot. Z. Syst. 2: 33 (1896).
Eleocharis boliviana Palla ex Svenson, Rhodora 31: 179 (1929). *Eleocharis albibracteata* var. *boliviana* (Palla ex Svenson) S.González & P.M.Peterson, Sida 17: 414 (1996).

Eleocharis crispovaginata Boeck., Bot. Jahrb. Syst. 8: 206 (1887).
Ecuador. 83 ECU. Hemicr.

Eleocharis cubangensis H.E.Hess, Ber. Schweiz. Bot. Ges. 62: 339 (1953).
Angola. 26 ANG 27? Hemicr.

Eleocharis cylindrica Buckley, Proc. Acad. Nat. Sci. Philadelphia 1862: 10 (1863). *Trichophyllum cylindricum* (Buckley) House, Amer. Midl. Naturalist 6: 204 (1920).
SC. U.S.A., Paraguay to N. Argentina. 77 NWM? TEX 85 AGE PAR. Hemicr.
Eleocharis texana Britton, Bull. Torrey Bot. Club 11: 87 (1884).
Eleocharis spegazzinii Barros, Anales Mus. Nac. Buenos Aires 34: 474 (1928).

Eleocharis cylindrostachys Boeck., Flora 58: 108 (1875).
E. Australia. 50 NSW QLD. Hemicr.

Eleocharis debilis Kunth, Enum. Pl. 2: 143 (1837). *Scirpus debilis* (Kunth) Kuntze, Revis. Gen. Pl. 2: 757 (1891), nom. illeg.
Trop. America. 80 BLZ HON 81 CUB PUE 82 GUY VEN 84 BZL BZS 85 PAR. Hemicr. or rhizome geophyte.
Eleocharis macra Kunth, Enum. Pl. 2: 142 (1837). *Eleocharis debilis* f. *macra* (Kunth) Boeck., Repert. Spec. Nov. Regni Veg. 23: 192 (1926).
Eleocharis shaferi Britton, Mem. Torrey Bot. Club 16: 59 (1920).
Eleocharis yunquensis Britton, Sci. Surv. Porto Rico & Virgin Islands 5: 92 (1923).
Eleocharis ekmanii Kük., Repert. Spec. Nov. Regni Veg. 23: 192 (1926).
Eleocharis debilis f. *anomala* S.González & Reznicek, Novon 6: 358 (1996).

Eleocharis decoriglumis Berhaut, Bull. Soc. Bot. France 100: 174 (1953).
W. Trop. Africa to Tanzania. 22 MLI NGR SEN 23 CMN 24 CHA 25 TAN UGA. Ther.

Eleocharis decumbens C.B.Clarke, Kew Bull., Addit. Ser. 8: 23 (1908). *Eleocharis montevidensis* var. *decumbens* (C.B.Clarke) V.E.Grant, Aliso 4: 87 (1958).
Oregon to N. California. 73 ORE 76 CAL. Hemicr.

Eleocharis deightonii S.S.Hooper, Kew Bull. 26: 582 (1972).
Guinea to Sierra Leone. 22 GNB GUI SIE.

Eleocharis densa Benth., Pl. Hartw.: 27 (1840). *Eleocharis geniculata* var. *densa* (Benth.) Boeck. ex Svenson, Rhodora 39: 262 (1937).
Mexico to Guatemala. 79 MXC MXE MXG MXS 80 GUA. Hemicr. or rhizome geophyte.

Eleocharis diandra C.Wright, Bull. Torrey Bot. Club 10: 101 (1883). *Trichophyllum diandrum* (C.Wright) House, Amer. Midl. Naturalist 6: 204 (1920).
SE. Canada to NE. U.S.A. 72 ONT 75 CNT MAS NWH NWY VER. Ther.
Eleocharis diandra var. *depressa* Fernald, Rhodora 2: 60 (1900).

Eleocharis dietrichiana Boeck., Flora 58: 107 (1875).
E. Australia. 50 NSW QLD. Hemicr.

Eleocharis difformis S.T.Blake, Proc. Roy. Soc. Queensland 50: 99 (1938 publ. 1939).
Queensland. 50 QLD. Hemicr.

Eleocharis dombeyana Kunth, Enum. Pl. 2: 145 (1837). *Scirpus dombeyanus* (Kunth) Kuntze, Revis. Gen. Pl. 2: 757 (1891).
Mexico to C. America, W. & S. South America. 79 MXC MXE MXG MXN MXS MXT 80 GUA HON 83 BOL CLM ECU PER 85 AGE AGW CLC. Hemicr.
Chaetocyperus stoloniferus Nees, Linnaea 19: 695 (1847).
Eleocharis truncata Schltdl., Bot. Zeitung (Berlin) 7: 118 (1849). *Limnochloa truncata* (Schltdl.) Liebm., Mexic. Neldeagt. Pl., V, 2: 244 (1851).
Eleocharis bivaginata Steud., Syn. Pl. Glumac. 2: 77 (1854).
Eleocharis stolonifera Boeck., Linnaea 36: 424 (1870). *Scirpus stolonifer* (Boeck.) Kuntze, Revis. Gen. Pl. 2: 758 (1891).

Eleocharis dregeana Steud., Syn. Pl. Glumac. 2: 78 (1854).
S. Africa. 27 CPP LES NAT OFS SWZ TVL. Hemicr.

Eleocharis dulcis (Burm.f.) Trin. ex Hensch., Vita Rumphii: 186 (1833).
Trop. & Subtrop. Old World. 22 BKN GAM GHA GNB GUI IVO LBR NGA SEN SIE TOG 23 GAB 24 CHA 25 KEN TAN UGA 26 ZAM 27 BOT NAM 29 MAU MDG SEY 36 CHC CHH CHS 38 JAP NNS TAI 40 IND PAK SRL 41 THA VIE 42 JAW LSI MLY MOL PHI SUL SUM 43 NWG 50 NSW NTA QLD WAU 60 FIJ NWC SAM TON 62 CRL MRN. Hemicr. or tuber geophyte. Widely cultivated for its edible tubers.
Scirpus plantaginoides Rottb., Descr. Icon. Rar. Pl.: 45 (1773). *Eleocharis plantaginoidea* W.Wight, Contr. U. S. Natl. Herb. 9: 267 (1905). *Eleocharis plantaginoides* (Rottb.) Domin, Biblioth. Bot. 85: 445 (1915).
Scirpus plantagineus Retz., Observ. Bot. 5: 14 (1788). *Eleocharis plantaginea* (Retz.) Roem. & Schult., Syst. Veg. 2: 150 (1817). *Limnochloa plantaginea* (Retz.) Nees in R.Wight, Contr. Bot. India: 114 (1834).
Scirpus tuberosus Roxb., Pl. Coromandel 3: 25 (1815), nom. illeg. *Eleocharis tuberosa* (Roxb.) Schult. in J.J.Roemer & J.A.Schultes, Mant. 2: 86 (1824). *Eleocharis dulcis* var. *tuberosa* (Roxb.) T.Koyama, J. Fac. Sci. Univ. Tokyo, Sect. 3, Bot. 8: 97 (1961).
Scirpus dubius Roxb., Fl. Ind. 1: 215 (1820).
Scirpus tumidus Roxb., Fl. Ind. 1: 212 (1820). *Eleocharis tumida* (Roxb.) Schult. in J.J.Roemer & J.A.Schultes, Mant. 2: 86 (1824). *Limnochloa tumida* (Roxb.) Nees in R.Wight, Contr. Bot. India: 114 (1834).
Eleocharis equisetina C.Presl, Reliq. Haenk. 1: 195 (1828).
Carex tuberosa Blanco, Fl. Filip.: 35 (1837), nom. illeg.
Scirpus spiralis Willd. ex Kunth, Enum. Pl. 2: 153 (1837).
Eleocharis austrocaledonica Vieill., Ann. Sci. Nat., Bot., IV, 16: 38 (1861).
Eleocharis esculenta Vieill., Ann. Sci. Nat., Bot., IV, 16: 37 (1862).
Eleocharis plantaginea var. *stolonifera* Boeck., Beitr. Cyper. 2: 14 (1890).
Eleocharis indica (Lour.) Druce, Bot. Soc. Exch. Club Brit. Isles 1916: 621 (1917).
Eleocharis plantagineiformis Tang & F.T.Wang, in Fl. Reipubl. Popul. Sin. 11: 225 (1961).

Eleocharis dunensis Kük., Repert. Spec. Nov. Regni Veg. 12: 94 (1913).
NE. Argentina to Uruguay. 85 AGE URU. Hemicr.

Eleocharis eglerioides S.González & Reznicek, Novon 6: 359 (1996).
S. Venezuela, Bolivia. 82 VEN 83 BOL. Hydrohemicr.

Eleocharis elegans (Kunth) Roem. & Schult., Syst. Veg. 2: 150 (1817).
Trop. America to N. Argentina. 79 MXC MXE MXG MXS MXT 80 BLZ COS ELS GUA HON NIC PAN 81 JAM LEE PUE TRT WIN 82 FRG GUY SUR VEN 83 BOL CLM ECU PER 84 BZC BZL BZN BZS 85 AGE AGW PAR URU. Hemicr.
**Scirpus elegans* Kunth in F.W.H.von Humboldt, A.J.A. Bonpland & C.S.Kunth, Nov. Gen. Sp. 1: 226 (1816).
Eleocharis constricta Schult. in J.J.Roemer & J.A.Schultes, Mant. 2: 87 (1824).
Limnochloa crassiculmis Nees in C.F.P.von Martius & auct. suc. (eds.), Fl. Bras. 2(1): 99 (1842). *Eleocharis crassiculmis* (Nees) Steud., Syn. Pl. Glumac. 2: 81 (1855). *Scirpus crassiculmis* (Nees) Schrad. ex Griseb., Symb. Fl. Argent.: 312 (1879).
Eleocharis viridis Boeck., Beitr. Cyper. 1: 17 (1888). *Scirpus viridis* (Boeck.) Kuntze, Revis. Gen. Pl. 2: 758 (1891).
Eleocharis mexicana Palla, Oesterr. Bot. Z. 63: 402 (1913), nom. illeg.

Eleocharis elliptica Kunth, Enum. Pl. 2: 146 (1837).
C. & E. Canada, N. & EC. U.S.A. 70 NWT YUK 71 ABT BRC MAN SAS 72 NFL NSC ONT QUE 73 IDA MNT WAS WYO 74 ILL IOW MIN NDA WIS 75 CNT INI MAI MAS MIC NWH NWJ NWY OHI PEN VER WVA 78 GEO LOU TEN. Hemicr.
Eleocharis capitata var. *borealis* Svenson, Rhodora 34: 200 (1932). *Eleocharis tenuis* var. *borealis* (Svenson) Gleason, Phytologia 4: 22 (1952). *Eleocharis compressa* var. *borealis* (Svenson) Drapalik & Mohlenbr., Amer. Midl. Naturalist 63: 147 (1960).
Eleocharis compressa var. *atrata* Svenson, Rhodora 34: 218 (1932). *Eleocharis elliptica* f. *atrata* (Svenson) Drapalik & Mohlenbr., Amer. Midl. Naturalist 64: 502 (1960). *Eleocharis tenuis* var. *atrata* (Svenson) B.Boivin, Naturaliste Canad. 94: 524 (1967). *Eleocharis elliptica* var. *atrata* (Svenson) S.G.Sm., Novon 11: 243 (2001).

Eleocharis elongata Chapm., Fl. South. U.S.: 515 (1860). *Trichophyllum elongatum* (Chapm.) House, Amer. Midl. Naturalist 6: 204 (1920).
SE. U.S.A. to C. America, W. Jamaica. 77 TEX 78 ALA FLA NCA 80 BLZ NIC 81 JAM 84? Hemicr. or rhizome geophyte.

Eleocharis emarginata (Nees) Klotzsch ex Boeck., Linnaea 36: 443 (1870).
Brazil to N. Argentina. 83 BOL 84 BZL 85 AGE PAR. Hemicr.
**Chaetocyperus emarginatus* Nees in C.F.P.von Martius & auct. suc. (eds.), Fl. Bras. 2(1): 96 (1842). *Scirpus emarginatus* (Nees) Kuntze, Revis. Gen. Pl. 2: 757 (1891). *Chlorocharis emarginata* (Nees) Rikli, Jahrb. Wiss. Bot. 27: 564 (1895).
Eleocharis quinquangularis Boeck., Beitr. Cyper. 1: 15 (1888).

Eleocharis engelmannii Steud., Syn. Pl. Glumac. 2: 79 (1854). *Eleocharis ovata* var. *engelmannii* (Steud.) Britton, J. New York Microscop. Soc. 5: 103 (1889). *Trichophyllum engelmannii* (Steud.) House, Amer. Midl. Naturalist 6: 204 (1920). *Eleocharis obtusa* var. *engelmannii* (Steud.) Gilly, Iowa State Coll. J. Sci. 21: 92 (1946).
SC. Canada to U.S.A. 71 ABT BRC MAN 72 ONT 73 COL IDA MNT ORE WAS 74 ILL IOW KAN MIN MSO NDA NEB OKL WIS 75 CNT INI MAS MIC NWH NWJ NWY OHI PEN RHO WVA 76 ARI CAL 77 NWM TEX 78 ALA ARK DEL GEO KTY MRY TEN VRG WDC. Ther.
Eleocharis engelmannii var. *detonsa* A.Gray in H.N.Patterson, Cat. Pl. Illinois: 46 (1876). *Eleocharis engelmannii* f. *detonsa* (A.Gray) Svenson, Rhodora 31: 208 (1929).
Eleocharis engelmannii var. *robusta* Fernald, Proc. Amer. Acad. Arts 34: 496 (1899).
Eleocharis monticola Fernald, Proc. Amer. Acad. Arts 34: 496 (1899). *Trichophyllum monticola* (Fernald) House, Amer. Midl. Naturalist 6: 205 (1920). *Eleocharis engelmannii* var. *monticola* (Fernald) Svenson, Rhodora 31: 209 (1929).
Eleocharis monticola var. *laeviseta* Fernald, Proc. Amer. Acad. Arts 34: 496 (1899). *Eleocharis engelmannii* f. *laeviseta* (Fernald) Svenson, Rhodora 31: 210 (1929).
Eleocharis monticola var. *pallida* H.St.John, N. W. Sci. 2(3): 81 (1928).

Eleocharis equisetoides (Elliott) Torr., Ann. Lyceum Nat. Hist. New York 3: 296 (1836).
S. Ontario to C. & E. U.S.A. 72 ONT 74 ILL† MSO† WIS† 75 CNT INI MAS MIC NWY RHO 76 CAL 77 TEX 78 ALA ARK? FLA GEO LOU MRY MSI NCA SCA TEN? VRG. Hemicr.
**Scirpus equisetoides* Elliott, Sketch Bot. S. Carolina 1: 79 (1816). *Eleocharis elliottii* A.Dietr., Sp. Pl. 2: 82 (1832), nom. illeg. *Limnochloa equisetoides* (Elliott) Walp., Ann. Bot. Syst. 3: 684 (1852).

Eleocharis erhaiensis Y.D.Chen, Bull. Bot. Res., Harbin 7(2): 122 (1987).
Qinghai. 36CHQ. Hemicr.

Eleocharis erythropoda Steud., Syn. Pl. Glumac. 2: 76 (1854).
Hawaiian Is., N. America. 63 HAW 70 ASK NWT 71 ABT BRC MAN SAS 72 NBR ONT QUE 73 COL IDA MNT ORE WAS WYO 74 ILL IOW KAN MIN MSO NDA NEB OKL SDA WIS 75 CNT INI MAI MAS MIC NWH NWJ NWY OHI PEN RHO VER WVA 76 ARI UTA 77 NWM TEX 78 ARK DEL GEO KTY LOU MRY MSI NCA TEN VRG. Hemicr.
Scirpus glaucus Torr., Fl. N. Middle United States 1: 44 (1823), nom. illeg.
Eleocharis palustris var. *australis* Nees, Nov. Actorum Acad. Caes. Leop.-Carol. Nat. Cur. 19(Suppl. 1): 96 (1843). *Eleocharis calva* var. *australis* (Nees) H.St.John, Pacific Sci. 13: 159 (1959).
Eleocharis palustris var. *calva* A.Gray, Manual: 522 (1848). *Eleocharis palustris* subvar. *calva* (A.Gray) Farw., Amer. Midl. Naturalist 12: 50 (1930).
Scirpus nudissimus Steud. & Jard., Bull. Soc. Linn. Normandie, II, 9: 280 (1875).
Eleocharis calva (A.Gray) Torr. ex Fernald & Brackett, Rhodora 31: 61 (1929).

Eleocharis fallax Weath., Rhodora 24: 23 (1922).
SE. Canada to E. U.S.A., Caribbean. 72 NSC QUE 75 CNT MAS NWJ RHO 77 TEX 78 FLA GEO LOU MSI NCA VRG 81 CUB PUE. Hemicr. or rhizome geophyte.

Eleocharis nervosa Kük., Repert. Spec. Nov. Regni Veg. 23: 192 (1926).
Eleocharis ambigens Fernald, Rhodora 37: 394 (1935).

Eleocharis filiculmis Kunth, Enum. Pl. 2: 144 (1837).
Scirpidium sulcatum Nees in C.F.P.von Martius & auct. suc. (eds.), Fl. Bras. 2(1): 98 (1842), nom. superfl.
Mexico to Trop. America. 79 MXE MXG MXS MXT 80 BLZ COS ELS GUA HON NIC PAN 81 CUB DOM TRT 82 FRG GUY SUR VEN 83 BOL CLM ECU PER 84 BZC BZE BZN 85 AGE AGW PAR URU. Ther. or hemicr.
Scirpus sulcatus Roth, Nov. Pl. Sp.: 30 (1821), nom. illeg. *Fimbristylis sulcata* Schult. in J.J.Roemer & J.A.Schultes, Mant. 2: 52 (1824). *Eleocharis sulcata* (Schult.) Nees, Linnaea 36: 445 (1870). *Eleocharis rothiana* Boeck., Flora 43: 3 (1860), nom. illeg.
Scirpus tenuiculus Schrad. ex Schult. in J.J.Roemer & J.A.Schultes, Mant. 2: 74 (1824). *Baeothryon tenuiculum* (Schrad. ex Schult.) A.Dietr., Sp. Pl. 2: 93 (1832). *Chaetocyperus tenuiculus* (Schrad. ex Schult.) Nees in C.F.P.von Martius & auct. suc. (eds.), Fl. Bras. 2(1): 96 (1842).
Scirpidium piauhiense Nees in C.F.P.von Martius & auct. suc. (eds.), Fl. Bras. 2(1): 97 (1842). *Eleocharis piauhiensis* (Nees) Steud., Syn. Pl. Glumac. 2: 77 (1854).
Limnochloa calyptrata Liebm., Mexic. Neldeagt. Pl., V, 2: 244 (1851). *Eleocharis calyptrata* (Liebm.) Steud., Syn. Pl. Glumac. 2: 81 (1855).
Eleocharis punctata Hochst. ex Steud., Syn. Pl. Glumac. 2: 75 (1854).
Eleocharis stylosa Steud., Syn. Pl. Glumac. 2: 76 (1854).
Eleocharis balansaeana Boeck., Flora 62: 159 (1879). *Chlorocharis balansaeana* (Boeck.) Rikli, Jahrb. Wiss. Bot. 27: 564 (1895).
Scirpus filiculmis Schrad. ex Griseb., Abh. Königl. Ges. Wiss. Göttingen 24: 311 (1879), nom. inval.
Eleocharis costaricensis Boeck., Allg. Bot. Z. Syst. 2: 34 (1896).
Eleocharis purpureovaginata Boeck., Allg. Bot. Z. Syst. 2: 34 (1896).
Eleocharis filiculmis var. *douradae* D.A.Simpson, Kew Bull. 43: 133 (1988).

Eleocharis flavescens (Poir.) Urb., Symb. Antill. 4: 116 (1903).
America. 72 NBR NCS ONT QUE 73 IDA MNT WYO 74 ILL MIN WIS 75 CNT INI MAI MAS MIC NWH NWJ NWY OHI PEN RHO VER 76 ARI CAL UTA 77 TEX 78 ALA ARK DEL FLA GEO LOU MRY MSI NCA SCA VRG 79 MXE MXG MXS MXT 80 COS 81 BAH BER CUB JAM LEE PUE TRT WIN 82 FRG GUY SUR VEN 83 BOL CLM ECU 84 BZL 85 AGE PAR URU. Hemicr.
**Scirpus flavescens* Poir. in J.B.A.M.de Lamarck, Encycl. 6: 756 (1805). *Baeothryon flavescens* (Poir.) A.Dietr., Sp. Pl. 2: 91 (1832).

var. ***flavescens***
Trop. & Subtrop. America. 73 IDA MNT WYO 74 ILL MIN WIS 75 CNT INI MAI MAS MIC NWH NWJ NWY OHI PEN RHO VER 76 ARI CAL UTA 77 TEX 78 ALA ARK DEL FLA GEO LOU MRY MSI NCA SCA VRG 79 MXS MXT 80 COS 81 BAH BER CUB JAM LEE PUE TRT WIN 82 FRG GUY SUR VEN 83 BOL CLM ECU 84 BZL 85 AGE PAR URU. Hemicr.
Scirpus flaccidus Rchb. ex A.Spreng., Tent. Suppl.: 3 (1828). *Eleocharis flaccida* (Rchb. ex A.Spreng.) Urb., Symb. Antill. 2: 165 (1900). *Eleocharis ochreata* var. *flaccida* (Rchb. ex A.Spreng.) Boeck. in I.Urban, Symb. Antill. 2: 63 (1900). *Eleocharis flaccida* var. *genuina* Osten, Anales Mus. Hist. Nat. Montevideo, II, 3: 166 (1931), nom. inval.
Eleocharis polymorpha Nees, Linnaea 9: 294 (1834), nom. inval.
Eleocharis sulciculmis Rchb. ex Kunth, Enum. Pl. 2: 150 (1837).
Scirpus gaudichaudianus Kunth, Enum. Pl. 2: 157 (1837). *Eleocharis albovaginata* var. *stricta* Boeck., Vidensk. Meddel. Dansk Naturhist. Foren. Kjøbenhavn 1869: 133 (1869).
Scirpus palmaris Willd. ex Kunth, Enum. Pl. 2: 150 (1837).
Eleocharis biocreata Nees, Linnaea 9: 294 (1838), nom. inval.
Eleogenus ochreatus Nees in C.F.P.von Martius & auct. suc. (eds.), Fl. Bras. 2(1): 102 (1842). *Eleocharis ochreata* (Nees) Steud., Syn. Pl. Glumac. 2: 79 (1854). *Scirpus ochreatus* (Nees) Griseb., Mem. Amer. Acad. Arts, n.s., 8: 535 (1863). *Trichophyllum ochreatum* (Nees) House, Amer. Midl. Naturalist 6: 205 (1920).
Scirpus alboater Schrad. ex Nees in C.F.P.von Martius & auct. suc. (eds.), Fl. Bras. 2(1): 120 (1842).
Scirpus melanospermus Salzm. ex Steud., Syn. Pl. Glumac. 2: 78 (1854).
Scirpus caribaeus Griseb., Abh. Königl. Ges. Wiss. Göttingen 7: 269 (1857), nom. illeg.
Eleocharis albovaginata Boeck., Linnaea 36: 436 (1870). *Scirpus albovaginatus* (Boeck.) Kuntze, Revis. Gen. Pl. 2: 757 (1891).
Eleocharis albovaginata var. *flaccida* Boeck., Linnaea 36: 437 (1870).
Eleocharis albovaginata var. *humilis* Boeck., Linnaea 36: 438 (1870).
Eleocharis albovaginata var. *macrostachya* Boeck., Linnaea 36: 438 (1870).
Eleocharis albovaginata var. *tenuis* Boeck., Linnaea 36: 437 (1870).
Scirpus anisochaetus C.Wright, Anales Acad. Ci. Méd. Habana 8: 78 (1871).
Eleocharis arechavaletae Boeck., Beitr. Cyper. 1: 17 (1888).
Eleocharis dussiana Boeck., Allg. Bot. Z. Syst. 2: 54 (1896). *Eleocharis flavescens* var. *dussiana* (Boeck.) Stehlé, Notul. Syst. (Paris) 13: 91 (1947).
Eleocharis thermalis Rydb., Mem. New York Bot. Gard. 1: 69 (1900). *Trichophyllum thermale* (Rydb.) House, Amer. Midl. Naturalist 6: 205 (1920). *Eleocharis flavescens* var. *thermalis* (Rydb.) Cronquist, Vasc. Pl. Pacific Northw. 1: 355 (1969).
Eleocharis praticola Britton in J.K.Small, Fl. S.E. U.S.: 182 (1903). *Trichophyllum praticola* (Britton) House, Amer. Midl. Naturalist 6: 205 (1920).
Eleocharis flaccida var. *capitata* H.Pfeiff., Herbarium 2: 53 (1921).
Eleocharis flaccida var. *fuscescens* Kük., Repert. Spec. Nov. Regni Veg. 23: 191 (1926). *Eleocharis flavescens* var. *fuscescens* (Kük.) Svenson, Rhodora 41: 48 (1939).

Eleocharis flaccida var. *arechavaletae* Osten, Anales Mus. Hist. Nat. Montevideo, II, 3: 167 (1931).

Eleocharis flaccida var. *humilis* Kük. ex Osten, Anales Mus. Hist. Nat. Montevideo, II, 3: 167 (1931).

Eleocharis flaccida var. *macrostachya* Osten, Anales Mus. Hist. Nat. Montevideo, II, 3: 168 (1931).

Eleocharis flaccida var. *olivacea* Fernald & Griscom, Rhodora 37: 155 (1935).

var. ***olivacea*** (Torr.) Gleason, Phytologia 4: 22 (1952).
SE. Canada to C. & E. U.S.A., Caribbean. 72 NBR NCS ONT QUE 74 ILL MIN WIS 75 CNT INI MAI MAS MIC NWH NWJ NWY OHI PEN RHO VER 77 TEX 78 ALA DEL FLA GEO LOU MRY NCA SCA VRG 81 PUE. Hemicr.

**Eleocharis olivacea* Torr., Ann. Lyceum Nat. Hist. New York 3: 300 (1836). *Scirpus olivaceus* (Torr.) Kuntze, Revis. Gen. Pl. 2: 758 (1891). *Trichophyllum olivaceum* (Torr.) House, Amer. Midl. Naturalist 6: 205 (1920).

Scirpus intermedius A.Gray ex Torr., Ann. Lyceum Nat. Hist. New York 3: 300 (1836).

Eleocharis olivacea f. *reductiseta* Schuyler & Ferren, Bartonia 43: 46 (1975). *Eleocharis olivacea* var. *reductiseta* (Schuyler & Ferren) Schuyler & Ferren, Proc. Acad. Nat. Sci. Philadelphia 132: 103 (1980).

Eleocharis fuscopurpurea (Steud.) H.Pfeiff., Repert. Spec. Nov. Regni Veg. 28: 19 (1930).
Chile to W. Argentina. 85 AGE AGW CLC CLS JNF. Hemicr.

**Isolepis fuscopurpurea* Steud., Syn. Pl. Glumac. 2: 99 (1855). *Eleocharis maculosa* subsp. *fuscopurpurea* (Steud.) Kük., Repert. Spec. Nov. Regni Veg. 16: 432 (1920).

Eleocharis melanocarpa Phil., Linnaea 29: 85 (1857), nom. illeg. *Eleocharis vincentina* Phil., Anales Univ. Chile 93: 348 (1896).

Eleocharis univaginata Boeck., Beitr. Cyper. 1: 14 (1888).

Eleocharis hyalinovaginata Boeck., Allg. Bot. Z. Syst. 2: 52 (1896).

Scirpus desvauxii Phil., Anales Univ. Chile 93: 482 (1896).

Eleocharis maculosa var. *irritans* Kük., Repert. Spec. Nov. Regni Veg. 16: 432 (1920).

Eleocharis geniculata (L.) Roem. & Schult., Syst. Veg. 2: 150 (1817).
Trop. & Subtrop. 13 SIC? 20 ALG EGY 22 BEN GAM GHA GNB GUI IVO LBR MTN NGA NGR SEN SIE 23 EQG GAB GGI 24 SOC SOM 27 BOT CPP NAT TVL 29 MDG 34 AFG IRN IRQ LBS 35 GST SAU YEM 36 CHH CHS 38 NNS TAI 40 BAN IND PAK SRL 41 THA VIE 42 BOR CKI JAW LSI MLY MOL PHI SUL SUM XMS 50 NFK NSW NTA QLD SOA TAS VIC WAU (60) fij GIL ton 61 COO SCI 62 CRL MRN (63) haw 72 ONT 74 ILL NEB OKL 75 INI MIC NWJ OHI PEN 76 ARI CAL NEV 77 NWM TEX 78 ALA FLA GEO LOU 79 MXC MXE MXG MXN MXS MXT 80 BLZ COS ELS GUA HON NIC PAN 81 ARU BAH BER CAY CUB HAI JAM LEE NLA PUE TRT VNA WIN 82 FRG GUY SUR VEN 83 ECU PER 84 BZC. Ther. or hemicr.

Scirpus capitatus L., Sp. Pl.: 48 (1753). *Eleocharis capitata* (L.) R.Br., Prodr.: 225 (1810). *Scirpus brownii* Spreng., Syst. Veg. 1: 204 (1824). *Bulbostylis capitata* (L.) Steven, Mém. Soc. Imp. Naturalistes Moscou 5: 355 (1814). *Megadenus capitatus* (L.) Raf., Neogenyton: 4 (1825). *Eleogenus capitatus* (L.) Nees in R.Wight, Contr. Bot. India: 112 (1834). *Cyperus capitatus* (L.) Burm.f. ex B.D.Jacks., Index Kew. 1: 691 (1893), nom. inval. *Chlorocharis capitata* (L.) Rikli, Jahrb. Wiss. Bot. 27: 564 (1895). *Trichophyllum capitatum* (L.) House, Amer. Midl. Naturalist 6: 204 (1920). *Eleocharis capitata* var. *typica* Svenson, Rhodora 34: 199 (1932), nom. inval.

**Scirpus geniculatus* L., Sp. Pl.: 48 (1753). *Bulbostylis geniculatus* (L.) Steven, Mém. Soc. Imp. Naturalistes Moscou 5: 355 (1814). *Limnochloa geniculata* (L.) Nees in C.F.P.von Martius & auct. suc. (eds.), Fl. Bras. 2(1): 99 (1842). *Chlorocharis geniculata* (L.) Rikli, Jahrb. Wiss. Bot. 27: 564 (1895).

Scirpus caribaeus Rottb., Descr. Icon. Rar. Pl.: 46 (1773). *Eleocharis caribaea* (Rottb.) S.F.Blake, Rhodora 20: 24 (1918).

Scirpus geniculatus var. *minor* Vahl, Enum. Pl. 2: 251 (1805). *Eleocharis geniculata* var. *minor* (Vahl) Roem. & Schult., Syst. Veg. 2: 150 (1817).

Eleocharis setacea R.Br., Prodr.: 225 (1810), nom. illeg.

Scirpus repens Willd. ex Schult. in J.J.Roemer & J.A.Schultes, Mant. 2: 84 (1824).

Eleocharis riparia Nees ex Spreng., Syst. Veg. 4(2): 27 (1827).

Scirpus interstinctus Poepp. ex C.Presl, Isis (Oken) 21: 269 (1828).

Scirpus depressus Vell., Fl. Flumin. 1: 35, t. 83 (1829).

Scirpus constrictus Schrad. ex Kunth, Enum. Pl. 2: 153 (1837). *Limnochloa constricta* (Schrad. ex Kunth) Nees in C.F.P.von Martius & auct. suc. (eds.), Fl. Bras. 2(1): 99 (1842).

Eleogenus capitatus var. *major* Nees in C.F.P.von Martius & auct. suc. (eds.), Fl. Bras. 2(1): 101 (1842).

Eleogenus capitatus var. *minor* Nees in C.F.P.von Martius & auct. suc. (eds.), Fl. Bras. 2(1): 101 (1842).

Limnochloa densa Liebm., Mexic. Neldeagt. Pl., V, 2: 244 (1851).

Eleocharis brizantha Steud., Syn. Pl. Glumac. 2: 79 (1854).

Eleocharis melanosperma Steud., Syn. Pl. Glumac. 2: 78 (1854).

Eleocharis densesquamata Steud., Syn. Pl. Glumac. 2: 82 (1855).

Eleocharis singularis Steud., Syn. Pl. Glumac. 2: 81 (1855).

Scirpus atrofactus Steud., Syn. Pl. Glumac. 2: 317 (1855).

Eleocharis mexicana Peyr., Linnaea 30: 14 (1859).

Eleocharis microformis Buckley, Proc. Acad. Nat. Sci. Philadelphia 1862: 10 (1863).

Eleocharis dispar E.J.Hill, Bot. Gaz. 7: 3 (1882). *Eleocharis capitata* var. *dispar* (E.J.Hill) Fernald, Rhodora 8: 129 (1906). *Eleocharis caribaea* var. *dispar* (E.J.Hill) S.F.Blake, Rhodora 20: 24 (1918).

Eleocharis valida Boeck., Flora 65: 350 (1882). *Scirpus validus* (Boeck.) Kuntze, Revis. Gen. Pl. 2: 758 (1891), nom. illeg.

Eleocharis sintenisii Boeck., Beitr. Cyper. 1: 16 (1888).

Eleocharis caribaea var. *achlamydea* Zavaro & Pabón, Fontqueria 42: 269 (1995).

Eleocharis geniculata f. *brunnea* S.González & Reznicek, Novon 6: 361 (1996).

Eleocharis glabella Y.D.Chen, Bull. Bot. Res., Harbin 7(2): 118 (1987).
China. 36+. Hemicr.

Eleocharis glauca Boeck., Vidensk. Meddel. Naturhist. Foren. Kjøbenhavn 1871: 150 (1871).
N. Brazil. 82 GUY? VEN? 84 BZN. Hemicr.
Eleocharis arenaria Benth., Hooker's J. Bot. Kew Gard. Misc. 2: 244 (1850).

Eleocharis glaucovirens Boeck., Beitr. Cyper. 1: 13 (1888). *Scirpus glaucovirens* (Boeck.) Kuntze, Revis. Gen. Pl. 2: 757 (1891).
S. Brazil. 84 BZS. Hemicr.

Eleocharis gossweileri H.E.Hess, Ber. Schweiz. Bot. Ges. 63: 327 (1953).
Zaïre. 23 ZAI. Hemicr.

Eleocharis gracilis R.Br., Prodr.: 224 (1810). *Scirpus gracilis* (R.Br.) Kuntze, Revis. Gen. Pl. 2: 757 (1891), nom. illeg.
E. & S. Australia, New Zealand. 50 NSW QLD SOA TAS VIC WAU 51 CTM NZN NZS. Hemicr.
Scirpus tener Spreng., Syst. Veg. 1: 204 (1824).
Eleocharis gracilis var. *gracillima* Hook.f., Fl. Nov.-Zel. 1: 270 (1853). *Eleocharis gracillima* (Hook.f.) Hook.f., Handb. N. Zeal. Fl. 2: 745 (1867).
Eleocharis gracilis var. *radicans* Hook.f., Fl. Nov.-Zel. 1: 270 (1853).
Eleocharis cunninghamii Boeck., Flora 41: 412 (1858). *Scirpus cunninghamii* (Boeck.) Kuntze, Revis. Gen. Pl. 2: 757 (1891).
Eleocharis hookeri Boeck., Linnaea 36: 430 (1870). *Scirpus hookeri* (Boeck.) Kuntze, Revis. Gen. Pl. 2: 757 (1891).

Eleocharis grandirostris (Lindm.) Mereles, Candollea 43: 416 (1988).
Paraguay. 85 PAR. Ther. or hemicr.
Eleocharis intermedia Maury ex Micheli, Mém. Soc. Phys. Genève 31(1): 138 (1890), nom. illeg.
**Eleocharis sulcata* var. *grandirostris* Lindm., Bih. Kongl. Svenska Vetensk.-Akad. Handl. 26(9): 13 (1900).

Eleocharis grisea Kük., Repert. Spec. Nov. Regni Veg. 23: 194 (1926).
Cuba (I. de la Juventud). 81 CUB. Hemicr.

Eleocharis grossimucronata Mereles, Novon 14: 190 (2004).
Peru (Cajamarca). 83 PER.

Eleocharis halmaturina J.M.Black, Trans. & Proc. Roy. Soc. South Australia 51: 378 (1927). *Tetraria halmaturina* (J.M.Black) J.M.Black, Fl. S. Austral., ed. 2: 153 (1943).
South Australia (Kangaroo I.). 50 SOA. Hemicr.

Eleocharis haumaniana Barros, Anales Mus. Nac. Buenos Aires 34: 482 (1928).
Brazil to Argentina. 84 BZN 85 AGE AGS URU. Hemicr.

Eleocharis ignota S.González & Reznicek, Contr. Univ. Michigan Herb. 24: 109 (2005).
Mexico. 79 MXE MXS.

Eleocharis intermedia Schult. in J.J.Roemer & J.A. Schultes, Mant. 2: 91 (1824). *Trichophyllum intermedium* (Schult.) House, Amer. Midl. Naturalist 6: 204 (1920).
SE. Canada to NC. & E. U.S.A. 72 NBR ONT QUE 73 MNT 74 ILL MIN WIS 75 CNT INI MAI MAS MIC NWH NWJ NWY OHI PEN VER WVA 78 MRY TEN VRG. Hemicr.
**Scirpus intermedius* Muhl., Descr. Gram.: 31 (1817), nom. illeg.
Eleocharis reclinata Kunth, Enum. Pl. 2: 143 (1837).
Scirpus pusillus Willd. ex Kunth, Enum. Pl. 2: 143 (1837).
Scirpus reclinatus Beyr. ex Kunth, Enum. Pl. 2: 143 (1837).
Eleocharis intermedia var. *habereri* Fernald, Rhodora 8: 130 (1906). *Eleocharis reclinata* var. *habereri* (Fernald) House, Bull. New York State Mus. Nat. Hist. 243–244: 43 (1923). *Eleocharis reclinata* f. *habereri* (Fernald) Svenson, Rhodora 39: 265 (1937).

Eleocharis interstincta (Vahl) Roem. & Schult., Syst. Veg. 2: 149 (1817).
Trop. & Subtrop. America. 74 OKL 77 TEX 78 FLA 79 MXE MXG MXS MXT 80 BLZ COS ELS GUA HON NIC PAN 81 BER CAY CUB HAI JAM LEE PUE TRT WIN 82 FRG GUY SUR VEN 83 BOL CLM ECU 84 BZC BZN 85 AGE PAR. Hemicr. or rhizome geophyte.
Scirpus plantagineus Sw., Fl. Ind. Occid. 1: 123 (1797), nom. illeg. *Eleocharis plantaginea* Boeck., Linnaea 36: 474 (1870), nom. illeg.
**Scirpus interstinctus* Vahl, Enum. Pl. 2: 149 (1805). *Trichophyllum interstinctum* (Vahl) House, Amer. Midl. Naturalist 6: 204 (1920).
Limnochloa articulata Nees, Linnaea 9: 294 (1834), nom. inval.
Eleocharis articulata Kunth, Enum. Pl. 2: 157 (1837), nom. inval.
Limnochloa obsoleta Nees in C.F.P.von Martius & auct. suc. (eds.), Fl. Bras. 2(1): 100 (1842). *Eleocharis obsoleta* (Nees) Steud., Syn. Pl. Glumac. 2: 81 (1855).
Eleocharis septata Miq., Linnaea 17: 58 (1843).
Eleocharis cognata Steud., Syn. Pl. Glumac. 2: 81 (1855).

Eleocharis jelskiana Boeck., Linnaea 38: 376 (1874).
Trinidad, Colombia to N. South America. 81 TRT 82 FRG GUY SUR VEN 83 CLM PER. Hemicr.
Eleocharis sagotii C.B.Clarke, Kew Bull., Addit. Ser. 8: 20 (1908).
Eleocharis sagotii var. *glochidiata* C.B.Clarke, Bull. Misc. Inform. Kew, Addit. Ser. 8: 21 (1908).

Eleocharis kamtschatica (C.A.Mey.) Kom., Fl. Kamtschatka 1: 207 (1927).
NE. China to Alaska, E. Canada. 31 KAM KUR PRM SAK 36 CHM 38 JAP KOR 70 ASK 71 BRC MAN 72 LAB QUE. Hemicr.
**Scirpus kamtschaticus* C.A.Mey., Mém. Sav. Étr. Acad. St. Pétersbourg 1: 198 (1831).
Scirpus camtschaticus Kunth, Enum. Pl. 2: 158 (1837).
Scirpus mitratus Franch. & Sav., Enum. Pl. Jap. 2: 544 (1878), nom. illeg. *Eleocharis mitrata* Makino, Bot. Mag. (Tokyo) 8: 380 (1894). *Eleocharis kamtschatica* var. *reducta* Ohwi, Bot. Mag. (Tokyo) 45: 185 (1931).
Scirpus sachalinensis Meinsh., Trudy Imp. S.-Peterburgsk. Bot. Sada 18: 280 (1901). *Eleocharis sachalinensis* (Meinsh.) Kom. in V.L.Komarov & E.N.Alissova-Klobukova, Key Pl. Far East. USSR 1: 266 (1931).
Eleocharis savatieri C.B.Clarke ex H.Lév., Bull. Acad. Int. Géogr. Bot. 14: 203 (1904).
Eleocharis triflora Kom., Repert. Spec. Nov. Regni Veg. 13: 162 (1914), nom. illeg. *Eleocharis komarovii* Zinserl. in V.L.Komarov (ed.), Fl. URSS 3: 588

(1935). *Eleocharis kamtschatica* var. *komarovii* (Zinserl.) Vorosch. in A.K.Skvortsov (ed.), Florist. Issl. Razn. Raionakh SSSR: 153 (1985).
Eleocharis glehnii Kom., Trudy Glavn. Bot. Sada 39: 34 (1923).
Eleocharis kamtschatica f. *reducta* Ohwi, Mem. Coll. Sci. Kyoto Imp. Univ., Ser. B, Biol. 18: 45 (1944).

Eleocharis keigheryi K.L.Wilson, Nuytsia 11: 273 (1997).
SW. Australia. 50 WAU. Hemicr.

Eleocharis kirkii C.B.Clarke in D.Oliver & auct. suc. (eds.), Fl. Trop. Afr. 8: 410 (1902).
S. Trop. Africa. 26 ZIM. Ther.

Eleocharis kleinii Barros, Sellowia 18: 49 (1966).
S. Brazil. 84 BZS. Hemicr.

Eleocharis knutei Pabón & Zavaro, Fontqueria 42: 269 (1995).
Cuba. 81 CUB. Hemicr.

Eleocharis kuoi Y.D.Chen, Bull. Bot. Res., Harbin 7(2): 120 (1987).
Qinghai. 36 CHQ. Hemicr.

Eleocharis kuroguwai Ohwi, J. Jap. Bot. 12: 654 (1936).
C. & S. Japan, Korea. 38 JAP KOR. Hemicr.

Eleocharis laeviglumis R.Trevis. & Boldrini, Novon 16: 155 (2006).
Brazil (Rio Grande do Sul). 84 BZS. Hemicr. or rhizome geophyte

Eleocharis laeviseta Nakai, Repert. Spec. Nov. Regni Veg. 13: 246 (1914).
Korea. 38 KOR. Hemicr.

Eleocharis lanceolata Fernald, Proc. Amer. Acad. Arts 34: 493 (1899). *Trichophyllum lanceolatum* (Fernald) House, Amer. Midl. Naturalist 6: 205 (1920). *Eleocharis obtusa* var. *lanceolata* (Fernald) Gilly, Iowa State Coll. J. Sci. 21: 92 (1946).
EC. U.S.A. to Texas. 74 KAN MSO OKL (76) cal 77 TEX 78 ARK LOU TEN. Ther.

Eleocharis lankana T.Koyama, Bull. Natl. Sci. Mus. Tokyo, n.s., 17: 67 (1974).
India, SW. Sri Lanka. 40 IND SRL. Hemicr.

subsp. ***lankana***
SW. Sri Lanka. 40 SRL. Hemicr.

subsp. ***mohamadii*** M.A.Wadood Khan, J. Econ. Taxon. Bot. 22: 560 (1998 publ. 1999).
India (Maharashtra). 40 IND. Hemicr.

Eleocharis lechleri Boeck., Linnaea 36: 422 (1870). *Scirpus lechleri* (Boeck.) Kuntze, Revis. Gen. Pl. 2: 757 (1891).
W. & S. South America. 83 BOL ECU 85 AGS AGW CLC CLN CLS. Hemicr.

Eleocharis lepta C.B.Clarke in W.H.Harvey & auct. suc. (eds.), Fl. Cap. 7: 758 (1900). *Scirpus lepta* (C.B.Clarke) Levyns, J. S. African Bot. 14: 84 (1948).
Cape Prov. 27 CPP. Hemicr.

Eleocharis liesneri S.González & Reznicek, Novon 6: 361 (1996).
Venezuela (Bolívar). 82 VEN. Hemicr.

Eleocharis limosa (Schrad.) Schult. in J.J.Roemer & J.A.Schultes, Mant. 2: 87 (1824).
S. Africa, C. Madagascar. 27 CPP NAT 29 MDG. Hemicr.
**Scirpus limosus* Schrad., Gött. Gel. Anz. 3: 2069 (1821). *Limnochloa limosa* (Schrad.) Nees, Linnaea 9: 294 (1834).
Eleocharis sororia Kunth, Enum. Pl. 2: 148 (1837).
Limnochloa capensis C.Krauss, Flora 28: 759 (1845).
Eleocharis baronii Baker, J. Linn. Soc., Bot. 20: 297 (1883).

Eleocharis liogieri T.Koyama, Brittonia 29: 237 (1977).
Caribbean. 81 DOM LEE. Hemicr.

Eleocharis liouana Tang & F.T.Wang, in Fl. Reipubl. Popul. Sin. 11: 226 (1961).
China (Yunnan). 36 CHC. Hemicr.

Eleocharis loefgreniana Boeck., Beitr. Cyper. 2: 12 (1890).
SE. Brazil. 84 BZL. Hemicr.
Eleocharis leptostachys Lindm., Bih. Kongl. Svenska Vetensk.-Akad. Handl. 26(9): 13 (1900).

Eleocharis lundellii Svenson, Amer. Midl. Naturalist 29: 470 (1943).
Texas. 77 TEX. Hemicr.

Eleocharis macbarronii K.L.Wilson, Telopea 5: 599 (1994).
E. & SE. Australia. 50 NSW QLD VIC. Hemicr.

Eleocharis × macounii Fernald, Proc. Amer. Acad. Arts 34: 387 (1899). *E. intermedia* × *E. obtusa*.
E. Canada. 72 QUE. Hemicr.

Eleocharis macrantha Boeck., Linnaea 36: 453 (1870). *Scirpus macranthus* (Boeck.) Kuntze, Revis. Gen. Pl. 2: 757 (1891).
Syria. 34 LBS. Hemicr.
Scirpus hexatrichus Ehrenb. ex Boiss., Fl. Orient. 5: 387 (1882).

Eleocharis macrostachya Britton in J.K.Small, Fl. S.E. U.S.: 184 (1903).
Temp. & Subalpine America. 70 ASK YUK 71 ABT BRC MAN SAS 72 ONT QUE 73 COL IDA MNT ORE WAS WYO 74 ILL IOW KAN MIN MSO NDA NEB OKL SDA WIS 76 ARI CAL NEV UTA 77 NWM TEX 78 ARK LOU MSI 79 MXC MXE MXN MXS MXT 83 PER 85 AGE AGS AGW CLC URU. Hemicr.
Eleocharis perlonga Fernald & Brackett, Rhodora 31: 70 (1929).
Eleocharis xyridiformis Fernald & Brackett, Rhodora 31: 76 (1929).

Eleocharis maculosa (Vahl) Roem. & Schult., Syst. Veg. 2: 154 (1817).
Mexico to Trop. America. 79 MXE MXT 80 BLZ GUA HON 81 LEE WIN 82 GUY VEN 83 BOL CLM ECU GAL PER 84 BZE BZL BZN BZS 85 AGE AGS AGW CLC PAR URU. Hemicr.
**Scirpus maculosus* Vahl, Enum. Pl. 2: 247 (1805). *Trichophyllum maculosum* (Vahl) House, Amer. Midl. Naturalist 6: 205 (1920). *Eleocharis maculosa* var. *genuina* Osten, Anales Mus. Hist. Nat. Montevideo, II, 3: 169 (1931), nom. inval.
Bulbostylis maculosus Steven, Mém. Soc. Imp. Naturalistes Moscou 5: 355 (1814).
Eleocharis arcuata Kunze ex Kunth, Enum. Pl. 2: 147 (1837). *Eleocharis vincentina* var. *arcuata* (Kunze ex Kunth) C.B.Clarke, Bot. Jahrb. Syst. 30(68): 19 (1901).
Eleogenus schottianus Nees in C.F.P.von Martius & auct. suc. (eds.), Fl. Bras. 2(1): 102 (1842). *Eleocharis schottiana* (Nees) Steud., Syn. Pl. Glumac. 2: 79 (1854).

Eleocharis gracillima Boeck., Linnaea 36: 459 (1870), nom. illeg.
Eleocharis glazioviana Boeck., Vidensk. Meddel. Naturhist. Foren. Kjøbenhavn 1871: 150 (1871).
Eleocharis lehmanniana Boeck., Bot. Jahrb. Syst. 8: 205 (1887).
Eleocharis mendoncae Boeck., Beitr. Cyper. 2: 12 (1890).
Eleocharis maculosa f. *arcuata* Osten, Anales Mus. Hist. Nat. Montevideo, II, 3: 170 (1931).
Eleocharis maculosa f. *depressa* Kük. ex Osten, Anales Mus. Hist. Nat. Montevideo, II, 3: 170 (1931).
Eleocharis maculosa var. *maritima* Osten, Anales Mus. Hist. Nat. Montevideo, II, 3: 169 (1931).

Eleocharis maidenii Kük., Repert. Spec. Nov. Regni Veg. 13: 135 (1914).
New South Wales. 50 NSW. Hemicr.
Eleocharis maidenii var. *subaquatica* Kük., Repert. Spec. Nov. Regni Veg. 13: 136 (1914).

Eleocharis mamillata (H.Lindb.) H.Lindb., Exsicc. (Herb. Norm.) 44: 108 (1902).
Europe to China, Subarctic America to N. U.S.A. 10 FIN GRB NOR SWE 11 AUT CZE GER HUN POL SWI 12 FRA SPA 13 BUL ITA ROM YUG 14 BLR BLT RUC RUN RUS RUW UKR 30+ 31+ 32+ 33 NCS 36 CHM 70 ASK NWT YUK 71 ABT BRC SAS 72 ONT QUE 73 WAS 74 MIN WIS. Hemicr.
**Scirpus mamillatus* H.Lindb., Acta Soc. Fauna Fl. Fenn. 33(7): 7 (1902). *Eleocharis palustris* subsp. *mamillata* (H.Lindb.) Beauverd, Bull. Soc. Bot. Genève, II, 13: 263 (1921).

subsp. ***austriaca*** (Hayek) Strandh., Opera Bot. 9(2): 9 (1965).
Europe to Caucasus. 10 GRB NOR 11 AUT CZE GER HUN POL SWI 12 FRA SPA 13 ITA ROM YUG 14 RUS RUW UKR 30 33 NCS. Hemicr.
**Eleocharis austriaca* Hayek, Sched. Fl. Stiriac. 19–20: 8 (1910). *Eleocharis palustris* subsp. *austriaca* (Hayek) Podp., Práce Morav. Přír. Spolecn. 5: 50 (1929).
Eleocharis benedicta Beauverd, Bull. Soc. Bot. Genève, II, 13: 245 (1923 publ. 1922).
Eleocharis leptostylopodiata Zinserl. in V.L. Komarov (ed.), Fl. URSS 3: 581 (1935).

subsp. ***mamillata***
Europe to China, Subarctic America to N. U.S.A. 10 FIN NOR SWE 11 AUT CZE GER HUN POL SWI 12 FRA 13 BUL ROM 14 BLR BLT RUC RUN RUW 30+ 31+ 32+ 36 CHM 70 ASK NWT YUK 71 ABT BRC SAS 72 ONT QUE 73 WAS 74 MIN WIS. Hemicr.
Eleocharis satoi Ohwi, Acta Phytotax. Geobot. 2: 28 (1933).

Eleocharis margaritacea (Hultén) Miyabe & Kudô, J. Fac. Agric. Hokkaido Imp. Univ. 26: 210 (1931).
Russian Far East to N. & C. Japan. 31 KAM KUR 38 JAP. Hemicr.
**Scirpus margaritaceus* Hultén, Kongl. Svenska Vetensk. Acad. Handl., III, 5(1): 167 (1927).

Eleocharis marginulata Hochst. ex Steud., Syn. Pl. Glumac. 2: 78 (1854). *Scirpus marginulatus* (Hochst. ex Steud.) Kuntze, Revis. Gen. Pl. 2: 757 (1891).
Ethiopia to Tanzania, Arabian Pen. 24 ETH 25 KEN TAN 35 SAU. Hemicr.
Eleocharis striata Hochst. ex Steud., Syn. Pl. Glumac. 2: 78 (1854).

Eleocharis maximowiczii Zinserl. in V.L.Komarov (ed.), Fl. URSS 3: 588 (1935).
Russian Far East. 31+. Hemicr.

Eleocharis melanocarpa Torr., Ann. Lyceum Nat. Hist. New York 3: 311 (1836). *Scirpus melanocarpus* (Torr.) Kuntze, Revis. Gen. Pl. 2: 757 (1891).
E. U.S.A. to Texas. 75 INI MAS MIC NWY RHO 77 TEX 78 ALA FLA GEO NCA SCA VRG. Hemicr.

Eleocharis melanocephala C.B.Clarke, Bot. Jahrb. Syst. 30(68): 24 (1902).
C. Chile to Argentina. 85 AGE AGS AGW CLC. Hemicr.

Eleocharis melanostachys (d'Urv.) C.B.Clarke, Bot. Jahrb. Syst. 30(68): 20 (1902).
Peru to Falkland Is. 83 PER 85 AGS CLC CLS 90 FAL. Hemicr.
**Scirpus melanostachys* d'Urv., Mém. Soc. Linn. Paris 4: 600 (1826). *Fimbristylis melanostachys* (d'Urv.) Brongn. in L.I.Duperrey, Voy. Monde, Phan.: 188 (1834).
Isolepis heteromorpha Steud., Syn. Pl. Glumac. 2: 100 (1855).
Eleocharis litoralis Phil., Linnaea 29: 77 (1857). *Scirpus litoralis* (Phil.) Kuntze, Revis. Gen. Pl. 2: 757 (1891), nom. illeg.
Eleocharis valdiviana Phil., Linnaea 29: 77 (1857). *Scirpus valdivianus* (Phil.) Kuntze, Revis. Gen. Pl. 2: 758 (1891).
Eleocharis macrorhiza Boeck., Flora 41: 413 (1858).
Scirpus andinus Phil., Anales Univ. Chile 1873: 554 (1873). *Eleocharis andina* (Phil.) H.Pfeiff., Herbarium 56: 54 (1921).
Eleocharis appendiculata C.B.Clarke, Bot. Jahrb. Syst. 30(68): 20 (1902), nom. illeg.
Eleocharis funebris Speg., Anales Mus. Nac. Buenos Aires 7: 177 (1902).
Eleocharis simulans C.B.Clarke, Bot. Jahrb. Syst. 30(68): 20 (1902).

Eleocharis mendocina Phil., Anales Univ. Chile 1873: 553 (1873). *Scirpus mendocinus* (Phil.) Kuntze, Revis. Gen. Pl. 2: 757 (1891).
NW. Argentina. 85 AGW. Hemicr.

Eleocharis microcarpa Torr., Ann. Lyceum Nat. Hist. New York 3: 312 (1836). *Scirpus microcarpus* (Torr.) Kuntze, Revis. Gen. Pl. 2: 758 (1891).
S. U.S.A. to Texas, Caribbean. 75 INI MAS MIC NWJ PEN 77 TEX 78 ALA FLA GEO LOU MRY MSI NCA SCA TEN VRG 81 CUB JAM LEE PUE 82 VEN. Ther. or hemicr.

var. ***filiculmis*** Torr., Ann. Lyceum Nat. Hist. New York 3: 312 (1836).
E. U.S.A. to Texas. 75 INI MAS MIC NWJ PEN 77 TEX 78 ALA FLA GEO LOU MRY MSI NCA SCA TEN VRG. Ther. or hemicr.
Eleocharis torreyana Boeck., Linnaea 36: 440 (1870). *Scirpus torreyanus* (Boeck.) Kuntze, Revis. Gen. Pl. 2: 758 (1891). *Trichophyllum torreyanum* (Boeck.) House, Amer. Midl. Naturalist 6: 205 (1920).

var. ***microcarpa***
SE. U.S.A., Caribbean. 78 FLA GEO LOU SCA 81 CUB JAM LEE PUE 82 VEN. Ther. or hemicr.
Eleocharis paracicularis Boeck., Bot. Jahrb. Syst. 7: 275 (1886). *Scirpus paracicularis* (Boeck.) C.Wright ex C.B.Clarke in I.Urban, Symb. Antill. 2: 71 (1900).

Eleocharis cubensis Boeck., Beitr. Cyper. 2: 10 (1890).

Eleocharis microlepis (Griseb.) D.A.Simpson, Kew Bull. 43: 429 (1988).
Cuba to Hispaniola. 81 CUB DOM HAI. Ther.
**Scirpus microlepis* Griseb., Cat. Pl. Cub.: 239 (1866).

Eleocharis migoana Ohwi & T.Koyama, Bull. Natl. Sci. Mus. Tokyo, n.s., 3(38): 27 (1956).
SE. China. 36 CHS. Hemicr.

Eleocharis minarum Boeck., Beitr. Cyper. 2: 12 (1890).
SE. Brazil. 84 BZL. Hemicr.

Eleocharis minima Kunth, Enum. Pl. 2: 139 (1837).
Trop. & Subtrop. America. 77 TEX 78 FLA GEO MRY MSI 79 MXE MXG MXN MXS MXT 80 BLZ COS ELS GUA HON NIC PAN 81 CAY CUB JAM LEE TRT 82 FRG? GUY VEN 83 BOL CLM ECU PER 84 BZC BZL BZN BZS 85 AGE AGW PAR. Ther. or hemicr.

var. ***bicolor*** (Chapm.) Svenson, Rhodora 39: 219 (1937).
Trop. & Subtrop. America. 78 FLA GEO MSI 80 BLZ COS HON NIC 81 TRT 82 GUY VEN 84 BZS 85 AGE PAR. Hemicr.
**Eleocharis bicolor* Chapm., Fl. South. U.S.: 517 (1860). *Trichophyllum bicolor* (Chapm.) House, Amer. Midl. Naturalist 6: 204 (1920).

var. ***minima***
Trop. & Subtrop. America. 77 TEX 78 FLA MRY 79 MXE MXG MXN MXS MXT 80 BLZ COS ELS GUA HON NIC PAN 81 CAY CUB JAM LEE TRT 82 FRG? GUY VEN 83 BOL CLM ECU PER 84 BZC BZL BZN BZS 85 AGE AGW PAR. Ther. or hemicr.
Eleocharis uncialis Chapm. ex Small in ?
Chaetocyperus viviparus Nees in C.F.P.von Martius & auct. suc. (eds.), Fl. Bras. 2(1): 93 (1842).
Scirpus tenerrimus Schrad. ex Nees in C.F.P.von Martius & auct. suc. (eds.), Fl. Bras. 2(1): 94 (1842).
Scirpus viviparus Schrad. ex Nees in C.F.P.von Martius & auct. suc. (eds.), Fl. Bras. 2(1): 93 (1842), nom. inval.
Chaetocyperus jamesonii Steud., Syn. Pl. Glumac. 2: 74 (1854). *Eleocharis jamesonii* (Steud.) N.E.Br., Bull. Misc. Inform. Kew 1922: 256 (1922).
Chaetocyperus reptans Boeck., Flora 40: 34 (1857).
Eleocharis subtilis Boeck., Linnaea 36: 426 (1870). *Scirpus subtilis* (Boeck.) Kuntze, Revis. Gen. Pl. 2: 758 (1891).
Eleocharis tenuissima Boeck., Linnaea 36: 419 (1870). *Scirpus tenuissimus* (Boeck.) Kuntze, Revis. Gen. Pl. 2: 758 (1891). *Eleocharis minima* var. *tenuissima* (Boeck.) D.A.Simpson, Kew Bull. 43: 131 (1988).
Eleocharis wrightiana Boeck., Beitr. Cyper. 1: 12 (1888).
Eleocharis villaricensis Maury ex Micheli, Mém. Soc. Phys. Genève 31(1): 138 (1890).
Eleocharis durandii Boeck., Allg. Bot. Z. Syst. 2: 34 (1896).
Eleocharis ambigua Urb., Symb. Antill. 2: 166 (1900), nom. illeg.
Eleocharis oropuchensis Britton, Bull. Torrey Bot. Club 48: 327 (1921 publ. 1922). *Eleocharis minima* var. *oropuchensis* (Britton) C.D.Adams, Fl. Trinidad & Tobago 3(5): 403 (1992).
Eleocharis savannarum Britton, Bull. Torrey Bot. Club 48: 327 (1921 publ. 1922).
Helonema estrellensis Suess., Bot. Jahrb. Syst. 73: 118 (1943).

Eleocharis minuta Boeck., Bot. Jahrb. Syst. 5: 503 (1884).
E. Trop. Africa, W. Indian Ocean. 25 KEN TAN UGA 29 MAU MDG REU (50) nsw qld vic. Ther. or hemicr.

Eleocharis minutissima Britton, Mem. Torrey Bot. Club 16: 60 (1920).
SE. Mexico to C. America, W. Cuba. 79 MXT 80 BLZ COS HON PAN 81 CUB. Ther.

Eleocharis mitracarpa Steud., Syn. Pl. Glumac. 2: 77 (1854).
Ukraine to China. 14 KRY RUE RUS UKR 32 KAZ KGZ TKM UZB 33 TCS 34 AFG IRN 36 CHC CHN 40 EHM PAK WHM. Hemicr.
Scirpus equisetiformis Meinsh., Trudy Imp. S.-Peterburgsk. Bot. Sada 18: 261 (1901). *Eleocharis equisetiformis* (Meinsh.) B.Fedtsch., Rastit. Turkest.: 165 (1915).
Eleocharis argyrolepidoides Zinserl. in V.L.Komarov (ed.), Fl. URSS 3: 584 (1935).
Eleocharis turcomanica Zinserl. in V.L.Komarov (ed.), Fl. URSS 3: 583 (1935).

Eleocharis monantha Nelmes, Kew Bull. 7: 290 (1952).
Central African Rep. 23 CAF. Hemicr.

Eleocharis montana (Kunth) Roem. & Schult., Syst. Veg. 2: 153 (1817).
Trop. & Subtrop. America. 76 ARI 77 TEX 78 ALA FLA GEO LOU 79 MXC MXE MXG MXN MXS MXT 80 BLZ COS ELS GUA HON NIC PAN 81 CUB DOM HAI JAM LEE PUE 82 SUR VEN 83 BOL CLM ECU GAL PER 84 BZC BZL BZN 85 AGE AGW CLC PAR URU. Hemicr.
**Scirpus montanus* Kunth in F.W.H.von Humboldt, A.J.A.Bonpland & C.S.Kunth, Nov. Gen. Sp. 1: 226 (1816).
Scirpus nodulosus Roth, Nov. Pl. Sp.: 29 (1821). *Eleocharis nodulosa* (Roth) Schult. in J.J.Roemer & J.A.Schultes, Mant. 2: 87 (1824). *Eleogenus nodulosus* (Roth) Nees in C.F.P.von Martius & auct. suc. (eds.), Fl. Bras. 2(1): 104 (1842). *Trichophyllum nodulosum* (Roth) House, Amer. Midl. Naturalist 6: 205 (1920). *Eleocharis montana* var. *nodulosa* (Roth) Svenson, J. Arnold Arbor. 25: 47 (1944).
Eleocharis consanguinea Kunth, Enum. Pl. 2: 148 (1837).
Eleocharis mutata Schltdl., Bot. Zeitung (Berlin) 7: 134 (1849), nom. illeg.
Eleocharis subnodulosa Steud., Syn. Pl. Glumac. 2: 81 (1855). *Eleocharis nodulosa* var. *subnodulosa* (Steud.) Kük., Repert. Spec. Nov. Regni Veg. 23: 192 (1926).
Eleocharis chrysocarpa Boeck., Vidensk. Meddel. Naturhist. Foren. Kjøbenhavn 1869: 134 (1869).
Eleocharis nodulosa var. *humilis* Boeck., Linnaea 36: 469 (1870).
Eleocharis nodulosa var. *tenuis* Boeck., Flora 62: 160 (1879).
Scirpus heteromorphus Steud. ex F.Phil., Cat. Pl. Vasc. Chil.: 311 (1881).
Eleocharis vulcani Boeck., Bot. Jahrb. Syst. 8: 206 (1887).
Eleocharis andesica C.B.Clarke, Kew Bull., Addit. Ser. 8: 23 (1908).

Eleocharis nodulosa var. *angulata* Svenson, Rhodora 39: 258 (1937).
Eleocharis nodulosa f. *trigyna* Barros in H.R.Descole (ed.), Gen. Sp. Pl. Argent. 4(1): 149 (1947).

Eleocharis montevidensis Kunth, Enum. Pl. 2: 144 (1837). *Limnochloa montevidensis* (Kunth) Nees in C.F.P.von Martius & auct. suc. (eds.), Fl. Bras. 2(1): 99 (1842). *Scirpus montevidensis* (Kunth) Kuntze, Revis. Gen. Pl. 2: 758 (1891). *Eleocharis montana* subsp. *montevidensis* (Kunth) Osten, Anales Mus. Hist. Nat. Montevideo, II, 3: 183 (1931).
C. & S. U.S.A. to Guatemala, Peru to S. Brazil. 74 KAN OKL 76 ARI CAL 77 NWM TEX 78 ALA FLA GEO LOU MSI NCA SCA 79 MXC MXE MXG MXN MXS MXT 80 GUA HON 83 PER 84 BZN BZS 85 AGE AGW URU. Hemicr.
Eleocharis arenicola Torr. ex Engelm. & A.Gray, Boston J. Nat. Hist. 5: 237 (1845). *Scirpus arenicolus* (Torr. ex Engelm. & A.Gray) Kuntze, Revis. Gen. Pl. 2: 757 (1891). *Trichophyllum arenicola* (Torr. ex Engelm. & A.Gray) House, Amer. Midl. Naturalist 6: 204 (1920).
Eleocharis palmeri Svenson, Rhodora 34: 223 (1932).
Eleocharis montevidensis f. *angustior* S.González & McVaugh, in Fl. Novo-Galiciana 13: 362 (1993).

Eleocharis moorei M.T.Strong & S.González, Sida 19: 118 (2000).
Mexico (Hidalgo, Veracruz). 79 MXE MXG. Hemicr. or rhizome geophyte.

Eleocharis morroi D.A.Simpson, Kew Bull. 43: 131 (1988).
NE. Brazil. 84 BZE. Hemicr.

Eleocharis multicaulis (Sm.) Desv., Observ. Pl. Angers: 74 (1818).
Azores, Europe to N. Africa. 10 DEN GRB IRE NOR SWE 11 BGM CZE GER NET POL 12 COR FRA POR SAR SPA 13 ITA KRI SIC YUG 14 UKR 20 ALG MOR TUN 21 AZO. Hemicr.
**Scirpus multicaulis* Sm., Fl. Brit. 1: 48 (1800). *Clavula multicaulis* (Sm.) Dumort., Fl. Belg.: 143 (1827). *Limnochloa multicaulis* (Sm.) Rchb., Fl. Germ. Excurs.: 140(11) (1831). *Cyperus multicaulis* (Sm.) Missbach & E.H.L.Krause in J.Sturm, Deutschl. Fl., ed. 2, 2: 24 (1900).
Clavula multicaulis var. *vivipara* Dumort., Fl. Belg.: 143 (1827). *Eleocharis multicaulis* var. *vivipara* (Dumort.) Lej. & Courtois, Comp. Fl. Belg. 1: 39 (1828). *Eleocharis multicaulis* f. *vivipara* (Dumort.) P.Silva, Bol. Soc. Brot., II, 60: 157 (1987).
Limnochloa multiculmis Rchb. in J.C.Mössler, Handb. Gewächsk. ed. 2, 3: 1808 (1830).
Scirpus multicalamis Bubani, Fl. Pyren. 4: 204 (1902).
Scirpus multiculmeus Bubani, Fl. Pyren. 4: 204 (1902).
Eleocharis multiculmis A.Terracc. ex Lojac., Fl. Sicul. 3: 210 (1909).

Eleocharis mutata (L.) Roem. & Schult., Syst. Veg. 2: 155 (1817).
Trop. America, W. & WC. Trop. Africa. 22 GHA GNB GUI LBR NGA SEN SIE TOG 23 EQG GAB GGI 25 TAN 77 TEX 78 ALA del 79 MXE MXG MXI MXS MXT 80 BLZ COS GUA HON NIC PAN 81 ARU CAY CUB DOM JAM LEE NLA PUE TRT WIN 82 FRG GUY SUR VEN 83 BOL CLM ECU GAL PER 84 BZE BZL 85 PAR. Hemicr.
**Scirpus mutatus* L., Syst. Nat. ed. 10: 867 (1759). *Limnochloa mutata* (L.) Nees in C.F.P.von Martius & auct. suc. (eds.), Fl. Bras. 2(1): 101 (1842). *Trichophyllum mutatum* (L.) House, Amer. Midl. Naturalist 6: 205 (1920).
Scirpus marginatus Muhl., Descr. Gram.: 28 (1817), nom. illeg.
Eleocharis scariosa Steud., Syn. Pl. Glumac. 2: 80 (1854).
Scirpus obtuse-trigonus Salzm. ex Steud., Syn. Pl. Glumac. 2: 80 (1854).
Eleocharis spiralis Boeck., Linnaea 36: 473 (1870), nom. illeg.

Eleocharis nana Kunth, Enum. Pl. 2: 140 (1837). *Scirpus nanus* (Kunth) Kuntze, Revis. Gen. Pl. 2: 758 (1891), nom. illeg.
Florida, Cuba, N. South America to Brazil. 78 FLA 81 CUB 82 GUY SUR VEN 84 BZL BZS. Ther. or hemicr.
Scirpus capillaceus Michx., Fl. Bor.-Amer. 1: 30 (1803). *Chaetocyperus polymorphus* Nees & Lindl., Linnaea 9: 289 (1834). *Chaetocyperus polymorphus* var. *capillaceus* (Michx.) Nees in C.F.P.von Martius & auct. suc. (eds.), Fl. Bras. 2(1): 94 (1842), nom. illeg.
Scirpus pusillus Vahl, Enum. Pl. 2: 246 (1805). *Baeothryon pusillum* (Vahl) A.Dietr., Sp. Pl. 2: 92 (1832).
Scirpus humilis Wallr., Sched. Crit.: 27 (1822), nom. illeg.
Scirpus floridanus Michx. ex Kunth, Enum. Pl. 2: 158 (1837).
Chaetocyperus polymorphus var. *sphagnicola* Nees in C.F.P.von Martius & auct. suc. (eds.), Fl. Bras. 2(1): 94 (1842).
Chaetocyperus punctatus Nees in C.F.P.von Martius & auct. suc. (eds.), Fl. Bras. 2(1): 93 (1842).
Scirpus punctatus Schrad. ex Nees in C.F.P.von Martius & auct. suc. (eds.), Fl. Bras. 2(1): 93 (1842), nom. inval.
Scirpus translucens Le Gall ex J.Lloyd, Fl. Loire-Inf.: 275 (1844).
Eleocharis punctata Boeck., Vidensk. Meddel. Naturhist. Foren. Kjøbenhavn 1869: 132 (1869), nom. illeg. *Scirpus punctatus* (Boeck.) Kuntze, Revis. Gen. Pl. 2: 758 (1891).
Scirpus camptotrichus C.Wright in F.A.Sauvalle, Fl. Cub.: 173 (1871). *Eleocharis camptotricha* (C. Wright) C.B.Clarke in I.Urban, Symb. Antill. 2: 69 (1900). *Eleocharis prolifera* var. *camptotricha* (C.Wright) Urb., Symb. Antill. 2: 165 (1900).
Scirpus leptos C.Wright in F.A.Sauvalle, Fl. Cub.: 176 (1871).
Eleocharis punctata Boeck. ex Tonduz, Bull. Herb. Boissier 3: 464 (1895), nom. illeg.
Eleocharis nana f. *capillaris* Boeck. in I.Urban, Symb. Antill. 2: 69 (1900).

Eleocharis naumanniana Boeck., Bot. Jahrb. Syst. 5: 92 (1884).
W. Trop. Africa to Botswana. 22 GHA GNB GUI IVO LBR MLI NGA SEN SIE 23 CAF GAB ZAI 26? 27 BOT. Hemicr.

var. ***caillei*** (Hutch. ex Nelmes) S.S.Hooper, Kew Bull. 26: 582 (1971).
W. Trop. Africa. 22 GUI LBR MLI SIE. Hemicr.
**Eleocharis caillei* Hutch. ex Nelmes, Kew Bull. 6: 165 (1951).

var. ***naumanniana***
W. Trop. Africa to Botswana. 22 GHA GNB GUI IVO LBR NGA SEN SIE 23 CAF GAB ZAI 26? 27 BOT. Hemicr.

Eleocharis neozelandica C.B.Clarke ex T.Kirk., Trans. & Proc. New Zealand Inst. 26: 260 (1893 publ. 1894).
N. & C. New Zealand. 51 NZN NZS. Hemicr.

Eleocharis niederleinii Boeck., Beitr. Cyper. 1: 13 (1888).
NE. Argentina to Uruguay. 85 AGE URU. Hemicr.

Eleocharis nigrescens (Nees) Steud., Syn. Pl. Glumac. 2: 77 (1855).
Chad to E. Trop. Africa, SE. U.S.A., S. Venezuela to Brazil. 24 CHA 25 KEN TAN UGA 78 FLA SCA 82 VEN 84 BZE BZN. Ther.
**Scirpidium nigrescens* Nees in C.F.P.von Martius & auct. suc. (eds.), Fl. Bras. 2(1): 97 (1842). *Isolepis nigrescens* (Nees) Steud., Syn. Pl. Glumac. 2: 91 (1855). *Scirpus nigrescens* (Nees) Salzm. ex Steud., Syn. Pl. Glumac. 2: 91 (1855).
Eleocharis hildebrandtii Boeck., Flora 61: 34 (1878).
Eleocharis perrieri Cherm., Bull. Soc. Bot. France 73: 554 (1926 publ. 1927).

Eleocharis nipponica Makino, Bot. Mag. (Tokyo) 18: 110 (1904).
Russian Far East to Japan. 31+ 38 JAP. Hemicr.

Eleocharis nitida Fernald, Rhodora 8: 129 (1906). *Scirpus nitidus* (Fernald) Hultén, Fl. Alaska Yukon: 293 (1942).
Alaska to N. U.S.A. 70 ASK 71 SAS 72 NFL NSC ONT QUE 74 MIN WIS 75 MIC NWH. Hemicr.

Eleocharis nuda C.B.Clarke, Kew Bull., Addit. Ser. 8: 21 (1908).
N. Australia. 50 NTA QLD. Hemicr.

Eleocharis nudipes (Kunth) Palla, Denkschr. Kaiserl. Akad. Wiss., Wien. Math.-Naturwiss. Kl. 79: 171 (1908).
Brazil to N. Argentina. 84 BZC BZL BZS 85 AGE PAR URU. Hemicr.
**Isolepis nudipes* Kunth, Enum. Pl. 2: 206 (1837).
Scirpidium grande Nees in C.F.P.von Martius & auct. suc. (eds.), Fl. Bras. 2(1): 97 (1842). *Eleocharis grandis* (Nees) Boeck., Vidensk. Meddel. Naturhist. Foren. Kjøbenhavn 1869: 134 (1869). *Scirpus grandis* (Nees) Kuntze, Revis. Gen. Pl. 2: 757 (1891).

Eleocharis nupeensis Hutch. in J.Hutchinson & J.M.Dalziel, Fl. W. Trop. Afr. 2: 467 (1936).
W. Trop. Africa to Tanzania. 22 GHA IVO NGA SEN SIE 24 CHA 25 TAN. Hemicr.
**Eleocharis mitrata* var. *africana* C.B.Clarke in D.Oliver & auct. suc. (eds.), Fl. Trop. Afr. 8: 406 (1902).
Eleocharis fistulosa var. *micrantha* Cherm., Arch. Bot. Bull. Mens. 7(4): 25 (1936).

Eleocharis obicis L.A.S.Johnson & O.D.Evans, Contr. New South Wales Natl. Herb. 4: 71 (1968).
New South Wales. 50 NSW. Hemicr.

Eleocharis obpyriformis D.A.Simpson, Kew Bull. 48: 705 (1993).
Brazil (Bahia). 84 BZE. Hemicr.

Eleocharis obtusa (Willd.) Schult. in J.J.Roemer & J.A.Schultes, Mant. 2: 89 (1824).
N. America, NE. Argentina. (12) por (13) ita (63) haw 71 BRC 72 NSC ONT PEI QUE 73 COL IDA ORE WAS WYO 74 ILL IOW MIN MSO NEB OKL SDA WIS 75 CNT INI MAI MAS MIC NWH NWJ NWY OHI PEN RHO VER WVA 76 CAL NEV 77 TEX 78 ALA ARK DEL FLA GEO KTY LOU MRY MSI NCA SCA TEN VRG WDC 85 AGE. Ther.
**Scirpus obtusus* Willd., Enum. Pl.: 76 (1809). *Trichophyllum obtusum* (Willd.) House, Amer. Midl. Naturalist 6: 205 (1920).
Eleocharis obtusa var. *jejuna* Fernald, Proc. Amer. Acad. Arts 34: 492 (1899).
Eleocharis obtusa var. *ellipsoidalis* Fernald ex Svenson, Rhodora 31: 216 (1929).
Eleocharis obtusa var. *gigantea* Svenson, Rhodora 31: 216 (1929).
Eleocharis obtusa var. *peasei* Svenson, Rhodora 31: 217 (1929).

Eleocharis obtusitrigona (Lindl. & Nees) Steud., Syn. Pl. Glumac. 2: 80 (1854).
SE. Texas to C. America, Brazil to N. Argentina. 77 TEX 79 MXG 80 COS HON NIC 84 BZE BZL 85 AGE PAR. Hemicr.
**Limnochloa obtusitrigona* Lindl. & Nees in C.F.P.von Martius & auct. suc. (eds.), Fl. Bras. 2(1): 100 (1842). *Eleocharis mutata* var. *obtusitrigona* (Lindl. & Nees) C.B.Clarke, Bull. Herb. Boissier 6(App. 1): 20 (1898). *Eleocharis fistulosa* var. *obtusitrigona* (Lindl. & Nees) M.Barros, Sellowia 12: 262 (1960).
Eleocharis strobilacea Pedersen, Darwiniana 12: 243 (1961).

Eleocharis occidentalis Mereles, Candollea 45: 653 (1990).
Paraguay. 85 PAR. Hemicr.

Eleocharis occulta S.G.Sm., Novon 11: 247 (2001).
Oklahoma to Texas. 74 OKL 77 TEX. Hemicr. or rhizome geophyte.

Eleocharis ochrostachys Steud., Syn. Pl. Glumac. 2: 80 (1854). *Scirpus ochrostachys* (Steud.) Kuntze, Revis. Gen. Pl. 2: 758 (1891).
Trop. & Subtrop. Asia to W. Pacific. 36 CHS 38 NNS TAI 40 IND SRL 41 THA VIE 42 BOR JAW MLY MOL PHI SUL SUM 43 BIS NWG SOL 50 NTA QLD 60 FIJ 62 CRL MRN. Hemicr.
Eleocharis subulata Boeck., Flora 41: 412 (1858).
Scirpus laxiflorus Thwaites, Enum. Pl. Zeyl.: 435 (1864). *Eleocharis variegata* var. *laxiflora* (Thwaites) C.B.Clarke, J. Linn. Soc., Bot. 36: 229 (1903). *Eleocharis laxiflora* (Thwaites) H.Pfeiff., Mitt. Inst. Allg. Bot. Hamburg 169 (1928).

Eleocharis oligantha C.B.Clarke in I.Urban, Symb. Antill. 2: 69 (1900).
Caribbean. 81 CUB JAM PUE. Ther.
Eleocharis wrightii C.B.Clarke in I.Urban, Symb. Antill. 2: 69 (1900).

Eleocharis olivaceonux D.A.Simpson, Kew Bull. 48: 705 (1993).
Brazil (Bahia). 84 BZE. Hemicr.

Eleocharis onthitensis H.E.Hess, Ber. Schweiz. Bot. Ges. 63: 336 (1953).
Angola to Namibia. 26 ANG 27 NAM. Hemicr.

Eleocharis ovata (Roth) Roem. & Schult., Syst. Veg. 2: 152 (1817).
Temp. Northern Hemisphere. 11 AUT BGM CZE GER HUN NET POL SWI 12 FRA 13 ROM 14 BLR BLT RUC RUE RUN RUS RUW UKR 30+ 31 PRM 32 KAZ 33 TCS 36 CHC CHM CHX 38 JAP 71 ABT BRC 72 LAB NBR NFL NSC ONT PEI QUE 73 ORE WAS 74 ILL IOW MIN OKL WIS 75 CNT INI MAI MAS MIC NWH NWJ NWY OHI PEN RHO VER WVA 76 ARI 78 KTY. Ther.

Scirpus capitatus Schreb., Spic. Fl. Lips.: 60 (1771), nom. illeg.
**Scirpus ovatus* Roth, Tent. Fl. Germ. 2(2): 562 (1793). *Bulbostylis ovatus* (Roth) Steven, Mém. Soc. Imp. Naturalistes Moscou 5: 355 (1814). *Clavula ovata* (Roth) Dumort., Fl. Belg.: 143 (1827). *Eleogenus ovatus* (Roth) Nees, Linnaea 9: 294 (1834). *Cyperus ovatus* (Roth) Missbach & E.H.L.Krause in J.Sturm, Deutschl. Fl., ed. 2, 2: 25 (1900), nom. illeg. *Trichophyllum ovatum* (Roth) House, Amer. Midl. Naturalist 6: 205 (1920).
Scirpus compressus Moench, Methodus: 849 (1794).
Scirpus annuus Thuill., Fl. Env. Paris, ed. 2: 22 (1799). *Eleocharis annua* House, Bull. New York State Mus. Nat. Sci. 243–244: 58 (1923).
Scirpus multicaulis C.C.Gmel., Fl. Bad. 1: 96 (1805).
Scirpus turgidus Pers., Syn. Pl. 1: 66 (1805), nom. illeg.
Scirpus soloniensis Dubois, Méth. Éprouv., ed. 2: 249 (1833). *Eleocharis soloniensis* (Dubois) Hara, J. Jap. Bot. 14: 338 (1938).
Scirpus elegantulus Steud., Syn. Pl. Glumac. 2: 317 (1855).
Eleocharis ovata var. *humifusa* A.Terracc., Malpighia 2: 311 (1889).
Eleocharis ovata var. *italica* A.Terracc., Malpighia 2: 310 (1889).
Eleocharis ovata f. *sessitensis* A.Terracc., Malpighia 2: 311 (1889).
Eleocharis ovata var. *aphanactis* J.W.Moore, Rhodora 52: 54 (1950).

Eleocharis oxylepis (Meinsh.) B.Fedtsch., Rastit. Turkest.: 165 (1915).
Ukraine to Kazakhstan. 14 RUS UKR 32 KAZ. Hemicr.
**Scirpus oxylepis* Meinsh., Trudy Imp. S.-Peterburgsk. Bot. Sada 18: 263 (1901).

Eleocharis pachycarpa É.Desv. in C.Gay, Fl. Chil. 6: 174 (1854). *Scirpus pachycarpus* (É.Desv.) Kuntze, Revis. Gen. Pl. 2: 758 (1891).
S. South America. (50) nsw (76) cal nev 85 AGS CLC PAR URU. Hemicr.
Eleocharis leptocaulis Steud., Syn. Pl. Glumac. 2: 77 (1854). *Eleocharis pachycarpa* var. *leptocaulis* (Steud.) C.B.Clarke, Bull. Herb. Boissier, II, 3: 1016 (1903).
Eleocharis dubia Phil., Linnaea 29: 77 (1857).
Eleocharis fuscosanguinea Boeck., Linnaea 36: 425 (1870). *Scirpus fuscosanguineus* (Boeck.) Kuntze, Revis. Gen. Pl. 2: 757 (1891).
Eleocharis liocarpa Phil., Anaes Univ. Chile 1873: 553 (1873). *Scirpus liocarpus* (Phil.) Kuntze, Revis. Gen. Pl. 2: 757 (1891).
Eleocharis lepida Phil., Anales Univ. Chile 93: 349 (1896).

Eleocharis pachystyla (C.Wright) C.B.Clarke in I.Urban, Symb. Antill. 2: 72 (1900).
Trop. America. 79 MXT 80 BLZ COS HON NIC PAN 81 CUB DOM HAI PUE TRT 82 FRG GUY SUR VEN 83 CLM. Hemicr.
Scirpus melanocarpus Griseb., Cat. Pl. Cub.: 239 (1866).
**Scirpus pachystylus* C.Wright in F.A.Sauvalle, Fl. Cub.: 174 (1871).
Eleocharis pachystyla var. *angustostachya* H.Pfeiff., Herbarium 2: 55 (1921).
Eleocharis pachystyla var. *macrostachya* H.Pfeiff., Herbarium 2: 55 (1921).

Eleocharis pallens S.T.Blake, Proc. Roy. Soc. Queensland 49: 154 (1937 publ. 1938).
Australia. 50 NSW NTA QLD SOA VIC. Hemicr.

Eleocharis palustris (L.) Roem. & Schult., Syst. Veg. 2: 151 (1817).
Temp. Northern Hemisphere to Himalaya. 11 AUT BGM CZE DEN FIN FOR GER GRB HUN ICE IRE NET NOR POL SWE SWI 12 BAL COR FRA POR SAR SPA 13 ALB BUL GRC ITA KRI ROM SIC TUE YUG 14 BLR BLT KRY RUC RUE RUN RUS RUW UKR 20 ALG EGY LBY MOR TUN 21 AZO CNY MDR 30 ALT BRY CTA IRK KRA TVA WSB YAK 31 KUR SAK 32 KAZ 33 NCS TCS 34 AFG CYP IRN IRQ LBS PAL TUR 35 YEM 36 CHM 37 MON 38 JAP 40 NEP PAK 70 ASK GNL NUN NWT YUK 71 ABT BRC MAN SAS 72 LAB NBR NFL NSC ONT PEI QUE 73 COL IDA MNT ORE WAS WYO 74 ILL IOW KAN MIN MSO NDA NEB OKL SDA WIS 75 CNT INI MAI MAS MIC NWH NWJ NWY OHI PEN RHO VER WVA 76 ARI CAL NEV UTA 77 NWM TEX 78 ALA ARK DEL KTY LOU MRY MSI NCA SCA TEN VRG 79 MXE. Hemicr.
**Scirpus palustris* L., Sp. Pl.: 47 (1753). *Schoenus palustris* (L.) Bernh., Syst. Verz.: 26 (1800). *Bulbostylis palustris* (L.) Steven, Mém. Soc. Imp. Naturalistes Moscou 5: 355 (1814). *Megadenus palustris* (L.) Raf., Neogenyton: 4 (1825). *Clavula palustris* (L.) Dumort., Fl. Belg.: 143 (1827). *Cyperus palustris* (L.) Sessé & Moç., Fl. Mexic.: 16 (1893). *Chlorocharis palustris* (L.) Rikli, Jahrb. Wiss. Bot. 27: 564 (1895). *Trichophyllum palustre* (L.) House, Amer. Midl. Naturalist 6: 205 (1920).

subsp. ***iranica*** Kukkonen, Ann. Bot. Fenn. 32: 156 (1995).
E. Medit. to Pakistan. 20 EGY 34 AFG IRN IRQ TUR 40 PAK. Hemicr.

subsp. ***palustris***
Temp. Northern Hemisphere to Himalaya. 11 AUT BGM CZE DEN FIN FOR GER GRB HUN ICE IRE NET NOR POL SWE SWI 12 BAL COR FRA POR SAR SPA 13 ALB BUL GRC ITA KRI ROM SIC TUE YUG 14 BLR BLT KRY RUC RUE RUN RUS RUW UKR 20 ALG LBY MOR TUN 21 AZO CNY MDR 30 ALT BRY CTA IRK KRA TVA WSB YAK 31 KUR SAK 32 33 NCS TCS 34 AFG CYP IRN IRQ 35 YEM 36 CHM 37 MON 38 JAP 40 NEP 70 ASK GNL NWT YUK 71 ABT BRC MAN SAS 72 LAB NBR NFL NSC ONT PEI QUE 73 COL IDA MNT ORE WAS WYO 74 ILL IOW KAN MIN MSO NDA NEB OKL SDA WIS 75 CNT INI MAI MAS MIC NWH NWJ NWY OHI PEN RHO VER WVA 76 ARI CAL NEV UTA 77 NWM TEX 78 ALA ARK DEL KTY LOU MRY MSI NCA SCA TEN VRG 79 MXE. Hemicr.
Scirpus acicularis Oeder in G.C.Oeder & al. (eds.), Fl. Dan.: t. 167 (1764), nom. illeg.
Scirpus ovatus Gilib., Excerc. Phyt. 2: 512 (1792), nom. illeg.
Scirpus conicus C.Presl ex Mert. & W.D.J.Koch in J.C.Röhling, Deutschl. Fl. 1: 427 (1796), nom. inval.
Scirpus intermedius Thuill., Fl. Env. Paris, ed. 2: 21 (1799).
Scirpus reptans Thuill., Fl. Env. Paris, ed. 2: 22 (1799), nom. illeg. *Scirpus varius* Schreb. ex Schweigg. & Körte, Fl. Erlang.: 11 (1811).

Scirpus baeothryon Schrad., Fl. Germ. 1: 127 (1806), nom. illeg.
Scirpus glaucescens Willd., Enum. Pl.: 76 (1809). *Eleocharis glaucescens* (Willd.) Schult. in J.J.Roemer & J.A.Schultes, Mant. 2: 89 (1824). *Eleocharis palustris* var. *glaucescens* (Willd.) Coss. & Germ., Fl. Descr. Anal. Paris: 607 (1845). *Trichophyllum palustre* var. *glaucescens* (Willd.) House, Amer. Midl. Naturalist 6: 205 (1920).
Scirpus tenuis Schreb. ex Schweigg. & Körte, Fl. Erlang.: 12 (1811).
Eleocharis palustris var. *limburgensis* Lej., Fl. Spa 2: 289 (1813).
Scirpus limburgensis Lej., Fl. Spa 2: 289 (1813).
Eleocharis minor Mazziari, Ionios Anthol. 2: 460 (1834).
Limnochloa capensis Nees, Linnaea 10: 185 (1836), nom. illeg.
Scirpus caespitosus Willd. ex Kunth, Enum. Pl. 2: 148 (1837).
Scirpus campestris Willd. ex Kunth, Enum. Pl. 2: 148 (1837), nom. inval.
Eleocharis palustris var. *degenerata* Hook., Fl. Bor.-Amer. 2: 228 (1839).
Eleocharis palustris var. *minor* Hook., Fl. Bor.-Amer. 2: 228 (1839).
Eleocharis limosa Drège, Linnaea 20: 246 (1847), nom. illeg.
Eleocharis palustris var. *major* Sond., Fl. Hamburg.: 22 (1850). *Eleocharis smallii* var. *major* (Sond.) F.Seym., Fl. New England, ed. 2: 113 (1982).
Eleocharis nebrodensis Parl., Fl. Ital. 2: 63 (1852). *Eleocharis palustris* subsp. *nebrodensis* (Parl.) Nyman, Consp. Fl. Eur.: 767 (1882). *Eleocharis palustris* var. *nebrodensis* (Parl.) A.Terracc., Malpighia 2: 304 (1889). *Eleocharis lereschii* subsp. *nebrodensis* (Parl.) Nyman, Consp. Fl. Eur., Suppl. 2: 319 (1890).
Eleocharis haematolepis Steud., Syn. Pl. Glumac. 2: 79 (1854).
Eleocharis appendiculata Phil., Linnaea 29: 85 (1857). *Scirpus appendiculatus* (Phil.) Kuntze, Revis. Gen. Pl. 2: 757 (1891).
Eleocharis pileata A.Gray, Mem. Amer. Acad. Arts, n.s., 6: 417 (1859). *Scirpus pileatus* (A.Gray) Franch. & Sav., Enum. Pl. Jap. 2: 110 (1877).
Scirpus bailii Kohts, Oesterr. Bot. Z. 19: 834 (1869). *Scirpus parvulus* var. *bailii* (Kohts) Nyman, Consp. Fl. Eur.: 765 (1882).
Eleocharis palustris f. *italica* A.Terracc., Malpighia 2: 304 (1889).
Eleocharis striata Hochst. ex Engl., Abh. Königl. Akad. Wiss. Berlin 1891: 149 (1892), nom. illeg.
Scirpus lanceiglumis C.B.Clarke in T.A.Durand & H. Schinz, Consp. Fl. Afric. 5: 634 (1894), nom. inval.
Cyperus paluster Missbach & E.H.L.Krause in J.Sturm, Deutschl. Fl., ed. 2, 2: 24 (1900).
Eleocharis boissieri Podp., Verh. K. K. Zool.-Bot. Ges. Wien 52: 677 (1902).
Eleocharis eupalustris H.Lindb., Acta Soc. Fauna Fl. Fenn. 27(7): 5 (1902).
Cyperus paluster var. *uniglumoides* E.H.L.Krause, Beih. Bot. Centralbl. 32(2): 331 (1914).
Trichophyllum palustre var. *calvum* (A.Gray) House, Amer. Midl. Naturalist 6: 205 (1920).
Eleocharis minima f. *major* Boeck., Herbarium 2: 55 (1921).
Eleocharis minima var. *mexicensis* H.Pfeiff., Herbarium 2: 55 (1921).
Eleocharis crassa Fisch. & C.A.Mey. ex Zinserl. in V.L.Komarov (ed.), Fl. URSS 3: 582 (1935).
Eleocharis ecarinata Zinserl. in V.L.Komarov (ed.), Fl. URSS 3: 583 (1935).
Eleocharis globularis Zinserl. in V.L.Komarov (ed.), Fl. URSS 3: 582 (1935). *Eleocharis palustris* subsp. *globularis* (Zinserl.) T.V.Egorova, Novosti Sist. Vyssh. Rast. 17: 79 (1980). *Eleocharis palustris* var. *globularis* (Zinserl.) A.E.Kozhevn., Sosud. Rast. Sovet. Dal'nego Vostoka 3: 210 (1988).
Eleocharis intersita Zinserl. in V.L.Komarov (ed.), Fl. URSS 3: 581 (1935). *Eleocharis palustris* subsp. *intersita* (Zinserl.) Tzvelev, Novosti Sist. Vyssh. Rast. 32: 183 (2000).
Eleocharis kasakstanica Zinserl. in V.L.Komarov (ed.), Fl. URSS 3: 583 (1935).
Eleocharis levinae Zoz, in Fl. RSS Ucr. 2: 426 (1940).
Eleocharis palustris subsp. *microcarpa* Walters, J. Econ. Taxon. Bot. 37: 194 (1949). *Eleocharis palustromicrocarpa* Á.Löve, Caryologia 3: 270 (1951). *Scirpus palustromicrocarpus* (Á.Löve) Á.Löve, Caryologia 3: 271 (1951).
Eleocharis kitamurana T.Koyama, Acta Phytotax. Geobot. 17: 48 (1957).
Eleocharis palustris var. *lindbergii* Strandh., Bot. Not. 114: 424 (1961). *Eleocharis lindbergii* (Strandh.) Tzvelev, Novosti Sist. Vyssh. Rast. 30: 163 (1996). *Eleocharis palustris* subsp. *lindbergii* (Strandh.) Tzvelev, Novosti Sist. Vyssh. Rast. 32: 183 (2000).
Eleocharis intersita f. *acetosa* Tang & F.T.Wang ex Y.L.Chang, in Fl. Plant. Herb. Chinae Bor.-Or. 11: 204 (1976).
Eleocharis oxystachys Sakalo, Novosti Sist. Vyssh. Nizsh. Rast. 1976: 50 (1977).
Eleocharis palustris var. *kurilensis* A.E.Kozhevn., Sosud. Rast. Sovet. Dal'nego Vostoka 3: 210 (1988).

var. ***vigens*** L.H.Bailey, J. New York Microscop. Soc. 5: 104 (1889). *Trichophyllum palustre* var. *vigens* (L.H.Bailey) House, Amer. Midl. Naturalist 6: 205 (1920).
N. America. 70 ASK GNL NUN NWT YUK 71 ABT BRC MAN SAS 72 LAB NBR NFL NSC ONT PEI QUE 73 COL IDA MNT ORE WAS WYO 74 ILL IOW KAN MIN MSO NDA NEB OKL SDA WIS 75 CNT INI MAI MAS MIC NWH NWJ NWY OHI PEN RHO VER WVA 76 ARI CAL NEV UTA 77 NWM TEX 78 ALA ARK DEL KTY LOU MRY MSI NCA SCA TEN VRG 79 MXE. Hemicr.
Eleocharis smallii Britton, Torreya 3: 23 (1903).

subsp. ***vulgaris*** Walters, J. Econ. Taxon. Bot. 37: 194 (1949). *Eleocharis vulgaris* (Walters) Á.Löve & D.Löve, Folia Geobot. Phytotax. 10: 275 (1975).
Europe. 10 DEN FIN FOR GRB ICE IRE NOR SWE 11 AUT BGM CZE GER HUN NET POL SWI 12 BAL COR FRA POR SAR SPA 13 ALB BUL GRC ITA KRI ROM SIC TUE YUG 14 BLT. Hemicr.
Eleocharis filiculmis Schur, Enum. Pl. Transsilv.: 690 (1866), nom. illeg.
Eleocharis gracilis Hayek, Sched. Fl. Stiriac. 19–20: 9 (1910), nom. illeg.

Eleocharis papillosa Latz, J. Adelaide Bot. Gard. 1: 343 (1979).
Northern Territory. 50 NTA. Hemicr.

Eleocharis paradoxa Y.D.Chen, Bull. Bot. Res., Harbin 7(2): 121 (1987).
Qinghai. 36 CHQ. Hemicr.

Eleocharis parishii Britton, J. New York Microscop. Soc. 5: 111 (1889). *Eleocharis montevidensis* var. *parishii* (Britton) V.E.Grant, Aliso 4: 87 (1958).
Oregon to N. Mexico. 73 ORE 74 KAN 76 ARI CAL NEV UTA 77 NWM TEX 79 MXE MXN. Ther.
Eleocharis disciformis Parish in A.Davidson & G.L. Moxley, Fl. S. Calif.: 70 (1923). *Eleocharis montevidensis* var. *disciformis* (Parish) V.E.Grant, Aliso 4: 87 (1958).

Eleocharis parodii Barros, Anales Mus. Nac. Buenos Aires 34: 480 (1928).
NE. Argentina to Uruguay. (50) nsw 85 AGE URU. Hemicr.

Eleocharis parvinux Ohwi, Mem. Coll. Sci. Kyoto Imp. Univ., Ser. B, Biol. 18(1): 47 (1944).
Japan (E. Honshu). 38 JAP. Hemicr.

Eleocharis parvula (Roem. & Schult.) Palla, Bot. Jahrb. Syst. 10: 299 (1888).
Temp. Northern Hemisphere to W. Malesia and Venezuela. 10 DEN FIN GRB IRE NOR SWE 11 GER POL SWI† 12 FRA POR SAR SPA 13 GRC ITA ROM YUG 14 BLT RUS RUW UKR 20 EGY 30 31 32 KAZ UZB 33 TCS 38 JAP 41 VIE 42 JAW 71 BRC 72 NBR NFL NSC QUE 73 ORE WAS 74 ILL KAN MSO 75 CNT INI MAI MAS MIC NWJ NWY OHI 76 CAL 78 ALA ARK FLA GEO LOU MRY MSI NCA SCA VRG 79 MXN MXT 80 GUA NIC 81 CUB 82 VEN 84 BZL. Tuber or rhizome geophyte.
Scirpus nanus Spreng., Pl. Min. Cogn. Pug. 1: 4 (1813), nom. illeg. *Limnochloa parvula* (Spreng.) Rchb. in J.C.Mössler, Handb. Gewächsk. ed. 2, 3: 1808 (1830). *Baeothryon nanum* (Spreng.) A.Dietr., Sp. Pl. 2: 91 (1832).
Scirpus pollicaris Delile, Descr. Egypte, Hist. Nat.: 50 (1813), nom. nud.
**Scirpus parvulus* Roem. & Schult., Syst. Veg. 2: 124 (1817). *Eleogiton parvula* (Roem. & Schult.) Link, Hort. Berol. 1: 285 (1827). *Cyperus parvulus* (Roem. & Schult.) Missbach & E.H.L.Krause in J.Sturm, Deutschl. Fl., ed. 2, 2: 26 (1900), nom. illeg. *Trichophorum parvulum* (Roem. & Schult.) Pignatti, Arch. Bot. (Forlì) 34: 7 (1958).
Clavula comosa Dumort., Fl. Belg.: 143 (1827). *Scirpus comosus* (Dumort.) B.D.Jacks., Index Kew. 2: 839 (1894), nom. illeg.
Eleocharis pygmaea Torr., Ann. Lyceum Nat. Hist. New York 3: 313 (1836). *Chaetocyperus pygmaeus* (Torr.) Torr. ex A.Gray, Manual: 524 (1848).
Chaetocyperus pusillus Steud., Syn. Pl. Glumac. 2: 74 (1854).
Isolepis leptos Steud., Syn. Pl. Glumac. 2: 91 (1855). *Scirpus leptos* (Steud.) C.Wright, Anales Acad. Ci. Méd. Habana 8: 80 (1871). *Eleocharis leptos* (Steud.) Svenson, Rhodora 31: 176 (1929), nom. illeg.
Chaetocyperus membranaceus Buckley, Proc. Acad. Nat. Sci. Philadelphia 1862: 10 (1863). *Eleocharis membranacea* (Buckley) Gilly, Amer. Midl. Naturalist 26: 66 (1941).
Eleocharis leptos var. *johnstonii* Svenson, Rhodora 31: 177 (1929).
Eleocharis parvula f. *spongiosa* Fassett, Rhodora 39: 273 (1937).

Eleocharis pellucida C.Presl, Reliq. Haenk. 1: 196 (1828).
Russian Far East to Malesia. 31+ 36 CHC CHM CHS 38 JAP KOR NNS TAI 40 EHM IND SRL 41 42 BOR JAW LSI MLY PHI SUL SUM. Hemicr.
Eleocharis afflata Steud., Syn. Pl. Glumac. 2: 76 (1854). *Scirpus afflatus* (Steud.) Benth., Fl. Hongk.: 394 (1861). *Scirpus afflatus* (Steud.) Kuntze, Revis. Gen. Pl. 2: 757 (1891).
Eleocharis subprolifera Steud., Syn. Pl. Glumac. 2: 80 (1854).
Eleocharis japonica Miq., Ann. Mus. Bot. Lugduno-Batavi 2: 142 (1865). *Scirpus japonicus* (Miq.) Franch. & Sav., Enum. Pl. Jap. 2: 109 (1877). *Eleocharis pellucida* var. *japonica* (Miq.) Tang & F.T.Wang, in Fl. Reipubl. Popul. Sin. 11: 58 (1961). *Eleocharis congesta* subsp. *japonica* (Miq.) T.Koyama, in Fl. Taiwan 5: 220 (1978).
Eleocharis japonica Boeck., Linnaea 36: 422 (1870), nom. illeg.
Eleocharis ochrostachys Boeck., Linnaea 36: 452 (1870), nom. illeg.
Eleocharis thomsonii Boeck., Linnaea 36: 451 (1870). *Scirpus thomsonii* (Boeck.) Kuntze, Revis. Gen. Pl. 2: 758 (1891).
Eleocharis chlorocarpa Boeck., Flora 61: 34 (1878).
Eleocharis kuntzei Boeck., Bot. Jahrb. Syst. 10: 14 (1888).
Scirpus zollingeri Kuntze, Revis. Gen. Pl. 2: 758 (1891).
Chlorocharis subprolifera Rikli, Jahrb. Wiss. Bot. 27: 564 (1895).
Eleocharis gableana Boeck., Allg. Bot. Z. Syst. 2: 54 (1896).
Eleocharis shimadai Hayata, Icon. Pl. Formosan. 6: 107 (1916).
Scirpus japonicus var. *thermalis* Hultén, Fl. Kamtchatka 1: 165 (1927). *Eleocharis congesta* var. *thermalis* (Hultén) T.Koyama, J. Fac. Sci. Univ. Tokyo, Sect. 3, Bot. 8: 90 (1961). *Eleocharis thermalis* (Hultén) T.V.Egorova, Novosti Sist. Vyssh. Rast. 17: 74 (1980), nom. illeg.
Eleocharis pellucida var. *sanguinolenta* Tang & F.T.Wang, in Fl. Reipubl. Popul. Sin. 11: 226 (1961).
Eleocharis pellucida var. *spongiosa* Tang & F.T.Wang, in Fl. Reipubl. Popul. Sin. 11: 226 (1961).
Eleocharis congesta f. *soranumaensis* Koji Ito, J. Geobot. 19: 6 (1971).

Eleocharis penchaoi Y.D.Chen, Bull. Bot. Res., Harbin 7(2): 119 (1987).
Qinghai. 36 CHQ. Hemicr.

Eleocharis philippinensis Svenson, Rhodora 31: 155 (1929).
Trop. Asia to E. Australia. 36 CHH 40 IND 41 THA VIE 42 JAW LSI MLY PHI 43 NWG 50 NSW QLD 60 NWC. Ther. or hemicr.

Eleocharis plana S.T.Blake, Proc. Roy. Soc. Queensland 49: 155 (1937 publ. 1938).
E. & SE. Australia. 50 NSW QLD SOA VIC. Hemicr.

Eleocharis platypus C.B.Clarke, Bot. Jahrb. Syst. 34(78): 3 (1904).
Ecuador. 83 ECU. Hemicr.

Eleocharis plicarhachis (Griseb.) Svenson, Rhodora 31: 158 (1929).

Trop. America. 79 MXT 80 COS GUA NIC PAN 81 CUB TRT 82 FRG GUY SUR VEN 83 BOL CLM ECU PER 84 BZC BZL BZN 85 AGE AGW PAR. Hemicr.

**Scirpus plicarhachis* Griseb., Cat. Pl. Cub.: 239 (1866).

Eleocharis elata Boeck., Vidensk. Meddel. Naturhist. Foren. Kjøbenhavn 1871: 151 (1871).

Eleocharis pseudofistulosa H.E.Hess, Ber. Schweiz. Bot. Ges. 63: 329 (1953).
Angola to Namibia. 26 ANG 27 NAM. Hemicr.

Eleocharis punctulata Boeck. ex Duss, Ann. Inst. Bot.-Géol. Colon. Marseille 3: 544 (1896 publ. 1897).
Martinique. 81 WIN. Hemicr.

Eleocharis punctata f. *major* Duss, Ann. Inst. Bot.-Géol. Colon. Marseille 3: 544 (1896 publ. 1897).

Eleocharis pusilla R.Br., Prodr.: 225 (1810). *Scirpus pusillus* (R.Br.) Poir. in J.B.A.M.de Lamarck, Encycl., Suppl. 5: 103 (1817).
Australia, New Zealand. 50 NSW NTA QLD SOA TAS VIC WAU 51 NZN NZS. Hemicr.

Scirpus pumilio Spreng., Syst. Veg. 1: 204 (1824).

Eleocharis acicularis var. *elongata* Benth., Fl. Austral. 7: 297 (1878).

Eleocharis qinghaiensis Y.D.Chen, Bull. Bot. Res., Harbin 7(2): 117 (1987).
Qinghai. 36 CHQ. Hemicr.

Eleocharis quadrangulata (Michx.) Roem. & Schult., Syst. Veg. 2: 155 (1817).
S. Ontario to C. & E. U.S.A., Mexico (Durango to Jalisco). 72 ONT 74 ILL KAN MSO OKL WIS 75 CNT INI MAS MIC NWJ NWY OHI PEN WVA (76) cal 77 TEX 78 ALA ARK DEL DEL FLA GEO KTY LOU MRY MSI NCA SCA TEN VRG WDC 79 MXE MXS. Hemicr.

**Scirpus quadrangulatus* Michx., Fl. Bor.-Amer. 1: 30 (1803). *Limnochloa quadrangulata* (Michx.) Nees, Linnaea 9: 294 (1834). *Eleocharis quadrangulata* var. *typica* Fernald, Rhodora 37: 393 (1935), nom. inval.

Eleocharis quadrangulata var. *crassior* Fernald, Rhodora 37: 393 (1935).

Eleocharis quinqueflora (Hartmann) O.Schwarz, Mitt. Thüring. Bot. Ges. 1: 89 (1949).
Temp. Northern Hemisphere, S. South America. 10 DEN FIN FOR GRB ICE IRE NOR SWE 11 AUT BGM CZE GER HUN NET POL SWI 12 COR FRA SPA? 13 ALB BUL GRC ITA YUG 14 BLR BLT KRY RUC RUE RUN RUS RUW UKR 20 EGY MOR 30 WSB 31 KAM 32 KAZ KGZ TKM UZB 33 TCS 34 AFG IRN IRQ TUR 36 CHX 37 MON 40 NEP PAK WHM 70 ASK GNL NWT YUK 71 ABT BRC MAN SAS 72 LAB NBR NFL NSC ONT PEI QUE 73 COL IDA MNT ORE WAS WYO 74 ILL IOW MIN NDA NEB WIS 75 INI MAI MAS MIC NWH NWJ NWY OHI PEN VER 76 ARI CAL NEV UTA 77 NWM 85 AGS AGW CLN. Hemicr.

**Scirpus quinqueflorus* Hartmann, Primae Lin. Inst. Bot., ed. 2: 85 (1767).

Scirpus cespitosus Pollich, Hist. Pl. Palat. 1: 39 (1776), nom. illeg.

Scirpus pauciflorus Lightf., Fl. Scot. 2: 1078 (1777). *Eleocharis pauciflora* (Lightf.) Link, Hort. Berol. 1: 284 (1827). *Baeothryon pauciflorum* (Lightf.) A.Dietr., Sp. Pl. 2: 90 (1832). *Limnochloa pauciflora* (Lightf.) Peterm., Fl. Bienitz: 21 (1841). *Cyperus pauciflorus* (Lightf.) Missbach & E.H.L.Krause in J.Sturm, Deutschl. Fl., ed. 2, 2: 27 (1900), nom. illeg. *Trichophorum pauciflorum* (Lightf.) Pignatti, Arch. Bot. (Forlì) 34: 7 (1958).

Scirpus baeothryon L.f., Suppl. Pl.: 103 (1782). *Clavula boeotryon* (L.f.) Dumort., Fl. Belg.: 143 (1827). *Limnochloa baeothryon* (L.f.) Rchb. in J.C.Mössler, Handb. Gewächsk. ed. 2, 3: 1808 (1830). *Eleocharis baeothryon* (L.f.) Nees, Linnaea 9: 294 (1834), nom. illeg.

Scirpus sepium Honck., Verz. Gew. Teutschl.: 78 (1782).

Scirpus halleri Vill., Hist. Pl. Dauphiné 2: 188 (1787). *Baeothryon halleri* (Vill.) T.Nees, Gen. Fl. Germ. 2: 17 (1816).

Scirpus campestris Rottb., Ann. Bot. (Usteri) 14: 18 (1795).

Eleocharis baeothryon Schult. in J.J.Roemer & J.A.Schultes, Mant. 2: 92 (1824).

Eleocharis atacamensis Phil., Fl. Atacam.: 53 (1860). *Scirpus atacamensis* (Phil.) Kuntze, Revis. Gen. Pl. 2: 757 (1891).

Isolepis andina Phil., Anales Mus. Nac. Santiago de Chile 1: 79 (1891).

Eleocharis vierhapperi Bojko, Verh. Zool.-Bot. Ges. Wien 79: 300 (1929 publ. 1930). *Trichophorum vierhapperi* (Bojko) Pignatti, Arch. Bot. (Forlì) 34: 7 (1958).

Eleocharis pauciflora var. *fernaldii* Svenson, Rhodora 36: 380 (1934). *Scirpus pauciflorus* var. *fernaldii* (Svenson) Hiitonen, Memoranda Soc. Fauna Fl. Fenn. 25: 74 (1950). *Eleocharis fernaldii* (Svenson) Á.Löve, Svensk Bot. Tidskr. 48: 218 (1954). *Eleocharis quinqueflora* subsp. *fernaldii* (Svenson) Hultén, Ark. Bot., a.s., 7(1): 24 (1968).

Eleocharis meridionalis Zinserl. in V.L.Komarov (ed.), Fl. URSS 3: 580 (1935). *Eleocharis quinqueflora* subsp. *meridionalis* (Zinserl.) T.V.Egorova, Novosti Sist. Vyssh. Rast. 18: 102 (1981).

Eleocharis czernjajevi Zoz, Trudy Inst. Bot. Kharkivs'k. Derzhavn. Univ. 1: 55 (1936).

Eleocharis obscura T.Koyama, Acta Phytotax. Geobot. 17: 102 (1958).

Eleocharis rabenii Boeck., Vidensk. Meddel. Naturhist. Foren. Kjøbenhavn 1871: 149 (1871).
S. Brazil to Uruguay. 84 BZS 85 URU. Hemicr.

Eleocharis radicans Kunth, Enum. Pl. 2: 142 (1837).
C. U.S.A. to S. South America. (63) haw 74 OKL 75 MIC 76 ARI CAL 77 TEX 78 FLA LOU VRG 79 MXC MXE MXG MXN MXS MXT 80 ELS GUA NIC (81) pue 83 BOL CLM ECU PER 85 AGE AGS AGW CLC CLS PAR URU. Hemicr.

**Scirpus radicans* Poir. in J.B.A.M.de Lamarck, Encycl. 6: 751 (1805), nom. illeg. *Eleogiton radicans* (Poir.) A.Dietr., Sp. Pl. 2: 97 (1832). *Scirpus radiatus* Steud., Nomencl. Bot., ed. 2, 2: 541 (1841). *Chaetocyperus radicans* (Poir.) Steud., Syn. Pl. Glumac. 2: 74 (1854). *Eleocharis acicularis* var. *radicans* (Poir.) Britton, J. New York Microscop. Soc. 1889: 105 (1889).

Scirpus exiguus Kunth in F.W.H.von Humboldt, A.J.A.Bonpland & C.S.Kunth, Nov. Gen. Sp. 1: 225 (1816). *Eleocharis exigua* (Kunth) Roem. & Schult., Syst. Veg. 2: 154 (1817).

Eleocharis costulata Nees & Meyen ex Kunth, Enum. Pl. 2: 142 (1837). *Chaetocyperus costulatus* (Nees & Meyen ex Kunth) Nees, Nov. Actorum Acad.

Caes. Leop.-Carol. Nat. Cur. 19(Suppl. 1): 96 (1843). *Scirpus costulatus* (Nees & Meyen ex Kunth) Kuntze, Revis. Gen. Pl. 2: 757 (1891).
Eleocharis rivularis Phil., Linnaea 33: 270 (1865).
Eleocharis acicularis var. *lindheimeri* C.B.Clarke, Rhodora 31: 199 (1929). *Eleocharis lindheimeri* Svenson, Rhodora 31: 199 (1929).

Eleocharis ramboii H.E.Hess, Ber. Schweiz. Bot. Ges. 67: 93 (1957).
S. Brazil. 84 BZS. Hemicr.

Eleocharis ravenelii Britton in J.K.Small, Fl. S.E. U.S.: 184 (1903). *Trichophyllum ravenelii* (Britton) House, Amer. Midl. Naturalist 6: 205 (1920).
Texas to Mexico (San Luis Potosí). 77 TEX 79 MXE. Hemicr.
Eleocharis austrotexana M.C.Johnst., S. W. Naturalist 9: 312 (1964).

Eleocharis reichei Boeck., Allg. Bot. Z. Syst. 2: 54 (1896).
Chile. 85 CLC. Hemicr.

Eleocharis retroflexa (Poir.) Urb., Symb. Antill. 2: 165 (1900).
Trop. & Subtrop. 22 GUI LBR MLI SIE 23 ZAI 25 TAN 26? 27 BOT 29 MAU 36 CHS 38 NNS 40 EHM IND NEP SRL 41 CBD THA VIE 42 BOR JAW LSI MLY MOL PHI SUL SUM 43 NWG 50 QLD 78 ALA 79 MXG MXT 80 BLZ COS ELS GUA HON NIC PAN 81 CUB DOM JAM LEE PUE TRT WIN 82 FRG GUY SUR VEN 83 CLM ECU PER? 84 BZN. Ther.
**Scirpus retroflexus* Poir. in J.B.A.M.de Lamarck, Encycl. 6: 753 (1805). *Baeothryon retroflexum* (Poir.) A.Dietr., Sp. Pl. 2: 93 (1832).

subsp. ***chaetaria*** (Roem. & Schult.) T.Koyama, Bull. Natl. Sci. Mus. Tokyo, n.s., 17: 68 (1974).
E. Trop. Africa to NE. Australia. 25 TAN 29 MAU 36 CHS 38 NNS 40 EHM IND NEP SRL 41 CBD THA VIE 42 BOR JAW LSI MLY MOL PHI SUL SUM 43 NWG 50 QLD. Ther.
Cyperus setaceus Retz., Observ. Bot. 5: 10 (1788).
Scirpus pygmaeus Lam., Tabl. Encycl. 1: 139 (1791). *Baeothryon pygmaeum* (Lam.) A.Dietr., Sp. Pl. 2: 92 (1832).
Eleocharis setacea R.Br., Prodr.: 224 (1810). *Chaetocyperus setaceus* (R.Br.) Nees in C.F.P.von Martius & auct. suc. (eds.), Fl. Bras. 2(1): 94 (1842).
**Eleocharis chaetaria* Roem. & Schult., Syst. Veg. 2: 154 (1817). *Scirpus chaetarius* (Roem. & Schult.) Spreng., Syst. Veg. 1: 203 (1824). *Trichophyllum chaetaria* (Roem. & Schult.) House, Amer. Midl. Naturalist 6: 204 (1920).
Chaetocyperus limnocharis Nees in R.Wight, Contr. Bot. India: 96 (1834).
Eleocharis recurvata Nees, Linnaea 9: 294 (1834), nom. inval.
Chaetocyperus arenicola Steud., Syn. Pl. Glumac. 2: 73 (1854).
Chaetocyperus depauperatus Steud., Syn. Pl. Glumac. 2: 73 (1854).
Isolepis ambigua Steud., Syn. Pl. Glumac. 2: 91 (1855). *Eleocharis minima* var. *ambigua* (Steud.) Kük., Repert. Spec. Nov. Regni Veg. 23: 194 (1926).

subsp. ***depressa*** Zavaro & Pabón, Fontqueria 42: 270 (1995).
Cuba. 81 CUB. Ther.

subsp. ***retroflexa***
SE. U.S.A. to Trop America. 22 GUI LBR MLI SIE 23 ZAI 25 TAN 26? 27 BOT 29 MAU 36 CHS 38 NNS 40 EHM IND NEP SRL 41 CBD THA VIE 78 ALA 79 MXG MXT 80 BLZ COS ELS GUA HON NIC PAN 81 CUB DOM JAM LEE PUE TRT WIN 82 FRG GUY SUR VEN 83 CLM ECU PER? 84 BZN. Ther.
Cyperus depauperatus Vahl, Enum. Pl. 2: 305 (1805). *Eleocharis depauperata* (Vahl) Kunth, Enum. Pl. 2: 140 (1837). *Chaetocyperus polymorphus* var. *depauperatus* (Vahl) Nees in C.F.P.von Martius & auct. suc. (eds.), Fl. Bras. 2(1): 94 (1842).
Chaetocyperus niveus Liebm., Mexic. Neldeagt. Pl., V, 2: 242 (1851).
Chaetocyperus rugulosus Nees, Bonplandia (Hannover) 3: 86 (1855).
Eleocharis triflora Boeck., Flora 63: 437 (1880).

subsp. ***subtilissima*** (Nelmes) Lye, Nordic J. Bot. 3: 243 (1983).
W. Trop. Africa to Botswana. 22 GUI LBR MLI SIE 23 ZAI 26? 27 BOT. Ther.
**Eleocharis subtilissima* Nelmes, Kew Bull. 6: 22 (1951 publ. 1952).

Eleocharis reverchonii Svenson, Rhodora 31: 203 (1929).
Texas. 77 TEX. Hemicr.

Eleocharis robbinsii Oakes, Mag. Hort. Bot. 7: 178 (1841). *Trichophyllum robbinsii* (Oakes) House, Amer. Midl. Naturalist 6: 205 (1920).
E. Canada to NC. & E. U.S.A. 72 NBR NCS ONT QUE 74 MIN WIS 75 CNT INI MAI MAS MIC NWH NWJ OHI PEN RHO 78 ALA DEL FLA GEO NCA SCA VRG. Hemicr.

Eleocharis robusta (Boeck.) H.E.Hess, Ber. Schweiz. Bot. Ges. 63: 331 (1953).
W. Trop. Africa. 22 SEN. Hemicr.
**Eleocharis fistulosa* var. *robusta* Boeck., Flora 62: 563 (1876).

Eleocharis rostellata (Torr.) Torr., Fl. New York 2: 347 (1843).
N. & C. America to Caribbean, N. Argentina. 71 ABT? BRC 72 NSC ONT 73 IDA MNT WAS WYO 74 ILL KAN MIN NEB OKL SDA WIS 75 CNT INI MAI MAS MIC NWJ NWY OHI PEN RHO WVA 76 ARI CAL NEV UTA 77 NWM TEX 78 DEL FLA LOU MRY NCA SCA VRG 79 MXC MXE MXN MXS MXT 80 GUA 81 BER CAY CUB HAI PUE 85 AGW. Hemicr. or rhizome geophyte.
**Scirpus rostellatus* Torr., Ann. Lyceum Nat. Hist. New York 3: 318 (1836). *Trichophyllum rostellatum* (Torr.) Farw., Rep. (Annual) Michigan Acad. Sci. 19: 253 (1917).
Eleocharis rostellata var. *occidentalis* S.Watson in S.Watson & al., Bot. California 2: 222 (1880). *Eleocharis rostellata* f. *occidentalis* (S.Watson) Beetle, Publ. Univ. Wyoming 5: 22 (1938).
Eleocharis rostellata var. *congdonii* Jeps., Fl. Calif.: 196 (1922).

Eleocharis rugosa D.A.Simpson, Kew Bull. 42: 900 (1987).
Brazil. 84 BZE. Hemicr.

Eleocharis rzedowskii S.González, Phytologia 57: 381 (1985).
Mexico (Nuevo León). 79 MXE. Hemicr.

Eleocharis schaffneri Boeck., Bot. Jahrb. Syst. 7: 274 (1886). *Scirpus schaffneri* (Boeck.) Kuntze, Revis. Gen. Pl. 2: 758 (1891), nom. illeg.
Mexico to C. America. 79 MXE MXS MXT 80 COS GUA HON NIC. Ther. or hemicr.
Eleocharis exilis Boeck., Beitr. Cyper. 1: 16 (1888).

Eleocharis schenckii Boeck., Beitr. Cyper. 2: 13 (1890).
Brazil. 84+. Hemicr. or rhizome geophyte.

Eleocharis schlechteri C.B.Clarke in W.H.Harvey & auct. suc. (eds.), Fl. Cap. 7: 758 (1900).
S. Africa. 27 CPP NAM. Hemicr.
Eleocharis seydeliana Podlech, Mitt. Bot. Staatssamml. München 3: 527 (1960).

Eleocharis sellowiana Kunth, Enum. Pl. 2: 149 (1837). *Eleogenus sellovianus* (Kunth) Nees in C.F.P.von Martius & auct. suc. (eds.), Fl. Bras. 2(1): 103 (1842). *Scirpus sellowianus* (Kunth) Griseb., Abh. Königl. Ges. Wiss. Göttingen 24: 312 (1879).
SE. Mexico to S. Trop. America. 79 MXT 80 COS ELS GUA HON NIC PAN 82 FRG GUY VEN 83 BOL CLM ECU GAL 84 BZC 85 AGE PAR URU. Hemicr.

var. ***homonyma*** (Steud.) H.Pfeiff., Exsicc. 56: 54 (1921).
Guyana. 82 GUY. Hemicr.
**Eleocharis homonyma* Steud., Syn. Pl. Glumac. 2: 79 (1854).

var. ***sellowiana***
SE. Mexico to S. Trop. America. 79 MXT 80 COS ELS GUA HON NIC PAN 82 FRG GUY VEN 83 BOL CLM ECU GAL 84 BZC 85 AGE PAR URU. Hemicr.
Eleocharis salzmanniana Steud., Syn. Pl. Glumac. 2: 78 (1854).
Scirpus bahiensis Steud., Syn. Pl. Glumac. 2: 83 (1855).
Eleocharis mesopotamica Boeck., Beitr. Cyper. 1: 17 (1888).
Eleocharis tuerckheimii Boeck., Beitr. Cyper. 1: 16 (1888).
Eleocharis pittieri Boeck., Allg. Bot. Z. Syst. 2: 35 (1896).
Eleocharis galapagensis Svenson, Rhodora 31: 233 (1929).

Eleocharis setifolia (A.Rich.) J.Raynal, Adansonia, n.s., 7: 318 (1967).
Trop. 22 BEN BKN GHA GUI IVO MLI NGA NGR SEN SIE TOG 23 CAF ZAI 24 CHA ETH SUD 25 KEN 26 ZAM 29 MDG 41 THA 42 PHI 50 NTA QLD SOA 78 FLA SCA 79 MXG MXS MXT 80 BLZ COS HON NIC PAN 81 CUB DOM JAM LEE TRT 82 GUY VEN 84 BZE BZN. Ther. or tuber geophyte.
**Isolepis setifolia* A.Rich., Tent. Fl. Abyss. 2: 498 (1850).

subsp. ***schweinfurthiana*** (Boeck.) D.A.Simpson, Kew Bull. 43: 428 (1988).
W. Trop. Africa to Sudan. 22 BKN GHA IVO MLI SEN SIE 23 ZAI 24 SUD. Ther. or tuber geophyte.
**Eleocharis schweinfurthiana* Boeck., Flora 62: 562 (1879).
Eleocharis helenae Buscal. & Muschl., Bot. Jahrb. Syst. 49: 461 (1913).

subsp. ***setifolia***
Trop. 22 BEN BKN GHA GUI IVO MLI NGA NGR SEN SIE TOG 23 CAF ZAI 24 CHA ETH 25 KEN 26 ZAM 29 MDG 41 THA 42 PHI 50 NTA QLD SOA 78 FLA SCA 79 MXS MXT 80 BLZ COS HON NIC PAN 81 CUB DOM JAM LEE TRT 82 GUY 84 BZE BZN. Ther. or tuber geophyte.
Eleocharis minutiflora Boeck., Bot. Jahrb. Syst. 7: 274 (1886). *Eleocharis nigrescens* var. *minutiflora* (Boeck.) Svenson, Rhodora 39: 226 (1937).
Eleocharis carolina Small, Man. S.E. Fl.: 165 (1933).

Eleocharis setulosa P.C.Li, Acta Bot. Yunnan. 12: 11 (1990).
China (Yunnan). 36 CHC. Hemicr.

Eleocharis sphacelata R.Br., Prodr.: 224 (1810). *Scirpus sphacelatus* (R.Br.) Poir. in J.B.A.M.de Lamarck, Encycl., Suppl. 5: 102 (1817).
New Guinea, Australia, New Zealand. 43 NWG 50 NSW QLD SOA TAS VIC WAU 51 NZN NZS. Hemicr.
Eleocharis biseptata Steud., Syn. Pl. Glumac. 2: 82 (1855).
Eleocharis subsphacelata Steud., Syn. Pl. Glumac. 2: 317 (1855).

Eleocharis spiralis (Rottb.) Roem. & Schult., Syst. Veg. 2: 155 (1817).
W. Indian Ocean to SW. Pacific. 29 MAU MDG 36 CHH CHS 40 BAN IND SRL 41 MYA THA VIE 42 BOR JAW MLY PHI 43 NWG 50 NTA QLD WAU 60 NWC. Hemicr.
**Scirpus spiralis* Rottb., Descr. Icon. Rar. Pl.: 42 (1773). *Limnochloa spiralis* (Rottb.) Nees in R.Wight, Contr. Bot. India: 114 (1834).
Eleocharis compacta R.Br., Prodr.: 224 (1810). *Scirpus compactus* (R.Br.) Spreng., Syst. Veg. 1: 203 (1824), nom. illeg.
Scirpus mutatus Roxb. ex C.B.Clarke in J.D.Hooker, Fl. Brit. India 6: 627 (1893).
Eleocharis spiralis f. *tenuis* Boeck. in I.Urban, Symb. Antill. 2: 62 (1900).

Eleocharis spongostyla H.E.Hess, Ber. Schweiz. Bot. Ges. 63: 343 (1953).
Angola. 26 ANG. Hemicr.

Eleocharis squamigera Svenson, Rhodora 36: 389 (1934).
S. Venezuela to Brazil. 82 VEN 84 BZS. Hemicr.

subsp. ***squamigera***
Brazil. 84 BZS. Hemicr.
Eleocharis squamata Boeck., Beitr. Cyper. 2: 11 (1890).

subsp. ***straminea*** S.González & Reznicek, Novon 6: 362 (1996).
S. Venezuela. 82 VEN. Hemicr.

Eleocharis stenocarpa Svenson, Rhodora 31: 205 (1929).
Colombia to Venezuela. 82 VEN 83 CLM. Hemicr.

Eleocharis steyermarkii S.González & Reznicek, Novon 6: 363 (1996).
Venezuela (Bolívar). 82 VEN. Hemicr.

Eleocharis × subangulata T.Koyama, J. Jap. Bot. 46: 66 (1971). *E. pellucina* × *E. tetraquetra*.
Japan to Nansei-shoto. 38 JAP NNS. Hemicr.

Eleocharis subarticulata (Nees) Boeck., Linnaea 36: 455 (1870).
SE. Brazil. 84 BZL. Hemicr.

**Chaetocyperus subarticulatus* Nees in C.F.P.von Martius & auct. suc. (eds.), Fl. Bras. 2(1): 96 (1842). *Scirpus subarticulatus* (Nees) Kuntze, Revis. Gen. Pl. 2: 758 (1891).

Eleocharis subcancellata C.B.Clarke, Kew Bull., Addit. Ser. 8: 21 (1908).
SW. Mexico (Jalisco). 79 MXS. Hemicr.

Eleocharis subfoliata C.B.Clarke, Kew Bull., Addit. Ser. 8: 22 (1908).
N. South America to Brazil. 82 GUY SUR VEN 84 BZN. Hemicr.
Eleocharis leucocarpa Boeck., Vidensk. Meddel. Naturhist. Foren. Kjøbenhavn 1869: 132 (1869).
Eleocharis spadicea C.B.Clarke, Kew Bull., Addit. Ser. 8: 21 (1908).

Eleocharis suksdorfiana Beauverd, Bull. Soc. Bot. Genève, II, 13: 267 (1922). *Eleocharis pauciflora* var. *suksdorfiana* (Beauverd) Svenson, Rhodora 31: 174 (1929). *Eleocharis quinqueflora* subsp. *suksdorfiana* (Beauverd) Hultén, Ark. Bot., a.s., 7(1): 24 (1968). *Eleocharis quinqueflora* var. *suksdorfiana* (Beauverd) J.T.Howell, Marin Fl., ed. 2: 363 (1969 publ. 1970).
W. Canada to W. U.S.A. 71 ABT BRC 73 COL IDA MNT ORE WAS 76 CAL NEV UTA. Hemicr. or rhizome geophyte.

Eleocharis sundaica J.Kern, Blumea 9: 220 (1958).
Lesser Sunda Is. (Kep. Alor), N. Australia. 42 LSI 50 NTA WAU. Hemicr.

Eleocharis svensoniana S.González, Phytologia 57: 383 (1985).
Mexico (Durango to Jalisco). 79 MXE MXN MXS. Hemicr.

Eleocharis swamyi Govind., Proc. Indian Acad. Sci., Pl. Sci. 94: 13 (1985).
India. 40 IND. Hemicr.

Eleocharis tenuis (Willd.) Schult. in J.J.Roemer & J.A.Schultes, Mant. 2: 89 (1824).
E. Canada to NC. & E. U.S.A., California. 72 NBR NSC PEI QUE 74 ILL IOW KAN MSO NEB OKL SDA WIS 75 CNT INI MAS NWH NWJ NWY OHI PEN RHO VER WVA 76 CAL 78 ALA ARK DEL GEO KTY LOU MRY MSI NCA SCA TEN VRG WDC. Hel.
**Scirpus tenuis* Willd., Enum. Pl.: 76 (1809). *Scirpidium tenue* (Willd.) Nees, Linnaea 9: 293 (1834). *Trichophyllum tenue* (Willd.) Farw., Rep. (Annual) Michigan Acad. Sci. 21: 350 (1919 publ. 1920).

var. ***pseudoptera*** (Weath. ex Svenson) Svenson, Rhodora 41: 65 (1939).
E. Canada to E. U.S.A. 72 NSC QUE 75 CNT INI NWJ NWY PEN VER 78 DEL GEO KTY LOU MRY MSI NCA TEN VRG WDC. Hemicr. or rhizome geophyte.
**Eleocharis capitata* var. *pseudoptera* Weath. ex Svenson, Rhodora 34: 202 (1932). *Eleocharis elliptica* var. *pseudoptera* (Weath. ex Svenson) L.J.Harms, Amer. J. Bot. 59: 487 (1972).

var. ***tenuis***
E. Canada to NC. & E. U.S.A., California. 72 NBR NSC PEI QUE 74 OKL 75 CNT MAS NWH NWJ NWY OHI PEN RHO VER WVA 76 CAL 78 DEL KTY LOU MRY NCA SCA TEN VRG. Hel.
Scirpus filiformis Lam., Tabl. Encycl. 1: 138 (1791), nom. illeg. *Isolepis filiformis* Roem. & Schult., Syst. Veg. 2: 106 (1817). *Eleogiton filiformis* (Roem. & Schult.) A.Dietr., Sp. Pl. 2: 96 (1832). *Eleocharis filiformis* (Roem. & Schult.) Kunth, Enum. Pl. 2: 146 (1837).
Scirpus capitatus Willd. ex Kunth, Enum. Pl. 2: 145 (1837), nom. illeg.
Scirpus ellipticus Willd. ex Kunth, Enum. Pl. 2: 146 (1837).
Scirpus multicaulis Vahl ex Kunth, Enum. Pl. 2: 146 (1837).

var. ***verrucosa*** (Svenson) Svenson, Rhodora 41: 66 (1939).
C. & E. U.S.A. 74 ILL IOW KAN MSO NEB OKL SDA WIS 75 INI PEN WVA 77 TEX 78 ALA ARK GEO KTY LOU MRY TEN VRG WDC. Hemicr. or hel.
**Eleocharis capitata* var. *verrucosa* Svenson, Rhodora 34: 202 (1932). *Eleocharis verrucosa* (Svenson) L.J.Harms, Amer. J. Bot. 59: 486 (1972).

Eleocharis testui Cherm., Bull. Soc. Bot. France 77: 276 (1930).
Gabon. 23 GAB. Hemicr.

Eleocharis tetraquetra Nees in R.Wight, Contr. Bot. India: 113 (1834). *Scirpus tetraquetrus* (Nees) Kuntze, Revis. Gen. Pl. 2: 757 (1891).
Afghanistan to E. Australia. 31 34 AFG 36 CHH CHS 38 JAP NNS TAI 40 EHM IND NEP PAK SRL 41 THA VIE 42 JAW LSI PHI SUL SUM 43 NWG 50 NSW QLD. Hemicr.
Limnochloa tetraquetra Nees in R.Wight, Contr. Bot. India: 113 (1834).
Eleocharis erythrochlamys Miq., Fl. Ned. Ind. 3: 300 (1856).
Scirpus tetraqueter Thwaites, Enum. Pl. Zeyl.: 434 (1864).
Scirpus hakonensis Franch. & Sav., Enum. Pl. Jap. 2: 110 (1877).
Scirpus onaei Franch. & Sav., Enum. Pl. Jap. 2: 544 (1878).
Eleocharis alta Boeck., Beitr. Cyper. 1: 17 (1888).
Scirpus yokuhamensis Kuntze, Revis. Gen. Pl. 2: 758 (1891).
Eleocharis tsurumachii Ohwi, Acta Phytotax. Geobot. 2: 29 (1933).
Eleocharis tetraquetra var. *micranthera* Ohwi, Bot. Mag. (Tokyo) 56: 201 (1942).

Eleocharis tiarata Gómez-Laur., Monogr. Syst. Bot. Missouri Bot. Gard. 92: 505 (2003).
Caribbean, C. & S. Trop. America. 80 BLZ COS GUA NIC 81 DOM TRT 82 FRG GUY SUR VEN 84 BZN. Hemicr. or hel.
**Scirpus mitratus* Griseb., Fl. Brit. W. I.: 570 (1864). *Eleocharis mitrata* (Griseb.) C.B.Clarke in I.Urban, Symb. Antill. 2: 62 (1900), nom. illeg.

Eleocharis torticulmis S.G.Sm., Novon 11: 250 (2001).
California (Plumas Co.). 76 CAL. Hemicr. or rhizome geophyte.

Eleocharis tortilis (Link) Schult. in J.J.Roemer & J.A.Schultes, Mant. 2: 92 (1824).
E. U.S.A. to Texas. 75 NWJ 77 TEX 78 ALA ARK FLA GEO LOU MRY MSI NCA SCA TEN VRG. Hemicr.
**Scirpus tortilis* Link, Jahrb. Gewächsk. 1(3): 78 (1820).
Scirpus spiralis Bosc ex Kunth, Enum. Pl. 2: 145 (1837).
Scirpus tuberculatus Beyr. ex Kunth, Enum. Pl. 2: 145 (1837), nom. inval.

Eleocharis tricostata Torr., Ann. Lyceum Nat. Hist. New York 3: 310 (1836). *Scirpus tricostatus* (Torr.) Kuntze, Revis. Gen. Pl. 2: 758 (1891). *Trichophyllum tricostatum* (Torr.) House, Amer. Midl. Naturalist 6: 205 (1920).
E. U.S.A. 75 MAS MIC NWJ NWY 78 ALA FLA GEO LOU NCA SCA VRG. Hemicr.

Eleocharis trilateralis Tang & F.T.Wang, in Fl. Reipubl. Popul. Sin. 11: 225 (1961).
China (Yunnan). 36 CHC. Hemicr.

Eleocharis trilophus C.B.Clarke in D.Oliver & auct. suc. (eds.), Fl. Trop. Afr. 8: 409 (1902).
Senegal, Chad. 22 SEN 24 CHA. Ther.

Eleocharis tuberculosa (Michx.) Roem. & Schult., Syst. Veg. 2: 152 (1817).
E. Canada, E. U.S.A. to Texas. 72 NSC 75 MAS NWJ 77 TEX 78 ALA ARK FLA GEO LOU MSI NCA SCA TEN. Hemicr.
**Scirpus tuberculosus* Michx., Fl. Bor.-Amer. 1: 30 (1803). *Bulbostylis tuberculosa* (Michx.) Steven, Mém. Soc. Imp. Naturalistes Moscou 5: 355 (1814). *Chlorocharis tuberculosa* (Michx.) Rikll, Jahrb. Wiss. Bot. 27: 564 (1895). *Trichophyllum tuberculosum* (Michx.) House, Amer. Midl. Naturalist 6: 205 (1920).
Scirpus tuberculatus Elliott, Sketch Bot. S. Carolina 1: 78 (1816). *Megadenus tuberculatus* (Elliott) Raf., Neogenyton: 4 (1825).
Rhynchospora monostachya Steud., Syn. Pl. Glumac. 2: 140 (1855).
Eleocharis tuberculosa var. *pubnicoensis* Fernald, Rhodora 23: 233 (1921). *Eleocharis tuberculosa* f. *pubnicoensis* (Fernald) Svenson, Rhodora 39: 250 (1937).
Eleocharis tuberculosa f. *retrorsa* Svenson, Rhodora 39: 250 (1937).

Eleocharis tucumanensis Barros, Lilloa 12: 7 (1947).
Bolivia to NW. Argentina. 83 BOL 85 AGW CLN. Hemicr.

Eleocharis tuvinica Bubnova, Bot. Zhurn. (Moscow & Leningrad) 71: 1405 (1986).
Siberia to C. Asia. 30 TVA 32+. Hemicr.

Eleocharis uniglumis (Link) Schult. in J.J.Roemer & J.A.Schultes, Mant. 2: 88 (1824).
Europe to Russian Far East and Pakistan, Subarctic America to N. U.S.A. 10 DEN FIN FOR GRB ICE IRE NOR SWE 11 AUT BGM CZE GER HUN NET POL SWI 12 FRA SAR SPA 13 ALB BUL GRC ITA ROM YUG 14 BLR BLT KRY RUC RUE RUN RUS RUW UKR 20 ALG MOR 30 31 32 KAZ KGZ UZB 33 TCS 34 AFG IRN IRQ TUR 35 SAU 36 CHM CHX 37 MON 40 EHM? PAK 70 ASK GNL NWT YUK 72 ABT BRC LAB MAN NBR NFL NSC ONT PEI QUE 73 COL 74 NDA NEB SDA 75 MAI MAS NWH NWJ NWY RHO 76 NEV UTA 77 NWM 78 DEL NCA VRG. Hemicr.
**Scirpus uniglumis* Link, Jahrb. Gewächsk. 1(3): 77 (1820). *Clavula uniglumis* (Link) Dumort., Fl. Belg.: 143 (1827). *Eleocharis palustris* subsp. *uniglumis* (Link) Hartm., Sv. Norsk. Exc.-Fl.: 10 (1846). *Cyperus uniglumis* (Link) Missbach & E.H.L.Krause in J.Sturm, Deutschl. Fl., ed. 2, 2: 25 (1900). *Eleocharis uniglumis* var. *typica* Fernald & Brackett, Rhodora 31: 72 (1929), nom. inval.

subsp. ***fennica*** (Palla ex Kneuck.) Vestergr. in C.A.M. Lindman, Sv. Fanerogamfl., ed. 2: 126 (1926).
Europe to China. 10 FIN SWE 11 GER 14 BLT RUW 32 KAZ 36 CHM. Hemicr.
**Eleocharis fennica* Palla ex Kneuck., Allg. Bot. Z. Syst. 7: 212 (1901). *Eleocharis uniglumis* var. *fennica* (Palla ex Kneuck.) E.Holmb., Skand. Fl. 2: 308 (1926).
Eleocharis sareptana Zinserl., Trudy Glavn. Bot. Sada 40: 279 (1929). *Eleocharis fennica* f. *sareptana* (Zinserl.) Tang & F.T.Wang, in Fl. Reipubl. Popul. Sin. 11: 69 (1961).

subsp. ***septentrionalis*** (Zinserl.) T.V.Egorova, in Fl. Evropeiskoi Chasti SSSR 2: 116 (1976).
N. & NE. Europe. 10 FIN NOR 14 RUN. Hemicr.
**Eleocharis septentrionalis* Zinserl. in V.L.Komarov (ed.), Fl. URSS 3: 586 (1935).

subsp. ***sterneri*** Strandh., Bot. Not. 114: 433 (1961).
Sweden (Gotland, Öland). 10 SWE. Hemicr.

subsp. ***uniglumis***
Europe to Russian Far East and Pakistan, Subarctic America to N. U.S.A. 10 DEN FIN FOR GRB ICE IRE NOR SWE 11 AUT BGM CZE GER HUN NET POL SWI 12 FRA SAR SPA 13 ALB? BUL GRC ITA ROM YUG 14 BLR BLT KRY RUC RUE RUN RUS RUW UKR 20 ALG MOR 30 31 32 KAZ KGZ UZB 33 TCS 34 AFG IRN IRQ TUR 35 SAU 36 CHX 37 MON 40 EHM? PAK 70 ASK GNL NWT YUK 72 ABT BRC LAB MAN NBR NFL NSC ONT PEI QUE 73 COL 74 NDA NEB SDA 75 MAI MAS NWH NWJ NWY RHO 76 NEV UTA 77 NWM 78 DEL NCA VRG. Hemicr.
Eleocharis conica J.Presl & C.Presl, Delic. Prag.: 143 (1822).
Eleocharis polycaula Wender., Schriften Ges. Beförd. Gesammten Naturwiss. Marburg 1: 132 (1823).
Eleocharis affinis C.A.Mey. ex Claus, Beitr. Pflanzenk. Russ. Reiches 8: 261 (1847). *Eleocharis uniglumis* var. *affinis* (C.A.Mey. ex Claus) Nyman, Consp. Fl. Eur.: 767 (1882).
Eleocharis watsonii Bab., Ann. Mag. Nat. Hist., II, 10: 20 (1852). *Eleocharis uniglumis* var. *watsonii* (Bab.) Nyman, Consp. Fl. Eur.: 767 (1882).
Scirpus klingei Meinsh., Trudy Imp. S.-Peterburgsk. Bot. Sada 18: 262 (1901). *Eleocharis klingei* (Meinsh.) B.Fedtsch., Rastit. Turkest.: 165 (1915).
Eleocharis korshinskyana Zinserl., Trudy Glavn. Bot. Sada 40: 280 (1929).
Eleocharis scythica Zinserl., Trudy Glavn. Bot. Sada 40: 279 (1929).
Eleocharis uniglumis var. *halophila* Fernald & Brackett, Rhodora 31: 72 (1929). *Eleocharis halophila* (Fernald & Brackett) Fernald & Brackett, Rhodora 37: 395 (1935).
Eleocharis euuniglumis Zinserl. in V.L.Komarov (ed.), Fl. URSS 3: 584 (1935).
Eleocharis multiseta Zinserl. in V.L.Komarov (ed.), Fl. URSS 3: 586 (1935).
Eleocharis paucidentata Zinserl. in V.L.Komarov (ed.), Fl. URSS 3: 588 (1935).
Eleocharis transcaucasica Zinserl. in V.L.Komarov (ed.), Fl. URSS 3: 585 (1935).
Eleocharis macrocarpa Zoz, Trudy Inst. Bot. Kharkivs'k. Derzhavn. Univ. 1: 55 (1936).
Eleocharis zinserlingi Zoz, Trudy Inst. Bot. Kharkivs'k. Derzhavn. Univ. 1: 55 (1936).
Eleocharis carinata Sakalo, Novosti Sist. Vyssh. Nizsh. Rast. 1976: 48 (1977).

Eleocharis urbanii Boeck., Allg. Bot. Z. Syst. 2: 20 (1896).
Brazil. 84+. Hemicr.

Eleocharis urceolata (Liebm.) Svenson, Rhodora 39: 219 (1937).
Mexico, Nicaragua. 79 MXG MXT 80 NIC. Ther.
**Chaetocyperus urceolatus* Liebm., Mexic. Neldeagt. Pl., V, 2: 243 (1851). *Eleocharis liebmanniana* Boeck., Linnaea 36: 439 (1870), nom. illeg. *Scirpus liebmannianus* (Boeck.) Kuntze, Revis. Gen. Pl. 2: 757 (1891).

Eleocharis ussuriensis Zinserl. in V.L.Komarov (ed.), Fl. URSS 3: 581 (1935). *Eleocharis mamillata* f. *ussuriensis* (Zinserl.) Y.L.Chang, in Fl. Plant. Herb. Chinae Bor.-Or. 11: 32 (1976). *Eleocharis mamillata* subsp. *ussuriensis* (Zinserl.) T.V.Egorova, in Fl. Evropeiskoi Chasti SSSR 2: 113 (1976). *Eleocharis austriaca* subsp. *ussuriensis* (Zinserl.) T.V.Egorova, Novosti Sist. Vyssh. Rast. 33: 71 (2001).
Siberia to Japan. 30+ 31 PRM 36 CHM 38 JAP KOR. Hemicr.
Eleocharis mamillata var. *cyclocarpa* Kitag., Lin. Fl. Manshur.: 119 (1939).

Eleocharis usteri Palla, Oesterr. Bot. Z. 58: 60 (1908).
Brazil. 84+. Hemicr.

Eleocharis valleculosa Ohwi, Acta Phytotax. Geobot. 2: 29 (1933).
N. China to Japan. 36 CHM CHN 38 JAP KOR. Hemicr.
Eleocharis valleculosa var. *setosa* Ohwi, Acta Phytotax. Geobot. 2: 29 (1933). *Eleocharis valleculosa* f. *setosa* (Ohwi) Kitag., Lin. Fl. Manshur.: 121 (1939).
Eleocharis truncatovaginata T.Koyama, J. Jap. Bot. 32: 148 (1957).

Eleocharis variegata (Poir.) C.Presl, Isis (Oken) 21: 269 (1828).
Trop. & S. Africa to New Guinea. 22 BEN GHA GUI IVO LBR NGA SEN SIE 23 BUR CAF CON GAB ZAI 24 CHA 25 TAN UGA 26 ANG 27 BOT 29 MAU MDG REU SEY 42 SUM 43 NWG. Hemicr.
**Scirpus variegatus* Poir. in J.B.A.M.de Lamarck, Encycl. 6: 749 (1805). *Baeothryon variegatum* (Poir.) A.Dietr., Sp. Pl. 2: 92 (1832).
Eleocharis sieberi Kunth, Enum. Pl. 2: 153 (1837). *Scirpus sieberi* (Kunth) Kuntze, Revis. Gen. Pl. 2: 758 (1891).
Scirpus laxifolius Thwaites, Enum. Pl. Zeyl.: 435 (1864).
Eleocharis graeffeana Boeck., Flora 58: 108 (1875).
Eleocharis calocarpa Cherm., Arch. Bot. Bull. Mens. 4(7): 41 (1931).

Eleocharis venezuelensis S.González & Reznicek, Novon 6: 363 (1996).
S. Venezuela. 82 VEN. Hemicr. or hel.

Eleocharis viridans Kük. ex Osten, Anales Mus. Hist. Nat. Montevideo, II, 3(2): 175 (1931).
Brazil to N. Argentina. 84 BZC 85 AGE AGW PAR URU. Hemicr.

Eleocharis vivipara Link, Hort. Berol. 1: 283 (1827). *Scirpus viviparus* (Link) Kuntze, Revis. Gen. Pl. 2: 758 (1891). *Chlorocharis vivipara* (Link) Rikli, Jahrb. Wiss. Bot. 27: 564 (1895). *Trichophyllum viviparum* (Link) House, Amer. Midl. Naturalist 6: 205 (1920).
SE. U.S.A. to Texas. 77 TEX 78 ALA FLA GEO LOU NCA SCA VRG. Ther.
Eleocharis curtisii Small, Man. S.E. Fl.: 165 (1933).

Eleocharis welwitschii Nelmes, Kew Bull. 13: 282 (1958).
Angola to Namibia. 26 ANG 27 NAM. Hemicr.

Eleocharis wichurai Boeck., Linnaea 36: 448 (1870).
Siberia to Japan. 30+ 31 PRM 36 CHM CHN 38 JAP KOR NNS. Hemicr.
Eleocharis hakonensis Franch. & Sav., Enum. Pl. Jap. 2: 110 (1879).
Scirpus petasatus Maxim., Bull. Soc. Imp. Naturalistes Moscou 54(1): 64 (1879). *Eleocharis petasata* (Maxim.) Zinserl. in V.L.Komarov (ed.), Fl. URSS 3: 589 (1935).
Eleocharis liukiuensis Makino, Bot. Mag. (Tokyo) 18: 111 (1904).
Eleocharis wichurai var. *teres* Hara, Bot. Mag. (Tokyo) 52: 398 (1938). *Eleocharis wichurai* f. *teres* (Hara) Ohwi, Mem. Coll. Sci. Kyoto Imp. Univ., Ser. B, Biol. 13: 38 (1944).
Eleocharis wichurai f. *vivipara* Ueno, J. Jap. Bot. 51: 198 (1976).

Eleocharis widgrenii Boeck., Bot. Jahrb. Syst. 5: 503 (1884). *Scirpus widgrenii* (Boeck.) Kuntze, Revis. Gen. Pl. 2: 758 (1891).
Brazil. 84+. Hemicr.

Eleocharis wilkensii Schuyler, Notul. Nat. Acad. Nat. Sci. Philadelphia 423: 4 (1969).
NE. Mexico. 79 MXE.

Eleocharis wolfii (A.Gray) A.Gray ex Britton, J. New York Microscop. Soc. 5: 105 (1889).
C. & E. U.S.A. 73 COL 74 ILL IOW KAN MIN MSO NDA OKL WIS 75 INI NWY OHI 77 TEX 78 ALA ARK LOU MSI TEN. Hemicr.
**Scirpus wolfii* A.Gray, Proc. Amer. Acad. Arts 10: 77 (1874).

Eleocharis yecorensis Roalson, Aliso 18: 57 (1999).
Mexico. 79 MXC MXN. Hemicr.

Eleocharis* ×*yezoensis Hara, Bot. Mag. (Tokyo) 52: 396 (1938). *E. congesta* × *E. wichurae*.
N. Japan. 38 JAP. Hemicr.

Eleocharis yunnanensis Svenson, Rhodora 41: 559 (1939).
China (Yunnan). 36 CHC. Hemicr.
Eleocharis pauciflora var. *rhizomatosa* Hand.-Mazz., Symb. Sin. 7: 1250 (1936).

Synonyms:
Eleocharis aciculariformis Greenm. = ***Eleocharis bonariensis*** Nees
Eleocharis acicularis var. *bella* Piper = ***Eleocharis bella*** (Piper) Svenson
Eleocharis acicularis subsp. *bonariensis* (Nees) Osten = ***Eleocharis bonariensis*** Nees
Eleocharis acicularis var. *elongata* Benth. = ***Eleocharis pusilla*** R.Br.
Eleocharis acicularis f. *filiformis* Osten = ***Eleocharis bonariensis*** Nees
Eleocharis acicularis var. *fluitans* Döll = ***Eleocharis acicularis*** (L.) Roem. & Schult. subsp. ***acicularis***
Eleocharis acicularis f. *fluitans* (Döll) Svenson = ***Eleocharis acicularis*** (L.) Roem. & Schult. subsp. ***acicularis***
Eleocharis acicularis var. *gracilescens* Svenson = ***Eleocharis acicularis*** (L.) Roem. & Schult. subsp. ***acicularis***
Eleocharis acicularis f. *inundata* Svenson = ***Eleocharis acicularis*** (L.) Roem. & Schult. subsp. ***acicularis***
Eleocharis acicularis var. *lilliputiana* Speg. = ***Eleocharis bonariensis*** Nees

Eleocharis acicularis var. *lindheimeri* C.B.Clarke = **Eleocharis radicans** Kunth
Eleocharis acicularis var. *longiseta* Svenson = **Eleocharis acicularis** subsp. **yokoscensis** (Franch. & Sav.) T.V.Egorova
Eleocharis acicularis var. *minima* Torr. = **Eleocharis bella** (Piper) Svenson
Eleocharis acicularis var. *occidentalis* Svenson = **Eleocharis acicularis** (L.) Roem. & Schult. subsp. **acicularis**
Eleocharis acicularis f. *occidentalis* (Svenson) Beetle = **Eleocharis acicularis** (L.) Roem. & Schult. subsp. **acicularis**
Eleocharis acicularis var. *radicans* (Poir.) Britton = **Eleocharis radicans** Kunth
Eleocharis acicularis f. *rigidula* (Rchb.) Svenson = **Eleocharis acicularis** (L.) Roem. & Schult. subsp. **acicularis**
Eleocharis acicularis f. *submersa* (Nilsson) Norman = **Eleocharis acicularis** (L.) Roem. & Schult. subsp. **acicularis**
Eleocharis acicularis f. *submersa* Druce = **Eleocharis acicularis** (L.) Roem. & Schult. subsp. **acicularis**
Eleocharis acicularis var. *submersa* (Nilsson) Svenson = **Eleocharis acicularis** (L.) Roem. & Schult. subsp. **acicularis**
Eleocharis acicularis f. *triangularis* (Reinsch) Svenson = **Eleocharis acicularis** (L.) Roem. & Schult. subsp. **acicularis**
Eleocharis acicularis var. *typica* Svenson = **Eleocharis acicularis** (L.) Roem. & Schult.
Eleocharis acuminata (Muhl.) Nees = **Eleocharis compressa** Sull. var. **compressa**
Eleocharis acuminata var. *tenerrima* Boeck. = **Eleocharis compressa** Sull. var. **compressa**
Eleocharis acuta var. *pallens* Benth. = **Eleocharis acuta** R.Br.
Eleocharis acuta var. *platylepis* Hook.f. = **Eleocharis acuta** R.Br.
Eleocharis acuta var. *tenuis* Carse = **Eleocharis acuta** R.Br.
Eleocharis acutisquamata Buckley = **Eleocharis compressa** var. **acutisquamata** (Buckley) S.G.Sm.
Eleocharis affinis C.A.Mey. ex Claus = **Eleocharis uniglumis** (Link) Schult. subsp. **uniglumis**
Eleocharis afflata Steud. = **Eleocharis pellucida** C.Presl
Eleocharis albibracteata var. *boliviana* (Palla ex Svenson) S.González & P.M.Peterson = **Eleocharis crinalis** (Griseb.) C.B.Clarke
Eleocharis albida var. *berlandieri* (C.B.Clarke) Britton = **Eleocharis albida** Torr.
Eleocharis albovaginata Boeck. = **Eleocharis flavescens** (Poir.) Urb. var. **flavescens**
Eleocharis albovaginata var. *flaccida* Boeck. = **Eleocharis flavescens** (Poir.) Urb. var. **flavescens**
Eleocharis albovaginata var. *humilis* Boeck. = **Eleocharis flavescens** (Poir.) Urb. var. **flavescens**
Eleocharis albovaginata var. *macrostachya* Boeck. = **Eleocharis flavescens** (Poir.) Urb. var. **flavescens**
Eleocharis albovaginata var. *stricta* Boeck. = **Eleocharis flavescens** (Poir.) Urb. var. **flavescens**
Eleocharis albovaginata var. *tenuis* Boeck. = **Eleocharis flavescens** (Poir.) Urb. var. **flavescens**
Eleocharis alpina (L.) Bluff, Nees & Schauer = **Trichophorum alpinum** (L.) Pers.
Eleocharis alta Boeck. = **Eleocharis tetraquetra** Nees
Eleocharis ambigens Fernald = **Eleocharis fallax** Weath.
Eleocharis ambigua Kirk & Buchanan = **Eleocharis acuta** R.Br.
Eleocharis ambigua Urb. = **Eleocharis minima** Kunth var. **minima**
Eleocharis amphibia Durand = **Eleocharis bonariensis** Nees
Eleocharis anceps Ridl. = **Eleocharis complanata** Boeck.
Eleocharis andesica C.B.Clarke = **Eleocharis montana** (Kunth) Roem. & Schult.
Eleocharis andina (Phil.) H.Pfeiff. = **Eleocharis melanostachys** (d'Urv.) C.B.Clarke
Eleocharis annua House = **Eleocharis ovata** (Roth) Roem. & Schult.
Eleocharis appendiculata C.B.Clarke = **Eleocharis melanostachys** (d'Urv.) C.B.Clarke
Eleocharis appendiculata Phil. = **Eleocharis palustris** (L.) Roem. & Schult. subsp. **palustris**
Eleocharis arcuata Kunze ex Kunth = **Eleocharis maculosa** (Vahl) Roem. & Schult.
Eleocharis arechavaletae Boeck. = **Eleocharis flavescens** (Poir.) Urb. var. **flavescens**
Eleocharis arenaria Benth. = **Eleocharis glauca** Boeck.
Eleocharis arenicola Torr. ex Engelm. & A.Gray = **Eleocharis montevidensis** Kunth
Eleocharis argyrolepidoides Zinserl. = **Eleocharis mitracarpa** Steud.
Eleocharis articulata Kunth = **Eleocharis interstincta** (Vahl) Roem. & Schult.
Eleocharis atacamensis Phil. = **Eleocharis quinqueflora** (Hartmann) O.Schwarz
Eleocharis atricha Benth. = **Eleocharis acicularis** (L.) Roem. & Schult. subsp. **acicularis**
Eleocharis atropurpurea var. *albivaginata* Boeck. = **Eleocharis atropurpurea** (Retz.) Kunth
Eleocharis atropurpurea var. *setiformis* Benth. = **Eleocharis atropurpurea** (Retz.) Kunth
Eleocharis atropurpurea var. *trigyna* Boeck. = **Eleocharis atropurpurea** (Retz.) Kunth
Eleocharis atropurpurea var. *zanardinii* (Parl.) A.Terracc. = **Eleocharis atropurpurea** (Retz.) Kunth
Eleocharis attenuata var. *erhizomatosa* Tang & F.T.Wang = **Eleocharis attenuata** (Franch. & Sav.) Palla
Eleocharis attenuata var. *major* (Hara) Hara = **Eleocharis attenuata** (Franch. & Sav.) Palla
Eleocharis austriaca Hayek = **Eleocharis mamillata** subsp. **austriaca** (Hayek) Strandh.
Eleocharis austriaca subsp. *ussuriensis* (Zinserl.) T.V.Egorova = **Eleocharis ussuriensis** Zinserl.
Eleocharis austrocaledonica Vieill. = **Eleocharis dulcis** (Burm.f.) Trin. ex Hensch.
Eleocharis austrotexana M.C.Johnst. = **Eleocharis ravenelii** Britton
Eleocharis baeothryon Schult. = **Eleocharis quinqueflora** (Hartmann) O.Schwarz
Eleocharis baeothryon (L.f.) Nees = **Eleocharis quinqueflora** (Hartmann) O.Schwarz
Eleocharis bahamensis f. *minor* Boeck. = **Eleocharis bahamensis** Boeck.
Eleocharis balansaeana Boeck. = **Eleocharis filiculmis** Kunth
Eleocharis baronii Baker = **Eleocharis limosa** (Schrad.) Schult.
Eleocharis bartoliana De Not. = **Eleocharis carniolica** W.D.J.Koch
Eleocharis benedicta Beauverd = **Eleocharis mamillata** subsp. **austriaca** (Hayek) Strandh.
Eleocharis berlandieri C.B.Clarke = **Eleocharis albida** Torr.
Eleocharis bermudiana Britton = **Eleocharis albida** Torr.

Eleocharis bicolor Chapm. = ***Eleocharis minima*** var. ***bicolor*** (Chapm.) Svenson
Eleocharis biocreata Nees = ***Eleocharis flavescens*** (Poir.) Urb. var. ***flavescens***
Eleocharis biseptata Steud. = ***Eleocharis sphacelata*** R.Br.
Eleocharis bivaginata Steud. = ***Eleocharis dombeyana*** Kunth
Eleocharis boissieri Podp. = ***Eleocharis palustris*** (L.) Roem. & Schult. subsp. ***palustris***
Eleocharis boliviana Palla ex Svenson = ***Eleocharis crinalis*** (Griseb.) C.B.Clarke
Eleocharis brehmeriana Boeck. = ***Eleocharis crinalis*** (Griseb.) C.B.Clarke
Eleocharis brizantha Steud. = ***Eleocharis geniculata*** (L.) Roem. & Schult.
Eleocharis caduca var. *gracilescens* A.Terracc. = ***Eleocharis caduca*** (Delile) Schult.
Eleocharis caillei Hutch. ex Nelmes = ***Eleocharis naumanniana*** var. ***caillei*** (Hutch. ex Nelmes) S.S. Hooper
Eleocharis calocarpa Cherm. = ***Eleocharis variegata*** (Poir.) C.Presl
Eleocharis calva (A.Gray) Torr. ex Fernald & Brackett = ***Eleocharis erythropoda*** Steud.
Eleocharis calva var. *australis* (Nees) H.St.John = ***Eleocharis erythropoda*** Steud.
Eleocharis calyptrata (Liebm.) Steud. = ***Eleocharis filiculmis*** Kunth
Eleocharis camptotricha (C.Wright) C.B.Clarke = ***Eleocharis nana*** Kunth
Eleocharis camptotricha var. *schweinitzii* C.B.Clarke = ***Eleocharis bahamensis*** Boeck.
Eleocharis capitata (L.) R.Br. = ***Eleocharis geniculata*** (L.) Roem. & Schult.
Eleocharis capitata var. *borealis* Svenson = ***Eleocharis elliptica*** Kunth
Eleocharis capitata var. *dispar* (E.J.Hill) Fernald = ***Eleocharis geniculata*** (L.) Roem. & Schult.
Eleocharis capitata var. *pseudoptera* Weath. ex Svenson = ***Eleocharis tenuis*** var. ***pseudoptera*** (Weath. ex Svenson) Svenson
Eleocharis capitata var. *typica* Svenson = ***Eleocharis geniculata*** (L.) Roem. & Schult.
Eleocharis capitata var. *verrucosa* Svenson = ***Eleocharis tenuis*** var. ***verrucosa*** (Svenson) Svenson
Eleocharis caribaea (Rottb.) S.F.Blake = ***Eleocharis geniculata*** (L.) Roem. & Schult.
Eleocharis caribaea var. *achlamydea* Zavaro & Pabón = ***Eleocharis geniculata*** (L.) Roem. & Schult.
Eleocharis caribaea var. *dispar* (E.J.Hill) S.F.Blake = ***Eleocharis geniculata*** (L.) Roem. & Schult.
Eleocharis carinata Sakalo = ***Eleocharis uniglumis*** (Link) Schult. subsp. ***uniglumis***
Eleocharis carniolica var. *macrostachya* Boeck. = ***Eleocharis carniolica*** W.D.J.Koch
Eleocharis carolina Small = ***Eleocharis setifolia*** (A.Rich.) J.Raynal subsp. ***setifolia***
Eleocharis cespitosa (L.) Link = ***Trichophorum cespitosum*** (L.) Hartm.
Eleocharis chaetaria Hance = ***Eleocharis acicularis*** (L.) Roem. & Schult. subsp. ***acicularis***
Eleocharis chaetaria Roem. & Schult. = ***Eleocharis retroflexa*** subsp. ***chaetaria*** (Roem. & Schult.) T.Koyama
Eleocharis chlorocarpa Boeck. = ***Eleocharis pellucida*** C.Presl
Eleocharis chrysocarpa Boeck. = ***Eleocharis montana*** (Kunth) Roem. & Schult.
Eleocharis cognata Steud. = ***Eleocharis interstincta*** (Vahl) Roem. & Schult.
Eleocharis comosa K.Richt. = ***Eleocharis acicularis*** (L.) Roem. & Schult. subsp. ***acicularis***
Eleocharis compacta R.Br. = ***Eleocharis spiralis*** (Rottb.) Roem. & Schult.
Eleocharis compressa var. *atrata* Svenson = ***Eleocharis elliptica*** Kunth
Eleocharis compressa var. *borealis* (Svenson) Drapalik & Mohlenbr. = ***Eleocharis elliptica*** Kunth
Eleocharis confervoides (Poir.) Steud. = ***Websteria confervoides*** (Poir.) S.S.Hooper
Eleocharis congesta subsp. *japonica* (Miq.) T.Koyama = ***Eleocharis pellucida*** C.Presl
Eleocharis congesta f. *soranumaensis* Koji Ito = ***Eleocharis pellucida*** C.Presl
Eleocharis congesta subsp. *subvivipara* (Boeck.) T.Koyama = ***Eleocharis congesta*** D.Don
Eleocharis congesta var. *thermalis* (Hultén) T.Koyama = ***Eleocharis pellucida*** C.Presl
Eleocharis conica J.Presl & C.Presl = ***Eleocharis uniglumis*** (Link) Schult. subsp. ***uniglumis***
Eleocharis consanguinea Kunth = ***Eleocharis montana*** (Kunth) Roem. & Schult.
Eleocharis constricta Schult. = ***Eleocharis elegans*** (Kunth) Roem. & Schult.
Eleocharis costaricensis Boeck. = ***Eleocharis filiculmis*** Kunth
Eleocharis costata J.Presl & C.Presl = ***Eleocharis acicularis*** (L.) Roem. & Schult. subsp. ***acicularis***
Eleocharis costulata Nees & Meyen ex Kunth = ***Eleocharis radicans*** Kunth
Eleocharis crassa Fisch. & C.A.Mey. ex Zinserl. = ***Eleocharis palustris*** (L.) Roem. & Schult. subsp. ***palustris***
Eleocharis crassiculmis (Nees) Steud. = ***Eleocharis elegans*** (Kunth) Roem. & Schult.
Eleocharis crinalis f. *humilis* Boeck. ex C.B.Clarke = ***Eleocharis albibracteata*** var. ***nubigena*** (C.B.Clarke) Boros
Eleocharis cubensis Boeck. = ***Eleocharis microcarpa*** Torr. var. ***microcarpa***
Eleocharis cunninghamii Boeck. = ***Eleocharis gracilis*** R.Br.
Eleocharis curtisii Small = ***Eleocharis vivipara*** Link
Eleocharis cyperina Roem. & Schult. = ***Cyperus hermaphroditus*** (Jacq.) Standl.
Eleocharis czernjajevi Zoz = ***Eleocharis quinqueflora*** (Hartmann) O.Schwarz
Eleocharis debilis f. *anomala* S.González & Reznicek = ***Eleocharis debilis*** Kunth
Eleocharis debilis f. *macra* (Kunth) Boeck. = ***Eleocharis debilis*** Kunth
Eleocharis densesquamata Steud. = ***Eleocharis geniculata*** (L.) Roem. & Schult.
Eleocharis depauperata (Vahl) Kunth = ***Eleocharis retroflexa*** (Poir.) Urb. subsp. ***retroflexa***
Eleocharis diandra var. *depressa* Fernald = ***Eleocharis diandra*** C.Wright
Eleocharis dichotoma (L.) H.Karst. = ***Fimbristylis dichotoma*** (L.) Vahl
Eleocharis disciformis Parish = ***Eleocharis parishii*** Britton
Eleocharis dispar E.J.Hill = ***Eleocharis geniculata*** (L.) Roem. & Schult.
Eleocharis dubia Phil. = ***Eleocharis pachycarpa*** É.Desv.
Eleocharis dulcis var. *tuberosa* (Roxb.) T.Koyama = ***Eleocharis dulcis*** (Burm.f.) Trin. ex Hensch.
Eleocharis durandii Boeck. = ***Eleocharis minima*** Kunth var. ***minima***

Eleocharis melanocarpa Phil. = ***Eleocharis fuscopurpurea*** (Steud.) H.Pfeiff.
Eleocharis melanocephala É.Desv. = ***Eleocharis albibracteata*** Nees & Meyen ex Kunth var. ***albibracteata***
Eleocharis melanosperma Steud. = ***Eleocharis geniculata*** (L.) Roem. & Schult.
Eleocharis membranacea (Buckley) Gilly = ***Eleocharis parvula*** (Roem. & Schult.) Palla
Eleocharis mendoncae Boeck. = ***Eleocharis maculosa*** (Vahl) Roem. & Schult.
Eleocharis meridionalis Zinserl. = ***Eleocharis quinqueflora*** (Hartmann) O.Schwarz
Eleocharis mesopotamica Boeck. = ***Eleocharis sellowiana*** Kunth var. ***sellowiana***
Eleocharis mexicana Palla = ***Eleocharis elegans*** (Kunth) Roem. & Schult.
Eleocharis mexicana Peyr. = ***Eleocharis geniculata*** (L.) Roem. & Schult.
Eleocharis microcarpa var. *brittonii* (Svenson ex Small) Svenson = ***Eleocharis brittonii*** Svenson ex Small
Eleocharis microformis Buckley = ***Eleocharis geniculata*** (L.) Roem. & Schult.
Eleocharis minima var. *ambigua* (Steud.) Kük. = ***Eleocharis retroflexa*** subsp. ***chaetaria*** (Roem. & Schult.) T.Koyama
Eleocharis minima f. *major* Boeck. = ***Eleocharis palustris*** (L.) Roem. & Schult. subsp. ***palustris***
Eleocharis minima var. *mexicensis* H.Pfeiff. = ***Eleocharis palustris*** (L.) Roem. & Schult. subsp. ***palustris***
Eleocharis minima var. *oropuchensis* (Britton) C.D.Adams = ***Eleocharis minima*** Kunth var. ***minima***
Eleocharis minima var. *tenuissima* (Boeck.) D.A.Simpson = ***Eleocharis minima*** Kunth var. ***minima***
Eleocharis minor Mazziari = ***Eleocharis palustris*** (L.) Roem. & Schult. subsp. ***palustris***
Eleocharis minutiflora Boeck. = ***Eleocharis setifolia*** (A.Rich.) J.Raynal subsp. ***setifolia***
Eleocharis mitrata Makino = ***Eleocharis kamtschatica*** (C.A.Mey.) Kom.
Eleocharis mitrata (Griseb.) C.B.Clarke = ***Eleocharis tiarata*** Gómez-Laur.
Eleocharis mitrata var. *africana* C.B.Clarke = ***Eleocharis nupeensis*** Hutch.
Eleocharis monandra Hochst. ex Steud. = ***Eleocharis atropurpurea*** (Retz.) Kunth
Eleocharis montana subsp. *montevidensis* (Kunth) Osten = ***Eleocharis montevidensis*** Kunth
Eleocharis montana var. *nodulosa* (Roth) Svenson = ***Eleocharis montana*** (Kunth) Roem. & Schult.
Eleocharis montevidensis f. *angustior* S.González & McVaugh = ***Eleocharis montevidensis*** Kunth
Eleocharis montevidensis var. *bolanderi* (A.Gray) V.E.Grant = ***Eleocharis bolanderi*** A.Gray
Eleocharis montevidensis var. *decumbens* (C.B.Clarke) V.E.Grant = ***Eleocharis decumbens*** C.B.Clarke
Eleocharis montevidensis var. *disciformis* (Parish) V.E.Grant = ***Eleocharis parishii*** Britton
Eleocharis montevidensis var. *parishii* (Britton) V.E.Grant = ***Eleocharis parishii*** Britton
Eleocharis monticola Fernald = ***Eleocharis engelmannii*** Steud.
Eleocharis monticola var. *laeviseta* Fernald = ***Eleocharis engelmannii*** Steud.
Eleocharis monticola var. *pallida* H.St.John = ***Eleocharis engelmannii*** Steud.
Eleocharis mucronulata Nees = ***Eleocharis acuta*** R.Br.
Eleocharis mucronulata var. *minor* Nees = ***Eleocharis acuta*** R.Br.
Eleocharis muehlenbergiana (Schult.) Steud. =
Eleocharis multicaulis var. *vivipara* (Dumort.) Lej. & Courtois = ***Eleocharis multicaulis*** (Sm.) Desv.
Eleocharis multicaulis f. *vivipara* (Dumort.) P.Silva = ***Eleocharis multicaulis*** (Sm.) Desv.
Eleocharis multiculmis A.Terracc. ex Lojac. = ***Eleocharis multicaulis*** (Sm.) Desv.
Eleocharis multiflora Chapm. = ***Eleocharis atropurpurea*** (Retz.) Kunth
Eleocharis multiseta Zinserl. = ***Eleocharis uniglumis*** (Link) Schult. subsp. ***uniglumis***
Eleocharis mutata Schltdl. = ***Eleocharis montana*** (Kunth) Roem. & Schult.
Eleocharis mutata var. *obtusitrigona* (Lindl. & Nees) C.B.Clarke = ***Eleocharis obtusitrigona*** (Lindl. & Nees) Steud.
Eleocharis nana f. *capillaris* Boeck. = ***Eleocharis nana*** Kunth
Eleocharis nebrodensis Parl. = ***Eleocharis palustris*** (L.) Roem. & Schult. subsp. ***palustris***
Eleocharis nervata Svenson = ***Eleocharis acicularis*** (L.) Roem. & Schult. subsp. ***acicularis***
Eleocharis nervosa Kük. = ***Eleocharis fallax*** Weath.
Eleocharis nigrescens var. *minutiflora* (Boeck.) Svenson = ***Eleocharis setifolia*** (A.Rich.) J.Raynal subsp. ***setifolia***
Eleocharis nodulosa (Roth) Schult. = ***Eleocharis montana*** (Kunth) Roem. & Schult.
Eleocharis nodulosa var. *angulata* Svenson = ***Eleocharis montana*** (Kunth) Roem. & Schult.
Eleocharis nodulosa var. *humilis* Boeck. = ***Eleocharis montana*** (Kunth) Roem. & Schult.
Eleocharis nodulosa var. *subnodulosa* (Steud.) Kük. = ***Eleocharis montana*** (Kunth) Roem. & Schult.
Eleocharis nodulosa var. *tenuis* Boeck. = ***Eleocharis montana*** (Kunth) Roem. & Schult.
Eleocharis nodulosa f. *trigyna* Barros = ***Eleocharis montana*** (Kunth) Roem. & Schult.
Eleocharis nubigena C.B.Clarke = ***Eleocharis albibracteata*** var. ***nubigena*** (C.B.Clarke) Boros
Eleocharis obscura T.Koyama = ***Eleocharis quinqueflora*** (Hartmann) O.Schwarz
Eleocharis obsoleta (Nees) Steud. = ***Eleocharis interstincta*** (Vahl) Roem. & Schult.
Eleocharis obtusa var. *ellipsoidalis* Fernald ex Svenson = ***Eleocharis obtusa*** (Willd.) Schult.
Eleocharis obtusa var. *engelmannii* (Steud.) Gilly = ***Eleocharis engelmannii*** Steud.
Eleocharis obtusa var. *gigantea* Svenson = ***Eleocharis obtusa*** (Willd.) Schult.
Eleocharis obtusa var. *jejuna* Fernald = ***Eleocharis obtusa*** (Willd.) Schult.
Eleocharis obtusa var. *lanceolata* (Fernald) Gilly = ***Eleocharis lanceolata*** Fernald
Eleocharis obtusa var. *peasei* Svenson = ***Eleocharis obtusa*** (Willd.) Schult.
Eleocharis ochreata (Nees) Steud. = ***Eleocharis flavescens*** (Poir.) Urb. var. ***flavescens***
Eleocharis ochreata var. *flaccida* (Rchb. ex A.Spreng.) Boeck. = ***Eleocharis flavescens*** (Poir.) Urb. var. ***flavescens***
Eleocharis ochrostachys Boeck. = ***Eleocharis pellucida*** C.Presl
Eleocharis olivacea Torr. = ***Eleocharis flavescens*** var. ***olivacea*** (Torr.) Gleason
Eleocharis olivacea f. *reductiseta* Schuyler & Ferren = ***Eleocharis flavescens*** var. ***olivacea*** (Torr.) Gleason

Eleocharis olivacea var. *reductiseta* (Schuyler & Ferren) Schuyler & Ferren = ***Eleocharis flavescens*** var. ***olivacea*** (Torr.) Gleason
Eleocharis oropuchensis Britton = ***Eleocharis minima*** Kunth var. ***minima***
Eleocharis ovata var. *aphanactis* J.W.Moore = ***Eleocharis ovata*** (Roth) Roem. & Schult.
Eleocharis ovata var. *engelmannii* (Steud.) Britton = ***Eleocharis engelmannii*** Steud.
Eleocharis ovata var. *gaetula* Maire = ***Eleocharis caduca*** (Delile) Schult.
Eleocharis ovata var. *humifusa* A.Terracc. = ***Eleocharis ovata*** (Roth) Roem. & Schult.
Eleocharis ovata var. *italica* A.Terracc. = ***Eleocharis ovata*** (Roth) Roem. & Schult.
Eleocharis ovata f. *sessitensis* A.Terracc. = ***Eleocharis ovata*** (Roth) Roem. & Schult.
Eleocharis oxyoneura Durand = ***Eleocharis bonariensis*** Nees
Eleocharis oxystachys Sakalo = ***Eleocharis palustris*** (L.) Roem. & Schult. subsp. ***palustris***
Eleocharis pachycarpa var. *leptocaulis* (Steud.) C.B.Clarke = ***Eleocharis pachycarpa*** É.Desv.
Eleocharis pachystyla var. *angustostachya* H.Pfeiff. = ***Eleocharis pachystyla*** (C.Wright) C.B.Clarke
Eleocharis pachystyla var. *macrostachya* H.Pfeiff. = ***Eleocharis pachystyla*** (C.Wright) C.B.Clarke
Eleocharis pallida Shuttlew. ex Steud. = ***Trichophorum cespitosum*** (L.) Hartm. subsp. ***cespitosum***
Eleocharis palmeri Svenson = ***Eleocharis montevidensis*** Kunth
Eleocharis palustris var. *australis* Nees = ***Eleocharis erythropoda*** Steud.
Eleocharis palustris subsp. *austriaca* (Hayek) Podp. = ***Eleocharis mamillata*** subsp. ***austriaca*** (Hayek) Strandh.
Eleocharis palustris var. *calva* A.Gray = ***Eleocharis erythropoda*** Steud.
Eleocharis palustris subvar. *calva* (A.Gray) Farw. = ***Eleocharis erythropoda*** Steud.
Eleocharis palustris var. *crassa* (Fisch. & C.A.Mey. ex Zinserl.) Nyman = ***Eleocharis palustris*** (L.) Roem. & Schult. subsp. ***palustris***
Eleocharis palustris var. *degenerata* Hook. = ***Eleocharis palustris*** (L.) Roem. & Schult. subsp. ***palustris***
Eleocharis palustris var. *glaucescens* (Willd.) Coss. & Germ. = ***Eleocharis palustris*** (L.) Roem. & Schult. subsp. ***palustris***
Eleocharis palustris subsp. *globularis* (Zinserl.) T.V.Egorova = ***Eleocharis palustris*** (L.) Roem. & Schult. subsp. ***palustris***
Eleocharis palustris var. *globularis* (Zinserl.) A.E.Kozhevn. = ***Eleocharis palustris*** (L.) Roem. & Schult. subsp. ***palustris***
Eleocharis palustris subsp. *intersita* (Zinserl.) Tzvelev = ***Eleocharis palustris*** (L.) Roem. & Schult. subsp. ***palustris***
Eleocharis palustris f. *italica* A.Terracc. = ***Eleocharis palustris*** (L.) Roem. & Schult. subsp. ***palustris***
Eleocharis palustris var. *kurilensis* A.E.Kozhevn. = ***Eleocharis palustris*** (L.) Roem. & Schult. subsp. ***palustris***
Eleocharis palustris var. *limburgensis* Lej. = ***Eleocharis palustris*** (L.) Roem. & Schult. subsp. ***palustris***
Eleocharis palustris var. *lindbergii* Strandh. = ***Eleocharis palustris*** (L.) Roem. & Schult. subsp. ***palustris***
Eleocharis palustris subsp. *lindbergii* (Strandh.) Tzvelev = ***Eleocharis palustris*** (L.) Roem. & Schult. subsp. ***palustris***
Eleocharis palustris var. *major* Sond. = ***Eleocharis palustris*** (L.) Roem. & Schult. subsp. ***palustris***
Eleocharis palustris subsp. *mamillata* (H.Lindb.) Beauverd = ***Eleocharis mamillata*** (H.Lindb.) H.Lindb.
Eleocharis palustris subsp. *microcarpa* Walters = ***Eleocharis palustris*** (L.) Roem. & Schult. subsp. ***palustris***
Eleocharis palustris var. *minor* Hook. = ***Eleocharis palustris*** (L.) Roem. & Schult. subsp. ***palustris***
Eleocharis palustris var. *mucronulata* Boeck. = ***Eleocharis acuta*** R.Br.
Eleocharis palustris subsp. *nebrodensis* (Parl.) Nyman = ***Eleocharis palustris*** (L.) Roem. & Schult. subsp. ***palustris***
Eleocharis palustris var. *nebrodensis* (Parl.) A.Terracc. = ***Eleocharis palustris*** (L.) Roem. & Schult. subsp. ***palustris***
Eleocharis palustris subsp. *uniglumis* (Link) Hartm. = ***Eleocharis uniglumis*** (Link) Schult.
Eleocharis palustris var. *watsonii* (Bab.) C.B.Clarke = ***Eleocharis uniglumis*** (Link) Schult.
Eleocharis palustromicrocarpa Á.Löve = ***Eleocharis palustris*** (L.) Roem. & Schult. subsp. ***palustris***
Eleocharis paracicularis Boeck. = ***Eleocharis microcarpa*** Torr. var. ***microcarpa***
Eleocharis paraguayensis Maury ex Micheli = [85 PAR]
Eleocharis parvula var. *anachaeta* (Torr.) Svenson = ***Eleocharis coloradoensis*** (Britton) Gilly
Eleocharis parvula var. *coloradoensis* (Britton) Beetle = ***Eleocharis coloradoensis*** (Britton) Gilly
Eleocharis parvula f. *spongiosa* Fassett = ***Eleocharis parvula*** (Roem. & Schult.) Palla
Eleocharis paucidentata Zinserl. = ***Eleocharis uniglumis*** (Link) Schult. subsp. ***uniglumis***
Eleocharis pauciflora (Lightf.) Link = ***Eleocharis quinqueflora*** (Hartmann) O.Schwarz
Eleocharis pauciflora var. *bernardina* (Munz & I.M.Johnst.) Svenson = ***Eleocharis bernardina*** (Munz & I.M.Johnst.) Munz & I.M.Johnst.
Eleocharis pauciflora var. *fernaldii* Svenson = ***Eleocharis quinqueflora*** (Hartmann) O.Schwarz
Eleocharis pauciflora var. *rhizomatosa* Hand.-Mazz. = ***Eleocharis yunnanensis*** Svenson
Eleocharis pauciflora var. *suksdorfiana* (Beauverd) Svenson = ***Eleocharis suksdorfiana*** Beauverd
Eleocharis pellucida f. *attenuata* (Franch. & Sav.) Ohwi = ***Eleocharis attenuata*** (Franch. & Sav.) Palla
Eleocharis pellucida var. *japonica* (Miq.) Tang & F.T.Wang = ***Eleocharis pellucida*** C.Presl
Eleocharis pellucida var. *sanguinolenta* Tang & F.T.Wang = ***Eleocharis pellucida*** C.Presl
Eleocharis pellucida var. *spongiosa* Tang & F.T.Wang = ***Eleocharis pellucida*** C.Presl
Eleocharis perlonga Fernald & Brackett = ***Eleocharis macrostachya*** Britton
Eleocharis perrieri Cherm. = ***Eleocharis nigrescens*** (Nees) Steud.
Eleocharis petasata (Maxim.) Zinserl. = ***Eleocharis wichurai*** Boeck.
Eleocharis piauhiensis (Nees) Steud. = ***Eleocharis filiculmis*** Kunth
Eleocharis pileata A.Gray = ***Eleocharis palustris*** (L.) Roem. & Schult. subsp. ***palustris***
Eleocharis pittieri Boeck. = ***Eleocharis sellowiana*** Kunth var. ***sellowiana***
Eleocharis planiculmis Steud. = ***Eleocharis acutangula*** (Roxb.) Schult.
Eleocharis planifolia (Spreng.) Nees = ***Trichophorum planifolium*** (Spreng.) Palla

Eleocharis plantaginea (Retz.) Roem. & Schult. = ***Eleocharis dulcis*** (Burm.f.) Trin. ex Hensch.
Eleocharis plantaginea Boeck. = ***Eleocharis interstincta*** (Vahl) Roem. & Schult.
Eleocharis plantaginea var. *stolonifera* Boeck. = ***Eleocharis dulcis*** (Burm.f.) Trin. ex Hensch.
Eleocharis plantagineiformis Tang & F.T.Wang = ***Eleocharis dulcis*** (Burm.f.) Trin. ex Hensch.
Eleocharis plantaginoidea W.Wight = ***Eleocharis dulcis*** (Burm.f.) Trin. ex Hensch.
Eleocharis plantaginoides (Rottb.) Domin = ***Eleocharis dulcis*** (Burm.f.) Trin. ex Hensch.
Eleocharis polycaula Wender. = ***Eleocharis uniglumis*** (Link) Schult. subsp. ***uniglumis***
Eleocharis polymorpha Nees = ***Eleocharis flavescens*** (Poir.) Urb. var. ***flavescens***
Eleocharis posadensis Kük. = ***Eleocharis contracta*** Maury ex Micheli
Eleocharis praticola Britton = ***Eleocharis flavescens*** (Poir.) Urb. var. ***flavescens***
Eleocharis prolifera Torr. = ***Eleocharis baldwinii*** (Torr.) Chapm.
Eleocharis prolifera var. *camptotricha* (C.Wright) Urb. = ***Eleocharis nana*** Kunth
Eleocharis punctata Hochst. ex Steud. = ***Eleocharis filiculmis*** Kunth
Eleocharis punctata Boeck. = ***Eleocharis nana*** Kunth
Eleocharis punctata Boeck. ex Tonduz = ***Eleocharis nana*** Kunth
Eleocharis punctata f. *major* Duss = ***Eleocharis punctulata*** Boeck. ex Duss
Eleocharis purpurascens Boeck. = ***Eleocharis congesta*** D.Don
Eleocharis purpureovaginata Boeck. = ***Eleocharis filiculmis*** Kunth
Eleocharis pygmaea (Suess.) L.T.Eiten = ***Eleocharis chamaegyne*** L.T.Eiten
Eleocharis pygmaea Torr. = ***Eleocharis parvula*** (Roem. & Schult.) Palla
Eleocharis pygmaea var. *anachaeta* Torr. = ***Eleocharis coloradoensis*** (Britton) Gilly
Eleocharis quadrangulata var. *crassior* Fernald = ***Eleocharis quadrangulata*** (Michx.) Roem. & Schult.
Eleocharis quadrangulata var. *typica* Fernald = ***Eleocharis quadrangulata*** (Michx.) Roem. & Schult.
Eleocharis quaesita Kitag. = [38 JAP]
Eleocharis quinquangularis Boeck. = ***Eleocharis emarginata*** (Nees) Klotzsch ex Boeck.
Eleocharis quinqueflora var. *bernardina* (Munz & I.M.Johnst.) S.González & P.M.Peterson = ***Eleocharis bernardina*** (Munz & I.M.Johnst.) Munz & I.M.Johnst.
Eleocharis quinqueflora subsp. *fernaldii* (Svenson) Hultén = ***Eleocharis quinqueflora*** (Hartmann) O.Schwarz
Eleocharis quinqueflora subsp. *meridionalis* (Zinserl.) T.V.Egorova = ***Eleocharis quinqueflora*** (Hartmann) O.Schwarz
Eleocharis quinqueflora subsp. *suksdorfiana* (Beauverd) Hultén = ***Eleocharis suksdorfiana*** Beauverd
Eleocharis quinqueflora var. *suksdorfiana* (Beauverd) J.T.Howell = ***Eleocharis suksdorfiana*** Beauverd
Eleocharis reclinata Kunth = ***Eleocharis intermedia*** Schult.
Eleocharis reclinata var. *habereri* (Fernald) House = ***Eleocharis intermedia*** Schult.
Eleocharis reclinata f. *habereri* (Fernald) Svenson = ***Eleocharis intermedia*** Schult.
Eleocharis recurvata Nees = ***Eleocharis retroflexa*** subsp. ***chaetaria*** (Roem. & Schult.) T.Koyama
Eleocharis riparia Nees ex Spreng. = ***Eleocharis geniculata*** (L.) Roem. & Schult.
Eleocharis rivularis Phil. = ***Eleocharis radicans*** Kunth
Eleocharis rostellata var. *congdonii* Jeps. = ***Eleocharis rostellata*** (Torr.) Torr.
Eleocharis rostellata var. *occidentalis* S.Watson = ***Eleocharis rostellata*** (Torr.) Torr.
Eleocharis rostellata f. *occidentalis* (S.Watson) Beetle = ***Eleocharis rostellata*** (Torr.) Torr.
Eleocharis rothiana Boeck. = ***Eleocharis filiculmis*** Kunth
Eleocharis rothiana var. *nuda* Maury =
Eleocharis sachalinensis (Meinsh.) Kom. = ***Eleocharis kamtschatica*** (C.A.Mey.) Kom.
Eleocharis sagotii C.B.Clarke = ***Eleocharis jelskiana*** Boeck.
Eleocharis sagotii var. *glochidiata* C.B.Clarke = ***Eleocharis jelskiana*** Boeck.
Eleocharis salzmanniana Steud. = ***Eleocharis sellowiana*** Kunth var. ***sellowiana***
Eleocharis sanguinea Maury ex Micheli = [85 PAR]
Eleocharis sareptana Zinserl. = ***Eleocharis uniglumis*** subsp. ***fennica*** (Palla ex Kneuck.) Vestergr.
Eleocharis satoi Ohwi = ***Eleocharis mamillata*** subsp. ***mamillata***
Eleocharis savannarum Britton = ***Eleocharis minima*** Kunth var. ***minima***
Eleocharis savatieri C.B.Clarke ex H.Lév. = ***Eleocharis kamtschatica*** (C.A.Mey.) Kom.
Eleocharis scariosa Steud. = ***Eleocharis mutata*** (L.) Roem. & Schult.
Eleocharis schottiana (Nees) Steud. = ***Eleocharis maculosa*** (Vahl) Roem. & Schult.
Eleocharis schweinfurthiana Boeck. = ***Eleocharis setifolia*** subsp. ***schweinfurthiana*** (Boeck.) D.A.Simpson
Eleocharis scythica Zinserl. = ***Eleocharis uniglumis*** (Link) Schult. subsp. ***uniglumis***
Eleocharis septata Miq. = ***Eleocharis interstincta*** (Vahl) Roem. & Schult.
Eleocharis septentrionalis Zinserl. = ***Eleocharis uniglumis*** subsp. ***septentrionalis*** (Zinserl.) T.V.Egorova
Eleocharis setacea R.Br. = ***Eleocharis geniculata*** (L.) Roem. & Schult.
Eleocharis setacea R.Br. = ***Eleocharis retroflexa*** subsp. ***chaetaria*** (Roem. & Schult.) T.Koyama
Eleocharis seydeliana Podlech = ***Eleocharis schlechteri*** C.B.Clarke
Eleocharis shaferi Britton = ***Eleocharis debilis*** Kunth
Eleocharis shimadai Hayata = ***Eleocharis pellucida*** C.Presl
Eleocharis sieberi Kunth = ***Eleocharis variegata*** (Poir.) C.Presl
Eleocharis simplex (Elliott) A.Dietr. = ***Eleocharis albida*** Torr.
Eleocharis simulans C.B.Clarke = ***Eleocharis melanostachys*** (d'Urv.) C.B.Clarke
Eleocharis singularis Steud. = ***Eleocharis geniculata*** (L.) Roem. & Schult.
Eleocharis sintenisii Boeck. = ***Eleocharis geniculata*** (L.) Roem. & Schult.
Eleocharis smallii Britton = ***Eleocharis palustris*** var. ***vigens*** L.H.Bailey
Eleocharis smallii var. *major* (Sond.) F.Seym. = ***Eleocharis palustris*** (L.) Roem. & Schult. subsp. ***palustris***
Eleocharis soloniensis (Dubois) Hara = ***Eleocharis ovata*** (Roth) Roem. & Schult.

Eleocharis sororia Kunth = ***Eleocharis limosa*** (Schrad.) Schult.
Eleocharis spadicea C.B.Clarke = ***Eleocharis subfoliata*** C.B.Clarke
Eleocharis spegazzinii Barros = ***Eleocharis cylindrica*** Buckley
Eleocharis spiralis Boeck. = ***Eleocharis mutata*** (L.) Roem. & Schult.
Eleocharis spiralis f. *tenuis* Boeck. = ***Eleocharis spiralis*** (Rottb.) Roem. & Schult.
Eleocharis squamata Boeck. = ***Eleocharis squamigera*** Svenson subsp. ***squamigera***
Eleocharis stolonifera Boeck. = ***Eleocharis dombeyana*** Kunth
Eleocharis striata Hochst. ex Steud. = ***Eleocharis marginulata*** Hochst. ex Steud.
Eleocharis striata Hochst. ex Engl. = ***Eleocharis palustris*** (L.) Roem. & Schult. subsp. ***palustris***
Eleocharis striatula É.Desv. = ***Eleocharis bonariensis*** Nees
Eleocharis strobilacea Pedersen = ***Eleocharis obtusitrigona*** (Lindl. & Nees) Steud.
Eleocharis stylosa Steud. = ***Eleocharis filiculmis*** Kunth
Eleocharis submersa Miq. = ***Websteria confervoides*** (Poir.) S.S.Hooper
Eleocharis subnodulosa Steud. = ***Eleocharis montana*** (Kunth) Roem. & Schult.
Eleocharis subprolifera Steud. = ***Eleocharis pellucida*** C.Presl
Eleocharis subsphacelata Steud. = ***Eleocharis sphacelata*** R.Br.
Eleocharis subtilis Boeck. = ***Eleocharis minima*** Kunth var. ***minima***
Eleocharis subtilissima Nelmes = ***Eleocharis retroflexa*** subsp. ***subtilissima*** (Nelmes) Lye
Eleocharis subulata Boeck. = ***Eleocharis ochrostachys*** Steud.
Eleocharis subvivipara Boeck. = ***Eleocharis congesta*** D.Don
Eleocharis sulcata (Schult.) Nees = ***Eleocharis filiculmis*** Kunth
Eleocharis sulcata var. *grandirostris* Lindm. = ***Eleocharis grandirostris*** (Lindm.) Mereles
Eleocharis sulciculmis Rchb. ex Kunth = ***Eleocharis flavescens*** (Poir.) Urb. var. ***flavescens***
Eleocharis svensonii Zinserl. = ***Eleocharis acicularis*** subsp. ***yokoscensis*** (Franch. & Sav.) T.V.Egorova
Eleocharis tenerrima Peter = [25 TAN]
Eleocharis tenuis var. *atrata* (Svenson) B.Boivin = ***Eleocharis elliptica*** Kunth
Eleocharis tenuis var. *borealis* (Svenson) Gleason = ***Eleocharis elliptica*** Kunth
Eleocharis tenuissima Boeck. = ***Eleocharis minima*** Kunth var. ***minima***
Eleocharis tetraquetra var. *micranthera* Ohwi = ***Eleocharis tetraquetra*** Nees
Eleocharis texana Britton = ***Eleocharis cylindrica*** Buckley
Eleocharis thermalis Rydb. = ***Eleocharis flavescens*** (Poir.) Urb. var. ***flavescens***
Eleocharis thermalis (Hultén) T.V.Egorova = ***Eleocharis pellucida*** C.Presl
Eleocharis thomsonii Boeck. = ***Eleocharis pellucida*** C.Presl
Eleocharis tibestica Quézel = ***Eleocharis caduca*** (Delile) Schult.
Eleocharis torreyana Boeck. = ***Eleocharis microcarpa*** var. ***filiculmis*** Torr.
Eleocharis transcaucasica Zinserl. = ***Eleocharis uniglumis*** (Link) Schult. subsp. ***uniglumis***
Eleocharis triangularis Reinsch = ***Eleocharis acicularis*** (L.) Roem. & Schult. subsp. ***acicularis***
Eleocharis triangulata (Roxb.) Sieber ex C.Presl = ***Schoenoplectus mucronatus*** (L.) Palla subsp. ***mucronatus***
Eleocharis trichoides (Kunth) Kuntze = ***Eleocharis acicularis*** (L.) Roem. & Schult. subsp. ***acicularis***
Eleocharis triflora Kom. = ***Eleocharis kamtschatica*** (C.A.Mey.) Kom.
Eleocharis triflora Boeck. = ***Eleocharis retroflexa*** (Poir.) Urb. subsp. ***retroflexa***
Eleocharis truncata Schltdl. = ***Eleocharis dombeyana*** Kunth
Eleocharis truncatovaginata T.Koyama = ***Eleocharis valleculosa*** Ohwi
Eleocharis tsurumachii Ohwi = ***Eleocharis tetraquetra*** Nees
Eleocharis tuberculosa var. *pubnicoensis* Fernald = ***Eleocharis tuberculosa*** (Michx.) Roem. & Schult.
Eleocharis tuberculosa f. *pubnicoensis* (Fernald) Svenson = ***Eleocharis tuberculosa*** (Michx.) Roem. & Schult.
Eleocharis tuberculosa f. *retrorsa* Svenson = ***Eleocharis tuberculosa*** (Michx.) Roem. & Schult.
Eleocharis tuberosa (Roxb.) Schult. = ***Eleocharis dulcis*** (Burm.f.) Trin. ex Hensch.
Eleocharis tuerckheimii Boeck. = ***Eleocharis sellowiana*** Kunth var. ***sellowiana***
Eleocharis tumida (Roxb.) Schult. = ***Eleocharis dulcis*** (Burm.f.) Trin. ex Hensch.
Eleocharis turcomanica Zinserl. = ***Eleocharis mitracarpa*** Steud.
Eleocharis uncialis Chapm. ex Small = ***Eleocharis minima*** Kunth var. ***minima***
Eleocharis uniflora O.Seberg = ***Chillania pusilla*** Roiv.
Eleocharis uniglumis var. *affinis* (C.A.Mey. ex Claus) Nyman = ***Eleocharis uniglumis*** (Link) Schult. subsp. ***uniglumis***
Eleocharis uniglumis var. *fennica* (Palla ex Kneuck.) E.Holmb. = ***Eleocharis uniglumis*** subsp. ***fennica*** (Palla ex Kneuck.) Vestergr.
Eleocharis uniglumis var. *halophila* Fernald & Brackett = ***Eleocharis uniglumis*** (Link) Schult. subsp. ***uniglumis***
Eleocharis uniglumis var. *typica* Fernald & Brackett = ***Eleocharis uniglumis*** (Link) Schult.
Eleocharis uniglumis var. *watsonii* (Bab.) Nyman = ***Eleocharis uniglumis*** (Link) Schult. subsp. ***uniglumis***
Eleocharis univaginata Boeck. = ***Eleocharis fuscopurpurea*** (Steud.) H.Pfeiff.
Eleocharis valdiviana Phil. = ***Eleocharis melanostachys*** (d'Urv.) C.B.Clarke
Eleocharis valida Boeck. = ***Eleocharis geniculata*** (L.) Roem. & Schult.
Eleocharis valleculosa var. *setosa* Ohwi = ***Eleocharis valleculosa*** Ohwi
Eleocharis valleculosa f. *setosa* (Ohwi) Kitag. = ***Eleocharis valleculosa*** Ohwi
Eleocharis variegata var. *laxiflora* (Thwaites) C.B.Clarke = ***Eleocharis ochrostachys*** Steud.
Eleocharis verrucosa (Svenson) L.J.Harms = ***Eleocharis tenuis*** var. ***verrucosa*** (Svenson) Svenson
Eleocharis vierhapperi Bojko = ***Eleocharis quinqueflora*** (Hartmann) O.Schwarz
Eleocharis villaricensis Maury ex Micheli = ***Eleocharis minima*** Kunth var. ***minima***
Eleocharis vincentina Phil. = ***Eleocharis fuscopurpurea*** (Steud.) H.Pfeiff.

Eleocharis vincentina var. *arcuata* (Kunze ex Kunth) C.B.Clarke = ***Eleocharis maculosa*** (Vahl) Roem. & Schult.
Eleocharis viridis Boeck. = ***Eleocharis elegans*** (Kunth) Roem. & Schult.
Eleocharis vulcani Boeck. = ***Eleocharis montana*** (Kunth) Roem. & Schult.
Eleocharis vulgaris (Walters) Á.Löve & D.Löve = ***Eleocharis palustris*** subsp. ***vulgaris*** Walters
Eleocharis watsonii Bab. = ***Eleocharis uniglumis*** (Link) Schult. subsp. ***uniglumis***
Eleocharis wichurai var. *teres* Hara = ***Eleocharis wichurai*** Boeck.
Eleocharis wichurai f. *teres* (Hara) Ohwi = ***Eleocharis wichurai*** Boeck.
Eleocharis wichurai f. *vivipara* Ueno = ***Eleocharis wichurai*** Boeck.
Eleocharis wrightiana Boeck. = ***Eleocharis minima*** Kunth var. ***minima***
Eleocharis wrightii C.B.Clarke = ***Eleocharis oligantha*** C.B.Clarke
Eleocharis xyridiformis Fernald & Brackett = ***Eleocharis macrostachya*** Britton
Eleocharis yokoscensis (Franch. & Sav.) Tang & F.T.Wang = ***Eleocharis acicularis*** subsp. ***yokoscensis*** (Franch. & Sav.) T.V.Egorova
Eleocharis yunquensis Britton = ***Eleocharis debilis*** Kunth
Eleocharis zanardinii Parl. = ***Eleocharis atropurpurea*** (Retz.) Kunth
Eleocharis zinserlingi Zoz = ***Eleocharis uniglumis*** (Link) Schult. subsp. ***uniglumis***

***Unplaced Names*:**
Eleocharis lechleri C.B.Clarke, Bot. Jahrb. Syst. 30(68): 24 (1902), nom. illeg. = [85 CLS]
Eleocharis leptocaulos Phil., Anales Univ. Chile 93: 349 (1896), nom. illeg. = [85 CLC]
Eleocharis macrorrhiza T.Koyama, Brittonia 31: 285 (1979), nom. illeg. = [41 THA]
Eleocharis paraguayensis Maury ex Micheli, Mém. Soc. Phys. Genève 31(1): 137 (1890). = [85 PAR]
Eleocharis quaesita Kitag., J. Jap. Bot. 27: 202 (1952). = [38 JAP]
Eleocharis sanguinea Maury ex Micheli, Mém. Soc. Phys. Genève 31(1): 135 (1890). = [85 PAR]
Eleocharis tenerrima Peter, Abh. Preuss. Akad. Wiss., Phys.-Math. Kl., n.f., 13(2): 52 (1928). = [25 TAN]

Eleogenus

Eleogenus Nees = ***Eleocharis*** R.Br.
Eleogenus atropurpureus (Retz.) Nees = ***Eleocharis atropurpurea*** (Retz.) Kunth
Eleogenus capitatus (L.) Nees = ***Eleocharis geniculata*** (L.) Roem. & Schult.
Eleogenus capitatus var. *major* Nees = ***Eleocharis geniculata*** (L.) Roem. & Schult.
Eleogenus capitatus var. *minor* Nees = ***Eleocharis geniculata*** (L.) Roem. & Schult.
Eleogenus laetevirens Nees = ***Eleocharis atropurpurea*** (Retz.) Kunth
Eleogenus laetevirens var. *tenuis* Nees = ***Eleocharis atropurpurea*** (Retz.) Kunth
Eleogenus laetovirens var. *umbonatus* Nees = ***Eleocharis atropurpurea*** (Retz.) Kunth
Eleogenus lereschii (Shuttlew.) Bluff, Nees & Schauer = ***Eleocharis atropurpurea*** (Retz.) Kunth
Eleogenus nodulosus (Roth) Nees = ***Eleocharis montana*** (Kunth) Roem. & Schult.
Eleogenus ochreatus Nees = ***Eleocharis flavescens*** (Poir.) Urb. var. ***flavescens***
Eleogenus ovatus (Roth) Nees = ***Eleocharis ovata*** (Roth) Roem. & Schult.
Eleogenus schottianus Nees = ***Eleocharis maculosa*** (Vahl) Roem. & Schult.
Eleogenus sellovianus (Kunth) Nees = ***Eleocharis sellowiana*** Kunth

Eleogiton

Eleogiton Link = ***Isolepis*** R.Br.
Eleogiton atropurpurea (Retz.) A.Dietr. = ***Eleocharis atropurpurea*** (Retz.) Kunth
Eleogiton beccarii (Boeck.) Soják = ***Isolepis beccarii*** (Boeck.) Goetgh. & D.A.Simpson
Eleogiton caricoides (Steud.) A.Dietr. = ***Ficinia angustifolia*** (Schrad.) C.B.Clarke
Eleogiton cernua (Vahl) A.Dietr. = ***Isolepis cernua*** (Vahl) Roem. & Schult.
Eleogiton crassiusculum (Hook.f.) Soják = ***Isolepis crassiuscula*** Hook.f.
Eleogiton curvula Nees = ***Isolepis fluitans*** (L.) R.Br. var. ***fluitans***
Eleogiton digitata (Nees ex Schrad.) Nees = ***Isolepis digitata*** Nees ex Schrad.
Eleogiton digitata var. *dissoluta* Nees = ***Isolepis digitata*** Nees ex Schrad.
Eleogiton duvalii Fourr. = ***Schoenoplectus lacustris*** (L.) Palla subsp. ***lacustris***
Eleogiton elongantha A.Dietr. = ***Trichophorum pumilum*** (Vahl) Schinz & Thell. subsp. ***pumilum***
Eleogiton exigua A.Dietr. = ***Eleocharis acicularis*** (L.) Roem. & Schult. subsp. ***acicularis***
Eleogiton fascicularis Nees = ***Isolepis fluitans*** (L.) R.Br. var. ***fluitans***
Eleogiton filiformis (Roem. & Schult.) A.Dietr. = ***Eleocharis tenuis*** (Willd.) Schult. var. ***tenuis***
Eleogiton fluitans (L.) Link = ***Isolepis fluitans*** (L.) R.Br.
Eleogiton fluitans var. *fasciculata* Miq. = ***Isolepis fluitans*** (L.) R.Br. var. ***fluitans***
Eleogiton lacustris (L.) Fourr. = ***Schoenoplectus lacustris*** (L.) Palla
Eleogiton litoralis (Schrad.) Fourr. = ***Schoenoplectus litoralis*** (Schrad.) Palla
Eleogiton longifolia Nees = ***Isolepis digitata*** Nees ex Schrad.
Eleogiton monostachya (J.König ex Roxb.) A.Dietr. = ***Fimbristylis schoenoides*** (Retz.) Vahl
Eleogiton parvula (Roem. & Schult.) Link = ***Eleocharis parvula*** (Roem. & Schult.) Palla
Eleogiton planifolia (Spreng.) A.Dietr. = ***Trichophorum planifolium*** (Spreng.) Palla
Eleogiton radicans (Poir.) A.Dietr. = ***Eleocharis radicans*** Kunth
Eleogiton rothii Feurr. = ***Schoenoplectus pungens*** (Vahl) Palla var. ***pungens***
Eleogiton rubicunda Nees = ***Isolepis ludwigii*** (Steud.) Kunth
Eleogiton scabra (Roxb.) A.Dietr. = ***Fimbristylis acuminata*** Vahl
Eleogiton scariosa (Thunb.) A.Dietr. = ***Ficinia deusta*** (P.J.Bergius) Levyns
Eleogiton spadicea (L.) A.Dietr. = ***Fimbristylis spadicea*** (L.) Vahl
Eleogiton striata Nees = ***Isolepis striata*** (Nees) Kunth
Eleogiton tabernaemontani (C.C.Gmel.) Fourr. = ***Schoenoplectus tabernaemontani*** (C.C.Gmel.) Palla
Eleogiton tenuissima Nees = ***Isolepis tenuissima*** (Nees) Kunth

Eleogiton triqueter (L.) Fourr. = ***Schoenoplectus triqueter*** (L.) Palla
Eleogiton verruculosa (Steud.) A.Dietr. = ***Isolepis cernua*** var. ***setiformis*** (Benth.) Muasya

Elogiton

Elogiton Link ex Steud. = ***Isolepis*** R.Br.

Elyna

Elyna Schrad. = ***Kobresia*** Willd.
Elyna bellardii (All.) K.Koch = ***Kobresia myosuroides*** (Vill.) Fiori & Paol.
Elyna bellardii Hartm. = ***Kobresia simpliciuscula*** (Wahlenb.) Mack. subsp. ***simpliciuscula***
Elyna capillifolia Decne. = ***Kobresia capillifolia*** (Decne.) C.B.Clarke
Elyna caricina Mert. & W.D.J.Koch = ***Kobresia simpliciuscula*** (Wahlenb.) Mack. subsp. ***simpliciuscula***
Elyna filifolia Turcz. = ***Kobresia filifolia*** (Turcz.) C.B.Clarke
Elyna filiformis Steud. = ***Kobresia myosuroides*** (Vill.) Fiori & Paol.
Elyna humilis C.A.Mey. ex Trautv. = ***Kobresia humilis*** (C.A.Mey. ex Trautv.) Serg.
Elyna kokanica Regel = ***Kobresia royleana*** (Nees) Boeck. subsp. ***royleana***
Elyna laxa (Nees) Kunth = ***Kobresia laxa*** Nees
Elyna myosuroides (Vill.) Fritsch ex Janch. = ***Kobresia myosuroides*** (Vill.) Fiori & Paol.
Elyna paniculata Meinsh. ex K.V.Ivanova = ***Kobresia royleana*** (Nees) Boeck. subsp. ***royleana***
Elyna schoenoides C.A.Mey. = ***Kobresia schoenoides*** (C.A.Mey.) Steud.
Elyna scirpina (Willd.) Pax = ***Kobresia myosuroides*** (Vill.) Fiori & Paol.
Elyna sibirica Turcz. = ***Kobresia sibirica*** (Turcz.) Boeck.
Elyna spicata Boiss. = ***Kobresia capillifolia*** (Decne.) C.B.Clarke
Elyna spicata Schrad. = ***Kobresia myosuroides*** (Vill.) Fiori & Paol.
Elyna stenocarpa Kar. & Kir. = ***Kobresia royleana*** (Nees) Boeck. subsp. ***royleana***
Elyna stricta Hoppe ex Steud. = ***Kobresia myosuroides*** (Vill.) Fiori & Paol.

Elynanthus

Elynanthus P.Beauv. ex Lestib. = ?
Elynanthus Nees = ***Tetraria*** P.Beauv.
Elynanthus antarcticus (Hook.f.) Macloskie = ***Schoenus antarcticus*** (Hook.f.) Dusén
Elynanthus aphyllus F.Muell. ex Boeck. = ***Schoenus subaphyllus*** Kük.
Elynanthus arenarius (Schrad.) Nees = ***Tetraria compar*** (L.) P.Beauv.
Elynanthus aristatus (Thunb.) Boeck. = ***Tetraria flexuosa*** (Thunb.) C.B.Clarke
Elynanthus auritus Nees = ***Tetraria sylvatica*** (Nees) C.B.Clarke
Elynanthus australis Nees = ***Gahnia australis*** (Nees) K.L.Wilson
Elynanthus bifidus Nees = ***Schoenus bifidus*** (Nees) Boeck.
Elynanthus borbonicus Steud. = ***Carpha borbonica*** (Steud.) C.B.Clarke
Elynanthus capillaceus Benth. = ***Tetraria capillaris*** (F.Muell.) J.M.Black
Elynanthus capitatus Nees = ***Schoenus subbulbosus*** Benth.
Elynanthus compar (L.) Nees = ***Tetraria compar*** (L.) P.Beauv.
Elynanthus cuspidatus (Rottb.) Nees = ***Tetraria cuspidata*** (Rottb.) C.B.Clarke
Elynanthus dregeanus Boeck. = ***Epischoenus dregeanus*** (Boeck.) Levyns
Elynanthus ensifolius Boeck. = ***Machaerina iridifolia*** (Bory) T.Koyama
Elynanthus gracilis Nees = ***Tetraria cuspidata*** (Rottb.) C.B.Clarke
Elynanthus grandiflorus Nees = ***Schoenus grandiflorus*** (Nees) F.Muell.
Elynanthus kraussii Hochst. = ***Cyathocoma hexandra*** (Nees) Browning
Elynanthus laxus (Hook.f.) Macloskie = ***Schoenus rhynchosporoides*** (Steud.) Kük.
Elynanthus ligulatus Boeck. = ***Tetraria ligulata*** (Boeck.) C.B.Clarke
Elynanthus loreus Nees = ***Tetraria cuspidata*** (Rottb.) C.B.Clarke
Elynanthus microstachyus Boeck. = ***Tetraria fasciata*** (Rottb.) C.B.Clarke
Elynanthus obtusifolius Nees = ***Schoenus obtusifolius*** (Nees) Boeck.
Elynanthus octandrus Nees = ***Tetraria octandra*** (Nees) Kük.
Elynanthus pictus Boeck. = ***Tetraria picta*** (Boeck.) C.B.Clarke
Elynanthus punctorius (Vahl) Benth. & Hook.f. = ***Neesenbeckia punctoria*** (Vahl) Levyns
Elynanthus revolutus Nees = ***Tetraria octandra*** (Nees) Kük.
Elynanthus sculptus Nees = ***Schoenus sculptus*** (Nees) Boeck.
Elynanthus sodalium (Har.) Planch. = ***Schoenus andinus*** (Phil.) H.Pfeiff.
Elynanthus spathaceus Nees = ***Schoenus nigricans*** L.
Elynanthus sylvaticus Nees = ***Tetraria sylvatica*** (Nees) C.B.Clarke
Elynanthus tenerrimus Boeck. = [27]
Elynanthus ustulatus (L.) Nees = ***Tetraria ustulata*** (L.) C.B.Clarke
Elynanthus viscosus (Schrad.) Nees = ***Tetraria compar*** (L.) P.Beauv.

Unplaced Names:
Elynanthus P.Beauv. ex Lestib., Essai Cypér.: 32 (1819). = ?
Elynanthus tenerrimus Boeck., Allg. Bot. Z. Syst. 2: 122 (1896). = [27]

Elytrospermum

Elytrospermum C.A.Mey. = ***Schoenoplectus*** (Rchb.) Palla
Elytrospermum californicum C.A.Mey. = ***Schoenoplectus californicus*** (C.A.Mey.) Soják

Ephippiorhynchium

Ephippiorhynchium Nees = ***Rhynchospora*** Vahl
Ephippiorhynchium amazonicum (Poepp. ex Kunth) Nees ex Boeck. = ***Rhynchospora amazonica*** Poepp. ex Kunth
Ephippiorhynchium longirostre (Nees) Nees = ***Rhynchospora amazonica*** subsp. ***guianensis*** (Kük.) T.Koyama
Ephippiorhynchium longirostre var. *amazonicum* (Poepp. ex Kunth) Nees = ***Rhynchospora amazonica*** subsp. ***guianensis*** (Kük.) T.Koyama
Ephippiorhynchium polycephalum (Nees) Nees = ***Rhynchospora holoschoenoides*** (Rich.) Herter

Ephippiorhynchium stenocarpum (Kunth) Nees = ***Rhynchospora stenocarpa*** Kunth
Ephippiorhynchium tenuirostre Nees = ***Rhynchospora holoschoenoides*** (Rich.) Herter
Ephippiorhynchium triflorum (Vahl) Nees = ***Rhynchospora triflora*** Vahl
Ephippiorhynchium trispicatum Nees = ***Rhynchospora trispicata*** (Nees) Schrad. ex Steud.

Epiandria

Epiandria C.Presl = ***Gahnia*** J.R.Forst. & G.Forst.
Epiandria teretifolia C.Presl = ***Gahnia filifolia*** (C.Presl) Kük. ex Benl

Epiphystis

Epiphystis Trin. = ***Cyperus*** L.
Epiphystis ophiuroides Trin. = ***Cyperus filiformis*** Sw.

Epischoenus

Epischoenus C.B.Clarke in W.H.Harvey & auct. suc. (eds.), Fl. Cap. 7: 273 (1898).
Cape Prov. 27 CPP.
8 Species

Epischoenus adnatus Levyns, J. S. African Bot. 13: 54 (1947).
SW. & S. Cape Prov. 27 CPP. Hemicr.

Epischoenus cernuus Levyns, J. S. African Bot. 25: 80 (1958).
WSW. Cape Prov. 27 CPP. Hemicr.

Epischoenus complanatus Levyns, J. S. African Bot. 25: 79 (1958).
SW. Cape Prov. 27 CPP. Hemicr.

Epischoenus dregeanus (Boeck.) Levyns, J. S. African Bot. 25: 76 (1958).
SW. Cape Prov. 27 CPP. Hemicr.
**Elynanthus dregeanus* Boeck., Linnaea 38: 262 (1874). *Schoenus dregeanus* (Boeck.) Kuntze, Revis. Gen. Pl. 2: 756 (1891). *Tetraria dregeana* (Boeck.) C.B.Clarke in T.A.Durand & H.Schinz, Consp. Fl. Afric. 5: 661 (1894).

Epischoenus gracilis Levyns, J. S. African Bot. 7: 81 (1941).
SW. Cape Prov. 27 CPP. Hemicr.

Epischoenus lucidus (C.B.Clarke) Levyns, J. S. African Bot. 25: 78 (1958).
SW. Cape Prov. 27 CPP. Hemicr.
**Tetraria lucida* C.B.Clarke in W.H.Harvey & auct. suc. (eds.), Fl. Cap. 7: 759 (1900).
Epischoenus eriophorus Levyns, J. S. African Bot. 13: 56 (1947).

Epischoenus quadrangularis (Boeck.) C.B.Clarke in W.H.Harvey & auct. suc. (eds.), Fl. Cap. 7: 273 (1898).
SW. & S. Cape Prov. 27 CPP. Hemicr.
**Schoenus quadrangularis* Boeck., Linnaea 38: 274 (1874). *Tetraria quadrangularis* (Boeck.) Kük., Repert. Spec. Nov. Regni Veg. 44: 191 (1938).

Epischoenus villosus Levyns, J. S. African Bot. 13: 56 (1947).
SW. Cape Prov. 27 CPP. Hemicr.

***Synonyms*:**
Epischoenus eriophorus Levyns = ***Epischoenus lucidus*** (C.B.Clarke) Levyns

Eriochaeta

Eriochaeta Torr. ex Steud. = ***Rhynchospora*** Vahl

Eriophorella

Eriophorella Holub = ***Trichophorum*** Pers.
Eriophorella alpina (L.) Holub = ***Trichophorum alpinum*** (L.) Pers.
Eriophorella pumila (Vahl) Kit Tan = ***Trichophorum pumilum*** (Vahl) Schinz & Thell.

Eriophoropsis

Eriophoropsis Palla = ***Eriophorum*** L.
Eriophoropsis virginica (L.) Palla = ***Eriophorum virginicum*** L.

Eriophorum

Eriophorum L., Sp. Pl.: 53 (1753).
Subarctic and Temp. Northern Hemisphere. 10 DEN FIN FOR GRB ICE IRE NOR SVA SWE SWI 11 AUT BGM CZE GER HUN NET POL SWI 12 FRA POR SPA 13 ALB BUL GRC ITA ROM YUG 14 BLR BLT RUC RUE RUN RUS RUW UKR 30 ALT TVA WSB YAK 31 AMU KRA MAG SAK 32 KAZ KGZ 33 NCS TCS 34 AFG IRN TUR 36 CHC CHI CHM 37 MON 38 JAP KOR 40 BAN EHM NEP PAK WHM 41 MYA VIE 70 ALU ASK GNL NUN NWT YUK 71 ABT BRC LAB MAN QUE SAS 72 LAB NBR NFL NSC ONT PEI QUE 73 COL IDA MNT ORE WAS WYO 74 ILL IOW MIN NCA NDA NEB SDA WIS 75 CNT INI MAI MAS MIC NWH NWJ NWY OHI PEN RHO VER WVA 76 CAL NEV UTA 77 NWM 78 DEL MRY NCA SCA TEN VRG WDC.
18 Species
Linagrostis Guett., Hist. Acad. Roy. Sci. Mém. Math. Phys. (Paris, 4to) 1750: 187 (1754).
Linagrostis Hill, Brit. Herb.: 506 (1757).
Plumaria Heist. ex Fabr., Enum.: 207 (1759).
Leucoma Ehrh., Beitr. Naturk. 4: 146 (1789).
Eriophoropsis Palla, Bot. Zeitung (Berlin) 54(1): 151 (1896).
Erioscirpus Palla, Bot. Zeitung (Berlin) 54(1): 151 (1896).
Plumaria Bubani, Fl. Pyren. 4: 195 (1902).

Eriophorum angustifolium Honck., Verz. Gew. Teutschl. 1: 153 (1782). *Eriophorum polystachyon* var. *angustifolium* (Honck.) A.Gray, Manual, ed. 2: 502 (1856). *Plumaria angustifolia* (Honck.) Bubani, Fl. Pyren. 4: 195 (1902). *Scirpus angustifolius* (Honck.) T.Koyama, J. Fac. Sci. Univ. Tokyo, Sect. 3, Bot. 7: 356 (1958).
Subarctic and Temp. Northern Hemisphere. 10 DEN FIN FOR GRB ICE IRE NOR SVA SWE 11 AUT BGM CZE GER HUN NET POL SWI 12 FRA POR SPA 13 ALB BUL GRC ITA ROM YUG 14 BLR BLT RUC RUE RUN RUS RUW 30 WSB 31+ 32 KAZ 33 NCS TCS 34 TUR 36 CHI CHM 37 MON 70 ASK GNL NUN NWT YUK 71 ABT BRC MAN SAS 72 LAB NBR NFL NSC ONT PEI QUE 73 COL IDA MNT ORE WAS WYO 74 ILL IOW MIN NDA NEB SDA WIS 75 INI MAI MIC NWY 77 NWM. Hemicr.

subsp. ***angustifolium***
Subarctic & Temp. Northern Hemisphere. 10 DEN FIN FOR GRB ICE IRE NOR SVA SWE 11 AUT BGM CZE GER HUN NET POL SWI 12

FRA POR SPA 13 ALB BUL GRC ITA ROM YUG 14 BLR BLT RUC RUE RUN RUS RUW 30 WSB 31+ 32+ 33 NCS TCS 34 TUR 36 CHI CHM 37 MON 70 ASK GNL NUN NWT YUK 71 ABT BRC MAN SAS 72 LAB NBR NFL NSC ONT PEI QUE 73 COL IDA MNT ORE WAS WYO 74 ILL IOW MIN NDA NEB SDA WIS 75 INI MAI MIC NWY 77 NWM. Hemicr.

Eriophorum monostachion Groschke in A.F.M.von Keyserling & Dersch., Beschr. Prov. Kurl.: 116.

Eriophorum polystachion L., Sp. Pl.: 52 (1753), nom. rejic. *Linagrostis polystachia* (L.) Scop., Fl. Carniol., ed. 2, 1: 48 (1771).

Linagrostis vulgaris Hill, Brit. Herb.: 506 (1757). *Eriophorum vulgare* (Hill) Druce, Rep. Bot. Exch. Club Soc. Brit. Isles 3: 439 (1913 publ. 1914), nom. illeg.

Linagrostis paniculata Lam., Fl. Franç. 3: 555 (1778), nom. superfl. *Eriophorum paniculatum* (Lam.) Gaterau, Descr. Pl. Montauban: 32 (1789).

Scirpus speciosus Salisb., Prodr. Stirp. Chap. Allerton: 31 (1796).

Eriophorum vaillantii Poit. & Turpin, Fl. Paris.: t. 52 (1809).

Eriophorum gracile Sm., Engl. Bot.: t. 2402 (1812), nom. illeg. *Eriophorum angustifolium* var. *gracile* Nyman, Consp. Fl. Eur.: 762 (1882).

Eriophorum intermedium Bastard, J. Bot. (Desvaux) 3: 19 (1814).

Eriophorum angustifolium var. *congestum* Mert. & W.D.J.Koch in J.C.Röhling, Deutschl. Fl., ed. 3, 1: 546 (1823).

Eriophorum angustifolium var. *megastachyon* Nutt. in W.H.Keating, Narr. Exp. St. Peter's River 2: 381 (1824).

Eriophorum angustifolium var. *brevifolium* Torr., Ann. Lyceum Nat. Hist. New York 3: 340 (1836).

Eriophorum dubium Hegetschw., Fl. Schweiz: 52 (1838).

Eriophorum alpicola Schur, Verh. Mitth. Siebenbürg. Vereins Naturwiss. Hermannstadt 4: 78 (1853).

Eriophorum angustifolium var. *minus* Bellynck, Fl. Namur: 280 (1855).

Eriophorum angustifolium f. *uniceps* Th.Fr., Öfvers. Kongl. Vetensk.-Akad. Förh. 26(2): 136 (1869 publ. 1870).

Eriophorum minus Dalla Torre in A.Hartinger & K.W.von Dalla Torre, Atlas Alpenfl.: 216 (1883). *Eriophorum angustifolium* subsp. *minus* (Dalla Torre) Holub, Folia Geobot. Phytotax. 23: 413 (1988).

Eriophorum polystachyon f. *elegans* (Bab.) Ostenf. & Lundager, Meddel. Grønland 43: 9 (1910).

Eriophorum polystachyon var. *viridicarinatum* Farw., Rep. (Annual) Michigan Acad. Sci. 21: 361 (1920).

Eriophorum angustifolium f. *rubrovaginatum* Raymond, Naturaliste Canad. 77: 67 (1950).

Eriophorum subarcticum V.N.Vassil., Bot. Mater. Gerb. Bot. Inst. Komarova Akad. Nauk S.S.S.R. 13: 58 (1950). *Eriophorum angustifolium* subsp. *subarcticum* (V.N.Vassil.) Hultén ex Kartesz & Gandhi, Phytologia 72: 22 (1992).

Eriophorum angustifolium f. *scammanianum* Raymond, Rhodora 53: 208 (1951).

Eriophorum angustifolium var. *coloratum* Hultén, Ark. Bot., a.s., 7(1): 23 (1968).

Eriophorum angustifolium var. *giganteum* Hultén, Ark. Bot., a.s., 7(1): 23 (1968).

subsp. ***komarovii*** (V.N.Vassil.) Vorosch. in A.K. Skvortsov (ed.), Florist. Issl. Razn. Raionakh SSSR: 152 (1985).

Siberia to Russian Far East. 30+ 31+. Hemicr. or rhizome geophyte.

**Eriophorum komarovii* V.N.Vassil., Bot. Mater. Gerb. Bot. Inst. Komarova Akad. Nauk S.S.S.R. 8: 102 (1940). *Eriophorum polystachion* subsp. *komarovii* (V.N.Vassil.) M.S.Novos., Bot. Zhurn. (Moscow & Leningrad) 79(11): 87 (1994 publ. 1995). *Eriophorum angustifolium* subsp. *komarovii* (V.N.Vassil.) M.S.Novos., Novosti Sist. Vyssh. Rast. 33: 48 (2001).

subsp. ***triste*** (T.C.E.Fr.) Hultén, Kongl. Svenska Vetensk. Acad. Handl. 8(5): 58 (1962).

Subarctic. 10 30+ 31+ 70 ASK GNL NUN NWT YUK 72 QUE. Hemicr. or rhizome geophyte.

**Eriophorum angustifolium* var. *triste* T.C.E.Fr., Öfvers. Kongl. Vetensk.-Akad. Förh. 1869(2): 135 (1869). *Eriophorum triste* (T.C.E.Fr.) Hadac & Á.Löve, Bot. Not. 1950: 34 (1950).

Eriophorum angustifolium subsp. *scabriusculum* Hultén, Svensk Bot. Tidskr. 30: 518 (1936).

Eriophorum* ×*beringianum Raymond, Naturaliste Canad. 84: 184 (1957). *E. angustifolium* × *E. chamissonis*.

Russian Far East to Alaska. 31+ 70 ALU ASK. Hemicr.

Eriophorum brachyantherum Trautv. & C.A.Mey., Fl. Ochot. 1(2), 3: 98 (1856). *Scirpus brachyantherus* (Trautv. & C.A.Mey.) T.Koyama, J. Fac. Sci. Univ. Tokyo, Sect. 3, Bot. 7: 359 (1958).

Subarctic Northern Hemisphere. 10 FIN NOR SWE 14 RUE RUN 30 WSB 31 MAG 37 MON 38 KOR 70 ASK NUN NWT YUK 71 ABT BRC MAN SAS 72 LAB NFL ONT QUE. Hemicr.

Eriophorum vaginatum subsp. *opacum* Björnstr., Grunddr. Pitea Lappm. Vaxtfys.: 35 (1856). *Eriophorum opacum* (Björnstr.) Fernald, Rhodora 7: 85 (1905).

Eriophorum opacum var. *cinnamomeum* A.E.Porsild, Canad. Field-Naturalist 56: 112 (1943).

Eriophorum brachyantherum var. *pellucidum* Lepage, Naturaliste Canad. 79: 177 (1952).

Eriophorum callitrix Cham. ex C.A.Mey., Mém. Sav. Étr. Acad. St. Pétersbourg 1: 203 (1831).

Siberia to Canada. 30+ 31 MAG 70 ASK GNL NUN NWT YUK 71 ABT BRC MAN SAS 72 LAB NFL ONT QUE 73 MNT WYO. Hemicr.

Eriophorum callitrix var. *moravium* Raymond, Bull. Soc. Bot. France 98: 7 (1951). *Eriophorum callitrix* f. *moravium* (Raymond) B.Boivin, Naturaliste Canad. 94: 524 (1967).

Eriophorum callitrix var. *pallidum* Hultén, Ark. Bot., a.s., 7(1): 23 (1968).

Eriophorum chamissonis C.A.Mey., Mém. Sav. Étr. Acad. St. Pétersbourg 1: 204 (1831). *Scirpus chamissonis* (C.A.Mey.) T.Koyama, J. Fac. Sci. Univ. Tokyo, Sect. 3, Bot. 7: 296 (1958).

Subarctic to WC. U.S.A. 10 FIN ICE? NOR SWE 14 RUN 30 ALT YAK 31 MAG SAK 36 CHM 37 MON 70 ALU ASK GNL NUN NWT YUK 71 ABT BRC MAN SAS 72 LAB NBR NFL NSC ONT PEI QUE 73 COL IDA MNT ORE WAS WYO 74 MIN NDA NEB WIS 76 UTA. Hemicr. or rhizome geophyte.

Eriophorum intermedium Cham. ex C.A.Mey., Mém. Sav. Étr. Acad. St. Pétersbourg 1: 204 (1831).
Eriophorum russeolum Fr. ex Hartm., Novit. Fl. Suec. Mant. 3: 67 (1843). *Scirpus russeolus* (Fr. ex Hartm.) T.Koyama, J. Fac. Sci. Univ. Tokyo, Sect. 3, Bot. 7: 359 (1958).
Eriophorum russeolum var. *albidum* F.Nyl., Acta Soc. Sci. Fenn. 3: 17 (1852). *Eriophorum chamissonis* var. *albidum* (F.Nyl.) Fernald, Rhodora 7: 84 (1905). *Eriophorum chamissonis* subsp. *albidum* (F.Nyl.) Piper & Beattie, Fl. N.W. Coast: 85 (1915). *Eriophorum chamissonis* f. *albidum* (F.Nyl.) Fernald, Rhodora 23: 131 (1921).
Eriophorum rufescens E.S.Anderson, Bot. Not. 1857: 79 (1857).
Eriophorum aquatile Norman, Skr. Vidensk.-Selsk. Christiana, Math.-Naturvidensk. Kl. 1893(Extr.): 43 (1893).
Eriophorum russeolum var. *majus* Sommier, Nuovo Giorn. Bot. Ital. 25(2): 95 (1893). *Scirpus russeolus* var. *major* (Sommier) T.Koyama, J. Fac. Sci. Univ. Tokyo, Sect. 3, Bot. 7: 360 (1958).
Eriophorum mandshuricum Meinsh., Trudy Imp. S.-Peterburgsk. Bot. Sada 18: 268 (1901).
Eriophorum strigosum Miyabe & Kudô, Trans. Sapporo Nat. Hist. Soc. 5: 67 (1913).
Eriophorum chamissonis subsp. *albidum* (F.Nyl.) Piper & Beattie, Fl. N.W. Coast: 85 (1915).
Eriophorum chamissonis var. *aquatile* (Norman) Fernald, Rhodora 27: 207 (1926).
Eriophorum russeolum f. *leucothrix* Blomgr., Skand. Fl. 2: 294 (1926). *Eriophorum russeolum* var. *leucothrix* (Blomgr.) Hultén, Acta Univ. Lund. 2(38; 1): 283 (1942).
Eriophorum eximium V.N.Vassil., Bot. Mater. Gerb. Bot. Inst. Komarova Akad. Nauk S.S.S.R. 8: 107 (1940).
Eriophorum chamissonis f. *turneri* Raymond, Svensk Bot. Tidskr. 48: 79 (1954).
Eriophorum altaicum var. *neogaeum* J.Raynal, Contr. Inst. Bot. Univ. Montréal 70: 103 (1957).
Eriophorum mandshuricum subsp. *sibiricum* M.S. Novos., Bot. Zhurn. (Moscow & Leningrad) 78(8): 87 (1993).
Eriophorum russeolum subsp. *leiocarpum* M.S.Novos., Bot. Zhurn. (Moscow & Leningrad) 78(8): 86 (1993).

Eriophorum comosum (Wall.) Nees in R.Wight, Contr. Bot. India: 110 (1834).
S. Iran to Indo-China. 34 AFG IRN 36 CHC 40 BAN EHM NEP PAK WHM 41 MYA VIE. Hemicr.
**Scirpus comosus* Wall. in W.Roxburgh, Fl. Ind. 1: 234 (1820). *Trichophorum comosum* (Wall.) A.Dietr., Sp. Pl. 2: 202 (1832). *Erioscirpus comosus* (Wall.) Palla, Bot. Zeitung (Berlin) 54(1): 151 (1896).
Scirpus elongatus Buch.-Ham. ex D.Don, Prodr. Fl. Nepal.: 40 (1825).
Eriophorum arundinaceum Nees in R.Wight, Contr. Bot. India: 110 (1834).
Eriophorum cannabina Royle, Ill. Bot. Himal. Mts.: 415 (1840).

Eriophorum crinigerum (A.Gray) Beetle, Leafl. W. Bot. 3: 165 (1942).
Oregon to California. 73 ORE 76 CAL. Hemicr.
**Scirpus criniger* A.Gray, Proc. Amer. Acad. Arts 7: 392 (1868).

Eriophorum* × *fellowsii (Fernald) M.S.Novos., Bot. Zhurn. (Moscow & Leningrad) 79(12): 73 (1994 publ. 1995) *E. virginicum* × *E. viridicarinatum*.
SE. Canada to NE. U.S.A. 72 ONT 75 MAI MAS. Hemicr.
**Eriophorum viridicarinatum* var. *fellowsii* Fernald, Rhodora 7: 91 (1905). *Eriophorum viridicarinatum* f. *fellowsii* (Fernald) Fernald, Manual, ed. 8: 279 (1950).

Eriophorum gracile Koch in A.W.Roth, Catal. Bot. 2(Add.): 259 (1799). *Scirpus ardea* T.Koyama, J. Fac. Sci. Univ. Tokyo, Sect. 3, Bot. 7: 354 (1958).
Temp. Northern Hemisphere. 10 DEN FIN IRE NOR SWI 11 AUT BGM CZE GER HUN NET POL SWI 12 FRA 13 BUL ITA ROM YUG 14 BLT RUC RUE RUN RUS RUW UKR 30 WSB 31+ 32 KAZ 36 CHM 38 KOR 70 ASK NUN NWT YUK 71 ABT BRC MAN SAS 72 LAB NBR NFL NSC ONT PEI QUE 73 COL IDA MNT ORE WAS WYO 74 ILL IOW MIN NDA NEB SDA WIS 75 CNT INI MAI MAS MIC NWH NWJ NWY OHI PEN RHO VER 76 CAL NEV UTA 78 DEL. Hemicr.
Eriophorum triquetrum Hoppe, Bot. Taschenb. Anfänger Wiss. Apothekerkunst 1800: 106 (1800).
Eriophorum pallidum Nees, Linnaea 9: 293 (1834), nom. inval.
Eriophorum gracile var. *paucinervium* Engelm., Amer. J. Sci. Arts 46: 103 (1844).
Eriophorum hoefftii F.Nyl., Acta Soc. Sci. Fenn. 3: 18 (1849). *Eriophorum gracile* subsp. *hoefftii* (F.Nyl.) Nyman, Consp. Fl. Eur., Suppl. 2: 318 (1890).
Eriophorum gracile var. *caurianum* Fernald, Rhodora 7: 87 (1905).
Eriophorum coreanum Palla, Oesterr. Bot. Z. 59: 190 (1909).
Eriophorum asiaticum V.N.Vassil., Bot. Mater. Gerb. Bot. Inst. Komarova Akad. Nauk S.S.S.R. 8: 104 (1940). *Eriophorum gracile* subsp. *asiaticum* (V.N.Vassil.) M.S.Novos., Bot. Zhurn. (Moscow & Leningrad) 79(11): 84 (1994 publ. 1995).

Eriophorum* ×*gracilifolium M.S.Novos., Bot. Zhurn. (Moscow & Leningrad) 79(5): 100 (1994). *E. gracile* × *E. latifolium*.
European Russia. 14 RUW. Hemicr.

Eriophorum humile Turcz., Bull. Soc. Imp. Naturalistes Moscou 1838: 103 (1838).
Siberia to C. Asia. 30 ALT TVA 31 AMU 32 KAZ 37 MON. Hemicr.

Eriophorum latifolium Hoppe, Bot. Taschenb. Anfänger Wiss. Apothekerkunst 1800: 108 (1800). *Eriophorum polystachyon* var. *latifolium* (Hoppe) A.Gray, Manual, ed. 2: 502 (1856). *Plumaria latifolia* (Hoppe) Bubani, Fl. Pyren. 4: 195 (1902).
Europe to Caucasus. 10 DEN FIN GRB IRE NOR 11 AUT BGM CZE GER HUN NET POL SWI 12 FRA SPA 13 ALB BUL ITA ROM YUG 14 BLR BLT RUC RUE RUN RUS RUW UKR 33 NCS TCS 34 TUR 37 MON. Hemicr.
Eriophorum vulgare Pers., Syn. Pl. 1: 70 (1805).
Carex alopecurus Lapeyr., Hist. Pl. Pyrénées, Suppl.: 141 (1818).
Eriophorum pubescens Sm., Engl. Fl. 1: 68 (1824).
Eriophorum latifolium var. *congestum* Baguet, Bull. Soc. Roy. Bot. Belgique 22: 89 (1883).

Eriophorum* ×*medium Andersson, Bot. Not. 1857: 62 (1857). *E. chamissonis* × *E. scheuchzeri*.
Subarctic Northern Hemisphere. 10 FIN NOR 14 RUN 30 WSB 31+ 70 ASK 71 LAB QUE. Hemicr.
Eriophorum ×*gauthieri* Boivin, Provancheria 25: 43 (1992).

Eriophorum microstachyum Boeck., Linnaea 38: 399 (1874). *Erioscirpus microstachyus* (Boeck.) Palla, Bot. Zeitung (Berlin) 54(1): 151 (1896). *Scirpus microstachyus* (Boeck.) T.Koyama, J. Fac. Sci. Univ. Tokyo, Sect. 3, Bot. 7: 294 (1958).
Himalaya. 40 EHM NEP WHM. Hemicr.

Eriophorum* ×*polystachiovaginatum Beauverd, Bull. Soc. Bot. Genève, II, 27: 90 (1937). *E. angustifolium* × *E. vaginatum*. *Eriophorum* ×*beauverdii* Soó, Acta Bot. Acad. Sci. Hung. 16: 369 (1970 publ. 1971), nom. illeg.
Europe. 12 FRA. Hemicr.
Eriophorum ×*churchillianum* Lepage, Naturaliste Canad. 84: 39 (1957).

Eriophorum* ×*pylaieanum Raymond, Svensk Bot. Tidskr. 45: 523 (1951). *E. chamissonis* × *E. vaginatum*.
Alaska to Canada. 70 ASK 72 LAB NFL NSC ONT QUE. Hemicr.
Eriophorum ×*porsildii* Raymond, Svensk Bot. Tidskr. 45: 530 (1951).

Eriophorum* ×*rousseauianum Raymond, Naturaliste Canad. 77: 67 (1950). *E. angustifolium* × *E. scheuchzeri*.
Alaska to Canada. 70 ASK 72 QUE. Hemicr.
Eriophorum ×*sorensenii* Raymond, Naturaliste Canad. 77: 69 (1950).

Eriophorum scabriculme (Beetle) Raymond, Naturaliste Canad. 84: 147 (1957).
Indo-China. 41 VIE. Hemicr.
**Scirpus scabriculmis* Beetle, Amer. J. Bot. 33: 665 (1946).

Eriophorum scheuchzeri Hoppe, Bot. Taschenb. Anfänger Wiss. Apothekerkunst 1800: 104 (1800). *Scirpus scheuchzeri* (Hoppe) Vitman, Summa Pl. 1: 150 (1789), nom. illeg. *Plumaria scheuchzeri* (Hoppe) Bubani, Fl. Pyren. 4: 247 (1902).
Europe to W. Himalaya, Suarctic N. America to Canada. 10 FIN ICE NOR SVA SWE 11 AUT CZE? GER SWI 12 FRA SPA 13 ITA ROM YUG 14 RUN UKR 30 WSB YAK 31 KRA MAG 32 KAZ KGZ 37 MON 40 PAK WHM 70 ASK GNL NUN NWT YUK 71 ABT BRC MAN SAS 72 LAB NFL ONT QUE 73 COL MNT WAS WYO 76 UTA. Hemicr.
Scirpus scheuchzeri Hartmann, Primae Lin. Inst. Bot.: 84 (1767).
Eriophorum capitatum Host, Icon. Descr. Gram. Austriac. 1: 30 (1801).
Eriophorum leucocephalum Boeck., Flora 41: 419 (1858). *Scirpus leucocephalus* (Boeck.) T.Koyama, J. Fac. Sci. Univ. Tokyo, Sect. 3, Bot. 7: 360 (1958).
Eriophorum sphaerocephalum Wulfen, Fl. Norica Phan.: 62 (1858).
Eriophorum altaicum Meinsh., Trudy Imp. S.-Peterburgsk. Bot. Sada 18: 267 (1901). *Eriophorum scheuchzeri* subsp. *altaicum* (Meinsh.) N.V. Bondareva, in Fl. Sibir. 3: 15 (1990).
Eriophorum scheuchzeri subsp. *arcticum* M.S.Novos., Bot. Zhurn. (Moscow & Leningrad) 79(4): 112 (1994).

Eriophorum tenellum Nutt., Gen. N. Amer. Pl.: 2 add. (1818).
E. Canada to NC. & NE. U.S.A. 70 NUN 72 LAB NBR NFL NSC ONT PEI QUE 75 CNT MAI MAS MIC NWH NWJ NWY PEN RHO VER. Hemicr.
Eriophorum strictum R.Br. in J.Richardson, Bot. App.: 3 (1823).
Eriophorum tenellum var. *monticola* Fernald, Rhodora 10: 47 (1908).
Eriophorum tenellum f. *axillare* Lakela, Rhodora 52: 46 (1950).
Eriophorum tenellum f. *gorhami* Raymond, Rhodora 53: 208 (1951).

Eriophorum tolmatchevii M.S.Novos., Bot. Zhurn. (Moscow & Leningrad) 79(4): 114 (1994).
Siberia. 30 YAK. Hemicr.

Eriophorum transiens Raymond, Naturaliste Canad. 86: 240 (1959).
SC. China. 36 CHC. Hemicr.

Eriophorum vaginatum L., Sp. Pl.: 52 (1753). *Linagrostis vaginata* (L.) Scop., Fl. Carniol., ed. 2, 1: 47 (1771). *Scirpus vaginatus* (L.) Salisb., Prodr. Stirp. Chap. Allerton: 31 (1796), nom. illeg. *Plumaria vaginata* (L.) Bubani, Fl. Pyren. 4: 196 (1902).
Subarctic & Temp. Northern Hemisphere. 10 DEN FIN FOR GRB IRE NOR SWE 11 AUT BGM CZE GER HUN NET POL SWI 12 FRA SPA 13 GRC ITA ROM YUG 14 BLR BLT RUC RUE RUN RUS RUW UKR 30 31 32 33 NCS TCS 34 TUR 36 CHM 38 JAP KOR 70 ASK GNL NUN NWT YUK 71 ABT BRC MAN SAS 72 LAB NBR NFL NSC ONT PEI QUE 74 MIN WIS 75 CNT INI MAI MAS MIC NWH NWJ NWY PEN RHO VER. Hemicr.
Eriophorum caespitosum Host, Icon. Descr. Gram. Austriac. 1: 30 (1801).
Eriophorum caespitosum var. *humilius* E.Mey., Pl. Labrador.: 21 (1830).
Eriophorum callithrix Lange in G.C.Oeder & al. (eds.), Fl. Dan. Suppl.: t. 122 (1874).
Eriophorum callitrix var. *erubescens* Fernald, Rhodora 7: 85 (1905). *Eriophorum spissum* var. *erubescens* (Fernald) Fernald, Rhodora 27: 209 (1926). *Eriophorum vaginatum* f. *erubescens* (Fernald) B.Boivin, Provancheria 25: 45 (1992).
Eriophorum fauriei E.G.Camus, Notul. Syst. (Paris) 1: 249 (1910). *Scirpus fauriei* (E.G.Camus) T.Koyama, J. Fac. Sci. Univ. Tokyo, Sect. 3, Bot. 7: 358 (1958). *Eriophorum vaginatum* var. *fauriei* (E.G.Camus) Kitag., Neo-Lineam. Fl. Manshur.: 150 (1979).
Eriophorum spissum Fernald, Rhodora 27: 208 (1925 publ. 1926). *Eriophorum vaginatum* subsp. *spissum* (Fernald) Hultén, Acta Univ. Lund., n.s., 38(1): 286 (1942). *Eriophorum vaginatum* var. *spissum* (Fernald) B.Boivin, Naturaliste Canad. 94: 524 (1967).
Eriophorum scabridum Ohwi, Bot. Mag. (Tokyo) 45: 187 (1931).

Eriophorum virginicum L., Sp. Pl.: 52 (1753). *Eriophoropsis virginica* (L.) Palla, Bot. Zeitung (Berlin) 54(1): 151 (1896). *Scirpus virginicus* (L.) T.Koyama, J. Fac. Sci. Univ. Tokyo, Sect. 3, Bot. 7: 294 (1958).
E. Canada to NC. & E. U.S.A. 72 LAB NBR NFL NSC ONT PEI QUE 74 ILL MIN WIS 75 CNT INI MAI MAS MIC NWH NWJ NWY OHI PEN RHO VER WVA 78 MRY NCA SCA TEN VRG WDC. Hemicr.
Eriophorum virginianum Houtt., Nat. Hist. 13: 127 (1782).
Eriophorum virginicum var. *gracile* Torr., Fl. N. Middle United States 1: 66 (1824).

Eriophorum confertissimum Alph.Wood, Class-book Bot., ed. 2: 572 (1847). *Eriophorum virginicum* var. *confertissimum* (Alph.Wood) Alph.Wood, Class-book Bot.: 740 (1861).
Eriophorum virginicum f. *album* (A.Gray) Wiegand, Rhodora 26: 2 (1924).

Eriophorum viridicarinatum (Engelm.) Fernald, Rhodora 7: 89 (1905).
Subarctic and N. U.S.A. 70 ASK NUN NWT 71 ABT BRC MAN SAS 72 LAB NBR NFL NSC ONT PEI QUE 73 COL IDA WYO 74 ILL IOW MIN NCA WIS 75 CNT INI MAI MAS MIC NWH NWY OHI PEN RHO VER. Hemicr.
**Eriophorum latifolium* var. *viridicarinatum* Engelm., Amer. J. Sci. Arts 46: 103 (1844).

***Synonyms*:**
Eriophorum alpicola Schur = ***Eriophorum angustifolium*** Honck. subsp. ***angustifolium***
Eriophorum alpinum L. = ***Trichophorum alpinum*** (L.) Pers.
Eriophorum altaicum Meinsh. = ***Eriophorum scheuchzeri*** Hoppe
Eriophorum altaicum var. *neogaeum* J.Raynal = ***Eriophorum chamissonis*** C.A.Mey.
Eriophorum angustifolium var. *brevifolium* Torr. = ***Eriophorum angustifolium*** Honck. subsp. ***angustifolium***
Eriophorum angustifolium var. *coloratum* Hultén = ***Eriophorum angustifolium*** Honck. subsp. ***angustifolium***
Eriophorum angustifolium var. *congestum* Mert. & W.D.J.Koch = ***Eriophorum angustifolium*** Honck. subsp. ***angustifolium***
Eriophorum angustifolium var. *giganteum* Hultén = ***Eriophorum angustifolium*** Honck. subsp. ***angustifolium***
Eriophorum angustifolium var. *gracile* Nyman = ***Eriophorum angustifolium*** Honck. subsp. ***angustifolium***
Eriophorum angustifolium subsp. *komarovii* (V.N.Vassil.) M.S.Novos. = ***Eriophorum angustifolium*** subsp. ***komarovii*** (V.N.Vassil.) Vorosch.
Eriophorum angustifolium var. *megastachyon* Nutt. = ***Eriophorum angustifolium*** Honck. subsp. ***angustifolium***
Eriophorum angustifolium var. *minus* Bellynck = ***Eriophorum angustifolium*** Honck. subsp. ***angustifolium***
Eriophorum angustifolium subsp. *minus* (Dalla Torre) Holub = ***Eriophorum angustifolium*** Honck. subsp. ***angustifolium***
Eriophorum angustifolium f. *rubrovaginatum* Raymond = ***Eriophorum angustifolium*** Honck. subsp. ***angustifolium***
Eriophorum angustifolium subsp. *scabriusculum* Hultén = ***Eriophorum angustifolium*** subsp. ***triste*** (T.C.E.Fr.) Hultén
Eriophorum angustifolium f. *scammanianum* Raymond = ***Eriophorum angustifolium*** Honck. subsp. ***angustifolium***
Eriophorum angustifolium subsp. *subarcticum* (V.N. Vassil.) Hultén ex Kartesz & Gandhi = ***Eriophorum angustifolium*** Honck. subsp. ***angustifolium***
Eriophorum angustifolium var. *triste* T.C.E.Fr. = ***Eriophorum angustifolium*** subsp. ***triste*** (T.C.E.Fr.) Hultén
Eriophorum angustifolium f. *uniceps* Th.Fr. = ***Eriophorum angustifolium*** Honck. subsp. ***angustifolium***
Eriophorum aquatile Norman = ***Eriophorum chamissonis*** C.A.Mey.
Eriophorum arundinaceum Nees = ***Eriophorum comosum*** (Wall.) Nees
Eriophorum asiaticum V.N.Vassil. = ***Eriophorum gracile*** Koch
Eriophorum ×*beauverdii* Soó = ***Eriophorum* ×*polystachiovaginatum*** Beauverd
Eriophorum bellardii Colla = [13 ITA]
Eriophorum brachyantherum var. *pellucidum* Lepage = ***Eriophorum brachyantherum*** Trautv. & C.A.Mey.
Eriophorum caespitosum Host = ***Eriophorum vaginatum*** L.
Eriophorum caespitosum var. *humilius* E.Mey. = ***Eriophorum vaginatum*** L.
Eriophorum callithrix Lange = ***Eriophorum vaginatum*** L.
Eriophorum callitrix var. *erubescens* Fernald = ***Eriophorum vaginatum*** L.
Eriophorum callitrix var. *moravium* Raymond = ***Eriophorum callitrix*** Cham. ex C.A.Mey.
Eriophorum callitrix f. *moravium* (Raymond) B.Boivin = ***Eriophorum callitrix*** Cham. ex C.A.Mey.
Eriophorum callitrix var. *pallidum* Hultén = ***Eriophorum callitrix*** Cham. ex C.A.Mey.
Eriophorum cannabina Royle = ***Eriophorum comosum*** (Wall.) Nees
Eriophorum capitatum Host = ***Eriophorum scheuchzeri*** Hoppe
Eriophorum chamissonis var. *albidum* (F.Nyl.) Fernald = ***Eriophorum chamissonis*** C.A.Mey.
Eriophorum chamissonis subsp. *albidum* (F.Nyl.) Piper & Beattie = ***Eriophorum chamissonis*** C.A.Mey.
Eriophorum chamissonis subsp. *albidum* (F.Nyl.) Piper & Beattie = ***Eriophorum chamissonis*** C.A.Mey.
Eriophorum chamissonis f. *albidum* (F.Nyl.) Fernald = ***Eriophorum chamissonis*** C.A.Mey.
Eriophorum chamissonis var. *aquatile* (Norman) Fernald = ***Eriophorum chamissonis*** C.A.Mey.
Eriophorum chamissonis f. *turneri* Raymond = ***Eriophorum chamissonis*** C.A.Mey.
Eriophorum ×*churchillianum* Lepage = ***Eriophorum* ×*polystachiovaginatum*** Beauverd
Eriophorum confertissimum Alph.Wood = ***Eriophorum virginicum*** L.
Eriophorum coreanum Palla = ***Eriophorum gracile*** Koch
Eriophorum cyperinum L. = ***Scirpus cyperinus*** (L.) Kunth
Eriophorum cyperinum var. *atrocinctus* Farw. =
Eriophorum cyperinum var. *brachypodum* (Fernald) Farw. = ***Scirpus atrocinctus*** Fernald
Eriophorum cyperinum f. *brachypodum* (Fernald) Farw. = ***Scirpus atrocinctus*** Fernald
Eriophorum cyperinum f. *condensatum* Peck = ***Scirpus atrocinctus*** Fernald
Eriophorum cyperinum var. *eriophorum* (Michx.) Farw. = ***Scirpus cyperinus*** (L.) Kunth
Eriophorum cyperinum f. *grande* (Fernald) Farw. = ***Scirpus cyperinus*** (L.) Kunth
Eriophorum cyperinum var. *laxum* (A.Gray) Britton, Sterns & Poggenb. = ***Scirpus cyperinus*** (L.) Kunth
Eriophorum cyperinum var. *pedicellatum* (Fernald) Farw. = ***Scirpus pedicellatus*** Fernald
Eriophorum cyperinum var. *pelium* (Fernald) Farw. = ***Scirpus cyperinus*** (L.) Kunth
Eriophorum depauperatum Arv.-Touv. = [12 FRA]
Eriophorum dubium Hegetschw. = ***Eriophorum angustifolium*** Honck. subsp. ***angustifolium***
Eriophorum dubium Steud. = ?
Eriophorum eximium V.N.Vassil. = ***Eriophorum chamissonis*** C.A.Mey.

Eriophorum fauriei E.G.Camus = ***Eriophorum vaginatum*** L.
Eriophorum filamentosum Boeck. = [42 MLY]
Eriophorum ×fulvum Bach.Pyl. ex Raymond = [72 NFL]
Eriophorum × gauthieri Boivin = ***Eriophorum ×medium*** Andersson
Eriophorum gracile Sm. = ***Eriophorum angustifolium*** Honck. subsp. ***angustifolium***
Eriophorum gracile subsp. *asiaticum* (V.N.Vassil.) M.S.Novos. = ***Eriophorum gracile*** Koch
Eriophorum gracile var. *caurianum* Fernald = ***Eriophorum gracile*** Koch
Eriophorum gracile subsp. *hoefftii* (F.Nyl.) Nyman = ***Eriophorum gracile*** Koch
Eriophorum gracile var. *paucinervium* Engelm. = ***Eriophorum gracile*** Koch
Eriophorum hoefftii F.Nyl. = ***Eriophorum gracile*** Koch
Eriophorum hudsonianum Michx. = ***Trichophorum alpinum*** (L.) Pers.
Eriophorum intercedens H.Lindb. = [10 FIN]
Eriophorum intermedium Bastard = ***Eriophorum angustifolium*** Honck. subsp. ***angustifolium***
Eriophorum intermedium Cham. ex C.A.Mey. = ***Eriophorum chamissonis*** C.A.Mey.
Eriophorum japonicum Maxim. = ***Scirpus maximowiczii*** C.B.Clarke
Eriophorum kerneri Ullep. = [11 HUN]
Eriophorum komarovii V.N.Vassil. = ***Eriophorum angustifolium*** subsp. ***komarovii*** (V.N.Vassil.) Vorosch.
Eriophorum latifolium var. *congestum* Baguet = ***Eriophorum latifolium*** Hoppe
Eriophorum latifolium var. *viridicarinatum* Engelm. = ***Eriophorum viridicarinatum*** (Engelm.) Fernald
Eriophorum leucocephalum Boeck. = ***Eriophorum scheuchzeri*** Hoppe
Eriophorum lineatum (Michx.) Benth. & Hook.f. = ***Scirpus lineatus*** Michx.
Eriophorum lockhartii (G.Don) Steud. =
Eriophorum mandshuricum Meinsh. = ***Eriophorum chamissonis*** C.A.Mey.
Eriophorum mandshuricum subsp. *sibiricum* M.S.Novos. = ***Eriophorum chamissonis*** C.A.Mey.
Eriophorum maximowiczii (C.B.Clarke) Beetle = ***Scirpus maximowiczii*** C.B.Clarke
Eriophorum ×medium nothosubsp. *album* J.Cay. = *E. russeolum* subsp. *leiocarpum* × *E. scheuchzeri*
Eriophorum minus Dalla Torre = ***Eriophorum angustifolium*** Honck. subsp. ***angustifolium***
Eriophorum monostachion Groschke = ***Eriophorum angustifolium*** Honck. subsp. ***angustifolium***
Eriophorum montevidense Link = ***Androtrichum trigynum*** (Spreng.) H.Pfeiff.
Eriophorum ocreatum A.Nelson = [7]
Eriophorum opacum (Björnstr.) Fernald = ***Eriophorum brachyantherum*** Trautv. & C.A.Mey.
Eriophorum opacum var. *cinnamomeum* A.E.Porsild = ***Eriophorum brachyantherum*** Trautv. & C.A.Mey.
Eriophorum pallidum Nees = ***Eriophorum gracile*** Koch
Eriophorum paniculatum (Lam.) Gaterau = ***Eriophorum angustifolium*** Honck. subsp. ***angustifolium***
Eriophorum paucinervium Eaton = [7]
Eriophorum paucinervum Eaton = [7]
Eriophorum polystachion L. = ***Eriophorum angustifolium*** Honck. subsp. ***angustifolium***
Eriophorum polystachion subsp. *komarovii* (V.N.Vassil.) M.S.Novos. = ***Eriophorum angustifolium*** subsp. ***komarovii*** (V.N.Vassil.) Vorosch.
Eriophorum polystachyon var. *angustifolium* (Honck.) A.Gray = ***Eriophorum angustifolium*** Honck.
Eriophorum polystachyon f. *elegans* (Bab.) Ostenf. & Lundager = ***Eriophorum angustifolium*** Honck. subsp. ***angustifolium***
Eriophorum polystachyon var. *latifolium* (Hoppe) A.Gray = ***Eriophorum latifolium*** Hoppe
Eriophorum polystachyon var. *viridicarinatum* Farw. = ***Eriophorum angustifolium*** Honck. subsp. ***angustifolium***
Eriophorum ×porsildii Raymond = ***Eriophorum ×pylaieanum*** Raymond
Eriophorum pubescens Sm. = ***Eriophorum latifolium*** Hoppe
Eriophorum rufescens E.S.Anderson = ***Eriophorum chamissonis*** C.A.Mey.
Eriophorum russeolum Fr. ex Hartm. = ***Eriophorum chamissonis*** C.A.Mey.
Eriophorum russeolum var. *albidum* F.Nyl. = ***Eriophorum chamissonis*** C.A.Mey.
Eriophorum russeolum subsp. *leiocarpum* M.S.Novos. = ***Eriophorum chamissonis*** C.A.Mey.
Eriophorum russeolum f. *leucothrix* Blomgr. = ***Eriophorum chamissonis*** C.A.Mey.
Eriophorum russeolum var. *leucothrix* (Blomgr.) Hultén = ***Eriophorum chamissonis*** C.A.Mey.
Eriophorum russeolum var. *majus* Sommier = ***Eriophorum chamissonis*** C.A.Mey.
Eriophorum russeolum subsp. *rufescens* (E.S.Anderson) Hyl. = ***Eriophorum chamissonis*** C.A.Mey.
Eriophorum scabridum Ohwi = ***Eriophorum vaginatum*** L.
Eriophorum scheuchzeri subsp. *altaicum* (Meinsh.) N.V.Bondareva = ***Eriophorum scheuchzeri*** Hoppe
Eriophorum scheuchzeri subsp. *arcticum* M.S.Novos. = ***Eriophorum scheuchzeri*** Hoppe
Eriophorum ×sorensenii Raymond = ***Eriophorum ×rousseauianum*** Raymond
Eriophorum sphaerocephalum Wulfen = ***Eriophorum scheuchzeri*** Hoppe
Eriophorum spissum Fernald = ***Eriophorum vaginatum*** L.
Eriophorum spissum var. *erubescens* (Fernald) Fernald = ***Eriophorum vaginatum*** L.
Eriophorum strictum R.Br. = ***Eriophorum tenellum*** Nutt.
Eriophorum strigosum Miyabe & Kudô = ***Eriophorum chamissonis*** C.A.Mey.
Eriophorum subarcticum V.N.Vassil. = ***Eriophorum angustifolium*** Honck. subsp. ***angustifolium***
Eriophorum tenellum f. *axillare* Lakela = ***Eriophorum tenellum*** Nutt.
Eriophorum tenellum f. *gorhami* Raymond = ***Eriophorum tenellum*** Nutt.
Eriophorum tenellum var. *monticola* Fernald = ***Eriophorum tenellum*** Nutt.
Eriophorum textile Larrañaga = [85 URU]
Eriophorum triquetrum Hoppe = ***Eriophorum gracile*** Koch
Eriophorum triste (T.C.E.Fr.) Hadac & Á.Löve = ***Eriophorum angustifolium*** subsp. ***triste*** (T.C.E.Fr.) Hultén
Eriophorum vaginatum f. *erubescens* (Fernald) B.Boivin = ***Eriophorum vaginatum*** L.
Eriophorum vaginatum var. *fauriei* (E.G.Camus) Kitag. = ***Eriophorum vaginatum*** L.
Eriophorum vaginatum subsp. *opacum* Björnstr. = ***Eriophorum brachyantherum*** Trautv. & C.A.Mey.
Eriophorum vaginatum subsp. *spissum* (Fernald) Hultén = ***Eriophorum vaginatum*** L.
Eriophorum vaginatum var. *spissum* (Fernald) B.Boivin = ***Eriophorum vaginatum*** L.

Eriophorum vaillantii Poit. & Turpin = ***Eriophorum angustifolium*** Honck. subsp. ***angustifolium***
Eriophorum virginianum Houtt. = ***Eriophorum virginicum*** L.
Eriophorum virginicum f. *album* (A.Gray) Wiegand = ***Eriophorum virginicum*** L.
Eriophorum virginicum var. *confertissimum* (Alph. Wood) Alph.Wood = ***Eriophorum virginicum*** L.
Eriophorum virginicum var. *gracile* Torr. = ***Eriophorum virginicum*** L.
Eriophorum viridicarinatum var. *fellowsii* Fernald = ***Eriophorum ×fellowsii*** (Fernald) M.S.Novos.
Eriophorum viridicarinatum f. *fellowsii* (Fernald) Fernald = ***Eriophorum ×fellowsii*** (Fernald) M.S.Novos.
Eriophorum vulgare (Hill) Druce = ***Eriophorum angustifolium*** Honck. subsp. ***angustifolium***
Eriophorum vulgare Pers. = ***Eriophorum latifolium*** Hoppe

***Unplaced Names*:**
Eriophorum bellardii Colla, Herb. Pedem. 5: 561 (1836). = [13 ITA]
Eriophorum depauperatum Arv.-Touv., Ess. Pl. Dauph.: 66 (1871). = [12 FRA]
Eriophorum dubium Steud., Syn. Pl. Glumac. 2: 128 (1855). = ?
Eriophorum filamentosum Boeck., Bot. Jahrb. Syst. 5: 506 (1843). = [42 MLY]
Eriophorum ×fulvum Bach.Pyl. ex Raymond, Svensk Bot. Tidskr. 45: 527 (1951). = [72 NFL]
Eriophorum intercedens H.Lindb., Exsicc. (Sched. Pl. Finland.) 1906: 32 (1906). = [10 FIN]
Eriophorum kerneri Ullep., Oesterr. Bot. Z. 43: 421 (1893). = [11 HUN]
Eriophorum ocreatum A.Nelson, Bull. Torrey Bot. Club 29: 401 (1902). = [7]
Eriophorum paucinervium Eaton in N.L.Britton, Man. Fl. N. States: 182 (1901). = [7]
Eriophorum paucinervum Eaton, Bull. Torrey Bot. Club 25: 341 (1898). = [7]
Eriophorum textile Larrañaga, Escritos D. A. Larrañaga 2: 26 (1923). = [85 URU]

Erioscirpus

Erioscirpus Palla = ***Eriophorum*** L.
Erioscirpus comosus (Wall.) Palla = ***Eriophorum comosum*** (Wall.) Nees
Erioscirpus microstachyus (Boeck.) Palla = ***Eriophorum microstachyum*** Boeck.

Eriospora

Eriospora Hochst. ex A.Rich. = ***Coleochloa*** Gilly
Eriospora abyssinica Hochst. ex A.Rich. = ***Coleochloa abyssinica*** (Hochst. ex A.Rich.) Gilly
Eriospora abyssinica var. *castanea* C.B.Clarke = ***Coleochloa abyssinica*** var. ***castanea*** (C.B.Clarke) Pic.Serm.
Eriospora oliveri (Boeck.) C.B.Clarke = ***Coleochloa setifera*** (Ridl.) Gilly
Eriospora pilosa (Boeck.) Benth. = ***Afrotrilepis pilosa*** (Boeck.) J.Raynal
Eriospora rehmanniana C.B.Clarke = ***Coleochloa setifera*** (Ridl.) Gilly
Eriospora schweinfurthiana (Boeck.) Benth. ex C.B.Clarke = ***Coleochloa schweinfurthiana*** (Boeck.) Nelmes
Eriospora setifera (Ridl.) C.B.Clarke = ***Coleochloa setifera*** (Ridl.) Gilly
Eriospora villosula C.B.Clarke = ***Coleochloa setifera*** (Ridl.) Gilly
Eriospora virgata K.Schum. = ***Coleochloa virgata*** (K.Schum.) Nelmes

Eucyperus

Eucyperus Rikli = ***Cyperus*** L.
Eucyperus alternifolius (L.) Rikli = ***Cyperus alternifolius*** L.
Eucyperus bruneovaginatus (Boeck.) Rikli = ***Cyperus marginatus*** Thunb.
Eucyperus diffusus (Vahl) Rikli = ***Cyperus diffusus*** Vahl
Eucyperus fuscus (L.) Rikli = ***Cyperus fuscus*** L.
Eucyperus gracilis (R.Br.) Rikli = ***Cyperus gracilis*** R.Br.
Eucyperus longifolius (Poir.) Rikli = ***Cyperus longifolius*** Poir.
Eucyperus longus (L.) Rikli = ***Cyperus longus*** L.
Eucyperus luzulae (L.) Rikli = ***Cyperus luzulae*** (L.) Retz.
Eucyperus martianus Rikli = ***Cyperus friburgensis*** Boeck.
Eucyperus pungens Rikli = ***Cyperus textilis*** Thunb.
Eucyperus schaffneri Rikli = ***Karinia mexicana*** (C.B.Clarke ex Britton) Reznicek & McVaugh
Eucyperus tenellus (L.f.) Palla = ***Cyperus tenellus*** L.f.
Eucyperus textilis (Thunb.) Rikli = ***Cyperus textilis*** Thunb.

Evandra

Evandra R.Br., Prodr.: 239 (1810).
SW. Australia. 50 WAU.
2 Species

Evandra aristata R.Br., Prodr.: 239 (1810).
SW. Australia. 50 WAU. Hemicr.

Evandra pauciflora R.Br., Prodr.: 239 (1810).
SW. Australia. 50 WAU. Hemicr.

Everardia

Everardia Ridl. ex Oliv., Trans. Linn. Soc. London, Bot. 2: 287 (1887).
S. Trop. America. 82 GUY SUR VEN 83 CLM PER 84 BZN.
11 Species
Pseudoeverardia Gilly, Fieldiana, Bot. 28(1): 59 (1951).

Everardia angusta N.E.Br., Trans. Linn. Soc. London, Bot. 6: 73 (1901).
SE. Venezuela to Guyana (Mt. Roraima). 82 GUY VEN. Cham.
Everardia gracilis Gilly, Bull. Torrey Bot. Club 68: 26 (1940).
Everardia steyermarkii Gilly, Fieldiana, Bot. 28(1): 55 (1951).

Everardia debilis T.Koyama & Maguire, Mem. New York Bot. Gard. 12(3): 32 (1965).
S. Venezuela (Cerro Yutaje, Cerro Guanay). 82 VEN. Cham.

Everardia diffusa T.Koyama & Maguire, Mem. New York Bot. Gard. 12(3): 21 (1965).
S. Venezuela. 82 VEN. Cham.

Everardia disticha T.Koyama & Maguire, Mem. New York Bot. Gard. 12(3): 25 (1965).
SE. Venezuela to Guyana (Mt. Roraima), Brazil (Serra Aracá). 82 GUY VEN 84 BZN. Cham.

Everardia erectolaxa T.Koyama & Maguire, Mem. New York Bot. Gard. 12(3): 27 (1965).
S. Venezuela (Ilu-tepuí). 82 VEN. Cham.

Everardia flexifolia (Gilly) T.Koyama & Maguire, Mem. New York Bot. Gard. 12(3): 34 (1965).
S. Venezuela (Cerro Duida). 82 VEN. Cham.
**Didymiandrum flexifolium* Gilly, Bull. Torrey Bot. Club 68: 331 (1941). *Pseudoeverardia flexifolia* (Gilly) Gilly, Fieldiana, Bot. 28(1): 59 (1951).

Everardia longifolia Gilly, Brittonia 3: 153 (1939).
Venezuela (SE. Bolívar). 82 VEN. Cham.

Everardia maguireana T.Koyama, Mem. New York Bot. Gard. 18(2): 23 (1969).
N. Brazil (Serra da Neblina). 84 BZN. Cham.

Everardia montana Ridl., Trans. Linn. Soc. London, Bot. 2: 287 (1887).
Venezuela to Suriname and Peru. 82 GUY SUR VEN 83 CLM PER 84 BZN. Cham.

subsp. ***duidae*** (Gilly) T.Koyama & Maguire, Mem. New York Bot. Gard. 12(3): 30 (1965).
S. Venezuela. 82 VEN. Cham.
**Everardia duidae* Gilly, Bull. Torrey Bot. Club 68: 30 (1940).

subsp. ***glaucifolia*** (Gilly) T.Koyama & Maguire, Mem. New York Bot. Gard. 12(3): 31 (1965).
SE. Colombia to S. Venezuela. 82 VEN 83 CLM. Cham.
**Everardia glaucifolia* Gilly, Bull. Torrey Bot. Club 68: 27 (1940).

subsp. ***guaiquinimae*** (Schnee) T.Koyama & Maguire, Mem. New York Bot. Gard. 12(3): 31 (1965).
S. Venezuela (Cerro Guaiquinima). 82 VEN. Cham.
**Everardia guaiquinimae* Schnee, Bol. Soc. Venez. Ci. Nat. 9: 83 (1944).

subsp. ***montana***
Venezuela to Suriname, N. Peru. 82 GUY SUR VEN 83 PER 84 BZN. Cham.
Everardia revoluta Gilly, Bull. Torrey Bot. Club 68: 28 (1940).
Everardia surinamensis Gilly, Bull. Torrey Bot. Club 76: 292 (1949).
Everardia hitchcockii Lasser & Maguire, Brittonia 7: 76 (1950).
Everardia montana var. *pilosa* Gilly, Fieldiana, Bot. 28(1): 54 (1951).

subsp. ***ptariensis*** (Gilly) T.Koyama & Maguire, Mem. New York Bot. Gard. 12(3): 23 (1965).
Venezuela (Bolívar). 82 VEN. Cham.
**Everardia ptariensis* Gilly, Fieldiana, Bot. 28(1): 55 (1951).

subsp. ***velutina*** T.Koyama, Mem. New York Bot. Gard. 18(2): 26 (1969).
Venezuela (Sierra de la Neblina), Brazil (Serra da Neblina). 82 VEN 84 BZN. Cham.

Everardia recurvigluma T.Koyama & Maguire, Mem. New York Bot. Gard. 12(3): 23 (1965).
Venezuela (Bolívar). 82 VEN. Cham.

Everardia vareschii Maguire, Acta Biol. Venez. 2(6): 43 (1957).
Venezuela (Bolívar). 82 VEN. Cham.

Synonyms:
Everardia duidae Gilly = ***Everardia montana*** subsp. ***duidae*** (Gilly) T.Koyama & Maguire
Everardia glaucifolia Gilly = ***Everardia montana*** subsp. ***glaucifolia*** (Gilly) T.Koyama & Maguire
Everardia gracilis Gilly = ***Everardia angusta*** N.E.Br.
Everardia guaiquinimae Schnee = ***Everardia montana*** subsp. ***guaiquinimae*** (Schnee) T.Koyama & Maguire
Everardia hitchcockii Lasser & Maguire = ***Everardia montana*** Ridl. subsp. ***montana***
Everardia lanata T.Koyama & Maguire = ***Lagenocarpus lanatus*** (T.Koyama & Maguire) T.Koyama
Everardia montana var. *pilosa* Gilly = ***Everardia montana*** Ridl. subsp. ***montana***
Everardia ptariensis Gilly = ***Everardia montana*** subsp. ***ptariensis*** (Gilly) T.Koyama & Maguire
Everardia revoluta Gilly = ***Everardia montana*** Ridl. subsp. ***montana***
Everardia steyermarkii Gilly = ***Everardia angusta*** N.E.Br.
Everardia surinamensis Gilly = ***Everardia montana*** Ridl. subsp. ***montana***

Exocarya

Exocarya Benth., Hooker's Icon. Pl. 13: t. 1206 (1877).
New Guinea to NE. New South Wales. 43 NWG 50 NSW QLD.
1 Species

Exocarya sclerioides (F.Muell.) Benth., Hooker's Icon. Pl. 13: t. 1206 (1877).
New Guinea to NE. New South Wales. 43 NWG 50 NSW QLD. Hemicr.
**Cladium sclerioides* F.Muell., Fragm. 9: 12 (1875).
Scleria ustulata F.M.Bailey, Syn. Queensl. Fl., Suppl. 3: 81 (1898).
Exocarya montivaga Domin, Biblioth. Bot. 85: 484 (1915).

Synonyms:
Exocarya montivaga Domin = ***Exocarya sclerioides*** (F.Muell.) Benth.

Exochogyne

Exochogyne C.B.Clarke = ***Lagenocarpus*** Nees
Exochogyne amazonica C.B.Clarke = ***Lagenocarpus amazonicus*** (C.B.Clarke) H.Pfeiff.
Exochogyne amazonica var. *esmeraldensis* Gilly = ***Lagenocarpus amazonicus*** (C.B.Clarke) H.Pfeiff.
Exochogyne decandra Tutin = ***Lagenocarpus amazonicus*** (C.B.Clarke) H.Pfeiff.
Exochogyne megalorrhyncha Tutin = ***Lagenocarpus amazonicus*** (C.B.Clarke) H.Pfeiff.
Exochogyne steyermarkii Gilly = ***Lagenocarpus amazonicus*** (C.B.Clarke) H.Pfeiff.

Facolos

Facolos Raf. = ***Carex*** L.
Facolos brunnescens (Pers.) Raf. = ***Carex brunnescens*** (Pers.) Poir.
Facolos complanata (Torr. & Hook.) Raf. = ***Carex complanata*** Torr. & Hook.
Facolos curta (Gooden.) Raf. = ***Carex canescens*** L. subsp. ***canescens***
Facolos limosa (L.) Raf. = ***Carex limosa*** L.

Ficinia

Ficinia Schrad., Commentat. Soc. Regiae Sci. Gott. Recent. 7: 143 (1832).
St. Helena, Ethiopia to S. Africa, Madagascar, S. & E. Australia, New Zealand, St. Paul I., Juan Fernández Is. 24 ETH 25 KEN TAN UGA 26+ MOZ ZIM 27 CPP LES NAM NAT OFS TVL 28 STH 29 MDG 50 NFK NSW SOA TAS VIC WAU 51 CTM KER NZN NZS 85 JNF 90 ASP.

74 Species
Melancranis Vahl, Enum. Pl. 2: 239 (1805).
Hypolepis P.Beauv. ex Lestib., Essai Cypér.: 33 (1819).
Hemichlaena Schrad., Gött. Gel. Anz. 3: 2066 (1821).
Acrolepis Schrad., Commentat. Soc. Regiae Sci. Gott. Recent. 7: 142 (1832).
Hypophialium Nees, Linnaea 7: 530 (1832).
Pleurachne Schrad., Commentat. Soc. Regiae Sci. Gott. Recent. 7: 147 (1832).
Schoenidium Nees, Linnaea 9: 291 (1834).
Sickmannia Nees, Linnaea 9: 292 (1834).
Chamaexiphium Hochst. ex Steud., Syn. Pl. Glumac. 2: 59 (1854), nom. illeg.

Ficinia acrostachys (Steud.) C.B.Clarke in T.A.Durand & H.Schinz, Consp. Fl. Afric. 5: 635 (1894).
Cape Prov. 27 CPP. Hemicr.
**Scirpus acrostachys* Steud., Flora 12: 148 (1829). *Isolepis acrostachya* (Steud.) A.Dietr., Sp. Pl. 2: 116 (1832).

Ficinia acuminata (Nees) Nees, Linnaea 9: 292 (1834).
Cape Prov. 27 CPP. Hemicr.
Scirpus acuminatus Steud., Flora 12: 146 (1829), nom. illeg. **Isolepis acuminata* Nees, Linnaea 7: 501 (1832). *Melancranis acuminata* (Nees) Kuntze, Revis. Gen. Pl. 2: 756 (1891).
Ficinia acuminata Kunth, Enum. Pl. 2: 256 (1837), nom. illeg. *Ficinia kunthiana* Boeck., Linnaea 37: 67 (1871). *Melancranis kunthiana* (Boeck.) Kuntze, Revis. Gen. Pl. 2: 756 (1891).
Ficinia elongata Boeck., Linnaea 37: 65 (1871). *Melancranis elongata* (Boeck.) Kuntze, Revis. Gen. Pl. 2: 756 (1891).
Ficinia ignorata Boeck., Beitr. Cyper. 2: 19 (1890).

Ficinia albicans Nees, Linnaea 10: 175 (1836).
Cape Prov. 27 CPP. Hemicr.

Ficinia anceps Nees, Linnaea 10: 179 (1836). *Melancranis anceps* (Nees) Kuntze, Revis. Gen. Pl. 2: 756 (1891).
SW. Cape Prov. 27 CPP. Hemicr.

Ficinia angustifolia (Schrad.) C.B.Clarke in T.A.Durand & H.Schinz, Consp. Fl. Afric. 5: 635 (1894).
S. Africa. 27 CPP TVL. Hemicr.
**Hemichlaena angustifolia* Schrad., Gött. Gel. Anz. 3: 2066 (1821).
Schoenus caricoides Steud., Flora 12: 137 (1829). *Eleogiton caricoides* (Steud.) A.Dietr., Sp. Pl. 2: 99 (1832). *Ficinia caricoidea* (Steud.) H.Pfeiff., Revis. Ficinia: 51 (1921).
Hemichlaena longifolia Nees, Linnaea 10: 129 (1836). *Ficinia longifolia* (Nees) C.B.Clarke in T.A.Durand & H.Schinz, Consp. Fl. Afric. 5: 640 (1894).
Fimbristylis vexata Steud., Syn. Pl. Glumac. 2: 318 (1855).
Isolepis podocarpa Boeck., Flora 43: 179 (1860).
Ficinia micrantha Boeck. ex C.B.Clarke in T.A.Durand & H.Schinz, Consp. Fl. Afric. 5: 635 (1894), nom. inval.
Ficinia rehmanniana Boeck. ex C.B.Clarke in T.A.Durand & H.Schinz, Consp. Fl. Afric. 5: 635 (1894), pro syn.

Ficinia anysbergensis Muasya, S. African J. Bot. 71: 197 (2005).
SW. Cape Prov. 27 CPP. Hemicr.

Ficinia arenicola T.H.Arnold & Gordon-Gray, Bothalia 12: 414 (1978).
S. Cape Prov. 27 CPP. Hemicr.

var. ***arenicola***
S. Cape Prov. 27 CPP. Hemicr.

var. ***erecta*** T.H.Arnold & Gordon-Gray, Bothalia 12: 416 (1978).
S. Cape Prov. 27 CPP. Hemicr.

Ficinia argyropus Nees, Linnaea 10: 177 (1836). *Melancranis argyropus* (Nees) Kuntze, Revis. Gen. Pl. 2: 756 (1891). *Ficinia paradoxa* var. *argyropus* (Nees) C.B.Clarke in T.A.Durand & H.Schinz, Consp. Fl. Afric. 5: 641 (1894).
Cape Prov. 27 CPP. Hemicr.
Ficinia argyropus var. *minor* Boeck., Linnaea 37: 71 (1871).

Ficinia atrostachya H.Pfeiff., Herbarium 54: 33 (1920).
KwaZulu-Natal. 27 NAT. Hemicr.

Ficinia borealis Lye, Lidia 3: 205 (1996).
Ethiopia. 24 ETH. Hemicr.

Ficinia brevifolia Nees ex Kunth, Enum. Pl. 2: 260 (1837). *Melancranis brevifolia* (Nees ex Kunth) Kuntze, Revis. Gen. Pl. 2: 756 (1891).
Cape Prov. 27 CPP. Hemicr.
Melancranis radiata Vahl, Enum. Pl. 2: 239 (1805).
Ficinia brevifolia f. *atrocastanea* C.B.Clarke in W.H. Harvey & auct. suc. (eds.), Fl. Cap. 7: 250 (1898).
Ficinia brevifolia f. *debilior* C.B.Clarke ex H.Pfeiff., Revis. Ficinia: 34 (1921).

Ficinia bulbosa (L.) Nees, Linnaea 8: 91 (1833).
Cape Prov. 27 CPP. Hemicr.
**Schoenus bulbosus* L., Mant. Pl. 2: 178 (1771). *Isolepis bulbosa* (L.) Nees, Linnaea 7: 507 (1832). *Melancranis bulbosa* (L.) Kuntze, Revis. Gen. Pl. 2: 756 (1891).
Scirpus capensis Rottb., Descr. Icon. Rar. Pl.: 53 (1773).
Schoenus spicatus Thunb., Prodr. Pl. Cap.: 16 (1794).
Scirpus vaginatus Thunb., Prodr. Pl. Cap.: 17 (1794).
Scirpus bicapitatus Poir. in J.B.A.M.de Lamarck, Encycl. 6: 761 (1805).
Scirpus biceps Roem. & Schult., Syst. Veg. 2: 134 (1817).

Ficinia capillifolia (Schrad.) C.B.Clarke in T.A.Durand & H.Schinz, Consp. Fl. Afric. 5: 637 (1894).
Cape Prov. 27 CPP. Hemicr.
**Hemichlaena capillifolia* Schrad., Gött. Gel. Anz. 3: 2066 (1821). *Hypophialium capillifolium* (Schrad.) Nees, Linnaea 7: 530 (1832).

Ficinia capitella (Thunb.) Nees, Linnaea 10: 168 (1836).
SW. Cape Prov. 27 CPP. Hemicr.
**Schoenus capitellum* Thunb., Prodr. Pl. Cap.: 16 (1794). *Chaetospora capitellum* (Thunb.) A.Dietr., Sp. Pl. 2: 27 (1832). *Hypolepis capitellum* (Thunb.) Nees, Linnaea 7: 522 (1832). *Ficinia indica* var. *capitella* (Thunb.) H.Pfeiff., Revis. Ficinia: 39 (1921).

Ficinia cedarbergensis T.H.Arnold & Gordon-Gray, Bothalia 12: 405 (1978).
WSW. Cape Prov. 27 CPP. Hemicr.

Ficinia ciliata Boeck., Abh. Naturwiss. Vereine Bremen 7: 38 (1880). *Ficinia laevis* var. *ciliata* (Boeck.) H.Pfeiff., Revis. Ficinia: 33 (1921).
C. Madagascar. 29 MDG. Hemicr. or rhizome geophyte.

Ficinia cinnamomea C.B.Clarke in W.H.Harvey & auct. suc. (eds.), Fl. Cap. 7: 246 (1898).
Namibia to Cape Prov. 27 CPP NAM. Hemicr.

Ficinia clandestina (Steud.) Boeck., Linnaea 37: 82 (1871).
Ethiopia. 24 ETH. Hemicr.
**Cyperus clandestinus* Steud., Flora 25: 594 (1842). *Chamaexiphium clandestinum* (Steud.) Hochst., Flora 18: 102 (1844). *Melancranis clandestina* (Steud.) Kuntze, Revis. Gen. Pl. 2: 756 (1891).
Ficinia dregeana var. *abyssinica* H.Pfeiff., Revis. Ficinia: 45 (1920).

Ficinia compasbergensis Drège ex Steud., Syn. Pl. Glumac. 2: 124 (1855). *Melancranis comparbergensis* (Drège ex Steud.) Kuntze, Revis. Gen. Pl. 2: 756 (1891).
Cape Prov. 27 CPP. Hemicr.

Ficinia crinita (Poir.) B.L Burtt, Notes Roy. Bot. Gard. Edinburgh 43: 358 (1986).
Cape Prov. 27 CPP. Hemicr.
**Cyperus crinitus* Poir. in J.B.A.M.de Lamarck, Encycl. 6: 752 (1805). *Scirpus crinitus* (Poir.) Poir. in J.B.A.M.de Lamarck, Encycl. 6: 752 (1805). *Baeothryon crinitum* (Poir.) A.Dietr., Sp. Pl. 2: 94 (1832). *Ficinia oligantha* var. *crinita* (Poir.) J.Raynal, Adansonia, n.s., 14: 210 (1974), nom. illeg.

Ficinia dasystachys C.B.Clarke in W.H.Harvey & auct. suc. (eds.), Fl. Cap. 7: 251 (1898).
S. Africa. 27 CPP NAT TVL. Hemicr.

Ficinia deusta (P.J.Bergius) Levyns, J. S. African Bot. 13: 69 (1947).
SW. & S. Cape Prov. 27 CPP. Hemicr.
**Schoenus deustus* P.J.Bergius, Descr. Pl. Cap.: 10 (1767).
Scirpus trigynus L., Mant. Pl. 2: 180 (1771). *Melancranis trigyna* (L.) Kuntze, Revis. Gen. Pl. 2: 756 (1891). *Ficinia trigyna* (L.) Druce, Bot. Soc. Exch. Club Brit. Isles 1916: 623 (1917).
Scirpus bulbosus Rottb., Descr. Icon. Rar. Pl.: 46 (1773).
Schoenus scariosus Thunb., Prodr. Pl. Cap.: 16 (1794). *Melancranis scariosa* (Thunb.) Vahl, Enum. Pl. 2: 239 (1805). *Eleogiton scariosa* (Thunb.) A.Dietr., Sp. Pl. 2: 98 (1832). *Hypolepis scariosa* (Thunb.) Nees, Linnaea 7: 521 (1832). *Isolepis scariosa* (Thunb.) Nees, Linnaea 7: 501 (1832). *Ficinia scariosa* (Thunb.) Nees, Linnaea 9: 292 (1834).
Ficinia conifera Nees, Linnaea 10: 181 (1836).

Ficinia distans C.B.Clarke, Kew Bull., Addit. Ser. 8: 31 (1908).
Cape Prov. 27 CPP. Hemicr.

Ficinia dunensis Levyns, J. S. African Bot. 13: 68 (1947).
SW. & S. Cape Prov. 27 CPP. Hemicr.

Ficinia dura Turrill, Bull. Misc. Inform. Kew 1925: 70 (1925).
Cape Prov. 27 CPP. Hemicr.

Ficinia ecklonea (Steud.) Nees, Linnaea 9: 292 (1834).
Cape Prov. 27 CPP. Hemicr.
Scirpus gracilis Poir. in J.B.A.M.de Lamarck, Encycl. 6: 763 (1805). *Trichelostylis gracilis* (Poir.) Nees, Linnaea 7: 492 (1832). *Isolepis gracilis* (Poir.) Nees, Linnaea 8: 91 (1833), nom. illeg. *Melancranis gracilis* (Poir.) Nees, Linnaea 9: 287 (1834). *Ficinia poiretii* Kunth, Enum. Pl. 2: 255 (1837). *Isolepis poiretii* (Kunth) Steud., Syn. Pl. Glumac. 2: 98 (1855). *Scirpus wightianus* Boeck., Linnaea 36: 765 (1870). *Fimbristylis wightiana* (Boeck.) Beetle, Amer. Midl. Naturalist 41: 493 (1949), nom. illeg.
**Scirpus eckloneus* Steud., Flora 12: 148 (1829). *Isolepis steudelii* Schrad., Anal. Fl. Cap. 1: 19 (1832), nom. illeg. *Ficinia steudelii* (Schrad.) Nees ex Kunth, Enum. Pl. 2: 235 (1837), nom. illeg. *Melancranis ecklonea* (Steud.) Kuntze, Revis. Gen. Pl. 2: 756 (1891).
Isolepis ecklonii A.Dietr., Sp. Pl. 2: 116 (1832).
Isolepis laciniata Eckl. ex Kunth, Enum. Pl. 2: 255 (1837).

Ficinia elatior Levyns, J. S. African Bot. 13: 71 (1947).
SW. Cape Prov. 27 CPP. Hemicr.

Ficinia esterhuyseniae Muasya, S. African J. Bot. 71: 198 (2005).
SW. Cape Prov. 27 CPP. Hemicr.

Ficinia fascicularis Nees, Linnaea 10: 178 (1836). *Melancranis fascicularis* (Nees) Kuntze, Revis. Gen. Pl. 2: 756 (1891).
Cape Prov. to KwaZulu-Natal. 27 CPP NAT. Hemicr.

Ficinia fastigiata (Thunb.) Nees, Linnaea 10: 170 (1836).
Cape Prov. 27 CPP. Hemicr.
Schoenus filiformis Thunb., Prodr. Pl. Cap.: 16 (1794).
**Scirpus fastigiatus* Thunb., Prodr. Pl. Cap.: 18 (1794). *Melancranis fastigiata* (Thunb.) Kuntze, Revis. Gen. Pl. 2: 756 (1891).

Ficinia filiculmea B.L.Burtt, Notes Roy. Bot. Gard. Edinburgh 43: 358 (1986).
Cape Prov. to KwaZulu-Natal. 27 CPP NAT. Hemicr.

Ficinia filiformis (Lam.) Schrad., Anal. Fl. Cap. 1: 46 (1832).
SW. Cape Prov. 27 CPP. Hemicr.
**Schoenus filiformis* Lam., Tabl. Encycl. 1: 135 (1791). *Ficinia filiformis* var. *capillaris* Nees, Linnaea 10: 173 (1836), nom. illeg. *Ficinia capillaris* (Nees) Levyns, J. S. African Bot. 13: 69 (1947), nom. illeg. *Melancranis filiformis* (Lam.) Kuntze, Revis. Gen. Pl. 2: 756 (1891).
Ficinia filamentosa Nees, Linnaea 9: 292 (1834), nom. inval.
Isolepis fibrosa Nees, Linnaea 9: 292 (1834), nom. inval.
Ficinia filiformis f. *albovaginata* Nees, Linnaea 10: 173 (1836).
Ficinia filiformis f. *brevifolia* Nees, Linnaea 10: 173 (1836).
Ficinia filiformis f. *rubiginosa* Nees, Linnaea 10: 173 (1836).
Ficinia tenuifolia Kunth, Enum. Pl. 2: 257 (1837). *Isolepis ficinioides* Steud., Syn. Pl. Glumac. 2: 98 (1855), nom. illeg. *Melancranis tenuifolia* (Kunth) Kuntze, Revis. Gen. Pl. 2: 756 (1891).
Schoenus aggregatus Steud. ex Kunth, Enum. Pl. 2: 202 (1837).
Scirpus elatus Boeck., Linnaea 36: 694 (1870), nom. illeg.
Schoenus setifolius Benth., Fl. Austral. 7: 359 (1878).
Ficinia ludwigii Boeck., Flora 65: 15 (1882).
Ficinia capillaris Nees ex C.B.Clarke in T.A.Durand & H.Schinz, Consp. Fl. Afric. 5: 635 (1894), nom. inval.
Scirpus aggregatus Steud. ex C.B.Clarke in T.A.Durand & H.Schinz, Consp. Fl. Afric. 5: 644 (1894), nom. inval.
Ficinia latinux C.B.Clarke in W.H.Harvey & auct. suc. (eds.), Fl. Cap. 7: 252 (1898). *Ficinia tenuifolia* var. *latinux* (C.B.Clarke) H.Pfeiff., Revis. Ficinia: 25 (1921).

Ficinia rugulosa C.B.Clarke in W.H.Harvey & auct. suc. (eds.), Fl. Cap. 7: 241 (1898). *Ficinia filiformis* var. *rugulosa* (C.B.Clarke) H.Pfeiff., Revis. Ficinia: 19 (1921).
Ficinia filiformis f. *micrantha* H.Pfeiff., Revis. Ficinia: 18 (1921).

Ficinia gracilis Schrad., Anal. Fl. Cap. 1: 44 (1832).
Kenya to S. Africa. 25 KEN TAN 26+ 27 CPP LES NAT OFS TVL. Hemicr.
Isolepis commutata Nees, Linnaea 10: 161 (1836). *Ficinia commutata* (Nees) Kunth, Enum. Pl. 2: 256 (1837). *Melancranis commutata* (Nees) Kuntze, Revis. Gen. Pl. 2: 756 (1891).
Isolepis lineata Nees, Linnaea 10: 160 (1836), nom. illeg.
Scirpus gracilis Zeyh. ex Kunth, Enum. Pl. 2: 256 (1837), nom. inval.

Ficinia grandiflora T.H.Arnold & Gordon-Gray, Bothalia 12: 408 (1978).
SW. Cape Prov. 27 CPP. Hemicr.

Ficinia gydomontana T.H.Arnold & Gordon-Gray, Bothalia 12: 410 (1978).
SW. Cape Prov. 27 CPP. Hemicr.

Ficinia indica (Lam.) H.Pfeiff., Revis. Ficinia: 38 (1921).
SW. & S. Cape Prov. 27 CPP. Hemicr.
**Schoenus indicus* Lam., Encycl. 1: 740 (1785). *Melancranis indica* (Lam.) Kuntze, Revis. Gen. Pl. 2: 756 (1891).
Schoenus striatus Thunb., Prodr. Pl. Cap.: 16 (1794). *Chaetospora striata* (Thunb.) A.Dietr., Sp. Pl. 2: 27 (1832). *Ficinia striata* (Thunb.) Kunth, Enum. Pl. 2: 257 (1837).
Schoenus atratus Schrad., Gött. Gel. Anz. 3: 2069 (1821). *Hypolepis atrata* (Schrad.) Nees, Linnaea 8: 93 (1833). *Ficinia atrata* (Schrad.) Nees, Linnaea 10: 167 (1836).
Schoenus coronatus Steud., Flora 12: 136 (1829).
Schoenus subserratus Steud., Flora 12: 136 (1829).
Scirpus coronatus Steud., Flora: 136 (1829).
Ficinia setiformis Schrad., Anal. Fl. Cap. 1: 45 (1832).
Ficinia pallida Nees, Linnaea 9: 292 (1834), nom. inval.
Ficinia varia Boeck., Verh. Bot. Vereins Prov. Brandenburg 39: 47 (1888). *Ficinia indica* var. *boeckeleriana* H.Pfeiff., Revis. Ficinia: 39 (1921).
Ficinia lithosperma var. *compacta* C.B.Clarke in W.H. Harvey & auct. suc. (eds.), Fl. Cap. 7: 253 (1898).

Ficinia involuta Nees, Linnaea 10: 174 (1836). *Ficinia tristachya* var. *involuta* (Nees) H.Pfeiff., Revis. Ficinia: 21 (1921).
Cape Prov. 27 CPP. Hemicr.

Ficinia ixioides Nees, Linnaea 10: 180 (1836). *Melancranis ixioides* (Nees) Kuntze, Revis. Gen. Pl. 2: 756 (1891).
SW. & S. Cape Prov. 27 CPP. Hemicr.

subsp. ***glabra*** T.H.Arnold & Gordon-Gray, Bothalia 12: 417 (1978).
WSW. Cape Prov. 27 CPP. Hemicr.

subsp. ***ixioides***
SW. & S. Cape Prov. 27 CPP. Hemicr.

Ficinia laciniata (Thunb.) Nees, Linnaea 10: 177 (1836).
Cape Prov. to KwaZulu-Natal. 27 CPP. Hemicr.
**Scirpus laciniatus* Thunb., Prodr. Pl. Cap.: 17 (1794). *Hypolepis composita* Nees, Linnaea 7: 525 (1832), nom. illeg. *Ficinia composita* (Nees) Nees, Linnaea 9: 292 (1834), nom. illeg. *Melancranis laciniata* (Thunb.) Kuntze, Revis. Gen. Pl. 2: 756 (1891).

Ficinia laevis Nees, Linnaea 10: 171 (1836). *Melancranis laevis* (Nees) Kuntze, Revis. Gen. Pl. 2: 756 (1891).
Cape Prov. 27 CPP. Hemicr.
Ficinia praeusta Nees, Linnaea 9: 292 (1834), nom. inval. *Isolepis praeusta* C.B.Clarke in T.A.Durand & H.Schinz, Consp. Fl. Afric. 5: 639 (1894).
Ficinia dregeana Kunth, Enum. Pl. 2: 259 (1837).

Ficinia lateralis (Vahl) Kunth, Enum. Pl. 2: 254 (1837).
Cape Prov. 27 CPP. Hemicr.
**Schoenus lateralis* Vahl, Enum. Pl. 2: 211 (1805). *Schoenidium laterale* (Vahl) Nees, Linnaea 9: 292 (1834). *Schoenidium laterale* (Vahl) Nees, Linnaea 9: 291 (1834). *Melancranis lateralis* (Vahl) Kuntze, Revis. Gen. Pl. 2: 756 (1891).
Ficinia aphylla Nees in S.L.Endlicher, Atacta Bot.: 3 (1833).
Ficinia aphylla Nees in S.L.Endlicher, Prodr. Fl. Norfolk.: 23 (1833).
Scirpus truncatus Sieber ex Nees in S.L.Endlicher, Prodr. Fl. Norfolk.: 23 (1833), nom. inval.
Isolepis arcuata Eckl. ex Kunth, Enum. Pl. 2: 254 (1837).
Scirpus truncatus Sieber ex Kunth, Enum. Pl. 2: 254 (1837), nom. inval.

Ficinia latifolia T.H.Arnold & Gordon-Gray, Bothalia 12: 412 (1978).
SW. Cape Prov. 27 CPP. Hemicr.

Ficinia leiocarpa Nees, Linnaea 10: 171 (1836). *Melancranis leiocarpa* (Nees) Kuntze, Revis. Gen. Pl. 2: 756 (1891).
Cape Prov. 27 CPP. Hemicr.
Ficinia semibracteata Nees, Linnaea 9: 292 (1834), nom. nud.
Ficinia punctata Hochst., Flora 28: 758 (1845).

Ficinia levynsiae T.H.Arnold & Gordon-Gray, Bothalia 12: 407 (1978).
SW. Cape Prov. 27 CPP. Hemicr.

Ficinia × lucida C.B.Clarke in W.H.Harvey & auct. suc. (eds.), Fl. Cap. 7: 254 (1898). *F. cedarbergensis* × *F. ixioides* subsp. *glabra*. *Ficinia lithosperma* var. *lucida* (C.B.Clarke) H.Pfeiff., Revis. Ficinia: 40 (1921).
Cape Prov. 27 CPP. Hemicr.

Ficinia macowanii C.B.Clarke in W.H.Harvey & auct. suc. (eds.), Fl. Cap. 7: 247 (1898). *Melancranis macowanii* (C.B.Clarke) Kuntze, Revis. Gen. Pl. 3(2): 335 (1898).
Cape Prov. 27 CPP. Hemicr.

Ficinia micrantha C.B.Clarke in W.H.Harvey & auct. suc. (eds.), Fl. Cap. 7: 240 (1898). *Ficinia tristachya* subvar. *micrantha* (C.B.Clarke) H.Pfeiff., Revis. Ficinia: 21 (1921).
Cape Prov. 27 CPP. Hemicr.

Ficinia minutiflora C.B.Clarke in W.H.Harvey & auct. suc. (eds.), Fl. Cap. 7: 759 (1900). *Ficinia tristachya* var. *minutiflora* (C.B.Clarke) H.Pfeiff., Revis. Ficinia: 21 (1921).
Cape Prov. 27 CPP. Hemicr.

Ficinia mucronata C.B.Clarke in É.A.J.de Wildeman, Pl. Nov. Hort. Then. 1: 29 (1904).
Cape Prov. 27 CPP. Hemicr.

Ficinia nana B.L.Burtt, Notes Roy. Bot. Gard. Edinburgh 43: 360 (1986).
Lesotho to KwaZulu-Natal. 27 LES NAT. Hemicr.

Ficinia nigrescens (Schrad.) J.Raynal, Adansonia, n.s., 14: 211 (1974).
SW. Namibia to Cape Prov. 27 CPP NAM. Hemicr.
Scirpus filiformis Burm.f., Fl. Indica, Prodr. Fl. Cap.: 3 (1768). *Schoenus filiformis* (Burm.f.) Schult. in J.J.Roemer & J.A.Schultes, Mant. 2: 528 (1824). *Ficinia filiformis* (Burm.f.) Beetle, Amer. Midl. Naturalist 41: 467 (1949), nom. illeg.
Scirpus scariosus Roem. & Schult., Syst. Veg. 2: 61 (1817).
**Melancranis nigrescens* Schrad., Gött. Gel. Anz. 3: 2072 (1821). *Hypolepis nigrescens* (Schrad.) Nees, Linnaea 7: 522 (1832). *Ficinia bracteata* Boeck., Linnaea 37: 83 (1871), nom. illeg.
Hypolepis nigrescens var. *filiformis* Nees, Linnaea 10: 141 (1836).
Melancranis rigidula Nees, Linnaea 10: 141 (1836). *Ficinia bracteata* f. *rigidula* (Nees) H.Pfeiff., Revis. Ficinia: 42 (1921).
Schoenus filiformis Willd. ex Kunth, Enum. Pl. 2: 265 (1837).
Ficinia bracteata f. *nigratofuscata* H.Pfeiff., Revis. Ficinia: 42 (1921).

Ficinia nodosa (Rottb.) Goetgh., Muasya & D.A.Simpson, Novon 10: 133 (2000).
St. Helena, S. Africa, S. & E. Australia, New Zealand, St. Paul I., Juan Fernández Is. 27 CPP NAT 28 STH 50 NFK NSW SOA TAS VIC WAU 51 CTM KER NZN NZS 85 JNF 90 ASP. Hemicr.
**Scirpus nodosus* Rottb., Descr. Icon. Rar. Pl.: 52 (1773). *Isolepis nodosa* (Rottb.) R.Br., Prodr.: 221 (1810). *Holoschoenus nodosus* (Rottb.) A.Dietr., Sp. Pl. 2: 165 (1832). *Scirpoides nodosus* (Rottb.) L.C.Br. ex R.T.F. Clifton, St. Helena: A Highly Critical Flora: 81 (1996).
Scirpus gracilis Rudge, Trans. Linn. Soc. London 10: 289 (1811), nom. illeg. *Isolepis gracilis* (Rudge) Sweet, Hort. Brit.: 438 (1826), nom. illeg.
Fimbristylis textilis Roxb. in A.Beatson, Tracts St. Helena: 309 (1816).
Scirpus globosus Spreng., Syst. Veg. 4(2): 27 (1827).
Ficinia guttata Endl., Prodr. Fl. Norfolk.: 23 (1833).
Scirpus glutinosus Banks & Sol. ex Hook.f., Fl. Nov.-Zel. 1: 270 (1853).
Isolepis monocephala Steud., Syn. Pl. Glumac. 2: 97 (1855).
Isolepis holoschoenus Melliss, St. Helena: 343 (1875).
Scirpus nodosus var. *macrostachya* Benth., Fl. Austral. 7: 332 (1878).
Scirpus atropurpureovaginatus Boeck., Flora 65: 14 (1882).
Fimbristylis schinzii Boeck., Verh. Bot. Vereins Prov. Brandenburg 29: 47 (1888).

Ficinia oligantha (Steud.) J.Raynal, Adansonia, n.s., 14: 210 (1974).
Kenya to S. Africa. 25 KEN TAN UGA 26+ 27 CPP LES TVL. Hemicr.
**Scirpus oliganthus* Steud., Flora 12: 146 (1829). *Isolepis oligantha* (Steud.) A.Dietr., Sp. Pl. 2: 103 (1832). *Schoenus oliganthos* (Steud.) Kunth, Enum. Pl. 2: 253 (1837).
Isolepis oliganthes Nees, Linnaea 7: 503 (1832).

Ficinia pallens (Schrad.) Nees, Linnaea 10: 169 (1836).
SW. Cape Prov. 27 CPP. Hemicr.
**Schoenus pallens* Schrad., Anal. Fl. Cap. 1: 25 (1832).

var. ***lithosperma*** (Boeck.) T.H.Arnold & Gordon-Gray, Bothalia 12: 418 (1978).
SW. Cape Prov. 27 CPP. Hemicr.
**Ficinia lithosperma* Boeck., Linnaea 37: 72 (1871). *Melancranis lithosperma* (Boeck.) Kuntze, Revis. Gen. Pl. 2: 756 (1891).
Ficinia bolusii Boeck., Bot. Jahrb. Syst. 5: 506 (1884). *Melancranis bolusii* (Boeck.) Kuntze, Revis. Gen. Pl. 2: 756 (1891). *Ficinia lithosperma* f. *major* H.Pfeiff., Revis. Ficinia: 40 (1921).

var. ***pallens***
SW. Cape Prov. 27 CPP. Hemicr.

Ficinia paradoxa (Schrad.) Nees, Linnaea 10: 178 (1836).
Cape Prov. 27 CPP. Hemicr.
**Isolepis paradoxa* Schrad., Gött. Gel. Anz. 3: 2068 (1821). *Ficinia laciniata* var. *paradoxa* (Schrad.) Nees, Linnaea 10: 178 (1836). *Melancranis paradoxa* (Schrad.) Kuntze, Revis. Gen. Pl. 2: 756 (1891).
Isolepis eckloniana Nees, Linnaea 7: 506 (1832).
Ficinia fimbriata C.A.Mey. ex C.B.Clarke in T.A.Durand & H.Schinz, Consp. Fl. Afric. 5: 641 (1894), nom. inval.

Ficinia petrophylla T.H.Arnold & Gordon-Gray, Bothalia 12: 408 (1978).
SSW. Cape Prov. 27 CPP. Hemicr.

Ficinia pinguior C.B.Clarke in W.H.Harvey & auct. suc. (eds.), Fl. Cap. 7: 254 (1898). *Ficinia lithosperma* var. *pinguior* (C.B.Clarke) H.Pfeiff., Revis. Ficinia: 40 (1921).
SW. Cape Prov. 27 CPP. Hemicr.

Ficinia polystachya Levyns, J. S. African Bot. 13: 66 (1947).
Cape Prov. 27 CPP. Hemicr.
**Hemichlaena angustifolia* var. *fascicularis* Nees, Linnaea 10: 129 (1836). *Hemichlaena fascicularis* (Nees) Steud., Syn. Pl. Glumac. 2: 2 (1854).
Hemichlaena angustifolia var. *polystachya* Kunth, Enum. Pl. 2: 330 (1837).

Ficinia praemorsa Nees, Linnaea 10: 167 (1836). *Melancranis praemorsa* (Nees) Kuntze, Revis. Gen. Pl. 2: 756 (1891).
Cape Prov. 27 CPP. Hemicr.

Ficinia pusilla C.B.Clarke in W.H.Harvey & auct. suc. (eds.), Fl. Cap. 7: 239 (1898).
Cape Prov. 27 CPP. Hemicr.
Ficinia bergiana var. *capillaris* C.B.Clarke in T.A. Durand & H.Schinz, Consp. Fl. Afric. 5: 636 (1894).

Ficinia pygmaea Boeck., Linnaea 37: 82 (1871).
SW. Cape Prov. 27 CPP. Hemicr.
Chamaexiphium dregeanum Steud., Syn. Pl. Glumac. 2: 59 (1854). *Melancranis dregeana* (Steud.) Kuntze, Revis. Gen. Pl. 2: 756 (1891). *Ficinia dregeana* (Steud.) H.Pfeiff., Revis. Ficinia: 45 (1921), nom. illeg.
Ficinia limosa Levyns, J. S. African Bot. 13: 68 (1947).

Ficinia quinquangularis Boeck., Linnaea 37: 63 (1871). *Melancranis quinquangularis* (Boeck.) Kuntze, Revis. Gen. Pl. 2: 756 (1891).
Cape Prov. 27 CPP. Hemicr.

Ficinia radiata (L.f.) Kunth, Enum. Pl. 2: 260 (1837).
SW. Cape Prov. 27 CPP. Hemicr.
**Schoenus radiatus* L.f., Suppl. Pl.: 101 (1782). *Scirpus radiatus* (L.f.) Thunb., Prodr. Pl. Cap.: 18

(1794). *Isolepis radiata* (L.f.) Roem. & Schult., Syst. Veg. 2: 113 (1817). *Sickmannia radiata* (L.f.) Nees, Linnaea 9: 292 (1834). *Melancranis radiata* (L.f.) Kunth, Enum. Pl. 2: 260 (1837), nom. illeg.

Ficinia ramosissima Kunth, Enum. Pl. 2: 262 (1837). *Acrolepis ramosissima* (Kunth) Boeck., Linnaea 37: 89 (1871).
Cape Prov. 27 CPP. Hemicr.

Ficinia repens Kunth, Enum. Pl. 2: 255 (1837). *Melancranis repens* (Kunth) Kuntze, Revis. Gen. Pl. 2: 756 (1891).
Cape Prov. 27 CPP. Hemicr.
Isolepis repens Nees, Linnaea 10: 158 (1836), nom. nud.

Ficinia rigida Levyns, J. S. African Bot. 13: 66 (1947).
Cape Prov. 27 CPP. Hemicr.

Ficinia secunda (Vahl) Kunth, Enum. Pl. 2: 262 (1837).
Cape Prov. 27 CPP. Hemicr.
Schoenus bulbosus Thunb., Prodr. Pl. Cap.: 16 (1794).
**Schoenus secundus* Vahl, Enum. Pl. 2: 215 (1805). *Pleurachne secunda* (Vahl) Schrad., Commentat. Soc. Regiae Sci. Gott. Recent. 7: 147 (1832). *Pleurachne sieberi* Schrad., Commentat. Soc. Regiae Sci. Gott. Recent. 7: 148 (1832). *Melancranis secunda* (Vahl) Kuntze, Revis. Gen. Pl. 2: 756 (1891).
Schoenus dispar Spreng., Neue Entdeck. Pflanzenk. 3: 8 (1822). *Chaetospora dispar* (Spreng.) A.Dietr., Sp. Pl. 2: 32 (1832). *Ficinia dispar* (Spreng.) Fourc., Trans. Roy. Soc. South Africa 21: 76 (1932).

Ficinia stolonifera Boeck., Linnaea 37: 60 (1871).
S. Trop. & S. Africa. 26+ 27 CPP LES NAT OFS TVL. Hemicr.
Ficinia filiformis Nees, Linnaea 10: 173 (1836), nom. illeg.
Ficinia filiformis f. *contorta* Nees, Linnaea 10: 173 (1836). *Melancranis contorta* (Nees) Kuntze, Revis. Gen. Pl. 2: 756 (1891). *Ficinia contorta* (Nees) H.Pfeiff., Revis. Ficinia: 19 (1921).
Scirpus leucocoleus K.Schum. in H.G.A.Engler, Pflanzenw. Ost-Afrikas, C: 125 (1895).

Ficinia sylvatica Kunth, Enum. Pl. 2: 254 (1837). *Melancranis sylvatica* (Kunth) Kuntze, Revis. Gen. Pl. 2: 756 (1891).
Cape Prov. 27 CPP. Hemicr.
Scirpus trispicatus L.f., Suppl. Pl.: 103 (1782). *Melancranis trispicata* (L.f.) Kuntze, Revis. Gen. Pl. 3(2): 335 (1898). *Ficinia trispicata* (L.f.) Druce, Bot. Soc. Exch. Club Brit. Isles 1916: 623 (1917).
Scirpus sparmannii Lam., Tabl. Encycl. 1: 140 (1791).
Ficinia leiocarpa Zeyh. ex C.B.Clarke in T.A.Durand & H.Schinz, Consp. Fl. Afric. 5: 643 (1894), nom. inval.

Ficinia trichodes (Schrad.) Benth. & Hook.f., Gen. Pl. 3: 1053 (1883).
Cape Prov. 27 CPP. Hemicr.
**Acrolepis trichodes* Schrad., Commentat. Soc. Regiae Sci. Gott. Recent. 7: 142 (1832).
Acrolepis ferruginea Nees ex Boeck., Linnaea 37: 91 (1871). *Ficinia ferruginea* (Nees ex Boeck.) C.B.Clarke in T.A.Durand & H.Schinz, Consp. Fl. Afric. 5: 638 (1894).

Ficinia tristachya (Rottb.) Nees, Linnaea 9: 292 (1834).
Cape Prov. 27 CPP. Hemicr.
**Scirpus tristachyos* Rottb., Descr. Icon. Rar. Pl.: 48 (1773). *Isolepis tristachya* (Rottb.) Roem. & Schult., Syst. Veg. 2: 110 (1817). *Melancranis tristachya* (Rottb.) Kuntze, Revis. Gen. Pl. 2: 756 (1891).
Scirpus pseudoschoenus Steud., Flora 12: 147 (1829). *Isolepis pseudoschoenus* (Steud.) A.Dietr., Sp. Pl. 2: 103 (1832). *Ficinia pseudoschoenus* (Steud.) Fourc., Trans. Roy. Soc. South Africa 21: 76 (1932).
Ficinia bergiana Kunth, Enum. Pl. 2: 254 (1837). *Ficinia filiformis* var. *bergiana* (Kunth) H.Pfeiff., Revis. Ficinia: 18 (1921).
Ficinia tribracteata Boeck., Linnaea 37: 62 (1871). *Melancranis tribracteata* (Boeck.) Kuntze, Revis. Gen. Pl. 2: 756 (1891). *Ficinia tristachya* var. *tribracteata* (Boeck.) H.Pfeiff., Revis. Ficinia: 21 (1921).
Ficinia tristachya var. *glaucescens* Boeck., Linnaea 37: 61 (1871).
Ficinia subacuta C.B.Clarke in W.H.Harvey & auct. suc. (eds.), Fl. Cap. 7: 244 (1898).
Ficinia tristachya var. *bibracteata* H.Pfeiff., Revis. Ficinia: 21 (1921).

Ficinia trollii (Kük.) Muasya & D.A.Simpson, Novon 10: 133 (2000).
Tanzania to S. Trop. Africa. 25 TAN 26 MOZ ZIM. Hemicr.
**Scirpus trollii* Kük., Feddes Repert. Spec. Nov. Regni Veg. 53: 72 (1944). *Isolepis trollii* (Kük.) Lye, Bot. Not. 130: 313 (1977).

Ficinia truncata (Thunb.) Schrad., Anal. Fl. Cap. 1: 43 (1832).
Cape Prov. 27 CPP. Hemicr.
**Scirpus truncatus* Thunb., Prodr. Pl. Cap.: 17 (1794). *Isolepis truncata* (Thunb.) Nees, Linnaea 7: 508 (1832). *Melancranis truncata* (Thunb.) Kuntze, Revis. Gen. Pl. 2: 756 (1891).

Ficinia undosa B.L.Burtt, Notes Roy. Bot. Gard. Edinburgh 43: 360 (1986).
S. Africa. 27 LES NAT OFS. Hemicr.

Ficinia zeyheri Boeck., Linnaea 37: 58 (1871). *Melancranis zeyheri* (Boeck.) Kuntze, Revis. Gen. Pl. 2: 756 (1891).
Cape Prov. 27 CPP. Hemicr.

Synonyms:

Ficinia acuminata Kunth = ***Ficinia acuminata*** (Nees) Nees
Ficinia ambigua Steud. = ***Fimbristylis dichotoma*** (L.) Vahl subsp. ***dichotoma***
Ficinia anomala (Retz.) Steud. = ***Hypolytrum nemorum*** (Vahl) Spreng. subsp. ***nemorum***
Ficinia antarctica Nees = ***Isolepis diabolica*** (Steud.) Schrad.
Ficinia aphylla Nees = ***Ficinia lateralis*** (Vahl) Kunth
Ficinia aphylla Nees = ***Ficinia lateralis*** (Vahl) Kunth
Ficinia argyropus var. *minor* Boeck. = ***Ficinia argyropus*** Nees
Ficinia atrata (Schrad.) Nees = ***Ficinia indica*** (Lam.) H.Pfeiff.
Ficinia bergiana Kunth = ***Ficinia tristachya*** (Rottb.) Nees
Ficinia bergiana var. *capillaris* C.B.Clarke = ***Ficinia pusilla*** C.B.Clarke
Ficinia bicolor (Carmich.) H.Pfeiff. = ***Isolepis bicolor*** Carmich.
Ficinia bolusii Boeck. = ***Ficinia pallens*** var. ***lithosperma*** (Boeck.) T.H.Arnold & Gordon-Gray
Ficinia bracteata Boeck. = ***Ficinia nigrescens*** (Schrad.) J.Raynal
Ficinia bracteata f. *nigratofuscata* H.Pfeiff. = ***Ficinia nigrescens*** (Schrad.) J.Raynal

Ficinia bracteata f. *rigidula* (Nees) H.Pfeiff. = ***Ficinia nigrescens*** (Schrad.) J.Raynal
Ficinia brevifolia f. *atrocastanea* C.B.Clarke = ***Ficinia brevifolia*** Nees ex Kunth
Ficinia brevifolia f. *debilior* C.B.Clarke ex H.Pfeiff. = ***Ficinia brevifolia*** Nees ex Kunth
Ficinia canaliculata H.Pfeiff. = ***Hellmuthia membranacea*** (Thunb.) R.W.Haines & Lye
Ficinia capillaris Nees ex C.B.Clarke = ***Ficinia filiformis*** (Lam.) Schrad.
Ficinia capillaris (Nees) Levyns = ***Ficinia filiformis*** (Lam.) Schrad.
Ficinia caricoidea (Steud.) H.Pfeiff. = ***Ficinia angustifolia*** (Schrad.) C.B.Clarke
Ficinia commutata (Nees) Kunth = ***Ficinia gracilis*** Schrad.
Ficinia composita (Nees) Nees = ***Ficinia laciniata*** (Thunb.) Nees
Ficinia conifera Nees = ***Ficinia deusta*** (P.J.Bergius) Levyns
Ficinia contexta (Nees) Nees = ***Bulbostylis contexta*** (Nees) Bodard
Ficinia contorta (Nees) H.Pfeiff. = ***Ficinia stolonifera*** Boeck.
Ficinia costata (Hochst. ex A.Rich.) H.Pfeiff. = ***Isolepis costata*** Hochst. ex A.Rich.
Ficinia decidua H.Pfeiff. = [27 CPP]
Ficinia dispar (Spreng.) Fourc. = ***Ficinia secunda*** (Vahl) Kunth
Ficinia distans C.B.Clarke = [27]
Ficinia dregeana Kunth = ***Ficinia laevis*** Nees
Ficinia dregeana (Steud.) H.Pfeiff. = ***Ficinia pygmaea*** Boeck.
Ficinia dregeana var. *abyssinica* H.Pfeiff. = ***Ficinia clandestina*** (Steud.) Boeck.
Ficinia elongata Boeck. = ***Ficinia acuminata*** (Nees) Nees
Ficinia ferruginea (Nees ex Boeck.) C.B.Clarke = ***Ficinia trichodes*** (Schrad.) Benth. & Hook.f.
Ficinia filamentosa Nees = ***Ficinia filiformis*** (Lam.) Schrad.
Ficinia filiformis (Burm.f.) Beetle = ***Ficinia nigrescens*** (Schrad.) J.Raynal
Ficinia filiformis Nees = ***Ficinia stolonifera*** Boeck.
Ficinia filiformis f. *albovaginata* Nees = ***Ficinia filiformis*** (Lam.) Schrad.
Ficinia filiformis var. *bergiana* (Kunth) H.Pfeiff. = ***Ficinia tristachya*** (Rottb.) Nees
Ficinia filiformis f. *brevifolia* Nees = ***Ficinia filiformis*** (Lam.) Schrad.
Ficinia filiformis var. *capillaris* Nees = ***Ficinia filiformis*** (Lam.) Schrad.
Ficinia filiformis f. *contorta* Nees = ***Ficinia stolonifera*** Boeck.
Ficinia filiformis f. *micrantha* H.Pfeiff. = ***Ficinia filiformis*** (Lam.) Schrad.
Ficinia filiformis f. *rubiginosa* Nees = ***Ficinia filiformis*** (Lam.) Schrad.
Ficinia filiformis var. *rugulosa* (C.B.Clarke) H.Pfeiff. = ***Ficinia filiformis*** (Lam.) Schrad.
Ficinia fimbriata C.A.Mey. ex C.B.Clarke = ***Ficinia paradoxa*** (Schrad.) Nees
Ficinia foliaceobracteata H.Pfeiff. = ***Cyperus cephalotes*** Vahl
Ficinia guttata Endl. = ***Ficinia nodosa*** (Rottb.) Goetgh., Muasya & D.A.Simpson
Ficinia ignorata Boeck. = ***Ficinia acuminata*** (Nees) Nees
Ficinia indica var. *boeckeleriana* H.Pfeiff. = ***Ficinia indica*** (Lam.) H.Pfeiff.
Ficinia indica var. *capitella* (Thunb.) H.Pfeiff. = ***Ficinia capitella*** (Thunb.) Nees
Ficinia kunthiana Boeck. = ***Ficinia acuminata*** (Nees) Nees
Ficinia laciniata var. *paradoxa* (Schrad.) Nees = ***Ficinia paradoxa*** (Schrad.) Nees
Ficinia laevis var. *ciliata* (Boeck.) H.Pfeiff. = ***Ficinia ciliata*** Boeck.
Ficinia latinux C.B.Clarke = ***Ficinia filiformis*** (Lam.) Schrad.
Ficinia leiocarpa Zeyh. ex C.B.Clarke = ***Ficinia sylvatica*** Kunth
Ficinia ligulata C.B.Clarke = [27]
Ficinia limosa Levyns = ***Ficinia pygmaea*** Boeck.
Ficinia lipocarphioides Kük. = ***Alinula lipocarphioides*** (Kük.) J.Raynal
Ficinia lithosperma Boeck. = ***Ficinia pallens*** var. ***lithosperma*** (Boeck.) T.H.Arnold & Gordon-Gray
Ficinia lithosperma var. *compacta* C.B.Clarke = ***Ficinia indica*** (Lam.) H.Pfeiff.
Ficinia lithosperma var. *lucida* (C.B.Clarke) H.Pfeiff. = ***Ficinia × lucida*** C.B.Clarke
Ficinia lithosperma f. *major* H.Pfeiff. = ***Ficinia pallens*** var. ***lithosperma*** (Boeck.) T.H.Arnold & Gordon-Gray
Ficinia lithosperma var. *pinguior* (C.B.Clarke) H.Pfeiff. = ***Ficinia pinguior*** C.B.Clarke
Ficinia longifolia (Nees) C.B.Clarke = ***Ficinia angustifolia*** (Schrad.) C.B.Clarke
Ficinia ludwigii Boeck. = ***Ficinia filiformis*** (Lam.) Schrad.
Ficinia marginata (Thunb.) Fourc. = ***Isolepis marginata*** (Thunb.) A.Dietr.
Ficinia membranacea (Thunb.) Kunth = ***Hellmuthia membranacea*** (Thunb.) R.W.Haines & Lye
Ficinia micrantha Boeck. ex C.B.Clarke = ***Ficinia angustifolia*** (Schrad.) C.B.Clarke
Ficinia monticola Kunth = [27 CPP]
Ficinia nuda Boeck. = [27 CPP]
Ficinia oligantha var. *crinita* (Poir.) J.Raynal = ***Ficinia crinita*** (Poir.) B.L Burtt
Ficinia pallida Nees = ***Ficinia indica*** (Lam.) H.Pfeiff.
Ficinia paradoxa var. *argyropus* (Nees) C.B.Clarke = ***Ficinia argyropus*** Nees
Ficinia poiretii Kunth = ***Ficinia ecklonea*** (Steud.) Nees
Ficinia praeusta Nees = ***Ficinia laevis*** Nees
Ficinia pseudoschoenus (Steud.) Fourc. = ***Ficinia tristachya*** (Rottb.) Nees
Ficinia pulchella Kunth = ***Schoenoplectus pulchellus*** (Kunth) J.Raynal
Ficinia punctata Hochst. = ***Ficinia leiocarpa*** Nees
Ficinia rehmanniana Boeck. ex C.B.Clarke = ***Ficinia angustifolia*** (Schrad.) C.B.Clarke
Ficinia rugulosa C.B.Clarke = ***Ficinia filiformis*** (Lam.) Schrad.
Ficinia scariosa (Thunb.) Nees = ***Ficinia deusta*** (P.J.Bergius) Levyns
Ficinia schinziana Boeck. = ***Isolepis diabolica*** (Steud.) Schrad.
Ficinia schlechteriana H.Pfeiff. = [27]
Ficinia semibracteata Nees = ***Ficinia leiocarpa*** Nees
Ficinia setiformis Schrad. = ***Ficinia indica*** (Lam.) H.Pfeiff.
Ficinia steudelii (Schrad.) Nees ex Kunth = ***Ficinia ecklonea*** (Steud.) Nees
Ficinia striata (Thunb.) Kunth = ***Ficinia indica*** (Lam.) H.Pfeiff.
Ficinia subacuta C.B.Clarke = ***Ficinia tristachya*** (Rottb.) Nees

Ficinia tenuifolia Kunth = ***Ficinia filiformis*** (Lam.) Schrad.
Ficinia tenuifolia var. *latinux* (C.B.Clarke) H.Pfeiff. = ***Ficinia filiformis*** (Lam.) Schrad.
Ficinia tenuis (Eckl. ex Kunth) C.B.Clarke =
Ficinia thyrsoidea H.Pfeiff. = [27 CPP]
Ficinia tribracteata Boeck. = ***Ficinia tristachya*** (Rottb.) Nees
Ficinia trigyna (L.) Druce = ***Ficinia deusta*** (P.J.Bergius) Levyns
Ficinia trinkleriana H.Pfeiff. = [27 CPP]
Ficinia trispicata (L.f.) Druce = ***Ficinia sylvatica*** Kunth
Ficinia tristachya var. *bibracteata* H.Pfeiff. = ***Ficinia tristachya*** (Rottb.) Nees
Ficinia tristachya var. *glaucescens* Boeck. = ***Ficinia tristachya*** (Rottb.) Nees
Ficinia tristachya var. *involuta* (Nees) H.Pfeiff. = ***Ficinia involuta*** Nees
Ficinia tristachya subvar. *micrantha* (C.B.Clarke) H.Pfeiff. = ***Ficinia micrantha*** C.B.Clarke
Ficinia tristachya var. *minutiflora* (C.B.Clarke) H.Pfeiff. = ***Ficinia minutiflora*** C.B.Clarke
Ficinia tristachya var. *tribracteata* (Boeck.) H.Pfeiff. = ***Ficinia tristachya*** (Rottb.) Nees
Ficinia varia Boeck. = ***Ficinia indica*** (Lam.) H.Pfeiff.

***Unplaced Names*:**
Ficinia decidua H.Pfeiff., Herbarium 54: 34 (1920). = [27 CPP]
Ficinia distans C.B.Clarke in É.A.J.de Wildeman, Pl. Nov. Hort. Then. 1: 30 (1904). = [27]
Ficinia ligulata C.B.Clarke, Kew Bull., Addit. Ser. 8: 31 (1908). = [27]
Ficinia monticola Kunth, Enum. Pl. 2: 261 (1837). = [27 CPP]
Ficinia nuda Boeck., Linnaea 38: 383 (1874). = [27 CPP]
Ficinia schlechteriana H.Pfeiff., Herbarium 54: 33 (1920). = [27]
Ficinia thyrsoidea H.Pfeiff., Herbarium 54: 33 (1920). = [27 CPP]
Ficinia trinkleriana H.Pfeiff., Herbarium 54: 34 (1920). = [27 CPP]

Fimbristylis

Fimbristylis Vahl, Enum. Pl. 2: 285 (1805), nom. cons.
Cosmopolitan. 11 SWI 12 FRA POR SPA 13 ALB BUL GRE ITA KRI ROM SIC TUE YUG 14 UKR 20 ALG EGY LBY MOR 21 CNY 22 BEN BKN GAM GHA GNB GUI IVO LBR MGA MLI MTN NGA NGR SEN SIE TOG 23 BUR CAF CMN CON EQG GAB GGI RWA ZAI 24 CHA SOM SUD 25 KEN TAN UGA 26 MOZ ZAM ZIM 27 BOT CPP LES NAM NAT OFS SWZ TVL 29 ALD COM MAU MDG REU SEY 31+ KAM KHA PRM 32 KAZ KGZ NWG TKM TZK UZB 33 TCS 34 AFG CYP IRN IRQ LBS PAL SIN TUR 35 GST SAU YEM 36 CHC CHH CHM CHN CHS CHT 38 JAP KOR KZN NNS OGA TAI 40 ASS BAN EHM IND NEP PAK SRL WHM 41 AND CBD LAO MYA NCB THA VIE 42 BOR CKI JAW LSI MLY MOL PHI SUL SUM XMS 43 BIS NWG SOL 50 NFK NSW NTA QLD SOA TAS VIC WAU 51 NZN 60 FIJ GIL HBI NUE NWC TOK TON 61 COO MRQ SCI TUA 62 CRL MRN MRS WAK 63 HAW 72 NBR ONT QUE 73 COL 74 ILL IOW KAN MIN MSO NEB OKL SDA WIS 75 CNT INI MAI MAS MIC NWH NWJ NWY OHI PEN RHO VER WVA 76 ARI CAL NEV UTA 77 NWM TEX 78 ALA ARK DEL FLA GEO KTY LOU MRY MSI NCA NCS SCA TEN VRG WDC 79 MXC MXE MXG MXN MXS MXT 80 BLZ COS CPI ELS GUA HON NIC PAN 81 BAH BER CAY CUB DOM HAI JAM LEE NLA PUE TRT WIN 82 FRG GUY SUR VEN 83 BOL CLM ECU PER 84+ BZC BZE BZL BZN 85 AGE AGW PAR URU.
303 Species
Abildgaardia Vahl, Enum. Pl. 2: 296 (1805).
Iria R.Hedw., Gen. Pl.: 360 (1806).
Echinolytrum Desv., J. Bot. (Desvaux) 1: 20 (1808).
Aplostemon Raf., J. Phys. Chim. Hist. Nat. Arts 89: 105 (1819).
Dichostylis P.Beauv. in T.G.Lestiboudois, Essai Cypér.: 39 (1819).
Trichelostylis T.Lestib., Essai Cypér.: 40 (1819).
Abildgardia Rchb., Consp. Regn. Veg.: 56 (1828), orth. var.
Gussonea J.Presl & C.Presl, Reliq. Haenk. 1: 183 (1828).
Pogonostylis Bertol., Fl. Ital. 1: 312 (1833).
Campylostachys E.Mey. in J.F.Drège, Zwei Pflanzengeogr. Dokum.: 83 (1843).
Pseudocyperus Steud., Syn. Pl. Glumac. 2: 108 (1855).
Mischospora Boeck., Flora 43: 113 (1860).
Iriha Kuntze, Revis. Gen. Pl. 2: 751 (1891), orth. var.
Tylocarya Nelmes, Kew Bull. 4: 139 (1949).

Fimbristylis acicularis R.Br., Prodr.: 226 (1810). *Iria acicularis* (R.Br.) Kuntze, Revis. Gen. Pl. 2: 753 (1891).
Philippines to N. Australia. 42 MOL PHI SUL 43 NWG SOL 50 NTA QLD WAU. Hemicr.
Eriocaulon setaceum Lour., Fl. Cochinch.: 60 (1790).
Fimbristylis setacea Benth., London J. Bot. 2: 239 (1843).
Abildgaardia brevifolia Steud., Syn. Pl. Glumac. 2: 72 (1854).
Fimbristylis acuminata var. *pumila* Nees, Hooker's J. Bot. Kew Gard. Misc. 6: 29 (1854).
Isolepis cochleata Steud., Syn. Pl. Glumac. 2: 100 (1855).
Fimbristylis acuminata var. *minor* Nees ex Boeck., Linnaea 37: 4 (1871).
Fimbristylis australica Boeck., Linnaea 38: 384 (1874).
Fimbristylis decumbens Boeck., Flora 58: 110 (1875).
Fimbristylis bursifolia S.Vidal, Phan. Cuming. Philipp.: 14, 156 (1885).
Fimbristylis setacea var. *brevifolia* C.B.Clarke in J.D.Hooker, Fl. Brit. India 6: 632 (1893).

Fimbristylis acuminata Vahl, Enum. Pl. 2: 285 (1805). *Iria acuminata* (Vahl) Kuntze, Revis. Gen. Pl. 2: 751 (1891).
Trop. & Subtrop. Asia to N. Australia. 36 CHS 38 NNS TAI 40 EHM IND SRL 41 THA VIE 42 BOR JAW MLY PHI SUM 43 NWG 50 NTA QLD WAU. Hemicr.
Scirpus scaber Roxb., Fl. Ind. 1: 220 (1820). *Fimbristylis scabra* (Roxb.) Schult. in J.J.Roemer & J.A.Schultes, Mant. 2: 59 (1824). *Eleogiton scabra* (Roxb.) A.Dietr., Sp. Pl. 2: 99 (1832).
Fimbristylis nutans var. *minor* Camus in H.Lecomte, Fl. Indo-Chine 7: 96 (1912).

Fimbristylis adenolepis J.Kern, Blumea 8: 123 (1955).
Indo-China, Jawa (Kep. Kangean). 41 THA VIE 42 JAW. Ther.

Fimbristylis adjuncta S.T.Blake, Proc. Roy. Soc. Queensland 58: 43 (1947).
N. Queensland. 50 QLD. Hemicr. or rhizome geophyte.

Fimbristylis* × *adventitia Ces., Linnaea 32: 261 (1863). *F. bisumbellata* × *F. squarrosa*. *Fimbristylis squarrosa* var. *adventitia* (Ces.) Nyman, Consp. Fl. Eur.: 763 (1882).
S. Europe. 13 ITA. Hemicr. or rhizome geophyte.

Fimbristylis aestivalis Vahl, Enum. Pl. 2: 288 (1805). *Scirpus aestivalis* (Vahl) Retz., Observ. Bot. 4: 12 (1786). *Iria aestivalis* (Vahl) Kuntze, Revis. Gen. Pl. 2: 751 (1891).
Trop. & Subtrop. Asia to Australia. 31+ 36 CHC CHS 38 JAP TAI 40 ASS EHM IND NEP SRL 41 THA VIE 42 JAW PHI SUL SUM 43 NWG 50 NSW NTA QLD SOA VIC 60 NWC (63) haw (84) bzc. Ther.

var. ***aestivalis***
Trop. & Subtrop. Asia to Australia. 31+ 36 CHC CHS 38 JAP TAI 40 ASS EHM IND NEP SRL 41 THA VIE 42 JAW PHI SUL SUM 43 NWG 50 NSW NTA QLD SOA VIC 60 NWC (63) haw (84) bzc. Ther.
Cyperus amane Siebold, Verh. Batav. Genootsch. Kunsten 12: 384 (1830).
Isolepis schomburghii Steud., Syn. Pl. Glumac. 2: 102 (1855).
Fimbristylis tricholepis Miq., Fl. Ned. Ind. 3: 319 (1856).
Fimbristylis leiocarpa Maxim., Prim. Fl. Amur.: 301 (1859). *Scirpus leiocarpus* (Maxim.) Meinsh., Trudy Imp. S.-Peterburgsk. Bot. Sada 18: 249 (1901).
Fimbristylis tokyoensis Makino, Bot. Mag. (Tokyo) 6: 47 (1892).
Fimbristylis trimenii Hook.f. in H.Trimen, Handb. Fl. Ceylon 5: 52 (1900).
Fimbristylis aestivalis subsp. *major* Trimen ex T.Koyama, Bot. Mag. (Tokyo) 87: 325 (1974).

var. ***macrostachya*** Benth., Fl. Austral. 7: 310 (1878).
Philippines, Queensland. 42 PHI 50 QLD. Ther.

var. ***trichopoda*** J.Kern, Blumea 12: 27 (1963).
SW. India, NE. New Guinea. 40 IND 43 NWG. Ther.

Fimbristylis aggregata C.E.C.Fisch., Bull. Misc. Inform. Kew 1931: 44 (1931).
S. India. 40 IND. Hemicr. or rhizome geophyte.

Fimbristylis alata E.G.Camus, Notul. Syst. (Paris) 1: 244 (1910).
Indo-China. 41 LAO THA VIE. Ther.

Fimbristylis albicans Nees in R.Wight, Contr. Bot. India: 100 (1834).
S. India. 40 IND. Hemicr. or rhizome geophyte.

Fimbristylis alboviridis (C.B.Clarke ex Scott-Elliot) C.B.Clarke in J.D.Hooker, Fl. Brit. India 6: 638 (1893).
W. & WC. Trop. Africa, Trop. Asia. 22 BKN GHA GUI NGA SEN SIE 23 ZAI 40 ASS BAN IND 41 AND NCB THA VIE 42 JAW MLY PHI. Ther.
**Cyperus alboviridis* C.B.Clarke ex Scott-Elliot, J. Linn. Soc., Bot. 29: 62 (1891).
Fimbristylis diphylla var. *tuberculata* Cherm., Arch. Bot. Bull. Mens. 4(7): 31 (1931).

Fimbristylis ambavanensis V.P.Prasad & N.P.Singh, J. Bombay Nat. Hist. Soc. 96: 454 (1999).
India (Maharashtra). 40 IND. Ther.

Fimbristylis ammobia Latz, J. Adelaide Bot. Gard. 1: 345 (1979).
Western Australia to Northern Territory. 50 NTA WAU. Hemicr. or rhizome geophyte.

Fimbristylis amplocarpa Govind., Proc. Indian Acad. Sci., B 80: 41 (1974).
S. India. 40 IND. Hemicr. or rhizome geophyte.

Fimbristylis angamoozhiensis Ravi & Anil Kumar, Rheedea 3: 108 (1993).
India (Kerala). 40 IND. Hemicr. or rhizome geophyte.

Fimbristylis anisoclada Ohwi, Blumea 8: 97 (1955).
Indo-China to Malesia. 41 THA VIE 42 JAW SUL. Hemicr.

Fimbristylis aphylla Steud., Syn. Pl. Glumac. 2: 114 (1855).
W. & WC. Trop. Africa, S. China to Trop. Asia. 22 GHA NGA SIE 23 BUR CAF ZAI 27 NAT 36 CHC 40 IND 41 THA VIE 42 JAW PHI SUM. Hemicr.
Fimbristylis quadrangularis var. *crassa* C.B.Clarke in J.D.Hooker, Fl. Brit. India 6: 644 (1893).
Fimbristylis vanderystii De Wild., Pl. Bequaert. 4: 205 (1927).
Fimbristylis testui Cherm., Arch. Bot. Bull. Mens. 4(7): 33 (1931).
Fimbristylis aphylla var. *gracilis* Tang & F.T.Wang, in Fl. Reipubl. Popul. Sin. 11: 227 (1961).

Fimbristylis argentea (Rottb.) Vahl, Enum. Pl. 2: 294 (1805).
India to Pen. Malaysia. 40 IND SRL 41 CBD MYA THA VIE 42 MLY. Hemicr.
**Scirpus argenteus* Rottb., Descr. Icon. Rar. Pl.: 51 (1773).
Scirpus monander Rottb., Descr. Icon. Rar. Pl.: 50 (1773). *Fimbristylis monandra* (Rottb.) F.Muell., Fragm. 1: 195 (1859). *Iria monandra* (Rottb.) Kuntze, Revis. Gen. Pl. 2: 753 (1891). *Iria muelleriana* Kuntze, Revis. Gen. Pl. 2: 753 (1891).
Scirpus nanus Poir. in J.B.A.M.de Lamarck, Encycl. 6: 759 (1805).
Aplostemon triquetrum Raf., J. Phys. Chim. Hist. Nat. Arts 89: 105 (1819).

Fimbristylis argillicola Kral, Sida 4: 128 (1971).
Mexico. 79 MXC MXE MXS. Hemicr.

Fimbristylis arnhemensis Latz, Nuytsia 7: 161 (1990).
NW. Northern Territory. 50 NTA. Ther.

Fimbristylis arnottiana Boeck., Linnaea 37: 28 (1871). *Iria arnottiana* (Boeck.) Kuntze, Revis. Gen. Pl. 2: 753 (1891).
S. India. 40 IND. Hemicr. or rhizome geophyte.

Fimbristylis arthrostyloides W.Fitzg., J. Proc. Roy. Soc. Western Australia 3: 121 (1918).
N. Western Australia. 50 WAU. Hemicr. or rhizome geophyte.

Fimbristylis aspera (Schrad.) Boeck., Linnaea 37: 29 (1871).
Guyana to Brazil. 82 GUY 84 BZN. Hemicr. or rhizome geophyte.
**Schoenus asper* Schrad. in J.J.Roemer & J.A.Schultes, Mant. 2: 41 (1824). *Trichelostylis aspera* (Schrad.) Nees in C.F.P.von Martius & auct. suc. (eds.), Fl. Bras. 2(1): 80 (1842).
Fimbristylis aspera var. *glaucovirens* Boeck., Linnaea 38: 394 (1874).

Fimbristylis assamica C.B.Clarke ex Guha Bakshi, Bull. Bot. Soc. Bengal 24: 111 (1970 publ. 1971).
Assam. 40 ASS. Hemicr. or rhizome geophyte.

Fimbristylis autumnalis (L.) Roem. & Schult. in J.J.Roemer & J.A.Schultes, Syst. Veg. 2: 97 (1817).
America. (34) lbs 72 NBR ONT QUE 74 ILL IOW KAN MIN MSO NEB OKL SDA WIS 75 CNT INI MAI MAS MIC NWH NWJ NWY OHI PEN RHO VER WVA 77 TEX 78 ALA ARK DEL FLA GEO KTY LOU MRY MSI NCA SCA TEN VRG WDC 79 MXE MXG MXS MXT 80 BLZ COS ELS GUA HON NIC 81 CUB TRT 82 FRG GUY? SUR VEN 83 ECU 84 BZC BZN 85 AGE AGW PAR URU. Ther.
**Scirpus autumnalis* L., Mant. Pl. 2: 180 (1771). *Isolepis autumnalis* (L.) J.Presl & C.Presl, Reliq. Haenk. 1: 189 (1828). *Trichelostylis autumnalis* (L.) Nees, Linnaea 9: 290 (1834). *Iria autumnalis* (L.) Kuntze, Revis. Gen. Pl. 2: 752 (1891).
Scirpus mucronulatus Michx., Fl. Bor.-Amer. 1: 31 (1803). *Trichelostylis mucronulata* (Michx.) Torr., Ann. Lyceum Nat. Hist. New York 3: 355 (1836). *Fimbristylis mucronulata* (Michx.) S.F.Blake, Rhodora 20: 25 (1918). *Fimbristylis autumnalis* var. *mucronulata* (Michx.) Fernald, Rhodora 37: 398 (1935).
Scirpus michauxii Pers., Syn. Pl. 1: 68 (1805).
Isolepis gracilis Kunth in F.W.H.von Humboldt, A.J.A.Bonpland & C.S.Kunth, Nov. Gen. Sp. 1: 223 (1816). *Fimbristylis autumnalis* var. *gracilis* (Kunth) Kük., Repert. Spec. Nov. Regni Veg. 23: 196 (1926).
Fimbristylis geminata Kunth, Enum. Pl. 2: 247 (1837). *Trichelostylis geminata* (Kunth) Nees in C.F.P.von Martius & auct. suc. (eds.), Fl. Bras. 2(1): 80 (1842). *Iria autumnalis* var. *geminata* (Kunth) Farw., Rep. (Annual) Michigan Acad. Sci. 21: 359 (1920).
Scirpus castaneus Willd. ex Kunth, Enum. Pl. 2: 227 (1837), nom. inval.
Fimbristylis firmula Steud., Syn. Pl. Glumac. 2: 111 (1855).
Fimbristylis frankii Steud., Syn. Pl. Glumac. 2: 111 (1855).
Fimbristylis leptostachya Steud., Syn. Pl. Glumac. 2: 118 (1855).
Fimbristylis microstachya Steud., Syn. Pl. Glumac. 2: 113 (1855).
Iria autumnalis var. *caespitosa* Kuntze, Revis. Gen. Pl. 2: 752 (1891).
Fimbristylis frankii var. *brachyactis* Fernald, Rhodora 11: 180 (1909). *Fimbristylis autumnalis* f. *brachyactis* (Fernald) S.F.Blake, Rhodora 20: 25 (1918).

Fimbristylis bahiensis Steud., Syn. Pl. Glumac. 2: 108 (1855).
Brazil. 84 BZC BZE BZL. Ther. or hemicr.
Cyperus geminatus Schrad. ex Schult. in J.J.Roemer & J.A.Schultes, Mant. 2: 95 (1824), nom. illeg.
Abildgaardia baeothryon A.St.-Hil., Voy. Distr. Diam. 1: 389 (1833), provisional synonym.
Abildgaardia aphylla Kunth, Enum. Pl. 2: 248 (1837). *Abildgaardia scirpoides* Nees in C.F.P.von Martius & auct. suc. (eds.), Fl. Bras. 2(1): 71 (1842), nom. illeg. *Iria aphylla* (Kunth) Kuntze, Revis. Gen. Pl. 2: 753 (1891).
Cyperus scirpinus Salzm. ex Steud., Syn. Pl. Glumac. 2: 108 (1855), nom. inval.
Pseudocyperus scirpinus Steud., Syn. Pl. Glumac. 2: 108 (1855).
Fimbristylis martii Boeck., Linnaea 37: 24 (1871). *Iria martii* (Boeck.) Kuntze, Revis. Gen. Pl. 2: 753 (1891).

Fimbristylis barteri Boeck., Linnaea 37: 33 (1871). *Iria barteri* (Boeck.) Kuntze, Revis. Gen. Pl. 2: 753 (1891).
W. Trop. Africa to Chad. 22 GHA GUI NGA SEN TOG 23 GAB 24 CHA. Hemicr.
Fimbristylis chevalieri Kük., Bull. Mus. Natl. Hist. Nat., II, 3: 547 (1931).

Fimbristylis benthamiana Govind., Rheedea 7: 115 (1997).
India (West Bengal). 40 IND. Hemicr. or rhizome geophyte.

Fimbristylis bispicula Govind., Rheedea 6: 59 (1996).
India (Karnataka). 40 IND. Hemicr. or rhizome geophyte.

Fimbristylis bisumbellata (Forssk.) Bubani, Dodecanthea: 30 (1850).
Old World. 11 SWI 12 FRA POR SPA 13 ALB BUL GRE ITA ROM SIC TUE YUG 14 UKR 20 EGY MOR 21 CNY 22 GHA IVO MLI MTN NGA SEN 23 CAF GAB 24 CHA SOM 25 KEN TAN 27 BOT CPP NAM NAT TVL 29 MDG 32 TKM 33 TCS 34 AFG IRN IRQ LBS PAL TUR 35 SAU YEM 36 CHC CHN CHS 38 NNS TAI 40 EHM IND NEP PAK SRL 41 THA VIE 42 JAW PHI 50 NTA QLD WAU. Ther.
**Scirpus bisumbellatus* Forssk., Fl. Aegypt.-Arab.: 15 (1775). *Iria bisumbellata* (Forssk.) Kuntze, Revis. Gen. Pl. 2: 752 (1891). *Fimbristylis dichotoma* subsp. *bisumbellata* (Forssk.) Luceño, Anales Jard. Bot. Madrid 57: 176 (1999).
Fimbristylis dichotoma var. *villosa* Vahl, Enum. Pl. 2: 287 (1805).
Scirpus pallescens Roxb., Fl. Ind. 1: 229 (1820). *Fimbristylis pallescens* (Roxb.) Nees in R.Wight, Contr. Bot. India: 101 (1834).
Fimbristylis liukiuensis Tuyama, J. Jap. Bot. 11: 250 (1935).
Fimbristylis bisumbellata var. *elata* Täckh., Publ. Cairo Univ. Herb. 5: 41 (1972 publ. 1974).

Fimbristylis bivalvis (Lam.) Lye, Lidia 3: 144 (1995).
E. Zaïre, E. Tanzania, KwaZulu-Natal, E. Madagascar. 23 ZAI 25 TAN 27 NAT 29 MDG. Hemicr.
**Scirpus bivalvis* Lam., Tabl. Encycl. 1: 141 (1791). *Isolepis bivalvis* (Lam.) Steud., Syn. Pl. Glumac. 2: 101 (1855).
Fimbristylis longiculmis Steud., Syn. Pl. Glumac. 2: 110 (1855).
Fimbristylis sansibariensis Boeck., Flora 63: 437 (1880).

Fimbristylis blakei Latz, Nuytsia 7: 162 (1990).
Western Australia to Northern Territory. 50 NTA WAU. Hemicr.

Fimbristylis blepharolepis J.Kern, Blumea 12: 25 (1963).
New Guinea. 43 NWG. Ther.

Fimbristylis borbonica Steud., Syn. Pl. Glumac. 2: 107 (1855).
Réunion. 29 REU. Hemicr. or rhizome geophyte.

Fimbristylis brevivaginata Kral, Sida 15: 318 (1992).
SE. U.S.A. 78 ALA GEO. Hemicr.

Fimbristylis brunneoides J.Kern, Blumea 8: 146 (1955).
NE. Thailand. 41 THA. Hemicr.

Fimbristylis caesia Miq., Fl. Ned. Ind. 3: 315 (1856).
Vietnam, Philippines, Jawa. 41 VIE 42 JAW PHI. Ther.

Fimbristylis caespitosa R.Br., Prodr.: 228 (1810). *Scirpus cespitosus* (R.Br.) Poir. in J.B.A.M.de Lamarck, Encycl., Suppl. 5: 99 (1817), nom. inval. *Iria caespitosa* (R.Br.) Kuntze, Revis. Gen. Pl. 2: 753 (1891).
N. Australia. 50 NTA QLD WAU. Hemicr. or rhizome geophyte.
Fimbristylis brachylaena F.Muell., Fragm. 1: 199 (1859).

Fimbristylis calcicola J.Kern, Blumea 8: 129 (1955).
Pen. Malaysia (P. Langkawi). 42 MLY. Ther.

Fimbristylis caloptera Latz, Nuytsia 7: 163 (1990).
Northern Territory. 50 NTA. Hemicr. or rhizome geophyte.

Fimbristylis cancellata Cherm., Bull. Soc. Bot. France 68: 418 (1921 publ. 1922).
W. Madagascar. 29 MDG. Ther.

Fimbristylis capilliculmis Ohwi, Blumea 8: 99 (1955).
New Guinea. 43 NWG. Ther.

Fimbristylis cardiocarpa F.Muell., Fragm. 1: 194 (1859). *Iria cardiocarpa* (F.Muell.) Kuntze, Revis. Gen. Pl. 2: 753 (1891).
N. Australia. 50 NTA QLD WAU. Hemicr. or rhizome geophyte.

Fimbristylis caroliniana (Lam.) Fernald, Rhodora 42: 246 (1940).
E. U.S.A. to Mexico, Cuba. 75 NWJ 77 TEX 78 ALA DEL FLA GEO LOU MRY MSI NCA SCA VRG 79 MXE MXG MXT 81 CUB. Hemicr.
**Scirpus carolinianus* Lam., Tabl. Encycl. 1: 142 (1791).
Fimbristylis harperi Britton in J.K.Small, Fl. Miami: 29 (1913).
Fimbristylis puberula f. *eucycla* Fernald, Rhodora 37: 396 (1935). *Fimbristylis caroliniana* f. *eucycla* (Fernald) Fernald, Rhodora 42: 247 (1940).
Fimbristylis puberula f. *pycnostachya* Fernald, Rhodora 37: 396 (1935). *Fimbristylis caroliniana* f. *pycnostachya* (Fernald) Fernald, Rhodora 42: 247 (1940).

Fimbristylis carolinii Latz, Nuytsia 7: 164 (1990).
Queensland. 50 QLD. Ther.

Fimbristylis carpopoda Govind., Proc. Indian Acad. Sci., B 79: 160 (1974).
Assam. 40 ASS. Hemicr. or rhizome geophyte.

Fimbristylis celebica Ohwi, Blumea 8: 103 (1955).
Sulawesi. 42 SUL. Ther.

Fimbristylis cephalophora F.Muell., Fragm. 1: 196 (1859).
Western Australia to Northern Territory. 50 NTA WAU. Hemicr. or rhizome geophyte.
Fimbristylis oligocephala W.Fitzg., J. Proc. Roy. Soc. Western Australia 3: 120 (1918).

Fimbristylis cephalotes Steud., Syn. Pl. Glumac. 2: 120 (1855).
Brazil. 84+. Hemicr. or rhizome geophyte.

Fimbristylis chingmaiensis S.M.Huang, in Fl. Hainan. 4: 536 (1977).
Hainan. 36 CHH. Hemicr. or rhizome geophyte.

Fimbristylis cinnamometorum (Vahl) Kunth, Enum. Pl. 2: 229 (1837).
Trop. & Subtrop. Asia to N. & E. Australia. 36 CHS 38 TAI 40 SRL 41 MYA NCB THA VIE 42 PHI SUM 43 NWG 50 NSW QLD WAU. Ther. or hemicr.
**Scirpus cinnamometorum* Vahl, Enum. Pl. 2: 278 (1805). *Isolepis cinnamometorum* (Vahl) Roem. & Schult., Syst. Veg. 2: 119 (1817). *Trichelostylis cinnamometorum* (Vahl) Nees, Linnaea 9: 290 (1834). *Abildgaardia cinnamometorum* (Vahl) Thwaites, Enum. Pl. Zeyl.: 347 (1864). *Iria cinnamometorum* (Vahl) Kuntze, Revis. Gen. Pl. 2: 753 (1891).
Fimbristylis cyperoides R.Br., Prodr.: 229 (1810).
Fimbristylis biflora Boeck., Linnaea 38: 393 (1874). *Iria biflora* (Boeck.) Kuntze, Revis. Gen. Pl. 2: 753 (1891).
Fimbristylis kamphoeveneri Boeck., Bot. Jahrb. Syst. 5: 505 (1884).

Fimbristylis circumciliata Govind., Proc. Indian Acad. Sci., B 79: 163 (1974).
Assam. 40 ASS. Hemicr. or rhizome geophyte.

Fimbristylis clavata S.T.Blake, Pap. Dept. Biol. Univ. Queensland 1(13): 7 (1940).
N. Australia. 50 NTA QLD WAU. Hemicr. or rhizome geophyte.

Fimbristylis compacta Turrill, Bull. Misc. Inform. Kew 1915: 118 (1915).
N. Northern Territory. 50 NTA. Hemicr. or rhizome geophyte.

Fimbristylis complanata (Retz.) Link, Hort. Berol. 1: 292 (1827).
Trop. & Subtrop. 22 LBR NGA NGR 23 GAB RWA ZAI 24 SOM 25 KEN TAN UGA 27 BOT CPP NAM NAT OFS SWZ TVL 29 SEY 35 YEM 36 CHC CHS 38 JAP KOR NNS TAI 40 EHM IND NEP PAK SRL 41 THA VIE 42 BOR JAW LSI MLY MOL PHI SUL SUM 43 NWG 50 NTA QLD (60) fij NWC TON 62 CRL MRN 79 MXC MXE MXG MXN MXS MXT 80 BLZ COS GUA HON NIC PAN 81 CUB DOM HAI JAM LEE PUE TRT WIN 82 FRG SUR VEN 83 BOL CLM ECU PER 84 BZE BZL BZN 85 AGE AGW PAR URU. Hemicr.
**Scirpus complanatus* Retz., Observ. Bot. 5: 14 (1788). *Cyperus complanatus* (Retz.) Willd., Sp. Pl. 1: 270 (1797), nom. illeg. *Isolepis complanata* (Retz.) Roem. & Schult., Syst. Veg. 2: 119 (1817). *Trichelostylis complanata* (Retz.) Nees in R.Wight, Contr. Bot. India: 103 (1834). *Fimbristylis autumnalis* var. *complanata* (Retz.) Kük., Bot. Jahrb. Syst. 59: 50 (1924).

var. ***complanata***
Trop. & Subtrop. 22 LBR NGA NGR 23 GAB RWA ZAI 24 SOM 25 KEN TAN UGA 27 BOT CPP NAM NAT OFS SWZ TVL 29 SEY 35 YEM 36 CHC CHS 38 JAP KOR NNS TAI 40 EHM IND NEP PAK SRL 41 THA VIE 42 BOR JAW LSI MLY MOL PHI SUL SUM 43 NWG 50 NTA QLD (60) fij NWC TON 62 CRL MRN 79 MXC MXE MXG MXN MXS MXT 80 BLZ COS GUA HON NIC PAN 81 CUB DOM HAI JAM LEE PUE TRT WIN 82 FRG SUR VEN 83 BOL CLM ECU PER 84 BZE BZL BZN 85 AGE AGW PAR URU. Hemicr.
Cyperus amentaceus Rudge, Pl. Guian.: 16 (1805). *Scirpus amentaceus* (Rudge) Griseb., Abh. Königl. Ges. Wiss. Göttingen 7: 270 (1857).
Scirpus anceps Willd., Mag. Neuesten Entdeck. Gesammten Naturk. Ges. Naturf. Freunde Berlin

2: 287 (1808), nom. illeg. *Isolepis willdenowii* Roem. & Schult., Syst. Veg. 2: 120 (1817). *Isolepis anceps* A.Dietr., Sp. Pl. 2: 125 (1832), nom. superfl. *Cyperus anceps* Steud., Syn. Pl. Glumac. 2: 22 (1854), nom. illeg. *Fimbristylis anceps* Steud., Syn. Pl. Glumac. 2: 112 (1855), nom. superfl.

Trichelostylis scabra Nees in R.Wight, Contr. Bot. India: 103 (1834). *Fimbristylis scabra* (Nees) Kunth, Enum. Pl. 2: 245 (1837), nom. illeg.

Scirpus kielmeyeri Schimp. ex Steud., Nomencl. Bot., ed. 2, 2: 539 (1841).

Trichelostylis rudgeana Nees in C.F.P.von Martius & auct. suc. (eds.), Fl. Bras. 2(1): 79 (1842). *Fimbristylis rudgeana* (Nees) Steud., Syn. Pl. Glumac. 2: 115 (1855).

Fimbristylis conferta A.Rich. in R.de la Sagra, Hist. Fis. Cuba, Bot. 11: 291 (1850).

Fimbristylis amblyphylla Steud., Syn. Pl. Glumac. 2: 116 (1855).

Fimbristylis boeckeleri Steud., Syn. Pl. Glumac. 2: 113 (1855).

Fimbristylis cylindrostachya Steud., Syn. Pl. Glumac. 2: 119 (1855).

Fimbristylis kielmaieri Steud., Syn. Pl. Glumac. 2: 113 (1855).

Trichelostylis kielmeyeri Nees ex Boeck., Linnaea 37: 38 (1871).

Fimbristylis subaphylla Boeck., Flora 62: 565 (1879). *Fimbristylis complanata* var. *subaphylla* (Boeck.) Lye, Bot. Not. 127: 499 (1974).

Fimbristylis obscura Fernald, Proc. Amer. Acad. Arts 36: 492 (1901).

Fimbristylis complanata var. *macrocarya* Domin, Biblioth. Bot. 85: 463 (1915).

Fimbristylis autumnalis f. *hemisphaerica* Kük., Bot. Jahrb. Syst. 59: 6 (1924).

Fimbristylis bequaertii De Wild., Pl. Bequaert. 4: 200 (1927).

Fimbristylis complanata f. *exaltata* T.Koyama, Bull. Arts Sci. Div. Ryukyu Univ. 3: 70 (1959).

var. ***keniaeensis*** (Kük.) Lye, Bot. Not. 127: 499 (1974).
Kenya. 25 KEN. Hemicr.

**Fimbristylis keniaeensis* Kük., Notizbl. Bot. Gart. Berlin-Dahlem 9: 307 (1925). *Fimbristylis complanata* subsp. *keniaeensis* (Kük.) Lye, Nordic J. Bot. 2: 334 (1982).

Fimbristylis composita Latz, Nuytsia 7: 165 (1990).
Northern Territory. 50 NTA. Hemicr.

Fimbristylis consanguinea Kunth, Enum. Pl. 2: 228 (1837). *Iria consanguinea* (Kunth) Kuntze, Revis. Gen. Pl. 2: 753 (1891) (1891).
S. Africa, Madagascar, India, Sri Lanka, C. & S. China, W. Jawa. 27 CPP 29 MDG 36 CHC CHN CHS 40 IND SRL 42 JAW. Hemicr.

Fimbristylis connectens Thwaites, Enum. Pl. Zeyl.: 349 (1864).

Fimbristylis complanata var. *kraussiana* C.B.Clarke in J.D.Hooker, Fl. Brit. India 6: 646 (1893).

Fimbristylis kraussiana Hochst. ex Hook.f. in H.Trimen, Handb. Fl. Ceylon 5: 63 (1900).

Fimbristylis horsfieldii C.B.Clarke, Kew Bull., Addit. Ser. 8: 25 (1908).

Fimbristylis contorta C.E.C.Fisch., Bull. Misc. Inform. Kew 1931: 45 (1931).
S. India. 40 IND. Hemicr. or rhizome geophyte.

Fimbristylis corynocarya F.Muell., Fragm. 1: 197 (1859). *Iria corynocarya* (F.Muell.) Kuntze, Revis. Gen. Pl. 2: 753 (1891).
N. Australia. 50 NTA QLD WAU. Hemicr. or rhizome geophyte.

Fimbristylis costiglumis Domin, Biblioth. Bot. 85: 451 (1915).
Northern Territory to Queensland. 50 NTA QLD. Hemicr. or rhizome geophyte.

Fimbristylis crystallina Govind., Proc. Indian Acad. Sci., B 79: 166 (1974).
S. India. 40 IND. Hemicr. or rhizome geophyte.

Fimbristylis cuneata Govind. & S.K.Varma, J. Econ. Taxon. Bot. 21: 377 (1997).
India (Bihar). 40 IND. Hemicr. or rhizome geophyte.

Fimbristylis cymosa R.Br., Prodr.: 228 (1810). *Scirpus cymosus* (R.Br.) Poir. in J.B.A.M.de Lamarck, Encycl., Suppl. 5: 99 (1816), nom. illeg. *Iria cymosa* (R.Br.) Kuntze, Revis. Gen. Pl. 2: 753 (1891). *Fimbristylis cymosa* var. *typica* Domin, Biblioth. Bot. 85: 463 (1915), nom. inval.
Trop. & Subtrop. 22 BEN GHA IVO LBR NGA NGR SEN SIE TOG 23 EQG GAB ZAI 24 SOM 25 KEN TAN 27 BOT CPP NAT 29 ALD MAU MDG REU SEY 34 IRN 35 GST SAU YEM 36 CHH CHS 38 JAP NNS TAI 40 IND NEP PAK SRL 41 THA VIE 42 BOR CKI JAW LSI MLY MOL PHI SUL SUM XMS 43 BIS NWG SOL 50 NTA QLD WAU 60 FIJ GIL HBI NUE NWC TOK TON 61 SCI TUA 62 CRL MRN MRS WAK 63 HAW 78 FLA 79 MXE MXG MXN MXS MXT 80 BLZ COS GUA HON NIC PAN 81 BAH CAY CUB DOM JAM LEE NLA PUE TRT WIN 82 FRG GUY SUR VEN 83 CLM ECU 84 BZE BZN. Hemicr.

subsp. ***cymosa***
Trop. & Subtrop. 22 BEN GHA IVO LBR NGA NGR SEN SIE TOG 23 EQG GAB ZAI 24 SOM 25 KEN TAN 27 BOT CPP NAT 29 ALD MAU MDG REU SEY 34 IRN 35 GST SAU YEM 36 CHH CHS 38 JAP TAI 40 IND NEP PAK SRL 41 THA VIE 42 BOR CKI JAW LSI MLY MOL PHI SUL SUM XMS 43 BIS NWG SOL 50 NTA QLD WAU 60 FIJ GIL NUE NWC TOK TON 61 SCI TUA 62 CRL MRN MRS WAK 63 HAW 78 FLA 79 MXE MXG MXN MXS MXT 80 BLZ COS GUA HON NIC PAN 81 BAH CAY CUB DOM JAM LEE NLA PUE TRT WIN 82 FRG GUY SUR VEN 83 CLM ECU 84 BZE BZN. Hemicr.

Kyllinga triceps Rottb., Descr. Icon. Rar. Pl.: 14 (1773). *Cyperus triceps* (Rottb.) Endl., Cat. Horti Vindob. 1: 94 (1842).

Scirpus glomeratus Retz., Observ. Bot. 4: 11 (1786), nom. illeg. *Isolepis glomerata* Schrad. ex Schult. in J.J.Roemer & J.A.Schultes, Mant. 2: 67 (1824). *Fimbristylis glomerata* (Schrad. ex Schult.) Nees, Linnaea 9: 290 (1834). *Iria glomerata* (Schrad. ex Schult.) Kuntze, Revis. Gen. Pl. 2: 753 (1891). *Fimbristylis glomerata* (Schrad. ex Schult.) Urb., Symb. Antill. 2: 166 (1900), nom. illeg. *Fimbristylis rigida* Kunth, Enum. Pl. 2: 231 (1837).

Scirpus cymosus Lam., Tabl. Encycl. 1: 141 (1791).

Scirpus strictus Roxb., Fl. Ind. 1: 226 (1820). *Isolepis stricta* (Roxb.) A.Dietr., Sp. Pl. 2: 121 (1832).

Fimbristylis spathacea Roth, Nov. Pl. Sp.: 24 (1821). *Fimbristylis cymosa* subsp. *spathacea* (Roth) T.Koyama, Micronesica 1: 83 (1964).

Fimbristylis cymosa var. *spathacea* (Roth) T.Koyama, J. Jap. Bot. 46: 66 (1971).
Scirpus rigescens Willd. ex Schult. in J.J.Roemer & J.A.Schultes, Mant. 2: 84 (1824). *Isolepis rigescens* (Willd. ex Schult.) C.Presl, Isis (Oken) 21: 268 (1828).
Fimbristylis capitata Sieber ex C.Presl, Isis (Oken) 21: 268 (1828), nom. illeg.
Isolepis haenkei J.Presl & C.Presl, Reliq. Haenk. 1: 187 (1828).
Isolepis schraderi A.Dietr., Sp. Pl. 2: 120 (1832).
Fimbristylis wightiana Nees in R.Wight, Contr. Bot. India: 99 (1834).
Fimbristylis ciliolata Steud., Syn. Pl. Glumac. 2: 109 (1855).
Fimbristylis laevissima Steud., Syn. Pl. Glumac. 2: 117 (1855).
Fimbristylis monini Steud., Syn. Pl. Glumac. 2: 120 (1855).
Fimbristylis reunionis Steud., Syn. Pl. Glumac. 2: 114 (1855).
Fimbristylis umbellatocapitata Steud., Syn. Pl. Glumac. 2: 120 (1855).
Fimbristylis biumbellulata Boeck., Flora 41: 603 (1858), nom. illeg.
Scirpus ciliolatus Boeck., Vidensk. Meddel. Naturhist. Foren. Kjøbenhavn 1871: 152 (1871).
Fimbristylis multifolia Boeck., Linnaea 38: 397 (1874). *Iria multifolia* (Boeck.) Kuntze, Revis. Gen. Pl. 2: 753 (1891).
Fimbristylis sintenisii Boeck., Bot. Jahrb. Syst. 7: 276 (1886).
Fimbristylis warburgii K.Schum. ex Warb., Bot. Jahrb. Syst. 13: 265 (1891).
Fimbristylis cymosa var. *subcapitata* C.B.Clarke, Bot. Tidsskr. 24: 90 (1901).
Fimbristylis melanospora Fernald, Proc. Amer. Acad. Arts 36: 491 (1901).
Fimbristylis capitulifera Merr., Philipp. J. Sci., C 9: 265 (1914).
Fimbristylis cymosa var. *multifolia* Domin, Biblioth. Bot. 85: 463 (1915).
Fimbristylis kankaoensis Hayata, Icon. Pl. Formosan. 6: 111 (1916).
Fimbristylis atollensis H.St.John, Pacific Sci. 6: 145 (1952).
Fimbristylis spathacea var. *depauperata* T.Koyama, J. Jap. Bot. 30: 129 (1955).

subsp. ***umbellatocapitata*** (Hillebr.) T.Koyama, Micronesica 1: 82 (1964).
S. China to Pacific. 36 CHS 38 NNS TAI 60 HBI NUE TON 61 SCI 62 CRL MRN 63 HAW. Hemicr.
**Fimbristylis cymosa* var. *umbellatocapitata* Hillebr., Fl. Hawaiian Isl.: 473 (1888). *Fimbristylis spathacea* var. *umbellatocapitata* (Hillebr.) T.Koyama, J. Fac. Sci. Univ. Tokyo, Sect. 3, Bot. 8: 107 (1961).
Fimbristylis pycnocephala Hillebr., Fl. Hawaiian Isl.: 473 (1888). *Iria pycnocephala* (Hillebr.) Kuntze, Revis. Gen. Pl. 2: 753 (1891). *Fimbristylis cymosa* var. *pycnocephala* (Hillebr.) Kük., Bernice P. Bishop Mus. Bull. 128: 20 (1935).
Fimbristylis cymosa var. *microcephala* F.Br., Fl. SE. Polyn. 1: 104 (1931).

Fimbristylis dauciformis Govind., Proc. Indian Acad. Sci., B 76: 181 (1972).
SW. India. 40 IND. Hemicr. or rhizome geophyte.

Fimbristylis debilis Steud., Syn. Pl. Glumac. 2: 109 (1855).
W. & WC. Trop. Africa. 22 BEN BKN GAM GHA IVO MLI NGA NGR SEN SIE 23 GAB. Ther.

Fimbristylis decipiens Kral, Sida 4: 120 (1971).
Texas to SE. U.S.A., Mexico. 77 TEX 78 ALA FLA GEO LOU MSI NCA SCA 79 MXN. Ther.

Fimbristylis densa S.T.Blake, Pap. Dept. Biol. Univ. Queensland 1(13): 11 (1940).
N. Australia. 50 NTA QLD WAU. Hemicr. or rhizome geophyte.

Fimbristylis denudata R.Br., Prodr.: 227 (1810). *Scirpus denudatus* (R.Br.) Poir. in J.B.A.M.de Lamarck, Encycl., Suppl. 5: 99 (1817), nom. inval. *Iria denutata* (R.Br.) Kuntze, Revis. Gen. Pl. 2: 753 (1891).
N. Australia. 50 NTA QLD WAU. Hemicr. or rhizome geophyte.
Fimbristylis elata R.Br., Prodr.: 227 (1810). *Scirpus elatus* (R.Br.) Poir. in J.B.A.M.de Lamarck, Encycl., Suppl. 5: 99 (1817), nom. inval. *Iria elata* (R.Br.) Kuntze, Revis. Gen. Pl. 2: 753 (1891).

Fimbristylis dichotoma (L.) Vahl, Enum. Pl. 2: 287 (1805).
Trop. & Subtrop. (11) swi (12) fra (13) ita 20 ALG MOR 22 BEN BKN GAM GHA GNB GUI IVO MLI NGA NGR SEN SIE TOG 23 BUR CMN EQG GAB GGI RWA ZAI 24 CHA 25 KEN TAN UGA 27 BOT CPP NAM NAT OFS SWZ TVL 29 COM MDG REU SEY 31 KAM 32 KAZ KGZ UZB 33 TCS 34 AFG 35 GST SAU YEM 36 CHC CHM CHN CHS 38 JAP KOR KZN NNS OGA TAI 40 ASS EHM IND NEP PAK SRL 41 THA VIE 42 43 BIS NWG 50 NFK NSW NTA QLD SOA TAS VIC WAU 60 FIJ GIL NWC TON 61 COO SCI 62 CRL MRN wak 63 HAW 74 ILL KAN MSO OKL 75 INI MIC NWJ NWY OHI PEN 76 ARI 77 TEX 78 ALA ARK DEL FLA GEO KTY LOU MRY MSI NCA SCA TEN VRG WDC 79 MXC MXE MXG MXN MXS MXT 80 BLZ COS CPI ELS GUA HON NIC PAN 81 BAH CAY CUB DOM HAI JAM LEE PUE TRT WIN 82 FRG GUY SUR VEN 83 BOL CLM ECU PER 84 BZC BZE BZL BZN 85 AGE PAR URU. Ther. or hemicr.
**Scirpus dichotomus* L., Sp. Pl.: 50 (1753). *Isolepis dichotoma* (L.) Kunth in F.W.H.von Humboldt, A.J.A.Bonpland & C.S.Kunth, Nov. Gen. Sp. 1: 223 (1816). *Eleocharis dichotoma* (L.) H.Karst., Deut. Fl.: 353 (1881).

subsp. ***depauperata*** (R.Br.) J.Kern, in Fl. Males. 7: 576 (1974).
S. China, S. Malesia to N. Australia. 36 CHS 42 LSI MOL 43 NWG 50 NTA QLD 62+. Ther.
**Fimbristylis depauperata* R.Br., Prodr.: 227 (1810). *Isolepis depauperata* (R.Br.) Link, Jahrb. Gewächsk. 1(3): 80 (1820). *Scirpus depauperatus* (R.Br.) Kunth, Enum. Pl. 2: 233 (1837). *Iria depauperata* (R.Br.) Kuntze, Revis. Gen. Pl. 2: 753 (1891). *Fimbristylis diphylla* var. *depauperata* (R.Br.) C.B.Clarke in J.D.Hooker, Fl. Brit. India 6: 637 (1893). *Fimbristylis depauperata* var. *typica* Domin, Biblioth. Bot. 89: 456 (1915), nom. inval. *Fimbristylis dichotoma* f. *depauperata* (R.Br.) Ohwi, J. Jap. Bot. 14: 578 (1938).
Fimbristylis spirostachya F.Muell. ex Benth., Fl. Austral. 7: 311 (1878). *Fimbristylis diphylla* var.

spirostachya (F.Muell. ex Benth.) C.B.Clarke in J.D.Hooker, Fl. Brit. India 6: 637 (1893).
Fimbristylis depauperata var. *chillagoensis* Domin, Biblioth. Bot. 89: 456 (1915).
Fimbristylis depauperata var. *plurispiculata* Domin, Biblioth. Bot. 89: 456 (1915).
Fimbristylis depauperata var. *polyphylla* Domin, Biblioth. Bot. 89: 456 (1915).

subsp. ***dichotoma***
Trop. & Subtrop. (11) swi (12) fra (13) ita 20 ALG MOR 22 BEN BKN GAM GHA GNB GUI IVO MLI NGA NGR SEN SIE TOG 23 BUR CMN EQG GAB GGI RWA ZAI 24 CHA 25 KEN TAN UGA 27 BOT CPP NAM NAT OFS SWZ TVL 29 COM MDG REU SEY 32 KGZ UZB 33 TCS 34 AFG 35 GST SAU YEM 36 CHC CHM CHN CHS 38 JAP KOR KZN NNS OGA TAI 40 ASS EHM IND NEP PAK SRL 41 THA VIE 42 43 BIS NWG 50 NFK NSW NTA QLD SOA TAS VIC WAU 60 FIJ GIL NWC TON 61 COO SCI 62 CRL MRN 63 HAW 74 ILL KAN MSO OKL 75 INI MIC NWJ NWY OHI PEN 76 ARI 77 TEX 78 ALA ARK DEL FLA GEO KTY LOU MRY MSI NCA SCA TEN VRG WDC 79 MXC MXE MXG MXN MXS MXT 80 BLZ COS CPI ELS GUA HON NIC PAN 81 BAH CAY CUB DOM HAI JAM LEE PUE TRT WIN 82 FRG GUY SUR VEN 83 BOL CLM ECU PER 84 BZC BZE BZL BZN 85 AGE PAR URU. Ther. or hemicr.
Scirpus annuus All., Fl. Pedem. 2: 277 (1785). *Fimbristylis annua* (All.) Roem. & Schult. in J.J.Roemer & J.A.Schultes, Syst. Veg. 2: 95 (1817). *Fimbristylis annua* f. *typica* Kük., Repert. Spec. Nov. Regni Veg. 23: 195 (1926), nom. inval. *Fimbristylis dichotoma* f. *annua* (All.) Ohwi, Mem. Coll. Sci. Kyoto Imp. Univ., Ser. B, Biol. 18: 81 (1944). *Fimbristylis dichotoma* var. *annua* (All.) T.Koyama, J. Fac. Sci. Univ. Tokyo, Sect. 3, Bot. 8: 111 (1961).
Scirpus diphyllus Retz., Observ. Bot. 5: 15 (1788). *Fimbristylis diphylla* (Retz.) Vahl, Enum. Pl. 2: 289 (1805).
Scirpus niloticus J.F.Gmel., Syst. Nat. 2: 126 (1791).
Scirpus obtusifolius Lam., Tabl. Encycl. 1: 141 (1791). *Isolepis obtusifolia* (Lam.) Roem. & Schult., Syst. Veg. 2: 118 (1817). *Trichelostylis obtusifolia* (Lam.) Nees, Linnaea 9: 290 (1834). *Fimbristylis obtusifolia* (Lam.) Kunth, Enum. Pl. 2: 240 (1837), nom. illeg. *Iria obtusifolia* (Lam.) Hitchc., Rep. (Annual) Missouri Bot. Gard. 4: 141 (1893).
Scirpus puberulus Michx., Fl. Bor.-Amer. 1: 31 (1803). *Fimbristylis castanea* var. *puberula* (Michx.) Britton ex House, Bull. New York State Mus. Nat. Hist. 254: 141 (1924).
Fimbristylis hirtella Vahl, Enum. Pl. 2: 286 (1805).
Fimbristylis laxa Vahl, Enum. Pl. 2: 292 (1805). *Scirpus laxus* (Vahl) Poir. in J.B.A.M.de Lamarck, Encycl., Suppl. 5: 101 (1817). *Fimbristylis dichotoma* var. *laxa* (Vahl) Napper, Kew Bull. 25: 436 (1971).
Fimbristylis serrulata Vahl, Enum. Pl. 2: 285 (1805). *Scirpus serrulatus* (Vahl) Vahl, Eclog. Amer. 3: 1 (1807).
Scirpus fimbriatus Poir. in J.B.A.M.de Lamarck, Encycl. 6: 779 (1805).
Scirpus pubescens Pers., Syn. Pl. 1: 68 (1805), nom. illeg.
Scirpus villosus Poir. in J.B.A.M.de Lamarck, Encycl. 6: 780 (1805). *Fimbristylis villosa* (Poir.) Roem. & Schult. in J.J.Roemer & J.A.Schultes, Syst. Veg. 2: 98 (1817).
Scirpus hirtellus Vahl, Eclog. Amer. 3: 1 (1807).
Fimbristylis elongata R.Br., Prodr.: 228 (1810). *Scirpus elongatus* (R.Br.) Poir. in J.B.A.M.de Lamarck, Encycl., Suppl. 5: 99 (1817), nom. inval.
Fimbristylis gracilis R.Br., Prodr.: 227 (1810). *Fimbristylis diphylla* var. *gracilis* (R.Br.) Benth., Fl. Austral. 7: 312 (1878).
Fimbristylis parviflora R.Br., Prodr.: 227 (1810). *Scirpus parviflorus* (R.Br.) Poir. in J.B.A.M.de Lamarck, Encycl., Suppl. 5: 99 (1817), nom. inval.
Fimbristylis stricta R.Br., Prodr.: 228 (1810). *Scirpus strictus* (R.Br.) Poir. in J.B.A.M.de Lamarck, Encycl., Suppl. 5: 99 (1817), nom. inval. *Trichelostylis stricta* (R.Br.) Nees, J. Bot. (Hooker) 2: 394 (1840). *Iria stricta* (R.Br.) Kuntze, Revis. Gen. Pl. 2: 753 (1891).
Fimbristylis variabilis R.Br., Prodr.: 228 (1810). *Scirpus variabilis* (R.Br.) Poir. in J.B.A.M.de Lamarck, Encycl., Suppl. 5: 99 (1817), nom. inval.
Scirpus sulcatus Elliott, Sketch Bot. S. Carolina 1: 56 (1816), nom. illeg.
Scirpus tomentosus Poir. in J.B.A.M.de Lamarck, Encycl., Suppl. 5: 101 (1817).
Fimbristylis candelabrum Link, Jahrb. Gewächsk. 1(3): 80 (1820).
Fimbristylis pubescens Link, Jahrb. Gewächsk. 1(3): 80 (1820).
Scirpus arvensis Roxb., Fl. Ind. 1: 224 (1820).
Scirpus brevifolius Roxb., Fl. Ind. 1: 229 (1820). *Fimbristylis brevifolia* (Roxb.) Wall. ex Schult. in J.J.Roemer & J.A.Schultes, Mant. 2: 54 (1824), nom. illeg.
Scirpus fuscescens Link, Jahrb. Gewächsk. 1(3): 81 (1820). *Fimbristylis fuscescens* (Link) Steud., Nomencl. Bot., ed. 2, 1: 610 (1840).
Scirpus glomeratus Roxb., Fl. Ind. 1: 244 (1820), nom. illeg.
Scirpus miliaceus Roxb., Fl. Ind. 1: 231 (1820), nom. illeg.
Fimbristylis nitidula Schult. in J.J.Roemer & J.A.Schultes, Mant. 2: 81 (1824).
Fimbristylis striata Labill., Sert Austro-Caledon.: t. 16, f. 2 (1824).
Isolepis brachyphylla Schult. in J.J.Roemer & J.A.Schultes, Mant. 2: 64 (1824). *Fimbristylis brachyphylla* (Schult.) Schult. in J.J.Roemer, J.A.Schultes & J.J.Schultes, Mant. 3: 530 (1827).
Isolepis curvifolia Schrad. in J.J.Roemer & J.A. Schultes, Mant. 2: 70 (1824). *Trichelostylis curvifolia* (Schrad.) Nees in R.Wight, Contr. Bot. India: 105 (1834). *Fimbristylis curvifolia* (Schrad.) Steud., Syn. Pl. Glumac. 2: 116 (1855).
Scirpus baldwinianus Schult. in J.J.Roemer & J.A.Schultes, Mant. 2: 85 (1824). *Fimbristylis baldwiniana* (Schult.) Torr., Ann. Lyceum Nat. Hist. New York 3: 344 (1836).
Fimbristylis foliosa Link, Hort. Berol. 1: 288 (1827).
Fimbristylis mezostachya Link, Hort. Berol. 1: 290 (1827).
Fimbristylis pallescens Link, Hort. Berol. 1: 291 (1827). *Isolepis pallescens* (Link) A.Dietr., Sp. Pl. 2: 121 (1832).

Fimbristylis pubigera Link, Hort. Berol. 1: 289 (1827).
Scirpus elliottii Spreng., Syst. Veg. 4(2): 28 (1827). *Isolepis elliottii* (Spreng.) A.Dietr., Sp. Pl. 2: 131 (1832).
Scirpus pallescens Willd. ex Link, Hort. Berol. 1: 292 (1827).
Scirpus pubigerus Bernh. ex Link, Hort. Berol. 1: 289 (1827).
Fimbristylis humboldtii J.Presl & C.Presl, Reliq. Haenk. 1: 190 (1828).
Fimbristylis juncifolia J.Presl & C.Presl, Reliq. Haenk. 1: 190 (1828).
Fimbristylis obtusifolia Nees ex C.Presl, Isis (Oken) 21: 268 (1828).
Fimbristylis verrucosa J.Presl & C.Presl, Reliq. Haenk. 1: 190 (1828).
Fimbristylis verrucosa var. *glabrescens* C.Presl, Reliq. Haenk. 1: 190 (1828).
Fimbristylis verrucosa var. *hirsuta* C.Presl, Reliq. Haenk. 1: 190 (1828).
Fimbristylis affinis Hook. & Arn., Bot. Beechey Voy.: 72 (1832). *Fimbristylis hookeri* Endl., Ann. Wiener Mus. Naturgesch. 1(160): 675 (1836).
Fimbristylis carinata Nees, Linnaea 9: 290 (1834), nom. inval.
Fimbristylis ovalis Nees in R.Wight, Contr. Bot. India: 98 (1834).
Fimbristylis royeniana Nees in R.Wight, Contr. Bot. India: 99 (1834). *Fimbristylis diphylla* f. *royeniana* (Nees) C.B.Clarke, Bull. Herb. Boissier, II, 3: 1017 (1903).
Cyperus annuus Willd. ex Kunth, Enum. Pl. 2: 233 (1837), nom. inval.
Fimbristylis communis Kunth, Enum. Pl. 2: 234 (1837).
Fimbristylis dregeana Kunth, Enum. Pl. 2: 232 (1837).
Fimbristylis mauritiana Nees ex Kunth, Enum. Pl. 2: 235 (1837).
Fimbristylis sororia Kunth, Enum. Pl. 2: 246 (1837).
Scirpus candelabrum Willd. ex Kunth, Enum. Pl. 2: 232 (1837), nom. inval.
Scirpus nitidulus Willd. ex Kunth, Enum. Pl. 2: 232 (1837).
Fimbristylis brizoides var. *humboldtiana* Nees in C.F.P.von Martius & auct. suc. (eds.), Fl. Bras. 2(1): 74 (1842).
Fimbristylis brizoides var. *pallens* Nees in C.F.P.von Martius & auct. suc. (eds.), Fl. Bras. 2(1): 75 (1842).
Fimbristylis brizoides var. *tenuifolia* Nees in C.F.P.von Martius & auct. suc. (eds.), Fl. Bras. 2(1): 75 (1842).
Fimbristylis trifurca Nees in C.F.P.von Martius & auct. suc. (eds.), Fl. Bras. 2(1): 74 (1842).
Isolepis varia Schrad. ex Nees in C.F.P.von Martius & auct. suc. (eds.), Fl. Bras. 2(1): 75 (1842).
Fimbristylis brizoides Sm. ex Nees & Meyen, Nov. Actorum Acad. Caes. Leop.-Carol. Nat. Cur. 19(Suppl. 1): 78 (1843). *Fimbristylis annua* f. *brizoides* (Sm. ex Nees & Meyen) Kük., Repert. Spec. Nov. Regni Veg. 23: 196 (1926). *Fimbristylis annua* var. *brizoides* (Sm. ex Nees & Meyen) A.Gross & Brade, Arq. Mus. Nac. Rio de Janeiro 34: 58 (1932).
Fimbristylis scherardii Bubani, Dodecanthea: 30 (1850).
Ficinia ambigua Steud., Syn. Pl. Glumac. 2: 125 (1855).
Fimbristylis ambigua Steud., Syn. Pl. Glumac. 2: 117 (1855).
Fimbristylis boiviniana Steud., Syn. Pl. Glumac. 2: 110 (1855).
Fimbristylis calocarpa Steud., Syn. Pl. Glumac. 2: 117 (1855).
Fimbristylis caripensis Steud., Syn. Pl. Glumac. 2: 112 (1855).
Fimbristylis circinnata Steud., Syn. Pl. Glumac. 2: 116 (1855).
Fimbristylis confinis Steud., Syn. Pl. Glumac. 2: 115 (1855).
Fimbristylis diffusula Steud., Syn. Pl. Glumac. 2: 116 (1855).
Fimbristylis fuscata Steud., Syn. Pl. Glumac. 2: 112 (1855).
Fimbristylis goeringiana Steud., Syn. Pl. Glumac. 2: 118 (1855).
Fimbristylis metzii Steud., Syn. Pl. Glumac. 2: 117 (1855).
Fimbristylis philippica Steud., Syn. Pl. Glumac. 2: 116 (1855).
Fimbristylis pohliana Steud., Syn. Pl. Glumac. 2: 119 (1855).
Fimbristylis similis Steud., Syn. Pl. Glumac. 2: 112 (1855).
Fimbristylis heynei Boeck., Flora 43: 244 (1860).
Fimbristylis diphylla var. *floribunda* Miq., Ann. Mus. Bot. Lugduno-Batavi 2: 144 (1865). *Fimbristylis dichotoma* var. *floribunda* (Miq.) T.Koyama, J. Jap. Bot. 63: 90 (1988).
Fimbristylis polymorpha Boeck., Vidensk. Meddel. Naturhist. Foren. Kjøbenhavn 1869: 141 (1869). *Iria polymorpha* (Boeck.) Kuntze, Revis. Gen. Pl. 2: 752 (1891). *Iria polymorpha* var. *normalis* Kuntze, Revis. Gen. Pl. 2: 752 (1891), nom. inval.
Fimbristylis dichotoma var. *glaberrima* Boeck., Linnaea 37: 13 (1871).
Fimbristylis squamulosa Hochst. ex Boeck., Linnaea 37: 15 (1871).
Fimbristylis sechellensis Baker, Fl. Mauritius: 418 (1877).
Fimbristylis novae-britanniae Boeck., Bot. Jahrb. Syst. 5: 93 (1884).
Fimbristylis eggersii Boeck., Beitr. Cyper. 2: 17 (1890).
Iria spirostachya Kuntze, Revis. Gen. Pl. 2: 753 (1891).
Fimbristylis diphylla var. *nilagirica* C.B.Clarke in J.D.Hooker, Fl. Brit. India 6: 637 (1893). *Fimbristylis dichotoma* var. *nilagirica* (C.B. Clarke) Karthik. in S.Karthikeyan & al., Fl. Ind. Enumerat. – Monocot.: 51 (1989).
Fimbristylis alamosana Fernald, Proc. Amer. Acad. Arts 36: 491 (1901).
Fimbristylis holwayana Fernald, Proc. Amer. Acad. Arts 36: 492 (1901).
Fimbristylis germainii E.G.Camus, Notul. Syst. (Paris) 1: 246 (1910).
Fimbristylis diphylla var. *laxa* E.G.Camus in H.Lecomte, Fl. Indo-Chine 7: 104 (1912).
Fimbristylis diphylla f. *coarctata* Domin, Biblioth. Bot. 85: 457 (1915).
Fimbristylis diphylla f. *glabra* Domin, Biblioth. Bot. 85: 457 (1915).
Fimbristylis diphylla var. *polystachya* Domin, Biblioth. Bot. 89: 457 (1915).

Fimbristylis diphylla f. *subpubescens* Domin, Biblioth. Bot. 85: 457 (1915).
Fimbristylis luzuliformis Domin, Biblioth. Bot. 85: 457 (1915).
Fimbristylis tikushiensis Hayata, Icon. Pl. Formosan. 6: 113 (1916). *Fimbristylis dichotoma* var. *tikushiensis* (Hayata) T.Koyama, J. Jap. Bot. 63: 91 (1988).
Fimbristylis diphylla var. *paucispiculata* Cherm., Bull. Soc. Bot. France 68: 418 (1921).
Fimbristylis annua var. *pseudoferruginea* Kük., Bot. Jahrb. Syst. 59: 5 (1924).
Fimbristylis annua f. *oblonga* Kük., Repert. Spec. Nov. Regni Veg. 23: 196 (1926).
Fimbristylis tashiroana Ohwi, Bot. Mag. (Tokyo) 45: 377 (1931). *Fimbristylis dichotoma* var. *tashiroana* (Ohwi) Ohwi, J. Jap. Bot. 14: 578 (1938).
Fimbristylis tashiroana var. *cincta* Ohwi, Bot. Mag. (Tokyo) 45: 377 (1931). *Fimbristylis dichotoma* f. *cincta* (Ohwi) T.Koyama, J. Jap. Bot. 63: 93 (1988).
Fimbristylis darlingtoniana Pennell, Bartonia 15: 30 (1933).
Fimbristylis annua var. *paucispiculata* Blatt. & McCann, J. Bombay Nat. Hist. Soc. 37: 544 (1934). *Fimbristylis dichotoma* var. *paucispiculata* (Blatt. & McCann) Panigrahi, Bull. Bot. Surv. India 11: 111 (1969 publ. 1971).
Fimbristylis ramosii Kük., Mitth. Thüring. Bot. Vereins, n.s., 50: 110 (1943).
Fimbristylis arenicola Wiggins, Contr. Dudley Herb. 4: 15 (1950).
Fimbristylis diphylla subsp. *diffusa* D.B.Ward, Castanea 33: 127 (1968).
Fimbristylis dichotoma f. *gracillima* T.Koyama, J. Jap. Bot. 63: 92 (1988).
Fimbristylis dichotoma var. *tentsuki* T.Koyama, J. Jap. Bot. 63: 91 (1988).
Fimbristylis dichotoma f. *tentsuki* T.Koyama, J. Jap. Bot. 63: 91 (1988).

subsp. ***glauca*** (Vahl) T.Koyama, Bot. Mag. (Tokyo) 87: 324 (1974).
S. India, Sri Lanka. 40 IND SRL. Ther. or hemicr.
**Fimbristylis glauca* Vahl, Enum. Pl. 2: 288 (1805).
Fimbristylis royeniana var. *microstachya* Nees in R.Wight, Contr. Bot. India: 99 (1834).

var. ***ochotensis*** (Meinsh.) T.Koyama, J. Jap. Bot. 63: 92 (1988).
Kamchatka. 31 KAM. Ther.
**Scirpus ochotensis* Meinsh., Trudy Imp. S.-Peterburgsk. Bot. Sada 18: 248 (1900). *Fimbristylis ochotensis* (Meinsh.) Kom., Fl. Kamtschatka 1: 212 (1927).

subsp. ***ophiticola*** (Britton) Zavaro, Fontqueria 42: 199 (1995).
Cuba. 81 CUB. Ther. or hemicr.
**Fimbristylis ophiticola* Britton, Mem. Torrey Bot. Club 16: 60 (1920).

subsp. ***podocarpa*** (Nees) T.Koyama, Micronesica 1: 87 (1964).
Trop. & Subtrop. Old World to NW. Pacific. 22 GAM GHA GUI MLI NGA SEN SIE TOG 23 BUR ZAI 29 MAU MDG 36 CHS 38 NNS TAI 40 ASS EHM IND NEP SRL WHM 41 THA VIE 42 JAW LSI MLY MOL PHI SUL SUM 43 BIS NWG 50 NTA QLD 62 CRL MRN (77) tex (78) ala ark fla geo lou msi nca sca. Ther. or hemicr.
Fimbristylis tomentosa Vahl, Enum. Pl. 2: 290 (1805). *Fimbristylis annua* f. *tomentosa* (Vahl) Kük., Repert. Spec. Nov. Regni Veg. 23: 196 (1926). *Fimbristylis dichotoma* f. *tomentosa* (Vahl) Ohwi, J. Jap. Bot. 14: 578 (1938). *Fimbristylis diphylla* var. *tomentosa* (Vahl) Barros, Anales Mus. Argent. Ci. Nat. "Bernardino Rivadavia" 41: 328 (1945).
Fimbristylis affinis C.Presl, Reliq. Haenk. 1: 191 (1828).
Fimbristylis cincta Nees in R.Wight, Contr. Bot. India: 98 (1834).
**Fimbristylis podocarpa* Nees in R.Wight, Contr. Bot. India: 98 (1834).
Fimbristylis subtristachya Steud., Syn. Pl. Glumac. 2: 109 (1855).
Fimbristylis diphylla var. *pluristriata* C.B.Clarke in J.D.Hooker, Fl. Brit. India 6: 637 (1893). *Fimbristylis pluristriata* (C.B.Clarke) Berhaut, Bull. Soc. Bot. France 101: 376 (1955). *Fimbristylis dichotoma* var. *pluristriata* (C.B.Clarke) Napper, Kew Bull. 25: 437 (1971).

Fimbristylis dichotomoides Tang & F.T.Wang, in Fl. Reipubl. Popul. Sin. 11: 228 (1961).
SE. China. 36 CHS. Hemicr. or rhizome geophyte.

Fimbristylis dictyocolea S.T.Blake, J. Arnold Arbor. 35: 209 (1954).
Indo-China, New Guinea to N. Australia. 41 CBD THA 43 NWG 50 NTA WAU. Hemicr.

Fimbristylis diglumoides Govind. & S.K.Varma, J. Econ. Taxon. Bot. 21: 380 (1997).
India (Bihar). 40 IND. Hemicr. or rhizome geophyte.

Fimbristylis dimorphonucifera Govind., Rheedea 7: 122 (1997).
India (Karnataka). 40 IND. Hemicr. or rhizome geophyte.

Fimbristylis diphylloides Makino, Fl. Japan: 1389 (1925).
S. China, S. Korea, C. & S. Japan. 36 CHC CHS 38 JAP KOR. Hemicr. or rhizome geophyte.
Fimbristylis campylophylla Tuyama, J. Jap. Bot. 11: 248 (1935).
Fimbristylis brevicollis Kük., Mitth. Thüring. Bot. Vereins, n.s., 50: 9 (1943).
Fimbristylis diphylloides var. *straminea* Tang & F.T. Wang, in Fl. Reipubl. Popul. Sin. 11: 227 (1961).

Fimbristylis dipsacea (Rottb.) C.B.Clarke in J.D.Hooker, Fl. Brit. India 6: 635 (1893).
W. Trop. Africa to Tanzania, Russian Far East to Trop. Asia, N. Australia, S. Venezuela to Brazil. 22 MGA MLI SIE 23 ZAI 25 TAN 31 KHA PRM 36 CHC CHM CHS 38 JAP 40 ASS IND SRL 41 MYA THA VIE 42 BOR PHI SUL SUM 43 NWG 50 NTA 82 VEN 84 BZN. Ther.
**Scirpus dipsaceus* Rottb., Descr. Icon. Rar. Pl.: 56 (1773). *Echinolytrum dipsaceum* (Rottb.) Desf., J. Bot. (Desvaux) 1: 20 (1808). *Isolepis dipsacea* (Rottb.) Roem. & Schult., Syst. Veg. 2: 119 (1817).
Scirpus minimus Roxb., Fl. Ind. 1: 219 (1820).
Isolepis elachista Schult. in J.J.Roemer & J.A. Schultes, Mant. 2: 61 (1824).
Isolepis verrucifera Maxim., Prim. Fl. Amur.: 300 (1859). *Fimbristylis verrucifera* (Maxim.) Makino,

Bot. Mag. (Tokyo) 9: 259 (1895). *Scirpus verruciferus* (Maxim.) Meinsh., Trudy Imp. S.-Peterburgsk. Bot. Sada 18: 246 (1901). *Echinolytrum verruciferum* (Maxim.) Makino, Bot. Mag. (Tokyo) 20: 9 (1906). *Fimbristylis dipsacea* var. *verrucifera* (Maxim.) T.Koyama, J. Fac. Sci. Univ. Tokyo, Sect. 3, Bot. 8: 118 (1961).
Fimbristylis nipponensis Makino, Bot. Mag. (Tokyo) 6: 47 (1892).

Fimbristylis disticha Boeck., Linnaea 38: 393 (1874). *Iria disticha* (Boeck.) Kuntze, Revis. Gen. Pl. 2: 753 (1891).
S. China to N. Sumatera. 36 CHS 41 AND MYA THA VIE 42 SUM. Ther.

Fimbristylis distincta S.T.Blake, Pap. Dept. Biol. Univ. Queensland 1(13): 12 (1940).
N. Australia. 50 NTA QLD. Hemicr. or rhizome geophyte.

Fimbristylis dolera S.T.Blake, Pap. Dept. Biol. Univ. Queensland 1(13): 7 (1940).
N. Australia. 50 NTA QLD. Hemicr. or rhizome geophyte.

Fimbristylis doliiformis Govind., Rheedea 8: 89 (1998).
India (Tamil Nadu). 40 IND. Hemicr. or rhizome geophyte.

Fimbristylis dunlopii Latz, Nuytsia 7: 166 (1990).
Northern Territory. 50 NTA. Ther.

Fimbristylis dura (Zoll. & Moritzi) Merr., Philipp. J. Sci., C 11: 53 (1916).
Trop. Asia. 40 IND SRL 41 MYA THA VIE 42 BOR JAW MLY SUM. Hemicr.
**Isolepis dura* Zoll. & Moritzi, Syst. Verz.: 97 (1846).
Fimbristylis asperrima Boeck., Linnaea 37: 40 (1871).
Trichelostylis asperrima Nees ex Boeck., Linnaea 37: 41 (1871).

Fimbristylis eichleriana Govind., Beitr. Biol. Pflanzen 70: 79 (1998).
S. India. 40 IND. Hemicr. or rhizome geophyte.

Fimbristylis elegans S.T.Blake, Pap. Dept. Biol. Univ. Queensland 1(13): 3 (1940).
N. Australia. 50 NTA QLD WAU. Hemicr. or rhizome geophyte.

Fimbristylis eligulata Govind., Proc. Indian Acad. Sci., B 76: 184 (1972).
S. India. 40 IND. Hemicr. or rhizome geophyte.

Fimbristylis engleriana Buscal. & Muschl., Bot. Jahrb. Syst. 49: 461 (1913).
Zimbabwe. 26 ZIM. Hemicr. or rhizome geophyte.

Fimbristylis eragrostis (Nees) Hance, J. Linn. Soc., Bot. 13: 132 (1873).
Trop. & Subtrop. Asia to N. queensland. 36 CHC CHH CHS 38 TAI 40 ASS IND SRL 41 NCB THA VIE 42 LSI MLY 43 NWG 50 QLD. Hemicr.
**Abildgaardia eragrostis* Nees in R.Wight, Contr. Bot. India: 95 (1834). *Iria eragrostis* (Nees) Kuntze, Revis. Gen. Pl. 2: 753 (1891).
Fimbristylis nigrobrunnea Thwaites, Enum. Pl. Zeyl.: 434 (1864). *Fimbristylis eragrostis* var. *nigrobrunnea* (Thwaites) S.S.Hooper in C.J.Saldanha & D.H. Nicolson, Fl. Hassan Distr. Karnataka: 678 (1976).
Fimbristylis subtetrastachya Boeck., Linnaea 37: 50 (1871). *Iria subtetrastachya* (Boeck.) Kuntze, Revis. Gen. Pl. 2: 753 (1891).
Fimbristylis pycnostachya Hance, J. Bot. 15: 338 (1877).
Fimbristylis lepidota E.G.Camus, Notul. Syst. (Paris) 1: 247 (1910).
Fimbristylis tortispica Turrill, Bull. Misc. Inform. Kew 1911: 348 (1911).
Fimbristylis schlechteri Kük., Bot. Jahrb. Syst. 59: 50 (1924).

Fimbristylis eremophila Latz, J. Adelaide Bot. Gard. 1: 345 (1979).
Western Australia to Northern Territory. 50 NTA WAU. Hemicr. or rhizome geophyte.

Fimbristylis falcata (Vahl) Kunth, Enum. Pl. 2: 239 (1837).
W. Indian Ocean, Trop. Asia. 29 COM MDG 40 ASS EHM IND NEP PAK SRL WHM 41 THA VIE 42 PHI 43 BIS NWG. Hemicr.
**Scirpus falcatus* Vahl, Enum. Pl. 2: 275 (1805). *Isolepis falcata* (Vahl) Roem. & Schult., Syst. Veg. 2: 118 (1817). *Iria falcata* (Vahl) Kuntze, Revis. Gen. Pl.: 753 (1891).

var. ***abbreviata*** (Boeck.) Karthik. in S.Karthikeyan & al., Fl. Ind. Enumerat. – Monocot.: 52 (1989).
Comoros, Madagascar, India to Indo-China. 29 COM MDG 40 IND 41+. Hemicr.
**Fimbristylis abbreviata* Boeck., Flora 12: 601 (1858). *Fimbristylis junciformis* var. *abbreviata* (Boeck.) C.B.Clarke in J.D.Hooker, Fl. Brit. India 6: 648 (1893).

var. ***falcata***
Trop. Asia. 40 ASS EHM IND NEP PAK SRL WHM 41 THA VIE 42 PHI 43 BIS NWG. Hemicr.
Fimbristylis brevifolia J.Presl & C.Presl, Reliq. Haenk. 1: 192 (1828), nom. illeg.
Fimbristylis brachyphylla J.Presl & C.Presl, Reliq. Haenk. 1: 351 (1830), nom. illeg.
Trichelostylis chaetorrhiza Nees in R.Wight, Contr. Bot. India: 106 (1834). *Fimbristylis chaetorrhiza* (Nees) Kunth, Enum. Pl. 2: 240 (1837).
Trichelostylis junciformis Nees in R.Wight, Contr. Bot. India: 106 (1834). *Fimbristylis junciformis* (Nees) Kunth, Enum. Pl. 2: 239 (1837). *Iria junciformis* (Nees) Kuntze, Revis. Gen. Pl. 2: 752 (1891).
Trichelostylis latifolia Nees in R.Wight, Contr. Bot. India: 106 (1834). *Fimbristylis latifolia* (Nees) Kunth, Enum. Pl. 2: 239 (1837). *Fimbristylis falcata* var. *latifolia* (Nees) H.B.Naithani & M.B.Raizada, Indian Forester 103: 414 (1977).
Trichelostylis torta Nees in R.Wight, Contr. Bot. India: 105 (1834). *Fimbristylis torta* (Nees) Kunth, Enum. Pl. 2: 241 (1837). *Iria torta* (Nees) Kuntze, Revis. Gen. Pl. 2: 753 (1891).
Scirpus obtusifolius B.Heyne ex Kunth, Enum. Pl. 2: 239 (1837).
Fimbristylis haenkei D.Dietr., Syn. Pl. 2: 161 (1840).
Fimbristylis crassipes Boeck., Flora 41: 602 (1858). *Iria crassipes* (Boeck.) Kuntze, Revis. Gen. Pl. 2: 753 (1891).
Fimbristylis junciformis var. *conostachya* C.B.Clarke, J. Bot. (Morot) 3: 197 (1889).

Fimbristylis falcifolia Boeck., Linnaea 37: 25 (1871). *Iria falcifolia* (Boeck.) Kuntze, Revis. Gen. Pl. 2: 753 (1891).
NE. Trop. Africa. 24 SUD. Hemicr. or rhizome geophyte.

Fimbristylis fenestrata Kük., Bot. Jahrb. Syst. 59: 49 (1924).
Vietnam, NE. New Guinea. 41 VIE 43 NWG. Ther.

Fimbristylis ferruginea (L.) Vahl, Enum. Pl. 2: 291 (1805).
Trop. & Subtrop. 20 ALG EGY LBY 22 BEN GAM GHA GNB GUI IVO LBR MTN NGA NGR SEN SIE TOG 23 EQG GAB GGI ZAI 24 CHA SOM 25 KEN TAN 27 BOT CPP NAM NAT SWZ TVL 29 ALD MAU MDG REU 34 LBS PAL SIN 35 GST SAU YEM 36 CHH CHS 38 JAP NNS OGA TAI 40 ASS IND SRL 41 AND MYA NCB VIE 42 BOR JAW LSI MLY MOL PHI SUL SUM 43 BIS NWG 50 NSW NTA QLD WAU 60 NWC 61 MRQ SCI 79 MXT 80 BLZ 81 BAH CAY CUB DOM JAM LEE NLA PUE TRT WIN 82 FRG GUY SUR VEN. Hemicr.
**Scirpus ferrugineus* L., Sp. Pl.: 50 (1753). *Isolepis ferruginea* (L.) Schltdl., Linnaea 6: 27 (1831). *Iria ferruginea* (L.) Kuntze, Revis. Gen. Pl. 2: 752 (1891). *Fimbristylis spadicea* var. *ferruginea* (L.) Alph.Wood, Class-book Bot.: 741 (1861).

var. ***anpinensis*** (Hayata) H.Y.Liu, in Fl. Taiwan, ed. 2, 5: 269 (2000).
SE. China to Temp. E. Asia. 36 CHS 38 NNS OGA TAI. Hemicr.
**Fimbristylis anpinensis* Hayata, Icon. Pl. Formosan. 6: 108 (1916). *Fimbristylis sieboldii* subsp. *anpinensis* (Hayata) T.Koyama, in Fl. Taiwan 5: 242 (1978).

var. ***ferruginea***
Trop. & Subtrop. 20 ALG EGY LBY 22 BEN GAM GHA GNB GUI IVO LBR MTN NGA NGR SEN SIE TOG 23 GAB ZAI 24 CHA SOM 25 KEN TAN 27 BOT CPP NAM NAT SWZ TVL 29 ALD MAU MDG REU 34 LBS PAL SIN 35 GST SAU YEM 36 CHH CHS 38 JAP NNS TAI 40 ASS IND SRL 41 AND VIE 42 BOR JAW LSI MLY MOL PHI SUL SUM 43 BIS NWG 50 NSW NTA QLD WAU 60 NWC 61 MRQ SCI 79 MXT 80 BLZ 81 BAH CAY CUB DOM JAM LEE NLA PUE TRT WIN 82 FRG GUY SUR VEN. Hemicr.
Schoenus polymorphus Rottb., Descr. Icon. Rar. Pl.: 67 (1773).
Scirpus miliaceus Thunb., Fl. Jap.: 37 (1784), nom. illeg.
Scirpus arvensis Retz., Observ. Bot. 4: 11 (1786). *Fimbristylis arvensis* (Retz.) Vahl, Enum. Pl. 2: 291 (1805).
Scirpus debilis Lam., Tabl. Encycl. 1: 141 (1791). *Scirpus ferrugineus* var. *debilis* (Lam.) Poir. in J.B.A.M.de Lamarck, Encycl. 6: 780 (1804).
Scirpus bonariensis Poir. in J.B.A.M.de Lamarck, Encycl. 6: 763 (1805).
Scirpus tranquebariensis Roth, Catal. Bot. 3: 6 (1806).
Fimbristylis brevifolia R.Br., Prodr.: 228 (1810).
Fimbristylis stans Rich. ex Spreng., Pl. Min. Cogn. Pug. 1: 6 (1813).
Fimbristylis compressa Roem. & Schult., Syst. Veg. 2: 100 (1817).
Scirpus brevifolius Poir. in J.B.A.M.de Lamarck, Encycl., Suppl. 5: 99 (1817), nom. inval.
Scirpus globulosus Roxb., Fl. Ind. 1: 217 (1820).
Scirpus tristachyos Roxb., Fl. Ind. 1: 221 (1820), nom. illeg.
Eleocharis juncoides Schult. in J.J.Roemer & J.A.Schultes, Mant. 2: 90 (1824).
Fimbristylis marginata Labill., Sert Austro-Caledon.: t. 16 (1824).
Scirpus stans Ham., Prodr. Pl. Ind. Occid.: 14 (1825).
Fimbristylis mauritiana Tausch ex C.Presl, Isis (Oken) 21: 268 (1828).
Fimbristylis roxburghii A.Dietr., Sp. Pl. 2: 162 (1832).
Fimbristylis sieberiana Kunth, Enum. Pl. 2: 237 (1837). *Fimbristylis ferruginea* var. *sieberiana* (Kunth) Boeck., Linnaea 37: 17 (1871). *Fimbristylis ferruginea* subsp. *sieberiana* (Kunth) Lye, Nordic J. Bot. 2: 335 (1982).
Scirpus cinereofuscus Willd. ex Kunth, Enum. Pl. 2: 236 (1837).
Scirpus pallescens Willd. ex Kunth, Enum. Pl. 2: 237 (1837).
Campylostachys abbreviata E.Mey. in J.F.Drège, Zwei Pflanzengeogr. Dokum.: 83 (1843), nom. illeg.
Fimbristylis lomentocarpa Hochst. ex A.Rich., Tent. Fl. Abyss. 2: 504 (1850).
Fimbristylis sublateralis Steud., Syn. Pl. Glumac. 2: 114 (1855).
Fimbristylis trispicata Steud., Syn. Pl. Glumac. 2: 107 (1855).
Fimbristylis cyrtophylla Miq., Fl. Ned. Ind. 3: 325 (1856).
Fimbristylis ochreata Boeck., Flora 41: 599 (1858).
Fimbristylis paucispicata F.Muell., Fragm. 1: 197 (1859).
Fimbristylis leiocarpa Miq., Ann. Mus. Bot. Lugduno-Batavi 2: 144 (1865), nom. illeg.
Fimbristylis andamandica Kurz, Rep. Veg. Andaman Isl., App. B: 21 (1870).
Scirpus habessinicus Ehrenb. ex Boeck., Linnaea 37: 17 (1871).
Fimbristylis sieboldii Miq. ex Franch. & Sav., Enum. Pl. Jap. 2: 118 (1877).
Fimbristylis ferruginea var. *foliata* Benth., Fl. Austral. 7: 312 (1878).
Fimbristylis ferruginea f. *decomposita* Domin, Biblioth. Bot. 85: 459 (1915).
Fimbristylis aginkotensis Hayata, Icon. Pl. Formosan. 6: 108 (1916).
Fimbristylis ciliata C.B.Clarke ex Drobow, Opred. Rast. Tashk. 1: 49 (1923).
Fimbristylis ferruginea var. *compacta* Kük., Repert. Spec. Nov. Regni Veg. 23: 196 (1926).

Fimbristylis fibrillosa Goetgh., Bull. Jard. Bot. Natl. Belg. 54: 79 (1984).
Zaïre. 23 ZAI. Hemicr. or rhizome geophyte.

Fimbristylis filifolia Boeck., Linnaea 37: 32 (1871). *Iria filifolia* (Boeck.) Kuntze, Revis. Gen. Pl. 2: 753 (1891).
SW. Darjiling to Assam. 40 ASS EHM. Hemicr. or rhizome geophyte.

Fimbristylis fimbristyloides (F.Muell.) Druce, Bot. Soc. Exch. Club Brit. Isles 1916: 623 (1917).
Trop. & Subtrop. Asia to Queensland. 38 KOR NNS 40 EHM IND NEP 41 MYA THA 42 BOR JAW LSI SUL SUM 43 NWG 50 QLD. Ther.
**Abildgaardia fimbristyloides* F.Muell., Fragm. 8: 273 (1874). *Iria fimbristyloides* (F.Muell.) Kuntze, Revis. Gen. Pl. 2: 753 (1891).
Fimbristylis dallachyi F.Muell. ex Benth., Fl. Austral. 7: 309 (1878). *Fimbristylis disticha* var. *dallachyi* (F.Muell. ex Benth.) Domin, Biblioth. Bot. 85: 454 (1915).

Fimbristylis lacei Turrill, Bull. Misc. Inform. Kew 1911: 348 (1911).
Fimbristylis nanofusca Tang & F.T.Wang, in Fl. Reipubl. Popul. Sin. 11: 229 (1961).

Fimbristylis fordii C.B.Clarke, J. Linn. Soc., Bot. 36: 236 (1903).
SE. China to Hainan. 36 CHH CHS. Hemicr. or rhizome geophyte.

Fimbristylis fuchsiana Govind., Beitr. Biol. Pflanzen 70: 81 (1998).
India (Uttar Pradesh). 40 IND. Hemicr. or rhizome geophyte.

Fimbristylis fulvescens (Thwaites) Thwaites, Enum. Pl. Zeyl.: 434 (1864).
Sri Lanka, Myanmar. 40 SRL 41 MYA. Hemicr.
**Abildgaardia fulvescens* Thwaites, Enum. Pl. Zeyl.: 347 (1864). *Fimbristylis fusca* subsp. *fulvescens* (Thwaites) T.Koyama, Bot. Mag. (Tokyo) 87: 307 (1974).
Abildgaardia fusca var. *longifolia* Boeck., Linnaea 37: 55 (1871).

Fimbristylis furva R.Br., Prodr.: 228 (1810). *Scirpus furva* (R.Br.) Poir. in J.B.A.M.de Lamarck, Encycl., Suppl. 5: 99 (1817), nom. inval. *Iria furva* (R.Br.) Kuntze, Revis. Gen. Pl. 2: 753 (1891).
New Guinea to N. Australia. 43 NWG 50 NTA QLD WAU. Hemicr.
Fimbristylis capitata R.Br., Prodr.: 228 (1810). *Trichelostylis capitata* (R.Br.) Nees, Linnaea 9: 290 (1834). *Iria capitata* (R.Br.) Kuntze, Revis. Gen. Pl. 2: 753 (1891).
Fimbristylis allenii Turrill, Bull. Misc. Inform. Kew 1915: 117 (1915).

Fimbristylis fusca (Nees) Benth. ex C.B.Clarke in J.D.Hooker, Fl. Brit. India 6: 649 (1893).
Trop. & Subtrop. Asia to N. Australia. 36 CHH CHS 38 JAP 40 NEP 41 THA VIE 42 BOR JAW MLY MOL PHI SUL SUM 43 NWG 50 NTA. Hemicr.
Gussonea cyperoides J.Presl & C.Presl, Reliq. Haenk. 1: 183 (1828). *Abildgaardia cyperoides* (J.Presl & C.Presl) Nees & Meyen in R.Wight, Contr. Bot. India: 95 (1834). *Fimbristylis cyperoides* (J.Presl & C.Presl) F.Muell., Fragm. 9: 11 (1875), nom. illeg.
Schoenus puberulus C.A.Mey., Mém. Sav. Étr. Acad. St. Pétersbourg 1: 196 (1831).
**Abildgaardia fusca* Nees in R.Wight, Contr. Bot. India: 95 (1834). *Iria fusca* (Nees) Kuntze, Revis. Gen. Pl. 2: 753 (1891).
Gussonea pauciflora Brongn. in L.I.Duperrey, Voy. Monde, Phan.: t. 34 (1834). *Abildgaardia pauciflora* (Brongn.) Kunth, Enum. Pl. 2: 249 (1837).
Isolepis longispica Steud., Syn. Pl. Glumac. 2: 104 (1855).
Rhynchospora anomala Steud., Syn. Pl. Glumac. 2: 149 (1855).
Fimbristylis subfusca E.G.Camus, Notul. Syst. (Paris) 1: 248 (1910).
Fimbristylis rigidifolia Ridl., J. Straits Branch Roy. Asiat. Soc. 59: 223 (1911).
Fimbristylis stenochlaena Kük., Mitth. Thüring. Bot. Vereins, n.s., 50: 11 (1943).

Fimbristylis fuscinux C.B.Clarke in J.D.Hooker, Fl. Brit. India 6: 638 (1893).
N. India to Darjiling. 40 EHM IND. Hemicr. or rhizome geophyte.

Fimbristylis fuscoides C.B.Clarke, Bull. Herb. Boissier, II, 5: 719 (1905).
Indo-China to W. Malesia. 41 CBD THA VIE 42 BOR MLY SUM. Hemicr.
Fimbristylis erythradenia E.G.Camus, Notul. Syst. (Paris) 1: 247 (1910).
Fimbristylis angustifolia Ridl., J. Straits Branch Roy. Asiat. Soc. 59: 223 (1911).

Fimbristylis gabonica Cherm., Bull. Soc. Bot. France 77: 276 (1930).
WC. Trop. Africa to Uganda. 23 GAB ZAI 25 UGA. Hemicr.
Fimbristylis robusta Lye, Bot. Not. 127: 498 (1974).

Fimbristylis gambleana Boeck., Beitr. Cyper. 2: 40 (1890).
S. India. 40 IND. Hemicr. or rhizome geophyte.

Fimbristylis gentarea Govind., Beitr. Biol. Pflanzen 70: 83 (1998).
India (Tamil Nadu). 40 IND. Hemicr. or rhizome geophyte.

Fimbristylis gigantea Kük. in R.E.Fries, Wiss. Erg. Schwed. Rhod.-Kongo Exped. 1: 5 (1921).
Zambia. 26 ZAM. Hemicr. or rhizome geophyte.

Fimbristylis glaucophylla (Boeck.) Beetle, Amer. Midl. Naturalist 41: 468 (1949).
Brazil. 84+. Hemicr. or rhizome geophyte.
**Scirpus glaucophyllus* Boeck., Vidensk. Meddel. Naturhist. Foren. Kjøbenhavn 1871: 153 (1871).

Fimbristylis glazioviana Boeck., Allg. Bot. Z. Syst. 2: 77 (1896).
SE. Brazil. 84 BZL. Hemicr. or rhizome geophyte.

Fimbristylis gracilenta Hance, J. Bot. 6: 89 (1868).
S. China to Sumatera. 36 CHS 41 THA VIE 42 SUM. Ther.

var. ***gracilenta***
S. China to Sumatera. 36 CHS 41 THA VIE 42 SUM. Ther.
Fimbristylis thorelii E.G.Camus, Notul. Syst. (Paris) 1: 246 (1910).

var. ***psilopoda*** J.Kern, Reinwardtia 6: 49 (1961).
NE. Thailand. 41 THA. Ther.

Fimbristylis griffithii Boeck., Flora 43: 241 (1860).
Trop. Asia. 32 NWG 40+ 41 AND MYA THA VIE 42 BOR JAW MLY MOL SUL SUM. Ther.
Fimbristylis aestivalis var. *glaberrima* Boeck., Linnaea 37: 11 (1871).

Fimbristylis hainanensis Tang & F.T.Wang, in Fl. Reipubl. Popul. Sin. 11: 228 (1961).
SE. China to Hainan. 36 CHH CHS. Hemicr. or rhizome geophyte.

Fimbristylis hawaiiensis Hillebr., Fl. Hawaiian Isl.: 472 (1888). *Iria hawaiiensis* (Hillebr.) Kuntze, Revis. Gen. Pl. 2: 753 (1891).
Hawaiian Is. 63 HAW. Hemicr. or rhizome geophyte.

Fimbristylis henryi C.B.Clarke, J. Linn. Soc., Bot. 36: 237 (1903).
C. & S. China. 36 CHC CHN CHS. Hemicr. or rhizome geophyte.
Fimbristylis unicolor Ohwi & T.Koyama, Bull. Natl. Sci. Mus. Tokyo, n.s., 3(38): 30 (1956).

Fimbristylis hirsutifolia Govind., Proc. Indian Acad. Sci., Pl. Sci. 100: 79 (1990).
India. 40 IND. Ther.

Fimbristylis hookeriana Boeck., Linnaea 37: 22 (1871). *Iria hookeriana* (Boeck.) Kuntze, Revis. Gen. Pl. 2: 753 (1891).
Assam to S. China and Indo-China. 36 CHS 40 ASS 41 LAO THA VIE. Ther. or hemicr.
Fimbristylis chalarocephala Ohwi & T.Koyama, Bull. Natl. Sci. Mus. Tokyo, n.s., 3(38): 27 (1956).

Fimbristylis humerosa Govind., Rheedea 7: 117 (1997).
India (Kerala). 40 IND. Hemicr. or rhizome geophyte.

Fimbristylis hyalina Govind. & Sasidh., J. Econ. Taxon. Bot. 21: 373 (1997).
India (Kerala). 40 IND. Hemicr. or rhizome geophyte.

Fimbristylis hygrophila Gordon-Gray, J. S. African Bot. 32: 129 (1966). *Abildgaardia hygrophila* (Gordon-Gray) Lye, Mitt. Bot. Staatssamml. München 10: 547 (1971).
S. Zaïre to S. Africa. 23 ZAI 26 ZAM 27 CPP NAT SWZ. Hemicr. or rhizome geophyte.

Fimbristylis inaguensis Britton, Torreya 13: 215 (1913).
Bahamas to Cuba (Cays), Anegada. 81 BAH CUB LEE. Hemicr. or rhizome geophyte.

Fimbristylis insignis Thwaites, Enum. Pl. Zeyl.: 349 (1864). *Iria insignis* (Thwaites) Kuntze, Revis. Gen. Pl. 2: 753 (1891).
S. China, Trop. Asia to N. Queensland. 36 CHS 40 SRL 41 LAO THA VIE 42 BOR MLY PHI SUL 43 NWG 50 QLD. Hemicr.
Fimbristylis thwaitesii Boeck., Linnaea 37: 34 (1871).
Fimbristylis petrogena Ohwi, Bot. Mag. (Tokyo) 56: 201 (1942).

Fimbristylis intonsa S.T.Blake, J. Arnold Arbor. 35: 221 (1954).
E. Himalaya, Sumatera, New Guinea. 40 EHM 42 SUM 43 NWG. Hemicr. or rhizome geophyte.

Fimbristylis × itaru-itoana T.Koyama, J. Jap. Bot. 30: 130 (1955). *F. microcarpa* × *F. stauntonii* var. *tonensis*.
Japan. 38 JAP. Hemicr. or rhizome geophyte.

Fimbristylis jucunda (C.B.Clarke) J.Kern, Blumea 10: 637 (1960).
Indo-China. 41 LAO THA VIE. Ther. or hemicr.
**Cyperus jucundus* C.B.Clarke, Kew Bull., Addit. Ser. 8: 3 (1908).
Fimbristylis brunnea C.B.Clarke ex E.G.Camus, Notul. Syst. (Paris) 1: 248 (1910).

Fimbristylis juncea (G.Forst.) R.Br. ex Roem. & Schult., Syst. Veg. 2: 102 (1817).
Marquesas, Society Is. 61 MRQ SCI. Hemicr. or rhizome geophyte.
**Scirpus junceus* G.Forst., Fl. Ins. Austr.: 6 (1786). *Iria juncea* (G.Forst.) Kuntze, Revis. Gen. Pl. 2: 753 (1891).

var. ***juncea***
Society Is. 61 SCI. Hemicr. or rhizome geophyte.

var. ***marquesana*** (Steud.) Fosberg, J. Jap. Bot. 63: 83 (1988).
Marquesas. 61 MRQ. Hemicr. or rhizome geophyte.
**Fimbristylis marquesana* Steud., Syn. Pl. Glumac. 2: 107 (1855).

var. ***nukahivensis*** (Steud.) Fosberg, J. Jap. Bot. 63: 82 (1988).
Marquesas. 61 MRQ. Hemicr. or rhizome geophyte.
**Fimbristylis nukahivensis* Steud., Syn. Pl. Glumac. 2: 117 (1855).
Fimbristylis separanda Steud. ex Jard., Mém. Soc. Sci. Nat. Cherbourg 5: 299 (1857), nom. inval.

var. ***separanda*** (Steud. ex Jard.) Fosberg, J. Jap. Bot. 63: 81 (1988).
Marquesas. 61 MRQ. Hemicr. or rhizome geophyte.
**Fimbristylis separanda* Steud. ex Jard., Mém. Soc. Sci. Nat. Cherbourg 5: 324 (1858).

var. ***tertia*** (Steud. ex Jard.) Fosberg, J. Jap. Bot. 63: 81 (1988).
Marquesas. 61 MRQ. Hemicr. or rhizome geophyte.
**Fimbristylis tertia* Steud. ex Jard., Mém. Soc. Sci. Nat. Cherbourg 5: 324 (1858).

Fimbristylis juncocephala Boeck., Flora 41: 600 (1858).
Brazil. 84+. Hemicr. or rhizome geophyte.

Fimbristylis kadzusana Ohwi, Repert. Spec. Nov. Regni Veg. 36: 43 (1934).
Japan (Honshu). 38 JAP. Hemicr. or rhizome geophyte.

Fimbristylis kernii T.Koyama, Brittonia 30: 102 (1978).
Indo-China. 41 CBD THA. Ther.

Fimbristylis kingii Gamble ex Boeck., Beitr. Cyper. 2: 40 (1890).
S. India. 40 IND. Hemicr. or rhizome geophyte.

Fimbristylis kwantungensis C.B.Clarke, J. Linn. Soc., Bot. 36: 238 (1903).
SE. China. 36 CHS. Hemicr. or rhizome geophyte.

Fimbristylis lanata Roem. & Schult., Syst. Veg. 2: 92 (1817).
Venezuela. 82 VEN. Hemicr. or rhizome geophyte.
Xyris eriophora Klotzsch in M.R.Schomburgk, Reis. Br.-Guiana: 1117 (1849).

Fimbristylis lanceolata C.B.Clarke, Kew Bull., Addit. Ser. 8: 25 (1908).
SE. Sulawesi to N. Australia. 42 MOL SUL 50 NTA QLD. Hemicr.

Fimbristylis lasiophylla J.Kern, Blumea 8: 115 (1955).
S. Vietnam. 41 VIE. Hemicr. or rhizome geophyte.

Fimbristylis latiglumifera Govind., J. Bombay Nat. Hist. Soc. 69: 161 (1972).
S. India. 40 IND. Hemicr. or rhizome geophyte.

Fimbristylis latinucifera Govind., J. Bombay Nat. Hist. Soc. 69: 159 (1972).
S. India. 40 IND. Hemicr. or rhizome geophyte.

Fimbristylis lawiana (Boeck.) J.Kern, Reinwardtia 4: 96 (1956).
W. India. 40 IND. Hemicr. or rhizome geophyte.
**Scirpus lawianus* Boeck., Linnaea 36: 497 (1870).
Fimbristylis digitata Boeck., Flora 61: 35 (1878).
Trichelostylis digitata Dalzell ex Boeck., Flora 61: 35 (1878).

Fimbristylis laxiglumis Latz, Nuytsia 7: 167 (1990).
N. Australia. 50 NTA QLD. Ther.

Fimbristylis leptoclada Benth., Fl. Hongk.: 893 (1861). *Iria leptoclada* (Benth.) Kuntze, Revis. Gen. Pl. 2: 753 (1891). *Fimbristylis vaginata* f. *leptoclada* (Benth.) Domin, Biblioth. Bot. 85: 460 (1915).
S. China to Trop. Asia. 36 CHS 38 NNS 40 SRL 41 THA VIE 42 BOR LSI MLY PHI SUM 43 NWG. Ther. or hemicr.

var. ***etuberculata*** J.Kern, Blumea 10: 637 (1960).
W. New Guinea. 43 NWG. Ther.

var. ***leptoclada***
S. China to Trop. Asia. 36 CHS 40 SRL 41 THA VIE 42 BOR LSI MLY PHI SUM. Ther. or hemicr.
Fimbristylis retusa Thwaites, Enum. Pl. Zeyl.: 349 (1864). *Iria retusa* (Thwaites) Kuntze, Revis. Gen. Pl. 2: 753 (1891).

var. ***takamineana*** (Ohwi) T.Koyama, J. Fac. Sci. Univ. Tokyo, Sect. 3, Bot. 8: 110 (1961).
Nansei-shoto. 38 NNS. Hemicr.
**Fimbristylis takamineana* Ohwi, J. Jap. Bot. 13: 332 (1937).

Fimbristylis leucocolea Benth., Fl. Austral. 7: 304 (1878). *Iria leucocolea* (Benth.) Kuntze, Revis. Gen. Pl. 2: 753 (1891).
N. Australia. 50 NTA WAU. Hemicr. or rhizome geophyte.

Fimbristylis leucostachya Roem. & Schult., Syst. Veg. 2: 92 (1817). *Iria leucostachya* (Roem. & Schult.) Kuntze, Revis. Gen. Pl. 2: 753 (1891).
Venezuela. 82 VEN. Hemicr. or rhizome geophyte.

Fimbristylis ligulata Govind., Proc. Indian Acad. Sci., B 76: 187 (1972).
S. India. 40 IND. Hemicr. or rhizome geophyte.

Fimbristylis limosa Poepp. & Kunth in C.S.Kunth, Enum. Pl. 2: 225 (1837).
Guyana. 82 GUY. Hemicr. or rhizome geophyte.

Fimbristylis lineatisquama Ohwi, Blumea 8: 109 (1955).
Philippines. 42 PHI. Hemicr.

Fimbristylis lithophila Govind., Beitr. Biol. Pflanzen 70: 87 (1998).
India (Tamil Nadu). 40 IND. Hemicr. or rhizome geophyte.

Fimbristylis littoralis Gaudich., Voy. Uranie: 413 (1829).
Trop. & Subtrop. 22 BEN GHA GNB IVO NGA NGR SEN 23 EQG GAB ZAI 25 TAN 29 MAU MDG REU SEY 34 AFG IRN IRQ 36 CHN CHS CHT 38 JAP KOR NNS TAI 40 ASS EHM IND NEP PAK SRL 41 VIE 42 BOR JAW LSI MLY MOL PHI SUL SUM 43 NWG 50 NTA QLD WAU (60) fij 61 SCI 62 CRL 63 HAW 74 MSO 77 TEX 78 ALA ARK FLA GEO KTY LOU MSI NCS SCA TEN 79 MXE MXG MXN MXS MXT 80 BLZ COS ELS GUA HON NIC PAN 81 BAH CUB DOM JAM JAM PUE TRT WIN 82 FRG GUY SUR VEN 83 BOL CLM ECU PER 84 BZC BZN. Ther. or hemicr.

var. ***koidzumiana*** (Ohwi) T.Koyama, J. Fac. Sci. Univ. Tokyo, Sect. 3, Bot. 8: 108 (1961).
Indo-China to Taiwan, NW. Pacific. 36 CHS 38 JAP NNS TAI 41 VIE 62+. Hemicr.
**Fimbristylis koidzumiana* Ohwi, Acta Phytotax. Geobot. 1: 78 (1932). *Fimbristylis miliacea* subsp. *koidzumiana* (Ohwi) T.Koyama, in Fl. Taiwan 5: 236 (1978).

var. ***littoralis***
Trop. & Subtrop. 22 BEN GHA GNB IVO NGA NGR SEN 23 EQG GAB ZAI 25 TAN 29 MAU MDG REU SEY 34 AFG IRN IRQ 36 CHN CHT 38 JAP KOR TAI 40 ASS EHM IND NEP PAK SRL 41 VIE 42 BOR JAW LSI MLY MOL PHI SUL SUM 43 NWG 50 NTA QLD WAU (60) fij 61 SCI 62 CRL 63 HAW 74 MSO 77 TEX 78 ALA ARK FLA GEO KTY LOU MSI NCS SCA TEN 79 MXE MXG MXN MXS MXT 80 BLZ COS ELS GUA HON NIC PAN 81 BAH CUB DOM JAM PUE TRT WIN 82 FRG GUY SUR VEN 83 BOL CLM ECU PER 84 BZC BZN. Ther. or hemicr.
Fimbristylis emarginata Wight in N.Wallich, Numer. List: 3500 (1831).
Fimbristylis tetragona A.Dietr., Sp. Pl. 2: 152 (1832).
Fimbristylis flaccidula Zoll., Syst. Verz. 1: 61 (1854), nom. inval.
Fimbristylis flaccida Steud., Syn. Pl. Glumac. 2: 113 (1855).
Fimbristylis quadrangularis A.Dietr. ex Steud., Syn. Pl. Glumac. 2: 114 (1855).
Fimbristylis hatsusimae Ohwi, Acta Phytotax. Geobot. 8: 68 (1939).

var. ***macrostachya*** (J.Kern) J.Kern, in Fl. Males. 7: 552 (1974).
Malesia. 42 JAW PHI. Ther. or hemicr.
**Fimbristylis miliacea* var. *macrostachya* J.Kern, Blumea 8: 117 (1955).

Fimbristylis longibracteata Pires de Lima, Bol. Soc. Brot., II, 2: 134 (1924).
Mozambique. 26 MOZ. Hemicr. or rhizome geophyte.

Fimbristylis longispica Steud., Syn. Pl. Glumac. 2: 118 (1855). *Fimbristylis dichotoma* subsp. *longispica* (Steud.) T.Koyama, J. Fac. Sci. Univ. Tokyo, Sect. 3, Bot. 8: 112 (1961).
China to Temp. E. Asia. 36 CHN CHS 38 JAP KOR KZN OGA 42 PHI?. Hemicr. or rhizome geophyte.

var. ***boninensis*** (Hayata) Ohwi, Mem. Coll. Sci. Kyoto Imp. Univ., Ser. B, Biol. 18: 80 (1944).
Ogasawara-shoto, Kazan-retto. 38 KZN OGA 42 PHI?. Ther. or hemicr.
**Fimbristylis boninensis* Hayata, Icon. Pl. Formosan. 6: 109 (1916). *Fimbristylis dichotoma* var. *boninensis* (Hayata) T.Koyama, J. Fac. Sci. Univ. Tokyo, Sect. 3, Bot. 8: 113 (1961).

var. ***hahajimensis*** (Tuyama) Ohwi, Mem. Coll. Sci. Kyoto Imp. Univ., Ser. B, Biol. 18: 80 (1944).
Ogasawara-shoto (Haha-jima). 38 OGA. Ther. or hemicr.
**Fimbristylis hahajimensis* Tuyama, Bot. Mag. (Tokyo) 49: 368 (1935). *Fimbristylis dichotoma* var. *hahajimensis* (Tuyama) Ohwi, J. Jap. Bot. 18: 134 (1942).

var. ***longispica***
China to Japan. 36 CHN CHS 38 JAP KOR. Hemicr. or rhizome geophyte.
Fimbristylis buergeri Miq., Ann. Mus. Bot. Lugduno-Batavi 2: 144 (1865).
Fimbristylis dietrichsenii Boeck., Bot. Jahrb. Syst. 5: 505 (1884).
Fimbristylis koreensis H.Lév., Bull. Acad. Int. Géogr. Bot. 14: 199 (1904).
Fimbristylis crassispica Palla, Oesterr. Bot. Z. 59: 192 (1909).

Fimbristylis longistigmata Govind., Proc. Indian Acad. Sci., B 76: 190 (1972).
S. India. 40 IND. Hemicr. or rhizome geophyte.

Fimbristylis longistipitata Tang & F.T.Wang, in Fl. Reipubl. Popul. Sin. 11: 229 (1961).
Hainan. 36 CHH. Hemicr. or rhizome geophyte.

Fimbristylis macassarensis Steud., Syn. Pl. Glumac. 2: 109 (1855).
Jawa (Madura) to N. Australia. 42 JAW PHI SUL 50 NTA WAU. Ther.
Fimbristylis corniculata Merr., Philipp. J. Sci., C 7: 231 (1912).

Fimbristylis macrantha Boeck., Linnaea 38: 388 (1874). *Iria macrantha* (Boeck.) Kuntze, Revis. Gen. Pl. 2: 753 (1891).
N. Australia. 50 NTA QLD WAU. Hemicr. or rhizome geophyte.

Fimbristylis madagascariensis Boeck., Abh. Naturwiss. Vereine Bremen 7: 38 (1880).
EC. & E. Trop. Africa, Madagascar. 23 BUR RWA 25 KEN TAN UGA 29 MDG. Hemicr.

Fimbristylis malayana Ohwi, Blumea 8: 96 (1955).
Pen. Malaysia (P. Langkawi). 42 MLY. Hemicr.

Fimbristylis mangorensis Cherm., Bull. Soc. Bot. France 72: 617 (1925).
C. Madagascar. 29 MDG. Hemicr.

Fimbristylis manilaliana Govind., Rheedea 8: 87 (1998).
India (Kerala). 40 IND. Hemicr. or rhizome geophyte.

Fimbristylis maracandica Zakirov, Bot. Mater. Gerb. Bot. Inst. Komarova Akad. Nauk S.S.S.R. 12: 64 (1950).
C. Asia. 32+. Hemicr. or rhizome geophyte.

Fimbristylis merguensis C.B.Clarke in J.D.Hooker, Fl. Brit. India 6: 642 (1893).
Pen. Malaysia. 42 MLY. Hemicr. or rhizome geophyte.

Fimbristylis merrillii J.Kern, Blumea 8: 135 (1955).
S. China, Indo-China to N. Australia. 36 CHC 41 THA VIE 42 JAW PHI SUL SUM 43 NWG 50+. Ther.

Fimbristylis mexicana Palla, Oesterr. Bot. Z. 58: 390 (1908). *Abildgaardia mexicana* (Palla) Kral, Sida 4: 71 (1971).
Mexico. 79 MXC MXE MXS MXT. Hemicr. or rhizome geophyte.
Fimbristylis crassipes Boeck., Linnaea 38: 392 (1874), nom. illeg.

Fimbristylis micans S.T.Blake, Pap. Dept. Biol. Univ. Queensland 1(13): 6 (1940).
Queensland. 50 QLD. Hemicr. or rhizome geophyte.

Fimbristylis microcarya F.Muell., Fragm. 1: 200 (1859). *Iria microcarya* (F.Muell.) Kuntze, Revis. Gen. Pl. 2: 753 (1891). *Fimbristylis complanata* var. *microcarya* (F.Muell.) C.B.Clarke in J.D.Hooker, Fl. Brit. India 6: 646 (1893). *Fimbristylis autumnalis* var. *microcarya* (F.Muell.) Kük., Bot. Jahrb. Syst. 69: 258 (1938).
W. & C. Trop. Africa to Namibia, Taiwan to Trop. Asia. 22 GHA IVO LBR NGA SIE TOG 23 BUR 27 NAM 38 TAI 40 EHM IND WHM 41 VIE 42 JAW LSI PHI 43 NWG 50 NTA QLD WAU. Ther.

var. ***microcarya***
W. & C. Trop. Africa to Namibia, Taiwan to Trop. Asia. 22 GHA IVO LBR NGA SIE TOG 23 BUR 27 NAM 38 TAI 40 EHM IND WHM 41 VIE 42 JAW LSI PHI 43 NWG 50 NTA QLD WAU. Ther.
Fimbristylis thonningiana Boeck., Linnaea 38: 395 (1874). *Iria thonningiana* (Boeck.) Kuntze, Revis. Gen. Pl. 2: 753 (1891).
Fimbristylis taiwanica Ohwi, J. Jap. Bot. 14: 574 (1938). *Fimbristylis autumnalis* subsp. *taiwanica* (Ohwi) T.Koyama, in Fl. Taiwan 5: 230 (1978).
Fimbristylis quinquangularis f. *abludens* Backer, Bekn. Fl. Java, em. ed. 10(246): 26 (1949).

var. ***tainanensis*** (Ohwi) H.Y.Liu, in Fl. Taiwan, ed. 2, 5: 272 (2000).
Taiwan. 38 TAI. Ther.
**Fimbristylis tainanensis* Ohwi, J. Jap. Bot. 14: 574 (1938). *Fimbristylis autumnalis* subsp. *tainanensis* (Ohwi) T.Koyama, Micronesica 1: 81 (1964).

Fimbristylis modesta S.T.Blake, Pap. Dept. Biol. Univ. Queensland 1(13): 11 (1940).
N. Australia. 50 NTA QLD WAU. Hemicr. or rhizome geophyte.

Fimbristylis monospicula Govind., Proc. Indian Acad. Sci., B 79: 169 (1974).
S. India. 40 IND. Hemicr. or rhizome geophyte.

Fimbristylis monticola Hochst. ex Steud., Syn. Pl. Glumac. 2: 111 (1855).
S. India, Sri Lanka. 40 IND SRL. Hemicr.
Fimbristylis tenuifolia Thwaites, Enum. Pl. Zeyl.: 434 (1864).

Fimbristylis mozambicensis Gand., Bull. Soc. Bot. France 66: 297 (1919 publ. 1920).
Mozambique. 26 MOZ. Hemicr. or rhizome geophyte.

Fimbristylis multicephala Govind., Proc. Indian Acad. Sci., B 80: 44 (1974).
N. India. 40 IND. Hemicr. or rhizome geophyte.

Fimbristylis multinervia Govind., J. Bombay Nat. Hist. Soc. 69: 162 (1972).
Assam. 40 ASS. Hemicr. or rhizome geophyte.

Fimbristylis mycosa Govind., Rheedea 7: 124 (1997).
India (Tamil Nadu). 40 IND. Hemicr. or rhizome geophyte.

Fimbristylis nagpurensis V.P.Prasad & N.P.Singh, J. Econ. Taxon. Bot. 21: 671 (1997 publ. 1998).
India (Maharashtra). 40 IND. Hemicr. or rhizome geophyte.

Fimbristylis nanningensis Tang & F.T.Wang, in Fl. Reipubl. Popul. Sin. 11: 229 (1961).
China (Guangxi). 36 CHS. Hemicr. or rhizome geophyte.

Fimbristylis narayanii C.E.C.Fisch., Bull. Misc. Inform. Kew 1931: 46 (1931).
S. India. 40 IND. Hemicr. or rhizome geophyte.

Fimbristylis neilsonii F.Muell., Fragm. 9: 70 (1875). *Iria neilsonii* (F.Muell.) Kuntze, Revis. Gen. Pl. 2: 753 (1891).
N. & E. Australia. 50 NSW NTA QLD WAU. Hemicr. or rhizome geophyte.

Fimbristylis nelmesii J.Kern, Acta Bot. Neerl. 7: 800 (1958).
Indo-China. 41 LAO THA VIE. Hemicr.
**Tylocarya cylindrostachya* Nelmes, Kew Bull. 4: 139 (1949).

Fimbristylis neocaledonica C.B.Clarke, Bot. Jahrb. Syst. 39: 24 (1906).
New Caledonia. 60 NWC. Hemicr. or rhizome geophyte.

Fimbristylis nigritana C.B.Clarke in D.Oliver & auct. suc. (eds.), Fl. Trop. Afr. 8: 418 (1902).
Togo to Nigeria. 22 NGA TOG. Hemicr.

Fimbristylis nuda Boeck., Flora 59: 110 (1875).
N. Australia. 50 NTA QLD WAU. Hemicr. or rhizome geophyte.

Fimbristylis nutans (Retz.) Vahl, Enum. Pl. 2: 285 (1805).
Trop. & Subtrop. Asia to N. & E. Australia. 36 CHH CHS 38 NNS TAI 40 ASS IND SRL 41 MYA NCB THA VIE 42 BOR MLY SUM 43 NWG 50 NSW NTA QLD WAU. Hemicr.
**Scirpus nutans* Retz., Observ. Bot. 4: 12 (1786). *Iria nutans* (Retz.) Kuntze, Revis. Gen. Pl. 2: 753 (1891). *Fimbristylis nutans* var. *typica* Domin, Biblioth. Bot. 85: 450 (1915), nom. inval.

Fimbristylis oblonga T.Koyama, Contr. Inst. Bot. Univ. Montréal 70: 41 (1957).
S. Vietnam. 41 VIE. Hemicr. or rhizome geophyte.

Fimbristylis obtusata (C.B.Clarke) Ridl., Fl. Malay Penins. 5: 157 (1925).
Bangladesh to Sulawesi. 40 BAN 41 MYA THA VIE 42 BOR MLY SUL. Ther.
**Fimbristylis tenera* var. *obtusata* C.B.Clarke in J.D.Hooker, Fl. Brit. India 6: 642 (1893).
Fimbristylis nigrobrunnea var. *thorelii* Camus, Notul. Syst. (Paris) 1: 248 (1910).

Fimbristylis odontocarpa S.T.Blake, Pap. Dept. Biol. Univ. Queensland 1(13): 13 (1940).
N. Australia. 50 NTA QLD WAU. Hemicr. or rhizome geophyte.

Fimbristylis onchnidiocarpa J.Kern, Adansonia, n.s., 7: 181 (1967).
Vietnam. 41 VIE. Hemicr. or rhizome geophyte.

Fimbristylis ovata (Burm.f.) J.Kern, Blumea 15: 126 (1967).
Trop. & Subtrop. 22 BEN BKN GHA IVO NGA TOG 23 BUR GGI RWA ZAI 25 KEN TAN UGA 27 CPP LES NAT OFS SWZ TVL 29 COM MAU MDG 35 YEM 36 CHC CHH CHS 38 JAP KOR NNS TAI 40 EHM IND PAK SRL 41 THA VIE 42 BOR JAW LSI MLY MOL PHI SUL SUM 43 BIS NWG 60 FIJ nue NWC TON 62 CRL 78 FLA 79 MXE MXG MXT 80 BLZ COS GUA HON NIC PAN 81 BAH CAY CUB DOM HAI JAM LEE NLA PUE TRT WIN 82 GUY VEN 83 BOL CLM ECU 84 BZN 85 AGE AGW PAR URU. Hemicr.
**Carex ovata* Burm.f., Fl. Indica: 194 (1768). *Abildgaardia ovata* (Burm.f.) Kral, Sida 4: 72 (1971).
Cyperus monostachyos L., Mant. Pl. 2: 180 (1771). *Abildgaardia monostachya* (L.) Vahl, Enum. Pl. 2: 296 (1805). *Fimbristylis monostachya* (L.) Hassk., Pl. Jav. Rar.: 61 (1848). *Iria monostachya* (L.) Kuntze, Revis. Gen. Pl. 2: 752 (1891). *Scirpus monostachyus* (L.) Kuntze, Revis. Gen. Pl. 3(2): 337 (1898).
Cyperus caribaeus Pers., Syn. Pl. 1: 65 (1805), nom. illeg.
Cyperus indicus Rich. ex Pers., Syn. Pl. 1: 65 (1805). *Iria indica* (Rich. ex Pers.) Schult. in J.J.Roemer & J.A.Schultes, Mant. 2: 94 (1824). *Abildgaardia indica* (Rich. ex Pers.) Nees, Linnaea 9: 289 (1834).
Xyris brasiliensis Spreng., Novi Provent.: 48 (1818).
Scirpus schoenoides Roxb., Fl. Ind. 1: 221 (1820), nom. illeg.
Iria caribea Steud., Nomencl. Bot.: 437 (1821).
Abildgaardia compressa J.Presl & C.Presl, Reliq. Haenk. 1: 179 (1828).
Abildgaardia rottboelliana Nees ex Hook. & Arn., Bot. Beechey Voy.: 272 (1838).
Fimbristylis compressa Fern.-Vill. in F.M.Blanco, Fl. Filip., ed. 3, 4(13A): 307 (1880).
Abildgaardia javanensis Gand., Bull. Soc. Bot. France 66: 296 (1919 publ. 1920).

Fimbristylis oxystachya F.Muell., Fragm. 1: 195 (1859). *Iria oxystachya* (F.Muell.) Kuntze, Revis. Gen. Pl. 2: 753 (1891).
N. Australia. 50 NTA QLD WAU. Hemicr. or rhizome geophyte.

Fimbristylis pachyptera S.T.Blake, Pap. Dept. Biol. Univ. Queensland 1(13): 14 (1940).
N. Australia. 50 NTA QLD WAU. Hemicr. or rhizome geophyte.

Fimbristylis palauensis Ohwi, J. Jap. Bot. 18: 134 (1942).
Caroline Is. 62 CRL. Hemicr. or rhizome geophyte.
Fimbristylis fosbergiana T.Koyama, Bull. Arts Sci. Div. Ryukyu Univ. 3: 66 (1959).

Fimbristylis pallida S.T.Blake, Proc. Roy. Soc. Queensland 58: 42 (1947).
Northern Territory. 50 NTA. Hemicr. or rhizome geophyte.
**Fimbristylis leucostachya* Boeck., Linnaea 38: 385 (1874), nom. illeg. *Fimbristylis nutans* var. *leucostachya* Domin, Biblioth. Bot. 85: 450 (1915).

Fimbristylis pandurata Govind., Rheedea 7: 124 (1997).
India (Tamil Nadu). 40 IND. Hemicr. or rhizome geophyte.

Fimbristylis parvilenta T.Koyama, Brittonia 31: 287 (1979).
Indo-China. 41 THA VIE. Hemicr.

Fimbristylis pauciflora R.Br., Prodr.: 225 (1810). *Scirpus pauciflorus* (R.Br.) Poir. in J.B.A.M.de Lamarck, Encycl., Suppl. 5: 98 (1817), nom. inval. *Iria pauciflora* (R.Br.) Kuntze, Revis. Gen. Pl. 2: 752 (1891).
Indo-China to N. Australia. 38 NNS 41 MYA THA VIE 42 BOR MLY MOL SUM 43 NWG 50 NTA QLD WAU 62 CRL. Hemicr.
Trichelostylis filiformis Nees in R.Wight, Contr. Bot. India: 102 (1834). *Fimbristylis filiformis* (Nees) Kunth, Enum. Pl. 2: 221 (1837).
Fimbristylis pumila Benth., London J. Bot. 2: 239 (1843).
Fimbristylis malaccana Boeck., Flora 12: 507 (1858).

Fimbristylis paupercula Boeck., Linnaea 38: 396 (1874). *Iria paupercula* (Boeck.) Kuntze, Revis. Gen. Pl. 2: 753 (1891).
S. India. 40 IND. Hemicr.

Fimbristylis pentastachya Boeck., Flora 40: 36 (1857).
Mexico. 79 MXE MXG MXN MXS. Hemicr.

Fimbristylis perlaxa Ohwi, Blumea 8: 101 (1955).
Maluku to New Guinea. 42 MOL 43 NWG. Ther.

Fimbristylis perpusilla R.M.Harper. ex Small & Britton in J.K.Small, Fl. S.E. U.S.: 188 (1903).
SE. U.S.A. 78 DEL GEO MRY NCA SCA TEN VRG. Ther.

Fimbristylis perspicua Govind. & Sasidh., J. Econ. Taxon. Bot. 21: 376 (1997).
India (Kerala). 40 IND. Hemicr. or rhizome geophyte.

Fimbristylis phaeolepis J.Kern, Adansonia, n.s., 7: 183 (1967).
Indo-China. 41 MYA THA. Ther.

Fimbristylis phaeoleuca S.T.Blake, Pap. Dept. Biol. Univ. Queensland 1(13): 4 (1940).
N. Australia. 50 NTA QLD WAU. Hemicr. or rhizome geophyte.

Fimbristylis pierotii Miq., Ann. Mus. Bot. Lugduno-Batavi 2: 145 (1865). *Iria pierotii* (Miq.) Kuntze, Revis. Gen. Pl. 2: 753 (1891).
Himalaya to Japan and Philippines. 36 CHC CHS 38 JAP KOR 40 EHM NEP WHM 41 VIE 42 PHI. Hemicr.
Fimbristylis pinetorum Merr., Philipp. J. Sci., C 9: 266 (1914).

Fimbristylis pilifera W.Fitzg., J. Proc. Roy. Soc. Western Australia 3: 118 (1918).
N. Australia. 50 NTA WAU. Hemicr. or rhizome geophyte.

Fimbristylis pilosa Vahl, Enum. Pl. 2: 290 (1805). *Iria pilosa* (Vahl) Kuntze, Revis. Gen. Pl. 2: 753 (1891). *Fimbristylis subpilosa* Kük., Repert. Spec. Nov. Regni Veg. Beih. 40(1): 405 (1937).
Trop. Africa. 22 BEN BKN GHA GUI IVO NGA SEN SIE TOG 23 BUR CON EQG ZAI 24 CHA 25 KEN TAN UGA. Ther. or hemicr.
Fimbristylis elongata Sieber ex C.Presl, Isis (Oken) 21: 268 (1828), nom. inval.
Scirpus fimbriatus Willd. ex Kunth, Enum. Pl. 2: 235 (1837).
Cyperus lanatus Ruiz ex Boeck., Linnaea 36: 750 (1870).

Fimbristylis polytrichoides (Retz.) R.Br., Prodr.: 226 (1810).
E. Trop. Africa, Madagascar, Trop. & Subtrop. Asia to N. & E. Australia. 25 KEN TAN 29 MDG 36 CHH CHS 38 TAI 40 BAN IND SRL 41 MYA THA VIE 42 JAW MLY PHI SUM 43 NWG 50 NSW NTA QLD WAU. Hemicr.
**Scirpus polytrichoides* Retz., Observ. Bot. 4: 11 (1786). *Aplostemon polytrichoides* (Retz.) Raf., J. Phys. Chim. Hist. Nat. Arts 89: 105 (1819). *Iria polytrichoides* (Retz.) Kuntze, Revis. Gen. Pl. 2: 753 (1891).
Abildgaardia javana Nees, Linnaea 9: 289 (1834), nom. inval.
Abildgaardia javanica Steud., Syn. Pl. Glumac. 2: 72 (1854).
Fimbristylis albescens Steud., Syn. Pl. Glumac. 2: 107 (1855).
Fimbristylis subbulbosa Boeck., Flora 41: 598 (1858).
Fimbristylis juncea Boeck., Linnaea 37: 4 (1871), nom. illeg.
Fimbristylis subbulbosa Boeck., Beitr. Cyper. 2: 39 (1890).
Fimbristylis takaoensis Hayata, Icon. Pl. Formosan. 6: 112 (1916). *Fimbristylis polytrichoides* var. *takaoensis* (Hayata) T.Koyama, J. Fac. Sci. Univ. Tokyo, Sect. 3, Bot. 8: 110 (1961).

Fimbristylis prabatensis D.A.Simpson, Kew Bull. 53: 225 (1998).
NE. Thailand. 41 THA. Hemicr.

Fimbristylis psammocola Tang & F.T.Wang, in Fl. Reipubl. Popul. Sin. 11: 227 (1961).
China (Yunnan). 36 CHC. Hemicr. or rhizome geophyte.

Fimbristylis psammophila J.Kern, Reinwardtia 4: 96 (1956).
NE. Thailand. 41 THA. Hemicr.
**Fimbristylis arenicola* J.Kern, Blumea 8: 146 (1955), nom. illeg.

Fimbristylis pseudomicrocarya Govind., Proc. Indian Acad. Sci., Pl. Sci. 100: 77 (1990).
India (Kerala). 40 IND. Ther.

Fimbristylis pseudonarayanii Ravi & Anil Kumar, Rheedea 3: 110 (1993).
India (Kerala). 40 IND. Hemicr. or rhizome geophyte.

Fimbristylis pterygosperma R.Br., Prodr.: 226 (1810). *Scirpus pterygosperma* (R.Br.) Poir. in J.B.A.M.de Lamarck, Encycl., Suppl. 5: 98 (1817), nom. inval. *Iria pterygosperma* (R.Br.) Kuntze, Revis. Gen. Pl. 2: 753 (1891).
N. Australia. 50 NTA QLD WAU. Hemicr. or rhizome geophyte.

Fimbristylis puberula Vahl, Enum. Pl. 2: 289 (1805).
SE. Canada to U.S.A. 72 ONT 73 COL 74 ILL KAN MSO NEB OKL 75 INI MIC NWJ NWY OHI 76 ARI UTA 77 NWM TEX 78 ALA ARK DEL FLA GEO KTY LOU MRY MSI NCA SCA TEN VRG. Hemicr.

var. ***interior*** (Britton) Kral, Sida 4: 135 (1971).
WC. & C. U.S.A. 73 COL 74 KAN NEB OKL 76 ARI UTA 77 NWM TEX. Hemicr.
**Fimbristylis interior* Britton in N.L.Britton & A.Brown, Ill. Fl. N. U.S., ed. 2, 1: 320 (1913).

var. ***puberula***
SE. Canada to C. & E. U.S.A. 72 ONT 74 ILL KAN MSO NEB OKL 75 INI MIC NWJ NWY OHI 76 UTA 77 TEX 78 ALA ARK DEL FLA GEO KTY LOU MRY MSI NCA SCA TEN VRG. Hemicr.
Isolepis drummondii Torr. & Hook., Ann. Lyceum Nat. Hist. New York 3: 350 (1836). *Fimbristylis drummondii* (Torr. & Hook.) Boeck., Flora 41: 603 (1858). *Iria drummondii* (Torr. & Hook.) Kuntze, Revis. Gen. Pl. 2: 753 (1891).
Fimbristylis anomala Boeck., Flora 43: 240 (1860).
Fimbristylis multistriata Boeck., Flora 43: 243 (1860).

Fimbristylis pubisquama J.Kern, Blumea 8: 131 (1955).
S. India to Indo-China, N. Australia. 40 IND SRL 41 CBD MYA VIE 50 NTA. Ther.
**Fimbristylis compressa* Boeck., Linnaea 38: 387 (1874), nom. illeg. *Iria compressa* Kuntze, Revis. Gen. Pl. 2: 753 (1891).

Fimbristylis punctata R.Br., Prodr.: 226 (1810). *Scirpus punctatus* (R.Br.) Poir. in J.B.A.M.de Lamarck, Encycl., Suppl. 5: 98 (1817), nom. inval. *Iria punctata* (R.Br.) Kuntze, Revis. Gen. Pl. 2: 753 (1891).
N. Australia. 50 NTA QLD WAU. Hemicr. or rhizome geophyte.

Fimbristylis pustulosa Govind., Proc. Indian Acad. Sci., Pl. Sci. 90: 444 (1981).
S. India. 40 IND. Hemicr. or rhizome geophyte.

Fimbristylis quadriflora (Boeck.) Beetle, Amer. Midl. Naturalist 41: 483 (1949).
Brazil. 84+. Hemicr. or rhizome geophyte.
Scirpus subquadriflorus Boeck., Vidensk. Meddel. Naturhist. Foren. Kjøbenhavn 1869: 139 (1869).
**Scirpus quadriflorus* Boeck., Linnaea 36: 761 (1870).

Fimbristylis quinquangularis (Vahl) Kunth, Enum. Pl. 2: 229 (1837).
Trop. Africa, Iraq to Trop. & Subtrop. Asia and N. Australia. 22 BKN IVO NGA SEN TOG 23 BUR EQG GAB ZAI 25 KEN TAN UGA 29 COM MAU MDG REU 32 KAZ UZB 34 AFG IRQ 36 CHC CHS 38 OGA TAI 40 ASS EHM IND NEP PAK SRL 41 THA VIE 42 JAW LSI MOL PHI SUL SUM 43 NWG 50 NTA QLD WAU (60) fij (61) sci (63) haw (81) lee pue (82) frg guy sur ven. Ther. or hemicr.
Scirpus miliaceus L., Syst. Nat. ed. 10: 868 (1759), nom. rejic. *Fimbristylis miliacea* (L.) Vahl, Enum. Pl. 2: 267 (1805), nom. rejic. *Isolepis miliacea* (L.) J.Presl & C.Presl, Reliq. Haenk. 1: 188 (1828). *Trichelostylis miliacea* (L.) Nees in R.Wight, Contr. Bot. India: 103 (1834). *Iria miliacea* (L.) Kuntze, Revis. Gen. Pl. 2: 752 (1891).
Fimbristylis mucronata Vahl, Enum. Pl. 2: 293 (1805).
Scirpus benghalensis Pers., Syn. Pl. 1: 68 (1805). *Fimbristylis benghalensis* (Pers.) Roem. & Schult. in J.J.Roemer & J.A.Schultes, Syst. Veg. 2: 94 (1817).
**Scirpus quinquangularis* Vahl, Enum. Pl. 2: 279 (1805). *Trichelostylis quinquangularis* (Vahl) Nees in R.Wight, Contr. Bot. India: 104 (1834). *Iria quinquangularis* (Vahl) Kuntze, Revis. Gen. Pl. 2: 752 (1891).
Scirpus pentagonus Roxb., Fl. Ind. 1: 218 (1820). *Isolepis pentagona* (Roxb.) Schult. in J.J.Roemer & J.A.Schultes, Mant. 2: 69 (1824).
Scirpus plantagineus Roxb., Fl. Ind. 1: 212 (1820), nom. illeg.
Fimbristylis angularis Link, Enum. Hort. Berol. Alt. 1: 289 (1821). *Trichelostylis angularis* (Link) Nees in R.Wight, Contr. Bot. India: 105 (1834). *Iria angularis* (Link) Kuntze, Revis. Gen. Pl. 2: 753 (1891).
Scirpus quadrangularis Thouars, Cat. Hort. Maurit.: 4 (1822).
Isolepis angularis Schrad. ex Schult. in J.J.Roemer & J.A.Schultes, Mant. 2: 69 (1824).
Isolepis miliacea var. *major* C.Presl, Reliq. Haenk. 1: 189 (1828).
Scirpus salbundius Buch.-Ham. ex Wall., Numer. List: 3499 (1831).
Trichelostylis pentagona Nees in R.Wight, Contr. Bot. India: 105 (1834).
Scirpus niloticus Blanco, Fl. Filip.: 33 (1837).
Scirpus parviflorus Willd. ex Kunth, Enum. Pl. 2: 230 (1837).
Fimbristylis fauriei Ohwi, Acta Phytotax. Geobot. 1: 77 (1932).
Fimbristylis quinquangularis var. *bistaminifera* Tang & F.T.Wang, in Fl. Reipubl. Popul. Sin. 11: 227 (1961).
Fimbristylis quinquangularis var. *elata* Tang & F.T. Wang, in Fl. Reipubl. Popul. Sin. 11: 227 (1961).
Fimbristylis miliacea subsp. *macroglumis* Lye, Nordic J. Bot. 2: 333 (1982).
Fimbristylis miliacea subsp. *pallescens* Lye, Nordic J. Bot. 2: 333 (1982).

Fimbristylis rara R.Br., Prodr.: 227 (1810). *Scirpus rarus* (R.Br.) Poir. in J.B.A.M.de Lamarck, Encycl., Suppl. 5: 99 (1817), nom. inval. *Iria rara* (R.Br.) Kuntze, Revis. Gen. Pl. 2: 753 (1891). *Fimbristylis rara* var. *typica* Domin, Biblioth. Bot. 85: 461 (1915), nom. inval.
N. Australia. 50 NTA QLD WAU. Hemicr. or rhizome geophyte.
Fimbristylis obtusangula F.Muell., Fragm. 1: 198 (1859). *Iria obtusangula* (F.Muell.) Kuntze, Revis. Gen. Pl. 2: 753 (1891). *Fimbristylis rara* var. *obtusangula* (F.Muell.) Domin, Biblioth. Bot. 85: 461 (1915).
Fimbristylis rufa Boeck., Bot. Jahrb. Syst. 5: 93 (1884).

Fimbristylis ratnagirica V.P.Prasad & N.P.Singh, J. Econ. Taxon. Bot. 21: 673 (1997 publ. 1998).
India (Maharashtra). 40 IND. Hemicr. or rhizome geophyte.

Fimbristylis raymondii T.Koyama, Contr. Inst. Bot. Univ. Montréal 70: 48 (1957).
N. Vietnam. 41 VIE. Hemicr. or rhizome geophyte.

Fimbristylis recta F.M.Bailey, Syn. Queensl. Fl., Suppl. 3: 80 (1890).
New Guinea to N. Australia. 43 NWG 50 NTA QLD. Hemicr.
Fimbristylis xyridis var. *rigidula* Benth., Fl. Austral. 7: 307 (1878).
Fimbristylis macgillivrayi C.B.Clarke, Kew Bull., Addit. Ser. 8: 24 (1908).
Fimbristylis stricticulmis Domin, Biblioth. Bot. 85: 452 (1915).

Fimbristylis rectifolia Govind., Proc. Indian Acad. Sci., B 80: 46 (1974).
S. India. 40 IND. Hemicr. or rhizome geophyte.

Fimbristylis rhizomatosa Pires de Lima, Bol. Soc. Brot., II, 2: 134 (1924).
Mozambique. 26 MOZ. Hemicr. or rhizome geophyte.

Fimbristylis rhodesiana Rendle, J. Linn. Soc., Bot. 40: 222 (1911).
Zimbabwe. 26 ZIM. Hemicr. or rhizome geophyte.

Fimbristylis rhyticarya F.Muell., Fragm. 1: 215 (1859). *Iria rhyticarya* (F.Muell.) Kuntze, Revis. Gen. Pl. 2: 753 (1891). *Fimbristylis nutans* var. *rhyticarya* (F.Muell.) Domin, Biblioth. Bot. 85: 450 (1915).
N. Australia. 50 NTA QLD WAU. Hemicr. or rhizome geophyte.

Fimbristylis rigidiuscula Govind., Proc. Indian Acad. Sci., B 78: 45 (1973).
S. India. 40 IND. Hemicr. or rhizome geophyte.

Fimbristylis rigidula Nees in R.Wight, Contr. Bot. India: 99 (1834).
Himalaya to Philippines. 36 CHC CHS 40 ASS BAN EHM NEP PAK WHM 41 THA VIE 42 PHI. Hemicr.
Fimbristylis hanceana Boeck., Linnaea 38: 894 (1874). *Iria hanceana* (Boeck.) Kuntze, Revis. Gen. Pl. 2: 753 (1891) (1891).

Fimbristylis rufoglumosa Tang & F.T.Wang, in Fl. Reipubl. Popul. Sin. 11: 228 (1961).
China (Guangdong). 36 CHS. Hemicr. or rhizome geophyte.

Fimbristylis rugosa Govind., Proc. Indian Acad. Sci., B 78: 48 (1973).
S. India. 40 IND. Hemicr. or rhizome geophyte.

Fimbristylis rupestris Latz, Nuytsia 7: 167 (1990).
N. Australia. 50 NTA WAU. Ther.

Fimbristylis sachetiana Fosberg, J. Jap. Bot. 63: 83 (1988).
Marquesas (Nuku Hiva). 61 MRQ. Hemicr. or rhizome geophyte.

Fimbristylis salbundia (Nees) Kunth, Enum. Pl. 2: 230 (1837).
Trop. Asia. 40 ASS IND SRL 41 MYA THA VIE 42 PHI SUM 43 NWG. Hemicr.
Trichelostylis pentaptera Nees in R.Wight, Contr. Bot. India: 105 (1834). *Fimbristylis pentaptera* (Nees) Kunth, Enum. Pl. 2: 229 (1837). *Iria pentaptera* (Nees) Kuntze, Revis. Gen. Pl. 2: 753 (1891). *Fimbristylis salbundia* subsp. *pentaptera* (Nees) T.Koyama, Bot. Mag. (Tokyo) 87: 317 (1974).
**Trichelostylis salbundia* Nees in R.Wight, Contr. Bot. India: 105 (1834). *Iria salbundia* (Nees) Kuntze, Revis. Gen. Pl. 2: 753 (1891).
Fimbristylis paludosa Merr., Philipp. J. Sci., C 9: 265 (1914).
Fimbristylis haspaniformis T.Koyama, Contr. Inst. Bot. Univ. Montréal 70: 45 (1957).

Fimbristylis savannicola J.Kern, Blumea 10: 635 (1960).
NE. Thailand. 41 THA. Hemicr.

Fimbristylis scaberrima Nees in R.Wight, Contr. Bot. India: 102 (1834). *Iria scaberrima* (Nees) Kuntze, Revis. Gen. Pl. 2: 753 (1891).
Assam to W. Malesia. 40 ASS 41 CBD VIE 42 BOR JAW SUM. Hemicr. or rhizome geophyte.

Fimbristylis scabrida Schumach., Beskr. Guin. Pl.: 32 (1827).
Trop. Africa, Madagascar. 22 BEN GHA GUI IVO NGA SIE TOG 23 BUR CAF ZAI 24 CHA 25 UGA 29 MDG. Hemicr.
Fimbristylis muriculata Benth. in W.J.Hooker, Niger Fl.: 554 (1849).
Fimbristylis mucronata Boeck., Linnaea 37: 41 (1871), nom. illeg. *Iria mucronata* Kuntze, Revis. Gen. Pl. 2: 753 (1891).
Fimbristylis pachystachys Cherm., Bull. Soc. Bot. France 68: 418 (1921 publ. 1922).
Fimbristylis pachystylis Cherm., Bull. Soc. Bot. France 68: 418 (1921 publ. 1922).

Fimbristylis scabrisquama Govind., Proc. Indian Acad. Sci., Pl. Sci. 91: 43 (1982).
S. India. 40 IND. Hemicr. or rhizome geophyte.

Fimbristylis schoenoides (Retz.) Vahl, Enum. Pl. 2: 286 (1805).
W. Trop. Africa to Tanzania, Trop. & Subtrop. Asia to N. Australia. 22 GHA GNB NGA SEN SIE 25 TAN 36 CHC CHS 38 TAI 40 ASS BAN EHM IND NEP PAK SRL 41 THA VIE 42 BOR JAW LSI MLY PHI SUM 50 NTA QLD (63) haw (78) ala fla geo lou msi. Hemicr.
**Scirpus schoenoides* Retz., Observ. Bot. 5: 14 (1788). *Iria schoenodes* (Retz.) Kuntze, Revis. Gen. Pl. 2: 752 (1891).
Scirpus bispicatus J.König ex Roxb., Fl. Ind. 1: 220 (1820). *Isolepis bispicata* (J.König ex Roxb.) Schult. in J.J.Roemer & J.A.Schultes, Mant. 2: 61 (1824). *Fimbristylis bispicata* (J.König ex Roxb.) Nees & Meyen in R.Wight, Contr. Bot. India: 97 (1834). *Iria bispicata* (J.König ex Roxb.) Kuntze, Revis. Gen. Pl. 2: 753 (1891).
Scirpus monostachyos J.König ex Roxb., Fl. Ind. 1: 219 (1820). *Isolepis monostachya* (J.König ex Roxb.) Spreng., Neue Entdeck. Pflanzenk. 3: 11 (1822). *Eleogiton monostachya* (J.König ex Roxb.) A.Dietr., Sp. Pl. 2: 97 (1832).
Isolepis rariflora Schrad. ex Schult. in J.J.Roemer & J.A.Schultes, Mant. 2: 61 (1824).
Abildgaardia nervosa J.Presl & C.Presl, Reliq. Haenk. 1: 180 (1828).
Fimbristylis inconstans Steud., Syn. Pl. Glumac. 2: 107 (1855).
Fimbristylis kunthiana Ridl., Trans. Linn. Soc. London, Bot. 2: 151 (1884).
Fimbristylis kagiensis Hayata, Icon. Pl. Formosan. 6: 110 (1916).
Fimbristylis longifolia S.T.Blake, Pap. Dept. Biol. Univ. Queensland 1(13): 9 (1940).

Fimbristylis schultzii Boeck., Linnaea 38: 391 (1874). *Fimbristylis platystachys* var. *schultzii* (Boeck.) Domin, Biblioth. Bot. 85: 464 (1915).
Lesser Sunda Is. to N. Australia. 42 LSI 50 NTA QLD WAU. Ther.
Fimbristylis platystachys Boeck., Linnaea 38: 390 (1874). *Iria platystachya* (Boeck.) Kuntze, Revis. Gen. Pl. 2: 753 (1891). *Fimbristylis platystachys* var. *typica* Domin, Biblioth. Bot. 85: 464 (1915), nom. inval.
Fimbristylis stellata S.T.Blake, Proc. Roy. Soc. Queensland 58: 46 (1947).

Fimbristylis schweinfurthiana Boeck., Flora 62: 565 (1879).
W. Trop. Africa to Sudan. 22 GUI MLI SIE 24 CHA SUD. Hemicr.

Fimbristylis semarangensis Ohwi, Blumea 8: 106 (1955).
SE. Vietnam C. Jawa. 41 VIE 42 JAW. Ther.

Fimbristylis semidisticha Govind., Proc. Indian Acad. Sci., B 78: 51 (1973).
S. India. 40 IND. Hemicr. or rhizome geophyte.

Fimbristylis sericea (Poir.) R.Br., Prodr.: 228 (1810).
E. India to Japan and N. Australia. 36 CHH CHS 38 JAP NNS? TAI 40 IND 41 THA VIE 42 BOR JAW MLY SUM 50 NTA QLD WAU. Hemicr.
**Scirpus sericeus* Poir. in J.B.A.M.de Lamarck, Encycl., Suppl. 5: 99 (1817), nom. inval. *Iria sericea* (Poir.) Kuntze, Revis. Gen. Pl. 2: 753 (1891).
Fimbristylis decora Nees & Meyen in R.Wight, Contr. Bot. India: 101 (1834).
Fimbristylis dasyphylla Miq., Fl. Ned. Ind. 3: 327 (1856).
Fimbristylis junciformis var. *latifolia* Camus in H.Lecomte, Fl. Indo-Chine 7: 120 (1912).
Fimbristylis velutina Franch., Bull. Soc. Bot. France 26: 88 (1979).

Fimbristylis shimadana Ohwi, Acta Phytotax. Geobot. 5: 184 (1936).
Taiwan. 38 TAI. Hemicr.

Fimbristylis signata S.T.Blake, J. Arnold Arbor. 35: 214 (1954).
New Guinea to N. Australia. 43 NWG 50 NTA QLD WAU. Ther.
**Fimbristylis debilis* F.Muell., Fragm. 1: 198 (1859), nom. illeg. *Iria debilis* Kuntze, Revis. Gen. Pl. 2: 753 (1891).

Fimbristylis simaoensis Y.Y.Qian, Bull. Bot. Res., Harbin 17: 128 (1997).
China (Yunnan). 36 CHC. Hemicr. or rhizome geophyte.

Fimbristylis simplex S.T.Blake, Pap. Dept. Biol. Univ. Queensland 1(13): 10 (1940).
N. Australia. 50 NTA QLD WAU. Hemicr. or rhizome geophyte.

Fimbristylis simpsonii V.P.Prasad & N.P.Singh, J. Bombay Nat. Hist. Soc. 96: 456 (1999).
India (Karnataka). 40 IND. Ther.

Fimbristylis simulans Latz, Nuytsia 7: 168 (1990).
N. Australia. 50 NTA WAU. Ther.

Fimbristylis singularis Govind., Beitr. Biol. Pflanzen 70: 88 (1998).
India (Karnataka). 40 IND. Hemicr. or rhizome geophyte.

Fimbristylis sleumeri J.Kern, Blumea 15: 437 (1968).
NE. Thailand. 41 THA. Hemicr.

Fimbristylis smitinandii T.Koyama, Brittonia 30: 106 (1978).
NE. Thailand. 41 THA. Ther.

Fimbristylis solidifolia F.Muell., Fragm. 1: 198 (1859). *Iria solidifolia* (F.Muell.) Kuntze, Revis. Gen. Pl. 2: 753 (1891).
N. Australia. 50 NTA WAU. Hemicr. or rhizome geophyte.

Fimbristylis spadicea (L.) Vahl, Enum. Pl. 2: 294 (1805).
E. U.S.A. to Trop. & Subtrop. America. 75 NWJ NWY 77 TEX 78 ALA DEL FLA GEO LOU MRY MSI NCA SCA VRG WDC 79 MXC MXE MXG MXN MXS MXT 80 BLZ COS ELS GUA HON NIC PAN 81 BAH BER CAY CUB DOM JAM LEE NLA PUE TRT WIN 82 FRG GUY SUR VEN 83 CLM ECU PER 84 BZC 85 AGE PAR. Hemicr.
**Scirpus spadiceus* L., Sp. Pl.: 51 (1753). *Schoenus spadiceus* (L.) Vahl, Enum. Pl. 2: 210 (1805). *Eleogiton spadicea* (L.) A.Dietr., Sp. Pl. 2: 99 (1832). *Iria spadicea* (L.) Kuntze, Revis. Gen. Pl. 2: 752 (1891). *Iria spadicea* var. *normalis* Kuntze, Revis. Gen. Pl. 2: 752 (1891), nom. inval.
Scirpus castaneus Michx., Fl. Bor.-Amer. 1: 31 (1803). *Fimbristylis castanea* (Michx.) Vahl, Enum. Pl. 2: 292 (1805). *Fimbristylis spadicea* var. *castanea* (Michx.) A.Gray, Manual, ed. 5: 566 (1867). *Iria castanea* (Michx.) Farw., Rep. (Annual) Michigan Acad. Sci. 21: 359 (1919 publ. 1920).
Fimbristylis cylindrica Vahl, Enum. Pl. 2: 293 (1805).
Scirpus domingensis Pers., Syn. Pl. 1: 67 (1805). *Fimbristylis spadicea* f. *domingensis* (Pers.) Kük., Repert. Spec. Nov. Regni Veg. 23: 196 (1926).
Fimbristylis speciosa Rohde ex Spreng., Pl. Min. Cogn. Pug. 1: 5 (1813).
Scirpus dichotomus G.Mey., Prim. Fl. Esseq.: 43 (1818), sensu auct.
Fimbristylis sterilis Nees, Linnaea 9: 290 (1834), nom. inval.
Scirpus cylindraceus Willd. ex Kunth, Enum. Pl. 2: 238 (1837).
Fimbristylis umbellata Schrad. ex Nees in C.F.P.von Martius & auct. suc. (eds.), Fl. Bras. 2(1): 73 (1842). *Iria umbellata* (Schrad. ex Nees) Kuntze, Revis. Gen. Pl. 2: 753 (1891).
Fimbristylis riehleana Steud., Syn. Pl. Glumac. 2: 114 (1855).
Scirpus sterilis Salzm. ex Steud., Syn. Pl. Glumac. 2: 118 (1855).
Iria spadicea var. *nigra* Kuntze, Revis. Gen. Pl. 2: 752 (1891).
Iria spadicea var. *pallida* Kuntze, Revis. Gen. Pl. 2: 752 (1891).
Fimbristylis pallidula Kral, Sida 4: 130 (1971).
Fimbristylis spadicea var. *longestigmata* Zavaro, Fontqueria 42: 199 (1995).

Fimbristylis sphaerocephala Benth., Fl. Austral. 7: 306 (1878). *Iria sphaerocephala* (Benth.) Kuntze, Revis. Gen. Pl. 2: 753 (1891).
N. Australia. 50 NTA WAU. Hemicr. or rhizome geophyte.
Fimbristylis semilevis F.Muell. ex C.B.Clarke, Kew Bull., Addit. Ser. 8: 24 (1908).

Fimbristylis spicigera J.Kern, Blumea 8: 144 (1955).
NE. Thailand to Laos. 41 LAO THA. Hemicr.

Fimbristylis spiralis R.Br., Prodr.: 226 (1810). *Iria spiralis* (R.Br.) Kuntze, Revis. Gen. Pl. 2: 753 (1891).
Northern Territory. 50 NTA. Hemicr. or rhizome geophyte.

Fimbristylis splendida C.B.Clarke in D.Oliver & auct. suc. (eds.), Fl. Trop. Afr. 8: 527 (1902).
WC. Trop. Africa. 23 CON EQG ZAI. Hemicr.

Fimbristylis squarrosa Vahl, Enum. Pl. 2: 289 (1805). *Scirpus squarrosus* (Vahl) Poir. in J.B.A.M.de Lamarck, Encycl., Suppl. 5: 100 (1817), nom. illeg. *Pogonostylis squarrosus* (Vahl) Bertol., Fl. Ital. 1: 312 (1833). *Iria squarrosa* (Vahl) Kuntze, Revis. Gen. Pl. 2: 753 (1891). *Fimbristylis aestivalis* var. *squarrosa* (Vahl) T.Koyama, J. Fac. Sci. Univ. Tokyo, Sect. 3, Bot. 8: 116 (1961).
Trop. & Subtrop. Old World. (13) ita 20 ALG 22 MLI NGA NGR SEN SIE 23 BUR CMN GAB ZAI 25 TAN 27 BOT CPP NAM TVL 29 MDG 31 PRM 33 TCS 34 IRN 36 CHC CHM CHN CHS 38 JAP KOR TAI 40 EHM NEP WHM 41 MYA THA VIE 42 JAW PHI SUM 50 NSW NTA QLD SOA VIC WAU 51 NZN 60 FIJ (75) nwj (80) hon (81) cub (82) ven (83) bol per (84) bzn (85) age par uru. Ther.

var. ***esquarrosa*** Makino, Bot. Mag. (Tokyo) 17: 47 (1903). *Fimbristylis aestivalis* var. *esquarrosa* (Makino) T.Koyama, J. Fac. Sci. Univ. Tokyo, Sect. 3, Bot. 8: 116 (1961). *Fimbristylis squarrosa* subsp. *esquarrosa* (Makino) T.Koyama, in Fl. Taiwan 5: 247 (1978).
Trop. & Subtrop. Old World. 31 PRM 36 CHM CHN CHS 38 JAP KOR TAI 41 THA VIE 42 JAW PHI SUM 50 NSW NTA QLD SOA VIC WAU 51 NZN 60 FIJ. Ther.
Fimbristylis velata R.Br., Prodr.: 227 (1810). *Scirpus velatus* (R.Br.) Poir. in J.B.A.M.de Lamarck, Encycl., Suppl. 5: 99 (1817), nom. inval. *Iria velata* (R.Br.) Kuntze, Revis. Gen. Pl. 2: 753 (1891). *Fimbristylis squarrosa* var. *velata* (R.Br.) C.B.Clarke ex Cheeseman, Man. New Zealand Fl.: 770 (1906).
Fimbristylis makinoana Ohwi, J. Jap. Bot. 14: 578 (1938).

var. ***squarrosa***
Trop. & Subtrop. Old World. (13) ita 20 ALG 22 MLI NGA NGR SEN SIE 23 BUR CMN GAB ZAI 25 TAN 27 BOT CPP NAM TVL 29 MDG 31 PRM 33 TCS 34 IRN 36 CHC CHN CHS 38 JAP KOR TAI 40 EHM NEP WHM 41 MYA 42 JAW PHI SUM 50 NSW NTA QLD SOA VIC WAU 51 NZN (60) fij (75) nwj (80) hon (81) cub (82) ven (83) bol per (84) bzn (85) age par uru. Ther.
Cyperus hirtus Thunb., Phytogr. Blätt. 1: 6 (1803). *Bulbostylis hirta* (Thunb.) Svenson, Contr. Ocas. Mus. Hist. Nat. Colegio "De La Salle" 4: 11 (1946).

Fimbristylis propinqua R.Br., Prodr.: 227 (1810). *Scirpus propinqua* (R.Br.) Poir. in J.B.A.M. de Lamarck, Encycl., Suppl. 5: 99 (1817), nom. inval.
Isolepis hirta Kunth in F.W.H.von Humboldt, A.J.A. Bonpland & C.S.Kunth, Nov. Gen. Sp. 1: 224 (1816). *Fimbristylis hirta* (Kunth) Roem. & Schult. in J.J.Roemer & J.A.Schultes, Syst. Veg. 2: 99 (1817). *Scirpus hirtus* (Kunth) Poir. in J.B.A.M. de Lamarck, Encycl., Suppl. 5: 106 (1817).
Scirpus aestivalis Wall. in W.Roxburgh, Fl. Ind. 1: 230 (1820).
Fimbristylis nana Roth, Nov. Pl. Sp.: 25 (1821).
Isolepis glaucescens Schrad. & Schult. in J.J.Roemer & J.A.Schultes, Mant. 2: 68 (1824).
Fimbristylis comata Nees in R.Wight, Contr. Bot. India: 102 (1834).
Fimbristylis ecklonii Nees, Linnaea 10: 145 (1836).
Fimbristylis squarrosa var. *robustior* Kunth, Enum. Pl. 2: 255 (1837).
Fimbristylis michelii Bubani, Dodecanthea: 34 (1850).
Fimbristylis hispidula var. *canina* Schltdl., Linnaea 26: 129 (1853). *Fimbristylis hirta* var. *canina* (Schltdl.) R.Knuth, Repert. Spec. Nov. Regni Veg. Beih. 43: 143 (1926).
Scirpus cornatus C.Wright in F.A.Sauvalle, Fl. Cub.: 177 (1871).
Scirpus cornatus C.Wright, Anales Acad. Ci. Méd. Habana 8: 81 (1871).

Fimbristylis squarrulosa F.Muell., Fragm. 1: 216 (1859).
N. Australia. 50 NTA QLD WAU. Hemicr. or rhizome geophyte.
*_Abildgaardia schoenoides_ R.Br., Prodr.: 229 (1810). *Cyperus schoenoides* (R.Br.) Poir. in J.B.A.M.de Lamarck, Encycl., Suppl. 5: 185 (1817).

Fimbristylis stauntonii Debeaux & Franch., Actes Soc. Linn. Bordeaux 31: t. 3 (1877).
China to Japan. 36 CHN CHS 38 JAP KOR. Hemicr. or rhizome geophyte.

var. ***stauntonii***
China to S. Japan. 36 CHN CHS 38 JAP KOR. Hemicr. or rhizome geophyte.

var. ***tonensis*** (Makino) T.Koyama, J. Fac. Sci. Univ. Tokyo, Sect. 3, Bot. 8: 119 (1961).
C. Japan. 38 JAP. Hemicr. or rhizome geophyte.
*_Fimbristylis tonensis_ Makino, Bot. Mag. (Tokyo) 9: 260 (1895).

Fimbristylis stenostachya S.T.Blake, Pap. Dept. Biol. Univ. Queensland 1(13): 8 (1940).
N. Australia. 50 NTA QLD. Hemicr. or rhizome geophyte.

Fimbristylis stigmatotecta Govind., Rheedea 6: 61 (1996).
India (Kerala). 40 IND. Hemicr. or rhizome geophyte.

Fimbristylis stolonifera C.B.Clarke in J.D.Hooker, Fl. Brit. India 6: 637 (1893).
Nepal to SC. China. 36 CHC 40 ASS NEP. Hemicr. or rhizome geophyte.

Fimbristylis straminea Turrill, Bull. Misc. Inform. Kew 1911: 192 (1911).
Indo-China. 41 MYA THA VIE. Ther.

Fimbristylis strigosa Govind., Proc. Indian Acad. Sci., B 78: 54 (1973).
S. India. 40 IND. Hemicr. or rhizome geophyte.

Fimbristylis striolata Napper, Kew Bull. 25: 438 (1971).
W. Trop. Africa. 22 GUI SEN SIE. Ther.

Fimbristylis subalata J.Kern, Blumea 8: 133 (1955).
Indo-China, New Guinea. 41 CBD LAO THA VIE 43 NWG. Hemicr.

Fimbristylis subaristata Benth., Fl. Austral. 7: 314 (1878). *Iria subaristata* (Benth.) Kuntze, Revis. Gen. Pl. 2: 753 (1891).
N. Australia. 50 NTA QLD WAU. Hemicr. or rhizome geophyte.
Fimbristylis subaristata var. *glaucina* Domin, Biblioth. Bot. 85: 460 (1915).

Fimbristylis subdura Ohwi, Blumea 8: 101 (1955).
Vietnam, C. Jawa. 41 VIE 42 JAW. Hemicr.

Fimbristylis subinclinata T.Koyama, Bot. Mag. (Tokyo) 80: 353 (1967).
Taiwan. 38 TAI. Hemicr.

Fimbristylis subtrabeculata C.B.Clarke in J.D.Hooker, Fl. Brit. India 6: 633 (1893).
S. India. 40 IND. Hemicr. or rhizome geophyte.

Fimbristylis subtricephala T.Koyama, Brittonia 31: 291 (1979).
Vietnam. 41 VIE. Hemicr. or rhizome geophyte.

Fimbristylis sumbaensis Ohwi, Blumea 8: 106 (1955).
Lesser Sunda Is. (Sumba). 42 LSI. Hemicr. or rhizome geophyte.

Fimbristylis swamyi Govind., Proc. Indian Acad. Sci., Pl. Sci. 94: 15 (1985).
S. India. 40 IND. Hemicr. or rhizome geophyte.

Fimbristylis tamaensis Steyerm., Fieldiana, Bot. 28(1): 39 (1951).
Venezuela. 82 VEN. Hemicr. or rhizome geophyte.

Fimbristylis tenera Schult. in J.J.Roemer & J.A.Schultes, Mant. 2: 57 (1824). *Iria tenera* (Schult.) Kuntze, Revis. Gen. Pl. 2: 753 (1891).
Pakistan to Sri Lanka. 40 IND PAK SRL. Ther. or hemicr.
*_Scirpus tenellus_ Roxb., Fl. Ind. 1: 224 (1820), nom. illeg. *Trichelostylis tenella* Nees in R.Wight, Contr. Bot. India: 103 (1834).
Fimbristylis glabra Hochst. ex Steud., Syn. Pl. Glumac. 2: 111 (1855).
Fimbristylis muricatula Steud., Syn. Pl. Glumac. 2: 113 (1855).
Fimbristylis oxylepis Steud., Syn. Pl. Glumac. 2: 110 (1855).

Fimbristylis tenuicula Boeck., Linnaea 38: 385 (1874). *Iria tenuicula* (Boeck.) Kuntze, Revis. Gen. Pl. 2: 753 (1891).
Assam to Sumatera. 40 ASS 41 MYA THA VIE 42 SUM. Ther.

Fimbristylis tenuinervia J.Kern, Blumea 8: 137 (1955).
Philippines, New Guinea. 42 PHI 43 NWG. Ther.

Fimbristylis tenuis Roem. & Schult. in J.J.Roemer & J.A.Schultes, Syst. Veg. 2: 94 (1817).
Venezuela. 82 VEN. Hemicr. or rhizome geophyte.
Scirpus gracilis Willd. ex Kunth, Enum. Pl. 2: 227 (1837), nom. inval. *Iria gracilis* Kuntze, Revis. Gen. Pl. 2: 753 (1891).

Fimbristylis tetragona R.Br., Prodr.: 226 (1810). *Scirpus tetragonus* (R.Br.) Poir. in J.B.A.M.de Lamarck, Encycl. 6: 767 (1805). *Scirpus tetragonus* (R.Br.) Poir.

in J.B.A.M.de Lamarck, Encycl., Suppl. 5: 98 (1817), nom. inval. *Isolepis tetragona* (R.Br.) Schult. in J.J.Roemer & J.A.Schultes, Mant. 2: 69 (1824). *Trichelostylis tetragona* (R.Br.) Nees in R.Wight, Contr. Bot. India: 104 (1834). *Iria tetragona* (R.Br.) Kuntze, Revis. Gen. Pl. 2: 753 (1891).
Trop. & Subtrop. Asia to N. Australia. 36 CHH CHS 38 TAI 40 IND SRL 41 MYA THA VIE 42 BOR JAW LSI MLY PHI SUL SUM 50 NTA QLD WAU. Hemicr.
Fimbristylis cylindrocarpa Kunth, Enum. Pl. 2: 222 (1837).
Fimbristylis abjiciens Steud., Syn. Pl. Glumac. 2: 107 (1855).
Trichelostylis xyroides Arn. ex Hook.f., Fl. Tasman.: xlviii (1855).
Mischospora efoliata Boeck., Flora 43: 113 (1860).
Fimbristylis oxyrhachis Miq., Fl. Ned. Ind., Eerste Bijv.: 601 (1861).
Fimbristylis arnottii Thwaites, Enum. Pl. Zeyl.: 348 (1864).
Fimbristylis xyrioides Arn. ex Thwaites, Enum. Pl. Zeyl.: 348 (1864).

Fimbristylis thermalis S.Watson, Botany [Fortieth Parallel]: 360 (1871). *Iria thermalis* (S.Watson) Kuntze, Revis. Gen. Pl. 2: 753 (1891).
SW. U.S.A. to Mexico (Baja California to Coahuila). 76 ARI CAL NEV UTA 79 MXE MXN. Hemicr.

Fimbristylis thomsonii Boeck., Linnaea 37: 37 (1871). *Iria thomsonii* (Boeck.) Kuntze, Revis. Gen. Pl. 2: 753 (1891).
Trop. & Subtrop. Asia. 36 CHC CHS 38 NNS TAI 40 ASS EHM 41 THA VIE 42 MLY PHI SUM. Hemicr.

Fimbristylis tortifolia Govind., Proc. Indian Acad. Sci., Pl. Sci. 91: 46 (1982).
S. India. 40 IND. Hemicr. or rhizome geophyte.

Fimbristylis trachycarya F.Muell., Fragm. 1: 199 (1859).
N. Australia. 50 NTA WAU. Hemicr. or rhizome geophyte.

Fimbristylis trichocaulis C.B.Clarke, Kew Bull., Addit. Ser. 8: 25 (1908).
Myanmar. 41 MYA. Hemicr. or rhizome geophyte.

Fimbristylis trichoides J.Kern, Blumea 13: 119 (1965).
E. Thailand. 41 THA. Ther.

Fimbristylis trichophylla Ridl., Fl. Malay Penins. 5: 155 (1925).
Indo-China to Pen. Malaysia. 41 THA VIE 42 MLY. Hemicr.

var. ***erecta*** Holttum ex J.Kern, Blumea 8: 41 (1955).
Pen. Thailand to Pen. Malaysia. 41 THA 42 MLY. Hemicr.

var. ***trichophylla***
Vietnam, Pen. Malaysia (P. Langkawi). 41 VIE 42 MLY. Hemicr.

Fimbristylis triflora (L.) K.Schum. ex Engl., Abh. Königl. Akad. Wiss. Berlin 1894: 14 (1894).
Kenya to S. Africa, Madagascar, S. India, Sri Lanka. 23 ZAI 25 KEN TAN 27 CPP NAT SWZ 29 MDG 40 IND SRL. Hemicr.
**Cyperus triflorus* L., Mant. Pl. 2: 180 (1771). *Iria triflora* (L.) Kuntze, Revis. Gen. Pl. 2: 753 (1891). *Abildgaardia triflora* (L.) Lye in R.W.Haines & K.A.Lye, Sedges & Rushes E. Afr.: 400 (1983).
Abildgaardia tristachya Vahl, Enum. Pl. 2: 297 (1805).
Fimbristylis tristachya (Vahl) Thwaites, Enum. Pl. Zeyl.: 434 (1864), nom. illeg.

Fimbristylis trigastrocarya F.Muell., Fragm. 1: 194 (1859). *Iria trigastrocarya* (F.Muell.) Kuntze, Revis. Gen. Pl. 2: 753 (1891) (1891).
N. Australia. 50 NTA WAU. Hemicr. or rhizome geophyte.

Fimbristylis tristachya R.Br., Prodr.: 227 (1810). *Fimbristylis ferruginea* var. *tristachya* (R.Br.) Domin, Biblioth. Bot. 85: 459 (1915).
Trop. Asia to Japan and Pacific. 31+ 36 CHN CHS 38 JAP KOR NNS TAI 40 BAN 41 VIE 42 BOR JAW LSI MLY PHI SUL SUM 43 NWG 50 QLD 62 MRN. Hemicr. or rhizome geophyte.

var. ***pacifica*** (Ohwi) T.Koyama, J. Fac. Sci. Univ. Tokyo, Sect. 3, Bot. 8: 114 (1961).
Japan to Nansei-shoto. 38 JAP NNS. Hemicr.
**Fimbristylis pacifica* Ohwi, J. Jap. Bot. 14: 576 (1938).

var. ***subbispicata*** (Nees) T.Koyama, J. Fac. Sci. Univ. Tokyo, Sect. 3, Bot. 8: 114 (1961).
China to Temp. E. Asia. 31+ 36 CHN CHS 38 JAP KOR NNS TAI 41 VIE. Hemicr.
**Fimbristylis subbispicata* Nees, Nov. Actorum Acad. Caes. Leop.-Carol. Nat. Cur. 19(Suppl. 1): 75 (1843).
Fimbristylis japonica Siebold & Zucc. ex Steud., Syn. Pl. Glumac. 2: 107 (1855).
Fimbristylis gynophora C.B.Clarke, Bull. Acad. Int. Géogr. Bot. 16: 60 (1906).

var. ***tristachya***
Trop. Asia to Pacific. 40 41 VIE 42 BOR JAW LSI MLY PHI SUL SUM 43 NWG 50 QLD 62 MRN. Hemicr. or rhizome geophyte.
Schoenus cyperoides Retz., Observ. Bot. 4: 8 (1786).
Abildgaardia lanceolata Schumach., Beskr. Guin. Pl.: 33 (1827).
Fimbristylis mariana Gaudich., Voy. Uranie: 418 (1829).
Fimbristylis bispicata var. *monostachya* Nees, Hooker's J. Bot. Kew Gard. Misc. 6: 29 (1854).
Fimbristylis macrostachya Boeck., Linnaea 38: 386 (1874). *Iria macrostachya* (Boeck.) Kuntze, Revis. Gen. Pl. 2: 753 (1891).
Fimbristylis subbulbosa Benth., Fl. Austral. 7: 305 (1878). *Iria subbulbosa* (Benth.) Kuntze, Revis. Gen. Pl. 2: 753 (1891).
Fimbristylis maxima K.Schum. in K.M.Schumann & U.M.Hollrung, Fl. Kais. Wilh. Land: 24 (1889).
Scirpus triflorus Kuntze, Revis. Gen. Pl. 3(2): 337 (1898).
Fimbristylis alleizettei E.G.Camus, Notul. Syst. (Paris) 1: 291 (1910).
Fimbristylis annamica E.G.Camus, Notul. Syst. (Paris) 1: 245 (1910).
Fimbristylis mariana var. *foenea* Kük., Repert. Spec. Nov. Regni Veg. 16: 432 (1920).

Fimbristylis tumida Govind., Rheedea 8: 91 (1998).
S. India. 40 IND. Hemicr. or rhizome geophyte.

Fimbristylis tunquinensis Boeck., Allg. Bot. Z. Syst. 2: 56 (1896).
N. Vietnam. 41 VIE. Hemicr. or rhizome geophyte.

Fimbristylis turkestanica (Regel) B.Fedtsch., Trudy Glavn. Bot. Sada 38: 181 (1924).
Medit. to India. 12 SPA 13 KRI 20 EGY 32 KGZ TZK UZB 33 TCS 34 AFG CYP IRN IRQ LBS PAL TUR 35 GST SAU 40 IND PAK. Hemicr.
**Isolepis turkestanica* Regel, Trudy Imp. S.-Peterburgsk. Bot. Sada 7: 561 (1880). *Scirpus turkestanicus* (Regel) Meinsh., Trudy Imp. S.-Peterburgsk. Bot. Sada 18: 247 (1900).
Fimbristylis vitiata C.B.Clarke, Kew Bull., Addit. Ser. 8: 24 (1908).
Fimbristylis schischkinii Pobed., Bot. Mater. Gerb. Bot. Inst. Komarova Akad. Nauk S.S.S.R. 12: 18 (1950).

Fimbristylis uliginosa Hochst. ex Steud., Syn. Pl. Glumac. 2: 109 (1855). *Iria uliginosa* (Hochst. ex Steud.) Kuntze, Revis. Gen. Pl. 2: 753 (1891) (1891).
S. India. 40 IND. Hemicr. or rhizome geophyte.

Fimbristylis ultragluma Govind., Rheedea 7: 119 (1997).
India (Maharashtra). 40 IND. Hemicr. or rhizome geophyte.

Fimbristylis umbellaris (Lam.) Vahl, Enum. Pl. 2: 291 (1805).
Trop. & Subtrop. Asia to W. Pacific. 36 CHC CHS 38 NNS TAI 40 ASS IND NEP PAK SRL 41 THA VIE 42 BOR JAW MLY PHI SUL SUM 43 NWG 60+ 62 CRL MRN. Hemicr.
Scirpus globulosus Retz., Observ. Bot. 6: 19 (1791). *Isolepis globulosa* (Retz.) Schult. in J.J.Roemer & J.A.Schultes, Mant. 2: 68 (1824). *Trichelostylis globulosa* (Retz.) Nees in R.Wight, Contr. Bot. India: 105 (1834). *Fimbristylis globulosa* (Retz.) Kunth, Enum. Pl. 2: 231 (1837). *Iria globulosa* (Retz.) Kuntze, Revis. Gen. Pl. 2: 752 (1891).
**Scirpus umbellaris* Lam., Tabl. Encycl. 1: 141 (1791).
Dichostylis torresiana Gaudich., Voy. Uranie: 413 (1829).
Fimbristylis torresiana Gaudich., Voy. Uranie: 413 (1829).
Fimbristylis efoliata Steud., Syn. Pl. Glumac. 2: 109 (1855).
Fimbristylis globulosa var. *vicaryi* C.B.Clarke in J.D.Hooker, Fl. Brit. India 6: 645 (1893). *Fimbristylis umbellaris* var. *vicaryi* (C.B.Clarke) Karthik. in S.Karthikeyan & al., Fl. Ind. Enumerat. - Monocot.: 56 (1989).
Fimbristylis utilis Elmer, Leafl. Philipp. Bot. 3: 855 (1910).
Fimbristylis hypsocolea Hosok., Trans. Nat. Hist. Soc. Taiwan 28: 152 (1938).

Fimbristylis unispicularis Govind. & Hemadri, Proc. Indian Acad. Sci., B 82: 205 (1975).
India (Maharashtra). 40 IND. Hemicr. or rhizome geophyte.

Fimbristylis urakasiana Kük., Bot. Jahrb. Syst. 59: 5 (1924).
NW. Pacific. 62 CRL MRN. Hemicr.

Fimbristylis vagans S.T.Blake, Pap. Dept. Biol. Univ. Queensland 1(13): 5 (1940).
Queensland. 50 QLD.

Fimbristylis vaginata (R.Br.) Domin, Biblioth. Bot. 95: 460 (1915).
N. Australia. 50 NTA QLD. Hemicr. or rhizome geophyte.
**Abildgaardia vaginata* R.Br., Prodr. 1: 229 (1810). *Fimbristylis brownii* Benth., Fl. Austral. 7: 308 (1878), nom. illeg. *Iria vaginata* (R.Br.) Kuntze, Revis. Gen. Pl. 2: 753 (1891).

Fimbristylis vahlii (Lam.) Link, Hort. Berol. 1: 287 (1827).
C. & S. U.S.A. to Trop. America. 74 ILL KAN MSO NEB OKL 76 ARI CAL 77 TEX 78 ALA ARK FLA GEO KTY LOU MSI NCA SCA TEN 79 MXE MXN MXT 80 BLZ NIC 81 LEE 82 FRG GUY SUR VEN 83 PER 84 BZN 85 AGE PAR. Ther.
**Scirpus vahlii* Lam., Tabl. Encycl. 1: 139 (1791). *Isolepis vahlii* (Lam.) Kunth in F.W.H.von Humboldt, A.J.A.Bonpland & C.S.Kunth, Nov. Gen. Sp. 1: 221 (1816). *Iria vahlii* (Lam.) Kuntze, Revis. Gen. Pl. 2: 753 (1891).
Fimbristylis congesta Torr., Ann. New York Acad. Sci. 3: 345 (1836). *Dichostylis congesta* (Torr.) Palla, Bot. Jahrb. Syst. 10: 299 (1888).
Scirpus trichophyllus Willd. ex Kunth, Enum. Pl. 2: 224 (1837).
Fimbristylis vincentii Steud., Syn. Pl. Glumac. 2: 109 (1855).
Isolepis hartmannii Steud., Syn. Pl. Glumac. 2: 100 (1855).
Isolepis subacaulis Steud., Syn. Pl. Glumac. 2: 318 (1855). *Fimbristylis vahlii* var. *subacaulis* (Steud.) Boeck., Linnaea 37: 9 (1871).
Isolepis vincentii Steud., Syn. Pl. Glumac. 2: 102 (1855).
Fimbristylis breviculmis Boeck., Flora 41: 643 (1858).
Scirpus apus A.Gray, Proc. Amer. Acad. Arts 10: 78 (1874). *Fimbristylis apus* (A.Gray) S.Watson in S.Watson & al., Bot. California 2: 224 (1880). *Iria apus* (A.Gray) Kuntze, Revis. Gen. Pl. 2: 753 (1891).

Fimbristylis vanoverberghii Kük. in H.G.A.Engler (ed.), Pflanzenr., IV, 20(101): 631 (1936).
Malesia to New Guinea. 42 BOR PHI SUM 43 NWG. Hemicr.
**Cladium cyperoides* Merr., Philipp. J. Sci., C 7: 74 (1912). *Mariscus fallax* Fernald, Rhodora 25: 53 (1923), nom. illeg. *Machaerina cyperoides* (Merr.) T.Koyama, Bot. Mag. (Tokyo) 69: 63 (1956).
Fimbristylis fusca var. *hispidissima* Kük., Mitth. Thüring. Bot. Vereins, n.s., 50: 11 (1943).

Fimbristylis variegata Gordon-Gray, J. S. African Bot. 32: 134 (1966). *Abildgaardia variegata* (Gordon-Gray) Lye, Mitt. Bot. Staatssamml. München 10: 547 (1971).
KwaZulu-Natal. 27 NAT. Hemicr. or rhizome geophyte.

Fimbristylis villosissima Steud., Syn. Pl. Glumac. 2: 119 (1855).
Brazil. 84+. Hemicr. or rhizome geophyte.

Fimbristylis virella Govind., Rheedea 8: 93 (1998).
SW. India. 40 IND. Hemicr. or rhizome geophyte.

Fimbristylis warmingii (Boeck.) Malme, Ark. Bot. 29A(5): 8 (1937).
Brazil. 84 BZC. Hemicr. or rhizome geophyte.
**Scirpus warmingii* Boeck., Vidensk. Meddel. Naturhist. Foren. Kjøbenhavn 1869: 136 (1969).

Fimbristylis wetarensis Ohwi, Blumea 8: 103 (1955).
Lesser Sunda Is. (Wetar). 42 LSI. Hemicr.

Fimbristylis willdenowiana Govind., Beitr. Biol. Pflanzen 70: 90 (1998).

India (Andhra Pradesh). 40 IND. Hemicr. or rhizome geophyte.

Fimbristylis woodrowii C.B.Clarke, Bull. Misc. Inform. Kew 1898: 227 (1898).
SE. Pakistan to W. India. 40 IND PAK. Ther.

Fimbristylis wukungshanensis Tang & F.T.Wang, in Fl. Reipubl. Popul. Sin. 11: 226 (1961).
China (Guangxi). 36 CHS. Hemicr. or rhizome geophyte.

Fimbristylis xyridis R.Br., Prodr.: 226 (1810). *Scirpus xyridis* (R.Br.) Poir. in J.B.A.M.de Lamarck, Encycl., Suppl. 5: 98 (1817), nom. inval. *Iria xyridis* (R.Br.) Kuntze, Revis. Gen. Pl. 2: 753 (1891).
N. Australia. 50 NTA WAU. Hemicr. or rhizome geophyte.

Fimbristylis yunnanensis C.B.Clarke, J. Linn. Soc., Bot. 36: 247 (1903).
Assam to SC. China. 36 CHC 40 ASS 41 THA. Hemicr.

Fimbristylis zeylanica T.Koyama, Bot. Mag. (Tokyo) 83: 188 (1970).
NC. Sri Lanka. 40 SRL. Ther.

***Synonyms*:**
Fimbristylis abbreviata Boeck. = ***Fimbristylis falcata*** var. ***abbreviata*** (Boeck.) Karthik.
Fimbristylis abjiciens Steud. = ***Fimbristylis tetragona*** R.Br.
Fimbristylis abortiva Steud. = ***Bulbostylis abortiva*** (Steud.) C.B.Clarke
Fimbristylis actinoschoenus C.B.Clarke = ***Actinoschoenus thouarsii*** (Kunth) Benth.
Fimbristylis acuminata var. *minor* Nees ex Boeck. = ***Fimbristylis acicularis*** R.Br.
Fimbristylis acuminata var. *pumila* Nees = ***Fimbristylis acicularis*** R.Br.
Fimbristylis aestivalis var. *esquarrosa* (Makino) T. Koyama = ***Fimbristylis squarrosa*** var. ***esquarrosa*** Makino
Fimbristylis aestivalis var. *glaberrima* Boeck. = ***Fimbristylis griffithii*** Boeck.
Fimbristylis aestivalis subsp. *major* Trimen ex T.Koyama = ***Fimbristylis aestivalis*** Vahl var. ***aestivalis***
Fimbristylis aestivalis var. *squarrosa* (Vahl) T.Koyama = ***Fimbristylis squarrosa*** Vahl
Fimbristylis affinis Hook. & Arn. = ***Fimbristylis dichotoma*** (L.) Vahl subsp. ***dichotoma***
Fimbristylis affinis C.Presl = ***Fimbristylis dichotoma*** subsp. ***podocarpa*** (Nees) T.Koyama
Fimbristylis africana C.B.Clarke = ***Bulbostylis pilosa*** (Willd.) Cherm.
Fimbristylis aginkotensis Hayata = ***Fimbristylis ferruginea*** (L.) Vahl var. ***ferruginea***
Fimbristylis alamosana Fernald = ***Fimbristylis dichotoma*** (L.) Vahl subsp. ***dichotoma***
Fimbristylis albescens Steud. = ***Fimbristylis polytrichoides*** (Retz.) R.Br.
Fimbristylis alleizettei E.G.Camus = ***Fimbristylis tristachya*** R.Br. var. ***tristachya***
Fimbristylis allenii Turrill = ***Fimbristylis furva*** R.Br.
Fimbristylis alpestris (Urb.) Alain = ***Bulbostylis alpestris*** Urb.
Fimbristylis ambigua Steud. = ***Fimbristylis dichotoma*** (L.) Vahl subsp. ***dichotoma***
Fimbristylis amblyphylla Steud. = ***Fimbristylis complanata*** (Retz.) Link var. ***complanata***
Fimbristylis anceps Steud. = ***Fimbristylis complanata*** (Retz.) Link var. ***complanata***
Fimbristylis andamandica Kurz = ***Fimbristylis ferruginea*** (L.) Vahl var. ***ferruginea***
Fimbristylis andongensis Ridl. = ***Bulbostylis andongensis*** (Ridl.) C.B.Clarke
Fimbristylis androgyna R.Br. = ***Cyperus denudatus*** L.f.
Fimbristylis angularis Link = ***Fimbristylis quinquangularis*** (Vahl) Kunth
Fimbristylis angustifolia Ridl. = ***Fimbristylis fuscoides*** C.B.Clarke
Fimbristylis annamica E.G.Camus = ***Fimbristylis tristachya*** R.Br. var. ***tristachya***
Fimbristylis annua (All.) Roem. & Schult. = ***Fimbristylis dichotoma*** (L.) Vahl subsp. ***dichotoma***
Fimbristylis annua f. *brizoides* (Sm. ex Nees & Meyen) Kük. = ***Fimbristylis dichotoma*** (L.) Vahl subsp. ***dichotoma***
Fimbristylis annua var. *brizoides* (Sm. ex Nees & Meyen) A.Gross & Brade = ***Fimbristylis dichotoma*** (L.) Vahl subsp. ***dichotoma***
Fimbristylis annua var. *diphylla* (Retz.) Kük. =
Fimbristylis annua f. *oblonga* Kük. = ***Fimbristylis dichotoma*** (L.) Vahl subsp. ***dichotoma***
Fimbristylis annua var. *paucispiculata* Blatt. & McCann = ***Fimbristylis dichotoma*** (L.) Vahl subsp. ***dichotoma***
Fimbristylis annua var. *pseudoferruginea* Kük. = ***Fimbristylis dichotoma*** (L.) Vahl subsp. ***dichotoma***
Fimbristylis annua f. *tomentosa* (Vahl) Kük. = ***Fimbristylis dichotoma*** subsp. ***podocarpa*** (Nees) T.Koyama
Fimbristylis annua f. *typica* Kük. = ***Fimbristylis dichotoma*** (L.) Vahl subsp. ***dichotoma***
Fimbristylis anomala Boeck. = ***Fimbristylis puberula*** Vahl var. ***puberula***
Fimbristylis anpinensis Hayata = ***Fimbristylis ferruginea*** var. ***anpinensis*** (Hayata) H.Y.Liu
Fimbristylis aphylla (R.Br.) F.Muell. = ***Arthrostylis aphylla*** R.Br.
Fimbristylis aphylla var. *gracilis* Tang & F.T.Wang = ***Fimbristylis aphylla*** Steud.
Fimbristylis aphyllanthoides Welw. ex Ridl. = ***Bulbostylis pilosa*** (Willd.) Cherm.
Fimbristylis apus (A.Gray) S.Watson = ***Fimbristylis vahlii*** (Lam.) Link
Fimbristylis arenaria Nees = ***Bulbostylis humilis*** (Kunth) C.B.Clarke
Fimbristylis arenicola Wiggins = ***Fimbristylis dichotoma*** (L.) Vahl subsp. ***dichotoma***
Fimbristylis arenicola J.Kern = ***Fimbristylis psammophila*** J.Kern
Fimbristylis argentina (Palla) Molfino = ***Bulbostylis juncoides*** (Vahl) Kük. ex Herter
Fimbristylis armerioides (Miq.) Beetle = ***Bulbostylis barbata*** (Rottb.) C.B.Clarke subsp. ***barbata***
Fimbristylis arnottii Thwaites = ***Fimbristylis tetragona*** R.Br.
Fimbristylis arvensis (Retz.) Vahl = ***Fimbristylis ferruginea*** (L.) Vahl var. ***ferruginea***
Fimbristylis aspera var. *glaucovirens* Boeck. = ***Fimbristylis aspera*** (Schrad.) Boeck.
Fimbristylis asperrima Boeck. = ***Fimbristylis dura*** (Zoll. & Moritzi) Merr.
Fimbristylis atollensis H.St.John = ***Fimbristylis cymosa*** R.Br. subsp. ***cymosa***
Fimbristylis atrosanguinea (Boeck.) K.Schum. = ***Bulbostylis atrosanguinea*** (Boeck.) C.B.Clarke
Fimbristylis australica Boeck. = ***Fimbristylis acicularis*** R.Br.
Fimbristylis autumnalis f. *brachyactis* (Fernald) S.F. Blake = ***Fimbristylis autumnalis*** (L.) Roem. & Schult.

Fimbristylis autumnalis var. *complanata* (Retz.) Kük. = ***Fimbristylis complanata*** (Retz.) Link
Fimbristylis autumnalis var. *gracilis* (Kunth) Kük. = ***Fimbristylis autumnalis*** (L.) Roem. & Schult.
Fimbristylis autumnalis f. *hemisphaerica* Kük. = ***Fimbristylis complanata*** (Retz.) Link var. ***complanata***
Fimbristylis autumnalis var. *microcarya* (F.Muell.) Kük. = ***Fimbristylis microcarya*** F.Muell.
Fimbristylis autumnalis var. *mucronulata* (Michx.) Fernald = ***Fimbristylis autumnalis*** (L.) Roem. & Schult.
Fimbristylis autumnalis subsp. *tainanensis* (Ohwi) T.Koyama = ***Fimbristylis microcarya*** var. ***tainanensis*** (Ohwi) H.Y.Liu
Fimbristylis autumnalis subsp. *taiwanica* (Ohwi) T.Koyama = ***Fimbristylis microcarya*** F.Muell. var. ***microcarya***
Fimbristylis baldwiniana (Schult.) Torr. = ***Fimbristylis dichotoma*** (L.) Vahl subsp. ***dichotoma***
Fimbristylis barbata (Rottb.) Benth. = ***Bulbostylis barbata*** (Rottb.) C.B.Clarke
Fimbristylis benghalensis (Pers.) Roem. & Schult. = ***Fimbristylis quinquangularis*** (Vahl) Kunth
Fimbristylis bequaertii De Wild. = ***Fimbristylis complanata*** (Retz.) Link var. ***complanata***
Fimbristylis biflora Boeck. = ***Fimbristylis cinnamometorum*** (Vahl) Kunth
Fimbristylis bispicata (J.König ex Roxb.) Nees & Meyen = ***Fimbristylis schoenoides*** (Retz.) Vahl
Fimbristylis bispicata var. *monostachya* Nees = ***Fimbristylis tristachya*** R.Br. var. ***tristachya***
Fimbristylis bisumbellata var. *elata* Täckh. = ***Fimbristylis bisumbellata*** (Forssk.) Bubani
Fimbristylis biumbellulata Boeck. = ***Fimbristylis cymosa*** R.Br. subsp. ***cymosa***
Fimbristylis boeckeleri Steud. = ***Fimbristylis complanata*** (Retz.) Link var. ***complanata***
Fimbristylis boiviniana Steud. = ***Fimbristylis dichotoma*** (L.) Vahl subsp. ***dichotoma***
Fimbristylis boninensis Hayata = ***Fimbristylis longispica*** var. ***boninensis*** (Hayata) Ohwi
Fimbristylis brachylaena F.Muell. = ***Fimbristylis caespitosa*** R.Br.
Fimbristylis brachyphylla (Schult.) Schult. = ***Fimbristylis dichotoma*** (L.) Vahl subsp. ***dichotoma***
Fimbristylis brachyphylla J.Presl & C.Presl = ***Fimbristylis falcata*** (Vahl) Kunth var. ***falcata***
Fimbristylis brevicollis Kük. = ***Fimbristylis diphylloides*** Makino
Fimbristylis breviculma Govind. = [40 IND]
Fimbristylis breviculmis Boeck. = ***Fimbristylis vahlii*** (Lam.) Link
Fimbristylis brevifolia (Roxb.) Wall. ex Schult. = ***Fimbristylis dichotoma*** (L.) Vahl subsp. ***dichotoma***
Fimbristylis brevifolia J.Presl & C.Presl = ***Fimbristylis falcata*** (Vahl) Kunth var. ***falcata***
Fimbristylis brevifolia R.Br. = ***Fimbristylis ferruginea*** (L.) Vahl var. ***ferruginea***
Fimbristylis brizoides Sm. ex Nees & Meyen = ***Fimbristylis dichotoma*** (L.) Vahl subsp. ***dichotoma***
Fimbristylis brizoides var. *humboldtiana* Nees = ***Fimbristylis dichotoma*** (L.) Vahl subsp. ***dichotoma***
Fimbristylis brizoides var. *pallens* Nees = ***Fimbristylis dichotoma*** (L.) Vahl subsp. ***dichotoma***
Fimbristylis brizoides var. *tenuifolia* Nees = ***Fimbristylis dichotoma*** (L.) Vahl subsp. ***dichotoma***
Fimbristylis bromeliifolia A.Spreng. = ***Diplasia karatifolia*** Rich. ex Pers.
Fimbristylis brownii Benth. = ***Fimbristylis vaginata*** (R.Br.) Domin
Fimbristylis brunnea C.B.Clarke ex E.G.Camus = ***Fimbristylis jucunda*** (C.B.Clarke) J.Kern
Fimbristylis brunneovaginata (Boeck.) H.Pfeiff. = ***Bulbostylis sphaerocephala*** (Boeck.) Lindm.
Fimbristylis buergeri Miq. = ***Fimbristylis longispica*** Steud. var. ***longispica***
Fimbristylis bufonia (Kunth) Alain = ***Bulbostylis tenuifolia*** (Rudge) J.F.Macbr.
Fimbristylis burchellii Ficalho & Hiern = ***Bulbostylis burchellii*** (Ficalho & Hiern) C.B.Clarke
Fimbristylis bursifolia S.Vidal = ***Fimbristylis acicularis*** R.Br.
Fimbristylis calocarpa Steud. = ***Fimbristylis dichotoma*** (L.) Vahl subsp. ***dichotoma***
Fimbristylis campylophylla Tuyama = ***Fimbristylis diphylloides*** Makino
Fimbristylis candelabrum Link = ***Fimbristylis dichotoma*** (L.) Vahl subsp. ***dichotoma***
Fimbristylis capillacea Hochst. ex Steud. = ***Bulbostylis densa*** (Wall.) Hand.-Mazz. subsp. ***densa***
Fimbristylis capillaris (L.) A.Gray = ***Bulbostylis capillaris*** (L.) Kunth ex C.B.Clarke
Fimbristylis capillaris var. *coarctata* (Elliott) Britton = ***Bulbostylis ciliatifolia*** var. ***coarctata*** (Elliott) Kral
Fimbristylis capillaris f. *major* Maury = ***Bulbostylis capillaris*** (L.) Kunth ex C.B.Clarke subsp. ***capillaris***
Fimbristylis capillaris f. *minima* Maury = ***Bulbostylis capillaris*** (L.) Kunth ex C.B.Clarke subsp. ***capillaris***
Fimbristylis capillaris var. *pilosa* Britton = ***Bulbostylis juncoides*** (Vahl) Kük. ex Herter
Fimbristylis capitata Sieber ex C.Presl = ***Fimbristylis cymosa*** R.Br. subsp. ***cymosa***
Fimbristylis capitata R.Br. = ***Fimbristylis furva*** R.Br.
Fimbristylis capitulifera Merr. = ***Fimbristylis cymosa*** R.Br. subsp. ***cymosa***
Fimbristylis cardiocarpa Ridl. = ***Bulbostylis filamentosa*** (Vahl) C.B.Clarke
Fimbristylis carinata Nees = ***Fimbristylis dichotoma*** (L.) Vahl subsp. ***dichotoma***
Fimbristylis caripensis Steud. = ***Fimbristylis dichotoma*** (L.) Vahl subsp. ***dichotoma***
Fimbristylis caroliniana f. *eucycla* (Fernald) Fernald = ***Fimbristylis caroliniana*** (Lam.) Fernald
Fimbristylis caroliniana f. *pycnostachya* (Fernald) Fernald = ***Fimbristylis caroliniana*** (Lam.) Fernald
Fimbristylis carteri (Britton ex Small) Alain = ***Bulbostylis ciliatifolia*** var. ***coarctata*** (Elliott) Kral
Fimbristylis castanea (Michx.) Vahl = ***Fimbristylis spadicea*** (L.) Vahl
Fimbristylis castanea var. *puberula* (Michx.) Britton ex House = ***Fimbristylis dichotoma*** (L.) Vahl subsp. ***dichotoma***
Fimbristylis chaetorrhiza (Nees) Kunth = ***Fimbristylis falcata*** (Vahl) Kunth var. ***falcata***
Fimbristylis chalarocephala Ohwi & T.Koyama = ***Fimbristylis hookeriana*** Boeck.
Fimbristylis chevalieri Kük. = ***Fimbristylis barteri*** Boeck.
Fimbristylis chinensis (Benth.) Tang & F.T.Wang = ***Actinoschoenus thouarsii*** (Kunth) Benth.
Fimbristylis chirigota (C.Wright) Alain = ***Bulbostylis hispidula*** (Vahl) R.W.Haines subsp. ***hispidula***
Fimbristylis ciliata C.B.Clarke ex Drobow = ***Fimbristylis ferruginea*** (L.) Vahl var. ***ferruginea***
Fimbristylis ciliolata Steud. = ***Fimbristylis cymosa*** R.Br. subsp. ***cymosa***
Fimbristylis cincta Nees = ***Fimbristylis dichotoma*** subsp. ***podocarpa*** (Nees) T.Koyama

Fimbristylis cinerea Ridl. = ***Bulbostylis trichobasis*** (Baker) C.B.Clarke
Fimbristylis cinnamomea (Boeck.) K.Schum. = ***Bulbostylis cinnamomea*** (Boeck.) C.B.Clarke
Fimbristylis cioniana Savi = ***Bulbostylis cioniana*** (Savi) Lye
Fimbristylis cioniana var. *microcarpa* Chiov. = ***Bulbostylis somaliensis*** subsp. ***microcarpa*** (Chiov.) Lye
Fimbristylis circinnata Steud. = ***Fimbristylis dichotoma*** (L.) Vahl subsp. ***dichotoma***
Fimbristylis clavinux C.B.Clarke = ***Bulbostylis hispidula*** (Vahl) R.W.Haines subsp. ***hispidula***
Fimbristylis coarctata Schwein. = ***Bulbostylis capillaris*** (L.) Kunth ex C.B.Clarke subsp. ***capillaris***
Fimbristylis coleotricha Hochst. ex A.Rich. = ***Bulbostylis coleotricha*** (Hochst. ex A.Rich.) C.B.Clarke
Fimbristylis collina Ridl. = ***Bulbostylis scabricaulis*** Cherm.
Fimbristylis comata Nees = ***Fimbristylis squarrosa*** Vahl var. ***squarrosa***
Fimbristylis communis Kunth = ***Fimbristylis dichotoma*** (L.) Vahl subsp. ***dichotoma***
Fimbristylis complanata f. *exaltata* T.Koyama = ***Fimbristylis complanata*** (Retz.) Link var. ***complanata***
Fimbristylis complanata subsp. *keniaeensis* (Kük.) Lye = ***Fimbristylis complanata*** var. ***keniaeensis*** (Kük.) Lye
Fimbristylis complanata var. *kraussiana* C.B.Clarke = ***Fimbristylis consanguinea*** Kunth
Fimbristylis complanata var. *macrocarya* Domin = ***Fimbristylis complanata*** (Retz.) Link var. ***complanata***
Fimbristylis complanata var. *microcarya* (F.Muell.) C.B.Clarke = ***Fimbristylis microcarya*** F.Muell.
Fimbristylis complanata var. *subaphylla* (Boeck.) Lye = ***Fimbristylis complanata*** (Retz.) Link var. ***complanata***
Fimbristylis compressa Roem. & Schult. = ***Fimbristylis ferruginea*** (L.) Vahl var. ***ferruginea***
Fimbristylis compressa Fern.-Vill. = ***Fimbristylis ovata*** (Burm.f.) J.Kern
Fimbristylis compressa Boeck. = ***Fimbristylis pubisquama*** J.Kern
Fimbristylis conferta A.Rich. = ***Fimbristylis complanata*** (Retz.) Link var. ***complanata***
Fimbristylis confinis Steud. = ***Fimbristylis dichotoma*** (L.) Vahl subsp. ***dichotoma***
Fimbristylis congesta Torr. = ***Fimbristylis vahlii*** (Lam.) Link
Fimbristylis conifera Rchb. ex Kunth = ***Bulbostylis conifera*** (Kunth) Beetle
Fimbristylis connectens Thwaites = ***Fimbristylis consanguinea*** Kunth
Fimbristylis conspicua Boeck. = ***Bulbostylis conspicua*** (Boeck.) H.Pfeiff.
Fimbristylis contexta (Nees) Kunth = ***Bulbostylis contexta*** (Nees) Bodard
Fimbristylis corniculata Merr. = ***Fimbristylis macassarensis*** Steud.
Fimbristylis crassipes Boeck. = ***Fimbristylis falcata*** (Vahl) Kunth var. ***falcata***
Fimbristylis crassipes Boeck. = ***Fimbristylis mexicana*** Palla
Fimbristylis crassispica Palla = ***Fimbristylis longispica*** Steud. var. ***longispica***
Fimbristylis cubensis C.B.Clarke = ***Bulbostylis setacea*** (Griseb.) Svenson
Fimbristylis cumingii (Steud.) Náves = ***Bulbostylis barbata*** (Rottb.) C.B.Clarke subsp. ***barbata***
Fimbristylis curassavica (Britton) Alain = ***Bulbostylis curassavica*** (Britton) Kük. & Ekman
Fimbristylis curvifolia (Schrad.) Steud. = ***Fimbristylis dichotoma*** (L.) Vahl subsp. ***dichotoma***
Fimbristylis cylindrica Vahl = ***Fimbristylis spadicea*** (L.) Vahl
Fimbristylis cylindrocarpa Kunth = ***Fimbristylis tetragona*** R.Br.
Fimbristylis cylindrostachya Steud. = ***Fimbristylis complanata*** (Retz.) Link var. ***complanata***
Fimbristylis cymosa var. *microcephala* F.Br. = ***Fimbristylis cymosa*** subsp. ***umbellatocapitata*** (Hillebr.) T.Koyama
Fimbristylis cymosa var. *multifolia* Domin = ***Fimbristylis cymosa*** R.Br. subsp. ***cymosa***
Fimbristylis cymosa var. *pycnocephala* (Hillebr.) Kük. = ***Fimbristylis cymosa*** subsp. ***umbellatocapitata*** (Hillebr.) T.Koyama
Fimbristylis cymosa subsp. *spathacea* (Roth) T.Koyama = ***Fimbristylis cymosa*** R.Br. subsp. ***cymosa***
Fimbristylis cymosa var. *spathacea* (Roth) T.Koyama = ***Fimbristylis cymosa*** R.Br. subsp. ***cymosa***
Fimbristylis cymosa var. *subcapitata* C.B.Clarke = ***Fimbristylis cymosa*** R.Br. subsp. ***cymosa***
Fimbristylis cymosa var. *typica* Domin = ***Fimbristylis cymosa*** R.Br.
Fimbristylis cymosa var. *umbellatocapitata* Hillebr. = ***Fimbristylis cymosa*** subsp. ***umbellatocapitata*** (Hillebr.) T.Koyama
Fimbristylis cyperoides R.Br. = ***Fimbristylis cinnamometorum*** (Vahl) Kunth
Fimbristylis cyperoides (J.Presl & C.Presl) F.Muell. = ***Fimbristylis fusca*** (Nees) Benth. ex C.B.Clarke
Fimbristylis cyrtophylla Miq. = ***Fimbristylis ferruginea*** (L.) Vahl var. ***ferruginea***
Fimbristylis dallachyi F.Muell. ex Benth. = ***Fimbristylis fimbristyloides*** (F.Muell.) Druce
Fimbristylis darlingtoniana Pennell = ***Fimbristylis dichotoma*** (L.) Vahl subsp. ***dichotoma***
Fimbristylis dasyphylla Miq. = ***Fimbristylis sericea*** (Poir.) R.Br.
Fimbristylis debilis F.Muell. = ***Fimbristylis signata*** S.T.Blake
Fimbristylis decora Nees & Meyen = ***Fimbristylis sericea*** (Poir.) R.Br.
Fimbristylis decumbens Boeck. = ***Fimbristylis acicularis*** R.Br.
Fimbristylis densa (Wall.) T.Koyama & T.I.Chuang = ***Bulbostylis densa*** (Wall.) Hand.-Mazz.
Fimbristylis depauperata R.Br. = ***Fimbristylis dichotoma*** subsp. ***depauperata*** (R.Br.) J.Kern
Fimbristylis depauperata var. *chillagoensis* Domin = ***Fimbristylis dichotoma*** subsp. ***depauperata*** (R.Br.) J.Kern
Fimbristylis depauperata var. *plurispiculata* Domin = ***Fimbristylis dichotoma*** subsp. ***depauperata*** (R.Br.) J.Kern
Fimbristylis depauperata var. *polyphylla* Domin = ***Fimbristylis dichotoma*** subsp. ***depauperata*** (R.Br.) J.Kern
Fimbristylis depauperata var. *typica* Domin = ***Fimbristylis dichotoma*** subsp. ***depauperata*** (R.Br.) J.Kern
Fimbristylis dichotoma f. *annua* (All.) Ohwi = ***Fimbristylis dichotoma*** (L.) Vahl subsp. ***dichotoma***
Fimbristylis dichotoma var. *annua* (All.) T.Koyama = ***Fimbristylis dichotoma*** (L.) Vahl subsp. ***dichotoma***
Fimbristylis dichotoma subsp. *bisumbellata* (Forssk.) Luceño = ***Fimbristylis bisumbellata*** (Forssk.) Bubani

Fimbristylis dichotoma var. *boninensis* (Hayata) T.Koyama = ***Fimbristylis longispica*** var. ***boninensis*** (Hayata) Ohwi
Fimbristylis dichotoma f. *cincta* (Ohwi) T.Koyama = ***Fimbristylis dichotoma*** (L.) Vahl subsp. ***dichotoma***
Fimbristylis dichotoma f. *depauperata* (R.Br.) Ohwi = ***Fimbristylis dichotoma*** subsp. ***depauperata*** (R.Br.) J.Kern
Fimbristylis dichotoma var. *floribunda* (Miq.) T.Koyama = ***Fimbristylis dichotoma*** (L.) Vahl subsp. ***dichotoma***
Fimbristylis dichotoma var. *glaberrima* Boeck. = ***Fimbristylis dichotoma*** (L.) Vahl subsp. ***dichotoma***
Fimbristylis dichotoma f. *gracillima* T.Koyama = ***Fimbristylis dichotoma*** (L.) Vahl subsp. ***dichotoma***
Fimbristylis dichotoma var. *hahajimensis* (Tuyama) Ohwi = ***Fimbristylis longispica*** var. ***hahajimensis*** (Tuyama) Ohwi
Fimbristylis dichotoma var. *laxa* (Vahl) Napper = ***Fimbristylis dichotoma*** (L.) Vahl subsp. ***dichotoma***
Fimbristylis dichotoma subsp. *longispica* (Steud.) T.Koyama = ***Fimbristylis longispica*** Steud.
Fimbristylis dichotoma var. *nilagirica* (C.B.Clarke) Karthik. = ***Fimbristylis dichotoma*** (L.) Vahl subsp. ***dichotoma***
Fimbristylis dichotoma var. *paucispiculata* (Blatt. & McCann) Panigrahi = ***Fimbristylis dichotoma*** (L.) Vahl subsp. ***dichotoma***
Fimbristylis dichotoma var. *pluristriata* (C.B.Clarke) Napper = ***Fimbristylis dichotoma*** subsp. ***podocarpa*** (Nees) T.Koyama
Fimbristylis dichotoma var. *tashiroana* (Ohwi) Ohwi = ***Fimbristylis dichotoma*** (L.) Vahl subsp. ***dichotoma***
Fimbristylis dichotoma var. *tentsuki* T.Koyama = ***Fimbristylis dichotoma*** (L.) Vahl subsp. ***dichotoma***
Fimbristylis dichotoma f. *tentsuki* T.Koyama = ***Fimbristylis dichotoma*** (L.) Vahl subsp. ***dichotoma***
Fimbristylis dichotoma var. *tikushiensis* (Hayata) T.Koyama = ***Fimbristylis dichotoma*** (L.) Vahl subsp. ***dichotoma***
Fimbristylis dichotoma f. *tomentosa* (Vahl) Ohwi = ***Fimbristylis dichotoma*** subsp. ***podocarpa*** (Nees) T.Koyama
Fimbristylis dichotoma var. *villosa* Vahl = ***Fimbristylis bisumbellata*** (Forssk.) Bubani
Fimbristylis dietrichsenii Boeck. = ***Fimbristylis longispica*** Steud. var. ***longispica***
Fimbristylis diffusula Steud. = ***Fimbristylis dichotoma*** (L.) Vahl subsp. ***dichotoma***
Fimbristylis digitata Boeck. = ***Fimbristylis lawiana*** (Boeck.) J.Kern
Fimbristylis diphylla (Retz.) Vahl = ***Fimbristylis dichotoma*** (L.) Vahl subsp. ***dichotoma***
Fimbristylis diphylla f. *coarctata* Domin = ***Fimbristylis dichotoma*** (L.) Vahl subsp. ***dichotoma***
Fimbristylis diphylla var. *depauperata* (R.Br.) C.B.Clarke = ***Fimbristylis dichotoma*** subsp. ***depauperata*** (R.Br.) J.Kern
Fimbristylis diphylla subsp. *diffusa* D.B.Ward = ***Fimbristylis dichotoma*** (L.) Vahl subsp. ***dichotoma***
Fimbristylis diphylla var. *floribunda* Miq. = ***Fimbristylis dichotoma*** (L.) Vahl subsp. ***dichotoma***
Fimbristylis diphylla f. *glabra* Domin = ***Fimbristylis dichotoma*** (L.) Vahl subsp. ***dichotoma***
Fimbristylis diphylla var. *gracilis* (R.Br.) Benth. = ***Fimbristylis dichotoma*** (L.) Vahl subsp. ***dichotoma***
Fimbristylis diphylla var. *laxa* E.G.Camus = ***Fimbristylis dichotoma*** (L.) Vahl subsp. ***dichotoma***
Fimbristylis diphylla var. *nilagirica* C.B.Clarke = ***Fimbristylis dichotoma*** (L.) Vahl subsp. ***dichotoma***
Fimbristylis diphylla var. *paucispiculata* Cherm. = ***Fimbristylis dichotoma*** (L.) Vahl subsp. ***dichotoma***
Fimbristylis diphylla var. *pluristriata* C.B.Clarke = ***Fimbristylis dichotoma*** subsp. ***podocarpa*** (Nees) T.Koyama
Fimbristylis diphylla var. *polystachya* Domin = ***Fimbristylis dichotoma*** (L.) Vahl subsp. ***dichotoma***
Fimbristylis diphylla f. *royeniana* (Nees) C.B.Clarke = ***Fimbristylis dichotoma*** (L.) Vahl subsp. ***dichotoma***
Fimbristylis diphylla var. *spirostachya* (F.Muell. ex Benth.) C.B.Clarke = ***Fimbristylis dichotoma*** subsp. ***depauperata*** (R.Br.) J.Kern
Fimbristylis diphylla f. *subpubescens* Domin = ***Fimbristylis dichotoma*** (L.) Vahl subsp. ***dichotoma***
Fimbristylis diphylla var. *tomentosa* (Vahl) Barros = ***Fimbristylis dichotoma*** subsp. ***podocarpa*** (Nees) T.Koyama
Fimbristylis diphylla var. *tuberculata* Cherm. = ***Fimbristylis alboviridis*** (C.B.Clarke ex Scott-Elliot) C.B.Clarke
Fimbristylis diphylloides var. *straminea* Tang & F.T.Wang = ***Fimbristylis diphylloides*** Makino
Fimbristylis dipsacea var. *verrucifera* (Maxim.) T.Koyama = ***Fimbristylis dipsacea*** (Rottb.) C.B.Clarke
Fimbristylis disticha var. *dallachyi* (F.Muell. ex Benth.) Domin = ***Fimbristylis fimbristyloides*** (F.Muell.) Druce
Fimbristylis divaricata Ham. = ***Rhynchospora trichodes*** C.B.Clarke
Fimbristylis donticola Hochst. ex Steud. = [40]
Fimbristylis dregeana Kunth = ***Fimbristylis dichotoma*** (L.) Vahl subsp. ***dichotoma***
Fimbristylis drummondii (Torr. & Hook.) Boeck. = ***Fimbristylis puberula*** Vahl var. ***puberula***
Fimbristylis ecklonii Nees = ***Fimbristylis squarrosa*** Vahl var. ***squarrosa***
Fimbristylis efoliata Steud. = ***Fimbristylis umbellaris*** (Lam.) Vahl
Fimbristylis eggersii Boeck. = ***Fimbristylis dichotoma*** (L.) Vahl subsp. ***dichotoma***
Fimbristylis elata R.Br. = ***Fimbristylis denudata*** R.Br.
Fimbristylis elongata R.Br. = ***Fimbristylis dichotoma*** (L.) Vahl subsp. ***dichotoma***
Fimbristylis elongata Sieber ex C.Presl = ***Fimbristylis pilosa*** Vahl
Fimbristylis elongata Pires de Lima = [26 MOZ]
Fimbristylis emarginata Wight = ***Fimbristylis littoralis*** Gaudich. var. ***littoralis***
Fimbristylis eragrostis var. *nigrobrunnea* (Thwaites) S.S.Hooper = ***Fimbristylis eragrostis*** (Nees) Hance
Fimbristylis erythradenia E.G.Camus = ***Fimbristylis fuscoides*** C.B.Clarke
Fimbristylis exigua Boeck. = ***Isolepis sepulcralis*** Steud.
Fimbristylis exilis (Kunth) Roem. & Schult. = ***Bulbostylis hispidula*** (Vahl) R.W.Haines subsp. ***hispidula***
Fimbristylis exilis var. *brachyphylla* Cherm. = ***Bulbostylis hensii*** (C.B.Clarke) R.W.Haines
Fimbristylis exilis var. *hirsutissima* Cherm. = ***Bulbostylis hispidula*** (Vahl) R.W.Haines subsp. ***hispidula***
Fimbristylis exilis var. *senegalensis* Cherm. = ***Bulbostylis hispidula*** subsp. ***senegalensis*** (Cherm.) Van den Berghen
Fimbristylis falcata var. *latifolia* (Nees) H.B.Naithani & M.B.Raizada = ***Fimbristylis falcata*** (Vahl) Kunth var. ***falcata***
Fimbristylis faulensis Beck = [51 NZS]
Fimbristylis fauriei Ohwi = ***Fimbristylis quinquangularis*** (Vahl) Kunth

Fimbristylis ferruginea var. *compacta* Kük. = **Fimbristylis ferruginea** (L.) Vahl var. **ferruginea**
Fimbristylis ferruginea f. *decomposita* Domin = **Fimbristylis ferruginea** (L.) Vahl var. **ferruginea**
Fimbristylis ferruginea var. *foliata* Benth. = **Fimbristylis ferruginea** (L.) Vahl var. **ferruginea**
Fimbristylis ferruginea var. *sieberiana* (Kunth) Boeck. = **Fimbristylis ferruginea** (L.) Vahl var. **ferruginea**
Fimbristylis ferruginea subsp. *sieberiana* (Kunth) Lye = **Fimbristylis ferruginea** (L.) Vahl var. **ferruginea**
Fimbristylis ferruginea var. *tristachya* (R.Br.) Domin = **Fimbristylis tristachya** R.Br.
Fimbristylis ferruginea var. *typica* Domin = **Fimbristylis ferruginea** (L.) Vahl
Fimbristylis filamentosa (Vahl) K.Schum. = **Bulbostylis filamentosa** (Vahl) C.B.Clarke
Fimbristylis filiformis (Nees) Kunth = **Fimbristylis pauciflora** R.Br.
Fimbristylis filiformis (Thwaites) Druce = **Actinoschoenus thouarsii** (Kunth) Benth.
Fimbristylis firmula Steud. = **Fimbristylis autumnalis** (L.) Roem. & Schult.
Fimbristylis flaccida Steud. = **Fimbristylis littoralis** Gaudich. var. **littoralis**
Fimbristylis flaccidula Zoll. = **Fimbristylis littoralis** Gaudich. var. **littoralis**
Fimbristylis flexuosa Ridl. = **Bulbostylis flexuosa** (Ridl.) Goetgh.
Fimbristylis floccosa (Griseb.) Alain = **Bulbostylis floccosa** (Griseb.) C.B.Clarke
Fimbristylis foliosa Link = **Fimbristylis dichotoma** (L.) Vahl subsp. **dichotoma**
Fimbristylis fosbergiana T.Koyama = **Fimbristylis palauensis** Ohwi
Fimbristylis frankii Steud. = **Fimbristylis autumnalis** (L.) Roem. & Schult.
Fimbristylis frankii var. *brachyactis* Fernald = **Fimbristylis autumnalis** (L.) Roem. & Schult.
Fimbristylis fusca subsp. *fulvescens* (Thwaites) T.Koyama = **Fimbristylis fulvescens** (Thwaites) Thwaites
Fimbristylis fusca var. *hispidissima* Kük. = **Fimbristylis vanoverberghii** Kük.
Fimbristylis fuscata Steud. = **Fimbristylis dichotoma** (L.) Vahl subsp. **dichotoma**
Fimbristylis fuscescens (Link) Steud. = **Fimbristylis dichotoma** (L.) Vahl subsp. **dichotoma**
Fimbristylis gaimardioides (É.Desv.) Steud. = **Zameioscirpus gaimardioides** (É.Desv.) Dhooge & Goetgh.
Fimbristylis geminata Kunth = **Fimbristylis autumnalis** (L.) Roem. & Schult.
Fimbristylis germainii E.G.Camus = **Fimbristylis dichotoma** (L.) Vahl subsp. **dichotoma**
Fimbristylis glabra Hochst. ex Steud. = **Fimbristylis tenera** Schult.
Fimbristylis glauca Vahl = **Fimbristylis dichotoma** subsp. **glauca** (Vahl) T.Koyama
Fimbristylis globulosa (Retz.) Kunth = **Fimbristylis umbellaris** (Lam.) Vahl
Fimbristylis globulosa var. *vicaryi* C.B.Clarke = **Fimbristylis umbellaris** (Lam.) Vahl
Fimbristylis glomerata (Schrad. ex Schult.) Nees = **Fimbristylis cymosa** R.Br. subsp. **cymosa**
Fimbristylis glomerata (Schrad. ex Schult.) Urb. = **Fimbristylis cymosa** R.Br. subsp. **cymosa**
Fimbristylis goeringiana Steud. = **Fimbristylis dichotoma** (L.) Vahl subsp. **dichotoma**
Fimbristylis gracilis R.Br. = **Fimbristylis dichotoma** (L.) Vahl subsp. **dichotoma**
Fimbristylis grisebachii Greenm. ex Combs = **Bulbostylis setacea** (Griseb.) Svenson
Fimbristylis gynophora C.B.Clarke = **Fimbristylis tristachya** var. **subbispicata** (Nees) T.Koyama
Fimbristylis haenkei D.Dietr. = **Fimbristylis falcata** (Vahl) Kunth var. **falcata**
Fimbristylis hahajimensis Tuyama = **Fimbristylis longispica** var. **hahajimensis** (Tuyama) Ohwi
Fimbristylis hamiltonii Steud. = [81 PUE]
Fimbristylis hanceana Boeck. = **Fimbristylis rigidula** Nees
Fimbristylis harperi Britton = **Fimbristylis caroliniana** (Lam.) Fernald
Fimbristylis harrisii (Britton) C.D.Adams = **Bulbostylis subaphylla** C.B.Clarke
Fimbristylis haspaniformis T.Koyama = **Fimbristylis salbundia** (Nees) Kunth
Fimbristylis hatsusimae Ohwi = **Fimbristylis littoralis** Gaudich. var. **littoralis**
Fimbristylis hensii C.B.Clarke = **Bulbostylis hensii** (C.B.Clarke) R.W.Haines
Fimbristylis heynei Boeck. = **Fimbristylis dichotoma** (L.) Vahl subsp. **dichotoma**
Fimbristylis hildebrandtii (Boeck.) Ridl. = **Bulbostylis hispidula** (Vahl) R.W.Haines subsp. **hispidula**
Fimbristylis hirsuta Hochst. ex Steud. = **Rhynchospora trichodes** C.B.Clarke
Fimbristylis hirta (Kunth) Roem. & Schult. = **Fimbristylis squarrosa** Vahl var. **squarrosa**
Fimbristylis hirta var. *canina* (Schltdl.) R.Knuth = **Fimbristylis squarrosa** Vahl var. **squarrosa**
Fimbristylis hirtella Vahl = **Fimbristylis dichotoma** (L.) Vahl subsp. **dichotoma**
Fimbristylis hirtella Ham. =
Fimbristylis hispaniolica (Kük. & Ekman) Alain = **Bulbostylis hispaniolica** Kük. & Ekman
Fimbristylis hispidula (Vahl) Kunth = **Bulbostylis hispidula** (Vahl) R.W.Haines
Fimbristylis hispidula subsp. *brachyphylla* (Cherm.) Napper = **Bulbostylis hensii** (C.B.Clarke) R.W.Haines
Fimbristylis hispidula var. *canina* Schltdl. = **Fimbristylis squarrosa** Vahl var. **squarrosa**
Fimbristylis hispidula subsp. *senegalensis* (Cherm.) Napper = **Bulbostylis hispidula** subsp. **senegalensis** (Cherm.) Van den Berghen
Fimbristylis holwayana Fernald = **Fimbristylis dichotoma** (L.) Vahl subsp. **dichotoma**
Fimbristylis hookeri Endl. = **Fimbristylis dichotoma** (L.) Vahl subsp. **dichotoma**
Fimbristylis horsfieldii C.B.Clarke = **Fimbristylis consanguinea** Kunth
Fimbristylis huillensis Ridl. = [26 ANG]
Fimbristylis humboldtii J.Presl & C.Presl = **Fimbristylis dichotoma** (L.) Vahl subsp. **dichotoma**
Fimbristylis humilis S.T.Blake = [50 QLD]
Fimbristylis humilis Peter = **Bulbostylis schimperiana** (Hochst. ex A.Rich.) C.B.Clarke
Fimbristylis hypsocolea Hosok. = **Fimbristylis umbellaris** (Lam.) Vahl
Fimbristylis inconstans Steud. = **Fimbristylis schoenoides** (Retz.) Vahl
Fimbristylis interior Britton = **Fimbristylis puberula** var. **interior** (Britton) Kral
Fimbristylis jacobinae Steud. = **Bulbostylis jacobinae** (Steud.) Lindm.
Fimbristylis japonica Siebold & Zucc. ex Steud. = **Fimbristylis tristachya** var. **subbispicata** (Nees) T.Koyama
Fimbristylis juncea Boeck. = **Fimbristylis polytrichoides** (Retz.) R.Br.

Fimbristylis juncifolia J.Presl & C.Presl = ***Fimbristylis dichotoma*** (L.) Vahl subsp. ***dichotoma***
Fimbristylis junciformis (Nees) Kunth = ***Fimbristylis falcata*** (Vahl) Kunth var. ***falcata***
Fimbristylis junciformis (Kunth) Maury = ***Bulbostylis junciformis*** (Kunth) C.B.Clarke
Fimbristylis junciformis var. *abbreviata* (Boeck.) C.B.Clarke = ***Fimbristylis falcata*** var. ***abbreviata*** (Boeck.) Karthik.
Fimbristylis junciformis var. *conostachya* C.B.Clarke = ***Fimbristylis falcata*** (Vahl) Kunth var. ***falcata***
Fimbristylis junciformis var. *latifolia* Camus = ***Fimbristylis sericea*** (Poir.) R.Br.
Fimbristylis juncoides (Vahl) Alain = ***Bulbostylis juncoides*** (Vahl) Kük. ex Herter
Fimbristylis kagiensis Hayata = ***Fimbristylis schoenoides*** (Retz.) Vahl
Fimbristylis kamphoeveneri Boeck. = ***Fimbristylis cinnamometorum*** (Vahl) Kunth
Fimbristylis kankaoensis Hayata = ***Fimbristylis cymosa*** R.Br. subsp. ***cymosa***
Fimbristylis keniaeensis Kük. = ***Fimbristylis complanata*** var. ***keniaeensis*** (Kük.) Lye
Fimbristylis kielmaieri Steud. = ***Fimbristylis complanata*** (Retz.) Link var. ***complanata***
Fimbristylis koidzumiana Ohwi = ***Fimbristylis littoralis*** var. ***koidzumiana*** (Ohwi) T.Koyama
Fimbristylis koreensis H.Lév. = ***Fimbristylis longispica*** Steud. var. ***longispica***
Fimbristylis kraussiana Hochst. ex Hook.f. = ***Fimbristylis consanguinea*** Kunth
Fimbristylis kunthiana Ridl. = ***Fimbristylis schoenoides*** (Retz.) Vahl
Fimbristylis lacei Turrill = ***Fimbristylis fimbristyloides*** (F.Muell.) Druce
Fimbristylis laevissima Steud. = ***Fimbristylis cymosa*** R.Br. subsp. ***cymosa***
Fimbristylis laniceps (C.B.Clarke) K.Schum. = ***Bulbostylis laniceps*** C.B.Clarke
Fimbristylis lanifera (Boeck.) K.Schum. = ***Bulbostylis lanifera*** (Boeck.) Kük.
Fimbristylis latifolia (Nees) Kunth = ***Fimbristylis falcata*** (Vahl) Kunth var. ***falcata***
Fimbristylis laxa Vahl = ***Fimbristylis dichotoma*** (L.) Vahl subsp. ***dichotoma***
Fimbristylis laxiflora C.B.Clarke = ***Actinoschoenus thouarsii*** (Kunth) Benth.
Fimbristylis leiocarpa Maxim. = ***Fimbristylis aestivalis*** Vahl var. ***aestivalis***
Fimbristylis leiocarpa Miq. = ***Fimbristylis ferruginea*** (L.) Vahl var. ***ferruginea***
Fimbristylis lepidota E.G.Camus = ***Fimbristylis eragrostis*** (Nees) Hance
Fimbristylis leptostachya Steud. = ***Fimbristylis autumnalis*** (L.) Roem. & Schult.
Fimbristylis leucostachya Boeck. = ***Fimbristylis pallida*** S.T.Blake
Fimbristylis lichtensteiniana (Kunth) Hemsl. = ***Bulbostylis lichtensteiniana*** (Kunth) C.B.Clarke
Fimbristylis liukiuensis Tuyama = ***Fimbristylis bisumbellata*** (Forssk.) Bubani
Fimbristylis lomentocarpa Hochst. ex A.Rich. = ***Fimbristylis ferruginea*** (L.) Vahl var. ***ferruginea***
Fimbristylis longiculmis Steud. = ***Fimbristylis bivalvis*** (Lam.) Lye
Fimbristylis longifolia S.T.Blake = ***Fimbristylis schoenoides*** (Retz.) Vahl
Fimbristylis ludwigii Steud. = ***Isolepis ludwigii*** (Steud.) Kunth
Fimbristylis luzuliformis Domin = ***Fimbristylis dichotoma*** (L.) Vahl subsp. ***dichotoma***
Fimbristylis macgillivrayi C.B.Clarke = ***Fimbristylis recta*** F.M.Bailey
Fimbristylis macra Ridl. = ***Bulbostylis macra*** (Ridl.) C.B.Clarke
Fimbristylis macrostachya Boeck. = ***Fimbristylis tristachya*** R.Br. var. ***tristachya***
Fimbristylis magnifica C.B.Clarke = [2]
Fimbristylis makinoana Ohwi = ***Fimbristylis squarrosa*** var. ***esquarrosa*** Makino
Fimbristylis malaccana Boeck. = ***Fimbristylis pauciflora*** R.Br.
Fimbristylis marginata Labill. = ***Fimbristylis ferruginea*** (L.) Vahl var. ***ferruginea***
Fimbristylis mariana Gaudich. = ***Fimbristylis tristachya*** R.Br. var. ***tristachya***
Fimbristylis mariana var. *foenea* Kük. = ***Fimbristylis tristachya*** R.Br. var. ***tristachya***
Fimbristylis marquesana Steud. = ***Fimbristylis juncea*** var. ***marquesana*** (Steud.) Fosberg
Fimbristylis marrana Miré & Quézel = ***Bulbostylis densa*** subsp. ***afromontana*** (Lye) R.W.Haines
Fimbristylis martii Boeck. = ***Fimbristylis bahiensis*** Steud.
Fimbristylis mauritiana Nees ex Kunth = ***Fimbristylis dichotoma*** (L.) Vahl subsp. ***dichotoma***
Fimbristylis mauritiana Tausch ex C.Presl = ***Fimbristylis ferruginea*** (L.) Vahl var. ***ferruginea***
Fimbristylis maxima K.Schum. = ***Fimbristylis tristachya*** R.Br. var. ***tristachya***
Fimbristylis megastachys Ridl. = ***Bulbostylis megastachys*** (Ridl.) C.B.Clarke
Fimbristylis melanocephala Ridl. = ***Bulbostylis melanocephala*** (Ridl.) C.B.Clarke
Fimbristylis melanospora Fernald = ***Fimbristylis cymosa*** R.Br. subsp. ***cymosa***
Fimbristylis melanostachys (d'Urv.) Brongn. = ***Eleocharis melanostachys*** (d'Urv.) C.B.Clarke
Fimbristylis metzii Steud. = ***Fimbristylis dichotoma*** (L.) Vahl subsp. ***dichotoma***
Fimbristylis mezostachya Link = ***Fimbristylis dichotoma*** (L.) Vahl subsp. ***dichotoma***
Fimbristylis micheliana (L.) Rchb. = ***Cyperus michelianus*** (L.) Delile
Fimbristylis michelii Bubani = ***Fimbristylis squarrosa*** Vahl var. ***squarrosa***
Fimbristylis microstachya Steud. = ***Fimbristylis autumnalis*** (L.) Roem. & Schult.
Fimbristylis miliacea (L.) Vahl = ***Fimbristylis quinquangularis*** (Vahl) Kunth
Fimbristylis miliacea subsp. *koidzumiana* (Ohwi) T.Koyama = ***Fimbristylis littoralis*** var. ***koidzumiana*** (Ohwi) T.Koyama
Fimbristylis miliacea subsp. *macroglumis* Lye = ***Fimbristylis quinquangularis*** (Vahl) Kunth
Fimbristylis miliacea var. *macrostachya* J.Kern = ***Fimbristylis littoralis*** var. ***macrostachya*** (J.Kern) J.Kern
Fimbristylis miliacea subsp. *pallescens* Lye = ***Fimbristylis quinquangularis*** (Vahl) Kunth
Fimbristylis minima Hochst. ex Boeck. = ***Bulbostylis oligostachys*** (Hochst. ex A.Rich.) C.B.Clarke
Fimbristylis minutissima Maire = ***Bulbostylis densa*** subsp. ***afromontana*** (Lye) R.W.Haines
Fimbristylis mollis (Wall.) Kunth = ***Rhynchospora mollis*** (Wall.) Schult.
Fimbristylis monandra (Rottb.) F.Muell. = ***Fimbristylis argentea*** (Rottb.) Vahl

Fimbristylis monandra Schult. = ***Bulbostylis barbata*** (Rottb.) C.B.Clarke subsp. ***barbata***
Fimbristylis monini Steud. = ***Fimbristylis cymosa*** R.Br. subsp. ***cymosa***
Fimbristylis monostachya (L.) Hassk. = ***Fimbristylis ovata*** (Burm.f.) J.Kern
Fimbristylis mucronata Vahl = ***Fimbristylis quinquangularis*** (Vahl) Kunth
Fimbristylis mucronata Boeck. = ***Fimbristylis scabrida*** Schumach.
Fimbristylis mucronulata (Michx.) S.F.Blake = ***Fimbristylis autumnalis*** (L.) Roem. & Schult.
Fimbristylis multifolia Boeck. = ***Fimbristylis cymosa*** R.Br. subsp. ***cymosa***
Fimbristylis multistriata Boeck. = ***Fimbristylis puberula*** Vahl var. ***puberula***
Fimbristylis muricatula Steud. = ***Fimbristylis tenera*** Schult.
Fimbristylis muriculata Benth. = ***Fimbristylis scabrida*** Schumach.
Fimbristylis nana Roth = ***Fimbristylis squarrosa*** Vahl var. ***squarrosa***
Fimbristylis nanofusca Tang & F.T.Wang = ***Fimbristylis fimbristyloides*** (F.Muell.) Druce
Fimbristylis neglecta Hemsl. = ***Bulbostylis neglecta*** (Hemsl.) C.B.Clarke
Fimbristylis nesiotis Hemsl. = ***Bulbostylis nesiotis*** (Hemsl.) C.B.Clarke
Fimbristylis nigrescens Steud. = [82]
Fimbristylis nigrobrunnea Thwaites = ***Fimbristylis eragrostis*** (Nees) Hance
Fimbristylis nigrobrunnea var. *thorelii* Camus = ***Fimbristylis obtusata*** (C.B.Clarke) Ridl.
Fimbristylis nipponensis Makino = ***Fimbristylis dipsacea*** (Rottb.) C.B.Clarke
Fimbristylis nitidula Schult. = ***Fimbristylis dichotoma*** (L.) Vahl subsp. ***dichotoma***
Fimbristylis novae-britanniae Boeck. = ***Fimbristylis dichotoma*** (L.) Vahl subsp. ***dichotoma***
Fimbristylis nukahivensis Steud. = ***Fimbristylis juncea*** var. ***nukahivensis*** (Steud.) Fosberg
Fimbristylis nutans var. *leucostachya* Domin = ***Fimbristylis pallida*** S.T.Blake
Fimbristylis nutans var. *minor* Camus = ***Fimbristylis acuminata*** Vahl
Fimbristylis nutans var. *rhyticarya* (F.Muell.) Domin = ***Fimbristylis rhyticarya*** F.Muell.
Fimbristylis nutans var. *typica* Domin = ***Fimbristylis nutans*** (Retz.) Vahl
Fimbristylis obscura Fernald = ***Fimbristylis complanata*** (Retz.) Link var. ***complanata***
Fimbristylis obtusangula F.Muell. = ***Fimbristylis rara*** R.Br.
Fimbristylis obtusifolia Nees ex C.Presl = ***Fimbristylis dichotoma*** (L.) Vahl subsp. ***dichotoma***
Fimbristylis obtusifolia (Lam.) Kunth = ***Fimbristylis dichotoma*** (L.) Vahl subsp. ***dichotoma***
Fimbristylis ochotensis (Meinsh.) Kom. = ***Fimbristylis dichotoma*** var. ***ochotensis*** (Meinsh.) T.Koyama
Fimbristylis ochreata Boeck. = ***Fimbristylis ferruginea*** (L.) Vahl var. ***ferruginea***
Fimbristylis oligocephala W.Fitzg. = ***Fimbristylis cephalophora*** F.Muell.
Fimbristylis oligostachya (Hochst. ex A.Rich.) K.Schum. = ***Bulbostylis oligostachys*** (Hochst. ex A.Rich.) C.B.Clarke
Fimbristylis oligostachys Hochst. ex A.Rich. = ***Bulbostylis oligostachys*** (Hochst. ex A.Rich.) C.B.Clarke
Fimbristylis ophiticola Britton = ***Fimbristylis dichotoma*** subsp. ***ophiticola*** (Britton) Zavaro
Fimbristylis oritrephes Ridl. = ***Bulbostylis oritrephes*** (Ridl.) C.B.Clarke
Fimbristylis ovalis Nees = ***Fimbristylis dichotoma*** (L.) Vahl subsp. ***dichotoma***
Fimbristylis oxylepis Steud. = ***Fimbristylis tenera*** Schult.
Fimbristylis oxyrhachis Miq. = ***Fimbristylis tetragona*** R.Br.
Fimbristylis pachystachys Cherm. = ***Fimbristylis scabrida*** Schumach.
Fimbristylis pachystylis Cherm. = ***Fimbristylis scabrida*** Schumach.
Fimbristylis pacifica Ohwi = ***Fimbristylis tristachya*** var. ***pacifica*** (Ohwi) T.Koyama
Fimbristylis pallescens (Roxb.) Nees = ***Fimbristylis bisumbellata*** (Forssk.) Bubani
Fimbristylis pallescens Link = ***Fimbristylis dichotoma*** (L.) Vahl subsp. ***dichotoma***
Fimbristylis pallidula Kral = ***Fimbristylis spadicea*** (L.) Vahl
Fimbristylis paludosa Merr. = ***Fimbristylis salbundia*** (Nees) Kunth
Fimbristylis papillosa (Kük.) Alain = ***Bulbostylis junciformis*** (Kunth) C.B.Clarke
Fimbristylis paradoxa (Spreng.) Alain = ***Bulbostylis paradoxa*** (Spreng.) Lindm.
Fimbristylis parva Ridl. = ***Bulbostylis pusilla*** (Hochst. ex A.Rich.) C.B.Clarke
Fimbristylis parviflora R.Br. = ***Fimbristylis dichotoma*** (L.) Vahl subsp. ***dichotoma***
Fimbristylis paucispicata F.Muell. = ***Fimbristylis ferruginea*** (L.) Vahl var. ***ferruginea***
Fimbristylis pentaptera (Nees) Kunth = ***Fimbristylis salbundia*** (Nees) Kunth
Fimbristylis petrogena Ohwi = ***Fimbristylis insignis*** Thwaites
Fimbristylis philippica Steud. = ***Fimbristylis dichotoma*** (L.) Vahl subsp. ***dichotoma***
Fimbristylis pilosa (Willd.) K.Schum. = ***Bulbostylis pilosa*** (Willd.) Cherm.
Fimbristylis pinetorum Merr. = ***Fimbristylis pierotii*** Miq.
Fimbristylis planifolia Steud. = [62 CRL]
Fimbristylis platystachys Boeck. = ***Fimbristylis schultzii*** Boeck.
Fimbristylis platystachys var. *schultzii* (Boeck.) Domin = ***Fimbristylis schultzii*** Boeck.
Fimbristylis platystachys var. *typica* Domin = ***Fimbristylis schultzii*** Boeck.
Fimbristylis pluristriata (C.B.Clarke) Berhaut = ***Fimbristylis dichotoma*** subsp. ***podocarpa*** (Nees) T.Koyama
Fimbristylis podocarpa Nees = ***Fimbristylis dichotoma*** subsp. ***podocarpa*** (Nees) T.Koyama
Fimbristylis pohliana Steud. = ***Fimbristylis dichotoma*** (L.) Vahl subsp. ***dichotoma***
Fimbristylis polymorpha Boeck. = ***Fimbristylis dichotoma*** (L.) Vahl subsp. ***dichotoma***
Fimbristylis polytrichoides var. *takaoensis* (Hayata) T.Koyama = ***Fimbristylis polytrichoides*** (Retz.) R.Br.
Fimbristylis portoricensis (Britton) Alain = ***Bulbostylis pauciflora*** (Liebm.) C.B.Clarke
Fimbristylis preslii Kunth = ***Bulbostylis hispidula*** (Vahl) R.W.Haines subsp. ***hispidula***
Fimbristylis prolifera Steud. = [82]
Fimbristylis propinqua R.Br. = ***Fimbristylis squarrosa*** Vahl var. ***squarrosa***
Fimbristylis puberula var. *drummondii* (Torr. & Hook.) D.B.Ward = ***Fimbristylis puberula*** Vahl

Fimbristylis puberula f. *eucycla* Fernald = ***Fimbristylis caroliniana*** (Lam.) Fernald
Fimbristylis puberula f. *pycnostachya* Fernald = ***Fimbristylis caroliniana*** (Lam.) Fernald
Fimbristylis pubescens Link = ***Fimbristylis dichotoma*** (L.) Vahl subsp. ***dichotoma***
Fimbristylis pubigera Link = ***Fimbristylis dichotoma*** (L.) Vahl subsp. ***dichotoma***
Fimbristylis pulchella (Thwaites) Trimen = ***Bulbostylis barbata*** subsp. ***pulchella*** (Thwaites) T.Koyama
Fimbristylis pumila Benth. = ***Fimbristylis pauciflora*** R.Br.
Fimbristylis purpureoatra (Boeck.) C.B.Clarke ex Engl. = ***Bulbostylis oligostachys*** (Hochst. ex A.Rich.) C.B.Clarke
Fimbristylis pusilla Hochst. ex A.Rich. = ***Bulbostylis pusilla*** (Hochst. ex A.Rich.) C.B.Clarke
Fimbristylis pycnocephala Hillebr. = ***Fimbristylis cymosa*** subsp. ***umbellatocapitata*** (Hillebr.) T.Koyama
Fimbristylis pycnostachya Hance = ***Fimbristylis eragrostis*** (Nees) Hance
Fimbristylis pygmaea Vahl = ***Isolepis cernua*** (Vahl) Roem. & Schult. var. ***cernua***
Fimbristylis quadrangularis A.Dietr. ex Steud. = ***Fimbristylis littoralis*** Gaudich. var. ***littoralis***
Fimbristylis quadrangularis var. *crassa* C.B.Clarke = ***Fimbristylis aphylla*** Steud.
Fimbristylis quaternella Ridl. = ***Bulbostylis quaternella*** (Ridl.) Goetgh.
Fimbristylis quinquangularis f. *abludens* Backer = ***Fimbristylis microcarya*** F.Muell. var. ***microcarya***
Fimbristylis quinquangularis var. *bistaminifera* Tang & F.T.Wang = ***Fimbristylis quinquangularis*** (Vahl) Kunth
Fimbristylis quinquangularis var. *elata* Tang & F.T.Wang = ***Fimbristylis quinquangularis*** (Vahl) Kunth
Fimbristylis ramosii Kük. = ***Fimbristylis dichotoma*** (L.) Vahl subsp. ***dichotoma***
Fimbristylis rara var. *obtusangula* (F.Muell.) Domin = ***Fimbristylis rara*** R.Br.
Fimbristylis rara var. *typica* Domin = ***Fimbristylis rara*** R.Br.
Fimbristylis retusa Thwaites = ***Fimbristylis leptoclada*** Benth. var. ***leptoclada***
Fimbristylis reunionis Steud. = ***Fimbristylis cymosa*** R.Br. subsp. ***cymosa***
Fimbristylis riehleana Steud. = ***Fimbristylis spadicea*** (L.) Vahl
Fimbristylis rigida Kunth = ***Fimbristylis cymosa*** R.Br. subsp. ***cymosa***
Fimbristylis rigida Beetle = ***Schoenoplectus californicus*** (C.A.Mey.) Soják
Fimbristylis rigidifolia Ridl. = ***Fimbristylis fusca*** (Nees) Benth. ex C.B.Clarke
Fimbristylis robusta Lye = ***Fimbristylis gabonica*** Cherm.
Fimbristylis rotundata Kük. = ***Bulbostylis rotundata*** (Kük.) R.W.Haines
Fimbristylis roxburghii A.Dietr. = ***Fimbristylis ferruginea*** (L.) Vahl var. ***ferruginea***
Fimbristylis royeniana Nees = ***Fimbristylis dichotoma*** (L.) Vahl subsp. ***dichotoma***
Fimbristylis royeniana var. *microstachya* Nees = ***Fimbristylis dichotoma*** subsp. ***glauca*** (Vahl) T. Koyama
Fimbristylis rudgeana (Nees) Steud. = ***Fimbristylis complanata*** (Retz.) Link var. ***complanata***
Fimbristylis rufa Boeck. = ***Fimbristylis rara*** R.Br.
Fimbristylis salbundia subsp. *pentaptera* (Nees) T.Koyama = ***Fimbristylis salbundia*** (Nees) Kunth
Fimbristylis sansibariensis Boeck. = ***Fimbristylis bivalvis*** (Lam.) Lye
Fimbristylis savannarum Alain = ***Bulbostylis juncoides*** (Vahl) Kük. ex Herter
Fimbristylis scabra (Roxb.) Schult. = ***Fimbristylis acuminata*** Vahl
Fimbristylis scabra (Nees) Kunth = ***Fimbristylis complanata*** (Retz.) Link var. ***complanata***
Fimbristylis scabra (J.Presl & C.Presl) Hauman & Vanderveken = ***Bulbostylis scabra*** (J.Presl & C.Presl) C.B.Clarke
Fimbristylis scherardii Bubani = ***Fimbristylis dichotoma*** (L.) Vahl subsp. ***dichotoma***
Fimbristylis schimperiana (Hochst. ex A.Rich.) K.Schum = ***Bulbostylis schimperiana*** (Hochst. ex A.Rich.) C.B.Clarke
Fimbristylis schinzii Boeck. = ***Ficinia nodosa*** (Rottb.) Goetgh., Muasya & D.A.Simpson
Fimbristylis schischkinii Pobed. = ***Fimbristylis turkestanica*** (Regel) B.Fedtsch.
Fimbristylis schlechteri Kük. = ***Fimbristylis eragrostis*** (Nees) Hance
Fimbristylis schoenoides (Kunth) K.Schum. = ***Bulbostylis schoenoides*** (Kunth) C.B.Clarke
Fimbristylis schomburgkiana Steud. = ***Bulbostylis schomburgkiana*** (Steud.) M.T.Strong
Fimbristylis sechellensis Baker = ***Fimbristylis dichotoma*** (L.) Vahl subsp. ***dichotoma***
Fimbristylis sellowiana (Kunth) Hauman & Vanderveken = ***Bulbostylis sellowiana*** (Kunth) Palla
Fimbristylis semihirsuta Boeck. = [36]
Fimbristylis semilevis F.Muell. ex C.B.Clarke = ***Fimbristylis sphaerocephala*** Benth.
Fimbristylis separanda Steud. ex Jard. = ***Fimbristylis juncea*** var. ***nukahivensis*** (Steud.) Fosberg
Fimbristylis separanda Steud. ex Jard. = ***Fimbristylis juncea*** var. ***separanda*** (Steud. ex Jard.) Fosberg
Fimbristylis serrulata Vahl = ***Fimbristylis dichotoma*** (L.) Vahl subsp. ***dichotoma***
Fimbristylis setacea Benth. = ***Fimbristylis acicularis*** R.Br.
Fimbristylis setacea var. *brevifolia* C.B.Clarke = ***Fimbristylis acicularis*** R.Br.
Fimbristylis setifolia A.Rich. = ***Bulbostylis atrosanguinea*** (Boeck.) C.B.Clarke
Fimbristylis sieberiana Kunth = ***Fimbristylis ferruginea*** (L.) Vahl var. ***ferruginea***
Fimbristylis sieboldii Miq. ex Franch. & Sav. = ***Fimbristylis ferruginea*** (L.) Vahl var. ***ferruginea***
Fimbristylis sieboldii subsp. *anpinensis* (Hayata) T.Koyama = ***Fimbristylis ferruginea*** var. ***anpinensis*** (Hayata) H.Y.Liu
Fimbristylis similis Steud. = ***Fimbristylis dichotoma*** (L.) Vahl subsp. ***dichotoma***
Fimbristylis sintenisii Boeck. = ***Fimbristylis cymosa*** R.Br. subsp. ***cymosa***
Fimbristylis sororia Kunth = ***Fimbristylis dichotoma*** (L.) Vahl subsp. ***dichotoma***
Fimbristylis spadicea var. *castanea* (Michx.) A.Gray = ***Fimbristylis spadicea*** (L.) Vahl
Fimbristylis spadicea f. *domingensis* (Pers.) Kük. = ***Fimbristylis spadicea*** (L.) Vahl
Fimbristylis spadicea var. *drummondii* (Torr. & Hook.) Mohlenbr. = ***Fimbristylis puberula*** Vahl
Fimbristylis spadicea var. *ferruginea* (L.) Alph.Wood = ***Fimbristylis spadicea*** (L.) Vahl
Fimbristylis spadicea var. *longestigmata* Zavaro = ***Fimbristylis spadicea*** (L.) Vahl
Fimbristylis spartium Ham. = [81 PUE]

Fimbristylis spathacea Roth = ***Fimbristylis cymosa*** R.Br. subsp. ***cymosa***
Fimbristylis spathacea var. *depauperata* T.Koyama = ***Fimbristylis cymosa*** R.Br. subsp. ***cymosa***
Fimbristylis spathacea var. *umbellatocapitata* (Hillebr.) T.Koyama = ***Fimbristylis cymosa*** subsp. ***umbellatocapitata*** (Hillebr.) T.Koyama
Fimbristylis speciosa Rohde ex Spreng. = ***Fimbristylis spadicea*** (L.) Vahl
Fimbristylis sphaerocarpa (Boeck.) K.Schum. = ***Bulbostylis sphaerocarpa*** (Boeck.) C.B.Clarke
Fimbristylis sphaerocephala (Boeck.) Hauman & Vanderveken = ***Bulbostylis sphaerocephala*** (Boeck.) Lindm.
Fimbristylis spirostachya F.Muell. ex Benth. = ***Fimbristylis dichotoma*** subsp. ***depauperata*** (R.Br.) J.Kern
Fimbristylis squamulosa Hochst. ex Boeck. = ***Fimbristylis dichotoma*** (L.) Vahl subsp. ***dichotoma***
Fimbristylis squarrosa var. *adventitia* (Ces.) Nyman = ***Fimbristylis* × *adventitia*** Ces.
Fimbristylis squarrosa subsp. *esquarrosa* (Makino) T. Koyama = ***Fimbristylis squarrosa*** var. ***esquarrosa*** Makino
Fimbristylis squarrosa var. *robustior* Kunth = ***Fimbristylis squarrosa*** Vahl var. ***squarrosa***
Fimbristylis squarrosa var. *velata* (R.Br.) C.B.Clarke ex Cheeseman = ***Fimbristylis squarrosa*** var. ***esquarrosa*** Makino
Fimbristylis stans Rich. ex Spreng. = ***Fimbristylis ferruginea*** (L.) Vahl var. ***ferruginea***
Fimbristylis stellata S.T.Blake = ***Fimbristylis schultzii*** Boeck.
Fimbristylis stenochlaena Kük. = ***Fimbristylis fusca*** (Nees) Benth. ex C.B.Clarke
Fimbristylis stenophylla (Elliott) Alain = ***Bulbostylis stenophylla*** (Elliott) C.B.Clarke
Fimbristylis sterilis Nees = ***Fimbristylis spadicea*** (L.) Vahl
Fimbristylis stradbrokensis (Domin) J.Kern = ***Trachystylis stradbrokensis*** (Domin) Kük.
Fimbristylis striata Labill. = ***Fimbristylis dichotoma*** (L.) Vahl subsp. ***dichotoma***
Fimbristylis stricta R.Br. = ***Fimbristylis dichotoma*** (L.) Vahl subsp. ***dichotoma***
Fimbristylis stricticulmis Domin = ***Fimbristylis recta*** F.M.Bailey
Fimbristylis strictior Steud. = ***Bulbostylis schomburgkiana*** (Steud.) M.T.Strong
Fimbristylis subaphylla Boeck. = ***Fimbristylis complanata*** (Retz.) Link var. ***complanata***
Fimbristylis subaristata var. *glaucina* Domin = ***Fimbristylis subaristata*** Benth.
Fimbristylis subbispicata Nees = ***Fimbristylis tristachya*** var. ***subbispicata*** (Nees) T.Koyama
Fimbristylis subbulbosa Boeck. = ***Fimbristylis polytrichoides*** (Retz.) R.Br.
Fimbristylis subbulbosa Boeck. = ***Fimbristylis polytrichoides*** (Retz.) R.Br.
Fimbristylis subbulbosa Benth. = ***Fimbristylis tristachya*** R.Br. var. ***tristachya***
Fimbristylis subfusca E.G.Camus = ***Fimbristylis fusca*** (Nees) Benth. ex C.B.Clarke
Fimbristylis sublateralis Steud. = ***Fimbristylis ferruginea*** (L.) Vahl var. ***ferruginea***
Fimbristylis subpilosa Kük. = ***Fimbristylis pilosa*** Vahl
Fimbristylis subtetrastachya Boeck. = ***Fimbristylis eragrostis*** (Nees) Hance
Fimbristylis subtristachya Steud. = ***Fimbristylis dichotoma*** subsp. ***podocarpa*** (Nees) T.Koyama
Fimbristylis subumbellata K.Schum. = ***Bulbostylis schimperiana*** (Hochst. ex A.Rich.) C.B.Clarke
Fimbristylis sulcata Schult. = ***Eleocharis filiculmis*** Kunth
Fimbristylis tainanensis Ohwi = ***Fimbristylis microcarya*** var. ***tainanensis*** (Ohwi) H.Y.Liu
Fimbristylis taiwanica Ohwi = ***Fimbristylis microcarya*** F.Muell. var. ***microcarya***
Fimbristylis takamineana Ohwi = ***Fimbristylis leptoclada*** var. ***takamineana*** (Ohwi) T.Koyama
Fimbristylis takaoensis Hayata = ***Fimbristylis polytrichoides*** (Retz.) R.Br.
Fimbristylis tashiroana Ohwi = ***Fimbristylis dichotoma*** (L.) Vahl subsp. ***dichotoma***
Fimbristylis tashiroana var. *cincta* Ohwi = ***Fimbristylis dichotoma*** (L.) Vahl subsp. ***dichotoma***
Fimbristylis taylorii K.Schum. = ***Bulbostylis taylorii*** (K.Schum.) C.B.Clarke
Fimbristylis tenella (Link) Schult. = ***Bulbostylis tenella*** (Link) C.B.Clarke
Fimbristylis tenera var. *obtusata* C.B.Clarke = ***Fimbristylis obtusata*** (C.B.Clarke) Ridl.
Fimbristylis tenuifolia Thwaites = ***Fimbristylis monticola*** Hochst. ex Steud.
Fimbristylis tenuissima Steud. = ***Bulbostylis barbata*** (Rottb.) C.B.Clarke subsp. ***barbata***
Fimbristylis tertia Steud. ex Jard. = ***Fimbristylis juncea*** var. ***tertia*** (Steud. ex Jard.) Fosberg
Fimbristylis testui Cherm. = ***Fimbristylis aphylla*** Steud.
Fimbristylis tetragona A.Dietr. = ***Fimbristylis littoralis*** Gaudich. var. ***littoralis***
Fimbristylis textilis Roxb. = ***Ficinia nodosa*** (Rottb.) Goetgh., Muasya & D.A.Simpson
Fimbristylis thonningiana Boeck. = ***Fimbristylis microcarya*** F.Muell. var. ***microcarya***
Fimbristylis thorelii E.G.Camus = ***Fimbristylis gracilenta*** Hance var. ***gracilenta***
Fimbristylis thouarsii (Kunth) Merr. = ***Actinoschoenus thouarsii*** (Kunth) Benth.
Fimbristylis thwaitesii Boeck. = ***Fimbristylis insignis*** Thwaites
Fimbristylis thwaitesii (Boeck.) Beetle =
Fimbristylis tikushiensis Hayata = ***Fimbristylis dichotoma*** (L.) Vahl subsp. ***dichotoma***
Fimbristylis tisserantii Cherm. = ***Bulbostylis viridicarinata*** (De Wild.) Goetgh.
Fimbristylis tokyoensis Makino = ***Fimbristylis aestivalis*** Vahl var. ***aestivalis***
Fimbristylis tomentosa Vahl = ***Fimbristylis dichotoma*** subsp. ***podocarpa*** (Nees) T.Koyama
Fimbristylis tonensis Makino = ***Fimbristylis stauntonii*** var. ***tonensis*** (Makino) T.Koyama
Fimbristylis torresiana Gaudich. = ***Fimbristylis umbellaris*** (Lam.) Vahl
Fimbristylis torta (Nees) Kunth = ***Fimbristylis falcata*** (Vahl) Kunth var. ***falcata***
Fimbristylis tortispica Turrill = ***Fimbristylis eragrostis*** (Nees) Hance
Fimbristylis transiens K.Schum. = ***Bulbostylis boeckeleriana*** var. ***transiens*** (K.Schum.) R.W.Haines
Fimbristylis tricholepis Miq. = ***Fimbristylis aestivalis*** Vahl var. ***aestivalis***
Fimbristylis trichophora Steud. ex C.B.Clarke = ***Bulbostylis hispidula*** (Vahl) R.W.Haines subsp. ***hispidula***
Fimbristylis trifida (Nees) Trin. = ***Bulbostylis densa*** subsp. ***afromontana*** (Lye) R.W.Haines
Fimbristylis trifurca Nees = ***Fimbristylis dichotoma*** (L.) Vahl subsp. ***dichotoma***

Fimbristylis trimenii Hook.f. = ***Fimbristylis aestivalis*** Vahl var. ***aestivalis***
Fimbristylis trispicata Steud. = ***Fimbristylis ferruginea*** (L.) Vahl var. ***ferruginea***
Fimbristylis tristachya (Vahl) Thwaites = ***Fimbristylis triflora*** (L.) K.Schum. ex Engl.
Fimbristylis umbellaris var. *vicaryi* (C.B.Clarke) Karthik. = ***Fimbristylis umbellaris*** (Lam.) Vahl
Fimbristylis umbellata Schrad. ex Nees = ***Fimbristylis spadicea*** (L.) Vahl
Fimbristylis umbellatocapitata Steud. = ***Fimbristylis cymosa*** R.Br. subsp. ***cymosa***
Fimbristylis unicolor Ohwi & T.Koyama = ***Fimbristylis henryi*** C.B.Clarke
Fimbristylis urvilleana Steud. = ***Bulbostylis scabra*** (J.Presl & C.Presl) C.B.Clarke
Fimbristylis utilis Elmer = ***Fimbristylis umbellaris*** (Lam.) Vahl
Fimbristylis vaginata f. *leptoclada* (Benth.) Domin = ***Fimbristylis leptoclada*** Benth.
Fimbristylis vahlii var. *subacaulis* (Steud.) Boeck. = ***Fimbristylis vahlii*** (Lam.) Link
Fimbristylis vanderystii De Wild. = ***Fimbristylis aphylla*** Steud.
Fimbristylis variabilis R.Br. = ***Fimbristylis dichotoma*** (L.) Vahl subsp. ***dichotoma***
Fimbristylis velata R.Br. = ***Fimbristylis squarrosa*** var. ***esquarrosa*** Makino
Fimbristylis velutina Franch. = ***Fimbristylis sericea*** (Poir.) R.Br.
Fimbristylis vermoesenii De Wild. = ***Bulbostylis cioniana*** (Savi) Lye
Fimbristylis verrucifera (Maxim.) Makino = ***Fimbristylis dipsacea*** (Rottb.) C.B.Clarke
Fimbristylis verrucosa J.Presl & C.Presl = ***Fimbristylis dichotoma*** (L.) Vahl subsp. ***dichotoma***
Fimbristylis verrucosa var. *glabrescens* C.Presl = ***Fimbristylis dichotoma*** (L.) Vahl subsp. ***dichotoma***
Fimbristylis verrucosa var. *hirsuta* C.Presl = ***Fimbristylis dichotoma*** (L.) Vahl subsp. ***dichotoma***
Fimbristylis vestita Steud. = ***Bulbostylis hispidula*** (Vahl) R.W.Haines subsp. ***hispidula***
Fimbristylis vestita (Kunth) Hemsl. = ***Bulbostylis vestita*** (Kunth) C.B.Clarke
Fimbristylis vexata Steud. = ***Ficinia angustifolia*** (Schrad.) C.B.Clarke
Fimbristylis villosa (Poir.) Roem. & Schult. = ***Fimbristylis dichotoma*** (L.) Vahl subsp. ***dichotoma***
Fimbristylis vincentii Steud. = ***Fimbristylis vahlii*** (Lam.) Link
Fimbristylis viridicarinata De Wild. = ***Bulbostylis viridicarinata*** (De Wild.) Goetgh.
Fimbristylis vitiata C.B.Clarke = ***Fimbristylis turkestanica*** (Regel) B.Fedtsch.
Fimbristylis warburgii K.Schum. ex Warb. = ***Fimbristylis cymosa*** R.Br. subsp. ***cymosa***
Fimbristylis wightiana Nees = ***Fimbristylis cymosa*** R.Br. subsp. ***cymosa***
Fimbristylis wightiana (Boeck.) Beetle = ***Ficinia ecklonea*** (Steud.) Nees
Fimbristylis wombaliensis De Wild. = ***Bulbostylis wombaliensis*** (De Wild.) R.W.Haines
Fimbristylis xyridis var. *rigidula* Benth. = ***Fimbristylis recta*** F.M.Bailey
Fimbristylis xyrioides Arn. ex Thwaites = ***Fimbristylis tetragona*** R.Br.
Fimbristylis zambesica K.Schum. = ***Bulbostylis macra*** (Ridl.) C.B.Clarke

***Unplaced Names*:**
Fimbristylis breviculma Govind., Rheedea 6: 61 (1996), nom. illeg. = [40 IND]
Fimbristylis donticola Hochst. ex Steud., Syn. Pl. Glumac. 2: 111 (1855). = [40]
Fimbristylis elongata Pires de Lima, Bol. Soc. Brot., II, 2: 133 (1924), nom. illeg. = [26 MOZ]
Fimbristylis faulensis Beck, Ann. K. K. Naturhist. Hofmus. 3: 252 (1888). = [51 NZS]
Fimbristylis hamiltonii Steud., Syn. Pl. Glumac. 2: 119 (1855). = [81 PUE]
Fimbristylis huillensis Ridl., Trans. Linn. Soc. London, Bot. 2: 154 (1884). = [26 ANG]
Fimbristylis humilis S.T.Blake, Proc. Roy. Soc. Queensland 48: 91 (1936 publ. 1937), nom. illeg. = [50 QLD]
Fimbristylis magnifica C.B.Clarke, Bull. Soc. Bot. France 54(8): 28 (1907). = [2]
Fimbristylis nigrescens Steud., Syn. Pl. Glumac. 2: 118 (1855). = [82]
Fimbristylis planifolia Steud., Syn. Pl. Glumac. 2: 118 (1855). = [62 CRL]
Fimbristylis prolifera Steud., Syn. Pl. Glumac. 2: 120 (1855). = [82]
Fimbristylis semihirsuta Boeck., Beitr. Cyper. 2: 18 (1890). = [36]
Fimbristylis spartium Ham., Prodr. Pl. Ind. Occid.: 14 (1825). = [81 PUE]

Fintelmannia

Fintelmannia Kunth = ***Trilepis*** Nees
Fintelmannia eximia C.B.Clarke = ***Trilepis eximia*** (C.B.Clarke) H.Pfeiff.
Fintelmannia microstachya C.B.Clarke = ***Trilepis microstachya*** (C.B.Clarke) H.Pfeiff.
Fintelmannia restioides Kunth = ***Trilepis lhotzkiana*** Nees
Fintelmannia robusta C.B.Clarke = ***Trilepis lhotzkiana*** Nees
Fintelmannia setifera Ridl. = ***Coleochloa setifera*** (Ridl.) Gilly

Forexeta

Forexeta Raf. = ***Carex*** L.
Forexeta nepalensis (Spreng.) Raf. = ***Carex longipes*** D.Don ex Tilloch & Taylor
Forexeta phyllostachys (C.A.Mey.) Raf. = ***Carex phyllostachys*** C.A.Mey.

Froelichia

Froelichia Wulfen = ***Kobresia*** Willd.
Froelichia caricoides Wulfen = ***Kobresia myosuroides*** (Vill.) Fiori & Paol.

Fuirena

Fuirena Rottb., Descr. Icon. Rar. Pl.: 70 (1773).
Cosmopolitan. 12 BAL COR POR SPA 13 ITA KRI 20 ALG EGY MOR TUN 22 BEN BKN GAM GHA GNB GUI IVO LBR MLI NGA NGR SEN SIE TOG 23 BUR CAF CMN EQG GAB GGI RWA ZAI 24 CHA ETH SOM 25 KEN TAN UGA 26+ ANG MLW MOZ ZAM ZIM 27 BOT CPP LES NAM NAT OFS SWZ TVL 29 COM MAU MDG REU 34 AFG EAI LBS PAL TUR 35 YEM 36 CHC CHH CHS 38 JAP KOR NNS TAI 40 ASS BAN EHM IND NEP PAK SRL WHM 41 THA VIE 42

BOR JAW LSI MLY PHI SUL SUM 43 NWG 50 NSW NTA QLD WAU 62 CRL MRN 72 ONT 74 ILL? KAN MSO NEB OKL 75 CNT INI MAS MIC NWJ NWY RHO 77 NWM TEX 78 ALA ARK DEL FLA GEO KTY LOU MRY MSI NCA SCA TEN VRG WDC 79 MXC MXE MXG MXN MXS MXT 80 BLZ COS ELS GUA HON NIC PAN 81 CUB DOM JAM LEE PUE TRT WIN 82 FRG GUY SUR VEN 83 BOL CLM ECU PER 84+ BZC BZE BZL BZN BZS 85 AGE PAR URU.
59 Species
Vaginaria Pers., Syn. Pl. 1: 70 (1805).
Pentasticha Turcz., Bull. Soc. Imp. Naturalistes Moscou 1862(2): 330 (1862).

Fuirena abnormalis C.B.Clarke in D.Oliver & auct. suc. (eds.), Fl. Trop. Afr. 8: 462 (1902). *Scirpus abnormalis* (C.B.Clarke) T.Koyama, J. Fac. Sci. Univ. Tokyo, Sect. 3, Bot. 7: 287 (1958).
Tanzania to S. Trop. Africa. 25 TAN 26+. Ther.

Fuirena angolensis (C.B.Clarke) Lye ex J.Raynal & Roessler, Mitt. Bot. Staatssamml. München 13: 354 (1977).
Kenya to Namibia. 25 KEN TAN 26 ANG 27 NAM. Ther.
**Fuirena glomerata* var. *angolensis* C.B.Clarke in D. Oliver & auct. suc. (eds.), Fl. Trop. Afr. 8: 466 (1902).

Fuirena appendiculata Peter, Abh. Preuss. Akad. Wiss., Phys.-Math. Kl., n.f., 13(2): 32 (1928).
Tanzania. 25 TAN. Hemicr. or rhizome geophyte.

Fuirena arenosa R.Br., Prodr.: 220 (1810).
Queensland. 50 QLD. Hemicr. or rhizome geophyte.
Fuirena repanda S.T.Blake, Proc. Roy. Soc. Queensland 52: 59 (1941), nom. illeg.

Fuirena bernieri Cherm., Bull. Soc. Bot. France 68: 424 (1921 publ. 1922).
E. Madagascar. 29 MDG. Hemicr. or rhizome geophyte.

Fuirena boreocoerulescens Lye, Candollea 51: 1 (1996).
N. Somalia. 24 SOM. Hemicr. or rhizome geophyte.

Fuirena brachylepis Peter, Abh. Preuss. Akad. Wiss., Phys.-Math. Kl., n.f., 13(2): 50 (1928).
Tanzania. 25 TAN. Hemicr. or rhizome geophyte.

Fuirena brasiliensis Palla, Denkschr. Kaiserl. Akad. Wiss., Wien. Math.-Naturwiss. Kl. 79: 168 (1908).
Brazil. 84+. Hemicr. or rhizome geophyte.

Fuirena breviseta (Coville) Coville, Bull. Torrey Bot. Club 28: 466 (1908).
Texas to SE. U.S.A., Cuba. 77 TEX 78 ALA FLA GEO LOU MSI NCA SCA VRG 81 CUB. Hemicr. or rhizome geophyte.
**Fuirena squarrosa* var. *breviseta* Coville, Bull. Torrey Bot. Club 17: 98 (1890).

Fuirena bullifera J.Raynal & Roessler, Mitt. Bot. Staatssamml. München 13: 255 (1977).
S. Africa. 27 BOT NAM TVL. Hemicr. or rhizome geophyte.

Fuirena bushii Kral, Sida 7: 331 (1978).
Oklahoma to Louisiana. 74 OKL 77 TEX 78 ARK LOU. Hemicr.
**Fuirena ciliata* Bush, Rep. (Annual) Missouri Bot. Gard. 1905: 97 (1905), nom. illeg.

Fuirena camptotricha C.Wright, Anales Acad. Ci. Méd. Habana 8: 82 (1871).
Cuba, SE. Mexico to C. America. 79 MXG MXT 80 BLZ COS GUA HON NIC PAN 81 CUB. Hemicr.
Fuirena bulbipes S.F.Blake, Contr. U. S. Natl. Herb. 24: 2 (1922).

Fuirena ciliaris (L.) Roxb., Hort. Bengal.: 81 (1814).
Trop. & Subtrop. Old World. 20 EGY 22 BEN GHA GNB GUI IVO NGA NGR SEN SIE TOG 23 CMN GAB 25 KEN TAN UGA 26 ZAM 27 BOT NAM NAT SWZ TVL 36 CHC CHH CHS 38 JAP KOR NNS TAI 40 ASS BAN EHM IND NEP PAK SRL WHM 41 THA VIE 42 BOR JAW LSI MLY PHI SUL SUM 43 NWG 50 NSW NTA QLD WAU. Ther.
**Scirpus ciliaris* L., Mant. Pl. 2: 182 (1771).

f. ***apetala*** (Wingf.) Lye, Nordic J. Bot. 3: 241 (1983).
E. Tanzania. 25 TAN. Ther.
**Fuirena ciliaris* var. *apetala* Wingf., Bot. Not. 130: 319 (1977).

f. ***ciliaris***
Trop. & Subtrop. Old World. 20 EGY 22 BEN GHA GNB GUI IVO NGA NGR SEN SIE TOG 23 CMN GAB 25 KEN TAN UGA 26 ZAM 27 BOT NAM NAT SWZ TVL 36 CHC CHH CHS 38 JAP KOR NNS TAI 40 ASS BAN EHM IND NEP PAK SRL WHM 41 THA VIE 42 BOR JAW LSI MLY PHI SUL SUM 43 NWG 50 NSW NTA QLD WAU. Ther.
Scirpus ciliatus Rottb., Descr. Pl. Rar.: 26 (1772), orth. var.
Fuirena glomerata Lam., Tabl. Encycl. 1: 150 (1791).
Scirpus pilosus Retz., Observ. Bot. 6: 19 (1791), nom. illeg. *Scirpus aristatus* Willd., Sp. Pl. 1: 300 (1797).
Fuirena rottboellii Nees in R.Wight, Contr. Bot. India: 94 (1834).
Fuirena striata Llanos, Fragm. Pl. Filip.: 21 (1851).

Fuirena claviseta Peter, Abh. Preuss. Akad. Wiss., Phys.-Math. Kl., n.f., 13(2): 50 (1928).
E. Tanzania. 25 TAN. Ther. or hemicr.

Fuirena coerulescens Steud., Flora 12: 153 (1829). *Scirpus coerulescens* (Steud.) Kuntze, Revis. Gen. Pl. 3(2): 336 (1898).
S. Africa. 27 CPP NAM NAT OFS TVL (78) vrg. Hemicr. or rhizome geophyte.
Fuirena mollicula Kunth, Enum. Pl. 2: 182 (1837).
Fuirena glauca Boeck. ex C.B.Clarke in T.A.Durand & H.Schinz, Consp. Fl. Afric. 5: 646 (1894), nom. inval.
Fuirena reticulata Kük., Repert. Spec. Nov. Regni Veg. 41: 271 (1937).

Fuirena cuspidata (Roth) Kunth, Enum. Pl. 2: 187 (1837).
India. 40 IND. Hemicr. or rhizome geophyte.
**Scirpus cuspidatus* Roth, Nov. Pl. Sp.: 31 (1821). *Isolepis cuspidata* (Roth) A.Dietr., Sp. Pl. 2: 120 (1832).
Fuirena wallichiana Kunth, Enum. Pl. 2: 182 (1837).
Fuirena wallichiana var. *evoluta* C.B.Clarke in J.D.Hooker, Fl. Brit. India 6: 666 (1893).

Fuirena ecklonii Nees, Linnaea 10: 143 (1836).
Cape Prov. to KwaZulu-Natal. 27 CPP NAT. Hemicr. or rhizome geophyte.

Fuirena enodis C.B.Clarke in W.H.Harvey & auct. suc. (eds.), Fl. Cap. 7: 263 (1898). *Scirpus enodis* (C.B.Clarke) T.Koyama, J. Fac. Sci. Univ. Tokyo, Sect. 3, Bot. 7: 287 (1958).
Cape Prov. 27 CPP. Hemicr. or rhizome geophyte.

Fuirena felicis S.S.Hooper, Publ. Cairo Univ. Herb. 78: 323 (1977).
Yemen. 35 YEM. Hemicr. or rhizome geophyte.

Fuirena friesii Kük. in R.E.Fries, Wiss. Erg. Schwed. Rhod.-Kongo Exped. 1: 8 (1921).
S. Trop. Africa. 26 ZAM. Hemicr. or rhizome geophyte.

Fuirena gracilis Kunth, Enum. Pl. 2: 181 (1837).
Cape Prov. 27 CPP. Hemicr. or rhizome geophyte.
Fuirena microlepis Kunth, Enum. Pl. 2: 182 (1837).
Fuirena glabra Hochst., Flora 28: 757 (1845), nom. illeg.

Fuirena hirsuta (P.J.Bergius) P.L.Forbes, J. S. African Bot. 35: 83 (1969).
S. Africa. 27 CPP NAT SWZ TVL. Hemicr. or rhizome geophyte.
**Cyperus hirsutus* P.J.Bergius, Descr. Pl. Cap.: 11 (1767).
Scirpus hottentottus L., Mant. Pl. 2: 182 (1771). *Fuirena hottentotta* (L.) Druce, Bot. Soc. Exch. Club Brit. Isles 1916: 623 (1917).
Fuirena hirta Vahl, Enum. Pl. 2: 387 (1805).
Fuirena cephalotes Schrad., Gött. Gel. Anz. 1821(3): 2071 (1821).
Fuirena erioloma Nees, Linnaea 10: 142 (1836).
Fuirena glabra Kunth, Enum. Pl. 2: 182 (1837). *Scirpus glaber* (Kunth) T.Koyama, J. Fac. Sci. Univ. Tokyo, Sect. 3, Bot. 7: 287 (1958).
Fuirena intermedia Kunth, Enum. Pl. 2: 181 (1837).

Fuirena incompleta Nees in C.F.P.von Martius & auct. suc. (eds.), Fl. Bras. 2(1): 107 (1842). *Scirpus incompletus* (Nees) T.Koyama, J. Fac. Sci. Univ. Tokyo, Sect. 3, Bot. 7: 287 (1958).
Mexico to Trop. America. 79 MXC MXE MXG MXN MXS MXT 80 BLZ GUA HON PAN 82 VEN 83 CLM ECU PER 84 BZC BZL 85 AGE PAR URU. Hemicr. or rhizome geophyte.
Fuirena hexachaeta Schltdl., Linnaea 19: 69 (1846).
Fuirena incompleta f. *elatior* H.Pfeiff., Repert. Spec. Nov. Regni Veg. 33: 204 (1933).
Fuirena incompleta var. *obliterata* Kral, Sida 7: 350 (1978).

Fuirena incrassata S.T.Blake, Proc. Roy. Soc. Queensland 52: 58 (1941).
C. & E. Australia. 50 NSW NTA QLD. Hemicr. or rhizome geophyte.

Fuirena lainzii Luceño & M.Alves, Anales Jard. Bot. Madrid 54: 415 (1996).
Brazil (Pernambuco). 84 BZE. Hemicr. or rhizome geophyte.

Fuirena leptostachya Oliv., Trans. Linn. Soc. London 29: 168 (1875).
Trop. & S. Africa. 22 BEN BKN MLI NGA 23 RWA ZAI 24 CHA 25 KEN TAN UGA 26 ANG MLW 27 BOT NAM SWZ TVL. Ther.

var. ***leptostachya***
Trop. & S. Africa. 22 BKN MLI NGA 23 RWA ZAI 24 CHA 25 KEN TAN UGA 26 MLW 27 BOT NAM TVL. Ther.
Fuirena schweinfurthiana Boeck., Flora 63: 438 (1880).
Fuirena pygmaea Welw. ex Ridl., Trans. Linn. Soc. London, Bot. 2: 160 (1884).
Fuirena moiseri Turrill, Bull. Misc. Inform. Kew 1925: 71 (1925).

var. ***nudiflora*** K.Schum. in H.G.A.Engler, Pflanzenw. Ost-Afrikas, C: 126 (1895).
Benin, Uganda to S. Africa. 22 BEN 25 KEN TAN UGA 26 ANG 27 BOT NAM SWZ TVL. Ther.
Fuirena leptostachya f. *nudiflora* Lye, Nordic J. Bot. 3: 241 (1983), nom. illeg.

Fuirena longa Chapm., Fl. South. U.S., ed. 2(Suppl. 2): 698 (1892).
SE. U.S.A. to Texas. 77 TEX 78 ALA FLA GEO LOU MSI. Hemicr. or rhizome geophyte.

Fuirena microcarpa Lye, Bot. Not. 127: 111 (1974).
E. Tanzania. 25 TAN. Ther.

Fuirena moritziana Boeck., Allg. Bot. Z. Syst. 2: 77 (1896).
Venezuela. 82 VEN. Hemicr. or rhizome geophyte.

Fuirena multiflora Peter, Abh. Preuss. Akad. Wiss., Phys.-Math. Kl., n.f., 13(2): 50 (1928).
Tanzania. 25 TAN. Hemicr. or rhizome geophyte.

Fuirena nudiflora S.T.Blake, Proc. Roy. Soc. Queensland 52: 57 (1941).
N. Australia. 50 NTA QLD WAU. Hemicr. or rhizome geophyte.

Fuirena nyasensis Nelmes, Mem. New York Bot. Gard. 9: 98 (1954).
Malawi. 26 MLW. Hemicr. or rhizome geophyte.

Fuirena obcordata P.L.Forbes, S. African J. Bot. 53: 185 (1987).
Mozambique to S. Africa. 26 MOZ 27 NAM NAT TVL. Hemicr. or rhizome geophyte.

Fuirena ochreata Nees ex Kunth, Enum. Pl. 2: 184 (1837).
Kenya to Angola, Madagascar. 25 KEN TAN 26 ANG 29 MDG. Hemicr.
Fuirena calolepis K.Schum., Abh. Königl. Akad. Wiss. Berlin 1894: 20 (1894). *Scirpus calolepis* (K.Schum.) Kuntze, Revis. Gen. Pl. 3(2): 336 (1898).
Fuirena cinerascens Bojer ex C.B.Clarke in D.Oliver & auct. suc. (eds.), Fl. Trop. Afr. 8: 467 (1902).
Fuirena cristata Turrill, Bull. Misc. Inform. Kew 1914: 170 (1914).

Fuirena oedipus C.B.Clarke, J. Linn. Soc., Bot. 37: 478 (1906).
S. Trop. Africa. 26 ZIM. Hemicr. or rhizome geophyte.

Fuirena ponmudiensis Ravi & Anil Kumar, Rheedea 4: 102 (1994).
SW. India. 40 IND. Hemicr. or rhizome geophyte.

Fuirena pubescens (Poir.) Kunth, Enum. Pl. 2: 182 (1837).
Africa, Medit. to NW. India. 12 BAL COR POR SPA 13 ITA KRI 20 ALG EGY MOR TUN 22 NGA NGR 23 RWA ZAI 24 CHA ETH 25 KEN TAN UGA 26 ANG MLW 27 BOT CPP LES NAM NAT SWZ TVL 29 MDG 34 AFG EAI LBS PAL TUR 35 YEM 40 IND PAK WHM. Hemicr.
**Carex pubescens* Poir., Voy. Barbarie 2: 254 (1789). *Scirpus pubescens* (Poir.) Lam., Tabl. Encycl. 1: 139 (1791). *Isolepis pubescens* (Poir.) Roem. & Schult., Syst. Veg. 2: 118 (1817).

var. ***abbreviata*** Lye, Bot. Not. 127: 112 (1974).
Uganda. 25 UGA. Hemicr.

var. ***buchananii*** (Boeck.) C.B.Clarke in T.A.Durand & H.Schinz, Consp. Fl. Afric. 5: 648 (1894).
S. Tanzania to S. Trop. Africa. 25 TAN 26 ANG MLW. Hemicr.
Fuirena welwitschii Ridl., Trans. Linn. Soc. London, Bot. 2: 161 (1884).
**Fuirena buchananii* Boeck., Beitr. Cyper. 1: 20 (1888).

var. ***major*** Lye, Bot. Not. 127: 112 (1974).
Uganda to S. Africa. 25 KEN TAN UGA 26 ANG 27 BOT SWZ TVL. Hemicr. or rhizome geophyte.
Fuirena macrostachya Boeck., Bot. Jahrb. Syst. 5: 507 (1884).
**Fuirena pachyrrhiza* Ridl., Trans. Linn. Soc. London, Bot. 2: 161 (1884).
Fuirena macrostachya Boeck. in H.G.A.Engler, Pflanzenw. Ost-Afrikas, A: 67 (1895).

var. ***pergamentacea*** C.E.C.Fisch. in J.S.Gamble, Fl. Madras: 1669 (1931).
S. India. 40 IND. Hemicr. or rhizome geophyte.

var. ***pubescens***
Africa, Medit. to NW. India. 12 BAL COR POR SPA 13 ITA KRI 20 ALG EGY MOR TUN 22 NGA NGR 23 RWA ZAI 24 CHA ETH 25 KEN TAN UGA 26 ANG 27 BOT CPP LES NAM NAT SWZ TVL 29 MDG 34 AFG EAI LBS PAL TUR 35 YEM 40 IND PAK WHM. Hemicr.
Carex tenuifolia Poir., Voy. Barbarie 2: 254 (1789).
Sparganium pubescens Poir., Voy. Barbarie 2: 254 (1789), nom. inval.
Carex poiretii J.F.Gmel., Syst. Nat. 2: 140 (1791).
Scirpus ciliaris Pers., Syn. Pl. 1: 77 (1805), nom. illeg.
Fuirena annua Royle, Ill. Bot. Himal. Mts.: 411 (1840).

Fuirena pumila (Torr.) Spreng., Syst. Veg. 1: 237 (1824).
SE. Canada to EC. & E. U.S.A. 72 ONT 74 ILL? 75 INI MAS MIC NWJ NWY RHO 77 TEX 78 ALA DEL FLA GEO LOU MRY NCA SCA VRG. Ther.
**Fuirena squarrosa* var. *pumila* Torr., Fl. N. Middle United States 1: 68 (1824). *Fuirena simplex* var. *pumila* (Torr.) C.B.Clarke ex Coville, Bull. Torrey Bot. Club 17: 6 (1890).
Fuirena torreyana L.C.Beck, Bot. North. Middle States: 429 (1833).

Fuirena quercina Cherm., Bull. Soc. Bot. France 68: 424 (1921 publ. 1922).
Madagascar. 29 MDG. Hemicr. or rhizome geophyte.

Fuirena repens Boeck., Bot. Jahrb. Syst. 7: 277 (1886).
Mexico (San Luis Potosí). 79 MXE. Hemicr. or rhizome geophyte.

Fuirena rhizomatifera Tang & F.T.Wang, in Fl. Reipubl. Popul. Sin. 11: 225 (1961).
SC. China. 36 CHC. Hemicr. or rhizome geophyte.
Fuirena umbellata var. *angustifolia* Kük. in H.F.von Handel-Mazzetti, Symb. Sin. 7: 1248 (1936).

Fuirena robusta Kunth, Enum. Pl. 2: 185 (1837).
Trop. America. 79 MXT 80 GUA HON NIC PAN 81 CUB PUE 82 FRG GUY SUR VEN 84 BZE BZL BZN BZS 85 AGE PAR. Hemicr. or rhizome geophyte.
Fuirena bahiensis Lindl. & Nees in C.F.P.von Martius & auct. suc. (eds.), Fl. Bras. 2(1): 108 (1842).
Fuirena glabrata Steud., Syn. Pl. Glumac. 2: 127 (1855).
Fuirena latifolia Steud., Syn. Pl. Glumac. 2: 126 (1855).
Fuirena schizophylla C.Wright, Anales Acad. Ci. Méd. Habana 8: 82 (1871).

Fuirena sagittata Lye, Bot. Not. 127: 110 (1974).
Tanzania. 25 TAN. Ther.

Fuirena scirpoidea Michx., Fl. Bor.-Amer. 1: 38 (1803).
Scirpus scirpoideus (Michx.) T.Koyama, J. Fac. Sci. Univ. Tokyo, Sect. 3, Bot. 7: 287 (1858).
SE. U.S.A. to Texas, C. America, Cuba. 77 TEX 78 ALA FLA GEO LOU MSI 80 NIC 81 CUB. Hemicr.
Vaginaria richardii Pers., Syn. Pl. 1: 70 (1805).

Fuirena simplex Vahl, Eclog. Amer. 2: 8 (1798).
C. & SE. U.S.A. to C. America, Cuba, N. South America. 74 KAN MSO NEB OKL 77 NWM NWM TEX TEX 78 ARK LOU 79 MXC MXE MXG MXN MXS MXT 80 BLZ ELS GUA HON NIC 81 CUB 82 FRG. Hemicr.

var. ***aristulata*** (Torr.) Kral, Sida 7: 336 (1978).
C. & SE. U.S.A. to NW. Mexico. 74 KAN MSO NEB OKL 77 NWM TEX 78 ARK LOU 79 MXN. Hemicr.
**Fuirena squarrosa* var. *aristulata* Torr., Ann. Lyceum Nat. Hist. New York 3: 291 (1836).

var. ***simplex***
C. & SE. U.S.A. to C. America, Caribbean, N. South America. 74 KAN OKL 77 NWM TEX 78 ARK LOU 79 MXC MXE MXG MXN MXS MXT 80 BLZ ELS GUA HON NIC 81 CUB PUE 82 FRG. Hemicr.
Fuirena obtusiflora Vahl, Eclog. Amer. 2: 8 (1798).
Fuirena schiedeana Kunth, Enum. Pl. 2: 183 (1837).
Fuirena aristulata Steud., Syn. Pl. Glumac. 2: 127 (1855).
Fuirena simplex f. *macrostachya* Coville, Bull. Torrey Bot. Club 17: 4 (1890).
Fuirena cylindrica Bush, Rep. (Annual) Missouri Bot. Gard. 1905: 91 (1905).
Fuirena zacapana Bartlett, Proc. Amer. Acad. Arts 43: 50 (1907).
Fuirena primiera M.E.Jones, Contr. U. S. Natl. Herb. 18: 25 (1933).

Fuirena simpsonii Ravi, N.Mohanan & Shaju, Nordic J. Bot. 22: 329 (2002 publ. 2003).
India (Kerala). 40 IND.

Fuirena somaliensis Lye in ?
S. Somalia. 24 SOM. Ther.

Fuirena squarrosa Michx., Fl. Bor.-Amer. 1: 37 (1803).
C. & E. U.S.A., Caribbean. 74 OKL 75 CNT NWJ NWY RHO 77 TEX 78 ALA ARK DEL FLA GEO KTY LOU MRY MSI NCA SCA TEN VRG WDC 81 CUB JAM. Ther. or hemicr.
Fuirena hispida Elliott, Sketch Bot. S. Carolina 1: 579 (1821).

Fuirena stephani Ramos & Diego, Acta Bot. Mex. 58: 52 (2002).
Mexico (Campeche). 79 MXT. Hemicr. or rhizome geophyte.

Fuirena striatella Lye, Candollea 51: 4 (1996).
N. Somalia. 24 SOM. Ther.

Fuirena stricta Steud., Syn. Pl. Glumac. 2: 128 (1855).
Trop. & S. Africa, Comoros, Madagascar. 22 BEN BKN GHA GNB GUI IVO LBR MLI NGA NGR SEN SIE TOG 23 BUR CMN RWA ZAI 24 CHA 25 KEN TAN UGA 26 ANG 27 BOT NAM NAT SWZ TVL 29 COM MDG. Hel. or hemicr.

subsp. ***chlorocarpa*** (Ridl.) Lye, Nordic J. Bot. 3: 241 (1983).
Nigeria to Ethiopia and Angola, Madagascar. 22 NGA 23 CMN ZAI 24 ETH 26 ANG 29 MDG. Hemicr.
**Fuirena chlorocarpa* Ridl., Trans. Linn. Soc. London, Bot. 2: 159 (1884).

subsp. ***stricta***
Trop. & S. Africa, Comoros, Madagascar. 22 BEN BKN GHA GNB GUI IVO LBR MLI NGA NGR SEN SIE TOG 23 BUR RWA ZAI 24 CHA 25 KEN TAN UGA 27 BOT NAM NAT SWZ TVL 29 COM MDG. Hel. or hemicr.
Rhynchospora senegalensis Steud., Syn. Pl. Glumac. 2: 149 (1855).
Pentasticha madagascariensis Turcz., Bull. Soc. Imp. Naturalistes Moscou 1862(2): 330 (1862).

Fuirena subdigitata C.B.Clarke, J. Linn. Soc., Bot. 37: 477 (1906).
S. Trop. Africa. 26 ZIM. Hemicr. or rhizome geophyte.

Fuirena swamyi Govind., Proc. Indian Acad. Sci., Pl. Sci. 94: 17 (1985).
India (Tamil Nadu). 40 IND. Hemicr.

Fuirena tenuis P.L.Forbes, S. African J. Bot. 3: 359 (1984).
Cape Prov. to Lesotho. 27 CPP LES. Hemicr. or rhizome geophyte.

Fuirena trilobites C.B.Clarke in J.D.Hooker, Fl. Brit. India 6: 666 (1893).
S. India. 40 IND. Hemicr. or rhizome geophyte.

Fuirena tuwensis M.B.Deshp. & Shah, Bull. Bot. Surv. India 10: 239 (1969).
C. & W. India. 40 IND. Hemicr. or rhizome geophyte.

Fuirena umbellata Rottb., Descr. Icon. Rar. Pl.: 70 (1773). *Scirpus umbellatus* (Rottb.) Kuntze, Revis. Gen. Pl. 3(2): 337 (1898). *Fuirena umbellata* var. *typica* Domin, Biblioth. Bot. 85: 467 (1915), nom. inval. *Scirpus fuirena* T.Koyama, J. Fac. Sci. Univ. Tokyo, Sect. 3, Bot. 7: 361 (1958).
Trop. & Subtrop. 22 BEN BKN GAM GHA GNB GUI IVO LBR MLI NGA NGR SEN SIE TOG 23 BUR CAF CMN EQG GAB GGI ZAI 24 CHA 25 KEN TAN UGA 26 ANG MLW 27 BOT NAT 29 MAU MDG REU 36 CHC CHH 38 NNS TAI 40 ASS BAN EHM IND NEP PAK SRL WHM 41 THA VIE 42 BOR JAW MLY PHI SUL SUM 43 NWG 50 NTA QLD WAU 62 CRL MRN 79 MXG MXS MXT 80 BLZ COS ELS GUA HON NIC PAN 81 CUB DOM JAM LEE PUE TRT WIN 82 FRG GUY SUR VEN 83 BOL CLM ECU PER 84 BZC BZL BZN BZS 85 AGE PAR. Hel.
Fuirena paniculata L.f., Suppl. Pl.: 105 (1782).
Fuirena canescens Vahl, Enum. Pl. 2: 385 (1805).
Fuirena pentagona Schumach., Beskr. Guin. Pl.: 42 (1827).
Fuirena tereticulmis C.Presl, Reliq. Haenk. 1: 186 (1828).
Fuirena pentagona Wight & Arn. ex Nees in R.Wight, Contr. Bot. India: 93 (1834), nom. illeg.
Fuirena rivularis Nees, Linnaea 9: 288 (1834), nom. inval.
Scirpus rivularis D.Don. ex Nees, Linnaea 9: 288 (1834), nom. inval.
Fuirena mauritiana Nees ex Kunth, Enum. Pl. 2: 186 (1837).
Fuirena mauritiana var. *angustifolia* Kunth, Enum. Pl. 2: 186 (1837).
Fuirena thouarsiana Kunth, Enum. Pl. 2: 186 (1837).
Fuirena quinguangularis Hassk., Flora 25(2 Beibl.): 3 (1842).
Fuirena guianensis Klotzsch in M.R.Schomburgk, Reis. Br.-Guiana: 895 (1849).
Fuirena haemaglottis Steud., Syn. Pl. Glumac. 2: 126 (1855).
Fuirena complotricha C.Wright, Fl. Cub.: 177 (1871).
Fuirena hildebrandtii Boeck., Flora 65: 15 (1882).
Fuirena seriata C.B.Clarke, Bull. Soc. Bot. France 54(8): 28 (1907).
Fuirena umbellata var. *pilosa* Domin, Biblioth. Bot. 85: 467 (1915).
Fuirena philippinensis Gand., Bull. Soc. Bot. France 66: 297 (1919 publ. 1920).
Fuirena umbellata var. *unguiculata* Kük., Repert. Spec. Nov. Regni Veg. 23: 200 (1926).
Fuirena umbellata var. *americana* H.Pfeiff., Repert. Spec. Nov. Regni Veg. 33: 204 (1933).
Fuirena mahouxii Cherm., Bull. Soc. Bot. France 81: 264 (1934).

Fuirena uncinata (Willd.) Kunth, Enum. Pl. 2: 184 (1837).
W. & S. India, Sri Lanka. 40 IND SRL. Ther. or hemicr.
Scirpus capitatus Burm.f., Fl. Indica: 21 (1768), nom. illeg. *Fuirena capitata* T.Koyama in M.D. Daddanayake & al. (eds.), Revised Handb. Fl. Ceylon 5: 151 (1985).
**Scirpus uncinatus* Willd., Sp. Pl. 1: 300 (1797). *Isolepis uncinata* (Willd.) Roem. & Schult., Syst. Veg. 2: 111 (1817).

Fuirena zambesiaca Lye, Bot. Not. 127: 109 (1974).
Kenya to Mozambique. 25 KEN TAN 26 MOZ. Ther.

Synonyms:
Fuirena annua Royle = ***Fuirena pubescens*** (Poir.) Kunth var. ***pubescens***
Fuirena aristulata Steud. = ***Fuirena simplex*** Vahl var. ***simplex***
Fuirena bahiensis Lindl. & Nees = ***Fuirena robusta*** Kunth
Fuirena buchananii Boeck. = ***Fuirena pubescens*** var. ***buchananii*** (Boeck.) C.B.Clarke
Fuirena bulbipes S.F.Blake = ***Fuirena camptotricha*** C.Wright
Fuirena calolepis K.Schum. = ***Fuirena ochreata*** Nees ex Kunth
Fuirena canescens Vahl = ***Fuirena umbellata*** Rottb.
Fuirena capitata T.Koyama = ***Fuirena uncinata*** (Willd.) Kunth
Fuirena cephalotes Schrad. = ***Fuirena hirsuta*** (P.J.Bergius) P.L.Forbes
Fuirena chlorocarpa Ridl. = ***Fuirena stricta*** subsp. ***chlorocarpa*** (Ridl.) Lye
Fuirena ciliaris var. *apetala* Wingf. = ***Fuirena ciliaris*** f. ***apetala*** (Wingf.) Lye

Fuirena ciliata Bush = ***Fuirena bushii*** Kral
Fuirena cinerascens Bojer ex C.B.Clarke = ***Fuirena ochreata*** Nees ex Kunth
Fuirena complotricha C.Wright = ***Fuirena umbellata*** Rottb.
Fuirena cristata Turrill = ***Fuirena ochreata*** Nees ex Kunth
Fuirena cylindrica Bush = ***Fuirena simplex*** Vahl var. ***simplex***
Fuirena erioloma Nees = ***Fuirena hirsuta*** (P.J.Bergius) P.L.Forbes
Fuirena filifolia Rchb. ex Kunth = ***Tetraria cuspidata*** (Rottb.) C.B.Clarke
Fuirena glabra Hochst. = ***Fuirena gracilis*** Kunth
Fuirena glabra Kunth = ***Fuirena hirsuta*** (P.J.Bergius) P.L.Forbes
Fuirena glabrata Steud. = ***Fuirena robusta*** Kunth
Fuirena glauca Boeck. ex C.B.Clarke = ***Fuirena coerulescens*** Steud.
Fuirena glomerata Lam. = ***Fuirena ciliaris*** (L.) Roxb. f. ***ciliaris***
Fuirena glomerata var. *angolensis* C.B.Clarke = ***Fuirena angolensis*** (C.B.Clarke) Lye ex J.Raynal & Roessler
Fuirena gracilis Spreng. = ***Rhynchospora tenuis*** subsp. ***tenuis***
Fuirena guianensis Klotzsch = ***Fuirena umbellata*** Rottb.
Fuirena haemaglottis Steud. = ***Fuirena umbellata*** Rottb.
Fuirena hexachaeta Schltdl. = ***Fuirena incompleta*** Nees
Fuirena hildebrandtii Boeck. = ***Fuirena umbellata*** Rottb.
Fuirena hirta Vahl = ***Fuirena hirsuta*** (P.J.Bergius) P.L.Forbes
Fuirena hispida Elliott = ***Fuirena squarrosa*** Michx.
Fuirena hottentotta (L.) Druce = ***Fuirena hirsuta*** (P.J.Bergius) P.L.Forbes
Fuirena incompleta f. *elatior* H.Pfeiff. = ***Fuirena incompleta*** Nees
Fuirena incompleta var. *obliterata* Kral = ***Fuirena incompleta*** Nees
Fuirena intermedia Kunth = ***Fuirena hirsuta*** (P.J.Bergius) P.L.Forbes
Fuirena latifolia Steud. = ***Fuirena robusta*** Kunth
Fuirena leptostachya f. *nudiflora* Lye = ***Fuirena leptostachya*** var. ***nudiflora*** K.Schum.
Fuirena macrostachya Boeck. = ***Fuirena pubescens*** var. ***major*** Lye
Fuirena macrostachya Boeck. = ***Fuirena pubescens*** var. ***major*** Lye
Fuirena mahouxii Cherm. = ***Fuirena umbellata*** Rottb.
Fuirena mauritiana Nees ex Kunth = ***Fuirena umbellata*** Rottb.
Fuirena mauritiana var. *angustifolia* Kunth = ***Fuirena umbellata*** Rottb.
Fuirena microlepis Kunth = ***Fuirena gracilis*** Kunth
Fuirena microstachya Boeck. = [82]
Fuirena moiseri Turrill = ***Fuirena leptostachya*** Oliv. var. ***leptostachya***
Fuirena mollicula Kunth = ***Fuirena coerulescens*** Steud.
Fuirena nana A.Rich. = ***Lipocarpha nana*** (A.Rich.) Cherm.
Fuirena obtusiflora Vahl = ***Fuirena simplex*** Vahl var. ***simplex***
Fuirena pachyrrhiza Ridl. = ***Fuirena pubescens*** var. ***major*** Lye
Fuirena paniculata L.f. = ***Fuirena umbellata*** Rottb.
Fuirena pentagona Schumach. = ***Fuirena umbellata*** Rottb.
Fuirena pentagona Wight & Arn. ex Nees = ***Fuirena umbellata*** Rottb.
Fuirena philippinensis Gand. = ***Fuirena umbellata*** Rottb.
Fuirena primiera M.E.Jones = ***Fuirena simplex*** Vahl var. ***simplex***
Fuirena pygmaea Welw. ex Ridl. = ***Fuirena leptostachya*** Oliv. var. ***leptostachya***
Fuirena quinguangularis Hassk. = ***Fuirena umbellata*** Rottb.
Fuirena repanda S.T.Blake = ***Fuirena arenosa*** R.Br.
Fuirena reticulata Kük. = ***Fuirena coerulescens*** Steud.
Fuirena rivularis Nees = ***Fuirena umbellata*** Rottb.
Fuirena rottboellii Nees = ***Fuirena ciliaris*** (L.) Roxb. f. ***ciliaris***
Fuirena rubiginosa Spreng. = ***Machaerina rubiginosa*** (Spreng.) T.Koyama
Fuirena schiedeana Kunth = ***Fuirena simplex*** Vahl var. ***simplex***
Fuirena schizophylla C.Wright = ***Fuirena robusta*** Kunth
Fuirena schweinfurthiana Boeck. = ***Fuirena leptostachya*** Oliv. var. ***leptostachya***
Fuirena seriata C.B.Clarke = ***Fuirena umbellata*** Rottb.
Fuirena simplex f. *macrostachya* Coville = ***Fuirena simplex*** Vahl var. ***simplex***
Fuirena simplex var. *pumila* (Torr.) C.B.Clarke ex Coville = ***Fuirena simplex*** Vahl var. ***simplex***
Fuirena squarrosa var. *aristulata* Torr. = ***Fuirena simplex*** var. ***aristulata*** (Torr.) Kral
Fuirena squarrosa var. *breviseta* Coville = ***Fuirena breviseta*** (Coville) Coville
Fuirena squarrosa var. *pumila* Torr. = ***Fuirena pumila*** (Torr.) Spreng.
Fuirena striata Llanos = ***Fuirena ciliaris*** (L.) Roxb. f. ***ciliaris***
Fuirena tereticulmis C.Presl = ***Fuirena umbellata*** Rottb.
Fuirena thouarsiana Kunth = ***Fuirena umbellata*** Rottb.
Fuirena torreyana L.C.Beck = ***Fuirena pumila*** (Torr.) Spreng.
Fuirena umbellata var. *americana* H.Pfeiff. = ***Fuirena umbellata*** Rottb.
Fuirena umbellata var. *angustifolia* Kük. = ***Fuirena rhizomatifera*** Tang & F.T.Wang
Fuirena umbellata var. *pilosa* Domin = ***Fuirena umbellata*** Rottb.
Fuirena umbellata var. *typica* Domin = ***Fuirena umbellata*** Rottb.
Fuirena umbellata var. *unguiculata* Kük. = ***Fuirena umbellata*** Rottb.
Fuirena verticillata Spreng. = ***Lagenocarpus verticillatus*** (Spreng.) T.Koyama & Maguire
Fuirena wallichiana Kunth = ***Fuirena cuspidata*** (Roth) Kunth
Fuirena wallichiana var. *evoluta* C.B.Clarke = ***Fuirena cuspidata*** (Roth) Kunth
Fuirena weigeltii A.Spreng. = ***Lagenocarpus verticillatus*** (Spreng.) T.Koyama & Maguire
Fuirena welwitschii Ridl. = ***Fuirena pubescens*** var. ***buchananii*** (Boeck.) C.B.Clarke
Fuirena zacapana Bartlett = ***Fuirena simplex*** Vahl var. ***simplex***

***Unplaced Names*:**
Fuirena microstachya Boeck., Linnaea 38: 398 (1874). = [82]

Gahnia

Gahnia J.R.Forst. & G.Forst., Char. Gen. Pl.: 26 (1775).
Trop. & Subtrop. Asia to Pacific and New Zealand. 36 CHC CHH CHS 38 NNS OGA TAI 41 THA VIE 42 BOR JAW MLY MOL PHI SUL SUM 43 NWG SOL 50 NFK NSW QLD SOA TAS VIC WAU 51 NZN NZS 60 FIJ NWC VAN 61 MRQ SCI TUB 63 HAW.

41 Species
Lampocarya R.Br., Prodr.: 238 (1810).
Epiandria C.Presl, Isis (Oken) 21: 270 (1828), nom. inval.
Didymonema C.Presl, Didymonema: 5 (1829).
Melachne Schrad. ex Schult. & Schult.f. in J.J.Roemer & J.A.Schultes, Syst. Veg. 7(2): lxxv (1830).
Psittacoschoenus Nees in J.G.C.Lehmann, Pl. Preiss. 2: 87 (1846).
Phacellanthus Steud. in H.Zollinger, Syst. Verz. 1: 61 (1854), nom. illeg. *Syziganthus* Steud., Syn. Pl. Glumac. 2: 153 (1855), nom. illeg.

Gahnia ancistrophylla Benth., Fl. Austral. 7: 415 (1878). *Mariscus ancistrophyllus* (Benth.) Kuntze, Revis. Gen. Pl. 2: 755 (1891).
S. Australia. 50 SOA VIC WAU. Hemicr. or rhizome geophyte.
Cladium ancistrophyllum F.Muell. ex Benth., Fl. Austral. 7: 415 (1878).

Gahnia aristata Benth., Fl. Austral. 7: 416 (1878). *Mariscus aristatus* (Benth.) Kuntze, Revis. Gen. Pl. 2: 755 (1891). *Gahnia lanigera* var. *aristata* (Benth.) Kük., Feddes Repert. Spec. Nov. Regni Veg. 52: 108 (1943).
Western Australia. 50 WAU. Hemicr. or rhizome geophyte.
Cladium aristatum F.Muell. ex Benth., Fl. Austral. 7: 416 (1878).

Gahnia aspera (R.Br.) Spreng., Syst. Veg. 2: 114 (1825).
Maluku to Pacific. 38 OGA 42 MOL 43 NWG 50 NSW QLD 60 FIJ NWC VAN 61 SCI TUB 63 HAW. Hemicr. or rhizome geophyte.
**Lampocarya aspera* R.Br., Prodr.: 238 (1810). *Cladium asperum* (R.Br.) F.Muell., Fragm. 9: 12 (1875). *Mariscus asper* (R.Br.) Kuntze, Revis. Gen. Pl. 2: 755 (1891), nom. illeg.

subsp. ***aspera***
Maluku to S. Pacific. 38 OGA 42 MOL 43 NWG 50 NSW QLD 60 FIJ NWC VAN 61 SCI TUB. Hemicr. or rhizome geophyte.
Lampocarya rawacensis Kunth, Enum. Pl. 2: 333 (1837). *Gahnia rawacensis* (Kunth) Steud., Syn. Pl. Glumac. 2: 164 (1855).
Hexalepis scabrifolia Boeck., Flora 58: 118 (1875).
Gahnia boninsimae Maxim., Bull. Acad. Imp. Sci. Saint-Pétersbourg 31: 112 (1887).
Gahnia aspera f. *distans* Domin, Biblioth. Bot. 85: 481 (1915).
Gahnia stokesii F.Br., Bernice P. Bishop Mus. Bull. 84: 113 (1931).
Gahnia aspera var. *angustinus* Kük., Feddes Repert. Spec. Nov. Regni Veg. 52: 94 (1943).

subsp. ***globosa*** (H.Mann) J.Kern, in Fl. Males. 7: 710 (1974).
Hawaiian Is. (Oahu). 63 HAW. Hemicr. or rhizome geophyte.
**Gahnia globosa* H.Mann, Proc. Amer. Acad. Arts 7: 210 (1867). *Mariscus globosus* (H.Mann) Kuntze, Revis. Gen. Pl. 2: 755 (1891).
Gahnia congesta Boeck., Linnaea 38: 353 (1874). *Mariscus congestus* (Boeck.) Kuntze, Revis. Gen. Pl. 2: 755 (1891).
Gahnia mucronata Boeck., Linnaea 38: 342 (1874).

Gahnia australis (Nees) K.L.Wilson, Telopea 1: 464 (1980).
SW. Australia. 50 WAU. Hemicr. or rhizome geophyte.
**Elynanthus australis* Nees, Ann. Mag. Nat. Hist. 6: 48 (1841). *Cladium elynanthoides* F.Muell., Fragm. 9: 31 (1875), nom. illeg. *Machaerina elynanthoides* (F.Muell.) T.Koyama, Bot. Mag. (Tokyo) 69: 63 (1956). *Mariscus australis* (Nees) Kuntze, Revis. Gen. Pl. 2: 755 (1891). *Cladium australe* (Nees) H.Pfeiff., Repert. Spec. Nov. Regni Veg. 23: 350 (1927), nom. illeg. *Baumea australis* (Nees) S.T.Blake, Contr. Queensland Herb. 8: 28 (1969).

Gahnia baniensis Benl, Repert. Spec. Nov. Regni Veg. 44: 197 (1938).
Hainan to W. Malesia. 36 CHH 41 VIE 42 BOR MLY SUM. Hemicr. or rhizome geophyte.
Gahnia javanica var. *penangensis* C.B.Clarke in J.D.Hooker, Fl. Brit. India 6: 677 (1894). *Gahnia penangensis* (C.B.Clarke) Kük., Bull. Jard. Bot. Buitenzorg, III, 16: 308 (1940).

Gahnia beecheyi H.Mann, Proc. Amer. Acad. Arts 7: 210 (1867). *Mariscus beecheyi* (H.Mann) Kuntze, Revis. Gen. Pl. 2: 755 (1891).
Hawaiian Is. 63 HAW. Hemicr. or rhizome geophyte.
Gahnia leptostachya Boeck., Linnaea 38: 354 (1874).
Lampocarya leptostachya Schrad. ex Boeck., Linnaea 38: 354 (1874).
Gahnia mannii Hillebr., Fl. Hawaiian Isl.: 482 (1888). *Mariscus mannii* (Hillebr.) Kuntze, Revis. Gen. Pl. 2: 755 (1891).
Mariscus mucronatus Kuntze, Revis. Gen. Pl. 2: 755 (1891).

Gahnia decomposita (R.Br.) Benth., Fl. Austral. 7: 417 (1878).
SW. Australia. 50 WAU. Hemicr. or rhizome geophyte.
**Cladium decompositum* R.Br., Prodr.: 237 (1810). *Mariscus decompositus* (R.Br.) Kuntze, Revis. Gen. Pl. 2: 755 (1891), nom. illeg.
Gahnia preissii Nees in J.G.C.Lehmann, Pl. Preiss. 2: 87 (1846). *Cladium preissii* (Nees) F.Muell., Fragm. 9: 13 (1875). *Mariscus preissii* (Nees) Kuntze, Revis. Gen. Pl. 2: 755 (1891).

Gahnia deusta (R.Br.) Benth., Fl. Austral. 7: 416 (1878).
S. Australia. 50 SOA VIC WAU. Hemicr. or rhizome geophyte.
**Cladium deustum* R.Br., Prodr.: 237 (1810). *Mariscus deustus* (R.Br.) Kuntze, Revis. Gen. Pl. 2: 755 (1891).

Gahnia drummondii (Steud.) K.L.Wilson, Telopea 1: 465 (1980).
SW. Australia. 50 WAU. Hemicr. or rhizome geophyte.
**Schoenus drummondii* Steud., Syn. Pl. Glumac. 2: 165 (1855).
Gahnia polyphylla Benth., Fl. Austral. 7: 415 (1878). *Cladium polyphyllum* (Benth.) F.Muell., Syst. Census Austral. Pl. 1: 129 (1882). *Mariscus polyphyllus* (Benth.) Kuntze, Revis. Gen. Pl. 2: 755 (1891), nom. illeg.

Gahnia erythrocarpa R.Br., Prodr.: 239 (1810). *Cladium erythrocarpum* (R.Br.) F.Muell., Key Vict. Pl. 1: 454 (1888). *Gahnia psittacorum* var. *erythrocarpa* (R.Br.) C.B.Clarke, Ill. Cyper.: t. 97 (1909).
CE. New South Wales. 50 NSW. Hemicr. or rhizome geophyte.
Gahnia urvilleana Kunth, Enum. Pl. 2: 332 (1837).
Gahnia gonyocarpa Steud., Syn. Pl. Glumac. 2: 164 (1855).

Gahnia filifolia (C.Presl) Kük. ex Benl, Bot. Arch. 40: 207 (1940).

EC. New South Wales. 50 NSW. Hemicr. or rhizome geophyte.
Epiandria teretifolia C.Presl, Isis (Oken) 21: 270 (1828), nom. inval.
Gahnia psittacorum Sieber ex C.Presl, Isis (Oken) 21: 270 (1828), nom. illeg.
**Didymonema filifolia* C.Presl, Didymonema: 5 (1829). *Gahnia teretifolia* Domin, Biblioth. Bot. 85: 481 (1915), nom. illeg. *Caustis filifolia* (C.Presl) H.Pfeiff., Repert. Spec. Nov. Regni Veg. 21: 364 (1925).
Melachne sieberi Schrad. ex Schult. & Schult.f. in J.J.Roemer & J.A.Schultes, Syst. Veg. 7(2): lxxv (1830). *Caustis sieberi* (Schrad. ex Schult. & Schult.f.) Kunth, Enum. Pl. 2: 307 (1837). *Gahnia sieberi* (Schrad. ex Schult. & Schult.f.) Boeck., Linnaea 38: 343 (1874). *Cladium sieberi* (Schrad. ex Schult. & Schult.f.) F.Muell., Fragm. 9: 14 (1875).

Gahnia filum (Labill.) F.Muell., Key Vict. Pl. 1: 456 (1888).
S. & SE. Australia. 50 NSW SOA TAS VIC WAU. Hemicr. or rhizome geophyte.
**Schoenus filum* Labill., Nov. Holl. Pl. 1: 18 (1805). *Cladium filum* (Labill.) R.Br., Prodr.: 237 (1810). *Mariscus filum* (Labill.) Kuntze, Revis. Gen. Pl. 2: 755 (1891).
Baumea longifolia Boeck., Linnaea 38: 244 (1874).

Gahnia graminifolia Rodway, Pap. & Proc. Roy. Soc. Tasmania 1893: 182 (1894).
Tasmania. 50 TAS. Hemicr. or rhizome geophyte.
Gahnia rodwayi F.Muell. ex Rodway, Pap. & Proc. Roy. Soc. Tasmania 1892: 93 (1893), nom. inval.

Gahnia grandis (Labill.) S.T.Blake, Contr. Queensland Herb. 8: 33 (1969).
SE. Australia. 50 NSW TAS VIC. Hemicr. or rhizome geophyte.
**Scleria grandis* Labill., Voy. Rech. Pérouse: 146 (1800).
Gahnia leucocarpa R.Br., Prodr.: 239 (1810).
Gahnia psilocaulon Boeck., Linnaea 38: 345 (1874). *Gahnia psittacorum* f. *psilocaulon* (Boeck.) Benl, Bot. Arch. 40: 202 (1940). *Gahnia psittacorum* var. *psilocaulon* (Boeck.) Kük., Feddes Repert. Spec. Nov. Regni Veg. 52: 67 (1943).
Gahnia psittacorum var. *oxylepis* Benth., Fl. Austral. 7: 419 (1878).
Gahnia psittacorum f. *leucosperma* Benl, Bot. Arch. 40: 202 (1940).

Gahnia howeana R.O.Gardner, New Zealand J. Bot. 35: 155 (1997).
Lord Howe I. 50 NFK. Hemicr. or rhizome geophyte.

Gahnia hystrix J.M.Black, Trans. & Proc. Roy. Soc. South Australia 51: 29 (1927).
South Australia (Kangaroo I.). 50 SOA. Hemicr. or rhizome geophyte.

Gahnia insignis S.T.Blake, Proc. Roy. Soc. Queensland 68: 39 (1957).
SE. Queensland to NE. New South Wales. 50 NSW QLD. Hemicr. or rhizome geophyte.

Gahnia javanica Moritzi, Syst. Verz.: 98 (1846). *Mariscus javanicus* (Moritzi) Kuntze, Revis. Gen. Pl. 2: 755 (1891).
SC. China to Jawa and Solomon Is. 36 CHC 41 VIE 42 BOR JAW MLY PHI SUL SUM 43 NWG SOL. Hemicr. or rhizome geophyte.
Schoenus paniculatus Hassk., Cat. Hort. Bot. Bogor.: 296 (1844), nom. illeg. *Schoenus hasskarlii* Steud., Syn. Pl. Glumac. 2: 166 (1855).
Phacellanthus multiflorus Steud. in H.Zollinger, Syst. Verz. 1: 61 (1854). *Syziganthus multiflorus* (Steud.) Steud., Syn. Pl. Glumac. 2: 153 (1855).
Gahnia castanea Ridl., J. Fed. Malay States Mus. 6: 60 (1915).
Gahnia javanica var. *paupercula* Kük., Bot. Jahrb. Syst. 64: 260 (1938).
Gahnia javanica f. *borneensis* Benl, Bot. Arch. 40: 173 (1940).
Gahnia javanica var. *philippinensis* Benl, Bot. Arch. 40: 174 (1940).
Gahnia javanica f. *sinensis* Benl, Bot. Arch. 40: 173 (1940).
Gahnia javanica subf. *subcapitata* Kük. ex Benl, Bot. Arch. 40: 173 (1940).
Gahnia javanica f. *sumatrensis* Benl, Bot. Arch. 40: 172 (1940).

Gahnia lacera (R.Lesson ex A.Rich.) Steud., Syn. Pl. Glumac. 2: 164 (1855).
New Zealand North I. 51 NZN. Hemicr. or rhizome geophyte.
**Lampocarya lacera* R.Lesson ex A.Rich. in J.S.C. Dumont d'Urville, Voy. Astrolabe: 109 (1832).

Gahnia lanaiensis O.Deg., I.Deg. & J.Kern, Blumea 12: 349 (1964).
Hawaiian Is. (Lanai). 63 HAW. Hemicr. or rhizome geophyte.

Gahnia lanigera (R.Br.) Benth., Fl. Austral. 7: 415 (1878).
S. & SE. Australia. 50 NSW SOA VIC WAU. Hemicr. or rhizome geophyte.
**Cladium lanigerum* R.Br., Prodr.: 237 (1810). *Mariscus laniger* (R.Br.) Kuntze, Revis. Gen. Pl. 2: 755 (1891).
Cladium medium R.Br., Prodr.: 237 (1810).

Gahnia marquisensis F.Br., Bernice P. Bishop Mus. Bull. 84: 112 (1931).
Marquesas. 61 MRQ. Hemicr. or rhizome geophyte.

Gahnia melanocarpa R.Br., Prodr.: 239 (1810). *Cladium melanocarpum* (R.Br.) F.Muell., Fragm. 9: 13 (1875). *Mariscus melanocarpus* (R.Br.) Kuntze, Revis. Gen. Pl. 2: 755 (1891).
SE. Queensland to Victoria. 50 NSW QLD VIC. Hemicr. or rhizome geophyte.

Gahnia microcarpa Guillaumin, Bull. Mus. Natl. Hist. Nat., II, 21: 379 (1949).
New Caledonia. 60 NWC. Hemicr. or rhizome geophyte.

Gahnia microstachya Benth., Fl. Austral. 7: 414 (1878). *Cladium microstachyum* (Benth.) F.Muell., Syst. Census Austral. Pl. 1: 129 (1882). *Mariscus microstachyus* (Benth.) Kuntze, Revis. Gen. Pl. 2: 755 (1891).
SE. Australia. 50 NSW TAS VIC. Hemicr. or rhizome geophyte.
Gahnia fitzgeraldii Rodway, Pap. & Proc. Roy. Soc. Tasmania 1892: 82 (1893). *Gahnia microstachya* var. *fitzgeraldii* (Rodway) Kük., Feddes Repert. Spec. Nov. Regni Veg. 52: 111 (1943).

Gahnia novocaledonensis Benl, Bot. Arch. 40: 225 (1940).
New Caledonia. 60 NWC. Hemicr. or rhizome geophyte.

Gahnia pauciflora Kirk, Trans. & Proc. New Zealand Inst. 1: 94 (1868).
New Zealand. 51 NZN NZS. Hemicr. or rhizome geophyte.
Gahnia hectori Kirk, Trans. & Proc. New Zealand Inst. 9: 551 (1877).
Gahnia exigua Colenso, Trans. & Proc. New Zealand Inst. 18: 279 (1885 publ. 1886).
Gahnia multiglumis Colenso, Trans. & Proc. New Zealand Inst. 18: 280 (1885 publ. 1886).
Gahnia parviflora Colenso, Trans. & Proc. New Zealand Inst. 18: 280 (1885 publ. 1886).
Gahnia scaberula Colenso, Trans. & Proc. New Zealand Inst. 18: 278 (1885 publ. 1886).

Gahnia procera J.R.Forst. & G.Forst., Char. Gen. Pl.: 26 (1775). *Mariscus procerus* (J.R.Forst. & G.Forst.) Kuntze, Revis. Gen. Pl. 2: 755 (1891), nom. illeg.
New Zealand. 51 NZN NZS. Hemicr. or rhizome geophyte.

Gahnia psittacorum Labill., Nov. Holl. Pl. 1: 89 (1805). *Cladium psittacorum* (Labill.) F.Muell., Fragm. 9: 13 (1875). *Mariscus psittacorum* (Labill.) Kuntze, Revis. Gen. Pl. 2: 755 (1891). *Gahnia psittacorum* var. *typica* Domin, Biblioth. Bot. 85: 482 (1915), nom. inval.
Tasmania. 50 TAS. Hemicr. or rhizome geophyte.
Psittacoschoenus erythrocarpus F.Muell. ex Boeck., Linnaea 38: 341 (1874).

Gahnia radula (R.Br.) Benth., Fl. Austral. 7: 417 (1878).
SE. Australia. 50 NSW SOA TAS VIC. Hemicr. or rhizome geophyte.
**Cladium radula* R.Br., Prodr.: 237 (1810). *Mariscus radula* (R.Br.) Kuntze, Revis. Gen. Pl. 2: 755 (1891).
Gahnia melanocarpa Hook.f., Fl. Tasman. 2: 97 (1858).

Gahnia rigida Kirk, Trans. & Proc. New Zealand Inst. 9: 551 (1877).
New Zealand. 51 NZN NZS. Hemicr. or rhizome geophyte.
Gahnia robusta Kirk, Trans. & Proc. New Zealand Inst. 26: 261 (1893 publ. 1894).

Gahnia schoenoides G.Forst., Fl. Ins. Austr.: 26 (1786). *Lampocarya schoenoides* (G.Forst.) R.Br., Prodr.: 238 (1810). *Mariscus schoenodes* (G.Forst.) Kuntze, Revis. Gen. Pl. 2: 755 (1891).
Society Is. 61 SCI. Hemicr. or rhizome geophyte.
Schoenus spilocarpus Sol. ex Seem., Fl. Vit.: 316 (1868).

Gahnia sclerioides K.L.Wilson, Nuytsia 11: 275 (1997).
SW. Australia. 50 WAU. Hemicr. or rhizome geophyte.

Gahnia setifolia Hook.f., Fl. New Zeal. 1: 278 (1853). *Mariscus setifolius* (Hook.f.) Kuntze, Revis. Gen. Pl. 2: 755 (1891).
New Zealand. 51 NZN NZS. Hemicr. or rhizome geophyte.
Lampocarya setifolia R.Lesson & A.Rich. ex Hook.f., Handb. N. Zeal. Fl. 1: 306 (1864), nom. inval.

Gahnia sieberiana Kunth, Enum. Pl. 2: 332 (1837).
Sulawesi to E. & SE. Australia. 42 SUL 43 NWG SOL 50 NSW QLD SOA TAS VIC 60 NWC. Hemicr. or rhizome geophyte.
Gahnia tetragonocarpa Boeck., Enum. Pl. 2: 347 (1837). *Cladium tetragonocarpum* (Boeck.) F.Muell., Syst. Census Austral. Pl. 1: 129 (1882). *Mariscus tetragonocarpus* (Boeck.) Kuntze, Revis. Gen. Pl. 2: 755 (1891).
Gahnia arbuscula Domin, Biblioth. Bot. 85: 482 (1915). *Gahnia sieberiana* var. *arbuscula* (Domin) Kük., Feddes Repert. Spec. Nov. Regni Veg. 52: 81 (1943).
Gahnia radula var. *oreogenes* Domin, Biblioth. Bot. 85: 481 (1915).
Gahnia breviaristata Benl, Repert. Spec. Nov. Regni Veg. 44: 199 (1938).
Gahnia clarkei Benl, Repert. Spec. Nov. Regni Veg. 44: 196 (1938).

Gahnia sinuosa J.Raynal, Adansonia, n.s., 15: 190 (1975).
New Caledonia. 60 NWC. Hemicr. or rhizome geophyte.

Gahnia subaequiglumis S.T.Blake, Contr. Queensland Herb. 8: 37 (1969).
SE. Queensland to Victoria. 50 NSW QLD VIC. Hemicr. or rhizome geophyte.

Gahnia trifida Labill., Nov. Holl. Pl. 1: 89 (1805). *Cladium trifidum* (Labill.) F.Muell., Pap. & Proc. Roy. Soc. Tasmania 1878: 117 (1878). *Mariscus trifidus* (Labill.) Kuntze, Revis. Gen. Pl. 2: 755 (1891).
S. Australia. 50 SOA TAS VIC WAU. Hemicr. or rhizome geophyte.
Lampocarya hexandra R.Br., Prodr.: 238 (1810).
Gahnia trifida var. *effusa* Benth., Fl. Austral. 7: 413 (1878).

Gahnia tristis Nees ex Hook. & Arn., Bot. Beechey Voy.: 228 (1837). *Mariscus tristis* (Nees ex Hook. & Arn.) Kuntze, Revis. Gen. Pl. 2: 755 (1891).
S. China to Nansei-shoto and W. Malesia. 36 CHS 38 NNS TAI 41 THA VIE 42 BOR MLY SUM. Hemicr. or rhizome geophyte.
Gahnia wichurae Boeck., Linnaea 38: 348 (1874). *Mariscus wichurae* (Boeck.) Kuntze, Revis. Gen. Pl. 2: 755 (1891).
Gahnia boniana Boeck., Allg. Bot. Z. Syst. 2: 141 (1896).
Gahnia stricta Boeck., Allg. Bot. Z. Syst. 2: 142 (1896).

Gahnia vitiensis Rendle, J. Linn. Soc., Bot. 39: 179 (1909).
Fiji, Hawaiian Is. 60 FIJ 63 HAW. Hemicr. or rhizome geophyte.

subsp. ***kauaiensis*** (Benl) T.Koyama, Bishop Mus. Occas. Pap. 29: 126 (1989).
Hawaiian Is. (Kauai). 63 HAW. Hemicr. or rhizome geophyte.
Gahnia javanica var. *ellipsoidea* Kük., Repert. Spec. Nov. Regni Veg. 16: 433 (1920).
**Gahnia kauaiensis* Benl, Bot. Arch. 40: 190 (1940).

subsp. ***vitiensis***
Fiji. 60 FIJ. Hemicr. or rhizome geophyte.

Gahnia xanthocarpa (Hook.f.) Hook.f., Handb. N. Zeal. Fl. 1: 306 (1864).
New Zealand North I. 51 NZN NZS. Hemicr. or rhizome geophyte.
**Lampocarya xanthocarpa* Hook.f., Fl. Nov.-Zel. 1: 278 (1853). *Cladium xanthocarpum* (Hook.f.) F.Muell., Fragm. 9: 13 (1875). *Mariscus xanthocarpus* (Hook.f.) Kuntze, Revis. Gen. Pl. 2: 755 (1891).
Gahnia ebenocarpa Hook.f. ex Kirk, Trans. & Proc. New Zealand Inst. 1, ed. 2: 94 (1875). *Gahnia setifolia* var. *xanthocarpa* (F.Muell.) Kük., Feddes Repert. Spec. Nov. Regni Veg. 52: 70 (1943).

***Synonyms*:**

Gahnia affinis (Brongn.) Steud. = ***Morelotia affinis*** (Brongn.) S.T.Blake
Gahnia arbuscula Domin = ***Gahnia sieberiana*** Kunth
Gahnia arenaria Hook.f. = ***Morelotia affinis*** (Brongn.) S.T.Blake
Gahnia articulata (R.Br.) F.Muell. = ***Machaerina articulata*** (R.Br.) T.Koyama
Gahnia aspera var. *angustinus* Kük. = ***Gahnia aspera*** (R.Br.) Spreng. subsp. ***aspera***
Gahnia aspera f. *distans* Domin = ***Gahnia aspera*** (R.Br.) Spreng. subsp. ***aspera***
Gahnia boniana Boeck. = ***Gahnia tristis*** Nees ex Hook. & Arn.
Gahnia boninsimae Maxim. = ***Gahnia aspera*** (R.Br.) Spreng. subsp. ***aspera***
Gahnia breviaristata Benl = ***Gahnia sieberiana*** Kunth
Gahnia castanea Ridl. = ***Gahnia javanica*** Moritzi
Gahnia clarkei Benl = ***Gahnia sieberiana*** Kunth
Gahnia congesta Boeck. = ***Gahnia aspera*** subsp. ***globosa*** (H.Mann) J.Kern
Gahnia ebenocarpa Hook.f. ex Kirk = ***Gahnia xanthocarpa*** (Hook.f.) Hook.f.
Gahnia exigua Colenso = ***Gahnia pauciflora*** Kirk
Gahnia fitzgeraldii Rodway = ***Gahnia microstachya*** Benth.
Gahnia gahniiformis (Gaudich.) A.Heller = ***Morelotia gahniiformis*** Gaudich.
Gahnia gaudichaudii (Brongn.) Steud. = ***Morelotia gahniiformis*** Gaudich.
Gahnia globosa H.Mann = ***Gahnia aspera*** subsp. ***globosa*** (H.Mann) J.Kern
Gahnia glomerata (R.Br.) F.Muell. = ***Machaerina rubiginosa*** (Spreng.) T.Koyama
Gahnia gonyocarpa Steud. = ***Gahnia erythrocarpa*** R.Br.
Gahnia gunnii (Hook.f.) F.Muell. = ***Machaerina gunnii*** (Hook.f.) J.Kern
Gahnia hectori Kirk = ***Gahnia pauciflora*** Kirk
Gahnia insularis (Benth.) F.Muell. = ***Machaerina insularis*** (Benth.) T.Koyama
Gahnia javanica f. *borneensis* Benl = ***Gahnia javanica*** Moritzi
Gahnia javanica var. *ellipsoidea* Kük. = ***Gahnia vitiensis*** subsp. ***kauaiensis*** (Benl) T.Koyama
Gahnia javanica var. *paupercula* Kük. = ***Gahnia javanica*** Moritzi
Gahnia javanica var. *penangensis* C.B.Clarke = ***Gahnia baniensis*** Benl
Gahnia javanica var. *philippinensis* Benl = ***Gahnia javanica*** Moritzi
Gahnia javanica f. *sinensis* Benl = ***Gahnia javanica*** Moritzi
Gahnia javanica subf. *subcapitata* Kük. ex Benl = ***Gahnia javanica*** Moritzi
Gahnia javanica f. *sumatrensis* Benl = ***Gahnia javanica*** Moritzi
Gahnia juncea (R.Br.) F.Muell. = ***Machaerina juncea*** (R.Br.) T.Koyama
Gahnia kauaiensis Benl = ***Gahnia vitiensis*** subsp. ***kauaiensis*** (Benl) T.Koyama
Gahnia lanigera var. *aristata* (Benth.) Kük. = ***Gahnia aristata*** Benth.
Gahnia leptostachya Boeck. = ***Gahnia beecheyi*** H.Mann
Gahnia leucocarpa R.Br. = ***Gahnia grandis*** (Labill.) S.T.Blake
Gahnia mannii Hillebr. = ***Gahnia beecheyi*** H.Mann
Gahnia melanocarpa Hook.f. = ***Gahnia radula*** (R.Br.) Benth.
Gahnia microstachya var. *fitzgeraldii* (Rodway) Kük. = ***Gahnia microstachya*** Benth.
Gahnia mucronata Boeck. = ***Gahnia aspera*** subsp. ***globosa*** (H.Mann) J.Kern
Gahnia multiglumis Colenso = ***Gahnia pauciflora*** Kirk
Gahnia parviflora Colenso = ***Gahnia pauciflora*** Kirk
Gahnia penangensis (C.B.Clarke) Kük. = ***Gahnia baniensis*** Benl
Gahnia polyphylla Benth. = ***Gahnia drummondii*** (Steud.) K.L.Wilson
Gahnia preissii Nees = ***Gahnia decomposita*** (R.Br.) Benth.
Gahnia psilocaulon Boeck. = ***Gahnia grandis*** (Labill.) S.T.Blake
Gahnia psittacorum Sieber ex C.Presl = ***Gahnia filifolia*** (C.Presl) Kük. ex Benl
Gahnia psittacorum var. *erythrocarpa* (R.Br.) C.B. Clarke = ***Gahnia erythrocarpa*** R.Br.
Gahnia psittacorum f. *leucosperma* Benl = ***Gahnia grandis*** (Labill.) S.T.Blake
Gahnia psittacorum var. *oxylepis* Benth. = ***Gahnia grandis*** (Labill.) S.T.Blake
Gahnia psittacorum f. *psilocaulon* (Boeck.) Benl = ***Gahnia grandis*** (Labill.) S.T.Blake
Gahnia psittacorum var. *psilocaulon* (Boeck.) Kük. = ***Gahnia grandis*** (Labill.) S.T.Blake
Gahnia psittacorum var. *typica* Domin = ***Gahnia psittacorum*** Labill.
Gahnia radula var. *oreogenes* Domin = ***Gahnia sieberiana*** Kunth
Gahnia rawacensis (Kunth) Steud. = ***Gahnia aspera*** (R.Br.) Spreng. subsp. ***aspera***
Gahnia robusta Kirk = ***Gahnia rigida*** Kirk
Gahnia rodwayi F.Muell. ex Rodway = ***Gahnia graminifolia*** Rodway
Gahnia scaberula Colenso = ***Gahnia pauciflora*** Kirk
Gahnia schoenoides (R.Br.) F.Muell. = ***Machaerina acuta*** (Labill.) J.Kern
Gahnia setifolia var. *xanthocarpa* (F.Muell.) Kük. = ***Gahnia xanthocarpa*** (Hook.f.) Hook.f.
Gahnia sieberi (Schrad. ex Schult. & Schult.f.) Boeck. = ***Gahnia filifolia*** (C.Presl) Kük. ex Benl
Gahnia sieberiana var. *arbuscula* (Domin) Kük. = ***Gahnia sieberiana*** Kunth
Gahnia stokesii F.Br. = ***Gahnia aspera*** (R.Br.) Spreng. subsp. ***aspera***
Gahnia stricta Boeck. = ***Gahnia tristis*** Nees ex Hook. & Arn.
Gahnia teretifolia Domin = ***Gahnia filifolia*** (C.Presl) Kük. ex Benl
Gahnia tetragonocarpa Boeck. = ***Gahnia sieberiana*** Kunth
Gahnia trifida var. *effusa* Benth. = ***Gahnia trifida*** Labill.
Gahnia urvilleana Kunth = ***Gahnia erythrocarpa*** R.Br.
Gahnia wichurae Boeck. = ***Gahnia tristis*** Nees ex Hook. & Arn.

Galilea

Galilea Parl. = ***Cyperus*** L.
Galilea mucronata (L.) Parl. = ***Cyperus capitatus*** Vand.

Genersichia

Genersichia Heuff. = ***Carex*** L.
Genersichia obtusata (Lilj.) Heuff. = ***Carex obtusata*** Lilj.

Gussonea

Gussonea J.Presl & C.Presl = ***Fimbristylis*** Vahl
Gussonea cyperoides J.Presl & C.Presl = ***Fimbristylis fusca*** (Nees) Benth. ex C.B.Clarke
Gussonea pauciflora Brongn. = ***Fimbristylis fusca*** (Nees) Benth. ex C.B.Clarke

Gymnochaeta

Gymnochaeta Steud. = ***Schoenus*** L.
Gymnochaeta drummondii Steud. = ***Schoenus submicrostachyus*** Kük.

Gymnoschoenus

Gymnoschoenus Nees, Ann. Mag. Nat. Hist. 6: 47 (1841).
50 NSW SOA TAS VIC WAU.
2 Species

Gymnoschoenus anceps (R.Br.) C.B.Clarke, Bull. Misc. Inform. Kew, Addit. Ser. 8: 92 (1908).
SW. Western Australia. 50 WAU. Hemicr. or rhizome geophyte.
**Chaetospora anceps* R.Br., Prodr.: 233 (1810). *Schoenus anceps* (R.Br.) Poir. in J.B.A.M.de Lamarck, Encycl., Suppl. 2: 252 (1811). *Mesomelaena anceps* (R.Br.) Benth., Fl. Austral. 7: 380 (1878).
Chaetospora elongata Nees in J.G.C.Lehmann, Pl. Preiss. 2: 275 (1848). *Schoenus elongatus* (Nees) F.Muell., Fragm. 9: 30 (1975).

Gymnoschoenus sphaerocephalus (R.Br.) Hook.f., Fl. Tasman. 2: 83 (1858).
SE. Australia. 50 NSW SOA TAS VIC. Hemicr. or rhizome geophyte.
**Chaetospora sphaerocephala* R.Br., Prodr.: 233 (1810). *Schoenus sphaerocephalus* (R.Br.) Poir. in J.B.A.M.de Lamarck, Encycl., Suppl. 2: 252 (1811). *Mesomelaena sphaerocephala* (R.Br.) Benth., Fl. Austral. 7: 380 (1878).
Gymnoschoenus adustus Nees, Ann. Mag. Nat. Hist. 6: 47 (1841).
Xyris laevis Nees ex Benth., Fl. Austral. 7: 380 (1878).

Synonyms:
Gymnoschoenus adustus Nees = ***Gymnoschoenus sphaerocephalus*** (R.Br.) Hook.f.

Haloschoenus

Haloschoenus Nees = ***Rhynchospora*** Vahl
Haloschoenus albidus Nees = ***Rhynchospora albida*** (Nees) Boeck.
Haloschoenus capillaris Nees = ***Rhynchospora tenuis*** subsp. ***tenuis***
Haloschoenus capillaris var. *congestus* Nees = ***Rhynchospora riparia*** (Nees) Boeck.
Haloschoenus capillaris var. *descendens* Nees = ***Rhynchospora tenuis*** subsp. ***tenuis***
Haloschoenus capillaris var. *minor* Nees = ***Rhynchospora tenuis*** subsp. ***austrobrasiliensis*** T. Koyama
Haloschoenus capillaris var. *ustilaginosa* Nees = ***Rhynchospora tenuis*** subsp. ***tenuis***
Haloschoenus caracasanus Liebm. = ***Rhynchospora tenuis*** subsp. ***tenuis***
Haloschoenus contractus Nees = ***Rhynchospora contracta*** (Nees) J.Raynal
Haloschoenus elatior (Kunth) Nees = ***Rhynchospora stricta*** Boeck.
Haloschoenus emaciatus Nees = ***Rhynchospora emaciata*** (Nees) Boeck.
Haloschoenus hispidulus (Vahl) Nees = ***Rhynchospora trichodes*** C.B.Clarke
Haloschoenus nardifolius (Kunth) Nees = ***Rhynchospora nardifolia*** (Kunth) Boeck.
Haloschoenus pygmaeus Nees = ***Rhynchospora contracta*** (Nees) J.Raynal
Haloschoenus rigidus Nees ex Kunth = ***Rhynchospora polyphylla*** (Vahl) Vahl
Haloschoenus riparius Nees = ***Rhynchospora riparia*** (Nees) Boeck.
Haloschoenus sabulosus Nees = ***Rhynchospora tenuis*** subsp. ***tenuis***
Haloschoenus sparsus Nees = ***Rhynchospora rariflora*** (Michx.) Elliott
Haloschoenus stoloniferus Nees = ***Rhynchospora stolonifera*** (Nees) Boeck.
Haloschoenus tenellus Nees = ***Rhynchospora tenella*** (Nees) Boeck.

Haplostylis

Haplostylis Nees = ***Rhynchospora*** Vahl
Haplostylis albiceps (Kunth) Nees = ***Rhynchospora albiceps*** Kunth
Haplostylis armeriifolia Nees = ***Rhynchospora armerioides*** J.Presl & C.Presl
Haplostylis bahiensis Nees = ***Rhynchospora barbata*** (Vahl) Kunth
Haplostylis barbata (Vahl) Nees = ***Rhynchospora barbata*** (Vahl) Kunth
Haplostylis barbata var. *humboldtii* (Nees) Nees = ***Rhynchospora barbata*** (Vahl) Kunth
Haplostylis elatior Nees = ***Rhynchospora elatior*** Kunth
Haplostylis elatior var. *hirsuta* Nees = ***Rhynchospora elatior*** Kunth
Haplostylis hirta Nees = ***Rhynchospora hirta*** (Nees) Boeck.
Haplostylis humboldtii Nees = ***Rhynchospora barbata*** (Vahl) Kunth
Haplostylis mexicana Liebm. = ***Rhynchospora mexicana*** (Liebm.) Steud.
Haplostylis meyenii Nees = ***Rhynchospora rubra*** (Lour.) Makino subsp. ***rubra***
Haplostylis minor Nees = ***Rhynchospora leucostachys*** Boeck.
Haplostylis wightianus Nees = ***Rhynchospora wightiana*** (Nees) Steud.

Hedychloa

Hedychloa Raf. = ***Kyllinga*** Rottb.
Hedychloa fragrans Raf. = ***Kyllinga pumila*** Michx.

Heleocharis

Heleocharis T.Lestib. = ***Eleocharis*** R.Br.

Heleogiton

Heleogiton Schult. = ***Schoenoplectus*** (Rchb.) Palla
Heleogiton compressa (L.) Peterm. = ***Blysmus compressus*** (L.) Panz. ex Link
Heleogiton fluitans (L.) Link ex Rchb. = ***Isolepis fluitans*** (L.) R.Br. var. ***fluitans***
Heleogiton glaucum Rchb. = ***Schoenoplectus lacustris*** (L.) Palla subsp. ***lacustris***
Heleogiton lejeunei (Weihe) Rchb. = ***Schoenoplectus triqueter*** (L.) Palla

Heleogiton litorale (Schrad.) Rchb. = ***Schoenoplectus litoralis*** (Schrad.) Palla
Heleogiton pungens (Vahl) Rchb. = ***Schoenoplectus pungens*** (Vahl) Palla
Heleogiton rothii (Hoppe) Schur = ***Schoenoplectus pungens*** (Vahl) Palla var. ***pungens***
Heleogiton tabernaemontani (C.C.Gmel.) Peterm. = ***Schoenoplectus tabernaemontani*** (C.C.Gmel.) Palla
Heleogiton ×*trigonum* (Sm.) Rchb. = ***Schoenoplectus* × *carinatus*** (Sm.) Palla nothosubsp. ***carinatus***
Heleogiton triquetrum (L.) Rchb. = ***Schoenoplectus triqueter*** (L.) Palla

Heleonastes

Heleonastes Ehrh. = ***Carex*** L.

Heleophylax

Heleophylax P.Beauv. ex Lestib. = ***Schoenoplectus*** (Rchb.) Palla
Heleophylax americanus (Pers.) Schinz & Thell. = ***Schoenoplectus americanus*** (Pers.) Volkart
Heleophylax ×*carinatus* (Sm.) Schinz & Thell. = ***Schoenoplectus* × *carinatus*** (Sm.) Palla
Heleophylax lacustris (L.) Schinz & Thell. = ***Schoenoplectus lacustris*** (L.) Palla
Heleophylax mucronatus (L.) Schinz & Thell. = ***Schoenoplectus mucronatus*** (L.) Palla
Heleophylax supinus (L.) Schinz & Thell. = ***Schoenoplectiella supina*** (L.) Lye
Heleophylax tabernaemontani (C.C.Gmel.) Schinz & Thell. = ***Schoenoplectus tabernaemontani*** (C.C.Gmel.) Palla
Heleophylax triquetrus (L.) Schinz & Thell. = ***Schoenoplectus triqueter*** (L.) Palla

Heliocharis

Heliocharis Lindl. = ***Eleocharis*** R.Br.

Hellmuthia

Hellmuthia Steud., Syn. Pl. Glumac. 2: 90 (1855).
Cape Prov. 27 CPP.
1 Species

Hellmuthia membranacea (Thunb.) R.W.Haines & Lye, Bot. Not. 129: 66 (1976).
Cape Prov. 27 CPP. Hemicr. or rhizome geophyte.
**Scirpus membranaceus* Thunb., Prodr. Pl. Cap.: 17 (1794). *Isolepis membranacea* (Thunb.) Nees, Linnaea 9: 291 (1834). *Isolepis membranacea* (Thunb.) Nees, Linnaea 10: 159 (1836). *Ficinia membranacea* (Thunb.) Kunth, Enum. Pl. 2: 252 (1837).
Hellmuthia restioides Steud., Syn. Pl. Glumac. 2: 90 (1855).
Ficinia canaliculata H.Pfeiff., Herbarium 54: 34 (1920).

***Synonyms*:**
Hellmuthia restioides Steud. = ***Hellmuthia membranacea*** (Thunb.) R.W.Haines & Lye

Helonema

Helonema Suess. = ***Eleocharis*** R.Br.
Helonema estrellensis Suess. = ***Eleocharis minima*** Kunth var. ***minima***

Helothrix

Helothrix Nees = ***Schoenus*** L.
Helothrix axillaris (R.Br.) Palla = ***Schoenus maschalinus*** Roem. & Schult.
Helothrix imberbis (R.Br.) Palla = ***Schoenus apogon*** Roem. & Schult.
Helothrix paludosa (R.Br.) Palla = ***Schoenus paludosus*** (R.Br.) Roem. & Schult.
Helothrix philippinensis Palla = ***Schoenus maschalinus*** Roem. & Schult.
Helothrix pusilla Nees = ***Schoenus maschalinus*** Roem. & Schult.

Hemicarex

Hemicarex Benth. = ***Kobresia*** Willd.
Hemicarex curvirostris C.B.Clarke = ***Kobresia curvirostris*** (C.B.Clarke) C.B.Clarke
Hemicarex ecklonii (Nees) Benth. = ***Schoenoxiphium ecklonii*** Nees
Hemicarex filicina C.B.Clarke = ***Kobresia esenbeckii*** (Kunth) Noltie var. ***esenbeckii***
Hemicarex hookeri (Boeck.) Benth. = ***Kobresia esenbeckii*** (Kunth) Noltie var. ***esenbeckii***
Hemicarex laxa Benth. = ***Kobresia laxa*** Nees
Hemicarex linearis (Boott) Benth. = ***Kobresia nepalensis*** (Nees) Kük. var. ***nepalensis***
Hemicarex meyeriana Benth. = ***Schoenoxiphium lanceum*** (Thunb.) Kük.
Hemicarex pygmaea C.B.Clarke = ***Kobresia pygmaea*** (C.B.Clarke) C.B.Clarke
Hemicarex sickmanniana (Kunth) Benth. = ***Schoenoxiphium lanceum*** (Thunb.) Kük.
Hemicarex thunbergii (Nees) Benth. = ***Schoenoxiphium thunbergii*** Nees
Hemicarex trinervis (Boeck.) Benth. & Hook.f. ex C.B.Clarke = ***Kobresia esenbeckii*** (Kunth) Noltie var. ***esenbeckii***

Hemicarpha

Hemicarpha Nees & Arn. = ***Lipocarpha*** R.Br.
Hemicarpha aristulata (Coville) A.Nelson = ***Lipocarpha aristulata*** (Coville) G.C.Tucker
Hemicarpha axillaris (Lam.) B.D.Jacks. = ***Rhynchospora glomerata*** (L.) Vahl
Hemicarpha caespitula (Liebm.) Palla = ***Lipocarpha micrantha*** (Vahl) G.C.Tucker
Hemicarpha drummondii Nees = ***Lipocarpha drummondii*** (Nees) G.C.Tucker
Hemicarpha intermedia Piper = ***Lipocarpha aristulata*** (Coville) G.C.Tucker
Hemicarpha isolepis Nees = ***Lipocarpha hemisphaerica*** (Roth) Goetgh.
Hemicarpha micrantha (Vahl) Britton = ***Lipocarpha micrantha*** (Vahl) G.C.Tucker
Hemicarpha micrantha var. *aristulata* Coville = ***Lipocarpha aristulata*** (Coville) G.C.Tucker
Hemicarpha micrantha var. *drummondii* (Nees) Frieland = ***Lipocarpha drummondii*** (Nees) G.C.Tucker
Hemicarpha micrantha var. *humboldtii* (Boeck.) Beetle = ***Lipocarpha micrantha*** (Vahl) G.C.Tucker
Hemicarpha micrantha var. *minor* (Schrad.) Friedl. = ***Lipocarpha micrantha*** (Vahl) G.C.Tucker
Hemicarpha nuda Boeck. = ***Eleocharis acuta*** R.Br.
Hemicarpha occidentalis A.Gray = ***Lipocarpha occidentalis*** (A.Gray) G.C.Tucker
Hemicarpha schomburgkii Friedl. = ***Lipocarpha schomburgkii*** (Friedl.) G.C.Tucker
Hemicarpha schraderi Kunth = ***Lipocarpha hemisphaerica*** (Roth) Goetgh.

Hemicarpha senegalensis Steud. = ***Lipocarpha hemisphaerica*** (Roth) Goetgh.
Hemicarpha subsquarrosa (Muhl.) Nees = ***Lipocarpha micrantha*** (Vahl) G.C.Tucker
Hemicarpha subsquarrosa var. *drummondii* (Nees) A.Gray = ***Lipocarpha drummondii*** (Nees) G.C.Tucker

Hemichlaena

Hemichlaena Schrad. = ***Ficinia*** Schrad.
Hemichlaena angustifolia Schrad. = ***Ficinia angustifolia*** (Schrad.) C.B.Clarke
Hemichlaena angustifolia var. *fascicularis* Nees = ***Ficinia polystachya*** Levyns
Hemichlaena angustifolia var. *polystachya* Kunth = ***Ficinia polystachya*** Levyns
Hemichlaena bulbosa Hochst. ex A.Rich. = ***Cyperus blysmoides*** Hochst. ex C.B.Clarke
Hemichlaena capillifolia Schrad. = ***Ficinia capillifolia*** (Schrad.) C.B.Clarke
Hemichlaena fascicularis (Nees) Steud. = ***Ficinia polystachya*** Levyns
Hemichlaena longifolia Nees = ***Ficinia angustifolia*** (Schrad.) C.B.Clarke

Heuffelia

Heuffelia Opiz = ***Carex*** L.
Heuffelia pyrenaica (Wahlenb.) Opiz = ***Carex pyrenaica*** Wahlenb.

Hexalepis

Hexalepis scabrifolia Boeck. = ***Gahnia aspera*** (R.Br.) Spreng. subsp. ***aspera***

Holmia

Holmia Börner = ***Kobresia*** Willd.
Holmia seticulmis (Boeck.) Fedde & J.Schust. = ***Kobresia esenbeckii*** var. ***fissiglumis*** (C.B.Clarke) Noltie

Holoschoenus

Holoschoenus Link = ***Scirpoides*** Ség.
Holoschoenus atrovirens (Willd.) Link = ***Scirpus atrovirens*** Willd.
Holoschoenus australis (L.) Rchb. = ***Scirpoides holoschoenus*** (L.) Soják subsp. ***holoschoenus***
Holoschoenus brunneus (Muhl.) Steud. = ***Scirpus polyphyllus*** Vahl
Holoschoenus diphyllus Montandon = ***Scirpoides holoschoenus*** (L.) Soják subsp. ***holoschoenus***
Holoschoenus elatior Nees = ***Rhynchospora cajennensis*** Boeck.
Holoschoenus exserens Rchb. = ***Scirpoides holoschoenus*** (L.) Soják subsp. ***holoschoenus***
Holoschoenus filiformis Rchb. = ***Scirpoides holoschoenus*** (L.) Soják subsp. ***holoschoenus***
Holoschoenus globifer (L.f.) Rchb. = ***Scirpoides holoschoenus*** subsp. ***globiferus*** (L.f.) Soják
Holoschoenus globiferus (L.f.) A.Dietr. = ***Scirpoides holoschoenus*** subsp. ***globiferus*** (L.f.) Soják
Holoschoenus globuliferus (Boiss.) Grossh. = ***Scirpoides holoschoenus*** (L.) Soják subsp. ***holoschoenus***
Holoschoenus incurvatus (Roxb.) A.Dietr. = ***Schoenoplectiella articulata*** (L.) Lye
Holoschoenus laxus Opiz = ***Scirpoides holoschoenus*** (L.) Soják subsp. ***holoschoenus***
Holoschoenus linnaei Rchb. = ***Scirpoides holoschoenus*** (L.) Soják subsp. ***holoschoenus***
Holoschoenus melanocarpus Boeck. = [79]
Holoschoenus mexicanus (C.B.Clarke ex Britton) Palla = ***Karinia mexicana*** (C.B.Clarke ex Britton) Reznicek & McVaugh
Holoschoenus nodosus (Rottb.) A.Dietr. = ***Ficinia nodosa*** (Rottb.) Goetgh., Muasya & D.A.Simpson
Holoschoenus panormitanus (Parl.) Nyman = ***Scirpoides holoschoenus*** (L.) Soják subsp. ***holoschoenus***
Holoschoenus romanus (L.) Fritsch = ***Scirpoides holoschoenus*** (L.) Soják subsp. ***holoschoenus***
Holoschoenus romanus var. *australis* (L.) Soó = ***Scirpoides holoschoenus*** (L.) Soják subsp. ***holoschoenus***
Holoschoenus romanus f. *filiformis* (Rchb.) Soó = ***Scirpoides holoschoenus*** (L.) Soják subsp. ***holoschoenus***
Holoschoenus romanus f. *monocephalus* (Waisb.) Soó = ***Scirpoides holoschoenus*** (L.) Soják subsp. ***holoschoenus***
Holoschoenus subarticulatus (Roxb.) A.Dietr. = ***Schoenoplectiella articulata*** (L.) Lye
Holoschoenus thunbergii (Schrad.) A.Dietr. = ***Scirpoides holoschoenus*** subsp. ***thunbergii*** (Schrad.) Soják
Holoschoenus tuberculatus Rchb. = ***Scirpoides holoschoenus*** (L.) Soják subsp. ***holoschoenus***
Holoschoenus vulgaris Link = ***Scirpoides holoschoenus*** (L.) Soják
Holoschoenus vulgaris subvar. *albovittatus* Rchb. ex Nyman = ***Scirpoides holoschoenus*** (L.) Soják subsp. ***holoschoenus***
Holoschoenus vulgaris var. *australis* (L.) Nyman = ***Scirpoides holoschoenus*** (L.) Soják subsp. ***holoschoenus***
Holoschoenus vulgaris var. *filiformis* (Rchb.) Nyman = ***Scirpoides holoschoenus*** (L.) Soják subsp. ***holoschoenus***
Holoschoenus vulgaris subsp. *panormitanus* (Parl.) Nyman = ***Scirpoides holoschoenus*** (L.) Soják subsp. ***holoschoenus***
Holoschoenus vulgaris var. *romanus* (L.) Vis. = ***Scirpoides holoschoenus*** (L.) Soják subsp. ***holoschoenus***
Holoschoenus vulgaris var. *typicus* Halácsy = ***Scirpoides holoschoenus*** (L.) Soják

Unplaced Names:
Holoschoenus melanocarpus Boeck., Flora 40: 37 (1857). = [79]

Homalostachys

Homalostachys Boeck. = ***Carex*** L.
Homalostachys sinensis Boeck. = ***Carex moupinensis*** Franch.

Hoppia

Hoppia Nees = ***Bisboeckelera*** Kuntze
Hoppia angustifolia Boeck. = ***Bisboeckelera longifolia*** (Rudge) Kuntze
Hoppia berroi C.B.Clarke = ***Cyperus berroi*** (C.B.Clarke) Barros
Hoppia bicolor C.B.Clarke = ***Becquerelia clarkei*** T.Koyama
Hoppia irrigua Nees = ***Bisboeckelera irrigua*** (Nees) Kuntze
Hoppia microcephala Boeck. = ***Bisboeckelera microcephala*** (Boeck.) T.Koyama

Hydroschoenus

Hydroschoenus Zoll. & Moritzi = ***Cyperus*** L.
Hydroschoenus kyllingioides Zoll. & Moritzi = ***Cyperus cephalotes*** Vahl

Hygrocharis

Hygrocharis Nees = ***Rhynchospora*** Vahl

Hymenochaeta

Hymenochaeta P.Beauv. ex Lestib. = ***Actinoscirpus*** (Ohwi) R.W.Haines & Lye
Hymenochaeta grossa (L.f.) Nees = ***Actinoscirpus grossus*** (L.f.) Goetgh. & D.A.Simpson
Hymenochaeta haematodes Nees & Meyen = ***Actinoscirpus grossus*** (L.f.) Goetgh. & D.A.Simpson var. ***grossus***
Hymenochaeta juncoides (Roxb.) Nakai = ***Schoenoplectiella juncoides*** (Roxb.) Lye
Hymenochaeta lacustris (L.) Nakai = ***Schoenoplectus lacustris*** (L.) Palla
Hymenochaeta makinoi Nakai = ***Schoenoplectus lacustris*** (L.) Palla subsp. ***lacustris***
Hymenochaeta maxima (Roxb.) Nees = ***Actinoscirpus grossus*** (L.f.) Goetgh. & D.A.Simpson var. ***grossus***
Hymenochaeta okuyamae (Ohwi) Nakai = ***Schoenoplectus komarovii*** (Roshev.) Soják
Hymenochaeta preslii (A.Dietr.) Nakai = ***Schoenoplectus mucronatus*** (L.) Palla subsp. ***mucronatus***
Hymenochaeta tabernaemontani (C.C.Gmel.) Nakai = ***Schoenoplectus tabernaemontani*** (C.C.Gmel.) Palla
Hymenochaeta triquetra (L.) Nakai = ***Schoenoplectus triqueter*** (L.) Palla
Hymenochaeta wallichii (Nees) Nakai = ***Schoenoplectiella wallichii*** (Nees) Lye

Hymenolytrum

Hymenolytrum Schrad. ex Nees = ***Scleria*** P.J.Bergius
Hymenolytrum comosum Nees = ***Scleria comosa*** (Nees) Steud.
Hymenolytrum cyperinum (Kunth) Nees = ***Scleria cyperina*** Kunth
Hymenolytrum martii Nees = ***Scleria martii*** (Nees) Steud.
Hymenolytrum poeppigii Nees = ***Scleria poeppigii*** (Nees) Steud.
Hymenolytrum silvestris Schrad. ex Nees = ***Scleria stipularis*** Nees

Hypaelyptum

Hypaelyptum Vahl = ***Lipocarpha*** R.Br.
Hypaelyptum albiceps (Ridl.) K.Schum. = ***Lipocarpha albiceps*** Ridl.
Hypaelyptum argenteum Vahl = ***Lipocarpha chinensis*** (Osbeck) J.Kern
Hypaelyptum filiforme Vahl = ***Lipocarpha filiformis*** (Vahl) Kunth
Hypaelyptum microcephalum R.Br. = ***Lipocarpha microcephala*** (R.Br.) Kunth
Hypaelyptum nemorum (Vahl) P.Beauv. = ***Hypolytrum nemorum*** (Vahl) Spreng.
Hypaelyptum pulcherrimum (Ridl.) K.Schum. = ***Lipocarpha nana*** (A.Rich.) Cherm.
Hypaelyptum pungens Vahl = ***Hypolytrum pulchrum*** (Rudge) H.Pfeiff.
Hypaelyptum senegalense (Lam.) K.Schum. = ***Lipocarpha chinensis*** (Osbeck) J.Kern
Hypaelyptum sphacelatum Vahl = ***Lipocarpha gracilis*** (Rich. ex Pers.) Nees

Hypolepis

Hypolepis P.Beauv. ex Lestib. = ***Ficinia*** Schrad.
Hypolepis atrata (Schrad.) Nees = ***Ficinia indica*** (Lam.) H.Pfeiff.
Hypolepis capitellum (Thunb.) Nees = ***Ficinia capitella*** (Thunb.) Nees
Hypolepis composita Nees = ***Ficinia laciniata*** (Thunb.) Nees
Hypolepis nigrescens (Schrad.) Nees = ***Ficinia nigrescens*** (Schrad.) J.Raynal
Hypolepis nigrescens var. *filiformis* Nees = ***Ficinia nigrescens*** (Schrad.) J.Raynal
Hypolepis scariosa (Thunb.) Nees = ***Ficinia deusta*** (P.J.Bergius) Levyns

Hypolytrum

Hypolytrum Rich. ex Pers., Syn. Pl. 1: 70 (1805).
Trop. 22 GHA GUI IVO LBR NGA SEN SIE 23 CAB CAF CMN CON EQG GAB GGI ZAI 25 TAN UGA 26 ANG 29 MAU MDG REU 36 CHH 38 TAI 40 ASS EHM IND SRL 41 AND THA VIE 42 BOR JAW MLY MOL PHI SUL SUM 43 BIS NWG SOL 50 NTA QLD 60 FIJ 62 CRL 79 MXG MXT 80 BLZ COS CPI GUA HON NIC PAN 81 TRT 82 FRG GUY SUR VEN 83 CLM ECU PER 84 BZC BZE BZL BZN BZS.
58 Species
Beera P.Beauv. in T.G.Lestiboudois, Essai Cypér.: 43 (1819).
Tunga Roxb., Fl. Ind. 1: 187 (1820).
Albikia J.Presl & C.Presl, Reliq. Haenk. 1: 184 (1828).

Hypolytrum africanum Nees ex Steud., Syn. Pl. Glumac. 2: 132 (1855).
W. Trop. Africa. 22 GUI LBR SIE. Hemicr. or rhizome geophyte.
Hypolytrum longiscaposum C.B.Clarke in D.Oliver & auct. suc. (eds.), Fl. Trop. Afr. 8: 489 (1902).
Hypolytrum attiense A.Chev. ex Hutch. & Dalziel, Fl. W. Trop. Afr. 2: 472 (1936), pro syn.

Hypolytrum amorimii M.Alves & W.W.Thomas, Brittonia 54: 124 (2002).
Brazil (Espírito Santo). 84 BZL. Hemicr. or rhizome geophyte.

Hypolytrum amplum Kunth, Enum. Pl. 2: 272 (1837).
Cocos I., Venezuela to N. Brazil. 80 CPI 82 FRG GUY SUR VEN 83 PER 84 BZN. Hemicr. or rhizome geophyte.
Hypolytrum costatum Hochst. ex Steud., Syn. Pl. Glumac. 2: 183 (1855).
Hypolytrum amplissimum M.Alves & W.W.Thomas, Feddes Repert. 113: 261 (2002).

Hypolytrum angolense Nelmes, Kew Bull. 10: 72 (1955).
Angola. 26 ANG. Hemicr. or rhizome geophyte.

Hypolytrum bahiense M.Alves & W.W.Thomas, Brittonia 54: 126 (2002).
Brazil (Bahia). 84 BZE. Hemicr. or rhizome geophyte.

Hypolytrum balakrishnanii Noot., Blumea 25: 319 (1979).
Andaman Is. 41 AND. Hemicr. or rhizome geophyte.

Hypolytrum bullatum C.B.Clarke, Kew Bull., Addit. Ser. 8: 51 (1908).
NE. Brazil. 84 BZE. Hemicr. or rhizome geophyte.
Calyptrocarya bullata C.B.Clarke ex H.Pfeiff., Repert. Spec. Nov. Regni Veg. 18: 382 (1922), pro syn.

Hypolytrum cacuminum Nelmes, Kew Bull. 10: 74 (1955).
W. Trop. Africa. 22 GUI IVO LBR. Hemicr. or rhizome geophyte.

Hypolytrum capitulatum Suringar ex C.B.Clarke, Kew Bull., Addit. Ser. 8: 51 (1908).
W. Borneo. 42 BOR. Hemicr. or rhizome geophyte.

Hypolytrum chevalieri Nelmes, Kew Bull. 10: 69 (1955).
Gabon. 23 GAB. Hemicr. or rhizome geophyte.

Hypolytrum compactum Nees & Meyen ex Kunth, Enum. Pl. 2: 271 (1837).
Andaman Is. to N. Queensland. 41 AND VIE 42 BOR MOL PHI SUL 43 NWG 50 QLD. Hemicr. or rhizome geophyte.
Hypolytrum xerocarpum C.B.Clarke, Kew Bull., Addit. Ser. 8: 52 (1908).
Mapania thoreliana E.G.Camus, Notul. Syst. (Paris) 1: 250 (1910).

Hypolytrum dissitiflorum Steud., Syn. Pl. Glumac. 2: 132 (1855).
Caroline Is. 62 CRL. Hemicr. or rhizome geophyte.
Hypolytrum oligostachyum K.Schum. & Lauterb., Fl. Schutzgeb. Südsee: 190 (1900).

Hypolytrum elegans (E.G.Camus) Uittien, Recueil Trav. Bot. Néerl. 33: 155 (1936).
N. Vietnam. 41 VIE. Hemicr. or rhizome geophyte.
**Mapania elegans* E.G.Camus, Notul. Syst. (Paris) 1: 249 (1910).

Hypolytrum espiritosantense M.Alves & W.W.Thomas, Feddes Repert. 113: 263 (2002).
Brazil (Espírito Santo). 84 BZL. Hemicr. or rhizome geophyte.

Hypolytrum flavinux (T.Koyama) D.A.Simpson, Rev. Gen. Mapania: 169 (1992).
Caroline Is. 62 CRL. Hemicr. or rhizome geophyte.
**Mapania flavinux* T.Koyama, Micronesica 1: 65 (1964).

Hypolytrum formosanum Ohwi, Mem. Coll. Sci. Kyoto Imp. Univ., Ser. B, Biol. 18(1): 168 (1944).
Taiwan. 38 TAI. Hemicr. or rhizome geophyte.

Hypolytrum glaziovii Boeck., Flora 63: 438 (1880).
Brazil (Rio de Janeiro). 84 BZL. Hemicr. or rhizome geophyte.

Hypolytrum glomerulatum M.Alves & W.W.Thomas, Feddes Repert. 113: 265 (2002).
Brazil (Bahia). 84 BZE. Hemicr. or rhizome geophyte.

Hypolytrum hainanense (Merr.) Tang & F.T.Wang, in Fl. Reipubl. Popul. Sin. 11: 201 (1961).
Hainan. 36 CHH. Hemicr. or rhizome geophyte.
**Mapania hainanensis* Merr., Lingnan Sci. J. 9: 35 (1930).

Hypolytrum heteromorphum Nelmes, Kew Bull. 9: 522 (1954 publ. 1955).
W. Trop. Africa to Tanzania. 22 GHA GUI IVO LBR NGA SEN SIE 23 CAB CAF CMN EQG GAB GGI ZAI 25 TAN UGA. Hemicr. or rhizome geophyte.

Hypolytrum heterophyllum Boeck., Beitr. Cyper. 1: 22 (1888).
W. Trop. Africa to Angola. 22 GHA IVO LBR NGA SIE 23 CAB CAF CMN CON EQG GAB GGI ZAI 26 ANG. Hemicr. or rhizome geophyte.
Hypolytrum buchholzianum Boeck., Beitr. Cyper. 1: 22 (1888).
Hypolytrum congense C.B.Clarke ex De Wild. & T.Durand, Bull. Soc. Roy. Bot. Belgique, Compt. Rend. 38: 115 (1899).
Hypolytrum purpurascens Cherm., Bull. Soc. Bot. France 80: 508 (1933).

Hypolytrum humile (Hassk. ex Steud.) Boeck., Linnaea 37: 128 (1871).
W. Jawa. 42 JAW. Hemicr. or rhizome geophyte.
**Pandanophyllum humile* Hassk. ex Steud., Syn. Pl. Glumac. 2: 134 (1855). *Lepironia humilis* (Hassk. ex Steud.) Miq., Ill. Fl. Archip. Ind.: 61 (1870). *Mapania humilis* (Hassk. ex Steud.) Fern.-Vill. in F.M.Blanco, Fl. Filip., ed. 3, 4(13A): 309 (1880).

Hypolytrum jardimii M.Alves & W.W.Thomas, Brittonia 54: 128 (2002).
Brazil (Bahia). 84 BZE. Hemicr. or rhizome geophyte.

Hypolytrum jenmanii C.B.Clarke, Kew Bull., Addit. Ser. 8: 49 (1908).
N. South America to N. Brazil. 82 FRG GUY SUR VEN 84 BZN. Hemicr. or rhizome geophyte.

subsp. ***jenmanii***
N. South America to N. Brazil. 82 FRG GUY SUR VEN 84 BZN. Hemicr. or rhizome geophyte.
Hypolytrum hoppioides Huber ex C.B.Clarke, Kew Bull., Addit. Ser. 8: 51 (1908).

subsp. ***serratifolium*** (Uittien) T.Koyama, Mem. New York Bot. Gard. 17: 70 (1967).
N. South America. 82 FRG GUY SUR. Hemicr. or rhizome geophyte.
**Hypolytrum jenmanii* var. *serratifolium* Uittien, Recueil Trav. Bot. Néerl. 22: 344 (1924).

Hypolytrum lancifolium C.B.Clarke in D.Oliver & auct. suc. (eds.), Fl. Trop. Afr. 8: 488 (1902).
WC. Trop. Africa. 23 CAB CMN EQG GAB. Hemicr. or rhizome geophyte.

Hypolytrum laxum Kunth, Enum. Pl. 2: 270 (1837).
SE. Colombia to S. Venezuela. 82 VEN 83 CLM ECU 84 BZN. Hemicr. or rhizome geophyte.
Beera laxa Schrad. ex Kunth, Enum. Pl. 2: 270 (1837).
Hypolytrum caespitosum Schrad. ex Nees in C.F.P.von Martius & auct. suc. (eds.), Fl. Bras. 2(1): 68 (1842).

Hypolytrum leptocalamum M.Alves & W.W.Thomas, Brittonia 54: 131 (2002).
N. South America to N. Brazil. 82 GUY VEN 84 BZN. Hemicr. or rhizome geophyte.

Hypolytrum longifolium (Rich.) Nees, Linnaea 9: 288 (1834).
SE. Mexico to Brazil. 79 MXG MXT 80 BLZ COS GUA HON NIC PAN 81 TRT 82 FRG GUY SUR VEN 83 CLM ECU PER 84 BZC BZE BZN. Hemicr. or rhizome geophyte.
**Scirpus longifolius* Rich., Actes Soc. Hist. Nat. Paris 1: 106 (1792).

subsp. ***longifolium***
Costa Rica to Brazil. 80 COS PAN 81 TRT 82 FRG GUY SUR VEN 83 CLM PER 84 BZC BZE BZN. Hemicr. or rhizome geophyte.
Beera silvatica Schrad. ex Nees in C.F.P.von Martius & auct. suc. (eds.), Fl. Bras. 2(1): 67 (1842).

Hypolytrum fuscum Nees in C.F.P.von Martius & auct. suc. (eds.), Fl. Bras. 2(1): 67 (1842).
Hypolytrum irriguum Nees in C.F.P.von Martius & auct. suc. (eds.), Fl. Bras. 2(1): 68 (1842).
Hypolytrum cariciforme Nees ex Steud., Syn. Pl. Glumac. 2: 132 (1855).

subsp. ***nicaraguense*** (Liebm.) T.Koyama, Mem. New York Bot. Gard. 17(1): 75 (1967).
SE. Mexico to N. Brazil. 79 MXG MXT 80 BLZ COS GUA HON NIC PAN 81+ 82 VEN 83 CLM ECU 84 BZN. Hemicr. or rhizome geophyte.
**Hypolytrum nicaraguense* Liebm., Mexic. Neldeagt. Pl., V, 2: 235 (1851).

subsp. ***rubescens*** (Huber ex C.B.Clarke) T.Koyama, Mem. New York Bot. Gard. 17(1): 74 (1967).
Panama to N. Brazil. 80 PAN 82 GUY VEN 83 CLM 84 BZN. Hemicr. or rhizome geophyte.
**Hypolytrum rubescens* Huber ex C.B.Clarke, Kew Bull., Addit. Ser. 8: 50 (1908).

subsp. ***sylvaticum*** (Poepp. ex Kunth) T.Koyama, Mem. New York Bot. Gard. 17(1): 74 (1967).
N. South America to Brazil. 82 FRG GUY SUR VEN 83 CLM ECU PER 84 BZN. Hemicr. or rhizome geophyte.
**Hypolytrum sylvaticum* Poepp. ex Kunth, Enum. Pl. 2: 270 (1837).
Hypolytrum punctatum Nees in C.F.P.von Martius & auct. suc. (eds.), Fl. Bras. 2(1): 66 (1842).
Hypolytrum corymbosum Rich. ex Boeck., Linnaea 37: 125 (1871).

Hypolytrum longirostre Thwaites, Enum. Pl. Zeyl.: 346 (1864).
SW. Sri Lanka. 40 SRL. Hemicr. or rhizome geophyte.

Hypolytrum lucennoi M.Alves & W.W.Thomas, Feddes Repert. 113: 267 (2002).
Brazil (Bahia). 84 BZE. Hemicr. or rhizome geophyte.

Hypolytrum mauritianum Nees ex Kunth, Enum. Pl. 2: 272 (1837).
W. Indian Ocean. 29 MAU MDG REU. Hemicr. or rhizome geophyte.
Rhynchospora binervis Sieber ex C.Presl, Isis (Oken) 21: 269 (1828).
Hypolytrum iridifolium Willd. ex Kunth, Enum. Pl. 2: 272 (1837).
Hypolytrum mauritianum var. *contractum* Nees, Bull. Soc. Bot. France 69: 724 (1922).
Hypolytrum mauritianum var. *testaceum* Cherm., Bull. Soc. Bot. France 69: 724 (1922).

Hypolytrum nemorum (Vahl) Spreng., Syst. Veg. 1: 233 (1825).
Taiwan, Trop. Asia to W. Pacific. 38 TAI 40 ASS EHM IND SRL 41 AND THA VIE 42 BOR JAW MLY MOL PHI SUL SUM 43 BIS NWG SOL 50 NTA QLD 60 FIJ 62 CRL. Hemicr.
**Schoenus nemorum* Vahl, Symb. Bot. 3: 8 (1794). *Hypaelyptum nemorum* (Vahl) P.Beauv., Fl. Oware 2: 13 (1810).

subsp. ***fuscorubens*** (T.Koyama) T.Koyama in M.D.Daddanayake & al. (eds.), Revised Handb. Fl. Ceylon 5: 140 (1985).
S. India, Sri Lanka. 40 IND SRL. Hemicr. or rhizome geophyte.
**Hypolytrum fuscorubens* T.Koyama, Phytologia 20: 449 (1970).

subsp. ***nemorum***
Taiwan, Trop. Asia to N. Australia. 38 TAI 40 ASS EHM IND SRL 41 AND THA VIE 42 BOR JAW MLY MOL PHI SUL SUM 43 BIS NWG SOL 50 NTA QLD. Hemicr.
Scirpus anomalus Retz., Observ. Bot. 5: 15 (1788). *Ficinia anomala* (Retz.) Steud., Syn. Pl. Glumac. 2: 125 (1855). *Hypolytrum anomalum* (Retz.) Domin, Biblioth. Bot. 85: 484 (1915), nom. illeg.
Hypolytrum latifolium Rich. ex Pers., Syn. Pl. 1: 70 (1805).
Tunga diandra Roxb., Fl. Ind. 1: 188 (1820). *Hypolytrum diandrum* (Roxb.) A.Dietr., Sp. Pl. 2: 365 (1832).
Albikia schoenoides C.Presl, Reliq. Haenk. 1: 184 (1828). *Hypolytrum schoenoides* (C.Presl) Nees, Linnaea 9: 288 (1834).
Albikia scirpoides C.Presl, Reliq. Haenk. 1: 184 (1828). *Hypolytrum scirpoides* (C.Presl) Merr., Enum. Philipp. Fl. Pl. 1: 103 (1923).
Hypolytrum giganteum Wall. ex Nees in R.Wight, Contr. Bot. India: 93 (1834).
Hypolytrum ensifolium Willd. ex Kunth, Enum. Pl. 2: 271 (1837).
Hypolytrum trinervium Kunth, Enum. Pl. 2: 272 (1837).
Hypolytrum myrianthum Miq., Fl. Ned. Ind. 3: 333 (1856).
Hypolytrum latifolium var. *minus* Thwaites, Enum. Pl. Zeyl.: 346 (1864).
Hypolytrum wightianum Boeck., Linnaea 37: 130 (1871).
Hypolytrum penangense C.B.Clarke in J.D.Hooker, Fl. Brit. India 6: 679 (1894). *Hypolytrum latifolium* var. *penangense* (C.B.Clarke) Ridl., Fl. Malay Penins. 5: 170 (1925).
Hypolytrum costato-nux C.B.Clarke, J. Straits Branch Roy. Asiat. Soc. 46: 225 (1906).
Hypolytrum philippense C.B.Clarke, Philipp. J. Sci., C 2: 109 (1907).
Hypolytrum viridinux C.B.Clarke, Philipp. J. Sci., C 2: 109 (1907).
Hypolytrum amplectens J.V.Suringar, Nova Guinea 8: 708 (1912).
Hypolytrum quadriglumatum J.V.Suringar, Nova Guinea 8: 709 (1912).
Hypolytrum anomalum var. *refractum* Domin, Biblioth. Bot. 85: 484 (1915).
Hypolytrum minus Ridl., Trans. Linn. Soc. London, Bot. 9: 244 (1916).
Hypolytrum scabrum Uittien, J. Arnold Arbor. 20: 215 (1939).
Hypolytrum latifolium var. *depauperatum* Ohwi, Bot. Mag. (Tokyo) 56: 269 (1942). *Hypolytrum depauperatum* (Ohwi) Ohwi & T.Koyama, Acta Phytotax. Geobot. 18: 25 (1959).
Hypolytrum microcarpum S.T.Blake, J. Arnold Arbor. 35: 235 (1954).

var. ***proliferum*** (Boeck.) J.Kern, in Fl. Males. 7: 492 (1974).
Vietnam to N. Queensland. 41 VIE 42 BOR MLY SUM 50 QLD. Hemicr. or rhizome geophyte.
**Hypolytrum proliferum* Boeck., Linnaea 37: 126 (1871).

subsp. ***vitiense*** (C.B.Clarke) T.Koyama, Micronesica 1: 68 (1964).
Caroline Is., Fiji. 60 FIJ 62 CRL. Hemicr. or rhizome geophyte.

**Hypolytrum vitiense* C.B.Clarke, Kew Bull., Addit. Ser. 8: 51 (1908).

Hypolytrum nudicaule Juss. ex Cherm., Bull. Soc. Bot. France 69: 724 (1922 publ. 1923).
E. Madagascar. 29 MDG. Hemicr. or rhizome geophyte.

Hypolytrum nudum C.B.Clarke, Kew Bull., Addit. Ser. 8: 49 (1908).
Colombia to Guyana. 82 GUY VEN 83 CLM 84 BZN. Hemicr. or rhizome geophyte.

Hypolytrum ohwianum T.Koyama, Jap. J. Bot. 15: 183 (1956).
Hainan. 36 CHH. Hemicr. or rhizome geophyte.

Hypolytrum pallidiceps S.S.Hooper & T.Koyama, Bot. Mag. (Tokyo) 97: 413 (1984).
Guyana. 82 GUY. Hemicr. or rhizome geophyte.

Hypolytrum paraense M.Alves & W.W.Thomas, Brittonia 54: 133 (2002).
Brazil (Pará). 84 BZN. Hemicr. or rhizome geophyte.

Hypolytrum paucistrobiliferum Tang & F.T.Wang, in Fl. Reipubl. Popul. Sin. 11: 200 (1961).
Hainan. 36 CHH. Hemicr. or rhizome geophyte.

Hypolytrum poecilolepis Nelmes, Kew Bull. 10: 77 (1955).
W. Trop. Africa. 22 GHA IVO LBR SIE. Hemicr. or rhizome geophyte.

Hypolytrum polystachyum Cherm., Bull. Soc. Bot. France 80: 507 (1933).
WC. Trop. Africa. 23 CON GAB. Hemicr. or rhizome geophyte.
Hypolytrum polystachyum var. *depauperatum* Cherm., Bull. Soc. Bot. France 80: 507 (1933).

Hypolytrum pseudomapanioides D.A.Simpson & Lye, Kew Bull. 59: 613 (2004).
Cameroon. 23 CMN. Hemicr. or rhizome geophyte.

Hypolytrum pulchrum (Rudge) H.Pfeiff., Bot. Arch. 12: 450 (1925).
N. South America to Brazil. 82 FRG GUY SUR VEN 84 BZE BZN. Hemicr. or rhizome geophyte.
Hypaelyptum pungens Vahl, Enum. Pl. 2: 283 (1805). *Scirpus pungens* (Vahl) Poir. in J.B.A.M. de Lamarck, Encycl., Suppl. 5: 102 (1817), nom. illeg. *Hypolytrum pungens* (Vahl) Kunth, Enum. Pl. 2: 269 (1837).
**Scirpus pulcher* Rudge, Pl. Guian.: 18 (1805).
Hypolytrum umbellatum Willd. ex Link, Jahrb. Gewächsk. 1(3): 91 (1820).
Hypolytrum glomeratum Rich. ex Kunth, Enum. Pl. 2: 270 (1837).

Hypolytrum pynaertii (De Wild.) Nelmes, Kew Bull. 10: 81 (1955).
WC. Trop. Africa. 23 CMN GAB ZAI. Hemicr. or rhizome geophyte.
**Mapania pynaertii* De Wild., Ann. Mus. Congo Belge, Bot., V, 3: 166 (1910).
Hypolytrum gabonicum Cherm., Bull. Soc. Bot. France 77: 277 (1930).
Hypolytrum gabonicum var. *plicatum* Cherm., Bull. Soc. Bot. France 83: 494 (1936). *Hypolytrum pynaertii* var. *plicatum* (Cherm.) Nelmes, Kew Bull. 10: 82 (1955).

Hypolytrum rigens Nees in C.F.P.von Martius & auct. suc. (eds.), Fl. Bras. 2(1): 67 (1842).
Brazil. 84 BZC BZE BZL. Hemicr. or rhizome geophyte.

Hypolytrum scaberrimum Boeck., Flora 65: 26 (1882).
Mapania scaberrima (Boeck.) C.B.Clarke in T.A. Durand & H.Schinz, Consp. Fl. Afric. 5: 667 (1894).
WC. Trop. Africa. 23 CMN CON EQG GAB. Hemicr. or rhizome geophyte.
Hypolytrum rhizomatanthum var. *elatum* Cherm., Bull. Soc. Bot. France 83: 494 (1936). *Hypolytrum elatum* (Cherm.) Nelmes, Kew Bull. 10: 80 (1955).

Hypolytrum schnellianum Lorougnon, Bull. Jard. Bot. Natl. Belg. 45: 181 (1975).
Ivory Coast (near Djiroutou). 22 IVO. Hel. or hemicr.

Hypolytrum schraderianum Nees in C.F.P.von Martius & auct. suc. (eds.), Fl. Bras. 2(1): 65 (1842).
S. Trop. America. 82 FRG GUY VEN 83 CLM ECU PER 84 BZC BZE BZL BZN BZS. Hemicr. or rhizome geophyte.
Hypolytrum selloianum Boeck., Linnaea 37: 134 (1871).
Hypolytrum microstachyum Boeck., Beitr. Cyper. 1: 23 (1888).
Hypolytrum macrophyllum Boeck., Allg. Bot. Z. Syst. 2: 78 (1896), nom. illeg.

Hypolytrum secans (K.Schum.) J.Raynal, Adansonia, n.s., 8: 429 (1968).
WC. Trop. Africa. 23 CMN GAB. Hemicr. or rhizome geophyte.
**Mapania secans* K.Schum., Notizbl. Königl. Bot. Gart. Berlin 3: 105 (1901).
Hypolytrum rhizomatanthum Cherm., Bull. Soc. Bot. France 80: 508 (1933).

Hypolytrum senegalense Rich. ex Pers., Syn. Pl. 1: 70 (1805).
Sierra Leone to Guinea, Gabon. 22 GUI SIE 23 GAB. Hemicr. or rhizome geophyte.
Hypolytrum senegalense C.B.Clarke in D.Oliver & auct. suc. (eds.), Fl. Trop. Afr. 8: 488 (1902), nom. illeg.

Hypolytrum sphaerostachyum Boeck., Vidensk. Meddel. Naturhist. Foren. Kjøbenhavn 1871: 155 (1871).
S. Venezuela to N. Peru. 82 VEN 83 PER 84 BZN. Hemicr. or rhizome geophyte.

Hypolytrum spongiosum T.Koyama, Mem. New York Bot. Gard. 17(1): 76 (1967).
SE. Colombia to Guyana. 82 GUY VEN 83 CLM 84 BZN. Hemicr. or rhizome geophyte.

Hypolytrum stemonifolium T.Koyama, Jap. J. Bot. 20: 129 (1969).
Guyana to Brazil. 82 GUY 84 BZC BZN. Hemicr. or rhizome geophyte.

Hypolytrum strictum Poepp. ex Kunth, Enum. Pl. 2: 270 (1837).
S. Venezuela to N. Peru. 82 VEN 83 PER 84 BZN. Hemicr. or rhizome geophyte.
Beera caespitosa Schrad. ex Kunth, Enum. Pl. 2: 270 (1837).
Hypolytrum scapigerum Nees in C.F.P.von Martius & auct. suc. (eds.), Fl. Bras. 2(1): 69 (1842).
Hypolytrum micranthum Boeck., Vidensk. Meddel. Naturhist. Foren. Kjøbenhavn 1871: 154 (1871).

Hypolytrum supervacuum C.B.Clarke, Kew Bull., Addit. Ser. 8: 50 (1908).
N. Brazil. 84 BZN. Hemicr. or rhizome geophyte.

Hypolytrum testui Cherm., Bull. Soc. Bot. France 77: 277 (1930).
WC. Trop. Africa to Tanzania. 23 CAB GAB ZAI 25 TAN. Hemicr.

Hypolytrum turgidum C.B.Clarke in J.D.Hooker, Fl. Brit. India 6: 679 (1894). *Hypolytrum latifolium* var. *turgidum* (C.B.Clarke) Hook.f. ex Trimen, Handb. Fl. Ceylon 5: 90 (1900).
Sri Lanka. 40 SRL. Hemicr. or rhizome geophyte.

Hypolytrum unispicatum Sosef & D.A.Simpson, Blumea 50: 523 (2005).
Cameroon. 23 CMN. Hemicr. or rhizome geophyte

Hypolytrum verticillatum T.Koyama, Darwiniana 16: 75 (1970).
Brazil (Bahia to Rio de Janeiro). 84 BZE BZL. Hemicr. or rhizome geophyte.

***Synonyms*:**
Hypolytrum albidum Willd. ex Kunth = ***Lipocarpha chinensis*** (Osbeck) J.Kern
Hypolytrum amplectens J.V.Suringar = ***Hypolytrum nemorum*** (Vahl) Spreng. subsp. ***nemorum***
Hypolytrum amplissimum M.Alves & W.W.Thomas = ***Hypolytrum amplum*** Kunth
Hypolytrum anomalum (Retz.) Domin = ***Hypolytrum nemorum*** (Vahl) Spreng. subsp. ***nemorum***
Hypolytrum anomalum Steud. = ***Scleria anomala*** (Steud.) J.Raynal
Hypolytrum anomalum var. *refractum* Domin = ***Hypolytrum nemorum*** (Vahl) Spreng. subsp. ***nemorum***
Hypolytrum anomalum var. *typicum* Domin = ***Hypolytrum nemorum*** (Vahl) Spreng.
Hypolytrum argenteum (Vahl) Kunth = ***Lipocarpha chinensis*** (Osbeck) J.Kern
Hypolytrum aschersonianum Boeck. = ***Mapania soyauxii*** (Boeck.) H.Pfeiff.
Hypolytrum attiense A.Chev. ex Hutch. & Dalziel = ***Hypolytrum africanum*** Nees ex Steud.
Hypolytrum berteroanum Schult. = ***Rhynchospora berteroi*** (Spreng.) C.B.Clarke
Hypolytrum berteroi Spreng. = ***Rhynchospora berteroi*** (Spreng.) C.B.Clarke
Hypolytrum borneense Kurz = ***Mapania bancana*** (Miq.) T.Koyama
Hypolytrum buchholzianum Boeck. = ***Hypolytrum heterophyllum*** Boeck.
Hypolytrum caespitosum Schrad. ex Nees = ***Hypolytrum laxum*** Kunth
Hypolytrum capillare Schrad. ex Nees = ***Lipocarpha micrantha*** (Vahl) G.C.Tucker
Hypolytrum cariciforme Nees ex Steud. = ***Hypolytrum longifolium*** (Rich.) Nees subsp. ***longifolium***
Hypolytrum ceylanicum B.Heyne ex Nees = ***Lipocarpha gracilis*** (Rich. ex Pers.) Nees
Hypolytrum condensatum C.B.Clarke = ***Mapania pycnostachya*** (Benth.) T.Koyama
Hypolytrum congense C.B.Clarke ex De Wild. & T.Durand = ***Hypolytrum heterophyllum*** Boeck.
Hypolytrum corymbosum Rich. ex Boeck. = ***Hypolytrum longifolium*** subsp. ***sylvaticum*** (Poepp. ex Kunth) T.Koyama
Hypolytrum costato-nux C.B.Clarke = ***Hypolytrum nemorum*** (Vahl) Spreng. subsp. ***nemorum***
Hypolytrum costatum Hochst. ex Steud. = ***Hypolytrum amplum*** Kunth
Hypolytrum costatum Nelmes = [22 LBR]
Hypolytrum costatum Thwaites = ***Scirpodendron ghaeri*** (Gaertn.) Merr.
Hypolytrum depauperatum (Ohwi) Ohwi & T.Koyama = ***Hypolytrum nemorum*** (Vahl) Spreng. subsp. ***nemorum***
Hypolytrum diandrum (Roxb.) A.Dietr. = ***Hypolytrum nemorum*** (Vahl) Spreng. subsp. ***nemorum***
Hypolytrum elatum (Cherm.) Nelmes = ***Hypolytrum scaberrimum*** Boeck.
Hypolytrum ensifolium Willd. ex Kunth = ***Hypolytrum nemorum*** (Vahl) Spreng. subsp. ***nemorum***
Hypolytrum floribundum Nees ex Steud. = ***Mapania floribunda*** (Nees ex Steud.) T.Koyama
Hypolytrum fuscorubens T.Koyama = ***Hypolytrum nemorum*** subsp. ***fuscorubens*** (T.Koyama) T.Koyama
Hypolytrum fuscum Nees = ***Hypolytrum longifolium*** (Rich.) Nees subsp. ***longifolium***
Hypolytrum gabonicum Cherm. = ***Hypolytrum pynaertii*** (De Wild.) Nelmes
Hypolytrum gabonicum var. *plicatum* Cherm. = ***Hypolytrum pynaertii*** (De Wild.) Nelmes
Hypolytrum giganteum Wall. ex Nees = ***Hypolytrum nemorum*** (Vahl) Spreng. subsp. ***nemorum***
Hypolytrum glomeratum Rich. ex Kunth = ***Hypolytrum pulchrum*** (Rudge) H.Pfeiff.
Hypolytrum glomeratum Schrad. ex Nees = ***Lipocarpha maculata*** (Michx.) Torr.
Hypolytrum gracile Rich. ex Pers. = ***Lipocarpha gracilis*** (Rich. ex Pers.) Nees
Hypolytrum grande (Uittien) T.Koyama = ***Principina grandis*** Uittien
Hypolytrum hoppioides Huber ex C.B.Clarke = ***Hypolytrum jenmanii*** C.B.Clarke subsp. ***jenmanii***
Hypolytrum imeriense R.Gross = ***Mapania imeriensis*** (R.Gross) T.Koyama
Hypolytrum iridifolium Willd. ex Kunth = ***Hypolytrum mauritianum*** Nees ex Kunth
Hypolytrum iridifolium Link = ***Diplasia karatifolia*** Rich. ex Pers.
Hypolytrum irriguum Nees = ***Hypolytrum longifolium*** (Rich.) Nees subsp. ***longifolium***
Hypolytrum jenmanii var. *serratifolium* Uittien = ***Hypolytrum jenmanii*** subsp. ***serratifolium*** (Uittien) T.Koyama
Hypolytrum kuntzeanum Boeck. = ***Cladium mariscus*** subsp. ***jamaicense*** (Crantz) Kük.
Hypolytrum laevigatum (Roxb.) Spreng. = ***Lipocarpha chinensis*** (Osbeck) J.Kern
Hypolytrum latifolium Rich. ex Pers. = ***Hypolytrum nemorum*** (Vahl) Spreng. subsp. ***nemorum***
Hypolytrum latifolium var. *depauperatum* Ohwi = ***Hypolytrum nemorum*** (Vahl) Spreng. subsp. ***nemorum***
Hypolytrum latifolium var. *minus* Thwaites = ***Hypolytrum nemorum*** (Vahl) Spreng. subsp. ***nemorum***
Hypolytrum latifolium var. *penangense* (C.B.Clarke) Ridl. = ***Hypolytrum nemorum*** (Vahl) Spreng. subsp. ***nemorum***
Hypolytrum latifolium var. *turgidum* (C.B.Clarke) Hook. f. ex Trimen = ***Hypolytrum turgidum*** C.B.Clarke
Hypolytrum leprieurii Nees ex Steud. = [22 GUI] Mapania sp. ?
Hypolytrum loefgrenii Boeck. = ***Pleurostachys loefgrenii*** (Boeck.) T.Koyama
Hypolytrum longiscaposum C.B.Clarke = ***Hypolytrum africanum*** Nees ex Steud.
Hypolytrum macranthum Boeck. = ***Mapania macrantha*** (Boeck.) H.Pfeiff.
Hypolytrum macrocephalum Gaudich. = ***Mapania macrocephala*** (Gaudich.) K.Schum.
Hypolytrum macrophyllum Boeck. = ***Hypolytrum schraderianum*** Nees

Hypolytrum macrophyllum Boeck. = ***Mapania macrophylla*** (Boeck.) H.Pfeiff.
Hypolytrum maculatum (Michx.) Kuntze = ***Lipocarpha maculata*** (Michx.) Torr.
Hypolytrum mauritianum var. *contractum* Nees = ***Hypolytrum mauritianum*** Nees ex Kunth
Hypolytrum mauritianum var. *testaceum* Cherm. = ***Hypolytrum mauritianum*** Nees ex Kunth
Hypolytrum micranthum Boeck. = ***Hypolytrum strictum*** Poepp. ex Kunth
Hypolytrum microcarpum S.T.Blake = ***Hypolytrum nemorum*** (Vahl) Spreng. subsp. ***nemorum***
Hypolytrum microstachyum Boeck. = ***Hypolytrum schraderianum*** Nees
Hypolytrum minus Ridl. = ***Hypolytrum nemorum*** (Vahl) Spreng. subsp. ***nemorum***
Hypolytrum multinerve Hochst. ex Steud. = ***Calyptrocarya glomerulata*** (Brongn.) Urb.
Hypolytrum myrianthum Miq. = ***Hypolytrum nemorum*** (Vahl) Spreng. subsp. ***nemorum***
Hypolytrum nicaraguense Liebm. = ***Hypolytrum longifolium*** subsp. ***nicaraguense*** (Liebm.) T.Koyama
Hypolytrum oligostachyum K.Schum. & Lauterb. = ***Hypolytrum dissitiflorum*** Steud.
Hypolytrum pandanophyllum F.Muell. = ***Mapania sumatrana*** subsp. ***pandanophylla*** (F.Muell.) D.A. Simpson
Hypolytrum parvibracteum C.B.Clarke = ***Paramapania parvibractea*** (C.B.Clarke) Uittien
Hypolytrum penangense C.B.Clarke = ***Hypolytrum nemorum*** (Vahl) Spreng. subsp. ***nemorum***
Hypolytrum philippense C.B.Clarke = ***Hypolytrum nemorum*** (Vahl) Spreng. subsp. ***nemorum***
Hypolytrum polystachyum var. *depauperatum* Cherm. = ***Hypolytrum polystachyum*** Cherm.
Hypolytrum proliferum Boeck. = ***Hypolytrum nemorum*** var. ***proliferum*** (Boeck.) J.Kern
Hypolytrum punctatum Nees = ***Hypolytrum longifolium*** subsp. ***sylvaticum*** (Poepp. ex Kunth) T.Koyama
Hypolytrum pungens (Vahl) Kunth = ***Hypolytrum pulchrum*** (Rudge) H.Pfeiff.
Hypolytrum purpurascens Cherm. = ***Hypolytrum heterophyllum*** Boeck.
Hypolytrum pycnocephalum Benth. = ***Mapania pycnocephala*** (Benth.) Benth.
Hypolytrum pynaertii var. *plicatum* (Cherm.) Nelmes = ***Hypolytrum pynaertii*** (De Wild.) Nelmes
Hypolytrum quadriglumatum J.V.Suringar = ***Hypolytrum nemorum*** (Vahl) Spreng. subsp. ***nemorum***
Hypolytrum radians (C.B.Clarke) Ridl. = ***Paramapania radians*** (C.B.Clarke) Uittien
Hypolytrum rhizomatanthum Cherm. = ***Hypolytrum secans*** (K.Schum.) J.Raynal
Hypolytrum rhizomatanthum var. *elatum* Cherm. = ***Hypolytrum scaberrimum*** Boeck.
Hypolytrum rubescens Huber ex C.B.Clarke = ***Hypolytrum longifolium*** subsp. ***rubescens*** (Huber ex C.B.Clarke) T.Koyama
Hypolytrum sandwithii Gilly = ***Mapania imeriensis*** (R.Gross) T.Koyama
Hypolytrum scabrum Uittien = ***Hypolytrum nemorum*** (Vahl) Spreng. subsp. ***nemorum***
Hypolytrum scapigerum Nees = ***Hypolytrum strictum*** Poepp. ex Kunth
Hypolytrum schoenoides (C.Presl) Nees = ***Hypolytrum nemorum*** (Vahl) Spreng. subsp. ***nemorum***
Hypolytrum scirpoides (C.Presl) Merr. = ***Hypolytrum nemorum*** (Vahl) Spreng. subsp. ***nemorum***
Hypolytrum selloianum Boeck. = ***Hypolytrum schraderianum*** Nees
Hypolytrum sellowiana (Kunth) Kuntze = ***Lipocarpha maculata*** (Michx.) Torr.
Hypolytrum senegalense C.B.Clarke = ***Hypolytrum senegalense*** Rich. ex Pers.
Hypolytrum seychellense C.B.Clarke = ***Mapania floribunda*** (Nees ex Steud.) T.Koyama
Hypolytrum soyauxii Boeck. = ***Mapania soyauxii*** (Boeck.) H.Pfeiff.
Hypolytrum sylvaticum Poepp. ex Kunth = ***Hypolytrum longifolium*** subsp. ***sylvaticum*** (Poepp. ex Kunth) T.Koyama
Hypolytrum tepuianum Steyerm. = ***Mapania tepuiana*** (Steyerm.) T.Koyama
Hypolytrum triceps (Roxb.) A.Dietr. = ***Lipocarpha gracilis*** (Rich. ex Pers.) Nees
Hypolytrum trinervium Kunth = ***Hypolytrum nemorum*** (Vahl) Spreng. subsp. ***nemorum***
Hypolytrum umbellatum Willd. ex Link = ***Hypolytrum pulchrum*** (Rudge) H.Pfeiff.
Hypolytrum viridinux C.B.Clarke = ***Hypolytrum nemorum*** (Vahl) Spreng. subsp. ***nemorum***
Hypolytrum vitiense C.B.Clarke = ***Hypolytrum nemorum*** subsp. ***vitiense*** (C.B.Clarke) T.Koyama
Hypolytrum wightianum Boeck. = ***Hypolytrum nemorum*** (Vahl) Spreng. subsp. ***nemorum***
Hypolytrum xerocarpum C.B.Clarke = ***Hypolytrum compactum*** Nees & Meyen ex Kunth

Unplaced Names:
Hypolytrum costatum Nelmes, Kew Bull. 10: 74 (1955), nom. illeg. = [22 LBR]
Hypolytrum leprieurii Nees ex Steud., Syn. Pl. Glumac. 2: 133 (1855). = [22 GUI] Mapania sp. ?

Hypophialium

Hypophialium Nees = ***Ficinia*** Schrad.
Hypophialium capillifolium (Schrad.) Nees = ***Ficinia capillifolia*** (Schrad.) C.B.Clarke

Hypoporum

Hypoporum Nees = ***Scleria*** P.J.Bergius
Hypoporum annulare Nees = ***Scleria annularis*** Steud.
Hypoporum baldwinii Torr. = ***Scleria baldwinii*** (Torr.) Steud.
Hypoporum capillare (R.Br.) Nees = ***Scleria lithosperma*** (L.) Sw. var. ***lithosperma***
Hypoporum capitatum Nees = ***Scleria neesii*** Kunth
Hypoporum compositum Nees = ***Scleria composita*** (Nees) Boeck.
Hypoporum diffusum Nees = ***Scleria verticillata*** Muhl. ex Willd.
Hypoporum gracile (Rich.) Torr. = ***Scleria lithosperma*** (L.) Sw. var. ***lithosperma***
Hypoporum hirtellum (Sw.) Nees = ***Scleria hirtella*** Sw.
Hypoporum humile Nees = ***Scleria distans*** Poir. var. ***distans***
Hypoporum interruptum Torr. = ***Scleria hirtella*** Sw.
Hypoporum interruptum (Rich.) Nees = ***Scleria interrupta*** Rich.
Hypoporum leptostachyum (Kunth) Nees = ***Scleria leptostachya*** Kunth
Hypoporum lithospermum (L.) Nees = ***Scleria lithosperma*** (L.) Sw.
Hypoporum micrococcum Liebm. = ***Scleria tenella*** Kunth
Hypoporum nutans (Willd. ex Kunth) Nees = ***Scleria distans*** Poir. var. ***distans***

Hypoporum pergracile Nees = ***Scleria pergracilis*** (Nees) Kunth
Hypoporum pleostachyum (Kunth) Nees = ***Scleria spicata*** (Spreng.) J.F.Macbr.
Hypoporum purpurascens Nees = ***Scleria lithosperma*** (L.) Sw. var. ***lithosperma***
Hypoporum purpurascens Liebm. = ***Scleria tenella*** Kunth
Hypoporum pygmaeum (R.Br.) Nees = ***Diplacrum pygmaeum*** (R.Br.) Nees ex Boeck.
Hypoporum sieberi Nees = ***Scleria lithosperma*** (L.) Sw. var. ***lithosperma***
Hypoporum tenellum (Kunth) Nees = ***Scleria tenella*** Kunth
Hypoporum variegatum Nees = ***Scleria variegata*** (Nees) Steud.
Hypoporum verticillatum Nees = ***Scleria tenella*** Kunth
Hypoporum verticillatum (Muhl. ex Willd.) Nees = ***Scleria verticillata*** Muhl. ex Willd.
Hypoporum virgatum Nees = ***Scleria virgata*** (Nees) Steud.

Ideleria

Ideleria Kunth = ***Tetraria*** P.Beauv.
Ideleria capensis Kunth = ***Cyathocoma hexandra*** (Nees) Browning
Ideleria hexandra (Nees) H.Pfeiff. = ***Cyathocoma hexandra*** (Nees) Browning
Ideleria neesii Kunth = ***Cyathocoma hexandra*** (Nees) Browning

Indocourtoisia

Indocourtoisia Bennet & Raizada = ***Courtoisina*** Soják
Indocourtoisia assimilis (Steud.) Bennet & Raizada = ***Courtoisina assimilis*** (Steud.) Maquet
Indocourtoisia cyperoides (Roxb.) Bennet & Raizada = ***Courtoisina cyperoides*** (Roxb.) Soják

Iria

Iria Kuntze = ***Cyperus*** L.
Iria R.Hedw. = ***Fimbristylis*** Vahl
Iria acicularis (R.Br.) Kuntze = ***Fimbristylis acicularis*** R.Br.
Iria acuminata (Vahl) Kuntze = ***Fimbristylis acuminata*** Vahl
Iria aestivalis (Vahl) Kuntze = ***Fimbristylis aestivalis*** Vahl
Iria androgyna (R.Br.) Kuntze = ***Cyperus denudatus*** L.f.
Iria angularis (Link) Kuntze = ***Fimbristylis quinquangularis*** (Vahl) Kunth
Iria aphylla (Kunth) Kuntze = ***Fimbristylis bahiensis*** Steud.
Iria apus (A.Gray) Kuntze = ***Fimbristylis vahlii*** (Lam.) Link
Iria arnottiana (Boeck.) Kuntze = ***Fimbristylis arnottiana*** Boeck.
Iria autumnalis (L.) Kuntze = ***Fimbristylis autumnalis*** (L.) Roem. & Schult.
Iria autumnalis var. *caespitosa* Kuntze = ***Fimbristylis autumnalis*** (L.) Roem. & Schult.
Iria autumnalis var. *geminata* (Kunth) Farw. = ***Fimbristylis autumnalis*** (L.) Roem. & Schult.
Iria barbata (Rottb.) Kuntze = ***Bulbostylis barbata*** (Rottb.) C.B.Clarke
Iria barteri (Boeck.) Kuntze = ***Fimbristylis barteri*** Boeck.
Iria biflora (Boeck.) Kuntze = ***Fimbristylis cinnamometorum*** (Vahl) Kunth
Iria bispicata (J.König ex Roxb.) Kuntze = ***Fimbristylis schoenoides*** (Retz.) Vahl
Iria bisumbellata (Forssk.) Kuntze = ***Fimbristylis bisumbellata*** (Forssk.) Bubani
Iria caespitosa (R.Br.) Kuntze = ***Fimbristylis caespitosa*** R.Br.
Iria capillaris (L.) Kuntze = ***Bulbostylis capillaris*** (L.) Kunth ex C.B.Clarke
Iria capitata (R.Br.) Kuntze = ***Fimbristylis furva*** R.Br.
Iria cardiocarpa (F.Muell.) Kuntze = ***Fimbristylis cardiocarpa*** F.Muell.
Iria caribea Steud. = ***Fimbristylis ovata*** (Burm.f.) J.Kern
Iria castanea (Michx.) Farw. = ***Fimbristylis spadicea*** (L.) Vahl
Iria cinnamometorum (Vahl) Kuntze = ***Fimbristylis cinnamometorum*** (Vahl) Kunth
Iria compressa Kuntze = ***Fimbristylis pubisquama*** J.Kern
Iria consanguinea (Kunth) Kuntze = ***Fimbristylis consanguinea*** Kunth
Iria corynocarya (F.Muell.) Kuntze = ***Fimbristylis corynocarya*** F.Muell.
Iria crassipes (Boeck.) Kuntze = ***Fimbristylis falcata*** (Vahl) Kunth var. ***falcata***
Iria cymosa (R.Br.) Kuntze = ***Fimbristylis cymosa*** R.Br.
Iria debilis Kuntze = ***Fimbristylis signata*** S.T.Blake
Iria denutata (R.Br.) Kuntze = ***Fimbristylis denudata*** R.Br.
Iria depauperata (R.Br.) Kuntze = ***Fimbristylis dichotoma*** subsp. ***depauperata*** (R.Br.) J.Kern
Iria disticha (Boeck.) Kuntze = ***Fimbristylis disticha*** Boeck.
Iria drummondii (Torr. & Hook.) Kuntze = ***Fimbristylis puberula*** Vahl var. ***puberula***
Iria elata (R.Br.) Kuntze = ***Fimbristylis denudata*** R.Br.
Iria eragrostis (Nees) Kuntze = ***Fimbristylis eragrostis*** (Nees) Hance
Iria falcata (Vahl) Kuntze = ***Fimbristylis falcata*** (Vahl) Kunth
Iria falcifolia (Boeck.) Kuntze = ***Fimbristylis falcifolia*** Boeck.
Iria ferruginea (L.) Kuntze = ***Fimbristylis ferruginea*** (L.) Vahl
Iria filifolia (Boeck.) Kuntze = ***Fimbristylis filifolia*** Boeck.
Iria fimbristyloides (F.Muell.) Kuntze = ***Fimbristylis fimbristyloides*** (F.Muell.) Druce
Iria furva (R.Br.) Kuntze = ***Fimbristylis furva*** R.Br.
Iria fusca (Nees) Kuntze = ***Fimbristylis fusca*** (Nees) Benth. ex C.B.Clarke
Iria globulosa (Retz.) Kuntze = ***Fimbristylis umbellaris*** (Lam.) Vahl
Iria glomerata (Schrad. ex Schult.) Kuntze = ***Fimbristylis cymosa*** R.Br. subsp. ***cymosa***
Iria gracilis Kuntze = ***Fimbristylis tenuis*** Roem. & Schult.
Iria hanceana (Boeck.) Kuntze = ***Fimbristylis rigidula*** Nees
Iria hawaiiensis (Hillebr.) Kuntze = ***Fimbristylis hawaiiensis*** Hillebr.
Iria hookeriana (Boeck.) Kuntze = ***Fimbristylis hookeriana*** Boeck.
Iria indica (Rich. ex Pers.) Schult. = ***Fimbristylis ovata*** (Burm.f.) J.Kern
Iria insignis (Thwaites) Kuntze = ***Fimbristylis insignis*** Thwaites
Iria juncea (G.Forst.) Kuntze = ***Fimbristylis juncea*** (G.Forst.) R.Br. ex Roem. & Schult.
Iria junciformis (Nees) Kuntze = ***Fimbristylis falcata*** (Vahl) Kunth var. ***falcata***

Iria leptoclada (Benth.) Kuntze = ***Fimbristylis leptoclada*** Benth.
Iria leucocolea (Benth.) Kuntze = ***Fimbristylis leucocolea*** Benth.
Iria leucostachya (Roem. & Schult.) Kuntze = ***Fimbristylis leucostachya*** Roem. & Schult.
Iria macrantha (Boeck.) Kuntze = ***Fimbristylis macrantha*** Boeck.
Iria macrostachya (Boeck.) Kuntze = ***Fimbristylis tristachya*** R.Br. var. ***tristachya***
Iria martii (Boeck.) Kuntze = ***Fimbristylis bahiensis*** Steud.
Iria microcarya (F.Muell.) Kuntze = ***Fimbristylis microcarya*** F.Muell.
Iria miliacea (L.) Kuntze = ***Fimbristylis quinquangularis*** (Vahl) Kunth
Iria monandra (Rottb.) Kuntze = ***Fimbristylis argentea*** (Rottb.) Vahl
Iria monostachya (L.) Kuntze = ***Fimbristylis ovata*** (Burm.f.) J.Kern
Iria mucronata Kuntze = ***Fimbristylis scabrida*** Schumach.
Iria muelleriana Kuntze = ***Fimbristylis argentea*** (Rottb.) Vahl
Iria multifolia (Boeck.) Kuntze = ***Fimbristylis cymosa*** R.Br. subsp. ***cymosa***
Iria neilsonii (F.Muell.) Kuntze = ***Fimbristylis neilsonii*** F.Muell.
Iria nutans (Retz.) Kuntze = ***Fimbristylis nutans*** (Retz.) Vahl
Iria obtusangula (F.Muell.) Kuntze = ***Fimbristylis rara*** R.Br.
Iria obtusifolia (Lam.) Hitchc. = ***Fimbristylis dichotoma*** (L.) Vahl subsp. ***dichotoma***
Iria oxystachya (F.Muell.) Kuntze = ***Fimbristylis oxystachya*** F.Muell.
Iria pauciflora (R.Br.) Kuntze = ***Fimbristylis pauciflora*** R.Br.
Iria paupercula (Boeck.) Kuntze = ***Fimbristylis paupercula*** Boeck.
Iria pentaptera (Nees) Kuntze = ***Fimbristylis salbundia*** (Nees) Kunth
Iria pierotii (Miq.) Kuntze = ***Fimbristylis pierotii*** Miq.
Iria pilosa (Vahl) Kuntze = ***Fimbristylis pilosa*** Vahl
Iria platystachya (Boeck.) Kuntze = ***Fimbristylis schultzii*** Boeck.
Iria polymorpha (Boeck.) Kuntze = ***Fimbristylis dichotoma*** (L.) Vahl subsp. ***dichotoma***
Iria polymorpha var. *normalis* Kuntze = ***Fimbristylis dichotoma*** (L.) Vahl subsp. ***dichotoma***
Iria polytrichoides (Retz.) Kuntze = ***Fimbristylis polytrichoides*** (Retz.) R.Br.
Iria pterygosperma (R.Br.) Kuntze = ***Fimbristylis pterygosperma*** R.Br.
Iria punctata (R.Br.) Kuntze = ***Fimbristylis punctata*** R.Br.
Iria pycnocephala (Hillebr.) Kuntze = ***Fimbristylis cymosa*** subsp. ***umbellatocapitata*** (Hillebr.) T.Koyama
Iria quinquangularis (Vahl) Kuntze = ***Fimbristylis quinquangularis*** (Vahl) Kunth
Iria rara (R.Br.) Kuntze = ***Fimbristylis rara*** R.Br.
Iria retusa (Thwaites) Kuntze = ***Fimbristylis leptoclada*** Benth. var. ***leptoclada***
Iria rhyticarya (F.Muell.) Kuntze = ***Fimbristylis rhyticarya*** F.Muell.
Iria salbundia (Nees) Kuntze = ***Fimbristylis salbundia*** (Nees) Kunth
Iria scaberrima (Nees) Kuntze = ***Fimbristylis scaberrima*** Nees
Iria schoenodes (Retz.) Kuntze = ***Fimbristylis schoenoides*** (Retz.) Vahl
Iria sericea (Poir.) Kuntze = ***Fimbristylis sericea*** (Poir.) R.Br.
Iria solidifolia (F.Muell.) Kuntze = ***Fimbristylis solidifolia*** F.Muell.
Iria spadicea (L.) Kuntze = ***Fimbristylis spadicea*** (L.) Vahl
Iria spadicea var. *nigra* Kuntze = ***Fimbristylis spadicea*** (L.) Vahl
Iria spadicea var. *normalis* Kuntze = ***Fimbristylis spadicea*** (L.) Vahl
Iria spadicea var. *pallida* Kuntze = ***Fimbristylis spadicea*** (L.) Vahl
Iria sphaerocephala (Benth.) Kuntze = ***Fimbristylis sphaerocephala*** Benth.
Iria spiralis (R.Br.) Kuntze = ***Fimbristylis spiralis*** R.Br.
Iria spirostachya Kuntze = ***Fimbristylis dichotoma*** (L.) Vahl subsp. ***dichotoma***
Iria squarrosa (Vahl) Kuntze = ***Fimbristylis squarrosa*** Vahl
Iria stricta (R.Br.) Kuntze = ***Fimbristylis dichotoma*** (L.) Vahl subsp. ***dichotoma***
Iria subaristata (Benth.) Kuntze = ***Fimbristylis subaristata*** Benth.
Iria subbulbosa (Benth.) Kuntze = ***Fimbristylis tristachya*** R.Br. var. ***tristachya***
Iria subtetrastachya (Boeck.) Kuntze = ***Fimbristylis eragrostis*** (Nees) Hance
Iria tenera (Schult.) Kuntze = ***Fimbristylis tenera*** Schult.
Iria tenuicula (Boeck.) Kuntze = ***Fimbristylis tenuicula*** Boeck.
Iria tetragona (R.Br.) Kuntze = ***Fimbristylis tetragona*** R.Br.
Iria thermalis (S.Watson) Kuntze = ***Fimbristylis thermalis*** S.Watson
Iria thomsonii (Boeck.) Kuntze = ***Fimbristylis thomsonii*** Boeck.
Iria thonningiana (Boeck.) Kuntze = ***Fimbristylis microcarya*** F.Muell. var. ***microcarya***
Iria torta (Nees) Kuntze = ***Fimbristylis falcata*** (Vahl) Kunth var. ***falcata***
Iria triflora (L.) Kuntze = ***Fimbristylis triflora*** (L.) K.Schum. ex Engl.
Iria trigastrocarya (F.Muell.) Kuntze = ***Fimbristylis trigastrocarya*** F.Muell.
Iria uliginosa (Hochst. ex Steud.) Kuntze = ***Fimbristylis uliginosa*** Hochst. ex Steud.
Iria umbellata (Schrad. ex Nees) Kuntze = ***Fimbristylis spadicea*** (L.) Vahl
Iria vaginata (R.Br.) Kuntze = ***Fimbristylis vaginata*** (R.Br.) Domin
Iria vahlii (Lam.) Kuntze = ***Fimbristylis vahlii*** (Lam.) Link
Iria velata (R.Br.) Kuntze = ***Fimbristylis squarrosa*** var. ***esquarrosa*** Makino
Iria xyridis (R.Br.) Kuntze = ***Fimbristylis xyridis*** R.Br.

Iriha

Iriha Kuntze = ***Fimbristylis*** Vahl

Isolepis

Isolepis R.Br., Prodr.: 221 (1810).
Cosmopolitan. 10 DEN FOR GRB ICE IRE NOR SWE 11 AUT BGM CZE GER HUN NET POL SWI 12 BAL COR FRA POR SAR SPA 13 ALB? BUL GRC ITA KRI ROM SIC TUE YUG 14 BLR

BLT RUC RUE RUS RUW UKR 20 ALG EGY LBY MOR TUN 21 AZO CNY CVI MDR 22 MLI NGA 23 BUR CMN RWA ZAI 24 ETH SUD 25 KEN TAN UGA 26 ANG MLW MOZ ZAM ZIM 27 BOT CPP LES NAM NAT OFS SWZ TVL 28 STH 29 COM MDG REU 30 32 KAZ KGZ TZK UZB 33 TCS 34 AFG CYP EAI IRN LBS PAL TUR 35 SAU YEM 36 CHC CHT 38 JAP 40 ASS EHM IND NEP PAK SRL WHM 41 THA VIE 42 BOR JAW PHI SUL SUM 43 NWG 50 NFK NSW NTA NZN QLD SOA TAS VIC WAU 51 ATP CTM NZN NZS (60) fij 71 BRC 73 ORE WAS 74 ILL KAN MSO OKL (75) nwj pen 76 CAL 77 TEX 78 ALA ARK FLA GEO KTY LOU MSI NCA SCA TEN vrg 79 MXN 80 COS GUA PAN (81) win 82 VEN 83 BOL CLM ECU PER 84 BZS 85 AGE AGS AGW CLC CLN CLS JNF URU 90 ASP FAL MAQ TDC.

74 Species

Eleogiton Link, Hort. Berol. 1: 293 (1827), nom. illeg. *Scirpidiella* Rauschert, Feddes Repert. 94: 299 (1983).

Elogiton Link ex Steud., Syn. Pl. Glumac. 2: 291 (1855).

Isolepis alpina Hook.f., Fl. Tasman. 2: 86 (1858). *Scirpus cartilagineus* var. *alpinus* (Hook.f.) Benth., Fl. Austral. 7: 328 (1878).

Victoria to Tasmania. 50 TAS VIC. Ther. or hemicr.

Scirpus gunnii Boeck., Linnaea 36: 493 (1870).

Scirpus cartilagineus Benth., Fl. Austral. 7: 328 (1878), nom. illeg.

Isolepis angelica B.L.Burtt, Notes Roy. Bot. Gard. Edinburgh 43: 362 (1986).

S. Africa. 27 CPP LES. Ther. or hemicr.

Isolepis antarctica (L.) Roem. & Schult., Syst. Veg. 2: 112 (1817).

S. Africa, New Zealand. 27 CPP (28) sth 51 NZN NZS. Ther. or hemicr.

**Scirpus antarcticus* L., Mant. Pl. 2: 181 (1771).

Scirpus setaceus Thunb., Prodr. Pl. Cap.: 17 (1794), nom. illeg.

Cyperus minutus Roth, Nov. Pl. Sp.: 32 (1821), nom. illeg.

Scirpus bergianus Spreng., Syst. Veg. 1: 212 (1824).

Isolepis sesslerioides Kunth, Enum. Pl. 2: 195 (1837).

Scirpus minimus Sieber ex Kunth, Enum. Pl. 2: 195 (1837).

Isolepis bergiana Drège, Zwei Pflanzengeogr. Dokum.: 120 (1843).

Scirpus dregeanus C.B.Clarke in W.H.Harvey & auct. suc. (eds.), Fl. Cap. 7: 224 (1898).

Scirpus antarcticus var. *alpinus* Ewart, Fl. Victoria: 219 (1931).

Scirpus antarcticus var. *propinquus* Ewart, Fl. Victoria: 219 (1931).

Isolepis aucklandica Hook.f., Fl. Antarct.: 88 (1844). *Scirpus aucklandicus* (Hook.f.) Boeck., Linnaea 36: 491 (1870).

New Guinea (Highlands), Amsterdam I., SE. Australia to New Zealand, Andes. 43 NWG 50 NSW TAS VIC 51 ATP NZN NZS 83 CLM ECU PER 85 AGS 90 ASP MAQ. Hemicr. or rhizome geophyte.

Isolepis subcucullata Berggr., Minneskr. Fisiog. Sallks. Lund 1877: 22 (1877). *Scirpus subcucullatus* (Berggr.) Berggr., Nya Art. Nyzeel. Fan.: 12 (1878).

Isolepis cartiliganea var. *rigida* Bergg., Minneskr. Fisiog. Sallks. Lund 8: 23 (1878).

Scirpus muscosus Kirk, Trans. & Proc. New Zealand Inst. 17: 224 (1885).

Isolepis australiensis (Maiden & Betche) K.L.Wilson, Telopea 2: 165 (1981).

Australia. 50 NSW NTA QLD SOA VIC. Ther. or hemicr.

**Scirpus cernuus* var. *australiensis* Maiden & Betche, Proc. Linn. Soc. New South Wales 33: 316 (1908). *Scirpus australiensis* (Maiden & Betche) S.T.Blake, Proc. Roy. Soc. Queensland 51: 179 (1939 publ. 1940).

Isolepis basilaris Hook.f., Handb. N. Zeal. Fl. 1: 302 (1864). *Scirpus basilaris* (Hook.f.) C.B.Clarke in T.F.Cheeseman, Man. New Zealand Fl.: 772 (1906).

New Zealand. 51 NZN NZS. Ther. or hemicr.

Isolepis novae-zealandiae Colenso, Trans. & Proc. New Zealand Inst. 21: 102 (1888 publ. 1889).

Isolepis beccarii (Boeck.) Goetgh. & D.A.Simpson, Kew Bull. 46: 174 (1991).

Sumatera. 42 SUM. Ther. or hemicr.

**Scirpus beccarii* Boeck., Bot. Jahrb. Syst. 7: 275 (1886). *Eleogiton beccarii* (Boeck.) Soják, Cas. Nár. Mus., Odd. Prír. 148: 193 (1979 publ. 1980). *Scirpidiella beccarii* (Boeck.) Rauschert, Feddes Repert. 94: 299 (1983).

Isolepis bicolor Carmich., Trans. Linn. Soc. London 12: 503 (1819). *Scirpus bicolor* (Carmich.) Spreng., Syst. Veg. 4(2): 28 (1827). *Scirpus thouarsianus* var. *bicolor* (Carmich.) Hemsl., Rep. Challenger, Bot. 1(3): 156 (1885). *Ficinia bicolor* (Carmich.) H.Pfeiff., Revis. Ficinia: 27 (1921).

Tristan da Cunha. 90 TDC. Ther. or hemicr.

Scirpus prolifer Thouars, Esquisse Fl. Tristan d'Acugna: 36 (1808), nom. illeg. *Isolepis prolifera* Carmich., Trans. Linn. Soc. London 12: 503 (1819), nom. illeg. *Scirpus thouarsianus* Schult. in J.J.Roemer & J.A.Schultes, Mant. 2: 84 (1824). *Isolepis thouarsii* A.Dietr., Sp. Pl. 2: 109 (1832).

Isolepis squarrosa Carmich., Trans. Linn. Soc. London 12: 503 (1819), nom. illeg. *Scirpus squarrosus* Spreng., Syst. Veg. 4(2): 28 (1827), nom. illeg.

Isolepis acugnana Schult. in J.J.Roemer & J.A.Schultes, Mant. 2: 582 (1824).

Scirpus oliveri Boeck., Flora 58: 260 (1875).

Scirpus proliferoramosus Boeck., Flora 58: 261 (1875).

Scirpus virens Boeck., Flora 58: 261 (1875). *Scirpus thouarsianus* var. *virens* (Boeck.) Hemsl., Rep. Challenger, Bot. 1(3): 157 (1885).

Scirpus pallescens Boeck. ex Hemsl., Rep. Challenger, Bot. 1(2): 158 (1884), nom. inval.

Scirpus thouarsianus var. *pallescens* Hemsl., Rep. Challenger, Bot. 1(3): 158 (1885).

Isolepis brevicaulis (Levyns) J.Raynal, Adansonia, n.s., 17: 55 (1977).

SW. Cape Prov. 27 CPP. Ther. or hemicr.

Scirpus burchellii C.B.Clarke in W.H.Harvey & auct. suc. (eds.), Fl. Cap. 7: 216 (1898).

**Scirpus brevicaulis* Levyns, J. S. African Bot. 10: 31 (1944).

Isolepis bulbifera (Boeck.) Muasya, Kew Bull. 57: 313 (2002).

SW. Cape Prov. 27 CPP. Ther. or hemicr.

**Scirpus bulbiferus* Boeck., Linnaea 36: 482 (1870).

Scirpus delicatulus Levyns, J. S. African Bot. 10: 30 (1944).

Isolepis caligenis (V.J.Cook) Soják, Cas. Nár. Mus., Odd. Prír. 148: 194 (1979 publ. 1980).
New Zealand. 51 NZS. Ther. or hemicr.
**Scirpus caligenis* V.J.Cook, Trans. & Proc. Roy. Soc. New Zealand 81: 158 (1953).
Scirpus caligenis var. *tristigmatosa* V.J.Cook, Trans. & Proc. Roy. Soc. New Zealand 81: 158 (1953).

Isolepis capensis Muasya, Kew Bull. 57: 305 (2002).
Cape Prov. 27 CPP. Ther. or hemicr.

Isolepis carinata Hook. & Arn. ex Torr., Ann. Lyceum Nat. Hist. New York 3: 349 (1836). *Trichelostylis carinata* (Hook. & Arn. ex Torr.) Alph.Wood, Class-book Bot.: 742 (1861). *Scirpus carinatus* (Hook. & Arn. ex Torr.) A.Gray, Proc. Amer. Acad. Arts 7: 392 (1868), nom. illeg.
C. & SE. U.S.A. 74 ILL KAN MSO OKL 76 CAL 77 TEX 78 ALA ARK FLA GEO KTY LOU MSI NCA SCA TEN. Ther.
Isolepis koilolepis Steud., Syn. Pl. Glumac. 2: 318 (1855). *Scirpus koilolepis* (Steud.) Gleason, Rhodora 42: 479 (1942).
Isolepis hookeri Nees ex Boeck., Linnaea 36: 498 (1870), nom. inval.

Isolepis cernua (Vahl) Roem. & Schult., Syst. Veg. 2: 106 (1817).
Temp. to Andes. 10 GRB IRE 12 BAL COR FRA POR SAR SPA 13 ALB? GRC ITA KRI SIC TUE YUG 20 ALG LBY MOR TUN 21 AZO CNY CVI MDR 24 ETH 25 KEN TAN 26 ZIM 27 CPP LES NAT OFS TVL (28) sth 34 CYP EAI LBS PAL TUR 50 NFK NSW QLD SOA TAS VIC WAU 51 ATP CTM NZN NZS 71 BRC 73 ORE WAS 76 CAL 77 TEX 79 MXN 83 BOL ECU PER 84 BZS 85 AGE AGS AGW CLC CLN CLS JNF URU 90 FAL TDC. Ther.
**Scirpus cernuus* Vahl, Enum. Pl. 2: 245 (1805). *Eleogiton cernua* (Vahl) A.Dietr., Sp. Pl. 2: 96 (1832). *Schoenoplectus cernuus* (Vahl) Hayek, Repert. Spec. Nov. Regni Veg. Beih. 30(3): 153 (1932).

var. ***andina*** Muasya, Kew Bull. 57: 300 (2002).
Ecuador to Bolivia. 83 BOL ECU. Ther. or hemicr.

var. ***cernua***
Temp. 10 GRB IRE 12 BAL COR FRA POR SAR SPA 13 ALB? GRC ITA KRI SIC TUE YUG 20 ALG LBY MOR TUN 21 AZO CNY CVI MDR 24 ETH 25 KEN 26 ZIM 27 CPP LES NAT OFS TVL (28) sth 34 CYP EAI LBS PAL TUR 50 NFK NSW QLD SOA TAS VIC WAU 51 ATP CTM NZN NZS 71 BRC 73 ORE WAS 76 CAL 77 TEX 79 MXN 83 BOL ECU PER 84 BZS 85 AGE AGS AGW CLC CLN CLS JNF URU 90 FAL TDC. Ther.
Scirpus savii var. *vahlii* Lange in ?. *Scirpus cernuus* var. *vahlii* (Lange) O.Bolòs & Vigo, Fl. Man. Paisos Catalans 4: 212 (2001).
Scirpus acicularis Ucria, Hort. Panhorm.: 52 (1789), nom. illeg.
Scirpus filiformis Savi, Fl. Pis. 1: 46 (1798), nom. illeg. *Scirpus savii* Sebast. & Mauri, Fl. Roman. Prodr.: 22 (1818). *Isolepis saviana* Schult. in J.J.Roemer & J.A.Schultes, Mant. 2: 63 (1824), nom. illeg. *Isolepis savii* (Sebast. & Mauri) Fourr., Ann. Soc. Linn. Lyon, n.s., 17: 174 (1869). *Scirpus setaceus* var. *savii* (Sebast. & Mauri) Firoi in A.Fiori & al., Fl. Italia 1: 120 (1896). *Schoenoplectus savii* (Sebast. & Mauri) Palla in W.D.J.Koch, Syn. Deut. Schweiz. Fl., ed. 3: 2539 (1905).
Fimbristylis pygmaea Vahl, Enum. Pl. 2: 285 (1805). *Isolepis pygmaea* (Vahl) Kunth, Enum. Pl. 2: 191 (1837). *Scirpus pygmaeus* (Kunth) A.Gray, Proc. Amer. Acad. Arts 7: 392 (1868), nom. illeg. *Scirpus cernuus* f. *pygmaeus* (Kunth) Lindm., Bih. Kongl. Svenska Vetensk.-Akad. Handl. 26(3; 9): 21 (1900). *Scirpus cernuus* var. *pygmaeus* (Kunth) Dusén, Ergebn. Schwed. Exp. Magell. 3(5): 212 (1900). *Scirpus setaceus* var. *pygmaeus* (Vahl) Boeck., Linnaea 36: 501 (1870).
Scirpus numidianus Vahl, Enum. Pl. 2: 254 (1805). *Isolepis numidiana* (Vahl) Roem. & Schult., Syst. Veg. 2: 110 (1817).
Isolepis riparia R.Br., Prodr.: 222 (1810). *Scirpus riparius* (R.Br.) Poir. in J.B.A.M.de Lamarck, Encycl., Suppl. 5: 103 (1817), nom. illeg.
Scirpus gracilis Koch in P.Salzmann, Enum. Pl. Rar.: 9 (1818), nom. illeg.
Isolepis sicula C.Presl, Cyper. Gramin. Sicul.: 13 (1820). *Scirpus nervosus* var. *siculus* (C.Presl) A.Terracc., Malpighia 2: 422 (1889). *Isolepis nervosa* var. *sicula* (C.Presl) Nyman, Consp. Fl. Eur., Suppl. 2: 319 (1890).
Isolepis tenuipes J.Presl, Delic. Prag. 1: 144 (1822).
Isolepis tenuis J.Presl, Delic. Prag.: 144 (1822). *Isolepis saviana* var. *tenuis* (J.Presl) Nyman, Consp. Fl. Eur.: 766 (1882).
Scirpus leptaleus Koch, Syll. Pl. Nov. 1: 8 (1824).
Scirpus brevis d'Urv., Mém. Soc. Linn. Paris 4: 600 (1826). *Isolepis brevis* (d'Urv.) Brongn. in L.I.Duperrey, Voy. Monde, Phan.: 180 (1834). *Isolepis pygmaea* var. *brevis* (d'Urv.) Hook.f., Fl. Antarct. 2: 361 (1846). *Scirpus cernuus* var. *brevis* (d'Urv.) Skottsb., Wiss. Ergebn. Schwed. Sudpol.-Exped. 4(10): 41 (1909).
Isolepis pumila Link, Hort. Berol. 1: 286 (1827).
Scirpus savii var. *major* Guss., Fl. Sicul. Prodr. 1: (1827).
Isolepis brachyphylla J.Presl & C.Presl, Reliq. Haenk. 1: 187 (1828).
Isolepis chaetodes Link in C.L.von Buch, Phys. Beschr. Canar. Ins.: 138 (1828).
Scirpus chaetodes Link in C.L.von Buch, Phys. Beschr. Canar. Ins.: 138 (1828).
Isolepis magellanica Gaudich., Voy. Uranie: 414 (1829).
Isolepis brevifolia J.Presl & C.Presl, Reliq. Haenk. 1: 351 (1830).
Isolepis setosa Raoul ex A.Rich. in J.S.C.Dumont d'Urville, Voy. Astrolabe: 104 (1832).
Scirpus minimus Hook. & Arn., Bot. Beechey Voy.: 49 (1832).
Cyperus pumilio Nees, Linnaea 9: 284 (1834), nom. inval.
Isolepis chlorostachya Nees, Linnaea 10: 149 (1836).
Isolepis microcarpa Nees, Linnaea 10: 150 (1836).
Isolepis meyeniana Nees, Linnaea 11: 291 (1837).
Isolepis rupestris Kunth, Enum. Pl. 2: 193 (1837).
Scirpus hookeri Kunth, Enum. Pl. 2: 173 (1837).
Scirpus pumilus Willd. ex Kunth, Enum. Pl. 2: 191 (1837).
Isolepis erubescens Steud., Flora 25: 604 (1842).
Isolepis heterolepis Steud., Flora 25: 602 (1842).
Isolepis heterophylla Steud., Flora 25: 603 (1842).
Isolepis furcata Nees & Meyen, Nov. Actorum Acad. Caes. Leop.-Carol. Nat. Cur. 19(Suppl. 1): 88 (1843).

Scirpus minaae Tod., Rar. Pl. Sicil. Prov., n.s., 1: 15 (1845). *Isolepis minaae* (Tod.) Parl., Giorn. Bot. Ital. 2(1): 146 (1846). *Scirpus savii* f. *todaroanus* A.Terracc., Malpighia 2: 426 (1889). *Scirpus savii* subsp. *minaae* (Tod.) Arcang., Camp. Fl. Ital., ed. 2: 80 (1894). *Scirpus savii* var. *minaae* (Tod.) Firoi in A.Fiori & al., Fl. Italia 1: 120 (1896).
Isolepis pygmaea var. *elongata* Hook.f., Fl. Antarct. 2: 351 (1846).
Isolepis multicaulis Schltdl., Linnaea 20: 562 (1847).
Cyperus ambiguus Steud., Syn. Pl. Glumac. 2: 16 (1854), nom. illeg.
Isolepis pygmaea var. *pallida* É.Desv. in C.Gay, Fl. Chil. 6: 186 (1854).
Isolepis pygmaea var. *sanguinea* É.Desv. in C.Gay, Fl. Chil. 6: 186 (1854).
Isolepis controversa Steud., Syn. Pl. Glumac. 2: 92 (1855). *Scirpus savii* f. *controversus* (Steud.) A.Terracc., Malpighia 2: 426 (1889).
Isolepis kochii Steud., Syn. Pl. Glumac. 2: 93 (1855).
Isolepis pholiodes Steud., Syn. Pl. Glumac. 2: 93 (1855).
Isolepis punctulata Steud., Syn. Pl. Glumac. 2: 92 (1855).
Isolepis purpurascens Steud., Syn. Pl. Glumac. 2: 92 (1855).
Isolepis striatella Steud., Syn. Pl. Glumac. 2: 92 (1855).
Isolepis leptocaulis Torr., Pacif. Railr. Rep. Parke, Bot. 4: 153 (1857).
Isolepis modesta Phil., Linnaea 29: 79 (1857). *Scirpus modestus* (Phil.) F.Phil., Cat. Pl. Vasc. Chil.: 311 (1881).
Isolepis pumilo Boeck., Flora 42: 448 (1859).
Isolepis fuscata Meyen ex Boeck., Linnaea 36: 500 (1870).
Isolepis subprolifer Boeck., Linnaea 36: 492 (1870).
Isolepis trachycarpa Boeck., Linnaea 36: 501 (1870), nom. inval.
Isolepis trigyna Kunze ex Boeck., Linnaea 36: 502 (1870), nom. inval.
Scirpus pictus Boeck., Linnaea 36: 503 (1870).
Scirpus subprolifer Boeck., Linnaea 36: 492 (1870).
Isolepis pygmaea var. *californica* Torr. in C.Wilkes, U.S. Expl. Exped. 17: 476 (1874). *Scirpus cernuus* var. *californicus* (Torr.) Beetle, Rhodora 46: 145 (1944). *Scirpus cernuus* subsp. *californicus* (Torr.) Thorne, Aliso 9: 195 (1978).
Isolepis prolifera Melliss, St. Helena: 344 (1875), nom. illeg.
Scirpus nudipes Griseb., Abh. Königl. Ges. Wiss. Göttingen 24: 312 (1879).
Scirpus aphyllus Boeck., Bot. Jahrb. Syst. 7: 275 (1886). *Scirpus cernuus* f. *aphyllus* (Boeck.) Barros, Anales Mus. Argent. Ci. Nat. "Bernardino Rivadavia" 38: 144 (1935). *Isolepis cernua* f. *aphylla* (Boeck.) Vegetti, Hickenia 2: 138 (1995).
Scirpus arechavaletae Boeck., Beitr. Cyper. 1: 18 (1888).
Scirpus nervosus var. *campanus* A.Terracc., Malpighia 2: 422 (1889). *Isolepis nervosa* var. *campana* (A.Terracc.) Nyman, Consp. Fl. Eur., Suppl. 2: 319 (1890). *Scirpus savii* subsp. *campanus* (A.Terracc.) Arcang., Camp. Fl. Ital., ed. 2: 80 (1894).
Scirpus savii var. *filiformis* A.Terracc., Malpighia 2: 425 (1889).
Scirpus savii var. *fuscescens* A.Terracc., Malpighia 2: 425 (1889).
Isolepis monostachya Phil., Anales Mus. Nac. Santiago de Chile 1: 79 (1891), nom. illeg.
Scirpus chloroticus Phil., Anales Univ. Chile 93: 480 (1896).
Scirpus microstachys Phil., Anales Univ. Chile 93: 478 (1896).
Scirpus subtilis Boeck., Allg. Bot. Z. Syst. 2: 54 (1896).
Scirpus terminalis Phil., Anales Univ. Chile 93: 479 (1896).
Scirpus cernuus var. *dura* C.B.Clarke, Bot. Jahrb. Syst. 30(68): 29 (1901).
Scirpus cernuus f. *macrostachyus* Maire & Weiller, in Fl. Afr. Nord 4: 58 (1957).

var. ***meruensis*** (R.W.Haines & Lye) Muasya, Kew Bull. 57: 299 (2002).
Tanzania (Mt. Meru). 25 TAN. Ther.
**Isolepis meruensis* R.W.Haines & Lye, Bot. Not. 130: 311 (1977).

var. ***platycarpa*** (S.T.Blake) Muasya, Kew Bull. 57: 300 (2002).
SE. Australia, New Zealand. 50 NSW SOA TAS VIC 51 NZN NZS. Ther. or hemicr.
**Scirpus platycarpus* S.T.Blake, Proc. Roy. Soc. Queensland 51: 180 (1939 publ. 1940). *Isolepis platycarpa* (S.T.Blake) Soják, Cas. Nár. Mus., Odd. Prír. 148: 194 (1979 publ. 1980).

var. ***setiformis*** (Benth.) Muasya, Kew Bull. 57: 299 (2002).
Cape Prov. S. Australia, Norfolk I., Tristan da Cunha. 27 CPP 50 NFK SOA WAU 90 TDC. Ther. or hemicr.
Scirpus verrucosulus Steud., Flora 12: 145 (1829). *Eleogiton verruculosa* (Steud.) A.Dietr., Sp. Pl. 2: 98 (1832). *Isolepis verrucosula* (Steud.) Nees, Linnaea 7: 495 (1832).
Isolepis eckloniana Schrad., Anal. Fl. Cap. 1: 15 (1832), nom. illeg.
Isolepis ptycholeptos Steud., Syn. Pl. Glumac. 2: 93 (1855).
Scirpus arenarius Benth., Fl. Austral. 7: 325 (1878), nom. illeg. *Scirpus psammophilus* S.T. Blake, Proc. Roy. Soc. Queensland 51: 178 (1939 publ. 1940).
**Scirpus arenarius* var. *setiformis* Benth., Fl. Austral. 7: 326 (1878). *Scirpus setiformis* (Benth.) S.T.Blake, Proc. Roy. Soc. Queensland 51: 178 (1939 publ. 1940). *Isolepis setiformis* (Benth.) K.L.Wilson, Telopea 2: 169 (1981).

Isolepis congrua Nees in J.G.C.Lehmann, Pl. Preiss. 2: 75 (1846). *Scirpus congruus* (Nees) S.T.Blake, Proc. Roy. Soc. Queensland 48: 90 (1936 publ. 1937).
S. & SE. Australia. 50 NSW SOA VIC WAU. Ther. or hemicr.
Scirpus kochii Maiden & Betche, Proc. Linn. Soc. New South Wales 33: 318 (1908).
Scirpus multicaulis F.Muell. ex C.B.Clarke, Kew Bull., Addit. Ser. 8: 29 (1908).

Isolepis costata Hochst. ex A.Rich., Tent. Fl. Abyss. 2: 499 (1850). *Scirpus costatus* (Hochst. ex A.Rich.) Boeck., Linnaea 36: 511 (1870). *Ficinia costata* (Hochst. ex A.Rich.) H.Pfeiff., Revis. Ficinia: 22 (1921).
Ethiopia to S. Africa. 23 BUR RWA ZAI 24 ETH SUD 25 KEN TAN UGA 26 MLW ZAM ZIM 27

BOT CPP LES NAM NAT OFS SWZ TVL 29 MDG. Hemicr.

Isolepis setacea var. *abyssinica* Boeck., Flora 41: 419 (1858).

Scirpus multicostatus Baker, J. Linn. Soc., Bot. 20: 298 (1883).

Scirpus macer Boeck., Bot. Jahrb. Syst. 5: 503 (1884). *Scirpus costatus* var. *macer* (Boeck.) Cherm., Bull. Soc. Bot. France 74: 607 (1927). *Isolepis costata* var. *macra* (Boeck.) B.L.Burtt, Notes Roy. Bot. Gard. Edinburgh 43: 363 (1986).

Isolepis crassiuscula Hook.f., Fl. Tasman. 2: 86 (1858). *Scirpus crassiusculus* (Hook.f.) Benth., Fl. Austral. 7: 326 (1878). *Eleogiton crassiusculum* (Hook.f.) Soják, Cas. Nár. Mus., Odd. Prír. 148: 193 (1979 publ. 1980). *Scirpidiella crassiuscula* (Hook.f.) Rauschert, Feddes Repert. 94: 299 (1983).

Japan (Honshu), New Guinea, SE. Australia to New Zealand. 38 JAP 43 NWG 50 NSW QLD TAS VIC 51 NZN. Hemicr.

Scirpus pseudofluitans Makino, Bot. Mag. (Tokyo) 19: 28 (1905). *Scirpus fluitans* var. *pseudofluitans* (Makino) T.Koyama, J. Fac. Sci. Univ. Tokyo, Sect. 3, Bot. 3(7): 326 (1958).

Isolepis cyperoides R.Br., Prodr.: 222 (1810). *Scirpus cyperoides* (R.Br.) Spreng., Syst. Veg. 1: 208 (1824), nom. illeg. *Scirpus brunonianus* S.T.Blake, Proc. Roy. Soc. Queensland 51: 179 (1939 publ. 1940).

SW. Australia. 50 WAU. Ther. or hemicr.

Isolepis diabolica (Steud.) Schrad., Anal. Fl. Cap. 1: 21 (1832).

S. Africa. 27 CPP LES NAM NAT. Ther. or hemicr.

**Scirpus diabolicus* Steud., Flora 12: 147 (1829). *Ficinia antarctica* Nees, Linnaea 9: 292 (1834).

Isolepis antarctica Nees, Linnaea 7: 505 (1832), nom. illeg.

Ficinia schinziana Boeck., Verh. Bot. Vereins Prov. Brandenburg 29: 47 (1888).

Isolepis digitata Nees ex Schrad., Anal. Fl. Cap. 1: 20 (1832). *Eleogiton digitata* (Nees ex Schrad.) Nees, Linnaea 9: 291 (1834). *Scirpus digitatus* (Nees ex Schrad.) Boeck., Linnaea 36: 691 (1870).

Cape Prov. 27 CPP. Ther. or hemicr.

Eleogiton digitata var. *dissoluta* Nees, Linnaea 10: 163 (1836). *Isolepis dissoluta* (Nees) Kunth, Enum. Pl. 2: 219 (1837).

Eleogiton longifolia Nees, Linnaea 10: 164 (1836).

Isolepis dubia Kunth, Enum. Pl. 2: 216 (1837).

Rhynchospora glomerata Vahl ex Kunth, Enum. Pl. 2: 216 (1837), nom. inval.

Rhynchospora oligosantha Endl. ex Kunth, Enum. Pl. 2: 216 (1837).

Schoenus aggregatus Spreng. ex Kunth, Enum. Pl. 2: 202 (1837).

Isolepis martiana Schrad. ex Nees in C.F.P.von Martius & auct. suc. (eds.), Fl. Bras. 2(1): 84 (1842).

Scirpus flaccifolius Steud., Syn. Pl. Glumac. 2: 83 (1855).

Eleocharis flexifolia Rchb. ex C.B.Clarke in T.A.Durand & H.Schinz, Consp. Fl. Afric. 5: 622 (1894), nom. inval.

Isolepis distigmatosa (C.B.Clarke) Edgar, New Zealand J. Bot. 24: 159 (1987).

New Zealand. 51 CTM NZN NZS. Ther. or hemicr.

Isolepis prolifera Hook.f., Fl. Nov.-Zel. 1: 271 (1853), nom. illeg.

**Scirpus sulcatus* var. *distigmatosus* C.B.Clarke in T.F. Cheeseman, Man. New Zealand Fl.: 775 (1906).

Isolepis expallescens Kunth, Enum. Pl. 2: 196 (1837). *Scirpus expallescens* (Kunth) Boeck., Linnaea 36: 509 (1870).

Cape Prov. 27 CPP. Hemicr. or rhizome geophyte.

Isolepis fluitans (L.) R.Br., Prodr.: 221 (1810).

Old World. 10 DEN GRB NOR SWE 11 BGM CZE GER NET POL? 12 FRA POR SPA 13 ITA YUG? 21 AZO 22 MLI NGA 23 BUR CMN RWA ZAI 24 ETH SUD 25 KEN TAN UGA 26 ANG MLW MOZ ZAM ZIM 27 CPP LES NAM NAT OFS SWZ TVL 29 COM MDG REU 40 ASS IND SRL 41 VIE 42 JAW 50 NSW QLD SOA TAS VIC WAU 51 NZN NZS. Hel. or ther.

**Scirpus fluitans* L., Sp. Pl.: 48 (1763). *Clavula fluitans* (L.) Dumort., Fl. Belg.: 143 (1827). *Eleogiton fluitans* (L.) Link, Hort. Berol. 1: 285 (1827). *Dichostylis fluitans* (L.) Rchb., Fl. Germ. Excurs.: 75 (1830). *Eleocharis fluitans* (L.) Hook., Brit. Fl.: 24 (1830). *Cyperus fluitans* (L.) Missbach & E.H.L.Krause in J.Sturm, Deutschl. Fl., ed. 2, 2: 22 (1900). *Schoenoplectus fluitans* (L.) Palla in W.D.J.Koch, Syn. Deut. Schweiz. Fl., ed. 3: 2539 (1905). *Scirpidiella fluitans* (L.) Rauschert, Feddes Repert. 94: 299 (1983). *Heleogiton fluitans* (L.) Link ex Rchb., Icon. Fl. Germ. Helv. 8: t. 298 (1846).

var. ***fluitans***

Old World. 10 DEN GRB NOR SWE 11 BGM CZE GER NET POL? 12 FRA POR SPA 13 ITA YUG? 21 AZO 22 MLI NGA 23 BUR CMN RWA ZAI 24 ETH SUD 25 KEN TAN UGA 26 ANG MLW MOZ ZAM ZIM 27 CPP LES NAM NAT OFS SWZ TVL 29 COM MDG REU 40 ASS IND SRL 41 VIE 42 JAW 50 NSW QLD SOA TAS VIC WAU 51 NZN NZS. Hel. or ther.

Scirpus stolonifer Roth, Ann. Bot. (Usteri) 10: 36 (1794).

Eleogiton curvula Nees in R.Wight, Contr. Bot. India: 110 (1834). *Isolepis curvula* (Nees) Kunth, Enum. Pl. 2: 189 (1837). *Isolepis curvata* (Nees) Zoll., Syst. Verz. 2: 62 (1854). *Scirpus fluitans* var. *curvulus* (Nees) C.B.Clarke in J.D.Hooker, Fl. Brit. India 6: 653 (1894).

Eleogiton fascicularis Nees, Linnaea 10: 165 (1836). *Isolepis fascicularis* (Nees) Kunth, Enum. Pl. 2: 188 (1837). *Scirpus fluitans* var. *fascicularis* (Nees) Boeck., Linnaea 36: 486 (1870).

Isolepis fascicularis Kunth, Enum. Pl. 2: 188 (1837).

Isolepis giraudyi Steud., Syn. Pl. Glumac. 2: 91 (1855).

Isolepis intricata Steud., Syn. Pl. Glumac. 2: 318 (1855).

Eleogiton fluitans var. *fasciculata* Miq., Fl. Ned. Ind. 3: 313 (1856).

Isolepis fluitans var. *terrestris* Hook.f., Fl. Tasman. 2: 86 (1859).

Scirpus giraudyi Boeck., Linnaea 36: 483 (1870).

Scirpus fluitans var. *microstachyus* Benth., Fl. Austral. 7: 325 (1878).

Scirpus fluitans var. *terrestris* Benth., Fl. Austral. 7: 325 (1878).

Scirpus fluitans var. *spadiceus* C.B.Clarke in J.D.Hooker, Fl. Brit. India 6: 653 (1893).

Scirpus fluitans var. *emergens* H.Pfeiff., Repert. Spec. Nov. Regni Veg. 33: 202 (1933). *Scirpus amazonicus* Beetle, Leafl. W. Bot. 4: 46 (1944).

Isolepis fluitans var. *major* Lye, Bot. Not. 127: 522 (1974).

var. ***lenticularis*** (R.Br.) Muasya, Kew Bull. 57: 281 (2002).
SE. Australia, New Zealand. 50 SOA TAS VIC 51 NZN. Ther. or hemicr.
**Isolepis lenticularis* R.Br., Prodr.: 222 (1810). *Scirpus lenticularis* (R.Br.) Spreng., Syst. Veg. 1: 208 (1824).
Scirpus carsei Kük., Trans. & Proc. New Zealand Inst. 48: 241 (1915 publ. 1916).

var. ***nervosa*** (Hochst. ex A.Rich.) Lye, Lidia 3: 132 (1994).
Ethiopia. 24 ETH. Ther. or hemicr.
**Isolepis nervosa* Hochst. ex A.Rich., Tent. Fl. Abyss. 2: 499 (1850). *Scirpus nervosus* (Hochst. ex A.Rich.) Boeck., Linnaea 36: 484 (1870).
Isolepis fuscescens Steud., Syn. Pl. Glumac. 2: 92 (1855).
Scirpus ramosus Boeck., Linnaea 38: 377 (1874).

Isolepis gaudichaudiana Kunth, Enum. Pl. 2: 201 (1837). *Scirpus gaudichaudii* Boeck., Linnaea 36: 511 (1870).
SE. Australia. 50 NSW VIC. Ther. or hemicr.
Scirpus inundatus var. *floribundus* Benth., Fl. Austral. 7: 330 (1878).

Isolepis graminoides (R.W.Haines & Lye) Lye, Bot. Not. 127: 525 (1974).
E. Trop. Africa (Mt. Elgon). 25 KEN UGA. Hemicr.
**Scirpus graminoides* R.W.Haines & Lye, Bot. Not. 123: 430 (1970). *Scirpidiella graminoides* (R.W.Haines & Lye) Rauschert, Feddes Repert. 94: 300 (1983).

Isolepis habra (Edgar) Soják, Cas. Nár. Mus., Odd. Prír. 148: 194 (1979 publ. 1980).
N. Borneo, Philippines, New Guinea, SE. Australia, New Zealand. 42 BOR PHI 43 NWG 50 NSW TAS VIC 51 ATP CTM NZN NZS. Ther. or hemicr.
**Scirpus habrus* Edgar, New Zealand J. Bot. 4: 199 (1966).
Isolepis limbata W.M.Curtis, Brunonia 7: 300 (1985).

Isolepis hemiuncialis (C.B.Clarke) J.Raynal, Adansonia, n.s., 17: 55 (1977).
Namibia to W. Cape Prov. 27 CPP NAM. Ther.
**Scirpus hemiuncialis* C.B.Clarke in É.A.J.de Wildeman, Pl. Nov. Hort. Then. 1: 23 (1904).
Scirpus aciformis B.Nord., Dinteria 11: 53 (1974). *Isolepis aciformis* (B.Nord.) J.Raynal, Adansonia, n.s., 17: 55 (1977).

Isolepis hookeriana Boeck., Flora 41: 418 (1858). *Scirpus hookerianus* (Boeck.) S.T.Blake, Contr. Queensland Herb. 8: 19 (1969).
S. & E. Australia. 50 NSW QLD SOA TAS VIC WAU. Ther. or hemicr.
Isolepis hookeriana var. *elatior* Boeck., Flora 41: 418 (1858).
Scirpus calocarpus S.T.Blake, Proc. Roy. Soc. Queensland 51: 179 (1939 publ. 1940).

Isolepis humbertii (Cherm.) J.Raynal, Adansonia, n.s., 17: 56 (1977).
C. Madagascar. 29 MDG. Hemicr.
**Scirpus humbertii* Cherm., Bull. Soc. Bot. France 68: 422 (1921 publ. 1922).

Isolepis humillima (Benth.) K.L.Wilson, Telopea 2: 167 (1981).
Australia. 50 NTA QLD WAU. Ther. or hemicr.
Isolepis acaulis F.Muell. ex Benth., Fl. Austral. 7: 324 (1878).
**Scirpus humillimus* Benth., Fl. Austral. 7: 324 (1878).

Isolepis hystrix (Thunb.) Schrad., Anal. Fl. Cap. 1: 23 (1832).
Cape Prov. 27 CPP (50) nsw soa vic (78) vrg. Ther.
**Scirpus hystrix* Thunb., Prodr. Pl. Cap.: 17 (1794).
Isolepis dregeana Kunth, Enum. Pl. 2: 204 (1837).
Scirpus dipsaceus Vahl ex Kunth, Enum. Pl. 2: 204 (1837).
Scirpus natalensis Boeck. ex C.B.Clarke in T.A. Durand & H.Schinz, Consp. Fl. Afric. 5: 623 (1894), nom. inval.

Isolepis incomtula Nees, Linnaea 10: 154 (1836). *Scirpus incomtulus* (Nees) Boeck., Linnaea 36: 693 (1870).
Cape Prov. 27 CPP. Ther. or hemicr.
Isolepis echinidium Nees, Linnaea 10: 155 (1836).
Isolepis exilis Nees, Linnaea 10: 147 (1836).
Isolepis kunthiana Steud., Syn. Pl. Glumac. 2: 95 (1855).

Isolepis inconspicua (Levyns) J.Raynal, Adansonia, n.s., 17: 56 (1977).
Cape Prov. 27 CPP. Ther. or hemicr.
**Scirpus inconspicuus* Levyns, J. S. African Bot. 10: 30 (1944).

Isolepis inundata R.Br., Prodr.: 222 (1810). *Scirpus inundatus* (R.Br.) Poir. in J.B.A.M.de Lamarck, Encycl., Suppl. 5: 103 (1817).
New Guinea to New Zealand. 43 NWG 50 NFK NSW QLD SOA TAS VIC 51 CTM NZN NZS (60) fij. Ther. or hemicr.
Isolepis propinqua R.Br., Prodr.: 22 (1810).
Isolepis conspersa Nees in S.L.Endlicher, Prodr. Fl. Norfolk.: 23 (1833). *Scirpus conspersus* (Nees) Laing, Trans. & Proc. New Zealand Inst. 47: 19 (1914 publ. 1915).
Isolepis propinqua Nees, Ann. Mag. Nat. Hist. 6: 46 (1841), nom. illeg.
Isolepis gunnii Steud., Syn. Pl. Glumac. 2: 94 (1855).
Isolepis urvillei Steud., Syn. Pl. Glumac. 2: 94 (1855).
Isolepis multinervosa Boeck., Flora 41: 418 (1858).
Scirpus urvillei Boeck., Linnaea 36: 510 (1870).
Scirpus nanus F.Phil., Cat. Pl. Vasc. Chil.: 311 (1881), nom. illeg.
Scirpus phaeocarpus F.Phil., Cat. Pl. Vasc. Chil.: 312 (1881).
Scirpus inundatus var. *major* C.B.Clarke in T.F. Cheeseman, Man. New Zealand Fl.: 775 (1906).

Isolepis inyangensis Muasya & Goetgh., Kew Bull. 57: 282 (2002).
Zimbabwe to KwaZulu-Natal. 26 ZIM 27 NAT SWZ. Ther. or hemicr.

Isolepis karroica (C.B.Clarke) J.Raynal, Adansonia, n.s., 17: 56 (1977).
Cape Prov. 27 CPP. Ther. or hemicr.
**Scirpus karroicus* C.B.Clarke in W.H.Harvey & auct. suc. (eds.), Fl. Cap. 7: 221 (1898).

Isolepis keniaensis Lye, Nordic J. Bot. 2: 565 (1983).
Kenya (Mt. Kenya). 25 KEN. Hemicr.

Isolepis kilimanjarica R.W.Haines & Lye, Bot. Not. 127: 522 (1974). *Scirpidiella kilimanjarica* (R.W. Haines & Lye) Rauschert, Feddes Repert. 94: 300 (1983).
Tanzania (Mt. Kilimanjaro). 25 TAN. Hemicr. or rhizome geophyte.

Isolepis leptostachya Kunth, Enum. Pl. 2: 190 (1837).
Scirpus leptostachyus (Kunth) Boeck., Linnaea 36: 485 (1870).
SW. Cape Prov. 27 CPP. Ther.

Isolepis leucoloma (Nees) C.Archer, Bothalia 28: 42 (1998).
Cape Prov. 27 CPP. Ther.
**Cyperus leucoloma* Nees, Linnaea 10: 133 (1836).

Isolepis ludwigii (Steud.) Kunth, Enum. Pl. 2: 189 (1837).
Cape Prov. 27 CPP. Ther. or hemicr.
**Fimbristylis ludwigii* Steud., Flora 12: 139 (1829). *Trichelostylis ludwigii* (Steud.) Nees, Linnaea 9: 290 (1834). *Dichostylis ludwigii* (Steud.) Nees ex Kunth, Enum. Pl. 2: 189 (1837). *Scirpus ludwigii* (Steud.) Boeck., Linnaea 36: 486 (1870).
Eleogiton rubicunda Nees, Linnaea 10: 164 (1836).
Scirpus fascicularis Willd. ex Kunth, Enum. Pl. 2: 188 (1837).
Isolepis tenuior Steud., Syn. Pl. Glumac. 2: 91 (1855).

Isolepis marginata (Thunb.) A.Dietr., Sp. Pl. 2: 110 (1832).
Cape Prov., S. & SE. Australia. 27 CPP 50 NSW SOA TAS VIC WAU. Ther. or hemicr.
**Scirpus marginatus* Thunb., Prodr. Pl. Cap.: 17 (1794). *Ficinia marginata* (Thunb.) Fourc., Trans. Roy. Soc. South Africa 21: 76 (1932).
Isolepis cartiliganea R.Br., Prodr.: 222 (1810). *Scirpus cartilagineus* (R.Br.) Spreng., Syst. Veg. 1: 208 (1824).
Cyperus rothianus Schult. in J.J.Roemer & J.A.Schultes, Mant. 2: 95 (1824).
Isolepis chrysocarpa Nees, Linnaea 7: 499 (1832).
Isolepis plebeia Schrad., Anal. Fl. Cap. 1: 18 (1832).
Isolepis plebeia var. *maior* Schrad., Anal. Fl. Cap. 1: 18 (1832).
Isolepis phaeocarpa Nees, Linnaea 10: 153 (1836), nom. nud.
Isolepis sphaerocarpa Nees, Linnaea 10: 153 (1836).
Isolepis notata Nees in J.G.C.Lehmann, Pl. Preiss. 2: 74 (1846).
Isolepis cartilaginea var. *elatior* Nees in J.G.C.Lehmann, Pl. Preiss. 2: 74 (1848).
Isolepis cartilaginea subvar. *minor* Nees in J.G.C.Lehmann, Pl. Preiss. 2: 74 (1848).
Isolepis cartilaginea var. *parva* Nees in J.G.C.Lehmann, Pl. Preiss. 2: 73 (1848).
Isolepis cartilaginea subvar. *pauciflora* Nees in J.G.C.Lehmann, Pl. Preiss. 2: 74 (1848).
Cyperus modestulus Steud., Syn. Pl. Glumac. 2: 16 (1854).
Isolepis semipedunculata Boeck., Flora 41: 417 (1858).
Scirpus cartilagineus var. *propinquus* Benth., Fl. Austral. 7: 328 (1878).
Scirpus ebenocarpus Kirk, Trans. & Proc. New Zealand Inst. 17: 224 (1885).

Isolepis minuta (Turrill) J.Raynal, Adansonia, n.s., 17: 56 (1977).
Cape Prov. 27 CPP. Ther. or hemicr.
**Scirpus minutus* Turrill, Bull. Misc. Inform. Kew 1925: 69 (1925).

Isolepis montivaga (S.T.Blake) K.L.Wilson, Telopea 2: 168 (1981).
SE. Australia. 50 NSW TAS VIC. Ther. or hemicr.
**Scirpus montivagus* S.T.Blake, Proc. Roy. Soc. Queensland 60: 46 (1950).

Isolepis moseleyana (Boeck.) Muasya, Kew Bull. 57: 321 (2002).
Tristan da Cunda. 90 TDC. Ther. or hemicr.
**Scirpus moseleyanus* Boeck., Flora 58: 262 (1875).

Isolepis natans (Thunb.) A.Dietr., Sp. Pl. 2: 106 (1832).
S. Trop. & S. Africa. 26 ANG ZIM 27 CPP NAT TVL. Hemicr.
**Scirpus natans* Thunb., Prodr. Pl. Cap.: 17 (1794).
Isolepis palustris Schrad., Anal. Fl. Cap. 1: 17 (1832).
Isolepis rivularis Schrad., Anal. Fl. Cap. 1: 19 (1832).
Scirpus rivularis (Schrad.) Boeck., Linnaea 36: 504 (1870).
Isolepis pallida Nees, Linnaea 8: 81 (1833).

Isolepis nigricans Kunth in F.W.H.von Humboldt, A.J.A.Bonpland & C.S.Kunth, Nov. Gen. Sp. 1: 220 (1816). *Scirpus nigricans* (Kunth) Poir. in J.B.A.M.de Lamarck, Encycl., Suppl. 5: 104 (1817).
C. America to Argentina. 80 COS GUA PAN 82 VEN 83 BOL CLM ECU PER 85 AGE AGS CLC. Ther. or hemicr.
Isolepis psilocarpa Kunth, Enum. Pl. 2: 193 (1837).
Cyperus urvilleanus Steud., Syn. Pl. Glumac. 2: 16 (1854).
Isolepis albescens É.Desv. in C.Gay, Fl. Chil. 6: 188 (1854), nom. illeg. *Scirpus albescens* (É.Desv.) F.Phil., Cat. Pl. Vasc. Chil.: 310 (1881).
Isolepis vivipara Schrad. ex É.Desv. in C.Gay, Fl. Chil. 6: 189 (1854).
Isolepis angachillensis Steud., Syn. Pl. Glumac. 2: 94 (1855). *Scirpus angachillensis* (Steud.) Boeck., Linnaea 36: 496 (1870).
Isolepis conceptionis Steud., Syn. Pl. Glumac. 2: 94 (1855).
Isolepis nana Phil., Linnaea 29: 79 (1857).
Isolepis ramulosa Steud. in W.Lechler, Berberid. Amer. Austral.: 50 (1857).
Isolepis tristachya Phil., Linnaea 30: 203 (1860).
Scirpus bridgesii Boeck., Linnaea 36: 508 (1870).
Scirpus gaudichaudii var. *tenuis* Boeck., Linnaea 36: 512 (1870).
Scirpus constitutionis Boeck. ex R.Phil., Anales Univ. Chile 93: 483 (1896).
Scirpus didymostachyus Phil., Anales Univ. Chile 93: 475 (1896).
Scirpus trichocaulos Phil., Anales Univ. Chile 93: 478 (1896).
Scirpus philippii C.B.Clarke, Bot. Jahrb. Syst. 30(68): 32 (1902), nom. illeg.

Isolepis oldfieldiana (S.T.Blake) K.L.Wilson, Telopea 2: 168 (1981).
SW. Australia. 50 WAU. Ther. or hemicr.
Scirpus brizoides Benth., Fl. Austral. 7: 326 (1878), nom. illeg. **Scirpus oldfieldianus* S.T.Blake, Proc. Roy. Soc. Queensland 51: 179 (1939 publ. 1940).

Isolepis omissa J.Raynal, Adansonia, n.s., 17: 54 (1977).
Ethiopia. 24 ETH. Hemicr.

Isolepis pellocolea B.L.Burtt, Notes Roy. Bot. Gard. Edinburgh 43: 363 (1986).
S. Africa (Drakensberg). 27 LES NAT. Ther. or hemicr.

Isolepis pottsii (V.J.Cook) Soják, Cas. Nár. Mus., Odd. Prír. 148: 194 (1979 publ. 1980).
New Zealand. 51 NZN NZS. Ther. or hemicr.
**Scirpus pottsii* V.J.Cook, Trans. & Proc. Roy. Soc. New Zealand 81: 157 (1953).

Isolepis praetextata (Edgar) Soják, Cas. Nár. Mus., Odd. Prír. 148: 194 (1979 publ. 1980).

New Zealand. 51 ATP NZN NZS. Ther. or hemicr.
**Scirpus praetextatus* Edgar, New Zealand J. Bot. 4: 196 (1966).

Isolepis producta (C.B.Clarke) K.L.Wilson, Telopea 2: 168 (1981).
SE. Australia. 50 NSW SOA TAS VIC. Ther. or hemicr.
**Scirpus productus* C.B.Clarke, Kew Bull., Addit. Ser. 8: 28 (1908).

Isolepis prolifera (Rottb.) R.Br., Prodr.: 223 (1810).
St. Helena, SW. & SE. Australia to New Zealand, Tristan da Cunha. (12) fra 27 CPP NAT 28 STH 50 NSW VIC WAU 51 NZN NZS 90 TDC. Hemicr.
**Scirpus prolifer* Rottb., Descr. Icon. Rar. Pl.: 55 (1773).
Cyperus punctatus Lam., Tabl. Encycl. 1: 144 (1791).
Cyperus aitonii Spreng., Mant. Prim. Fl. Hal. 1: 29 (1807).
Isolepis erythronegma Steud., Syn. Pl. Glumac. 2: 96 (1855).
Isolepis globosa Buchanan, Trans. & Proc. New Zealand Inst. 3: 211 (1871).

Isolepis pseudosetacea (Daveau) Vasc., Bol. Soc. Brot., II, 44: 83 (1970).
C. & SE. U.S.A., W. Medit. 12 POR SPA 20 ALG MOR 74 MSO 77 TEX 78 ALA ARK GEO LOU MSI. Ther. or hemicr.
**Scirpus pseudosetaceus* Daveau, Bol. Soc. Brot. 9: 85 (1891).
Isolepis gracilis Pau, Broteria, Sér. Trimestr. 2: 50 (1933), nom. illeg.
Scirpus molestus M.C.Johnst., S. W. Naturalist 9: 310 (1964). *Isolepis molesta* (M.C.Johnst.) S.G.Sm., Novon 5: 101 (1995).

Isolepis pusilla Kunth, Enum. Pl. 2: 190 (1837).
Cape Prov. 27 CPP. Ther. or hemicr.
Scirpus tenuis Spreng., Flora 12(1 Beibl.): 12 (1829), nom. illeg. *Isolepis tenuis* Schrad., Anal. Fl. Cap. 1: 15 (1832), nom. illeg. *Scirpus africanus* Beetle, Leafl. W. Bot. 4: 45 (1944). *Scirpus nanodes* Levyns, J. S. African Bot. 10: 29 (1944).
Isolepis atropurpurea Nees, Linnaea 7: 495 (1832).

Isolepis ranko (Steud.) Vegetti, Hickenia 2: 138 (1995).
Chile to S. Argentina. 85 AGS CLS. Ther. or hemicr.
**Cyperus ranko* Steud., Syn. Pl. Glumac. 2: 315 (1855). *Scirpus ranko* (Steud.) Gunckel, Revista Univ. (Santiago) 33(1): 24 (1948).
Scirpus perpusillus Boeck., Linnaea 36: 496 (1870). *Isolepis perpusilla* (Boeck.) Phil., Anales Univ. Chile 1873: 555 (1873).

Isolepis reticularis Colenso, Trans. & Proc. New Zealand Inst. 18: 277 (1885 publ. 1886). *Scirpus reticularis* (Colenso) Edgar in Moore & E.Edgar, Fl. N. Zeal. 2: 178 (1970).
New Zealand. 51 CTM NZN NZS. Ther. or hemicr.
Scirpus inundatus var. *gracillimus* C.B.Clarke in T.F.Cheeseman, Man. New Zealand Fl.: 775 (1906).

Isolepis rubicunda Kunth, Enum. Pl. 2: 188 (1837). *Scirpus rubicundus* (Kunth) Parl., Fl. Ital. 2: 83 (1852).
Cape Prov. 27 CPP. Ther. or hemicr.
Scirpus globiceps C.B.Clarke in W.H.Harvey & auct. suc. (eds.), Fl. Cap. 7: 214 (1898), nom. nud.

Isolepis ruwenzoriensis R.W.Haines & Lye, Bot. Not. 127: 524 (1974). *Scirpidiella ruwenzoriensis* (R.W.Haines & Lye) Rauschert, Feddes Repert. 94: 300 (1983).
Uganda (Ruwenzori). 25 UGA. Hemicr. or rhizome geophyte.

Isolepis sepulcralis Steud., Syn. Pl. Glumac. 2: 94 (1855).
Kenya to S. Africa, Madagascar, Tristan da Cunha. 25 KEN TAN 26 ANG ZIM 27 BOT CPP LES NAT OFS TVL (28) sth 29 MDG (50) nsw (51) nzn (81) win 90 TDC. Ther.
Fimbristylis exigua Boeck., Bot. Jahrb. Syst. 5: 506 (1884).
Scirpus griquensium C.B.Clarke in W.H.Harvey & auct. suc. (eds.), Fl. Cap. 7: 222 (1898).
Scirpus antipodus V.J.Cook, Trans. & Proc. Roy. Soc. New Zealand 81: 159 (1953).

Isolepis setacea (L.) R.Br., Prodr.: 222 (1810).
Temp. & Subtrop. Eurasia, Africa. 10 DEN FOR GRB ICE IRE NOR SWE 11 AUT BGM CZE GER HUN NET POL SWI 12 COR FRA POR SPA 13 ALB? BUL GRC ITA ROM SIC YUG 14 BLR BLT RUC RUE RUS RUW UKR 20 ALG EGY MOR 21 AZO MDR 23 CMN RWA ZAI 24 ETH 25 KEN TAN UGA 27 CPP LES NAM NAT OFS TVL 30 32 KAZ KGZ TZK UZB 33 TCS 34 AFG IRN LBS TUR 35 SAU YEM 36 CHC CHT 40 ASS EHM NEP PAK WHM 41 THA (50) tas (51) nzn (71) brc (73) ore was (75) nwj pen (76) cal (80) cos (83) bol ecu. Ther.
**Scirpus setaceus* L., Sp. Pl.: 49 (1753). *Mariscus setaceus* (L.) Moench, Methodus: 350 (1794). *Cyperus setaceus* (L.) Missbach & E.H.L.Krause in J.Sturm, Deutschl. Fl., ed. 2, 2: 21 (1900), nom. illeg. *Schoenoplectus setaceus* (L.) Palla in W.D.J.Koch, Syn. Deut. Schweiz. Fl., ed. 3: 2538 (1905).
Isolepis monandra Link, Jahrb. Gewächsk. 1(3): 78 (1820).
Scirpus setaceus var. *major* Lej., Rev. Fl. Spa: 12 (1824).
Scirpus clathratus Rchb., Icon. Fl. Germ. Helv. 8: 39 (1846). *Scirpus setaceus* var. *clathratus* (Rchb.) A.Terracc., Malpighia 2: 424 (1889).
Scirpus gracillimus Kohts, Oesterr. Bot. Z. 19: 333 (1869). *Isolepis setacea* subsp. *gracillima* (Kohts) Nyman, Consp. Fl. Eur.: 766 (1882). *Scirpus setaceus* var. *gracillimus* (Kohts) A.Terracc., Malpighia 2: 424 (1889).
Scirpus setaceus var. *savii* Boeck., Linnaea 36: 502 (1870).
Isolepis pleurocarpa Hochst. ex Boiss., Fl. Orient. 5: 379 (1882).
Isolepis setacea var. *aberdarica* R.W.Haines & Lye, Bot. Not. 130: 311 (1977).

Isolepis sororia Kunth, Enum. Pl. 2: 192 (1837). *Scirpus sororius* (Kunth) C.B.Clarke in T.A.Durand & H.Schinz, Consp. Fl. Afric. 5: 630 (1894).
S. Africa. 27 CPP LES OFS TVL. Ther. or hemicr.
Scirpus minutissimus Boeck., Verh. Bot. Vereins Prov. Brandenburg 29: 46 (1888), nom. nud.

Isolepis stellata (C.B.Clarke) K.L.Wilson, Telopea 2: 169 (1981).
S. & SE. Australia. 50 NSW SOA TAS VIC WAU. Ther. or hemicr.
**Scirpus stellatus* C.B.Clarke, Kew Bull., Addit. Ser. 8: 29 (1908).

Isolepis striata (Nees) Kunth, Enum. Pl. 2: 189 (1837).
Cape Prov. 27 CPP. Ther. or hemicr.
**Eleogiton striata* Nees, Linnaea 10: 165 (1836). *Scirpus striatus* (Nees) Fourc., Trans. Roy. Soc. South Africa 21: 76 (1932).

Scirpus capillifolius Parl., Fl. Ital. 2: 83 (1852).
Isolepis robustula Steud., Syn. Pl. Glumac. 2: 90 (1855).
Scirpus fluitans var. *robustus* Boeck., Linnaea 36: 486 (1870).

Isolepis subtilissima Boeck., Flora 41: 416 (1858). *Scirpus subtilissimus* (Boeck.) S.T.Blake, Contr. Queensland Herb. 8: 20 (1969).
Borneo to New Zealand. 42 BOR PHI SUL 43 NWG 50 NSW NZN QLD SOA TAS VIC. Ther. or hemicr.
Schoenoplectus merrillii Palla, Allg. Bot. Z. Syst. 17(Beih.): 3 (1911). *Scirpus merrillii* (Palla) Kük. ex Merr., Enum. Philipp. Fl. Pl. 1: 117 (1923).

Isolepis sulcata (Thouars) Carmich., Trans. Linn. Soc. London 12: 503 (1819).
Tristan da Cunha. 90 TDC. Ther. or hemicr.
**Scirpus sulcatus* Thouars, Esquisse Fl. Tristan d'Acugna: 36 (1808). *Isolepis carmichaelii* A.Dietr., Sp. Pl. 2: 107 (1832), nom. illeg.
Scirpus thuarsii Spreng., Syst. Veg. 4(2): 27 (1827).

Isolepis tasmanica (S.T.Blake) K.L.Wilson, Telopea 2: 170 (1981).
Tasmania. 50 TAS. Ther. or hemicr.
**Scirpus tasmanicus* S.T.Blake, Contr. Queensland Herb. 8: 21 (1969).

Isolepis tenuissima (Nees) Kunth, Enum. Pl. 2: 190 (1837).
Cape Prov. to KwaZulu-Natal. 27 CPP NAT. Hemicr. or rhizome geophyte.
**Eleogiton tenuissima* Nees, Linnaea 10: 163 (1836). *Scirpus tenuissimus* (Nees) Boeck., Linnaea 36: 481 (1870).
Isolepis aquatilis Kunth, Enum. Pl. 2: 215 (1837), nom. illeg.
Isolepis subtilis Kunth, Enum. Pl. 2: 191 (1837). *Scirpus cernuus* var. *subtilis* (Kunth) C.B.Clarke in T.A.Durand & H.Schinz, Consp. Fl. Afric. 5: 619 (1894). *Scirpus chlorostachyus* Levyns, J. S. African Bot. 10: 29 (1944). *Isolepis chlorostachya* (Levyns) Soják, Cas. Nár. Mus., Odd. Prír. 148: 194 (1979 publ. 1980).
Schoenus natans Thunb. ex C.B.Clarke, Revis. Gen. Pl. 2: 633 (1891).

Isolepis trachysperma Nees, Linnaea 10: 152 (1836). *Scirpus trachyspermus* (Nees) C.B.Clarke in T.A.Durand & H.Schinz, Consp. Fl. Afric. 5: 633 (1894).
Cape Prov. 27 CPP (50) soa vic. Ther. or hemicr.

Isolepis varians Steud., Syn. Pl. Glumac. 2: 94 (1855). *Scirpus varians* (Steud.) Boeck., Linnaea 36: 698 (1870).
Chile to Argentina. 85 AGE AGS CLC CLS. Ther. or hemicr.
Isolepis valdiviae Steud., Syn. Pl. Glumac. 2: 94 (1855).

Isolepis venustula Kunth, Enum. Pl. 2: 192 (1837). *Scirpus venustulus* (Kunth) Boeck., Linnaea 36: 479 (1870).
Cape Prov. 27 CPP. Hemicr. or rhizome geophyte.
Isolepis bicolor Nees, Linnaea 10: 148 (1836), nom. illeg.
Isolepis perpusilla Nees ex C.B.Clarke in T.A.Durand & H.Schinz, Consp. Fl. Afric. 5: 634 (1894), nom. inval.
Scirpus bachmannianus Boeck. ex C.B.Clarke in T.A.Durand & H.Schinz, Consp. Fl. Afric. 5: 634 (1894), nom. inval.

Isolepis victoriensis (N.A.Wakef.) K.L.Wilson, Telopea 2: 170 (1981).
SE. Australia. 50 NSW VIC. Ther. or hemicr.
**Scirpus victoriensis* N.A.Wakef., Victoria Naturalist 73: 163 (1957).

Isolepis wakefieldiana (S.T.Blake) K.L.Wilson, Telopea 2: 170 (1981).
Victoria to Tasmania. 50 TAS VIC. Ther. or hemicr.
Scirpus costatus var. *gracilis* Boeck., Linnaea 36: 511 (1870).
Scirpus inundatus var. *pseudosetaceus* Kük., Candollea 6: 427 (1936). **Scirpus wakefieldianus* S.T.Blake, Proc. Roy. Soc. Queensland 60: 45 (1950).

***Synonyms*:**
Isolepis abyssinica Hochst. ex C.B.Clarke = ***Bulbostylis hispidula*** (Vahl) R.W.Haines subsp. ***hispidula***
Isolepis acaulis F.Muell. ex Benth. = ***Isolepis humillima*** (Benth.) K.L.Wilson
Isolepis acaulis (Phil.) Boeck. = ***Phylloscirpus acaulis*** (Phil.) Goetgh. & D.A.Simpson
Isolepis acicularis (L.) Schltdl. = ***Eleocharis acicularis*** (L.) Roem. & Schult.
Isolepis aciformis (B.Nord.) J.Raynal = ***Isolepis hemiuncialis*** (C.B.Clarke) J.Raynal
Isolepis acrostachya (Steud.) A.Dietr. = ***Ficinia acrostachys*** (Steud.) C.B.Clarke
Isolepis acugnana Schult. = ***Isolepis bicolor*** Carmich.
Isolepis acuminata Nees = ***Ficinia acuminata*** (Nees) Nees
Isolepis albescens É.Desv. = ***Isolepis nigricans*** Kunth
Isolepis albescens Miq. = ***Rhynchospora albescens*** (Miq.) Kük.
Isolepis allochroa Steud. = ***Eleocharis atropurpurea*** (Retz.) Kunth
Isolepis ambigua Steud. = ***Eleocharis retroflexa*** subsp. ***chaetaria*** (Roem. & Schult.) T.Koyama
Isolepis ambigua Zoll. = ***Schoenoplectiella lateriflora*** (J.F.Gmel.) Lye
Isolepis anceps A.Dietr. = ***Fimbristylis complanata*** (Retz.) Link var. ***complanata***
Isolepis andina Phil. = ***Eleocharis quinqueflora*** (Hartmann) O.Schwarz
Isolepis angachillensis Steud. = ***Isolepis nigricans*** Kunth
Isolepis angularis Schrad. ex Schult. = ***Fimbristylis quinquangularis*** (Vahl) Kunth
Isolepis antarctica Nees = ***Isolepis diabolica*** (Steud.) Schrad.
Isolepis aphylla Boeck. = ***Eleocharis albida*** Torr.
Isolepis aquatilis Kunth = ***Isolepis tenuissima*** (Nees) Kunth
Isolepis arcuata Eckl. ex Kunth = ***Ficinia lateralis*** (Vahl) Kunth
Isolepis arenaria Nees = ***Bulbostylis juncoides*** (Vahl) Kük. ex Herter
Isolepis aristata J.Presl = ***Schoenoplectus pungens*** (Vahl) Palla var. ***pungens***
Isolepis armerioides Miq. = ***Bulbostylis barbata*** (Rottb.) C.B.Clarke subsp. ***barbata***
Isolepis articulata (L.) Nees = ***Schoenoplectiella articulata*** (L.) Lye
Isolepis ascolepis A.Rich. = ***Ascolepis eriocauloides*** (Steud.) Nees ex Steud.
Isolepis asperiuscula Kunth = ***Bulbostylis tenuifolia*** (Rudge) J.F.Macbr.
Isolepis asperula Kunth = ***Bulbostylis tenuifolia*** (Rudge) J.F.Macbr.
Isolepis atacamensis Phil. = ***Zameioscirpus atacamensis*** (Phil.) Dhooge & Goetgh.

Isolepis atropurpurea Nees = ***Isolepis pusilla*** Kunth
Isolepis atropurpurea (Retz.) Roem. & Schult. = ***Eleocharis atropurpurea*** (Retz.) Kunth
Isolepis australis (L.) Roem. & Schult. ex G.Don = ***Scirpoides holoschoenus*** (L.) Soják subsp. ***holoschoenus***
Isolepis australis (L.) Bluff, Nees & Schauer = ***Scirpoides holoschoenus*** (L.) Soják subsp. ***holoschoenus***
Isolepis autumnalis (L.) J.Presl & C.Presl = ***Fimbristylis autumnalis*** (L.) Roem. & Schult.
Isolepis barbata (Rottb.) R.Br. = ***Bulbostylis barbata*** (Rottb.) C.B.Clarke
Isolepis bellula Steud. = ***Lipocarpha hemisphaerica*** (Roth) Goetgh.
Isolepis bergiana (Spreng.) Schult. = ***Isolepis antarctica*** (L.) Roem. & Schult.
Isolepis bergiana Drège = ***Isolepis antarctica*** (L.) Roem. & Schult.
Isolepis berlandieri Steud. = ***Bulbostylis junciformis*** (Kunth) C.B.Clarke
Isolepis beyrichii Schrad. ex B.D.Jacks. = ***Bulbostylis juncoides*** (Vahl) Kük. ex Herter
Isolepis bicolor Nees = ***Isolepis venustula*** Kunth
Isolepis bispicata (J.König ex Roxb.) Schult. = ***Fimbristylis schoenoides*** (Retz.) Vahl
Isolepis bivalvis (Lam.) Steud. = ***Fimbristylis bivalvis*** (Lam.) Lye
Isolepis boeckeleri Oliv. = ***Cyperus dubius*** Rottb. subsp. ***dubius***
Isolepis brachyphylla J.Presl & C.Presl = ***Isolepis cernua*** (Vahl) Roem. & Schult. var. ***cernua***
Isolepis brachyphylla Schult. = ***Fimbristylis dichotoma*** (L.) Vahl subsp. ***dichotoma***
Isolepis breviculmis Kunth = ***Bulbostylis humilis*** (Kunth) C.B.Clarke
Isolepis brevifolia J.Presl & C.Presl = ***Isolepis cernua*** (Vahl) Roem. & Schult. var. ***cernua***
Isolepis brevis (d'Urv.) Brongn. = ***Isolepis cernua*** (Vahl) Roem. & Schult. var. ***cernua***
Isolepis bufonia Kunth = ***Bulbostylis tenuifolia*** (Rudge) J.F.Macbr.
Isolepis bulbosa (L.) Nees = ***Ficinia bulbosa*** (L.) Nees
Isolepis caespitula Liebm. = ***Lipocarpha micrantha*** (Vahl) G.C.Tucker
Isolepis capillaris (L.) Roem. & Schult. = ***Bulbostylis capillaris*** (L.) Kunth ex C.B.Clarke
Isolepis capillaris f. *borealiamericanae* Kunth = ***Bulbostylis capillaris*** (L.) Kunth ex C.B.Clarke subsp. ***capillaris***
Isolepis capillaris var. *capitata* Miq. = ***Bulbostylis densa*** var. ***capitata*** (Miq.) Ohwi
Isolepis capillaris f. *europaea* Kunth = ***Bulbostylis capillaris*** (L.) Kunth ex C.B.Clarke subsp. ***capillaris***
Isolepis capillaris f. *indicae* Kunth = ***Bulbostylis capillaris*** (L.) Kunth ex C.B.Clarke subsp. ***capillaris***
Isolepis capillaris f. *meridionaliamericanae* Kunth = ***Bulbostylis capillaris*** (L.) Kunth ex C.B.Clarke subsp. ***capillaris***
Isolepis carmichaelii A.Dietr. = ***Isolepis sulcata*** (Thouars) Carmich.
Isolepis cartilaginea var. *elatior* Nees = ***Isolepis marginata*** (Thunb.) A.Dietr.
Isolepis cartilaginea subvar. *minor* Nees = ***Isolepis marginata*** (Thunb.) A.Dietr.
Isolepis cartilaginea var. *parva* Nees = ***Isolepis marginata*** (Thunb.) A.Dietr.
Isolepis cartilaginea subvar. *pauciflora* Nees = ***Isolepis marginata*** (Thunb.) A.Dietr.
Isolepis cartiliganea R.Br. = ***Isolepis marginata*** (Thunb.) A.Dietr.
Isolepis cartiliganea var. *rigida* Bergg. = ***Isolepis aucklandica*** Hook.f.
Isolepis cernua f. *aphylla* (Boeck.) Vegetti = ***Isolepis cernua*** (Vahl) Roem. & Schult. var. ***cernua***
Isolepis chaetodes Link = ***Isolepis cernua*** (Vahl) Roem. & Schult. var. ***cernua***
Isolepis chlorostachya Nees = ***Isolepis cernua*** (Vahl) Roem. & Schult. var. ***cernua***
Isolepis chlorostachya (Levyns) Soják = ***Isolepis tenuissima*** (Nees) Kunth
Isolepis chrysocarpa Nees = ***Isolepis marginata*** (Thunb.) A.Dietr.
Isolepis ciliata C.Presl = ***Bulbostylis ciliata*** (C.Presl) C.Presl ex Nees
Isolepis ciliatifolia (Elliott) Torr. = ***Bulbostylis ciliatifolia*** (Elliott) Fernald
Isolepis ciliifolia Steud. = ***Bulbostylis conifera*** (Kunth) Beetle
Isolepis cinnamometorum (Vahl) Roem. & Schult. = ***Fimbristylis cinnamometorum*** (Vahl) Kunth
Isolepis circinata Kunth = ***Bulbostylis circinata*** (Kunth) C.B.Clarke
Isolepis coarctata Torr. = ***Bulbostylis capillaris*** (L.) Kunth ex C.B.Clarke subsp. ***capillaris***
Isolepis cochleata Steud. = ***Fimbristylis acicularis*** R.Br.
Isolepis collina Kunth = ***Bulbostylis contexta*** (Nees) Bodard
Isolepis commutata Nees = ***Ficinia gracilis*** Schrad.
Isolepis complanata (Retz.) Roem. & Schult. = ***Fimbristylis complanata*** (Retz.) Link
Isolepis compressa Nees = [27]
Isolepis conceptionis Steud. = ***Isolepis nigricans*** Kunth
Isolepis congesta Schrad. ex Schult. = ***Bulbostylis junciformis*** (Kunth) C.B.Clarke
Isolepis conifera Kunth = ***Bulbostylis conifera*** (Kunth) Beetle
Isolepis conostachya Boeck. = ***Bulbostylis conostachya*** (Boeck.) Beetle
Isolepis consanguinea Kunth = ***Bulbostylis consanguinea*** (Kunth) C.B.Clarke
Isolepis consocialis Steud. = ***Bulbostylis hispidula*** (Vahl) R.W.Haines subsp. ***hispidula***
Isolepis conspersa Nees = ***Isolepis inundata*** R.Br.
Isolepis controversa Steud. = ***Isolepis cernua*** (Vahl) Roem. & Schult. var. ***cernua***
Isolepis coronaria Roem. & Schult. = ***Cyperus leucocephalus*** Retz.
Isolepis corymbosa J.Presl & C.Presl =
Isolepis corymbosa Roth ex Roem. & Schult. = ***Schoenoplectus corymbosus*** (Roth ex Roem. & Schult.) J.Raynal
Isolepis costata var. *macra* (Boeck.) B.L.Burtt = ***Isolepis costata*** Hochst. ex A.Rich.
Isolepis crinita Schrad. ex Nees = ***Bulbostylis circinata*** (Kunth) C.B.Clarke
Isolepis cumingii Steud. = ***Bulbostylis barbata*** (Rottb.) C.B.Clarke subsp. ***barbata***
Isolepis curvata (Nees) Zoll. = ***Isolepis fluitans*** (L.) R.Br. var. ***fluitans***
Isolepis curvifolia Schrad. = ***Fimbristylis dichotoma*** (L.) Vahl subsp. ***dichotoma***
Isolepis curvula (Nees) Kunth = ***Isolepis fluitans*** (L.) R.Br. var. ***fluitans***
Isolepis cuspidata (Roth) A.Dietr. = ***Fuirena cuspidata*** (Roth) Kunth
Isolepis decipiens Nees = ***Schoenoplectus decipiens*** (Nees) J.Raynal

Isolepis densa (Wall.) Schult. = ***Bulbostylis densa*** (Wall.) Hand.-Mazz.
Isolepis depauperata (R.Br.) Link = ***Fimbristylis dichotoma*** subsp. ***depauperata*** (R.Br.) J.Kern
Isolepis deserticola Phil. = ***Zameioscirpus gaimardioides*** (É.Desv.) Dhooge & Goetgh.
Isolepis dichotoma (L.) Kunth = ***Fimbristylis dichotoma*** (L.) Vahl
Isolepis dichroa Steud. = ***Eleocharis atropurpurea*** (Retz.) Kunth
Isolepis dioeca Kunth = ***Scirpoides dioecus*** (Kunth) Browning
Isolepis dipsacea (Rottb.) Roem. & Schult. = ***Fimbristylis dipsacea*** (Rottb.) C.B.Clarke
Isolepis dissoluta (Nees) Kunth = ***Isolepis digitata*** Nees ex Schrad.
Isolepis disticha Boeck. = ***Bulbostylis pauciflora*** (Liebm.) C.B.Clarke
Isolepis divaricata (Elliott) A.Dietr. = ***Scirpus divaricatus*** Elliott
Isolepis dregeana Kunth = ***Isolepis hystrix*** (Thunb.) Schrad.
Isolepis drummondii Torr. & Hook. = ***Fimbristylis puberula*** Vahl var. ***puberula***
Isolepis dubia Kunth = ***Isolepis digitata*** Nees ex Schrad.
Isolepis dura Zoll. & Moritzi = ***Fimbristylis dura*** (Zoll. & Moritzi) Merr.
Isolepis echinidium Nees = ***Isolepis incomtula*** Nees
Isolepis echinocephala Oliv. = ***Oxycaryum cubense*** (Poepp. & Kunth) Palla
Isolepis echinulata Kunth = ***Cyperus squarrosus*** L.
Isolepis eckloniana Schrad. = ***Isolepis cernua*** var. ***setiformis*** (Benth.) Muasya
Isolepis eckloniana Nees = ***Ficinia paradoxa*** (Schrad.) Nees
Isolepis ecklonii A.Dietr. = ***Ficinia ecklonea*** (Steud.) Nees
Isolepis elachista Schult. = ***Fimbristylis dipsacea*** (Rottb.) C.B.Clarke
Isolepis elliottii (Spreng.) A.Dietr. = ***Fimbristylis dichotoma*** (L.) Vahl subsp. ***dichotoma***
Isolepis elongata C.A.Mey. = ***Trichophorum pumilum*** (Vahl) Schinz & Thell. subsp. ***pumilum***
Isolepis erubescens Steud. = ***Isolepis cernua*** (Vahl) Roem. & Schult. var. ***cernua***
Isolepis erythronegma Steud. = ***Isolepis prolifera*** (Rottb.) R.Br.
Isolepis exigua Hook. & Arn. = ***Eleocharis acicularis*** (L.) Roem. & Schult. subsp. ***acicularis***
Isolepis exilis Nees = ***Isolepis incomtula*** Nees
Isolepis exilis Kunth = ***Bulbostylis hispidula*** (Vahl) R.W.Haines subsp. ***hispidula***
Isolepis exserens (Rchb.) Bluff, Nees & Schauer = ***Scirpoides holoschoenus*** (L.) Soják subsp. ***holoschoenus***
Isolepis falcata (Vahl) Roem. & Schult. = ***Fimbristylis falcata*** (Vahl) Kunth
Isolepis fascicularis (Nees) Kunth = ***Isolepis fluitans*** (L.) R.Br. var. ***fluitans***
Isolepis fascicularis Kunth = ***Isolepis fluitans*** (L.) R.Br. var. ***fluitans***
Isolepis fascicularis Sieber ex Kunth = ***Bulbostylis barbata*** (Rottb.) C.B.Clarke subsp. ***barbata***
Isolepis ferruginea (L.) Schltdl. = ***Fimbristylis ferruginea*** (L.) Vahl
Isolepis festucoides (Poir.) Kunth = ***Bulbostylis festucoides*** (Poir.) C.B.Clarke
Isolepis fibrosa Nees = ***Ficinia filiformis*** (Lam.) Schrad.
Isolepis ficinioides Steud. = ***Ficinia filiformis*** (Lam.) Schrad.
Isolepis filamentosa (Vahl) Roem. & Schult. = ***Bulbostylis filamentosa*** (Vahl) C.B.Clarke
Isolepis filiformis Roem. & Schult. = ***Eleocharis tenuis*** (Willd.) Schult. var. ***tenuis***
Isolepis filiformis (Rchb.) Bluff, Nees & Schauer = ***Scirpoides holoschoenus*** (L.) Soják subsp. ***holoschoenus***
Isolepis fimbriata (Nees) Schrad. ex B.D.Jacks. = ***Bulbostylis fimbriata*** (Nees) C.B.Clarke
Isolepis fistulosa (Forssk.) Delile = ***Schoenoplectiella articulata*** (L.) Lye
Isolepis fluitans var. *major* Lye = ***Isolepis fluitans*** (L.) R.Br. var. ***fluitans***
Isolepis fluitans var. *terrestris* Hook.f. = ***Isolepis fluitans*** (L.) R.Br. var. ***fluitans***
Isolepis foenicularis Schrad. ex Boeck. = ***Bulbostylis barbata*** (Rottb.) C.B.Clarke subsp. ***barbata***
Isolepis fontana Steud. = ***Scirpoides holoschoenus*** (L.) Soják subsp. ***holoschoenus***
Isolepis funckii Steud. = ***Bulbostylis funckii*** (Steud.) C.B.Clarke
Isolepis furcata Nees & Meyen = ***Isolepis cernua*** (Vahl) Roem. & Schult. var. ***cernua***
Isolepis fusca Link = ***Bulbostylis capillaris*** (L.) Kunth ex C.B.Clarke subsp. ***capillaris***
Isolepis fuscata Meyen ex Boeck. = ***Isolepis cernua*** (Vahl) Roem. & Schult. var. ***cernua***
Isolepis fuscescens Steud. = ***Isolepis fluitans*** var. ***nervosa*** (Hochst. ex A.Rich.) Lye
Isolepis fuscopurpurea Steud. = ***Eleocharis fuscopurpurea*** (Steud.) H.Pfeiff.
Isolepis gardneriana Steud. = ***Bulbostylis gardneriana*** (Steud.) Beetle
Isolepis giraudyi Steud. = ***Isolepis fluitans*** (L.) R.Br. var. ***fluitans***
Isolepis glaucescens Schrad. & Schult. = ***Fimbristylis squarrosa*** Vahl var. ***squarrosa***
Isolepis globosa Buchanan = ***Isolepis prolifera*** (Rottb.) R.Br.
Isolepis globulifera Steud. = ***Scirpoides holoschoenus*** subsp. ***globiferus*** (L.f.) Soják
Isolepis globulosa (Retz.) Schult. = ***Fimbristylis umbellaris*** (Lam.) Vahl
Isolepis glomerata Schrad. ex Schult. = ***Fimbristylis cymosa*** R.Br. subsp. ***cymosa***
Isolepis gracilis Pau = ***Isolepis pseudosetacea*** (Daveau) Vasc.
Isolepis gracilis Nees = ***Bulbostylis puberula*** C.B.Clarke
Isolepis gracilis (Poir.) Nees = ***Ficinia ecklonea*** (Steud.) Nees
Isolepis gracilis (Rudge) Sweet = ***Ficinia nodosa*** (Rottb.) Goetgh., Muasya & D.A.Simpson
Isolepis gracilis Kunth = ***Fimbristylis autumnalis*** (L.) Roem. & Schult.
Isolepis gracillima Hochst. ex C.B.Clarke = ***Bulbostylis capillaris*** (L.) Kunth ex C.B.Clarke subsp. ***capillaris***
Isolepis grandispica Steud. = ***Bolboschoenus maritimus*** (L.) Palla subsp. ***maritimus***
Isolepis gunnii Steud. = ***Isolepis inundata*** R.Br.
Isolepis haenkei J.Presl & C.Presl = ***Fimbristylis cymosa*** R.Br. subsp. ***cymosa***
Isolepis hamulosa (M.Bieb.) Kunth = ***Cyperus hamulosus*** M.Bieb.
Isolepis hartmannii Steud. = ***Fimbristylis vahlii*** (Lam.) Link
Isolepis hemisphaerica (Roth) A.Dietr. = ***Lipocarpha hemisphaerica*** (Roth) Goetgh.

Isolepis heterolepis Steud. = ***Isolepis cernua*** (Vahl) Roem. & Schult. var. ***cernua***
Isolepis heterolepis Steud. = ***Carex setifolia*** Kunze var. ***setifolia***
Isolepis heteromorpha Steud. = ***Eleocharis melanostachys*** (d'Urv.) C.B.Clarke
Isolepis heterophylla Steud. = ***Isolepis cernua*** (Vahl) Roem. & Schult. var. ***cernua***
Isolepis hirta Kunth = ***Fimbristylis squarrosa*** Vahl var. ***squarrosa***
Isolepis hirtella Schrad. ex Schult. = ***Bulbostylis juncoides*** (Vahl) Kük. ex Herter
Isolepis hispida Steud. = ***Bulbostylis hispida*** (Steud.) Beetle
Isolepis hispidula Nees = [40 IND]
Isolepis holoschoenus Melliss = ***Ficinia nodosa*** (Rottb.) Goetgh., Muasya & D.A.Simpson
Isolepis holoschoenus (L.) Roem. & Schult. = ***Scirpoides holoschoenus*** (L.) Soják
Isolepis holoschoenus var. *australis* Roem. & Schult. = ***Scirpoides holoschoenus*** (L.) Soják
Isolepis holoschoenus var. *romanus* (L.) Roem. & Schult. = ***Scirpoides holoschoenus*** (L.) Soják
Isolepis hookeri Nees ex Boeck. = ***Isolepis carinata*** Hook. & Arn. ex Torr.
Isolepis hookeriana var. *elatior* Boeck. = ***Isolepis hookeriana*** Boeck.
Isolepis humboldtii Roem. & Schult. = ***Lipocarpha micrantha*** (Vahl) G.C.Tucker
Isolepis humilis Kunth = ***Bulbostylis humilis*** (Kunth) C.B.Clarke
Isolepis humillima Hochst. ex C.B.Clarke = ***Bulbostylis humilis*** (Kunth) C.B.Clarke
Isolepis inclinata Delile = ***Schoenoplectus corymbosus*** (Roth ex Roem. & Schult.) J.Raynal
Isolepis incurvata (Roxb.) Nees = ***Schoenoplectiella articulata*** (L.) Lye
Isolepis intricata Steud. = ***Isolepis fluitans*** (L.) R.Br. var. ***fluitans***
Isolepis involucellata Steud. = ***Bulbostylis barbata*** (Rottb.) C.B.Clarke subsp. ***barbata***
Isolepis junciformis Kunth = ***Bulbostylis junciformis*** (Kunth) C.B.Clarke
Isolepis juncoides Miq. = ***Schoenoplectiella lateriflora*** (J.F.Gmel.) Lye
Isolepis kernii (Raymond) Lye = ***Lipocarpha kernii*** (Raymond) Goetgh.
Isolepis kochii Steud. = ***Isolepis cernua*** (Vahl) Roem. & Schult. var. ***cernua***
Isolepis koilolepis Steud. = ***Isolepis carinata*** Hook. & Arn. ex Torr.
Isolepis krausei Phil. = ***Schoenus nitens*** (R.Br.) Roem. & Schult. var. ***nitens***
Isolepis kunthiana Steud. = ***Isolepis incomtula*** Nees
Isolepis kyllingioides A.Rich. = ***Kyllingiella microcephala*** (Steud.) R.W.Haines & Lye
Isolepis labillardierei Steud. = ***Bulbostylis labillardierei*** (Steud.) Beetle
Isolepis laciniata Eckl. ex Kunth = ***Ficinia ecklonea*** (Steud.) Nees
Isolepis lanata Kunth = ***Bulbostylis lanata*** (Kunth) Lindm.
Isolepis langsdorffiana Kunth = ***Bulbostylis ciliata*** (C.Presl) C.Presl ex Nees
Isolepis latifolia (Bertol.) A.Dietr. = ***Pleurostachys tenuiflora*** Brongn.
Isolepis lenticularis R.Br. = ***Isolepis fluitans*** var. ***lenticularis*** (R.Br.) Muasya
Isolepis leptalea (Koch) Schult. = ***Isolepis cernua*** (Vahl) Roem. & Schult. var. ***cernua***
Isolepis leptocaulis Torr. = ***Isolepis cernua*** (Vahl) Roem. & Schult. var. ***cernua***
Isolepis leptos Steud. = ***Eleocharis parvula*** (Roem. & Schult.) Palla
Isolepis lereschii Shuttlew. ex Parl. = ***Eleocharis atropurpurea*** (Retz.) Kunth
Isolepis leucostachya Kunth = ***Bulbostylis leucostachya*** (Kunth) Kunth
Isolepis lichtensteiniana Kunth = ***Bulbostylis lichtensteiniana*** (Kunth) C.B.Clarke
Isolepis limbata W.M.Curtis = ***Isolepis habra*** (Edgar) Soják
Isolepis lineata Nees = ***Ficinia gracilis*** Schrad.
Isolepis lineata (Michx.) Roem. & Schult. = ***Scirpus lineatus*** Michx.
Isolepis linnaei (Rchb.) Bluff, Nees & Schauer = ***Scirpoides holoschoenus*** (L.) Soják subsp. ***holoschoenus***
Isolepis longifolia Steud. = ***Eleocharis acicularis*** (L.) Roem. & Schult. subsp. ***acicularis***
Isolepis longispica Steud. = ***Fimbristylis fusca*** (Nees) Benth. ex C.B.Clarke
Isolepis lupulina Nees = ***Schoenoplectiella roylei*** (Nees) Lye
Isolepis magellanica Gaudich. = ***Isolepis cernua*** (Vahl) Roem. & Schult. var. ***cernua***
Isolepis margaritifera Nees = ***Schoenus apogon*** Roem. & Schult. var. ***apogon***
Isolepis martiana Schrad. ex Nees = ***Isolepis digitata*** Nees ex Schrad.
Isolepis martii Roem. & Schult. = ***Cladium mariscus*** subsp. ***martii*** (Roem. & Schult.) T.V.Egorova
Isolepis maxima A.Dietr. = ***Actinoscirpus grossus*** (L.f.) Goetgh. & D.A.Simpson var. ***grossus***
Isolepis megapotamica A.Dietr. = ***Pycreus megapotamicus*** (A.Dietr.) Nees
Isolepis membranacea (Thunb.) Nees = ***Hellmuthia membranacea*** (Thunb.) R.W.Haines & Lye
Isolepis membranacea (Thunb.) Nees = ***Hellmuthia membranacea*** (Thunb.) R.W.Haines & Lye
Isolepis meruensis R.W.Haines & Lye = ***Isolepis cernua*** var. ***meruensis*** (R.W.Haines & Lye) Muasya
Isolepis mesantha Nees = ***Schoenoplectiella lateriflora*** (J.F.Gmel.) Lye
Isolepis meyeniana Nees = ***Isolepis cernua*** (Vahl) Roem. & Schult. var. ***cernua***
Isolepis micheliana (L.) Roem. & Schult. = ***Cyperus michelianus*** (L.) Delile
Isolepis micrantha (Vahl) Roem. & Schult. = ***Lipocarpha micrantha*** (Vahl) G.C.Tucker
Isolepis microcarpa Nees = ***Isolepis cernua*** (Vahl) Roem. & Schult. var. ***cernua***
Isolepis microcephala (Steud.) Lye = ***Kyllingiella microcephala*** (Steud.) R.W.Haines & Lye
Isolepis miliacea (L.) J.Presl & C.Presl = ***Fimbristylis quinquangularis*** (Vahl) Kunth
Isolepis miliacea var. *major* C.Presl = ***Fimbristylis quinquangularis*** (Vahl) Kunth
Isolepis minaae (Tod.) Parl. = ***Isolepis cernua*** (Vahl) Roem. & Schult. var. ***cernua***
Isolepis minima Schrad. = ***Lipocarpha hemisphaerica*** (Roth) Goetgh.
Isolepis modesta Phil. = ***Isolepis cernua*** (Vahl) Roem. & Schult. var. ***cernua***
Isolepis molesta (M.C.Johnst.) S.G.Sm. = ***Isolepis pseudosetacea*** (Daveau) Vasc.
Isolepis monandra Link = ***Isolepis setacea*** (L.) R.Br.
Isolepis monocephala Steud. = ***Ficinia nodosa*** (Rottb.) Goetgh., Muasya & D.A.Simpson

Isolepis monostachya Phil. = ***Isolepis cernua*** (Vahl) Roem. & Schult. var. ***cernua***
Isolepis monostachya (J.König ex Roxb.) Spreng. = ***Fimbristylis schoenoides*** (Retz.) Vahl
Isolepis mucronata (L.) Fourr. = ***Schoenoplectus mucronatus*** (L.) Palla
Isolepis muehlenbergii A.Dietr. = ***Bulbostylis capillaris*** (L.) Kunth ex C.B.Clarke subsp. ***capillaris***
Isolepis multicaulis Schltdl. = ***Isolepis cernua*** (Vahl) Roem. & Schult. var. ***cernua***
Isolepis multinervosa Boeck. = ***Isolepis inundata*** R.Br.
Isolepis nana Phil. = ***Isolepis nigricans*** Kunth
Isolepis nervosa Hochst. ex A.Rich. = ***Isolepis fluitans*** var. ***nervosa*** (Hochst. ex A.Rich.) Lye
Isolepis nervosa var. *campana* (A.Terracc.) Nyman = ***Isolepis cernua*** (Vahl) Roem. & Schult. var. ***cernua***
Isolepis nervosa var. *sicula* (C.Presl) Nyman = ***Isolepis cernua*** (Vahl) Roem. & Schult. var. ***cernua***
Isolepis nigrescens (Nees) Steud. = ***Eleocharis nigrescens*** (Nees) Steud.
Isolepis nitens (Vahl) Roem. & Schult. = ***Rhynchospora nitens*** (Vahl) A.Gray
Isolepis nodosa (Rottb.) R.Br. = ***Ficinia nodosa*** (Rottb.) Goetgh., Muasya & D.A.Simpson
Isolepis notata Nees = ***Isolepis marginata*** (Thunb.) A.Dietr.
Isolepis novae-zealandiae Colenso = ***Isolepis basilaris*** Hook.f.
Isolepis nudipes Kunth = ***Eleocharis nudipes*** (Kunth) Palla
Isolepis numidiana (Vahl) Roem. & Schult. = ***Isolepis cernua*** (Vahl) Roem. & Schult. var. ***cernua***
Isolepis obtusifolia (Lam.) Roem. & Schult. = ***Fimbristylis dichotoma*** (L.) Vahl subsp. ***dichotoma***
Isolepis oligantha (Steud.) A.Dietr. = ***Ficinia oligantha*** (Steud.) J.Raynal
Isolepis oligantha C.A.Mey. = ***Trichophorum pumilum*** (Vahl) Schinz & Thell. subsp. ***pumilum***
Isolepis oliganthes Nees = ***Ficinia oligantha*** (Steud.) J.Raynal
Isolepis oreophila Phil. = ***Amphiscirpus nevadensis*** (S.Watson) Oteng-Yeb.
Isolepis oryzetorum Steud. = ***Schoenoplectiella lateriflora*** (J.F.Gmel.) Lye
Isolepis ottonis Boeck. = ***Bulbostylis ottonis*** (Boeck.) Beetle
Isolepis pallescens (Link) A.Dietr. = ***Fimbristylis dichotoma*** (L.) Vahl subsp. ***dichotoma***
Isolepis pallida Nees = ***Isolepis natans*** (Thunb.) A.Dietr.
Isolepis paludicola Kunth = ***Schoenoplectus paludicola*** (Kunth) Palla
Isolepis palustris Schrad. = ***Isolepis natans*** (Thunb.) A.Dietr.
Isolepis panamensis J.Presl & C.Presl = [80 PAN] Rhynchospora sp ?
Isolepis paniculata Gray = ***Scirpoides holoschoenus*** (L.) Soják subsp. ***holoschoenus***
Isolepis panormitana Parl. = ***Scirpoides holoschoenus*** (L.) Soják subsp. ***holoschoenus***
Isolepis paradoxa (Spreng.) Kunth = ***Bulbostylis paradoxa*** (Spreng.) Lindm.
Isolepis paradoxa Schrad. = ***Ficinia paradoxa*** (Schrad.) Nees
Isolepis paraensis Schrad. ex Nees = ***Bulbostylis paraensis*** C.B.Clarke
Isolepis pauciflora (Liebm.) Steud. = ***Bulbostylis pauciflora*** (Liebm.) C.B.Clarke
Isolepis pentagona (Roxb.) Schult. = ***Fimbristylis quinquangularis*** (Vahl) Kunth
Isolepis pentasticha Boeck. = ***Schoenoplectiella supina*** (L.) Lye
Isolepis perpusilla (Boeck.) Phil. = ***Isolepis ranko*** (Steud.) Vegetti
Isolepis perpusilla Nees ex C.B.Clarke = ***Isolepis venustula*** Kunth
Isolepis perrottetii Steud. = ***Bulbostylis hispidula*** (Vahl) R.W.Haines subsp. ***hispidula***
Isolepis phaeocarpa Nees = ***Isolepis marginata*** (Thunb.) A.Dietr.
Isolepis pholiodes Steud. = ***Isolepis cernua*** (Vahl) Roem. & Schult. var. ***cernua***
Isolepis pilosa (Willd.) Steud. = ***Bulbostylis pilosa*** (Willd.) Cherm.
Isolepis planifolia Spreng. = ***Trichophorum planifolium*** (Spreng.) Palla
Isolepis platycarpa (S.T.Blake) Soják = ***Isolepis cernua*** var. ***platycarpa*** (S.T.Blake) Muasya
Isolepis plebeia Schrad. = ***Isolepis marginata*** (Thunb.) A.Dietr.
Isolepis plebeia var. *maior* Schrad. = ***Isolepis marginata*** (Thunb.) A.Dietr.
Isolepis pleurocarpa Hochst. ex Boiss. = ***Isolepis setacea*** (L.) R.Br.
Isolepis podocarpa Boeck. = ***Ficinia angustifolia*** (Schrad.) C.B.Clarke
Isolepis pohliana Boeck. = ***Bulbostylis pohliana*** (Boeck.) Beetle
Isolepis poiretii (Kunth) Steud. = ***Ficinia ecklonea*** (Steud.) Nees
Isolepis poiretii Roem. & Schult. = ***Scirpoides holoschoenus*** (L.) Soják subsp. ***holoschoenus***
Isolepis polycolea De Not. = ***Schoenoplectiella supina*** (L.) Lye
Isolepis polyphylla A.Rich. = ***Kyllingiella polyphylla*** (A.Rich.) Lye
Isolepis praeusta C.B.Clarke = ***Ficinia laevis*** Nees
Isolepis preslii A.Dietr. =
Isolepis prolifera Carmich. = ***Isolepis bicolor*** Carmich.
Isolepis prolifera Melliss = ***Isolepis cernua*** (Vahl) Roem. & Schult. var. ***cernua***
Isolepis prolifera Hook.f. = ***Isolepis distigmatosa*** (C.B.Clarke) Edgar
Isolepis prolongata Nees = ***Schoenoplectiella articulata*** (L.) Lye
Isolepis propinqua R.Br. = ***Isolepis inundata*** R.Br.
Isolepis propinqua Nees = ***Isolepis inundata*** R.Br.
Isolepis proxima Steud. = ***Schoenoplectiella proxima*** (Steud.) Lye
Isolepis pseudojunciformis Boeck. = ***Bulbostylis pseudojunciformis*** (Boeck.) Beetle
Isolepis pseudoschoenus (Steud.) A.Dietr. = ***Ficinia tristachya*** (Rottb.) Nees
Isolepis psilocarpa Kunth = ***Isolepis nigricans*** Kunth
Isolepis ptycholeptos Steud. = ***Isolepis cernua*** var. ***setiformis*** (Benth.) Muasya
Isolepis puberula (Poir.) Kunth = ***Bulbostylis puberula*** C.B.Clarke
Isolepis pubescens (Poir.) Roem. & Schult. = ***Fuirena pubescens*** (Poir.) Kunth
Isolepis pubiculmis Boeck. = ***Bulbostylis hispidula*** (Vahl) R.W.Haines subsp. ***hispidula***
Isolepis pubigera Rchb. ex Kunth = ***Bulbostylis hispidula*** (Vahl) R.W.Haines subsp. ***hispidula***
Isolepis pubigera (Link) Schrad. = ***Fimbristylis dichotoma*** (L.) Vahl subsp. ***dichotoma***
Isolepis pulchella Thwaites = ***Bulbostylis barbata*** subsp. ***pulchella*** (Thwaites) T.Koyama
Isolepis pumila Link = ***Isolepis cernua*** (Vahl) Roem.

& Schult. var. ***cernua***

Isolepis pumila (Vahl) Roem. & Schult. = ***Trichophorum pumilum*** (Vahl) Schinz & Thell.

Isolepis pumilo Boeck. = ***Isolepis cernua*** (Vahl) Roem. & Schult. var. ***cernua***

Isolepis punctulata Steud. = ***Isolepis cernua*** (Vahl) Roem. & Schult. var. ***cernua***

Isolepis purpurascens Steud. = ***Isolepis cernua*** (Vahl) Roem. & Schult. var. ***cernua***

Isolepis pygmaea (Vahl) Kunth = ***Isolepis cernua*** (Vahl) Roem. & Schult. var. ***cernua***

Isolepis pygmaea var. *brevis* (d'Urv.) Hook.f. = ***Isolepis cernua*** (Vahl) Roem. & Schult. var. ***cernua***

Isolepis pygmaea var. *californica* Torr. = ***Isolepis cernua*** (Vahl) Roem. & Schult. var. ***cernua***

Isolepis pygmaea var. *elongata* Hook.f. = ***Isolepis cernua*** (Vahl) Roem. & Schult. var. ***cernua***

Isolepis pygmaea var. *pallida* É.Desv. = ***Isolepis cernua*** (Vahl) Roem. & Schult. var. ***cernua***

Isolepis pygmaea var. *sanguinea* É.Desv. = ***Isolepis cernua*** (Vahl) Roem. & Schult. var. ***cernua***

Isolepis racemosa (Spreng.) Schult. =

Isolepis radiata (L.f.) Roem. & Schult. = ***Ficinia radiata*** (L.f.) Kunth

Isolepis radiciflora Steud. = ***Bulbostylis capillaris*** (L.) Kunth ex C.B.Clarke subsp. ***capillaris***

Isolepis ramulosa Steud. = ***Isolepis nigricans*** Kunth

Isolepis rariflora Schrad. ex Schult. = ***Fimbristylis schoenoides*** (Retz.) Vahl

Isolepis rehmannii (Ridl.) Lye = ***Lipocarpha rehmannii*** (Ridl.) Goetgh.

Isolepis repens Nees = ***Ficinia repens*** Kunth

Isolepis rigescens (Willd. ex Schult.) C.Presl = ***Fimbristylis cymosa*** R.Br. subsp. ***cymosa***

Isolepis rigida Steud. = ***Trichophorum rigidum*** (Steud.) Goetgh., Muasya & D.A.Simpson

Isolepis riparia R.Br. = ***Isolepis cernua*** (Vahl) Roem. & Schult. var. ***cernua***

Isolepis rivularis Schrad. = ***Isolepis natans*** (Thunb.) A.Dietr.

Isolepis robustula Steud. = ***Isolepis striata*** (Nees) Kunth

Isolepis romana (L.) Roem. & Schult. ex G.Don = ***Scirpoides holoschoenus*** (L.) Soják subsp. ***holoschoenus***

Isolepis roylei Nees = ***Schoenoplectiella roylei*** (Nees) Lye

Isolepis rubescens J.Presl = [84] Bulbostylsis sp.

Isolepis rugosa Boeck. = ***Schoenoplectiella articulata*** (L.) Lye

Isolepis rupestris Kunth = ***Isolepis cernua*** (Vahl) Roem. & Schult. var. ***cernua***

Isolepis saviana Schult. = ***Isolepis cernua*** (Vahl) Roem. & Schult. var. ***cernua***

Isolepis saviana var. *tenuis* (J.Presl) Nyman = ***Isolepis cernua*** (Vahl) Roem. & Schult. var. ***cernua***

Isolepis savii (Sebast. & Mauri) Fourr. = ***Isolepis cernua*** (Vahl) Roem. & Schult. var. ***cernua***

Isolepis scabra J.Presl & C.Presl = ***Bulbostylis scabra*** (J.Presl & C.Presl) C.B.Clarke

Isolepis scariosa (Thunb.) Nees = ***Ficinia deusta*** (P.J.Bergius) Levyns

Isolepis schimperiana Hochst. ex A.Rich. = ***Bulbostylis schimperiana*** (Hochst. ex A.Rich.) C.B.Clarke

Isolepis schoenoides Kunth = ***Bulbostylis schoenoides*** (Kunth) C.B.Clarke

Isolepis schomburghii Steud. = ***Fimbristylis aestivalis*** Vahl var. ***aestivalis***

Isolepis schraderi A.Dietr. = ***Fimbristylis cymosa*** R.Br. subsp. ***cymosa***

Isolepis schraderiana Steud. = ***Bulbostylis schraderiana*** (Steud.) Beetle

Isolepis schweinfurthiana Oliv. = ***Bulbostylis hispidula*** (Vahl) R.W.Haines subsp. ***hispidula***

Isolepis sellowiana Kunth = ***Bulbostylis sellowiana*** (Kunth) Palla

Isolepis semipedunculata Boeck. = ***Isolepis marginata*** (Thunb.) A.Dietr.

Isolepis senegalensis Steud. = ***Schoenoplectiella senegalensis*** (Steud.) Lye

Isolepis sesslerioides Kunth = ***Isolepis antarctica*** (L.) Roem. & Schult.

Isolepis setacea var. *aberdarica* R.W.Haines & Lye = ***Isolepis setacea*** (L.) R.Br.

Isolepis setacea var. *abyssinica* Boeck. = ***Isolepis costata*** Hochst. ex A.Rich.

Isolepis setacea subsp. *gracillima* (Kohts) Nyman = ***Isolepis setacea*** (L.) R.Br.

Isolepis setifolia A.Rich. = ***Eleocharis setifolia*** (A.Rich.) J.Raynal

Isolepis setiformis (Benth.) K.L.Wilson = ***Isolepis cernua*** var. ***setiformis*** (Benth.) Muasya

Isolepis setosa Raoul ex A.Rich. = ***Isolepis cernua*** (Vahl) Roem. & Schult. var. ***cernua***

Isolepis sicula C.Presl = ***Isolepis cernua*** (Vahl) Roem. & Schult. var. ***cernua***

Isolepis sieberi Schrad. = ***Bulbostylis puberula*** C.B.Clarke

Isolepis simillima Steud. = ***Schoenoplectiella supina*** (L.) Lye

Isolepis sphaerocarpa Nees = ***Isolepis marginata*** (Thunb.) A.Dietr.

Isolepis sphaerophora Ehrenb. ex Boeck. = ***Scirpoides holoschoenus*** (L.) Soják subsp. ***holoschoenus***

Isolepis spiralis A.Rich. = ***Desmoschoenus spiralis*** (A.Rich.) Hook.f.

Isolepis squarrosa Carmich. = ***Isolepis bicolor*** Carmich.

Isolepis squarrosa Kunth = ***Lipocarpha micrantha*** (Vahl) G.C.Tucker

Isolepis squarrosa (L.) Roem. & Schult. = ***Lipocarpha squarrosa*** (L.) Goetgh.

Isolepis stenophylla (Elliott) Torr. = ***Bulbostylis stenophylla*** (Elliott) C.B.Clarke

Isolepis steudelii Schrad. = ***Ficinia ecklonea*** (Steud.) Nees

Isolepis striatella Steud. = ***Isolepis cernua*** (Vahl) Roem. & Schult. var. ***cernua***

Isolepis stricta (Roxb.) A.Dietr. = ***Fimbristylis cymosa*** R.Br. subsp. ***cymosa***

Isolepis striolata Nees ex Boeck. = ***Schoenoplectiella supina*** (L.) Lye

Isolepis subacaulis Steud. = ***Fimbristylis vahlii*** (Lam.) Link

Isolepis subcucullata Berggr. = ***Isolepis aucklandica*** Hook.f.

Isolepis subprolifer Boeck. = ***Isolepis cernua*** (Vahl) Roem. & Schult. var. ***cernua***

Isolepis subsquarrosa (Muhl.) Schrad. ex Schult. = ***Lipocarpha micrantha*** (Vahl) G.C.Tucker

Isolepis subsquarrosa var. *minor* Roem. & Schult. = ***Lipocarpha micrantha*** (Vahl) G.C.Tucker

Isolepis subtilis Kunth = ***Isolepis tenuissima*** (Nees) Kunth

Isolepis subtristachya Hochst. ex Boeck. = ***Bulbostylis barbata*** (Rottb.) C.B.Clarke subsp. ***barbata***

Isolepis subtristachya Hochst. ex C.B.Clarke = ***Bulbostylis barbata*** (Rottb.) C.B.Clarke subsp. ***barbata***

Isolepis supina (L.) R.Br. = ***Schoenoplectiella supina*** (L.) Lye

Isolepis tenella (Link) Kunth = ***Bulbostylis tenella*** (Link) C.B.Clarke
Isolepis tenerrima Fisch. & C.A.Mey. ex Ledeb. = ***Bulbostylis tenerrima*** (Fisch. & C.A.Mey. ex Ledeb.) Palla
Isolepis tenuifolia (Rudge) A.Dietr. = ***Bulbostylis tenuifolia*** (Rudge) J.F.Macbr.
Isolepis tenuior Steud. = ***Isolepis ludwigii*** (Steud.) Kunth
Isolepis tenuipes J.Presl = ***Isolepis cernua*** (Vahl) Roem. & Schult. var. ***cernua***
Isolepis tenuis J.Presl = ***Isolepis cernua*** (Vahl) Roem. & Schult. var. ***cernua***
Isolepis tenuis Schrad. = ***Isolepis pusilla*** Kunth
Isolepis tenuissima D.Don = ***Bulbostylis densa*** (Wall.) Hand.-Mazz. subsp. ***densa***
Isolepis tetragona (R.Br.) Schult. = ***Fimbristylis tetragona*** R.Br.
Isolepis thouarsii A.Dietr. = ***Isolepis bicolor*** Carmich.
Isolepis thouarsii (Roem. & Schult.) Nees = ***Bulbostylis puberula*** C.B.Clarke
Isolepis thunbergiana Nees = ***Scirpoides holoschoenus*** subsp. ***thunbergii*** (Schrad.) Soják
Isolepis thunbergii Schrad. = ***Scirpoides holoschoenus*** subsp. ***thunbergii*** (Schrad.) Soják
Isolepis trachycarpa Boeck. = ***Isolepis cernua*** (Vahl) Roem. & Schult. var. ***cernua***
Isolepis trichoides Schrad. ex Schult. = ***Bulbostylis trichoides*** (Schrad. ex Schult.) Beetle
Isolepis trichokolea Steud. = ***Bulbostylis densa*** (Wall.) Hand.-Mazz. subsp. ***densa***
Isolepis trifida Nees = ***Bulbostylis densa*** subsp. ***afromontana*** (Lye) R.W.Haines
Isolepis trigyna Kunze ex Boeck. = ***Isolepis cernua*** (Vahl) Roem. & Schult. var. ***cernua***
Isolepis tristachya Phil. = ***Isolepis nigricans*** Kunth
Isolepis tristachya (Rottb.) Roem. & Schult. = ***Ficinia tristachya*** (Rottb.) Nees
Isolepis trollii (Kük.) Lye = ***Ficinia trollii*** (Kük.) Muasya & D.A.Simpson
Isolepis truncata Schrad. ex Nees = [84]
Isolepis truncata (Thunb.) Nees = ***Ficinia truncata*** (Thunb.) Schrad.
Isolepis turkestanica Regel = ***Fimbristylis turkestanica*** (Regel) B.Fedtsch.
Isolepis uliginosa Steud. = ***Eleocharis acicularis*** (L.) Roem. & Schult. subsp. ***acicularis***
Isolepis uncinata (Willd.) Roem. & Schult. = ***Fuirena uncinata*** (Willd.) Kunth
Isolepis uninodis Delile = ***Schoenoplectiella supina*** (L.) Lye
Isolepis urvillei Steud. = ***Isolepis inundata*** R.Br.
Isolepis vahlii (Lam.) Kunth = ***Fimbristylis vahlii*** (Lam.) Link
Isolepis valdiviae Steud. = ***Isolepis varians*** Steud.
Isolepis varia Schrad. ex Nees = ***Fimbristylis dichotoma*** (L.) Vahl subsp. ***dichotoma***
Isolepis variegata Steud. = ***Carex setifolia*** Kunze var. ***setifolia***
Isolepis verrucifera Maxim. = ***Fimbristylis dipsacea*** (Rottb.) C.B.Clarke
Isolepis verrucosula (Steud.) Nees = ***Isolepis cernua*** var. ***setiformis*** (Benth.) Muasya
Isolepis vestita Kunth = ***Bulbostylis vestita*** (Kunth) C.B.Clarke
Isolepis vincentii Steud. = ***Fimbristylis vahlii*** (Lam.) Link
Isolepis vivipara Schrad. ex É.Desv. = ***Isolepis nigricans*** Kunth
Isolepis wallichiana Schult. = ***Bulbostylis barbata*** (Rottb.) C.B.Clarke subsp. ***barbata***
Isolepis warei Torr. = ***Bulbostylis warei*** (Torr.) C.B.Clarke
Isolepis willdenowii Kunth = ***Bulbostylis barbata*** (Rottb.) C.B.Clarke subsp. ***barbata***
Isolepis willdenowii Roem. & Schult. = ***Fimbristylis complanata*** (Retz.) Link var. ***complanata***

***Unplaced Names*:**
Isolepis compressa Nees, Linnaea 10: 157 (1836). = [27]
Isolepis hispidula Nees, Ann. Nat. Hist. 1: 284 (1838). = [40 IND]
Isolepis panamensis J.Presl & C.Presl, Reliq. Haenk. 1: 351 (1830). = [80 PAN] Rhynchospora sp ?
Isolepis rubescens J.Presl, Delic. Prag.: 157 (1822). = [84] Bulbostylsis sp.
Isolepis truncata Schrad. ex Nees in C.F.P.von Martius & auct. suc. (eds.), Fl. Bras. 2(1): 83 (1842), nom. illeg. = [84]

Isoschoenus

Isoschoenus Nees = ***Schoenus*** L.
Isoschoenus acuminatus Nees = ***Schoenus acuminatus*** R.Br.
Isoschoenus armeria Nees = ***Schoenus armeria*** (Nees) Boeck.
Isoschoenus barbatus (Vahl) Nees = ***Rhynchospora barbata*** (Vahl) Kunth
Isoschoenus drummondii Steud. = ***Schoenus subflavus*** Kük.
Isoschoenus flavus Nees = ***Schoenus subflavus*** Kük.

Itheta

Itheta Raf. = ***Carex*** L.
Itheta breviseta (Torr.) Raf. = ***Carex filifolia*** Nutt. var. ***filifolia***

Juncellus

Juncellus C.B.Clarke = ***Cyperus*** L.
Juncellus alopecuroides (Rottb.) C.B.Clarke = ***Cyperus alopecuroides*** Rottb.
Juncellus altus Turrill = ***Pycreus altus*** (Turrill) Lye
Juncellus ater C.B.Clarke = ***Pycreus melas*** (Ridl.) C.B.Clarke
Juncellus calanthus Peter = ***Cyperus laevigatus*** subsp. ***distachyos*** (All.) Maire & Weiller
Juncellus distachyos (All.) Turrill = ***Cyperus laevigatus*** subsp. ***distachyos*** (All.) Maire & Weiller
Juncellus inundatus C.B.Clarke = ***Cyperus serotinus*** Rottb.
Juncellus kuekenthalii (Merr.) Backer =
Juncellus laevigatus (L.) C.B.Clarke = ***Cyperus laevigatus*** L.
Juncellus laevigatus subsp. *distachyos* (All.) P.H.Davis = ***Cyperus laevigatus*** subsp. ***distachyos*** (All.) Maire & Weiller
Juncellus leucolepis C.B.Clarke = ***Pycreus pumilus*** (L.) Nees
Juncellus limosus (Maxim.) C.B.Clarke = ***Cyperus limosus*** Maxim.
Juncellus michelianus (L.) Blatt. & McCann = ***Cyperus michelianus*** (L.) Delile
Juncellus minutus C.B.Clarke = [25 TAN]
Juncellus nipponicus (Franch. & Sav.) C.B.Clarke = ***Cyperus nipponicus*** Franch. & Sav.
Juncellus pallidiflorus Peter = ***Cyperus alopecuroides*** Rottb.
Juncellus pannonicus (Jacq.) C.B.Clarke = ***Cyperus pannonicus*** Jacq.

Juncellus pervillei (Boeck.) C.B.Clarke = ***Pycreus pervillei*** (Boeck.) C.B.Clarke
Juncellus pumilus Peter = ***Pycreus pumilus*** (L.) Nees
Juncellus pustulatus (Vahl) C.B.Clarke = ***Cyperus pustulatus*** Vahl
Juncellus pygmaeus (Rottb.) C.B.Clarke = ***Cyperus michelianus*** subsp. ***pygmaeus*** (Rottb.) Asch. & Graebn.
Juncellus serotinus (Rottb.) C.B.Clarke = ***Cyperus serotinus*** Rottb.
Juncellus serotinus var. *capitatus* D.Z.Ma = ***Cyperus serotinus*** Rottb.
Juncellus serotinus f. *contortus* (Ces. ex Asch. & Graebn.) Soó = ***Cyperus serotinus*** Rottb.
Juncellus serotinus f. *grandis* (Kük.) Soó = ***Cyperus serotinus*** Rottb.
Juncellus stylosus (C.B.Clarke) C.B.Clarke = ***Cyperus serotinus*** Rottb.

***Unplaced Names*:**
Juncellus minutus C.B.Clarke, J. Bot. 34: 224 (1896). = [25 TAN]

Karinia

Karinia Reznicek & McVaugh, Sida 16: 207 (1994).
Mexico. 79 MXC MXE MXG MXS.
1 Species

Karinia mexicana (C.B.Clarke ex Britton) Reznicek & McVaugh, Sida 16: 207 (1994).
Mexico. 79 MXC MXE MXG MXS. Hemicr. or rhizome geophyte.
Scirpus cyperoides Hemsl., Biol. Cent.-Amer., Bot. 3: 461 (1885), nom. illeg. **Scirpus mexicanus* C.B.Clarke ex Britton, Trans. New York Acad. Sci. 11: 77 (1892). *Holoschoenus mexicanus* (C.B. Clarke ex Britton) Palla, Oesterr. Bot. Z. 63: 401 (1913).
Scirpus orbicephalus Beetle, Madroño 7: 160 (1944). *Cyperus orbicephalus* (Beetle) T.Koyama & McVaugh, Bull. Torrey Bot. Club 90: 231 (1963).
Cyperus schaffneri Boeck., Bot. Jahrb. Syst. 7: 273 (1886), nom. illeg. *Eucyperus schaffneri* Rikli, Jahrb. Wiss. Bot. 27: 568 (1895).

Khaosokia

Khaosokia D.A.Simpson, Chayam. & J.Parn., Bot. J. Linn. Soc. 149: 358 (2005).
Thailand. 41 THA.
1 Species

Khaosokia caricoides D.A.Simpson, Chayam. & J.Parn., Bot. J. Linn. Soc. 149: 358 (2005).
Thailand. 41 THA. Hemicr. or rhizome geophyte.

Killinga

Killinga T.Lestib. = ***Kyllinga*** Rottb.

Kleistrocalyx

Kleistrocalyx Steud. = ***Rhynchospora*** Vahl
Kleistrocalyx junciformis Steud. = ***Rhynchospora oligantha*** A.Gray var. ***oligantha***

Kobresia

Kobresia Willd., Sp. Pl. 4: 205 (1805).
Temp. Northern Hemisphere to Sumatera. 10 FIN GRB ICA NOR SWE 11 AUT CZE GER POL SWI 12 FRA ITA ROM SPA 13 BUL ITA ROM YUG 14 RUE RUN 30 ALT IRK KRA TVA WSB YAK 31+ KUR MAG 32 KAZ KGZ TZK UZB 33 TCS 34 AFG IRN IRQ TUR 36 CHC CHI CHM CHN CHQ CHT CHX 37 MON 38 KOR 40 ASS EHM NEP PAK WHM 41 MYA THA 42 SUM 70 ASK GNL NUN NWT YUK 71 ABT BRC MAN 72 LAB NFL ONT QUE 73 COL IDA MNT? ORE WYO 76 CAL UTA 77 NWM.
73 Species
Elyna Schrad., Fl. Germ. 1: 155 (1806).
Kobresia Pers., Syn. Pl. 2: 534 (1807), orth. var.
Froelichia Wulfen, Fl. Norica Phan.: 729 (1858), nom. illeg.
Hemicarex Benth., J. Linn. Soc., Bot. 18: 367 (1881).
Kobria St.-Lag., Ann. Soc. Bot. Lyon 8: 175 (1881).
Holmia Börner, Abh. Naturwiss. Vereine Bremen 22: 262 (1913).
Blysmocarex N.A.Ivanova, Bot. Zhurn. S.S.S.R. 24: 502 (1939).

Kobresia brunnescens Boeck., Beitr. Cyper. 1: 45 (1888).
Tibet. 36 CHT. Hemicr. or rhizome geophyte.

Kobresia burangensis Y.C.Yang, in Fl. Xizangica 5: 374 (1987).
Tibet. 36 CHT. Hemicr. or rhizome geophyte.

Kobresia capillifolia (Decne.) C.B.Clarke, J. Linn. Soc., Bot. 20: 378 (1883).
Afghanistan to Himalaya. 34 AFG 36 CHC CHI CHN CHT CHX 40 EHM NEP PAK WHM. Hemicr.
**Elyna capillifolia* Decne. in V.Jacquemont, Voy. Inde 4: 173 (1844).
Elyna spicata Boiss., Fl. Orient. 5: 394 (1882).
Kobresia thomsonii Maxim. ex K.V.Ivanova, Bot. Zhurn. S.S.S.R. 24: 486 (1939).

Kobresia capilliformis K.V.Ivanova, Bot. Zhurn. S.S.S.R. 24: 484 (1939).
S. Siberia to Mongolia. 30 TVA 32 KAZ KGZ 37 MON. Hemicr. or rhizome geophyte.

Kobresia cercostachys (Franch.) C.B.Clarke, J. Linn. Soc., Bot. 36: 267 (1903).
Himalaya to Tibet. 36 CHC CHQ CHT 40 EHM WHM. Hemicr. or rhizome geophyte.
**Carex cercostachys* Franch., Bull. Soc. Philom. Paris, VIII, 7: 27 (1895).
Kobresia stiebritziana Hand.-Mazz., Anz. Akad. Wiss. Wien, Math.-Naturwiss. Kl. 57: 54 (1920). *Kobresia nepalensis* var. *stiebritziana* (Hand.-Mazz.) R.C.Srivast., Novon 8: 203 (1998).

Kobresia coninux F.T.Wang & Tang, Acta Phytotax. Sin. 1: 182 (1951).
China (Hubei). 36 CHC. Hemicr. or rhizome geophyte.

Kobresia cuneata Kük., Acta Horti Gothob. 5: 39 (1930).
Tibet to C. China. 36 CHC CHN CHQ CHT. Hemicr. or rhizome geophyte.

Kobresia curticeps (C.B.Clarke) Kük. in H.G.A.Engler (ed.), Pflanzenr., IV, 20(38): 47 (1909).
C. & E. Himalaya to Tibet. 36 CHT 40 EHM NEP. Hemicr. or rhizome geophyte.
**Carex curticeps* C.B.Clarke in J.D.Hooker, Fl. Brit. India 6: 729 (1894).

Kobresia curvata C.B.Clarke, Kew Bull., Addit. Ser. 8: 68 (1908).
Sikkim to Tibet. 36 CHT 40 EHM. Hemicr.

Kobresia curvirostris (C.B.Clarke) C.B.Clarke in J.D.Hooker, Fl. Brit. India 6: 699 (1894).

E. Himalaya to N. Thailand. 40 EHM 41 THA. Hemicr. or rhizome geophyte.
**Hemicarex curvirostris* C.B.Clarke, J. Linn. Soc., Bot. 20: 384 (1883). *Kobresia laxa* var. *curvirostris* (C.B. Clarke) R.C.Srivast., in Fl. Sikkim 1: 224 (1996).
Kobresia siamensis Ohwi, Acta Phytotax. Geobot. 23: 109 (1968).

Kobresia daqingshanica X.Y.Mao, Acta Sci. Nat. Univ. Intramongol. 19: 341 (1988).
Inner Mongolia. 36 CHI. Hemicr. or rhizome geophyte.

Kobresia duthiei C.B.Clarke in J.D.Hooker, Fl. Brit. India 6: 697 (1894).
Himalaya to SC. China. 36 CHC CHT 40 EHM NEP WHM. Hemicr. or rhizome geophyte.

Kobresia elata Boeck., Beitr. Cyper. 2: 32 (1890).
Tibet. 36 CHT. Hemicr. or rhizome geophyte.

Kobresia esbirajbhandarii Rajbh. & H.Ohba, J. Jap. Bot. 62: 272 (1987).
Nepal. 40 NEP. Hemicr. or rhizome geophyte.

Kobresia esenbeckii (Kunth) Noltie, Edinburgh J. Bot. 50: 43 (1993).
Himalaya to SC. China. 36 CHC CHT 40 EHM NEP WHM 41 MYA. Hemicr. or rhizome geophyte.
**Carex esenbeckii* Kunth, Enum. Pl. 2: 522 (1837).

var. ***esenbeckii***
Himalaya to SC. China. 36 CHC CHT 40 EHM NEP WHM 41 MYA. Hemicr. or rhizome geophyte.
Carex trinervis Nees in R.Wight, Contr. Bot. India: 120 (1834), nom. illeg. *Kobresia trinervis* Boeck., Linnaea 39: 4 (1875). *Hemicarex trinervis* (Boeck.) Benth. & Hook.f. ex C.B.Clarke, J. Linn. Soc., Bot. 20: 382 (1883).
Kobresia hookeri Boeck., Linnaea 39: 4 (1875). *Hemicarex hookeri* (Boeck.) Benth., J. Linn. Soc., Bot. 18: 367 (1881).
Carex mutans Boott ex C.B.Clarke, J. Linn. Soc., Bot. 20: 383 (1883).
Hemicarex filicina C.B.Clarke, J. Linn. Soc., Bot. 20: 384 (1883). *Kobresia filicina* (C.B.Clarke) C.B.Clarke in J.D.Hooker, Fl. Brit. India 6: 696 (1894).
Kobresia mutans Boott ex C.B.Clarke, J. Linn. Soc., Bot. 20: 383 (1883).
Carex polygyna Boeck., Beitr. Cyper. 1: 40 (1888).
Kobresia angusta C.B.Clarke in J.D.Hooker, Fl. Brit. India 6: 695 (1894).
Kobresia foliosa C.B.Clarke in J.D.Hooker, Fl. Brit. India 6: 696 (1894).
Kobresia filicina var. *subfilicinoides* P.C.Li, Acta Phytotax. Sin. 37: 155 (1999).

var. ***fissiglumis*** (C.B.Clarke) Noltie, Edinburgh J. Bot. 50: 43 (1993).
Nepal to SC. China. 36 CHC CHT 40 EHM NEP. Hemicr. or rhizome geophyte.
Kobresia seticulmis Boeck., Linnaea 39: 3 (1875). *Holmia seticulmis* (Boeck.) Fedde & J.Schust., Just's Bot. Jahresber. 41(2): 10 (1913 publ. 1918).
**Kobresia fissiglumis* C.B.Clarke in J.D.Hooker, Fl. Brit. India 6: 696 (1894).

Kobresia falcata F.T.Wang & Tang ex P.C.Li, Acta Bot. Yunnan. 12: 18 (1990).
China (Sichuan). 36 CHC. Hemicr. or rhizome geophyte.

Kobresia filifolia (Turcz.) C.B.Clarke, J. Linn. Soc., Bot. 20: 381 (1883).
Siberia to Mongolia. 30 ALT IRK KRA TVA YAK 31 MAG 36 CHI CHN 37 MON. Hemicr. or rhizome geophyte.
**Elyna filifolia* Turcz., Bull. Soc. Imp. Naturalistes Moscou 28(1): 353 (1855).
Kobresia gracilis Meinsh., Trudy Imp. S.-Peterburgsk. Bot. Sada 18: 276 (1901).

Kobresia fragilis C.B.Clarke, J. Linn. Soc., Bot. 36: 267 (1903). *Schoenoxiphium fragile* (C.B.Clarke) C.B.Clarke, Kew Bull., Addit. Ser. 8: 67 (1908).
Nepal to SC. China. 36 CHC CHT 40 EHM NEP. Hemicr. or rhizome geophyte.
Schoenoxiphium clarkeanum Kük., Bull. Herb. Boissier, II, 4: 49 (1904). *Kobresia clarkeana* (Kük.) Kük. in H.G.A.Engler (ed.), Pflanzenr., IV, 38: 48 (1909).
Kobresia bonatiana Kük., Bull. Acad. Int. Géogr. Bot. 22: 250 (1912). *Carex bonatiana* (Kük.) K.V. Ivanova, Bot. Zhurn. S.S.S.R. 24: 501 (1939).
Kobresia yunnanensis Hand.-Mazz., Symb. Sin. 7: 1256 (1936).
Kobresia curticeps var. *gyirongensis* Y.C.Yang, in Fl. Xizangica 5: 391 (1987).

Kobresia gammiei C.B.Clarke, Kew Bull., Addit. Ser. 8: 68 (1908).
C. & E. Himalaya. 36 CHT 40 EHM NEP. Hemicr. or rhizome geophyte.
Kobresia williamsii T.Koyama, Bot. Mag. (Tokyo) 86: 279 (1973).

Kobresia gandakiensis Rajbh. & H.Ohba, Bull. Univ. Mus. Univ. Tokyo 34: 132 (1991).
Nepal. 40 NEP. Hemicr. or rhizome geophyte.

Kobresia graminifolia C.B.Clarke, J. Linn. Soc., Bot. 36: 268 (1903).
Tibet to C. China. 36 CHN CHT. Hemicr. or rhizome geophyte.

Kobresia handel-mazzettii K.V.Ivanova, Bot. Zhurn. S.S.S.R. 24: 494 (1939).
Tibet to SC. China. 36 CHC CHT. Hemicr.

Kobresia harae Rajbh. & H.Ohba, J. Jap. Bot. 62: 193 (1987).
Nepal. 40 NEP. Hemicr. or rhizome geophyte.

Kobresia helanshanica W.Z.Di & M.J.Zhong, Acta Bot. Boreal.-Occid. Sin. 5: 311 (1985).
Inner Mongolia. 36 CHI. Hemicr. or rhizome geophyte.

Kobresia hispida Kük., Acta Horti Gothob. 5: 39 (1930).
China (Sichuan). 36 CHC. Hemicr. or rhizome geophyte.

Kobresia hohxilensis R.F.Huang in S.G.Wu & Z.J.Feng (eds.), Biol. & Human Physiol. Hoh Xil Region: 101 (1996).
Qinghai. 36 CHQ. Hemicr. or rhizome geophyte.

Kobresia humilis (C.A.Mey. ex Trautv.) Serg. in V.L.Komarov (ed.), Fl. URSS 3: 111 (1935).
Caucasus to Mongolia and Himalaya. 32 KAZ KGZ TZK UZB 33 TCS 34 AFG IRN IRQ TUR 36 CHT CHX 37 MON 40 EHM NEP PAK WHM. Hemicr. or rhizome geophyte.
**Elyna humilis* C.A.Mey. ex Trautv., Trudy Imp. S.-Peterburgsk. Bot. Sada 1: 21 (1871). *Kobresia royleana* var. *humilis* (C.A.Mey. ex Trautv.) Kük. in

H.G.A.Engler (ed.), Pflanzenr., IV, 38: 46 (1909).
Kobresia persica Kük. & Bornm., Oesterr. Bot. Z. 47: 133 (1897).
Kobresia royleana var. *parvinux* T.Koyama, Acta Phytotax. Geobot. 16: 168 (1956).
Kobresia karakorumensis Dickoré, Stapfia 39: 77 (1995).

Kobresia hyalinolepis Boeck., Beitr. Cyper. 1: 39 (1888).
Tibet. 36 CHT. Hemicr. or rhizome geophyte.

Kobresia inflata P.C.Li ex S.R.Zhang, Acta Phytotax. Sin. 42: 202 (2004).
Tibet. 36 CHT. Hemicr.

Kobresia kanaii Rajbh. & H.Ohba, Bull. Univ. Mus. Univ. Tokyo 34: 135 (1991).
Nepal. 40 NEP. Hemicr. or rhizome geophyte.

Kobresia kansuensis Kük., Acta Horti Gothob. 5: 38 (1930).
Tibet to C. China. 36 CHC CHN CHQ CHT. Hemicr. or rhizome geophyte.

Kobresia kashgarica Dickoré, Stapfia 39: 79 (1995).
Xinjiang. 36 CHX. Hemicr. or rhizome geophyte.

Kobresia kobresioidea (Kük.) J.Kern, Acta Bot. Neerl. 7: 795 (1958).
N. Sumatera. 42 SUM. Hemicr. or rhizome geophyte.
**Schoenoxiphium kobresioideum* Kük., Bull. Jard. Bot. Buitenzorg, III, 16: 312 (1940).

Kobresia kuekenthaliana Hand.-Mazz., Anz. Akad. Wiss. Wien, Math.-Naturwiss. Kl. 57: 290 (1920). *Schoenoxiphium kuekenthalianum* (Hand.-Mazz.) K.V.Ivanova, Bot. Zhurn. S.S.S.R. 24: 501 (1939).
SC. China. 36 CHC. Hemicr. or rhizome geophyte.

Kobresia lacustris P.C.Li, Acta Bot. Yunnan. 12: 14 (1990).
China (Sichuan). 36 CHC. Hemicr. or rhizome geophyte.

Kobresia laxa Nees in R.Wight, Contr. Bot. India: 119 (1834). *Elyna laxa* (Nees) Kunth, Enum. Pl. 2: 534 (1837). *Schoenoxiphium laxum* (Nees) K.V.Ivanova, Bot. Zhurn. S.S.S.R. 24: 501 (1939).
Afghanistan to Himalaya. 32 TZK 34 AFG 36 CHT 40 EHM NEP PAK WHM. Hemicr.
Hemicarex laxa Benth., J. Linn. Soc., Bot. 18: 367 (1881).
Kobresia pseudolaxa C.B.Clarke, J. Linn. Soc., Bot. 20: 381 (1883).
Schoenoxiphium hissaricum Pissjauk., Bot. Mater. Gerb. Bot. Inst. Komarova Akad. Nauk S.S.S.R. 12: 72 (1950). *Kobresia hissarica* (Pissjauk.) Soják, Cas. Nár. Mus., Odd. Prír. 148: 194 (1979 publ. 1980). *Kobresia laxa* subsp. *hissarica* (Pissjauk.) Kukkonen, Ann. Naturhist. Mus. Wien, B 98B(Suppl.): 91 (1996).
Kobresia afghanica Raymond, Biol. Skr. 14(4): 17 (1965).

Kobresia lepidochlamys F.T.Wang & Tang ex P.C.Li, Acta Bot. Yunnan. 12: 15 (1990).
China (Yunnan). 36 CHC. Hemicr. or rhizome geophyte.

Kobresia littledalei C.B.Clarke, Kew Bull., Addit. Ser. 8: 67 (1908).
Tibet. 36 CHT. Hemicr. or rhizome geophyte.

Kobresia loliacea F.T.Wang & Tang ex P.C.Li, Acta Bot. Yunnan. 12: 13 (1990).
Tibet to SC. China. 36 CHC CHT. Hemicr. or rhizome geophyte.

Kobresia longiaristita P.C.Li, Acta Bot. Yunnan. 12: 16 (1990).
China (Sichuan). 36 CHC. Hemicr. or rhizome geophyte.

Kobresia macrantha Boeck., Beitr. Cyper. 1: 39 (1888). *Blysmocarex macrantha* (Boeck.) K.V.Ivanova, Bot. Zhurn. S.S.S.R. 24: 502 (1939).
Himalaya to C. China. 36 CHC CHN CHQ CHT 40 NEP WHM. Hemicr. or rhizome geophyte.
Blysmocarex nudicarpa Y.C.Yang, Acta Bot. Yunnan. 4: 325 (1982). *Kobresia nudicarpa* (Y.C.Yang) S.R.Zhang, Acta Phytotax. Sin. 33: 160 (1995). *Kobresia macrantha* var. *nudicarpa* (Y.C.Yang) P.C.Li, in Fl. Reipubl. Popul. Sin. 12: 26 (2000). *Blysmocarex macrantha* subsp. *nudicarpa* (Y.C.Yang) D.S.Deng, Guihaia 22: 120 (2002).

Kobresia macrolepis Meinsh., Trudy Imp. S.-Peterburgsk. Bot. Sada 18: 276 (1901).
Caucasus. 33 TCS. Hemicr. or rhizome geophyte.

Kobresia macroprophylla (Y.C.Yang) P.C.Li, in Fl. Reipubl. Popul. Sin. 12: 17 (2000).
Qinghai to NC. China. 36 CHN CHQ. Hemicr. or rhizome geophyte.
**Kobresia filifolia* var. *macroprophylla* Y.C.Yang, Acta Biol. Plateau Sin. 2: 8 (1984).

Kobresia mallae Rajbh. & H.Ohba, J. Jap. Bot. 62: 270 (1987).
Nepal. 40 NEP. Hemicr. or rhizome geophyte.

Kobresia maquensis Y.C.Yang, Acta Biol. Plateau Sin. 2: 4 (1984).
C. China. 36 CHC CHN. Hemicr. or rhizome geophyte.

Kobresia menyuanica Y.C.Yang, Acta Biol. Plateau Sin. 2: 3 (1984).
Qinghai. 36 CHQ. Hemicr. or rhizome geophyte.

Kobresia myosuroides (Vill.) Fiori & Paol. in A.Fiori & al., Fl. Italia 1: 125 (1896).
Temp. Northern Hemisphere. 10 FIN ICA NOR SWE 11 AUT CZE GER SWI 12 FRA SPA 13 BUL ITA ROM YUG 14 RUE RUN 30 ALT KRA TVA WSB YAK 31 KUR MAG 32 KAZ 36 CHC CHI CHM CHN CHQ CHT CHX 37 MON 38 KOR 70 ASK GNL NUN NWT YUK 71 ABT BRC MAN 72 LAB NFL ONT QUE 73 COL IDA ORE WYO 76 CAL UTA 77 NWM. Hemicr.
**Carex myosuroides* Vill., Prosp. Hist. Pl. Dauphiné: 17 (1779). *Elyna myosuroides* (Vill.) Fritsch ex Janch., Mitt. Naturwiss. Vereins Univ. Wien 5: 110 (1907).
Carex bellardii All., Fl. Pedem. 2: 264 (1785). *Kobresia bellardii* (All.) Degl. ex Loisel., Fl. Gall.: 626 (1807). *Scirpus bellardii* (All.) Wahlenb., Fl. Lapp.: 15 (1812). *Elyna bellardii* (All.) K.Koch, Linnaea 21: 616 (1848). *Froelichia caricoides* Wulfen, Fl. Norica Phan.: 729 (1858), nom. illeg.
Carex hermaphrodita J.F.Gmel., Syst. Nat. 2: 138 (1791).
Kobresia scirpina Willd., Sp. Pl. 4: 205 (1805). *Elyna scirpina* (Willd.) Pax in H.G.A.Engler & K.A.E. Prantl, Nat. Pflanzenfam. 2(2): 122 (1887). *Carex scirpina* (Willd.) Missbach & E.H.L.Krause in J. Sturm, Deutschl. Fl., ed. 2, 2: 63 (1900), nom. illeg.
Elyna spicata Schrad., Fl. Germ. 1: 155 (1806).
Kobresia nardina Hornem. in G.C.Oeder & al. (eds.), Fl. Dan.: 74 (1827).
Kobresia filiformis Dewey, Amer. J. Sci. Arts 29: 253 (1836).

Elyna stricta Hoppe ex Steud., Nomencl. Bot., ed. 2, 1: 551 (1840), nom. inval.
Elyna filiformis Steud., Syn. Pl. Glumac. 2: 245 (1855).
Carex vulcanicola Nakai, Bot. Mag. (Tokyo) 28: 327 (1914).
Kobresia bistaminata W.Z.Di & M.J.Zhong, Acta Bot. Boreal.-Occid. Sin. 6: 275 (1986). *Kobresia myosuroides* subsp. *bistaminata* (W.Z.Di & M.J.Zhong) S.R.Zhang, Novon 9: 453 (1999).

Kobresia nepalensis (Nees) Kük. in H.G.A.Engler (ed.), Pflanzenr., IV, 20(38)(38): 40 (1909).
N. Pakistan to Himalaya. 36 CHC CHT 40 EHM NEP PAK WHM. Hemicr.
**Uncinia nepalensis* Nees in R.Wight, Contr. Bot. India: 129 (1834).

var. ***nepalensis***
N. Pakistan to Himalaya. 36 CHC CHT 40 EHM NEP PAK WHM. Hemicr.
Carex essenbeckii Boott, Ill. Carex 1: 52 (1858).
Carex linearis Boott, Ill. Carex 1: 51 (1858). *Hemicarex linearis* (Boott) Benth., J. Linn. Soc., Bot. 18: 367 (1881).
Carex unciniiformis Boeck., Beitr. Cyper. 1: 40 (1888).
Carex linearis var. *elachista* C.B.Clarke in J.D. Hooker, Fl. Brit. India 6: 713 (1894). *Kobresia nepalensis* var. *elachista* (C.B.Clarke) Kük. in H.G.A.Engler (ed.), Pflanzenr., IV, 38: 40 (1909).

var. ***vaginosa*** (C.B.Clarke) R.C.Srivast., in Fl. Sikkim 1: 225 (1996).
Himalaya to Tibet. 36 CHT 40 EHM NEP WHM. Hemicr. or rhizome geophyte.
**Kobresia vaginosa* C.B.Clarke in J.D.Hooker, Fl. Brit. India 6: 695 (1894). *Kobresia nepalensis* subsp. *vaginosa* (C.B.Clarke) T.Koyama, Enum. Fl. Pl. Nepal 1: 113 (1978).
Kobresia cercostachys var. *capillacea* P.C.Li, Acta Bot. Yunnan. 12: 17 (1990).

Kobresia nitens C.B.Clarke, J. Linn. Soc., Bot. 20: 379 (1883).
N. Pakistan to Nepal. 40 NEP PAK WHM. Hemicr. or rhizome geophyte.

Kobresia ovczinnikovii T.V.Egorova, Trudy Bot. Inst. Akad. Nauk S.S.S.R., Rast. Tsentral. Azii 3: 35 (1967).
C. Asia. 32 TZK. Hemicr. or rhizome geophyte.

Kobresia pinetorum F.T.Wang & Tang ex P.C.Li, Acta Bot. Yunnan. 12: 14 (1990).
China (Yunnan). 36 CHC. Hemicr. or rhizome geophyte.

Kobresia prainii Kük., Bull. Herb. Boissier, II, 4: 50 (1904).
E. Himalaya to Tibet. 36 CHT 40 EHM. Hemicr. or rhizome geophyte.
Kobresia utriculata C.B.Clarke, Kew Bull., Addit. Ser. 8: 67 (1908).
Kobresia prainii var. *elliptica* Y.C.Yang, in Fl. Xizangica 5: 387 (1987).

Kobresia pseuduncinoides Noltie, Edinburgh J. Bot. 50: 47 (1993).
Nepal to Bhutan and Tibet. 36 CHT 40 EHM NEP. Hemicr. or rhizome geophyte.

Kobresia pusilla K.V.Ivanova, Bot. Zhurn. S.S.S.R. 24: 296 (1939).
China. 36 CHC CHI CHN CHQ CHT. Hemicr. or rhizome geophyte.

Kobresia pygmaea (C.B.Clarke) C.B.Clarke in J.D.Hooker, Fl. Brit. India 6: 696 (1894).
Himalaya to China. 36 CHC CHI CHN CHQ CHT CHX 40 EHM NEP PAK WHM 41 MYA. Hemicr. or rhizome geophyte.
**Hemicarex pygmaea* C.B.Clarke, J. Linn. Soc., Bot. 20: 383 (1883).

var. ***filiculmis*** Kük., Acta Horti Gothob. 5: 37 (1930).
SC. China. 36 CHC. Hemicr. or rhizome geophyte.
Kobresia microstachya K.V.Ivanova, Bot. Zhurn. S.S.S.R. 24: 488 (1939).

var. ***pygmaea***
Himalaya to China. 36 CHC CHI CHN CHQ CHT CHX 40 EHM NEP PAK WHM 41 MYA. Hemicr. or rhizome geophyte.

Kobresia robusta Maxim., Bull. Acad. Imp. Sci. Saint-Pétersbourg 29: 218 (1883).
Mongolia to China. 36 CHN CHQ CHT 37 MON. Hemicr. or rhizome geophyte.
Kobresia sargentiana Hemsl., J. Linn. Soc., Bot. 30: 139 (1894).

Kobresia royleana (Nees) Boeck., Linnaea 39: 8 (1875).
Afghanistan to China. 30 32 KAZ KGZ TZK 34 AFG 36 CHC CHN CHQ CHT CHX 40 EHM NEP PAK WHM. Hemicr. or rhizome geophyte.
**Trilepis royleana* Nees, Edinburgh New Philos. J. 17: 267 (1834).

subsp. ***minshanica*** (Y.C.Yang) S.R.Zhang, Novon 9: 453 (1999).
China (Sichuan, Gansu). 36 CHC CHN. Hemicr. or rhizome geophyte.
**Kobresia minshanica* Y.C.Yang, Acta Biol. Plateau Sin. 2: 1 (1984).

subsp. ***royleana***
Afghanistan to SC. China. 30 32 KAZ TZK 34 AFG 36 CHC CHQ CHT CHX 40 EHM NEP PAK WHM. Hemicr. or rhizome geophyte.
Elyna stenocarpa Kar. & Kir., Bull. Soc. Imp. Naturalistes Moscou 15: 526 (1842). *Kobresia stenocarpa* (Kar. & Kir.) Steud., Syn. Pl. Glumac. 2: 246 (1855).
Elyna kokanica Regel, Trudy Imp. S.-Peterburgsk. Bot. Sada 7: 563 (1880).
Kobresia paniculata Meinsh., Trudy Imp. S.-Peterburgsk. Bot. Sada 18: 279 (1901).
Elyna paniculata Meinsh. ex K.V.Ivanova, Bot. Zhurn. S.S.S.R. 24: 492 (1939), nom. inval.
Kobresia stenocarpa var. *simplex* Y.C.Yang, in Fl. Xizangica 5: 395 (1987).
Kobresia royleana var. *himalaica* Rajbh. & H.Ohba, Bull. Univ. Mus. Univ. Tokyo 34: 150 (1991).

Kobresia sanguinea (Boott) Raymond, Biol. Skr. 14(4): 19 (1965).
E. Afghanistan to W. Himalaya. 34 AFG 40 PAK WHM. Hemicr. or rhizome geophyte.
**Carex sanguinea* Boott, Trans. Linn. Soc. London 20: 137 (1846).

Kobresia schoenoides (C.A.Mey.) Steud., Syn. Pl. Glumac. 2: 246 (1855).
Caucasus to Himalaya. 32 KAZ KGZ TZK UZB 33 TCS 34 AFG IRN 36 CHC CHT CHX 40 EHM NEP PAK WHM. Hemicr.

Elyna schoenoides C.A.Mey. in C.F.von Ledebour, Fl. Altaic. 4: 235 (1833).
Kobresia deasyi C.B.Clarke, Kew Bull., Addit. Ser. 8: 68 (1908).
Kobresia pamiroalaica K.V.Ivanova, Bot. Zhurn. S.S.S.R. 24: 481 (1939).
Kobresia pamiralaica K.V.Ivanova, in Fl. Uzbekist. 1: 540 (1941).
Kobresia septatonodosa T.Koyama, Acta Phytotax. Geobot. 16: 168 (1956).
Kobresia glaucifolia F.T.Wang & Tang ex P.C.Li, Acta Phytotax. Sin. 37: 153 (1999).

Kobresia setschwanensis Hand.-Mazz., Symb. Sin. 7: 1254 (1936).
Tibet to C. China. 36 CHC CHN CHT. Hemicr. or rhizome geophyte.

subsp. ***setschwanensis***
Tibet to C. China. 36 CHC CHN CHT. Hemicr. or rhizome geophyte.

subsp. ***squamiformis*** (Y.C.Yang) S.R.Zhang, Novon 9: 453 (1999).
China (Gansu). 36 CHN. Hemicr. or rhizome geophyte.
Kobresia squamiformis Y.C.Yang, Acta Biol. Plateau Sin. 2: 9 (1984).

Kobresia sibirica (Turcz.) Boeck., Linnaea 39: 7 (1875).
Subarctic to NW. U.S.A. 14 RUE 30 KRA WSB YAK 31 MAG 37 MON 70 ASK NUN NWT YUK 73 COL MNT? WYO. Hemicr.
Elyna sibirica Turcz. in C.F.von Ledebour, Fl. Ross. 4: 262 (1852).
Kobresia macrocarpa Clokey in N.L.Britton & al. (eds.), N. Amer. Fl. 18: 5 (1931). *Kobresia bellardii* var. *macrocarpa* (Clokey) H.D.Harr., Man. Pl. Colorado: 641 (1954).
Kobresia arctica A.E.Porsild, Sargentia 4: 15 (1943), nom. illeg. *Kobresia hyperborea* A.E.Porsild, Bull. Natl. Mus. Canada 121: 103 (1951).
Kobresia hyperborea var. *alaskana* Duman, Bull. Torrey Bot. Club 83: 194 (1956).
Kobresia hyperborea var. *lepagei* Duman, Bull. Torrey Bot. Club 83: 193 (1956). *Kobresia schoenoides* var. *lepagei* (Duman) Boivin, Naturaliste Canad. 94: 525 (1967).

Kobresia sikkimensis Kük. in H.G.A.Engler (ed.), Pflanzenr., IV, 20(38): 47 (1909).
Nepal to Assam. 40 ASS EHM NEP. Hemicr. or rhizome geophyte.
Carex prainii Kük., Bull. Herb. Boissier, II, 55: 51 (1903).

Kobresia simpliciuscula (Wahlenb.) Mack., Bull. Torrey Bot. Club 50: 349 (1923).
Europe to Caucasus, Subartic America to W. U.S.A. 10 FIN GRB NOR SWE 11 AUT CZE GER POL SWI 12 FRA ITA ROM SPA 14 RUN 30 KRA WSB YAK 31 MAG 33 TCS 34 TUR 36 CHT 37 MON 70 ASK GNL NUN NWT YUK 71 ABT BRC MAN 72 LAB NFL ONT QUE 73 COL IDA ORE WYO 76 UTA. Hemicr.
Carex simpliciuscula Wahlenb., Kongl. Vetensk. Acad. Nya Handl. 1803: 141 (1803).

subsp. ***simpliciuscula***
Europe to Caucasus, Subartic America to W. U.S.A. 10 FIN GRB NOR SWE 11 AUT CZE GER POL SWI 12 FRA ITA ROM SPA 33 TCS 34 TUR 36 CHT 37 MON 70 ASK GNL NUN NWT YUK 71 ABT BRC MAN 72 LAB NFL ONT QUE 73 COL IDA ORE WYO 76 UTA. Hemicr.
Carex bipartita All., Fl. Pedem. 2: 265 (1785), nom. rejic. *Kobresia bipartita* (All.) Dalla Torre in A.Hartinger & K.W.von Dalla Torre, Atlas Alpenfl.: 216 (1883).
Elyna caricina Mert. & W.D.J.Koch in J.C.Röhling, Deutschl. Fl. 1: 458 (1796).
Kobresia caricina Willd., Sp. Pl. 4: 206 (1805). *Carex kobresia* Missbach & E.H.L.Krause in J.Sturm, Deutschl. Fl., ed. 2, 2: 64 (1900).
Schoenus monoicus Sm., Engl. Bot.: t. 1410 (1805).
Carex mirabilis Host, Icon. Descr. Gram. Austriac. 4: t. 78 (1809).
Elyna bellardii Hartm., Kongl. Vetensk. Acad. Handl. 1814: 107 (1814).
Carex hybrida Schkuhr ex Kunth, Enum. Pl. 2: 533 (1837).
Kobresia simpliciuscula var. *americana* Duman, Bull. Torrey Bot. Club 83: 194 (1956).

subsp. ***subfilifolia*** (T.V.Egorova, Jurtzev & V.V. Petrovsky) T.V.Egorova, Novosti Sist. Vyssh. Rast. 20: 84 (1983).
Siberia to Russian Far East. 30+ 31+. Hemicr.
Kobresia filifolia subsp. *subfilifolia* T.V.Egorova, Jurtzev & V.V.Petrovsky, Bot. Zhurn. (Moscow & Leningrad) 66: 1042 (1981).

subsp. ***subholarctica*** T.V.Egorova, Novosti Sist. Vyssh. Rast. 20: 83 (1983). *Kobresia subholarctica* (T.V.Egorova) T.V.Egorova, Bot. Zhurn. (Moscow & Leningrad) 76: 1736 (1991 publ. 1992). *Kobresia simpliciuscula* var. *subholarctica* (T.V.Egorova) A.E.Kozhevn., Sosud. Rast. Sovet. Dal'nego Vostoka 3: 229 (1998).
NE. Europe to Subarctic Asia. 14 RUN 30 KRA WSB YAK 31 MAG. Hemicr.

Kobresia smirnovii K.V.Ivanova, Bot. Zhurn. S.S.S.R. 24: 480 (1939).
Siberia to Mongolia. 30 ALT 32 KAZ 37 MON. Hemicr.

Kobresia stolonifera Y.C.Tang ex P.C.Li, Acta Phytotax. Sin. 37: 154 (1999). *Blysmocarex macrantha* subsp. *stolonifera* (Y.C.Tang ex P.C.Li) D.S.Deng, Guihaia 22: 120 (2002).
NW. China to Tibet. 36 CHT CHX. Hemicr. or rhizome geophyte.

Kobresia tibetica Maxim., Bull. Acad. Imp. Sci. Saint-Pétersbourg 29: 219 (1883). *Kobresia capillifolia* var. *tibetica* (Maxim.) Kük. in H.G.A.Engler (ed.), Pflanzenr., IV, 38: 36 (1909).
Himalaya to C. China. 36 CHN CHT 40 WHM. Hemicr.

Kobresia tunicata Hand.-Mazz., Symb. Sin. 7: 1254 (1936).
NW. & C. China. 36 CHC CHN CHX. Hemicr. or rhizome geophyte.
Kobresia capillifolia var. *condensata* Kük., Notes Roy. Bot. Gard. Edinburgh 7: 134 (1912).

Kobresia uncinioides (Boott) C.B.Clarke in J.D.Hooker, Fl. Brit. India 6: 698 (1894).
C. & E. Himalaya to SC. China. 36 CHC CHT 40 EHM NEP. Hemicr. or rhizome geophyte.
Carex uncinioides Boott, Ill. Carex 1: 8 (1858).

Kobresia vidua (Boott ex C.B.Clarke) Kük. in H.G.A.Engler (ed.), Pflanzenr., IV, 20(38): 40 (1909).

Himalaya to C. China. 36 CHC CHN CHQ CHT 40 EHM NEP. Hemicr. or rhizome geophyte.
**Carex vidua* Boott ex C.B.Clarke in J.D.Hooker, Fl. Brit. India 6: 713 (1894).
Kobresia prattii C.B.Clarke, J. Linn. Soc., Bot. 36: 268 (1903).
Kobresia harrysmithii Kük., Acta Horti Gothob. 5: 37 (1930).

Kobresia woodii Noltie, Edinburgh J. Bot. 50: 48 (1993).
C. Blutan to S. Tibet. 36 CHT 40 EHM. Hemicr. or rhizome geophyte.

Kobresia yadongensis Y.C.Yang, in Fl. Xizangica 5: 387 (1987).
Tibet. 36 CHT. Hemicr. or rhizome geophyte.

Kobresia yangii S.R.Zhang, Acta Phytotax. Sin. 33: 160 (1995).
China (Sichuan). 36 CHC. Hemicr. or rhizome geophyte.
**Kobresia gracilis* Y.C.Yang, Acta Biol. Plateau Sin. 2: 10 (1984), nom. illeg.

Kobresia yushuensis Y.C.Yang, Acta Biol. Plateau Sin. 2: 6 (1984).
Qinghai. 36 CHQ. Hemicr. or rhizome geophyte.

***Synonyms*:**
Kobresia Pers. = ***Kobresia*** Willd.
Kobresia afghanica Raymond = ***Kobresia laxa*** Nees
Kobresia angusta C.B.Clarke = ***Kobresia esenbeckii*** (Kunth) Noltie var. ***esenbeckii***
Kobresia arctica A.E.Porsild = ***Kobresia sibirica*** (Turcz.) Boeck.
Kobresia arctica K.V.Ivanova = ***Carex nardina*** Fr. var. ***nardina***
Kobresia bellardii (All.) Degl. ex Loisel. = ***Kobresia myosuroides*** (Vill.) Fiori & Paol.
Kobresia bellardii var. *macrocarpa* (Clokey) H.D.Harr. = ***Kobresia sibirica*** (Turcz.) Boeck.
Kobresia bipartita (All.) Dalla Torre = ***Kobresia simpliciuscula*** (Wahlenb.) Mack. subsp. ***simpliciuscula***
Kobresia bistaminata W.Z.Di & M.J.Zhong = ***Kobresia myosuroides*** (Vill.) Fiori & Paol.
Kobresia bonatiana Kük. = ***Kobresia fragilis*** C.B.Clarke
Kobresia buchananii (C.B.Clarke) T.Koyama = ***Schoenoxiphium rufum*** Nees
Kobresia bucharica (Kük.) K.V.Ivanova = ***Carex bucharica*** Kük.
Kobresia capillifolia var. *condensata* Kük. = ***Kobresia tunicata*** Hand.-Mazz.
Kobresia capillifolia var. *tibetica* (Maxim.) Kük. = ***Kobresia tibetica*** Maxim.
Kobresia caricina Willd. = ***Kobresia simpliciuscula*** (Wahlenb.) Mack. subsp. ***simpliciuscula***
Kobresia cercostachys var. *capillacea* P.C.Li = ***Kobresia nepalensis*** var. ***vaginosa*** (C.B.Clarke) R.C.Srivast.
Kobresia clarkeana (Kük.) Kük. = ***Kobresia fragilis*** C.B.Clarke
Kobresia curticeps var. *gyirongensis* Y.C.Yang = ***Kobresia fragilis*** C.B.Clarke
Kobresia curvula (All.) K.V.Ivanova = ***Carex curvula*** All.
Kobresia cyperina Willd. = ***Cyperus hermaphroditus*** (Jacq.) Standl.
Kobresia deasyi C.B.Clarke = ***Kobresia schoenoides*** (C.A.Mey.) Steud.
Kobresia dregeana (Kunth) T.Koyama = ***Schoenoxiphium dregeanum*** Kunth
Kobresia ecklonii (Nees) T.Koyama = ***Schoenoxiphium ecklonii*** Nees
Kobresia elachycarpa (Fernald) Fernald = ***Carex sterilis*** Willd.
Kobresia filicina (C.B.Clarke) C.B.Clarke = ***Kobresia esenbeckii*** (Kunth) Noltie var. ***esenbeckii***
Kobresia filicina var. *subfilicinoides* P.C.Li = ***Kobresia esenbeckii*** (Kunth) Noltie var. ***esenbeckii***
Kobresia filifolia var. *macroprophylla* Y.C.Yang = ***Kobresia macroprophylla*** (Y.C.Yang) P.C.Li
Kobresia filifolia subsp. *subfilifolia* T.V.Egorova, Jurtzev & V.V.Petrovsky = ***Kobresia simpliciuscula*** subsp. ***subfilifolia*** (T.V.Egorova, Jurtzev & V.V.Petrovsky) T.V.Egorova
Kobresia filiformis Dewey = ***Kobresia myosuroides*** (Vill.) Fiori & Paol.
Kobresia fissiglumis C.B.Clarke = ***Kobresia esenbeckii*** var. ***fissiglumis*** (C.B.Clarke) Noltie
Kobresia foliosa C.B.Clarke = ***Kobresia esenbeckii*** (Kunth) Noltie var. ***esenbeckii***
Kobresia glaucifolia F.T.Wang & Tang ex P.C.Li = ***Kobresia schoenoides*** (C.A.Mey.) Steud.
Kobresia globularis Dewey = ***Carex filifolia*** Nutt. var. ***filifolia***
Kobresia gracilis Meinsh. = ***Kobresia filifolia*** (Turcz.) C.B.Clarke
Kobresia gracilis Y.C.Yang = ***Kobresia yangii*** S.R.Zhang
Kobresia harrysmithii Kük. = ***Kobresia vidua*** (Boott ex C.B.Clarke) Kük.
Kobresia hepburnii (Boott) K.V.Ivanova = ***Carex nardina*** var. ***hepburnii*** (Boott) Kük.
Kobresia hissarica (Pissjauk.) Soják = ***Kobresia laxa*** Nees
Kobresia hookeri Boeck. = ***Kobresia esenbeckii*** (Kunth) Noltie var. ***esenbeckii***
Kobresia hyperborea A.E.Porsild = ***Kobresia sibirica*** (Turcz.) Boeck.
Kobresia hyperborea var. *alaskana* Duman = ***Kobresia sibirica*** (Turcz.) Boeck.
Kobresia hyperborea var. *lepagei* Duman = ***Kobresia sibirica*** (Turcz.) Boeck.
Kobresia karakorumensis Dickoré = ***Kobresia humilis*** (C.A.Mey. ex Trautv.) Serg.
Kobresia kunthiana (Kük.) T.Koyama = ***Schoenoxiphium sparteum*** (Wahlenb.) C.B.Clarke
Kobresia lancea (Thunb.) T.Koyama = ***Schoenoxiphium lanceum*** (Thunb.) Kük.
Kobresia laxa var. *curvirostris* (C.B.Clarke) R.C.Srivast. = ***Kobresia curvirostris*** (C.B.Clarke) C.B.Clarke
Kobresia laxa subsp. *hissarica* (Pissjauk.) Kukkonen = ***Kobresia laxa*** Nees
Kobresia lehmannii (Nees) T.Koyama = ***Schoenoxiphium lehmannii*** (Nees) Kunth ex Steud.
Kobresia lolonum Hand.-Mazz. = ***Carex parva*** Nees
Kobresia macrantha var. *nudicarpa* (Y.C.Yang) P.C.Li = ***Kobresia macrantha*** Boeck.
Kobresia macrocarpa Clokey = ***Kobresia sibirica*** (Turcz.) Boeck.
Kobresia microglochin (Wahlenb.) Tang & W.T.Wang = ***Carex microglochin*** Wahlenb.
Kobresia microstachya K.V.Ivanova = ***Kobresia pygmaea*** var. ***filiculmis*** Kük.
Kobresia minshanica Y.C.Yang = ***Kobresia royleana*** subsp. ***minshanica*** (Y.C.Yang) S.R.Zhang
Kobresia mutans Boott ex C.B.Clarke = ***Kobresia esenbeckii*** (Kunth) Noltie var. ***esenbeckii***
Kobresia myosuroides subsp. *bistaminata* (W.Z.Di & M.J.Zhong) S.R.Zhang = ***Kobresia myosuroides*** (Vill.) Fiori & Paol.
Kobresia nardina Hornem. = ***Kobresia myosuroides*** (Vill.) Fiori & Paol.

Kobresia nepalensis var. *elachista* (C.B.Clarke) Kük. = ***Kobresia nepalensis*** (Nees) Kük. var. ***nepalensis***
Kobresia nepalensis var. *stiebritziana* (Hand.-Mazz.) R.C.Srivast. = ***Kobresia cercostachys*** (Franch.) C.B.Clarke
Kobresia nepalensis subsp. *vaginosa* (C.B.Clarke) T.Koyama = ***Kobresia nepalensis*** var. ***vaginosa*** (C.B.Clarke) R.C.Srivast.
Kobresia nudicarpa (Y.C.Yang) S.R.Zhang = ***Kobresia macrantha*** Boeck.
Kobresia pamiralaica K.V.Ivanova = ***Kobresia schoenoides*** (C.A.Mey.) Steud.
Kobresia pamiroalaica K.V.Ivanova = ***Kobresia schoenoides*** (C.A.Mey.) Steud.
Kobresia paniculata Meinsh. = ***Kobresia royleana*** (Nees) Boeck. subsp. ***royleana***
Kobresia persica Kük. & Bornm. = ***Kobresia humilis*** (C.A.Mey. ex Trautv.) Serg.
Kobresia prainii var. *elliptica* Y.C.Yang = ***Kobresia prainii*** Kük.
Kobresia pratensis Freyn = [30]
Kobresia prattii C.B.Clarke = ***Kobresia vidua*** (Boott ex C.B.Clarke) Kük.
Kobresia pseudolaxa C.B.Clarke = ***Kobresia laxa*** Nees
Kobresia royleana var. *himalaica* Rajbh. & H.Ohba = ***Kobresia royleana*** (Nees) Boeck. subsp. ***royleana***
Kobresia royleana var. *humilis* (C.A.Mey. ex Trautv.) Kük. = ***Kobresia humilis*** (C.A.Mey. ex Trautv.) Serg.
Kobresia royleana var. *parvinux* T.Koyama = ***Kobresia humilis*** (C.A.Mey. ex Trautv.) Serg.
Kobresia rufa (Nees) T.Koyama = ***Schoenoxiphium rufum*** Nees
Kobresia sargentiana Hemsl. = ***Kobresia robusta*** Maxim.
Kobresia schoenoides var. *lepagei* (Duman) Boivin = ***Kobresia sibirica*** (Turcz.) Boeck.
Kobresia scirpina Willd. = ***Kobresia myosuroides*** (Vill.) Fiori & Paol.
Kobresia septatonodosa T.Koyama = ***Kobresia schoenoides*** (C.A.Mey.) Steud.
Kobresia seticulmis Boeck. = ***Kobresia esenbeckii*** var. ***fissiglumis*** (C.B.Clarke) Noltie
Kobresia siamensis Ohwi = ***Kobresia curvirostris*** (C.B.Clarke) C.B.Clarke
Kobresia simpliciuscula var. *americana* Duman = ***Kobresia simpliciuscula*** (Wahlenb.) Mack. subsp. ***simpliciuscula***
Kobresia simpliciuscula var. *subholarctica* (T.V.Egorova) A.E.Kozhevn. = ***Kobresia simpliciuscula*** subsp. ***subholarctica*** T.V.Egorova
Kobresia spartea (Wahlenb.) T.Koyama = ***Schoenoxiphium sparteum*** (Wahlenb.) C.B.Clarke
Kobresia squamiformis Y.C.Yang = ***Kobresia setschwanensis*** subsp. ***squamiformis*** (Y.C.Yang) S.R.Zhang
Kobresia stenocarpa (Kar. & Kir.) Steud. = ***Kobresia royleana*** (Nees) Boeck. subsp. ***royleana***
Kobresia stenocarpa var. *simplex* Y.C.Yang = ***Kobresia royleana*** (Nees) Boeck. subsp. ***royleana***
Kobresia stiebritziana Hand.-Mazz. = ***Kobresia cercostachys*** (Franch.) C.B.Clarke
Kobresia subholarctica (T.V.Egorova) T.V.Egorova = ***Kobresia simpliciuscula*** subsp. ***subholarctica*** T.V.Egorova
Kobresia thomsonii Maxim. ex K.V.Ivanova = ***Kobresia capillifolia*** (Decne.) C.B.Clarke
Kobresia tibetica subsp. *littledalei* (C.B.Clarke) P.C.Li = ***Kobresia schoenoides*** (C.A.Mey.) Steud.
Kobresia trinervis Boeck. = ***Kobresia esenbeckii*** (Kunth) Noltie var. ***esenbeckii***
Kobresia utriculata C.B.Clarke = ***Kobresia prainii*** Kük.
Kobresia vaginosa C.B.Clarke = ***Kobresia nepalensis*** var. ***vaginosa*** (C.B.Clarke) R.C.Srivast.
Kobresia williamsii T.Koyama = ***Kobresia gammiei*** C.B.Clarke
Kobresia yunnanensis Hand.-Mazz. = ***Kobresia fragilis*** C.B.Clarke

Unplaced Names:
Kobresia pratensis Freyn, Oesterr. Bot. Z. 40: 266 (1890). = [30]

Kobria

Kobria St.-Lag. = ***Kobresia*** Willd.

Kolerma

Kolerma Raf. = ***Carex*** L.
Kolerma alata (Torr.) Raf. = ***Carex alata*** Torr.

Koyamaea

Koyamaea W.W.Thomas & G.Davidse, Syst. Bot. 14: 189 (1989).
S. Venezuela to N. Brazil. 82 VEN 84 BZN.
1 Species

Koyamaea neblinensis W.W.Thomas & G.Davidse, Syst. Bot. 14: 189 (1989).
Venezuela (Sierra de la Neblina), Brazil (Serra da Neblina). 82 VEN 84 BZN. Hemicr.

Kreczetoviczia

Kreczetoviczia Tzvelev = ***Trichophorum*** Pers.
Kreczetoviczia cespitosa (L.) Tzvelev = ***Trichophorum cespitosum*** (L.) Hartm.
Kreczetoviczia pumila (Vahl) Tzvelev = ***Trichophorum pumilum*** (Vahl) Schinz & Thell.
Kreczetoviczia uniflora (Trautv.) Tzvelev = ***Trichophorum uniflorum*** (Trautv.) Malyschev & Lukitsch.

Kuekenthalia

Kuekenthalia Börner = ***Carex*** L.
Kuekenthalia secalina (Willd. ex Wahlenb.) Fedde & J.Schust. = ***Carex secalina*** Willd. ex Wahlenb.

Kyllinga

Kyllinga Rottb., Descr. Icon. Rar. Pl.: 12 (1773), nom. cons.
Trop. & Subtrop. (21) azo CVI mdr 22 BEN BKN GAM GHA GNB GUI IVO LBR MLI MTN NGA NGR SEN SIE TOG 23 BUR CAF CMN CON EQG GAB GGI RWA ZAI 24 CHA ERI ETH SOC SOM SUD TAN 25 KEN TAN UGA 26 ANG MLW MOZ ZAM ZIM 27 BOT CPP LES NAM NAT OFS SWZ TVL 28 ASC STH 29 ALD CGS COM MAU MCI MDG REU ROD SEY 31 KAM 33 TCS 34 AFG 35 SAU YEM 36 CHC CHH CHM CHS 38 JAP KOR KZN NNS OGA TAI 40 ASS BAN EHM IND NEP PAK SRL WHM 41 AND MYA THA VIE 42 BOR CKI JAW LSI MLY MOL PHI SUL SUM XMS 43 BIS NWG SOL 50 NFK NSW NTA QLD SOA WAU 51 NZN (60) fij GIL nue NWC SAM TON VAN 61 COO EAS MRQ PIT SCI TUA 62 CRL MRN MRS (63) haw 74 ILL KAN MSO OKL (75) cnt INI NWJ NWY OHI PEN WVA (76) cal 77 NWM TEX 78 ALA ARK DEL FLA GEO KTY LOU MRY MSI NCA SCA

TEN VRG 79 MXC MXE MXG MXN MXS MXT 80 BLZ COS CPI ELS GUA HON NIC PAN 81 BER CAY CUB DOM HAI JAM LEE PUE TRT WIN 82 FRG GUY SUR VEN 83 BOL CLM ECU GAL PER 84 BZC BZE BZL BZN BZS 85 AGE AGW PAR URU.

73 Species

Thryocephalon J.R.Forst. & G.Forst., Char. Gen. Pl.: 65 (1775).

Killinga T.Lestib., Essai Cypér.: 28 (1819).

Hedychloa Raf., Ann. Nat. 1: 16 (1820).

Cyprolepis Steud., Flora 33: 229 (1850), nom. nud.

Lyprolepis Steud., Syn. Pl. Glumac. 2: 130 (1855).

Kyllinga afro-occidentalis Lye, Nordic J. Bot. 1: 744 (1981 publ. 1982).
Ivory Coast. 22 IVO. Hemicr. or rhizome geophyte.

Kyllinga afropumila Lye, Nordic J. Bot. 1: 741 (1981 publ. 1982). *Cyperus afropumilus* (Lye) Lye in R.W.Haines & K.A.Lye, Sedges & Rushes E. Afr., App. 3: 2 (1983).
S. Tanzania. 25 TAN. Hemicr.

Kyllinga alba Nees, Linnaea 10: 140 (1836). *Cyperus alatus* subsp. *albus* (Nees) Lye, Lidia 3: 172 (1995).
Trop. & S. Africa, Seychelles. 22 GHA IVO NGA TOG 23 CON EQG GAB ZAI 24 CHA ERI ETH SOM SUD 25 KEN TAN UGA 26 ANG MLW MOZ ZAM ZIM 27 BOT CPP LES NAM NAT OFS SWZ TVL 29 SEY. Hemicr. or rhizome geophyte.

subsp. ***alba***
Trop. & S. Africa, Seychelles. 22 GHA IVO NGA TOG 23 GAB ZAI 24 CHA ERI ETH SOM SUD 25 KEN TAN UGA 26 ANG MLW MOZ ZAM ZIM 27 BOT CPP LES NAM NAT OFS SWZ TVL 29 SEY. Hemicr. or rhizome geophyte.

Kyllinga alata Nees, Linnaea 10: 139 (1836). *Cyperus alatus* (Nees) F.Muell., Fragm. 8: 272 (1874). *Kyllinga alba* var. *alata* (Nees) C.B.Clarke in T.A.Durand & H.Schinz, Consp. Fl. Afric. 5: 526 (1894).

Kyllinga cristata Kunth, Enum. Pl. 2: 136 (1837). *Cyperus cristatus* (Kunth) Mattf. & Kük. in H.G.A.Engler (ed.), Pflanzenr., IV, 20(101): 609 (1936).

Kyllinga nigritana C.B.Clarke in D.Oliver & auct. suc. (eds.), Fl. Trop. Afr. 8: 272 (1901). *Cyperus cristatus* var. *nigritanus* (C.B.Clarke) Kük. in H.G.A.Engler (ed.), Pflanzenr., IV, 20(101): 610 (1936). *Kyllinga alba* subsp. *nigritana* (C.B.Clarke) J.-P.Lebrun & Stork, Énum. Pl. Fl. Afr. Trop. 3: 190 (1995).

Cyperus alatus var. *serratus* Peter & Kük. in H.G.A.Engler (ed.), Pflanzenr., IV, 20(101): 611 (1936).

subsp. ***ascolepidioides*** (Cherm.) Lye, Nordic J. Bot. 1: 746 (1981 publ. 1982).
WC. Trop. Africa. 23 CON ZAI. Hemicr. or rhizome geophyte.

**Kyllinga ascolepidioides* Cherm., Arch. Bot. Bull. Mens. 4(7): 4 (1931). *Cyperus ascolepidioides* (Cherm.) Kük. in H.G.A.Engler (ed.), Pflanzenr., IV, 20(101): 610 (1936). *Cyperus cristatus* subsp. *ascolepidioides* (Cherm.) Lye in R.W.Haines & K.A.Lye, Sedges & Rushes E. Afr., App. 3: 2 (1983). *Cyperus alatus* subsp. *ascolepidioides* (Cherm.) Lye, Lidia 3: 172 (1995).

Kyllinga alba-purpurea Lye, Nordic J. Bot. 1: 743 (1981 publ. 1982). *Cyperus alba-purpureus* (Lye) Lye in R.W.Haines & K.A.Lye, Sedges & Rushes E. Afr., App. 3: 2 (1983).
Tanzania (Sumbawanga Distr.). 25 TAN. Hemicr. or rhizome geophyte.

Kyllinga albiceps (Ridl.) Rendle in W.P.Hiern, Cat. Afr. Pl. 2: 106 (1899).
Trop. & S. Africa. 22 NGA TOG 23 CMN CON ZAI 25 TAN UGA 26 ANG MLW MOZ 27 BOT NAM. Hemicr. or rhizome geophyte.

**Cyperus albiceps* Ridl., J. Bot. 22: 16 (1884).

Kyllinga merxmuelleri Podlech, Mitt. Bot. Staatssamml. München 3: 525 (1960). *Cyperus merxmuelleri* (Podlech) Lye in R.W.Haines & K.A.Lye, Sedges & Rushes E. Afr., App. 3: 2 (1983).

Kyllinga albogracilis Lye, Nordic J. Bot. 1: 742 (1981 publ. 1982). *Cyperus albogracilis* (Lye) Lye in R.W.Haines & K.A.Lye, Sedges & Rushes E. Afr., App. 3: 2 (1983).
S. Tanzania to Zambia. 25 TAN 26 ZAM. Hemicr. or rhizome geophyte.

Kyllinga aureovillosa Lye, Nordic J. Bot. 1: 743 (1981 publ. 1982). *Cyperus aureovillosus* (Lye) Lye in R.W.Haines & K.A.Lye, Sedges & Rushes E. Afr., App. 3: 2 (1983).
Malawi (Nyika plateau). 26 MLW. Hemicr. or rhizome geophyte.

Kyllinga bigibbosa (Fosberg) Lye, Nordic J. Bot. 1: 746 (1981 publ. 1982).
Aldabra. 29 ALD. Hemicr.

**Cyperus bigibbosa* Fosberg, Kew Bull. 31: 832 (1977).

Kyllinga brevifolia Rottb., Descr. Icon. Rar. Pl.: 13 (1773). *Cyperus brevifolius* (Rottb.) Hassk., Cat. Hort. Bot. Bogor.: 24 (1844).
Trop. & Subtrop. (21) azo mdr 22 LBR NGA TOG 23 BUR CMN EQG RWA ZAI 24 ETH SOC SUD 25 KEN TAN UGA 27 BOT NAM NAT 28 ASC STH 29 ALD CGS COM MAU MCI MDG REU ROD SEY 34 AFG 36 CHC CHM 38 JAP KOR KZN NNS OGA TAI 40 ASS BAN EHM IND NEP PAK SRL 41 THA VIE 42 BOR CKI JAW LSI MLY MOL PHI SUL SUM XMS 43 NWG (50) nfk NSW NTA QLD SOA WAU 51 NZN (60) fij GIL nue ton 61 EAS MRQ PIT SCI TUA (63) haw 74 OKL (76) cal 77 TEX 78 ALA FLA GEO LOU MSI NCA SCA 79 MXC MXG MXN MXS MXT 80 BLZ COS CPI ELS GUA HON NIC PAN 81 BER CAY CUB DOM JAM LEE PUE TRT WIN 82 FRG GUY SUR VEN 83 BOL CLM ECU GAL PER 84 BZN 85 AGE AGW PAR URU. Hemicr. or rhizome geophyte.

subsp. ***brevifolia***
Trop. & Subtrop. (21) azo mdr 22 LBR NGA TOG 23 CMN 24 SOC 25 KEN TAN UGA 27 NAT 28 ASC STH 29 ALD CGS COM MAU MCI MDG REU ROD SEY 34 AFG 36 CHC CHM 38 JAP KOR KZN NNS OGA TAI 40 ASS BAN EHM IND NEP PAK SRL 41 THA VIE 42 BOR CKI JAW LSI MLY MOL PHI SUL SUM XMS 43 NWG (50) nfk NSW NTA QLD SOA WAU 51 NZN (60) fij GIL nue ton 61 EAS MRQ PIT SCI TUA (63) haw 74 OKL (76) cal 77 TEX 78 ALA FLA GEO LOU MSI NCA SCA 79 MXC MXG MXN MXS MXT 80 BLZ COS CPI ELS GUA

HON NIC PAN 81 BER CAY CUB DOM JAM LEE PUE TRT WIN 82 FRG GUY SUR VEN 83 BOL CLM ECU GAL PER 84 BZN 85 AGE AGW PAR URU. Hemicr. or rhizome geophyte.

Schoenus capitatus Crantz, Inst. Rei Herb. 1: 408 (1766).

Kyllinga monocephala L.f., Suppl. Pl.: 104 (1782), nom. illeg.

Kyllinga monocephala Thunb., Fl. Jap.: 35 (1784), nom. illeg.

Kyllinga tricephala Salisb., Prodr. Stirp. Chap. Allerton: 32 (1796).

Kyllinga nivea Pers., Syn. Pl. 1: 57 (1805).

Kyllinga intermedia R.Br., Prodr.: 219 (1810).

Kyllinga elongata Kunth in F.W.H.von Humboldt, A.J.A.Bonpland & C.S.Kunth, Nov. Gen. Sp. 1: 211 (1816).

Kyllinga cruciformis Schrad. ex Schult. in J.J.Roemer & J.A.Schultes, Mant. 2: 107 (1824). *Cyperus cruciformis* (Schrad. ex Schult.) Endl., Cat. Horti Vindob. 1: 94 (1842). *Kyllinga brevifolia* var. *cruciformis* (Schrad. ex Schult.) Cherm., Ann. Inst. Bot.-Géol. Colon. Marseille 27(2): 38 (1919).

Kyllinga tenuis Baldwin, Trans. Amer. Philos. Soc., n.s., 2: 168 (1825).

Kyllinga cruciata Nees, Linnaea 9: 286 (1834), nom. inval.

Kyllinga gracilis Kunth, Enum. Pl. 2: 134 (1837).

Kyllinga sororia Kunth, Enum. Pl. 2: 131 (1837).

Kyllinga laxa Schrad. ex Nees in C.F.P.von Martius & auct. suc. (eds.), Fl. Bras. 2(1): 14 (1842).

Kyllinga odorata Liebm., Mexic. Neldeagt. Pl., V, 2: 233 (1851), nom. illeg.

Kyllinga hohenackeri Hochst. ex Steud., Syn. Pl. Glumac. 2: 60 (1854).

Kyllinga pumilio Steud., Syn. Pl. Glumac. 2: 67 (1854).

Kyllinga tenuissima Steud., Syn. Pl. Glumac. 2: 67 (1854).

Mariscus kyllingioides Steud., Syn. Pl. Glumac. 2: 60 (1854).

Kyllinga fuscata Miq., Fl. Ned. Ind. 3: 294 (1856).

Kyllinga longiculmis Miq., Fl. Ned. Ind. 3: 292 (1856).

Kyllinga brevifolia var. *longifolia* Boeck., Linnaea 35: 426 (1868).

Kyllinga oligostachya Boeck., Linnaea 35: 407 (1868).

Kyllinga honolulu Steud. ex Jard., Bull. Soc. Linn. Normandie, II, 9: 278 (1875).

Kyllinga brevifolia f. *laxa* H.Pfeiff., Repert. Spec. Nov. Regni Veg. 27: 95 (1929).

Cyperus brevifolius f. *firmiculmis* Kük. in H.G.A. Engler (ed.), Pflanzenr., IV, 20(101): 602 (1936).

subsp. ***intricata*** (Cherm.) Lye, Nordic J. Bot. 1: 747 (1981 publ. 1982).

Ethiopia to S. Africa, Madagascar. 23 BUR RWA ZAI 24 ETH SUD 25 KEN TAN UGA 27 BOT NAM NAT TVL 29 MDG. Hemicr. or rhizome geophyte.

Kyllinga aurata Nees, Linnaea 10: 139 (1836). *Cyperus erectus* var. *auratus* (Nees) Kük. in H.G.A.Engler (ed.), Pflanzenr., IV, 20: 589 (1926). *Kyllinga colorata* var. *aurata* (Nees) Lye, Bot. Not. 125: 211 (1972).

Kyllinga consanguinea Kunth, Enum. Pl. 2: 135 (1837).

Kyllinga sojauxii Boeck., Flora 62: 515 (1879).

**Kyllinga intricata* Cherm., Bull. Mus. Natl. Hist. Nat. 25: 211 (1919). *Cyperus erectus* var. *intricatus* (Cherm.) Kük. in H.G.A.Engler (ed.), Pflanzenr., IV, 20(101): 590 (1936). *Kyllinga erecta* var. *intricata* (Cherm.) Kük. in H.G.A.Engler (ed.), Pflanzenr., IV, 20(101): 590 (1936). *Cyperus brevifolius* subsp. *intricatus* (Cherm.) Lye in R.W.Haines & K.A.Lye, Sedges & Rushes E. Afr., App. 3: 2 (1983).

subsp. ***lurida*** (Kük.) Lye, Nordic J. Bot. 1: 747 (1981 publ. 1982).

E. Trop. Africa. 25 KEN TAN. Hemicr. or rhizome geophyte.

**Kyllinga erecta* var. *lurida* Kük., Notizbl. Bot. Gart. Berlin-Dahlem 9: 300 (1925). *Cyperus erectus* var. *luridus* (Kük.) Kük. in H.G.A.Engler (ed.), Pflanzenr., IV, 20(101): 590 (1936). *Kyllinga aurata* var. *lurida* (Kük.) Napper, J. E. Africa Nat. Hist. Soc. Natl. Mus. 28: 21 (1971). *Kyllinga colorata* var. *lurida* (Kük.) Lye, Bot. Not. 125: 218 (1972). *Cyperus brevifolius* subsp. *luridus* (Kük.) Lye in R.W.Haines & K.A.Lye, Sedges & Rushes E. Afr., App. 3: 2 (1983).

var. ***stellulata*** (J.V.Suringar) Hooper ex Ohwi in S.Karthikeyan & al., Fl. Ind. Enumerat. – Monocot.: 60 (1989).

S. India, Malesia to New Guinea. 40 IND 42 JAW LSI PHI 43 NWG. Hemicr. or rhizome geophyte.

**Cyperus brevifolius* f. *stellulatus* J.V.Suringar, Sur. Gesl. Cyp. Mal. Arch.: 48 (1898). *Cyperus brevifolius* var. *stellulatus* (J.V.Suringar) Kük. in H.G.A.Engler (ed.), Pflanzenr., IV, 20(101)(101): 603 (1936).

Kyllinga brunneoalata Cherm., Bull. Soc. Bot. France 83: 492 (1936).

Gabon. 23 GAB. Hemicr. or rhizome geophyte.

Kyllinga brunneoalba Lye, Nordic J. Bot. 1: 741 (1981 publ. 1982). *Cyperus brunneoalbus* (Lye) Lye in R.W.Haines & K.A.Lye, Sedges & Rushes E. Afr., App. 3: 2 (1983).

Kenya (Mandera Distr.). 25 KEN. Hemicr.

Kyllinga buchananii C.B.Clarke in W.H.Harvey & auct. suc. (eds.), Fl. Cap. 7: 155 (1897). *Cyperus sublaevicarinatus* Mattf. & Kük. in H.G.A.Engler (ed.), Pflanzenr., IV, 20(101): 572 (1936).

S. Trop. & S. Africa. 26 MLW 27 NAT. Hemicr. or rhizome geophyte.

Kyllinga bulbosa P.Beauv., Fl. Oware 1: 11 (1805).

Trop. Africa to Indo-China. 22 BEN BKN GAM GHA GUI IVO MLI NGA SEN SIE TOG 23 BUR CAF CMN CON EQG GAB GGI RWA ZAI 24 CHA ETH SUD 25 KEN TAN UGA 26 ANG MLW MOZ 36 CHS 40 ASS BAN EHM IND NEP PAK SRL 41 MYA THA VIE (42) mly (50) nsw qld. Hemicr. or rhizome geophyte.

Schoenus niveus L., Syst. Veg. ed. 13: 81 (1774). *Mariscus niveus* (L.) Merr., Philipp. J. Sci. 14: 369 (1919).

Kyllinga monocephala Nees in R.Wight, Contr. Bot. India: 91 (1834), nom. illeg.

Kyllinga nana Nees ex Hook. & Arn., Bot. Beechey Voy.: 224 (1837).

Kyllinga pumila Steud., Flora 25: 596 (1842), nom. illeg.

Kyllinga macrocephala A.Rich., Tent. Fl. Abyss. 2: 491 (1850). *Cyperus richardii* Steud., Syn. Pl. Glumac. 2: 8 (1854).

Kyllinga geminiflora Steud., Syn. Pl. Glumac. 2: 70 (1854). *Cyperus geminiflora* (Steud.) Wickens, Bull. Republ. Sudan Min. Agric. & Forest., n.s., 14: 40 (1969).
Kyllinga triceps var. *obtusiflora* Boeck., Linnaea 35: 414 (1868). *Cyperus triceps* var. *obtusiflorus* (Boeck.) Kük. in H.G.A.Engler (ed.), Pflanzenr., IV, 20(101): 579 (1936).
Kyllinga leucantha Boeck., Linnaea 38: 356 (1874).
Kyllinga schimperi Hochst. ex Engl., Abh. Königl. Akad. Wiss. Berlin 1891: 145 (1892), pro syn.
Kyllinga macrocephala var. *angustior* C.B.Clarke in T.A.Durand & H.Schinz, Études Fl. Congo: 279 (1896). *Cyperus richardii* var. *angustior* (C.B.Clarke) Kük. in H.G.A.Engler (ed.), Pflanzenr., IV, 20(101): 570 (1936).
Kyllinga parvula C.B.Clarke ex Rendle in W.P.Hiern, Cat. Afr. Pl. 2: 103 (1899).
Kyllinga macrocephala var. *oligantha* Cherm., Bull. Soc. Bot. France 82: 334 (1935). *Cyperus richardii* var. *oliganthus* (Cherm.) Kük. in H.G.A.Engler (ed.), Pflanzenr., IV, 20(101): 570 (1936).
Cyperus triceps var. *angustifolius* Kük. in H.G.A. Engler (ed.), Pflanzenr., IV, 20(101): 579 (1936).

Kyllinga cardosii Meneses, Garcia de Orta 4: 240 (1956).
Angola. 26 ANG. Hemicr. or rhizome geophyte.

Kyllinga cartilaginea K.Schum. in H.G.A.Engler, Pflanzenw. Ost-Afrikas, C: 123 (1895). *Cyperus cartilagineus* (K.Schum.) Mattf. & Kük. in H.G.A. Engler (ed.), Pflanzenr., IV, 20(101): 608 (1936).
Tanzania to S. Trop. Africa, Madagascar. 23 ZAI 24 TAN 26 ANG MOZ 29 MDG. Hemicr. or rhizome geophyte.
Kyllinga alba var. *laevissima* Cherm., Bull. Soc. Bot. France 82: 334 (1935). *Cyperus cartilagineus* var. *laevissimus* (Cherm.) Kük. in H.G.A.Engler (ed.), Pflanzenr., IV, 20(101): 609 (1936).
Cyperus cartilagineus var. *serratangulus* Peter & Kük. in H.G.A.Engler (ed.), Pflanzenr., IV, 20(101): 609 (1936).

Kyllinga chlorotropis Steud., Flora 25: 598 (1842). *Cyperus chlorotropis* (Steud.) Mattf. & Kük. in H.G.A. Engler (ed.), Pflanzenr., IV, 20(101): 576 (1936).
NE. & E. Trop. Africa, Arabian Pen. 24 ETH 25 KEN TAN UGA 35 SAU YEM. Hemicr.

Kyllinga chrysantha K.Schum. in H.G.A.Engler, Pflanzenw. Ost-Afrikas, C: 123 (1895). *Cyperus aureostramineus* Mattf. & Kük. in H.G.A.Engler (ed.), Pflanzenr., IV, 20(101): 573 (1936).
Somalia to Tanzania. 23 RWA 24 SOM 25 KEN TAN UGA. Hemicr.

Kyllinga chrysanthoides Mtot., Bot. Jahrb. Syst. 111: 292 (1990).
Burundi. 23 BUR. Hemicr. or rhizome geophyte.

Kyllinga comosipes (Mattf. & Kük.) Napper, J. E. Africa Nat. Hist. Soc. Natl. Mus. 28: 24 (1971).
Ethiopia to Zambia. 24 ETH SOM 25 KEN TAN UGA 26 ZAM. Hemicr.

subsp. ***comosipes***
Ethiopia to Zambia. 24 ETH SOM 25 KEN TAN UGA 26 ZAM. Hemicr.
Kyllinga aurea Thomson in J.H.Speke, J. Discov. Source Nile: 654 (1863), nom. nud.
Kyllinga leucocephala Boeck., Flora 58: 257 (1875), nom. illeg.
Kyllinga eximia var. *kelleri* C.B.Clarke in D.Oliver & auct. suc. (eds.), Fl. Trop. Afr. 8: 288 (1901). *Cyperus eximius* var. *kelleri* (C.B.Clarke) Kük. in H.G.A.Engler (ed.), Pflanzenr., IV, 20(101): 568 (1936).
Cyperus cartilagineus var. *angustatus* Peter & Kük. in H.G.A.Engler (ed.), Pflanzenr., IV, 20(101): 609 (1936). *Kyllinga comosipes* var. *angustata* (Peter & Kük.) Napper, J. E. Africa Nat. Hist. Soc. Natl. Mus. 28: 24 (1971).

subsp. ***decolorans*** (Kük.) Lye, Nordic J. Bot. 1: 747 (1981 publ. 1982).
E. Trop. Africa. 25 KEN TAN UGA. Hemicr.
**Kyllinga chrysantha* var. *decolorans* Kük., Repert. Spec. Nov. Regni Veg. 12: 92 (1913). *Cyperus aureostramineus* var. *decolorans* (Kük.) Kük. in H.G.A.Engler (ed.), Pflanzenr., IV, 20(101): 574 (1936). *Cyperus comosipes* subsp. *decolorans* (Kük.) Lye in R.W.Haines & K.A.Lye, Sedges & Rushes E. Afr., App. 3: 2 (1983). *Kyllinga decolorans* (Kük.) Mtot., Bot. Jahrb. Syst. 111: 291 (1990).

Kyllinga controversa Steud., Syn. Pl. Glumac. 2: 70 (1854). *Cyperus controversus* (Steud.) Mattf. & Kük. in H.G.A.Engler (ed.), Pflanzenr., IV, 20(101): 611 (1936). *Cyperus alatus* subsp. *controversus* (Steud.) Lye, Lidia 3: 172 (1995).
Trop. Africa. 22 NGA SEN 24 ERI ETH SUD 26 ANG MOZ. Hemicr. or rhizome geophyte.
Kyllinga alba Steud., Flora 16: 597 (1842), nom. illeg.
Kyllinga dorsocrena Nees, Linnaea 35: 430 (1868).

Kyllinga coriacea Cherm., Bull. Mus. Natl. Hist. Nat. 25: 208 (1919).
W. Madagascar. 29 MDG. Hemicr. or rhizome geophyte.

Kyllinga crassipes Boeck., Flora 42: 444 (1859). *Cyperus bulbipes* Mattf. & Kük. in H.G.A.Engler (ed.), Pflanzenr., IV, 20(101): 587 (1936).
E. & S. Trop. Africa. 25 KEN TAN UGA 26 MOZ. Hemicr. or rhizome geophyte.
Cyperus bulbipes var. *pallescens* Kük. in H.G.A.Engler (ed.), Pflanzenr., IV, 20(101): 588 (1936).

Kyllinga curvispiculosa (T.Koyama) P.H.Hô, Cayco Vietnam 3: 699 (1993).
S. Vietnam. 41 VIE. Hemicr. or rhizome geophyte.
**Cyperus curvispiculosus* T.Koyama, Naturaliste Canad. 82: 215 (1955).

Kyllinga debilis C.B.Clarke, Bull. Soc. Bot. France 54(8): 26 (1907). *Cyperus leptorhachis* Mattf. & Kük. in H.G.A.Engler (ed.), Pflanzenr., IV, 20(101): 595 (1936).
W. & WC. Trop. Africa. 22 BEN BKN GUI IVO MLI NGA SEN TOG 23 CAF CON. Ther.

Kyllinga echinata S.S.Hooper, Kew Bull. 26: 580 (1972).
Burkina Faso, Ghana, Gabon. 22 BKN GHA 23 GAB. Hemicr.

Kyllinga eglandulosa Govind. & Ramani, J. Econ. Taxon. Bot. 18: 335 (1994).
India (Karnataka). 40 IND. Hemicr. or rhizome geophyte.

Kyllinga elatior Kunth, Enum. Pl. 2: 135 (1837). *Cyperus aromaticus* var. *elatior* (Kunth) Kük. in H.G.A.Engler (ed.), Pflanzenr., IV, 20(101): 582 (1936).
Trop. & S. Africa, Madagascar. 23 CMN GGI RWA

ZAI 24 ETH 25 KEN TAN UGA 26 ZIM 27 CPP NAT 29 MDG. Hemicr. or rhizome geophyte.

Kyllinga pinguis C.B.Clarke, Bot. Jahrb. Syst. 38: 131 (1906). *Cyperus pinguis* (C.B.Clarke) Mattf. & Kük. in H.G.A.Engler (ed.), Pflanzenr., IV, 20(101): 583 (1936).

Kyllinga erecta Schumach., Beskr. Guin. Pl.: 42 (1827). *Cyperus erectus* (Schumach.) Mattf. & Kük. in H.G.A. Engler (ed.), Pflanzenr., IV, 20(101): 588 (1936).

Trop. & S. Africa, W. Indian Ocean. 22 BEN BKN GHA GUI IVO LBR MLI NGA SEN SIE TOG 23 CAF CMN CON EQG GAB GGI RWA ZAI 24 CHA ETH 25 KEN TAN UGA 26 ANG MLW MOZ ZIM 27 CPP LES NAT OFS SWZ TVL 29 MAU MDG. Hemicr. or rhizome geophyte.

var. ***africana*** (Kük.) S.S.Hooper, Kew Bull. 26: 580 (1972).

W. & WC. Trop. Africa. 22 BEN GHA IVO MLI NGA TOG 23 CAF CMN ZAI. Hemicr. or rhizome geophyte.

**Cyperus obtusatus* var. *africanus* Kük. in H.G.A.Engler (ed.), Pflanzenr., IV, 20(101): 586 (1936).

subsp. ***albescens*** Lye, Nordic J. Bot. 1: 745 (1981 publ. 1982). *Cyperus erectus* subsp. *albescens* (Lye) Lye in R.W.Haines & K.A.Lye, Sedges & Rushes E. Afr., App. 3: 2 (1983).

WC. Trop. Africa. 23 GAB ZAI. Hemicr. or rhizome geophyte.

subsp. ***erecta***

Trop. & S. Africa, W. Indian Ocean. 22 BEN BKN GHA GUI IVO LBR MLI NGA SEN SIE TOG 23 CMN CON GAB RWA ZAI 24 CHA ETH 25 KEN TAN UGA 26 ANG MLW MOZ ZIM 27 CPP LES NAT OFS SWZ TVL 29 MAU MDG. Hemicr. or rhizome geophyte.

Kyllinga consanguinea Kunth, Enum. Pl. 2: 135 (1837).

Cyperus monocephalus F.Muell., Fragm. 8: 271 (1874), nom. illeg.

Kyllinga erecta var. *intercedens* Kük., Repert. Spec. Nov. Regni Veg. 12: 91 (1913). *Cyperus erectus* var. *intercedens* (Kük.) Kük. in H.G.A.Engler (ed.), Pflanzenr., IV, 20(101): 589 (1936).

Kyllinga erecta var. *pleiocarpa* Kük., Repert. Spec. Nov. Regni Veg. 12: 91 (1913). *Cyperus erectus* var. *pleiocarpus* (Kük.) Kük. in H.G.A.Engler (ed.), Pflanzenr., IV, 20(101): 589 (1936).

Kyllinga erecta var. *schlechteri* Kük., Repert. Spec. Nov. Regni Veg. 12: 91 (1913). *Cyperus erectus* var. *schlechteri* (Kük.) Kük. in H.G.A.Engler (ed.), Pflanzenr., IV, 20(101): 589 (1936).

Cyperus erectus f. *minor* Kük. in H.G.A.Engler (ed.), Pflanzenr., IV, 20(101): 590 (1936).

Cyperus erectus f. *pallescens* Kük. in H.G.A.Engler (ed.), Pflanzenr., IV, 20(101): 590 (1936).

Kyllinga exigua Boeck., Abh. Naturwiss. Vereine Bremen 7: 36 (1880). *Cyperus luteostramineus* Mattf. & Kük. in H.G.A.Engler (ed.), Pflanzenr., IV, 20(101): 571 (1936).

W. Indian Ocean. 24 SOM? 29 MAU MDG REU. Hemicr. or rhizome geophyte.

Kyllinga eximia C.B.Clarke in D.Oliver & auct. suc. (eds.), Fl. Trop. Afr. 8: 287 (1901). *Cyperus eximius* (C.B.Clarke) Mattf. & Kük. in H.G.A.Engler (ed.), Pflanzenr., IV, 20(101): 567 (1936).

Ethiopia to Kenya. 24 ETH SOM 25 KEN. Hemicr. or rhizome geophyte.

Kyllinga filicula C.B.Clarke in D.Oliver & auct. suc. (eds.), Fl. Trop. Afr. 8: 526 (1902).

Zaïre. 23 ZAI. Hemicr. or rhizome geophyte.

Kyllinga gracillima Miq., Ann. Mus. Bot. Lugduno-Batavi 2: 142 (1865). *Cyperus brevifolioides* Thieret & Delahouss., Sida 3: 129 (1967). *Kyllinga brevifolioides* (Thieret & Delahouss.) G.C.Tucker, J. Arnold Arbor. 68: 408 (1987).

Temp. Asia. 31 33 36 CHC CHS 38 JAP KOR 40 EHM NEP (75) cnt nwj nwy pen (78) ala ark del geo kty mry msi nca ten vrg. Hemicr. or rhizome geophyte.

Kyllinga monocephala var. *leiolepis* Franch. & Sav., Enum. Pl. Jap. 2: 108 (1877). *Kyllinga brevifolia* subsp. *leiolepis* (Franch. & Sav.) T.Koyama, Enum. Fl. Pl. Nepal 1: 115 (1978). *Kyllinga brevifolia* var. *leiolepis* (Franch. & Sav.) Karthik. in S.Karthikeyan & al., Fl. Ind. Enumerat. – Monocot.: 60 (1989).

Kyllinga inaurata Nees ex Boeck., Linnaea 35: 406 (1868). *Cyperus inauratus* (Nees ex Boeck.) Mattf. & Kük. in H.G.A.Engler (ed.), Pflanzenr., IV, 20(101): 573 (1936).

Tanzania to S .Africa. 25 TAN 27 CPP TVL. Hemicr. or rhizome geophyte.

Kyllinga tetragona Spreng. ex C.B.Clarke in W.H. Harvey & auct. suc. (eds.), Fl. Cap. 7: 154 (1897).

Cyperus inauratus var. *laevicarinatus* Kük. in H.G.A. Engler (ed.), Pflanzenr., IV, 20(101): 573 (1936).

Kyllinga jubensis Chiov., Fl. Somala 2: 433 (1932). *Cyperus erectus* subsp. *jubensis* (Chiov.) Lye, Candollea 51: 424 (1996).

S. Somalia. 24 SOM. Hemicr. or rhizome geophyte.

Kyllinga kamschatica Meinsh., Trudy Imp. S.-Peterburgsk. Bot. Sada 18: 229 (1901). *Kyllinga brevifolia* subsp. *kamtschatica* (Meinsh.) Vorosch. in A.K.Skvortsov (ed.), Florist. Issl. Razn. Raionakh SSSR: 152 (1985).

Russian Far East. 31 KAM. Hemicr. or rhizome geophyte.

Kyllinga riederiana Meinsh. ex Kom., Fl. Kamtschatka 1: 200 (1927), pro syn.

Kyllinga kilianii Muasya & D.A.Simpson, Kew Bull. 51: 183 (1996).

Kenya. 25 KEN. Hemicr. or rhizome geophyte.

Kyllinga lehmannii Nees, Linnaea 10: 139 (1836). *Cyperus lehmannii* (Nees) F.Muell., Fragm. 8: 272 (1874).

Cape Prov. 27 CPP. Hemicr. or rhizome geophyte.

Kyllinga ciliata Kunth, Enum. Pl. 2: 136 (1837).

Kyllinga melanosperma Nees in R.Wight, Contr. Bot. India: 91 (1834). *Cyperus melanospermus* (Nees) Suringar, Cyperus Mal. Arch.: 50 (1898).

Trop. Old World. 22 NGA 23 CMN GAB GGI ZAI 25 KEN TAN UGA 26 ANG 27 BOT CPP NAT SWZ TVL 29 COM MDG SEY 36 CHC CHH 40 IND SRL 41 THA VIE 42 JAW LSI MLY PHI SUL SUM 43 NWG 50 QLD (60) fij. Hemicr. or rhizome geophyte.

subsp. ***bifolia*** (Miq.) Karthik. in S.Karthikeyan & al., Fl. Ind. Enumerat. – Monocot.: 60 (1989).

India to New Guinea. 40 IND 41 THA 42 JAW LSI MLY 43 NWG. Hemicr. or rhizome geophyte.

Kyllinga bifolia Miq., Fl. Ned. Ind. 3: 293 (1856). *Cyperus melanospermus* subsp. *bifolius* (Miq.) J.Kern, Reinwardtia 3: 62 (1954).
Cyperus brevifolius f. *vaginatus* J.V.Suringar, Sur. Gesl. Cyp. Mal. Arch.: 47 (1898).

subsp. ***elata*** (Steud.) Lye, Nordic J. Bot. 1: 747 (1981 publ. 1982).
E. Trop. Africa to Seychelles. 25 KEN TAN 29 COM SEY. Hemicr. or rhizome geophyte.
Kyllinga elata Steud., Syn. Pl. Glumac. 2: 70 (1854). *Cyperus aromaticus* var. *elatus* (Steud.) Kük. in H.G.A.Engler (ed.), Pflanzenr., IV, 20(101): 582 (1936). *Kyllinga polyphylla* var. *elata* (Steud.) Lye, Bot. Not. 125: 218 (1972). *Cyperus melanospermus* subsp. *elatus* (Steud.) Lye in R.W.Haines & K.A.Lye, Sedges & Rushes E. Afr., App. 3: 2 (1983). *Kyllinga melanosperma* var. *elata* (Steud.) J.-P.Lebrun & Stork, Énum. Pl. Fl. Afr. Trop. 3: 191 (1995).

var. ***hexalata*** Lye, Nordic J. Bot. 1: 746 (1981 publ. 1982). *Cyperus melanospermus* var. *hexalatus* (Lye) Lye in R.W.Haines & K.A.Lye, Sedges & Rushes E. Afr., App. 3: 2 (1983).
Uganda. 25 UGA. Hemicr. or rhizome geophyte.

subsp. ***melanosperma***
Trop. Old World. 22 NGA 23 CMN GAB GGI ZAI 25 KEN TAN UGA 26 ANG 27 BOT CPP NAT SWZ TVL 29 MDG 36 CHC CHH 40 IND SRL 41 THA VIE 42 JAW LSI MLY PHI SUL SUM 43 NWG 50 QLD (60) fij. Hemicr. or rhizome geophyte.
Kyllinga vaginata Zoll., Syst. Verz.: 63 (1845).
Kyllinga fuscescens Boeck., Linnaea 35: 421 (1868).
Kyllinga imerinensis Cherm., Bull. Mus. Natl. Hist. Nat. 25: 210 (1919). *Cyperus melanospermus* var. *imerinensis* (Cherm.) Kük. in H.G.A.Engler (ed.), Pflanzenr., IV, 20(101): 584 (1936).
Kyllinga perrieri Cherm., Bull. Mus. Natl. Hist. Nat. 25: 211 (1919). *Kyllinga imerinensis* var. *perrieri* (Cherm.) Cherm., Bull. Soc. Bot. France 74: 603 (1927). *Cyperus melanospermus* var. *perrieri* (Cherm.) Kük. in H.G.A.Engler (ed.), Pflanzenr., IV, 20(101): 584 (1936).

var. ***plurifoliata*** Kük., Repert. Spec. Nov. Regni Veg. 12: 92 (1913). *Kyllinga plurifoliata* (Kük.) Cherm. in T.A.Durand & H.Schinz, Consp. Fl. Afric. 5: 531 (1894). *Cyperus melanospermus* var. *plurifoliatus* (Kük.) Kük. in H.G.A.Engler (ed.), Pflanzenr., IV, 20(101): 585 (1936).
C. Madagascar. 29 MDG. Hemicr. or rhizome geophyte.

Kyllinga microbracteata Lye, Nordic J. Bot. 1: 744 (1981 publ. 1982). *Cyperus microbracteatus* (Lye) Lye in R.W.Haines & K.A.Lye, Sedges & Rushes E. Afr., App. 3: 2 (1983).
S. Tanzania. 25 TAN. Hemicr. or tuber geophyte.

Kyllinga microbulbosa Lye, Bot. Not. 125: 217 (1972). *Cyperus microbulbosus* (Lye) Lye in R.W.Haines & K.A.Lye, Sedges & Rushes E. Afr., App. 3: 2 (1983).
Tanzania (Kaitokoi pools). 25 TAN. Hemicr.

Kyllinga microstyla C.B.Clarke, Bull. Misc. Inform. Kew 1895: 229 (1895). *Cyperus microstylus* (C.B.Clarke) Mattf. & Kük. in H.G.A.Engler (ed.), Pflanzenr., IV, 20(101): 578 (1936).
Ethiopia to Kenya. 24 ETH SOM 25 KEN. Hemicr.

Kyllinga nemoralis (J.R.Forst. & G.Forst.) Dandy ex Hutch. & Dalziel, Fl. W. Trop. Afr. 2: 487 (1936).
Trop. & Subtrop. Old World. 22 BEN GHA GUI IVO LBR NGA TOG 23 CMN CON GAB GGI 24 SOC 25 KEN TAN UGA 26 MOZ 27 NAT 28 STH 29 COM MAU MDG REU SEY 33 TCS 36 CHH CHS 38 KZN NNS OGA TAI 40 ASS EHM IND NEP PAK SRL WHM 41 AND MYA THA VIE 42 BOR JAW LSI MLY MOL PHI SUL SUM xms 43 BIS NWG SOL 50 NFK QLD (60) fij GIL nue NWC SAM TON VAN 61 COO SCI TUA 62 CRL MRN MRS (63) haw (81) jam lee pue trt win (82) sur (84) bze. Hemicr. or rhizome geophyte.
Scirpus cephalotes Jacq., Hort. Bot. Vindob. 1: 42 (1771), nom. illeg. *Kyllinga cephalotes* Druce, Bot. Soc. Exch. Club Brit. Isles 1916: 630 (1917).
Kyllinga monocephala Rottb., Descr. Icon. Rar. Pl.: 13 (1773), excl. synon. *Cyperus kyllingia* Endl., Cat. Horti Vindob. 1: 94 (1842).
**Thryocephalon nemorale* J.R.Forst. & G.Forst., Char. Gen. Pl.: 65 (1775).
Kyllinga monocephala Stokes, Bot. Mat. Med. 1: 120 (1812), nom. illeg.
Kyllinga monocephala var. *subtriceps* Kunth, Enum. Pl. 2: 130 (1837). *Cyperus kyllingia* f. *subtriceps* (Kunth) Kük. in H.G.A.Engler (ed.), Pflanzenr., IV, 20(101): 608 (1936).
Kyllinga gracilis Zoll., Syst. Verz.: 63 (1845), nom. illeg.
Cyperus leucocephalus Hassk., Pl. Jav. Rar.: 87 (1848).
Kyllinga mindorensis Steud., Syn. Pl. Glumac. 2: 67 (1854).
Kyllinga intermedia Seem., Bonplandia (Hannover) 9: 261 (1861).
Kyllinga monocephala var. *humilis* Boeck., Linnaea 35: 429 (1868). *Cyperus kyllingia* f. *humilis* (Boeck.) Kük. in H.G.A.Engler (ed.), Pflanzenr., IV, 20(101): 608 (1936).
Kyllinga monocephala var. *tenuis* Boeck., Linnaea 35: 429 (1868). *Cyperus kyllingia* f. *tenuis* (Boeck.) Kük. in H.G.A.Engler (ed.), Pflanzenr., IV, 20(101): 608 (1936).
Kyllinga planiculmis var. *mucronata* Cherm., Bull. Soc. Bot. France 72: 613 (1925).

Kyllinga nervosa Steud., Flora 25: 597 (1842). *Cyperus costatus* Mattf. & Kük. in H.G.A.Engler (ed.), Pflanzenr., IV, 20(101): 575 (1936). *Cyperus oblongus* subsp. *nervosus* (Steud.) Lye in R.W.Haines & K.A.Lye, Sedges & Rushes E. Afr., App. 3: 2 (1983).
Ethiopia to Tanzania. 23 RWA ZAI 24 ERI ETH SOM 25 KEN TAN UGA. Hemicr.

subsp. ***flava*** (C.B.Clarke) Lye, Nordic J. Bot. 1: 747 (1981 publ. 1982).
E. Trop. Africa. 25 KEN TAN. Hemicr.
**Kyllinga flava* C.B.Clarke in D.Oliver & auct. suc. (eds.), Fl. Trop. Afr. 8: 281 (1901). *Kyllinga nervosa* var. *flava* (C.B.Clarke) Lye, Bot. Not. 125: 218 (1972). *Cyperus oblongus* subsp. *flavus* (C.B.Clarke) Lye in R.W.Haines & K.A.Lye, Sedges & Rushes E. Afr., App. 3: 2 (1983).

subsp. ***jubensis*** (Mtot.) J.-P.Lebrun & Stork, Énum. Pl. Fl. Afr. Trop. 4: 602 (1997).
S. Somalia. 24 SOM. Hemicr.
**Kyllinga flava* subsp. *jubensis* Mtot., Sinet 13: 39 (1990). *Cyperus oblongus* subsp. *jubensis* (Mtot.) Lye, Lidia 3: 132 (1994).

subsp. ***nervosa***
Ethiopia to Tanzania. 23 RWA ZAI 24 ERI ETH SOM 25 KEN TAN UGA. Hemicr.

subsp. ***oblonga*** (C.B.Clarke) J.-P.Lebrun & Stork, Énum. Pl. Fl. Afr. Trop. 3: 192 (1995).
E. Kenya. 25 KEN. Hemicr.
**Kyllinga oblonga* C.B.Clarke in D.Oliver & auct. suc. (eds.), Fl. Trop. Afr. 8: 284 (1901). *Cyperus oblongus* (C.B.Clarke) Kük. in H.G.A.Engler (ed.), Pflanzenr., IV, 20(101): 580 (1936).

var. ***ruwenzoriensis*** (C.B.Clarke) Lye, Bot. Not. 125: 218 (1972).
W. Uganda. 25 UGA. Hemicr.
**Kyllinga ruwenzoriensis* C.B.Clarke in D.Oliver & auct. suc. (eds.), Fl. Trop. Afr. 8: 283 (1901). *Cyperus oblongus* var. *ruwenzoriensis* (C.B.Clarke) Kük. in H.G.A.Engler (ed.), Pflanzenr., IV, 20(101): 580 (1936).

subsp. ***sidamoensis*** Mtot., Sinet 13: 37 (1990). *Cyperus costatus* subsp. *sidamoensis* (Mtot.) Lye, Lidia 3: 132 (1994).
Ethiopia to Kenya. 24 ETH 25 KEN. Hemicr. or rhizome geophyte.

Kyllinga ngothe Mtot., Bot. Jahrb. Syst. 111: 294 (1990).
E. Trop. Africa. 25 KEN TAN. Hemicr. or rhizome geophyte.

Kyllinga nigripes C.B.Clarke in D.Oliver & auct. suc. (eds.), Fl. Trop. Afr. 8: 285 (1901). *Cyperus nigripes* (C.B.Clarke) Kük. in H.G.A.Engler (ed.), Pflanzenr., IV, 20(101): 572 (1936).
Tanzania to Malawi. 25 TAN 26 MLW. Hemicr. or rhizome geophyte.
Cyperus nigripes var. *grandiceps* Kük. in H.G.A. Engler (ed.), Pflanzenr., IV, 20(101): 572 (1936).

Kyllinga nudiceps C.B.Clarke ex Standl., Publ. Field Columbian Mus., Bot. Ser. 4: 199 (1929). *Cyperus nudiceps* (C.B.Clarke ex Standl.) O'Neill, Leafl. W. Bot. 4: 38 (1944).
Cocos I. 80 CPI. Hemicr. or rhizome geophyte.

Kyllinga odorata Vahl, Enum. Pl. 2: 382 (1805). *Kyllinga odorata* var. *genuina* Osten, Anales Mus. Hist. Nat. Montevideo, II, 3: 120 (1931), nom. inval.
Trop. & Subtrop. 22 BEN GNB GUI IVO LBR NGA SEN SIE TOG 23 CMN CON EQG GAB GGI RWA ZAI 24 CHA ETH SOM 25 KEN TAN UGA 26 ANG 27 BOT CPP NAT SWZ TVL 29 MDG 35 YEM 36 CHC CHS 38 NNS TAI 40 ASS EHM IND NEP SRL 41 THA VIE 42 BOR JAW LSI MOL PHI SUM 43 NWG 50 NSW NTA QLD 74 OKL 77 NWM TEX 78 ALA ARK FLA GEO LOU MSI NCA SCA 79 MXC MXE MXG MXN MXS MXT 80 BLZ COS ELS GUA HON NIC PAN 81 CUB JAM LEE PUE TRT WIN 82 FRG GUY SUR VEN 83 BOL ECU PER 84 BZC BZN 85 AGE AGW PAR URU. Hemicr. or rhizome geophyte.

subsp. ***appendiculata*** (K.Schum.) Lye, Nordic J. Bot. 1: 746 (1981 publ. 1982).
Togo to NE. & E. Trop. Africa. 22 TOG 23 CMN EQG GGI RWA ZAI 24 ETH SOM 25 KEN TAN. Hemicr. or rhizome geophyte.
Kyllinga macrocephala Hook.f., J. Proc. Linn. Soc., Bot. 7: 225 (1864), nom. illeg.
Kyllinga cylindrica var. *major* C.B.Clarke in T.A.Durand & H.Schinz, Consp. Fl. Afric. 5: 528 (1894). *Kyllinga odorata* var. *major* (C.B.Clarke) Chiov., Fl. Somala 2: 432 (1932). *Cyperus sesquiflorus* var. *major* (C.B.Clarke) Kük. in H.G.A.Engler (ed.), Pflanzenr., IV, 20(101): 594 (1936).
**Kyllinga appendiculata* K.Schum., Bot. Jahrb. Syst. 24: 338 (1897). *Cyperus sesquiflorus* subsp. *appendiculatus* (K.Schum.) Lye in R.W. Haines & K.A.Lye, Sedges & Rushes E. Afr., App. 3: 2 (1983).

subsp. ***cylindrica*** (Nees) T.Koyama, Gard. Bull. Singapore 30: 161 (1977).
Trop. & Subtrop. Old World. 22 NGA SIE TOG 23 CMN CON GGI ZAI 24 ETH 25 KEN TAN UGA 26 ANG 29 MDG 36 CHS 38 NNS TAI 40 ASS EHM IND NEP SRL 41 THA VIE 42 BOR JAW LSI MOL PHI SUM 43 NWG. Hemicr. or rhizome geophyte.
**Kyllinga cylindrica* Nees in R.Wight, Contr. Bot. India: 91 (1834). *Kyllinga odorata* var. *cylindrica* (Nees) Kük., J. Straits Branch Roy. Asiat. Soc. 76: 80 (1917). *Cyperus sesquiflorus* var. *cylindricus* (Nees) Kük. in H.G.A.Engler (ed.), Pflanzenr., IV, 20(101): 593 (1936). *Cyperus sesquiflorus* subsp. *cylindricus* (Nees) T.Koyama, Bot. Mag. (Tokyo) 83: 187 (1970). *Kyllinga sesquiflora* subsp. *cylindrica* (Nees) T.Koyama, in Fl. Taiwan 5: 288 (1978).
Kyllinga cylindrica var. *subtriceps* Nees in R.Wight, Contr. Bot. India: 91 (1834). *Cyperus sesquiflorus* var. *subtriceps* (Nees) T.Koyama, Quart. J. Taiwan Mus. 14: 191 (1961).
Kyllinga viridula Hochst. ex A.Rich., Tent. Fl. Abyss. 2: 492 (1850).
Kyllinga cylindrica f. *latifolia* (Benth.) Domin, Biblioth. Bot. 85: 414 (1915).
Kyllinga madagascariensis Gand., Bull. Soc. Bot. France 66: 298 (1919 publ. 1920).

subsp. ***odorata***
Trop. & Subtrop. 22 BEN GNB GUI IVO LBR NGA SEN SIE TOG 23 CMN GAB GGI RWA ZAI 24 CHA SOM 25 KEN TAN UGA 26 ANG 27 BOT CPP NAT SWZ TVL 29 MDG 35 YEM 36 CHC 38 NNS 40 ASS EHM IND NEP SRL SRL 41 THA VIE VIE 42 BOR JAW LSI MOL PHI SUM 43 NWG 50 NSW NTA QLD 74 OKL 77 NWM TEX 78 ALA ARK FLA GEO LOU MSI NCA SCA 79 MXC MXE MXG MXN MXS MXT 80 BLZ COS ELS GUA HON NIC PAN 81 CUB JAM LEE PUE TRT WIN 82 FRG GUY SUR VEN 83 BOL ECU PER 84 BZC BZN 85 AGE AGW PAR URU. Hemicr. or rhizome geophyte.
Schoenus odoratus Aubl., Hist. Pl. Guiane 1: 44 (1775).
Kyllinga triceps Sw., Observ. Bot.: 33 (1791), nom. illeg.
Kyllinga monocephala Muhl., Descr. Gram.: 3 (1817), nom. illeg.
Kyllinga leucocephala Baldwin, Trans. Amer. Philos. Soc., n.s., 2: 170 (1825).
Kyllinga pumila Sieber ex C.Presl, Isis (Oken) 21: 270 (1828).
Kyllinga sesquiflora Torr., Ann. Lyceum Nat. Hist. New York 3: 287 (1836). *Cyperus sesquiflorus* (Torr.) Mattf. & Kük. in H.G.A.Engler (ed.), Pflanzenr., IV, 20(101): 19 (1935).
Kyllinga martiana Schrad. ex Nees in C.F.P.von Martius & auct. suc. (eds.), Fl. Bras. 2(1): 14 (1842).

Kyllinga tuberosa Scheele, Linnaea 17: 342 (1843).
Kyllinga campestris Salzm. ex Steud., Syn. Pl. Glumac. 2: 68 (1854).
Kyllinga odorata var. *elongata* Boeck., Linnaea 35: 415 (1868). *Cyperus sesquiflorus* f. *elongatus* (Boeck.) Kük. in H.G.A.Engler (ed.), Pflanzenr., IV, 20(101): 594 (1936).
Kyllinga odorata var. *gracilis* Boeck., Linnaea 35: 411 (1868). *Cyperus sesquiflorus* f. *gracilis* (Boeck.) Kük. in H.G.A.Engler (ed.), Pflanzenr., IV, 20(101): 593 (1936).
Kyllinga odorata var. *minor* Boeck., Linnaea 35: 411 (1868). *Cyperus sesquiflorus* var. *minor* (Boeck.) Kük. in H.G.A.Engler (ed.), Pflanzenr., IV, 20(101): 593 (1936).
Kyllinga odorata var. *latifolia* Benth., Fl. Austral. 7: 252 (1878).
Kyllinga pellucidoalbida Boeck., Beitr. Cyper. 1: 1 (1888).
Kyllinga pellucidolucida Boeck., Beitr. Cyper. 1: 1 (1888).
Kyllinga odorata var. *rigida* Boeck., Vidensk. Meddel. Dansk Naturhist. Foren. Kjøbenhavn 1894: 237 (1895).
Kyllinga leucocephala var. *pluriceps* Kük., Repert. Spec. Nov. Regni Veg. 12: 92 (1913). *Kyllinga odorata* var. *pluriceps* (Kük.) Kük., Notizbl. Bot. Gart. Berlin-Dahlem 9: 299 (1925). *Cyperus sesquiflorus* var. *pluriceps* (Kük.) Kük. in H.G.A.Engler (ed.), Pflanzenr., IV, 20(101): 595 (1936).
Kyllinga odorata var. *bulbifera* Kük., Repert. Spec. Nov. Regni Veg. 12: 91 (1913). *Cyperus sesquiflorus* f. *bulbiferus* (Kük.) Kük. in H.G.A.Engler (ed.), Pflanzenr., IV, 20(101): 593 (1936).
Kyllinga odorata var. *fallax* Kük., Notizbl. Bot. Gart. Berlin-Dahlem 9: 299 (1925). *Cyperus sesquiflorus* var. *fallax* (Kük.) Kük. in H.G.A.Engler (ed.), Pflanzenr., IV, 20(101): 595 (1936).
Kyllinga odorata var. *dunensis* Osten, Anales Mus. Hist. Nat. Montevideo, IIa, 3(2): 120 (1931). *Cyperus sesquiflorus* f. *dunensis* (Osten) Kük. in H.G.A.Engler (ed.), Pflanzenr., IV, 20(101): 593 (1936).
Cyperus sesquiflorus f. *denticulatus* Kük. in H.G.A.Engler (ed.), Pflanzenr., IV, 20(101): 595 (1936).
Cyperus sesquiflorus f. *globosus* Kük. in H.G.A.Engler (ed.), Pflanzenr., IV, 20(101): 594 (1936).
Cyperus sesquiflorus f. *spinulosus* Kük. in H.G.A.Engler (ed.), Pflanzenr., IV, 20(101): 593 (1936).
Cyperus kernianus Ohwi & T.Koyama, J. Jap. Bot. 30: 126 (1955).

Kyllinga pachystyla Kük., Bull. Misc. Inform. Kew 1911: 153 (1911). *Cyperus pachystylus* (Kük.) Kük. in H.G.A.Engler (ed.), Pflanzenr., IV, 20(101): 591 (1936).
Angola. 26 ANG. Hemicr. or rhizome geophyte.

Kyllinga pauciflora Ridl., Trans. Linn. Soc. London, Bot. 2: 147 (1884). *Cyperus ridleyi* Mattf. & Kük. in H.G.A.Engler (ed.), Pflanzenr., IV, 20(101): 599 (1936).
SW. Tanzania S. Africa. 25 TAN 26 ANG MLW ZIM 27 NAT. Hemicr. or rhizome geophyte.

Kyllinga peteri (Kük.) Lye, Nordic J. Bot. 1: 746 (1981 publ. 1982).
Tanzania to Zambia. 25 TAN 26 ZAM. Hemicr. or rhizome geophyte.
**Cyperus peteri* Kük. in H.G.A.Engler (ed.), Pflanzenr., IV, 20(101): 575 (1936).

Kyllinga planiculmis C.B.Clarke ex Cherm., Bull. Mus. Natl. Hist. Nat. 25: 209 (1919).
W. Indian Ocean. 29 COM MAU MDG. Hemicr. or rhizome geophyte.
**Kyllinga monocephala* var. *latifolia* Boeck., Linnaea 35: 429 (1868). *Cyperus kyllingia* var. *latifolius* (Boeck.) Kük. in H.G.A.Engler (ed.), Pflanzenr., IV, 20(101): 608 (1936). *Cyperus sesquiflorus* f. *latifolius* (Boeck.) Kük. in H.G.A.Engler (ed.), Pflanzenr., IV, 20(101): 593 (1936).

Kyllinga platyphylla K.Schum., Bot. Jahrb. Syst. 30: 270 (1901). *Cyperus ciliatopilosus* Mattf. & Kük. in H.G.A.Engler (ed.), Pflanzenr., IV, 20(101): 7 (1935).
S. Tanzania to S. Trop. Africa. 25 TAN 26 ANG MLW MOZ ZAM. Hemicr. or rhizome geophyte.
Kyllinga platyphylla var. *longifolia* Kük., Repert. Spec. Nov. Regni Veg. 12: 92 (1913). *Cyperus ciliatopilosus* var. *longifolius* (Kük.) Kük. in H.G.A. Engler (ed.), Pflanzenr., IV, 20(101): 571 (1936).
Cyperus ciliatopilosus var. *rhizomatosus* Kük. in H.G.A. Engler (ed.), Pflanzenr., IV, 20(101): 571 (1936).

Kyllinga pluristaminea Govind. & Ramani, J. Econ. Taxon. Bot. 18: 363 (1994).
India (Tamil Nadu). 40 IND. Hemicr. or rhizome geophyte.

Kyllinga polyphylla Willd. ex Kunth, Enum. Pl. 2: 134 (1837). *Kyllinga erecta* var. *polyphylla* (Willd. ex Kunth) S.S.Hooper, Kew Bull. 26: 580 (1972).
Trop. & S. Africa, W. Indian Ocean. 22 GHA GUI NGA TOG 23 BUR CAF CMN CON GAB ZAI 24 SOM 25 KEN TAN UGA 26 ANG 27 NAT 29 COM MAU MDG REU SEY (40) ind srl (41) tha vie (42) mly xms (43) sol (50) qld (60) fij sam (81) win. Hemicr. or rhizome geophyte.
Kyllinga macrantha Boeck., Linnaea 35: 420 (1868).
Kyllinga involucrata Bojer ex Baker, Fl. Mauritius: 415 (1877).
Kyllinga aromatica Ridl., Trans. Linn. Soc. London, Bot. 2: 146 (1884). *Cyperus aromaticus* (Ridl.) Mattf. & Kük. in H.G.A.Engler (ed.), Pflanzenr., IV, 20(101): 581 (1936).
Kyllinga planiceps C.B.Clarke in T.A.Durand & H.Schinz, Consp. Fl. Afric. 5: 531 (1894), nom. nud.
Kyllinga teres C.B.Clarke in D.Oliver & auct. suc. (eds.), Fl. Trop. Afr. 8: 276 (1901). *Cyperus aromaticus* var. *teres* (C.B.Clarke) Kük. in H.G.A. Engler (ed.), Pflanzenr., IV, 20(101): 582 (1936).
Cyperus aromaticus var. *brachyrhizomatosus* Kük. in H.G.A.Engler (ed.), Pflanzenr., IV, 20(101): 583 (1936).
Cyperus aromaticus var. *repens* Kük. in H.G.A.Engler (ed.), Pflanzenr., IV, 20(101): 583 (1936).

Kyllinga pseudobulbosa Mtot., Nordic J. Bot. 9: 638 (1990).
Tanzania. 25 TAN. Hemicr. or rhizome geophyte.

Kyllinga pulchella Kunth, Enum. Pl. 2: 137 (1837). *Cyperus teneristolon* Mattf. & Kük. in H.G.A.Engler (ed.), Pflanzenr., IV, 20(101): 574 (1936).
Ethiopia to S. Africa. 24 ETH 25 KEN TAN 27 CPP LES NAT OFS TVL. Hemicr. or rhizome geophyte.

Kyllinga atrosanguinea Steud., Flora 25: 599 (1842).
Kyllinga bracheilema Steud., Flora 25: 598 (1842). *Cyperus bracheilema* (Steud.) Mattf. & Kük. in H.G.A.Engler (ed.), Pflanzenr., IV, 20(101): 576 (1936).
Kyllinga pulchella f. *robustior* Kük., Repert. Spec. Nov. Regni Veg. 12: 92 (1913). *Cyperus teneristolon* var. *robustior* (Kük.) Kük. in H.G.A. Engler (ed.), Pflanzenr., IV, 20(101): 575 (1936).
Cyperus transitorius Kük. in H.G.A.Engler (ed.), Pflanzenr., IV, 20(101): 35 (1935). *Kyllinga transitoria* (Kük.) T.Koyama, J. Jap. Bot. 51: 317 (1976).
Kyllinga anomala Peter & Kük. in H.G.A.Engler (ed.), Pflanzenr., IV, 20(101): 35 (1935), nom. inval.

Kyllinga pumila Michx., Fl. Bor.-Amer. 1: 28 (1803). *Kyllinga caespitosa* var. *pumila* (Michx.) Boeck., Linnaea 35: 412 (1868). *Thryocephalon pumilum* (Michx.) Nieuwl., Amer. Midl. Naturalist 4: 94 (1915). *Cyperus densicaespitosus* Mattf. & Kük. in H.G.A. Engler (ed.), Pflanzenr., IV, 20(101): 597 (1936).
Trop. & S. Africa, Madagascar, E. U.S.A. to N. Argentina. 21 CVI 22 BEN BKN GAM GHA GNB GUI IVO LBR MLI NGA NGR SEN SIE TOG 23 BUR CMN CON EQG GAB GGI RWA ZAI 24 CHA ETH 25 KEN TAN UGA 26 ANG MLW 29 MDG 74 ILL KAN MSO OKL 75 INI NWJ NWY OHI PEN WVA 77 TEX 78 ALA ARK DEL FLA GEO KTY LOU MRY MSI NCA SCA TEN VRG 79 MXC MXE MXG MXS MXT 80 BLZ COS CPI ELS GUA HON NIC PAN 81 CUB JAM LEE PUE TRT WIN 82 FRG GUY SUR VEN 83 BOL CLM ECU GAL PER 84 BZC BZE BZL BZN BZS 85 AGE AGW PAR URU. Ther. or hemicr.
Kyllinga viridiflora Roxb. ex Spreng., Index Seminum (HAL).
Kyllinga odorata Kunth in F.W.H.von Humboldt, A.J.A.Bonpland & C.S.Kunth, Nov. Gen. Sp. 1: 211 (1816), nom. illeg.
Hedychloa fragrans Raf., Ann. Nat. 1: 16 (1820).
Kyllinga uncinata Link, Enum. Hort. Berol. Alt. 1: 47 (1821).
Kyllinga viridiflora Link, Hort. Berol. 2: 322 (1827).
Kyllinga pumila var. *elatior* Kunth, Enum. Pl. 2: 132 (1837).
Kyllinga monocephala Sieber ex Steud., Nomencl. Bot., ed. 2, 1: 851 (1840), nom. inval.
Kyllinga caespitosa Nees in C.F.P.von Martius & auct. suc. (eds.), Fl. Bras. 2(1): 12 (1842).
Kyllinga caespitosa var. *major* Nees in C.F.P.von Martius & auct. suc. (eds.), Fl. Bras. 2(1): 12 (1842). *Cyperus densicaespitosus* var. *major* (Nees) Kük. in H.G.A. Engler (ed.), Pflanzenr., IV, 20(101): 599 (1936).
Kyllinga hortensis Salzm. ex Schltdl., Bot. Zeitung (Berlin) 7: 117 (1849).
Kyllinga rigidula Steud., Syn. Pl. Glumac. 2: 71 (1854). *Cyperus densicaespitosus* var. *rigidulus* (Steud.) Kük. in H.G.A.Engler (ed.), Pflanzenr., IV, 20(101): 599 (1936).
Kyllinga tenuifolia Steud., Syn. Pl. Glumac. 2: 69 (1854). *Cyperus tenuifolius* (Steud.) Dandy in A.W.Exell, Cat. Vasc. Pl. S. Tome: 363 (1944).
Kyllinga fraterna Steud. in W.Lechler, Berberid. Amer. Austral.: 56 (1857).
Kyllinga caespitosa var. *elatior* (Kunth) Boeck., Linnaea 35: 413 (1868).
Kyllinga robusta Boeck., Linnaea 35: 409 (1868). *Kyllinga brevifolia* var. *robusta* (Boeck.) H.Pfeiff., Repert. Spec. Nov. Regni Veg. 27: 96 (1929).
Kyllinga naumanniana Boeck., Flora 62: 516 (1879).
Kyllinga flexuosa Boeck., Beitr. Cyper. 2: 1 (1890).
Kyllinga caespitosa subvar. *debilis* Kuntze, Revis. Gen. Pl. 2: 754 (1891).
Kyllinga blepharinota Hochst. ex Engl., Abh. Königl. Akad. Wiss. Berlin 1891: 145 (1892).

Kyllinga rhizomafragilis Lye, Nordic J. Bot. 1: 744 (1981 publ. 1982). *Cyperus rhizomafragilis* (Lye) Lye in R.W.Haines & K.A.Lye, Sedges & Rushes E. Afr., App. 3: 2 (1983).
Zambia. 26 ZAM. Hemicr. or rhizome geophyte.

Kyllinga robinsoniana Mtot., Nordic J. Bot. 9: 637 (1990).
Kenya to S. Trop. Africa. 25 KEN TAN 26 ANG MOZ ZAM. Hemicr. or rhizome geophyte.

Kyllinga senegalensis C.B.Clarke in D.Oliver & auct. suc. (eds.), Fl. Trop. Afr. 8: 276 (1901). *Cyperus senegalensis* (C.B.Clarke) Mattf. & Kük. in H.G.A. Engler (ed.), Pflanzenr., IV, 20(101): 581 (1936).
W. Trop. Africa. 22 SEN TOG. Hemicr. or rhizome geophyte.

Kyllinga songeensis Lye, Bot. Not. 125: 218 (1972). *Cyperus songeensis* (Lye) Lye in R.W.Haines & K.A.Lye, Sedges & Rushes E. Afr., App. 3: 2 (1983).
E. Trop. Africa. 25 KEN TAN. Hemicr. or tuber geophyte.

Kyllinga sphaerocephala Boeck., Flora 58: 258 (1875). *Cyperus purpureoglandulosus* Mattf. & Kük. in H.G.A.Engler (ed.), Pflanzenr., IV, 20(101): 570 (1936).
S. Somalia to Angola. 23 BUR RWA 24 SOM 25 KEN TAN UGA 26 ANG. Hemicr. or rhizome geophyte.

Kyllinga squamulata Vahl, Enum. Pl. 2: 381 (1805).
Africa to Indo-China, Queensland. 21 CVI 22 BEN BKN GAM GHA GNB GUI IVO LBR MLI MTN NGA NGR SEN TOG 23 CMN CON ZAI 24 CHA ETH SUD 25 TAN UGA 29 MDG 33 TCS 40 ASS EHM IND NEP PAK 41 VIE 50 QLD (78) fla (81) win. Ther.
Kyllinga squamulosa Kunth, Enum. Pl. 2: 135 (1837).
Kyllinga cristata Afzel. ex A.Rich., Tent. Fl. Abyss. 2: 493 (1850).
Kyllinga dentata Hochst. ex A.Rich., Tent. Fl. Abyss. 2: 493 (1850).
Kyllinga metzii Hochst. ex Steud., Syn. Pl. Glumac. 2: 70 (1854). *Cyperus metzii* (Hochst. ex Steud.) Mattf. & Kük. in H.G.A.Engler (ed.), Pflanzenr., IV, 20(101): 25 (1935).

Kyllinga stenophylla K.Schum. ex C.B.Clarke, Bot. Jahrb. Syst. 38: 131 (1906). *Cyperus densicaespitosus* var. *stenophyllus* (K.Schum. ex C.B.Clarke) Kük. in H.G.A. Engler (ed.), Pflanzenr., IV, 20(101): 599 (1936).
WC. Trop. Africa. 23 CMN RWA. Hemicr. or rhizome geophyte.

Kyllinga tanzaniae Lye, Bot. Not. 125: 217 (1972). *Cyperus tanzaniae* (Lye) Lye in R.W.Haines & K.A.Lye, Sedges & Rushes E. Afr., App. 3: 2 (1983).
Tanzania. 25 TAN. Hemicr. or rhizome geophyte.

Kyllinga tibialis Poit. ex Ledeb., Diss. Bot. Pl. Doming.: 6 (1805). *Cyperus tibialis* (Poit. ex Ledeb.) Govaerts, World Checklist Seed Pl. 3(1): 21 (1999).
Trop. America, W. Trop. Africa. 22 NGA SEN 80 BLZ COS ELS GUA HON NIC PAN 81 DOM JAM PUE TRT 83 CLM ECU. Hemicr. or rhizome geophyte.

Mariscus aphyllus Vahl, Enum. Pl. 2: 373 (1805). *Kyllinga aphylla* (Vahl) Kunth, Enum. Pl. 2: 127 (1837). *Cyperus aphyllus* (Kunth) F.Muell., Fragm. 8: 272 (1874), nom. illeg. *Cyperus aphyllus* (Vahl) Hassk., Cat. Hort. Bot. Bogor.: 24 (1844), nom. illeg.
Kyllinga peruviana var. *foliata* Kük., Repert. Spec. Nov. Regni Veg. 12: 92 (1913). *Cyperus peruvianus* var. *foliatus* (Kük.) Kük. in H.G.A.Engler (ed.), Pflanzenr., IV, 20(101): 587 (1936).

Kyllinga tisserantii Cherm., Arch. Bot. Bull. Mens. 4(7): 6 (1931).
W. Trop. Africa to Chad. 22 IVO NGA SIE TOG 23 CAF CMN 24 CHA. Hemicr. or rhizome geophyte.

Kyllinga tisserantioides Mtot., Nordic J. Bot. 9: 639 (1990).
Benin to Nigeria. 22 BEN NGA. Hemicr. or rhizome geophyte.

Kyllinga ugogensis (Peter & Kük.) Lye, Bot. Not. 125: 218 (1972).
Tanzania. 25 TAN. Hemicr. or rhizome geophyte.
**Cyperus ugogensis* Peter & Kük. in H.G.A.Engler (ed.), Pflanzenr., IV, 20(101): 572 (1936).
Kyllinga ferruginea Peter ex Kük. in H.G.A.Engler (ed.), Pflanzenr., IV, 20(101): 572 (1936).

Kyllinga uniflora Mtot., Nordic J. Bot. 9: 640 (1990).
Tanzania. 25 TAN. Hemicr. or rhizome geophyte.

Kyllinga urbanii Kük., Repert. Spec. Nov. Regni Veg. 23: 183 (1926). *Cyperus neourbanii* Kük. in H.G.A.Engler (ed.), Pflanzenr., IV, 20(101): 604 (1936).
W. Cuba. 81 CUB. Hemicr. or rhizome geophyte.

Kyllinga vaginata Lam., Tabl. Encycl. 1: 148 (1791). *Kyllinga peruviana* Lam., Encycl. 3: 366 (1792), nom. superfl. *Cyperus peruvianus* (Lam.) F.N.Williams, Bull. Herb. Boissier, II, 7: 90 (1907).
Trop. America, Trop. Africa. 21 CVI 22 BEN NGA TOG 23 CMN CON EQG GAB GGI RWA ZAI 25 TAN 26 ANG MLW 80 BLZ COS ELS GUA HON NIC PAN 81 CAY DOM HAI JAM LEE PUE 82 FRG GUY SUR VEN 83 BOL ECU PER 84 BZC BZE BZL BZN BZS 85 AGE AGW PAR URU. Hemicr. or rhizome geophyte.
Kyllinga capitata P.Beauv., Fl. Oware 1: t. 31 (1806).
Kyllinga globosa P.Beauv., Fl. Oware 1: t. 31 (1806).
Kyllinga rigida Baldwin, Trans. Amer. Philos. Soc., n.s., 2: 169 (1825).
Kyllinga pungens Link, Hort. Berol. 1: 326 (1827).
Kyllinga obtusata J.Presl & C.Presl, Reliq. Haenk. 1: 183 (1828). *Cyperus obtusatus* (J.Presl & C.Presl) Mattf. & Kük. in H.G.A.Engler (ed.), Pflanzenr., IV, 20(101): 585 (1936).
Kyllinga vaginata Rchb.f. ex Kunth, Enum. Pl. 2: 128 (1837). *Kyllinga obtusata* var. *aphylla* Boeck., Vidensk. Meddel. Dansk Naturhist. Foren. Kjøbenhavn 1869: 159 (1869).
Schoenus angularis Spreng. ex Kunth, Enum. Pl. 2: 128 (1837).
Kyllinga stricta Schrad. ex Nees in C.F.P.von Martius & auct. suc. (eds.), Fl. Bras. 2(1): 13 (1842).
Cyprolepis denudata Steud., Flora 33: 229 (1850).
Kyllinga repens Salzm. ex Steud., Syn. Pl. Glumac. 2: 67 (1854).
Lyprolepis denudata Steud., Syn. Pl. Glumac. 2: 130 (1855).
Kyllinga obtusata var. *cylindrostachya* Boeck., Linnaea 35: 419 (1868). *Cyperus obtusatus* var. *cylindrostachyus* (Boeck.) Kük. in H.G.A. Engler (ed.), Pflanzenr., IV, 20(101): 586 (1936).
Kyllinga obtusata var. *subaphylla* Boeck., Linnaea 35: 418 (1868).
Kyllinga tenuis Boeck., Linnaea 35: 423 (1868), nom. illeg. *Cyperus obtusatus* var. *tenuis* (Boeck.) Kük. in H.G.A.Engler (ed.), Pflanzenr., IV, 20(101): 586 (1936).

Kyllinga welwitschii Ridl., Trans. Linn. Soc. London, Bot. 2: 147 (1884). *Cyperus welwitschii* (Ridl.) Lye in R.W.Haines & K.A.Lye, Sedges & Rushes E. Afr., App. 3: 2 (1983).
Dry Trop. Africa to Namibia. 22 BKN IVO MTN NGA NGR SEN 23 CMN 24 CHA ETH SOM 25 KEN UGA 26 ANG 27 NAM. Hemicr.
Kyllinga triceps var. *ciliata* Boeck. in W.C.H.Peters, Naturw. Reise Mossambique: 535 (1864). *Cyperus triceps* var. *ciliatus* (Boeck.) Kük. in H.G.A.Engler (ed.), Pflanzenr., IV, 20(101): 579 (1936).
Kyllinga controversa var. *subexalata* C.B.Clarke in D.Oliver & auct. suc. (eds.), Fl. Trop. Afr. 8: 271 (1902). *Cyperus controversus* var. *subexalatus* (C.B.Clarke) Kük. in H.G.A.Engler (ed.), Pflanzenr., IV, 20(101): 612 (1936).

***Synonyms*:**
Kyllinga alata Nees = ***Kyllinga alba*** Nees subsp. ***alba***
Kyllinga alba Steud. = ***Kyllinga controversa*** Steud.
Kyllinga alba var. *alata* (Nees) C.B.Clarke = ***Kyllinga alba*** Nees subsp. ***alba***
Kyllinga alba var. *laevissima* Cherm. = ***Kyllinga cartilaginea*** K.Schum.
Kyllinga alba subsp. *nigritana* (C.B.Clarke) J.-P.Lebrun & Stork = ***Kyllinga alba*** Nees subsp. ***alba***
Kyllinga albescens Steud. = ***Lipocarpha chinensis*** (Osbeck) J.Kern
Kyllinga ambigua Steud. = [24 CHA ETH] Isolepis ?
Kyllinga anomala Peter & Kük. = ***Kyllinga pulchella*** Kunth
Kyllinga aphylla (Vahl) Kunth = ***Kyllinga tibialis*** Poit. ex Ledeb.
Kyllinga appendiculata K.Schum. = ***Kyllinga odorata*** subsp. ***appendiculata*** (K.Schum.) Lye
Kyllinga aromatica Ridl. = ***Kyllinga polyphylla*** Willd. ex Kunth
Kyllinga ascolepidioides Cherm. = ***Kyllinga alba*** subsp. ***ascolepidioides*** (Cherm.) Lye
Kyllinga atrosanguinea Steud. = ***Kyllinga pulchella*** Kunth
Kyllinga aurata Nees = ***Kyllinga brevifolia*** subsp. ***intricata*** (Cherm.) Lye
Kyllinga aurata var. *lurida* (Kük.) Napper = ***Kyllinga brevifolia*** subsp. ***lurida*** (Kük.) Lye
Kyllinga aurea Thomson = ***Kyllinga comosipes*** (Mattf. & Kük.) Napper subsp. ***comosipes***
Kyllinga baoulensis A.Chev. = ***Lipocarpha barteri*** C.B.Clarke
Kyllinga bifolia Miq. = ***Kyllinga melanosperma*** subsp. ***bifolia*** (Miq.) Karthik.
Kyllinga biglumis (Gaertn.) Steud. = ***Cyperus cyperoides*** (L.) Kuntze subsp. ***cyperoides***
Kyllinga blepharinota Hochst. ex Engl. = ***Kyllinga pumila*** Michx.
Kyllinga bracheilema Steud. = ***Kyllinga pulchella*** Kunth
Kyllinga brevifolia Roem. & Schult. = ***Rhynchospora berteroi*** (Spreng.) C.B.Clarke
Kyllinga brevifolia var. *cruciformis* (Schrad. ex Schult.) Cherm. = ***Kyllinga brevifolia*** Rottb. subsp. ***brevifolia***
Kyllinga brevifolia subsp. *kamtschatica* (Meinsh.) Vorosch. = ***Kyllinga kamschatica*** Meinsh.

Kyllinga brevifolia f. *laxa* H.Pfeiff. = ***Kyllinga brevifolia*** Rottb. subsp. ***brevifolia***
Kyllinga brevifolia subsp. *leiolepis* (Franch. & Sav.) T.Koyama = ***Kyllinga gracillima*** Miq.
Kyllinga brevifolia var. *leiolepis* (Franch. & Sav.) Karthik. = ***Kyllinga gracillima*** Miq.
Kyllinga brevifolia var. *longifolia* Boeck. = ***Kyllinga brevifolia*** Rottb. subsp. ***brevifolia***
Kyllinga brevifolia var. *robusta* (Boeck.) H.Pfeiff. = ***Kyllinga pumila*** Michx.
Kyllinga brevifolioides (Thieret & Delahouss.) G.C.Tucker = ***Kyllinga gracillima*** Miq.
Kyllinga bulbocaulis Boeck. = ***Cyperus mollipes*** (C.B.Clarke) K.Schum.
Kyllinga bulbosa J.König ex Vahl = ***Cyperus cyperinus*** (Retz.) Suringar
Kyllinga bulbosa Steud. = ***Cyperus plateilema*** (Steud.) Kük.
Kyllinga caespitosa Nees = ***Kyllinga pumila*** Michx.
Kyllinga caespitosa subvar. *debilis* Kuntze = ***Kyllinga pumila*** Michx.
Kyllinga caespitosa var. *elatior* (Kunth) Boeck. = ***Kyllinga pumila*** Michx.
Kyllinga caespitosa var. *major* Nees = ***Kyllinga pumila*** Michx.
Kyllinga caespitosa var. *pumila* (Michx.) Boeck. = ***Kyllinga pumila*** Michx.
Kyllinga campestris Salzm. ex Steud. = ***Kyllinga odorata*** Vahl subsp. ***odorata***
Kyllinga capensis Steud. = ***Cyperus capensis*** (Steud.) Endl.
Kyllinga capitata P.Beauv. = ***Kyllinga vaginata*** Lam.
Kyllinga cayennensis Lam. = ***Cyperus aggregatus*** (Willd.) Endl. var. ***aggregatus***
Kyllinga cephalotes Druce = ***Kyllinga nemoralis*** (J.R.Forst. & G.Forst.) Dandy ex Hutch. & Dalziel
Kyllinga chrysantha var. *comosipes* (Mattf. & Kük.) J.-P.Lebrun & Stork = ***Kyllinga comosipes*** (Mattf. & Kük.) Napper
Kyllinga chrysantha var. *decolorans* Kük. = ***Kyllinga comosipes*** subsp. ***decolorans*** (Kük.) Lye
Kyllinga ciliata Kunth = ***Kyllinga lehmannii*** Nees
Kyllinga colorata (L.) Druce = ***Rhynchospora colorata*** (L.) H.Pfeiff.
Kyllinga colorata var. *aurata* (Nees) Lye = ***Kyllinga brevifolia*** subsp. ***intricata*** (Cherm.) Lye
Kyllinga colorata var. *lurida* (Kük.) Lye = ***Kyllinga brevifolia*** subsp. ***lurida*** (Kük.) Lye
Kyllinga comosipes var. *angustata* (Peter & Kük.) Napper = ***Kyllinga comosipes*** (Mattf. & Kük.) Napper subsp. ***comosipes***
Kyllinga consanguinea Kunth = ***Kyllinga brevifolia*** subsp. ***intricata*** (Cherm.) Lye
Kyllinga consanguinea Kunth = ***Kyllinga erecta*** Schumach. subsp. ***erecta***
Kyllinga controversa var. *subexalata* C.B.Clarke = ***Kyllinga welwitschii*** Ridl.
Kyllinga cristata Kunth = ***Kyllinga alba*** Nees subsp. ***alba***
Kyllinga cristata Afzel. ex A.Rich. = ***Kyllinga squamulata*** Vahl
Kyllinga cruciata Nees = ***Kyllinga brevifolia*** Rottb. subsp. ***brevifolia***
Kyllinga cruciformis Schrad. ex Schult. = ***Kyllinga brevifolia*** Rottb. subsp. ***brevifolia***
Kyllinga cylindrica Nees = ***Kyllinga odorata*** subsp. ***cylindrica*** (Nees) T.Koyama
Kyllinga cylindrica f. *latifolia* (Benth.) Domin = ***Kyllinga odorata*** subsp. ***cylindrica*** (Nees) T.Koyama
Kyllinga cylindrica var. *major* C.B.Clarke = ***Kyllinga odorata*** subsp. ***appendiculata*** (K.Schum.) Lye
Kyllinga cylindrica var. *subtriceps* Nees = ***Kyllinga odorata*** subsp. ***cylindrica*** (Nees) T.Koyama
Kyllinga cyperina Retz. = ***Cyperus cyperinus*** (Retz.) Suringar
Kyllinga cyperoides Roxb. = ***Courtoisina cyperoides*** (Roxb.) Soják
Kyllinga decolorans (Kük.) Mtot. = ***Kyllinga comosipes*** subsp. ***decolorans*** (Kük.) Lye
Kyllinga decora Steud. = ***Ascolepis brasiliensis*** (Kunth) Benth. ex C.B.Clarke
Kyllinga dentata Hochst. ex A.Rich. = ***Kyllinga squamulata*** Vahl
Kyllinga dipsacoides Schumach. = ***Ascolepis dipsacoides*** (Schumach.) J.Raynal
Kyllinga dorsocrena Nees = ***Kyllinga controversa*** Steud.
Kyllinga elata Steud. = ***Kyllinga melanosperma*** subsp. ***elata*** (Steud.) Lye
Kyllinga elongata Kunth = ***Kyllinga brevifolia*** Rottb. subsp. ***brevifolia***
Kyllinga erecta var. *intercedens* Kük. = ***Kyllinga erecta*** Schumach. subsp. ***erecta***
Kyllinga erecta var. *intricata* (Cherm.) Kük. = ***Kyllinga brevifolia*** subsp. ***intricata*** (Cherm.) Lye
Kyllinga erecta var. *lurida* Kük. = ***Kyllinga brevifolia*** subsp. ***lurida*** (Kük.) Lye
Kyllinga erecta var. *pleiocarpa* Kük. = ***Kyllinga erecta*** Schumach. subsp. ***erecta***
Kyllinga erecta var. *polyphylla* (Willd. ex Kunth) S.S.Hooper = ***Kyllinga polyphylla*** Willd. ex Kunth
Kyllinga erecta var. *schlechteri* Kük. = ***Kyllinga erecta*** Schumach. subsp. ***erecta***
Kyllinga eriocauloides Steud. = ***Ascolepis eriocauloides*** (Steud.) Nees ex Steud.
Kyllinga eximia var. *kelleri* C.B.Clarke = ***Kyllinga comosipes*** (Mattf. & Kük.) Napper subsp. ***comosipes***
Kyllinga ferruginea Peter ex Kük. = ***Kyllinga ugogensis*** (Peter & Kük.) Lye
Kyllinga filiformis Sw. = ***Cyperus swartzii*** (A.Dietr.) Boeck. ex Kük.
Kyllinga filiformis var. *capillaris* (Sw.) Griseb. = ***Cyperus nanus*** Willd. var. ***nanus***
Kyllinga filiformis var. *junciformis* C.Wright ex Griseb. = ***Cyperus swartzii*** (A.Dietr.) Boeck. ex Kük.
Kyllinga flava C.B.Clarke = ***Kyllinga nervosa*** subsp. ***flava*** (C.B.Clarke) Lye
Kyllinga flava subsp. *jubensis* Mtot. = ***Kyllinga nervosa*** subsp. ***jubensis*** (Mtot.) J.-P.Lebrun & Stork
Kyllinga flexuosa Boeck. = ***Kyllinga pumila*** Michx.
Kyllinga fraterna Steud. = ***Kyllinga pumila*** Michx.
Kyllinga fuscata Miq. = ***Kyllinga brevifolia*** Rottb. subsp. ***brevifolia***
Kyllinga fuscescens Boeck. = ***Kyllinga melanosperma*** Nees subsp. ***melanosperma***
Kyllinga geminiflora Steud. = ***Kyllinga bulbosa*** P.Beauv.
Kyllinga globosa P.Beauv. = ***Kyllinga vaginata*** Lam.
Kyllinga gracilis Kunth = ***Kyllinga brevifolia*** Rottb. subsp. ***brevifolia***
Kyllinga gracilis Zoll. = ***Kyllinga nemoralis*** (J.R.Forst. & G.Forst.) Dandy ex Hutch. & Dalziel
Kyllinga granularis Desf. ex Link = ***Cyperus swartzii*** (A.Dietr.) Boeck. ex Kük.
Kyllinga hohenackeri Hochst. ex Steud. = ***Kyllinga brevifolia*** Rottb. subsp. ***brevifolia***
Kyllinga honolulu Steud. ex Jard. = ***Kyllinga brevifolia*** Rottb. subsp. ***brevifolia***
Kyllinga hortensis Salzm. ex Schltdl. = ***Kyllinga pumila*** Michx.

Kyllinga hyalina (Vahl) T.Koyama = ***Queenslandiella hyalina*** (Vahl) Ballard
Kyllinga hymenopoda Peter = ***Cyperus pseudovestitus*** (C.B.Clarke) Kük.
Kyllinga imerinensis Cherm. = ***Kyllinga melanosperma*** Nees subsp. ***melanosperma***
Kyllinga imerinensis var. *perrieri* (Cherm.) Cherm. = ***Kyllinga melanosperma*** Nees subsp. ***melanosperma***
Kyllinga incompleta Jacq. = ***Cyperus hermaphroditus*** (Jacq.) Standl.
Kyllinga intermedia R.Br. = ***Kyllinga brevifolia*** Rottb. subsp. ***brevifolia***
Kyllinga intermedia Seem. = ***Kyllinga nemoralis*** (J.R.Forst. & G.Forst.) Dandy ex Hutch. & Dalziel
Kyllinga intricata Cherm. = ***Kyllinga brevifolia*** subsp. ***intricata*** (Cherm.) Lye
Kyllinga involucrata Bojer ex Baker = ***Kyllinga polyphylla*** Willd. ex Kunth
Kyllinga laxa Schrad. ex Nees = ***Kyllinga brevifolia*** Rottb. subsp. ***brevifolia***
Kyllinga leucantha Boeck. = ***Kyllinga bulbosa*** P.Beauv.
Kyllinga leucocephala Boeck. = ***Kyllinga comosipes*** (Mattf. & Kük.) Napper subsp. ***comosipes***
Kyllinga leucocephala Baldwin = ***Kyllinga odorata*** Vahl subsp. ***odorata***
Kyllinga leucocephala var. *pluriceps* Kük. = ***Kyllinga odorata*** Vahl subsp. ***odorata***
Kyllinga liebmannii Steud. = ***Cyperus nanus*** var. ***liebmannii*** (Steud.) Kük.
Kyllinga longiculmis Miq. = ***Kyllinga brevifolia*** Rottb. subsp. ***brevifolia***
Kyllinga macrantha Boeck. = ***Kyllinga polyphylla*** Willd. ex Kunth
Kyllinga macrocephala A.Rich. = ***Kyllinga bulbosa*** P.Beauv.
Kyllinga macrocephala Hook.f. = ***Kyllinga odorata*** subsp. ***appendiculata*** (K.Schum.) Lye
Kyllinga macrocephala var. *angustior* C.B.Clarke = ***Kyllinga bulbosa*** P.Beauv.
Kyllinga macrocephala var. *oligantha* Cherm. = ***Kyllinga bulbosa*** P.Beauv.
Kyllinga maculata Michx. = ***Lipocarpha maculata*** (Michx.) Torr.
Kyllinga madagascariensis Gand. = ***Kyllinga odorata*** subsp. ***cylindrica*** (Nees) T.Koyama
Kyllinga mariae Steud. = ***Cyperus dubius*** Rottb. subsp. ***dubius***
Kyllinga martiana Schrad. ex Nees = ***Kyllinga odorata*** Vahl subsp. ***odorata***
Kyllinga melanosperma var. *elata* (Steud.) J.-P.Lebrun & Stork = ***Kyllinga melanosperma*** subsp. ***elata*** (Steud.) Lye
Kyllinga merxmuelleri Podlech = ***Kyllinga albiceps*** (Ridl.) Rendle
Kyllinga metzii Hochst. ex Steud. = ***Kyllinga squamulata*** Vahl
Kyllinga microcephala Liebm. = ***Cyperus swartzii*** (A.Dietr.) Boeck. ex Kük.
Kyllinga microcephala Steud. = ***Kyllingiella microcephala*** (Steud.) R.W.Haines & Lye
Kyllinga mindorensis Steud. = ***Kyllinga nemoralis*** (J.R.Forst. & G.Forst.) Dandy ex Hutch. & Dalziel
Kyllinga monocephala L.f. = ***Kyllinga brevifolia*** Rottb. subsp. ***brevifolia***
Kyllinga monocephala Thunb. = ***Kyllinga brevifolia*** Rottb. subsp. ***brevifolia***
Kyllinga monocephala Nees = ***Kyllinga bulbosa*** P.Beauv.
Kyllinga monocephala Rottb. = ***Kyllinga nemoralis*** (J.R.Forst. & G.Forst.) Dandy ex Hutch. & Dalziel
Kyllinga monocephala Stokes = ***Kyllinga nemoralis*** (J.R.Forst. & G.Forst.) Dandy ex Hutch. & Dalziel
Kyllinga monocephala Muhl. = ***Kyllinga odorata*** Vahl subsp. ***odorata***
Kyllinga monocephala Sieber ex Steud. = ***Kyllinga pumila*** Michx.
Kyllinga monocephala var. *humilis* Boeck. = ***Kyllinga nemoralis*** (J.R.Forst. & G.Forst.) Dandy ex Hutch. & Dalziel
Kyllinga monocephala var. *latifolia* Boeck. = ***Kyllinga planiculmis*** C.B.Clarke ex Cherm.
Kyllinga monocephala var. *leiolepis* Franch. & Sav. = ***Kyllinga gracillima*** Miq.
Kyllinga monocephala var. *subtriceps* Kunth = ***Kyllinga nemoralis*** (J.R.Forst. & G.Forst.) Dandy ex Hutch. & Dalziel
Kyllinga monocephala var. *tenuis* Boeck. = ***Kyllinga nemoralis*** (J.R.Forst. & G.Forst.) Dandy ex Hutch. & Dalziel
Kyllinga multinervia Steud. = ***Cyperus dubius*** Rottb. subsp. ***dubius***
Kyllinga nana Nees ex Hook. & Arn. = ***Kyllinga bulbosa*** P.Beauv.
Kyllinga naumanniana Boeck. = ***Kyllinga pumila*** Michx.
Kyllinga nervosa var. *flava* (C.B.Clarke) Lye = ***Kyllinga nervosa*** subsp. ***flava*** (C.B.Clarke) Lye
Kyllinga nigritana C.B.Clarke = ***Kyllinga alba*** Nees subsp. ***alba***
Kyllinga nivea Pers. = ***Kyllinga brevifolia*** Rottb. subsp. ***brevifolia***
Kyllinga oblonga C.B.Clarke = ***Kyllinga nervosa*** subsp. ***oblonga*** (C.B.Clarke) J.-P.Lebrun & Stork
Kyllinga obtusata J.Presl & C.Presl = ***Kyllinga vaginata*** Lam.
Kyllinga obtusata var. *aphylla* Boeck. = ***Kyllinga vaginata*** Lam.
Kyllinga obtusata var. *cylindrostachya* Boeck. = ***Kyllinga vaginata*** Lam.
Kyllinga obtusata var. *subaphylla* Boeck. = ***Kyllinga vaginata*** Lam.
Kyllinga odorata Liebm. = ***Kyllinga brevifolia*** Rottb. subsp. ***brevifolia***
Kyllinga odorata Kunth = ***Kyllinga pumila*** Michx.
Kyllinga odorata var. *bulbifera* Kük. = ***Kyllinga odorata*** Vahl subsp. ***odorata***
Kyllinga odorata var. *cylindrica* (Nees) Kük. = ***Kyllinga odorata*** subsp. ***cylindrica*** (Nees) T.Koyama
Kyllinga odorata var. *dunensis* Osten = ***Kyllinga odorata*** Vahl subsp. ***odorata***
Kyllinga odorata var. *elongata* Boeck. = ***Kyllinga odorata*** Vahl subsp. ***odorata***
Kyllinga odorata var. *fallax* Kük. = ***Kyllinga odorata*** Vahl subsp. ***odorata***
Kyllinga odorata var. *genuina* Osten = ***Kyllinga odorata*** Vahl
Kyllinga odorata var. *gracilis* Boeck. = ***Kyllinga odorata*** Vahl subsp. ***odorata***
Kyllinga odorata var. *latifolia* Benth. = ***Kyllinga odorata*** Vahl subsp. ***odorata***
Kyllinga odorata var. *major* (C.B.Clarke) Chiov. = ***Kyllinga odorata*** subsp. ***appendiculata*** (K.Schum.) Lye
Kyllinga odorata var. *minor* Boeck. = ***Kyllinga odorata*** Vahl subsp. ***odorata***
Kyllinga odorata var. *pluriceps* (Kük.) Kük. = ***Kyllinga odorata*** Vahl subsp. ***odorata***
Kyllinga odorata var. *rigida* Boeck. = ***Kyllinga odorata*** Vahl subsp. ***odorata***

Kyllinga oligostachya Boeck. = ***Kyllinga brevifolia*** Rottb. subsp. ***brevifolia***
Kyllinga ovularis Michx. = ***Cyperus echinatus*** (L.) Alph.Wood
Kyllinga panicea Rottb. = ***Cyperus paniceus*** (Rottb.) Boeck.
Kyllinga parvula C.B.Clarke ex Rendle = ***Kyllinga bulbosa*** P.Beauv.
Kyllinga pellucidoalbida Boeck. = ***Kyllinga odorata*** Vahl subsp. ***odorata***
Kyllinga pellucidolucida Boeck. = ***Kyllinga odorata*** Vahl subsp. ***odorata***
Kyllinga perrieri Cherm. = ***Kyllinga melanosperma*** Nees subsp. ***melanosperma***
Kyllinga peruviana Lam. = ***Kyllinga vaginata*** Lam.
Kyllinga peruviana var. *foliata* Kük. = ***Kyllinga tibialis*** Poit. ex Ledeb.
Kyllinga pierreana E.G.Camus = ***Cyperus leucocephalus*** Retz.
Kyllinga pinguis C.B.Clarke = ***Kyllinga elatior*** Kunth
Kyllinga planiceps C.B.Clarke = ***Kyllinga polyphylla*** Willd. ex Kunth
Kyllinga planiculmis var. *mucronata* Cherm. = ***Kyllinga nemoralis*** (J.R.Forst. & G.Forst.) Dandy ex Hutch. & Dalziel
Kyllinga platyphylla var. *longifolia* Kük. = ***Kyllinga platyphylla*** K.Schum.
Kyllinga plurifoliata (Kük.) Cherm. = ***Kyllinga melanosperma*** var. ***plurifoliata*** Kük.
Kyllinga polycephala (Lam.) Link = ***Cyperus luzulae*** (L.) Retz.
Kyllinga polyphylla var. *elata* (Steud.) Lye = ***Kyllinga melanosperma*** subsp. ***elata*** (Steud.) Lye
Kyllinga pulchella f. *robustior* Kük. = ***Kyllinga pulchella*** Kunth
Kyllinga pumila Steud. = ***Kyllinga bulbosa*** P.Beauv.
Kyllinga pumila Sieber ex C.Presl = ***Kyllinga odorata*** Vahl subsp. ***odorata***
Kyllinga pumila var. *elatior* Kunth = ***Kyllinga pumila*** Michx.
Kyllinga pumilio Steud. = ***Kyllinga brevifolia*** Rottb. subsp. ***brevifolia***
Kyllinga pungens Link = ***Kyllinga vaginata*** Lam.
Kyllinga punicea J.F.Gmel. = ***Cyperus paniceus*** (Rottb.) Boeck.
Kyllinga ramosa Sessé & Moç. = [81 PUE]
Kyllinga repens Salzm. ex Steud. = ***Kyllinga vaginata*** Lam.
Kyllinga riederiana Meinsh. ex Kom. = ***Kyllinga kamschatica*** Meinsh.
Kyllinga rigida Baldwin = ***Kyllinga vaginata*** Lam.
Kyllinga rigida Link = ***Cyperus cyperinus*** (Retz.) Suringar
Kyllinga rigidifolia Boeck. = ***Cyperus swartzii*** (A.Dietr.) Boeck. ex Kük.
Kyllinga rigidula Steud. = ***Kyllinga pumila*** Michx.
Kyllinga robusta Boeck. = ***Kyllinga pumila*** Michx.
Kyllinga ruwenzoriensis C.B.Clarke = ***Kyllinga nervosa*** var. ***ruwenzoriensis*** (C.B.Clarke) Lye
Kyllinga schimperi Hochst. ex Engl. = ***Kyllinga bulbosa*** P.Beauv.
Kyllinga scirpina Rchb. ex Steud. = ***Cyperus luzulae*** (L.) Retz.
Kyllinga scirpina Rchb. ex C.B.Clarke = ***Oxycaryum cubense*** (Poepp. & Kunth) Palla
Kyllinga sesquiflora Torr. = ***Kyllinga odorata*** Vahl subsp. ***odorata***
Kyllinga sesquiflora subsp. *cylindrica* (Nees) T.Koyama = ***Kyllinga odorata*** subsp. ***cylindrica*** (Nees) T.Koyama
Kyllinga sojauxii Boeck. = ***Kyllinga brevifolia*** subsp. ***intricata*** (Cherm.) Lye
Kyllinga sororia Kunth = ***Kyllinga brevifolia*** Rottb. subsp. ***brevifolia***
Kyllinga squamulosa Kunth = ***Kyllinga squamulata*** Vahl
Kyllinga squarrosa Baldwin = ***Cyperus aggregatus*** (Willd.) Endl. var. ***aggregatus***
Kyllinga squarrosa Steud. = ***Lipocarpha microcephala*** (R.Br.) Kunth
Kyllinga stricta Schrad. ex Nees = ***Kyllinga vaginata*** Lam.
Kyllinga sumatrensis Retz. = ***Cyperus cyperoides*** (L.) Kuntze subsp. ***cyperoides***
Kyllinga tenuifolia Steud. = ***Kyllinga pumila*** Michx.
Kyllinga tenuis Baldwin = ***Kyllinga brevifolia*** Rottb. subsp. ***brevifolia***
Kyllinga tenuis Boeck. = ***Kyllinga vaginata*** Lam.
Kyllinga tenuissima Steud. = ***Kyllinga brevifolia*** Rottb. subsp. ***brevifolia***
Kyllinga teres C.B.Clarke = ***Kyllinga polyphylla*** Willd. ex Kunth
Kyllinga tetragona Spreng. ex C.B.Clarke = ***Kyllinga inaurata*** Nees ex Boeck.
Kyllinga transitoria (Kük.) T.Koyama = ***Kyllinga pulchella*** Kunth
Kyllinga tricephala Salisb. = ***Kyllinga brevifolia*** Rottb. subsp. ***brevifolia***
Kyllinga triceps Sw. = ***Kyllinga odorata*** Vahl subsp. ***odorata***
Kyllinga triceps Rottb. = ***Fimbristylis cymosa*** R.Br. subsp. ***cymosa***
Kyllinga triceps var. *ciliata* Boeck. = ***Kyllinga welwitschii*** Ridl.
Kyllinga triceps var. *obtusiflora* Boeck. = ***Kyllinga bulbosa*** P.Beauv.
Kyllinga tuberosa Scheele = ***Kyllinga odorata*** Vahl subsp. ***odorata***
Kyllinga umbellata Rottb. = ***Cyperus cylindrostachyus*** Boeck.
Kyllinga uncinata Link = ***Kyllinga pumila*** Michx.
Kyllinga vaginata Zoll. = ***Kyllinga melanosperma*** Nees subsp. ***melanosperma***
Kyllinga vaginata Rchb.f. ex Kunth = ***Kyllinga vaginata*** Lam.
Kyllinga viridiflora Roxb. ex Spreng. = ***Kyllinga pumila*** Michx.
Kyllinga viridiflora Link = ***Kyllinga pumila*** Michx.
Kyllinga viridula Hochst. ex A.Rich. = ***Kyllinga odorata*** subsp. ***cylindrica*** (Nees) T.Koyama

Unplaced Names:
Kyllinga ambigua Steud., Syn. Pl. Glumac. 2: 70 (1854). = [24 CHA ETH] Isolepis ?
Kyllinga ramosa Sessé & Moç., Fl. Mexic.: 16 (1893). = [81 PUE]

Kyllingiella

Kyllingiella R.W.Haines & Lye, Bot. Not. 131: 176 (1978).
Trop. & S. Africa, Indian Subcontinent. 22 BKN GHA GUI MLI NGA NGR SEN 23 GGI ZAI 24 CHA ETH 25 KEN TAN UGA 26 ZAM 27 BOT NAM TVL 40 IND WHM.
4 Species

Kyllingiella microcephala (Steud.) R.W.Haines & Lye, Bot. Not. 131: 176 (1978).
Trop. & S. Africa, Indian Subcontinent. 22 BKN GHA GUI MLI NGA NGR SEN 23 GGI 24 CHA ETH 25 KEN TAN UGA 27 BOT NAM TVL 40 IND WHM. Hemicr.

**Kyllinga microcephala* Steud., Flora 25: 597 (1842). *Scirpus microcephalus* (Steud.) Dandy in F.W.Andrews, Fl. Pl. Sudan 3: 366 (1956). *Isolepis microcephala* (Steud.) Lye, Bot. Not. 124: 480 (1971).
Isolepis kyllingioides A.Rich., Tent. Fl. Abyss. 2: 502 (1850). *Scirpus kyllingioides* (A.Rich.) Boeck., Linnaea 36: 733 (1870).

Kyllingiella polyphylla (A.Rich.) Lye in R.W.Haines & K.A.Lye, Sedges & Rushes E. Afr.: 143 (1983).
Ethiopia to E. Trop. Africa. 24 ETH 25 KEN TAN. Hemicr.
**Isolepis polyphylla* A.Rich., Tent. Fl. Abyss. 2: 503 (1850).

Kyllingiella simpsonii Muasya, Kew Bull. 57: 997 (2002).
Tanzania to Zambia. 23 ZAI 25 TAN 26 ZAM.

Kyllingiella ugandensis R.W.Haines & Lye, Bot. Not. 131: 177 (1978).
E. Trop. Africa. 25 KEN TAN UGA. Hemicr.

Lagenocarpus

Lagenocarpus Nees, Linnaea 9: 304 (1834).
Trop. America. 80 BLZ HON NIC 81 CUB PUE TRT 82 FRG GUY SUR VEN 83 BOL CLM 84+ BZC BZE BZL BZN BZS.
31 Species
Anogyna Nees, J. Bot. (Hooker) 2: 395 (1840).
Acrocarpus Nees in C.F.P.von Martius & auct. suc. (eds.), Fl. Bras. 2(1): 158 (1842), nom. illeg.
Adamantogeton Schrad. ex Nees in C.F.P.von Martius & auct. suc. (eds.), Fl. Bras. 2(1): 165 (1842).
Cryptangium Schrad. ex Nees in C.F.P.von Martius & auct. suc. (eds.), Fl. Bras. 2(1): 163 (1842).
Microlepis Schrad. ex Nees in C.F.P.von Martius & auct. suc. (eds.), Fl. Bras. 2(1): 157 (1842).
Orobium Schrad. ex Nees in C.F.P.von Martius & auct. suc. (eds.), Fl. Bras. 2(1): 164 (1842).
Phaenopyrum Schrad. ex Nees in C.F.P.von Martius & auct. suc. (eds.), Fl. Bras. 2(1): 166 (1842).
Lerisca Schltdl., Bot. Zeitung (Berlin) 3: 476 (1845).
Cladotheca Steud., Syn. Pl. Glumac. 2: 178 (1855).
Exochogyne C.B.Clarke, Verh. Bot. Vereins Prov. Brandenburg 47: 101 (1905).
Neo-Senaea K.Schum. ex H.Pfeiff., Repert. Spec. Nov. Regni Veg. 21: 34 (1925).
Ulea-flos C.B.Clarke ex H.Pfeiff., Repert. Spec. Nov. Regni Veg. 21: 35 (1925), nom. inval.
Didymiandrum Gilly, Bull. Torrey Bot. Club 68: 330 (1941).

Lagenocarpus adamantinus Nees in C.F.P.von Martius & auct. suc. (eds.), Fl. Bras. 2(1): 165 (1842). *Scleria adamantina* (Nees) Steud., Syn. Pl. Glumac. 2: 177 (1855).
Brazil (Minas Gerais). 84 BZL. Hemicr. or rhizome geophyte.

Lagenocarpus alboniger (A.St.-Hil.) C.B.Clarke, Kew Bull., Addit. Ser. 8: 64 (1908).
E. Brazil. 84 BZE BZL. Hemicr. or rhizome geophyte.
**Scleria albonigra* A.St.-Hil., Voy. Distr. Diam. 1: 371 (1833).
Scleria tristis A.St.-Hil., Voy. Distr. Diam. 1: 370 (1833).
Cryptangium insigne Boeck., Allg. Bot. Z. Syst. 2: 143 (1896). *Lagenocarpus insignis* (Boeck.) H.Pfeiff., Ber. Deutsch. Bot. Ges. 39: 132 (1921).
Lagenocarpus alboniger var. *subglabra* H.Pfeiff., Repert. Spec. Nov. Regni Veg. 18: 85 (1922).

Lagenocarpus amazonicus (C.B.Clarke) H.Pfeiff., Repert. Spec. Nov. Regni Veg. 21: 35 (1925).
SE. Colombia to N. South America. 82 FRG GUY SUR VEN 83 CLM 84 BZC BZN.
**Exochogyne amazonica* C.B.Clarke, Verh. Bot. Vereins Prov. Brandenburg 47: 101 (1905).
Exochogyne decandra Tutin, Hooker's Icon. Pl. 33: t. 3275 (1935).
Exochogyne megalorrhyncha Tutin, Hooker's Icon. Pl. 33: t. 3274 (1935).
Exochogyne amazonica var. *esmeraldensis* Gilly, Fieldiana, Bot. 28(1): 53 (1951).
Exochogyne steyermarkii Gilly, Fieldiana, Bot. 28(1): 53 (1951).

Lagenocarpus bracteosus C.B.Clarke, Repert. Spec. Nov. Regni Veg. 2: 145 (1906). *Lagenocarpus adamantinus* var. *bracteosus* (C.B.Clarke) H.Pfeiff., Repert. Spec. Nov. Regni Veg. 18: 85 (1922).
Brazil (Minas Gerais). 84 BZL. Hemicr. or rhizome geophyte.
Lagenocarpus ciponensis H.Pfeiff., Repert. Spec. Nov. Regni Veg. 21: 34 (1925).

Lagenocarpus celiae T.Koyama & Maguire, Mem. New York Bot. Gard. 12(3): 46 (1965).
SE. Colombia to Brazil. 82 VEN 83 CLM 84 BZC BZN. Hemicr. or rhizome geophyte.

Lagenocarpus clarkei H.Pfeiff., Repert. Spec. Nov. Regni Veg. 18: 92 (1922).
SE. Brazil. 84 BZL. Hemicr. or rhizome geophyte.
**Cryptangium strictum* C.B.Clarke, Kew Bull., Addit. Ser. 8: 66 (1908). *Cephalocarpus clarkei* H.Pfeiff., Repert. Spec. Nov. Regni Veg. 18: 72 (1922), nom. illeg.

Lagenocarpus comatus (Boeck.) H.Pfeiff., Ber. Deutsch. Bot. Ges. 39: 131 (1921).
SE. Brazil. 84 BZL. Hemicr. or rhizome geophyte.
**Cryptangium comatum* Boeck., Flora 65: 351 (1882). *Cephalocarpus comatus* (Boeck.) H.Pfeiff., Repert. Spec. Nov. Regni Veg. 18: 72 (1922).

Lagenocarpus compactus D.A.Simpson, Kew Bull. 48: 707 (1993).
Brazil (Bahia). 84 BZE. Hemicr. or rhizome geophyte.

Lagenocarpus cubensis Kük., Repert. Spec. Nov. Regni Veg. 23: 219 (1926).
E. Cuba (Sierra del Cristal). 81 CUB. Hemicr. or rhizome geophyte.

Lagenocarpus distichophyllus (Boeck.) H.Pfeiff., Ber. Deutsch. Bot. Ges. 39: 131 (1921).
Brazil. 84+. Hemicr. or rhizome geophyte.
**Cryptangium distichophyllum* Boeck., Beitr. Cyper. 2: 28 (1890).

Lagenocarpus eriopodus T.Koyama & Maguire, Mem. New York Bot. Gard. 12(3): 47 (1965).
Venezuela (Amazonas). 82 VEN. Hemicr. or rhizome geophyte.

Lagenocarpus glomerulatus Gilly, Fieldiana, Bot. 28(1): 57 (1951).
S. Venezuela to Guyana. 82 GUY VEN 84 BZN. Hemicr. or rhizome geophyte.

Lagenocarpus griseus (Boeck.) H.Pfeiff., Ber. Deutsch. Bot. Ges. 39: 131 (1921).

Brazil (Minas Gerais). 84 BZL. Hemicr. or rhizome geophyte.
**Cryptangium griseum* Boeck., Allg. Bot. Z. Syst. 2: 143 (1896).
Lagenocarpus vestitus C.B.Clarke, Kew Bull., Addit. Ser. 8: 64 (1908).

Lagenocarpus guianensis Nees, Linnaea 9: 304 (1834). *Scleria guianensis* (Nees) Steud., Syn. Pl. Glumac. 2: 177 (1855). *Lagenocarpus guianensis* var. *typicus* H.Pfeiff., Repert. Spec. Nov. Regni Veg. 18: 75 (1922), nom. inval.
Trop. America. 80 BLZ HON NIC 81 CUB PUE TRT 82 FRG GUY SUR VEN 83 BOL 84 BZE BZL BZN. Hemicr. or rhizome geophyte.

subsp. ***guianensis***
Trop. America. 80 BLZ HON NIC 81 CUB PUE TRT 82 FRG GUY SUR VEN 83 BOL 84 BZE BZL BZN. Hemicr. or rhizome geophyte.
Phaenopyrum paniculatum Schrad. ex Nees in C.F.P.von Martius & auct. suc. (eds.), Fl. Bras. 2(1): 166 (1842), nom. inval.
Lagenocarpus riedelianus C.B.Clarke, Kew Bull., Addit. Ser. 8: 63 (1908).
Lagenocarpus guianensis var. *paraensis* H.Pfeiff., Repert. Spec. Nov. Regni Veg. 18: 75 (1922).
Lagenocarpus sprucei H.Pfeiff., Repert. Spec. Nov. Regni Veg. 18: 74 (1922).
Lagenocarpus portoricensis Britton, Bull. Torrey Bot. Club 50: 55 (1923).
Lagenocarpus guianensis f. *gracilior* H.Pfeiff., Repert. Spec. Nov. Regni Veg. 20: 42 (1924).

subsp. ***hypochaetus*** T.Koyama, Mem. New York Bot. Gard. 12(3): 44 (1965).
S. Venezuela to N. Brazil. 82 VEN 84 BZN. Hemicr. or rhizome geophyte.

Lagenocarpus humilis (Nees) Kuntze, Revis. Gen. Pl. 2: 754 (1891).
Brazil (Minas Gerais). 84 BZL. Hemicr. or rhizome geophyte.
**Acrocarpus humilis* Nees in C.F.P.von Martius & auct. suc. (eds.), Fl. Bras. 2(1): 161 (1842). *Scleria humilis* (Nees) Steud., Syn. Pl. Glumac. 2: 178 (1855). *Cryptangium humile* (Nees) Boeck., Linnaea 38: 413 (1874). *Cephalocarpus humilis* (Nees) H.Pfeiff., Repert. Spec. Nov. Regni Veg. 18: 72 (1922).

Lagenocarpus junciformis (Kunth) Kuntze, Revis. Gen. Pl. 2: 754 (1891).
SE. & S. Brazil. 84 BZL BZS. Hemicr. or rhizome geophyte.
**Scleria junciformis* Kunth, Enum. Pl. 2: 356 (1837). *Acrocarpus junciformis* (Kunth) Nees in C.F.P.von Martius & auct. suc. (eds.), Fl. Bras. 2(1): 158 (1842). *Cryptangium junciforme* (Kunth) Boeck., Linnaea 38: 415 (1874). *Lagenocarpus strictus* var. *junciformis* (Kunth) H.Pfeiff., Repert. Spec. Nov. Regni Veg. 18: 89 (1922).
Cryptangium pauciflorum Boeck., Allg. Bot. Z. Syst. 2: 142 (1896). *Lagenocarpus pauciflorus* (Boeck.) H.Pfeiff., Ber. Deutsch. Bot. Ges. 39: 132 (1921). *Lagenocarpus minarum* var. *paucifolius* (Boeck.) H.Pfeiff., Repert. Spec. Nov. Regni Veg. 18: 90 (1922).
Cryptangium nudipes C.B.Clarke, Kew Bull., Addit. Ser. 8: 65 (1908). *Lagenocarpus nudipes* (C.B. Clarke) H.Pfeiff., Repert. Spec. Nov. Regni Veg. 18: 93 (1922).

Lagenocarpus lanatus (T.Koyama & Maguire) T.Koyama, Makinoa, n.s., 4: 39 (2004).
Colombia (Cerro Isibokuri), Venezuela (Macizo del Chimantá). 82 VEN 83 CLM. Cham.
**Everardia lanata* T.Koyama & Maguire, Mem. New York Bot. Gard. 12(3): 34 (1965).

Lagenocarpus minarum (Nees) Kuntze, Revis. Gen. Pl. 2: 754 (1891).
Brazil. 84 BZC BZL BZS. Hemicr. or rhizome geophyte.
Acrocarpus campestris Nees in C.F.P.von Martius & auct. suc. (eds.), Fl. Bras. 2(1): 160 (1842). *Scleria campestris* (Nees) Steud., Syn. Pl. Glumac. 2: 178 (1855). *Cryptangium campestre* (Nees) Boeck., Linnaea 38: 416 (1874). *Lagenocarpus campestris* (Nees) Kuntze, Revis. Gen. Pl. 2: 754 (1891).
Acrocarpus densifolius Nees in C.F.P.von Martius & auct. suc. (eds.), Fl. Bras. 2(1): 159 (1842). *Scleria densifolia* (Nees) Steud., Syn. Pl. Glumac. 2: 178 (1855). *Cryptangium densifolium* (Nees) Boeck., Linnaea 38: 418 (1874). *Lagenocarpus densifolius* (Nees) Kuntze, Revis. Gen. Pl. 2: 754 (1891). *Lagenocarpus minarum* var. *densifolius* (Nees) H.Pfeiff., Repert. Spec. Nov. Regni Veg. 18: 90 (1922).
**Acrocarpus minarum* Nees in C.F.P.von Martius & auct. suc. (eds.), Fl. Bras. 2(1): 159 (1842). *Scleria minarum* (Nees) Steud., Syn. Pl. Glumac. 2: 178 (1855). *Cryptangium minarum* (Nees) Boeck., Linnaea 38: 417 (1874).
Microlepis minarum-generalium Schrad. in C.F.P.von Martius & auct. suc. (eds.), Fl. Bras. 2(1): 157 (1842).
Cryptangium glaziovii Boeck., Vidensk. Meddel. Naturhist. Foren. Kjøbenhavn 1879-1880: 29 (1880). *Lagenocarpus glaziovii* (Boeck.) H.Pfeiff., Ber. Deutsch. Bot. Ges. 39: 131 (1921).
Cryptangium paucifolium Boeck., Vidensk. Meddel. Naturhist. Foren. Kjøbenhavn 1879-1880: 28 (1880).
Cryptangium melanocarpum Boeck., Beitr. Cyper. 2: 28 (1890). *Lagenocarpus melanocarpus* (Boeck.) H.Pfeiff., Ber. Deutsch. Bot. Ges. 39: 132 (1921).
Cryptangium claussenii C.B.Clarke, Kew Bull., Addit. Ser. 8: 65 (1908). *Lagenocarpus claussenii* (C.B.Clarke) H.Pfeiff., Repert. Spec. Nov. Regni Veg. 18: 91 (1922).
Lagenocarpus compositus H.Pfeiff., Repert. Spec. Nov. Regni Veg. 18: 87 (1922).
Lagenocarpus glaziovii var. *pyramidalis* H.Pfeiff., Repert. Spec. Nov. Regni Veg. 18: 88 (1922).

Lagenocarpus parvulus (C.B.Clarke) H.Pfeiff., Repert. Spec. Nov. Regni Veg. 18: 91 (1922).
Brazil (Minas Gerais). 84 BZL. Hemicr. or rhizome geophyte.
**Cryptangium parvulum* C.B.Clarke, Repert. Spec. Nov. Regni Veg. 2: 145 (1906).
Lagenocarpus ciliatus f. *humilis* H.Pfeiff., Repert. Spec. Nov. Regni Veg. 18: 91 (1922).

Lagenocarpus pendulus T.Koyama, Mem. New York Bot. Gard. 12(3): 44 (1965).
SE. Colombia to N. Brazil. 82 VEN 83 CLM 84 BZN. Hemicr. or rhizome geophyte.

Lagenocarpus polyphyllus (Nees) Kuntze, Revis. Gen. Pl. 2: 754 (1891).
Brazil (Rio de Janeiro). 84 BZL. Hemicr. or rhizome geophyte.
**Acrocarpus polyphyllus* Nees in C.F.P.von Martius & auct. suc. (eds.), Fl. Bras. 2(1): 158 (1842). *Scleria*

polyphylla (Nees) Steud., Syn. Pl. Glumac. 2: 178 (1855). *Cryptangium polyphyllum* (Nees) Boeck., Linnaea 38: 414 (1874). *Cephalocarpus polyphyllus* (Nees) H.Pfeiff., Repert. Spec. Nov. Regni Veg. 18: 72 (1922). *Lagenocarpus polyphyllus* var. *typicus* H.Pfeiff., Repert. Spec. Nov. Regni Veg. 18: 92 (1922).
Lagenocarpus polyphyllus var. *ganglioneus* H.Pfeiff., Repert. Spec. Nov. Regni Veg. 18: 92 (1922).
Lagenocarpus polyphyllus var. *itatiayus* H.Pfeiff., Repert. Spec. Nov. Regni Veg. 18: 92 (1922).

Lagenocarpus rigidus (Kunth) Nees in C.F.P.von Martius & auct. suc. (eds.), Fl. Bras. 2(1): 167 (1842).
Caribbean, S. Trop. America. 81 CUB TRT 82 FRG GUY SUR VEN 83 CLM 84 BZC BZE BZL BZN BZS. Hemicr. or rhizome geophyte.
**Scleria rigida* Kunth, Enum. Pl. 2: 355 (1837).

subsp. ***rigidus***
Caribbean, S. Trop. America. 81 CUB TRT 82 FRG GUY SUR VEN 83 CLM 84 BZC BZE BZL BZN BZS. Hemicr. or rhizome geophyte.
Anogyna tremula Nees, J. Bot. (Hooker) 2: 395 (1840). *Lagenocarpus tremulus* (Nees) Nees in C.F.P.von Martius & auct. suc. (eds.), Fl. Bras. 2(1): 168 (1842). *Scleria tremula* (Nees) Steud., Syn. Pl. Glumac. 2: 177 (1855). *Cryptangium tremulum* (Nees) Boeck., Allg. Bot. Z. Syst. 2: 143 (1896). *Lagenocarpus rigidus* subsp. *tremulus* (Nees) T.Koyama & Maguire, Mem. New York Bot. Gard. 12(3): 41 (1965).
Lagenocarpus martii Nees in C.F.P.von Martius & auct. suc. (eds.), Fl. Bras. 2(1): 169 (1842). *Scleria viridifolia* Steud., Syn. Pl. Glumac. 2: 177 (1855).
Lagenocarpus topazinus Nees in C.F.P.von Martius & auct. suc. (eds.), Fl. Bras. 2(1): 168 (1842). *Scleria topazina* (Nees) Steud., Syn. Pl. Glumac. 2: 177 (1855).
Microlepis bahiensis Schrad. ex Nees in C.F.P.von Martius & auct. suc. (eds.), Fl. Bras. 2(1): 157 (1842).
Microlepis elatior Schrad. in C.F.P.von Martius & auct. suc. (eds.), Fl. Bras. 2(1): 157 (1842).
Microlepis rigidifolia Schrad. ex Nees in C.F.P.von Martius & auct. suc. (eds.), Fl. Bras. 2(1): 157 (1842).
Lagenocarpus neesii Boeck., Linnaea 38: 426 (1874). *Lagenocarpus rigidus* f. *neesii* (Boeck.) H.Pfeiff., Repert. Spec. Nov. Regni Veg. 18: 80 (1922).
Lagenocarpus crassipes Boeck., Flora 63: 453 (1880).
Cryptangium arundinaceum Boeck., Flora 65: 351 (1882). *Lagenocarpus rigidus* var. *arundinaceus* (Boeck.) H.Pfeiff., Repert. Spec. Nov. Regni Veg. 18: 80 (1922).
Cryptangium kuntzeanum Boeck., Beitr. Cyper. 1: 32 (1888). *Lagenocarpus kuntzeanus* (Boeck.) Kuntze, Revis. Gen. Pl. 2: 754 (1891).
Cryptangium dioicum Boeck., Allg. Bot. Z. Syst. 2: 142 (1896). *Lagenocarpus dioicus* (Boeck.) H.Pfeiff., Ber. Deutsch. Bot. Ges. 39: 131 (1921). *Lagenocarpus martii* f. *dioicus* (Boeck.) H.Pfeiff., Repert. Spec. Nov. Regni Veg. 18: 79 (1922).
Lagenocarpus mendoncae C.B.Clarke, Kew Bull., Addit. Ser. 8: 63 (1908).
Lagenocarpus oocarpus C.B.Clarke, Kew Bull., Addit. Ser. 8: 63 (1908).
Lagenocarpus pauloensis Palla, Denkschr. Kaiserl. Akad. Wiss., Wien. Math.-Naturwiss. Kl. 79: 198 (1908).
Lagenocarpus pedicellatus C.B.Clarke, Kew Bull., Addit. Ser. 8: 63 (1908).
Lagenocarpus dusenii H.Pfeiff., Repert. Spec. Nov. Regni Veg. 18: 80 (1922).
Lagenocarpus giganteus H.Pfeiff., Repert. Spec. Nov. Regni Veg. 18: 81 (1922).
Lagenocarpus giganteus f. *polystachyus* H.Pfeiff., Repert. Spec. Nov. Regni Veg. 18: 82 (1922).
Lagenocarpus martii f. *intermixtus* H.Pfeiff., Repert. Spec. Nov. Regni Veg. 18: 79 (1922).
Lagenocarpus martii f. *longinux* H.Pfeiff., Repert. Spec. Nov. Regni Veg. 18: 79 (1922).
Lagenocarpus martii f. *longirostris* H.Pfeiff., Repert. Spec. Nov. Regni Veg. 18: 79 (1922).
Lagenocarpus pseudocladius H.Pfeiff., Repert. Spec. Nov. Regni Veg. 18: 83 (1922).
Lagenocarpus rigidus f. *elatior* H.Pfeiff., Repert. Spec. Nov. Regni Veg. 18: 80 (1922).
Lagenocarpus rigidus f. *gracilior* H.Pfeiff., Repert. Spec. Nov. Regni Veg. 18: 80 (1922).
Lagenocarpus rigidus f. *robustior* H.Pfeiff., Repert. Spec. Nov. Regni Veg. 18: 80 (1922).
Lagenocarpus scalariformis H.Pfeiff., Repert. Spec. Nov. Regni Veg. 18: 77 (1922).
Lagenocarpus tremulus f. *multigyna* H.Pfeiff., Repert. Spec. Nov. Regni Veg. 18: 75 (1922).
Lagenocarpus warmingii H.Pfeiff., Repert. Spec. Nov. Regni Veg. 18: 76 (1922).
Lagenocarpus bifidus Gilly, Fieldiana, Bot. 28(1): 56 (1951).
Lagenocarpus diffusus Gilly, Fieldiana, Bot. 28(1): 56 (1951).
Lagenocarpus steyermarkii Gilly, Fieldiana, Bot. 28(1): 58 (1951).

subsp. ***tenuifolius*** (Boeck.) T.Koyama & Maguire, Mem. New York Bot. Gard. 12(3): 41 (1965).
Brazil. 84 BZC BZE BZL. Hemicr. or rhizome geophyte.
**Cryptangium tenuifolium* Boeck., Beitr. Cyper. 1: 32 (1888).
Cryptangium brevifolium Boeck., Allg. Bot. Z. Syst. 2: 143 (1896). *Lagenocarpus brevifolius* (Boeck.) H.Pfeiff., Ber. Deutsch. Bot. Ges. 39: 131 (1921). *Lagenocarpus tremulus* var. *brevifolius* (Boeck.) H.Pfeiff., Repert. Spec. Nov. Regni Veg. 18: 75 (1922).
Lagenocarpus brevirostris C.B.Clarke, Kew Bull., Addit. Ser. 8: 65 (1908).
Lagenocarpus inversus C.B.Clarke, Kew Bull., Addit. Ser. 8: 64 (1908).
Lagenocarpus tenuifolius var. *riedelii* C.B.Clarke, Kew Bull., Addit. Ser. 8: 64 (1908).
Lagenocarpus lapaensis H.Pfeiff., Repert. Spec. Nov. Regni Veg. 18: 86 (1922).

Lagenocarpus sabanensis Gilly, Fieldiana, Bot. 28(1): 57 (1951).
E. Colombia to W. Venezuela and Brazil. 82 VEN 83 CLM 84 BZC BZN. Hemicr. or rhizome geophyte.

Lagenocarpus sericeus H.Pfeiff., Repert. Spec. Nov. Regni Veg. 18: 83 (1922).
SE. Brazil. 84 BZL. Hemicr. or rhizome geophyte.

Lagenocarpus stellatus (Boeck.) Kuntze, Revis. Gen. Pl. 2: 754 (1891).
Venezuela (Bolívar) to S. Guyana, N. Brazil. 82 GUY VEN 84 BZN. Hemicr. or rhizome geophyte.

Acrocarpus stellatus Nees ex Boeck., Linnaea 38: 422 (1874), nom. inval.

**Cryptangium stellatum* Boeck., Linnaea 38: 421 (1874). *Didymiandrum stellatum* (Boeck.) Gilly, Bull. Torrey Bot. Club 68: 331 (1941).

Didymiandrum guaiquinimae Schnee, Bol. Soc. Venez. Ci. Nat. 9: 81 (1944).

Lagenocarpus subaphyllus T.Koyama, Brittonia 24: 279 (1972).
Brazil (Espírito Santo). 84 BZL. Hemicr. or rhizome geophyte.

Lagenocarpus tenuifolius (Boeck.) Kuntze, Revis. Gen. Pl. 2: 754 (1891).
Brazil. 84 BZC BZE BZL BZS. Hemicr. or rhizome geophyte.

Lagenocarpus triqueter (Boeck.) Kuntze, Revis. Gen. Pl. 2: 754 (1891).
Brazil (Rio de Janeiro to Santa Catarina). 84 BZL BZS. Hemicr. or rhizome geophyte.

**Cryptangium triquetrum* Boeck., Linnaea 38: 413 (1874).

Cryptangium schenckianum Boeck., Beitr. Cyper. 2: 27 (1890). *Lagenocarpus schenckianus* (Boeck.) H.Pfeiff., Ber. Deutsch. Bot. Ges. 39: 132 (1921).

Lagenocarpus velutinus Nees in C.F.P.von Martius & auct. suc. (eds.), Fl. Bras. 2(1): 166 (1842). *Scleria velutina* (Nees) Steud., Syn. Pl. Glumac. 2: 177 (1855).
E. Brazil. 84 BZE BZL. Hemicr. or rhizome geophyte.

Lagenocarpus venezuelensis Davidse, Novon 3: 244 (1993).
Venezuela (Amazonas). 82 VEN. Hemicr. or rhizome geophyte.

**Lagenocarpus amazonicus* Davidse, Novon 2: 319 (1992), nom. illeg.

Lagenocarpus verticillatus (Spreng.) T.Koyama & Maguire, Mem. New York Bot. Gard. 12(3): 49 (1965).
N. South America to Brazil. 82 GUY SUR VEN 83 CLM 84 BZC BZE BZL BZN. Hemicr. or rhizome geophyte.

**Fuirena verticillata* Spreng., Novi Provent.: 47 (1818).

Fuirena weigeltii A.Spreng., Tent. Suppl.: 4 (1828). *Lagenocarpus weigeltii* (A.Spreng.) Uittien in A.A. Pulle, Fl. Suriname 1(1): 501 (1943).

Scleria leptocladia Poepp. ex Kunth, Enum. Pl. 2: 355 (1837). *Cryptangium leptocladium* (Poepp. ex Kunth) Boeck., Linnaea 38: 420 (1874). *Lagenocarpus leptocladius* (Poepp. ex Kunth) Kuntze, Revis. Gen. Pl. 2: 754 (1891). *Lagenocarpus strictus* var. *leptocladius* (Poepp. ex Kunth) H.Pfeiff., Repert. Spec. Nov. Regni Veg. 18: 89 (1922).

Scleria stricta Kunth, Enum. Pl. 2: 355 (1837). *Cryptangium uliginosum* Schrad. ex Nees in C.F.P.von Martius & auct. suc. (eds.), Fl. Bras. 2(1): 164 (1842). *Lagenocarpus uliginosus* (Schrad. ex Nees) Kuntze, Revis. Gen. Pl. 2: 754 (1891). *Lagenocarpus strictus* (Kunth) Kuntze, Revis. Gen. Pl. 2: 754 (1891). *Cryptangium strictum* (Kunth) H.Pfeiff., Ber. Deutsch. Bot. Ges. 39: 133 (1921), nom. illeg.

Scleria kunthii Miq., Sert. Exot.: 4 (1843). *Lagenocarpus kunthii* (Miq.) Uittien in A.A.Pulle, Fl. Suriname 1(1): 132 (1934).

Cladotheca vaginans Steud., Syn. Pl. Glumac. 2: 178 (1855).

Scleria hostmanniana Steud., Syn. Pl. Glumac. 2: 178 (1855).

Cryptangium leptocladium var. *longifolium* Boeck., Linnaea 38: 421 (1874).

Cryptangium ciliatum Boeck., Beitr. Cyper. 2: 28 (1890). *Lagenocarpus ciliatus* (Boeck.) H.Pfeiff., Ber. Deutsch. Bot. Ges. 39: 131 (1921).

Cryptangium uliginosum f. *humilis* C.B.Clarke, Repert. Spec. Nov. Regni Veg. 18: 91 (1922).

***Synonyms*:**

Lagenocarpus adamantinus var. *bracteosus* (C.B.Clarke) H.Pfeiff. = ***Lagenocarpus bracteosus*** C.B.Clarke

Lagenocarpus alboniger var. *subglabra* H.Pfeiff. = ***Lagenocarpus alboniger*** (A.St.-Hil.) C.B.Clarke

Lagenocarpus amazonicus Davidse = ***Lagenocarpus venezuelensis*** Davidse

Lagenocarpus bifidus Gilly = ***Lagenocarpus rigidus*** (Kunth) Nees subsp. ***rigidus***

Lagenocarpus brevifolius (Boeck.) H.Pfeiff. = ***Lagenocarpus rigidus*** subsp. ***tenuifolius*** (Boeck.) T.Koyama & Maguire

Lagenocarpus brevirostris C.B.Clarke = ***Lagenocarpus rigidus*** subsp. ***tenuifolius*** (Boeck.) T.Koyama & Maguire

Lagenocarpus campestris (Nees) Kuntze = ***Lagenocarpus minarum*** (Nees) Kuntze

Lagenocarpus ciliatus (Boeck.) H.Pfeiff. = ***Lagenocarpus verticillatus*** (Spreng.) T.Koyama & Maguire

Lagenocarpus ciliatus f. *humilis* H.Pfeiff. = ***Lagenocarpus parvulus*** (C.B.Clarke) H.Pfeiff.

Lagenocarpus ciponensis H.Pfeiff. = ***Lagenocarpus bracteosus*** C.B.Clarke

Lagenocarpus claussenii (C.B.Clarke) H.Pfeiff. = ***Lagenocarpus minarum*** (Nees) Kuntze

Lagenocarpus compositus H.Pfeiff. = ***Lagenocarpus minarum*** (Nees) Kuntze

Lagenocarpus crassipes Boeck. = ***Lagenocarpus rigidus*** (Kunth) Nees subsp. ***rigidus***

Lagenocarpus densifolius (Nees) Kuntze = ***Lagenocarpus minarum*** (Nees) Kuntze

Lagenocarpus diffusus Gilly = ***Lagenocarpus rigidus*** (Kunth) Nees subsp. ***rigidus***

Lagenocarpus dioicus (Boeck.) H.Pfeiff. = ***Lagenocarpus rigidus*** (Kunth) Nees subsp. ***rigidus***

Lagenocarpus dracaenula (Nees) H.Pfeiff. = ***Cephalocarpus dracaenula*** Nees

Lagenocarpus dusenii H.Pfeiff. = ***Lagenocarpus rigidus*** (Kunth) Nees subsp. ***rigidus***

Lagenocarpus giganteus H.Pfeiff. = ***Lagenocarpus rigidus*** (Kunth) Nees subsp. ***rigidus***

Lagenocarpus giganteus f. *polystachyus* H.Pfeiff. = ***Lagenocarpus rigidus*** (Kunth) Nees subsp. ***rigidus***

Lagenocarpus glaziovii (Boeck.) H.Pfeiff. = ***Lagenocarpus minarum*** (Nees) Kuntze

Lagenocarpus glaziovii var. *pyramidalis* H.Pfeiff. = ***Lagenocarpus minarum*** (Nees) Kuntze

Lagenocarpus guianensis f. *gracilior* H.Pfeiff. = ***Lagenocarpus guianensis*** Nees subsp. ***guianensis***

Lagenocarpus guianensis var. *paraensis* H.Pfeiff. = ***Lagenocarpus guianensis*** Nees subsp. ***guianensis***

Lagenocarpus guianensis var. *typicus* H.Pfeiff. = ***Lagenocarpus guianensis*** Nees

Lagenocarpus insignis (Boeck.) H.Pfeiff. = ***Lagenocarpus alboniger*** (A.St.-Hil.) C.B.Clarke

Lagenocarpus inversus C.B.Clarke = ***Lagenocarpus rigidus*** subsp. ***tenuifolius*** (Boeck.) T.Koyama & Maguire

Lagenocarpus kunthii (Miq.) Uittien = ***Lagenocarpus verticillatus*** (Spreng.) T.Koyama & Maguire

Lagenocarpus kuntzeanus (Boeck.) Kuntze = ***Lagenocarpus rigidus*** (Kunth) Nees subsp. ***rigidus***

Lagenocarpus laniger C.B.Clarke ex H.Pfeiff. = ?
Lagenocarpus lapaensis H.Pfeiff. = ***Lagenocarpus rigidus*** subsp. ***tenuifolius*** (Boeck.) T.Koyama & Maguire
Lagenocarpus leptocladius (Poepp. ex Kunth) Kuntze = ***Lagenocarpus verticillatus*** (Spreng.) T.Koyama & Maguire
Lagenocarpus martii Nees = ***Lagenocarpus rigidus*** (Kunth) Nees subsp. ***rigidus***
Lagenocarpus martii f. *dioicus* (Boeck.) H.Pfeiff. = ***Lagenocarpus rigidus*** (Kunth) Nees subsp. ***rigidus***
Lagenocarpus martii f. *intermixtus* H.Pfeiff. = ***Lagenocarpus rigidus*** (Kunth) Nees subsp. ***rigidus***
Lagenocarpus martii f. *longinux* H.Pfeiff. = ***Lagenocarpus rigidus*** (Kunth) Nees subsp. ***rigidus***
Lagenocarpus martii f. *longirostris* H.Pfeiff. = ***Lagenocarpus rigidus*** (Kunth) Nees subsp. ***rigidus***
Lagenocarpus martii f. *typicus* H.Pfeiff. ex Luetzelb. = ***Lagenocarpus rigidus*** (Kunth) Nees
Lagenocarpus melanocarpus (Boeck.) H.Pfeiff. = ***Lagenocarpus minarum*** (Nees) Kuntze
Lagenocarpus mendoncae C.B.Clarke = ***Lagenocarpus rigidus*** (Kunth) Nees subsp. ***rigidus***
Lagenocarpus minarum var. *densifolius* (Nees) H.Pfeiff. = ***Lagenocarpus minarum*** (Nees) Kuntze
Lagenocarpus minarum var. *paucifolius* (Boeck.) H.Pfeiff. = ***Lagenocarpus minarum*** (Nees) Kuntze
Lagenocarpus neesii Boeck. = ***Lagenocarpus rigidus*** (Kunth) Nees subsp. ***rigidus***
Lagenocarpus nudipes (C.B.Clarke) H.Pfeiff. = ***Lagenocarpus junciformis*** (Kunth) Kuntze
Lagenocarpus oocarpus C.B.Clarke = ***Lagenocarpus rigidus*** (Kunth) Nees subsp. ***rigidus***
Lagenocarpus pauciflorus (Boeck.) H.Pfeiff. = ***Lagenocarpus junciformis*** (Kunth) Kuntze
Lagenocarpus pauciflorus var. *triqueter* (Boeck.) H.Pfeiff. =
Lagenocarpus pauloensis Palla = ***Lagenocarpus rigidus*** (Kunth) Nees subsp. ***rigidus***
Lagenocarpus pedicellatus C.B.Clarke = ***Lagenocarpus rigidus*** (Kunth) Nees subsp. ***rigidus***
Lagenocarpus polyphyllus var. *ganglioneus* H.Pfeiff. = ***Lagenocarpus polyphyllus*** (Nees) Kuntze
Lagenocarpus polyphyllus var. *itatiayus* H.Pfeiff. = ***Lagenocarpus polyphyllus*** (Nees) Kuntze
Lagenocarpus polyphyllus var. *typicus* H.Pfeiff. = ***Lagenocarpus polyphyllus*** (Nees) Kuntze
Lagenocarpus portoricensis Britton = ***Lagenocarpus guianensis*** Nees subsp. ***guianensis***
Lagenocarpus pseudocladius H.Pfeiff. = ***Lagenocarpus rigidus*** (Kunth) Nees subsp. ***rigidus***
Lagenocarpus riedelianus C.B.Clarke = ***Lagenocarpus guianensis*** Nees subsp. ***guianensis***
Lagenocarpus rigidus var. *arundinaceus* (Boeck.) H.Pfeiff. = ***Lagenocarpus rigidus*** (Kunth) Nees subsp. ***rigidus***
Lagenocarpus rigidus f. *elatior* H.Pfeiff. = ***Lagenocarpus rigidus*** (Kunth) Nees subsp. ***rigidus***
Lagenocarpus rigidus f. *gracilior* H.Pfeiff. = ***Lagenocarpus rigidus*** (Kunth) Nees subsp. ***rigidus***
Lagenocarpus rigidus f. *neesii* (Boeck.) H.Pfeiff. = ***Lagenocarpus rigidus*** (Kunth) Nees subsp. ***rigidus***
Lagenocarpus rigidus f. *robustior* H.Pfeiff. = ***Lagenocarpus rigidus*** (Kunth) Nees subsp. ***rigidus***
Lagenocarpus rigidus subsp. *tremulus* (Nees) T.Koyama & Maguire = ***Lagenocarpus rigidus*** (Kunth) Nees subsp. ***rigidus***
Lagenocarpus scalariformis H.Pfeiff. = ***Lagenocarpus rigidus*** (Kunth) Nees subsp. ***rigidus***
Lagenocarpus schenckianus (Boeck.) H.Pfeiff. = ***Lagenocarpus triqueter*** (Boeck.) Kuntze
Lagenocarpus sprucei H.Pfeiff. = ***Lagenocarpus guianensis*** Nees subsp. ***guianensis***
Lagenocarpus steyermarkii Gilly = ***Lagenocarpus rigidus*** (Kunth) Nees subsp. ***rigidus***
Lagenocarpus strictus (Kunth) Kuntze = ***Lagenocarpus verticillatus*** (Spreng.) T.Koyama & Maguire
Lagenocarpus strictus var. *junciformis* (Kunth) H.Pfeiff. = ***Lagenocarpus junciformis*** (Kunth) Kuntze
Lagenocarpus strictus var. *leptocladius* (Poepp. ex Kunth) H.Pfeiff. = ***Lagenocarpus verticillatus*** (Spreng.) T.Koyama & Maguire
Lagenocarpus tenuifolius var. *riedelii* C.B.Clarke = ***Lagenocarpus rigidus*** subsp. ***tenuifolius*** (Boeck.) T.Koyama & Maguire
Lagenocarpus topazinus Nees = ***Lagenocarpus rigidus*** (Kunth) Nees subsp. ***rigidus***
Lagenocarpus tremulus (Nees) Nees = ***Lagenocarpus rigidus*** (Kunth) Nees subsp. ***rigidus***
Lagenocarpus tremulus var. *brevifolius* (Boeck.) H.Pfeiff. = ***Lagenocarpus rigidus*** subsp. ***tenuifolius*** (Boeck.) T.Koyama & Maguire
Lagenocarpus tremulus f. *longinux* H.Pfeiff. =
Lagenocarpus tremulus f. *multigyna* H.Pfeiff. = ***Lagenocarpus rigidus*** (Kunth) Nees subsp. ***rigidus***
Lagenocarpus uliginosus (Schrad. ex Nees) Kuntze = ***Lagenocarpus verticillatus*** (Spreng.) T.Koyama & Maguire
Lagenocarpus vestitus C.B.Clarke = ***Lagenocarpus griseus*** (Boeck.) H.Pfeiff.
Lagenocarpus warmingii H.Pfeiff. = ***Lagenocarpus rigidus*** (Kunth) Nees subsp. ***rigidus***
Lagenocarpus weigeltii (A.Spreng.) Uittien = ***Lagenocarpus verticillatus*** (Spreng.) T.Koyama & Maguire

***Unplaced Names*:**
Lagenocarpus laniger C.B.Clarke ex H.Pfeiff., Beih. Bot. Centralbl. 39(2): 441 (1923). = ?

Lampocarya

Lampocarya R.Br. = ***Gahnia*** J.R.Forst. & G.Forst.
Lampocarya affinis Brongn. = ***Morelotia affinis*** (Brongn.) S.T.Blake
Lampocarya aspera R.Br. = ***Gahnia aspera*** (R.Br.) Spreng.
Lampocarya gaudichaudii Brongn. = ***Morelotia gahniiformis*** Gaudich.
Lampocarya hexandra R.Br. = ***Gahnia trifida*** Labill.
Lampocarya lacera R.Lesson ex A.Rich. = ***Gahnia lacera*** (R.Lesson ex A.Rich.) Steud.
Lampocarya leptostachya Schrad. ex Boeck. = ***Gahnia beecheyi*** H.Mann
Lampocarya rawacensis Kunth = ***Gahnia aspera*** (R.Br.) Spreng. subsp. ***aspera***
Lampocarya schoenoides (G.Forst.) R.Br. = ***Gahnia schoenoides*** G.Forst.
Lampocarya setifolia R.Lesson & A.Rich. ex Hook.f. = ***Gahnia setifolia*** Hook.f.
Lampocarya tenax Hook.f. = ***Machaerina tenax*** (Hook.f.) T.Koyama
Lampocarya xanthocarpa Hook.f. = ***Gahnia xanthocarpa*** (Hook.f.) Hook.f.

Lamprochlaenia

Lamprochlaenia Börner = ***Carex*** L.
Lamprochlaenia nitida Fedde & J.Schust. = ***Carex liparocarpos*** Gaudin subsp. ***liparocarpos***

Langevinia

Langevinia Jacq.-Fél. = ***Mapania*** Aubl.
Langevinia monosperma Jacq.-Fél. = ***Mapania amplivaginata*** K.Schum.

Leiophyllum

Leiophyllum Ehrh. = ***Scirpus*** Tourn. ex L.

Lepidosperma

Lepidosperma Labill., Nov. Holl. Pl. 1: 14 (1804).
S. China, Indo-China to New Caledonia and New Zealand. 36 CHS 41 VIE 42 BOR MLY SUM 43 NWG 50 NSW QLD SOA TAS VIC WAU 51 CTM NZN NZS 60 NWC.
57 Species
Lepidotosperma Roem. & Schult., Syst. Veg. 2: 78 (1817).
Vauthiera A.Rich. in J.S.C.Dumont d'Urville, Voy. Astrolabe 1: 106 (1832).
Machaerina Nees in J.G.C.Lehmann, Nov. Stirp. Pug. 8: 54 (1844), nom. illeg.

Lepidosperma angustatum R.Br., Prodr.: 235 (1810).
SW. Australia. 50 WAU. Hemicr. or rhizome geophyte.
Lepidosperma fimbriatum Nees in J.G.C.Lehmann, Pl. Preiss. 2: 91 (1846).
Lepidosperma muelleri Boeck., Linnaea 38: 320 (1874).

Lepidosperma aphyllum R.Br., Prodr.: 235 (1810).
S. Western Australia. 50 WAU. Hemicr. or rhizome geophyte.

Lepidosperma australe (A.Rich.) Hook.f., Fl. Nov.-Zel. 1: 279 (1853).
New Zealand. 51 CTM NZN NZS. Hemicr. or rhizome geophyte.
**Vauthiera australis* A.Rich. in J.S.C.Dumont d'Urville, Voy. Astrolabe 1: 107 (1832). *Cladium vauthiera* C.B.Clarke in T.F.Cheeseman, Man. New Zealand Fl.: 788 (1906). *Mariscus vauthiera* (C.B.Clarke) Fernald, Rhodora 25: 54 (1923). *Cladium australe* (A.Rich.) Druce, Rep. Bot. Exch. Club Brit. Isles 1916: 615 (1917). *Machaerina australis* (A.Rich.) T.Koyama, Bot. Mag. (Tokyo) 69: 62 (1956).

Lepidosperma avium K.L.Wilson, Telopea 5: 603 (1994).
South Australia. 50 SOA. Hemicr. or rhizome geophyte.

Lepidosperma benthamianum C.B.Clarke, Kew Bull., Addit. Ser. 8: 91 (1908).
SW. Australia. 50 WAU. Hemicr. or rhizome geophyte.

Lepidosperma brunonianum Nees in J.G.C.Lehmann, Pl. Preiss. 2: 92 (1846).
SW. Australia. 50 WAU. Hemicr. or rhizome geophyte.
Lepidosperma brunonianum var. *binuciferum* Kük., Repert. Spec. Nov. Regni Veg. 50: 43 (1941).

Lepidosperma canescens Boeck., Linnaea 38: 330 (1874).
S. Australia. 50 SOA VIC WAU. Hemicr. or rhizome geophyte.

Lepidosperma carphoides F.Muell. ex Benth., Fl. Austral. 7: 400 (1878).
S. Australia. 50 SOA VIC WAU. Hemicr. or rhizome geophyte.

Lepidosperma chinense Nees & Meyen ex Kunth, Enum. Pl. 2: 320 (1837).
S. China, Indo-China to New Guinea. 36 CHS 41 VIE 42 BOR MLY SUM 43 NWG. Hemicr. or rhizome geophyte.
Cladium borneense C.B.Clarke ex Stapf, Trans. Linn. Soc. London, Bot. 4: 245 (1894). *Mariscus borneensis* (C.B.Clarke ex Stapf) Fernald, Rhodora 25: 52 (1923).

Lepidosperma clipeicola K.L.Wilson, Telopea 5: 604 (1994).
E. Australia. 50 NSW QLD. Hemicr. or rhizome geophyte.

Lepidosperma concavum R.Br., Prodr.: 234 (1810).
Lepidosperma squamatum var. *concavum* (R.Br.) Kük., Repert. Spec. Nov. Regni Veg. 50: 33 (1941).
SE. Australia. 50 SOA TAS VIC. Hemicr. or rhizome geophyte.
Chaetospora concava Nees, Ann. Mag. Nat. Hist. 6: 47 (1841).

Lepidosperma congestum R.Br., Prodr.: 234 (1810).
South Australia to Victoria. 50 SOA VIC. Hemicr. or rhizome geophyte.

Lepidosperma costale Nees in J.G.C.Lehmann, Pl. Preiss. 2: 92 (1846).
SW. Australia. 50 WAU. Hemicr. or rhizome geophyte.
Lepidosperma costale var. *densispicatum* Kük., Repert. Spec. Nov. Regni Veg. 50: 41 (1941).

Lepidosperma curtisiae K.L.Wilson & D.I.Morris, Telopea 5: 423 (1993).
SE. Australia. 50 NSW SOA TAS VIC. Hemicr. or rhizome geophyte.

Lepidosperma drummondii Benth., Fl. Austral. 7: 391 (1878).
SW. Australia. 50 WAU. Hemicr. or rhizome geophyte.
Lepidosperma drummondii var. *angustisquamellosum* Kük., Repert. Spec. Nov. Regni Veg. 50: 42 (1941).
Lepidosperma drummondii var. *floribundum* Kük., Repert. Spec. Nov. Regni Veg. 50: 43 (1941).

Lepidosperma effusum Benth., Fl. Austral. 7: 387 (1878).
SW. Australia. 50 WAU. Hemicr. or rhizome geophyte.

Lepidosperma elatius Labill., Nov. Holl. Pl. 1: 15 (1805).
SE. Queensland, Victoria to Tasmania. 50 QLD TAS VIC. Hemicr. or rhizome geophyte.
Lepidosperma elatius var. *planoconvexum* Kük., Repert. Spec. Nov. Regni Veg. 50: 35 (1941).

Lepidosperma ensiforme (Rodway) D.I.Morris in W.M.Curtis & D.I.Morris, Stud. Fl. Tasman. 4B: 425 (1994).
Tasmania. 50 TAS. Hemicr. or rhizome geophyte.
**Lepidosperma elatius* var. *ensiforme* Rodway, Tasman. Fl.: 246 (1903).

Lepidosperma evansianum K.L.Wilson, Telopea 5: 604 (1994).
New South Wales. 50 NSW. Hemicr. or rhizome geophyte.

Lepidosperma exaltatum R.Br., Prodr.: 234 (1810).
S. Western Australia. 50 WAU. Hemicr. or rhizome geophyte.

Lepidosperma filiforme Labill., Nov. Holl. Pl. 1: 17 (1804).
SE. Australia, New Zealand. 50 NSW TAS VIC 51 NZN NZS. Hemicr. or rhizome geophyte.

Lepidosperma flexuosum R.Br., Prodr.: 235 (1810).
E. New South Wales. 50 NSW. Hemicr. or rhizome geophyte.

Lepidosperma forsythii A.A.Ham., Proc. Linn. Soc. New South Wales 35: 411 (1910).
SE. Australia. 50 NSW TAS VIC. Hemicr. or rhizome geophyte.

Lepidosperma gladiatum Labill., Nov. Holl. Pl. 1: 15 (1805).
S. & SE. Australia. 50 NSW SOA TAS VIC WAU. Hemicr. or rhizome geophyte.
Lepidosperma ensatum Nees, Ann. Mag. Nat. Hist. 6: 47 (1841).

Lepidosperma globosum Labill., Nov. Holl. Pl. 1: 16 (1805).
Victoria to Tasmania. 50 TAS VIC. Hemicr. or rhizome geophyte.
Lepidosperma laeve R.Br., Prodr.: 235 (1810).

Lepidosperma gracile R.Br., Prodr.: 235 (1810). *Lepidosperma lineare* var. *gracile* (R.Br.) Kük., Repert. Spec. Nov. Regni Veg. 50: 49 (1941).
SW. Australia. 50 WAU. Hemicr. or rhizome geophyte.

Lepidosperma gunnii Boeck., Linnaea 38: 325 (1874).
SE. Queensland to Tasmania. 50 NSW QLD TAS VIC. Hemicr. or rhizome geophyte.

Lepidosperma inops F.Muell. ex Rodway, Pap. & Proc. Roy. Soc. Tasmania 1892: 93 (1894). *Lepidosperma lineare* var. *inops* (F.Muell. ex Rodway) Rodway, Pap. & Proc. Roy. Soc. Tasmania 1893: 182 (1894).
Tasmania. 50 TAS. Hemicr. or rhizome geophyte.

Lepidosperma latens K.L.Wilson, Telopea 5: 610 (1994).
New South Wales. 50 NSW. Hemicr. or rhizome geophyte.

Lepidosperma laterale R.Br., Prodr.: 234 (1810). *Lepidosperma laterale* var. *typicum* Domin, Biblioth. Bot. 85: 478 (1915), nom. inval.
E. & SE. Australia, New Zealand North I. 50 NSW QLD SOA TAS VIC 51 NZN. Hemicr. or rhizome geophyte.
Lepidosperma lineare R.Br., Prodr.: 235 (1810).
Lepidosperma longitudinale R.Br., Prodr.: 234 (1810).
Lepidosperma tetragynum R.Br., Prodr.: 234 (1810). *Lepidosperma laterale* var. *majus* Benth., Fl. Austral. 7: 394 (1878). *Lepidosperma laterale* var. *tetragynum* (R.Br.) Domin, Biblioth. Bot. 85: 478 (1915), nom. superfl.
Lepidosperma lineare var. *humile* Nees in J.G.C.Lehmann, Pl. Preiss. 2: 90 (1848).
Lepidosperma angustifolium Hook.f., Fl. Tasman. 2: 92 (1858).
Lepidosperma concavum Hook.f., Fl. Tasman. 2: 91 (1858), nom. illeg.
Lepidosperma humile (Nees) Boeck., Linnaea 38: 325 (1874).
Lepidosperma laterale var. *angustum* Benth., Fl. Austral. 7: 394 (1878).
Lepidosperma lineare var. *depauperatum* Benth., Fl. Austral. 7: 395 (1878).

Lepidosperma leptophyllum Benth., Fl. Austral. 7: 398 (1878).
SW. Australia. 50 WAU. Hemicr. or rhizome geophyte.

Lepidosperma leptostachyum Benth., Fl. Austral. 7: 397 (1878).
SW. Australia. 50 WAU. Hemicr. or rhizome geophyte.
Lepidosperma leptostachyum var. *asperatum* Kük., Repert. Spec. Nov. Regni Veg. 50: 117 (1941).

Lepidosperma limicola N.A.Wakef., Victoria Naturalist 70: 75 (1953).
SE. Queensland to SE. Australia. 50 NSW QLD SOA TAS VIC. Hemicr. or rhizome geophyte.

Lepidosperma longitudinale Labill., Nov. Holl. Pl. 1: 16 (1805).
S. & E. Australia. 50 NSW QLD SOA TAS VIC WAU. Hemicr. or rhizome geophyte.

Lepidosperma neesii Kunth, Enum. Pl. 2: 319 (1837).
New South Wales to Victoria. 50 NSW VIC. Hemicr. or rhizome geophyte.

Lepidosperma oldfieldii Hook.f., Fl. Tasman. 2: 91 (1858). *Lepidosperma elatius* var. *oldfieldii* (Hook.f.) Rodway, Tasman. Fl.: 246 (1903).
Tasmania. 50 TAS. Hemicr. or rhizome geophyte.

Lepidosperma pauperum Kük., Repert. Spec. Nov. Regni Veg. 50: 125 (1941).
New Caledonia. 60 NWC. Hemicr. or rhizome geophyte.

Lepidosperma perplanum Guillaumin, Bull. Soc. Bot. France 85: 44 (1938).
New Caledonia. 60 NWC. Hemicr. or rhizome geophyte.

Lepidosperma persecans S.T.Blake, Proc. Roy. Soc. Queensland 60: 50 (1950).
SW. Western Australia. 50 WAU. Hemicr. or rhizome geophyte.

Lepidosperma perteres C.B.Clarke, Kew Bull., Addit. Ser. 8: 47 (1908).
New Caledonia. 60 NWC. Hemicr. or rhizome geophyte.

Lepidosperma pruinosum Kük., Repert. Spec. Nov. Regni Veg. 50: 115 (1941).
SW. Australia. 50 WAU. Hemicr. or rhizome geophyte.
Lepidosperma pruinosum var. *rigidulum* Kük., Repert. Spec. Nov. Regni Veg. 50: 115 (1941).

Lepidosperma pubisquameum Steud., Syn. Pl. Glumac. 2: 158 (1855).
SW. Australia. 50 WAU. Hemicr. or rhizome geophyte.

Lepidosperma quadrangulatum A.A.Ham., Proc. Linn. Soc. New South Wales 45: 261 (1920). *Lepidosperma neesii* var. *quadrangulatum* (A.A.Ham.) Kük., Repert. Spec. Nov. Regni Veg. 50: 122 (1941).
E. New South Wales. 50 NSW. Hemicr. or rhizome geophyte.

Lepidosperma resinosum (Nees) Benth., Fl. Austral. 7: 392 (1878).
SW. Australia. 50 WAU. Hemicr. or rhizome geophyte.
**Machaerina resinosa* Nees in J.G.C.Lehmann, Nov. Stirp. Pug. 8: 54 (1844).
Lepidosperma sieberi Nees in J.G.C.Lehmann, Pl. Preiss. 2: 90 (1846).
Lepidosperma resinosum var. *pleianthemum* Kük., Repert. Spec. Nov. Regni Veg. 50: 45 (1941).

Lepidosperma rostratum S.T.Blake, Proc. Roy. Soc. Queensland 60: 51 (1950).
WSW. Western Australia. 50 WAU. Hemicr. or rhizome geophyte.

Lepidosperma rupestre Benth., Fl. Austral. 7: 388 (1878).
WSW. Western Australia. 50 WAU. Hemicr. or rhizome geophyte.

Lepidosperma scabrum Nees in J.G.C.Lehmann, Pl. Preiss. 2: 92 (1846).
SW. Australia. 50 WAU. Hemicr. or rhizome geophyte.
Lepidosperma scabrum var. *effusum* Benth., Fl. Austral. 7: 397 (1878).

Lepidosperma semiteres Boeck., Linnaea 38: 327 (1874). *Lepidosperma gracile* var. *semiteres* (Boeck.) Domin, Biblioth. Bot. 85: 478 (1915).
SE. Australia. 50 NSW SOA VIC. Hemicr. or rhizome geophyte.

Lepidosperma squamatum Labill., Nov. Holl. Pl. 1: 17 (1805).
SW. Australia. 50 WAU. Hemicr. or rhizome geophyte.
Lepidosperma sieberi Kunth, Enum. Pl. 2: 320 (1837).
Lepidosperma angustatum var. *curvispiculum* Benth., Fl. Austral. 7: 391 (1878). *Lepidosperma squamatum* var. *curvispiculum* (Benth.) Domin, Biblioth. Bot. 85: 477 (1915).
Lepidosperma gladiatum Nees ex Benth., Fl. Austral. 7: 390 (1878), nom. inval.

Lepidosperma striatum R.Br., Prodr.: 235 (1810).
SW. Australia. 50 WAU. Hemicr. or rhizome geophyte.
Lepidosperma confine Nees in J.G.C.Lehmann, Pl. Preiss. 2: 93 (1846).
Lepidosperma obtusum Kük., Repert. Spec. Nov. Regni Veg. 50: 120 (1941).

Lepidosperma tenue Benth., Fl. Austral. 7: 397 (1878).
SW. Australia. 50 WAU. Hemicr. or rhizome geophyte.

Lepidosperma tetraquetrum Nees in J.G.C.Lehmann, Pl. Preiss. 2: 90 (1846).
SW. Australia. 50 WAU. Hemicr. or rhizome geophyte.

Lepidosperma tortuosum F.Muell., Fragm. 9: 23 (1875).
SE. Australia. 50 NSW SOA? TAS VIC. Hemicr. or rhizome geophyte.
Lepidosperma falcatum Rodway, Pap. & Proc. Roy. Soc. Tasmania 1895: 103 (1896).

Lepidosperma tuberculatum Nees in J.G.C.Lehmann, Pl. Preiss. 2: 90 (1846).
SE. Queensland, SW. Australia. 50 QLD WAU. Hemicr. or rhizome geophyte.

var. ***grande*** Kük., Repert. Spec. Nov. Regni Veg. 50: 44 (1941).
SE. Queensland. 50 QLD. Hemicr. or rhizome geophyte.

var. ***tuberculatum***
SW. Australia. 50 WAU. Hemicr. or rhizome geophyte.

Lepidosperma urophorum N.A.Wakef., Victoria Naturalist 70: 76 (1953).
SE. Queensland to Victoria. 50 NSW QLD VIC. Hemicr. or rhizome geophyte.

Lepidosperma ustulatum Steud., Syn. Pl. Glumac. 2: 157 (1855). *Lepidosperma angustatum* var. *ustulatum* (Steud.) Benth., Fl. Austral. 7: 391 (1878).
SW. Australia. 50 WAU. Hemicr. or rhizome geophyte.

Lepidosperma viscidum R.Br., Prodr.: 234 (1810).
S. Australia. 50 SOA VIC WAU. Hemicr. or rhizome geophyte.
Lepidosperma viscidum var. *flaccidum* Kük., Repert. Spec. Nov. Regni Veg. 50: 46 (1941).
Lepidosperma viscidum var. *subpyramidale* Kük., Repert. Spec. Nov. Regni Veg. 50: 46 (1941).

***Synonyms*:**

Lepidosperma anceps (Poir.) Willd. ex Link = ***Machaerina anceps*** (Poir.) Bojer
Lepidosperma angustatum var. *curvispiculum* Benth. = ***Lepidosperma squamatum*** Labill.
Lepidosperma angustatum var. *ustulatum* (Steud.) Benth. = ***Lepidosperma ustulatum*** Steud.
Lepidosperma angustifolium Hook.f. = ***Lepidosperma laterale*** R.Br.
Lepidosperma brehmeri Boeck. = ***Tetraria bromoides*** (Lam.) H.Pfeiff.
Lepidosperma brunonianum var. *binuciferum* Kük. = ***Lepidosperma brunonianum*** Nees
Lepidosperma burmanii (Vahl) Spreng. = ***Tetraria burmanii*** (Vahl) C.B.Clarke
Lepidosperma colensoi Boeck. = ***Machaerina juncea*** (R.Br.) T.Koyama
Lepidosperma concavum Hook.f. = ***Lepidosperma laterale*** R.Br.
Lepidosperma concavum var. *pyramidatum* Benth. =
Lepidosperma confine Nees = ***Lepidosperma striatum*** R.Br.
Lepidosperma costale var. *densispicatum* Kük. = ***Lepidosperma costale*** Nees
Lepidosperma dregei Boeck. = ***Tetraria nigrovaginata*** (Nees) C.B.Clarke
Lepidosperma drummondii var. *angustisquamellosum* Kük. = ***Lepidosperma drummondii*** Benth.
Lepidosperma drummondii var. *floribundum* Kük. = ***Lepidosperma drummondii*** Benth.
Lepidosperma elatius var. *ensiforme* Rodway = ***Lepidosperma ensiforme*** (Rodway) D.I.Morris
Lepidosperma elatius var. *oldfieldii* (Hook.f.) Rodway = ***Lepidosperma oldfieldii*** Hook.f.
Lepidosperma elatius var. *planoconvexum* Kük. = ***Lepidosperma elatius*** Labill.
Lepidosperma ensatum Nees = ***Lepidosperma gladiatum*** Labill.
Lepidosperma exsul C.B.Clarke = ***Tricostularia neesii*** Nees
Lepidosperma falcatum Rodway = ***Lepidosperma tortuosum*** F.Muell.
Lepidosperma fimbriatum Nees = ***Lepidosperma angustatum*** R.Br.
Lepidosperma gladiatum Nees ex Benth. = ***Lepidosperma squamatum*** Labill.
Lepidosperma gracile var. *semiteres* (Boeck.) Domin = ***Lepidosperma semiteres*** Boeck.
Lepidosperma humile (Nees) Boeck. = ***Lepidosperma laterale*** R.Br.
Lepidosperma involucratum (Rottb.) Roem. & Schult. = ***Tetraria involucrata*** (Rottb.) C.B.Clarke
Lepidosperma iridifolium (Bory) Willd. ex Link = ***Machaerina iridifolia*** (Bory) T.Koyama
Lepidosperma laeve R.Br. = ***Lepidosperma globosum*** Labill.
Lepidosperma laterale var. *angustum* Benth. = ***Lepidosperma laterale*** R.Br.
Lepidosperma laterale var. *majus* Benth. = ***Lepidosperma laterale*** R.Br.
Lepidosperma laterale var. *tetragynum* (R.Br.) Domin = ***Lepidosperma laterale*** R.Br.

Lepidosperma laterale var. *typicum* Domin = ***Lepidosperma laterale*** R.Br.
Lepidosperma leptostachyum var. *asperatum* Kük. = ***Lepidosperma leptostachyum*** Benth.
Lepidosperma lineare R.Br. = ***Lepidosperma laterale*** R.Br.
Lepidosperma lineare var. *depauperatum* Benth. = ***Lepidosperma laterale*** R.Br.
Lepidosperma lineare var. *gracile* (R.Br.) Kük. = ***Lepidosperma gracile*** R.Br.
Lepidosperma lineare var. *humile* Nees = ***Lepidosperma laterale*** R.Br.
Lepidosperma lineare var. *inops* (F.Muell. ex Rodway) Rodway = ***Lepidosperma inops*** F.Muell. ex Rodway
Lepidosperma longitudinale R.Br. = ***Lepidosperma laterale*** R.Br.
Lepidosperma muelleri Boeck. = ***Lepidosperma angustatum*** R.Br.
Lepidosperma neesii var. *quadrangulatum* (A.A.Ham.) Kük. = ***Lepidosperma quadrangulatum*** A.A.Ham.
Lepidosperma obtusum Kük. = ***Lepidosperma striatum*** R.Br.
Lepidosperma pauciflorum (R.Br.) F.Muell. = ***Tricostularia pauciflora*** (R.Br.) Benth.
Lepidosperma pruinosum var. *rigidulum* Kük. = ***Lepidosperma pruinosum*** Kük.
Lepidosperma resinosum var. *pleianthemum* Kük. = ***Lepidosperma resinosum*** (Nees) Benth.
Lepidosperma rottboellii Schrad. = ***Tetraria bromoides*** (Lam.) H.Pfeiff.
Lepidosperma scabrum var. *effusum* Benth. = ***Lepidosperma scabrum*** Nees
Lepidosperma sieberi Nees = ***Lepidosperma resinosum*** (Nees) Benth.
Lepidosperma sieberi Kunth = ***Lepidosperma squamatum*** Labill.
Lepidosperma squamatum var. *concavum* (R.Br.) Kük. = ***Lepidosperma concavum*** R.Br.
Lepidosperma squamatum var. *curvispiculum* (Benth.) Domin = ***Lepidosperma squamatum*** Labill.
Lepidosperma squamatum var. *pyramidatum* (Benth.) Kük. = ***Lepidosperma squamatum*** Labill.
Lepidosperma teres Nees ex C.B.Clarke = ***Tetraria bromoides*** (Lam.) H.Pfeiff.
Lepidosperma tetragonum Labill. = ***Machaerina tetragona*** (Labill.) T.Koyama
Lepidosperma tetragynum R.Br. = ***Lepidosperma laterale*** R.Br.
Lepidosperma thermale (L.) Schrad. = ***Tetraria thermalis*** (L.) C.B.Clarke
Lepidosperma triangulare Boeck. = ***Tetraria triangularis*** (Boeck.) C.B.Clarke
Lepidosperma uncinatum Nees = ***Mesomelaena stygia*** (R.Br.) Nees subsp. ***stygia***
Lepidosperma viscidum var. *flaccidum* Kük. = ***Lepidosperma viscidum*** R.Br.
Lepidosperma viscidum var. *subpyramidale* Kük. = ***Lepidosperma viscidum*** R.Br.
Lepidosperma waigiuensis Steud. = ***Machaerina glomerata*** (Gaudich.) T.Koyama
Lepidosperma zeylanicum Lindl. ex Boeck. = ***Tricostularia undulata*** (Thwaites) J.Kern

Lepidospora

Lepidospora (F.Muell.) F.Muell. = ***Schoenus*** L.
Lepidospora tenuissima F.Muell. = ***Schoenus lepidosperma*** (F.Muell.) K.L.Wilson subsp. ***lepidosperma***

Lepidotosperma

Lepidotosperma Roem. & Schult. = ***Lepidosperma*** Labill.

Lepironia

Lepironia Rich. ex Pers., Syn. Pl. 1: 70 (1805).
Madagascar, Sri Lanka to W. Pacific. 29 MDG 36 CHS 38 NNS TAI 40 IND SRL 41 THA VIE 42 BOR MLY MOL SUL SUM 43 NWG 50 NSW NTA QLD 60 FIJ NWC TON 62 CRL.
1 Species
Chondrachne R.Br., Prodr.: 220 (1810).
Lepyronia T.Lestib., Essai Cypér.: 25 (1819).
Choricarpha Boeck., Flora 41: 19 (1858).

Lepironia articulata (Retz.) Domin, Biblioth. Bot. 85: 486 (1915).
Madagascar, Sri Lanka to W. Pacific. 29 MDG 36 CHS 38 NNS TAI 40 IND SRL 41 THA VIE 42 BOR MLY MOL SUL SUM 43 NWG 50 NSW NTA QLD 60 FIJ NWC TON 62 CRL. Hemicr. or rhizome geophyte.
**Restio articulatus* Retz., Observ. Bot. 4: 14 (1786). *Chondrachne articulata* (Retz.) R.Br., Prodr.: 220 (1810).
Lepironia mucronata Rich. ex Pers., Syn. Pl. 1: 70 (1805).
Scirpus coniferus Poir. in J.B.A.M.de Lamarck, Encycl. 6: 756 (1805). *Lepironia conifera* (Poir.) Druce, Bot. Soc. Exch. Club Brit. Isles 1916: 631 (1917).
Choricarpha aphylla Boeck., Flora 41: 20 (1858).
Lepironia mucronata var. *capitata* F.Muell., Fragm. 9: 17 (1875). *Lepironia articulata* var. *capitata* (F.Muell.) Domin, Biblioth. Bot. 85: 486 (1915).
Lepironia compressa Boeck., Allg. Bot. Z. Syst. 2: 78 (1896).

Synonyms:
Lepironia articulata var. *capitata* (F.Muell.) Domin = ***Lepironia articulata*** (Retz.) Domin
Lepironia bancana Miq. = ***Mapania bancana*** (Miq.) T.Koyama
Lepironia compressa Boeck. = ***Lepironia articulata*** (Retz.) Domin
Lepironia conifera (Poir.) Druce = ***Lepironia articulata*** (Retz.) Domin
Lepironia cuspidata Miq. = ***Mapania cuspidata*** (Miq.) Uittien
Lepironia cymbaria (R.Br.) T.Koyama = ***Chorizandra cymbaria*** R.Br.
Lepironia enodis Miq. = ***Mapania enodis*** (Miq.) C.B.Clarke
Lepironia foliosa Miq. = ***Mapania enodis*** (Miq.) C.B.Clarke
Lepironia humilis (Hassk. ex Steud.) Miq. = ***Hypolytrum humile*** (Hassk. ex Steud.) Boeck.
Lepironia macrocephala (Gaudich.) Miq. = ***Mapania macrocephala*** (Gaudich.) K.Schum.
Lepironia mucronata Rich. ex Pers. = ***Lepironia articulata*** (Retz.) Domin
Lepironia mucronata var. *capitata* F.Muell. = ***Lepironia articulata*** (Retz.) Domin
Lepironia multiarticulata (Nees) T.Koyama = ***Chorizandra multiarticulata*** Nees
Lepironia neesii T.Koyama = ***Chorizandra enodis*** Nees
Lepironia palustris (Hassk. ex Steud.) Miq. = ***Mapania palustris*** (Hassk. ex Steud.) Fern.-Vill.
Lepironia sphaerocephala (R.Br.) T.Koyama = ***Chorizandra sphaerocephala*** R.Br.

Lepironia squamata (Kurz) Miq. = ***Mapania squamata*** (Kurz) C.B.Clarke
Lepironia sumatrana Miq. = ***Mapania sumatrana*** (Miq.) Benth.
Lepironia zeylanica (Thwaites) Miq. = ***Mapania zeylanica*** (Thwaites) Trimen

Lepisia

Lepisia C.Presl = ***Tetraria*** P.Beauv.
Lepisia ustulata (L.) C.Presl = ***Tetraria ustulata*** (L.) C.B.Clarke

Lepistachya

Lepistachya Zipp. ex Miq. = ***Mapania*** Aubl.
Lepistachya praemorsa Zipp. ex Miq. = ***Mapania cuspidata*** (Miq.) Uittien var. ***cuspidata***

Leptolepis

Leptolepis Boeck. = ***Blysmus*** Panz. ex Schult.
Leptolepis tibetica Boeck. = ***Blysmus compressus*** var. ***brevifolius*** (Decne.) Karth.

Leptoschoenus

Leptoschoenus Nees = ***Rhynchospora*** Vahl
Leptoschoenus prolifer Nees = ***Rhynchospora schomburgkiana*** (Boeck.) T.Koyama

Leptostachys

Leptostachys Ehrh. = ***Carex*** L.

Leptovignea

Leptovignea Börner = ***Carex*** L.
Leptovignea elongata (L.) Fedde & J.Schust. = ***Carex elongata*** L.
Leptovignea loliacea (L.) Fedde & J.Schust. = ***Carex loliacea*** L.

Lepyronia

Lepyronia T.Lestib. = ***Lepironia*** Rich. ex Pers.

Lerisca

Lerisca Schltdl. = ***Lagenocarpus*** Nees

Leucocoma

Leucocoma Rydb. = ***Trichophorum*** Pers.
Leucocoma alpina (L.) Rydb. = ***Trichophorum alpinum*** (L.) Pers.
Leucocoma alpina (L.) Rydb. = ***Trichophorum alpinum*** (L.) Pers.

Leucoglochin

Leucoglochin Heuffl. = ***Carex*** L.
Leucoglochin pauciflorus (Lightf.) Heuff. = ***Carex pauciflora*** Lightf.

Leucoma

Leucoma Ehrh. = ***Eriophorum*** L.

Limivasculum

Limivasculum Börner = ***Carex*** L.
Limivasculum atropictum (Steud.) Fedde & J.Schust. = ***Carex atropicta*** Steud.
Limivasculum cespitosum (L.) Fedde & J.Schust. = ***Carex cespitosa*** L.

Limnocharis

Limnocharis Kunth = ***Eleocharis*** R.Br.

Limnochloa

Limnochloa P.Beauv. ex Lestib. = ***Eleocharis*** R.Br.
Limnochloa acicularis (L.) Rchb. = ***Eleocharis acicularis*** (L.) Roem. & Schult.
Limnochloa acutangula (Roxb.) Nees = ***Eleocharis acutangula*** (Roxb.) Schult.
Limnochloa albibracteata Nees ex Kunth = ***Eleocharis albibracteata*** Nees & Meyen ex Kunth var. ***albibracteata***
Limnochloa alpina (L.) Drejer = ***Trichophorum alpinum*** (L.) Pers.
Limnochloa alpina (Schleich.) Rchb. = ***Trichophorum pumilum*** (Vahl) Schinz & Thell. subsp. ***pumilum***
Limnochloa articulata Nees = ***Eleocharis interstincta*** (Vahl) Roem. & Schult.
Limnochloa baeothryon (L.f.) Rchb. = ***Eleocharis quinqueflora*** (Hartmann) O.Schwarz
Limnochloa caduciflora Turcz. ex Trin. = (Poaceae)
Limnochloa calyptrata Liebm. = ***Eleocharis filiculmis*** Kunth
Limnochloa capensis C.Krauss = ***Eleocharis limosa*** (Schrad.) Schult.
Limnochloa capensis Nees = ***Eleocharis palustris*** (L.) Roem. & Schult. subsp. ***palustris***
Limnochloa cespitosa (L.) Rchb. = ***Trichophorum cespitosum*** (L.) Hartm.
Limnochloa constricta (Schrad. ex Kunth) Nees = ***Eleocharis geniculata*** (L.) Roem. & Schult.
Limnochloa costulata Nees ex Kunth = ***Eleocharis acicularis*** (L.) Roem. & Schult. subsp. ***acicularis***
Limnochloa crassiculmis Nees = ***Eleocharis elegans*** (Kunth) Roem. & Schult.
Limnochloa densa Liebm. = ***Eleocharis geniculata*** (L.) Roem. & Schult.
Limnochloa equisetoides (Elliott) Walp. = ***Eleocharis equisetoides*** (Elliott) Torr.
Limnochloa fistulosa (Link) Nees = ***Eleocharis acutangula*** (Roxb.) Schult.
Limnochloa geniculata (L.) Nees = ***Eleocharis geniculata*** (L.) Roem. & Schult.
Limnochloa limosa (Schrad.) Nees = ***Eleocharis limosa*** (Schrad.) Schult.
Limnochloa media (Roxb.) Nees = ***Eleocharis acutangula*** (Roxb.) Schult.
Limnochloa montevidensis (Kunth) Nees = ***Eleocharis montevidensis*** Kunth
Limnochloa multicaulis (Sm.) Rchb. = ***Eleocharis multicaulis*** (Sm.) Desv.
Limnochloa multiculmis Rchb. = ***Eleocharis multicaulis*** (Sm.) Desv.
Limnochloa mutata (L.) Nees = ***Eleocharis mutata*** (L.) Roem. & Schult.
Limnochloa obsoleta Nees = ***Eleocharis interstincta*** (Vahl) Roem. & Schult.
Limnochloa obtusitrigona Lindl. & Nees = ***Eleocharis obtusitrigona*** (Lindl. & Nees) Steud.
Limnochloa parvula (Spreng.) Rchb. = ***Eleocharis parvula*** (Roem. & Schult.) Palla
Limnochloa pauciflora (Lightf.) Peterm. = ***Eleocharis quinqueflora*** (Hartmann) O.Schwarz
Limnochloa plantaginea (Retz.) Nees = ***Eleocharis dulcis*** (Burm.f.) Trin. ex Hensch.

Limnochloa quadrangulata (Michx.) Nees = ***Eleocharis quadrangulata*** (Michx.) Roem. & Schult.
Limnochloa spiralis (Rottb.) Nees = ***Eleocharis spiralis*** (Rottb.) Roem. & Schult.
Limnochloa tetraquetra Nees = ***Eleocharis tetraquetra*** Nees
Limnochloa truncata (Schltdl.) Liebm. = ***Eleocharis dombeyana*** Kunth
Limnochloa tumida (Roxb.) Nees = ***Eleocharis dulcis*** (Burm.f.) Trin. ex Hensch.

Limonaetes

Limonaetes Ehrh. = ***Carex*** L.

Linagrostis

Linagrostis Guett. = ***Eriophorum*** L.
Linagrostis Hill = ***Eriophorum*** L.
Linagrostis alpina (L.) Scop. = ***Trichophorum alpinum*** (L.) Pers.
Linagrostis paniculata Lam. = ***Eriophorum angustifolium*** Honck. subsp. ***angustifolium***
Linagrostis polystachia (L.) Scop. = ***Eriophorum angustifolium*** Honck. subsp. ***angustifolium***
Linagrostis vaginata (L.) Scop. = ***Eriophorum vaginatum*** L.
Linagrostis vulgaris Hill = ***Eriophorum angustifolium*** Honck. subsp. ***angustifolium***

Lipocarpha

Lipocarpha R.Br. in J.H.Tuckey, Narr. Exped. Zaire: 459 (1818).
Trop. & Subtrop. to N. America. 22 BEN BKN GHA GNB GUI IVO LBR MLI NGA NGR SEN SIE TOG 23 BUR CAF CMN CON EQG GAB RWA ZAI 24 CHA ETH SUD 25 KEN TAN UGA 26 ANG MLW MOZ ZAM ZIM 27 BOT LES NAM NAT OFS SWZ TVL 29 MAU MDG 36 CHH CHS 38 JAP KOR NNS TAI 40 ASS EHM IND NEP PAK SRL WHM 41 CBD MYA THA VIE 42 BOR JAW MLY MOL PHI SUL SUM 43 BIS NWG SOL 50 NSW NTA QLD SOA VIC WAU 61 TUA 62 CRL 71 BRC 72 ONT QUE 73 COL ORE WAS WYO 74 ILL IOW KAN MIN MSO NDA NEB OKL WIS 75 CNT INI MAI MAS MIC NWH NWJ NWY OHI PEN RHO 76 ARI CAL 77 NWM TEX 78 ALA ARK FLA GEO KTY MRY NCA SCA VRG WDC 79 MXC MXE MXG MXN MXS MXT 80 BLZ COS CPI ELS GUA HON NIC PAN 81 ARU CUB JAM LEE NLA PUE TRT WIN 82 FRG GUY SUR VEN 83 BOL CLM ECU GAL PER 84 BZC BZE BZL BZN BZS 85 AGE PAR URU.
36 Species
Hypaelyptum Vahl, Enum. Pl. 2: 283 (1805), nom. rejic.
Hemicarpha Nees & Arn., Edinburgh New Philos. J. 17: 263 (1834).
Rikliella J.Raynal, Adansonia, n.s., 13: 154 (1973).

Lipocarpha abietina Goetgh., Wageningen Agric. Univ. Pap. 89(1): 19 (1989).
Trop. Africa to Botswana. 22 BKN IVO MLI SEN 23 BUR CAF CMN ZAI 24 CHA 25 TAN 26 ANG 27 BOT. Hemicr. or rhizome geophyte.

Lipocarpha albiceps Ridl., Trans. Linn. Soc. London, Bot. 2: 163 (1884). *Hypaelyptum albiceps* (Ridl.) K.Schum. in H.G.A.Engler, Pflanzenw. Ost-Afrikas, C: 127 (1895). *Cyperus echinolepis* T.Koyama, Bot. Mag. (Tokyo) 73: 438 (1960).
Trop. Africa. 22 BEN BKN GHA IVO MLI NGA SEN TOG 23 BUR CMN ZAI 24 CHA 25 KEN TAN UGA 26 ANG MLW ZAM ZIM. Hemicr. or rhizome geophyte.
Lipocarpha purpureolutea Ridl., Trans. Linn. Soc. London, Bot. 2: 163 (1884).

Lipocarpha aristulata (Coville) G.C.Tucker, J. Arnold Arbor. 68: 410 (1987).
W. & C. U.S.A. 73 COL ORE WAS WYO 74 KAN MIN MSO NEB OKL 76 ARI CAL 77 NWM TEX. Ther.
**Hemicarpha micrantha* var. *aristulata* Coville, Bull. Torrey Bot. Club 21: 36 (1894). *Hemicarpha aristulata* (Coville) A.Nelson, Bull. Torrey Bot. Club 29: 400 (1902).
Hemicarpha intermedia Piper in C.V.Piper & R.K.Beattie, Fl. Palouse Reg.: 36 (1901).

Lipocarpha atra Ridl., Trans. Linn. Soc. London, Bot. 2: 162 (1884).
Trop. Africa. 22 BKN SEN 23 CAF CMN ZAI 24 CHA 25 TAN 26 ANG MOZ ZAM ZIM. Hemicr. or rhizome geophyte.

Lipocarpha barteri C.B.Clarke in D.Oliver & auct. suc. (eds.), Fl. Trop. Afr. 8: 472 (1902). *Cyperus neobarteri* T.Koyama, Bot. Mag. (Tokyo) 73: 438 (1960). *Lipocarpha atra* var. *barteri* (C.B.Clarke) J.Raynal, Adansonia, n.s., 7: 85 (1967).
W. Trop. Africa to Cameroon. 22 BEN GHA IVO NGA TOG 23 CMN. Hemicr. or rhizome geophyte.
Kyllinga baoulensis A.Chev., Explor. Bot. Afrique Occ. Franç. 1: 698 (1920), nom. inval.

Lipocarpha bawangensis R.H.Miau, Acta Sci. Nat. Univ. Sunyatseni 30: 86 (1991).
Hainan. 36 CHH. Hemicr. or rhizome geophyte.

Lipocarpha chinensis (Osbeck) J.Kern, Blumea, Suppl. 4: 167 (1958).
Trop. & Subtrop. Old World. 22 BEN BKN GHA GNB GUI IVO LBR MLI NGA NGR SEN SIE TOG 23 BUR CAF CMN CON EQG GAB RWA ZAI 24 CHA ETH SUD 25 KEN TAN UGA 26 ANG MLW MOZ ZAM ZIM 27 BOT NAT SWZ TVL 29 MAU MDG 36 CHS 38 NNS TAI 40 ASS EHM IND NEP SRL WHM 41 THA VIE 42 BOR JAW MLY MOL PHI SUL SUM 43 NWG 50 QLD. Hemicr.
**Scirpus chinensis* Osbeck, Dagb. Ostind. Resa: 220 (1757). *Cyperus lipocarpha* T.Koyama, Bot. Mag. (Tokyo) 73: 438 (1960).
Scirpus senegalensis Lam., Tabl. Encycl. 1: 140 (1791). *Hypaelyptum argenteum* Vahl, Enum. Pl. 2: 283 (1805), nom. illeg. *Hypolytrum argenteum* (Vahl) Kunth in F.W.H.von Humboldt, A.J.A.Bonpland & C.S.Kunth, Nov. Gen. Sp. 1: 218 (1816). *Lipocarpha humboldtiana* Nees, Linnaea 9: 287 (1834). *Lipocarpha argentea* (Vahl) R.Br. in J.H.Tuckey, Narr. Exped. Zaire: 477 (1818). *Cyperus submaculatus* (Vahl) T.Koyama, Bot. Mag. (Tokyo) 73: 438 (1960). *Hypaelyptum senegalense* (Lam.) K.Schum. in H.G.A.Engler, Pflanzenw. Ost-Afrikas, C: 127 (1895). *Lipocarpha senegalensis* (Lam.) T.Durand & H.Durand, Syll. Fl. Congol.: 619 (1909).
Tunga laevigata Roxb., Fl. Ind. 1: 188 (1820). *Hypolytrum laevigatum* (Roxb.) Spreng., Syst. Veg. 1: 233 (1824). *Lipocarpha laevigata* (Roxb.) Nees in R.Wight, Contr. Bot. India: 92 (1834).

Hypolytrum albidum Willd. ex Kunth, Enum. Pl. 2: 266 (1837).
Kyllinga albescens Steud., Syn. Pl. Glumac. 2: 68 (1854).
Lipocarpha debilis Ridl., Trans. Linn. Soc. London, Bot. 9: 243 (1916).

Lipocarpha comosa J.Raynal, Bull. Mus. Natl. Hist. Nat., II, 41: 974 (1970).
Tanzania to Zambia. 23 ZAI 25 TAN 26 MLW ZAM. Hemicr.

Lipocarpha constricta Goetgh., Wageningen Agric. Univ. Pap. 89(1): 31 (1989).
Ethiopia to Burundi. 23 BUR 24 ETH. Hemicr. or rhizome geophyte.

Lipocarpha crassicuspis (J.Raynal) Goetgh., Wageningen Agric. Univ. Pap. 89(1): 32 (1989).
Senegal. 22 SEN. Ther.
**Lipocarpha prieuriana* var. *crassicuspis* J.Raynal, Adansonia, n.s., 7: 86 (1967).

Lipocarpha drummondii (Nees) G.C.Tucker, J. Arnold Arbor. 68: 410 (1987).
EC. & S. U.S.A. 74 ILL KAN MSO NDA NEB OKL WIS 75 INI OHI 76 ARI 77 NWM TEX 78 ARK. Ther.
**Hemicarpha drummondii* Nees in C.F.P.von Martius & auct. suc. (eds.), Fl. Bras. 2(1): 62 (1842). *Hemicarpha subsquarrosa* var. *drummondii* (Nees) A.Gray, Manual, ed. 2: 495 (1856). *Hemicarpha micrantha* var. *drummondii* (Nees) Frieland, Amer. J. Bot. 28: 860 (1941). *Scirpus micranthus* var. *drummondii* (Nees) Mohlenbr., Amer. Midl. Naturalist 70: 22 (1963).

Lipocarpha echinus J.Raynal, Adansonia, n.s., 13: 159 (1973).
Zambia. 26 ZAM. Ther.

Lipocarpha filiformis (Vahl) Kunth, Enum. Pl. 2: 267 (1837).
W. Trop. Africa to Chad and Angola. 22 BEN BKN GHA GNB GUI IVO LBR MLI NGA SEN SIE TOG 23 CMN GAB ZAI 24 CHA 26 ANG. Ther.
**Hypaelyptum filiforme* Vahl, Enum. Pl. 2: 284 (1805).

Lipocarpha gracilis (Rich. ex Pers.) Nees, Linnaea 9: 287 (1834).
Indian Subcontinent to Indo-China. 40 ASS IND NEP SRL 41 MYA THA (84) bzc. Ther.
Scirpus monander Willd., Sp. Pl. 1: 311 (1797), nom. illeg.
Hypaelyptum sphacelatum Vahl, Enum. Pl. 2: 283 (1805). *Lipocarpha sphacelata* (Vahl) Kunth, Enum. Pl. 2: 267 (1837). *Cyperus ceylanicus* (Vahl) T.Koyama, Bot. Mag. (Tokyo) 73: 438 (1960).
**Hypolytrum gracile* Rich. ex Pers., Syn. Pl. 1: 70 (1805).
Tunga triceps Roxb., Fl. Ind. 1: 187 (1820). *Hypolytrum triceps* (Roxb.) A.Dietr., Sp. Pl. 2: 363 (1832). *Lipocarpha triceps* (Roxb.) Nees in R.Wight, Contr. Bot. India: 92 (1834).
Hypolytrum ceylanicum B.Heyne ex Nees, Linnaea 9: 288 (1834), nom. nud.

Lipocarpha hemisphaerica (Roth) Goetgh., Wageningen Agric. Univ. Pap. 89(1): 37 (1989).
Trop. & Subtrop. Old World. 22 GHA NGA SEN TOG 23 CAF CMN ZAI 24 ETH 25 KEN TAN UGA 26 ANG MLW ZAM ZIM 27 BOT NAM OFS 40 IND 41 THA. Ther.
Isolepis minima Schrad., Gött. Gel. Anz. 3: 2068 (1821). *Hemicarpha schraderi* Kunth, Enum. Pl. 2: 268 (1837), nom. illeg.
**Scirpus hemisphaericus* Roth, Nov. Pl. Sp.: 29 (1821). *Isolepis hemisphaerica* (Roth) A.Dietr., Sp. Pl. 2: 109 (1832).
Hemicarpha isolepis Nees, Edinburgh New Philos. J. 17: 263 (1834). *Scirpus isolepis* (Nees) Boeck., Linnaea 36: 498 (1870). *Lipocarpha isolepis* (Nees) R.W.Haines, Bot. Not. 124: 476 (1971).
Scirpus minimus Willd. ex Kunth, Enum. Pl. 2: 268 (1837).
Hemicarpha senegalensis Steud., Syn. Pl. Glumac. 2: 130 (1855).
Isolepis bellula Steud., Syn. Pl. Glumac. 2: 318 (1855).
Lipocarpha rautanenii Boeck. ex Schinz, Verh. Bot. Vereins Prov. Brandenburg 31: 179 (1890).
Lipocarpha monocephala Turrill, Bull. Misc. Inform. Kew 1913: 307 (1913).

Lipocarpha kernii (Raymond) Goetgh., Wageningen Agric. Univ. Pap. 89(1): 42 (1989).
Trop. Africa, India. 22 BEN BKN GHA MLI NGA NGR SEN TOG 23 CAF CMN 24 CHA ETH SUD 25 TAN 26 MLW ZAM ZIM 40 IND. Ther.
**Scirpus kernii* Raymond, Naturaliste Canad. 86: 230 (1959). *Isolepis kernii* (Raymond) Lye, Bot. Not. 124: 479 (1971). *Rikliella kernii* (Raymond) J.Raynal, Adansonia, n.s., 13: 155 (1973).

Lipocarpha leucaspis J.Raynal, Bull. Mus. Natl. Hist. Nat., II, 41: 978 (1970).
Nigeria, Burundi to Zambia. 22 NGA 23 BUR ZAI 26 ZAM. Ther. or hemicr.

Lipocarpha maculata (Michx.) Torr., Ann. Lyceum Nat. Hist. New York 3: 288 (1836). *Cyperus neotropicalis* Alain, Bull. Torrey Bot. Club 92: 291 (1965).
Trop. & Subtrop. America. 74 ILL 78 ALA FLA GEO NCA SCA VRG WDC 79 MXG MXN MXS MXT 80 COS HON PAN 81 CUB TRT 82 GUY SUR VEN 83 BOL CLM PER 84 BZC BZL BZN BZS 85 AGE PAR URU. Ther.
Scirpus cephalotes Walter, Fl. Carol.: 71 (1788), nom. illeg. *Dichromena cephalotes* Britton, Bull. Torrey Bot. Club 15: 100 (1888).
**Kyllinga maculata* Michx., Fl. Bor.-Amer. 1: 29 (1803). *Mariscus maculatus* (Michx.) Roem. & Schult., Syst. Veg. 2: 243 (1817). *Hypolytrum maculatum* (Michx.) Kuntze, Revis. Gen. Pl. 3(2): 335 (1898).
Lipocarpha sellowiana Kunth, Enum. Pl. 2: 267 (1837). *Hypolytrum sellowiana* (Kunth) Kuntze, Revis. Gen. Pl. 3(2): 335 (1898). *Cyperus sellowianus* (Kunth) T.Koyama, Bot. Mag. (Tokyo) 73: 438 (1960).
Hypolytrum glomeratum Schrad. ex Nees in C.F.P.von Martius & auct. suc. (eds.), Fl. Bras. 2(1): 64 (1842), nom. inval.
Lipocarpha glomerata Nees in C.F.P.von Martius & auct. suc. (eds.), Fl. Bras. 2(1): 64 (1842).
Scirpus pycnostachys Salzm. ex Steud., Syn. Pl. Glumac. 2: 129 (1855).
Ascolepis venezuelensis Schnee, Bol. Soc. Venez. Ci. Nat. 9: 5 (1944).

Lipocarpha mangarevica H.St.John, Pacific Pl. Stud. 43: 9 (1988).
Tuamotu (Mangareva). 61 TUA. Hemicr.

Lipocarpha mexicana Liebm., Mexic. Neldeagt. Pl., V, 2: 235 (1851).

Mexico to Brazil, Madagascar. 29 MDG 79 MXE MXG MXS 80 ELS GUA HON NIC PAN 82 VEN 83 BOL CLM 84 BZC BZL. Ther.

Lipocarpha micrantha (Vahl) G.C.Tucker, J. Arnold Arbor.: 68: 410 (1987).
America, Senegal, S. Trop. & S. Africa, Madagascar. 22 SEN 26 ANG ZIM 27 NAM NAT TVL 29 MDG 71 BRC 72 ONT QUE 73 ORE 74 ILL IOW KAN MIN MSO NEB OKL WIS 75 CNT INI MAI MAS MIC NWH NWJ NWY OHI PEN RHO 76 ARI CAL 77 NWM TEX 78 ALA FLA KTY MRY NCA VRG 79 MXC MXE MXG MXN MXS MXT 80 BLZ COS CPI ELS GUA HON NIC PAN 81 ARU CUB JAM LEE NLA PUE TRT WIN 82 FRG GUY SUR VEN 83 BOL CLM ECU GAL PER 84 BZE BZN 85 PAR URU. Ther.
**Scirpus micranthus* Vahl, Enum. Pl. 2: 254 (1805). *Isolepis micrantha* (Vahl) Roem. & Schult., Syst. Veg. 2: 110 (1817). *Hemicarpha micrantha* (Vahl) Britton, Bull. Torrey Bot. Club 21: 35 (1894).
Scirpus minimus Pursh, Fl. Amer. Sept. 1: 55 (1813), nom. illeg.
Isolepis squarrosa Kunth in F.W.H.von Humboldt, A.J.A.Bonpland & C.S.Kunth, Nov. Gen. Sp. 1: 221 (1816).
Isolepis humboldtii Roem. & Schult., Syst. Veg. 2: 112 (1817). *Scirpus humboldtii* (Roem. & Schult.) A.Dietr., Sp. Pl. 2: 112 (1832).
Scirpus subsquarrosus Muhl., Descr. Gram.: 39 (1817). *Isolepis subsquarrosa* (Muhl.) Schrad. ex Schult. in J.J.Roemer & J.A.Schultes, Mant. 2: 64 (1824). *Hemicarpha subsquarrosa* (Muhl.) Nees in C.F.P.von Martius & auct. suc. (eds.), Fl. Bras. 2(1): 61 (1842).
Isolepis subsquarrosa var. *minor* Roem. & Schult., Mant. 2: 64 (1824). *Scirpus micranthus* var. *minor* (Roem. & Schult.) B.Boivin, Provancheria 25: 55 (1992).
Scirpus sesquipollicaris Willd. ex Kunth, Enum. Pl. 2: 203 (1837).
Hypolytrum capillare Schrad. ex Nees in C.F.P.von Martius & auct. suc. (eds.), Fl. Bras. 2(1): 61 (1842).
Isolepis caespitula Liebm., Mexic. Neldeagt. Pl., V, 2: 237 (1851). *Hemicarpha caespitula* (Liebm.) Palla, Oesterr. Bot. Z. 58: 417 (1908).
Scirpus micranthus var. *humboldtii* Boeck., Linnaea 36: 500 (1870). *Hemicarpha micrantha* var. *humboldtii* (Boeck.) Beetle, Amer. Midl. Naturalist 41: 476 (1949).
Hemicarpha micrantha var. *minor* (Schrad.) Friedl., Amer. J. Bot. 28: 860 (1942).

Lipocarpha microcephala (R.Br.) Kunth, Enum. Pl. 2: 268 (1837).
Trop. & Subtrop. Asia to Australia. 36 CHH CHS 38 JAP KOR NNS TAI 41 MYA THA VIE 42 JAW MLY MOL PHI SUL 43 BIS NWG SOL 50 NSW NTA QLD SOA VIC WAU 62 CRL (78) ala. Ther.
**Hypaelyptum microcephalum* R.Br., Prodr.: 220 (1810).
Kyllinga squarrosa Steud., Syn. Pl. Glumac. 2: 68 (1854).
Ascolepis kyllingioides Steud., Syn. Pl. Glumac. 2: 105 (1855). *Lipocarpha zollingeriana* Boeck., Flora 42: 100 (1859), nom. illeg. *Cyperus zollingerianus* (Boeck.) T.Koyama, Bot. Mag. (Tokyo) 73: 438 (1960).
Scirpus leptocarpus F.Muell., Trans. Philos. Soc. Victoria 1: 109 (1855).
Scirpus dietrichiae Boeck., Flora 58: 109 (1875). *Scirpus squarrosus* var. *dietrichiae* (Boeck.) Benth., Fl. Austral. 7: 329 (1878).
Rikliella australiensis J.Raynal, Adansonia, n.s., 16: 220 (1976).

Lipocarpha monostachya R.Gross & Mattf., Notizbl. Bot. Gart. Berlin-Dahlem 14: 189 (1938).
Tanzania to Zimbabwe. 23 ZAI 25 TAN 26 MLW ZIM. Ther.

Lipocarpha nana (A.Rich.) Cherm., Bull. Soc. Bot. France 71: 142 (1924).
Trop. & S. Africa, Madagascar. 22 GUI NGA 23 BUR CMN ZAI 24 ETH SUD 25 KEN TAN UGA 26 ANG MLW MOZ ZAM ZIM 27 LES NAM NAT OFS SWZ TVL 29 MDG. Ther.
**Fuirena nana* A.Rich., Tent. Fl. Abyss. 2: 497 (1850).
Lipocarpha pulcherrima Ridl., Trans. Linn. Soc. London, Bot. 2: 162 (1884). *Hypaelyptum pulcherrimum* (Ridl.) K.Schum. in H.G.A.Engler, Pflanzenw. Ost-Afrikas, C: 127 (1895). *Cyperus persquarrosus* T.Koyama, Bot. Mag. (Tokyo) 73: 438 (1960).
Lipocarpha atropurpurea Boeck., Beitr. Cyper. 1: 21 (1888).
Lipocarpha tenera Boeck., Beitr. Cyper. 1: 21 (1888).
Lipocarpha minima Cherm., Bull. Soc. Bot. France 68: 425 (1921 publ. 1922). *Cyperus unistamen* T.Koyama, Bot. Mag. (Tokyo) 73: 438 (1960).

Lipocarpha occidentalis (A.Gray) G.C.Tucker, J. Arnold Arbor. 68: 410 (1987).
W. U.S.A. 73 ORE WAS 76 CAL. Ther.
**Hemicarpha occidentalis* A.Gray, Proc. Amer. Acad. Arts 7: 391 (1868). *Scirpus occidentalis* (A.Gray) C.B.Clarke, Kew Bull., Addit. Ser. 8: 30 (1908), nom. illeg.

Lipocarpha perspicua S.S.Hooper, Kew Bull. 41: 424 (1986).
Angola. 26 ANG. Ther.

Lipocarpha prieuriana Steud., Syn. Pl. Glumac. 2: 130 (1855). *Cyperus prieurianus* (Steud.) T.Koyama, Bot. Mag. (Tokyo) 73: 438 (1960).
Trop. Africa. 22 BEN BKN GHA GUI MLI NGA NGR SEN TOG 23 CAF CMN 24 CHA ETH SUD 25 TAN 26 MLW ZAM ZIM. Ther. or hemicr.
Lipocarpha schweinfurthiana Boeck., Flora 62: 567 (1879).

Lipocarpha pygmaea J.Kern, Blumea 10: 638 (1960).
Indo-China. 41 CBD MYA THA. Ther.

Lipocarpha raynaleana Govind., Adansonia, n.s., 20: 369 (1981).
SW. India. 40 IND. Ther.

Lipocarpha reddyi S.S.Hooper, Kew Bull. 41: 427 (1986).
S. India. 40 IND. Ther.

Lipocarpha rehmannii (Ridl.) Goetgh., Wageningen Agric. Univ. Pap. 89(!): 64 (1989).
Kenya to S. Africa, Madagascar. 23 ZAI 25 KEN TAN 26 ANG MLW MOZ ZAM ZIM 27 BOT LES NAM NAT TVL 29 MDG (78) vrg. Ther.
**Scirpus rehmannii* Ridl., Trans. Linn. Soc. London, Bot. 2: 159 (1884). *Isolepis rehmannii* (Ridl.) Lye, Bot. Not. 124: 479 (1971). *Rikliella rehmannii* (Ridl.) J.Raynal, Adansonia, n.s., 13: 155 (1973).
Scirpus hystricoides B.Nord., Dinteria 11: 55 (1974).

Lipocarpha robinsonii J.Raynal, Adansonia, n.s., 7: 81 (1967).
S. Trop. Africa. 26 ANG ZAM. Hemicr. or rhizome geophyte.

Lipocarpha salzmanniana Steud., Syn. Pl. Glumac. 2: 129 (1855).
NC. Mexico to Brazil, Cuba. 79 MXG 80 BLZ COS HON NIC PAN 81 CUB 82 GUY VEN 83 CLM 84 BZC BZE BZN. Ther.
Lipocarpha cochleata Griseb., Cat. Pl. Cub.: 241 (1866).

Lipocarpha schomburgkii (Friedl.) G.C.Tucker, J. Arnold Arbor. 68: 410 (1987).
S. Venezuela to Guyana. 82 GUY VEN. Ther.
**Hemicarpha schomburgkii* Friedl., Amer. J. Bot. 28: 858 (1941).

Lipocarpha squarrosa (L.) Goetgh., Wageningen Agric. Univ. Pap. 89(1): 71 (1989).
Indian Subcontinent to S. China and Pen. Malaysia. 36 CHS 40 ASS EHM IND NEP PAK SRL WHM 41 CBD MYA THA VIE 42 MLY (78) fla. Ther.
**Scirpus squarrosus* L., Mant. Pl. 2: 181 (1771). *Isolepis squarrosa* (L.) Roem. & Schult., Syst. Veg. 2: 111 (1817). *Rikliella squarrosa* (L.) J.Raynal, Adansonia, n.s., 13: 154 (1973).
Scirpus squamosus Link, Jahrb. Gewächsk. 1(3): 79 (1820), orth. var.
Scirpus sinensis Kunth, Enum. Pl. 2: 202 (1837).
Ascolepis tenuior Steud., Syn. Pl. Glumac. 2: 105 (1855).

Lipocarpha thermalis J.Raynal ex Goetgh., Wageningen Agric. Univ. Pap. 89(1): 73 (1989).
S. Zaïre. 23 ZAI. Ther.

Synonyms:
Lipocarpha argentea (Vahl) R.Br. = ***Lipocarpha chinensis*** (Osbeck) J.Kern
Lipocarpha atra var. *barteri* (C.B.Clarke) J.Raynal = ***Lipocarpha barteri*** C.B.Clarke
Lipocarpha atropurpurea Boeck. = ***Lipocarpha nana*** (A.Rich.) Cherm.
Lipocarpha cochleata Griseb. = ***Lipocarpha salzmanniana*** Steud.
Lipocarpha debilis Ridl. = ***Lipocarpha chinensis*** (Osbeck) J.Kern
Lipocarpha foliosa Miq. = ***Remirea maritima*** Aubl.
Lipocarpha glomerata Nees = ***Lipocarpha maculata*** (Michx.) Torr.
Lipocarpha humboldtiana Nees = ***Lipocarpha maculata*** (Michx.) Torr.
Lipocarpha isolepis (Nees) R.W.Haines = ***Lipocarpha hemisphaerica*** (Roth) Goetgh.
Lipocarpha laevigata (Roxb.) Nees = ***Lipocarpha chinensis*** (Osbeck) J.Kern
Lipocarpha minima Cherm. = ***Lipocarpha nana*** (A.Rich.) Cherm.
Lipocarpha monocephala Turrill = ***Lipocarpha hemisphaerica*** (Roth) Goetgh.
Lipocarpha multibracteata C.B.Clarke = ***Ascolepis pusilla*** Ridl. var. ***pusilla***
Lipocarpha paradoxa Cherm. = ***Alinula paradoxa*** (Cherm.) Goetgh. & Vorster
Lipocarpha prieuriana var. *crassicuspis* J.Raynal = ***Lipocarpha crassicuspis*** (J.Raynal) Goetgh.
Lipocarpha pulcherrima Ridl. = ***Lipocarpha nana*** (A.Rich.) Cherm.
Lipocarpha purpureolutea Ridl. = ***Lipocarpha albiceps*** Ridl.
Lipocarpha rautanenii Boeck. ex Schinz = ***Lipocarpha hemisphaerica*** (Roth) Goetgh.
Lipocarpha schweinfurthiana Boeck. = ***Lipocarpha prieuriana*** Steud.
Lipocarpha sellowiana Kunth = ***Lipocarpha maculata*** (Michx.) Torr.
Lipocarpha senegalensis (Lam.) T.Durand & H.Durand = ***Lipocarpha chinensis*** (Osbeck) J.Kern
Lipocarpha sphacelata (Vahl) Kunth = ***Lipocarpha gracilis*** (Rich. ex Pers.) Nees
Lipocarpha sphacelata var. *gracilis* (Rich. ex Pers.) Boeck. =
Lipocarpha tenera Boeck. = ***Lipocarpha nana*** (A.Rich.) Cherm.
Lipocarpha triceps (Roxb.) Nees = ***Lipocarpha gracilis*** (Rich. ex Pers.) Nees
Lipocarpha zollingeriana Boeck. = ***Lipocarpha microcephala*** (R.Br.) Kunth

Lonchostylis

Lonchostylis Torr. = ***Rhynchospora*** Vahl

Loncoperis

Loncoperis Raf. = ***Carex*** L.
Loncoperis bromoides (Willd.) Raf. = ***Carex bromoides*** Willd.
Loncoperis scoparia (Willd.) Raf. = ***Carex scoparia*** Willd.
Loncoperis stipata (Muhl. ex Willd.) Raf. = ***Carex stipata*** Muhl. ex Willd.

Lophocarpus

Lophocarpus Boeck. = ***Schoenus*** L.
Lophocarpus bonia Boeck. = ***Schoenus falcatus*** R.Br.
Lophocarpus tonquinensis Boeck. = ***Schoenus falcatus*** R.Br.

Lophoschoenus

Lophoschoenus Stapf = ***Costularia*** C.B.Clarke
Lophoschoenus arundinaceus (Sol. ex Vahl) Stapf = ***Costularia arundinacea*** (Sol. ex Vahl) Kük.
Lophoschoenus chamaedendron Guillaumin = ***Costularia chamaedendron*** (Guillaumin) Kük.
Lophoschoenus conosus (C.B.Clarke) Stapf = ***Costularia comosa*** (C.B.Clarke) Kük.
Lophoschoenus elongatus (Kunth) H.Pfeiff. = ***Costularia elongata*** (Kunth) Kük.
Lophoschoenus falcifolius Däniker = ***Costularia xyridioides*** (Däniker) Kük.
Lophoschoenus fragilis Däniker = ***Costularia fragilis*** (Däniker) Kük.
Lophoschoenus hornei (C.B.Clarke) Stapf = ***Costularia hornei*** (C.B.Clarke) Kük.
Lophoschoenus montis-fontium Guillaumin = ***Costularia neocaledonica*** (C.B.Clarke) Rendle
Lophoschoenus neocaledonicus (C.B.Clarke) H.Pfeiff. = ***Costularia neocaledonica*** (C.B.Clarke) Rendle
Lophoschoenus stagnalis Däniker = ***Costularia stagnalis*** (Däniker) Kük.
Lophoschoenus urvilleanus (Gaudich. ex Boeck.) Stapf = ***Costularia pilisepala*** (Steud.) F.Kern
Lophoschoenus xyridioides (Däniker) Guillaumin = ***Costularia xyridioides*** (Däniker) Kük.

Loxanisa

Loxanisa Raf. = ***Carex*** L.
Loxanisa atrata (L.) Raf. = ***Carex atrata*** L.
Loxanisa gracillima (Schwein.) Raf. = ***Carex gracillima*** Schwein.

Loxotrema

Loxotrema Raf. = ***Carex*** L.
Loxotrema alpestris Raf. = ***Carex albicans*** Willd. ex Spreng. var. ***albicans***
Loxotrema capillaris (L.) Raf. = ***Carex capillaris*** L.
Loxotrema castanea (Wahlenb.) Raf. = ***Carex castanea*** Wahlenb.
Loxotrema cuspidata (Host) Raf. = ***Carex flacca*** subsp. ***erythrostachys*** (Hoppe) Holub
Loxotrema davisii (Schwein. & Torr.) Raf. = ***Carex davisii*** Schwein. & Torr.
Loxotrema halseyana (Dewey) Raf. = ***Carex polymorpha*** Muhl.
Loxotrema hitchcockiana (Dewey) Raf. = ***Carex hitchcockiana*** Dewey
Loxotrema leptostachys (Ehrh. ex L.f.) Raf. = ***Carex strigosa*** Huds.
Loxotrema lusitanica (Willd.) Raf. = ***Carex paniculata*** subsp. ***lusitanica*** (Willd.) Maire
Loxotrema pilosa (Scop.) Raf. = ***Carex pilosa*** Scop.
Loxotrema triflora (Schkuhr) Raf. = ***Carex depauperata*** Curtis ex With
Loxotrema tuberosa (Degl.) Raf. = ***Carex distachya*** Desf. var. ***distachya***
Loxotrema vestita (Willd.) Raf. = ***Carex vestita*** Willd.

Lyprolepis

Lyprolepis Steud. = ***Kyllinga*** Rottb.
Lyprolepis denudata Steud. = ***Kyllinga vaginata*** Lam.

Machaerina

Machaerina Vahl, Enum. Pl. 2: 238 (1805).
Tanzania, W. Indian Ocean to Pacific, Trop. America. 25 TAN 29 COM MAU MDG REU SEY 36+ CHH 38 NNS OGA 40 ASS BAN IND SRL 41 VIE 42 BOR JAW MLY MOL PHI SUL SUM 43 BIS NWG SOL 50 NFK NSW NTA QLD SOA TAS VIC WAU 51 CTM NZN NZS 60 FIJ NWC SAM VAN 61 MRQ SCI TUB 62 CRL MRN 63 HAW 81 CUB HAI LEE PUE WIN 82 GUY 84+ BZL BZS 85 JNF.
51 Species
Baumea Gaudich., Voy. Uranie: 416 (1826).
Vincentia Gaudich., Voy. Uranie: 417 (1829).
Chapelliera Nees, Linnaea 9: 298 (1834).
Trachyrhynchium Nees, Nov. Actorum Acad. Caes. Leop.-Carol. Nat. Cur. 19(Suppl. 1): 113 (1843).
Terobera Steud., Syn. Pl. Glumac. 2: 164 (1855).
Agylla Phil., Anales Univ. Chile 1865(1): 643 (1865).

Machaerina acuta (Labill.) J.Kern, Acta Bot. Neerl. 8: 266 (1959).
E. & S. Australia. 50 NSW QLD SOA TAS VIC WAU. Hemicr. or rhizome geophyte.
**Schoenus acutus* Labill., Nov. Holl. Pl. 1: 18 (1805). *Cladium schoenoides* R.Br., Prodr.: 237 (1810), nom. illeg. *Baumea schoenoides* (R.Br.) Boeck., Linnaea 38: 246 (1874). *Gahnia schoenoides* (R.Br.) F.Muell., Key Vict. Pl. 1: 456 (1888). *Machaerina schoenoides* (R.Br.) T.Koyama, Bot. Mag. (Tokyo) 69: 65 (1956). *Cladium acutum* (Labill.) Poir. in G.-F.Cuvier, Dict. Sci. Nat., ed. 2, 9: 344 (1817). *Mariscus acutus* (Labill.) Kuntze, Revis. Gen. Pl. 2: 755 (1891). *Baumea acuta* (Labill.) Palla, Allg. Bot. Z. Syst. 8: 69 (1902).
Schoenus falcatus Nees, Linnaea 9: 298 (1834), nom. inval.

Machaerina anceps (Poir.) Bojer, Hortus Maurit.: 385 (1837).
W. Indian Ocean. 29 MAU MDG REU. Hemicr. or rhizome geophyte.
**Scirpus anceps* Poir. in J.B.A.M.de Lamarck, Encycl. 6: 782 (1805). *Lepidosperma anceps* (Poir.) Willd. ex Link, Jahrb. Gewächsk. 1(3): 75 (1820). *Vincentia anceps* (Poir.) Kunth, Enum. Pl. 2: 315 (1837). *Mariscus anceps* (Poir.) Kuntze, Revis. Gen. Pl. 2: 755 (1891).

Machaerina angustifolia (Gaudich.) T.Koyama, Bot. Mag. (Tokyo) 69: 62 (1956).
C. Pacific. 61 SCI 63 HAW. Hemicr. or rhizome geophyte.
Machaerina restioides Gaudich., Voy. Uranie: 104 (1827), nom. illeg.
**Vincentia angustifolia* Gaudich., Voy. Uranie: 417 (1829). *Mariscus angustifolius* (Gaudich.) Kuntze, Revis. Gen. Pl. 2: 755 (1891). *Cladium angustifolium* (Gaudich.) Benth. & Hook.f. ex Drake, Ill. Fl. Ins. Pacif.: 334 (1892). *Cladium vincentia* C.B.Clarke, Ill. Cyper.: t. 84 (1909).

Machaerina arthrophylla (Nees) T.Koyama, Bot. Mag. (Tokyo) 69: 62 (1956).
Australia. 50 NSW NTA QLD SOA TAS VIC WAU. Hemicr. or rhizome geophyte.
**Chapelliera arthrophylla* Nees in J.G.C.Lehmann, Pl. Preiss. 2: 77 (1846). *Baumea arthrophylla* (Nees) Boeck., Linnaea 38: 242 (1874). *Cladium arthrophyllum* (Nees) F.Muell., Fragm. 9: 14 (1875). *Mariscus arthrophyllus* (Nees) Kuntze, Revis. Gen. Pl. 2: 755 (1891).

Machaerina articulata (R.Br.) T.Koyama, Bot. Mag. (Tokyo) 69: 62 (1956).
New Guinea to New Zealand. 43 NWG 50 NSW QLD SOA TAS VIC WAU 51 NZN 60 NWC VAN. Hemicr. or rhizome geophyte.
**Cladium articulatum* R.Br., Prodr.: 237 (1810). *Chapelliera articulata* (R.Br.) Nees in J.G.C.Lehmann, Pl. Preiss. 2: 76 (1846). *Gahnia articulata* (R.Br.) F.Muell., Key Vict. Pl. 1: 456 (1888). *Mariscus articulatus* (R.Br.) Kuntze, Revis. Gen. Pl. 2: 755 (1891). *Baumea articulata* (R.Br.) S.T.Blake, Contr. Queensland Herb. 8: 28 (1969).
Baumea loculata Boeck., Linnaea 38: 243 (1874).

Machaerina aspericaulis (Kük.) T.Koyama, Bot. Mag. (Tokyo) 69: 62 (1956).
N. Borneo (Mt. Kinabalu). 42 BOR. Hemicr. or rhizome geophyte.
**Cladium aspericaule* Kük., Bull. Jard. Bot. Buitenzorg, III, 16: 309 (1940). *Baumea aspericaulis* (Kük.) S.T.Blake, Contr. Queensland Herb. 8: 28 (1969).

Machaerina austrobrasiliensis M.T.Strong, Novon 7: 317 (1997).
S. Brazil. 84 BZS. Hemicr. or rhizome geophyte.

Machaerina ayangannensis M.T.Strong, Novon 7: 313 (1997).
Guyana. 82 GUY. Hemicr. or rhizome geophyte.

Machaerina bidwellii (Stapf ex Setch.) T.Koyama, Bot. Mag. (Tokyo) 69: 62 (1956).
Society Is. 61 SCI. Hemicr. or rhizome geophyte.
**Vincentia bidwellii* Stapf ex Setch., Dept. Marine Biol. Carnegie Inst. Washington 20: 111 (1924). *Mariscus bidwellii* (Stapf ex Setch.) J.W.Moore,

Bernice P. Bishop Mus. Bull. 102: 20 (1933). *Cladium bidwellii* (Stapf ex Setch.) Kük., Repert. Spec. Nov. Regni Veg. 51: 148 (1942).

Machaerina complanata (Berggr.) T.Koyama, Bot. Mag. (Tokyo) 69: 63 (1956).
New Zealand North I. 51 NZN. Hemicr.
**Cladium complanatum* Berggr., Minneskr. Fisiog. Sallks. Lund 8: 23 (1877). *Mariscus complanatus* (Berggr.) Fernald, Rhodora 25: 53 (1923). *Baumea complanata* (Berggr.) S.T.Blake, Contr. Queensland Herb. 8: 29 (1969).

Machaerina cubensis (Kük.) T.Koyama, Bot. Mag. (Tokyo) 69: 63 (1956).
E. Cuba. 81 CUB. Hemicr. or rhizome geophyte.
**Cladium cubense* Kük., Repert. Spec. Nov. Regni Veg. 23: 213 (1926).

Machaerina deplanchei (Boeck.) T.Koyama, Bot. Mag. (Tokyo) 69: 63 (1956).
New Caledonia. 60 NWC. Hemicr. or rhizome geophyte.
**Baumea deplanchei* Boeck., Flora 61: 143 (1878). *Cladium deplanchei* (Boeck.) C.B.Clarke, Bot. Jahrb. Syst. 39: 26 (1906). *Mariscus deplanchei* (Boeck.) Schinz & Guillaumin in C.F.Sarasin & J.Roux, Nova Caledonia, Bot. 1: 121 (1920).

Machaerina disticha (C.B.Clarke) T.Koyama, Bot. Mag. (Tokyo) 69: 63 (1956).
Borneo to NW. New Guinea. 42 BOR MOL PHI SUL 43 NWG. Hemicr. or rhizome geophyte.
**Cladium distichum* C.B.Clarke, Philipp. J. Sci., C 2: 102 (1907). *Mariscus distichus* (C.B.Clarke) Fernald, Rhodora 25: 53 (1923). *Baumea disticha* (C.B.Clarke) S.T.Blake, Contr. Queensland Herb. 8: 29 (1969).
Cladium micranthes C.B.Clarke, Kew Bull., Addit. Ser. 8: 46 (1908). *Mariscus micranthes* (C.B. Clarke) Fernald, Rhodora 25: 54 (1923). *Machaerina micranthes* (C.B.Clarke) T.Koyama, Bot. Mag. (Tokyo) 69: 64 (1956).
Cladium philippinense Merr., Philipp. J. Sci., C 5: 171 (1910). *Mariscus geniculatus* Fernald, Rhodora 25: 53 (1923). *Machaerina philippinensis* (Merr.) T.Koyama, Bot. Mag. (Tokyo) 69: 65 (1956).

Machaerina ekmanii (Kük.) T.Koyama, Bot. Mag. (Tokyo) 69: 63 (1956).
Haiti. 81 HAI. Hemicr. or rhizome geophyte.
**Cladium ekmanii* Kük., Repert. Spec. Nov. Regni Veg. 23: 213 (1926).

Machaerina ensigera (Hance) T.Koyama, Bot. Mag. (Tokyo) 69: 63 (1956).
China. 36+. Hemicr. or rhizome geophyte.
**Cladium ensigerum* Hance, J. Bot. 23: 80 (1885). *Mariscus ensiger* (Hance) Fernald, Rhodora 25: 53 (1923). *Baumea ensigera* (Hance) S.T.Blake, Contr. Queensland Herb. 8: 29 (1969).

Machaerina falcata (Nees) T.Koyama, Bot. Mag. (Tokyo) 69: 63 (1956).
Indo-China to Pacific. 41 VIE 42 BOR PHI SUL SUM 43 NWG SOL 60 FIJ SAM VAN 62 CRL 63 HAW. Hemicr. or rhizome geophyte.
**Baumea falcata* Nees, Hooker's J. Bot. Kew Gard. Misc. 6: 29 (1854). *Cladium falcatum* (Nees) C.B.Clarke, Kew Bull., Addit. Ser. 8: 46 (1908). *Vincentia falcata* (Nees) Stapf, J. Linn. Soc., Bot. 42: 179 (1914). *Mariscus falcatus* (Nees) Fernald, Rhodora 25: 53 (1923).
Cladium samoense C.B.Clarke ex Stapf, Trans. Linn. Soc. London, Bot. 4: 245 (1894). *Vincentia samoensis* (C.B.Clarke ex Stapf) Stapf, J. Linn. Soc., Bot. 42: 178 (1914). *Mariscus samoensis* (C.B. Clarke ex Stapf) Fernald, Rhodora 25: 54 (1923).
Cladium samoense var. *uniseta* C.B.Clarke ex Stapf, Trans. Linn. Soc. London, Bot. 4: 246 (1894).
Cladium latifolium Merr., Philipp. J. Sci., C 2: 262 (1907), nom. illeg. *Mariscus platyphyllus* Fernald, Rhodora 25: 54 (1923). *Machaerina latifolia* T.Koyama, Bot. Mag. (Tokyo) 69: 64 (1956), nom. illeg.
Cladium falcatum var. *sumatrense* C.B.Clarke, Kew Bull., Addit. Ser. 8: 46 (1908).
Vincentia malesiaca Stapf, J. Linn. Soc., Bot. 42: 178 (1914). *Machaerina malesiaca* (Stapf) T.Koyama, Bot. Mag. (Tokyo) 69: 65 (1956).
Vincentia dissoluta Stapf, Dept. Marine Biol. Carnegie Inst. Washington 20: 111 (1924).
Cladium falcatum var. *celebicum* Kük., Repert. Spec. Nov. Regni Veg. 51: 146 (1942).
Cladium iris Ohwi, Bot. Mag. (Tokyo) 56: 208 (1942). *Machaerina iris* (Ohwi) T.Koyama, Bot. Mag. (Tokyo) 69: 64 (1956).
Cladium ponapense Ohwi, J. Jap. Bot. 18: 136 (1942). *Machaerina ponapensis* (Ohwi) Ohwi & T.Koyama, Bot. Mag. (Tokyo) 69: 65 (1956).

Machaerina ficticia (Hemsl.) T.Koyama, Bot. Mag. (Tokyo) 69: 63 (1956).
Brazil. 84 BZL. Hemicr. or rhizome geophyte.
**Cladium ficticium* Hemsl., Rep. Challenger, Bot. 3: 59 (1884). *Mariscus ficticius* (Hemsl.) Fernald, Rhodora 25: 53 (1923). *Machaerina scirpoidea* subsp. *ficticia* (Hemsl.) T.Koyama, Mem. New York Bot. Gard. 23: 89 (1972).

Machaerina filifolia Griseb., Cat. Pl. Cub.: 242 (1866). *Cladium filifolium* (Griseb.) Benth. & Hook.f., Gen. Pl. 3: 1066 (1883). *Mariscus filifolius* (Griseb.) Kuntze, Revis. Gen. Pl. 2: 755 (1891).
E. Cuba. 81 CUB. Hemicr. or rhizome geophyte.

Machaerina flexuosa (Boeck.) J.Kern, Acta Bot. Neerl. 8: 266 (1959).
Tanzania, Comoros, Madagascar. 25 TAN 29 COM MDG. Hemicr.
**Baumea flexuosa* Boeck., Abh. Naturwiss. Vereine Bremen 7: 39 (1880). *Cladium flexuosum* (Boeck.) C.B.Clarke in T.A.Durand & H.Schinz, Consp. Fl. Afric. 5: 665 (1894).

subsp. ***flexuosa***
Comoros, Madagascar. 29 COM MDG. Hemicr.
Chapelliera iridifolia Nees, Linnaea 9: 298 (1834). *Trachyrhynchium iridifolium* (Nees) Nees, Nov. Actorum Acad. Caes. Leop.-Carol. Nat. Cur. 19(Suppl. 1): 114 (1843). *Baumea iridifolia* (Nees) Boeck., Linnaea 38: 244 (1874).

subsp. ***laevinux*** (J.Raynal) Lye, Nordic J. Bot. 3: 243 (1983).
Madagascar. 29 MDG. Hemicr. or rhizome geophyte.
**Baumea iridifolia* subsp. *laevinux* J.Raynal, Adansonia, n.s., 12: 108 (1972).

subsp. ***polyanthemum*** (Kük.) Lye, Nordic J. Bot. 3: 243 (1983).
Tanzania (Rondo Plateau). 25 TAN. Hemicr.
**Cladium flexuosum* var. *polyanthemum* Kük., Repert. Spec. Nov. Regni Veg. 51: 160 (1942).

Machaerina glomerata (Gaudich.) T.Koyama, Bot. Mag. (Tokyo) 69: 63 (1956).
Malesia to Ogasawara-shoto. 38 OGA 42 BOR MOL PHI SUL 43 BIS NWG SOL 62 CRL. Hemicr. or rhizome geophyte.
**Baumea glomerata* Gaudich., Voy. Uranie: 416 (1829). *Cladium glomeratum* (Gaudich.) H.Pfieff., Repert. Spec. Nov. Regni Veg. 23: 349 (1927), nom. illeg.
Lepidosperma waigiuensis Steud., Syn. Pl. Glumac. 2: 158 (1855), provisional synonym.
Cladium colpodes Lauterb. in K.M.Schumann & C.A.G.Lauterbach, Fl. Schutzgeb. Südsee, Nachtr.: 59 (1905). *Mariscus colpodes* (Lauterb.) Fernald, Rhodora 25: 53 (1923). *Machaerina mariscoides* subsp. *colpodes* (Lauterb.) T.Koyama, Micronesica 1: 103 (1964).
Cladium gaudichaudii W.Wight, Contr. U. S. Natl. Herb. 9: 230 (1905). *Mariscus gaudichaudii* (W.Wight) Fernald, Rhodora 25: 53 (1923). *Machaerina gaudichaudii* (W.Wight) T.Koyama, Bot. Mag. (Tokyo) 69: 64 (1956).
Cladium globiceps C.B.Clarke, Kew Bull., Addit. Ser. 8: 46 (1908). *Mariscus globiceps* (C.B.Clarke) Fernald, Rhodora 25: 53 (1923).
Cladium juncoides Elmer, Leafl. Philipp. Bot. 3: 854 (1910). *Machaerina juncoides* (Elmer) T.Koyama, Bot. Mag. (Tokyo) 69: 64 (1956).
Cladium boninsimae Nakai, Bot. Mag. (Tokyo) 25: 223 (1911). *Mariscus boninsimae* (Nakai) Fernald, Rhodora 25: 51 (1923). *Machaerina boninsimae* (Nakai) T.Koyama, Bot. Mag. (Tokyo) 69: 62 (1956).
Cladium sinuatum Ridl., Trans. Linn. Soc. London, Bot. 9: 243 (1916). *Machaerina sinuata* (Ridl.) T.Koyama, Bot. Mag. (Tokyo) 69: 65 (1956).

Machaerina gunnii (Hook.f.) J.Kern, Acta Bot. Neerl. 8: 266 (1959).
E. New Guinea, SE. Queensland to Tasmania. 43 NWG 50 NSW QLD SOA TAS VIC. Hemicr. or rhizome geophyte.
**Cladium gunnii* Hook.f., Fl. Tasman. 2: 95 (1858). *Gahnia gunnii* (Hook.f.) F.Muell., Key Vict. Pl. 1: 456 (1888). *Baumea gunnii* (Hook.f.) S.T.Blake, Contr. Queensland Herb. 8: 27 (1969).
Cladium laxiflorum Hook.f., Fl. Tasman. 2: 95 (1858).
Cladium gunnii var. *brevipaniculatum* Kük., Bot. Jahrb. Syst. 69: 260 (1938). *Cladium brevipaniculatum* (Kük.) Kük., Repert. Spec. Nov. Regni Veg. 51: 176 (1942). *Machaerina brevipaniculata* (Kük.) T.Koyama, Bot. Mag. (Tokyo) 69: 62 (1956).

Machaerina hirta Boeck., Flora 58: 117 (1875). *Cladium hirtum* (Boeck.) C.B.Clarke, Kew Bull., Addit. Ser. 8: 46 (1908). *Mariscus hirtus* (Boeck.) Fernald, Rhodora 25: 53 (1923).
Samoa. 60 SAM. Hemicr. or rhizome geophyte.

Machaerina huttonii (Kirk) T.Koyama, Bot. Mag. (Tokyo) 69: 64 (1956).
New Zealand. 51 NZN NZS. Hemicr. or rhizome geophyte.
**Cladium huttonii* Kirk, Trans. & Proc. New Zealand Inst. 9: 551 (1877). *Mariscus huttonii* (Kirk) Fernald, Rhodora 25: 53 (1923). *Baumea huttonii* (Kirk) S.T.Blake, Contr. Queensland Herb. 8: 29 (1969).

Machaerina insularis (Benth.) T.Koyama, Bot. Mag. (Tokyo) 69: 64 (1956).
Lord Howe I. (Mt. Gower, Mt. Lidgbird). 50 NFK. Hemicr.
**Cladium insulare* Benth., Fl. Austral. 7: 403 (1878). *Gahnia insularis* (Benth.) F.Muell., Syst. Census Austral. Pl. 2: 216 (1889). *Mariscus insularis* (Benth.) Kuntze, Revis. Gen. Pl. 2: 755 (1891).

Machaerina involuta H.St.John, Nordic J. Bot. 4: 57 (1984).
Tubuai Is. 61 TUB. Hemicr. or rhizome geophyte.

Machaerina iridifolia (Bory) T.Koyama, Bot. Mag. (Tokyo) 69: 64 (1956).
Mascarenes. 29 MAU. Hemicr. or rhizome geophyte.
**Scirpus iridifolius* Bory, Voy. îles Afrique 2: 94 (1804). *Lepidosperma iridifolium* (Bory) Willd. ex Link, Jahrb. Gewächsk. 1(3): 75 (1820). *Cladium iridifolium* (Bory) Baker, Fl. Mauritius: 424 (1877). *Mariscus iridifolius* (Bory) Kuntze, Revis. Gen. Pl. 2: 755 (1891).
Vincentia latifolia Kunth, Enum. Pl. 2: 314 (1837). *Cladium latifolium* (Kunth) Benth. & Hook.f. ex Drake, Ill. Fl. Ins. Pacif.: 334 (1892).
Elynanthus ensifolius Boeck., Linnaea 38: 264 (1874). *Cladium ensifolium* (Boeck.) Benth. & Hook.f. ex B.D.Jacks., Index Kew. 1: 551 (1893). *Baumea ensifolia* (Boeck.) Palla, Denkschr. Kaiserl. Akad. Wiss., Wien. Math.-Naturwiss. Kl. 79: 194 (1908). *Machaerina ensifolia* (Boeck.) T.Koyama, Bot. Mag. (Tokyo) 69: 63 (1956).
Vincentia ensifolia Boeck. ex C.B.Clarke in T.A.Durand & H.Schinz, Consp. Fl. Afric. 5: 665 (1894), nom. inval.

Machaerina juncea (R.Br.) T.Koyama, Bot. Mag. (Tokyo) 69: 64 (1956).
New Caledonia, E. & S. Australia to New Zealand. 50 NFK NSW QLD SOA TAS VIC WAU 51 NZN NZS 60 NWC. Hemicr. or rhizome geophyte.
**Cladium junceum* R.Br., Prodr.: 237 (1810). *Chapelliera juncea* (R.Br.) Nees in J.G.C.Lehmann, Pl. Preiss. 2: 76 (1846). *Gahnia juncea* (R.Br.) F.Muell., Key Vict. Pl. 1: 456 (1888). *Mariscus junceus* (R.Br.) Kuntze, Revis. Gen. Pl. 2: 755 (1891). *Baumea juncea* (R.Br.) Palla, Allg. Bot. Z. Syst. 15: 113 (1909).
Lepidosperma colensoi Boeck., Linnaea 38: 328 (1874).
Mariscus ouveanus Däniker, Vierteljahrsschr. Naturf. Ges. Zürich 77(19): 83 (1932). *Cladium ouveanum* (Däniker) Guillaumin, Bull. Soc. Bot. France 85: 45 (1938).

Machaerina lamii (Kük.) J.Kern, Acta Bot. Neerl. 8: 266 (1959).
W. New Guinea (Mt. Doorman). 43 NWG. Hemicr. or rhizome geophyte.
**Cladium lamii* Kük., Bull. Jard. Bot. Buitenzorg, III, 16: 309 (1940).

Machaerina lavarum (Poir.) Bojer, Hortus Maurit.: 385 (1837).
W. Indian Ocean. 29 MAU MDG REU SEY. Hemicr. or rhizome geophyte.
**Scirpus lavarum* Poir. in J.B.A.M.de Lamarck, Encycl. 6: 783 (1805). *Mariscus lavarum* (Poir.) Kuntze, Revis. Gen. Pl. 2: 755 (1891). *Cladium lavarum* (Poir.) Cordem., Fl. Réunion: 133 (1895).

Machaerina laxa (Nees) T.Koyama, Bot. Mag. (Tokyo) 69: 64 (1956).
S. Australia. 50 SOA VIC WAU. Hemicr. or rhizome geophyte.

Chapelliera laxa Nees in J.G.C.Lehmann, Pl. Preiss. 2: 76 (1846). *Baumea laxa* (Nees) Boeck., Linnaea 38: 245 (1874). *Cladium laxum* (Nees) Benth., Fl. Austral. 7: 405 (1878). *Mariscus laxus* (Nees) Kuntze, Revis. Gen. Pl. 2: 755 (1891). *Cladium preissii* var. *laxum* (Nees) Kük., Repert. Spec. Nov. Regni Veg. 51: 152 (1942).

Cladium gracile J.M.Black, Fl. S. Austral. 4: 678 (1929). *Machaerina gracilis* (J.M.Black) T.Koyama, Bot. Mag. (Tokyo) 69: 63 (1956).

Machaerina macrophylla (Boeck.) T.Koyama, Bot. Mag. (Tokyo) 69: 64 (1956).
Brazil. 84+. Hemicr. or rhizome geophyte.
Vincentia macrophylla Boeck., Allg. Bot. Z. Syst. 2: 112 (1896). *Cladium ficticium* var. *macrophyllum* (Boeck.) Kük., Repert. Spec. Nov. Regni Veg. 51: 143 (1942).

Machaerina maingayi (C.B.Clarke) T.Koyama, Bot. Mag. (Tokyo) 69: 64 (1956).
Vietnam to Pen. Malaysia. 41 VIE 42 MLY. Hemicr. or rhizome geophyte.
Cladium maingayi C.B.Clarke in J.D.Hooker, Fl. Brit. India 6: 674 (1894). *Vincentia maingayi* (C.B.Clarke) Stapf, J. Linn. Soc., Bot. 42: 179 (1914). *Mariscus maingayi* (C.B.Clarke) Fernald, Rhodora 25: 53 (1923).

Machaerina mariscoides (Gaudich.) J.Kern, Acta Bot. Neerl. 8: 266 (1959).
New Guinea to Pacific. 43 NWG SOL 61 MRQ SCI 62 CRL MRN 63 HAW. Hemicr. or rhizome geophyte.
Baumea mariscoides Gaudich., Voy. Uranie: 417 (1829). *Cladium mariscoides* (Gaudich.) Fern.-Vill. in F.M.Blanco, Fl. Filip., ed. 3, 4(13A): 309 (1880), nom. illeg.
Baumea meyenii Kunth, Enum. Pl. 2: 314 (1837). *Mariscus meyenii* (Kunth) Kuntze, Revis. Gen. Pl. 2: 755 (1891). *Cladium meyenii* (Kunth) Benth. & Hook.f. ex Drake, Ill. Fl. Ins. Pacif.: 334 (1892). *Machaerina mariscoides* subsp. *meyenii* (Kunth) T.Koyama, Micronesica 1: 104 (1964).
Mariscus marianus Kuntze, Revis. Gen. Pl. 2: 755 (1891).
Mariscus societensis J.W.Moore, Bernice P. Bishop Mus. Bull. 102: 21 (1933).

Machaerina milnei (C.B.Clarke) T.Koyama, Bot. Mag. (Tokyo) 69: 64 (1956).
Vanuatu. 60 VAN. Hemicr. or rhizome geophyte.
Cladium milnei C.B.Clarke, Kew Bull., Addit. Ser. 8: 46 (1908). *Mariscus milnei* (C.B.Clarke) Fernald, Rhodora 25: 54 (1923). *Baumea milnei* (C.B.Clarke) S.T.Blake, Contr. Queensland Herb. 8: 29 (1969).

Machaerina montana (J.Raynal) Lye, Nordic J. Bot. 3: 243 (1983).
Madagascar. 29 MDG. Hemicr. or rhizome geophyte.
Baumea montana J.Raynal, Adansonia, n.s., 12: 110 (1972).

Machaerina monticola (Guillaumin) T.Koyama, Bot. Mag. (Tokyo) 69: 64 (1956).
Vanuatu. 60 VAN. Hemicr. or rhizome geophyte.
Vincentia monticola Guillaumin, Bull. Mus. Natl. Hist. Nat., II, 9: 304 (1937).

Machaerina muelleri (C.B.Clarke) T.Koyama, Bot. Mag. (Tokyo) 69: 64 (1956).
SE. Queensland to NE. New South Wales. 50 NSW QLD. Hemicr. or rhizome geophyte.
Cladium schoenoides var. *elongatum* Benth., Fl. Austral. 7: 407 (1878).
Cladium muelleri C.B.Clarke, Kew Bull., Addit. Ser. 8: 47 (1908). *Mariscus muelleri* (C.B.Clarke) Fernald, Rhodora 25: 54 (1923). *Baumea muelleri* (C.B.Clarke) S.T.Blake, Contr. Queensland Herb. 8: 29 (1969).

Machaerina myriantha (Chun & F.C.How) Y.C.Tang, in Fl. Hainan. 4: 537 (1977).
Hainan. 36 CHH. Hemicr. or rhizome geophyte.
Cladium myrianthum Chun & F.C.How, Acta Phytotax. Sin. 7: 83 (1958).

Machaerina nuda (Steud.) J.Kern, Acta Bot. Neerl. 8: 266 (1959).
E. Australia. 50 NSW QLD. Hemicr. or rhizome geophyte.
Schoenus nudus Steud., Syn. Pl. Glumac. 2: 165 (1855). *Cladium nudum* (Steud.) Boeck., Linnaea 38: 236 (1874). *Mariscus nudus* (Steud.) Kuntze, Revis. Gen. Pl. 2: 755 (1891). *Baumea nuda* (Steud.) S.T.Blake, Contr. Queensland Herb. 8: 28 (1969).
Cladium nudum var. *rigidulum* Kük., Repert. Spec. Nov. Regni Veg. 51: 180 (1942).

Machaerina nukuhivensis (F.Br.) T.Koyama, Bot. Mag. (Tokyo) 69: 65 (1956).
Marquesas. 61 MRQ. Hemicr. or rhizome geophyte.
Cladium nukuhivense F.Br., Bernice P. Bishop Mus. Bull. 84: 111 (1931).

Machaerina preissii (Nees) L.A.S.Johnson & Koyama, Naturaliste Canad. 88: 249 (1961).
SW. Australia. 50 WAU. Hemicr. or rhizome geophyte.
Baumea preissii Nees in J.G.C.Lehmann, Pl. Preiss. 2: 75 (1846).
Cladium latissimum F.Muell., Fragm. 9: 15 (1875). *Machaerina latissima* (F.Muell.) T.Koyama, Bot. Mag. (Tokyo) 69: 64 (1956).
Cladium drummondii C.B.Clarke, Kew Bull., Addit. Ser. 8: 91 (1908). *Mariscus xiphioides* Fernald, Rhodora 25: 54 (1923).

Machaerina raiateensis (J.W.Moore) S.L.Welsh, Fl. Societensis: 320 (1998).
Society Is. (Raiatea). 61 SCI. Hemicr.
Mariscus raiateensis J.W.Moore, Bernice P. Bishop Mus. Bull. 226: 4 (1963).

Machaerina restioides (Sw.) Vahl, Enum. Pl. 2: 238 (1805).
Caribbean. 81 CUB HAI LEE PUE WIN. Hemicr. or rhizome geophyte.
Schoenus restioides Sw., Prodr.: 19 (1788). *Mariscus restiodes* (Sw.) Kuntze, Revis. Gen. Pl. 2: 755 (1891). *Cladium restioides* (Sw.) Benth. ex C.B.Clarke in I.Urban, Symb. Antill. 2: 135 (1900).

subsp. ***effusa*** (Griseb.) T.Koyama, Mem. New York Bot. Gard. 23: 89 (1972).
Cuba to Hispaniola. 81 CUB HAI. Hemicr. or rhizome geophyte.
Machaerina restioides var. *effusa* Griseb., Mem. Amer. Acad. Arts, n.s., 8: 535 (1860). *Mariscus restioides* var. *effusus* (Griseb.) Fernald, Rhodora 25: 54 (1923).

subsp. ***restioides***
Caribbean. 81 LEE PUE WIN. Hemicr. or rhizome geophyte.

Machaerina riparia (Nees) T.Koyama, Bot. Mag. (Tokyo) 69: 65 (1956).
SW. Australia. 50 WAU. Hemicr. or rhizome geophyte.
**Chapelliera riparia* Nees in J.G.C.Lehmann, Pl. Preiss. 2: 76 (1846). *Baumea riparia* (Nees) Boeck., Linnaea 38: 246 (1874). *Cladium riparium* (Nees) Benth., Fl. Austral. 7: 405 (1878). *Mariscus riparius* (Nees) Kuntze, Revis. Gen. Pl. 2: 755 (1891), nom. illeg. *Cladium arthrophyllum* var. *riparium* (Nees) Kük., Repert. Spec. Nov. Regni Veg. 51: 167 (1942).

Machaerina rubiginosa (Spreng.) T.Koyama, J. Fac. Sci. Univ. Tokyo, Sect. 3, Bot. 8: 123 (1961).
Trop. Asia to Nansei-shoto and New Zealand. 38 NNS OGA 40 ASS BAN IND SRL 41 VIE 42 JAW MLY MOL PHI SUM 43 NWG 50 NSW NTA QLD SOA TAS VIC WAU 51 CTM NZN NZS 60 NWC. Hemicr. or rhizome geophyte.
Schoenus rubiginosus Sol. ex G.Forst., Fl. Ins. Austr.: 89 (1786), nom. nud.
**Fuirena rubiginosa* Spreng., Mant. Prim. Fl. Hal.: 29 (1807). *Melancranis rubiginosa* (Spreng.) Spreng., Syst. Veg. 1: 236 (1824). *Baumea rubiginosa* (Spreng.) Boeck., Linnaea 38: 241 (1874). *Cladium rubiginosum* (Spreng.) Domin, Biblioth. Bot. 85: 476 (1915).
Cladium glomeratum R.Br., Prodr.: 237 (1810). *Chapelliera glomerata* (R.Br.) Nees in J.G.C.Lehmann, Pl. Preiss. 2: 76 (1846). *Baumea brownii* Boeck., Linnaea 38: 242 (1874). *Gahnia glomerata* (R.Br.) F.Muell., Key Vict. Pl. 1: 456 (1888). *Mariscus glomeratus* (R.Br.) Kuntze, Revis. Gen. Pl. 2: 755 (1891), nom. illeg.
Cladium dubium Spreng., Syst. Veg. 4(2): 21 (1827).
Baumea crassa Thwaites, Enum. Pl. Zeyl.: 353 (1864). *Mariscus crassus* (Thwaites) Kuntze, Revis. Gen. Pl. 2: 755 (1891). *Cladium crassum* (Thwaites) Kük., Bull. Jard. Bot. Buitenzorg, III, 16: 311 (1940). *Machaerina crassa* (Thwaites) T.Koyama, Bot. Mag. (Tokyo) 69: 63 (1956). *Machaerina rubiginosa* subsp. *crassa* (Thwaites) T.Koyama, Bot. Mag. (Tokyo) 93: 352 (1980).
Cladium glomeratum var. *subseptatum* Benth., Fl. Austral. 7: 404 (1878).
Cladium rubiginosum var. *subseptatum* (Benth.) Domin, Biblioth. Bot. 89: 476 (1915).
Cladium glomeratum f. *major* Kük. ex Carse, Trans. & Proc. New Zealand Inst. 48: 240 (1916).
Cladium brevistigma Nakai ex Tuyama, Bot. Mag. (Tokyo) 49: 509 (1935). *Machaerina brevistigma* (Nakai ex Tuyama) T.Koyama, Bot. Mag. (Tokyo) 69: 63 (1956).
Cladium rubiginosum var. *subriparium* Kük., Repert. Spec. Nov. Regni Veg. 51: 173 (1942).
Cladium nipponense Ohwi, Mem. Coll. Sci. Kyoto Imp. Univ., Ser. B, Biol. 18(1): 11 (1944). *Machaerina nipponense* (Ohwi) Ohwi & T.Koyama, Bot. Mag. (Tokyo) 69: 65 (1956).

Machaerina scirpoidea (Steud.) T.Koyama, Bot. Mag. (Tokyo) 69: 65 (1956).
Juan Fernández Is. (Robinson Crusoe I.). 85 JNF. Hemicr.
**Terobera scirpoidea* Steud., Syn. Pl. Glumac. 2: 164 (1855). *Vincentia scirpoidea* (Steud.) Boeck., Linnaea 38: 250 (1874). *Cladium scirpoideum* (Steud.) Benth. & Hook.f. ex Hemsl., Rep. Challenger, Bot. 3: 59 (1884). *Mariscus scirpodeus* (Steud.) Kuntze, Revis. Gen. Pl. 2: 755 (1891).
Agylla ensifolia Phil., Anales Univ. Chile 1865(1): 643 (1865).

Machaerina sinclairii (Hook.f.) T.Koyama, Bot. Mag. (Tokyo) 69: 65 (1956).
Malesia, New Zealand North I. 42 MOL PHI SUL SUM 51 NZN. Hemicr. or rhizome geophyte.
Cladium anceps Hook.f., Handb. N. Zeal. Fl. 1: 305 (1864).
**Cladium sinclairii* Hook.f., Handb. N. Zeal. Fl. 1: 304 (1864). *Vincentia gladiata* (Hook.f.) Boeck., Linnaea 38: 250 (1874).
Cladium gahnoides Colenso, Trans. & Proc. New Zealand Inst. 16: 340 (1884).
Vincentia robinsonii Merr., Philipp. J. Sci., C 11: 258 (1916). *Machaerina robinsonii* (Merr.) T.Koyama, Bot. Mag. (Tokyo) 69: 65 (1956).
Vincentia crinita Stapf, Philipp. J. Sci. 19: 65 (1921). *Cladium crinitum* (Stapf) Merr., Enum. Philipp. Fl. Pl. 1: 129 (1923). *Machaerina crinita* (Stapf) T.Koyama, Bot. Mag. (Tokyo) 69: 63 (1956).
Cladium maingayi var. *subsetosum* Kük., Repert. Spec. Nov. Regni Veg. 51: 12 (1942).

Machaerina tenax (Hook.f.) T.Koyama, Bot. Mag. (Tokyo) 69: 66 (1956).
New Zealand. 51 CTM NZN NZS. Hemicr. or rhizome geophyte.
**Lampocarya tenax* Hook.f., Fl. Nov.-Zel. 1: 277 (1853). *Cladium tenax* (Hook.f.) Druce, Rep. Bot. Exch. Club Brit. Isles 1916: 615 (1917). *Baumea tenax* (Hook.f.) S.T.Blake, Contr. Queensland Herb. 8: 27 (1969).

Machaerina teretifolia (R.Br.) T.Koyama, Bot. Mag. (Tokyo) 69: 66 (1956).
W. New Guinea, E. Australia, New Zealand. 43 NWG 50 NSW QLD 51 NZN NZS. Hemicr. or rhizome geophyte.
**Cladium teretifolium* R.Br., Prodr.: 237 (1810). *Chapelliera teretifolia* (R.Br.) Nees in J.G.C. Lehmann, Pl. Preiss. 2: 76 (1846). *Mariscus teretifolius* (R.Br.) Kuntze, Revis. Gen. Pl. 2: 755 (1891). *Baumea teretifolia* (R.Br.) Palla, Allg. Bot. Z. Syst. 8: 69 (1902). *Cladium teretifolium* var. *typicum* Domin, Biblioth. Bot. 89: 473 (1915), nom. inval.
Cladium rigidifolium C.Presl, Isis (Oken) 21: 270 (1828).
Cladium arfakense Rendle in L.S.Gibbs, Fl. Arfak Mts.: 90 (1917). *Machaerina arfakensis* (Rendle) T.Koyama, Bot. Mag. (Tokyo) 69: 62 (1956).

Machaerina tetragona (Labill.) T.Koyama, Bot. Mag. (Tokyo) 69: 66 (1956).
SE. Queensland to SE. Australia. 50 NSW QLD SOA TAS VIC. Hemicr. or rhizome geophyte.
**Lepidosperma tetragonum* Labill., Nov. Holl. Pl. 1: 17 (1805). *Cladium tetragonum* (Labill.) J.M.Black, Fl. S. Austral.: 95 (1922). *Baumea tetragona* (Labill.) S.T.Blake, Contr. Queensland Herb. 8: 30 (1969).
Cladium tetraquetrum Hook.f., Fl. Tasman. 2: 95 (1858). *Mariscus tetraquetrus* (Hook.f.) Kuntze, Revis. Gen. Pl. 2: 755 (1891). *Cladium teretifolium* var. *tetraquetrum* (Hook.f.) Domin, Biblioth. Bot. 89: 473 (1915).
Cladium teretifolium var. *asperrimum* Kük., Repert. Spec. Nov. Regni Veg. 51: 169 (1942).

Machaerina vaginalis (Benth.) T.Koyama, Bot. Mag. (Tokyo) 69: 66 (1956).
SW. Australia. 50 WAU. Hemicr. or rhizome geophyte.

**Cladium vaginale* Benth., Fl. Austral. 7: 408 (1878). *Mariscus vaginalis* (Benth.) Kuntze, Revis. Gen. Pl. 2: 755 (1891). *Baumea vaginalis* (Benth.) S.T.Blake, Contr. Queensland Herb. 8: 30 (1969).

***Synonyms*:**
Machaerina Nees = ***Lepidosperma*** Labill.
Machaerina affinis (Brongn.) J.Kern = ***Morelotia affinis*** (Brongn.) S.T.Blake
Machaerina arfakensis (Rendle) T.Koyama = ***Machaerina teretifolia*** (R.Br.) T.Koyama
Machaerina aromatica (Merr.) T.Koyama = ***Schoenus punctatus*** R.Br.
Machaerina australis (A.Rich.) T.Koyama = ***Lepidosperma australe*** (A.Rich.) Hook.f.
Machaerina boninsimae (Nakai) T.Koyama = ***Machaerina glomerata*** (Gaudich.) T.Koyama
Machaerina brevipaniculata (Kük.) T.Koyama = ***Machaerina gunnii*** (Hook.f.) J.Kern
Machaerina brevistigma (Nakai ex Tuyama) T.Koyama = ***Machaerina rubiginosa*** (Spreng.) T.Koyama
Machaerina capillacea (Benth.) T.Koyama = ***Tetraria capillaris*** (F.Muell.) J.M.Black
Machaerina crassa (Thwaites) T.Koyama = ***Machaerina rubiginosa*** (Spreng.) T.Koyama
Machaerina crinita (Stapf) T.Koyama = ***Machaerina sinclairii*** (Hook.f.) T.Koyama
Machaerina cyperoides (Merr.) T.Koyama = ***Fimbristylis vanoverberghii*** Kük.
Machaerina elynanthoides (F.Muell.) T.Koyama = ***Gahnia australis*** (Nees) K.L.Wilson
Machaerina ensifolia (Boeck.) T.Koyama = ***Machaerina iridifolia*** (Bory) T.Koyama
Machaerina gahniiformis (Gaudich.) J.Kern = ***Morelotia gahniiformis*** Gaudich.
Machaerina gaudichaudii (W.Wight) T.Koyama = ***Machaerina glomerata*** (Gaudich.) T.Koyama
Machaerina gracilis (J.M.Black) T.Koyama = ***Machaerina laxa*** (Nees) T.Koyama
Machaerina iris (Ohwi) T.Koyama = ***Machaerina falcata*** (Nees) T.Koyama
Machaerina juncoides (Elmer) T.Koyama = ***Machaerina glomerata*** (Gaudich.) T.Koyama
Machaerina latifolia T.Koyama = ***Machaerina falcata*** (Nees) T.Koyama
Machaerina latissima (F.Muell.) T.Koyama = ***Machaerina preissii*** (Nees) L.A.S.Johnson & Koyama
Machaerina malesiaca (Stapf) T.Koyama = ***Machaerina falcata*** (Nees) T.Koyama
Machaerina mariscoides subsp. *colpodes* (Lauterb.) T.Koyama = ***Machaerina glomerata*** (Gaudich.) T.Koyama
Machaerina mariscoides subsp. *meyenii* (Kunth) T.Koyama = ***Machaerina mariscoides*** (Gaudich.) J.Kern
Machaerina melicoides (Poir.) Bojer = ***Costularia melicoides*** (Poir.) C.B.Clarke
Machaerina melleri (Baker) T.Koyama = ***Costularia melleri*** (Baker) C.B.Clarke
Machaerina meyenii (Kunth) O.Deg. & I.Deg. = ***Machaerina mariscoides*** (Gaudich.) J.Kern
Machaerina micranthes (C.B.Clarke) T.Koyama = ***Machaerina disticha*** (C.B.Clarke) T.Koyama
Machaerina monocarpa (J.M.Black) T.Koyama = ***Schoenus carsei*** Cheeseman
Machaerina nipponense (Ohwi) Ohwi & T.Koyama = ***Machaerina rubiginosa*** (Spreng.) T.Koyama
Machaerina pantopoda (Baker) T.Koyama = ***Costularia pantopoda*** (Baker) C.B.Clarke
Machaerina philippinensis (Merr.) T.Koyama = ***Machaerina disticha*** (C.B.Clarke) T.Koyama
Machaerina ponapensis (Ohwi) Ohwi & T.Koyama = ***Machaerina falcata*** (Nees) T.Koyama
Machaerina procera (S.T.Blake) T.Koyama = ***Cladium mariscus*** subsp. ***intermedium*** Kük.
Machaerina pulchra (Ridl.) T.Koyama = ***Tricostularia undulata*** (Thwaites) J.Kern
Machaerina punctata (R.Br.) T.Koyama = ***Schoenus punctatus*** R.Br.
Machaerina resinosa Nees = ***Lepidosperma resinosum*** (Nees) Benth.
Machaerina restioides Gaudich. = ***Machaerina angustifolia*** (Gaudich.) T.Koyama
Machaerina restioides var. *effusa* Griseb. = ***Machaerina restioides*** subsp. ***effusa*** (Griseb.) T.Koyama
Machaerina robinsonii (Merr.) T.Koyama = ***Machaerina sinclairii*** (Hook.f.) T.Koyama
Machaerina rubiginosa subsp. *crassa* (Thwaites) T.Koyama = ***Machaerina rubiginosa*** (Spreng.) T.Koyama
Machaerina schoenoides (R.Br.) T.Koyama = ***Machaerina acuta*** (Labill.) J.Kern
Machaerina scirpoidea subsp. *ficticia* (Hemsl.) T.Koyama = ***Machaerina ficticia*** (Hemsl.) T.Koyama
Machaerina sinuata (Ridl.) T.Koyama = ***Machaerina glomerata*** (Gaudich.) T.Koyama
Machaerina stradbrokensis (Domin) T.Koyama = ***Trachystylis stradbrokensis*** (Domin) Kük.
Machaerina sucinonux T.Koyama = [38 OGA]
Machaerina undulata (Thwaites) T.Koyama = ***Tricostularia undulata*** (Thwaites) J.Kern

***Unplaced Names*:**
Machaerina sucinonux T.Koyama, Bot. Mag. (Tokyo) 69: 66 (1956). = [38 OGA]

Macrochaetium

Macrochaetium Steud. = ***Tetraria*** P.Beauv.
Macrochaetium dregei Steud. = ***Cyathocoma hexandra*** (Nees) Browning
Macrochaetium ecklonii (Nees) Levyns = ***Cyathocoma ecklonii*** Nees
Macrochaetium hexandrum (Nees) H.Pfeiff. = ***Cyathocoma hexandra*** (Nees) Browning
Macrochaetium neesii (Kunth) Kük. = ***Cyathocoma hexandra*** (Nees) Browning

Macrolomia

Macrolomia Schrad. ex Nees = ***Scleria*** P.J.Bergius
Macrolomia bracteata (Cav.) Schrad. ex Nees = ***Scleria bracteata*** Cav.

Malacochaete

Malacochaete Nees = ***Schoenoplectus*** (Rchb.) Palla
Malacochaete aemulans Arn. ex C.B.Clarke = ***Schoenoplectus corymbosus*** (Roth ex Roem. & Schult.) J.Raynal
Malacochaete assimilis Ces. = ***Schoenoplectus californicus*** (C.A.Mey.) Soják
Malacochaete chilensis Nees & Meyen ex Boeck. = ***Schoenoplectus californicus*** (C.A.Mey.) Soják
Malacochaete corymbosa (Roth ex Roem. & Schult.) Arn. = ***Schoenoplectus corymbosus*** (Roth ex Roem. & Schult.) J.Raynal
Malacochaete lehmannii Nees ex Boeck. = ***Schoenoplectus litoralis*** (Schrad.) Palla subsp. ***litoralis***

Malacochaete litoralis (Schrad.) Nees = ***Schoenoplectus litoralis*** (Schrad.) Palla
Malacochaete oligostachya Phil. = ***Schoenoplectus californicus*** (C.A.Mey.) Soják
Malacochaete pectinata (Roxb.) Nees = ***Schoenoplectus litoralis*** subsp. ***thermalis*** (Trab.) S.S.Hooper
Malacochaete pterolepis Nees = ***Schoenoplectus litoralis*** subsp. ***thermalis*** (Trab.) S.S.Hooper
Malacochaete riparia Nees & Meyen = ***Schoenoplectus californicus*** (C.A.Mey.) Soják
Malacochaete riparia var. *chamissoi* Nees = ***Schoenoplectus californicus*** (C.A.Mey.) Soják
Malacochaete sanguinolenta Nees & Meyen ex Boeck. = ***Schoenoplectus californicus*** (C.A.Mey.) Soják
Malacochaete tatora (Kunth) Meyen = ***Schoenoplectus californicus*** (C.A.Mey.) Soják

Manochlaenia

Manochlaenia Börner = ***Carex*** L.
Manochlaenia grisea (Wahlenb.) Fedde & J.Schust. = ***Carex grisea*** Wahlenb.
Manochlaenia pendula (Huds.) Fedde & J.Schust. = ***Carex pendula*** Huds.

Mapania

Mapania Aubl., Hist. Pl. Guiane 1: 47 (1775).
Trop. 22 GHA GUI IVO LBR NGA SIE 23 CAB CMN CON EQG GAB GGI ZAI 29 SEY 36 CHH CHS 40 ASS BAN EHM IND SRL 41 AND CBD THA VIE 42 BOR JAW LSI MLY MOL PHI SUL SUM 43 BIS NWG SOL 50 NTA QLD 60 FIJ SAM SCZ VAN 62 CRL MRN 80 COS HON NIC PAN 82 FRG GUY SUR VEN 83 CLM ECU PER 84 BZN.
84 Species
Pandanophyllum Hassk., Tijdschr. Natuurl. Gesch. Physiol. 10: 118 (1843).
Cephaloscirpus Kurz, J. Asiat. Soc. Bengal, Pt. 2, Nat. Hist. 38(2): 83 (1869).
Thoracostachyum Kurz, J. Asiat. Soc. Bengal, Pt. 2, Nat. Hist. 38(2): 75 (1869).
Lepistachya Zipp. ex Miq., Ill. Fl. Archip. Ind.: 61 (1870).
Mapaniopsis C.B.Clarke, Kew Bull., Addit. Ser. 8: 55 (1908).
Apartea Pellegr., Bull. Soc. Bot. France 78: 180 (1931).
Langevinia Jacq.-Fél., Bull. Mus. Natl. Hist. Nat., II, 19: 88 (1947).

Mapania africana Boeck., Linnaea 37: 137 (1871).
WC. Trop. Africa. 23 CAB CMN EQG GAB. Hemicr. or rhizome geophyte.

subsp. ***africana***
WC. Trop. Africa. 23 CAB CMN EQG GAB. Hemicr. or rhizome geophyte.
Mapania dolichostachya K.Schum., Notizbl. Königl. Bot. Gart. Berlin 3: 106 (1901).

subsp. ***filipes*** (J.Raynal) D.A.Simpson, Rev. Gen. Mapania: 127 (1992).
WC. Trop. Africa. 23 CMN GAB. Hemicr. or rhizome geophyte.
**Mapania africana* var. *filipes* J.Raynal, Adansonia, n.s., 8: 412 (1968).

Mapania amplivaginata K.Schum., Notizbl. Königl. Bot. Gart. Berlin 3: 105 (1901).
S. Nigeria to WC. Trop. Africa. 22 NGA 23 CAB CMN EQG GAB GGI. Hemicr. or rhizome geophyte.
Mapania oblonga C.B.Clarke in D.Oliver & auct. suc. (eds.), Fl. Trop. Afr. 8: 491 (1902), nom. nud.
Mapania oblonga var. *elliptica* C.B.Clarke in D.Oliver & auct. suc. (eds.), Fl. Trop. Afr. 8: 492 (1902), nom. nud.
Mapania mildbraedii Graebn. in G.W.J.Mildbread (ed.), Wiss. Erg. Deut. Zentr.-Afr. Exped., Bot. 2: 93 (1922), nom. nud.
Apartea le-testui Pellegr., Bull. Soc. Bot. France 78: 180 (1931).
Langevinia monosperma Jacq.-Fél., Bull. Mus. Natl. Hist. Nat., II, 19: 88 (1947). *Mapania monosperma* (Jacq.-Fél.) Maguire & Koyama, Mem. New York Bot. Gard. 17(1): 51 (1967).

Mapania angustifolia Uittien, Recueil Trav. Bot. Néerl. 32: 197 (1935).
Borneo (Sarawak: Pegunungan Dulit). 42 BOR. Hemicr.

Mapania archboldii Uittien, J. Arnold Arbor. 20: 214 (1939).
New Guinea. 43 NWG. Hemicr. or rhizome geophyte.
Mapania dictyophlebia S.T.Blake, J. Arnold Arbor. 28: 211 (1947).

Mapania arunachalensis G.D.Pal, J. Bombay Nat. Hist. Soc. 91: 438 (1994 publ. 1995).
Arunachal Pradesh. 40 EHM. Hemicr. or rhizome geophyte.

Mapania assimilis T.Koyama, Mem. New York Bot. Gard. 17(1): 53 (1967).
C. America to N. Ecuador. 80 COS HON NIC PAN 82 FRG 83 CLM ECU. Hemicr. or rhizome geophyte.

subsp. ***assimilis***
C. America to N. Ecuador. 80 COS HON NIC PAN 83 CLM ECU. Hemicr. or rhizome geophyte.

subsp. ***guianensis*** D.A.Simpson, Kew Bull. 44: 141 (1989).
French Guiana. 82 FRG. Hemicr. or rhizome geophyte.

Mapania aturensis D.A.Simpson, Kew Bull. 51: 730 (1996).
S. Venezuela (Cerro Cuao). 82 VEN. Hemicr. or rhizome geophyte.

Mapania baccifera C.B.Clarke, Kew Bull., Addit. Ser. 8: 53 (1908).
Papuasia. 43 BIS NWG SOL. Hemicr. or rhizome geophyte.

subsp. ***baccifera***
Papuasia. 43 NWG SOL. Hemicr. or rhizome geophyte.
Mapania papuana Ridl., Trans. Linn. Soc. London, Bot. 9: 246 (1916).
Mapania radulosa Ridl., Trans. Linn. Soc. London, Bot. 9: 245 (1916).

subsp. ***manusensis*** D.A.Simpson, Rev. Gen. Mapania: 103 (1993).
Bismarck Arch. (Manus I.). 43 BIS. Hemicr. or rhizome geophyte.

Mapania balansae (E.G.Camus) T.Koyama, Bull. Arts Sci. Div. Ryukyu Univ. 3: 73 (1959).
Vietnam. 41 VIE. Hemicr. or rhizome geophyte.
**Thoracostachyum balansae* E.G.Camus, Notul. Syst. (Paris) 1: 252 (1910).
Thoracostachyum balansae var. *latifolium* K.K.Nguyen, Bot. Zhurn. (Moscow & Leningrad) 84: 856 (1979).

Mapania baldwinii Nelmes, Kew Bull. 6: 420 (1951 publ. 1952).
W. Trop. Africa to Equatorial Guinea. 22 GHA IVO LBR 23 EQG. Hemicr. or rhizome geophyte.
Mapania comoensis A.Chev. ex Hutch. & Dalziel, Fl. W. Trop. Afr. 2: 471 (1936), no latin descr.

Mapania bancana (Miq.) T.Koyama, Bull. Arts Sci. Div. Ryukyu Univ. 3: 74 (1959).
S. Indo-China to New Guinea. 41 THA VIE 42 BOR MLY MOL SUL SUM 43 NWG. Hemicr. or rhizome geophyte.
**Lepironia bancana* Miq., Fl. Ned. Ind., Eerste Bijv.: 604 (1861). *Thoracostachyum bancanum* (Miq.) Kurz, J. Asiat. Soc. Bengal, Pt. 2, Nat. Hist. 38(2): 75 (1869).
Hypolytrum borneense Kurz, J. Asiat. Soc. Bengal, Pt. 2, Nat. Hist. 38(2): 74 (1869).
Mapania bancana var. *longispica* C.B.Clarke in J.D.Hooker, Fl. Brit. India 6: 680 (1894).
Thoracostachyum ridleyi C.B.Clarke, J. Straits Branch Roy. Asiat. Soc. 46: 226 (1906).
Thoracostachyum subcapitatum J.V.Suringar, Nova Guinea 8: 710 (1912).

Mapania borneensis Merr., J. Straits Branch Roy. Asiat. Soc. 76: 78 (1917).
Borneo. 42 BOR. Hemicr. or rhizome geophyte.

Mapania bougainvillensis D.A.Simpson, Rev. Gen. Mapania: 91 (1992).
Bismarck Arch. to Solomon Is. 43 BIS SOL. Hemicr. or rhizome geophyte.

Mapania caudata Kük., Repert. Spec. Nov. Regni Veg. 29: 201 (1931).
Pen. Malaysia, Borneo. 42 BOR MLY. Hemicr. or rhizome geophyte.
Mapania petiolata var. *pumila* Uittien, Recueil Trav. Bot. Néerl. 33: 283 (1936). *Mapania cuspidata* var. *pumila* (Uittien) Uittien, J. Arnold Arbor. 20: 214 (1939).

Mapania coodei D.A.Simpson, Kew Bull. 55: 997 (2000).
W. New Guinea. 43 NWG. Hemicr. or rhizome geophyte.

Mapania coriandrum Nelmes, Kew Bull. 6: 419 (1951 publ. 1952).
W. Trop. Africa. 22 GHA GUI IVO LBR SIE. Hemicr. or rhizome geophyte.

Mapania cuatrecasasii T.Koyama, Mem. New York Bot. Gard. 17(1): 59 (1967).
Costa Rica to Guyana. 80 COS PAN 82 GUY 83 CLM. Hemicr. or rhizome geophyte.
Mapania herrerae Gómez-Laur., Phytologia 60: 73 (1986).

Mapania cuspidata (Miq.) Uittien, J. Arnold Arbor. 20: 213 (1939).
Thailand to New Guinea. 41 THA 42 BOR MLY MOL PHI SUL SUM 43 NWG. Cham.
**Lepironia cuspidata* Miq., Fl. Ned. Ind., Eerste Bijv.: 603 (1861).

var. ***angustifolia*** (Uittien) Uittien, J. Arnold Arbor. 20: 214 (1939).
Malesia to New Guinea. 42 BOR MOL PHI SUL 43 NWG. Cham.
**Mapania petiolata* var. *angustifolia* Uittien, Recueil Trav. Bot. Néerl. 33: 282 (1936).

var. ***cuspidata***
Thailand to W. Malesia. 41 THA 42 BOR MLY SUM. Cham.
Pandanophyllum zippelianum Kurz, Natuurk. Tijdschr. Ned.-Indië 27: 244 (1864), nom. nud.
Lepistachya praemorsa Zipp. ex Miq., Ill. Fl. Archip. Ind.: 61 (1870).
Pandanophyllum wendlandii auct., Gard. Chron., n.s., 21: 711 (1884).
Mapania lucida N.E.Br., Ill. Hort. 32: 77 (1885).
Mapania triquetra Ridl., J. Straits Branch Roy. Asiat. Soc. 41: 51 (1904).
Mapania petiolata C.B.Clarke, Kew Bull., Addit. Ser. 8: 54 (1908). *Mapania humilis* var. *petiolata* (C.B.Clarke) Ridl., Fl. Malay Penins. 5: 174 (1925). *Mapania cuspidata* var. *petiolata* (C.B. Clarke) Uittien, J. Arnold Arbor. 20: 213 (1939).
Mapania platyphylla Merr., Philipp. J. Sci., C 11: 54 (1916).
Mapania inopinata Uittien, Recueil Trav. Bot. Néerl. 33: 151 (1936).
Mapania stolonifera Uittien, Recueil Trav. Bot. Néerl. 33: 279 (1936).

Mapania debilis C.B.Clarke, J. Straits Branch Roy. Asiat. Soc. 46: 227 (1906).
Borneo. 42 BOR. Hemicr.
Mapania gracilipes Merr., J. Straits Branch Roy. Asiat. Soc. 85: 158 (1922).
Mapania flagellaris Uittien, Recueil Trav. Bot. Néerl. 32: 198 (1935).

Mapania effusa (C.B.Clarke) T.Koyama, J. Fac. Sci. Univ. Tokyo, Sect. 3, Bot. 8: 68 (1961).
N. South America to N. Brazil. 82 FRG VEN 84 BZN. Hemicr. or rhizome geophyte.
**Mapaniopsis effusa* C.B.Clarke, Kew Bull., Addit. Ser. 8: 55 (1908).

Mapania enodis (Miq.) C.B.Clarke, Kew Bull., Addit. Ser. 8: 53 (1908).
S. Thailand to W. Malesia. 41 THA 42 BOR MLY SUM. Hemicr. or rhizome geophyte.
**Lepironia enodis* Miq., Fl. Ned. Ind., Eerste Bijv.: 603 (1861). *Pandanophyllum miquelianum* Kurz, J. Asiat. Soc. Bengal, Pt. 2, Nat. Hist. 38(2): 81 (1869), nom. illeg.
Lepironia foliosa Miq., Fl. Ned. Ind., Eerste Bijv.: 603 (1861).
Mapania longa Ridl. ex C.B.Clarke in J.D.Hooker, Fl. Brit. India 6: 683 (1894).
Mapania tumida Uittien, Gard. Bull. Straits Settlem. 10: 182 (1939).

Mapania ferruginea Ridl., Bol. Soc. Brot. 5: 209 (1887).
São Tomé, Príncipe. 23 GGI. Hemicr. or rhizome geophyte.
Mapania subcomposita C.B.Clarke in D.Oliver & auct. suc. (eds.), Fl. Trop. Afr. 8: 490 (1902).

Mapania floribunda (Nees ex Steud.) T.Koyama, J. Fac. Sci. Univ. Tokyo, Sect. 3, Bot. 8: 69 (1961).
Seychelles. 29 SEY. Hemicr.
**Hypolytrum floribundum* Nees ex Steud., Syn. Pl. Glumac. 2: 132 (1855). *Thoracostachyum floribundum* (Nees ex Steud.) C.B.Clarke in T.A.Durand & H.Schinz, Consp. Fl. Afric. 5: 667 (1894).
Hypolytrum seychellense C.B.Clarke, Kew Bull., Addit. Ser. 8: 52 (1908).
Thoracostachyum angustifolium C.B.Clarke, Kew Bull., Addit. Ser. 8: 52 (1908).

Mapania foxworthyi Merr., Philipp. J. Sci., C 11: 53 (1916).
Borneo. 42 BOR. Hemicr.

Mapania graminea Uittien, Recueil Trav. Bot. Néerl. 33: 147 (1936).
Borneo. 42 BOR. Rhizome geophyte.
Mapania graminea var. *stipitata* Uittien, Recueil Trav. Bot. Néerl. 33: 147 (1936).

Mapania hispida D.A.Simpson, Rev. Gen. Mapania: 109 (1992).
NW. Borneo. 42 BOR. Hemicr. or rhizome geophyte.

Mapania holttumii J.Kern, Blumea 9: 215 (1958).
Pen. Malaysia. 42 MLY. Hemicr.
**Mapania insignis* Holttum, Gard. Bull. Singapore 11: 293 (1947), nom. illeg.

Mapania imeriensis (R.Gross) T.Koyama, Mem. New York Bot. Gard. 17(1): 56 (1967).
S. Guyana to Suriname. 82 GUY SUR VEN. Hemicr. or rhizome geophyte.
**Hypolytrum imeriense* R.Gross, Repert. Spec. Nov. Regni Veg. 50: 210 (1941).
Hypolytrum sandwithii Gilly, Bull. Torrey Bot. Club 76: 291 (1949).

Mapania immersa (Thwaites) Benth. ex C.B.Clarke in J.D.Hooker, Fl. Brit. India 6: 682 (1894).
Sri Lanka. 40 SRL. Hel.
**Pandanophyllum immersum* Thwaites, Enum. Pl. Zeyl.: 483 (1864).

Mapania insignis Sandwith, Bull. Misc. Inform. Kew 1933: 496 (1933).
Guyana. 82 GUY. Hemicr. or rhizome geophyte.

Mapania ivorensis (J.Raynal) J.Raynal, Adansonia, n.s., 8: 415 (1968).
Liberia to Ivory Coast. 22 IVO LBR. Hemicr. or rhizome geophyte.
**Mapania macrantha* subsp. *ivorensis* J.Raynal, Adansonia, n.s., 5: 278 (1965).

Mapania kurzii C.B.Clarke in J.D.Hooker, Fl. Brit. India 6: 681 (1894).
Pen. Thailand to Sumatera. 41 THA 42 MLY SUM. Hemicr. or rhizome geophyte.
Mapania multispicata C.B.Clarke in J.D.Hooker, Fl. Brit. India 6: 682 (1894).
Mapania longispica Ridl., J. Straits Branch Roy. Asiat. Soc. 44: 205 (1905).

Mapania latifolia Uittien, Recueil Trav. Bot. Néerl. 32: 199 (1935).
Borneo. 42 BOR. Hemicr. or rhizome geophyte.
Mapania montana Ridl., J. Straits Branch Roy. Asiat. Soc. 44: 206 (1905), sensu auct.

Mapania liberiensis D.A.Simpson, Rev. Gen. Mapania: 133 (1992).
Liberia. 22 LBR. Hemicr. or rhizome geophyte.

Mapania linderi Hutch. ex Nelmes, Kew Bull. 7: 421 (1952).
W. Trop. Africa. 22 GUI IVO LBR SIE. Hemicr. or rhizome geophyte.

Mapania longiflora C.B.Clarke, Kew Bull., Addit. Ser. 8: 54 (1908).
Pen. Malaysia, Borneo. 42 BOR MLY. Hemicr.

Mapania lorea Uittien, Recueil Trav. Bot. Néerl. 33: 150 (1936).
W. Malesia. 42 BOR MLY SUM. Hemicr. or rhizome geophyte.

Mapania macrantha (Boeck.) H.Pfeiff., Bot. Arch. 12: 450 (1925).
WC. Trop. Africa. 23 CMN EQG GAB. Hemicr. or rhizome geophyte.
**Hypolytrum macranthum* Boeck., Bot. Jahrb. Syst. 5: 507 (1884). *Mapania superba* C.B.Clarke in T.A.Durand & H.Schinz, Consp. Fl. Afric. 5: 667 (1894), nom. illeg.

Mapania macrocephala (Gaudich.) K.Schum., Bot. Jahrb. Syst. 13: 265 (1891).
C. Malesia to W. Pacific. 42 LSI MOL PHI SUL 43 BIS NWG SOL 50 QLD 60 SAM 62 CRL. Hemicr. or rhizome geophyte.
**Hypolytrum macrocephalum* Gaudich., Voy. Uranie: 415 (1829). *Cephaloscirpus macrocephalus* (Gaudich.) Kurz, J. Asiat. Soc. Bengal, Pt. 2, Nat. Hist. 38(2): 84 (1869). *Lepironia macrocephala* (Gaudich.) Miq., Ill. Fl. Archip. Ind.: 64 (1870). *Pandanophyllum macrocephalum* (Gaudich.) Boeck., Flora 58: 112 (1875).

subsp. ***longifolia*** (Boeck.) D.A.Simpson, Rev. Gen. Mapania: 63 (1992).
Bismarck Arch. to Samoa. 43 BIS SOL 60 SAM. Hemicr. or rhizome geophyte.
**Pandanophyllum longifolium* Boeck., Flora 58: 112 (1875). *Mapania longifolia* (Boeck.) C.B.Clarke, Kew Bull., Addit. Ser. 8: 54 (1908).

subsp. ***macrocephala***
C. Malesia to N. Queensland. 42 LSI MOL PHI SUL 43 BIS NWG 50 QLD 62 CRL. Hemicr. or rhizome geophyte.
Mapania pandanacea Ridl., Trans. Linn. Soc. London, Bot. 9: 246 (1916).
Mapania lactea Kük., Bot. Jahrb. Syst. 59: 56 (1924).
Mapania moseleyi f. *latifolia* Uittien, J. Arnold Arbor. 20: 213 (1939).
Mapania geelvinkensis Ohwi, Bot. Mag. (Tokyo) 56: 210 (1942).
Mapania margaritae Ohwi, Bot. Mag. (Tokyo) 56: 211 (1942).
Mapania palauensis Hosok., Trans. Nat. Hist. Soc. Taiwan 32: 7 (1942).
Mapania yapensis Hosok., Trans. Nat. Hist. Soc. Taiwan 32: 8 (1942).

Mapania macrophylla (Boeck.) H.Pfeiff., Bot. Arch. 12: 449 (1925).
N. South America to Brazil. 82 FRG GUY SUR VEN 84 BZN. Hemicr. or rhizome geophyte.
**Hypolytrum macrophyllum* Boeck., Flora 61: 142 (1878). *Mapania schomburgkii* C.B.Clarke, Kew Bull., Addit. Ser. 8: 54 (1908), nom. illeg.

Mapania maguireana T.Koyama & Steyerm., Mem. New York Bot. Gard. 17(1): 63 (1967).
Guyana to Brazil (Serra da Neblina). 82 GUY 84 BZN. Hemicr. or rhizome geophyte.

Mapania mangenotiana Lorougnon, Bull. Jard. Bot. État 34: 297 (1964).
Liberia, Ivory Coast, Gabon. 22 IVO LBR 23 GAB. Hemicr. or rhizome geophyte.
Mapania africana subsp. *occidentalis* J.Raynal, Adansonia, n.s., 5: 277 (1965).

Mapania mannii C.B.Clarke in D.Oliver & auct. suc. (eds.), Fl. Trop. Afr. 8: 491 (1902).

WC. Trop. Africa. 23 CMN CON GAB GGI ZAI. Hemicr. or rhizome geophyte.

subsp. ***bieleri*** (De Wild.) J.Raynal ex D.A.Simpson, Rev. Gen. Mapania: 114 (1992).
WC. Trop. Africa. 23 CON GAB ZAI. Hemicr. or rhizome geophyte.
**Mapania bieleri* De Wild., Ann. Mus. Congo Belge, Bot., V, 3: 165 (1910).

subsp. ***mannii***
WC. Trop. Africa. 23 CMN CON GAB GGI ZAI. Hemicr. or rhizome geophyte.

Mapania maschalina J.Kern, Blumea 12: 23 (1963).
Borneo (Sarawak, Sabah). 42 BOR. Rhizome geophyte.

Mapania meditensis D.A.Simpson, Rev. Gen. Mapania: 118 (1992).
Pen. Thailand, Pen. Malaysia, Borneo. 41 THA 42 BOR MLY. Rhizome geophyte.

Mapania micrococca (T.Koyama) D.A.Simpson, Kew Bull. 51: 737 (1996).
S. Venezuela to Guyana (Mt. Roraima). 82 GUY VEN. Hemicr. or rhizome geophyte.
**Mapaniopsis micrococca* T.Koyama, Jap. J. Bot. 20: 130 (1969).

Mapania micropandanus Holttum, Gard. Bull. Singapore 11: 294 (1947).
Pen. Malaysia (SE. Johor). 42 MLY. Rhizome geophyte.

Mapania minor (Nelmes) J.Raynal, Adansonia, n.s., 5: 278 (1965).
Liberia to Ivory Coast. 22 IVO LBR. Hemicr. or rhizome geophyte.
**Mapania macrantha* var. *minor* Nelmes, Kew Bull. 6: 420 (1951).

Mapania monostachya Uittien, Recueil Trav. Bot. Néerl. 32: 194 (1935).
Pen. Malaysia, Borneo. 42 BOR MLY. Rhizome geophyte.

Mapania moseleyi C.B.Clarke, Kew Bull., Addit. Ser. 8: 55 (1908).
Papuasia. 43 BIS NWG SOL. Hemicr. or rhizome geophyte.
Mapania ledermannii Kük., Bot. Jahrb. Syst. 59: 57 (1924).

Mapania neblina D.A.Simpson, Rev. Gen. Mapania: 159 (1992).
S. Venezuela (Sierra de la Neblina). 82 VEN. Hemicr. or rhizome geophyte.

Mapania nudispica T.Koyama, Contr. Inst. Bot. Univ. Montréal 70: 54 (1957).
Vietnam. 41 VIE. Hemicr. or rhizome geophyte.

Mapania obscuriflora D.A.Simpson, Rev. Gen. Mapania: 100 (1992).
Borneo (Kalimantan). 42 BOR. Hemicr. or rhizome geophyte.

Mapania pacifica (Hosok.) T.Koyama, Bull. Arts Sci. Div. Ryukyu Univ. 3: 74 (1959).
Caroline Is. 62 CRL. Hemicr. or rhizome geophyte.
Thoracostachyum hypolytroides var. *immensum* Kük., Bot. Jahrb. Syst. 59: 8 (1924). *Mapania pandanophylla* subsp. *immensa* (Kük.) T.Koyama, Micronesica 1: 65 (1964).
**Thoracostachyum pacificum* Hosok., Trans. Nat. Hist. Soc. Taiwan 32: 6 (1942).

Mapania palustris (Hassk. ex Steud.) Fern.-Vill. in F.M.Blanco, Fl. Filip., ed. 3, 4(13A): 309 (1880).
Andaman Is. to Vanuatu. 41 AND THA 42 BOR JAW MLY MOL PHI SUM 43 BIS NWG SOL 60 VAN. Hemicr.
**Pandanophyllum palustre* Hassk. ex Steud., Syn. Pl. Glumac. 2: 134 (1855). *Lepironia palustris* (Hassk. ex Steud.) Miq., Ill. Fl. Archip. Ind.: 63 (1870).

var. ***andamanica*** (C.B.Clarke) D.A.Simpson, Rev. Gen. Mapania: 69 (1992).
Andaman Is., Pen. Thailand to N. Pen. Malaysia. 41 AND THA 42 MLY. Hemicr.
**Mapania andamanica* C.B.Clarke in J.D.Hooker, Fl. Brit. India 6: 681 (1894).

var. ***palustris***
Pen. Thailand to Vanuatu. 41 THA 42 BOR JAW MLY MOL PHI SUM 43 BIS NWG SOL 60 VAN. Hemicr.
Mapania albescens C.B.Clarke, Kew Bull., Addit. Ser. 8: 54 (1908).
Mapania banahaensis Elmer, Leafl. Philipp. Bot. 2: 574 (1909).
Mapania affinis Merr., J. Straits Branch Roy. Asiat. Soc. 85: 157 (1922).
Mapania grandiceps Kük., Bot. Jahrb. Syst. 59: 55 (1924).
Mapania javana Uittien, Recueil Trav. Bot. Néerl. 33: 287 (1936).

Mapania paradoxa J.Raynal, Adansonia, n.s., 15: 532 (1976).
Franch Guiana. 82 FRG. Hemicr. or rhizome geophyte.

Mapania pedunculata D.A.Simpson, Kew Bull. 51: 732 (1996).
S. Venezuela (Cerro Aracamuni). 82 VEN. Hemicr. or rhizome geophyte.

Mapania pubisquama Cherm., Bull. Soc. Bot. France 77: 278 (1930).
WC. Trop. Africa. 23 CAB CMN EQG GAB. Hemicr. or rhizome geophyte.

Mapania purpuriceps (C.B.Clarke) J.Raynal, Adansonia, n.s., 8: 411 (1968).
Gabon. 23 GAB. Hemicr. or rhizome geophyte.
**Mapania subcomposita* var. *purpuriceps* C.B.Clarke in D.Oliver & auct. suc. (eds.), Fl. Trop. Afr. 8: 491 (1902).

Mapania pycnocephala (Benth.) Benth., J. Linn. Soc., Bot. 15: 512 (1877).
Nicaragua to S. Trop. America. 80 COS NIC PAN 82 FRG GUY VEN 83 CLM ECU 84 BZN. Hemicr. or rhizome geophyte.
**Hypolytrum pycnocephalum* Benth., Bot. Voy. Sulphur: 177 (1846).

subsp. ***fluviatilis*** (Sandwith) T.Koyama, Mem. New York Bot. Gard. 17(1): 66 (1967).
N. South America to Brazil. 82 FRG GUY VEN 84 BZN. Hemicr. or rhizome geophyte.
**Mapania fluviatilis* Sandwith, Bull. Misc. Inform. Kew 1933: 495 (1933).

subsp. ***pycnocephala***
Nicaragua to Venezuela and N. Ecuador. 80 COS NIC PAN 82 VEN 83 CLM ECU. Hemicr. or rhizome geophyte.

Mapania pycnostachya (Benth.) T.Koyama, Mem. New York Bot. Gard. 17(1): 60 (1967).

S. Venezuela (Amazonas) to N. Brazil. 82 VEN 84 BZN. Hemicr. or rhizome geophyte.
Diplasia pycnostachya Benth., J. Linn. Soc., Bot. 15: 512 (1877). *Hypolytrum condensatum* C.B.Clarke, Kew Bull., Addit. Ser. 8: 50 (1908), nom. illeg.

Mapania raynaliana D.A.Simpson, Rev. Gen. Mapania: 131 (1992).
WC. Trop. Africa. 23 CMN GAB. Hemicr. or rhizome geophyte.

Mapania rhynchocarpa Lorougnon & Raynal, Adansonia, n.s., 8: 419 (1968).
W. Trop. Africa. 22 GHA IVO LBR NGA SIE. Hemicr. or rhizome geophyte.

Mapania richardsii Uittien, Recueil Trav. Bot. Néerl. 32: 195 (1935).
Borneo. 42 BOR. Hemicr.

Mapania rionegrensis D.A.Simpson, Kew Bull. 51: 732 (1996).
S. Venezuela to N. Peru. 82 VEN 83 PER. Hemicr. or rhizome geophyte.

Mapania sessilis Merr., J. Straits Branch Roy. Asiat. Soc. 85: 156 (1922). *Mapania longiflora* var. *sessilis* (Merr.) Uittien, Recueil Trav. Bot. Néerl. 33: 146 (1936).
W. Malesia. 42 BOR MLY SUM. Hemicr. or rhizome geophyte.

Mapania seychellaria D.A.Simpson, Rev. Gen. Mapania: 58 (1992).
Seychelles (Praslin). 29 SEY. Hemicr. or rhizome geophyte.

Mapania silhetensis C.B.Clarke in J.D.Hooker, Fl. Brit. India 6: 681 (1894).
Assam to Hainan. 36 CHH CHS 40 ASS BAN 41 VIE. Hemicr. or rhizome geophyte.
Pandanophyllum palustre var. *silhetana* Kurz, J. Asiat. Soc. Bengal, Pt. 2, Nat. Hist. 38(2): 78 (1869).
Mapania sinensis Uittien, Recueil Trav. Bot. Néerl. 33: 288 (1936).
Mapania dolichopoda Tang & F.T.Wang, in Fl. Reipubl. Popul. Sin. 11: 233 (1961).

Mapania soyauxii (Boeck.) H.Pfeiff., Repert. Spec. Nov. Regni Veg. 28: 22 (1930).
WC. Trop. Africa. 23 CAB CMN GAB ZAI. Hemicr. or rhizome geophyte.
Hypolytrum aschersonianum Boeck., Flora 65: 26 (1882). *Mapania aschersoniana* (Boeck.) H.Pfeiff., Bot. Arch. 12: 449 (1925).
**Hypolytrum soyauxii* Boeck., Flora 65: 25 (1882).
Mapania dwanensis Cherm., Bull. Soc. Bot. France 83: 495 (1936).

Mapania spadicea Uittien, Recueil Trav. Bot. Néerl. 33: 278 (1936).
Borneo. 42 BOR. Hemicr. or rhizome geophyte.

Mapania squamata (Kurz) C.B.Clarke, Kew Bull., Addit. Ser. 8: 53 (1908).
W. Malesia. 42 BOR JAW MLY SUM. Hemicr.
**Pandanophyllum squamatum* Kurz, J. Asiat. Soc. Bengal, Pt. 2, Nat. Hist. 38(2): 80 (1869). *Lepironia squamata* (Kurz) Miq., Ill. Fl. Archip. Ind.: 64 (1870).
Mapania heterocephala Merr., J. Straits Branch Roy. Asiat. Soc. 85: 158 (1922).

Mapania steyermarkii T.Koyama, Mem. New York Bot. Gard. 17(1): 61 (1967).
SE. Venezuela to Guyana. 82 GUY VEN. Hemicr. or rhizome geophyte.

Mapania sumatrana (Miq.) Benth., Fl. Austral. 7: 341 (1878).
Malesia to NW. Pacific. 42 BOR JAW MLY SUL SUM 43 NWG 50 NTA QLD 62 CRL MRN. Hemicr. or rhizome geophyte.
**Lepironia sumatrana* Miq., Fl. Ned. Ind., Eerste Bijv.: 604 (1861). *Thoracostachyum sumatranum* (Miq.) Kurz, J. Asiat. Soc. Bengal, Pt. 2, Nat. Hist. 38(2): 75 (1869).

subsp. ***pandanophylla*** (F.Muell.) D.A.Simpson, Rev. Gen. Mapania: 53 (1992).
Malesia to NW. Pacific. 42 BOR MLY SUL 43 NWG 50 NTA QLD 62 CRL MRN. Hemicr. or rhizome geophyte.
**Hypolytrum pandanophyllum* F.Muell., Fragm. 9: 16 (1875). *Mapania pandanophylla* (F.Muell.) K.Schum. in K.M.Schumann & U.M.Hollrung, Fl. Kais. Wilh. Land: 25 (1889). *Thoracostachyum pandanophyllum* (F.Muell.) Domin, Biblioth. Bot. 85: 484 (1915).
Pandanophyllum hypolytroides F.Muell., Fragm. 9: 16 (1875). *Mapania hypolytroides* (F.Muell.) F.Muell. ex Benth., Fl. Austral. 7: 341 (1878). *Thoracostachyum hypolytroides* (F.Muell.) C.B.Clarke in J.D.Hooker, Fl. Brit. India 6: 680 (1894).
Mapania heyneana Backer, Bull. Jard. Bot. Buitenzorg, III, 2: 328 (1920). *Thoracostachyum heyneanum* (Backer) Uittien, Recueil Trav. Bot. Néerl. 32: 200 (1935).

subsp. ***sumatrana***
W. Malesia. 42 BOR JAW SUM. Hemicr. or rhizome geophyte.
Thoracostachyum dichromenoides Ridl., Bot. Jahrb. Syst. 44: 525 (1910).

Mapania surinamensis Uittien in A.A.Pulle, Fl. Suriname 1(1): 128 (1934).
Suriname. 82 SUR. Hemicr. or rhizome geophyte.
**Mapania montana* Uittien, Recueil Trav. Bot. Néerl. 22: 339 (1925), nom. illeg.

Mapania sylvatica Aubl., Hist. Pl. Guiane 1: 47 (1775).
N. South America to N. Brazil, Gabon. 23 GAB 82 FRG GUY SUR VEN 84 BZN. Hemicr. or rhizome geophyte.

subsp. ***gabonica*** (Cherm.) D.A.Simpson, Kew Bull. 44: 140 (1989).
Gabon. 23 GAB. Hemicr. or rhizome geophyte.
**Mapania gabonica* Cherm., Bull. Soc. Bot. France 77: 277 (1930).

subsp. ***sylvatica***
N. South America to N. Brazil. 82 FRG GUY SUR VEN 84 BZN. Hemicr. or rhizome geophyte.
Mapania sylvestris Vahl, Enum. Pl. 2: 391 (1805).

Mapania tamdaoensis N.K.Khoi, J. Biol. (Vietnam) 18(4): 4 (1996).
Vietnam. 41 VIE. Hemicr. or rhizome geophyte.

Mapania tenuiscapa C.B.Clarke in J.D.Hooker, Fl. Brit. India 6: 683 (1894).
Indo-China to Philippines. 41 CBD THA VIE 42 BOR MLY PHI. Hemicr. or rhizome geophyte.
Mapania valida Ridl., J. Straits Branch Roy. Asiat. Soc. 44: 205 (1905).

Mapania tepuiana (Steyerm.) T.Koyama, Mem. New York Bot. Gard. 17(1): 54 (1967).
N. South America. 82 FRG? GUY VEN. Hemicr. or rhizome geophyte.
**Hypolytrum tepuianum* Steyerm., Fieldiana, Bot. 28(1): 51 (1951).

Mapania testui Cherm., Bull. Soc. Bot. France 77: 278 (1930).
Gabon. 23 GAB. Hemicr. or rhizome geophyte.

Mapania theobromina D.A.Simpson, Kew Bull. 51: 738 (1996).
S. Venezuela (Sierra de la Neblina). 82 VEN. Hemicr. or rhizome geophyte.

Mapania tonkinensis (E.G.Camus) T.Koyama, Bull. Arts Sci. Div. Ryukyu Univ. 3: 74 (1959).
Vietnam. 41 VIE. Hemicr. or rhizome geophyte.
**Diplasia tonkinensis* E.G.Camus, Notul. Syst. (Paris) 1: 250 (1910). *Thoracostachyum tonkinense* (E.G.Camus) Uittien, Recueil Trav. Bot. Néerl. 33: 140 (1936).

Mapania vitiensis (Uittien) T.Koyama, Allertonia 1: 341 (1978).
Solomon Is. to Fiji (Vanua Levu). 43 SOL 60 FIJ SCZ. Hemicr. or rhizome geophyte.
**Thoracostachyum vitiense* Uittien, Bernice P. Bishop Mus. Bull. 141: 16 (1936).

Mapania wallichii C.B.Clarke in J.D.Hooker, Fl. Brit. India 6: 682 (1894).
W. Malesia. 42 BOR MLY SUM. Hemicr.
Pandanophyllum palustre var. *malesica* Kurz, J. Asiat. Soc. Bengal, Pt. 2, Nat. Hist. 38(2): 78 (1869).

Mapania zeylanica (Thwaites) Trimen, Syst. Cat. Fl. Pl. Ceylon: 102 (1885).
SE. India, Sri Lanka. 40 IND SRL. Hemicr. or rhizome geophyte.
**Pandanophyllum zeylanicum* Thwaites, Enum. Pl. Zeyl.: 345 (1864). *Lepironia zeylanica* (Thwaites) Miq., Ill. Fl. Archip. Ind.: 61 (1870).

Synonyms:
Mapania affinis Merr. = ***Mapania palustris*** (Hassk. ex Steud.) Fern.-Vill. var. ***palustris***
Mapania africana var. *filipes* J.Raynal = ***Mapania africana*** subsp. ***filipes*** (J.Raynal) D.A.Simpson
Mapania africana subsp. *occidentalis* J.Raynal = ***Mapania mangenotiana*** Lorougnon
Mapania albescens C.B.Clarke = ***Mapania palustris*** (Hassk. ex Steud.) Fern.-Vill. var. ***palustris***
Mapania alpina Elmer ex Merr. = [42 PHI]
Mapania andamanica C.B.Clarke = ***Mapania palustris*** var. ***andamanica*** (C.B.Clarke) D.A.Simpson
Mapania aschersoniana (Boeck.) H.Pfeiff. = ***Mapania soyauxii*** (Boeck.) H.Pfeiff.
Mapania banahaensis Elmer = ***Mapania palustris*** (Hassk. ex Steud.) Fern.-Vill. var. ***palustris***
Mapania bancana var. *longispica* C.B.Clarke = ***Mapania bancana*** (Miq.) T.Koyama
Mapania bieleri De Wild. = ***Mapania mannii*** subsp. ***bieleri*** (De Wild.) J.Raynal ex D.A.Simpson
Mapania comoensis A.Chev. ex Hutch. & Dalziel = ***Mapania baldwinii*** Nelmes
Mapania cuspidata var. *petiolata* (C.B.Clarke) Uittien = ***Mapania cuspidata*** (Miq.) Uittien var. ***cuspidata***
Mapania cuspidata var. *pumila* (Uittien) Uittien = ***Mapania caudata*** Kük.
Mapania deistelii K.Schum. = [23 CMN]
Mapania dictyophlebia S.T.Blake = ***Mapania archboldii*** Uittien
Mapania dolichopoda Tang & F.T.Wang = ***Mapania silhetensis*** C.B.Clarke
Mapania dolichostachya K.Schum. = ***Mapania africana*** Boeck. subsp. ***africana***
Mapania dwanensis Cherm. = ***Mapania soyauxii*** (Boeck.) H.Pfeiff.
Mapania elegans E.G.Camus = ***Hypolytrum elegans*** (E.G.Camus) Uittien
Mapania flagellaris Uittien = ***Mapania debilis*** C.B.Clarke
Mapania flavinux T.Koyama = ***Hypolytrum flavinux*** (T.Koyama) D.A.Simpson
Mapania fluviatilis Sandwith = ***Mapania pycnocephala*** subsp. ***fluviatilis*** (Sandwith) T.Koyama
Mapania gabonica Cherm. = ***Mapania sylvatica*** subsp. ***gabonica*** (Cherm.) D.A.Simpson
Mapania geelvinkensis Ohwi = ***Mapania macrocephala*** (Gaudich.) K.Schum. subsp. ***macrocephala***
Mapania gracilipes Merr. = ***Mapania debilis*** C.B.Clarke
Mapania gracillima Kük. & Merr. = ***Paramapania gracillima*** (Kük. & Merr.) Uittien
Mapania graminea var. *stipitata* Uittien = ***Mapania graminea*** Uittien
Mapania grandiceps Kük. = ***Mapania palustris*** (Hassk. ex Steud.) Fern.-Vill. var. ***palustris***
Mapania grandis (Uittien) T.Koyama = ***Principina grandis*** Uittien
Mapania hainanensis Merr. = ***Hypolytrum hainanense*** (Merr.) Tang & F.T.Wang
Mapania herrerae Gómez-Laur. = ***Mapania cuatrecasasii*** T.Koyama
Mapania heterocephala Merr. = ***Mapania squamata*** (Kurz) C.B.Clarke
Mapania heyneana Backer = ***Mapania sumatrana*** subsp. ***pandanophylla*** (F.Muell.) D.A.Simpson
Mapania humilis (Hassk. ex Steud.) Fern.-Vill. = ***Hypolytrum humile*** (Hassk. ex Steud.) Boeck.
Mapania humilis var. *petiolata* (C.B.Clarke) Ridl. = ***Mapania cuspidata*** (Miq.) Uittien var. ***cuspidata***
Mapania hypolytroides (F.Muell.) F.Muell. ex Benth. = ***Mapania sumatrana*** subsp. ***pandanophylla*** (F.Muell.) D.A.Simpson
Mapania inopinata Uittien = ***Mapania cuspidata*** (Miq.) Uittien var. ***cuspidata***
Mapania insignis Holttum = ***Mapania holttumii*** J.Kern
Mapania javana Uittien = ***Mapania palustris*** (Hassk. ex Steud.) Fern.-Vill. var. ***palustris***
Mapania johorensis (Uittien) T.Koyama = ***Paramapania parvibractea*** (C.B.Clarke) Uittien
Mapania lactea Kük. = ***Mapania macrocephala*** (Gaudich.) K.Schum. subsp. ***macrocephala***
Mapania ledermannii Kük. = ***Mapania moseleyi*** C.B.Clarke
Mapania longa Ridl. ex C.B.Clarke = ***Mapania enodis*** (Miq.) C.B.Clarke
Mapania longiflora var. *sessilis* (Merr.) Uittien = ***Mapania sessilis*** Merr.
Mapania longifolia (Boeck.) C.B.Clarke = ***Mapania macrocephala*** subsp. ***longifolia*** (Boeck.) D.A.Simpson
Mapania longirostris Kük. = ***Paramapania longirostris*** (Kük.) Uittien
Mapania longispica Ridl. = ***Mapania kurzii*** C.B.Clarke
Mapania lucbanensis Elmer = ***Paramapania parvibractea*** (C.B.Clarke) Uittien
Mapania lucida N.E.Br. = ***Mapania cuspidata*** (Miq.) Uittien var. ***cuspidata***

Mapania macilentula (Ohwi) T.Koyama = ***Paramapania parvibractea*** (C.B.Clarke) Uittien
Mapania macrantha subsp. *ivorensis* J.Raynal = ***Mapania ivorensis*** (J.Raynal) J.Raynal
Mapania macrantha var. *minor* Nelmes = ***Mapania minor*** (Nelmes) J.Raynal
Mapania margaritae Ohwi = ***Mapania macrocephala*** (Gaudich.) K.Schum. subsp. ***macrocephala***
Mapania mildbraedii Graebn. = ***Mapania amplivaginata*** K.Schum.
Mapania monosperma (Jacq.-Fél.) Maguire & Koyama = ***Mapania amplivaginata*** K.Schum.
Mapania montana Ridl. = ***Mapania latifolia*** Uittien
Mapania montana Uittien = ***Mapania surinamensis*** Uittien
Mapania montana Lauterb. & K.Schum. = ***Paramapania parvibractea*** (C.B.Clarke) Uittien
Mapania moseleyi f. *latifolia* Uittien = ***Mapania macrocephala*** (Gaudich.) K.Schum. subsp. ***macrocephala***
Mapania multispicata C.B.Clarke = ***Mapania kurzii*** C.B.Clarke
Mapania oblonga C.B.Clarke = ***Mapania amplivaginata*** K.Schum.
Mapania oblonga var. *elliptica* C.B.Clarke = ***Mapania amplivaginata*** K.Schum.
Mapania palauensis Hosok. = ***Mapania macrocephala*** (Gaudich.) K.Schum. subsp. ***macrocephala***
Mapania pandanacea Ridl. = ***Mapania macrocephala*** (Gaudich.) K.Schum. subsp. ***macrocephala***
Mapania pandanifolia Sander = ?
Mapania pandanophylla (F.Muell.) K.Schum. = ***Mapania sumatrana*** subsp. ***pandanophylla*** (F.Muell.) D.A.Simpson
Mapania pandanophylla subsp. *immensa* (Kük.) T.Koyama = ***Mapania pacifica*** (Hosok.) T.Koyama
Mapania papuana Ridl. = ***Mapania baccifera*** C.B.Clarke subsp. ***baccifera***
Mapania parvibractea (C.B.Clarke) T.Koyama = ***Paramapania parvibractea*** (C.B.Clarke) Uittien
Mapania petiolata C.B.Clarke = ***Mapania cuspidata*** (Miq.) Uittien var. ***cuspidata***
Mapania petiolata var. *angustifolia* Uittien = ***Mapania cuspidata*** var. ***angustifolia*** (Uittien) Uittien
Mapania petiolata var. *pumila* Uittien = ***Mapania caudata*** Kük.
Mapania platyphylla Merr. = ***Mapania cuspidata*** (Miq.) Uittien var. ***cuspidata***
Mapania pynaertii De Wild. = ***Hypolytrum pynaertii*** (De Wild.) Nelmes
Mapania radians C.B.Clarke = ***Paramapania radians*** (C.B.Clarke) Uittien
Mapania radulosa Ridl. = ***Mapania baccifera*** C.B.Clarke subsp. ***baccifera***
Mapania rostrata Elmer ex Uittien = ***Paramapania rostrata*** Uittien
Mapania scaberrima (Boeck.) C.B.Clarke = ***Hypolytrum scaberrimum*** Boeck.
Mapania schomburgkii C.B.Clarke = ***Mapania macrophylla*** (Boeck.) H.Pfeiff.
Mapania secans K.Schum. = ***Hypolytrum secans*** (K.Schum.) J.Raynal
Mapania simplex (Ridl.) T.Koyama = ***Paramapania simplex*** (Ridl.) Uittien
Mapania sinensis Uittien = ***Mapania silhetensis*** C.B.Clarke
Mapania stolonifera Uittien = ***Mapania cuspidata*** (Miq.) Uittien var. ***cuspidata***
Mapania subcomposita C.B.Clarke = ***Mapania ferruginea*** Ridl.
Mapania subcomposita var. *purpuriceps* C.B.Clarke = ***Mapania purpuriceps*** (C.B.Clarke) J.Raynal
Mapania superba C.B.Clarke = ***Mapania macrantha*** (Boeck.) H.Pfeiff.
Mapania sylvatica Pursh = ***Cymophyllus fraserianus*** (Ker Gawl.) Kartesz & Gandhi
Mapania sylvestris Vahl = ***Mapania sylvatica*** Aubl. subsp. ***sylvatica***
Mapania thoreliana E.G.Camus = ***Hypolytrum compactum*** Nees & Meyen ex Kunth
Mapania triquetra Ridl. = ***Mapania cuspidata*** (Miq.) Uittien var. ***cuspidata***
Mapania tumida Uittien = ***Mapania enodis*** (Miq.) C.B.Clarke
Mapania valida Ridl. = ***Mapania tenuiscapa*** C.B.Clarke
Mapania yapensis Hosok. = ***Mapania macrocephala*** (Gaudich.) K.Schum. subsp. ***macrocephala***

Unplaced Names:
Mapania alpina Elmer ex Merr., Enum. Philipp. Fl. Pl. 1: 132 (1922). = [42 PHI]
Mapania deistelii K.Schum., Notizbl. Königl. Bot. Gart. Berlin 3: 106 (1901). = [23 CMN]
Mapania pandanifolia Sander, New Pl.: 14 (1897). = ?

Mapaniopsis

Mapaniopsis C.B.Clarke = ***Mapania*** Aubl.
Mapaniopsis effusa C.B.Clarke = ***Mapania effusa*** (C.B.Clarke) T.Koyama
Mapaniopsis micrococca T.Koyama = ***Mapania micrococca*** (T.Koyama) D.A.Simpson

Mariscopsis

Mariscopsis Cherm. = ***Queenslandiella*** Domin
Mariscopsis hyalinus (Vahl) Ballard = ***Queenslandiella hyalina*** (Vahl) Ballard
Mariscopsis suaveolens Cherm. = ***Queenslandiella hyalina*** (Vahl) Ballard

Marisculus

Marisculus Goetgh. = ***Alinula*** J.Raynal
Marisculus peteri (Kük.) Goetgh. = ***Alinula peteri*** (Kük.) Goetgh. & Vorster

Mariscus

Mariscus Scop. = ***Cladium*** P.Browne
Mariscus Gaertn. = ***Cyperus*** L.
Mariscus Vahl = ***Cyperus*** L.
Mariscus acicularis (L.) Moench = ***Eleocharis acicularis*** (L.) Roem. & Schult.
Mariscus acutus (R.Br.) C.B.Clarke = ***Cyperus lucidus*** R.Br.
Mariscus acutus (Labill.) Kuntze = ***Machaerina acuta*** (Labill.) J.Kern
Mariscus affinis (Brongn.) Kuntze = ***Morelotia affinis*** (Brongn.) S.T.Blake
Mariscus aggregatus Willd. = ***Cyperus aggregatus*** (Willd.) Endl.
Mariscus alatus Schult. = ***Cyperus coriifolius*** Boeck.
Mariscus albescens Gaudich. = ***Cyperus javanicus*** Houtt.
Mariscus albomarginatus C.B.Clarke = ***Cyperus vestitus*** Hochst. ex C.Krauss
Mariscus albopilosus C.B.Clarke = ***Cyperus albopilosus*** (C.B.Clarke) Kük.
Mariscus albosanguineus (Kük.) Napper = ***Cyperus albosanguineus*** Kük.

Mariscus albus (L.) Gilib. = ***Rhynchospora alba*** (L.) Vahl
Mariscus alopecuroides Steud. = [84]
Mariscus alpestris (K.Schum.) C.B.Clarke = ***Cyperus tomaiophyllus*** K.Schum.
Mariscus alpinus (Liebm.) C.B.Clarke = ***Cyperus hermaphroditus*** (Jacq.) Standl.
Mariscus alterniflorus (R.Br.) C.B.Clarke = ***Cyperus alterniflorus*** R.Br.
Mariscus alternifolius Vahl = ***Cyperus cyperoides*** (L.) Kuntze subsp. ***cyperoides***
Mariscus amauropus (Steud.) Cufod. = ***Cyperus amauropus*** Steud.
Mariscus ambiguus Liebm. = ***Cyperus hermaphroditus*** (Jacq.) Standl.
Mariscus amomodorus (K.Schum.) Cufod. = ***Cyperus amomodorus*** K.Schum.
Mariscus anceps (Poir.) Kuntze = ***Machaerina anceps*** (Poir.) Bojer
Mariscus ancistrophyllus (Benth.) Kuntze = ***Gahnia ancistrophylla*** Benth.
Mariscus andringitrensis Cherm. = ***Cyperus aster*** (C.B.Clarke ex Cherm.) Kük. var. ***aster***
Mariscus angularis Turrill = ***Cyperus chersinus*** (N.E.Br.) Kük.
Mariscus angustatus (R.Br.) C.B.Clarke = ***Cyperus angustatus*** R.Br.
Mariscus angustifolius (Gaudich.) Kuntze = ***Machaerina angustifolia*** (Gaudich.) T.Koyama
Mariscus aphyllus Vahl = ***Kyllinga tibialis*** Poit. ex Ledeb.
Mariscus apiculatus (Liebm.) T.Koyama = ***Cyperus manimae*** var. ***apiculatus*** (Liebm.) C.D.Adams
Mariscus appendiculatus Brongn. = ***Cyperus appendiculatus*** (Brongn.) Kunth
Mariscus arcuatoreflexus Cherm. = ***Cyperus varicus*** (C.B.Clarke ex Cherm.) Kük.
Mariscus aridicola Domin = ***Cyperus xerophilus*** Cherm.
Mariscus aristatus (Rottb.) Cherm. = ***Cyperus squarrosus*** L.
Mariscus aristatus (Benth.) Kuntze = ***Gahnia aristata*** Benth.
Mariscus armstrongii (Benth.) C.B.Clarke = ***Cyperus armstrongii*** Benth.
Mariscus aromaticus (Merr.) Fernald = ***Schoenus punctatus*** R.Br.
Mariscus arthrophyllus (Nees) Kuntze = ***Machaerina arthrophylla*** (Nees) T.Koyama
Mariscus articulatus (R.Br.) Kuntze = ***Machaerina articulata*** (R.Br.) T.Koyama
Mariscus asper Liebm. = ***Cyperus mutisii*** (Kunth) Andersson
Mariscus asper (R.Br.) Kuntze = ***Gahnia aspera*** (R.Br.) Spreng.
Mariscus assimilis (Steud.) Podlech = ***Courtoisina assimilis*** (Steud.) Maquet
Mariscus aster C.B.Clarke ex Cherm. = ***Cyperus aster*** (C.B.Clarke ex Cherm.) Kük.
Mariscus atrosanguineus Hochst. ex A.Rich. = ***Cyperus plateilema*** (Steud.) Kük.
Mariscus atrosanguineus Liebm. = ***Cyperus sanguineoater*** Boeck.
Mariscus aureus Bertol. = ***Cyperus retrorsus*** Chapm.
Mariscus australis (Nees) Kuntze = ***Gahnia australis*** (Nees) K.L.Wilson
Mariscus avenicola C.B.Clarke ex Britton = ***Cyperus retrorsus*** Chapm.
Mariscus aximensis C.B.Clarke = ***Cyperus tenuis*** Sw.
Mariscus badius Willd. ex Kunth = ***Cyperus subbadius*** Kük.
Mariscus balansae (Maury) C.B.Clarke = ***Cyperus impolitus*** Kunth
Mariscus baoulensis Hutch. = ***Cyperus baoulensis*** Kük.
Mariscus beecheyi (H.Mann) Kuntze = ***Gahnia beecheyi*** H.Mann
Mariscus bequaertii Cherm. = ***Cyperus ferrugineoviridis*** (C.B.Clarke) Kük.
Mariscus beyrichii Schrad. ex Nees = ***Cyperus beyrichii*** (Schrad. ex Nees) Steud.
Mariscus bidentatus (Vahl) J.Raynal = ***Cyperus confertus*** Sw.
Mariscus bidwellii (Stapf ex Setch.) J.W.Moore = ***Machaerina bidwellii*** (Stapf ex Setch.) T.Koyama
Mariscus biglumis Gaertn. = ***Cyperus cyperoides*** (L.) Kuntze subsp. ***cyperoides***
Mariscus binucifer C.B.Clarke = ***Cyperus vestitus*** Hochst. ex C.Krauss
Mariscus blatteri McCann = [40 IND]
Mariscus blodgettii (Britton) T.Koyama = ***Cyperus retrorsus*** Chapm.
Mariscus boeckeleri C.B.Clarke = ***Cyperus amomodorus*** K.Schum.
Mariscus bojeri C.B.Clarke = ***Cyperus longibracteatus*** (Cherm.) Kük.
Mariscus bolusii C.B.Clarke = ***Cyperus owanii*** Boeck.
Mariscus boninsimae (Nakai) Fernald = ***Machaerina glomerata*** (Gaudich.) T.Koyama
Mariscus borneensis (C.B.Clarke ex Stapf) Fernald = ***Lepidosperma chinense*** Nees & Meyen ex Kunth
Mariscus bowmanii (F.Muell. ex Benth.) C.B.Clarke = ***Cyperus bowmanii*** F.Muell. ex Benth.
Mariscus bracheilema Steud. = ***Cyperus echinatus*** (L.) Alph.Wood
Mariscus brachystachys Hook.f. = ***Cyperus anderssonii*** Boeck.
Mariscus bracteosus Schltdl. = ***Cyperus pohlii*** (Nees) Steud. subsp. ***pohlii***
Mariscus brevibracteatus (Domin) Domin = ***Cyperus dietrichiae*** var. ***brevibracteatus*** (Domin) Kük.
Mariscus brittonii C.B.Clarke = ***Cyperus retrorsus*** Chapm.
Mariscus brizaeus C.B.Clarke = ***Cyperus planifolius*** Rich.
Mariscus brunneus (Sw.) C.B.Clarke = ***Cyperus planifolius*** Rich.
Mariscus bulbocaulis A.Rich. = ***Cyperus plateilema*** (Steud.) Kük.
Mariscus bulbosus C.B.Clarke = ***Cyperus clarkei*** T.Cooke
Mariscus bulbosus Steud. = ***Cyperus plateilema*** (Steud.) Kük.
Mariscus bullatus (Kük.) Podlech = ***Cyperus chersinus*** (N.E.Br.) Kük.
Mariscus caesius (Boeck.) C.B.Clarke = ***Cyperus caesius*** Boeck.
Mariscus californicus (S.Watson) Fernald = ***Cladium mariscus*** subsp. ***californicum*** (S.Watson) Govaerts
Mariscus camphoratus (Liebm.) T.Koyama = ***Cyperus camphoratus*** Liebm.
Mariscus capensis (Steud.) Schrad. = ***Cyperus capensis*** (Steud.) Endl.
Mariscus capillaceus (Benth.) Fernald = ***Tetraria capillaris*** (F.Muell.) J.M.Black
Mariscus capillaris (Sw.) Vahl = ***Cyperus nanus*** Willd. var. ***nanus***
Mariscus capitatus Steud. = ***Remirea maritima*** Aubl.
Mariscus cariciformis F.Muell. = ***Cyperus paniceus*** (Rottb.) Boeck.

Mariscus carinatus (R.Br.) C.B.Clarke = ***Cyperus carinatus*** R.Br.
Mariscus castaneus Kunze ex Kunth = ***Cyperus poeppigii*** Kunth
Mariscus cayennensis (Lam.) Urb. = ***Cyperus aggregatus*** (Willd.) Endl. var. ***aggregatus***
Mariscus cayennensis var. *humilis* (Benth. ex C.B. Clarke) Stehlé = ***Cyperus redolens*** Maury ex Micheli
Mariscus cearaensis (R.Gross ex Kük.) T.Koyama = ***Cyperus cearaensis*** R.Gross ex Kük.
Mariscus cephalanthus (Torr. & Hook.) C.B.Clarke = ***Cyperus cephalanthus*** Torr. & Hook.
Mariscus cephalotes Poepp. ex Kunth = ***Cyperus humilis*** Kunth
Mariscus chaetophyllus Chiov. = ***Cyperus chaetophyllus*** (Chiov.) Kük.
Mariscus chersinus N.E.Br. = ***Cyperus chersinus*** (N.E.Br.) Kük.
Mariscus chionocephalus Chiov. = ***Cyperus chionocephalus*** (Chiov.) Chiov. ex Chiarugi
Mariscus chiuensis (Nees ex Hook. & Arn.) Fernald = ***Cladium mariscus*** subsp. ***chinense*** (Nees ex Hook. & Arn.) T.Koyama
Mariscus chlorocephalus C.B.Clarke = ***Cyperus chlorocephalus*** (C.B.Clarke) Kük.
Mariscus chrysocephalus K.Schum. = ***Cyperus chrysocephalus*** (K.Schum.) Kük.
Mariscus ciliatus (Jungh.) T.Koyama = ***Cyperus ciliatus*** Jungh.
Mariscus circumclusus C.B.Clarke = ***Cyperus circumclusus*** (C.B.Clarke) Schweinf.
Mariscus cladium (Sw.) Kuntze = ***Cladium mariscus*** (L.) Pohl subsp. ***mariscus***
Mariscus clarkei (T.Cooke) T.Koyama = ***Cyperus clarkei*** T.Cooke
Mariscus clarkei Turrill ex Craib = ***Cyperus cyperoides*** (L.) Kuntze subsp. ***cyperoides***
Mariscus coloratus (Vahl) Nees = ***Cyperus dubius*** Rottb. subsp. ***dubius***
Mariscus coloratus var. *macrocephalus* C.B.Clarke = ***Cyperus dubius*** subsp. ***macrocephalus*** (C.B.Clarke) Lye
Mariscus colpodes (Lauterb.) Fernald = ***Machaerina glomerata*** (Gaudich.) T.Koyama
Mariscus compactus (Retz.) Bold. = ***Cyperus compactus*** Retz.
Mariscus compactus var. *macrostachys* (Boeck.) H.B.Naithani & S.Biswas = ***Cyperus compactus*** Retz.
Mariscus complanatus (Berggr.) Fernald = ***Machaerina complanata*** (Berggr.) T.Koyama
Mariscus conceptionis (Steud.) C.B.Clarke = ***Cyperus rigens*** J.Presl & C.Presl subsp. ***rigens***
Mariscus concinnus C.B.Clarke = ***Cyperus concinniformis*** Kük.
Mariscus concinnus Schrad. ex Nees = ***Cyperus cyperoides*** (L.) Kuntze subsp. ***cyperoides***
Mariscus concinnus var. *evolutior* (C.B.Clarke) Panigrahi = ***Cyperus cyperoides*** (L.) Kuntze subsp. ***cyperoides***
Mariscus concinnus var. *khasianus* (C.B.Clarke) Panigrahi = ***Cyperus cyperoides*** (L.) Kuntze subsp. ***cyperoides***
Mariscus concinnus var. *subcompositus* (C.B.Clarke) Panigrahi = ***Cyperus cyperoides*** (L.) Kuntze subsp. ***cyperoides***
Mariscus confertospicatus Steud. = ***Cyperus aggregatus*** (Willd.) Endl. var. ***aggregatus***
Mariscus confertus Kunth = ***Cyperus aggregatus*** (Willd.) Endl. var. ***aggregatus***
Mariscus confertus (Sw.) C.B.Clarke = ***Cyperus confertus*** Sw.
Mariscus conflexus Link = ***Cyperus hermaphroditus*** (Jacq.) Standl.
Mariscus congestus (Vahl) C.B.Clarke = ***Cyperus congestus*** Vahl
Mariscus congestus (Boeck.) Kuntze = ***Gahnia aspera*** subsp. ***globosa*** (H.Mann) J.Kern
Mariscus congestus var. *glanduliferus* C.B.Clarke = ***Cyperus congestus*** Vahl
Mariscus conicus R.Br. = ***Cyperus conicus*** (R.Br.) Boeck.
Mariscus cooperi C.B.Clarke = ***Cyperus congestus*** Vahl
Mariscus coriaceus G.Mey. = ***Cyperus ligularis*** L.
Mariscus cornelii-ostenii (Kük.) T.Koyama = ***Cyperus cornelii-ostenii*** Kük.
Mariscus cornutus Andersson = ***Cyperus elegans*** subsp. ***rubiginosus*** (Hook.f.) Eliasson
Mariscus corymbosus Boeck. = ***Cyperus cyperoides*** (L.) Kuntze subsp. ***cyperoides***
Mariscus crassus (Thwaites) Kuntze = ***Machaerina rubiginosa*** (Spreng.) T.Koyama
Mariscus cufodontii Chiov. = [24 ETH]
Mariscus cunninghamii C.B.Clarke = ***Cyperus cunninghamii*** (C.B.Clarke) C.A.Gardner
Mariscus cupreus Hochst. ex Boeck. = ***Cyperus impubes*** Steud. var. ***impubes***
Mariscus curtisii C.B.Clarke = ***Cyperus ovatus*** Baldwin
Mariscus cylindricus Elliott = ***Cyperus retrorsus*** Chapm.
Mariscus cylindricus var. *australis* Lindm. = ***Cyperus aggregatus*** var. ***gigas*** (Lindm.) Guagl.
Mariscus cylindristachyus Steud. = ***Cyperus cyperoides*** (L.) Kuntze subsp. ***cyperoides***
Mariscus cyperiformis (Muhl.) Torr. = ***Cyperus lupulinus*** subsp. ***macilentus*** (Fernald) Marcks
Mariscus cyperinus (Retz.) Vahl = ***Cyperus cyperinus*** (Retz.) Suringar
Mariscus cyperinus var. *bengalensis* C.B.Clarke = ***Cyperus cyperinus*** (Retz.) Suringar
Mariscus cyperinus subsp. *laxatus* (C.B.Clarke) T.Koyama = ***Cyperus cyperinus*** (Retz.) Suringar
Mariscus cyperinus var. *maximus* C.B.Clarke = ***Cyperus cyperinus*** (Retz.) Suringar
Mariscus cyperinus var. *pictus* (Nees) Karthik. = ***Cyperus cyperinus*** (Retz.) Suringar
Mariscus cyperoides (Roxb.) A.Dietr. = ***Courtoisina cyperoides*** (Roxb.) Soják
Mariscus cyperoides (L.) Urb. = ***Cyperus cyperoides*** (L.) Kuntze
Mariscus cyperoides var. *capitatus* Domin = ***Cyperus cyperoides*** (L.) Kuntze subsp. ***cyperoides***
Mariscus dactyliformis (Boeck.) C.B.Clarke = ***Cyperus dactyliformis*** Boeck.
Mariscus decaryi Cherm. = ***Cyperus dubius*** Rottb. subsp. ***dubius***
Mariscus decaryi var. *pauciflorus* Cherm. = ***Cyperus dubius*** Rottb. subsp. ***dubius***
Mariscus deciduus (Boeck.) C.B.Clarke = ***Cyperus deciduus*** Boeck.
Mariscus decipiens (Hillebr.) C.B.Clarke = ***Cyperus hillebrandii*** var. ***decipiens*** (Hillebr.) Kük.
Mariscus decompositus R.Br. = ***Cyperus decompositus*** (R.Br.) F.Muell.
Mariscus decompositus (R.Br.) Kuntze = ***Gahnia decomposita*** (R.Br.) Benth.
Mariscus delicatulus Fernald = ***Schoenus delicatulus*** (Fernald) J.Kern
Mariscus deplanchei (Boeck.) Schinz & Guillaumin = ***Machaerina deplanchei*** (Boeck.) T.Koyama

Mariscus detersus C.B.Clarke ex Cherm. = ***Cyperus dubius*** Rottb. subsp. ***dubius***
Mariscus deustus (R.Br.) Kuntze = ***Gahnia deusta*** (R.Br.) Benth.
Mariscus diamantinus D.A.Simpson = ***Cyperus diamantinus*** (D.A.Simpson) Govaerts & D.A.Simpson
Mariscus dicigerus (Liebm.) T.Koyama = ***Cyperus tenuis*** Sw.
Mariscus dietrichiae (Boeck.) C.B.Clarke = ***Cyperus dietrichiae*** Boeck.
Mariscus dietrichiae subsp. *bowmannii* (F.Muell. ex Benth.) Domin = ***Cyperus bowmanii*** F.Muell. ex Benth.
Mariscus dietrichiae subsp. *brevibracteatus* Domin = ***Cyperus dietrichiae*** var. ***brevibracteatus*** (Domin) Kük.
Mariscus dietrichiae var. *brevibracteatus* Domin = ***Cyperus dietrichiae*** var. ***brevibracteatus*** (Domin) Kük.
Mariscus dilutus Nees = ***Cyperus compactus*** Retz.
Mariscus dipsaceus (Liebm.) C.B.Clarke = ***Cyperus dipsaceus*** Liebm.
Mariscus dissitiflorus (Nees ex Torr.) C.B.Clarke = ***Cyperus thyrsiflorus*** Jungh. ex Schltdl.
Mariscus dissitiflorus var. *lucidior* C.B.Clarke = ***Cyperus thyrsiflorus*** Jungh. ex Schltdl.
Mariscus dissolutus (Kunth) T.Koyama = ***Cyperus dissolutus*** Kunth
Mariscus distichus (C.B.Clarke) Fernald = ***Machaerina disticha*** (C.B.Clarke) T.Koyama
Mariscus diurensis (Boeck.) C.B.Clarke = ***Cyperus diurensis*** Boeck.
Mariscus drakensbergensis Vorster = ***Cyperus drakensbergensis*** (Vorster) Govaerts
Mariscus dregeanus Kunth = ***Cyperus dubius*** Rottb. subsp. ***dubius***
Mariscus dregeanus var. *buchananii* C.B.Clarke = ***Cyperus dubius*** Rottb. subsp. ***dubius***
Mariscus drummondii Steud. = ***Cyperus echinatus*** (L.) Alph.Wood
Mariscus dubius (Rottb.) Kük. ex G.E.C.Fischer = ***Cyperus dubius*** Rottb.
Mariscus dubius var. *coloratus* (Vahl) Kük. = ***Cyperus dubius*** Rottb. subsp. ***dubius***
Mariscus dubius subsp. *macrocephalus* (C.B.Clarke) J.-P.Lebrun & Stork = ***Cyperus dubius*** subsp. ***macrocephalus*** (C.B.Clarke) Lye
Mariscus dunensis Cherm. = ***Cyperus dunensis*** (Cherm.) Kük.
Mariscus duripes (I.M.Johnst.) Conz. = ***Cyperus duripes*** I.M.Johnst.
Mariscus durus (Kunth) C.B.Clarke = ***Cyperus durus*** Kunth
Mariscus echinatus (L.) Elliott = ***Cyperus echinatus*** (L.) Alph.Wood
Mariscus ehrenbergianus (Boeck.) C.B.Clarke = ***Cyperus mutisii*** (Kunth) Andersson
Mariscus ekmanii (Kük.) T.Koyama = ***Cyperus ekmanii*** Kük.
Mariscus elatior (Boeck.) C.B.Clarke = ***Cyperus dactyliformis*** Boeck.
Mariscus elatus Kunth = ***Cyperus aggregatus*** (Willd.) Endl. var. ***aggregatus***
Mariscus elatus Vahl = ***Cyperus coriifolius*** Boeck.
Mariscus elephantinus C.B.Clarke = ***Cyperus elephantinus*** (C.B.Clarke) C.B.Clarke
Mariscus elevatus J.W.Moore = ***Cladium mariscus*** subsp. ***jamaicense*** (Crantz) Kük.
Mariscus ensiger (Hance) Fernald = ***Machaerina ensigera*** (Hance) T.Koyama
Mariscus eurystachys (Ridl.) C.B.Clarke = ***Cyperus tenuis*** Sw.
Mariscus falcatus (Nees) Fernald = ***Machaerina falcata*** (Nees) T.Koyama
Mariscus fallax Cherm. = ***Cyperus impubes*** var. ***fallax*** (Cherm.) Kük.
Mariscus fallax Fernald = ***Fimbristylis vanoverberghii*** Kük.
Mariscus fallax var. *major* Cherm. = ***Cyperus impubes*** var. ***fallax*** (Cherm.) Kük.
Mariscus familiaris (Steud.) T.Koyama = ***Cyperus familiaris*** Steud.
Mariscus fasciculigerus (F.Muell.) Domin = ***Cyperus cunninghamii*** (C.B.Clarke) C.A.Gardner subsp. ***cunninghamii***
Mariscus fauriei (Kük.) T.Koyama = ***Cyperus fauriei*** Kük.
Mariscus fendlerianus (Boeck.) T.Koyama = ***Cyperus fendlerianus*** Boeck.
Mariscus ferax (Rich.) C.B.Clarke = ***Cyperus odoratus*** L. subsp. ***odoratus***
Mariscus ferrugineoviridis (C.B.Clarke) Cherm. = ***Cyperus ferrugineoviridis*** (C.B.Clarke) Kük.
Mariscus ficticius (Hemsl.) Fernald = ***Machaerina ficticia*** (Hemsl.) T.Koyama
Mariscus filiculmis (Vahl) T.Koyama = ***Cyperus filiculmis*** Vahl
Mariscus filifolius (Griseb.) Kuntze = ***Machaerina filifolia*** Griseb.
Mariscus filiformis Kunth = ***Cyperus filifolius*** Willd. ex Kunth
Mariscus filiformis (Sw.) Roem. & Schult. = ***Cyperus filiformis*** Sw.
Mariscus filiformis (Sw.) Spreng. = ***Cyperus swartzii*** (A.Dietr.) Boeck. ex Kük.
Mariscus filum (Labill.) Kuntze = ***Gahnia filum*** (Labill.) F.Muell.
Mariscus firmipes C.B.Clarke = ***Cyperus firmipes*** (C.B.Clarke) Kük.
Mariscus flabelliformis Kunth = ***Cyperus tenuis*** Sw.
Mariscus flabelliformis var. *aximensis* (C.B.Clarke) S.S.Hooper = ***Cyperus tenuis*** Sw.
Mariscus flavus Vahl = ***Cyperus aggregatus*** (Willd.) Endl. var. ***aggregatus***
Mariscus flavus Seem. = ***Cyperus seemannianus*** Boeck.
Mariscus flavus var. *gigas* Lindm. = ***Cyperus aggregatus*** var. ***gigas*** (Lindm.) Guagl.
Mariscus flavus var. *humilis* Benth. ex C.B.Clarke = ***Cyperus redolens*** Maury ex Micheli
Mariscus flavus f. *umbellatoflavus* C.B.Clarke = ***Cyperus aggregatus*** (Willd.) Endl.
Mariscus floridensis C.B.Clarke = ***Cyperus floridanus*** Britton
Mariscus foliosissimus Steud. = ***Oxycaryum cubense*** (Poepp. & Kunth) Palla
Mariscus foliosus C.B.Clarke = ***Cyperus luteus*** Boeck.
Mariscus fuligineus (Chapm.) C.B.Clarke = ***Cyperus fuligineus*** Chapm.
Mariscus fulvus (R.Br.) C.B.Clarke = ***Cyperus fulvus*** R.Br.
Mariscus fulvus var. *canescens* C.B.Clarke ex Domin = ***Cyperus fulvus*** R.Br.
Mariscus fulvus var. *confusus* C.B.Clarke ex Domin = ***Cyperus fulvus*** R.Br.
Mariscus fulvus var. *densispiculatus* Domin = ***Cyperus fulvus*** R.Br.
Mariscus fulvus var. *jucundus* Domin = ***Cyperus fulvus*** R.Br.
Mariscus gaudichaudii (W.Wight) Fernald = ***Machaerina glomerata*** (Gaudich.) T.Koyama

Mariscus gaudichaudii (Brongn.) Kuntze = ***Morelotia gahniiformis*** Gaudich.
Mariscus gayi C.B.Clarke = ***Cyperus gayi*** (C.B.Clarke) Kük.
Mariscus geniculatus Fernald = ***Machaerina disticha*** (C.B.Clarke) T.Koyama
Mariscus giganteus Boeck. = ***Cyperus compactus*** Retz.
Mariscus glandulosus Bojer = ***Cyperus ligularis*** L.
Mariscus globiceps (C.B.Clarke) Fernald = ***Machaerina glomerata*** (Gaudich.) T.Koyama
Mariscus globifer C.B.Clarke = ***Cyperus globifer*** (C.B.Clarke) Lye
Mariscus globosus (H.Mann) Kuntze = ***Gahnia aspera*** subsp. ***globosa*** (H.Mann) J.Kern
Mariscus globulosus Urb. = ***Cyperus echinatus*** (L.) Alph.Wood
Mariscus glomeratus W.Barton = ***Cyperus lupulinus*** subsp. ***macilentus*** (Fernald) Marcks
Mariscus glomeratus (R.Br.) Kuntze = ***Machaerina rubiginosa*** (Spreng.) T.Koyama
Mariscus goniobolbus Cherm. = ***Cyperus pseudovestitus*** (C.B.Clarke) Kük.
Mariscus goniobolbus var. *angustifolius* Cherm. = ***Cyperus pseudovestitus*** (C.B.Clarke) Kük.
Mariscus goniobolbus var. *perrieri* (Cherm.) Cherm. = ***Cyperus pseudovestitus*** (C.B.Clarke) Kük.
Mariscus grabowskianus (Boeck.) Merr. = ***Cyperus compactus*** Retz.
Mariscus gracilis Vahl = ***Cyperus swartzii*** (A.Dietr.) Boeck. ex Kük.
Mariscus grantii C.B.Clarke = ***Cyperus vorsteri*** K.L.Wilson
Mariscus grayi (Torr.) C.B.Clarke = ***Cyperus grayi*** Torr.
Mariscus gregorii C.B.Clarke = ***Cyperus hemisphaericus*** Boeck.
Mariscus guatemalensis C.B.Clarke = ***Cyperus manimae*** var. ***asperrimus*** (Liebm.) Kük.
Mariscus gueinzii C.B.Clarke = ***Cyperus dactyliformis*** Boeck.
Mariscus guianensis C.B.Clarke = ***Cyperus guianensis*** (C.B.Clarke) Kük.
Mariscus gunnii (Hook.f.) C.B.Clarke = ***Cyperus gunnii*** Hook.f.
Mariscus gunnii var. *novae-hollandiae* (Boeck.) Domin = ***Cyperus gunnii*** var. ***novae-hollandiae*** (Boeck.) Kük.
Mariscus gunnii var. *typicus* Domin = ***Cyperus gunnii*** Hook.f.
Mariscus haematodes (Endl.) C.B.Clarke = ***Cyperus haematodes*** Endl.
Mariscus haenkei J.Presl & C.Presl = ***Cyperus regiomontanus*** Britton
Mariscus hainanensis Chun & F.C.How = [36 CHH]
Mariscus hamulosus (M.Bieb.) S.S.Hooper = ***Cyperus hamulosus*** M.Bieb.
Mariscus hassleri C.B.Clarke = ***Cyperus impolitus*** Kunth
Mariscus havanensis Kunth = ***Cyperus filiformis*** Sw.
Mariscus hemisphaericus (Boeck.) C.B.Clarke = ***Cyperus hemisphaericus*** Boeck.
Mariscus hermaphroditus (Jacq.) Urb. = ***Cyperus hermaphroditus*** (Jacq.) Standl.
Mariscus hermaphroditus var. *peduncularis* (Britton) Stehlé = ***Cyperus dentoniae*** G.C.Tucker
Mariscus hillebrandii (Boeck.) T.Koyama = ***Cyperus hillebrandii*** Boeck.
Mariscus hillebrandii subsp. *decipiens* (Hillebr.) T.Koyama = ***Cyperus hillebrandii*** var. ***decipiens*** (Hillebr.) Kük.
Mariscus hilsenbergii C.B.Clarke = ***Cyperus pseudobrunneus*** (C.B.Clarke ex Cherm.) Kük.
Mariscus hirtellus Chiov. = ***Cyperus hirtellus*** (Chiov.) Kük.
Mariscus hirtus (Boeck.) Fernald = ***Machaerina hirta*** Boeck.
Mariscus hochstetteri Walp. = ***Cyperus plateilema*** (Steud.) Kük.
Mariscus holoschoenus (R.Br.) C.B.Clarke = ***Cyperus holoschoenus*** R.Br.
Mariscus hookerianus C.B.Clarke = ***Cyperus sikkimensis*** Kük.
Mariscus houghtonii (Torr.) T.Koyama = ***Cyperus houghtonii*** Torr.
Mariscus huarmensis Kunth = ***Cyperus odoratus*** L. subsp. ***odoratus***
Mariscus humbertii Cherm. = ***Cyperus macrocarpus*** (Kunth) Boeck.
Mariscus humilis Zeyh. ex Schrad. = ***Cyperus tabularis*** Schrad.
Mariscus humilis Nees ex C.B.Clarke. = ***Cyperus tabularis*** Schrad.
Mariscus huttonii (Kirk) Fernald = ***Machaerina huttonii*** (Kirk) T.Koyama
Mariscus hypochlorus (Hillebr.) C.B.Clarke = ***Cyperus hypochlorus*** Hillebr.
Mariscus hypochlorus subsp. *brevior* (Kük.) T.Koyama = ***Cyperus hypochlorus*** var. ***brevior*** Kük.
Mariscus impolitus (Kunth) J.Raynal = ***Cyperus impolitus*** Kunth
Mariscus impubes (Steud.) Napper = ***Cyperus impubes*** Steud.
Mariscus incompletus Urb. = ***Cyperus coriifolius*** Boeck.
Mariscus indecorus (Kunth) Podlech = ***Cyperus indecorus*** Kunth
Mariscus inflatus C.B.Clarke = ***Cyperus vestitus*** Hochst. ex C.Krauss
Mariscus infucatus (Kunth) T.Koyama = ***Cyperus infucatus*** Kunth
Mariscus insularis (Benth.) Kuntze = ***Machaerina insularis*** (Benth.) T.Koyama
Mariscus intricatus (L.) Cufod. = ***Cyperus squarrosus*** L.
Mariscus involutus C.B.Clarke = ***Cyperus dactyliformis*** Boeck.
Mariscus iridifolius (Bory) Kuntze = ***Machaerina iridifolia*** (Bory) T.Koyama
Mariscus irroratus Nees = ***Cyperus dubius*** Rottb. subsp. ***dubius***
Mariscus ischnos (Schltdl.) C.B.Clarke = ***Cyperus ischnos*** Schltdl.
Mariscus ivohibensis Cherm. = ***Cyperus ivohibensis*** (Cherm.) Kük.
Mariscus jacquinii Kunth = ***Cyperus hermaphroditus*** (Jacq.) Standl.
Mariscus jacquinii var. *angustior* C.B.Clarke = ***Cyperus hermaphroditus*** (Jacq.) Standl.
Mariscus jacquinii subsp. *angustior* C.B.Clarke = ***Cyperus hermaphroditus*** (Jacq.) Standl.
Mariscus jamaicensis (Crantz) Britton = ***Cladium mariscus*** subsp. ***jamaicense*** (Crantz) Kük.
Mariscus javanicus (Houtt.) Merr. & F.P.Metcalf = ***Cyperus javanicus*** Houtt.
Mariscus javanicus (Moritzi) Kuntze = ***Gahnia javanica*** Moritzi
Mariscus junceus (R.Br.) Kuntze = ***Machaerina juncea*** (R.Br.) T.Koyama
Mariscus karisimbiensis Cherm. = ***Cyperus karisimbiensis*** (Cherm.) Kük.

Mariscus karwinskianus Kunth = ***Cyperus mutisii*** (Kunth) Andersson
Mariscus keniensis (Kük.) S.S.Hooper = ***Cyperus keniensis*** Kük.
Mariscus kerstenii (Boeck.) C.B.Clarke = ***Cyperus kerstenii*** Boeck.
Mariscus kraussii Hochst. = ***Cyperus dubius*** Rottb. subsp. ***dubius***
Mariscus kraussii var. *capitatus* Cherm. = ***Cyperus dubius*** Rottb. subsp. ***dubius***
Mariscus kunthianus Gaudich. = ***Cyperus neokunthianus*** Kük.
Mariscus kyllingiiformis Boeck. = ***Cyperus dubius*** Rottb. subsp. ***dubius***
Mariscus kyllingioides Steud. = ***Kyllinga brevifolia*** Rottb. subsp. ***brevifolia***
Mariscus laeteflorens C.B.Clarke = ***Cyperus laeteflorens*** (C.B.Clarke) Kük.
Mariscus laetus (J.Presl & C.Presl) Lindm. = ***Cyperus laetus*** J.Presl & C.Presl
Mariscus laevigatus Roem. & Schult. = ***Cyperus aggregatus*** (Willd.) Endl. var. ***aggregatus***
Mariscus laevigatus Hook. & Arn. = ***Cyperus seemannianus*** Boeck.
Mariscus laevis Kunth = ***Cyperus aggregatus*** (Willd.) Endl. var. ***aggregatus***
Mariscus laevis R.Br. = ***Cyperus leiocaulon*** Benth.
Mariscus lancastriensis (Porter) C.B.Clarke = ***Cyperus lancastriensis*** Porter
Mariscus laniger (R.Br.) Kuntze = ***Gahnia lanigera*** (R.Br.) Benth.
Mariscus latibracteatus Palla = ***Cyperus aggregatus*** (Willd.) Endl. var. ***aggregatus***
Mariscus latifolius Schrad. ex Nees = ***Cyperus coriifolius*** Boeck.
Mariscus lavarum (Poir.) Kuntze = ***Machaerina lavarum*** (Poir.) Bojer
Mariscus laxiflorus Turrill = ***Cyperus turrillii*** Kük.
Mariscus laxus (Nees) Kuntze = ***Machaerina laxa*** (Nees) T.Koyama
Mariscus lentiginosus (Millsp. & Chase) Conz. = ***Cyperus lentiginosus*** Millsp. & Chase
Mariscus leptochloides Steud. = ***Cyperus imbricatus*** Retz.
Mariscus leptophyllus (Hochst. ex Steud.) C.B.Clarke = ***Cyperus amauropus*** Steud.
Mariscus leptostachyus (Nees) Kuntze = ***Cladium mariscus*** subsp. ***jamaicense*** (Crantz) Kük.
Mariscus ligularis (L.) Urb. = ***Cyperus ligularis*** L.
Mariscus litoreus C.B.Clarke = ***Cyperus ovatus*** Baldwin
Mariscus littoralis G.Mey. = ***Cyperus aggregatus*** (Willd.) Endl. var. ***aggregatus***
Mariscus longibracteatus Cherm. = ***Cyperus longibracteatus*** (Cherm.) Kük.
Mariscus longibracteatus var. *keniensis* (Kük.) Maquet = ***Cyperus keniensis*** Kük.
Mariscus longiradiatus Liebm. = ***Cyperus hermaphroditus*** (Jacq.) Standl.
Mariscus longus Schrad. ex Nees = ***Cyperus infucatus*** Kunth
Mariscus lucidus (R.Br.) C.B.Clarke = ***Cyperus lucidus*** R.Br.
Mariscus luridus T.Durand & De Wild. = ***Cyperus tenuis*** Sw.
Mariscus luteus (Boeck.) C.B.Clarke = ***Cyperus luteus*** Boeck.
Mariscus luzuliformis (Boeck.) C.B.Clarke = ***Cyperus indecorus*** Kunth
Mariscus macer Kunth = ***Cyperus cyperoides*** subsp. ***pseudoflavus*** (C.B.Clarke) Lye
Mariscus macrocarpus Kunth = ***Cyperus macrocarpus*** (Kunth) Boeck.
Mariscus macrophyllus Brongn. = ***Cyperus macrophyllus*** (Brongn.) Boeck.
Mariscus macropus C.B.Clarke = ***Cyperus circumclusus*** (C.B.Clarke) Schweinf.
Mariscus maculatus (Michx.) Roem. & Schult. = ***Lipocarpha maculata*** (Michx.) Torr.
Mariscus maderaspatanus (Willd.) Napper = ***Cyperus maderaspatanus*** Willd.
Mariscus magnus C.B.Clarke = ***Cyperus tomaiophyllus*** K.Schum.
Mariscus maingayi (C.B.Clarke) Fernald = ***Machaerina maingayi*** (C.B.Clarke) T.Koyama
Mariscus malawicus J.Raynal = ***Alinula malawica*** (J.Raynal) Goetgh. & Vorster
Mariscus manimae (Kunth) C.B.Clarke = ***Cyperus manimae*** Kunth
Mariscus manimae var. *apiculatus* (Liebm.) C.B.Clarke = ***Cyperus manimae*** var. ***apiculatus*** (Liebm.) C.D.Adams
Mariscus manimae var. *divergens* (Kunth) C.B.Clarke = ***Cyperus manimae*** Kunth var. ***manimae***
Mariscus manimae var. *hartwegianus* C.B.Clarke = ***Cyperus manimae*** Kunth var. ***manimae***
Mariscus mannii (Hillebr.) Kuntze = ***Gahnia beecheyi*** H.Mann
Mariscus manongarivensis Cherm. = ***Cyperus luteus*** Boeck.
Mariscus marianus Kuntze = ***Machaerina mariscoides*** (Gaudich.) J.Kern
Mariscus mariscoides (Muhl.) Kuntze = ***Cladium mariscoides*** (Muhl.) Torr.
Mariscus mariscoides f. *congestus* (Fernald) Fernald = ***Cladium mariscoides*** (Muhl.) Torr.
Mariscus mariscus (L.) Borbás = ***Cladium mariscus*** (L.) Pohl
Mariscus maritimus Miq. = ***Cyperus dubius*** Rottb. subsp. ***dubius***
Mariscus maritimus C.B.Clarke = ***Cyperus karisimbiensis*** (Cherm.) Kük. var. ***karisimbiensis***
Mariscus marlothii (Boeck.) C.B.Clarke = ***Cyperus marlothii*** Boeck.
Mariscus marlothii var. *globospica* C.B.Clarke = ***Cyperus capensis*** (Steud.) Endl.
Mariscus martii (Dufour ex Roem. & Schult.) Fernald = ***Cladium mariscus*** (L.) Pohl subsp. ***mariscus***
Mariscus mayottensis Cherm. = [29 COM]
Mariscus melanocarpus (R.Br.) Kuntze = ***Gahnia melanocarpa*** R.Br.
Mariscus merrillii C.B.Clarke = ***Cyperus dubius*** Rottb. subsp. ***dubius***
Mariscus mexicanus Willd. ex Link = ***Cyperus aggregatus*** (Willd.) Endl. var. ***aggregatus***
Mariscus mexicanus Steud. = ***Cyperus mutisii*** (Kunth) Andersson
Mariscus meyenianus (Kunth) Nees = ***Cyperus meyenianus*** Kunth
Mariscus meyenii (Kunth) Kuntze = ***Machaerina mariscoides*** (Gaudich.) J.Kern
Mariscus micans Nees = ***Cyperus mutisii*** (Kunth) Andersson
Mariscus micranthes (C.B.Clarke) Fernald = ***Machaerina disticha*** (C.B.Clarke) T.Koyama
Mariscus microbolbos (C.B.Clarke) Vorster = ***Cyperus microbolbos*** C.B.Clarke

Mariscus microcephalus J.Presl & C.Presl = ***Cyperus compactus*** Retz.
Mariscus microcephalus var. *pauciflorus* H.Pfeiff. = ***Cyperus compactus*** Retz.
Mariscus microstachyus (Benth.) Kuntze = ***Gahnia microstachya*** Benth.
Mariscus milnei (C.B.Clarke) Fernald = ***Machaerina milnei*** (C.B.Clarke) T.Koyama
Mariscus mohrii (Britton ex C.B.Clarke) T.Koyama = ***Cyperus spectabilis*** Link var. ***spectabilis***
Mariscus mollipes C.B.Clarke = ***Cyperus mollipes*** (C.B.Clarke) K.Schum.
Mariscus moniliferus Chiov. = ***Cyperus impubes*** Steud. var. ***impubes***
Mariscus monospermus S.M.Huang = [36 CHH]
Mariscus mucronatus (L.) Gaertn. = ***Cyperus capitatus*** Vand.
Mariscus mucronatus C.Presl = ***Cyperus capitatus*** Vand.
Mariscus mucronatus Kuntze = ***Gahnia beecheyi*** H.Mann
Mariscus muelleri (C.B.Clarke) Fernald = ***Machaerina muelleri*** (C.B.Clarke) T.Koyama
Mariscus multifolius (Kunth) T.Koyama = ***Cyperus multifolius*** Kunth
Mariscus mutisii Hook.f. = ***Cyperus elegans*** subsp. ***rubiginosus*** (Hook.f.) Eliasson
Mariscus mutisii Kunth = ***Cyperus mutisii*** (Kunth) Andersson
Mariscus myrmecias (Ridl.) C.B.Clarke = ***Cyperus myrmecias*** Ridl.
Mariscus nanus (Willd.) T.Koyama = ***Cyperus nanus*** Willd.
Mariscus neglectus Schult. = ***Cyperus retrorsus*** Chapm.
Mariscus niveus (L.) Merr. = ***Kyllinga bulbosa*** P.Beauv.
Mariscus nogalensis Chiov. = ***Cyperus grandibulbosus*** C.B.Clarke
Mariscus nossibeensis Steud. = ***Cyperus cyperoides*** (L.) Kuntze subsp. ***cyperoides***
Mariscus nudus (Steud.) Kuntze = ***Machaerina nuda*** (Steud.) J.Kern
Mariscus nyassensis Podlech = ***Cyperus nyassensis*** (Podlech) Lye
Mariscus oaxacensis Steud. = ***Cyperus mutisii*** (Kunth) Andersson
Mariscus oblonginux C.B.Clarke = ***Cyperus amomodorus*** K.Schum.
Mariscus obsoletenervosus (Peter & Kük.) Greenway = ***Cyperus obsoletenervosus*** Peter & Kük.
Mariscus oligostachys Schrad. ex Nees = ***Cyperus meyenianus*** Kunth
Mariscus ouveanus Däniker = ***Machaerina juncea*** (R.Br.) T.Koyama
Mariscus ovularis (Michx.) Vahl = ***Cyperus echinatus*** (L.) Alph.Wood
Mariscus ovularis Kunth = ***Cyperus echinatus*** (L.) Alph.Wood
Mariscus ovularis Torr. = ***Cyperus retrorsus*** Chapm.
Mariscus ovularis var. *tenellus* Torr. = ***Cyperus echinatus*** (L.) Alph.Wood
Mariscus owanii (Boeck.) C.B.Clarke = ***Cyperus owanii*** Boeck.
Mariscus pachystachyus Steud. = ***Cyperus cyperoides*** (L.) Kuntze subsp. ***cyperoides***
Mariscus pallens Liebm. = ***Cyperus thyrsiflorus*** Jungh. ex Schltdl.
Mariscus palustris Schrad. = ***Cyperus meyenianus*** Kunth
Mariscus panamensis C.B.Clarke = ***Cyperus panamensis*** (C.B.Clarke) Britton ex Standl.
Mariscus panamensis var. *deformis* C.B.Clarke = ***Cyperus panamensis*** (C.B.Clarke) Britton ex Standl.
Mariscus paniceus (Rottb.) Vahl = ***Cyperus paniceus*** (Rottb.) Boeck.
Mariscus paniceus var. *queenslandicus* Domin = ***Cyperus paniceus*** (Rottb.) Boeck.
Mariscus paniceus var. *roxburghianus* C.B.Clarke = ***Cyperus paniceus*** (Rottb.) Boeck.
Mariscus pantopodus (Baker) Fernald = ***Costularia pantopoda*** (Baker) C.B.Clarke
Mariscus paradoxus (Cherm.) Cherm. = ***Alinula paradoxa*** (Cherm.) Goetgh. & Vorster
Mariscus parviflorus Nees = ***Cyperus luzulae*** (L.) Retz.
Mariscus patulus Schrad. = ***Cyperus hermaphroditus*** (Jacq.) Standl.
Mariscus pedunculatus (R.Br.) T.Koyama = ***Remirea maritima*** Aubl.
Mariscus pennatiformis (Kük.) T.Koyama = ***Cyperus pennatiformis*** Kük.
Mariscus pennatiformis subsp. *bryanii* (Kük.) T.Koyama = ***Cyperus pennatiformis*** var. ***bryanii*** Kük.
Mariscus pennatus (Lam.) Domin = ***Cyperus javanicus*** Houtt.
Mariscus pennatus var. *ventricosus* (R.Br.) Domin = ***Cyperus ventricosus*** R.Br.
Mariscus perrieri Cherm. = ***Cyperus pseudovestitus*** (C.B.Clarke) Kük.
Mariscus phaeocephalus (Griseb.) C.B.Clarke = ***Cyperus manimae*** Kunth var. ***manimae***
Mariscus philippensis Steud. = ***Cyperus cyperoides*** (L.) Kuntze subsp. ***cyperoides***
Mariscus phillipsiae C.B.Clarke = ***Cyperus phillipsiae*** (C.B.Clarke) Kük.
Mariscus phleoides Nees ex Kunth = ***Cyperus phleoides*** (Nees ex Kunth) H.Mann
Mariscus phleoides subsp. *hawaiiensis* (H.Mann) T.Koyama = ***Cyperus phleoides*** var. ***hawaiiensis*** (H.Mann) Kük.
Mariscus picardae (Boeck.) T.Koyama = ***Cyperus picardae*** Boeck.
Mariscus pictus Nees = ***Cyperus cyperinus*** (Retz.) Suringar
Mariscus pilosulus C.B.Clarke = ***Cyperus pilosulus*** (C.B.Clarke) K.Schum. ex Kük.
Mariscus piluliferus G.Bertol. = ***Rhynchospora holoschoenoides*** (Rich.) Herter
Mariscus planifolius (Rich.) Urb. = ***Cyperus planifolius*** Rich.
Mariscus plateilema Steud. = ***Cyperus plateilema*** (Steud.) Kük.
Mariscus platyphyllus Fernald = ***Machaerina falcata*** (Nees) T.Koyama
Mariscus poeppigianus Kunth = ***Cyperus mutisii*** (Kunth) Andersson
Mariscus pohlianus Nees = ***Cyperus odoratus*** L. subsp. ***odoratus***
Mariscus pohlii (Nees) D.A.Simpson = ***Cyperus pohlii*** (Nees) Steud.
Mariscus pohlii subsp. *bahiensis* D.A.Simpson = ***Cyperus pohlii*** subsp. ***bahiensis*** (D.A.Simpson) D.A.Simpson
Mariscus polycephalus A.Rich. = ***Cyperus appendiculatus*** (Brongn.) Kunth var. ***appendiculatus***
Mariscus polycephalus (Lam.) Link = ***Cyperus luzulae*** (L.) Retz.
Mariscus polyphyllus Steud. = ***Cyperus cyperoides*** (L.) Kuntze subsp. ***cyperoides***
Mariscus polyphyllus Kunth = ***Cyperus multifolius*** Kunth

Mariscus polyphyllus (Benth.) Kuntze = ***Gahnia drummondii*** (Steud.) K.L.Wilson
Mariscus polystachyus Brongn. ex A.Rich. = ***Cyperus appendiculatus*** (Brongn.) Kunth var. ***appendiculatus***
Mariscus praelongatus (Steud.) C.B.Clarke = ***Cyperus strigosus*** L.
Mariscus pratensis (Boeck.) Cufod. = ***Cyperus pratensis*** Boeck.
Mariscus preissii (Nees) Kuntze = ***Gahnia decomposita*** (R.Br.) Benth.
Mariscus procerus Schrad. ex Nees = ***Cyperus coriifolius*** Boeck.
Mariscus procerus A.Rich. = ***Cyperus impubes*** Steud. var. ***impubes***
Mariscus procerus (J.R.Forst. & G.Forst.) Kuntze = ***Gahnia procera*** J.R.Forst. & G.Forst.
Mariscus pseudobrunneus C.B.Clarke ex Cherm. = ***Cyperus pseudobrunneus*** (C.B.Clarke ex Cherm.) Kük.
Mariscus pseudoflavus C.B.Clarke = ***Cyperus cyperoides*** subsp. ***pseudoflavus*** (C.B.Clarke) Lye
Mariscus pseudopilosus T.Durand & De Wild. = ***Cyperus pseudopilosus*** (T.Durand & De Wild.) Govaerts
Mariscus pseudovestitus C.B.Clarke = ***Cyperus pseudovestitus*** (C.B.Clarke) Kük.
Mariscus psilostachys C.B.Clarke = ***Cyperus pluribracteatus*** (Kük.) Govaerts
Mariscus psittacorum (Labill.) Kuntze = ***Gahnia psittacorum*** Labill.
Mariscus pubens (Kük.) Podlech = ***Cyperus pubens*** Kük.
Mariscus pubescens C.Presl = ***Cyperus dentoniae*** G.C.Tucker
Mariscus pulcher (Ridl.) Fernald = ***Tricostularia undulata*** (Thwaites) J.Kern
Mariscus pullu Steud. = ***Cyperus paniceus*** (Rottb.) Boeck.
Mariscus pungens Steud. = ***Remirea maritima*** Aubl.
Mariscus purpurascens (Vahl) C.B.Clarke = ***Cyperus planifolius*** Rich.
Mariscus pycnostachyus Kunth = ***Cyperus pycnostachyus*** (Kunth) Kunth
Mariscus quarrei Cherm. = ***Cyperus cyperoides*** (L.) Kuntze subsp. ***cyperoides***
Mariscus radians (Nees & Meyen ex Kunth) Tang & F.T.Wang = ***Cyperus radians*** Nees & Meyen ex Kunth
Mariscus radians var. *floribundus* (E.G.Camus) S.M.Huang = ***Cyperus radians*** Nees & Meyen ex Kunth
Mariscus radiatus Hochst. = ***Cyperus macrocarpus*** (Kunth) Boeck.
Mariscus radula (R.Br.) Kuntze = ***Gahnia radula*** (R.Br.) Benth.
Mariscus raiateensis J.W.Moore = ***Machaerina raiateensis*** (J.W.Moore) S.L.Welsh
Mariscus ramosus (Benth.) Domin = ***Cyperus ramosus*** (Benth.) Kük.
Mariscus rechingeri Palla = ***Cyperus stenophyllus*** J.V.Suringar
Mariscus rehmannianus (Boeck.) C.B.Clarke = ***Cyperus vestitus*** Hochst. ex C.Krauss
Mariscus remotus C.B.Clarke = ***Cyperus remotus*** (C.B.Clarke) Kük.
Mariscus restiodes (Sw.) Kuntze = ***Machaerina restioides*** (Sw.) Vahl
Mariscus restioides var. *effusus* (Griseb.) Fernald = ***Machaerina restioides*** subsp. ***effusa*** (Griseb.) T.Koyama
Mariscus retroflexus Willd. ex Kunth = ***Cyperus echinatus*** (L.) Alph.Wood
Mariscus retrofractus Gaertn. = ***Cyperus retroflexus*** Buckley var. ***retroflexus***
Mariscus retrofractus (L.) Vahl = ***Cyperus retrofractus*** (L.) Torr.
Mariscus retrorsus (Chapm.) C.B.Clarke = ***Cyperus retrorsus*** Chapm.
Mariscus rhaphiostachys Liebm. = ***Cyperus odoratus*** L. subsp. ***odoratus***
Mariscus rhodesicus Podlech = ***Cyperus hirtellus*** (Chiov.) Kük.
Mariscus richardii Steud. = ***Cyperus impubes*** Steud. var. ***impubes***
Mariscus rigens (J.Presl & C.Presl) T.Koyama = ***Cyperus rigens*** J.Presl & C.Presl
Mariscus rigidellus (Benth.) C.B.Clarke = ***Cyperus rigidellus*** (Benth.) J.M.Black
Mariscus rigidus Spreng. = ***Cyperus aggregatus*** (Willd.) Endl. var. ***aggregatus***
Mariscus riparius Schrad. = ***Cyperus thunbergii*** Vahl
Mariscus riparius (Nees) Kuntze = ***Machaerina riparia*** (Nees) T.Koyama
Mariscus riparius var. *gillii* C.B.Clarke = ***Cyperus thunbergii*** Vahl
Mariscus riparius var. *robustior* C.B.Clarke = ***Cyperus thunbergii*** Vahl
Mariscus riparius var. *trisumbellatus* C.B.Clarke = ***Cyperus thunbergii*** Vahl
Mariscus rivularis Walp. = ***Cyperus echinatus*** (L.) Alph.Wood
Mariscus rockii (Kük.) T.Koyama = ***Cyperus rockii*** Kük.
Mariscus rohlfsii (Boeck.) C.B.Clarke = ***Cyperus rohlfsii*** Boeck.
Mariscus roxburghianus (C.B.Clarke) Alston = ***Cyperus paniceus*** (Rottb.) Boeck.
Mariscus rubrotinctus Cherm. = ***Cyperus longibracteatus*** (Cherm.) Kük.
Mariscus rufus Kunth = ***Cyperus ligularis*** L.
Mariscus rutilans C.B.Clarke = ***Cyperus lhotskyanus*** Beck
Mariscus samoensis (C.B.Clarke ex Stapf) Fernald = ***Machaerina falcata*** (Nees) T.Koyama
Mariscus sandwicensis (Kük.) T.Koyama = ***Cyperus sandwicensis*** Kük.
Mariscus sanguineofuscus (Nees) C.B.Clarke = ***Cyperus lucidus*** R.Br.
Mariscus saturatus Donn.Sm. = ***Cyperus miliifolius*** Poepp. & Kunth
Mariscus scaber R.Br. = ***Cyperus scaber*** (R.Br.) Boeck.
Mariscus schaffnerianus Schltdl. = ***Cyperus pallidicolor*** (Kük.) G.C.Tucker
Mariscus schimperi Hochst. ex A.Rich. = ***Cyperus cruentus*** Rottb.
Mariscus schoenodes (G.Forst.) Kuntze = ***Gahnia schoenoides*** G.Forst.
Mariscus schweinfurthii Chiov. = ***Cyperus schweinfurthii*** (Chiov.) Kük.
Mariscus schweinitzii (Torr.) T.Koyama = ***Cyperus schweinitzii*** Torr.
Mariscus scirpodeus (Steud.) Kuntze = ***Machaerina scirpoidea*** (Steud.) T.Koyama
Mariscus scleropodus (Chiov.) Cufod. = ***Cyperus scleropodus*** Chiov.
Mariscus seemannianus (Boeck.) Palla = ***Cyperus seemannianus*** Boeck.
Mariscus serratus Gilib. = ***Cladium mariscus*** (L.) Pohl subsp. ***mariscus***

Mariscus sertularinus (Liebm.) T.Koyama = ***Cyperus hermaphroditus*** (Jacq.) Standl.
Mariscus setaceus (L.) Moench = ***Isolepis setacea*** (L.) R.Br.
Mariscus setifolius (Hook.f.) Kuntze = ***Gahnia setifolia*** Hook.f.
Mariscus setiglumis C.B.Clarke ex S.Moore = ***Cyperus pohlii*** (Nees) Steud. subsp. ***pohlii***
Mariscus sieberianus Nees ex C.B.Clarke = ***Cyperus cyperoides*** (L.) Kuntze subsp. ***cyperoides***
Mariscus sieberianus var. *evolutior* C.B.Clarke = ***Cyperus cyperoides*** (L.) Kuntze subsp. ***cyperoides***
Mariscus sieberianus var. *khasianus* C.B.Clarke = ***Cyperus cyperoides*** (L.) Kuntze subsp. ***cyperoides***
Mariscus sieberianus var. *nossibeensis* (Steud.) Cherm. = ***Cyperus cyperoides*** (L.) Kuntze subsp. ***cyperoides***
Mariscus sieberianus var. *subcompositus* C.B.Clarke = ***Cyperus cyperoides*** (L.) Kuntze subsp. ***cyperoides***
Mariscus socialis (C.B.Clarke) S.S.Hooper = ***Cyperus pseudopilosus*** (T.Durand & De Wild.) Govaerts
Mariscus societensis J.W.Moore = ***Machaerina mariscoides*** (Gaudich.) J.Kern
Mariscus solidus (Kunth) Vorster = ***Cyperus solidus*** Kunth
Mariscus somaliensis C.B.Clarke = ***Cyperus somaliensis*** C.B.Clarke
Mariscus soyauxii (Boeck.) C.B.Clarke = ***Cyperus soyauxii*** Boeck.
Mariscus spectabilis (Link) T.Koyama = ***Cyperus spectabilis*** Link
Mariscus sphaericus Willd. ex Kunth = ***Cyperus echinatus*** (L.) Alph.Wood
Mariscus sphaerocephalus Steud. = ***Cyperus echinatus*** (L.) Alph.Wood
Mariscus splendens Cherm. = ***Cyperus splendens*** (Cherm.) Kük.
Mariscus squarrosus (L.) C.B.Clarke = ***Cyperus squarrosus*** L.
Mariscus stenolepis (Torr.) C.B.Clarke = ***Cyperus strigosus*** L.
Mariscus steudelianus (Boeck.) Cufod. = ***Cyperus cyperoides*** (L.) Kuntze subsp. ***cyperoides***
Mariscus stolonifer C.B.Clarke = ***Cyperus chevalieri*** Kük.
Mariscus stramineoferrugineus (Kük.) Napper = ***Cyperus stramineoferrugineus*** Kük.
Mariscus strigosus (L.) C.B.Clarke = ***Cyperus strigosus*** L.
Mariscus stuppeus (G.Forst.) Merr. = ***Cyperus javanicus*** Houtt.
Mariscus subcaracasanus (Kük.) T.Koyama = ***Cyperus subcaracasanus*** Kük.
Mariscus sublimis C.B.Clarke = ***Cyperus cyperoides*** (L.) Kuntze subsp. ***cyperoides***
Mariscus subulatus Nees = ***Cyperus meyenianus*** Kunth
Mariscus subuniflorus (Britton) T.Koyama = ***Cyperus retroflexus*** var. ***pumilus*** (Britton) R.Carter & S.D.Jones
Mariscus sumatrensis (Retz.) J.Raynal = ***Cyperus cyperoides*** (L.) Kuntze subsp. ***cyperoides***
Mariscus sumatrensis var. *evolutior* (C.B.Clarke) Karthik. = ***Cyperus cyperoides*** (L.) Kuntze subsp. ***cyperoides***
Mariscus sumatrensis var. *khasianus* (C.B.Clarke) Karthik. = ***Cyperus cyperoides*** (L.) Kuntze subsp. ***cyperoides***
Mariscus sumatrensis var. *macrocarpus* (Kunth) Maquet = ***Cyperus macrocarpus*** (Kunth) Boeck.
Mariscus sumatrensis var. *subcompositus* (C.B.Clarke) Karthik. = ***Cyperus cyperoides*** (L.) Kuntze subsp. ***cyperoides***
Mariscus sundaicus Miq. = ***Cyperus cyperinus*** (Retz.) Suringar
Mariscus swartzii A.Dietr. = ***Cyperus swartzii*** (A.Dietr.) Boeck. ex Kük.
Mariscus sylvestris Kunth = ***Cyperus aggregatus*** (Willd.) Endl. var. ***aggregatus***
Mariscus tabina (Steud. ex Boeck.) C.B.Clarke = ***Cyperus tabina*** Steud. ex Boeck.
Mariscus tabularis (Schrad.) C.B.Clarke = ***Cyperus tabularis*** Schrad.
Mariscus tabularis var. *humilis* C.B.Clarke = ***Cyperus tabularis*** Schrad.
Mariscus tabularis var. *zeyheri* C.B.Clarke = ***Cyperus tabularis*** Schrad.
Mariscus tacnensis (Nees & Meyen) J.Raynal = ***Cyperus tacnensis*** Nees & Meyen
Mariscus tamakugu Masam. & Syozi = [36 CHH 38 TAI]
Mariscus tanyphyllus (Ridl.) C.B.Clarke = ***Cyperus tanyphyllus*** Ridl.
Mariscus taylorii C.B.Clarke = ***Cyperus oblongoincrassatus*** Kük.
Mariscus taylorii var. *groteanus* (Kük.) Napper = ***Cyperus oblongoincrassatus*** Kük.
Mariscus taylorii var. *udigensis* (Kük.) Napper = ***Cyperus oblongoincrassatus*** Kük.
Mariscus tenuifolius Schrad. ex Nees = ***Cyperus cyperinus*** (Retz.) Suringar
Mariscus tenuis C.B.Clarke =
Mariscus tenuis (Sw.) Nelmes = ***Cyperus tenuis*** Sw.
Mariscus teretifolius (R.Br.) Kuntze = ***Machaerina teretifolia*** (R.Br.) T.Koyama
Mariscus tetragonocarpus (Boeck.) Kuntze = ***Gahnia sieberiana*** Kunth
Mariscus tetragonus (Elliott) C.B.Clarke = ***Cyperus tetragonus*** Elliott
Mariscus tetraquetrus (Hook.f.) Kuntze = ***Machaerina tetragona*** (Labill.) T.Koyama
Mariscus thomensis C.B.Clarke = ***Cyperus cylindrostachyus*** Boeck.
Mariscus thunbergii (Vahl) Schrad. = ***Cyperus thunbergii*** Vahl
Mariscus thwaitesii Livera = ***Cyperus procerus*** Rottb.
Mariscus thyrsiflorus (Jungh. ex Schltdl.) T.Koyama = ***Cyperus thyrsiflorus*** Jungh. ex Schltdl.
Mariscus toluccensis Steud. = ***Cyperus aggregatus*** (Willd.) Endl. var. ***aggregatus***
Mariscus tomaiophyllus (K.Schum.) C.B.Clarke = ***Cyperus tomaiophyllus*** K.Schum.
Mariscus tovarii Kunth = ***Cyperus hermaphroditus*** (Jacq.) Standl.
Mariscus trialatus (Boeck.) Tang & F.T.Wang = ***Cyperus trialatus*** (Boeck.) J.Kern
Mariscus tribrachiatus Liebm. = ***Cyperus thyrsiflorus*** Jungh. ex Schltdl.
Mariscus trifidus (Labill.) Kuntze = ***Gahnia trifida*** Labill.
Mariscus trinervis C.B.Clarke = ***Cyperus pseudopilosus*** (T.Durand & De Wild.) Govaerts
Mariscus tristis (Nees ex Hook. & Arn.) Kuntze = ***Gahnia tristis*** Nees ex Hook. & Arn.
Mariscus uitenhagensis Steud. = ***Cyperus capensis*** (Steud.) Endl.
Mariscus uliginosus Schrad. ex Nees = ***Cyperus meyenianus*** Kunth
Mariscus umbellatus (Rottb.) Vahl = ***Cyperus cylindrostachyus*** Boeck.
Mariscus umbellatus J.Presl & C.Presl = ***Cyperus cyperoides*** (L.) Kuntze subsp. ***cyperoides***
Mariscus umbellatus Pursh = ***Cyperus retrorsus*** Chapm.
Mariscus umbellatus Zoll. = ***Rhynchospora rubra*** (Lour.) Makino subsp. ***rubra***

Mariscus umbellatus var. *procerior* Zoll. = ***Rhynchospora rubra*** (Lour.) Makino subsp. ***rubra***
Mariscus umbilensis C.B.Clarke ex W.Watson = ***Cyperus owanii*** Boeck.
Mariscus uniflorus Steud. = ***Cyperus retroflexus*** Buckley var. ***retroflexus***
Mariscus unifolius (Boeck.) T.Koyama = ***Cyperus unifolius*** Boeck.
Mariscus urbanii (Boeck.) C.B.Clarke = ***Cyperus urbanii*** Boeck.
Mariscus ustulatus (A.Rich.) C.B.Clarke = ***Cyperus ustulatus*** A.Rich.
Mariscus vaginalis (Benth.) Kuntze = ***Machaerina vaginalis*** (Benth.) T.Koyama
Mariscus varicus C.B.Clarke ex Cherm. = ***Cyperus varicus*** (C.B.Clarke ex Cherm.) Kük.
Mariscus vauthiera (C.B.Clarke) Fernald = ***Lepidosperma australe*** (A.Rich.) Hook.f.
Mariscus vestitus (Hochst. ex C.Krauss) C.B.Clarke = ***Cyperus vestitus*** Hochst. ex C.Krauss
Mariscus vestitus var. *decurvatus* C.B.Clarke = ***Cyperus vestitus*** Hochst. ex C.Krauss
Mariscus viguieri Cherm. = ***Cyperus impubes*** var. ***viguieri*** (Cherm.) Kük.
Mariscus viguieri var. *contractus* Cherm. = ***Cyperus impubes*** var. ***viguieri*** (Cherm.) Kük.
Mariscus viridis Hochst. ex Schweinf. = ***Cyperus cruentus*** Rottb.
Mariscus wacketii (Kük.) T.Koyama = ***Cyperus impolitus*** Kunth
Mariscus wallichianus Kunth = ***Cyperus paniceus*** (Rottb.) Boeck.
Mariscus whitmeei C.B.Clarke = ***Cyperus whitmeei*** (C.B.Clarke) Kük.
Mariscus wichurae (Boeck.) Kuntze = ***Gahnia tristis*** Nees ex Hook. & Arn.
Mariscus wrightii (Britton) T.Koyama = ***Cyperus dipsaceus*** Liebm.
Mariscus xanthocarpus (Hook.f.) Kuntze = ***Gahnia xanthocarpa*** (Hook.f.) Hook.f.
Mariscus xerophilus Domin = [50 NTA QLD]
Mariscus xiphioides Fernald = ***Machaerina preissii*** (Nees) L.A.S.Johnson & Koyama

***Unplaced Names*:**
Mariscus alopecuroides Steud., Syn. Pl. Glumac. 2: 65 (1854). = [84]
Mariscus blatteri McCann, J. Bombay Nat. Hist. Soc. 36: 532 (1934). = [40 IND]
Mariscus cufodontii Chiov., Miss. Biol. Borana, Racc. Bot., Angiosp.-Gymnosp.: 294 (1939). = [24 ETH]
Mariscus hainanensis Chun & F.C.How, Acta Phytotax. Sin. 7: 87 (1958). = [36 CHH]
Mariscus mayottensis Cherm., Notul. Syst. (Paris) 5: 270 (1936). = [29 COM]
Mariscus monospermus S.M.Huang, in Fl. Hainan. 4: 537 (1977). = [36 CHH]
Mariscus tamakugu Masam. & Syozi, Acta Phytotax. Geobot. 14: 89 (1951). = [36 CHH 38 TAI]
Mariscus xerophilus Domin, Biblioth. Bot. 85: 438 (1915). = [50 NTA QLD]

Mastigloscleria

Mastigloscleria Nees = ***Scleria*** P.J.Bergius
Mastigloscleria reflexa (Kunth) Nees = ***Scleria secans*** (L.) Urb.
Mastigloscleria reflexa var. *styligera* Nees = ***Scleria secans*** (L.) Urb.

Maukschia

Maukschia Heuff. = ***Carex*** L.
Maukschia laevis (Hoppe) Heuff. = ***Carex dioica*** L.
Maukschia scabra (Hoppe) Heuff. = ***Carex davalliana*** Sm.

Maximoviczia

Maximoviczia A.P.Khokhr. = ***Scirpus*** Tourn. ex L.
Maximoviczia japonica (Maxim.) A.P.Khokhr. = ***Scirpus maximowiczii*** C.B.Clarke

Maximowicziella

Maximowicziella A.P.Khokhr. = ***Scirpus*** Tourn. ex L.
Maximowicziella japonica (Maxim.) M.S.Novos. = ***Scirpus maximowiczii*** C.B.Clarke

Megadenus

Megadenus Raf. = ***Eleocharis*** R.Br.
Megadenus capitatus (L.) Raf. = ***Eleocharis geniculata*** (L.) Roem. & Schult.
Megadenus palustris (L.) Raf. = ***Eleocharis palustris*** (L.) Roem. & Schult.
Megadenus tuberculatus (Elliott) Raf. = ***Eleocharis tuberculosa*** (Michx.) Roem. & Schult.

Megarrhena

Megarrhena Schrad. ex Nees = ***Androtrichum*** (Brongn.) Brongn.

Melachne

Melachne Schrad. ex Schult. & Schult.f. = ***Gahnia*** J.R.Forst. & G.Forst.
Melachne sieberi Schrad. ex Schult. & Schult.f. = ***Gahnia filifolia*** (C.Presl) Kük. ex Benl

Melancranis

Melancranis Vahl = ***Ficinia*** Schrad.
Melancranis acuminata (Nees) Kuntze = ***Ficinia acuminata*** (Nees) Nees
Melancranis anceps (Nees) Kuntze = ***Ficinia anceps*** Nees
Melancranis argyropus (Nees) Kuntze = ***Ficinia argyropus*** Nees
Melancranis bolusii (Boeck.) Kuntze = ***Ficinia pallens*** var. ***lithosperma*** (Boeck.) T.H.Arnold & Gordon-Gray
Melancranis brevifolia (Nees ex Kunth) Kuntze = ***Ficinia brevifolia*** Nees ex Kunth
Melancranis bulbosa (L.) Kuntze = ***Ficinia bulbosa*** (L.) Nees
Melancranis clandestina (Steud.) Kuntze = ***Ficinia clandestina*** (Steud.) Boeck.
Melancranis commutata (Nees) Kuntze = ***Ficinia gracilis*** Schrad.
Melancranis comparbergensis (Drège ex Steud.) Kuntze = ***Ficinia compasbergensis*** Drège ex Steud.
Melancranis contorta (Nees) Kuntze = ***Ficinia stolonifera*** Boeck.
Melancranis dregeana (Steud.) Kuntze = ***Ficinia pygmaea*** Boeck.
Melancranis ecklonea (Steud.) Kuntze = ***Ficinia ecklonea*** (Steud.) Nees
Melancranis elongata (Boeck.) Kuntze = ***Ficinia acuminata*** (Nees) Nees
Melancranis fascicularis (Nees) Kuntze = ***Ficinia fascicularis*** Nees

Melancranis fastigiata (Thunb.) Kuntze = ***Ficinia fastigiata*** (Thunb.) Nees
Melancranis filiformis (Lam.) Kuntze = ***Ficinia filiformis*** (Lam.) Schrad.
Melancranis gracilis (Poir.) Nees = ***Ficinia ecklonea*** (Steud.) Nees
Melancranis indica (Lam.) Kuntze = ***Ficinia indica*** (Lam.) H.Pfeiff.
Melancranis ixioides (Nees) Kuntze = ***Ficinia ixioides*** Nees
Melancranis kunthiana (Boeck.) Kuntze = ***Ficinia acuminata*** (Nees) Nees
Melancranis laciniata (Thunb.) Kuntze = ***Ficinia laciniata*** (Thunb.) Nees
Melancranis laevis (Nees) Kuntze = ***Ficinia laevis*** Nees
Melancranis lateralis (Vahl) Kuntze = ***Ficinia lateralis*** (Vahl) Kunth
Melancranis leiocarpa (Nees) Kuntze = ***Ficinia leiocarpa*** Nees
Melancranis lithosperma (Boeck.) Kuntze = ***Ficinia pallens*** var. ***lithosperma*** (Boeck.) T.H.Arnold & Gordon-Gray
Melancranis macowanii (C.B.Clarke) Kuntze = ***Ficinia macowanii*** C.B.Clarke
Melancranis monticola (Kunth) Kuntze =
Melancranis nigrescens Schrad. = ***Ficinia nigrescens*** (Schrad.) J.Raynal
Melancranis nuda (Boeck.) Kuntze =
Melancranis paradoxa (Schrad.) Kuntze = ***Ficinia paradoxa*** (Schrad.) Nees
Melancranis praemorsa (Nees) Kuntze = ***Ficinia praemorsa*** Nees
Melancranis quinquangularis (Boeck.) Kuntze = ***Ficinia quinquangularis*** Boeck.
Melancranis radiata Vahl = ***Ficinia brevifolia*** Nees ex Kunth
Melancranis radiata (L.f.) Kunth = ***Ficinia radiata*** (L.f.) Kunth
Melancranis repens (Kunth) Kuntze = ***Ficinia repens*** Kunth
Melancranis rigidula Nees = ***Ficinia nigrescens*** (Schrad.) J.Raynal
Melancranis rubiginosa (Spreng.) Spreng. = ***Machaerina rubiginosa*** (Spreng.) T.Koyama
Melancranis scariosa (Thunb.) Vahl = ***Ficinia deusta*** (P.J.Bergius) Levyns
Melancranis secunda (Vahl) Kuntze = ***Ficinia secunda*** (Vahl) Kunth
Melancranis sylvatica (Kunth) Kuntze = ***Ficinia sylvatica*** Kunth
Melancranis tenuifolia (Kunth) Kuntze = ***Ficinia filiformis*** (Lam.) Schrad.
Melancranis tenuis Eckl. ex Kunth = [27]
Melancranis tribracteata (Boeck.) Kuntze = ***Ficinia tristachya*** (Rottb.) Nees
Melancranis trigyna (L.) Kuntze = ***Ficinia deusta*** (P.J.Bergius) Levyns
Melancranis trispicata (L.f.) Kuntze = ***Ficinia sylvatica*** Kunth
Melancranis tristachya (Rottb.) Kuntze = ***Ficinia tristachya*** (Rottb.) Nees
Melancranis truncata (Thunb.) Kuntze = ***Ficinia truncata*** (Thunb.) Schrad.
Melancranis zeyheri (Boeck.) Kuntze = ***Ficinia zeyheri*** Boeck.

Unplaced Names:
Melancranis tenuis Eckl. ex Kunth, Enum. Pl. 2: 265 (1837). = [27]

Melanoschoenos

Melanoschoenos Ség. = ***Schoenus*** L.

Meltrema

Meltrema Raf. = ***Carex*** L.

Mesomelaena

Mesomelaena Nees in J.G.C.Lehmann, Pl. Preiss. 2: 88 (1846).
SW. Australia. 50 WAU.
5 Species

Mesomelaena graciliceps (C.B.Clarke) K.L.Wilson, Telopea 2: 186 (1981).
SW. Australia. 50 WAU. Hemicr.
**Carpha graciliceps* C.B.Clarke, Bot. Jahrb. Syst. 35: 81 (1904). *Ptilanthelium graciliceps* (C.B.Clarke) Kük., Repert. Spec. Nov. Regni Veg. 47: 216 (1939).

Mesomelaena preissii Nees in J.G.C.Lehmann, Pl. Preiss. 2: 88 (1846). *Mesomelaena stygia* var. *preissii* (Nees) Kük., Repert. Spec. Nov. Regni Veg. 48: 54 (1940).
SW. Australia. 50 WAU. Hemicr.

Mesomelaena pseudostygia (Kük.) K.L.Wilson, Telopea 2: 190 (1981).
WSW. Western Australia. 50 WAU. Hemicr.
**Mesomelaena stygia* var. *pseudostygia* Kük., Repert. Spec. Nov. Regni Veg. 48: 53 (1940).

Mesomelaena stygia (R.Br.) Nees in J.G.C.Lehmann, Pl. Preiss. 2: 89 (1846).
SW. Australia. 50 WAU. Hemicr.
**Chaetospora stygia* R.Br., Prodr.: 233 (1810). *Schoenus stygius* (R.Br.) Poir. in J.B.A.M.de Lamarck, Encycl., Suppl. 2: 252 (1811).

subsp. ***deflexa*** (Kük.) K.L.Wilson, Telopea 2: 192 (1981).
WSW. Western Australia. 50 WAU. Hemicr.
**Mesomelaena stygia* var. *deflexa* Kük., Repert. Spec. Nov. Regni Veg. 48: 53 (1940).

subsp. ***stygia***
SW. Australia. 50 WAU. Hemicr.
Lepidosperma uncinatum Nees in J.G.C.Lehmann, Pl. Preiss. 2: 93 (1846). *Mesomelaena uncinata* (Nees) C.B.Clarke, Kew Bull., Addit. Ser. 8: 48 (1908).

Mesomelaena tetragona (R.Br.) Benth., Fl. Austral. 7: 379 (1878).
SW. Australia. 50 WAU. Hemicr.
**Chaetospora tetragona* R.Br., Prodr.: 233 (1810). *Schoenus tetragonus* (R.Br.) Poir. in J.B.A.M.de Lamarck, Encycl., Suppl. 2: 252 (1811).
Mesomelaena bicapitata Steud., Syn. Pl. Glumac. 2: 163 (1855).

Synonyms:
Mesomelaena anceps (R.Br.) Benth. = ***Gymnoschoenus anceps*** (R.Br.) C.B.Clarke
Mesomelaena bicapitata Steud. = ***Mesomelaena tetragona*** (R.Br.) Benth.
Mesomelaena deusta (R.Br.) Benth. = ***Ptilothrix deusta*** (R.Br.) K.L.Wilson
Mesomelaena sphaerocephala (R.Br.) Benth. = ***Gymnoschoenus sphaerocephalus*** (R.Br.) Hook.f.
Mesomelaena stygia var. *deflexa* Kük. = ***Mesomelaena stygia*** subsp. ***deflexa*** (Kük.) K.L.Wilson
Mesomelaena stygia var. *preissii* (Nees) Kük. = ***Mesomelaena preissii*** Nees

Mesomelaena stygia var. *pseudostygia* Kük. = ***Mesomelaena pseudostygia*** (Kük.) K.L.Wilson
Mesomelaena uncinata (Nees) C.B.Clarke = ***Mesomelaena stygia*** (R.Br.) Nees subsp. ***stygia***

Michrochaeta

Michrochaeta Rchb. = ***Rhynchospora*** Vahl

Microdracoides

Microdracoides Hua, Bull. Mus. Hist. Nat. (Paris) 12: 421 (1906).
W. & WC. Trop. Africa. 22 GUI NGA SIE 23 CMN EQG?
1 Species
Schoenodendron Engl., Bot. Jahrb. Syst. 44(101): 34 (1910).

Microdracoides squamosus Hua, Bull. Mus. Hist. Nat. (Paris) 12: 422 (1906).
W. Trop. Africa to Cameroon. 22 GUI NGA SIE 23 CMN EQG?. Cham.
Schoenodendron buecheri Engl., Bot. Jahrb. Syst. 44(101): 34 (1910).

Microlepis

Microlepis Schrad. ex Nees = ***Lagenocarpus*** Nees
Microlepis bahiensis Schrad. ex Nees = ***Lagenocarpus rigidus*** (Kunth) Nees subsp. ***rigidus***
Microlepis elatior Schrad. = ***Lagenocarpus rigidus*** (Kunth) Nees subsp. ***rigidus***
Microlepis minarum-generalium Schrad. = ***Lagenocarpus minarum*** (Nees) Kuntze
Microlepis rigidifolia Schrad. ex Nees = ***Lagenocarpus rigidus*** (Kunth) Nees subsp. ***rigidus***

Micropapyrus

Micropapyrus Suess. = ***Rhynchospora*** Vahl
Micropapyrus viviparoides Suess. = ***Rhynchospora schomburgkiana*** (Boeck.) T.Koyama

Mischospora

Mischospora Boeck. = ***Fimbristylis*** Vahl
Mischospora efoliata Boeck. = ***Fimbristylis tetragona*** R.Br.

Mitrospora

Mitrospora Nees = ***Rhynchospora*** Vahl
Mitrospora cephalophora Nees = ***Rhynchospora exaltata*** Kunth
Mitrospora ciliata Nees = ***Rhynchospora neesii*** Steud.
Mitrospora pallida Nees = ***Rhynchospora patuligluma*** C.B.Clarke ex Lindm.
Mitrospora polyphylla (Vahl) Nees = ***Rhynchospora polyphylla*** (Vahl) Vahl
Mitrospora recurvata Nees = ***Rhynchospora recurvata*** (Nees) Steud.
Mitrospora wageneriana Schltdl. = ***Rhynchospora wageneriana*** (Schltdl.) Schnee

Mondo

Mondo Adans. = ***Carex*** L.

Morelotia

Morelotia Gaudich., Voy. Uranie: 416 (1829).
New Zealand, Hawaiian Is. 51 NZN NZS 63 HAW.
2 Species

Morelotia affinis (Brongn.) S.T.Blake, Contr. Queensland Herb. 8: 38 (1969).
New Zealand. 51 NZN NZS. Hemicr. or rhizome geophyte.
Morelotia gahniiformis var. *minor* A.Rich. in J.S.C. Dumont d'Urville, Voy. Astrolabe 1: 115 (1832).
**Lampocarya affinis* Brongn. in L.I.Duperrey, Voy. Monde, Phan.: 166 (1833). *Gahnia affinis* (Brongn.) Steud., Syn. Pl. Glumac. 2: 164 (1855). *Gahnia arenaria* Hook.f., Handb. N. Zeal. Fl. 1: 306 (1864), nom. illeg. *Mariscus affinis* (Brongn.) Kuntze, Revis. Gen. Pl. 2: 755 (1891). *Machaerina affinis* (Brongn.) J.Kern, Acta Bot. Neerl. 11: 223 (1962).

Morelotia gahniiformis Gaudich., Voy. Uranie: 416 (1829). *Gahnia gahniiformis* (Gaudich.) A.Heller, Minnesota Bot. Stud. 1: 802 (1896). *Machaerina gahniiformis* (Gaudich.) J.Kern, Acta Bot. Neerl. 11: 223 (1962).
Hawaiian Is. 63 HAW. Hemicr. or rhizome geophyte.
Lampocarya gaudichaudii Brongn. in L.I.Duperrey, Voy. Monde, Phan.: 166 (1833). *Gahnia gaudichaudii* (Brongn.) Steud., Syn. Pl. Glumac. 2: 164 (1855). *Mariscus gaudichaudii* (Brongn.) Kuntze, Revis. Gen. Pl. 2: 755 (1891).
Cladium quadrangulare Nees, Linnaea 9: 301 (1834), nom. inval.

***Synonyms*:**
Morelotia gahniiformis var. *minor* A.Rich. = ***Morelotia affinis*** (Brongn.) S.T.Blake

Morisia

Morisia Nees = ***Rhynchospora*** Vahl
Morisia capitata Nees = ***Rhynchospora rubra*** (Lour.) Makino subsp. ***rubra***
Morisia wallichii Nees = ***Rhynchospora rubra*** (Lour.) Makino subsp. ***rubra***

Neesenbeckia

Neesenbeckia Levyns, J. S. African Bot. 13: 74 (1947).
Cape Prov. 27 CPP.
1 Species
**Buekia* Nees, Linnaea 9: 300 (1834), nom. illeg.

Neesenbeckia punctoria (Vahl) Levyns, J. S. African Bot. 13: 74 (1947).
Cape Prov. 27 CPP. Hemicr. or rhizome geophyte.
**Schoenus punctorius* Vahl, Enum. Pl. 2: 217 (1805). *Chaetospora punctoria* (Vahl) A.Dietr., Sp. Pl. 2: 28 (1832). *Buekia punctoria* (Vahl) Nees, Linnaea 9: 300 (1834). *Elynanthus punctorius* (Vahl) Benth. & Hook.f., Gen. Pl. 3: 1064 (1883). *Tetraria punctoria* (Vahl) C.B.Clarke in T.A.Durand & H.Schinz, Consp. Fl. Afric. 5: 662 (1894).

Neilreichia

Neilreichia Kotula = ***Carex*** L.
Neilreichia umbrosa Kotula = ***Carex praecox*** Schreb.

Nelmesia

Nelmesia Van der Veken, Bull. Jard. Bot. État 25: 143 (1955).
WC. Trop. Africa. 23 ZAI.
1 Species

Nelmesia melanostachya Van der Veken, Bull. Jard. Bot. État 25: 143 (1955).
Zaïre. 23 ZAI. Hemicr. or rhizome geophyte.

Nemocharis

Nemocharis Beurl. = ***Scirpus*** Tourn. ex L.
Nemocharis radicans Burl. = ***Scirpus radicans*** Schkuhr
Nemocharis sylvatica (L.) Beurl. = ***Scirpus sylvaticus*** L.

Nemochloa

Nemochloa Nees = ***Pleurostachys*** Brongn.
Nemochloa beyrichii Nees = ***Pleurostachys beyrichii*** (Nees) Steud.
Nemochloa elegans (Kunth) Nees = ***Pleurostachys tenuiflora*** Brongn.
Nemochloa extenuata Nees = ***Pleurostachys millegrana*** (Nees) Steud.
Nemochloa foliosa (Kunth) Nees = ***Pleurostachys foliosa*** Kunth
Nemochloa gaudichaudii (Brongn.) Nees = ***Pleurostachys gaudichaudii*** Brongn.
Nemochloa macrantha (Kunth) Nees = ***Pleurostachys orbignyana*** Brongn.
Nemochloa martiana Nees = ***Pleurostachys martiana*** (Nees) Steud.
Nemochloa millegrana Nees = ***Pleurostachys millegrana*** (Nees) Steud.
Nemochloa orbignyana (Brongn.) Nees ex Steud. = ***Pleurostachys orbignyana*** Brongn.
Nemochloa sellovii (Kunth) Nees = ***Pleurostachys sellowii*** Kunth
Nemochloa sparsiflora (Kunth) Nees = ***Pleurostachys sparsiflora*** Kunth
Nemochloa stricta (Kunth) Nees = ***Pleurostachys stricta*** Kunth
Nemochloa tenuiflora (Brongn.) Nees = ***Pleurostachys tenuiflora*** Brongn.
Nemochloa turbinata Nees = ***Pleurostachys turbinata*** (Nees) H.Pfeiff.
Nemochloa urvillei (Brongn.) Nees = ***Pleurostachys urvillei*** Brongn.

Nemum

Nemum Desv. ex Ham., Prodr. Pl. Ind. Occid.: 13 (1825).
Trop. Africa. 22 BEN BKN GHA GNB GUI IVO LBR MLI NGA SEN SIE 23 CAF CMN ZAI 25 UGA 26 ANG ZAM.
5 Species

Nemum bulbostyloides (S.S.Hooper) J.Raynal, Adansonia, n.s., 13: 150 (1973).
W. Trop. Africa. 22 GUI IVO SIE. Hemicr. or rhizome geophyte.
**Scirpus bulbostyloides* S.S.Hooper, Kew Bull. 26: 581 (1972).

Nemum equitans (Kük.) J.Raynal, Adansonia, n.s., 13: 150 (1973).
S. Zaïre to Angola. 23 ZAI 26 ANG ZAM. Hemicr. or rhizome geophyte.
**Scirpus equitans* Kük. in R.E.Fries, Wiss. Erg. Schwed. Rhod.-Kongo Exped. 1: 7 (1921). *Bulbostylis equitans* (Kük.) Raymond, Mém. Jard. Bot. Montréal 55: 38 (1962).

Nemum megastachyum (Cherm.) J.Raynal, Adansonia, n.s., 13: 150 (1973).
Central African Rep. to N. Zaïre. 23 CAF ZAI. Hemicr. or rhizome geophyte.
**Scirpus angolensis* var. *megastachyum* Cherm., Arch. Bot. Bull. Mens. 4(7): 25 (1931).

Nemum parviflorum Lye, Lidia 2: 33 (1989).
Guinea. 22 GUI. Ther.

Nemum spadiceum (Lam.) Desv. in W.Hamilton, Prodr. Pl. Ind. Occid.: 13 (1825).
Trop. Africa. 22 BEN BKN GHA GNB GUI IVO LBR MLI NGA SEN SIE 23 CAF CMN ZAI 25 UGA 26 ANG ZAM. Ther.
**Eriocaulon spadiceum* Lam., Encycl. 3: 217 (1789). *Scirpus spadiceus* (Lam.) Boeck., Linnaea 36: 493 (1870), nom. illeg. *Scirpus briziformis* Hutch. in J.Hutchinson & J.M.Dalziel, Fl. W. Trop. Afr. 2: 466 (1936). *Scirpus angolensis* var. *briziformis* (Hutch.) S.S.Hooper, Kew Bull. 26: 580 (1972). *Scirpus ustulatus* Podlech, Mitt. Bot. Staatssamml. München 4: 118 (1961).
Scirpus spadiceus var. *ciliatus* Ridl., Trans. Linn. Soc. London, Bot. 2: 156 (1884). *Scirpus angolensis* C.B.Clarke in T.A.Durand & H.Schinz, Consp. Fl. Afric. 5: 617 (1894).

Neo-Senaea

Neo-Senaea K.Schum. ex H.Pfeiff. = ***Lagenocarpus*** Nees

Neolophocarpus

Neolophocarpus E.G.Camus = ***Schoenus*** L.
Neolophocarpus tonquinensis (Boeck.) E.G.Camus = ***Schoenus falcatus*** R.Br.

Neskiza

Neskiza Raf. = ***Carex*** L.
Neskiza aquatilis (Wahlenb.) Raf. = ***Carex aquatilis*** Wahlenb.
Neskiza aurea (Nutt.) Raf. = ***Carex aurea*** Nutt.
Neskiza cespitosa (L.) Raf. = ***Carex cespitosa*** L.
Neskiza crinita (Lam.) Raf. = ***Carex crinita*** Lam.
Neskiza glareosa (Schkuhr ex Wahlenb.) Raf. = ***Carex glareosa*** Schkuhr ex Wahlenb.
Neskiza loliacea (L.) Raf. = ***Carex loliacea*** L.
Neskiza salina (Wahlenb.) Raf. = ***Carex salina*** Wahlenb.
Neskiza saxatilis Raf. = ***Carex bigelowii*** Torr. ex Schwein. subsp. ***bigelowii***
Neskiza trisperma (Dewey) Raf. = ***Carex trisperma*** Dewey

Nomochloa

Nomochloa P.Beauv. = ***Blysmus*** Panz. ex Schult.
Nomochloa Nees = ***Pleurostachys*** Brongn.
Nomochloa compressa (L.) Beetle = ***Blysmus compressus*** (L.) Panz. ex Link
Nomochloa elegans (Kunth) Nees = ***Pleurostachys tenuiflora*** Brongn.
Nomochloa orbignyana (Brongn.) Nees = ***Pleurostachys orbignyana*** Brongn.
Nomochloa rufa (Huds.) Beetle = ***Blysmus rufus*** (Huds.) Link
Nomochloa rufa var. *neogaea* (Fernald) Beetle = ***Blysmus rufus*** (Huds.) Link
Nomochloa stricta (Kunth) Nees = ***Pleurostachys stricta*** Kunth
Nomochloa tenuiflora (Brongn.) Nees = ***Pleurostachys tenuiflora*** Brongn.
Nomochloa turbinata Nees = ***Pleurostachys turbinata*** (Nees) H.Pfeiff.

Olamblis

Olamblis Raf. = ***Carex*** L.
Olamblis barrattii (Torr. ex Schwein.) Raf. = ***Carex barrattii*** Torr. ex Schwein.

Olamblis elliottii (Schwein. & Torr.) Raf. = ***Carex elliottii*** Schwein. & Torr.
Olamblis fraseri (Andrews) Raf. = ***Cymophyllus fraserianus*** (Ker Gawl.) Kartesz & Gandhi
Olamblis hirsuta Raf. = ***Carex hirsutella*** Mack.
Olamblis laxa (Wahlenb.) Raf. = ***Carex laxa*** Wahlenb.
Olamblis miliacea Raf. = ***Carex prasina*** Wahlenb.
Olamblis oakesiana (Torr.) Raf. = ***Carex oligosperma*** Michx.
Olamblis verna (Chaix & Vill.) Raf. = ***Carex caryophyllea*** Latourr. var. ***caryophyllea***
Olamblis virescens (Muhl. ex Willd.) Raf. = ***Carex virescens*** Muhl. ex Willd.

Olotrema

Olotrema Raf. = ***Carex*** L.
Olotrema acuminata (Willd.) Raf. = ***Carex flacca*** Schreb. subsp. ***flacca***
Olotrema brunnea (Thunb.) Raf. = ***Carex brunnea*** Thunb.
Olotrema collinsii (Nutt.) Raf. = ***Carex collinsii*** Nutt.
Olotrema filifolia (Nutt.) Raf. = ***Carex filifolia*** Nutt.
Olotrema gebhardii (Willd.) Raf. = ***Carex elongata*** L.
Olotrema geminata (Schkuhr) Raf. = ***Carex geminata*** Schkuhr
Olotrema gynobasis (Vill.) Raf. = ***Carex halleriana*** Asso
Olotrema juncifolia (All.) Raf. = ***Carex maritima*** Gunnerus
Olotrema leporina (L.) Raf. = ***Carex leporina*** L.
Olotrema oligocarpa (Willd.) Raf. = ***Carex oligocarpa*** Willd.
Olotrema tetanica (Schkuhr) Raf. = ***Carex tetanica*** Schkuhr
Olotrema umbellata (Willd.) Raf. = ***Carex umbellata*** Willd.
Olotrema washingtoniana (Dewey) Raf. = ***Carex aquatilis*** var. ***minor*** Boott

Omoscleria

Omoscleria Nees = ***Scleria*** P.J.Bergius
Omoscleria flagellum (J.F.Gmel.) Nees = ***Scleria secans*** (L.) Urb.
Omoscleria tenacissima Nees = ***Scleria tenacissima*** (Nees) Steud.

Oncostylis

Oncostylis Nees = ***Bulbostylis*** Kunth
Oncostylis arenaria Nees = ***Bulbostylis juncoides*** (Vahl) Kük. ex Herter
Oncostylis bivalvis Nees = [84 BZL]
Oncostylis ciliata Nees = ***Bulbostylis capillaris*** (L.) Kunth ex C.B.Clarke subsp. ***capillaris***
Oncostylis circinata (Kunth) Nees = ***Bulbostylis circinata*** (Kunth) C.B.Clarke
Oncostylis conifera (Kunth) Nees = ***Bulbostylis conifera*** (Kunth) Beetle
Oncostylis dubia Nees = ***Bulbostylis capillaris*** (L.) Kunth ex C.B.Clarke subsp. ***capillaris***
Oncostylis fimbriata Nees = ***Bulbostylis fimbriata*** (Nees) C.B.Clarke
Oncostylis hirtella (Schrad. ex Schult.) Nees = ***Bulbostylis juncoides*** (Vahl) Kük. ex Herter
Oncostylis hirtella var. *hispidula* Nees = ***Bulbostylis juncoides*** (Vahl) Kük. ex Herter
Oncostylis hispida Liebm. = ***Bulbostylis vestita*** (Kunth) C.B.Clarke
Oncostylis junciformis (Kunth) Nees = ***Bulbostylis junciformis*** (Kunth) C.B.Clarke
Oncostylis junciformis var. *ambigua* Nees = ***Bulbostylis junciformis*** (Kunth) C.B.Clarke
Oncostylis junciformis var. *congesta* (Schrad. ex Schult.) Nees = ***Bulbostylis junciformis*** (Kunth) C.B.Clarke
Oncostylis junciformis var. *humboldtiana* Nees = ***Bulbostylis junciformis*** (Kunth) C.B.Clarke
Oncostylis junciformis f. *laxa* Nees = ***Bulbostylis junciformis*** (Kunth) C.B.Clarke
Oncostylis junciformis var. *speciosior* Nees = ***Bulbostylis junciformis*** (Kunth) C.B.Clarke
Oncostylis junciformis var. *stricta* Nees = ***Bulbostylis junciformis*** (Kunth) C.B.Clarke
Oncostylis kunthiana Nees = ***Bulbostylis consanguinea*** (Kunth) C.B.Clarke
Oncostylis kunthiana var. *laxa* Nees = ***Bulbostylis consanguinea*** (Kunth) C.B.Clarke
Oncostylis paradoxa (Spreng.) Nees = ***Bulbostylis paradoxa*** (Spreng.) Lindm.
Oncostylis pauciflora Liebm. = ***Bulbostylis pauciflora*** (Liebm.) C.B.Clarke
Oncostylis tenella (Link) Nees = ***Bulbostylis tenella*** (Link) C.B.Clarke
Oncostylis tenuifolia (Rudge) Nees = ***Bulbostylis tenuifolia*** (Rudge) J.F.Macbr.
Oncostylis truncata Nees = ***Bulbostylis truncata*** (Nees) M.T.Strong
Oncostylis vestita (Kunth) Nees = ***Bulbostylis vestita*** (Kunth) C.B.Clarke

Unplaced Names:
Oncostylis bivalvis Nees in C.F.P.von Martius & auct. suc. (eds.), Fl. Bras. 2(1): 205 (1842). = [84 BZL]

Onkerma

Onkerma Raf. = ***Carex*** L.
Onkerma badia (Pers.) Raf. = ***Carex canescens*** L. subsp. ***canescens***
Onkerma globularis (L.) Raf. = ***Carex globularis*** L.
Onkerma oederi (Retz.) Raf. = ***Carex pilulifera*** L. subsp. ***pilulifera***
Onkerma rigida Raf. = ***Carex bigelowii*** Torr. ex Schwein. subsp. ***bigelowii***
Onkerma rotundata (Wahlenb.) Raf. = ***Carex rotundata*** Wahlenb.
Onkerma sphaerocarpa Raf. = ***Carex supina*** Willd. ex Wahlenb. var. ***supina***

Opetiola

Opetiola Gaertn. = ***Cyperus*** L.
Opetiola myosuroides Gaertn. = ***Cyperus cyperinus*** (Retz.) Suringar

Ophryoscleria

Ophryoscleria Nees = ***Scleria*** P.J.Bergius
Ophryoscleria asperrima Liebm. = ***Scleria macrophylla*** J.Presl & C.Presl
Ophryoscleria lucida Nees = ***Scleria mitis*** P.J.Bergius
Ophryoscleria microcarpa (Nees ex Kunth) Nees = ***Scleria microcarpa*** Nees ex Kunth
Ophryoscleria mitis (P.J.Bergius) Nees = ***Scleria mitis*** P.J.Bergius
Ophryoscleria paludosa (Poepp. & Kunth) Nees = ***Scleria macrophylla*** J.Presl & C.Presl
Ophryoscleria racemosa (Poir.) Nees = ***Scleria racemosa*** Poir.
Ophryoscleria riparia (Poepp. & Kunth) Nees = ***Scleria mitis*** P.J.Bergius

Ophryoscleria schiedeana (Schltdl.) Liebm. = ***Scleria schiedeana*** Schltdl.
Ophryoscleria sororia (Kunth) Nees = ***Scleria sororia*** Kunth

Oreobolopsis

Oreobolopsis T.Koyama & Guagl., Darwiniana 28: 79 (1987 publ. 1988).
California, W. South America. 76 CAL 83 BOL ECU PER.
3 Species

Oreobolopsis clementis (M.E.Jones) Dhooge & Goetgh., Novon 12: 341 (2002).
California. 76 CAL. Hemicr. or rhizome geophyte.
**Scirpus clementis* M.E.Jones, Contr. U. S. Natl. Herb. 14: 21 (1912). *Baeothryon clementis* (M.E.Jones) Á.Löve & D.Löve, Univ. Colorado Stud. Ser. Biol. 17: 14 (1965). *Trichophorum clementis* (M.E.Jones) S.G.Sm., Novon 5: 102 (1995).
Scirpus yosemitanus Smiley, Univ. Calif. Publ. Bot. 9: 108 (1921).

Oreobolopsis inversa Dhooge & Goetgh., Novon 12: 338 (2002).
Ecuador to Peru. 83 ECU PER. Hemicr. or rhizome geophyte.

Oreobolopsis tepalifera T.Koyama & Guagl., Darwiniana 28: 80 (1987 publ. 1988).
SC. Ecuador to Bolivia. 83 BOL ECU PER. Hemicr. or rhizome geophyte.

Oreobolus

Oreobolus R.Br., Prodr.: 235 (1810).
Malesia to Australasia, Hawaiian Is., Costa Rica to Falkland Is. 42 BOR MLY SUL SUM 43 NWG 50 NSW TAS VIC 51 ATP NZN NZS 63 HAW 80 COS PAN 82 VEN 83 CLM ECU PER 85 AGS CLS JNF 90 FAL.
17 Species
Voladeria Benoist, Bull. Soc. Bot. France 84: 637 (1938).
Schoenoides O.Seberg, Willdenowia 16: 181 (1986).

Oreobolus acutifolius S.T.Blake, Contr. Queensland Herb. 8: 39 (1969).
Tasmania. 50 TAS. Hemicr. or rhizome geophyte.

Oreobolus ambiguus Kük. & Steenis, Bull. Jard. Bot. Buitenzorg, III, 14: 48 (1936).
N. Borneo, SW. Sulawesi, New Guinea. 42 BOR SUL 43 NWG. Hemicr. or cham.

Oreobolus cleefii L.E.Moro, Acad. Colomb. Cienc., Ser. J.A. Lleras 1: 165 (1987).
Colombia to Ecuador. 83 CLM ECU. Hemicr. or rhizome geophyte.

Oreobolus distichus F.Muell., Trans. Philos. Soc. Victoria 1: 109 (1855).
SE. Australia. 50 NSW TAS VIC. Hemicr. or rhizome geophyte.

Oreobolus ecuadorensis T.Koyama, Bot. Mag. (Tokyo) 97: 415 (1984).
Ecuador to Peru. 83 ECU PER. Hemicr. or rhizome geophyte.

Oreobolus furcatus H.Mann, Proc. Amer. Acad. Arts 7: 210 (1867).
Hawaiian Is. 63 HAW. Hemicr. or rhizome geophyte.

Oreobolus goeppingeri Suess., Bot. Jahrb. Syst. 72: 289 (1942).
Costa Rica to Peru. 80 COS PAN 82 VEN 83 CLM ECU PER. Hemicr.
Oreobolus obtusangulus var. *rubrovaginatus* Kük., Repert. Spec. Nov. Regni Veg. 48: 69 (1940). *Oreobolus obtusangulus* subsp. *rubrovaginatus* (Kük.) T.Koyama, Mem. New York Bot. Gard. 18(2): 27 (1969).
Oreobolus goeppingeri var. *sumapacii* L.E.Moro, Acad. Colomb. Cienc., Ser. J.A. Lleras 1: 180 (1987).
Oreobolus goeppingeri var. *tuquerrensis* L.E.Moro, Acad. Colomb. Cienc., Ser. J.A. Lleras 1: 181 (1987).

Oreobolus impar Edgar, New Zealand J. Bot. 2: 456 (1964).
New Zealand. 51 NZN NZS. Hemicr. or rhizome geophyte.

Oreobolus kuekenthalii Steenis ex Kük., Repert. Spec. Nov. Regni Veg. 44: 187 (1938).
C. Pen. Malaysia to N. Sumatera. 42 MLY SUM. Hemicr. or cham.
**Schoenus distichus* Ridl., J. Fed. Malay States Mus. 6: 194 (1915). *Oreobolus distichus* (Ridl.) Kük. & Steenis, Bull. Jard. Bot. Buitenzorg, III, 13: 292 (1935), nom. illeg.

Oreobolus obtusangulus Gaudich., Ann. Sci. Nat. (Paris) 5: 99 (1825).
Colombia to Falkland Is. 83 CLM ECU PER 85 AGS CLS JNF 90 FAL. Hemicr.

subsp. ***obtusangulus***
Peru to Falkland Is. 83 PER 85 AGS CLS JNF 90 FAL. Hemicr.
Oreobolus clandestinus Phil., Linnaea 29: 81 (1857).
Oreobolus obtusangulus f. *borealis* Skottsb., Veröff. Geobot. Inst. Rübel Zürich 3: 88 (1925).
Voladeria carchiensis Benoist, Bull. Soc. Bot. France 84: 637 (1938).

subsp. ***unispicus*** O.Seberg, Bot. J. Linn. Soc. 96: 163 (1988).
Colombia to Peru. 83 CLM ECU PER. Hemicr. or cham.

Oreobolus oligocephalus W.M.Curtis, Brunonia 7: 301 (1985). *Schoenoides oligocephalus* (W.M.Curtis) O.Seberg, Willdenowia 16: 182 (1986).
Tasmania. 50 TAS. Hemicr. or rhizome geophyte.

Oreobolus oxycarpus S.T.Blake, Contr. Queensland Herb. 8: 40 (1969).
SE. Australia. 50 NSW TAS VIC. Hemicr. or rhizome geophyte.

subsp. ***brownii*** O.Seberg, Bot. J. Linn. Soc. 96: 178 (1988).
Tasmania. 50 TAS. Hemicr. or rhizome geophyte.

subsp. ***oxycarpus***
SE. Australia. 50 NSW TAS VIC. Hemicr. or rhizome geophyte.

Oreobolus pectinatus Hook.f., Fl. Antarct.: 87 (1844). *Oreobolus pumilio* var. *pectinatus* (Hook.f.) C.B.Clarke ex Cheeseman, Man. New Zealand Fl.: 797 (1906).
New Zealand. 51 ATP NZN NZS. Hemicr. or rhizome geophyte.
Oreobolus serrulata Colenso, Trans. & Proc. New Zealand Inst. 22: 492 (1889 publ. 1890).

Oreobolus pumilio R.Br., Prodr.: 236 (1810).
New Guinea, SE. Australia. 43 NWG 50 NSW TAS VIC. Hemicr. or cham.
Oreobolus pfeifferianus Barros ex H.Pfeiff., Repert. Spec. Nov. Regni Veg. 23: 352 (1927).
Oreobolus clemensiae Kük., Repert. Spec. Nov. Regni Veg. 48: 69 (1940). *Oreobolus pumilio* subsp. *clemensiae* (Kük.) O.Seberg, Bot. J. Linn. Soc.: 96: 170 (1988).

Oreobolus strictus Berggr., Minneskr. Fisiog. Sallks. Lund 8: 25 (1877).
New Zealand. 51 NZN NZS. Hemicr. or rhizome geophyte.

Oreobolus tholicarpus D.I.Morris, Muelleria 15: 28 (2001).
Tasmania. 50 TAS. Hemicr. or rhizome geophyte.

Oreobolus venezuelensis Steyerm., Bol. Soc. Venez. Ci. Nat. 11: 308 (1949).
Costa Rica to Peru. 80 COS PAN 82 VEN 83 ECU PER. Hemicr.

Synonyms:
Oreobolus clandestinus Phil. = ***Oreobolus obtusangulus*** Gaudich. subsp. ***obtusangulus***
Oreobolus clemensiae Kük. = ***Oreobolus pumilio*** R.Br.
Oreobolus distichus (Ridl.) Kük. & Steenis = ***Oreobolus kuekenthalii*** Steenis ex Kük.
Oreobolus goeppingeri var. *sumapacii* L.E.Moro = ***Oreobolus goeppingeri*** Suess.
Oreobolus goeppingeri var. *tuquerrensis* L.E.Moro = ***Oreobolus goeppingeri*** Suess.
Oreobolus obtusangulus f. *borealis* Skottsb. = ***Oreobolus obtusangulus*** Gaudich. subsp. ***obtusangulus***
Oreobolus obtusangulus var. *rubrovaginatus* Kük. = ***Oreobolus goeppingeri*** Suess.
Oreobolus obtusangulus subsp. *rubrovaginatus* (Kük.) T.Koyama = ***Oreobolus goeppingeri*** Suess.
Oreobolus pfeifferianus Barros ex H.Pfeiff. = ***Oreobolus pumilio*** R.Br.
Oreobolus pumilio subsp. *clemensiae* (Kük.) O.Seberg = ***Oreobolus pumilio*** R.Br.
Oreobolus pumilio var. *pectinatus* (Hook.f.) C.B.Clarke ex Cheeseman = ***Oreobolus pectinatus*** Hook.f.
Oreobolus serrulata Colenso = ***Oreobolus pectinatus*** Hook.f.

Oreograstis

Oreograstis K.Schum. = ***Carpha*** Banks & Sol. ex R.Br.
Oreograstis eminii K.Schum. = ***Carpha eminii*** (K.Schum.) C.B.Clarke

Orobium

Orobium Schrad. ex Nees = ***Lagenocarpus*** Nees

Osculisa

Osculisa Raf. = ***Carex*** L.
Osculisa acuta (L.) Raf. = ***Carex acuta*** L.
Osculisa sudetica (Opiz) Raf. = ***Carex atrata*** L. subsp. ***atrata***
Osculisa trinervis (Degl.) Raf. = ***Carex trinervis*** Degl.

Oxycaryum

Oxycaryum Nees in C.F.P.von Martius & auct. suc. (eds.), Fl. Bras. 2(1): 90 (1842).
Trop. & Subtrop. Africa, Trop. & Subtrop. America. 22 BEN BKN GHA MLI NGA NGR SEN TOG 23 RWA ZAI 25 TAN UGA 27 BOT NAM 29 MDG 77 TEX 78 ALA FLA LOU 79 MXS MXT 80 BLZ COS ELS GUA HON NIC PAN 81 CUB DOM JAM TRT 82 FRG GUY SUR VEN 83 BOL ECU PER 84 BZC BZN 85 AGE PAR URU.
1 Species
Crepidocarpus Klotzsch ex Boeck, Linnaea 36: 414 (1870).

Oxycaryum cubense (Poepp. & Kunth) Palla, Denkschr. Kaiserl. Akad. Wiss., Wien. Math.-Naturwiss. Kl. 79: 169 (1908).
Trop. & Subtrop. Africa, Trop. & Subtrop. America. 22 BEN BKN GHA MLI NGA NGR SEN TOG 23 RWA ZAI 25 TAN UGA 27 BOT NAM 29 MDG 77 TEX 78 ALA FLA LOU 79 MXS MXT 80 BLZ COS ELS GUA HON NIC PAN 81 CUB DOM JAM TRT 82 FRG GUY SUR VEN 83 BOL ECU PER 84 BZC BZN 85 AGE PAR URU. Hel.
**Scirpus cubensis* Poepp. & Kunth, Enum. Pl. 2: 172 (1837). *Anosporum cubense* (Poepp. & Kunth) Boeck., Bot. Zeitung (Berlin) 27: 26 (1869). *Crepidocarpus cubensis* (Poepp. & Kunth) Klotzsch ex Boeck., Linnaea 36: 414 (1870). *Kyllinga scirpina* Rchb. ex C.B.Clarke in T.A.Durand & H.Schinz, Consp. Fl. Afric. 5: 615 (1894).
Oxycaryum schomburgkianum Nees in C.F.P.von Martius & auct. suc. (eds.), Fl. Bras. 2(1): 90 (1842).
Cyperus blepharoleptos Steud., Syn. Pl. Glumac. 2: 28 (1854).
Mariscus foliosissimus Steud., Syn. Pl. Glumac. 2: 65 (1854).
Scirpus ablepharus Griseb., Cat. Pl. Cub.: 240 (1866). *Anosporum ablepharum* (Griseb.) Maury ex Micheli, Mém. Soc. Phys. Genève 31: 125 (1890).
Anosporum cubense var. *gracile* Boeck., Linnaea 36: 414 (1870). *Scirpus cubensis* var. *gracilis* (Boeck.) Beetle, Rhodora 46: 146 (1944).
Scirpus cubensis var. *minor* C.Wright & Sauvalle, Anales Acad. Ci. Méd. Habana 8: 177 (1871).
Isolepis echinocephala Oliv., Trans. Linn. Soc. London 29: 167 (1875).
Anosporum schinzii Boeck., Verh. Bot. Vereins Prov. Brandenburg 29(Abh.): 46 (1888). *Oxycaryum schinzii* (Boeck.) Palla, Denkschr. Kaiserl. Akad. Wiss., Wien. Math.-Naturwiss. Kl. 79: 169 (1908).
Crepidocarpus schinzii Klotzsch ex Boeck., Verh. Bot. Vereins Prov. Brandenburg 29: 46 (1888).
Anosporum paraguayense Maury, Mém. Soc. Phys. Genève 31: 124 (1890). *Oxycaryum paraguayense* (Maury) Palla, Denkschr. Kaiserl. Akad. Wiss., Wien. Math.-Naturwiss. Kl. 79: 169 (1908). *Scirpus cubensis* var. *paraguayensis* (Maury) Kük. ex Barros, Anales Mus. Argent. Ci. Nat. "Bernardino Rivadavia" 38: 150 (1935). *Scirpus paraguayensis* (Maury) Herter, Fl. Illustr. Urug. 1: 161 (1942). *Oxycaryum cubense* f. *paraguayense* (Maury) Pedersen, Hickenia 2: 138 (1995).
Anosporum piliferum Maury, Mém. Soc. Phys. Genève 31: 124 (1890). *Oxycaryum piliferum* (Maury) Palla, Denkschr. Kaiserl. Akad. Wiss., Wien. Math.-Naturwiss. Kl. 79: 169 (1908). *Scirpus piliferus* (Maury) Pickel, Bol. Mus. Nac. Rio de Janeiro 13(1-2): 124 (1937 publ. 1938).
Oxycaryum guianense Palla, Denkschr. Kaiserl. Akad. Wiss., Wien. Math.-Naturwiss. Kl. 79: 169 (1908).

Synonyms:
Oxycaryum cubense f. *paraguayense* (Maury) Pedersen = ***Oxycaryum cubense*** (Poepp. & Kunth) Palla

Phaeocephalum microcarpum (Baldwin ex A.Gray) House = ***Rhynchospora microcarpa*** Baldwin ex A.Gray
Phaeocephalum miliaceum (Lam.) House = ***Rhynchospora miliacea*** (Lam.) A.Gray
Phaeocephalum mixtum (Britton) House = ***Rhynchospora mixta*** Britton
Phaeocephalum pallidum (M.A.Curtis) House = ***Rhynchospora pallida*** M.A.Curtis
Phaeocephalum patulum (A.Gray) House = ***Rhynchospora caduca*** Elliott
Phaeocephalum perplexum (Britton) House = ***Rhynchospora perplexa*** Britton
Phaeocephalum plankii (Britton) House = ***Rhynchospora harveyi*** W.Boott var. ***harveyi***
Phaeocephalum plumosum (Elliott) House = ***Rhynchospora plumosa*** Elliott
Phaeocephalum proliferum (Nees) House = ***Rhynchospora schomburgkiana*** (Boeck.) T.Koyama
Phaeocephalum punctatum (Elliott) House = ***Rhynchospora punctata*** Elliott
Phaeocephalum pusillum (Chapm. ex M.A.Curtis) House = ***Rhynchospora pusilla*** Chapm. ex M.A.Curtis
Phaeocephalum rariflorum (Michx.) House = ***Rhynchospora rariflora*** (Michx.) Elliott
Phaeocephalum schoenoides (Alph.Wood) House = ***Rhynchospora elliottii*** A.Dietr.
Phaeocephalum solitare (R.M.Harper) House = ***Rhynchospora solitaria*** R.M.Harper
Phaeocephalum stipitatum (Chapm.) House = ***Rhynchospora odorata*** C.Wright ex Griseb.
Phaeocephalum torreyanum (A.Gray) House = ***Rhynchospora torreyana*** A.Gray
Phaeocephalum tracyi (Britton) House = ***Rhynchospora tracyi*** Britton

Phaeolorum

Phaeolorum Ehrh. = ***Carex*** L.

Phylloscirpus

Phylloscirpus C.B.Clarke, Kew Bull., Addit. Ser. 8: 44 (1908).
Ecuador to W. Argentina. 83 BOL CLM ECU PER 85 AGS AGW CLN.
3 Species

Phylloscirpus acaulis (Phil.) Goetgh. & D.A.Simpson, Kew Bull. 46: 175 (1991).
Ecuador to W. Argentina. 83 BOL ECU PER 85 AGS AGW CLN. Hemicr. or rhizome geophyte.
**Scirpus acaulis* Phil., Linnaea 29: 78 (1857). *Isolepis acaulis* (Phil.) Boeck., Linnaea 36: 440 (1870).
Scirpus nubigenus Phil., Anales Univ. Chile 1873: 554 (1873).
Scirpus sessiliflorus Phil., Anales Univ. Chile 93: 479 (1896).
Scirpus atacamensis var. *exscapus* Kurtz, Bol. Acad. Nac. Ci. 15: 520 (1897).
Phylloscirpus andesinus C.B.Clarke, Kew Bull., Addit. Ser. 8: 45 (1908).

Phylloscirpus boliviensis (Barros) Dhooge & Goetgh., Novon 14: 282 (2004).
W. South America to N. Chile. 83 BOL CLM ECU PER 85 CLN. Hemicr. or rhizome geophyte.
**Scirpus boliviensis* Barros, Darwiniana 11: 764 (1959).

Phylloscirpus deserticola (Phil.) Dhooge & Goetgh., Novon 14: 278 (2004).
Ecuador to NW. Argentina. 83 BOL ECU PER 85 AGW CLN. Hemicr.
**Scirpus deserticola* Phil., Fl. Atacam.: 53 (1860).
Scirpus semisubterraneus Boeck., Linnaea 36: 495 (1870). *Scirpus deserticola* var. *semisubterraneus* (Boeck.) Barros, Anales Mus. Argent. Ci. Nat. "Bernardino Rivadavia" 38: 156 (1935).

***Synonyms*:**
Phylloscirpus andesinus C.B.Clarke = ***Phylloscirpus acaulis*** (Phil.) Goetgh. & D.A.Simpson

Phyllostachys

Phyllostachys Torr. = ***Carex*** L.

Physiglochis

Physiglochis Neck. = ***Carex*** L.
Physiglochis buxbaumii (Wahlenb.) Raf. = ***Carex buxbaumii*** Wahlenb.
Physiglochis dioica (L.) Raf. = ***Carex dioica*** L.
Physiglochis paniculata (L.) Raf. = ***Carex paniculata*** L.
Physiglochis teretiuscula (Gooden.) Raf. = ***Carex diandra*** Schrank
Physiglochis wormskioldii Raf. = ***Carex scirpoidea*** Michx. var. ***scirpoidea***

Platylepis

Platylepis Kunth = ***Ascolepis*** Nees ex Steud.
Platylepis brasiliensis Kunth = ***Ascolepis brasiliensis*** (Kunth) Benth. ex C.B.Clarke
Platylepis capensis Kunth = ***Ascolepis capensis*** (Kunth) Ridl.
Platylepis dioica Steud. = ***Ascolepis capensis*** (Kunth) Ridl.
Platylepis guianensis Nees = ***Ascolepis brasiliensis*** (Kunth) Benth. ex C.B.Clarke
Platylepis leucocephala Nees = ***Ascolepis brasiliensis*** (Kunth) Benth. ex C.B.Clarke
Platylepis xanthocephala Nees = ***Ascolepis brasiliensis*** (Kunth) Benth. ex C.B.Clarke

Pleurachne

Pleurachne Schrad. = ***Ficinia*** Schrad.
Pleurachne secunda (Vahl) Schrad. = ***Ficinia secunda*** (Vahl) Kunth
Pleurachne sieberi Schrad. = ***Ficinia secunda*** (Vahl) Kunth

Pleuranthus

Pleuranthus Rich. ex Pers. = ***Dulichium*** Pers.

Pleurostachys

Pleurostachys Brongn. in L.I.Duperrey, Voy. Monde, Phan.: 172 (1833). *Nemochloa* Nees in F.J.F.Meyen, Observ. Bot. 1: 108 (1834), nom. illeg.
Trinidad to S. Trop. America. 81 TRT 82 FRG GUY SUR VEN 83 BOL CLM ECU PER 84 BZE BZL BZN BZS.
33 Species
Nomochloa Nees, Linnaea 9: 299 (1834), nom. illeg.

Pleurostachys angustifolia Boeck., Beitr. Cyper. 2: 18 (1890).
Brazil (Rio de Janeiro). 84 BZL. Hemicr. or rhizome geophyte.

Pleurostachys beyrichii (Nees) Steud., Syn. Pl. Glumac. 2: 139 (1855).
SE. & S. Brazil. 84 BZL BZS. Hemicr. or rhizome geophyte.
Nemochloa beyrichii Nees in C.F.P.von Martius & auct. suc. (eds.), Fl. Bras. 2(1): 152 (1842). *Rhynchospora fluminensis* L.B.Sm., Phytologia 1: 81 (1934).
Pleurostachys paniculata Boeck., Beitr. Cyper. 2: 20 (1890).
Pleurostachys beyrichii var. *panicoides* H.Pfeiff., Repert. Spec. Nov. Regni Veg. 17: 234 (1921). *Pleurostachys panicoides* (H.Pfeiff.) H.Pfeiff., Bot. Arch. 9: 240 (1925).

Pleurostachys bracteolata C.B.Clarke, Kew Bull., Addit. Ser. 8: 41 (1908).
Brazil (Rio de Janeiro). 84 BZL. Hemicr. or rhizome geophyte.

Pleurostachys bradei R.Gross, Repert. Spec. Nov. Regni Veg. 29: 245 (1931).
Brazil (Rio de Janeiro). 84 BZL. Hemicr. or rhizome geophyte.
Pleurostachys bradei var. *alata* Kük. & R.Gross, Bot. Jahrb. Syst. 75: 472 (1952).

Pleurostachys calyptrocaryoides R.Gross, Repert. Spec. Nov. Regni Veg. 42: 175 (1937).
Brazil (Rio de Janeiro). 84 BZL. Hemicr. or rhizome geophyte.

Pleurostachys densifoliata H.Pfeiff., Repert. Spec. Nov. Regni Veg. 17: 233 (1921).
Brazil (Paraná). 84 BZS. Hemicr. or rhizome geophyte.

Pleurostachys distichophylla (Boeck.) C.B.Clarke, Kew Bull., Addit. Ser. 8: 41 (1908).
Brazil (Rio de Janeiro). 84 BZL. Hemicr. or rhizome geophyte.
Rhynchospora distichophylla Boeck., Vidensk. Meddel. Naturhist. Foren. Kjøbenhavn 41-42: 26 (1879).

Pleurostachys douglasii C.B.Clarke, Kew Bull., Addit. Ser. 8: 41 (1908).
Brazil (Rio de Janeiro). 84 BZL. Hemicr. or rhizome geophyte.

Pleurostachys foliosa Kunth, Enum. Pl. 2: 284 (1837). *Nemochloa foliosa* (Kunth) Nees in C.F.P.von Martius & auct. suc. (eds.), Fl. Bras. 2(1): 152 (1842). *Rhynchospora foliosa* (Kunth) L.B.Sm., Phytologia 1: 81 (1934).
SE. Brazil. 84 BZL. Hemicr. or rhizome geophyte.
Pleurostachys pauloensis Palla, Denkschr. Kaiserl. Akad. Wiss., Wien. Math.-Naturwiss. Kl. 79: 187 (1908).
Pleurostachys longa Lindm., Bih. Kongl. Svenska Vetensk.-Akad. Handl. 26(9): 29 (190o).

Pleurostachys gaudichaudii Brongn. in L.I.Duperrey, Voy. Monde, Phan.: 174 (1833). *Nemochloa gaudichaudii* (Brongn.) Nees in C.F.P.von Martius & auct. suc. (eds.), Fl. Bras. 2(1): 152 (1842). *Pleurostachys gaudichaudii* f. *genuina* H.Pfeiff., Bot. Arch. 9: 233 (1925), nom. inval. *Rhynchospora gaudichaudii* (Brongn.) L.B.Sm., Phytologia 1: 81 (1934).
SE. & S. Brazil. 84 BZL BZS. Hemicr. or rhizome geophyte.
Pleurostachys gaudichaudii f. *minor* H.Pfeiff., Bot. Arch. 9: 233 (1925).

Pleurostachys geraldiana Davie, J. Bot. 55: 222 (1917).
Brazil (São Paulo). 84 BZL. Hemicr. or rhizome geophyte.

Pleurostachys graminifolia Brongn. in L.I.Duperrey, Voy. Monde, Phan.: 175 (1833). *Rhynchospora graminifolia* (Brongn.) L.B.Sm., Phytologia 1: 81 (1934).
SE. & S. Brazil. 84 BZL BZS. Hemicr. or rhizome geophyte.
Pleurostachys gracilis Boeck., Allg. Bot. Z. Syst. 2: 111 (1896). *Rhynchospora hunnewellii* L.B.Sm., Rhodora 50: 132 (1948). *Pleurostachys graminifolia* var. *gracilis* (Boeck.) Kük. & R.Gross, Bot. Jahrb. Syst. 75: 480 (1952).
Pleurostachys foliosa var. *gaudichaudii* H.Pfeiff., Repert. Spec. Nov. Regni Veg. 17: 232 (1921).
Pleurostachys graminifolia var. *glabra* Kük. & R.Gross, Bot. Jahrb. Syst. 75: 480 (1952).

Pleurostachys guianensis Uittien, Recueil Trav. Bot. Néerl. 22: 337 (1925).
Suriname. 82 SUR. Hemicr. or rhizome geophyte.

Pleurostachys hoehneana R.Gross, Repert. Spec. Nov. Regni Veg. 42: 175 (1937).
SE. Brazil. 84 BZL. Hemicr. or rhizome geophyte.

Pleurostachys kuhlmannii Kük., Bot. Jahrb. Syst. 75: 476 (1952).
N. Brazil. 84 BZN. Hemicr. or rhizome geophyte.

Pleurostachys loefgrenii (Boeck.) T.Koyama, Darwiniana 16: 89 (1970).
E. Brazil. 84 BZE BZL. Hemicr. or rhizome geophyte.
**Hypolytrum loefgrenii* Boeck., Vidensk. Meddel. Naturhist. Foren. Kjøbenhavn 1894: 239 (1895).
Pleurostachys arrojadii H.Pfeiff., Repert. Spec. Nov. Regni Veg. 19: 295 (1924).
Pleurostachys arrojadii f. *latifolia* Kük. & R.Gross, Bot. Jahrb. Syst. 75: 470 (1952).
Pleurostachys arrojadii f. *rigida* Kük., Bot. Jahrb. Syst. 75: 470 (1952).

Pleurostachys martiana (Nees) Steud., Syn. Pl. Glumac. 2: 139 (1855).
SE. Brazil. 84 BZL. Hemicr. or rhizome geophyte.
**Nemochloa martiana* Nees in C.F.P.von Martius & auct. suc. (eds.), Fl. Bras. 2(1): 150 (1842). *Rhynchospora martiana* (Nees) L.B.Sm., Phytologia 1: 81 (1934).

Pleurostachys millegrana (Nees) Steud., Syn. Pl. Glumac. 2: 139 (1855).
Peru to Bolivia, SE. Brazil. 83 BOL PER 84 BZL. Hemicr. or rhizome geophyte.
Nemochloa extenuata Nees in C.F.P.von Martius & auct. suc. (eds.), Fl. Bras. 2(1): 152 (1842). *Pleurostachys extenuata* (Nees) Steud., Syn. Pl. Glumac. 2: 139 (1855).
**Nemochloa millegrana* Nees in C.F.P.von Martius & auct. suc. (eds.), Fl. Bras. 2(1): 150 (1842). *Rhynchospora millegrana* (Nees) L.B.Sm., Phytologia 1: 81 (1934).
Pleurostachys grandifolia Boeck., Vidensk. Meddel. Naturhist. Foren. Kjøbenhavn 41-42: 25 (1879). *Pleurostachys millegrana* var. *grandifolia* (Boeck.) H.Pfeiff., Repert. Spec. Nov. Regni Veg. 17: 235 (1921).
Pleurostachys millegrana var. *minuscula* Kük., Repert. Spec. Nov. Regni Veg. 26: 254 (1929).

Pleurostachys muelleri Boeck., Beitr. Cyper. 2: 19 (1890). *Pleurostachys muelleri* f. *normalis* Boeck., Allg. Bot. Z. Syst. 2: 111 (1896), nom. inval.
S. Brazil. 84 BZS. Hemicr. or rhizome geophyte.

Pleurostachys orbignyana Brongn. in L.I.Duperrey, Voy. Monde, Phan.: 175 (1833). *Nomochloa orbignyana* (Brongn.) Nees, Linnaea 9: 299 (1834). *Nemochloa orbignyana* (Brongn.) Nees ex Steud., Nomencl. Bot., ed. 2, 2: 189 (1841). *Rhynchospora orbignyana* (Brongn.) L.B.Sm., Phytologia 1: 81 (1934).
Brazil (Rio de Janeiro). 84 BZL. Hemicr. or rhizome geophyte.
Pleurostachys macrantha Kunth, Enum. Pl. 2: 286 (1837). *Nemochloa macrantha* (Kunth) Nees in C.F.P.von Martius & auct. suc. (eds.), Fl. Bras. 2(1): 151 (1842).

Pleurostachys pearcei C.B.Clarke, Kew Bull., Addit. Ser. 8: 41 (1908). *Dichromena pearcei* (C.B.Clarke) J.F.Macbr., Publ. Field Mus. Nat. Hist., Bot. Ser. 11: 6 (1931).
Peru. 83 PER. Hemicr. or rhizome geophyte.

Pleurostachys peruviana C.B.Clarke, Kew Bull., Addit. Ser. 8: 42 (1908). *Dichromena peruviana* (C.B.Clarke) J.F.Macbr., Publ. Field Mus. Nat. Hist., Bot. Ser. 11: 6 (1931).
Peru. 83 PER. Hemicr. or rhizome geophyte.

Pleurostachys puberula Boeck., Flora 63: 453 (1880). *Rhynchospora puberula* (Boeck.) L.B.Sm., Phytologia 1: 82 (1934).
Trinidad to S. Trop. America. 81 TRT 82 FRG GUY VEN 83 BOL CLM ECU 84 BZL. Hemicr. or rhizome geophyte.
Pleurostachys montana Palla, Denkschr. Kaiserl. Akad. Wiss., Wien. Math.-Naturwiss. Kl. 79: 184 (1908). *Pleurostachys puberula* var. *montana* (Palla) Kük. & R.Gross, Bot. Jahrb. Syst. 75: 478 (1952).
Pleurostachys buchtienii Kük., Repert. Spec. Nov. Regni Veg. 26: 254 (1929). *Pleurostachys puberula* var. *buchtienii* (Kük.) Kük., Bot. Jahrb. Syst. 75: 477 (1952).

Pleurostachys regnellii C.B.Clarke, Kew Bull., Addit. Ser. 8: 42 (1908).
SE. Brazil. 84 BZL. Hemicr. or rhizome geophyte.

Pleurostachys scaposa C.B.Clarke, Kew Bull., Addit. Ser. 8: 42 (1908).
Brazil (Acre). 84 BZN. Hemicr. or rhizome geophyte.

Pleurostachys sellowii Kunth, Enum. Pl. 2: 285 (1837). *Nemochloa sellovii* (Kunth) Nees in C.F.P.von Martius & auct. suc. (eds.), Fl. Bras. 2(1): 152 (1842).
SE. & S. Brazil. 84 BZL BZS. Hemicr. or rhizome geophyte.
Pleurostachys longiradiata Palla, Denkschr. Kaiserl. Akad. Wiss., Wien. Math.-Naturwiss. Kl. 79: 187 (1908). *Pleurostachys sellowii* var. *longiradiata* (Palla) Kük. & R.Gross, Bot. Jahrb. Syst. 75: 471 (1952).
Pleurostachys cimaensis H.Pfeiff., Repert. Spec. Nov. Regni Veg. 17: 233 (1921).

Pleurostachys sparsiflora Kunth, Enum. Pl. 2: 286 (1837). *Nemochloa sparsiflora* (Kunth) Nees in C.F.P.von Martius & auct. suc. (eds.), Fl. Bras. 2(1): 152 (1842). *Rhynchospora sparsiflora* (Kunth) L.B.Sm., Phytologia 1: 82 (1934).
N. Brazil. 84 BZN. Hemicr. or rhizome geophyte.
Chaetospora effusa Schrad. ex Kunth, Enum. Pl. 2: 286 (1837).

Pleurostachys stricta Kunth, Enum. Pl. 2: 286 (1837). *Nomochloa stricta* (Kunth) Nees, J. Bot. (Hooker) 2: 398 (1840). *Nemochloa stricta* (Kunth) Nees in C.F.P.von Martius & auct. suc. (eds.), Fl. Bras. 2(1): 152 (1842).
SE. & S. Brazil. 84 BZL BZS. Hemicr. or rhizome geophyte.
Pleurostachys rabenii Boeck., Beitr. Cyper. 2: 19 (1890). *Pleurostachys stricta* var. *rabenii* (Boeck.) H.Pfeiff., Repert. Spec. Nov. Regni Veg. 38: 98 (1935).
Pleurostachys spicata Boeck., Allg. Bot. Z. Syst. 2: 112 (1896).
Pleurostachys stricta var. *angustocarpa* H.Pfeiff., Repert. Spec. Nov. Regni Veg. 17: 234 (1921).
Pleurostachys stricta var. *dissoluta* R.Gross & Kük., Bot. Jahrb. Syst. 75: 482 (1952).

Pleurostachys tenuiflora Brongn. in L.I.Duperrey, Voy. Monde, Phan.: 175 (1833). *Nomochloa tenuiflora* (Brongn.) Nees, Linnaea 9: 299 (1834). *Nemochloa tenuiflora* (Brongn.) Nees in C.F.P.von Martius & auct. suc. (eds.), Fl. Bras. 2(1): 149 (1842). *Rhynchospora tenuiflora* (Brongn.) L.B.Sm., Phytologia 1: 82 (1934).
Brazil (Rio de Janeiro, São Paulo). 84 BZL. Hemicr. or rhizome geophyte.
Scirpus latifolius Bertol., Opusc. Sci. 3: 406 (1819). *Isolepis latifolia* (Bertol.) A.Dietr., Sp. Pl. 2: 132 (1832).
Pleurostachys elegans Kunth, Enum. Pl. 2: 285 (1837). *Nemochloa elegans* (Kunth) Nees in C.F.P.von Martius & auct. suc. (eds.), Fl. Bras. 2(1): 149 (1842). *Nomochloa elegans* (Kunth) Nees, Nov. Actorum Acad. Caes. Leop.-Carol. Nat. Cur. 19(Suppl. 1): 113 (1843). *Pleurostachys tenuiflora* var. *elegans* (Kunth) H.Pfeiff., Repert. Spec. Nov. Regni Veg. 17: 231 (1921).
Pleurostachys floribunda Kunth, Enum. Pl. 2: 285 (1837). *Pleurostachys tenuiflora* var. *floribunda* (Kunth) H.Pfeiff., Bot. Arch. 9: 228 (1925).
Pleurostachys latifolia H.Pfeiff., Repert. Spec. Nov. Regni Veg. 19: 295 (1924).
Pleurostachys luetzelburgiana H.Pfeiff., Repert. Spec. Nov. Regni Veg. 19: 294 (1924).
Pleurostachys tenuiflora var. *longifolia* Kük. & R.Gross, Bot. Jahrb. Syst. 75: 460 (1952).

Pleurostachys turbinata (Nees) H.Pfeiff., Repert. Spec. Nov. Regni Veg. 19: 295 (1924).
Brazil (Rio de Janeiro). 84 BZL. Hemicr. or rhizome geophyte.
**Nemochloa turbinata* Nees in F.J.F.Meyen, Observ. Bot. 1: 108 (1834). *Pleurostachys kunthiana* C.B.Clarke, Kew Bull., Addit. Ser. 8: 42 (1908).
Nomochloa turbinata Nees, Nov. Actorum Acad. Caes. Leop.-Carol. Nat. Cur. 19(Suppl. 1): 112 (1843).

Pleurostachys ulei Boeck., Beitr. Cyper. 2: 19 (1890).
S. Brazil. 84 BZS. Hemicr. or rhizome geophyte.

Pleurostachys undulatifolia R.Gross, Repert. Spec. Nov. Regni Veg. 42: 174 (1937).
Brazil (Rio de Janeiro). 84 BZL. Hemicr. or rhizome geophyte.

Pleurostachys urvillei Brongn. in L.I.Duperrey, Voy. Monde, Phan.: 178 (1834). *Nemochloa urvillei* (Brongn.) Nees in C.F.P.von Martius & auct. suc. (eds.), Fl. Bras. 2(1): 150 (1842). *Rhynchospora scalaris* L.B.Sm., Phytologia 1: 82 (1934).
S. Brazil (to Minas Gerais). 84 BZL BZS. Hemicr. or rhizome geophyte.
Pleurostachys paranensis Palla, Denkschr. Kaiserl. Akad. Wiss., Wien. Math.-Naturwiss. Kl. 79: 185

(1908). *Pleurostachys urvillei* var. *paranensis* (Palla) Kük. & R.Gross, Bot. Jahrb. Syst. 75: 463 (1952).

Pleurostachys polygonatum Palla, Denkschr. Kaiserl. Akad. Wiss., Wien. Math.-Naturwiss. Kl. 79: 186 (1908). *Pleurostachys urvillei* f. *polygonatum* (Palla) Kük., Bot. Jahrb. Syst. 75: 464 (1952).

Pleurostachys paranensis var. *dusenii* H.Pfeiff., Repert. Spec. Nov. Regni Veg. 17: 231 (1921). *Pleurostachys dusenii* (H.Pfeiff.) H.Pfeiff., Bot. Arch. 9: 299 (1925). *Pleurostachys urvillei* f. *dusenii* (H.Pfeiff.) Kük. & R.Gross, Bot. Jahrb. Syst. 75: 464 (1952).

Pleurostachys paranensis var. *ypirangensis* H.Pfeiff., Repert. Spec. Nov. Regni Veg. 17: 231 (1921). *Pleurostachys ypirangensis* (H.Pfeiff.) H.Pfeiff., Bot. Arch. 9: 229 (1925). *Pleurostachys urvillei* f. *ypirangensis* (H.Pfeiff.) Kük. & R.Gross, Bot. Jahrb. Syst. 75: 464 (1952).

Synonyms:

Pleurostachys arrojadii H.Pfeiff. = ***Pleurostachys loefgrenii*** (Boeck.) T.Koyama

Pleurostachys arrojadii f. *latifolia* Kük. & R.Gross = ***Pleurostachys loefgrenii*** (Boeck.) T.Koyama

Pleurostachys arrojadii f. *rigida* Kük. = ***Pleurostachys loefgrenii*** (Boeck.) T.Koyama

Pleurostachys beyrichii var. *panicoides* H.Pfeiff. = ***Pleurostachys beyrichii*** (Nees) Steud.

Pleurostachys bradei var. *alata* Kük. & R.Gross = ***Pleurostachys bradei*** R.Gross

Pleurostachys buchtienii Kük. = ***Pleurostachys puberula*** Boeck.

Pleurostachys cimaensis H.Pfeiff. = ***Pleurostachys sellowii*** Kunth

Pleurostachys dusenii (H.Pfeiff.) H.Pfeiff. = ***Pleurostachys urvillei*** Brongn.

Pleurostachys elegans Kunth = ***Pleurostachys tenuiflora*** Brongn.

Pleurostachys extenuata (Nees) Steud. = ***Pleurostachys millegrana*** (Nees) Steud.

Pleurostachys floribunda Kunth = ***Pleurostachys tenuiflora*** Brongn.

Pleurostachys foliosa var. *gaudichaudii* H.Pfeiff. = ***Pleurostachys graminifolia*** Brongn.

Pleurostachys gaudichaudii f. *genuina* H.Pfeiff. = ***Pleurostachys gaudichaudii*** Brongn.

Pleurostachys gaudichaudii f. *minor* H.Pfeiff. = ***Pleurostachys gaudichaudii*** Brongn.

Pleurostachys gracilis Boeck. = ***Pleurostachys graminifolia*** Brongn.

Pleurostachys graminifolia var. *glabra* Kük. & R.Gross = ***Pleurostachys graminifolia*** Brongn.

Pleurostachys graminifolia var. *gracilis* (Boeck.) Kük. & R.Gross = ***Pleurostachys graminifolia*** Brongn.

Pleurostachys grandifolia Boeck. = ***Pleurostachys millegrana*** (Nees) Steud.

Pleurostachys kunthiana C.B.Clarke = ***Pleurostachys turbinata*** (Nees) H.Pfeiff.

Pleurostachys latifolia H.Pfeiff. = ***Pleurostachys tenuiflora*** Brongn.

Pleurostachys longa Lindm. = ***Pleurostachys foliosa*** Kunth

Pleurostachys longiradiata Palla = ***Pleurostachys sellowii*** Kunth

Pleurostachys luetzelburgiana H.Pfeiff. = ***Pleurostachys tenuiflora*** Brongn.

Pleurostachys macrantha Kunth = ***Pleurostachys orbignyana*** Brongn.

Pleurostachys millegrana var. *grandifolia* (Boeck.) H.Pfeiff. = ***Pleurostachys millegrana*** (Nees) Steud.

Pleurostachys millegrana var. *minuscula* Kük. = ***Pleurostachys millegrana*** (Nees) Steud.

Pleurostachys montana Palla = ***Pleurostachys puberula*** Boeck.

Pleurostachys muelleri f. *normalis* Boeck. = ***Pleurostachys muelleri*** Boeck.

Pleurostachys panicoides (H.Pfeiff.) H.Pfeiff. = ***Pleurostachys beyrichii*** (Nees) Steud.

Pleurostachys paniculata Boeck. = ***Pleurostachys beyrichii*** (Nees) Steud.

Pleurostachys paranensis Palla = ***Pleurostachys urvillei*** Brongn.

Pleurostachys paranensis var. *dusenii* H.Pfeiff. = ***Pleurostachys urvillei*** Brongn.

Pleurostachys paranensis var. *ypirangensis* H.Pfeiff. = ***Pleurostachys urvillei*** Brongn.

Pleurostachys pauloensis Palla = ***Pleurostachys foliosa*** Kunth

Pleurostachys polygonatum Palla = ***Pleurostachys urvillei*** Brongn.

Pleurostachys puberula var. *buchtienii* (Kük.) Kük. = ***Pleurostachys puberula*** Boeck.

Pleurostachys puberula var. *montana* (Palla) Kük. & R.Gross = ***Pleurostachys puberula*** Boeck.

Pleurostachys rabenii Boeck. = ***Pleurostachys stricta*** Kunth

Pleurostachys schottmuelleri (Boeck.) Kük. = ***Rhynchospora schottmuelleri*** Boeck.

Pleurostachys sellowii var. *longiradiata* (Palla) Kük. & R.Gross = ***Pleurostachys sellowii*** Kunth

Pleurostachys spicata Boeck. = ***Pleurostachys stricta*** Kunth

Pleurostachys stricta var. *angustocarpa* H.Pfeiff. = ***Pleurostachys stricta*** Kunth

Pleurostachys stricta var. *dissoluta* R.Gross & Kük. = ***Pleurostachys stricta*** Kunth

Pleurostachys stricta var. *rabenii* (Boeck.) H.Pfeiff. = ***Pleurostachys stricta*** Kunth

Pleurostachys tenuiflora var. *elegans* (Kunth) H.Pfeiff. = ***Pleurostachys tenuiflora*** Brongn.

Pleurostachys tenuiflora var. *floribunda* (Kunth) H.Pfeiff. = ***Pleurostachys tenuiflora*** Brongn.

Pleurostachys tenuiflora var. *longifolia* Kük. & R.Gross = ***Pleurostachys tenuiflora*** Brongn.

Pleurostachys urvillei f. *dusenii* (H.Pfeiff.) Kük. & R.Gross = ***Pleurostachys urvillei*** Brongn.

Pleurostachys urvillei var. *paranensis* (Palla) Kük. & R.Gross = ***Pleurostachys urvillei*** Brongn.

Pleurostachys urvillei f. *polygonatum* (Palla) Kük. = ***Pleurostachys urvillei*** Brongn.

Pleurostachys urvillei f. *ypirangensis* (H.Pfeiff.) Kük. & R.Gross = ***Pleurostachys urvillei*** Brongn.

Pleurostachys ypirangensis (H.Pfeiff.) H.Pfeiff. = ***Pleurostachys urvillei*** Brongn.

Plumaria

Plumaria Heist. ex Fabr. = ***Eriophorum*** L.

Plumaria Bubani = ***Eriophorum*** L.

Plumaria alpina (L.) Bubani = ***Trichophorum alpinum*** (L.) Pers.

Plumaria angustifolia (Honck.) Bubani = ***Eriophorum angustifolium*** Honck.

Plumaria latifolia (Hoppe) Bubani = ***Eriophorum latifolium*** Hoppe

Plumaria scheuchzeri (Hoppe) Bubani = ***Eriophorum scheuchzeri*** Hoppe

Plumaria vaginata (L.) Bubani = ***Eriophorum vaginatum*** L.

Pogonostylis

Pogonostylis Bertol. = ***Fimbristylis*** Vahl
Pogonostylis squarrosus (Vahl) Bertol. = ***Fimbristylis squarrosa*** Vahl

Polyglochin

Polyglochin Ehrh. = ***Carex*** L.

Principina

Principina Uittien, Recueil Trav. Bot. Néerl. 32: 282 (1835).
Príncipe. 23 GGI.
1 Species

Principina grandis Uittien, Recueil Trav. Bot. Néerl. 32: 282 (1935). *Hypolytrum grande* (Uittien) T.Koyama, J. Fac. Sci. Univ. Tokyo, Sect. 3, Bot. 8: 68 (1961). *Mapania grandis* (Uittien) T.Koyama, Curr. Topics Pl. Sci.: 226 (1969).
Príncipe. 23 GGI. Hemicr. or rhizome geophyte.

Proteocarpus

Proteocarpus Börner = ***Carex*** L.
Proteocarpus atrofuscus (Schkuhr) Fedde & J.Schust. = ***Carex atrofusca*** Schkuhr
Proteocarpus flavus (L.) Fedde & J.Schust. = ***Carex flava*** L.
Proteocarpus sylvaticus (Huds.) Fedde & J.Schust. = ***Carex sylvatica*** Huds.

Pseudocarex

Pseudocarex Miq. = ***Carex*** L.
Pseudocarex plantaginea Miq. = ***Carex siderosticta*** Hance var. ***siderosticta***

Pseudocyperus

Pseudocyperus Steud. = ***Fimbristylis*** Vahl
Pseudocyperus scirpinus Steud. = ***Fimbristylis bahiensis*** Steud.

Pseudoeverardia

Pseudoeverardia Gilly = ***Everardia*** Ridl. ex Oliv.
Pseudoeverardia flexifolia (Gilly) Gilly = ***Everardia flexifolia*** (Gilly) T.Koyama & Maguire

Pseudomariscus

Pseudomariscus Rauschert = ***Courtoisina*** Soják
Pseudomariscus cyperoides (Roxb.) Rauschert = ***Courtoisina cyperoides*** (Roxb.) Soják
Pseudomariscus olivaceus (Boeck.) Rauschert =

Pseudoschoenus

Pseudoschoenus (C.B.Clarke) Oteng-Yeb., Notes Roy. Bot. Gard. Edinburgh 33: 308 (1974).
S. Africa. 27 CPP LES NAM OFS.
1 Species

Pseudoschoenus inanis (Thunb.) Oteng-Yeb., Notes Roy. Bot. Gard. Edinburgh 33: 309 (1974).
S. Africa. 27 CPP LES NAM OFS. Hemicr. or rhizome geophyte.
**Schoenus inanis* Thunb., Prodr. Pl. Cap.: 16 (1794). *Scirpus inanis* (Thunb.) Steud., Syn. Pl. Glumac. 2: 86 (1855).
Scirpus spathaceus Hochst., Flora 28: 759 (1845).

Psilocarya

Psilocarya Torr. = ***Rhynchospora*** Vahl
Psilocarya candida Nees = ***Rhynchospora candida*** (Nees) Boeck.
Psilocarya conferta Nees = ***Rhynchospora robusta*** (Kunth) Boeck.
Psilocarya corymbifera (C.Wright) Benth. = ***Rhynchospora ignorata*** Boeck.
Psilocarya corymbiformis Benth. & Hook.f. = ***Rhynchospora scirpoides*** (Torr.) Griseb.
Psilocarya divergens Nees = ***Rhynchospora velutina*** (Kunth) Boeck.
Psilocarya mexicana Liebm. = ***Rhynchospora velutina*** (Kunth) Boeck.
Psilocarya nitens (Vahl) Alph.Wood = ***Rhynchospora nitens*** (Vahl) A.Gray
Psilocarya pauloensis Boeck. = ***Rhynchospora velutina*** (Kunth) Boeck.
Psilocarya pohlii Nees = ***Rhynchospora velutina*** (Kunth) Boeck.
Psilocarya portoricensis Britton = ***Rhynchospora nitens*** (Vahl) A.Gray var. ***nitens***
Psilocarya rhynchosporoides Torr. = ***Rhynchospora nitens*** (Vahl) A.Gray var. ***nitens***
Psilocarya robusta (Kunth) Nees = ***Rhynchospora robusta*** (Kunth) Boeck.
Psilocarya robusta var. *major* Nees = ***Rhynchospora robusta*** (Kunth) Boeck.
Psilocarya robusta var. *minor* Nees = ***Rhynchospora robusta*** (Kunth) Boeck.
Psilocarya rufa Nees = ***Rhynchospora velutina*** (Kunth) Boeck.
Psilocarya schiedeana (Kunth) Liebm. = ***Rhynchospora eximia*** (Nees) Boeck.
Psilocarya scirpoides Torr. = ***Rhynchospora scirpoides*** (Torr.) Griseb.
Psilocarya scirpoides var. *grimesii* Fernald & Griscom = ***Rhynchospora scirpoides*** (Torr.) Griseb.
Psilocarya sellowiana (Kunth) Nees = ***Rhynchospora velutina*** (Kunth) Boeck.
Psilocarya teneriffae Steud. = ***Rhynchospora contracta*** (Nees) J.Raynal
Psilocarya texensis Torr. = ***Rhynchospora nitens*** (Vahl) A.Gray var. ***nitens***
Psilocarya velutina (Kunth) Nees = ***Rhynchospora velutina*** (Kunth) Boeck.
Psilocarya velutina var. *glabrescens* Boeck. = ***Rhynchospora velutina*** (Kunth) Boeck.

Psittacoschoenus

Psittacoschoenus Nees = ***Gahnia*** J.R.Forst. & G.Forst.
Psittacoschoenus erythrocarpus F.Muell. ex Boeck. = ***Gahnia psittacorum*** Labill.

Psyllophora

Psyllophora Ehrh. = ***Carex*** L.
Psyllophora capitata (L.) Schur = ***Carex capitata*** L.
Psyllophora davalliana (Sm.) Schur = ***Carex davalliana*** Sm.
Psyllophora dioica (L.) Schur = ***Carex dioica*** L.
Psyllophora obtusata (Lilj.) Schur = ***Carex obtusata*** Lilj.
Psyllophora pauciflora (Lightf.) Schur = ***Carex pauciflora*** Lightf.
Psyllophora pulicaris (L.) Schur = ***Carex pulicaris*** L.
Psyllophora puliciformis Montandon = ***Carex pulicaris*** L.

Psyllophora pyrenaica (Wahlenb.) Schur = ***Carex pyrenaica*** Wahlenb.
Psyllophora sieberiana (Opiz) Opiz = ***Carex davalliana*** Sm.
Psyllophora vulgaris Heuff. = ***Carex pulicaris*** L.

Ptacoseia

Ptacoseia Ehrh. = ***Carex*** L.

Pterachne

Pterachne Schrad. ex Nees = ***Ascolepis*** Nees ex Steud.

Pterochaete

Pterochaete Arn. ex Boeck. = ***Rhynchospora*** Vahl
Pterochaete drummondii Arn. ex Boeck. = ***Rhynchospora rariflora*** (Michx.) Elliott

Pterocyperus

Pterocyperus Opiz = ***Cyperus*** L.
Pterocyperus esculentus (L.) Opiz = ***Cyperus esculentus*** L.

Pterogyne

Pterogyne Schrad. ex Nees = ***Ascolepis*** Nees ex Steud.

Pterolepis

Pterolepis Schrad. = ***Schoenoplectus*** (Rchb.) Palla
Pterolepis scirpoides Schrad. = ***Schoenoplectus litoralis*** subsp. ***thermalis*** (Trab.) S.S.Hooper

Pteroscleria

Pteroscleria Nees = ***Diplacrum*** R.Br.
Pteroscleria guianensis Nees = ***Diplacrum guianense*** (Nees) T.Koyama
Pteroscleria longifolia Griseb. = ***Diplacrum capitatum*** (Willd.) Boeck.

Pterotheca

Pterotheca C.Presl = ***Rhynchospora*** Vahl
Pterotheca vahlii C.Presl = ***Rhynchospora barbata*** (Vahl) Kunth

Pterygolepis

Pterygolepis Rchb. = ***Schoenoplectus*** (Rchb.) Palla

Ptilanthelium

Ptilanthelium Steud. = ***Schoenus*** L.
Ptilanthelium chauvinii Steud. = ***Schoenus turbinatus*** (R.Br.) Roem. & Schult.
Ptilanthelium deustum (R.Br.) Kük. = ***Ptilothrix deusta*** (R.Br.) K.L.Wilson
Ptilanthelium graciliceps (C.B.Clarke) Kük. = ***Mesomelaena graciliceps*** (C.B.Clarke) K.L.Wilson

Ptilochaeta

Ptilochaeta Nees = ***Rhynchospora*** Vahl
Ptilochaeta diodon Nees = ***Rhynchospora diodon*** (Nees) Griseb.

Ptilosciadium

Ptilosciadium Steud. = ***Rhynchospora*** Vahl
Ptilosciadium lindenii Steud. = ***Rhynchospora hirta*** (Nees) Boeck.

Ptilothrix

Ptilothrix K.L.Wilson, Telopea 5: 612 (1994).
E. Australia. 50 NSW QLD.
1 Species

Ptilothrix deusta (R.Br.) K.L.Wilson, Telopea 5: 613 (1994).
E. Australia. 50 NSW QLD. Hemicr. or rhizome geophyte.
**Carpha deusta* R.Br., Prodr.: 230 (1810). *Rhynchospora deusta* (R.Br.) Spreng., Syst. Veg. 1: 195 (1824). *Chaetospora deusta* (R.Br.) F.Muell., Fragm. 9: 39 (1875). *Mesomelaena deusta* (R.Br.) Benth., Fl. Austral. 7: 379 (1878). *Schoenus deustus* (R.Br.) F.Muell., Syst. Census Austral. Pl. 1: 128 (1882), nom. illeg. *Ptilanthelium deustum* (R.Br.) Kük., Repert. Spec. Nov. Regni Veg. 47: 215 (1939).

Ptychocaryum

Ptychocaryum R.Br. ex Kuntze = ***Scirpodendron*** Zipp. ex Kurz
Ptychocaryum ghaeri (Gaertn.) H.Pfeiff. = ***Scirpodendron ghaeri*** (Gaertn.) Merr.

Pycreus

Pycreus P.Beauv., Fl. Oware 2: 48 (1816).
Cosmopolitan. 11 BGM† CZE GER HUN NET POL SWI 12 COR FRA POR SAR SPA 13 ALB BUL GRC ITA KRI ROM SIC TUE YUG 14 BLR BLT KRY RUC RUE RUS RUW UKR 20 ALG EGY LBY MOR TUN 21 AZO CNY CVI MDR 22 BEN BKN GAM GHA GNB GUI IVO LBR MLI MTN NGA NGR SEN SIE TOG 23 BUR CAF CMN CON EQG GAB GGI RWA ZAI 24 CHA ERI ETH SOM SUD 25 KEN TAN UGA 26 ANG MLW MOZ ZAM ZIM 27 BOT CPP LES NAM NAT OFS SWZ TVL 28 ASC STH 29 ALD COM MAU MDG REU SEY 31 AMU PRM 32 KAZ KGZ TKM TZK UZB 33 NCS TCS 34 AFG IRN IRQ LBS PAL TUR 35 OMA SAU YEM 36 CHC CHH CHM CHN CHS CHT 38 JAP KOR KZN NNS OGA TAI 40 ASS BAN EHM IND NEP PAK SRL WHM 41 AND CBD LAO MYA NCB SCS THA VIE 42 BOR CKI JAW LSI MLY MOL PHI SUL SUM 43 BIS NWG SOL (50) nfk NSW NTA QLD SOA TAS VIC WAU (60) fij gil sam ton (61) coo EAS sci tua 62 CRL MRN wak 63 HAW 72 NBR NSC ONT QUE 73 COL ORE WAS 74 ILL IOW KAN MIN MSO NDA NEB OKL SDA WIS 75 CNT INI MAI MAS MIC NWH NWJ NWY OHI PEN RHO VER WVA 76 ARI CAL NEV UTA 77 NWM TEX 78 ALA ARK DEL FLA GEO KTY LOU MRY MSI NCA SCA TEN VRG WDC 79 MXC MXE MXG MXN MXS MXT 80 BLZ COS CPI ELS GUA HON NIC PAN 81 BER CAY CUB DOM HAI JAM LEE PUE TRT WIN 82 FRG GUY SUR VEN 83 BOL CLM ECU GAL PER 84 BZC BZE BZL BZN BZS 85 AGE AGW BZE BZN CLC CLN PAR URU.
118 Species
Distimus Raf., J. Phys. Chim. Hist. Nat. Arts 89: 105 (1819).
Torreya Raf., J. Phys. Chim. Hist. Nat. Arts 89: 105 (1819).

Pycreus acaulis Nelmes, Kew Bull. 10: 91 (1955).
Malawi. 26 MLW. Hemicr. or rhizome geophyte.

Pycreus acuticarinatus (Kük.) Cherm., Bull. Soc. Bot. France 81: 262 (1934).

W. Trop. Africa to Cameroon. 22 GHA GNB GUI IVO MLI NGA SEN TOG 23 CMN. Hemicr. or rhizome geophyte.
***Cyperus acuticarinatus* Kük., Repert. Spec. Nov. Regni Veg. 12: 93 (1913). *Cyperus macranthus* f. *acuticarinatus* (Kük.) Kük. in H.G.A.Engler (ed.), Pflanzenr., IV, 20(101): 389 (1936).
Cyperus crustaceus Raymond, Garcia de Orta 11: 374 (1963).

Pycreus aethiops (Welw. ex Ridl.) C.B.Clarke in T.A.Durand & H.Schinz, Consp. Fl. Afric. 5: 534 (1894).
Trop. & S. Africa. 22 IVO 24 CHA ETH SUD 25 TAN 26 ANG 27 BOT. Hemicr. or rhizome geophyte.
**Cyperus aethiops* Welw. ex Ridl., Trans. Linn. Soc. London, Bot. 2: 129 (1884).
Cyperus aberdarensis Kük., Notizbl. Bot. Gart. Berlin-Dahlem 9: 301 (1925). *Cyperus aethiops* var. *aberdarensis* (Kük.) Kük. in H.G.A.Engler (ed.), Pflanzenr., IV, 20(101): 367 (1936).
Cyperus aethiops var. *tessmannii* Kük. in H.G.A. Engler (ed.), Pflanzenr., IV, 20(101): 366 (1936).

Pycreus afrozonatus Lye, Nordic J. Bot. 1: 622 (1981).
WC. Trop. Africa. 23 CON ZAI. Ther.

Pycreus alleizettei Cherm., Bull. Mus. Natl. Hist. Nat. 25: 140 (1919). *Cyperus alleizettei* (Cherm.) Kük. in H.G.A.Engler (ed.), Pflanzenr., IV, 20(101): 390 (1936).
C. Madagascar. 29 MDG. Ther.

Pycreus altus (Turrill) Lye, Nordic J. Bot. 1: 622 (1981).
S. Trop. Africa. 26 ANG ZAM. Hemicr.
**Juncellus altus* Turrill, Bull. Misc. Inform. Kew 1814: 338 (1914). *Cyperus praealtus* Kük. in H.G.A. Engler (ed.), Pflanzenr., IV, 20(101): 319 (1936).

Pycreus apiculatus Govind., Proc. Indian Acad. Sci., Pl. Sci. 100: 415 (1990).
India. 40 IND. Ther.

Pycreus aschenbornianus (Boeck.) C.B.Clarke, Kew Bull., Addit. Ser. 8: 2 (1908).
Mexico to C. America. (41) vie 79 MXC MXG MXS 80 PAN. Hemicr. or rhizome geophyte.
**Cyperus aschenbornianus* Boeck., Linnaea 35: 460 (1868).

Pycreus atribulbus (Kük.) Napper, J. E. Africa Nat. Hist. Soc. Natl. Mus. 28(124): 5 (1971).
E. Tanzania to KwaZulu-Natal. 25 TAN 26 MOZ 27 NAT. Hemicr. or bulb geophyte.
**Cyperus atribulbus* Kük. in H.G.A.Engler (ed.), Pflanzenr., IV, 20(101): 363 (1936).

Pycreus atrobrunneus (Baker) C.B.Clarke in T.A. Durand & H.Schinz, Consp. Fl. Afric. 5: 535 (1894).
Madagascar. 29 MDG. Hemicr. or rhizome geophyte.
**Cyperus atrobrunneus* Baker, J. Linn. Soc., Bot. 20: 281 (1883).

Pycreus atroglumosus (Govind.) P.Singh & V.Singh, J. Econ. Taxon. Bot. 5: 467 (1984).
SW. India. 40 IND. Hemicr. or rhizome geophyte.
**Cyperus atroglumosa* Govind., Proc. Indian Acad. Sci., B 81: 187 (1975).

Pycreus atronervatus (Boeck.) C.B.Clarke in T.A.Durand & H.Schinz, Consp. Fl. Afric. 5: 535 (1894).
Ethiopia. 24 ETH. Hemicr. or rhizome geophyte.
**Cyperus atronervatus* Boeck., Linnaea 38: 358 (1874).

Pycreus atropurpureus C.B.Clarke in T.A.Durand & H.Schinz, Consp. Fl. Afric. 5: 535 (1894).
Madagascar. 29 MDG. Hemicr. or rhizome geophyte.
**Cyperus atropurpureus* Boeck., Bot. Jahrb. Syst. 5: 497 (1884), nom. illeg. *Cyperus chermezonii* Kük. in H.G.A.Engler (ed.), Pflanzenr., IV, 20(101): 9 (1935).
Pycreus mundtii var. *perrieri* Cherm., Bull. Soc. Bot. France 72: 19 (1925).

Pycreus atrorubidus Nelmes, Kew Bull. 6: 320 (1951 publ. 1952). *Cyperus atrorubidus* (Nelmes) Raymond, Naturaliste Canad. 91: 127 (1964).
W. Trop. Africa to Zambia. 22 GUI IVO LBR SIE 26 ZAM. Hemicr. or rhizome geophyte.

Pycreus betschuanus (Boeck.) C.B.Clarke in T.A.Durand & H.Schinz, Consp. Fl. Afric. 5: 535 (1894).
S. Africa. 27 CPP NAM NAT OFS TVL. Hemicr. or rhizome geophyte.
**Cyperus betschuanus* Boeck., Bot. Jahrb. Syst. 11: 406 (1890).

Pycreus bipartitus (Torr.) C.B.Clarke, Kew Bull., Addit. Ser. 8: 2 (1908).
SE. Canada to N. Argentina. 72 ONT QUE 73 COL ORE WAS 74 ILL IOW KAN MIN MSO NDA NEB OKL SDA WIS 75 CNT INI MAI MAS MIC NWH NWJ NWY OHI PEN RHO VER WVA 76 ARI CAL NEV UTA 77 NWM TEX 78 ALA ARK DEL GEO KTY LOU MRY MSI NCA SCA TEN VRG WDC 79 MXE MXG MXT 80 COS ELS GUA HON NIC PAN 82 VEN 83 BOL ECU PER 85 AGE AGW CLN. Ther.
Cyperus bicolor Barton, Comp. Fl. Philadelph. 1: 27 (1818), nom. illeg. *Cyperus flavescens* var. *bicolor* Farw., Amer. Midl. Naturalist 12: 118 (1930).
Cyperus angustifolius A.Dietr., Sp. Pl. 2: 322 (1832), nom. illeg.
**Cyperus bipartitus* Torr., Ann. Lyceum Nat. Hist. New York 3: 257 (1836). *Cyperus diandrus* var. *bipartitus* (Torr.) Kük. in H.G.A.Engler (ed.), Pflanzenr., IV, 20(101): 384 (1936).
Cyperus rivularis Kunth, Enum. Pl. 2: 6 (1837). *Chlorocyperus rivularis* (Kunth) Rikli, Jahrb. Wiss. Bot. 27: 564 (1895). *Pycreus rivularis* (Kunth) Palla, Allg. Bot. Z. Syst. 13: 43 (1907). *Cyperus niger* var. *rivularis* (Kunth) V.E.Grant, Aliso 4: 87 (1958).
Cyperus tenuiculus Kunth, Enum. Pl. 2: 106 (1837).
Cyperus castaneus Bigelow, Fl. Boston., ed. 3: 19 (1840), nom. illeg.
Cyperus bigelowii Steud., Syn. Pl. Glumac. 2: 5 (1854).
Cyperus lagunetto Steud., Syn. Pl. Glumac. 2: 5 (1854). *Pycreus lagunetto* (Steud.) C.B.Clarke, Bot. Jahrb. Syst. 30(68): 8 (1902). *Cyperus rivularis* subsp. *lagunetto* (Steud.) Kük. in H.G.A.Engler (ed.), Pflanzenr., IV, 20(101): 383 (1936). *Cyperus rivularis* var. *lagunetto* (Steud.) O'Neill, Rhodora 44: 86 (1942).
Cyperus rivularis var. *elongatus* Boeck., Linnaea 35: 453 (1868). *Cyperus bipartitus* f. *elongatus* (Boeck.) Mohlenbr., Ill. Fl. Illinois, Sedges: Cyperus to Scleria, ed. 2: 184 (2001).
Cyperus pauper Phil., Anales Univ. Chile 1873: 553 (1873), nom. illeg.
Cyperus argentinus C.B.Clarke, J. Linn. Soc., Bot. 21: 64 (1884), nom. illeg.
Cyperus rivularis var. *acutatus* C.B.Clarke, J. Linn. Soc., Bot. 21: 66 (1884). *Cyperus rivularis* f.

acutatus (C.B.Clarke) Kük. in H.G.A.Engler (ed.), Pflanzenr., IV, 20(101): 383 (1936).
Cyperus rivularis var. *depauperatus* C.B.Clarke, J. Linn. Soc., Bot. 21: 65 (1884). *Cyperus rivularis* f. *depauperatus* (C.B.Clarke) Kük. in H.G.A. Engler (ed.), Pflanzenr., IV, 20(101): 383 (1936).
Cyperus rivularis var. *elutus* C.B.Clarke, J. Linn. Soc., Bot. 21: 65 (1884). *Cyperus rivularis* f. *elutus* (C.B.Clarke) Kük. in H.G.A.Engler (ed.), Pflanzenr., IV, 20(101): 383 (1936). *Cyperus bipartitus* f. *elutus* (C.B.Clarke) Mohlenbr., Ill. Fl. Illinois, Sedges: Cyperus to Scleria, ed. 2: 184 (2001).
Cyperus rivularis f. *subacaulis* Kük. in H.G.A. Engler (ed.), Pflanzenr., IV, 20(101): 384 (1936).

Pycreus blastophorus Cherm., Bull. Soc. Bot. France 72: 614 (1925). *Cyperus blastophorus* (Cherm.) Kük. in H.G.A.Engler (ed.), Pflanzenr., IV, 20(101): 390 (1936).
C. Madagascar. 29 MDG. Hemicr. or rhizome geophyte.

Pycreus bolei S.M.Almeida, J. Bombay Nat. Hist. Soc. 83: 180 (1986).
India (Maharashtra). 40 IND. Ther.

Pycreus camagueyensis (Britton) T.Koyama, Phytologia 29: 74 (1974).
EC. Cuba. 81 CUB. Hemicr. or rhizome geophyte.
**Cyperus camagueyensis* Britton, Mem. Torrey Bot. Club 16: 59 (1926).

Pycreus capillifolius (A.Rich.) C.B.Clarke in T.A. Durand & H.Schinz, Consp. Fl. Afric. 5: 535 (1894).
E. Brazil, Trop. Africa, Madagascar. 22 BEN BKN GHA GUI IVO LBR MLI NGA SEN SIE 23 CAF CON RWA ZAI 24 CHA ETH SUD 25 KEN TAN UGA 26 ANG 29 MDG 82 GUY? 84 BZE BZL. Ther.
**Cyperus capillifolius* A.Rich., Tent. Fl. Abyss. 2: 475 (1850).
Cyperus afzelii Boeck., Linnaea 35: 475 (1868).
Pycreus setaceus C.B.Clarke, Kew Bull., Addit. Ser. 8: 2 (1908).
Pycreus capillifolius var. *major* Cherm., Arch. Bot. (Forlì) 4(7): 12 (1931). *Cyperus capillifolius* var. *major* (Cherm.) Kük. in H.G.A.Engler (ed.), Pflanzenr., IV, 20(101): 359 (1936).

Pycreus cataractarum C.B.Clarke, Bot. Jahrb. Syst. 38: 132 (1906). *Cyperus cataractarum* (C.B.Clarke) K.Schum. ex Engl. in H.G.A.Engler & C.G.O.Drude, Veg. Erde 9: 200 (1908).
S. Nigeria to WC. Trop. Africa. 22 NGA 23 CMN EQG GAB ZAI. Hemicr. or rhizome geophyte.

Pycreus chekiangensis Tang & F.T.Wang, in Fl. Reipubl. Popul. Sin. 11: 232 (1961).
SE. China. 36 CHS. Hemicr. or rhizome geophyte.

Pycreus chrysanthus (Boeck.) C.B.Clarke in T.A.Durand & H.Schinz, Consp. Fl. Afric. 5: 535 (1894).
S. Trop. & S. Africa. 26 ZAM 27 BOT CPP LES NAM SWZ TVL. Hemicr. or rhizome geophyte.
**Cyperus chrysanthus* Boeck., Linnaea 35: 476 (1868).
Cyperus chrysanthus var. *occidentalis* Kük. in H.G.A. Engler (ed.), Pflanzenr., IV, 20(101): 337 (1936).

Pycreus compressiformis Cherm., Bull. Soc. Bot. France 73: 552 (1926 publ. 1827). *Cyperus compressiformis* (Cherm.) Kük. in H.G.A.Engler (ed.), Pflanzenr., IV, 20(101): 375 (1936).
W. Madagascar. 29 MDG. Ther.

Pycreus cooperi C.B.Clarke in W.H.Harvey & auct. suc. (eds.), Fl. Cap. 7: 160 (1897). *Cyperus cooperi* (C.B.Clarke) Kük., Bot. Not. 1934: 68 (1934), nom. illeg.
S. Trop. & S. Africa. 26 ANG ZIM 27 CPP LES NAT SWZ TVL. Hemicr. or rhizome geophyte.

Pycreus cuanzensis (Ridl.) C.B.Clarke in T.A.Durand & H.Schinz, Consp. Fl. Afric. 5: 536 (1894).
Angola. 26 ANG. Hemicr. or rhizome geophyte.
**Cyperus cuanzensis* Ridl., Trans. Linn. Soc. London, Bot. 2: 128 (1884).

Pycreus curvibracteatus (Govind.) P.Singh & V.Singh, J. Econ. Taxon. Bot. 5: 468 (1984).
S. India. 40 IND. Ther.
**Cyperus curvibracteatus* Govind., Proc. Indian Acad. Sci., B 88: 231 (1979).

Pycreus decumbens T.Koyama, Bull. Natl. Sci. Mus., Tokyo, B 2: 167 (1976).
Brazil. 84 BZL BZS. Hemicr. or rhizome geophyte.

Pycreus delavayi C.B.Clarke, J. Linn. Soc., Bot. 36: 203 (1903). *Cyperus delavayi* (C.B.Clarke) Kük. in H.G.A. Engler (ed.), Pflanzenr., IV, 20(101): 331 (1936).
SC. China. 36 CHC. Hemicr. or rhizome geophyte.

Pycreus demangei J.Raynal, Kew Bull. 23: 314 (1969). *Cyperus demangei* (J.Raynal) Lye, Nordic J. Bot. 3: 231 (1983).
Mali, Chad, S. Tanzania. 22 MLI 24 CHA 25 TAN. Ther.

Pycreus dewildeorum J.Raynal, Kew Bull. 23: 313 (1969). *Cyperus dewildeorum* (J.Raynal) Lye, Nordic J. Bot. 3: 231 (1983).
Cameroon. 23 CMN. Hemicr. or rhizome geophyte.

Pycreus diander (Torr.) C.B.Clarke, Kew Bull., Addit. Ser. 8: 2 (1908).
E. Canada to NC. & E. U.S.A. 72 NBR NSC ONT QUE 74 ILL IOW KAN MIN MSO NEB WIS 75 CNT INI MAI MAS MIC NWH NWJ NWY PEN RHO VER 78 DEL MRY VRG. Ther.
**Cyperus diandrus* Torr., Cat. Pl. New York: 89 (1819).
Torreya maritima Raf., J. Phys. Chim. Hist. Nat. Arts 89: 105 (1819).
Cyperus diandrus var. *pauciflorus* Alph.Wood, Class-book Bot.: 732 (1861).
Cyperus diandrus var. *elongatus* Britton, Bull. Torrey Bot. Club 19: 226 (1892). *Cyperus diandrus* f. *elongatus* (Britton) Fernald, Rhodora 41: 423 (1939).

Pycreus diaphanus (Schrad. ex Roem. & Schult.) S.S.Hooper & T.Koyama, J. Jap. Bot. 51: 316 (1976).
Trop. & Subtrop. Asia. 31 PRM 36 CHC CHM 38 JAP KOR 40 ASS BAN EHM IND NEP WHM 41 THA VIE 42 PHI SUL SUM. Ther.
**Cyperus diaphanus* Schrad. ex Roem. & Schult. in J.J.Roemer & J.A.Schultes, Mant. 2: 477 (1824). *Cyperus latispicatus* var. *diaphanus* (Schrad. ex Roem. & Schult.) Kük. in H.G.A.Engler (ed.), Pflanzenr., IV, 20(101): 393 (1936).
Cyperus latispicatus Boeck., Flora 42: 433 bis (1859). *Pycreus latispicatus* (Boeck.) C.B.Clarke in J.D. Hooker, Fl. Brit. India 6: 696 (1893). *Chlorocyperus latispicatus* (Boeck.) Rikli, Jahrb. Wiss. Bot. 27: 563 (1895). *Cyperus diaphanus* var. *latispicatus* (Boeck.) J.Kern, in Fl. Males. 7(3): 653 (1974).
Cyperus aphyllus Boeck., Linnaea 35: 450 (1868).
Cyperus pseudobromoides Boeck., Linnaea 35: 464 (1868).

Cyperus latispicatus var. *acaulis* C.B.Clarke, J. Linn. Soc., Bot. 21: 41 (1884). *Pycreus diaphanus* var. *acaulis* (C.B.Clarke) Karthik. in S.Karthikeyan & al., Fl. Ind. Enumerat. – Monocot.: 64 (1989).
Cyperus setiformis Korsh., Trudy Imp. S.-Peterburgsk. Bot. Sada 12: 405 (1892). *Pycreus setiformis* (Korsh.) Nakai, Bot. Mag. (Tokyo) 26: 201 (1912). *Cyperus diaphanus* subsp. *setiformis* (Korsh.) J.Kern, in Fl. Males. 7(3): 653 (1974).
Cyperus latispicatus Boeck. ex C.B.Clarke, J. Linn. Soc., Bot. 36: 202 (1903).
Cyperus latispicatus var. *gracilescens* Kük. in H.G.A.Engler (ed.), Pflanzenr., IV, 20(101): 393 (1936). *Cyperus diaphanus* var. *gracilescens* (Kük.) H.O.Saxena, Indian Forester 99: 505 (1973). *Pycreus diaphanus* var. *gracilescens* (Kük.) S.S.Hooper in C.J.Saldanha & D.H. Nicolson, Fl. Hassan Distr. Karnataka: 692 (1976).
Cyperus latispicatus var. *laxus* Kük. in H.G.A.Engler (ed.), Pflanzenr., IV, 20(101): 393 (1936). *Pycreus diaphanus* var. *laxus* (Kük.) H.B.Naithani & S.Biswas in H.B.Naithani, Flow. Pl. India, Nepal & Bhutan: 471 (1990).

Pycreus diloloensis Kük. ex Cherm., Rev. Zool. Bot. Africaines 24: 296 (1934). *Cyperus diloloensis* (Kük. ex Cherm.) Kük. in H.G.A.Engler (ed.), Pflanzenr., IV, 20(101): 340 (1936).
Zaïre. 23 ZAI. Hemicr. or rhizome geophyte.

Pycreus divulsus (Ridl.) C.B.Clarke in T.A.Durand & H.Schinz, Consp. Fl. Afric. 5: 536 (1894).
Trop. Africa, C. Madagascar. 22 NGA SIE 23 CMN 24 ETH 26 ZAM 29 MDG. Ther.
**Cyperus divulsus* Ridl., J. Bot. 22: 15 (1884).

subsp. ***africanus*** S.S.Hooper, Kew Bull. 26: 579 (1972).
Trop. Africa. 22 NGA SIE 23 CMN 24 ETH 26 ZAM. Ther.

subsp. ***divulsus***
C. Madagascar. 29 MDG. Ther.

Pycreus dwarkensis (K.C.Sahni & H.B.Naithani) S.S.Hooper, Kew Bull. 40: 467 (1985).
C. Somalia to NW. India. 24 SOM 35 OMA 40 IND PAK. Ther.
**Cyperus dwarkensis* K.C.Sahni & H.B.Naithani, Indian Forester 102: 357 (1976).

Pycreus elegantulus (Steud.) C.B.Clarke in T.A.Durand & H.Schinz, Consp. Fl. Afric. 5: 536 (1894).
Trop. Africa to Arabian Pen. 22 NGA 23 BUR CMN GGI RWA ZAI 24 ERI ETH 25 KEN TAN UGA 26 ANG MLW 27 NAT TVL 35 SAU YEM. Hemicr.
Cyperus melanocephalus R.Br. in H.Salt, Voy. Abyss., App.: lxii (1814), nom. nud.
Cyperus atronitens Hochst., Flora 24(1): 20 (1841).
**Cyperus elegantulus* Steud., Flora 25: 583 (1842). *Pycreus niger* subsp. *elegantulus* (Steud.) Lye, Nordic J. Bot. 1: 622 (1981). *Cyperus niger* subsp. *elegantulus* (Steud.) Lye in R.W.Haines & K.A.Lye, Sedges & Rushes E. Afr.: 271 (1983).
Cyperus ischnocormus Steud., Syn. Pl. Glumac. 2: 11 (1854).
Cyperus elegantulus var. *submelanostachyus* Kük. in H.G.A.Engler (ed.), Pflanzenr., IV, 20(101): 343 (1936).

Pycreus fasciculatus Govind., J. Econ. Taxon. Bot. 20: 301 (1996).
S. India. 40 IND. Ther.

Pycreus felicis J.Raynal, Kew Bull. 26: 568 (1972).
Guinea. 22 GUI. Hemicr.

Pycreus fibrillosus (Kük.) Cherm., Rev. Zool. Bot. Africaines 22: 63 (1932).
WC. Trop. Africa to Zambia. 23 CMN CON GAB ZAI 25 TAN 26 ZAM. Hemicr.
**Cyperus fibrillosus* Kük. in R.E.Fries, Wiss. Erg. Schwed. Rhod.-Kongo Exped. 1: 1 (1921).
Pycreus katangensis Cherm. in É.A.J.De Wildeman, Contr. Fl. Katanga, Suppl. 5: 11 (1933), nom. nud.
Pycreus scaettae Cherm., Rev. Zool. Bot. Africaines 24: 295 (1934). *Cyperus fibrillosus* var. *scaettae* (Cherm.) Kük. in H.G.A.Engler (ed.), Pflanzenr., IV, 20(101): 348 (1936).
Pycreus vanderystii Cherm., Rev. Zool. Bot. Africaines 24: 296 (1934). *Cyperus fibrillosus* var. *vanderystii* (Cherm.) Kük. in H.G.A.Engler (ed.), Pflanzenr., IV, 20(101): 348 (1936).
Pycreus scaettae var. *katangensis* Cherm., Bull. Jard. Bot. État 13: 279 (1935). *Cyperus fibrillosus* f. *katangensis* (Cherm.) Kük. in H.G.A.Engler (ed.), Pflanzenr., IV, 20(101): 348 (1936).

Pycreus filicinus (Vahl) T.Koyama, Phytologia 29: 74 (1974).
E. U.S.A. 74 ILL 75 CNT MAI MAS NWH NWJ NWY RHO 78 GEO LOU MSI NCA SCA VRG. Ther. or hemicr.
**Cyperus filicinus* Vahl, Enum. Pl. 2: 332 (1805). *Cyperus polystachyos* var. *filicinus* (Vahl) C.B.Clarke, J. Linn. Soc., Bot. 21: 55 (1884).
Cyperus brizaeus Pursh, Fl. Amer. Sept. 1: 51 (1813), nom. illeg.
Torreya caespitosa Raf., J. Phys. Chim. Hist. Nat. Arts 89: 105 (1819).
Cyperus nuttallii Eddy ex Spreng., Neue Entdeck. Pflanzenk. 1: 240 (1820).
Cyperus torreyanus Schult. in J.J.Roemer & J.A.Schultes, Mant. 2: 101 (1824).
Cyperus caricinus D.Don, Prodr. Fl. Nepal.: 39 (1825).
Cyperus cleaveri Torr., Ann. Lyceum Nat. Hist. New York 3: 258 (1836). *Cyperus polystachyos* var. *cleaveri* (Torr.) C.B.Clarke, J. Linn. Soc., Bot. 21: 56 (1884). *Pycreus cleaveri* (Torr.) C.B.Clarke, Kew Bull., Addit. Ser. 8: 2 (1908). *Cyperus filicinus* f. *cleaveri* (Torr.) Kük. in H.G.A.Engler (ed.), Pflanzenr., IV, 20(101): 374 (1936).
Cyperus tenellus Torr., Ann. Lyceum Nat. Hist. New York 3: 257 (1836), nom. illeg.
Cyperus louisianae Steud., Syn. Pl. Glumac. 2: 10 (1854).
Cyperus nuttallii var. *minimus* Alph.Wood, Class-book Bot.: 733 (1861).
Cyperus filicinus f. *splendens* Kük. in H.G.A.Engler (ed.), Pflanzenr., IV, 20(101): 374 (1936).

Pycreus flavescens (L.) P.Beauv. ex Rchb., Fl. Germ. Excurs.: 72 (1830).
Cosmopolitan. 11 BGM† CZE GER HUN NET POL SWI 12 COR FRA POR SAR SPA 13 ALB BUL GRC ITA KRI ROM SIC TUE YUG 14 BLR BLT KRY RUC RUE RUS RUW UKR 20 ALG EGY MOR TUN 21 AZO MDR 22 BEN BKN GHA GNB GUI IVO MLI NGA NGR SEN SIE TOG 23 BUR CMN CON RWA ZAI 24 CHA ETH SUD 25 KEN TAN UGA 26 ANG MLW MOZ ZAM ZIM 27 BOT NAM NAT SWZ TVL 29 MDG 32 KAZ TZK UZB 33 TCS 34 AFG IRN IRQ LBS PAL TUR 35 YEM 40 IND PAK WHM 41 VIE (50) 72 ONT 74 ILL KAN MSO OKL 75 INI

MAS MIC NWJ NWY OHI PEN WVA 76 CAL 77 TEX 78 ALA ARK DEL FLA GEO KTY LOU MRY MSI NCA SCA TEN VRG WDC 79 MXC MXE MXG MXN MXS MXT 80 BLZ COS CPI ELS GUA HON NIC PAN 81 BER CUB DOM PUE TRT WIN 82 GUY VEN 83 CLM 84 BZN BZS. Ther.

**Cyperus flavescens* L., Sp. Pl.: 46 (1753). *Distimus flavescens* (L.) Raf., J. Phys. Chim. Hist. Nat. Arts 89: 105 (1819). *Chlorocyperus flavescens* (L.) Rikli, Jahrb. Wiss. Bot. 27: 563 (1895).

subsp. ***fallaciosus*** (Cherm.) Lye, Nordic J. Bot. 1: 622 (1981).

Trop. Africa. 22 GHA GUI MLI NGA NGR SEN 23 CAF 24 CHA 25 KEN TAN UGA. Ther.

**Pycreus fallaciosus* Cherm., Arch. Bot. Bull. Mens. 7(4): 7 (1936). *Cyperus fallaciosus* (Cherm.) Raymond, Naturaliste Canad. 91: 129 (1964). *Cyperus flavescens* subsp. *fallaciosus* (Cherm.) Lye in R.W.Haines & K.A.Lye, Sedges & Rushes E. Afr., App. 3: 2 (1983).

subsp. ***flavescens***

Cosmopolitan. 11 BGM† CZE GER HUN NET POL SWI 12 COR FRA POR SAR SPA 13 ALB BUL GRC ITA KRI ROM SIC TUE YUG 14 BLR BLT KRY RUC RUE RUS RUW UKR 20 ALG EGY MOR TUN 21 AZO MDR 22 BEN BKN GHA GNB GUI IVO MLI NGA NGR SEN SIE TOG 23 BUR CMN CON RWA ZAI 24 ETH SUD 25 KEN TAN UGA 26 ANG MLW MOZ ZAM ZIM 27 BOT NAM NAT SWZ TVL 32 KAZ TZK UZB 33 TCS 34 AFG IRN IRQ LBS PAL TUR 35 YEM 40 IND PAK WHM 41 VIE (50) 72 ONT 74 ILL KAN MSO OKL 75 INI MAS MIC NWJ NWY OHI PEN WVA 76 CAL 77 TEX 78 ALA ARK DEL FLA GEO KTY LOU MRY MSI NCA SCA TEN VRG WDC 79 MXC MXE MXG MXN MXS MXT 80 BLZ COS CPI ELS GUA HON NIC PAN 81 BER CUB DOM PUE TRT WIN 82 GUY VEN 83 CLM 84 BZN BZS. Ther.

Cyperus flavens Pall., Tabl. Taur.: 45 (1795).

Cyperus poiformis Pursh, Fl. Amer. Sept. 1: 50 (1813). *Cyperus flavescens* var. *poiformis* (Pursh) Fernald, Rhodora 41: 529 (1939).

Cyperus stramineus Link, Jahrb. Gewächsk. 1(3): 84 (1818).

Cyperus flavescens f. *minor* Gaudin, Fl. Helv. 1: 134 (1828). *Pycreus flavescens* f. *minor* (Gaudin) Soó, Acta Bot. Acad. Sci. Hung. 16: 369 (1970 publ. 1971).

Cyperus xanthinus C.Presl, Isis (Oken) 21: 271 (1828).

Cyperus polystachyus Jungh., Linnaea 6: 23 (1831), nom. illeg.

Cyperus fontanesii Kunth, Enum. Pl. 2: 5 (1837). *Cyperus flavescens* var. *fontanesii* (Kunth) C.B.Clarke, J. Linn. Soc., Bot. 21: 38 (1884). *Cyperus flavescens* f. *fontanesii* (Kunth) Kük. in H.G.A.Engler (ed.), Pflanzenr., IV, 20(101): 400 (1936).

Cyperus paraensis Mart. & Schrad. ex Nees in C.F.P.von Martius & auct. suc. (eds.), Fl. Bras. 2(1): 8 (1842).

Pycreus paraensis Nees in C.F.P.von Martius & auct. suc. (eds.), Fl. Bras. 2(1): 8 (1842). *Cyperus flavescens* var. *paraensis* (Nees) Boeck., Linnaea 35: 439 (1868). *Cyperus flavescens* f. *paraensis* (Nees) Kük. in H.G.A.Engler (ed.), Pflanzenr., IV, 20(101): 401 (1936).

Cyperus gussonei Gasp. ex Guss., Fl. Sicul. Syn. 2: 779 (1844). *Cyperus flavescens* var. *gussonei* (Gasp. ex Guss.) Nyman, Consp. Fl. Eur.: 760 (1882).

Cyperus flavescens f. *caucasicus* K.Koch, Linnaea 21: 622 (1848). *Pycreus flavescens* f. *caucasicus* (K.Koch) Soó, Acta Bot. Acad. Sci. Hung. 16: 369 (1970 publ. 1971).

Cyperus abyssinicus Hochst. ex A.Rich., Tent. Fl. Abyss. 2: 474 (1850). *Cyperus flavescens* var. *abyssinicus* (Hochst. ex A.Rich.) C.B.Clarke, J. Linn. Soc., Bot. 21: 38 (1884). *Chlorocyperus abyssinicus* (Hochst. ex A.Rich.) Rikli, Jahrb. Wiss. Bot. 27: 564 (1895). *Cyperus flavescens* f. *abyssinicus* (Hochst. ex A.Rich.) Kük. in H.G.A.Engler (ed.), Pflanzenr., IV, 20(101): 401 (1936).

Cyperus piceus Liebm., Mexic. Halvgr.: 200 (1850). *Pycreus piceus* (Liebm.) C.B.Clarke, Kew Bull., Addit. Ser. 8: 1 (1908). *Cyperus flavescens* var. *piceus* (Liebm.) Fernald, Rhodora 41: 529 (1939).

Cyperus squalidus Liebm., Mexic. Halvgr.: 199 (1850). *Cyperus piceus* f. *squalidus* (Liebm.) Kük. in H.G.A.Engler (ed.), Pflanzenr., IV, 20(101): 397 (1936).

Cyperus melanostachyus Walp., Ann. Bot. Syst. 3: 667 (1852), nom. illeg.

Cyperus setaceus Wulfen, Fl. Norica Phan.: 49 (1858), nom. illeg. *Cyperus flavescens* f. *setaceus* Kük. in H.G.A.Engler (ed.), Pflanzenr., IV, 20(101): 400 (1936). *Pycreus flavescens* f. *setaceus* (Kük.) Soó, Acta Bot. Acad. Sci. Hung. 16: 369 (1970 publ. 1971).

Cyperus pallidus Kit. ex Kanitz, Verh. K. K. Zool.-Bot. Ges. Wien 13: 505 (1863), nom. illeg.

Cyperus tristachyus Boeck., Linnaea 35: 454 (1868). *Pycreus tristachyus* (Boeck.) C.B.Clarke, Kew Bull., Addit. Ser. 8: 1 (1908).

Cyperus tristachyus var. *elongatus* Boeck., Linnaea 35: 455 (1868).

Cyperus tristachyus var. *minor* Boeck., Linnaea 35: 455 (1868).

Cyperus flavescens f. *monostachyus* Sacc., Atti Reale Ist. Veneto Sci. Lett. Arti, III, 14: 305 (1869). *Pycreus flavescens* f. *monostachyus* (Sacc.) Soó, Acta Bot. Acad. Sci. Hung. 16: 369 (1970 publ. 1971).

Cyperus flavescens var. *pygmaeus* Goiran, Spec. Morphogr. Veg.: 19 (1875).

Cyperus durandii Boeck., Allg. Bot. Z. Syst. 1: 185 (1895).

Cyperus tonduzianus Boeck., Allg. Bot. Z. Syst. 1: 187 (1895).

Cyperus flavescens f. *compactus* Asch. & Graebn., Syn. Mitteleur. Fl. 2(2): 268 (1903).

Cyperus flavescens f. *gracillimus* Asch. & Graebn., Syn. Mitteleur. Fl. 2(2): 268 (1903). *Pycreus flavescens* f. *gracillimus* (Asch. & Graebn.) Soó, Acta Bot. Acad. Sci. Hung. 16: 369 (1970 publ. 1971), without basionym date.

Pycreus potosiensis C.B.Clarke, Kew Bull., Addit. Ser. 8: 1 (1908). *Cyperus piceus* f. *potosiensis* (C.B.Clarke) Kük. in H.G.A.Engler (ed.), Pflanzenr., IV, 20(101): 397 (1936).

Cyperus flavescens f. *major* Lojac., Fl. Sicul. 3: 200 (1909).

Cyperus flavescens subvar. *mohrii* Farw., Amer. Midl. Naturalist 12: 118 (1930).
Cyperus nieuwlandii Geise, Amer. Midl. Naturalist 15: 245 (1934).

subsp. ***intermedius*** (Rikli) Lye, Nordic J. Bot. 1: 622 (1981).
Senegal, Ethiopia to Angola, Madagascar. 22 SEN 24 ETH 26 ANG MLW 29 MDG. Ther.
Cyperus intermedius Steud., Flora 25: 581 (1842), nom. illeg. **Chlorocyperus intermedius* Rikli, Jahrb. Wiss. Bot. 27: 563 (1895). *Pycreus intermedius* (Rikli) C.B.Clarke in D.Oliver & auct. suc. (eds.), Fl. Trop. Afr. 8: 290 (1901). *Cyperus flavescens* subsp. *intermedius* (Rikli) Lye, Lidia 3: 132 (1994). *Cyperus subintermedius* Kük. in H.G.A.Engler (ed.), Pflanzenr., IV, 20(101): 390 (1936).
Cyperus eragrostis C.Krauss, Flora 28: 754 (1845), nom. illeg.
Pycreus vicinus Cherm., Bull. Soc. Bot. France 67: 328 (1920 publ. 1921). *Cyperus subintermedius* var. *vicinus* (Cherm.) Kük. in H.G.A.Engler (ed.), Pflanzenr., IV, 20(101): 392 (1936).
Cyperus subintermedius var. *angustisquamatus* Kük. in H.G.A.Engler (ed.), Pflanzenr., IV, 20(101): 392 (1936).

subsp. ***microglumis*** Lye, Nordic J. Bot. 1: 621 (1981). *Cyperus flavescens* subsp. *microglumis* (Lye) Lye in R.W.Haines & K.A.Lye, Sedges & Rushes E. Afr., App. 3: 2 (1983).
Uganda. 25 UGA. Ther.

var. ***rehmannianus*** (C.B.Clarke) Govaerts in R.H.A. Govaerts & D.A.Simpson, World Checklist of Cyperaceae: 604 (2007).
Kenya to S. Africa, Madagascar. 25 KEN TAN UGA 26 MLW ZAM 27 CPP LES NAT SWZ TVL 29 MDG. Ther.
**Pycreus rehmannianus* C.B.Clarke in W.H.Harvey & auct. suc. (eds.), Fl. Cap. 7: 156 (1897). *Cyperus rehmannianus* (C.B.Clarke) Boeck. ex Kuntze, Revis. Gen. Pl. 3(2): 334 (1898). *Cyperus flavescens* var. *rehmannianus* (C.B.Clarke) Kük., Wiss. Erg. Schwed. Rhod.-Kongo Exped. 1: 1 (1921).
Pycreus bathiei Cherm., Bull. Soc. Bot. France 69: 719 (1922 publ. 1923). *Cyperus rehmannianus* var. *bathiei* (Cherm.) Kük. in H.G.A.Engler (ed.), Pflanzenr., IV, 20(101): 398 (1936).
Cyperus rehmannianus f. *minor* Kük. in H.G.A. Engler (ed.), Pflanzenr., IV, 20(101): 397 (1936).
Cyperus rehmannianus var. *rigidiculmis* Kük. in H.G.A.Engler (ed.), Pflanzenr., IV, 20(101): 398 (1936).
Pycreus flavescens var. *castaneus* Lye, Nordic J. Bot. 1: 621 (1981). *Cyperus flavescens* var. *castaneus* (Lye) Lye in R.W.Haines & K.A.Lye, Sedges & Rushes E. Afr., App. 3: 2 (1983).

subsp. ***tanaensis*** (Kük.) Lye, Nordic J. Bot. 1: 622 (1981).
Kenya (Tana River). 25 KEN. Ther.
Pycreus debilissimus C.B.Clarke, J. Bot. 34: 224 (1896).
**Cyperus tanaensis* Kük. in H.G.A.Engler (ed.), Pflanzenr., IV, 20(101): 397 (1936). *Cyperus flavescens* subsp. *tanaensis* (Kük.) Lye in R.W.Haines & K.A.Lye, Sedges & Rushes E. Afr., App. 3: 2 (1983).

Pycreus flavidus (Retz.) T.Koyama, J. Jap. Bot. 51: 316 (1976).
S. Europe to Japan, Madagascar to E. Australia. 12 FRA SPA 13 BUL GRC ITA KRI 20 ALG EGY MOR 27 CPP NAT SWZ 29 MAU MDG SEY 32 KAZ KGZ TKM TZK UZB 33 NCS TCS 34 AFG IRN IRQ LBS PAL TUR 36 CHC CHM CHN CHS 38 JAP KOR KZN NNS OGA TAI 40 ASS BAN EHM IND NEP PAK SRL WHM 41 MYA THA VIE 42 BOR JAW LSI MOL PHI SUL SUM 43 NWG 50 NSW QLD VIC. Ther. or hemicr.
Cyperus globosus All., Fl. Pedem. 1: 49 (1785), nom. illeg. *Pycreus globosus* Rchb., Fl. Germ. Excurs.: 140(10) (1831). *Chlorocyperus globosus* (Rchb.) Palla, Allg. Bot. Z. Syst. 6: 61 (1900). *Pycreus globosus* var. *typicus* Domin, Biblioth. Bot. 85: 418 (1915), nom. inval.
**Cyperus flavidus* Retz., Observ. Bot. 5: 13 (1788).
Cyperus confertus Lam., Tabl. Encycl. 1: 145 (1791), nom. illeg.
Cyperus divaricatus Lam., Tabl. Encycl. 1: 145 (1791).
Cyperus strictus Lam., Tabl. Encycl. 1: 146 (1791). *Cyperus globosus* var. *strictus* (Lam.) C.B.Clarke, J. Linn. Soc., Bot. 20: 280 (1883). *Pycreus globosus* var. *strictus* (Lam.) Domin, Biblioth. Bot. 85: 419 (1915). *Pycreus strictus* (Lam.) Alston in H.Trimen, Handb. Fl. Ceylon 4(Suppl.): 307 (1931). *Pycreus flavidus* var. *strictus* (Lam.) Karthik. in S.Karthikeyan & al., Fl. Ind. Enumerat. – Monocot.: 65 (1989).
Cyperus curvatus Vahl, Enum. Pl. 2: 338 (1805).
Cyperus fascicularis DC. in J.B.A.M.de Lamarck & A.P. de Candolle, Fl. Franç., ed. 3, 3: 722 (1805), nom. illeg.
Cyperus humifusus Clemente, Ens. Var. Vid. Com.: 284 (1807). *Cyperus globosus* var. *humifusus* (Clemente) Nyman, Consp. Fl. Eur.: 760 (1882).
Cyperus capillaris J.König ex Roxb., Fl. Ind. 1: 194 (1820). *Pycreus capillaris* (J.König ex Roxb.) Nees ex C.B.Clarke in J.D.Hooker, Fl. Brit. India 6: 591 (1893).
Cyperus tortuosus J.König ex Roxb., Fl. Ind. 1: 201 (1820).
Cyperus lamarckianus Schult. in J.J.Roemer & J.A.Schultes, Mant. 2: 108 (1824).
Cyperus complanatus C.Presl, Isis (Oken) 21: 271 (1828), nom. illeg.
Cyperus horizontalis Willd. ex Kunth, Enum. Pl. 2: 5 (1837), nom. illeg.
Cyperus linnaeanus Kunth, Enum. Pl. 2: 115 (1837).
Cyperus vulgaris Sieber ex Kunth, Enum. Pl. 2: 4 (1837).
Cyperus atroferrugineus Steud., Syn. Pl. Glumac. 2: 2 (1854).
Cyperus jungendus Steud., Syn. Pl. Glumac. 2: 3 (1854).
Cyperus mucronatus Steud. in H.Zollinger, Syst. Verz. 1: 63 (1854), nom. illeg.
Cyperus nilagiricus Hochst. ex Steud., Syn. Pl. Glumac. 2: 2 (1854). *Cyperus globosus* var. *nilagiricus* (Hochst. ex Steud.) C.B.Clarke, J. Linn. Soc., Bot. 21: 49 (1884). *Chlorocyperus nilagiricus* (Hochst. ex Steud.) Rikli, Jahrb. Wiss. Bot. 27: 564 (1895). *Pycreus nilagiricus* (Hochst. ex Steud.) Schischk. in V.L.Komarov (ed.), Fl. URSS 3: 6 (1935). *Pycreus flavidus* var. *nilagiricus* (Hochst. ex Steud.) Karthik. in S.Karthikeyan & al., Fl. Ind. Enumerat. – Monocot.: 65 (1989).
Cyperus trachyrhachis Steud., Syn. Pl. Glumac. 2: 3 (1854).

Cyperus junghuhnii Miq., Fl. Ned. Ind. 3: 260 (1856).
Cyperus ater Dalzell & A.Gibson, Bombay Fl.: 283 (1861), nom. illeg.
Cyperus globosus var. *submonocephalus* Boeck., Linnaea 35: 459 (1868).
Cyperus nilagiricus var. *pauperior* Boeck., Linnaea 35: 458 (1868). *Cyperus globosus* f. *pauperior* (Boeck.) Kük. in H.G.A.Engler (ed.), Pflanzenr., IV, 20(101): 355 (1936).
Cyperus globosus var. *bubanii* Debeaux, Rec. Med. Pharm.: s.p. (1874).
Cyperus globosus var. *khasianus* C.B.Clarke, J. Linn. Soc., Bot. 21: 48 (1884). *Pycreus flavidus* var. *khasianus* (C.B.Clarke) Karthik. in S.Karthikeyan & al., Fl. Ind. Enumerat. – Monocot.: 65 (1989).
Cyperus semiteres B.Heyne ex C.B.Clarke, J. Linn. Soc., Bot. 21: 49 (1884).
Cyperus fusco-ater Meinsh. ex Korsh., Trudy Imp. S.-Peterburgsk. Bot. Sada 12: 406 (1892). *Cyperus globosus* f. *fusco-ater* (Meinsh. ex Korsh.) Kük. in H.G.A.Engler (ed.), Pflanzenr., IV, 20(101): 355 (1936).
Pycreus globosus var. *erectus* C.B.Clarke, J. Linn. Soc., Bot. 34: 15 (1898). *Cyperus globosus* var. *erectus* (C.B.Clarke) Kük. in H.G.A.Engler (ed.), Pflanzenr., IV, 20(101): 356 (1936). *Pycreus flavidus* var. *erectus* (C.B.Clarke) Karthik. in S.Karthikeyan & al., Fl. Ind. Enumerat. – Monocot.: 65 (1989).
Pycreus globosus var. *nilagiricus* Domin, Biblioth. Bot. 85: 419 (1915).
Pycreus reductus Cherm., Bull. Mus. Natl. Hist. Nat. 27: 552 (1921).
Cyperus globosus f. *minimus* Kük., Acta Horti Gothob. 5: 108 (1929).
Cyperus globosus f. *depauperatus* Kük. in H.G.A. Engler (ed.), Pflanzenr., IV, 20(101): 354 (1936).

Pycreus fluminalis (Ridl.) Rendle in W.P.Hiern, Cat. Afr. Pl. 2: 106 (1899).
Trop. Africa, Madagascar. 23 CMN CON 25 TAN UGA 26 ANG 29 MDG. Hemicr.
**Cyperus fluminalis* Ridl., Trans. Linn. Soc. London, Bot. 2: 127 (1884).
Cyperus monocephalus Baker, J. Linn. Soc., Bot. 22: 531 (1887), nom. illeg. *Pycreus monocephalus* C.B.Clarke in T.A.Durand & H.Schinz, Consp. Fl. Afric. 5: 538 (1894).
Cyperus spiculosus F.N.Williams, Bull. Herb. Boissier, II, 4: 224 (1904), nom. illeg.
Cyperus fluminalis var. *longifolius* Cherm. ex Kük. in H.G.A.Engler (ed.), Pflanzenr., IV, 20(101): 402 (1936).

Pycreus fontinalis Cherm., Bull. Soc. Bot. France 67: 327 (1920 publ. 1921). *Cyperus fontinalis* (Cherm.) Kük. in H.G.A.Engler (ed.), Pflanzenr., IV, 20(101): 341 (1936).
C. Madagascar. 29 MDG. Hemicr.

Pycreus fugax (Liebm.) C.D.Adams, Ann. Missouri Bot. Gard. 78: 254 (1991).
Texas to Trop. America. 77 TEX 78 LOU 79 MXE MXG MXN MXS MXT 80 BLZ COS ELS GUA HON NIC PAN 81 CUB JAM 82 VEN 83 ECU GAL 85 BZE BZN. Ther.
Cyperus tenellus J.Presl & C.Presl, Reliq. Haenk. 1: 176 (1828), nom. illeg.
**Cyperus fugax* Liebm., Mexic. Halvgr.: 196 (1850). *Cyperus polystachyos* f. *fugax* (Liebm.) Kük., Repert. Spec. Nov. Regni Veg. 23: 184 (1926).
Cyperus fugax var. *humilis* Boeck., Linnaea 35: 475 (1868).

Pycreus govindarajalui V.S.Raju, Taxon 40: 103 (1991).
India (Maharashtra). 40 IND. Ther.
**Cyperus decumbens* Govind., J. Indian Bot. Soc. 52: 72 (1973 publ. 1974). *Pycreus decumbens* (Govind.) P.Singh & V.Singh, J. Econ. Taxon. Bot. 5: 467 (1984), nom. illeg.

Pycreus gracillimus Chiov., Ann. Bot. (Rome) 13: 58 (1914). *Cyperus gracillimus* (Chiov.) Kük. in H.G.A. Engler (ed.), Pflanzenr., IV, 20(101): 393 (1936).
S. Zaïre. 23 ZAI. Hemicr. or rhizome geophyte.

Pycreus grammicus (Kunze ex Kunth) C.B.Clarke, Bot. Jahrb. Syst. 30(68): 6 (1902).
Ecuador to Chile. 83 ECU 85 CLC. Hemicr. or rhizome geophyte.
**Cyperus grammicus* Kunze ex Kunth, Enum. Pl. 2: 6 (1837). *Chlorocyperus grammicus* (Kunze ex Kunth) Rikli, Jahrb. Wiss. Bot. 27: 563 (1895).
Cyperus gramineus Steud., Syn. Pl. Glumac. 2: 5 (1854), nom. illeg.
Cyperus parvulus Steud., Syn. Pl. Glumac. 2: 4 (1854).

Pycreus gratissimus Kitag., Bot. Mag. (Tokyo) 49: 222 (1935).
NE. China. 36 CHM. Hemicr. or rhizome geophyte.

Pycreus heterochrous Nelmes, Kew Bull. 7: 321 (1952).
Zambia. 26 ZAM. Hemicr. or rhizome geophyte.

Pycreus hildebrandtii C.B.Clarke in D.Oliver & auct. suc. (eds.), Fl. Trop. Afr. 8: 299 (1901). *Cyperus hildebrandtii* (C.B.Clarke) K.Schum. ex T.Durand & B.D.Jacks., Index Kew., Suppl. 1: 122 (1902), nom. illeg. *Cyperus pseudohildebrandtii* Kük. in H.G.A. Engler (ed.), Pflanzenr., IV, 20(101): 366 (1936).
E. Kenya to E. Tanzania. 25 KEN TAN. Ther. or hemicr.
Cyperus minutulus K.Schum., Bot. Jahrb. Syst. 24: 339 (1897), nom. illeg.
Pycreus minimus C.B.Clarke in D.Oliver & auct. suc. (eds.), Fl. Trop. Afr. 8: 302 (1901).

Pycreus intactus (Vahl) J.Raynal, Adansonia, n.s., 17: 46 (1977).
S. Africa, Madagascar, S. India. (22) gnb sen 27 CPP NAT SWZ TVL 29 MDG 40 IND. Hemicr. or rhizome geophyte.
**Cyperus intactus* Vahl, Enum. Pl. 2: 332 (1805).
Cyperus ferrugineus Poir. in J.B.A.M.de Lamarck, Encycl. 7: 261 (1806), nom. illeg. *Cyperus polystachyos* var. *ferrugineus* Boeck., Linnaea 35: 479 (1868). *Pycreus ferrugineus* (Boeck.) C.B. Clarke in J.D.Hooker, Fl. Brit. India 6: 595 (1893).
Cyperus foliosus Willd. ex Kunth, Enum. Pl. 2: 11 (1837).
Cyperus micans Kunth, Enum. Pl. 2: 12 (1837). *Cyperus polystachyos* var. *micans* (Kunth) C.B.Clarke, J. Linn. Soc., Bot. 21: 54 (1884). *Pycreus micans* (Kunth) Karthik. in S.Karthikeyan & al., Fl. Ind. Enumerat. – Monocot.: 65 (1989).
Cyperus elongatus Steud., Syn. Pl. Glumac. 2: 11 (1854), nom. illeg.
Cyperus minor Steud., Syn. Pl. Glumac. 2: 4 (1854).
Cyperus polystachyos var. *baronii* C.B.Clarke, J. Linn. Soc., Bot. 20: 280 (1883). *Cyperus ferrugineus* var. *baronii* (C.B.Clarke) Kük. in H.G.A.Engler (ed.), Pflanzenr., IV, 20(101): 373 (1936).

Pycreus kanarensis V.P.Prasad & N.P.Singh, J. Econ. Taxon. Bot. 21: 667 (1997 publ. 1998).
India (Karnataka). 40 IND. Ther.

Pycreus lancelotii S.M.Almeida, J. Bombay Nat. Hist. Soc. 83: 182 (1986).
India (Maharashtra). 40 IND. Ther.

Pycreus lanceolatus (Poir.) C.B.Clarke in T.A.Durand & H.Schinz, Consp. Fl. Afric. 5: 538 (1894).
Trop. Africa, W. Indian Ocean, Trop. & Subtrop. America. 22 BEN BKN GAM GHA GNB GUI IVO LBR MLI NGA NGR SEN SIE TOG 23 CMN CON GAB ZAI 24 CHA ETH SUD 25 KEN TAN UGA 26 ANG MLW 29 MAU MDG 77 TEX 78 ALA FLA GEO LOU 79 MXC MXE MXG MXN MXS MXT 80 BLZ COS ELS GUA HON NIC PAN 81 CUB DOM HAI JAM 83 BOL CLM 84 BZE BZL BZS 85 AGE PAR URU. Hemicr. or rhizome geophyte.
Cyperus lanceolatus Poir. in J.B.A.M.de Lamarck, Encycl. 7: 245 (1806).
Cyperus fasciculatus Elliott, Sketch Bot. S. Carolina 1: 63 (1816). *Cyperus elliottianus* Schult. in J.J. Roemer & J.A.Schultes, Mant. 2: 100 (1824), nom. illeg. *Pycreus elliottianus* Nees, Linnaea 9: 283 (1834).
Cyperus densus Link, Jahrb. Gewächsk. 1(3): 83 (1818). *Pycreus densus* (Link) C.B.Clarke, Symb. Antill. 2: 164 (1900).
Cyperus humboldtianus Schult. in J.J.Roemer & J.A.Schultes, Mant. 2: 100 (1824). *Pycreus humboldtianus* (Schult.) Cufod., Bull. Jard. Bot. Natl. Belg. 40(Suppl.): 1442 (1970).
Cyperus lanceolatus var. *compositus* C.Presl, Reliq. Haenk. 1: 167 (1828).
Cyperus flavus J.Presl & C.Presl, Reliq. Haenk. 1: 176 (1830).
Cyperus olfersianus Kunth, Enum. Pl. 2: 10 (1837). *Pycreus olfersianus* (Kunth) Nees in C.F.P.von Martius & auct. suc. (eds.), Fl. Bras. 2(1): 8 (1842).
Cyperus maximiliani Schrad. ex Nees in C.F.P.von Martius & auct. suc. (eds.), Fl. Bras. 2(1): 7 (1842), pro syn.
Pycreus infirmus Nees in C.F.P.von Martius & auct. suc. (eds.), Fl. Bras. 2(1): 8 (1842).
Pycreus maximiliani Nees in C.F.P.von Martius & auct. suc. (eds.), Fl. Bras. 2(1): 7 (1842). *Cyperus olfersianus* var. *maximiliani* (Nees) Boeck., Linnaea 35: 440 (1868).
Pycreus propinquus Nees in C.F.P.von Martius & auct. suc. (eds.), Fl. Bras. 2(1): 7 (1842).
Cyperus ambiguus Liebm., Mexic. Halvgr.: 199 (1850).
Cyperus helvus Liebm., Mexic. Halvgr.: 197 (1850). *Pycreus helvus* (Liebm.) C.B.Clarke in I.Urban, Symb. Antill. 2: 19 (1900). *Cyperus lanceolatus* var. *helvus* (Liebm.) Kük., Repert. Spec. Nov. Regni Veg. 32: 74 (1933).
Cyperus correctus Steud., Syn. Pl. Glumac. 2: 7 (1854).
Cyperus pycnocephalus Steud., Syn. Pl. Glumac. 2: 5 (1854).
Pycreus elliottianus var. *humilis* Nees in B.Seemann, Bot. Voy. Herald: 221 (1854).
Cyperus variegatus Griseb., Fl. Brit. W. I.: 562 (1864), nom. illeg.
Cyperus jacquemontii Boeck., Linnaea 35: 442 (1868).
Cyperus olfersianus var. *elatior* Boeck., Linnaea 35: 440 (1868).
Cyperus lepidus Phil., Anales Univ. Chile 93: 347 (1896), nom. illeg.
Pycreus olfersianus Urb., Symb. Antill. 2: 164 (1900), nom. illeg.
Cyperus lanceolatus var. *subunioloides* Kük. in H.G.A. Engler (ed.), Pflanzenr., IV, 20(101): 351 (1936).

Pycreus latovaginatus (Govind.) P.Singh & V.Singh, J. Econ. Taxon. Bot. 5: 467 (1984).
S. India. 40 IND. Ther.
Cyperus latovaginata Govind., J. Indian Bot. Soc. 52: 74 (1973 publ. 1974).

Pycreus lijiangensis L.K.Dai, in Fl. Reipubl. Popul. Sin. 11: 233 (1961).
China (Yunnan). 36 CHC. Hemicr. or rhizome geophyte.

Pycreus longistolon (Peter & Kük.) Napper, J. E. Africa Nat. Hist. Soc. Natl. Mus. 28: 6 (1971).
Tanzania. 25 TAN. Hemicr. or rhizome geophyte.
Cyperus longistolon Peter & Kük. in H.G.A.Engler (ed.), Pflanzenr., IV, 20(101): 333 (1936).

subsp. ***atrofuscus*** Lye, Nordic J. Bot. 1: 618 (1981).
Cyperus longistolon subsp. *atrofuscus* (Lye) Lye in R.W.Haines & K.A.Lye, Sedges & Rushes E. Afr., App. 3: 2 (1983).
S. Tanzania. 25 TAN. Hemicr. or rhizome geophyte.

subsp. ***longistolon***
W. Tanzania. 25 TAN. Hemicr. or rhizome geophyte.

Pycreus longivaginans (Kük.) Cherm., Bull. Soc. Bot. France 74: 604 (1927 publ. 1828)).
C. Madagascar. 29 MDG. Hemicr.
Cyperus longivaginans Kük., Repert. Spec. Nov. Regni Veg. 12: 93 (1913).
Pycreus solidifolius Cherm., Bull. Mus. Natl. Hist. Nat. 25: 139 (1919).

Pycreus lurida (Govind.) P.Singh & V.Singh, J. Econ. Taxon. Bot. 5: 467 (1984).
S. India. 40 IND. Ther.
Cyperus lurida Govind., Proc. Indian Acad. Sci., B 81: 189 (1975).

Pycreus macranthus (Boeck.) C.B.Clarke in T.A.Durand & H.Schinz, Consp. Fl. Afric. 5: 538 (1894).
Trop. & S. Africa. 22 IVO 24 ETH 25 KEN TAN UGA 26 ANG MLW 27 BOT CPP LES NAT OFS SWZ TVL. Hemicr.
Cyperus lanceus var. *macrostachya* Kunth, Enum. Pl. 2: 8 (1837).
Cyperus lanceus var. *mucronatus* Kunth, Enum. Pl. 2: 8 (1837). *Cyperus macranthus* var. *mucronatus* (Kunth) Kük. in H.G.A.Engler (ed.), Pflanzenr., IV, 20(101): 389 (1936).
Cyperus macranthus Boeck., Linnaea 35: 462 (1868).
Cyperus permutatus Boeck., Linnaea 35: 477 (1868). *Pycreus permutatus* (Boeck.) Napper, J. E. Africa Nat. Hist. Soc. Natl. Mus. 28: 6 (1971).
Cyperus lanceus var. *angustifolius* Ridl., Trans. Linn. Soc. London, Bot. 7: 126 (1884). *Cyperus macranthus* var. *angustifolius* (Ridl.) Kük. in H.G.A. Engler (ed.), Pflanzenr., IV, 20(101): 389 (1936).
Pycreus segmentatus C.B.Clarke, Kew Bull., Addit. Ser. 8: 1 (1908).

Pycreus macrostachyos (Lam.) J.Raynal, Kew Bull. 23: 314 (1969).
Trop. & Subtrop. 22 BEN BKN GAM GHA GNB GUI IVO MLI NGA NGR SEN SIE TOG 23 BUR CMN CON RWA ZAI 24 CHA ETH SOM SUD 25 KEN TAN UGA 26 ANG 27 BOT NAM NAT SWZ TVL 29 MAU MDG REU 41 MYA 50 NTA QLD WAU 79 MXE MXN MXS MXT 80 COS ELS HON PAN 82 VEN 83 BOL ECU PER 84 BZC BZE BZN 85 AGE AGW PAR. Ther.

**Cyperus macrostachyos* Lam., Tabl. Encycl. 1: 147 (1791).

subsp. ***macrostachyos***

Trop. & Subtrop. 22 BEN BKN GAM GHA GNB GUI IVO MLI NGA NGR SEN SIE TOG 23 CMN CON RWA ZAI 24 CHA ETH SOM SUD 25 KEN TAN UGA 26 ANG 27 BOT NAM NAT SWZ TVL 29 MAU MDG 40 IND 41 MYA 50 NTA QLD WAU 79 MXE MXN MXS MXT 80 COS ELS HON PAN 82 VEN 83 BOL ECU PER 84 BZC BZE BZN 85 AGE AGW PAR. Ther.

Cyperus flavicomus Michx., Fl. Bor.-Amer. 1: 27 (1803). *Pycreus flavicomus* (Michx.) C.D.Adams, Ann. Missouri Bot. Gard. 78: 254 (1991).

Cyperus flavicomus Vahl, Enum. Pl. 2: 360 (1805), nom. illeg. *Distimus flavicomus* Raf., J. Phys. Chim. Hist. Nat. Arts 89: 105 (1819).

Pycreus albomarginatus Mart. & Schrad. ex Nees in C.F.P.von Martius & auct. suc. (eds.), Fl. Bras. 2(1): 9 (1842). *Cyperus albomarginatus* (Mart. & Schrad. ex Nees) Steud., Syn. Pl. Glumac. 2: 10 (1854).

Pycreus sabulosus Nees in C.F.P.von Martius & auct. suc. (eds.), Fl. Bras. 2(1): 10 (1842). *Cyperus sabulosus* (Nees) Boeck., Linnaea 35: 468 (1868). *Cyperus albomarginatus* var. *sabulosus* (Nees) Kük. in H.G.A.Engler (ed.), Pflanzenr., IV, 20(101): 361 (1936).

Cyperus hochstetteri Nees ex Krauss, Flora 29: 755 (1846).

Cyperus retusus A.Rich., Tent. Fl. Abyss. 2: 485 (1850).

Cyperus argyraeus Steud., Syn. Pl. Glumac. 2: 11 (1854).

Cyperus patuliflorus Boeck., Linnaea 35: 473 (1868). *Chlorocyperus patuliflorus* (Boeck.) Rikli, Jahrb. Wiss. Bot. 27: 563 (1895).

Cyperus ornatissimus F.Muell., Fragm. 8: 265 (1874).

Cyperus hochstetteri var. *tenuis* Boeck., Flora 62: 546 (1879). *Cyperus albomarginatus* var. *tenuis* (Boeck.) Kük. in H.G.A.Engler (ed.), Pflanzenr., IV, 20(101): 361 (1936). *Pycreus macrostachyos* var. *tenuis* (Boeck.) Wickens, Kew Bull., Addit. Ser. 5: 165 (1976).

Pycreus albomarginatus var. *bromoides* C.B.Clarke, Bull. Herb. Boissier 6(App. 1): 19 (1898).

Cyperus albomarginatus var. *pachyanthemos* Kük. in H.G.A.Engler (ed.), Pflanzenr., IV, 20(101): 361 (1936).

subsp. ***tremulus*** (Poir.) Lye, Nordic J. Bot. 1: 622 (1981).

Trop. Africa, W. Indian Ocean. 22 SEN 23 BUR CON ZAI 24 CHA SUD 25 TAN 26 MLW MOZ 29 MDG REU. Ther.

**Cyperus tremulus* Poir. in J.B.A.M.de Lamarck, Encycl. 7: 264 (1806). *Pycreus tremulus* (Poir.) C.B.Clarke in T.A.Durand & H.Schinz, Consp. Fl. Afric. 5: 542 (1894). *Cyperus macrostachyos* subsp. *tremulus* (Poir.) Lye in R.W.Haines & K.A.Lye, Sedges & Rushes E. Afr., App. 3: 2 (1983).

Cyperus retusus Nees ex Steud., Syn. Pl. Glumac. 2: 22 (1854). *Pycreus retusus* (Nees ex Steud.) Nees ex C.B.Clarke, J. Linn. Soc., Bot. 21: 71 (1884).

Cyperus tremulus var. *opulentus* Boeck. in W.C.H.Peters, Naturw. Reise Mossambique: 535 (1864).

Cyperus tremulus var. *minor* Boeck., Linnaea 35: 470 (1868).

Pycreus woronowii Palla, Vestn. Tiflissk. Bot. Sada 21: 22 (1912). *Cyperus tremulus* var. *woronowii* (Palla) Kük. in H.G.A.Engler (ed.), Pflanzenr., IV, 20(101): 362 (1936).

Pycreus mahadevanii Govind., J. Econ. Taxon. Bot. 20: 299 (1996).

India (Karnataka). 40 IND. Ther.

Pycreus malabaricus C.B.Clarke, J. Linn. Soc., Bot. 34: 12 (1898). *Cyperus malabaricus* (C.B.Clarke) T.Cooke, Fl. Bombay 2: 856 (1908).

S. India. 40 IND. Ther.

Pycreus malangensis Meneses, Garcia de Orta 4: 242 (1956).

Angola. 26 ANG. Hemicr. or rhizome geophyte.

Pycreus megapotamicus (A.Dietr.) Nees in C.F.P.von Martius & auct. suc. (eds.), Fl. Bras. 2(1): 6 (1842).

Brazil to Chile. 82 SUR? 84 BZC BZL BZS 85 AGE CLC CLN PAR URU. Hemicr. or rhizome geophyte.

**Isolepis megapotamica* A.Dietr., Sp. Pl. 2: 132 (1832). *Cyperus megapotamicus* (A.Dietr.) Kunth, Enum. Pl. 2: 10 (1837). *Chlorocyperus megapotamicus* (A.Dietr.) Rikli, Jahrb. Wiss. Bot. 27: 563 (1895). *Cyperus megapotamicus* var. *genuinus* Osten, Anales Mus. Hist. Nat. Montevideo, II, 3: 130 (1931), nom. inval.

Cyperus nematodes Schrad. ex Nees in C.F.P.von Martius & auct. suc. (eds.), Fl. Bras. 2(1): 6 (1842), nom. inval.

Cyperus jaeggii Boeck., Beitr. Cyper. 1: 2 (1888). *Cyperus megapotamicus* var. *jaeggii* (Boeck.) Kük., Physis (Buenos Aires) 9: 360 (1929). *Pycreus megapotamicus* var. *jaeggii* (Boeck.) Guagl., Hickenia 2: 137 (1995).

Cyperus mauryi Kuntze, Revis. Gen. Pl. 3(2): 333 (1898).

Pycreus nematodes C.B.Clarke, Bot. Jahrb. Syst. 30(68): 5 (1902).

Cyperus megapotamicus var. *nematodes* Kük. ex Osten, Anales Mus. Hist. Nat. Montevideo, II, 3: 131 (1931).

Pycreus melanacme Nelmes, Kew Bull. 10: 91 (1955). *Cyperus melanacme* (Nelmes) Raymond, Naturaliste Canad. 91: 129 (1964).

S. Zaïre to Zambia. 23 ZAI 26 ZAM. Hemicr. or rhizome geophyte.

Pycreus melas (Ridl.) C.B.Clarke in T.A.Durand & H.Schinz, Consp. Fl. Afric. 5: 538 (1894).

Trop. Africa. 22 BEN GHA NGA TOG 23 CMN CON 24 CHA SUD 25 TAN 26 ANG ZAM. Ther.

**Cyperus melas* Ridl., Trans. Linn. Soc. London, Bot. 2: 127 (1884).

Juncellus ater C.B.Clarke, Bull. Soc. Bot. France 54(8): 26 (1907). *Pycreus ater* (C.B.Clarke) Cherm., Arch. Bot. Bull. Mens. 4(7): 11 (1931).

Pycreus membranaceus (Vahl) Govind., Proc. Indian Acad. Sci., Pl. Sci. 100: 424 (1990).

Indian Subcontinent to Indo-China. 40 ASS IND SRL 41 THA. Ther.

**Cyperus membranaceus* Vahl, Enum. Pl. 2: 330 (1805). *Cyperus pumilus* var. *membranaceus* (Vahl) C.B.Clarke, J. Linn. Soc., Bot. 21: 44 (1884). *Pycreus pumilus* subsp. *membranaceus* (Vahl) T.Koyama, Gard. Bull. Singapore 30: 151

(1977). *Pycreus pumilus* var. *membranaceus* (Vahl) Karthik. in S.Karthikeyan & al., Fl. Ind. Enumerat. – Monocot.: 66 (1989).

Pycreus micromelas Lye, Nordic J. Bot. 1: 617 (1981). *Cyperus micromelas* (Lye) Lye in R.W.Haines & K.A. Lye, Sedges & Rushes E. Afr., App. 3: 2 (1983).
N. Zambia. 26 ZAM. Ther.

Pycreus mortonii S.S.Hooper, Kew Bull. 23: 313 (1969). *Cyperus mortonii* (S.S.Hooper) Lye, Nordic J. Bot. 3: 231 (1983).
W. Trop. Africa to Uganda. 22 BEN GHA NGA SEN 23 CMN 24 CHA 25 UGA. Ther. or hemicr.
Pycreus lanceolatus subsp. *ugandensis* Lye, Nordic J. Bot. 1: 619 (1981).

Pycreus mundtii Nees, Linnaea 10: 131 (1836). *Cyperus mundtii* (Nees) Kunth, Enum. Pl. 2: 17 (1837). *Chlorocyperus mundtii* (Nees) Rikli, Jahrb. Wiss. Bot. 27: 563 (1895).
SW. Spain, Africa, India. 12 SPA 20 ALG EGY LBY MOR 21 CVI 22 BEN GHA GNB GUI IVO MTN NGA NGR SEN 23 BUR CON RWA ZAI 24 CHA ETH SUD 25 KEN TAN UGA 26 ANG MOZ ZAM ZIM 27 BOT CPP LES NAT OFS SWZ TVL 29 MDG REU 40 IND (81) cub. Hemicr. or rhizome geophyte.
Cyperus densifolius Kunth, Enum. Pl. 2: 114 (1837). *Pycreus densifolius* (Kunth) Nees ex Cherm., Ann. Inst. Bot.-Géol. Colon. Marseille 27: 77 (1919).
Cyperus turfosus Salzm. ex Kunth, Enum. Pl. 2: 7 (1837).
Cyperus distichophyllus Steud., Flora 25: 582 (1842). *Cyperus mundtii* f. *distichophyllus* (Steud.) Kük. in R.E.Fries, Wiss. Erg. Schwed. Rhod.-Kongo Exped. 1: 2 (1921). *Cyperus mundtii* var. *distichophyllus* (Steud.) Kük. in H.G.A.Engler (ed.), Pflanzenr., IV, 20(101): 381 (1936).
Cyperus densifolius Steud., Syn. Pl. Glumac. 2: 6 (1854), nom. illeg.
Cyperus eragrostis Clementi ex Willk. & Lange, Prodr. Fl. Hispan. 1: 137 (1861), nom. illeg.
Cyperus mundtii var. *glaucus* Boeck., Linnaea 35: 449 (1868).
Pycreus sanguinolentus var. *uniceps* C.B.Clarke, Bot. Jahrb. Syst. 38: 132 (1906). *Cyperus mundtii* var. *uniceps* (C.B.Clarke) Kük., Bot. Not. 1934: 69 (1934). *Pycreus mundtii* var. *uniceps* (C.B.Clarke) Napper, J. E. Africa Nat. Hist. Soc. Natl. Mus. 28: 3 (1971).
Pycreus mundtii var. *gracilis* Cherm., Ann. Inst. Bot.-Géol. Colon. Marseille 27: 77 (1919).
Cyperus mundtii var. *densispiculosus* Kük. in H.G.A. Engler (ed.), Pflanzenr., IV, 20(101): 381 (1936).

Pycreus muricatus (Kük.) Napper, J. E. Africa Nat. Hist. Soc. Natl. Mus. 28: 6 (1971).
S. Tanzania to Northern Prov. 25 TAN 26 MLW ZIM 27 TVL. Hemicr.
**Cyperus muricatus* Kük., Repert. Spec. Nov. Regni Veg. 12: 92 (1913).

Pycreus niederleinianus (Boeck.) Lindm., Bih. Kongl. Svenska Vetensk.-Akad. Handl. 26(9): 6 (1900).
Ecuador to NE. Argentina. 83 ECU 84 BZL 85 AGE PAR. Hemicr. or rhizome geophyte.
**Cyperus niederleinianus* Boeck., Beitr. Cyper. 1: (1888).

Pycreus niger (Ruiz & Pav.) Cufod., Bull. Jard. Bot. Natl. Belg. 40(Suppl.): 1443 (1970).
Trop. & Subtrop. America. 74 OKL 76 ARI CAL 77 NWM TEX 79 MXC MXE MXG MXN MXS MXT 80 COS GUA HON NIC PAN 82 VEN 83 BOL CLM ECU PER 84 BZS 85 AGE AGW. Hemicr.
**Cyperus niger* Ruiz & Pav., Fl. Peruv. 1: 47 (1798).
Cyperus melanostachyus Kunth in F.W.H.von Humboldt, A.J.A.Bonpland & C.S.Kunth, Nov. Gen. Sp. 1: 207 (1817). *Chlorocyperus melanostachyus* (Kunth) Rikli, Jahrb. Wiss. Bot. 27: 564 (1895). *Pycreus melanostachyus* (Kunth) C.B.Clarke, Bot. Jahrb. Syst. 30(68): 6 (1902).
Cyperus variegatus Kunth in F.W.H.von Humboldt, A.J.A.Bonpland & C.S.Kunth, Nov. Gen. Sp. 1: 208 (1817). *Pycreus melanostachyus* var. *variegatus* (Kunth) C.B.Clarke, Contr. U. S. Natl. Herb. 10: 446 (1908).
Cyperus adustus J.Presl & C.Presl, Reliq. Haenk. 1: 167 (1828).
Cyperus cimicinus J.Presl & C.Presl, Reliq. Haenk. 1: 166 (1828). *Chlorocyperus cimicinus* (J.Presl & C.Presl) Rikli, Jahrb. Wiss. Bot. 27: 563 (1895). *Pycreus cimicinus* (J.Presl & C.Presl) H.Pfeiff., Repert. Spec. Nov. Regni Veg. 28: 17 (1930).
Cyperus melanostachyus var. *gracilis* Liebm., Mexic. Halvgr.: 14 (1850).
Cyperus melanostachyus var. *robustus* Liebm., Mexic. Halvgr.: 201 (1850). *Cyperus niger* var. *robustus* (Liebm.) Kük. in H.G.A.Engler (ed.), Pflanzenr., IV, 20(101): 344 (1936).
Cyperus longicruris Boeck., Linnaea 38: 358 (1874).
Cyperus lorentzianus Boeck., Linnaea 38: 357 (1874). *Cyperus niger* var. *lorentzianus* (Boeck.) Kük. in H.G.A.Engler (ed.), Pflanzenr., IV, 20(101): 345 (1936).
Cyperus diandrus var. *castaneus* S.Watson in S.Watson & al., Bot. California 2: 214 (1880). *Cyperus niger* var. *castaneus* (S.Watson) Kük. in H.G.A.Engler (ed.), Pflanzenr., IV, 20(101): 344 (1936).
Cyperus intricatoramosus Boeck., Allg. Bot. Z. Syst. 1: 186 (1885). *Cyperus niger* var. *intricatoramosus* (Boeck.) Kük. in H.G.A.Engler (ed.), Pflanzenr., IV, 20(101): 345 (1936).
Cyperus diandrus var. *capitatus* Britton, Bull. Torrey Bot. Club 13: 205 (1886). *Cyperus flavescens* subvar. *capitatus* (Britton) Farw., Amer. Midl. Naturalist 12: 118 (1930). *Cyperus niger* var. *capitatus* (Britton) O'Neill, Rhodora 44: 86 (1942).
Cyperus lorentzianus var. *compressus* Kük., Bot. Jahrb. Syst. 7: 273 (1886).
Cyperus lorentzianus var. *elatus* Kük., Bot. Jahrb. Syst. 7: 273 (1886).
Cyperus melanostachyus var. *caespitosus* Boeck., Bot. Jahrb. Syst. 8: 205 (1887).
Cyperus pergracilis Boeck., Beitr. Cyper. 2: 1 (1890).
Cyperus biradiatus Boeck., Allg. Bot. Z. Syst. 1: 186 (1895).
Cyperus niger var. *pseudoelegantulus* Kük. in H.G.A. Engler (ed.), Pflanzenr., IV, 20(101): 345 (1936).
Cyperus piceus f. *densior* Kük. in H.G.A.Engler (ed.), Pflanzenr., IV, 20(101): 397 (1936).

Pycreus nigricans (Steud.) C.B.Clarke, Trans. Linn. Soc. London, Bot. 4: 53 (1894).
Ethiopia to Zimbabwe, Madagascar. 23 BUR RWA ZAI 24 ETH 25 KEN TAN UGA 26 MLW ZIM 29 MDG. Hemicr.
**Cyperus nigricans* Steud., Flora 25: 584 (1842).
Pycreus nigrescens C.B.Clarke in J.Buchan, Nyassaland: 20 (1889).

Pycreus nyasensis C.B.Clarke in D.Oliver & auct. suc. (eds.), Fl. Trop. Afr. 8: 304 (1901).
Cyperus nigricans var. *firmior* Kük., Repert. Spec. Nov. Regni Veg. 12: 94 (1913).
Pycreus simulans Cherm., Bull. Mus. Natl. Hist. Nat. 25: 139 (1919). *Pycreus nigricans* var. *simulans* (Cherm.) Cherm., Bull. Soc. Bot. France 74: 605 (1927). *Cyperus nigricans* var. *simulans* (Cherm.) Kük. in H.G.A.Engler (ed.), Pflanzenr., IV, 20(101): 337 (1936).
Pycreus vavavatensis Cherm., Bull. Mus. Natl. Hist. Nat. 25: 138 (1919).

Pycreus nitidus (Lam.) J.Raynal, Kew Bull. 23: 314 (1969).
Africa to Israel. 22 BEN NGA NGR SEN 23 BUR CMN RWA ZAI 24 CHA ETH 25 KEN TAN UGA 27 BOT CPP NAM NAT OFS SWZ TVL 29 MDG 34 PAL. Hemicr. or rhizome geophyte.
**Cyperus nitidus* Lam., Tabl. Encycl. 1: 145 (1791).
Cyperus lanceus Thunb., Prodr. Pl. Cap.: 18 (1794). *Pycreus lanceus* (Thunb.) Turrill, Bull. Misc. Inform. Kew 1925: 67 (1925).
Cyperus ater Vahl, Enum. Pl. 2: 335 (1805).
Pycreus umbrosus Nees, Linnaea 10: 130 (1836).
Cyperus lanceus var. *humilis* Kunth, Enum. Pl. 2: 8 (1837).
Cyperus umbraticola Kunth, Enum. Pl. 2: 13 (1837).
Cyperus melanopus Boeck., Flora 62: 545 (1879). *Cyperus lanceus* var. *melanopus* (Boeck.) Kük. in H.G.A.Engler (ed.), Pflanzenr., IV, 20(101): 335 (1936).
Cyperus brunneoater Boeck., Bot. Jahrb. Syst. 5: 498 (1884).
Cyperus fulvus Ridl., Trans. Linn. Soc. London, Bot. 2: 126 (1884), nom. illeg.
Cyperus lanceus var. *grantii* C.B.Clarke, J. Linn. Soc., Bot. 21: 66 (1884).
Cyperus lanceus var. *divaricatus* Kük., Repert. Spec. Nov. Regni Veg. 12: 93 (1913).
Pycreus antsirabensis Cherm., Bull. Mus. Natl. Hist. Nat. 25: 138 (1919).
Pycreus antsirabensis var. *densior* Cherm., Bull. Soc. Bot. France 72: 171 (1925). *Cyperus lanceus* f. *densior* (Cherm.) Kük. in H.G.A.Engler (ed.), Pflanzenr., IV, 20(101): 336 (1936).
Cyperus lanceus var. *palaestinensis* Kük. in H.G.A.Engler (ed.), Pflanzenr., IV, 20(101): 336 (1936).
Cyperus lanceus var. *ramosus* Kük. in H.G.A.Engler (ed.), Pflanzenr., IV, 20(101): 336 (1936).

Pycreus nuerensis (Boeck.) S.S.Hooper, Kew Bull. 26: 579 (1972).
Trop. Africa. 22 BEN GUI IVO NGA SIE 23 RWA ZAI 24 SUD 25 KEN UGA 26 ZIM. Hemicr. or rhizome geophyte.
**Cyperus nuerensis* Boeck., Flora 62: 555 (1879). *Cyperus globosus* var. *nuerensis* (Boeck.) Kük. in H.G.A.Engler (ed.), Pflanzenr., IV, 20(101): 356 (1936).

Pycreus oakfortensis (Boeck. ex C.B.Clarke) C.B.Clarke in W.H.Harvey & auct. suc. (eds.), Fl. Cap. 7: 159 (1897).
S. Africa. 27 CPP NAT SWZ TVL. Hemicr. or rhizome geophyte.
**Cyperus oakfortensis* Boeck. ex C.B.Clarke in W.H.Harvey & auct. suc. (eds.), Fl. Cap. 7: 159 (1897).

Pycreus okavangensis Podlech, Mitt. Bot. Staatssamml. München 3: 522 (1960).
Botswana to Namibia. 27 BOT NAM. Hemicr. or rhizome geophyte.

Pycreus opulentus Govind., Proc. Indian Acad. Sci., Pl. Sci. 100: 417 (1990).
India. 40 IND. Ther.

Pycreus overlaetii Cherm. ex S.S.Hooper & J.Raynal, Kew Bull. 23: 314 (1969). *Cyperus overlaetii* (Cherm. ex S.S.Hooper & J.Raynal) Lye, Nordic J. Bot. 3: 231 (1983).
EC. Trop. Africa. 23 ZAI 25 TAN. Ther.
Pycreus flavescens subsp. *laevinux* Lye, Nordic J. Bot. 1: 620 (1981).

Pycreus pagotii J.Raynal, Kew Bull. 23: 314 (1969). *Cyperus pagotii* (J.Raynal) Lye, Nordic J. Bot. 3: 231 (1983).
S. Chad. 24 CHA. Hemicr. or rhizome geophyte.

Pycreus palghattensis Govind., Proc. Indian Acad. Sci., Pl. Sci. 100: 427 (1990).
SW. India. 40 IND. Ther.

Pycreus pauper (Hochst. ex A.Rich.) C.B.Clarke in T.A.Durand & H.Schinz, Consp. Fl. Afric. 5: 540 (1894).
Trop. Africa. 22 NGA 23 CAF CMN 24 ETH 25 TAN UGA 26 ANG ZAM. Ther.
**Cyperus pauper* Hochst. ex A.Rich., Tent. Fl. Abyss. 2: 478 (1850). *Chlorocyperus pauper* (Hochst. ex A.Rich.) Rikli, Jahrb. Wiss. Bot. 27: 563 (1894).
Cyperus angulatus W.Watson in E.T.Atkinson, Gaz. N. W. Ind. 10: 393 (1876), nom. illeg.

Pycreus pelophilus (Ridl.) C.B.Clarke in T.A.Durand & H.Schinz, Consp. Fl. Afric. 5: 540 (1894).
Trop. & S. Africa, SW. Madagascar. 22 GUI 23 BUR CON 25 TAN UGA 26 ANG MLW MOZ 27 BOT NAM SWZ TVL 29 MDG 83 ECU. Ther.
**Cyperus pelophilus* Ridl., Trans. Linn. Soc. London, Bot. 2: 129 (1884).
Pycreus chorisanthus C.B.Clarke in D.Oliver & auct. suc. (eds.), Fl. Trop. Afr. 8: 526 (1902).
Pycreus decaryi Cherm., Bull. Soc. Bot. France 72: 19 (1925).
Cyperus pelophilus f. *nanus* Kük. in H.G.A.Engler (ed.), Pflanzenr., IV, 20(101): 364 (1936).

Pycreus pervillei (Boeck.) C.B.Clarke in T.A.Durand & H.Schinz, Consp. Fl. Afric. 5: 540 (1894).
Madagascar. 29 MDG. Hemicr.
**Cyperus pervillei* Boeck., Linnaea 38: 359 (1874). *Juncellus pervillei* (Boeck.) C.B.Clarke, Kew Bull., Addit. Ser. 8: 3 (1908).

Pycreus plicatus Govind., Proc. Indian Acad. Sci., Pl. Sci. 100: 417 (1990).
India. 40 IND. Ther.

Pycreus plumbeonucea (Govind.) P.Singh & V.Singh, J. Econ. Taxon. Bot. 5: 467 (1984).
S. India. 40 IND. Ther.
**Cyperus plumbeonucea* Govind., J. Indian Bot. Soc. 52: 76 (1973 publ. 1974).

Pycreus plurinodosus (Govind.) P.Singh & V.Singh, J. Econ. Taxon. Bot. 5: 467 (1984).
S. India. 40 IND. Ther.
**Cyperus plurinodosa* Govind., Proc. Indian Acad. Sci., B 81: 192 (1975).

Pycreus poikilostachys Nelmes, Kew Bull. 6: 320 (1951 publ. 1952).
Zambia. 26 ZAM. Hemicr. or rhizome geophyte.

Pycreus polystachyos (Rottb.) P.Beauv., Fl. Oware 2: 48 (1816).
Trop. & Subtrop. 13 ITA 20 ALG EGY LBY MOR 21 CNY CVI 22 BEN BKN GAM GHA GNB GUI IVO LBR NGA NGR SEN SIE TOG 23 BUR EQG GAB GGI ZAI 24 ETH 25 KEN TAN UGA 26 ANG MLW MOZ ZAM ZIM 27 BOT CPP NAM NAT SWZ TVL 28 ASC STH 29 COM MAU MDG REU SEY 31 34 IRQ PAL 35 YEM 36 CHH CHS 38 JAP KOR KZN NNS OGA TAI 40 ASS EHM IND PAK SRL 41 AND CBD LAO MYA NCB SCS THA VIE 42 BOR CKI JAW LSI MLY MOL PHI SUL SUM 43 BIS NWG SOL (50) nfk NSW NTA QLD VIC WAU (60) fij gil sam ton (61) coo EAS sci tua 62 CRL MRN 63 HAW 79 MXE MXG MXN MXS MXT 80 BLZ COS ELS GUA HON NIC PAN 81 BER CAY CUB DOM JAM LEE PUE TRT WIN 82 FRG GUY SUR VEN 83 BOL CLM ECU GAL PER 84 BZC BZE BZL BZN BZS 85 AGE AGW PAR URU. Ther. or hemicr.
**Cyperus polystachyos* Rottb., Descr. Icon. Rar. Pl.: 39 (1773). *Chlorocyperus polystachyus* (Rottb.) Rikli, Jahrb. Wiss. Bot. 27: 563 (1895). *Pycreus polystachyos* var. *typicus* Domin, Biblioth. Bot. 85: 416 (1915), nom. inval.

var. ***microdontus*** (Torr.) Govaerts in R.H.A.Govaerts & D.A.Simpson, World Checklist of Cyperaceae: 610 (2007).
Trop. & Subtrop. 21 CVI 22 GAM GNB GUI SEN SIE 24 ETH 25 KEN TAN UGA 26 ANG 27 NAM 40 ASS IND SRL 41 THA VIE 42 BOR MLY PHI 50 NTA 63 HAW 74 MSO OKL 75 MAI MAS NWJ NWY PEN WVA 77 TEX 78 ALA ARK DEL FLA GEO KTY LOU MRY MSI NCA SCA TEN VRG WDC 79 MXG MXS 80 BLZ COS GUA PAN 81 CUB JAM 82 VEN 83 ECU 84 BZE 85 PAR. Hemicr. or rhizome geophyte.
Cyperus holosericeus Link, Hort. Berol. 1: 317 (1827). *Cyperus polystachyos* var. *holosericeus* (Link) C.B.Clarke, J. Linn. Soc., Bot. 21: 55 (1884). *Cyperus polystachyos* subsp. *holosericeus* (Link) T.Koyama, Madroño 20: 253 (1970). *Pycreus polystachyos* subsp. *holosericeus* (Link) T.Koyama, Bishop Mus. Occas. Pap. 29: 125 (1989).
**Cyperus microdontus* Torr., Ann. Lyceum Nat. Hist. New York 3: 255 (1836).
Cyperus microdontus var. *texensis* Torr., Ann. Lyceum Nat. Hist. New York 3: 430 (1836). *Cyperus polystachyos* var. *texensis* (Torr.) Fernald, Rhodora 41: 530 (1939).
Cyperus inconspicuus Liebm., Mexic. Halvgr.: 197 (1850). *Cyperus polystachyos* f. *inconspicuus* (Liebm.) Kük. in H.G.A.Engler (ed.), Pflanzenr., IV, 20(101): 372 (1936).
Cyperus polystachyos var. *leptostachyus* Boeck., Linnaea 35: 478 (1868).
Cyperus polystachyos var. *laxiflorus* Benth., Fl. Austral. 7: 261 (1878). *Pycreus polystachyos* var. *laxiflorus* (Benth.) C.B.Clarke in J.D.Hooker, Fl. Brit. India 6: 592 (1893). *Cyperus polystachyos* subsp. *laxiflorus* (Benth.) Lye in R.W.Haines & K.A.Lye, Sedges & Rushes E. Afr., App. 3: 2 (1983).
Pycreus holosericeus Merr., Philipp. J. Sci., C 12: 231 (1917).

subsp. ***polystachyos***
Trop. & Subtrop. 13 ITA 20 ALG EGY LBY MOR 21 CNY CVI 22 BEN BKN GAM GHA GNB GUI IVO LBR NGA NGR SEN SIE TOG 23 BUR EQG GAB GGI ZAI 24 ETH 25 KEN TAN UGA 26 ANG MLW MOZ ZAM ZIM 27 BOT CPP NAM NAT SWZ TVL 28 ASC STH 29 COM MAU MDG REU SEY 31 34 IRQ PAL 35 YEM 36 CHH CHS 38 JAP KOR KZN NNS OGA TAI 40 ASS EHM IND PAK SRL 41 AND CBD LAO MYA NCB SCS THA VIE 42 BOR CKI JAW LSI MLY MOL PHI SUL SUM 43 BIS NWG SOL (50) nfk NSW NTA QLD VIC WAU (60) fij gil sam ton (61) coo EAS sci tua 62 CRL MRN 63 HAW 79 MXG MXN MXS MXT 80 BLZ COS ELS GUA HON NIC PAN 81 BER CAY CUB DOM JAM LEE PUE TRT WIN 82 FRG GUY SUR VEN 83 BOL CLM ECU GAL PER 84 BZC BZE BZL BZN BZS 85 AGE AGW PAR URU. Ther. or hemicr.
Cyperus paniculatus Rottb., Descr. Icon. Rar. Pl.: 40 (1773). *Pycreus paniculatus* (Rottb.) Nees, Linnaea 9: 283 (1834). *Cyperus polystachyos* var. *paniculatus* (Rottb.) C.B.Clarke, J. Linn. Soc., Bot. 21: 55 (1884). *Pycreus polystachyos* var. *paniculatus* (Rottb.) Donn.Sm., Enum. Pl. Guatem. 3: 86 (1893).
Cyperus fascicularis Poir., Voy. Barbarie 2: 88 (1789). *Cyperus polystachyos* var. *fascicularis* (Poir.) Kuntze, Revis. Gen. Pl. 2: 750 (1891).
Cyperus scopellatus Rich., Actes Soc. Hist. Nat. Paris 1: 106 (1792).
Cyperus olidus Vahl, Enum. Pl. 2: 333 (1805).
Cyperus olivaceus Vahl, Enum. Pl. 2: 332 (1805).
Cyperus caespitosus Poir. in J.B.A.M.de Lamarck, Encycl. 7: 250 (1806). *Pycreus caespitosus* (Poir.) C.B.Clarke in T.A.Durand & H.Schinz, Consp. Fl. Afric. 5: 535 (1894). *Cyperus polystachyos* var. *caespitosus* (Poir.) Kük. in H.G.A.Engler (ed.), Pflanzenr., IV, 20(101): 372 (1936).
Cyperus tenuis Muhl., Descr. Gram.: 22 (1817), nom. illeg.
Cyperus litos Schult. in J.J.Roemer & J.A.Schultes, Mant. 2: 101 (1824).
Cyperus gatesii Torr., Ann. Lyceum Nat. Hist. New York 3: 255 (1836).
Cyperus incrassatus Beyr. ex Kunth, Enum. Pl. 2: 15 (1837).
Cyperus thouarsii Kunth, Enum. Pl. 2: 12 (1837). *Cyperus polystachyos* var. *thouarsii* (Kunth) Kük. in H.G.A.Engler (ed.), Pflanzenr., IV, 20(101): 373 (1936).
Cyperus sonderi J.A.Schmidt, Beitr. Fl. Cap Verd. Ins.: 162 (1852).
Cyperus liebmannii Steud., Syn. Pl. Glumac. 2: 7 (1854).
Cyperus teretifructus Steud., Syn. Pl. Glumac. 2: 3 (1854).
Cyperus texensis Steud., Syn. Pl. Glumac. 2: 9 (1854).
Cyperus chlorostachys Boeck. in W.C.H.Peters, Naturw. Reise Mossambique: 540 (1864). *Cyperus polystachyos* var. *chlorostachys* (Boeck.) Kük. in H.G.A.Engler (ed.), Pflanzenr., IV, 20(101): 371 (1936).
Cyperus boivinii Boeck., Linnaea 35: 481 (1868).

Cyperus polystachyos f. *glaucescens* Boeck., Linnaea 35: 479 (1868).
Cyperus polystachyos var. *macrostachyos* Boeck., Linnaea 35: 479 (1868).
Cyperus polystachyos var. *macrostachyus* Boeck., Linnaea 35: 479 (1868).
Cyperus aquatilis F.Muell., Fragm. 8: 270 (1874), nom. illeg.
Cyperus hahnianus Boeck., Flora 61: 138 (1878). *Pycreus polystachyos* var. *hahnianus* (Boeck.) C.B.Clarke in I.Urban, Symb. Antill. 2: 18 (1900). *Cyperus polystachyos* var. *hahnianus* (Boeck.) Kük. in H.G.A.Engler (ed.), Pflanzenr., IV, 20(101): 370 (1936). *Pycreus odoratus* var. *hahnianus* (Boeck.) Stehlé, Caribbean Forester 5: 184 (1944).
Cyperus hookerianus Arn. ex C.B.Clarke, J. Linn. Soc., Bot. 20: 280 (1883), nom. inval.
Cyperus strictus Rottler ex C.B.Clarke, J. Linn. Soc., Bot. 21: 54 (1884), nom. inval.
Cyperus polystachyos var. *pallidus* Hillebr., Fl. Hawaiian Isl.: 463 (1888).
Cyperus circinatus Ridl., J. Linn. Soc., Bot. 27: 64 (1890). *Cyperus polystachyos* var. *circinatus* (Ridl.) Kük. in H.G.A.Engler (ed.), Pflanzenr., IV, 20(101): 370 (1936). *Pycreus polystachyos* var. *circinatus* (Ridl.) H.B.Naithani & S.Biswas in H.B.Naithani, Flow. Pl. India, Nepal & Bhutan: 472 (1990).
Cyperus polystachyos var. *longispiculatus* Suringar, Cyperus Mal. Arch.: 62 (1898).
Pycreus odoratus Urb., Symb. Antill. 2: 164 (1900).
Cyperus polystachyos f. *minor* Cherm. ex Kük. in H.G.A.Engler (ed.), Pflanzenr., IV, 20(101): 373 (1936).
Cyperus polystachyos var. *miser* Kük. in H.G.A. Engler (ed.), Pflanzenr., IV, 20(101): 370 (1936).
Cyperus polystachyos f. *pernanus* Kük. in H.G.A.Engler (ed.), Pflanzenr., IV, 20(101): 371 (1936).
Cyperus polystachyos var. *sanguineus* Kük. in H.G.A.Engler (ed.), Pflanzenr., IV, 20(101): 371 (1936).
Pycreus caespitosus var. *minor* Cherm., Bull. Soc. Bot. France 83: 71 (1936).

subsp. ***riograndensis*** D.A.Simpson, Kew Bull. 44: 281 (1989).
SE. & S. Brazil. 84 BZL BZS. Hemicr. or rhizome geophyte.

Pycreus pratorum (Korotky) Schischk. in V.L.Komarov (ed.), Fl. URSS 3: 8 (1935).
Amur. 31 AMU. Hemicr. or rhizome geophyte.
**Cyperus pratorum* Korotky, Repert. Spec. Nov. Regni Veg. 13: 292 (1914). *Cyperus sanguinolentus* var. *pratorum* (Korotky) Kük. in H.G.A.Engler (ed.), Pflanzenr., IV, 20(101): 388 (1936).

Pycreus pseudodiaphanus S.S.Hooper, Kew Bull. 23: 313 (1969). *Cyperus pseudodiaphanus* (S.S.Hooper) Lye, Nordic J. Bot. 3: 231 (1983).
W. Trop. Africa. 22 BKN GHA GNB GUI MLI NGA SEN TOG. Ther.

var. ***occidentalis*** S.S.Hooper, Kew Bull. 23: 313 (1969).
W. Trop. Africa. 22 GHA GUI SEN. Ther.

var. ***pseudodiaphanus***
W. Trop. Africa. 22 BKN GHA GNB MLI NGA SEN TOG. Ther.

Pycreus pseudolatespicatus L.K.Dai, in Fl. Reipubl. Popul. Sin. 11: 233 (1961).
China (Sichuan). 36 CHC. Hemicr. or rhizome geophyte.

Pycreus pubescens Turrill, Bull. Misc. Inform. Kew 1914: 339 (1914). *Cyperus gossweileri* Kük. in H.G.A. Engler (ed.), Pflanzenr., IV, 20(101): 7 (1935).
S. Angola. 26 ANG. Hemicr. or rhizome geophyte.

Pycreus pumilus (L.) Nees, Linnaea 9: 283 (1834).
Trop. & Subtrop. Old World. 22 BEN BKN GHA GUI IVO MLI NGA NGR SEN SIE TOG 23 CON ZAI 24 CHA ETH SOM 25 KEN TAN UGA 26 ANG MLW MOZ ZAM ZIM 27 BOT NAM NAT SWZ TVL 29 ALD COM MDG 36 CHS 38 TAI 40 ASS BAN EHM IND NEP PAK SRL WHM 41 MYA THA VIE 42 BOR JAW LSI MLY MOL PHI SUL SUM 43 NWG 50 NTA QLD VIC WAU (62) wak (78) fla geo (81) trt win (82) frg. Ther.
**Cyperus pumilus* L., Cent. Pl. II: 6 (1756).
Cyperus pygmaeus Retz., Observ. Bot. 4: 9 (1786), nom. illeg.
Cyperus nitens Retz., Observ. Bot. 5: 13 (1788). *Pycreus nitens* (Retz.) Nees, Nov. Actorum Acad. Caes. Leop.-Carol. Nat. Cur. 19(Suppl. 1): 53 (1843). *Dichostylis nitens* (Retz.) Palla, Bot. Jahrb. Syst. 10: 296 (1888).
Cyperus nitens Vahl, Enum. Pl. 2: 331 (1805), nom. illeg.
Cyperus patens Vahl, Enum. Pl. 2: 334 (1805). *Cyperus nitens* var. *patens* (Vahl) Boeck., Linnaea 35: 484 (1868). *Dichostylis patens* (Vahl) Palla, Bot. Jahrb. Syst. 10: 296 (1888). *Dichostylis patens* (Vahl) Rikli, Jahrb. Wiss. Bot. 27: 565 (1895), nom. illeg. *Pycreus patens* (Vahl) Cherm., Arch. Bot. (Forlì) 4(7): 11 (1931). *Cyperus pumilus* var. *patens* (Vahl) Kük. in H.G.A.Engler (ed.), Pflanzenr., IV, 20(101): 378 (1936).
Cyperus pusillus Vahl, Enum. Pl. 2: 303 (1805).
Cyperus punctatus Roxb., Fl. Ind. 1: 197 (1820), nom. illeg. *Cyperus pumilus* var. *punctatus* C.B.Clarke, J. Linn. Soc., Bot. 21: 46 (1884). *Pycreus pumilus* var. *punctatus* Domin, Biblioth. Bot. 85: 417 (1915). *Pycreus punctatus* Govind., Proc. Indian Acad. Sci., Pl. Sci. 100: 426 (1990).
Cyperus pulvinatus Nees & Meyen in R.Wight, Contr. Bot. India: 74 (1834). *Pycreus pulvinatus* (Nees & Meyen) Nees, Nov. Actorum Acad. Caes. Leop.-Carol. Nat. Cur. 19(Suppl. 1): 53 (1843).
Cyperus truncatulus Steud., Flora 25: 581 (1842).
Cyperus commutatus Steud., Syn. Pl. Glumac. 2: 10 (1854).
Cyperus gymnoleptus Steud., Syn. Pl. Glumac. 2: 3 (1854).
Cyperus laxiusculus Steud., Syn. Pl. Glumac. 2: 14 (1854).
Cyperus obstinatus Steud., Syn. Pl. Glumac. 2: 10 (1854).
Cyperus terminalis Marsh ex Steud., Syn. Pl. Glumac. 2: 5 (1854).
Cyperus divergens Chapm., Fl. South. U.S.: 512 (1860), nom. illeg. *Cyperus leucolepis* Carey ex C.B.Clarke, J. Linn. Soc., Bot. 21: 61 (1884), nom. illeg. *Juncellus leucolepis* C.B.Clarke, Kew Bull., Addit. Ser. 8: 3 (1908). *Cyperus pumilus* var. *leucolepis* (C.B.Clarke) Kük. in H.G.A.Engler (ed.), Pflanzenr., IV, 20(101): 378 (1936).
Cyperus breviculmis F.Muell., Fragm. 8: 267 (1874), nom. illeg.

Cyperus strictus Wight ex C.B.Clarke, J. Linn. Soc., Bot. 21: 46 (1884), nom. inval.
Pycreus nitens f. *gracilis* E.G.Camus, Notul. Syst. (Paris) 1: 241 (1910). *Cyperus pumilus* f. *gracilis* (E.G.Camus) Kük. in H.G.A.Engler (ed.), Pflanzenr., IV, 20(101): 376 (1936).
Juncellus pumilus Peter, Abh. Preuss. Akad. Wiss., Phys.-Math. Kl., n.f., 13(2): 48 (1928).
Cyperus pumilus var. *nervulosus* Kük. in H.G.A. Engler (ed.), Pflanzenr., IV, 20(101): 378 (1936). *Cyperus nervulosus* (Kük.) S.T.Blake, Proc. Roy. Soc. Queensland 51: 41 (1939 publ. 1940).
Cyperus pumilus var. *prorepens* Kük. in H.G.A. Engler (ed.), Pflanzenr., IV, 20(101): 378 (1936).

Pycreus puncticulatus (Vahl) Nees in C.F.P.von Martius & auct. suc. (eds.), Fl. Bras. 2(1): 10 (1842).
S. India to China and Pen. Malaysia. 36 CHC 40 IND SRL 41 VIE 42 MLY. Ther.
Cyperus puncticulatus Vahl, Enum. Pl. 2: 348 (1805).
Cyperus baccha Nees, Linnaea 9: 283 (1834), nom. inval. *Pycreus baccha* Nees in C.F.P.von Martius & auct. suc. (eds.), Fl. Bras. 2(1): 10 (1842).
Cyperus inundatus Nees in R.Wight, Contr. Bot. India: 76 (1834), nom. illeg.
Pycreus inundatus Nees in C.F.P.von Martius & auct. suc. (eds.), Fl. Bras. 2(1): 10 (1842).
Cyperus puncticulatus var. *quinquagintiflorus* C.B. Clarke, J. Linn. Soc., Bot. 21: 69 (1884). *Pycreus puncticulatus* var. *quinquagintiflorus* (C.B.Clarke) C.B.Clarke in J.D.Hooker, Fl. Brit. India 6: 594 (1893).

Pycreus pyramidalis Govind., J. Econ. Taxon. Bot. 20: 301 (1996).
S. India. 40 IND. Ther.

Pycreus rhizomatosus C.B.Clarke, Kew Bull., Addit. Ser. 8: 2 (1908). *Cyperus rhizomatosus* (C.B.Clarke) Kük. in H.G.A.Engler (ed.), Pflanzenr., IV, 20(101): 328 (1936).
Madagascar. 29 MDG. Hemicr.
Cyperus commersonii C.B.Clarke, Kew Bull., Addit. Ser. 8: 6 (1908). *Pycreus commersonii* (C.B.Clarke) Cherm., Ann. Inst. Bot.-Géol. Colon. Marseille, III, 7(2): 69 (1919).

Pycreus rubriglumosus (Govind.) P.Singh & V.Singh, J. Econ. Taxon. Bot. 5: 468 (1984).
S. India. 40 IND. Hemicr. or rhizome geophyte.
Cyperus rubriglumosus Govind., Proc. Indian Acad. Sci., B 88: 229 (1979).

Pycreus sanguineosquamatus Van der Veken, Bull. Jard. Bot. État 24: 145 (1955).
Zaïre. 23 ZAI. Hemicr. or rhizome geophyte.

Pycreus sanguinolentus (Vahl) Nees, Linnaea 9: 283 (1834).
Old World. 24 ERI ETH 25 KEN TAN 28 STH 31 PRM 32 KAZ KGZ TKM TZK UZB 33 TCS 34 AFG TUR 35 SAU YEM 36 CHC CHH CHM CHN CHS CHT 38 JAP KOR KZN NNS OGA TAI 40 ASS BAN EHM IND NEP PAK SRL WHM 41 MYA THA VIE 42 BOR JAW LSI MLY MOL PHI SUL SUM 43 NWG 50 NSW QLD SOA TAS VIC (63) haw (78) ala geo lou msi. Hemicr. or rhizome geophyte.
Cyperus sanguinolentus Vahl, Enum. Pl. 2: 351 (1805), nom. cons.

var. ***cyrtostachys*** (Miq.) L.K.Dai, in Fl. Hainan. 4: 538 (1977).
Malesia. 42 BOR JAW LSI MLY MOL PHI SUL SUM 43 NWG. Ther. or hemicr.
Cyperus eragrostis var. *cyrtostachys* Miq., Fl. Ned. Ind. 3: 257 (1856). *Cyperus sanguinolentus* f. *cyrtostachys* (Miq.) Kük. in H.G.A.Engler (ed.), Pflanzenr., IV, 20(101): 387 (1936). *Cyperus sanguinolentus* subsp. *cyrtostachys* (Miq.) A.Kern., Reinwardtia 3: 57 (1954). *Pycreus sanguinolentus* subsp. *cyrtostachys* (Miq.) Karthik. in S.Karthikeyan & al., Fl. Ind. Enumerat. – Monocot.: 66 (1989).

subsp. ***melanocephalus*** (Miq.) Karthik. in S.Karthikeyan & al., Fl. Ind. Enumerat. – Monocot.: 66 (1989).
Trop. & Subtrop. Asia. 36 CHC 40 ASS EHM 42 JAW PHI 43 NWG. Ther. or hemicr.
Cyperus melanocephalus Miq., Fl. Ned. Ind. 3: 257 (1856). *Cyperus sanguinolentus* f. *melanocephalus* (Miq.) Kük. in H.G.A.Engler (ed.), Pflanzenr., IV, 20(101): 387 (1936). *Cyperus sanguinolentus* subsp. *melanocephalus* (Miq.) A.Kern., Reinwardtia 3: 55 (1954).

subsp. ***nairobiensis*** Lye, Nordic J. Bot. 1: 617 (1981). *Cyperus sanguinolentus* subsp. *nairobiensis* (Lye) Lye in R.W.Haines & K.A.Lye, Sedges & Rushes E. Afr., App. 3: 2 (1983).
Kenya. 25 KEN. Ther.

subsp. ***sanguinolentus***
Old World. 24 ERI ETH 25 KEN TAN 28 STH 31 PRM 32 KAZ KGZ TKM TZK UZB 33 TCS 34 AFG TUR 35 SAU YEM 36 CHC CHH CHM CHN CHS CHT 38 JAP KOR KZN NNS OGA TAI 40 ASS BAN EHM IND NEP PAK SRL WHM 41 MYA THA VIE 42 JAW LSI MLY MOL PHI SUL SUM 50 NSW QLD SOA TAS VIC (63) haw (78) ala geo lou msi. Hemicr. or rhizome geophyte.
Cyperus pumilus Rottb., Descr. Icon. Rar. Pl.: 29 (1773), nom. illeg.
Cyperus albidus Lam., Tabl. Encycl. 1: 146 (1791), nom. illeg.
Cyperus eragrostis Vahl, Enum. Pl. 2: 322 (1805), nom. illeg. *Pycreus eragrostis* Palla, Ann. K. K. Naturhist. Hofmus. 23: 204 (1909).
Cyperus areolatus R.Br., Prodr.: 216 (1810). *Pycreus sanguinolentus* var. *typicus* Domin, Biblioth. Bot. 85: 418 (1915), nom. inval. *Cyperus sanguinolentus* var. *areolatus* (R.Br.) Kük. in H.G.A.Engler (ed.), Pflanzenr., IV, 20(101): 388 (1936).
Cyperus erythraeus Schrad. ex Schult. in J.J.Roemer & J.A.Schultes, Mant. 2: 477 (1824).
Cyperus flavescens var. *rubromarginatus* Schrenk in F.E.L.von Fischer & C.A.von Meyer, Enum. Pl. Nov. 1: 3 (1841). *Cyperus sanguinolentus* f. *rubromarginatus* (Schrenk) Kük. in H.G.A. Engler (ed.), Pflanzenr., IV, 20(101): 386 (1936).
Cyperus neurotropis Steud., Flora 25: 583 (1842). *Cyperus eragrostis* var. *microstachyus* Boeck., Linnaea 35: 445 (1868). *Cyperus sanguinolentus* f. *neurotropis* (Steud.) Kük. in H.G.A.Engler (ed.), Pflanzenr., IV, 20(101): 386 (1936).
Cyperus atratus Steud., Syn. Pl. Glumac. 2: 12 (1854).
Cyperus concolor Steud., Syn. Pl. Glumac. 2: 6 (1854).
Cyperus eragrostis var. *humilis* Miq., Fl. Ned. Ind. 3: 257 (1856). *Cyperus sanguinolentus* f. *humilis*

(Miq.) Kük. in H.G.A.Engler (ed.), Pflanzenr., IV, 20(101): 386 (1936).
Cyperus eragrostis var. *flaccidulus* Boeck., Linnaea 35: 446 (1868). *Cyperus sanguinolentus* f. *flaccidulus* (Boeck.) Kük. in H.G.A.Engler (ed.), Pflanzenr., IV, 20(101): 387 (1936).
Cyperus eragrostis var. *melanocephalus* Boeck., Linnaea 35: 445 (1868).
Cyperus teysmannii Boeck., Flora 58: 259 (1875). *Cyperus sanguinolentus* var. *teysmannii* (Boeck.) Kük. in H.G.A.Engler (ed.), Pflanzenr., IV, 20(101): 387 (1936). *Cyperus sanguinolentus* subsp. *teysmannii* (Boeck.) A.Kern., Reinwardtia 3: 57 (1954).
Cyperus rehmannii Boiss., Fl. Orient. 5: 364 (1882).
Cyperus eragrostis var. *micronux* C.B.Clarke, J. Linn. Soc., Bot. 21: 59 (1884). *Cyperus sanguinolentus* var. *micronux* (C.B.Clarke) Kük. in H.G.A.Engler (ed.), Pflanzenr., IV, 20(101): 387 (1936). *Pycreus sanguinolentus* var. *micronux* (C.B.Clarke) Karthik. in S.Karthikeyan & al., Fl. Ind. Enumerat. – Monocot.: 66 (1989). *Pycreus sanguinolentus* subsp. *micronux* (C.B.Clarke) H.B.Naithani & S.Biswas in H.B.Naithani, Flow. Pl. India, Nepal & Bhutan: 472 (1990).
Cyperus levis F.Muell. ex C.B.Clarke, J. Linn. Soc., Bot. 21: 59 (1884).
Cyperus korshinskyi Meinsh., Trudy Imp. S.-Peterburgsk. Bot. Sada 18: 235 (1900). *Cyperus sanguinolentus* var. *korshinskyi* (Meinsh.) Kük. in H.G.A.Engler (ed.), Pflanzenr., IV, 20(101): 387 (1936). *Pycreus korshinskyi* (Meinsh.) V.I.Krecz., Bot. Mater. Gerb. Bot. Inst. Komarova Akad. Nauk S.S.S.R. 7: 27 (1937).
Pycreus rubromarginatus E.G.Camus, Notul. Syst. (Paris) 1: 240 (1910). *Cyperus rubromarginatus* (E.G.Camus) Drobow in A.I.Vvedensky & al., Key Fl. Tashkent: 46 (1923), nom. illeg.
Cyperus eragrostis var. *pauperita* J.M.Black, Trans. & Proc. Roy. Soc. South Australia 48: 253 (1924).
Cyperus eragrostis var. *minor* Ewart, Fl. Victoria: 212 (1931).
Cyperus sanguinolentus f. *setaceus* Kük. in H.G.A.Engler (ed.), Pflanzenr., IV, 20(101): 388 (1936).
Cyperus sanguinolentus subsp. *sacer* Korlahalli, Bull. Bot. Surv. India 9: 235 (1967 publ. 1968). *Pycreus sanguinolentus* subsp. *sacer* (Korlahalli) Karthik. in S.Karthikeyan & al., Fl. Ind. Enumerat. – Monocot.: 66 (1989).
Cyperus louisianensis Thieret, Proc. Louisiana Acad. Sci. 40: 23 (1977).

Pycreus similinervulosus Govind., Proc. Indian Acad. Sci., Pl. Sci. 100: 430 (1990).
SW. India. 40 IND. Ther.

Pycreus smithianus (Ridl.) C.B.Clarke in T.A.Durand & H.Schinz, Consp. Fl. Afric. 5: 542 (1894).
W. & WC. Trop. Africa. 22 BEN BKN GUI IVO NGA SIE 23 BUR CMN EQG GAB GGI ZAI. Hemicr.
**Cyperus smithianus* Ridl., J. Bot. 22: 15 (1884).

Pycreus spissiflorus C.B.Clarke, Trans. Linn. Soc. London, Bot. 4: 53 (1894 publ. 1895). *Cyperus spissiflorus* (C.B.Clarke) K.Schum. in H.G.A.Engler, Pflanzenw. Ost-Afrikas, C: 117 (1895).
Malawi (Mt. Mulanje). 26 MLW. Hemicr. or rhizome geophyte.

Pycreus squarrosulus Cherm., Bull. Mus. Natl. Hist. Nat. 25: 137 (1919). *Cyperus squarrosulus* (Cherm.) Kük. in H.G.A.Engler (ed.), Pflanzenr., IV, 20(101): 375 (1936).
W. Madagascar. 29 MDG. Ther.

Pycreus stramineus C.B.Clarke in J.D.Hooker, Fl. Brit. India 6: 589 (1893).
Nepal to Pen. Malaysia. 40 ASS BAN EHM IND NEP SRL 41 MYA THA VIE 42 MLY. Ther.
**Cyperus stramineus* Nees in R.Wight, Contr. Bot. India: 74 (1834), nom. illeg. *Cyperus substramineus* Kük. in H.G.A.Engler (ed.), Pflanzenr., IV, 20(101): 10 (1935). *Pycreus substramineus* (Kük.) N.K.Khoi, Novosti Sist. Vyssh. Rast. 18: 140 (1981), nom. illeg.
Cyperus capillaris Hochst. ex Steud., Syn. Pl. Glumac. 2: 4 (1854), nom. illeg.
Cyperus coromandelinus Boeck., Linnaea 35: 480 (1868), nom. illeg.

Pycreus stricticulmis (Govind.) P.Singh & V.Singh, J. Econ. Taxon. Bot. 5: 467 (1984).
S. India. 40 IND. Hemicr. or rhizome geophyte.
**Cyperus stricticulmis* Govind., Proc. Indian Acad. Sci., B 81: 194 (1975).

Pycreus subtrigonus C.B.Clarke in T.A.Durand & H.Schinz, Études Fl. Congo 1: 282 (1896). *Cyperus subtrigonus* (C.B.Clarke) Kük. in H.G.A.Engler (ed.), Pflanzenr., IV, 20(101): 17 (1935).
WC. Trop. Africa. 23 CAF CON ZAI. Hemicr. or rhizome geophyte.

Pycreus sulcinux (C.B.Clarke) C.B.Clarke in J.D.Hooker, Fl. Brit. India 6: 593 (1893).
Trop. & Subtrop. Asia to Queensland. 36 CHS 38 TAI 40 ASS BAN EHM IND 41 MYA THA VIE 42 BOR JAW LSI MOL PHI 43 NWG 50 QLD 62 CRL. Ther.
**Cyperus sulcinux* C.B.Clarke, J. Linn. Soc., Bot. 21: 56 (1884).
Cyperus bonianus Boeck., Allg. Bot. Z. Syst. 1: 186 (1895).
Pycreus substellatus E.G.Camus, Notul. Syst. (Paris) 1: 240 (1910).

Pycreus tener C.B.Clarke, Bull. Herb. Boissier, II, 3: 940 (1903). *Cyperus lanceolatus* var. *tener* (C.B.Clarke) Kük., Physis (Buenos Aires) 9: 362 (1929). *Cyperus barrosianus* Herter, Revista Sudamer. Bot. 9: 145 (1953).
Brazil to NE. Argentina. 84 85 AGE PAR URU. Hemicr. or rhizome geophyte.

Pycreus testui Cherm., Arch. Bot. Bull. Mens. 4(7): 13 (1931). *Cyperus laxespicatus* var. *testui* (Cherm.) Kük. in H.G.A.Engler (ed.), Pflanzenr., IV, 20(101): 332 (1936).
W. & WC. Trop. Afrca. 22 GHA GNB GUI IVO MLI NGA SEN TOG 23 CAF CMN CON. Hemicr. or rhizome geophyte.

Pycreus unioloides (R.Br.) Urb., Symb. Antill. 2: 164 (1900).
Trop. & Subtrop. 22 BKN GAM IVO MLI NGA 23 CAF 24 ETH 25 TAN UGA 27 BOT CPP LES NAT SWZ TVL 29 MDG 36 CHC CHS 38 JAP TAI 40 ASS EHM IND NEP 41 MYA THA VIE 42 JAW PHI SUM 43 NWG 50 NSW QLD 76 CAL 79 MXC MXE MXG MXS MXT 80 BLZ COS GUA HON PAN 81 CUB DOM HAI JAM PUE 82 SUR VEN 83 BOL CLM PER 84 BZC BZL BZN BZS 85 AGE PAR URU. Hemicr. or rhizome geophyte.

Cyperus unioloides R.Br., Prodr.: 216 (1810). *Pycreus unioloides* var. *typicus* Domin, Biblioth. Bot. 85: 417 (1915), nom. inval.
Cyperus bromoides Link, Jahrb. Gewächsk. 1(3): 85 (1818). *Cyperus unioloides* var. *bromoides* (Link) C.B.Clarke, J. Linn. Soc., Bot. 21: 61 (1884). *Pycreus angulatus* f. *bromoides* (Link) Lindm., Bih. Kongl. Svenska Vetensk.-Akad. Handl. 26(3; 9): 6 (1900).
Cyperus angulatus Nees in R.Wight, Contr. Bot. India: 73 (1834). *Pycreus angulatus* (Nees) Nees ex C.B.Clarke in J.D.Hooker, Fl. Brit. India 6: 593 (1893). *Pycreus unioloides* var. *angulatus* (Nees) Domin, Biblioth. Bot. 85: 417 (1915).
Cyperus selloanus Boeck., Linnaea 35: 446 (1868).
Cyperus spinuliferus Boeck., Linnaea 35: 465 (1868).
Cyperus lanceus F.Muell., Fragm. 8: 259 (1874), nom. illeg.
Cyperus luteolus Boeck., Flora 58: 82 (1875).
Cyperus tosaensis Makino, Bot. Mag. (Tokyo) 6: 47 (1892).
Cyperus unioloides var. *wightii* C.B.Clarke in J.D.Hooker, Fl. Brit. India 6: 593 (1893).
Pycreus angulatus var. *wightii* Nees ex C.B.Clarke in J.D.Hooker, Fl. Brit. India 6: 593 (1893). *Pycreus unioloides* var. *wightii* (Nees ex C.B.Clarke) Karthik. in S.Karthikeyan & al., Fl. Ind. Enumerat. – Monocot.: 67 (1989).
Cyperus taubertii Boeck., Allg. Bot. Z. Syst. 1: 185 (1895).
Pycreus unioloides var. *insularis* Domin, Biblioth. Bot. 85: 417 (1915). *Cyperus unioloides* f. *insularis* (Domin) Kük. in H.G.A.Engler (ed.), Pflanzenr., IV, 20(101): 340 (1936).
Pycreus umbrosiformis Cherm., Bull. Soc. Bot. France 67: 326 (1920 publ. 1921).
Cyperus unioloides f. *pallescens* Kük., Repert. Spec. Nov. Regni Veg. 23: 184 (1926).
Cyperus unioloides var. *fissivaginatus* Kük. in H.G.A. Engler (ed.), Pflanzenr., IV, 20(101): 340 (1936).
Cyperus unioloides f. *reductus* Kük. in H.G.A.Engler (ed.), Pflanzenr., IV, 20(101): 340 (1936).
Cyperus unioloides var. *compositus* T.Koyama, Naturaliste Canad. 82: 213 (1955). *Pycreus unioloides* var. *compositus* (T.Koyama) N.K.Khoi, Novosti Sist. Vyssh. Rast. 18: 137 (1981).

Pycreus waillyi Cherm., Bull. Soc. Bot. France 85: 366 (1938). *Cyperus waillyi* (Cherm.) Lye in R.W.Haines & K.A.Lye, Sedges & Rushes E. Afr., App. 3: 2 (1983).
Mali, Tanzania, Zambia. 22 MLI 25 TAN 26 ZAM. Ther.

Pycreus xantholepis Nelmes, Kew Bull. 6: 319 (1951 publ. 1952).
Zaïre to Angola. 23 ZAI 26 ANG. Hemicr. or rhizome geophyte.

Pycreus zonatus Cherm., Bull. Soc. Bot. France 67: 328 (1920 publ. 1921). *Pycreus zonatissimus* Cherm., Bull. Soc. Bot. France 74: 605 (1927 publ. 1928), nom. illeg. *Cyperus zonatissimus* Kük. in H.G.A.Engler (ed.), Pflanzenr., IV, 20(101): 395 (1936).
E. Trop. Africa, C. Madagascar. 25 KEN TAN 29 MDG. Ther.

Synonyms:
Pycreus albomarginatus Mart. & Schrad. ex Nees = ***Pycreus macrostachyos*** (Lam.) J.Raynal subsp. ***macrostachyos***
Pycreus albomarginatus var. *bromoides* C.B.Clarke = ***Pycreus macrostachyos*** (Lam.) J.Raynal subsp. ***macrostachyos***
Pycreus angulatus (Nees) Nees ex C.B.Clarke = ***Pycreus unioloides*** (R.Br.) Urb.
Pycreus angulatus f. *bromoides* (Link) Lindm. = ***Pycreus unioloides*** (R.Br.) Urb.
Pycreus angulatus var. *wightii* Nees ex C.B.Clarke = ***Pycreus unioloides*** (R.Br.) Urb.
Pycreus antsirabensis Cherm. = ***Pycreus nitidus*** (Lam.) J.Raynal
Pycreus antsirabensis var. *densior* Cherm. = ***Pycreus nitidus*** (Lam.) J.Raynal
Pycreus ater (C.B.Clarke) Cherm. = ***Pycreus melas*** (Ridl.) C.B.Clarke
Pycreus baccha Nees = ***Pycreus puncticulatus*** (Vahl) Nees
Pycreus badius (Desf.) Hayek = ***Cyperus longus*** subsp. ***badius*** (Desf.) Murb.
Pycreus bathiei Cherm. = ***Pycreus flavescens*** var. ***rehmannianus*** (C.B.Clarke) Govaerts
Pycreus caespitosus (Poir.) C.B.Clarke = ***Pycreus polystachyos*** (Rottb.) P.Beauv. subsp. ***polystachyos***
Pycreus caespitosus var. *minor* Cherm. = ***Pycreus polystachyos*** (Rottb.) P.Beauv. subsp. ***polystachyos***
Pycreus capillaris (J.König ex Roxb.) Nees ex C.B.Clarke = ***Pycreus flavidus*** (Retz.) T.Koyama
Pycreus capillifolius var. *major* Cherm. = ***Pycreus capillifolius*** (A.Rich.) C.B.Clarke
Pycreus chorisanthus C.B.Clarke = ***Pycreus pelophilus*** (Ridl.) C.B.Clarke
Pycreus cimicinus (J.Presl & C.Presl) H.Pfeiff. = ***Pycreus niger*** (Ruiz & Pav.) Cufod.
Pycreus cleaveri (Torr.) C.B.Clarke = ***Pycreus filicinus*** (Vahl) T.Koyama
Pycreus colchicus (K.Koch) Schischk. ex Grossh. = ***Cyperus serotinus*** Rottb.
Pycreus commersonii (C.B.Clarke) Cherm. = ***Pycreus rhizomatosus*** C.B.Clarke
Pycreus congestus (Vahl) Hayek = ***Cyperus congestus*** Vahl
Pycreus debilissimus C.B.Clarke = ***Pycreus flavescens*** subsp. ***tanaensis*** (Kük.) Lye
Pycreus decaryi Cherm. = ***Pycreus pelophilus*** (Ridl.) C.B.Clarke
Pycreus decumbens (Govind.) P.Singh & V.Singh = ***Pycreus govindarajalui*** V.S.Raju
Pycreus densifolius (Kunth) Nees ex Cherm. = ***Pycreus mundtii*** Nees
Pycreus densus (Link) C.B.Clarke = ***Pycreus lanceolatus*** (Poir.) C.B.Clarke
Pycreus diaphanus var. *acaulis* (C.B.Clarke) Karthik. = ***Pycreus diaphanus*** (Schrad. ex Roem. & Schult.) S.S.Hooper & T.Koyama
Pycreus diaphanus var. *gracilescens* (Kük.) S.S.Hooper = ***Pycreus diaphanus*** (Schrad. ex Roem. & Schult.) S.S.Hooper & T.Koyama
Pycreus diaphanus var. *laxus* (Kük.) H.B.Naithani & S.Biswas = ***Pycreus diaphanus*** (Schrad. ex Roem. & Schult.) S.S.Hooper & T.Koyama
Pycreus djalonis A.Chev. = ***Cyperus pustulatus*** Vahl
Pycreus elliottianus Nees = ***Pycreus lanceolatus*** (Poir.) C.B.Clarke
Pycreus elliottianus var. *humilis* Nees = ***Pycreus lanceolatus*** (Poir.) C.B.Clarke
Pycreus eragrostis Palla = ***Pycreus sanguinolentus*** (Vahl) Nees subsp. ***sanguinolentus***
Pycreus esculentus (L.) Hayek = ***Cyperus esculentus*** L.
Pycreus fallaciosus Cherm. = ***Pycreus flavescens*** subsp. ***fallaciosus*** (Cherm.) Lye

Pycreus fernandezianus C.B.Clarke = ***Cyperus reflexus*** Vahl
Pycreus ferrugineus (Boeck.) C.B.Clarke = ***Pycreus intactus*** (Vahl) J.Raynal
Pycreus flavescens var. *castaneus* Lye = ***Pycreus flavescens*** var. ***rehmannianus*** (C.B.Clarke) Govaerts
Pycreus flavescens f. *caucasicus* (K.Koch) Soó = ***Pycreus flavescens*** (L.) P.Beauv. ex Rchb. subsp. ***flavescens***
Pycreus flavescens f. *gracillimus* (Asch. & Graebn.) Soó = ***Pycreus flavescens*** (L.) P.Beauv. ex Rchb. subsp. ***flavescens***
Pycreus flavescens subsp. *laevinux* Lye = ***Pycreus overlaetii*** Cherm. ex S.S.Hooper & J.Raynal
Pycreus flavescens f. *minor* (Gaudin) Soó = ***Pycreus flavescens*** (L.) P.Beauv. ex Rchb. subsp. ***flavescens***
Pycreus flavescens f. *monostachyus* (Sacc.) Soó = ***Pycreus flavescens*** (L.) P.Beauv. ex Rchb. subsp. ***flavescens***
Pycreus flavescens f. *setaceus* (Kük.) Soó = ***Pycreus flavescens*** (L.) P.Beauv. ex Rchb. subsp. ***flavescens***
Pycreus flavicomus (Michx.) C.D.Adams = ***Pycreus macrostachyos*** (Lam.) J.Raynal subsp. ***macrostachyos***
Pycreus flavidus var. *erectus* (C.B.Clarke) Karthik. = ***Pycreus flavidus*** (Retz.) T.Koyama
Pycreus flavidus var. *khasianus* (C.B.Clarke) Karthik. = ***Pycreus flavidus*** (Retz.) T.Koyama
Pycreus flavidus var. *nilagiricus* (Hochst. ex Steud.) Karthik. = ***Pycreus flavidus*** (Retz.) T.Koyama
Pycreus flavidus var. *strictus* (Lam.) Karthik. = ***Pycreus flavidus*** (Retz.) T.Koyama
Pycreus glaber (L.) Hayek = ***Cyperus glaber*** L.
Pycreus globosus Rchb. = ***Pycreus flavidus*** (Retz.) T.Koyama
Pycreus globosus var. *erectus* C.B.Clarke = ***Pycreus flavidus*** (Retz.) T.Koyama
Pycreus globosus var. *nilagiricus* Domin = ***Pycreus flavidus*** (Retz.) T.Koyama
Pycreus globosus var. *strictus* (Lam.) Domin = ***Pycreus flavidus*** (Retz.) T.Koyama
Pycreus globosus var. *typicus* Domin = ***Pycreus flavidus*** (Retz.) T.Koyama
Pycreus glomeratus (L.) Hayek = ***Cyperus glomeratus*** L.
Pycreus helvus (Liebm.) C.B.Clarke = ***Pycreus lanceolatus*** (Poir.) C.B.Clarke
Pycreus holosericeus Merr. = ***Pycreus polystachyos*** var. ***microdontus*** (Torr.) Govaerts
Pycreus humboldtianus (Schult.) Cufod. = ***Pycreus lanceolatus*** (Poir.) C.B.Clarke
Pycreus hyalinus (Vahl) Domin = ***Queenslandiella hyalina*** (Vahl) Ballard
Pycreus infirmus Nees = ***Pycreus lanceolatus*** (Poir.) C.B.Clarke
Pycreus intermedius (Rikli) C.B.Clarke = ***Pycreus flavescens*** subsp. ***intermedius*** (Rikli) Lye
Pycreus inundatus Nees = ***Pycreus puncticulatus*** (Vahl) Nees
Pycreus katangensis Cherm. = ***Pycreus fibrillosus*** (Kük.) Cherm.
Pycreus korshinskyi (Meinsh.) V.I.Krecz. = ***Pycreus sanguinolentus*** (Vahl) Nees subsp. ***sanguinolentus***
Pycreus laevigatus (L.) Nees = ***Cyperus laevigatus*** L.
Pycreus lagunetto (Steud.) C.B.Clarke = ***Pycreus bipartitus*** (Torr.) C.B.Clarke
Pycreus lanceolatus subsp. *ugandensis* Lye = ***Pycreus mortonii*** S.S.Hooper
Pycreus lanceus (Thunb.) Turrill = ***Pycreus nitidus*** (Lam.) J.Raynal
Pycreus lateralis (Forssk.) Nees = ***Cyperus laevigatus*** L. subsp. ***laevigatus***
Pycreus latispicatus (Boeck.) C.B.Clarke = ***Pycreus diaphanus*** (Schrad. ex Roem. & Schult.) S.S.Hooper & T.Koyama
Pycreus limosus (Maxim.) Schischk. = ***Cyperus limosus*** Maxim.
Pycreus longus (L.) Hayek = ***Cyperus longus*** L.
Pycreus macranthus var. *angustifolius* (Ridl.) C.B.Clarke ex Rendle = ***Cyperus laevigatus*** L. subsp. ***laevigatus***
Pycreus macrostachyos var. *tenuis* (Boeck.) Wickens = ***Pycreus macrostachyos*** (Lam.) J.Raynal subsp. ***macrostachyos***
Pycreus maximiliani Nees = ***Pycreus lanceolatus*** (Poir.) C.B.Clarke
Pycreus megapotamicus var. *jaeggii* (Boeck.) Guagl. = ***Pycreus megapotamicus*** (A.Dietr.) Nees
Pycreus melanostachyus (Kunth) C.B.Clarke = ***Pycreus niger*** (Ruiz & Pav.) Cufod.
Pycreus melanostachyus var. *variegatus* (Kunth) C.B.Clarke = ***Pycreus niger*** (Ruiz & Pav.) Cufod.
Pycreus micans (Kunth) Karthik. = ***Pycreus intactus*** (Vahl) J.Raynal
Pycreus minimus C.B.Clarke = ***Pycreus hildebrandtii*** C.B.Clarke
Pycreus monocephalus C.B.Clarke = ***Pycreus fluminalis*** (Ridl.) Rendle
Pycreus monti (L.f.) P.Beauv. ex Rchb. = ***Cyperus serotinus*** Rottb.
Pycreus mucronatus (Rottb.) Nees = ***Cyperus laevigatus*** subsp. ***distachyos*** (All.) Maire & Weiller
Pycreus mundtii var. *gracilis* Cherm. = ***Pycreus mundtii*** Nees
Pycreus mundtii var. *perrieri* Cherm. = ***Pycreus atropurpureus*** C.B.Clarke
Pycreus mundtii var. *uniceps* (C.B.Clarke) Napper = ***Pycreus mundtii*** Nees
Pycreus nematodes C.B.Clarke = ***Pycreus megapotamicus*** (A.Dietr.) Nees
Pycreus niger subsp. *elegantulus* (Steud.) Lye = ***Pycreus elegantulus*** (Steud.) C.B.Clarke
Pycreus nigrescens C.B.Clarke = ***Pycreus nigricans*** (Steud.) C.B.Clarke
Pycreus nigricans var. *simulans* (Cherm.) Cherm. = ***Pycreus nigricans*** (Steud.) C.B.Clarke
Pycreus nilagiricus (Hochst. ex Steud.) Schischk. = ***Pycreus flavidus*** (Retz.) T.Koyama
Pycreus nitens (Retz.) Nees = ***Pycreus pumilus*** (L.) Nees
Pycreus nitens f. *gracilis* E.G.Camus = ***Pycreus pumilus*** (L.) Nees
Pycreus nyasensis C.B.Clarke = ***Pycreus nigricans*** (Steud.) C.B.Clarke
Pycreus odoratus Urb. = ***Pycreus polystachyos*** (Rottb.) P.Beauv. subsp. ***polystachyos***
Pycreus odoratus var. *hahnianus* (Boeck.) Stehlé = ***Pycreus polystachyos*** (Rottb.) P.Beauv. subsp. ***polystachyos***
Pycreus olfersianus (Kunth) Nees = ***Pycreus lanceolatus*** (Poir.) C.B.Clarke
Pycreus olfersianus Urb. = ***Pycreus lanceolatus*** (Poir.) C.B.Clarke
Pycreus paniculatus (Rottb.) Nees = ***Pycreus polystachyos*** (Rottb.) P.Beauv. subsp. ***polystachyos***
Pycreus pannonicus (Jacq.) P.Beauv. ex Rchb. = ***Cyperus pannonicus*** Jacq.
Pycreus paraensis Nees = ***Pycreus flavescens*** (L.) P.Beauv. ex Rchb. subsp. ***flavescens***
Pycreus patens (Vahl) Cherm. = ***Pycreus pumilus*** (L.) Nees

Pycreus permutatus (Boeck.) Napper = ***Pycreus macranthus*** (Boeck.) C.B.Clarke
Pycreus piceus (Liebm.) C.B.Clarke = ***Pycreus flavescens*** (L.) P.Beauv. ex Rchb. subsp. ***flavescens***
Pycreus polystachyos var. *circinatus* (Ridl.) H.B. Naithani & S.Biswas = ***Pycreus polystachyos*** (Rottb.) P.Beauv. subsp. ***polystachyos***
Pycreus polystachyos var. *hahnianus* (Boeck.) C.B.Clarke = ***Pycreus polystachyos*** (Rottb.) P.Beauv. subsp. ***polystachyos***
Pycreus polystachyos subsp. *holosericeus* (Link) T.Koyama = ***Pycreus polystachyos*** var. ***microdontus*** (Torr.) Govaerts
Pycreus polystachyos var. *laxiflorus* (Benth.) C.B.Clarke = ***Pycreus polystachyos*** var. ***microdontus*** (Torr.) Govaerts
Pycreus polystachyos var. *paniculatus* (Rottb.) Donn.Sm. = ***Pycreus polystachyos*** (Rottb.) P.Beauv. subsp. ***polystachyos***
Pycreus polystachyos var. *typicus* Domin = ***Pycreus polystachyos*** (Rottb.) P.Beauv.
Pycreus potosiensis C.B.Clarke = ***Pycreus flavescens*** (L.) P.Beauv. ex Rchb. subsp. ***flavescens***
Pycreus propinquus Nees = ***Pycreus lanceolatus*** (Poir.) C.B.Clarke
Pycreus pulvinatus (Nees & Meyen) Nees = ***Pycreus pumilus*** (L.) Nees
Pycreus pumilus subsp. *membranaceus* (Vahl) T.Koyama = ***Pycreus membranaceus*** (Vahl) Govind.
Pycreus pumilus var. *membranaceus* (Vahl) Karthik. = ***Pycreus membranaceus*** (Vahl) Govind.
Pycreus pumilus var. *punctatus* Domin = ***Pycreus pumilus*** (L.) Nees
Pycreus pumilus var. *substerilis* E.G.Camus = ***Queenslandiella hyalina*** (Vahl) Ballard
Pycreus punctatus Govind. = ***Pycreus pumilus*** (L.) Nees
Pycreus puncticulatus var. *quinquagintiflorus* (C.B. Clarke) C.B.Clarke = ***Pycreus puncticulatus*** (Vahl) Nees
Pycreus pygmaeus (Rottb.) Nees = ***Cyperus michelianus*** subsp. ***pygmaeus*** (Rottb.) Asch. & Graebn.
Pycreus reductus Cherm. = ***Pycreus flavidus*** (Retz.) T.Koyama
Pycreus rehmannianus C.B.Clarke = ***Pycreus flavescens*** var. ***rehmannianus*** (C.B.Clarke) Govaerts
Pycreus retusus (Nees ex Steud.) Nees ex C.B.Clarke = ***Pycreus macrostachyos*** subsp. ***tremulus*** (Poir.) Lye
Pycreus rivularis (Kunth) Palla = ***Pycreus bipartitus*** (Torr.) C.B.Clarke
Pycreus rotundus (L.) Hayek = ***Cyperus rotundus*** L.
Pycreus rubromarginatus E.G.Camus = ***Pycreus sanguinolentus*** (Vahl) Nees subsp. ***sanguinolentus***
Pycreus sabulosus Nees = ***Pycreus macrostachyos*** (Lam.) J.Raynal subsp. ***macrostachyos***
Pycreus sanguinolentus subsp. *cyrtostachys* (Miq.) Karthik. = ***Pycreus sanguinolentus*** var. ***cyrtostachys*** (Miq.) L.K.Dai
Pycreus sanguinolentus var. *micronux* (C.B.Clarke) Karthik. = ***Pycreus sanguinolentus*** (Vahl) Nees subsp. ***sanguinolentus***
Pycreus sanguinolentus subsp. *micronux* (C.B.Clarke) H.B.Naithani & S.Biswas = ***Pycreus sanguinolentus*** (Vahl) Nees subsp. ***sanguinolentus***
Pycreus sanguinolentus subsp. *sacer* (Korlahalli) Karthik. = ***Pycreus sanguinolentus*** (Vahl) Nees subsp. ***sanguinolentus***
Pycreus sanguinolentus var. *typicus* Domin = ***Pycreus sanguinolentus*** (Vahl) Nees subsp. ***sanguinolentus***
Pycreus sanguinolentus var. *uniceps* C.B.Clarke = ***Pycreus mundtii*** Nees
Pycreus scaettae Cherm. = ***Pycreus fibrillosus*** (Kük.) Cherm.
Pycreus scaettae var. *katangensis* Cherm. = ***Pycreus fibrillosus*** (Kük.) Cherm.
Pycreus segmentatus C.B.Clarke = ***Pycreus macranthus*** (Boeck.) C.B.Clarke
Pycreus serotinus (Rottb.) Hayek = ***Cyperus serotinus*** Rottb.
Pycreus setaceus C.B.Clarke = ***Pycreus capillifolius*** (A.Rich.) C.B.Clarke
Pycreus setiformis (Korsh.) Nakai = ***Pycreus diaphanus*** (Schrad. ex Roem. & Schult.) S.S.Hooper & T.Koyama
Pycreus simulans Cherm. = ***Pycreus nigricans*** (Steud.) C.B.Clarke
Pycreus solidifolius Cherm. = ***Pycreus longivaginans*** (Kük.) Cherm.
Pycreus squarrosus (L.) Nees = ***Cyperus squarrosus*** L.
Pycreus strictus (Lam.) Alston = ***Pycreus flavidus*** (Retz.) T.Koyama
Pycreus substellatus E.G.Camus = ***Pycreus sulcinux*** (C.B.Clarke) C.B.Clarke
Pycreus substramineus (Kük.) N.K.Khoi = ***Pycreus stramineus*** C.B.Clarke
Pycreus tremulus (Poir.) C.B.Clarke = ***Pycreus macrostachyos*** subsp. ***tremulus*** (Poir.) Lye
Pycreus tristachyus (Boeck.) C.B.Clarke = ***Pycreus flavescens*** (L.) P.Beauv. ex Rchb. subsp. ***flavescens***
Pycreus umbrosiformis Cherm. = ***Pycreus unioloides*** (R.Br.) Urb.
Pycreus umbrosus Nees = ***Pycreus nitidus*** (Lam.) J.Raynal
Pycreus unioloides var. *angulatus* (Nees) Domin = ***Pycreus unioloides*** (R.Br.) Urb.
Pycreus unioloides var. *compositus* (T.Koyama) N.K.Khoi = ***Pycreus unioloides*** (R.Br.) Urb.
Pycreus unioloides var. *insularis* Domin = ***Pycreus unioloides*** (R.Br.) Urb.
Pycreus unioloides var. *typicus* Domin = ***Pycreus unioloides*** (R.Br.) Urb.
Pycreus unioloides var. *wightii* (Nees ex C.B.Clarke) Karthik. = ***Pycreus unioloides*** (R.Br.) Urb.
Pycreus vanderystii Cherm. = ***Pycreus fibrillosus*** (Kük.) Cherm.
Pycreus vavavatensis Cherm. = ***Pycreus nigricans*** (Steud.) C.B.Clarke
Pycreus vicinus Cherm. = ***Pycreus flavescens*** subsp. ***intermedius*** (Rikli) Lye
Pycreus woronowii Palla = ***Pycreus macrostachyos*** subsp. ***tremulus*** (Poir.) Lye
Pycreus zonatissimus Cherm. = ***Pycreus zonatus*** Cherm.

Queenslandiella

Queenslandiella Domin, Biblioth. Bot. 85: 415 (1915).
Trop. Old World. 24 SOM 25 KEN TAN 26 MOZ ZAM 29 MAU MDG 40 IND LDV SRL 41 VIE 42 BOR CKI JAW LSI MOL 50 QLD (78) fla.
1 Species
Mariscopsis Cherm., Bull. Mus. Natl. Hist. Nat. 25: 60 (1919).

Queenslandiella hyalina (Vahl) Ballard, Hooker's Icon. Pl. 33: t. 3208 (1933).
S. Somalia to S. Trop. Africa and Queensland. 24 SOM 25 KEN TAN 26 MOZ ZAM 29 MAU MDG 40 IND LDV SRL 41 VIE 42 BOR CKI JAW LSI MOL 50 QLD (78) fla. Ther.

*Cyperus hyalinus Vahl, Enum. Pl. 2: 320 (1805). *Pycreus hyalinus* (Vahl) Domin, Biblioth. Bot. 85: 417 (1916). *Mariscopsis hyalinus* (Vahl) Ballard, Bull. Misc. Inform. Kew 1932: 458 (1932). *Kyllinga hyalina* (Vahl) T.Koyama, J. Jap. Bot. 51: 317 (1976).

Pycreus pumilus var. *substerilis* E.G.Camus, Notul. Syst. (Paris) 1: 241 (1910). *Cyperus hyalinus* var. *substerilis* (E.G.Camus) Kük. in H.G.A.Engler (ed.), Pflanzenr., IV, 20(101): 499 (1936).

Queenslandiella mira Domin, Biblioth. Bot. 85: 415 (1915).

Cyperus suaveolens Boivin ex Cherm., Bull. Mus. Natl. Hist. Nat. 25: 61 (1919), nom. inval.

Mariscopsis suaveolens Cherm., Bull. Mus. Natl. Hist. Nat. 25: 61 (1919).

***Synonyms*:**

Queenslandiella mira Domin = ***Queenslandiella hyalina*** (Vahl) Ballard

×*Raymondiella*

×*Raymondiella* Boivin = *Dulichium* ×*Rhynchospora*

×*Raymondiella* ×*weatherbiana* (Fernald) Boivin =

***Unplaced Names*:**

×*Raymondiella* Boivin, Taxon 22: 278 (1973), not accordance with Art. H. 7. = *Dulichium* ×*Rhynchospora*

Raynalia

Raynalia Soják = ***Alinula*** J.Raynal

Raynalia lipocarphioides (Kük.) Soják = ***Alinula lipocarphioides*** (Kük.) J.Raynal

Reedia

Reedia F.Muell., Fragm. 1: 239 (1859).
SW. Australia. 50 WAU.
1 Species

Reedia spathacea F.Muell., Fragm. 1: 240 (1859).
SW. Western Australia. 50 WAU. Hemicr. or rhizome geophyte.

Reigera

Reigera Opiz = ***Bolboschoenus*** (Asch.) Palla

Reigera maritima (L.) Opiz = ***Bolboschoenus maritimus*** (L.) Palla

Remirea

Remirea Aubl., Hist. Pl. Guiane 1: 44 (1775).
Trop. 22 BEN GHA IVO LBR NGA SIE TOG 23 EQG GAB GGI 25 TAN 29 MDG SEY 38 NNS TAI 40 IND SRL 41 AND MYA NCB THA VIE 42 BOR JAW LSI MLY MOL PHI SUL SUM 43 BIS NWG SOL 50 NTA QLD 78 FLA 79 MXG MXT 80 BLZ COS GUA HON NIC PAN 81 CAY DOM JAM LEE PUE TRT 82 FRG SUR.
1 Species

Remirea maritima Aubl., Hist. Pl. Guiane 1: 45 (1775). *Duval-jouvea maritima* (Aubl.) Palla, Denkschr. Kaiserl. Akad. Wiss., Wien. Math.-Naturwiss. Kl. 89: 500 (1913).
Trop. 22 BEN GHA IVO LBR NGA SIE TOG 23 EQG GAB GGI 25 TAN 29 MDG SEY 38 NNS TAI 40 IND SRL 41 AND MYA NCB THA VIE 42 BOR JAW LSI MLY MOL PHI SUL SUM 43 BIS NWG SOL 50 NTA QLD 78 FLA 79 MXG MXT 80 BLZ COS GUA HON NIC PAN 81 CAY DOM JAM LEE PUE TRT 82 FRG SUR. Hemicr. or rhizome geophyte.

Remirea pedunculata R.Br., Prodr.: 236 (1810). *Remirea maritima* var. *pedunculata* (R.Br.) Benth., Fl. Austral. 7: 347 (1878). *Cyperus pedunculatus* (R.Br.) J.Kern, Acta Bot. Neerl. 7: 798 (1958). *Mariscus pedunculatus* (R.Br.) T.Koyama, Gard. Bull. Singapore 30: 159 (1977).

Remirea wightiana Nees in R.Wight, Contr. Bot. India: 92 (1834).

Cyperus kegelianus Steud., Syn. Pl. Glumac. 2: 29 (1854).

Mariscus capitatus Steud., Syn. Pl. Glumac. 2: 60 (1854).

Mariscus pungens Steud., Syn. Pl. Glumac. 2: 60 (1854).

Remirea rigidissima Steud., Syn. Pl. Glumac. 2: 317 (1855).

Lipocarpha foliosa Miq., Fl. Ned. Ind. 3: 332 (1856).

Remirea distichophylla Boeck., Flora 41: 410 (1858).

***Synonyms*:**

Remirea diffusa Sieber ex Steud. = (Poaceae)

Remirea distichophylla Boeck. = ***Remirea maritima*** Aubl.

Remirea maritima var. *pedunculata* (R.Br.) Benth. = ***Remirea maritima*** Aubl.

Remirea pedunculata R.Br. = ***Remirea maritima*** Aubl.

Remirea rigidissima Steud. = ***Remirea maritima*** Aubl.

Remirea wightiana Nees = ***Remirea maritima*** Aubl.

Rhaptocalymma

Rhaptocalymma Borrer = ***Carex*** L.

Rhaptocalymna

Rhaptocalymna incurva (Lightf.) Fedde & J.Schust. = ***Carex maritima*** Gunnerus

Rhynchocladium

Rhynchocladium T.Koyama, Mem. New York Bot. Gard. 23: 86 (1972).
S. Venezuela to Guyana. 82 GUY VEN.
1 Species

Rhynchocladium steyermarkii (T.Koyama) T.Koyama, Mem. New York Bot. Gard. 23: 88 (1972).
S. Venezuela to Guyana. 82 GUY VEN. Hemicr. or rhizome geophyte.
*Cladium steyermarkii T.Koyama, Mem. New York Bot. Gard. 18(2): 28 (1969).

Rhynchopera

Rhynchopera Börner = ***Carex*** L.

Rhynchopera paniculata (L.) Fedde & J.Schust. = ***Carex paniculata*** L.

Rhynchospora

Rhynchospora Vahl, Enum. Pl. 2: 229 (1805), nom. & orth. cons.
Cosmopolitan. 10 DEN FIN GRB IRE NOR SWE 11 AUT BGM CZE GER HUN† NET POL SWI 12 FRA POR SPA 13 ITA ROM YUG 14 BLR BLT RUC RUE RUN UKR 20 ALG MOR TUN 22 BEN BKN GAM GHA GNB GUI IVO LBR MLI NGA NGR SEN SIE TOG 23 BUR CAF CMN EQG GAB GGI RWA ZAI 24 CHA ETH SUD 25 KEN

MOZ TAN UGA 26 ANG MLW MOZ ZAM ZIM 27 BOT CPP LES NAM NAT SWZ TVL 29 MAU MDG REU 30 IRK 31 KAM KUR 32 KAZ 33 TCS 34 TUR 36 CHC CHH CHS 38 JAP KOR NNS OGA TAI 40 ASS BAN EHM IND NEP SRL 41 AND CBD LAO MYA NCB THA VIE 42 BOR JAW LSI MLY MOL PHI SUL SUM 43 BIS NWG 50 NSW NTA QLD VIC WAU 60 FIJ SAM TON VAN 61 MRQ SCI TUB 62 CRL MRN 63 HAW 70 ABT ASK MAN NWT SAS 71 ABT BRC MAN SAS 72 LAB NBR NFL NSC ONT PEI QUE 73 IDA ORE WAS 74 ILL IOW KAN MIN MSO NDA OKL SDA WIS 75 CNT INI MAI MAS MIC NWH NWJ NWY OHI PEN RHO VER WVA 76 CAL 77 TEX 78 ALA ARK DEL FLA GEO KTY LOU MRY MSI NCA SCA TEN VRG WDC 79 MXC MXE MXG MXN MXS MXT 80 BLZ COS CPI ELS GUA HON NIC PAN 81 BAH BER CAY CUB DOM HAI JAM LEE PUE TRT VNA WIN 82 FRG GUY SUR VEN 83 BOL CLM ECU GAL PER 84 AGE BZC BZE BZL BZN BZS PAR 85 AGE AGS AGW CLC CLN PAR URU.

351 Species

Phaeocephalum Ehrh., Beitr. Naturk. 4: 146 (1789).
Dichromena Michx., Fl. Bor.-Amer. 1: 37 (1803).
Triodon Rich. ex Pers., Syn. Pl. 1: 60 (1805).
Chaetospora Kunth in F.W.H.von Humboldt, A.J.A.Bonpland & C.S.Kunth, Nov. Gen. Sp. 1: 229 (1816), nom. illeg.
Spermodon P.Beauv. ex Lestib., Essai Cypér.: 27 (1819).
Zosterospermum P.Beauv. in T.G.Lestiboudois, Essai Cypér.: 28 (1819).
Dichroma Ham., Prodr. Pl. Ind. Occid.: 15 (1825), nom. illeg.
Pterotheca C.Presl, Symb. Bot. 1: 55 (1831).
Calyptrostylis Nees, Linnaea 9: 295 (1834).
Cephaloschoenus Nees, Linnaea 9: 295 (1834).
Ceratoschoenus Nees, Linnaea 9: 296 (1834).
Diplochaete Nees, Linnaea 9: 296 (1834).
Echinoschoenus Nees & Meyen in F.J.F.Meyen, Observ. Bot. 1: 116 (1834).
Haloschoenus Nees, Linnaea 9: 296 (1834).
Haplostylis Nees, Linnaea 9: 295 (1834).
Mitrospora Nees, Linnaea 9: 295 (1834).
Morisia Nees, Edinburgh New Philos. J., Oct. 1834: 265 (1834).
Psilocarya Torr., Ann. Lyceum Nat. Hist. New York 3: 359 (1836).
Eriochaeta Torr. ex Steud., Nomencl. Bot., ed. 2, 1: 586 (1840).
Leptoschoenus Nees, J. Bot. (Hooker) 2: 393 (1840).
Michrochaeta Rchb., Deut. Bot. Herb.-Buch: 192 (1841).
Asteroschoenus Nees in C.F.P.von Martius & auct. suc. (eds.), Fl. Bras. 2(1): 124 (1842).
Ephippiorhynchium Nees in C.F.P.von Martius & auct. suc. (eds.), Fl. Bras. 2(1): 134 (1842).
Hygrocharis Nees in C.F.P.von Martius & auct. suc. (eds.), Fl. Bras. 2(1): 147 (1842).
Pachymitra Nees in C.F.P.von Martius & auct. suc. (eds.), Fl. Bras. 2(1): 115 (1842).
Ptilochaeta Nees in C.F.P.von Martius & auct. suc. (eds.), Fl. Bras. 2(1): 147 (1842).
Sphaeroschoenus Nees, Nov. Actorum Acad. Caes. Leop.-Carol. Nat. Cur. 19(Suppl. 1): 97 (1843).
Astroschoenus Lindl., Veg. Kingd., ed. 2.: 119 (1847).
Kleistrocalyx Steud., Flora 33: 229 (1850).
Calyptrolepis Steud., Syn. Pl. Glumac. 2: 151 (1855).
Cleistocalyx Steud., Syn. Pl. Glumac. 2: 151 (1855).
Ptilosciadium Steud., Syn. Pl. Glumac. 2: 149 (1855).
Trichochaeta Steud., Syn. Pl. Glumac. 2: 155 (1855).
Pterochaete Arn. ex Boeck., Linnaea 37: 600 (1873).
Lonchostylis Torr., Ann. Lyceum Nat. Hist. New York 3: 370 (1886).
Syntrinema H.Pfeiff., Repert. Spec. Nov. Regni Veg. 21: 238 (1925).
Micropapyrus Suess., Bot. Jahrb. Syst. 73: 115 (1943).

Rhynchospora aberrans C.B.Clarke, Kew Bull., Addit. Ser. 8: 35 (1908). *Dichromena aberrans* (C.B.Clarke) J.F.Macbr., Publ. Field Mus. Nat. Hist., Bot. Ser. 11: 6 (1931).
Peru, E. Brazil. 83 PER 84 BZE. Hemicr. or rhizome geophyte.

Rhynchospora affinis W.Fitzg., J. Proc. Roy. Soc. Western Australia 3: 124 (1918).
N. Australia. 50 NSW QLD WAU. Hemicr. or rhizome geophyte.

Rhynchospora agostiniana T.Koyama, Mem. New York Bot. Gard. 23: 25 (1972).
Venezuela (Amazonas). 82 VEN. Hemicr. or rhizome geophyte.

Rhynchospora alba (L.) Vahl, Enum. Pl. 2: 236 (1805).
Temp. Northern Hemisphere. 10 DEN FIN GRB IRE NOR SWE 11 AUT BGM CZE GER HUN† NET POL SWI 12 FRA POR SPA 13 ITA ROM YUG 14 BLR BLT RUC RUE RUN UKR 30 IRK 31 KAM KUR 32 KAZ 33 TCS 34 TUR 38 JAP KOR TAI 70 ASK NWT 71 ABT BRC MAN SAS 72 NBR NFL NSC ONT PEI QUE 73 IDA ORE WAS 74 ILL MIN WIS 75 CNT INI MAI MAS MIC NWH NWJ NWY OHI PEN RHO VER WVA 76 CAL 78 DEL GEO MRY NCA SCA TEN VRG 81 PUE. Hemicr.
**Schoenus albus* L., Sp. Pl.: 44 (1753). *Mariscus albus* (L.) Gilib., Excerc. Phyt. 2: 512 (1792). *Scirpus albus* (L.) Salisb., Prodr. Stirp. Chap. Allerton: 31 (1796). *Triodon albus* (L.) Farw., Rep. (Annual) Michigan Acad. Sci. 19: 253 (1917). *Phaeocephalum album* (L.) House, Amer. Midl. Naturalist 6: 201 (1920). *Dichromena alba* (L.) J.F.Macbr., Publ. Field Columbian Mus., Bot. Ser. 4: 166 (1929).
Rhynchospora alba var. *kiusiana* Makino, Bot. Mag. (Tokyo) 17: 191 (1903).
Phaeocephalum album var. *marcum* Farw., Rep. (Annual) Michigan Acad. Sci. 21: 361 (1920).
Rhynchospora luquillensis Britton, Bull. Torrey Bot. Club 50: 56 (1923).
Rhynchospora alba f. *laeviseta* Gale, Rhodora 46: 123 (1944).

Rhynchospora albescens (Miq.) Kük., Bot. Jahrb. Syst. 75: 304 (1951).
Costa Rica to S. Trop. America. 80 COS 82 FRG GUY SUR VEN 83 CLM 84 BZN. Ther. or hemicr.
**Isolepis albescens* Miq., Linnaea 19: 226 (1847). *Bulbostylis albescens* (Miq.) Beetle, Amer. Midl. Naturalist 34: 724 (1945).
Rhynchospora monostachya Boeck., Allg. Bot. Z. Syst. 2: 78 (1896), nom. illeg. *Dichromena monostachya* C.B.Clarke, Kew Bull., Addit. Ser. 8: 32 (1908).

Rhynchospora albiceps Kunth, Enum. Pl. 2: 289 (1837). *Haplostylis albiceps* (Kunth) Nees in C.F.P.von Martius & auct. suc. (eds.), Fl. Bras. 2(1): 129 (1842).
Brazil to Argentina (Misiones). 84 BZC BZL BZS 85 AGE PAR. Hemicr. or rhizome geophyte.

Rhynchospora auriculata Boeck., Vidensk. Meddel. Naturhist. Foren. Kjøbenhavn 1869: 144 (1869).
Dichromena foliosa Hochst. ex Boeck., Linnaea 37: 534 (1872).
Dichromena canescens Maury ex Micheli, Mém. Soc. Phys. Genève 31(1): 143 (1890). *Rhynchospora canescens* (Maury ex Micheli) H.Pfeiff., Repert. Spec. Nov. Regni Veg. 38: 96 (1936).
Rhynchospora leucantha Boeck., Beitr. Cyper. 2: 22 (1890).

Rhynchospora albida (Nees) Boeck., Linnaea 37: 561 (1873).
Colombia to Guyana. 82 GUY VEN 83 CLM 84 BZN. Hemicr. or rhizome geophyte.
**Haloschoenus albidus* Nees in C.F.P.von Martius & auct. suc. (eds.), Fl. Bras. 2(1): 122 (1842). *Dichromena albida* (Nees) Steud., Syn. Pl. Glumac. 2: 136 (1855).

Rhynchospora albomarginata Kük., Bot. Jahrb. Syst. 56(125): 16 (1921).
N. South America to N. Brazil. 82 FRG GUY VEN 84 BZN. Hemicr. or rhizome geophyte.
Rhynchospora baileyi Britton, Gentes Herb. 2: 200 (1930).

Rhynchospora albotuberculata Kük., Boissiera 7: 101 (1943).
Brazil (Goiás). 84 BZS. Hemicr. or rhizome geophyte.

Rhynchospora almensis D.A.Simpson, Kew Bull. 50: 151 (1995).
Brazil (Bahia). 84 BZE. Hemicr. or rhizome geophyte.

Rhynchospora amazonica Poepp. ex Kunth, Enum. Pl. 2: 292 (1837). *Ephippiorhynchium amazonicum* (Poepp. ex Kunth) Nees ex Boeck., Linnaea 37: 625 (1873). *Dichromena amazonica* (Poepp. ex Kunth) J.F.Macbr., Publ. Field Mus. Nat. Hist., Bot. Ser. 11: 6 (1931). *Rhynchospora corniculata* subsp. *amazonica* (Poepp. ex Kunth) H.Pfeiff., Repert. Spec. Nov. Regni Veg. 49: 76 (1940).
S. Trop. America. 82 FRG GUY SUR VEN 83 BOL CLM ECU PER 84 BZN. Hemicr. or rhizome geophyte.

subsp. ***amazonica***
S. Trop. America. 82 FRG GUY SUR VEN 83 BOL CLM ECU PER 84 BZN. Hemicr. or rhizome geophyte.

subsp. ***guianensis*** (Kük.) T.Koyama, Mem. New York Bot. Gard. 23: 28 (1972).
S. Venezuela to Guyana and N. Brazil. 82 GUY VEN 83 ECU 84 BZN. Hemicr. or rhizome geophyte.
Calyptrostylis longirostris Nees, J. Bot. (Hooker) 2: 394 (1840). *Ephippiorhynchium longirostre* (Nees) Nees in C.F.P.von Martius & auct. suc. (eds.), Fl. Bras. 2(1): 136 (1842). *Cephaloschoenus longirostris* (Nees) Nees, Nov. Actorum Acad. Caes. Leop.-Carol. Nat. Cur. 19(Suppl. 1): 101 (1843). *Rhynchospora longirostris* (Nees) Steud., Syn. Pl. Glumac. 2: 145 (1855), nom. illeg.

Rhynchospora andresii Gómez-Laur., Phytologia 51: 477 (1982).
Costa Rica to Panama. 80 COS PAN. Hemicr. or rhizome geophyte.

Rhynchospora angolensis Turrill, Bull. Misc. Inform. Kew 1914: 136 (1914).
Trop. Africa, Madagascar. 22 BEN 23 CAF 25 TAN UGA 26 ANG 29 MDG. Hemicr. or rhizome geophyte.
Rhynchospora africana Cherm., Arch. Bot. Bull. Mens. 4(7): 44 (1931).

Rhynchospora angosturensis W.W.Thomas, Mem. New York Bot. Gard. 37: 75 (1984).
Mexico (San Luis Potosí). 79 MXE. Hemicr.

Rhynchospora angustifolia Palla, Denkschr. Kaiserl. Akad. Wiss., Wien. Math.-Naturwiss. Kl. 79: 193 (1908).
SE. Brazil. 84 BZL. Hemicr. or rhizome geophyte.

Rhynchospora angustipaniculata M.T.Strong, Novon 11: 261 (2001).
Guyana (Pakaraima Mts.). 82 GUY. Hemicr. or rhizome geophyte.

Rhynchospora arenicola Uittien, Recueil Trav. Bot. Néerl. 22: 335 (1925). *Rhynchospora pilosa* subsp. *arenicola* (Uittien) T.Koyama, Mem. New York Bot. Gard. 23: 61 (1972).
N. South America to N. Brazil. 82 FRG GUY SUR VEN 84 BZN. Hemicr. or rhizome geophyte.

Rhynchospora argentea Standl., Contr. U. S. Natl. Herb. 18: 87 (1916).
E. Panama to NW. Colombia. 80 PAN 83 CLM. Hemicr. or rhizome geophyte.

Rhynchospora aripoensis Britton, Bull. Torrey Bot. Club 48: 327 (1921 publ. 1922).
N. Trinidad. 81 TRT. Hemicr. or rhizome geophyte.

Rhynchospora aristata Boeck., Flora 40: 36 (1857).
Mexico to Trop. America. 79 MXC MXE MXG MXN MXS MXT 80 COS GUA HON NIC PAN 81 CUB DOM HAI JAM 82 GUY? VEN 83 BOL CLM ECU PER. Hemicr. or rhizome geophyte.

var. ***aristata***
Mexico to Trop. America. 79 MXC MXE MXG MXN MXS MXT 80 COS GUA HON NIC 81 DOM HAI JAM 82 GUY 83 BOL CLM ECU PER. Hemicr. or rhizome geophyte.
Chaetospora aurea Cham. & Schltdl., Linnaea 6: 28 (1831), nom. illeg.
Rhynchospora fuertesii Urb., Repert. Spec. Nov. Regni Veg. 15: 98 (1917). *Rhynchospora aristata* var. *fuertesii* (Urb.) Kük., Repert. Spec. Nov. Regni Veg. 23: 205 (1926).
Rhynchospora aristata var. *latilaminata* Kük., Repert. Spec. Nov. Regni Veg. 26: 253 (1929).

var. ***brevifoliata*** Kük., Repert. Spec. Nov. Regni Veg. 23: 205 (1926).
E. Cuba. 81 CUB. Hemicr. or rhizome geophyte.

var. ***moritziana*** (Boeck.) Kük., Bot. Jahrb. Syst. 74: 401 (1949).
Venezuela to Ecuador. 82 VEN 83 CLM ECU. Hemicr. or rhizome geophyte.
**Rhynchospora moritziana* Boeck., Linnaea 37: 640 (1873).

var. ***suberecta*** Kük., Repert. Spec. Nov. Regni Veg. 23: 204 (1926).
SW. Mexico (to S. Durango). 79 MXC MXE MXS 83 ECU. Hemicr. or rhizome geophyte.

Rhynchospora armerioides J.Presl & C.Presl, Reliq. Haenk. 1: 197 (1828).
Costa Rica to Brazil. 80 COS PAN 82 GUY SUR VEN

83 BOL CLM 84 BZC BZE BZL BZN. Hemicr.
Haplostylis armeriifolia Nees in C.F.P.von Martius & auct. suc. (eds.), Fl. Bras. 2(1): 126 (1842).

Rhynchospora asperula (Nees) Steud., Syn. Pl. Glumac. 2: 144 (1855).
Brazil to NE. Argentina. 83 BOL 84 BZC BZE BZL BZS 85 AGE PAR URU. Hemicr. or rhizome geophyte.
**Calyptrostylis asperula* Nees in C.F.P.von Martius & auct. suc. (eds.), Fl. Bras. 2(1): 139 (1842). *Rhynchospora corymbosa* var. *asperula* (Nees) Kük., Bot. Jahrb. Syst. 74: 416 (1949).

Rhynchospora ayangannensis M.T.Strong, Novon 11: 263 (2001).
Guyana (Pakaraima Mts.). 82 GUY. Hemicr. or rhizome geophyte.

Rhynchospora bakhuisensis M.T.Strong, Novon 11: 266 (2001).
Suriname (Bakhuis Mts.). 82 SUR. Hemicr. or rhizome geophyte.

Rhynchospora baldwinii A.Gray, Ann. Lyceum Nat. Hist. New York 3: 210 (1835). *Phaeocephalum baldwinii* (A.Gray) House, Amer. Midl. Naturalist 6: 201 (1920).
SE. U.S.A. 78 ALA FLA GEO LOU MSI NCA SCA. Ther. or hemicr.
Rhynchospora glauca Baldwin ex A.Gray, Ann. Lyceum Nat. Hist. New York 3: 210 (1835), nom. illeg.

Rhynchospora barbata (Vahl) Kunth, Enum. Pl. 2: 290 (1837).
Trop. America. 79 MXG MXT 80 BLZ COS GUA HON NIC PAN 81 CUB DOM TRT 82 FRG GUY SUR VEN 83 BOL CLM 84 BZC BZE BZN. Hemicr.
**Schoenus barbatus* Vahl, Eclog. Amer. 2: 4 (1798). *Haplostylis barbata* (Vahl) Nees, Linnaea 9: 295 (1834). *Isoschoenus barbatus* (Vahl) Nees in J.G.C.Lehmann, Pl. Preiss. 2: 80 (1846). *Dichromena barbata* (Vahl) J.F.Macbr., Publ. Field Mus. Nat. Hist., Bot. Ser. 11: 5 (1931).
Schoenus globosus Rudge, Pl. Guian.: 14 (1805). *Rhynchospora globosa* (Rudge) Desv. ex Ham., Prodr. Pl. Ind. Occid.: 15 (1825), nom. illeg. *Rhynchospora hamiltonii* Kunth, Enum. Pl. 2: 291 (1837).
Chaetospora pterocarpa Kunth in F.W.H.von Humboldt, A.J.A.Bonpland & C.S.Kunth, Nov. Gen. Sp. 1: 230 (1816). *Rhynchospora pterocarpa* (Kunth) Roem. & Schult., Syst. Veg. 2: 89 (1817). *Schoenus pterocarpus* (Kunth) Poir. in J.B.A.M.de Lamarck, Encycl., Suppl. 5: 617 (1817). *Haplostylis humboldtii* Nees, Linnaea 9: 295 (1834), nom. illeg. *Haplostylis barbata* var. *humboldtii* (Nees) Nees in C.F.P.von Martius & auct. suc. (eds.), Fl. Bras. 2(1): 128 (1842). *Rhynchospora barbata* var. *glabra* Maury, J. Bot. (Morot) 3: 209 (1889).
Chaetospora pterosperma Kunth in F.W.H.von Humboldt, A.J.A.Bonpland & C.S.Kunth, Nov. Gen. Sp. 1: 230 (1816).
Schoenus hemisphaericus Humb. ex Link, Jahrb. Gewächsk. 1(3): 75 (1820).
Eleocharis macrostylis Spreng., Neue Entdeck. Pflanzenk. 3: 11 (1822). *Scirpus macrostylis* (Spreng.) Spreng., Syst. Veg. 1: 208 (1824).
Pterotheca vahlii C.Presl, Symb. Bot. 1: 55 (1831).
Rhynchospora macrostylis A.Dietr., Sp. Pl. 2: 74 (1832).
Diplochaete guianensis Nees, Linnaea 9: 296 (1834).
Haplostylis bahiensis Nees, Linnaea 9: 295 (1834), nom. inval.
Chaetospora pterosperma Nees in C.F.P.von Martius & auct. suc. (eds.), Fl. Bras. 2(1): 128 (1842), orth. var.
Rhynchospora bahiensis Steud., Syn. Pl. Glumac. 2: 143 (1855). *Rhynchospora barbata* var. *bahiensis* (Steud.) Boeck., Linnaea 37: 546 (1873).
Rhynchospora ciliatifolia Steud., Syn. Pl. Glumac. 2: 143 (1855).
Schoenus monocephalus Salzm. ex Steud., Syn. Pl. Glumac. 2: 143 (1855).
Rhynchospora heterocaulis var. *heterolepis* Böckel, Allg. Bot. Z. Syst. 2: 94 (1896).
Rhynchospora heterocaulis C.B.Clarke, Kew Bull., Addit. Ser. 8: 32 (1908).

Rhynchospora barrosiana Guagl., Darwiniana 22: 287 (1979).
Brazil to NE. Argentina, S. Africa. 27 CPP NAT 84 BZL BZS 85 AGE PAR URU. Hemicr. or rhizome geophyte.

Rhynchospora berteroi (Spreng.) C.B.Clarke in I.Urban, Symb. Antill. 2: 119 (1900).
Cuba to Leeward Is., Belize. 80 BLZ 81 CUB DOM HAI JAM LEE PUE. Ther. or hemicr.
Schoenus pusillus Sw., Prodr.: 20 (1788). *Dichromena pusilla* (Sw.) Kunth, Enum. Pl. 2: 275 (1837). *Rhynchospora pusilla* (Sw.) Griseb., Abh. Königl. Ges. Wiss. Göttingen 7: 271 (1857), nom. illeg.
**Hypolytrum berteroi* Spreng., Neue Entdeck. Pflanzenk. 1: 241 (1820).
Hypolytrum berteroanum Schult. in J.J.Roemer & J.A.Schultes, Mant. 2: 133 (1824).
Kyllinga brevifolia Roem. & Schult., Mant. 2: 134 (1824), nom. illeg.
Scleria microdiscus Steud., Syn. Pl. Glumac. 2: 175 (1855).
Rhynchospora berteroana Boeck., Flora 41: 645 (1858), orth. var.
Dichromena filiformis C.B.Clarke in I.Urban, Symb. Antill. 2: 102 (1900), nom. illeg. *Rhynchospora subfiliformis* H.Pfeiff., Repert. Spec. Nov. Regni Veg. 49: 80 (1940). *Rhynchospora nervosa* var. *subfiliformis* (H.Pfeiff.) Kük., Bot. Jahrb. Syst. 75: 295 (1951).

Rhynchospora biflora Boeck., Vidensk. Meddel. Naturhist. Foren. Kjøbenhavn 1871: 156 (1871).
S. Venezuela, SE. & S. Brazil. 82 VEN 84 BZL BZS. Hemicr. or rhizome geophyte.
Rhynchospora alpina Palla, Denkschr. Kaiserl. Akad. Wiss., Wien. Math.-Naturwiss. Kl. 79: 191 (1908), nom. illeg.
Rhynchospora cryptantha C.B.Clarke, Kew Bull., Addit. Ser. 8: 37 (1908). *Rhynchospora biflora* var. *cryptantha* (C.B.Clarke) Kük., Bot. Jahrb. Syst. 74: 449 (1949). *Rhynchospora biflora* var. *cryptantha* (C.B.Clarke) Kük., Bot. Jahrb. Syst. 74: 449 (1949).
Rhynchospora sororopana Steyerm., Fieldiana, Bot. 28(1): 46 (1951).

Rhynchospora blepharophora (J.Presl & C.Presl) H.Pfeiff., Repert. Spec. Nov. Regni Veg. 38: 94 (1935).
W. South America to W. Venezuela. 82 VEN 83 CLM ECU PER. Hemicr.
Dichromena ciliata J.Presl & C.Presl in C.B.Presl, Reliq. Haenk. 1: 197 (1828), nom. illeg. *Dichromena preslii* A.Dietr., Sp. Pl. 2: 170 (1832).

Dichromena blepharophora J.Presl & C.Presl in C.B.Presl, Reliq. Haenk. 1: 351 (1830). *Rhynchospora nervosa* var. *blepharophora* (J.Presl & C.Presl) Kük., Bot. Jahrb. Syst. 75: 298 (1951).
Dichromena pulchella Poepp. & Kunth in C.S.Kunth, Enum. Pl. 2: 277 (1837).
Dichromena pittieri Standl., Contr. U. S. Natl. Herb. 17: 427 (1914). *Rhynchospora pittieri* (Standl.) H.Pfeiff., Repert. Spec. Nov. Regni Veg. 17: 238 (1921), nom. illeg. *Rhynchospora standleyana* H.Pfeiff., Repert. Spec. Nov. Regni Veg. 38: 92 (1935). *Rhynchospora nervosa* var. *standleyana* (H.Pfeiff.) Kük., Bot. Jahrb. Syst. 75: 297 (1951).

Rhynchospora boivinii Boeck., Linnaea 38: 403 (1874).
Réunion. 29 REU. Hemicr. or rhizome geophyte.

Rhynchospora bolivarana Steyerm., Fieldiana, Bot. 28(1): 40 (1951).
S. Venezuela to W. Guyana. 82 GUY VEN. Hemicr. or rhizome geophyte.

Rhynchospora boliviensis C.B.Clarke, Kew Bull., Addit. Ser. 8: 37 (1908).
S. Mexico to Bolivia. 79 MXS 80 HON 83 BOL CLM. Hemicr.

Rhynchospora boninensis Nakai ex Tuyama, Bot. Mag. (Tokyo) 49: 511 (1935).
Ogasawara-shoto. 38 OGA. Hemicr. or rhizome geophyte.

Rhynchospora brachychaeta C.Wright, Anales Acad. Ci. Méd. Habana 8: 85 (1871).
SE. U.S.A. to Guyana. 78 ALA FLA MSI 80 BLZ NIC 81 CUB DOM PUE 82 GUY. Hemicr.
Phaeocephalum brachychaetum House, Amer. Midl. Naturalist 6: 201 (1920).
Rhynchospora blauneri Britton, Bull. Torrey Bot. Club 50: 56 (1923).

Rhynchospora bracteovillosa A.C.Araujo & W.W.Thomas, Brittonia 55: 31 (2003).
Brazil (Goiás, Mato Grosso). 84 BZC.

Rhynchospora brasiliensis Boeck., Vidensk. Meddel. Naturhist. Foren. Kjøbenhavn 41-42: 26 (1879).
S. Trop. America. 82 GUY VEN 83 CLM PER 84 BZC BZL BZS. Hemicr. or rhizome geophyte.

Rhynchospora brevirostris Griseb., Cat. Pl. Cub.: 246 (1866).
Trop. Africa, Mexico to Trop. America. 22 GNB GUI MLI NGA SEN SIE 23 CAF CMN ZAI 25 TAN 26 ZAM 79 MXC MXS MXT 80 BLZ COS HON NIC PAN 81 CUB DOM TRT 82 FRG GUY SUR VEN 83 BOL CLM 84 BZC BZL BZN. Ther. or hemicr.
Rhynchospora pringlei C.B.Clarke, Kew Bull., Addit. Ser. 8: 89 (1908), nom. illeg.
Rhynchospora brevirostris var. *truncata* Kük., Bot. Jahrb. Syst. 75: 287 (1951).

Rhynchospora breviuscula H.Pfeiff., Repert. Spec. Nov. Regni Veg. 38: 95 (1935).
Brazil (São Paulo, Paraná). 84 BZL BZS. Hemicr.
**Dichromena gracilis* C.B.Clarke, Kew Bull., Addit. Ser. 8: 32 (1908), nom. illeg.

Rhynchospora brittonii Gale, Rhodora 46: 241 (1944).
SE. Mexico, W. Cuba, S. Brazil to Argentina (Corrientes). 79 MXT 81 CUB 84 BZS 85 AGE URU. Hemicr.

Rhynchospora brunnea Boeck., Linnaea 37: 593 (1873).
N. Peru. 83 PER. Hemicr. or rhizome geophyte.

Rhynchospora bucherorum León, Contr. Ocas. Mus. Hist. Nat. Colegio "De La Salle" 8: 226 (1946).
E. Cuba. 81 CUB. Hemicr. or rhizome geophyte.

Rhynchospora cabecarae Gómez-Laur., Novon 5: 270 (1995).
Costa Rica. 80 COS. Hemicr. or rhizome geophyte.

Rhynchospora cacuminicola Gale, Rhodora 46: 275 (1944).
Dominican Rep. 81 DOM. Hemicr. or rhizome geophyte.

Rhynchospora caduca Elliott, Sketch Bot. S. Carolina 1: 62 (1816). *Phaeocephalum caducum* (Elliott) House, Amer. Midl. Naturalist 6: 201 (1920).
C. & SE. U.S.A. (63) haw 74 OKL 77 TEX 78 ALA ARK FLA GEO LOU MSI NCA SCA TEN VRG. Hemicr. or rhizome geophyte.
Rhynchospora patula A.Gray, Ann. Lyceum Nat. Hist. New York 3: 201 (1835). *Phaeocephalum patulum* (A.Gray) House, Amer. Midl. Naturalist 6: 202 (1920).
Rhynchospora longiseta Steud., Syn. Pl. Glumac. 2: 140 (1855).

Rhynchospora caesionux T.Koyama, Brittonia 24: 281 (1972).
Brazil (Mato Grosso). 84 BZC. Hemicr. or rhizome geophyte.

Rhynchospora cajennensis Boeck., Linnaea 38: 404 (1874).
N. South America. 82 FRG GUY VEN. Hemicr. or rhizome geophyte.
Holoschoenus elatior Nees, J. Bot. (Hooker) 2: 394 (1840). *Rhynchospora demerarensis* C.B.Clarke, Kew Bull., Addit. Ser. 8: 40 (1908). *Rhynchospora stricta* var. *demerarensis* (C.B.Clarke) Kük., Bot. Jahrb. Syst. 75: 194 (1950).

Rhynchospora calderana D.A.Simpson, Kew Bull. 42: 891 (1987).
Brazil (Bahia). 84 BZE. Hemicr.

Rhynchospora californica Gale, Rhodora 46: 272 (1944).
California. 76 CAL. Hemicr. or rhizome geophyte.

Rhynchospora candida (Nees) Boeck., Linnaea 37: 605 (1873).
Trop. & S. Africa Madagascar, S. Trop. America. 22 BEN BKN GNB GUI IVO LBR MLI NGA SEN SIE 23 BUR CAF CMN EQG GAB ZAI 24 CHA 25 TAN UGA 26 ANG MLW MOZ ZAM 27 BOT 29 MDG 82 GUY SUR VEN 83 BOL CLM 84 BZN. Hemicr. or rhizome geophyte.
**Psilocarya candida* Nees in C.F.P.von Martius & auct. suc. (eds.), Fl. Bras. 2(1): 117 (1842). *Pachymitra candida* (Nees) Nees ex Boeck., Linnaea 37: 606 (1873). *Dichromena candida* (Nees) Ridl., Trans. Linn. Soc. London, Bot. 2: 149 (1884). *Rhynchospora adscendens* C.B.Clarke in T.A.Durand & H.Schinz, Consp. Fl. Afric. 5: 65 (1894), nom. illeg.
Rhynchospora candida var. *pilosa* Schnee, Bol. Soc. Venez. Ci. Nat. 9: 62 (1944).

Rhynchospora capillacea Torr., Fl. N. Middle United States 1: 55 (1823). *Phaeocephalum capillaceum* (Torr.) Farw., Rep. (Annual) Michigan Acad. Sci. 21: 361 (1919 publ. 1920).

Canada to C. & E. U.S.A. 70 ABT MAN SAS 72 NBR NFL ONT QUE 74 ILL IOW MIN MSO NDA OKL SDA WIS 75 CNT INI MAI MAS MIC NWH NWJ NWY OHI PEN RHO VER 77 TEX 78 ALA ARK TEN VRG. Hemicr. or rhizome geophyte.

Schoenus setaceus Muhl., Descr. Gram.: 6 (1817), nom. illeg. *Rhynchospora setacea* MacMill., Metasp. Minnesota Valley: 104 (1892).

Rhynchospora smallii Britton in J.K.Small, Fl. S.E. U.S.: 1321 (1903).

Phaeocephalum capillaceum var. *levisetum* Farw., Rep. (Annual) Michigan Acad. Sci. 21: 361 (1920).

Rhynchospora capillacea f. *leviseta* Fernald, Rhodora 37: 252 (1935).

Rhynchospora capillifolia Boeck., Beitr. Cyper. 2: 22 (1890).

SE. Brazil. 84 BZL. Hemicr. or rhizome geophyte.

Rhynchospora capitata (Kunth) Roem. & Schult., Syst. Veg. 2: 88 (1817).

Colombia (Meta) to Venezuela (Amazonas). 82 VEN 83 CLM. Hemicr. or rhizome geophyte.

**Chaetospora capitata* Kunth in F.W.H.von Humboldt, A.J.A.Bonpland & C.S.Kunth, Nov. Gen. Sp. 1: 229 (1816). *Schoenus capitatus* (Kunth) Poir. in J.B.A.M.de Lamarck, Encycl., Suppl. 5: 617 (1817), nom. illeg.

Rhynchospora lanceolata Willd. ex Kunth, Enum. Pl. 2: 288 (1837).

Rhynchospora capitellata (Michx.) Vahl, Enum. Pl. 2: 235 (1805).

E. Canada to U.S.A. 72 NBR NSC ONT QUE 73 ORE 74 ILL KAN MIN MSO WIS 75 CNT INI MAI MAS MIC NWH NWJ NWY OHI PEN RHO VER WVA 76 CAL 77 TEX 78 ALA ARK DEL FLA GEO KTY LOU MRY MSI NCA SCA TEN VRG WDC. Hemicr. or rhizome geophyte.

**Schoenus capitellatus* Michx., Fl. Bor.-Amer. 1: 36 (1803). *Rhynchospora glomerata* var. *capitellata* (Michx.) Kük., Bot. Jahrb. Syst. 75: 98 (1950).

Rhynchospora glomerata var. *leptocarpa* Chapm. ex Britton, Trans. New York Acad. Sci. 11: 88 (1892). *Rhynchospora leptocarpa* (Chapm. ex Britton) Small, Man. S.E. Fl.: 181 (1933). *Rhynchospora glomerata* f. *leptocarpa* (Chapm. ex Britton) Kük., Bot. Jahrb. Syst. 75: 99 (1950).

Rhynchospora capitellata var. *controversa* S.F.Blake, Rhodora 20: 28 (1918). *Rhynchospora capitellata* f. *controversa* (S.F.Blake) Gale, Rhodora 46: 119 (1944).

Rhynchospora capitellata var. *discutiens* S.F.Blake, Rhodora 20: 28 (1918). *Rhynchospora capitellata* f. *discutiens* (S.F.Blake) Gale, Rhodora 46: 119 (1944).

Rhynchospora capitellata var. *leptocarpa* S.F.Blake, Rhodora 20: 28 (1918).

Rhynchospora caracasana (Kunth) Boeck., Linnaea 37: 607 (1873).

N. South America to Bolivia. 82 GUY SUR VEN 83 BOL 84 BZN. Hemicr. or rhizome geophyte.

**Dichromena caracasana* Kunth, Enum. Pl. 2: 281 (1837).

Rhynchospora filiformis Willd. ex Kunth, Enum. Pl. 2: 281 (1837), nom. inval.

Rhynchospora andina Kük., Feddes Repert. Spec. Nov. Regni Veg. 53: 73 (1944).

Rhynchospora careyana Fernald, Rhodora 20: 140 (1918).

New Jersey to SE. U.S.A., Cuba, W. Jamaica. 75 NWJ 78 ALA DEL FLA GEO LOU MSI NCA SCA 81 CUB JAM. Hemicr. or rhizome geophyte.

**Ceratoschoenus macrostachys* var. *patulus* Chapm., Fl. South. U.S.: 529 (1860). *Rhynchospora macrostachya* var. *patula* (Chapm.) Chapm., Fl. South. U.S., ed. 3: 556 (1897).

Rhynchospora corniculata var. *patula* (Chapm.) Britton, Trans. New York Acad. Sci. 11: 84 (1892).

Rhynchospora cariciformis Nees in C.F.P.von Martius & auct. suc. (eds.), Fl. Bras. 2(1): 145 (1842). *Dichromena cariciformis* (Nees) J.F.Macbr., Publ. Field Mus. Nat. Hist., Bot. Ser. 11: 42 (1931).

Colombia to Venezuela and Brazil. 82 VEN 83 CLM PER 84 BZC BZN. Hemicr. or rhizome geophyte.

Rhynchospora igarapicola T.Koyama, Mem. New York Bot. Gard. 23: 51 (1972).

Rhynchospora carrillensis Gómez-Laur., Brenesia 22: 353 (1984).

SE. Mexico to C. America. 79 MXT 80 COS PAN. Hemicr.

Rhynchospora castanea T.Koyama, Bull. Natl. Sci. Mus., Tokyo, B 2: 169 (1976).

Brazil. 84 BZL. Hemicr. or rhizome geophyte.

Rhynchospora catharinensis Barros, Sellowia 19: 57 (1967).

S. Brazil. 84 BZS. Hemicr. or rhizome geophyte.

Rhynchospora caucasica Palla, Vestn. Tiflissk. Bot. Sada 30: 26 (1913).

Caucasus. 33 TCS. Hemicr. or rhizome geophyte.

Rhynchospora cephalantha A.Gray, Ann. Lyceum Nat. Hist. New York 3: 218 (1835). *Rhynchospora cephalantha* var. *typica* Fernald & Gale, Rhodora 42: 423 (1940), nom. inval.

E. U.S.A. to Texas. 75 NWJ NWY 77 TEX 78 ALA DEL FLA GEO LOU MRY MSI NCA SCA VRG. Hemicr. or rhizome geophyte.

Rhynchospora biceps Torr. ex A.Gray, Ann. Lyceum Nat. Hist. New York 3: 219 (1835).

Rhynchospora intermedia Beyr. ex Kunth, Enum. Pl. 2: 296 (1837), nom. inval.

Rhynchospora cephalantha var. *luxurians* Boeck., Linnaea 37: 572 (1873).

Rhynchospora cephalantha var. *pleiocephala* Fernald & Gale, Rhodora 42: 424 (1940).

Rhynchospora cephalantha f. *antrorsa* Gale, Rhodora 46: 109 (1944).

Rhynchospora cephalantha var. *attenuata* Gale, Rhodora 46: 110 (1944).

Rhynchospora cephalantha f. *controversa* Gale, Rhodora 46: 110 (1944).

Rhynchospora cephalotes (L.) Vahl, Enum. Pl. 2: 237 (1805).

S. Mexico to Trop. America. 79 MXG MXS MXT 80 BLZ COS GUA HON NIC PAN 81 CUB JAM TRT 82 FRG GUY SUR VEN 83 BOL CLM ECU PER 84 BZC BZE BZN. Hemicr. or rhizome geophyte.

**Scirpus cephalotes* L., Sp. Pl. ed. 2: 76 (1762). *Schoenus cephalotes* (L.) Rottb., Descr. Icon. Rar. Pl.: 61 (1773). *Dichromena schoenus* J.F.Macbr., Publ. Field Mus. Nat. Hist., Bot. Ser. 11: 6 (1931).

Rhynchospora conica Desv. ex Ham., Prodr. Pl. Ind. Occid.: 15 (1825).

Scleria fasciculata Willd. ex Kunth, Enum. Pl. 2: 301 (1837).

Rhynchospora cephalotes var. *pseudocomata* Kük., Bot. Jahrb. Syst. 75: 123 (1950).

Rhynchospora cernua Griseb., Cat. Pl. Cub.: 248 (1866).
Cuba. 81 CUB. Hemicr. or rhizome geophyte.

Rhynchospora chalarocephala Fernald & Gale, Rhodora 42: 426 (1940).
EC. & SE. U.S.A. 74 OKL 75 NWJ 77 TEX 78 ALA DEL FLA GEO LOU MRY MSI NCA SCA TEN VRG WDC. Hemicr. or rhizome geophyte.

Rhynchospora chapmanii M.A.Curtis, Amer. J. Sci. Arts, II, 7: 409 (1849). *Phaeocephalum chapmanii* (M.A.Curtis) House, Amer. Midl. Naturalist 6: 201 (1920).
SE. U.S.A., C. America. 78 ALA FLA GEO LOU MSI NCA SCA 80 BLZ NIC. Hemicr.
Rhynchospora grayana Chapm. ex M.A.Curtis, Amer. J. Sci. Arts, II, 7: 409 (1849).
Dichromena heterophylla Boeck., Flora 41: 644 (1858). *Rhynchospora heterophylla* (Boeck.) Boeck., Linnaea 37: 553 (1873).

Rhynchospora chimantensis W.W.Thomas, Contr. Univ. Michigan Herb. 15: 311 (1982).
Venezuela (Bolívar). 84 BZN. Hemicr. or rhizome geophyte.

Rhynchospora chinensis Nees & Meyen, Nov. Actorum Acad. Caes. Leop.-Carol. Nat. Cur. 19(Suppl. 1): 108 (1843). *Rhynchospora glauca* var. *chinensis* (Nees & Meyen) C.B.Clarke in J.D.Hooker, Fl. Brit. India 6: 672 (1893).
W. Indian Ocean, Indo-China to Hawaiian Is. 29 MAU MDG REU 36+ 38 JAP KOR NNS OGA TAI 40 ASS IND SRL 41 MYA THA VIE 42+ 63 HAW. Hemicr. or rhizome geophyte.

subsp. ***chinensis***
W. Indian Ocean, Indo-China to Japan. 29 MAU MDG REU 36 38 JAP KOR NNS TAI 40 ASS IND SRL 41 MYA THA VIE 42 (63) haw. Hemicr. or rhizome geophyte.
Rhynchospora japonica Makino, Bot. Mag. (Tokyo) 17: 184 (1903).
Rhynchospora longisetigera Hayata, Icon. Pl. Formosan. 6: 116 (1916).

var. ***curvoaristata*** (Tuyama) Ohwi, Mem. Coll. Sci. Kyoto Imp. Univ., Ser. B, Biol. 18(1): 19 (1944).
Ogasawara-shoto (Chichi-jima). 38 OGA. Hemicr. or rhizome geophyte.
**Rhynchospora japonica* var. *curvoaristata* Tuyama, Bot. Mag. (Tokyo) 52: 567 (1938).

subsp. ***spiciformis*** (Hillebr.) T.Koyama, Bishop Mus. Occas. Pap. 29: 127 (1989).
Hawaiian Is. 63 HAW. Hemicr. or rhizome geophyte.
**Rhynchospora spiciformis* Hillebr., Fl. Hawaiian Isl.: 477 (1888).

Rhynchospora ciliaris (Michx.) C.Mohr, Contr. U. S. Natl. Herb. 6: 408 (1901).
SE. U.S.A. 78 ALA FLA GEO LOU MSI NCA SCA. Hemicr. or rhizome geophyte.
**Schoenus ciliaris* Michx., Fl. Bor.-Amer. 1: 36 (1803). *Rhynchospora ciliata* Vahl, Enum. Pl. 2: 235 (1805), nom. superfl. *Phaeocephalum ciliare* (Michx.) House, Amer. Midl. Naturalist 6: 201 (1920).
Rhynchospora rappiana Small, Man. S.E. Fl.: 179 (1933).

Rhynchospora ciliolata Boeck., Linnaea 37: 563 (1873).
SE. Brazil. 84 BZL. Hemicr. or rhizome geophyte.

Rhynchospora colorata (L.) H.Pfeiff., Repert. Spec. Nov. Regni Veg. 38: 89 (1935).
SE. U.S.A. to N. South America. 77 TEX 78 ALA ARK FLA GEO LOU MSI NCA SCA VRG 79 MXC MXE MXG MXS MXT 80 BLZ COS GUA 81 BAH BER CAY CUB JAM LEE PUE WIN 82 FRG VEN. Hemicr. or rhizome geophyte.
**Schoenus coloratus* L., Sp. Pl.: 43 (1753). *Dichromena colorata* (L.) C.L.Hitchc., Rep. (Annual) Missouri Bot. Gard. 4: 141 (1893). *Kyllinga colorata* (L.) Druce, Bot. Soc. Exch. Club Brit. Isles 1916: 630 (1917).
Schoenus stellatus Lam., Encycl. 1: 741 (1785). *Dichromena ciliata* Pers., Syn. Pl. 1: 58 (1805). *Dichromena persooniana* Nees in C.F.P.von Martius & auct. suc. (eds.), Fl. Bras. 2(1): 112 (1842), nom. superfl. *Rhynchospora persooniana* (Nees) Griseb., Mem. Amer. Acad. Arts, n.s., 8: 536 (1863). *Dichroma ciliatum* Ham., Prodr. Pl. Ind. Occid.: 15 (1825), nom. superfl. *Rhynchospora stellata* (Lam.) Griseb., Abh. Königl. Ges. Wiss. Göttingen 7: 271 (1857).
Dichromena leucocephala Michx., Fl. Bor.-Amer. 1: 37 (1803). *Dichromena leucocephalum* (Michx.) Pers., Syn. Pl. 1: 57 (1805). *Rhynchospora leucocephala* (Michx.) Boeck., Vidensk. Meddel. Naturhist. Foren. Kjøbenhavn 1869: 144 (1869).
Dichromena glabra Pers., Syn. Pl. 1: 57 (1805).
Dichromena drummondiana Steud., Syn. Pl. Glumac. 2: 135 (1855).
Rhynchospora drummondiana Steud., Syn. Pl. Glumac. 2: 135 (1855).
Rhynchospora stellata f. *intercedens* Kük., Bot. Jahrb. Syst. 74: 300 (1951).

Rhynchospora comata (Link) Schult. in J.J.Roemer & J.A.Schultes, Mant. 2: 50 (1824).
Trop. America. 80 COS PAN 81 CUB JAM TRT 82 FRG GUY SUR VEN 83 CLM PER 84 BZC BZE BZL BZN. Hemicr. or rhizome geophyte.
**Schoenus comatus* Link, Enum. Hort. Berol. Alt. 1: 41 (1821). *Rhynchospora cephalotes* var. *comata* (Link) Kük., Repert. Spec. Nov. Regni Veg. 23: 210 (1926).
Rhynchospora sylvatica Nees, Linnaea 9: 297 (1834), nom. inval.
Rhynchospora bromoides Kunth, Enum. Pl. 2: 300 (1837). *Rhynchospora comata* f. *bromoides* (Kunth) A.Gross & Brade, Arq. Mus. Nac. Rio de Janeiro 34: 51 (1932).
Schoenus sylvaticus Salzm. ex Schltdl., Bot. Zeitung (Berlin) 7: 167 (1849).
Rhynchospora cephalotoides Griseb., Cat. Pl. Cub.: 242 (1866).
Rhynchospora diedrichsenii Boeck., Beitr. Cyper. 2: 25 (1890).
Rhynchospora umbraticola var. *broadwayi* Kük., Bot. Jahrb. Syst. 74: 392 (1949). *Rhynchospora broadwayi* (Kük.) Raymond, Naturaliste Canad. 99: 220 (1972).

Rhynchospora compressa J.Carey ex Chapm., Fl. South. U.S.: 525 (1860). *Rhynchospora cymosa* var. *compressa* (J.Carey ex Chapm.) C.B.Clarke, Trans. New York Acad. Sci. 11: 91 (1892). *Phaeocephalum compressum* (J.Carey ex Chapm.) House, Amer. Midl. Naturalist 6: 201 (1920).
SE. U.S.A. 78 ALA FLA GEO LOU MSI SCA. Ther. or hemicr.

Rhynchospora confinis (Nees) C.B.Clarke, Kew Bull., Addit. Ser. 8: 40 (1908).
Brazil to N. Argentina. 83 BOL 84 BZC 85 AGE AGW PAR. Hemicr. or rhizome geophyte.
**Spermodon confinis* Nees in C.F.P.von Martius & auct. suc. (eds.), Fl. Bras. 2(1): 119 (1842).
Rhynchospora lundii Boeck., Vidensk. Meddel. Naturhist. Foren. Kjøbenhavn 1869: 147 (1869).

Rhynchospora confusa Ballard, Hooker's Icon. Pl. 33: t. 3250 (1934).
Brazil. 84 BZC BZE. Hemicr. or rhizome geophyte.
**Syntrinema brasiliense* Radlk. & H.Pfeiff., Repert. Spec. Nov. Regni Veg. 21: 238 (1925).

Rhynchospora consanguinea (Kunth) Boeck., Linnaea 37: 536 (1873).
Brazil to Paraguay. 84 BZC BZE BZL BZN BZS 85 PAR. Hemicr. or rhizome geophyte.
Dichromena compressa Spreng., Neue Entdeck. Pflanzenk. 1: 239 (1820).
**Dichromena consanguinea* Kunth, Enum. Pl. 2: 278 (1837). *Asteroschoenus consanguineus* (Kunth) Nees in C.F.P.von Martius & auct. suc. (eds.), Fl. Bras. 2(1): 125 (1842).
Dichromena rigida Kunth, Enum. Pl. 2: 278 (1837). *Asteroschoenus rigidus* (Kunth) Nees in C.F.P.von Martius & auct. suc. (eds.), Fl. Bras. 2(1): 126 (1842). *Rhynchospora rigida* (Kunth) Boeck., Vidensk. Meddel. Naturhist. Foren. Kjøbenhavn 1869: 145 (1869), nom. illeg. *Rhynchospora consanguinea* var. *rigida* (Kunth) Kük., Bot. Jahrb. Syst. 74: 477 (1949).
Dichromena speciosa Kunth, Enum. Pl. 2: 278 (1837). *Asteroschoenus speciosus* (Kunth) Nees in C.F.P. von Martius & auct. suc. (eds.), Fl. Bras. 2(1): 125 (1842). *Rhynchospora speciosa* (Kunth) Boeck., Linnaea 27: 536 (1873). *Rhynchospora consanguinea* var. *speciosa* (Kunth) Kük., Bot. Jahrb. Syst. 74: 477 (1949).
Asteroschoenus consanguineus var. *angustifolius* Nees in C.F.P.von Martius & auct. suc. (eds.), Fl. Bras. 2(1): 125 (1842).
Dichromena hispida Schrad. ex Nees in C.F.P.von Martius & auct. suc. (eds.), Fl. Bras. 2(1): 111 (1842).
Rhynchospora consanguinea f. *subrigida* Kük., Bot. Jahrb. Syst. 74: 477 (1949).
Rhynchospora consanguinea f. *subspeciosa* Kük., Bot. Jahrb. Syst. 74: 477 (1949).

Rhynchospora contracta (Nees) J.Raynal, Adansonia, n.s., 17: 277 (1978).
Mexico to Trop. America, W. Trop. Africa. 22 GNB 78 GEO 79 MXC MXC MXG MXS MXT 80 BLZ COS ELS GUA HON NIC PAN 81 JAM LEE PUE TRT WIN 82 FRG VEN 83 CLM ECU GAL 84 BZE. Ther.
Juncus parviflorus Poir. in J.B.A.M.de Lamarck, Encycl., Suppl. 3: 160 (1813), nom. illeg.
Scleria micrantha Poir. in J.B.A.M.de Lamarck, Encycl., Suppl. 5: 108 (1817).
Zosterospermum gracile Desv. ex Ham., Prodr. Pl. Ind. Occid.: 14 (1825).
Scleria minutiflora Rich. ex Spreng., Syst. Veg. 3: 831 (1826), nom. inval. *Rhynchospora minutiflora* C.D.Adams, Phytologia 21: 70 (1971), nom. inval.
Rhynchospora sparsa Sieber ex C.Presl, Isis (Oken) 21: 268 (1828), nom. illeg.
Haloschoenus pygmaeus Nees, Linnaea 9: 296 (1834), nom. inval.
Schoenus teneriffae Vent. ex Torr., Ann. Lyceum Nat. Hist. New York 3: 362 (1836), nom. inval.
Rhynchospora parviflora Vahl ex Kunth, Enum. Pl. 2: 279 (1837), nom. inval.
Rhynchospora polyphylla Balb. ex Kunth, Enum. Pl. 2: 279 (1837), nom. inval.
**Haloschoenus contractus* Nees in C.F.P.von Martius & auct. suc. (eds.), Fl. Bras. 2(1): 123 (1842). *Dichromena contracta* (Nees) Steud., Syn. Pl. Glumac. 2: 136 (1855). *Rhynchospora micrantha* var. *contracta* (Nees) Kük., Bot. Jahrb. Syst. 75: 276 (1951).
Psilocarya teneriffae Steud., Syn. Pl. Glumac. 2: 151 (1855).
Scleria cincta Steud., Syn. Pl. Glumac. 2: 177 (1855).

Rhynchospora cordatachenia M.T.Strong, Novon 15: 480 (2005).
N. French Guiana. 82 FRG. Hemicr.

Rhynchospora coriifolia Boeck., Linnaea 38: 406 (1874).
Brazil (Rio de Janeiro). 84 BZL. Hemicr. or rhizome geophyte.

Rhynchospora corniculata (Lam.) A.Gray, Ann. Lyceum Nat. Hist. New York 3: 205 (1835).
C. & E. U.S.A., Cuba. 74 ILL MSO OKL 75 INI 77 TEX 78 ALA ARK DEL FLA GEO KTY LOU MRY MSI NCA SCA TEN VRG 81 CUB. Hemicr. or rhizome geophyte.
Schoenus umbellatus Walter, Fl. Carol.: 70 (1788), nom. illeg.
**Schoenus corniculatus* Lam., Tabl. Encycl. 1: 137 (1791). *Ceratoschoenus corniculatus* (Lam.) Nees, Linnaea 9: 296 (1834). *Rhynchospora corniculata* var. *typica* Waterf., Rhodora 52: 22 (1950), nom. inval.
Schoenus longirostris Michx., Fl. Bor.-Amer. 1: 35 (1803). *Rhynchospora longirostris* (Michx.) Elliott, Sketch Bot. S. Carolina 1: 59 (1816). *Ceratoschoenus longirostris* (Michx.) A.Gray, Ann. Lyceum Nat. Hist. New York 3: 369 (1836).
Rhynchospora laxa Vahl, Enum. Pl. 2: 231 (1805).
Rhynchospora corniculata var. *interior* Fernald, Rhodora 20: 140 (1918).

Rhynchospora corymbosa (L.) Britton, Trans. New York Acad. Sci. 11: 84 (1892).
Trop. & Subtrop. 22 BEN BKN GAM GHA GNB GUI IVO LBR MLI NGA SEN SIE TOG 23 CAF CMN EQG GAB GGI ZAI 24 CHA 25 TAN UGA 27 CPP NAT 29 MAU MDG REU 36 CHC CHH CHS 38 NNS TAI 40 ASS BAN EHM IND SRL 41 AND MYA NCB THA VIE 42 BOR JAW LSI MLY MOL PHI SUL SUM 43 BIS NWG 50 NSW NTA QLD 60 FIJ SAM TON VAN 61 SCI 62 CRL MRN 79 MXC MXG MXS MXT 80 BLZ COS GUA HON NIC PAN 81 CUB DOM JAM LEE PUE TRT WIN 82 FRG GUY SUR VEN 83 BOL CLM ECU GAL PER 84 BZC BZE BZL BZN BZS 85 AGE AGS AGW PAR URU. Hemicr. or rhizome geophyte.
**Scirpus corymbosus* L., Cent. Pl. II: 7 (1756). *Schoenus corymbosus* (L.) Pers., Syn. Pl. 1: 59 (1805). *Dichromena corymbosa* (L.) J.F.Macbr., Publ. Field Mus. Nat. Hist., Bot. Ser. 11: 6 (1931).
Schoenus surinamensis Rottb., Descr. Icon. Rar. Pl.: 68 (1773). *Rhynchospora surinamensis* (Rottb.) Nees, Linnaea 9: 297 (1834).
Rhynchospora aurea Vahl, Enum. Pl. 2: 229 (1805). *Chaetospora aurea* (Vahl) Kunth in F.W.H.von

Humboldt, A.J.A.Bonpland & C.S.Kunth, Nov. Gen. Sp. 1: 231 (1816).
Schoenus floridus Rudge, Pl. Guian.: 15 (1805). *Rhynchospora florida* (Rudge) Schult. in J.J.Roemer, J.A.Schultes & J.J.Schultes, Mant. 2: 45 (1824). *Calyptrostylis rudgei* Nees, Linnaea 9: 295 (1834). *Calyptrostylis florida* (Rudge) Nees in C.F.P.von Martius & auct. suc. (eds.), Fl. Bras. 2(1): 138 (1842). *Rhynchospora aurea* f. *florida* (Rudge) Lindm., Bih. Kongl. Svenska Vetensk.-Akad. Handl. 26(3; 9): 27 (1900). *Rhynchospora corymbosa* var. *florida* (Rudge) Kük., Bot. Jahrb. Syst. 74: 417 (1949).
Schoenus articulatus Roxb., Fl. Ind. 1: 184 (1820). *Rhynchospora articulata* (Roxb.) Schult. in J.J.Roemer & J.A.Schultes, Mant. 2: 49 (1824). *Cephaloschoenus articulatus* (Roxb.) Nees, Edinburgh New Philos. J. 17: 266 (1834). *Calyptrostylis articulata* (Roxb.) Nees, J. Bot. (Hooker) 2: 394 (1840).
Scirpus bangalorensis B.Heyne ex Wall., Numer. List: 3464 (1831), nom. inval.
Calyptrostylis gaudichaudii Nees, Linnaea 9: 295 (1834), nom. inval.
Cephaloschoenus divergens Nees, Linnaea 9: 296 (1834), nom. inval.
Scirpus umbellatus Roxb. ex Kunth, Enum. Pl. 2: 294 (1837).
Calyptrostylis asperula var. *minor* Nees in C.F.P.von Martius & auct. suc. (eds.), Fl. Bras. 2(1): 139 (1842). *Rhynchospora corymbosa* var. *minor* (Nees) Kük., Bot. Jahrb. Syst. 74: 416 (1949).
Calyptrostylis divergens Nees in C.F.P.von Martius & auct. suc. (eds.), Fl. Bras. 2(1): 139 (1842).
Calyptrostylis fascicularis Nees in C.F.P.von Martius & auct. suc. (eds.), Fl. Bras. 2(1): 139 (1842).
Rhynchospora corymbifera Nees, Nov. Actorum Acad. Caes. Leop.-Carol. Nat. Cur. 19(Suppl. 1): 106 (1843).
Rhynchospora schraderiana Steud., Syn. Pl. Glumac. 2: 146 (1855).
Rhynchospora subulirostris Steud., Syn. Pl. Glumac. 2: 149 (1855).
Rhynchospora macrocarpa Boeck., Linnaea 37: 629 (1873).
Rhynchospora corymbosa f. *chacoensis* Barros, Trans. New York Acad. Sci. 11: 84 (1892). *Rhynchospora corymbosa* var. *chacoensis* (Barros) Svenson, Lilloa 14: 73 (1949).
Rhynchospora surinamensis K.Schum. in H.G.A. Engler, Pflanzenw. Ost-Afrikas, C: 127 (1895).
Rhynchospora puncticulata var. *pleiocephala* Osten, Anales Mus. Hist. Nat. Montevideo, II, 3(2): 214 (1931).
Rhynchospora corymbosa var. *bonariensis* Barros ex Cabrera & Dawson, Revista Mus. La Plata, Secc. Bot. 5: 314 (1944).
Rhynchospora corymbosa f. *angustirostris* Barros, Anales Mus. Argent. Ci. Nat. "Bernardino Rivadavia" 41: 388 (1945). *Rhynchospora corymbosa* var. *angustirostris* (Barros) Guagl., Darwiniana 39: 308 (2001).
Rhynchospora corymbosa var. *singularis* Kük., Bot. Jahrb. Syst. 74: 417 (1949).

Rhynchospora crinigera Boeck., Beitr. Cyper. 1: 28 (1888).
S. Brazil to S. Uruguay. 84 BZS 85 URU. Hemicr. or rhizome geophyte.
Rhynchospora crinigera var. *spicata* Kük. ex Osten, Anales Mus. Hist. Nat. Montevideo, II, 3: 221 (1931).

Rhynchospora crinipes Gale, Rhodora 46: 173 (1944). *Rhynchospora filifolia* var. *crinipes* (Gale) Kük., Bot. Jahrb. Syst. 75: 106 (1950).
SE. U.S.A. 78 ALA FLA GEO MSI NCA. Hemicr. or rhizome geophyte.

Rhynchospora crispa Gale, Rhodora 46: 213 (1944).
E. Cuba. 81 CUB. Hemicr. or rhizome geophyte.

Rhynchospora cubensis A.Rich. in R.de la Sagra, Hist. Fis. Cuba, Bot. 11: 294 (1850).
Cuba, Hispaniola. 81 CUB DOM HAI. Hemicr. or rhizome geophyte.
Rhynchospora deflexa Griseb., Cat. Pl. Cub.: 243 (1866).
Rhynchospora cubensis var. *stenophylloidea* Kük., Repert. Spec. Nov. Regni Veg. 32: 78 (1933). *Rhynchospora stenophylloidea* (Kük.) Gale, Rhodora 46: 226 (1944).
Rhynchospora cubensis var. *brevifoliata* Kük., Bot. Jahrb. Syst. 75: 141 (1950).

Rhynchospora curtissii Britton in J.K.Small, Fl. S.E. U.S.: 195 (1903). *Phaeocephalum curtissii* (Britton) House, Amer. Midl. Naturalist 6: 201 (1920).
SE. U.S.A. 78 ALA FLA MSI. Ther. or hemicr.
Rhynchospora filifolia var. *ellipsoidea* Kük., Bot. Jahrb. Syst. 75: 107 (1950).

Rhynchospora curvula Griseb., Fl. Brit. W. I.: 574 (1864).
Trinidad, C. & S. Trop. America. 80 BLZ NIC 81 TRT 82 FRG GUY SUR VEN 83 CLM 84 BZN. Hemicr.
Rhynchospora scapigera Boeck., Linnaea 38: 402 (1874).

Rhynchospora davidsei W.W.Thomas, Brittonia 44: 37 (1992).
Panama. 80 PAN. Hemicr.

Rhynchospora debilis Gale, Rhodora 46: 194 (1944). *Rhynchospora fascicularis* var. *debilis* (Gale) Kük., Bot. Jahrb. Syst. 75: 111 (1950).
SE. U.S.A. to Texas. 77 TEX 78 ALA FLA GEO LOU MSI NCA SCA VRG. Hemicr. or rhizome geophyte.

Rhynchospora decurrens Chapm., Fl. South. U.S.: 525 (1860). *Phaeocephalum decurrens* (Chapm.) House, Amer. Midl. Naturalist 6: 201 (1920).
SE. U.S.A. 78 ALA FLA GEO LOU MSI NCA SCA. Ther. or hemicr.

Rhynchospora dentinux C.B.Clarke, Kew Bull., Addit. Ser. 8: 33 (1908).
Colombia to N. South America. 82 FRG VEN 83 CLM. Hemicr. or rhizome geophyte.

Rhynchospora depauperata Palla, Denkschr. Kaiserl. Akad. Wiss., Wien. Math.-Naturwiss. Kl. 79: 189 (1908).
Brazil. 84 BZL. Hemicr. or rhizome geophyte.

Rhynchospora depressa (Kük.) Gale, Rhodora 46: 217 (1944).
E. Cuba. 81 CUB. Hemicr. or rhizome geophyte.
**Rhynchospora lindeniana* var. *depressa* Kük., Repert. Spec. Nov. Regni Veg. 23: 211 (1926). *Rhynchospora cernua* var. *depressa* (Kük.) Kük., Bot. Jahrb. Syst. 75: 165 (1950).

Rhynchospora depressirostris M.T.Strong, Brittonia 52: 241 (2000).
Puerto Rico. 81 PUE. Hemicr. or rhizome geophyte.

Rhynchospora dilectissimi-patris Luceño, Anales Jard. Bot. Madrid 55: 309 (1997).
Brazil (Piauí). 84 BZE. Hemicr. or rhizome geophyte.

Rhynchospora diodon (Nees) Griseb., Cat. Pl. Cub.: 245 (1866).
C. America to Brazil, Cuba. 80 HON 81 CUB 83 CLM 84. Ther.
Chaetospora diodon Schrad. ex Nees in C.F.P.von Martius & auct. suc. (eds.), Fl. Bras. 2(1): 148 (1842).
**Ptilochaeta diodon* Nees in C.F.P.von Martius & auct. suc. (eds.), Fl. Bras. 2(1): 147 (1842).
Rhynchospora lunata Griseb., Cat. Pl. Cub.: 244 (1866).

Rhynchospora dissitiflora Steud. ex Boeck., Linnaea 37: 598 (1873).
Costa Rica to W. South America. 80 COS PAN 82 VEN 83 BOL CLM ECU PER. Hemicr.
Rhynchospora culmenicola Steyerm., Fieldiana, Bot. 28(1): 41 (1951).

Rhynchospora dissitispicula T.Koyama, Bull. Natl. Sci. Mus., Tokyo, B 2: 169 (1976).
Brazil (Goiás). 84 BZC. Hemicr. or rhizome geophyte.

Rhynchospora divergens Chapm. ex M.A.Curtis, Amer. J. Sci. Arts, II, 7: 409 (1849). *Phaeocephalum divergens* (Chapm. ex M.A.Curtis) House, Amer. Midl. Naturalist 6: 202 (1920).
SE. U.S.A., SE. Mexico to Belize, Caribbean. 77 TEX 78 ALA FLA GEO LOU MSI NCA SCA 79 MXT 80 BLZ 81 BAH CUB DOM. Ther. or hemicr.
Rhynchospora ravenelii M.A.Curtis, Amer. J. Sci. Arts, II, 7: 410 (1849).
Rhynchospora lamprosperma C.Wright, Anales Acad. Ci. Méd. Habana 8: 87 (1871).
Rhynchospora divergens var. *expansa* Kük., Bot. Jahrb. Syst. 75: 283 (1951).

Rhynchospora dives Standl., Publ. Field Mus. Nat. Hist., Bot. Ser. 17: 315 (1938).
SE. Mexico to C. America. 79 MXG MXT 80 COS GUA HON. Hemicr. or rhizome geophyte.
Rhynchospora orizabensis C.B.Clarke ex Britton, Trans. New York Acad. Sci. 11: 85 (1892), nom. inval.
Rhynchospora orizabensis C.B.Clarke ex Kük., Bot. Jahrb. Syst. 74: 406 (1949).

Rhynchospora domingensis Urb., Symb. Antill. 7: 170 (1912). *Rhynchospora longiflora* var. *domingensis* (Urb.) Kük., Bot. Jahrb. Syst. 74: 396 (1949).
E. Cuba to WC. Puerto Rico. 81 CUB DOM PUE. Hemicr. or rhizome geophyte.

Rhynchospora donselaarii M.T.Strong, Novon 11: 268 (2001).
Suriname. 82 SUR. Hemicr. or rhizome geophyte.

Rhynchospora duckei R.Gross, Repert. Spec. Nov. Regni Veg. 50: 212 (1941).
SE. Colombia to N. Brazil. 82 VEN 83 CLM 84 BZN. Hemicr. or rhizome geophyte.
Dichromena brevicaulis Suess., Bot. Jahrb. Syst. 73: 199 (1943).

Rhynchospora duidae Steyerm., Fieldiana, Bot. 28(1): 43 (1951).
S. Venezuela. 82 VEN. Hemicr. or rhizome geophyte.

Rhynchospora durangensis Kral & W.W.Thomas, Brittonia 38: 214 (1986).
Mexico (Durango). 79 MXE. Hemicr. or rhizome geophyte.

Rhynchospora ebracteata (Standl.) H.Pfeiff., Repert. Spec. Nov. Regni Veg. 17: 238 (1921).
Tobago. 81 TRT. Hemicr. or rhizome geophyte.
**Dichromena ebracteata* Standl., Contr. U. S. Natl. Herb. 17: 428 (1914).

Rhynchospora eburnea Kral & W.W.Thomas, Brittonia 40: 32 (1988).
S. Venezuela to N. Brazil. 82 VEN 84 BZN. Hemicr. or rhizome geophyte.

Rhynchospora edwalliana Boeck., Vidensk. Meddel. Dansk Naturhist. Foren. Kjøbenhavn 1894: 239 (1895).
SE. & S. Brazil. 84 BZL BZS. Hemicr. or rhizome geophyte.
Rhynchospora flexuosa C.B.Clarke, Kew Bull., Addit. Ser. 8: 90 (1908).

Rhynchospora elatior Kunth, Enum. Pl. 2: 289 (1837).
Brazil. 84 BZC BZL BZN BZS. Hemicr. or rhizome geophyte.
Chaetospora ciliata Schrad. ex Nees in C.F.P.von Martius & auct. suc. (eds.), Fl. Bras. 2(1): 128 (1842).
Haplostylis elatior Nees in C.F.P.von Martius & auct. suc. (eds.), Fl. Bras. 2(1): 128 (1842).
Haplostylis elatior var. *hirsuta* Nees in C.F.P.von Martius & auct. suc. (eds.), Fl. Bras. 2(1): 128 (1842).
Rhynchospora elatior var. *hirsuta* Nees in C.F.P.von Martius & auct. suc. (eds.), Fl. Bras. 2(1): 128 (1842).
Rhynchospora macrocephala Boeck., Beitr. Cyper. 2: 21 (1890).
Rhynchospora burchellii C.B.Clarke, Kew Bull., Addit. Ser. 8: 34 (1908). *Rhynchospora elatior* var. *burchellii* (C.B.Clarke) Kük., Bot. Jahrb. Syst. 74: 470 (1949). *Rhynchospora elatior* var. *burchellii* (C.B.Clarke) Kük., Bot. Jahrb. Syst. 74: 470 (1949).
Rhynchospora elatior var. *variegatopes* H.Pfeiff., Repert. Spec. Nov. Regni Veg. 33: 207 (1933).
Rhynchospora elatior var. *variegatopes* H.Pfeiff., Repert. Spec. Nov. Regni Veg. 33: 207 (1933).

Rhynchospora elegantula Maury, J. Bot. (Morot) 3: 209 (1889).
S. Venezuela. 82 VEN. Hemicr. or rhizome geophyte.

Rhynchospora elliottii A.Dietr., Sp. Pl. 2: 69 (1832).
SE. U.S.A. to Texas. 77 TEX 78 ALA FLA GEO LOU MSI NCA SCA. Hemicr.
**Scirpus schoenoides* Elliott, Sketch Bot. S. Carolina 1: 89 (1816), nom. illeg. *Rhynchospora schoenoides* Alph.Wood, Class-book Bot.: 744 (1861). *Phaeocephalum schoenoides* (Alph.Wood) House, Amer. Midl. Naturalist 6: 202 (1920).
Rhynchospora multiflora A.Gray, Ann. Lyceum Nat. Hist. New York 3: 200 (1835).
Rhynchospora pendula A.Gray, Ann. Lyceum Nat. Hist. New York 3: 201 (1835).

Rhynchospora emaciata (Nees) Boeck., Vidensk. Meddel. Naturhist. Foren. Kjøbenhavn 31: 149 (1869 publ. 1870).
Trinidad, C. & S. Trop. America. 80 COS NIC 81 TRT 82 GUY SUR VEN VEN 83 BOL CLM 84 AGE BZC BZE BZL BZN BZS PAR. Hemicr.
**Haloschoenus emaciatus* Nees in C.F.P.von Martius & auct. suc. (eds.), Fl. Bras. 2(1): 121 (1842). *Dichromena emaciata* (Nees) Steud., Syn. Pl. Glumac. 2: 136 (1855). *Rhynchospora tenuis* var. *emaciata* (Nees) Lindm., Bih. Kongl. Svenska Vetensk.-Akad. Handl. 26(3; 9): 28 (1900).

Dichromena pohliana Steud., Syn. Pl. Glumac. 2: 319 (1855).
Rhynchospora leptostachya Boeck., Linnaea 37: 591 (1873).

Rhynchospora enmanuelis Luceño & Rocha, Hoehnea 29: 197 (2002).
Brazil (São Paulo). 84 BZL.

Rhynchospora eurycarpa A.C.Araujo & Longhi-Wagner, Novon 14: 6 (2004).
Brazil (Mato Grosso). 84 BZC.

Rhynchospora exaltata Kunth, Enum. Pl. 2: 291 (1837). *Dichromena exaltata* (Kunth) J.F.Macbr., Publ. Field Columbian Mus., Bot. Ser. 8: 113 (1930).
Trop. America to Paraguay. 80 BLZ 81 CUB DOM 82 GUY VEN 83 BOL PER 84 BZC BZE BZL BZS 85 PAR. Hemicr. or rhizome geophyte.
Chaetospora triceps Cham. & Schltdl., Linnaea 6: 29 (1831). *Echinoschoenus triceps* (Cham. & Schltdl.) Nees & Meyen in F.J.F.Meyen, Observ. Bot. 1: 116 (1834). *Rhynchospora triceps* (Cham. & Schltdl.) Steud., Nomencl. Bot., ed. 2, 2: 456 (1841), nom. illeg.
Echinoschoenus beyrichii Nees in C.F.P.von Martius & auct. suc. (eds.), Fl. Bras. 2(1): 132 (1842). *Rhynchospora beyrichii* (Nees) Steud., Syn. Pl. Glumac. 2: 144 (1855). *Rhynchospora exaltata* var. *beyrichii* (Nees) Kük., Bot. Jahrb. Syst. 74: 440 (1949).
Mitrospora cephalophora Nees in C.F.P.von Martius & auct. suc. (eds.), Fl. Bras. 2(1): 133 (1842). *Rhynchospora cephalophora* (Nees) Steud., Syn. Pl. Glumac. 2: 146 (1855). *Rhynchospora exaltata* var. *cephalophora* (Nees) Kük., Bot. Jahrb. Syst. 74: 440 (1949).
Echinoschoenus sparganioides Lindl. ex Nees, Nov. Actorum Acad. Caes. Leop.-Carol. Nat. Cur. 19(Suppl. 1): 103 (1843). *Rhynchospora sparganoides* (Lindl. ex Nees) Steud., Syn. Pl. Glumac. 2: 144 (1855).
Rhynchospora arundinacea Boeck., Vidensk. Meddel. Naturhist. Foren. Kjøbenhavn 41-42: 27 (1879).
Rhynchospora schenckiana Boeck., Beitr. Cyper. 1: 22 (1888).
Rhynchospora exaltata f. *angusta* Kük., Ark. Bot. 22A(17): 8 (1929).
Rhynchospora exaltata var. *ovalis* Kük., Repert. Spec. Nov. Regni Veg. 26: 253 (1929).
Rhynchospora exaltata var. *microcephala* Kük., Bot. Jahrb. Syst. 74: 440 (1949).
Rhynchospora exaltata f. *monocephala* Kük., Bot. Jahrb. Syst. 74: 440 (1949).
Rhynchospora simplex var. *angusta* Kük., Bot. Jahrb. Syst. 74: 441 (1949).

Rhynchospora eximia (Nees) Boeck., Linnaea 37: 601 (1873).
Florida to Trop. America, Trop. Africa. 22 BEN GHA GNB GUI IVO MLI NGA SEN SIE TOG 23 CAF CMN GAB 24 CHA 26 MLW 78 FLA 79 MXG MXS MXT 80 BLZ COS GUA HON NIC PAN 81 CUB DOM 82 VEN 83 BOL 84 BZN. Ther.
Chaetospora ferruginea Cham. & Schltdl., Linnaea 6: 28 (1831).
Dichromena schiedeana Kunth, Enum. Pl. 2: 282 (1837). *Psilocarya schiedeana* (Kunth) Liebm., Mexic. Neldeagt. Pl., V, 2: 248 (1851). *Rhynchospora grosserugosa* Kük., Bot. Jahrb. Syst. 75: 280 (1951).
**Spermodon eximius* Nees in B.Seemann, Bot. Voy. Herald: 222 (1854).
Rhynchospora psilocaryoides Griseb., Cat. Pl. Cub.: 247 (1866).
Rhynchospora oxycephala C.Wright, Anales Acad. Ci. Méd. Habana 8: 152 (1871).
Rhynchospora schroederi K.Schum. ex C.B.Clarke, Bot. Jahrb. Syst. 38: 135 (1906).
Rhynchospora elegans Kük., Bot. Jahrb. Syst. 56(125): 18 (1921).
Rhynchospora testui var. *pleiantha* Cherm., Bull. Soc. Bot. France 81: 267 (1934).
Rhynchospora eximia var. *multiflorens* Kük., Bot. Jahrb. Syst. 75: 182 (1950).

Rhynchospora exserta C.B.Clarke, Kew Bull., Addit. Ser. 8: 33 (1908). *Rhynchospora longisetis* subsp. *exserta* (C.B.Clarke) Kük., Bull. Jard. Bot. Buitenzorg, III, 16: 303 (1940).
N. Queensland. 50 QLD. Hemicr. or rhizome geophyte.

Rhynchospora faberi C.B.Clarke, J. Linn. Soc., Bot. 36: 259 (1903).
China to Russian Far East. 31 36 CHS 38 JAP KOR. Hemicr. or rhizome geophyte.
Rhynchospora fauriei var. *leviseta* C.B.Clarke ex H.Lév., Bull. Soc. Philom. Paris, VII, 10: 104 (1886).
Rhynchospora hattoriana Makino, Bot. Mag. (Tokyo) 17: 189 (1903).
Rhynchospora miyakeana Makino, Bot. Mag. (Tokyo) 17: 184 (1903).
Rhynchospora umemurae Makino, Bot. Mag. (Tokyo) 17: 187 (1903).
Rhynchospora breviseta Palla, Oesterr. Bot. Z. 59: 187 (1909).
Rhynchospora ouensanensis Koidz., Bot. Mag. (Tokyo) 43: 389 (1929).
Rhynchospora umemurae var. *yakusimensis* Masam., Trans. Nat. Hist. Soc. Taiwan 23: 210 (1933). *Rhynchospora yakusimensis* (Masam.) Masam., Mem. Fac. Sci. Taihoku Imp. Univ. 11(4): 522 (1934).

Rhynchospora fallax Uittien, Recueil Trav. Bot. Néerl. 22: 336 (1925). *Rhynchospora setacea* var. *fallax* (Uittien) Kük., Bot. Jahrb. Syst. 75: 290 (1951). *Rhynchospora tenerrima* subsp. *fallax* (Uittien) T.Koyama, Mem. New York Bot. Gard. 23: 70 (1972).
S. Trop. America. 82 FRG SUR VEN 83 CLM 84 BZC. Ther.
Schoenus setaceus Rottb., Descr. Pl. Rar.: 14 (1772).
Rhynchospora setacea f. *filicula* Kük., Bot. Jahrb. Syst. 75: 289 (1951).

Rhynchospora fascicularis (Michx.) Vahl, Enum. Pl. 2: 234 (1805).
SE. U.S.A. to N. South America. 77 TEX 78 ALA FLA GEO LOU MSI NCA SCA VRG 79 MXG MXT 80 BLZ GUA HON NIC 81 BER CUB DOM JAM LEE PUE 82 GUY SUR. Hemicr.
**Schoenus fascicularis* Michx., Fl. Bor.-Amer. 1: 37 (1803). *Phaeocephalum fasciculare* (Michx.) House, Amer. Midl. Naturalist 6: 202 (1920). *Rhynchospora distans* var. *fascicularis* (Michx.) Kük., Repert. Spec. Nov. Regni Veg. 23: 208 (1926). *Rhynchospora fascicularis* var. *typica* Gale, Rhodora 46: 188 (1944), nom. inval.
Rhynchospora fascicularis var. *trichoides* Chapm., Fl. South. U.S.: 527 (1860).

subsp. ***fascicularis***
SE. U.S.A. to C. America, Caribbean. 77 TEX 78 ALA FLA GEO LOU MSI NCA SCA VRG 79 MXG MXT 80 BLZ HON NIC 81 BER CUB DOM JAM LEE PUE. Hemicr.
Schoenus distans Michx., Fl. Bor.-Amer. 1: 36 (1803). *Rhynchospora distans* (Michx.) Vahl, Enum. Pl. 2: 235 (1805). *Rhynchospora fascicularis* var. *distans* (Michx.) Chapm., Enum. Pl. 2: 296 (1837). *Phaeocephalum distans* (Michx.) House, Amer. Midl. Naturalist 6: 201 (1920). *Dichromena distans* (Michx.) J.F.Macbr., Publ. Field Mus. Nat. Hist., Bot. Ser. 11: 5 (1931).
Rhynchospora setacea Vahl, Enum. Pl. 2: 233 (1805).
Rhynchospora filiformis Balb. ex Kunth, Enum. Pl. 2: 297 (1837), nom. inval.
Rhynchospora dommucensis A.H.Moore, List Pl. Bermuda: 6 (1906).

subsp. ***sierrensis*** W.W.Thomas, Brittonia 44: 29 (1992).
Guatemala. 80 GUA. Hemicr.

Rhynchospora fauriei Franch., Bull. Soc. Philom. Paris, VII, 10: 104 (1886).
Korea, Japan. 38 JAP KOR. Hemicr. or rhizome geophyte.

Rhynchospora fernaldii Gale, Rhodora 46: 182 (1944). *Rhynchospora fascicularis* var. *fernaldii* (Gale) Kük., Bot. Jahrb. Syst. 75: 111 (1950).
SE. U.S.A. 78 ALA FLA GEO MSI. Hemicr. or rhizome geophyte.

Rhynchospora filifolia A.Gray, Ann. Lyceum Nat. Hist. New York 3: 366 (1836). *Phaeocephalum filifolium* (A.Gray) House, Amer. Midl. Naturalist 6: 202 (1920).
New Jersey to Texas, SE. Mexico to C. America, Cuba. 75 NWJ 77 TEX 78 ALA DEL FLA GEO LOU MRY MSI NCA SCA VRG 79 MXT 80 BLZ NIC 81 CUB. Hemicr.
Rhynchospora fuscoides C.B.Clarke ex Britton, Trans. New York Acad. Sci. 11: 89 (1892). *Phaeocephalum fuscoides* (C.B.Clarke ex Britton) House, Amer. Midl. Naturalist 6: 202 (1920).

Rhynchospora filiformis Vahl, Enum. Pl. 2: 232 (1805).
Trop. America. 79 MXT 80 BLZ COS GUA HON NIC 81 CUB DOM LEE PUE TRT 82 FRG GUY SUR VEN 83 BOL CLM 84 BZC BZE BZL BZN. Hemicr.
Schoenus distans Poir. in J.B.A.M.de Lamarck, Encycl., Suppl. 2: 249 (1811), nom. illeg.
Dichromena filiformis Kunth, Enum. Pl. 2: 281 (1837). *Spermodon filiformis* (Kunth) Nees in C.F.P.von Martius & auct. suc. (eds.), Fl. Bras. 2(1): 118 (1842).
Rhynchospora inexpansa Sieber ex Steud., Nomencl. Bot., ed. 2, 2: 456 (1841).
Rhynchospora cubensis Griseb., Cat. Pl. Cub.: 246 (1866), nom. illeg.
Rhynchospora podosperma C.Wright ex Sauvalle, Anales Acad. Ci. Méd. Habana 8: 87 (1871).
Rhynchospora longispicata Boeck., Linnaea 37: 600 (1873).

Rhynchospora floridensis (Britton ex Small) H.Pfeiff., Repert. Spec. Nov. Regni Veg. 49: 82 (1940).
S. Florida, Bahamas, SE. Mexico to Belize. 78 FLA 79 MXT 80 BLZ 81 BAH. Hemicr.
**Dichromena floridensis* Britton ex Small, Fl. S.E. U.S.: 190 (1903). *Rhynchospora stellata* f. *floridensis* (Britton ex Small) Kük., Bot. Jahrb. Syst. 75: 300 (1951).
Dichromena inaguensis Britton in N.L.Britton & C.F.Millspaugh, Bahama Fl.: 53 (1920).

Rhynchospora fujiana Makino, Bot. Mag. (Tokyo) 17: 183 (1903).
Russian Far East to Korea. 31 38 JAP KOR. Hemicr. or rhizome geophyte.
Rhynchospora fujiana var. *scabrista* Makino, Bot. Mag. (Tokyo) 17: 184 (1903).
Rhynchospora yasudana Makino, Bot. Mag. (Tokyo) 17: 188 (1903).
Rhynchospora franchetiana C.B.Clarke ex H.Lév., Bull. Acad. Int. Géogr. Bot. 14: 197 (1904).
Rhynchospora coreana Palla, Oesterr. Bot. Z. 59: 186 (1909).

Rhynchospora fusca (L.) W.T.Aiton, Hortus Kew. 1: 127 (1810).
Temp. Northern Hemisphere. 10 DEN FIN GRB IRE NOR SWE 11 AUT BGM CZE GER NET POL SWI 12 FRA 13 ITA YUG 14 BLR BLT RUN 71 SAS 72 LAB NBR NFL NSC ONT QUE 74 ILL MIN WIS 75 CNT INI MAI MAS MIC NWH NWJ NWY PEN RHO VER 78 DEL MRY. Hemicr.
**Schoenus fuscus* L., Sp. Pl. ed. 2: 1664 (1763). *Rhynchospora alba* var. *fusca* (L.) Pursh, Fl. Amer. Sept. 1: 49 (1813). *Phaeocephalum fuscum* (L.) House, Amer. Midl. Naturalist 6: 202 (1920).
Schoenus setaceus Thuill., Fl. Env. Paris, ed. 2: 19 (1799), nom. illeg.
Rhynchospora capillacea Steud., Syn. Pl. Glumac. 2: 141 (1855), nom. illeg.

Rhynchospora gageri Britton, Contr. New York Bot. Gard. 194: 12 (1917). *Rhynchospora joveroensis* var. *gageri* (Britton) Kük., Bot. Jahrb. Syst. 75: 108 (1950).
Cuba. 81 CUB. Hemicr. or rhizome geophyte.
Rhynchospora longifrons var. *reducta* Kük., Repert. Spec. Nov. Regni Veg. 23: 209 (1926).

Rhynchospora gigantea Link, Jahrb. Gewächsk. 1(3): 76 (1820). *Calyptrostylis gigantea* (Link) Nees in C.F.P.von Martius & auct. suc. (eds.), Fl. Bras. 2(1): 137 (1842). *Dichromena gigantea* (Link) J.F.Macbr., Publ. Field Mus. Nat. Hist., Bot. Ser. 11: 6 (1931).
Trop. America. 79 MXG MXT 80 NIC 81 CUB JAM PUE 82 FRG GUY SUR VEN 84 BZC BZE BZL BZN BZS 85 PAR. Hemicr. or rhizome geophyte.
Cephaloschoenus pseudomariscus Nees, Linnaea 9: 296 (1834), nom. inval.
Rhynchospora pseudomariscus Kunth, Enum. Pl. 2: 303 (1837).
Rhynchospora scaberrima Boeck., Flora 63: 452 (1880). *Rhynchospora gigantea* var. *scaberrima* (Boeck.) Kük., Bot. Jahrb. Syst. 74: 419 (1949).

Rhynchospora glaziovii Boeck., Linnaea 38: 405 (1874).
SE. Brazil. 84 BZL. Hemicr. or rhizome geophyte.
Rhynchospora glaziovii var. *compacta* Kük., Bot. Jahrb. Syst. 74: 445 (1949).

Rhynchospora globosa (Kunth) Roem. & Schult., Syst. Veg. 2: 89 (1817).
Trop. America. 79 MXG MXS MXT 80 BLZ COS ELS GUA HON NIC PAN 81 CUB TRT 82 FRG GUY SUR VEN 83 BOL CLM PER 84 BZC BZE BZL BZN BZS 85 AGE PAR URU. Hemicr. or rhizome geophyte.

**Chaetospora globosa* Kunth, Nov. Gen. Sp. 1: 135 (1816). *Cephaloschoenus globosus* (Kunth) Nees, Linnaea 9: 296 (1834). *Schoenus globosus* (Kunth) Willd. ex Kunth, Enum. Pl. 2: 288 (1837).

var. ***globosa***
Trop. America. 79 MXS 80 BLZ COS ELS GUA HON NIC PAN 81 CUB TRT 82 FRG GUY SUR VEN 83 BOL CLM PER 84 BZC BZE BZL BZN BZS 85 AGE PAR URU. Hemicr. or rhizome geophyte.
Schoenus cerasiformis Poir. in J.B.A.M.de Lamarck, Encycl., Suppl. 5: 617 (1817).
Cephaloschoenus armeria Nees, Linnaea 9: 296 (1834), nom. inval. *Cephaloschoenus globosus* var. *armeria* Nees in C.F.P.von Martius & auct. suc. (eds.), Fl. Bras. 2(1): 130 (1842).
Cephaloschoenus globosus var. *minor* Nees in C.F.P.von Martius & auct. suc. (eds.), Fl. Bras. 2(1): 130 (1842).
Cephaloschoenus pohlianus Nees in C.F.P.von Martius & auct. suc. (eds.), Fl. Bras. 2(1): 131 (1842). *Rhynchospora pohliana* (Nees) Steud., Syn. Pl. Glumac. 2: 142 (1855).
Rhynchospora cephalantha A.Rich. in R.de la Sagra, Hist. Fis. Cuba, Bot. 11: 293 (1850), nom. illeg.
Cephaloschoenus marginatus Liebm., Mexic. Neldeagt. Pl., V, 2: 250 (1851). *Rhynchospora marginata* (Liebm.) Steud., Syn. Pl. Glumac. 2: 142 (1855).
Rhynchospora presleana Steud., Syn. Pl. Glumac. 2: 142 (1855).
Rhynchospora alta Boeck., Beitr. Cyper. 2: 20 (1890). *Dichromena alta* (Boeck.) J.F.Macbr., Publ. Field Columbian Mus., Bot. Ser. 4: 166 (1929).
Rhynchospora loefgrenii Boeck., Beitr. Cyper. 2: 21 (1890). *Rhynchospora globosa* var. *loefgrenii* (Boeck.) Kük., Bot. Jahrb. Syst. 74: 465 (1949).
Xyris triquetra Kuntze, Revis. Gen. Pl. 2: 719 (1891).
Rhynchospora epiglobosa C.B.Clarke, Kew Bull., Addit. Ser. 8: 34 (1908). *Rhynchospora globosa* var. *epiglobosa* (C.B.Clarke) Kük., Bot. Jahrb. Syst. 74: 465 (1949).
Rhynchospora globosa var. *stenocarpa* Kük., Bot. Jahrb. Syst. 74: 465 (1949).

var. ***tenuifolia*** León, Fl. Cuba 1: 219 (1946).
Cuba (I. de la Juventud). 81 CUB. Hemicr. or rhizome geophyte.

Rhynchospora globularis (Chapm.) Small, Man. S.E. Fl.: 184 (1933).
C. & E. U.S.A., Caribbean. (63) haw 74 MSO OKL 75 INI NWJ OHI PEN (76) cal 77 TEX 78 ALA ARK FLA GEO LOU MRY MSI NCA SCA TEN VRG 81 CUB JAM. Hemicr. or rhizome geophyte.
**Rhynchospora cymosa* var. *globularis* Chapm., Fl. South. U.S.: 525 (1860). *Rhynchospora globularis* var. *typica* Gale, Rhodora 46: 244 (1944), nom. inval.

var. ***globularis***
C. & E. U.S.A. (63) haw 74 MSO OKL 75 INI NWJ OHI PEN (76) cal 77 TEX 78 ALA ARK FLA GEO LOU MRY MSI NCA SCA TEN VRG. Hemicr. or rhizome geophyte.
Rhynchospora obliterata Gale, Rhodora 46: 242 (1944).

var. ***pinetorum*** (Britton & Small) Gale, Rhodora 46: 248 (1944).
SE. U.S.A. to Texas, Caribbean. 77 TEX 78 ALA FLA GEO LOU MSI NCA SCA 81 CUB JAM. Hemicr. or rhizome geophyte.
**Rhynchospora pinetorum* Britton & Small in J.K.Small, Man. S.E. Fl.: 183 (1933).

var. ***saxicola*** (Small) Kük., Bot. Jahrb. Syst. 75: 159 (1950).
SE. U.S.A. 78 ALA GEO SCA. Ther. or hemicr.
**Rhynchospora saxicola* Small, Man. S.E. Fl.: 185 (1933).

Rhynchospora glomerata (L.) Vahl, Enum. Pl. 2: 234 (1805).
C. & E. U.S.A. 74 ILL OKL 75 NWJ WVA 77 TEX 78 ALA ARK DEL FLA GEO KTY LOU MRY MSI NCA SCA TEN VRG. Hemicr. or rhizome geophyte.
**Schoenus glomeratus* L., Sp. Pl.: 44 (1753). *Phaeocephalum glomeratum* (L.) House, Amer. Midl. Naturalist 6: 202 (1920). *Rhynchospora glomerata* var. *typica* Fernald, Rhodora 37: 401 (1935), nom. inval.
Schoenus axillaris Lam., Tabl. Encycl. 1: 137 (1791). *Rhynchospora axillaris* (Lam.) Britton, Prelim. Cat.: 61 (1888). *Hemicarpha axillaris* (Lam.) B.D.Jacks., Index Kew., Suppl. 1: 201 (1902). *Phaeocephalum axillare* (Lam.) House, Amer. Midl. Naturalist 6: 201 (1920).
Schoenus capitatus Pers., Syn. Pl. 1: 60 (1805), nom. illeg.
Rhynchospora capitellata Elliott, Sketch Bot. S. Carolina 1: 61 (1816), nom. illeg.
Rhynchospora filiformis Sieber ex C.Presl, Isis (Oken) 21: 269 (1828), nom. illeg.
Rhynchospora pauciflora C.Presl, Isis (Oken) 21: 269 (1828).
Rhynchospora teres Eaton, Man. Bot., ed. 5: 364 (1829).
Rhynchospora paniculata A.Gray, Ann. Lyceum Nat. Hist. New York 3: 211 (1835), nom. illeg. *Rhynchospora glomerata* var. *paniculata* Chapm., Fl. South. U.S.: 528 (1860).
Rhynchospora distans Willd. ex Kunth, Enum. Pl. 2: 296 (1837).
Rhynchospora humilis Beyr. ex Kunth, Enum. Pl. 2: 297 (1837).
Rhynchospora torreyana Griseb., Cat. Pl. Cub.: 243 (1866).
Rhynchospora glomerata var. *discutiens* C.B.Clarke ex Britton, Trans. New York Acad. Sci. 11: 89 (1892). *Rhynchospora glomerata* f. *discutiens* (C.B.Clarke ex Britton) Fernald, Rhodora 37: 402 (1935).
Rhynchospora glomerata var. *minor* Britton, Trans. New York Acad. Sci. 11: 89 (1892). *Phaeocephalum glomeratum* var. *minus* (Britton) Farw., Rep. (Annual) Michigan Acad. Sci. 21: 361 (1920).
Rhynchospora glomerata f. *controversa* Fernald, Rhodora 37: 402 (1935).
Rhynchospora glomerata var. *angusta* Gale, Rhodora 46: 114 (1944).

Rhynchospora gollmeri Boeck., Linnaea 37: 556 (1873).
Venezuela. 82 VEN. Hemicr. or rhizome geophyte.

Rhynchospora gracilenta A.Gray, Ann. Lyceum Nat. Hist. New York 3: 215 (1835). *Phaeocephalum gracilentum* (A.Gray) House, Amer. Midl. Naturalist 6: 202 (1920).

C. & E. U.S.A. to C. America, Cuba. 74 OKL 75 NWJ PEN 77 TEX 78 ALA ARK DEL FLA GEO LOU MRY MSI NCA SCA TEN VRG 79 MXT 80 BLZ NIC 81 CUB. Hemicr.

Rhynchospora fusca A.Gray, Ann. Lyceum Nat. Hist. New York 3: 216 (1835), nom. illeg.

Rhynchospora drummondiana Boeck., Flora 41: 644 (1858), nom. illeg.

Rhynchospora gracilenta var. *diversifolia* Fernald, Rhodora 37: 399 (1935).

Rhynchospora trichophylla Fernald, Rhodora 39: 389 (1937).

Rhynchospora gracillima Thwaites & Hook., Enum. Pl. Zeyl.: 435 (1864).

Trop. & Subtrop. Old World. 22 BEN BKN GHA GUI IVO LBR MLI NGA SEN TOG 23 CMN ZAI 25 TAN UGA 26 ZAM ZIM 27 NAT SWZ 29 MDG 36+ 40 ASS SRL 41 NCB THA 42 SUL SUM 43 NWG 50 QLD. Ther. or hemicr.

subsp. ***gracillima***

Trop. Asia to N. Queensland. 36 40 ASS SRL 41 NCB THA 42 SUL SUM 43 NWG 50 QLD. Ther. or hemicr.

Rhynchospora kamphoeveneri Boeck., Bot. Jahrb. Syst. 5: 508 (1884).

subsp. ***subquadrata*** (Cherm.) J.Raynal, Adansonia, n.s., 7: 321 (1967).

Trop. & S. Africa, Madagascar. 22 BEN BKN GHA GUI IVO LBR MLI NGA SEN TOG 23 CMN ZAI 25 TAN UGA 26 ZAM ZIM 27 NAT SWZ 29 MDG. Ther. or hemicr.

**Rhynchospora subquadrata* Cherm., Bull. Soc. Bot. France 69: 720 (1922 publ. 1923).

Rhynchospora testui Cherm., Arch. Bot. Bull. Mens. 4(7): 42 (1931).

Rhynchospora grandifolia Boeck., Bot. Jahrb. Syst. 25: 587 (1898).

Samoa. 60 SAM. Hemicr. or rhizome geophyte.

Rhynchospora grayi Kunth, Enum. Pl. 2: 539 (1837). *Phaeocephalum grayi* (Kunth) House, Amer. Midl. Naturalist 6: 202 (1920).

SE. U.S.A. to Texas, W. Cuba. 77 TEX 78 ALA FLA GEO LOU MSI NCA SCA VRG 81 CUB. Hemicr. or rhizome geophyte.

Rhynchospora distans Elliott, Sketch Bot. S. Carolina 1: 59 (1816), nom. illeg. *Schoenus distans* Muhl., Descr. Gram.: 11 (1817), nom. illeg. **Rhynchospora elliottii* A.Gray, Ann. Lyceum Nat. Hist. New York 3: 204 (1835), nom. illeg.

Schoenus fuscus Muhl., Descr. Gram.: 6 (1817).

Rhynchospora suberostris Boeck., Linnaea 37: 579 (1873).

Rhynchospora* × *hakkodensis Mochizuki, J. Geobot. 24: 58 (1977). *R. alba* × *R. yasudana*.

Japan. 38 JAP. Hemicr. or rhizome geophyte.

Rhynchospora harperi Small, Man. S.E. Fl.: 182 (1933). *Rhynchospora fascicularis* var. *harperi* (Small) Kük., Bot. Jahrb. Syst. 75: 111 (1950).

SE. U.S.A., Belize. 78 ALA DEL FLA GEO MRY MSI NCA SCA 80 BLZ. Hemicr.

Rhynchospora harveyi W.Boott, Bot. Gaz. 9: 85 (1884). *Rhynchospora grayi* var. *harveyi* (W.Boott) Kük., Bot. Jahrb. Syst. 75: 120 (1950).

C. & SE. U.S.A. 74 KAN MSO OKL 77 TEX 78 ALA ARK FLA GEO LOU MSI NCA SCA TEN VRG. Hemicr.

var. ***culixa*** (Gale) Kral, Novon 9: 206 (1999).

Georgia to Florida. 78 FLA GEO. Hemicr.

**Rhynchospora culixa* Gale, Rhodora 46: 235 (1944). *Rhynchospora grayi* var. *culixa* (Gale) Kük., Bot. Jahrb. Syst. 75: 120 (1950).

var. ***harveyi***

C. & SE. U.S.A. 74 KAN MSO OKL 77 TEX 78 ALA ARK FLA GEO LOU MSI NCA SCA TEN VRG. Hemicr.

Rhynchospora earlei Britton in J.K.Small, Fl. S.E. U.S.: 197 (1903). *Phaeocephalum earlei* (Britton) House, Amer. Midl. Naturalist 6: 202 (1920).

Rhynchospora plankii Britton in J.K.Small, Fl. S.E. U.S.: 196 (1903). *Phaeocephalum plankii* (Britton) House, Amer. Midl. Naturalist 6: 202 (1920).

Rhynchospora hassleri C.B.Clarke, Bull. Herb. Boissier, II, 3: 1023 (1903).

C. & S. Trop. America. 80 BLZ COS HON NIC 82 FRG GUY SUR 83 BOL PER 84 BZN 85 PAR. Hemicr. or rhizome geophyte.

Rhynchospora viridilutea C.B.Clarke, Kew Bull., Addit. Ser. 8: 36 (1908).

Rhynchospora hatschbachii T.Koyama, Bull. Natl. Sci. Mus., Tokyo, B 2: 172 (1976).

Brazil (Paraná). 84 BZS. Hemicr. or rhizome geophyte.

Rhynchospora heterochaeta S.T.Blake, Proc. Roy. Soc. Queensland 51: 47 (1939 publ. 1940).

Vietnam to N. Australia. 41 VIE 42 JAW LSI PHI 43 NWG 50 NTA QLD. Ther.

Rhynchospora heterochaeta var. *irregularis* Kük., Bot. Jahrb. Syst. 74: 497 (1949).

Rhynchospora heterolepis Boeck., Allg. Bot. Z. Syst. 2: 94 (1896).

Brazil. 84+. Hemicr. or rhizome geophyte.

Rhynchospora hieronymi Boeck., Beitr. Cyper. 1: 30 (1888).

C. & S. Trop. America. 80 COS GUA PAN 82 GUY? VEN 83 BOL ECU 84 BZS 85 AGE AGW URU. Hemicr. or rhizome geophyte.

var. ***floribunda*** (Boeck.) Kük., Bot. Jahrb. Syst. 74: 397 (1949).

Venezuela to Argentina (Misiones). 82 VEN 83 ECU 84+ 85 AGE. Hemicr. or rhizome geophyte.

**Rhynchospora floribunda* Boeck., Beitr. Cyper. 2: 27 (1890).

var. ***hieronymi***

C. & S. Trop. America. 80 COS GUA PAN 82 GUY VEN 83 BOL 84 BZS 85 AGW. Hemicr. or rhizome geophyte.

Rhynchospora hieronymi f. *compacta* Kük., Bot. Jahrb. Syst. 74: 397 (1949).

var. ***montevidensis*** Barros, Anales Mus. Argent. Ci. Nat. "Bernardino Rivadavia" 41: 398 (1945). *Rhynchospora montevidensis* (Barros) Herter, Revista Sudamer. Bot. 9: 158 (1953). *Rhynchospora hieronymi* subsp. *montevidensis* (Barros) Guagl., Darwiniana 23: 499 (1981).

S. Brazil to NE. Argentina. 84 BZS 85 AGE URU. Hemicr. or rhizome geophyte.

Rhynchospora hildebrandtii Boeck., Bot. Jahrb. Syst. 5: 508 (1884).

C. Madagascar. 29 MDG. Hemicr.

Rhynchospora hirsuta (Vahl) Vahl, Enum. Pl. 2: 231 (1805).
Trop. America. 79 MXT 80 BLZ COS GUA HON NIC PAN 81 CUB DOM TRT 82 FRG GUY SUR VEN 83 CLM 84 BZC BZE BZN. Ther.
**Schoenus hirsutus* Vahl, Eclog. Amer. 1: 6 (1797). *Dichromena hirsuta* (Vahl) Kunth, Enum. Pl. 2: 281 (1837).
Rhynchospora venezolana Kük., Feddes Repert. Spec. Nov. Regni Veg. 53: 74 (1944).

Rhynchospora hirta (Nees) Boeck., Vidensk. Meddel. Naturhist. Foren. Kjøbenhavn 1869: 146 (1869).
Brazil to Bolivia. 82 VEN? 83 BOL 84 BZC BZL. Hemicr. or rhizome geophyte.
**Haplostylis hirta* Nees in C.F.P.von Martius & auct. suc. (eds.), Fl. Bras. 2(1): 128 (1842).
Ptilosciadium lindenii Steud., Syn. Pl. Glumac. 2: 150 (1855).
Rhynchospora monadelpha C.B.Clarke, Kew Bull., Addit. Ser. 8: 33 (1908).
Rhynchospora hirta var. *tenuior* Kük., Bot. Jahrb. Syst. 74: 467 (1949).

Rhynchospora hirticeps (Kük.) T.Koyama, Brittonia 31: 288 (1979).
Thailand to New Guinea. 41 THA 42 PHI SUL 43 NWG. Hemicr. or rhizome geophyte.
**Rhynchospora rubra* var. *hirticeps* Kük., Bot. Jahrb. Syst. 74: 495 (1949).

Rhynchospora hispidula Griseb., Cat. Pl. Cub.: 245 (1866). *Rhynchospora grisebachii* Boeck. ex C.B.Clarke in I.Urban, Symb. Antill. 2: 166 (1900), nom. illeg.
W. Cuba. 81 CUB. Hemicr. or rhizome geophyte.

Rhynchospora holoschoenoides (Rich.) Herter, Revista Sudamer. Bot. 9: 157 (1953).
Mexico to Trop. America, Trop. & S. Africa, Madagascar. 22 BEN GHA GNB GUI IVO LBR NGR SEN SIE 23 EQG GAB 25 TAN 26 ANG ZAM ZIM 27 BOT CPP NAM NAT 29 MAU MDG REU 79 MXG MXS MXT 80 BLZ COS ELS GUA HON NIC PAN 81 BAH CUB DOM JAM LEE PUE TRT WIN 82 FRG GUY SUR VEN 83 BOL 84 BZC BZE BZL BZN BZS 85 AGE PAR URU. Hemicr. or hel.
Schoenus cyperoides Sw., Prodr.: 19 (1788), nom. illeg. *Rhynchospora cyperoides* Mart., Denkschr. Königl. Akad. Wiss. München 6: 149 (1824). *Dichromena cyperoides* (Mart.) J.F.Macbr., Publ. Field Mus. Nat. Hist., Bot. Ser. 11: 6 (1931).
Schoenus cyperinus J.F.Gmel., Syst. Nat. 2: 1621 (1792), orth. var.
**Schoenus holoschoenoides* Rich., Actes Soc. Hist. Nat. Paris 1: 106 (1792).
Schoenus triceps Vahl, Eclog. Amer. 2: 4 (1798). *Rhynchospora triceps* (Vahl) Hochst., Flora 28: 759 (1845), nom. illeg.
Schoenus fragiferus Rudge, Pl. Guian.: 15 (1805).
Schoenus polycephalus Pers., Syn. Pl. 1: 59 (1805), nom. illeg. *Cephaloschoenus polycephalus* Nees, Linnaea 9: 296 (1834). *Rhynchospora polycephala* (Nees) Wydler ex Kunth, Enum. Pl. 2: 291 (1837). *Ephippiorhynchium polycephalum* (Nees) Nees in C.F.P.von Martius & auct. suc. (eds.), Fl. Bras. 2(1): 134 (1842).
Scirpus muricatus Poir. in J.B.A.M.de Lamarck, Encycl. 6: 765 (1805).
Rhynchospora globulifera Link, Jahrb. Gewächsk. 1(3): 76 (1820). *Schoenus globuliferus* (Link) Willd. ex Kunth, Enum. Pl. 2: 291 (1837).
Cephaloschoenus tenuirostris Nees, Linnaea 9: 296 (1834), nom. inval. *Rhynchospora tenuirostris* (Nees) Kunth, Enum. Pl. 2: 303 (1837).
Rhynchospora sparganoides Bojer, Hortus Maurit.: 386 (1837), nom. inval.
Schoenus sparganoides Lindl. ex Kunth, Enum. Pl. 2: 291 (1837), pro syn.
Ephippiorhynchium tenuirostre Nees in C.F.P.von Martius & auct. suc. (eds.), Fl. Bras. 2(1): 135 (1842).
Cephaloschoenus oligocephalus Hochst., Flora 28: 759 (1845).
Mariscus piluliferus G.Bertol., Mem. Reale Accad. Sci. Ist. Bologna 5: 466 (1854).
Rhynchospora mauritii Steud., Syn. Pl. Glumac. 2: 149 (1855).
Rhynchospora urvillei Steud., Syn. Pl. Glumac. 2: 149 (1855).
Rhynchospora arechavaletae Boeck., Beitr. Cyper. 1: 24 (1888).
Rhynchospora sphaerocephala Boeck., Beitr. Cyper. 1: 24 (1888).
Rhynchospora schoenoides Britton, Trans. New York Acad. Sci. 11: 92 (1892).
Rhynchospora dolichostyla K.Schum. in O.Warburg (ed.), Kunene-Sambesi Exped.: 179 (1903).
Rhynchospora ostenii Kük., Repert. Spec. Nov. Regni Veg. 13: 136 (1914). *Rhynchospora arechavaletae* var. *ostenii* (Kük.) Kük., Bot. Jahrb. Syst. 74: 436 (1949).
Rhynchospora riedeliana Gand., Bull. Soc. Bot. France 66: 298 (1919 publ. 1920), nom. illeg.
Rhynchospora arechavaletae f. *pleiocephala* Osten, Anales Mus. Hist. Nat. Montevideo, II, 3: 213 (1931).
Rhynchospora cyperoides var. *triquetrifolia* Kük., Bot. Jahrb. Syst. 74: 435 (1949).

Rhynchospora hookeri Boeck., Linnaea 37: 621 (1873).
E. Himalaya to W. Malesia. 40 ASS BAN EHM IND 41 MYA THA VIE 42 JAW MLY SUL?. Hemicr. or rhizome geophyte.

Rhynchospora iberae Guagl., Darwiniana 24: 469 (1982).
Brazil to Argentina (Corrientes). 84 BZS 85 AGE PAR. Hemicr. or rhizome geophyte.

Rhynchospora ierensis C.D.Adams, Phytologia 63: 298 (1987).
SW. Trinidad. 81 TRT. Hemicr. or rhizome geophyte.

Rhynchospora ignorata Boeck., Bot. Jahrb. Syst. 5: 508 (1884).
W. Cuba. 81 CUB. Ther.
Scirpus corymbiferus C.Wright, Anales Acad. Ci. Méd. Habana 8: 80 (1871). *Psilocarya corymbifera* (C.Wright) Benth., Hooker's Icon. Pl. 15: 25 (1881). *Rhynchospora corymbifera* (C.Wright) Kük., Repert. Spec. Nov. Regni Veg. 23: 205 (1926), nom. illeg.
Rhynchospora corymbifera var. *pseudeximia* Kük., Repert. Spec. Nov. Regni Veg. 23: 206 (1926).

Rhynchospora imeriensis (Kük.) W.W.Thomas, Brittonia 48: 481 (1996 publ. 1997).
Venezuela (Amazonas). 82 VEN. Hemicr. or rhizome geophyte.
**Rhynchospora cephalotes* var. *imeriensis* Kük., Bot. Jahrb. Syst. 75: 124 (1950).

Rhynchospora immensa Kük., Bot. Jahrb. Syst. 56(125): 17 (1921). *Rhynchospora aristata* var. *immensa* (Kük.) Kük., Bot. Jahrb. Syst. 74: 401 (1949).
C. America to Brazil (Roraima), Trinidad. 80 COS GUA PAN 81 TRT 82 GUY VEN 83 CLM 84 BZN. Hemicr. or rhizome geophyte.

Rhynchospora inexpansa (Michx.) Vahl, Enum. Pl. 2: 232 (1805).
SE. U.S.A. to Texas. 77 TEX 78 ALA ARK FLA GEO LOU MSI NCA SCA VRG. Hemicr. or rhizome geophyte.
**Schoenus inexpansus* Michx., Fl. Bor.-Amer. 1: 35 (1803). *Phaeocephalum inexpansum* (Michx.) House, Amer. Midl. Naturalist 6: 202 (1920).

Rhynchospora intermedia (Chapm.) Britton, Trans. New York Acad. Sci. 11: 87 (1892).
Florida, Cuba. 78 FLA 81 CUB. Hemicr. or rhizome geophyte.
**Rhynchospora plumosa* var. *intermedia* Chapm., Fl. South. U.S.: 524 (1860). *Phaeocephalum intermedium* (Chapm.) House, Amer. Midl. Naturalist 6: 201 (1920).
Rhynchospora pineticola C.B.Clarke, Kew Bull., Addit. Ser. 8: 40 (1908).

Rhynchospora inundata (Oakes) Fernald, Rhodora 20: 139 (1918).
E. U.S.A. 75 MAS NWJ NWY RHO 78 ALA DEL FLA GEO LOU MRY MSI NCA SCA VRG. Hel.
**Ceratoschoenus macrostachys* var. *inundatus* Oakes, Mag. Hort. Bot. 7: 185 (1841). *Rhynchospora macrostachya* var. *inundata* (Oakes) Fernald, Rhodora 8: 164 (1906). *Ceratoschoenus inundatus* (Oakes) House, Bull. New York State Mus. Nat. Sci. 254: 152 (1924).

Rhynchospora jaliscensis McVaugh, Acta Bot. Mex. 11: 15 (1990).
Mexico (Jalisco). 79 MXS. Hemicr. or rhizome geophyte.

Rhynchospora jamaicensis Britton, Bull. Torrey Bot. Club 41: 1 (1914).
Caribbean. 81 JAM PUE. Hemicr. or rhizome geophyte.

Rhynchospora joveroensis Britton, Contr. New York Bot. Gard. 194: 11 (1917).
W. Cuba. 81 CUB. Hemicr. or rhizome geophyte.
Rhynchospora longifrons Kük., Repert. Spec. Nov. Regni Veg. 23: 209 (1926).
Rhynchospora longifrons var. *breviseta* Kük., Repert. Spec. Nov. Regni Veg. 23: 109 (1926).
Rhynchospora joveroensis var. *breviseta* Kük., Bot. Jahrb. Syst. 75: 108 (1950).

Rhynchospora jubata Liebm., Mexic. Neldeagt. Pl., V, 2: 254 (1851).
SE. Mexico (to Veracruz). 79 MXG MXT. Hemicr.

Rhynchospora junciformis (Kunth) Boeck., Flora 41: 646 (1858).
N. South America to Brazil. 82 FRG GUY SUR VEN 84 BZC BZL BZN BZS. Hemicr. or rhizome geophyte.
Spermodon edentulatus Nees, Linnaea 9: 296 (1834), nom. inval.
**Dichromena junciformis* Kunth, Enum. Pl. 2: 279 (1837). *Rhynchospora edentula* H.Pfeiff., Repert. Spec. Nov. Regni Veg. 23: 345 (1927), nom. superfl.
Rhynchospora junciformis var. *monocarpa* Kük., Bot. Jahrb. Syst. 56(125): 19 (1921).

Rhynchospora killipii R.Gross, Publ. Field Mus. Nat. Hist., Bot. Ser. 11: 40 (1931). *Dichromena killipii* (R.Gross) J.F.Macbr., Publ. Field Mus. Nat. Hist., Bot. Ser. 13(1): 301 (1936).
C. Peru. 83 PER. Hemicr. or rhizome geophyte.

Rhynchospora knieskernii J.Carey, Amer. J. Sci. Arts, II, 4: 25 (1847). *Phaeocephalum knieskernii* (J.Carey) House, Amer. Midl. Naturalist 6: 202 (1920).
New Jersey, Delaware. 75 NWJ 78 DEL. Hemicr. or rhizome geophyte.

Rhynchospora kunthii Nees ex Kunth, Enum. Pl. 2: 296 (1837).
California to W. South America. 76 CAL 79 MXC MXE MXG MXS MXT 80 COS GUA PAN 83 CLM ECU PER. Hemicr. or rhizome geophyte.
Rhynchospora kunthii var. *composita* Nees in C.F.P.von Martius & auct. suc. (eds.), Fl. Bras. 2(1): 147 (1842).
Rhynchospora schaffneri Boeck., Linnaea 37: 575 (1873).

Rhynchospora kuntzei C.B.Clarke ex Kuntze, Revis. Gen. Pl. 3(2): 336 (1898). *Dichromena kuntzei* (C.B.Clarke ex Kuntze) J.F.Macbr., Publ. Field Mus. Nat. Hist., Bot. Ser. 11: 42 (1931). *Rhynchospora umbraticola* var. *kuntzei* (C.B.Clarke ex Kuntze) Kük., Bot. Jahrb. Syst. 74: 391 (1949).
E. Peru to Bolivia. 83 BOL PER. Hemicr. or rhizome geophyte.
Rhynchospora uleana Kük., Bot. Jahrb. Syst. 56(125): 17 (1921), nom. illeg.

Rhynchospora lapensis C.B.Clarke, Kew Bull., Addit. Ser. 8: 38 (1908).
Brazil (Minas Gerais). 84 BZL. Hemicr. or rhizome geophyte.
Rhynchospora schwackei C.B.Clarke, Kew Bull., Addit. Ser. 8: 35 (1908).
Rhynchospora campanulata Kük., Repert. Spec. Nov. Regni Veg. 12: 95 (1913). *Rhynchospora lapensis* var. *campanulata* (Kük.) Kük., Bot. Jahrb. Syst. 74: 457 (1949).

Rhynchospora latibracteata Guagl., Darwiniana 39: 318 (2001).
S. Brazil to Uruguay. 84 BZS 85 URU. Hemicr. or rhizome geophyte.

Rhynchospora latifolia (Baldwin ex Elliott) W.W.Thomas, Mem. New York Bot. Gard. 37: 86 (1984).
SE. U.S.A. to SE. Texas. 77 TEX 78 ALA FLA GEO LOU MSI NCA SCA. Hemicr. or rhizome geophyte.
**Dichromena latifolia* Baldwin ex Elliott, Sketch Bot. S. Carolina 1: 90 (1816). *Rhynchospora stellata* var. *latifolia* (Baldwin ex Elliott) Kük., Bot. Jahrb. Syst. 74: 301 (1951).

Rhynchospora leae C.B.Clarke, Kew Bull., Addit. Ser. 8: 34 (1908).
N. & NE. Queensland. 50 QLD. Hemicr. or rhizome geophyte.

Rhynchospora legrandii Kük. ex Barros, Darwiniana 5: 182 (1941). *Rhynchospora corymbosa* subsp. *legrandii* (Kük. ex Barros) Luceño & Vanzela, Canad. J. Bot. 76: 448 (1998). *Rhynchospora corymbosa* var. *legrandii* (Kük. ex Barros) Guagl., Darwiniana 39: 314 (2001).
Brazil to NE. Argentina. 84 BZS 85 AGE URU. Hemicr. or rhizome geophyte.

Rhynchospora leptorhyncha C.Wright, Anales Acad. Ci. Méd. Habana 8: 84 (1971). *Phaeocephalum leptorhynchum* (C.Wright) House, Amer. Midl. Naturalist 6: 202 (1920).
W. Cuba. 81 CUB. Hemicr. or rhizome geophyte.
Rhynchospora leptorhyncha var. *laevirostris* Kük., Repert. Spec. Nov. Regni Veg. 23: 209 (1926).

Rhynchospora leucoloma A.C.Araujo & Longhi-Wagner, Brittonia 55: 33 (2003).
Venezuela to Brazil. 82 VEN 84 BZC BZN.

Rhynchospora leucostachys Boeck., Allg. Bot. Z. Syst. 2: 79 (1896).
Brazil to Paraguay. 84 BZC BZL 85 PAR. Hemicr. or rhizome geophyte.
Haplostylis minor Nees in C.F.P.von Martius & auct. suc. (eds.), Fl. Bras. 2(1): 129 (1842). *Rhynchospora minor* (Nees) Steud., Syn. Pl. Glumac. 2: 143 (1855), nom. illeg. *Rhynchospora diamantina* var. *brevispiculosa* Kük., Feddes Repert. Spec. Nov. Regni Veg. 53: 73 (1944). *Rhynchospora elatior* var. *brevispiculosa* (Kük.) A.C.Araujo, Novon 14: 10 (2004).
Rhynchospora diamantina C.B.Clarke ex Kük., Feddes Repert. Spec. Nov. Regni Veg. 53: 72 (1944).

Rhynchospora lindeniana Griseb., Cat. Pl. Cub.: 244 (1866). *Rhynchospora lindeniana* var. *typica* Gale, Rhodora 46: 222 (1944), nom. inval.
SE. Mexico to Guatemala, Caribbean. 79 MXT 80 GUA 81 BAH CUB DOM JAM PUE. Hemicr. or rhizome geophyte.
Rhynchospora bahamensis Britton, Torreya 13: 217 (1913). *Rhynchospora lindeniana* var. *bahamensis* (Britton) Gale, Rhodora 46: 223 (1944).

Rhynchospora lindmanii H.Pfeiff., Repert. Spec. Nov. Regni Veg. 49: 83 (1940).
Brazil (Mato Grosso). 84 BZC. Hemicr. or rhizome geophyte.

Rhynchospora locuples C.B.Clarke, Bot. Jahrb. Syst. 34(78): 5 (1904). *Dichromena locuples* (C.B.Clarke) J.F.Macbr., Publ. Field Columbian Mus., Bot. Ser. 8: 114 (1930). *Rhynchospora schiedeana* var. *locuples* (C.B.Clarke) Kük., Bot. Jahrb. Syst. 74: 398 (1949).
Costa Rica to W. South America. 80 COS PAN 82 VEN 83 BOL CLM ECU PER. Hemicr.

Rhynchospora longa (Lindm.) H.Pfeiff., Repert. Spec. Nov. Regni Veg. 38: 94 (1935).
Ecuador to WC. Brazil. 83 ECU PER 84 BZC. Hemicr.
**Dichromena longa* Lindm., Bih. Kongl. Svenska Vetensk.-Akad. Handl. 26(9): 23 (1901).

Rhynchospora longibracteata Boeck., Linnaea 37: 534 (1873).
Trinidad, S. Trop. America. 81 TRT 82 GUY VEN 83 CLM 84 BZS. Hemicr. or rhizome geophyte.
Rhynchospora setigera Griseb., Fl. Brit. W. I.: 577 (1864), nom. illeg.
Rhynchospora longibracteata var. *minor* H.Pfeiff., Repert. Spec. Nov. Regni Veg. 17: 238 (1921).

Rhynchospora longiflora C.Presl, Symb. Bot. 1: 41 (1831).
Mexico, Caribbean. 79 MXG 81 JAM LEE TRT WIN. Hemicr. or rhizome geophyte.
Rhynchospora laxa Schult. in J.J.Roemer & J.A.Schultes, Mant. 2: 46 (1824), nom. illeg.
Schoenus longirostris Sieber ex C.Presl, Isis (Oken) 21: 269 (1828).
Rhynchospora florida Griseb., Fl. Brit. W. I.: 575 (1864), nom. illeg.
Rhynchospora corymbifera Boeck., Linnaea 37: 619 (1873), nom. illeg.
Rhynchospora cruegeriana Boeck., Beitr. Cyper. 2: 26 (1890). *Rhynchospora aristata* var. *cruegeriana* (Boeck.) Kük., Repert. Spec. Nov. Regni Veg. 23: 205 (1926).
Rhynchospora eggersiana Boeck., Beitr. Cyper. 2: 25 (1890).
Rhynchospora corymbifera var. *subcostata* C.B.Clarke in I.Urban, Symb. Antill. 2: 113 (1900). *Rhynchospora longiflora* var. *subcostata* (C.B.Clarke) Kük., Bot. Jahrb. Syst. 74: 296 (1949).

Rhynchospora longifolia Boeck., Allg. Bot. Z. Syst. 2: 79 (1896).
Brazil. 84+. Hemicr. or rhizome geophyte.

Rhynchospora longisetis R.Br., Prodr.: 230 (1810). *Schoenus longisetis* (R.Br.) Poir. in J.B.A.M.de Lamarck, Encycl., Suppl. 2: 252 (1811). *Cephaloschoenus longisetis* (R.Br.) Nees, Linnaea 9: 296 (1834).
Trop. Asia to N. Australia. 40 IND 41 CBD LAO MYA THA VIE 43 NWG 50 NTA QLD. Ther.
Rhynchospora prescottiana Wall., Numer. List: 3423 (1831), nom. inval.
Rhynchospora massieana E.G.Camus, Notul. Syst. (Paris) 1: 249 (1910).

Rhynchospora luetzelburgiana Kük., Feddes Repert. Spec. Nov. Regni Veg. 53: 74 (1944).
Brazil (Rio de Janeiro). 84 BZL. Hemicr. or rhizome geophyte.

Rhynchospora macra (C.B.Clarke ex Britton) Small, Man. S.E. Fl.: 180 (1933).
SE. U.S.A. to Texas, C. America. 77 TEX 78 ALA FLA GEO LOU MSI NCA SCA 80 NIC. Hemicr.
**Rhynchospora alba* var. *macra* C.B.Clarke ex Britton, Trans. New York Acad. Sci. 11: 88 (1892).
Triodon albus var. *macer* (C.B.Clarke) Farw., Rep. (Annual) Michigan Acad. Sci. 19: 253 (1917).

Rhynchospora macrochaeta Steud. ex Boeck., Linnaea 37: 632 (1873). *Dichromena macrochaeta* (Steud. ex Boeck.) J.F.Macbr., Publ. Field Columbian Mus., Bot. Ser. 4: 167 (1929).
Costa Rica to Guyana and Bolivia. 80 COS 82 GUY? VEN 83 BOL CLM PER. Hemicr. or rhizome geophyte.
Rhynchospora macrochaeta var. *jamesonii* C.B.Clarke, Bot. Jahrb. Syst. 34(78): 5 (1904).
Rhynchospora macrochaeta var. *colombiensis* (Boeck.) Kük., Bot. Jahrb. Syst. 74: 394 (1949).

var. ***macrochaeta***
Costa Rica to Guyana and Bolivia. 80 COS 82 GUY VEN 83 BOL CLM PER. Hemicr. or rhizome geophyte.
Rhynchospora columbiensis Boeck., Beitr. Cyper. 1: 29 (1888).
Rhynchospora pittieri Boeck., Allg. Bot. Z. Syst. 2: 109 (1896).
Rhynchospora tamana Steyerm., Fieldiana, Bot. 28(1): 47 (1951).

var. ***triceps*** Kük., Bot. Jahrb. Syst. 74: 394 (1949).
C. Peru. 83 PER. Hemicr. or rhizome geophyte.
**Rhynchospora triceps* Boeck., Linnaea 37: 642 (1873), nom. illeg. *Rhynchospora ruiziana* var. *triceps* K.Schum. in I.Urban, Symb. Antill. 2:

115 (1900). *Dichromena ruiziana* var. *triceps* J.F.Macbr., Publ. Field Mus. Nat. Hist., Bot. Ser. 4: 167 (1929).

Rhynchospora macrostachya Torr. ex A.Gray, Ann. Lyceum Nat. Hist. New York 3: 206 (1835). *Ceratoschoenus macrostachys* (Torr. ex A.Gray) A.Gray, Ann. Lyceum Nat. Hist. New York 3: 369 (1836). *Rhynchospora corniculata* var. *macrostachya* (Torr. ex A.Gray) Britton, Trans. New York Acad. Sci. 11: 84 (1892).
C. & E. U.S.A. 74 KAN MSO OKL 75 CNT INI MAI MAS MIC NWJ NWY RHO 77 TEX 78 ALA ARK DEL FLA GEO KTY LOU MRY NCA SCA TEN VRG WDC. Hemicr. or rhizome geophyte.
Ceratoschoenus macrophyllus Tuck., Amer. J. Sci. Arts, II, 6: 232 (1848).
Rhynchospora macrostachya var. *colpophila* Fernald & Gale, Rhodora 42: 421 (1940).

Rhynchospora maguireana T.Koyama, Mem. New York Bot. Gard. 23: 54 (1972).
Venezuela to SW. Guyana. 82 GUY VEN. Hemicr. or rhizome geophyte.

Rhynchospora malasica C.B.Clarke in J.D.Hooker, Fl. Brit. India 6: 670 (1893).
Temp. E. Asia to W. Malesia. 38 JAP KOR NNS TAI 42 BOR MLY SUM. Hemicr. or rhizome geophyte.
Rhynchospora nipponica Makino, Bot. Mag. (Tokyo) 18: 145 (1904).

Rhynchospora mandonii C.B.Clarke, Kew Bull., Addit. Ser. 8: 38 (1908). *Dichromena mandonii* (C.B.Clarke) J.F.Macbr., Publ. Field Columbian Mus., Bot. Ser. 4: 167 (1929).
Peru to Bolivia. 83 BOL PER. Hemicr. or rhizome geophyte.
Rhynchospora polyphylla var. *laxa* (P.Beauv.) H.Pfeiff., Repert. Spec. Nov. Regni Veg. 49: 78 (1940).

Rhynchospora marcelo-guerrae Luceño & M.Alves, Candollea 52: 192 (1997).
Brazil (Tocantins). 84 BZN. Hemicr. or rhizome geophyte.

Rhynchospora marisculus Lindl. & Nees in C.F.P.von Martius & auct. suc. (eds.), Fl. Bras. 2(1): 142 (1842). *Dichromena marisculus* (Lindl. & Nees) J.F.Macbr., Publ. Field Mus. Nat. Hist., Bot. Ser. 11: 5 (1931).
Trop. America to N. Argentina. 79 MXG MXT 80 BLZ COS ELS GUA HON NIC PAN 81 CUB DOM JAM LEE PUE WIN 82 GUY SUR VEN 83 BOL CLM ECU PER 84 BZC BZE BZL BZS 85 AGE AGW PAR URU. Hemicr. or rhizome geophyte.
Rhynchospora rigida Schrad. ex Kunth, Enum. Pl. 2: 297 (1837), nom. inval.
Schoenus marisculus Salzm. ex Steud., Syn. Pl. Glumac. 2: 147 (1855), nom. inval.
Rhynchospora marisculus var. *elatior* Boeck., Vidensk. Meddel. Dansk Naturhist. Foren. Kjøbenhavn 1869: 149 (1870).
Rhynchospora tenuiseta C.Wright, Anales Acad. Ci. Méd. Habana 8: 83 (1871).
Rhynchospora uleana Boeck., Allg. Bot. Z. Syst. 2: 110 (1896).
Rhynchospora weberbaueri C.B.Clarke, Bot. Jahrb. Syst. 37: 518 (1906). *Dichromena weberbaueri* (C.B.Clarke) J.F.Macbr., Publ. Field Columbian Mus., Bot. Ser. 4: 167 (1929).
Rhynchospora borinquensis Britton, Bull. Torrey Bot. Club 42: 387 (1915).
Rhynchospora rigidula Schrad. ex Knuth, Repert. Spec. Nov. Regni Veg. Beih. 43: 148 (1926).

Rhynchospora marquisensis F.Br., Bernice P. Bishop Mus. Bull. 84: 108 (1931). *Rhynchospora sclerioides* var. *marquisensis* (F.Br.) Kük., Bot. Jahrb. Syst. 74: 504 (1949).
Marquesas. 61 MRQ. Hemicr. or rhizome geophyte.

Rhynchospora mayarensis León, Contr. Ocas. Mus. Hist. Nat. Colegio "De La Salle" 8: 233 (1946).
E. Cuba. 81 CUB. Hemicr. or rhizome geophyte.
**Rhynchospora nipensis* Kük., Repert. Spec. Nov. Regni Veg. 23: 207 (1926), nom. illeg.
Rhynchospora subnipensis Kük., Bot. Jahrb. Syst. 75: 193 (1950), nom. illeg.

Rhynchospora megalocarpa A.Gray, Ann. Lyceum Nat. Hist. New York 3: 208 (1835).
SE. U.S.A. 78 ALA FLA GEO MSI NCA SCA. Hemicr. or rhizome geophyte.
Rhynchospora dodecandra Baldwin ex A.Gray., Ann. Lyceum Nat. Hist. New York 3: 207 (1835).
Rhynchospora pycnocarpa A.Gray, Ann. Lyceum Nat. Hist. New York 3: 208 (1835).
Phaeocephalum dodecandrum House, Amer. Midl. Naturalist 6: 202 (1920).

Rhynchospora megaplumosa E.L.Bridges & Orzell, Lundellia 2: 30 (2000).
C. Florida. 78 FLA. Hemicr.

Rhynchospora megapotamica (A.Spreng.) H.Pfeiff., Revista Sudamer. Bot. 7: 129 (1943).
Bolivia to S. South America. 83 BOL 84 BZS 85 AGE AGW CLN PAR URU. Hemicr. or rhizome geophyte.
**Scirpus megapotamicus* A.Spreng., Tent. Suppl.: 4 (1828).
Rhynchospora luzuliformis Boeck., Linnaea 38: 632 (1873).
Rhynchospora maculata Maury ex Micheli, Denkschr. Schweiz. Naturf. Ges. 31(1): 146 (1890).
Rhynchospora spicata Boeck., Allg. Bot. Z. Syst. 2: 95 (1896).
Rhynchospora luzuliformis var. *elongata* Kuntze, Revis. Gen. Pl. 3(2): 336 (1898).
Rhynchospora luzuliformis var. *subcapitata* Kuntze, Revis. Gen. Pl. 3(2): 336 (1898).
Rhynchospora luzuliformis var. *pusilla* Osten, Anales Mus. Hist. Nat. Montevideo, II, 3: 219 (1931).
Rhynchospora luzuliformis var. *spicata* Kük., Anales Mus. Hist. Nat. Montevideo, II, 3: 221 (1931).
Rhynchospora megapotamica var. *spicata* Kük., Revista Sudamer. Bot. 7: 130 (1943).

Rhynchospora melanocarpa A.C.Araujo & W.W.Thomas, Brittonia 55: 33 (2003).
Bolivia. 83 BOL.

Rhynchospora mendoncana Boeck., Beitr. Cyper. 2: 23 (1890).
Brazil. 84+. Hemicr. or rhizome geophyte.

Rhynchospora metralis C.B.Clarke, Kew Bull., Addit. Ser. 8: 90 (1908).
N. Brazil. 84 BZN. Hemicr. or rhizome geophyte.

Rhynchospora mexicana (Liebm.) Steud., Syn. Pl. Glumac. 2: 144 (1855).
S. Mexico to Venezuela, Caribbean. 79 MXS MXT 80 BLZ COS GUA HON PAN 81 DOM HAI 82 VEN 84. Ther. or hemicr.

Haplostylis mexicana Liebm., Mexic. Neldeagt. Pl., V, 2: 249 (1851).
Rhynchospora buchii Urb., Symb. Antill. 7: 169 (1912).

Rhynchospora microcarpa Baldwin ex A.Gray, Ann. Lyceum Nat. Hist. New York 3: 202 (1835). *Phaeocephalum microcarpum* (Baldwin ex A.Gray) House, Amer. Midl. Naturalist 6: 202 (1920).
SE. U.S.A. to Texas, Caribbean, Belize. 77 TEX 78 ALA FLA GEO LOU MSI NCA SCA 80 BLZ 81 CUB PUE. Hemicr.
Rhynchospora parviflora Beyr. ex Kunth, Enum. Pl. 2: 297 (1837).
Rhynchospora torreyana var. *microrhyncha* Griseb., Cat. Pl. Cub.: 243 (1866).
Rhynchospora microcarpa var. *remotiflora* Kük., Repert. Spec. Nov. Regni Veg. 23: 210 (1926).
Rhynchospora edisoniana Britton ex Small, Man. S.E. Fl.: 184 (1933).
Rhynchospora sulcata Gale, Rhodora 46: 240 (1944).

Rhynchospora microcephala (Britton) Britton in J.K.Small, Fl. S.E. U.S.: 195 (1903).
New Jersey to SE. U.S.A., W. Cuba, Jamaica. 75 NWJ 78 ALA DEL FLA GEO LOU MRY MSI NCA SCA VRG WDC 81 CUB JAM. Hemicr. or rhizome geophyte.
**Rhynchospora axillaris* var. *microcephala* Britton, Trans. New York Acad. Sci. 11: 89 (1892). *Rhynchospora cephalantha* var. *microcephala* (Britton) Kük., Bot. Jahrb. Syst. 75: 101 (1950).

Rhynchospora miliacea (Lam.) A.Gray, Ann. Lyceum Nat. Hist. New York 3: 198 (1835).
SE. U.S.A. to Texas, Caribbean. 77 TEX 78 ALA FLA GEO LOU MSI NCA SCA VRG 81 CUB JAM. Hemicr. or rhizome geophyte.
**Schoenus miliaceus* Lam., Tabl. Encycl. 1: 137 (1791). *Phaeocephalum miliaceum* (Lam.) House, Amer. Midl. Naturalist 6: 202 (1920).
Schoenus sparsus Michx., Fl. Bor.-Amer. 1: 35 (1803). *Rhynchospora sparsa* (Michx.) Vahl, Enum. Pl. 2: 230 (1805).

Rhynchospora minor Nees in C.F.P.von Martius & auct. suc. (eds.), Fl. Bras. 2(1): 143 (1842).
Brazil. 84+. Hemicr. or rhizome geophyte.

Rhynchospora mixta Britton in J.K.Small, Fl. S.E. U.S.: 197 (1903). *Phaeocephalum mixtum* (Britton) House, Amer. Midl. Naturalist 6: 202 (1920).
SE. U.S.A. to Texas. 77 TEX 78 ALA FLA GEO LOU MSI NCA SCA. Hemicr. or rhizome geophyte.
Rhynchospora prolifera Small, Fl. S.E. U.S.: 198 (1903).

Rhynchospora modesti-lucennoi Castrov., Nordic J. Bot. 15: 569 (1995 publ. 1996).
Spain. 12 SPA. Hemicr. or rhizome geophyte.

Rhynchospora mollis (Wall.) Schult. in J.J.Roemer & J.A.Schultes, Mant. 2: 49 (1824).
Nepal. 40 NEP. Hemicr. or rhizome geophyte. Provisionally accepted.
**Scirpus mollis* Wall. in W.Roxburgh, Fl. Ind. 1: 227 (1820). *Fimbristylis mollis* (Wall.) Kunth, Enum. Pl. 2: 246 (1837).

Rhynchospora montana (Uittien) H.Pfeiff., Repert. Spec. Nov. Regni Veg. 38: 91 (1935).
E. Suriname. 82 SUR. Hemicr.
**Dichromena montana* Uittien, Recueil Trav. Bot. Néerl. 22: 334 (1925).

Rhynchospora nanuzae Rocha & Luceño, Hoehnea 29: 201 (2002).
Brazil (Minas Gerais). 84 BZL.

Rhynchospora nardifolia (Kunth) Boeck., Linnaea 37: 560 (1873).
Brazil. 84 BZC BZE BZL BZS. Hemicr. or rhizome geophyte.
**Dichromena nardifolia* Kunth, Enum. Pl. 2: 280 (1837). *Haloschoenus nardifolius* (Kunth) Nees in C.F.P. von Martius & auct. suc. (eds.), Fl. Bras. 2(1): 122 (1842).

Rhynchospora neesii Steud., Syn. Pl. Glumac. 2: 146 (1855).
Brazil. 84+. Hemicr. or rhizome geophyte.
**Mitrospora ciliata* Nees in C.F.P.von Martius & auct. suc. (eds.), Fl. Bras. 2(1): 133 (1842).

Rhynchospora nervosa (Vahl) Boeck., Vidensk. Meddel. Naturhist. Foren. Kjøbenhavn 1869: 143 (1869).
Mexico to Trop. America. 79 MXC MXE MXG MXN MXS MXT 80 BLZ COS CPI ELS GUA HON NIC PAN 81 CUB DOM JAM LEE TRT WIN 82 FRG GUY SUR VEN 83 BOL CLM ECU PER 84 BZC BZE BZL BZN BZS 85 AGE PAR. Hemicr. or rhizome geophyte.
**Dichromena nervosa* Vahl, Enum. Pl. 2: 241 (1805).

subsp. ***ciliata*** (G.Mey.) T.Koyama, Madroño 20: 254 (1970).
Trop. America. 79 MXG MXT 80 BLZ COS ELS GUA HON NIC PAN 81 CUB LEE PUE TRT WIN 82 FRG GUY SUR VEN 83 BOL CLM ECU PER 84 BZE BZL BZN BZS. Hemicr. or rhizome geophyte.
Dichromena ciliata Vahl, Enum. Pl. 2: 240 (1805), nom. illeg. **Schoenus ciliatus* G.Mey., Prim. Fl. Esseq.: 23 (1818). *Rhynchospora ciliata* (G.Mey.) Kük., Bot. Jahrb. Syst. 56(125): 16 (1921), nom. illeg. *Rhynchospora nervosa* var. *ciliata* (G.Mey.) Kük., Bot. Jahrb. Syst. 75: 295 (1951). *Rhynchospora vahliana* Griseb., Fl. Brit. W. I.: 577 (1864).
Schoenus globosus Kunth in F.W.H.von Humboldt, A.J.A.Bonpland & C.S.Kunth, Nov. Gen. Sp. 1: 229 (1816). *Dichromena globosa* (Kunth) Roem. & Schult., Syst. Veg. 2: 90 (1817).
Dichromena leucophylla Willk. ex Kunth, Enum. Pl. 2: 276 (1837), pro syn.
Dichromena obtusiflora Schrad. ex Nees in C.F.P. von Martius & auct. suc. (eds.), Fl. Bras. 2(1): 113 (1842).
Dichromena pura Nees in C.F.P.von Martius & auct. suc. (eds.), Fl. Bras. 2(1): 112 (1842). *Rhynchospora pura* (Nees) Griseb., Fl. Brit. W. I.: 577 (1864).
Dichromena obtusifolia Steud., Syn. Pl. Glumac. 2: 135 (1855).
Rhynchospora microseta Steud., Syn. Pl. Glumac. 2: 142 (1855).
Rhynchospora hirsuta Griseb., Fl. Brit. W. I.: 576 (1864), nom. illeg.
Rhynchospora jelskiana Boeck., Linnaea 38: 401 (1874). *Rhynchospora nervosa* var. *jelskiana* (Boeck.) Kük., Bot. Jahrb. Syst. 75: 298 (1951).
Dichromena ciliata var. *angustifolia* Cherm., Bull. Soc. Bot. France 71: 1051 (1925).
Dichromena ciliata var. *cinnamomea* Kük., Repert. Spec. Nov. Regni Veg. 23: 200 (1926). *Rhynchospora nervosa* var. *cinnamomea* (Kük.) Kük., Bot. Jahrb. Syst. 75: 297 (1951).

subsp. ***nervosa***
Mexico to Trop. America. 79 MXC MXE MXG MXN MXS MXT 80 COS CPI ELS GUA HON NIC PAN 81 CUB DOM JAM LEE PUE TRT 82 FRG GUY SUR VEN 83 BOL CLM ECU PER 84 BZC BZE BZL BZN BZS 85 AGE PAR. Hemicr. or rhizome geophyte.
Schoenus tenuifolius Kunth in F.W.H.von Humboldt, A.J.A.Bonpland & C.S.Kunth, Nov. Gen. Sp. 1: 228 (1816). *Dichromena tenuifolia* (Kunth) Spreng., Syst. Veg. 1: 203 (1824). *Rhynchospora nervosa* var. *tenuifolia* (Kunth) H.Pfeiff., Repert. Spec. Nov. Regni Veg. 38: 92 (1935).
Dichromena pilosa Spreng., Syst. Veg. 1: 203 (1824).
Rhynchospora exigua Boeck., Vidensk. Meddel. Naturhist. Foren. Kjøbenhavn, V, 5: 238 (1894).
Dichromena ciliata var. *vahliana* C.B.Clarke in I.Urban, Symb. Antill. 2: 101 (1900).

Rhynchospora nipensis Britton, Contr. New York Bot. Gard. 194: 13 (1917).
E. Cuba (Sierra de Nipe). 81 CUB. Hemicr. or rhizome geophyte.
Rhynchospora fibrillosa Kük., Repert. Spec. Nov. Regni Veg. 23: 212 (1926).

Rhynchospora nitens (Vahl) A.Gray, Manual, ed. 5: 568 (1867).
E. U.S.A. to C. America, Caribbean, Guyana. 75 INI MAS MIC NWJ NWY 77 TEX 78 ALA DEL FLA GEO LOU MSI NCA SCA VRG 80 BLZ NIC 81 CUB DOM JAM PUE 82 GUY. Ther.
**Scirpus nitens* Vahl, Enum. Pl. 2: 272 (1805). *Isolepis nitens* (Vahl) Roem. & Schult., Syst. Veg. 2: 117 (1817). *Dichromena nitens* (Vahl) Kunth, Enum. Pl. 2: 282 (1837). *Psilocarya nitens* (Vahl) Alph.Wood, Amer. Bot. Fl.: 364 (1870). *Dichostylis nitens* (Vahl) Rikli, Jahrb. Wiss. Bot. 27: 565 (1895), nom. illeg.

var. ***hispaniolica*** Kük., Repert. Spec. Nov. Regni Veg. 32: 76 (1933).
Caribbean. 81 DOM JAM. Ther.

var. ***nitens***
E. U.S.A. to C. America, Caribbean, Guyana. 75 INI MAS MIC NWJ NWY 77 TEX 78 ALA DEL FLA GEO LOU MSI NCA SCA VRG 80 BLZ NIC 81 CUB PUE 82 GUY. Ther.
Psilocarya rhynchosporoides Torr., Ann. Lyceum Nat. Hist. New York 3: 361 (1836).
Psilocarya texensis Torr., Ann. Lyceum Nat. Hist. New York 3: 361 (1836). *Rhynchospora texensis* (Torr.) Boeck., Linnaea 37: 602 (1873).
Scirpus carolinianus Panz. ex Kunth, Enum. Pl. 2: 282 (1837).
Psilocarya portoricensis Britton, Bull. Torrey Bot. Club 43: 387 (1915).

Rhynchospora nivea Boeck., Linnaea 37: 527 (1873). *Dichromena nivea* (Boeck.) Britton, Bull. Torrey Bot. Club 15: 101 (1888). *Dichromena nivea* (Boeck.) Boeck., Contr. Herb. Franklin Marshall Coll. 1: 20 (1895), nom. illeg.
Oklahoma to Texas. 74 OKL 77 TEX. Hemicr. or rhizome geophyte.
Dichromena reverchonii S.H.Wright, Bull. Torrey Bot. Club 9: 86 (1882).

Rhynchospora nuda Gale, Rhodora 46: 167 (1944). *Rhynchospora chapmanii* var. *nuda* (Gale) Kük., Bot. Jahrb. Syst. 75: 117 (1950).
Cuba (I. de la Juventud). 81 CUB. Hemicr. or rhizome geophyte.

Rhynchospora oaxacana Kral & W.W.Thomas, Brittonia 38: 210 (1986).
S. Mexico (to Veracruz). 79 MXG MXS MXT. Hemicr. or bulb geophyte.

Rhynchospora odorata C.Wright ex Griseb., Cat. Pl. Cub.: 242 (1866).
SE. U.S.A., Caribbean. 78 ALA FLA GEO NCA SCA 81 BAH BER CUB DOM JAM PUE. Hemicr. or rhizome geophyte.
Rhynchospora stipitata Chapm., Fl. South. U.S. 2(Suppl.): 660 (1883). *Phaeocephalum stipitatum* (Chapm.) House, Amer. Midl. Naturalist 6: 202 (1920).

Rhynchospora oligantha A.Gray, Ann. Lyceum Nat. Hist. New York 3: 212 (1835). *Rhynchospora oligantha* var. *typica* Gale, Rhodora 46: 129 (1944), nom. inval.
New Jersey to Texas, C. America, Caribbean. 75 NWJ 77 TEX 78 ALA DEL FLA GEO LOU MSI NCA SCA VRG WDC 80 BLZ NIC 81 CUB DOM JAM PUE. Hemicr.

var. ***breviseta*** Gale, Rhodora 46: 129 (1944). *Rhynchospora breviseta* (Gale) Channell, Rhodora 58: 336 (1957), nom. illeg.
SE. U.S.A., Caribbean. 78 ALA FLA GEO LOU MSI NCA SCA VRG 81 CUB DOM JAM PUE. Hemicr.

var. ***oligantha***
New Jersey to Texas, C. America. 75 NWJ 77 TEX 78 ALA DEL FLA GEO LOU MSI NCA SCA VRG WDC 80 BLZ NIC. Hemicr.
Rhynchospora triflora M.A.Curtis, Boston J. Nat. Hist. 1: 140 (1837), nom. illeg.
Kleistrocalyx junciformis Steud., Flora 33: 229 (1850). *Calyptrolepis junciformis* (Steud.) Steud., Syn. Pl. Glumac. 2: 151 (1855). *Cleistocalyx junciformis* (Steud.) Steud., Syn. Pl. Glumac. 2: 151 (1855).

Rhynchospora oreoboloidea Gómez-Laur., Phytologia 50: 459 (1982).
Costa Rica to Ecuador. 80 COS PAN 83 CLM ECU. Hemicr.
Rhynchospora paramora Mora, Mutisia 29: 4 (1966), nom. illeg.

Rhynchospora organensis C.B.Clarke, Kew Bull., Addit. Ser. 8: 37 (1908). *Rhynchospora corymbosa* var. *organensis* (C.B.Clarke) Kük., Bot. Jahrb. Syst. 74: 416 (1949).
Brazil to NE. Argentina. 84 BZL 85 AGE PAR URU. Hemicr. or rhizome geophyte.

Rhynchospora pallida M.A.Curtis, Amer. J. Sci. Arts, II, 7: 409 (1849). *Rhynchospora curtisii* Steud., Syn. Pl. Glumac. 2: 141 (1855), nom. illeg. *Phaeocephalum pallidum* (M.A.Curtis) House, Amer. Midl. Naturalist 6: 202 (1920).
E. U.S.A. 75 NWJ NWY 78 DEL MRY NCA SCA VRG. Hemicr. or rhizome geophyte.

Rhynchospora palustris Boeck., Beitr. Cyper. 2: 24 (1890).
Brazil. 84+. Hemicr. or rhizome geophyte.

Rhynchospora panicifolia Maury, J. Bot. (Morot) 2: 422 (1888).
Colombia. 83 CLM. Hemicr. or rhizome geophyte.

Rhynchospora papillosa W.W.Thomas, Mem. New York Bot. Gard. 37: 100 (1984).

Venezuela to SW. Guyana. 82 GUY VEN 84 BZN. Hemicr. or rhizome geophyte.

Rhynchospora paraensis Schrad. ex Kunth, Enum. Pl. 1: 300 (1837).
Brazil (Pará). 84 BZN. Hemicr. or rhizome geophyte.

Rhynchospora paramora Steyerm., Fieldiana, Bot. 28(1): 44 (1951).
Venezuela. 82 VEN. Hemicr. or rhizome geophyte.

Rhynchospora paranaensis A.C.Araujo & W.W.Thomas, Novon 14: 8 (2004).
Brazil (Paraná). 84 BZS. Hemicr.

Rhynchospora patuligluma C.B.Clarke ex Lindm., Bih. Kongl. Svenska Vetensk.-Akad. Handl. 26(9): 25 (1900).
S. Venezuela to Brazil. 82 VEN 84 BZC BZL BZS. Hemicr. or rhizome geophyte.
**Mitrospora pallida* Nees in C.F.P.von Martius & auct. suc. (eds.), Fl. Bras. 2(1): 133 (1842). *Rhynchospora pallida* (Nees) Steud., Syn. Pl. Glumac. 2: 146 (1855), nom. illeg.

Rhynchospora pedersenii Guagl., Darwiniana 39: 321 (2001).
Trop. America. 79 MXG MXT 80 COS NIC PAN 81 PUE 82 VEN 83 BOL 84 BZL BZS 85 PAR. Hemicr. or rhizome geophyte.
**Rhynchospora gigantea* var. *latifolia* H.Pfeiff., Repert. Spec. Nov. Regni Veg. 17: 236 (1921).

Rhynchospora perplexa Britton in J.K.Small, Fl. S.E. U.S.: 197 (1903). *Phaeocephalum perplexum* (Britton) House, Amer. Midl. Naturalist 6: 202 (1920). *Rhynchospora microcarpa* var. *perplexa* (Britton) Kük., Bot. Jahrb. Syst. 75: 133 (1950).
SE. U.S.A. to Texas, Caribbean. 77 TEX 78 ALA FLA GEO LOU MSI NCA SCA TEN VRG 81 BAH CUB DOM JAM. Hemicr. or rhizome geophyte.
Rhynchospora perplexa var. *virginiana* Fernald, Rhodora 49: 125 (1947).

Rhynchospora perrieri Cherm., Bull. Soc. Bot. France 69: 721 (1923).
Trop. & S. Africa, Madagascar. 22 BKN GNB GUI IVO LBR NGA SEN SIE TOG 23 CAF CMN 25 KEN TAN 26 ZAM ZIM 27 BOT NAT 29 MDG. Ther.
Rhynchospora deightonii Hutch. in J.Hutchinson & J.M.Dalziel, Fl. W. Trop. Afr. 2: 468 (1936), no latin descr.

Rhynchospora pilosa Boeck., Linnaea 37: 537 (1873).
N. South America to Brazil. 82 GUY VEN 84 BZC BZE BZL BZS. Hemicr. or rhizome geophyte.
**Dichromena pilosa* Kunth, Enum. Pl. 2: 279 (1837), nom. illeg.
Schoenus hirtus Schrad. ex Nees in C.F.P.von Martius & auct. suc. (eds.), Fl. Bras. 2(1): 114 (1842).
Scleria nana Boeck., Allg. Bot. Z. Syst. 2: 158 (1896). *Dichromena nana* (Boeck.) Core, Brittonia 1: 243 (1934). *Rhynchospora nana* (Boeck.) H.Pfeiff., Repert. Spec. Nov. Regni Veg. 49: 82 (1940).
Dichromena minarum C.B.Clarke, Kew Bull., Addit. Ser. 8: 32 (1908). *Rhynchospora graminosa* H.Pfeiff., Repert. Spec. Nov. Regni Veg. 38: 95 (1935).

Rhynchospora pilulifera Bertol., Opusc. Sci. 3: 406 (1819).
Brazil. 84+. Hemicr. or rhizome geophyte.

Rhynchospora pirrensis W.W.Thomas, Brittonia 38: 314 (1986).
Panama (Cerro Pirre). 80 PAN. Hemicr. or rhizome geophyte.

Rhynchospora planifolia (T.Koyama) W.W.Thomas, Brittonia 48: 485 (1996 publ. 1997).
SE. Colombia to SW. Venezuela. 82 VEN 83 CLM. Hemicr. or rhizome geophyte.
**Rhynchospora rigidifolia* subsp. *planifolia* T.Koyama, Mem. New York Bot. Gard. 23: 63 (1972).

Rhynchospora pleiantha (Kük.) Gale, Rhodora 46: 171 (1944).
SE. U.S.A., Cuba. 78 ALA FLA GEO NCA 81 CUB. Hemicr. or rhizome geophyte.
**Rhynchospora filifolia* var. *pleiantha* Kük., Repert. Spec. Nov. Regni Veg. 23: 208 (1926).

Rhynchospora plumosa Elliott, Sketch Bot. S. Carolina 1: 58 (1816). *Phaeocephalum plumosum* (Elliott) House, Amer. Midl. Naturalist 6: 202 (1920).
SE. U.S.A. to C. America, Caribbean. 77 TEX 78 ALA DEL FLA GEO LOU MSI NCA SCA 80 BLZ HON NIC 81 CUB PUE. Hemicr.
Schoenus ciliaris Muhl., Descr. Gram.: 11 (1817), nom. illeg.
Rhynchospora semiplumosa A.Gray, Ann. Lyceum Nat. Hist. New York 3: 213 (1835). *Rhynchospora plumosa* var. *semiplumosa* (A.Gray) Alph.Wood, Class-book Bot.: 743 (1861).
Rhynchospora penniseta Griseb., Cat. Pl. Cub.: 244 (1866).
Rhynchospora plumosa var. *interrupta* Kük., Bot. Jahrb. Syst. 74: 508 (1949).

Rhynchospora pluricarpa Pilg., Bot. Jahrb. Syst. 30: 143 (1901).
Brazil (Mato Grosso). 84 BZC. Hemicr. or rhizome geophyte.

Rhynchospora plusquamrobusta Luceño & M.Martins, Anales Jard. Bot. Madrid 57: 174 (1999).
Brazil (Espírito Santo). 84 BZL. Hemicr. or rhizome geophyte.

Rhynchospora polyantha Steud., Syn. Pl. Glumac. 2: 147 (1855).
Peru, SE. Brazil. 83 PER 84 BZL. Hemicr. or rhizome geophyte.
**Rhynchospora multiflora* Nees in C.F.P.von Martius & auct. suc. (eds.), Fl. Bras. 2(1): 143 (1842), nom. illeg.

Rhynchospora polyphylla (Vahl) Vahl, Enum. Pl. 2: 230 (1805).
Trop. America. 79 MXG MXT 80 BLZ COS CPI GUA HON NIC PAN 81 CUB DOM JAM LEE TRT WIN 82 FRG GUY SUR VEN 83 BOL CLM ECU PER 84 BZN. Hemicr. or rhizome geophyte.
**Schoenus polyphyllus* Vahl, Eclog. Amer. 2: 5 (1798). *Mitrospora polyphylla* (Vahl) Nees, Linnaea 9: 295 (1834). *Dichromena polyphylla* (Vahl) J.F.Macbr., Publ. Field Columbian Mus., Bot. Ser. 4: 167 (1929).
Rhynchospora rigida Spreng., Syst. Veg. 1: 196 (1824).
Rhynchospora pyramidata Desv. ex Ham., Prodr. Pl. Ind. Occid.: 15 (1825).
Rhynchospora ferruginea Sieber ex C.Presl, Isis (Oken) 21: 269 (1828), nom. inval.

Rhynchospora polycephala C.Presl, Isis (Oken) 21: 269 (1828), nom. inval.
Rhynchospora paniculata C.Presl, Symb. Bot. 1: 42 (1831).
Haloschoenus rigidus Nees ex Kunth, Enum. Pl. 2: 300 (1837).
Rhynchospora adulta Schltdl., Bot. Zeitung (Berlin) 7: 166 (1849).
Rhynchospora costaricensis Boeck., Allg. Bot. Z. Syst. 2: 110 (1896).
Rhynchospora polyphylla var. *brunnescens* Kük., Repert. Spec. Nov. Regni Veg. 23: 204 (1926).
Rhynchospora polyphylla var. *contracta* Kük., Repert. Spec. Nov. Regni Veg. 23: 204 (1926).
Rhynchospora polyphylla var. *longispiculosa* Kük., Bot. Jahrb. Syst. 74: 406 (1949).

Rhynchospora polystachys (Turrill) H.Pfeiff., Repert. Spec. Nov. Regni Veg. 38: 93 (1935).
Honduras to Peru. 80 COS HON 83 CLM ECU PER. Hemicr.
**Dichromena polystachys* Turrill, Mém. Soc. Sci. Nat. Neuchâtel 5: 348 (1913). *Rhynchospora nervosa* var. *polystachys* (Turrill) Kük., Bot. Jahrb. Syst. 75: 298 (1951).

Rhynchospora praecincta Maury ex Micheli, Denkschr. Schweiz. Naturf. Ges. 31(1): 146 (1890).
Paraguay to Uruguay. 85 AGE PAR URU. Hemicr. or rhizome geophyte.
Rhynchospora hyalinolepis Boeck., Beitr. Cyper. 2: 23 (1890).

Rhynchospora prenteloupiana Boeck., Beitr. Cyper. 1: 26 (1888).
Dominican Rep. 81 DOM. Hemicr. or rhizome geophyte.

Rhynchospora pruinosa Griseb., Mem. Amer. Acad. Arts, n.s., 8: 535 (1863).
E. Cuba. 81 CUB. Hemicr. or rhizome geophyte.

Rhynchospora pseudomacrostachya Gerry Moore, Guagl. & Zartman, Brittonia 54: 340 (2002 publ. 2003).
S. Brazil. 84 BZS.

Rhynchospora pterochaeta F.Muell., Fragm. 9: 17 (1875).
N. & NE. Australia. 50 NTA QLD. Hemicr. or rhizome geophyte.

Rhynchospora pubera (Vahl) Boeck., Linnaea 37: 528 (1873).
Costa Rica to S. Trop. America. 80 COS PAN 82 FRG GUY SUR VEN 83 BOL CLM PER 84 BZC BZN. Ther. or hemicr.
**Dichromena pubera* Vahl, Enum. Pl. 2: 241 (1805).

subsp. ***parvula*** W.W.Thomas, Mem. New York Bot. Gard. 37: 50 (1984).
Costa Rica to N. South America. 80 COS PAN 82 GUY SUR VEN 83 CLM 84 BZN. Ther. or hemicr.

subsp. ***pubera***
S. Trop America. 82 FRG GUY SUR VEN 83 BOL CLM PER 84 BZC BZN. Ther. or hemicr.
Schoenus pubescens Kunth in F.W.H.von Humboldt, A.J.A.Bonpland & C.S.Kunth, Nov. Gen. Sp. 1: 228 (1816). *Dichromena humboldtiana* Nees in C.F.P.von Martius & auct. suc. (eds.), Fl. Bras. 2(1): 111 (1842), nom. illeg. *Rhynchospora humboldtiana* (Nees) Griseb., Fl. Brit. W. I.: 577 (1864), nom. illeg.
Dichromena erostris Nees, Linnaea 9: 291 (1834), nom. inval.
Dichromena tenera Schrad. ex Nees in C.F.P.von Martius & auct. suc. (eds.), Fl. Bras. 2(1): 113 (1842).
Dichromena unicolor Salzm. ex Steud., Syn. Pl. Glumac. 2: 136 (1855).
Rhynchospora pubera var. *multispiculosa* Kük., Bot. Jahrb. Syst. 75: 308 (1951).

Rhynchospora pubisquama M.T.Strong, Novon 11: 270 (2001).
Guianas. 82 FRG SUR. Hemicr. or rhizome geophyte.
**Rhynchospora cephalotes* var. *angustifolia* Uittien, Recueil Trav. Bot. Néerl. 30: 188 (1933).

Rhynchospora punctata Elliott, Sketch Bot. S. Carolina 1: 60 (1816). *Phaeocephalum punctatum* (Elliott) House, Amer. Midl. Naturalist 6: 202 (1920).
Georgia to Florida. 78 FLA GEO. Hemicr.

Rhynchospora pusilla Chapm. ex M.A.Curtis, Amer. J. Sci. Arts, II, 7: 409 (1849). *Phaeocephalum pusillum* (Chapm. ex M.A.Curtis) House, Amer. Midl. Naturalist 6: 202 (1920).
SE. U.S.A. to C. America, Caribbean. 77 TEX 78 ALA FLA GEO LOU MSI NCA SCA 79 MXT 80 BLZ GUA NIC 81 CUB DOM HAI JAM PUE. Hemicr.
Rhynchospora intermixta C.Wright, Anales Acad. Ci. Méd. Habana 8: 88 (1871).
Rhynchospora bruneri Britton, Sci. Surv. Porto Rico & Virgin Islands 5: 103 (1923).
Rhynchospora pusilla var. *ferruginescens* Kük., Repert. Spec. Nov. Regni Veg. 23: 207 (1926).

Rhynchospora racemosa C.Wright ex Sauvalle, Anales Acad. Ci. Méd. Habana 8: 86 (1871).
Caribbean, Colombia. 81 CUB DOM HAI JAM PUE 83 CLM. Hemicr. or rhizome geophyte.
Schoenus gracilis Sw., Prodr.: 19 (1788).
Rhynchospora polyphylla f. *umbrosa* Boeck., Linnaea 37: 636 (1873).
Rhynchospora racemosa f. *setifera* Kük., Repert. Spec. Nov. Regni Veg. 23: 203 (1926).
Rhynchospora racemosa f. *setifolia* Kük., Repert. Spec. Nov. Regni Veg. 23: 203 (1926).

Rhynchospora radicans (Schltdl. & Cham.) H.Pfeiff., Repert. Spec. Nov. Regni Veg. 38: 93 (1935).
Mexico to Trop. America. 79 MXC MXE MXG MXS MXT 80 BLZ COS ELS GUA HON NIC PAN 81 CUB DOM HAI JAM LEE PUE TRT WIN 82 FRG GUY SUR VEN 83 BOL CLM ECU PER 84 BZN. Hemicr. or rhizome geophyte.
**Dichromena radicans* Schltdl. & Cham., Linnaea 6: 28 (1831). *Dichromena radicans* var. *typica* A.Gross & Brade, Arq. Mus. Nac. Rio de Janeiro 34: 58 (1932), nom. inval.

subsp. ***microcephala*** (Bertero ex Spreng.) W.W. Thomas, Mem. New York Bot. Gard. 37: 60 (1984).
Trop. America. 79 MXG MXS MXT 80 BLZ COS ELS GUA HON NIC PAN 81 CUB DOM HAI JAM LEE PUE TRT WIN 82 FRG GUY SUR VEN 83 BOL CLM ECU PER 84 BZN. Hemicr. or rhizome geophyte.
**Dichromena microcephala* Bertero ex Spreng., Syst. Veg. 1: 202 (1824). *Rhynchospora microcephala* (Bertero ex Spreng.) Kük., Bot. Jahrb. Syst. 75: 311 (1951), nom. illeg.
Rhynchospora pubera var. *microcarpa* Boeck., Linnaea 37: 529 (1872).

Rhynchospora pubera f. *elatior* Duss, Ann. Inst. Bot.-Géol. Colon. Marseille 3: 552 (1896).

subsp. ***radicans***
Mexico to Brazil, Trinidad. 79 MXC MXE MXG MXS MXT 80 BLZ COS GUA HON PAN 81 TRT 82 VEN 83 CLM 84 BZN. Hemicr. or rhizome geophyte.
Dichromena tenuis Steud., Syn. Pl. Glumac. 2: 135 (1855).

Rhynchospora rariflora (Michx.) Elliott, Sketch Bot. S. Carolina 1: 58 (1816).
New Jersey to C. America, Caribbean. 74 OKL 75 NWJ 77 TEX 78 ALA ARK FLA GEO LOU MRY MSI NCA SCA TEN VRG 80 BLZ HON NIC 81 CUB DOM JAM PUE. Ther. or hemicr.
**Schoenus rariflorus* Michx., Fl. Bor.-Amer. 1: 35 (1803). *Rhynchospora micrantha* Vahl, Enum. Pl. 2: 231 (1805), nom. illeg. *Dichromena micrantha* (Vahl) Kunth, Enum. Pl. 2: 278 (1837). *Rhynchospora micrantha* Pursh, Fl. Amer. Sept. 1: 48 (1813), nom. illeg. *Phaeocephalum rariflorum* (Michx.) House, Amer. Midl. Naturalist 6: 202 (1920).
Schoenus glaucus Poir. in J.B.A.M.de Lamarck, Encycl., Suppl. 2: 249 (1811).
Haloschoenus sparsus Nees, Linnaea 9: 296 (1834), nom. inval.
Rhynchospora setacea Willd. ex Kunth, Enum. Pl. 2: 297 (1837).
Schoenus setaceus Vahl ex Kunth., Enum. Pl. 2: 297 (1837).
Pterochaete drummondii Arn. ex Boeck., Linnaea 37: 600 (1873).

Rhynchospora recognita (Gale) Kral, Novon 9: 205 (1999).
California, C. & E. U.S.A. to C. America, Caribbean. 74 ILL KAN MSO OKL 75 INI MIC NWJ OHI PEN WVA 76 CAL 77 TEX 78 ALA ARK DEL FLA GEO KTY LOU LOU MRY MSI NCA SCA TEN VRG WDC 79 MXT 80 BLZ GUA HON NIC 81 CUB JAM PUE WIN. Hemicr. or rhizome geophyte.
Schoenus cymosus Muhl., Descr. Gram.: 8 (1817), nom. illeg.
**Rhynchospora globularis* var. *recognita* Gale, Rhodora 46: 245 (1944).

Rhynchospora recurvata (Nees) Steud., Syn. Pl. Glumac. 2: 145 (1855).
E. Brazil. 84 BZE BZL. Hemicr. or rhizome geophyte.
**Mitrospora recurvata* Nees in C.F.P.von Martius & auct. suc. (eds.), Fl. Bras. 2(1): 132 (1842).
Schoenus recurvatus Schrad. ex Nees in C.F.P.von Martius & auct. suc. (eds.), Fl. Bras. 2(1): 132 (1842), nom. inval.
Rhynchospora hemicephala Boeck., Beitr. Cyper. 1: 24 (1888). *Rhynchospora recurvata* var. *hemicephala* (Boeck.) Kük., Bot. Jahrb. Syst. 74: 459 (1949). *Rhynchospora recurvata* var. *hemicephala* (Boeck.) Kük., Bot. Jahrb. Syst. 74: 459 (1949).

Rhynchospora reptans (Rich.) Kük., Bot. Jahrb. Syst. 75: 304 (1951).
N. South America to Brazil. 82 FRG GUY SUR 84 BZC BZN. Hemicr. or rhizome geophyte.
**Scirpus reptans* Rich., Actes Soc. Hist. Nat. Paris 1: 106 (1792). *Dichromena reptans* (Rich.) Pers., Syn. Pl. 1: 57 (1805).
Dichromena repens Vahl, Enum. Pl. 2: 241 (1805). *Rhynchospora repens* (Vahl) Boeck., Linnaea 37: 531 (1873).
Rhynchospora sprucei Boeck., Allg. Bot. Z. Syst. 2: 79 (1896).

Rhynchospora ridleyi C.B.Clarke, Kew Bull., Addit. Ser. 8: 36 (1908).
NE. Brazil. 84 BZE. Hemicr.

Rhynchospora riedeliana C.B.Clarke, Kew Bull., Addit. Ser. 8: 34 (1908).
Brazil (Minas Gerais). 84 BZL. Hemicr. or rhizome geophyte.

Rhynchospora rigidifolia (Gilly) T.Koyama, Mem. New York Bot. Gard. 23: 62 (1972).
S. Venezuela. 82 VEN. Hemicr. or rhizome geophyte.
**Dichromena rigidifolia* Gilly, Brittonia 3: 154 (1939).

subsp. ***capillacea*** T.Koyama, Mem. New York Bot. Gard. 23: 63 (1972). *Rhynchospora capillifolia* W.W.Thomas, Brittonia 48: 484 (1996 publ. 1997), nom. illeg.
Venezuela (Bolívar) to W. Guyana. 82 GUY VEN. Hemicr. or rhizome geophyte.

subsp. ***rigidifolia***
S. Venezuela. 82 VEN. Hemicr. or rhizome geophyte.

Rhynchospora riparia (Nees) Boeck., Linnaea 37: 561 (1873).
N. South America to Uruguay. 82 FRG GUY SUR VEN 83 CLM 84 BZC BZE BZL BZN 85 URU. Hemicr. or rhizome geophyte.
Haloschoenus capillaris var. *congestus* Nees in C.F.P.von Martius & auct. suc. (eds.), Fl. Bras. 2(1): 121 (1842). *Rhynchospora tenuis* var. *congesta* (Nees) Kük., Bot. Jahrb. Syst. 75: 191 (1950).
**Haloschoenus riparius* Nees in C.F.P.von Martius & auct. suc. (eds.), Fl. Bras. 2(1): 120 (1842). *Dichromena riparia* (Nees) Steud., Syn. Pl. Glumac. 2: 136 (1855). *Rhynchospora tenuis* subsp. *riparia* (Nees) T.Koyama, Mem. New York Bot. Gard. 23: 78 (1972).
Dichromena canaliculata Steud., Syn. Pl. Glumac. 2: 136 (1855). *Rhynchospora canaliculata* (Steud.) Boeck., Flora 63: 451 (1880).
Schoenus maritimus Salzm. ex Steud., Syn. Pl. Glumac. 2: 137 (1855), nom. illeg. *Rhynchospora tenuis* var. *maritima* Boeck., Linnaea 37: 612 (1873).
Rhynchospora rudioi Boeck., Flora 65: 27 (1882).

Rhynchospora robusta (Kunth) Boeck., Linnaea 37: 616 (1973).
Mexico to Trop. America. 79 MXC MXG MXT 80 BLZ GUA HON PAN 81 CUB JAM 82 FRG? GUY? VEN 83 CLM 84 BZC BZL BZN BZS 85 AGE PAR. Hemicr.
**Dichromena robusta* Kunth, Enum. Pl. 2: 283 (1837). *Psilocarya robusta* (Kunth) Nees in C.F.P.von Martius & auct. suc. (eds.), Fl. Bras. 2(1): 116 (1842).
Psilocarya conferta Nees in C.F.P.von Martius & auct. suc. (eds.), Fl. Bras. 2(1): 116 (1842). *Rhynchospora conferta* (Nees) Boeck., Linnaea 37: 615 (1873).
Psilocarya robusta var. *major* Nees in C.F.P.von Martius & auct. suc. (eds.), Fl. Bras. 2(1): 116 (1842).
Psilocarya robusta var. *minor* Nees in C.F.P.von Martius & auct. suc. (eds.), Fl. Bras. 2(1): 116 (1842).
Schoenus pubigerus Schrad. ex Nees in C.F.P.von Martius & auct. suc. (eds.), Fl. Bras. 2(1): 116 (1842).

Rhynchospora robusta var. *platyrrhyncha* C.B.Clarke ex Chodat & Hassl., Bull. Herb. Boissier, II, 3: 1024 (1903).
Rhynchospora robusta var. *polyantha* Kük., Bot. Jahrb. Syst. 75: 171 (1950).

Rhynchospora roraimae Kük., Bot. Jahrb. Syst. 56(125): 19 (1921).
S. Venezuela to W. Guyana. 82 GUY VEN. Hemicr. or rhizome geophyte.

Rhynchospora rosae W.W.Thomas, Brittonia 44: 31 (1992).
Mexico (Chiapas). 79 MXT. Hemicr.

Rhynchospora rosemariana D.A.Simpson, Kew Bull. 44: 289 (1989). *Rhynchospora terminalis* var. *rosemariana* (D.A.Simpson) A.C.Araujo, Novon 14: 11 (2004).
Brazil (Minas Gerais, Mato Grosso). 84 BZC BZL. Hemicr. or rhizome geophyte.

Rhynchospora rostrata Lindm., Bih. Kongl. Svenska Vetensk.-Akad. Handl. 26(9): 27 (1900).
Brazil. 84 BZC. Hemicr. or rhizome geophyte.

Rhynchospora rubra (Lour.) Makino, Bot. Mag. (Tokyo) 17: 180 (1903).
Trop. & Subtrop. Old World. 22 BEN GHA IVO LBR SEN SIE 23 EQG GAB 25 MOZ TAN 27 BOT NAT 29 MDG 36 CHC CHH CHS 38 JAP KOR NNS TAI 40 ASS EHM IND NEP SRL 41 AND NCB THA VIE 42 BOR JAW LSI MLY MOL PHI SUL SUM 43 BIS NWG 50 NSW NTA QLD 62 CRL MRN. Ther. or hemicr.
**Schoenus ruber* Lour., Fl. Cochinch.: 52 (1790).

subsp. ***africana*** J.Raynal, Adansonia, n.s., 7: 323 (1967).
Trop. & S. Africa, Madagascar. 22 BEN GHA IVO LBR SEN SIE 23 EQG GAB 25 MOZ TAN 27 BOT NAT 29 MDG. Ther. or hemicr.
Chaetospora madagascariensis Steud., Syn. Pl. Glumac. 2: 161 (1855). *Rhynchospora madagascariensis* (Steud.) Cherm., Mém. Acad. Malagache 10: 38 (1931).
Rhynchospora minor Nelmes, Kew Bull. 11: 533 (1956 publ. 1957), nom. illeg.

subsp. ***rubra***
Trop. & Subtrop. Asia to NW. Pacific. 36 CHC CHH CHS 38 JAP KOR NNS TAI 40 ASS EHM IND NEP SRL 41 AND NCB THA VIE 42 BOR JAW LSI MLY MOL PHI SUL SUM 43 BIS NWG 50 NSW NTA QLD 62 CRL MRN. Ther. or hemicr.
Rhynchospora haenkei J.Presl & C.Presl, Reliq. Haenk. 1: 199 (1828).
Scirpus retusus J.König ex Wall., Numer. List: 3428 (1831).
Cephaloschoenus parvus Nees, Linnaea 9: 296 (1834), nom. inval. *Rhynchospora parva* (Nees) Kunth, Enum. Pl. 2: 302 (1837).
Haplostylis meyenii Nees, Linnaea 9: 295 (1834), nom. inval.
Morisia capitata Nees, Linnaea 9: 295 (1834), nom. inval.
Morisia wallichii Nees, Edinburgh New Philos. J., Oct. 1834: 265 (1834). *Sphaeroschoenus wallichii* (Nees) Nees, Nov. Actorum Acad. Caes. Leop.-Carol. Nat. Cur. 19(Suppl. 1): 97 (1843). *Rhynchospora wallichii* (Nees) K.Schum. in K.M.Schumann & U.M.Hollrung, Fl. Kais. Wilh. Land: 25 (1889).
Rhynchospora wallichiana Kunth, Enum. Pl. 2: 289 (1837).
Mariscus umbellatus Zoll., Syst. Verz. 1: 63 (1854), nom. illeg.
Mariscus umbellatus var. *procerior* Zoll., Syst. Verz. 1: 63 (1854).
Rhynchospora rubra f. *elata* Domin, Biblioth. Bot. 85: 468 (1915).
Rhynchospora mairei H.Lév., Cat. Pl. Yunnan: 80 (1916).

subsp. ***senegalensis*** J.Raynal, Adansonia, n.s., 7: 323 (1967).
Senegal. 22 SEN. Ther. or hemicr.

Rhynchospora rudis C.B.Clarke, Kew Bull., Addit. Ser. 8: 39 (1908).
Venezuela. 82 VEN. Hemicr. or rhizome geophyte.

Rhynchospora rugosa (Vahl) Gale, Rhodora 46: 275 (1944).
Trop. & Subtrop. 12 POR SPA 20 ALG MOR TUN 22 IVO NGA 23 CMN GAB RWA ZAI 24 ETH 25 TAN UGA 26 ANG MLW MOZ ZAM ZIM 27 BOT CPP LES NAT SWZ TVL 29 MAU MDG REU 36 CHS 38 JAP NNS TAI 40 ASS EHM IND NEP SRL 41 THA VIE 42 BOR JAW MLY PHI SUL SUM 43 NWG 50 NSW QLD VIC 62 CRL 63 HAW 79 MXC MXG MXS MXT 80 BLZ COS GUA HON NIC PAN 81 CUB DOM JAM LEE PUE TRT WIN 82 FRG GUY SUR VEN 83 BOL CLM ECU PER 84 BZC BZE BZL BZN BZS 85 AGE AGW PAR URU. Hemicr.
**Schoenus rugosus* Vahl, Eclog. Amer. 2: 5 (1798).

subsp. ***americana*** (Guagl.) Govaerts in R.H.A. Govaerts & D.A.Simpson, World Checklist of Cyperaceae: 640 (2007).
SE. Mexico to S. South America. 79 MXT 80 GUA HON NIC PAN 82 VEN 83 BOL CLM ECU PER 84 BZC BZL BZS 85 AGE CLC PAR URU. Hemicr. or rhizome geophyte.
**Rhynchospora brownii* subsp. *americana* Guagl., Darwiniana 22: 272 (1979).

subsp. ***brownii*** (Roem. & Schult.) T.Koyama, Enum. Fl. Pl. Nepal 1: 118 (1978).
Trop. & Subtrop. Old World. 12 POR SPA 20 ALG MOR TUN 22 IVO NGA 23 CMN GAB RWA ZAI 24 ETH 25 TAN UGA 26 ANG MLW MOZ ZAM ZIM 27 BOT CPP LES NAT SWZ TVL 29 MAU MDG REU 36 CHS 38 JAP NNS TAI 40 ASS EHM IND NEP SRL 41 THA VIE 42 BOR JAW MLY PHI SUL SUM 43 NWG 50 NSW QLD VIC. Hemicr. or rhizome geophyte.
Rhynchospora glauca Vahl, Enum. Pl. 2: 233 (1805). *Dichromena glauca* (Vahl) J.F.Macbr., Publ. Field Columbian Mus., Bot. Ser. 4: 166 (1929).
Rhynchospora gracilis Vahl, Enum. Pl. 2: 234 (1805).
Rhynchospora laxa R.Br., Prodr.: 230 (1810), nom. illeg. *Schoenus laxus* Poir. in J.B.A.M.de Lamarck, Encycl., Suppl. 2: 252 (1811). **Rhynchospora brownii* Roem. & Schult., Syst. Veg. 2: 86 (1817). *Rhynchospora sinensis* Steud., Nomencl. Bot., ed. 2, 2: 456 (1841).
Chaetospora ferruginea Kunth in F.W.H.von Humboldt, A.J.A.Bonpland & C.S.Kunth, Nov. Gen. Sp. 2: 85 (1817).
Chaetospora vicozensis Schrad. ex Schult., Mant. 2: 46 (1824).

Rhynchospora lavarum Hook. & Arn., Bot. Beechey Voy.: 98 (1832).
Rhynchospora thouarsii Nees, Linnaea 9: 297 (1834), nom. inval.
Rhynchospora castanea Nees & Meyen in F.J.F.Meyen, Observ. Bot. 2: 119 (1835).
Rhynchospora brunnea Willd. ex Kunth, Enum. Pl. 2: 297 (1837).
Rhynchospora distans Beyr. ex Kunth, Enum. Pl. 2: 297 (1837).
Rhynchospora juncea Willd. ex Kunth, Enum. Pl. 2: 298 (1837).
Schoenus pauciflorus Willd. ex Kunth, Enum. Pl. 2: 298 (1837).
Rhynchospora effusa Schrad. ex Nees in C.F.P.von Martius & auct. suc. (eds.), Fl. Bras. 2(1): 142 (1842).
Rhynchospora glauca var. *strobilacea* Nees in C.F.P.von Martius & auct. suc. (eds.), Fl. Bras. 2(1): 141 (1842).
Rhynchospora glauca var. *ustilaginosa* Nees in C.F.P.von Martius & auct. suc. (eds.), Fl. Bras. 2(1): 141 (1842).
Rhynchospora petraea Nees in C.F.P.von Martius & auct. suc. (eds.), Fl. Bras. 2(1): 142 (1842).
Rhynchospora laevis Schltdl., Linnaea 25: 166 (1852).
Chaetospora mauritii Steud., Syn. Pl. Glumac. 2: 161 (1855).
Rhynchospora laxa var. *minor* Thwaites, Enum. Pl. Zeyl.: 352 (1864).
Rhynchospora griffithii Boeck., Linnaea 38: 404 (1874). *Rhynchospora rugosa* var. *griffithii* (Boeck.) D.M.Verma & V.Chandra, Bull. Bot. Surv. India 22: 128 (1980 publ. 1982).
Rhynchospora sikkimensis C.B.Clarke in J.D.Hooker, Fl. Brit. India 6: 672 (1893). *Rhynchospora rugosa* var. *sikkimensis* (C.B. Clarke) D.M.Verma & V.Chandra, Bull. Bot. Surv. India 22: 130 (1980 publ. 1982).
Rhynchospora glauca var. *altior* Kük. ex O.C.Schmidt, Repert. Spec. Nov. Regni Veg. 32: 77 (1933).
Rhynchospora glauca var. *condensata* Kük., Bot. Jahrb. Syst. 69: 259 (1938).

subsp. ***lavarum*** (Gaudich.) T.Koyama, Bishop Mus. Occas. Pap. 29: 127 (1989).
Hawaiian Is. 63 HAW. Hemicr.
**Rhynchospora lavarum* Gaudich., Voy. Uranie: 415 (1829).

f. ***ponapensis*** (Hosok.) T.Koyama, Micronesica 1: 107 (1964).
Caroline Is. (Pohnpei). 62 CRL. Hemicr. or rhizome geophyte.
**Rhynchospora ponapensis* Hosok., Trans. Nat. Hist. Soc. Taiwan 25: 263 (1935).

subsp. ***rugosa***
Mexico to Trop. America. 79 MXC MXG MXS MXT 80 BLZ COS GUA HON NIC PAN 81 CUB DOM JAM LEE PUE TRT WIN 82 FRG GUY SUR VEN 83 BOL CLM ECU PER 84 BZC BZE BZL BZN BZS 85 AGE AGW PAR URU. Hemicr.
Rhynchospora ferruginea Roem. & Schult., Syst. Veg. 2: 85 (1817).
Rhynchospora vicozensis Schult. in J.J.Roemer & J.A.Schultes, Mant. 2: 46 (1824).
Rhynchospora pungens Liebm., Mexic. Neldeagt. Pl., V, 2: 252 (1851).
Rhynchospora laevigata Steud., Syn. Pl. Glumac. 2: 145 (1855).
Rhynchospora pauloensis Boeck., Beitr. Cyper. 2: 24 (1890).
Rhynchospora durandiana Boeck., Allg. Bot. Z. Syst. 2: 94 (1896).

Rhynchospora ruiziana Boeck., Linnaea 37: 641 (1873). *Dichromena ruiziana* (Boeck.) J.F.Macbr., Publ. Field Columbian Mus., Bot. Ser. 4: 167 (1929). *Rhynchospora macrochaeta* var. *ruiziana* (Boeck.) Kük., Bot. Jahrb. Syst. 74: 393 (1949).
S. Mexico to W. South America and Guyana. 79 MXS MXT 80 COS PAN 82 GUY VEN 83 BOL CLM ECU PER. Hemicr. or rhizome geophyte.
Rhynchospora stuebelii Boeck., Linnaea 37: 641 (1873).
Rhynchospora quadrispicata Boeck., Beitr. Cyper. 1: 28 (1888). *Rhynchospora macrochaeta* var. *quadrispicata* (Boeck.) Kük., Bot. Jahrb. Syst. 74: 395 (1949).
Rhynchospora quinquespicata Boeck., Beitr. Cyper. 1: 29 (1888). *Rhynchospora macrochaeta* var. *quinquespicata* (Boeck.) Kük., Bot. Jahrb. Syst. 74: 394 (1949).
Rhynchospora infuscata Boeck., Beitr. Cyper. 2: 31 (1890).
Rhynchospora macrochaeta f. *condensata* Kük., Bot. Jahrb. Syst. 74: 394 (1949).

Rhynchospora rupicola M.T.Strong, Novon 11: 270 (2001).
Guianas. 82 GUY SUR. Hemicr. or rhizome geophyte.

Rhynchospora sampaioana R.Gross, Repert. Spec. Nov. Regni Veg. 29: 244 (1931).
Brazil (Rio de Janeiro). 84 BZL. Hemicr. or rhizome geophyte.

Rhynchospora sanariapensis Steyerm., Fieldiana, Bot. 28(1): 46 (1951).
SE. Colombia to SW. Guyana. 82 VEN 83 CLM. Ther. or hemicr.

Rhynchospora sartoriana Boeck., Flora 40: 36 (1857).
Mexico. 79 MXG. Hemicr. or rhizome geophyte.

Rhynchospora saxisavannicola M.T.Strong, Novon 15: 482 (2005).
Guianas. 82 FRG SUR. Hemicr. or rhizome geophyte.

Rhynchospora scabrata Griseb., Cat. Pl. Cub.: 243 (1866). *Rhynchospora scabrata* var. *typica* Gale, Rhodora 46: 220 (1944).
E. Cuba (Sierra de Nipe). 81 CUB. Hemicr. or rhizome geophyte.
Rhynchospora randii Britton, Contr. New York Bot. Gard. 194: 14 (1917). *Rhynchospora pruinosa* var. *randii* (Britton) Kük., Bot. Jahrb. Syst. 75: 162 (1950).

Rhynchospora schiedeana (Schltdl.) Kunth, Enum. Pl. 2: 300 (1837). *Calyptrostylis schiedeana* (Kunth) Liebm., Mexic. Neldeagt. Pl., V, 2: 252 (1851).
Mexico to W. South America. 79 MXC MXG MXS MXT 80 COS ELS GUA PAN 83 BOL CLM ECU PER. Hemicr. or rhizome geophyte.
**Chaetospora schiedeana* Schltdl., Linnaea 6: 28 (1831).
Calyptrostylis paniculata Liebm., Mexic. Neldeagt. Pl., V, 2: 252 (1851).

Rhynchospora paniculifera Steud., Syn. Pl. Glumac. 2: 146 (1855).
Rhynchospora schiedeana subsp. *varica* C.B. Clarke, Contr. U. S. Natl. Herb. 10: 463 (1908).

Rhynchospora schmidtii Kük., Repert. Spec. Nov. Regni Veg. 23: 202 (1926). *Rhynchospora trispicata* subsp. *schmidtii* (Kük.) T.Koyama, Mem. New York Bot. Gard. 23: 81 (1972).
W. Cuba. 81 CUB. Hemicr. or rhizome geophyte.

Rhynchospora schomburgkiana (Boeck.) T.Koyama, Mem. New York Bot. Gard. 23: 67 (1972).
Colombia to Guyana. 82 GUY VEN 83 CLM 84 BZN. Hemicr. or rhizome geophyte.
Leptoschoenus prolifer Nees, J. Bot. (Hooker) 2: 393 (1840). *Rhynchospora prolifera* (Nees) C.B.Clarke, Kew Bull., Addit. Ser. 8: 90 (1908), nom. illeg. *Phaeocephalum proliferum* (Nees) House, Amer. Midl. Naturalist 6: 202 (1920).
**Scirpus schomburgkianus* Boeck., Linnaea 36: 757 (1870).
Micropapyrus viviparoides Suess., Bot. Jahrb. Syst. 73: 116 (1943).

Rhynchospora schottmuelleri Boeck., Flora 65: 27 (1882). *Pleurostachys schottmuelleri* (Boeck.) Kük., Bot. Jahrb. Syst. 75: 313 (1951).
Brazil. 84+. Hemicr. or rhizome geophyte.

Rhynchospora scirpoides (Torr.) Griseb., Cat. Pl. Cub.: 247 (1866).
EC. & E. U.S.A., Cuba. 74 WIS 75 CNT INI MAS MIC NWJ NWY RHO 77 TEX 78 ALA DEL FLA GEO LOU MRY MSI NCA SCA VRG 81 CUB. Hemicr. or rhizome geophyte.
**Psilocarya scirpoides* Torr., Ann. Lyceum Nat. Hist. New York 3: 360 (1836).
Psilocarya corymbiformis Benth. & Hook.f., Gen. Pl. 3: 1048 (1883).
Rhynchospora corymbiformis Britton, Trans. New York Acad. Sci. 11: 86 (1892).
Psilocarya scirpoides var. *grimesii* Fernald & Griscom, Rhodora 37: 154 (1935).

Rhynchospora sclerioides Hook. & Arn., Bot. Beechey Voy.: 99 (1832).
C. Pacific. 61 SCI 63 HAW. Hemicr. or rhizome geophyte.
Rhynchospora thyrsoidea Nees & Meyen ex Kunth, Enum. Pl. 2: 204 (1837).

Rhynchospora scutellata Griseb., Cat. Pl. Cub.: 246 (1866).
Texas to Belize, W. Cuba, Ecuador to Uruguay. 77 TEX 79 MXE MXS MXT 80 BLZ 81 CUB 83 ECU 84 BZC 85 AGE PAR URU. Hemicr. or rhizome geophyte.
Rhynchospora indianolensis Small, Fl. S.E. U.S.: 193 (1903).
Rhynchospora pringlei Greenm., Proc. Amer. Acad. Arts 39: 69 (1903).
Rhynchospora puncticulata Kük. ex Osten, Anales Mus. Hist. Nat. Montevideo, II, 3(2): 210 (1931). *Rhynchospora corymbosa* var. *puncticulata* (Kük. ex Osten) Kük., Bot. Jahrb. Syst. 74: 416 (1949).

Rhynchospora sellowiana Steud., Syn. Pl. Glumac. 2: 145 (1855). *Rhynchospora pseudosellowiana* Kük., Bot. Jahrb. Syst. 75: 155 (1950), nom. illeg.
Brazil. 84 BZC BZL. Hemicr. or rhizome geophyte.
Rhynchospora filifolia Kunth, Enum. Pl. 2: 299 (1837), nom. illeg.
Rhynchospora pallens Schrad. ex Nees in C.F.P.von Martius & auct. suc. (eds.), Fl. Bras. 2(1): 146 (1842), pro syn.
Rhynchospora pseudosellowiana var. *castanea* Kük., Bot. Jahrb. Syst. 75: 156 (1950).

Rhynchospora semihirsuta Boeck., Linnaea 37: 614 (1873).
S. Brazil. 84 BZS. Hemicr. or rhizome geophyte.

Rhynchospora semiinvolucrata J.Presl & C.Presl, Reliq. Haenk. 1: 198 (1828). *Dichromena semiinvolucrata* (J.Presl & C.Presl) J.F.Macbr., Publ. Field Columbian Mus., Bot. Ser. 4: 167 (1929).
Peru. 83 PER. Hemicr. or rhizome geophyte.

Rhynchospora seslerioides Griseb., Cat. Pl. Cub.: 247 (1866). *Dichromena seslerioides* (Griseb.) Benth. ex C.B.Clarke in I.Urban, Symb. Antill. 2: 100 (1900).
Cuba. 81 CUB. Hemicr. or rhizome geophyte.

Rhynchospora setigera (Kunth) Boeck., Vidensk. Meddel. Naturhist. Foren. Kjøbenhavn 1869: 145 (1869).
Brazil to NE. Argentina. 84 BZE BZL 85 AGE PAR URU. Hemicr. or rhizome geophyte.
**Dichromena setigera* Kunth, Enum. Pl. 2: 277 (1837).
Rhynchospora stelligera Herter, Revista Sudamer. Bot. 9: 159 (1953), nom. illeg.

Rhynchospora shaferi Britton, Contr. New York Bot. Gard. 194: 13 (1917).
E. Cuba (Sierra de Nipe). 81 CUB. Hemicr. or rhizome geophyte.
Rhynchospora lingulata Kük., Repert. Spec. Nov. Regni Veg. 23: 211 (1926).

Rhynchospora siguaneana Britton, Contr. New York Bot. Gard. 194: 6 (1917).
Cuba (I. de la Juventud). 81 CUB. Hemicr. or rhizome geophyte.

Rhynchospora simplex (Kük.) Kük., Bot. Jahrb. Syst. 74: 441 (1949).
Cuba to Hispaniola. 81 CUB DOM HAI. Hemicr. or rhizome geophyte.
**Rhynchospora exaltata* var. *simplex* Kük., Repert. Spec. Nov. Regni Veg. 23: 261 (1926).

Rhynchospora smithii W.W.Thomas, Mem. New York Bot. Gard. 37: 37 (1984).
Brazil (Santa Catarina). 84 BZS. Hemicr. or rhizome geophyte.

Rhynchospora sola Gale, Rhodora 46: 162 (1944).
W. Cuba. 81 CUB. Hemicr. or rhizome geophyte.

Rhynchospora solitaria R.M.Harper, Bull. Torrey Bot. Club 28: 468 (1901). *Phaeocephalum solitare* (R.M.Harper) House, Amer. Midl. Naturalist 6: 202 (1920).
Georgia. 78 GEO. Hemicr. or rhizome geophyte.

Rhynchospora spectabilis Hochst. ex C.Krauss, Flora 28: 760 (1845).
Sudan to S. Africa. 23 ZAI 24 SUD 26 ZAM 27 NAT. Hemicr. or rhizome geophyte.
Rhynchospora corymbosa var. *grandispiculosa* Kük., Bot. Jahrb. Syst. 74: 417 (1949).

Rhynchospora splendens Lindm., Bih. Kongl. Svenska Vetensk.-Akad. Handl. 26(9): 26 (1900).
Venezuela to W. Guyana, Bolivia, SE. & S. Brazil. 82 VEN 83 BOL 84 BZL BZS. Hemicr. or rhizome geophyte.

Rhynchospora macrostachya Palla, Denkschr. Kaiserl. Akad. Wiss., Wien. Math.-Naturwiss. Kl. 79: 191 (1908), nom. illeg.
Rhynchospora splendens f. *composita* Kük., Bot. Jahrb. Syst. 74: 442 (1949).
Rhynchospora splendens f. *laetevirens* Kük., Bot. Jahrb. Syst. 74: 443 (1949).

Rhynchospora spruceana C.B.Clarke, Kew Bull., Addit. Ser. 8: 40 (1908).
N. South America to N. Brazil. 82 FRG GUY SUR VEN 84 BZN. Hemicr. or rhizome geophyte.
Rhynchospora filiformis var. *latifolia* Uittien, Recueil Trav. Bot. Néerl. 22: 336 (1926).
Rhynchospora graminea Uittien in A.A.Pulle, Fl. Suriname 1(1): 105 (1934).

Rhynchospora squamulosa Kük., Repert. Spec. Nov. Regni Veg. 23: 201 (1926).
W. Cuba. 81 CUB. Hemicr. or rhizome geophyte.

Rhynchospora stenocarpa Kunth, Enum. Pl. 2: 294 (1837). *Ephippiorhynchium stenocarpum* (Kunth) Nees in C.F.P.von Martius & auct. suc. (eds.), Fl. Bras. 2(1): 136 (1842).
SE. Brazil. 84 BZL. Hemicr. or rhizome geophyte.

Rhynchospora stenophylla Chapm., Fl. South. U.S.: 525 (1860).
SE. U.S.A. to Texas. 77 TEX 78 ALA FLA GEO LOU MSI NCA SCA. Ther. or hemicr.

Rhynchospora steyermarkii T.Koyama, Mem. New York Bot. Gard. 23: 70 (1972).
I. Margarita, Venezuela to W. Guyana. 81 VNA 82 GUY VEN. Hemicr. or rhizome geophyte.

Rhynchospora stokesii F.Br., Bernice P. Bishop Mus. Bull. 84: 109 (1931). *Rhynchospora sclerioides* var. *stokesii* (F.Br.) Kük., Bot. Jahrb. Syst. 74: 505 (1949).
Tubuai Is. (Rapa-Iti). 61 TUB. Hemicr. or rhizome geophyte.
Rhynchospora stokesii f. *saxicola* F.Br., Bernice P. Bishop Mus. Bull. 84: 110 (1931). *Rhynchospora sclerioides* f. *saxicola* (F.Br.) Kük., Bot. Jahrb. Syst. 74: 505 (1949).

Rhynchospora stolonifera (Nees) Boeck., Linnaea 37: 562 (1873).
SE. Brazil. 84 BZL. Hemicr. or rhizome geophyte.
**Haloschoenus stoloniferus* Nees in C.F.P.von Martius & auct. suc. (eds.), Fl. Bras. 2(1): 120 (1842). *Dichromena stolonifera* (Nees) Steud., Syn. Pl. Glumac. 2: 136 (1855).

Rhynchospora stricta Boeck., Linnaea 37: 603 (1873).
S. Brazil. 84 BZS. Hemicr. or rhizome geophyte.
**Dichromena elatior* Kunth, Enum. Pl. 2: 282 (1837). *Haloschoenus elatior* (Kunth) Nees in C.F.P.von Martius & auct. suc. (eds.), Fl. Bras. 2(1): 120 (1842).

Rhynchospora subdicephala T.Koyama, Bull. Natl. Sci. Mus., Tokyo, B 2: 173 (1976).
S. Venezuela to Brazil (Amapá). 82 FRG VEN 84 BZN. Hemicr. or rhizome geophyte.

Rhynchospora subimberbis Griseb., Cat. Pl. Cub.: 245 (1866).
W. Cuba. 81 CUB. Hemicr. or rhizome geophyte.

Rhynchospora sublanata Lindeman & Donsel., Acta Bot. Neerl. 20: 444 (1971).
Guianas. 82 GUY SUR. Hemicr. or rhizome geophyte.

Rhynchospora submarginata Kük., Bot. Jahrb. Syst. 74: 498 (1949).
Trop. Asia to N. Australia. 40 IND 41 THA VIE 42 MLY SUM 43 NWG 50 NTA. Ther.
**Rhynchospora marginata* C.B.Clarke, Kew Bull., Addit. Ser. 8: 89 (1908), nom. illeg.
Rhynchospora submarginata var. *glabrinux* Kük., Bot. Jahrb. Syst. 74: 499 (1949).
Rhynchospora submarginata var. *tonkinensis* Kük., Bot. Jahrb. Syst. 74: 499 (1949).

Rhynchospora subplumosa C.B.Clarke, Kew Bull., Addit. Ser. 8: 34 (1908).
N. South America to Brazil (Roraima). 82 GUY SUR VEN 83 CLM 84 BZN. Hemicr. or rhizome geophyte.
Rhynchospora trichochaeta var. *venezuelensis* Steyerm., Fieldiana, Bot. 28: 49 (1951).

Rhynchospora subsetigera H.Pfeiff., Repert. Spec. Nov. Regni Veg. 49: 81 (1940).
E. Cuba. 81 CUB. Hemicr. or rhizome geophyte.
**Rhynchospora setigera* Griseb., Cat. Pl. Cub.: 248 (1866), nom. illeg. *Dichromena grisebachii* C.B.Clarke in I.Urban, Symb. Antill. 2: 102 (1900).

Rhynchospora subsetosa C.B.Clarke, Kew Bull., Addit. Ser. 8: 36 (1908).
SE. & S. Brazil. 84 BZL BZS. Hemicr. or rhizome geophyte.
**Rhynchospora sellowiana* Boeck., Linnaea 37: 620 (1873).

Rhynchospora subtenuifolia Kük., Bot. Jahrb. Syst. 74: 499 (1949).
New Guinea to N. Australia. 43 NWG 50 NTA QLD WAU. Ther.
Rhynchospora longisetis F.Muell., Fragm. 9: 17 (1875), nom. illeg.
**Rhynchospora tenuifolia* Benth., Fl. Austral. 7: 350 (1878), nom. illeg.

Rhynchospora subtilis Boeck., Beitr. Cyper. 1: 26 (1888).
Brazil (Rio de Janeiro to Santa Catarina). 84 BZL BZS. Hemicr. or rhizome geophyte.

Rhynchospora talamancensis Gómez-Laur. & W.W.Thomas, Brittonia 44: 37 (1992).
Costa Rica to Colombia. 80 COS PAN 83 CLM. Hemicr.

Rhynchospora tenella (Nees) Boeck., Linnaea 37: 595 (1873).
N. South America to Brazil. 82 FRG GUY SUR VEN 83 CLM 84 BZL BZN BZS. Hemicr. or rhizome geophyte.
**Haloschoenus tenellus* Nees in C.F.P.von Martius & auct. suc. (eds.), Fl. Bras. 2(1): 123 (1842). *Dichromena tenella* (Nees) Steud., Syn. Pl. Glumac. 2: 136 (1855).
Rhynchospora tenella f. *haplostylis* Uittien, Recueil Trav. Bot. Néerl. 22: 337 (1926). *Rhynchospora tenella* var. *haplostylis* (Uittien) Kük., Bot. Jahrb. Syst. 75: 276 (1951).

Rhynchospora tenerrima Nees ex Spreng., Syst. Veg. 4(2): 26 (1827).
W. Trop. Africa, Trop. America. 22 GNB GUI LBR SEN SIE 79 MXT 80 BLZ COS GUA HON NIC PAN 81 CUB JAM LEE PUE TRT WIN 82 FRG GUY SUR VEN 83 BOL CLM 84 BZC BZE BZL BZN. Ther. or hemicr.

subsp. ***microcarpa*** J.Raynal, Adansonia, n.s., 7: 325 (1967).
W. Trop. Africa. 22 GNB GUI LBR SEN SIE. Hemicr. or rhizome geophyte.

subsp. ***tenerrima***
Trop. America. 79 MXT 80 BLZ COS GUA HON NIC PAN 81 CUB DOM JAM LEE PUE TRT WIN 82 FRG GUY SUR VEN 83 BOL CLM 84 BZC BZE BZL BZN. Ther. or hemicr.
Schoenus setaceus P.J.Bergius, Acta Helv. Phys.-Math. 7: 130 (1772). *Triodon setaceus* (P.J. Bergius) Rich. ex Pers., Syn. Pl. 1: 60 (1805), pro syn. *Spermodon setaceus* (P.J. Bergius) Nees, Linnaea 9: 296 (1834). *Dichromena setacea* (P.J.Bergius) Kunth, Enum. Pl. 2: 279 (1837). *Rhynchospora setacea* (P.J.Bergius) Boeck., Vidensk. Meddel. Naturhist. Foren. Kjøbenhavn 1869: 159 (1869), nom. illeg.
Scleria setacea Poir. in J.B.A.M.de Lamarck, Encycl. 7: 4 (1806).
Rhynchospora spermodon Griseb., Fl. Brit. W. I.: 576 (1864).

Rhynchospora tenuifolia Griseb., Cat. Pl. Cub.: 244 (1866).
Cuba to Hispaniola. 81 CUB DOM. Hemicr. or rhizome geophyte.
Rhynchospora stenophylla var. *albescens* Kük., Repert. Spec. Nov. Regni Veg. 23: 211 (1926).

Rhynchospora tenuis Link, Jahrb. Gewächsk. 1(3): 76 (1818). *Dichromena linkii* J.F.Macbr., Publ. Field Mus. Nat. Hist., Bot. Ser. 11: 5 (1931), nom. illeg.
Trop. America. 79 MXG MXS MXT 80 BLZ GUA HON NIC PAN 81 BAH CUB DOM JAM LEE TRT WIN 82 FRG GUY SUR VEN 83 BOL CLM ECU GAL 84 BZC BZE BZL BZN BZS 85 AGE AGW PAR URU. Hemicr.

subsp. ***austrobrasiliensis*** T.Koyama, Mem. New York Bot. Gard. 23: 78 (1972).
Guyana to Argentina (Misiones). 82 GUY 83 BOL 84 BZL 85 AGE PAR. Hemicr. or rhizome geophyte.
Haloschoenus capillaris var. *minor* Nees in C.F.P.von Martius & auct. suc. (eds.), Fl. Bras. 2(1): 121 (1842).

subsp. ***tenuis***
Trop. America. 79 MXG MXS MXT 80 BLZ GUA HON NIC PAN 81 BAH CUB DOM JAM LEE TRT WIN 82 FRG GUY SUR VEN 83 BOL CLM ECU GAL 84 BZC BZE BZL BZN BZS 85 AGE AGW PAR URU. Hemicr.
Fuirena gracilis Spreng., Novi Provent.: 46 (1818). *Dichromena gracilis* (Spreng.) Kunth, Enum. Pl. 2: 280 (1837). *Rhynchospora gracilis* (Spreng.) Kuntze, Revis. Gen. Pl. 3(2): 335 (1898), nom. illeg.
Haloschoenus sabulosus Nees, Linnaea 9: 296 (1834), nom. inval.
Dichromena capillaris Kunth, Enum. Pl. 2: 283 (1837), nom. inval.
Dichromena cubensis Poepp. & Kunth in C.S. Kunth, Enum. Pl. 2: 281 (1837). *Rhynchospora cubensis* (Poepp. & Kunth) Griseb., Fl. Brit. W. I.: 576 (1864), nom. illeg.
Dichromena sabulosa Kunth, Enum. Pl. 2: 283 (1837).
Haloschoenus capillaris Nees in C.F.P.von Martius & auct. suc. (eds.), Fl. Bras. 2(1): 121 (1842).
Haloschoenus capillaris var. *descendens* Nees in C.F.P.von Martius & auct. suc. (eds.), Fl. Bras. 2(1): 121 (1842).
Haloschoenus capillaris var. *ustilaginosa* Nees in C.F.P.von Martius & auct. suc. (eds.), Fl. Bras. 2(1): 121 (1842).
Spermodon curvulus Nees in C.F.P.von Martius & auct. suc. (eds.), Fl. Bras. 2(1): 119 (1842). *Rhynchospora curvula* (Nees) Boeck., Linnaea 37: 596 (1873), nom. illeg.
Haloschoenus caracasanus Liebm., Mexic. Neldeagt. Pl., V, 2: 249 (1851).
Rhynchospora pseudolunata Boeck., Beitr. Cyper. 1: 25 (1888). *Rhynchospora tenuis* var. *pseudolunata* (Boeck.) Kük., Repert. Spec. Nov. Regni Veg. 23: 206 (1926).
Rhynchospora juncellus C.B.Clarke in I.Urban, Symb. Antill. 2: 117 (1900).
Rhynchospora tenuis f. *major* C.B.Clarke, Bull. Herb. Boissier, II, 3: 1025 (1903).
Rhynchospora tenuis f. *minor* C.B.Clarke, Bull. Herb. Boissier, II, 3: 1025 (1903).
Rhynchospora tenuis var. *antillana* Kük., Repert. Spec. Nov. Regni Veg. 23: 206 (1926).
Rhynchospora tenuis var. *brevinux* Kük., Repert. Spec. Nov. Regni Veg. 23: 207 (1926).
Rhynchospora tenuis var. *prorepens* Kük., Repert. Spec. Nov. Regni Veg. 23: 206 (1926).

Rhynchospora tepuiana Steyerm., Fieldiana, Bot. 28(1): 48 (1951). *Rhynchospora flexuosa* subsp. *tepuiana* (Steyerm.) T.Koyama, Mem. New York Bot. Gard. 23: 48 (1972).
S. Venezuela to W. Guyana. 82 GUY VEN. Hemicr. or rhizome geophyte.
Rhynchospora tenerrima Boeck., Linnaea 37: 554 (1873), nom. illeg. *Rhynchospora flexuosa* var. *tenerrima* Kük., Bot. Jahrb. Syst. 75: 153 (1950).

Rhynchospora terminalis (Nees) Steud., Syn. Pl. Glumac. 2: 143 (1855).
Brazil. 84 BZC BZL BZS. Hemicr. or rhizome geophyte.
Calyptrostylis capitata Nees in C.F.P.von Martius & auct. suc. (eds.), Fl. Bras. 2(1): 140 (1842).
**Calyptrostylis terminalis* Nees in C.F.P.von Martius & auct. suc. (eds.), Fl. Bras. 2(1): 140 (1842).
Calyptrostylis terminalis var. *pallida* Nees in C.F.P.von Martius & auct. suc. (eds.), Fl. Bras. 2(1): 140 (1842). *Rhynchospora terminalis* f. *pallida* (Nees) Kük., Bot. Jahrb. Syst. 74: 471 (1949).
Rhynchospora minarum Steud., Syn. Pl. Glumac. 2: 143 (1855).
Rhynchospora crassipes Boeck., Vidensk. Meddel. Naturhist. Foren. Kjøbenhavn 1869: 146 (1869). *Rhynchospora terminalis* var. *crassipes* (Boeck.) Kük., Bot. Jahrb. Syst. 74: 471 (1949). *Rhynchospora terminalis* var. *crassipes* (Boeck.) Kük., Bot. Jahrb. Syst. 74: 471 (1949).
Rhynchospora subcapitata Boeck., Linnaea 37: 547 (1873).
Rhynchospora exilis Boeck., Flora 63: 440 (1880).

Rhynchospora testacea Boeck., Vidensk. Meddel. Naturhist. Foren. Kjøbenhavn 1869: 148 (1869).
Brazil. 84+. Hemicr. or rhizome geophyte.

Rhynchospora thornei Kral, Sida 7: 42 (1977).
SE. U.S.A. 78 ALA FLA GEO NCA SCA. Hemicr. or rhizome geophyte.

Rhynchospora tomentosa Steyerm., Fieldiana, Bot.

28(1): 49 (1951).
Venezuela. 82 VEN. Hemicr. or rhizome geophyte.

Rhynchospora torresiana Britton & Standl., J. Wash. Acad. Sci. 15: 473 (1925).
Costa Rica (Cord. de Talamanca). 80 COS. Hemicr.

Rhynchospora torreyana A.Gray, Ann. Lyceum Nat. Hist. New York 3: 197 (1835). *Phaeocephalum torreyanum* (A.Gray) House, Amer. Midl. Naturalist 6: 202 (1920).
E. U.S.A. 75 CNT MAS NWJ NWY RHO 78 ALA DEL FLA GEO MRY NCA SCA VRG. Hemicr. or rhizome geophyte.
Rhynchospora micrantha A.Gray, Ann. Lyceum Nat. Hist. New York 3: 197 (1835), nom. illeg.

Rhynchospora tracyi Britton, Trans. New York Acad. Sci. 11: 84 (1892). *Phaeocephalum tracyi* (Britton) House, Amer. Midl. Naturalist 6: 202 (1920).
SE. U.S.A. to Texas, Belize, Bahamas to Cuba. 77 TEX 78 ALA FLA GEO LOU MSI NCA SCA 80 BLZ 81 BAH CUB. Hemicr. or rhizome geophyte.
**Ceratoschoenus capitatus* Chapm., Fl. South. U.S.: 529 (1860).
Rhynchospora hitchcockii Gand., Bull. Soc. Bot. France 66: 298 (1919 publ. 1920).

Rhynchospora trichochaeta C.B.Clarke, Kew Bull., Addit. Ser. 8: 89 (1908). *Rhynchospora mexicana* var. *trichochaeta* (C.B.Clarke) Kük., Bot. Jahrb. Syst. 56(125): 16 (1921).
SE. Colombia to Guyana and Brazil. 82 FRG GUY VEN 83 CLM 84 BZC BZL BZN. Hemicr. or rhizome geophyte.
**Trichochaeta tenuis* Steud., Syn. Pl. Glumac. 2: 156 (1855).
Rhynchospora flavida Boeck., Flora 63: 439 (1880). *Rhynchospora trichochaeta* var. *flavida* (Boeck.) Kük., Bot. Jahrb. Syst. 74: 482 (1949). *Rhynchospora trichochaeta* var. *flavida* (Boeck.) Kük., Bot. Jahrb. Syst. 74: 482 (1949).
Rhynchospora denticulata Huber, Bol. Mus. Goeldi Hist. Nat. Ethnogr. 5: 317 (1909).

Rhynchospora trichodes C.B.Clarke in I.Urban, Symb. Antill. 2: 116 (1900).
Trop. America. 80 BLZ COS NIC PAN 81 PUE TRT 82 FRG GUY SUR VEN 83 BOL CLM 84 BZC BZN. Ther. or hemicr.
**Schoenus hispidulus* Vahl, Enum. Pl. 2: 219 (1805). *Dichromena hispidula* (Vahl) Kunth, Enum. Pl. 2: 279 (1837). *Haloschoenus hispidulus* (Vahl) Nees in C.F.P.von Martius & auct. suc. (eds.), Fl. Bras. 2(1): 124 (1842). *Rhynchospora hispidula* (Vahl) Boeck., Linnaea 37: 604 (1873), nom. illeg.
Fimbristylis divaricata Ham., Prodr. Pl. Ind. Occid.: 14 (1825).
Fimbristylis hirsuta Hochst. ex Steud., Syn. Pl. Glumac. 2: 116 (1855).
Rhynchospora hispidula var. *major* Kük., Bot. Jahrb. Syst. 75: 179 (1950).

Rhynchospora trichophora Nees in C.F.P.von Martius & auct. suc. (eds.), Fl. Bras. 2(1): 145 (1842). *Rhynchospora cariciformis* var. *trichophora* (Nees) Kük., Bot. Jahrb. Syst. 75: 126 (1950).
S. Venezuela to Peru. 82 VEN 83 CLM PER 84 BZN. Hemicr. or rhizome geophyte.

Rhynchospora triflora Vahl, Enum. Pl. 2: 231 (1805). *Scirpus triflorus* (Vahl) Poir. in J.B.A.M.de Lamarck, Encycl. 2: 249 (1786). *Ephippiorhynchium triflorum* (Vahl) Nees in C.F.P.von Martius & auct. suc. (eds.), Fl. Bras. 2(1): 136 (1842). *Dichromena triflora* (Vahl) J.F.Macbr., Publ. Field Mus. Nat. Hist., Bot. Ser. 11: 6 (1931).
Trop. 22 BEN GHA GNB GUI IVO MLI NGA SEN SIE TOG 23 CAF CMN 24 CHA 26 ANG MOZ ZAM 40 IND SRL 41 VIE 42 MLY 43 NWG 79 MXT 80 BLZ COS GUA NIC PAN 81 CUB DOM TRT 82 FRG GUY SUR VEN 83 BOL CLM 84 BZC BZL BZN 85 PAR. Hemicr. or rhizome geophyte.
Cephaloschoenus ceylanicus Nees, Edinburgh New Philos. J. 17: 265 (1834). *Rhynchospora ceylanica* (Nees) Kunth, Enum. Pl. 2: 294 (1837). *Calyptrostylis ceylanica* (Nees) Nees, J. Bot. (Hooker) 2: 394 (1840).
Rhynchospora stenorhyncha Griseb., Fl. Brit. W. I.: 575 (1864).
Rhynchospora triflora var. *papuana* Kük., Bot. Jahrb. Syst. 74: 427 (1949).
Rhynchospora karuaiana Steyerm., Fieldiana, Bot. 28(1): 43 (1951).

Rhynchospora trispicata (Nees) Schrad. ex Steud., Syn. Pl. Glumac. 2: 145 (1855).
Trop. America. 79 MXG MXT 80 BLZ COS HON NIC 81 CUB 82 FRG GUY SUR VEN 83 BOL CLM 84 BZC BZE BZN 85 PAR. Hemicr. or rhizome geophyte.
**Ephippiorhynchium trispicatum* Nees in C.F.P.von Martius & auct. suc. (eds.), Fl. Bras. 2(1): 136 (1842).
Rhynchospora guianensis Lindeman & Donsel., Acta Bot. Neerl. 20: 440 (1971).

Rhynchospora tuerckheimii C.B.Clarke ex Kük., Bot. Jahrb. Syst. 74: 402 (1949).
S. Mexico to N. South America. 79 MXG MXS MXT 80 COS GUA HON NIC PAN 82 GUY VEN. Hemicr. or rhizome geophyte.

Rhynchospora umbraticola Poepp. & Kunth in C.S.Kunth, Enum. Pl. 2: 300 (1837). *Dichromena umbraticola* (Poepp. & Kunth) J.F.Macbr., Publ. Field Columbian Mus., Bot. Ser. 4: 167 (1929).
Peru. 83 PER. Hemicr. or rhizome geophyte.

Rhynchospora uniflora Boeck., Flora 63: 439 (1880).
Caribbean to Venezuela, SE. Brazil. 81 CUB DOM HAI JAM PUE 82 VEN 84 BZL. Hemicr. or rhizome geophyte.
Rhynchospora elongata Boeck., Beitr. Cyper. 1: 26 (1888).
Rhynchospora ekmanii Urb., Ark. Bot. 17(7): 15 (1922). *Rhynchospora elongata* var. *ekmanii* (Urb.) Kük., Repert. Spec. Nov. Regni Veg. 23: 203 (1926). *Rhynchospora uniflora* var. *ekmanii* (Urb.) Kük., Bot. Jahrb. Syst. 74: 450 (1949). *Rhynchospora uniflora* var. *ekmanii* (Urb.) Kük., Bot. Jahrb. Syst. 74: 450 (1949).
Rhynchospora uniflora var. *uberior* Kük., Bot. Jahrb. Syst. 74: 450 (1949).
Rhynchospora uniflora var. *uberior* Kük., Bot. Jahrb. Syst. 74: 450 (1949).

Rhynchospora unisetosa T.Koyama, Mem. New York Bot. Gard. 23: 82 (1972).
Colombia (Meta) to Venezuela. 82 VEN 83 CLM. Hemicr. or rhizome geophyte.

Rhynchospora urbanii Boeck., Beitr. Cyper. 1: 26 (1888).
Brazil (Rio de Janeiro). 84 BZL. Hemicr. or rhizome geophyte.

Rhynchospora velloziiformis T.Koyama, Jap. J. Bot. 20: 131 (1969).
WC. Brazil. 84 BZC. Hemicr. or rhizome geophyte.

Rhynchospora velutina (Kunth) Boeck., Vidensk. Meddel. Naturhist. Foren. Kjøbenhavn 1869: 149 (1869).
Trop. America. 79 MXG MXT 80 COS GUA HON NIC 81 CUB DOM PUE 82 FRG GUY SUR VEN 83 BOL CLM 84 BZC BZL BZN BZS 85 AGE PAR URU. Hemicr. or rhizome geophyte.
Dichromena sellowiana Kunth, Enum. Pl. 2: 282 (1837). *Psilocarya sellowiana* (Kunth) Nees in C.F.P.von Martius & auct. suc. (eds.), Fl. Bras. 2(1): 117 (1842). *Rhynchospora kuekenthalii* Herter, Revista Sudamer. Bot. 9: 158 (1953), nom. illeg.
**Dichromena velutina* Kunth, Enum. Pl. 2: 282 (1837). *Pachymitra velutina* (Kunth) Nees in C.F.P.von Martius & auct. suc. (eds.), Fl. Bras. 2(1): 115 (1842), pro syn. *Psilocarya velutina* (Kunth) Nees in C.F.P.von Martius & auct. suc. (eds.), Fl. Bras. 2(1): 115 (1842).
Schoenus velutinus Schrad. ex Kunth, Enum. Pl. 2: 282 (1837), nom. inval.
Psilocarya divergens Nees in C.F.P.von Martius & auct. suc. (eds.), Fl. Bras. 2(1): 116 (1842).
Psilocarya pohlii Nees in C.F.P.von Martius & auct. suc. (eds.), Fl. Bras. 2(1): 115 (1842).
Psilocarya rufa Nees in C.F.P.von Martius & auct. suc. (eds.), Fl. Bras. 2(1): 117 (1842). *Rhynchospora rufa* (Nees) Boeck., Vidensk. Meddel. Naturhist. Foren. Kjøbenhavn 1869: 149 (1869). *Pachymitra rufa* (Nees) Nees ex Boeck., Linnaea 37: 616 (1873), pro syn.
Psilocarya mexicana Liebm., Mexic. Neldeagt. Pl., V, 2: 248 (1851). *Rhynchospora mexicana* (Liebm.) C.Wright, Anales Acad. Ci. Méd. Habana 8: 87 (1871), nom. illeg. *Psilocarya velutina* var. *glabrescens* Boeck., Bot. Jahrb. Syst. 1: 364 (1881). *Rhynchospora rufa* var. *liebmanniana* C.B.Clarke in I.Urban, Symb. Antill. 2: 117 (1900), nom. superfl.
Pachymitra conferta Nees ex Boeck., Linnaea 37: 615 (1873).
Rhynchospora niederleiniana Boeck., Beitr. Cyper. 1: 27 (1888).
Psilocarya pauloensis Boeck., Beitr. Cyper. 2: 24 (1890).
Rhynchospora velutina f. *glabrescens* Pilg., Bot. Jahrb. Syst. 30: 143 (1901).
Rhynchospora aureiformis C.B.Clarke, Ill. Cyper.: t. 71, f. 3 (1909).

Rhynchospora vulcani Boeck., Linnaea 37: 638 (1873).
SE. Mexico to W. South America. 79 MXT 80 COS GUA NIC PAN 82 VEN 83 BOL CLM ECU PER. Hemicr. or rhizome geophyte.
Rhynchospora hoffmannii Boeck., Linnaea 37: 637 (1873). *Rhynchospora vulcani* var. *hoffmannii* (Boeck.) Kük., Bot. Jahrb. Syst. 74: 402 (1949).
Rhynchospora lechleri Steud. ex Boeck., Linnaea 37: 633 (1873). *Dichromena lechleri* (Steud. ex Boeck.) J.F.Macbr., Publ. Field Columbian Mus., Bot. Ser. 4: 167 (1929).
Rhynchospora ecuadorensis C.B.Clarke, Bot. Jahrb. Syst. 34(78): 5 (1904). *Rhynchospora vulcani* var. *ecuadorensis* (C.B.Clarke) Kük., Bot. Jahrb. Syst. 74: 402 (1949).
Rhynchospora ecuadorensis C.B.Clarke, Kew Bull., Addit. Ser. 8: 39 (1908).
Rhynchospora ptaritepuiana Steyerm., Fieldiana, Bot. 28(1): 45 (1951).

Rhynchospora wageneriana (Schltdl.) Schnee in H.F.Pittier & al., Cat. Fl. Venez. 1: 129 (1945).
Colombia to Venezuela. 82 VEN 83 CLM. Hemicr. or rhizome geophyte.
**Mitrospora wageneriana* Schltdl., Linnaea 26: 129 (1853).

Rhynchospora warmingii Boeck., Vidensk. Meddel. Naturhist. Foren. Kjøbenhavn 1869: 143 (1869).
Brazil. 84 BZC BZL BZN. Hemicr. or rhizome geophyte.

Rhynchospora waspamensis Kral & W.W.Thomas, Brittonia 40: 34 (1988).
Nicaragua. 80 NIC. Hemicr.

Rhynchospora watsonii (Britton) Davidse, Ann. Missouri Bot. Gard. 61: 529 (1974).
S. Mexico to Ecuador and Martinique. 79 MXG MXS MXT 80 BLZ COS GUA NIC PAN 81 WIN 82 FRG GUY SUR VEN 83 CLM ECU. Hemicr.
Dichromena fasciata Maury, J. Bot. (Morot) 2: 396 (1888).
**Dichromena watsonii* Britton, Bull. Torrey Bot. Club 15: 101 (1888). *Rhynchospora radicans* var. *watsonii* (Britton) Kük., Bot. Jahrb. Syst. 75: 311 (1951).
Rhynchospora dussiana Boeck., Beitr. Cyper. 2: 40 (1890).
Rhynchospora dussii Boeck. ex Duss, Ann. Inst. Bot.-Géol. Colon. Marseille 3: 553 (1896 publ. 1897).

Rhynchospora widgrenii Boeck., Beitr. Cyper. 2: 25 (1890).
Brazil (Minas Gerais). 84 BZL. Hemicr. or rhizome geophyte.

Rhynchospora wightiana (Nees) Steud., Syn. Pl. Glumac. 2: 148 (1855).
S. India, Sri Lanka, Vietnam, Jawa (Madura I.), New Guinea. 40 IND SRL 41 VIE 42 JAW 43 NWG. Ther.
**Haplostylis wightianus* Nees, Nov. Actorum Acad. Caes. Leop.-Carol. Nat. Cur. 19(Suppl. 1): 101 (1843).
Rhynchospora discolor Hochst. ex Steud., Syn. Pl. Glumac. 2: 148 (1855).

Rhynchospora wrightiana Boeck., Beitr. Cyper. 2: 25 (1890).
SE. U.S.A., Caribbean. 78 ALA FLA GEO MSI NCA SCA VRG 81 CUB PUE. Hemicr. or rhizome geophyte.
Rhynchospora tenuis Baldwin ex A.Gray, Ann. Lyceum Nat. Hist. New York 3: 217 (1835), nom. illeg. *Rhynchospora distans* var. *tenuis* Britton, Trans. New York Acad. Sci. 11: 90 (1892).
**Rhynchospora gracillima* C.Wright, Anales Acad. Ci. Méd. Habana 8: 85 (1871), nom. illeg. *Rhynchospora distans* var. *gracillima* Kük., Repert. Spec. Nov. Regni Veg. 23: 208 (1926).

Rhynchospora zacualtipanensis M.T.Strong, Sida 19: 115 (2000).
Mexico (Hidalgo). 79 MXE. Hemicr. or rhizome geophyte.

Synonyms:
Rhynchospora adscendens C.B.Clarke = ***Rhynchospora candida*** (Nees) Boeck.
Rhynchospora adulta Schltdl. = ***Rhynchospora polyphylla*** (Vahl) Vahl
Rhynchospora africana Cherm. = ***Rhynchospora angolensis*** Turrill

Rhynchospora alba var. *fusca* (L.) Pursh = ***Rhynchospora fusca*** (L.) W.T.Aiton
Rhynchospora alba var. *kiusiana* Makino = ***Rhynchospora alba*** (L.) Vahl
Rhynchospora alba f. *laeviseta* Gale = ***Rhynchospora alba*** (L.) Vahl
Rhynchospora alba var. *macra* C.B.Clarke ex Britton = ***Rhynchospora macra*** (C.B.Clarke ex Britton) Small
Rhynchospora alba var. *meridiana* H.Pfeiff. = [79 MXG 84 BZN]
Rhynchospora alpina Palla = ***Rhynchospora biflora*** Boeck.
Rhynchospora alpina (R.Br.) Spreng. = ***Carpha alpina*** R.Br.
Rhynchospora alta Boeck. = ***Rhynchospora globosa*** (Kunth) Roem. & Schult. var. ***globosa***
Rhynchospora amazonica var. *guianensis* Kük. = ***Rhynchospora amazonica*** Poepp. ex Kunth
Rhynchospora andina Kük. = ***Rhynchospora caracasana*** (Kunth) Boeck.
Rhynchospora anomala Steud. = ***Fimbristylis fusca*** (Nees) Benth. ex C.B.Clarke
Rhynchospora arechavaletae Boeck. = ***Rhynchospora holoschoenoides*** (Rich.) Herter
Rhynchospora arechavaletae var. *ostenii* (Kük.) Kük. = ***Rhynchospora holoschoenoides*** (Rich.) Herter
Rhynchospora arechavaletae f. *pleiocephala* Osten = ***Rhynchospora holoschoenoides*** (Rich.) Herter
Rhynchospora aristata var. *cruegeriana* (Boeck.) Kük. = ***Rhynchospora longiflora*** C.Presl
Rhynchospora aristata var. *fuertesii* (Urb.) Kük. = ***Rhynchospora aristata*** Boeck. var. ***aristata***
Rhynchospora aristata var. *immensa* (Kük.) Kük. = ***Rhynchospora immensa*** Kük.
Rhynchospora aristata var. *latilaminata* Kük. = ***Rhynchospora aristata*** Boeck. var. ***aristata***
Rhynchospora articulata (Roxb.) Schult. = ***Rhynchospora corymbosa*** (L.) Britton
Rhynchospora arundinacea Boeck. = ***Rhynchospora exaltata*** Kunth
Rhynchospora aurea Vahl = ***Rhynchospora corymbosa*** (L.) Britton
Rhynchospora aurea f. *florida* (Rudge) Lindm. = ***Rhynchospora corymbosa*** (L.) Britton
Rhynchospora aureiformis C.B.Clarke = ***Rhynchospora velutina*** (Kunth) Boeck.
Rhynchospora auriculata Boeck. = ***Rhynchospora albiceps*** Kunth
Rhynchospora avenacea (R.Br.) Spreng. = ***Cyathochaeta avenacea*** (R.Br.) Benth.
Rhynchospora axillaris (Lam.) Britton = ***Rhynchospora glomerata*** (L.) Vahl
Rhynchospora axillaris var. *microcephala* Britton = ***Rhynchospora microcephala*** (Britton) Britton
Rhynchospora bahamensis Britton = ***Rhynchospora lindeniana*** Griseb.
Rhynchospora bahiensis Steud. = ***Rhynchospora barbata*** (Vahl) Kunth
Rhynchospora baileyi Britton = ***Rhynchospora albomarginata*** Kük.
Rhynchospora barbata var. *bahiensis* (Steud.) Boeck. = ***Rhynchospora barbata*** (Vahl) Kunth
Rhynchospora barbata var. *glabra* Maury = ***Rhynchospora barbata*** (Vahl) Kunth
Rhynchospora berteroana Boeck. = ***Rhynchospora berteroi*** (Spreng.) C.B.Clarke
Rhynchospora beyrichii (Nees) Steud. = ***Rhynchospora exaltata*** Kunth
Rhynchospora biceps Torr. ex A.Gray = ***Rhynchospora cephalantha*** A.Gray
Rhynchospora biflora var. *cryptantha* (C.B.Clarke) Kük. = ***Rhynchospora biflora*** Boeck.
Rhynchospora biflora var. *cryptantha* (C.B.Clarke) Kük. = ***Rhynchospora biflora*** Boeck.
Rhynchospora binervis Sieber ex C.Presl = ***Hypolytrum mauritianum*** Nees ex Kunth
Rhynchospora blauneri Britton = ***Rhynchospora brachychaeta*** C.Wright
Rhynchospora borinquensis Britton = ***Rhynchospora marisculus*** Lindl. & Nees
Rhynchospora brevirostris var. *clarkei* (Rose) Kük. = ***Rhynchospora brevirostris*** Griseb.
Rhynchospora brevirostris var. *truncata* Kük. = ***Rhynchospora brevirostris*** Griseb.
Rhynchospora breviseta Palla = ***Rhynchospora faberi*** C.B.Clarke
Rhynchospora breviseta (Gale) Channell = ***Rhynchospora oligantha*** var. ***breviseta*** Gale
Rhynchospora broadwayi (Kük.) Raymond = ***Rhynchospora comata*** (Link) Schult.
Rhynchospora bromoides Kunth = ***Rhynchospora comata*** (Link) Schult.
Rhynchospora brownii Roem. & Schult. = ***Rhynchospora rugosa*** subsp. ***brownii*** (Roem. & Schult.) T.Koyama
Rhynchospora brownii subsp. *americana* Guagl. = ***Rhynchospora rugosa*** subsp. ***americana*** (Guagl.) Govaerts
Rhynchospora bruneri Britton = ***Rhynchospora pusilla*** Chapm. ex M.A.Curtis
Rhynchospora brunnea Willd. ex Kunth = ***Rhynchospora rugosa*** subsp. ***brownii*** (Roem. & Schult.) T.Koyama
Rhynchospora buchii Urb. = ***Rhynchospora mexicana*** (Liebm.) Steud.
Rhynchospora bulbocaulis (Boeck.) Boeck. = ***Cyperus mollipes*** (C.B.Clarke) K.Schum.
Rhynchospora burchellii C.B.Clarke = ***Rhynchospora elatior*** Kunth
Rhynchospora campanulata Kük. = ***Rhynchospora lapensis*** C.B.Clarke
Rhynchospora canaliculata (Steud.) Boeck. = ***Rhynchospora riparia*** (Nees) Boeck.
Rhynchospora candida var. *pilosa* Schnee = ***Rhynchospora candida*** (Nees) Boeck.
Rhynchospora canescens (Maury ex Micheli) H.Pfeiff. = ***Rhynchospora albiceps*** Kunth
Rhynchospora capillacea Steud. = ***Rhynchospora fusca*** (L.) W.T.Aiton
Rhynchospora capillacea f. *leviseta* Fernald = ***Rhynchospora capillacea*** Torr.
Rhynchospora capillifolia W.W.Thomas = ***Rhynchospora rigidifolia*** subsp. ***capillacea*** T.Koyama
Rhynchospora capitellata Elliott = ***Rhynchospora glomerata*** (L.) Vahl
Rhynchospora capitellata var. *controversa* S.F.Blake = ***Rhynchospora capitellata*** (Michx.) Vahl
Rhynchospora capitellata f. *controversa* (S.F.Blake) Gale = ***Rhynchospora capitellata*** (Michx.) Vahl
Rhynchospora capitellata var. *discutiens* S.F.Blake = ***Rhynchospora capitellata*** (Michx.) Vahl
Rhynchospora capitellata f. *discutiens* (S.F.Blake) Gale = ***Rhynchospora capitellata*** (Michx.) Vahl
Rhynchospora capitellata var. *leptocarpa* S.F.Blake = ***Rhynchospora capitellata*** (Michx.) Vahl
Rhynchospora capitellata var. *minor* (Britton) S.F.Blake = ***Rhynchospora capitellata*** (Michx.) Vahl

Rhynchospora cariciformis var. *trichophora* (Nees) Kük. = ***Rhynchospora trichophora*** Nees
Rhynchospora castanea Nees & Meyen = ***Rhynchospora rugosa*** subsp. ***brownii*** (Roem. & Schult.) T.Koyama
Rhynchospora cephalantha A.Rich. = ***Rhynchospora globosa*** (Kunth) Roem. & Schult. var. ***globosa***
Rhynchospora cephalantha f. *antrorsa* Gale = ***Rhynchospora cephalantha*** A.Gray
Rhynchospora cephalantha var. *attenuata* Gale = ***Rhynchospora cephalantha*** A.Gray
Rhynchospora cephalantha f. *controversa* Gale = ***Rhynchospora cephalantha*** A.Gray
Rhynchospora cephalantha var. *luxurians* Boeck. = ***Rhynchospora cephalantha*** A.Gray
Rhynchospora cephalantha var. *microcephala* (Britton) Kük. = ***Rhynchospora microcephala*** (Britton) Britton
Rhynchospora cephalantha var. *pleiocephala* Fernald & Gale = ***Rhynchospora cephalantha*** A.Gray
Rhynchospora cephalantha var. *typica* Fernald & Gale = ***Rhynchospora cephalantha*** A.Gray
Rhynchospora cephalophora (Nees) Steud. = ***Rhynchospora exaltata*** Kunth
Rhynchospora cephalotes var. *angustifolia* Uittien = ***Rhynchospora pubisquama*** M.T.Strong
Rhynchospora cephalotes var. *comata* (Link) Kük. = ***Rhynchospora comata*** (Link) Schult.
Rhynchospora cephalotes var. *imeriensis* Kük. = ***Rhynchospora imeriensis*** (Kük.) W.W.Thomas
Rhynchospora cephalotes var. *pseudocomata* Kük. = ***Rhynchospora cephalotes*** (L.) Vahl
Rhynchospora cephalotoides Griseb. = ***Rhynchospora comata*** (Link) Schult.
Rhynchospora cernua var. *depressa* (Kük.) Kük. = ***Rhynchospora depressa*** (Kük.) Gale
Rhynchospora ceylanica (Nees) Kunth = ***Rhynchospora triflora*** Vahl
Rhynchospora chapmanii var. *nuda* (Gale) Kük. = ***Rhynchospora nuda*** Gale
Rhynchospora ciliata Vahl = ***Rhynchospora ciliaris*** (Michx.) C.Mohr
Rhynchospora ciliata (G.Mey.) Kük. = ***Rhynchospora nervosa*** subsp. ***ciliata*** (G.Mey.) T.Koyama
Rhynchospora ciliatifolia Steud. = ***Rhynchospora barbata*** (Vahl) Kunth
Rhynchospora cinnamomea Boeck. = [7]
Rhynchospora clandestina (R.Br.) Spreng. = ***Cyathochaeta clandestina*** (R.Br.) Benth.
Rhynchospora clarkei Rose = ***Rhynchospora brevirostris*** Griseb.
Rhynchospora columbiensis Boeck. = ***Rhynchospora macrochaeta*** Steud. ex Boeck. var. ***macrochaeta***
Rhynchospora comata f. *bromoides* (Kunth) A.Gross & Brade = ***Rhynchospora comata*** (Link) Schult.
Rhynchospora conferta (Nees) Boeck. = ***Rhynchospora robusta*** (Kunth) Boeck.
Rhynchospora conica Desv. ex Ham. = ***Rhynchospora cephalotes*** (L.) Vahl
Rhynchospora consanguinea var. *rigida* (Kunth) Kük. = ***Rhynchospora consanguinea*** (Kunth) Boeck.
Rhynchospora consanguinea var. *speciosa* (Kunth) Kük. = ***Rhynchospora consanguinea*** (Kunth) Boeck.
Rhynchospora consanguinea f. *subrigida* Kük. = ***Rhynchospora consanguinea*** (Kunth) Boeck.
Rhynchospora consanguinea f. *subspeciosa* Kük. = ***Rhynchospora consanguinea*** (Kunth) Boeck.
Rhynchospora coreana Palla = ***Rhynchospora fujiana*** Makino
Rhynchospora corniculata subsp. *amazonica* (Poepp. ex Kunth) H.Pfeiff. = ***Rhynchospora amazonica*** Poepp. ex Kunth
Rhynchospora corniculata var. *interior* Fernald = ***Rhynchospora corniculata*** (Lam.) A.Gray
Rhynchospora corniculata var. *macrostachya* (Torr. ex A.Gray) Britton = ***Rhynchospora macrostachya*** Torr. ex A.Gray
Rhynchospora corniculata var. *patula* (Chapm.) Britton = ***Rhynchospora careyana*** Fernald
Rhynchospora corniculata var. *typica* Waterf. = ***Rhynchospora corniculata*** (Lam.) A.Gray
Rhynchospora corymbifera Nees = ***Rhynchospora corymbosa*** (L.) Britton
Rhynchospora corymbifera (C.Wright) Kük. = ***Rhynchospora ignorata*** Boeck.
Rhynchospora corymbifera Boeck. = ***Rhynchospora longiflora*** C.Presl
Rhynchospora corymbifera var. *pseudeximia* Kük. = ***Rhynchospora ignorata*** Boeck.
Rhynchospora corymbifera var. *subcostata* C.B.Clarke = ***Rhynchospora longiflora*** C.Presl
Rhynchospora corymbiformis Britton = ***Rhynchospora scirpoides*** (Torr.) Griseb.
Rhynchospora corymbosa f. *angustirostris* Barros = ***Rhynchospora corymbosa*** (L.) Britton
Rhynchospora corymbosa var. *angustirostris* (Barros) Guagl. = ***Rhynchospora corymbosa*** (L.) Britton
Rhynchospora corymbosa var. *asperula* (Nees) Kük. = ***Rhynchospora asperula*** (Nees) Steud.
Rhynchospora corymbosa var. *bonariensis* Barros ex Cabrera & Dawson = ***Rhynchospora corymbosa*** (L.) Britton
Rhynchospora corymbosa f. *chacoensis* Barros = ***Rhynchospora corymbosa*** (L.) Britton
Rhynchospora corymbosa var. *chacoensis* (Barros) Svenson = ***Rhynchospora corymbosa*** (L.) Britton
Rhynchospora corymbosa var. *florida* (Rudge) Kük. = ***Rhynchospora corymbosa*** (L.) Britton
Rhynchospora corymbosa var. *grandispiculosa* Kük. = ***Rhynchospora spectabilis*** Hochst. ex C.Krauss
Rhynchospora corymbosa subsp. *legrandii* (Kük. ex Barros) Luceño & Vanzela = ***Rhynchospora legrandii*** Kük. ex Barros
Rhynchospora corymbosa var. *legrandii* (Kük. ex Barros) Guagl. = ***Rhynchospora legrandii*** Kük. ex Barros
Rhynchospora corymbosa var. *minor* (Nees) Kük. = ***Rhynchospora corymbosa*** (L.) Britton
Rhynchospora corymbosa var. *organensis* (C.B.Clarke) Kük. = ***Rhynchospora organensis*** C.B.Clarke
Rhynchospora corymbosa var. *puncticulata* (Kük. ex Osten) Kük. = ***Rhynchospora scutellata*** Griseb.
Rhynchospora corymbosa var. *singularis* Kük. = ***Rhynchospora corymbosa*** (L.) Britton
Rhynchospora costaricensis Boeck. = ***Rhynchospora polyphylla*** (Vahl) Vahl
Rhynchospora crassipes Boeck. = ***Rhynchospora terminalis*** (Nees) Steud.
Rhynchospora crinigera var. *spicata* Kük. ex Osten = ***Rhynchospora crinigera*** Boeck.
Rhynchospora cruegeriana Boeck. = ***Rhynchospora longiflora*** C.Presl
Rhynchospora cryptantha C.B.Clarke = ***Rhynchospora biflora*** Boeck.
Rhynchospora cubensis Griseb. = ***Rhynchospora filiformis*** Vahl
Rhynchospora cubensis (Poepp. & Kunth) Griseb. = ***Rhynchospora tenuis*** subsp. ***tenuis***
Rhynchospora cubensis var. *brevifoliata* Kük. = ***Rhynchospora cubensis*** A.Rich.

Rhynchospora cubensis var. *stenophylloidea* Kük. = ***Rhynchospora cubensis*** A.Rich.
Rhynchospora culixa Gale = ***Rhynchospora harveyi*** var. ***culixa*** (Gale) Kral
Rhynchospora culmenicola Steyerm. = ***Rhynchospora dissitiflora*** Steud. ex Boeck.
Rhynchospora curtisii Steud. = ***Rhynchospora pallida*** M.A.Curtis
Rhynchospora curvula (Nees) Boeck. = ***Rhynchospora tenuis*** subsp. ***tenuis***
Rhynchospora cymosa (Willd.) Muhl. ex Elliott = ***Scirpus polyphyllus*** Vahl
Rhynchospora cymosa var. *compressa* (J.Carey ex Chapm.) C.B.Clarke = ***Rhynchospora compressa*** J.Carey ex Chapm.
Rhynchospora cymosa var. *globularis* Chapm. = ***Rhynchospora globularis*** (Chapm.) Small
Rhynchospora cyperoides Mart. = ***Rhynchospora holoschoenoides*** (Rich.) Herter
Rhynchospora cyperoides var. *triquetrifolia* Kük. = ***Rhynchospora holoschoenoides*** (Rich.) Herter
Rhynchospora deflexa Griseb. = ***Rhynchospora cubensis*** A.Rich.
Rhynchospora deightonii Hutch. = ***Rhynchospora perrieri*** Cherm.
Rhynchospora demerarensis C.B.Clarke = ***Rhynchospora cajennensis*** Boeck.
Rhynchospora denticulata Huber = ***Rhynchospora trichochaeta*** C.B.Clarke
Rhynchospora deusta (R.Br.) Spreng. = ***Ptilothrix deusta*** (R.Br.) K.L.Wilson
Rhynchospora diamantina C.B.Clarke ex Kük. = ***Rhynchospora leucostachys*** Boeck.
Rhynchospora diamantina var. *brevispiculosa* Kük. = ***Rhynchospora leucostachys*** Boeck.
Rhynchospora diandra (R.Br.) Spreng. = ***Cyathochaeta diandra*** (R.Br.) Nees
Rhynchospora diedrichsenii Boeck. = ***Rhynchospora comata*** (Link) Schult.
Rhynchospora discolor Hochst. ex Steud. = ***Rhynchospora wightiana*** (Nees) Steud.
Rhynchospora distans (Michx.) Vahl = ***Rhynchospora fascicularis*** (Michx.) Vahl subsp. ***fascicularis***
Rhynchospora distans Willd. ex Kunth = ***Rhynchospora glomerata*** (L.) Vahl
Rhynchospora distans Elliott = ***Rhynchospora grayi*** Kunth
Rhynchospora distans Beyr. ex Kunth = ***Rhynchospora rugosa*** subsp. ***brownii*** (Roem. & Schult.) T. Koyama
Rhynchospora distans var. *fascicularis* (Michx.) Kük. = ***Rhynchospora fascicularis*** (Michx.) Vahl
Rhynchospora distans var. *gracillima* Kük. = ***Rhynchospora wrightiana*** Boeck.
Rhynchospora distans var. *tenuis* Britton = ***Rhynchospora wrightiana*** Boeck.
Rhynchospora distichophylla Boeck. = ***Pleurostachys distichophylla*** (Boeck.) C.B.Clarke
Rhynchospora divergens var. *expansa* Kük. = ***Rhynchospora divergens*** Chapm. ex M.A.Curtis
Rhynchospora dodecandra Baldwin ex A.Gray. = ***Rhynchospora megalocarpa*** A.Gray
Rhynchospora dolichostyla K.Schum. = ***Rhynchospora holoschoenoides*** (Rich.) Herter
Rhynchospora dommucensis A.H.Moore = ***Rhynchospora fascicularis*** (Michx.) Vahl subsp. ***fascicularis***
Rhynchospora drummondiana Steud. = ***Rhynchospora colorata*** (L.) H.Pfeiff.
Rhynchospora drummondiana Boeck. = ***Rhynchospora gracilenta*** A.Gray
Rhynchospora durandiana Boeck. = ***Rhynchospora rugosa*** (Vahl) Gale subsp. ***rugosa***
Rhynchospora dussiana Boeck. = ***Rhynchospora watsonii*** (Britton) Davidse
Rhynchospora dussii Boeck. ex Duss = ***Rhynchospora watsonii*** (Britton) Davidse
Rhynchospora earlei Britton = ***Rhynchospora harveyi*** W.Boott var. ***harveyi***
Rhynchospora echinata Willd. ex Kunth = ***Actinoschoenus thouarsii*** (Kunth) Benth.
Rhynchospora ecuadorensis C.B.Clarke = ***Rhynchospora vulcani*** Boeck.
Rhynchospora ecuadorensis C.B.Clarke = ***Rhynchospora vulcani*** Boeck.
Rhynchospora edentula H.Pfeiff. = ***Rhynchospora junciformis*** (Kunth) Boeck.
Rhynchospora edisoniana Britton ex Small = ***Rhynchospora microcarpa*** Baldwin ex A.Gray
Rhynchospora effusa Schrad. ex Nees = ***Rhynchospora rugosa*** subsp. ***brownii*** (Roem. & Schult.) T.Koyama
Rhynchospora eggersiana Boeck. = ***Rhynchospora longiflora*** C.Presl
Rhynchospora ekmanii Urb. = ***Rhynchospora uniflora*** Boeck.
Rhynchospora elatior var. *brevispiculosa* (Kük.) A.C.Araujo = ***Rhynchospora leucostachys*** Boeck.
Rhynchospora elatior var. *burchellii* (C.B.Clarke) Kük. = ***Rhynchospora elatior*** Kunth
Rhynchospora elatior var. *burchellii* (C.B.Clarke) Kük. = ***Rhynchospora elatior*** Kunth
Rhynchospora elatior var. *hirsuta* Nees = ***Rhynchospora elatior*** Kunth
Rhynchospora elatior var. *variegatopes* H.Pfeiff. = ***Rhynchospora elatior*** Kunth
Rhynchospora elatior var. *variegatopes* H.Pfeiff. = ***Rhynchospora elatior*** Kunth
Rhynchospora elegans Kük. = ***Rhynchospora eximia*** (Nees) Boeck.
Rhynchospora elliottii A.Gray = ***Rhynchospora grayi*** Kunth
Rhynchospora elongata Boeck. = ***Rhynchospora uniflora*** Boeck.
Rhynchospora elongata var. *ekmanii* (Urb.) Kük. = ***Rhynchospora uniflora*** Boeck.
Rhynchospora epiglobosa C.B.Clarke = ***Rhynchospora globosa*** (Kunth) Roem. & Schult. var. ***globosa***
Rhynchospora erinacea (Ridl.) C.B.Clarke = ***Sphaerocyperus erinaceus*** (Ridl.) Lye
Rhynchospora etuberculata Steud. = ***Schoenoplectus etuberculatus*** (Steud.) Soják
Rhynchospora exaltata f. *angusta* Kük. = ***Rhynchospora exaltata*** Kunth
Rhynchospora exaltata var. *beyrichii* (Nees) Kük. = ***Rhynchospora exaltata*** Kunth
Rhynchospora exaltata var. *cephalophora* (Nees) Kük. = ***Rhynchospora exaltata*** Kunth
Rhynchospora exaltata var. *microcephala* Kük. = ***Rhynchospora exaltata*** Kunth
Rhynchospora exaltata f. *monocephala* Kük. = ***Rhynchospora exaltata*** Kunth
Rhynchospora exaltata var. *ovalis* Kük. = ***Rhynchospora exaltata*** Kunth
Rhynchospora exaltata var. *simplex* Kük. = ***Rhynchospora simplex*** (Kük.) Kük.
Rhynchospora exigua Boeck. = ***Rhynchospora nervosa*** (Vahl) Boeck. subsp. ***nervosa***
Rhynchospora exilis Boeck. = ***Rhynchospora terminalis*** (Nees) Steud.

Rhynchospora eximia var. *multiflorens* Kük. = ***Rhynchospora eximia*** (Nees) Boeck.
Rhynchospora fabri (Rottb.) Steud. = (Poaceae)
Rhynchospora fascicularis var. *debilis* (Gale) Kük. = ***Rhynchospora debilis*** Gale
Rhynchospora fascicularis var. *distans* (Michx.) Chapm. = ***Rhynchospora fascicularis*** (Michx.) Vahl subsp. ***fascicularis***
Rhynchospora fascicularis var. *fernaldii* (Gale) Kük. = ***Rhynchospora fernaldii*** Gale
Rhynchospora fascicularis var. *harperi* (Small) Kük. = ***Rhynchospora harperi*** Small
Rhynchospora fascicularis var. *trichoides* Chapm. = ***Rhynchospora fascicularis*** (Michx.) Vahl
Rhynchospora fascicularis var. *typica* Gale = ***Rhynchospora fascicularis*** (Michx.) Vahl
Rhynchospora fauriei C.B.Clarke ex H.Lév. = [36]
Rhynchospora fauriei var. *leviseta* C.B.Clarke ex H.Lév. = ***Rhynchospora faberi*** C.B.Clarke
Rhynchospora ferruginea Sieber ex C.Presl = ***Rhynchospora polyphylla*** (Vahl) Vahl
Rhynchospora ferruginea Roem. & Schult. = ***Rhynchospora rugosa*** (Vahl) Gale subsp. ***rugosa***
Rhynchospora fibrillosa Kük. = ***Rhynchospora nipensis*** Britton
Rhynchospora filifolia Kunth = ***Rhynchospora sellowiana*** Steud.
Rhynchospora filifolia var. *crinipes* (Gale) Kük. = ***Rhynchospora crinipes*** Gale
Rhynchospora filifolia var. *ellipsoidea* Kük. = ***Rhynchospora curtissii*** Britton
Rhynchospora filifolia var. *pleiantha* Kük. = ***Rhynchospora pleiantha*** (Kük.) Gale
Rhynchospora filiformis Willd. ex Kunth = ***Rhynchospora caracasana*** (Kunth) Boeck.
Rhynchospora filiformis Balb. ex Kunth = ***Rhynchospora fascicularis*** (Michx.) Vahl subsp. ***fascicularis***
Rhynchospora filiformis Sieber ex C.Presl = ***Rhynchospora glomerata*** (L.) Vahl
Rhynchospora filiformis var. *latifolia* Uittien = ***Rhynchospora spruceana*** C.B.Clarke
Rhynchospora flavida Boeck. = ***Rhynchospora trichochaeta*** C.B.Clarke
Rhynchospora flexuosa C.B.Clarke = ***Rhynchospora edwalliana*** Boeck.
Rhynchospora flexuosa var. *tenerrima* Kük. = ***Rhynchospora tepuiana*** Steyerm.
Rhynchospora flexuosa subsp. *tepuiana* (Steyerm.) T.Koyama = ***Rhynchospora tepuiana*** Steyerm.
Rhynchospora floribunda Boeck. = ***Rhynchospora hieronymi*** var. ***floribunda*** (Boeck.) Kük.
Rhynchospora florida (Rudge) Schult. = ***Rhynchospora corymbosa*** (L.) Britton
Rhynchospora florida Griseb. = ***Rhynchospora longiflora*** C.Presl
Rhynchospora fluminensis L.B.Sm. = ***Pleurostachys beyrichii*** (Nees) Steud.
Rhynchospora foliata Steud. = [7]
Rhynchospora foliosa (Kunth) L.B.Sm. = ***Pleurostachys foliosa*** Kunth
Rhynchospora franchetiana C.B.Clarke ex H.Lév. = ***Rhynchospora fujiana*** Makino
Rhynchospora fuertesii Urb. = ***Rhynchospora aristata*** Boeck. var. ***aristata***
Rhynchospora fujiana var. *scabrista* Makino = ***Rhynchospora fujiana*** Makino
Rhynchospora fusca A.Gray = ***Rhynchospora gracilenta*** A.Gray
Rhynchospora fuscoides C.B.Clarke ex Britton = ***Rhynchospora filifolia*** A.Gray
Rhynchospora gaudichaudii (Brongn.) L.B.Sm. = ***Pleurostachys gaudichaudii*** Brongn.
Rhynchospora gigantea var. *latifolia* H.Pfeiff. = ***Rhynchospora pedersenii*** Guagl.
Rhynchospora gigantea var. *scaberrima* (Boeck.) Kük. = ***Rhynchospora gigantea*** Link
Rhynchospora glauca Baldwin ex A.Gray = ***Rhynchospora baldwinii*** A.Gray
Rhynchospora glauca Vahl = ***Rhynchospora rugosa*** subsp. ***brownii*** (Roem. & Schult.) T.Koyama
Rhynchospora glauca var. *altior* Kük. ex O.C.Schmidt = ***Rhynchospora rugosa*** subsp. ***brownii*** (Roem. & Schult.) T.Koyama
Rhynchospora glauca var. *chinensis* (Nees & Meyen) C.B.Clarke = ***Rhynchospora chinensis*** Nees & Meyen
Rhynchospora glauca var. *condensata* Kük. = ***Rhynchospora rugosa*** subsp. ***brownii*** (Roem. & Schult.) T.Koyama
Rhynchospora glauca var. *strobilacea* Nees = ***Rhynchospora rugosa*** subsp. ***brownii*** (Roem. & Schult.) T.Koyama
Rhynchospora glauca var. *ustilaginosa* Nees = ***Rhynchospora rugosa*** subsp. ***brownii*** (Roem. & Schult.) T.Koyama
Rhynchospora glaziovii var. *compacta* Kük. = ***Rhynchospora glaziovii*** Boeck.
Rhynchospora globosa (Rudge) Desv. ex Ham. = ***Rhynchospora barbata*** (Vahl) Kunth
Rhynchospora globosa var. *epiglobosa* (C.B.Clarke) Kük. = ***Rhynchospora globosa*** (Kunth) Roem. & Schult. var. ***globosa***
Rhynchospora globosa var. *loefgrenii* (Boeck.) Kük. = ***Rhynchospora globosa*** (Kunth) Roem. & Schult. var. ***globosa***
Rhynchospora globosa var. *stenocarpa* Kük. = ***Rhynchospora globosa*** (Kunth) Roem. & Schult. var. ***globosa***
Rhynchospora globularis var. *obliterata* (Gale) Kük. = ***Rhynchospora globularis*** (Chapm.) Small
Rhynchospora globularis var. *recognita* Gale = ***Rhynchospora recognita*** (Gale) Kral
Rhynchospora globularis var. *typica* Gale = ***Rhynchospora globularis*** (Chapm.) Small
Rhynchospora globulifera Link = ***Rhynchospora holoschoenoides*** (Rich.) Herter
Rhynchospora glomerata Vahl ex Kunth = ***Isolepis digitata*** Nees ex Schrad.
Rhynchospora glomerata var. *angusta* Gale = ***Rhynchospora glomerata*** (L.) Vahl
Rhynchospora glomerata var. *capitellata* (Michx.) Kük. = ***Rhynchospora capitellata*** (Michx.) Vahl
Rhynchospora glomerata f. *controversa* Fernald = ***Rhynchospora glomerata*** (L.) Vahl
Rhynchospora glomerata var. *discutiens* C.B.Clarke ex Britton = ***Rhynchospora glomerata*** (L.) Vahl
Rhynchospora glomerata f. *discutiens* (C.B.Clarke ex Britton) Fernald = ***Rhynchospora glomerata*** (L.) Vahl
Rhynchospora glomerata var. *leptocarpa* Chapm. ex Britton = ***Rhynchospora capitellata*** (Michx.) Vahl
Rhynchospora glomerata f. *leptocarpa* (Chapm. ex Britton) Kük. = ***Rhynchospora glomerata*** (L.) Vahl
Rhynchospora glomerata var. *minor* Britton = ***Rhynchospora glomerata*** (L.) Vahl
Rhynchospora glomerata var. *paniculata* Chapm. = ***Rhynchospora glomerata*** (L.) Vahl

Rhynchospora glomerata var. *typica* Fernald = ***Rhynchospora glomerata*** (L.) Vahl
Rhynchospora gracilenta var. *diversifolia* Fernald = ***Rhynchospora gracilenta*** A.Gray
Rhynchospora gracilis Vahl = ***Rhynchospora rugosa*** subsp. ***brownii*** (Roem. & Schult.) T.Koyama
Rhynchospora gracilis (Spreng.) Kuntze = ***Rhynchospora tenuis*** subsp. ***tenuis***
Rhynchospora gracillima C.Wright = ***Rhynchospora wrightiana*** Boeck.
Rhynchospora graminea Uittien = ***Rhynchospora spruceana*** C.B.Clarke
Rhynchospora graminifolia (Brongn.) L.B.Sm. = ***Pleurostachys graminifolia*** Brongn.
Rhynchospora graminosa H.Pfeiff. = ***Rhynchospora pilosa*** Boeck.
Rhynchospora grayana Chapm. ex M.A.Curtis = ***Rhynchospora chapmanii*** M.A.Curtis
Rhynchospora grayi var. *culixa* (Gale) Kük. = ***Rhynchospora harveyi*** var. ***culixa*** (Gale) Kral
Rhynchospora grayi var. *harveyi* (W.Boott) Kük. = ***Rhynchospora harveyi*** W.Boott
Rhynchospora griffithii Boeck. = ***Rhynchospora rugosa*** subsp. ***brownii*** (Roem. & Schult.) T.Koyama
Rhynchospora grisebachii Boeck. ex C.B.Clarke = ***Rhynchospora hispidula*** Griseb.
Rhynchospora grosserugosa Kük. = ***Rhynchospora eximia*** (Nees) Boeck.
Rhynchospora guianensis Lindeman & Donsel. = ***Rhynchospora trispicata*** (Nees) Schrad. ex Steud.
Rhynchospora haenkei J.Presl & C.Presl = ***Rhynchospora rubra*** (Lour.) Makino subsp. ***rubra***
Rhynchospora hamiltonii Kunth = ***Rhynchospora barbata*** (Vahl) Kunth
Rhynchospora hattoriana Makino = ***Rhynchospora faberi*** C.B.Clarke
Rhynchospora hemicephala Boeck. = ***Rhynchospora recurvata*** (Nees) Steud.
Rhynchospora heterocaulis C.B.Clarke = ***Rhynchospora barbata*** (Vahl) Kunth
Rhynchospora heterocaulis var. *heterolepis* Böckel = ***Rhynchospora barbata*** (Vahl) Kunth
Rhynchospora heterochaeta var. *irregularis* Kük. = ***Rhynchospora heterochaeta*** S.T.Blake
Rhynchospora heterophylla (Boeck.) Boeck. = ***Rhynchospora chapmanii*** M.A.Curtis
Rhynchospora hieronymi f. *compacta* Kük. = ***Rhynchospora hieronymi*** Boeck. var. ***hieronymi***
Rhynchospora hieronymi subsp. *montevidensis* (Barros) Guagl. = ***Rhynchospora hieronymi*** var. ***montevidensis*** Barros
Rhynchospora hirsuta Griseb. = ***Rhynchospora nervosa*** subsp. ***ciliata*** (G.Mey.) T.Koyama
Rhynchospora hirta var. *tenuior* Kük. = ***Rhynchospora hirta*** (Nees) Boeck.
Rhynchospora hispidula (Vahl) Boeck. = ***Rhynchospora trichodes*** C.B.Clarke
Rhynchospora hispidula var. *major* Kük. = ***Rhynchospora trichodes*** C.B.Clarke
Rhynchospora hitchcockii Gand. = ***Rhynchospora tracyi*** Britton
Rhynchospora hoffmannii Boeck. = ***Rhynchospora vulcani*** Boeck.
Rhynchospora humboldtiana (Nees) Griseb. = ***Rhynchospora pubera*** (Vahl) Boeck. subsp. ***pubera***
Rhynchospora humilis Beyr. ex Kunth = ***Rhynchospora glomerata*** (L.) Vahl
Rhynchospora hunnewellii L.B.Sm. = ***Pleurostachys graminifolia*** Brongn.
Rhynchospora hyalinolepis Boeck. = ***Rhynchospora praecincta*** Maury ex Micheli
Rhynchospora igarapicola T.Koyama = ***Rhynchospora cariciformis*** Nees
Rhynchospora indianolensis Small = ***Rhynchospora scutellata*** Griseb.
Rhynchospora inexpansa Sieber ex Steud. = ***Rhynchospora filiformis*** Vahl
Rhynchospora infuscata Boeck. = ***Rhynchospora ruiziana*** Boeck.
Rhynchospora intermedia Beyr. ex Kunth = ***Rhynchospora cephalantha*** A.Gray
Rhynchospora intermixta C.Wright = ***Rhynchospora pusilla*** Chapm. ex M.A.Curtis
Rhynchospora japonica Makino = ***Rhynchospora chinensis*** Nees & Meyen subsp. ***chinensis***
Rhynchospora japonica var. *curvoaristata* Tuyama = ***Rhynchospora chinensis*** var. ***curvoaristata*** (Tuyama) Ohwi
Rhynchospora jelskiana Boeck. = ***Rhynchospora nervosa*** subsp. ***ciliata*** (G.Mey.) T.Koyama
Rhynchospora joveroensis var. *breviseta* Kük. = ***Rhynchospora joveroensis*** Britton
Rhynchospora joveroensis var. *gageri* (Britton) Kük. = ***Rhynchospora gageri*** Britton
Rhynchospora juncea Willd. ex Kunth = ***Rhynchospora rugosa*** subsp. ***brownii*** (Roem. & Schult.) T. Koyama
Rhynchospora juncellus C.B.Clarke = ***Rhynchospora tenuis*** subsp. ***tenuis***
Rhynchospora junciformis var. *monocarpa* Kük. = ***Rhynchospora junciformis*** (Kunth) Boeck.
Rhynchospora kamphoeveneri Boeck. = ***Rhynchospora gracillima*** Thwaites & Hook. subsp. ***gracillima***
Rhynchospora karuaiana Steyerm. = ***Rhynchospora triflora*** Vahl
Rhynchospora kuekenthalii H.Pfeiff. = ***Rhynchospora kuntzei*** C.B.Clarke ex Kuntze
Rhynchospora kuekenthalii Herter = ***Rhynchospora velutina*** (Kunth) Boeck.
Rhynchospora kunthii var. *composita* Nees = ***Rhynchospora kunthii*** Nees ex Kunth
Rhynchospora kysoor (Roxb.) A.Dietr. = ***Actinoscirpus grossus*** var. ***kysoor*** (Roxb.) Noltie
Rhynchospora laevigata Steud. = ***Rhynchospora rugosa*** (Vahl) Gale subsp. ***rugosa***
Rhynchospora laevis Schltdl. = ***Rhynchospora rugosa*** subsp. ***brownii*** (Roem. & Schult.) T.Koyama
Rhynchospora lamprosperma C.Wright = ***Rhynchospora divergens*** Chapm. ex M.A.Curtis
Rhynchospora lanceolata Willd. ex Kunth = ***Rhynchospora capitata*** (Kunth) Roem. & Schult.
Rhynchospora lapensis var. *campanulata* (Kük.) Kük. = ***Rhynchospora lapensis*** C.B.Clarke
Rhynchospora lavarum Hook. & Arn. = ***Rhynchospora rugosa*** subsp. ***brownii*** (Roem. & Schult.) T.Koyama
Rhynchospora lavarum Gaudich. = ***Rhynchospora rugosa*** subsp. ***lavarum*** (Gaudich.) T.Koyama
Rhynchospora laxa Vahl = ***Rhynchospora corniculata*** (Lam.) A.Gray
Rhynchospora laxa Schult. = ***Rhynchospora longiflora*** C.Presl
Rhynchospora laxa R.Br. = ***Rhynchospora rugosa*** subsp. ***brownii*** (Roem. & Schult.) T.Koyama
Rhynchospora laxa var. *minor* Thwaites = ***Rhynchospora rugosa*** subsp. ***brownii*** (Roem. & Schult.) T.Koyama
Rhynchospora lechleri Steud. ex Boeck. = ***Rhynchospora vulcani*** Boeck.

Rhynchospora leptocarpa (Chapm. ex Britton) Small = ***Rhynchospora capitellata*** (Michx.) Vahl
Rhynchospora leptorhyncha var. *laevirostris* Kük. = ***Rhynchospora leptorhyncha*** C.Wright
Rhynchospora leptostachya Boeck. = ***Rhynchospora emaciata*** (Nees) Boeck.
Rhynchospora leucantha Boeck. = ***Rhynchospora albiceps*** Kunth
Rhynchospora leucocarpa Ridl. = ***Costularia leucocarpa*** (Ridl.) H.Pfeiff.
Rhynchospora leucocephala (Michx.) Boeck. = ***Rhynchospora colorata*** (L.) H.Pfeiff.
Rhynchospora lindeniana var. *bahamensis* (Britton) Gale = ***Rhynchospora lindeniana*** Griseb.
Rhynchospora lindeniana var. *depressa* Kük. = ***Rhynchospora depressa*** (Kük.) Gale
Rhynchospora lindeniana var. *typica* Gale = ***Rhynchospora lindeniana*** Griseb.
Rhynchospora lingulata Kük. = ***Rhynchospora shaferi*** Britton
Rhynchospora loefgrenii Boeck. = ***Rhynchospora globosa*** (Kunth) Roem. & Schult. var. ***globosa***
Rhynchospora longibracteata var. *minor* H.Pfeiff. = ***Rhynchospora longibracteata*** Boeck.
Rhynchospora longiflora var. *domingensis* (Urb.) Kük. = ***Rhynchospora domingensis*** Urb.
Rhynchospora longiflora var. *subcostata* (C.B.Clarke) Kük. = ***Rhynchospora longiflora*** C.Presl
Rhynchospora longifrons Kük. = ***Rhynchospora joveroensis*** Britton
Rhynchospora longifrons var. *breviseta* Kük. = ***Rhynchospora joveroensis*** Britton
Rhynchospora longifrons var. *reducta* Kük. = ***Rhynchospora gageri*** Britton
Rhynchospora longirostris (Nees) Steud. = ***Rhynchospora amazonica*** subsp. ***guianensis*** (Kük.) T.Koyama
Rhynchospora longirostris (Michx.) Elliott = ***Rhynchospora corniculata*** (Lam.) A.Gray
Rhynchospora longiseta Steud. = ***Rhynchospora caduca*** Elliott
Rhynchospora longisetigera Hayata = ***Rhynchospora chinensis*** Nees & Meyen subsp. ***chinensis***
Rhynchospora longisetis F.Muell. = ***Rhynchospora subtenuifolia*** Kük.
Rhynchospora longisetis subsp. *exserta* (C.B.Clarke) Kük. = ***Rhynchospora exserta*** C.B.Clarke
Rhynchospora longispicata Boeck. = ***Rhynchospora filiformis*** Vahl
Rhynchospora lunata Griseb. = ***Rhynchospora diodon*** (Nees) Griseb.
Rhynchospora lundii Boeck. = ***Rhynchospora confinis*** (Nees) C.B.Clarke
Rhynchospora luquillensis Britton = ***Rhynchospora alba*** (L.) Vahl
Rhynchospora luzuliformis Boeck. = ***Rhynchospora megapotamica*** (A.Spreng.) H.Pfeiff.
Rhynchospora luzuliformis var. *elongata* Kuntze = ***Rhynchospora megapotamica*** (A.Spreng.) H.Pfeiff.
Rhynchospora luzuliformis var. *pusilla* Osten = ***Rhynchospora megapotamica*** (A.Spreng.) H.Pfeiff.
Rhynchospora luzuliformis var. *spicata* Kük. = ***Rhynchospora megapotamica*** (A.Spreng.) H.Pfeiff.
Rhynchospora luzuliformis var. *subcapitata* Kuntze = ***Rhynchospora megapotamica*** (A.Spreng.) H.Pfeiff.
Rhynchospora macrocarpa Boeck. = ***Rhynchospora corymbosa*** (L.) Britton
Rhynchospora macrocephala Boeck. = ***Rhynchospora elatior*** Kunth
Rhynchospora macrochaeta var. *colombiensis* (Boeck.) Kük. = ***Rhynchospora macrochaeta*** Steud. ex Boeck.
Rhynchospora macrochaeta f. *condensata* Kük. = ***Rhynchospora ruiziana*** Boeck.
Rhynchospora macrochaeta var. *jamesonii* C.B.Clarke = ***Rhynchospora macrochaeta*** Steud. ex Boeck.
Rhynchospora macrochaeta var. *quadrispicata* (Boeck.) Kük. = ***Rhynchospora ruiziana*** Boeck.
Rhynchospora macrochaeta var. *quinquespicata* (Boeck.) Kük. = ***Rhynchospora ruiziana*** Boeck.
Rhynchospora macrochaeta var. *ruiziana* (Boeck.) Kük. = ***Rhynchospora ruiziana*** Boeck.
Rhynchospora macrostachya Palla = ***Rhynchospora splendens*** Lindm.
Rhynchospora macrostachya var. *colpophila* Fernald & Gale = ***Rhynchospora macrostachya*** Torr. ex A.Gray
Rhynchospora macrostachya var. *inundata* (Oakes) Fernald = ***Rhynchospora inundata*** (Oakes) Fernald
Rhynchospora macrostachya var. *patula* (Chapm.) Chapm. = ***Rhynchospora careyana*** Fernald
Rhynchospora macrostylis A.Dietr. = ***Rhynchospora barbata*** (Vahl) Kunth
Rhynchospora maculata Maury ex Micheli = ***Rhynchospora megapotamica*** (A.Spreng.) H.Pfeiff.
Rhynchospora madagascariensis (Steud.) Cherm. = ***Rhynchospora rubra*** subsp. ***africana*** J.Raynal
Rhynchospora mairei H.Lév. = ***Rhynchospora rubra*** (Lour.) Makino subsp. ***rubra***
Rhynchospora marginata (Liebm.) Steud. = ***Rhynchospora globosa*** (Kunth) Roem. & Schult. var. ***globosa***
Rhynchospora marginata C.B.Clarke = ***Rhynchospora submarginata*** Kük.
Rhynchospora marisculus var. *elatior* Boeck. = ***Rhynchospora marisculus*** Lindl. & Nees
Rhynchospora martiana (Nees) L.B.Sm. = ***Pleurostachys martiana*** (Nees) Steud.
Rhynchospora massieana E.G.Camus = ***Rhynchospora longisetis*** R.Br.
Rhynchospora mauritii Steud. = ***Rhynchospora holoschoenoides*** (Rich.) Herter
Rhynchospora megapotamica var. *spicata* Kük. = ***Rhynchospora megapotamica*** (A.Spreng.) H.Pfeiff.
Rhynchospora mexicana (Liebm.) C.Wright = ***Rhynchospora velutina*** (Kunth) Boeck.
Rhynchospora mexicana var. *trichochaeta* (C.B.Clarke) Kük. = ***Rhynchospora trichochaeta*** C.B.Clarke
Rhynchospora micrantha Vahl = ***Rhynchospora rariflora*** (Michx.) Elliott
Rhynchospora micrantha Pursh = ***Rhynchospora rariflora*** (Michx.) Elliott
Rhynchospora micrantha A.Gray = ***Rhynchospora torreyana*** A.Gray
Rhynchospora micrantha var. *contracta* (Nees) Kük. = ***Rhynchospora contracta*** (Nees) J.Raynal
Rhynchospora microcarpa var. *perplexa* (Britton) Kük. = ***Rhynchospora perplexa*** Britton
Rhynchospora microcarpa var. *remotiflora* Kük. = ***Rhynchospora microcarpa*** Baldwin ex A.Gray
Rhynchospora microcephala (Bertero ex Spreng.) Kük. = ***Rhynchospora radicans*** subsp. ***microcephala*** (Bertero ex Spreng.) W.W.Thomas
Rhynchospora microseta Steud. = ***Rhynchospora nervosa*** subsp. ***ciliata*** (G.Mey.) T.Koyama
Rhynchospora millegrana Schrad. ex Steud. = [8] Check if the same as the other millegrana.
Rhynchospora millegrana (Nees) L.B.Sm. = ***Pleurostachys millegrana*** (Nees) Steud.

Rhynchospora minarum Steud. = ***Rhynchospora terminalis*** (Nees) Steud.
Rhynchospora minor (Nees) Steud. = ***Rhynchospora leucostachys*** Boeck.
Rhynchospora minor Nelmes = ***Rhynchospora rubra*** subsp. ***africana*** J.Raynal
Rhynchospora minutiflora C.D.Adams = ***Rhynchospora contracta*** (Nees) J.Raynal
Rhynchospora miyakeana Makino = ***Rhynchospora faberi*** C.B.Clarke
Rhynchospora monadelpha C.B.Clarke = ***Rhynchospora hirta*** (Nees) Boeck.
Rhynchospora monostachya Boeck. = ***Rhynchospora albescens*** (Miq.) Kük.
Rhynchospora monostachya Steud. = ***Eleocharis tuberculosa*** (Michx.) Roem. & Schult.
Rhynchospora montevidensis (Barros) Herter = ***Rhynchospora hieronymi*** var. ***montevidensis*** Barros
Rhynchospora moritziana Boeck. = ***Rhynchospora aristata*** var. ***moritziana*** (Boeck.) Kük.
Rhynchospora multiflora A.Gray = ***Rhynchospora elliottii*** A.Dietr.
Rhynchospora multiflora Nees = ***Rhynchospora polyantha*** Steud.
Rhynchospora nana (Boeck.) H.Pfeiff. = ***Rhynchospora pilosa*** Boeck.
Rhynchospora nervosa var. *blepharophora* (J.Presl & C.Presl) Kük. = ***Rhynchospora blepharophora*** (J.Presl & C.Presl) H.Pfeiff.
Rhynchospora nervosa var. *ciliata* (G.Mey.) Kük. = ***Rhynchospora nervosa*** subsp. ***ciliata*** (G.Mey.) T.Koyama
Rhynchospora nervosa var. *cinnamomea* (Kük.) Kük. = ***Rhynchospora nervosa*** subsp. ***ciliata*** (G.Mey.) T.Koyama
Rhynchospora nervosa var. *jelskiana* (Boeck.) Kük. = ***Rhynchospora nervosa*** subsp. ***ciliata*** (G.Mey.) T.Koyama
Rhynchospora nervosa var. *polystachys* (Turrill) Kük. = ***Rhynchospora polystachys*** (Turrill) H.Pfeiff.
Rhynchospora nervosa var. *standleyana* (H.Pfeiff.) Kük. = ***Rhynchospora blepharophora*** (J.Presl & C.Presl) H.Pfeiff.
Rhynchospora nervosa var. *subfiliformis* (H.Pfeiff.) Kük. = ***Rhynchospora berteroi*** (Spreng.) C.B.Clarke
Rhynchospora nervosa var. *tenuifolia* (Kunth) H.Pfeiff. = ***Rhynchospora nervosa*** (Vahl) Boeck. subsp. ***nervosa***
Rhynchospora niederleiniana Boeck. = ***Rhynchospora velutina*** (Kunth) Boeck.
Rhynchospora nipensis Kük. = ***Rhynchospora mayarensis*** León
Rhynchospora nipponica Makino = ***Rhynchospora malasica*** C.B.Clarke
Rhynchospora nitida Spreng. = ***Tetraria compar*** (L.) P.Beauv.
Rhynchospora obliterata Gale = ***Rhynchospora globularis*** (Chapm.) Small var. ***globularis***
Rhynchospora ochrocephala Boeck. = ***Cyperus rhynchosporoides*** Kük.
Rhynchospora oligantha var. *typica* Gale = ***Rhynchospora oligantha*** A.Gray
Rhynchospora oligosantha Endl. ex Kunth = ***Isolepis digitata*** Nees ex Schrad.
Rhynchospora orbignyana (Brongn.) L.B.Sm. = ***Pleurostachys orbignyana*** Brongn.
Rhynchospora orizabensis C.B.Clarke ex Britton = ***Rhynchospora dives*** Standl.
Rhynchospora orizabensis C.B.Clarke ex Kük. = ***Rhynchospora dives*** Standl.
Rhynchospora ostenii Kük. = ***Rhynchospora holoschoenoides*** (Rich.) Herter
Rhynchospora ouensanensis Koidz. = ***Rhynchospora faberi*** C.B.Clarke
Rhynchospora oxycephala C.Wright = ***Rhynchospora eximia*** (Nees) Boeck.
Rhynchospora pallens Schrad. ex Nees = ***Rhynchospora sellowiana*** Steud.
Rhynchospora pallida (Nees) Steud. = ***Rhynchospora patuligluma*** C.B.Clarke ex Lindm.
Rhynchospora paniculata A.Gray = ***Rhynchospora glomerata*** (L.) Vahl
Rhynchospora paniculata C.Presl = ***Rhynchospora polyphylla*** (Vahl) Vahl
Rhynchospora paniculifera Steud. = ***Rhynchospora schiedeana*** (Schltdl.) Kunth
Rhynchospora paramora Mora = ***Rhynchospora oreoboloidea*** Gómez-Laur.
Rhynchospora parva (Nees) Kunth = ***Rhynchospora rubra*** (Lour.) Makino subsp. ***rubra***
Rhynchospora parviflora Vahl ex Kunth = ***Rhynchospora contracta*** (Nees) J.Raynal
Rhynchospora parviflora Beyr. ex Kunth = ***Rhynchospora microcarpa*** Baldwin ex A.Gray
Rhynchospora patula A.Gray = ***Rhynchospora caduca*** Elliott
Rhynchospora pauciflora C.Presl = ***Rhynchospora glomerata*** (L.) Vahl
Rhynchospora pauciflora Willd. ex Kunth = ***Actinoschoenus thouarsii*** (Kunth) Benth.
Rhynchospora pauloensis Boeck. = ***Rhynchospora rugosa*** (Vahl) Gale subsp. ***rugosa***
Rhynchospora pendula A.Gray = ***Rhynchospora elliottii*** A.Dietr.
Rhynchospora penniseta Griseb. = ***Rhynchospora plumosa*** Elliott
Rhynchospora perplexa var. *virginiana* Fernald = ***Rhynchospora perplexa*** Britton
Rhynchospora perrigida Boeck. = ***Bulbostylis paradoxa*** (Spreng.) Lindm.
Rhynchospora persooniana (Nees) Griseb. = ***Rhynchospora colorata*** (L.) H.Pfeiff.
Rhynchospora petraea Nees = ***Rhynchospora rugosa*** subsp. ***brownii*** (Roem. & Schult.) T.Koyama
Rhynchospora pfeifferi Herter = ?
Rhynchospora pilosa subsp. *arenicola* (Uittien) T.Koyama = ***Rhynchospora arenicola*** Uittien
Rhynchospora pineticola C.B.Clarke = ***Rhynchospora intermedia*** (Chapm.) Britton
Rhynchospora pinetorum Britton & Small = ***Rhynchospora globularis*** var. ***pinetorum*** (Britton & Small) Gale
Rhynchospora pittieri (Standl.) H.Pfeiff. = ***Rhynchospora blepharophora*** (J.Presl & C.Presl) H.Pfeiff.
Rhynchospora pittieri Boeck. = ***Rhynchospora macrochaeta*** Steud. ex Boeck. var. ***macrochaeta***
Rhynchospora plankii Britton = ***Rhynchospora harveyi*** W.Boott var. ***harveyi***
Rhynchospora plumosa var. *intermedia* Chapm. = ***Rhynchospora intermedia*** (Chapm.) Britton
Rhynchospora plumosa var. *interrupta* Kük. = ***Rhynchospora plumosa*** Elliott
Rhynchospora plumosa var. *semiplumosa* (A.Gray) Alph.Wood = ***Rhynchospora plumosa*** Elliott
Rhynchospora podosperma C.Wright ex Sauvalle = ***Rhynchospora filiformis*** Vahl
Rhynchospora pohliana (Nees) Steud. = ***Rhynchospora globosa*** (Kunth) Roem. & Schult. var. ***globosa***

Rhynchospora polycephala (Nees) Wydler ex Kunth = ***Rhynchospora holoschoenoides*** (Rich.) Herter
Rhynchospora polycephala C.Presl = ***Rhynchospora polyphylla*** (Vahl) Vahl
Rhynchospora polyphylla Balb. ex Kunth = ***Rhynchospora contracta*** (Nees) J.Raynal
Rhynchospora polyphylla var. *brunnescens* Kük. = ***Rhynchospora polyphylla*** (Vahl) Vahl
Rhynchospora polyphylla var. *contracta* Kük. = ***Rhynchospora polyphylla*** (Vahl) Vahl
Rhynchospora polyphylla var. *laxa* (P.Beauv.) H.Pfeiff. = ***Rhynchospora mandonii*** C.B.Clarke
Rhynchospora polyphylla var. *longispiculosa* Kük. = ***Rhynchospora polyphylla*** (Vahl) Vahl
Rhynchospora polyphylla f. *umbrosa* Boeck. = ***Rhynchospora racemosa*** C.Wright ex Sauvalle
Rhynchospora ponapensis Hosok. = ***Rhynchospora rugosa*** f. ***ponapensis*** (Hosok.) T.Koyama
Rhynchospora prescottiana Wall. = ***Rhynchospora longisetis*** R.Br.
Rhynchospora presleana Steud. = ***Rhynchospora globosa*** (Kunth) Roem. & Schult. var. ***globosa***
Rhynchospora pringlei C.B.Clarke = ***Rhynchospora brevirostris*** Griseb.
Rhynchospora pringlei Greenm. = ***Rhynchospora scutellata*** Griseb.
Rhynchospora prolifera Small = ***Rhynchospora mixta*** Britton
Rhynchospora prolifera (Nees) C.B.Clarke = ***Rhynchospora schomburgkiana*** (Boeck.) T.Koyama
Rhynchospora pruinosa var. *randii* (Britton) Kük. = ***Rhynchospora scabrata*** Griseb.
Rhynchospora pseudolunata Boeck. = ***Rhynchospora tenuis*** subsp. ***tenuis***
Rhynchospora pseudomariscus Kunth = ***Rhynchospora gigantea*** Link
Rhynchospora pseudosellowiana Kük. = ***Rhynchospora sellowiana*** Steud.
Rhynchospora pseudosellowiana var. *castanea* Kük. = ***Rhynchospora sellowiana*** Steud.
Rhynchospora psilocaryoides Griseb. = ***Rhynchospora eximia*** (Nees) Boeck.
Rhynchospora ptaritepuiana Steyerm. = ***Rhynchospora vulcani*** Boeck.
Rhynchospora pterocarpa (Kunth) Roem. & Schult. = ***Rhynchospora barbata*** (Vahl) Kunth
Rhynchospora pubera f. *elatior* Duss = ***Rhynchospora radicans*** subsp. ***microcephala*** (Bertero ex Spreng.) W.W.Thomas
Rhynchospora pubera var. *microcarpa* Boeck. = ***Rhynchospora radicans*** subsp. ***microcephala*** (Bertero ex Spreng.) W.W.Thomas
Rhynchospora pubera var. *multispiculosa* Kük. = ***Rhynchospora pubera*** (Vahl) Boeck. subsp. ***pubera***
Rhynchospora puberula (Boeck.) L.B.Sm. = ***Pleurostachys puberula*** Boeck.
Rhynchospora punctata Boeck. = [79 82 VEN]
Rhynchospora puncticulata Kük. ex Osten = ***Rhynchospora scutellata*** Griseb.
Rhynchospora puncticulata var. *pleiocephala* Osten = ***Rhynchospora corymbosa*** (L.) Britton
Rhynchospora pungens Liebm. = ***Rhynchospora rugosa*** (Vahl) Gale subsp. ***rugosa***
Rhynchospora pura (Nees) Griseb. = ***Rhynchospora nervosa*** subsp. ***ciliata*** (G.Mey.) T.Koyama
Rhynchospora pusilla (Sw.) Griseb. = ***Rhynchospora berteroi*** (Spreng.) C.B.Clarke
Rhynchospora pusilla var. *ferruginescens* Kük. = ***Rhynchospora pusilla*** Chapm. ex M.A.Curtis
Rhynchospora pycnocarpa A.Gray = ***Rhynchospora megalocarpa*** A.Gray
Rhynchospora pyramidata Desv. ex Ham. = ***Rhynchospora polyphylla*** (Vahl) Vahl
Rhynchospora quadrispicata Boeck. = ***Rhynchospora ruiziana*** Boeck.
Rhynchospora quinquespicata Boeck. = ***Rhynchospora ruiziana*** Boeck.
Rhynchospora racemosa f. *setifera* Kük. = ***Rhynchospora racemosa*** C.Wright ex Sauvalle
Rhynchospora racemosa f. *setifolia* Kük. = ***Rhynchospora racemosa*** C.Wright ex Sauvalle
Rhynchospora radicans var. *watsonii* (Britton) Kük. = ***Rhynchospora watsonii*** (Britton) Davidse
Rhynchospora randii Britton = ***Rhynchospora scabrata*** Griseb.
Rhynchospora rappiana Small = ***Rhynchospora ciliaris*** (Michx.) C.Mohr
Rhynchospora ravenelii M.A.Curtis = ***Rhynchospora divergens*** Chapm. ex M.A.Curtis
Rhynchospora recurvata var. *hemicephala* (Boeck.) Kük. = ***Rhynchospora recurvata*** (Nees) Steud.
Rhynchospora recurvata var. *hemicephala* (Boeck.) Kük. = ***Rhynchospora recurvata*** (Nees) Steud.
Rhynchospora repens (Vahl) Boeck. = ***Rhynchospora reptans*** (Rich.) Kük.
Rhynchospora riedeliana Gand. = ***Rhynchospora holoschoenoides*** (Rich.) Herter
Rhynchospora rigida (Kunth) Boeck. = ***Rhynchospora consanguinea*** (Kunth) Boeck.
Rhynchospora rigida Schrad. ex Kunth = ***Rhynchospora marisculus*** Lindl. & Nees
Rhynchospora rigida Spreng. = ***Rhynchospora polyphylla*** (Vahl) Vahl
Rhynchospora rigidifolia subsp. *planifolia* T.Koyama = ***Rhynchospora planifolia*** (T.Koyama) W.W. Thomas
Rhynchospora rigidula Schrad. ex Knuth = ***Rhynchospora marisculus*** Lindl. & Nees
Rhynchospora robusta var. *platyrrhyncha* C.B.Clarke ex Chodat & Hassl. = ***Rhynchospora robusta*** (Kunth) Boeck.
Rhynchospora robusta var. *polyantha* Kük. = ***Rhynchospora robusta*** (Kunth) Boeck.
Rhynchospora rubra f. *elata* Domin = ***Rhynchospora rubra*** (Lour.) Makino subsp. ***rubra***
Rhynchospora rubra var. *hirticeps* Kük. = ***Rhynchospora hirticeps*** (Kük.) T.Koyama
Rhynchospora rudioi Boeck. = ***Rhynchospora riparia*** (Nees) Boeck.
Rhynchospora rufa (Nees) Boeck. = ***Rhynchospora velutina*** (Kunth) Boeck.
Rhynchospora rufa var. *liebmanniana* C.B.Clarke = ***Rhynchospora velutina*** (Kunth) Boeck.
Rhynchospora rugosa var. *griffithii* (Boeck.) D.M.Verma & V.Chandra = ***Rhynchospora rugosa*** subsp. ***brownii*** (Roem. & Schult.) T.Koyama
Rhynchospora rugosa var. *sikkimensis* (C.B.Clarke) D.M.Verma & V.Chandra = ***Rhynchospora rugosa*** subsp. ***brownii*** (Roem. & Schult.) T.Koyama
Rhynchospora ruiziana var. *triceps* K.Schum. = ***Rhynchospora macrochaeta*** var. ***triceps*** Kük.
Rhynchospora ruppioides Benth. = ***Websteria confervoides*** (Poir.) S.S.Hooper
Rhynchospora saxicola Small = ***Rhynchospora globularis*** var. ***saxicola*** (Small) Kük.
Rhynchospora scaberrima Boeck. = ***Rhynchospora gigantea*** Link
Rhynchospora scabrata var. *typica* Gale = ***Rhynchospora***

scabrata Griseb.
Rhynchospora scalaris L.B.Sm. = ***Pleurostachys urvillei*** Brongn.
Rhynchospora scapigera Boeck. = ***Rhynchospora curvula*** Griseb.
Rhynchospora schaffneri Boeck. = ***Rhynchospora kunthii*** Nees ex Kunth
Rhynchospora schenckiana Boeck. = ***Rhynchospora exaltata*** Kunth
Rhynchospora schiedeana var. *locuples* (C.B.Clarke) Kük. = ***Rhynchospora locuples*** C.B.Clarke
Rhynchospora schiedeana subsp. *varica* C.B.Clarke = ***Rhynchospora schiedeana*** (Schltdl.) Kunth
Rhynchospora schoenoides Alph.Wood = ***Rhynchospora elliottii*** A.Dietr.
Rhynchospora schoenoides Britton = ***Rhynchospora holoschoenoides*** (Rich.) Herter
Rhynchospora schraderiana Steud. = ***Rhynchospora corymbosa*** (L.) Britton
Rhynchospora schroederi K.Schum. ex C.B.Clarke = ***Rhynchospora eximia*** (Nees) Boeck.
Rhynchospora schwackei C.B.Clarke = ***Rhynchospora lapensis*** C.B.Clarke
Rhynchospora sclerioides var. *marquisensis* (F.Br.) Kük. = ***Rhynchospora marquisensis*** F.Br.
Rhynchospora sclerioides f. *saxicola* (F.Br.) Kük. = ***Rhynchospora stokesii*** F.Br.
Rhynchospora sclerioides var. *stokesii* (F.Br.) Kük. = ***Rhynchospora stokesii*** F.Br.
Rhynchospora sellowiana Boeck. = ***Rhynchospora subsetosa*** C.B.Clarke
Rhynchospora semiplumosa A.Gray = ***Rhynchospora plumosa*** Elliott
Rhynchospora senegalensis Steud. = ***Fuirena stricta*** Steud. subsp. ***stricta***
Rhynchospora setacea MacMill. = ***Rhynchospora capillacea*** Torr.
Rhynchospora setacea Vahl = ***Rhynchospora fascicularis*** (Michx.) Vahl subsp. ***fascicularis***
Rhynchospora setacea Willd. ex Kunth = ***Rhynchospora rariflora*** (Michx.) Elliott
Rhynchospora setacea (P.J.Bergius) Boeck. = ***Rhynchospora tenerrima*** Nees ex Spreng. subsp. ***tenerrima***
Rhynchospora setacea var. *fallax* (Uittien) Kük. = ***Rhynchospora fallax*** Uittien
Rhynchospora setacea f. *filicula* Kük. = ***Rhynchospora fallax*** Uittien
Rhynchospora setigera Griseb. = ***Rhynchospora longibracteata*** Boeck.
Rhynchospora setigera Griseb. = ***Rhynchospora subsetigera*** H.Pfeiff.
Rhynchospora sikkimensis C.B.Clarke = ***Rhynchospora rugosa*** subsp. ***brownii*** (Roem. & Schult.) T.Koyama
Rhynchospora simplex var. *angusta* Kük. = ***Rhynchospora exaltata*** Kunth
Rhynchospora sinensis Steud. = ***Rhynchospora rugosa*** subsp. ***brownii*** (Roem. & Schult.) T.Koyama
Rhynchospora smallii Britton = ***Rhynchospora capillacea*** Torr.
Rhynchospora sororopana Steyerm. = ***Rhynchospora biflora*** Boeck.
Rhynchospora sparganoides (Lindl. ex Nees) Steud. = ***Rhynchospora exaltata*** Kunth
Rhynchospora sparganoides Bojer = ***Rhynchospora holoschoenoides*** (Rich.) Herter
Rhynchospora sparsa Sieber ex C.Presl = ***Rhynchospora contracta*** (Nees) J.Raynal
Rhynchospora sparsa (Michx.) Vahl = ***Rhynchospora miliacea*** (Lam.) A.Gray
Rhynchospora sparsiflora (Kunth) L.B.Sm. = ***Pleurostachys sparsiflora*** Kunth
Rhynchospora speciosa (Kunth) Boeck. = ***Rhynchospora consanguinea*** (Kunth) Boeck.
Rhynchospora spermodon Griseb. = ***Rhynchospora tenerrima*** Nees ex Spreng. subsp. ***tenerrima***
Rhynchospora sphaerocephala Boeck. = ***Rhynchospora holoschoenoides*** (Rich.) Herter
Rhynchospora spicata Boeck. = ***Rhynchospora megapotamica*** (A.Spreng.) H.Pfeiff.
Rhynchospora spicata Spreng. = ***Scleria spicata*** (Spreng.) J.F.Macbr.
Rhynchospora spiciformis Hillebr. = ***Rhynchospora chinensis*** subsp. ***spiciformis*** (Hillebr.) T.Koyama
Rhynchospora splendens f. *composita* Kük. = ***Rhynchospora splendens*** Lindm.
Rhynchospora splendens f. *laetevirens* Kük. = ***Rhynchospora splendens*** Lindm.
Rhynchospora sprucei Boeck. = ***Rhynchospora reptans*** (Rich.) Kük.
Rhynchospora standleyana H.Pfeiff. = ***Rhynchospora blepharophora*** (J.Presl & C.Presl) H.Pfeiff.
Rhynchospora stellata (Lam.) Griseb. = ***Rhynchospora colorata*** (L.) H.Pfeiff.
Rhynchospora stellata f. *floridensis* (Britton ex Small) Kük. = ***Rhynchospora floridensis*** (Britton ex Small) H.Pfeiff.
Rhynchospora stellata f. *intercedens* Kük. = ***Rhynchospora colorata*** (L.) H.Pfeiff.
Rhynchospora stellata var. *latifolia* (Baldwin ex Elliott) Kük. = ***Rhynchospora latifolia*** (Baldwin ex Elliott) W.W.Thomas
Rhynchospora stelligera Herter = ***Rhynchospora setigera*** (Kunth) Boeck.
Rhynchospora stenophylla var. *albescens* Kük. = ***Rhynchospora tenuifolia*** Griseb.
Rhynchospora stenophylloidea (Kük.) Gale = ***Rhynchospora cubensis*** A.Rich.
Rhynchospora stenorhyncha Griseb. = ***Rhynchospora triflora*** Vahl
Rhynchospora stipitata Chapm. = ***Rhynchospora odorata*** C.Wright ex Griseb.
Rhynchospora stokesii f. *saxicola* F.Br. = ***Rhynchospora stokesii*** F.Br.
Rhynchospora stricta var. *demerarensis* (C.B.Clarke) Kük. = ***Rhynchospora cajennensis*** Boeck.
Rhynchospora stuebelii Boeck. = ***Rhynchospora ruiziana*** Boeck.
Rhynchospora subcapitata Boeck. = ***Rhynchospora terminalis*** (Nees) Steud.
Rhynchospora suberostris Boeck. = ***Rhynchospora grayi*** Kunth
Rhynchospora subfiliformis H.Pfeiff. = ***Rhynchospora berteroi*** (Spreng.) C.B.Clarke
Rhynchospora sublateralis Steud. = ***Schoenus sublateralis*** (Steud.) C.B.Clarke
Rhynchospora submarginata var. *glabrinux* Kük. = ***Rhynchospora submarginata*** Kük.
Rhynchospora submarginata var. *tonkinensis* Kük. = ***Rhynchospora submarginata*** Kük.
Rhynchospora subnipensis Kük. = ***Rhynchospora mayarensis*** León
Rhynchospora subquadrata Cherm. = ***Rhynchospora gracillima*** subsp. ***subquadrata*** (Cherm.) J.Raynal
Rhynchospora subulirostris Steud. = ***Rhynchospora corymbosa*** (L.) Britton
Rhynchospora sulcata Gale = ***Rhynchospora microcarpa*** Baldwin ex A.Gray

Rhynchospora surinamensis (Rottb.) Nees = ***Rhynchospora corymbosa*** (L.) Britton
Rhynchospora surinamensis K.Schum. = ***Rhynchospora corymbosa*** (L.) Britton
Rhynchospora sylvatica Nees = ***Rhynchospora comata*** (Link) Schult.
Rhynchospora tamana Steyerm. = ***Rhynchospora macrochaeta*** Steud. ex Boeck. var. ***macrochaeta***
Rhynchospora tenella f. *haplostylis* Uittien = ***Rhynchospora tenella*** (Nees) Boeck.
Rhynchospora tenella var. *haplostylis* (Uittien) Kük. = ***Rhynchospora tenella*** (Nees) Boeck.
Rhynchospora tenerrima Boeck. = ***Rhynchospora tepuiana*** Steyerm.
Rhynchospora tenerrima subsp. *fallax* (Uittien) T.Koyama = ***Rhynchospora fallax*** Uittien
Rhynchospora tenerrima var. *flexuosa* (C.B.Clarke) Boeck. =
Rhynchospora tenuiflora (Brongn.) L.B.Sm. = ***Pleurostachys tenuiflora*** Brongn.
Rhynchospora tenuiflora var. *albescens* (Kük.) Kük. =
Rhynchospora tenuifolia Benth. = ***Rhynchospora subtenuifolia*** Kük.
Rhynchospora tenuirostris (Nees) Kunth = ***Rhynchospora holoschoenoides*** (Rich.) Herter
Rhynchospora tenuis Baldwin ex A.Gray = ***Rhynchospora wrightiana*** Boeck.
Rhynchospora tenuis var. *antillana* Kük. = ***Rhynchospora tenuis*** subsp. ***tenuis***
Rhynchospora tenuis var. *brevinux* Kük. = ***Rhynchospora tenuis*** subsp. ***tenuis***
Rhynchospora tenuis var. *congesta* (Nees) Kük. = ***Rhynchospora riparia*** (Nees) Boeck.
Rhynchospora tenuis var. *emaciata* (Nees) Lindm. = ***Rhynchospora emaciata*** (Nees) Boeck.
Rhynchospora tenuis f. *major* C.B.Clarke = ***Rhynchospora tenuis*** subsp. ***tenuis***
Rhynchospora tenuis var. *maritima* Boeck. = ***Rhynchospora riparia*** (Nees) Boeck.
Rhynchospora tenuis f. *minor* C.B.Clarke = ***Rhynchospora tenuis*** subsp. ***tenuis***
Rhynchospora tenuis var. *prorepens* Kük. = ***Rhynchospora tenuis*** subsp. ***tenuis***
Rhynchospora tenuis var. *pseudolunata* (Boeck.) Kük. = ***Rhynchospora tenuis*** subsp. ***tenuis***
Rhynchospora tenuis subsp. *riparia* (Nees) T.Koyama = ***Rhynchospora riparia*** (Nees) Boeck.
Rhynchospora tenuiseta C.Wright = ***Rhynchospora marisculus*** Lindl. & Nees
Rhynchospora teres Eaton = ***Rhynchospora glomerata*** (L.) Vahl
Rhynchospora terminalis var. *crassipes* (Boeck.) Kük. = ***Rhynchospora terminalis*** (Nees) Steud.
Rhynchospora terminalis var. *crassipes* (Boeck.) Kük. = ***Rhynchospora terminalis*** (Nees) Steud.
Rhynchospora terminalis f. *pallida* (Nees) Kük. = ***Rhynchospora terminalis*** (Nees) Steud.
Rhynchospora terminalis var. *rosemariana* (D.A. Simpson) A.C.Araujo = ***Rhynchospora rosemariana*** D.A.Simpson
Rhynchospora testui Cherm. = ***Rhynchospora gracillima*** subsp. ***subquadrata*** (Cherm.) J.Raynal
Rhynchospora testui var. *pleiantha* Cherm. = ***Rhynchospora eximia*** (Nees) Boeck.
Rhynchospora texensis (Torr.) Boeck. = ***Rhynchospora nitens*** (Vahl) A.Gray var. ***nitens***
Rhynchospora thouarsii Nees = ***Rhynchospora rugosa*** subsp. ***brownii*** (Roem. & Schult.) T.Koyama
Rhynchospora thyrsoidea Nees & Meyen ex Kunth = ***Rhynchospora sclerioides*** Hook. & Arn.
Rhynchospora torreyana Griseb. = ***Rhynchospora glomerata*** (L.) Vahl
Rhynchospora torreyana var. *microrhyncha* Griseb. = ***Rhynchospora microcarpa*** Baldwin ex A.Gray
Rhynchospora triceps (Cham. & Schltdl.) Steud. = ***Rhynchospora exaltata*** Kunth
Rhynchospora triceps (Vahl) Hochst. = ***Rhynchospora holoschoenoides*** (Rich.) Herter
Rhynchospora triceps Boeck. = ***Rhynchospora macrochaeta*** var. ***triceps*** Kük.
Rhynchospora trichochaeta var. *flavida* (Boeck.) Kük. = ***Rhynchospora trichochaeta*** C.B.Clarke
Rhynchospora trichochaeta var. *flavida* (Boeck.) Kük. = ***Rhynchospora trichochaeta*** C.B.Clarke
Rhynchospora trichochaeta var. *venezuelensis* Steyerm. = ***Rhynchospora subplumosa*** C.B.Clarke
Rhynchospora trichophylla Fernald = ***Rhynchospora gracilenta*** A.Gray
Rhynchospora triflora M.A.Curtis = ***Rhynchospora oligantha*** A.Gray var. ***oligantha***
Rhynchospora triflora var. *papuana* Kük. = ***Rhynchospora triflora*** Vahl
Rhynchospora trigyna Hochst. = [24 ETH]
Rhynchospora trispicata subsp. *schmidtii* (Kük.) T.Koyama = ***Rhynchospora schmidtii*** Kük.
Rhynchospora uleana Kük. = ***Rhynchospora kuntzei*** C.B.Clarke ex Kuntze
Rhynchospora uleana Boeck. = ***Rhynchospora marisculus*** Lindl. & Nees
Rhynchospora umbraticola var. *broadwayi* Kük. = ***Rhynchospora comata*** (Link) Schult.
Rhynchospora umbraticola var. *kuntzei* (C.B.Clarke ex Kuntze) Kük. = ***Rhynchospora kuntzei*** C.B.Clarke ex Kuntze
Rhynchospora umemurae Makino = ***Rhynchospora faberi*** C.B.Clarke
Rhynchospora umemurae var. *yakusimensis* Masam. = ***Rhynchospora faberi*** C.B.Clarke
Rhynchospora uniflora var. *ekmanii* (Urb.) Kük. = ***Rhynchospora uniflora*** Boeck.
Rhynchospora uniflora var. *ekmanii* (Urb.) Kük. = ***Rhynchospora uniflora*** Boeck.
Rhynchospora uniflora var. *uberior* Kük. = ***Rhynchospora uniflora*** Boeck.
Rhynchospora uniflora var. *uberior* Kük. = ***Rhynchospora uniflora*** Boeck.
Rhynchospora urvillei Steud. = ***Rhynchospora holoschoenoides*** (Rich.) Herter
Rhynchospora vahliana Griseb. = ***Rhynchospora nervosa*** subsp. ***ciliata*** (G.Mey.) T.Koyama
Rhynchospora velutina f. *glabrescens* Pilg. = ***Rhynchospora velutina*** (Kunth) Boeck.
Rhynchospora venezolana Kük. = ***Rhynchospora hirsuta*** (Vahl) Vahl
Rhynchospora vicozensis Schult. = ***Rhynchospora rugosa*** (Vahl) Gale subsp. ***rugosa***
Rhynchospora viridilutea C.B.Clarke = ***Rhynchospora hassleri*** C.B.Clarke
Rhynchospora vulcani var. *ecuadorensis* (C.B.Clarke) Kük. = ***Rhynchospora vulcani*** Boeck.
Rhynchospora vulcani var. *hoffmannii* (Boeck.) Kük. = ***Rhynchospora vulcani*** Boeck.
Rhynchospora wallichiana Kunth = ***Rhynchospora rubra*** (Lour.) Makino subsp. ***rubra***
Rhynchospora wallichii (Nees) K.Schum. = ***Rhynchospora rubra*** (Lour.) Makino subsp. ***rubra***
Rhynchospora weberbaueri C.B.Clarke = ***Rhynchospora marisculus*** Lindl. & Nees

Rhynchospora yakusimensis (Masam.) Masam. = ***Rhynchospora faberi*** C.B.Clarke
Rhynchospora yasudana Makino = ***Rhynchospora fujiana*** Makino

***Unplaced Names*:**
Rhynchospora alba var. *meridiana* H.Pfeiff., Repert. Spec. Nov. Regni Veg. 33: 210 (1933). = [79 MXG 84 BZN]
Rhynchospora cinnamomea Boeck., Flora 41: 645 (1858). = [7]
Rhynchospora fauriei C.B.Clarke ex H.Lév., Bull. Acad. Int. Géogr. Bot. 14: 197 (1904), nom. illeg. = [36]
Rhynchospora foliata Steud., Syn. Pl. Glumac. 2: 142 (1855). = [7]
Rhynchospora millegrana Schrad. ex Steud., Syn. Pl. Glumac. 2: 139 (1855). = [8] Check if the same as the other millegrana.
Rhynchospora pfeifferi Herter, Revista Sudamer. Bot. 9: 159 (1953). = ?
Rhynchospora punctata Boeck., Linnaea 37: 580 (1873), nom. illeg. = [79 82 VEN]
Rhynchospora trigyna Hochst., Flora 24(1 Intell.): 21 (1841). = [24 ETH]

Rikliella

Rikliella J.Raynal = ***Lipocarpha*** R.Br.
Rikliella australiensis J.Raynal = ***Lipocarpha microcephala*** (R.Br.) Kunth
Rikliella kernii (Raymond) J.Raynal = ***Lipocarpha kernii*** (Raymond) Goetgh.
Rikliella rehmannii (Ridl.) J.Raynal = ***Lipocarpha rehmannii*** (Ridl.) Goetgh.
Rikliella squarrosa (L.) J.Raynal = ***Lipocarpha squarrosa*** (L.) Goetgh.

Schelhammeria

Schelhammeria Moench = ***Carex*** L.
Schelhammeria capitata Moench = ***Carex bohemica*** Schreb.
Schelhammeria cyperoides (L.) Dumort. = ***Carex bohemica*** Schreb.

Schizolepis

Schizolepis Schrad. ex Nees = ***Scleria*** P.J.Bergius
Schizolepis abortiva Nees = ***Scleria abortiva*** Nees ex Kunth
Schizolepis arguta Nees = ***Scleria arguta*** (Nees) Steud.
Schizolepis arundinacea (Kunth) Palla = ***Scleria latifolia*** Sw.
Schizolepis foliosa Nees = ***Scleria plusiophylla*** Steud.
Schizolepis latifolia (Sw.) Nees = ***Scleria latifolia*** Sw.
Schizolepis myricocarpa (Kunth) Nees = ***Scleria myricocarpa*** Kunth
Schizolepis olyroides (Kunth) Nees = ***Scleria latifolia*** Sw.
Schizolepis panicoides (Kunth) Nees = ***Scleria panicoides*** Kunth
Schizolepis paranensis Palla = ***Scleria schiedeana*** Schltdl.
Schizolepis rufescens Schrad. ex Nees = ***Scleria latifolia*** Sw.
Schizolepis sieberi (Nees) Nees = ***Scleria sieberi*** Nees
Schizolepis sylvestris (Poepp. & Kunth) Nees = ***Scleria latifolia*** Sw.
Schizolepis trigonocarpa Nees = ***Scleria latifolia*** Sw.
Schizolepis trigonocarpa var. *plena* Nees = ***Scleria latifolia*** Sw.

Schoenidium

Schoenidium Nees = ***Ficinia*** Schrad.
Schoenidium laterale (Vahl) Nees = ***Ficinia lateralis*** (Vahl) Kunth
Schoenidium laterale (Vahl) Nees = ***Ficinia lateralis*** (Vahl) Kunth

Schoenodendron

Schoenodendron Engl. = ***Microdacoides*** Hua
Schoenodendron buecheri Engl. = ***Microdacoides squamosus*** Hua

Schoenoides

Schoenoides O.Seberg = ***Oreobolus*** R.Br.
Schoenoides oligocephalus (W.M.Curtis) O.Seberg = ***Oreobolus oligocephalus*** W.M.Curtis

Schoenoplectiella

Schoenoplectiella Lye, Lidia 6: 20 (2003).
Cosmopolitan. 11 AUT CZE GER HUN POL SWI 12 FRA por SPA 13 BUL GRC ITA ROM TUE YUG 14 KRY RUC RUS UKR 20 ALG EGY 22 BEN BKN GHA GNB GUI IVO MLI MTN NGA NGR SEN SIE TOG 23 BUR CAF CMN EQG RWA ZAI 24 CHA ETH SOM SUD 25 KEN TAN UGA 26 ANG ZIM 27 BOT CPP NAM NAT OFS SWZ TVL 29 MAU MDG REU SEY 30 (31) 32 KAZ TZK UZB 33 TCS 34 AFG IRN IRQ PAL 35 YEM 36 CHM CHS 38 JAP KOR NNS TAI 40 ASS EHM IND NEP PAK SRL WHM 41 MYA THA VIE 42 BOR JAW LSI MLY PHI SUL SUM 43 NWG 50 NSW NTA QLD SOA VIC WAU 60 FIJ 62 CRL MRN 63 HAW 71 BRC 72 ONT QUE 73 COL WYO 74 ILL IOW† KAN MIN MSO NEB OKL SDA WIS 75 CNT INI MAI MAS MIC NWH NWJ NWY OHI PEN RHO WVA 76 CAL UTA 77 TEX 78 ALA DEL FLA GEO KTY MRY MSI NCA SCA TEN VRG WDC 79 MXE MXT 84 85 AGE PAR.
26 Species

Schoenoplectiella aberrans (Cherm.) Lye, Lidia 6: 24 (2003).
C. Madagascar. 29 MDG. Hemicr. or rhizome geophyte.
**Scirpus aberrans* Cherm., Bull. Soc. Bot. France 67: 423 (1921 publ. 1923). *Schoenoplectus aberrans* (Cherm.) J.Raynal, Adansonia, n.s., 16: 152 (1976).

Schoenoplectiella articulata (L.) Lye, Lidia 6: 20 (2003).
Trop. & Subtrop. Old World. 20 EGY 22 GHA NGA NGR TOG 23 BUR CMN ZAI 24 CHA SOM SUD 25 KEN TAN UGA 27 BOT CPP NAM NAT SWZ TVL 29 MAU MDG 36 CHS 40 ASS EHM IND NEP SRL 41 THA VIE 42 JAW LSI PHI SUL 43 NWG 50 NTA QLD. Ther.
**Scirpus articulatus* L., Sp. Pl.: 47 (1753). *Isolepis articulata* (L.) Nees in R.Wight, Contr. Bot. India: 108 (1834). *Schoenoplectus articulatus* (L.) Palla, Bot. Jahrb. Syst. 10: 299 (1888).
Scirpus fistulosus Forssk., Fl. Aegypt.-Arab.: 14 (1775). *Isolepis fistulosa* (Forssk.) Delile, Descr. Egypte, Hist. Nat.: 50 (1813). *Schoenoplectus fistulosus* (Forssk.) Soják, Cas. Nár. Mus., Odd. Přír. 148: 194 (1979 publ. 1980).
Scirpus incurvatus Roxb., Fl. Ind. 1: 216 (1820). *Eleocharis incurvata* (Roxb.) Schult. in J.J.Roemer

& J.A.Schultes, Mant. 2: 92 (1824). *Holoschoenus incurvatus* (Roxb.) A.Dietr., Sp. Pl. 2: 165 (1832). *Isolepis incurvata* (Roxb.) Nees in R.Wight, Contr. Bot. India: 108 (1834).
Scirpus subarticulatus Roxb., Fl. Ind. 1: 215 (1820). *Holoschoenus subarticulatus* (Roxb.) A.Dietr., Sp. Pl. 2: 166 (1832).
Isolepis prolongata Nees in R.Wight, Contr. Bot. India: 108 (1834).
Scirpus prolongatus Poir. ex Nees, Linnaea 9: 291 (1834), nom. inval.
Isolepis rugosa Boeck., Linnaea 36: 702 (1870).
Scirpus articulatus var. *major* Boeck., Linnaea 36: 702 (1870).

Schoenoplectiella dissachantha (S.T.Blake) Lye, Lidia 6: 24 (2003).
Australia. 50 NSW NTA QLD SOA VIC WAU. Hemicr. or rhizome geophyte.
**Scirpus dissachanthus* S.T.Blake, Victoria Naturalist 63: 116 (1946). *Schoenoplectus dissachanthus* (S.T.Blake) J.Raynal, Adansonia, n.s., 16: 139 (1976).

Schoenoplectiella erecta (Poir.) Lye, Lidia 6: 25 (2003).
Tanzania to S. Africa, Madagascar, Indo-China, SE. U.S.A. to Mexico, S. South America. 25 TAN UGA 26 ZIM 27 BOT NAM NAT TVL 29 MAU MDG 41 VIE (50) nsw qld vic 77 TEX 78 ALA FLA GEO SCA 79 MXE MXT 85 PAR. Ther.
Scirpus lateralis Retz., Observ. Bot. 4: 11 (1786), nom. illeg.
**Scirpus erectus* Poir. in J.B.A.M.de Lamarck, Encycl. 6: 761 (1805). *Schoenoplectus erectus* (Poir.) Palla ex J.Raynal, Adansonia, n.s., 16: 141 (1976).
Scirpus ternatus Wall., Numer. List: 3462 (1831).
Scirpus supinus Boeck., Linnaea 36: 700 (1870).
Scirpus ternatus Ham. ex Hook.f., Fl. Brit. India 6: 656 (1893), nom. inval.
Scirpus erismanae Schuyler, Notul. Nat. Acad. Nat. Sci. Philadelphia 423: 3 (1969).
Scirpus wilkensii Schuyler, Notul. Nat. Acad. Nat. Sci. Philadelphia 423: 4 (1969).
Scirpus raynalii Schuyler, Notul. Nat. Acad. Nat. Sci. Philadelphia 438: 1 (1971). *Schoenoplectus erectus* subsp. *raynalii* (Schuyler) Lye, Nordic J. Bot. 3: 243 (1983). *Scirpus erectus* subsp. *raynalii* (Schuyler) B.F.Hansen & Wunderlin, Bot. Explor. (Florida) 3: 38 (2003).
Scirpus sinuatus Schuyler, Notul. Nat. Acad. Nat. Sci. Philadelphia 438: 2 (1971). *Schoenoplectus erectus* subsp. *sinuatus* (Schuyler) Lye, Nordic J. Bot. 3: 243 (1983).

Schoenoplectiella hallii (A.Gray) Lye, Lidia 6: 25 (2003).
NC. & E. U.S.A. 74 ILL IOW† KAN MSO NEB OKL WIS 75 INI MAS† MIC 77 TEX 78 GEO KTY. Hemicr. or rhizome geophyte.
**Scirpus hallii* A.Gray, Manual, ed. 5: 563 (1867). *Scirpus supinus* var. *hallii* (A.Gray) A.Gray, Manual, ed. 5: 563 (1867). *Scirpus uninodis* var. *hallii* (A.Gray) Beetle, Amer. J. Bot. 29: 656 (1942). *Schoenoplectus supinus* subsp. *hallii* (A.Gray) Soják, Cas. Nár. Mus., Odd. Prír. 141: 62 (1972). *Schoenoplectus hallii* (A.Gray) S.G.Sm., Novon 5: 101 (1995).

Schoenoplectiella heterophylla (Schuyler) Lye, Lidia 6: 25 (2003).
Madagascar. 29 MDG. Hemicr. or rhizome geophyte.
**Scirpus heterophyllus* Schuyler, Brittonia 22: 151 (1970).

Schoenoplectiella hooperae (J.Raynal) Lye, Lidia 6: 25 (2003).
Tanzania. 25 TAN. Ther.
**Schoenoplectus hooperae* J.Raynal, Adansonia, n.s., 16: 146 (1976).

Schoenoplectiella juncea (Willd.) Lye, Lidia 6: 25 (2003).
Trop. Africa. 22 BEN GHA GNB GUI NGA SEN TOG 24 SOM 25 KEN TAN UGA. Ther.
**Schoenus junceus* Willd., Phytographia: 2 (1794). *Scirpus aureiglumis* S.S.Hooper, Kew Bull. 26: 581 (1972). *Schoenoplectus junceus* (Willd.) J.Raynal, Adansonia, n.s., 16: 139 (1976).

Schoenoplectiella juncoides (Roxb.) Lye, Lidia 6: 25 (2003).
W. Indian Ocean, Asia to Pacific. (12) por 29 MDG REU SEY 32 UZB 33 TCS 34 IRN 36 CHM CHS 38 JAP KOR NNS TAI 40 ASS EHM IND NEP PAK SRL WHM 41 THA VIE 42 BOR JAW LSI MLY PHI SUL SUM 43 NWG 60 FIJ 62 CRL MRN 63 HAW. Ther. or hemicr.
**Scirpus juncoides* Roxb., Fl. Ind. 1: 218 (1820). *Schoenoplectus juncoides* (Roxb.) Palla, Bot. Jahrb. Syst. 10: 299 (1888). *Hymenochaeta juncoides* (Roxb.) Nakai, Bull. Natl. Sci. Mus. Tokyo 31: 133 (1952).
Scirpus quadrangulus D.Don, Prodr. Fl. Nepal.: 40 (1825).
Scirpus donianus Spreng., Syst. Veg. 4(2): 28 (1827).
Scirpus luzonensis J.Presl & C.Presl, Reliq. Haenk. 1: 193 (1828).
Scirpus junciformis Nees in R.Wight, Contr. Bot. India: 112 (1834), nom. illeg.
Scirpus timorensis Kunth, Enum. Pl. 2: 162 (1837).
Scirpus supinus var. *elatior* Boeck., Linnaea 36: 700 (1870).
Scirpus rockii Kük., Repert. Spec. Nov. Regni Veg. 16: 432 (1920). *Schoenoplectus juncoides* subsp. *rockii* (Kük.) Soják, Cas. Nár. Mus., Odd. Prír. 141: 62 (1972).
Scirpus subquadrangularis Makino, J. Jap. Bot. 7: 8 (1931).
Scirpus ohwianus T.Koyama, Bot. Mag. (Tokyo) 69: 212 (1956). *Schoenoplectus juncoides* subsp. *ohwianus* (T.Koyama) Soják, Cas. Nár. Mus., Odd. Prír. 141: 62 (1972). *Schoenoplectus ohwianus* (T.Koyama) Holub, Folia Geobot. Phytotax. 11: 83 (1976).

Schoenoplectiella laevis (S.T.Blake) Lye, Lidia 6: 25 (2003).
Australia. 50 NSW NTA QLD WAU. Hemicr. or rhizome geophyte.
**Scirpus laevis* S.T.Blake, Proc. Roy. Soc. Queensland 51: 177 (1939 publ. 1940). *Schoenoplectus laevis* (S.T.Blake) J.Raynal, Adansonia, n.s., 16: 151 (1976).

Schoenoplectiella lateriflora (J.F.Gmel.) Lye, Lidia 6: 25 (2003).
Trop. & Subtrop. Old World to N. Australia. 22 BEN BKN GHA MLI NGA NGR SEN TOG 23 EQG 24 SOM 25 KEN TAN UGA 27 CPP 29 MDG 30 32 TZK 34 AFG 36 CHS 38 TAI 40 ASS IND NEP PAK SRL 41 THA 42 JAW LSI MLY PHI 50 NTA QLD WAU. Ther.
**Scirpus lateriflorus* J.F.Gmel., Syst. Nat. 2: 127 (1791). *Scirpus supinus* var. *lateriflorus* (J.F.Gmel.)

T.Koyama, J. Fac. Sci. Univ. Tokyo, Sect. 3, Bot. 7: 302 (1958). *Schoenoplectus lateriflorus* (J.F.Gmel.) Lye, Bot. Not. 124: 290 (1971). *Schoenoplectus supinus* subsp. *lateriflorus* (J.F.Gmel.) Soják, Cas. Nár. Mus., Odd. Prír. 141: 62 (1972). *Schoenoplectus supinus* subsp. *lateriflorus* (J.F.Gmel.) T.Koyama, Enum. Fl. Pl. Nepal 1: 119 (1978).

Isolepis mesantha Nees, Linnaea 9: 291 (1834), nom. inval.

Isolepis ambigua Zoll., Syst. Verz. 2: 62 (1854), nom. inval.

Isolepis oryzetorum Steud., Syn. Pl. Glumac. 2: 96 (1855). *Scirpus oryzetorum* (Steud.) Ohwi, Mem. Coll. Sci. Kyoto Imp. Univ., Ser. B, Biol. 18(1): 112 (1944). *Schoenoplectus oryzetorum* (Steud.) V.I.Krecz., in Fl. Tadzhikskoi SSR 2: 41 (1963).

Isolepis juncoides Miq., Fl. Ned. Ind. 3: 312 (1856).

Schoenoplectus lateriflorus subsp. *laevinux* Lye, Nordic J. Bot. 3: 242 (1983).

Schoenoplectiella leucantha (Boeck.) Lye, Lidia 6: 25 (2003).

S. Africa. 27 CPP OFS TVL. Hemicr. or rhizome geophyte.

**Scirpus leucanthus* Boeck., Verh. Bot. Vereins Prov. Brandenburg 29: 46 (1888). *Schoenoplectus leucanthus* (Boeck.) J.Raynal, Adansonia, n.s., 16: 143 (1976).

Schoenoplectiella microglumis (Lye) Lye, Lidia 6: 26 (2003).

Rwanda to E. Trop. Africa. 23 RWA 25 KEN TAN UGA. Ther.

Scirpus tenerrimus Peter, Repert. Spec. Nov. Regni Veg. Beih. 40(1): 124 (1936).

**Schoenoplectus microglumis* Lye, Bot. Not. 124: 287 (1971).

Schoenoplectiella oxyjulos (S.S.Hooper) Lye, Lidia 6: 26 (2003).

W. Trop. Africa to Sudan. 22 GUI NGA SIE 23 CAF CMN 24 SUD. Hemicr.

**Scirpus oxyjulos* S.S.Hooper, Kew Bull. 26: 581 (1972). *Schoenoplectus oxyjulos* (S.S.Hooper) J.Raynal, Adansonia, n.s., 16: 151 (1976).

Schoenoplectiella perrieri (Cherm.) Lye, Lidia 6: 26 (2003).

W. Madagascar. 29 MDG. Hemicr.

**Scirpus perrieri* Cherm., Arch. Bot. Bull. Mens. 3: 194 (1929 publ. 1931). *Schoenoplectus perrieri* (Cherm.) J.Raynal, Adansonia, n.s., 16: 152 (1976).

Schoenoplectiella praelongata (Poir.) Lye, Lidia 6: 26 (2003).

Egypt to Chad, S. Africa, Trop. Asia to N. Australia. 20 EGY 24 CHA 27 BOT NAM NAT SWZ TVL 40 ASS IND PAK WHM 41 VIE 42 PHI 50 NTA QLD WAU. Ther.

**Scirpus praelongatus* Poir. in J.B.A.M.de Lamarck, Encycl. 6: 764 (1805). *Schoenoplectus praelongatus* (Poir.) J.Raynal, Adansonia, n.s., 16: 148 (1976).

Schoenoplectiella proxima (Steud.) Lye, Lidia 6: 26 (2003).

Tanzania, Botswana, SW. Arabian Pen. 25 TAN 27 BOT 35 YEM. Ther.

**Isolepis proxima* Steud., Syn. Pl. Glumac. 2: 95 (1855). *Schoenoplectus proximus* (Steud.) J.Raynal, Adansonia, n.s., 16: 144 (1976).

Schoenoplectus proximus var. *botswanensis* Hayas., J. Jap. Bot. 50: 161 (2005).

Schoenoplectiella purshiana (Fernald) Lye, Lidia 6: 26 (2003).

E. Canada to NC. & E. U.S.A. 72 ONT QUE 74 ILL MIN MSO WIS 75 CNT INI MAI MAS MIC NWH NWJ NWY OHI PEN RHO WVA 78 ALA DEL GEO KTY MRY MSI NCA SCA TEN VRG WDC. Ther.

Scirpus debilis Pursh, Fl. Amer. Sept. 1: 55 (1813), nom. illeg. **Scirpus purshianus* Fernald, Rhodora 44: 479 (1942). *Schoenoplectus juncoides* subsp. *purshianus* (Fernald) Soják, Cas. Nár. Mus., Odd. Prír. 141: 62 (1972). *Schoenoplectus purshianus* (Fernald) M.T.Strong, Novon 3: 202 (1993).

Scirpus supinus var. *digynus* Boeck., Linnaea 36: 700 (1870). *Scirpus juncoides* var. *digynus* (Boeck.) T.Koyama, Phytologia 8: 457 (1963).

Scirpus debilis var. *williamsii* Fernald, Rhodora 3: 252 (1901). *Scirpus purshianus* f. *williamsii* (Fernald) Fernald, Rhodora 44: 479 (1942). *Scirpus smithii* var. *williamsii* (Fernald) Beetle, Amer. J. Bot. 29: 655 (1942). *Scirpus juncoides* var. *williamsii* (Fernald) T.Koyama, Canad. J. Bot. 40: 914 (1962). *Schoenoplectus smithii* subsp. *williamsii* (Fernald) Soják, Cas. Nár. Mus., Odd. Prír. 148: 194 (1979 publ. 1980). *Schoenoplectus purshianus* var. *williamsii* (Fernald) S.G.Sm., Novon 12: 107 (2002).

Schoenoplectiella raynaliana (U.Scholz) Lye, Lidia 6: 26 (2003).

Togo. 22 TOG. Hemicr. or rhizome geophyte.

**Schoenoplectus raynalianus* U.Scholz, Willdenowia 11: 91 (1981).

Schoenoplectiella reductus (Cherm.) Lye, Lidia 6: 26 (2003).

SW. Madagascar. 29 MDG. Hemicr.

**Scirpus reductus* Cherm., Bull. Soc. Bot. France 68: 423 (1921 publ. 1923). *Schoenoplectus reductus* (Cherm.) J.Raynal, Adansonia, n.s., 16: 151 (1976).

Schoenoplectiella roylei (Nees) Lye, Lidia 6: 26 (2003).

Trop. & S. Africa, Iran to Indian Subcontinent. 22 BKN GHA GNB MLI MTN NGA SEN 23 ZAI 24 CHA SOM 25 KEN TAN UGA 26 ANG 27 BOT NAM 32 KAZ TZK UZB 34 AFG IRN IRQ 40 ASS IND PAK WHM. Ther.

Isolepis lupulina Nees in R.Wight, Contr. Bot. India: 107 (1834). *Scirpus lupulinus* (Nees) Roshev. in V.L.Komarov (ed.), Fl. URSS 3: 53 (1935), nom. illeg. *Schoenoplectus lupulinus* (Nees) V.I.Krecz., in Fl. Uzbekist. 1: 330 (1941).

**Isolepis roylei* Nees in R.Wight, Contr. Bot. India: 107 (1834). *Scirpus roylei* (Nees) R.Parker in J.F.Duthie, Fl. Gangetic Plain 3: 361 (1929). *Schoenoplectus roylei* (Nees) Ovcz. & Czukav., in Fl. Tadzhikskoi SSR 2: 40 (1963).

Scirpus quinquefarius Buch.-Ham. ex Boeck., Linnaea 36: 701 (1870). *Schoenoplectus quinquefarius* (Buch.-Ham. ex Boeck.) Palla, Bot. Jahrb. Syst. 10: 299 (1888).

Scirpus rehmannianus Boeck. ex C.B.Clarke in T.A.Durand & H.Schinz, Consp. Fl. Afric. 5: 629 (1894), nom. inval.

Schoenoplectiella saximontana (Fernald) Lye, Lidia 6: 27 (2003).

W. Canada to WC. & EC. U.S.A., NE. Mexico. 71 BRC 73 COL WYO 74 ILL KAN MSO NEB OKL SDA 75 OHI 76 CAL UTA 77 TEX 79 MXE. Ther. or hemicr.

Scirpus saximontanus Fernald, Rhodora 3: 251 (1901). *Scirpus supinus* var. *saximontanus* (Fernald) T.Koyama, Canad. J. Bot. 40: 920 (1962). *Schoenoplectus supinus* subsp. *saximontanus* (Fernald) Soják, Cas. Nár. Mus., Odd. Prír. 141: 62 (1972). *Schoenoplectus saximontanus* (Fernald) J.Raynal, Adansonia, n.s., 16: 141 (1976).
Scirpus bergsonii Schuyler, Notul. Nat. Acad. Nat. Sci. Philadelphia 423: 6 (1969).

Schoenoplectiella senegalensis (Steud.) Lye, Lidia 6: 27 (2003).
Africa, S. India. 20 EGY 22 BEN BKN GHA GUI IVO MLI MTN NGA NGR SEN TOG 23 RWA ZAI 24 ETH SOM 25 KEN TAN UGA 27 BOT NAM 40 IND. Ther.
**Isolepis senegalensis* Steud., Syn. Pl. Glumac. 2: 96 (1855). *Schoenoplectus senegalensis* (Steud.) Palla, Bot. Jahrb. Syst. 10: 299 (1888).
Scirpus jacobi C.E.C.Fisch., Bull. Misc. Inform. Kew 1931: 103 (1931). *Schoenoplectus jacobi* (C.E.C. Fisch.) Lye, Bot. Not. 124: 290 (1971).

Schoenoplectiella supina (L.) Lye, Lidia 6: 27 (2003).
Europe to Himalaya, Africa, Brazil to NE. Argentina. 11 AUT CZE GER HUN POL SWI 12 FRA SPA 13 BUL GRC ITA ROM TUE YUG 14 KRY RUC RUS UKR 20 ALG EGY 22 MLI NGA SEN TOG 24 SUD 29 MDG 30 (31) 32 KAZ UZB 33 TCS 34 IRN PAL 40 ASS IND PAK WHM 41 VIE 84 85 AGE. Ther.
**Scirpus supinus* L., Sp. Pl.: 49 (1753). *Isolepis supina* (L.) R.Br., Prodr.: 221 (1810). *Schoenoplectus supinus* (L.) Palla, Bot. Jahrb. Syst. 10: 299 (1888). *Cyperus supinus* (L.) Missbach & E.H.L.Krause in J.Sturm, Deutschl. Fl., ed. 2, 2: 20 (1900). *Heleophylax supinus* (L.) Schinz & Thell., Vierteljahrsschr. Naturf. Ges. Zürich 53: 587 (1908).
Scirpus lateralis Forssk., Fl. Aegypt.-Arab.: 15 (1775).
Scirpus halleri Vitman, Summa Pl. 1: 150 (1789).
Isolepis uninodis Delile, Descr. Egypte, Hist. Nat.: 152 (1813). *Scirpus uninodis* (Delile) Coss. & Durieu, Expl. Sci. Algérie 2: 310 (1868).
Scirpus mucronatus Roxb., Fl. Ind. 1: 219 (1820), nom. illeg.
Scirpus melanospermus C.A.Mey., Mém. Sav. Étr. Acad. St. Pétersbourg 1: 199 (1831). *Schoenoplectus melanospermus* (C.A.Mey.) Grossh., Fl. Kavkaza 1: 146 (1928).
Scirpus adscendens Willd. ex Kunth, Enum. Pl. 2: 198 (1837).
Scirpus natans Bojer, Hortus Maurit.: 383 (1837), nom. illeg.
Isolepis polycolea De Not., Index Seminum (G) 1847: s.p. (1847). *Scirpus polycoleus* (De Not.) De Not., Ann. Sci. Nat., Bot., III, 9: 326 (1848).
Isolepis simillima Steud., Syn. Pl. Glumac. 2: 95 (1855).
Scirpus tristachyos Zoll. ex Steud., Syn. Pl. Glumac. 2: 96 (1855), nom. inval.
Isolepis pentasticha Boeck., Flora 42: 446 (1859).
Isolepis striolata Nees ex Boeck., Linnaea 36: 700 (1870).
Scirpus supinus var. *glomeratus* A.Terracc., Malpighia 2: 425 (1889).
Scirpus guaraniticus Pedersen, Bot. Tidsskr. 57: 42 (1961).

Schoenoplectiella vohemarensis (Cherm.) Lye, Lidia 6: 27 (2003).
Madagascar. 29 MDG. Ther. or hemicr.
**Scirpus vohemarensis* Cherm., Bull. Soc. Bot. France 68: 423 (1921 publ. 1922). *Schoenoplectus vohemarensis* (Cherm.) J.Raynal, Adansonia, n.s., 16: 150 (1976).

Schoenoplectiella wallichii (Nees) Lye, Lidia 6: 27 (2003).
Assam to Japan and Philippines. 36 38 JAP KOR TAI 40 ASS 41 MYA VIE 42 MLY PHI. Hemicr. or rhizome geophyte.
**Scirpus wallichii* Nees in R.Wight, Contr. Bot. India: 112 (1834). *Hymenochaeta wallichii* (Nees) Nakai, Bull. Natl. Sci. Mus. Tokyo 31: 134 (1952). *Schoenoplectus wallichii* (Nees) T.Koyama, in Fl. Taiwan 5: 210 (1978).
Scirpus sasakii Hayata, Icon. Pl. Formosan. 6: 115 (1916).
Scirpus erectus var. *angustus* Makino, J. Jap. Bot. 8: 46 (1933).

Schoenoplectus

Schoenoplectus (Rchb.) Palla, Verh. K. K. Zool.-Bot. Ges. Wien 38(Sitzungsber.): 49 (1888), nom. cons.
Cosmopolitan. 10 DEN FIN GRB IRE NOR SWE 11 AUT BGM CZE GER HUN NET POL SWI 12 BAL COR FRA POR SAR SPA 13 ALB BUL GRC ITA KRI ROM SIC TUE YUG 14 BLR BLT KRY RUC RUE RUN RUS RUW UKR 20 ALG EGY LBY MOR TUN 21 AZO CNY MDR 22 BEN BKN GHA GNB GUI MLI NGA NGR SEN SIE 23 BUR CAF CMN GAB GGI RWA ZAI 24 CHA ETH SOM 25 KEN TAN UGA 26 MOZ ZAM ZIM 27 BOT CPP LES NAM NAT OFS SWZ TVL 29 COM MAU MDG REU 30 ALT BRY CTA IRK KRA TVA WSB YAK 31 AMU KAM KHA KUR MAG PRM SAK 32 KAZ KGZ TKM TZK UZB 33 NCS TCS 34 AFG EAI IRN IRQ LBS PAL TUR 35 OMA SAU YEM 36 CHC CHM CHN CHS CHT CHX 37 MON 38 JAP KOR NNS OGA TAI 40 ASS BAN EHM IND NEP PAK SRL WHM 41 NCB THA VIE 42 BOR JAW LSI MLY PHI SUL SUM 43 NWG 50 NFK NSW NTA QLD SOA TAS VIC WAU 51 NZN NZS 60 NWC VAN (61) coo EAS 62 MRN 63 HAW 70 ASK NWT YUK 71 ABT BRC MAN NBR NSC ONT PEI QUE SAS 72 LAB NBR NFL NSC ONT PEI QUE 73 CAL COL IDA MNT ORE WAS WYO 74 ILL IOW KAN MIN MSO NDA NEB OKL SDA WIS 75 CNT INI MAI MAS MIC NWH NWJ NWY OHI PEN RHO VER WVA 76 ARI CAL NEV UTA 77 NWM TEX 78 ALA ARK DEL FLA GEO KTY LOU MRY MSI NCA SCA TEN VRG WDC 79 MXC MXE MXG MXN MXS MXT 80 BLZ COS ELS GUA HON NIC 81 BER CAY CUB DOM JAM LEE PUE 83 BOL CLM ECU PER 84 BZS 85 AGE AGS AGW CLC CLN CLS PAR URU 90 FAL.
39 Species
Heleophylax P.Beauv. ex Lestib., Essai Cypér.: 41 (1819). *Heleogiton* Schult. in J.J.Roemer & J.A. Schultes, Mant. 2: 2 (1824), nom. illeg.
Pterolepis Schrad., Gött. Gel. Anz. 3: 2071 (1821).
Elytrospermum C.A.Mey., Mém. Sav. Étr. Acad. St. Pétersbourg 1: 200 (1830).
Malacochaete Nees, Linnaea 9: 292 (1834).
Pterygolepis Rchb., Deut. Bot. Herb.-Buch: 236 (1841).

Schoenoplectus acutus (Muhl. ex Bigelow) Á.Löve & D.Löve, Bull. Torrey Bot. Club 81: 33 (1954).
Alaska to U.S.A. 70 ASK NWT YUK 71 ABT BRC MAN NBR NSC ONT PEI QUE SAS 73 COL IDA

MNT ORE WAS WYO 74 ILL IOW KAN MIN MSO NDA NEB OKL SDA WIS 75 CNT INI MAI MAS MIC NWH NWJ NWY OHI PEN RHO VER WVA 76 ARI CAL NEV UTA 77 NWM TEX 78 ARK DEL KTY MRY NCA VRG 79 MXN. Hemicr. or rhizome geophyte.

**Scirpus acutus* Muhl. ex Bigelow, Fl. Boston.: 15 (1814). *Schoenoplectus lacustris* subsp. *acutus* (Muhl. ex Bigelow) Á.Löve & D.Löve, Taxon 30: 849 (1981).

var. ***acutus***

Alaska to U.S.A. 70 ASK NWT YUK 71 ABT BRC MAN NBR NSC ONT PEI QUE SAS 73 COL IDA MNT WAS WYO 74 ILL IOW KAN MIN MSO NDA NEB OKL SDA WIS 75 CNT INI MAI MAS MIC NWH NWJ NWY OHI PEN RHO VER WVA 77 TEX 78 ARK DEL KTY MRY NCA VRG. Hemicr. or rhizome geophyte.

Scirpus malheurensis L.F.Hend., Rhodora 32: 20 (1930).

var. ***occidentalis*** (S.Watson) S.G.Sm., Novon 5: 97 (1995).

W. U.S.A. to Texas. 73 COL IDA MNT ORE WAS WYO 76 ARI CAL NEV UTA 77 NWM TEX 79 MXN. Hemicr. or rhizome geophyte.

**Scirpus lacustris* var. *occidentalis* S.Watson in S.Watson & al., Bot. California 2: 218 (1880). *Scirpus occidentalis* (S.Watson) Chase, Rhodora 6: 68 (1904). *Scirpus acutus* var. *occidentalis* (S.Watson) Beetle, Publ. Univ. Wyoming 13: 5 (1948).

Scirpus occidentalis var. *congestus* Farw., Rep. (Annual) Michigan Acad. Sci. 19: 247 (1917). *Scirpus acutus* f. *congestus* (Farw.) Fernald, Rhodora 23: 131 (1921). *Schoenoplectus acutus* f. *congestus* (Farw.) Mohlenbr., Ill. Fl. Illinois, Sedges: Cyperus to Scleria, ed. 2: 201 (2001).

Scirpus rubiginosus Beetle, Amer. J. Bot. 28: 697 (1941). *Schoenoplectus rubiginosus* (Beetle) Soják, Cas. Nár. Mus., Odd. Prír. 140: 127 (1972).

Schoenoplectus americanus (Pers.) Volkart in H.Schinz & R.Keller, Fl. Schweiz, ed. 2: 75 (1905).

New Zealand, N. & C. America, Caribbean, Peru to Argentina. 51 NZN NZS 70 ASK 71 BRC 72 NSC 73 IDA ORE WAS WYO 74 ILL KAN MSO OKL 75 CNT MAI MAS MIC NWH NWJ NWY RHO VER 76 ARI CAL NEV UTA 77 NWM TEX 78 ALA DEL FLA GEO LOU MRY MSI NCA SCA VRG 79 MXC MXE MXG MXN MXS MXT 80 BLZ COS GUA NIC 81 BER CUB DOM JAM LEE PUE 83 BOL PER 85 AGE AGS AGW CLC CLN URU. Hemicr. or rhizome geophyte.

**Scirpus americanus* Pers., Syn. Pl. 1: 68 (1805). *Heleophylax americanus* (Pers.) Schinz & Thell., Vierteljahrsschr. Naturf. Ges. Zürich 53: 587 (1908).

Scirpus triqueter var. *triangularis* Pers., Syn. Pl. 1: 91 (1805). *Scirpus americanus* var. *triangularis* (Pers.) Beetle, Amer. J. Bot. 30: 400 (1943). *Schoenoplectus americanus* subsp. *triangularis* (Pers.) Soó, Acta Bot. Acad. Sci. Hung. 16: 369 (1970 publ. 1971).

Scirpus chilensis Nees & Meyen, Nova Acta Acad. Caes. Leop.-Carol. German. Nat. Cur. 19(Suppl. 1): 93 (1843). *Schoenoplectus chilensis* (Nees & Meyen) Soják, Cas. Nár. Mus., Odd. Prír. 140: 127 (1972).

Scirpus olneyi A.Gray, Boston J. Nat. Hist. 5: 238 (1845). *Schoenoplectus olneyi* (A.Gray) Palla, Bot. Jahrb. Syst. 10: 299 (1888).

Scirpus triqueter Torr., Pacif. Railr. Rep. Parke, Bot. 4: 153 (1857), nom. illeg.

Scirpus novae-zealandiae Colenso, Trans. & Proc. New Zealand Inst. 18: 277 (1885 publ. 1886).

Scirpus solispicatus Lunell, Amer. Midl. Naturalist 4: 230 (1915).

Scirpus olneyi var. *longisetis* Kük., Ark. Bot. 22A(17): 8 (1929). *Scirpus chilensis* var. *longisetis* (Kük.) Beetle, Amer. J. Bot. 33: 663 (1946).

Schoenoplectus annamicus (Raymond) T.Koyama, Brittonia 31: 291 (1979).

Vietnam. 41 VIE. Hemicr.

**Scirpus annamicus* Raymond, Naturaliste Canad. 84: 137 (1957).

Schoenoplectus bucharicus (Roshev.) Grossh., Raspr. Kavk. Odnod. Prish.: 46 (1939).

Caucasus to C. Asia. 32 TZK UZB 33 TCS 34 AFG IRN IRQ. Hemicr. or rhizome geophyte.

**Scirpus bucharicus* Roshev. in V.L.Komarov (ed.), Fl. URSS 3: 579 (1935).

Schoenoplectus californicus (C.A.Mey.) Soják, Cas. Nár. Mus., Odd. Prír. 140: 127 (1972).

Easter I., C. & S. U.S.A. to America, Ecuador to Argentina. (51) nzn (61) coo EAS (63) haw 74 KAN OKL 76 ARI CAL NEV 77 NWM TEX 78 ALA ARK FLA GEO LOU MSI NCA SCA TEN 79 MXC MXE MXG MXN MXS MXT 80 COS ELS GUA HON 83 BOL CLM ECU 84 BZS 85 AGE AGS AGW CLC CLN CLS PAR URU 90 FAL. Hemicr. or rhizome geophyte.

Scirpus riparius J.Presl & C.Presl, Reliq. Haenk. 1: 193 (1828), nom. illeg. *Malacochaete riparia* Nees & Meyen, Linnaea 9: 292 (1834). *Schoenoplectus riparius* (Nees & Meyen) Palla, Bot. Jahrb. Syst. 10: 299 (1888).

**Elytrospermum californicum* C.A.Mey., Mém. Sav. Étr. Acad. St. Pétersbourg 1: 200 (1830). *Scirpus californicus* (C.A.Mey.) Steud., Nomencl. Bot., ed. 2, 2: 538 (1841).

Scirpus decipiens A.St.-Hil., Voy. Distr. Diam. 2: 424 (1833).

Scirpus tatora Kunth, Enum. Pl. 2: 166 (1837). *Malacochaete tatora* (Kunth) Meyen, Nov. Actorum Acad. Caes. Leop.-Carol. Nat. Cur. 19(Suppl. 1): 90 (1843). *Schoenoplectus tatora* (Kunth) Palla, Bot. Jahrb. Syst. 10: 299 (1888). *Scirpus californicus* var. *tatora* (Kunth) Barros, Darwiniana 6: 126 (1942). *Scirpus californicus* subsp. *tatora* (Kunth) T.Koyama, Canad. J. Bot. 41: 1129 (1963).

Scirpus validus Benth., Pl. Hartw.: 27 (1840), nom. illeg.

Malacochaete riparia var. *chamissoi* Nees in C.F.P.von Martius & auct. suc. (eds.), Fl. Bras. 2(1): 107 (1842). *Schoenoplectus chamissoi* (Nees) Soják, Cas. Nár. Mus., Odd. Prír. 141: 62 (1972).

Scirpus chamissoi Schrad. ex Nees in C.F.P.von Martius & auct. suc. (eds.), Fl. Bras. 2(1): 107 (1842).

Scirpus decipiens A.St.-Hil. ex Nees in C.F.P.von Martius & auct. suc. (eds.), Fl. Bras. 2(1): 208 (1842).

Scirpus rigidus Schrad. ex Nees in C.F.P.von Martius & auct. suc. (eds.), Fl. Bras. 2(1): 107 (1842).

Dichromena atrosanguinea É.Desv. in C.Gay, Fl. Chil. 6: 190 (1854). *Scirpus atrosanguineus*

(É.Desv.) Skottsb., Gefässpfl.-Fl. Westpatag.: 6 (1924), nom. illeg.
Scirpus pseudotriqueter Steud., Syn. Pl. Glumac. 2: 86 (1855). *Schoenoplectus pseudotriqueter* (Steud.) Soják, Cas. Nár. Mus., Odd. Prír. 141: 62 (1972).
Scirpus tereticulmis Steud., Syn. Pl. Glumac. 2: 85 (1855). *Scirpus riparius* var. *tereticulmis* (Steud.) C.B.Clarke, Bot. Jahrb. Syst. 30(68): 36 (1901). *Schoenoplectus californicus* subsp. *tereticulmis* (Steud.) Soják, Cas. Nár. Mus., Odd. Prír. 148: 194 (1979 publ. 1980). *Schoenoplectus californicus* var. *tereticulmis* (Steud.) Vegetti, Hickenia 2: 138 (1995).
Malacochaete oligostachya Phil., Linnaea 29: 78 (1857).
Malacochaete chilensis Nees & Meyen ex Boeck., Linnaea 36: 717 (1870).
Malacochaete sanguinolenta Nees & Meyen ex Boeck., Linnaea 36: 718 (1870).
Malacochaete assimilis Ces., Atti Accad. Sci. Fis. 5(7): 17 (1873).
Scirpus paschalis Kük., Monde Pl. 16: 15 (1914), nom. nud.
Scirpus riparius var. *paschalis* Kük., Repert. Spec. Nov. Regni Veg. 16: 432 (1920). *Scirpus californicus* var. *paschalis* (Kük.) Beetle, Amer. Midl. Naturalist 41: 459 (1949).
Scirpus californicus var. *spoliatus* Barros, Lilloa 1: 69 (1937). *Schoenoplectus californicus* var. *spoliatus* (Barros) Vegetti, Hickenia 2: 138 (1995).
Scirpus californicus var. *chamissoi* (Nees ex Mart) Beetle, Amer. Midl. Naturalist 41: 459 (1949).

Schoenoplectus* × *carinatus (Sm.) Palla, Bot. Jahrb. Syst. 10: 299 (1888). *S. lacustris* × *S. triqueter*.
Europe to W. Himalaya. 10 GRB 11 GER NET 13 ITA 32 KAZ KGZ TKM TZK UZB 34 AFG 40 PAK WHM. Hemicr.
**Scirpus* × *carinatus* Sm., Engl. Fl. 1: 60 (1824). *Heleophylax* × *carinatus* (Sm.) Schinz & Thell., Vierteljahrsschr. Naturf. Ges. Zürich 53: 587 (1908). *Heleogiton* × *trigonum* (Sm.) Rchb. in J.C. Mössler, Handb. Gewächsk. ed. 2, 3: 1807 (1830).

nothosubsp. ***carinatus***. *S. lacustris* × *S. triqueter*.
Europe. 10 GRB 11 GER NET 13 ITA. Hemicr.
Scirpus × *duvalii* Hoppe in J.Sturm, Deutschl. Fl. 36: (1814), provisional synonym.
Scirpus × *rosellinii* Ces., Pass. & Gibelli, Comp. Fl. Ital.: 99 (1884).
Scirpus × *kalmusii* Asch., Abrom. & Graebn. in P.F.A. Ascherson & K.O.R.Graebner, Fl. Nordostdeut. Flachl.: 133 (1898). *Schoenoplectus* × *kalmussii* (Asch., Abrom. & Graebn.) Palla, Allg. Bot. Z. Syst. 6: 215 (1900).

nothosubsp. ***triquetriformis*** (V.I.Krecz., Ovcz. & Czukav.) Govaerts in R.H.A.Govaerts & D.A.Simpson, World Checklist of Cyperaceae: 662 (2007). *S. lacustris* subsp. *hyppoliti* × *S. triqueter*.
C. Asia to W. Himalaya. 32 KAZ KGZ TKM TZK UZB 34 AFG 40 PAK WHM. Hemicr. or rhizome geophyte.
**Schoenoplectus* × *triquetriformis* V.I.Krecz., Ovcz. & Czukav., in Fl. Tadzhikskoi SSR 2: 423 (1963). *Scirpus triquetriformis* (V.I.Krecz.) T.V.Egorova, Opred. Rast. Sred. Azii 5: 15 (1976).

Schoenoplectus confusus (N.E.Br.) Lye, Bot. Not. 124: 290 (1971).
Ethiopia to S. Africa. 24 ETH 25 KEN TAN UGA 26 ZIM 27 NAT. Rhizome geophyte.
**Scirpus confusus* N.E.Br., Bull. Misc. Inform. Kew 1921: 300 (1921).

subsp. ***confusus***
Ethiopia to E. Trop. Africa. 24 ETH 25 KEN TAN UGA. Rhizome geophyte.

subsp. ***natalitius*** Browning, S. African J. Bot. 57: 258 (1991).
KwaZulu-Natal. 27 NAT. Rhizome geophyte.

var. ***rogersii*** (N.E.Br.) Lye, Nordic J. Bot. 3: 242 (1983).
E. & S. Trop. Africa. 25 KEN TAN 26 ZIM. Rhizome geophyte.
**Scirpus rogersii* N.E.Br., Bull. Misc. Inform. Kew 1921: 301 (1921). *Schoenoplectus rogersii* (N.E.Br.) Lye, Bot. Not. 124: 290 (1971).

Schoenoplectus* × *contortus (Eames) S.G.Sm., Novon 5: 101 (1995). *S. americanus* × *S. pungens*.
U.S.A. 75 CNT MAS MIC NWJ 76 CAL 77 NWM 78 DEL FLA VRG. Hemicr. or rhizome geophyte.
**Scirpus olneyi* var. *contortus* Eames, Rhodora 9: 220 (1907). *Scirpus* × *contortus* (Eames) T.Koyama, Canad. J. Bot. 41: 1123 (1963).

Schoenoplectus corymbosus (Roth ex Roem. & Schult.) J.Raynal in B.Peyre de Fabregues & J.P.Lebrun, Cat. Pl. Vasc. Niger: 343 (1976).
Africa to Indian Subcontinent. 20 ALG EGY 22 BKN MLI NGA NGR 23 CMN GGI RWA ZAI 24 CHA ETH SOM 25 KEN TAN UGA 27 BOT CPP NAM NAT OFS SWZ TVL 29 MDG 35 SAU YEM 40 IND PAK. Hel.
Isolepis inclinata Delile, Descr. Egypte, Hist. Nat.: 50 (1813). *Scirpus inclinatus* (Delile) Asch. & Schweinf. ex Boiss., Fl. Orient. 5: 381 (1882). *Schoenoplectus inclinatus* (Delile) Lye, Bot. Not. 124: 290 (1971).
**Isolepis corymbosa* Roth ex Roem. & Schult., Syst. Veg. 2: 110 (1817). *Scirpus corymbosus* (Roth ex Roem. & Schult.) B.Heyne in A.W.Roth, Nov. Pl. Sp.: 28 (1821). *Malacochaete corymbosa* (Roth ex Roem. & Schult.) Arn. in R.Wight, Cat. Ind. Pl.: 143 (1837).
Scirpus brachyceras Hochst. ex A.Rich., Tent. Fl. Abyss. 2: 496 (1850). *Schoenoplectus brachyceras* (Hochst. ex A.Rich.) Lye, Bot. Not. 124: 290 (1971). *Schoenoplectus corymbosus* var. *brachyceras* (Hochst. ex A.Rich.) Lye, Nordic J. Bot. 3: 242 (1983).
Scirpus madagascariensis Boeck., Abh. Naturwiss. Vereine Bremen 7: 37 (1880).
Scirpus lyallii Baker, J. Linn. Soc., Bot. 20: 297 (1883).
Malacochaete aemulans Arn. ex C.B.Clarke in T.A.Durand & H.Schinz, Consp. Fl. Afric. 5: 620 (1894), nom. inval.

Schoenoplectus decipiens (Nees) J.Raynal, Adansonia, n.s., 15: 540 (1976).
S. Africa. 27 BOT CPP LES NAT OFS TVL. Hemicr. or rhizome geophyte.
**Isolepis decipiens* Nees, Linnaea 10: 157 (1836).

Schoenoplectus deltarum (Schuyler) Soják, Cas. Nár. Mus., Odd. Prír. 141: 62 (1972).
C. & SE. U.S.A. 74 KAN MSO 77 TEX 78 ALA FLA LOU MSI. Hemicr. or rhizome geophyte.
**Scirpus deltarum* Schuyler, Notul. Nat. Acad. Nat. Sci. Philadelphia 427: 1 (1970).

Schoenoplectus ehrenbergii (Boeck.) Soják, Cas. Nár. Mus., Odd. Prír. 140: 127 (1972).
E. European Russia to N. China. 14 RUE 32 KAZ 36 CHN CHX. Hemicr.
**Scirpus ehrenbergii* Boeck., Linnaea 36: 712 (1870).
Scirpus ewersmannii Fisch. ex Litw., Trudy Glavn. Bot. Sada 40: 271 (1929), nom. inval.

Schoenoplectus etuberculatus (Steud.) Soják, Cas. Nár. Mus., Odd. Prír. 140: 127 (1972).
Rhode Island to Texas. 74 MSO 75 RHO 77 TEX 78 ALA DEL FLA GEO LOU MRY MSI NCA SCA VRG. Hemicr. or rhizome geophyte.
**Rhynchospora etuberculata* Steud., Syn. Pl. Glumac. 2: 142 (1855). *Scirpus etuberculatus* (Steud.) Kuntze, Revis. Gen. Pl. 2: 758 (1891).
Scirpus macranthus Boeck., Flora 41: 415 (1858).
Scirpus leptolepis Chapm., Fl. South. U.S.: 520 (1860).
Scirpus canbyi A.Gray, Proc. Acad. Nat. Sci. Philadelphia 1864: 18 (1864).
Scirpus erythropodus Kuntze, Revis. Gen. Pl. 2: 757 (1891).

Schoenoplectus fuscorubens (T.Koyama) T.Koyama, Enum. Fl. Pl. Nepal 1: 118 (1978).
Nepal to SC. China. 36 CHC 40 EHM NEP. Hemicr. or rhizome geophyte.
**Scirpus fuscorubens* T.Koyama, Willdenowia 5: 491 (1969).

Schoenoplectus heterochaetus (Chase) Soják, Cas. Nár. Mus., Odd. Prír. 140: 127 (1927).
Canada to N. & NC. U.S.A. 71 ABT MAN 72 ONT QUE 73 IDA MNT ORE WYO 74 ILL IOW KAN MIN MSO NDA NEB OKL SDA WIS 75 MAS MIC NWY PEN VER 76 CAL. Hemicr. or rhizome geophyte.
Scirpus lacustris var. *tenuiculmis* E.Sheld., Bull. Geol. Nat. Hist. Surv. 9: 583 (1896). *Schoenoplectus lacustris* subsp. *tenuiculmis* (E.Sheld.) Soják, Cas. Nár. Mus., Odd. Prír. 141: 62 (1972).
**Scirpus heterochaetus* Chase, Rhodora 6: 70 (1904).

Schoenoplectus hondoensis (Ohwi) Soják, Cas. Nár. Mus., Odd. Prír. 141: 62 (1971).
Japan (N. & C. Honshu). 38 JAP. Hemicr. or rhizome geophyte.
**Scirpus hondoensis* Ohwi, Bot. Mag. (Tokyo) 45: 189 (1931).

Schoenoplectus hotarui (Ohwi) Holub, Folia Geobot. Phytotax. 11: 83 (1976).
S. Russian Far East to Japan. 31 AMU PRM 38 JAP. Hemicr. or rhizome geophyte.
**Scirpus hotarui* Ohwi, Repert. Spec. Nov. Regni Veg. 36: 44 (1934). *Scirpus juncoides* var. *hotarui* (Ohwi) Ohwi, Mem. Coll. Sci. Kyoto Imp. Univ., Ser. B, Biol. 18(1): 114 (1944). *Schoenoplectus juncoides* subsp. *hotarui* (Ohwi) Soják, Cas. Nár. Mus., Odd. Prír. 141: 62 (1972).

Schoenoplectus komarovii (Roshev.) Soják, Cas. Nár. Mus., Odd. Prír. 140: 127 (1972).
Russian Far East to Japan. 31 AMU KHA PRM 36 CHM 38 JAP KOR. Ther.
Scirpus supinus var. *leiocarpus* Kom., Fl. Manshur. 1: 345 (1901). *Schoenoplectus smithii* subsp. *leiocarpus* (Kom.) Soják, Cas. Nár. Mus., Odd. Prír. 141: 62 (1972).
**Scirpus komarovii* Roshev. in V.L.Komarov (ed.), Fl. URSS 3: 579 (1935).
Scirpus okuyamae Ohwi, Acta Phytotax. Geobot. 4: 64 (1935). *Hymenochaeta okuyamae* (Ohwi) Nakai, Bull. Natl. Sci. Mus. Tokyo 31: 133 (1952).

Schoenoplectus* × *kuekenthalianus (Junge) D.H.Kent, Watsonia 18: 213 (1990). *S. tabernaemontana* × *C. triqueter.*
E. Canada to N. U.S.A. 72 ONT QUE 73 ORE 74 MIN 75 NWY VER. Hemicr. or rhizome geophyte.
**Scirpus* × *kuekenthalianus* Junge, Jahrb. Hamburg. Wiss. Anst. 22(Beih.): 73 (1905). *Schoenoplectus* × *carinatus* nothosubsp. *kuekenthalianus* (Junge) P.D.Sell in P.D.Sell & G.Murrell, Fl. Great Britain Ireland 5: 363 (1996).
Scirpus × *arunensis* Druce, Bot. Soc. Exch. Club Brit. Isles 7: 68 (1923 publ. 1924), minimal descr.

Schoenoplectus lacustris (L.) Palla, Bot. Jahrb. Syst. 10: 299 (1888).
Europe, Medit. to Mongolia, S. Africa. 10 DEN FIN GRB IRE NOR SWE 11 AUT BGM CZE GER HUN NET POL SWI 12 BAL COR FRA POR SAR SPA 13 ALB BUL GRC ITA KRI ROM SIC TUE YUG 14 BLR BLT KRY RUC RUE RUN RUS RUW UKR 20 ALG LBY MOR TUN 27 CPP LES OFS TVL 30 ALT BRY CTA IRK KRA TVA WSB YAK 31 AMU KHA PRM SAK 32 KAZ KGZ TKM TZK UZB 33 NCS TCS 34 AFG IRN PAL TUR 35 YEM 37 MON 40 IND WHM. Hemicr.
**Scirpus lacustris* L., Sp. Pl.: 48 (1753). *Eleogiton lacustris* (L.) Fourr., Ann. Soc. Linn. Lyon, n.s., 17: 173 (1869). *Cyperus lacustris* (L.) Missbach & E.H.L.Krause in J.Sturm, Deutschl. Fl., ed. 2, 2: 30 (1900), nom. illeg. *Heleophylax lacustris* (L.) Schinz & Thell., Vierteljahrsschr. Naturf. Ges. Zürich 53: 587 (1908). *Hymenochaeta lacustris* (L.) Nakai, Bull. Natl. Sci. Mus. Tokyo 31: 133 (1952). *Schoenus lacustris* (L.) Bernh., Syst. Verz.: 26 (1800).

subsp. ***hippolyti*** (V.I.Krecz.) Kukkonen, Ann. Naturhist. Mus. Wien, B 98B(Suppl.): 89 (1996).
Caucasus to Mongolia. 14 RUC RUS 30 ALT BRY CTA KRA TVA WSB 31 AMU KHA PRM SAK 32 KAZ KGZ TKM TZK UZB 33 NCS TCS 34 AFG IRN 37 MON 40 IND WHM. Hemicr. or rhizome geophyte.
**Scirpus hippolyti* V.I.Krecz., Bot. Mater. Gerb. Bot. Inst. Komarova Akad. Nauk S.S.S.R. 7: 28 (1937). *Schoenoplectus hippolyti* (V.I.Krecz.) V.I.Krecz. ex Grossh., Fl. Kavkaza, ed. 2, 2: 16 (1940).
Schoenoplectus grossheimii Pobed., Bot. Mater. Gerb. Bot. Inst. Komarova Akad. Nauk S.S.S.R. 12: 17 (1950). *Scirpus grossheimii* (Pobed.) Czerep., Sosud. Rast. SSSR: 207 (1981).

subsp. ***lacustris***
Europe, Medit. to Afghanistan, S. Africa. 10 DEN FIN GRB IRE NOR SWE 11 AUT BGM CZE GER HUN NET POL SWI 12 BAL COR FRA POR SAR SPA 13 ALB BUL GRC ITA KRI ROM SIC TUE YUG 14 BLR BLT KRY RUC RUE RUN RUS RUW UKR 20 ALG LBY MOR TUN 27 CPP LES OFS TVL 30 IRK KRA WSB YAK 32 KAZ KGZ TZK UZB 33 NCS TCS 34 AFG IRN PAL TUR 35 YEM. Hemicr.
Scirpus lacustris var. *carinatus* Nakai in ?
Scirpus lacustris var. *medius* Schrad. ex Lej., Rev. Fl. Spa: 11 (1824).
Cyperus americanus Garsault, Fig. Pl. Méd.: t. 13c (1764), opus utique oppr.

Scirpus holoschoenus Oeder, Fl. Dan.: t. 454 (1769).
Scirpus altissimus Gilib., Excerc. Phyt. 2: 514 (1792).
Scirpus turgidus Thuill., Fl. Env. Paris, ed. 2: 22 (1799).
Scirpus lacustris f. *minor* Baumg., Enum. Stirp. Transsilv. 3: 320 (1815). *Schoenoplectus lacustris* f. *minor* (Baumg.) Soó, Acta Bot. Acad. Sci. Hung. 16: 369 (1970 publ. 1971).
Scirpus brayi Hoppe ex Roem. & Schult., Syst. Veg. 2: 137 (1817).
Scirpus orgyalis Raf., Ann. Nat. 1: 16 (1820).
Scirpus medius Gray, Nat. Arr. Brit. Pl. 2: 75 (1821).
Scirpus lacustris f. *palustris* Schltdl., Fl. Berol. 1: (1823). *Schoenoplectus lacustris* f. *palustris* (Schltdl.) Soó, Acta Bot. Acad. Sci. Hung. 16: 368 (1970 publ. 1971), without basionym page.
Scirpus andrzejowscii Besser ex Schult. in J.J.Roemer & J.A.Schultes, Mant. 2: 536 (1824).
Scirpus janii Besser ex Schult. in J.J.Roemer & J.A.Schultes, Mant. 2: 536 (1824).
Scirpus lithuanicus Besser ex Schult. in J.J.Roemer & J.A.Schultes, Mant. 2: 535 (1824).
Scirpus macrophyllus Besser ex Schult. in J.J.Roemer & J.A.Schultes, Mant. 2: 535 (1824).
Scirpus wolfgangii Besser ex Schult. in J.J.Roemer & J.A.Schultes, Mant. 2: 536 (1824).
Scirpus trigonus Nolte, Novit. Fl. Holsat.: 9 (1826).
Heleogiton glaucum Rchb. in J.C.Mössler, Handb. Gewächsk. ed. 2, 3: 1807 (1830).
Scirpus barbatulus Kunth, Enum. Pl. 2: 583 (1837).
Scirpus barbulatus Kunth, Enum. Pl. 2: 165 (1837).
Scirpus pungens Willd. ex Kunth, Enum. Pl. 2: 164 (1837), nom. illeg.
Scirpus custoris Hegetschw, Fl. Schweiz: 49 (1839). *Scirpus lacustris* var. *custoris* (Hegetschw) Nyman, Consp. Fl. Eur.: 764 (1882).
Scirpus meyenii Nees in J.G.C.Lehmann, Pl. Preiss. 2: 75 (1846).
Scirpus ciliatus Steud., Syn. Pl. Glumac. 2: 96 (1855).
Scirpus globifer Welw. ex Steud., Syn. Pl. Glumac. 2: 87 (1855).
Scirpus lacustris f. *foliosus* Des Moul., Cat. Rais. Phan. Dordogne, Suppl. Fin.: 306 (1859). *Schoenoplectus lacustris* f. *foliosus* (Des Moul.) Soó, Acta Bot. Acad. Sci. Hung. 16: 368 (1970 publ. 1971).
Scirpus lacustris f. *confertus* Schur, Enum. Pl. Transsilv.: 692 (1866). *Schoenoplectus lacustris* f. *confertus* (Schur) Soó, Acta Bot. Acad. Sci. Hung. 16: 369 (1970 publ. 1971).
Scirpus lacustris f. *laxiflorus* Kuntze, Taschen-Fl. Leipzig: 25 (1867). *Schoenoplectus lacustris* f. *laxiflorus* (Kuntze) Soó, Acta Bot. Acad. Sci. Hung. 16: 369 (1970 publ. 1971).
Eleogiton duvalii Fourr., Ann. Soc. Linn. Lyon, n.s., 17: 173 (1869).
Scirpus jardinei Steud. & Jard., Bull. Soc. Linn. Normandie, II, 9: 280 (1875).
Scirpus lacustris f. *compactus* Beckh., Fl. Westfalen: 936 (1893). *Schoenoplectus lacustris* f. *compactus* (Beckh.) Soó, Acta Bot. Acad. Sci. Hung. 16: 369 (1970 publ. 1971).
Scirpus lacustris var. *condensatus* Peck, Rep. (Annual) New York State Mus. Nat. Hist. 53: 853 (1900). *Scirpus validus* var. *condensatus* (Peck) Beetle, Publ. Univ. Wyoming 13: 6 (1948).
Scirpus lacustris f. *conglomeratus* Junge, Jahrb. Hamburg. Wiss. Anst. 25(3): 255 (1908). *Schoenoplectus lacustris* f. *conglomeratus* (Junge) Soó, Acta Bot. Acad. Sci. Hung. 16: 369 (1970 publ. 1971).
Scirpus drepanensis Lojac., Fl. Sicul. 3: 215 (1909).
Scirpus lacustris f. *radiatus* Vollm., Fl. Bayern: 125 (1914). *Schoenoplectus lacustris* f. *radiatus* (Vollm.) Soó, Acta Bot. Acad. Sci. Hung. 16: 368 (1970 publ. 1971).
Scirpus lacustris f. *terrestris* Glück, Beih. Bot. Centralbl. 39B: 313 (1922). *Schoenoplectus lacustris* f. *terrestris* (Glück) Soó, Acta Bot. Acad. Sci. Hung. 16: 368 (1970 publ. 1971), without basionym date.
Scirpus validus var. *creber* Fernald, Rhodora 45: 283 (1943). *Scirpus validus* f. *creber* (Fernald) Beetle, Publ. Univ. Wyoming 13: 6 (1948). *Schoenoplectus validus* subsp. *creber* (Fernald) Á.Löve & D.Löve, Bull. Torrey Bot. Club 81: 33 (1954). *Scirpus lacustris* subsp. *creber* (Fernald) T.Koyama, Phytologia 8: 457 (1963). *Schoenoplectus lacustris* subsp. *creber* (Fernald) Á.Löve & D.Löve, Taxon 30: 849 (1981).
Scirpus validus f. *megastachyus* Fernald, Rhodora 45: 283 (1943).
Hymenochaeta makinoi Nakai, Bull. Natl. Sci. Mus. Tokyo 31: 133 (1952).
Scirpus validus f. *dutillyanus* Lepage, Naturaliste Canad. 86: 68 (1959).
Scirpus lacustris f. *sparganiifolia* Volobaev, Izv. Sibirsk. Otd. Akad. Nauk SSSR, Ser. Biol. Nauk 1990(2): 82 (1990).

Schoenoplectus lineolatus (Franch. & Sav.) T.Koyama, in Fl. Taiwan 5: 215 (1978).
Russian Far East to Temp. E. Asia. 31 PRM 38 JAP NNS TAI (50) nsw vic. Hemicr. or rhizome geophyte.
**Scirpus lineolatus* Franch. & Sav., Enum. Pl. Jap. 2: 545 (1878).
Scirpus forsythii Kük., Repert. Spec. Nov. Regni Veg. 12: 94 (1913).

Schoenoplectus litoralis (Schrad.) Palla, Bot. Jahrb. Syst. 10: 299 (1888).
Africa, Europe to Himalaya, Australia. 11 AUT? HUN 12 COR SPA 13 ALB BUL GRC ITA KRI ROM SIC TUE YUG 14 KRY UKR 20 ALG EGY LBY MOR TUN 22 BEN GHA GNB GUI NGR SEN SIE 23 BUR GAB RWA ZAI 24 CHA SOM 25 KEN TAN UGA 27 BOT CPP NAM NAT 29 COM MDG 32 KAZ KGZ TKM TZK UZB 33 NCS TCS 34 AFG EAI IRN IRQ LBS PAL TUR 35 OMA SAU YEM 36 37 MON 40 ASS IND PAK SRL WHM 41 NCB THA VIE 42 JAW LSI PHI 43 NWG 50 NSW NTA QLD SOA WAU 60 NWC 62 MRN. Hemicr. or rhizome geophyte.
**Scirpus litoralis* Schrad., Fl. Germ. 1: 142 (1806). *Heleogiton litorale* (Schrad.) Rchb. in J.C.Mössler, Handb. Gewächsk. ed. 2, 3: 1808 (1830). *Malacochaete litoralis* (Schrad.) Nees, Linnaea 9: 292 (1834). *Eleogiton litoralis* (Schrad.) Fourr., Ann. Soc. Linn. Lyon, n.s., 17: 173 (1869).

subsp. ***kasachstanicus*** (Dobrochot.) Soják, Cas. Nár. Mus., Odd. Přír. 148: 194 (1979 publ. 1980).
C. Asia. 32 KAZ UZB. Hemicr. or rhizome geophyte.

Scirpus kasachstanicus Dobrochot., Bot. Mater. Gerb. Bot. Inst. Komarova Akad. Nauk S.S.S.R. 12: 61 (1950).

subsp. ***litoralis***

Africa, Europe to Himalaya, Australia. 11 AUT? HUN 12 COR SPA 13 ALB BUL GRC ITA KRI ROM SIC TUE YUG 14 KRY UKR 20 ALG LBY MOR TUN 22 BEN GHA GNB GUI SEN SIE 23 BUR GAB RWA ZAI 24 CHA SOM 25 KEN TAN UGA 27 BOT CPP NAM NAT 29 COM MDG 32 KAZ KGZ TKM TZK UZB 33 NCS TCS 34 AFG EAI IRN IRQ LBS PAL TUR 35 OMA SAU YEM 36 37 MON 40 ASS IND PAK WHM 41 NCB THA VIE 42 JAW LSI PHI 43 NWG 50 NSW NTA QLD SOA WAU 60 NWC 62 MRN. Hemicr. or rhizome geophyte.

Scirpus plumosus R.Br., Prodr.: 223 (1810).

Scirpus fimbrisetus Delile, Descr. Egypte, Hist. Nat.: 155 (1813).

Scirpus philippii Tineo, Pl. Rar. Sicil.: 36 (1817). *Scirpus litoralis* var. *philippii* (Tineo) Nyman, Consp. Fl. Eur.: 764 (1882). *Schoenoplectus philippii* (Tineo) Pignatti, Arch. Bot. (Forlì) 34: 8 (1958).

Scirpus aegyptiacus Decne., Ann. Sci. Nat., Bot., II, 4: 196 (1835).

Scirpus balearicus Willd. ex Kunth, Enum. Pl. 2: 166 (1837).

Scirpus desprauxii Steud., Syn. Pl. Glumac. 2: 86 (1855).

Scirpus triqueter Gren. & Godr., Fl. France 3: 373 (1855), nom. illeg.

Malacochaete lehmannii Nees ex Boeck., Linnaea 36: 617 (1870).

Scirpus arabicus Ehrenb. ex Boeck., Linnaea 36: 716 (1870).

Scirpus tabernaemontani Salzm. ex Ball, J. Linn. Soc., Bot. 16: 702 (1878), nom. inval.

Scirpus arrhenopogon Hochst. ex C.B.Clarke in T.A.Durand & H.Schinz, Consp. Fl. Afric. 5: 625 (1894), nom. inval.

subsp. ***thermalis*** (Trab.) S.S.Hooper in C.J.Saldanha & D.H. Nicolson, Fl. Hassan Distr. Karnataka: 698 (1976).

Africa to W. Pacific. 20 ALG EGY MOR 22 NGR SEN 23 BUR GAB RWA ZAI 24 CHA SOM 25 KEN TAN UGA 29 COM MDG 34 AFG IRN 35 38 NNS 40 IND PAK SRL 41 NCB THA VIE 42 JAW LSI PHI 43 NWG 60 NWC 62 MRN. Hel.

Scirpus subulatus Vahl, Enum. Pl. 2: 268 (1805). *Scirpus litoralis* var. *subulatus* (Vahl) Chiov., Pl. Nov. Aethiop.: 16 (1928). *Schoenoplectus subulatus* (Vahl) Lye, Bot. Not. 124: 290 (1971). *Schoenoplectus litoralis* subsp. *subulatus* (Vahl) Soják, Cas. Nár. Mus., Odd. Prír. 141: 62 (1972). *Scirpus litoralis* subsp. *subulatus* (Vahl) M.A.Wadood Khan, Rheedea 10: 28 (2000).

Scirpus pectinatus Roxb., Fl. Ind. 1: 220 (1820). *Malacochaete pectinata* (Roxb.) Nees, Linnaea 9: 292 (1834).

Pterolepis scirpoides Schrad., Gött. Gel. Anz. 3: 2071 (1821). *Schoenoplectus scirpoides* (Schrad.) Browning, S. African J. Bot. 60: 172 (1994).

Malacochaete pterolepis Nees, Linnaea 10: 184 (1836). *Scirpus pterolepis* (Nees) Kunth, Enum. Pl. 2: 166 (1837). *Scirpus litoralis* var. *pterolepis* (Nees) C.B.Clarke in T.A.Durand & H.Schinz, Consp. Fl. Afric. 5: 625 (1894).

Scirpus thermalis Trab. in J.A.Battandier & al., Fl. Algérie, Monocot.: 99 (1895). *Scirpus litoralis* subsp. *thermalis* (Trab.) Murb., Acta Univ. Lund., n.s., 2(18) 3: 18 (1922).

Scirpus wardianus J.R.Drumm., J. Asiat. Soc. Bengal, Pt. 2, Nat. Hist. 73: 147 (1904).

Schoenoplectus mucronatus (L.) Palla, Bot. Jahrb. Syst. 10: 299 (1888).

Old World to Australia. 11 AUT CZE GER HUN POL SWI 12 COR FRA POR SPA 13 BUL GRC ITA ROM SIC YUG 14 UKR 20 EGY 21 AZO 22 GNB GUI NGA 23 CAF CMN GAB 25 TAN UGA 26 ZAM 29 MAU MDG REU 31 PRM 32 KAZ KGZ TZK UZB 33 NCS TCS 34 AFG IRN IRQ TUR 35 YEM 37 MON 38 JAP NNS TAI 40 ASS BAN EHM IND NEP PAK SRL 41 THA VIE 42 BOR JAW LSI MLY PHI SUL SUM 43 NWG 50 NSW NTA QLD WAU (73) ore (74) ill iow mso (75) nwj (76) cal (78) kty ten. Hel. or rhizome geophyte.

Scirpus mucronatus L., Sp. Pl.: 50 (1753). *Isolepis mucronata* (L.) Fourr., Ann. Soc. Linn. Lyon, n.s., 17: 173 (1869). *Heleophylax mucronatus* (L.) Schinz & Thell., Vierteljahrsschr. Naturf. Ges. Zürich 53: 587 (1908).

subsp. ***clemensii*** (Kük.) Soják, Cas. Nár. Mus., Odd. Prír. 148: 194 (1979 publ. 1980).

New Guinea. 43 NWG. Hemicr. or rhizome geophyte.

Scirpus mucronatus subsp. *clemensii* Kük., Bot. Jahrb. Syst. 69: 259 (1938). *Scirpus clemensii* (Kük.) Ohwi, Bot. Mag. (Tokyo) 56: 203 (1942).

subsp. ***mucronatus***

Old World to Australia. 11 AUT CZE GER HUN POL SWI 12 COR FRA POR SPA 13 BUL GRC ITA ROM SIC YUG 14 UKR 20 EGY 21 AZO 22 GNB GUI NGA 23 CAF CMN GAB 25 TAN UGA 26 ZAM 29 MAU MDG REU 31 PRM 32 KAZ KGZ TZK UZB 33 NCS TCS 34 AFG IRN IRQ TUR 35 YEM 37 MON 38 JAP NNS TAI 40 ASS BAN EHM IND NEP PAK SRL 41 THA VIE 42 BOR JAW LSI MLY PHI SUL SUM 43 NWG 50 NSW NTA QLD WAU (73) ore (74) ill iow mso (75) nwj (76) cal (78) kty ten. Hel. or rhizome geophyte.

Scirpus glomeratus Scop., Fl. Carniol., ed. 2, 1: 47 (1771), nom. illeg.

Scirpus tricarinatus Pers., Syn. Pl. 1: 68 (1805).

Scirpus triangulatus Roxb., Fl. Ind. 1: 219 (1820). *Eleocharis triangulata* (Roxb.) Sieber ex C.Presl, Isis (Oken) 21: 268 (1828). *Schoenoplectus triangulatus* (Roxb.) Soják, Cas. Nár. Mus., Odd. Prír. 141: 62 (1972).

Scirpus muticus D.Don, Prodr. Fl. Nepal.: 41 (1825).

Scirpus acutus J.Presl & C.Presl, Reliq. Haenk. 1: 192 (1828), nom. illeg. *Scirpus preslii* A.Dietr., Sp. Pl. 2: 175 (1832). *Hymenochaeta preslii* (A.Dietr.) Nakai, Bull. Natl. Sci. Mus. Tokyo 31: 133 (1952).

Scirpus mysurensis B.Heyne ex Wall., Numer. List: 3467 (1831).

Scirpus javanus Nees in R.Wight, Contr. Bot. India: 112 (1834). *Schoenoplectus javanus* (Nees) Palla, Bot. Jahrb. Syst. 10: 299 (1888).

Scirpus mauritanicus Steud., Syn. Pl. Glumac. 2: 84 (1855).

Scirpus sundanus Miq., Fl. Ned. Ind. 3: 304 (1856). *Scirpus cognatus* Hance, Ann. Sci. Nat., Bot., IV, 15: 228 (1861).
Scirpus mucronatus var. *robustus* Miq., Ann. Mus. Bot. Lugduno-Batavi 2: 143 (1865). *Schoenoplectus mucronatus* subsp. *robustus* (Miq.) T.Koyama, in Fl. Taiwan 5: 214 (1978). *Schoenoplectus mucronatus* f. *robustus* (Miq.) Hayas. & H.Ohashi, J. Jap. Bot. 75: 319 (2000).
Scirpus abactus Ohwi, Bot. Mag. (Tokyo) 45: 186 (1931).
Scirpus mucronatus var. *tataranus* Honda, Bot. Mag. (Tokyo) 46: 373 (1932). *Schoenoplectus mucronatus* var. *tataranus* (Honda) Kohno, Iokawa & Daigobo, J. Jap. Bot. 76: 229 (2001).
Scirpus triangulatus var. *brevibracteatus* T.Koyama, J. Fac. Sci. Univ. Tokyo, Sect. 3, Bot. 7: 320 (1958).
Scirpus triangulatus var. *sanguineus* Tang & F.T.Wang, in Fl. Reipubl. Popul. Sin. 11: 222 (1961). *Schoenoplectus mucronatus* var. *sanguineus* (Tang & F.T.Wang) P.C.Li, Vasc. Pl. Hengduan Mount. 2: 2323 (1994).
Schoenoplectus mucronatus f. *hosoiri* Hayas. & H.Ohashi, J. Jap. Bot. 75: 319 (2000).
Schoenoplectus mucronatus var. *ishizawae* Kohno, Iokawa & Daigobo, J. Jap. Bot. 76: 228 (2001).
Schoenoplectus mucronatus f. *brevisetaceus* Horiuchi, J. Jap. Bot. 78: 255 (2003).
Schoenoplectus mucronatus var. *antrorsispinulosus* Iokawa, Kohno & Daigobo, J. Jap. Bot. 79: 1 (2004).

Schoenoplectus multisetus Hayas. & C.Sato, J. Jap. Bot. 79: 322 (2004).
Japan (Honshu, Kyushu). 38 JAP. Hemicr. or rhizome geophyte.

Schoenoplectus muricinux (C.B.Clarke) J.Raynal, Adansonia, n.s., 15: 538 (1976).
Rwanda to S. Africa. 23 RWA 26 ZAM ZIM 27 BOT CPP LES NAM NAT OFS SWZ TVL. Hemicr.
**Scirpus muricinux* C.B.Clarke, Bot. Jahrb. Syst. 38: 135 (1906).

Schoenoplectus muriculatus (Kük.) Browning, S. African J. Bot. 57: 254 (1991).
S. Trop. & S. Africa. 26 ZIM 27 CPP LES NAT SWZ TVL. Hemicr. or rhizome geophyte.
**Scirpus muriculatus* Kük., Bot. Not. 1934: 75 (1934).

Schoenoplectus nipponicus (Makino) Soják, Cas. Nár. Mus., Odd. Přír. 140: 127 (1972).
Russian Far East to NE. China, Korea (Cheju-do), Japan. 31 AMU KHA PRM 36 CHM 38 JAP KOR. Hemicr. or rhizome geophyte.
**Scirpus nipponicus* Makino, Bot. Mag. (Tokyo) 9: 311 (1895).
Scirpus depauperatus Kom., Trudy Imp. S.-Peterburgsk. Bot. Sada 20: 345 (1903), nom. illeg.

Schoenoplectus novae-angliae (Britton) M.T.Strong, Novon 3: 203 (1993).
E. U.S.A. 75 CNT MAI MAS NWH NWJ NWY RHO 78 DEL GEO MRY NCA VRG. Hemicr. or rhizome geophyte.
Scirpus maritimus var. *cylindricus* Torr., Ann. Lyceum Nat. Hist. New York 3: 325 (1836). *Scirpus cylindricus* (Torr.) Britton, Trans. New York Acad. Sci. 11: 79 (1892).
**Scirpus novae-angliae* Britton in N.L.Britton & A.Brown, Ill. Fl. N. U.S. 3: 509 (1898). *Scirpus campestris* var. *novae-angliae* (Britton) Fernald, Rhodora 8: 163 (1906). *Scirpus robustus* var. *novae-angliae* (Britton) Beetle, Amer. J. Bot. 29: 87 (1942). *Bolboschoenus novae-angliae* (Britton) S.G.Sm., Brittonia 47: 434 (1995).

Schoenoplectus × oblongus (T.Koyama) Soják, Cas. Nár. Mus., Odd. Přír. 141: 62 (1972). *S. acutus* × *S. tabernaemontani*.
E. Canada to U.S.A. 72 ONT QUE 73 ORE 74 MIN 75 NWY VER. Hemicr. or rhizome geophyte.
**Scirpus* × *oblongus* T.Koyama, Canad. J. Bot. 40: 928 (1962).

Schoenoplectus orthorhizomatus (Kats.Arai & Miyam.) Hayas. & H.Ohashi, J. Jap. Bot. 75: 224 (2000).
Japan. 38 JAP. Hemicr. or rhizome geophyte.
**Scirpus orthorhizomatus* Kats.Arai & Miyam., J. Jap. Bot. 72: 299 (1997).

Schoenoplectus × osoreyamensis (M.Kikuchi) Hayas., J. Jap. Bot. 80: 308 (2005). *S. hondoensis* × *S. mucronatus*
**Scirpus* ×*osoreyamensis* M.Kikuchi
Japan. 38 JAP. Hemicr. or rhizome geophyte

Schoenoplectus paludicola (Kunth) Palla, Bot. Jahrb. Syst. 10: 299 (1888).
S. Africa. 27 CPP NAT TVL. Hemicr. or rhizome geophyte.
Isolepis paludicola Kunth, Enum. Pl. 2: 198 (1837).
**Scirpus paludicola* Kunth, Enum. Pl. 2: 163 (1837).

Schoenoplectus patentiglumis Hayashi, J. Jap. Bot. 78: 65 (2003).
S. Trop. Africa to KwaZulu-Natal. 26 MOZ ZIM 27 NAT.

Schoenoplectus pulchellus (Kunth) J.Raynal, Adansonia, n.s., 15: 542 (1976).
S. Afrrica. 27 LES NAT TVL. Hemicr. or rhizome geophyte.
**Ficinia pulchella* Kunth, Enum. Pl. 2: 261 (1837).
Scirpus pulchellus (Kunth) Boeck., Linnaea 36: 698 (1870).

Schoenoplectus pungens (Vahl) Palla, Bot. Jahrb. Syst. 10: 299 (1888).
Europe, E. Canada to Mexico, Caribbean, Peru to Argentina, Australia to New Zealand. (10) grb 11 AUT BGM† CZE? GER HUN NET POL SWI 12 FRA POR SPA 13 ITA 14 BLT† 50 NSW SOA TAS VIC WAU 72 QUE 74 ILL KAN MIN MSO WIS 75 CNT INI MAI MAS MIC NWH NWJ NWY OHI PEN RHO VER WVA 78 ALA ARK DEL FLA GEO KTY LOU MRY MSI NCA SCA TEN VRG WDC 79 MXE MXN MXS 81 PUE 83 BOL PER 85 AGE AGS AGW CLC CLN. Hemicr. or rhizome geophyte.
**Scirpus pungens* Vahl, Enum. Pl. 2: 255 (1805). *Heleogiton pungens* (Vahl) Rchb. in J.C.Mössler, Handb. Gewächsk. ed. 2, 3: 1808 (1830). *Cyperus pungens* (Vahl) Missbach & E.H.L.Krause in J.Sturm, Deutschl. Fl., ed. 2, 2: 32 (1900), nom. illeg. *Scirpus americanus* var. *pungens* (Vahl) Barros & Osten, Anales Mus. Hist. Nat. Montevideo, II, 3: 203 (1931).

var. ***badius*** (C.Presl) S.G.Sm., Novon 5: 100 (1995).
W. Canada to W. U.S.A., Mexico (Baja California), S. South America, Australia to New Zealand. 50 NSW SOA TAS VIC WAU 71 BRC 73 CAL ORE WAS 79 MXN 85 AGE AGS AGW CLC URU. Hemicr. or rhizome geophyte.

**Scirpus badius* C.Presl, Reliq. Haenk. 1: 193 (1828).
Scirpus monophyllus C.Presl, Reliq. Haenk. 1: 193 (1828). *Scirpus americanus* subsp. *monophyllus* (C.Presl) T.Koyama, Canad. J. Bot. 41: 1118 (1963). *Schoenoplectus monophyllus* (C.Presl) Soják, Cas. Nár. Mus., Odd. Prír. 141: 62 (1972). *Scirpus pungens* subsp. *monophyllus* (C.Presl) R.L.Taylor & MacBryde, Canad. J. Bot. 56: 190 (1978).
Scirpus pungens var. *longisetis* Benth., Fl. Austral. 7: 333 (1878).
Scirpus pungens var. *nanus* Benth., Fl. Austral. 7: 333 (1878).
Scirpus olneyi f. *australis* Barros & Osten, Anales Mus. Hist. Nat. Montevideo, II, 3: 204 (1931).
Scirpus olneyi f. *minor* Barros in H.R.Descole (ed.), Gen. Sp. Pl. Argent. 4(1): 201 (1949).

var. ***longispicatus*** (Britton) S.G.Sm., Novon 5: 100 (1995).
Canada to W. & EC. U.S.A., NE. Argentina. 71 MAN 72 ONT 73 COL IDA MNT ORE WAS WYO 74 ILL IOW KAN MIN MSO NDA NEB OKL WIS 75 INI 76 ARI CAL NEV UTA 77 NWM TEX 78 MSI 85 AGE. Hemicr. or rhizome geophyte.
Scirpus pungens var. *polyphyllus* Boeck., Linnaea 36: 709 (1870). *Scirpus americanus* var. *polyphyllus* (Boeck.) Beetle, Amer. J. Bot. 30: 399 (1943). *Schoenoplectus pungens* var. *polyphyllus* (Boeck.) Dorn, Vasc. Pl. Wyoming, ed. 3: 376 (2001).
**Scirpus americanus* var. *longispicatus* Britton, Trans. New York Acad. Sci. 11: 78 (1892). *Schoenoplectus americanus* subsp. *longispicatus* (Britton) Soják, Cas. Nár. Mus., Odd. Prír. 141: 62 (1972). *Scirpus pungens* var. *longispicatus* (Britton) Cronquist in A.Cronquist & al, Intermountain Fl. 6: 74 (1977). *Schoenoplectus pungens* subsp. *longispicatus* (Britton) Á.Löve & D.Löve, Taxon 30: 849 (1981).
Scirpus americanus var. *longibracteatus* Barros & Osten, Anales Mus. Nac. Montevideo, II, 3: 203 (1931).
Scirpus americanus var. *longisetis* (Benth. & F.Muell.) Beetle, Amer. J. Bot. 30: 400 (1943).

var. ***pungens***
Europe, E. Canada to U.S.A., Caribbean, Peru to Argentina. (10) grb 11 AUT BGM† CZE? GER HUN NET POL SWI 12 FRA POR SPA 13 ITA 14 BLT† 50 NSW SOA TAS VIC WAU 72 QUE 74 ILL KAN MIN MSO WIS 75 CNT INI MAI MAS MIC NWH NWJ NWY OHI PEN RHO VER WVA 78 ALA ARK DEL FLA GEO KTY LOU MRY MSI NCA SCA TEN VRG WDC 79 MXN MXS 81 PUE 83 BOL PER 85 AGE AGS AGW CLC CLN. Hemicr. or rhizome geophyte.
Scirpus mucronatus Roth, Tent. Fl. Germ. 1: 23 (1788), nom. illeg.
Scirpus tenuifolius DC. in J.B.A.M.de Lamarck & A.P.de Candolle, Fl. Franç., ed. 3, Suppl.: 300 (1805).
Scirpus triqueter R.Br., Prodr.: 223 (1810), nom. illeg.
Scirpus rothii Hoppe in J.Sturm, Deutschl. Fl. 36: (1814). *Heleogiton rothii* (Hoppe) Schur, Enum. Pl. Transsilv.: 693 (1866).
Scirpus conglomeratus Kunth in F.W.H.von Humboldt, A.J.A.Bonpland & C.S.Kunth, Nov. Gen. Sp. 1: 227 (1816). *Scirpus americanus* var. *conglomeratus* (Kunth) T.Koyama, Canad. J. Bot. 41: 1119 (1963). *Schoenoplectus monophyllus* subsp. *conglomeratus* (Kunth) Soják, Cas. Nár. Mus., Odd. Prír. 141: 62 (1972). *Scirpus pungens* var. *conglomeratus* (Kunth) R.L.Taylor & MacBryde, Canad. J. Bot. 56: 190 (1978).
Scirpus triangularis Steud., Nomencl. Bot.: 752 (1821).
Isolepis aristata J.Presl, Delic. Prag.: 144 (1822). *Scirpus aristatus* (J.Presl) K.Richt., Pl. Eur. 1: 138 (1890), nom. illeg.
Eleocharis leptophylla Schult. in J.J.Roemer & J.A.Schultes, Mant. 2: 88 (1824).
Scirpus longifolius Hook. & Arn., Bot. Beechey Voy.: 49 (1832).
Scirpus melas Kunze ex Kunth, Enum. Pl. 2: 169 (1837).
Scirpus subappendiculatus Steud., Syn. Pl. Glumac. 2: 84 (1855).
Scirpus brevivaginatus Boeck., Flora 41: 413 (1858).
Eleogiton rothii Feurr., Ann. Soc. Linn. Lyon, n.s., 17: 173 (1869).
Scirpus aggregatus Ruiz ex Boeck., Linnaea 36: 709 (1870).
Scirpus melanorrhizus Boeck., Bot. Jahrb. Syst. 5: 504 (1884).
Scirpus triangularis MacMill., Metasp. Minnesota Valley: 99 (1892).
Scirpus limensis C.B.Clarke, Kew Bull., Addit. Ser. 8: 29 (1908).
Scirpus lomanus H.Pfeiff., Mitt. Inst. Allg. Bot. Hamburg 8: 39 (1929).
Scirpus americanus f. *cordilleranus* Kük. & Barros, Anales Mus. Argent. Ci. Nat. "Bernardino Rivadavia" 38: 159 (1935).

Schoenoplectus rechingeri Kukkonen, Ann. Bot. Fenn. 32: 154 (1995).
Pakistan (Swat). 40 PAK. Hemicr. or rhizome geophyte.

Schoenoplectus rhodesicus (Podlech) Lye, Nordic J. Bot. 3: 242 (1983).
Tanzania to Zambia. 25 TAN 26 ZAM. Hel.
**Scirpus rhodesicus* Podlech, Mitt. Bot. Staatssamml. München 4: 117 (1961).

Schoenoplectus schoofii (Beetle) Soják, Cas. Nár. Mus., Odd. Prír. 148: 194 (1979 publ. 1980).
China (Yunnan). 36 CHC. Hemicr. or rhizome geophyte.
**Scirpus schoofii* Beetle, Amer. J. Bot. 29: 654 (1942).

Schoenoplectus smithii (A.Gray) Soják, Cas. Nár. Mus., Odd. Prír. 141: 62 (1972).
E. Canada to NC. & E. U.S.A. 72 NBR ONT QUE 74 ILL MIN WIS 75 CNT INI MAI MAS MIC NWH NWJ NWY OHI PEN RHO VER 78 DEL MRY NCA VRG. Ther.
**Scirpus smithii* A.Gray, Manual, ed. 5: 563 (1867).

var. ***levisetus*** (Fernald) S.G.Sm., Novon 12: 108 (2002).
E. Canada to Virginia. 72 NBR QUE 75 MAI 78 VRG. Ther.
**Scirpus smithii* var. *levisetus* Fernald, Rhodora 23: 42 (1921).
Scirpus smithii f. *levisetus* (Fassett) Fernald, Rhodora 44: 479 (1942).

var. ***setosus*** (Fernald) S.G.Sm., Novon 12: 107 (2002).
E. Canada to NC. & E. U.S.A. 72 ONT QUE 74 ILL MIN WIS 75 CNT INI MAI MAS MIC NWH NWY OHI 78 DEL MRY NCA. Ther.
**Scirpus smithii* var. *setosus* Fernald, Rhodora 3: 252 (1901). *Scirpus smithii* f. *setosus* (Fernald) Fernald, Rhodora 44: 479 (1942).

var. ***smithii***
E. Canada to NC. & NE. U.S.A. 72 ONT QUE 74 MIN WIS 75 CNT INI MAS MIC NWJ NWY OHI PEN RHO VER 78 DEL. Ther.

Schoenoplectus* × *steinmetzii (Fernald) S.G.Sm., Novon 5: 99 (1995). *S. heterochaetus* × *S. tabernaemontani.*
NC. & NE. U.S.A. 74 WIS 75 MAI. Hemicr. or rhizome geophyte.
**Scirpus* ×*steinmetzii* Fernald, Rhodora 45: 286 (1943). *Scirpus validus* f. *steinmetzii* (Fernald) Beetle, Publ. Univ. Wyoming 13: 6 (1948).

Schoenoplectus subbisetosus (T.Koyama) Soják, Cas. Nár. Mus., Odd. Prír. 148: 194 (1979 publ. 1980).
Tibet. 36 CHT. Hemicr. or rhizome geophyte.
**Scirpus subbisetosus* T.Koyama, Bot. Mag. (Tokyo) 86: 281 (1973).

Schoenoplectus subterminalis (Torr.) Soják, Cas. Nár. Mus., Odd. Prír. 140: 127 (1972).
Alaska to N. & EC. U.S.A. 70 ASK 71 BRC SAS 72 LAB NBR NFL NSC ONT PEI QUE 73 IDA MNT ORE WAS 74 ILL† MIN MSO WIS 75 CNT INI MAI MAS MIC NWH NWJ NWY OHI PEN RHO VER 76 CAL 78 ALA DEL FLA GEO MRY NCA SCA VRG. Hemicr. or rhizome geophyte.
**Scirpus subterminalis* Torr., Fl. N. Middle United States 1: 47 (1823).
Scirpus subterminalis f. *terrestris* (Paine) Fernald, Rhodora 45: 288 (1943).

Schoenoplectus tabernaemontani (C.C.Gmel.) Palla, Bot. Jahrb. Syst. 10: 299 (1888).
Cosmopolitan. 10 DEN FIN GRB IRE NOR SWE 11 AUT BGM CZE GER HUN NET POL SWI 12 BAL COR FRA POR SAR SPA 13 ALB BUL GRC ITA KRI ROM SIC TUE YUG 14 BLR BLT KRY RUC RUE RUN RUS RUW UKR 20 ALG MOR 30 ALT BRY CTA IRK KRA TVA WSB YAK 31 AMU KAM KHA KUR MAG PRM SAK 32 KAZ KGZ TKM TZK UZB 33 NCS TCS 34 AFG EAI IRN LBS PAL TUR 38 JAP NNS TAI 40 NEP PAK WHM 41 VIE 42 PHI 43 NWG 50 NFK NSW QLD SOA TAS VIC WAU 51 NZN NZS 60 NWC VAN 63 HAW 70 ASK NWT YUK 71 ABT BRC MAN SAS 72 LAB NBR NFL NSC ONT PEI QUE 73 COL IDA MNT ORE WAS WYO 74 ILL IOW KAN MIN MSO NDA NEB OKL SDA WIS 75 CNT INI MAI MAS MIC NWH NWJ NWY OHI PEN RHO VER WVA 76 ARI CAL NEV UTA 77 NWM TEX 78 ALA ARK DEL FLA GEO KTY LOU MRY MSI NCA SCA TEN VRG WDC 79 MXC MXE MXG MXS MXT 80 COS GUA HON NIC 81 BER CAY CUB DOM JAM PUE 84 85 AGE AGW PAR URU. Hemicr. or rhizome geophyte.
**Scirpus tabernaemontani* C.C.Gmel., Fl. Bad. 1: 101 (1805). *Heleogiton tabernaemontani* (C.C.Gmel.) Peterm., Fl. Bienitz: 22 (1841). *Eleogiton tabernaemontani* (C.C.Gmel.) Fourr., Ann. Soc. Linn. Lyon, n.s., 17: 173 (1869). *Scirpus lacustris* var. *tabernaemontani* (C.C.Gmel.) Nyman, Consp. Fl. Eur.: 764 (1882). *Scirpus lacustris* var. *tabernaemontani* (C.C.Gmel.) Crép., Man. Fl. Belgique, ed. 5: 409 (1884). *Scirpus maritimus* subsp. *tabernaemontani* (C.C.Gmel.) Nyman, Consp. Fl. Eur., Suppl. 2: 318 (1890). *Cyperus tabernaemontani* (C.C.Gmel.) Missbach & E.H.L.Krause in J.Sturm, Deutschl. Fl., ed. 2, 2: 31 (1900). *Heleophylax tabernaemontani* (C.C.Gmel.) Schinz & Thell., Vierteljahrsschr. Naturf. Ges. Zürich 53: 587 (1908). *Hymenochaeta tabernaemontani* (C.C.Gmel.) Nakai, Bull. Natl. Sci. Mus. Tokyo 31: 133 (1952). *Schoenoplectus lacustris* subsp. *tabernaemontani* (C.C.Gmel.) Á.Löve & D.Löve, Folia Geobot. Phytotax. 10: 275 (1975).
Scirpus validus Vahl, Enum. Pl. 2: 268 (1805). *Scirpus lacustris* var. *validus* (Vahl) Kük., Repert. Spec. Nov. Regni Veg. 23: 200 (1926). *Schoenoplectus validus* (Vahl) Á.Löve & D.Löve, Bull. Torrey Bot. Club 81: 33 (1954). *Scirpus lacustris* subsp. *validus* (Vahl) T.Koyama, Canad. J. Bot. 40: 927 (1962). *Schoenoplectus lacustris* subsp. *validus* (Vahl) T.Koyama, Bishop Mus. Occas. Pap. 29: 128 (1989).
Scirpus glaucus Sm., Engl. Bot.: t. 2321 (1812). *Scirpus lacustris* f. *glaucus* (Sm.) Peterm., Anal. Pfl.-Schlüss.: 480 (1846). *Scirpus lacustris* subsp. *glaucus* (Sm.) Hartm., Sv. Norsk. Exc.-Fl.: 10 (1846). *Schoenoplectus lacustris* f. *glaucus* (Sm.) Soó, Acta Bot. Acad. Sci. Hung. 16: 368 (1970 publ. 1971). *Schoenoplectus lacustris* subsp. *glaucus* (Sm.) Luceño & Marín, Anales Jard. Bot. Madrid 59: 352 (2001 publ. 2002).
Scirpus tabernaemontani f. *paludosus* Boenn., Prodr. Fl. Monast. Westphal. (1824). *Schoenoplectus tabernaemontani* f. *paludosus* (Boenn.) Soó, Acta Bot. Acad. Sci. Hung. 16: 369 (1970 publ. 1971), without basionym page.
Scirpus uliginosus Kar. & Kir., Bull. Soc. Imp. Naturalistes Moscou 15: 520 (1842).
Scirpus tabernaemontani f. *luxurians* Miq., Ann. Mus. Bot. Lugduno-Batavi 2: 143 (1886). *Schoenoplectus validus* subsp. *luxurians* (Miq.) Soják, Cas. Nár. Mus., Odd. Prír. 141: 62 (1972).
Scirpus tabernaemontani f. *sub-duvalii* Beckh., Fl. Westfalen: 936 (1893). *Schoenoplectus tabernaemontani* f. *sub-duvalii* (Beckh.) Soó, Acta Bot. Acad. Sci. Hung. 16: 369 (1970 publ. 1971).
Scirpus siculus Lojac., Fl. Sicul. 3: 214 (1909).
Scirpus validus var. *laeviglumis* Tang & F.T.Wang, in Fl. Reipubl. Popul. Sin. 11: 221 (1961).

Schoenoplectus torreyi (Olney) Palla, Allg. Bot. Z. Syst. 17(Beih.): 3 (1911).
C. & E. Canada to NC. & NE. U.S.A. 71 MAN 72 NBR ONT QUE 74 ILL IOW MIN MSO WIS 75 CNT INI MAI MAS MIC NWH NWJ NWY OHI PEN RHO VER WVA 78 DEL KTY MRY VRG. Hemicr. or rhizome geophyte.
**Scirpus torreyi* Olney, Proc. Providence Franklin Soc. 1: 32 (1847).
Scirpus subterminalis var. *cylindricus* (Torr.) T.Koyama, Canad. J. Bot. 40: 930 (1962).

Schoenoplectus* × *trapezoideus (Koidz.) Hayas. & H. Ohashi, J. Jap. Bot. 75: 244 (2000). *Schoenoplectiella juncoides* × *S. mucronatus.*
Japan to Nansei-shoto. 38 JAP NNS. Hemicr. or rhizome geophyte.
**Scirpus* ×*trapezoideus* Koidz., Bot. Mag. (Tokyo) 39: 26 (1925).

Scirpus erectus var. *triangulatus* Honda, Bot. Mag. (Tokyo) 45: 45 (1931). *Scirpus juncoides* var. *triangulatus* (Honda) Ohwi, Mem. Coll. Sci. Kyoto Imp. Univ., Ser. B, Biol. 18: 115 (1944). *Schoenoplectus juncoides* subsp. *triangulatus* (Honda) Soják, Cas. Nár. Mus., Odd. Prír. 148: 194 (1979 publ. 1980).

Schoenoplectus triqueter (L.) Palla, Bot. Jahrb. Syst. 10: 299 (1888).

Temp. Eurasia to Himalaya. 10 GRB IRE 11 AUT BGM CZE GER HUN NET SWI 12 FRA POR SAR SPA 13 BUL ITA ROM SIC TUE YUG 14 RUE RUS UKR 20 EGY 21 CNY MDR 27 CPP TVL 30 ALT 31 PRM 32 KAZ KGZ TZK UZB 33 NCS TCS 34 AFG IRN TUR 36 CHS 38 JAP NNS OGA TAI (40) ehm IND PAK WHM (73) ore was. Hemicr.

**Scirpus triqueter* L., Mant. Pl. 1: 29 (1767). *Heleogiton triquetrum* (L.) Rchb. in J.C.Mössler, Handb. Gewächsk. ed. 2, 3: 1808 (1830). *Eleogiton triqueter* (L.) Fourr., Ann. Soc. Linn. Lyon, n.s., 17: 173 (1869). *Cyperus triqueter* (L.) Missbach & E.H.L.Krause in J.Sturm, Deutschl. Fl., ed. 2, 2: 31 (1900), nom. illeg. *Heleophylax triquetrus* (L.) Schinz & Thell., Vierteljahrsschr. Naturf. Ges. Zürich 53: 587 (1908). *Hymenochaeta triquetra* (L.) Nakai, Bull. Natl. Sci. Mus. Tokyo 31: 134 (1952).

Scirpus trigonus Roth, Neue Beytr. Bot. 1: 90 (1802).

Scirpus mucronatus Host, Icon. Descr. Gram. Austriac. 3: 44 (1805), nom. illeg.

Scirpus triqueter var. *conglomeratus* Dumort., Fl. Belg.: 144 (1827), nom. nud.

Scirpus hoppei Weihe, Flora 11: 118 (1828). *Scirpus triqueter* var. *hoppei* (Weihe) Nyman, Consp. Fl. Eur.: 764 (1882). *Schoenoplectus triqueter* f. *hoppei* (Weihe) Soó, Acta Bot. Acad. Sci. Hung. 16: 369 (1970 publ. 1971).

Scirpus lejeunei Weihe, Flora 11: 116 (1828). *Heleogiton lejeunei* (Weihe) Rchb., Fl. Germ. Excurs.: 77 (1830). *Scirpus triqueter* var. *lejeunei* (Weihe) Nyman, Consp. Fl. Eur.: 764 (1882).

Scirpus triqueter f. *conglomeratus* Döll, Rhein. Fl.: 163 (1843). *Schoenoplectus triqueter* f. *conglomeratus* (Döll) Soó, Acta Bot. Acad. Sci. Hung. 16: 369 (1970 publ. 1971).

Scirpus triqueter f. *pallidus* Döll, Rhein. Fl.: 163 (1843). *Schoenoplectus triqueter* f. *pallidus* (Döll) Soó, Acta Bot. Acad. Sci. Hung. 16: 369 (1970 publ. 1971).

Scirpus pollichii Gren. & Godr., Fl. France 3: 374 (1855).

Scirpus triqueter var. *segregatus* C.B.Clarke in J.D.Hooker, Fl. Brit. India 6: 658 (1893). *Schoenoplectus triqueter* var. *segregatus* (C.B.Clarke) Karthik. in S.Karthikeyan & al., Fl. Ind. Enumerat. - Monocot.: 70 (1989).

Scirpus triqueter f. *monostachys* Junge, Jahrb. Hamburg. Wiss. Anst. 22(3): 72 (1905). *Schoenoplectus triqueter* f. *monostachys* (Junge) Soó, Acta Bot. Acad. Sci. Hung. 16: 369 (1970 publ. 1971).

Scirpus triqueter f. *macrostachys* Junge, Jahrb. Hamburg. Wiss. Anst. 25(3): 257 (1908). *Schoenoplectus triqueter* f. *macrostachys* (Junge) Soó, Acta Bot. Acad. Sci. Hung. 16: 369 (1970 publ. 1971).

Schoenoplectus* × *uzenensis (Ohwi ex T.Koyama) Hayas. & H.Ohashi, J. Jap. Bot. 75: 224 (2000). *S. lineolatus* × *S. mucronatus*.

Japan. 38 JAP. Hemicr. or rhizome geophyte.

**Scirpus* × *uzenensis* Ohwi ex T.Koyama, J. Fac. Sci. Univ. Tokyo, Sect. 3, Bot. 7: 363 (1958).

Synonyms:

Schoenoplectus aberrans (Cherm.) J.Raynal = ***Schoenoplectiella aberrans*** (Cherm.) Lye

Schoenoplectus acutus f. *congestus* (Farw.) Mohlenbr. = ***Schoenoplectus acutus*** var. ***occidentalis*** (S.Watson) S.G.Sm.

Schoenoplectus americanus subsp. *longispicatus* (Britton) Soják = ***Schoenoplectus pungens*** var. ***longispicatus*** (Britton) S.G.Sm.

Schoenoplectus americanus subsp. *triangularis* (Pers.) Soó = ***Schoenoplectus americanus*** (Pers.) Volkart

Schoenoplectus articulatus (L.) Palla = ***Schoenoplectiella articulata*** (L.) Lye

Schoenoplectus brachyceras (Hochst. ex A.Rich.) Lye = ***Schoenoplectus corymbosus*** (Roth ex Roem. & Schult.) J.Raynal

Schoenoplectus californicus var. *spoliatus* (Barros) Vegetti = ***Schoenoplectus californicus*** (C.A.Mey.) Soják

Schoenoplectus californicus subsp. *tereticulmis* (Steud.) Soják = ***Schoenoplectus californicus*** (C.A.Mey.) Soják

Schoenoplectus californicus var. *tereticulmis* (Steud.) Vegetti = ***Schoenoplectus californicus*** (C.A.Mey.) Soják

Schoenoplectus × *carinatus* nothosubsp. *kuekenthalianus* (Junge) P.D.Sell = ***Schoenoplectus* × *kuekenthalianus*** (Junge) D.H.Kent

Schoenoplectus cernuus (Vahl) Hayek = ***Isolepis cernua*** (Vahl) Roem. & Schult.

Schoenoplectus chamissoi (Nees) Soják = ***Schoenoplectus californicus*** (C.A.Mey.) Soják

Schoenoplectus chilensis (Nees & Meyen) Soják = ***Schoenoplectus americanus*** (Pers.) Volkart

Schoenoplectus corymbosus var. *brachyceras* (Hochst. ex A.Rich.) Lye = ***Schoenoplectus corymbosus*** (Roth ex Roem. & Schult.) J.Raynal

Schoenoplectus dissachanthus (S.T.Blake) J.Raynal = ***Schoenoplectiella dissachantha*** (S.T.Blake) Lye

Schoenoplectus erectus (Poir.) Palla ex J.Raynal = ***Schoenoplectiella erecta*** (Poir.) Lye

Schoenoplectus erectus subsp. *raynalii* (Schuyler) Lye = ***Schoenoplectiella erecta*** (Poir.) Lye

Schoenoplectus erectus subsp. *sinuatus* (Schuyler) Lye = ***Schoenoplectiella erecta*** (Poir.) Lye

Schoenoplectus fistulosus (Forssk.) Soják = ***Schoenoplectiella articulata*** (L.) Lye

Schoenoplectus fluitans (L.) Palla = ***Isolepis fluitans*** (L.) R.Br.

Schoenoplectus fluviatilis (Torr.) M.T.Strong = ***Bolboschoenus fluviatilis*** (Torr.) Soják

Schoenoplectus grossheimii Pobed. = ***Schoenoplectus lacustris*** subsp. ***hippolyti*** (V.I.Krecz.) Kukkonen

Schoenoplectus grossus (L.f.) Palla = ***Actinoscirpus grossus*** (L.f.) Goetgh. & D.A.Simpson

Schoenoplectus hallii (A.Gray) S.G.Sm. = ***Schoenoplectiella hallii*** (A.Gray) Lye

Schoenoplectus hippolyti (V.I.Krecz.) V.I.Krecz. ex Grossh. = ***Schoenoplectus lacustris*** subsp. ***hippolyti*** (V.I.Krecz.) Kukkonen

Schoenoplectus hooperae J.Raynal = ***Schoenoplectiella hooperae*** (J.Raynal) Lye

Schoenoplectus × *igaensis* (T.Koyama) Hayas. & H.Ohashi = [38 JAP] *Schoenoplectiella juncoides* × *S. lineolatus*

Schoenoplectus inclinatus (Delile) Lye = ***Schoenoplectus corymbosus*** (Roth ex Roem. & Schult.) J.Raynal

Schoenoplectus jacobi (C.E.C.Fisch.) Lye = ***Schoenoplectiella senegalensis*** (Steud.) Lye

Schoenoplectus javanus (Nees) Palla = ***Schoenoplectus mucronatus*** (L.) Palla subsp. ***mucronatus***
Schoenoplectus junceus (Willd.) J.Raynal = ***Schoenoplectiella juncea*** (Willd.) Lye
Schoenoplectus ×*juncohotarui* Yashiro = [38 JAP] *Schoenoplectiella juncoides* × *Schoenoplectus hotarui*
Schoenoplectus juncoides (Roxb.) Palla = ***Schoenoplectiella juncoides*** (Roxb.) Lye
Schoenoplectus juncoides subsp. *hotarui* (Ohwi) Soják = ***Schoenoplectus hotarui*** (Ohwi) Holub
Schoenoplectus juncoides subsp. *ohwianus* (T.Koyama) Soják = ***Schoenoplectiella juncoides*** (Roxb.) Lye
Schoenoplectus juncoides subsp. *purshianus* (Fernald) Soják = ***Schoenoplectiella purshiana*** (Fernald) Lye
Schoenoplectus juncoides subsp. *rockii* (Kük.) Soják = ***Schoenoplectiella juncoides*** (Roxb.) Lye
Schoenoplectus juncoides subsp. *triangulatus* (Honda) Soják = ***Schoenoplectus*** ×***trapezoideus*** (Koidz.) Hayas. & H.Ohashi
Schoenoplectus ×*kalmussii* (Asch., Abrom. & Graebn.) Palla = ***Schoenoplectus*** ×***carinatus*** (Sm.) Palla nothosubsp. ***carinatus***
Schoenoplectus lacustris subsp. *acutus* (Muhl. ex Bigelow) Á.Löve & D.Löve = ***Schoenoplectus acutus*** (Muhl. ex Bigelow) Á.Löve & D.Löve
Schoenoplectus lacustris f. *compactus* (Beckh.) Soó = ***Schoenoplectus lacustris*** (L.) Palla subsp. ***lacustris***
Schoenoplectus lacustris f. *confertus* (Schur) Soó = ***Schoenoplectus lacustris*** (L.) Palla subsp. ***lacustris***
Schoenoplectus lacustris f. *conglomeratus* (Junge) Soó = ***Schoenoplectus lacustris*** (L.) Palla subsp. ***lacustris***
Schoenoplectus lacustris subsp. *creber* (Fernald) Á.Löve & D.Löve = ***Schoenoplectus lacustris*** (L.) Palla subsp. ***lacustris***
Schoenoplectus lacustris f. *foliosus* (Des Moul.) Soó = ***Schoenoplectus lacustris*** (L.) Palla subsp. ***lacustris***
Schoenoplectus lacustris f. *glaucus* (Sm.) Soó = ***Schoenoplectus tabernaemontani*** (C.C.Gmel.) Palla
Schoenoplectus lacustris subsp. *glaucus* (Sm.) Luceño & Marín = ***Schoenoplectus tabernaemontani*** (C.C. Gmel.) Palla
Schoenoplectus lacustris f. *laxiflorus* (Kuntze) Soó = ***Schoenoplectus lacustris*** (L.) Palla subsp. ***lacustris***
Schoenoplectus lacustris f. *minor* (Baumg.) Soó = ***Schoenoplectus lacustris*** (L.) Palla subsp. ***lacustris***
Schoenoplectus lacustris f. *palustris* (Schltdl.) Soó = ***Schoenoplectus lacustris*** (L.) Palla subsp. ***lacustris***
Schoenoplectus lacustris f. *radiatus* (Vollm.) Soó = ***Schoenoplectus lacustris*** (L.) Palla subsp. ***lacustris***
Schoenoplectus lacustris subsp. *tabernaemontani* (C.C.Gmel.) Á.Löve & D.Löve = ***Schoenoplectus tabernaemontani*** (C.C.Gmel.) Palla
Schoenoplectus lacustris subsp. *tenuiculmis* (E.Sheld.) Soják = ***Schoenoplectus heterochaetus*** (Chase) Soják
Schoenoplectus lacustris f. *terrestris* (Glück) Soó = ***Schoenoplectus lacustris*** (L.) Palla subsp. ***lacustris***
Schoenoplectus lacustris subsp. *validus* (Vahl) T.Koyama = ***Schoenoplectus tabernaemontani*** (C.C.Gmel.) Palla
Schoenoplectus laevis (S.T.Blake) J.Raynal = ***Schoenoplectiella laevis*** (S.T.Blake) Lye
Schoenoplectus lateriflorus (J.F.Gmel.) Lye = ***Schoenoplectiella lateriflora*** (J.F.Gmel.) Lye
Schoenoplectus lateriflorus subsp. *laevinux* Lye = ***Schoenoplectiella lateriflora*** (J.F.Gmel.) Lye
Schoenoplectus leucanthus (Boeck.) J.Raynal = ***Schoenoplectiella leucantha*** (Boeck.) Lye
Schoenoplectus litoralis subsp. *subulatus* (Vahl) Soják = ***Schoenoplectus litoralis*** subsp. ***thermalis*** (Trab.) S.S.Hooper
Schoenoplectus lupulinus (Nees) V.I.Krecz. = ***Schoenoplectiella roylei*** (Nees) Lye
Schoenoplectus maritimus (L.) Lye = ***Bolboschoenus maritimus*** (L.) Palla
Schoenoplectus melanospermus (C.A.Mey.) Grossh. = ***Schoenoplectiella supina*** (L.) Lye
Schoenoplectus merrillii Palla = ***Isolepis subtilissima*** Boeck.
Schoenoplectus microglumis Lye = ***Schoenoplectiella microglumis*** (Lye) Lye
Schoenoplectus monophyllus (C.Presl) Soják = ***Schoenoplectus pungens*** var. ***badius*** (C.Presl) S.G.Sm.
Schoenoplectus monophyllus subsp. *conglomeratus* (Kunth) Soják = ***Schoenoplectus pungens*** (Vahl) Palla var. ***pungens***
Schoenoplectus mucronatus var. *antrorsispinulosus* Iokawa, Kohno & Daigobo = ***Schoenoplectus mucronatus*** (L.) Palla subsp. ***mucronatus***
Schoenoplectus mucronatus f. *brevisetaceus* Horiuchi = ***Schoenoplectus mucronatus*** (L.) Palla subsp. ***mucronatus***
Schoenoplectus mucronatus f. *hosoiri* Hayas. & H.Ohashi = ***Schoenoplectus mucronatus*** (L.) Palla subsp. ***mucronatus***
Schoenoplectus mucronatus var. *ishizawae* Kohno, Iokawa & Daigobo = ***Schoenoplectus mucronatus*** (L.) Palla subsp. ***mucronatus***
Schoenoplectus mucronatus subsp. *robustus* (Miq.) T.Koyama = ***Schoenoplectus mucronatus*** (L.) Palla subsp. ***mucronatus***
Schoenoplectus mucronatus f. *robustus* (Miq.) Hayas. & H.Ohashi = ***Schoenoplectus mucronatus*** (L.) Palla subsp. ***mucronatus***
Schoenoplectus mucronatus var. *sanguineus* (Tang & F.T.Wang) P.C.Li = ***Schoenoplectus mucronatus*** (L.) Palla subsp. ***mucronatus***
Schoenoplectus mucronatus var. *tataranus* (Honda) Kohno, Iokawa & Daigobo = ***Schoenoplectus mucronatus*** (L.) Palla subsp. ***mucronatus***
Schoenoplectus nevadensis (S.Watson) Soják = ***Amphiscirpus nevadensis*** (S.Watson) Oteng-Yeb.
Schoenoplectus nobilis (Ridl.) Lye = ***Bolboschoenus nobilis*** (Ridl.) Goetgh. & D.A.Simpson
Schoenoplectus ×*oguraensis* (T.Koyama) Hayas. & H.Ohashi = [38 JAP] *S. mucronatus* × *Schoenoplectiella wallichii*
Schoenoplectus ohwianus (T.Koyama) Holub = ***Schoenoplectiella juncoides*** (Roxb.) Lye
Schoenoplectus olneyi (A.Gray) Palla = ***Schoenoplectus americanus*** (Pers.) Volkart
Schoenoplectus oryzetorum (Steud.) V.I.Krecz. = ***Schoenoplectiella lateriflora*** (J.F.Gmel.) Lye
Schoenoplectus oxyjulos (S.S.Hooper) J.Raynal = ***Schoenoplectiella oxyjulos*** (S.S.Hooper) Lye
Schoenoplectus perrieri (Cherm.) J.Raynal = ***Schoenoplectiella perrieri*** (Cherm.) Lye
Schoenoplectus philippii (Tineo) Pignatti = ***Schoenoplectus litoralis*** (Schrad.) Palla subsp. ***litoralis***
Schoenoplectus praelongatus (Poir.) J.Raynal = ***Schoenoplectiella praelongata*** (Poir.) Lye
Schoenoplectus proximus (Steud.) J.Raynal = ***Schoenoplectiella proxima*** (Steud.) Lye
Schoenoplectus proximus var. *botswanensis* Hayas. = ***Schoenoplectiella proxima*** (Steud.) Lye
Schoenoplectus pseudotriqueter (Steud.) Soják = ***Schoenoplectus californicus*** (C.A.Mey.) Soják

Schoenoplectus pungens subsp. *longispicatus* (Britton) Á.Löve & D.Löve = ***Schoenoplectus pungens*** var. ***longispicatus*** (Britton) S.G.Sm.
Schoenoplectus pungens var. *polyphyllus* (Boeck.) Dorn = ***Schoenoplectus pungens*** var. ***longispicatus*** (Britton) S.G.Sm.
Schoenoplectus purshianus (Fernald) M.T.Strong = ***Schoenoplectiella purshiana*** (Fernald) Lye
Schoenoplectus purshianus var. *williamsii* (Fernald) S.G. Sm. = ***Schoenoplectiella purshiana*** (Fernald) Lye
Schoenoplectus quinquefarius (Buch.-Ham. ex Boeck.) Palla = ***Schoenoplectiella roylei*** (Nees) Lye
Schoenoplectus raynalianus U.Scholz = ***Schoenoplectiella raynaliana*** (U.Scholz) Lye
Schoenoplectus reductus (Cherm.) J.Raynal = ***Schoenoplectiella reductus*** (Cherm.) Lye
Schoenoplectus riparius (Nees & Meyen) Palla = ***Schoenoplectus californicus*** (C.A.Mey.) Soják
Schoenoplectus robustus (Pursh) M.T.Strong = ***Bolboschoenus robustus*** (Pursh) Soják
Schoenoplectus rogersii (N.E.Br.) Lye = ***Schoenoplectus confusus*** var. ***rogersii*** (N.E.Br.) Lye
Schoenoplectus roylei (Nees) Ovcz. & Czukav. = ***Schoenoplectiella roylei*** (Nees) Lye
Schoenoplectus rubiginosus (Beetle) Soják = ***Schoenoplectus acutus*** var. ***occidentalis*** (S.Watson) S.G.Sm.
Schoenoplectus savii (Sebast. & Mauri) Palla = ***Isolepis cernua*** (Vahl) Roem. & Schult. var. ***cernua***
Schoenoplectus saximontanus (Fernald) J.Raynal = ***Schoenoplectiella saximontana*** (Fernald) Lye
Schoenoplectus scirpoides (Schrad.) Browning = ***Schoenoplectus litoralis*** subsp. ***thermalis*** (Trab.) S.S.Hooper
Schoenoplectus senegalensis (Steud.) Palla = ***Schoenoplectiella senegalensis*** (Steud.) Lye
Schoenoplectus setaceus (L.) Palla = ***Isolepis setacea*** (L.) R.Br.
Schoenoplectus smithii subsp. *leiocarpus* (Kom.) Soják = ***Schoenoplectus komarovii*** (Roshev.) Soják
Schoenoplectus smithii subsp. *williamsii* (Fernald) Soják = ***Schoenoplectiella purshiana*** (Fernald) Lye
Schoenoplectus subulatus (Vahl) Lye = ***Schoenoplectus litoralis*** subsp. ***thermalis*** (Trab.) S.S.Hooper
Schoenoplectus supinus (L.) Palla = ***Schoenoplectiella supina*** (L.) Lye
Schoenoplectus supinus subsp. *hallii* (A.Gray) Soják = ***Schoenoplectiella hallii*** (A.Gray) Lye
Schoenoplectus supinus subsp. *lateriflorus* (J.F.Gmel.) Soják = ***Schoenoplectiella lateriflora*** (J.F.Gmel.) Lye
Schoenoplectus supinus subsp. *lateriflorus* (J.F.Gmel.) T.Koyama = ***Schoenoplectiella lateriflora*** (J.F.Gmel.) Lye
Schoenoplectus supinus subsp. *saximontanus* (Fernald) Soják = ***Schoenoplectiella saximontana*** (Fernald) Lye
Schoenoplectus tabernaemontani f. *paludosus* (Boenn.) Soó = ***Schoenoplectus tabernaemontani*** (C.C.Gmel.) Palla
Schoenoplectus tabernaemontani f. *sub-duvalii* (Beckh.) Soó = ***Schoenoplectus tabernaemontani*** (C.C.Gmel.) Palla
Schoenoplectus tatora (Kunth) Palla = ***Schoenoplectus californicus*** (C.A.Mey.) Soják
Schoenoplectus triangulatus (Roxb.) Soják = ***Schoenoplectus mucronatus*** (L.) Palla subsp. ***mucronatus***
Schoenoplectus triqueter f. *conglomeratus* (Döll) Soó = ***Schoenoplectus triqueter*** (L.) Palla
Schoenoplectus triqueter f. *hoppei* (Weihe) Soó = ***Schoenoplectus triqueter*** (L.) Palla
Schoenoplectus triqueter f. *macrostachys* (Junge) Soó = ***Schoenoplectus triqueter*** (L.) Palla
Schoenoplectus triqueter f. *monostachys* (Junge) Soó = ***Schoenoplectus triqueter*** (L.) Palla
Schoenoplectus triqueter f. *pallidus* (Döll) Soó = ***Schoenoplectus triqueter*** (L.) Palla
Schoenoplectus triqueter var. *segregatus* (C.B.Clarke) Karthik. = ***Schoenoplectus triqueter*** (L.) Palla
Schoenoplectus ×*triquetriformis* V.I.Krecz., Ovcz. & Czukav. = ***Schoenoplectus* ×*carinatus*** nothosubsp. ***triquetriformis*** (V.I.Krecz., Ovcz. & Czukav.) Govaerts
Schoenoplectus validus (Vahl) Á.Löve & D.Löve = ***Schoenoplectus tabernaemontani*** (C.C.Gmel.) Palla
Schoenoplectus validus subsp. *creber* (Fernald) Á.Löve & D.Löve = ***Schoenoplectus lacustris*** (L.) Palla subsp. ***lacustris***
Schoenoplectus validus subsp. *luxurians* (Miq.) Soják = ***Schoenoplectus tabernaemontani*** (C.C.Gmel.) Palla
Schoenoplectus vohemarensis (Cherm.) J.Raynal = ***Schoenoplectiella vohemarensis*** (Cherm.) Lye
Schoenoplectus wallichii (Nees) T.Koyama = ***Schoenoplectiella wallichii*** (Nees) Lye

***Unplaced Names*:**
Schoenoplectus ×*igaensis* (T.Koyama) Hayas. & H.Ohashi, J. Jap. Bot. 75: 224 (2000). = [38 JAP] *Schoenoplectiella juncoides* × *S. lineolatus*
Schoenoplectus ×*juncohotarui* Yashiro, J. Jap. Bot. 79: 97 (2004). = [38 JAP] *Schoenoplectiella juncoides* × *Schoenoplectus hotarui*
Schoenoplectus ×*oguraensis* (T.Koyama) Hayas. & H.Ohashi, J. Jap. Bot. 75: 224 (2000). = [38 JAP] *S. mucronatus* × *Schoenoplectiella wallichii*

Schoenopsis

Schoenopsis P.Beauv. = ***Tetraria*** P.Beauv.
Schoenopsis burmanii (Vahl) Nees = ***Tetraria burmanii*** (Vahl) C.B.Clarke
Schoenopsis flexuosa (Thunb.) Nees = ***Tetraria flexuosa*** (Thunb.) C.B.Clarke

Schoenoxiphium

Schoenoxiphium Nees, Linnaea 7: 581 (1832).
Ethiopia to S. Africa, Madagascar, Arabian Pen., Tibet. 24 ETH 25 KEN TAN UGA 26 MLW 27 CPP LES NAT OFS SWZ TVL 29 MDG 35 YEM 36 CHT.
20 Species
Archaeocarex Börner, Abh. Naturwiss. Vereine Bremen 21: 262 (1913).

Schoenoxiphium altum Kukkonen, Notes Roy. Bot. Gard. Edinburgh 43: 365 (1986).
Cape Prov. 27 CPP. Hemicr. or rhizome geophyte.

Schoenoxiphium basutorum Turrill, Bull. Misc. Inform. Kew 1914: 19 (1914).
Free State to Lesotho. 27 LES OFS. Hemicr. or rhizome geophyte.

Schoenoxiphium bracteosum Kukkonen, Notes Roy. Bot. Gard. Edinburgh 43: 365 (1986).
S. Africa. 27 CPP LES NAT. Hemicr. or rhizome geophyte.

Schoenoxiphium burttii Kukkonen, Notes Roy. Bot. Gard. Edinburgh 43: 365 (1986).
KwaZulu-Natal. 27 NAT. Hemicr. or rhizome geophyte.

Schoenoxiphium caricinum Kük., Bull. Herb. Boissier, II, 4: 49 (1904).
Tibet. 36 CHT. Hemicr. or rhizome geophyte.

Schoenoxiphium distinctum Kukkonen, Bot. Not. 131: 263 (1978).
Lesotho to KwaZulu-Natal. 27 LES NAT. Hemicr. or rhizome geophyte.

Schoenoxiphium dregeanum Kunth, Enum. Pl. 2: 529 (1837). *Archaeocarex dregeana* (Kunth) Pissjauk., Bot. Mater. Gerb. Bot. Inst. Komarova Akad. Nauk S.S.S.R. 12: 83 (1950). *Kobresia dregeana* (Kunth) T.Koyama, J. Fac. Sci. Univ. Tokyo, Sect. 3, Bot. 8: 80 (1961).
Cape Prov. 27 CPP. Hemicr. or rhizome geophyte.
Schoenoxiphium burkei C.B.Clarke, J. Linn. Soc., Bot. 20: 386 (1883).

Schoenoxiphium ecklonii Nees, Linnaea 10: 200 (1836). *Hemicarex ecklonii* (Nees) Benth., J. Linn. Soc., Bot. 18: 367 (1881). *Archaeocarex ecklonii* (Nees) Pissjauk., Bot. Mater. Gerb. Bot. Inst. Komarova Akad. Nauk S.S.S.R. 12: 83 (1950). *Kobresia ecklonii* (Nees) T.Koyama, J. Fac. Sci. Univ. Tokyo, Sect. 3, Bot. 8: 80 (1961).
Cape Prov. 27 CPP. Hemicr. or rhizome geophyte.
Carex capensis Schkuhr, Beschr. Riedgräs. 2: 39 (1806).
Carex bisexualis C.B.Clarke in W.H.Harvey & auct. suc. (eds.), Fl. Cap. 7: 302 (1898).
Carex zeyheri C.B.Clarke in W.H.Harvey & auct. suc. (eds.), Fl. Cap. 7: 303 (1898).
Schoenoxiphium ecklonii var. *unisexuale* Kük. in H.G.A.Engler (ed.), Pflanzenr., IV, 20(38): 33 (1909).

Schoenoxiphium filiforme Kük., Bull. Misc. Inform. Kew 1910: 129 (1910).
S. Africa. 27 CPP LES NAT. Hemicr. or rhizome geophyte.

Schoenoxiphium gracile Cherm., Bull. Soc. Bot. France 70: 300 (1923).
C. Madagascar (Mt. Tsaratanana). 29 MDG. Hemicr.

Schoenoxiphium lanceum (Thunb.) Kük. in H.G.A.Engler (ed.), Pflanzenr., IV, 20(38): 28 (1909).
S. Africa. 27 CPP NAT OFS TVL. Hemicr. or rhizome geophyte.
**Schoenus lanceus* Thunb., Prodr. Pl. Cap.: 17 (1794). *Carex lancea* (Thunb.) Baill., Hist. Pl. 12: 341 (1894). *Kobresia lancea* (Thunb.) T.Koyama, J. Fac. Sci. Univ. Tokyo, Sect. 3, Bot. 8: 80 (1961).
Schoenoxiphium capense Nees, Linnaea 7: 533 (1832).
Carex ramosa Eckl. ex Kunth, Enum. Pl. 2: 531 (1837).
Schoenoxiphium meyerianum Kunth, Enum. Pl. 2: 530 (1837).
Schoenoxiphium sickmannianum Kunth, Enum. Pl. 2: 530 (1837). *Hemicarex sickmanniana* (Kunth) Benth., J. Linn. Soc., Bot. 18: 367 (1881).
Carex schimperiana Boeck., Linnaea 40: 373 (1876). *Schoenoxiphium schimperianum* (Boeck.) C.B.Clarke, Kew Bull., Addit. Ser. 8: 67 (1908).
Hemicarex meyeriana Benth., J. Linn. Soc., Bot. 18: 367 (1881).

Schoenoxiphium lehmannii (Nees) Kunth ex Steud., Syn. Pl. Glumac. 2: 245 (1855).
Ethiopia to S. Africa. 24 ETH 25 KEN TAN UGA 26 27 CPP NAT OFS TVL. Hemicr.
**Uncinia lehmannii* Nees, Linnaea 10: 206 (1836). *Kobresia lehmannii* (Nees) T.Koyama, J. Fac. Sci. Univ. Tokyo, Sect. 3, Bot. 8: 80 (1961).
Carex uhligii K.Schum. ex C.B.Clarke, Bot. Jahrb. Syst. 38: 136 (1906).

Schoenoxiphium madagascariense Cherm., Bull. Soc. Bot. France 70: 299 (1923).
S. Africa, Madagascar. 27 NAT OFS TVL 29 MDG. Hemicr.

Schoenoxiphium molle Kukkonen, Notes Roy. Bot. Gard. Edinburgh 43: 366 (1986).
KwaZulu-Natal. 27 NAT. Hemicr. or rhizome geophyte.

Schoenoxiphium perdensum Kukkonen, Bot. Not. 131: 265 (1978).
S. Africa. 27 CPP NAT OFS. Hemicr. or rhizome geophyte.

Schoenoxiphium rufum Nees, Linnaea 10: 201 (1836). *Carex rufa* (Nees) Baill., Hist. Pl. 12: 340 (1894), nom. illeg. *Archaeocarex rufus* (Nees) Fedde & J.Schust., Just's Bot. Jahresber. 41(2): 7 (1913 publ. 1918). *Kobresia rufa* (Nees) T.Koyama, J. Fac. Sci. Univ. Tokyo, Sect. 3, Bot. 8: 80 (1961).
S. Tanzania to S. Africa. 25 TAN 26 MLW 27 CPP LES NAT SWZ TVL. Hemicr.
Schoenoxiphium ludwigii Hochst., Flora 28: 764 (1845).
Schoenoxiphium buchananii C.B.Clarke in T.A.Durand & H.Schinz, Consp. Fl. Afric. 5: 676 (1894), nom. nud. *Kobresia buchananii* (C.B.Clarke) T.Koyama, J. Fac. Sci. Univ. Tokyo, Sect. 3, Bot. 8: 80 (1961).
Carex buchananii C.B.Clarke in W.H.Harvey & auct. suc. (eds.), Fl. Cap. 7: 305 (1898), nom. illeg.
Schoenoxiphium rufum var. *pondoense* Kük. in H.G.A. Engler (ed.), Pflanzenr., IV, 20(38): 31 (1909).

Schoenoxiphium schweickerdtii Merxm. & Podlech, Mitt. Bot. Staatssamml. München 3: 529 (1960).
S. Africa. 27 CPP NAT TVL. Hemicr. or rhizome geophyte.

Schoenoxiphium sparteum (Wahlenb.) C.B.Clarke, Kew Bull., Addit. Ser. 8: 67 (1908).
Ethiopia to S. Africa, Arabian Pen. 24 ETH 25 KEN TAN UGA 26 27 CPP LES NAT OFS SWZ TVL 35 YEM. Hemicr.
Carex indica Schkuhr, Beschr. Riedgräs. 1: 37 (1801), nom. illeg.
**Carex spartea* Wahlenb., Kongl. Vetensk. Acad. Nya Handl. 1803: 149 (1803). *Uncinia spartea* (Wahlenb.) Nees, Linnaea 10: 205 (1836). *Archaeocarex spartea* (Wahlenb.) Pissjauk., Bot. Mater. Gerb. Bot. Inst. Komarova Akad. Nauk S.S.S.R. 12: 83 (1950). *Kobresia spartea* (Wahlenb.) T.Koyama, J. Fac. Sci. Univ. Tokyo, Sect. 3, Bot. 8: 80 (1961).
Carex dregeana Kunth, Enum. Pl. 2: 511 (1837).
Carex esenbeckiana Boeck., Linnaea 40: 372 (1876).
Carex sprengelii Boeck., Linnaea 40: 371 (1876), nom. illeg.
Carex bolusii C.B.Clarke in W.H.Harvey & auct. suc. (eds.), Fl. Cap. 7: 304 (1898).
Schoenoxiphium caricoides C.B.Clarke, Kew Bull., Addit. Ser. 8: 67 (1908).
Schoenoxiphium kunthianum Kük. in H.G.A.Engler (ed.), Pflanzenr., IV, 20(38): 31 (1909). *Archaeocarex kunthiana* (Kük.) Pissjauk., Bot. Mater. Gerb. Bot. Inst. Komarova Akad. Nauk S.S.S.R. 12: 83 (1950). *Kobresia kunthiana* (Kük.) T.Koyama, J. Fac. Sci. Univ. Tokyo, Sect. 3, Bot. 8: 80 (1961).

Schoenoxiphium strictum Kukkonen, Notes Roy. Bot. Gard. Edinburgh 43: 366 (1986).
KwaZulu-Natal. 27 NAT. Hemicr. or rhizome geophyte.

Schoenoxiphium thunbergii Nees, Linnaea 9: 305 (1834). *Hemicarex thunbergii* (Nees) Benth., J. Linn. Soc., Bot. 18: 367 (1881). *Archaeocarex thunbergii* (Nees) Pissjauk., Bot. Mater. Gerb. Bot. Inst. Komarova Akad. Nauk S.S.S.R. 12: 83 (1950).
Swaziland. 27 SWZ. Hemicr. or rhizome geophyte.
**Carex capensis* Thunb., Prodr. Pl. Cap.: 14 (1794).

***Synonyms*:**
Schoenoxiphium buchananii C.B.Clarke = ***Schoenoxiphium rufum*** Nees
Schoenoxiphium burkei C.B.Clarke = ***Schoenoxiphium dregeanum*** Kunth
Schoenoxiphium capense Nees = ***Schoenoxiphium lanceum*** (Thunb.) Kük.
Schoenoxiphium caricoides C.B.Clarke = ***Schoenoxiphium sparteum*** (Wahlenb.) C.B.Clarke
Schoenoxiphium clarkeanum Kük. = ***Kobresia fragilis*** C.B.Clarke
Schoenoxiphium ecklonii var. *unisexuale* Kük. = ***Schoenoxiphium ecklonii*** Nees
Schoenoxiphium fragile (C.B.Clarke) C.B.Clarke = ***Kobresia fragilis*** C.B.Clarke
Schoenoxiphium hissaricum Pissjauk. = ***Kobresia laxa*** Nees
Schoenoxiphium kobresioideum Kük. = ***Kobresia kobresioidea*** (Kük.) J.Kern
Schoenoxiphium kuekenthalianum (Hand.-Mazz.) K.V. Ivanova = ***Kobresia kuekenthaliana*** Hand.-Mazz.
Schoenoxiphium kunthianum Kük. = ***Schoenoxiphium sparteum*** (Wahlenb.) C.B.Clarke
Schoenoxiphium laxum (Nees) K.V.Ivanova = ***Kobresia laxa*** Nees
Schoenoxiphium leucocephalum Thorel ex E.G.Camus = ***Carex tricephala*** Boeck.
Schoenoxiphium ludwigii Hochst. = ***Schoenoxiphium rufum*** Nees
Schoenoxiphium meyerianum Kunth = ***Schoenoxiphium lanceum*** (Thunb.) Kük.
Schoenoxiphium rufum var. *pondoense* Kük. = ***Schoenoxiphium rufum*** Nees
Schoenoxiphium schimperianum (Boeck.) C.B.Clarke = ***Schoenoxiphium lanceum*** (Thunb.) Kük.
Schoenoxiphium sickmannianum Kunth = ***Schoenoxiphium lanceum*** (Thunb.) Kük.

Schoenus

Schoenus L., Sp. Pl.: 42 (1753).
Cosmopolitan. 10 DEN FIN GRB IRE NOR SVA SWE 11 AUT BGM CZE GER HUN NET POL SWI 12 BAL COR FRA POR SAR SPA 13 ALB BUL GRC ITA KRI ROM SIC TUE YUG 14 BLR BLT RUC RUE RUS RUW UKR 20 ALG LBY MOR TUN 24 ERI ETH SOM 27 CPP NAT OFS TVL 32 KGZ TKM TZK UZB 33 TCS 34 EAI IRN IRQ LBS PAL SIN TUR 35 SAU YEM 36 CHC CHS 38 JAP NNS OGA TAI 40 PAK WHM 41 THA VIE 42 BOR JAW LSI MLY MOL PHI SUL SUM 43 BIS NWG SOL 50 NSW NTA NZN QLD SOA TAS VIC WAU 51 ATP CTM NZN NZS 60 FIJ NWC 62 CRL MRN 76 CAL NEV 77 TEX 78 FLA 79 MXE MXS MXT 80 BLZ HON 81 BAH CUB DOM JAM 82 84 BZS 85 AGS CLC CLS URU.
108 Species
Melanoschoenos Ség., Pl. Veron. 3: 70 (1754).
Chaetospora R.Br., Prodr.: 232 (1810).
Streblidia Link, Hort. Berol. 1: 276 (1827).
Helothrix Nees, Ann. Mag. Nat. Hist. 6: 45 (1841).
Isoschoenus Nees, Ann. Mag. Nat. Hist. 6: 49 (1841).
Cyclocampe Steud., Syn. Pl. Glumac. 2: 156 (1855).
Gymnochaeta Steud., Syn. Pl. Glumac. 2: 156 (1855).
Ptilanthelium Steud., Syn. Pl. Glumac. 2: 166 (1855).
Lepidospora (F.Muell.) F.Muell., Fragm. 9: 34 (1875).
Lophocarpus Boeck., Allg. Bot. Z. Syst. 2: 110 (1896).
Neolophocarpus E.G.Camus in H.Lecomte, Fl. Indo-Chine 7: 148 (1912).

Schoenus absconditus Kük., Repert. Spec. Nov. Regni Veg. 44: 99 (1938).
Tasmania. 50 TAS. Hemicr. or rhizome geophyte.

Schoenus achaetus (T.Koyama) T.Koyama, Allertonia 1: 341 (1978).
Caroline Is. (Pohnpei), Fiji (Viti Levu). 60 FIJ 62 CRL. Hemicr. or rhizome geophyte.
**Schoenus tendo* subsp. *achaetus* T.Koyama, Micronesica 1: 104 (1964).

Schoenus acuminatus R.Br., Prodr.: 231 (1810).
SW. Western Australia. 50 WAU. Hemicr. or rhizome geophyte.
Isoschoenus acuminatus Nees, Ann. Mag. Nat. Hist. 6: 49 (1841).
Schoenus acuminatus var. *evolutus* Kük., Repert. Spec. Nov. Regni Veg. 44: 75 (1938).

Schoenus andinus (Phil.) H.Pfeiff., Repert. Spec. Nov. Regni Veg. 23: 348 (1927).
Chile to S. Argentina. 85 AGS CLC CLS. Hemicr. or rhizome geophyte.
**Carpha andina* Phil., Anales Univ. Chile 1873: 555 (1873).
Schoenus sodalium Har., Bull. Soc. Bot. France 31: 154 (1884). *Elynanthus sodalium* (Har.) Planch., Miss. Sci. Cap Horn 5: 357 (1889).

Schoenus andrewsii W.Fitzg., Proc. Linn. Soc. New South Wales 28: 112 (1904).
WSW. Western Australia. 50 WAU. Hemicr. or rhizome geophyte.

Schoenus antarcticus (Hook.f.) Dusén, Ergebn. Schwed. Exp. Magell. 3(5): 213 (1900).
Magellan Reg. 85 AGS CLS. Hemicr. or rhizome geophyte.
**Chaetospora antarctica* Hook.f., Fl. Antarct. 2: 361 (1846). *Carpha antarctica* (Hook.f.) C.B.Clarke, Bot. Jahrb. Syst. 30(68): 40 (1902). *Elynanthus antarcticus* (Hook.f.) Macloskie, Rep. Princeton Univ. Exp. Patagonia, Botany 8: 267 (1905).

Schoenus apogon Roem. & Schult., Syst. Veg. 2: 77 (1817).
Temp. E. Asia to Indo-China, E. & SE. Australia to New Zealand. 38 JAP NNS TAI 41 VIE 50 NSW QLD SOA TAS VIC 51 NZN NZS. Ther.
**Chaetospora imberbis* R.Br., Prodr.: 233 (1810). *Helothrix imberbis* (R.Br.) Palla, Allg. Bot. Z. Syst. 8: 68 (1902). *Schoenus brownii* Hook.f., Handb. N. Zeal. Fl. 1: 298 (1864).

var. ***apogon***
Temp. E. Asia to Indo-China, E. & SE. Australia to New Zealand. 38 JAP NNS TAI 41 VIE 50 NSW QLD SOA TAS VIC 51 NZN NZS. Ther.
Isolepis margaritifera Nees, Ann. Mag. Nat. Hist. 6: 46 (1841). *Scirpus margaritiferus* (Nees) Boeck., Linnaea 36: 697 (1870).

Chaetospora tenuissima Steud., Syn. Pl. Glumac. 2: 319 (1855).
Schoenus laxiflorus Steud., Syn. Pl. Glumac. 2: 166 (1855). *Schoenus apogon* var. *laxiflorus* (Steud.) C.B.Clarke ex Cheeseman, Man. New Zealand Fl.: 783 (1906).
Chaetospora albescens Franch. & Sav., Enum. Pl. Jap. 2: 548 (1878).
Chaetospora japonica Franch. & Sav., Enum. Pl. Jap. 2: 548 (1878).
Schoenus vacillans Kirk, Trans. & Proc. New Zealand Inst. 10: 421 (1878).
Chaetospora umbellulifera Boeck., Flora 65: 28 (1882).
Schoenus apogon var. *reductus* Kük., Repert. Spec. Nov. Regni Veg. 44: 93 (1938).

var. ***caespitans*** (Petrie) Edgar, Fl. New Zeal. 2: 196 (1970).
New Zealand North I. 51 NZN. Hemicr. or rhizome geophyte.
**Schoenus caespitans* Petrie, Trans. & Proc. New Zealand Inst. 56: 9 (1926).

Schoenus armeria (Nees) Boeck., Linnaea 38: 279 (1874).
SW. Australia. 50 WAU. Hemicr. or rhizome geophyte.
**Isoschoenus armeria* Nees, Ann. Mag. Nat. Hist. 6: 49 (1841).

Schoenus asperocarpus F.Muell., Fragm. 9: 29 (1875).
SW. Australia. 50 WAU. Hemicr. or rhizome geophyte.

Schoenus badius Rye, Nuytsia 11: 263 (1997).
Western Australia. 50 WAU. Hemicr. or rhizome geophyte.

Schoenus benthamii F.Muell., Syst. Census Austral. Pl. 1: 127 (1882).
WSW. Western Australia. 50 WAU. Hemicr. or rhizome geophyte.
Chaetospora compressa Nees in J.G.C.Lehmann, Pl. Preiss. 2: 85 (1846), nom. illeg. *Schoenus compressus* Benth., Fl. Austral. 7: 357 (1878), nom. illeg.

Schoenus bifidus (Nees) Boeck., Linnaea 38: 282 (1874).
SW. Australia. 50 WAU. Hemicr. or rhizome geophyte.
**Elynanthus bifidus* Nees, Ann. Mag. Nat. Hist. 6: 48 (1841).
Chaetospora oligostachya F.Muell., Fragm. 9: 38 (1875).
Schoenus oligostachyus F.Muell., Fragm. 9: 38 (1875).
Schoenus jamesonianus W.Fitzg., Proc. Linn. Soc. New South Wales 27: 245 (1903). *Schoenus bifidus* var. *jamesonianus* (W.Fitzg.) Kük., Repert. Spec. Nov. Regni Veg. 44: 79 (1938).

Schoenus biglumis Kük., Repert. Spec. Nov. Regni Veg. 44: 164 (1938).
Tasmania. 50 TAS. Hemicr. or rhizome geophyte.

Schoenus breviculmis Benth., Fl. Austral. 7: 364 (1878).
S. Australia. 50 SOA VIC WAU. Hemicr. or rhizome geophyte.

Schoenus brevifolius R.Br., Prodr.: 231 (1810).
Ogasawara-shoto to New Zealand. 38 OGA 50 NSW QLD TAS VIC WAU 51 NZN NZS 60 NWC. Hemicr. or rhizome geophyte.
Chaetospora tenax Hook.f., Fl. Nov.-Zel. 1: 273 (1853). *Schoenus tenax* (Hook.f.) Hook.f., Handb. N. Zeal. Fl. 1: 298 (1864). *Schoenus brevifolius* subsp. *tenax* (Hook.f.) Kük., Repert. Spec. Nov. Regni Veg. 44: 69 (1938).

Schoenus brevisetis (R.Br.) Roem. & Schult., Syst. Veg. 2: 77 (1817).
SW. Australia. 50 WAU. Hemicr. or rhizome geophyte.
**Chaetospora brevisetis* R.Br., Prodr.: 232 (1810).
Schoenus flaviculmis Nees in J.G.C.Lehmann, Pl. Preiss. 2: 81 (1846).
Schoenus brevisetis var. *subimberbis* Benth., Fl. Austral. 7: 361 (1878).
Schoenus fuscescens C.B.Clarke, Bot. Jahrb. Syst. 35: 79 (1904). *Schoenus brevisetis* var. *fuscescens* (C.B.Clarke) Kük., Repert. Spec. Nov. Regni Veg. 44: 174 (1938).

Schoenus caespititius W.Fitzg., Proc. Linn. Soc. New South Wales 28: 110 (1904).
SW. Australia. 50 WAU. Hemicr. or rhizome geophyte.

Schoenus calcatus K.L.Wilson, Nuytsia 11: 276 (1997).
Western Australia. 50 WAU. Hemicr. or rhizome geophyte.

Schoenus calostachyus (R.Br.) Poir. in J.B.A.M.de Lamarck, Encycl., Suppl. 2: 251 (1811).
Nansei-shoto to E. Australia. 38 NNS 41 THA VIE 42 BOR MLY SUM 43 NWG 50 NSW NTA QLD 62 CRL. Hemicr. or rhizome geophyte.
**Chaetospora calostachys* R.Br., Prodr.: 233 (1810).
Cyclocampe waigiouensis Steud., Syn. Pl. Glumac. 2: 156 (1855).
Schoenus triangularis Volkens, Bot. Jahrb. Syst. 31: 458 (1901).

Schoenus calyptratus Kük., Repert. Spec. Nov. Regni Veg. 48: 248 (1940).
SE. Australia. 50 NSW TAS VIC. Hemicr. or rhizome geophyte.

Schoenus capillifolius D.A.Cooke, Muelleria 4: 299 (1981).
WSW. Western Australia. 50 WAU. Hemicr. or rhizome geophyte.

Schoenus carsei Cheeseman, Man. New Zealand Fl.: 781 (1906).
SE. Australia, New Zealand. 50 NZN SOA TAS VIC. Hemicr. or rhizome geophyte.
Schoenus monocarpus J.M.Black, Trans. & Proc. Roy. Soc. South Australia 52: 225 (1928). *Tetraria monocarpa* (J.M.Black) J.M.Black, Trans. & Proc. Roy. Soc. South Australia 58: 168 (1934). *Schoenus carsei* var. *monocarpus* (J.M.Black) Kük., Repert. Spec. Nov. Regni Veg. 44: 69 (1938). *Machaerina monocarpa* (J.M.Black) T.Koyama, Bot. Mag. (Tokyo) 69: 64 (1956).
Cladium monocarpum J.M.Black, Trans. & Proc. Roy. Soc. South Australia 53: 261 (1929).

Schoenus centralis Latz, J. Adelaide Bot. Gard. 1: 348 (1979).
C. Australia. 50 NSW NTA QLD WAU. Ther.

Schoenus clandestinus S.T.Blake, Proc. Roy. Soc. Queensland 60: 49 (1950).
WSW. Western Australia. 50 WAU. Hemicr. or rhizome geophyte.

Schoenus cruentus (Nees) Benth., Fl. Austral. 7: 357 (1878).
SW. Western Australia. 50 WAU. Hemicr. or rhizome geophyte.
**Chaetospora cruenta* Nees in J.G.C.Lehmann, Pl. Preiss. 2: 85 (1846).

Schoenus curvifolius (R.Br.) Roem. & Schult., Syst. Veg. 2: 76 (1817).
SW. Australia. 50 WAU. Hemicr. or rhizome geophyte.
**Chaetospora curvifolia* R.Br., Prodr.: 232 (1810).
Chaetospora aurata Nees, Ann. Mag. Nat. Hist. 6: 49 (1841).

Schoenus curvulus F.Muell., Trans. Roy. Soc. Victoria, n.s., 1(2): 36 (1889).
N. Borneo, SW. Sulawesi, New Guinea. 42 BOR SUL 43 NWG. Hemicr. or rhizome geophyte.
Schoenus kinabaluensis Stapf, J. Linn. Soc., Bot. 42: 176 (1914).

Schoenus cygneus (Nees) Nees in J.G.C.Lehmann, Pl. Preiss. 2: 81 (1846).
SW. Australia. 50 WAU. Hemicr. or rhizome geophyte.
**Chaetospora cygnea* Nees, Ann. Mag. Nat. Hist. 6: 49 (1841). *Schoenus brevisetis* var. *cygneus* (Nees) Kük., Repert. Spec. Nov. Regni Veg. 44: 175 (1938).

Schoenus deformis (R.Br.) Roem. & Schult., Syst. Veg. 2: 76 (1817).
South Australia to Victoria. 50 SOA VIC. Hemicr. or rhizome geophyte.
**Chaetospora deformis* R.Br., Prodr.: 232 (1810).

Schoenus delicatulus (Fernald) J.Kern, Blumea, Suppl. 4: 167 (1958).
N. Borneo, Philippines (Palawan). 42 BOR PHI. Hemicr. or rhizome geophyte.
Cladium filiforme Merr., Philipp. J. Sci., C 5: 172 (1910). **Mariscus delicatulus* Fernald, Rhodora 25: 53 (1923).

Schoenus discifer Tate, Trans. & Proc. Roy. South Australia 12: 131 (1889).
SW. Western Australia, SE. South Australia. 50 SOA WAU. Hemicr. or rhizome geophyte.

Schoenus efoliatus F.Muell., Fragm. 9: 32 (1875).
SW. Western Australia. 50 WAU. Hemicr. or rhizome geophyte.

Schoenus elegans S.T.Blake, Proc. Roy. Soc. Queensland 52: 60 (1941).
WSW. Western Australia. 50 WAU. Hemicr. or rhizome geophyte.

Schoenus ericetorum R.Br., Prodr.: 231 (1810).
SE. Queensland to New South Wales. 50 NSW QLD. Hemicr. or rhizome geophyte.

Schoenus evansianus K.L.Wilson, Telopea 5: 615 (1994).
New South Wales. 50 NSW. Hemicr. or rhizome geophyte.

Schoenus falcatus R.Br., Prodr.: 232 (1810).
Nansei-shoto to Indo-China and N. Australia. 38 NNS TAI 41 THA VIE 42 JAW LSI MLY 43 NWG SOL 50 NTA QLD WAU. Hemicr. or rhizome geophyte.
Schoenus elatus Boeck., Flora 58: 117 (1875).
Lophocarpus bonia Boeck., Allg. Bot. Z. Syst. 2: 110 (1896).
Lophocarpus tonquinensis Boeck., Allg. Bot. Z. Syst. 2: 111 (1896). *Neolophocarpus tonquinensis* (Boeck.) E.G.Camus in H.Lecomte, Fl. Indo-Chine 7: 149 (1912).

Schoenus ferrugineus L., Sp. Pl.: 43 (1753). *Streblidia ferruginea* (L.) Link, Hort. Berol. 1: 276 (1827). *Chaetospora ferruginea* (L.) Rchb., Fl. Germ. Excurs.: 74 (1830).
Europe. 10 DEN FIN GRB† NOR SWE 11 AUT CZE GER HUN? SWI 12 FRA 13 ALB bul ITA ROM YUG 14 BLR BLT RUC RUE RUS RUW UKR. Hemicr.
Schoenus nigricans Hoppe, Bot. Taschenb. Anfänger Wiss. Apothekerkunst 1800: 135 (1800), sensu auct.
Schoenus karpatii Pénzes, Acta Bot. Acad. Sci. Hung. 8: 333 (1962).

Schoenus fluitans Hook.f., Fl. Tasman. 2: 81 (1858).
SE. Australia, New Zealand North I. 50 SOA TAS VIC 51 NZN. Hemicr. or rhizome geophyte.

Schoenus globifer Nees in J.G.C.Lehmann, Pl. Preiss. 2: 81 (1846). *Schoenus armeria* var. *globifer* (Nees) Kük., Repert. Spec. Nov. Regni Veg. 44: 178 (1938).
Western Australia. 50 WAU. Hemicr. or rhizome geophyte.

Schoenus grammatophyllus F.Muell., Fragm. 9: 31 (1875).
WSW. Western Australia. 50 WAU. Hemicr. or rhizome geophyte.

Schoenus grandiflorus (Nees) F.Muell., Fragm. 9: 30 (1875).
SW. Australia. 50 WAU. Hemicr. or rhizome geophyte.
**Elynanthus grandiflorus* Nees in J.G.C.Lehmann, Pl. Preiss. 2: 78 (1846).

Schoenus griffinianus K.L.Wilson, Nuytsia 11: 278 (1997).
SW. Australia. 50 WAU. Hemicr. or rhizome geophyte.

Schoenus hexandrus F.Muell. & Tate, Trans. & Proc. Rep. Roy. Soc. South Australia 16: 380 (1896).
SC. & SW. Western Australia. 50 WAU. Hemicr. or rhizome geophyte.

Schoenus humilis Benth., Fl. Austral. 7: 374 (1878).
SW. Australia. 50 WAU. Hemicr. or rhizome geophyte.

Schoenus imberbis R.Br., Prodr.: 231 (1810). *Schoenus ericetorum* var. *imberbis* (R.Br.) Kük., Repert. Spec. Nov. Regni Veg. 44: 179 (1938).
New South Wales to Victoria. 50 NSW VIC. Hemicr. or rhizome geophyte.

Schoenus indutus Benth., Fl. Austral. 7: 372 (1878).
WSW. Western Australia. 50 WAU. Hemicr. or rhizome geophyte.

Schoenus insolitus K.L.Wilson, Nuytsia 11: 280 (1997).
Western Australia. 50 WAU. Hemicr. or rhizome geophyte.

Schoenus kennyi (F.M.Bailey) S.T.Blake, Proc. Roy. Soc. Queensland 51: 48 (1939 publ. 1940).
E. Australia. 50 NSW QLD. Hemicr. or rhizome geophyte.
**Arthrostylis kennyi* F.M.Bailey, Queensland Agric. J. 28: 278 (1912). *Schoenus subaphyllus* var. *kennyi* (F.M.Bailey) Kük., Repert. Spec. Nov. Regni Veg. 48: 250 (1940).

Schoenus laevigatus W.Fitzg., Proc. Linn. Soc. New South Wales 28: 111 (1904).
S. Australia. 50 SOA VIC WAU. Hemicr. or rhizome geophyte.
Schoenus brachyphyllus F.Muell. ex J.M.Black, Fl. S. Austral. 4: 677 (1929).

Schoenus laevinux (Kük.) Ohwi, J. Jap. Bot. 18: 136 (1942).
Maluku to Caroline Is. 42 MOL 43 BIS NWG 62 CRL. Hemicr. or rhizome geophyte.
**Schoenus melanostachyus* var. *laevinux* Kük., Bot. Jahrb. Syst. 59: 52 (1924).

Schoenus lanatus Labill., Nov. Holl. Pl. 1: 19 (1805). *Chaetospora lanata* (Labill.) R.Br., Prodr.: 232 (1810).
SW. Australia. 50 WAU. Hemicr. or rhizome geophyte.

Schoenus latilaminatus Kük., Repert. Spec. Nov. Regni Veg. 44: 88 (1938).
SE. Queensland to SE. Australia. 50 NSW QLD SOA TAS VIC. Hemicr. or rhizome geophyte.

Schoenus latitans S.T.Blake, Proc. Roy. Soc. Queensland 60: 50 (1950).
WSW. Western Australia. 50 WAU. Hemicr. or rhizome geophyte.

Schoenus lepidosperma (F.Muell.) K.L.Wilson, Telopea 5: 619 (1994).
SE. Queensland to SE. Australia. 50 NSW QLD SOA TAS VIC. Hemicr.
**Chaetospora lepidosperma* F.Muell., Monthly Not. Pap. & Proc. Roy. Soc. Tasmania: 12 (1869).

subsp. ***lepidosperma***
SE. Australia. 50 NSW SOA TAS VIC. Hemicr.
Chaetospora tenuissima Hook.f., Fl. Tasman. 2: 81 (1858), nom. illeg. *Lepidospora tenuissima* F.Muell., Fragm. 9: 34 (1875).

subsp. ***pachylepis*** (S.T.Blake) K.L.Wilson, Telopea 5: 619 (1994).
SE. Queensland to Victoria. 50 NSW QLD VIC. Hemicr. or rhizome geophyte.
**Schoenus pachylepis* S.T.Blake, Proc. Roy. Soc. Queensland 54: 69 (1943).

Schoenus loliaceus Kük., Repert. Spec. Nov. Regni Veg. 44: 86 (1938).
SW. Australia. 50 WAU. Hemicr. or rhizome geophyte.

Schoenus longibracteatus Kük., Repert. Spec. Nov. Regni Veg. 44: 70 (1938).
N. Borneo, W. New Guinea. 42 BOR 43 NWG. Hemicr.
Schoenus neoguineensis Kük., Repert. Spec. Nov. Regni Veg. 44: 77 (1938).

Schoenus lymansmithii M.T.Strong, Harvard Pap. Bot. 4: 275 (1999).
Brazil (Santa Catarina). 84 BZS. Hemicr. or rhizome geophyte.

Schoenus maschalinus Roem. & Schult., Syst. Veg. 2: 77 (1817).
Malesia to New Zealand and Marianas. 42 PHI SUM 43 NWG 50 NSW QLD SOA TAS VIC 51 CTM NZN NZS 62 MRN. Hemicr.
Chaetospora axillaris R.Br., Prodr.: 233 (1810). *Schoenus axillaris* (R.Br.) Hook.f., Handb. N. Zeal. Fl.: 298 (1864), nom. illeg. *Helothrix axillaris* (R.Br.) Palla, Allg. Bot. Z. Syst. 8: 68 (1902). *Schoenus subaxillaris* Kük., Repert. Spec. Nov. Regni Veg. 44: 89 (1938).
Helothrix pusilla Nees, Ann. Mag. Nat. Hist. 6: 45 (1841).
Scirpus foliatus Hook.f., London J. Bot. 3: 414 (1844). *Schoenus foliatus* (Hook.f.) S.T.Blake, Proc. Roy. Soc. Queensland 51: 48 (1939 publ. 1940).
Scleria setifera Boeck., Flora 41: 648 (1858).
Chaetospora axillaris f. *terrestris* Boeck., Linnaea 38: 289 (1874). *Schoenus subaxillaris* f. *terrestris* (Boeck.) Kük., Repert. Spec. Nov. Regni Veg. 44: 90 (1938).
Helothrix philippinensis Palla, Allg. Bot. Z. Syst. 17(Beil.): 1 (1911). *Schoenus philippinensis* (Palla) Kük. ex Merr., Enum. Philipp. Fl. Pl. 1: 128 (1923).
Schoenus mariannae Hosok., Trans. Nat. Hist. Soc. Taiwan 25: 263 (1935).
Schoenus erythrosiphon Ohwi, Bot. Mag. (Tokyo) 56: 205 (1942).

Schoenus melanostachyus R.Br., Prodr.: 231 (1810).
N. Borneo, Philippines (Mindoro), E. Australia. 42 BOR PHI 50 NSW QLD VIC. Hemicr. or rhizome geophyte.
Schoenus melanostachyus var. *altissimus* F.Muell., Fragm. 9: 29 (1875).
Schoenus halconensis Kük., Repert. Spec. Nov. Regni Veg. 44: 65 (1938).

Schoenus minutulus F.Muell., Fragm. 9: 32 (1875).
S. Western Australia. 50 WAU. Hemicr. or rhizome geophyte.

Schoenus moorei Benth., Fl. Austral. 7: 367 (1878).
New South Wales. 50 NSW. Hemicr. or rhizome geophyte.

Schoenus multiglumis Benth., Fl. Austral. 7: 368 (1878).
SW. Australia. 50 WAU. Hemicr. or rhizome geophyte.

Schoenus nanus (Nees) Benth., Fl. Austral. 7: 364 (1878).
S. Australia. 50 SOA VIC WAU. Hemicr. or rhizome geophyte.
**Chaetospora nana* Nees in J.G.C.Lehmann, Pl. Preiss. 2: 85 (1846).

Schoenus natans (F.Muell.) Benth., Fl. Austral. 7: 375 (1878).
SW. Australia. 50 WAU. Hemicr. or rhizome geophyte.
**Chaetospora natans* F.Muell., Fragm. 9: 38 (1875).

Schoenus nigricans L., Sp. Pl.: 43 (1753). *Chaetospora nigricans* (L.) Kunth, Enum. Pl. 2: 323 (1837).
Cosmopolitan. 10 DEN GRB IRE NOR† SVA SWE 11 AUT BGM CZE GER HUN NET POL SWI 12 BAL COR FRA POR SAR SPA 13 ALB BUL GRC ITA KRI ROM SIC TUE YUG 14 BLR BLT RUS UKR 20 ALG LBY MOR TUN 24 ERI ETH SOM 27 CPP NAT OFS TVL 32 KGZ TKM TZK UZB 33 TCS 34 EAI IRN IRQ LBS PAL SIN TUR 35 SAU YEM 40 PAK WHM 76 CAL NEV 77 TEX 78 FLA 79 MXE MXS MXT 80 BLZ HON 81 BAH CUB DOM JAM 82 85 URU. Hemicr.
Schoenus aggregatus Thunb., Phytogr. Blätt. 1: 5 (1803).
Schoenus compressus C.Presl, Cyper. Gramin. Sicul.: 12 (1820), nom. illeg.
Schoenus hypomelas Spreng., Neue Entdeck. Pflanzenk. 3: 8 (1822).

Schoenus bonannii C.Presl, Fl. Sicul.: xliii (1826). *Blysmus bonannii* (C.Presl) Steud., Nomencl. Bot., ed. 2, 1: 212 (1840).
Elynanthus spathaceus Nees, Linnaea 10: 186 (1836).
Chaetospora oligostachya Boeck., Flora 42: 101 (1859).
Schoenus nigricans var. *ambiguus* Kük., Repert. Spec. Nov. Regni Veg. 44: 171 (1938).
Schoenus nigricans var. *plurisquamatus* Kük., Repert. Spec. Nov. Regni Veg. 44: 171 (1938).

Schoenus nitens (R.Br.) Roem. & Schult., Syst. Veg. 2: 77 (1817).
W. New Guinea, Australia, New Zealand. 43 NWG 50 NSW QLD SOA TAS VIC WAU 51 CTM NZN NZS. Hemicr. or rhizome geophyte.
**Chaetospora nitens* R.Br., Prodr.: 233 (1810). *Scirpus nitens* (R.Br.) Boeck., Linnaea 36: 696 (1870), nom. illeg.

var. ***concinnus*** (Hook.f.) Cheeseman, Man. New Zealand Fl.: 784 (1906).
New Zealand. 51 NZN NZS. Hemicr. or rhizome geophyte.
**Chaetospora concinna* Hook.f., Fl. Nov.-Zel. 1: 274 (1853). *Schoenus concinnus* (Hook.f.) Hook.f., Handb. N. Zeal. Fl. 1: 299 (1864).
Schoenus moorei Kirk, Trans. & Proc. New Zealand Inst. 13: 384 (1881), nom. illeg.

var. ***nitens***
W. New Guinea, Australia, New Zealand. 43 NWG 50 NSW QLD SOA TAS VIC WAU 51 CTM NZN NZS. Hemicr. or rhizome geophyte.
Chaetospora lucens Poir. in G.-F.Cuvier, Dict. Sci. Nat. 8: 58 (1817).
Scirpus spadiceus Phil., Linnaea 29: 80 (1857), nom. illeg.
Isolepis krausei Phil., Linnaea 30: 203 (1860). *Schoenus krausei* (Phil.) Gunckel, Anales Mus. Hist. Nat. Valparaiso 4: 42 (1971).
Schoenus nitens var. *major* Ewart & Jean White, Proc. Roy. Soc. Victoria, n.s., 22: 326 (1910).

Schoenus nudifructus C.Chen, in Fl. Reipubl. Popul. Sin. 11: 230 (1961).
China (Yunnan). 36 CHC. Hemicr. or rhizome geophyte.

Schoenus obtusifolius (Nees) Boeck., Linnaea 38: 281 (1874).
SW. Australia. 50 WAU. Hemicr. or rhizome geophyte.
**Elynanthus obtusifolius* Nees in J.G.C.Lehmann, Pl. Preiss. 2: 78 (1846).

Schoenus odontocarpus F.Muell., Fragm. 9: 32 (1875).
SW. Australia. 50 WAU. Hemicr. or rhizome geophyte.

Schoenus ornithopodioides (Kük.) S.T.Blake, Contr. Queensland Herb. 8: 41 (1969).
SE. Queensland. 50 QLD. Hemicr. or rhizome geophyte.
**Schoenus ericetorum* var. *ornithopodioides* Kük., Repert. Spec. Nov. Regni Veg. 44: 179 (1938).

Schoenus paludosus (R.Br.) Roem. & Schult., Syst. Veg. 2: 77 (1817).
SE. Queensland to New South Wales. 50 NSW QLD. Hemicr. or rhizome geophyte.
**Chaetospora paludosa* R.Br., Prodr.: 233 (1810). *Tricostularia paludosa* (R.Br.) Benth., Fl. Austral. 7: 382 (1878). *Costularia paludosa* (R.Br.) C.B.Clarke, Bull. Misc. Inform. Kew 1899: 114 (1899). *Helothrix paludosa* (R.Br.) Palla, Allg. Bot. Z. Syst. 8: 68 (1902).

Schoenus pauciflorus (Hook.f.) Hook.f., Handb. N. Zeal. Fl. 1: 298 (1864).
New Zealand. 51 ATP CTM NZN NZS. Hemicr. or rhizome geophyte.
**Chaetospora pauciflora* Hook.f., Fl. Nov.-Zel. 1: 273 (1853).

Schoenus pedicellatus (R.Br.) Roem. & Schult., Syst. Veg. 2: 76 (1817).
WSW. Western Australia. 50 WAU. Hemicr. or rhizome geophyte.
**Chaetospora pedicellata* R.Br., Prodr.: 232 (1810).

Schoenus pennisetis S.T.Blake, Proc. Roy. Soc. Queensland 60: 48 (1950).
WSW. Western Australia. 50 WAU. Hemicr. or rhizome geophyte.

Schoenus pleiostemoneus F.Muell., Fragm. 9: 52 (1875).
SW. Australia. 50 WAU. Hemicr. or rhizome geophyte.

Schoenus plumosus Rye, Nuytsia 11: 264 (1997).
SW. Australia. 50 WAU. Hemicr. or rhizome geophyte.

Schoenus punctatus R.Br., Prodr.: 232 (1810). *Cladium punctatum* (R.Br.) H.Pfieff., Repert. Spec. Nov. Regni Veg. 23: 350 (1927). *Machaerina punctata* (R.Br.) T.Koyama, Bot. Mag. (Tokyo) 69: 65 (1956).
Pen. Thailand, New Guinea to N. Australia, Marianas (Guam). 41 THA 43 NWG 50 NTA QLD WAU 62 MRN. Hemicr. or rhizome geophyte.
Cladium aromaticum Merr., Philipp. J. Sci., C 9: 59 (1914). *Mariscus aromaticus* (Merr.) Fernald, Rhodora 25: 51 (1923). *Machaerina aromatica* (Merr.) T.Koyama, Bot. Mag. (Tokyo) 69: 62 (1956).

Schoenus pygmaeus S.T.Blake, Contr. Queensland Herb. 8: 41 (1969).
Tasmania. 50 TAS. Hemicr. or rhizome geophyte.

Schoenus racemosus J.M.Black, Trans. Roy. Soc. South Australia 63: 242 (1939).
South Australia. 50 SOA. Hemicr. or rhizome geophyte.

Schoenus rhynchosporoides (Steud.) Kük., Repert. Spec. Nov. Regni Veg. 44: 19 (1938).
Chile to S. Argentina. 85 AGS CLS. Hemicr. or rhizome geophyte.
Chaetospora laxa Hook.f., Fl. Antarct. 2: 361 (1846). *Elynanthus laxus* (Hook.f.) Macloskie, Rep. Princeton Univ. Exp. Patagonia, Botany 8: 267 (1905). *Carpha laxa* (Hook.f.) H.Pfieff., Repert. Spec. Nov. Regni Veg. 23: 345 (1927).
**Chaetospora rhynchosporoides* Steud., Syn. Pl. Glumac. 2: 162 (1855).
Carpha paniculata Phil., Linnaea 29: 80 (1857).
Chaetospora rhynchospermoides F.Phil., Cat. Pl. Vasc. Chil.: 306 (1881).

Schoenus rigens S.T.Blake, Proc. Roy. Soc. Queensland 60: 49 (1950).
WSW. Western Australia. 50 WAU. Hemicr. or rhizome geophyte.
**Chaetospora distans* F.Muell., Fragm. 9: 35 (1875).

Schoenus rodwayanus W.Fitzg., Proc. Linn. Soc. New South Wales 27: 243 (1903).
SW. Australia. 50 WAU. Hemicr. or rhizome geophyte.

Schoenus scabripes Benth., Fl. Austral. 7: 368 (1878).
SE. Queensland to NE. New South Wales. 50 NSW QLD. Hemicr. or rhizome geophyte.

Schoenus* × *scheuchzeri Brügger, Jahresber. Naturf. Ges. Graubündens, II, 23–24: 119 (1880). *S. ferrugineus* × *S. nigricans*.
Europe. 10 SWE 11 AUT CZE GER SWI 12 FRA. Hemicr. or rhizome geophyte.
Schoenus × *intermedius* Brügger, Jahresber. Naturf. Ges. Graubündens, II, 25: 108 (1882). *Chaetospora* × *intermedia* (Brügger) Beck, Fl. Nieder-Österreich 1: 128 (1896).

Schoenus sculptus (Nees) Boeck., Linnaea 38: 286 (1874).
S. Australia. 50 SOA VIC WAU. Hemicr. or rhizome geophyte.
**Elynanthus sculptus* Nees in J.G.C.Lehmann, Pl. Preiss. 2: 79 (1846).

Schoenus sesquispicula C.B.Clarke, Bot. Jahrb. Syst. 35: 80 (1904).
S. Western Australia. 50 WAU. Hemicr. or rhizome geophyte.

Schoenus setiformis S.T.Blake, J. Arnold Arbor. 29: 92 (1948).
W. New Guinea. 43 NWG. Hemicr. or rhizome geophyte.

Schoenus sinensis Hand.-Mazz., Anz. Akad. Wiss. Wien, Math.-Naturwiss. Kl. 62: 150 (1925).
S. China. 36 CHC CHS. Hemicr. or rhizome geophyte.

Schoenus smitinandii T.Koyama, Brittonia 31: 289 (1979).
E. Thailand. 41 THA. Hemicr. or rhizome geophyte.

Schoenus sparteus R.Br., Prodr.: 231 (1810).
Lesser Sunda Is. to N. Australia. 42 LSI 43 NWG 50 NTA QLD. Hemicr. or rhizome geophyte.
Schoenus fuscoguttatus Ohwi, Bot. Mag. (Tokyo) 56: 206 (1942).

Schoenus subaphyllus Kük., Repert. Spec. Nov. Regni Veg. 44: 179 (1938).
E. & S. Australia. 50 NSW QLD SOA VIC WAU. Hemicr. or rhizome geophyte.
Elynanthus aphyllus F.Muell. ex Boeck., Linnaea 38: 281 (1874), nom. inval.
**Schoenus aphyllus* Boeck., Linnaea 38: 280 (1874), nom. illeg.
Schoenus subaphyllus var. *barbulatus* Kük., Repert. Spec. Nov. Regni Veg. 48: 250 (1940).

Schoenus subbarbatus Kük., Repert. Spec. Nov. Regni Veg. 44: 176 (1938).
SW. Australia. 50 WAU. Hemicr. or rhizome geophyte.
**Schoenus barbatus* Boeck., Linnaea 38: 277 (1874), nom. illeg.

Schoenus subbulbosus Benth., Fl. Austral. 7: 358 (1878).
SW. Australia. 50 WAU. Hemicr. or rhizome geophyte.
Elynanthus capitatus Nees, Ann. Mag. Nat. Hist. 6: 48 (1841). *Schoenus capitatus* (Nees) F.Muell., Fragm. 9: 58 (1875), nom. illeg.
Chaetospora brevisetis F.Muell., Fragm. 9: 37 (1875), nom. illeg.
Schoenus subbulbosus var. *junceus* Benth., Fl. Austral. 7: 358 (1878).

Schoenus subfascicularis Kük., Repert. Spec. Nov. Regni Veg. 44: 32 (1938).
SW. Australia. 50 WAU. Hemicr. or rhizome geophyte.
**Schoenus fascicularis* Nees, Ann. Mag. Nat. Hist. 6: 48 (1841), nom. illeg.

Schoenus subflavus Kük., Repert. Spec. Nov. Regni Veg. 44: 175 (1938).
SW. Australia. 50 WAU. Hemicr. or rhizome geophyte.
**Isoschoenus flavus* Nees, Ann. Mag. Nat. Hist. 6: 49 (1841). *Schoenus flavus* (Nees) Boeck., Linnaea 38: 278 (1874), nom. illeg.
Isoschoenus drummondii Steud., Syn. Pl. Glumac. 2: 167 (1855).

Schoenus sublateralis (Steud.) C.B.Clarke, Kew Bull., Addit. Ser. 8: 44 (1908).
SW. Australia. 50 WAU. Hemicr. or rhizome geophyte.
**Rhynchospora sublateralis* Steud., Syn. Pl. Glumac. 2: 148 (1855).

Schoenus sublaxus Kük., Repert. Spec. Nov. Regni Veg. 44: 77 (1938).
SW. Australia. 50 WAU. Hemicr. or rhizome geophyte.
**Schoenus laxus* W.Fitzg., Proc. Linn. Soc. New South Wales 28: 111 (1904), nom. illeg.

Schoenus submicrostachyus Kük., Repert. Spec. Nov. Regni Veg. 44: 164 (1938).
S. Western Australia. 50 WAU. Hemicr. or rhizome geophyte.
**Chaetospora microstachya* Nees in J.G.C.Lehmann, Pl. Preiss. 2: 84 (1846). *Schoenus microstachyus* (Nees) H.Pfeiff., Repert. Spec. Nov. Regni Veg. 23: 348 (1927), nom. illeg.
Gymnochaeta drummondii Steud., Syn. Pl. Glumac. 2: 156 (1855). *Schoenus drummondii* (Steud.) Benth., Fl. Austral. 7: 259 (1878), nom. illeg.

Schoenus tendo (Hook.f.) Banks & Sol. ex Hook.f., Handb. N. Zeal. Fl. 1: 298 (1864).
New Zealand North I. 51 NZN. Hemicr. or rhizome geophyte.
**Chaetospora tendo* Hook.f., Fl. Nov.-Zel. 1: 273 (1853).
Cyperus tendo Banks & Sol. ex Hook.f., Fl. Nov.-Zel. 1: 273 (1853), nom. inval.

Schoenus tenellus Benth., Fl. Austral. 7: 375 (1878).
Schoenus fluitans var. *tenellus* (Benth.) Kük., Repert. Spec. Nov. Regni Veg. 44: 86 (1938).
SW. Australia. 50 WAU. Hemicr. or rhizome geophyte.

Schoenus tenuissimus Benth., Fl. Austral. 7: 365 (1878).
S. Australia. 50 SOA TAS VIC WAU. Hemicr. or rhizome geophyte.

Schoenus tesquorum J.M.Black, Trans. & Proc. Roy. Soc. South Australia 46: 565 (1922).
SE. Australia. 50 SOA TAS VIC. Hemicr. or rhizome geophyte.

Schoenus trachycarpus F.Muell., Fragm. 9: 33 (1875).
SW. Australia. 50 WAU. Hemicr. or rhizome geophyte.
Schoenus tepperi F.Muell., Fragm. 11: 106 (1881). *Schoenus breviculmis* var. *tepperi* (F.Muell.) Kük., Repert. Spec. Nov. Regni Veg. 44: 99 (1938).

Schoenus turbinatus (R.Br.) Roem. & Schult., Syst. Veg. 2: 76 (1817).
SE. Queensland to Tasmania. 50 NSW QLD TAS VIC. Hemicr. or rhizome geophyte.
**Chaetospora turbinata* R.Br., Prodr.: 232 (1810).
Ptilanthelium chauvinii Steud., Syn. Pl. Glumac. 2: 167 (1855).

Schoenus unispiculatus F.Muell. ex Benth., Fl. Austral. 7: 365 (1878).
SW. Australia. 50 WAU. Hemicr. or rhizome geophyte.

Schoenus vaginatus Benth., Fl. Austral. 7: 371 (1878).
SE. Queensland. 50 QLD. Hemicr. or rhizome geophyte.

Schoenus variicellae Rye, Nuytsia 11: 266 (1997).
SW. Australia. 50 WAU. Hemicr. or rhizome geophyte.

Schoenus villosus R.Br., Prodr.: 231 (1810). *Chaetospora villosa* (R.Br.) Nees ex Steud., Nomencl. Bot., ed. 2, 1: 341 (1840).
SE. Queensland to New South Wales. 50 NSW QLD. Hemicr. or rhizome geophyte.

Schoenus yarrabensis Domin, Biblioth. Bot. 85: 472 (1915). *Costularia yarrabensis* (Domin) H.Pfeiff., Repert. Spec. Nov. Regni Veg. 23: 347 (1927). *Schoenus apogon* var. *yarrabensis* (Domin) Kük., Repert. Spec. Nov. Regni Veg. 44: 93 (1938).
Queensland. 50 QLD. Hemicr. or rhizome geophyte.

***Synonyms*:**
Schoenus aculeatus L. = (Poaceae)
Schoenus acuminatus var. *evolutus* Kük. = ***Schoenus acuminatus*** R.Br.
Schoenus acutus Labill. = ***Machaerina acuta*** (Labill.) J.Kern
Schoenus aggregatus Thunb. = ***Schoenus nigricans*** L.
Schoenus aggregatus Steud. ex Kunth = ***Ficinia filiformis*** (Lam.) Schrad.
Schoenus aggregatus Spreng. ex Kunth = ***Isolepis digitata*** Nees ex Schrad.
Schoenus albus L. = ***Rhynchospora alba*** (L.) Vahl
Schoenus altissimus Salzm. ex Steud. = ***Cladium mariscus*** (L.) Pohl subsp. ***mariscus***
Schoenus anceps (R.Br.) Poir. = ***Gymnoschoenus anceps*** (R.Br.) C.B.Clarke
Schoenus angularis Spreng. ex Kunth = ***Kyllinga vaginata*** Lam.
Schoenus angustifolius Vahl = ***Dulichium arundinaceum*** (L.) Britton var. ***arundinaceum***
Schoenus aphyllus Boeck. = ***Schoenus subaphyllus*** Kük.
Schoenus aphyllus Vahl = ***Actinoschoenus thouarsii*** (Kunth) Benth.
Schoenus apogon var. *laxiflorus* (Steud.) C.B.Clarke ex Cheeseman = ***Schoenus apogon*** Roem. & Schult. var. ***apogon***
Schoenus apogon var. *reductus* Kük. = ***Schoenus apogon*** Roem. & Schult. var. ***apogon***
Schoenus apogon var. *yarrabensis* (Domin) Kük. = ***Schoenus yarrabensis*** Domin
Schoenus arenarius Schrad. = ***Tetraria compar*** (L.) P.Beauv.
Schoenus aristatus Nees ex Kunth = ***Tetraria cuspidata*** (Rottb.) C.B.Clarke
Schoenus aristatus Thunb. = ***Tetraria flexuosa*** (Thunb.) C.B.Clarke
Schoenus armeria var. *globifer* (Nees) Kük. = ***Schoenus globifer*** Nees
Schoenus articulatus Roxb. = ***Rhynchospora corymbosa*** (L.) Britton
Schoenus arundinaceus Sol. ex Vahl = ***Costularia arundinacea*** (Sol. ex Vahl) Kük.
Schoenus asper Schrad. = ***Fimbristylis aspera*** (Schrad.) Boeck.
Schoenus atratus Schrad. = ***Ficinia indica*** (Lam.) H.Pfeiff.
Schoenus attenuatus Gand. = [12 FRA]
Schoenus axillaris (R.Br.) Hook.f. = ***Schoenus maschalinus*** Roem. & Schult.
Schoenus axillaris Lam. = ***Rhynchospora glomerata*** (L.) Vahl
Schoenus barbatus Boeck. = ***Schoenus subbarbatus*** Kük.
Schoenus barbatus Vahl = ***Rhynchospora barbata*** (Vahl) Kunth
Schoenus bifidus var. *jamesonianus* (W.Fitzg.) Kük. = ***Schoenus bifidus*** (Nees) Boeck.
Schoenus bobartiae Retz. = ***Cyperus dubius*** Rottb. subsp. ***dubius***
Schoenus bonannii C.Presl = ***Schoenus nigricans*** L.
Schoenus brachyphyllus F.Muell. ex J.M.Black = ***Schoenus laevigatus*** W.Fitzg.
Schoenus breviculmis var. *tepperi* (F.Muell.) Kük. = ***Schoenus trachycarpus*** F.Muell.
Schoenus brevifolius subsp. *tenax* (Hook.f.) Kük. = ***Schoenus brevifolius*** R.Br.
Schoenus brevisetis var. *cygneus* (Nees) Kük. = ***Schoenus cygneus*** (Nees) Nees
Schoenus brevisetis var. *fuscescens* (C.B.Clarke) Kük. = ***Schoenus brevisetis*** (R.Br.) Roem. & Schult.
Schoenus brevisetis var. *subimberbis* Benth. = ***Schoenus brevisetis*** (R.Br.) Roem. & Schult.
Schoenus bromoides Lam. = ***Tetraria bromoides*** (Lam.) H.Pfeiff.
Schoenus brownii Hook.f. = ***Schoenus apogon*** Roem. & Schult. var. ***apogon***
Schoenus brunneus Sol. ex Roem. & Schult. = ***Blysmus rufus*** (Huds.) Link
Schoenus bulbosus L. = ***Ficinia bulbosa*** (L.) Nees
Schoenus bulbosus Thunb. = ***Ficinia secunda*** (Vahl) Kunth
Schoenus burmanii Vahl = ***Tetraria burmanii*** (Vahl) C.B.Clarke
Schoenus caespitans Petrie = ***Schoenus apogon*** var. ***caespitans*** (Petrie) Edgar
Schoenus calostachyus var. *distans* (F.Muell.) Benth. = ***Schoenus calostachyus*** (R.Br.) Poir.
Schoenus capillaceus Thunb. = ***Tetraria capillacea*** (Thunb.) C.B.Clarke
Schoenus capillaris Sw. = ***Cyperus nanus*** Willd. var. ***nanus***
Schoenus capillaris (F.Muell.) F.Muell. = ***Tetraria capillaris*** (F.Muell.) J.M.Black
Schoenus capitatus (Nees) F.Muell. = ***Schoenus subbulbosus*** Benth.
Schoenus capitatus Dubois = ***Cyperus michelianus*** (L.) Delile subsp. ***michelianus***
Schoenus capitatus Crantz = ***Kyllinga brevifolia*** Rottb. subsp. ***brevifolia***
Schoenus capitatus (Kunth) Poir. = ***Rhynchospora capitata*** (Kunth) Roem. & Schult.
Schoenus capitatus Pers. = ***Rhynchospora glomerata*** (L.) Vahl
Schoenus capitellatus Michx. = ***Rhynchospora capitellata*** (Michx.) Vahl
Schoenus capitellum Thunb. = ***Ficinia capitella*** (Thunb.) Nees
Schoenus caricoides Steud. = ***Ficinia angustifolia*** (Schrad.) C.B.Clarke

Schoenus carsei var. *monocarpus* (J.M.Black) Kük. = ***Schoenus carsei*** Cheeseman
Schoenus castaneus Willd. ex Kunth = ***Cladium mariscus*** (L.) Pohl subsp. ***mariscus***
Schoenus cephalotes (L.) Rottb. = ***Rhynchospora cephalotes*** (L.) Vahl
Schoenus cerasiformis Poir. = ***Rhynchospora globosa*** (Kunth) Roem. & Schult. var. ***globosa***
Schoenus ciliaris Michx. = ***Rhynchospora ciliaris*** (Michx.) C.Mohr
Schoenus ciliaris Muhl. = ***Rhynchospora plumosa*** Elliott
Schoenus ciliatus G.Mey. = ***Rhynchospora nervosa*** subsp. ***ciliata*** (G.Mey.) T.Koyama
Schoenus circinalis (Schrad.) Schult. = ***Tetraria microstachys*** (Vahl) H.Pfeiff.
Schoenus cladium Sw. = ***Cladium mariscus*** (L.) Pohl subsp. ***mariscus***
Schoenus coarctatus Seenus = (Juncaceae)
Schoenus coloratus L. = ***Rhynchospora colorata*** (L.) H.Pfeiff.
Schoenus comatus Link = ***Rhynchospora comata*** (Link) Schult.
Schoenus comosus C.B.Clarke = ***Costularia comosa*** (C.B.Clarke) Kük.
Schoenus compar L. = ***Tetraria compar*** (L.) P.Beauv.
Schoenus compressus Benth. = ***Schoenus benthamii*** F.Muell.
Schoenus compressus C.Presl = ***Schoenus nigricans*** L.
Schoenus compressus L. = ***Blysmus compressus*** (L.) Panz. ex Link
Schoenus concinnus (Hook.f.) Hook.f. = ***Schoenus nitens*** var. ***concinnus*** (Hook.f.) Cheeseman
Schoenus confervoides (Poir.) Willd. ex Kunth = ***Websteria confervoides*** (Poir.) S.S.Hooper
Schoenus congestus Willd. ex Link = ***Cladium mariscus*** (L.) Pohl subsp. ***mariscus***
Schoenus corniculatus Lam. = ***Rhynchospora corniculata*** (Lam.) A.Gray
Schoenus coronatus Steud. = ***Ficinia indica*** (Lam.) H.Pfeiff.
Schoenus corymbosus (L.) Pers. = ***Rhynchospora corymbosa*** (L.) Britton
Schoenus cuspidatus Rottb. = ***Tetraria cuspidata*** (Rottb.) C.B.Clarke
Schoenus cymosus Muhl. = ***Rhynchospora recognita*** (Gale) Kral
Schoenus cymosus Willd. = ***Scirpus polyphyllus*** Vahl
Schoenus cyperinus J.F.Gmel. = ***Rhynchospora holoschoenoides*** (Rich.) Herter
Schoenus cyperoides Retz. = ***Fimbristylis tristachya*** R.Br. var. ***tristachya***
Schoenus cyperoides Sw. = ***Rhynchospora holoschoenoides*** (Rich.) Herter
Schoenus dactyloides Vahl = ***Carpha glomerata*** Nees
Schoenus dactyloides Nees ex Kunth = ***Tetraria thermalis*** (L.) C.B.Clarke
Schoenus deustus P.J.Bergius = ***Ficinia deusta*** (P.J. Bergius) Levyns
Schoenus deustus (R.Br.) F.Muell. = ***Ptilothrix deusta*** (R.Br.) K.L.Wilson
Schoenus dispar Spreng. = ***Ficinia secunda*** (Vahl) Kunth
Schoenus distans (F.Muell.) F.Muell. = ***Schoenus calostachyus*** (R.Br.) Poir.
Schoenus distans Michx. = ***Rhynchospora fascicularis*** (Michx.) Vahl subsp. ***fascicularis***
Schoenus distans Poir. = ***Rhynchospora filiformis*** Vahl
Schoenus distans Muhl. = ***Rhynchospora grayi*** Kunth
Schoenus distichus Ridl. = ***Oreobolus kuekenthalii*** Steenis ex Kük.
Schoenus dregeanus (Boeck.) Kuntze = ***Epischoenus dregeanus*** (Boeck.) Levyns
Schoenus drummondii (Steud.) Benth. = ***Schoenus submicrostachyus*** Kük.
Schoenus drummondii Steud. = ***Gahnia drummondii*** (Steud.) K.L.Wilson
Schoenus effusus Sw. = ***Cladium mariscus*** (L.) Pohl subsp. ***mariscus***
Schoenus elatus Boeck. = ***Schoenus falcatus*** R.Br.
Schoenus elevatus Sol. ex G.Forst. = ***Cladium mariscus*** (L.) Pohl subsp. ***mariscus***
Schoenus elongatus Willd. ex Kunth = ***Costularia elongata*** (Kunth) Kük.
Schoenus elongatus (Nees) F.Muell. = ***Gymnoschoenus anceps*** (R.Br.) C.B.Clarke
Schoenus ericetorum var. *imberbis* (R.Br.) Kük. = ***Schoenus imberbis*** R.Br.
Schoenus ericetorum var. *ornithopodioides* Kük. = ***Schoenus ornithopodioides*** (Kük.) S.T.Blake
Schoenus erinaceus Ridl. = ***Sphaerocyperus erinaceus*** (Ridl.) Lye
Schoenus erraticus Hook.f. = ***Bulbostylis erratica*** (Hook.f.) C.B.Clarke
Schoenus erythrosiphon Ohwi = ***Schoenus maschalinus*** Roem. & Schult.
Schoenus fabri Rottb. = (Poaceae)
Schoenus falcatus Nees = ***Machaerina acuta*** (Labill.) J.Kern
Schoenus fasciatus Rottb. = ***Tetraria fasciata*** (Rottb.) C.B.Clarke
Schoenus fascicularis Nees = ***Schoenus subfascicularis*** Kük.
Schoenus fascicularis Michx. = ***Rhynchospora fascicularis*** (Michx.) Vahl
Schoenus ferrugineus Krock. = (Juncaceae)
Schoenus ferrugineus Huds. = ***Blysmus rufus*** (Huds.) Link
Schoenus festucoides (Poir.) Willd. ex Kunth = ***Bulbostylis festucoides*** (Poir.) C.B.Clarke
Schoenus filiformis Thunb. = ***Ficinia fastigiata*** (Thunb.) Nees
Schoenus filiformis Lam. = ***Ficinia filiformis*** (Lam.) Schrad.
Schoenus filiformis (Burm.f.) Schult. = ***Ficinia nigrescens*** (Schrad.) J.Raynal
Schoenus filiformis Willd. ex Kunth = ***Ficinia nigrescens*** (Schrad.) J.Raynal
Schoenus filum Labill. = ***Gahnia filum*** (Labill.) F.Muell.
Schoenus fimbristyloides (Benth.) F.Muell. = ***Tricostularia undulata*** (Thwaites) J.Kern
Schoenus flaviculmis Nees = ***Schoenus brevisetis*** (R.Br.) Roem. & Schult.
Schoenus flavus (Nees) Boeck. = ***Schoenus subflavus*** Kük.
Schoenus flexuosus Steud. ex Kunth = ***Tetraria cuspidata*** (Rottb.) C.B.Clarke
Schoenus flexuosus Thunb. = ***Tetraria flexuosa*** (Thunb.) C.B.Clarke
Schoenus floribundus Willd. ex C.Presl = ***Cladium mariscus*** (L.) Pohl subsp. ***mariscus***
Schoenus floridus Rudge = ***Rhynchospora corymbosa*** (L.) Britton
Schoenus fluitans var. *tenellus* (Benth.) Kük. = ***Schoenus tenellus*** Benth.
Schoenus foliatus (Hook.f.) S.T.Blake = ***Schoenus maschalinus*** Roem. & Schult.
Schoenus fragiferus Rudge = ***Rhynchospora holoschoenoides*** (Rich.) Herter
Schoenus fuscescens C.B.Clarke = ***Schoenus brevisetis*** (R.Br.) Roem. & Schult.

Schoenus fuscoguttatus Ohwi = ***Schoenus sparteus*** R.Br.
Schoenus fuscus L. = ***Rhynchospora fusca*** (L.) W.T.Aiton
Schoenus fuscus Muhl. = ***Rhynchospora grayi*** Kunth
Schoenus glaucus Sieber ex C.B.Clarke = ***Cyperus conglomeratus*** Rottb. subsp. ***conglomeratus***
Schoenus glaucus Poir. = ***Rhynchospora rariflora*** (Michx.) Elliott
Schoenus globosus Rudge = ***Rhynchospora barbata*** (Vahl) Kunth
Schoenus globosus (Kunth) Willd. ex Kunth = ***Rhynchospora globosa*** (Kunth) Roem. & Schult.
Schoenus globosus Kunth = ***Rhynchospora nervosa*** subsp. ***ciliata*** (G.Mey.) T.Koyama
Schoenus globuliferus (Link) Willd. ex Kunth = ***Rhynchospora holoschoenoides*** (Rich.) Herter
Schoenus glomeratus Thunb. = ***Carpha glomerata*** Nees
Schoenus glomeratus L. = ***Rhynchospora glomerata*** (L.) Vahl
Schoenus gracilis Sw. = ***Rhynchospora racemosa*** C.Wright ex Sauvalle
Schoenus guillauminii Kük. = ***Tricostularia guillauminii*** (Kük.) J.Raynal
Schoenus halconensis Kük. = ***Schoenus melanostachyus*** R.Br.
Schoenus halleri Honck. = [1]
Schoenus hasskarlii Steud. = ***Gahnia javanica*** Moritzi
Schoenus hattorianus Nakai = [38 JAP]
Schoenus hemisphaericus Humb. ex Link = ***Rhynchospora barbata*** (Vahl) Kunth
Schoenus hirsutus Vahl = ***Rhynchospora hirsuta*** (Vahl) Vahl
Schoenus hirtus Schrad. ex Nees = ***Rhynchospora pilosa*** Boeck.
Schoenus hispidulus Vahl = ***Rhynchospora trichodes*** C.B.Clarke
Schoenus holoschoenoides Rich. = ***Rhynchospora holoschoenoides*** (Rich.) Herter
Schoenus hornei C.B.Clarke = ***Costularia hornei*** (C.B.Clarke) Kük.
Schoenus hypomelas Spreng. = ***Schoenus nigricans*** L.
Schoenus inanis Thunb. = ***Pseudoschoenus inanis*** (Thunb.) Oteng-Yeb.
Schoenus incanus Forssk. = [35 YEM]
Schoenus indicus Lam. = ***Ficinia indica*** (Lam.) H.Pfeiff.
Schoenus inermis Willd. ex Kunth = ***Scleria triglomerata*** Michx.
Schoenus inexpansus Michx. = ***Rhynchospora inexpansa*** (Michx.) Vahl
Schoenus ×intermedius Brügger = ***Schoenus ×scheuchzeri*** Brügger
Schoenus involucratus Rottb. = ***Tetraria involucrata*** (Rottb.) C.B.Clarke
Schoenus jamesonianus W.Fitzg. = ***Schoenus bifidus*** (Nees) Boeck.
Schoenus junceus Willd. = ***Schoenoplectiella juncea*** (Willd.) Lye
Schoenus juncoides Vahl = ***Bulbostylis juncoides*** (Vahl) Kük. ex Herter
Schoenus juvenis C.B.Clarke = ***Costularia neocaledonica*** (C.B.Clarke) Rendle
Schoenus karpatii Pénzes = ***Schoenus ferrugineus*** L.
Schoenus kinabaluensis Stapf = ***Schoenus curvulus*** F.Muell.
Schoenus krausei (Phil.) Gunckel = ***Schoenus nitens*** (R.Br.) Roem. & Schult. var. ***nitens***
Schoenus lacustris (L.) Bernh. = ***Schoenoplectus lacustris*** (L.) Palla subsp. ***lacustris***
Schoenus laevis Thunb. = [27 CPP]
Schoenus lanatus (Kunth) Willd. ex Kunth = ***Bulbostylis lanata*** (Kunth) Lindm.
Schoenus lanceus Thunb. = ***Schoenoxiphium lanceum*** (Thunb.) Kük.
Schoenus lateralis Vahl = ***Ficinia lateralis*** (Vahl) Kunth
Schoenus latifolius Vahl = ***Scleria melaleuca*** Rchb. ex Schltdl. & Cham.
Schoenus laxiflorus Steud. = ***Schoenus apogon*** Roem. & Schult. var. ***apogon***
Schoenus laxus W.Fitzg. = ***Schoenus sublaxus*** Kük.
Schoenus laxus Poir. = ***Rhynchospora rugosa*** subsp. ***brownii*** (Roem. & Schult.) T.Koyama
Schoenus ligulatus (Boeck.) Kuntze = ***Tetraria ligulata*** (Boeck.) C.B.Clarke
Schoenus lithospermus L. = ***Scleria flagellum-nigrorum*** P.J.Bergius
Schoenus longifolius Rudge = ***Bisboeckelera longifolia*** (Rudge) Kuntze
Schoenus longirostris Michx. = ***Rhynchospora corniculata*** (Lam.) A.Gray
Schoenus longirostris Sieber ex C.Presl = ***Rhynchospora longiflora*** C.Presl
Schoenus longisetis (R.Br.) Poir. = ***Rhynchospora longisetis*** R.Br.
Schoenus longus Gand. = [12 FRA]
Schoenus loreus (Nees) Kuntze = ***Tetraria cuspidata*** (Rottb.) C.B.Clarke
Schoenus macrocephalus Boeck. = [27 CPP]
Schoenus macrorrhizus Rich. ex C.B.Clarke = ***Cyperus conglomeratus*** Rottb. subsp. ***conglomeratus***
Schoenus macrostachyus Noë ex Rchb. = ***Bolboschoenus maritimus*** (L.) Palla subsp. ***maritimus***
Schoenus marginatus Kunze ex Kunth = ***Carex aphylla*** Kunth
Schoenus mariannae Hosok. = ***Schoenus maschalinus*** Roem. & Schult.
Schoenus mariscoides Muhl. = ***Cladium mariscoides*** (Muhl.) Torr.
Schoenus marisculus Salzm. ex Steud. = ***Rhynchospora marisculus*** Lindl. & Nees
Schoenus mariscus L. = ***Cladium mariscus*** (L.) Pohl
Schoenus maritimus Lam. = ***Cyperus capitatus*** Vand.
Schoenus maritimus Salzm. ex Steud. = ***Rhynchospora riparia*** (Nees) Boeck.
Schoenus medwedewii Meinsh. = ***Cladium mariscus*** (L.) Pohl subsp. ***mariscus***
Schoenus megalocephalus Gand. = [12 FRA]
Schoenus melanostachyus var. *altissimus* F.Muell. = ***Schoenus melanostachyus*** R.Br.
Schoenus melanostachyus var. *laevinux* Kük. = ***Schoenus laevinux*** (Kük.) Ohwi
Schoenus microcephalus J.Kern = [60 NWC]
Schoenus microstachys Vahl = ***Tetraria microstachys*** (Vahl) H.Pfeiff.
Schoenus microstachyus (Nees) H.Pfeiff. = ***Schoenus submicrostachyus*** Kük.
Schoenus microstachyus (Boeck.) Kuntze = ***Tetraria fasciata*** (Rottb.) C.B.Clarke
Schoenus miliaceus Lam. = ***Rhynchospora miliacea*** (Lam.) A.Gray
Schoenus moniliferus Sparrm. ex Spreng. = ***Scleria flagellum-nigrorum*** P.J.Bergius
Schoenus monocarpus J.M.Black = ***Schoenus carsei*** Cheeseman
Schoenus monocephalus Salzm. ex Steud. = ***Rhynchospora barbata*** (Vahl) Kunth
Schoenus monoicus Sm. = ***Kobresia simpliciuscula*** (Wahlenb.) Mack. subsp. ***simpliciuscula***

Schoenus moorei Kirk = ***Schoenus nitens*** var. ***concinnus*** (Hook.f.) Cheeseman
Schoenus mucronatus L. = ***Cyperus capitatus*** Vand.
Schoenus natans Thunb. ex C.B.Clarke = ***Isolepis tenuissima*** (Nees) Kunth
Schoenus neesii (Nees) F.Muell. = ***Tricostularia neesii*** Nees
Schoenus nemorum Vahl = ***Hypolytrum nemorum*** (Vahl) Spreng.
Schoenus neocaledonicus C.B.Clarke = ***Costularia neocaledonica*** (C.B.Clarke) Rendle
Schoenus neoguineensis Kük. = ***Schoenus longibracteatus*** Kük.
Schoenus nigricans Hoppe = ***Schoenus ferrugineus*** L.
Schoenus nigricans var. *ambiguus* Kük. = ***Schoenus nigricans*** L.
Schoenus nigricans var. *plurisquamatus* Kük. = ***Schoenus nigricans*** L.
Schoenus nitens Willd. ex Kunth = ***Carpha nitens*** (Kunth) Kük.
Schoenus nitens subsp. *krausei* (Phil.) Kük. = ***Schoenus nitens*** (R.Br.) Roem. & Schult.
Schoenus nitens var. *major* Ewart & Jean White = ***Schoenus nitens*** (R.Br.) Roem. & Schult. var. ***nitens***
Schoenus niveus L. = ***Kyllinga bulbosa*** P.Beauv.
Schoenus nudus Steud. = ***Machaerina nuda*** (Steud.) J.Kern
Schoenus octandrus (Nees) F.Muell. = ***Tetraria octandra*** (Nees) Kük.
Schoenus odoratus Aubl. = ***Kyllinga odorata*** Vahl subsp. ***odorata***
Schoenus oliganthos (Steud.) Kunth = ***Ficinia oligantha*** (Steud.) J.Raynal
Schoenus oligostachyus F.Muell. = ***Schoenus bifidus*** (Nees) Boeck.
Schoenus pachylepis S.T.Blake = ***Schoenus lepidosperma*** subsp. ***pachylepis*** (S.T.Blake) K.L.Wilson
Schoenus pallens Schrad. = ***Ficinia pallens*** (Schrad.) Nees
Schoenus pallescens Gand. = [12 FRA]
Schoenus palustris (L.) Bernh. = ***Eleocharis palustris*** (L.) Roem. & Schult.
Schoenus paniculatus Hassk. = ***Gahnia javanica*** Moritzi
Schoenus paniculatus Burm.f. = ***Scleria terrestris*** (L.) Fassett var. ***terrestris***
Schoenus paradoxus Spreng. = ***Bulbostylis paradoxa*** (Spreng.) Lindm.
Schoenus pauciflorus Willd. ex Kunth = ***Rhynchospora rugosa*** subsp. ***brownii*** (Roem. & Schult.) T.Koyama
Schoenus pauciflorus (F.Muell.) F.Muell. = ***Tricostularia pauciflora*** (R.Br.) Benth.
Schoenus philippinensis (Palla) Kük. ex Merr. = ***Schoenus maschalinus*** Roem. & Schult.
Schoenus pictus (Boeck.) Kuntze = ***Tetraria picta*** (Boeck.) C.B.Clarke
Schoenus pilosus Willd. = ***Bulbostylis pilosa*** (Willd.) Cherm.
Schoenus polycephalus Pers. = ***Rhynchospora holoschoenoides*** (Rich.) Herter
Schoenus polymorphus Rottb. = ***Fimbristylis ferruginea*** (L.) Vahl var. ***ferruginea***
Schoenus polyphyllus Vahl = ***Rhynchospora polyphylla*** (Vahl) Vahl
Schoenus pterocarpus (Kunth) Poir. = ***Rhynchospora barbata*** (Vahl) Kunth
Schoenus puberulus C.A.Mey. = ***Fimbristylis fusca*** (Nees) Benth. ex C.B.Clarke
Schoenus pubescens Kunth = ***Rhynchospora pubera*** (Vahl) Boeck. subsp. ***pubera***
Schoenus pubigerus Schrad. ex Nees = ***Rhynchospora robusta*** (Kunth) Boeck.
Schoenus punctorius Vahl = ***Neesenbeckia punctoria*** (Vahl) Levyns
Schoenus pungens Willd. ex Kunth = ***Tetraria compar*** (L.) P.Beauv.
Schoenus pusillus Sw. = ***Rhynchospora berteroi*** (Spreng.) C.B.Clarke
Schoenus quadrangularis Boeck. = ***Epischoenus quadrangularis*** (Boeck.) C.B.Clarke
Schoenus radiatus J.R.Forst. ex Vahl = ***Costularia arundinacea*** (Sol. ex Vahl) Kük.
Schoenus radiatus L.f. = ***Ficinia radiata*** (L.f.) Kunth
Schoenus rariflorus Michx. = ***Rhynchospora rariflora*** (Michx.) Elliott
Schoenus recurvatus Schrad. ex Nees = ***Rhynchospora recurvata*** (Nees) Steud.
Schoenus restioides Sw. = ***Machaerina restioides*** (Sw.) Vahl
Schoenus rottboellii (Schrad.) Kuntze = ***Tetraria bromoides*** (Lam.) H.Pfeiff.
Schoenus ruber Lour. = ***Rhynchospora rubra*** (Lour.) Makino
Schoenus rubiginosus Sol. ex G.Forst. = ***Machaerina rubiginosa*** (Spreng.) T.Koyama
Schoenus rufus Huds. = ***Blysmus rufus*** (Huds.) Link
Schoenus rugosus Vahl = ***Rhynchospora rugosa*** (Vahl) Gale
Schoenus scariosus Thunb. = ***Ficinia deusta*** (P.J. Bergius) Levyns
Schoenus secans L. = ***Scleria secans*** (L.) Urb.
Schoenus secundus Vahl = ***Ficinia secunda*** (Vahl) Kunth
Schoenus setaceus Muhl. = ***Rhynchospora capillacea*** Torr.
Schoenus setaceus Rottb. = ***Rhynchospora fallax*** Uittien
Schoenus setaceus Thuill. = ***Rhynchospora fusca*** (L.) W.T.Aiton
Schoenus setaceus Vahl ex Kunth. = ***Rhynchospora rariflora*** (Michx.) Elliott
Schoenus setaceus P.J.Bergius = ***Rhynchospora tenerrima*** Nees ex Spreng. subsp. ***tenerrima***
Schoenus setifolius Benth. = ***Ficinia filiformis*** (Lam.) Schrad.
Schoenus sodalium Har. = ***Schoenus andinus*** (Phil.) H.Pfeiff.
Schoenus spadiceus Kunth = ***Bulbostylis paradoxa*** (Spreng.) Lindm.
Schoenus spadiceus (L.) Vahl = ***Fimbristylis spadicea*** (L.) Vahl
Schoenus sparganoides Lindl. ex Kunth = ***Rhynchospora holoschoenoides*** (Rich.) Herter
Schoenus sparsus Michx. = ***Rhynchospora miliacea*** (Lam.) A.Gray
Schoenus spathaceus L. = ***Dulichium arundinaceum*** (L.) Britton var. ***arundinaceum***
Schoenus sphaerocephalus (R.Br.) Poir. = ***Gymnoschoenus sphaerocephalus*** (R.Br.) Hook.f.
Schoenus spicatus Burm.f. = [27 CPP]
Schoenus spicatus Thunb. = ***Ficinia bulbosa*** (L.) Nees
Schoenus spilocarpus Sol. ex Seem. = ***Gahnia schoenoides*** G.Forst.
Schoenus squamellatus Kük. = [50 WAU]
Schoenus stellatus Lam. = ***Rhynchospora colorata*** (L.) H.Pfeiff.
Schoenus striatus Thunb. = ***Ficinia indica*** (Lam.) H.Pfeiff.
Schoenus stygius (R.Br.) Poir. = ***Mesomelaena stygia*** (R.Br.) Nees
Schoenus subaphyllus var. *barbulatus* Kük. = ***Schoenus subaphyllus*** Kük.

Schoenus subaphyllus var. *kennyi* (F.M.Bailey) Kük. = ***Schoenus kennyi*** (F.M.Bailey) S.T.Blake
Schoenus subaxillaris Kük. = ***Schoenus maschalinus*** Roem. & Schult.
Schoenus subaxillaris f. *terrestris* (Boeck.) Kük. = ***Schoenus maschalinus*** Roem. & Schult.
Schoenus subbulbosus var. *junceus* Benth. = ***Schoenus subbulbosus*** Benth.
Schoenus subcontortus Gand. = [12 FRA]
Schoenus subserratus Steud. = ***Ficinia indica*** (Lam.) H.Pfeiff.
Schoenus surinamensis Rottb. = ***Rhynchospora corymbosa*** (L.) Britton
Schoenus sylvaticus Salzm. ex Schltdl. = ***Rhynchospora comata*** (Link) Schult.
Schoenus sylvaticus (L.) Bernh. = ***Scirpus sylvaticus*** L.
Schoenus sylvaticus (Nees) Kuntze = ***Tetraria sylvatica*** (Nees) C.B.Clarke
Schoenus tenax (Hook.f.) Hook.f. = ***Schoenus brevifolius*** R.Br.
Schoenus tendo subsp. *achaetus* T.Koyama = ***Schoenus achaetus*** (T.Koyama) T.Koyama
Schoenus tener Spreng. = [27 CPP]
Schoenus teneriffae Vent. ex Torr. = ***Rhynchospora contracta*** (Nees) J.Raynal
Schoenus tenuiflorus Willd. ex Link = [27 CPP]
Schoenus tenuifolius Kunth = ***Rhynchospora nervosa*** (Vahl) Boeck. subsp. ***nervosa***
Schoenus tenuis Kirk = ***Tetraria capillaris*** (F.Muell.) J.M.Black
Schoenus tepperi F.Muell. = ***Schoenus trachycarpus*** F.Muell.
Schoenus tetragonus (R.Br.) Poir. = ***Mesomelaena tetragona*** (R.Br.) Benth.
Schoenus thermalis L. = ***Tetraria thermalis*** (L.) C.B.Clarke
Schoenus triangularis Volkens = ***Schoenus calostachyus*** (R.Br.) Poir.
Schoenus triceps Vahl = ***Rhynchospora holoschoenoides*** (Rich.) Herter
Schoenus tricostularia F.Muell. = ***Tricostularia compressa*** Nees
Schoenus tristachyus Thunb. = [27 CPP]
Schoenus tuberosus Burm.f. = ***Cyperus rotundus*** L. subsp. ***rotundus***
Schoenus umbellatus Jacq. = ***Cyperus echinatus*** (L.) Alph.Wood
Schoenus umbellatus Walter = ***Rhynchospora corniculata*** (Lam.) A.Gray
Schoenus ustulatus L. = ***Tetraria ustulata*** (L.) C.B.Clarke
Schoenus vacillans Kirk = ***Schoenus apogon*** Roem. & Schult. var. ***apogon***
Schoenus velutinus Schrad. ex Kunth = ***Rhynchospora velutina*** (Kunth) Boeck.
Schoenus vendeanus Gand. = [12 FRA]
Schoenus viscosus Schrad. = ***Tetraria compar*** (L.) P.Beauv.
Schoenus xipholepis (Baker) Summerh. = ***Costularia elongata*** (Kunth) Kük.
Schoenus xyridioides Däniker = ***Costularia xyridioides*** (Däniker) Kük.

***Unplaced Names*:**
Schoenus attenuatus Gand., Dec. Pl. Nov. 2: 35 (1876). = [12 FRA]
Schoenus halleri Honck., Verz. Gew. Teutschl.: 127 (1782). = [1]
Schoenus hattorianus Nakai, Bot. Mag. (Tokyo) 26: 79 (1912). = [38 JAP]
Schoenus incanus Forssk., Fl. Aegypt.-Arab.: 12 (1775). = [35 YEM]
Schoenus laevis Thunb., Prodr. Pl. Cap.: 17 (1794). = [27 CPP]
Schoenus longus Gand., Dec. Pl. Nov. 2: 35 (1876). = [12 FRA]
Schoenus macrocephalus Boeck., Flora 61: 144 (1878). = [27 CPP]
Schoenus megalocephalus Gand., Dec. Pl. Nov. 2: 36 (1876). = [12 FRA]
Schoenus microcephalus J.Kern, Blumea 12: 28 (1963). = [60 NWC]
Schoenus pallescens Gand., Dec. Pl. Nov. 2: 35 (1876). = [12 FRA]
Schoenus spicatus Burm.f., Fl. Indica, Prodr. Fl. Cap.: 3 (1768). = [27 CPP]
Schoenus squamellatus Kük., Repert. Spec. Nov. Regni Veg. 44: 71 (1938). = [50 WAU]
Schoenus subcontortus Gand., Dec. Pl. Nov. 2: 36 (1876). = [12 FRA]
Schoenus tener Spreng., Neue Entdeck. Pflanzenk. 3: 9 (1822). = [27 CPP]
Schoenus tenuiflorus Willd. ex Link, Jahrb. Gewächsk. 1(3): 75 (1820). = [27 CPP]
Schoenus tristachyus Thunb., Prodr. Pl. Cap.: 16 (1794). = [27 CPP]
Schoenus vendeanus Gand., Dec. Pl. Nov. 2: 34 (1876). = [12 FRA]

Scirpidiella

Scirpidiella Rauschert = ***Isolepis*** R.Br.
Scirpidiella beccarii (Boeck.) Rauschert = ***Isolepis beccarii*** (Boeck.) Goetgh. & D.A.Simpson
Scirpidiella crassiuscula (Hook.f.) Rauschert = ***Isolepis crassiuscula*** Hook.f.
Scirpidiella fluitans (L.) Rauschert = ***Isolepis fluitans*** (L.) R.Br.
Scirpidiella graminoides (R.W.Haines & Lye) Rauschert = ***Isolepis graminoides*** (R.W.Haines & Lye) Lye
Scirpidiella kilimanjarica (R.W.Haines & Lye) Rauschert = ***Isolepis kilimanjarica*** R.W.Haines & Lye
Scirpidiella ruwenzoriensis (R.W.Haines & Lye) Rauschert = ***Isolepis ruwenzoriensis*** R.W.Haines & Lye

Scirpidium

Scirpidium Nees = ***Eleocharis*** R.Br.
Scirpidium aciculare (L.) Nees = ***Eleocharis acicularis*** (L.) Roem. & Schult.
Scirpidium grande Nees = ***Eleocharis nudipes*** (Kunth) Palla
Scirpidium nigrescens Nees = ***Eleocharis nigrescens*** (Nees) Steud.
Scirpidium piauhiense Nees = ***Eleocharis filiculmis*** Kunth
Scirpidium sulcatum Nees = ***Eleocharis filiculmis*** Kunth
Scirpidium sulcatum var. *lituratum* Nees =
Scirpidium tenue (Willd.) Nees = ***Eleocharis tenuis*** (Willd.) Schult.

Scirpocyperus

Scirpocyperus Ség. = ***Bolboschoenus*** (Asch.) Palla
Scirpocyperus septentrionalis Montandon = ***Bolboschoenus maritimus*** (L.) Palla subsp. ***maritimus***

Scirpodendron

Scirpodendron Zipp. ex Kurz, J. Asiat. Soc. Bengal, Pt. 2, Nat. Hist. 38(2): 84 (1869). *Ptychocaryum* R.Br. ex Kuntze in T.E.von Post & C.E.O.Kuntze, Lex. Gen. Phan.: 470 (1903), nom. illeg.
India to W. Pacific. 40 IND SRL 41 AND THA VIE 42 BOR JAW MLY MOL PHI SUL SUM 43 NWG 50 QLD 60 FIJ SAM VAN 62 CRL.
2 Species

Scirpodendron bogneri S.S.Hooper, Kew Bull. 52: 683 (1997).
Pen. Malaysia. 42 MLY. Rhizome geophyte or hel.

Scirpodendron ghaeri (Gaertn.) Merr., Philipp. J. Sci., C 9: 268 (1914).
E. India, SW. Sri Lanka to W. Pacific. 40 IND SRL 41 AND THA VIE 42 BOR JAW MLY MOL PHI SUL SUM 43 NWG 50 QLD 60 FIJ SAM VAN 62 CRL. Rhizome geophyte or hel.
Scleria macrocarpa Wall., Numer. List: 3538 (1831), nom. inval.
Hypolytrum costatum Thwaites, Enum. Pl. Zeyl.: 346 (1864). *Scirpodendron costatum* (Thwaites) Kurz, J. Asiat. Soc. Bengal, Pt. 2, Nat. Hist. 38(2): 85 (1869).
Scirpodendron pandaniforme Zipp. ex Kurz, J. Asiat. Soc. Bengal, Pt. 2, Nat. Hist. 38(2): 85 (1869), pro syn.
Scirpodendron sulcatum Miq., Ill. Fl. Archip. Ind.: 65 (1870), orth. var.
Ptychocaryum ghaeri (Gaertn.) H.Pfeiff., Repert. Spec. Nov. Regni Veg. 21: 240 (1925).

Synonyms:
Scirpodendron costatum (Thwaites) Kurz = ***Scirpodendron ghaeri*** (Gaertn.) Merr.
Scirpodendron pandaniforme Zipp. ex Kurz = ***Scirpodendron ghaeri*** (Gaertn.) Merr.
Scirpodendron sulcatum Miq. = ***Scirpodendron ghaeri*** (Gaertn.) Merr.

Scirpoides

Scirpoides Ség., Pl. Veron. 3: 73 (1754).
Canary Is. to W. Himalaya, S. Africa. 10 GRB 11 AUT BGM† CZE GER HUN POL SWI 12 BAL COR FRA POR SAR SPA 13 ALB BUL GRC ITA KRI ROM SIC TUE YUG 14 BLR KRY RUS UKR 20 ALG EGY LBY MOR TUN WSA 21 CNY 24 CHA 27 BOT CPP LES NAM NAT OFS TVL 30 32 KAZ KGZ TKM TZK UZB 33 TCS 34 AFG EAI IRN IRQ LBS PAL SIN TUR 35 SAU YEM 40 PAK WHM (75) nwy (78) kty.
3 Species
Holoschoenus Link, Hort. Berol. 1: 293 (1827).

Scirpoides burkei (C.B.Clarke) Goetgh., Muasya & D.A.Simpson, Novon 10: 132 (2000).
S. Africa. 27 CPP LES NAT OFS TVL. Hemicr. or rhizome geophyte.
Scirpus schinzianus Boeck. ex C.B.Clarke in T.A.Durand & H.Schinz, Consp. Fl. Afric. 5: 618 (1894), nom. inval.
**Scirpus burkei* C.B.Clarke in W.H.Harvey & auct. suc. (eds.), Fl. Cap. 7: 227 (1898).

Scirpoides dioecus (Kunth) Browning, S. African J. Bot. 60: 318 (1994).
S. Africa. 27 BOT CPP NAM OFS TVL. Hemicr. or rhizome geophyte.
**Isolepis dioeca* Kunth, Enum. Pl. 2: 199 (1837). *Scirpus dioecus* (Kunth) Boeck., Linnaea 36: 719 (1870).
Scirpus schinzii Boeck., Verh. Bot. Vereins Prov. Brandenburg 29: 47 (1888).

Scirpoides holoschoenus (L.) Soják, Cas. Nár. Mus., Odd. Prír. 140: 127 (1972).
Canary Is., Europe to W. Himalaya, Chad, Cape Prov. 10 GRB 11 AUT BGM† CZE GER HUN POL SWI 12 BAL COR FRA POR SAR SPA 13 ALB BUL GRC ITA KRI ROM SIC TUE YUG 14 BLR KRY RUS UKR 20 ALG EGY LBY MOR TUN WSA 21 CNY 24 CHA 27 CPP 30+ 32 KAZ KGZ TKM TZK UZB 33 TCS 34 AFG EAI IRN IRQ LBS PAL SIN TUR 35 SAU YEM 40 PAK WHM (75) nwy (78) kty. Hemicr.
**Scirpus holoschoenus* L., Sp. Pl.: 49 (1753). *Isolepis holoschoenus* (L.) Roem. & Schult., Syst. Veg. 2: 115 (1817). *Isolepis holoschoenus* var. *australis* Roem. & Schult., Syst. Veg. 2: 115 (1817), nom. inval. *Holoschoenus vulgaris* Link, Hort. Berol. 1: 293 (1827). *Cyperus holoschoenus* (L.) Missbach & E.H.L.Krause in J.Sturm, Deutschl. Fl., ed. 2, 2: 20 (1900), nom. illeg. *Scirpus holoschoenus* subsp. *euholoschoenus* Briq., Prodr. Fl. Corse 1: 231 (1910), nom. inval.

subsp. ***globiferus*** (L.f.) Soják, Cas. Nár. Mus., Odd. Prír. 141: 61 (1972).
Canary Is. to Iran. 12 SAR 20 ALG MOR TUN 21 CNY 34 IRN. Hemicr. or rhizome geophyte.
**Scirpus globiferus* L.f., Suppl. Pl.: 104 (1782). *Holoschoenus globifer* (L.f.) Rchb., Flora 13(2): 518 (1830). *Holoschoenus globiferus* (L.f.) A.Dietr., Sp. Pl. 2: 164 (1832). *Scirpus holoschoenus* var. *globiferus* (L.f.) Parl., Fl. Ital. 2: 96 (1852). *Scirpus holoschoenus* subsp. *globiferus* (L.f.) Husn., Cypéracées: 67 (1906). *Scirpoides globiferus* (L.f.) Soják, Cas. Nár. Mus., Odd. Prír. 140: 127 (1972).
Isolepis globulifera Steud., Syn. Pl. Glumac. 2: 97 (1855).

subsp. ***holoschoenus***
Europe to W. Himalaya, Chad. 10 GRB 11 AUT BGM† CZE GER HUN POL SWI 12 BAL COR FRA POR SAR SPA 13 ALB BUL GRC ITA KRI ROM SIC TUE YUG 14 BLR KRY RUS UKR 20 ALG EGY LBY MOR TUN WSA 24 CHA 30 32 KAZ KGZ TKM TZK UZB 33 TCS 34 AFG EAI IRN IRQ LBS PAL SIN TUR 35 SAU YEM 40 PAK WHM (75) nwy (78) kty. Hemicr.
Scirpus romanus L., Sp. Pl.: 49 (1753). *Isolepis romana* (L.) Roem. & Schult. ex G.Don in J.C.Loudon, Hort. Brit.: 21 (1830). *Holoschoenus vulgaris* var. *romanus* (L.) Vis., Fl. Dalmat. 1: 111 (1842). *Scirpus holoschoenus* var. *romanus* (L.) W.D.J.Koch, Syn. Fl. Germ. Helv., ed. 2, 2: 857 (1844). *Holoschoenus romanus* (L.) Fritsch, Excursionsfl. Oesterreich: 80 (1897). *Scirpoides romanus* (L.) Soják, Cas. Nár. Mus., Odd. Prír. 140: 127 (1972). *Scirpus holoschoenus* subsp. *romanus* (L.) Mateo & Figuerola, Fl. Analit. Valencia: 369 (1987).
Scirpus australis L., Syst. Veg. ed. 13: 85 (1774). *Holoschoenus australis* (L.) Rchb., Flora 13: 499 (1830). *Isolepis australis* (L.) Roem. & Schult. ex G.Don in J.C.Loudon, Hort. Brit.: 21 (1830). *Isolepis australis* (L.) Bluff, Nees & Schauer,

Comp. Fl. German. 1(1): 82 (1836). *Scirpus holoschoenus* var. *australis* (L.) W.D.J.Koch, Syn. Fl. Germ. Helv., ed. 2, 2: 857 (1844). *Holoschoenus vulgaris* var. *australis* (L.) Nyman, Consp. Fl. Eur.: 765 (1882). *Holoschoenus romanus* var. *australis* (L.) Soó, Acta Bot. Acad. Sci. Hung. 16: 368 (1970 publ. 1971). *Scirpoides holoschoenus* subsp. *australis* (L.) Soják, Cas. Nár. Mus., Odd. Prír. 141: 61 (1972).

Scirpus ellychniarius Molina, Sag. Stor. Nat. Chili: 153 (1782).

Scirpus intermedius Poir. in J.B.A.M.de Lamarck, Encycl. 6: 767 (1805), nom. illeg.

Isolepis poiretii Roem. & Schult., Syst. Veg. 2: 116 (1817).

Isolepis paniculata Gray, Nat. Arr. Brit. Pl. 2: 78 (1821).

Seidlia atrovirens Opiz, Naturalientausch 11: 349 (1826).

Holoschoenus exserens Rchb., Flora 13: 501 (1830). *Isolepis exserens* (Rchb.) Bluff, Nees & Schauer, Comp. Fl. German. 1(1): 81 (1836).

Holoschoenus filiformis Rchb., Flora 13: 501 (1830). *Isolepis filiformis* (Rchb.) Bluff, Nees & Schauer, Comp. Fl. German. 1(1): 81 (1836). *Holoschoenus vulgaris* var. *filiformis* (Rchb.) Nyman, Consp. Fl. Eur.: 765 (1882). *Holoschoenus romanus* f. *filiformis* (Rchb.) Soó, Acta Bot. Acad. Sci. Hung. 16: 368 (1970 publ. 1971).

Holoschoenus linnaei Rchb., Fl. Germ. Excurs.: 76 (1830). *Isolepis linnaei* (Rchb.) Bluff, Nees & Schauer, Comp. Fl. German. 1(1): 82 (1836). *Scirpus holoschoenus* prol. *linnaei* (Rchb.) Asch. & Graebn., Syn. Mitteleur. Fl. 2(2): 322 (1903), nom. illeg.

Holoschoenus tuberculatus Rchb., Flora 12: 498 (1830).

Scirpus longibracteatus Salzm. ex Rchb., Flora 13: 497 (1830).

Isolepis panormitana Parl., Giorn. Bot. Ital. 2(1): 146 (1846). *Scirpus panormitanus* (Parl.) Parl., Fl. Ital. 2: 99 (1852). *Holoschoenus panormitanus* (Parl.) Nyman, Consp. Fl. Eur.: 765 (1882). *Holoschoenus vulgaris* subsp. *panormitanus* (Parl.) Nyman, Consp. Fl. Eur.: 765 (1882). *Scirpoides holoschoenus* subsp. *panormitanus* (Parl.) Soják, Cas. Nár. Mus., Odd. Prír. 152: 21 (1983).

Holoschoenus laxus Opiz, Seznam: 23 (1852).

Scirpus parlatoris Biv. ex Parl., Fl. Ital. 2: 96 (1852). *Scirpus holoschoenus* var. *macrostachyus* Husn., Cypéracées: 67 (1906).

Isolepis fontana Steud., Syn. Pl. Glumac. 2: 97 (1855).

Holoschoenus diphyllus Montandon in F.Friche-Joset, Syn. Fl. Jura: 346 (1856).

Scirpus albovittatus Schur, Enum. Pl. Transsilv.: 694 (1866), nom. inval.

Scirpus polymorphus Dulac, Fl. Hautes -Pyrénées: 60 (1867).

Isolepis sphaerophora Ehrenb. ex Boeck., Linnaea 36: 721 (1870).

Holoschoenus vulgaris subvar. *albovittatus* Rchb. ex Nyman, Consp. Fl. Eur.: 765 (1882), nom. nud.

Scirpus holoschoenus var. *globuliferus* Boiss., Fl. Orient. 5: 381 (1882). *Holoschoenus globuliferus* (Boiss.) Grossh., Beih. Bot. Centralbl. 44(2): 204 (1927).

Scirpus romanus f. *monocephalus* Waisb., Magyar Bot. Lapok 4: 65 (1905). *Holoschoenus romanus* f. *monocephalus* (Waisb.) Soó, Acta Bot. Acad. Sci. Hung. 16: 368 (1970 publ. 1971).

subsp. ***thunbergii*** (Schrad.) Soják, Cas. Nár. Mus., Odd. Prír. 141: 61 (1972).

Cape Prov. 27 CPP. Hemicr. or rhizome geophyte.

Scirpus holoschoenus Thunb., Fl. Cap.: 368 (1811), nom. illeg.

**Isolepis thunbergii* Schrad., Gött. Gel. Anz. 3: 2068 (1821). *Scirpus thunbergii* (Schrad.) A.Spreng., Tent. Suppl.: 4 (1828). *Holoschoenus thunbergii* (Schrad.) A.Dietr., Sp. Pl. 2: 164 (1832). *Scirpus holoschoenus* var. *thunbergii* (Schrad.) C.B.Clarke in T.A.Durand & H.Schinz, Consp. Fl. Afric. 5: 623 (1894). *Scirpoides thunbergii* (Schrad.) Soják, Cas. Nár. Mus., Odd. Prír. 140: 127 (1972).

Isolepis thunbergiana Nees, Linnaea 7: 508 (1832). *Scirpus thunbergianus* (Nees) Levyns, J. S. African Bot. 10: 28 (1944).

Synonyms:

Scirpoides globiferus (L.f.) Soják = ***Scirpoides holoschoenus*** subsp. ***globiferus*** (L.f.) Soják

Scirpoides holoschoenus subsp. *australis* (L.) Soják = ***Scirpoides holoschoenus*** (L.) Soják subsp. ***holoschoenus***

Scirpoides holoschoenus subsp. *panormitanus* (Parl.) Soják = ***Scirpoides holoschoenus*** (L.) Soják subsp. ***holoschoenus***

Scirpoides nodosus (Rottb.) L.C.Br. ex R.T.F.Clifton = ***Ficinia nodosa*** (Rottb.) Goetgh., Muasya & D.A.Simpson

Scirpoides romanus (L.) Soják = ***Scirpoides holoschoenus*** (L.) Soják subsp. ***holoschoenus***

Scirpoides thunbergii (Schrad.) Soják = ***Scirpoides holoschoenus*** subsp. ***thunbergii*** (Schrad.) Soják

Scirpus

Scirpus Tourn. ex L., Sp. Pl.: 51 (1753), nom. cons.

Temp. Northern Hemisphere to S. South America, S. Africa. 10 DEN FIN GRB IRE NOR SWE 11 AUT BGM CZE GER HUN NET POL SWI 12 FRA SPA 13 ALB BUL GRC ITA ROM TUE YUG 14 BLR BLT KRY RUC RUE RUN RUS RUW UKR 27 CPP LES NAT OFS SWZ TVL 30 ALT BRY CTA IRK KRA TVA WSB YAK 31 AMU KAM KHA KUR MAG PRM SAK 32 KAZ 33 NCS TCS 34 TUR 36 CHC CHH CHM CHN CHS 37 MON 38 JAP KOR NNS OGA TAI 40 ASS EHM IND NEP WHM 41 LAO THA VIE 42 BOR JAW MOL PHI SUL SUM 43 NWG 50 NSW QLD VIC (51) 70 ASK NWT YUK 71 ABT BRC MAN SAS 72 LAB NBR NFL NSC ONT PEI QUE 73 COL IDA IOW KAN MIN MNT MSO NDA NEB OKL ORE SDA WAS WIS WYO 74 ILL IOW KAN MIN MSO NDA NEB OKL SDA WIS 75 CNT INI MAI MAS MIC NWH NWJ NWY OHI PEN RHO VER WVA 76 ARI CAL NEV UTA 77 NWM TEX 78 ALA ARK DEL FLA GEO KTY LOU MRY MSI NCA SCA TEN VER VRG WDC 79 MXE MXN 83 BOL PER 85 AGE AGS AGW CLC CLN CLS.

66 Species

Chamaeschoenus Ehrh., Beitr. Naturk. 4: 147 (1789).

Leiophyllum Ehrh., Beitr. Naturk. 4: 146 (1789).

Dichismus Raf., J. Phys. Chim. Hist. Nat. Arts 89: 106 (1819).
Diplarinus Raf., J. Phys. Chim. Hist. Nat. Arts 89: 106 (1819).
Seidlia Opiz, Naturalientausch 11: 349 (1826).
Actaeogeton Steud., Nomencl. Bot., ed. 2, 1: 22 (1840).
Blepharolepis Nees in C.F.P.von Martius & auct. suc. (eds.), Fl. Bras. 2(1): 91 (1842).
Nemocharis Beurl., Bot. Not. 1853: 52 (1853).
Taphrogiton Montandon in F.Friche-Joset, Syn. Fl. Jura, ed. 2: 346 (1868).
Maximoviczia A.P.Khokhr., Fl. Magadansk. Obl.: 73 (1985), nom. illeg. *Maximowicziella* A.P.Khokhr., Analiz Fl. Kolymskogo Nagor'ya: 15 (1989).

Scirpus ancistrochaetus Schuyler, Rhodora 64: 44 (1962).
SE. Canada to Virginia. 72 QUE 75 MAS NWH NWY PEN VER WVA 78 MRY VRG. Hemicr. or rhizome geophyte.

Scirpus angustisquamis Beetle, Amer. J. Bot. 33: 661 (1946).
S. Chile. 85 CLS. Hemicr.

Scirpus asper J.Presl & C.Presl, Reliq. Haenk. 1: 194 (1828).
Peru to Argentina. 83 BOL PER 85 AGE AGS AGW CLC CLN. Hemicr. or rhizome geophyte.
Scirpus glaucus Nees & Meyen ex Kunth, Enum. Pl. 2: 160 (1837), nom. illeg. *Scirpus subasper* Beetle, Revista Univ. (Cuzco) 33(87): 139 (1945).
Scirpus leptopus Boeck., Flora 41: 414 (1858).
Scirpus asper var. *polystachyus* C.B.Clarke, Bot. Jahrb. Syst. 30(68): 36 (1901). *Scirpus subasper* var. *polystachyus* (C.B.Clarke) Beetle ex J.F.Macbr., Revista Univ. (Cuzco) 33: 139 (1945).
Scirpus subasper var. *diffusus* Beetle, Amer. J. Bot. 33: 661 (1946). *Scirpus asper* var. *diffusus* (Beetle) Beetle, Bol. Soc. Argent. Bot. 5: 82 (1953).

Scirpus atrocinctus Fernald, Proc. Amer. Acad. Arts 34: 502 (1899). *Trichophorum atrocinctum* (Fernald) Á.Löve & D.Löve, Taxon 30: 849 (1981).
Canada to N. U.S.A. 71 ABT BRC MAN SAS 72 LAB NBR NFL NSC ONT PEI QUE 73 MNT WAS WYO 74 ILL IOW MIN SDA WIS 75 CNT MAI MAS MIC NWH NWJ NWY PEN RHO VER WVA. Hemicr. or rhizome geophyte.
Eriophorum cyperinum f. *condensatum* Peck, Rep. (Annual) Regents Univ. State New York State Mus. 46: 130 (1893). *Scirpus atrocinctus* var. *condensatus* (Peck) House, Bull. New York State Mus. Nat. Hist. 233-234: 62 (1922).
Scirpus atrocinctus var. *brachypodus* Fernald, Proc. Amer. Acad. Arts 34: 503 (1899). *Scirpus atrocinctus* f. *brachypodus* (Fernald) S.F.Blake, Rhodora 15: 161 (1913).
Eriophorum cyperinum var. *brachypodum* (Fernald) Farw., Rep. (Annual) Michigan Acad. Sci. 6: 204 (1904).
Eriophorum cyperinum f. *brachypodum* (Fernald) Farw., Rep. (Annual) Michigan Acad. Sci. 21: 360 (1920).
Scirpus cyperinus var. *brachypodus* (Fernald) Gilly, Iowa State Coll. J. Sci. 21: 84 (1946).

Scirpus atrovirens Willd., Enum. Pl.: 79 (1809). *Holoschoenus atrovirens* (Willd.) Link, Hort. Berol. 1: 293 (1827). *Scirpus sylvaticus* var. *atrovirens* (Willd.) A.Gray, Manual, ed. 2: 500 (1856).
Canada to U.S.A. (11) ger swi (12) fra (13) ita 71 MAN 72 NFL ONT QUE 73 MNT 74 ILL IOW KAN MIN MSO NDA NEB OKL SDA WIS 75 CNT INI MAI MAS MIC NWH NWJ NWY OHI PEN RHO VER WVA 76 ARI 77 TEX 78 ALA DEL GEO KTY LOU MRY MSI NCA SCA TEN VRG WDC. Hemicr.
Scirpus sylvaticus var. *sychnocephalus* S.N.Cowles, Amer. Midl. Naturalist 3: 101 (1869). *Scirpus atrovirens* f. *sychnocephalus* (S.N.Cowles) S.F.Blake, Rhodora 15: 161 (1913).
Scirpus atrovirens var. *pycnocephalus* Fernald, Rhodora 8: 163 (1906).
Scirpus atrovirens f. *viviparus* Vict., Proc. & Trans. Roy. Soc. Canada, III, 23(2; 5): 267 (1929).
Scirpus atrovirens f. *proliferus* F.J.Herm., Rhodora 40: 77 (1938).
Scirpus atrovirens f. *angustispicatus* Fernald, Rhodora 45: 296 (1943).
Scirpus atrovirens f. *cephalanthus* Fernald, Rhodora 45: 296 (1943).

Scirpus chen-moui Tang & F.T.Wang, in Fl. Reipubl. Popul. Sin. 11: 223 (1961).
China (Yunnan). 36 CHC. Hemicr. or rhizome geophyte.

Scirpus chuanus Tang & F.T.Wang, in Fl. Reipubl. Popul. Sin. 11: 222 (1961).
China (Jiangsu). 36 CHS.

Scirpus chunianus Tang & F.T.Wang, in Fl. Reipubl. Popul. Sin. 11: 221 (1961).
Hainan. 36 CHH. Hemicr. or rhizome geophyte.

Scirpus colchicus Kimer., Soobshch. Akad. Nauk Gruzinsk. S.S.R. 80: 441 (1975).
Caucasus. 33 NCS TCS. Hemicr. or rhizome geophyte.

Scirpus congdonii Britton, Torreya 18: 36 (1918).
W. U.S.A. 73 ORE 76 CAL NEV. Hemicr. or rhizome geophyte.
Scirpus congdonii var. *minor* L.F.Hend., Rhodora 32: 21 (1930).

Scirpus cyperinus (L.) Kunth, Enum. Pl. 2: 170 (1837).
SC. & E. Canada to U.S.A., Mexico. 71 MAN 72 LAB NBR NFL NSC ONT PEI QUE 73 IDA MNT ORE WAS 74 ILL IOW MIN MSO OKL SDA WIS 75 CNT INI MAI MAS MIC NWH NWJ NWY OHI PEN RHO VER WVA 77 TEX 78 ALA ARK DEL FLA GEO KTY LOU MRY MSI NCA SCA TEN VRG 79 MXE. Hemicr. or rhizome geophyte.
**Eriophorum cyperinum* L., Sp. Pl. ed. 2: 77 (1762). *Trichophorum cyperinum* (L.) Pers., Syn. Pl. 1: 69 (1805). *Scirpus cyperinus* var. *normalis* Kuntze, Revis. Gen. Pl. 2: 757 (1891), nom. inval.
Scirpus eriophorum Michx., Fl. Bor.-Amer. 1: 33 (1803). *Scirpus cyperinus* var. *eriophorum* (Michx.) Kuntze, Revis. Gen. Pl. 2: 757 (1891). *Eriophorum cyperinum* var. *eriophorum* (Michx.) Farw., Rep. (Annual) Michigan Acad. Sci. 2: 39 (1901).
Scirpus thyrsiflorus Willd., Enum. Pl.: 78 (1809).
Scirpus eriophorum var. *cyperinus* A.Gray, Manual, ed. 2: 501 (1856).
Scirpus eriophorum var. *laxus* A.Gray, Manual, ed. 5: 501 (1856). *Eriophorum cyperinum* var. *laxum* (A.Gray) Britton, Sterns & Poggenb., Prelim. Cat.: 61 (1888).
Scirpus sylvaticus f. *cephaloideus* E.Sheld., Bull. Geol. Nat. Hist. Surv. 9: 68 (1894). *Scirpus*

cyperinus f. *cephaloideus* (E.Sheld.) Beetle, Rhodora 46: 146 (1944).
Scirpus eriophorum var. *andrewsii* Fernald, Proc. Amer. Acad. Arts 34: 501 (1899). *Scirpus cyperinus* var. *andrewsii* (Fernald) Fernald, Rhodora 2: 16 (1900). *Scirpus cyperinus* f. *andrewsii* (Fernald) D.S.Carp., Fl. Vermont, ed. 3: 74 (1937).
Scirpus eriophorum var. *condensatus* Fernald, Proc. Amer. Acad. Arts 34: 501 (1899). *Scirpus cyperinus* var. *condensatus* (Fernald) Fernald, Rhodora 2: 16 (1900). *Scirpus cyperinus* f. *condensatus* (Fernald) S.F.Blake, Rhodora 15: 162 (1913).
Scirpus atrocinctus var. *grandis* Fernald, Rhodora 2: 17 (1900). *Eriophorum cyperinum* f. *grande* (Fernald) Farw., Rep. (Annual) Michigan Acad. Sci. 21: 360 (1920). *Scirpus atrocinctus* f. *grandis* (Fernald) D.S.Carp., Fl. Vermont, ed. 3: 74 (1937).
Scirpus atrocinctus Fernald in N.L.Britton, Man. Fl. N. States: 1045 (1901), nom. illeg.
Scirpus cyperinus var. *pelius* Fernald, Rhodora 8: 164 (1906). *Eriophorum cyperinum* var. *pelium* (Fernald) Farw., Rep. (Annual) Michigan Acad. Sci. 19: 253 (1917).
Scirpus cyperinus var. *congestus* House, Bull. New York State Mus. Nat. Hist. 233-234: 62 (1922). *Scirpus cyperinus* f. *congestus* (House) House, Bull. New York State Mus. Nat. Hist. 254: 150 (1924).
Scirpus eriophorum f. *praelongus* Fernald, Rhodora 44: 383 (1942). *Scirpus rubricosus* f. *praelongus* (Fernald) Fernald, Rhodora 47: 124 (1945).
Scirpus rubricosus Fernald, Rhodora 47: 124 (1945). *Scirpus cyperinus* var. *rubricosus* (Fernald) Gilly, Iowa State Coll. J. Sci. 21: 82 (1946).
Scirpus cyperinus f. *confertus* Beetle, Amer. J. Bot. 33: 663 (1946).

Scirpus dialgamensis Majeed Kak & Javeid, J. Bombay Nat. Hist. Soc. 79: 172 (1982).
W. Himalaya. 40 WHM. Hemicr. or rhizome geophyte.

Scirpus diffusus Schuyler, Brittonia 18: 140 (1966).
California. 76 CAL. Hemicr. or rhizome geophyte.

Scirpus divaricatus Elliott, Sketch Bot. S. Carolina 1: 88 (1816). *Isolepis divaricata* (Elliott) A.Dietr., Sp. Pl. 2: 131 (1832).
Missouri to SE. U.S.A. 74 MSO 77 TEX 78 ALA ARK FLA GEO LOU NCA SCA TEN VER. Hemicr. or rhizome geophyte.
Scirpus ambiguus Schult. in J.J.Roemer & J.A. Schultes, Mant. 2: 85 (1824).

Scirpus expansus Fernald, Rhodora 45: 293 (1943).
E. Canada to E. U.S.A. 72 NSC ONT 75 CNT MAI MAS MIC NWH NWJ NWY OHI PEN RHO VER WVA 78 ALA DEL GEO KTY MRY NCA SCA TEN VRG WDC. Hemicr. or rhizome geophyte.
Scirpus sylvaticus var. *bissellii* Fernald, Rhodora 2: 21 (1900). *Scirpus microcarpus* var. *bissellii* (Fernald) House, Bull. New York State Mus. Nat. Hist. 243-244: 18 (1923). *Scirpus sylvaticus* f. *bissellii* (Fernald) D.S.Carp., Fl. Vermont, ed. 3: 76 (1937). *Scirpus expansus* f. *bissellii* (Fernald) Fernald, Rhodora 45: 294 (1943).
Scirpus expansus f. *globulosus* Fernald, Rhodora 45: 294 (1943).

Scirpus falsus C.B.Clarke in W.H.Harvey & auct. suc. (eds.), Fl. Cap. 7: 230 (1898).
S. Africa. 27 CPP LES NAT OFS TVL. Hemicr. or rhizome geophyte.

Scirpus ficinioides Kunth, Enum. Pl. 2: 172 (1837).
S. Africa. 27 CPP LES NAT OFS SWZ TVL. Hemicr. or rhizome geophyte.
Scirpus kunthii Boeck., Linnaea 36: 508 (1870).

Scirpus filipes C.B.Clarke, J. Linn. Soc., Bot. 36: 249 (1903).
China (Fujian). 36 CHS. Hemicr. or rhizome geophyte.
Scirpus filipes var. *paucispiculatus* Tang & F.T.Wang, in Fl. Reipubl. Popul. Sin. 11: 221 (1961).

Scirpus flaccidifolius (Fernald) Schuyler, Rhodora 69: 198 (1967).
Virginia to North Carolina. 78 NCA VRG. Hemicr. or rhizome geophyte.
**Scirpus atrovirens* var. *flaccidifolius* Fernald, Rhodora 40: 396 (1938).

Scirpus fohaiensis Tang & F.T.Wang, in Fl. Reipubl. Popul. Sin. 11: 222 (1961).
China (Yunnan). 36 CHC. Hemicr. or rhizome geophyte.

Scirpus fragrans Ruiz & Pav., Fl. Peruv. 1: 47 (1798).
Peru (Huánuco). 83 PER. Hemicr. or rhizome geophyte.

Scirpus fuirenoides Maxim., Bull. Acad. Imp. Sci. Saint-Pétersbourg 31: 109 (1887).
Japan (Honshu, Kyushu). 38 JAP. Hemicr. or rhizome geophyte.

Scirpus georgianus R.M.Harper, Bull. Torrey Bot. Club 27: 331 (1900). *Scirpus atrovirens* var. *georgianus* (R.M.Harper) Fernald, Rhodora 23: 134 (1921). *Scirpus atrovirens* f. *georgianus* (R.M.Harper) F.Seym., Fl. New England: 119 (1969).
E. Canada to C. & E. U.S.A. (51) 72 ONT PEI 74 ILL IOW KAN MIN MSO NEB WIS 75 CNT INI MAI MAS NWH NWJ NWY OHI PEN RHO VER WVA 77 TEX 78 ALA ARK GEO KTY LOU MRY NCA SCA TEN VRG. Hemicr.

Scirpus hainanensis S.M.Huang, in Fl. Hainan. 4: 538 (1977).
Hainan. 36 CHH. Hemicr. or rhizome geophyte.

Scirpus hattorianus Makino, J. Jap. Bot. 8: 44 (1933).
E. Canada to NC. & NE. U.S.A., introduced in C. Japan. (38) jap 72 LAB NBR NFL NSC ONT QUE 74 ILL MIN WIS 75 CNT INI MAI MAS MIC NWH NWJ NWY OHI PEN RHO VER WVA 78 MRY NCA VRG. Hemicr. or rhizome geophyte.
Scirpus atrovirens subvar. *viviparus* Farw., Amer. Midl. Naturalist 11: 72 (1928).

Scirpus heleocharidioides F.T.Wang & Tang, Acta Phytotax. Sin. 1: 181 (1951).
China (Hebei). 36 CHN. Hemicr. or rhizome geophyte.

Scirpus huae T.Koyama, Naturaliste Canad. 88: 239 (1961).
China (Guangxi). 36 CHS. Hemicr. or rhizome geophyte.

Scirpus jingmenensis Tang & F.T.Wang, in Fl. Reipubl. Popul. Sin. 11: 222 (1961).
China (Hubei). 36 CHC.

Scirpus karuisawensis Makino, Bot. Mag. (Tokyo) 18: 119 (1904). *Scirpus fuirenoides* var. *karuisawensis* (Makino) Hara, J. Jap. Bot. 9: 125 (1933). *Scirpus cyperinus* f. *karuisawensis* (Makino) Makino, Ill. Fl. Jap.: 812 (1940).

NE. China to C. Japan. 36 CHM 38 JAP KOR. Hemicr. or rhizome geophyte.
Scirpus fuirenoides var. *jaluanus* Kom., Fl. Manshur. 1: 341 (1901). *Scirpus jaluanus* (Kom.) Nakai in T.Mori, Enum. Pl. Corea: 77 (1922).
Scirpus coreanus Palla, Oesterr. Bot. Z. 59: 188 (1909).

Scirpus kimsonensis N.K.Khoi, J. Biol. (Vietnam) 18: 5 (1996).
Vietnam. 41 VIE. Hemicr. or rhizome geophyte.

Scirpus kunkelii Barros, Darwiniana 12: 521 (1962).
C. Peru. 83 PER. Hemicr. or rhizome geophyte.

Scirpus lineatus Michx., Fl. Bor.-Amer. 1: 32 (1803). *Trichophorum lineatum* (Michx.) Pers., Syn. Pl. 1: 69 (1805). *Isolepis lineata* (Michx.) Roem. & Schult., Syst. Veg. 2: 117 (1817). *Eriophorum lineatum* (Michx.) Benth. & Hook.f., Gen. Pl. 3: 1052 (1883).
SE. U.S.A. 78 ALA FLA GEO LOU MSI NCA SCA VRG. Hemicr. or rhizome geophyte.
Trichophorum pendulum Muhl., Descr. Gram.: 44 (1817).
Scirpus brizoides Willd. ex Link, Jahrb. Gewächsk. 1(3): 80 (1820).
Scirpus drummondii Steud., Syn. Pl. Glumac. 2: 318 (1855).
Scirpus fontinalis R.M.Harper, Bull. Torrey Bot. Club 30: 322 (1903).
Scirpus lineatus var. *habereri* House, Bull. New York State Mus. Nat. Hist. 176: 38 (1915).
Scirpus fontinalis var. *virginiana* Fernald, Rhodora 41: 532 (1939).

Scirpus longii Fernald, Rhodora 13: 6 (1911).
SE. Canada to NE. U.S.A. 72 NSC 75 CNT† MAI MAS NWH NWY† RHO. Hemicr. or rhizome geophyte.

Scirpus lushanensis Ohwi, Acta Phytotax. Geobot. 7: 134 (1938). *Scirpus lineatus* var. *lushanensis* (Ohwi) T.Koyama, J. Fac. Sci. Univ. Tokyo, Sect. 3, Bot. 7(6): 341 (1958).
Primorye to Sumatera. 31 PRM 36 CHC CHN CHS 38 KOR 40 ASS EHM 41 THA VIE 42 SUM. Hemicr. or rhizome geophyte.
Scirpus asiaticus Beetle, Amer. J. Bot. 33: 662 (1946). *Scirpus lineatus* var. *asiaticus* (Beetle) T.Koyama, J. Fac. Sci. Univ. Tokyo, Sect. 3, Bot. 7(6): 342 (1958). *Scirpus wichurae* var. *asiaticus* (Beetle) T.Koyama ex W.T.Lee, Lin. Fl. Koreae: 1532 (1996).

Scirpus macrolepis Phil., Anales Univ. Chile 93: 481 (1896).
Chile to W. & S. Argentina. 85 AGS AGW CLC. Hemicr. or rhizome geophyte.

Scirpus mattfeldianus Kük., Repert. Spec. Nov. Regni Veg. 27: 108 (1929). *Baeothryon mattfeldianum* (Kük.) Á.Löve & D.Löve, Univ. Colorado Stud. Ser. Biol. 17: 15 (1965).
SE. China to Vietnam. 36 CHS 41 VIE. Hemicr. or rhizome geophyte.

Scirpus maximowiczii C.B.Clarke, Kew Bull., Addit. Ser. 8: 30 (1908). *Eriophorum maximowiczii* (C.B.Clarke) Beetle, Amer. J. Bot. 33: 663 (1946).
Siberia to Japan. 30 BRY CTA IRK YAK 31 AMU KHA KUR MAG SAK 38 JAP KOR. Hemicr. or rhizome geophyte.
**Eriophorum japonicum* Maxim., Bull. Acad. Imp. Sci. Saint-Pétersbourg 31: 111 (1887). *Scirpus japonicus* (Maxim.) Kuntze, Revis. Gen. Pl. 2: 757 (1891), nom. illeg. *Maximoviczia japonica* (Maxim.) A.P.Khokhr., Fl. Magadansk. Obl.: 73 (1985). *Maximowicziella japonica* (Maxim.) M.S.Novos., Bot. Zhurn. (Moscow & Leningrad) 79(12): 74 (1994 publ. 1995).

Scirpus melanocaulos Phil., Anales Univ. Chile 93: 476 (1896).
Chile. 85 CLC. Hemicr. or rhizome geophyte.

Scirpus microcarpus J.Presl & C.Presl, Reliq. Haenk. 1: 195 (1828). *Scirpus sylvaticus* var. *microcarpus* (J.Presl & C.Presl) MacMill., Metasp. Minnesota Valley: 97 (1892).
Russian Far East, Alaska to W. & N. U.S.A., Mexico (Baja California). 31 KAM 70 ASK NWT YUK 71 ABT BRC MAN SAS 72 LAB NBR NFL NSC ONT PEI QUE 73 COL IDA MNT ORE WAS WYO 74 ILL IOW MIN NDA NEB SDA WIS 75 CNT MAI MAS MIC NWH NWJ NWY PEN RHO VER WVA 76 ARI CAL NEV UTA 77 NWM 78 KTY 79 MXN. Hemicr. or rhizome geophyte.
Scirpus lenticularis Torr., Ann. Lyceum Nat. Hist. New York 3: 328 (1836).
Scirpus rubrotinctus Fernald, Rhodora 2: 20 (1900). *Scirpus microcarpus* var. *rubrotinctus* (Fernald) M.E.Jones, Bull. Montana State Univ., Biol. Ser. 15: 20 (1910). *Scirpus microcarpus* subsp. *rubrotinctus* (Fernald) Á.Löve & D.Löve, Taxon 30: 849 (1981).
Scirpus rubrotinctus var. *confertus* Fernald, Rhodora 2: 21 (1900). *Scirpus microcarpus* var. *confertus* (Fernald) House, Bull. New York State Mus. Nat. Hist. 233-234: 62 (1922). *Scirpus rubrotinctus* f. *confertus* (Fernald) Weath., Bull. Connecticut State Geol. Surv. 48: 29 (1931).
Scirpus macounii T.Holm, Amer. J. Sci., IV, 18: 21 (1904).
Scirpus avatshensis Kom., Repert. Spec. Nov. Regni Veg. 13: 163 (1914).
Scirpus microcarpus var. *longispicatus* M.Peck, Proc. Biol. Soc. Wash. 47: 185 (1934).
Scirpus rubrotinctus f. *radiosus* Fernald, Rhodora 45: 295 (1943). *Scirpus microcarpus* f. *radiosus* (Fernald) C.F.Reed, Phytologia 67: 453 (1989).

Scirpus mitsukurianus Makino, Bot. Mag. (Tokyo) 17: 7 (1903).
C. & S. Japan. 38 JAP. Hemicr. or rhizome geophyte.

Scirpus molinianus Beetle, Bol. Soc. Argent. Bot. 5: 85 (1953).
S. Chile. 85 CLS. Hemicr. or rhizome geophyte.

Scirpus monocephalus J.Q.He, Acta Phytotax. Sin. 37: 291 (1999).
China (Anhui). 36 CHS. Hemicr. or rhizome geophyte.

Scirpus naikianus M.A.Wadood Khan, Rheedea 8: 71 (1998).
India (Maharashtra). 40 IND. Hemicr. or rhizome geophyte.

Scirpus neochinensis Tang & F.T.Wang, in Fl. Reipubl. Popul. Sin. 11: 223 (1961).
China (Guangdong). 36 CHS. Hemicr. or rhizome geophyte.

Scirpus oligosetus A.E.Kozhevn., Bot. Zhurn. (Moscow & Leningrad) 72: 1255 (1987).
Russian Far East. 31 PRM. Hemicr. or rhizome geophyte.

Scirpus orientalis Ohwi, Acta Phytotax. Geobot. 1: 76 (1932). *Scirpus sylvaticus* subsp. *orientalis* (Ohwi) Vorosch. in A.K.Skvortsov (ed.), Florist. Issl. Razn. Raionakh SSSR: 152 (1985).
Siberia to Japan. 30 ALT BRY CTA IRK KRA TVA WSB YAK 31 AMU KHA PRM SAK 36 CHM 37 MON 38 JAP KOR. Hemicr. or rhizome geophyte.
Scirpus sylvaticus var. *maximowiczii* Regel, Tent. Fl. Ussur.: 161 (1861). *Scirpus sylvaticus* subsp. *maximowiczii* (Regel) N.R.Cui, Claves Pl. Xinjiang. 1: 293 (1982).
Scirpus sylvaticus var. *subradicans* Kük. ex Tang, Bull. Fan Mem. Inst. Biol. 3: 136 (1932).

Scirpus pallidus (Britton) Fernald, Rhodora 8: 163 (1906).
S. Canada to U.S.A. 71 ABT BRC MAN SAS 72 ONT 73 COL IDA IOW KAN MIN MNT MSO NDA NEB OKL ORE SDA WAS WIS WYO 75 PEN 76 ARI UTA 77 NWM TEX. Hemicr. or rhizome geophyte.
**Scirpus atrovirens* var. *pallidus* Britton, Trans. New York Acad. Sci. 9: 14 (1889). *Scirpus atrovirens* subsp. *pallidus* (Britton) D.Löve & J.-P.Bernard, Svensk Bot. Tidskr. 53: 379 (1959).

Scirpus paniculatocorymbosus Kük., Acta Horti Gothob. 5: 35 (1930).
China (Sichuan). 36 CHC. Hemicr. or rhizome geophyte.

Scirpus peckii Britton, Trans. New York Acad. Sci. 11: 82 (1892).
SW. Québec to NE. U.S.A. 72 QUE 75 CNT MAI MAS NWH NWY RHO VER. Hemicr. or rhizome geophyte.
Scirpus atratus Fernald, Rhodora 2: 18 (1900).

Scirpus pedicellatus Fernald, Rhodora 2: 16 (1900). *Eriophorum cyperinum* var. *pedicellatum* (Fernald) Farw., Rep. (Annual) Michigan Acad. Sci. 21: 360 (1920). *Scirpus cyperinus* var. *pedicellatus* (Fernald) Schuyler, Michigan Botanist 1: 80 (1962).
E. Canada to NC. & NE. U.S.A. 72 NBR NSC ONT QUE 74 ILL IOW MIN MSO WIS 75 CNT INI MAI MAS MIC NWH NWJ NWY OHI PEN RHO VER. Hemicr. or rhizome geophyte.
Scirpus pedicellatus var. *pullus* Fernald, Rhodora 2: 17 (1900).
Scirpus pedicellatus f. *viviparus* F.G.Bernard, Naturaliste Canad. 95: 225 (1968).

Scirpus pendulus Muhl., Descr. Gram.: 44 (1817).
E. Canada to U.S.A., Mexico (Coahuila). (50) vic 72 NBR ONT QUE 73 COL ORE 74 ILL IOW KAN MIN MSO NEB OKL SDA WIS 75 CNT INI MAI MAS MIC NWH NWJ NWY OHI PEN VER WVA 76 ARI CAL 77 NWM TEX 78 ALA ARK DEL FLA GEO KTY MRY MSI NCA SCA TEN VRG WDC 79 MXE. Hemicr. or rhizome geophyte.
Scirpus lineatus Muhl., Descr. Gram.: 45 (1817), nom. illeg.
Scirpus lineatus f. *elongatus* Eames, Rhodora 33: 168 (1931). *Scirpus pendulus* f. *elongatus* (Eames) C.F.Reed, Phytologia 67: 453 (1989).

Scirpus petelotii R.Gross, Notizbl. Bot. Gart. Berlin-Dahlem 14: 193 (1938).
Indo-China. 41 LAO VIE. Hemicr. or rhizome geophyte.

Scirpus pinguiculus C.B.Clarke in W.H.Harvey & auct. suc. (eds.), Fl. Cap. 7: 222 (1898).
S. Africa. 27 CPP LES NAT. Hemicr. or rhizome geophyte.

Scirpus polyphyllus Vahl, Enum. Pl. 2: 274 (1805).
EC. & E. U.S.A. 74 ILL MSO 75 CNT INI MAS NWH NWJ NWY OHI PEN VER WVA 78 ALA ARK DEL GEO KTY MRY MSI NCA SCA TEN VRG WDC. Hemicr. or rhizome geophyte.
Schoenus cymosus Willd., Sp. Pl. 1: 265 (1797). *Rhynchospora cymosa* (Willd.) Muhl. ex Elliott, Sketch Bot. S. Carolina 1: 58 (1816). *Scirpus cymosus* (Willd.) Spreng., Syst. Veg. 1: 209 (1824), nom. illeg. *Phaeocephalum cymosum* (Willd.) House, Amer. Midl. Naturalist 6: 201 (1920). *Dichromena cymosa* (Willd.) J.F.Macbr., Publ. Field Mus. Nat. Hist., Bot. Ser. 11: 5 (1931).
Scirpus exaltatus Pursh, Fl. Amer. Sept. 1: 56 (1813), nom. illeg.
Scirpus exaltatus var. *viviparus* Pursh, Fl. Amer. Sept. 1: 56 (1813). *Scirpus brunneus* var. *viviparus* (Pursh) Torr., Ann. Lyceum Nat. Hist. New York 3: 327 (1836). *Scirpus polyphyllus* f. *viviparus* (Pursh) Beetle, Amer. Midl. Naturalist 41: 482 (1949).
Scirpus brunneus Muhl., Descr. Gram.: 42 (1817). *Holoschoenus brunneus* (Muhl.) Steud., Nomencl. Bot., ed. 2, 1: 773 (1840).
Scirpus brunneus var. *crispus* Torr., Ann. Lyceum Nat. Hist. New York 3: 327 (1836).
Scirpus polyphyllus var. *macrostachys* Boeck., Linnaea 36: 731 (1870). *Scirpus polyphyllus* f. *macrostachys* (Boeck.) Fernald, Rhodora 45: 296 (1943).

Scirpus polystachyus F.Muell., Trans. Philos. Soc. Victoria 1: 108 (1855).
SE. Queensland to Victoria. 50 NSW QLD VIC. Hemicr. or rhizome geophyte.

Scirpus pseudoarticulatus L.K.Dai & S.M.Huang, in Fl. Hainan. 4: 538 (1977).
Hainan. 36 CHH. Hemicr. or rhizome geophyte.

Scirpus radicans Schkuhr, Ann. Bot. (Usteri) 4: 49 (1793). *Scirpus sylvaticus* var. *radicans* (Schkuhr) Willd., Sp. Pl. 1: 308 (1798).
Europe to Japan. 10 FIN NOR SWE 11 AUT CZE GER HUN POL 12 FRA 13 ITA ROM YUG 14 BLR BLT RUC RUE RUN RUW UKR 30 ALT BRY CTA IRK KRA WSB YAK 31 AMU KHA PRM SAK 32 KAZ 37 MON 38 JAP KOR. Hemicr. or rhizome geophyte.
Scirpus sylvaticus Vahl, Enum. Pl. 2: 271 (1805), sensu auct.
Nemocharis radicans Burl., Bot. Not. 1853: 52 (1853).
Taphrogiton radicans Montandon in F.Friche-Joset, Syn. Fl. Jura, ed. 2: 346 (1868).
Cyperus radicans Missbach & E.H.L.Krause in J.Sturm, Deutschl. Fl., ed. 2, 2: 35 (1900).
Scirpus hokkaidoensis Beetle, Amer. J. Bot. 33: 662 (1946).

Scirpus reichei Boeck., Allg. Bot. Z. Syst. 2: 55 (1896).
Chile. 85 CLC. Hemicr. or rhizome geophyte.

Scirpus rongchengensis F.Z.Li, Acta Phytotax. Sin. 25: 71 (1987).
China (Shandong). 36 CHN. Hemicr. or rhizome geophyte.

Scirpus rosthornii Diels, Bot. Jahrb. Syst. 29: 228 (1900).
E. Nepal to Japan (S. Kyushu). 36 CHC CHS 38 JAP 40 EHM NEP. Hemicr. or rhizome geophyte.

Scirpus kiushuensis Ohwi, Repert. Spec. Nov. Regni Veg. 36: 44 (1934). *Scirpus ternatanus* var. *kiushuensis* (Ohwi) T.Koyama, J. Fac. Sci. Univ. Tokyo, Sect. 3, Bot. 7(6): 349 (1958).

Scirpus spegazzinianus Barros, Bol. Soc. Argent. Bot. 5: 155 (1954).
S. Argentina (Santa Cruz). 85 AGS. Hemicr. or rhizome geophyte.

Scirpus sylvaticus L., Sp. Pl.: 49 (1753). *Schoenus sylvaticus* (L.) Bernh., Syst. Verz.: 27 (1800). *Seidlia sylvatica* (L.) Opiz, Naturalientausch 11: 349 (1826). *Nemocharis sylvatica* (L.) Beurl., Bot. Not. 1853: 52 (1853). *Taphrogiton sylvaticum* (L.) Montandon in F.Friche-Joset, Syn. Fl. Jura, ed. 2: 346 (1868). *Cyperus sylvaticus* (L.) Missbach & E.H.L.Krause in J.Sturm, Deutschl. Fl., ed. 2, 2: 34 (1900).
Europe to C. Asia. 10 DEN FIN GRB IRE NOR SWE 11 AUT BGM CZE GER HUN NET POL SWI 12 FRA SPA 13 ALB BUL GRC ITA ROM TUE YUG 14 BLR BLT KRY RUC RUE RUN RUS RUW UKR 30 ALT BRY IRK KRA WSB 32 KAZ 33 NCS TCS 34 TUR. Hemicr.
Scirpus gramineus Neck., Delic. Gallo-Belg.: 27 (1773).
Scirpus latifolius Gilib., Excerc. Phyt. 2: 515 (1792), nom. inval.
Seidlia radicans Opiz, Naturalientausch 11: 349 (1826).
Scirpus sylvaticus var. *dissitiflorus* Sond., Fl. Hamburg.: 38 (1850).
Scirpus sylvaticus var. *compactus* C.Klinggr., Fl. Preussen 2: 132 (1866).
Scirpus sylvaticus var. *digynus* Boeck., Linnaea 36: 727 (1870). *Scirpus sylvaticus* subsp. *digynus* (Boeck.) T.Koyama, Canad. J. Bot. 43: 1392 (1965).
Scirpus intermedius Èelak., Arch. Naturwiss. Landesdurchf. Böhmen 4: 742 (1881). *Scirpus celakovskyanus* Holub, Folia Geobot. Phytotax. 18: 206 (1983).
Scirpus sylvaticus var. *jechlii* Opiz ex Nyman, Consp. Fl. Eur.: 763 (1882), nom. nud.
Scirpus sylvaticus f. *jungei* Soó, Acta Bot. Acad. Sci. Hung. 16: 368 (1970 publ. 1971).

Scirpus ternatanus Reinw. ex Miq., Fl. Ned. Ind. 3: 307 (1856).
Trop. & Subtrop. Asia. 36 CHC CHS 38 JAP NNS OGA TAI 40 ASS EHM WHM 41 THA VIE 42 BOR JAW MOL PHI SUL SUM 43 NWG. Hemicr.
Scirpus chinensis Munro in B.Seemann, Bot. Voy. Herald: 423 (1857), nom. illeg.

Scirpus trachycaulos Phil., Anales Univ. Chile 93: 482 (1896).
Chile. 85 CLC. Hemicr. or rhizome geophyte.

Scirpus trisetosus Tang & F.T.Wang, in Fl. Reipubl. Popul. Sin. 11: 221 (1961).
China. 36 CHN.

Scirpus wichurae Boeck., Linnaea 36: 729 (1870). *Scirpus cyperinus* var. *wichurae* (Boeck.) Makino, Bot. Mag. (Tokyo) 18: 120 (1904). *Scirpus lineatus* f. *wichurae* (Boeck.) T.Koyama, J. Fac. Sci. Univ. Tokyo, Sect. 3, Bot. 7(6): 339 (1958). *Scirpus lineatus* subsp. *wichurae* (Boeck.) T.Koyama, J. Fac. Sci. Univ. Tokyo, Sect. 3, Bot. 7(6): 339 (1958). *Scirpus lineatus* var. *wichurae* (Boeck.) T.Koyama, J. Fac. Sci. Univ. Tokyo, Sect. 3, Bot. 7(6): 339 (1958).
Kuril Is. to Japan. 31 KUR 38 JAP. Hemicr. or rhizome geophyte.
Scirpus eriophorum var. *nipponicum* Franch. & Sav, Enum. Pl. Jap. 2: 114 (1877).
Scirpus concolor Maxim., Bull. Acad. Imp. Sci. Saint-Pétersbourg 31: 110 (1887). *Scirpus cyperinus* var. *concolor* (Maxim.) Makino, J. Jap. Bot. 8: 46 (1932). *Scirpus wichurae* f. *concolor* (Maxim.) Ohwi, Mem. Coll. Sci. Kyoto Imp. Univ., Ser. B, Biol. 18(1): 107 (1944). *Scirpus lineatus* f. *concolor* (Maxim.) T.Koyama, J. Fac. Sci. Univ. Tokyo, Sect. 3, Bot. 7(6): 341 (1958).
Scirpus fuirenoides Maxim. ex C.B.Clarke, J. Linn. Soc., Bot. 36: 250 (1903), nom. illeg.
Scirpus fuirenoides f. *minor* C.B.Clarke, Bull. Acad. Int. Géogr. Bot. 14: 200 (1904).
Scirpus wichurae var. *borealis* Ohwi in K.Mayebara, Fl. Austro-Higo.: 84 (1931). *Scirpus wichurae* f. *borealis* (Ohwi) Ohwi, Mem. Coll. Sci. Kyoto Imp. Univ., Ser. B, Biol. 18(1): 107 (1944). *Scirpus borealis* (Ohwi) T.Koyama, J. Jap. Bot. 32: 152 (1957).
Scirpus cyperinus f. *cylindricus* Makino, J. Jap. Bot. 8: 46 (1932). *Scirpus wichurae* f. *cylindricus* (Makino) Nemoto, Fl. Japan, Suppl.: 1023 (1936). *Scirpus lineatus* f. *cylindricus* (Makino) T.Koyama, J. Fac. Sci. Univ. Tokyo, Sect. 3, Bot. 7(6): 341 (1958).

Synonyms:
Scirpus abactus Ohwi = ***Schoenoplectus mucronatus*** (L.) Palla subsp. ***mucronatus***
Scirpus aberrans Cherm. = ***Schoenoplectiella aberrans*** (Cherm.) Lye
Scirpus ablepharus Griseb. = ***Oxycaryum cubense*** (Poepp. & Kunth) Palla
Scirpus abnormalis (C.B.Clarke) T.Koyama = ***Fuirena abnormalis*** C.B.Clarke
Scirpus acaulis Phil. = ***Phylloscirpus acaulis*** (Phil.) Goetgh. & D.A.Simpson
Scirpus acicularis L. = ***Eleocharis acicularis*** (L.) Roem. & Schult.
Scirpus acicularis Zanardini ex Parl. = ***Eleocharis atropurpurea*** (Retz.) Kunth
Scirpus acicularis Oeder = ***Eleocharis palustris*** (L.) Roem. & Schult. subsp. ***palustris***
Scirpus acicularis Ucria = ***Isolepis cernua*** (Vahl) Roem. & Schult. var. ***cernua***
Scirpus acicularis var. *fluitans* Döll = ***Eleocharis acicularis*** (L.) Roem. & Schult. subsp. ***acicularis***
Scirpus acicularis var. *rigidula* Rchb. = ***Eleocharis acicularis*** (L.) Roem. & Schult. subsp. ***acicularis***
Scirpus acicularis var. *submersa* Nilsson = ***Eleocharis acicularis*** (L.) Roem. & Schult. subsp. ***acicularis***
Scirpus acicularis var. *submersus* (Porsild) Nilsson = ***Eleocharis acicularis*** (L.) Roem. & Schult. subsp. ***acicularis***
Scirpus acicularis f. *submersus* (Osten) Porsild = ***Eleocharis acicularis*** (L.) Roem. & Schult. subsp. ***acicularis***
Scirpus aciformis B.Nord. = ***Isolepis hemiuncialis*** (C.B.Clarke) J.Raynal
Scirpus acrostachys Steud. = ***Ficinia acrostachys*** (Steud.) C.B.Clarke
Scirpus acuminatus Muhl. = ***Eleocharis compressa*** Sull. var. ***compressa***
Scirpus acuminatus Steud. = ***Ficinia acuminata*** (Nees) Nees
Scirpus acutangulus Roxb. = ***Eleocharis acutangula*** (Roxb.) Schult.
Scirpus acutus (R.Br.) Spreng. = ***Eleocharis acuta*** R.Br.
Scirpus acutus Muhl. ex Bigelow = ***Schoenoplectus acutus*** (Muhl. ex Bigelow) Á.Löve & D.Löve

Scirpus acutus J.Presl & C.Presl = ***Schoenoplectus mucronatus*** (L.) Palla subsp. ***mucronatus***
Scirpus acutus f. *congestus* (Farw.) Fernald = ***Schoenoplectus acutus*** var. ***occidentalis*** (S.Watson) S.G.Sm.
Scirpus acutus var. *malheurensis* (L.F.Hend.) Beetle = ***Schoenoplectus acutus*** (Muhl. ex Bigelow) Á.Löve & D.Löve
Scirpus acutus var. *occidentalis* (S.Watson) Beetle = ***Schoenoplectus acutus*** var. ***occidentalis*** (S.Watson) S.G.Sm.
Scirpus adscendens Willd. ex Kunth = ***Schoenoplectiella supina*** (L.) Lye
Scirpus aegialites Phil. = [85 CLC] Isolepis sp. ?
Scirpus aegyptiacus Poir. = ***Bolboschoenus maritimus*** (L.) Palla subsp. ***maritimus***
Scirpus aegyptiacus Decne. = ***Schoenoplectus litoralis*** (Schrad.) Palla subsp. ***litoralis***
Scirpus aemulans Steud. = ***Actinoscirpus grossus*** (L.f.) Goetgh. & D.A.Simpson var. ***grossus***
Scirpus aestivalis (Vahl) Retz. = ***Fimbristylis aestivalis*** Vahl
Scirpus aestivalis Wall. = ***Fimbristylis squarrosa*** Vahl var. ***squarrosa***
Scirpus affinis Roth = ***Bolboschoenus maritimus*** subsp. ***affinis*** (Roth) T.Koyama
Scirpus afflatus (Steud.) Benth. = ***Eleocharis pellucida*** C.Presl
Scirpus afflatus (Steud.) Kuntze = ***Eleocharis pellucida*** C.Presl
Scirpus africanus Beetle = ***Isolepis pusilla*** Kunth
Scirpus aggregatus Steud. ex C.B.Clarke = ***Ficinia filiformis*** (Lam.) Schrad.
Scirpus aggregatus Ruiz ex Boeck. = ***Schoenoplectus pungens*** (Vahl) Palla var. ***pungens***
Scirpus albescens (É.Desv.) F.Phil. = ***Isolepis nigricans*** Kunth
Scirpus albibracteatus (Nees & Meyen ex Kunth) Kuntze = ***Eleocharis albibracteata*** Nees & Meyen ex Kunth
Scirpus albidus (Torr.) Kuntze = ***Eleocharis albida*** Torr.
Scirpus alboater Schrad. ex Nees = ***Eleocharis flavescens*** (Poir.) Urb. var. ***flavescens***
Scirpus albovaginatus (Boeck.) Kuntze = ***Eleocharis flavescens*** (Poir.) Urb. var. ***flavescens***
Scirpus albovittatus Schur = ***Scirpoides holoschoenus*** (L.) Soják subsp. ***holoschoenus***
Scirpus albus (L.) Salisb. = ***Rhynchospora alba*** (L.) Vahl
Scirpus alpinus (L.) Dalla Torre & Sarnth. = ***Trichophorum alpinum*** (L.) Pers.
Scirpus alpinus Schleich. = ***Trichophorum pumilum*** (Vahl) Schinz & Thell. subsp. ***pumilum***
Scirpus altissimus Gilib. = ***Schoenoplectus lacustris*** (L.) Palla subsp. ***lacustris***
Scirpus amazonicus Beetle = ***Isolepis fluitans*** (L.) R.Br. var. ***fluitans***
Scirpus ambiguus Schult. = ***Scirpus divaricatus*** Elliott
Scirpus amentaceus (Rudge) Griseb. = ***Fimbristylis complanata*** (Retz.) Link var. ***complanata***
Scirpus americanus Pers. = ***Schoenoplectus americanus*** (Pers.) Volkart
Scirpus americanus var. *conglomeratus* (Kunth) T.Koyama = ***Schoenoplectus pungens*** (Vahl) Palla var. ***pungens***
Scirpus americanus f. *cordilleranus* Kük. & Barros = ***Schoenoplectus pungens*** (Vahl) Palla var. ***pungens***
Scirpus americanus var. *longibracteatus* Barros & Osten = ***Schoenoplectus pungens*** var. ***longispicatus*** (Britton) S.G.Sm.
Scirpus americanus var. *longisetis* (Benth. & F.Muell.) Beetle = ***Schoenoplectus pungens*** var. ***longispicatus*** (Britton) S.G.Sm.
Scirpus americanus var. *longispicatus* Britton = ***Schoenoplectus pungens*** var. ***longispicatus*** (Britton) S.G.Sm.
Scirpus americanus subsp. *monophyllus* (C.Presl) T.Koyama = ***Schoenoplectus pungens*** var. ***badius*** (C.Presl) S.G.Sm.
Scirpus americanus var. *polyphyllus* (Boeck.) Beetle = ***Schoenoplectus pungens*** var. ***longispicatus*** (Britton) S.G.Sm.
Scirpus americanus var. *pungens* (Vahl) Barros & Osten = ***Schoenoplectus pungens*** (Vahl) Palla
Scirpus americanus var. *triangularis* (Pers.) Beetle = ***Schoenoplectus americanus*** (Pers.) Volkart
Scirpus amphibius (Durand) Bonnier = ***Eleocharis bonariensis*** Nees
Scirpus analecti Beetle = ***Cypringlea analecta*** (Beetle) M.T.Strong
Scirpus analecti var. *evadens* C.D.Adams = ***Cypringlea analecta*** (Beetle) M.T.Strong
Scirpus anceps Willd. = ***Fimbristylis complanata*** (Retz.) Link var. ***complanata***
Scirpus anceps Poir. = ***Machaerina anceps*** (Poir.) Bojer
Scirpus andinus Phil. = ***Eleocharis melanostachys*** (d'Urv.) C.B.Clarke
Scirpus androgyna (R.Br.) Poir. = ***Cyperus denudatus*** L.f.
Scirpus andrzejowscii Besser ex Schult. = ***Schoenoplectus lacustris*** (L.) Palla subsp. ***lacustris***
Scirpus angachillensis (Steud.) Boeck. = ***Isolepis nigricans*** Kunth
Scirpus angolensis C.B.Clarke = ***Nemum spadiceum*** (Lam.) Desv.
Scirpus angolensis var. *briziformis* (Hutch.) S.S.Hooper = ***Nemum spadiceum*** (Lam.) Desv.
Scirpus angolensis var. *megastachyum* Cherm. = ***Nemum megastachyum*** (Cherm.) J.Raynal
Scirpus angulatus Willd. ex Kunth = ***Eleocharis acutangula*** (Roxb.) Schult.
Scirpus angustifolius (Honck.) T.Koyama = ***Eriophorum angustifolium*** Honck.
Scirpus anisochaetus C.Wright = ***Eleocharis flavescens*** (Poir.) Urb. var. ***flavescens***
Scirpus annamicus Raymond = ***Schoenoplectus annamicus*** (Raymond) T.Koyama
Scirpus annuus Thuill. = ***Eleocharis ovata*** (Roth) Roem. & Schult.
Scirpus annuus All. = ***Fimbristylis dichotoma*** (L.) Vahl subsp. ***dichotoma***
Scirpus anomalus Retz. = ***Hypolytrum nemorum*** (Vahl) Spreng. subsp. ***nemorum***
Scirpus antarcticus Vahl = ***Bulbostylis barbata*** (Rottb.) C.B.Clarke subsp. ***barbata***
Scirpus antarcticus L. = ***Isolepis antarctica*** (L.) Roem. & Schult.
Scirpus antarcticus var. *alpinus* Ewart = ***Isolepis antarctica*** (L.) Roem. & Schult.
Scirpus antarcticus var. *propinquus* Ewart = ***Isolepis antarctica*** (L.) Roem. & Schult.
Scirpus antipodus V.J.Cook = ***Isolepis sepulcralis*** Steud.
Scirpus aphyllus Boeck. = ***Isolepis cernua*** (Vahl) Roem. & Schult. var. ***cernua***
Scirpus appendiculatus (Phil.) Kuntze = ***Eleocharis palustris*** (L.) Roem. & Schult. subsp. ***palustris***
Scirpus apus A.Gray = ***Fimbristylis vahlii*** (Lam.) Link
Scirpus arabicus Ehrenb. ex Boeck. = ***Schoenoplectus litoralis*** (Schrad.) Palla subsp. ***litoralis***

Scirpus ardea T.Koyama = ***Eriophorum gracile*** Koch
Scirpus arechavaletae Boeck. = ***Isolepis cernua*** (Vahl) Roem. & Schult. var. ***cernua***
Scirpus arenarius (Nees) Boeck. = ***Bulbostylis humilis*** (Kunth) C.B.Clarke
Scirpus arenarius Benth. = ***Isolepis cernua*** var. ***setiformis*** (Benth.) Muasya
Scirpus arenarius var. *setiformis* Benth. = ***Isolepis cernua*** var. ***setiformis*** (Benth.) Muasya
Scirpus arenicolus (Torr. ex Engelm. & A.Gray) Kuntze = ***Eleocharis montevidensis*** Kunth
Scirpus argenteus Rottb. = ***Fimbristylis argentea*** (Rottb.) Vahl
Scirpus aristatus Willd. = ***Fuirena ciliaris*** (L.) Roxb. f. ***ciliaris***
Scirpus aristatus (J.Presl) K.Richt. = ***Schoenoplectus pungens*** (Vahl) Palla var. ***pungens***
Scirpus arrhenopogon Hochst. ex C.B.Clarke = ***Schoenoplectus litoralis*** (Schrad.) Palla subsp. ***litoralis***
Scirpus articulatus L. = ***Schoenoplectiella articulata*** (L.) Lye
Scirpus articulatus var. *major* Boeck. = ***Schoenoplectiella articulata*** (L.) Lye
Scirpus ×arunensis Druce = ***Schoenoplectus ×kuekenthalianus*** (Junge) D.H.Kent
Scirpus arvensis Roxb. = ***Fimbristylis dichotoma*** (L.) Vahl subsp. ***dichotoma***
Scirpus arvensis Retz. = ***Fimbristylis ferruginea*** (L.) Vahl var. ***ferruginea***
Scirpus asiaticus Beetle = ***Scirpus lushanensis*** Ohwi
Scirpus asper var. *diffusus* (Beetle) Beetle = ***Scirpus asper*** J.Presl & C.Presl
Scirpus asper var. *polystachyus* C.B.Clarke = ***Scirpus asper*** J.Presl & C.Presl
Scirpus asperulus (Kunth) Spreng. = ***Bulbostylis tenuifolia*** (Rudge) J.F.Macbr.
Scirpus atacamensis (Phil.) Kuntze = ***Eleocharis quinqueflora*** (Hartmann) O.Schwarz
Scirpus atacamensis (Phil.) Boeck. = ***Zameioscirpus atacamensis*** (Phil.) Dhooge & Goetgh.
Scirpus atacamensis var. *exscapus* Kurtz = ***Phylloscirpus acaulis*** (Phil.) Goetgh. & D.A.Simpson
Scirpus atratus Fernald = ***Scirpus peckii*** Britton
Scirpus atrichus (R.Br.) Poir. = ***Eleocharis atricha*** R.Br.
Scirpus atrichus (Palla) Dalla Torre & Sarnth. = ***Trichophorum pumilum*** (Vahl) Schinz & Thell. subsp. ***pumilum***
Scirpus atrocinctus Fernald = ***Scirpus cyperinus*** (L.) Kunth
Scirpus atrocinctus var. *brachypodus* Fernald = ***Scirpus atrocinctus*** Fernald
Scirpus atrocinctus f. *brachypodus* (Fernald) S.F.Blake = ***Scirpus atrocinctus*** Fernald
Scirpus atrocinctus var. *condensatus* (Peck) House = ***Scirpus atrocinctus*** Fernald
Scirpus atrocinctus var. *grandis* Fernald = ***Scirpus cyperinus*** (L.) Kunth
Scirpus atrocinctus f. *grandis* (Fernald) D.S.Carp. = ***Scirpus cyperinus*** (L.) Kunth
Scirpus atrofactus Steud. = ***Eleocharis geniculata*** (L.) Roem. & Schult.
Scirpus atropurpureovaginatus Boeck. = ***Ficinia nodosa*** (Rottb.) Goetgh., Muasya & D.A.Simpson
Scirpus atropurpureus Retz. = ***Eleocharis atropurpurea*** (Retz.) Kunth
Scirpus atrosanguineus Boeck. = ***Bulbostylis atrosanguinea*** (Boeck.) C.B.Clarke
Scirpus atrosanguineus (É.Desv.) Skottsb. = ***Schoenoplectus californicus*** (C.A.Mey.) Soják
Scirpus atrovirens f. *angustispicatus* Fernald = ***Scirpus atrovirens*** Willd.
Scirpus atrovirens f. *cephalanthus* Fernald = ***Scirpus atrovirens*** Willd.
Scirpus atrovirens var. *flaccidifolius* Fernald = ***Scirpus flaccidifolius*** (Fernald) Schuyler
Scirpus atrovirens var. *georgianus* (R.M.Harper) Fernald = ***Scirpus georgianus*** R.M.Harper
Scirpus atrovirens f. *georgianus* (R.M.Harper) F.Seym. = ***Scirpus georgianus*** R.M.Harper
Scirpus atrovirens var. *pallidus* Britton = ***Scirpus pallidus*** (Britton) Fernald
Scirpus atrovirens subsp. *pallidus* (Britton) D.Löve & J.-P.Bernard = ***Scirpus pallidus*** (Britton) Fernald
Scirpus atrovirens f. *proliferus* F.J.Herm. = ***Scirpus atrovirens*** Willd.
Scirpus atrovirens var. *pycnocephalus* Fernald = ***Scirpus atrovirens*** Willd.
Scirpus atrovirens f. *sychnocephalus* (S.N.Cowles) S.F.Blake = ***Scirpus atrovirens*** Willd.
Scirpus atrovirens f. *viviparus* Vict. = ***Scirpus atrovirens*** Willd.
Scirpus atrovirens subvar. *viviparus* Farw. = ***Scirpus hattorianus*** Makino
Scirpus attenuatus Franch. & Sav. = ***Eleocharis attenuata*** (Franch. & Sav.) Palla
Scirpus aturensis Maury = ***Bulbostylis aturensis*** (Maury) C.B.Clarke
Scirpus aucklandicus (Hook.f.) Boeck. = ***Isolepis aucklandica*** Hook.f.
Scirpus aureiglumis S.S.Hooper = ***Schoenoplectiella juncea*** (Willd.) Lye
Scirpus auronitens Nees & Ehrenb. ex Boeck. = ***Bolboschoenus maritimus*** (L.) Palla subsp. ***maritimus***
Scirpus australiensis (Maiden & Betche) S.T.Blake = ***Isolepis australiensis*** (Maiden & Betche) K.L.Wilson
Scirpus australis L. = ***Scirpoides holoschoenus*** (L.) Soják subsp. ***holoschoenus***
Scirpus autumnalis Rottb. = ***Cyperus haspan*** L.
Scirpus autumnalis L. = ***Fimbristylis autumnalis*** (L.) Roem. & Schult.
Scirpus avatshensis Kom. = ***Scirpus microcarpus*** J.Presl & C.Presl
Scirpus bachmannianus Boeck. ex C.B.Clarke = ***Isolepis venustula*** Kunth
Scirpus badius C.Presl = ***Schoenoplectus pungens*** var. ***badius*** (C.Presl) S.G.Sm.
Scirpus baeothryon Schrad. = ***Eleocharis palustris*** (L.) Roem. & Schult. subsp. ***palustris***
Scirpus baeothryon L.f. = ***Eleocharis quinqueflora*** (Hartmann) O.Schwarz
Scirpus bahiensis Steud. = ***Eleocharis sellowiana*** Kunth var. ***sellowiana***
Scirpus bailii Kohts = ***Eleocharis palustris*** (L.) Roem. & Schult. subsp. ***palustris***
Scirpus baldwinianus Schult. = ***Fimbristylis dichotoma*** (L.) Vahl subsp. ***dichotoma***
Scirpus balearicus Willd. ex Kunth = ***Schoenoplectus litoralis*** (Schrad.) Palla subsp. ***litoralis***
Scirpus bangalorensis B.Heyne ex Wall. = ***Rhynchospora corymbosa*** (L.) Britton
Scirpus barbatulus Kunth = ***Schoenoplectus lacustris*** (L.) Palla subsp. ***lacustris***
Scirpus barbatus Rottb. = ***Bulbostylis barbata*** (Rottb.) C.B.Clarke
Scirpus barbatus Vahl ex Kunth = ***Bulbostylis vestita*** (Kunth) C.B.Clarke

Scirpus barbatus var. *americanus* Boeck. = ***Bulbostylis barbata*** (Rottb.) C.B.Clarke
Scirpus barbulatus Kunth = ***Schoenoplectus lacustris*** (L.) Palla subsp. ***lacustris***
Scirpus basilaris (Hook.f.) C.B.Clarke = ***Isolepis basilaris*** Hook.f.
Scirpus beccarii Boeck. = ***Isolepis beccarii*** (Boeck.) Goetgh. & D.A.Simpson
Scirpus bellardii (All.) Wahlenb. = ***Kobresia myosuroides*** (Vill.) Fiori & Paol.
Scirpus benghalensis Pers. = ***Fimbristylis quinquangularis*** (Vahl) Kunth
Scirpus bergianus Spreng. = ***Isolepis antarctica*** (L.) Roem. & Schult.
Scirpus bergsonii Schuyler = ***Schoenoplectiella saximontana*** (Fernald) Lye
Scirpus berlandieri Steud. = ***Bulbostylis junciformis*** (Kunth) C.B.Clarke
Scirpus bernardinus Munz & I.M.Johnst. = ***Eleocharis bernardina*** (Munz & I.M.Johnst.) Munz & I.M.Johnst.
Scirpus bibracteatus Kük. = [27 CPP]
Scirpus bicapitatus Poir. = ***Ficinia bulbosa*** (L.) Nees
Scirpus biceps Roem. & Schult. = ***Ficinia bulbosa*** (L.) Nees
Scirpus bicolor (Carmich.) Spreng. = ***Isolepis bicolor*** Carmich.
Scirpus biconcavus Ohwi = ***Bolboschoenus planiculmis*** (F.Schmidt) E.M.Egorova
Scirpus bifolius Wallr. = ***Blysmus rufus*** (Huds.) Link
Scirpus bispicatus J.König ex Roxb. = ***Fimbristylis schoenoides*** (Retz.) Vahl
Scirpus bisumbellatus Forssk. = ***Fimbristylis bisumbellata*** (Forssk.) Bubani
Scirpus bivalvis Lam. = ***Fimbristylis bivalvis*** (Lam.) Lye
Scirpus boeckelerianus Schweinf. = ***Bulbostylis boeckeleriana*** (Schweinf.) Beetle
Scirpus boliviensis Barros = ***Phylloscirpus boliviensis*** (Barros) Dhooge & Goetgh.
Scirpus bonariensis (Nees) Kuntze = ***Eleocharis bonariensis*** Nees
Scirpus bonariensis Poir. = ***Fimbristylis ferruginea*** (L.) Vahl var. ***ferruginea***
Scirpus boraeanus Tourlet = ***Bolboschoenus maritimus*** (L.) Palla subsp. ***maritimus***
Scirpus borealis (Ohwi) T.Koyama = ***Scirpus wichurae*** Boeck.
Scirpus borneensis H.Pfeiff. = [42 BOR]
Scirpus brachyantherus (Trautv. & C.A.Mey.) T.Koyama = ***Eriophorum brachyantherum*** Trautv. & C.A.Mey.
Scirpus brachyceras Hochst. ex A.Rich. = ***Schoenoplectus corymbosus*** (Roth ex Roem. & Schult.) J.Raynal
Scirpus brachyphyllus Willd. ex Kunth = ***Bulbostylis barbata*** (Rottb.) C.B.Clarke subsp. ***barbata***
Scirpus bracteatus Bigelow = ***Trichophorum cespitosum*** (L.) Hartm. subsp. ***cespitosum***
Scirpus brayi Hoppe ex Roem. & Schult. = ***Schoenoplectus lacustris*** (L.) Palla subsp. ***lacustris***
Scirpus brevicaulis Levyns = ***Isolepis brevicaulis*** (Levyns) J.Raynal
Scirpus breviculmis (Kunth) Boeck. = ***Bulbostylis humilis*** (Kunth) C.B.Clarke
Scirpus brevifolius Decne. = ***Blysmus compressus*** var. ***brevifolius*** (Decne.) Karth.
Scirpus brevifolius Bernh. ex Kunth = ***Bulbostylis barbata*** (Rottb.) C.B.Clarke subsp. ***barbata***
Scirpus brevifolius Roxb. = ***Fimbristylis dichotoma*** (L.) Vahl subsp. ***dichotoma***
Scirpus brevifolius Poir. = ***Fimbristylis ferruginea*** (L.) Vahl var. ***ferruginea***
Scirpus brevis d'Urv. = ***Isolepis cernua*** (Vahl) Roem. & Schult. var. ***cernua***
Scirpus brevivaginatus Boeck. = ***Schoenoplectus pungens*** (Vahl) Palla var. ***pungens***
Scirpus bridgesii Boeck. = ***Isolepis nigricans*** Kunth
Scirpus brittonianus Piper = ***Bolboschoenus maritimus*** subsp. ***paludosus*** (A.Nelson) T.Koyama
Scirpus briziformis Hutch. = ***Nemum spadiceum*** (Lam.) Desv.
Scirpus brizoides Willd. ex Link = ***Scirpus lineatus*** Michx.
Scirpus brizoides Benth. = ***Isolepis oldfieldiana*** (S.T.Blake) K.L.Wilson
Scirpus bromeliifolius Rudge = ***Diplasia karatifolia*** Rich. ex Pers.
Scirpus brownii Spreng. = ***Eleocharis geniculata*** (L.) Roem. & Schult.
Scirpus brunneovaginatus Boeck. = ***Bulbostylis sphaerocephala*** (Boeck.) Lindm.
Scirpus brunneus Muhl. = ***Scirpus polyphyllus*** Vahl
Scirpus brunneus var. *crispus* Torr. = ***Scirpus polyphyllus*** Vahl
Scirpus brunneus var. *viviparus* (Pursh) Torr. = ***Scirpus polyphyllus*** Vahl
Scirpus brunonianus S.T.Blake = ***Isolepis cyperoides*** R.Br.
Scirpus bucharicus Roshev. = ***Schoenoplectus bucharicus*** (Roshev.) Grossh.
Scirpus buettnerianus Boeck. = [22 GUI]
Scirpus bufonius (Kunth) Spreng. = ***Bulbostylis tenuifolia*** (Rudge) J.F.Macbr.
Scirpus bulbiferus Boeck. = ***Isolepis bulbifera*** (Boeck.) Muasya
Scirpus bulbostyloides S.S.Hooper = ***Nemum bulbostyloides*** (S.S.Hooper) J.Raynal
Scirpus bulbosus Rottb. = ***Ficinia deusta*** (P.J.Bergius) Levyns
Scirpus burchellii C.B.Clarke = ***Isolepis brevicaulis*** (Levyns) J.Raynal
Scirpus burkei (C.B.Clarke) K.Schum. = ***Bulbostylis contexta*** (Nees) Bodard
Scirpus burkei C.B.Clarke = ***Scirpoides burkei*** (C.B.Clarke) Goetgh., Muasya & D.A.Simpson
Scirpus caducus Delile = ***Eleocharis caduca*** (Delile) Schult.
Scirpus caespitosus Willd. ex Kunth = ***Eleocharis palustris*** (L.) Roem. & Schult. subsp. ***palustris***
Scirpus caldwellii V.J.Cook = ***Bolboschoenus caldwellii*** (V.J.Cook) Soják
Scirpus californicus (C.A.Mey.) Steud. = ***Schoenoplectus californicus*** (C.A.Mey.) Soják
Scirpus californicus var. *chamissoi* (Nees ex Mart) Beetle = ***Schoenoplectus californicus*** (C.A.Mey.) Soják
Scirpus californicus var. *paschalis* (Kük.) Beetle = ***Schoenoplectus californicus*** (C.A.Mey.) Soják
Scirpus californicus var. *spoliatus* Barros = ***Schoenoplectus californicus*** (C.A.Mey.) Soják
Scirpus californicus var. *tatora* (Kunth) Barros = ***Schoenoplectus californicus*** (C.A.Mey.) Soják
Scirpus californicus subsp. *tatora* (Kunth) T.Koyama = ***Schoenoplectus californicus*** (C.A.Mey.) Soják
Scirpus caligenis V.J.Cook = ***Isolepis caligenis*** (V.J.Cook) Soják
Scirpus caligenis var. *tristigmatosa* V.J.Cook = ***Isolepis caligenis*** (V.J.Cook) Soják
Scirpus calocarpus S.T.Blake = ***Isolepis hookeriana*** Boeck.

Scirpus calolepis (K.Schum.) Kuntze = ***Fuirena ochreata*** Nees ex Kunth
Scirpus campestris Britton = ***Bolboschoenus maritimus*** subsp. ***paludosus*** (A.Nelson) T.Koyama
Scirpus campestris Willd. ex Kunth = ***Eleocharis palustris*** (L.) Roem. & Schult. subsp. ***palustris***
Scirpus campestris Rottb. = ***Eleocharis quinqueflora*** (Hartmann) O.Schwarz
Scirpus campestris var. *fernaldii* (E.P.Bicknell) Bartlett = ***Bolboschoenus maritimus*** (L.) Palla
Scirpus campestris var. *novae-angliae* (Britton) Fernald = ***Schoenoplectus novae-angliae*** (Britton) M.T.Strong
Scirpus campestris var. *paludosus* (A.Nelson) Fernald = ***Bolboschoenus maritimus*** subsp. ***paludosus*** (A.Nelson) T.Koyama
Scirpus camptotrichus C.Wright = ***Eleocharis nana*** Kunth
Scirpus camtschaticus Kunth = ***Eleocharis kamtschatica*** (C.A.Mey.) Kom.
Scirpus canaliculatotriqueter Steud. = ***Actinoscirpus grossus*** (L.f.) Goetgh. & D.A.Simpson var. ***grossus***
Scirpus canbyi A.Gray = ***Schoenoplectus etuberculatus*** (Steud.) Soják
Scirpus candelabrum Willd. ex Kunth = ***Fimbristylis dichotoma*** (L.) Vahl subsp. ***dichotoma***
Scirpus capensis Burm.f. = ***Bolboschoenus capensis*** (Burm.f.) Holub
Scirpus capensis Rottb. = ***Ficinia bulbosa*** (L.) Nees
Scirpus capillaceus (Kunth) Griseb. = ***Eleocharis capillacea*** Kunth
Scirpus capillaceus Michx. = ***Eleocharis nana*** Kunth
Scirpus capillaris L. = ***Bulbostylis capillaris*** (L.) Kunth ex C.B.Clarke
Scirpus capillaris var. *elatior* Griseb. = ***Bulbostylis capillaris*** (L.) Kunth ex C.B.Clarke subsp. ***capillaris***
Scirpus capillaris var. *sphaerolepis* Boeck. = ***Bulbostylis capillaris*** (L.) Kunth ex C.B.Clarke subsp. ***capillaris***
Scirpus capillaris var. *tenuifolius* (Rudge) Pilg. = ***Bulbostylis tenuifolia*** (Rudge) J.F.Macbr.
Scirpus capillifolius Parl. = ***Isolepis striata*** (Nees) Kunth
Scirpus capitatus Muhl. = ***Eleocharis albida*** Torr.
Scirpus capitatus L. = ***Eleocharis geniculata*** (L.) Roem. & Schult.
Scirpus capitatus Schreb. = ***Eleocharis ovata*** (Roth) Roem. & Schult.
Scirpus capitatus Willd. ex Kunth = ***Eleocharis tenuis*** (Willd.) Schult. var. ***tenuis***
Scirpus capitatus Burm.f. = ***Fuirena uncinata*** (Willd.) Kunth
Scirpus capsularis Lour. = (Juncaceae)
Scirpus caribaeus Griseb. = ***Eleocharis flavescens*** (Poir.) Urb. var. ***flavescens***
Scirpus caribaeus Rottb. = ***Eleocharis geniculata*** (L.) Roem. & Schult.
Scirpus cariciformis Vest = ***Blysmus compressus*** (L.) Panz. ex Link subsp. ***compressus***
Scirpus caricinus Schrad. = ***Blysmus compressus*** (L.) Panz. ex Link subsp. ***compressus***
Scirpus caricis Retz. = ***Blysmus compressus*** (L.) Panz. ex Link subsp. ***compressus***
Scirpus caricis var. *brevifolius* (Decne.) C.B.Clarke = ***Blysmus compressus*** var. ***brevifolius*** (Decne.) Karth.
Scirpus caricis var. *dissitus* C.B.Clarke = ***Blysmus compressus*** var. ***brevifolius*** (Decne.) Karth.
Scirpus caricis var. *sikkimensis* C.B.Clarke = ***Blysmus compressus*** var. ***brevifolius*** (Decne.) Karth.
Scirpus carinatus (Hook. & Arn. ex Torr.) A.Gray = ***Isolepis carinata*** Hook. & Arn. ex Torr.
Scirpus ×*carinatus* Sm. = ***Schoenoplectus ×carinatus*** (Sm.) Palla
Scirpus carniolicus (W.D.J.Koch) Ambrosi = ***Eleocharis carniolica*** W.D.J.Koch
Scirpus carolinianus Lam. = ***Fimbristylis caroliniana*** (Lam.) Fernald
Scirpus carolinianus Panz. ex Kunth = ***Rhynchospora nitens*** (Vahl) A.Gray var. ***nitens***
Scirpus caronarius Vahl = ***Cyperus coronarius*** (Vahl) Kunth
Scirpus carsei Kük. = ***Isolepis fluitans*** var. ***lenticularis*** (R.Br.) Muasya
Scirpus cartilagineus Benth. = ***Isolepis alpina*** Hook.f.
Scirpus cartilagineus (R.Br.) Spreng. = ***Isolepis marginata*** (Thunb.) A.Dietr.
Scirpus cartilagineus var. *alpinus* (Hook.f.) Benth. = ***Isolepis alpina*** Hook.f.
Scirpus cartilagineus var. *propinquus* Benth. = ***Isolepis marginata*** (Thunb.) A.Dietr.
Scirpus castaneus Muhl. = ***Bulbostylis capillaris*** (L.) Kunth ex C.B.Clarke subsp. ***capillaris***
Scirpus castaneus Willd. ex Kunth = ***Fimbristylis autumnalis*** (L.) Roem. & Schult.
Scirpus castaneus Michx. = ***Fimbristylis spadicea*** (L.) Vahl
Scirpus celakovskyanus Holub = ***Scirpus sylvaticus*** L.
Scirpus cellulosus (Torr.) Kuntze = ***Eleocharis cellulosa*** Torr.
Scirpus cephalotes B.Heyne ex Roth = ***Bolboschoenus maritimus*** (L.) Palla subsp. ***maritimus***
Scirpus cephalotes Jacq. = ***Kyllinga nemoralis*** (J.R.Forst. & G.Forst.) Dandy ex Hutch. & Dalziel
Scirpus cephalotes Walter = ***Lipocarpha maculata*** (Michx.) Torr.
Scirpus cephalotes L. = ***Rhynchospora cephalotes*** (L.) Vahl
Scirpus cernuus Vahl = ***Isolepis cernua*** (Vahl) Roem. & Schult.
Scirpus cernuus f. *aphyllus* (Boeck.) Barros = ***Isolepis cernua*** (Vahl) Roem. & Schult. var. ***cernua***
Scirpus cernuus var. *australiensis* Maiden & Betche = ***Isolepis australiensis*** (Maiden & Betche) K.L.Wilson
Scirpus cernuus var. *brevis* (d'Urv.) Skottsb. = ***Isolepis cernua*** (Vahl) Roem. & Schult. var. ***cernua***
Scirpus cernuus var. *californicus* (Torr.) Beetle = ***Isolepis cernua*** (Vahl) Roem. & Schult. var. ***cernua***
Scirpus cernuus subsp. *californicus* (Torr.) Thorne = ***Isolepis cernua*** (Vahl) Roem. & Schult. var. ***cernua***
Scirpus cernuus var. *dura* C.B.Clarke = ***Isolepis cernua*** (Vahl) Roem. & Schult. var. ***cernua***
Scirpus cernuus f. *macrostachyus* Maire & Weiller = ***Isolepis cernua*** (Vahl) Roem. & Schult. var. ***cernua***
Scirpus cernuus f. *pygmaeus* (Kunth) Lindm. = ***Isolepis cernua*** (Vahl) Roem. & Schult. var. ***cernua***
Scirpus cernuus var. *pygmaeus* (Kunth) Dusén = ***Isolepis cernua*** (Vahl) Roem. & Schult. var. ***cernua***
Scirpus cernuus var. *subtilis* (Kunth) C.B.Clarke = ***Isolepis tenuissima*** (Nees) Kunth
Scirpus cernuus var. *vahlii* (Lange) O.Bolòs & Vigo = ***Isolepis cernua*** (Vahl) Roem. & Schult. var. ***cernua***
Scirpus cespitosus Pollich = ***Eleocharis quinqueflora*** (Hartmann) O.Schwarz
Scirpus cespitosus (R.Br.) Poir. = ***Fimbristylis caespitosa*** R.Br.
Scirpus cespitosus L. = ***Trichophorum cespitosum*** (L.) Hartm.
Scirpus cespitosus var. *callosus* Bigelow ex Torr. = ***Trichophorum cespitosum*** (L.) Hartm. subsp. ***cespitosum***

Scirpus cespitosus var. *delicatulus* Fernald = ***Trichophorum cespitosum*** (L.) Hartm. subsp. ***cespitosum***
Scirpus cespitosus subsp. *germanicus* (Palla) Brodd. = ***Trichophorum cespitosum*** subsp. ***germanicum*** (Palla) Hegi
Scirpus chaeta Schult. = ***Eleocharis acicularis*** (L.) Roem. & Schult. subsp. ***acicularis***
Scirpus chaetarius (Roem. & Schult.) Spreng. = ***Eleocharis retroflexa*** subsp. ***chaetaria*** (Roem. & Schult.) T.Koyama
Scirpus chaetodes Link = ***Isolepis cernua*** (Vahl) Roem. & Schult. var. ***cernua***
Scirpus chamissoi Schrad. ex Nees = ***Schoenoplectus californicus*** (C.A.Mey.) Soják
Scirpus chamissonis (C.A.Mey.) T.Koyama = ***Eriophorum chamissonis*** C.A.Mey.
Scirpus chilensis Nees & Meyen = ***Schoenoplectus americanus*** (Pers.) Volkart
Scirpus chilensis var. *longisetis* (Kük.) Beetle = ***Schoenoplectus americanus*** (Pers.) Volkart
Scirpus chinensis Munro = ***Scirpus ternatanus*** Reinw. ex Miq.
Scirpus chinensis Osbeck = ***Lipocarpha chinensis*** (Osbeck) J.Kern
Scirpus chirigota C.Wright = ***Bulbostylis hispidula*** (Vahl) R.W.Haines subsp. ***hispidula***
Scirpus chlorostachyus Levyns = ***Isolepis tenuissima*** (Nees) Kunth
Scirpus chloroticus Phil. = ***Isolepis cernua*** (Vahl) Roem. & Schult. var. ***cernua***
Scirpus chubutensis C.B.Clarke = ***Amphiscirpus nevadensis*** (S.Watson) Oteng-Yeb.
Scirpus ciliaris L. = ***Fuirena ciliaris*** (L.) Roxb.
Scirpus ciliaris Pers. = ***Fuirena pubescens*** (Poir.) Kunth var. ***pubescens***
Scirpus ciliatifolius Elliott = ***Bulbostylis ciliatifolia*** (Elliott) Fernald
Scirpus ciliatus Rottb. = ***Fuirena ciliaris*** (L.) Roxb. f. ***ciliaris***
Scirpus ciliatus Steud. = ***Schoenoplectus lacustris*** (L.) Palla subsp. ***lacustris***
Scirpus ciliolatus Boeck. = ***Fimbristylis cymosa*** R.Br. subsp. ***cymosa***
Scirpus cinereofuscus Willd. ex Kunth = ***Fimbristylis ferruginea*** (L.) Vahl var. ***ferruginea***
Scirpus cinnamometorum Vahl = ***Fimbristylis cinnamometorum*** (Vahl) Kunth
Scirpus cinnamomeus Boeck. = ***Bulbostylis cinnamomea*** (Boeck.) C.B.Clarke
Scirpus circinatus (Kunth) Boeck. = ***Bulbostylis circinata*** (Kunth) C.B.Clarke
Scirpus clarkei Stapf = ***Trichophorum subcapitatum*** (Thwaites & Hook.) D.A.Simpson
Scirpus clathratus Rchb. = ***Isolepis setacea*** (L.) R.Br.
Scirpus clemensii (Kük.) Ohwi = ***Schoenoplectus mucronatus*** subsp. ***clemensii*** (Kük.) Soják
Scirpus clementis M.E.Jones = ***Oreobolopsis clementis*** (M.E.Jones) Dhooge & Goetgh.
Scirpus clintonii A.Gray = ***Trichophorum clintonii*** (A.Gray) S.G.Sm.
Scirpus coahuilensis Svenson = ***Cypringlea coahuilensis*** (Svenson) M.T.Strong
Scirpus coarctatus Elliott = ***Bulbostylis ciliatifolia*** var. ***coarctata*** (Elliott) Kral
Scirpus coerulescens (Steud.) Kuntze = ***Fuirena coerulescens*** Steud.
Scirpus cognatus Hance = ***Schoenoplectus mucronatus*** (L.) Palla subsp. ***mucronatus***
Scirpus coleotrichus (Hochst. ex A.Rich.) Boeck. = ***Bulbostylis coleotricha*** (Hochst. ex A.Rich.) C.B.Clarke
Scirpus collinus Boeck. = ***Bulbostylis contexta*** (Nees) Bodard
Scirpus coloradoensis Britton = ***Eleocharis coloradoensis*** (Britton) Gilly
Scirpus comosus (Dumort.) B.D.Jacks. = ***Eleocharis parvula*** (Roem. & Schult.) Palla
Scirpus comosus Wall. = ***Eriophorum comosum*** (Wall.) Nees
Scirpus compactus Hoffm. = ***Bolboschoenus maritimus*** (L.) Palla subsp. ***maritimus***
Scirpus compactus (R.Br.) Spreng. = ***Eleocharis spiralis*** (Rottb.) Roem. & Schult.
Scirpus compactus var. *orientalis* Litv. = ***Bolboschoenus planiculmis*** (F.Schmidt) E.M.Egorova
Scirpus complanatus Retz. = ***Fimbristylis complanata*** (Retz.) Link
Scirpus compressus (L.) Pers. = ***Blysmus compressus*** (L.) Panz. ex Link
Scirpus compressus Moench = ***Eleocharis ovata*** (Roth) Roem. & Schult.
Scirpus compressus var. *gracilis* Zapal. = ***Blysmus compressus*** (L.) Panz. ex Link subsp. ***compressus***
Scirpus concolor Maxim. = ***Scirpus wichurae*** Boeck.
Scirpus confervoides Poir. = ***Websteria confervoides*** (Poir.) S.S.Hooper
Scirpus confusus N.E.Br. = ***Schoenoplectus confusus*** (N.E.Br.) Lye
Scirpus congdonii var. *minor* L.F.Hend. = ***Scirpus congdonii*** Britton
Scirpus congestus Rchb. ex Kunth = ***Bulbostylis junciformis*** (Kunth) C.B.Clarke
Scirpus congestus (D.Don) Buch.-Ham. ex D.Don = ***Eleocharis congesta*** D.Don
Scirpus conglomeratus Kunth = ***Schoenoplectus pungens*** (Vahl) Palla var. ***pungens***
Scirpus congruus (Nees) S.T.Blake = ***Isolepis congrua*** Nees
Scirpus conicus C.Presl ex Mert. & W.D.J.Koch = ***Eleocharis palustris*** (L.) Roem. & Schult. subsp. ***palustris***
Scirpus coniferus Poir. = ***Lepironia articulata*** (Retz.) Domin
Scirpus conostachys (Boeck.) Boeck. = ***Bulbostylis conostachya*** (Boeck.) Beetle
Scirpus consanguineus (Kunth) Boeck. = ***Bulbostylis consanguinea*** (Kunth) C.B.Clarke
Scirpus conspersus (Nees) Laing = ***Isolepis inundata*** R.Br.
Scirpus constitutionis Boeck. ex R.Phil. = ***Isolepis nigricans*** Kunth
Scirpus constrictus Schrad. ex Kunth = ***Eleocharis geniculata*** (L.) Roem. & Schult.
Scirpus ×*contortus* (Eames) T.Koyama = ***Schoenoplectus* × *contortus*** (Eames) S.G.Sm.
Scirpus coreanus Palla = ***Scirpus karuisawensis*** Makino
Scirpus cornatus C.Wright = ***Fimbristylis squarrosa*** Vahl var. ***squarrosa***
Scirpus cornatus C.Wright = ***Fimbristylis squarrosa*** Vahl var. ***squarrosa***
Scirpus coronarius Vahl = ***Cyperus coronarius*** (Vahl) Kunth
Scirpus coronatus Steud. = ***Ficinia indica*** (Lam.) H.Pfeiff.
Scirpus corymbiferus C.Wright = ***Rhynchospora ignorata*** Boeck.
Scirpus corymbosus Forssk. = ***Bolboschoenus maritimus*** (L.) Palla subsp. ***maritimus***

Scirpus corymbosus L. = ***Rhynchospora corymbosa*** (L.) Britton
Scirpus corymbosus (Roth ex Roem. & Schult.) B.Heyne = ***Schoenoplectus corymbosus*** (Roth ex Roem. & Schult.) J.Raynal
Scirpus costatus (Hochst. ex A.Rich.) Boeck. = ***Isolepis costata*** Hochst. ex A.Rich.
Scirpus costatus var. *gracilis* Boeck. = ***Isolepis wakefieldiana*** (S.T.Blake) K.L.Wilson
Scirpus costatus var. *macer* (Boeck.) Cherm. = ***Isolepis costata*** Hochst. ex A.Rich.
Scirpus costulatus (Nees & Meyen ex Kunth) Kuntze = ***Eleocharis radicans*** Kunth
Scirpus crassiculmis (Nees) Schrad. ex Griseb. = ***Eleocharis elegans*** (Kunth) Roem. & Schult.
Scirpus crassiusculus (Hook.f.) Benth. = ***Isolepis crassiuscula*** Hook.f.
Scirpus crinalis Griseb. = ***Eleocharis crinalis*** (Griseb.) C.B.Clarke
Scirpus criniger A.Gray = ***Eriophorum crinigerum*** (A.Gray) Beetle
Scirpus crinitus (Poir.) Poir. = ***Ficinia crinita*** (Poir.) B.L Burtt
Scirpus cubensis Poepp. & Kunth = ***Oxycaryum cubense*** (Poepp. & Kunth) Palla
Scirpus cubensis var. *gracilis* (Boeck.) Beetle = ***Oxycaryum cubense*** (Poepp. & Kunth) Palla
Scirpus cubensis var. *minor* C.Wright & Sauvalle = ***Oxycaryum cubense*** (Poepp. & Kunth) Palla
Scirpus cubensis var. *paraguayensis* (Maury) Kük. ex Barros = ***Oxycaryum cubense*** (Poepp. & Kunth) Palla
Scirpus cumingii Boeck. = ***Carex setifolia*** Kunze var. ***setifolia***
Scirpus cunninghamii (Boeck.) Kuntze = ***Eleocharis gracilis*** R.Br.
Scirpus cuspidatus Roth = ***Fuirena cuspidata*** (Roth) Kunth
Scirpus custoris Hegetschw = ***Schoenoplectus lacustris*** (L.) Palla subsp. ***lacustris***
Scirpus cylindraceus Willd. ex Kunth = ***Fimbristylis spadicea*** (L.) Vahl
Scirpus cylindricus (Torr.) Britton = ***Schoenoplectus novae-angliae*** (Britton) M.T.Strong
Scirpus cymosus (Willd.) Spreng. = ***Scirpus polyphyllus*** Vahl
Scirpus cymosus (R.Br.) Poir. = ***Fimbristylis cymosa*** R.Br.
Scirpus cymosus Lam. = ***Fimbristylis cymosa*** R.Br. subsp. ***cymosa***
Scirpus cyperiformis Muhl. = ***Cyperus lupulinus*** subsp. ***macilentus*** (Fernald) Marcks
Scirpus cyperinus var. *andrewsii* (Fernald) Fernald = ***Scirpus cyperinus*** (L.) Kunth
Scirpus cyperinus f. *andrewsii* (Fernald) D.S.Carp. = ***Scirpus cyperinus*** (L.) Kunth
Scirpus cyperinus var. *brachypodus* (Fernald) Gilly = ***Scirpus atrocinctus*** Fernald
Scirpus cyperinus f. *cephaloideus* (E.Sheld.) Beetle = ***Scirpus cyperinus*** (L.) Kunth
Scirpus cyperinus var. *concolor* (Maxim.) Makino = ***Scirpus wichurae*** Boeck.
Scirpus cyperinus var. *condensatus* (Fernald) Fernald = ***Scirpus cyperinus*** (L.) Kunth
Scirpus cyperinus f. *condensatus* (Fernald) S.F.Blake = ***Scirpus cyperinus*** (L.) Kunth
Scirpus cyperinus f. *confertus* Beetle = ***Scirpus cyperinus*** (L.) Kunth
Scirpus cyperinus var. *congestus* House = ***Scirpus cyperinus*** (L.) Kunth
Scirpus cyperinus f. *congestus* (House) House = ***Scirpus cyperinus*** (L.) Kunth
Scirpus cyperinus f. *cylindricus* Makino = ***Scirpus wichurae*** Boeck.
Scirpus cyperinus var. *eriophorum* (Michx.) Kuntze = ***Scirpus cyperinus*** (L.) Kunth
Scirpus cyperinus f. *karuisawensis* (Makino) Makino = ***Scirpus karuisawensis*** Makino
Scirpus cyperinus var. *normalis* Kuntze = ***Scirpus cyperinus*** (L.) Kunth
Scirpus cyperinus var. *pedicellatus* (Fernald) Schuyler = ***Scirpus pedicellatus*** Fernald
Scirpus cyperinus var. *pelius* Fernald = ***Scirpus cyperinus*** (L.) Kunth
Scirpus cyperinus var. *rubricosus* (Fernald) Gilly = ***Scirpus cyperinus*** (L.) Kunth
Scirpus cyperinus var. *wichurae* (Boeck.) Makino = ***Scirpus wichurae*** Boeck.
Scirpus cyperoides Lam. = ***Bolboschoenus maritimus*** (L.) Palla subsp. ***maritimus***
Scirpus cyperoides L. = ***Cyperus cyperoides*** (L.) Kuntze
Scirpus cyperoides (R.Br.) Spreng. = ***Isolepis cyperoides*** R.Br.
Scirpus cyperoides Hemsl. = ***Karinia mexicana*** (C.B.Clarke ex Britton) Reznicek & McVaugh
Scirpus debilis (Kunth) Kuntze = ***Eleocharis debilis*** Kunth
Scirpus debilis Lam. = ***Fimbristylis ferruginea*** (L.) Vahl var. ***ferruginea***
Scirpus debilis Pursh = ***Schoenoplectiella purshiana*** (Fernald) Lye
Scirpus debilis var. *williamsii* Fernald = ***Schoenoplectiella purshiana*** (Fernald) Lye
Scirpus decipiens A.St.-Hil. = ***Schoenoplectus californicus*** (C.A.Mey.) Soják
Scirpus decipiens A.St.-Hil. ex Nees = ***Schoenoplectus californicus*** (C.A.Mey.) Soják
Scirpus decumanus Willd. ex Kunth = ***Bolboschoenus maritimus*** (L.) Palla subsp. ***maritimus***
Scirpus delicatulus Levyns = ***Isolepis bulbifera*** (Boeck.) Muasya
Scirpus deltarum Schuyler = ***Schoenoplectus deltarum*** (Schuyler) Soják
Scirpus densus Wall. = ***Bulbostylis densa*** (Wall.) Hand.-Mazz.
Scirpus denudatus (R.Br.) Poir. = ***Fimbristylis denudata*** R.Br.
Scirpus depauperatus (R.Br.) Kunth = ***Fimbristylis dichotoma*** subsp. ***depauperata*** (R.Br.) J.Kern
Scirpus depauperatus Kom. = ***Schoenoplectus nipponicus*** (Makino) Soják
Scirpus depressus Vell. = ***Eleocharis geniculata*** (L.) Roem. & Schult.
Scirpus deserticola Phil. = ***Phylloscirpus deserticola*** (Phil.) Dhooge & Goetgh.
Scirpus deserticola var. *semisubterraneus* (Boeck.) Barros = ***Phylloscirpus deserticola*** (Phil.) Dhooge & Goetgh.
Scirpus desprauxii Steud. = ***Schoenoplectus litoralis*** (Schrad.) Palla subsp. ***litoralis***
Scirpus desvauxii Phil. = ***Eleocharis fuscopurpurea*** (Steud.) H.Pfeiff.
Scirpus diabolicus Steud. = ***Isolepis diabolica*** (Steud.) Schrad.
Scirpus dichotomus L. = ***Fimbristylis dichotoma*** (L.) Vahl
Scirpus dichotomus G.Mey. = ***Fimbristylis spadicea*** (L.) Vahl
Scirpus dichromenoides C.B.Clarke = [82 VEN]

Scirpus dictyospermus C.Wright = ***Eleocharis cellulosa*** Torr.
Scirpus didymostachyus Phil. = ***Isolepis nigricans*** Kunth
Scirpus dietrichiae Boeck. = ***Lipocarpha microcephala*** (R.Br.) Kunth
Scirpus digitatus (Nees ex Schrad.) Boeck. = ***Isolepis digitata*** Nees ex Schrad.
Scirpus dioecus (Kunth) Boeck. = ***Scirpoides dioecus*** (Kunth) Browning
Scirpus diphyllus Retz. = ***Fimbristylis dichotoma*** (L.) Vahl subsp. ***dichotoma***
Scirpus dipsaceus Rottb. = ***Fimbristylis dipsacea*** (Rottb.) C.B.Clarke
Scirpus dipsaceus Vahl ex Kunth = ***Isolepis hystrix*** (Thunb.) Schrad.
Scirpus dissachanthus S.T.Blake = ***Schoenoplectiella dissachantha*** (S.T.Blake) Lye
Scirpus dissitus Duthie = ***Blysmus compressus*** var. ***dissitus*** (Duthie) Karth.
Scirpus distichophyllus Boeck. = ***Zameioscirpus gaimardioides*** (É.Desv.) Dhooge & Goetgh.
Scirpus distichus Peterm. = ***Blysmus compressus*** (L.) Panz. ex Link subsp. ***compressus***
Scirpus distigmaticus (Kük.) Tang & F.T.Wang = ***Trichophorum pumilum*** (Vahl) Schinz & Thell. subsp. ***pumilum***
Scirpus dombeyanus (Kunth) Kuntze = ***Eleocharis dombeyana*** Kunth
Scirpus domingensis Pers. = ***Fimbristylis spadicea*** (L.) Vahl
Scirpus donianus Spreng. = ***Schoenoplectiella juncoides*** (Roxb.) Lye
Scirpus dregeanus C.B.Clarke = ***Isolepis antarctica*** (L.) Roem. & Schult.
Scirpus drepanensis Lojac. = ***Schoenoplectus lacustris*** (L.) Palla subsp. ***lacustris***
Scirpus drummondii Steud. = ***Scirpus lineatus*** Michx.
Scirpus dubius Roxb. = ***Eleocharis dulcis*** (Burm.f.) Trin. ex Hensch.
Scirpus dussianus Boeck. ex Duss = ***Bulbostylis barbata*** (Rottb.) C.B.Clarke subsp. ***barbata***
Scirpus dussii Boeck. = ***Bulbostylis barbata*** (Rottb.) C.B.Clarke subsp. ***barbata***
Scirpus ×*duvalii* Hoppe = ***Schoenoplectus* ×*carinatus*** (Sm.) Palla nothosubsp. ***carinatus***
Scirpus ebenocarpus Kirk = ***Isolepis marginata*** (Thunb.) A.Dietr.
Scirpus echinatus L. = ***Cyperus echinatus*** (L.) Alph.Wood
Scirpus eckloneus Steud. = ***Ficinia ecklonea*** (Steud.) Nees
Scirpus edwallianus Boeck. = ***Bulbostylis edwalliana*** (Boeck.) Prata & M.G.López
Scirpus ehrenbergii Boeck. = ***Schoenoplectus ehrenbergii*** (Boeck.) Soják
Scirpus elatus Gueldenst. ex Ledeb. = [14]
Scirpus elatus Boeck. = ***Ficinia filiformis*** (Lam.) Schrad.
Scirpus elatus (R.Br.) Poir. = ***Fimbristylis denudata*** R.Br.
Scirpus elegans Salisb. = ***Cyperus luzulae*** (L.) Retz.
Scirpus elegans Kunth = ***Eleocharis elegans*** (Kunth) Roem. & Schult.
Scirpus elegantulus Steud. = ***Eleocharis ovata*** (Roth) Roem. & Schult.
Scirpus elliottii Spreng. = ***Fimbristylis dichotoma*** (L.) Vahl subsp. ***dichotoma***
Scirpus ellipticus Willd. ex Kunth = ***Eleocharis tenuis*** (Willd.) Schult. var. ***tenuis***
Scirpus ellychniarius Molina = ***Scirpoides holoschoenus*** (L.) Soják subsp. ***holoschoenus***
Scirpus elongatus Buch.-Ham. ex D.Don = ***Eriophorum comosum*** (Wall.) Nees
Scirpus elongatus (R.Br.) Poir. = ***Fimbristylis dichotoma*** (L.) Vahl subsp. ***dichotoma***
Scirpus emarginatus (Nees) Kuntze = ***Eleocharis emarginata*** (Nees) Klotzsch ex Boeck.
Scirpus emergens (Norman) Fernald = ***Trichophorum pumilum*** (Vahl) Schinz & Thell. subsp. ***pumilum***
Scirpus enodis (C.B.Clarke) T.Koyama = ***Fuirena enodis*** C.B.Clarke
Scirpus equisetiformis Meinsh. = ***Eleocharis mitracarpa*** Steud.
Scirpus equisetoides Elliott = ***Eleocharis equisetoides*** (Elliott) Torr.
Scirpus equitans Kük. = ***Nemum equitans*** (Kük.) J.Raynal
Scirpus erectogracilis Hayata = [38 TAI]
Scirpus erectus Poir. = ***Schoenoplectiella erecta*** (Poir.) Lye
Scirpus erectus var. *angustus* Makino = ***Schoenoplectiella wallichii*** (Nees) Lye
Scirpus erectus subsp. *raynalii* (Schuyler) B.F.Hansen & Wunderlin = ***Schoenoplectiella erecta*** (Poir.) Lye
Scirpus erectus var. *triangulatus* Honda = ***Schoenoplectus* ×*trapezoideus*** (Koidz.) Hayas. & H.Ohashi
Scirpus eriophorum Michx. = ***Scirpus cyperinus*** (L.) Kunth
Scirpus eriophorum var. *andrewsii* Fernald = ***Scirpus cyperinus*** (L.) Kunth
Scirpus eriophorum var. *condensatus* Fernald = ***Scirpus cyperinus*** (L.) Kunth
Scirpus eriophorum var. *cyperinus* A.Gray = ***Scirpus cyperinus*** (L.) Kunth
Scirpus eriophorum var. *laxus* A.Gray = ***Scirpus cyperinus*** (L.) Kunth
Scirpus eriophorum var. *nipponicum* Franch. & Sav = ***Scirpus wichurae*** Boeck.
Scirpus eriophorum f. *praelongus* Fernald = ***Scirpus cyperinus*** (L.) Kunth
Scirpus eriophorus Poepp. ex C.Presl =
Scirpus erismanae Schuyler = ***Schoenoplectiella erecta*** (Poir.) Lye
Scirpus erraticus Rota ex De Not. = ***Eleocharis atropurpurea*** (Retz.) Kunth
Scirpus erythropodus Kuntze = ***Schoenoplectus etuberculatus*** (Steud.) Soják
Scirpus etuberculatus (Steud.) Kuntze = ***Schoenoplectus etuberculatus*** (Steud.) Soják
Scirpus ewersmannii Fisch. ex Litw. = ***Schoenoplectus ehrenbergii*** (Boeck.) Soják
Scirpus exaltatus Pursh = ***Scirpus polyphyllus*** Vahl
Scirpus exaltatus var. *viviparus* Pursh = ***Scirpus polyphyllus*** Vahl
Scirpus excelsus Salisb. = ***Cladium mariscus*** (L.) Pohl subsp. ***mariscus***
Scirpus exiguus Kunth = ***Eleocharis radicans*** Kunth
Scirpus exilis (Kunth) Poir. = ***Bulbostylis hispidula*** (Vahl) R.W.Haines subsp. ***hispidula***
Scirpus expallescens (Kunth) Boeck. = ***Isolepis expallescens*** Kunth
Scirpus expansus f. *bissellii* (Fernald) Fernald = ***Scirpus expansus*** Fernald
Scirpus expansus f. *globulosus* Fernald = ***Scirpus expansus*** Fernald
Scirpus falcatus Vahl = ***Fimbristylis falcata*** (Vahl) Kunth

Scirpus fascicularis Willd. ex Kunth = ***Isolepis ludwigii*** (Steud.) Kunth
Scirpus fastigiatus Thunb. = ***Ficinia fastigiata*** (Thunb.) Nees
Scirpus fauriei (E.G.Camus) T.Koyama = ***Eriophorum vaginatum*** L.
Scirpus fernaldii E.P.Bicknell = ***Bolboschoenus maritimus*** (L.) Palla subsp. ***maritimus***
Scirpus ferrugineus L. = ***Fimbristylis ferruginea*** (L.) Vahl
Scirpus ferrugineus var. *debilis* (Lam.) Poir. = ***Fimbristylis ferruginea*** (L.) Vahl var. ***ferruginea***
Scirpus festucoides Poir. = ***Bulbostylis festucoides*** (Poir.) C.B.Clarke
Scirpus filamentosus Vahl = ***Bulbostylis filamentosa*** (Vahl) C.B.Clarke
Scirpus filiculmis Schrad. ex Griseb. = ***Eleocharis filiculmis*** Kunth
Scirpus filiformis Lam. = ***Eleocharis tenuis*** (Willd.) Schult. var. ***tenuis***
Scirpus filiformis Burm.f. = ***Ficinia nigrescens*** (Schrad.) J.Raynal
Scirpus filiformis Savi = ***Isolepis cernua*** (Vahl) Roem. & Schult. var. ***cernua***
Scirpus filipes var. *paucispiculatus* Tang & F.T.Wang = ***Scirpus filipes*** C.B.Clarke
Scirpus fimbriatus (Nees) Boeck. = ***Bulbostylis fimbriata*** (Nees) C.B.Clarke
Scirpus fimbriatus Poir. = ***Fimbristylis dichotoma*** (L.) Vahl subsp. ***dichotoma***
Scirpus fimbriatus Willd. ex Kunth = ***Fimbristylis pilosa*** Vahl
Scirpus fimbrisetus Delile = ***Schoenoplectus litoralis*** (Schrad.) Palla subsp. ***litoralis***
Scirpus fimbristyloides K.Schum. ex C.B.Clarke = ***Bulbostylis fimbristyloides*** C.B.Clarke
Scirpus fimbristyloides K.Schum. ex C.B.Clarke = ***Bulbostylis schimperiana*** (Hochst. ex A.Rich.) C.B.Clarke
Scirpus fistulosus Poir. = ***Eleocharis acutangula*** (Roxb.) Schult.
Scirpus fistulosus (Link) Kuntze = ***Eleocharis acutangula*** (Roxb.) Schult.
Scirpus fistulosus Forssk. = ***Schoenoplectiella articulata*** (L.) Lye
Scirpus flaccidus Rchb. ex A.Spreng. = ***Eleocharis flavescens*** (Poir.) Urb. var. ***flavescens***
Scirpus flaccifolius Steud. = ***Isolepis digitata*** Nees ex Schrad.
Scirpus flavescens Poir. = ***Eleocharis flavescens*** (Poir.) Urb.
Scirpus floccosus Griseb. = ***Bulbostylis floccosa*** (Griseb.) C.B.Clarke
Scirpus floridanus Michx. ex Kunth = ***Eleocharis nana*** Kunth
Scirpus fluitans L. = ***Isolepis fluitans*** (L.) R.Br.
Scirpus fluitans var. *curvulus* (Nees) C.B.Clarke = ***Isolepis fluitans*** (L.) R.Br. var. ***fluitans***
Scirpus fluitans var. *emergens* H.Pfeiff. = ***Isolepis fluitans*** (L.) R.Br. var. ***fluitans***
Scirpus fluitans var. *fascicularis* (Nees) Boeck. = ***Isolepis fluitans*** (L.) R.Br. var. ***fluitans***
Scirpus fluitans var. *microstachyus* Benth. = ***Isolepis fluitans*** (L.) R.Br. var. ***fluitans***
Scirpus fluitans var. *pseudofluitans* (Makino) T.Koyama = ***Isolepis crassiuscula*** Hook.f.
Scirpus fluitans var. *robustus* Boeck. = ***Isolepis striata*** (Nees) Kunth
Scirpus fluitans var. *spadiceus* C.B.Clarke = ***Isolepis fluitans*** (L.) R.Br. var. ***fluitans***
Scirpus fluitans var. *terrestris* Benth. = ***Isolepis fluitans*** (L.) R.Br. var. ***fluitans***
Scirpus fluviatilis (Torr.) A.Gray = ***Bolboschoenus fluviatilis*** (Torr.) Soják
Scirpus foliatus Hook.f. = ***Schoenus maschalinus*** Roem. & Schult.
Scirpus fontinalis R.M.Harper = ***Scirpus lineatus*** Michx.
Scirpus fontinalis var. *virginiana* Fernald = ***Scirpus lineatus*** Michx.
Scirpus forsythii Kük. = ***Schoenoplectus lineolatus*** (Franch. & Sav.) T.Koyama
Scirpus frondosus Banks & Sol. ex Boeck. = ***Desmoschoenus spiralis*** (A.Rich.) Hook.f.
Scirpus fuirena T.Koyama = ***Fuirena umbellata*** Rottb.
Scirpus fuirenoides Maxim. ex C.B.Clarke = ***Scirpus wichurae*** Boeck.
Scirpus fuirenoides var. *jaluanus* Kom. = ***Scirpus karuisawensis*** Makino
Scirpus fuirenoides var. *karuisawensis* (Makino) Hara = ***Scirpus karuisawensis*** Makino
Scirpus fuirenoides f. *minor* C.B.Clarke = ***Scirpus wichurae*** Boeck.
Scirpus ×fujimakii M.Kikuchi & Koyama = [38 JAP]
Scirpus furva (R.Br.) Poir. = ***Fimbristylis furva*** R.Br.
Scirpus fuscescens Link = ***Fimbristylis dichotoma*** (L.) Vahl subsp. ***dichotoma***
Scirpus fuscorubens T.Koyama = ***Schoenoplectus fuscorubens*** (T.Koyama) T.Koyama
Scirpus fuscosanguineus (Boeck.) Kuntze = ***Eleocharis pachycarpa*** É.Desv.
Scirpus fuscus Willd. ex Kunth = ***Bulbostylis tenuifolia*** (Rudge) J.F.Macbr.
Scirpus gaillardii Maury = [8]
Scirpus gaudichaudianus Kunth = ***Eleocharis flavescens*** (Poir.) Urb. var. ***flavescens***
Scirpus gaudichaudii Boeck. = ***Isolepis gaudichaudiana*** Kunth
Scirpus gaudichaudii var. *tenuis* Boeck. = ***Isolepis nigricans*** Kunth
Scirpus geniculatus L. = ***Eleocharis geniculata*** (L.) Roem. & Schult.
Scirpus geniculatus var. *minor* Vahl = ***Eleocharis geniculata*** (L.) Roem. & Schult.
Scirpus germanicus (Palla) W.F.Christ. = ***Trichophorum cespitosum*** subsp. ***germanicum*** (Palla) Hegi
Scirpus giganteus Kunth = ***Androtrichum giganteum*** (Kunth) H.Pfeiff.
Scirpus giraudyi Boeck. = ***Isolepis fluitans*** (L.) R.Br. var. ***fluitans***
Scirpus glaber (Kunth) T.Koyama = ***Fuirena hirsuta*** (P.J.Bergius) P.L.Forbes
Scirpus glaucescens Willd. = ***Eleocharis palustris*** (L.) Roem. & Schult. subsp. ***palustris***
Scirpus glaucophyllus Boeck. = ***Fimbristylis glaucophylla*** (Boeck.) Beetle
Scirpus glaucovirens (Boeck.) Kuntze = ***Eleocharis glaucovirens*** Boeck.
Scirpus glaucus Nees & Meyen ex Kunth = ***Scirpus asper*** J.Presl & C.Presl
Scirpus glaucus Lam. = ***Bolboschoenus glaucus*** S.G.Sm.
Scirpus glaucus Torr. = ***Eleocharis erythropoda*** Steud.
Scirpus glaucus Sm. = ***Schoenoplectus tabernaemontani*** (C.C.Gmel.) Palla
Scirpus glaziovii Boeck. = ***Bulbostylis juncoides*** (Vahl) Kük. ex Herter
Scirpus globiceps C.B.Clarke = ***Isolepis rubicunda*** Kunth
Scirpus globifer Welw. ex Steud. = ***Schoenoplectus lacustris*** (L.) Palla subsp. ***lacustris***

Scirpus globiferus L.f. = ***Scirpoides holoschoenus*** subsp. ***globiferus*** (L.f.) Soják
Scirpus globosus Spreng. = ***Ficinia nodosa*** (Rottb.) Goetgh., Muasya & D.A.Simpson
Scirpus globulosus Roxb. = ***Fimbristylis ferruginea*** (L.) Vahl var. ***ferruginea***
Scirpus globulosus Retz. = ***Fimbristylis umbellaris*** (Lam.) Vahl
Scirpus glomeratus B.Heyne ex Wall. = ***Cyperus arenarius*** Retz.
Scirpus glomeratus L. = ***Cyperus dubius*** Rottb. subsp. ***dubius***
Scirpus glomeratus Retz. = ***Fimbristylis cymosa*** R.Br. subsp. ***cymosa***
Scirpus glomeratus Roxb. = ***Fimbristylis dichotoma*** (L.) Vahl subsp. ***dichotoma***
Scirpus glomeratus Scop. = ***Schoenoplectus mucronatus*** (L.) Palla subsp. ***mucronatus***
Scirpus glutinosus Banks & Sol. ex Hook.f. = ***Ficinia nodosa*** (Rottb.) Goetgh., Muasya & D.A.Simpson
Scirpus gracilis (R.Br.) Kuntze = ***Eleocharis gracilis*** R.Br.
Scirpus gracilis Poir. = ***Ficinia ecklonea*** (Steud.) Nees
Scirpus gracilis Zeyh. ex Kunth = ***Ficinia gracilis*** Schrad.
Scirpus gracilis Rudge = ***Ficinia nodosa*** (Rottb.) Goetgh., Muasya & D.A.Simpson
Scirpus gracilis Willd. ex Kunth = ***Fimbristylis tenuis*** Roem. & Schult.
Scirpus gracilis Koch = ***Isolepis cernua*** (Vahl) Roem. & Schult. var. ***cernua***
Scirpus gracillimus Boeck. = [24 ETH]
Scirpus gracillimus Kohts = ***Isolepis setacea*** (L.) R.Br.
Scirpus graecus Quézel & Contandr. = [13 GRC]
Scirpus gramineus Neck. = ***Scirpus sylvaticus*** L.
Scirpus graminoides R.W.Haines & Lye = ***Isolepis graminoides*** (R.W.Haines & Lye) Lye
Scirpus grandis (Nees) Kuntze = ***Eleocharis nudipes*** (Kunth) Palla
Scirpus grandispicus (Steud.) Berhaut = ***Bolboschoenus maritimus*** (L.) Palla subsp. ***maritimus***
Scirpus granulatohirtellus Boeck. ex Schinz = ***Bulbostylis hispidula*** (Vahl) R.W.Haines subsp. ***hispidula***
Scirpus griffithii Boeck. = ***Actinoscirpus grossus*** (L.f.) Goetgh. & D.A.Simpson var. ***grossus***
Scirpus griquensium C.B.Clarke = ***Isolepis sepulcralis*** Steud.
Scirpus grisebachii Britton = ***Websteria confervoides*** (Poir.) S.S.Hooper
Scirpus grossheimii (Pobed.) Czerep. = ***Schoenoplectus lacustris*** subsp. ***hippolyti*** (V.I.Krecz.) Kukkonen
Scirpus grossus L.f. = ***Actinoscirpus grossus*** (L.f.) Goetgh. & D.A.Simpson
Scirpus grossus f. *kysoor* (Roxb.) Beetle = ***Actinoscirpus grossus*** var. ***kysoor*** (Roxb.) Noltie
Scirpus guaraniticus Pedersen = ***Schoenoplectiella supina*** (L.) Lye
Scirpus gunnii Boeck. = ***Isolepis alpina*** Hook.f.
Scirpus habessinicus Ehrenb. ex Boeck. = ***Fimbristylis ferruginea*** (L.) Vahl var. ***ferruginea***
Scirpus habrus Edgar = ***Isolepis habra*** (Edgar) Soják
Scirpus hahnii C.B.Clarke = ***Cyperus seslerioides*** Kunth
Scirpus hakonensis Franch. & Sav. = ***Eleocharis tetraquetra*** Nees
Scirpus halleri Vill. = ***Eleocharis quinqueflora*** (Hartmann) O.Schwarz
Scirpus halleri Vitman = ***Schoenoplectiella supina*** (L.) Lye
Scirpus hallii A.Gray = ***Schoenoplectiella hallii*** (A.Gray) Lye
Scirpus hamulosus (M.Bieb.) Steven = ***Cyperus hamulosus*** M.Bieb.
Scirpus helferi Beetle = [41 AND MYA]
Scirpus hemisphaericus Roth = ***Lipocarpha hemisphaerica*** (Roth) Goetgh.
Scirpus hemiuncialis C.B.Clarke = ***Isolepis hemiuncialis*** (C.B.Clarke) J.Raynal
Scirpus heterocarpus S.Watson = ***Bulbostylis funckii*** (Steud.) C.B.Clarke
Scirpus heterochaetus Chase = ***Schoenoplectus heterochaetus*** (Chase) Soják
Scirpus heteromorphus Steud. ex F.Phil. = ***Eleocharis montana*** (Kunth) Roem. & Schult.
Scirpus heterophyllus Schuyler = ***Schoenoplectiella heterophylla*** (Schuyler) Lye
Scirpus hexatrichus Ehrenb. ex Boiss. = ***Eleocharis macrantha*** Boeck.
Scirpus hieronymi Boeck. = ***Zameioscirpus atacamensis*** (Phil.) Dhooge & Goetgh.
Scirpus hildebrandtii Boeck. = ***Bulbostylis hispidula*** (Vahl) R.W.Haines subsp. ***hispidula***
Scirpus hippolyti V.I.Krecz. = ***Schoenoplectus lacustris*** subsp. ***hippolyti*** (V.I.Krecz.) Kukkonen
Scirpus hirsutus Willd. ex Kunth = ***Bulbostylis vestita*** (Kunth) C.B.Clarke
Scirpus hirtellus Vahl = ***Fimbristylis dichotoma*** (L.) Vahl subsp. ***dichotoma***
Scirpus hirtus Griseb. = ***Bulbostylis vestita*** (Kunth) C.B.Clarke
Scirpus hirtus (Kunth) Poir. = ***Fimbristylis squarrosa*** Vahl var. ***squarrosa***
Scirpus hispanicus Gueldenst. ex Ledeb. = [14]
Scirpus hispidulus Vahl = ***Bulbostylis hispidula*** (Vahl) R.W.Haines
Scirpus hochstetteri Boeck. = [24 ETH]
Scirpus hokkaidoensis Beetle = ***Scirpus radicans*** Schkuhr
Scirpus holoschoenus Oeder = ***Schoenoplectus lacustris*** (L.) Palla subsp. ***lacustris***
Scirpus holoschoenus L. = ***Scirpoides holoschoenus*** (L.) Soják
Scirpus holoschoenus Thunb. = ***Scirpoides holoschoenus*** subsp. ***thunbergii*** (Schrad.) Soják
Scirpus holoschoenus var. *australis* (L.) W.D.J.Koch = ***Scirpoides holoschoenus*** (L.) Soják subsp. ***holoschoenus***
Scirpus holoschoenus subsp. *euholoschoenus* Briq. = ***Scirpoides holoschoenus*** (L.) Soják
Scirpus holoschoenus var. *globiferus* (L.f.) Parl. = ***Scirpoides holoschoenus*** subsp. ***globiferus*** (L.f.) Soják
Scirpus holoschoenus subsp. *globiferus* (L.f.) Husn. = ***Scirpoides holoschoenus*** subsp. ***globiferus*** (L.f.) Soják
Scirpus holoschoenus var. *globuliferus* Boiss. = ***Scirpoides holoschoenus*** (L.) Soják subsp. ***holoschoenus***
Scirpus holoschoenus prol. *linnaei* (Rchb.) Asch. & Graebn. = ***Scirpoides holoschoenus*** (L.) Soják subsp. ***holoschoenus***
Scirpus holoschoenus var. *macrostachyus* Husn. = ***Scirpoides holoschoenus*** (L.) Soják subsp. ***holoschoenus***
Scirpus holoschoenus var. *romanus* (L.) W.D.J.Koch = ***Scirpoides holoschoenus*** (L.) Soják subsp. ***holoschoenus***
Scirpus holoschoenus subsp. *romanus* (L.) Mateo & Figuerola = ***Scirpoides holoschoenus*** (L.) Soják subsp. ***holoschoenus***

Scirpus holoschoenus var. *thunbergii* (Schrad.) C.B.Clarke = ***Scirpoides holoschoenus*** subsp. ***thunbergii*** (Schrad.) Soják
Scirpus hondoensis Ohwi = ***Schoenoplectus hondoensis*** (Ohwi) Soják
Scirpus hookeri (Boeck.) Kuntze = ***Eleocharis gracilis*** R.Br.
Scirpus hookeri Kunth = ***Isolepis cernua*** (Vahl) Roem. & Schult. var. ***cernua***
Scirpus hookerianus (Boeck.) S.T.Blake = ***Isolepis hookeriana*** Boeck.
Scirpus hoppei Weihe = ***Schoenoplectus triqueter*** (L.) Palla
Scirpus hotarui Ohwi = ***Schoenoplectus hotarui*** (Ohwi) Holub
Scirpus hottentottus L. = ***Fuirena hirsuta*** (P.J.Bergius) P.L.Forbes
Scirpus hudsonianus (Michx.) Fernald = ***Trichophorum alpinum*** (L.) Pers.
Scirpus humbertii Cherm. = ***Isolepis humbertii*** (Cherm.) J.Raynal
Scirpus humboldtii Spreng. = ***Bulbostylis junciformis*** (Kunth) C.B.Clarke
Scirpus humboldtii (Roem. & Schult.) A.Dietr. = ***Lipocarpha micrantha*** (Vahl) G.C.Tucker
Scirpus humilis Willd. ex Kunth = ***Bulbostylis scabra*** (J.Presl & C.Presl) C.B.Clarke
Scirpus humilis Link = ***Bulbostylis tenuifolia*** (Rudge) J.F.Macbr.
Scirpus humilis Wallr. = ***Eleocharis nana*** Kunth
Scirpus humillimus Benth. = ***Isolepis humillima*** (Benth.) K.L.Wilson
Scirpus hyalinolepis Steud. ex Jard. = ***Bolboschoenus maritimus*** (L.) Palla subsp. ***maritimus***
Scirpus hystricoides B.Nord. = ***Lipocarpha rehmannii*** (Ridl.) Goetgh.
Scirpus hystrix Thunb. = ***Isolepis hystrix*** (Thunb.) Schrad.
Scirpus ×*igaensis* T.Koyama =
Scirpus inanis (Thunb.) Steud. = ***Pseudoschoenus inanis*** (Thunb.) Oteng-Yeb.
Scirpus inclinatus (Delile) Asch. & Schweinf. ex Boiss. = ***Schoenoplectus corymbosus*** (Roth ex Roem. & Schult.) J.Raynal
Scirpus incompletus (Nees) T.Koyama = ***Fuirena incompleta*** Nees
Scirpus incomtulus (Nees) Boeck. = ***Isolepis incomtula*** Nees
Scirpus inconspicuus Levyns = ***Isolepis inconspicua*** (Levyns) J.Raynal
Scirpus incurvatus Roxb. = ***Schoenoplectiella articulata*** (L.) Lye
Scirpus interior Britton = ***Bolboschoenus maritimus*** subsp. ***paludosus*** (A.Nelson) T.Koyama
Scirpus intermedius Èelak. = ***Scirpus sylvaticus*** L.
Scirpus ×*intermedius* Tang & F.T.Wang = [36 CHC]
Scirpus intermedius A.Gray ex Torr. = ***Eleocharis flavescens*** var. ***olivacea*** (Torr.) Gleason
Scirpus intermedius Muhl. = ***Eleocharis intermedia*** Schult.
Scirpus intermedius Thuill. = ***Eleocharis palustris*** (L.) Roem. & Schult. subsp. ***palustris***
Scirpus intermedius Poir. = ***Scirpoides holoschoenus*** (L.) Soják subsp. ***holoschoenus***
Scirpus interstinctus Poepp. ex C.Presl = ***Eleocharis geniculata*** (L.) Roem. & Schult.
Scirpus interstinctus Vahl = ***Eleocharis interstincta*** (Vahl) Roem. & Schult.
Scirpus intricatus L. = ***Cyperus squarrosus*** L.
Scirpus inundatus (R.Br.) Poir. = ***Isolepis inundata*** R.Br.
Scirpus inundatus var. *floribundus* Benth. = ***Isolepis gaudichaudiana*** Kunth
Scirpus inundatus var. *gracillimus* C.B.Clarke = ***Isolepis reticularis*** Colenso
Scirpus inundatus var. *major* C.B.Clarke = ***Isolepis inundata*** R.Br.
Scirpus inundatus var. *pseudosetaceus* Kük. = ***Isolepis wakefieldiana*** (S.T.Blake) K.L.Wilson
Scirpus iridifolius Bory = ***Machaerina iridifolia*** (Bory) T.Koyama
Scirpus isdellensis W.Fitzg. = [50 WAU]
Scirpus iseensis T.Koyama & Shimizu = [38 JAP]
Scirpus isolepis (Nees) Boeck. = ***Lipocarpha hemisphaerica*** (Roth) Goetgh.
Scirpus jacobi C.E.C.Fisch. = ***Schoenoplectiella senegalensis*** (Steud.) Lye
Scirpus jaluanus (Kom.) Nakai = ***Scirpus karuisawensis*** Makino
Scirpus janii Besser ex Schult. = ***Schoenoplectus lacustris*** (L.) Palla subsp. ***lacustris***
Scirpus japonicus (Maxim.) Kuntze = ***Scirpus maximowiczii*** C.B.Clarke
Scirpus japonicus (Miq.) Franch. & Sav. = ***Eleocharis pellucida*** C.Presl
Scirpus japonicus var. *thermalis* Hultén = ***Eleocharis pellucida*** C.Presl
Scirpus jardinei Steud. & Jard. = ***Schoenoplectus lacustris*** (L.) Palla subsp. ***lacustris***
Scirpus javanus Nees = ***Schoenoplectus mucronatus*** (L.) Palla subsp. ***mucronatus***
Scirpus junceus G.Forst. = ***Fimbristylis juncea*** (G.Forst.) R.Br. ex Roem. & Schult.
Scirpus junciformis Retz. = [36+]
Scirpus junciformis (Kunth) Poir. = ***Bulbostylis junciformis*** (Kunth) C.B.Clarke
Scirpus junciformis Nees = ***Schoenoplectiella juncoides*** (Roxb.) Lye
Scirpus junciformis f. *robustus* Kuntze = ***Bulbostylis junciformis*** (Kunth) C.B.Clarke
Scirpus juncoides Willd. ex Kunth = ***Bulbostylis junciformis*** (Kunth) C.B.Clarke
Scirpus juncoides Roxb. = ***Schoenoplectiella juncoides*** (Roxb.) Lye
Scirpus juncoides var. *digynus* (Boeck.) T.Koyama = ***Schoenoplectiella purshiana*** (Fernald) Lye
Scirpus juncoides var. *hotarui* (Ohwi) Ohwi = ***Schoenoplectus hotarui*** (Ohwi) Holub
Scirpus juncoides var. *nanus* Griseb. = ***Bulbostylis juncoides*** (Vahl) Kük. ex Herter
Scirpus juncoides var. *triangulatus* (Honda) Ohwi = ***Schoenoplectus ×trapezoideus*** (Koidz.) Hayas. & H.Ohashi
Scirpus juncoides var. *williamsii* (Fernald) T.Koyama = ***Schoenoplectiella purshiana*** (Fernald) Lye
Scirpus junghuhnii Miq. = ***Sumatroscirpus junghuhnii*** (Miq.) Oteng-Yeb.
Scirpus junghuhnii var. *minor* Kük. = ***Sumatroscirpus junghuhnii*** (Miq.) Oteng-Yeb.
Scirpus kalli Forssk. = ***Cyperus capitatus*** Vand.
Scirpus ×*kalmusii* Asch., Abrom. & Graebn. = ***Schoenoplectus ×carinatus*** (Sm.) Palla nothosubsp. ***carinatus***
Scirpus kamerunensis K.Schum. ex C.B.Clarke = ***Bulbostylis erratica*** (Hook.f.) C.B.Clarke
Scirpus kamtschaticus C.A.Mey. = ***Eleocharis kamtschatica*** (C.A.Mey.) Kom.
Scirpus karroicus C.B.Clarke = ***Isolepis karroica*** (C.B.Clarke) J.Raynal

Scirpus kasachstanicus Dobrochot. = ***Schoenoplectus litoralis*** subsp. ***kasachstanicus*** (Dobrochot.) Soják
Scirpus kernii Raymond = ***Lipocarpha kernii*** (Raymond) Goetgh.
Scirpus kielmeyeri Schimp. ex Steud. = ***Fimbristylis complanata*** (Retz.) Link var. ***complanata***
Scirpus kirkii (C.B.Clarke) K.Schum. = ***Bulbostylis kirkii*** C.B.Clarke
Scirpus kiushuensis Ohwi = ***Scirpus rosthornii*** Diels
Scirpus klingei Meinsh. = ***Eleocharis uniglumis*** (Link) Schult. subsp. ***uniglumis***
Scirpus kochii Maiden & Betche = ***Isolepis congrua*** Nees
Scirpus koilolepis (Steud.) Gleason = ***Isolepis carinata*** Hook. & Arn. ex Torr.
Scirpus komarovii Roshev. = ***Schoenoplectus komarovii*** (Roshev.) Soják
Scirpus koshewnikowii Litv. = ***Bolboschoenus planiculmis*** (F.Schmidt) E.M.Egorova
Scirpus ×*kuekenthalianus* Junge = ***Schoenoplectus ×kuekenthalianus*** (Junge) D.H.Kent
Scirpus kunthii Boeck. = ***Scirpus ficinioides*** Kunth
Scirpus kyllingioides (A.Rich.) Boeck. = ***Kyllingiella microcephala*** (Steud.) R.W.Haines & Lye
Scirpus kysoor Roxb. = ***Actinoscirpus grossus*** var. ***kysoor*** (Roxb.) Noltie
Scirpus laciniatus Nees & Ehrenb. ex Boeck. = ***Bolboschoenus maritimus*** (L.) Palla subsp. ***maritimus***
Scirpus laciniatus Thunb. = ***Ficinia laciniata*** (Thunb.) Nees
Scirpus lacustris L. = ***Schoenoplectus lacustris*** (L.) Palla
Scirpus lacustris var. *carinatus* Nakai = ***Schoenoplectus lacustris*** (L.) Palla subsp. ***lacustris***
Scirpus lacustris f. *compactus* Beckh. = ***Schoenoplectus lacustris*** (L.) Palla subsp. ***lacustris***
Scirpus lacustris var. *condensatus* Peck = ***Schoenoplectus lacustris*** (L.) Palla subsp. ***lacustris***
Scirpus lacustris f. *confertus* Schur = ***Schoenoplectus lacustris*** (L.) Palla subsp. ***lacustris***
Scirpus lacustris f. *conglomeratus* Junge = ***Schoenoplectus lacustris*** (L.) Palla subsp. ***lacustris***
Scirpus lacustris subsp. *creber* (Fernald) T.Koyama = ***Schoenoplectus lacustris*** (L.) Palla subsp. ***lacustris***
Scirpus lacustris var. *custoris* (Hegetschw) Nyman = ***Schoenoplectus lacustris*** (L.) Palla subsp. ***lacustris***
Scirpus lacustris f. *foliosus* Des Moul. = ***Schoenoplectus lacustris*** (L.) Palla subsp. ***lacustris***
Scirpus lacustris f. *glaucus* (Sm.) Peterm. = ***Schoenoplectus tabernaemontani*** (C.C.Gmel.) Palla
Scirpus lacustris subsp. *glaucus* (Sm.) Hartm. = ***Schoenoplectus tabernaemontani*** (C.C.Gmel.) Palla
Scirpus lacustris f. *laxiflorus* Kuntze = ***Schoenoplectus lacustris*** (L.) Palla subsp. ***lacustris***
Scirpus lacustris var. *medius* Schrad. ex Lej. = ***Schoenoplectus lacustris*** (L.) Palla
Scirpus lacustris f. *minor* Baumg. = ***Schoenoplectus lacustris*** (L.) Palla subsp. ***lacustris***
Scirpus lacustris var. *occidentalis* S.Watson = ***Schoenoplectus acutus*** var. ***occidentalis*** (S.Watson) S.G.Sm.
Scirpus lacustris f. *palustris* Schltdl. = ***Schoenoplectus lacustris*** (L.) Palla subsp. ***lacustris***
Scirpus lacustris f. *radiatus* Vollm. = ***Schoenoplectus lacustris*** (L.) Palla subsp. ***lacustris***
Scirpus lacustris f. *sparganiifolia* Volobaev = ***Schoenoplectus lacustris*** (L.) Palla subsp. ***lacustris***
Scirpus lacustris var. *tabernaemontani* (C.C.Gmel.) Nyman = ***Schoenoplectus tabernaemontani*** (C.C.Gmel.) Palla
Scirpus lacustris var. *tabernaemontani* (C.C.Gmel.) Crép. = ***Schoenoplectus tabernaemontani*** (C.C. Gmel.) Palla
Scirpus lacustris var. *tenuiculmis* E.Sheld. = ***Schoenoplectus heterochaetus*** (Chase) Soják
Scirpus lacustris f. *terrestris* Glück = ***Schoenoplectus lacustris*** (L.) Palla subsp. ***lacustris***
Scirpus lacustris var. *validus* (Vahl) Kük. = ***Schoenoplectus tabernaemontani*** (C.C.Gmel.) Palla
Scirpus lacustris subsp. *validus* (Vahl) T.Koyama = ***Schoenoplectus tabernaemontani*** (C.C.Gmel.) Palla
Scirpus laetevirens Schrad. ex Nees = ***Eleocharis atropurpurea*** (Retz.) Kunth
Scirpus laevis S.T.Blake = ***Schoenoplectiella laevis*** (S.T.Blake) Lye
Scirpus lagoensis Boeck. = ***Bulbostylis lagoensis*** (Boeck.) Prata & M.G.López
Scirpus lanatus (Kunth) Poir. = ***Bulbostylis lanata*** (Kunth) Lindm.
Scirpus lanceiglumis C.B.Clarke = ***Eleocharis palustris*** (L.) Roem. & Schult. subsp. ***palustris***
Scirpus laniferus Boeck. = ***Bulbostylis lanifera*** (Boeck.) Kük.
Scirpus lappaceus Lam. = ***Cyperus squarrosus*** L.
Scirpus lateralis Retz. = ***Schoenoplectiella erecta*** (Poir.) Lye
Scirpus lateralis Forssk. = ***Schoenoplectiella supina*** (L.) Lye
Scirpus lateriflorus J.F.Gmel. = ***Schoenoplectiella lateriflora*** (J.F.Gmel.) Lye
Scirpus latifolius Gilib. = ***Scirpus sylvaticus*** L.
Scirpus latifolius Bertol. = ***Pleurostachys tenuiflora*** Brongn.
Scirpus lavarum Poir. = ***Machaerina lavarum*** (Poir.) Bojer
Scirpus lawianus Boeck. = ***Fimbristylis lawiana*** (Boeck.) J.Kern
Scirpus laxiflorus Thwaites = ***Eleocharis ochrostachys*** Steud.
Scirpus laxifolius Thwaites = ***Eleocharis variegata*** (Poir.) C.Presl
Scirpus laxus (Vahl) Poir. = ***Fimbristylis dichotoma*** (L.) Vahl subsp. ***dichotoma***
Scirpus lechleri (Boeck.) Kuntze = ***Eleocharis lechleri*** Boeck.
Scirpus leiocarpus (Maxim.) Meinsh. = ***Fimbristylis aestivalis*** Vahl var. ***aestivalis***
Scirpus lejeunei Weihe = ***Schoenoplectus triqueter*** (L.) Palla
Scirpus lenticularis Torr. = ***Scirpus microcarpus*** J.Presl & C.Presl
Scirpus lenticularis (R.Br.) Spreng. = ***Isolepis fluitans*** var. ***lenticularis*** (R.Br.) Muasya
Scirpus lepta (C.B.Clarke) Levyns = ***Eleocharis lepta*** C.B.Clarke
Scirpus leptaleus Koch = ***Isolepis cernua*** (Vahl) Roem. & Schult. var. ***cernua***
Scirpus leptocarpus F.Muell. = ***Lipocarpha microcephala*** (R.Br.) Kunth
Scirpus leptolepis Chapm. = ***Schoenoplectus etuberculatus*** (Steud.) Soják
Scirpus leptopus Boeck. = ***Scirpus asper*** J.Presl & C.Presl
Scirpus leptos C.Wright = ***Eleocharis nana*** Kunth
Scirpus leptos (Steud.) C.Wright = ***Eleocharis parvula*** (Roem. & Schult.) Palla
Scirpus leptostachyus (Kunth) Boeck. = ***Isolepis leptostachya*** Kunth
Scirpus lereschii (Shuttlew.) Thomas ex W.D.J.Koch = ***Eleocharis atropurpurea*** (Retz.) Kunth

Scirpus leucanthus Boeck. = ***Schoenoplectiella leucantha*** (Boeck.) Lye
Scirpus leucocephalus (Boeck.) T.Koyama = ***Eriophorum scheuchzeri*** Hoppe
Scirpus leucocoleus K.Schum. = ***Ficinia stolonifera*** Boeck.
Scirpus leucostachyus (Kunth) Poir. = ***Bulbostylis leucostachya*** (Kunth) Kunth
Scirpus libanoticus Post = [34 LBS]
Scirpus lichtensteinianus (Kunth) Boeck. = ***Bulbostylis lichtensteiniana*** (Kunth) C.B.Clarke
Scirpus liebmannianus (Boeck.) Kuntze = ***Eleocharis urceolata*** (Liebm.) Svenson
Scirpus limburgensis Lej. = ***Eleocharis palustris*** (L.) Roem. & Schult. subsp. ***palustris***
Scirpus limensis C.B.Clarke = ***Schoenoplectus pungens*** (Vahl) Palla var. ***pungens***
Scirpus limosus Schrad. = ***Eleocharis limosa*** (Schrad.) Schult.
Scirpus lineatus Muhl. = ***Scirpus pendulus*** Muhl.
Scirpus lineatus var. *asiaticus* (Beetle) T.Koyama = ***Scirpus wichurae*** Boeck.
Scirpus lineatus f. *concolor* (Maxim.) T.Koyama = ***Scirpus wichurae*** Boeck.
Scirpus lineatus f. *cylindricus* (Makino) T.Koyama = ***Scirpus wichurae*** Boeck.
Scirpus lineatus f. *elongatus* Eames = ***Scirpus pendulus*** Muhl.
Scirpus lineatus var. *habereri* House = ***Scirpus lineatus*** Michx.
Scirpus lineatus var. *lushanensis* (Ohwi) T.Koyama = ***Scirpus lushanensis*** Ohwi
Scirpus lineatus f. *wichurae* (Boeck.) T.Koyama = ***Scirpus wichurae*** Boeck.
Scirpus lineatus subsp. *wichurae* (Boeck.) T.Koyama = ***Scirpus wichurae*** Boeck.
Scirpus lineatus var. *wichurae* (Boeck.) T.Koyama = ***Scirpus wichurae*** Boeck.
Scirpus lineolatus Franch. & Sav. = ***Schoenoplectus lineolatus*** (Franch. & Sav.) T.Koyama
Scirpus liocarpus (Phil.) Kuntze = ***Eleocharis pachycarpa*** É.Desv.
Scirpus lithospermus L. = ***Scleria lithosperma*** (L.) Sw.
Scirpus lithuanicus Besser ex Schult. = ***Schoenoplectus lacustris*** (L.) Palla subsp. ***lacustris***
Scirpus litoralis (Phil.) Kuntze = ***Eleocharis melanostachys*** (d'Urv.) C.B.Clarke
Scirpus litoralis Schrad. = ***Schoenoplectus litoralis*** (Schrad.) Palla
Scirpus litoralis var. *philippii* (Tineo) Nyman = ***Schoenoplectus litoralis*** (Schrad.) Palla subsp. ***litoralis***
Scirpus litoralis var. *pterolepis* (Nees) C.B.Clarke = ***Schoenoplectus litoralis*** subsp. ***thermalis*** (Trab.) S.S.Hooper
Scirpus litoralis var. *subulatus* (Vahl) Chiov. = ***Schoenoplectus litoralis*** subsp. ***thermalis*** (Trab.) S.S.Hooper
Scirpus litoralis subsp. *subulatus* (Vahl) M.A.Wadood Khan = ***Schoenoplectus litoralis*** subsp. ***thermalis*** (Trab.) S.S.Hooper
Scirpus litoralis subsp. *thermalis* (Trab.) Murb. = ***Schoenoplectus litoralis*** subsp. ***thermalis*** (Trab.) S.S.Hooper
Scirpus littoralis Flüggé ex Rchb. = ***Blysmus rufus*** (Huds.) Link
Scirpus loefgrenii Boeck. = ***Bulbostylis loefgrenii*** (Boeck.) Prata & M.G.López
Scirpus lomanus H.Pfeiff. = ***Schoenoplectus pungens*** (Vahl) Palla var. ***pungens***
Scirpus longibracteatus Salzm. ex Rchb. = ***Scirpoides holoschoenus*** (L.) Soják subsp. ***holoschoenus***
Scirpus longifolius Rich. = ***Hypolytrum longifolium*** (Rich.) Nees
Scirpus longifolius Hook. & Arn. = ***Schoenoplectus pungens*** (Vahl) Palla var. ***pungens***
Scirpus longivaginatus Boeck. = [84]
Scirpus lorentzii Boeck. = ***Bulbostylis juncoides*** (Vahl) Kük. ex Herter
Scirpus lucidus Less. ex Kunth = ***Bolboschoenus maritimus*** (L.) Palla subsp. ***maritimus***
Scirpus ludwigii (Steud.) Boeck. = ***Isolepis ludwigii*** (Steud.) Kunth
Scirpus lugardii C.B.Clarke = ***Cyperus hamulosus*** M.Bieb.
Scirpus lupulinus Spreng. = ***Cyperus lupulinus*** (Spreng.) Marcks
Scirpus lupulinus (Nees) Roshev. = ***Schoenoplectiella roylei*** (Nees) Lye
Scirpus lusulinus Rchb. ex Kunth = ***Bulbostylis junciformis*** (Kunth) C.B.Clarke
Scirpus luteus Gilib. = ***Trichophorum cespitosum*** (L.) Hartm. subsp. ***cespitosum***
Scirpus luzonensis J.Presl & C.Presl = ***Schoenoplectiella juncoides*** (Roxb.) Lye
Scirpus luzulae L. = ***Cyperus luzulae*** (L.) Retz.
Scirpus lyallii Baker = ***Schoenoplectus corymbosus*** (Roth ex Roem. & Schult.) J.Raynal
Scirpus macer Boeck. = ***Isolepis costata*** Hochst. ex A.Rich.
Scirpus macounii T.Holm = ***Scirpus microcarpus*** J.Presl & C.Presl
Scirpus macranthus (Boeck.) Kuntze = ***Eleocharis macrantha*** Boeck.
Scirpus macranthus Boeck. = ***Schoenoplectus etuberculatus*** (Steud.) Soják
Scirpus macrolepis Boeck. ex C.B.Clarke = ***Bulbostylis kirkii*** C.B.Clarke
Scirpus macrophyllus Besser ex Schult. = ***Schoenoplectus lacustris*** (L.) Palla subsp. ***lacustris***
Scirpus macrostachyos Lam. = ***Bolboschoenus maritimus*** (L.) Palla subsp. ***maritimus***
Scirpus macrostachys Boeck. = [40 EHM]
Scirpus macrostachys Willd. = ***Bolboschoenus glaucus*** S.G.Sm.
Scirpus macrostylis (Spreng.) Spreng. = ***Rhynchospora barbata*** (Vahl) Kunth
Scirpus maculosus Vahl = ***Eleocharis maculosa*** (Vahl) Roem. & Schult.
Scirpus madagascariensis Boeck. = ***Schoenoplectus corymbosus*** (Roth ex Roem. & Schult.) J.Raynal
Scirpus malheurensis L.F.Hend. = ***Schoenoplectus acutus*** (Muhl. ex Bigelow) Á.Löve & D.Löve var. ***acutus***
Scirpus mamillatus H.Lindb. = ***Eleocharis mamillata*** (H.Lindb.) H.Lindb.
Scirpus margaritaceus Hultén = ***Eleocharis margaritacea*** (Hultén) Miyabe & Kudô
Scirpus margaritiferus (Nees) Boeck. = ***Schoenus apogon*** Roem. & Schult. var. ***apogon***
Scirpus marginatus Muhl. = ***Eleocharis mutata*** (L.) Roem. & Schult.
Scirpus marginatus Thunb. = ***Isolepis marginata*** (Thunb.) A.Dietr.
Scirpus marginulatus (Hochst. ex Steud.) Kuntze = ***Eleocharis marginulata*** Hochst. ex Steud.
Scirpus ×*mariqueter* Tang & F.T.Wang = [36 CHC]
Scirpus maritimus L. = ***Bolboschoenus maritimus*** (L.) Palla

Scirpus maritimus subsp. *affinis* (Roth) Norl. = ***Bolboschoenus maritimus*** subsp. ***affinis*** (Roth) T.Koyama
Scirpus maritimus var. *agonus* Beetle = ***Bolboschoenus maritimus*** (L.) Palla
Scirpus maritimus f. *agonus* (Beetle) Fernald = ***Bolboschoenus maritimus*** (L.) Palla
Scirpus maritimus var. *compactus* (Hoffm.) Lej. = ***Bolboschoenus maritimus*** (L.) Palla
Scirpus maritimus var. *cylindricus* Torr. = ***Schoenoplectus novae-angliae*** (Britton) M.T.Strong
Scirpus maritimus var. *cymosus* Rchb. = ***Bolboschoenus maritimus*** var. ***cymosus*** (Rchb.) Kit Tan & Oteng-Yeb.
Scirpus maritimus var. *digynus* Godr. = ***Bolboschoenus maritimus*** (L.) Palla subsp. ***maritimus***
Scirpus maritimus f. *digynus* Zapal. = ***Bolboschoenus maritimus*** (L.) Palla subsp. ***maritimus***
Scirpus maritimus var. *fernaldii* (E.P.Bicknell) Beetle = ***Bolboschoenus maritimus*** (L.) Palla
Scirpus maritimus var. *fluviatilis* Torr. = ***Bolboschoenus fluviatilis*** (Torr.) Soják
Scirpus maritimus f. *longifolius* Peterm. = ***Bolboschoenus maritimus*** (L.) Palla subsp. ***maritimus***
Scirpus maritimus var. *macrostachys* (Willd.) Dumort. = ***Bolboschoenus glaucus*** S.G.Sm.
Scirpus maritimus var. *macrostachyus* Michx. = ***Bolboschoenus robustus*** (Pursh) Soják
Scirpus maritimus var. *megastachyus* (Steud.) Nyman = ***Bolboschoenus maritimus*** (L.) Palla subsp. ***maritimus***
Scirpus maritimus var. *monostachyus* Lej. & Courtois = ***Bolboschoenus maritimus*** (L.) Palla subsp. ***maritimus***
Scirpus maritimus var. *ovatus* Nees = ***Bolboschoenus maritimus*** (L.) Palla
Scirpus maritimus var. *paludosus* (A.Nelson) Kük. = ***Bolboschoenus maritimus*** subsp. ***paludosus*** (A.Nelson) T.Koyama
Scirpus maritimus var. *robustus* (Pursh) Kük. = ***Bolboschoenus robustus*** (Pursh) Soják
Scirpus maritimus f. *sarmaticus* Zapal. = ***Bolboschoenus maritimus*** (L.) Palla subsp. ***maritimus***
Scirpus maritimus f. *subumbellatus* Schur = ***Bolboschoenus maritimus*** (L.) Palla subsp. ***maritimus***
Scirpus maritimus subsp. *tabernaemontani* (C.C.Gmel.) Nyman = ***Schoenoplectus tabernaemontani*** (C.C.Gmel.) Palla
Scirpus maritimus var. *tuberosus* (Desf.) Roem. & Schult. = ***Bolboschoenus glaucus*** S.G.Sm.
Scirpus martii Dufour ex Roem. & Schult. = ***Cladium mariscus*** (L.) Pohl subsp. ***mariscus***
Scirpus mauritanicus Steud. = ***Schoenoplectus mucronatus*** (L.) Palla subsp. ***mucronatus***
Scirpus maximus Roxb. = ***Actinoscirpus grossus*** (L.f.) Goetgh. & D.A.Simpson var. ***grossus***
Scirpus medianus V.J.Cook = ***Bolboschoenus medianus*** (V.J.Cook) Soják
Scirpus medius Roxb. = ***Eleocharis acutangula*** (Roxb.) Schult.
Scirpus medius Gray = ***Schoenoplectus lacustris*** (L.) Palla subsp. ***lacustris***
Scirpus megapotamicus A.Spreng. = ***Rhynchospora megapotamica*** (A.Spreng.) H.Pfeiff.
Scirpus megastachyus Steud. = ***Bolboschoenus maritimus*** (L.) Palla subsp. ***maritimus***
Scirpus melanocarpus (Boeck.) Hemsl. =
Scirpus melanocarpus (Torr.) Kuntze = ***Eleocharis melanocarpa*** Torr.
Scirpus melanocarpus Griseb. = ***Eleocharis pachystyla*** (C.Wright) C.B.Clarke
Scirpus melanocephalus (É.Desv.) Griseb. = ***Eleocharis albibracteata*** Nees & Meyen ex Kunth var. ***albibracteata***
Scirpus melanorrhizus Boeck. = ***Schoenoplectus pungens*** (Vahl) Palla var. ***pungens***
Scirpus melanospermus Salzm. ex Steud. = ***Eleocharis flavescens*** (Poir.) Urb. var. ***flavescens***
Scirpus melanospermus C.A.Mey. = ***Schoenoplectiella supina*** (L.) Lye
Scirpus melanostachys d'Urv. = ***Eleocharis melanostachys*** (d'Urv.) C.B.Clarke
Scirpus melas Kunze ex Kunth = ***Schoenoplectus pungens*** (Vahl) Palla var. ***pungens***
Scirpus membranaceus Thunb. = ***Hellmuthia membranacea*** (Thunb.) R.W.Haines & Lye
Scirpus mendocinus (Phil.) Kuntze = ***Eleocharis mendocina*** Phil.
Scirpus merrillii (Palla) Kük. ex Merr. = ***Isolepis subtilissima*** Boeck.
Scirpus mexicanus C.B.Clarke ex Britton = ***Karinia mexicana*** (C.B.Clarke ex Britton) Reznicek & McVaugh
Scirpus meyenii Nees = ***Schoenoplectus lacustris*** (L.) Palla subsp. ***lacustris***
Scirpus meyeri Trautv. = ***Trichophorum pumilum*** (Vahl) Schinz & Thell. subsp. ***pumilum***
Scirpus michauxii Pers. = ***Fimbristylis autumnalis*** (L.) Roem. & Schult.
Scirpus michelianus L. = ***Cyperus michelianus*** (L.) Delile
Scirpus michelianus f. *conglomeratus* Beck = ***Cyperus michelianus*** (L.) Delile subsp. ***michelianus***
Scirpus michelii Bubani = ***Cyperus michelianus*** (L.) Delile subsp. ***michelianus***
Scirpus micranthus Vahl = ***Lipocarpha micrantha*** (Vahl) G.C.Tucker
Scirpus micranthus var. *drummondii* (Nees) Mohlenbr. = ***Lipocarpha drummondii*** (Nees) G.C.Tucker
Scirpus micranthus var. *humboldtii* Boeck. = ***Lipocarpha micrantha*** (Vahl) G.C.Tucker
Scirpus micranthus var. *minor* (Roem. & Schult.) B.Boivin = ***Lipocarpha micrantha*** (Vahl) G.C.Tucker
Scirpus microcarpus (Torr.) Kuntze = ***Eleocharis microcarpa*** Torr.
Scirpus microcarpus var. *bissellii* (Fernald) House = ***Scirpus expansus*** Fernald
Scirpus microcarpus var. *confertus* (Fernald) House = ***Scirpus microcarpus*** J.Presl & C.Presl
Scirpus microcarpus var. *longispicatus* M.Peck = ***Scirpus microcarpus*** J.Presl & C.Presl
Scirpus microcarpus f. *radiosus* (Fernald) C.F.Reed = ***Scirpus microcarpus*** J.Presl & C.Presl
Scirpus microcarpus var. *rubrotinctus* (Fernald) M.E.Jones = ***Scirpus microcarpus*** J.Presl & C.Presl
Scirpus microcarpus subsp. *rubrotinctus* (Fernald) Á.Löve & D.Löve = ***Scirpus microcarpus*** J.Presl & C.Presl
Scirpus microcephalus (Steud.) Dandy = ***Kyllingiella microcephala*** (Steud.) R.W.Haines & Lye
Scirpus microlepis Griseb. = ***Eleocharis microlepis*** (Griseb.) D.A.Simpson
Scirpus microstachys Boeck. = ***Bulbostylis capillaris*** (L.) Kunth ex C.B.Clarke subsp. ***capillaris***
Scirpus microstachys Phil. = ***Isolepis cernua*** (Vahl) Roem. & Schult. var. ***cernua***

Scirpus microstachyus (Boeck.) T.Koyama = ***Eriophorum microstachyum*** Boeck.
Scirpus miliaceus Roxb. = ***Fimbristylis dichotoma*** (L.) Vahl subsp. ***dichotoma***
Scirpus miliaceus Thunb. = ***Fimbristylis ferruginea*** (L.) Vahl var. ***ferruginea***
Scirpus miliaceus L. = ***Fimbristylis quinquangularis*** (Vahl) Kunth
Scirpus minaae Tod. = ***Isolepis cernua*** (Vahl) Roem. & Schult. var. ***cernua***
Scirpus mindorensis Beetle = ***Bulbostylis barbata*** (Rottb.) C.B.Clarke subsp. ***barbata***
Scirpus minimus Vahl = ***Bulbostylis barbata*** (Rottb.) C.B.Clarke subsp. ***barbata***
Scirpus minimus Roxb. = ***Fimbristylis dipsacea*** (Rottb.) C.B.Clarke
Scirpus minimus Sieber ex Kunth = ***Isolepis antarctica*** (L.) Roem. & Schult.
Scirpus minimus Hook. & Arn. = ***Isolepis cernua*** (Vahl) Roem. & Schult. var. ***cernua***
Scirpus minimus Willd. ex Kunth = ***Lipocarpha hemisphaerica*** (Roth) Goetgh.
Scirpus minimus Pursh = ***Lipocarpha micrantha*** (Vahl) G.C.Tucker
Scirpus minutissimus Boeck. = ***Isolepis sororia*** Kunth
Scirpus minutus Turrill = ***Isolepis minuta*** (Turrill) J.Raynal
Scirpus mitratus Franch. & Sav. = ***Eleocharis kamtschatica*** (C.A.Mey.) Kom.
Scirpus mitratus Griseb. = ***Eleocharis tiarata*** Gómez-Laur.
Scirpus modestus (Phil.) F.Phil. = ***Isolepis cernua*** (Vahl) Roem. & Schult. var. ***cernua***
Scirpus molestus M.C.Johnst. = ***Isolepis pseudosetacea*** (Daveau) Vasc.
Scirpus mollis Wall. = ***Rhynchospora mollis*** (Wall.) Schult.
Scirpus monander Roxb. = ***Bulbostylis barbata*** (Rottb.) C.B.Clarke subsp. ***barbata***
Scirpus monander Rottb. = ***Fimbristylis argentea*** (Rottb.) Vahl
Scirpus monander Willd. = ***Lipocarpha gracilis*** (Rich. ex Pers.) Nees
Scirpus monophyllus C.Presl = ***Schoenoplectus pungens*** var. ***badius*** (C.Presl) S.G.Sm.
Scirpus monostachyos J.König ex Roxb. = ***Fimbristylis schoenoides*** (Retz.) Vahl
Scirpus monostachyus (L.) Kuntze = ***Fimbristylis ovata*** (Burm.f.) J.Kern
Scirpus montanus Kunth = ***Eleocharis montana*** (Kunth) Roem. & Schult.
Scirpus montevidensis (Link) C.B.Clarke = ***Androtrichum trigynum*** (Spreng.) H.Pfeiff.
Scirpus montevidensis (Kunth) Kuntze = ***Eleocharis montevidensis*** Kunth
Scirpus montivagus S.T.Blake = ***Isolepis montivaga*** (S.T.Blake) K.L.Wilson
Scirpus morrisonensis Hayata = ***Trichophorum subcapitatum*** (Thwaites & Hook.) D.A.Simpson
Scirpus moseleyanus Boeck. = ***Isolepis moseleyana*** (Boeck.) Muasya
Scirpus mucronatus Pollich = ***Bolboschoenus maritimus*** (L.) Palla subsp. ***maritimus***
Scirpus mucronatus Roxb. = ***Schoenoplectiella supina*** (L.) Lye
Scirpus mucronatus L. = ***Schoenoplectus mucronatus*** (L.) Palla
Scirpus mucronatus Roth = ***Schoenoplectus pungens*** (Vahl) Palla var. ***pungens***
Scirpus mucronatus Host = ***Schoenoplectus triqueter*** (L.) Palla
Scirpus mucronatus subsp. *clemensii* Kük. = ***Schoenoplectus mucronatus*** subsp. ***clemensii*** (Kük.) Soják
Scirpus mucronatus var. *robustus* Miq. = ***Schoenoplectus mucronatus*** (L.) Palla subsp. ***mucronatus***
Scirpus mucronatus var. *tataranus* Honda = ***Schoenoplectus mucronatus*** (L.) Palla subsp. ***mucronatus***
Scirpus mucronulatus (Nees) Kuntze = ***Eleocharis acuta*** R.Br.
Scirpus mucronulatus Michx. = ***Fimbristylis autumnalis*** (L.) Roem. & Schult.
Scirpus muehlenbergianus Schult. = [7]
Scirpus muehlenbergii Spreng. = ***Bulbostylis barbata*** (Rottb.) C.B.Clarke subsp. ***barbata***
Scirpus muelleri Boeck. = [50]
Scirpus multicalamis Bubani = ***Eleocharis multicaulis*** (Sm.) Desv.
Scirpus multicaulis Sm. = ***Eleocharis multicaulis*** (Sm.) Desv.
Scirpus multicaulis C.C.Gmel. = ***Eleocharis ovata*** (Roth) Roem. & Schult.
Scirpus multicaulis Vahl ex Kunth = ***Eleocharis tenuis*** (Willd.) Schult. var. ***tenuis***
Scirpus multicaulis F.Muell. ex C.B.Clarke = ***Isolepis congrua*** Nees
Scirpus multicostatus Baker = ***Isolepis costata*** Hochst. ex A.Rich.
Scirpus multiculmeus Bubani = ***Eleocharis multicaulis*** (Sm.) Desv.
Scirpus muricatus Poir. = ***Rhynchospora holoschoenoides*** (Rich.) Herter
Scirpus muricinux C.B.Clarke = ***Schoenoplectus muricinux*** (C.B.Clarke) J.Raynal
Scirpus muriculatus Kük. = ***Schoenoplectus muriculatus*** (Kük.) Browning
Scirpus muscosus Kirk = ***Isolepis aucklandica*** Hook.f.
Scirpus mutatus L. = ***Eleocharis mutata*** (L.) Roem. & Schult.
Scirpus mutatus Roxb. ex C.B.Clarke = ***Eleocharis spiralis*** (Rottb.) Roem. & Schult.
Scirpus muticus D.Don = ***Schoenoplectus mucronatus*** (L.) Palla subsp. ***mucronatus***
Scirpus mysurensis B.Heyne ex Wall. = ***Schoenoplectus mucronatus*** (L.) Palla subsp. ***mucronatus***
Scirpus nanodes Levyns = ***Isolepis pusilla*** Kunth
Scirpus nanus (Kunth) Kuntze = ***Eleocharis nana*** Kunth
Scirpus nanus Spreng. = ***Eleocharis parvula*** (Roem. & Schult.) Palla
Scirpus nanus Poir. = ***Fimbristylis argentea*** (Rottb.) Vahl
Scirpus nanus F.Phil. = ***Isolepis inundata*** R.Br.
Scirpus nanus var. *anachaetus* (Torr.) Britton = ***Eleocharis coloradoensis*** (Britton) Gilly
Scirpus natalensis Boeck. ex C.B.Clarke = ***Isolepis hystrix*** (Thunb.) Schrad.
Scirpus natans Thunb. = ***Isolepis natans*** (Thunb.) A.Dietr.
Scirpus natans Bojer = ***Schoenoplectiella supina*** (L.) Lye
Scirpus natans Griseb. = ***Websteria confervoides*** (Poir.) S.S.Hooper
Scirpus neesii Boeck. =
Scirpus nervosus (Hochst. ex A.Rich.) Boeck. = ***Isolepis fluitans*** var. ***nervosa*** (Hochst. ex A.Rich.) Lye
Scirpus nervosus var. *campanus* A.Terracc. = ***Isolepis cernua*** (Vahl) Roem. & Schult. var. ***cernua***
Scirpus nervosus var. *siculus* (C.Presl) A.Terracc. = ***Isolepis cernua*** (Vahl) Roem. & Schult. var. ***cernua***

Scirpus nevadensis S.Watson = ***Amphiscirpus nevadensis*** (S.Watson) Oteng-Yeb.
Scirpus nevadensis var. *remireoides* (Griseb.) Beetle = ***Amphiscirpus nevadensis*** (S.Watson) Oteng-Yeb.
Scirpus niederleinianus Boeck. = ***Bulbostylis scabra*** (J.Presl & C.Presl) C.B.Clarke
Scirpus nigrescens (Nees) Salzm. ex Steud. = ***Eleocharis nigrescens*** (Nees) Steud.
Scirpus nigricans (Kunth) Poir. = ***Isolepis nigricans*** Kunth
Scirpus niloticus J.F.Gmel. = ***Fimbristylis dichotoma*** (L.) Vahl subsp. ***dichotoma***
Scirpus niloticus Blanco = ***Fimbristylis quinquangularis*** (Vahl) Kunth
Scirpus nindensis Ficalho & Hiern = [2]
Scirpus nipponicus Makino = ***Schoenoplectus nipponicus*** (Makino) Soják
Scirpus nitens Vahl = ***Rhynchospora nitens*** (Vahl) A.Gray
Scirpus nitens (R.Br.) Boeck. = ***Schoenus nitens*** (R.Br.) Roem. & Schult.
Scirpus nitidulus Willd. ex Kunth = ***Fimbristylis dichotoma*** (L.) Vahl subsp. ***dichotoma***
Scirpus nitidus (Fernald) Hultén = ***Eleocharis nitida*** Fernald
Scirpus nobilis Ridl. = ***Bolboschoenus nobilis*** (Ridl.) Goetgh. & D.A.Simpson
Scirpus nodosus Rottb. = ***Ficinia nodosa*** (Rottb.) Goetgh., Muasya & D.A.Simpson
Scirpus nodosus var. *macrostachya* Benth. = ***Ficinia nodosa*** (Rottb.) Goetgh., Muasya & D.A.Simpson
Scirpus nodulosus Roth = ***Eleocharis montana*** (Kunth) Roem. & Schult.
Scirpus novae-angliae Britton = ***Schoenoplectus novae-angliae*** (Britton) M.T.Strong
Scirpus novae-zealandiae Colenso = ***Schoenoplectus americanus*** (Pers.) Volkart
Scirpus nubigenus Phil. = ***Phylloscirpus acaulis*** (Phil.) Goetgh. & D.A.Simpson
Scirpus nudifructus Kük. = [26 ZAM]
Scirpus nudipes Griseb. = ***Isolepis cernua*** (Vahl) Roem. & Schult. var. ***cernua***
Scirpus nudissimus Steud. & Jard. = ***Eleocharis erythropoda*** Steud.
Scirpus nudus Raf. = [7]
Scirpus numidianus Vahl = ***Isolepis cernua*** (Vahl) Roem. & Schult. var. ***cernua***
Scirpus nutans Retz. = ***Fimbristylis nutans*** (Retz.) Vahl
Scirpus ×*oblongus* T.Koyama = ***Schoenoplectus* ×*oblongus*** (T.Koyama) Soják
Scirpus obtuse-trigonus Salzm. ex Steud. = ***Eleocharis mutata*** (L.) Roem. & Schult.
Scirpus obtusifolius Lam. = ***Fimbristylis dichotoma*** (L.) Vahl subsp. ***dichotoma***
Scirpus obtusifolius B.Heyne ex Kunth = ***Fimbristylis falcata*** (Vahl) Kunth var. ***falcata***
Scirpus obtusus Willd. = ***Eleocharis obtusa*** (Willd.) Schult.
Scirpus occidentalis (A.Gray) C.B.Clarke = ***Lipocarpha occidentalis*** (A.Gray) G.C.Tucker
Scirpus occidentalis (S.Watson) Chase = ***Schoenoplectus acutus*** var. ***occidentalis*** (S.Watson) S.G.Sm.
Scirpus occidentalis var. *congestus* Farw. = ***Schoenoplectus acutus*** var. ***occidentalis*** (S.Watson) S.G.Sm.
Scirpus occultus C.B.Clarke = [2]
Scirpus ochotensis Meinsh. = ***Fimbristylis dichotoma*** var. ***ochotensis*** (Meinsh.) T.Koyama
Scirpus ochreatus (Nees) Griseb. = ***Eleocharis flavescens*** (Poir.) Urb. var. ***flavescens***
Scirpus ochrostachys (Steud.) Kuntze = ***Eleocharis ochrostachys*** Steud.
Scirpus ×*oguraensis* T.Koyama =
Scirpus ohwianus T.Koyama = ***Schoenoplectiella juncoides*** (Roxb.) Lye
Scirpus okuyamae Ohwi = ***Schoenoplectus komarovii*** (Roshev.) Soják
Scirpus oldfieldianus S.T.Blake = ***Isolepis oldfieldiana*** (S.T.Blake) K.L.Wilson
Scirpus oliganthus Steud. = ***Ficinia oligantha*** (Steud.) J.Raynal
Scirpus oligostachys (Hochst. ex A.Rich.) Boeck. = ***Bulbostylis oligostachys*** (Hochst. ex A.Rich.) C.B.Clarke
Scirpus olivaceus (Torr.) Kuntze = ***Eleocharis flavescens*** var. ***olivacea*** (Torr.) Gleason
Scirpus oliveri Boeck. = ***Isolepis bicolor*** Carmich.
Scirpus olneyi A.Gray = ***Schoenoplectus americanus*** (Pers.) Volkart
Scirpus olneyi f. *australis* Barros & Osten = ***Schoenoplectus pungens*** var. ***badius*** (C.Presl) S.G.Sm.
Scirpus olneyi var. *contortus* Eames = ***Schoenoplectus* ×*contortus*** (Eames) S.G.Sm.
Scirpus olneyi var. *longisetis* Kük. = ***Schoenoplectus americanus*** (Pers.) Volkart
Scirpus olneyi f. *minor* Barros = ***Schoenoplectus pungens*** var. ***badius*** (C.Presl) S.G.Sm.
Scirpus onaei Franch. & Sav. = ***Eleocharis tetraquetra*** Nees
Scirpus orbicephalus Beetle = ***Karinia mexicana*** (C.B.Clarke ex Britton) Reznicek & McVaugh
Scirpus orgyalis Raf. = ***Schoenoplectus lacustris*** (L.) Palla subsp. ***lacustris***
Scirpus orthorhizomatus Kats.Arai & Miyam. = ***Schoenoplectus orthorhizomatus*** (Kats.Arai & Miyam.) Hayas. & H.Ohashi
Scirpus oryzetorum (Steud.) Ohwi = ***Schoenoplectiella lateriflora*** (J.F.Gmel.) Lye
Scirpus ×*osoreyamensis* M.Kikuchi = *Schoenoplectus hondoensis* × *Schoenoplectus mucronatus*
Scirpus ovatus Roth = ***Eleocharis ovata*** (Roth) Roem. & Schult.
Scirpus ovatus Gilib. = ***Eleocharis palustris*** (L.) Roem. & Schult. subsp. ***palustris***
Scirpus oxyjulos S.S.Hooper = ***Schoenoplectiella oxyjulos*** (S.S.Hooper) Lye
Scirpus oxylepis Meinsh. = ***Eleocharis oxylepis*** (Meinsh.) B.Fedtsch.
Scirpus pachycarpus (É.Desv.) Kuntze = ***Eleocharis pachycarpa*** É.Desv.
Scirpus pachystylus C.Wright = ***Eleocharis pachystyla*** (C.Wright) C.B.Clarke
Scirpus pacificus Britton ex Parish = ***Bolboschoenus maritimus*** subsp. ***paludosus*** (A.Nelson) T.Koyama
Scirpus pakapakensis Stapf = ***Trichophorum subcapitatum*** (Thwaites & Hook.) D.A.Simpson
Scirpus pallescens Roxb. = ***Fimbristylis bisumbellata*** (Forssk.) Bubani
Scirpus pallescens Willd. ex Link = ***Fimbristylis dichotoma*** (L.) Vahl subsp. ***dichotoma***
Scirpus pallescens Willd. ex Kunth = ***Fimbristylis ferruginea*** (L.) Vahl var. ***ferruginea***
Scirpus pallescens Boeck. ex Hemsl. = ***Isolepis bicolor*** Carmich.
Scirpus palmaris Willd. ex Kunth = ***Eleocharis flavescens*** (Poir.) Urb. var. ***flavescens***

Scirpus paludicola Kunth = ***Schoenoplectus paludicola*** (Kunth) Palla
Scirpus paludosus A.Nelson = ***Bolboschoenus maritimus*** subsp. ***paludosus*** (A.Nelson) T.Koyama
Scirpus paludosus var. *atlanticus* Fernald = ***Bolboschoenus maritimus*** subsp. ***paludosus*** (A.Nelson) T.Koyama
Scirpus palustris L. = ***Eleocharis palustris*** (L.) Roem. & Schult.
Scirpus palustromicrocarpus (Á.Löve) Á.Löve = ***Eleocharis palustris*** (L.) Roem. & Schult. subsp. ***palustris***
Scirpus panormitanus (Parl.) Parl. = ***Scirpoides holoschoenus*** (L.) Soják subsp. ***holoschoenus***
Scirpus paracicularis (Boeck.) C.Wright ex C.B.Clarke = ***Eleocharis microcarpa*** Torr. var. ***microcarpa***
Scirpus paradoxus (Spreng.) Boeck. = ***Bulbostylis paradoxa*** (Spreng.) Lindm.
Scirpus paraguayensis (Maury) Herter = ***Oxycaryum cubense*** (Poepp. & Kunth) Palla
Scirpus parlatoris Biv. ex Parl. = ***Scirpoides holoschoenus*** (L.) Soják subsp. ***holoschoenus***
Scirpus parviflorus (R.Br.) Poir. = ***Fimbristylis dichotoma*** (L.) Vahl subsp. ***dichotoma***
Scirpus parviflorus Willd. ex Kunth = ***Fimbristylis quinquangularis*** (Vahl) Kunth
Scirpus parvinux (C.B.Clarke) K.Schum. = ***Bulbostylis parvinux*** C.B.Clarke
Scirpus parvulus Roem. & Schult. = ***Eleocharis parvula*** (Roem. & Schult.) Palla
Scirpus parvulus var. *bailii* (Kohts) Nyman = ***Eleocharis palustris*** (L.) Roem. & Schult. subsp. ***palustris***
Scirpus paschalis Kük. = ***Schoenoplectus californicus*** (C.A.Mey.) Soják
Scirpus pauciflorus Lightf. = ***Eleocharis quinqueflora*** (Hartmann) O.Schwarz
Scirpus pauciflorus (R.Br.) Poir. = ***Fimbristylis pauciflora*** R.Br.
Scirpus pauciflorus var. *fernaldii* (Svenson) Hiitonen = ***Eleocharis quinqueflora*** (Hartmann) O.Schwarz
Scirpus pectinatus Roxb. = ***Schoenoplectus litoralis*** subsp. ***thermalis*** (Trab.) S.S.Hooper
Scirpus pedicellatus var. *pullus* Fernald = ***Scirpus pedicellatus*** Fernald
Scirpus pedicellatus f. *viviparus* F.G.Bernard = ***Scirpus pedicellatus*** Fernald
Scirpus pendulus f. *elongatus* (Eames) C.F.Reed = ***Scirpus pendulus*** Muhl.
Scirpus pentagonus Roxb. = ***Fimbristylis quinquangularis*** (Vahl) Kunth
Scirpus perpusillus Boeck. = ***Isolepis ranko*** (Steud.) Vegetti
Scirpus perrieri Cherm. = ***Schoenoplectiella perrieri*** (Cherm.) Lye
Scirpus perviridis V.J.Cook = ***Bolboschoenus fluviatilis*** (Torr.) Soják
Scirpus petasatus Maxim. = ***Eleocharis wichurai*** Boeck.
Scirpus phaeocarpus F.Phil. = ***Isolepis inundata*** R.Br.
Scirpus philippii C.B.Clarke = ***Isolepis nigricans*** Kunth
Scirpus philippii Tineo = ***Schoenoplectus litoralis*** (Schrad.) Palla subsp. ***litoralis***
Scirpus pictus Boeck. = ***Isolepis cernua*** (Vahl) Roem. & Schult. var. ***cernua***
Scirpus pileatus (A.Gray) Franch. & Sav. = ***Eleocharis palustris*** (L.) Roem. & Schult. subsp. ***palustris***
Scirpus piliferus (Maury) Pickel = ***Oxycaryum cubense*** (Poepp. & Kunth) Palla
Scirpus pilosus Thunb. = [27 CPP]
Scirpus pilosus Larrañaga = [85 URU]
Scirpus pilosus Retz. = ***Fuirena ciliaris*** (L.) Roxb. f. ***ciliaris***
Scirpus pitardii Trab. ex Pit. = ***Cyperus hamulosus*** M.Bieb.
Scirpus planiculmis F.Schmidt = ***Bolboschoenus planiculmis*** (F.Schmidt) E.M.Egorova
Scirpus planifolius Grimm = ***Blysmus compressus*** (L.) Panz. ex Link subsp. ***compressus***
Scirpus planifolius Muhl. = ***Trichophorum planifolium*** (Spreng.) Palla
Scirpus planifolius var. *brevifolius* Torr. = ***Trichophorum planifolium*** (Spreng.) Palla
Scirpus plantagineus Retz. = ***Eleocharis dulcis*** (Burm.f.) Trin. ex Hensch.
Scirpus plantagineus Sw. = ***Eleocharis interstincta*** (Vahl) Roem. & Schult.
Scirpus plantagineus Roxb. = ***Fimbristylis quinquangularis*** (Vahl) Kunth
Scirpus plantaginoides Rottb. = ***Eleocharis dulcis*** (Burm.f.) Trin. ex Hensch.
Scirpus platycarpus S.T.Blake = ***Isolepis cernua*** var. ***platycarpa*** (S.T.Blake) Muasya
Scirpus plicarhachis Griseb. = ***Eleocharis plicarhachis*** (Griseb.) Svenson
Scirpus plumosus R.Br. = ***Schoenoplectus litoralis*** (Schrad.) Palla subsp. ***litoralis***
Scirpus pollicaris Delile = ***Eleocharis parvula*** (Roem. & Schult.) Palla
Scirpus pollichii Gren. & Godr. = ***Schoenoplectus triqueter*** (L.) Palla
Scirpus polycoleus (De Not.) De Not. = ***Schoenoplectiella supina*** (L.) Lye
Scirpus polymorphus Salzm. ex Steud. = ***Eleocharis acicularis*** (L.) Roem. & Schult. subsp. ***acicularis***
Scirpus polymorphus Dulac = ***Scirpoides holoschoenus*** (L.) Soják subsp. ***holoschoenus***
Scirpus polyopsis Bubani = ***Bolboschoenus maritimus*** (L.) Palla subsp. ***maritimus***
Scirpus polyphyllus var. *macrostachys* Boeck. = ***Scirpus polyphyllus*** Vahl
Scirpus polyphyllus f. *macrostachys* (Boeck.) Fernald = ***Scirpus polyphyllus*** Vahl
Scirpus polyphyllus f. *viviparus* (Pursh) Beetle = ***Scirpus polyphyllus*** Vahl
Scirpus polytrichoides Retz. = ***Fimbristylis polytrichoides*** (Retz.) R.Br.
Scirpus potosiensis C.B.Clarke = ***Cyperus elegans*** L. subsp. ***elegans***
Scirpus potosinus Britton = ***Cypringlea analecta*** (Beetle) M.T.Strong
Scirpus pottsii V.J.Cook = ***Isolepis pottsii*** (V.J.Cook) Soják
Scirpus praelongatus Poir. = ***Schoenoplectiella praelongata*** (Poir.) Lye
Scirpus praetextatus Edgar = ***Isolepis praetextata*** (Edgar) Soják
Scirpus preslii A.Dietr. = ***Schoenoplectus mucronatus*** (L.) Palla subsp. ***mucronatus***
Scirpus pretorianus Boeck. ex C.B.Clarke = ***Bulbostylis cinnamomea*** (Boeck.) C.B.Clarke
Scirpus pringlei Britton = ***Bulbostylis schaffneri*** (Boeck.) C.B.Clarke
Scirpus productus C.B.Clarke = ***Isolepis producta*** (C.B.Clarke) K.L.Wilson
Scirpus prolifer Thouars = ***Isolepis bicolor*** Carmich.
Scirpus prolifer Rottb. = ***Isolepis prolifera*** (Rottb.) R.Br.
Scirpus proliferoramosus Boeck. = ***Isolepis bicolor*** Carmich.
Scirpus ×*prolifulcus* Carse = [51 NZN] *Isolepis inundata* × *I. prolifera*

Scirpus ×*prolifundus* Carse = [51 NZN] *Isolepis distigmatosa* × *I. prolifera*
Scirpus prolongatus Poir. ex Nees = ***Schoenoplectiella articulata*** (L.) Lye
Scirpus propinqua (R.Br.) Poir. = ***Fimbristylis squarrosa*** Vahl var. ***squarrosa***
Scirpus psammophilus S.T.Blake = ***Isolepis cernua*** var. ***setiformis*** (Benth.) Muasya
Scirpus pseudofluitans Makino = ***Isolepis crassiuscula*** Hook.f.
Scirpus pseudoschoenus Steud. = ***Ficinia tristachya*** (Rottb.) Nees
Scirpus pseudosetaceus Daveau = ***Isolepis pseudosetacea*** (Daveau) Vasc.
Scirpus pseudotriqueter Steud. = ***Schoenoplectus californicus*** (C.A.Mey.) Soják
Scirpus pterolepis (Nees) Kunth = ***Schoenoplectus litoralis*** subsp. ***thermalis*** (Trab.) S.S.Hooper
Scirpus pterygosperma (R.Br.) Poir. = ***Fimbristylis pterygosperma*** R.Br.
Scirpus puberulus Poir. = ***Bulbostylis puberula*** C.B.Clarke
Scirpus puberulus Michx. = ***Fimbristylis dichotoma*** (L.) Vahl subsp. ***dichotoma***
Scirpus pubescens Pers. = ***Fimbristylis dichotoma*** (L.) Vahl subsp. ***dichotoma***
Scirpus pubescens (Poir.) Lam. = ***Fuirena pubescens*** (Poir.) Kunth
Scirpus pubigerus Bernh. ex Link = ***Fimbristylis dichotoma*** (L.) Vahl subsp. ***dichotoma***
Scirpus pulchellus (Kunth) Boeck. = ***Schoenoplectus pulchellus*** (Kunth) J.Raynal
Scirpus pulcher Rudge = ***Hypolytrum pulchrum*** (Rudge) H.Pfeiff.
Scirpus pulogensis Merr. = ***Trichophorum subcapitatum*** (Thwaites & Hook.) D.A.Simpson
Scirpus pumilio Spreng. = ***Eleocharis pusilla*** R.Br.
Scirpus pumilus Willd. ex Kunth = ***Isolepis cernua*** (Vahl) Roem. & Schult. var. ***cernua***
Scirpus pumilus Vahl = ***Trichophorum pumilum*** (Vahl) Schinz & Thell.
Scirpus pumilus subsp. *distigmaticus* Kük. = ***Trichophorum pumilum*** (Vahl) Schinz & Thell. subsp. ***pumilum***
Scirpus pumilus var. *rollandii* (Fernald) Beetle = ***Trichophorum pumilum*** subsp. ***rollandii*** (Fernald) R.L.Taylor & MacBryde
Scirpus pumilus subsp. *rollandii* (Fernald) Raymond = ***Trichophorum pumilum*** subsp. ***rollandii*** (Fernald) R.L.Taylor & MacBryde
Scirpus punctatus Schrad. ex Nees = ***Eleocharis nana*** Kunth
Scirpus punctatus (Boeck.) Kuntze = ***Eleocharis nana*** Kunth
Scirpus punctatus (R.Br.) Poir. = ***Fimbristylis punctata*** R.Br.
Scirpus pungens (Vahl) Poir. = ***Hypolytrum pulchrum*** (Rudge) H.Pfeiff.
Scirpus pungens Willd. ex Kunth = ***Schoenoplectus lacustris*** (L.) Palla subsp. ***lacustris***
Scirpus pungens Vahl = ***Schoenoplectus pungens*** (Vahl) Palla
Scirpus pungens var. *conglomeratus* (Kunth) R.L.Taylor & MacBryde = ***Schoenoplectus pungens*** (Vahl) Palla var. ***pungens***
Scirpus pungens var. *longisetis* Benth. = ***Schoenoplectus pungens*** var. ***badius*** (C.Presl) S.G.Sm.
Scirpus pungens var. *longispicatus* (Britton) Cronquist = ***Schoenoplectus pungens*** var. ***longispicatus*** (Britton) S.G.Sm.
Scirpus pungens subsp. *monophyllus* (C.Presl) R.L.Taylor & MacBryde = ***Schoenoplectus pungens*** var. ***badius*** (C.Presl) S.G.Sm.
Scirpus pungens var. *nanus* Benth. = ***Schoenoplectus pungens*** var. ***badius*** (C.Presl) S.G.Sm.
Scirpus pungens var. *polyphyllus* Boeck. = ***Schoenoplectus pungens*** var. ***longispicatus*** (Britton) S.G.Sm.
Scirpus purpurascens (Boeck.) Kuntze = ***Eleocharis congesta*** D.Don
Scirpus purpureoater Boeck. = ***Bulbostylis oligostachys*** (Hochst. ex A.Rich.) C.B.Clarke
Scirpus purshianus Fernald = ***Schoenoplectiella purshiana*** (Fernald) Lye
Scirpus purshianus f. *williamsii* (Fernald) Fernald = ***Schoenoplectiella purshiana*** (Fernald) Lye
Scirpus pusillus Kunth = ***Eleocharis acicularis*** (L.) Roem. & Schult. subsp. ***acicularis***
Scirpus pusillus Willd. ex Kunth = ***Eleocharis intermedia*** Schult.
Scirpus pusillus Vahl = ***Eleocharis nana*** Kunth
Scirpus pusillus (R.Br.) Poir. = ***Eleocharis pusilla*** R.Br.
Scirpus pycnostachys Salzm. ex Steud. = ***Lipocarpha maculata*** (Michx.) Torr.
Scirpus pygmaeus Lam. = ***Eleocharis retroflexa*** subsp. ***chaetaria*** (Roem. & Schult.) T.Koyama
Scirpus pygmaeus (Kunth) A.Gray = ***Isolepis cernua*** (Vahl) Roem. & Schult. var. ***cernua***
Scirpus quadrangularis Thouars = ***Fimbristylis quinquangularis*** (Vahl) Kunth
Scirpus quadrangulatus Michx. = ***Eleocharis quadrangulata*** (Michx.) Roem. & Schult.
Scirpus quadrangulus D.Don = ***Schoenoplectiella juncoides*** (Roxb.) Lye
Scirpus quadriflorus Boeck. = ***Fimbristylis quadriflora*** (Boeck.) Beetle
Scirpus quinquangularis Vahl = ***Fimbristylis quinquangularis*** (Vahl) Kunth
Scirpus quinquefarius Buch.-Ham. ex Boeck. = ***Schoenoplectiella roylei*** (Nees) Lye
Scirpus quinqueflorus Hartmann = ***Eleocharis quinqueflora*** (Hartmann) O.Schwarz
Scirpus racemosus Spreng. = [84]
Scirpus radiatus Steud. = ***Eleocharis radicans*** Kunth
Scirpus radiatus (L.f.) Thunb. = ***Ficinia radiata*** (L.f.) Kunth
Scirpus radicans Poir. = ***Eleocharis radicans*** Kunth
Scirpus radiciflorus Maury = ***Bulbostylis radiciflora*** (Maury) Beetle
Scirpus ramosus Boeck. = ***Isolepis fluitans*** var. ***nervosa*** (Hochst. ex A.Rich.) Lye
Scirpus ranko (Steud.) Gunckel = ***Isolepis ranko*** (Steud.) Vegetti
Scirpus rarus (R.Br.) Poir. = ***Fimbristylis rara*** R.Br.
Scirpus raynalii Schuyler = ***Schoenoplectiella erecta*** (Poir.) Lye
Scirpus reclinatus Beyr. ex Kunth = ***Eleocharis intermedia*** Schult.
Scirpus reductus Cherm. = ***Schoenoplectiella reductus*** (Cherm.) Lye
Scirpus rehmannianus Boeck. ex C.B.Clarke = ***Schoenoplectiella roylei*** (Nees) Lye
Scirpus rehmannii Ridl. = ***Lipocarpha rehmannii*** (Ridl.) Goetgh.
Scirpus reichenbachii Boeck. = ***Bulbostylis conifera*** (Kunth) Beetle
Scirpus remireoides Griseb. = ***Amphiscirpus nevadensis*** (S.Watson) Oteng-Yeb.
Scirpus renschii Boeck. = ***Bulbostylis renschii*** (Boeck.) C.B.Clarke

Scirpus repens Willd. ex Schult. = ***Eleocharis geniculata*** (L.) Roem. & Schult.
Scirpus reptans Thuill. = ***Eleocharis palustris*** (L.) Roem. & Schult. subsp. ***palustris***
Scirpus reptans Rich. = ***Rhynchospora reptans*** (Rich.) Kük.
Scirpus reticularis (Colenso) Edgar = ***Isolepis reticularis*** Colenso
Scirpus reticulatus Lam. = ***Cyperus virens*** Michx. var. ***virens***
Scirpus retroflexus Poir. = ***Eleocharis retroflexa*** (Poir.) Urb.
Scirpus retrofractus L. = ***Cyperus retrofractus*** (L.) Torr.
Scirpus retusus J.König ex Wall. = ***Rhynchospora rubra*** (Lour.) Makino subsp. ***rubra***
Scirpus rhodesicus Podlech = ***Schoenoplectus rhodesicus*** (Podlech) Lye
Scirpus rigescens Willd. ex Schult. = ***Fimbristylis cymosa*** R.Br. subsp. ***cymosa***
Scirpus rigidus Griseb. = [85 AGE]
Scirpus rigidus Schrad. ex Nees = ***Schoenoplectus californicus*** (C.A.Mey.) Soják
Scirpus rigidus (Steud.) Boeck. = ***Trichophorum rigidum*** (Steud.) Goetgh., Muasya & D.A.Simpson
Scirpus riparius Pers. = ***Bolboschoenus maritimus*** (L.) Palla subsp. ***maritimus***
Scirpus riparius (R.Br.) Poir. = ***Isolepis cernua*** (Vahl) Roem. & Schult. var. ***cernua***
Scirpus riparius J.Presl & C.Presl = ***Schoenoplectus californicus*** (C.A.Mey.) Soják
Scirpus riparius var. *paschalis* Kük. = ***Schoenoplectus californicus*** (C.A.Mey.) Soják
Scirpus riparius var. *tereticulmis* (Steud.) C.B.Clarke = ***Schoenoplectus californicus*** (C.A.Mey.) Soják
Scirpus rivularis D.Don. ex Nees = ***Fuirena umbellata*** Rottb.
Scirpus rivularis (Schrad.) Boeck. = ***Isolepis natans*** (Thunb.) A.Dietr.
Scirpus robustus Pursh = ***Bolboschoenus robustus*** (Pursh) Soják
Scirpus robustus var. *campestris* Fernald = ***Bolboschoenus maritimus*** subsp. ***paludosus*** (A.Nelson) T.Koyama
Scirpus robustus var. *compactus* Burtt Davy = ***Bolboschoenus robustus*** (Pursh) Soják
Scirpus robustus var. *novae-angliae* (Britton) Beetle = ***Schoenoplectus novae-angliae*** (Britton) M.T.Strong
Scirpus robustus var. *paludosus* (A.Nelson) Fernald = ***Bolboschoenus maritimus*** subsp. ***paludosus*** (A.Nelson) T.Koyama
Scirpus robustus f. *protrusus* Fernald = ***Bolboschoenus robustus*** (Pursh) Soják
Scirpus rockii Kük. = ***Schoenoplectiella juncoides*** (Roxb.) Lye
Scirpus rogersii N.E.Br. = ***Schoenoplectus confusus*** var. ***rogersii*** (N.E.Br.) Lye
Scirpus rollandii Fernald = ***Trichophorum pumilum*** subsp. ***rollandii*** (Fernald) R.L.Taylor & MacBryde
Scirpus romanus L. = ***Scirpoides holoschoenus*** (L.) Soják subsp. ***holoschoenus***
Scirpus romanus f. *monocephalus* Waisb. = ***Scirpoides holoschoenus*** (L.) Soják subsp. ***holoschoenus***
Scirpus ×*rosellinii* Ces., Pass. & Gibelli = ***Schoenoplectus* ×*carinatus*** (Sm.) Palla nothosubsp. ***carinatus***
Scirpus rostellatus Torr. = ***Eleocharis rostellata*** (Torr.) Torr.
Scirpus rothii Hoppe = ***Schoenoplectus pungens*** (Vahl) Palla var. ***pungens***
Scirpus roylei (Nees) R.Parker = ***Schoenoplectiella roylei*** (Nees) Lye
Scirpus rubicundus (Kunth) Parl. = ***Isolepis rubicunda*** Kunth
Scirpus rubiginosus Beetle = ***Schoenoplectus acutus*** var. ***occidentalis*** (S.Watson) S.G.Sm.
Scirpus rubricosus Fernald = ***Scirpus cyperinus*** (L.) Kunth
Scirpus rubricosus f. *praelongus* (Fernald) Fernald = ***Scirpus cyperinus*** (L.) Kunth
Scirpus rubrotinctus Fernald = ***Scirpus microcarpus*** J.Presl & C.Presl
Scirpus rubrotinctus var. *confertus* Fernald = ***Scirpus microcarpus*** J.Presl & C.Presl
Scirpus rubrotinctus f. *confertus* (Fernald) Weath. = ***Scirpus microcarpus*** J.Presl & C.Presl
Scirpus rubrotinctus f. *radiosus* Fernald = ***Scirpus microcarpus*** J.Presl & C.Presl
Scirpus rufescens Boeck. = ***Bulbostylis rufescens*** (Boeck.) Beetle
Scirpus rufus (Huds.) Schrad. = ***Blysmus rufus*** (Huds.) Link
Scirpus rufus subsp. *exilis* Printz = ***Blysmus rufus*** (Huds.) Link
Scirpus rufus var. *neogaeus* Fernald = ***Blysmus rufus*** (Huds.) Link
Scirpus ruppioides Thwaites ex C.Wright = ***Websteria confervoides*** (Poir.) S.S.Hooper
Scirpus russeolus (Fr. ex Hartm.) T.Koyama = ***Eriophorum chamissonis*** C.A.Mey.
Scirpus russeolus var. *major* (Sommier) T.Koyama = ***Eriophorum chamissonis*** C.A.Mey.
Scirpus sachalinensis Meinsh. = ***Eleocharis kamtschatica*** (C.A.Mey.) Kom.
Scirpus salbundius Buch.-Ham. ex Wall. = ***Fimbristylis quinquangularis*** (Vahl) Kunth
Scirpus salinus Schmidt ex Steud. = ***Bolboschoenus maritimus*** (L.) Palla subsp. ***maritimus***
Scirpus sasakii Hayata = ***Schoenoplectiella wallichii*** (Nees) Lye
Scirpus savii Sebast. & Mauri = ***Isolepis cernua*** (Vahl) Roem. & Schult. var. ***cernua***
Scirpus savii subsp. *campanus* (A.Terracc.) Arcang. = ***Isolepis cernua*** (Vahl) Roem. & Schult. var. ***cernua***
Scirpus savii f. *controversus* (Steud.) A.Terracc. = ***Isolepis cernua*** (Vahl) Roem. & Schult. var. ***cernua***
Scirpus savii var. *filiformis* A.Terracc. = ***Isolepis cernua*** (Vahl) Roem. & Schult. var. ***cernua***
Scirpus savii var. *fuscescens* A.Terracc. = ***Isolepis cernua*** (Vahl) Roem. & Schult. var. ***cernua***
Scirpus savii var. *major* Guss. = ***Isolepis cernua*** (Vahl) Roem. & Schult. var. ***cernua***
Scirpus savii subsp. *minaae* (Tod.) Arcang. = ***Isolepis cernua*** (Vahl) Roem. & Schult. var. ***cernua***
Scirpus savii var. *minaae* (Tod.) Firoi = ***Isolepis cernua*** (Vahl) Roem. & Schult. var. ***cernua***
Scirpus savii f. *todaroanus* A.Terracc. = ***Isolepis cernua*** (Vahl) Roem. & Schult. var. ***cernua***
Scirpus savii var. *vahlii* Lange = ***Isolepis cernua*** (Vahl) Roem. & Schult. var. ***cernua***
Scirpus saximontanus Fernald = ***Schoenoplectiella saximontana*** (Fernald) Lye
Scirpus scaber (J.Presl & C.Presl) Kuntze = ***Bulbostylis scabra*** (J.Presl & C.Presl) C.B.Clarke
Scirpus scaber Roxb. = ***Fimbristylis acuminata*** Vahl
Scirpus scaberrimus Boeck. = ***Actinoscirpus grossus*** (L.f.) Goetgh. & D.A.Simpson var. ***grossus***
Scirpus scabriculmis Beetle = ***Eriophorum scabriculme*** (Beetle) Raymond

Scirpus scariosus Roem. & Schult. = ***Ficinia nigrescens*** (Schrad.) J.Raynal
Scirpus schaffneri Boeck. = ***Bulbostylis schaffneri*** (Boeck.) C.B.Clarke
Scirpus schaffneri (Boeck.) Kuntze = ***Eleocharis schaffneri*** Boeck.
Scirpus schansiensis (Hand.-Mazz.) Tang & F.T.Wang = ***Trichophorum schansiense*** Hand.-Mazz.
Scirpus scheuchzeri Hartmann = ***Eriophorum scheuchzeri*** Hoppe
Scirpus scheuchzeri (Hoppe) Vitman = ***Eriophorum scheuchzeri*** Hoppe
Scirpus schimperianus (Hochst. ex A.Rich.) Boeck. = ***Bulbostylis schimperiana*** (Hochst. ex A.Rich.) C.B.Clarke
Scirpus schinzianus Boeck. ex C.B.Clarke = ***Scirpoides burkei*** (C.B.Clarke) Goetgh., Muasya & D.A.Simpson
Scirpus schinzii Boeck. = ***Scirpoides dioecus*** (Kunth) Browning
Scirpus schmidii Raymond = ***Bolboschoenus maritimus*** subsp. ***affinis*** (Roth) T.Koyama
Scirpus ×*schmidtianus* Junge = ?
Scirpus schoenoides Roxb. = ***Fimbristylis ovata*** (Burm.f.) J.Kern
Scirpus schoenoides Retz. = ***Fimbristylis schoenoides*** (Retz.) Vahl
Scirpus schoenoides Elliott = ***Rhynchospora elliottii*** A.Dietr.
Scirpus schomburgkianus Boeck. = ***Rhynchospora schomburgkiana*** (Boeck.) T.Koyama
Scirpus schoofii Beetle = ***Schoenoplectus schoofii*** (Beetle) Soják
Scirpus schwackeanus Boeck. = ***Bulbostylis sphaerocephala*** (Boeck.) Lindm.
Scirpus schweinfurthianus Boeck. = ***Bulbostylis abortiva*** (Steud.) C.B.Clarke
Scirpus scirpoideus (Michx.) T.Koyama = ***Fuirena scirpoidea*** Michx.
Scirpus scleropus (C.B.Clarke) K.Schum. = ***Bulbostylis scleropus*** C.B.Clarke
Scirpus sellowianus (Kunth) Boeck. = ***Bulbostylis sellowiana*** (Kunth) Palla
Scirpus sellowianus (Kunth) Griseb. = ***Eleocharis sellowiana*** Kunth
Scirpus semisubterraneus Boeck. = ***Phylloscirpus deserticola*** (Phil.) Dhooge & Goetgh.
Scirpus semiuncialis C.B.Clarke = [27]
Scirpus senegalensis Lam. = ***Lipocarpha chinensis*** (Osbeck) J.Kern
Scirpus seoanei Merino = [12 FRA]
Scirpus sepium Honck. = ***Eleocharis quinqueflora*** (Hartmann) O.Schwarz
Scirpus sericeus Poir. = ***Fimbristylis sericea*** (Poir.) R.Br.
Scirpus serrulatus (Vahl) Vahl = ***Fimbristylis dichotoma*** (L.) Vahl subsp. ***dichotoma***
Scirpus sesquipollicaris Willd. ex Kunth = ***Lipocarpha micrantha*** (Vahl) G.C.Tucker
Scirpus sessiliflorus Phil. = ***Phylloscirpus acaulis*** (Phil.) Goetgh. & D.A.Simpson
Scirpus setaceus Thunb. = ***Isolepis antarctica*** (L.) Roem. & Schult.
Scirpus setaceus L. = ***Isolepis setacea*** (L.) R.Br.
Scirpus setaceus var. *clathratus* (Rchb.) A.Terracc. = ***Isolepis setacea*** (L.) R.Br.
Scirpus setaceus var. *gracillimus* (Kohts) A.Terracc. = ***Isolepis setacea*** (L.) R.Br.
Scirpus setaceus var. *major* Lej. = ***Isolepis setacea*** (L.) R.Br.
Scirpus setaceus var. *pygmaeus* (Vahl) Boeck. = ***Isolepis cernua*** (Vahl) Roem. & Schult. var. ***cernua***
Scirpus setaceus var. *savii* (Sebast. & Mauri) Firoi = ***Isolepis cernua*** (Vahl) Roem. & Schult. var. ***cernua***
Scirpus setaceus var. *savii* Boeck. = ***Isolepis setacea*** (L.) R.Br.
Scirpus setifolius Boeck. = ***Bulbostylis sphaerocephala*** (Boeck.) Lindm.
Scirpus setiformis (Benth.) S.T.Blake = ***Isolepis cernua*** var. ***setiformis*** (Benth.) Muasya
Scirpus siamensis (C.B.Clarke) J.Kern = ***Ascolepis dipsacoides*** subsp. ***siamensis*** (C.B.Clarke) J.Raynal
Scirpus siculus Lojac. = ***Schoenoplectus tabernaemontani*** (C.C.Gmel.) Palla
Scirpus sieberi (Kunth) Kuntze = ***Eleocharis variegata*** (Poir.) C.Presl
Scirpus simplex Elliott = ***Eleocharis albida*** Torr.
Scirpus sinensis Kunth = ***Lipocarpha squarrosa*** (L.) Goetgh.
Scirpus sinuatus Schuyler = ***Schoenoplectiella erecta*** (Poir.) Lye
Scirpus smithii A.Gray = ***Schoenoplectus smithii*** (A.Gray) Soják
Scirpus smithii var. *levisetus* Fernald = ***Schoenoplectus smithii*** var. ***levisetus*** (Fernald) S.G.Sm.
Scirpus smithii f. *levisetus* (Fassett) Fernald = ***Schoenoplectus smithii*** var. ***levisetus*** (Fernald) S.G.Sm.
Scirpus smithii var. *setosus* Fernald = ***Schoenoplectus smithii*** var. ***setosus*** (Fernald) S.G.Sm.
Scirpus smithii f. *setosus* (Fernald) Fernald = ***Schoenoplectus smithii*** var. ***setosus*** (Fernald) S.G.Sm.
Scirpus smithii var. *williamsii* (Fernald) Beetle = ***Schoenoplectiella purshiana*** (Fernald) Lye
Scirpus solispicatus Lunell = ***Schoenoplectus americanus*** (Pers.) Volkart
Scirpus soloniensis Dubois = ***Eleocharis ovata*** (Roth) Roem. & Schult.
Scirpus sororius (Kunth) C.B.Clarke = ***Isolepis sororia*** Kunth
Scirpus spadiceus L. = ***Fimbristylis spadicea*** (L.) Vahl
Scirpus spadiceus (Lam.) Boeck. = ***Nemum spadiceum*** (Lam.) Desv.
Scirpus spadiceus Phil. = ***Schoenus nitens*** (R.Br.) Roem. & Schult. var. ***nitens***
Scirpus spadiceus var. *ciliatus* Ridl. = ***Nemum spadiceum*** (Lam.) Desv.
Scirpus sparmannii Lam. = ***Ficinia sylvatica*** Kunth
Scirpus spathaceus (L.) Michx. = ***Dulichium arundinaceum*** (L.) Britton var. ***arundinaceum***
Scirpus spathaceus Hochst. = ***Pseudoschoenus inanis*** (Thunb.) Oteng-Yeb.
Scirpus speciosus Salisb. = ***Eriophorum angustifolium*** Honck. subsp. ***angustifolium***
Scirpus sphacelatus (R.Br.) Poir. = ***Eleocharis sphacelata*** R.Br.
Scirpus sphaerocarpus Boeck. = ***Bulbostylis sphaerocarpa*** (Boeck.) C.B.Clarke
Scirpus sphaerocephalus Boeck. = ***Bulbostylis sphaerocephala*** (Boeck.) Lindm.
Scirpus sphaerocephalus var. *macrocephalus* Maury = ***Bulbostylis sphaerocephala*** (Boeck.) Lindm.
Scirpus sphaerolepis Boeck. = ***Bulbostylis scabra*** (J.Presl & C.Presl) C.B.Clarke
Scirpus spiralis (A.Rich.) Druce = ***Desmoschoenus spiralis*** (A.Rich.) Hook.f.
Scirpus spiralis Willd. ex Kunth = ***Eleocharis dulcis*** (Burm.f.) Trin. ex Hensch.

Scirpus spiralis Rottb. = ***Eleocharis spiralis*** (Rottb.) Roem. & Schult.
Scirpus spiralis Bosc ex Kunth = ***Eleocharis tortilis*** (Link) Schult.
Scirpus spiraloides Roxb. = ***Eleocharis acutangula*** (Roxb.) Schult.
Scirpus squamosus Link = ***Lipocarpha squarrosa*** (L.) Goetgh.
Scirpus squarrosulus Steud. = ***Bolboschoenus maritimus*** (L.) Palla subsp. ***maritimus***
Scirpus squarrosus (Vahl) Poir. = ***Fimbristylis squarrosa*** Vahl
Scirpus squarrosus Spreng. = ***Isolepis bicolor*** Carmich.
Scirpus squarrosus L. = ***Lipocarpha squarrosa*** (L.) Goetgh.
Scirpus squarrosus var. *dietrichiae* (Boeck.) Benth. = ***Lipocarpha microcephala*** (R.Br.) Kunth
Scirpus squarrosus var. *siamensis* C.B.Clarke = ***Ascolepis dipsacoides*** subsp. ***siamensis*** (C.B.Clarke) J.Raynal
Scirpus stagnicola Raymond = ***Bolboschoenus stagnicola*** (Raymond) Soják
Scirpus stans Ham. = ***Fimbristylis ferruginea*** (L.) Vahl var. ***ferruginea***
Scirpus stauntonii C.B.Clarke = ***Cyperus nipponicus*** Franch. & Sav.
Scirpus ×steinmetzii Fernald = ***Schoenoplectus ×steinmetzii*** (Fernald) S.G.Sm.
Scirpus stellatus C.B.Clarke = ***Isolepis stellata*** (C.B.Clarke) K.L.Wilson
Scirpus stenophyllus Elliott = ***Bulbostylis stenophylla*** (Elliott) C.B.Clarke
Scirpus sterilis Maiden & Betche = ***Bulbostylis barbata*** (Rottb.) C.B.Clarke subsp. ***barbata***
Scirpus sterilis Salzm. ex Steud. = ***Fimbristylis spadicea*** (L.) Vahl
Scirpus steudneri Boeck. =
Scirpus stolonifer (Boeck.) Kuntze = ***Eleocharis dombeyana*** Kunth
Scirpus stolonifer Roth = ***Isolepis fluitans*** (L.) R.Br. var. ***fluitans***
Scirpus stramineus C.Presl = [7]
Scirpus striatulus (É.Desv.) Griseb. = ***Eleocharis bonariensis*** Nees
Scirpus striatus (Nees) Fourc. = ***Isolepis striata*** (Nees) Kunth
Scirpus strictus Roxb. = ***Fimbristylis cymosa*** R.Br. subsp. ***cymosa***
Scirpus strictus (R.Br.) Poir. = ***Fimbristylis dichotoma*** (L.) Vahl subsp. ***dichotoma***
Scirpus strobiliferus Steud. = ***Bolboschoenus maritimus*** (L.) Palla subsp. ***maritimus***
Scirpus strobilinus Roxb. = ***Bolboschoenus maritimus*** subsp. ***affinis*** (Roth) T.Koyama
Scirpus stygius Ferber ex J.Fleisch. = (Juncaceae)
Scirpus subaphyllus Boeck. = ***Trichophorum subcapitatum*** (Thwaites & Hook.) D.A.Simpson
Scirpus subappendiculatus Steud. = ***Schoenoplectus pungens*** (Vahl) Palla var. ***pungens***
Scirpus subarticulatus (Nees) Kuntze = ***Eleocharis subarticulata*** (Nees) Boeck.
Scirpus subarticulatus Roxb. = ***Schoenoplectiella articulata*** (L.) Lye
Scirpus subasper Beetle = ***Scirpus asper*** J.Presl & C.Presl
Scirpus subasper var. *diffusus* Beetle = ***Scirpus asper*** J.Presl & C.Presl
Scirpus subasper var. *polystachyus* (C.B.Clarke) Beetle ex J.F.Macbr. = ***Scirpus asper*** J.Presl & C.Presl
Scirpus subbisetosus T.Koyama = ***Schoenoplectus subbisetosus*** (T.Koyama) Soják
Scirpus subcapitatus Thwaites & Hook. = ***Trichophorum subcapitatum*** (Thwaites & Hook.) D.A.Simpson
Scirpus subcapitatus subsp. *celebicus* J.Kern = ***Trichophorum subcapitatum*** (Thwaites & Hook.) D.A.Simpson
Scirpus subcapitatus subsp. *pulogensis* (Merr.) J.Kern = ***Trichophorum subcapitatum*** (Thwaites & Hook.) D.A.Simpson
Scirpus subcapitatus f. *rigidus* Kük. = ***Trichophorum subcapitatum*** (Thwaites & Hook.) D.A.Simpson
Scirpus subcapitatus var. *triangularis* Kük. = ***Trichophorum subcapitatum*** (Thwaites & Hook.) D.A.Simpson
Scirpus subcucullatus (Berggr.) Berggr. = ***Isolepis aucklandica*** Hook.f.
Scirpus subdistichus Boeck. = ***Bulbostylis pauciflora*** (Liebm.) C.B.Clarke
Scirpus submersus C.Wright = ***Websteria confervoides*** (Poir.) S.S.Hooper
Scirpus subprolifer Boeck. = ***Isolepis cernua*** (Vahl) Roem. & Schult. var. ***cernua***
Scirpus subquadrangularis Makino = ***Schoenoplectiella juncoides*** (Roxb.) Lye
Scirpus subquadriflorus Boeck. = ***Fimbristylis quadriflora*** (Boeck.) Beetle
Scirpus subsquarrosus Muhl. = ***Lipocarpha micrantha*** (Vahl) G.C.Tucker
Scirpus subterminalis Torr. = ***Schoenoplectus subterminalis*** (Torr.) Soják
Scirpus subterminalis var. *cylindricus* (Torr.) T.Koyama = ***Schoenoplectus torreyi*** (Olney) Palla
Scirpus subterminalis f. *terrestris* (Paine) Fernald = ***Schoenoplectus subterminalis*** (Torr.) Soják
Scirpus subtilis (Boeck.) Kuntze = ***Eleocharis minima*** Kunth var. ***minima***
Scirpus subtilis Boeck. = ***Isolepis cernua*** (Vahl) Roem. & Schult. var. ***cernua***
Scirpus subtilissimus (Boeck.) S.T.Blake = ***Isolepis subtilissima*** Boeck.
Scirpus subulatus Vahl = ***Schoenoplectus litoralis*** subsp. ***thermalis*** (Trab.) S.S.Hooper
Scirpus subviviparus (Boeck.) Kuntze = ***Eleocharis congesta*** D.Don
Scirpus sulcatus Roth = ***Eleocharis filiculmis*** Kunth
Scirpus sulcatus Elliott = ***Fimbristylis dichotoma*** (L.) Vahl subsp. ***dichotoma***
Scirpus sulcatus Thouars = ***Isolepis sulcata*** (Thouars) Carmich.
Scirpus sulcatus var. *distigmatosus* C.B.Clarke = ***Isolepis distigmatosa*** (C.B.Clarke) Edgar
Scirpus sulciculmis Schrad. ex Nees = ***Eleocharis atropurpurea*** (Retz.) Kunth
Scirpus sundanus Miq. = ***Schoenoplectus mucronatus*** (L.) Palla subsp. ***mucronatus***
Scirpus supinus Boeck. = ***Schoenoplectiella erecta*** (Poir.) Lye
Scirpus supinus L. = ***Schoenoplectiella supina*** (L.) Lye
Scirpus supinus var. *digynus* Boeck. = ***Schoenoplectiella purshiana*** (Fernald) Lye
Scirpus supinus var. *elatior* Boeck. = ***Schoenoplectiella juncoides*** (Roxb.) Lye
Scirpus supinus var. *glomeratus* A.Terracc. = ***Schoenoplectiella supina*** (L.) Lye
Scirpus supinus var. *hallii* (A.Gray) A.Gray = ***Schoenoplectiella hallii*** (A.Gray) Lye

Scirpus supinus var. *lateriflorus* (J.F.Gmel.) T.Koyama = ***Schoenoplectiella lateriflora*** (J.F.Gmel.) Lye
Scirpus supinus var. *leiocarpus* Kom. = ***Schoenoplectus komarovii*** (Roshev.) Soják
Scirpus supinus var. *saximontanus* (Fernald) T.Koyama = ***Schoenoplectiella saximontana*** (Fernald) Lye
Scirpus swampianus Bosc ex Kunth = ***Bolboschoenus maritimus*** (L.) Palla subsp. ***maritimus***
Scirpus sylvaticus Vahl = ***Scirpus radicans*** Schkuhr
Scirpus sylvaticus var. *atrovirens* (Willd.) A.Gray = ***Scirpus atrovirens*** Willd.
Scirpus sylvaticus var. *bissellii* Fernald = ***Scirpus expansus*** Fernald
Scirpus sylvaticus f. *bissellii* (Fernald) D.S.Carp. = ***Scirpus expansus*** Fernald
Scirpus sylvaticus f. *cephaloideus* E.Sheld. = ***Scirpus cyperinus*** (L.) Kunth
Scirpus sylvaticus var. *compactus* C.Klinggr. = ***Scirpus sylvaticus*** L.
Scirpus sylvaticus var. *digynus* Boeck. = ***Scirpus sylvaticus*** L.
Scirpus sylvaticus subsp. *digynus* (Boeck.) T.Koyama = ***Scirpus sylvaticus*** L.
Scirpus sylvaticus var. *dissitiflorus* Sond. = ***Scirpus sylvaticus*** L.
Scirpus sylvaticus var. *jechlii* Opiz ex Nyman = ***Scirpus sylvaticus*** L.
Scirpus sylvaticus f. *jungei* Soó = ***Scirpus sylvaticus*** L.
Scirpus sylvaticus var. *maximowiczii* Regel = ***Scirpus orientalis*** Ohwi
Scirpus sylvaticus subsp. *maximowiczii* (Regel) N.R.Cui = ***Scirpus orientalis*** Ohwi
Scirpus sylvaticus var. *microcarpus* (J.Presl & C.Presl) MacMill. = ***Scirpus microcarpus*** J.Presl & C.Presl
Scirpus sylvaticus subsp. *orientalis* (Ohwi) Vorosch. = ***Scirpus orientalis*** Ohwi
Scirpus sylvaticus var. *radicans* (Schkuhr) Willd. = ***Scirpus radicans*** Schkuhr
Scirpus sylvaticus var. *subradicans* Kük. ex Tang = ***Scirpus orientalis*** Ohwi
Scirpus sylvaticus var. *sychnocephalus* S.N.Cowles = ***Scirpus atrovirens*** Willd.
Scirpus tabernaemontani Salzm. ex Ball = ***Schoenoplectus litoralis*** (Schrad.) Palla subsp. ***litoralis***
Scirpus tabernaemontani C.C.Gmel. = ***Schoenoplectus tabernaemontani*** (C.C.Gmel.) Palla
Scirpus tabernaemontani f. *luxurians* Miq. = ***Schoenoplectus tabernaemontani*** (C.C.Gmel.) Palla
Scirpus tabernaemontani f. *paludosus* Boenn. = ***Schoenoplectus tabernaemontani*** (C.C.Gmel.) Palla
Scirpus tabernaemontani f. *sub-duvalii* Beckh. = ***Schoenoplectus tabernaemontani*** (C.C.Gmel.) Palla
Scirpus tasmanicus S.T.Blake = ***Isolepis tasmanica*** (S.T.Blake) K.L.Wilson
Scirpus tatora Kunth = ***Schoenoplectus californicus*** (C.A.Mey.) Soják
Scirpus tegetalis Burch. = ***Dulichium arundinaceum*** (L.) Britton var. ***arundinaceum***
Scirpus tenellus Link = ***Bulbostylis tenella*** (Link) C.B.Clarke
Scirpus tenellus Roxb. = ***Fimbristylis tenera*** Schult.
Scirpus tener Spreng. = ***Eleocharis gracilis*** R.Br.
Scirpus tenerrimus Schrad. ex Nees = ***Eleocharis minima*** Kunth var. ***minima***
Scirpus tenerrimus Peter = ***Schoenoplectiella microglumis*** (Lye) Lye
Scirpus tenuiculus Schrad. ex Schult. = ***Eleocharis filiculmis*** Kunth
Scirpus tenuifolius Rudge = ***Bulbostylis tenuifolia*** (Rudge) J.F.Macbr.
Scirpus tenuifolius DC. = ***Schoenoplectus pungens*** (Vahl) Palla var. ***pungens***
Scirpus tenuis (Gilli) Parsa = ***Blysmus compressus*** (L.) Panz. ex Link subsp. ***compressus***
Scirpus tenuis Schreb. ex Schweigg. & Körte = ***Eleocharis palustris*** (L.) Roem. & Schult. subsp. ***palustris***
Scirpus tenuis Willd. = ***Eleocharis tenuis*** (Willd.) Schult.
Scirpus tenuis Spreng. = ***Isolepis pusilla*** Kunth
Scirpus tenuispicatus Boeck. = ***Bulbostylis funckii*** (Steud.) C.B.Clarke
Scirpus tenuissimus (Boeck.) Kuntze = ***Eleocharis minima*** Kunth var. ***minima***
Scirpus tenuissimus (Nees) Boeck. = ***Isolepis tenuissima*** (Nees) Kunth
Scirpus tereticulmis Steud. = ***Schoenoplectus californicus*** (C.A.Mey.) Soják
Scirpus terminalis Phil. = ***Isolepis cernua*** (Vahl) Roem. & Schult. var. ***cernua***
Scirpus ternatanus var. *kiushuensis* (Ohwi) T.Koyama = ***Scirpus rosthornii*** Diels
Scirpus ternatus Wall. = ***Schoenoplectiella erecta*** (Poir.) Lye
Scirpus ternatus Ham. ex Hook.f. = ***Schoenoplectiella erecta*** (Poir.) Lye
Scirpus tetragonus Walker = [10 GRB]
Scirpus tetragonus (R.Br.) Poir. = ***Fimbristylis tetragona*** R.Br.
Scirpus tetragonus (R.Br.) Poir. = ***Fimbristylis tetragona*** R.Br.
Scirpus tetraqueter Thwaites = ***Eleocharis tetraquetra*** Nees
Scirpus tetraquetrus (Nees) Kuntze = ***Eleocharis tetraquetra*** Nees
Scirpus thermalis Trab. = ***Schoenoplectus litoralis*** subsp. ***thermalis*** (Trab.) S.S.Hooper
Scirpus thermalis Spruce ex Benth. & Hook.f. = ***Zameioscirpus gaimardioides*** (É.Desv.) Dhooge & Goetgh.
Scirpus thomsonii (Boeck.) Kuntze = ***Eleocharis pellucida*** C.Presl
Scirpus thouarsianus Schult. = ***Isolepis bicolor*** Carmich.
Scirpus thouarsianus var. *bicolor* (Carmich.) Hemsl. = ***Isolepis bicolor*** Carmich.
Scirpus thouarsianus var. *pallescens* Hemsl. = ***Isolepis bicolor*** Carmich.
Scirpus thouarsianus var. *virens* (Boeck.) Hemsl. = ***Isolepis bicolor*** Carmich.
Scirpus thouarsii Roem. & Schult. = ***Bulbostylis puberula*** C.B.Clarke
Scirpus thuarsii Spreng. = ***Isolepis sulcata*** (Thouars) Carmich.
Scirpus thunbergianus (Nees) Levyns = ***Scirpoides holoschoenus*** subsp. ***thunbergii*** (Schrad.) Soják
Scirpus thunbergii (Schrad.) A.Spreng. = ***Scirpoides holoschoenus*** subsp. ***thunbergii*** (Schrad.) Soják
Scirpus thwaitesii Boeck. = [40 IND]
Scirpus thyrsiflorus Willd. = ***Scirpus cyperinus*** (L.) Kunth
Scirpus timorensis Kunth = ***Schoenoplectiella juncoides*** (Roxb.) Lye
Scirpus tomentosus Poir. = ***Fimbristylis dichotoma*** (L.) Vahl subsp. ***dichotoma***
Scirpus torreyanus (Boeck.) Kuntze = ***Eleocharis microcarpa*** var. ***filiculmis*** Torr.
Scirpus torreyi Olney = ***Schoenoplectus torreyi*** (Olney) Palla

Scirpus elatus Gueldenst. ex Ledeb., Fl. Ross. 4: 252 (1852). = [14]
Scirpus erectogracilis Hayata, Icon. Pl. Formosan. 6: 114 (1916). = [38 TAI]
Scirpus ×fujimakii M.Kikuchi & Koyama, J. Jap. Bot. 32: 255 (1957). = [38 JAP]
Scirpus gaillardii Maury, J. Bot. (Morot) 3: 197 (1889). = [8]
Scirpus gracillimus Boeck., Vidensk. Meddel. Naturhist. Foren. Kjøbenhavn 1869: 140 (1869). = [24 ETH]
Scirpus graecus Quézel & Contandr., Taxon 16: 239 (1967). = [13 GRC]
Scirpus helferi Beetle, Amer. J. Bot. 33: 662 (1946). = [41 AND MYA]
Scirpus hispanicus Gueldenst. ex Ledeb., Fl. Ross. 4: 252 (1852). = [14]
Scirpus hochstetteri Boeck., Linnaea 36: 739 (1870). = [24 ETH]
Scirpus ×intermedius Tang & F.T.Wang, in Fl. Reipubl. Popul. Sin. 11: 224 (1961), nom. illeg. = [36 CHC]
Scirpus isdellensis W.Fitzg., J. Proc. Roy. Soc. Western Australia 3: 123 (1918). = [50 WAU]
Scirpus iseensis T.Koyama & Shimizu, J. Jap. Bot. 42: 177 (1967). = [38 JAP]
Scirpus junciformis Retz., Observ. Bot. 6: 19 (1791). = [36+]
Scirpus libanoticus Post, Fl. Syria: 833 (1896). = [34 LBS]
Scirpus longivaginatus Boeck., Beitr. Cyper. 2: 16 (1890). = [84]
Scirpus macrostachys Boeck., Beitr. Cyper. 1: 19 (1888), nom. illeg. = [40 EHM]
Scirpus ×mariqueter Tang & F.T.Wang, in Fl. Reipubl. Popul. Sin. 11: 223 (1961). = [36 CHC]
Scirpus muehlenbergianus Schult. in J.J.Roemer & J.A.Schultes, Mant. 2: 74 (1824). = [7]
Scirpus muelleri Boeck., Beitr. Cyper. 2: 14 (1890). = [50]
Scirpus nindensis Ficalho & Hiern, Trans. Linn. Soc. London, Bot. 2: 27 (1881). = [2]
Scirpus nudifructus Kük. in R.E.Fries, Wiss. Erg. Schwed. Rhod.-Kongo Exped. 1: 7 (1921). = [26 ZAM]
Scirpus nudus Raf., Atlantic J.: 18 (1833). = [7]
Scirpus occultus C.B.Clarke, Bull. Soc. Bot. France 54(8): 28 (1907). = [2]
Scirpus pilosus Thunb., Prodr. Pl. Cap.: 18 (1794). = [27 CPP]
Scirpus pilosus Larrañaga, Escritos D. A. Larrañaga 2: 25 (1923), nom. illeg. = [85 URU]
Scirpus ×prolifulcus Carse, Trans. & Proc. New Zealand Inst. 60: 572 (1930). = [51 NZN] *Isolepis inundata* × *I. prolifera*
Scirpus ×prolifundus Carse, Trans. & Proc. New Zealand Inst. 60: 572 (1930). = [51 NZN] *Isolepis distigmatosa* × *I. prolifera*
Scirpus racemosus Spreng., Syst. Veg. 1: 213 (1824). = [84]
Scirpus rigidus Griseb., Abh. Königl. Ges. Wiss. Göttingen 24: 312 (1879), nom. illeg. = [85 AGE]
Scirpus ×schmidtianus Junge, Jahrb. Hamburg. Wiss. Anst. 22(Beih.): 72 (1905). = ?
Scirpus semiuncialis C.B.Clarke, Kew Bull., Addit. Ser. 8: 29 (1908). = [27]
Scirpus seoanei Merino, Contr. Fl. Galicia: 173 (1897). = [12 FRA]
Scirpus stramineus C.Presl, Isis (Oken) 21: 269 (1828). = [7]
Scirpus tetragonus Walker, Essays Nat. Hist. Rural Econ.: 242 (1808). = [10 GRB]
Scirpus thwaitesii Boeck., Linnaea 38: 380 (1874). = [40 IND]
Scirpus typhinus Raf., Atlantic J.: 18 (1833). = [7]
Scirpus uleanus Boeck., Allg. Bot. Z. Syst. 2: 55 (1895). = [84]
Scirpus umbellatus Post, Fl. Syria: 21 (1896). = [34 PAL]
Scirpus varius Boeck. ex C.B.Clarke in W.H.Harvey & auct. suc. (eds.), Fl. Cap. 7: 229 (1898), nom. illeg. = [27 NAT SWZ TVL]
Scirpus welwitschii K.Richt., Pl. Eur. 1: 141 (1890). = [12 POR]
Scirpus xerophilus Pilg., Bot. Jahrb. Syst. 30: 142 (1901). = [84 BZC]

Scleria

Scleria P.J.Bergius, Kongl. Vetensk. Acad. Handl. 26: 142 (1765).
Trop. & Subtrop. to N. America. 22 BEN BKN GAM GHA GNB GUI IVO LBR MLI MLW MTN NGA NGR SEN SIE TOG 23 BUR CAF CMN CON EQG GAB GGI RWA ZAI 24 CHA ERI ETH SUD 25 KEN TAN UGA ZAM 26 ANG MLW MOZ ZAM ZIM 27 BOT CPP LES NAM NAT OFS SWZ TVL 29 COM MAU MDG SEY 35 YEM 36 CHC CHH CHS 38 JAP KOR NNS OGA TAI 40 ASS BAN EHM IND NEP SRL WHM 41 AND CBD LAO MYA NCB THA VIE 42 BOR JAW LSI MLY MOL PHI SUL SUM 43 BIS NWG SOL 50 NSW NTA QLD WAU 60 FIJ NWC SAM TON WAL 62 CRL MRN 63 HAW 72 ONT 74 ILL IOW KAN MIN MSO NEB OKL WIS 75 CNT INI MAS MIC NWH NWJ NWY OHI PEN RHO VER WVA 77 TEX 78 ALA ARK DEL FLA GEO KTY LOU MRY MSI NCA SCA TEN VRG WDC 79+ MXC MXE MXG MXN MXS MXT 80 BLZ COS ELS GUA HON NIC PAN 81 BAH CAY CUB DOM HAI JAM LEE PUE TRT VNA WIN 82 FRG GUY SUR VEN 83 BOL CLM ECU GAL PER 84+ BZC BZE BZL BZN BZS 85 AGE PAR URU.
263 Species
Diaphora Lour., Fl. Cochinch.: 578 (1790).
Catagyna P.Beauv. ex Lestib., Essai Cypér.: 26 (1819).
Cylindropus Nees, Linnaea 9: 303 (1834).
Hypoporum Nees, Linnaea 9: 202 (1834).
Anerma Schrad. ex Nees in C.F.P.von Martius & auct. suc. (eds.), Fl. Bras. 2(1): 170 (1842).
Chondrolomia Nees in C.F.P.von Martius & auct. suc. (eds.), Fl. Bras. 2(1): 173 (1842).
Cryptopodium Schrad. ex Nees in C.F.P.von Martius & auct. suc. (eds.), Fl. Bras. 2(1): 180 (1842), nom. inval.
Hymenolytrum Schrad. ex Nees in C.F.P.von Martius & auct. suc. (eds.), Fl. Bras. 2(1): 174 (1842).
Macrolomia Schrad. ex Nees in C.F.P.von Martius & auct. suc. (eds.), Fl. Bras. 2(1): 181 (1842).
Mastigloscleria Nees in C.F.P.von Martius & auct. suc. (eds.), Fl. Bras. 2(1): 177 (1842).
Omoscleria Nees in C.F.P.von Martius & auct. suc. (eds.), Fl. Bras. 2(1): 180 (1842).
Ophryoscleria Nees in C.F.P.von Martius & auct. suc. (eds.), Fl. Bras. 2(1): 182 (1842).
Schizolepis Schrad. ex Nees in C.F.P.von Martius & auct. suc. (eds.), Fl. Bras. 2(1): 186 (1842).
Trachylomia Nees in C.F.P.von Martius & auct. suc. (eds.), Fl. Bras. 2(1): 174 (1842).
Diploscyphus Liebm., Kongel. Danske Vidensk. Selsk. Skr., Naturvidensk. Math. Afd., V, 2: 262 (1850).
Sphaeropus Boeck., Flora 56: 89 (1873).

Acriulus Ridl., J. Linn. Soc., Bot. 20: 336 (1883).
Tonduzia Boeck. ex Tonduz, Bull. Herb. Boissier 3: 464 (1895).
Durandia Boeck., Allg. Bot. Z. Syst. 2: 160 (1896).

Scleria abortiva Nees ex Kunth, Enum. Pl. 2: 346 (1837).
Comoros, Madagascar. 29 COM MDG. Hemicr. or rhizome geophyte.
Schizolepis abortiva Nees in C.F.P.von Martius & auct. suc. (eds.), Fl. Bras. 2(1): 186 (1842).
Scleria abortiva var. *planifolia* Cherm., Bull. Soc. Bot. France 70: 299 (1923).

Scleria acanthocarpa Boeck., Vidensk. Meddel. Naturhist. Foren. Kjøbenhavn 1869: 154 (1869).
French Guiana to Brazil. 82 FRG 84+. Hemicr. or rhizome geophyte.
Scleria glazioviana Boeck., Flora 65: 352 (1882).

Scleria achtenii De Wild., Rev. Zool. Bot. Africaines 14(Suppl. Bot.): 16 (1926).
Trop. & S. Africa. 22 BEN GHA GNB GUI IVO LBR NGA SEN SIE TOG 23 GAB ZAI 25 TAN UGA 27 NAT. Hemicr. or rhizome geophyte.
Scleria substriatoalveolata De Wild., Rev. Zool. Bot. Africaines 14(Suppl. Bot.): 23 (1926).
Scleria subintegriloba De Wild., Pl. Bequaert. 4: 238 (1927). *Scleria achtenii* var. *subintegriloba* (De Wild.) Piérart, Lejeunia Mém. 13: 47 (1951 publ. 1953).

Scleria adpressohirta (Kük.) E.A.Rob., Kirkia 3: 10 (1962).
Tanzania to Zambia. 25 TAN 26 ZAM. Hemicr. or tuber geophyte.
**Scleria canaliculatotriquetra* var. *adpressohirta* Kük., Repert. Spec. Nov. Regni Veg. Beih. 40(1): 533 (1938).

Scleria afroreflexa Lye, Nordic J. Bot. 23: 431 (2005).
Cameroon. 23 CMN. Hemicr. or rhizome geophyte.

Scleria alpina Core, Brittonia 1: 241 (1934).
Brazil (Espírito Santo). 84 BZL. Hemicr. or rhizome geophyte.

Scleria amazonica Camelb., M.T.Strong & Goetgh., Novon 7: 98 (1997).
Venezuela (Amazonas). 82 VEN. Hemicr. or rhizome geophyte.

Scleria anceps Liebm., Mexic. Neldeagt. Pl., V, 2: 257 (1851).
SE. Mexico to S. Trop. America. 79 MXG MXT 80 COS GUA NIC PAN 82 GUY SUR 83 BOL. Hemicr. or rhizome geophyte.

Scleria andringitrensis Cherm., Bull. Soc. Bot. France 70: 297 (1923).
C. Madagascar. 29 MDG. Hemicr. or rhizome geophyte.

Scleria angusta Nees ex Kunth, Enum. Pl. 2: 346 (1837).
S. Africa, Madagascar. 27 NAT 29 MDG. Hemicr. or rhizome geophyte.
Scleria angustata Steud., Nomencl. Bot., ed. 2, 2: 542 (1841).

Scleria angustifolia E.A.Rob., Kirkia 3: 8 (1962).
Zambia. 26 ZAM. Hemicr. or rhizome geophyte.

Scleria annularis Steud., Syn. Pl. Glumac. 2: 176 (1855).
Indian Subcontinent to SC. China, New Guinea. 36 CHC 40 IND WHM 43 NWG. Ther.
Hypoporum annulare Nees, Linnaea 9: 303 (1834), nom. nud.

Scleria anomala (Steud.) J.Raynal, Adansonia, n.s., 17: 38 (1977).
Mauritius. 29 MAU. Hemicr.
**Hypolytrum anomalum* Steud., Syn. Pl. Glumac. 2: 133 (1855).

Scleria arcuata E.A.Rob., Kew Bull. 18: 541 (1966).
Zambia. 26 ZAM. Hemicr. or tuber geophyte.

Scleria arguta (Nees) Steud., Syn. Pl. Glumac. 2: 172 (1855).
Bolivia to Brazil. 83 BOL 84 BZL. Hemicr. or rhizome geophyte.
**Schizolepis arguta* Nees in C.F.P.von Martius & auct. suc. (eds.), Fl. Bras. 2(1): 189 (1842).

Scleria aromatica Core, Brittonia 1: 239 (1934).
Brazil. 84 BZC BZL. Hemicr. or rhizome geophyte.

Scleria assamica (C.B.Clarke) D.M.Verma, Bull. Bot. Surv. India 32: 71 (1990 publ. 1992).
Assam. 40 ASS. Hemicr. or rhizome geophyte.
**Scleria bracteata* var. *assamica* C.B.Clarke in J.D.Hooker, Fl. Brit. India 6: 694 (1894).

Scleria atroglumis D.A.Simpson, Kew Bull. 48: 709 (1993).
Brazil (Bahia to Minas Gerais). 84 BZE BZL. Hemicr. or rhizome geophyte.

Scleria balansae Maury ex Micheli, Denkschr. Schweiz. Naturf. Ges. 31(1): 148 (1890). *Scleria sellowiana* var. *balansae* (Maury ex Micheli) C.B.Clarke, Bull. Herb. Boissier, II, 3: 1028 (1903).
Bolivia to Uruguay. 83 BOL 84+ 85 AGE PAR URU. Hemicr. or rhizome geophyte.
Scleria sellowiana f. *glabriuscula* Osten in ?

Scleria baldwinii (Torr.) Steud., Syn. Pl. Glumac. 2: 175 (1855).
SE. U.S.A. to Texas, Cuba. 77 TEX 78 ALA FLA GEO LOU MSI SCA 81 BAH CUB. Hemicr. or rhizome geophyte.
**Hypoporum baldwinii* Torr., Ann. Lyceum Nat. Hist. New York 3: 382 (1836).
Scleria baldwinii var. *costata* Britton, Ann. New York Acad. Sci. 3: 229 (1885). *Scleria costata* (Britton) Small, Man. S.E. Fl.: 191 (1933).
Scleria dioica Leconte ex Britton, Ann. New York Acad. Sci. 3: 229 (1885).

Scleria bambariensis Cherm., Arch. Bot. Bull. Mens. 4(7): 48 (1931).
Trop. Africa, Madagascar. 23 CAF 25 KEN TAN 26 ZAM 29 MDG. Ther.

Scleria baroni-clarkei De Wild., Bull. Acad. Roy. Sci. Belgique, V, 13(10): 703 (1927).
WC. Trop. Africa. 23 GAB ZAI. Hemicr. or rhizome geophyte.
**Scleria clarkei* De Wild., Bull. Soc. Roy. Bot. Belgique 59: 130 (1927), nom. illeg.

Scleria baronii C.B.Clarke ex Cherm., Bull. Soc. Bot. France 70: 298 (1923).
Madagascar. 29 MDG. Hemicr. or rhizome geophyte.

Scleria benthamii C.B.Clarke, Kew Bull., Addit. Ser. 8: 58 (1908).
Trop. Asia to Queensland. 40 ASS EHM 41 THA VIE 42 PHI 43 NWG 50 QLD. Hemicr.
**Scleria tessellata* Benth., Fl. Hongk.: 399 (1861), nom. illeg.
Scleria khasiana C.B.Clarke ex Hook.f., Fl. Brit. India 6: 692 (1891), nom. illeg.

Scleria bequaertii De Wild., Rev. Zool. Bot. Africaines 14(Suppl. Bot.): 17 (1926).
Zaïre to Angola. 23 ZAI 26 ANG ZAM. Hemicr. or tuber geophyte.
Scleria bequaertii var. *laevis* Piérart, Lejeunia Mém. 13: 39 (1951 publ. 1953).

Scleria bicolor Nelmes, Kew Bull. 10: 444 (1955).
Angola. 26 ANG. Ther.
**Scleria remota* var. *hispida* Ridl., Trans. Linn. Soc. London, Bot. 2: 169 (1884).

Scleria biflora Roxb., Fl. Ind. ed. 1832, 3: 573 (1832).
Trop. & Subtrop. Asia. 36 CHH 38 NNS TAI 40 ASS IND NEP SRL WHM 41 MYA THA VIE 42 BOR JAW MLY PHI SUL SUM. Ther.
Scleria propinqua Steud., Syn. Pl. Glumac. 2: 169 (1855).
Scleria steudeliana Miq., Fl. Ned. Ind. 3: 344 (1856).
Scleria ferruginea Ohwi, Acta Phytotax. Geobot. 7: 37 (1938), nom. illeg. *Scleria biflora* subsp. *ferruginea* J.Kern, Reinwardtia 6: 76 (1961).

Scleria boivinii Steud., Syn. Pl. Glumac. 2: 173 (1855).
Trop. Africa, Comoros, Madagascar. 22 BEN GHA GNB GUI IVO LBR NGA SEN SIE TOG 23 CAF CMN EQG GAB GGI 25 TAN UGA 26 ANG 29 COM MDG. (Cl.) hemicr.
Scleria reflexa Benth. in W.J.Hooker, Niger Fl.: 555 (1849), nom. illeg.
Scleria barteri Boeck., Linnaea 38: 504 (1874).

Scleria boniana Boeck., Allg. Bot. Z. Syst. 2: 158 (1896).
Vietnam. 41 VIE. Hemicr. or rhizome geophyte.

Scleria borii D.M.Verma, Bull. Bot. Surv. India 29: 14 (1987 publ. 1989).
Assam. 40 ASS. Hemicr. or rhizome geophyte.

Scleria bourgeaui Boeck., Flora 65: 29 (1882).
Mexico. (78) ala 79 MXE MXG MXS MXT. Hemicr. or rhizome geophyte.
Scleria graminifolia Vasey, Bull. Torrey Bot. Club 15: 104 (1888).

Scleria bracteata Cav., Icon. 5: 34 (1799). *Macrolomia bracteata* (Cav.) Schrad. ex Nees in C.F.P.von Martius & auct. suc. (eds.), Fl. Bras. 2(1): 182 (1842).
Mexico to S. Trop. America, Trinidad. 79 MXG MXS MXT 80 BLZ COS GUA HON NIC PAN 81 TRT 82 FRG GUY SUR VEN 83 BOL CLM ECU PER 84 BZC BZE BZL BZN 85 PAR. Cl. cham.
Scleria floribunda Kunth in F.W.H.von Humboldt, A.J.A.Bonpland & C.S.Kunth, Nov. Gen. Sp. 1: 233 (1816). *Scleria bracteata* var. *floribunda* (Kunth) H.Pfeiff., Revista Sudamer. Bot. 5: 173 (1938).
Scleria papillata Willd. ex Kunth, Enum. Pl. 2: 345 (1837).
Scleria bracteata f. *simplicior* Kük., Repert. Spec. Nov. Regni Veg. 26: 253 (1929).
Scleria bracteata f. *supra-gynaecea* H.Pfeiff., Revista Sudamer. Bot. 5: 174 (1938).

Scleria bradei R.Gross, Repert. Spec. Nov. Regni Veg. 42: 173 (1937).
Brazil (Rio de Janeiro). 84 BZL. Hemicr. or rhizome geophyte.

Scleria brownii Kunth, Enum. Pl. 2: 349 (1837). *Scleria brownii* var. *typica* Domin, Biblioth. Bot. 85: 487 (1915), nom. inval.
N. Australia to SW. Pacific. 43 NWG 50 NTA QLD WAU 60 NWC TON. Hemicr.
**Scleria distans* R.Br., Prodr.: 240 (1810), nom. illeg.
Scleria brownii var. *glabra* Domin, Biblioth. Bot. 85: 487 (1915), nom. inval.
Scleria pallidiflora Boeck., Flora 58: 119 (1875).
Scleria setosoasperula Boeck., Flora 58: 120 (1875).

Scleria buekiana Schltdl., Bot. Zeitung (Berlin) 3: 291 (1845).
Brazil. 84+. Hemicr. or rhizome geophyte.

Scleria buettneri Boeck., Beitr. Cyper. 1: 36 (1888).
Gabon. 23 GAB. Hemicr. or rhizome geophyte.

Scleria bulbifera Hochst. ex A.Rich., Tent. Fl. Abyss. 2: 510 (1850).
Trop. & S. Africa, Madagascar, SW. Arabian Pen. 22 BEN BKN GHA GUI IVO MLI NGA SEN SIE TOG 23 BUR CAF CMN ZAI 24 CHA ETH SUD 25 KEN TAN UGA 26 ANG MLW MOZ ZAM ZIM 27 CPP NAT SWZ TVL 29 MDG 35 YEM. Hemicr. or rhizome geophyte.
Scleria atrosanguinea Hochst. ex Steud., Syn. Pl. Glumac. 2: 175 (1855).
Scleria schweinfurthiana Boeck., Flora 62: 570 (1879). *Scleria bulbifera* var. *schweinfurthiana* (Boeck.) Piérart, Lejeunia Mém. 13: 26 (1951 publ. 1953).
Scleria mechowiana Boeck., Bot. Jahrb. Syst. 5: 510 (1884). *Scleria bulbifera* var. *mechowiana* (Boeck.) Kük., Repert. Spec. Nov. Regni Veg. Beih. 40(1): 530 (1938).
Scleria buchananii Boeck., Beitr. Cyper. 1: 33 (1888).
Scleria cenchroides Hochst. ex C.B.Clarke in T.A.Durand & H.Schinz, Consp. Fl. Afric. 5: 669 (1894), nom. inval.
Scleria buchananii var. *latifolia* De Wild., Contr. Fl. Katanga: 20 (1919). *Scleria bulbifera* var. *latifolia* (De Wild.) Piérart, Lejeunia Mém. 13: 26 (1951 publ. 1953).
Scleria verdickii De Wild., Rev. Zool. Bot. Africaines 14(Suppl. Bot.): 26 (1926).
Scleria buchananii var. *laevinux* R.Gross, Notizbl. Bot. Gart. Berlin-Dahlem 11: 656 (1932).
Scleria schliebenii R.Gross, Notizbl. Bot. Gart. Berlin-Dahlem 11: 657 (1932).
Scleria schliebenii var. *ferruginea* Peter, Repert. Spec. Nov. Regni Veg. Beih. 40(1): 142 (1937).
Scleria thomasii Piérart, Bull. Soc. Roy. Bot. Belgique 83: 405 (1951).

Scleria burchellii C.B.Clarke, Kew Bull., Addit. Ser. 8: 56 (1908).
Brazil. 84 BZC BZL. Ther.

Scleria calcicola E.A.Rob., Kew Bull. 18: 496 (1966).
Zambia. 26 ZAM. Ther.

Scleria camaratensis Core, Mem. New York Bot. Gard. 12(3): 63 (1965).
Venezuela (Bolívar). 82 VEN. Hemicr. or rhizome geophyte.

Scleria canescens Boeck., Beitr. Cyper. 1: 37 (1888).
Puerto Rico. 81 PUE. Hemicr.

Scleria carphiformis Ridl., J. Fed. Malay States Mus. 6: 194 (1915).
Vietnam to N. Queensland. 41 VIE 42 BOR MLY SUL 50 QLD. Hemicr.

Scleria castanea Core, Brittonia 1: 239 (1934).
Peru (Cusco) to Bolivia. 83 BOL PER. Hemicr. or rhizome geophyte.

Scleria catharinensis Boeck., Allg. Bot. Z. Syst. 2: 158 (1896).

S. Brazil. 84 BZS. Hemicr. or rhizome geophyte.

Scleria catophylla C.B.Clarke in T.A.Durand & H.Schinz, Consp. Fl. Afric. 5: 670 (1894).
Trop. & S. Africa. 22 BEN GHA GNB GUI IVO LBR NGA SEN SIE TOG 23 BUR CAF CMN EQG GAB ZAI 24 SUD 25 KEN TAN UGA 26 ANG MLW ZAM 27 CPP NAT TVL. Hemicr.
**Scleria hirtella* var. *aterrima* Ridl., Trans. Linn. Soc. London, Bot. 2: 166 (1884). *Scleria aterrima* (Ridl.) Napper, Kew Bull. 25: 445 (1971), nom. illeg.
Scleria catophylla var. *chondrocarpa* Nelmes, Kew Bull. 10: 451 (1955).

Scleria chevalieri J.Raynal, Adansonia, n.s., 4: 150 (1964).
Senegal. 22 SEN. Hemicr.

Scleria chlorocalyx E.A.Rob., Kirkia 2: 184 (1961).
Zambia. 26 ZAM. Ther.

Scleria ciliaris Nees in R.Wight, Contr. Bot. India: 117 (1834).
Trop. & Subtrop. Asia to Caroline Is. 36 CHS 41 MYA THA VIE 42 BOR JAW MLY MOL PHI SUL SUM 43 BIS NWG SOL 50 NTA QLD 62 CRL. Hemicr.
Scleria chinensis Kunth, Enum. Pl. 2: 357 (1837).
Scleria scrobiculata Moritzi, Syst. Verz.: 98 (1846), nom. illeg.
Scleria bancana Miq., Fl. Ned. Ind., Eerste Bijv.: 602 (1861).
Scleria malaccensis Boeck., Linnaea 38: 507 (1874).
Scleria bancana var. *nana* Ridl., J. Straits Branch Roy. Asiat. Soc. 59: 225 (1911).

Scleria ciliata Michx., Fl. Bor.-Amer. 2: 167 (1803).
C. & SE. U.S.A. to C. America, Caribbean, Brazil to NE. Argentina. 74 MSO OKL 77 TEX 78 ALA FLA GEO LOU MSI NCA SCA TEN VRG 79 MXC MXE MXS MXT 80 BLZ COS GUA HON NIC 81 CUB DOM HAI JAM PUE 84+ 85 AGE PAR. Hemicr. or rhizome geophyte.

var. ***ciliata***
C. & SE. U.S.A. to C. America, Caribbean, Brazil to NE. Argentina. 74 MSO OKL 77 TEX 78 ALA FLA GEO LOU MSI SCA TEN VRG 79 MXC MXS 80 BLZ COS GUA HON NIC 81 CUB DOM HAI JAM PUE 84+ 85 AGE PAR. Hemicr. or rhizome geophyte.
Scleria hirtella Michx., Fl. Bor.-Amer. 2: 168 (1803), nom. illeg.
Acorus brasiliensius Schott ex Nees in C.F.P.von Martius & auct. suc. (eds.), Fl. Bras. 2(1): 170 (1842).
Scleria coriacea Liebm., Mexic. Neldeagt. Pl., V, 2: 259 (1851).
Scleria macrantha Boeck., Flora 41: 647 (1858).
Scleria ciliata var. *glabrescens* Maury, Mém. Soc. Phys. Genève 31(1; 1): 149 (1889).
Scleria ciliata var. *pauciflora* Kük., Repert. Spec. Nov. Regni Veg. 23: 215 (1926).

var. ***elliottii*** (Chapm.) Fernald, Rhodora 39: 392 (1937).
SE. U.S.A. to Texas, Cuba. 77 TEX 78 ALA FLA GEO LOU MSI NCA SCA VRG 81 CUB. Hemicr. or rhizome geophyte.
**Scleria elliottii* Chapm., Fl. South. U.S.: 531 (1860).

var. ***glabra*** (Chapm.) Fairey, Castanea 32: 48 (1967).
SE. U.S.A. to Texas, Cuba, Puerto Rico. 77 TEX 78 ALA FLA GEO LOU MSI NCA SCA 81 CUB PUE. Hemicr. or rhizome geophyte.
**Scleria pauciflora* var. *glabra* Chapm., Fl. South. U.S.: 532 (1860). *Scleria glabra* (Chapm.) Britton in J.K.Small, Fl. S.E. U.S.: 200 (1903), nom. illeg. *Scleria brittonii* Core ex Small, Man. S.E. Fl.: 190 (1933).

Scleria clarkei Lindm., Bih. Kongl. Svenska Vetensk.-Akad. Handl. 26(9): 32 (1900).
Brazil (Mato Grosso). 84 BZC. Hemicr. or rhizome geophyte.

Scleria clathrata Hochst. ex A.Rich., Tent. Fl. Abyss. 2: 510 (1850).
Ethiopia to Mozambique. 24 ETH 25 KEN TAN 26 MOZ ZAM. Ther.

Scleria colorata Core, Brittonia 1: 242 (1934).
S. Brazil. 84 BZS. Hemicr. or rhizome geophyte.

Scleria comosa (Nees) Steud., Syn. Pl. Glumac. 2: 171 (1855).
Bolivia to Brazil. 83 BOL 84 BZC BZN. Hemicr. or rhizome geophyte.
**Hymenolytrum comosum* Nees in C.F.P.von Martius & auct. suc. (eds.), Fl. Bras. 2(1): 174 (1842).

Scleria complanata Boeck., Flora 62: 571 (1879).
Sudan. 24 SUD. Hemicr. or rhizome geophyte.

Scleria composita (Nees) Boeck., Linnaea 38: 444 (1874).
Brazil to NE. Argentina. 83 BOL 84 BZC BZS 85 AGE PAR. Hemicr. or rhizome geophyte.
**Hypoporum compositum* Nees in C.F.P.von Martius & auct. suc. (eds.), Fl. Bras. 2(1): 171 (1842).
Scleria scabrosa Maury ex Micheli, Denkschr. Schweiz. Naturf. Ges. 31(1): 148 (1890). *Scleria composita* var. *scabrosa* (Maury ex Micheli) H.Pfeiff., Revista Sudamer. Bot. 10: 216 (1956).

Scleria corymbosa Roxb., Fl. Ind. ed. 1832, 3: 574 (1832).
SE. China to Trop. Asia. 36 CHS 40 ASS IND SRL 41 THA VIE 42 JAW MLY PHI SUM. Hemicr.
Scleria major Moon, Cat. Pl. Ceylon: 62 (1824), nom. nud.
Scleria androgyna Nees in R.Wight, Contr. Bot. India: 117 (1834).
Scleria corymbifera Boeck., Linnaea 38: 537 (1874).
Scleria ridleyi C.B.Clarke in J.D.Hooker, Fl. Brit. India 6: 686 (1894).

Scleria cuyabensis Pilg., Bot. Jahrb. Syst. 30: 144 (1901).
Brazil (Mato Grosso). 84 BZC. Ther.

Scleria cyathophora Holttum, Gard. Bull. Singapore 11: 294 (1947).
W. Malesia. 42 BOR MLY SUM. Hemicr.

Scleria cyperina Kunth, Enum. Pl. 2: 345 (1837). *Hymenolytrum cyperinum* (Kunth) Nees in C.F.P.von Martius & auct. suc. (eds.), Fl. Bras. 2(1): 174 (1842).
C. & S. Trop. America. 80 HON 82 FRG GUY SUR VEN 83 CLM CLM PER 84 BZN. Hemicr. or rhizome geophyte.
Scleria cyperina var. *dentata* H.Pfeiff., Repert. Spec. Nov. Regni Veg. 52: 169 (1943).

Scleria delicatula Nelmes, Kew Bull. 10: 448 (1955).
Tanzania to Zambia. 25 TAN 26 ZAM. Ther.
Scleria spondylogona Nelmes, Kew Bull. 10: 448 (1955).

Scleria densispicata (C.B.Clarke) J.Kern, Blumea 11: 159 (1961).
Philippines (Luzon). 42 PHI. Hemicr.
**Scleria motleyi* var. *densispicata* C.B.Clarke, Philipp. J. Sci. 1: 134 (1923).

Scleria depauperata Boeck., Flora 58: 118 (1875).
Samoa. 60 SAM. Hemicr. or rhizome geophyte.

Scleria depressa (C.B.Clarke) Nelmes, Amer. J. Bot. 39: 392 (1952).
W. Trop. Africa to Cameroon. 22 BEN BKN GAM GHA GNB GUI IVO LBR MLI MTN NGA SEN SIE TOG 23 CMN. Hemicr.
**Scleria racemosa* var. *depressa* C.B.Clarke in D.Oliver & auct. suc. (eds.), Fl. Trop. Afr. 8: 508 (1902). *Scleria racemosa* subsp. *depressa* (C.B.Clarke) J.Raynal, Adansonia, n.s., 4: 153 (1964).

Scleria distans Poir. in J.B.A.M.de Lamarck, Encycl. 7: 4 (1806).
Trop. & Subtrop. America, Trop. & S. Africa, Madagascar. 22 NGA TOG 23 CMN GAB RWA ZAI 24 SUD? 25 KEN TAN UGA 27 BOT CPP NAT TVL 29 MDG 77 TEX 78 ALA FLA GEO LOU MSI NCA 79 MXG MXS MXT 80 BLZ COS GUA HON NIC PAN 81 CUB DOM JAM PUE TRT WIN 82 FRG VEN 83 BOL CLM ECU GAL PER 84 BZC BZN 85 AGE PAR URU. Hemicr. or rhizome geophyte.

var. ***chondrocarpa*** (Nelmes) Lye, Nordic J. Bot. 3: 243 (1983).
Rwanda to E. Trop. Africa. 23 RWA 25 TAN UGA. Hemicr. or rhizome geophyte.
**Scleria hirtella* var. *chondrocarpa* Nelmes, Kew Bull. 10: 451 (1955).

var. ***distans***
Trop. & Subtrop. America, Trop. & S. Africa, Madagascar. 22 NGA TOG 23 CMN GAB RWA ZAI 25 KEN TAN UGA 27 BOT CPP NAT TVL 29 MDG 77 TEX 78 ALA FLA GEO LOU MSI NCA 79 MXG MXS MXT 80 BLZ COS GUA HON NIC PAN 81 CUB DOM JAM PUE TRT WIN 82 FRG VEN 83 BOL CLM ECU GAL PER 84 BZC BZN 85 AGE PAR URU. Hemicr. or rhizome geophyte.
Hypoporum humile Nees, Linnaea 9: 303 (1834), nom. inval.
Scleria nutans Willd. ex Kunth, Enum. Pl. 2: 351 (1837). *Hypoporum nutans* (Willd. ex Kunth) Nees in C.F.P.von Martius & auct. suc. (eds.), Fl. Bras. 2(1): 170 (1842).
Scleria tenella Griseb., Cat. Pl. Cub.: 249 (1866), nom. illeg.
Scleria hirtella Boeck., Linnaea 38: 439 (1874), nom. illeg.

var. ***glomerulata*** (Oliv.) Lye, Nordic J. Bot. 3: 243 (1983).
Uganda (Madi). 24 SUD? 25 UGA. Ther. or hemicr.
**Scleria glomerulata* Oliv., Trans. Linn. Soc. London 29: 170 (1875).

Scleria dregeana Kunth, Enum. Pl. 2: 354 (1837).
Uganda to S. Africa. 23 ZAI 25 TAN UGA 26 ANG MLW ZAM ZIM 27 BOT CPP LES NAT SWZ TVL. Hemicr. or rhizome geophyte.
Scleria holcoides Kunth, Enum. Pl. 2: 354 (1837).
Scleria meyeriana Kunth, Enum. Pl. 2: 354 (1837).
Scleria caespitosa Ridl., Trans. Linn. Soc. London, Bot. 2: 167 (1884).
Scleria setulosa Boeck., Beitr. Cyper. 1: 33 (1888).

Scleria dulungensis P.C.Li, Acta Bot. Yunnan. 12: 12 (1990).
China. 36 CHC. Hemicr. or rhizome geophyte.

Scleria eggersiana Boeck., Beitr. Cyper. 2: 41 (1890). *Scleria mitis* subsp. *eggersiana* (Boeck.) T.Koyama, Acta Amazon. 14: 112 (1984 publ. 1986).
Trop. America. 79 MXG MXT 80 BLZ COS GUA NIC PAN 81 CUB JAM LEE PUE TRT WIN 82 GUY SUR VEN 83 CLM ECU. Hemicr. or rhizome geophyte.
Scleria microcarpa var. *latifolia* Boeck., Linnaea 38: 517 (1874).
Scleria grisebachii C.B.Clarke in I.Urban, Symb. Antill. 2: 150 (1900).

Scleria elongatissima Piérart, Lejeunia Mém. 13: 68 (1953).
Zaïre. 23 ZAI. Hemicr. or rhizome geophyte.
**Scleria elongata* Piérart, Lejeunia Mém. 13: 45 (1953), nom. illeg.

Scleria erythrorrhiza Ridl., Trans. Linn. Soc. London, Bot. 2: 167 (1884).
S. Tanzania to Angola. 23 ZAI 25 TAN 26 ANG ZAM. Hemicr. or rhizome geophyte.

Scleria fauriei Ohwi, Acta Phytotax. Geobot. 1: 78 (1932).
Taiwan. 38 TAI. Hemicr. or rhizome geophyte.

Scleria filiculmis Boeck., Allg. Bot. Z. Syst. 2: 158 (1896). *Scleria georgiana* var. *filiculmis* (Boeck.) H.Pfeiff., Revista Sudamer. Bot. 7: 130 (1943).
S. Brazil. 84 BZS. Hemicr. or rhizome geophyte.

Scleria flagellum-nigrorum P.J.Bergius, Kongl. Vetensk. Acad. Handl. 26: 144 (1765).
SE. Mexico to S. Trop. America. 79 MXG MXT 80 COS HON 82 GUY SUR VEN 83 BOL CLM ECU PER 84 BZC BZL BZN. Hemicr. or rhizome geophyte.
Schoenus lithospermus L., Sp. Pl. ed. 2: 65 (1762). *Carex lithosperma* (L.) L., Syst. Nat. ed. 12, 2: 706 (1767).
Scleria margaritifera Gaertn., Fruct. Sem. Pl. 1: 13 (1788).
Scleria flabellum Poir. in J.B.A.M.de Lamarck, Encycl. 7: 1 (1806), orth. var.
Schoenus moniliferus Sparrm. ex Spreng., Syst. Veg. 3: 830 (1826).
Scleria nigricans J.Presl & C.Presl, Reliq. Haenk. 1: 201 (1828).
Scleria guahensis Schrad. ex Boeck., Linnaea 38: 512 (1874).

Scleria flexuosa Boeck., Beitr. Cyper. 1: 33 (1888).
Trop. & S. Africa. 22 GUI IVO SIE 25 TAN 26 ANG MLW ZAM 27 CPP LES NAT TVL. Hemicr. or rhizome geophyte.
Scleria remota Ridl., Trans. Linn. Soc. London, Bot. 2: 169 (1884).
Scleria dieterlenii Turrill, Bull. Misc. Inform. Kew 1914: 20 (1914).
Scleria lateritica Nelmes, Kew Bull. 10: 432 (1955).

Scleria foliosa Hochst. ex A.Rich., Tent. Fl. Abyss. 2: 509 (1850).
Trop. & S. Africa, Madagascar, India. 22 BEN BKN GHA IVO NGA NGR SEN TOG 23 BUR EQG?

24 ETH 25 KEN TAN UGA 26 27 BOT NAM SWZ TVL 29 MDG 40 IND. Ther.

Scleria dillonia Boeck., Flora 61: 38 (1878).

Scleria dumicola Ridl., Trans. Linn. Soc. London, Bot. 2: 169 (1884).

Scleria perrieri Cherm., Bull. Soc. Bot. France 70: 297 (1923).

Scleria foveolata Cav., Icon. 5: 35 (1799).
Panama (Cerro Ancón). 80 PAN. Hemicr. or rhizome geophyte.

Scleria friesii Kük. in R.E.Fries, Wiss. Erg. Schwed. Rhod.-Kongo Exped. 1: 9 (1921).
Zambia. 26 ZAM. Hemicr. or rhizome geophyte.

Scleria fujianensis G.P.Li, Fl. Fujianica 6: 689 (1995).
China (Fujian). 36 CHS. Hemicr. or rhizome geophyte.

Scleria fulvipilosa E.A.Rob., Kew Bull. 18: 515 (1966).
N. Zambia. 26 ZAM. Hemicr.

Scleria georgiana Core, Brittonia 1: 243 (1934).
SE. U.S.A. to C. America, Caribbean. 77 TEX 78 ALA FLA GEO LOU MSI NCA SCA 80 BLZ NIC 81 CUB DOM JAM PUE. Hemicr. or rhizome geophyte.

**Scleria gracilis* Elliott, Sketch Bot. S. Carolina 2: 557 (1824), nom. illeg.

Scleria glabra Boeck., Beitr. Cyper. 1: 35 (1888).
S. Chad to Mozambique. 23 BUR ZAI 24 CHA 25 TAN 26 MLW MOZ ZAM. Ther.

var. ***glabra***
Burundi to Mozambique. 23 BUR ZAI 25 TAN 26 MLW MOZ ZAM. Ther.

var. ***pallidior*** J.Raynal, Adansonia, n.s., 7: 244 (1967).
S. Chad. 24 CHA. Ther.

Scleria globonux C.B.Clarke in T.A.Durand & H.Schinz, Consp. Fl. Afric. 5: 671 (1894).
W. Trop. Africa to Uganda. 22 GHA GNB NGA SEN SIE 24 SUD 25 UGA. Ther.

Scleria goossensii De Wild., Rev. Zool. Bot. Africaines 14(Suppl. Bot.): 20 (1926).
WC. Trop. Africa. 23 GAB ZAI. Hemicr. or rhizome geophyte.

Scleria gracillima Boeck., Flora 42: 570 (1879).
Trop. Africa. 22 BEN GHA IVO NGA SEN 23 CAF 24 CHA SUD 25 TAN 26 MLW ZAM. Ther.

Scleria griegiifolia (Ridl.) C.B.Clarke in D.Oliver & auct. suc. (eds.), Fl. Trop. Afr. 8: 509 (1902).
Uganda to S. Africa, Madagascar. 23 ZAI 25 TAN UGA 26 ANG MLW ZAM ZIM 27 NAT 29 MDG. Hemicr. or rhizome geophyte.

**Acriulus griegiifolius* Ridl., J. Linn. Soc., Bot. 20: 336 (1883).

Acriulus madagascariensis Ridl., J. Linn. Soc., Bot. 20: 336 (1883). *Scleria acriulus* C.B.Clarke in D. Oliver & auct. suc. (eds.), Fl. Trop. Afr. 8: 509 (1902).

Scleria guineensis J.Raynal, Adansonia, n.s., 4: 148 (1964).
Guinea. 22 GUI. Ther.

Scleria harlandii Hance in Benth., Ann. Sci. Nat., Bot., V, 5: 248 (1866).
SE. China to Vietnam. 36 CHS 41 VIE. Hemicr. or rhizome geophyte.

**Scleria purpurascens* Benth., Fl. Hongk.: 400 (1861), nom. illeg.

Scleria havanensis Britton, Bull. Torrey Bot. Club 42: 492 (1915).
Cuba to Puerto Rico. 81 CUB DOM HAI PUE. Hemicr. or rhizome geophyte.

Scleria hildebrandtii Boeck., Flora 63: 454 (1880).
E. Trop. Africa, Madagascar. 25 KEN TAN 29 MDG. Ther.

Scleria hilsenbergii Ridl., J. Bot. 22: 16 (1884).
C. Madagascar. 29 MDG. Hemicr. or rhizome geophyte.

Scleria hirta Boeck., Flora 65: 29 (1882).
Mexico (Veracruz). 79 MXG. Hemicr. or rhizome geophyte.

Scleria hirtella Sw., Prodr.: 19 (1788). *Carex hirtella* (Sw.) J.F.Gmel., Syst. Nat. 2: 138 (1791). *Hypoporum hirtellum* (Sw.) Nees, Linnaea 9: 303 (1834).
Trop. & Subtrop. America, W. & WC. Trop. Africa, Madagascar. 22 GHA GUI IVO LBR NGA SEN SIE 23 BUR CMN 29 MDG 77 TEX 78 ALA FLA GEO LOU MSI 79 MXG MXS MXT 80 BLZ GUA HON PAN 81 CUB JAM LEE PUE TRT VNA WIN 82 FRG GUY SUR VEN 83 BOL CLM ECU PER 84 BZE BZL BZN BZS. Ther.

Scleria interrupta Michx., Fl. Bor.-Amer. 2: 168 (1803), nom. illeg. *Hypoporum interruptum* Torr., Ann. Lyceum Nat. Hist. New York 3: 382 (1836), nom. illeg.

Scleria hirtella var. *glabra* C.Presl, Reliq. Haenk. 1: 203 (1828).

Scleria hirtella var. *pubescens* C.Presl, Reliq. Haenk. 1: 203 (1828).

Scleria enchroides Kunth, Enum. Pl. 2: 352 (1837).

Scleria mollis Kunth, Enum. Pl. 2: 352 (1837).

Scleria michauxii Chapm., Fl. South. U.S.: 522 (1860).

Scleria hirtella var. *pauciciliata* Britton, Ann. New York Acad. Sci. 3: 236 (1884).

Scleria humilis Britton, Ann. New York Acad. Sci. 3: 235 (1885).

Scleria hirtella var. *tuberculata* Boeck. ex C.B.Clarke in W.H.Harvey & auct. suc. (eds.), Fl. Cap. 7: 294 (1898).

Scleria lindleyana C.B.Clarke, Kew Bull., Addit. Ser. 8: 56 (1908).

Scleria doradoensis Britton, Bull. Torrey Bot. Club 50: 55 (1923).

Scleria hispidior (C.B.Clarke) Nelmes, Kew Bull. 10: 435 (1955).
Ethiopia to Uganda. 24 ETH 25 UGA. Ther.

**Scleria hispidula* var. *hispidior* C.B.Clarke in D. Oliver & auct. suc. (eds.), Fl. Trop. Afr. 8: 497 (1902).

Scleria hispidula Hochst. ex A.Rich., Tent. Fl. Abyss. 2: 511 (1850). *Anerma hispidula* (Hochst. ex A.Rich.) Schrad. ex Nees in C.F.P.von Martius & auct. suc. (eds.), Fl. Bras. 2(1): 170 (1842).
Eritrea to Zambia. 24 ERI ETH 25 TAN ZAM. Ther.

Scleria interrupta Schltdl., Linnaea 20: 544 (1847), nom. illeg.

Scleria huberi C.B.Clarke, Kew Bull., Addit. Ser. 8: 60 (1908).
W. South America to Brazil. 83 BOL CLM PER 84 BZN. Hemicr. or rhizome geophyte.

Scleria peruviana Kük., Bot. Jahrb. Syst. 56(125): 22 (1921).

Scleria induta Turrill, Bull. Misc. Inform. Kew 1914: 137 (1914).
WC. & S. Trop. Afica. 23 GAB ZAI 26 ANG ZAM. Hemicr. or rhizome geophyte.
Scleria ovuligera Rchb. ex Nees, Linnaea 9: 303 (1834), nom. inval.
Scleria angolensis Turrill, Bull. Misc. Inform. Kew 1914: 136 (1914).
Scleria duvigneaudii Piérart, Bull. Séances Inst. Roy. Colon. Belge 22: 351 (1951).

Scleria interrupta Rich., Actes Soc. Hist. Nat. Paris 1: 112 (1792). *Hypoporum interruptum* (Rich.) Nees, Linnaea 9: 303 (1834). *Scleria distans* var. *interrupta* (Rich.) Kük., Repert. Spec. Nov. Regni Veg. 23: 214 (1926).
SE. Mexico to Trop. America, Trop. Africa. 22 BEN GUI LBR MLI SIE 23 CAF GAB ZAI 25 KEN 26 ZAM 79 MXT 80 BLZ COS GUA HON NIC PAN 81 CUB DOM 82 FRG GUY SUR VEN 83 ECU. Ther.
Scleria melanotricha var. *glabrior* C.B.Clarke in D.Oliver & auct. suc. (eds.), Fl. Trop. Afr. 8: 496 (1902).
Scleria pinetorum Britton, Bull. Torrey Bot. Club 42: 492 (1915).
Scleria schweinfurthiana var. *melanocarpa* Cherm., Arch. Bot. Bull. Mens. 4(7): 46 (1931).
Scleria tricholepis Nelmes, Kew Bull. 10: 447 (1955).

Scleria iostephana Nelmes, Kew Bull. 11: 94 (1956).
Trop. Africa. 22 BEN GHA IVO NGA TOG 23 CAF EQG GAB ZAI 24 CHA 25 TAN UGA 26 ZAM. Hemicr. or bulb geophyte.

Scleria jiangchengensis Y.Y.Qian, Bull. Bot. Res., Harbin 19: 241 (1999).
China (Yunnan). 36 CHC. Hemicr. or rhizome geophyte.

Scleria junghuhniana Boeck., Linnaea 38: 499 (1874).
Indo-China, W. & C. Jawa. 41 CBD VIE 42 JAW. Hemicr. or rhizome geophyte.

Scleria kerrii Turrill, Bull. Misc. Inform. Kew 1910: 384 (1910).
Indo-China. 41 THA VIE. Hemicr. or rhizome geophyte.

Scleria khasiana Boeck., Beitr. Cyper. 2: 29 (1890).
Assam. 40 ASS. Hemicr. or rhizome geophyte.

Scleria killipiana Britton, Brittonia 1: 240 (1934).
Colombia. 83 CLM. Hemicr. or rhizome geophyte.
Scleria hirtella var. *killipiana* (Britton ex Core) H.Pfeiff., Feddes Repert. Spec. Nov. Regni Veg. 55: 116 (1953).

Scleria kindtiana Graebn., Repert. Spec. Nov. Regni Veg. 16: 24 (1919).
Angola. 26 ANG. Hemicr. or rhizome geophyte.

Scleria kwangtungensis Chun & F.C.How, Acta Phytotax. Sin. 7: 88 (1958).
SE. China to Hainan. 36 CHH CHS. Hemicr. or rhizome geophyte.

Scleria lacustris C.Wright, Anales Acad. Ci. Méd. Habana 8: 163 (1871).
Trop. America, Trop. & S. Africa, Madagascar. 22 GHA GNB NGR SEN SIE 23 GAB ZAI 26 ZAM 27 BOT 29 MDG (78) fla 80 COS 81 CUB JAM 82 FRG GUY SUR VEN 83 BOL 84+ 85 PAR. Ther.
Scleria aquatica Cherm., Bull. Soc. Bot. France 77: 279 (1930).

Scleria laeviformis Tang & F.T.Wang, in Fl. Reipubl. Popul. Sin. 11: 235 (1961).
S. China to Taiwan. 36 CHC 38 TAI. Hemicr. or rhizome geophyte.

Scleria lagoensis Boeck., Vidensk. Meddel. Naturhist. Foren. Kjøbenhavn 1869: 151 (1869).
Trop. & S. Africa, Comoros, Madagascar, S. Trop. America. 22 BEN GHA GUI IVO NGA SEN SIE TOG 23 GAB 24 CHA 25 TAN UGA 27 SWZ 29 COM MDG 82 VEN 83 BOL CLM PER 84 BZL. Hemicr. or rhizome geophyte.

subsp. ***canaliculatotriquetra*** (Boeck.) Lye, Nordic J. Bot. 3: 243 (1983).
Trop. & S. Africa, Comoros, Madagascar. 22 BEN GHA GUI IVO NGA SEN SIE TOG 23 GAB 24 CHA 25 TAN UGA 27 SWZ 29 COM MDG. Hemicr. or tuber geophyte.
**Scleria canaliculatotriquetra* Boeck., Flora 62: 573 (1879).
Scleria mayottensis C.B.Clarke, Kew Bull., Addit. Ser. 8: 92 (1908).

subsp. ***lagoensis***
S. Trop. America. 82 VEN 83 BOL CLM PER 84 BZL. Hemicr. or rhizome geophyte.
Scleria moritziana Boeck., Linnaea 38: 460 (1874).
Scleria diurensis Boeck., Flora 62: 573 (1879).
Scleria cervina Ridl., Trans. Linn. Soc. London, Bot. 2: 171 (1884).
Scleria vanderystii De Wild., Rev. Zool. Bot. Africaines 14(Suppl. Bot.): 25 (1926).

Scleria latifolia Sw., Prodr.: 18 (1788). *Carex latifolia* (Sw.) J.F.Gmel., Syst. Nat. 2: 138 (1791). *Schizolepis latifolia* (Sw.) Nees in C.F.P.von Martius & auct. suc. (eds.), Fl. Bras. 2(1): 186 (1842).
Trop. America. 79 MXT 80 BLZ COS GUA HON NIC PAN 81 LEE TRT WIN 82 FRG GUY SUR VEN 83 BOL CLM ECU PER 84 BZC BZL BZN 85 AGE PAR. Hemicr. or rhizome geophyte.
Scleria nervosa Wikstr., Kongl. Vetensk. Acad. Handl. 1827: 75 (1827).
Scleria latifolia Nees, Flora 11: 303 (1828), nom. illeg.
Scleria arundinacea Kunth, Enum. Pl. 2: 347 (1837). *Schizolepis arundinacea* (Kunth) Palla, Denkschr. Kaiserl. Akad. Wiss., Wien. Math.-Naturwiss. Kl. 79: 196 (1908). *Scleria latifolia* var. *arundinacea* (Kunth) H.Pfeiff., Repert. Spec. Nov. Regni Veg. 52: 171 (1943).
Scleria cyanocarpa Kunth, Enum. Pl. 2: 347 (1837).
Scleria olyroides Kunth, Enum. Pl. 2: 348 (1837). *Schizolepis olyroides* (Kunth) Nees in C.F.P.von Martius & auct. suc. (eds.), Fl. Bras. 2(1): 186 (1842).
Scleria sylvestris Poepp. & Kunth in C.S.Kunth, Enum. Pl. 2: 346 (1837). *Schizolepis sylvestris* (Poepp. & Kunth) Nees in C.F.P.von Martius & auct. suc. (eds.), Fl. Bras. 2(1): 223 (1842).
Schizolepis rufescens Schrad. ex Nees in C.F.P.von Martius & auct. suc. (eds.), Fl. Bras. 2(1): 186 (1842).
Schizolepis trigonocarpa Nees in C.F.P.von Martius & auct. suc. (eds.), Fl. Bras. 2(1): 186 (1842).
Schizolepis trigonocarpa var. *plena* Nees in C.F.P.von Martius & auct. suc. (eds.), Fl. Bras. 2(1): 186 (1842).
Scleria grandifolia Miq., Linnaea 19: 230 (1847).

Scleria kappleriana Hochst. ex Steud, Syn. Pl. Glumac. 2: 172 (1855).
Scleria trigonocarpa Steud., Syn. Pl. Glumac. 2: 171 (1855).
Scleria lacunosa Boeck., Allg. Bot. Z. Syst. 2: 160 (1896).

Scleria laxa R.Br., Prodr.: 240 (1810).
Philippines to N. Australia. 42 PHI 43 NWG 50 NTA QLD WAU. Ther.
Scleria laxa var. *pseudotesselata* Domin, Biblioth. Bot. 85: 489 (1915).
Scleria filipendula S.T.Blake, Proc. Roy. Soc. Queensland 58: 49 (1947).

Scleria laxiflora R.Gross, Notizbl. Bot. Gart. Berlin-Dahlem 11: 658 (1932).
S. Tanzania to Zambia. 23 ZAI 25 TAN 26 ZAM. Hemicr. or tuber geophyte.

Scleria leptostachya Kunth, Enum. Pl. 2: 354 (1837). *Hypoporum leptostachyum* (Kunth) Nees in C.F.P.von Martius & auct. suc. (eds.), Fl. Bras. 2(1): 172 (1842).
Brazil to Argentina (Misiones). 83 BOL 84 BZC BZL BZS 85 AGE PAR URU. Hemicr. or rhizome geophyte.

Scleria levis Retz., Observ. Bot. 4: 13 (1786). *Carex laevis* (Retz.) J.F.Gmel., Syst. Nat. 2: 138 (1791).
Trop. & Subtrop. Asia to W. Pacific. 36 CHS 38 JAP NNS OGA TAI 40 ASS EHM IND NEP SRL 41 THA VIE 42 BOR JAW LSI MLY MOL PHI SUL SUM 43 NWG 50 NSW QLD 60 NWC 62 CRL. Hemicr.
Juncus zeilanicus Houtt., Nat. Hist. 2(13): t. 93 (1783).
Scleria zeylanica Poir. in J.B.A.M.de Lamarck, Encycl. 7: 3 (1806).
Scleria alata Moon, Cat. Pl. Ceylon: 62 (1824).
Scleria zelanica Moon, Cat. Pl. Ceylon: 62 (1824).
Scleria villosula Wall., Numer. List: 3414 (1831), nom. inval.
Scleria hebecarpa Nees in R.Wight, Contr. Bot. India: 117 (1834).
Scleria japonica Steud., Syn. Pl. Glumac. 2: 169 (1855).
Scleria pubescens Steud., Syn. Pl. Glumac. 2: 168 (1855). *Scleria hebecarpa* var. *pubescens* (Steud.) C.B.Clarke in J.D.Hooker, Fl. Brit. India 6: 689 (1894). *Scleria levis* var. *pubescens* (Steud.) C.Z.Zheng, in Fl. Zhejiang 7: 286 (1993).
Scleria sumatrensis var. *pubescens* Miq., Fl. Ned. Ind. 3: 344 (1856).
Scleria alta Boeck., Linnaea 38: 485 (1874).
Scleria melanosperma Nees & Arn. ex Boeck., Linnaea 38: 515 (1874).
Scleria vestita Boeck., Linnaea 38: 482 (1874).
Scleria dietrichiae Boeck., Flora 58: 131 (1875).
Scleria wichurai Boeck., Bot. Jahrb. Syst. 5: 510 (1884).
Scleria bulusanensis Elmer, Leafl. Philipp. Bot. 10: 3540 (1938).

Scleria lingulata C.B.Clarke, Kew Bull., Addit. Ser. 8: 58 (1908).
N. Northern Territory. 50 NTA. Hemicr. or rhizome geophyte.

Scleria lithosperma (L.) Sw., Prodr.: 18 (1788).
Trop. & Subtrop. 22 BEN GHA IVO NGA TOG 23 ZAI 25 KEN TAN 26 MOZ 29 COM MDG 38 TAI 40 ASS EHM IND SRL 41 AND THA VIE 42 BOR JAW LSI MLY MOL PHI SUL SUM 43 NWG 50 NTA QLD 60 FIJ TON WAL 62 CRL MRN 78 FLA LOU 79 MXE MXG MXS MXT 80 COS GUA HON NIC PAN 81 CAY CUB DOM JAM LEE PUE TRT WIN 82 FRG VEN 83 BOL 84 BZC. Hemicr. or rhizome geophyte.
**Scirpus lithospermus* L., Sp. Pl.: 51 (1753). *Hypoporum lithospermum* (L.) Nees in R.Wight, Contr. Bot. India: 117 (1834). *Scleria wightiana* Steud., Syn. Pl. Glumac. 2: 176 (1855), nom. illeg.

var. ***linearis*** Benth., Fl. Austral. 7: 430 (1878). *Scleria roxburghii* var. *australiensis* Domin, Biblioth. Bot. 85: 487 (1915), nom. illeg. *Scleria lithosperma* subsp. *linearis* (Benth.) T.Koyama in M.D.Daddanayake & al. (eds.), Revised Handb. Fl. Ceylon 5: 353 (1985).
Trop. Asia to Queensland. 40 IND SRL 41 THA VIE 42 LSI MLY PHI 43 NWG 50 QLD. Hemicr.
Scleria lithosperma var. *roxburghii* C.B.Clarke in J.D.Hooker, Fl. Brit. India 6: 686 (1894). *Scleria roxburghii* (C.B.Clarke) Domin, Biblioth. Bot. 85: 487 (1915).
Scleria lithosperma var. *multispiculata* Govind., Proc. Indian Acad. Sci., B 71: 221 (1970).
Scleria lithosperma var. *muricata* Govind., Proc. Indian Acad. Sci., B 71: 222 (1970).

var. ***lithosperma***
Trop. & Subtrop. 22 BEN GHA IVO NGA TOG 23 ZAI 25 KEN TAN 26 MOZ 29 COM MDG 38 TAI 40 ASS EHM IND SRL 41 AND 42 BOR JAW LSI MLY MOL PHI SUL SUM 43 NWG 50 NTA QLD 60 FIJ TON WAL 62 CRL MRN 78 FLA LOU 79 MXE MXS MXT 80 COS GUA HON NIC PAN 81 CAY CUB DOM JAM LEE PUE TRT WIN 82 FRG VEN 83 BOL 84 BZC. Hemicr. or rhizome geophyte.
Scleria tenuis Retz., Observ. Bot. 4: 13 (1786). *Carex tenuis* (Retz.) J.F.Gmel., Syst. Nat. 2: 138 (1791).
Scleria filiformis Sw., Prodr.: 19 (1788). *Carex subulata* J.F.Gmel., Syst. Nat. 2: 138 (1791). *Scleria subulata* (J.F.Gmel.) Steud., Nomencl. Bot., ed. 2, 2: 296 (1841). *Scleria lithosperma* var. *filiformis* (Sw.) Britton, Ann. New York Acad. Sci. 3: 231 (1884).
Scleria gracilis Rich., Actes Soc. Hist. Nat. Paris 1: 113 (1792). *Hypoporum gracile* (Rich.) Torr., Ann. Lyceum Nat. Hist. New York 3: 381 (1836).
Scleria purpurea Poir. in J.B.A.M.de Lamarck, Encycl. 7: 4 (1806). *Hypoporum purpurascens* Nees, Linnaea 9: 303 (1834), nom. illeg.
Scleria capillaris R.Br., Prodr.: 240 (1810). *Hypoporum capillare* (R.Br.) Nees, Linnaea 9: 303 (1834).
Scleria elongata J.Presl & C.Presl, Reliq. Haenk. 1: 202 (1828).
Scleria glaucescens J.Presl & C.Presl, Reliq. Haenk. 1: 202 (1828).
Hypoporum sieberi Nees, Linnaea 9: 303 (1834), nom. inval.
Scleria pauciflora Sauvalle, Fl. Cub.: 183 (1871), nom. illeg.
Scleria krugiana Boeck., Beitr. Cyper. 1: 35 (1888).
Scleria puzzolanea K.Schum., Abh. Königl. Akad. Wiss. Berlin 1894: 21 (1894).
Scleria lithosperma f. *ramosa* Domin, Biblioth. Bot. 85: 487 (1915).

Scleria longigluma Kük., Bot. Jahrb. Syst. 56(125): 22 (1921).
Guyana to Argentina (Misiones). 82 GUY 84 BZC BZN 85 AGE PAR. Hemicr. or rhizome geophyte.

Scleria longispiculata Nelmes, Kew Bull. 13: 150 (1958).
Tanzania to Namibia. 25 TAN 26 ZAM 27 NAM. Hemicr. or rhizome geophyte.

Scleria lucentinigricans E.A.Rob., Kew Bull. 18: 530 (1966).
Zambia. 26 ZAM. Ther.

Scleria macbrideana R.Gross, Publ. Field Mus. Nat. Hist., Bot. Ser. 11: 40 (1931).
Costa Rica to Peru. 80 COS 83 ECU PER. Hemicr. or rhizome geophyte.

Scleria mackaviensis Boeck., Flora 58: 119 (1875).
E. Australia. 50 NSW QLD. Hemicr. or rhizome geophyte.
Scleria mackaviensis var. *atrata* Domin, Biblioth. Bot. 85: 488 (1915).
Scleria mackaviensis var. *pallida* Domin, Biblioth. Bot. 85: 488 (1915).

Scleria macrogyne C.B.Clarke, Kew Bull., Addit. Ser. 8: 59 (1908).
S. Trop. America. 82 FRG GUY SUR VEN 83 PER 84 BZN. Hemicr. or rhizome geophyte.

Scleria macrolomioides H.Pfeiff., Repert. Spec. Nov. Regni Veg. 33: 214 (1933).
Mexico. 79+. Hemicr. or rhizome geophyte.

Scleria macrophylla J.Presl & C.Presl, Reliq. Haenk. 1: 200 (1828).
Mexico to Trop. America. 79 MXG MXN MXS MXT 80 BLZ COS GUA HON NIC PAN 81 CUB TRT 82 FRG GUY SUR VEN 83 BOL CLM ECU PER 84 BZC BZE BZN. Hemicr. or rhizome geophyte.
Scleria paludosa Poepp. & Kunth in C.S.Kunth, Enum. Pl. 2: 344 (1837). *Ophryoscleria paludosa* (Poepp. & Kunth) Nees in C.F.P.von Martius & auct. suc. (eds.), Fl. Bras. 2(1): 185 (1842).
Scleria palmifolia Hoffmanns. & Schltdl., Bot. Zeitung (Berlin) 3: 492 (1845).
Ophryoscleria asperrima Liebm., Mexic. Neldeagt. Pl., V, 2: 261 (1850). *Scleria asperrima* (Liebm.) Steud., Syn. Pl. Glumac. 2: 170 (1855).
Scleria phyllomacra Steud., Syn. Pl. Glumac. 2: 173 (1855).
Scleria macrocarpa Salzm. ex Boeck., Linnaea 38: 521 (1874).

Scleria madagascariensis Boeck., Bot. Jahrb. Syst. 5: 514 (1884).
Madagascar. 29 MDG. Hemicr. or rhizome geophyte.

Scleria martii (Nees) Steud., Syn. Pl. Glumac. 2: 171 (1855).
S. Trop. America. 82 FRG SUR VEN 83 BOL ECU PER 84 BZL BZN. Hemicr. or rhizome geophyte.
**Hymenolytrum martii* Nees in C.F.P.von Martius & auct. suc. (eds.), Fl. Bras. 2(1): 174 (1842).
Scleria pyramidalis Hochst. ex Steud., Syn. Pl. Glumac. 2: 171 (1855).
Scleria cyperinoides C.B.Clarke, Kew Bull., Addit. Ser. 8: 61 (1908).

Scleria melaleuca Rchb. ex Schltdl. & Cham., Linnaea 6: 29 (1831). *Scleria pterota* var. *melaleuca* (Rchb. ex Schltdl. & Cham.) Uittien in A.A.Pulle, Fl. Suriname 1(1): 140 (1934).
Trop. America, Trop. Africa, Madagascar. 22 BEN BKN GHA LBR MLI NGA SEN SIE TOG 23 GAB GGI 24 CHA 25 TAN 29 MDG 79 MXG MXS MXT 80 BLZ COS ELS GUA HON NIC PAN 81 CUB JAM LEE PUE TRT WIN 82 FRG GUY SUR VEN 83 BOL CLM ECU GAL PER 84 BZC BZE BZL BZN 85 AGE PAR. Hemicr. or rhizome geophyte.
Schoenus latifolius Vahl, Enum. Pl. 2: 226 (1805). *Scleria latifolia* (Vahl) C.Presl, Isis (Oken) 21: 268 (1828), nom. illeg. *Dichromena vahlii* A.Dietr., Sp. Pl. 2: 169 (1832).
Scleria asperata C.Presl, Isis (Oken) 21: 268 (1828), nom. inval.
Scleria pterota C.Presl, Isis (Oken) 21: 268 (1828), nom. nud.
Scleria communis Kunth, Enum. Pl. 2: 340 (1837).
Scleria affinis C.Presl ex Steud., Nomencl. Bot., ed. 2, 2: 542 (1841).
Scleria conspersa Sellow ex Nees in C.F.P.von Martius & auct. suc. (eds.), Fl. Bras. 2(1): 179 (1842).
Scleria pratensis var. *melanocarpa* Nees in C.F.P.von Martius & auct. suc. (eds.), Fl. Bras. 2(1): 179 (1842).
Scleria selloana Schrad. ex Nees in C.F.P.von Martius & auct. suc. (eds.), Fl. Bras. 2(1): 179 (1842).
Scleria pratensis Lindl. ex Nees, Nov. Actorum Acad. Caes. Leop.-Carol. Nat. Cur. 19(Suppl. 1): 121 (1843).
Scleria simplicior Steud., Syn. Pl. Glumac. 2: 169 (1855).
Scleria flagellata Sw. ex Boeck., Linnaea 38: 506 (1874).
Scleria ottonis Boeck., Linnaea 38: 490 (1874).
Scleria longifolia Boeck., Flora 65: 30 (1882).
Scleria pittieri Boeck., Allg. Bot. Z. Syst. 2: 159 (1896).
Scleria boliviana Palla in O.Buchtien, Contr. Fl. Bolivia: 90 (1910), nom. nud.
Scleria congolensis De Wild., Rev. Zool. Bot. Africaines 14(Suppl. Bot.): 19 (1926).
Scleria pterota var. *submelaleuca* Kük., Repert. Spec. Nov. Regni Veg. 23: 216 (1926).

Scleria melanomphala Kunth, Enum. Pl. 2: 345 (1837).
Trop. & S. Africa, Madagascar. 22 GHA GNB GUI IVO LBR NGA SIE TOG 23 CAF CMN EQG RWA ZAI 24 CHA 25 KEN TAN UGA 27 BOT CPP NAT SWZ TVL 29 MDG. Hemicr. or rhizome geophyte.
Scleria melanocephala Drège, Zwei Pflanzengeogr. Dokum.: 147 (1843).
Scleria macrantha Boeck., Flora 62: 572 (1879), nom. illeg. *Scleria centralis* Cherm., Arch. Bot. Bull. Mens. 4(7): 50 (1931).
Scleria tisserantii Cherm., Arch. Bot. Bull. Mens. 4(7): 50 (1931).

Scleria melanotricha Hochst. & A.Rich., Tent. Fl. Abyss. 2: 511 (1850).
Trop. Africa. 22 BEN BKN GHA GUI IVO MLI NGA SEN SIE 23 CAF CMN GAB ZAI 24 CHA ETH SUD 25 TAN 26 ZAM. Ther.
Scleria grata Nelmes, Kew Bull. 10: 453 (1955). *Scleria melanotricha* var. *grata* (Nelmes) Lye, Nordic J. Bot. 3: 243 (1983).

Scleria melicoides Schltdl., Linnaea 26: 127 (1853).
Venezuela. 82 VEN. Hemicr. or rhizome geophyte.

Scleria microcarpa Nees ex Kunth, Enum. Pl. 2: 341 (1837). *Ophryoscleria microcarpa* (Nees ex Kunth)

Nees in C.F.P.von Martius & auct. suc. (eds.), Fl. Bras. 2(1): 184 (1842).
Trop. America. 79 MXG MXT 80 BLZ COS GUA HON NIC PAN 81 CUB JAM LEE PUE TRT WIN 82 FRG GUY SUR VEN 83 BOL CLM ECU PER 84 BZE BZL BZN 85 PAR. Hemicr. or rhizome geophyte.
Scleria foliosa C.Wright, Anales Acad. Ci. Méd. Habana 8: 154 (1871), nom. illeg.
Scleria microcarpa var. *foliosa* C.B.Clarke in I.Urban, Symb. Antill. 2: 149 (1900).
Scleria microcarpa var. *subeggersiana* Kük. ex O.C. Schmidt, Repert. Spec. Nov. Regni Veg. 32: 78 (1933).

Scleria microstachya Boeck., Flora 63: 454 (1880).
Brazil. 84+. Hemicr. or rhizome geophyte.

Scleria mikawana Makino, Bot. Mag. (Tokyo) 27: 57 (1913).
Trop. & Subtrop. Old World. 22 GNB IVO SEN SIE 23 GAB ZAI 24 CHA 25 UGA 26 ANG ZAM 38 JAP 40 IND SRL 41 THA 43 NWG. Ther.
Scleria glabroreticulata De Wild., Pl. Bequaert. 4: 230 (1927).

Scleria millespicula T.Koyama, Bot. Mag. (Tokyo) 97: 418 (1984).
E. Brazil. 84 BZE BZL. Hemicr. or rhizome geophyte.

Scleria minima C.B.Clarke, Kew Bull., Addit. Ser. 8: 57 (1908).
Brazil. 84 BZC BZN. Hemicr. or rhizome geophyte.
Scleria minima f. *angustifolia* Kük., Bot. Jahrb. Syst. 56(125): 20 (1921).

Scleria minor (Britton) W.Stone, Pl. S. New Jersey: 283 (1911 publ. 1912).
E. U.S.A. to Texas. 75 NWJ NWY PEN 77 TEX 78 ALA DEL FLA GEO LOU MRY MSI NCA SCA TEN VRG WDC. Hemicr. or rhizome geophyte.
Scleria triglomerata var. *gracilis* Britton, Ann. New York Acad. Sci. 3: 230 (1884).
**Scleria triglomerata* var. *minor* Britton, Ill. Fl. N. U.S. 1: 282 (1896).

Scleria mitis P.J.Bergius, Kongl. Vetensk. Acad. Handl. 27: 145 (1765). *Carex mitis* (P.J.Bergius) J.F.Gmel., Syst. Nat. 2: 138 (1791). *Ophryoscleria mitis* (P.J.Bergius) Nees in C.F.P.von Martius & auct. suc. (eds.), Fl. Bras. 2(1): 185 (1842).
Trop. America. 80 BLZ COS GUA HON NIC PAN 81 CUB LEE TRT WIN 82 FRG GUY SUR VEN 83 BOL CLM ECU PER 84 BZC BZN 85 PAR. Hemicr. or rhizome geophyte.
Scleria riparia Poepp. & Kunth in C.S.Kunth, Enum. Pl. 2: 341 (1837). *Ophryoscleria riparia* (Poepp. & Kunth) Nees in C.F.P.von Martius & auct. suc. (eds.), Fl. Bras. 2(1): 184 (1842).
Ophryoscleria lucida Nees in C.F.P.von Martius & auct. suc. (eds.), Fl. Bras. 2(1): 183 (1842). *Scleria lucida* (Nees) Steud., Syn. Pl. Glumac. 2: 168 (1855).
Scleria praealta Salzm. ex Schltdl., Bot. Zeitung (Berlin) 3: 461 (1845).
Scleria trialata Bertero ex Boeck., Linnaea 38: 521 (1874).
Scleria trinitatis Boeck., Beitr. Cyper. 2: 31 (1890).
Scleria mitis var. *atrobrunnea* Kük., Repert. Spec. Nov. Regni Veg. 23: 219 (1926).

Scleria monticola Nelmes ex Napper, Kew Bull. 25: 444 (1971).
Sierra Leone. 22 SIE. Hemicr. or rhizome geophyte.

Scleria motemboensis Britton, Bull. Torrey Bot. Club 48: 341 (1921 publ. 1922).
EC. Cuba. 81 CUB. Hemicr. or rhizome geophyte.

Scleria motleyi C.B.Clarke, Philipp. J. Sci., C 2: 104 (1907).
Andaman Is. to New Guinea. 41 AND 42 BOR MLY PHI SUM 43 NWG. Hemicr.

subsp. ***motleyi***
Andaman Is. to New Guinea. 41 AND 42 BOR MLY PHI SUM 43 NWG. Hemicr.
Scleria trigonocarpa Ridl., Mat. Fl. Malay. Penins. 3: 110 (1907).
Scleria gonocarpa Ridl., Fl. Malay Penins. 5: 177 (1925).

subsp. ***rostrata*** J.Kern, Blumea 11: 158 (1961).
Philippines. 42 PHI. Hemicr.
Scleria trigona Merr., Philipp. J. Sci., C 8: 363 (1913).
Scleria sorsogonensis Elmer ex Merr., Enum. Philipp. Fl. Pl. 1: 134 (1922), no latin descr.
Scleria subrostrata Elmer, Leafl. Philipp. Bot. 10: 3542 (1938), nom. inval.

Scleria mucronata Poir. in J.B.A.M.de Lamarck, Encycl. 7: 3 (1806).
Caribbean. 81 CUB DOM JAM PUE. Hemicr. or rhizome geophyte.
Scleria cubensis Boeck., Beitr. Cyper. 2: 42 (1890).
Scleria microcarpa var. *subeciliata* C.B.Clarke in I.Urban, Symb. Antill. 2: 149 (1900). *Scleria mitis* var. *subeciliata* (C.B.Clarke) H.Pfeiff., Repert. Spec. Nov. Regni Veg. 52: 171 (1943).
Scleria catalinae Britton, Bull. Torrey Bot. Club 42: 489 (1915).

Scleria muehlenbergii Steud., Nomencl. Bot., ed. 2, 2: 543 (1841).
C. & E. U.S.A. to Trop. America. 74 ILL MSO OKL 75 INI NWJ NWY PEN 77 TEX 78 ALA ARK DEL FLA GEO KTY LOU MRY MSI NCA SCA TEN VRG WDC 79 MXC MXE MXG MXN MXS MXT 80 BLZ COS GUA HON NIC PAN 81 BAH CUB DOM PUE TRT 82 FRG GUY SUR VEN 83 BOL CLM ECU 84 BZC BZN 85 AGE. Hemicr. or rhizome geophyte.
**Scleria reticularis* Muhl., Descr. Gram.: 266 (1817), nom. illeg. *Scleria laxa* Torr., Ann. Lyceum Nat. Hist. New York 3: 376 (1836), nom. illeg. *Scleria torreyana* Walp., Ann. Bot. Syst. 3: 696 (1852), nom. superfl.
Scleria hemitaphra Steud., Syn. Pl. Glumac. 2: 169 (1855). *Scleria setacea* var. *hemitaphra* (Steud.) Kük., Repert. Spec. Nov. Regni Veg. 23: 215 (1926).
Scleria dictyocarpa C.Wright ex Griseb., Cat. Pl. Cub.: 249 (1866).
Scleria debilis C.Wright, Anales Acad. Ci. Méd. Habana 8: 154 (1871).
Scleria reticularis var. *pubescens* Britton, Ann. New York Acad. Sci. 3: 232 (1884).
Scleria latilacunosa Kük., Bot. Jahrb. Syst. 56(125): 21 (1921).
Scleria stevensiana Britton, Bull. Torrey Bot. Club 50: 55 (1923).

Scleria multilacunosa T.Koyama, Bull. Natl. Sci. Mus. Tokyo, n.s., 17: 71 (1974).
N. Sri Lanka. 40 SRL. Ther.

Scleria mutoensis Nakai, J. Jap. Bot. 18: 421 (1942).
Korea. 38 KOR. Hemicr. or rhizome geophyte.

Scleria myricocarpa Kunth, Enum. Pl. 2: 347 (1837). *Schizolepis myricocarpa* (Kunth) Nees in C.F.P.von Martius & auct. suc. (eds.), Fl. Bras. 2(1): 186 (1842).
SE. Brazil. 84 BZL. Hemicr. or rhizome geophyte.
Scleria atropurpurea Boeck., Vidensk. Meddel. Naturhist. Foren. Kjøbenhavn 1879-1880: 29 (1879).

Scleria nankingensis Tang & F.T.Wang, in Fl. Reipubl. Popul. Sin. 11: 234 (1961).
SE. China. 36 CHS. Hemicr. or rhizome geophyte.

Scleria natalensis Boeck. ex C.B.Clarke. in W.H.Harvey & auct. suc. (eds.), Fl. Cap. 7: 296 (1898).
Cape Prov. to KwaZulu-Natal. 27 CPP NAT. Hemicr. or rhizome geophyte.

Scleria naumanniana Boeck., Bot. Jahrb. Syst. 5: 94 (1884).
W. Trop. Africa to Angola. 22 BEN GHA GNB GUI IVO LBR MLI NGA SEN SIE TOG 23 CMN EQG? GAB ZAI 26 ANG. Hemicr.
Scleria flagellum Benth. in W.J.Hooker, Niger Fl.: 555 (1849), nom. illeg.
Scleria ovuligera Nees ex Boeck., Linnaea 38: 497 (1874), nom. illeg.

Scleria neesii Kunth, Enum. Pl. 2: 358 (1837).
SW. India, Sri Lanka, Indo-China to Pen. Malaysia (Perlis). 40 IND SRL 41 AND LAO NCB THA VIE 42 MLY. Hemicr.
**Hypoporum capitatum* Nees, Edinburgh New Philos. J. 17(34): 267 (1834). *Scleria capitata* (Nees) H.Pfeiff., Repert. Spec. Nov. Regni Veg. 26: 262 (1929), nom. illeg.
Scleria stricta Moon ex Pfeiff., Repert. Spec. Nov. Regni Veg. 26: 262 (1929), nom. illeg.

Scleria neocaledonica Rendle, J. Linn. Soc., Bot. 45: 262 (1921).
New Caledonia. 60 NWC. Hemicr. or rhizome geophyte.

Scleria neogranatensis C.B.Clarke, Kew Bull., Addit. Ser. 8: 62 (1908).
W. South America. 83 CLM ECU PER. Hemicr. or rhizome geophyte.

Scleria novae-hollandiae Boeck., Flora 58: 120 (1875).
Philippines to NW. Pacific and N. Australia. 42 PHI 43 NWG 50 NTA QLD WAU 62 CRL MRN. Ther.
Scleria merrillii Palla, Allg. Bot. Z. Syst. 17(Beil.): 8 (1911).

Scleria nyasensis C.B.Clarke in D.Oliver & auct. suc. (eds.), Fl. Trop. Afr. 8: 504 (1902).
Uganda to S. Trop. Africa. 23 BUR ZAI 25 TAN UGA 26 MLW ZAM ZIM. Hemicr. or rhizome geophyte.

Scleria oblata S.T.Blake ex J.Kern, Reinwardtia 6: 73 (1961).
S. China to Trop. Asia. 36 CHS 40 ASS BAN SRL 41 AND MYA THA VIE 42 BOR JAW MLY PHI SUL SUM. Hemicr.
Scleria laevis Willd., Sp. Pl. 4: 13 (1805), sensu auct.

Scleria obtusa Core, Brittonia 1: 240 (1934).
Bolivia to Paraguay. 83 BOL 84+ 85 PAR. Hemicr. or rhizome geophyte.

Scleria oligantha Michx., Fl. Bor.-Amer. 2: 167 (1803).
C. & E. U.S.A. to C. America. 74 ILL MSO OKL 75 INI OHI 77 TEX 78 ALA ARK FLA GEO KTY LOU MSI NCA SCA TEN VRG WDC 79 MXE MXG MXT 80 GUA HON. Hemicr. or rhizome geophyte.
Scleria purpureozonata Liebm., Mexic. Neldeagt. Pl., V, 2: 258 (1851).
Scleria leptoculmis Alph.Wood, Class-book Bot.: 746 (1861).

Scleria oligochondra Nelmes, Kew Bull. 11: 81 (1956).
Zaïre. 23 ZAI. Hemicr. or rhizome geophyte.

Scleria orchardii C.D.Adams, Phytologia 63: 299 (1987).
Trinidad. 81 TRT. Hemicr. or rhizome geophyte.

Scleria ovinux J.Raynal ex Fosberg, Bull. Mus. Natl. Hist. Nat., B, Adansonia 16: 225 (1994 publ. 1995).
New Caledonia. 60 NWC. Hemicr. or rhizome geophyte.

Scleria pachyrrhyncha Nelmes, Kew Bull. 11: 99 (1956).
Tanzania to Zimbabwe. 25 TAN 26 ZIM. Hemicr. or rhizome geophyte.

Scleria panicoides Kunth, Enum. Pl. 2: 348 (1837). *Schizolepis panicoides* (Kunth) Nees in C.F.P.von Martius & auct. suc. (eds.), Fl. Bras. 2(1): 186 (1842).
Brazil to Argentina (Misiones). 83 BOL 84 BZC BZL BZS 85 AGE PAR. Hemicr. or rhizome geophyte.

Scleria papuana J.Kern, Blumea 11: 160 (1961).
New Guinea. 43 NWG. Hemicr.

Scleria parallella C.B.Clarke, Kew Bull., Addit. Ser. 8: 59 (1908).
Colombia (Guianía) to S. Venezuela. 82 VEN 83 CLM. Hemicr. or rhizome geophyte.

Scleria parvula Steud., Syn. Pl. Glumac. 2: 174 (1855).
Trop. & Subtrop. Old World. 22 IVO SEN SIE 25 KEN TAN 36 CHS 38 JAP KOR 40 ASS EHM IND NEP SRL 41 THA VIE 42 PHI 43 NWG. Ther.
Scleria uliginosa Hochst. ex Boeck., Linnaea 38: 471 (1874).
Scleria fenestrata Franch. & Sav., Enum. Pl. Jap. 2: 549 (1878).
Scleria coreana Palla ex Nakai, Bot. Mag. (Tokyo) 30: 274 (1916).

Scleria patula E.A.Rob., Kirkia 2: 186 (1961).
Zambia. 26 ZAM. Ther.

Scleria pauciflora Muhl. ex Willd., Sp. Pl. 4: 318 (1805).
SE. Canada to C. & E. U.S.A., Cuba. 72 ONT 74 ILL KAN MSO OKL 75 CNT INI MAS NWH NWH NWJ NWY OHI PEN RHO VER 77 TEX 78 ALA ARK DEL FLA GEO KTY LOU MRY MSI NCA SCA TEN VRG WDC 79 MXE 81 CUB. Hemicr. or rhizome geophyte.

var. **caroliniana** (Willd.) Alph.Wood, Amer. Bot. Fl., ed. 2: 368 (1871).
NC. & E. U.S.A. to Mexico. 74 ILL MSO OKL 75 CNT INI MAS NWH NWJ NWY OHI PEN RHO 78 ALA ARK DEL FLA GEO KTY LOU MRY MSI NCA SCA TEN VRG WDC 79 MXE. Hemicr. or rhizome geophyte.
**Scleria caroliniana* Willd., Sp. Pl. 4: 318 (1805).

var. **curtissii** (Britton) Fairey, Castanea 34: 89 (1969).
SE. U.S.A., Cuba. 78 FLA GEO 81 CUB. Hemicr. or rhizome geophyte.
**Scleria curtissii* Britton in J.K.Small, Fl. S.E. U.S.: 200 (1903). *Scleria ciliata* var. *curtissii* (Britton) J.W.Kessler, Sida 12: 398 (1987).

var. ***pauciflora***
SE. Canada to C. & E. U.S.A., Cuba. 72 ONT 74 ILL KAN MSO OKL 75 MAS NWH NWJ NWY OHI PEN VER 77 TEX 78 ARK ARK DEL FLA GEO KTY LOU MRY MSI NCA SCA TEN VRG WDC 81 CUB. Hemicr. or rhizome geophyte.
Scleria pauciflora var. *elongata* Britton, Ann. New York Acad. Sci. 3: 235 (1884).
Scleria pauciflora var. *effusa* C.B.Clarke in I.Urban, Symb. Antill. 2: 143 (1900).
Scleria pauciflora var. *kansana* Fernald, Rhodora 8: 165 (1906).

Scleria paupercula E.A.Rob., Kew Bull. 18: 508 (1966).
Tanzania to Zimbabwe. 25 TAN 26 ZAM ZIM. Hemicr. or rhizome geophyte.

Scleria pergracilis (Nees) Kunth, Enum. Pl. 2: 354 (1837).
Trop. & Subtrop. Old World. 22 BEN GHA GNB GUI IVO MLI NGA SEN TOG 23 CAF CMN CON 24 CHA SUD 25 TAN 26 ZIM 27 NAT 36 CHC 40 ASS BAN IND NEP SRL WHM 41 THA VIE 42 PHI SUM 43 NWG 50 QLD. Ther.
**Hypoporum pergracile* Nees, Edinburgh New Philos. J. 1834: 267 (1834).
Scleria salebrosa Spreng. ex Boeck., Linnaea 38: 438 (1874).
Scleria salebrosa Boeck. ex C.B.Clarke. in T.A.Durand & H.Schinz, Consp. Fl. Afric. 5: 673 (1894).
Scleria pergracilis var. *brachystachys* Nelmes, Kew Bull. 10: 446 (1955).

Scleria pernambucana Luceño & M.Alves, Candollea 52: 196 (1997).
Brazil (Pernambuco). 84 BZE. Hemicr. or rhizome geophyte.

Scleria perpusilla Cherm., Bull. Soc. Bot. France 76: 557 (1929).
C. Madagascar. 29 MDG. Ther.

Scleria pilosa Boeck., Linnaea 38: 473 (1874).
SW. Sri Lanka. 40 SRL. Hemicr. or rhizome geophyte.
Scleria hirsuta Moon, Cat. Pl. Ceylon: 103 (1824), nom. inval.
**Cylindropus junciformis* Nees in R.Wight, Contr. Bot. India: 118 (1834). *Scleria junciformis* (Nees) Thwaites, Enum. Pl. Zeyl.: 354 (1864), nom. illeg.

Scleria pilosissima Britton, Bull. Torrey Bot. Club 42: 491 (1915).
E. Cuba (Sierra de Nipe). 81 CUB. Hemicr. or rhizome geophyte.

Scleria plusiophylla Steud., Syn. Pl. Glumac. 2: 172 (1855).
Brazil to Argentina (Misiones). 83 BOL 84 BZL 85 AGE PAR. Hemicr. or rhizome geophyte.
**Schizolepis foliosa* Nees in C.F.P.von Martius & auct. suc. (eds.), Fl. Bras. 2(1): 186 (1842). *Scleria latifolia* subsp. *foliosa* (Nees) T.Koyama, Acta Amazon. 14: 108 (1984 publ. 1986).

Scleria poaeoides Ridl., Trans. Linn. Soc. London, Bot. 2: 170 (1884).
Trop. Africa, Madagascar. 22 BEN GUI MLW NGA SIE 23 CAF ZAI 24 CHA 25 KEN TAN 26 ANG ZAM ZIM 29 MDG. Hemicr. or rhizome geophyte.
Scleria multispiculata Boeck., Beitr. Cyper. 1: 36 (1888).
Scleria multispiculosa Boeck., Beitr. Cyper. 1: 36 (1888).
Scleria prophyllata Nelmes, Kew Bull. 10: 434 (1955).

Scleria poeppigii (Nees) Steud., Syn. Pl. Glumac. 2: 171 (1855).
N. Brazil to Peru. 83 PER PER 84 BZN. Hemicr. or rhizome geophyte.
**Hymenolytrum poeppigii* Nees in C.F.P.von Martius & auct. suc. (eds.), Fl. Bras. 2(1): 174 (1842).

Scleria poiformis Retz., Observ. Bot. 4: 13 (1786). *Carex poiformis* (Retz.) J.F.Gmel., Syst. Nat. 2: 138 (1791).
Tanzania to S .Africa, Madagascar, Trop. & Subtrop. Asia to N. Australia. 25 TAN 26 MOZ 27 NAT 29 MDG 36 CHH 40 ASS IND SRL 41 THA VIE 42 BOR JAW MLY PHI SUL SUM 43 NWG 50 NTA QLD. Hemicr. or rhizome geophyte.
Scleria oryzoides J.Presl & C.Presl, Reliq. Haenk. 1: 201 (1828).
Scleria coriacea G.Bertol., Mem. Reale Accad. Sci. Ist. Bologna 5: 476 (1854), nom. illeg.
Scleria bertolonii M.Martens, Flora 40: 570 (1857).

Scleria poklei M.A.Wadood Khan, J. Econ. Taxon. Bot. 22: 559 (1998 publ. 1999).
W. India. 40 IND. Hemicr. or rhizome geophyte.

Scleria polycarpa Boeck., Linnaea 38: 509 (1874).
Maluku to W. Pacific. 42 MOL 43 BIS NWG 50 NTA QLD 60 FIJ SAM TON 62 CRL MRN. Hemicr.
Carex lithosperma G.Forst., Fl. Ins. Austr.: 65 (1786), nom. illeg.
Scleria margaritifera Willd., Sp. Pl. 4: 312 (1805), nom. illeg.
Scleria graeffeana Boeck., Flora 58: 121 (1875).
Scleria micrantha C.B.Clarke, Kew Bull., Addit. Ser. 8: 58 (1908), nom. illeg.
Scleria ternifolia Domin, Biblioth. Bot. 85: 490 (1915).

Scleria polyrrhiza E.A.Rob., Kew Bull. 18: 513 (1966).
Zambia. 26 ZAM. Hemicr.

Scleria porphyrocarpa E.A.Rob., Kew Bull. 18: 519 (1966).
S. Zaïre to Zambia. 23 ZAI 26 ZAM. (Cl.) hemicr.

Scleria procumbens E.A.Rob., Kew Bull. 18: 521 (1966).
Zambia. 26 ZAM. Hemicr. or tuber geophyte.

Scleria psilorrhiza C.B.Clarke in J.D.Hooker, Fl. Brit. India 6: 691 (1894).
Trop. Asia to N. Australia. 40 IND 41 CBD THA 42 JAW PHI 43 NWG 50 NTA QLD WAU. Hemicr. or rhizome geophyte.

Scleria pulchella Ridl., Trans. Linn. Soc. London, Bot. 2: 168 (1884).
S. Tanzania to S. Trop. Africa. 23 ZAI 25 TAN 26 ANG MLW ZAM. Ther.
Scleria ustulata Ridl., Trans. Linn. Soc. London, Bot. 2: 168 (1884).
Scleria hilsenbergii Boeck., Beitr. Cyper. 1: 34 (1888).

Scleria purdiei C.B.Clarke, Kew Bull., Addit. Ser. 8: 57 (1908).
SE. Mexico to Bolivia. 79 MXT 80 COS PAN 82 VEN 83 BOL CLM PER. Hemicr. or rhizome geophyte.
Scleria hitchcockii Standl., Contr. U. S. Natl. Herb. 18: 88 (1916).

Scleria purpurascens Steud., Syn. Pl. Glumac. 2: 169 (1855).
Indo-China to Malesia. 41 AND MYA NCB THA VIE 42 BOR JAW MLY PHI SUL SUM. Hemicr.

var. ***ophirensis*** (C.B.Clarke) J.Kern, Blumea 11: 188 (1961).
Pen. Malaysia to Sumatera. 42 MLY SUM. Hemicr.
**Scleria multifoliolata* var. *ophirensis* C.B.Clarke in J.D.Hooker, Fl. Brit. India 6: 693 (1894).

var. ***purpurascens***
Indo-China to Malesia. 41 AND MYA NCB THA VIE 42 BOR JAW MLY PHI SUL SUM. Hemicr.
Scleria multifoliata Boeck., Linnaea 38: 510 (1874).

Scleria pusilla Pilg., Bot. Jahrb. Syst. 30: 144 (1901).
Brazil (Mato Grosso). 84 BZC. Hemicr. or rhizome geophyte.

Scleria racemosa Poir. in J.B.A.M.de Lamarck, Encycl. 7: 6 (1806). *Ophryoscleria racemosa* (Poir.) Nees in C.F.P.von Martius & auct. suc. (eds.), Fl. Bras. 2(1): 183 (1842).
Sudan to S. Trop. Africa, Comores, Madagascar. 23 BUR EQG GAB ZAI 24 CHA SUD 25 KEN TAN UGA 26 ANG MLW MOZ ZAM ZIM 29 COM MDG. Hemicr. or rhizome geophyte.
Scleria macrosperma Nees ex Kunth, Enum. Pl. 2: 344 (1837).
Scleria macrocarpa Nees in C.F.P.von Martius & auct. suc. (eds.), Fl. Bras. 2(1): 183 (1842).
Scleria ciliolata Boeck., Flora 65: 31 (1882).

Scleria ramosa C.B.Clarke, Kew Bull., Addit. Ser. 8: 59 (1908).
Venezuela to Brazil. 82 VEN 83 PER 84 BZC. Hemicr. or rhizome geophyte.

Scleria rehmannii C.B.Clarke in W.H.Harvey & auct. suc. (eds.), Fl .Cap. 7: 295 (1898).
Trop. & S. Africa. 22 GNB GUI SEN TOG 23 CAF ZAI 25 TAN UGA 26 ANG MLW MOZ ZAM ZIM 27 NAM TVL. Hemicr. or rhizome geophyte.

Scleria reticularis Michx., Fl. Bor.-Amer. 2: 167 (1803).
EC. & E. U.S.A. 74 WIS 75 CNT INI MAS MIC NWH NWJ NWY PEN RHO 77 TEX 78 ALA DEL FLA GEO MRY MSI NCA SCA VRG. Ther. or hemicr.
Carex lithosperma Walter, Fl. Carol.: 227 (1788), nom. illeg.
Scleria muhlenbergiana Liebm., Mexic. Neldeagt. Pl., V, 2: 258 (1851).
Scleria jelskiana Boeck., Linnaea 38: 469 (1874).
Scleria reticularis var. *obscura* Britton, Ann. New York Acad. Sci. 3: 232 (1884).
Scleria reticularis var. *pumila* Britton, Ann. New York Acad. Sci. 3: 232 (1884).
Scleria trichopoda C.Wright ex Britton, Ann. New York Acad. Sci. 3: 232 (1885).

Scleria reticulata (Holttum) J.Kern, Reinwardtia 6: 71 (1961).
E. India to Pen. Malaysia. 40 BAN IND 41 MYA THA 42 MLY. Ther.
**Diplacrum reticulatum* Holttum, Gard. Bull. Singapore 11: 295 (1947).

Scleria retroserrata Kük., Bot. Jahrb. Syst. 56(125): 21 (1921).
N. Brazil. 84 BZN. Ther.

Scleria richardsiae E.A.Rob., Kirkia 3: 9 (1962).
Tanzania to Zambia. 25 TAN 26 MLW ZAM. Hemicr. or rhizome geophyte.

Scleria robinsoniana J.Raynal, Adansonia, n.s., 7: 239 (1967).
Sierra Leone to Guinea. 22 GUI SIE. Ther.

Scleria robusta Camelb. & Goetgh., Novon 10: 359 (2000).
S. Trop. America. 82 GUY? VEN 83 CLM PER 84 BZN. Hemicr. or rhizome geophyte.
**Scleria grandis* Core, J. Wash. Acad. Sci. 35: 322 (1945), nom. illeg.

Scleria rugosa R.Br., Prodr.: 240 (1810).
Trop. & Subtrop. Asia to N. Australia. 36 CHS 38 JAP NNS TAI 40 ASS IND SRL 41 AND MYA NCB THA VIE 42 BOR JAW MLY MOL PHI 43 NWG SOL 50 NTA QLD WAU 62 CRL. Ther.
Scleria lateriflora Boeck., Linnaea 38: 455 (1874).
Scleria onoei Franch. & Sav., Enum. Pl. Jap. 2: 549 (1878).
Scleria flaccida C.B.Clarke in J.D.Hooker, Fl. Brit. India 6: 688 (1894), nom. illeg.
Scleria pubigera Makino, Bot. Mag. (Tokyo) 27: 55 (1913). *Scleria fenestrata* var. *pubigera* (Makino) Ohwi, Bot. Mag. (Tokyo) 45: 190 (1931).
Scleria toksanensis Nakai, Bot. Mag. (Tokyo) 30: 274 (1916).
Scleria tokusanensis Nakai, Bot. Mag. (Tokyo) 30: 274 (1916), orth. var.

Scleria rutenbergiana Boeck., Abh. Naturwiss. Vereine Bremen 7: 40 (1880).
Madagascar. 29 MDG. Hemicr. or rhizome geophyte.

Scleria scabra Willd., Sp. Pl. 4: 315 (1805).
W. Cuba, S. Trop. America. 81 CUB 82 GUY SUR VEN 83 BOL CLM PER 84 BZC BZL BZN. Hemicr. or rhizome geophyte.
Scleria porphyrorhiza C.Wright, Anales Acad. Ci. Méd. Habana 8: 155 (1871).
Scleria lobulata Palla, Denkschr. Kaiserl. Akad. Wiss., Wien. Math.-Naturwiss. Kl. 79: 197 (1908).
Scleria cerradicola T.Koyama, Acta Amazon. 14: 112 (1984 publ. 1986).

Scleria scabriuscula Schltdl., Linnaea 20: 544 (1847).
S. Mexico to Guatemala. 79 MXG MXS MXT 80 GUA. Hemicr. or rhizome geophyte.
Scleria schiedeana Schltdl., Bot. Zeitung (Berlin) 3: 494 (1845), nom. illeg.
Scleria homonyma Steud., Syn. Pl. Glumac. 2: 179 (1855).

Scleria scandens Core, J. Wash. Acad. Sci. 35: 322 (1945).
N. Brazil. 84 BZN. Hemicr. or rhizome geophyte.

Scleria schenskyana Boeck., Beitr. Cyper. 2: 31 (1890).
Brazil (Minas Gerais). 84 BZL. Hemicr. or rhizome geophyte.

Scleria schiedeana Schltdl., Bot. Zeitung (Berlin) 3: 489 (1845). *Ophryoscleria schiedeana* (Schltdl.) Liebm., Mexic. Neldeagt. Pl., V, 2: 261 (1851).
Mexico to C. America, S. Brazil. 79 MXG MXS 80 GUA HON 84 BZS. Hemicr.
Schizolepis paranensis Palla, Denkschr. Kaiserl. Akad. Wiss., Wien. Math.-Naturwiss. Kl. 79: 196 (1908).

Scleria schimperiana Boeck., Linnaea 38: 466 (1874).
Trop. Africa. 22 NGA 23 RWA ZAI 24 ETH 25 TAN UGA 26 ZAM ZIM. Ther.

Scleria hypoxis Schweinf. ex Boeck., Linnaea 38: 465 (1874).

Scleria schmitzii Piérart, Lejeunia Mém. 13: 42 (1953). Zaïre. 23 ZAI. Hemicr. or rhizome geophyte.

Scleria schulzii Barros, Anales Mus. Argent. Ci. Nat. "Bernardino Rivadavia" 41: 439 (1945).
Argentina (Chaco, Corrientes). 85 AGE. Hemicr. or rhizome geophyte.

Scleria scindens Nees ex Kunth, Enum. Pl. 2: 343 (1837).
Caribbean. 81 CUB DOM LEE PUE WIN. Hemicr. or rhizome geophyte.
Scleria chlorantha Boeck., Linnaea 38: 506 (1874).
Scleria scaberrima Boeck., Beitr. Cyper. 2: 41 (1890).

Scleria scrobiculata Nees & Meyen in R.Wight, Contr. Bot. India: 117 (1834).
Taiwan, Indo-China to W. Pacific. 38 TAI 41 AND THA VIE 42 BOR JAW LSI MLY MOL PHI SUL SUM 43 NWG 50 NTA QLD 60 SAM 62 CRL. Hemicr. or rhizome geophyte.

subsp. ***discocarpa*** J.Kern, Blumea 11: 186 (1961).
Maluku to New Guinea. 42 MOL 43 NWG. Hemicr.

subsp. ***scrobiculata***
Taiwan, Indo-China to W. Pacific. 38 TAI 41 AND THA VIE 42 BOR JAW LSI MLY MOL PHI SUL SUM 43 NWG 50 NTA QLD 60 SAM 62 CRL. Hemicr. or rhizome geophyte.
Scleria tessellata Decne., Nouv. Ann. Mus. Hist. Nat. 3: 362 (1834), nom. illeg.
Scleria timorensis Nees, Linnaea 9: 303 (1834), nom. nud.
Scleria neesiana Hook. & Arn., Bot. Beechey Voy.: 299 (1838).
Scleria approximata Hassk., Tijdschr. Natuurl. Gesch. Physiol. 10: 118 (1843).
Scleria foveolata Llanos, Fragm. Pl. Filip.: 103 (1851).
Scleria waigiouensis Steud., Syn. Pl. Glumac. 2: 173 (1855).
Scleria doederleiniana Boeck., Bot. Jahrb. Syst. 5: 512 (1884).
Scleria purpureovaginata Boeck., Bot. Jahrb. Syst. 5: 513 (1884).
Scleria keyensis K.Schum. ex Warb., Bot. Jahrb. Syst. 13: 267 (1890).

Scleria secans (L.) Urb., Symb. Antill. 2: 169 (1900).
Trop. America. 79 MXG MXT 80 BLZ COS GUA HON NIC PAN 81 CUB JAM LEE PUE TRT WIN 82 FRG GUY SUR VEN 83 BOL CLM ECU PER 84 BZC BZE BZL BZN BZS 85 AGE PAR. Hemicr. or rhizome geophyte.
**Schoenus secans* L., Syst. Nat. ed. 10: 865 (1759).
Carex flagellum J.F.Gmel., Syst. Nat. 2: 138 (1791). *Scleria flagellum* (J.F.Gmel.) Sw., Fl. Ind. Occid. 1: 88 (1797). *Omoscleria flagellum* (J.F.Gmel.) Nees in C.F.P.von Martius & auct. suc. (eds.), Fl. Bras. 2(1): 180 (1842).
Scleria reflexa Kunth in F.W.H.von Humboldt, A.J.A.Bonpland & C.S.Kunth, Nov. Gen. Sp. 1: 232 (1816). *Mastigoscleria reflexa* (Kunth) Nees in C.F.P.von Martius & auct. suc. (eds.), Fl. Bras. 2(1): 177 (1842).
Carex margaritifera Steud., Nomencl. Bot., ed. 2, 1: 292 (1840).
Mastigoscleria reflexa var. *styligera* Nees in C.F.P.von Martius & auct. suc. (eds.), Fl. Bras. 2(1): 178 (1842).
Scleria renggeriana Steud., Syn. Pl. Glumac. 2: 173 (1855).
Scleria flagellum Griseb., Fl. Brit. W. I.: 578 (1864), nom. illeg.
Scleria reflexa var. *surinamensis* Boeck., Linnaea 38: 504 (1874).
Scleria weigeltiana Schrad. ex Boeck., Linnaea 38: 504 (1874).

Scleria sellowiana Kunth, Enum. Pl. 2: 350 (1837). *Chondrolomia sellowiana* (Kunth) Nees in C.F.P.von Martius & auct. suc. (eds.), Fl. Bras. 2(1): 174 (1842).
Brazil to NE. Argentina. 84 BZS 85 AGE PAR URU. Hemicr. or rhizome geophyte.

Scleria setulosociliata Boeck., Flora 65: 30 (1882).
Trop. America. 79 MXG MXS MXT 80 BLZ COS GUA HON NIC PAN 81 CUB JAM 82 VEN 83 ECU 84 BZN. Hemicr. or rhizome geophyte.

Scleria sheilae J.Raynal, Adansonia, n.s., 7: 243 (1967).
Cameroon. 23 CMN. Hemicr. or rhizome geophyte.

Scleria sieberi Nees, Linnaea 9: 303 (1834). *Schizolepis sieberi* (Nees) Nees in C.F.P.von Martius & auct. suc. (eds.), Fl. Bras. 2(1): 188 (1842) (1842).
Seychelles, Mascarenes. 29 MAU SEY. Hemicr.
**Scleria latifolia* Sieber ex C.Presl, Isis (Oken) 21: 268 (1828), nom. illeg.
Scleria racemosa Bojer, Hortus Maurit.: 385 (1837), nom. inval.

Scleria skutchii M.T.Strong & J.R.Grant, Novon 4: 296 (1994).
Costa Rica to Ecuador. 80 COS 83 CLM ECU. Hemicr. or rhizome geophyte.

Scleria sobolifer E.F.Franklin, Kew Bull. 38: 33 (1983).
KwaZulu-Natal. 27 NAT. Hemicr. or rhizome geophyte.

Scleria sororia Kunth, Enum. Pl. 2: 343 (1837). *Ophryoscleria sororia* (Kunth) Nees in C.F.P.von Martius & auct. suc. (eds.), Fl. Bras. 2(1): 184 (1842).
Mexico to Ecuador, Brasil. 79 MXG MXS MXT 80 GUA HON NIC 83 CLM ECU 84+. Hemicr. or rhizome geophyte.
Diploscyphus mexicanus Liebm., Kongel. Danske Vidensk. Selsk. Skr., Naturvidensk. Math. Afd., V, 2: 262 (1850). *Scleria mexicana* (Liebm.) Boeck., Linnaea 38: 500 (1874).
Scleria chimborazensis C.B.Clarke, Kew Bull., Addit. Ser. 8: 60 (1908).

Scleria sphacelata F.Muell., Fragm. 9: 20 (1875).
N. Northern Territory to Queensland. 50 NTA QLD. Hemicr. or rhizome geophyte.

Scleria sphaerocarpa (E.A.Rob.) Napper, Kew Bull. 25: 441 (1971).
Trop. Africa. 22 BEN BKN GHA IVO LBR NGA SEN SIE TOG 24 CHA SUD 26 ANG ZAM. Ther.
**Scleria tessellata* var. *sphaerocarpa* E.A.Rob., Kew Bull. 18: 526 (1966).

Scleria spicata (Spreng.) J.F.Macbr., Publ. Field Columbian Mus., Bot. Ser. 4: 168 (1929).
Peru, SE. Brazil. 83 PER 84 BZL. Hemicr. or rhizome geophyte.
**Rhynchospora spicata* Spreng., Syst. Veg. 1: 194 (1824).

Scleria pleostachya Kunth, Enum. Pl. 2: 355 (1837). *Hypoporum pleostachyum* (Kunth) Nees in C.F.P.von Martius & auct. suc. (eds.), Fl. Bras. 2(1): 170 (1842).

Scleria spiciformis Benth. in W.J.Hooker, Niger Fl.: 556 (1849).
W. & WC. Trop. Africa. 22 GUI IVO LBR SIE 23 EQG GAB ZAI. Hemicr.

Scleria splitgerberiana Henrard ex Uittien, Recueil Trav. Bot. Néerl. 30: 190 (1933).
N. & W. South America. 82 GUY SUR 83 BOL CLM ECU PER. Hemicr. or rhizome geophyte.

Scleria sprucei C.B.Clarke, Kew Bull., Addit. Ser. 8: 61 (1908).
Venezuela to Brazil. 82 VEN 83 BOL CLM 84 BZE BZN. Hemicr. or rhizome geophyte.

Scleria staheliana Uittien, Recueil Trav. Bot. Néerl. 30: 189 (1933).
Guianas. 82 FRG SUR. Hemicr. or rhizome geophyte.

Scleria stenophylla Core, Brittonia 1: 242 (1934).
Colombia. 83 CLM. Hemicr. or rhizome geophyte.

Scleria stereorrhiza C.Wright ex C.B.Clarke in I.Urban, Symb. Antill. 2: 147 (1900).
Cuba. 81 CUB. Hemicr. or rhizome geophyte.

var. ***major*** C.B.Clarke in I.Urban, Symb. Antill. 2: 147 (1900).
Cuba. 81 CUB. Hemicr. or rhizome geophyte.

var. ***stereorrhiza***
Cuba. 81 CUB. Hemicr. or rhizome geophyte.

Scleria stipitata Uittien, Recueil Trav. Bot. Néerl. 30: 189 (1933).
N. & W. South America. 82 GUY SUR 83 ECU PER. Hemicr. or rhizome geophyte.

Scleria stipularis Nees, J. Bot. (Hooker) 2: 394 (1840).
S. Trop. America. 82 FRG GUY SUR VEN 83 BOL CLM PER 84 BZN. Hemicr. or rhizome geophyte.
Hymenolytrum silvestris Schrad. ex Nees in C.F.P.von Martius & auct. suc. (eds.), Fl. Bras. 2(1): 174 (1842).

Scleria stocksiana Boeck., Linnaea 38: 474 (1874).
India. 40 IND. Hemicr. or rhizome geophyte.

Scleria suaveolens Nelmes, Kew Bull. 10: 442 (1955).
Zambia. 26 ZAM. Hemicr. or rhizome geophyte.

Scleria sumatrensis Retz., Observ. Bot. 5: 19 (1788).
Seychelles, Trop. & Subtrop. Asia to N. Australia. 29 SEY 36 CHS 38 TAI 40 ASS IND SRL 41 AND MYA NCB THA VIE 42 BOR JAW MLY PHI SUL SUM 50 NTA QLD 62 CRL. Hemicr.
Scleria setigera Roxb., Fl. Ind. ed. 1832, 3: 575 (1832).

Scleria swamyi Govind., Proc. Indian Acad. Sci., Pl. Sci. 94: 19 (1985).
S. India. 40 IND. Hemicr. or rhizome geophyte.

Scleria tenacissima (Nees) Steud., Syn. Pl. Glumac. 2: 175 (1855).
Venezuela to Bolivia. 82 VEN 83 BOL CLM 84 BZN. Hemicr. or rhizome geophyte.
**Omoscleria tenacissima* Nees in C.F.P.von Martius & auct. suc. (eds.), Fl. Bras. 2(1): 181 (1842).

Scleria tenella Kunth, Enum. Pl. 2: 353 (1837). *Hypoporum tenellum* (Kunth) Nees in C.F.P.von Martius & auct. suc. (eds.), Fl. Bras. 2(1): 171 (1842).
Mexico to Trop. America. 79 MXG MXS 80 BLZ COS ELS GUA HON NIC PAN 81 BAH CUB 82 FRG GUY SUR VEN 83 BOL 84 BZC. Ther.
Hypoporum micrococcum Liebm., Mexic. Neldeagt. Pl., V, 2: 255 (1851). *Scleria micrococca* (Liebm.) Steud., Syn. Pl. Glumac. 2: 179 (1855).
Hypoporum purpurascens Liebm., Mexic. Neldeagt. Pl., V, 2: 256 (1851).
Hypoporum verticillatum Nees, Bonplandia (Hannover) 3: 87 (1855), nom. illeg.
Scleria liebmannii Steud., Syn. Pl. Glumac. 2: 179 (1855).
Scleria luzuliformis C.Wright, Anales Acad. Ci. Méd. Habana 8: 156 (1871).
Scleria costaricensis Boeck., Allg. Bot. Z. Syst. 2: 157 (1896).
Scleria areolata Lundell, Amer. Midl. Naturalist 29: 471 (1943).

Scleria tepuiensis Core, Fieldiana, Bot. 28(1): 52 (1951).
N. South America. 82 FRG GUY VEN. Hemicr. or rhizome geophyte.

Scleria terrestris (L.) Fassett, Rhodora 26: 159 (1924).
Trop. & Subtrop. Asia to N. Australia. 36 CHS 38 NNS TAI 40 ASS EHM IND NEP SRL WHM 41 THA VIE 42 BOR JAW LSI MLY MOL PHI SUL SUM 43 NWG 50 NTA QLD. Hemicr.

var. ***hookeriana*** (Boeck.) D.M.Verma, Bull. Bot. Surv. India 32: 70 (1990 publ. 1992).
Assam. 40 ASS. Hemicr. or rhizome geophyte.
**Scleria hookeriana* Boeck., Linnaea 38: 498 (1874).

var. ***terrestris***
Trop. & Subtrop. Asia to N. Australia. 36 CHS 38 NNS TAI 40 ASS EHM IND NEP SRL WHM 41 THA VIE 42 BOR JAW LSI MLY MOL PHI SUL SUM 43 NWG 50 NTA QLD. Hemicr.
Schoenus paniculatus Burm.f., Fl. Indica: 19 (1768).
Diaphora cochinchinensis Lour., Fl. Cochinch.: 578 (1790). *Scleria cochinchinensis* (Lour.) Druce, Bot. Soc. Exch. Club Brit. Isles 1916: 646 (1917).
Scleria radula Hance, Ann. Sci. Nat., Bot., IV, 18: 232 (1862).
Scleria elata Thwaites, Enum. Pl. Zeyl.: 353 (1864).
Scleria hirsuta Boeck., Linnaea 38: 489 (1874).
Scleria melanostoma Boeck., Linnaea 38: 514 (1874).
Scleria exaltata Boeck., Bot. Jahrb. Syst. 5: 511 (1884).
Scleria haematostachys Boeck., Bot. Jahrb. Syst. 5: 512 (1884).
Scleria hasskarliana Boeck., Bot. Jahrb. Syst. 5: 511 (1884).
Scleria ploemii Boeck., Bot. Jahrb. Syst. 5: 513 (1884).
Scleria kuntzei Boeck., Beitr. Cyper. 1: 34 (1888).
Scleria rinkiana Boeck., Beitr. Cyper. 2: 30 (1890).
Scleria chinensis var. *biauriculata* C.B.Clarke in J.D.Hooker, Fl. Brit. India 6: 690 (1894).
Scleria elata var. *decolorans* C.B.Clarke in J.D.Hooker, Fl. Brit. India 6: 690 (1894).
Scleria elata var. *latior* C.B.Clarke in J.D.Hooker, Fl. Brit. India 6: 690 (1894).
Scleria luzonensis Palla, Allg. Bot. Z. Syst. 13: 49 (1907).
Scleria aspera Boeck., Linnaea 38: 482 (1974).

var. ***thomsoniana*** (Boeck.) D.M.Verma, Bull. Bot. Surv. India 32: 70 (1990 publ. 1992).
Assam. 40 ASS. Hemicr. or rhizome geophyte.
**Scleria thomsoniana* Boeck., Linnaea 38: 479 (1874).

Scleria tessellata Willd., Sp. Pl. 4: 315 (1805).
Trop. Africa, Madagascar, India to Thailand, N. Northern Territory. 22 BEN BKN GHA GUI IVO MLI NGA NGR SEN TOG 24 SUD 25 TAN 26 ANG ZAM 29 MDG 40 ASS IND 41 THA 50 NTA. Ther.
Carex indica J.König ex Willd., Sp. Pl. 4: 315 (1805), nom. illeg.
Scleria glandiformis Boeck., Flora 41: 647 (1858).

Scleria testacea Nees ex Kunth, Enum. Pl. 2: 341 (1837).
Hawaiian Is., Mexico to Paraguay. 63 HAW 79 MXT 80 NIC PAN 81 CUB 84 BZC 85 PAR. Hemicr. or rhizome geophyte.
Scleria phylloptera C.Wright ex Griseb., Cat. Pl. Cub.: 248 (1866). *Scleria microcarpa* var. *phylloptera* (C.Wright ex Griseb.) Kük., Repert. Spec. Nov. Regni Veg. 23: 218 (1926).

Scleria thwaitesiana Boeck., Linnaea 38: 454 (1874).
Sri Lanka, Thailand to Pen. Malaysia. 40 SRL 41 THA 42 MLY. Ther.

Scleria tonkinensis C.B.Clarke, Kew Bull., Addit. Ser. 8: 57 (1908).
S. China to Indo-China. 36 CHS 41 THA VIE. Hemicr. or rhizome geophyte.

Scleria transvaalensis E.F.Franklin, Kew Bull. 38: 35 (1983).
S. Africa. 27 NAT SWZ TVL. Hemicr. or rhizome geophyte.

Scleria trialata Poir. in J.B.A.M.de Lamarck, Encycl. 7: 6 (1806).
Comoros, Madagascar. 29 COM MDG. Hemicr. or rhizome geophyte.
Scleria rosea Cherm., Bull. Soc. Bot. France 70: 298 (1923). *Scleria trialata* var. *rosea* (Cherm.) Cherm., Bull. Soc. Bot. France 74: 608 (1927).

Scleria tricuspidata S.T.Blake, Blumea 11: 203 (1961).
New Guinea (Kep. Aru), NE. Australia. 43 NWG 50 NSW NTA QLD. Ther.
Scleria tessellata var. *debilis* Benth., Fl. Austral. 7: 430 (1878).

Scleria triglomerata Michx., Fl. Bor.-Amer. 2: 168 (1803). *Cladium triglomeratum* (Michx.) Nees, Linnaea 9: 301 (1834). *Trachylomia triglomerata* (Michx.) Nees in C.F.P.von Martius & auct. suc. (eds.), Fl. Bras. 2(1): 174 (1842).
S. Ontario to C. & E. U.S.A. 72 ONT 74 ILL IOW KAN MIN MSO NEB OKL WIS 75 CNT INI MAS MIC NWJ NWY OHI PEN RHO WVA 77 TEX 78 ALA ARK DEL FLA GEO KTY LOU MRY MSI NCA SCA TEN VRG WDC 81 PUE. Hemicr. or rhizome geophyte.
Schoenus inermis Willd. ex Kunth, Enum. Pl. 2: 304 (1837).
Scleria nitida Willd. ex Kunth, Enum. Pl. 3: 350 (1837).
Scleria flaccida Steud., Syn. Pl. Glumac. 2: 174 (1855).

Scleria triquetra M.T.Strong, Novon 4: 300 (1994).
French Guiana to Brazil (Pará). 82 FRG 84 BZN. Hemicr. or rhizome geophyte.

Scleria trisulcata G.P.Li, Fl. Fujianica 6: 689 (1995).
China (Fujian). 36 CHS. Hemicr. or rhizome geophyte.

Scleria tryonii Domin, Biblioth. Bot. 85: 487 (1915).
NE. Queensland. 50 QLD. Hemicr. or rhizome geophyte.

Scleria uleana Boeck., Allg. Bot. Z. Syst. 2: 159 (1896).
S. Brazil. 84 BZS. Hemicr. or rhizome geophyte.

Scleria unguiculata E.A.Rob., Kew Bull. 18: 536 (1966).
Trop. & S. Africa. 22 TOG 23 CAF RWA 25 TAN 26 ZAM 27 BOT. Hemicr.

Scleria vaginata Steud., Syn. Pl. Glumac. 2: 179 (1855).
C. & S. Trop. America. 80 COS PAN 82 SUR VEN 83 BOL CLM ECU PER 84 BZE. (Cl.) hemicr. or cham.
Tonduzia macrophylla Boeck., Bull. Herb. Boissier 3: 464 (1895). *Durandia macrophylla* (Boeck.) Boeck., Allg. Bot. Z. Syst. 2: 173 (1896).
Scleria tonduzii Boeck., Allg. Bot. Z. Syst. 2: 160 (1896).

Scleria valdemuricata Kük., Bot. Jahrb. Syst. 56(125): 20 (1921).
N. Brazil. 84 BZN. Hemicr. or rhizome geophyte.

Scleria variegata (Nees) Steud., Syn. Pl. Glumac. 2: 176 (1855).
SE. & S. Brazil. 84 BZL BZS. Hemicr. or rhizome geophyte.
**Hypoporum variegatum* Nees in C.F.P.von Martius & auct. suc. (eds.), Fl. Bras. 2(1): 173 (1842).

Scleria venezuelensis Core, Mem. New York Bot. Gard. 12(3): 62 (1965).
Venezuela. 82 VEN. Hemicr. or rhizome geophyte.

Scleria verrucosa Willd., Sp. Pl. 4: 313 (1805).
W. Trop. Africa to Tanzania. 22 BEN GHA LBR NGA SEN SIE TOG 23 CAF CMN EQG GAB ZAI 25 TAN UGA. Hemicr. or rhizome geophyte.
Scleria spinulosa Boeck., Beitr. Cyper. 2: 30 (1890).

Scleria verticillata Muhl. ex Willd., Sp. Pl. 4: 317 (1805). *Hypoporum verticillatum* (Muhl. ex Willd.) Nees, Linnaea 9: 303 (1834).
S. Ontario to C. & E. U.S.A., Caribbean, NE. Mexico, S. Trop. America. 72 ONT 74 ILL IOW MIN MSO OKL WIS 75 CNT INI MIC NWJ NWY OHI PEN 77 TEX 78 ALA ARK FLA GEO LOU MRY MSI NCA SCA TEN VRG 79 MXE 81 BAH CUB PUE 82 FRG GUY SUR VEN 83 BOL CLM 84 BZN. Ther.
Hypoporum diffusum Nees, Linnaea 9: 303 (1834), nom. inval.
Scleria diffusa Michx. ex Kunth, Enum. Pl. 2: 359 (1837).
Scleria tenuiflora Willd. ex Kunth, Enum. Pl. 2: 353 (1837).
Scleria kunthiana Steud., Syn. Pl. Glumac. 2: 176 (1855).
Scleria verticillata f. *brevis* Kük., Repert. Spec. Nov. Regni Veg. 23: 214 (1926).
Scleria verticillata f. *capillaris* Kük., Repert. Spec. Nov. Regni Veg. 23: 214 (1926).

Scleria veseyfitzgeraldii E.A.Rob., Kew Bull. 18: 503 (1966).
Tanzania to Namibia. 25 TAN 26 ZAM 27 BOT NAM. Ther. or hemicr.

Scleria vichadensis F.J.Herm., Caldasia 5: 38 (1948).
Colombia. 83 CLM. Hemicr. or rhizome geophyte.

Scleria violacea Pilg., Bot. Jahrb. Syst. 30: 144 (1901).
French Guiana to Brazil. 82 FRG 84 BZC. Hemicr. or rhizome geophyte.

Scleria virgata (Nees) Steud., Syn. Pl. Glumac. 2: 176 (1855).
Brazil to Argentina (Misiones). 84 BZE BZL 85 AGE. Hemicr. or rhizome geophyte.
**Hypoporum virgatum* Nees in C.F.P.von Martius & auct. suc. (eds.), Fl. Bras. 2(1): 172 (1842).
Scleria trigonocarpa Boeck., Beitr. Cyper. 1: 37 (1888), nom. illeg.
Scleria diffusa C.B.Clarke, Kew Bull., Addit. Ser. 8: 57 (1908), nom. illeg.

Scleria vogelii C.B.Clarke in D.Oliver & auct. suc. (eds.), Fl. Trop. Afr. 8: 508 (1902).
W. & WC. Trop. Africa. 22 GHA IVO LBR NGA 23 CMN EQG GAB GGI. Hemicr.

Scleria warmingiana Boeck., Vidensk. Meddel. Naturhist. Foren. Kjøbenhavn 1869: 152 (1869).
Brazil (Minas Gerais). 84 BZL. Hemicr. or rhizome geophyte.

Scleria welwitschii C.B.Clarke in T.A.Durand & H.Schinz, Consp. Fl. Afric. 5: 675 (1894).
Tanzania to S. Africa. 25 TAN 26 ANG ZAM ZIM 27 CPP NAT SWZ TVL. Hemicr. or rhizome geophyte.
**Scleria junciformis* Welw. ex Ridl., Trans. Linn. Soc. London, Bot. 2: 168 (1884), nom. illeg.

Scleria williamsii R.Gross, Publ. Field Mus. Nat. Hist., Bot. Ser. 11: 39 (1931).
N. Peru. 83 PER. Hemicr. or rhizome geophyte.

Scleria woodii C.B.Clarke in W.H.Harvey & auct. suc. (eds.), Fl. Cap. 7: 295 (1898).
Trop. & S. Africa. 22 GHA NGA 23 CAF CMN CON RWA ZAI 24 SUD 25 TAN UGA 26 ANG ZAM ZIM 27 BOT CPP NAT OFS SWZ TVL. Hemicr. or rhizome geophyte.

var. ***ornata*** (Cherm.) J.Schultze-Motel, Willdenowia 2: 504 (1960).
Trop. Africa. 22 GHA NGA 23 CAF CMN CON ZAI 24 SUD 25 KEN TAN UGA 26 ANG ZAM. Hemicr. or rhizome geophyte.
Scleria striatonux De Wild., Rev. Zool. Bot. Africaines 14(Suppl. Bot.): 22 (1926).
**Scleria rehmannii* var. *ornata* Cherm., Bull. Jard. Bot. État 13: 283 (1935).

var. ***woodii***
Uganda to S. Africa. 23 RWA ZAI 25 TAN UGA 26 ANG ZAM ZIM 27 BOT CPP NAT OFS SWZ TVL. Hemicr. or rhizome geophyte.
Scleria striatonux var. *lacunosa* Piérart, Lejeunia Mém. 13: 30 (1951 publ. 1953).

Scleria wrightiana Boeck., Flora 64: 79 (1881).
Cuba. 81 CUB. Hemicr. or rhizome geophyte.
Scleria auricoma Otto ex Sweet, Hort. Brit. 2: 549 (1830), nom. inval.
**Scleria elata* C.Wright, Anales Acad. Ci. Méd. Habana 8: 153 (1871), nom. illeg.

Scleria xerophila E.A.Rob., Kew Bull. 18: 540 (1966).
Zambia. 26 ZAM. Hemicr. or tuber geophyte.

Scleria zambesica E.A.Rob., Kew Bull. 18: 499 (1966).
Zambia. 26 ZAM. Ther.

Synonyms:

Scleria abortiva var. *planifolia* Cherm. = ***Scleria abortiva*** Nees ex Kunth
Scleria achtenii var. *subintegriloba* (De Wild.) Piérart = ***Scleria achtenii*** De Wild.
Scleria acriulus C.B.Clarke = ***Scleria griegiifolia*** (Ridl.) C.B.Clarke
Scleria adamantina (Nees) Steud. = ***Lagenocarpus adamantinus*** Nees
Scleria affinis C.Presl ex Steud. = ***Scleria melaleuca*** Rchb. ex Schltdl. & Cham.
Scleria africana Benth. = ***Diplacrum africanum*** (Benth.) C.B.Clarke
Scleria alata Moon = ***Scleria levis*** Retz.
Scleria albonigra A.St.-Hil. = ***Lagenocarpus alboniger*** (A.St.-Hil.) C.B.Clarke
Scleria alta Boeck. = ***Scleria levis*** Retz.
Scleria amphigaea Raymond = ***Diplacrum capitatum*** (Willd.) Boeck.
Scleria amphigaea var. *angustifolia* (Kük.) Raymond = ***Diplacrum capitatum*** (Willd.) Boeck.
Scleria androgyna Nees = ***Scleria corymbosa*** Roxb.
Scleria angolensis Turrill = ***Scleria induta*** Turrill
Scleria angustata Steud. = ***Scleria angusta*** Nees ex Kunth
Scleria approximata Hassk. = ***Scleria scrobiculata*** Nees & Meyen subsp. ***scrobiculata***
Scleria aquatica Cherm. = ***Scleria lacustris*** C.Wright
Scleria areolata Lundell = ***Scleria tenella*** Kunth
Scleria arundinacea Kunth = ***Scleria latifolia*** Sw.
Scleria aspera Boeck. = ***Scleria terrestris*** (L.) Fassett var. ***terrestris***
Scleria asperata C.Presl = ***Scleria melaleuca*** Rchb. ex Schltdl. & Cham.
Scleria asperrima (Liebm.) Steud. = ***Scleria macrophylla*** J.Presl & C.Presl
Scleria aterrima (Ridl.) Napper = ***Scleria catophylla*** C.B.Clarke
Scleria atropurpurea Boeck. = ***Scleria myricocarpa*** Kunth
Scleria atrosanguinea Hochst. ex Steud. = ***Scleria bulbifera*** Hochst. ex A.Rich.
Scleria auricoma Otto ex Sweet = ***Scleria wrightiana*** Boeck.
Scleria axillaris Moon = ***Diplacrum caricinum*** R.Br.
Scleria baldwinii var. *costata* Britton = ***Scleria baldwinii*** (Torr.) Steud.
Scleria bancana Miq. = ***Scleria ciliaris*** Nees
Scleria bancana var. *nana* Ridl. = ***Scleria ciliaris*** Nees
Scleria barteri Boeck. = ***Scleria boivinii*** Steud.
Scleria bequaertii var. *laevis* Piérart = ***Scleria bequaertii*** De Wild.
Scleria bertolonii M.Martens = ***Scleria poiformis*** Retz.
Scleria biflora subsp. *ferruginea* J.Kern = ***Scleria biflora*** Roxb.
Scleria boliviana Palla = ***Scleria melaleuca*** Rchb. ex Schltdl. & Cham.
Scleria bracteata var. *assamica* C.B.Clarke = ***Scleria assamica*** (C.B.Clarke) D.M.Verma
Scleria bracteata var. *floribunda* (Kunth) H.Pfeiff. = ***Scleria bracteata*** Cav.
Scleria bracteata f. *simplicior* Kük. = ***Scleria bracteata*** Cav.
Scleria bracteata f. *supra-gynaecea* H.Pfeiff. = ***Scleria bracteata*** Cav.
Scleria brittonii Core ex Small = ***Scleria ciliata*** var. ***glabra*** (Chapm.) Fairey
Scleria brownii var. *glabra* Domin = ***Scleria brownii*** Kunth

Scleria brownii var. *typica* Domin = ***Scleria brownii*** Kunth
Scleria buchananii Boeck. = ***Scleria bulbifera*** Hochst. ex A.Rich.
Scleria buchananii var. *laevinux* R.Gross = ***Scleria bulbifera*** Hochst. ex A.Rich.
Scleria buchananii var. *latifolia* De Wild. = ***Scleria bulbifera*** Hochst. ex A.Rich.
Scleria bulbifera var. *latifolia* (De Wild.) Piérart = ***Scleria bulbifera*** Hochst. ex A.Rich.
Scleria bulbifera var. *mechowiana* (Boeck.) Kük. = ***Scleria bulbifera*** Hochst. ex A.Rich.
Scleria bulbifera var. *schweinfurthiana* (Boeck.) Piérart = ***Scleria bulbifera*** Hochst. ex A.Rich.
Scleria bulusanensis Elmer = ***Scleria levis*** Retz.
Scleria caespitosa Ridl. = ***Scleria dregeana*** Kunth
Scleria campestris (Nees) Steud. = ***Lagenocarpus minarum*** (Nees) Kuntze
Scleria canaliculatotriquetra Boeck. = ***Scleria lagoensis*** subsp. ***canaliculatotriquetra*** (Boeck.) Lye
Scleria canaliculatotriquetra var. *adpressohirta* Kük. = ***Scleria adpressohirta*** (Kük.) E.A.Rob.
Scleria capillaris R.Br. = ***Scleria lithosperma*** (L.) Sw. var. ***lithosperma***
Scleria capitata (Nees) H.Pfeiff. = ***Scleria neesii*** Kunth
Scleria capitata Willd. = ***Diplacrum capitatum*** (Willd.) Boeck.
Scleria caricina (R.Br.) Benth. = ***Diplacrum caricinum*** R.Br.
Scleria caroliniana Willd. = ***Scleria pauciflora*** var. ***caroliniana*** (Willd.) Alph.Wood
Scleria catalinae Britton = ***Scleria mucronata*** Poir.
Scleria catophylla var. *chondrocarpa* Nelmes = ***Scleria catophylla*** C.B.Clarke
Scleria cenchroides Hochst. ex C.B.Clarke = ***Scleria bulbifera*** Hochst. ex A.Rich.
Scleria centralis Cherm. = ***Scleria melanomphala*** Kunth
Scleria cerradicola T.Koyama = ***Scleria scabra*** Willd.
Scleria cervina Ridl. = ***Scleria lagoensis*** Boeck. subsp. ***lagoensis***
Scleria chimborazensis C.B.Clarke = ***Scleria sororia*** Kunth
Scleria chinensis Kunth = ***Scleria ciliaris*** Nees
Scleria chinensis var. *biauriculata* C.B.Clarke = ***Scleria terrestris*** (L.) Fassett var. ***terrestris***
Scleria chlorantha Boeck. = ***Scleria scindens*** Nees ex Kunth
Scleria ciliata var. *curtissii* (Britton) J.W.Kessler = ***Scleria pauciflora*** var. ***curtissii*** (Britton) Fairey
Scleria ciliata var. *glabrescens* Maury = ***Scleria ciliata*** Michx. var. ***ciliata***
Scleria ciliata var. *pauciflora* Kük. = ***Scleria ciliata*** Michx. var. ***ciliata***
Scleria ciliolata Boeck. = ***Scleria racemosa*** Poir.
Scleria cincta Steud. = ***Rhynchospora contracta*** (Nees) J.Raynal
Scleria clandestina Rchb. ex Core = ***Calyptrocarya glomerulata*** (Brongn.) Urb.
Scleria clarkei De Wild. = ***Scleria baroni-clarkei*** De Wild.
Scleria cochinchinensis (Lour.) Druce = ***Scleria terrestris*** (L.) Fassett var. ***terrestris***
Scleria communis Kunth = ***Scleria melaleuca*** Rchb. ex Schltdl. & Cham.
Scleria composita var. *scabrosa* (Maury ex Micheli) H.Pfeiff. = ***Scleria composita*** (Nees) Boeck.
Scleria congolensis De Wild. = ***Scleria melaleuca*** Rchb. ex Schltdl. & Cham.
Scleria conspersa Sellow ex Nees = ***Scleria melaleuca*** Rchb. ex Schltdl. & Cham.
Scleria coreana Palla ex Nakai = ***Scleria parvula*** Steud.
Scleria coriacea Liebm. = ***Scleria ciliata*** Michx. var. ***ciliata***
Scleria coriacea G.Bertol. = ***Scleria poiformis*** Retz.
Scleria corymbifera Boeck. = ***Scleria corymbosa*** Roxb.
Scleria costaricensis Boeck. = ***Scleria tenella*** Kunth
Scleria costata (Britton) Small = ***Scleria baldwinii*** (Torr.) Steud.
Scleria cubensis Boeck. = ***Scleria mucronata*** Poir.
Scleria curtissii Britton = ***Scleria pauciflora*** var. ***curtissii*** (Britton) Fairey
Scleria cyanocarpa Kunth = ***Scleria latifolia*** Sw.
Scleria cymosa (Brongn.) Boeck. = ***Becquerelia cymosa*** Brongn.
Scleria cyperina var. *dentata* H.Pfeiff. = ***Scleria cyperina*** Kunth
Scleria cyperinoides C.B.Clarke = ***Scleria martii*** (Nees) Steud.
Scleria debilis C.Wright = ***Scleria muehlenbergii*** Steud.
Scleria densifolia (Nees) Steud. = ***Lagenocarpus minarum*** (Nees) Kuntze
Scleria dictyocarpa C.Wright ex Griseb. = ***Scleria muehlenbergii*** Steud.
Scleria dieterlenii Turrill = ***Scleria flexuosa*** Boeck.
Scleria dietrichiae Boeck. = ***Scleria levis*** Retz.
Scleria diffusa Michx. ex Kunth = ***Scleria verticillata*** Muhl. ex Willd.
Scleria diffusa C.B.Clarke = ***Scleria virgata*** (Nees) Steud.
Scleria dillonia Boeck. = ***Scleria foliosa*** Hochst. ex A.Rich.
Scleria dioica Leconte ex Britton = ***Scleria baldwinii*** (Torr.) Steud.
Scleria discolor (Kunth) Boeck. = ***Becquerelia discolor*** Kunth
Scleria distans R.Br. = ***Scleria brownii*** Kunth
Scleria distans var. *interrupta* (Rich.) Kük. = ***Scleria interrupta*** Rich.
Scleria diurensis Boeck. = ***Scleria lagoensis*** Boeck. subsp. ***lagoensis***
Scleria divaricata Nees = ***Becquerelia divaricata*** (Nees) Palla
Scleria doederleiniana Boeck. = ***Scleria scrobiculata*** Nees & Meyen subsp. ***scrobiculata***
Scleria doradoensis Britton = ***Scleria hirtella*** Sw.
Scleria dumicola Ridl. = ***Scleria foliosa*** Hochst. ex A.Rich.
Scleria duvigneaudii Piérart = ***Scleria induta*** Turrill
Scleria elata Thwaites = ***Scleria terrestris*** (L.) Fassett var. ***terrestris***
Scleria elata C.Wright = ***Scleria wrightiana*** Boeck.
Scleria elata var. *decolorans* C.B.Clarke = ***Scleria terrestris*** (L.) Fassett var. ***terrestris***
Scleria elata var. *latior* C.B.Clarke = ***Scleria terrestris*** (L.) Fassett var. ***terrestris***
Scleria elliottii Chapm. = ***Scleria ciliata*** var. ***elliottii*** (Chapm.) Fernald
Scleria elongata Piérart = ***Scleria elongatissima*** Piérart
Scleria elongata J.Presl & C.Presl = ***Scleria lithosperma*** (L.) Sw. var. ***lithosperma***
Scleria enchroides Kunth = ***Scleria hirtella*** Sw.
Scleria exaltata Boeck. = ***Scleria terrestris*** (L.) Fassett var. ***terrestris***
Scleria exigua J.Kern = ***Diplacrum exiguum*** (J.Kern) T.Koyama
Scleria fasciculata Willd. ex Kunth = ***Rhynchospora cephalotes*** (L.) Vahl
Scleria fenestrata Franch. & Sav. = ***Scleria parvula*** Steud.

Scleria fenestrata var. *pubigera* (Makino) Ohwi = ***Scleria rugosa*** R.Br.
Scleria ferruginea Ohwi = ***Scleria biflora*** Roxb.
Scleria ferruginea (Peter) Peter = ***Scleria bulbifera*** Hochst. ex A.Rich.
Scleria filiformis Sw. = ***Scleria lithosperma*** (L.) Sw. var. ***lithosperma***
Scleria filipendula S.T.Blake = ***Scleria laxa*** R.Br.
Scleria flabellum Poir. = ***Scleria flagellum-nigrorum*** P.J.Bergius
Scleria flaccida C.B.Clarke = ***Scleria rugosa*** R.Br.
Scleria flaccida Steud. = ***Scleria triglomerata*** Michx.
Scleria flagellata Sw. ex Boeck. = ***Scleria melaleuca*** Rchb. ex Schltdl. & Cham.
Scleria flagellum Benth. = ***Scleria naumanniana*** Boeck.
Scleria flagellum (J.F.Gmel.) Sw. = ***Scleria secans*** (L.) Urb.
Scleria flagellum Griseb. = ***Scleria secans*** (L.) Urb.
Scleria floribunda Kunth = ***Scleria bracteata*** Cav.
Scleria foliosa C.Wright = ***Scleria microcarpa*** Nees ex Kunth
Scleria foveolata Llanos = ***Scleria scrobiculata*** Nees & Meyen subsp. ***scrobiculata***
Scleria georgiana var. *filiculmis* (Boeck.) H.Pfeiff. = ***Scleria filiculmis*** Boeck.
Scleria glabra (Chapm.) Britton = ***Scleria ciliata*** var. ***glabra*** (Chapm.) Fairey
Scleria glabroreticulata De Wild. = ***Scleria mikawana*** Makino
Scleria glandiformis Boeck. = ***Scleria tessellata*** Willd.
Scleria glaucescens J.Presl & C.Presl = ***Scleria lithosperma*** (L.) Sw. var. ***lithosperma***
Scleria glazioviana Boeck. = ***Scleria acanthocarpa*** Boeck.
Scleria glomerulata Oliv. = ***Scleria distans*** var. ***glomerulata*** (Oliv.) Lye
Scleria gonocarpa Ridl. = ***Scleria motleyi*** C.B.Clarke subsp. ***motleyi***
Scleria gracilis Elliott = ***Scleria georgiana*** Core
Scleria gracilis Rich. = ***Scleria lithosperma*** (L.) Sw. var. ***lithosperma***
Scleria graeffeana Boeck. = ***Scleria polycarpa*** Boeck.
Scleria graminifolia Vasey = ***Scleria bourgeaui*** Boeck.
Scleria grandifolia Miq. = ***Scleria latifolia*** Sw.
Scleria grandis Core = ***Scleria robusta*** Camelb. & Goetgh.
Scleria grandis Labill. = ***Gahnia grandis*** (Labill.) S.T.Blake
Scleria grata Nelmes = ***Scleria melanotricha*** Hochst. & A.Rich.
Scleria grisebachii C.B.Clarke = ***Scleria eggersiana*** Boeck.
Scleria guahensis Schrad. ex Boeck. = ***Scleria flagellum-nigrorum*** P.J.Bergius
Scleria guianensis (Nees) Steud. = ***Lagenocarpus guianensis*** Nees
Scleria haematostachys Boeck. = ***Scleria terrestris*** (L.) Fassett var. ***terrestris***
Scleria hasskarliana Boeck. = ***Scleria terrestris*** (L.) Fassett var. ***terrestris***
Scleria hebecarpa Nees = ***Scleria levis*** Retz.
Scleria hebecarpa var. *pubescens* (Steud.) C.B.Clarke = ***Scleria levis*** Retz.
Scleria hemitaphra Steud. = ***Scleria muehlenbergii*** Steud.
Scleria hilsenbergii Boeck. = ***Scleria pulchella*** Ridl.
Scleria hirsuta Moon = ***Scleria pilosa*** Boeck.
Scleria hirsuta Boeck. = ***Scleria terrestris*** (L.) Fassett var. ***terrestris***
Scleria hirtella Michx. = ***Scleria ciliata*** Michx. var. ***ciliata***
Scleria hirtella Boeck. = ***Scleria distans*** Poir. var. ***distans***
Scleria hirtella var. *aterrima* Ridl. = ***Scleria catophylla*** C.B.Clarke
Scleria hirtella var. *chondrocarpa* Nelmes = ***Scleria distans*** var. ***chondrocarpa*** (Nelmes) Lye
Scleria hirtella var. *glabra* C.Presl = ***Scleria hirtella*** Sw.
Scleria hirtella var. *glabrescens* Boeck. = ***Scleria distans*** Poir.
Scleria hirtella var. *killipiana* (Britton ex Core) H.Pfeiff. = ***Scleria killipiana*** Britton
Scleria hirtella var. *pauciciliata* Britton = ***Scleria hirtella*** Sw.
Scleria hirtella var. *pubescens* C.Presl = ***Scleria hirtella*** Sw.
Scleria hirtella var. *tuberculata* Boeck. ex C.B.Clarke = ***Scleria hirtella*** Sw.
Scleria hispidula var. *hispidior* C.B.Clarke = ***Scleria hispidior*** (C.B.Clarke) Nelmes
Scleria hitchcockii Standl. = ***Scleria purdiei*** C.B.Clarke
Scleria holcoides Kunth = ***Scleria dregeana*** Kunth
Scleria homonyma Steud. = ***Scleria scabriuscula*** Schltdl.
Scleria hookeriana Boeck. = ***Scleria terrestris*** var. ***hookeriana*** (Boeck.) D.M.Verma
Scleria hostmanniana Steud. = ***Lagenocarpus verticillatus*** (Spreng.) T.Koyama & Maguire
Scleria humilis Britton = ***Scleria hirtella*** Sw.
Scleria humilis (Nees) Steud. = ***Lagenocarpus humilis*** (Nees) Kuntze
Scleria hypoxis Schweinf. ex Boeck. = ***Scleria schimperiana*** Boeck.
Scleria interrupta Michx. = ***Scleria hirtella*** Sw.
Scleria interrupta Schltdl. = ***Scleria hispidula*** Hochst. ex A.Rich.
Scleria japonica Steud. = ***Scleria levis*** Retz.
Scleria jelskiana Boeck. = ***Scleria reticularis*** Michx.
Scleria junciformis (Nees) Thwaites = ***Scleria pilosa*** Boeck.
Scleria junciformis Welw. ex Ridl. = ***Scleria welwitschii*** C.B.Clarke
Scleria junciformis Kunth = ***Lagenocarpus junciformis*** (Kunth) Kuntze
Scleria kappleriana Hochst. ex Steud = ***Scleria latifolia*** Sw.
Scleria keyensis K.Schum. ex Warb. = ***Scleria scrobiculata*** Nees & Meyen subsp. ***scrobiculata***
Scleria khasiana C.B.Clarke ex Hook.f. = ***Scleria benthamii*** C.B.Clarke
Scleria krugiana Boeck. = ***Scleria lithosperma*** (L.) Sw. var. ***lithosperma***
Scleria kunthiana Steud. = ***Scleria verticillata*** Muhl. ex Willd.
Scleria kunthii Miq. = ***Lagenocarpus verticillatus*** (Spreng.) T.Koyama & Maguire
Scleria kuntzei Boeck. = ***Scleria terrestris*** (L.) Fassett var. ***terrestris***
Scleria lacunosa Boeck. = ***Scleria latifolia*** Sw.
Scleria laevis Willd. = ***Scleria oblata*** S.T.Blake ex J.Kern
Scleria lateriflora Boeck. = ***Scleria rugosa*** R.Br.
Scleria lateritica Nelmes = ***Scleria flexuosa*** Boeck.
Scleria latifolia Nees = ***Scleria latifolia*** Sw.
Scleria latifolia (Vahl) C.Presl = ***Scleria melaleuca*** Rchb. ex Schltdl. & Cham.
Scleria latifolia Sieber ex C.Presl = ***Scleria sieberi*** Nees
Scleria latifolia var. *arundinacea* (Kunth) H.Pfeiff. =

Scleria latifolia Sw.
Scleria latifolia subsp. *foliosa* (Nees) T.Koyama = **Scleria plusiophylla** Steud.
Scleria latilacunosa Kük. = **Scleria muehlenbergii** Steud.
Scleria laxa Torr. = **Scleria muehlenbergii** Steud.
Scleria laxa var. *pseudotesselata* Domin = **Scleria laxa** R.Br.
Scleria leptocladia Poepp. ex Kunth = **Lagenocarpus verticillatus** (Spreng.) T.Koyama & Maguire
Scleria leptoculmis Alph.Wood = **Scleria oligantha** Michx.
Scleria levis var. *pubescens* (Steud.) C.Z.Zheng = **Scleria levis** Retz.
Scleria liebmannii Steud. = **Scleria tenella** Kunth
Scleria lindleyana C.B.Clarke = **Scleria hirtella** Sw.
Scleria lithosperma var. *filiformis* (Sw.) Britton = **Scleria lithosperma** (L.) Sw. var. **lithosperma**
Scleria lithosperma subsp. *linearis* (Benth.) T.Koyama = **Scleria lithosperma** var. **linearis** Benth.
Scleria lithosperma var. *multispiculata* Govind. = **Scleria lithosperma** var. **linearis** Benth.
Scleria lithosperma var. *muricata* Govind. = **Scleria lithosperma** var. **linearis** Benth.
Scleria lithosperma f. *ramosa* Domin = **Scleria lithosperma** (L.) Sw. var. **lithosperma**
Scleria lithosperma var. *roxburghii* C.B.Clarke = **Scleria lithosperma** var. **linearis** Benth.
Scleria lobulata Palla = **Scleria scabra** Willd.
Scleria longifolia Boeck. = **Scleria melaleuca** Rchb. ex Schltdl. & Cham.
Scleria longifolia (Griseb.) Roberty = **Diplacrum capitatum** (Willd.) Boeck.
Scleria lucida (Nees) Steud. = **Scleria mitis** P.J.Bergius
Scleria luzonensis Palla = **Scleria terrestris** (L.) Fassett var. **terrestris**
Scleria luzuliformis C.Wright = **Scleria tenella** Kunth
Scleria mackaviensis var. *atrata* Domin = **Scleria mackaviensis** Boeck.
Scleria mackaviensis var. *pallida* Domin = **Scleria mackaviensis** Boeck.
Scleria macrantha Boeck. = **Scleria ciliata** Michx. var. **ciliata**
Scleria macrantha Boeck. = **Scleria melanomphala** Kunth
Scleria macrocarpa Salzm. ex Boeck. = **Scleria macrophylla** J.Presl & C.Presl
Scleria macrocarpa Nees = **Scleria racemosa** Poir.
Scleria macrocarpa Wall. = **Scirpodendron ghaeri** (Gaertn.) Merr.
Scleria macrosperma Nees ex Kunth = **Scleria racemosa** Poir.
Scleria major Moon = **Scleria corymbosa** Roxb.
Scleria malaccensis Boeck. = **Scleria ciliaris** Nees
Scleria margaritifera Gaertn. = **Scleria flagellum-nigrorum** P.J.Bergius
Scleria margaritifera Willd. = **Scleria polycarpa** Boeck.
Scleria mayottensis C.B.Clarke = **Scleria lagoensis** subsp. **canaliculatotriquetra** (Boeck.) Lye
Scleria mechowiana Boeck. = **Scleria bulbifera** Hochst. ex A.Rich.
Scleria melanocephala Drège = **Scleria melanomphala** Kunth
Scleria melanosperma Nees & Arn. ex Boeck. = **Scleria levis** Retz.
Scleria melanostoma Boeck. = **Scleria terrestris** (L.) Fassett var. **terrestris**
Scleria melanotricha var. *glabrior* C.B.Clarke = **Scleria interrupta** Rich.
Scleria melanotricha var. *grata* (Nelmes) Lye = **Scleria melanotricha** Hochst. & A.Rich.
Scleria merkeliana (Nees) C.B.Clarke = **Becquerelia divaricata** (Nees) Palla
Scleria merrillii Palla = **Scleria novae-hollandiae** Boeck.
Scleria mexicana (Liebm.) Boeck. = **Scleria sororia** Kunth
Scleria meyeriana Kunth = **Scleria dregeana** Kunth
Scleria michauxii Chapm. = **Scleria hirtella** Sw.
Scleria micrantha C.B.Clarke = **Scleria polycarpa** Boeck.
Scleria micrantha Poir. = **Rhynchospora contracta** (Nees) J.Raynal
Scleria microcarpa var. *foliosa* C.B.Clarke = **Scleria microcarpa** Nees ex Kunth
Scleria microcarpa var. *latifolia* Boeck. = **Scleria eggersiana** Boeck.
Scleria microcarpa var. *phylloptera* (C.Wright ex Griseb.) Kük. = **Scleria testacea** Nees ex Kunth
Scleria microcarpa var. *subeciliata* C.B.Clarke = **Scleria mucronata** Poir.
Scleria microcarpa var. *subeggersiana* Kük. ex O.C.Schmidt = **Scleria microcarpa** Nees ex Kunth
Scleria micrococca (Liebm.) Steud. = **Scleria tenella** Kunth
Scleria microdiscus Steud. = **Rhynchospora berteroi** (Spreng.) C.B.Clarke
Scleria minarum (Nees) Steud. = **Lagenocarpus minarum** (Nees) Kuntze
Scleria minima f. *angustifolia* Kük. = **Scleria minima** C.B.Clarke
Scleria minutiflora Rich. ex Spreng. = **Rhynchospora contracta** (Nees) J.Raynal
Scleria mitis var. *atrobrunnea* Kük. = **Scleria mitis** P.J.Bergius
Scleria mitis subsp. *eggersiana* (Boeck.) T.Koyama = **Scleria eggersiana** Boeck.
Scleria mitis var. *subeciliata* (C.B.Clarke) H.Pfeiff. = **Scleria mucronata** Poir.
Scleria mitracarpoides Standl. & L.O.Williams = **Diplacrum mitracarpoides** (Standl. & L.O.Williams) C.D.Adams
Scleria mollis Kunth = **Scleria hirtella** Sw.
Scleria moritziana Boeck. = **Scleria lagoensis** Boeck. subsp. **lagoensis**
Scleria motleyi var. *densispicata* C.B.Clarke = **Scleria densispicata** (C.B.Clarke) J.Kern
Scleria muhlenbergiana Liebm. = **Scleria reticularis** Michx.
Scleria multifoliata Boeck. = **Scleria purpurascens** Steud. var. **purpurascens**
Scleria multifoliolata var. *ophirensis* C.B.Clarke = **Scleria purpurascens** var. **ophirensis** (C.B.Clarke) J.Kern
Scleria multispiculata Boeck. = **Scleria poaeoides** Ridl.
Scleria multispiculosa Boeck. = **Scleria poaeoides** Ridl.
Scleria muricata Boeck. = **Becquerelia muricata** (Boeck.) Nees
Scleria nana Boeck. = **Rhynchospora pilosa** Boeck.
Scleria neesiana Hook. & Arn. = **Scleria scrobiculata** Nees & Meyen subsp. **scrobiculata**
Scleria nervosa Wikstr. = **Scleria latifolia** Sw.
Scleria nigricans J.Presl & C.Presl = **Scleria flagellum-nigrorum** P.J.Bergius
Scleria nitida Willd. ex Kunth = **Scleria triglomerata** Michx.
Scleria notoptera Nees ex Steud. = **Diplacrum guianense** (Nees) T.Koyama

Scleria nutans Willd. ex Kunth = ***Scleria distans*** Poir. var. ***distans***
Scleria olyroides Kunth = ***Scleria latifolia*** Sw.
Scleria onoei Franch. & Sav. = ***Scleria rugosa*** R.Br.
Scleria oryzoides J.Presl & C.Presl = ***Scleria poiformis*** Retz.
Scleria ottonis Boeck. = ***Scleria melaleuca*** Rchb. ex Schltdl. & Cham.
Scleria ovuligera Rchb. ex Nees = ***Scleria induta*** Turrill
Scleria ovuligera Nees ex Boeck. = ***Scleria naumanniana*** Boeck.
Scleria pallidiflora Boeck. = ***Scleria brownii*** Kunth
Scleria palmetto Rchb. ex Core = ***Calyptrocarya glomerulata*** (Brongn.) Urb.
Scleria palmifolia Hoffmanns. & Schltdl. = ***Scleria macrophylla*** J.Presl & C.Presl
Scleria paludosa Poepp. & Kunth = ***Scleria macrophylla*** J.Presl & C.Presl
Scleria papillata Willd. ex Kunth = ***Scleria bracteata*** Cav.
Scleria pauciflora Sauvalle = ***Scleria lithosperma*** (L.) Sw. var. ***lithosperma***
Scleria pauciflora var. *effusa* C.B.Clarke = ***Scleria pauciflora*** Muhl. ex Willd. var. ***pauciflora***
Scleria pauciflora var. *elongata* Britton = ***Scleria pauciflora*** Muhl. ex Willd. var. ***pauciflora***
Scleria pauciflora var. *glabra* Chapm. = ***Scleria ciliata*** var. ***glabra*** (Chapm.) Fairey
Scleria pauciflora var. *kansana* Fernald = ***Scleria pauciflora*** Muhl. ex Willd. var. ***pauciflora***
Scleria pergracilis var. *brachystachys* Nelmes = ***Scleria pergracilis*** (Nees) Kunth
Scleria perrieri Cherm. = ***Scleria foliosa*** Hochst. ex A.Rich.
Scleria peruviana Kük. = ***Scleria huberi*** C.B.Clarke
Scleria phyllomacra Steud. = ***Scleria macrophylla*** J.Presl & C.Presl
Scleria phylloptera C.Wright ex Griseb. = ***Scleria testacea*** Nees ex Kunth
Scleria pinetorum Britton = ***Scleria interrupta*** Rich.
Scleria pittieri Boeck. = ***Scleria melaleuca*** Rchb. ex Schltdl. & Cham.
Scleria pleostachya Kunth = ***Scleria spicata*** (Spreng.) J.F.Macbr.
Scleria ploemii Boeck. = ***Scleria terrestris*** (L.) Fassett var. ***terrestris***
Scleria polyphylla (Nees) Steud. = ***Lagenocarpus polyphyllus*** (Nees) Kuntze
Scleria porphyrorhiza C.Wright = ***Scleria scabra*** Willd.
Scleria praealta Salzm. ex Schltdl. = ***Scleria mitis*** P.J.Bergius
Scleria pratensis Lindl. ex Nees = ***Scleria melaleuca*** Rchb. ex Schltdl. & Cham.
Scleria pratensis var. *melanocarpa* Nees = ***Scleria melaleuca*** Rchb. ex Schltdl. & Cham.
Scleria prophyllata Nelmes = ***Scleria poaeoides*** Ridl.
Scleria propinqua Steud. = ***Scleria biflora*** Roxb.
Scleria pterota C.Presl = ***Scleria melaleuca*** Rchb. ex Schltdl. & Cham.
Scleria pterota var. *melaleuca* (Rchb. ex Schltdl. & Cham.) Uittien = ***Scleria melaleuca*** Rchb. ex Schltdl. & Cham.
Scleria pterota var. *submelaleuca* Kük. = ***Scleria melaleuca*** Rchb. ex Schltdl. & Cham.
Scleria pubescens Steud. = ***Scleria levis*** Retz.
Scleria pubigera Makino = ***Scleria rugosa*** R.Br.
Scleria purpurascens Benth. = ***Scleria harlandii*** Hance
Scleria purpurea Poir. = ***Scleria lithosperma*** (L.) Sw. var. ***lithosperma***
Scleria purpureovaginata Boeck. = ***Scleria scrobiculata*** Nees & Meyen subsp. ***scrobiculata***
Scleria purpureozonata Liebm. = ***Scleria oligantha*** Michx.
Scleria puzzolanea K.Schum. = ***Scleria lithosperma*** (L.) Sw. var. ***lithosperma***
Scleria pygmaea R.Br. = ***Diplacrum pygmaeum*** (R.Br.) Nees ex Boeck.
Scleria pygmaeopsis J.Kern = ***Diplacrum pygmaeopsis*** (J.Kern) T.Koyama
Scleria pyramidalis Hochst. ex Steud. = ***Scleria martii*** (Nees) Steud.
Scleria racemosa Bojer = ***Scleria sieberi*** Nees
Scleria racemosa var. *depressa* C.B.Clarke = ***Scleria depressa*** (C.B.Clarke) Nelmes
Scleria racemosa subsp. *depressa* (C.B.Clarke) J.Raynal = ***Scleria depressa*** (C.B.Clarke) Nelmes
Scleria radula Hance = ***Scleria terrestris*** (L.) Fassett var. ***terrestris***
Scleria reflexa Benth. = ***Scleria boivinii*** Steud.
Scleria reflexa Kunth = ***Scleria secans*** (L.) Urb.
Scleria reflexa var. *surinamensis* Boeck. = ***Scleria secans*** (L.) Urb.
Scleria rehmannii var. *ornata* Cherm. = ***Scleria woodii*** var. ***ornata*** (Cherm.) J.Schultze-Motel
Scleria remota Ridl. = ***Scleria flexuosa*** Boeck.
Scleria remota var. *hispida* Ridl. = ***Scleria bicolor*** Nelmes
Scleria renggeriana Steud. = ***Scleria secans*** (L.) Urb.
Scleria reticularis Muhl. = ***Scleria muehlenbergii*** Steud.
Scleria reticularis var. *obscura* Britton = ***Scleria reticularis*** Michx.
Scleria reticularis var. *pubescens* Britton = ***Scleria muehlenbergii*** Steud.
Scleria reticularis var. *pumila* Britton = ***Scleria reticularis*** Michx.
Scleria ridleyi C.B.Clarke = ***Scleria corymbosa*** Roxb.
Scleria rigida Kunth = ***Lagenocarpus rigidus*** (Kunth) Nees
Scleria rinkiana Boeck. = ***Scleria terrestris*** (L.) Fassett var. ***terrestris***
Scleria riparia Poepp. & Kunth = ***Scleria mitis*** P.J.Bergius
Scleria rosea Cherm. = ***Scleria trialata*** Poir.
Scleria roxburghii (C.B.Clarke) Domin = ***Scleria lithosperma*** var. ***linearis*** Benth.
Scleria roxburghii var. *australiensis* Domin = ***Scleria lithosperma*** var. ***linearis*** Benth.
Scleria salebrosa Spreng. ex Boeck. = ***Scleria pergracilis*** (Nees) Kunth
Scleria salebrosa Boeck. ex C.B.Clarke. = ***Scleria pergracilis*** (Nees) Kunth
Scleria scaberrima Boeck. = ***Scleria scindens*** Nees ex Kunth
Scleria scabrosa Maury ex Micheli = ***Scleria composita*** (Nees) Boeck.
Scleria schiedeana Schltdl. = ***Scleria scabriuscula*** Schltdl.
Scleria schliebenii R.Gross = ***Scleria bulbifera*** Hochst. ex A.Rich.
Scleria schliebenii var. *ferruginea* Peter = ***Scleria bulbifera*** Hochst. ex A.Rich.
Scleria schweinfurthiana Boeck. = ***Scleria bulbifera*** Hochst. ex A.Rich.
Scleria schweinfurthiana var. *melanocarpa* Cherm. = ***Scleria interrupta*** Rich.
Scleria scrobiculata Moritzi = ***Scleria ciliaris*** Nees
Scleria selloana Schrad. ex Nees = ***Scleria melaleuca*** Rchb. ex Schltdl. & Cham.

Scleria sellowiana var. *balansae* (Maury ex Micheli) C.B.Clarke = ***Scleria balansae*** Maury ex Micheli
Scleria sellowiana f. *glabriuscula* Osten = ***Scleria balansae*** Maury ex Micheli
Scleria setacea Poir. = ***Rhynchospora tenerrima*** Nees ex Spreng. subsp. ***tenerrima***
Scleria setacea var. *hemitaphra* (Steud.) Kük. = ***Scleria reticularis*** Michx.
Scleria setifera Boeck. = ***Schoenus maschalinus*** Roem. & Schult.
Scleria setigera Roxb. = ***Scleria sumatrensis*** Retz.
Scleria setosoasperula Boeck. = ***Scleria brownii*** Kunth
Scleria setulosa Boeck. = ***Scleria dregeana*** Kunth
Scleria simplicior Steud. = ***Scleria melaleuca*** Rchb. ex Schltdl. & Cham.
Scleria sinensis (Boeck.) H.Pfeiff. = ***Carex moupinensis*** Franch.
Scleria sorsogonensis Elmer ex Merr. = ***Scleria motleyi*** subsp. ***rostrata*** J.Kern
Scleria spicata var. *castanea* (Core) H.Pfeiff. =
Scleria spinulosa Boeck. = ***Scleria verrucosa*** Willd.
Scleria spondylogona Nelmes = ***Scleria delicatula*** Nelmes
Scleria steudeliana Miq. = ***Scleria biflora*** Roxb.
Scleria stevensiana Britton = ***Scleria muehlenbergii*** Steud.
Scleria striatonux De Wild. = ***Scleria woodii*** var. ***ornata*** (Cherm.) J.Schultze-Motel
Scleria striatonux var. *lacunosa* Piérart = ***Scleria woodii*** C.B.Clarke var. ***woodii***
Scleria stricta Moon ex Pfeiff. = ***Scleria neesii*** Kunth
Scleria stricta Kunth = ***Lagenocarpus verticillatus*** (Spreng.) T.Koyama & Maguire
Scleria subintegriloba De Wild. = ***Scleria achtenii*** De Wild.
Scleria subrostrata Elmer = ***Scleria motleyi*** subsp. ***rostrata*** J.Kern
Scleria substriatoalveolata De Wild. = ***Scleria achtenii*** De Wild.
Scleria subulata (J.F.Gmel.) Steud. = ***Scleria lithosperma*** (L.) Sw. var. ***lithosperma***
Scleria suffulta C.B.Clarke = [50 QLD]
Scleria sumatrensis var. *pubescens* Miq. = ***Scleria levis*** Retz.
Scleria sylvestris Poepp. & Kunth = ***Scleria latifolia*** Sw.
Scleria tenella Griseb. = ***Scleria distans*** Poir. var. ***distans***
Scleria tenuiflora Willd. ex Kunth = ***Scleria verticillata*** Muhl. ex Willd.
Scleria tenuis Retz. = ***Scleria lithosperma*** (L.) Sw. var. ***lithosperma***
Scleria ternifolia Domin = ***Scleria polycarpa*** Boeck.
Scleria tessellata Benth. = ***Scleria benthamii*** C.B.Clarke
Scleria tessellata Decne. = ***Scleria scrobiculata*** Nees & Meyen subsp. ***scrobiculata***
Scleria tessellata var. *debilis* Benth. = ***Scleria tricuspidata*** S.T.Blake
Scleria tessellata var. *sphaerocarpa* E.A.Rob. = ***Scleria sphaerocarpa*** (E.A.Rob.) Napper
Scleria thomasii Piérart = ***Scleria bulbifera*** Hochst. ex A.Rich.
Scleria thomsoniana Boeck. = ***Scleria terrestris*** var. ***thomsoniana*** (Boeck.) D.M.Verma
Scleria timorensis Nees = ***Scleria scrobiculata*** Nees & Meyen subsp. ***scrobiculata***
Scleria tisserantii Cherm. = ***Scleria melanomphala*** Kunth
Scleria toksanensis Nakai = ***Scleria rugosa*** R.Br.
Scleria tokusanensis Nakai = ***Scleria rugosa*** R.Br.
Scleria tonduzii Boeck. = ***Scleria vaginata*** Steud.
Scleria topazina (Nees) Steud. = ***Lagenocarpus rigidus*** (Kunth) Nees subsp. ***rigidus***
Scleria torreyana Walp. = ***Scleria muehlenbergii*** Steud.
Scleria tremula (Nees) Steud. = ***Lagenocarpus rigidus*** (Kunth) Nees subsp. ***rigidus***
Scleria trialata Bertero ex Boeck. = ***Scleria mitis*** P.J.Bergius
Scleria trialata var. *rosea* (Cherm.) Cherm. = ***Scleria trialata*** Poir.
Scleria tricholepis Nelmes = ***Scleria interrupta*** Rich.
Scleria trichopoda C.Wright ex Britton = ***Scleria reticularis*** Michx.
Scleria triglomerata var. *gracilis* Britton = ***Scleria minor*** (Britton) W.Stone
Scleria triglomerata var. *minor* Britton = ***Scleria minor*** (Britton) W.Stone
Scleria trigona Merr. = ***Scleria motleyi*** subsp. ***rostrata*** J.Kern
Scleria trigonocarpa Steud. = ***Scleria latifolia*** Sw.
Scleria trigonocarpa Ridl. = ***Scleria motleyi*** C.B.Clarke subsp. ***motleyi***
Scleria trigonocarpa Boeck. = ***Scleria virgata*** (Nees) Steud.
Scleria trinitatis Boeck. = ***Scleria mitis*** P.J.Bergius
Scleria tristis A.St.-Hil. = ***Lagenocarpus alboniger*** (A.St.-Hil.) C.B.Clarke
Scleria tuberculata Boeck. = ***Becquerelia tuberculata*** (Boeck.) H.Pfeiff.
Scleria uliginosa Hochst. ex Boeck. = ***Scleria parvula*** Steud.
Scleria ustulata Ridl. = ***Scleria pulchella*** Ridl.
Scleria ustulata F.M.Bailey = ***Exocarya sclerioides*** (F.Muell.) Benth.
Scleria vanderystii De Wild. = ***Scleria lagoensis*** Boeck. subsp. ***lagoensis***
Scleria velutina (Nees) Steud. = ***Lagenocarpus velutinus*** Nees
Scleria verdickii De Wild. = ***Scleria bulbifera*** Hochst. ex A.Rich.
Scleria verticillata f. *brevis* Kük. = ***Scleria verticillata*** Muhl. ex Willd.
Scleria verticillata f. *capillaris* Kük. = ***Scleria verticillata*** Muhl. ex Willd.
Scleria verticillata var. *tenella* (Kunth) Kük. = ***Scleria verticillata*** Muhl. ex Willd.
Scleria vestita Boeck. = ***Scleria levis*** Retz.
Scleria villosula Wall. = ***Scleria levis*** Retz.
Scleria viridifolia Steud. = ***Lagenocarpus rigidus*** (Kunth) Nees subsp. ***rigidus***
Scleria waigiouensis Steud. = ***Scleria scrobiculata*** Nees & Meyen subsp. ***scrobiculata***
Scleria weigeltiana Schrad. ex Boeck. = ***Scleria secans*** (L.) Urb.
Scleria wichurai Boeck. = ***Scleria levis*** Retz.
Scleria wightiana Steud. = ***Scleria lithosperma*** (L.) Sw. var. ***lithosperma***
Scleria zelanica Moon = ***Scleria levis*** Retz.
Scleria zeylanica Poir. = ***Scleria levis*** Retz.

Unplaced Names:
Scleria suffulta C.B.Clarke, Kew Bull., Addit. Ser. 8: 58 (1908). = [50 QLD]

Sclerochaetium

Sclerochaetium Nees = ***Tetraria*** P.Beauv.
Sclerochaetium angustifolium Hochst. = ***Tetraria bromoides*** (Lam.) H.Pfeiff.
Sclerochaetium dactyloides (C.Presl) Nees ex C.B.Clarke = ***Tetraria involucrata*** (Rottb.) C.B.Clarke

Sclerochaetium giganteum Steud. = ***Tetraria thermalis*** (L.) C.B.Clarke
Sclerochaetium involucratum (Rottb.) Nees = ***Tetraria involucrata*** (Rottb.) C.B.Clarke
Sclerochaetium koenigii Nees ex Hochst. = ***Tetraria bromoides*** (Lam.) H.Pfeiff.
Sclerochaetium rottboellii (Schrad.) Nees = ***Tetraria bromoides*** (Lam.) H.Pfeiff.
Sclerochaetium spirale Hochst. = ***Tetraria spiralis*** (Hochst.) C.B.Clarke
Sclerochaetium teres Nees ex C.B.Clarke = ***Tetraria bromoides*** (Lam.) H.Pfeiff.
Sclerochaetium thermale (L.) Nees = ***Tetraria thermalis*** (L.) C.B.Clarke

Scuria

Scuria Raf. = ***Carex*** L.

Seidlia

Seidlia Opiz = ***Scirpus*** Tourn. ex L.
Seidlia atrovirens Opiz = ***Scirpoides holoschoenus*** (L.) Soják subsp. ***holoschoenus***
Seidlia radicans Opiz = ***Scirpus sylvaticus*** L.
Seidlia sylvatica (L.) Opiz = ***Scirpus sylvaticus*** L.

Sickmannia

Sickmannia Nees = ***Ficinia*** Schrad.
Sickmannia radiata (L.f.) Nees = ***Ficinia radiata*** (L.f.) Kunth

Sorostachys

Sorostachys Steud. = ***Cyperus*** L.
Sorostachys kyllingioides Steud. = ***Cyperus leucocephalus*** Retz.
Sorostachys leucocephalus (Retz.) Lye = ***Cyperus leucocephalus*** Retz.
Sorostachys pulchellus (R.Br.) Lye = ***Cyperus pulchellus*** R.Br.

Spermodon

Spermodon P.Beauv. ex Lestib. = ***Rhynchospora*** Vahl
Spermodon confinis Nees = ***Rhynchospora confinis*** (Nees) C.B.Clarke
Spermodon curvulus Nees = ***Rhynchospora tenuis*** subsp. ***tenuis***
Spermodon edentulatus Nees = ***Rhynchospora junciformis*** (Kunth) Boeck.
Spermodon eximius Nees = ***Rhynchospora eximia*** (Nees) Boeck.
Spermodon filiformis (Kunth) Nees = ***Rhynchospora filiformis*** Vahl
Spermodon setaceus (P.J.Bergius) Nees = ***Rhynchospora tenerrima*** Nees ex Spreng. subsp. ***tenerrima***

Sphaerocyperus

Sphaerocyperus Lye, Bot. Not. 125: 214 (1972).
Tanzania to Angola. 25 TAN 26 ANG.
1 Species

Sphaerocyperus erinaceus (Ridl.) Lye, Bot. Not. 125: 214 (1972).
Tanzania to Angola. 25 TAN 26 ANG. Hemicr. or rhizome geophyte.
**Schoenus erinaceus* Ridl., Trans. Linn. Soc. London, Bot. 2: 165 (1884). *Rhynchospora erinacea* (Ridl.) C.B.Clarke in T.A.Durand & H.Schinz, Consp. Fl. Afric. 5: 654 (1894). *Cyperus erinaceus* (Ridl.) Kük., Boissiera 7: 103 (1943). *Actinoschoenus erinaceus* (Ridl.) Raymond, Mitt. Bot. Staatssamml. München 10: 588 (1971).

Sphaeromariscus

Sphaeromariscus E.G.Camus = ***Cyperus*** L.
Sphaeromariscus microcephalus (J.Presl & C.Presl) E.G.Camus = ***Cyperus compactus*** Retz.

Sphaeropus

Sphaeropus Boeck. = ***Scleria*** P.J.Bergius
Sphaeropus pygmaeus (R.Br.) Boeck. = ***Diplacrum pygmaeum*** (R.Br.) Nees ex Boeck.

Sphaeroschoenus

Sphaeroschoenus Nees = ***Rhynchospora*** Vahl
Sphaeroschoenus wallichii (Nees) Nees = ***Rhynchospora rubra*** (Lour.) Makino subsp. ***rubra***

Stenophyllus

Stenophyllus Raf. = ***Bulbostylis*** Kunth
Stenophyllus alpestris (Urb.) Britton = ***Bulbostylis alpestris*** Urb.
Stenophyllus antillanus Britton = ***Bulbostylis capillaris*** subsp. ***antillana*** (Britton) T.Koyama
Stenophyllus arenarius (Nees) J.F.Macbr. = ***Bulbostylis juncoides*** (Vahl) Kük. ex Herter
Stenophyllus argentinus (Palla) Killip = ***Bulbostylis juncoides*** (Vahl) Kük. ex Herter
Stenophyllus barbatus (Rottb.) T.Cooke = ***Bulbostylis barbata*** (Rottb.) C.B.Clarke
Stenophyllus barbatus (Rottb.) Fyson = ***Bulbostylis barbata*** (Rottb.) C.B.Clarke
Stenophyllus barbatus f. *oligostachyus* Domin = ***Bulbostylis barbata*** (Rottb.) C.B.Clarke
Stenophyllus caespitosus Raf. = ***Bulbostylis stenophylla*** (Elliott) C.B.Clarke
Stenophyllus capillaris (L.) Britton = ***Bulbostylis capillaris*** (L.) Kunth ex C.B.Clarke
Stenophyllus capillaris var. *ciliatus* (C.Presl) J.F.Macbr. = ***Bulbostylis ciliata*** (C.Presl) C.Presl ex Nees
Stenophyllus capillaris var. *coarctatus* (Elliott) Britton =
Stenophyllus capillaris var. *cryptostachys* Fernald = ***Bulbostylis capillaris*** (L.) Kunth ex C.B.Clarke subsp. ***capillaris***
Stenophyllus capillaris var. *trifidus* (Kunth) Domin = ***Bulbostylis capillaris*** (L.) Kunth ex C.B.Clarke subsp. ***capillaris***
Stenophyllus carteri Britton ex Small = ***Bulbostylis ciliatifolia*** var. ***coarctata*** (Elliott) Kral
Stenophyllus ciliatifolius (Elliott) C.Mohr = ***Bulbostylis ciliatifolia*** (Elliott) Fernald
Stenophyllus circinatus (Kunth) H.Pfeiff. = ***Bulbostylis circinata*** (Kunth) C.B.Clarke
Stenophyllus coarctatus (Elliott) Britton = ***Bulbostylis ciliatifolia*** var. ***coarctata*** (Elliott) Kral
Stenophyllus collinus (Kunth) Chiov. = ***Bulbostylis contexta*** (Nees) Bodard
Stenophyllus coniferus (Kunth) Britton = ***Bulbostylis conifera*** (Kunth) Beetle
Stenophyllus conspicuus (Boeck.) H.Pfeiff. = ***Bulbostylis conspicua*** (Boeck.) H.Pfeiff.
Stenophyllus craspedotus (Chiov.) Chiov. = ***Bulbostylis craspedota*** Chiov.
Stenophyllus curassavicus Britton = ***Bulbostylis curassavica*** (Britton) Kük. & Ekman

Stenophyllus filiformis (C.B.Clarke) H.Pfeiff. = ***Bulbostylis hispidula*** subsp. ***filiformis*** (C.B.Clarke) R.W.Haines
Stenophyllus fimbriatus (Nees) Britton = ***Bulbostylis fimbriata*** (Nees) C.B.Clarke
Stenophyllus floridanus Nash = ***Bulbostylis barbata*** (Rottb.) C.B.Clarke subsp. ***barbata***
Stenophyllus francaensis (Palla) H.Pfeiff. = ***Bulbostylis edwalliana*** (Boeck.) Prata & M.G.López
Stenophyllus funckii (Steud.) Britton = ***Bulbostylis funckii*** (Steud.) C.B.Clarke
Stenophyllus harrisii Britton = ***Bulbostylis subaphylla*** C.B.Clarke
Stenophyllus hirtellus (Schrad. ex Schult.) H.Pfeiff. = ***Bulbostylis juncoides*** (Vahl) Kük. ex Herter
Stenophyllus hirtellus var. *elatior* H.Pfeiff. = ***Bulbostylis juncoides*** (Vahl) Kük. ex Herter
Stenophyllus hirtellus var. *humilis* H.Pfeiff. = ***Bulbostylis juncoides*** (Vahl) Kük. ex Herter
Stenophyllus junciformis (Kunth) Britton = ***Bulbostylis junciformis*** (Kunth) C.B.Clarke
Stenophyllus lanatus (Kunth) Britton = ***Bulbostylis lanata*** (Kunth) Lindm.
Stenophyllus leucostachyus (Kunth) Britton = ***Bulbostylis leucostachya*** (Kunth) Kunth
Stenophyllus nesioticus I.M.Johnst. = ***Bulbostylis nesiotica*** (I.M.Johnst.) Fernald
Stenophyllus paradoxus (Spreng.) Standl. = ***Bulbostylis paradoxa*** (Spreng.) Lindm.
Stenophyllus pauciflorus (Liebm.) Britton = ***Bulbostylis pauciflora*** (Liebm.) C.B.Clarke
Stenophyllus portoricensis Britton = ***Bulbostylis pauciflora*** (Liebm.) C.B.Clarke
Stenophyllus puberulus (Poir.) Killip = ***Bulbostylis puberula*** C.B.Clarke
Stenophyllus scaber (J.Presl & C.Presl) H.Pfeiff. = ***Bulbostylis scabra*** (J.Presl & C.Presl) C.B.Clarke
Stenophyllus sellowianus (Kunth) I.M.Johnst. = ***Bulbostylis sellowiana*** (Kunth) Palla
Stenophyllus sphaerolepis (Boeck.) Rusby = ***Bulbostylis scabra*** (J.Presl & C.Presl) C.B.Clarke
Stenophyllus stenophyllus (Elliott) Britton = ***Bulbostylis stenophylla*** (Elliott) C.B.Clarke
Stenophyllus subaphyllus (C.B.Clarke) Britton = ***Bulbostylis subaphylla*** C.B.Clarke
Stenophyllus tenellus (Link) H.Pfeiff. = ***Bulbostylis tenella*** (Link) C.B.Clarke
Stenophyllus tenuifolius (Rudge) Britton = ***Bulbostylis tenuifolia*** (Rudge) J.F.Macbr.
Stenophyllus tenuifolius var. *asperula* (Kunth) H.Pfeiff. = ***Bulbostylis tenuifolia*** (Rudge) J.F.Macbr.
Stenophyllus tenuifolius var. *latesquamata* H.Pfeiff. = ***Bulbostylis tenuifolia*** (Rudge) J.F.Macbr.
Stenophyllus tenuifolius var. *longinux* H.Pfeiff. = ***Bulbostylis tenuifolia*** (Rudge) J.F.Macbr.
Stenophyllus tuerckheimii (Urb.) Britton = ***Bulbostylis tuerckheimii*** Urb.
Stenophyllus vestitus (Kunth) Britton = ***Bulbostylis vestita*** (Kunth) C.B.Clarke
Stenophyllus warei (Torr.) Britton = ***Bulbostylis warei*** (Torr.) C.B.Clarke
Stenophyllus wilsonii Britton = ***Bulbostylis wilsonii*** (Britton) Urb.

Streblidia

Streblidia Link = ***Schoenus*** L.
Streblidia ferruginea (L.) Link = ***Schoenus ferrugineus*** L.

Sumatroscirpus

Sumatroscirpus Oteng-Yeb., Notes Roy. Bot. Gard. Edinburgh 33: 307 (1974).
Sumatera. 42 SUM.
1 Species

Sumatroscirpus junghuhnii (Miq.) Oteng-Yeb., Notes Roy. Bot. Gard. Edinburgh 33: 307 (1974).
N. & C. Sumatera. 42 SUM. Hemicr. or rhizome geophyte.
**Scirpus junghuhnii* Miq., Fl. Ned. Ind. 3: 307 (1856).
Scirpus junghuhnii var. *minor* Kük., Bull. Jard. Bot. Buitenzorg, III, 16: 301 (1940).

Syntrinema

Syntrinema H.Pfeiff. = ***Rhynchospora*** Vahl
Syntrinema brasiliense Radlk. & H.Pfeiff. = ***Rhynchospora confusa*** Ballard

Syziganthus

Syziganthus Steud. = ***Gahnia*** J.R.Forst. & G.Forst.
Syziganthus multiflorus (Steud.) Steud. = ***Gahnia javanica*** Moritzi

Taphrogiton

Taphrogiton Montandon = ***Scirpus*** Tourn. ex L.
Taphrogiton radicans Montandon = ***Scirpus radicans*** Schkuhr
Taphrogiton sylvaticum (L.) Montandon = ***Scirpus sylvaticus*** L.

Temnemis

Temnemis Raf. = ***Carex*** L.
Temnemis scouleri (Torr.) Raf. = ***Carex lyngbyei*** Hornem.

Terobera

Terobera Steud. = ***Machaerina*** Vahl
Terobera scirpoidea Steud. = ***Machaerina scirpoidea*** (Steud.) T.Koyama

Tetralepis

Tetralepis Steud. = ***Cyathochaeta*** Nees
Tetralepis australis Steud. = ***Cyathochaeta avenacea*** (R.Br.) Benth.

Tetraria

Tetraria P.Beauv., Mém. Cl. Sci. Math. Inst. Natl. France 1812(2): 54 (1816).
Tanzania to S. Africa, Borneo, Australia. 23 ZAI 25 TAN 26 MLW 27 CPP LES NAT SWZ TVL 42 BOR 50 NSW QLD SOA TAS VIC WAU 51 NZN NZS.
55 Species
Schoenopsis P.Beauv. in T.G.Lestiboudois, Essai Cypér.: 341 (1819).
Lepisia C.Presl, Lepisia: 1 (1829).
Trichoballia C.Presl, Symb. Bot. 1: 9 (1830).
Elynanthus Nees, Linnaea 7: 520 (1832), nom. illeg.
Sclerochaetium Nees, Linnaea 7: 511 (1832).
Aulacorhynchus Nees, Linnaea 9: 305 (1834).
Ideleria Kunth, Enum. Pl. 2: 310 (1837).
Macrochaetium Steud., Syn. Pl. Glumac. 2: 159 (1855).
Decalepis Boeck., Bot. Jahrb. Syst. 5: 509 (1884).

Boeckeleria T.Durand, Index Gen. Phan.: 504 (1888).
Tetrariopsis C.B.Clarke, Kew Bull., Addit. Ser. 8: 45 (1908).

Tetraria australiensis C.B.Clarke, Kew Bull., Addit. Ser. 8: 48 (1908).
WSW. Western Australia. 50 WAU. Hemicr. or rhizome geophyte.

Tetraria autumnalis Levyns, J. S. African Bot. 13: 88 (1947).
Cape Prov. 27 CPP. Hemicr. or rhizome geophyte.

Tetraria bachmannii Kük., Repert. Spec. Nov. Regni Veg. 29: 191 (1931).
SE. Cape Prov. to KwaZulu-Natal. 27 CPP NAT. Hemicr. or rhizome geophyte.

Tetraria bolusii C.B.Clarke in W.H.Harvey & auct. suc. (eds.), Fl. Cap. 7: 279 (1898).
Cape Prov. 27 CPP. Hemicr. or rhizome geophyte.

Tetraria borneensis J.Kern, Blumea 9: 222 (1958).
Borneo. 42 BOR. Hemicr. or rhizome geophyte.

Tetraria brachyphylla Levyns, J. S. African Bot. 13: 90 (1947).
Cape Prov. 27 CPP. Hemicr. or rhizome geophyte.

Tetraria bromoides (Lam.) H.Pfeiff., Repert. Spec. Nov. Regni Veg. 23: 347 (1927).
Cape Prov. 27 CPP. Hemicr. or rhizome geophyte.
**Schoenus bromoides* Lam., Tabl. Encycl. 1: 137 (1791).
Lepidosperma rottboellii Schrad., Anal. Fl. Cap. 1: 37 (1832). *Sclerochaetium rottboellii* (Schrad.) Nees, Linnaea 8: 89 (1833). *Tetraria rottboellii* (Schrad.) C.B.Clarke in T.A.Durand & H.Schinz, Consp. Fl. Afric. 5: 662 (1894). *Schoenus rottboellii* (Schrad.) Kuntze, Revis. Gen. Pl. 3(2): 336 (1898).
Sclerochaetium angustifolium Hochst., Flora 28: 762 (1845). *Tetraria rottboellioides* C.B.Clarke in T.A.Durand & H.Schinz, Consp. Fl. Afric. 5: 663 (1894). *Tetraria angustifolia* (Hochst.) H.Pfeiff., Repert. Spec. Nov. Regni Veg. 23: 348 (1927).
Sclerochaetium koenigii Nees ex Hochst., Flora 28: 763 (1845).
Lepidosperma teres Nees ex C.B.Clarke in T.A.Durand & H.Schinz, Consp. Fl. Afric. 5: 622 (1894), nom. inval.
Sclerochaetium teres Nees ex C.B.Clarke in T.A.Durand & H.Schinz, Consp. Fl. Afric. 5: 622 (1894), nom. inval.
Lepidosperma brehmeri Boeck., Allg. Bot. Z. Syst. 2: 113 (1896).

Tetraria burmanii (Vahl) C.B.Clarke in T.A.Durand & H.Schinz, Consp. Fl. Afric. 5: 659 (1894).
Cape Prov. 27 CPP. Hemicr. or rhizome geophyte.
**Schoenus burmanii* Vahl, Enum. Pl. 2: 227 (1805). *Lepidosperma burmanii* (Vahl) Spreng., Syst. Veg. 1: 194 (1824). *Chaetospora burmanii* (Vahl) Schrad., Anal. Fl. Cap. 1: 32 (1832). *Schoenopsis burmanii* (Vahl) Nees, Linnaea 7: 528 (1832).

Tetraria capillacea (Thunb.) C.B.Clarke in T.A.Durand & H.Schinz, Consp. Fl. Afric. 5: 659 (1894).
Cape Prov. 27 CPP. Hemicr. or rhizome geophyte.
**Schoenus capillaceus* Thunb., Prodr. Pl. Cap.: 16 (1794). *Chaetospora capillacea* (Thunb.) Nees, Linnaea 10: 192 (1836).

Tetraria capillaris (F.Muell.) J.M.Black, Trans. & Proc. Roy. Soc. South Australia 58: 169 (1934).
S. & E. Australia, New Zealand. 50 NSW QLD SOA TAS VIC WAU 51 NZN NZS. Hemicr. or rhizome geophyte.
Chaetospora capillacea Hook.f., Fl. Tasman. 2: 81 (1858), nom. illeg. *Elynanthus capillaceus* Benth., Fl. Austral. 7: 377 (1878). *Cladium capillaceum* (Benth.) C.B.Clarke in T.F.Cheeseman, Man. New Zealand Fl.: 789 (1906). *Mariscus capillaceus* (Benth.) Fernald, Rhodora 25: 52 (1923). *Machaerina capillacea* (Benth.) T.Koyama, Bot. Mag. (Tokyo) 69: 63 (1956).
Schoenus tenuis Kirk, Trans. & Proc. New Zealand Inst. 1(ed. 2): 94 (1871), nom. nud.
**Chaetospora capillaris* F.Muell., Fragm. 9: 34 (1875). *Schoenus capillaris* (F.Muell.) F.Muell., Syst. Census Austral. Pl. 1: 128 (1882), nom. illeg.
Cladium capillaceum var. *laminatum* Kük., Repert. Spec. Nov. Regni Veg. 51: 183 (1942).

Tetraria capitata Kük., Repert. Spec. Nov. Regni Veg. 29: 192 (1931).
Cape Prov. 27 CPP. Hemicr. or rhizome geophyte.

Tetraria compacta Levyns, J. S. African Bot. 13: 83 (1947).
Cape Prov. 27 CPP. Hemicr. or rhizome geophyte.

Tetraria compar (L.) P.Beauv. in T.G.Lestiboudois, Essai Cypér.: 36 (1819).
Cape Prov. 27 CPP. Hemicr. or rhizome geophyte.
**Schoenus compar* L., Mant. Pl. 2: 177 (1771). *Elynanthus compar* (L.) Nees, Linnaea 7: 520 (1832).
Tetraria thuarii P.Beauv., Mém. Cl. Sci. Math. Inst. Natl. France 1812(2): 54 (1816).
Rhynchospora nitida Spreng., Neue Entdeck. Pflanzenk. 3: 10 (1822).
Schoenus arenarius Schrad., Anal. Fl. Cap. 1: 27 (1832). *Elynanthus arenarius* (Schrad.) Nees, Linnaea 9: 298 (1834).
Schoenus viscosus Schrad., Anal. Fl. Cap. 1: 26 (1832). *Elynanthus viscosus* (Schrad.) Nees, Linnaea 10: 186 (1836).
Schoenus pungens Willd. ex Kunth, Enum. Pl. 2: 308 (1837).

Tetraria compressa Turrill, Bull. Misc. Inform. Kew 1925: 72 (1925).
Cape Prov. 27 CPP. Hemicr. or rhizome geophyte.

Tetraria crassa Levyns, J. S. African Bot. 13: 87 (1947).
Cape Prov. 27 CPP. Hemicr. or rhizome geophyte.

Tetraria crinifolia (Nees) C.B.Clarke in T.A.Durand & H.Schinz, Consp. Fl. Afric. 5: 660 (1894).
Cape Prov. 27 CPP. Hemicr. or rhizome geophyte.
**Aulacorhynchus crinifolius* Nees, Linnaea 9: 305 (1834).

Tetraria cuspidata (Rottb.) C.B.Clarke in T.A.Durand & H.Schinz, Consp. Fl. Afric. 5: 660 (1894).
S. Africa. 27 CPP LES NAT TVL. Hemicr. or rhizome geophyte.
**Schoenus cuspidatus* Rottb., Descr. Icon. Rar. Pl.: 66 (1773). *Chaetospora cuspidata* (Rottb.) Nees, Linnaea 7: 529 (1832). *Elynanthus cuspidatus* (Rottb.) Nees, Linnaea 10: 188 (1836).
Elynanthus gracilis Nees, Linnaea 10: 189 (1836).
Elynanthus loreus Nees, Linnaea 10: 188 (1836). *Schoenus loreus* (Nees) Kuntze, Revis. Gen. Pl. 2: 756 (1891).
Fuirena filifolia Rchb. ex Kunth, Enum. Pl. 2: 309 (1837).

Schoenus aristatus Nees ex Kunth, Enum. Pl. 2: 309 (1837), nom. inval.
Schoenus flexuosus Steud. ex Kunth, Enum. Pl. 2: 309 (1837).

Tetraria exilis Levyns, J. S. African Bot. 13: 88 (1947).
Cape Prov. 27 CPP. Hemicr. or rhizome geophyte.

Tetraria eximia C.B.Clarke in W.H.Harvey & auct. suc. (eds.), Fl .Cap. 7: 286 (1898).
Cape Prov. 27 CPP. Hemicr. or rhizome geophyte.

Tetraria fasciata (Rottb.) C.B.Clarke in T.A.Durand & H.Schinz, Consp. Fl. Afric. 5: 661 (1894).
Cape Prov. 27 CPP. Hemicr. or rhizome geophyte.
**Schoenus fasciatus* Rottb., Descr. Pl. Rar.: 13 (1772).
Elynanthus microstachyus Boeck., Linnaea 38: 261 (1874). *Schoenus microstachyus* (Boeck.) Kuntze, Revis. Gen. Pl. 2: 756 (1891).

Tetraria ferruginea C.B.Clarke in W.H.Harvey & auct. suc. (eds.), Fl. Cap. 7: 759 (1900).
Cape Prov. 27 CPP. Hemicr. or rhizome geophyte.

Tetraria fimbriolata (Nees) C.B.Clarke in T.A.Durand & H.Schinz, Consp. Fl. Afric. 5: 661 (1894).
Cape Prov. 27 CPP. Hemicr. or rhizome geophyte.
**Chaetospora fimbriolata* Nees, Linnaea 10: 191 (1836).

Tetraria flexuosa (Thunb.) C.B.Clarke in T.A.Durand & H.Schinz, Consp. Fl. Afric. 5: 661 (1894).
Cape Prov. 27 CPP. Hemicr. or rhizome geophyte.
Schoenus aristatus Thunb., Prodr. Pl. Cap.: 16 (1794). *Elynanthus aristatus* (Thunb.) Boeck., Linnaea 38: 260 (1874). *Tetraria aristata* (Thunb.) C.B.Clarke in T.A.Durand & H.Schinz, Consp. Fl. Afric. 5: 659 (1894).
**Schoenus flexuosus* Thunb., Prodr. Pl. Cap.: 16 (1794). *Chaetospora flexuosa* (Thunb.) Schrad., Anal. Fl. Cap. 1: 33 (1832). *Schoenopsis flexuosa* (Thunb.) Nees, Linnaea 7: 528 (1832).

Tetraria fourcadei Turrill & Schönland, Bull. Misc. Inform. Kew 1925: 73 (1925).
Cape Prov. 27 CPP. Hemicr. or rhizome geophyte.

Tetraria galpinii Schönland & Turrill, Bull. Misc. Inform. Kew 1925: 73 (1925).
Cape Prov. 27 CPP. Hemicr. or rhizome geophyte.

Tetraria gracilis Turrill, Bull. Misc. Inform. Kew 1925: 73 (1925).
Cape Prov. 27 CPP. Hemicr. or rhizome geophyte.

Tetraria graminifolia Levyns, J. S. African Bot. 13: 85 (1947).
Cape Prov. 27 CPP. Hemicr. or rhizome geophyte.

Tetraria involucrata (Rottb.) C.B.Clarke in T.A.Durand & H.Schinz, Consp. Fl. Afric. 5: 661 (1894).
Cape Prov. 27 CPP. Hemicr. or rhizome geophyte.
**Schoenus involucratus* Rottb., Descr. Icon. Rar. Pl.: 64 (1773). *Lepidosperma involucratum* (Rottb.) Roem. & Schult., Syst. Veg. 2: 81 (1817). *Trichoballia involucrata* (Rottb.) C.Presl, Symb. Bot. 1: 9 (1830). *Sclerochaetium involucratum* (Rottb.) Nees, Linnaea 7: 511 (1832).
Trichoballia dactyloides C.Presl, Symb. Bot. 1: 9 (1830). *Sclerochaetium dactyloides* (C.Presl) Nees ex C.B.Clarke in T.A.Durand & H.Schinz, Consp. Fl. Afric. 5: 661 (1894).

Tetraria ligulata (Boeck.) C.B.Clarke in T.A.Durand & H.Schinz, Consp. Fl. Afric. 5: 661 (1894).
Cape Prov. 27 CPP. Hemicr. or rhizome geophyte.
**Elynanthus ligulatus* Boeck., Linnaea 38: 262 (1874). *Schoenus ligulatus* (Boeck.) Kuntze, Revis. Gen. Pl. 2: 756 (1891).

Tetraria macowaniana B.L.Burtt, Notes Roy. Bot. Gard. Edinburgh 34: 282 (1976).
Cape Prov. to KwaZulu-Natal. 27 CPP NAT. Hemicr. or rhizome geophyte.
**Chaetospora hexandra* Boeck., Flora 61: 37 (1878). *Tetraria macowanii* C.B.Clarke in T.A.Durand & H.Schinz, Consp. Fl. Afric. 5: 661 (1894), nom. illeg.

Tetraria maculata Schönland & Turrill, Bull. Misc. Inform. Kew 1925: 74 (1925).
Cape Prov. 27 CPP. Hemicr. or rhizome geophyte.

Tetraria microcarpa S.T.Blake, Proc. Roy. Soc. Queensland 60: 52 (1950).
WSW. Western Australia. 50 WAU. Hemicr. or rhizome geophyte.

Tetraria microstachys (Vahl) H.Pfeiff., Repert. Spec. Nov. Regni Veg. 23: 347 (1927).
Cape Prov. 27 CPP. Hemicr. or rhizome geophyte.
**Schoenus microstachys* Vahl, Enum. Pl. 2: 220 (1805). *Chaetospora microstachys* (Vahl) Schrad., Anal. Fl. Cap. 1: 33 (1832).
Chaetospora circinalis Schrad., Gött. Gel. Anz. 3: 2070 (1821). *Schoenus circinalis* (Schrad.) Schult. in J.J.Roemer & J.A.Schultes, Mant. 2: 43 (1824). *Tetraria circinalis* (Schrad.) C.B.Clarke in T.A. Durand & H.Schinz, Consp. Fl. Afric. 5: 659 (1894).

Tetraria mlanjensis J.Raynal, Adansonia, n.s., 12: 213 (1972).
Malawi (Mt. Mulanje). 26 MLW. Hemicr.

Tetraria natalensis (C.B.Clarke) T.Koyama, J. Fac. Sci. Univ. Tokyo, Sect. 3, Bot. 8: 75 (1961).
S. Africa. 27 NAT SWZ TVL. Hemicr. or rhizome geophyte.
**Costularia natalensis* C.B.Clarke in W.H.Harvey & auct. suc. (eds.), Fl. Cap. 7: 274 (1898).

Tetraria nigrovaginata (Nees) C.B.Clarke in T.A.Durand & H.Schinz, Consp. Fl. Afric. 5: 662 (1894).
Cape Prov. 27 CPP. Hemicr. or rhizome geophyte.
**Cyathocoma nigrovaginata* Nees, Linnaea 10: 196 (1836). *Decalepis dregeana* (Nees) Boeck., Bot. Jahrb. Syst. 5: 509 (1884). *Boeckeleria nigrovaginata* (Nees) H.Pfiriff., Repert. Spec. Nov. Regni Veg. 23: 344 (1927).
Lepidosperma dregei Boeck., Allg. Bot. Z. Syst. 2: 113 (1896).

Tetraria octandra (Nees) Kük., Repert. Spec. Nov. Regni Veg. 29: 193 (1931).
SW. Australia. 50 WAU. Hemicr. or rhizome geophyte.
**Elynanthus octandrus* Nees, Ann. Mag. Nat. Hist. 6: 48 (1841). *Schoenus octandrus* (Nees) F.Muell., Fragm. 9: 31 (1875). *Tetrariopsis octandra* (Nees) C.B.Clarke, Kew Bull., Addit. Ser. 8: 45 (1908).
Elynanthus revolutus Nees in J.G.C.Lehmann, Pl. Preiss. 2: 78 (1846).

Tetraria paludosa Levyns, J. S. African Bot. 13: 85 (1947).
Cape Prov. 27 CPP. Hemicr. or rhizome geophyte.

Tetraria picta (Boeck.) C.B.Clarke in T.A.Durand & H.Schinz, Consp. Fl. Afric. 5: 662 (1894).

Cape Prov. 27 CPP. Hemicr. or rhizome geophyte.
Elynanthus pictus Boeck., Linnaea 38: 244 (1874). *Schoenus pictus* (Boeck.) Kuntze, Revis. Gen. Pl. 2: 633 (1891).

Tetraria pillansii Levyns, J. S. African Bot. 13: 93 (1947).
Cape Prov. 27 CPP. Hemicr. or rhizome geophyte.

Tetraria pleosticha C.B.Clarke in W.H.Harvey & auct. suc. (eds.), Fl. Cap. 7: 280 (1898).
Cape Prov. 27 CPP. Hemicr. or rhizome geophyte.

Tetraria pubescens Schönland & Turrill, Bull. Misc. Inform. Kew 1925: 75 (1925).
Cape Prov. 27 CPP. Hemicr. or rhizome geophyte.

Tetraria pygmaea Levyns, J. S. African Bot. 13: 90 (1947).
Cape Prov. 27 CPP. Hemicr. or rhizome geophyte.

Tetraria robusta (Kunth) C.B.Clarke in T.A.Durand & H.Schinz, Consp. Fl. Afric. 5: 662 (1894).
Cape Prov. 27 CPP. Hemicr. or rhizome geophyte.
Chaetospora robusta Kunth, Enum. Pl. 2: 325 (1837).

Tetraria scariosa Kük., Repert. Spec. Nov. Regni Veg. 29: 190 (1931).
Cape Prov. 27 CPP. Hemicr. or rhizome geophyte. Provisionally accepted.

Tetraria schonlandii Turrill, Bull. Misc. Inform. Kew 1925: 75 (1925).
Cape Prov. 27 CPP. Hemicr. or rhizome geophyte.

Tetraria secans C.B.Clarke in W.H.Harvey & auct. suc. (eds.), Fl. Cap. 7: 286 (1898).
Cape Prov. 27 CPP. Hemicr. or rhizome geophyte.

Tetraria spiralis (Hochst.) C.B.Clarke in T.A.Durand & H.Schinz, Consp. Fl. Afric. 5: 663 (1894).
Cape Prov. 27 CPP. Hemicr. or rhizome geophyte.
Sclerochaetium spirale Hochst., Flora 28: 761 (1845).

Tetraria sylvatica (Nees) C.B.Clarke in T.A.Durand & H.Schinz, Consp. Fl. Afric. 5: 663 (1894).
Cape Prov. 27 CPP. Hemicr. or rhizome geophyte.
Elynanthus auritus Nees, Linnaea 10: 187 (1836).
Elynanthus sylvaticus Nees, Linnaea 10: 188 (1836). *Schoenus sylvaticus* (Nees) Kuntze, Revis. Gen. Pl. 2: 633 (1891), nom. illeg.

Tetraria thermalis (L.) C.B.Clarke in T.A.Durand & H.Schinz, Consp. Fl. Afric. 5: 663 (1894).
Cape Prov. 27 CPP. Hemicr. or rhizome geophyte.
Schoenus thermalis L., Mant. Pl. 2: 172 (1771). *Lepidosperma thermale* (L.) Schrad., Gött. Gel. Anz. 3: 2070 (1821). *Trichoballia thermalis* (L.) C.Presl, Symb. Bot. 1: 9 (1830). *Sclerochaetium thermale* (L.) Nees, Linnaea 7: 512 (1832).
Schoenus dactyloides Nees ex Kunth, Enum. Pl. 2: 321 (1837).
Sclerochaetium giganteum Steud., Syn. Pl. Glumac. 2: 159 (1855).

Tetraria triangularis (Boeck.) C.B.Clarke in T.A.Durand & H.Schinz, Consp. Fl. Afric. 5: 663 (1894).
Cape Prov. 27 CPP. Hemicr. or rhizome geophyte.
Lepidosperma triangulare Boeck., Linnaea 38: 336 (1874).

Tetraria usambarensis K.Schum. in H.G.A.Engler, Pflanzenw. Ost-Afrikas, C: 128 (1895).
Zaïre, Tanzania (Usambara Mts.). 23 ZAI 25 TAN. Hemicr.

Tetraria ustulata (L.) C.B.Clarke in T.A.Durand & H.Schinz, Consp. Fl. Afric. 5: 664 (1894).
Cape Prov. 27 CPP. Hemicr. or rhizome geophyte.
Schoenus ustulatus L., Mant. Pl. 2: 178 (1771). *Lepisia ustulata* (L.) C.Presl, Lepisia: 1 (1829). *Elynanthus ustulatus* (L.) Nees, Linnaea 10: 198 (1836).

Tetraria vaginata Schönland & Turrill, Bull. Misc. Inform. Kew 1925: 76 (1925).
Cape Prov. 27 CPP. Hemicr. or rhizome geophyte.

Tetraria variabilis Levyns, J. S. African Bot. 13: 87 (1947).
Cape Prov. 27 CPP. Hemicr. or rhizome geophyte.

Synonyms:
Tetraria angustifolia (Hochst.) H.Pfeiff. = ***Tetraria bromoides*** (Lam.) H.Pfeiff.
Tetraria aristata (Thunb.) C.B.Clarke = ***Tetraria flexuosa*** (Thunb.) C.B.Clarke
Tetraria arundinacea (Sol. ex Vahl) T.Koyama = ***Costularia arundinacea*** (Sol. ex Vahl) Kük.
Tetraria brevicaulis C.B.Clarke = ***Capeobolus brevicaulis*** (C.B.Clarke) Browning
Tetraria brevifolia (Cherm.) T.Koyama = ***Costularia brevifolia*** Cherm.
Tetraria chamaedendron (Guillaumin) T.Koyama = ***Costularia chamaedendron*** (Guillaumin) Kük.
Tetraria circinalis (Schrad.) C.B.Clarke = ***Tetraria microstachys*** (Vahl) H.Pfeiff.
Tetraria comosa (C.B.Clarke) T.Koyama = ***Costularia comosa*** (C.B.Clarke) Kük.
Tetraria daenikeri (Kük.) T.Koyama = ***Costularia arundinacea*** (Sol. ex Vahl) Kük.
Tetraria dregeana (Boeck.) C.B.Clarke = ***Epischoenus dregeanus*** (Boeck.) Levyns
Tetraria elongata (Kunth) T.Koyama = ***Costularia elongata*** (Kunth) Kük.
Tetraria falcifolia (Däniker) T.Koyama = ***Costularia xyridioides*** (Däniker) Kük.
Tetraria fragilis (Däniker) T.Koyama = ***Costularia fragilis*** (Däniker) Kük.
Tetraria halmaturina (J.M.Black) J.M.Black = ***Eleocharis halmaturina*** J.M.Black
Tetraria hexandra (Nees) Kük. = ***Cyathocoma hexandra*** (Nees) Browning
Tetraria hornei (C.B.Clarke) T.Koyama = ***Costularia hornei*** (C.B.Clarke) Kük.
Tetraria laxa (Cherm.) T.Koyama = ***Costularia laxa*** Cherm.
Tetraria leucocarpa (Ridl.) T.Koyama = ***Costularia leucocarpa*** (Ridl.) H.Pfeiff.
Tetraria lucida C.B.Clarke = ***Epischoenus lucidus*** (C.B.Clarke) Levyns
Tetraria macowanii C.B.Clarke = ***Tetraria macowaniana*** B.L.Burtt
Tetraria melleri (Baker) T.Koyama = ***Costularia melleri*** (Baker) C.B.Clarke
Tetraria monocarpa (J.M.Black) J.M.Black = ***Schoenus carsei*** Cheeseman
Tetraria pantopoda (Baker) T.Koyama = ***Costularia pantopoda*** (Baker) C.B.Clarke
Tetraria punctoria (Vahl) C.B.Clarke = ***Neesenbeckia punctoria*** (Vahl) Levyns
Tetraria purpurea (Cherm.) T.Koyama = ***Costularia purpurea*** Cherm.
Tetraria quadrangularis (Boeck.) Kük. = ***Epischoenus quadrangularis*** (Boeck.) C.B.Clarke
Tetraria rottboellii (Schrad.) C.B.Clarke = ***Tetraria bromoides*** (Lam.) H.Pfeiff.

Tetraria rottboellioides C.B.Clarke = ***Tetraria bromoides*** (Lam.) H.Pfeiff.
Tetraria stagnalis (Däniker) T.Koyama = ***Costularia stagnalis*** (Däniker) Kük.
Tetraria thuarii P.Beauv. = ***Tetraria compar*** (L.) P.Beauv.
Tetraria thuarii var. *gracilior* C.B.Clarke = ***Cyathocoma bachmannii*** (Kük.) C.Archer
Tetraria urvilleana (Gaudich. ex Boeck.) T.Koyama = ***Costularia pilisepala*** (Steud.) F.Kern
Tetraria wallichiana C.B.Clarke = [27 CPP]
Tetraria xyridioides (Däniker) T.Koyama = ***Costularia xyridioides*** (Däniker) Kük.

Unplaced Names:
Tetraria wallichiana C.B.Clarke in W.H.Harvey & auct. suc. (eds.), Fl. Cap. 7: 280 (1898). = [27 CPP]

Tetrariopsis

Tetrariopsis C.B.Clarke = ***Tetraria*** P.Beauv.
Tetrariopsis octandra (Nees) C.B.Clarke = ***Tetraria octandra*** (Nees) Kük.

Thoracostachyum

Thoracostachyum Kurz = ***Mapania*** Aubl.
Thoracostachyum angustifolium C.B.Clarke = ***Mapania floribunda*** (Nees ex Steud.) T.Koyama
Thoracostachyum balansae E.G.Camus = ***Mapania balansae*** (E.G.Camus) T.Koyama
Thoracostachyum balansae var. *latifolium* K.K.Nguyen = ***Mapania balansae*** (E.G.Camus) T.Koyama
Thoracostachyum bancanum (Miq.) Kurz = ***Mapania bancana*** (Miq.) T.Koyama
Thoracostachyum dichromenoides Ridl. = ***Mapania sumatrana*** (Miq.) Benth. subsp. ***sumatrana***
Thoracostachyum floribundum (Nees ex Steud.) C.B.Clarke = ***Mapania floribunda*** (Nees ex Steud.) T.Koyama
Thoracostachyum heyneanum (Backer) Uittien = ***Mapania sumatrana*** subsp. ***pandanophylla*** (F.Muell.) D.A.Simpson
Thoracostachyum hypolytroides (F.Muell.) C.B.Clarke = ***Mapania sumatrana*** subsp. ***pandanophylla*** (F.Muell.) D.A.Simpson
Thoracostachyum hypolytroides var. *immensum* Kük. = ***Mapania pacifica*** (Hosok.) T.Koyama
Thoracostachyum longistylum Kük. = ***Paramapania parvibractea*** (C.B.Clarke) Uittien
Thoracostachyum longistylum f. *parvum* Kük. = ***Paramapania parvibractea*** (C.B.Clarke) Uittien
Thoracostachyum lucbanense (Elmer) Kük. ex Merr. = ***Paramapania parvibractea*** (C.B.Clarke) Uittien
Thoracostachyum macilentulum Ohwi = ***Paramapania parvibractea*** (C.B.Clarke) Uittien
Thoracostachyum montanum (Lauterb. & K.Schum.) J.V.Suringar = ***Paramapania parvibractea*** (C.B.Clarke) Uittien
Thoracostachyum pacificum Hosok. = ***Mapania pacifica*** (Hosok.) T.Koyama
Thoracostachyum pandanophyllum (F.Muell.) Domin = ***Mapania sumatrana*** subsp. ***pandanophylla*** (F.Muell.) D.A.Simpson
Thoracostachyum parvibracteum (C.B.Clarke) Kük. = ***Paramapania parvibractea*** (C.B.Clarke) Uittien
Thoracostachyum ridleyi C.B.Clarke = ***Mapania bancana*** (Miq.) T.Koyama
Thoracostachyum simplex Ridl. = ***Paramapania simplex*** (Ridl.) Uittien
Thoracostachyum subcapitatum J.V.Suringar = ***Mapania bancana*** (Miq.) T.Koyama
Thoracostachyum sumatranum (Miq.) Kurz = ***Mapania sumatrana*** (Miq.) Benth.
Thoracostachyum tonkinense (E.G.Camus) Uittien = ***Mapania tonkinensis*** (E.G.Camus) T.Koyama
Thoracostachyum vietnamense K.K.Nguyen = [41 VIE]
Thoracostachyum vitiense Uittien = ***Mapania vitiensis*** (Uittien) T.Koyama

Unplaced Names:
Thoracostachyum vietnamense K.K.Nguyen, Bot. Zhurn. (Moscow & Leningrad) 64: 855 (1979). = [41 VIE]

Thryocephalon

Thryocephalon J.R.Forst. & G.Forst. = ***Kyllinga*** Rottb.
Thryocephalon nemorale J.R.Forst. & G.Forst. = ***Kyllinga nemoralis*** (J.R.Forst. & G.Forst.) Dandy ex Hutch. & Dalziel
Thryocephalon pumilum (Michx.) Nieuwl. = ***Kyllinga pumila*** Michx.

Thysanocarex

Thysanocarex Börner = ***Carex*** L.
Thysanocarex cyperoides (L.) Fedde & J.Schust. = ***Carex bohemica*** Schreb.
Thysanocarex muskingumensis (Schwein.) Fedde & J.Schust. = ***Carex muskingumensis*** Schwein.

Tonduzia

Tonduzia Boeck. ex Tonduz = ***Scleria*** P.J.Bergius
Tonduzia macrophylla Boeck. = ***Scleria vaginata*** Steud.

Torreya

Torreya Raf. = ***Pycreus*** P.Beauv.
Torreya caespitosa Raf. = ***Pycreus filicinus*** (Vahl) T.Koyama
Torreya maritima Raf. = ***Pycreus diander*** (Torr.) C.B.Clarke

Torulinium

Torulinium Desv. = ***Cyperus*** L.
Torulinium angolense Turrill = ***Cyperus laxus*** subsp. ***buchholzii*** (Boeck.) Lye
Torulinium caucasicum Palla = ***Cyperus odoratus*** L. subsp. ***odoratus***
Torulinium confertum Desv. ex Ham. = ***Cyperus odoratus*** L. subsp. ***odoratus***
Torulinium correllii T.Koyama = ***Cyperus correllii*** (T.Koyama) G.C.Tucker
Torulinium eggersii (Boeck.) C.B.Clarke = ***Cyperus odoratus*** L. subsp. ***odoratus***
Torulinium ferax (Rich.) Ham. = ***Cyperus odoratus*** L. subsp. ***odoratus***
Torulinium filiforme (Sw.) C.B.Clarke = ***Cyperus filiformis*** Sw.
Torulinium flexuosum (Vahl) T.Koyama = ***Cyperus flexuosus*** Vahl
Torulinium harrisii (Kük.) T.Koyama = ***Cyperus harrisii*** Kük.
Torulinium hayesii C.B.Clarke = ***Cyperus hayesii*** (C.B.Clarke) Standl.
Torulinium macrocephalum (Liebm.) C.B.Clarke = ***Cyperus odoratus*** L. subsp. ***odoratus***
Torulinium macrocephalum var. *davidsei* C.D.Adams = ***Cyperus odoratus*** L. subsp. ***odoratus***

Torulinium macrocephalum var. *eggersii* (Boeck.) C.D.Adams = ***Cyperus odoratus*** L. subsp. ***odoratus***
Torulinium michauxianum C.B.Clarke = ***Cyperus odoratus*** L. subsp. ***odoratus***
Torulinium odoratum (L.) S.S.Hooper = ***Cyperus odoratus*** L.
Torulinium odoratum subsp. *auriculatum* (Nees & Meyen ex Kunth) T.Koyama = ***Cyperus auriculatus*** Nees & Meyen ex Kunth
Torulinium poeppigii (Kunth) C.B.Clarke = ***Cyperus poeppigii*** Kunth
Torulinium samoense (Boeck.) Palla = ***Cyperus cyperinus*** (Retz.) Suringar
Torulinium vahlii (Schrad. ex Nees) C.B.Clarke = ***Cyperus flexuosus*** Vahl

Trachylomia

Trachylomia Nees = ***Scleria*** P.J.Bergius
Trachylomia triglomerata (Michx.) Nees = ***Scleria triglomerata*** Michx.

Trachyrhynchium

Trachyrhynchium Nees = ***Machaerina*** Vahl
Trachyrhynchium iridifolium (Nees) Nees = ***Machaerina flexuosa*** (Boeck.) J.Kern subsp. ***flexuosa***

Trachystylis

Trachystylis S.T.Blake, Proc. Roy. Soc. Queensland 48: 89 (1936 publ. 1937).
Queensland. 50 QLD.
1 Species

Trachystylis stradbrokensis (Domin) Kük., Bot. Jahrb. Syst. 75: 496 (1952).
Queensland. 50 QLD. Hemicr. or rhizome geophyte.
**Cladium stradbrokense* Domin, Biblioth. Bot. 85: 476 (1915). *Machaerina stradbrokensis* (Domin) T.Koyama, Bot. Mag. (Tokyo) 69: 65 (1956). *Fimbristylis stradbrokensis* (Domin) J.Kern, Acta Bot. Neerl. 8: 267 (1959).
Trachystylis foliosa S.T.Blake, Proc. Roy. Soc. Queensland 48: 89 (1936 publ. 1937).

***Synonyms*:**
Trachystylis foliosa S.T.Blake = ***Trachystylis stradbrokensis*** (Domin) Kük.

Trasis

Trasis P.Beauv. = ***Cladium*** P.Browne

Trasus

Trasus Gray = ***Carex*** L.
Trasus ampullaceus (Gooden.) Gray = ***Carex rostrata*** Stokes var. ***rostrata***
Trasus atratus (L.) Gray = ***Carex atrata*** L.
Trasus binervis (Sm.) Gray = ***Carex binervis*** Sm.
Trasus capillaris (L.) Gray = ***Carex capillaris*** L.
Trasus chlorostachyos Gray = ***Carex pseudocyperus*** L. var. ***pseudocyperus***
Trasus clandestinus (Gooden.) Gray = ***Carex humilis*** Leyss. var. ***humilis***
Trasus depauperatus (Curtis ex With) Gray = ***Carex depauperata*** Curtis ex With
Trasus digitatus (L.) Gray = ***Carex digitata*** L.
Trasus distans (L.) Gray = ***Carex distans*** L.
Trasus erectus (Vill.) Gray = ***Carex sempervirens*** Vill.
Trasus extensus (Gooden.) Gray = ***Carex extensa*** Gooden.
Trasus filiformis (L.) Gray = ***Carex filiformis*** L.
Trasus flavus (L.) Gray = ***Carex flava*** L.
Trasus fulvus Gray = ***Carex hostiana*** DC.
Trasus glaucus (Scop.) Gray = ***Carex flacca*** Schreb. subsp. ***flacca***
Trasus hirtus (L.) Gray = ***Carex hirta*** L.
Trasus hostianus (DC.) Gray = ***Carex hostiana*** DC.
Trasus laevigatus (Sm.) Gray = ***Carex laevigata*** Sm.
Trasus limosus (L.) Gray = ***Carex limosa*** L.
Trasus oederi (Retz.) Gray = ***Carex pilulifera*** L. subsp. ***pilulifera***
Trasus pallescens (L.) Gray = ***Carex pallescens*** L.
Trasus paludosus (Gooden.) Gray = ***Carex acutiformis*** Ehrh.
Trasus paniceus (L.) Gray = ***Carex panicea*** L.
Trasus pauciflorus (Lightf.) Gray = ***Carex pauciflora*** Lightf.
Trasus pendulus (Huds.) Gray = ***Carex pendula*** Huds.
Trasus piluliferus (L.) Gray = ***Carex pilulifera*** L.
Trasus praecox (Schreb.) Gray = ***Carex praecox*** Schreb.
Trasus rariflorus (Wahlenb.) Gray = ***Carex rariflora*** (Wahlenb.) Sm.
Trasus riparius (Curtis) Gray = ***Carex riparia*** Curtis
Trasus strigosus (Huds.) Gray = ***Carex strigosa*** Huds.
Trasus sylvaticus (Huds.) Gray = ***Carex sylvatica*** Huds.
Trasus tomentosus Gray = ***Carex filiformis*** L.
Trasus ustulatus (Wahlenb.) Gray = ***Carex atrofusca*** Schkuhr subsp. ***atrofusca***
Trasus vesicarius (L.) Gray = ***Carex vesicaria*** L.

Trentepohlia

Trentepohlia Boeck. = ***Cyperus*** L.
Trentepohlia bifoliata Boeck. = ***Cyperus cephalotes*** Vahl

Trianoptiles

Trianoptiles Fenzl ex Endl., Gen. Pl.: 113 (1836).
Cape Prov. 27 CPP.
3 Species
**Ecklonea* Steud., Flora 12: 138 (1829), nom. illeg.

Trianoptiles capensis (Steud.) Harv., Gen. S. Afr. Pl., ed. 2: 422 (1868).
Cape Prov. 27 CPP. Hemicr. or rhizome geophyte.
**Ecklonea capensis* Steud., Flora 12: 138 (1829). *Carpha capensis* (Steud.) M.Martens ex Steud., Nomencl. Bot., ed. 2, 1: 300 (1840). *Carpha capensis* (Steud.) H.Pfeiff., Repert. Spec. Nov. Regni Veg. 29: 180 (1931), nom. illeg.

Trianoptiles solitaria (C.B.Clarke) Levyns, J. S. African Bot. 9: 26 (1943).
Cape Prov. 27 CPP. Hemicr. or rhizome geophyte.
**Ecklonea solitaria* C.B.Clarke in W.H.Harvey & auct. suc. (eds.), Fl. Cap. 7: 759 (1900). *Carpha solitaria* (C.B.Clarke) H.Pfeiff., Repert. Spec. Nov. Regni Veg. 29: 180 (1931).

Trianoptiles stipitata Levyns, J. S. African Bot. 9: 25 (1943).
Cape Prov. 27 CPP. Hemicr. or rhizome geophyte.

Trichelostylis

Trichelostylis T.Lestib. = ***Fimbristylis*** Vahl
Trichelostylis angularis (Link) Nees = ***Fimbristylis quinquangularis*** (Vahl) Kunth
Trichelostylis aspera (Schrad.) Nees = ***Fimbristylis aspera*** (Schrad.) Boeck.
Trichelostylis asperrima Nees ex Boeck. = ***Fimbristylis dura*** (Zoll. & Moritzi) Merr.

Trichelostylis autumnalis (L.) Nees = ***Fimbristylis autumnalis*** (L.) Roem. & Schult.
Trichelostylis borealis Alph.Wood = [74 ILL 75 MIC]
Trichelostylis capillaris (L.) Alph.Wood = ***Bulbostylis capillaris*** (L.) Kunth ex C.B.Clarke
Trichelostylis capitata (R.Br.) Nees = ***Fimbristylis furva*** R.Br.
Trichelostylis carinata (Hook. & Arn. ex Torr.) Alph.Wood = ***Isolepis carinata*** Hook. & Arn. ex Torr.
Trichelostylis chaetorrhiza Nees = ***Fimbristylis falcata*** (Vahl) Kunth var. ***falcata***
Trichelostylis ciliatifolia (Elliott) Alph.Wood = ***Bulbostylis ciliatifolia*** (Elliott) Fernald
Trichelostylis cinnamometorum (Vahl) Nees = ***Fimbristylis cinnamometorum*** (Vahl) Kunth
Trichelostylis coarctata (Elliott) Alph.Wood = ***Bulbostylis ciliatifolia*** var. ***coarctata*** (Elliott) Kral
Trichelostylis complanata (Retz.) Nees = ***Fimbristylis complanata*** (Retz.) Link
Trichelostylis contexta Nees = ***Bulbostylis contexta*** (Nees) Bodard
Trichelostylis curvifolia (Schrad.) Nees = ***Fimbristylis dichotoma*** (L.) Vahl subsp. ***dichotoma***
Trichelostylis digitata Dalzell ex Boeck. = ***Fimbristylis lawiana*** (Boeck.) J.Kern
Trichelostylis filiformis Nees = ***Fimbristylis pauciflora*** R.Br.
Trichelostylis geminata (Kunth) Nees = ***Fimbristylis autumnalis*** (L.) Roem. & Schult.
Trichelostylis globulosa (Retz.) Nees = ***Fimbristylis umbellaris*** (Lam.) Vahl
Trichelostylis gracilis (Poir.) Nees = ***Ficinia ecklonea*** (Steud.) Nees
Trichelostylis junciformis Nees = ***Fimbristylis falcata*** (Vahl) Kunth var. ***falcata***
Trichelostylis kielmeyeri Nees ex Boeck. = ***Fimbristylis complanata*** (Retz.) Link var. ***complanata***
Trichelostylis latifolia Nees = ***Fimbristylis falcata*** (Vahl) Kunth var. ***falcata***
Trichelostylis ludwigii (Steud.) Nees = ***Isolepis ludwigii*** (Steud.) Kunth
Trichelostylis miliacea (L.) Nees = ***Fimbristylis quinquangularis*** (Vahl) Kunth
Trichelostylis miliacea var. *microstachya* Nees =
Trichelostylis mucronulata (Michx.) Torr. = ***Fimbristylis autumnalis*** (L.) Roem. & Schult.
Trichelostylis obtusifolia (Lam.) Nees = ***Fimbristylis dichotoma*** (L.) Vahl subsp. ***dichotoma***
Trichelostylis pentagona Nees = ***Fimbristylis quinquangularis*** (Vahl) Kunth
Trichelostylis pentaptera Nees = ***Fimbristylis salbundia*** (Nees) Kunth
Trichelostylis pilosula Nees ex Boeck. = ***Bulbostylis hispidula*** (Vahl) R.W.Haines subsp. ***hispidula***
Trichelostylis quinquangularis (Vahl) Nees = ***Fimbristylis quinquangularis*** (Vahl) Kunth
Trichelostylis rudgeana Nees = ***Fimbristylis complanata*** (Retz.) Link var. ***complanata***
Trichelostylis salbundia Nees = ***Fimbristylis salbundia*** (Nees) Kunth
Trichelostylis scabra Nees = ***Fimbristylis complanata*** (Retz.) Link var. ***complanata***
Trichelostylis schimperiana (Hochst. ex A.Rich.) Hochst. ex Boeck. = ***Bulbostylis schimperiana*** (Hochst. ex A.Rich.) C.B.Clarke
Trichelostylis stenophylla (Elliott) Alph.Wood = ***Bulbostylis stenophylla*** (Elliott) C.B.Clarke
Trichelostylis stricta (R.Br.) Nees = ***Fimbristylis dichotoma*** (L.) Vahl subsp. ***dichotoma***
Trichelostylis tenella Nees = ***Fimbristylis tenera*** Schult.
Trichelostylis tetragona (R.Br.) Nees = ***Fimbristylis tetragona*** R.Br.
Trichelostylis torta Nees = ***Fimbristylis falcata*** (Vahl) Kunth var. ***falcata***
Trichelostylis warei (Torr.) Alph.Wood = ***Bulbostylis warei*** (Torr.) C.B.Clarke
Trichelostylis xyroides Arn. ex Hook.f. = ***Fimbristylis tetragona*** R.Br.

Unplaced Names:
Trichelostylis borealis Alph.Wood, Amer. Bot. Fl.: 364 (1870). = [74 ILL 75 MIC]

Trichoballia

Trichoballia C.Presl = ***Tetraria*** P.Beauv.
Trichoballia dactyloides C.Presl = ***Tetraria involucrata*** (Rottb.) C.B.Clarke
Trichoballia involucrata (Rottb.) C.Presl = ***Tetraria involucrata*** (Rottb.) C.B.Clarke
Trichoballia thermalis (L.) C.Presl = ***Tetraria thermalis*** (L.) C.B.Clarke

Trichochaeta

Trichochaeta Steud. = ***Rhynchospora*** Vahl
Trichochaeta tenuis Steud. = ***Rhynchospora trichochaeta*** C.B.Clarke

Trichophorum

Trichophorum Pers., Syn. Pl. 1: 69 (1805).
Temp. & Subtrop. Northern Hemisphere, W. South America to NW. Argentina. 10 DEN FIN FOR GRB ICE IRE NOR SWE 11 AUT BGM CZE GER HUN NET POL SWI 12 COR FRA POR SPA 13 BUL GRC? ITA ROM? YUG 14 BLR BLT RUC RUE RUN RUW 20 ALG 30 31 KAM KUR SAK 32 KAZ KGZ TZK UZB 33 TCS 34 AFG IRN 36 CHC CHN CHS CHX 37 MON 38 JAP KOR TAI 40 ASS IND NEP PAK SRL WHM 41 THA VIE 42 BOR MLY PHI SUL SUM 43 NWG 70 ALU ASK GNL NUN NWT YUK 71 ABT BRC MAN SAS 72 LAB NBR NFL NSC ONT PEI QUE 73 IDA MNT ORE WAS 74 ILL MIN MSO WIS 75 CNT MAI MAS MIC NWH NWJ NWY OHI PEN RHO VER WVA 76 CAL UTA 78 DEL MRY NCA SCA TEN VRG WDC 83 BOL ECU PER 85 AGW.
10 Species
Leucocoma Rydb., Fl. Rocky Mts.: 108 (1918).
Eriophorella Holub, Folia Geobot. Phytotax. 19: 97 (1984).
Kreczetoviczia Tzvelev, Bot. Zhurn. (Moscow & Leningrad) 84(7): 112 (1999).

Trichophorum alpinum (L.) Pers., Syn. Pl. 1: 70 (1805).
Temp. Northern Hemisphere. 10 DEN FIN GRB† NOR SWE 11 AUT CZE GER HUN† POL SWI 12 FRA SPA 13 ITA ROM? YUG 14 BLR RUC RUN 30 31 KAM KUR 38 JAP KOR 70 ASK NUN NWT YUK 71 ABT BRC MAN SAS 72 LAB NBR NFL NSC ONT QUE 73 MNT 74 MIN WIS 75 CNT MAI MAS MIC NWH NWJ NWY VER. Hemicr.
**Eriophorum alpinum* L., Sp. Pl.: 53 (1753). *Linagrostis alpina* (L.) Scop., Fl. Carniol., ed. 2, 1: 48 (1771). *Eleocharis alpina* (L.) Bluff, Nees & Schauer, Comp. Fl. German. 1(1): 92 (1836). *Limnochloa alpina* (L.) Drejer in C.J.Hartmann, Fl.

Excurs. Hafn.: 18 (1838), nom. illeg. *Plumaria alpina* (L.) Bubani, Fl. Pyren. 4: 247 (1902). *Scirpus alpinus* (L.) Dalla Torre & Sarnth., Fl. Tirol 6(1): 389 (1906). *Leucocoma alpina* (L.) Rydb., Fl. Rocky Mts.: 108 (1918). *Leucocoma alpina* (L.) Rydb., Fl. Rocky Mts.: 108 (1918). *Baeothryon alpinum* (L.) T.V.Egorova, Novosti Sist. Vyssh. Rast. 8: 85 (1971). *Eriophorella alpina* (L.) Holub, Folia Geobot. Phytotax. 19: 99 (1984).

Eriophorum hudsonianum Michx., Fl. Bor.-Amer. 1: 34 (1803). *Trichophorum hudsonianum* (Michx.) Pers., Syn. Pl. 1: 69 (1805). *Scirpus hudsonianus* (Michx.) Fernald, Rhodora 8: 161 (1906). *Baeothryon hudsonianum* (Michx.) Soják, Cas. Nár. Mus., Odd. Prír. 148: 193 (1979 publ. 1980).

Scirpus trichophorum Asch. & Graebn., Syn. Mitteleur. Fl. 2(2): 301 (1904).

Trichophorum cespitosum (L.) Hartm., Handb. Skand. Fl., ed. 5: 259 (1849).

Temp. Northern Hemisphere. 10 DEN FIN FOR GRB ICE IRE NOR SWE 11 AUT BGM CZE GER HUN? NET POL SWI 12 COR FRA POR SPA 13 BUL GRC? ITA YUG 14 BLR BLT RUN RUW 20 ALG 30 31 KUR SAK 32 KAZ 38 JAP 70 ALU ASK GNL NUN NWT 71 ABT BRC MAN SAS 72 LAB NBR NFL NSC ONT PEI QUE 73 IDA MNT ORE WAS 74 ILL MIN WIS 75 MAI NWH NWY VER 76 UTA 78 NCA SCA TEN. Hemicr.

**Scirpus cespitosus* L., Sp. Pl.: 48 (1753). *Clavula cespitosa* (L.) Dumort., Fl. Belg.: 143 (1827). *Eleocharis cespitosa* (L.) Link, Hort. Berol. 1: 283 (1827). *Limnochloa cespitosa* (L.) Rchb. in J.C.Mössler, Handb. Gewächsk. ed. 2, 3: 1808 (1830). *Baeothryon cespitosum* (L.) A.Dietr., Sp. Pl. 2: 89 (1832). *Kreczetoviczia cespitosa* (L.) Tzvelev, Bot. Zhurn. (Moscow & Leningrad) 84(7): 112 (1999).

subsp. ***cespitosum***

Temp. Northern Hemisphere. 10 DEN FIN FOR GRB ICE IRE NOR SWE 11 AUT BGM CZE GER HUN? NET POL SWI 12 COR FRA POR SPA 13 BUL GRC? ITA YUG 14 BLR BLT RUN RUW 20 ALG 30 31 KUR SAK 38 JAP 70 ALU ASK GNL NUN NWT 71 ABT BRC MAN SAS 72 LAB NBR NFL NSC ONT PEI QUE 73 IDA MNT ORE WAS 74 ILL MIN WIS 75 MAI NWH NWY VER 76 UTA 78 NCA SCA TEN. Hemicr.

Scirpus luteus Gilib., Excerc. Phyt. 2: 513 (1792).

Scirpus bracteatus Bigelow, New England J. Med. Surg. 5: 335 (1816). *Aplostemon bracteatum* (Bigelow) Raf., J. Phys. Chim. Hist. Nat. Arts 89: 105 (1819). *Trichophorum bracteatum* (Bigelow) V.I.Krecz. ex Kuseneva, Fl. Murm. Prov. 2: 26 (1954).

Scirpus cespitosus var. *callosus* Bigelow ex Torr., Fl. N. Middle United States 1: 47 (1823). *Trichophorum cespitosum* var. *callosum* (Bigelow ex Torr.) Mohlenbr., Ill. Fl. Illinois, Sedges: Cyperus to Scleria, ed. 2: 209 (2001), nom. illeg.

Eleocharis pallida Shuttlew. ex Steud., Syn. Pl. Glumac. 2: 76 (1854).

Trichophorum austriacum Palla, Ber. Deutsch. Bot. Ges. 55: 468 (1897).

Scirpus cespitosus var. *delicatulus* Fernald, Rhodora 23: 25 (1921). *Baeothryon cespitosum* subsp. *delicatulum* (Fernald) Á.Löve & D.Löve, Univ. Colorado Stud. Ser. Biol. 17: 15 (1965).

nothosubsp. ***foersteri*** Swan, Watsonia 22: 230 (1999). *T. cespitosum* subsp. *cespitosum* × *T. cespitosum* subsp. *subsp. germanicum*.

W. Europe. 10 GRB IRE NOR 11 BGM GER NET. Hemicr. or rhizome geophyte.

subsp. ***germanicum*** (Palla) Hegi, Ill. Fl. Mitt.-Eur. 2: 25 (1909).

W. Europe. 10 DEN GRB IRE NOR SWE 11 BGM GER NET 12 FRA. Hemicr.

**Trichophorum germanicum* Palla, Ber. Deutsch. Bot. Ges. 55: 468 (1897). *Scirpus cespitosus* subsp. *germanicus* (Palla) Brodd., Bot. Not. 1912: 82 (1912). *Scirpus germanicus* (Palla) W.F.Christ., Schriften Naturwiss. Vereins Schleswig-Holstein 21: 23 (1935). *Baeothryon germanicum* (Palla) Holub, Folia Geobot. Phytotax. 18: 204 (1983).

Trichophorum clintonii (A.Gray) S.G.Sm., Novon 5: 102 (1995).

Canada to NC. & NE. U.S.A. 71 ABT SAS 72 NBR ONT QUE 74 MIN WIS 75 MAI MIC NWY. Hemicr. or rhizome geophyte.

**Scirpus clintonii* A.Gray, Amer. J. Sci. Arts, II, 38: 290 (1864). *Baeothryon clintonii* (A.Gray) Á.Löve & D.Löve, Univ. Colorado Stud. Ser. Biol. 17: 14 (1965).

Trichophorum dolichocarpum Zakirov, Bot. Mater. Gerb. Bot. Inst. Komarova Akad. Nauk S.S.S.R. 13: 38 (1950). *Baeothryon dolichocarpum* (Zakirov) T.V.Egorova, Opred. Rast. Sred. Azii 5: 24 (1976).

C. Asia. 32 UZB. Hemicr. or rhizome geophyte.

Trichophorum planifolium (Spreng.) Palla, Oesterr. Bot. Z. 63: 402 (1913).

SE. Canada, EC. & NE. U.S.A. to Virginia. 72 ONT 74 ILL MSO 75 CNT MAS NWJ NWY OHI PEN RHO VER WVA 78 DEL MRY VRG WDC. Hemicr. or rhizome geophyte.

Scirpus planifolius Muhl., Descr. Gram.: 32 (1817), nom. illeg. **Isolepis planifolia* Spreng., Neue Entdeck. Pflanzenk. 3: 10 (1822). *Eleogiton planifolia* (Spreng.) A.Dietr., Sp. Pl. 2: 98 (1832). *Eleocharis planifolia* (Spreng.) Nees, Linnaea 9: 294 (1834). *Baeothryon planifolium* (Spreng.) Soják, Cas. Nár. Mus., Odd. Prír. 148: 193 (1979 publ. 1980). *Scirpus verecundus* Fernald, Rhodora 50: 284 (1948). *Baeothryon verecundum* (Fernald) Á.Löve & D.Löve, Univ. Colorado Stud. Ser. Biol. 17: 15 (1965).

Scirpus planifolius var. *brevifolius* Torr., Ann. Lyceum Nat. Hist. New York 3: 442 (1836).

Trichophorum pumilum (Vahl) Schinz & Thell., Vierteljahrsschr. Naturf. Ges. Zürich 66: 265 (1921).

Temp. Northern Hemisphere. 10 NOR 11 CZE SWI 12 FRA 13 ITA 14 RUE 30 32 KAZ KGZ TZK UZB 33 TCS 34 AFG IRN 36 CHC CHX 37 MON 40 NEP PAK WHM 70 ASK NWT YUK 71 ABT BRC SAS 72 QUE 73 MNT 76 CAL. Hemicr. or rhizome geophyte.

**Scirpus pumilus* Vahl, Enum. Pl. 2: 243 (1805). *Isolepis pumila* (Vahl) Roem. & Schult., Syst. Veg. 2: 106 (1817). *Baeothryon pumilum* (Vahl) Á.Löve & D.Löve, Univ. Colorado Stud. Ser. Biol. 17: 14 (1965). *Eriophorella pumila* (Vahl) Kit Tan in P.H.Davis (ed.), Fl. Turkey 9: 68 (1985). *Kreczetoviczia pumila* (Vahl) Tzvelev, Bot. Zhurn. (Moscow & Leningrad) 84(7): 112 (1999).

subsp. ***pumilum***
Europe to Himalaya. 10 NOR 11 CZE SWI 12 FRA 13 ITA 14 RUE 30 32 KAZ KGZ TZK UZB 33 TCS 34 AFG IRN 36 CHC CHX 37 MON 40 NEP PAK WHM 70 ASK NWT YUK 71 ABT BRC SAS 72 QUE 73 MNT 76 CAL. Hemicr. or rhizome geophyte.
Scirpus alpinus Schleich., Cat. Pl. Helv., ed. 4: 32 (1821). *Limnochloa alpina* (Schleich.) Rchb., Fl. Germ. Excurs.: 140(11) (1831).
Isolepis elongata C.A.Mey. in C.F.von Ledebour, Fl. Altaic. 1: 64 (1829).
Isolepis oligantha C.A.Mey., Mém. Sav. Étr. Acad. St. Pétersbourg 1: 197 (1831). *Trichophorum oliganthum* (C.A.Mey.) Fritsch, Excursionsfl. Oesterreich, ed. 2: 87 (1909).
Eleogiton elongantha A.Dietr., Sp. Pl. 2: 96 (1832).
Trichophorum emergens Norman, Soc. Reg. Sci. Norveg. 5: 319 (1868). *Scirpus emergens* (Norman) Fernald, Rhodora 45: 281 (1943).
Scirpus meyeri Trautv., Trudy Imp. S.-Peterburgsk. Bot. Sada 5: 121 (1877).
Trichophorum atrichum Palla, Bot. Jahrb. Syst. 10: 296 (1888). *Scirpus atrichus* (Palla) Dalla Torre & Sarnth., Fl. Tirol 6(1): 389 (1906), nom. illeg.
Scirpus pumilus subsp. *distigmaticus* Kük., Acta Horti Gothob. 5: 34 (1929). *Scirpus distigmaticus* (Kük.) Tang & F.T.Wang, in Fl. Reipubl. Popul. Sin. 11: 32 (1961). *Trichophorum distigmaticum* (Kük.) T.V.Egorova, Trudy Bot. Inst. Akad. Nauk S.S.S.R., Rast. Tsentral. Azii 3: 14 (1967). *Baeothryon distigmaticum* (Kük.) T.V.Egorova, Novosti Sist. Vyssh. Rast. 8: 86 (1971).

subsp. ***rollandii*** (Fernald) R.L.Taylor & MacBryde, Canad. J. Bot. 56: 191 (1978).
Alaska to W. U.S.A. 70 ASK NWT YUK 71 ABT BRC SAS 72 QUE 73 MNT 76 CAL. Hemicr. or rhizome geophyte.
**Scirpus rollandii* Fernald, Rhodora 45: 282 (1943). *Scirpus pumilus* var. *rollandii* (Fernald) Beetle, Amer. Midl. Naturalist 41: 483 (1949). *Scirpus pumilus* subsp. *rollandii* (Fernald) Raymond, Contr. Inst. Bot. Univ. Montréal 70: 91 (1957). *Trichophorum rollandii* (Fernald) Hultén, Kongl. Svenska Vetensk. Acad. Handl., IV, 7(1): 246 (1958).

Trichophorum rigidum (Steud.) Goetgh., Muasya & D.A.Simpson, Novon 10: 132 (2000).
W. South America to NW. Argentina. 83 BOL ECU PER 85 AGW. Hemicr. or rhizome geophyte.
**Isolepis rigida* Steud. in W.Lechler, Berberid. Amer. Austral.: 56 (1857). *Scirpus rigidus* (Steud.) Boeck., Linnaea 36: 492 (1870), nom. illeg. *Baeothryon rigidum* (Steud.) Soják, Cas. Nár. Mus., Odd. Prír. 148: 193 (1979 publ. 1980).

Trichophorum schansiense Hand.-Mazz., Oesterr. Bot. Z. 85: 224 (1936). *Scirpus schansiensis* (Hand.-Mazz.) Tang & F.T.Wang, in Fl. Reipubl. Popul. Sin. 11: 31 (1961).
NC. China. 36 CHN.

Trichophorum subcapitatum (Thwaites & Hook.) D.A.Simpson, Kew Bull. 53: 227 (1998).
Trop. & Subtrop. Asia. 36 CHS 38 TAI 40 ASS IND SRL 41 THA VIE 42 BOR MLY PHI SUL SUM 43 NWG. Hemicr. or rhizome geophyte.
**Scirpus subcapitatus* Thwaites & Hook., Enum. Pl. Zeyl.: 351 (1864). *Baeothryon subcapitatum* (Thwaites & Hook.) Á.Löve & D.Löve, Univ. Colorado Stud. Ser. Biol. 17: 15 (1965).
Scirpus subaphyllus Boeck., Linnaea 36: 705 (1870).
Scirpus clarkei Stapf, Trans. Linn. Soc. London, Bot. 4: 244 (1894).
Scirpus morrisonensis Hayata, J. Coll. Sci. Imp. Univ. Tokyo 25(19): 230 (1908).
Scirpus pulogensis Merr., Philipp. J. Sci., C 5: 333 (1910). *Scirpus subcapitatus* subsp. *pulogensis* (Merr.) J.Kern, Reinwardtia 4: 90 (1956). *Baeothryon subcapitatum* subsp. *pulogense* (Merr.) Soják, Cas. Nár. Mus., Odd. Prír. 148: 193 (1979 publ. 1980).
Scirpus pakapakensis Stapf, J. Linn. Soc., Bot. 42: 174 (1914).
Scirpus subcapitatus f. *rigidus* Kük., Bull. Jard. Bot. Buitenzorg, III, 16: 301 (1940).
Scirpus subcapitatus var. *triangularis* Kük., Bull. Jard. Bot. Buitenzorg, III, 16: 301 (1940).
Scirpus subcapitatus subsp. *celebicus* J.Kern, Reinwardtia 4: 90 (1956). *Baeothryon subcapitatum* subsp. *celebicum* (J.Kern) Soják, Cas. Nár. Mus., Odd. Prír. 149: 193 (1979 publ. 1980).

Trichophorum uniflorum (Trautv.) Malyschev & Lukitsch., Sched. Herb. Fl. URSS 15: 77 (1963).
Siberia to Russian Far East. 30 31. Hemicr. or rhizome geophyte.
**Scirpus uniflorus* Trautv., Trudy Imp. S.-Peterburgsk. Bot. Sada 5: 120 (1877). *Baeothryon uniflorum* (Trautv.) T.V.Egorova, Novosti Sist. Vyssh. Rast. 8: 85 (1971). *Kreczetoviczia uniflora* (Trautv.) Tzvelev, Bot. Zhurn. (Moscow & Leningrad) 84(7): 112 (1999).

Synonyms:
Trichophorum atrichum Palla = ***Trichophorum pumilum*** (Vahl) Schinz & Thell. subsp. ***pumilum***
Trichophorum atrocinctum (Fernald) Á.Löve & D.Löve = ***Scirpus atrocinctus*** Fernald
Trichophorum austriacum Palla = ***Trichophorum cespitosum*** (L.) Hartm. subsp. ***cespitosum***
Trichophorum bracteatum (Bigelow) V.I.Krecz. ex Kuseneva = ***Trichophorum cespitosum*** (L.) Hartm. subsp. ***cespitosum***
Trichophorum cespitosum var. *callosum* (Bigelow ex Torr.) Mohlenbr. = ***Trichophorum cespitosum*** (L.) Hartm. subsp. ***cespitosum***
Trichophorum clementis (M.E.Jones) S.G.Sm. = ***Oreobolopsis clementis*** (M.E.Jones) Dhooge & Goetgh.
Trichophorum comosum (Wall.) A.Dietr. = ***Eriophorum comosum*** (Wall.) Nees
Trichophorum cyperinum (L.) Pers. = ***Scirpus cyperinus*** (L.) Kunth
Trichophorum distigmaticum (Kük.) T.V.Egorova = ***Trichophorum pumilum*** (Vahl) Schinz & Thell. subsp. ***pumilum***
Trichophorum emergens Norman = ***Trichophorum pumilum*** (Vahl) Schinz & Thell. subsp. ***pumilum***
Trichophorum germanicum Palla = ***Trichophorum cespitosum*** subsp. ***germanicum*** (Palla) Hegi
Trichophorum hudsonianum (Michx.) Pers. = ***Trichophorum alpinum*** (L.) Pers.
Trichophorum lineatum (Michx.) Pers. = ***Scirpus lineatus*** Michx.
Trichophorum lockhartii G.Don = [81 TRT]
Trichophorum montevidense (Link) G.Don = ***Androtrichum trigynum*** (Spreng.) H.Pfeiff.

Trichophorum oliganthum (C.A.Mey.) Fritsch = ***Trichophorum pumilum*** (Vahl) Schinz & Thell. subsp. ***pumilum***
Trichophorum parvulum (Roem. & Schult.) Pignatti = ***Eleocharis parvula*** (Roem. & Schult.) Palla
Trichophorum pauciflorum (Lightf.) Pignatti = ***Eleocharis quinqueflora*** (Hartmann) O.Schwarz
Trichophorum pendulum Muhl. = ***Scirpus lineatus*** Michx.
Trichophorum rollandii (Fernald) Hultén = ***Trichophorum pumilum*** subsp. ***rollandii*** (Fernald) R.L.Taylor & MacBryde
Trichophorum vierhapperi (Bojko) Pignatti = ***Eleocharis quinqueflora*** (Hartmann) O.Schwarz

***Unplaced Names*:**
Trichophorum lockhartii G.Don in J.C.Loudon, Hort. Brit.: 468 (1830). = [81 TRT]

Trichophyllum

Trichophyllum Ehrh. = ***Eleocharis*** R.Br.
Trichophyllum aciculare (L.) House = ***Eleocharis acicularis*** (L.) Roem. & Schult.
Trichophyllum acuminatum (Muhl.) House = ***Eleocharis compressa*** Sull. var. ***compressa***
Trichophyllum albidum (Torr.) House = ***Eleocharis albida*** Torr.
Trichophyllum arenicola (Torr. ex Engelm. & A.Gray) House = ***Eleocharis montevidensis*** Kunth
Trichophyllum atropurpureum (Retz.) House = ***Eleocharis atropurpurea*** (Retz.) Kunth
Trichophyllum baldwinii (Torr.) House = ***Eleocharis baldwinii*** (Torr.) Chapm.
Trichophyllum bicolor (Chapm.) House = ***Eleocharis minima*** var. ***bicolor*** (Chapm.) Svenson
Trichophyllum capitatum (L.) House = ***Eleocharis geniculata*** (L.) Roem. & Schult.
Trichophyllum cellulosum (Torr.) House = ***Eleocharis cellulosa*** Torr.
Trichophyllum chaetaria (Roem. & Schult.) House = ***Eleocharis retroflexa*** subsp. ***chaetaria*** (Roem. & Schult.) T.Koyama
Trichophyllum cylindricum (Buckley) House = ***Eleocharis cylindrica*** Buckley
Trichophyllum diandrum (C.Wright) House = ***Eleocharis diandra*** C.Wright
Trichophyllum elongatum (Chapm.) House = ***Eleocharis elongata*** Chapm.
Trichophyllum engelmannii (Steud.) House = ***Eleocharis engelmannii*** Steud.
Trichophyllum intermedium (Schult.) House = ***Eleocharis intermedia*** Schult.
Trichophyllum intermedium var. *habereri* (Fernald) House =
Trichophyllum interstinctum (Vahl) House = ***Eleocharis interstincta*** (Vahl) Roem. & Schult.
Trichophyllum lanceolatum (Fernald) House = ***Eleocharis lanceolata*** Fernald
Trichophyllum maculosum (Vahl) House = ***Eleocharis maculosa*** (Vahl) Roem. & Schult.
Trichophyllum monticola (Fernald) House = ***Eleocharis engelmannii*** Steud.
Trichophyllum mutatum (L.) House = ***Eleocharis mutata*** (L.) Roem. & Schult.
Trichophyllum nodulosum (Roth) House = ***Eleocharis montana*** (Kunth) Roem. & Schult.
Trichophyllum obtusum (Willd.) House = ***Eleocharis obtusa*** (Willd.) Schult.
Trichophyllum ochreatum (Nees) House = ***Eleocharis flavescens*** (Poir.) Urb. var. ***flavescens***
Trichophyllum olivaceum (Torr.) House = ***Eleocharis flavescens*** var. ***olivacea*** (Torr.) Gleason
Trichophyllum ovatum (Roth) House = ***Eleocharis ovata*** (Roth) Roem. & Schult.
Trichophyllum palustre (L.) House = ***Eleocharis palustris*** (L.) Roem. & Schult.
Trichophyllum palustre var. *calvum* (A.Gray) House = ***Eleocharis palustris*** (L.) Roem. & Schult. subsp. ***palustris***
Trichophyllum palustre var. *glaucescens* (Willd.) House = ***Eleocharis palustris*** (L.) Roem. & Schult. subsp. ***palustris***
Trichophyllum palustre var. *vigens* (L.H.Bailey) House = ***Eleocharis palustris*** var. ***vigens*** L.H.Bailey
Trichophyllum praticola (Britton) House = ***Eleocharis flavescens*** (Poir.) Urb. var. ***flavescens***
Trichophyllum proliferum (Torr.) House = ***Eleocharis baldwinii*** (Torr.) Chapm.
Trichophyllum ravenelii (Britton) House = ***Eleocharis ravenelii*** Britton
Trichophyllum robbinsii (Oakes) House = ***Eleocharis robbinsii*** Oakes
Trichophyllum rostellatum (Torr.) Farw. = ***Eleocharis rostellata*** (Torr.) Torr.
Trichophyllum simplex (Elliott) House = ***Eleocharis tortilis*** (Link) Schult.
Trichophyllum tenue (Willd.) Farw. = ***Eleocharis tenuis*** (Willd.) Schult.
Trichophyllum thermale (Rydb.) House = ***Eleocharis flavescens*** (Poir.) Urb. var. ***flavescens***
Trichophyllum torreyanum (Boeck.) House = ***Eleocharis microcarpa*** var. ***filiculmis*** Torr.
Trichophyllum tricostatum (Torr.) House = ***Eleocharis tricostata*** Torr.
Trichophyllum tuberculosum (Michx.) House = ***Eleocharis tuberculosa*** (Michx.) Roem. & Schult.
Trichophyllum viviparum (Link) House = ***Eleocharis vivipara*** Link

Trichoschoenus

Trichoschoenus J.Raynal, Adansonia, n.s., 8: 223 (1968).
Madagascar. 29 MDG.
1 Species

Trichoschoenus bosseri J.Raynal, Adansonia, n.s., 8: 225 (1968).
Madagascar. 29 MDG. Hemicr. or rhizome geophyte.

Tricostularia

Tricostularia Nees in J.G.C.Lehmann, Nov. Stirp. Pug. 8: 50 (1844).
Sri Lanka to Australia. 40 SRL 41 THA VIE 42 BOR MLY SUM 43 NWG 50 NSW NTA QLD SOA TAS WAU 60 NWC.
5 Species
Discopodium Steud., Syn. Pl. Glumac. 2: 150 (1855).

Tricostularia compressa Nees in J.G.C.Lehmann, Nov. Stirp. Pug. 8: 51 (1844). *Schoenus tricostularia* F.Muell., Syst. Census Austral. Pl. 1: 128 (1882).
SW. Australia. 50 WAU. Hemicr. or rhizome geophyte.
Discopodium drummondii Steud., Syn. Pl. Glumac. 2: 150 (1855).
Chaetospora spicata Boeck., Linnaea 38: 297 (1874).

Tricostularia guillauminii (Kük.) J.Raynal, Adansonia, n.s., 14: 269 (1974).
New Caledonia. 60 NWC. Hemicr. or rhizome geophyte.

Schoenus guillauminii Kük., Repert. Spec. Nov. Regni Veg. 44: 95 (1938).
Tricostularia neocaledonica Rendle ex Kük., Repert. Spec. Nov. Regni Veg. 44: 95 (1938), nom. inval.

Tricostularia neesii Nees in J.G.C.Lehmann, Nov. Stirp. Pug. 8: 52 (1844). *Chaetospora neesii* (Nees) Boeck., Linnaea 38: 297 (1874). *Schoenus neesii* (Nees) F.Muell., Syst. Census Austral. Pl. 1: 128 (1882).
SW. Australia. 50 WAU. Hemicr. or rhizome geophyte.
Tricostularia neesii var. *elatior* Benth., Fl. Austral. 7: 384 (1878).
Lepidosperma exsul C.B.Clarke, Kew Bull., Addit. Ser. 8: 47 (1908).

Tricostularia pauciflora (R.Br.) Benth., Fl. Austral. 7: 383 (1878).
SE. Australia. 50 NSW SOA TAS. Hemicr. or rhizome geophyte.
**Cladium pauciflorum* R.Br., Prodr.: 237 (1810). *Chapelliera pauciflora* (R.Br.) Nees in J.G.C.Lehmann, Pl. Preiss. 2: 76 (1846). *Lepidosperma pauciflorum* (R.Br.) F.Muell., Fragm. 9: 23 (1875). *Schoenus pauciflorus* (F.Muell.) F.Muell., Syst. Census Austral. Pl. 1: 128 (1882). *Cladium junceum* var. *pauciflorum* (R.Br.) Kük., Repert. Spec. Nov. Regni Veg. 51: 182 (1942).

Tricostularia undulata (Thwaites) J.Kern, Acta Bot. Neerl. 8: 267 (1959).
Sri Lanka to N. Australia. 40 SRL 41 THA VIE 42 BOR MLY SUM 43 NWG 50 NTA QLD. Hemicr. or rhizome geophyte.
**Cladium undulatum* Thwaites, Enum. Pl. Zeyl.: 353 (1864). *Machaerina undulata* (Thwaites) T.Koyama, Bot. Mag. (Tokyo) 69: 66 (1956).
Carpha junciformis Boeck., Linnaea 38: 267 (1874).
Lepidosperma zeylanicum Lindl. ex Boeck., Linnaea 38: 332 (1874). *Tricostularia zeylanica* (Lindl. ex Boeck.) Benth., Fl. Austral. 7: 384 (1878).
Chaetospora fimbristyloides F.Muell., Fragm. 9: 34 (1875). *Cladium undulatum* var. *fimbristyloides* (F.Muell.) Domin, Biblioth. Bot. 89: 473 (1915).
Tricostularia fimbristyloides Benth., Fl. Austral. 7: 384 (1878). *Schoenus fimbristyloides* (Benth.) F.Muell., Syst. Census Austral. Pl. 1: 128 (1882).
Cladium pulchrum Ridl., J. Fed. Malay States Mus. 6: 192 (1915). *Mariscus pulcher* (Ridl.) Fernald, Rhodora 25: 54 (1923). *Machaerina pulchra* (Ridl.) T.Koyama, Bot. Mag. (Tokyo) 69: 65 (1956).

Synonyms:
Tricostularia fimbristyloides Benth. = ***Tricostularia undulata*** (Thwaites) J.Kern
Tricostularia neesii var. *elatior* Benth. = ***Tricostularia neesii*** Nees
Tricostularia neocaledonica Rendle ex Kük. = ***Tricostularia guillauminii*** (Kük.) J.Raynal
Tricostularia paludosa (R.Br.) Benth. = ***Schoenus paludosus*** (R.Br.) Roem. & Schult.
Tricostularia zeylanica (Lindl. ex Boeck.) Benth. = ***Tricostularia undulata*** (Thwaites) J.Kern

Trilepis

Trilepis Nees, Edinburgh New Philos. J. 17: 267 (1834).
N. South America to Brazil. 82 FRG GUY SUR VEN 84 BZE BZL.
7 Species
Fintelmannia Kunth, Enum. Pl. 2: 362 (1837).

Trilepis alcatrazensis Gilly, Brittonia 5: 17 (1943).
Brazil (São Paulo). 84 BZL. Hemicr. or rhizome geophyte.

Trilepis ciliatifolia T.Koyama, Jap. J. Bot. 20: 132 (1969).
Brazil (Rio de Janeiro). 84 BZL. Hemicr. or rhizome geophyte.

Trilepis eximia (C.B.Clarke) H.Pfeiff., Repert. Spec. Nov. Regni Veg. 18: 385 (1922).
Brazil (Rio de Janeiro). 84 BZL. Hemicr. or rhizome geophyte.
**Fintelmannia eximia* C.B.Clarke, Kew Bull., Addit. Ser. 8: 66 (1908).

Trilepis kanukuensis Gilly, Brittonia 5: 18 (1943).
N. South America. 82 FRG GUY SUR VEN. Cham.

Trilepis lhotzkiana Nees, Edinburgh New Philos. J. 17: 267 (1834). *Carex lhotzkiana* (Nees) Steud., Nomencl. Bot., ed. 2, 1: 292 (1840).
E. Brazil. 84 BZE BZL. Hemicr. or rhizome geophyte.
Fintelmannia restioides Kunth, Enum. Pl. 2: 363 (1837).
Fintelmannia robusta C.B.Clarke, Kew Bull., Addit. Ser. 8: 67 (1908). *Trilepis lhotzkiana* f. *robusta* (C.B.Clarke) H.Pfeiff., Repert. Spec. Nov. Regni Veg. 18: 384 (1922).

Trilepis microstachya (C.B.Clarke) H.Pfeiff., Repert. Spec. Nov. Regni Veg. 18: 285 (1922).
SE. Brazil. 84 BZL. Hemicr. or rhizome geophyte.
**Fintelmannia microstachya* C.B.Clarke, Kew Bull., Addit. Ser. 8: 66 (1908).

Trilepis tenuis Vitta, Brittonia 54: 120 (2002).
Brazil (Rio de Janeiro). 84 BZL. Hemicr. or rhizome geophyte.

Synonyms:
Trilepis abyssinica (Hochst. ex A.Rich.) Boeck. = ***Coleochloa abyssinica*** (Hochst. ex A.Rich.) Gilly
Trilepis lhotzkiana f. *robusta* (C.B.Clarke) H.Pfeiff. = ***Trilepis lhotzkiana*** Nees
Trilepis oliveri Boeck. = ***Coleochloa setifera*** (Ridl.) Gilly
Trilepis pilosa Boeck. = ***Afrotrilepis pilosa*** (Boeck.) J.Raynal
Trilepis rehmanniana Boeck. ex C.B.Clarke = ***Coleochloa setifera*** (Ridl.) Gilly
Trilepis royleana Nees = ***Kobresia royleana*** (Nees) Boeck.
Trilepis setifera (Ridl.) Kuntze = ***Coleochloa setifera*** (Ridl.) Gilly

Triodon

Triodon Rich. ex Pers. = ***Rhynchospora*** Vahl
Triodon albus (L.) Farw. = ***Rhynchospora alba*** (L.) Vahl
Triodon albus var. *macer* (C.B.Clarke) Farw. = ***Rhynchospora macra*** (C.B.Clarke ex Britton) Small
Triodon setaceus (P.J.Bergius) Rich. ex Pers. = ***Rhynchospora tenerrima*** Nees ex Spreng. subsp. ***tenerrima***

Triodus

Triodus Raf. = ***Carex*** L.

Triplima

Triplima Raf. = ***Carex*** L.

Tunga

Tunga Roxb. = ***Hypolytrum*** Rich. ex Pers.
Tunga diandra Roxb. = ***Hypolytrum nemorum*** (Vahl) Spreng. subsp. ***nemorum***
Tunga laevigata Roxb. = ***Lipocarpha chinensis*** (Osbeck) J.Kern
Tunga triceps Roxb. = ***Lipocarpha gracilis*** (Rich. ex Pers.) Nees

Tylocarya

Tylocarya Nelmes = ***Fimbristylis*** Vahl
Tylocarya cylindrostachya Nelmes = ***Fimbristylis nelmesii*** J.Kern

Ulea-flos

Ulea-flos C.B.Clarke ex H.Pfeiff. = ***Lagenocarpus*** Nees

Ulva

Ulva Adans. = ***Carex*** L.

Uncinia

Uncinia Pers., Syn. Pl. 2: 534 (1807).
Malesian Mts. to New Zealand and Pacific, Mexico to Trop. America, Subantarctic Is. 42 BOR PHI SUL 43 NWG 50 NFK NSW TAS VIC 51 ATP CTM NZN NZS 60 NWC 61 SCI TUB 63 HAW (78) ala 79 MXC MXE MXG MXS MXT 80 COS ELS GUA HON NIC PAN 81 CUB DOM JAM 82 VEN 83 BOL CLM ECU PER 84 BZN 85 AGE AGS AGW CHS CLC CLS JNF 90 ASP CRZ FAL KEG MAQ MPE SGE TDC.
67 Species
Agistron Raf., Good Book: 28 (1840).

Uncinia affinis (Colenso ex C.B.Clarke) Hamlin, Domin. Mus. Bull. 19: 30 (1959).
New Zealand. 51 NZN NZS. Hemicr.
**Uncinia riparia* var. *affinis* Colenso ex C.B.Clarke, J. Linn. Soc., Bot. 20: 392 (1883).
Uncinia purpurata var. *subcaespitosa* Kük. in H.G.A. Engler (ed.), Pflanzenr., IV, 38: 61 (1909).

Uncinia andina G.A.Wheeler, Hickenia 2: 218 (1997).
SC. Chile to SW. Argentina. 85 AGS CLC. Hemicr.

Uncinia angustifolia Hamlin, Domin. Mus. Bull. 19: 42 (1959).
New Zealand. 51 NZN NZS. Hemicr.
**Uncinia rupestris* var. *planifolia* Kük. in H.G.A. Engler (ed.), Pflanzenr., IV, 38: 64 (1909).

Uncinia araucana G.A.Wheeler, Aliso 15: 3 (1996 publ. 1997).
SC. Chile (La Araucaria). 85 CLC. Hemicr.

Uncinia astonii Hamlin, Domin. Mus. Bull. 19: 64 (1959).
New Zealand. 51 NZN NZS. Hemicr.

Uncinia aucklandica Hamlin, Domin. Mus. Bull. 19: 63 (1949).
S. New Zealand South I., Antipodean Is. 51 ATP NZS. Hemicr.

Uncinia austroamericana G.A.Wheeler, Darwiniana 43: 271 (2005).
Tierra del Fuego. 85 AGS. Hemicr.

Uncinia banksii Boott in J.D.Hooker, Fl. Nov.-Zel. 1: 287 (1853). *Uncinia riparia* var. *banksii* (Boott) C.B.Clarke, J. Linn. Soc., Bot. 20: 392 (1883).
New Zealand. 51 NZN NZS. Hemicr.
Uncinia banksii var. *major* Boott in J.D.Hooker, Fl. Nov.-Zel. 1: 287 (1853).
Uncinia capillaris Colenso, Trans. & Proc. New Zealand Inst. 20: 210 (1888).

Uncinia brevicaulis (Thouars) Kunth, Enum. Pl. 2: 528 (1837).
Hawaiian Is. (E. Maui), Juan Fernádez Is. to Subantarctic Is. 63 HAW 83 PER 85 AGS CLS JNF 90 ASP FAL TDC. Hemicr. or rhizome geophyte.
**Carex brevicaulis* Thouars, Esquisse Fl. Tristan d'Acugna: 35 (1808).
Uncinia breviculmis Carmich., Trans. Linn. Soc. London 12: 508 (1819).
Uncinia rigida Boeck., Flora 65: 64 (1882).
Uncinia cylindrica Franch., Miss. Sci. Cap Horn 5: 379 (1889). *Uncinia macloviana* var. *cylindrica* (Franch.) Kük., Bot. Centralbl. 76: 212 (1898).
Uncinia phleoides var. *laticarpa* Kük., Bot. Centralbl. 82: 130 (1900). *Uncinia brevicaulis* var. *laticarpa* (Kük.) Kük. in H.G.A.Engler (ed.), Pflanzenr., IV, 20(38): 52 (1909).
Uncinia brevicaulis f. *montana* Kük. in H.G.A.Engler (ed.), Pflanzenr., IV, 20(38): 52 (1909).

Uncinia caespitosa Colenso ex Boott in J.D.Hooker, Fl. Nov.-Zel. 1: 287 (1853).
New Zealand. 51 NZN NZS. Hemicr.
Uncinia compacta var. *caespitiformis* Kük. in L.Cockayne, Rep. Bot. Surv. Stewart Isl.: 42 (1909).
Uncinia caespitosa var. *collina* Petrie, Trans. & Proc. New Zealand Inst. 52: 19 (1920).

Uncinia chilensis G.A.Wheeler, Aliso 15: 1 (1996 publ. 1997).
SC. Chile to Argentina (Rio Negro). 85 AGS CLC. Hemicr.

Uncinia clavata (Kük.) Hamlin, Domin. Mus. Bull. 19: 68 (1959).
New Zealand. 51 NZN NZS. Hemicr.
**Uncinia australis* var. *clavata* Kük. in T.F.Cheeseman, Man. New Zealand Fl.: 802 (1906).

Uncinia compacta R.Br., Prodr.: 241 (1810). *Carex compacta* (R.Br.) Poir. in J.B.A.M.de Lamarck, Encycl., Suppl. 3: 282 (1813), nom. illeg.
Malesian Mts, SE. Australia, New Zealand, Subantarctic Is. 42 BOR PHI SUL 43 NWG 50 NSW TAS VIC 51 CTM NZN NZS 90 ASP CRZ KEG MPE TDC. Hemicr. or rhizome geophyte.

var. ***alpina*** Noot., Blumea 24: 519 (1978).
New Guinea (Mt. Wilhelm, Mt. Giluwe). 43 NWG. Hemicr. or rhizome geophyte.

var. ***compacta***
Malesian Mts, SE. Australia, New Zealand, Subantarctic Is. 42 BOR PHI SUL 43 NWG 50 NSW TAS VIC 51 CTM NZN NZS 90 ASP CRZ KEG MPE TDC. Hemicr. or rhizome geophyte.
Uncinia rupestris Raoul, Ann. Sci. Nat., Bot., II, 2: 117 (1844).
Uncinia moseleyana Boeck., Flora 61: 170 (1878).
Uncinia filiformis Boott ex C.B.Clarke, J. Linn. Soc., Bot. 20: 392 (1883).
Uncinia hookeri Colenso ex C.B.Clarke, J. Linn. Soc., Bot. 20: 395 (1883).
Uncinia horizontalis Colenso, Trans. & Proc. New Zealand Inst. 15: 334 (1883).

Uncinia caespitosa var. *minor* Kük. in T.F. Cheeseman, Man. New Zealand Fl.: 802 (1906).
Uncinia riparia var. *stolonifera* Kük. & Steen., Bull. Jard. Bot. Buitenzorg, III, 13: 213 (1934).
Uncinia subtrigona Nelmes, Kew Bull. 4: 144 (1949).

var. ***nervosa*** (Boott) C.B.Clarke, J. Linn. Soc., Bot. 20: 395 (1883).
New Guinea, SE. Australia, New Zealand. 43 NWG 50 TAS 51 NZN NZS. Hemicr. or rhizome geophyte.
**Uncinia nervosa* Boott in J.D.Hooker, Fl. Tasman. 2: 102 (1858).
Uncinia cheesemanniana Boeck., Bot. Jahrb. Syst. 5: 521 (1884).

Uncinia costata Kük., Repert. Spec. Nov. Regni Veg. 16: 433 (1920).
Juan Fernández Is. (Alejandro Selkirk I.). 85 JNF. Hemicr. or rhizome geophyte.

Uncinia dawsonii Hamlin, Trans. Roy. Soc. New Zealand, Bot. 2: 128 (1963).
New Caledonia. 60 NWC. Hemicr.

Uncinia debilior F.Muell., Fragm. 8: 151 (1874). *Uncinia filiformis* var. *debilior* (F.Muell.) W.R.B.Oliv., Trans. & Proc. New Zealand Inst. 49: 128 (1917).
Lord Howe I. 50 NFK. Hemicr.

Uncinia dikei Nelmes, Kew Bull. 4: 377 (1949).
Marion-Prince Edward Is. 90 MPE. Hemicr.

Uncinia distans Colenso ex Boott in J.D.Hooker, Fl. Nov.-Zel. 1: 286 (1853).
New Zealand North I. 51 NZN. Hemicr.
Uncinia nelmesii Hamlin, Trans. Roy. Soc. New Zealand, Bot. 2: 127 (1963).

Uncinia divaricata Boott in J.D.Hooker, Fl. Nov.-Zel. 1: 286 (1853). *Uncinia compacta* var. *divaricata* (Boott) Hook.f., Handb. N. Zeal. Fl. 1: 309 (1864).
New Zealand to Maquarie Is. 51 ATP NZN NZS 90 MAQ. Hemicr. or rhizome geophyte.
Uncinia clarkei Petrie, Trans. & Proc. New Zealand Inst. 20: 185 (1887 publ. 1888).
Uncinia rupestris var. *flavescens* Kük., Bot. Centralbl. 76: 211 (1898).
Uncinia compacta var. *petriei* C.B.Clarke in T.F.Cheeseman, Man. New Zealand Fl.: 800 (1906).
Uncinia compacta var. *clarkei* (Petrie) Kük. in H.G.A. Engler (ed.), Pflanzenr., IV, 20(38): 65 (1909).

Uncinia douglasii Boott in J.D.Hooker, Fl. Antarct. 2: 369 (1846). *Uncinia macloviana* var. *douglasii* (Boott) Kük., Bot. Centralbl. 82: 133 (1900).
Juan Fernández Is. 85 JNF. Hemicr. or rhizome geophyte.
Uncinia angusta Nees, Linnaea 9: 305 (1834), nom. inval.
Uncinia angustata Boeck., Linnaea 41: 347 (1877).
Uncinia fernandesiana Nees ex Boeck., Linnaea 41: 347 (1877).

Uncinia drucei Hamlin, Domin. Mus. Bull. 19: 58 (1959).
New Zealand. 51 NZN NZS. Hemicr. or rhizome geophyte.
Uncinia drucei var. *payciflora* Hamlin, Domin. Mus. Bull. 19: 59 (1959).

Uncinia ecuadorensis G.A.Wheeler & Goetgh., Aliso 15: 10 (1996 publ. 1997).
N. & C. Ecuador. 83 ECU. Hemicr. or rhizome geophyte.

Uncinia egmontiana Hamlin, Domin. Mus. Bull. 19: 33 (1959).
New Zealand. 51 NZN NZS. Hemicr.
Uncinia silvestris var. *squamata* Hamlin, Domin. Mus. Bull. 19: 28 (1959).

Uncinia elegans (Kük.) Hamlin, Domin. Mus. Bull. 19: 11 (1959).
Tasmania, New Zealand South I. 50 TAS 51 NZN. Hemicr. or rhizome geophyte.
**Uncinia sinclairii* var. *elegans* Kük. in T.F.Cheeseman, Man. New Zealand Fl.: 799 (1906).

Uncinia erinacea (Cav.) Pers., Syn. Pl. 2: 534 (1807).
Magellan Reg. 85 AGS CLS. Hemicr. or rhizome geophyte.
**Carex erinacea* Cav., Icon. 5: 40 (1799). *Agistron erinacea* (Cav.) Raf., Good Book: 28 (1840).
Uncinia longifolia Kunth, Enum. Pl. 2: 527 (1837).
Uncinia longiaristata Steud., Syn. Pl. Glumac. 2: 243 (1855).
Uncinia philippii Hohen. ex Steud., Syn. Pl. Glumac. 2: 243 (1855).
Uncinia macrotricha Franch., Miss. Sci. Cap Horn 5: 379 (1889).
Uncinia erinacea var. *angustata* Kük., Bot. Centralbl. 82: 101 (1900).

Uncinia ferruginea Boott in J.D.Hooker, Fl. Nov.-Zel. 1: 288 (1853).
New Zealand. 51 NZN NZS. Hemicr.
Uncinia nigra Colenso, Trans. & Proc. New Zealand Inst. 17: 253 (1885).
Uncinia variegata Colenso, Trans. & Proc. New Zealand Inst. 20: 211 (1887 publ. 1888).

Uncinia filiformis Colenso ex Boott in J.D.Hooker, Fl. Nov.-Zel. 1: 286 (1853).
New Zealand. 51 NZN NZS. Hemicr.
Uncinia rupestris var. *capillacea* Kük. in H.G.A. Engler (ed.), Pflanzenr., IV, 38: 64 (1909).

Uncinia flaccida S.T.Blake, Proc. Roy. Soc. Queensland 51: 49 (1939 publ. 1940).
SE. Australia. 50 NSW TAS VIC. Hemicr. or rhizome geophyte.

Uncinia fuscovaginata Kük., Bull. Herb. Boissier, II, 4: 50 (1904).
New Zealand. 51 NZN NZS. Hemicr. or rhizome geophyte.
Uncinia fuscovaginata var. *caespitans* Hamlin, Domin. Mus. Bull. 19: 21 (1959).

Uncinia gracilenta Hamlin, Domin. Mus. Bull. 19: 47 (1959).
New Zealand. 51 NZN NZS. Hemicr.

Uncinia hamata (Sw.) Urb., Symb. Antill. 2: 169 (1900).
Mexico to Trop. America. (78) ala 79 MXC MXE MXG MXS MXT 80 COS ELS GUA HON NIC PAN 81 CUB DOM JAM 82 VEN 83 BOL CLM ECU PER 84 BZN 85 AGE AGW. Hemicr. or rhizome geophyte.
**Carex hamata* Sw., Prodr.: 18 (1788).
Carex jamaicensis Poir. in J.B.A.M.de Lamarck, Encycl. 3: 246 (1789). *Uncinia jamaicensis* (Poir.) Pers., Syn. Pl. 2: 534 (1807).
Carex uncinata Schkuhr ex Steud., Nomencl. Bot., ed. 2, 1: 297 (1840).
Uncinia phleoides C.A.Mey., Bull. Acad. Roy. Sci. Bruxelles 9: 249 (1842), nom. illeg.
Uncinia jamaicensis Liebm., Mexic. Neldeagt. Pl., V, 2: 272 (1851), nom. illeg.

Uncinia mexicana Steud., Syn. Pl. Glumac. 2: 243 (1855). *Uncinia hamata* var. *mexicana* (Steud.) Kük. in H.G.A.Engler (ed.), Pflanzenr., IV, 20(38): 54 (1909).
Uncinia galeottii Boott ex C.B.Clarke, J. Linn. Soc., Bot. 20: 400 (1883).
Uncinia multifolia Boeck., Bot. Jahrb. Syst. 8: 207 (1887).
Uncinia hamata f. *angustifolia* Kük. in H.G.A.Engler (ed.), Pflanzenr., IV, 20(38): 54 (1909).

Uncinia hookeri Boott in J.D.Hooker, Fl. Antarct.: 91 (1844). *Uncinia riparia* var. *hookeri* (Boott) Kük. in H.G.A.Engler (ed.), Pflanzenr., IV, 38: 63 (1909).
Antipodean Is. to Maquarie Is. 51 ATP 90 MAQ. Hemicr.

Uncinia involuta Hamlin, Domin. Mus. Bull. 19: 49 (1959).
New Zealand. 51 NZN NZS. Hemicr.

Uncinia koyamae Gómez-Laur., Brenesia 18: 92 (1980).
Mexico (Chiapas), Costa Rica. 79 MXT 80 COS. Hemicr. or rhizome geophyte.

Uncinia lacustris G.A.Wheeler, Aliso 14: 141 (1995).
NC. Ecuador. 83 ECU. Hemicr. or rhizome geophyte.

Uncinia laxiflora Petrie, Trans. & Proc. New Zealand Inst. 17: 271 (1885).
New Zealand. 51 NZN NZS. Hemicr.

Uncinia lechleriana Steud., Syn. Pl. Glumac. 2: 244 (1855).
Chile to S. Argentina. 85 AGS CHS. Hemicr. or rhizome geophyte.

Uncinia leptostachya Raoul, Ann. Sci. Nat., Bot., II, 2: 116 (1844).
New Zealand. 51 NZN NZS. Hemicr.

Uncinia longifructus (Kük.) Petrie, Trans. & Proc. New Zealand Inst. 52: 17 (1919 publ. 1920).
New Zealand. 51 NZN NZS. Hemicr.
**Uncinia tenella* var. *longifructus* Kük. in H.G.A. Engler (ed.), Pflanzenr., IV, 38: 66 (1909).

Uncinia macloviana Gaudich., Voy. Uranie: 412 (1829). *Uncinia gracilis* var. *macloviana* (Gaudich.) C.B.Clarke, J. Linn. Soc., Bot. 20: 400 (1883). *Uncinia brevicaulis* var. *macloviana* (Gaudich.) Kük. in H.G.A.Engler (ed.), Pflanzenr., IV, 20(38): 52 (1909).
Chile to S. Argentina. 85 AGS CLS. Hemicr. or rhizome geophyte.
Uncinia de-la-costa Steud. in W.Lechler, Berberid. Amer. Austral.: 52 (1857). *Carex de-la-costa* (Steud.) Kuntze, Revis. Gen. Pl. 3(2): 332 (1898).
Uncinia montana Phil., Anales Univ. Chile 1865(2): 322 (1865). *Uncinia macloviana* var. *montana* (Phil.) Kük., Bot. Centralbl. 82: 132 (1900).

Uncinia macrolepis Decne. in J.S.C.Dumont d'Urville, Voy. Pôle Sud 2: 13 (1853).
Venezuela to Subantarctic Is. 82 VEN 83 BOL CLM ECU PER 85 AGS CLS 90 SGE TDC. Hemicr. or rhizome geophyte.
Uncinia meridensis Steyerm., Fieldiana, Bot. 28(1): 61 (1951).
Uncinia smithii Philcox, Kew Bull. 15: 229 (1961).

Uncinia macrophylla Steud., Syn. Pl. Glumac. 2: 244 (1855).
S. Chile. 85 CLS. Hemicr.
Uncinia bella Phil., Linnaea 30: 204 (1860).
Uncinia phalaroides Boott ex C.B.Clarke, J. Linn. Soc., Bot. 20: 396 (1883).
Uncinia bracteosa Phil., Anales Univ. Chile 93: 503 (1896).

Uncinia multifaria Nees ex Boott in J.D.Hooker, Fl. Antarct. 2: 369 (1846).
S. Chile. 85 CLS. Hemicr.
Uncinia macrostachya É.Desv. in C.Gay, Fl. Chil. 6: 229 (1854). *Uncinia multifaria* var. *macrostachya* (É.Desv.) Kük., Bot. Centralbl. 82: 102 (1900).

Uncinia negeri Kük., Bot. Centralbl. 76: 210 (1898).
Chile to SW. Argentina. 85 AGS CLS. Hemicr. or rhizome geophyte.
Uncinia negeri var. *araucana* Gunckel, Revista Univ. (Santiago) 30: 58 (1945).

Uncinia nemoralis K.L.Wilson, Telopea 5: 620 (1994).
SE. Australia. 50 NSW VIC. Hemicr.

Uncinia obtusifolia Heenan, New Zealand J. Bot. 34: 11 (1996).
New Zealand. 51 NZN NZS. Hemicr.

Uncinia paludosa G.A.Wheeler & Goetgh., Aliso 14: 142 (1995).
NE. Colombia to Peru. 83 CLM ECU PER. Hemicr. or rhizome geophyte.

Uncinia perplexa Heenan & de Lange, New Zealand J. Bot. 39: 376 (2001).
New Zealand North I. 51 NZN. Hemicr.

Uncinia phleoides (Cav.) Pers., Syn. Pl. 2: 534 (1807).
C. Mexico, W. & S. South America. 79 MXC 83 BOL CLM ECU PER 85 AGE AGS AGW CLC CLS JNF. Hemicr.
**Carex phleoides* Cav., Icon. 5: 40 (1799). *Agistron phleoides* (Cav.) Raf., Good Book: 28 (1840).
Uncinia trichocarpa C.A.Mey., Cyperac. Nov.: 11 (1831). *Uncinia phleoides* var. *trichocarpa* (C.A. Mey.) C.B.Clarke, J. Linn. Soc., Bot. 20: 399 (1883).
Uncinia cumingii Nees, Linnaea 9: 305 (1834), nom. inval.
Uncinia longifolia É.Desv. in C.Gay, Fl. Chil. 6: 226 (1854), nom. illeg.
Uncinia trichocarpa É.Desv. in C.Gay, Fl. Chil. 6: 227 (1854).
Uncinia durvillei Steud., Syn. Pl. Glumac. 2: 243 (1855).
Uncinia urvillei Steud., Syn. Pl. Glumac. 2: 243 (1855).
Uncinia longispica Boeck., Flora 41: 650 (1858). *Uncinia trichocarpa* var. *longispica* (Boeck.) Kük., Bot. Centralbl. 82: 131 (1900).
Uncinia montteana Phil., Linnaea 30: 205 (1860).
Uncinia chlorostachya Phil., Linnaea 33: 275 (1865).
Uncinia leptostachya Phil., Linnaea 33: 274 (1865).
Uncinia lasiocarpa Steud. ex Boeck., Linnaea 41: 349 (1877).
Uncinia longifolia Phil. ex C.B.Clarke, J. Linn. Soc., Bot. 20: 399 (1883).
Uncinia phleoides var. *nux-nigra* C.B.Clarke, J. Linn. Soc., Bot. 20: 399 (1883).
Uncinia phleoides f. *longispica* Franch., Miss. Sci. Cap Horn 5: 378 (1889).
Uncinia loliacea Phil., Anales Univ. Chile 93: 503 (1896).
Uncinia phleoides var. *brachytricha* Speg., Revista Fac. Agron. Univ. Nac. La Plata 3: 626 (1897).
Uncinia phleoides var. *krausei* Kük., Bot. Centralbl. 76: 211 (1898).

Uncinia purpurata Petrie, Trans. & Proc. New Zealand Inst. 17: 272 (1885).
New Zealand South I. 51 NZS. Hemicr.

Uncinia rapaensis H.St.John, Nordic J. Bot. 4: 60 (1984).
Tubuai Is. (Rapa-Iti). 61 TUB. Hemicr. or rhizome geophyte.

Uncinia riparia R.Br., Prodr.: 241 (1810). *Carex riparia* (R.Br.) Poir. in J.B.A.M.de Lamarck, Encycl., Suppl. 3: 282 (1813), nom. illeg.
New Guinea, SE. Australia, New Zealand South I. 43 NWG 50 TAS VIC 51 NZS. Hemicr. or rhizome geophyte.
Uncinia obtusata Colenso, Trans. & Proc. New Zealand Inst. 16: 341 (1884).
Uncinia sclerophylla Nelmes, Kew Bull. 4: 143 (1949).
Uncinia ohwiana T.Koyama, Bot. Mag. (Tokyo) 69: 214 (1956).

Uncinia rubra Colenso ex Boott in J.D.Hooker, Fl. Nov.-Zel. 1: 287 (1853).
New Zealand. 51 NZN NZS. Hemicr.
Uncinia rubra var. *fallax* Kük. in H.G.A.Engler (ed.), Pflanzenr., IV, 38: 64 (1909).

Uncinia ×rubrovaginata Hamlin, Domin. Mus. Bull. 19: 24 (1959). *U. fuscovaginata* × *U. rubra*.
New Zealand. 51 NZN. Hemicr.

Uncinia scabra Colenso ex Boott in J.D.Hooker, Fl. Nov.-Zel. 1: 285 (1853).
New Zealand. 51 NZN NZS. Hemicr.
Uncinia disticha Colenso, Trans. & Proc. New Zealand Inst. 20: 210 (1888).

Uncinia scabriuscula G.A.Wheeler, Hickenia 2: 215 (1997).
S. Chile to SW. Argentina. 85 AGS CLS. Hemicr.

Uncinia silvestris Hamlin, Domin. Mus. Bull. 19: 26 (1959).
New Zealand North I. 51 NZN. Hemicr.

Uncinia sinclairii Boott ex Hook.f., Handb. N. Zeal. Fl. 1: 309 (1864).
E. New Zealand South I. (50) nsw 51 NZS. Hemicr. or rhizome geophyte.

Uncinia strictissima (Kük.) Petrie, Trans. & Proc. New Zealand Inst. 47: 55 (1914 publ. 1915).
New Zealand. 51 ATP NZN NZS. Hemicr.
Uncinia rigida Petrie, Trans. & Proc. New Zealand Inst. 17: 271 (1885), nom. illeg.
**Uncinia rubra* var. *strictissima* Kük. in H.G.A.Engler (ed.), Pflanzenr., IV, 38: 64 (1909).

Uncinia subsacculata G.A.Wheeler & Goetgh., Aliso 14: 145 (1995).
Ecuador (Pichincha). 83 ECU. Hemicr. or rhizome geophyte.

Uncinia sulcata K.L.Wilson, Telopea 5: 620 (1994).
SE. Australia. 50 NSW VIC. Hemicr.

Uncinia tenella R.Br., Prodr.: 241 (1810).
SE. Australia. 50 NSW TAS VIC. Hemicr. or rhizome geophyte.
Carex tenella Poir. in J.B.A.M.de Lamarck, Encycl., Suppl. 3: 282 (1813), nom. illeg.
Uncinia tenella var. *robustior* Kük., Bot. Centralbl. 76: 211 (1898).

Uncinia tenuifolia G.A.Wheeler & Goetgh., Aliso 14: 144 (1995).
SE. Ecuador. 83 ECU. Hemicr. or rhizome geophyte.

Uncinia tenuis Poepp. ex Kunth, Enum. Pl. 2: 525 (1837).
Costa Rica, Colombia to Falkland Is. 80 COS 83 CLM ECU 85 AGS CLC CLS JNF 90 FAL. Hemicr. or rhizome geophyte.
Uncinia gracilis Decne. in J.S.C.Dumont d'Urville, Voy. Pôle Sud, Atlas: t. 6, f. B (1843).
Uncinia tenuis f. *firmula* Kük., Repert. Spec. Nov. Regni Veg. 16: 433 (1920).

Uncinia triquetra Kük., Bot. Centralbl. 82: 97 (1900). *Uncinia lechleriana* var. *triquetra* (Kük.) Kük. in H.G.A.Engler (ed.), Pflanzenr., IV, 20(38): 58 (1909).
Magellan Reg. 85 AGS CLS. Hemicr. or rhizome geophyte.

Uncinia uncinata (L.f.) Kük. in H.G.A.Engler (ed.), Pflanzenr., IV, 20(38): 62 (1909).
New Zealand, C. Pacific. 51 ATP CTM NZN NZS 61 SCI 63 HAW. Hemicr.
Uncinia uncinata var. *uliginosa* Skottsb. in ?
**Carex uncinata* L.f., Suppl. Pl.: 413 (1782). *Uncinia australis* Pers., Syn. Pl. 2: 534 (1807), nom. illeg. *Carex hamosa* Thouars, Esquisse Fl. Tristan d'Acugna: 35 (1808), nom. illeg.
Uncinia scaberrima Nees, Linnaea 9: 305 (1834), nom. inval.
Uncinia lindleyana Kunth, Enum. Pl. 2: 526 (1837).
Uncinia rigidula Steud., Syn. Pl. Glumac. 2: 245 (1855).
Uncinia alopecuroides Colenso, Trans. & Proc. New Zealand Inst. 15: 335 (1883).
Uncinia bractata Colenso, Trans. & Proc. New Zealand Inst. 16: 341 (1884).
Uncinia polyneura Colenso, Trans. & Proc. New Zealand Inst. 19: 270 (1886 publ. 1887).
Uncinia pedicellata Kük. in H.G.A.Engler (ed.), Pflanzenr., IV, 20(38): 61 (1909). *Uncinia uncinata* var. *pedicellata* (Kük.) Petrie, Trans. & Proc. New Zealand Inst. 47: 54 (1915).
Uncinia uncinata var. *laxior* Carse, Trans. & Proc. New Zealand Inst. 48: 240 (1916).

Uncinia viridis (C.B.Clarke) Edgar in Moore & E.Edgar, Fl. N. Zeal. 2: 299 (1970).
New Zealand. 51 NZN NZS. Hemicr. or rhizome geophyte.
**Uncinia compacta* var. *viridis* C.B.Clarke, J. Linn. Soc., Bot. 20: 395 (1883).

Uncinia zotovii Hamlin, Domin. Mus. Bull. 19: 37 (1959).
New Zealand. 51 CTM NZN NZS. Hemicr.

Synonyms:
Uncinia alopecuroides Colenso = ***Uncinia uncinata*** (L.f.) Kük.
Uncinia angusta Nees = ***Uncinia douglasii*** Boott
Uncinia angustata Boeck. = ***Uncinia douglasii*** Boott
Uncinia australis Pers. = ***Uncinia uncinata*** (L.f.) Kük.
Uncinia australis var. *clavata* Kük. = ***Uncinia clavata*** (Kük.) Hamlin
Uncinia banksii var. *major* Boott = ***Uncinia banksii*** Boott
Uncinia bella Phil. = ***Uncinia macrophylla*** Steud.
Uncinia bractata Colenso = ***Uncinia uncinata*** (L.f.) Kük.
Uncinia bracteosa Phil. = ***Uncinia macrophylla*** Steud.
Uncinia brevicaulis var. *laticarpa* (Kük.) Kük. = ***Uncinia brevicaulis*** (Thouars) Kunth
Uncinia brevicaulis var. *macloviana* (Gaudich.) Kük. = ***Uncinia macloviana*** Gaudich.

Uncinia brevicaulis f. *montana* Kük. = ***Uncinia brevicaulis*** (Thouars) Kunth
Uncinia breviculmis Carmich. = ***Uncinia brevicaulis*** (Thouars) Kunth
Uncinia breviseta Torr. = ***Carex filifolia*** Nutt. var. ***filifolia***
Uncinia caespitosa var. *collina* Petrie = ***Uncinia caespitosa*** Colenso ex Boott
Uncinia caespitosa var. *minor* Kük. = ***Uncinia compacta*** R.Br. var. ***compacta***
Uncinia capillaris Colenso = ***Uncinia banksii*** Boott
Uncinia cheesemanniana Boeck. = ***Uncinia compacta*** var. ***nervosa*** (Boott) C.B.Clarke
Uncinia chlorostachya Phil. = ***Uncinia phleoides*** (Cav.) Pers.
Uncinia clarkei Petrie = ***Uncinia divaricata*** Boott
Uncinia compacta var. *caespitiformis* Kük. = ***Uncinia caespitosa*** Colenso ex Boott
Uncinia compacta var. *clarkei* (Petrie) Kük. = ***Uncinia divaricata*** Boott
Uncinia compacta var. *divaricata* (Boott) Hook.f. = ***Uncinia divaricata*** Boott
Uncinia compacta var. *petriei* C.B.Clarke = ***Uncinia divaricata*** Boott
Uncinia compacta var. *viridis* C.B.Clarke = ***Uncinia viridis*** (C.B.Clarke) Edgar
Uncinia cumingii Nees = ***Uncinia phleoides*** (Cav.) Pers.
Uncinia cylindrica Franch. = ***Uncinia brevicaulis*** (Thouars) Kunth
Uncinia de-la-costa Steud. = ***Uncinia macloviana*** Gaudich.
Uncinia digyna Hochst. ex Steud. = ***Carex melanostachya*** M.Bieb. ex Willd.
Uncinia disticha Colenso = ***Uncinia scabra*** Colenso ex Boott
Uncinia drucei var. *payciflora* Hamlin = ***Uncinia drucei*** Hamlin
Uncinia durvillei Steud. = ***Uncinia phleoides*** (Cav.) Pers.
Uncinia erinacea var. *angustata* Kük. = ***Uncinia erinacea*** (Cav.) Pers.
Uncinia europaea J.Gay = ***Carex microglochin*** Wahlenb. subsp. ***microglochin***
Uncinia fernandesiana Nees ex Boeck. = ***Uncinia douglasii*** Boott
Uncinia filifolia (Nutt.) Nees = ***Carex filifolia*** Nutt.
Uncinia filiformis Boott ex C.B.Clarke = ***Uncinia compacta*** R.Br. var. ***compacta***
Uncinia filiformis var. *debilior* (F.Muell.) W.R.B.Oliv. = ***Uncinia debilior*** F.Muell.
Uncinia fuscovaginata var. *caespitans* Hamlin = ***Uncinia fuscovaginata*** Kük.
Uncinia galeottii Boott ex C.B.Clarke = ***Uncinia hamata*** (Sw.) Urb.
Uncinia gracilis Decne. = ***Uncinia tenuis*** Poepp. ex Kunth
Uncinia gracilis var. *macloviana* (Gaudich.) C.B.Clarke = ***Uncinia macloviana*** Gaudich.
Uncinia hamata f. *angustifolia* Kük. = ***Uncinia hamata*** (Sw.) Urb.
Uncinia hamata var. *mexicana* (Steud.) Kük. = ***Uncinia hamata*** (Sw.) Urb.
Uncinia hookeri Colenso ex C.B.Clarke = ***Uncinia compacta*** R.Br. var. ***compacta***
Uncinia horizontalis Colenso = ***Uncinia compacta*** R.Br. var. ***compacta***
Uncinia jamaicensis (Poir.) Pers. = ***Uncinia hamata*** (Sw.) Urb.
Uncinia jamaicensis Liebm. = ***Uncinia hamata*** (Sw.) Urb.
Uncinia kingii R.Br. ex Boott = ***Carex kingii*** (R.Br. ex Boott) Reznicek
Uncinia lasiocarpa Steud. ex Boeck. = ***Uncinia phleoides*** (Cav.) Pers.
Uncinia lechleriana var. *triquetra* (Kük.) Kük. = ***Uncinia triquetra*** Kük.
Uncinia lehmannii Nees = ***Schoenoxiphium lehmannii*** (Nees) Kunth ex Steud.
Uncinia leptostachya Phil. = ***Uncinia phleoides*** (Cav.) Pers.
Uncinia lindleyana Kunth = ***Uncinia uncinata*** (L.f.) Kük.
Uncinia loliacea Phil. = ***Uncinia phleoides*** (Cav.) Pers.
Uncinia longiaristata Steud. = ***Uncinia erinacea*** (Cav.) Pers.
Uncinia longifolia Kunth = ***Uncinia erinacea*** (Cav.) Pers.
Uncinia longifolia É.Desv. = ***Uncinia phleoides*** (Cav.) Pers.
Uncinia longifolia Phil. ex C.B.Clarke = ***Uncinia phleoides*** (Cav.) Pers.
Uncinia longispica Boeck. = ***Uncinia phleoides*** (Cav.) Pers.
Uncinia macloviana var. *cylindrica* (Franch.) Kük. = ***Uncinia brevicaulis*** (Thouars) Kunth
Uncinia macloviana var. *douglasii* (Boott) Kük. = ***Uncinia douglasii*** Boott
Uncinia macloviana var. *montana* (Phil.) Kük. = ***Uncinia macloviana*** Gaudich.
Uncinia macrostachya É.Desv. = ***Uncinia multifaria*** Nees ex Boott
Uncinia macrotricha Franch. = ***Uncinia erinacea*** (Cav.) Pers.
Uncinia meridensis Steyerm. = ***Uncinia macrolepis*** Decne.
Uncinia mexicana Steud. = ***Uncinia hamata*** (Sw.) Urb.
Uncinia microglochin (Wahlenb.) Spreng. = ***Carex microglochin*** Wahlenb.
Uncinia montana Phil. = ***Uncinia macloviana*** Gaudich.
Uncinia montteana Phil. = ***Uncinia phleoides*** (Cav.) Pers.
Uncinia moseleyana Boeck. = ***Uncinia compacta*** R.Br. var. ***compacta***
Uncinia multifaria var. *macrostachya* (É.Desv.) Kük. = ***Uncinia multifaria*** Nees ex Boott
Uncinia multifolia Boeck. = ***Uncinia hamata*** (Sw.) Urb.
Uncinia negeri var. *araucana* Gunckel = ***Uncinia negeri*** Kük.
Uncinia nelmesii Hamlin = ***Uncinia distans*** Colenso ex Boott
Uncinia nepalensis Nees = ***Kobresia nepalensis*** (Nees) Kük.
Uncinia nervosa Boott = ***Uncinia compacta*** var. ***nervosa*** (Boott) C.B.Clarke
Uncinia nigra Colenso = ***Uncinia ferruginea*** Boott
Uncinia obtusata Colenso = ***Uncinia riparia*** R.Br.
Uncinia ohwiana T.Koyama = ***Uncinia riparia*** R.Br.
Uncinia pedicellata Kük. = ***Uncinia uncinata*** (L.f.) Kük.
Uncinia phalaroides Boott ex C.B.Clarke = ***Uncinia macrophylla*** Steud.
Uncinia phalaroides (Kunth) Nees = ***Carex phalaroides*** Kunth
Uncinia philippii Hohen. ex Steud. = ***Uncinia erinacea*** (Cav.) Pers.
Uncinia phleoides C.A.Mey. = ***Uncinia hamata*** (Sw.) Urb.
Uncinia phleoides var. *brachytricha* Speg. = ***Uncinia phleoides*** (Cav.) Pers.

Uncinia phleoides var. *krausei* Kük. = ***Uncinia phleoides*** (Cav.) Pers.
Uncinia phleoides var. *laticarpa* Kük. = ***Uncinia brevicaulis*** (Thouars) Kunth
Uncinia phleoides f. *longispica* Franch. = ***Uncinia phleoides*** (Cav.) Pers.
Uncinia phleoides var. *nux-nigra* C.B.Clarke = ***Uncinia phleoides*** (Cav.) Pers.
Uncinia phleoides var. *trichocarpa* (C.A.Mey.) C.B.Clarke = ***Uncinia phleoides*** (Cav.) Pers.
Uncinia phyllostachya Nees = ***Carex seticulmis*** Boeck.
Uncinia polyneura Colenso = ***Uncinia uncinata*** (L.f.) Kük.
Uncinia purpurata var. *subcaespitosa* Kük. = ***Uncinia affinis*** (Colenso ex C.B.Clarke) Hamlin
Uncinia rigida Boeck. = ***Uncinia brevicaulis*** (Thouars) Kunth
Uncinia rigida Petrie = ***Uncinia strictissima*** (Kük.) Petrie
Uncinia rigidula Steud. = ***Uncinia uncinata*** (L.f.) Kük.
Uncinia riparia var. *affinis* Colenso ex C.B.Clarke = ***Uncinia affinis*** (Colenso ex C.B.Clarke) Hamlin
Uncinia riparia var. *banksii* (Boott) C.B.Clarke = ***Uncinia banksii*** Boott
Uncinia riparia var. *hookeri* (Boott) Kük. = ***Uncinia hookeri*** Boott
Uncinia riparia var. *stolonifera* Kük. & Steen. = ***Uncinia compacta*** R.Br. var. ***compacta***
Uncinia rubra var. *fallax* Kük. = ***Uncinia rubra*** Colenso ex Boott
Uncinia rubra var. *strictissima* Kük. = ***Uncinia strictissima*** (Kük.) Petrie
Uncinia runssoroensis (K.Schum.) Chiov. = ***Carex runssoroensis*** K.Schum.
Uncinia rupestris Raoul = ***Uncinia compacta*** R.Br. var. ***compacta***
Uncinia rupestris var. *capillacea* Kük. = ***Uncinia filiformis*** Colenso ex Boott
Uncinia rupestris var. *flavescens* Kük. = ***Uncinia divaricata*** Boott
Uncinia rupestris var. *planifolia* Kük. = ***Uncinia angustifolia*** Hamlin
Uncinia scaberrima Nees = ***Uncinia uncinata*** (L.f.) Kük.
Uncinia sclerophylla Nelmes = ***Uncinia riparia*** R.Br.
Uncinia sellowiana (Schltdl.) Nees = ***Carex sellowiana*** Schltdl.
Uncinia silvestris var. *squamata* Hamlin = ***Uncinia egmontiana*** Hamlin
Uncinia sinclairii var. *elegans* Kük. = ***Uncinia elegans*** (Kük.) Hamlin
Uncinia smithii Philcox = ***Uncinia macrolepis*** Decne.
Uncinia spartea (Wahlenb.) Nees = ***Schoenoxiphium sparteum*** (Wahlenb.) C.B.Clarke
Uncinia sprengelii (Dewey ex Spreng.) Nees = ***Carex sprengelii*** Dewey ex Spreng.
Uncinia subtrigona Nelmes = ***Uncinia compacta*** R.Br. var. ***compacta***
Uncinia tenella var. *longifructus* Kük. = ***Uncinia longifructus*** (Kük.) Petrie
Uncinia tenella var. *robustior* Kük. = ***Uncinia tenella*** R.Br.
Uncinia tenuis f. *firmula* Kük. = ***Uncinia tenuis*** Poepp. ex Kunth
Uncinia trichocarpa C.A.Mey. = ***Uncinia phleoides*** (Cav.) Pers.
Uncinia trichocarpa É.Desv. = ***Uncinia phleoides*** (Cav.) Pers.
Uncinia trichocarpa var. *longispica* (Boeck.) Kük. = ***Uncinia phleoides*** (Cav.) Pers.
Uncinia uncinata var. *laxior* Carse = ***Uncinia uncinata*** (L.f.) Kük.
Uncinia uncinata var. *pedicellata* (Kük.) Petrie = ***Uncinia uncinata*** (L.f.) Kük.
Uncinia uncinata var. *uliginosa* Skottsb. = ***Uncinia uncinata*** (L.f.) Kük.
Uncinia urvillei Steud. = ***Uncinia phleoides*** (Cav.) Pers.
Uncinia variegata Colenso = ***Uncinia ferruginea*** Boott

Ungeria

Ungeria Nees ex C.B.Clarke = ***Cyperus*** L.
Ungeria monocephala (Roxb.) Nees ex C.B.Clarke = ***Cyperus cephalotes*** Vahl

Vaginaria

Vaginaria Pers. = ***Fuirena*** Rottb.
Vaginaria richardii Pers. = ***Fuirena scirpoidea*** Michx.

Vauthiera

Vauthiera A.Rich. = ***Lepidosperma*** Labill.
Vauthiera australis A.Rich. = ***Lepidosperma australe*** (A.Rich.) Hook.f.

Vesicarex

Vesicarex Steyerm. = ***Carex*** L.
Vesicarex collumanthus Steyerm. = ***Carex collumanthus*** (Steyerm.) L.E.Mora

Vignantha

Vignantha Schur = ***Carex*** L.
Vignantha acuta (L.) Schur = ***Carex acuta*** L.
Vignantha dacica (Heuff.) Schur = ***Carex nigra*** subsp. ***dacica*** (Heuff.) Soó
Vignantha drejeri (Lange) Schur = ***Carex cespitosa*** L. var. ***cespitosa***
Vignantha drejeri var. *alpina* Schur = ***Carex cespitosa*** L. var. ***cespitosa***
Vignantha moenchiana (Wender.) Schur = ***Carex acuta*** L.
Vignantha mucronata (All.) Schur = ***Carex mucronata*** All.
Vignantha prolixa (Fr.) Fourr. = ***Carex acuta*** L.
Vignantha rigida (Raf.) Schur = ***Carex bigelowii*** Torr. ex Schwein. subsp. ***bigelowii***
Vignantha stricta (Lam.) Schur = ***Carex stricta*** Lam.
Vignantha touranginiana (Boreau) Fourr. = ***Carex acuta*** L.
Vignantha vulgaris (R.E.Fr.) Schur = ***Carex nigra*** (L.) Reichard subsp. ***nigra***
Vignantha vulgaris var. *subfastigiata* Schur = ***Carex nigra*** (L.) Reichard subsp. ***nigra***

Vignea

Vignea P.Beauv. = ***Carex*** L.
Vignea acroandra Schur =
Vignea acuta (L.) Rchb. = ***Carex acuta*** L.
Vignea ×*allensteiniensis* (R.Gross) Soják =
Vignea alta (Boott) Soják = ***Carex remota*** subsp. ***alta*** (Boott) Kük.
Vignea altissima Schur = ***Carex muricata*** L.
Vignea ammophila (Willd.) Rchb. = ***Carex divisa*** Huds.
Vignea angustior (Mack.) Soják = ***Carex echinata*** Murray subsp. ***echinata***
Vignea appressa (L.) Rchb. = ***Carex paniculata*** L. subsp. ***paniculata***
Vignea appropinquata (Schumach.) Soják = ***Carex appropinquata*** Schumach.

Vignea aquatilis (Wahlenb.) Rchb. = ***Carex aquatilis*** Wahlenb.
Vignea arenaria (L.) Rchb. = ***Carex arenaria*** L.
Vignea argyroglochin (Hornem.) Rchb. = ***Carex leporina*** L.
Vignea ×*arthuriana* (Beckm. & Figert) Dostál = ***Carex* ×*arthuriana*** Beckm. & Figert
Vignea austroafghanica (Raymond) Soják = ***Carex divisa*** Huds.
Vignea axillaris (L.) Rchb. = ***Carex remota*** L. subsp. ***remota***
Vignea ×*behurica* (Simonk.) Soják = ***Carex* ×*biharica*** Simonk.
Vignea bertolonii (Schkuhr) Rchb. = ***Carex divisa*** Huds.
Vignea bicolor (Bellardi ex All.) Rchb. = ***Carex bicolor*** Bellardi ex All.
Vignea ×*boenninghauseniana* (Weihe) Rchb. = ***Carex* ×*boenninghausiana*** Weihe
Vignea bohemica (Schreb.) Soják = ***Carex bohemica*** Schreb.
Vignea bonanzensis (Britton) Soják = ***Carex bonanzensis*** Britton
Vignea bracteosa (Kunze) Rchb. = ***Carex bracteosa*** Kunze
Vignea brizoides (L.) Rchb. = ***Carex brizoides*** L.
Vignea brunnescens (Pers.) Soják = ***Carex brunnescens*** (Pers.) Poir.
Vignea brunnescens subsp. *alaskana* (Kalela) Soják = ***Carex brunnescens*** (Pers.) Poir. subsp. ***brunnescens***
Vignea brunnescens subsp. *pacifica* (Kalela) Soják = ***Carex brunnescens*** (Pers.) Poir. subsp. ***brunnescens***
Vignea brunnescens subsp. *vitilis* (Fr.) Soják = ***Carex brunnescens*** subsp. ***sphaerostachya*** (Tuck.) Kalela
Vignea canariensis (Kük.) Soják = ***Carex canariensis*** Kük.
Vignea canescens (L.) Rchb. = ***Carex canescens*** L.
Vignea capitata (L.) Soják = ***Carex capitata*** L.
Vignea cephalophora (Muhl. ex Willd.) Rchb. = ***Carex cephalophora*** Muhl. ex Willd.
Vignea cerina Rchb. = ***Carex nigra*** (L.) Reichard subsp. ***nigra***
Vignea cespitosa (L.) Rchb. = ***Carex cespitosa*** L.
Vignea chlorocarpa Schur = ***Carex nigra*** (L.) Reichard subsp. ***nigra***
Vignea chordorrhiza (L.f.) Rchb. = ***Carex chordorrhiza*** L.f.
Vignea cinerea (Pollich) Dostál = ***Carex canescens*** L. subsp. ***canescens***
Vignea colchica (J.Gay) Soják = ***Carex colchica*** J.Gay
Vignea compacta Rchb. = ***Carex nigra*** (L.) Reichard subsp. ***nigra***
Vignea conjugata (Willd. ex Kunth) Rchb. = ***Carex brunnea*** Thunb. subsp. ***brunnea***
Vignea conjuncta (Boott) Soják = ***Carex conjuncta*** Boott
Vignea craspedotricha (Nelmes) Soják = ***Carex remota*** subsp. ***alta*** (Boott) Kük.
Vignea crassinervis Schur = ***Carex nigra*** (L.) Reichard subsp. ***nigra***
Vignea cuprina (I.Sándor ex Heuff.) Soják = ***Carex cuprina*** (I.Sándor ex Heuff.) Nendtv. ex A.Kern.
Vignea curaica (Kunth) Soják = ***Carex curaica*** Kunth
Vignea curta (Gooden.) Opiz = ***Carex canescens*** L. subsp. ***canescens***
Vignea curvata (Asch.) Soják = ***Carex praecox*** Schreb.
Vignea curvula Opiz = ***Carex praecox*** Schreb.
Vignea cyperoides (L.) Peterm. = ***Carex bohemica*** Schreb.
Vignea dacica (Heuff.) Fuss = ***Carex nigra*** subsp. ***dacica*** (Heuff.) Soó
Vignea davalliana (Sm.) Rchb. = ***Carex davalliana*** Sm.
Vignea deinbolliana (J.Gay) Soják = ***Carex stenophylla*** Wahlenb. subsp. ***stenophylla***
Vignea deweyana (Schwein.) Raf. = ***Carex deweyana*** Schwein.
Vignea diandra (Schrank) Soják = ***Carex diandra*** Schrank
Vignea dioica (L.) Rchb. = ***Carex dioica*** L.
Vignea disperma (Dewey) Raf. = ***Carex disperma*** Dewey
Vignea disticha (Huds.) Peterm. = ***Carex disticha*** Huds.
Vignea disticha f. *floribunda* Peterm. = ***Carex disticha*** Huds.
Vignea divisa (Huds.) Rchb. = ***Carex divisa*** Huds.
Vignea divulsa (Stokes) Rchb. = ***Carex divulsa*** Stokes
Vignea divulsa subsp. *leersiana* (Rauschert) Dostál = ***Carex divulsa*** subsp. ***leersii*** (Kneuck.) W.Koch
Vignea drejeri (Lange) Fuss = ***Carex cespitosa*** L. var. ***cespitosa***
Vignea duriuscula (C.A.Mey.) Soják = ***Carex duriuscula*** C.A.Mey.
Vignea echinata (Murray) Fourr. = ***Carex echinata*** Murray
Vignea ×*elanescens* (Cif. & Giacom.) Soják =
Vignea elongata (L.) Rchb. = ***Carex elongata*** L.
Vignea ×*emmae* (L.Gross) Soják = ***Carex* ×*emmae*** L.Gross
Vignea enervis (C.A.Mey.) Soják = ***Carex enervis*** C.A.Mey.
Vignea exilis (Dewey) Raf. = ***Carex exilis*** Dewey
Vignea festucacea (Schkuhr ex Willd.) Rchb. = ***Carex festucacea*** Schkuhr ex Willd.
Vignea ×*figertii* (Asch. & Graebn.) Soják = ***Carex* ×*figertii*** Asch. & Graebn.
Vignea ×*filkukae* (Podp.) Soják =
Vignea floridana (Schwein.) Raf. = ***Carex floridana*** Schwein.
Vignea fluviatilis (Boott) Soják = ***Carex fluviatilis*** Boott
Vignea foenea (Willd.) Rchb. = ***Carex foenea*** Willd.
Vignea foetida (All.) Rchb. = ***Carex pseudofoetida*** Kük. ex Ostenf. subsp. ***pseudofoetida***
Vignea ×*fussii* (Simonk.) Soják = ***Carex* ×*fussii*** Simonk.
Vignea ×*gaudiniana* (Guthnick) Pritz. = ***Carex* ×*gaudiniana*** Guthnick
Vignea ×*gaudiniana* (Guthnick) Soják = ***Carex* ×*gaudiniana*** Guthnick
Vignea gebhardii (Willd.) Rchb. = ***Carex elongata*** L.
Vignea ×*gerhaldtii* (Figert) Soják = ***Carex* ×*gerhaldtii*** Figert
Vignea ×*germanica* (K.Richt.) Soják = ***Carex* ×*germanica*** K.Richt.
Vignea glareosa (Schkuhr ex Wahlenb.) Rchb. = ***Carex glareosa*** Schkuhr ex Wahlenb.
Vignea grypos (Schkuhr) Rchb. = ***Carex echinata*** Murray subsp. ***echinata***
Vignea guestphalica Rchb. = ***Carex divulsa*** Stokes subsp. ***divulsa***
Vignea gynocrates (Wormsk.) Soják = ***Carex gynocrates*** Wormsk.
Vignea ×*hanseniana* (Junge) Soják =
Vignea ×*haussknechtii* Soják =
Vignea heleonastes (Ehrh. ex L.f.) Rchb. = ***Carex heleonastes*** Ehrh. ex L.f.
Vignea hepburnii (Boott) Soják = ***Carex nardina*** var. ***hepburnii*** (Boott) Kük.
Vignea hydrophila (Dumort.) Rchb. = ***Carex echinata*** Murray subsp. ***echinata***
Vignea ×*ilseana* (Ruhmer) Soják =

Vignea incurva (Lightf.) Rchb. = ***Carex maritima*** Gunnerus
Vignea intermedia Rchb. = ***Carex disticha*** Huds.
Vignea inversa (R.Br.) Soják = ***Carex inversa*** R.Br.
Vignea juncoides (J.Presl & C.Presl) Opiz = ***Carex melanostachya*** M.Bieb. ex Willd.
Vignea ×*kneuckeriana* (Zahn) Soják = ***Carex ×pseudoaxillaris*** K.Richt.
Vignea kobomugi (Ohwi) Soják = ***Carex kobomugi*** Ohwi
Vignea ×*kukenthalii* (Dörfl.) Soják =
Vignea lachenalii (Schkuhr) Soják = ***Carex lachenalii*** Schkuhr
Vignea lagopina (Wahlenb.) Rchb. = ***Carex lachenalii*** Schkuhr subsp. ***lachenalii***
Vignea lagopodioides (Willd.) Rchb. = ***Carex tribuloides*** Wahlenb. var. ***tribuloides***
Vignea leiorhyncha (C.A.Mey.) Soják = ***Carex leiorhyncha*** C.A.Mey.
Vignea leporina (L.) Rchb. = ***Carex leporina*** L.
Vignea leporina var. *alpina* (Asch. & Graebn.) Dostál = ***Carex leporina*** L.
Vignea ligerica (J.Gay) Soják = ***Carex colchica*** J.Gay
Vignea ×*limnogena* (Appel ex Kneuck.) Soják =
Vignea lithophila (Turcz.) Soják = ***Carex lithophila*** Turcz.
Vignea loliacea (L.) Rchb. = ***Carex loliacea*** L.
Vignea lucorum (Willd.) Rchb. = ***Carex lucorum*** Willd.
Vignea ×*ludibunda* (J.Gay) Soják = ***Carex ×ludibunda*** J.Gay
Vignea macloviana (d'Urv.) Soják = ***Carex macloviana*** d'Urv.
Vignea macrocephala (Willd. ex Spreng.) Soják = ***Carex macrocephala*** Willd. ex Spreng.
Vignea macrostylon (Lapeyr.) Soják = ***Carex macrostylon*** Lapeyr.
Vignea maritima (Gunnerus) Rchb. = ***Carex maritima*** Gunnerus
Vignea maritima subsp. *setina* (H.Christ) Soják = ***Carex maritima*** Gunnerus
Vignea ×*microstachya* (Ehrh.) Rchb. = ***Carex ×microstachya*** Ehrh.
Vignea ×*microstylis* (J.Gay ex Gaudin) Rchb. = ***Carex ×microstyla*** J.Gay ex Gaudin
Vignea ×*mithala* (Callier) Soják = ***Carex ×mithala*** Callier
Vignea moenchiana (Wender.) Rchb. = ***Carex acuta*** L.
Vignea monopleura (V.I.Krecz.) Soják = ***Carex remota*** subsp. ***rochebrunii*** (Franch. & Sav.) Kük.
Vignea mucronata (All.) Rchb. = ***Carex mucronata*** All.
Vignea muhlenbergii (Willd.) Raf. = ***Carex muehlenbergii*** Willd.
Vignea multiflora (Muhl. ex Willd.) Rchb. = ***Carex vulpinoidea*** Michx.
Vignea muricata (L.) Rchb. = ***Carex muricata*** L.
Vignea muricata subsp. *lamprocarpa* (Èelak.) Soják = ***Carex pairae*** F.W.Schultz
Vignea muskingumensis (Schwein.) Soják = ***Carex muskingumensis*** Schwein.
Vignea nardina (Fr.) Soják = ***Carex nardina*** Fr.
Vignea neglecta Peterm. = ***Carex cespitosa*** L. var. ***cespitosa***
Vignea nemorosa Rchb. = ***Carex vulpina*** L.
Vignea nemurensis (Franch.) Soják = ***Carex nemurensis*** Franch.
Vignea nenningiana Opiz =
Vignea norvegica (Retz.) Rchb. = ***Carex norvegica*** Retz.
Vignea nubigena (D.Don ex Tilloch & Taylor) Soják = ***Carex nubigena*** D.Don ex Tilloch & Taylor
Vignea ochrochlamys (Ohwi) Soják = ***Carex ochrochlamys*** Ohwi
Vignea ohmuelleriana (O.Lang) Pritz. = ***Carex arenaria*** L.
Vignea omiana (Franch. & Sav.) Soják = ***Carex omiana*** Franch. & Sav.
Vignea oreophila (C.A.Mey.) Soják = ***Carex oreophila*** C.A.Mey.
Vignea otrubae (Podp.) Soják = ***Carex cuprina*** (I.Sándor ex Heuff.) Nendtv. ex A.Kern.
Vignea ovalis (Gooden.) Soják = ***Carex leporina*** L.
Vignea pachystylis (J.Gay) Soják = ***Carex pachystylis*** J.Gay
Vignea pacifica (Drejer) Rchb. ex Nyman = ***Carex cespitosa*** L. var. ***cespitosa***
Vignea paniculata (L.) Rchb. = ***Carex paniculata*** L.
Vignea paniculata subsp. *lusitanica* (Willd.) Soják = ***Carex paniculata*** subsp. ***lusitanica*** (Willd.) Maire
Vignea ×*paponii* (Muret ex T.Durand & Pittier) Soják =
Vignea paradoxa Rchb. = ***Carex appropinquata*** Schumach.
Vignea parallela (Laest.) Soják = ***Carex parallela*** (Laest.) Sommerf.
Vignea perileia (S.T.Blake) Soják = ***Carex echinata*** Murray subsp. ***echinata***
Vignea persica (Nelmes) Soják = ***Carex divulsa*** Stokes subsp. ***divulsa***
Vignea persoonii (Láng) Schur = ***Carex canescens*** L. subsp. ***canescens***
Vignea phaeoleuca (Ohwi) Soják = ***Carex nubigena*** var. ***franchetiana*** Ohwi
Vignea planata (Franch. & Sav.) Soják = ***Carex planata*** Franch. & Sav.
Vignea pleistogyna (V.I.Krecz.) Soják = ***Carex nubigena*** D.Don ex Tilloch & Taylor subsp. ***nubigena***
Vignea ×*ploettneriana* (Beyer) Soják = ***Carex ×ploettneriana*** Beyer
Vignea polyphylla (Kar. & Kir.) Soják = ***Carex divulsa*** subsp. ***leersii*** (Kneuck.) W.Koch
Vignea praecox (Schreb.) Soják = ***Carex praecox*** Schreb.
Vignea praecox subsp. *curvata* (Asch.) Dostál = ***Carex praecox*** Schreb.
Vignea pribylovensis (Macoun) Soják = ***Carex glareosa*** subsp. ***pribylovensis*** (Macoun) G.Halliday & Chater
Vignea psammophila Schur = ***Carex brizoides*** L.
Vignea pseudoarenaria Rchb. = ***Carex colchica*** J.Gay
Vignea ×*pseudoaxillaris* (K.Richt.) Dostál = ***Carex ×pseudoaxillaris*** K.Richt.
Vignea pseudobrizoides (Chabaud) Schur = ***Carex pseudobrizoides*** Chabaud
Vignea pseudoschreberi Schur = ***Carex brizoides*** L.
Vignea ×*pseudovulpina* (K.Richt.) Soják =
Vignea pulicaris (L.) Rchb. = ***Carex pulicaris*** L.
Vignea pulla (Gooden.) Rchb. = ***Carex saxatilis*** L.
Vignea redowskiana (C.A.Mey.) Soják = ***Carex parallela*** subsp. ***redowskiana*** (C.A.Mey.) T.V.Egorova
Vignea reichenbachii (Bonnet) Soják = ***Carex pseudobrizoides*** Chabaud
Vignea remota (L.) Rchb. = ***Carex remota*** L.
Vignea remota var. *subloliacea* Schur = ***Carex remota*** L. subsp. ***remota***
Vignea remotiuscula (Wahlenb.) Soják = ***Carex remotiuscula*** Wahlenb.
Vignea repens (Bellardi) Rchb. = ***Carex repens*** Bellardi
Vignea repens (Bellardi) Schur = ***Carex repens*** Bellardi
Vignea reptabunda (Trautv.) Soják = ***Carex reptabunda*** (Trautv.) V.I.Krecz.
Vignea retroflexa (Muhl. ex Willd.) Rchb. = ***Carex retroflexa*** Muhl. ex Willd.
Vignea ×*rieseana* (Figert) Soják =
Vignea rigida Fuss = ***Carex nigra*** (L.) Reichard subsp. ***nigra***

Vignea rivularis (Schkuhr) Schur = ***Carex divisa*** Huds.
Vignea roborowskii (V.I.Krecz.) Soják = ***Carex pseudofoetida*** Kük. ex Ostenf. subsp. ***pseudofoetida***
Vignea rochebrunii (Franch. & Sav.) Soják = ***Carex remota*** subsp. ***rochebrunii*** (Franch. & Sav.) Kük.
Vignea rosea (Willd.) Rchb. = ***Carex rosea*** Willd.
Vignea rotae (De Not.) Soják = ***Carex paniculata*** L. subsp. ***paniculata***
Vignea salina (Wahlenb.) Rchb. = ***Carex salina*** Wahlenb.
Vignea saxatilis (L.) Rchb. = ***Carex saxatilis*** L.
Vignea schreberi (Schrank) Rchb. = ***Carex praecox*** Schreb.
Vignea ×*schuetzeana* (Figert) Dostál =
Vignea scirpoides Rchb. = ***Carex interior*** L.H.Bailey
Vignea scoparia (Willd.) Rchb. = ***Carex scoparia*** Willd.
Vignea setacea (Dewey) Raf. = ***Carex vulpinoidea*** Michx.
Vignea setifolia Fourr. = ***Carex divisa*** Huds.
Vignea siccata (Dewey) Raf. = ***Carex siccata*** Dewey
Vignea spicata (Huds.) Soják = ***Carex spicata*** Huds.
Vignea spicata f. *humilis* Peterm. = ***Carex spicata*** Huds.
Vignea squarrosa (L.) Rchb. = ***Carex squarrosa*** L.
Vignea stellulata (Gooden.) Rchb. = ***Carex echinata*** Murray subsp. ***echinata***
Vignea stenophylla (Wahlenb.) Rchb. = ***Carex stenophylla*** Wahlenb.
Vignea stenophylla f. *elata* Schur = ***Carex stenophylla*** Wahlenb. subsp. ***stenophylla***
Vignea stenophylla var. *exigua* Schur = ***Carex stenophylla*** Wahlenb. subsp. ***stenophylla***
Vignea stipata (Muhl. ex Willd.) Rchb. = ***Carex stipata*** Muhl. ex Willd.
Vignea straminea (Willd. ex Schkuhr) Raf. = ***Carex straminea*** Willd. ex Schkuhr
Vignea striata Rchb. = ***Carex brownii*** Tuck. subsp. ***brownii***
Vignea stricta (Lam.) Rchb. = ***Carex stricta*** Lam.
Vignea strictifolia Opiz = ***Carex acuta*** L.
Vignea ×*suziella* (Podp.) Soják =
Vignea tenuiflora (Wahlenb.) Soják = ***Carex tenuiflora*** Wahlenb.
Vignea tenuissima (Schur) Schur = ***Carex muricata*** L.
Vignea tereticaulis (F.Muell.) Soják = ***Carex tereticaulis*** F.Muell.
Vignea teretiuscula (Gooden.) Rchb. = ***Carex diandra*** Schrank
Vignea trinervis (Degl.) Rchb. = ***Carex trinervis*** Degl.
Vignea tripartita (All.) Rchb. = ***Carex lachenalii*** Schkuhr subsp. ***lachenalii***
Vignea ursina (Dewey) Soják = ***Carex ursina*** Dewey
Vignea virens Rchb. = ***Carex muricata*** L.
Vignea vulgaris (R.E.Fr.) Fuss = ***Carex nigra*** (L.) Reichard subsp. ***nigra***
Vignea vulpina (L.) Rchb. = ***Carex vulpina*** L.
Vignea vulpina f. *capitulata* Peterm. = ***Carex vulpina*** L.
Vignea vulpina var. *crassinervis* Schur = ***Carex vulpina*** L.
Vignea vulpinaris (Nees) Soják = ***Carex vulpinaris*** Nees
Vignea vulpinoidea (Michx.) Soják = ***Carex vulpinoidea*** Michx.
Vignea wallichiana (Spreng.) Soják = ***Carex foliosa*** D.Don ex Tilloch & Taylor
Vignea wendelboi (Nelmes) Soják = ***Carex vulpinaris*** Nees

Vignidula

Vignidula Börner = ***Carex*** L.
Vignidula nikoensis (Franch. & Sav.) Fedde & J.Schust. = ***Carex satsumensis*** Franch. & Sav

Vincentia

Vincentia Gaudich. = ***Machaerina*** Vahl
Vincentia anceps (Poir.) Kunth = ***Machaerina anceps*** (Poir.) Bojer
Vincentia angustifolia Gaudich. = ***Machaerina angustifolia*** (Gaudich.) T.Koyama
Vincentia bidwellii Stapf ex Setch. = ***Machaerina bidwellii*** (Stapf ex Setch.) T.Koyama
Vincentia crinita Stapf = ***Machaerina sinclairii*** (Hook.f.) T.Koyama
Vincentia dissoluta Stapf = ***Machaerina falcata*** (Nees) T.Koyama
Vincentia ensifolia Boeck. ex C.B.Clarke = ***Machaerina iridifolia*** (Bory) T.Koyama
Vincentia falcata (Nees) Stapf = ***Machaerina falcata*** (Nees) T.Koyama
Vincentia gladiata (Hook.f.) Boeck. = ***Machaerina sinclairii*** (Hook.f.) T.Koyama
Vincentia latifolia Kunth = ***Machaerina iridifolia*** (Bory) T.Koyama
Vincentia macrophylla Boeck. = ***Machaerina macrophylla*** (Boeck.) T.Koyama
Vincentia maingayi (C.B.Clarke) Stapf = ***Machaerina maingayi*** (C.B.Clarke) T.Koyama
Vincentia malesiaca Stapf = ***Machaerina falcata*** (Nees) T.Koyama
Vincentia monticola Guillaumin = ***Machaerina monticola*** (Guillaumin) T.Koyama
Vincentia robinsonii Merr. = ***Machaerina sinclairii*** (Hook.f.) T.Koyama
Vincentia samoensis (C.B.Clarke ex Stapf) Stapf = ***Machaerina falcata*** (Nees) T.Koyama
Vincentia scirpoidea (Steud.) Boeck. = ***Machaerina scirpoidea*** (Steud.) T.Koyama

Voladeria

Voladeria Benoist = ***Oreobolus*** R.Br.
Voladeria carchiensis Benoist = ***Oreobolus obtusangulus*** Gaudich. subsp. ***obtusangulus***

Volkiella

Volkiella Merxm. & Czech, Mitt. Bot. Staatssamml. München 8: 317 (1953).
Namibia. 27 NAM.
1 Species

Volkiella disticha Merxm. & Czech, Mitt. Bot. Staatssamml. München 8: 318 (1953).
Namibia. 27 NAM. Hemicr. or rhizome geophyte.

Websteria

Websteria S.H.Wright, Bull. Torrey Bot. Club 14: 135 (1887).
Trop. & Subtrop. America, Trop. & S. Africa to Sri Lanka. 22 IVO NGA 23 ZAI 25 TAN 26 ZAM 27 BOT 28 MDG 40 IND SRL 42 MLY 78 FLA GEO 80 COS GUA HON NIC 81 CUB TRT 82 FRG GUY SUR? VEN 83 ECU? 84 BZN 85 AGE PAR.
1 Species

Websteria confervoides (Poir.) S.S.Hooper, Kew Bull. 26: 582 (1972).
Trop. & Subtrop. America, Trop. & S. Africa to Sri Lanka. 22 IVO NGA 23 ZAI 25 TAN 26 ZAM 27 BOT 28 MDG 40 IND SRL 42 MLY 78 FLA GEO 80 COS GUA HON NIC 81 CUB TRT 82 FRG GUY SUR? VEN 83 ECU? 84 BZN 85 AGE PAR.
Hel. or hydrocham.

**Scirpus confervoides* Poir. in J.B.A.M.de Lamarck, Encycl. 6: 755 (1805). *Baeothryon confervoides* (Poir.) A.Dietr., Sp. Pl. 2: 94 (1832). *Schoenus confervoides* (Poir.) Willd. ex Kunth, Enum. Pl. 2: 173 (1837). *Eleocharis confervoides* (Poir.) Steud., Syn. Pl. Glumac. 2: 82 (1855). *Dulichium confervoides* (Poir.) Alston in H.Trimen, Handb. Fl. Ceylon 6(Suppl.): 310 (1931).
Eleocharis submersa Miq., Linnaea 19: 225 (1847).
Scirpus natans Griseb., Cat. Pl. Cub.: 238 (1866), nom. illeg.
Scirpus ruppioides Thwaites ex C.Wright, Fl. Cub.: 176 (1871).
Scirpus submersus C.Wright, Anales Acad. Ci. Méd. Habana 8: 79 (1871). *Websteria submersa* (C.Wright) Britton, Bull. Torrey Bot. Club 15: 99 (1888).
Rhynchospora ruppioides Benth., Hooker's Icon. Pl. 14: t. 1344 (1880).
Websteria limnophila C.Wright, Bull. Torrey Bot. Club 14: 135 (1887).
Websteria submersa var. *luetzelburgii* Suess., Bot. Jahrb. Syst. 73: 125 (1943).
Websteria submersa var. *negrensis* Suess., Bot. Jahrb. Syst. 73: 124 (1943).

***Synonyms*:**
Websteria limnophila C.Wright = ***Websteria confervoides*** (Poir.) S.S.Hooper
Websteria submersa (C.Wright) Britton = ***Websteria confervoides*** (Poir.) S.S.Hooper
Websteria submersa var. *luetzelburgii* Suess. = ***Websteria confervoides*** (Poir.) S.S.Hooper
Websteria submersa var. *negrensis* Suess. = ***Websteria confervoides*** (Poir.) S.S.Hooper

Zameioscirpus

Zameioscirpus Dhooge & Goetgh., Pl. Syst. Evol. 243: 75 (2003).
W. Peru to W. Argentina. 83 BOL PER 85 AGS AGW CLC CLN.
3 Species

Zameioscirpus atacamensis (Phil.) Dhooge & Goetgh., Pl. Syst. Evol. 243: 79 (2003).
S. Bolivia to NW. Argentina. 83 BOL 85 AGW CLN. Hemicr.
**Isolepis atacamensis* Phil., Fl. Atacam.: 53 (1860). *Scirpus atacamensis* (Phil.) Boeck., Linnaea 36: 482 (1870). *Baeothryon atacamense* (Phil.) Soják, Cas. Nár. Mus., Odd. Prír. 148: 193 (1979 publ. 1980).
Scirpus hieronymi Boeck., Beitr. Cyper. 1: 18 (1888).

Zameioscirpus gaimardioides (É.Desv.) Dhooge & Goetgh., Pl. Syst. Evol. 243: 82 (2003).
Peru, Chile (Coquimbo) to Argentina (San Juan, Neuquén). 83 PER 85 AGS AGW CLC. Hemicr.
**Carex gaimardioides* É.Desv. in C.Gay, Fl. Chil. 6: 202 (1854). *Fimbristylis gaimardioides* (É.Desv.) Steud. in W.Lechler, Berberid. Amer. Austral.: 56 (1857).
Isolepis deserticola Phil., Fl. Atacam.: 53 (1860).
Scirpus thermalis Spruce ex Benth. & Hook.f., Gen. Pl. 3: 1051 (1883).
Scirpus distichophyllus Boeck., Beitr. Cyper. 1: 19 (1888).

Zameioscirpus muticus Dhooge & Goetgh., Pl. Syst. Evol. 246: 133 (2004).
W. Peru to NW. Argentina. 83 BOL PER 85 AGW CLN. Hemicr.

Zosterospermum

Zosterospermum P.Beauv. = ***Rhynchospora*** Vahl
Zosterospermum gracile Desv. ex Ham. = ***Rhynchospora contracta*** (Nees) J.Raynal

Eriocaulon [Eriocaulaceae]

Eriocaulon setaceum Lour. = ***Fimbristylis acicularis*** R.Br.
Eriocaulon spadiceum Lam. = ***Nemum spadiceum*** (Lam.) Desv.
Eriocaulon sphacelatum Willd. ex Kunth = ***Bulbostylis paradoxa*** (Spreng.) Lindm.

Juncus [Juncaceae]

Juncus parviflorus Poir. = ***Rhynchospora contracta*** (Nees) J.Raynal
Juncus zeilanicus Houtt. = ***Scleria levis*** Retz.

Restio [Restionaceae]

Restio articulatus Retz. = ***Lepironia articulata*** (Retz.) Domin
Restio pilisepalus Steud. = ***Costularia pilisepala*** (Steud.) F.Kern

Sparganium [Typhaceae]

Sparganium pubescens Poir. = ***Fuirena pubescens*** (Poir.) Kunth var. ***pubescens***
Sparganium tenuifolium Poir. = ***Carex halleriana*** Asso
Sparganium trifidum Poir. = ***Dulichium arundinaceum*** (L.) Britton var. ***arundinaceum***

Willdenowia [Restionaceae]

Willdenowia compressa Steud. = ***Chrysitrix capensis*** L. var. ***capensis***

Xyris [Xyridaceae]

Xyris brasiliensis Spreng. = ***Fimbristylis ovata*** (Burm.f.) J.Kern
Xyris eriophora Klotzsch = ***Fimbristylis lanata*** Roem. & Schult.
Xyris laevis Nees ex Benth. = ***Gymnoschoenus sphaerocephalus*** (R.Br.) Hook.f.
Xyris triquetra Kuntze = ***Rhynchospora globosa*** (Kunth) Roem. & Schult. var. ***globosa***

Unplaced names

Abildgaardia abbreviata Lye, Lidia 1: 35 (1985). = [26 ZAM]
Abildgaardia acutispicata Lye, Lidia 1: 33 (1985). = [26 ZAM]
Abildgaardia angustespicata Lye, Lidia 1: 33 (1985). = [25 TAN]
Abildgaardia capitata Lye, Lidia 1: 37 (1985). = [26 ZAM]
Abildgaardia chinensis Gand., Bull. Soc. Bot. France 66: 296 (1919 publ. 1920). = ?
Abildgaardia densiflora Lye, Lidia 1: 31 (1985). = [26 ZAM]
Abildgaardia lacunosa Lye, Lidia 1: 33 (1985). = [26 ZAM]
Abildgaardia laxispicata Lye, Lidia 1: 37 (1985). = [26 ZAM]
Abildgaardia longispicata Lye, Lidia 1: 35 (1985). = [24 ETH]
Abildgaardia macroanthela Lye, Lidia 1: 35 (1985). = [26 ZAM]
Abildgaardia malawiensis Lye, Lidia 1: 35 (1985). = [26 MLW]
Abildgaardia metalliphila Lye, Lidia 1: 37 (1985). = [23 ZAI]
Abildgaardia microcephala Lye, Lidia 1: 33 (1985). = [26 ZAM]
Abildgaardia microrotundata Lye, Lidia 1: 33 (1985). = [25 TAN]
Abildgaardia natalensis Gand., Bull. Soc. Bot. France 66: 296 (1919 publ. 1920). = [27 NAT]
Abildgaardia nudiuscula Lye, Lidia 1: 37 (1985). = [26 ZAM]
Abildgaardia papillosa Kral & M.T.Strong, Sida 18: 842 (1999). = [84 BZE]
Abildgaardia pluricephala Lye, Lidia 1: 35 (1985). = [26 ZAM]
Abildgaardia polygama Spreng., Novi Provent.: 45 (1818). = [84]
Abildgaardia scrobiculata Lye, Lidia 1: 37 (1985). = [26 ZAM]
Abildgaardia squarrosa Lye, Lidia 1: 31 (1985). = [25 KEN]
Acriulus titan C.B.Clarke, Kew Bull., Addit. Ser. 8: 62 (1908). = [2]
Baumea acuminata Regel, Cat. Pl. Hort. Aksakov.: 15 (1860). = ?
Baumea johnsonii K.L.Wilson, Telopea 1: 457 (1980). = [50 NSW]
Baumea veillonis J.Raynal, Adansonia, n.s., 13: 469 (1973 publ. 1974). = [60 NWC]
Becquerelia gigantea Hochst. ex Steud., Syn. Pl. Glumac. 2: 180 (1855). = ?
Blepharolepis leucoblepharis Nees in C.F.P.von Martius & auct. suc. (eds.), Fl. Bras. 2(1): 91 (1842). = [84]
Bulbostylis tisserantii Cherm., Arch. Bot. Mém. 3: 11 (1936). = [23 CAF]
Calisto Néraud ex Gaudich., Voy. Uranie: 26 (1826). = [29 MAU]
Carex acroandra Schur, Enum. Pl. Transsilv.: 699 (1866). = ?
Carex acutiformis Brot., Fl. Lusit. 1: 66 (1804), nom. illeg. = ?
Carex aequalis Notø, Nyt Mag. Naturvidensk. 61: 223 (1924). = ?
Carex akiensis K.Okamoto, Bull. Okayama Coll. Sci. 1: 53 (1965). = [38 JAP]
Carex ×albertii H.Lév., Bull. Acad. Int. Géogr. Bot. 22: 48 (1912). = ?
Carex albirostris C.B.Clarke, Kew Bull., Addit. Ser. 8: 78 (1908). = ?
Carex ×allensteiniensis R.Gross, Jahrb. Preuss. Bot. Ver. 1927-9: 9 (1930). = ?
Carex ×alluvialis Figert, Allg. Bot. Z. Syst. 13: 4 (1907). = ?
Carex ×alluvialis Schube, Jahresber. Schles. Ges. Vaterl. Cult. 86(2): 51 (1909), nom. illeg. = ?
Carex alterna C.B.Clarke, Kew Bull., Addit. Ser. 8: 69 (1908). = ?
Carex andalusica H.Lindb., Acta Soc. Sci. Fenn., Ser. B, Opera Biol. 1(2): 28 (1932). = [12 SPA]
Carex ×anselmii Sennen, Treb. Inst. Catalana Hist. Nat. 15(1): 37 (1931). = ?
Carex ×appeliana Zahn, Oesterr. Bot. Z. 40: 364 (1890). = ?
Carex arcuata C.Laest., Bidr. Växtligh.: 26 (1860). = ?
Carex aristata D.Don ex Tilloch & Taylor, Philos. Mag. J. 62: 455 (Dec. 1823). = ?
Carex ×aschersonii H.Lév., Bull. Acad. Int. Géogr. Bot. 22: 48 (1912). = ?
Carex ×auroniensis L.C.Lamb., Bull. Acad. Int. Géogr. Bot. 18: 339 (1908). = ?
Carex ×barrei H.Lév., Bull. Acad. Int. Géogr. Bot. 22: 48 (1912). = ?
Carex ×bengyana H.Lév. & L.C.Lamb., Bull. Acad. Int. Géogr. Bot. 21: 267 (1911). = ?
Carex biconvexis Michx. in ?. = ?
Carex bicostata Olney, Car. Bar. Am.: 9 (1871). = ?
Carex bifida Roth ex Steud., Nomencl. Bot. 1: 155 (1821). = ?
Carex ×binderi Podp., Oesterr. Bot. Z. 1: 212 (1900). = ?
Carex binensis Franch., Bull. Soc. Philom. Paris, VIII, 7: 28 (1895). = ?
Carex binerviformis Holmboe, Stud. Veg. Cypr.: 42 (1914). = ?
Carex ×bogdanensis Zapal., Consp. Fl. Gallic. Crit. 3: 233 (1911). = ?
Carex bolanderi Gand., Bull. Soc. Bot. France 66: 296 (1919 publ. 1920), nom. illeg. = ?
Carex bordabensis Litv., Trudy Bot. Muz. Imp. Akad. Nauk 7: 100 (1910). = ?
Carex boscii Willd. ex Spreng., Syst. Veg. 3: 812 (1826). = ?
Carex ×botasgdensis Kük., Mitth. Thüring. Bot. Vereins, n.s., 5: 19 (1893). = ?
Carex brevicaulis Mack., Bull. Torrey Bot. Club 40: 547 (1913), nom. illeg. = [71 BRC 73 ORE WAS]
Carex breviglumis Le Gall, Fl. Morbihan: 658 (1852). = ?
Carex ×brevirostris Cederstr., Kongl. Vetensk. Acad. Handl. 14: 199 (1857). = ?
Carex brevirostris Blytt, Bot. Not. 1857: 133 (1857). = ?
Carex brotherorum H.Christ, Trudy Imp. S.-Peterburgsk. Bot. Sada 16: 434 (1900). = ?

Carex ×*brueggeri* K.Richt., Pl. Eur. 1: 170 (1890). = ?
Carex bufonia Gueldenst. ex Ledeb., Fl. Ross. 4: 320 (1852). = ?
Carex burjatorum Korotky, Repert. Spec. Nov. Regni Veg. 13: 293 (1914). = ?
Carex cantabrorum Gand., Bull. Soc. Bot. France 56: 134 (1909). = ?
Carex ×*castriferrei* Waisb., Magyar Bot. Lapok 4: 75 (1905). = ?
Carex ×*catalaunica* Sennen, Treb. Inst. Catalana Hist. Nat. 1917: 236, 182 (1917). = ?
Carex ×*catteyensis* A.Benn., Ann. Scott. Nat. Hist. 1911: 49 (1911). = [10 GRB]
Carex cernua Raeusch., Nomencl. Bot., ed. 3: 270 (1797). = ?
Carex cespitosa var. *ramosa* Dewey, Amer. J. Sci. Arts 12: 297 (1827). = [75 MAS]
Carex ×*cetica* Rech., Allg. Bot. Z. Syst. 20: 21 (1914). = ?
Carex chlamydea Norman, Forh. Vidensk.-Selsk. Kristiana 1893: 45 (1893). = ?
Carex ×*christii* Boeck., Bot. Centralbl. 42: 134 (1890). = ?
Carex ×*ciliatifructus* Ohwi, Acta Phytotax. Geobot. 1: 297 (1932). = ?
Carex ×*cokae* Podp., Vestn. Klubu Prír. v Prostejove 10: 7 (1907 publ. 1908). = [11 CZE]
Carex conspicua Boott ex C.B.Clarke, Kew Bull., Addit. Ser. 8: 82 (1908). = ?
Carex ×*corcoutica* Domin in K.Domin & J.Podpera, Klíc Kvet. Ceskoslov.: 961 (1928). = [11 CZE]
Carex coriophora subsp. *langtaodianensis* S.Yun Liang, in Fl. Reipubl. Popul. Sin. 12: 524 (2000). = [36 CHN]
Carex ×*corstorphinei* Druce, Rep. Bot. Exch. Club Brit. Isles 4: 216 (1915 publ. 1916). = ?
Carex ×*costei* Rouy, Ill. Pl. Eur.: t. 149 (1897). = ?
Carex crustacea Nelmes, Kew Bull. 1: 11 (1946). = ?
Carex ×*cuatrecasasii* Sennen, Treb. Inst. Catalana Hist. Nat. 15(1): 37 (1931). = [12 SPA]
Carex ×*dandoensis* T.Koyama, J. Jap. Bot. 31: 287 (1956). = ?
Carex ×*danielis* H.Lév., Bull. Acad. Int. Géogr. Bot. 22: 48 (1912). = ?
Carex deserta Merino, Fl. Galicia 3: 189 (1909). = ?
Carex displodens Printz, Veg. Siber.-Mongol. Front.: 153 (1921). = ?
Carex dissimilis J.F.Gmel., Syst. Nat. 2: 144 (1791). = ?
Carex dubia Missbach & E.H.L.Krause in J.Sturm, Deutschl. Fl., ed. 2, 2: 142 (1900), nom. illeg. = ?
Carex ×*ducellieri* Beauverd, Bull. Soc. Bot. Genève, II, 11: 6 (1919). = ?
Carex ×*duerrnbergeri* Ritzb., Oesterr. Bot. Z. 41: 246 (1891). = ?
Carex dyrrachiensis Nelmes, Kew Bull. 7: 67 (1952). = ?
Carex eboracensis Nelmes, J. Bot. 77: 114 (1939). = ?
Carex ×*elanescens* Cif. & Giacom., Nomencl. Fl. Ital. 1: 65 (1950). = ?
Carex ×*elisabeth* J.Andrés & al., Stud. Bot. Univ. Salamanca 7: 243 (1988 publ. 1989), contrary to Art. H.5.2 ICBN (1988). = ?
Carex eloyi C.B.Clarke, Kew Bull., Addit. Ser. 8: 88 (1908). = ?
Carex engleriana H.Lév. & Vaniot, Bull. Acad. Int. Géogr. Bot. 10: 274 (1901). = ?
Carex ×*enshuensis* T.Koyama, J. Jap. Bot. 30: 134 (1955). = ?
Carex erythraea Schrad., Linnaea 11: 87 (1837). = ?
Carex ×*ewingii* E.S.Marshall, J. Bot. 49: 197 (1911). = ?
Carex exaltata Kit., Linnaea 32: 322 (1863), nom. illeg. = ?
Carex ×*felixii* L.C.Lamb., Bull. Acad. Int. Géogr. Bot. 16(202): vii (1906). = ?
Carex ×*filkukae* Podp., Práce Morav. Prír. Spolecn. 6(3): 199 (1928). = ?
Carex ×*fontis-sancti* Podp., Spisy Prír. Fak. Masarykovy Univ. 12: 15 (1922). = ?
Carex ×*fragosoana* Pau, Bol. Soc. Ibér. Ci. Nat. 25: 33 (1926). = ?
Carex fuegiana Gand., Bull. Soc. Bot. France 66: 296 (1919 publ. 1920). = ?
Carex ×*fulva* Gooden., Trans. Linn. Soc. London 2: 177 (1794), nom. illeg. = ?
Carex furcata Lapeyr., Hist. Pl. Pyrénées: 568 (1813). = ?
Carex ×*furusei* T.Koyama, J. Jap. Bot. 30: 135 (1955). = [38 JAP] *C. brownii* × *C. brownii* subsp. transversa
Carex fuscotincta Merino, Mem. Real Soc. Esp. Hist. Nat. 2: 467 (1904). = ?
Carex gandogeri H.Lév. & Vaniot, Bull. Acad. Int. Géogr. Bot. 14: 184 (1905). = ?
Carex gandogeri H.Lév. ex Gand., Bull. Soc. Bot. France 55: 223 (1908). = ?
Carex gasparrinii Parl., Fl. Ital. 2: 149 (1852). = ?
Carex geihokuensis K.Okamoto, Bull. Okayama Coll. Sci. 1: 51 (1965). = [38 JAP]
Carex ×*ghristii* Boeck., Bot. Centralbl. 42: 134 (1890). = ?
Carex ×*gogelana* Podp., Práce Morav. Prír. Spolecn. 6(3): 375 (1928). = ?
Carex ×*goroi* T.Koyama, J. Jap. Bot. 30: 133 (1955). = ?
Carex ×*gotlandica* Englund, Memoranda Soc. Fauna Fl. Fenn. 6: 104 (1930). = ?
Carex ×*goyozanensis* M.Kikuchi & Mochizuki, J. Geobot. 18: 73 (1970). = ?
Carex ×*hageri* E.Baumann, Ber. Schweiz. Bot. Ges. 26-29: 180 (1920). = ?
Carex halei L.H.Bailey ex C.B.Clarke, Kew Bull., Addit. Ser. 8: 87 (1908), nom. illeg. = ?
Carex ×*hanasakensis* T.Koyama, J. Jap. Bot. 30: 133 (1955). = ?
Carex ×*hanseniana* Junge, Jahrb. Hamburg. Wiss. Anst. 25: 157, 232 (1908). = ?
Carex hausmannii Tapp. ex Kalela, Ann. Bot. Soc. Zool.-Bot. Fenn. "Vanamo" 19(3): 57 (1944). = ?
Carex herbariorum L.H.Bailey, Bot. Gaz. 17: 150 (1892). = ?
Carex hians Steud., Syn. Pl. Glumac. 2: 236 (1855). = ?
Carex ×*hibernica* A.Benn., J. Bot. 35: 250 (1897). = ?
Carex hirtiformis Dalla Torre & Sarnth., Fl. Tirol 6(2): 380 (1906). = ?
Carex hobkodensis Franch., Bull. Soc. Philom. Paris, VIII, 7: 28 (1895). = ?
Carex ×*hokariensis* T.Koyama, J. Jap. Bot. 31: 288 (1956). = ?
Carex hooker-grayi C.B.Clarke, Kew Bull., Addit. Ser. 8: 85 (1908). = ?
Carex ×*hosoii* T.Koyama, J. Jap. Bot. 31: 287 (1956). = ?
Carex hymenina Nees ex Boott, Ill. Carex 1: 67 (1858). = ?
Carex hystericina var. *angustior* L.H.Bailey, Contr. U. S. Natl. Herb. 1: 126 (1891). = [76 ARI]
Carex ×*ilseana* Ruhmer, Jahrb. Königl. Bot. Gart. Berlin 1: 258 (1881). = ?
Carex imbricata Drobow in M.G.Popov (ed.), Key Pl. Envir. Tashkent: 51 (1923). = [32]
Carex ×*inaensis* T.Koyama, J. Jap. Bot. 30: 133 (1955). = ?

Carex inaequalis J.F.Gmel., Syst. Nat. 2: 146 (1791), nom. illeg. = ?
Carex inandrensis Kihlm. ex Hjelt, Acta Soc. Sci. Fenn. 5: 298 (1895). = ?
Carex ×*indagata* Torges & Hausskn., Mitt. Geogr. Ges. (Thüringen) Jena 9: 26 (1891). = ?
Carex ×*innocua* Wallr., Erst. Beitr. Fl. Hercyn.: 299 (1840). = ?
Carex ×*insolita* T.Koyama, J. Jap. Bot. 30: 135 (1955). = ?
Carex intacta Samp., Bol. Soc. Brot. 24: 118 (1908). = ?
Carex isomadensis Simonk. ex K.Richt., Pl. Eur. 1: (1890). = ?
Carex jankae Gand., Contr. Fl. Terr. Slav. Merid. 1: 30 (1883). = ?
Carex kamagariensis K.Okamoto, Bull. Okayama Coll. Sci. 1: 66 (1965). = [38 JAP]
Carex karashidaniensis Akiyama, J. Jap. Bot. 13: 652 (1937). = ?
Carex ×*kohtsii* K.Richt., Pl. Eur. 1: 170 (1890). = ?
Carex komarovii V.I.Krecz., Bot. Mater. Gerb. Bot. Inst. Komarova Akad. Nauk S.S.S.R. 7: 33 (1937), nom. illeg. = ?
Carex ×*krajinae* Domin, Preslia 13-15: 30 (1935). = ?
Carex ×*krausei* P.Fourn., Fl. Compl. Plaine Franç.: 558 (1928), nom. illeg. = ?
Carex ×*kuekenthaliana* Appel & Brückn., Ber. Bayer. Bot. Ges. 1: 75 (1891). = ?
Carex kuekenthalii K.Schum. ex C.B.Clarke, Bot. Jahrb. Syst. 38: 136 (1906), nom. illeg. = ?
Carex ×*kukenthalii* Dörfl., Jahreskat. Wiener Bot. Tauschanst. 1900: 130 (1900). = ?
Carex ×*kyyhkynenii* Hiitonen in ?. = *C. canescens* × *C. tenuiflora*
Carex ×*lackowitziana* Paul, Allg. Bot. Z. Syst. 12: 160 (1906). = ?
Carex ×*laestadii* E.Holmb., Bot. Not. 1910: 87 (1910). = ?
Carex ×*lambertiana* H.Lév., Bull. Acad. Int. Géogr. Bot. 21: 268 (1911), nom. illeg. = ?
Carex lanescens J.Fellm., Bull. Soc. Imp. Naturalistes Moscou 8: 283 (1835). = ?
Carex ×*langii* Steud., Syn. Pl. Glumac. 2: 227 (1855). = ?
Carex lanosa Gueldenst. ex Ledeb., Fl. Ross. 4: 320 (1852). = ?
Carex ×*larminatii* H.Lév., Bull. Acad. Int. Géogr. Bot. 23: 286 (1914). = ?
Carex ×*lausii* Podp., Vestn. Klubu Prír. v Prostejove 10: 7 (1907 publ. 1908). = ?
Carex leveillei Husn., Cypéracées: 12 (1905). = [12 FRA]
Carex ×*lidii* Hadac, Stud. Bot. Cech. 5: 2 (1942). = ?
Carex ×*lidii* Flatberg, Norweg. J. Bot. 19: 93 (1972), nom. illeg. = *C. canescens* × *C. chordorrhiza*
Carex ×*ligniciensis* Figert, Allg. Bot. Z. Syst. 6: 38 (1900). = ?
Carex ×*limnogena* Appel ex Kneuck., Deutsche Bot. Monatsschr. 10: 168 (1892). = ?
Carex lineola C.A.Mey. ex Fr., Novit. Fl. Suec. Mant. 3: 145 (1842). = [10 SWE]
Carex lonchosacca C.B.Clarke, Kew Bull., Addit. Ser. 8: 77 (1908). = ?
Carex longiaristata Biv., Stirp. Rar. Sicilia 4: 8 (1818). = ?
Carex longibracteata Hornem., Hort. Bot. Hafn. 2: 881 (1815). = ?
Carex ×*loretii* Rouy in G.Rouy & J.Foucaud, Fl. France 13: 486 (1912). = ?
Carex lyonii Kurtz, Bot. Jahrb. Syst. 19: 419 (1894). = ?
Carex macrophylla K.Schum., Bot. Jahrb. Syst. 30: 381 (1901), nom. illeg. = ?
Carex macrostachyos Link, J. Bot. (Schrader) 2: 811 (1799). = ?
Carex maidenii Gand., Bull. Soc. Bot. France 46: 392 (1900). = ?
Carex marcida Raeusch., Nomencl. Bot., ed. 3: 270 (1797). = ?
Carex ×*massonii* Cay. & Lepage, Naturaliste Canad. 104: 567 (1977). = ?
Carex ×*mastnyana* Otruba & Podp., Sched. Fl. Exsicc. Reip. Bohem. Slov. Cent. 6: 21 (1931). = ?
Carex megalandra Gand., Bull. Soc. Bot. France 66: 295 (1919 publ. 1920). = ?
Carex melanocephala Kar. & Kir., Bull. Soc. Imp. Naturalistes Moscou 14: 436 (1841), nom. illeg. = ?
Carex melanostylis Desv. in C.Gay, Fl. Chil. 6: t. 73 (1854). = ?
Carex meridiana Akiyama, J. Jap. Bot. 13: 652 (1937). = ?
Carex mesomorpha Gand., Contr. Fl. Terr. Slav. Merid. 1: 30 (1883). = ?
Carex microsacca C.B.Clarke, Kew Bull., Addit. Ser. 8: 83 (1908). = [38 JAP]
Carex ×*mihashii* T.Koyama, J. Jap. Bot. 31: 288 (1956). = ?
Carex miliacea Schrank, Baier. Fl. 1: 300 (1789). = ?
Carex moscowensis C.B.Clarke ex B.Fedtsch., Bot. Zhurn. (St. Petersburg) 1: 191 (1906). = [14]
Carex ×*murrii* Appel, Oesterr. Bot. Z. 41: 124 (1891). = ?
Carex ×*naegelii* Brügger ex Sennholz, Verh. K. K. Zool.-Bot. Ges. Wien 37: 72 (1887). = ?
Carex nebrodensis Lojac., Fl. Sicul. 3: 231 (1909). = ?
Carex nenningiana Opiz, Seznam: 103 (1852). = ?
Carex neonelmesii Raymond, Naturaliste Canad. 82: 158 (1955). = ?
Carex nigricans Kit., Linnaea 32: 320 (1863), nom. illeg. = ?
Carex ×*nikaii* T.Koyama, J. Jap. Bot. 30: 132 (1955). = ?
Carex nitens Phil., Anales Univ. Chile 1873: 557 (1873). = ?
Carex oblonga J.F.Gmel., Syst. Nat. 2: 146 (1791). = ?
Carex obtusa Kit., Linnaea 32: 319 (1863). = ?
Carex pachyandra C.B.Clarke, Kew Bull., Addit. Ser. 8: 82 (1908). = ?
Carex ×*paczoskii* Zapal., Consp. Fl. Gallic. Crit. 3: 234 (1911). = ?
Carex pagesii H.Lév., Bull. Acad. Int. Géogr. Bot. 20: 208 (1910). = ?
Carex pairae var. *javanica* Nelmes, Kew Bull. 5: 208 (1950). = ?
Carex ×*palezieuxii* Kneuck., Allg. Bot. Z. Syst. 19: 1 (1913). = ?
Carex ×*paponii* Muret ex T.Durand & Pittier, Bull. Soc. Roy. Bot. Belgique 25: 336 (1886). = ?
Carex paradoxa J.F.Gmel., Syst. Nat. 2: 145 (1791). = ?
Carex pendulina Lojac., Fl. Sicul. 3: 224 (1909). = [13 SIC]
Carex ×*peraffinis* Appel, Ber. Deutsch. Bot. Ges. 10: 77 (1892). = ?
Carex personata C.B.Clarke, Kew Bull., Addit. Ser. 8: 71 (1908). = ?
Carex ×*petrae-fu* F.Murr, Progr. K. K. Oberrealschule Insbr.: 57 (1891). = ?
Carex pinguipes C.B.Clarke, Kew Bull., Addit. Ser. 8: 85 (1908). = ?

Carex platycarpos Rchb. in J.C.Mössler, Handb. Gewächsk. ed. 2, 3: 1656 (1830). = ?
Carex polygama J.F.Gmel., Syst. Nat. 2: 145 (1791). = ?
Carex ×*prahliana* Junge, Jahrb. Hamburg. Wiss. Anst. 22(Beih.): 71 (1904). = ?
Carex ×*prahliana* Junge, Verh. Vereins Naturwiss. Heimatf. Hamburg, III, 14: 118 (1906). = ?
Carex prolifera Kit., Linnaea 32: 320 (1863). = ?
Carex ×*pseudoaphanolepis* Ohwi, Mem. Coll. Sci. Kyoto Imp. Univ., Ser. B, Biol. 11(5): 512 (1936). = ?
Carex pseudodiandra Waisb., Magyar Bot. Lapok 4: 60, 72 (1905). = ?
Carex ×*pseudoheleonastes* Hand.-Mazz., Verh. K. K. Zool.-Bot. Ges. Wien 58: 102 (1908). = ?
Carex pseudomacrogyna Kük., Izv. Glavn. Bot. Sada S.S.S.R. 28: 220 (1929). = ?
Carex ×*pseudomairii* Camus, Bull. Soc. Bot. France 33: 479 (1886). = ?
Carex pseudonutans Boreau ex É.Martin, Cat. Pl. Romorantin: 309 (1875). = ?
Carex pseudothunbergii Akiyama, J. Jap. Bot. 16: 100 (1940). = ?
Carex pseudoturfosa Dalla Torre & Sarnth., Fl. Tirol 6(1): 330 (1906). = ?
Carex ×*pseudoturfosa* H.Lév. & L.C.Lamb., Bull. Acad. Int. Géogr. Bot. 21: 267 (1911), nom. illeg. = ?
Carex ×*pseudovulpina* K.Richt., Pl. Eur. 1: 168 (1890). = ?
Carex psilochloa Wall., Numer. List: 3534 (1831), nom. inval. = ?
Carex pulchella T.Holm, Amer. J. Sci., IV, 16: 457 (1903), nom. illeg. = ?
Carex pusilla Valbusa, Nuovo Giorn. Bot. Ital., n.s., 4: 174 (1897). = ?
Carex ×*raciborskii* Zapal., Consp. Fl. Gallic. Crit. 3: 233 (1911). = ?
Carex regelii C.B.Clarke ex O.Fedtsch., Fl. Pamir: 71 (1903). = [32]
Carex richmondii Boott ex C.B.Clarke, Kew Bull., Addit. Ser. 8: 83 (1908). = ?
Carex ×*rieseana* Figert, Allg. Bot. Z. Syst. 9: 112 (1903). = *C. paradoxa* × *C. remota*
Carex riloensis Stoeva & E.D.Popova, Fragm. Florist. Geobot. 38: 39 (1993). = [13 BUL]
Carex ×*rossiana* Degen, Fl. Veleb. 1: 585 (1936). = [13 YUG]
Carex rotoensis Petrie, Trans. & Proc. New Zealand Inst. 56: 9 (1926). = [51 CTM]
Carex rutilus Humn., Cat. Pl. Luxeuil: 25 (1876). = ?
Carex salina var. *mutica* Wahlenb., Fl. Lapp.: 246 (1812). = *C. aquatilis* × *C. flavicans*
Carex ×*salisiana* Brügger, Jahresber. Naturf. Ges. Graubündens, II, 23-24: 120 (1880). = ?
Carex sandwicensis Boeck., Flora 18: 265 (1835). = [63 HAW]
Carex ×*sanionis* K.Richt., Pl. Eur. 1: 170 (1890). = ?
Carex ×*sarntheinii* Vetter, Verh. K. K. Zool.-Bot. Ges. Wien 57: 236 (1907). = ?
Carex scaberrima Scheele, Linnaea 22: 345 (1849). = ?
Carex ×*schalkertii* Murr, Neue Übers. Bl.-Pfl. Voralberg: 57 (1923). = ?
Carex ×*schuetzeana* Figert. = ?
Carex scordiae Lojac., Fl. Sicul. 3: 230 (1909). = [13 SIC]
Carex seckeri C.B.Clarke, Kew Bull., Addit. Ser. 8: 88 (1908). = ?
Carex semicrinita C.B.Clarke, Kew Bull., Addit. Ser. 8: 70 (1908). = ?
Carex semicylindrica Kit., Linnaea 32: 319 (1863). = ?
Carex ×*senayana* Soó, Acta Bot. Acad. Sci. Hung. 16: 371 (1970 publ. 1971). = ?
Carex ×*serravalensis* Beauverd, Bull. Soc. Bot. Genève, II, 14: 39 (1922 publ. 1923). = ?
Carex setifolia Heer in J.Hegetschweiler-Bodmer, Fl. Schweiz: 929 (1840), nom. illeg. = ?
Carex skorpili Velen., Reliq. Mrkvic.: 31 (1922). = ?
Carex ×*solstitialis* Figert, Deutsche Bot. Monatsschr. 7: 86 (1889). = ?
Carex sonderiana Boott ex C.B.Clarke, Kew Bull., Addit. Ser. 8: 86 (1908), nom. illeg. = ?
Carex ×*sooi* Jákucs, Ann. Hist.-Nat. Mus. Natl. Hung., n.s., 3: 90 (1953). = ?
Carex spathulata Becc. in J.D.Hooker, Fl. Brit. India 6: 559 (1893). = ?
Carex sphaerica Lapeyr., Hist. Pl. Pyrénées: 570 (1813). = ?
Carex ×*stagnalis* L.C.Lamb., Bull. Acad. Int. Géogr. Bot. 21: 10 (1911). = [12 FRA]
Carex ×*stenolepis* f. *gracilior* Kük. in H.G.A.Engler (ed.), Pflanzenr., IV, 20(38): 715 (1909). = ?
Carex striata var. *japonica* Koidz., Bot. Mag. (Tokyo) 29: 388 (1915). = [38 JAP]
Carex ×*strigosula* Chatenier, Bull. Soc. Bot. France 58: 347 (1911). = [12 FRA]
Carex ×*subdecolorans* Nyár. in T.Savulescu, Fl. Reipubl. Popul. Roman. 11: 842 (1966). = ?
Carex sublivida Norrl., Not. Sällsk. Fauna Fl. Fenn. Förh. 10: 51 (1871). = ?
Carex subsabulosa Norman, Forh. Vidensk.-Selsk. Kristiania 1893: 48 (1893). = ?
Carex ×*superriparia* Hübl, Verh. Zool.-Bot. Ges. Wien 82: 22, 56 (1932). = ?
Carex suteri Kunth, Enum. Pl. 2: 523 (1837). = ?
Carex ×*suziella* Podp., Práce Morav. Prír. Spolecn. 6(3): 191 (1928). = ?
Carex taraspensis Brügger & Killias, Jahresber. Naturf. Ges. Graubündens, II, 31(Beibl.): 193 (1888). = ?
Carex tenuirostris Boeck., Flora 63: 456 (1880). = ?
Carex ×*teramotoi* T.Koyama, J. Jap. Bot. 30: 134 (1955). = ?
Carex thomsonii Gand., Bull. Soc. Bot. France 60: 27 (1913), nom. illeg. = ?
Carex ×*torgesiana* Kük., Mitth. Thüring. Bot. Vereins, n.s., 28: 62 (1911). = ?
Carex ×*tornabenii* Chiov., Ann. Bot. (Rome) 17: 83 (1927). = ?
Carex ×*toshironis* T.Koyama, J. Jap. Bot. 31: 287 (1956). = ?
Carex ×*tourletii* Gillies ex Tourlet, Bull. Soc. Bot. France 1: 313 (1854). = ?
Carex triceps Schrank, Baier. Fl. 1: 287 (1789). = ?
Carex truncatula Kit., Linnaea 32: 318 (1863). = ?
Carex tschonoskii V.I.Krecz., Bot. Mater. Gerb. Bot. Inst. Komarova Akad. Nauk S.S.S.R. 7: 32 (1937). = ?
Carex turuli Simonk., Magyar Bot. Lapok 3: 82 (1904). = ?
Carex ×*uechtritziana* K.Richt., Pl. Eur. 1: 170 (1896). = ?
Carex ulaviformis Hoppe in J.Sturm, Deutschl. Fl.: 61 (1833). = ?
Carex varica C.B.Clarke, Kew Bull., Addit. Ser. 8: 69 (1908). = ?
Carex vesiciformis Peterm., Flora 27: 339 (1844). = ?
Carex ×*viadrina* Figert, Allg. Bot. Z. Syst. 13: 3 (1907). = ?
Carex ×*viadrina* Schube, Jahresber. Schles. Ges. Vaterl. Cult. 86(2): 51 (1909). = ?
Carex ×*villacensis* Kük., Mitth. Thüring. Bot. Vereins, n.s., 38: 35 (1929). = ?

Carex ×*vimariensis* Hausskn. ex Berthold, Ber. Deutsch. Bot. Ges. 10: 131 (1890). = ?
Carex viridis Spenn., Fl. Friburg. 1: 48 (1825), nom. illeg. = ?
Carex volkensii K.Schum. in H.G.A.Engler, Pflanzenw. Ost-Afrikas, C: 130 (1895). = [25]
Carex ×*vratislaviensis* Figert, Allg. Bot. Z. Syst. 6: 39 (1900). = ?
Carex ×*xenostachya* T.Koyama, J. Jap. Bot. 31: 286 (1956). = ?
Carex yodoensis Boeck., Bot. Jahrb. Syst. 5: 515 (1884). = ?
Carex zallichiana Prescott ex Nees in R.Wight, Contr. Bot. India: 129 (1834). = ?
Chaetospora paniculata Steud., Syn. Pl. Glumac. 2: 161 (1855). = [29 REU]
Chaetospora tenella Rupr. in F.von der Osten-Saken, Sert. Tianschan.: 34 (1869). = [32]
Cladium durandoi Chabert, Bull. Soc. Bot. France 36: 321 (1889). = ?
Cladium tetraquetrum var. *planifolium* Benth., Fl. Austral. 7: 407 (1878). = [50 NSW QLD VIC]
Courtoisia olivacea Boeck., Flora 44: 331 (1861). = [82 SUR] Perhaps identical with Oxycaryum cubense.
Cyperus abnormis Steud., Syn. Pl. Glumac. 2: 43 (1854). = ?
Cyperus acutiusculus Lag., Gen. Sp. Pl.: 2 (1816). = [79]
Cyperus adscendens Steud., Syn. Pl. Glumac. 2: 14 (1854). = ?
Cyperus albidus Retz., Observ. Bot. 6: 21 (1791), nom. illeg. = [36]
Cyperus ×*amuricocompressus* T.Koyama, J. Jap. Bot. 33: 256 (1958). = [38 JAP]
Cyperus ×*amuricoides* T.Koyama, J. Jap. Bot. 30: 130 (1955). = ?
Cyperus apatris Steud., Syn. Pl. Glumac. 2: 47 (1854). = ?
Cyperus arenarius Salzm. ex Steud., Syn. Pl. Glumac. 2: 46 (1854), nom. illeg. = ?
Cyperus aristatus Boeck., Linnaea 35: 501 (1868), nom. illeg. = ?
Cyperus aureolus Steud., Syn. Pl. Glumac. 2: 23 (1854). = [81 WIN]
Cyperus ×*babae* T.Koyama, Quart. J. Taiwan Mus. 14: 194 (1961). = ?
Cyperus baldwinii Chapm., Fl. South. U.S.: 511 (1860), nom. illeg. = ?
Cyperus benghalensis C.B.Clarke, J. Linn. Soc., Bot. 21: 151 (1884), nom. illeg. = ?
Cyperus bifidus Pourr., Chlor. Hisp.: 174 (1788). = ?
Cyperus boeckleri Phil., Anales Univ. Chile 93: 344 (1896). = ?
Cyperus boninensis Ohwi, Mem. Coll. Sci. Kyoto Imp. Univ., Ser. B, Biol. 18(1): 161 (1944). = [38 OGA]
Cyperus ×*borealikiiensis* T.Koyama & Ogawa, J. Jap. Bot. 33: 256 (1958). = ?
Cyperus brizimorphus Steud., Syn. Pl. Glumac. 2: 5 (1854). = ?
Cyperus bromoides Britton ex Parish, Bull. S. Calif. Acad. Sci. 3: 49 (1904), nom. illeg. = ?
Cyperus caducus Steud., Syn. Pl. Glumac. 2: 315 (1855). = ?
Cyperus capensis Boeck., Linnaea 36: 378 (1870), nom. illeg. = ?
Cyperus carrii Kük., Bot. Jahrb. Syst. 69: 256 (1938). = ?
Cyperus catharinensis Boeck., Beitr. Cyper. 2: 9 (1890). = [84 BZS]
Cyperus coercens Rchb. ex Kunth, Enum. Pl. 2: 20 (1837). = ?
Cyperus ×*condensatus* T.Koyama, J. Jap. Bot. 28: 344 (1953). = ?
Cyperus convallium Steud., Syn. Pl. Glumac. 2: 5 (1854). = ?
Cyperus cylindricus Boeck., Flora 42: 437 bis (1859). = [84]
Cyperus cyrtostachys C.B.Clarke, J. Linn. Soc., Bot. 21: 59 (1884). = ?
Cyperus dentifer Kom., Izv. Bot. Sada Akad. Nauk S.S.S.R. 30: 198 (1931 publ. 1932). = ?
Cyperus depauperatus Steud., Syn. Pl. Glumac. 2: 34 (1854), nom. illeg. = ?
Cyperus donianus Boeck., Allg. Bot. Z. Syst. 1: 165 (1895), nom. illeg. = ?
Cyperus drummondii Nees ex Boeck., Linnaea 36: 314 (1870), nom. illeg. = ?
Cyperus dubius Roxb., Fl. Ind. 1: 188 (1820), nom. illeg. = ?
Cyperus elytropiptos Steud., Syn. Pl. Glumac. 2: 27 (1854). = ?
Cyperus filiculmis A.Rich. in R.de la Sagra, Hist. Fis. Cuba, Bot. 11: 285 (1850), nom. illeg. = ?
Cyperus flavicomus Schltdl., Linnaea 6: 24 (1831), nom. illeg. = ?
Cyperus fossarum Schltdl., Linnaea 31: 316 (1862), nom. illeg. = [82 SUR]
Cyperus frionensis Boeck., Beitr. Cyper. 2: 10 (1890). = ?
Cyperus fusciformis Drobow in M.G.Popov (ed.), Key Pl. Envir. Tashkent: 46 (1923). = ?
Cyperus gardneri Baker, Fl. Mauritius: 413 (1877), nom. illeg. = ?
Cyperus ginge Welw., Apont.: 586 (1859). = ?
Cyperus guatemalensis Steud., Syn. Pl. Glumac. 2: 47 (1854). = ?
Cyperus hangianus Gand., Bull. Soc. Bot. France 66: 297 (1919 publ. 1920). = ?
Cyperus heptaphyllus Larrañaga, Escritos D. A. Larrañaga 2: 24 (1923). = ?
Cyperus ×*heribaudii* Pau, Bol. Soc. Aragonesa Ci. Nat. 4: 328 (1905). = *Pycreus flavescens* × *C. fuscus* ?
Cyperus honoluluensis Steud. ex Jard., Bull. Soc. Linn. Normandie, II, 9: 287 (1875). = ?
Cyperus ×*hybridus* Hausskn., Mitth. Thüring. Bot. Vereins, n.s., 13-14: 34 (1899). = *C. longus* × *C. rotundus*
Cyperus jouannatii Pau, Bol. Soc. Aragonesa Ci. Nat. 4: 328 (1905). = ?
Cyperus ×*kadzusensis* T.Koyama, Quart. J. Taiwan Mus. 14: 193 (1961). = ?
Cyperus kerstingii Engl. in H.G.A.Engler & C.G.O.Drude, Veg. Erde 9: 204 (1908). = [24 SUD]
Cyperus kuekenthalii Merr., J. Arnold Arbor. 19: 322 (1938). = ?
Cyperus kuntzei Boeck., Beitr. Cyper. 1: 1 (1888). = ?
Cyperus laetus C.B.Clarke, Kew Bull., Addit. Ser. 8: 9 (1908), nom. illeg. = [50 QLD]
Cyperus lamprocarpus Boeck., Linnaea 35: 490 (1868), nom. illeg. = ?
Cyperus lapidicola Kük. in H.G.A.Engler (ed.), Pflanzenr., IV, 20(101): 17 (1935). = [27]
Cyperus laxus Willemet, Ann. Bot. (Usteri) 18: 6 (1796), nom. illeg. = [29 MAU]
Cyperus leonensis Steud., Nomencl. Bot., ed. 2, 1: 470 (1840). = [22 SIE]
Cyperus leptanthos Steud., Syn. Pl. Glumac. 2: 26 (1854). = ?
Cyperus leptophyes C.B.Clarke, Bot. Jahrb. Syst. 30(68): 14 (1902). = ?

Cyperus ligusticus Gand., Dec. Pl. Nov. 2: 28 (1876). = [13 ITA]
Cyperus lividus Steud., Syn. Pl. Glumac. 2: 35 (1854). = ?
Cyperus lomatophyllus Engl., Pflanzenw. Ost-Afrikas, A: 125 (1895). = ?
Cyperus lucidulus Link, Jahrb. Gewächsk. 1(3): 86 (1818). = ?
Cyperus macrostachys Link, J. Bot. (Schrader) 2: 310 (1799). = ?
Cyperus mariscoides Elliott, Sketch Bot. S. Carolina 1: 67 (1816). = ?
Cyperus mauritianus Willemet, Ann. Bot. (Usteri) 18: 4 (1796). = [29 MAU]
Cyperus ×*mihashii* T.Koyama, J. Jap. Bot. 28: 346 (1953). = ?
Cyperus minimus Forssk., Fl. Aegypt.-Arab.: 202 (1775), nom. illeg. = ?
Cyperus minutiflorus J.Presl & C.Presl, Reliq. Haenk. 1: 351 (1830). = [42 PHI]
Cyperus minutus J.F.Gmel., Syst. Nat. 2: 130 (1791). = ?
Cyperus navicularis Steud., Syn. Pl. Glumac. 2: 32 (1854). = ?
Cyperus nyassicus Chiov., Papiro Ital.: 72 (1931). = ?
Cyperus obtusiusculus Gand., Dec. Pl. Nov. 2: 29 (1876). = [12 FRA]
Cyperus ×*ogawai* T.Koyama, J. Jap. Bot. 31: 325 (1956). = ?
Cyperus oliganthus Gand., Dec. Pl. Nov. 2: 37 (1876). = [12 FRA]
Cyperus ostenii Kük. ex Osten, Anales Mus. Hist. Nat. Montevideo 3(2): 135 (1931). = ?
Cyperus pallidus Link, Jahrb. Gewächsk. 1(3): 86 (1818). = ?
Cyperus paniculatus Hook. & Arn., Bot. Beechey Voy.: 99 (1832), nom. illeg. = [63]
Cyperus paradoxus Steud., Syn. Pl. Glumac. 2: 8 (1854). = [29 REU]
Cyperus paramattensis Link, Enum. Pl.: 46 (1809). = ?
Cyperus parvus Hassk., Cat. Hort. Bot. Bogor.: 24 (1844). = ?
Cyperus patulus Willd. ex Link, Enum. Hort. Berol. Alt. 1: 46 (1821), nom. illeg. = ?
Cyperus pes-avium Bertol., Misc. Bot. 8: 30 (1849). = ?
Cyperus philippensis J.Presl & C.Presl, Reliq. Haenk. 1: 174 (1828). = [42 PHI]
Cyperus politus Welw. ex Ridl., Trans. Linn. Soc. London, Bot. 2: 125 (1884). = ?
Cyperus polyphyllus Larrañaga, Escritos D. A. Larrañaga 2: 24 (1923), nom. illeg. = ?
Cyperus pubescens J.Presl & C.Presl, Reliq. Haenk. 1: 181 (1830). = ?
Cyperus punctorius Schrad. ex Boeck., Linnaea 35: 567 (1868). = ?
Cyperus pungens Kunth, Enum. Pl. 2: 32 (1837). = ?
Cyperus rheedei F.Muell., Descr. Notes Papuan Pl. 7: 34 (1886). = ?
Cyperus rhenanus Schrad. ex Kunth, Enum. Pl. 2: 60 (1837). = ?
Cyperus rhigiophyllus Hochst. ex C.Krauss, Flora 28: 756 (1845). = [27 NAT]
Cyperus rigidulus Vahl, Enum. Pl. 2: 342 (1805). = ?
Cyperus riparius Salisb., Prodr. Stirp. Chap. Allerton: 30 (1796). = ?
Cyperus robustior Gand., Dec. Pl. Nov. 2: 37 (1876). = [12 FRA]
Cyperus rubromarginatus Meinsh., Trudy Imp. S.-Peterburgsk. Bot. Sada 18: 236 (1900). = ?
Cyperus rufescens Schrad. ex Steud., Syn. Pl. Glumac. 2: 27 (1854), nom. illeg. = ?
Cyperus sanguineofuscus Hook.f., Fl. Tasman. 2: 80 (1858), nom. illeg. = ?
Cyperus scoparius Poir. in J.B.A.M.de Lamarck, Encycl. 7: 253 (1806). = [29 MDG] Perhaps identical with Cyperus platyphyllus.
Cyperus seminudus Zoll. & Moritzi in H.Zollinger, Syst. Verz.: 96 (1854), nom. illeg. = ?
Cyperus st.-crucei Liebm., Mexic. Halvgr.: 223 (1850). = ?
Cyperus stachyphorus K.Koch, Linnaea 21: 623 (1848). = ?
Cyperus subanceps Boeck., Linnaea 38: 368 (1874). = [81]
Cyperus tener Vahl, Enum. Pl. 2: 299 (1805). = ?
Cyperus tenuifolius T.L.Dai, in Fl. Reipubl. Popul. Sin. 11: 231 (1961), nom. illeg. = [36]
Cyperus triceps F.N.Williams, Bull. Herb. Boissier, II, 7: 90 (1907), nom. illeg. = ?
Cyperus triphyllus Larrañaga, Escritos D. A. Larrañaga 2: 24 (1923). = ?
Cyperus tuberosus Roxb., Fl. Ind. 1: 199 (1820), nom. illeg. = ?
Cyperus ugandensis Chiov., Papiro Ital.: 73 (1931). = ?
Cyperus umbellatus Burm.f., Fl. Indica: 21 (1768). = ?
Cyperus umbellatus Hillebr., Fl. Hawaiian Isl.: 470 (1888), nom. illeg. = ?
Cyperus uncinatus Desv. ex Ham., Prodr. Pl. Ind. Occid.: 12 (1825), nom. illeg. = ?
Cyperus uniflorus Thunb., Fl. Jav.: 8 (1825). = ?
Cyperus venustus Kunth, Enum. Pl. 2: 68 (1837), nom. illeg. = ?
Cyperus ×*weatherbianus* Fernald, Rhodora 20: 190 (1918). = ?
Cyperus ×*yamamotoi* T.Koyama, J. Jap. Bot. 28: 344 (1953). = ?
Dichromena candida Baker, J. Linn. Soc., Bot. 21: 450 (1885), nom. illeg. = ?
Dichromena rufidula Steud., Syn. Pl. Glumac. 2: 137 (1855). = ?
Dichromena trachycarpa Steud., Syn. Pl. Glumac. 2: 136 (1855). = ?
Eleocharis lechleri C.B.Clarke, Bot. Jahrb. Syst. 30(68): 24 (1902), nom. illeg. = [85 CLS]
Eleocharis leptocaulos Phil., Anales Univ. Chile 93: 349 (1896), nom. illeg. = [85 CLC]
Eleocharis macrorrhiza T.Koyama, Brittonia 31: 285 (1979), nom. illeg. = [41 THA]
Eleocharis paraguayensis Maury ex Micheli, Mém. Soc. Phys. Genève 31(1): 137 (1890). = [85 PAR]
Eleocharis quaesita Kitag., J. Jap. Bot. 27: 202 (1952). = [38 JAP]
Eleocharis sanguinea Maury ex Micheli, Mém. Soc. Phys. Genève 31(1): 135 (1890). = [85 PAR]
Eleocharis tenerrima Peter, Abh. Preuss. Akad. Wiss., Phys.-Math. Kl., n.f., 13(2): 52 (1928). = [25 TAN]
Elynanthus P.Beauv. ex Lestib., Essai Cypér.: 32 (1819). = ?
Elynanthus tenerrimus Boeck., Allg. Bot. Z. Syst. 2: 122 (1896). = [27]
Eriophorum bellardii Colla, Herb. Pedem. 5: 561 (1836). = [13 ITA]
Eriophorum depauperatum Arv.-Touv., Ess. Pl. Dauph.: 66 (1871). = [12 FRA]
Eriophorum dubium Steud., Syn. Pl. Glumac. 2: 128 (1855). = ?
Eriophorum filamentosum Boeck., Bot. Jahrb. Syst. 5: 506 (1843). = [42 MLY]
Eriophorum ×*fulvum* Bach.Pyl. ex Raymond, Svensk Bot. Tidskr. 45: 527 (1951). = [72 NFL]

Eriophorum intercedens H.Lindb., Exsicc. (Sched. Pl. Finland.) 1906: 32 (1906). = [10 FIN]
Eriophorum kerneri Ullep., Oesterr. Bot. Z. 43: 421 (1893). = [11 HUN]
Eriophorum ocreatum A.Nelson, Bull. Torrey Bot. Club 29: 401 (1902). = [7]
Eriophorum paucinervium Eaton in N.L.Britton, Man. Fl. N. States: 182 (1901). = [7]
Eriophorum paucinervum Eaton, Bull. Torrey Bot. Club 25: 341 (1898). = [7]
Eriophorum textile Larrañaga, Escritos D. A. Larrañaga 2: 26 (1923). = [85 URU]
Ficinia decidua H.Pfeiff., Herbarium 54: 34 (1920). = [27 CPP]
Ficinia distans C.B.Clarke in É.A.J.de Wildeman, Pl. Nov. Hort. Then. 1: 30 (1904). = [27]
Ficinia ligulata C.B.Clarke, Kew Bull., Addit. Ser. 8: 31 (1908). = [27]
Ficinia monticola Kunth, Enum. Pl. 2: 261 (1837). = [27 CPP]
Ficinia nuda Boeck., Linnaea 38: 383 (1874). = [27 CPP]
Ficinia schlechteriana H.Pfeiff., Herbarium 54: 33 (1920). = [27]
Ficinia thyrsoidea H.Pfeiff., Herbarium 54: 33 (1920). = [27 CPP]
Ficinia trinkleriana H.Pfeiff., Herbarium 54: 34 (1920). = [27 CPP]
Fimbristylis breviculma Govind., Rheedea 6: 61 (1996), nom. illeg. = [40 IND]
Fimbristylis donticola Hochst. ex Steud., Syn. Pl. Glumac. 2: 111 (1855). = [40]
Fimbristylis elongata Pires de Lima, Bol. Soc. Brot., II, 2: 133 (1924), nom. illeg. = [26 MOZ]
Fimbristylis faulensis Beck, Ann. K. K. Naturhist. Hofmus. 3: 252 (1888). = [51 NZS]
Fimbristylis hamiltonii Steud., Syn. Pl. Glumac. 2: 119 (1855). = [81 PUE]
Fimbristylis huillensis Ridl., Trans. Linn. Soc. London, Bot. 2: 154 (1884). = [26 ANG]
Fimbristylis humilis S.T.Blake, Proc. Roy. Soc. Queensland 48: 91 (1936 publ. 1937), nom. illeg. = [50 QLD]
Fimbristylis magnifica C.B.Clarke, Bull. Soc. Bot. France 54(8): 28 (1907). = [2]
Fimbristylis nigrescens Steud., Syn. Pl. Glumac. 2: 118 (1855). = [82]
Fimbristylis planifolia Steud., Syn. Pl. Glumac. 2: 118 (1855). = [62 CRL]
Fimbristylis prolifera Steud., Syn. Pl. Glumac. 2: 120 (1855). = [82]
Fimbristylis semihirsuta Boeck., Beitr. Cyper. 2: 18 (1890). = [36]
Fimbristylis spartium Ham., Prodr. Pl. Ind. Occid.: 14 (1825). = [81 PUE]
Fuirena microstachya Boeck., Linnaea 38: 398 (1874). = [82]
Holoschoenus melanocarpus Boeck., Flora 40: 37 (1857). = [79]
Hypolytrum costatum Nelmes, Kew Bull. 10: 74 (1955), nom. illeg. = [22 LBR]
Hypolytrum leprieurii Nees ex Steud., Syn. Pl. Glumac. 2: 133 (1855). = [22 GUI] Mapania sp. ?
Isolepis compressa Nees, Linnaea 10: 157 (1836). = [27]
Isolepis hispidula Nees, Ann. Nat. Hist. 1: 284 (1838). = [40 IND]
Isolepis panamensis J.Presl & C.Presl, Reliq. Haenk. 1: 351 (1830). = [80 PAN] Rhynchospora sp ?
Isolepis rubescens J.Presl, Delic. Prag.: 157 (1822). = [84] Bulbostylsis sp.
Isolepis truncata Schrad. ex Nees in C.F.P.von Martius & auct. suc. (eds.), Fl. Bras. 2(1): 83 (1842), nom. illeg. = [84]
Juncellus minutus C.B.Clarke, J. Bot. 34: 224 (1896). = [25 TAN]
Kobresia pratensis Freyn, Oesterr. Bot. Z. 40: 266 (1890). = [30]
Kyllinga ambigua Steud., Syn. Pl. Glumac. 2: 70 (1854). = [24 CHA ETH] Isolepis ?
Kyllinga ramosa Sessé & Moç., Fl. Mexic.: 16 (1893). = [81 PUE]
Lagenocarpus laniger C.B.Clarke ex H.Pfeiff., Beih. Bot. Centralbl. 39(2): 441 (1923). = ?
Machaerina sucinonux T.Koyama, Bot. Mag. (Tokyo) 69: 66 (1956). = [38 OGA]
Mapania alpina Elmer ex Merr., Enum. Philipp. Fl. Pl. 1: 132 (1922). = [42 PHI]
Mapania deistelii K.Schum., Notizbl. Königl. Bot. Gart. Berlin 3: 106 (1901). = [23 CMN]
Mapania pandanifolia Sander, New Pl.: 14 (1897). = ?
Mariscus alopecuroides Steud., Syn. Pl. Glumac. 2: 65 (1854). = [84]
Mariscus blatteri McCann, J. Bombay Nat. Hist. Soc. 36: 532 (1934). = [40 IND]
Mariscus cufodontii Chiov., Miss. Biol. Borana, Racc. Bot., Angiosp.-Gymnosp.: 294 (1939). = [24 ETH]
Mariscus hainanensis Chun & F.C.How, Acta Phytotax. Sin. 7: 87 (1958). = [36 CHH]
Mariscus mayottensis Cherm., Notul. Syst. (Paris) 5: 270 (1936). = [29 COM]
Mariscus monospermus S.M.Huang, in Fl. Hainan. 4: 537 (1977). = [36 CHH]
Mariscus tamakugu Masam. & Syozi, Acta Phytotax. Geobot. 14: 89 (1951). = [36 CHH 38 TAI]
Mariscus xerophilus Domin, Biblioth. Bot. 85: 438 (1915). = [50 NTA QLD]
Melancranis tenuis Eckl. ex Kunth, Enum. Pl. 2: 265 (1837). = [27]
Oncostylis bivalvis Nees in C.F.P.von Martius & auct. suc. (eds.), Fl. Bras. 2(1): 205 (1842). = [84 BZL]
×*Raymondiella* Boivin, Taxon 22: 278 (1973), not accordance with Art. H. 7. = *Dulichium* × *Rhynchospora*
Rhynchospora alba var. *meridiana* H.Pfeiff., Repert. Spec. Nov. Regni Veg. 33: 210 (1933). = [79 MXG 84 BZN]
Rhynchospora cinnamomea Boeck., Flora 41: 645 (1858). = [7]
Rhynchospora fauriei C.B.Clarke ex H.Lév., Bull. Acad. Int. Géogr. Bot. 14: 197 (1904), nom. illeg. = [36]
Rhynchospora foliata Steud., Syn. Pl. Glumac. 2: 142 (1855). = [7]
Rhynchospora millegrana Schrad. ex Steud., Syn. Pl. Glumac. 2: 139 (1855). = [8] Check if the same as the other millegrana.
Rhynchospora pfeifferi Herter, Revista Sudamer. Bot. 9: 159 (1953). = ?
Rhynchospora punctata Boeck., Linnaea 37: 580 (1873), nom. illeg. = [79 82 VEN]
Rhynchospora trigyna Hochst., Flora 24(1 Intell.): 21 (1841). = [24 ETH]
Schoenoplectus ×*igaensis* (T.Koyama) Hayas. & H.Ohashi, J. Jap. Bot. 75: 224 (2000). = [38 JAP] *Schoenoplectiella juncoides* × *S. lineolatus*
Schoenoplectus ×*juncohotarui* Yashiro, J. Jap. Bot. 79: 97 (2004). = [38 JAP] *Schoenoplectiella juncoides* × *Schoenoplectus hotarui*
Schoenoplectus ×*oguraensis* (T.Koyama) Hayas. & H.Ohashi, J. Jap. Bot. 75: 224 (2000). = [38 JAP] *S. mucronatus* × *Schoenoplectiella wallichii*

Schoenus attenuatus Gand., Dec. Pl. Nov. 2: 35 (1876). = [12 FRA]
Schoenus halleri Honck., Verz. Gew. Teutschl.: 127 (1782). = [1]
Schoenus hattorianus Nakai, Bot. Mag. (Tokyo) 26: 79 (1912). = [38 JAP]
Schoenus incanus Forssk., Fl. Aegypt.-Arab.: 12 (1775). = [35 YEM]
Schoenus laevis Thunb., Prodr. Pl. Cap.: 17 (1794). = [27 CPP]
Schoenus longus Gand., Dec. Pl. Nov. 2: 35 (1876). = [12 FRA]
Schoenus macrocephalus Boeck., Flora 61: 144 (1878). = [27 CPP]
Schoenus megalocephalus Gand., Dec. Pl. Nov. 2: 36 (1876). = [12 FRA]
Schoenus microcephalus J.Kern, Blumea 12: 28 (1963). = [60 NWC]
Schoenus pallescens Gand., Dec. Pl. Nov. 2: 35 (1876). = [12 FRA]
Schoenus spicatus Burm.f., Fl. Indica, Prodr. Fl. Cap.: 3 (1768). = [27 CPP]
Schoenus squamellatus Kük., Repert. Spec. Nov. Regni Veg. 44: 71 (1938). = [50 WAU]
Schoenus subcontortus Gand., Dec. Pl. Nov. 2: 36 (1876). = [12 FRA]
Schoenus tener Spreng., Neue Entdeck. Pflanzenk. 3: 9 (1822). = [27 CPP]
Schoenus tenuiflorus Willd. ex Link, Jahrb. Gewächsk. 1(3): 75 (1820). = [27 CPP]
Schoenus tristachyus Thunb., Prodr. Pl. Cap.: 16 (1794). = [27 CPP]
Schoenus vendeanus Gand., Dec. Pl. Nov. 2: 34 (1876). = [12 FRA]
Scirpus aegialites Phil., Anales Univ. Chile 93: 477 (1896). = [85 CLC] Isolepis sp. ?
Scirpus bibracteatus Kük., Mitth. Thüring. Bot. Vereins, n.s., 50: 12 (1943). = [27 CPP]
Scirpus borneensis H.Pfeiff., Mitt. Inst. Allg. Bot. Hamburg 7: 172 (1928). = [42 BOR]
Scirpus buettnerianus Boeck., Beitr. Cyper. 1: 20 (1888). = [22 GUI]
Scirpus dichromenoides C.B.Clarke, Kew Bull., Addit. Ser. 8: 89 (1908). = [82 VEN]
Scirpus elatus Gueldenst. ex Ledeb., Fl. Ross. 4: 252 (1852). = [14]
Scirpus erectogracilis Hayata, Icon. Pl. Formosan. 6: 114 (1916). = [38 TAI]
Scirpus ×*fujimakii* M.Kikuchi & Koyama, J. Jap. Bot. 32: 255 (1957). = [38 JAP]
Scirpus gaillardii Maury, J. Bot. (Morot) 3: 197 (1889). = [8]
Scirpus gracillimus Boeck., Vidensk. Meddel. Naturhist. Foren. Kjøbenhavn 1869: 140 (1869). = [24 ETH]
Scirpus graecus Quézel & Contandr., Taxon 16: 239 (1967). = [13 GRC]
Scirpus helferi Beetle, Amer. J. Bot. 33: 662 (1946). = [41 AND MYA]
Scirpus hispanicus Gueldenst. ex Ledeb., Fl. Ross. 4: 252 (1852). = [14]
Scirpus hochstetteri Boeck., Linnaea 36: 739 (1870). = [24 ETH]
Scirpus ×*intermedius* Tang & F.T.Wang, in Fl. Reipubl. Popul. Sin. 11: 224 (1961), nom. illeg. = [36 CHC]
Scirpus isdellensis W.Fitzg., J. Proc. Roy. Soc. Western Australia 3: 123 (1918). = [50 WAU]
Scirpus iseensis T.Koyama & Shimizu, J. Jap. Bot. 42: 177 (1967). = [38 JAP]
Scirpus junciformis Retz., Observ. Bot. 6: 19 (1791). = [36+]
Scirpus libanoticus Post, Fl. Syria: 833 (1896). = [34 LBS]
Scirpus longivaginatus Boeck., Beitr. Cyper. 2: 16 (1890). = [84]
Scirpus macrostachys Boeck., Beitr. Cyper. 1: 19 (1888), nom. illeg. = [40 EHM]
Scirpus ×*mariqueter* Tang & F.T.Wang, in Fl. Reipubl. Popul. Sin. 11: 223 (1961). = [36 CHC]
Scirpus muehlenbergianus Schult. in J.J.Roemer & J.A.Schultes, Mant. 2: 74 (1824). = [7]
Scirpus muelleri Boeck., Beitr. Cyper. 2: 14 (1890). = [50]
Scirpus nindensis Ficalho & Hiern, Trans. Linn. Soc. London, Bot. 2: 27 (1881). = [2]
Scirpus nudifructus Kük. in R.E.Fries, Wiss. Erg. Schwed. Rhod.-Kongo Exped. 1: 7 (1921). = [26 ZAM]
Scirpus nudus Raf., Atlantic J.: 18 (1833). = [7]
Scirpus occultus C.B.Clarke, Bull. Soc. Bot. France 54(8): 28 (1907). = [2]
Scirpus pilosus Thunb., Prodr. Pl. Cap.: 18 (1794). = [27 CPP]
Scirpus pilosus Larrañaga, Escritos D. A. Larrañaga 2: 25 (1923), nom. illeg. = [85 URU]
Scirpus ×*prolifulcus* Carse, Trans. & Proc. New Zealand Inst. 60: 572 (1930). = [51 NZN] *Isolepis inundata* × *I. prolifera*
Scirpus ×*prolifundus* Carse, Trans. & Proc. New Zealand Inst. 60: 572 (1930). = [51 NZN] *Isolepis distigmatosa* × *I. prolifera*
Scirpus racemosus Spreng., Syst. Veg. 1: 213 (1824). = [84]
Scirpus rigidus Griseb., Abh. Königl. Ges. Wiss. Göttingen 24: 312 (1879), nom. illeg. = [85 AGE]
Scirpus ×*schmidtianus* Junge, Jahrb. Hamburg. Wiss. Anst. 22(Beih.): 72 (1905). = ?
Scirpus semiuncialis C.B.Clarke, Kew Bull., Addit. Ser. 8: 29 (1908). = [27]
Scirpus seoanei Merino, Contr. Fl. Galicia: 173 (1897). = [12 FRA]
Scirpus stramineus C.Presl, Isis (Oken) 21: 269 (1828). = [7]
Scirpus tetragonus Walker, Essays Nat. Hist. Rural Econ.: 242 (1808). = [10 GRB]
Scirpus thwaitesii Boeck., Linnaea 38: 380 (1874). = [40 IND]
Scirpus typhinus Raf., Atlantic J.: 18 (1833). = [7]
Scirpus uleanus Boeck., Allg. Bot. Z. Syst. 2: 55 (1895). = [84]
Scirpus umbellatus Post, Fl. Syria: 21 (1896). = [34 PAL]
Scirpus varius Boeck. ex C.B.Clarke in W.H.Harvey & auct. suc. (eds.), Fl. Cap. 7: 229 (1898), nom. illeg. = [27 NAT SWZ TVL]
Scirpus welwitschii K.Richt., Pl. Eur. 1: 141 (1890). = [12 POR]
Scirpus xerophilus Pilg., Bot. Jahrb. Syst. 30: 142 (1901). = [84 BZC]
Scleria suffulta C.B.Clarke, Kew Bull., Addit. Ser. 8: 58 (1908). = [50 QLD]
Tetraria wallichiana C.B.Clarke in W.H.Harvey & auct. suc. (eds.), Fl. Cap. 7: 280 (1898). = [27 CPP]
Thoracostachyum vietnamense K.K.Nguyen, Bot. Zhurn. (Moscow & Leningrad) 64: 855 (1979). = [41 VIE]
Trichelostylis borealis Alph.Wood, Amer. Bot. Fl.: 364 (1870). = [74 ILL 75 MIC]
Trichophorum lockhartii G.Don in J.C.Loudon, Hort. Brit.: 468 (1830). = [81 TRT]

www.ingramcontent.com/pod-product-compliance
Ingram Content Group UK Ltd.
Pitfield, Milton Keynes, MK11 3LW, UK
UKHW020129290926
15471UKWH00003B/372

9 781842 461990